BIOCHEMISTRY

4TH EDITION

BIOCHEMISTRY

DONALD VOET
University of Pennsylvania

JUDITH G. VOET
Swarthmore College

JOHN WILEY & SONS, INC.

VP & Publisher   *Kaye Pace*
Associate Publisher & Editor   *Petra Recta*
Sponsoring Editor   *Joan Kalkut*
Editorial Assistant   *Yelena Zolotorevskaya/Patrick White*
Marketing Manager   *Kristine Ruff*
Production Manager   *Dorothy Sinclair*
Production Editor   *Sandra Dumas*
Senior Designer   *Madelyn Lesure*
Senior Illustration Editor   *Anna Melhorn*
Executive Media Editor   *Thomas Kulesa*
Media Editor   *Marc Wedzecki*
Photo Department Manager   *Hilary Newman*
Photo Researcher   *Elyse Rieder*
Production Management Services   *Ingrao Associates*

Cover and part opening art: Illustrations, Irving Geis, Images from Irving Geis Collection/Howard Hughes Medical Institute. Rights owned by HHMI. Not to be reproduced without permission.

This book was typeset in 10/12 Times Ten Roman at Aptara®, Inc. and printed and bound by RR Donnelley. The cover was printed by RR Donnelley.

Founded in 1807, John Wiley & Sons, Inc. has been a valued source of knowledge and understanding for more than 200 years, helping people around the world meet their needs and fulfill their aspirations. Our company is built on a foundation of principles that include responsibility to the communities we serve and where we live and work. In 2008, we launched a Corporate Citizenship Initiative, a global effort to address the environmental, social, economic, and ethical challenges we face in our business. Among the issues we are addressing are carbon impact, paper specifications and procurement, ethical conduct within our business and among our vendors, and community and charitable support. For more information, please visit our website: www.wiley.com/go/citizenship.

The paper in this book was manufactured by a mill whose forest management programs include sustained yield -harvesting of its timberlands. Sustained yield harvesting principles ensure that the number of trees cut each year does not exceed the amount of new growth.

This book is printed on acid-free paper. ∞

ISBN 13   978-0470-57095-1
ISBN 13   978-0470-91745-9

Printed in the United States of America.
10 9 8 7 6

For our grandchildren: Maya, Leo, Cora, and Elisabeth

ABOUT THE COVER

The cover contains two paintings of horse heart cytochrome c. The upper painting, which was drawn by Irving Geis in collaboration with Richard Dickerson, was designed to show the influence of amino acid side chains on the protein's three-dimensional folding pattern. The lower painting, also made by Geis, is of cytochrome c illuminated by its single iron atom in which its hydrophilic side chains are drawn in green. These paintings were made in the 1970s, when only a handful of protein structures were known (around 70,000 are now known) and the personal computers that we presently use to visualize them were many years in the future. It reminds us that biochemistry is a process that is driven by the creativity of the human mind. Our visualization tools have developed from pen, ink, and colored pencils to sophisticated computers and software. Without creativity, however, these tools have little use.

PREFACE

Biochemistry is a field of enormous fascination and utility, arising, no doubt, from our own self-interest. Human welfare, particularly its medical and nutritional aspects, has been vastly improved by our rapidly growing understanding of biochemistry. Indeed, scarcely a day passes without the report of a biomedical discovery that benefits a significant portion of humanity. Further advances in this rapidly expanding field of knowledge will no doubt lead to even more spectacular gains in our ability to understand nature and to control our destinies. It is therefore essential that individuals embarking on a career in biomedical sciences be well versed in biochemistry.

This textbook is a distillation of our experiences in teaching undergraduate and graduate students at the University of Pennsylvania and Swarthmore College and is intended to provide such students with a thorough grounding in biochemistry. We assume that students who use this textbook have had the equivalent of one year of college chemistry and sufficient organic chemistry so that they are familiar with basic principles and nomenclature. We also assume that students have taken a one-year college course in general biology in which elementary biochemical concepts were discussed. Students who lack these prerequisites are advised to consult the appropriate introductory textbooks in those subjects.

NEW TO THIS EDITION

Since the third edition of *Biochemistry* was published in 2004, the field of biochemistry has continued its phenomenal and rapidly accelerating growth. This remarkable expansion of our knowledge, the work of many thousands of talented and dedicated scientists, has been characterized by numerous new paradigms, as well as an enormous enrichment of almost every aspect of the field. For example, the number of known protein and nucleic acid structures as determined by X-ray and NMR techniques has increased by over 3-fold. Moreover, the quality and complexity of these structures, which include numerous membrane proteins, has significantly improved, thereby providing enormous advances in our understanding of structural biochemistry. Bioinformatics, an only recently coined word, has come to dominate the way that many aspects of biochemistry are conceived and practiced. Since the third edition of *Biochemistry* was published, the number of known genome sequences has increased by over 10-fold and the goal of personalized medicine to determine the genome sequence of each individual seems to be within reach. Likewise, the state of knowledge has exploded in such subdisciplines as eukaryotic and prokaryotic molecular biology, metabolic control, protein folding, electron transport, membrane transport, immunology, signal transduction, etc. New and improved methodologies such as DNA microarrays, rapid DNA sequencing, RNAi, cryoelectron microscopy, mass spectrometry, single molecule techniques,

and robotic devices are now routinely used in the laboratory to answer questions that seemed entirely out of reach a decade ago. Indeed, these advances have affected our everyday lives in that they have changed the way that medicine is practiced, the way that we protect our own health, and the way in which food is produced.

THEMES

In writing this textbook we have emphasized several themes. First, biochemistry is a body of knowledge compiled by people through experimentation. In presenting what is known, we therefore stress how we have come to know it. The extra effort the student must make in following such a treatment, we believe, is handsomely repaid since it engenders the critical attitudes required for success in any scientific endeavor. Although science is widely portrayed as an impersonal subject, it is, in fact, a discipline shaped through the often idiosyncratic efforts of individual scientists. We therefore identify some of the major contributors to biochemistry (most of whom are still professionally active) and, in many cases, consider the approaches they have taken to solve particular biochemical puzzles. Students should realize, however, that most of the work described could not have been done without the dedicated and often indispensable efforts of numerous coworkers.

The unity of life and its variation through evolution is a second dominant theme running through the text. Certainly one of the most striking characteristics of life on earth is its enormous variety and adaptability. Yet, biochemical research has amply demonstrated that all living things are closely related at the molecular level. As a consequence, the molecular differences among the various species have provided intriguing insights into how organisms have evolved from one another and have helped delineate the functionally significant portions of their molecular machinery.

A third major theme is that biological processes are organized into elaborate and interdependent control networks. Such systems permit organisms to maintain relatively constant internal environments, to respond rapidly to external stimuli, and to grow and differentiate.

A fourth theme is that biochemistry has important medical consequences. We therefore frequently illustrate biochemical principles by examples of normal and abnormal human physiology and discuss the mechanisms of action of a variety drugs.

ORGANIZATION AND COVERAGE

As the information explosion in biochemistry has been occurring, teachers have been exploring more active learning methods such as problem-based learning, discovery-based learning, and cooperative learning. These new teaching and learning techniques involve more interaction among students and teachers and, most importantly, require more in-class time. In writing the fourth edition of this textbook, we have therefore been faced with the dual pressures of increased content and pedagogical innovation. We have responded to this challenge by presenting the subject matter of biochemistry as thoroughly and accurately as we can so as to provide students and instructors alike with this information as they explore various innovative learning strategies. In this way we deal with the widespread concern that these novel methods of stimulating student learning tend to significantly diminish course content. We have thus written a textbook that permits teachers to direct their students to areas of content that can be explored outside of class as well as providing material for in-class discussion.

We have reported many of the advances that have occurred in the last seven years in the fourth edition of *Biochemistry* and have thereby substantially enriched nearly all of its sections. Nevertheless, the basic organization of the fourth edition remains the same as that of the third edition.

The text is organized into five parts:

I. Introduction and Background: An introductory chapter followed by chapters that review the properties of aqueous solutions and the elements of thermodynamics.

II. Biomolecules: A description of the structures and functions of proteins, nucleic acids, carbohydrates, and lipids.

III. Mechanisms of Enzyme Action: An introduction to the properties, reaction kinetics, and catalytic mechanisms of enzymes.

IV. Metabolism: A discussion of how living things synthesize and degrade carbohydrates, lipids, amino acids, and nucleotides with emphasis on energy generation and consumption.

V. Expression and Transmission of Genetic Information: An expansion of the discussion of nucleic acid structure that is given in Part II followed by an exposition of both prokaryotic and eukaryotic molecular biology.

This organization permits us to cover the major areas of biochemistry in a logical and coherent fashion. Yet, modern biochemistry is a subject of such enormous scope that to maintain a relatively even depth of coverage throughout the text, we include more material than most one-year biochemistry courses will cover in detail. This depth of coverage, we feel, is one of the strengths of this book; it permits the instructor to teach a course of his/her own design and yet provide the student with a resource on biochemical subjects not emphasized in the course.

The order in which the subject matter of the text is presented more or less parallels that of most biochemistry courses. However, several aspects of the textbook's organization deserve comment:

1. Chapter 5 (Nucleic Acids, Gene Expression, and Recombinant DNA Technology) introduces molecular biology early in the narrative in response to the central role that recombinant DNA technology has come to play in modern biochemistry. Likewise, the burgeoning field of bioinformatics is discussed in a separate section of Chapter 7.

2. We have split our presentation of thermodynamics between two chapters. Basic thermodynamic principles—enthalpy, entropy, free energy, and equilibrium—are discussed in Chapter 3 because these subjects are prerequisites for understanding structural biochemistry, enzyme mechanisms, and kinetics. Metabolic aspects of thermodynamics—the thermodynamics of phosphate compounds and oxidation–reduction reactions—are presented in Chapter 16 since knowledge of these subjects is not required until the chapters that follow.

3. Techniques of protein purification are described in a separate chapter (Chapter 6) that precedes the discussions of protein structure and function. We have chosen this order so that students will not feel that proteins are somehow "pulled out of a hat." Nevertheless, Chapter 6 has been written as a resource chapter to be consulted repeatedly as the need arises. Techniques of nucleic acid purification are also discussed in that chapter for the above-described reasons.

4. Chapter 10 describes the properties of hemoglobin in detail so as to illustrate concretely the preceding discussions of protein structure and function. This chapter introduces allosteric theory to explain the cooperative nature of hemoglobin oxygen binding. The subsequent extension of allosteric theory to enzymology in Chapter 13 is then a straightforward matter.

5. Concepts of metabolic control are presented in the chapters on glycolysis (Chapter 17) and glycogen metabolism (Chapter 18) through the consideration of flux generation, allosteric regulation, substrate cycles, covalent enzyme modification, cyclic cascades, and a discussion of metabolic control analysis. We feel that these concepts are best understood when studied in metabolic context rather than as independent topics.

6. The rapid growth in our knowledge of biological signal transduction necessitates that this important subject have its own chapter, Chapter 19.

7. There is no separate chapter on coenzymes. These substances, we feel, are more logically studied in the context of the enzymatic reactions in which they participate.

8. Glycolysis (Chapter 17), glycogen metabolism (Chapter 18), the citric acid cycle (Chapter 21), and electron transport and oxidative phosphorylation (Chapter 22) are detailed as models of general metabolic pathways with emphasis placed on many of the catalytic and control mechanisms of the enzymes involved. The principles illustrated in these chapters are reiterated in somewhat less detail in the other chapters of Part IV.

9. Consideration of membrane transport (Chapter 20) precedes that of mitochondrially based metabolic pathways such as the citric acid cycle, electron transport, and oxidative phosphorylation. In this manner, the idea of the compartmentalization of biological processes can be easily assimilated. Chapter 20 also contains a discussion of neurotransmission because it is intimately involved with membrane transport.

10. Discussions of both the synthesis and the degradation of lipids have been placed in a single chapter (Chapter 25), as have the analogous discussions of amino acids (Chapter 26) and nucleotides (Chapter 28).

11. Energy metabolism is summarized and integrated in terms of organ specialization in Chapter 27, following the descriptions of carbohydrate, lipid, and amino acid metabolism.

12. The principles of both prokaryotic and eukaryotic molecular biology are expanded from their introduction in Chapter 5 in sequential chapters on DNA replication, repair and recombination (Chapter 30), transcription (Chapter 31), and translation (Chapter 32). Viruses (Chapter 33) are then considered as paradigms of more complex cellular functions, followed by discussions of eukaryotic gene expression (Chapter 34).

13. Chapter 35, the final chapter, is a series of minichapters that describe the biochemistry of a variety of well-characterized human physiological processes: blood clotting, the immune response, and muscle contraction.

14. Chapters 33, 34, and 35 are available on the Book Companion Site (www.wiley.com/college/voet) with the same appearance and level of detail as the chapters in the printed textbook.

The old adage that you learn a subject best by teaching it simply indicates that learning is an active rather than a passive process. The problems we provide at the end of each chapter are therefore designed to make students think rather than to merely regurgitate poorly assimilated and rapidly forgotten information. Few of the problems are trivial and some of them (particularly those marked with an asterisk) are quite difficult. Yet, successfully working out such problems can be one of the most rewarding aspects of the learning process. Only by thinking long and hard for themselves can students make a body of knowledge truly their own. The answers to the problems are worked out in detail in the *Solutions Manual* for this text. The manual can be an effective learning tool, however, only if the student makes a serious effort to solve a problem before looking up its answer.

We have included lists of references at the end of every chapter to provide students with starting points for independent biochemical explorations. The enormity of the biochemical research literature prevents us from giving all but a few of the most seminal research reports. Rather, we list what we have found to be the most useful reviews and monographs on the various subjects covered in each chapter.

Finally, although we have made every effort to make this text error free, we are under no illusions that we have done so. Thus, we are particularly grateful to the many readers of previous editions, students and faculty alike, who have taken the trouble to write us with suggestions on how to improve the textbook and to point out errors they have found. We earnestly hope that the readers of the fourth edition will continue this practice.

Donald Voet
Judith G. Voet

STUDENT AND INSTRUCTOR RESOURCES

The Book Companion Site for *Biochemistry* (www.wiley.com/college/voet) provides online resources for both students and instructors. These online resources are designed to enhance student understanding of biochemistry. They are all keyed to figures or sections in the text and called out in the text with a red mouse icon.

Bioinformatics Exercises: A set of exercises cover the contents and uses of databases related to nucleic acids, protein sequences, protein structures, enzyme inhibition, and other topics. These exercises, written by Paul Craig, Rochester Institute of Technology, Rochester, New York, use real data sets, pose specific questions, and prompt students to obtain information from online databases and to access the software tools for analyzing such data.

Guided Explorations: 30 self-contained presentations, many with narration, employ extensive animated computer graphics to enhance student understanding of key topics.

Interactive Exercises: 58 molecular structures from the text have been rendered in Jmol by Stephen Rouse. Jmol is a browser-independent interface for manipulating structures in three dimensions, and structures are paired with questions designed to facilitate comprehension of concepts. A Tutorial for using Jmol is also provided.

Kinemages: A set of 22 exercises comprising 55 three-dimensional images of selected protein and nucleic acids that can be manipulated by users as suggested by accompanying text.

Animated Figures: 67 figures from the text, illustrating various concepts, techniques, and processes, are presented as brief animations to facilitate learning.

Case Studies: A set of 33 case studies by Kathleen Cornely, Providence College, Providence, Rhode Island, use problem-based learning to promote understanding of biochemical concepts. Each case presents data from the literature and asks questions that require students to apply principles to novel situations often involving topics from multiple chapters in the textbook.

In addition, a printed *Solutions Manual* containing detailed solutions for all of the textbook's end-of-chapter problems is available for purchase.

FOR INSTRUCTORS

• **PowerPoint Slides** of all the figures and tables in the text are optimized with bold leader lines and large labels for classroom projection. The figures and tables are also available for importing individually as jpeg files from the *Wiley Image Gallery*.

• **Test Bank** by Marilee Benore, University of Michigan–Dearborn, Dearborn, Michigan, and Robert Kane, Baylor University, Waco, Texas, has over 1000 questions containing a variety of question types (multiple choice, matching, fill in the blank, and short answer). Each question is rated by difficulty.

• **Classroom Response Questions** ("clicker questions") by Rachel Milner and Adrienne Wright, University of Alberta, Edmonton, Alberta, Canada, are interactive questions for classroom response systems, to facilitate classroom participation and discussion. These questions can also be used by instructors as prelecture questions that help gauge students knowledge of overall concepts while addressing common misconceptions.

ACKNOWLEDGMENTS

This textbook is the result of the dedicated effort of numerous individuals, many of whom deserve special mention: Laura Ierardi cleverly combined text, figures, and tables in designing each of this textbook's pages. Suzanne Ingrao, our Production Coordinator, skillfully managed the production of the textbook. Madelyn Lesure designed the book's typography and cover. Joan Kalkut, our Editor, skillfully organized and managed the entire project. Hilary Newman and Elyse Rieder acquired many of the photographs in the textbook and kept track of all of them. Connie Parks, our copy editor, put the final polish on the manuscript and eliminated large numbers of grammatical and typographical errors. Special thanks to Alyson Rentrop, our Associate Editor, who coordinated and managed an exceptional supplements package, and to Tom Kulesa, Senior Media Editor, and Marc Wezdecki, Media Editor, who substantially improved and developed the media resources. Much of the art in this fourth edition of *Biochemistry* is the creative legacy of the drawings made for its first and second editions by John and Bette Woolsey and Patrick Lane of J/B Woolsey Associates. The late Irving Geis provided us with his extraordinary molecular art and gave freely of his wise counsel.

The atomic coordinates of most of the proteins and nucleic acids that we have drawn for use in this textbook were obtained from the Protein Data Bank (PDB), which is administered by the Research Collaboratory for Structural Bioinformatics (RCSB). We created these drawings using the molecular graphics programs PyMOL by Warren DeLano; RIBBONS by Mike Carson; and GRASP by Anthony Nicholls, Kim Sharp, and Barry Honig.

The interactive computer graphics diagrams that are presented on the website that accompanies this textbook are either Jmol images or Kinemages. Jmol is a free, open source, interactive, Web browser applet for manipulating molecules in three dimensions. It is based on the program RasMol by Roger Sayle, which was generously made publicly available. Kinemages are displayed by the program KiNG, which was written and generously provided by David C. Richardson, who also wrote and provided the program PREKIN, which we used to help generate the Kinemages. KiNG (Kinemage, Next Generation) is an interactive system for three-dimensional vector graphics that runs on Windows, Mac OS X, and Linux/Unix systems.

We wish especially to thank those colleagues who reviewed this textbook, in both its current and earlier editions, and provided us with their prudent advice:

Joseph Babitch, *Texas Christian University*

E.J. Berhman, *Ohio State University*

Karl D. Bishop, *Bucknell University*

Robert Blankenshop, *Arizona State University*

Charles L. Borders, Jr., *The College of Wooster*

Kenneth Brown, *University of Texas at Arlington*

Larry G. Butler, *Purdue University*

Carol Caparelli, *Fox Chase Cancer Center*

W. Scott Champney, *East Tennessee State University*

Paul F. Cook, *The University of Oklahoma*

Glenn Cunningham, *University of Central Florida*

Eugene Davidson, *Georgetown University*

Don Dennis, *University of Delaware*

Walter A. Deutsch, *Louisiana State University*

Kelsey R. Downum, *Florida International University*

William A. Eaton, *National Institutes of Health*

David Eisenberg, *University of California at Los Angeles*

Jeffrey Evans, *University of Southern Mississippi*

David Fahrney, *Colorado State University*

Paul Fitzpatrick, *Texas A&M University*

Robert Fletterick, *University of California at San Francisco*

Norbert C. Furumo, *Eastern Illinois University*

Scott Gilbert, *Swarthmore College*

Guido Guidotti, *Harvard University*

James H. Hageman, *New Mexico State University*

Lowell Hager, *University of Illinois at Urbana–Champaign*

James H. Hammons, *Swarthmore College*

Edward Harris, *Texas A&M University*

Angela Hoffman, *University of Portland*

Ralph A. Jacobson, *California Polytechnic State University*

Eileen Jaffe, *Fox Chase Cancer Center*

Jan G. Jaworski, *Miami University*

William P. Jencks, *Brandeis University*

Mary Ellen Jones, *University of North Carolina*

Jason D. Kahn, *University of Maryland*

Tokuji Kimura, *Wayne State University*

Barrie Kitto, *University of Texas at Austin*

Daniel J. Kosman, *State University of New York at Buffalo*

Robert D. Kuchta, *University of Colorado, Boulder*

Thomas Laue, *University of New Hampshire*

Albert Light, *Purdue University*

Dennis Lohr, *Arizona State University*

Larry Louters, *Calvin College*

Robert D. Lynch, *University of Lowell*

Harold G. Martinson, *University of California at Los Angeles*

Michael Mendenhall, *University of Kentucky*

Sabeeha Merchant, *University of California at Los Angeles*

Christopher R. Meyer, *California State University at Fullerton*

Ronald Montelaro, *Louisiana State University*

Scott Moore, *Boston University*

Harry F. Noller, *University of California at Santa Cruz*

John Ohlsson, *University of Colorado*

Gary L. Powell, *Clemson University*

Alan R. Price, *University of Michigan*

Paul Price, *University of California at San Diego*

Thomas I. Pynadath, *Kent State University*

Frank M. Raushel, *Texas A&M University*

Ivan Rayment, *University of Wisconsin*

Frederick Rudolph, *Rice University*

Raghupathy Sarma, *State University of New York at Stony Brook*

Paul R. Schimmel, *The Scripps Research Institute*

Thomas Schleich, *University of California at Santa Cruz*

Allen Scism, *Central Missouri State University*

Charles Shopsis, *Adelphi University*

Marvin A. Smith, *Brigham Young University*

Thomas Sneider, *Colorado State University*

Jochanan Stenish, *Western Michigan University*

Phyllis Strauss, *Northeastern University*

JoAnne Stubbe, *Massachusetts Institute of Technology*

William Sweeney, *Hunter College*

John Tooze, *European Molecular Biology Organization*

Mary Lynn Trawick, *Baylor University*

Francis Vella, *University of Saskatchewan*

Harold White, *University of Delaware*

William Widger, *University of Houston*

Ken Willeford, *Mississippi State University*

Lauren Williams, *Georgia Institute of Technology*

Jeffery T. Wong, *University of Toronto*

Beulah M. Woodfin, *The University of New Mexico*

James Zimmerman, *Clemson University*

D.V.
J.G.V.

BRIEF CONTENTS

CONTENTS

*Chapters 33, 34, and 35 are available on our book companion website, www.wiley.com/college/voet.

GUIDE TO MEDIA RESOURCES

The book website (www.wiley.com/college/voet) offers the following resources to enhance student understanding of biochemistry. These are all keyed to figures or sections in the text. They are called out in the text with a red mouse icon or margin note.

Chapter	Media Type	Title	Text Reference	Page
2 Acids, Bases, and Buffers	Animated Figure	Acid-base titration curves of 1-L solutions of $1M$ acetic acid, $H_2PO_4^-$, and NH_4^+ by a strong base	*Figure 2-11*	**47**
	Animated Figure	Titration curve of a 1-L solution of $1M$ H_3PO_4	*Figure 2-13*	**49**
4 Amino Acids	Animated Figure	Titration curve of glycine	*Figure 4-6*	**72**
5 Nucleic Acids, Gene Expression, and Recombinant DNA Technology	Kinemage Exercise	2.1. B-DNA	*Figure 5-11*	**89**
	Interactive Exercise	1. B-DNA	*Figure 5-11*	**89**
	Kinemage Exercise	2-2, 17-2. The Watson–Crick base pairs	*Figure 5-12*	**89**
	Animated Figure	Demonstration of the semiconservative nature of DNA replication in *E. coli* by density gradient ultracentrifugation	*Figure 5-13*	**91**
	Animated Figure	UV absorbance spectra of the nucleic acid bases and DNA	*Figure 5-15*	**92**
	Animated Figure	Example of a DNA melting curve	*Figure 5-16*	**93**
	Guided Exploration	1: Overview of transcription and translation	*Section 5-4*	**95**
	Guided Exploration	2: Regulation of gene expression by the *lac* repressor system	*Figure 5-25*	**97**
	Animated Figure	Construction of a recombinant DNA molecule	*Figure 5-44*	**109**
	Animated Figure	Cloning of foreign DNA in λ phages	*Figure 5-47*	**111**
	Guided Exploration	3: PCR and site-directed mutagenesis	*Section 5-5F*	**114**
	Animated Figure	Site-directed mutagenesis	*Figure 5-55*	**119**
6 Techniques of Protein and Nucleic Acid Purification	Animated Figure	An enzyme-linked immunosorbent assay (ELISA)	*Figure 6-1*	**132**
	Animated Figure	Ion exchange chromatography using stepwise elution	*Figure 6-6*	**136**
	Animated Figure	Gel filtration chromatography	*Figure 6-9*	**139**
7 Covalent Structures of Proteins and Nucleic Acids	Guided Exploration	4: Protein sequence determination	*Section 7-1*	**164**
	Animated Figure	The Edman degradation	*Figure 7-4*	**167**
	Animated Figure	The amino acid sequence of a polypeptide chain is determined by comparing the sequence of two sets of mutually overlapping peptide fragments	*Figure 7-6*	**171**
	Guided Exploration	5: DNA sequence determination by the chain-terminator method	*Section 7-2A*	**176**
	Guided Exploration	6: Bioinformatics	*Section 7-4*	**194**
8 Three-Dimensional Structures of Proteins	Kinemage Exercise	3-1. The peptide group	*Figure 8-1, 8-2, 8-4*	**221, 222, 223**
	Guided Exploration	7: Stable helices in proteins: the α helix	*Section 8-1B*	**225**
	Kinemage Exercise	3–2. The α helix	*Figure 8-11, 8-12*	**226, 227**

BIOCHEMISTRY

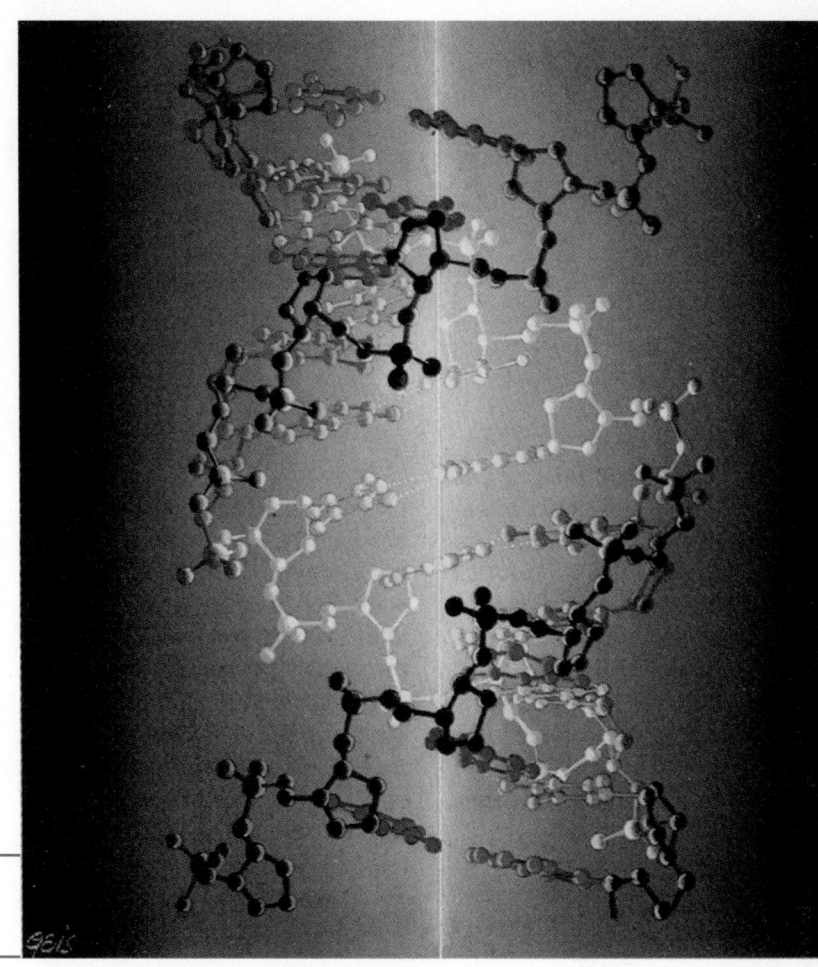

"Hot wire" DNA illuminated by its helix axis.

PART I

INTRODUCTION AND BACKGROUND

Life

CHAPTER 1

It is usually easy to decide whether or not something is alive. This is because living things share many common attributes, such as the capacity to extract energy from nutrients to drive their various functions, the power to actively respond to changes in their environment, and the ability to grow, to differentiate, and—perhaps most telling of all—to reproduce. Of course, a given organism may not have all of these traits. For example, mules, which are obviously alive, rarely reproduce. Conversely, inanimate matter may exhibit some lifelike properties. For instance, crystals may grow larger when immersed in a supersaturated solution of the crystalline material. Therefore, life, as are many other complex phenomena, is perhaps impossible to define in a precise fashion. Norman Horowitz, however, proposed a useful set of criteria for living systems: *Life possesses the properties of replication, catalysis, and mutability*. Much of this text is concerned with the manner in which living organisms exhibit these properties.

Biochemistry is the study of life on the molecular level. The significance of such studies is greatly enhanced if they are related to the biology of the corresponding organisms or even communities of such organisms. This introductory chapter therefore begins with a synopsis of the biological realm. This is followed by an outline of biochemistry, a review of genetics, a discussion of the origin of life, and finally, an introduction to the biochemical literature.

1 PROKARYOTES

It has long been recognized that life is based on morphological units known as **cells.** The formulation of this concept is generally attributed to an 1838 paper by Matthias Schleiden and Theodor Schwann, but its origins may be traced to the seventeenth century observations of early microscopists such as Robert Hooke. There are two major classifications of cells: the **eukaryotes** (Greek: *eu,* good or true + *karyon,* kernel or nut), which have a membrane-enclosed **nucleus** encapsulating their **DNA (deoxyribonucleic acid);** and the **prokaryotes** (Greek: *pro,* before), which lack this organelle. Prokaryotes, which comprise the various types of bacteria, have relatively simple structures and are invariably unicellular (although they may form filaments or colonies of independent cells). They are estimated to represent about half of Earth's biomass. Eukaryotes, which may be multicellular as well as unicellular, are vastly more complex than prokaryotes. (**Viruses,** which are much simpler entities than cells, are not classified as living because they lack the metabolic apparatus to reproduce outside their host cells. They are essentially large molecular aggregates.) This section is a discussion of prokaryotes. Eukaryotes are considered in the following section.

A. Form and Function

Prokaryotes are the most numerous and widespread organisms on Earth. This is because their varied and often highly adaptable metabolisms suit them to an enormous variety of habitats. Besides inhabiting our familiar temperate and aerobic environment, certain types of bacteria may thrive in or even require conditions that are hostile to eukaryotes such as unusual chemical environments, high temperatures (as high as 130°C), and lack of oxygen. Moreover, the rapid reproductive rate of prokaryotes (optimally <20 min per cell division for many species) permits them to take advantage of transiently favorable conditions, and conversely, the ability of many bacteria to form resistant **spores** allows them to survive adverse conditions.

a. Prokaryotes Have Relatively Simple Anatomies

Prokaryotes, which were first observed in 1683 by the inventor of the microscope, Antonie van Leeuwenhoek, have sizes that are mostly in the range 1 to 10 μm. They have one of three basic shapes (Fig. 1-1): spheroidal **(cocci),** rodlike **(bacilli),** and helically coiled **(spirilla),** but all have the same general design (Fig. 1-2). They are bounded, as are all cells, by an ~70-Å-thick **cell membrane (plasma membrane),** which consists of a lipid bilayer containing embedded proteins that control the passage of molecules in and out of the cell and catalyze a variety of reactions. The cells of most prokaryotic species are surrounded by a rigid, 30- to 250-Å-thick polysaccharide **cell wall** that mainly functions to protect the cell from mechanical injury and to prevent it from bursting in media more osmotically dilute than its contents. Some bacteria further encase themselves in a gelatinous polysaccharide **capsule** that protects them from the defenses of higher organisms. Although prokaryotes lack the membranous subcellular organelles characteristic of eukaryotes (Section 1-2), their plasma membranes may be infolded to form multilayered structures known as **mesosomes.** The mesosomes are thought to serve as the site of DNA replication and other specialized enzymatic reactions.

The prokaryotic **cytoplasm** (cell contents) is by no means a homogeneous soup. Its single **chromosome** (DNA molecule, several copies of which may be present in a rapidly growing cell) is condensed to form a body known as a **nucleoid.** The cytoplasm also contains numerous species of **RNA (ribonucleic acid),** a variety of soluble **enzymes** (proteins that catalyze specific reactions), and many thousands of 250-Å-diameter particles known as **ribosomes,** which are the sites of protein synthesis.

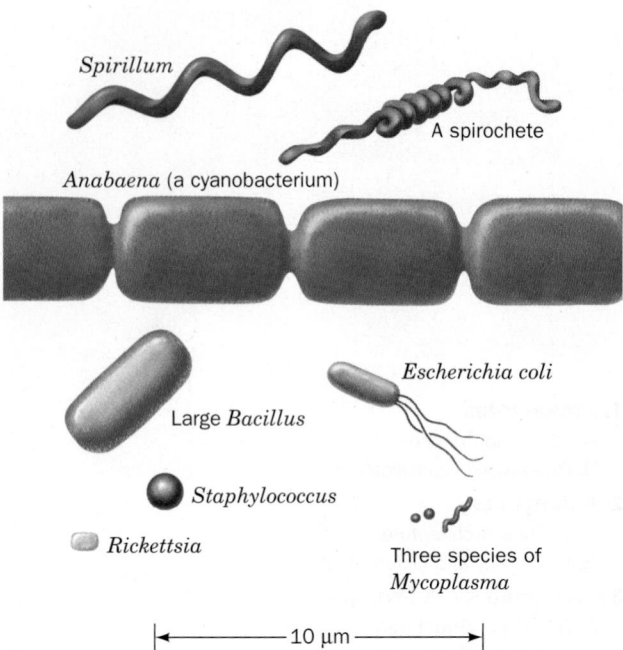

Figure 1-1 **Scale drawings of some prokaryotic cells.**

Many bacterial cells bear one or more whiplike appendages known as **flagella,** which are used for locomotion (Section 35-3I). Certain bacteria also have filamentous projections named **pili,** some types of which function as conduits for DNA during sexual conjugation (a process in which DNA is transferred from one cell to another; prokaryotes usually reproduce by binary fission) or aid in the attachment of the bacterium to a host organism's cells.

The bacterium ***Escherichia coli*** (abbreviated ***E. coli*** and named after its discoverer, Theodor Escherich) is the

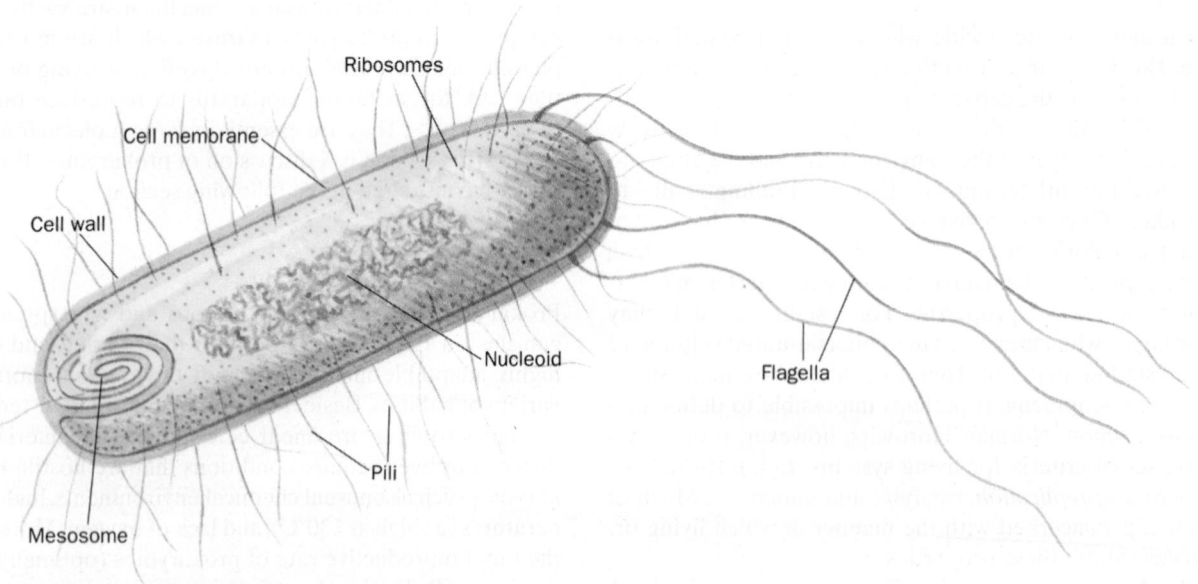

Figure 1-2 **Schematic diagram of a prokaryotic cell.**

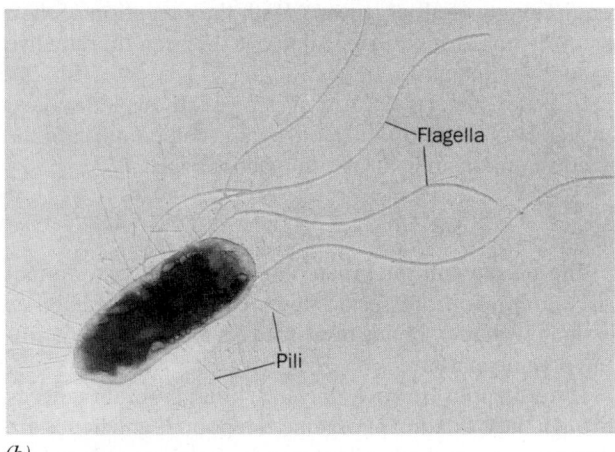

(a)   *(b)*

Figure 1-3   **Electron micrographs of *E. coli* cells.** (*a*) Stained to show internal structure. (*b*) Stained to reveal flagella and pili. [*a*: CNRI/Photo Researchers; *b*: Courtesy of Howard Berg, Harvard University.]

biologically most well-characterized organism as a result of its intensive biochemical and genetic study over the past 70 years. Indeed, much of the subject matter of this text deals with the biochemistry of *E. coli*. Cells of this normal inhabitant of the higher mammalian colon (Fig. 1-3) are typically 2-μm-long rods that are 1 μm in diameter and weigh ~2 × 10^{-12} g. Its DNA, which has a molecular mass of 2.5 × 10^9 **daltons (D),*** encodes ~4300 proteins (of which only ~60 to 70% have been identified), although, typically, only ~2600 different proteins are present in a cell at any given time. Altogether an *E. coli* cell contains 3 to 6 thousand different types of molecules, including proteins, nucleic acids, polysaccharides, lipids, and various small molecules and ions (Table 1-1).

b. Prokaryotes Employ a Wide Variety of Metabolic Energy Sources

The nutritional requirements of the prokaryotes are enormously varied. **Autotrophs** (Greek: *autos,* self + *trophikos,* to feed) can synthesize all their cellular constituents from simple molecules such as H_2O, CO_2, NH_3, and H_2S. Of course they need an energy source to do so as well as to power their other functions. **Chemolithotrophs** (Greek: *lithos,* stone) obtain their energy through the oxidation of inorganic compounds such as NH_3, H_2S, or even Fe^{2+}:

$$2\,NH_3 + 4\,O_2 \longrightarrow 2\,HNO_3 + 2\,H_2O$$
$$H_2S + 2\,O_2 \longrightarrow H_2SO_4$$
$$4\,FeCO_3 + O_2 + 6\,H_2O \longrightarrow 4\,Fe(OH)_3 + 4\,CO_2$$

Indeed, studies have revealed the existence of extensive al-

---

*The **molecular mass** of a particle may be expressed in units of daltons, which are defined as 1/12th the mass of a ^{12}C atom [atomic mass units (amu)]. Alternatively, this quantity may be expressed in terms of **molecular weight,** a dimensionless quantity defined as the ratio of the particle mass to 1/12th the mass of a ^{12}C atom and symbolized M_r (for relative molecular mass). In this text, we shall refer to the molecular mass of a particle rather than to its molecular weight.

Table 1-1   **Molecular Composition of *E. coli***

Component	Percentage by Weight
H_2O	70
Protein	15
Nucleic acids:	
DNA	1
RNA	6
Polysaccharides and precursors	3
Lipids and precursors	2
Other small organic molecules	1
Inorganic ions	1

Source: Watson, J.D., *Molecular Biology of the Gene* (3rd ed.), *p.* 69, Benjamin (1976).

beit extremely slow-growing colonies of chemolithotrophs that live as far as 5 kilometers underground and whose aggregate biomass appears to rival that of surface-dwelling organisms.

Photoautotrophs are autotrophs that obtain energy via **photosynthesis** (Chapter 24), a process in which light energy powers the transfer of electrons from inorganic donors to CO_2 yielding **carbohydrates** $[(CH_2O)_n]$. In the most widespread form of photosynthesis, the electron donor in the light-driven reaction sequence is H_2O.

$$n\,CO_2 + n\,H_2O \longrightarrow CH_2O_n + n\,O_2$$

This process is carried out by **cyanobacteria** (e.g., the green slimy organisms that grow on the walls of aquariums; cyanobacteria were formerly known as **blue-green algae**), as well as by plants. This form of photosynthesis is thought to have generated the O_2 in Earth's atmosphere. Some species of cyanobacteria have the ability to convert N_2 from the atmosphere to organic nitrogen compounds. This **nitrogen fixation** capacity gives them the simplest nutritional requirements of all organisms: With the exception of their need for small amounts of minerals, they can literally live on sunlight and air.

In a more primitive form of photosynthesis, substances such as H_2, H_2S, thiosulfate, or organic compounds are the electron donors in light-driven reactions such as

$$n\,CO_2 + 2n\,H_2S \longrightarrow CH_2O_n + n\,H_2O + 2n\,S$$

The **purple** and the **green photosynthetic bacteria** that carry out these processes occupy such oxygen-free habitats as shallow muddy ponds in which H_2S is generated by rotting organic matter.

Heterotrophs (Greek: *hetero,* other) obtain energy through the oxidation of organic compounds and hence are ultimately dependent on autotrophs for these substances. **Obligate aerobes** (which include animals) must utilize O_2, whereas **anaerobes** employ oxidizing agents such as sulfate **(sulfate-reducing bacteria)** or nitrate **(denitrifying bacteria).** Many organisms can partially metabolize various organic compounds in intramolecular oxidation–reduction processes known as **fermentation. Facultative anaerobes** such as *E. coli* can grow in either the presence or the absence of O_2. **Obligate anaerobes,** in contrast, are poisoned by the presence of O_2. Their metabolisms are thought to resemble those of the earliest life-forms (which arose over 3.8 billion years ago when Earth's atmosphere lacked O_2; Section 1-5B). At any rate, there are few organic compounds that cannot be metabolized by some prokaryotic organism.

B. *Prokaryotic Classification*

The traditional methods of **taxonomy** (the science of biological classification), which are based largely on the anatomical comparisons of both contemporary and fossil organisms, are essentially inapplicable to prokaryotes. This is because the relatively simple cell structures of prokaryotes, including those of ancient bacteria as revealed by their microfossil remnants, provide little indication of their phylogenetic relationships (**phylogenesis:** evolutionary development). Compounding this problem is the observation that prokaryotes exhibit little correlation between form and metabolic function. Moreover, the eukaryotic definition of a species as a population that can interbreed is meaningless for the asexually reproducing prokaryotes. Consequently, the conventional prokaryotic classification schemes are rather arbitrary and lack the implied evolutionary relationships of the eukaryotic classification scheme (Section 1-2B).

In the most widely used prokaryotic classification scheme, the **prokaryotae** (also known as **monera**) have two divisions: the cyanobacteria and the **bacteria.** The latter are further subdivided into 19 parts based on their various distinguishing characteristics, most notably cell structure, metabolic behavior, and staining properties.

A simpler classification scheme, which is based on cell wall properties, distinguishes three major types of prokaryotes: the **mycoplasmas,** the **gram-positive bacteria,** and the

gram-negative bacteria. Mycoplasmas lack the rigid cell wall of other prokaryotes. They are the smallest of all living cells (as small as 0.12 μm in diameter, Fig. 1-1) and possess ~20% of the DNA of an *E. coli.* Presumably this quantity of genetic information approaches the minimum amount necessary to specify the essential metabolic machinery required for cellular life. Gram-positive and gram-negative bacteria are distinguished according to whether or not they take up **gram stain** (a procedure developed in 1884 by Christian Gram in which heat-fixed cells are successively treated with the dye crystal violet and iodine and then destained with either ethanol or acetone). Gram-negative bacteria possess a complex **outer membrane** that surrounds their cell wall and excludes gram stain, whereas gram-positive bacteria lack such a membrane (Section 11-3B).

The development, in recent decades, of techniques for determining amino acid sequences in proteins (Section 7-1) and base sequences in nucleic acids (Section 7-2A) has provided abundant indications as to the genealogical relationships between organisms. Indeed, these techniques make it possible to place these relationships on a quantitative basis, and thus to construct a phylogenetically based classification system for prokaryotes.

By the analysis of ribosomal RNA sequences, Carl Woese showed that a group of prokaryotes he named the **Archaea** (also known as the **archaebacteria**) are as distantly related to the other prokaryotes, the **Bacteria** (also called the **eubacteria**), as both of these groups are to the **Eukarya** (the eukaryotes). The Archaea initially appeared to constitute three different kinds of unusual organisms: the **methanogens,** obligate anaerobes that produce methane (marsh gas) by the reduction of CO_2 with H_2; the **halobacteria,** which can live only in concentrated brine solutions ($>2M$ NaCl); and certain **thermoacidophiles,** organisms that inhabit acidic hot springs (~90°C and pH < 2). However, recent evidence indicates that ~40% of the microorganisms in the oceans are Archaea, and hence they may be the most common form of life on Earth.

On the basis of a number of fundamental biochemical traits that differ among the Archaea, the Bacteria, and the Eukarya, but that are common within each group, Woese proposed that these groups of organisms constitute the three primary **urkingdoms** or **domains** of evolutionary descent (rather than the traditional division into prokaryotes and eukaryotes). However, further sequence determinations have revealed that the Eukarya share sequence similarities with the Archaea that they do not share with the Bacteria. Evidently, the Archaea and the Bacteria diverged from some simple primordial life-form following which the Eukarya diverged from the Archaea, as the **phylogenetic tree** in Fig. 1-4 indicates.

2 EUKARYOTES

Eukaryotic cells are generally 10 to 100 μm in diameter and thus have a thousand to a million times the volume of typical prokaryotes. It is not size, however, but a profusion of membrane-enclosed organelles, each with a specialized

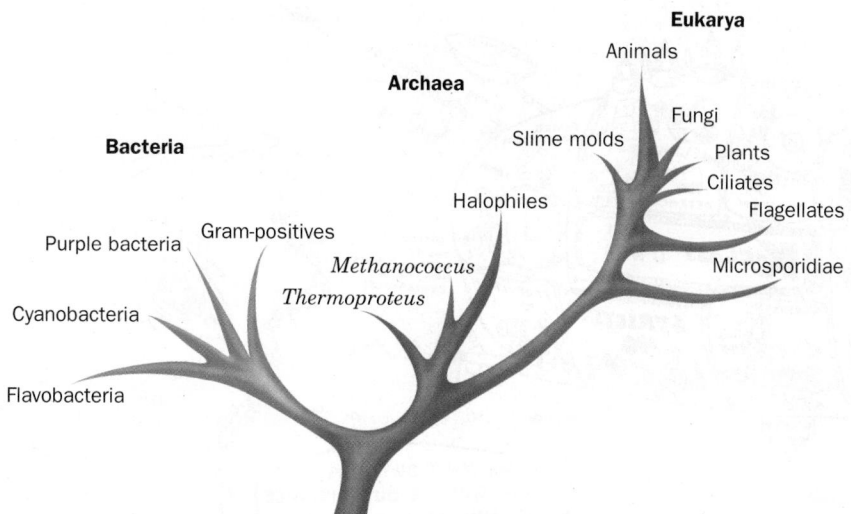

Figure 1-4 **Phylogenetic tree.** This "family tree" indicates the evolutionary relationships among the three domains of life. The root of the tree represents the last common ancestor of all life on Earth. [After Wheelis, M.L., Kandler, O., and Woese, C.R., *Proc. Natl. Acad. Sci.* **89,** 2931 (1992).]

function, that best characterizes eukaryotic cells (Fig. 1-5). In fact, *eukaryotic structure and function are more complex than those of prokaryotes at all levels of organization,* from the molecular level on up.

Eukaryotes and prokaryotes have developed according to fundamentally different evolutionary strategies.

Prokaryotes have exploited the advantages of simplicity and miniaturization: Their rapid growth rates permit them to occupy ecological niches in which there may be drastic fluctuations of the available nutrients. In contrast, the complexity of eukaryotes, which renders them larger and more slowly growing than prokaryotes, gives them the competitive

Figure 1-5 **Schematic diagram of an animal cell accompanied by electron micrographs of its organelles.** [Nucleus: Tektoff-RM, CNRI/Photo Researchers; rough endoplasmic reticulum: Pietro M. Motta & Tomonori Naguro/Photo Researchers, Inc. and Golgi apparatus: Secchi-Lecaque/Roussel-UCLAF/CNRI/Photo Researchers, Inc.; smooth endoplasmic reticulum: David M. Phillips/Visuals Unlimited; mitochondrion: CNRI/Photo Researchers; lysosome: Biophoto Associates/Photo Researchers.]

Figure 1-6 [Drawing by T.A. Bramley, in Carlile, M., *Trends Biochem. Sci.* **7,** 128 (1982). Copyright © Elsevier Biomedical Press, 1982. Used by permission.]

advantage in stable environments with limited resources (Fig. 1-6). It is therefore erroneous to consider prokaryotes as evolutionarily primitive with respect to eukaryotes. Both types of organisms are well adapted to their respective lifestyles.

The earliest known microfossils of eukaryotes date from ~1.4 billion years ago, some 2.4 billion years after life arose. This observation supports the classical notion that eukaryotes are descended from a highly developed prokaryote, possibly a mycoplasma. The differences between eukaryotes and modern prokaryotes, however, are so profound as to render this hypothesis improbable. Perhaps the early eukaryotes, which according to Woese's evidence evolved from a primordial life-form, were relatively unsuccessful and hence rare. Only after they had developed some of the complex organelles described in the following section did they become common enough to generate significant fossil remains.

A. Cellular Architecture

Eukaryotic cells, like prokaryotes, are bounded by a plasma membrane. The large size of eukaryotic cells results in their surface-to-volume ratios being much smaller than those of prokaryotes (the surface area of an object increases as the square of its radius, whereas volume does so as the cube). This geometrical constraint, coupled with the fact that many

essential enzymes are membrane associated, partially rationalizes the large amounts of intracellular membranes in eukaryotes (the plasma membrane typically constitutes <10% of the membrane in a eukaryotic cell). Since all the matter that enters or leaves a cell must somehow pass through its plasma membrane, the surface areas of many eukaryotic cells are increased by numerous projections and/or invaginations (Fig. 1-7). Moreover, portions of the plasma membrane often bud inward, in a process known as **endocytosis,** so that the cell surrounds portions of the external medium. Thus eukaryotic cells can engulf and digest food particles such as bacteria, whereas prokaryotes are limited to the absorption of individual nutrient molecules. The reverse of endocytosis, a process termed **exocytosis,** is a common eukaryotic secretory mechanism.

a. The Nucleus Contains the Cell's DNA

The nucleus, the eukaryotic cell's most conspicuous organelle, is the repository of its genetic information. This information is encoded in the base sequences of DNA molecules that form the discrete number of chromosomes characteristic of each species. The chromosomes consist of **chromatin,** a complex of DNA and protein. The amount of genetic information carried by eukaryotes is enormous; for example, a human cell has over 700 times the DNA of *E. coli* [in the terms commonly associated with computer memories, the **genome** (genetic complement) in each human

Figure 1-7 **Scanning electron micrograph of a fibroblast.**
[Courtesy of Guenther Albrecht-Buehler, Northwestern University.]

cell specifies around 800 megabytes of information—about 200 times the information content of this text]. Within the nucleus, the genetic information encoded by the DNA is transcribed into molecules of RNA (Chapter 31), which, after extensive processing, are transported to the cytoplasm (in eukaroytes, the cell contents exclusive of the nucleus), where they direct the ribosomal synthesis of proteins (Chapter 32). The nuclear envelope consists of a double membrane that is perforated by numerous ~90-Å-wide pores that regulate the flow of proteins and RNA between the nucleus and the cytoplasm.

The nucleus of most eukaryotic cells contains at least one dark-staining body known as the **nucleolus,** which is the site of ribosomal assembly. It contains chromosomal segments bearing multiple copies of genes specifying ribosomal RNA. These genes are transcribed in the nucleolus, and the resulting RNA is combined with ribosomal proteins that have been imported from their site of synthesis in the **cytosol** (the cytoplasm exclusive of its membrane-bound organelles). The resulting immature ribosomes are then exported to the cytosol, where their assembly is completed. Thus protein synthesis occurs almost entirely in the cytosol.

b. The Endoplasmic Reticulum and the Golgi Apparatus Function to Modify Membrane-Bound and Secretory Proteins

The most extensive membrane in the cell, which was discovered in 1945 by Keith Porter, forms a labyrinthine compartment named the **endoplasmic reticulum.** A large portion of this organelle, called the **rough endoplasmic reticulum,** is studded with ribosomes that are engaged in the synthesis of proteins that are either membrane-bound or destined for secretion. The **smooth endoplasmic reticulum,** which is devoid of ribosomes, is the site of lipid synthesis. Many of the products synthesized in the endoplasmic reticulum are eventually transported to the **Golgi**

apparatus (named after Camillo Golgi, who first described it in 1898), a stack of flattened membranous sacs in which these products are further processed (Section 23-3B).

c. Mitochondria Are the Site of Oxidative Metabolism

The **mitochondria** (Greek: *mitos,* thread + *chondros,* granule) are the site of cellular **respiration** (aerobic metabolism) in almost all eukaryotes. These cytoplasmic organelles, which are large enough to have been discovered by nineteenth century cytologists, vary in their size and shape but are often ellipsoidal with dimensions of around 1.0×2.0 μm—much like a bacterium. A eukaryotic cell typically contains on the order of 2000 mitochondria, which occupy roughly one-fifth of its total cell volume.

The mitochondrion, as the electron microscopic studies of George Palade and Fritjof Sjöstrand first revealed, has two membranes: a smooth outer membrane and a highly folded inner membrane whose invaginations are termed **cristae** (Latin: crests). Thus the mitochondrion contains two compartments, the **intermembrane space** and the internal **matrix space.** The enzymes that catalyze the reactions of respiration are components of either the gel-like **matrix** or the inner mitochondrial membrane. *These enzymes couple the energy-producing oxidation of nutrients to the energy-requiring synthesis of* **adenosine triphosphate** (**ATP;** Section 1-3B and Chapter 22). Adenosine triphosphate, after export to the rest of the cell, fuels its various energy-consuming processes.

Mitochondria are bacteria-like in more than size and shape. Their matrix space contains mitochondrion-specific DNA, RNA, and ribosomes that participate in the synthesis of several mitochondrial components. Moreover, they reproduce by binary fission, and the respiratory processes that they mediate bear a remarkable resemblance to those of modern aerobic bacteria. These observations led to the now widely accepted hypothesis championed by Lynn Margulis that mitochondria evolved from originally free-living gram-negative aerobic bacteria, which formed a symbiotic relationship with a primordial anaerobic eukaryote. The eukaryote-supplied nutrients consumed by the bacteria were presumably repaid severalfold by the highly efficient oxidative metabolism that the bacteria conferred on the eukaryote. This hypothesis is corroborated by the observation that the amoeba *Pelomyxa palustris,* one of the few eukaryotes that lack mitochondria, permanently harbors aerobic bacteria in such a symbiotic relationship.

d. Lysosomes and Peroxisomes Are Containers of Degradative Enzymes

Lysosomes, which were discovered in 1949 by Christian de Duve, are organelles bounded by a single membrane that are of variable size and morphology, although most have diameters in the range 0.1 to 0.8 μm. Lysosomes, which are essentially membranous bags containing a large variety of hydrolytic enzymes, function to digest materials ingested by endocytosis and to recycle cellular components (Section 32-6). Cytological investigations have revealed that lysosomes form by budding from the Golgi apparatus.

Peroxisomes (also known as **microbodies**) are membrane-enclosed organelles, typically 0.5 μm in diameter, that contain oxidative enzymes. They are so named because some peroxisomal reactions generate **hydrogen peroxide** (H_2O_2), a reactive substance that is either utilized in the enzymatic oxidation of other substances or degraded through a disproportionation reaction catalyzed by the enzyme **catalase:**

$$2 H_2O_2 \longrightarrow 2 H_2O + O_2$$

It is thought that peroxisomes function to protect sensitive cell components from oxidative attack by H_2O_2. Certain plants contain a specialized type of peroxisome, the **glyoxysome,** so named because it is the site of a series of reactions that are collectively termed the **glyoxylate pathway** (Section 23-2).

e. The Cytoskeleton Organizes the Cytosol

The cytosol, far from being a homogeneous solution, is a highly organized gel that can vary significantly in its composition throughout the cell. Much of its internal variability arises from the action of the **cytoskeleton,** an extensive array of filaments that gives the cell its shape and the ability to move and is responsible for the arrangement and internal motions of its organelles (Fig. 1-8).

The most conspicuous cytoskeletal components, the **microtubules,** are ~250-Å-diameter tubes that are composed of the protein **tubulin** (Section 35-3G). They form the supportive framework that guides the movements of organelles within a cell. For example, the **mitotic spindle** is an assembly of microtubules and associated proteins that participates in the separation of replicated chromosomes during cell division. Microtubules are also major constituents of **cilia,** the hairlike appendages extending from many cells, whose whiplike motions move the surrounding fluid past the cell or propel single cells through solution. Very long cilia, such as sperm tails, are termed **flagella** (prokaryotic flagella, which are composed of the protein **flagellin,** are quite different from and unrelated to those of eukaryotes).

The **microfilaments** are ~90-Å-diameter fibers that consist of the protein **actin.** Microfilaments, as do microtubules, have a mechanically supportive function. Furthermore, through their interactions with the protein **myosin,** microfilaments form contractile assemblies that are responsible for many types of intracellular movements such as cytoplasmic streaming and the formation of cellular protuberances or invaginations. More conspicuously, however, actin and myosin are the major protein components of muscle (Section 35-3A).

The third major cytoskeletal component, the **intermediate filaments,** are protein fibers that are 100 to 150 Å in diameter. Their prominence in parts of the cell that are subject to mechanical stress suggests that they have a load-bearing function. For example, skin in higher animals contains an extensive network of intermediate filaments made of the protein **keratin** (Section 8-2A), which is largely

(a) (b) (c) (d)

Figure 1-8 **Immunofluorescence micrographs showing cytoskeletal components.** Cells were treated with antibodies raised against (*a*) tubulin, (*b*) actin, (*c*) keratin, and (*d*) **vimentin** (a protein constituent of a type of intermediate filament) and then stained with fluorescently labeled antibodies that bound to the foregoing antibodies. [*a* and *d*: K.G. Murti/Visuals Unlimited; *b*: M. Schliwa/Visuals Unlimited; *c*: courtesy of Mary Osborn, Max Planck Institute for Biophysical Chemistry, Göttingen, Germany.]

responsible for the toughness of this protective outer covering. In contrast to the case with microtubules and microfilaments, the proteins forming intermediate filaments vary greatly in size and composition, both among the different cell types within a given organism and among the corresponding cell types in different organisms.

f. Plant Cells Are Enclosed by Rigid Cell Walls

Plant cells (Fig. 1-9) contain all of the previously described organelles. They also have several additional features, the most conspicuous of which is a rigid cell wall that lies outside the plasma membrane. These cell walls, whose major component is the fibrous polysaccharide **cellulose** (Section 11-2C), account for the structural strength of plants.

A **vacuole** is a membrane-enclosed space filled with fluid. Although vacuoles occur in animal cells, they are most prominent in plant cells, where they typically occupy 90% of the volume of a mature cell. Vacuoles function as

storage depots for nutrients, wastes, and specialized materials such as pigments. The relatively high concentration of solutes inside a plant vacuole causes it to take up water osmotically, thereby raising its internal pressure. This effect, combined with the cell walls' resistance to bursting, is largely responsible for the turgid rigidity of nonwoody plants.

g. Chloroplasts Are the Site of Photosynthesis in Plants

One of the definitive characteristics of plants is their ability to carry out photosynthesis. The site of photosynthesis is an organelle known as the **chloroplast,** which, although generally several times larger than a mitochondrion, resembles it in that both organelles have an inner and an outer membrane. Furthermore, the chloroplast's inner membrane space, the **stroma,** is similar to the mitochondrial matrix in that it contains many soluble enzymes. However, the inner chloroplast membrane is not folded

Figure 1-9 **Drawing of a plant cell accompanied by electron micrographs of its organelles.** [Plasmodesma: Courtesy of Hilton Mollenhauer, USDA; nucleus: Courtesy of Myron Ledbetter, Brookhaven National Laboratory; Golgi apparatus: Courtesy of W. Gordon Whaley, University of Texas; chloroplast: Courtesy of Lewis Shumway, College of Eastern Utah; amyloplast: Biophoto Associates; endoplasmic reticulum: Biophoto Associates/Photo Researchers.]

into cristae. Rather, the stroma encloses a third membrane system that forms interconnected stacks of disklike sacs called **thylakoids,** which contain the photosynthetic pigment **chlorophyll.** The thylakoid uses chlorophyll-trapped light energy to generate ATP, which is used in the stroma to drive biosynthetic reactions forming carbohydrates and other products (Chapter 24).

Chloroplasts, as do mitochondria, contain their own DNA, RNA, and ribosomes, and they reproduce by fission. Apparently chloroplasts, much like mitochondria, evolved from an ancient cyanobacterium that took up symbiotic residence in an ancestral nonphotosynthetic eukaryote. In fact, several modern nonphotosynthetic eukaryotes have just such a symbiotic relationship with authentic cyanobacteria. Hence *most modern eukaryotes are genetic "mongrels" in that they simultaneously have nuclear, mitochondrial, and in the case of plants, chloroplast lines of descent.*

B. *Phylogeny and Differentiation*

One of the most remarkable characteristics of eukaryotes is their enormous morphological diversity, on both the cellular and organismal levels. Compare, for example, the architectures of the various human cells drawn in Fig. 1-10. Similarly, recall the great anatomical differences among, say, an amoeba, an oak tree, and a human being.

Taxonometric schemes based on gross morphology as well as on protein and nucleic acid sequences (Sections 7-1 and 7-2) indicate that eukaryotes may be classified into three kingdoms: **Fungi, Plantae** (plants), and **Animalia** (animals). The relative structural simplicity of many unicellular eukaryotes, however, makes their classification under this scheme rather arbitrary. Consequently, these organisms are usually assigned a fourth eukaryotic kingdom, the **Protista.** (Note that biological classification schemes are for the convenience of biologists; nature is rarely neatly categorized.) Figure 1-11 is a phylogenetic tree for eukaryotes.

Anatomical comparisons among living and fossil organisms indicate that the various kingdoms of multicellular organisms independently evolved from Protista (Fig. 1-11). The programs of growth, differentiation, and development followed by multicellular animals (the **metazoa**) in their transformation from fertilized ova to adult organisms provide a remarkable indication of this evolutionary history. For example, all vertebrates exhibit gill-like pouches in their early embryonic stages, which presumably reflect their common fish ancestry (Fig. 1-12). Indeed, these early embryos are similar in size and anatomy even though their respective adult forms are vastly different in these characteristics. Such observations led Ernst Haeckel to formulate his famous (although overstated) dictum: *Ontogeny recapitulates phylogeny* (ontogeny: biological development).

Figure 1-10 **Drawings of some human cells.** (*a*) An osteocyte (bone cell), (*b*) a sperm, (*c*) a pancreatic acinar cell (which secretes digestive enzymes), and (*d*) a neuron (nerve cell).

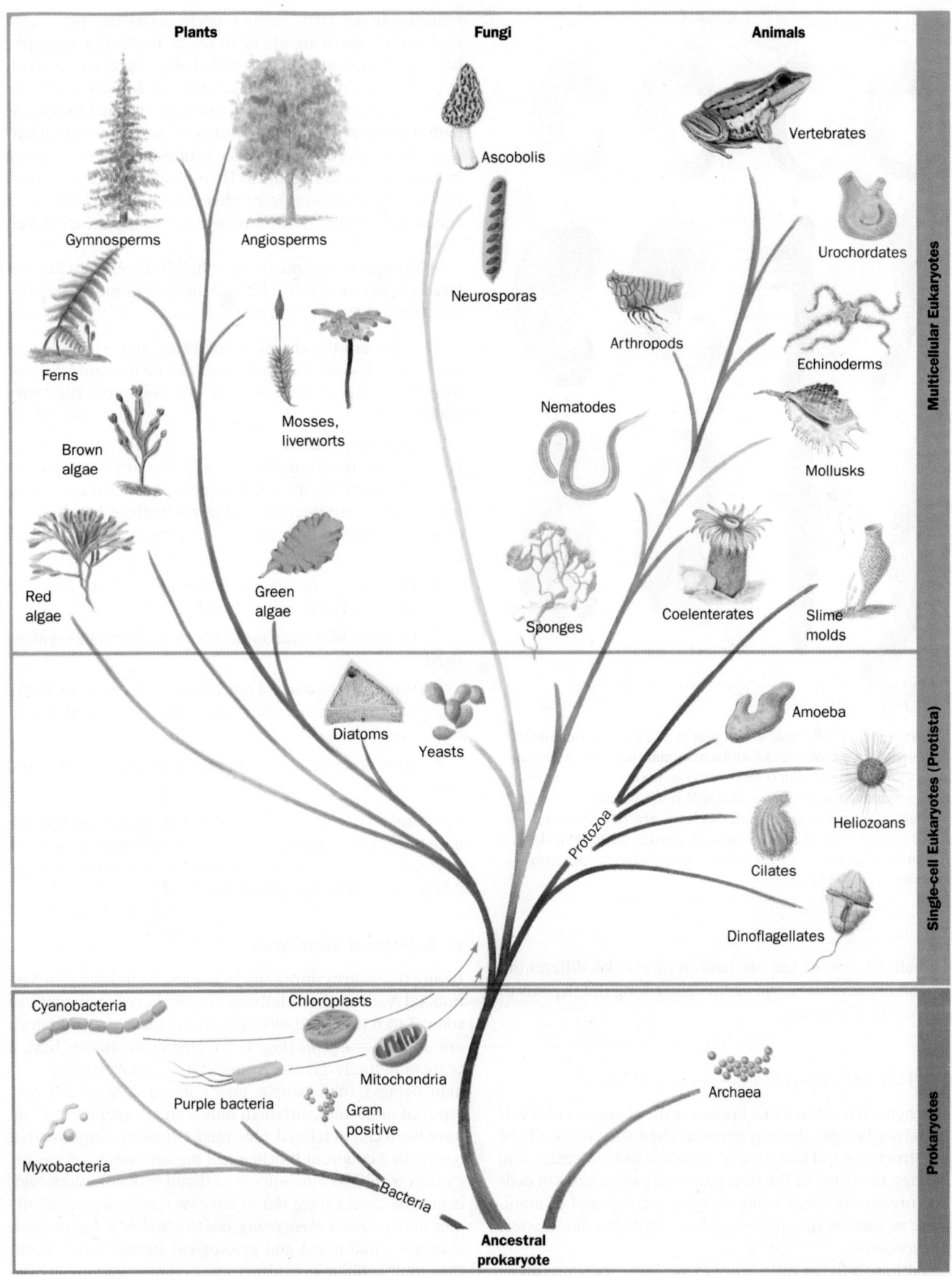

Figure 1-11 **Evolutionary tree indicating the lines of descent of cellular life on Earth.**

Gill pouches

Fish   Salamander   Chick   Human

Figure 1-12 **Embryonic development of a fish, an amphibian (salamander), a bird (chick), and a mammal (human).** At early stages they are similar in both size and anatomy (the top drawings have around the same scale), although it is now known that their similarites are not as great as these classic drawings indicate. Later they diverge in both of these properties. [After Haeckel, E., *Anthropogenie oder Entwickelungsgeschichte des Menschen,* Engelmann (1874).]

The elucidation of the mechanism of cellular differentiation in eukaryotes is one of the major long-range goals of modern biochemistry.

3 BIOCHEMISTRY: A PROLOGUE

Biochemistry, as the name implies, is the chemistry of life. It therefore bridges the gap between chemistry, the study of the structures and interactions of atoms and molecules, and biology, the study of the structures and interactions of cells and organisms. Since living things are composed of inanimate molecules, *life, at its most basic level, is a biochemical phenomenon.*

Although living organisms, as we have seen, are enormously diverse in their macroscopic properties, there is a remarkable similarity in their biochemistry that provides a unifying theme with which to study them. For example, hereditary information is encoded and expressed in an almost identical manner in all cellular life. Moreover, the series of biochemical reactions, which are termed **metabolic pathways,** as well as the structures of the enzymes that catalyze them are, for many basic processes, nearly identical from organism to organism. This strongly suggests that all known life-forms are descended from a single primordial ancestor in which these biochemical features first developed.

Although biochemistry is a highly diverse field, it is largely concerned with a limited number of interrelated issues. These are

1. What are the chemical and three-dimensional structures of biological molecules and assemblies, how do they form these structures, and how do their properties vary with them?

2. How do proteins work; that is, what are the molecular mechanisms of enzymatic catalysis, how do receptors recognize and bind specific molecules, and what are the intramolecular and intermolecular mechanisms by which receptors transmit information concerning their binding states?

3. How is genetic information expressed and how is it transmitted to future cell generations?

4. How are biological molecules and assemblies synthesized?

5. What are the control mechanisms that coordinate the myriad biochemical reactions that take place in cells and in organisms?

6. How do cells and organisms grow, differentiate, and reproduce?

These issues are previewed in this section and further illuminated in later chapters. However, as will become obvious as you read further, in all cases, our knowledge, extensive as it is, is dwarfed by our ignorance.

A. *Biological Structures*

Living things are enormously complex. As indicated in Section 1-1A, even the relatively simple *E. coli* cell contains some 3 to 6 thousand different compounds, most of which are unique to *E. coli* (Fig. 1-13). Higher organisms have a correspondingly greater complexity. ***Homo sapiens*** (human beings), for example, may contain 100,000 different types of molecules, although only a small fraction of them have been characterized. One might therefore suppose that to obtain a coherent biochemical understanding of any organism would be a hopelessly difficult task. This, however, is not the case. *Living things have an underlying regularity that derives from their being constructed in a hierarchical manner.* Anatomical and cytological studies have shown that multicellular organisms are organizations of organs, which are made of tissues consisting of cells, composed of

Figure 1-13 **Simulated cross section of an *E. coli* cell magnified around one millionfold.** The right side of the drawing shows the multilayered cell wall and membrane, decorated on its exterior surface with lipopolysaccharides (Section 11-3Bc). A flagellum *(lower right)* is driven by a motor anchored in the inner membrane (Section 35-3I). The cytoplasm, which occupies the middle region of the drawing, is predominantly filled with ribosomes engaged in protein synthesis (Section 32-3). The left side of the drawing contains a dense tangle of DNA in complex with specific proteins. Only the largest macromolecules and molecular assemblies are shown. In a living cell, the remaining space in the cytoplasm would be crowded with smaller molecules and water (a water molecule would be about the size of the period at the end of this sentence). [After a drawing by David Goodsell, UCLA.]

subcellular organelles (e.g., Fig. 1-14). At this point in our hierarchical descent, we enter the biochemical realm since organelles consist of **supramolecular assemblies,** such as membranes or fibers, that are organized clusters of **macromolecules** (polymeric molecules with molecular masses from thousands of daltons on up).

As Table 1-1 indicates, *E. coli,* and living things in general, contain only a few different types of macromolecules: **proteins** (Greek: *proteios,* of first importance; a term coined in 1838 by Jacob Berzelius), **nucleic acids,** and **polysaccharides** (Greek: *sakcharon,* sugar). *All of these substances have a modular construction; they consist of linked monomeric units that occupy the lowest level of our structural hierarchy.* Thus, as Fig. 1-15 indicates, proteins are polymers of amino acids (Section 4-1B), nucleic acids are polymers of nucleotides (Section 5-1), and polysaccharides are polymers of sugars (Section 11-2). **Lipids** (Greek: *lipos,*

fat), the fourth major class of biological molecules, are too small to be classified as macromolecules but also have a modular construction (Section 12-1).

The task of the biochemist has been vastly simplified by the finding that *there are relatively few species of monomeric units that occur in each class of biological macromolecule.* Proteins are all synthesized from the same 20 species of **amino acids,** nucleic acids are made from 8 types of **nucleotides** (4 each in DNA and RNA), and there are ~8 commonly occurring types of **sugars** in polysaccharides. The great variation in properties observed among macromolecules of each type largely arises from the enormous number of ways its monomeric units can be arranged and, in many cases, derivatized.

One of the central questions in biochemistry is how biological structures are formed. As is explained in later chapters, the monomeric units of macromolecules are either

(a) Organism: human being

├─────1 m─────┤

(b) Organ: skin   ├─1 mm─┤

(c) Tissue: epidermis   ├100 µm┤

(d) Cell: basal cell

(e) Organelle: mitochondrion   ├5 µm┤

├─1 µm─┤

(f) Supramolecular assembly:
inner mitochondrial membrane

Lipid

Protein

├─100 Å─┤

(g) Macromolecule:
cytochrome *c*

Polypeptide
chain

Heme

├─10 Å─┤

Figure 1-14   **Example of the hierarchical organization of biological structures.**

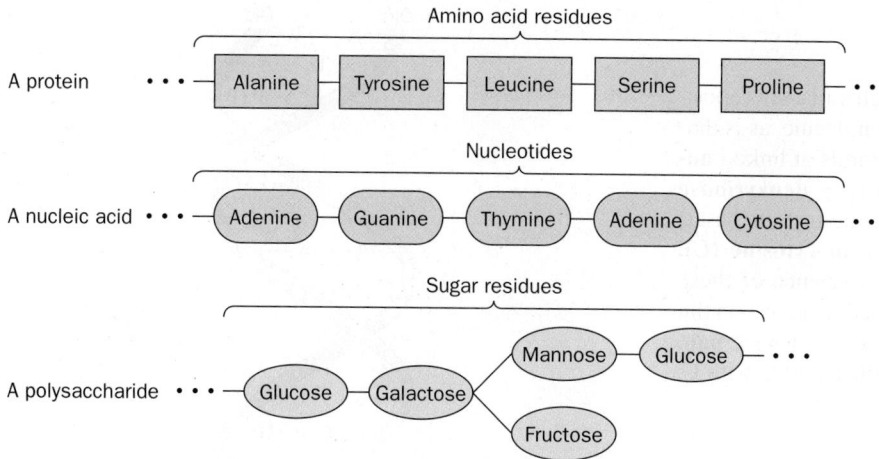

Figure 1-15 **Polymeric organization of proteins, nucleic acids, and polysaccharides.**

directly acquired by the cell as nutrients or enzymatically synthesized from simpler substances. Macromolecules are synthesized from their monomeric precursors in complex enzymatically mediated processes.

Newly synthesized proteins spontaneously fold to assume their native conformations (Section 9-1A); that is, they undergo **self-assembly.** Apparently their amino acid sequences specify their three-dimensional structures. Likewise, the structures of other types of macromolecules are specified by the sequences of their monomeric units. The principle of self-assembly extends at least to the level of supramolecular assemblies. However, the way in which higher levels of biological structures are generated is largely unknown. The elucidation of the mechanisms of cellular and organismal growth and differentiation is a major area of biological research.

B. *Metabolic Processes*

There is a bewildering array of chemical reactions that simultaneously occur in any living cell. Yet these reactions follow a pattern that organizes them into the coherent process we refer to as life. For instance, most biological reactions are members of a metabolic pathway; that is, they function as one of a sequence of reactions that produce one or more specific products. Moreover, one of the hallmarks of life is that the rates of its reactions are so tightly regulated that there is rarely an unsatisfied need for a reactant in a metabolic pathway or an unnecessary buildup of some product.

Metabolism has been traditionally (although not necessarily logically) divided into two major categories:

1. Catabolism or degradation, in which nutrients and cell constituents are broken down so as to salvage their components and/or to generate energy.

2. Anabolism or biosynthesis, in which biomolecules are synthesized from simpler components.

The energy required by anabolic processes is provided by catabolic processes largely in the form of **adenosine**

triphosphate (ATP). For instance, such energy-generating processes as photosynthesis and the biological oxidation of nutrients produce ATP from **adenosine diphosphate (ADP)** and a phosphate ion.

Adenosine diphosphate (ADP)

Adenosine triphosphate (ATP)

Conversely, such energy-consuming processes as biosynthesis, the transport of molecules against a concentration gradient, and muscle contraction are driven by the reverse of this reaction, the hydrolysis of ATP:

$$ATP + H_2O \rightleftharpoons ADP + HPO_4^{2-}$$

Thus, *anabolic and catabolic processes are coupled together through the mediation of the universal biological energy "currency," ATP.*

C. *Expression and Transmission of Genetic Information*

Deoxyribonucleic acid (DNA) is the cell's master repository of genetic information. This macromolecule, as is diagrammed in Fig. 1-16, consists of two strands of linked **nucleotides,** each of which is composed of a **deoxyribose** sugar residue, a phosphoryl group, and one of four bases: **adenine (A), thymine (T), guanine (G), or cytosine (C).** Genetic information is encoded in the sequence of these bases. Each DNA base is hydrogen bonded to a base on the opposite strand to form an entity known as a **base pair.** However, A can only hydrogen bond with T, and G with C,

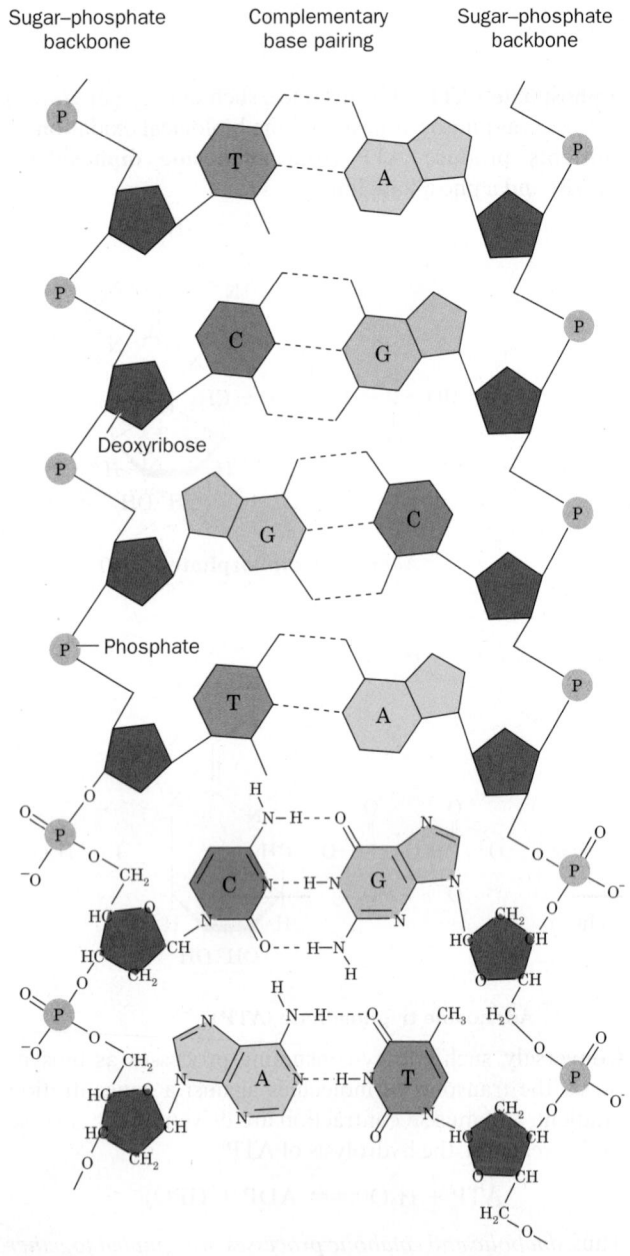

Sugar–phosphate backbone	Complementary base pairing	Sugar–phosphate backbone

Figure 1-16   Double-stranded DNA. The two polynucleotide chains associate through complementary base pairing. A pairs with T, and G pairs with C by forming specific hydrogen bonds.

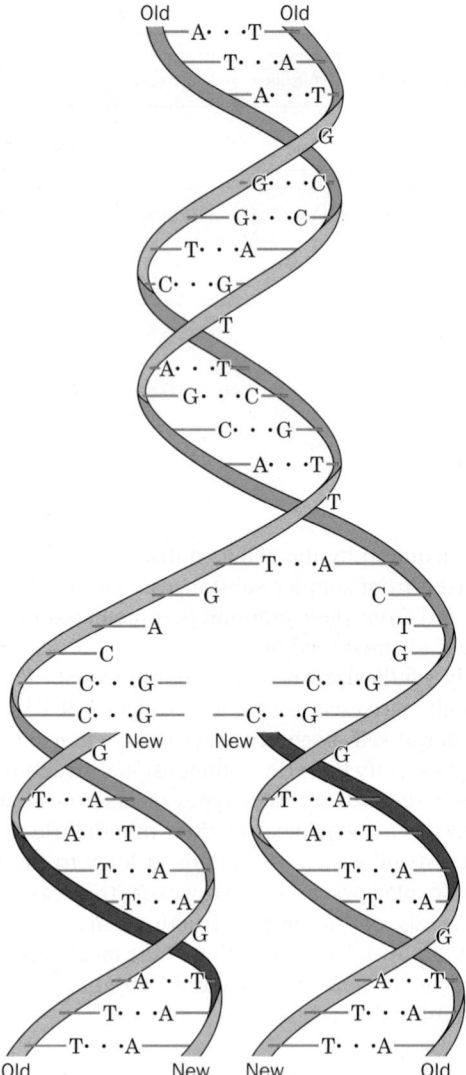

Figure 1-17   Schematic diagram of DNA replication. Each strand of parental DNA (*red*) acts as a template for the synthesis of a complementary daughter strand (*green*). Consequently, the resulting double-stranded molecules are identical.

so that the two strands are **complementary;** that is, the sequence of one strand implies the sequence of the other.

The division of a cell must be accompanied by the replication of its DNA. In this enzymatically mediated process, each DNA strand acts as a template for the formation of its complementary strand (Fig. 1-17; Section 5-4C). Consequently, every progeny cell contains a complete DNA molecule (or set of DNA molecules), each of which consists of one parental strand and one daughter strand. Mutations arise when, through rare copying errors or damage to a parental strand, one or more wrong bases are incorporated into a daughter strand. Most mutations are either innocuous or deleterious. Occasionally, however, one results in a new characteristic that confers some sort of selective advantage on its recipient. Individuals with such mutations, according to the tenets of the Darwinian theory of evolution, have an increased probability of reproducing. New species arise through a progression of such mutations.

The expression of genetic information is a two-stage process. In the first stage, which is termed **transcription,** a DNA strand serves as a template for the synthesis of a complementary strand of ribonucleic acid (RNA; Section 31-2). This nucleic acid, which is generally single stranded, differs chemically from DNA (Fig. 1-16) only in that it has **ribose** sugar residues in place of DNA's deoxyribose and **uracil (U)** replacing DNA's thymine base.

Ribose   **Uracil**

In the second stage of genetic expression, which is known as **translation,** ribosomes enzymatically link together amino acids to form proteins (Section 32-3). The order in which the amino acids are linked together is prescribed by the RNA's sequence of bases. Consequently, since proteins are self-assembling, the genetic information encoded by DNA serves, through the intermediacy of RNA, to specify protein structure and function. Just which genes are expressed in a given cell under a particular set of circumstances is controlled by complex regulatory systems whose workings are understood only in outline.

4 GENETICS: A REVIEW

One has only to note the resemblance between parent and child to realize that physical traits are inherited. Yet the mechanism of inheritance was unknown until the mid-twentieth century. The theory of **pangenesis,** which originated with the ancient Greeks, held that semen, which clearly has something to do with procreation, consists of representative particles from all over the body **(pangenes).** This idea was extended in the late eighteenth century by Jean Baptiste de Lamarck, who, in a theory known as **Lamarckism,** hypothesized that an individual's acquired characteristics, such as large muscles resulting from exercise, would be transmitted to his/her offspring. Pangenesis and at least some aspects of Lamarckism were accepted by most nineteenth century biologists, including Charles Darwin.

The realization, in the mid-nineteenth century, that all organisms are derived from single cells set the stage for the development of modern biology. In his **germ plasm theory,** August Weismann pointed out that sperm and ova, the **germ cells** (whose primordia are set aside early in embryonic development), are directly descended from the germ cells of the previous generation and that other cells of the body, the **somatic cells,** although derived from germ cells, do not give rise to them. He refuted the ideas of pangenesis and Lamarckism by demonstrating that the progeny of many successive generations of mice whose tails had been cut off had tails of normal length.

Figure 1-18 **Chromosomes.** A photomicrograph of a plant cell (*Scadoxus katherinae* Bak.) during anaphase of mitosis showing its chromosomes being pulled to opposite poles of the cell by the mitotic spindle. The microtubules forming the mitotic spindle are stained red and the chromosomes are blue. [Courtesy of Andrew S. Bajer, University of Oregon.]

A. *Chromosomes*

In the 1860s, eukaryotic cell nuclei were observed to contain linear bodies that were named chromosomes (Greek: *chromos,* color + *soma,* body) because they are strongly stained by certain basic dyes (Fig. 1-18). There are normally two copies of each chromosome **(homologous pairs)** present in every somatic cell. The number of unique chromosomes (N) in such a cell is known as its **haploid number,** and the total number of chromosomes ($2N$) is its **diploid number.** Different species differ in their haploid number of chromosomes (Table 1-2).

Table 1-2 **Number of Chromosomes ($2N$) in Some Eukaryotes**

Organism	Chromosomes
Human	46
Dog	78
Rat	42
Turkey	82
Frog	26
Fruit fly	8
Hermit crab	~254
Garden pea	14
Potato	48
Yeast	34
Green alga	~20

Source: Ayala, F.J. and Kiger, J.A., Jr., *Modern Genetics* (2nd ed.), *p.* 9, Benjamin/Cummings (1984).

a. Somatic Cells Divide by Mitosis

The division of somatic cells, a process known as **mitosis** (Fig. 1-19), is preceded by the duplication of each chromosome to form a cell with $4N$ chromosomes. During cell division, each chromosome attaches by its **centromere** to the **mitotic spindle** such that the members of each duplicate pair line up across the equatorial plane of the cell. The members of each duplicate pair are then pulled to opposite poles of the dividing cell by the action of the spindle to yield diploid daughter cells that each have the same $2N$ chromosomes as the parent cell.

b. Germ Cells Are Formed by Meiosis

The formation of germ cells, a process known as **meiosis** (Fig. 1-20), requires two consecutive cell divisions. Before the first meiotic division each chromosome replicates, but the resulting sister **chromatids** remain attached at their centromere. The homologous pairs of the doubled chromosomes then line up across the equatorial plane of the cell in zipperlike fashion, which permits an exchange of the corresponding sections of homologous chromosomes in a process known as **crossing-over.** The spindle then moves the members of each homologous pair to opposite poles of the cell so that, after the first meiotic division, each daughter cell contains N doubled chromosomes. In the second meiotic division, the sister chromatids separate to form chromosomes and move to opposite poles of the dividing cell to yield a total of four haploid cells that are known as **gametes.** Fertilization consists of the fusion of a male gamete (sperm) with a female gamete (ovum) to yield a diploid cell known as a **zygote** that has received N chromosomes from each of its parents.

B. *Mendelian Inheritance*

The basic laws of inheritance were reported in 1866 by Gregor Mendel. They were elucidated by the analysis of a series of **genetic crosses** between true-breeding strains (producing progeny that have the same characteristics as the parents) of garden peas, *Pisum sativum,* that differ in certain well-defined traits such as seed shape (round vs wrinkled), seed color (yellow vs green), or flower color (purple vs white). Mendel found that in crossing parents (P) that differ in a single trait, say seed shape, the progeny (F_1; first filial generation) all have the trait of only one of the parents, in this case round seeds (Fig. 1-21). The trait appearing in F_1 is said to be **dominant,** whereas the alternative trait is called **recessive.** In F_2, the progeny of F_1, three-quarters have the dominant trait and one-quarter have the recessive trait. Those peas with the recessive trait breed true; that is, self-crossing recessive F_2's results in progeny (F_3) that also have the recessive trait. The F_2's exhibiting the dominant trait, however, fall into two categories: One-third of them breed true, whereas the remainder have progeny with the same 3:1 ratio of dominant to recessive traits as do the members of F_2.

Mendel accounted for his observations by hypothesizing that *the various pairs of contrasting traits each result from a factor (now called a **gene**) that has alternative forms*

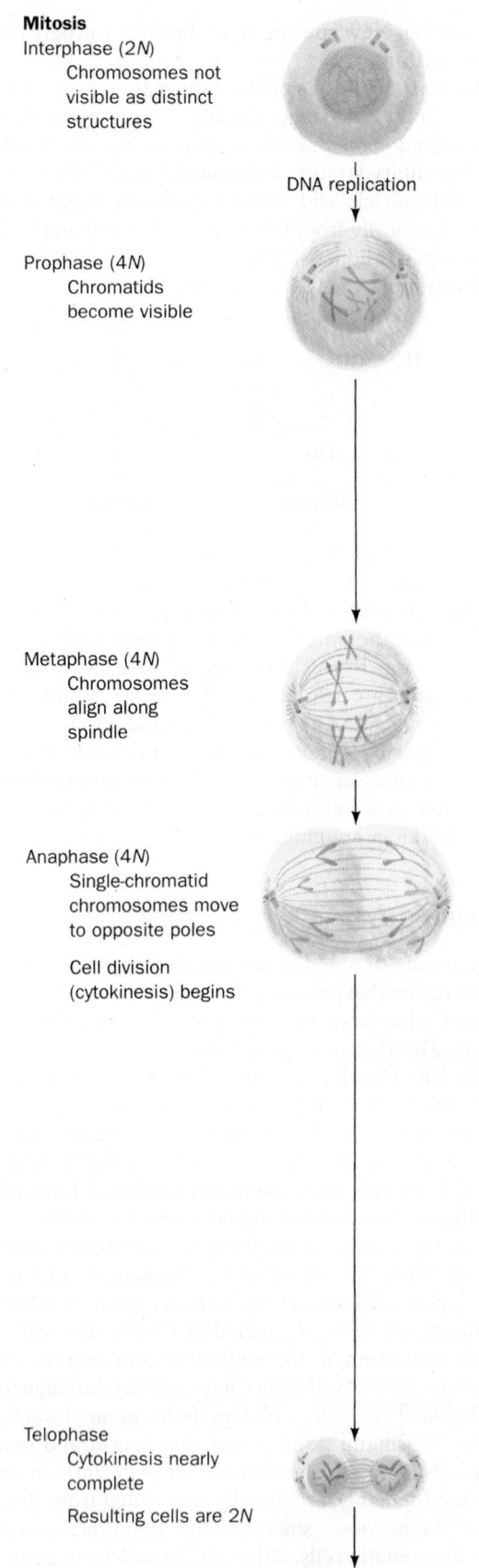

Mitosis

Interphase ($2N$)
    Chromosomes not visible as distinct structures

DNA replication

Prophase ($4N$)
    Chromatids become visible

Metaphase ($4N$)
    Chromosomes align along spindle

Anaphase ($4N$)
    Single-chromatid chromosomes move to opposite poles

    Cell division (cytokinesis) begins

Telophase
    Cytokinesis nearly complete

    Resulting cells are $2N$

Cell division

Figure 1-19 **Mitosis, the usual form of cell division in eukaryotes.** Mitosis yields two daughter cells, each with the same chromosomal complement as the parental cell.

Meiosis

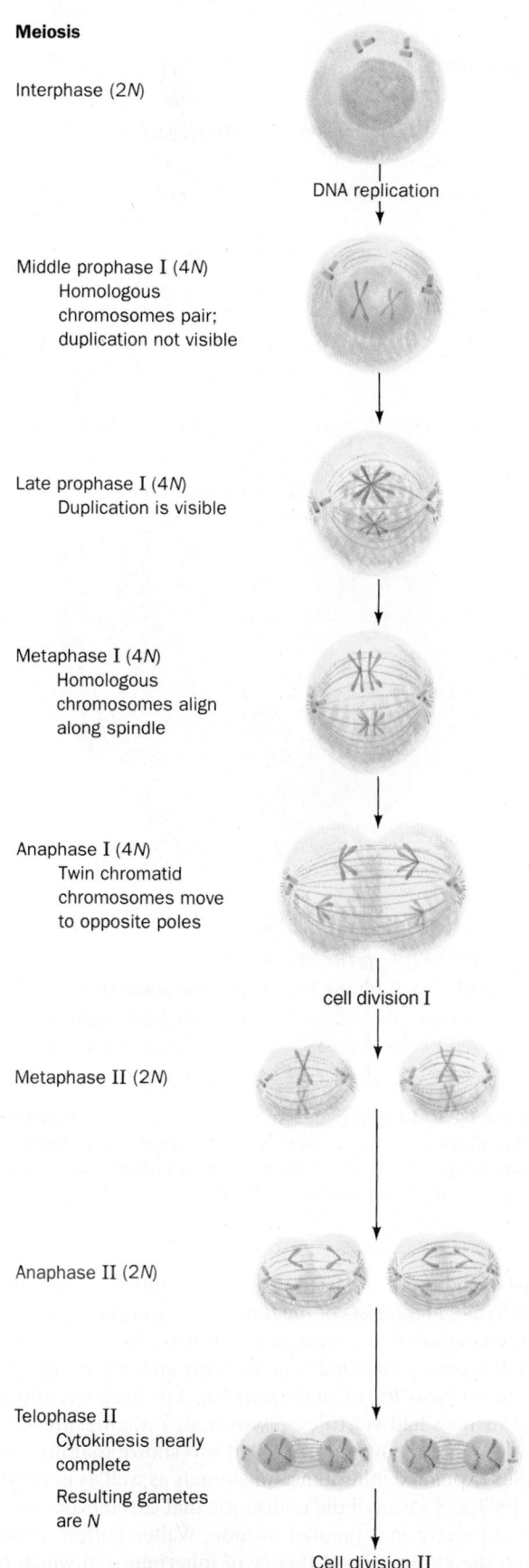

Interphase (2*N*)

DNA replication

Middle prophase I (4*N*)
Homologous
chromosomes pair;
duplication not visible

Late prophase I (4*N*)
Duplication is visible

Metaphase I (4*N*)
Homologous
chromosomes align
along spindle

Anaphase I (4*N*)
Twin chromatid
chromosomes move
to opposite poles

cell division I

Metaphase II (2*N*)

Anaphase II (2*N*)

Telophase II
Cytokinesis nearly
complete

Resulting gametes
are *N*

Cell division II

Figure 1-20 **Meiosis, which leads to the formation of gametes (sex cells).** Meiosis comprises two consecutive cell divisions to yield four daughter cells, each with half of the chromosomal complement of its parental cell.

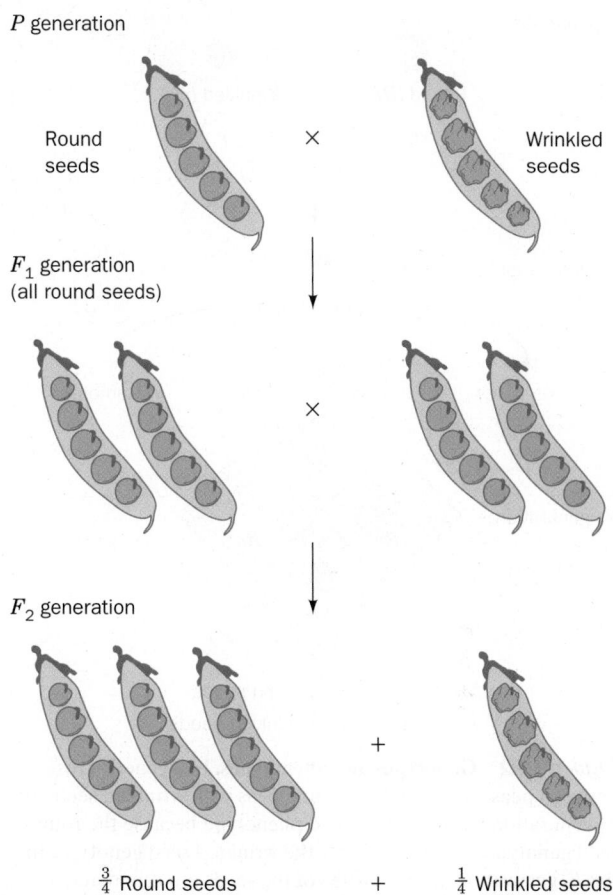

P generation

Round
seeds

×

Wrinkled
seeds

F$_1$ generation
(all round seeds)

×

F$_2$ generation

+

$\frac{3}{4}$ Round seeds + $\frac{1}{4}$ Wrinkled seeds

Figure 1-21 **Genetic crosses.** Crossing a pea plant that has round seeds with one that has wrinkled seeds yields *F*$_1$ progeny that all have round seeds. Crossing these *F*$_1$ peas yields an *F*$_2$ generation, of which three-quarters have round seeds and one-quarter have wrinkled seeds.

(alleles). Every plant contains a pair of genes governing a particular trait, one inherited from each of its parents. The alleles for seed shape are symbolized *R* for round seeds and *r* for wrinkled seeds (gene symbols are generally given in italics). The pure-breeding plants with round and wrinkled seeds, respectively, have *RR* and *rr* **genotypes** (genetic composition) and are both said to be **homozygous** in seed shape. Plants with the *Rr* genotype are **heterozygous** in seed shape and have the round seed **phenotype** (appearance or character) because *R* is dominant over *r*. *The two alleles do not blend or mix in any way in the plant and are independently transmitted through gametes to progeny (Fig. 1-22).*

Mendel also found that *different traits are independently inherited.* For example, crossing peas that have round yellow seeds (*RRYY*) with peas that have wrinkled green seeds (*rryy*) results in *F*$_1$ progeny (*RrYy*) that have round yellow seeds (yellow seeds are dominant over green seeds). The *F*$_2$ phenotypes appear in the ratio 9 round yellow : 3 round green : 3 wrinkled yellow : 1 wrinkled green. This result indicates that there is no tendency for the genes from

P generation

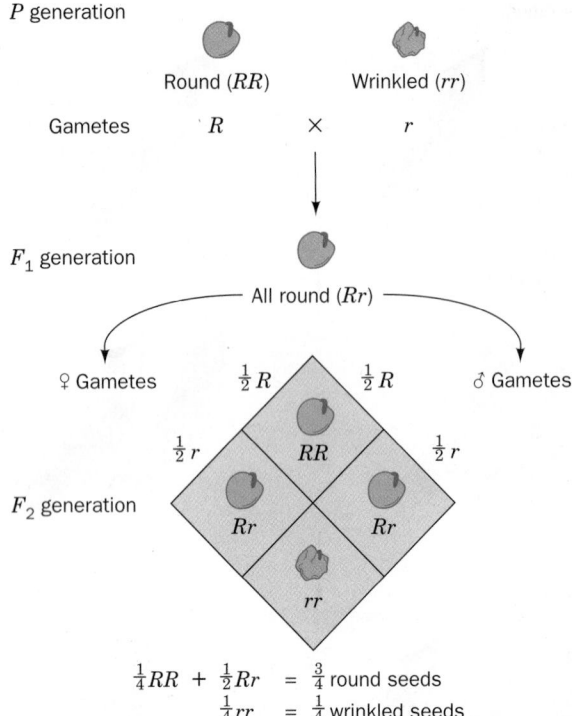

Round (*RR*)   Wrinkled (*rr*)

Gametes   *R*   ×   *r*

F_1 generation

All round (*Rr*)

♀ Gametes $\frac{1}{2}R$ $\frac{1}{2}R$ ♂ Gametes

F_2 generation

$$\frac{1}{4}RR + \frac{1}{2}Rr = \frac{3}{4} \text{ round seeds}$$
$$\frac{1}{4}rr = \frac{1}{4} \text{ wrinkled seeds}$$

Figure 1-22 **Genotypes and phenotypes.** In a genetic cross between peas with round seeds and peas with wrinkled seeds, the F_1 generation has the round seed phenotype because the round seed genotype is dominant over the wrinkled seed genotype. In the F_2 generation, three-fourths of the seeds are round and one-fourth are wrinkled because the genes for these alleles are independently transmitted by haploid gametes.

any parent to assort together (Fig. 1-23). It was later shown, however, *that only genes that occur on different chromosomes exhibit such independence.*

    The dominance of one trait over another is a common but not universal phenomenon. For example, crossing a pure-breeding red variety of the snapdragon *Antirrhinum* with a pure-breeding white variety results in pink-colored F_1 progeny. The F_2 progeny have red, pink, and white flowers in a 1:2:1 ratio because the flowers of homozygotes for the red color (*AA*) contain more red pigment than do the heterozygotes (*Aa*; Fig. 1-24). The red and white traits are therefore said to be **codominant.** In the case of codominance, the phenotype reveals the genotype.

    A given gene may have multiple alleles. A familiar example is the human **ABO blood group system** (Section 12-3E). A person may have type A, type B, type AB, or type O blood depending on whether his/her red blood cells bear A antigens, B antigens, both, or neither. The A and B antigens are specified by the codominant I^A and I^B alleles, respectively, and the O phenotype is homozygous for the recessive *i* allele.

C. *Chromosomal Theory of Inheritance*

Mendel's theory of inheritance was almost universally ignored by his contemporaries. This was partially because in

---

P generation

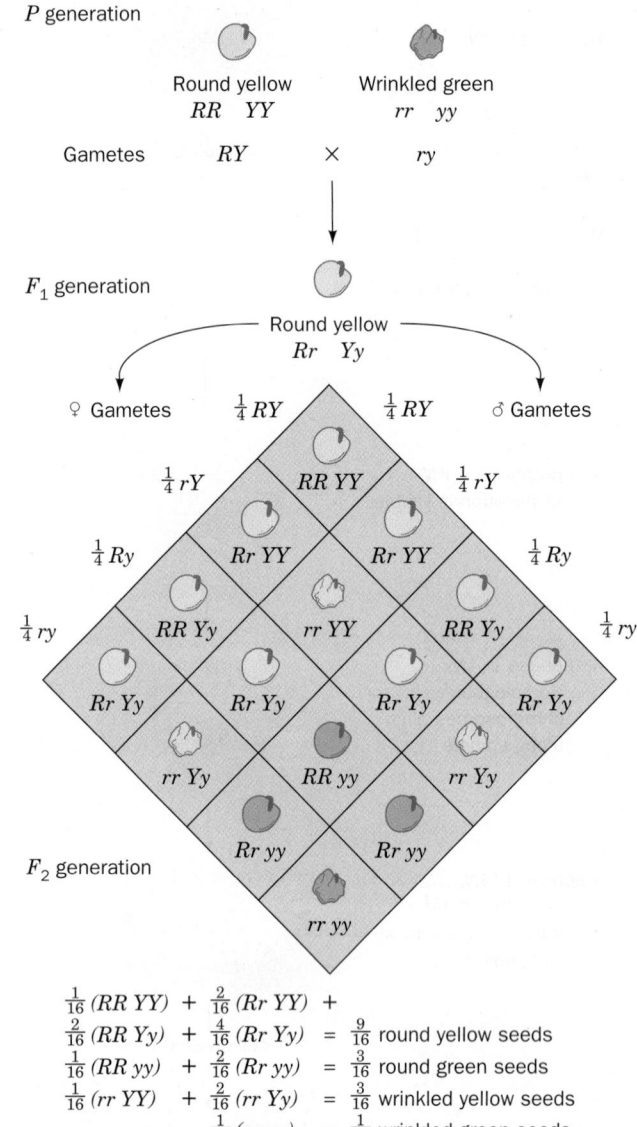

Round yellow   Wrinkled green
RR YY   *rr yy*

Gametes   *RY*   ×   *ry*

F_1 generation

Round yellow
Rr Yy

♀ Gametes $\frac{1}{4}RY$ $\frac{1}{4}RY$ ♂ Gametes

F_2 generation

$$\frac{1}{16}(RR\,YY) + \frac{2}{16}(Rr\,YY) +$$
$$\frac{2}{16}(RR\,Yy) + \frac{4}{16}(Rr\,Yy) = \frac{9}{16} \text{ round yellow seeds}$$
$$\frac{1}{16}(RR\,yy) + \frac{2}{16}(Rr\,yy) = \frac{3}{16} \text{ round green seeds}$$
$$\frac{1}{16}(rr\,YY) + \frac{2}{16}(rr\,Yy) = \frac{3}{16} \text{ wrinkled yellow seeds}$$
$$\frac{1}{16}(rr\,yy) = \frac{1}{16} \text{ wrinkled green seeds}$$

Figure 1-23 **Independent assortment.** The genes for round (*R*) versus wrinkled (*r*) and yellow (*Y*) versus green (*y*) pea seeds assort independently. The F_2 progeny consist of nine genotypes comprising the four possible phenotypes.

analyzing his data he used probability theory, an alien subject to most biologists of the time. The major reason his theory was ignored, however, is that it was ahead of its time: Contemporary knowledge of anatomy and physiology provided no basis for its understanding. For instance, mitosis and meiosis had yet to be discovered. Yet, after Mendel's work was rediscovered in 1900, it was shown that his principles explained inheritance in animals as well as in plants. In 1903, as a result of the realization that chromosomes and genes behave in a parallel fashion, Walter Sutton formulated the **chromosomal theory of inheritance** in which he hypothesized that genes are parts of chromosomes.

    The first trait to be assigned a chromosomal location was that of sex. *In most eukaryotes, the cells of females each contain two copies of the **X chromosome** (XX), whereas*

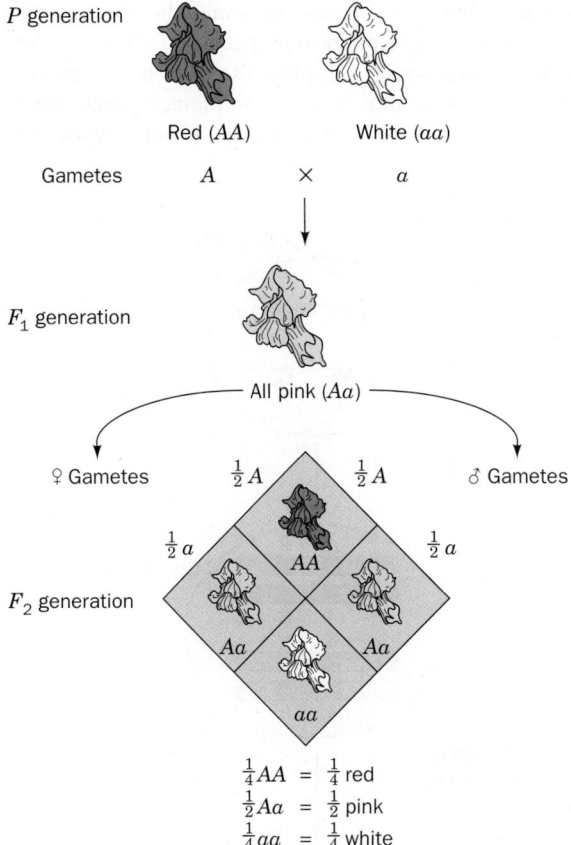

$$\tfrac{1}{4}AA \;=\; \tfrac{1}{4}\text{ red}$$
$$\tfrac{1}{2}Aa \;=\; \tfrac{1}{2}\text{ pink}$$
$$\tfrac{1}{4}aa \;=\; \tfrac{1}{4}\text{ white}$$

Figure 1-24 **Codominance.** In a cross between snapdragons with red (*AA*) and white (*aa*) flowers, the F_1 generation is pink (*Aa*), which demonstrates that the two alleles, *A* and *a*, are codominant. The F_2 flowers are red, pink, and white in a 1:2:1 ratio.

male cells contain one copy of X and a morphologically distinct **Y chromosome** *(XY; Fig. 1-25).* Ova must therefore contain a single X chromosome, and sperm contain either an X or a Y chromosome (Fig. 1-25). Fertilization by an X-bearing sperm therefore results in a female zygote and fertilization by a Y-bearing sperm yields a male zygote. This explains the observed 1:1 ratio of males to females in most species. The X and Y chromosomes are referred to as **sex chromosomes;** the others are known as **autosomes.**

a. Fruit Flies Are Favorite Genetic Subjects

The pace of genetic research greatly accelerated after Thomas Hunt Morgan began using the fruit fly *Drosophila melanogaster* as an experimental subject. This small prolific insect (Fig. 1-26), which is often seen hovering around ripe fruit in summer and fall, is easily maintained in the laboratory, where it produces a new generation every 14 days. With *Drosophila*, the results of genetic crosses can be determined some 25 times faster than they can with peas. *Drosophila* is presently the genetically best characterized higher organism.

The first known mutant strain of *Drosophila* had white eyes rather than the red eyes of the **wild type** (occurring in nature). Through genetic crosses of the white eye strain

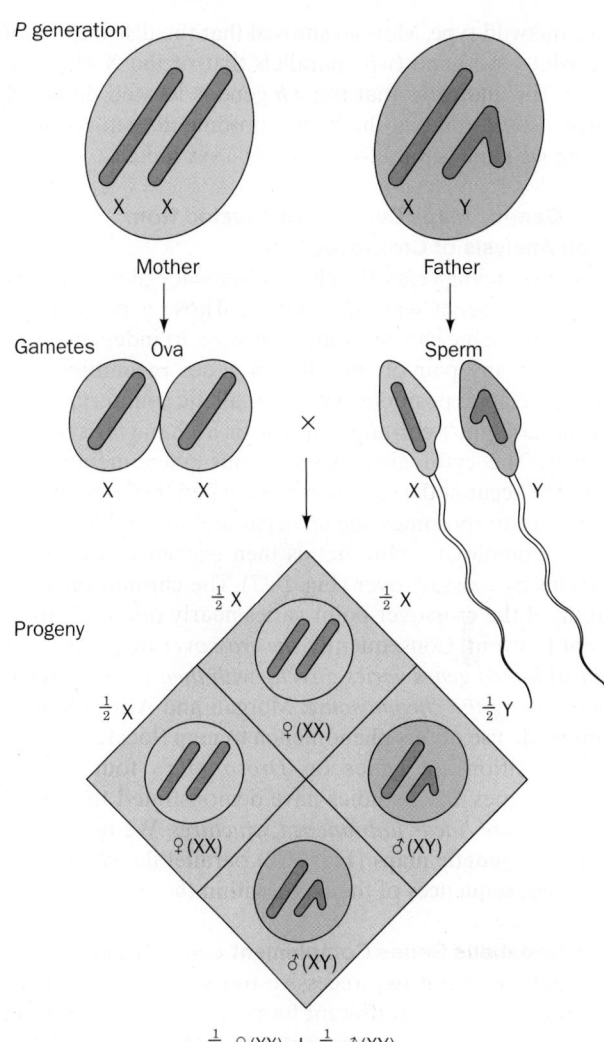

$$\tfrac{1}{2}\;\female(\text{XX}) \;+\; \tfrac{1}{2}\;\male(\text{XY})$$

Figure 1-25 **Independent segregation.** The independent segregation of the sex chromosomes, X and Y, results in a 1:1 ratio of males to females.

Figure 1-26 **The fruit fly *Drosophila melanogaster*.** The male (*left*) and the female (*right*) are shown in their relative sizes; they are actually ~2 mm long and weigh ~1 mg.

with the wild type, Morgan showed that the distribution of the white eye gene (*wh*) parallels that of the X chromosome. This indicates that the *wh* gene is located on the X chromosome and that the Y chromosome does not contain it. The *wh* gene is therefore said to be **sex linked.**

b. Genetic Maps Can Be Constructed from an Analysis of Crossover Rates

In succeeding years, the chromosomal locations of many *Drosophila* genes were determined. Those genes that reside on the same chromosome do not assort independently. However, any pair of such **linked** genes **recombine** (exchange relative positions with their allelic counterparts on the homologous chromosome) with a characteristic frequency. The cytological basis of this phenomenon was found to occur at the start of meiosis when the homologous doubled chromosomes line up in parallel (metaphase I; Fig. 1-20). Homologous chromatids then exchange equivalent sections by crossing-over (Fig. 1-27). The chromosomal location of the crossover point varies nearly randomly from event to event. Consequently, *the crossover frequency of a pair of linked genes varies directly with their physical separation along the chromosome.* Morgan and Alfred Sturtevant made use of this phenomenon to **map** (locate) the relative positions of genes on *Drosophila*'s four unique chromosomes. Such studies have demonstrated that *chromosomes are linear unbranched structures.* We now know that such **genetic maps** (Fig. 1-28) parallel the corresponding base sequences of the DNA within the chromosomes.

c. Nonallelic Genes Complement One Another

Whether or not two recessive traits that affect similar functions are allelic (different forms of the same gene) can be determined by a **complementation test.** In this test, a homozygote for one of the traits is crossed with a homozygote

for the other. If the two traits are nonallelic, the progeny will have the wild-type phenotype because each of the homologous chromosomes supplies the wild-type function that the other lacks; that is, they complement each other. For example, crossing a *Drosophila* that is homozygous for

Figure 1-27 **Crossing-over.** (*a*) An electron micrograph, together with an interpretive drawing, of two homologous pairs of chromatids during meiosis in the grasshopper *Chorthippus parallelus.* Nonsister chromatids (*different colors*) may recombine at any of the points where they cross over. [Courtesy of Bernard John, The Australian National University.] (*b*) A diagram showing the recombination of pairs of allelic genes (*A, B*) and (*a, b*) during crossover.

(*a*)   (*b*)

an eye color mutation known as purple (*pr*) with a homozygote for another eye color mutation known as brown (*bw*) yields progeny with wild-type eye color, thereby demonstrating that these two genes are not allelic (Fig. 1-29*a*).

In contrast, in crossing a female *Drosophila* that is homozygous for the sex-linked white eye color allele (*wh*) with a male carrying the sex-linked coffee eye color allele (*cf*), the female progeny do not have wild-type eye color (Fig. 1-29*b*). The *wh* and *cf* genes must therefore be allelic.

d. Genes Direct Protein Expression

The question of how genes control the characteristics of organisms took some time to be answered. Archibald Garrod was the first to suggest a specific connection between genes and enzymes. Individuals with **alkaptonuria** produce urine that darkens alarmingly on exposure to air, a consequence of the oxidation of the **homogentisic acid** they excrete (Section 16-3Ab). In 1902, Garrod showed that this

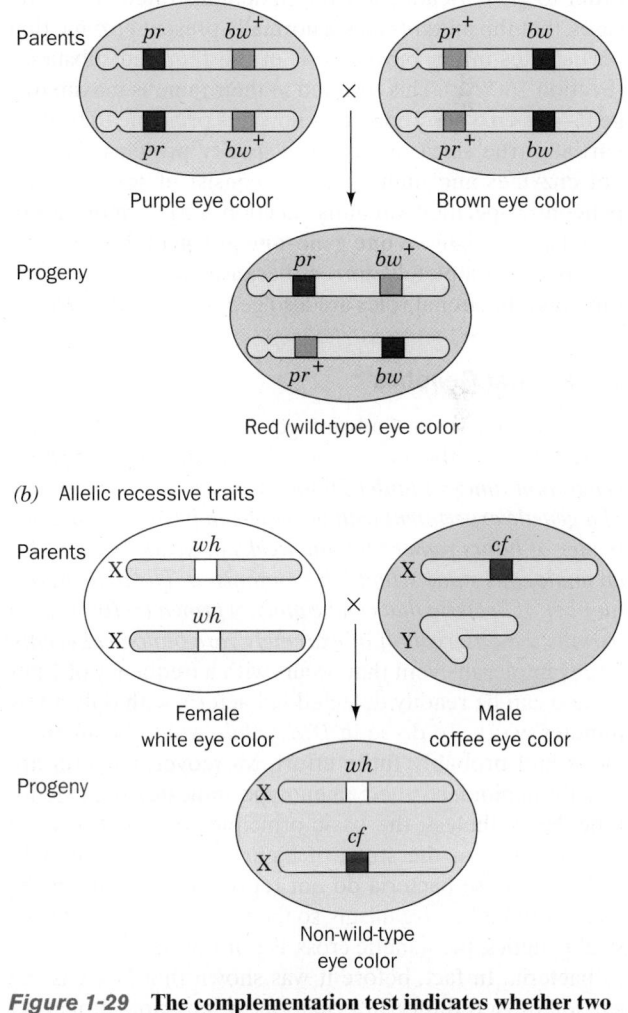

Figure 1-28 **Portion of the genetic map of chromosome 2 of *Drosophila*.** The positions of the genes are given in map units. Two genes separated by *m* map units recombine with a frequency of *m*%.

Figure 1-29 **The complementation test indicates whether two recessive traits are allelic.** Two examples in *Drosophila* are shown. (*a*) Crossing a homozygote for purple eye color (*pr*) with a homozygote for brown eye color (*bw*) yields progeny with wild-type eye color. This indicates that *pr* and *bw* are nonallelic. Here the superscript "+" indicates the wild-type allele. (*b*) In crossing a female that is homozygous for the sex-linked white eye color gene *wh* with a male bearing the sex-linked coffee eye color gene *cf*, the female progeny do not have the wild-type eye color. The *wh* and *cf* genes must therefore be allelic.

rather benign metabolic disorder (its only adverse effect is arthritis in later life) results from a recessive trait that is inherited in a Mendelian fashion. He further demonstrated that alkaptonurics are unable to metabolize the homogentisic acid fed to them and therefore concluded that *they lack an enzyme that metabolizes this substance.* Garrod described alkaptonuria and several other inherited human diseases he had studied as **inborn errors of metabolism.**

Beginning in 1940, George Beadle and Edward Tatum, in a series of investigations that mark the beginning of biochemical genetics, showed that *there is a one-to-one correspondence between a mutation and the lack of a specific enzyme.* The wild-type mold *Neurospora* grows on a "minimal medium" in which the only sources of carbon and nitrogen are glucose and NH_3. Certain mutant varieties of *Neurospora* that were generated by means of irradiation with X-rays, however, require an additional substance in order to grow. Beadle and Tatum demonstrated, in several cases, that the mutants lack a normally present enzyme that participates in the biosynthesis of the required substance (Section 16-3Ac). This resulted in their famous maxim **one gene–one enzyme.** Today we know this principle to be only partially true since many genes specify proteins that are not enzymes and many proteins consist of several independently specified subunits (Section 8-5). A more accurate dictum might be **one gene–one polypeptide.** Yet even that is not completely correct because RNAs with structural and functional roles are also genetically specified.

D. *Bacterial Genetics*

Bacteria offer several advantages for genetic study. Foremost of these is that *under favorable conditions, many have generation times of under 20 min. Consequently, the results of a genetic experiment with bacteria can be ascertained in a matter of hours rather than the weeks or years required for an analogous study with higher organisms. The tremendous number of bacteria that can be quickly grown (~10^{10} mL^{-1}) permits the observation of extremely rare biological events.* For example, an event that occurs with a frequency of 1 per million can be readily detected in bacteria with only a few minutes' work. To do so in *Drosophila* would be an enormous and probably futile effort. Moreover, bacteria are usually haploid, so their phenotype indicates their genotype. Nevertheless, the basic principles of genetics were elucidated from the study of higher plants and animals. This is because bacteria do not reproduce sexually in the manner of higher organisms, so the basic technique of classical genetics, the genetic cross, is not normally applicable to bacteria. In fact, before it was shown that DNA is the carrier of hereditary information, it was not altogether clear that bacteria had chromosomes.

The study of bacterial genetics effectively began in the 1940s when procedures were developed for isolating bacterial mutants. Since bacteria have few easily recognized morphological features, *their mutants are usually detected (selected for) by their ability or inability to grow under certain conditions.* For example, wild-type *E. coli* can grow on

1. Master plate with colonies grown on complete medium

Velvet   Handle

2. Velvet pressed to master plate and transferred to plate with different medium

Mutant colony missing

3. Colonies grow on replica plate

4. Replica and master plate are compared. Mutant colony is missing on replica plate

Figure 1-30   Replica plating. A technique for rapidly and conveniently transferring colonies from a "master" culture plate (Petri dish) to a different medium on another culture plate. Since the colonies on the master plate and on the replicas should have the same spatial distribution, it is easy to identify the desired mutants.

a medium in which glucose is the only carbon source. Mutants that are unable to synthesize the amino acid **leucine,** for instance, require the presence of leucine in their growth media. Mutants that are resistant to an antibiotic, say **ampicillin,** can grow in the presence of that antibiotic, whereas the wild type cannot. Mutants in which an essential protein has become temperature sensitive grow at 30°C but not at 42°C, whereas the wild type grows at either temperature. By using a suitable screening protocol, a bacterial colony containing a particular mutation or combination of mutations can be selected. This is conveniently done by the method of **replica plating** (Fig. 1-30).

E. *Viral Genetics*

*Viruses are infectious particles consisting of a nucleic acid molecule enclosed by a protective **capsid** (coat) that consists largely or entirely of protein. A virus specifically adsorbs to a susceptible cell into which it insinuates its nucleic acid.* Over the course of the infection (Fig. 1-31), the viral chromosome redirects the cell's metabolism so as to

Virus particle

Adsorption to host cell

Capsid   Viral chromosome

Host cell   Host chromosome

Injection of viral chromosome into the host cell

Host cell lysis releasing mature virus particles

Encapsulation of viral chromosomes by protective coat

Replication of viral chromosome

Figure 1-31 **The life cycle of a virus.**

produce new viruses. A viral infection usually culminates in the **lysis** (breaking open) of the host cell, thereby releasing large numbers (tens to thousands) of mature virus particles that can each initiate a new round of infection. Viruses, having no metabolism of their own, are the ultimate parasites. They are not living organisms since, in the absence of their host, they are as biologically inert as any other large molecule.

a. Viruses Are Subject to Complementation and Recombination

The genetics of viruses can be studied in much the same way as that of cellular organisms. Since viruses have no metabolism, however, their presence is usually detected by their ability to kill their host. The presence of viable **bacteriophages** (viruses infecting bacteria, **phages** for short; Greek: *phagein,* to eat) is conveniently indicated by **plaques** (clear spots) on a "lawn" of bacteria on a culture plate (Fig. 1-32). Plaques mark the spots where single phage particles had multiplied with the resulting lysis of the bacteria in the area. A mutant phage, which can produce progeny under certain **permissive conditions,** is detected by its inability to do so under other **restrictive conditions** in which the wild-type phage is viable. These conditions usually involve differences in the strain of the bacterial host employed or in the temperature.

Viruses are subject to complementation. Simultaneous infection of a bacterium by two different mutant varieties of a phage may yield progeny under conditions in which neither variety by itself can reproduce. If this occurs, then each mutant phage must have supplied a function that could not be supplied by the other. Each such mutation is

Figure 1-32 **Screening for viral mutants.** A culture plate covered with a lawn of bacteria on which bacteriophage have formed plaques. [Jack Bostrack/Visuals Unlimited.]

said to belong to a different **complementation group,** a term synonymous for gene.

Viral chromosomes are also subject to recombination. This occurs when a single cell is simultaneously infected by two mutant strains of a virus (Fig. 1-33). The dynamics of viral recombination differ from those in eukaryotes or bacteria because the viral chromosome undergoes recombination throughout the several rounds of DNA replication that occur during the viral life cycle. Recombinant viral

Figure 1-33   Viral recombination. Recombination of bacteriophage chromosomes occurs on simultaneous infection of a bacterial host by two phage strains carrying the genes *Ab* and *aB*.

progeny therefore consist of many if not all of the possible recombinant types.

b. The Recombinational Unit Is a Base Pair

The enormous rate at which bacteriophages reproduce permits the detection of recombinational events that occur with a frequency of as little as 1 in 10⁸. In the 1950s, Seymour Benzer carried out high-resolution genetic studies of the *rII* region of the **bacteriophage T4** chromosome. This ~4000-base pair **(bp)** region, which represents ~2% of the T4 chromosome, consists of two adjacent complementation groups designated *rIIA* and *rIIB*. In a permissive host, *E. coli* B, a mutation that inactivates the product of either gene causes the formation of plaques that are easily identified because they are much larger than those of the wild-type phage (the designation *r* stands for rapid lysis). However, only the wild-type phage will lyse the restrictive host, *E. coli* K12(λ). The presence of plaques in an *E. coli* K12(λ) culture plate that had been simultaneously infected with two different *rII* mutants in the same complementation group demonstrated that *recombination can take place within a gene.* This refuted a then widely held model of the chromosome in which genes were thought to be discrete entities, rather like beads on a string, such that recombination could take place only between intact genes. The genetic mapping of mutations at over 300 distinguishable sites in the *rIIA* and *rIIB* regions

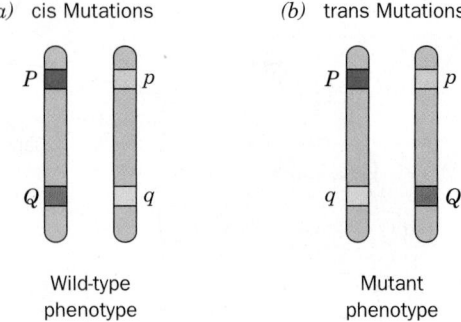

Figure 1-34   The cis–trans test. Consider a chromosome that is present in two copies in which two positions on the same gene, *P* and *Q*, have defective (recessive) mutants, *p* and *q*, respectively. (*a*) If the two mutations are cis (physically on the same chromosome), one gene will be wild type, so the organism will have a wild-type phenotype. (*b*) If the mutations are trans (on physically different chromosomes), both genes will be defective and the organism will have a mutant phenotype.

indicated that *genes, as are chromosomes, are linear unbranched structures.*

Benzer also demonstrated that a complementation test between two mutations in the same complementation group yields progeny in the restrictive host when the two mutations are in the **cis** configuration (on the same chromosome; Fig. 1-34*a*), but fails to do so when they are in the **trans** configuration (on physically different chromosomes; Fig. 1-34*b*). This is because only when both mutations physically occur in the same gene will the other gene be functionally intact. The term **cistron** was coined to mean a functional genetic unit defined according to this **cis–trans test.** This word has since become synonymous with gene or complementation group.

The recombination of pairs of *rII* mutants was observed to occur at frequencies as low as 0.01% (although frequencies as low as 0.0001% could, in principle, have been detected). Since a recombination frequency in T4 of 1% corresponds to a 240-bp separation of mutation sites, the unit of recombination can be no larger than $0.01 \times 240 = 2.4$ bp. For reasons having to do with the mechanism of recombination, this is an upper-limit estimate. On the basis of high-resolution genetic mapping, it was therefore concluded that *the unit of recombination is about the size of a single base pair.*

5 THE ORIGIN OF LIFE

People have always pondered the riddle of their existence. Indeed, all known cultures, past and present, primitive and sophisticated, have some sort of a creation myth that rationalizes how life arose. Only in the modern era, however, has it been possible to consider the origin of life in terms of a scientific framework, that is, in a manner subject to experimental verification. One of the first to do so was Charles

Darwin, the originator of the theory of evolution. In 1871, he wrote in a letter to a colleague:

> *It is often said that all the conditions for the first production of a living organism are now present, which could ever have been present. But if (and oh what a big if) we could conceive in some warm little pond, with all sorts of ammonia and phosphoric salts, light, heat, electricity, etc., present, that a protein compound was chemically formed ready to undergo still more complex changes, at the present day such matter would be instantly devoured, or absorbed, which would not have been the case before living creatures were formed.*

Radioactive dating studies indicate that Earth formed ~4.6 billion years ago but, due to the impacts of numerous large objects, its surface remained too hot to support life for several hundred million years thereafter. The earliest evidence of cellular life, microfossils of what appear to be organisms resembling modern cyanobacteria (Fig. 1-35), is ~3.5 billion years old. However, the oldest known sedimentary rocks on Earth, which are ~3.8 billion years old, have been subject to such extensive metamorphic forces (500°C and 5000 atm) that any microfossils they contained would have been obliterated. Nevertheless, geochemical analysis indicates (although not without dispute) that these rocks contain carbonaceous inclusions that are likely to be of biological origin and hence that life must have existed at the time these sedimentary rocks were laid down. If so, life on Earth must have arisen within a window of as little as a hundred million years that opened up ~4 billion years ago.

Since the prebiotic era left no direct record, *we cannot hope to determine exactly how life arose. Through laboratory experimentation, however, we can at least demonstrate what sorts of abiotic chemical reactions may have led to the formation of a living system.* Moreover, we are not entirely without traces of prebiotic development. The underlying biochemical and genetic unity of modern organisms suggests that life as we know it arose but once (if life arose more than once, the other forms must have rapidly died

out, possibly because they were "eaten" by the present form). Thus, by comparing the corresponding genetic messages of a wide variety of modern organisms it may be possible to derive reasonable models of the primordial messages from which they have descended.

It is generally accepted that the development of life occupied three stages (Fig. 1-36):

1. Chemical evolution, in which simple geologically occurring molecules reacted to form complex organic polymers.

2. The self-organization of collections of these polymers to form replicating entities. At some point in this process, the transition from a lifeless collection of reacting molecules to a living system occurred.

3. Biological evolution to ultimately form the complex web of modern life.

In this section, we outline what has been surmised about these processes. We precede this discussion by a consideration of why only carbon, of all the elements, is suitable as the basis of the complex chemistry required for life.

A. *The Unique Properties of Carbon*

Living matter, as Table 1-3 indicates, consists of a relatively small number of elements. C, H, O, N, P, and S, all of which readily form covalent bonds, comprise 92% of the dry weight of living things (most organisms are ~70% water). The balance consists of elements that are mainly present as ions and for the most part occur only in trace quantities (they usually carry out their functions at the active sites of enzymes). Note, however, that there is no known biological requirement for 64 of the 90 naturally occurring elements

Figure 1-35 Microfossil of what appears to be a cyanobacterium. This fossil, shown with its interpretive drawing, is from ~3.5-billion-year-old rock from western Australia. [Courtesy of J. William Schopf, UCLA.]

Table 1-3 **Elemental Composition of the Human Body**

Element	Dry Weight (%)[a]	Elements Present in Trace Amounts
C	61.7	B
N	11.0	F
O	9.3	Si
H	5.7	V
Ca	5.0	Cr
P	3.3	Mn
K	1.3	Fe
S	1.0	Co
Cl	0.7	Ni
Na	0.7	Cu
Mg	0.3	Zn
		Se
		Mo
		Sn
		I

[a]Calculated from Frieden, E., *Sci. Am.* **227**(1), 54–55 (1972).

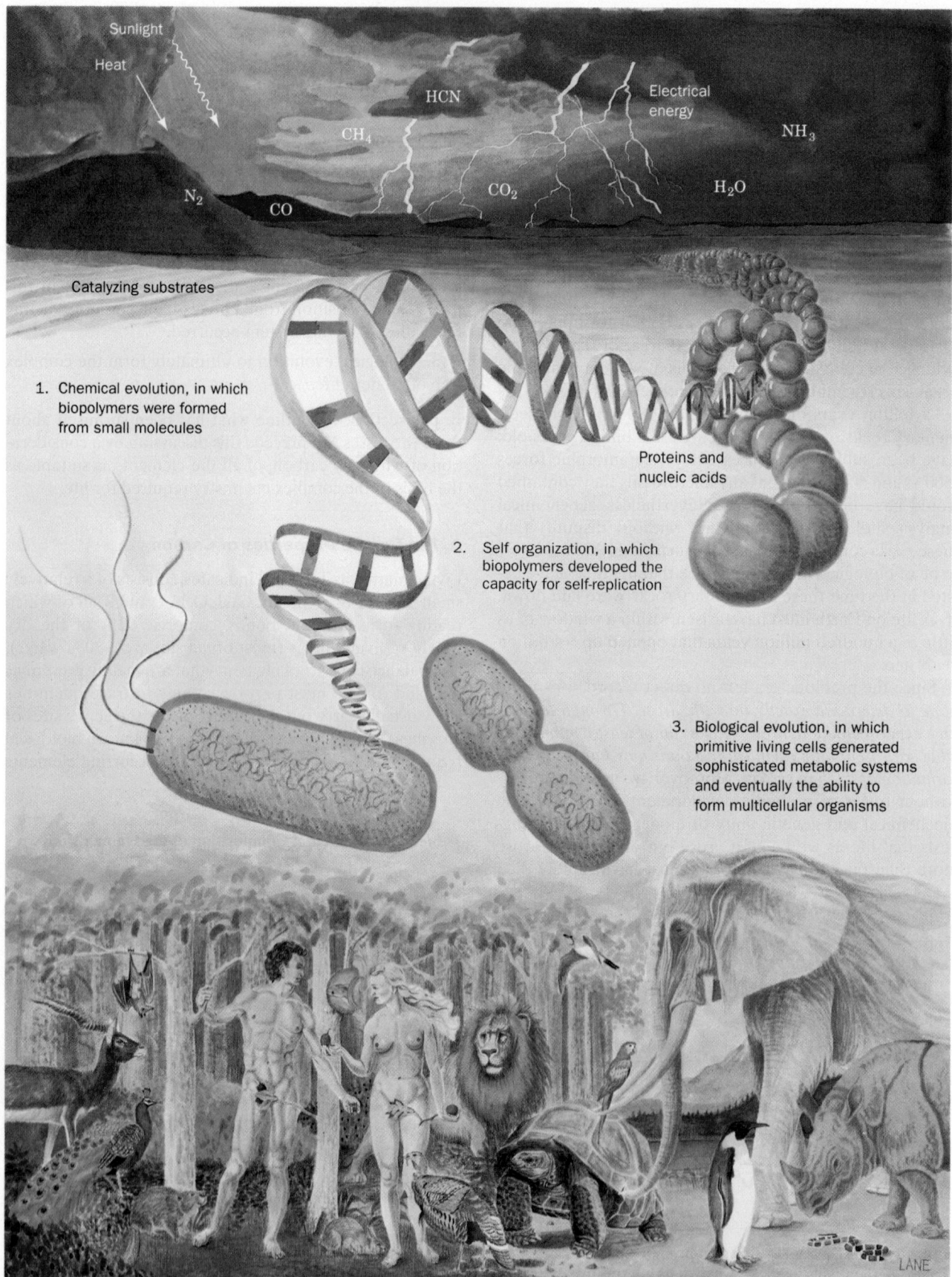

Sunlight

Heat

HCN

Electrical
energy

CH_4

NH_3

N_2 CO CO_2 H_2O

Catalyzing substrates

1. Chemical evolution, in which
   biopolymers were formed
   from small molecules

Proteins and
nucleic acids

2. Self organization, in which
   biopolymers developed the
   capacity for self-replication

3. Biological evolution, in which
   primitive living cells generated
   sophisticated metabolic systems
   and eventually the ability to
   form multicellular organisms

LANE

Figure 1-36 **The three stages in the evolution of life.**

Figure 1-37 **Periodic table in which the 26 elements utilized by living systems are highlighted in blue.**

(Fig. 1-37). Conversely, with the exceptions of oxygen and calcium, the biologically most abundant elements are but minor constituents of Earth's crust (the most abundant components of which are O, 47%; Si, 28%; Al, 7.9%; Fe, 4.5%; and Ca, 3.5%).

The predominance of carbon in living matter is no doubt a result of its tremendous chemical versatility compared with all the other elements. Carbon has the unique ability to form a virtually infinite number of compounds as a result of its capacity to make as many as four highly stable covalent bonds (including single, double, and triple bonds) combined with its ability to form covalently linked C—C *chains of unlimited extent.* Thus, of the nearly 40 million chemical compounds that are presently known, ~90% are organic (carbon-containing) substances. Let us examine the other elements in the periodic table to ascertain why they lack these combined properties.

Only five elements, B, C, N, Si, and P, have the capacity to make three or more bonds each and thus to form chains of covalently linked atoms that can also have pendant side chains. The other elements are either metals, which tend to form ions rather than covalent bonds; noble gases, which are essentially chemically inert; or atoms such as H or O that can each make only one or two covalent bonds. However, although B, N, Si, and P can each participate in at least three covalent bonds, they are, for reasons indicated below, unsuitable as the basis of a complex chemistry.

Boron, having fewer valence electrons (3) than valence orbitals (4), is electron deficient. This severely limits the types and stabilities of compounds that B can form. Nitrogen has the opposite problem; its 5 valence electrons make it electron rich. The repulsions between the lone pairs of electrons on covalently bonded N atoms serve to greatly reduce the bond energy of an N—N bond (171 kJ · mol^{-1} vs 348 kJ · mol^{-1} for a C—C single bond) relative to the unusually stable triple bond of the N_2 molecule (946 kJ · mol^{-1}). Even short chains of covalently bonded N atoms therefore tend to decompose, usually violently, to N_2. Silicon and carbon, being in the same column of the periodic table, might be expected to have similar chemistries. Sili-

con's large atomic radius, however, prevents two Si atoms from approaching each other closely enough to gain much orbital overlap. Consequently Si—Si bonds are weak (177 kJ · mol^{-1}) and the corresponding multiple bonds are rarely stable. Si—O bonds, in contrast, are so stable (369 kJ · mol^{-1}) that chains of alternating Si and O atoms are essentially inert (silicate minerals, whose frameworks consist of such bonds, form Earth's crust). Science fiction writers have speculated that **silicones,** which are oily or rubbery organosilicon compounds with backbones of linked Si—O units, for example, **methyl silicones,**

$$
\cdots -\underset{\underset{CH_3}{|}}{\overset{\overset{CH_3}{|}}{Si}}-O-\underset{\underset{CH_3}{|}}{\overset{\overset{CH_3}{|}}{Si}}-O-\underset{\underset{CH_3}{|}}{\overset{\overset{CH_3}{|}}{Si}}-O-\underset{\underset{CH_3}{|}}{\overset{\overset{CH_3}{|}}{Si}}-O-\cdots
$$

could form the chemical basis of extraterrestrial life-forms. Yet the very inertness of the Si—O bond makes this seem unlikely. Phosphorus, being below N in the periodic table, forms even less stable chains of covalently bonded atoms.

The foregoing does not imply that heteronuclear bonds are unstable. On the contrary, proteins contain C—N—C linkages, carbohydrates have C—O—C linkages, and nucleic acids possess C—O—P—O—C linkages. However, *these heteronuclear linkages are less stable than are* C—C *bonds. Indeed, they usually form the sites of chemical cleavage in the degradation of macromolecules and, conversely, are the bonds formed when monomer units are linked together to form macromolecules.* In the same vein, homonuclear linkages other than C—C bonds are so reactive that they are, with the exception of S—S bonds in proteins, extremely rare in biological systems.

B. Chemical Evolution

In the remainder of this section, we describe the most widely favored scenario for the origin of life. *Keep in mind, however, that there are valid scientific objections to this scenario as well as to the several others that have been seriously entertained, so that we are far from certain as to how life arose.*

The solar system is thought to have formed by the gravitationally induced collapse of a large interstellar cloud of dust and gas. The major portion of this cloud, which was composed mostly of hydrogen and helium, condensed to form the sun. The rising temperature and pressure at the center of the protosun eventually ignited the self-sustaining thermonuclear reaction that has since served as the sun's energy source. The planets, which formed from smaller clumps of dust, were not massive enough to support such a process. In fact the smaller planets, including Earth, consist of mostly heavier elements because their masses are too small to gravitationally retain much H_2 and He.

The primordial Earth's atmosphere was quite different from what it is today. It could not have contained signifi-

cant quantities of O_2, a highly reactive substance. Rather, in addition to the H_2O, N_2, and CO_2 that it presently has, the atmosphere probably contained smaller amounts of CO, CH_4, NH_3, SO_2, and possibly H_2, all molecules that have been spectroscopically detected in interstellar space. The chemical properties of such a gas mixture make it a **reducing atmosphere** in contrast to Earth's present atmosphere, which is an **oxidizing atmosphere.**

In the 1920s, Alexander Oparin and J.B.S. Haldane independently suggested that *ultraviolet (UV) radiation from the sun [which is presently largely absorbed by an ozone (O_3) layer high in the atmosphere] or lightning discharges caused the molecules of the primordial reducing atmosphere to react to form simple organic compounds such as amino acids, nucleic acid bases, and sugars.* That this process is possible was first experimentally demonstrated in 1953 by Stanley Miller and Harold Urey, who, in the apparatus diagrammed in Fig. 1-38, simulated effects of lightning storms in the primordial atmosphere by subjecting a refluxing mixture of H_2O, CH_4, NH_3, and H_2 to an electric discharge for about a week. (Although it now appears that Earth's primordial atmosphere did not have the strongly reducing composition assumed by Miller and Urey, localized reducing environments may have existed, particularly near volcanic plumes.) The resulting solution contained significant amounts of water-soluble organic compounds, the most abundant of which are listed in Table 1-4, together with a

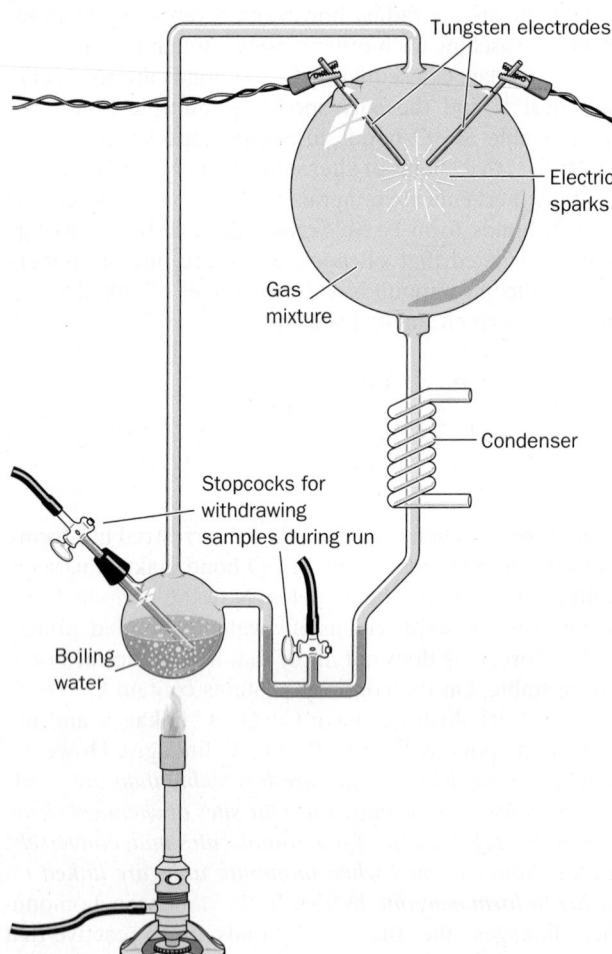

Figure 1-38 Apparatus for emulating the synthesis of organic compounds on the prebiotic Earth. A mixture of gases thought to resemble the primitive Earth's reducing atmosphere is subjected to an electric discharge, to simulate the effects of lightning, while the water in the flask is refluxed so that the newly formed compounds dissolve in the water and accumulate in the flask. [After Miller, S.L. and Orgel, L.E., *The Origins of Life on Earth*, p. 84, Prentice-Hall (1974).]

Table 1-4 **Yields from Sparking a Mixture of CH_4, NH_3, H_2O, and H_2**

Compound	Yield (%)
Glycine[a]	2.1
Glycolic acid	1.9
Sarcosine	0.25
Alanine[a]	1.7
Lactic acid	1.6
N-Methylalanine	0.07
α-Amino-n-butyric acid	0.34
α-Aminoisobutyric acid	0.007
α-Hydroxybutyric acid	0.34
β-Alanine	0.76
Succinic acid	0.27
Aspartic acid[a]	0.024
Glutamic acid[a]	0.051
Iminodiacetic acid	0.37
Iminoaceticpropionic acid	0.13
Formic acid	4.0
Acetic acid	0.51
Propionic acid	0.66
Urea	0.034
N-Methylurea	0.051

[a]Amino acid constituent of proteins.

Source: Miller, S.J. and Orgel, L.E., *The Origins of Life on Earth*, p. 85, Prentice-Hall (1974).

substantial quantity of insoluble tar (polymerized material). Several of the soluble compounds are amino acid components of proteins, and many of the others, as we shall see, are also of biochemical significance. Similar experiments in which the reaction conditions, the gas mixture, and/or the energy source were varied have resulted in the synthesis of many other amino acids. This, together with the observation that carbonaceous meteorites contain many of the same amino acids, strongly suggests that these substances were present in significant quantities on the primordial Earth. Indeed, it seems likely that large quantities of organic molecules were delivered to the primordial Earth by the meteorites and dust that so heavily bombarded it.

Nucleic acid bases can also be synthesized under supposed prebiotic conditions. In particular, adenine is formed by the condensation of HCN, a plentiful component of the prebiotic atmosphere, in a reaction catalyzed by NH_3 [note that the chemical formula of adenine is $(HCN)_5$]. The other bases have been synthesized in similar reactions involving HCN and H_2O. Sugars have been synthesized by the polymerization of formaldehyde (CH_2O) in reactions catalyzed by divalent cations, alumina, or clays. It is probably no accident that these compounds are the basic components of biological molecules. *They were apparently the most common organic substances in prebiotic times.*

The above described prebiotic reactions probably occurred over a period of hundreds of millions of years. Ultimately, it has been estimated, the oceans attained the organic consistency of a thin bouillon soup. Of course there must have been numerous places, such as tidal pools and shallow lakes, where the prebiotic soup became much more concentrated. In such environments its component organic molecules could have condensed to form, for example, polypeptides and polynucleotides (nucleic acids). Quite possibly these reactions were catalyzed by the adsorption of the reactants on minerals such as clays. If, however, life were to have formed, the rates of synthesis of these complex polymers would have had to be greater than their rates of hydrolysis. Therefore, the "pond" in which life arose may have been cold rather than warm, possibly even below 0°C (seawater freezes solidly only below −21°C), since hydrolysis reactions are greatly retarded at such low temperatures.

C. *The Rise of Living Systems*

Living systems have the ability to replicate themselves. The inherent complexity of such a process is such that no man-made device has even approached having this capacity. Clearly there is but an infinitesimal probability that a collection of molecules can simply gather at random to form a living entity (the likelihood of a living cell forming spontaneously from simple organic molecules has been said to be comparable to that of a modern jet aircraft being assembled by a tornado passing through a junkyard). How then did life arise? The answer, most probably, is that it was guided according to the Darwinian principle of the survival of the fittest as it applies at the molecular level.

a. **Life Probably Arose Through the Development of Self-Replicating RNA Molecules**

The primordial self-replicating system is widely believed to have been a collection of nucleic acid molecules because such molecules, as we have seen in Section 1-3C, can direct the synthesis of molecules complementary to themselves. RNA, as does DNA, can direct the synthesis of a complementary strand. In fact, RNA serves as the hereditary material of many viruses (Chapter 33). The polymerization of the progeny molecules would, at first, have been a simple chemical process and hence could hardly be expected to be accurate. The early progeny molecules would therefore have been only approximately complementary to their parents. Nevertheless, repeated cycles of nucleic acid synthesis would eventually exhaust the supply of free nucleotides so that the synthesis rate of new nucleic acid molecules would be ultimately limited by the hydrolytic degradation rate of old ones. Suppose, in this process, a nucleic acid molecule randomly arose that, through folding, was more resistant to degradation than its cousins. The progeny of this molecule, or at least its more faithful copies, could then propagate at the expense of the nonresistant molecules; that is, the resistant molecules would have a Darwinian advantage over their fellows. Theoretical studies suggest that such a system of molecules would evolve so as to optimize its replication efficiency under its inherent physical and chemical limitations.

In the next stage of the evolution of life, it is thought the dominant nucleic acids evolved the capacity to influence the efficiency and accuracy of their own replication. This process occurs in living systems through the nucleic acid–directed ribosomal synthesis of enzymes that catalyze nucleic acid synthesis. How nucleic acid–directed protein synthesis could have occurred before ribosomes arose is unknown because nucleic acids are not known to interact selectively with particular amino acids. This difficulty exemplifies the major problem in tracing the pathway of prebiotic evolution. Suppose some sort of rudimentary nucleic acid–influenced system arose that increased the efficiency of nucleic acid replication. This system must have eventually been replaced, presumably with almost no trace of its existence, by the much more efficient ribosomal system. Our hypothetical nucleic acid synthesis system is therefore analogous to the scaffolding used in the construction of a building. After the building has been erected the scaffolding is removed, leaving no physical evidence that it ever was there. *Most of the statements in this section must therefore be taken as educated guesses.* Without our having witnessed the event, it seems unlikely that we shall ever be certain of how life arose.

A plausible hypothesis for the evolution of self-replicating systems is that they initially consisted entirely of RNA, a scenario known as the **"RNA world."** This idea is based, in part, on the observation that certain species of RNA exhibit enzymelike catalytic properties (Section 31-4A). Moreover, since ribosomes are approximately two-thirds RNA and only one-third protein, it is plausible that the primordial ribosomes were entirely RNA. A cooperative relationship between RNA and protein might have arisen when these self-replicating protoribosomes evolved the ability to influence the synthesis of proteins that increased the efficiency and/or the accuracy of RNA synthesis. *From this point of view, RNA is the primary substance of life; the participation of DNA and proteins were later refinements that increased the Darwinian fitness of an already established self-replicating system.*

The types of systems that we have so far described were bounded only by the primordial "pond." A self-replicating system that developed a more efficient component would therefore have to share its benefits with all the "inhabitants" of the "pond," a situation that minimizes the improvement's selective advantage. Only through compartmentalization, that is, the generation of cells, could developing biological systems reap the benefits of any improvements that they might have acquired. Of course, cell formation would also hold together and protect any self-replicating system and therefore help it spread beyond its "pond" of origin. Indeed, the importance of compartmentalization is such that it may have preceded the development of self-replicating systems. The erection of cell boundaries is not without its price, however. Cells, as we shall see in later chapters, must expend much of their metabolic effort in selectively transporting substances across their cell membranes. How cell boundaries first arose, or even what they were made from, is presently unknown. However, one plausible theory holds that membranes first arose as empty vesicles whose exteriors could serve as attachment sites for such entities as enzymes and chromosomes in ways that facilitated their function. Evolution then flattened and folded these vesicles so that they enclosed their associated molecular assemblies, thereby defining the primordial cells.

b. Competition for Energy Resources Led to the Development of Metabolic Pathways, Photosynthesis, and Respiration

At this stage in their development, the entities we have been describing already fit Horowitz's criteria for life (exhibiting replication, catalysis, and mutability). The polymerization reactions through which these primitive organisms replicated were entirely dependent on the environment to supply the necessary monomeric units and the energy-rich compounds such as ATP or, more likely, just polyphosphates, that powered these reactions. As some of the essential components in the prebiotic soup became scarce, organisms developed the enzymatic systems that could synthesize these substances from simpler but more abundant precursors. As a consequence, energy-producing metabolic pathways arose. This latter development only postponed an "energy crisis," however, because these pathways consumed other preexisting energy-rich substances. The increasing scarcity of all such substances ultimately stimulated the development of photosynthesis to take advantage of a practically inexhaustible energy supply, the sun. Yet this process, as we saw in Section 1-1Ab, consumes reducing agents such as H_2S. The eventual exhaustion of these substances led to the refinement of the photosynthetic process so that it used the ubiquitous H_2O as its re-

ducing agent, thereby yielding O_2 as a by-product. The discovery, in ~3.5-billion-year-old rocks, of what appears to be fossilized cyanobacteria-like microorganisms (Fig. 1-35) suggests that oxygen-producing photosynthesis developed very early in the history of life.

The development of oxygen-producing photosynthesis led to yet another problem. The accumulation of the highly reactive O_2, which over the eons converted the reducing atmosphere of the prebiotic Earth to the modern oxidizing atmosphere (21% O_2), eventually interfered with the existing metabolic apparatus, which had evolved to operate under reducing conditions. The O_2 accumulation therefore stimulated the development of metabolic refinements that protected organisms from oxidative damage. More importantly, it led to the evolution of a much more efficient form of energy metabolism than had previously been possible, **respiration** (oxidative metabolism), which used the newly available O_2 as an oxidizing agent. [The availability of atmospheric O_2 is also responsible for the generation of the stratospheric ozone (O_3) layer that absorbs most of the biologically harmful ultraviolet radiation that strikes Earth.]

As previously outlined, the basic replicative and metabolic apparatus of modern organisms evolved quite early in the history of life on Earth. Indeed, many modern prokaryotes appear to resemble their very ancient ancestors. The rise of eukaryotes, as Section 1-2 indicates, occurred perhaps 2 billion years after prokaryotes had become firmly established. Multicellular organisms are a relatively recent evolutionary innovation, having not appeared, according to the fossil record, until ~700 million years ago.

6 THE BIOCHEMICAL LITERATURE

The biochemical literature contains the results of the work of tens of thousands of scientists extending well over a century. Consequently a biochemistry textbook can report only selected highlights of this vast amount of information. Moreover, the tremendous rate at which biochemical knowledge is presently being acquired, which is perhaps greater than that of any other intellectual endeavor, guarantees that there will have been significant biochemical advances even in the year or so that it took to produce this text from its final draft. A serious student of biochemistry must therefore regularly read the biochemical literature to flesh out the details of subjects covered in (or omitted from) this text, as well as to keep abreast of new developments. This section provides a few suggestions on how to do so.

A. Conducting a Literature Search

The primary literature of biochemistry, those publications that report the results of biochemical research, is presently being generated at a rate of tens of thousands of papers per year appearing in over 200 periodicals. An individual can therefore only read this voluminous literature in a highly selective fashion. Indeed, most biochemists tend to "read" only those publications that are likely to contain reports pertaining to their interests. By "read" it is meant that they scan the tables of contents of these journals for the titles of articles that seem of sufficient interest to warrant further perusal.

It is difficult to learn about a new subject by beginning with its primary literature. Instead, to obtain a general overview of a particular biochemical subject it is best to first peruse appropriate reviews and monographs. These usually present a synopsis of recent (at the time of their writing) developments in the area, often from the authors' particular point of view. There are more or less two types of reviews: those that are essentially a compilation of facts and those that critically evaluate the data and attempt to place them in some larger context. The latter type of review is of course more valuable, particularly for a novice in the field. Most reviews are published in specialized books and journals, although many journals that publish research reports also occasionally print reviews. Table 1-5 provides a list of many of the important biochemical review publications.

Monographs and reviews relevant to a subject of interest are usually easy to find through the use of a library catalog and the subject indexes of the major review publications (the chapter-end references of this text may also be helpful in this respect). An important part of any review is its reference list. It usually has previous reviews in the same or allied fields as well as indicating the most significant research reports in the area. Note the authors of these articles and the journals in which they tend to publish. When the most current reviews and research articles you have found tend to refer to the same group of earlier articles, you can be reasonably confident that your search for these earlier articles is largely complete. Finally, to familiarize yourself with the latest developments in the field, search the recent primary literature for the work of its most active research groups and visit the websites of these groups.

Academic libraries subscribe to Web-based reference search services such as MedLine, SciFinder Scholar, BIOSIS Previews, and Web of Science. MedLine can also be accessed free of charge through the National Center for Biotechnology Information (NCBI) (http://www.ncbi.nlm.nih.gov/PubMed). Google Scholar (http://scholar.google.com/) is a freely available search engine that indexes the scholarly literature across many disciplines. When used properly, these bibliographic search services are highly efficient tools for locating specific information. Wikipedia (http://wikipedia.org/) is also a valuable and easily available resource.

B. *Reading a Research Article*

Research reports more or less all have the same six-part format. They usually have a short abstract or summary located before the main body of the paper. The paper then continues with an introduction, which often contains a short synopsis of the field, the motivation for the research reported, and a preview of its conclusions. The next section contains a description of the methods used to obtain the experimental data. This is followed by a presentation of the results of the

Table 1-5 Some Biochemical Review Publications

Accounts of Chemical Research
Advances in Protein Chemistry and Structural Biology
Annual Review of Biochemistry
Annual Review of Biophysics
Annual Review of Cell and Developmental Biology
Annual Review of Genetics
Annual Review of Genomics and Human Genetics
Annual Review of Immunology
Annual Review of Medicine
Annual Review of Microbiology
Annual Review of Physiology
Annual Review of Plant Biology
Biochemical Journal[a]
Biochemistry and Molecular Biology Education[a]
Biochimica et Biophysica Acta[a]
BioEssays
Cell[a]
Chemistry and Biology
Critical Reviews in Biochemistry and Molecular Biology
Current Biology
Current Opinion in Biotechnology
Current Opinion in Cell Biology
Current Opinion in Chemical Biology
Current Opinion in Genetics and Development
Current Opinion in Structural Biology
FASEB Journal[a]
Journal of Biological Chemistry[a]
Methods in Enzymology
Molecular Cell[a]
Nature[a]
Nature Reviews Molecular Cell Biology
Nature Structural & Molecular Biology[a]
Proceedings of the National Academy of Sciences USA[a]
Progress in Biophysics and Molecular Biology
Progress in Nucleic Acid Research and Molecular Biology
Protein Science[a]
Quarterly Reviews of Biophysics
Science[a]
Scientific American
Structure[a]
Trends in Biochemical Sciences
Trends in Cell Biology
Trends in Genetics

[a]Periodicals that mainly publish research reports.

investigation and then by a discussion section wherein the conclusions of the investigation are set forth and placed in the context of other work in the field. Finally, there is a list of references. Most articles are "full papers," which may be tens of pages long. However, many journals also contain "communications" or "letters," which are usually only a few pages in length and are often published more quickly than are full papers. Many papers have accompanying supplementary material that is available on the journal's website.

It is by no means obvious how to read a scientific paper. Perhaps the worst way to do so is to read it from beginning to end as if it were some kind of a short story. In fact, most practicing scientists only occasionally read a research article in its entirety. It simply takes too long and is rarely productive. Rather, they scan selected parts of a paper and only dig deeper if it appears that to do so will be profitable. The following paragraph describes a reasonably efficient scheme for reading scientific papers. *This should be an active process in which the reader is constantly evaluating what is being read and relating it to his/her previous knowledge.* Moreover, the reader should maintain a healthy skepticism since there is a reasonable probability that any paper, particularly in its interpretation of experimental data and in its speculations, may be erroneous.

If the title of a paper indicates that it may be of interest, then this should be confirmed by a reading of its abstract. For many papers, even those containing useful informa-tion, it is unnecessary to read further. If you choose to con-tinue, it is probably best to do so by scanning the introduc-tion so as to obtain an overview of the work reported. At this point most experienced scientists scan the conclusions section of the paper to gain a better understanding of what was found. If further effort seems warranted, they scan the results section to ascertain whether the experimental data support the conclusions. The methods section (which in many journals is largely relegated to the supplementary materials) is usually not read in detail because it is often written in a condensed form that is only fully interpretable by an expert in the field. However, for such experts, the methods section may be the most valuable part of the pa-per. At this point, what to read next, if anything, is largely dictated by the remaining points of confusion. In many cases this confusion can only be eliminated by reading some of the references given in the paper. At any rate, un-less you plan to repeat or extend some of the work de-scribed, it is rarely necessary to read an article in detail. To do so in a critical manner, you will find, takes several hours for a paper of even moderate size.

CHAPTER SUMMARY

1 Prokaryotes   Prokaryotes are single-celled organisms that lack a membrane-enclosed nucleus. Most prokaryotes have similar anatomies: a rigid cell wall surrounding a cell membrane that encloses the cytoplasm. The cell's single chro-mosome is condensed to form a nucleoid. *Escherichia coli,* the biochemically most well-characterized organism, is a typical prokaryote. Prokaryotes have quite varied nutritional require-ments. The chemolithotrophs metabolize inorganic sub-stances. Photolithotrophs, such as cyanobacteria, carry out photosynthesis. Heterotrophs, which live by oxidizing organic substances, are classified as aerobes if they use oxygen in this process and as anaerobes if some other oxidizing agent serves as their terminal electron acceptor. Traditional prokaryotic classification schemes are rather arbitrary because of poor correlation between bacterial form and metabolism. Sequence comparisons of nucleic acids and proteins, however, have es-tablished that all life-forms can be classified into three do-mains of evolutionary descent: the Archaea (archaebacteria), the Bacteria (eubacteria), and the Eukarya (eukaryotes).

2 Eukaryotes   Eukaryotic cells, which are far more com-plex than those of prokaryotes, are characterized by having numerous membrane-enclosed organelles. The most conspicu-ous of these is the nucleus, which contains the cell's chromo-somes, and the nucleolus, where ribosomes are assembled. The endoplasmic reticulum is the site of synthesis of lipids and of proteins that are destined for secretion. Further processing of these products occurs in the Golgi apparatus. The mitochon-dria, wherein oxidative metabolism occurs, are thought to have evolved from a symbiotic relationship between an aero-bic bacterium and a primitive eukaryote. The chloroplast, the site of photosynthesis in plants, similarly evolved from a cyanobacterium. Other eukaryotic organelles include the lyso-some, which functions as an intracellular digestive chamber, and the peroxisome, which contains a variety of oxidative en-zymes including some that generate H_2O_2. The eukaryotic cy-toplasm is pervaded by a cytoskeleton whose components in-clude microtubules, which consist of tubulin; microfilaments, which are composed of actin; and intermediate filaments, which are made of different proteins in different types of cells. Eukaryotes have enormous morphological diversity on the cellular as well as on the organismal level. They have been classified into four kingdoms: Protista, Plantae, Fungi, and An-imalia. The pattern of embryonic development in multicellular organisms partially mirrors their evolutionary history.

3 Biochemistry: A Prologue   Organisms have a hier-archical structure that extends down to the submolecular level. They contain but three basic types of macromolecules: proteins, nucleic acids, and polysaccharides, as well as lipids, each of which are constructed from only a few different species of monomeric units. Macromolecules and supramolec-ular assemblies form their native biological structures through a process of self-assembly. The assembly mechanisms of higher biological structures are largely unknown. Metabolic processes are organized into a series of tightly regulated path-ways. These are classified as catabolic or anabolic depending on whether they participate in degradative or biosynthetic processes. The common energy "currency" in all these processes is ATP, whose synthesis is the product of many cata-bolic pathways and whose hydrolysis drives most anabolic pathways. DNA, the cell's hereditary molecule, encodes ge-netic information in its sequence of bases. The complementary base sequences of its two strands permit them to act as tem-plates for their own replication and for the synthesis of com-plementary strands of RNA. Ribosomes synthesize proteins by linking amino acids together in the order specified by the base sequences of RNAs.

4 Genetics: A Review   Eukaryotic cells contain a charac-teristic number of homologous pairs of chromosomes. In mitosis each daughter cell receives a copy of each of these chromosomes, but in meiosis each resulting gamete receives only one member of each homologous pair. Fertilization is the fusion of two haploid gametes to form a diploid zygote. The Mendelian laws of inheritance state that alternative forms of true-breeding traits are specified by different alleles of the same gene. Alleles may be dominant, codominant, or recessive depending on the phenotype of the heterozygote. Different genes assort independently unless they are on the same chromosome. The linkage between genes on the same chromosome, however, is never complete because of crossing-over among homologous chromosomes during meiosis. The rate at which genes recombine varies with their physical separation because crossing-over occurs essentially at random. This permits the construction of genetic maps. Whether two recessive traits are allelic may be determined by the complementation test. The nature of genes is largely defined by the dictum "one gene–one polypeptide." Mutant varieties of bacteriophages are detected by their ability to kill their host under various restrictive conditions. The fine structure analysis of the *rII* region of the bacteriophage T4 chromosome has revealed that recombination may take place within a gene, that genes are linear unbranched structures, and that the unit of mutation is ∼1 bp.

5 The Origin of Life   Life is carbon based because only carbon, among all the elements in the periodic table, has a sufficiently complex chemistry together with the ability to form virtually infinite stable chains of covalently bonded atoms. Reactions among the molecules in the reducing atmosphere of the prebiotic Earth are thought to have formed the simple organic precursors from which biological molecules developed. Eventually, in reactions that may have been catalyzed by minerals such as clays, polypeptides and polynucleotides formed. These evolved under the pressure of competition for the available monomeric units. Ultimately, a nucleic acid, most probably RNA, developed the capability of influencing its own replication by directing the synthesis of proteins that catalyze polynucleotide synthesis. This was followed by the development of cell membranes so as to form living entities. Subsequently, metabolic processes evolved to synthesize necessary intermediates from available precursors as well as the high-energy compounds required to power these reactions. Likewise, photosynthesis and respiration arose in response to environmental pressures brought about by the action of living organisms.

6 The Biochemical Literature   The sheer size and rate of increase of the biochemical literature requires that it be read to attain a thorough understanding of any aspect of biochemistry. The review literature provides an entrée into a given subspeciality. To remain current in any field, however, requires a regular perusal of its primary literature. This should be read in a critical and highly selective fashion.

REFERENCES

Prokaryotes and Eukaryotes

Becker, W.M., Kleinsmith, L.J., Hardin, J., and Bertoni, G.P., *The World of the Cell* (7th ed.), Benjamin Cummings (2009). [A highly readable cell biology text.]

Boone, D.R. and Castenholz, R.W. (Eds.), *Bergey's Manual of Systematic Bacteriology* (2nd ed.), Vol. I; *and* Brenner, D.J., Kreig, N.R., and Staley, J.T. (Eds.), *Bergey's Manual of Systematic Bacteriology* (2nd ed.), Vols. IIA, B, & C, Springer (2001 and 2005).

Campbell, N.A. and Reece, J.B., *Biology* (8th ed.) Benjamin Cummings (2008). [A comprehensive general biology text. There are several others available with similar content.]

Frieden, E., The chemical elements of life, *Sci. Am.* **227**(1), 52–60 (1972).

Goodsell, D.S., *The Machinery of Life,* Springer-Verlag (1998).

Jørgensen, B.B. and D'Hondt, S., A starving majority beneath the sea floor, *Science* **314**, 932–934 (2006). [Discusses the prokaryotes that live in the rocks deep below Earth's surface.]

Madigan, M.T., Martinko, J.M., Dunlap, P.V., and Clark, D.P., *Brock Biology of Microorganisms* (12th ed.), Pearson Benjamin Cummings (2009).

Margulis, L. and Schwartz, K.V., *Five Kingdoms. An Illustrated Guide to the Phyla of Life on Earth* (3rd ed.), Freeman (1998).

Pace, N.R., A molecular view of microbial diversity and the biosphere, *Science* **276**, 734–740 (1997).

Whitman, W.B., Coleman, D.C., and Wiebe, W.J., Prokaryotes: The unseen majority, *Proc. Natl. Acad. Sci.* **95**, 6578–6583 (1998). [Estimates the number of prokaryotes on Earth $(4-6 \times 10^{30}$ cells) and the aggregate mass of their cellular carbon $(3.5-5.5 \times 10^{14}$ kg, which therefore comprises 66–100% of the carbon in plants).]

Genetics

Benzer, S., The fine structure of the gene, *Sci. Am.* **206**(1), 70–84 (1962).

Cairns, J., Stent, G.S., and Watson, J. (Eds.), *Phage and the Origins of Molecular Biology, The Centennial Edition*, Cold Spring Harbor Laboratory (2007). [A series of scientific memoirs by many of the pioneers of molecular biology.]

Hartwell, L.H., Hood, L., Goldberg, M.L., Reynolds, A.E., Silver, L.M., and Veres, R.C., *Genetics. From Genes to Genomes* (3rd ed.), Chapters 1–5, McGraw-Hill (2008).

Snustad, D.P. and Simmons, M.J., *Principles of Genetics* (5th ed.), Wiley (2009).

Origin of Life

Berstein, M.P., Sandford, S.A., and Allamandola, S.A., Life's far-flung raw materials, *Sci. Am.* **281**(1), 42–49 (1999). [A discussion of the possibility that the complex organic molecules which provided the starting materials for life were delivered to the primordial Earth by meteorites and dust.]

Brack, A. (Ed.), *The Molecular Origins of Life,* Cambridge University Press (1998).

Doolittle, F.W., Phylogenetic classification and the universal tree, *Science* **284**, 2124–2128 (1999). [A discussion of how lateral gene transfer among the various forms of life may have confounded the ability to elucidate the "universal tree of life" if, in fact, such a tree is a reasonable model of the history of life.]

Dyson, F., *Origins of Life,* Cambridge University Press (1985). [A fascinating philosophical discourse on theories of life's origins by a respected theoretical physicist.]

Fraústo da Silva, J.R. and Williams, R.J.P., *The Biological Chemistry of the Elements,* Oxford (1991).

Gesteland, R.F., Cech, T.R., and Atkins, J.F. (Eds.), *The RNA World* (3rd ed.), Chapters 1–3, Cold Spring Harbor Laboratory Press (2006).

Herdewijn, P. and Kisakürek, M.V. (Eds.), *Origin of Life. Chemical Approach,* Wiley-VCH (2008).

Knoll, A.H., The early evolution of eukaryotes: A geological perspective, *Science* **256,** 622–627 (1992).

Lahav, N., *Biogenesis. Theories of Life's Origins,* Oxford University Press (1999).

Lazcano, A. and Miller, S.L., The origin and early evolution of life: Prebiotic chemistry, the pre-RNA world, and time, *Cell* **85,** 793–798 (1996); Bada, J.L. and Lazcano, A., Prebiotic soup—Revisiting the Miller experiment, *Science* **300,** 745–746 (2003); *and* Johnson, A.P., Cleaves, H.J., Dworkin, J.P., Glavin, D.P., Lazcano, A., and Bada, J.L., The Miller volcanic spark discharge experiment, *Science* **322,** 404 (2008).

Lifson, S., On the crucial stages in the origin of animate matter, *J. Mol. Evol.* **44,** 1–8 (1997).

Lurquin, P.F., *The Origins of Life and the Universe,* Columbia University Press (2003).

McNichol, J., Primordial soup, fool's gold, and spontaneous generation, *Biochem. Mol. Biol. Ed.* **36,** 255–261 (2008). [A brief introduction to the theory, history, and philosophy of the search for the origin of life.]

Mojzsis, S.J., Arrhenius, G., McKeegan, K.D., Harrison, T.M., Nutman, A.P., and Friend, C.R.L., Evidence for life on Earth before 3,800 million years ago, *Nature* **384,** 55–57 (1996).

Orgel, L.E., The origin of life—a review of facts and speculations, *Trends Biochem. Sci.* **23,** 491–495 (1998). [Reviews the most widely accepted hypotheses on the origin of life and discusses the evidence supporting them and their difficulties.]

Schopf, J.W., Fossil evidence of Archean life, *Philos. Trans. R. Soc. B* **361,** 869–885 (2006).

Shapiro, R., *Origins. A Skeptic's Guide to the Creation of Life on Earth,* Summit Books (1986). [An incisive and entertaining critique of the reigning theories of the origin of life.]

PROBLEMS

It is very difficult to learn something well without somehow participating in it. The chapter-end problems are therefore an important part of this book. They contain few problems of the regurgitation type. Rather they are designed to make you think and to offer insights not discussed in the text. Their difficulties range from those that require only a few moments' reflection to those that might take an hour or more of concentrated effort to work out. The more difficult problems are indicated by a leading asterisk (*). The answers to the problems are worked out in detail in the *Solutions Manual to Accompany Biochemistry* (4th ed.) by Donald Voet and Judith G. Voet. You should, of course, make every effort to work out a problem before consulting the *Solutions Manual.*

1. Under optimal conditions for growth, an *E. coli* cell will divide around every 20 min. If no cells died, how long would it take a single *E. coli* cell, under optimal conditions in a 10-L culture flask, to reach its maximum cell density of 10^{10} cells $\cdot$ mL^{-1} (a "saturated" culture)? Assuming that optimum conditions could be maintained, how long would it take for the total volume of the cells alone to reach 1 km^3? (Assume an *E. coli* cell to be a cylinder 2 μm long and 1 μm in diameter.)

2. Without looking them up, draw schematic diagrams of a bacterial cell and an animal cell. What are the functions of their various organelles? How many lines of descent might a typical animal cell have?

3. Compare the surface-to-volume ratios of a typical *E. coli* cell (its dimensions are given in Problem 1) and a spherical eukaryotic cell that is 20 μm in diameter. How does this difference affect the lifestyles of these two cell types? In order to improve their ability to absorb nutrients, the **brush border cells** of the intestinal epithelium have velvetlike patches of **microvilli** facing into the intestine. How does the surface-to-volume ratio of this eukaryotic cell change if 20% of its surface area is covered with cylindrical microvilli that are 0.1 μm in diameter, 1 μm in length, and occur on a square grid with 0.2-μm center-to-center spacing?

4. Many proteins in *E. coli* are normally present at concentrations of two molecules per cell. What is the molar concentration of such a protein? (The dimensions of *E. coli* are given in Problem 1.)

Conversely, how many glucose molecules does an *E. coli* cell contain if it has an internal glucose concentration of 1.0 mM?

5. The DNA of an *E. coli* chromosome measures 1.6 mm in length, when extended, and 20 Å in diameter. What fraction of an *E. coli* cell is occupied by its DNA? (The dimensions of *E. coli* are given in Problem 1.) A human cell has some 700 times the DNA of an *E. coli* cell and is typically spherical with a diameter of 20 μm. What fraction of such a human cell is occupied by its DNA?

*6. A new planet has been discovered that has approximately the same orbit about the sun as Earth but is invisible from Earth because it is always on the opposite side of the sun. Interplanetary probes have already established that this planet has a significant atmosphere. The National Aeronautics and Space Administration is preparing to launch a new unmanned probe that will land on the surface of the planet. Outline a simple experiment for this lander that will test for the presence of life on the surface of this planet (assume that the life-forms, if any, on the planet are likely to be microorganisms and therefore unable to walk up to the lander's video cameras and say "Hello").

7. It has been suggested that an all-out nuclear war would so enshroud Earth with clouds of dust and smoke that the entire surface of the planet would be quite dark and therefore intensely cold (well below 0°C) for several years (the so-called nuclear winter). In that case, it is thought, eukaryotic life would die out and bacteria would inherit Earth. Why?

8. One method that Mendel used to test his laws is known as a **testcross.** In it, F_1 hybrids are crossed with their recessive parent. What is the expected distribution of progeny and what are their phenotypes in a testcross involving peas with different-colored seeds? What is it for snapdragons with different flower colors (use the white parent in this testcross)?

9. The disputed paternity of a child can often be decided on the basis of blood tests. The M, N, and MN blood groups (Section 12-3E) result from two alleles, L^M and L^N; the Rh$^+$ blood group arises from a dominant allele, *R*. Both sets of alleles occur on a different chromosome from each other and from the alleles responsible for the ABO blood groups. The following table gives the blood types of three children, their mother, and the two possible fathers. Indicate, where possible, each child's paternity and justify your answer.

Child 1	B	M	Rh$^-$
Child 2	B	MN	Rh$^+$
Child 3	AB	MN	Rh$^+$
Mother	B	M	Rh$^+$
Male 1	B	MN	Rh$^+$
Male 2	AB	N	Rh$^+$

10. The most common form of color blindness, red–green color blindness, afflicts almost only males. What are the genotypes and phenotypes of the children and grandchildren of a red–green color-blind man and a woman with no genetic history of color blindness? Assume the children mate with individuals who also have no history of color blindness.

11. Green and purple photosynthetic bacteria are thought to resemble the first organisms that could carry out photosynthesis. Speculate on the composition of Earth's atmosphere when these organisms first arose.

12. Explore your local biochemistry library (it may be disguised as a biology, chemistry, or medical library). Locate where the current periodicals, the bound periodicals, and the books are kept. Browse through the contents of a current major biochemistry journal, such as *Biochemistry, Cell,* or *Proceedings of the National Academy of Sciences,* and pick a title that interests you. Scan the corresponding paper and note its organization. Likewise, peruse one of the articles in the latest volume of *Annual Review of Biochemistry.*

13. Using MedLine, look up the publications over the past 5 years of your favorite biomedical scientist. This person might be a recent Nobel prize winner or someone at your college/university. Note that even if the person you choose has an unusual name, it is likely that publications by other individuals with the same name will be included in your initial list.

Aqueous Solutions

CHAPTER 2

Life, as we know it, occurs in aqueous solution. Indeed, terrestrial life apparently arose in some primordial sea (Section 1-5B) and, as the fossil record indicates, did not venture onto dry land until comparatively recent times. Yet even those organisms that did develop the capacity to live out of water still carry the ocean with them: The compositions of their intracellular and extracellular fluids are remarkably similar to that of seawater. This is true even of organisms that live in such unusual environments as saturated brine, acidic hot sulfur springs, and petroleum.

Water is so familiar, we generally consider it to be a rather bland fluid of simple character. It is, however, a chemically reactive liquid with such extraordinary physical properties that, if chemists had discovered it in recent times, it would undoubtedly have been classified as an exotic substance.

The properties of water are of profound biological significance. *The structures of the molecules on which life is based—proteins, nucleic acids, lipids, and complex carbohydrates—result directly from their interactions with their aqueous environment. The combination of solvent properties responsible for the intramolecular and intermolecular associations of these substances is peculiar to water; no other solvent even resembles water in this respect.* Although the hypothesis that life could be based on organic polymers other than proteins and nucleic acids seems plausible, it is all but inconceivable that the complex structural organization and chemistry of living systems could exist in other than an aqueous medium. Indeed, direct observations on the surface of Mars, the only other planet in the solar system with temperatures compatible with life, indicate that it is presently devoid of both water and life.

Biological structures and processes can only be understood in terms of the physical and chemical properties of water. We therefore begin this chapter with a discussion of the molecular and solvent properties of water. In the following section we review its chemical behavior, that is, the nature of aqueous acids and bases.

1 PROPERTIES OF WATER

Water's peculiar physical and solvent properties stem largely from its extraordinary internal cohesiveness compared to that of almost any other liquid. In this section, we explore the physical basis of this phenomenon.

A. Structure and Interactions

The H_2O molecule has a bent geometry with an O—H bond distance of 0.958 Å and an H—O—H bond angle of 104.5° (Fig. 2-1). The large electronegativity differenc between H and O confers a 33% ionic character on the O—H bond as is indicated by water's dipole moment of 1.85 debye units. Water is clearly a highly polar molecule, a phenomenon with enormous implications for living systems.

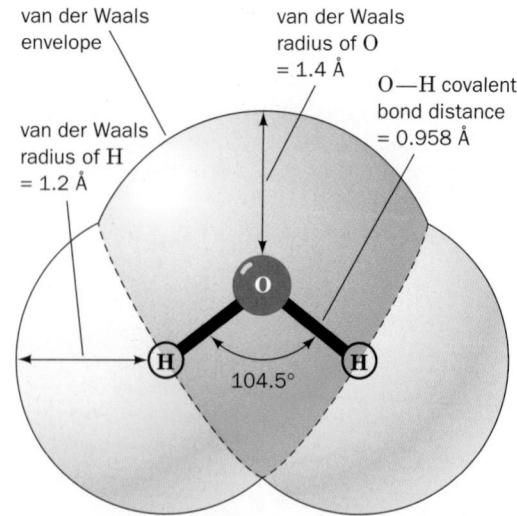

Figure 2-1   Structure of the water molecule. The outline represents the van der Waals envelope of the molecule (where the attractive components of the van der Waals interactions balance the repulsive components). The skeletal model of the molecule indicates its covalent bonds.

a. Water Molecules Associate Through Hydrogen Bonds

The electrostatic attractions between the dipoles of two water molecules tend to orient them such that the O—H bond on one water molecule points toward a lone-pair electron cloud on the oxygen atom of the other water molecule. This results in a directional intermolecular association known as a **hydrogen bond** (Fig. 2-2), an interaction that is crucial both to the properties of water itself and to its role as a biochemical solvent. In general, *a hydrogen bond may be represented as D—H··A , where D—H is a weakly acidic "donor group" such as N—H or O—H, and A is a lone-pair-bearing and thus weakly basic "acceptor atom" such as N or O.* Hence, a hydrogen bond is better represented as $^{\delta-}D$—$H^{\delta+}\cdots{}^{\delta-}A$, where the charge separation in the D—H bond arises from the greater electronegativity of D relative to H. The peculiar requirement of a central hydrogen atom rather than some other atom in a hydrogen bond stems from the hydrogen atom's small size: Only a hydrogen nucleus can approach the lone-pair electron cloud of an acceptor atom closely enough to permit an electrostatic association of significant magnitude. Moreover, as X-ray scattering measurements have revealed, hydrogen bonds are partially (~10%) covalent in character.

Hydrogen bonds are structurally characterized by an H···A distance that is at least 0.5 Å shorter than the calculated van der Waals distance (distance of closest approach between two nonbonded atoms) between the atoms. In water, for example, the O···H hydrogen bond distance is ~1.8 Å versus 2.6 Å for the corresponding van der Waals distance. The energy of a hydrogen bond (~20 kJ · mol^{-1} in H$_2$O) is small compared to covalent bond energies (for instance, 460 kJ · mol^{-1} for an O—H covalent bond). Nevertheless, most biological molecules have so many hydrogen bonding groups that hydrogen bonding is of paramount importance in determining their three-dimensional structures and their intermolecular associations. Hydrogen bonding is further discussed in Section 8-4B.

b. The Physical Properties of Ice and Liquid Water Largely Result from Intermolecular Hydrogen Bonding

The structure of ice provides a striking example of the cumulative strength of many hydrogen bonds. X-ray and neutron diffraction studies have established that water molecules in ice are arranged in an unusually open structure. Each water molecule is tetrahedrally surrounded by four nearest neighbors to which it is hydrogen bonded (Fig. 2-3). In two of these hydrogen bonds the central H$_2$O molecule is the "donor," and in the other two it is the "acceptor." As a consequence of its open structure, water is one of the very few substances that expands on freezing (at 0°C, liquid water has a density of 1.00 g · mL^{-1}, whereas ice has a density of 0.92 g · mL^{-1}).

The expansion of water on freezing has overwhelming consequences for life on Earth. Suppose that water contracted on freezing, that is, became more dense rather than less dense. Ice would then sink to the bottoms of lakes and oceans rather than float. This ice would be insulated from the

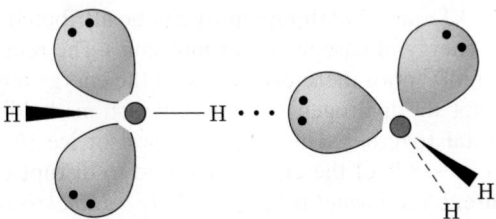

Figure 2-2   **Hydrogen bond between two water molecules.** The strength of this interaction is maximal when the O—H covalent bond points directly along a lone-pair electron cloud of the oxygen atom to which it is hydrogen bonded.

Figure 2-3   **Structure of ice.** The tetrahedral arrangement of the water molecules is a consequence of the roughly tetrahedral disposition of each oxygen atom's sp^3-hybridized bonding and lone-pair orbitals (Fig. 2-2). Oxygen and hydrogen atoms are represented, respectively, by red and white spheres, and hydrogen bonds are indicated by dashed lines. Note the open structure that gives ice its low density relative to liquid water. [After Pauling, L., *The Nature of the Chemical Bond* (3rd ed.), *p.* 465, Cornell University Press (1960).]

sun so that oceans, with the exception of a thin surface layer of liquid in warm weather, would be permanently frozen solid (the water at great depths even in tropical oceans is close to 4°C, its temperature of maximum density). The reflection of sunlight by these frozen oceans and their cooling effect on the atmosphere would ensure that land temperatures would also be much colder than at present; that is, Earth would have a permanent ice age. Furthermore, since life apparently evolved in the ocean, it seems unlikely that life could have developed at all if ice contracted on freezing.

Although the melting of ice is indicative of the cooperative collapse of its hydrogen bonded structure, hydrogen bonds between water molecules persist in the liquid state. The heat of sublimation of ice at 0°C is 46.9 kJ · mol^{-1}. Yet only ~6 kJ · mol^{-1} of this quantity can be attributed to the kinetic energy of gaseous water molecules. The remaining 41 kJ · mol^{-1} must therefore represent the energy required to disrupt the hydrogen bonding interactions holding an ice crystal together. The heat of fusion of ice (6.0 kJ · mol^{-1}) is ~15% of the energy required to disrupt the ice structure. *Liquid water is therefore only ~15% less hydrogen bonded than ice at 0°C.* Indeed, the boiling point of water is 264°C higher than that of methane (CH_4), a substance with nearly the same molecular mass as H_2O but which is incapable of hydrogen bonding (in the absence of intermolecular associations, substances with equal molecular masses should have similar boiling points). This reflects the extraordinary internal cohesiveness of liquid water resulting from its intermolecular hydrogen bonding.

c. Liquid Water Has a Rapidly Fluctuating Structure

X-ray and neutron scattering measurements of liquid water reveal a complex structure. Near 0°C, water exhibits an average nearest-neighbor O···O distance of 2.82 Å, which is slightly greater than the corresponding 2.76-Å distance in ice despite the greater density of the liquid. The X-ray data further indicate that each water molecule is surrounded by an average of about 4.4 nearest neighbors, which strongly suggests that the short-range structure of liquid water is predominantly tetrahedral in character. This picture is corroborated by the additional intermolecular distances in liquid water of around 4.5 and 7.0 Å, which are near the expected second and third nearest-neighbor distances in an icelike tetrahedral structure. Liquid water, however, also exhibits a 3.5-Å intermolecular distance, which cannot be rationalized in terms of an icelike structure. These average distances, moreover, become less sharply defined as the temperature increases into the physiologically significant range, thereby signaling the thermal breakdown of the short-range water structure.

The structure of liquid water is not simply described. This is because each water molecule reorients about once every 10^{-12} s, which makes the determination of water's instantaneous structure an experimentally and theoretically difficult problem (very few experimental techniques can make measurements over such short time spans). Indeed, only with the advent of modern computational methods have theoreticians felt that they are beginning to have a reasonable understanding of liquid water at the molecular level.

For the most part, molecules in liquid water are each hydrogen bonded to four nearest neighbors as they are in ice. These hydrogen bonds are distorted, however, so that the networks of linked molecules are irregular and varied, with the number of hydrogen bonds formed by each water molecule ranging from 3 to 6. Thus, for example, 3- to 7-membered rings of hydrogen bonded molecules commonly occur in liquid water (Fig. 2-4), in contrast to the cyclohexane-like 6-membered rings characteristic of ice (Fig. 2-3). Moreover, these networks are continually breaking up and re-forming over time periods on the order of 2×10^{-11} s. *Liquid water therefore consists of a rapidly fluctuating, space-filling network of hydrogen bonded H_2O molecules that, over short distances, resembles that of ice.*

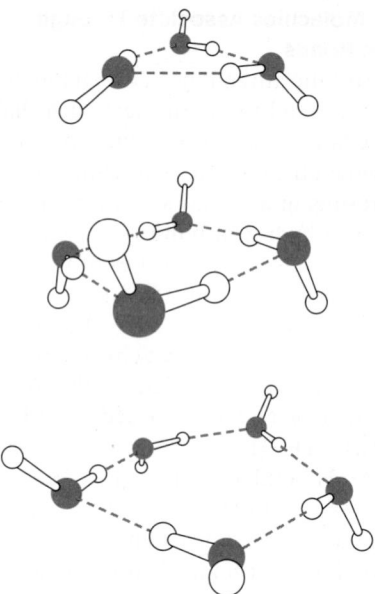

Figure 2-4   Theoretically predicted and spectroscopically confirmed structures of the water trimer, tetramer, and pentamer. Note that these rings are all essentially planar, with each water molecule acting as both a hydrogen bonding donor and acceptor and with the free hydrogens located above and below the planes of the rings. [After Liu, K., Cruzan, J.D., and Saykelly, R.J., *Science* **271**, 930 (1996).]

B. *Water as a Solvent*

Solubility depends on the ability of a solvent to interact with a solute more strongly than solute particles interact with each other. Water is said to be the "universal solvent." Although this statement cannot literally be true, water certainly dissolves more types of substances and in greater amounts than any other solvent. In particular, the polar character of water makes it an excellent solvent for polar and ionic materials, which are therefore said to be **hydrophilic** (Greek: *hydor*, water + *philos*, loving). On the other hand, nonpolar substances are virtually insoluble in water ("oil and water don't mix") and are consequently described as being **hydrophobic** (Greek: *phobos*, fear). Nonpolar substances, however, are soluble in nonpolar solvents such as CCl_4 or hexane. This information is summarized by another maxim, "like dissolves like."

Why do salts dissolve in water? Salts, such as NaCl or K_2HPO_4, are held together by ionic forces. The ions of a salt, as do any electrical charges, interact according to **Coulomb's law:**

$$F = \frac{kq_1q_2}{Dr^2} \qquad [2.1]$$

where F is the force between two electrical charges, q_1 and q_2, that are separated by the distance r, D is the **dielectric constant** of the medium between them, and k is a proportionality constant (8.99 × 10^9 J · m · C^{-2}). Thus, as the dielectric constant of a medium increases, the force between its embedded charges decreases; that is, the dielectric constant

Table 2-1   Dielectric Constants and Permanent Molecular Dipole Moments of Some Common Solvents

Substance	Dielectric Constant	Dipole Moment (debye)
Formamide	110.0	3.37
Water	78.5	1.85
Dimethyl sulfoxide	48.9	3.96
Methanol	32.6	1.66
Ethanol	24.3	1.68
Acetone	20.7	2.72
Ammonia	16.9	1.47
Chloroform	4.8	1.15
Diethyl ether	4.3	1.15
Benzene	2.3	0.00
Carbon tetrachloride	2.2	0.00
Hexane	1.9	0.00

Source: Brey, W.S., *Physical Chemistry and Its Biological Applications*, p. 26, Academic Press (1978).

of a solvent is a measure of its ability to keep opposite charges apart. In a vacuum, D is unity and in air, it is only negligibly larger. The dielectric constants of several common solvents, together with their permanent molecular dipole moments, are listed in Table 2-1. Note that these quantities tend to increase together, although not in a regular way.

The dielectric constant of water is among the highest of any pure liquid, whereas those of nonpolar substances, such as hydrocarbons, are relatively small. The force between two ions separated by a given distance in nonpolar liquids such as hexane or benzene is therefore 30 to 40 times greater than that in water. Consequently, in nonpolar solvents (low D), ions of opposite charge attract each other so strongly that they coalesce to form a salt, whereas the much weaker forces between ions in water solution (high D) permit significant quantities of the ions to remain separated.

An ion immersed in a polar solvent attracts the oppositely charged ends of the solvent dipoles, as is diagrammed in Fig. 2-5 for water. The ion is thereby surrounded by several concentric shells of oriented solvent molecules. Such ions are said to be **solvated** or, if water is the solvent, to be **hydrated.** The electric field produced by the solvent dipoles opposes that of the ion so that, in effect, the ionic charge is spread over the volume of the solvated complex. This arrangement greatly attenuates the coulombic forces between ions, which is why polar solvents have such high dielectric constants.

The orienting effect of ionic charges on dipolar molecules is opposed by thermal motions, which continually tend to randomly reorient all molecules. The dipoles in a solvated complex are therefore only partially oriented. The reason why the dielectric constant of water is so much greater than that of other liquids with comparable dipole moments is that liquid water's hydrogen bonded structure permits it to form oriented structures that resist thermal randomization, thereby more effectively distributing ionic charges. Indeed, ice under high pressure has a measured dielectric constant of 3 because its water molecules cannot reorient in response to an external electric field.

The bond dipoles of uncharged polar molecules make them soluble in aqueous solutions for the same reasons that ionic substances are water soluble. The solubilities of polar and ionic substances are enhanced if they carry functional groups, such as hydroxyl (—OH), keto (—C=O), carboxyl (—CO_2H or —COOH), or amino (—NH_2) groups, that can form hydrogen bonds with water, as is illustrated in Fig. 2-6. Indeed, water-soluble biomolecules such as proteins, nucleic acids, and carbohydrates bristle with just such groups. Nonpolar substances, in contrast, lack both hydrogen bonding donor and acceptor groups.

a. Amphiphiles Form Micelles and Bilayers

Most biological molecules have both polar (or ionically charged) and nonpolar segments and are therefore simultaneously hydrophilic and hydrophobic. Such molecules, for

Figure 2-6   Hydrogen bonding by functional groups. Hydrogen bonds form between water and (*a*) hydroxyl groups, (*b*) keto groups, (*c*) carboxyl groups, and (*d*) amino groups.

Figure 2-5   Solvation of ions by oriented water molecules.

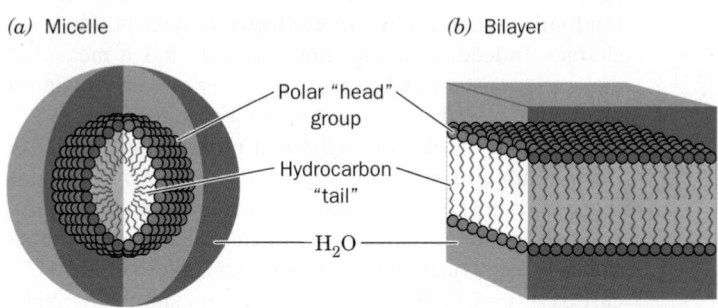

$$CH_3CH_2CH_2CH_2CH_2CH_2CH_2CH_2CH_2CH_2CH_2CH_2CH_2CH_2CH_2 - \overset{\overset{\displaystyle O}{\|}}{C} - O^-$$

Palmitate ($C_{15}H_{31}COO^-$)

$$CH_3CH_2CH_2CH_2CH_2CH_2CH_2CH_2 - \overset{\overset{\displaystyle H}{|}}{C} = \overset{\overset{\displaystyle H}{|}}{C} - CH_2CH_2CH_2CH_2CH_2CH_2CH_2 - \overset{\overset{\displaystyle O}{\|}}{C} - O^-$$

Oleate ($C_{17}H_{33}COO^-$)

Figure 2-7 **Examples of fatty acid anions.** They consist of a polar carboxylate group coupled to a long nonpolar hydrocarbon chain.

(a) Micelle

(b) Bilayer

Polar "head" group

Hydrocarbon "tail"

H₂O

Figure 2-8 **Associations of amphipathic molecules in aqueous solutions.** The polar "head" groups are hydrated, whereas the nonpolar "tails" aggregate so as to exclude the aqueous solution. (*a*) A spheroidal aggregate of amphipathic molecules known as a micelle. (*b*) An extended planar aggregate of amphipathic molecules called a **bilayer.** The bilayer may form a closed spheroidal shell, known as a vesicle, that encloses a small amount of aqueous solution.

example, **fatty acid** ions (soap ions; Fig. 2-7), are said to be **amphiphilic** or, synonymously, **amphipathic** (Greek: *amphi,* both + *pathos,* passion). How do amphiphiles interact with an aqueous solvent? Water, of course, tends to hydrate the hydrophilic portion of an amphiphile, but it also tends to exclude its hydrophobic portion. Amphiphiles consequently tend to form water-dispersed structurally ordered aggregates. Such aggregates may take the form of **micelles,** which are globules of up to several thousand amphiphiles arranged with their hydrophilic groups at the globule surface so that they can interact with the aqueous solvent while the hydrophobic groups associate at the center so as to exclude solvent (Fig. 2-8*a*). However, the model drawn in Fig. 2-8*a* is an oversimplification because it is geometrically impossible for all the hydrophobic groups to occupy the center of the micelle. Instead, the amphipilic molecules pack in a more disorganized and rapidly fluctuating fashion that largely buries their hydrophobic groups and exposes their polar groups (Fig. 2-9). Alternatively, amphiphiles may arrange themselves to form bilayered sheets or vesicles (Fig. 2-8*b*) in which the polar groups face the aqueous phase.

The interactions stabilizing a micelle or bilayer are collectively described as **hydrophobic forces** or **hydrophobic interactions** to indicate that they result from the tendency of water to exclude hydrophobic groups. Hydrophobic interactions are relatively weak compared to hydrogen bonds and lack directionality. Nevertheless, hydrophobic interactions are of pivotal biological importance because, as we shall see in later chapters, they are largely responsible for the structural integrity of biological macromolecules (Sections 8-4C and 29-2C), as well as that of supramolecular aggregates such as membranes. Note that hydrophobic interactions are peculiar to an aqueous environment. Other polar solvents do not promote such associations.

C. Proton Mobility

When an electrical current is passed through an ionic solution, the ions migrate toward the electrode of opposite polarity at a rate proportional to the electrical field and inversely proportional to the frictional drag experienced by the ion as it moves through the solution. This latter quantity, as Table 2-2 indicates, varies with the size of the ion. Note, however, that the ionic mobilities of both H_3O^+ and

Figure 2-9 **Model of a micelle.** Twenty molecules of octyl glucoside (an eight-carbon chain with a sugar head group) are shown in space-filling form in this computer-generated model. The molecules' polar O atoms are red and the C atoms are gray. H atoms have been omitted for clarity. Computer simulations indicate that such micelles have an irregular, rapidly fluctuating structure (unlike the symmetric aggregate pictured in Fig. 2-8*a*) such that portions of the hydrophobic tails are exposed on the micelle surface at any given instant. [Courtesy of Michael Garavito and Shelagh Ferguson-Miller, Michigan State University.]

Table 2.2 Ionic Mobilities^a in H$_2$O at 25°C

Ion	Mobility $\times 10^{-5}$ (cm$^2 \cdot$ V$^{-1} \cdot$ s^{-1})
H$_3$O$^+$	362.4
Li$^+$	40.1
Na$^+$	51.9
K$^+$	76.1
NH$_4^+$	76.0
Mg^{2+}	55.0
Ca^{2+}	61.6
OH$^-$	197.6
Cl$^-$	76.3
Br$^-$	78.3
CH$_3$COO$^-$	40.9
SO$_4^{2-}$	79.8

^{a} Ionic mobility is the distance an ion moves in 1 s under the influence of an electric field of 1 V $\cdot$ cm^{-1}.

Source: Brey, W.S., *Physical Chemistry and Its Biological Applications,* p. 172, Academic Press (1978).

OH$^-$ are anomalously large compared to those of other ions. For H$_3$O$^+$ (the **hydronium ion,** which is abbreviated H$^+$; a bare proton has no stable existence in aqueous solution), this high migration rate results from the ability of protons to jump rapidly from one water molecule to another, as is diagrammed in Fig. 2-10. Although a given hydronium ion can physically migrate through solution in the manner of, say, an Na$^+$ ion, the rapidity of the proton-jump mechanism makes the H$_3$O$^+$ ion's effective ionic mobility much greater than it otherwise would be (the mean lifetime of a given H$_3$O$^+$ ion is 10^{-12} s at 25°C). The anomalously high ionic mobility of the OH$^-$ ion is likewise accounted for by the proton-jump mechanism but, in this case, the apparent direction of ionic migration is opposite

Figure 2-10   Mechanism of hydronium ion migration in aqueous solution via proton jumps. Proton jumps, which mostly occur at random, take place rapidly compared with direct molecular migration, thereby accounting for the observed high ionic mobilities of hydronium and hydroxyl ions in aqueous solutions.

to the direction of proton jumping. Proton jumping is also responsible for the observation that *acid–base reactions are among the fastest reactions that take place in aqueous solutions,* and as we shall see (Section 23-3B), is of importance in biological proton-transfer reactions.

2 ACIDS, BASES, AND BUFFERS

Biological molecules, such as proteins and nucleic acids, bear numerous functional groups, such as carboxyl and amino groups, that can undergo acid–base reactions. Many properties of these molecules therefore vary with the acidities of the solutions in which they are immersed. In this section we discuss the nature of acid–base reactions and how acidities are controlled, both physiologically and in the laboratory.

A. *Acid–Base Reactions*

Acids and **bases,** in a definition coined in the 1880s by Svante Arrhenius, are, respectively, substances capable of donating protons and hydroxide ions. This definition is rather limited, because, for example, it does not account for the observation that NH$_3$, which lacks an OH$^-$ group, exhibits basic properties. In a more general definition, which was formulated in 1923 by Johannes Brønsted and Thomas Lowry, *an acid is a substance that can donate protons (as in the Arrhenius definition) and a base is a substance that can accept protons.* Under this definition, in every acid–base reaction,

$$\text{HA} + \text{H}_2\text{O} \rightleftharpoons \text{H}_3\text{O}^+ + \text{A}^-$$

a **Brønsted acid** (here HA) reacts with a **Brønsted base** (here H$_2$O) to form the **conjugate base** of the acid (A$^-$) and the **conjugate acid** of the base (H$_3$O$^+$) (this reaction is usually abbreviated HA $\rightleftharpoons$ H$^+$ + A$^-$ with the participation of H$_2$O implied). Accordingly, the acetate ion (CH$_3$COO$^-$) is the conjugate base of acetic acid (CH$_3$COOH) and the ammonium ion (NH$_4^+$) is the conjugate acid of ammonia (NH$_3$). (In a yet more general definition of acids and bases, Gilbert Lewis described a **Lewis acid** as a substance that can accept an electron pair and a **Lewis base** as a substance that can donate an electron pair. This definition, which is applicable to both aqueous and nonaqueous systems, is unnecessarily broad for describing most biochemical phenomena.)

a. The Strength of an Acid Is Specified by Its Dissociation Constant

The above acid dissociation reaction is characterized by its **equilibrium constant,** which, for acid–base reactions, is known as a **dissociation constant,**

$$K = \frac{[\text{H}_3\text{O}^+][\text{A}^-]}{[\text{HA}][\text{H}_2\text{O}]} \qquad [2.2]$$

a quantity that is a measure of the relative proton affinities of the HA/A$^-$ and H$_3$O$^+$/H$_2$O conjugate acid–base pairs. Here, as throughout the text, quantities in square brackets symbolize the molar concentrations of the enclosed substances. Since in dilute aqueous solutions the water concentration is essentially constant with $[H_2O] = 1000 \text{ g} \cdot L^{-1}/18.015 \text{ g} \cdot mol^{-1} = 55.5 \text{ M}$, this term is customarily combined with the dissociation constant, which then takes the form

$$K_a = K[H_2O] = \frac{[H^+][A^-]}{[HA]} \qquad [2.3]$$

For brevity, however, we shall henceforth omit the subscript "*a*." The dissociation constants for acids useful in preparing biochemical solutions are listed in Table 2-3.

Acids may be classified according to their relative strengths, that is, according to their abilities to transfer a proton to water. Acids with dissociation constants smaller than that of H_3O^+ (which, by definition, is unity in aqueous solutions) are only partially ionized in aqueous solutions and are known are **weak acids** ($K < 1$). Conversely, **strong acids** have dissociation constants larger than that of H_3O^+ so that they are almost completely ionized in aqueous solutions ($K > 1$). The acids listed in Table 2-3 are all weak acids. However, many of the so-called mineral acids, such as $HClO_4$, HNO_3, HCl, and H_2SO_4 (for its first ionization), are strong acids. Since strong acids rapidly transfer all their protons to H_2O, the strongest acid that can stably exist in aqueous solutions is H_3O^+. Likewise, there can be no stronger base in aqueous solutions than OH^-.

Table 2-3 **Dissociation Constants and pK's at 25°C of Some Acids in Common Laboratory Use as Biochemical Buffers**

Acid	K (M)	pK
Oxalic acid	5.37×10^{-2}	1.27 (pK_1)
H_3PO_4	7.08×10^{-3}	2.15 (pK_1)
Citric acid	7.41×10^{-4}	3.13 (pK_1)
Formic acid	1.78×10^{-4}	3.75
Succinic acid	6.17×10^{-5}	4.21 (pK_1)
Oxalate$^-$	5.37×10^{-5}	4.27 (pK_2)
Acetic acid	1.74×10^{-5}	4.76
Citrate$^-$	1.74×10^{-5}	4.76 (pK_2)
Citrate^{2-}	3.98×10^{-6}	5.40 (pK_3)
Succinate$^-$	2.29×10^{-6}	5.64 (pK_2)
2-(N-Morpholino)ethanesulfonic acid (MES)	8.13×10^{-7}	6.09
Cacodylic acid	5.37×10^{-7}	6.27
H_2CO_3	4.47×10^{-7}	6.35 (pK_1)
N-(2-Acetamido)iminodiacetic acid (ADA)	2.69×10^{-7}	6.57
Piperazine-N,N'-bis(2-ethanesulfonic acid) (PIPES)	1.74×10^{-7}	6.76
N-(2-Acetamido)-2-aminoethanesulfonic acid (ACES)	1.58×10^{-7}	6.80
$H_2PO_4^-$	1.51×10^{-7}	6.82 (pK_2)
3-(N-Morpholino)propanesulfonic acid (MOPS)	7.08×10^{-8}	7.15
N-2-Hydroxyethylpiperazine-N'-2-ethanesulfonic acid (HEPES)	3.39×10^{-8}	7.47
N-2-Hydroxyethylpiperazine-N'-3-propanesulfonic acid (HEPPS)	1.10×10^{-8}	7.96
N-[Tris(hydroxymethyl)methyl]glycine (Tricine)	8.91×10^{-9}	8.05
Tris(hydroxymethyl)aminomethane (Tris)	8.32×10^{-9}	8.08
Glycylglycine	5.62×10^{-9}	8.25
N,N-Bis(2-hydroxyethyl)glycine (Bicine)	5.50×10^{-9}	8.26
Boric acid	5.75×10^{-10}	9.24
NH_4^+	5.62×10^{-10}	9.25
Glycine	1.66×10^{-10}	9.78
HCO_3^-	4.68×10^{-11}	10.33 (pK_2)
Piperidine	7.58×10^{-12}	11.12
HPO_4^{2-}	4.17×10^{-13}	12.38 (pK_3)

Source: Mostly Dawson, R.M.C., Elliott, D.C., Elliott, W.H., and Jones, K.M., *Data for Biochemical Research* (3rd ed.), *pp.* 424–425, Oxford Science Publications (1986); *and* Good, N.E., Winget, G.D., Winter, W., Connolly, T.N., Izawa, S., and Singh, R.M.M., *Biochemistry* **5,** 467 (1966).

Water, being an acid, has a dissociation constant:

$$K = \frac{[H^+][OH^-]}{[H_2O]}$$

As above, the constant $[H_2O] = 55.5M$ can be incorporated into the dissociation constant to yield the expression for the ionization constant of water,

$$K_w = [H^+][OH^-] \qquad [2.4]$$

The value of K_w at 25°C is $10^{-14}\ M^2$. Pure water must contain equimolar amounts of H^+ and OH^- so that $[H^+] = [OH^-] = (K_w)^{1/2} = 10^{-7}\ M$. Since $[H^+]$ and $[OH^-]$ are reciprocally related by Eq. [2.4], if $[H^+]$ is greater than this value, $[OH^-]$ must be correspondingly less and vice versa. Solutions with $[H^+] = 10^{-7}\ M$ are said to be **neutral,** those with $[H^+] > 10^{-7}\ M$ are said to be **acidic,** and those with $[H^+] < 10^{-7}\ M$ are said to be **basic.** Most physiological solutions have hydrogen ion concentrations near neutrality. For example, human blood is normally slightly basic, with $[H^+] = 4.0 \times 10^{-8}\ M$.

The values of $[H^+]$ for most solutions are inconveniently small and difficult to compare. A more practical quantity, which was devised in 1909 by Søren Sørensen, is known as the **pH:**

$$pH = -\log[H^+] \qquad [2.5]$$

The pH of pure water is 7.0, whereas acidic solutions have pH < 7.0 and basic solutions have pH > 7.0. For a $1M$ solution of a strong acid, pH = 0 and for a $1M$ solution of a strong base, pH = 14. Note that if two solutions differ in pH by one unit, they differ in $[H^+]$ by a factor of 10. The pH of a solution may be accurately and easily determined through electrochemical measurements with a device known as a **pH meter.**

b. The pH of a Solution Is Determined by the Relative Concentrations of Acids and Bases

The relationship between the pH of a solution and the concentrations of an acid and its conjugate base can be easily derived by rearranging Eq. [2.3]

$$[H^+] = K\left(\frac{[HA]}{[A^-]}\right)$$

and substituting it into Eq. [2.5]

$$pH = -\log K + \log\left(\frac{[A^-]}{[HA]}\right)$$

Defining $pK = -\log K$ in analogy with Eq. [2.5], we obtain the **Henderson–Hasselbalch equation:**

$$pH = pK + \log\left(\frac{[A^-]}{[HA]}\right) \qquad [2.6]$$

This equation indicates that *the pK of an acid is numerically equal to the pH of the solution when the molar concentrations of the acid and its conjugate base are equal.* Table 2-3 lists the pK values of several acids.

B. *Buffers*

A 0.01-mL droplet of $1M$ HCl added to 1 L of pure water changes the water's pH from 7 to 5, which represents a 100-fold increase in $[H^+]$. Yet, since the properties of biological substances vary significantly with small changes in pH, they require environments in which the pH is insensitive to additions of acids or bases. To understand how this is possible, let us consider the titration of a weak acid with a strong base.

Figure 2-11 shows how the pH values of 1-L solutions of $1M$ acetic acid, $H_2PO_4^-$, and ammonium ion (NH_4^+), vary with the quantity of OH^- added. Titration curves such as those in Fig. 2-11, as well as distribution curves such as those

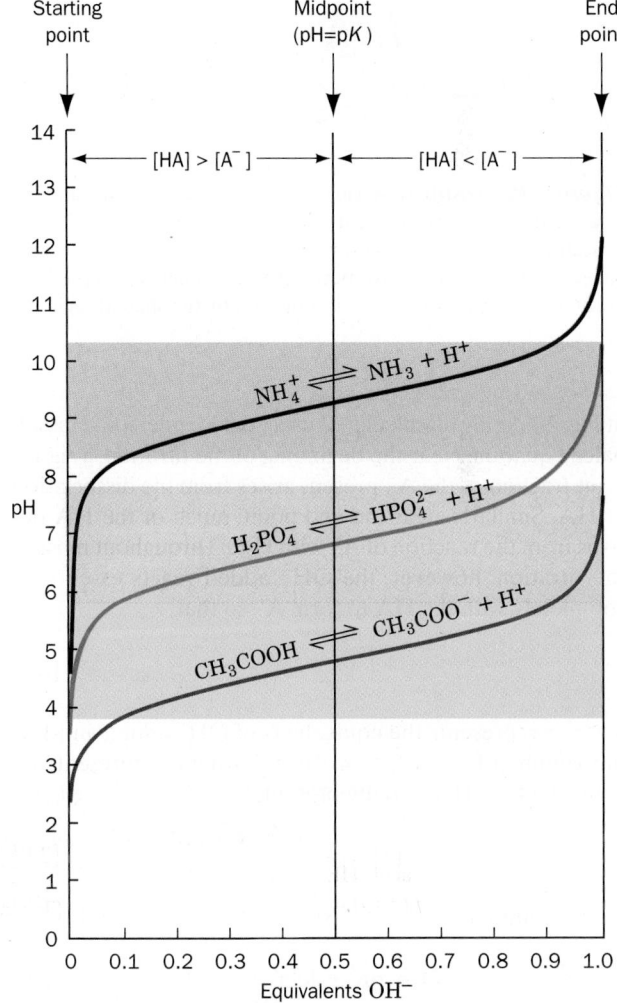

Figure 2-11 **Acid–base titration curves of 1-L solutions of $1M$ acetic acid, $H_2PO_4^-$, and NH_4^+ by a strong base.** At the starting point of each titration, the acid form of the conjugate acid–base pair overwhelmingly predominates. At the midpoint of the titration, where pH = pK, the concentration of the acid is equal to that of its conjugate base. Finally, at the end point of the titration, where the equivalents of strong base added equal the equivalents of acid at the starting point, the conjugate base is in great excess over acid. The shaded bands indicate the pH ranges over which the corresponding solution can function effectively as a buffer. **See the Animated Figures.**

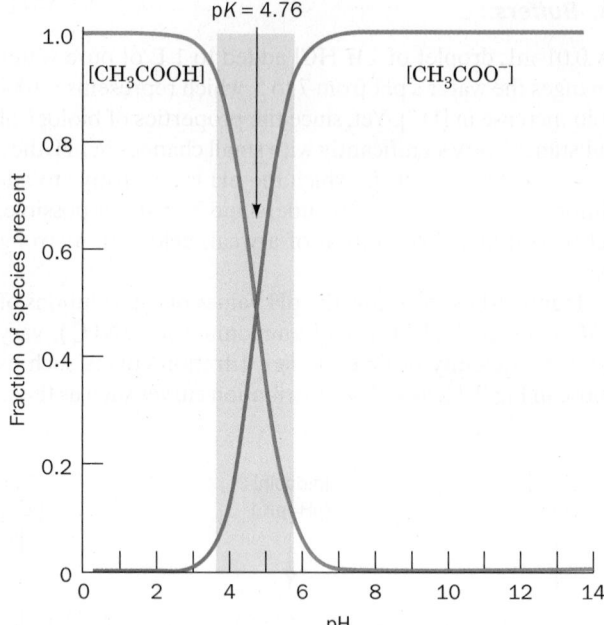

Figure 2-12   Distribution curves for acetic acid and acetate ion.
The fraction of species present is given as the ratio of the
concentration of CH_3COOH or CH_3COO^- to the total
concentrations of these two species. The customarily accepted
useful buffer range of $pK \pm 1$ is indicated by the shaded region.

in Fig. 2-12, may be calculated using the Henderson–Hasselbalch equation. Near the beginning of the titration, a significant fraction of the A^- present arises from the dissociation
of HA. Similarly, near the end point, much of the HA derives from the reaction of A^- with H_2O. Throughout most of
the titration, however, the OH^- added reacts essentially
completely with the HA to form A^- so that

$$[A^-] = \frac{x}{V} \qquad [2.7]$$

where x represents the equivalents of OH^- added and V is
the volume of the solution. Then, letting c_0 represent the
equivalents of HA initially present,

$$[HA] = \frac{c_0 - x}{V} \qquad [2.8]$$

Incorporating these relationships into Eq. [2.6] yields

$$pH = pK + \log\left(\frac{x}{c_0 - x}\right) \qquad [2.9]$$

which accurately describes a titration curve except near its
wings (these regions require more exact treatments that
take into account the ionizations of water).

Several details about the titration curves in Fig. 2-11
should be noted:

1. The curves have similar shapes but are shifted vertically along the pH axis.

2. The pH at the **equivalence point** of each titration
(where the equivalents of OH^- added equal the equiva-

lents of HA initially present) is >7 because of the reaction
of A^- with H_2O to form HA + OH^-; similarly, each initial
pH is <7.

3. The pH at the midpoint of each titration is numerically equal to the pK of its corresponding acid; here, according to the Henderson–Hasselbalch equation, [HA] = [A^-].

4. The slope of each titration curve is much less near its
midpoint than it is near its wings. This indicates that *when
[HA] $\approx$ [A^-], the pH of the solution is relatively insensitive
to the addition of strong base or strong acid. Such a solution,
which is known as an* **acid–base buffer,** *is resistant to pH
changes because small amounts of added H^+ or OH^-, respectively, react with the A^- or HA present without greatly
changing the value of log([A^-]/[HA]).*

a. Buffers Stabilize a Solution's pH

The ability of a buffer to resist pH changes with added
acid or base is directly proportional to the total concentration of the conjugate acid–base pair, [HA] + [A^-]. It is
maximal when pH = pK and decreases rapidly with a
change in pH from that point. A good rule of thumb is that
*a weak acid is in its useful buffer range within 1 pH unit of
its pK* (the shaded regions of Figs. 2-11 and 2-12). Above
this range, where the ratio [A^-]/[HA] > 10, the pH of the
solution changes rapidly with added strong base. A buffer
is similarly impotent with addition of strong acid when its
pK exceeds the pH by more than a unit.

Biological fluids, both those found intracellularly and
extracellularly, are heavily buffered. For example, the pH
of the blood in healthy individuals is closely controlled at
pH 7.4. The phosphate and carbonate ions that are components of most biological fluids are important in this respect
because they have pK's in this range (Table 2-3). Moreover, many biological molecules, such as proteins, nucleic
acids, and lipids, as well as numerous small organic molecules, bear multiple acid–base groups that are effective as
buffer components in the physiological pH range.

The concept that the properties of biological molecules
vary with the acidity of the solution in which they are dissolved was not fully appreciated before the beginning of
the twentieth century so that the acidities of biochemical
preparations made before that time were rarely controlled.
Consequently these early biochemical experiments yielded
poorly reproducible results. More recently, biochemical
preparations have been routinely buffered to simulate the
properties of naturally occurring biological fluids. Many of
the weak acids listed in Table 2-3 are commonly used as
buffers in biochemical preparations. In practice, the chosen
weak acid and one of its soluble salts are dissolved in the
(nearly equal) mole ratio necessary to provide the desired
pH and, with the aid of a pH meter, the resulting solution is
fine-tuned by titration with strong acid or base.

C. *Polyprotic Acids*

Substances that bear more than one acid–base group, such
as H_3PO_4 or H_2CO_3, as well as most biomolecules, are
known as **polyprotic acids.** The titration curves of such sub-

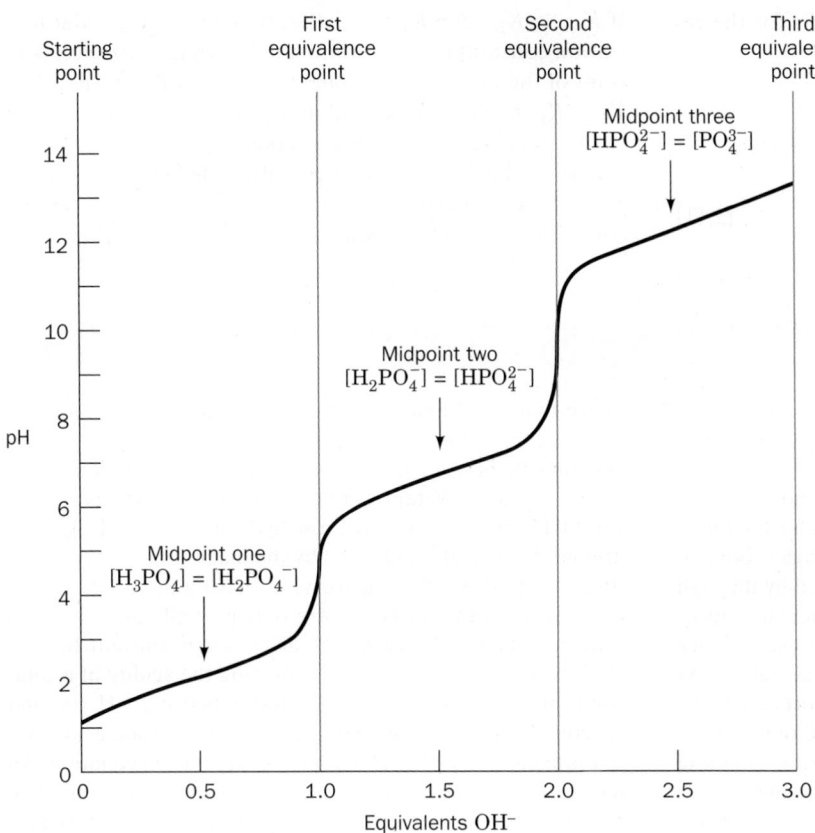

Starting point — First equivalence point — Second equivalence point — Third equivalence point

Midpoint three
$[HPO_4^{2-}] = [PO_4^{3-}]$

Midpoint two
$[H_2PO_4^-] = [HPO_4^{2-}]$

Midpoint one
$[H_3PO_4] = [H_2PO_4^-]$

pH

Equivalents OH⁻

Figure 2-13   Titration curve of a 1-L solution of $1M$ H_3PO_4. The two intermediate equivalence points occur at the steepest parts of the curve. Note the flatness of the curve near its starting points and end points in comparison with the curved ends of the titration curves in Fig. 2-11. This indicates that H_3PO_4 (pK_1 = 2.15) is verging on being a strong acid and PO_4^{3-} (pK_3 = 12.38) is verging on being a strong base. 🐾 **See the Animated Figures.**

stances, as is illustrated in Fig. 2-13 for H_3PO_4, are characterized by multiple pK's, one for each ionization step. Exact calculations of the concentrations of the various ionic species present at a given pH is clearly a more complex task than for a **monoprotic acid.**

The pK's of two closely associated acid–base groups are not independent. The ionic charge resulting from a proton dissociation electrostatically inhibits further proton dissociation from the same molecule, thereby increasing the values of the corresponding pK's. This effect, according to Coulomb's law, decreases as the distance between the ionizing groups increases. For example, the pK's of **oxalic acid's** two adjacent carboxyl groups differ by 3 pH units (Table 2-3), whereas those of **succinic acid,** in which the carboxyl groups are separated by two methylene groups, differ by 1.4 units.

$$H-O-\overset{\overset{O}{\|}}{C}-\overset{\overset{O}{\|}}{C}-O-H \qquad H-O-\overset{\overset{O}{\|}}{C}-CH_2CH_2-\overset{\overset{O}{\|}}{C}-O-H$$

Oxalic acid   **Succinic acid**

Likewise, successive ionizations from the same center, such as in H_3PO_4 or H_2CO_3, have pK's that differ by 4 to 5 pH units. If the pK's for successive ionizations of a polyprotic acid differ by at least 3 pH units, it can be accurately assumed that, at a given pH, only the members of the conjugate acid–base pair characterized by the nearest pK are present in significant concentrations. This, of course, greatly simplifies the calculations for determining the concentrations of the various ionic species present.

a. Polyprotic Acids with Closely Spaced pK's Have Molecular Ionization Constants

If the pK's of a polyprotic acid differ by less than ~2 pH units, as is true in perhaps the majority of biomolecules, the ionization constants measured by titration are not true group ionization constants but, rather, reflect the average ionization of the groups involved. The resulting ionization constants are therefore known as **molecular ionization constants.**

Consider the acid–base equilibria shown in Fig. 2-14 in which there are two nonequivalent protonation sites. Here, the quantities K_A, K_B, K_C, and K_D, the ionization constants for each group, are alternatively called **microscopic ionization constants.** The molecular ionization constant for the removal of the first proton from HAH is

$$K_1 = \frac{[H^+]([AH^-] + [HA^-])}{[HAH]} = K_A + K_B \quad [2.10]$$

Figure 2-14   Ionization of an acid that has two nonequivalent protonation sites.

Similarly, the molecular ionization constant K_2 for the removal of the second proton is

$$K_2 = \frac{[H^+][A^{2-}]}{[AH^-] + [HA^-]} = \frac{1}{(1/K_C) + (1/K_D)}$$

$$= \frac{K_C K_D}{K_C + K_D} \qquad [2.11]$$

If $K_A \gg K_B$, then $K_1 \approx K_A$; that is, the first molecular ionization constant is equal to the microscopic ionization constant of the more acidic group. Likewise, if $K_D \gg K_C$, then $K_2 \approx K_C$, so that the second molecular ionization constant is the microscopic ionization constant of the less acidic group. If the ionization steps differ sufficiently in their pK's, the molecular ionization constants, as expected, become identical to the microscopic ionization constants.

CHAPTER SUMMARY

1 Properties of Water Water is an extraordinary substance, the properties of which are of great biological importance. A water molecule can simultaneously participate in as many as four hydrogen bonds: two as a donor and two as an acceptor. These hydrogen bonds are responsible for the open, low-density structure of ice. Much of this hydrogen bonded structure exists in the liquid phase, as is evidenced by the high boiling point of water compared to those of other substances of similar molecular masses. Physical and theoretical evidence indicates that liquid water maintains a rapidly fluctuating, hydrogen bonded molecular structure that, over short ranges, resembles that of ice. The unique solvent properties of water derive from its polarity as well as its hydrogen bonding properties. In aqueous solutions, ionic and polar substances are surrounded by multiple concentric hydration shells of oriented water dipoles that act to attenuate the electrostatic interactions between the charges in the solution. The thermal randomization of the oriented water molecules is resisted by their hydrogen bonding associations, thereby accounting for the high dielectric constant of water. Nonpolar substances are essentially insoluble in water. However, amphipathic substances aggregate in aqueous solutions to form micelles and bilayers due to the combination of hydrophobic interactions among the nonpolar portions of these molecules and the hydrophilic interactions of their polar groups with the aqueous solvent. The H_3O^+ and OH^- ions have anomalously large ionic mobilities in aqueous solutions because the migration of these ions through solution occurs largely via proton jumping from one H_2O molecule to another.

2 Acids, Bases, and Buffers A Brønsted acid is a substance that can donate protons, whereas a Brønsted base can accept protons. On losing a proton, a Brønsted acid becomes its conjugate base. In an acid–base reaction, an acid donates its proton to a base. Water can react as an acid to form hydroxide ion, OH^-, or as a base to form hydronium ion, H_3O^+. The strength of an acid is indicated by the magnitude of its dissociation constant, K. Weak acids, which have a dissociation constant less than that of H_3O^+, are only partially dissociated in aqueous solution. Water has the dissociation constant 10^{-14} at 25°C. A practical quantity for expressing the acidity of a solution is pH $= -\log[H^+]$. The relationship between pH, pK, and the concentrations of the members of its conjugate acid–base pair is expressed by the Henderson–Hasselbalch equation. An acid–base buffer is a mixture of a weak acid with its conjugate base in a solution that has a pH near the pK of the acid. The ratio $[A^-]/[HA]$ in a buffer is not very sensitive to the addition of strong acids or bases, so that the pH of a buffer is not greatly affected by these substances. Buffers are operationally effective only in the pH range of p$K \pm 1$. Outside of this range, the pH of the solution changes rapidly with the addition of strong acid or base. Buffer capacity also depends on the total concentration of the conjugate acid–base pair. Biological fluids are generally buffered near neutrality. Many acids are polyprotic. However, unless the pK's of their various ionizations differ by less than 2 or 3 pH units, pH calculations can effectively treat them as if they were a mixture of separate weak acids. For polyprotic acids with pK's that differ by less than this amount, the observed molecular ionization constants are simply related to the microscopic ionization constants of the individual dissociating groups.

REFERENCES

Cooke, R. and Kuntz, I.D., The properties of water in biological systems, *Annu. Rev. Biophys. Bioeng.* **3,** 95–126 (1974).

Dill, K.A., Truskett, T.M., Vlachy, V., and Hribar-Lee, B., Modeling water, the hydrophobic effect, and ion solvation, *Annu. Rev. Biophys. Biomol. Struct.* **34,** 173–199 (2005).

Eisenberg, D. and Kauzman, W., *The Structure and Properties of Water,* Oxford University Press (1969). [A comprehensive monograph with a wealth of information.]

Finney, J.L., Water? What's so special about it? *Philos. Trans. R. Soc. Lond. B Biol. Sci.* **29,** 1145–1163 (2004). [Includes discussions of the structure of water molecules, hydrogen bonding, structures of ice and liquid water, and how these relate to biological function.]

Franks, F., *Water,* The Royal Society of Chemistry (1993).

Gestein, M. and Levitt, M., Simulating water and the molecules of life, *Sci. Am.* 279(5), 100–105 (1998).

Martin, T.W. and Derewenda, Z.S., The name is bond—H bond, *Nature Struct. Biol.* **6,** 403–406 (1999). [Reviews the history and nature of the hydrogen bond and describes the X-ray scattering experiments that demonstrated that hydrogen bonds have a partially covalent character.]

Mohammed, O.F., Pines, D., Dreyer, J., Pines, E., and Nibbering, E.T.J., Sequential proton transfer through water bridges in acid–base reactions, *Science* **310,** 83–86 (2005).

Stillinger, F.H., Water revisited, *Science* **209,** 451–457 (1980). [An outline of water structure on an elementary level.]

Tanford, C., *The Hydrophobic Effect: Formation of Micelles and Biological Membranes* (2nd ed.), Chapters 5 and 6, Wiley–

Interscience (1980). [Discussion of the structure of water and of micelles.]

Westhof, E., *Water and Biological Macromolecules,* CRC Press (1993).

Zumdahl, S.S., *Chemical Principles* (5th ed.), Chapters 7 and 8, Houghton Mifflin (2005). [Discusses acid–base chemistry. Most other general chemistry textbooks contain similar information.]

PROBLEMS

1. Draw the hydrogen bonding pattern that water forms with acetamide (CH_3CONH_2) and with pyridine (benzene with a CH group replaced by N).

2. Explain why the dielectric constants of the following pairs of liquids have the order given in Table 2-1: (a) carbon tetrachloride and chloroform; (b) ethanol and methanol; and (c) acetone and formamide.

3. "Inverted" micelles are made by dispersing amphipathic molecules in a nonpolar solvent, such as benzene, together with a small amount of water (counterions are also provided if the head groups are ionic). Draw the structure of an inverted micelle and describe the forces that stabilize it.

*4. Amphipathic molecules in aqueous solutions tend to concentrate at surfaces such as liquid–solid or liquid–gas interfaces. They are therefore said to be **surface-active molecules** or **surfactants.** Rationalize this behavior in terms of the properties of the amphiphiles and indicate the effect that surface-active molecules have on the surface tension of water (surface tension is a measure of the internal cohesion of a liquid as manifested by the force necessary to increase its surface area). Explain why surfactants such as soaps and detergents are effective in dispersing oily substances and oily dirt in aqueous solutions. Why do aqueous solutions of surfactants foam and why does the presence of oily substances reduce this foaming?

5. Indicate how hydrogen bonding forces and hydrophobic forces vary with the dielectric constant of the medium.

6. Using the data in Table 2-2, indicate the times it would take a K^+ and an H^+ ion to each move 1 cm in an electric field of $100\,V \cdot cm^{-1}$.

7. Explain why the mobility of H^+ in ice is only about an order of magnitude less than that in liquid water, whereas the mobility of Na^+ in solid NaCl is zero.

8. Calculate the pH of: (a) 0.1*M* HCl; (b) 0.1*M* NaOH; (c) $3 \times 10^{-5}\,M$ HNO_3; (d) $5 \times 10^{-10}\,M$ $HClO_4$; and (e) $2 \times 10^{-8}\,M$ KOH.

9. The volume of a typical bacterial cell is on the order of 1.0 μm^3. At pH 7, how many hydrogen ions are contained inside a bacterial cell? A bacterial cell contains thousands of macromolecules, such as proteins and nucleic acids, that each bear multiple ionizable groups. What does your result indicate about the common notion that ionizable groups are continuously bathed with H^+ and OH^- ions?

10. Using the data in Table 2-3, calculate the concentrations of all molecular and ionic species and the pH in aqueous solutions that have the following formal compositions: (a) 0.01*M* acetic acid; (b) 0.25*M* ammonium chloride; (c) 0.05*M* acetic acid +

0.10*M* sodium acetate; and (d) 0.20*M* boric acid [$B(OH)_3$] + 0.05*M* sodium borate [$NaB(OH)_4$].

11. **Acid–base indicators** are weak acids that change color on changing ionization states. When a small amount of an appropriately chosen indicator is added to a solution of an acid or base being titrated, the color change "indicates" the **end point** of the titration. **Phenolphthalein** is a commonly used acid–base indicator that, in aqueous solutions, changes from colorless to red-violet in the pH range between 8.2 and 10.0. Referring to Figs. 2-11 and 2-13, describe the effectiveness of phenolphthalein for accurately detecting the end point of a titration with strong base of: (a) acetic acid; (b) NH_4Cl; and (c) H_3PO_4 (at each of its three equivalence points).

*12. The formal composition of an aqueous solution is 0.12*M* K_2HPO_4 + 0.08*M* KH_2PO_4. Using the data in Table 2-3, calculate the concentrations of all ionic and molecular species in the solution and the pH of the solution.

13. Distilled water in equilibrium with air contains dissolved carbon dioxide at a concentration of $1.0 \times 10^{-5}\,M$. Using the data in Table 2-3, calculate the pH of such a solution.

14. Calculate the formal concentrations of acetic acid and sodium acetate necessary to prepare a buffer solution of pH 5 that is 0.20*M* in total acetate. The p*K* of acetic acid is given in Table 2-3.

15. In order to purify a certain protein, you require 0.1*M* glycine buffer at pH 9.4. Unfortunately, your stockroom has run out of glycine. However, you manage to find two 0.1*M* glycine buffer solutions, one at pH 9.0 and the other at pH 10.0. What volumes of these two solutions must you mix in order to obtain 200 mL of your required buffer?

16. An enzymatic reaction takes place in 10 mL of a solution that has a total citrate concentration of 120 m*M* and an initial pH of 7.00. During the reaction (which does not involve citrate), 0.2 milliequivalents of acid are produced. Using the data in Table 2-3, calculate the final pH of the solution. What would the final pH of the solution be in the absence of the citrate buffer assuming that the other components of the solution have no significant buffering capacity and that the solution is initially at pH 7?

*17. A solution's **buffer capacity** β is defined as the ratio of an incremental amount of base added, in equivalents, to the corresponding pH change. This is the reciprocal of the slope of the titration curve, Eq. [2.9]. Derive the equation for β and show that it is maximal at pH = p*K*.

18. Using the data in Table 2-3, calculate the microscopic ionization constants for oxalic acid and for succinic acid. How do these values compare with their corresponding molecular ionization constants?

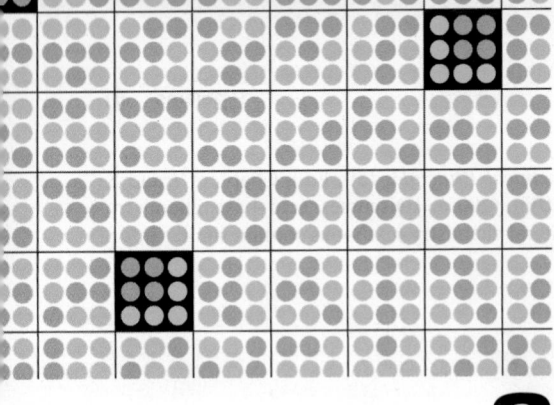

Thermodynamic Principles: A Review

CHAPTER 3

You can't win.
 First law of thermodynamics
You can't even break even.
 Second law of thermodynamics
You can't stay out of the game.
 Third law of thermodynamics

Living things require a continuous throughput of energy. For example, through photosynthesis, plants convert radiant energy from the sun, the primary energy source for life on Earth, to the chemical energy of carbohydrates and other organic substances. The plants, or the animals that eat them, then metabolize these substances to power such functions as the synthesis of biomolecules, the maintenance of concentration gradients, and the movement of muscles. These processes ultimately transform the energy to heat, which is dissipated to the environment. A considerable portion of the cellular biochemical apparatus must therefore be devoted to the acquisition and utilization of energy.

Thermodynamics (Greek: *therme*, heat + *dynamis*, power) is a marvelously elegant description of the relationships among the various forms of energy and how energy affects matter on the macroscopic as opposed to the molecular level; that is, it deals with amounts of matter large enough for their average properties, such as temperature

and pressure, to be well defined. Indeed, the basic principles of thermodynamics were developed in the nineteenth century before the atomic theory of matter had been generally accepted.

With a knowledge of thermodynamics we can determine whether a physical process is possible. Thermodynamics is therefore essential for understanding why macromolecules fold to their native conformations, how metabolic pathways are designed, why molecules cross biological membranes, how muscles generate mechanical force, and so on. The list is endless. Yet the reader should be cautioned that thermodynamics does not indicate the rates at which possible processes actually occur. For instance, although thermodynamics tells us that glucose and oxygen react with the release of copious amounts of energy, it does not indicate that this mixture is indefinitely stable at room temperature in the absence of the appropriate enzymes. The prediction of reaction rates requires, as we shall see in Section 14-1C, a mechanistic description of molecular processes. Yet thermodynamics is also an indispensable guide in formulating such mechanistic models because such models must conform to thermodynamic principles.

Thermodynamics, as it applies to biochemistry, is most frequently concerned with describing the conditions under which processes occur *spontaneously* (by themselves). We shall consequently review the elements of thermodynamics that enable us to predict chemical and biochemical spontaneity: the first and second laws of thermodynamics, the concept of free energy, and the nature of processes at equilibrium. Familiarity with these principles is indispensable for understanding many of the succeeding discussions in this text. We shall, however, postpone consideration of the thermodynamic aspects of metabolism until Sections 16-4 through 16-6.

1 FIRST LAW OF THERMODYNAMICS: ENERGY IS CONSERVED

In thermodynamics, a **system** is defined as that part of the universe that is of interest, such as a reaction vessel or an organism; the rest of the universe is known as the **surroundings.** A system is said to be **open, closed,** or **isolated** according to whether or not it can exchange matter and energy with its surroundings, only energy, or neither matter nor energy. Living organisms, which take up nutrients, release waste products, and generate work and heat, are examples of open systems; if an organism were sealed inside an uninsulated box, it would, together with the box, constitute a closed system, whereas if the box were perfectly insulated, the system would be isolated.

A. *Energy*

The **first law of thermodynamics** is a mathematical statement of the law of conservation of energy: *Energy can be neither created nor destroyed.*

$$\Delta U = U_{final} - U_{initial} = q + w \qquad [3.1]$$

Here U is energy, q represents the **heat** absorbed *by* the system *from* the surroundings, and w is the **work** done *on* the system *by* the surroundings. Heat is a reflection of random molecular motion, whereas work, which is defined as force times the distance moved under its influence, is associated with organized motion. Force may assume many different forms, including the gravitational force exerted by one mass on another, the expansional force exerted by a gas, the tensional force exerted by a spring or muscle fiber, the electrical force of one charge on another, or the dissipative forces of friction and viscosity. Processes in which the system releases heat, which by convention are assigned a negative q, are known as **exothermic processes** (Greek: *exo*, out of); those in which the system gains heat (positive q) are known as **endothermic processes** (Greek: *endon*, within). Under this convention, work done by the system against an external force is defined as a negative quantity.

The SI unit of energy, the **joule (J)**, is steadily replacing the **calorie (cal)** in modern scientific usage. The **large calorie (Cal**, with a capital C) is a unit favored by nutritionists. The relationships among these quantities and other units, as well as the values of constants that will be useful throughout this chapter, are collected in Table 3-1.

a. State Functions Are Independent of the Path a System Follows

Experiments have invariably demonstrated that the energy of a system depends only on its current properties or state, not on how it reached that state. For example, the state of a system composed of a particular gas sample is completely described by its pressure and temperature. The energy of this gas sample is a function only of these so-called **state functions** (quantities that depend only on the state of the system) and is therefore a state function itself. Consequently, there is no net change in energy ($\Delta U = 0$) for any process in which the system returns to its initial state (a **cyclic process**).

Neither heat nor work is separately a state function because each is dependent on the **path** followed by a system in changing from one state to another. For example, in the process of changing from an initial to a final state, a gas may do work by expanding against an external force, or do no work by following a path in which it encounters no external resistance. If Eq. [3.1] is to be obeyed, heat must also be path dependent. It is therefore meaningless to refer to

Table 3-1 **Thermodynamic Units and Constants**

Joule (J)
$1\,J = 1\,kg \cdot m^2 \cdot s^{-2}$ $1\,J = 1\,C \cdot V$ (coulomb volt)
$1\,J = 1\,N \cdot m$ (newton meter)
Calorie (cal)
1 cal heats 1 g of H_2O from 14.5 to 15.5°C
1 cal = 4.184 J
Large calorie (Cal)
1 Cal = 1 kcal   1 Cal = 4184 J
Avogadro's number (N)
$N = 6.0221 \times 10^{23}$ molecules $\cdot$ mol^{-1}
Coulomb (C)
$1\,C = 6.241 \times 10^{18}$ electron charges
Faraday ($\mathcal{F}$)
$1\,\mathcal{F} = N$ electron charges
$1\,\mathcal{F} = 96{,}485\,C \cdot mol^{-1} = 96{,}485\,J \cdot V^{-1} \cdot mol^{-1}$
Kelvin temperature scale (K)
0 K = absolute zero   273.15 K = 0°C
Boltzmann constant (k_B)
$k_B = 1.3807 \times 10^{-23}\,J \cdot K^{-1}$
Gas constant (R)
$R = Nk_B$ $R = 1.9872\,cal \cdot K^{-1} \cdot mol^{-1}$
$R = 8.3145\,J \cdot K^{-1} \cdot mol^{-1}$ $R = 0.08206\,L \cdot atm \cdot K^{-1} \cdot mol^{-1}$

the heat or work content of a system (in the same way that it is meaningless to refer to the number of one dollar bills and ten dollar bills in a bank account containing $85.00). To indicate this property, the heat or work produced during a change of state is never referred to as Δq or Δw but rather as just q or w.

B. *Enthalpy*

Any combination of only state functions must also be a state function. One such combination, which is known as **enthalpy** (Greek: *enthalpein*, to warm in), is defined

$$H = U + PV \qquad [3.2]$$

where V is the volume of the system and P is its pressure. Enthalpy is a particularly convenient quantity with which to describe biological systems because *under constant pressure, a condition typical of most biochemical processes, the enthalpy change between the initial and final states of a process, ΔH, is the easily measured heat that it generates or absorbs.* To show this, let us divide work into two categories: pressure–volume (P–V) work, which is work performed by expansion against an external pressure ($-P\,\Delta V$), and all other work (w'):

$$w = -P\,\Delta V + w' \qquad [3.3]$$

Then, by combining Eqs. [3.1], [3.2], and [3.3], we see that

$$\Delta H = \Delta U + P\,\Delta V = q_P + w + P\,\Delta V = q_P + w' \qquad [3.4]$$

where q_P is the heat transferred at constant pressure. Thus if $w' = 0$, as is often true of chemical reactions, $\Delta H = q_P$. Moreover, the volume changes in most biochemical processes are negligible, so that the differences between their ΔU and ΔH values are usually insignificant.

We are now in a position to understand the utility of state functions. For instance, suppose we wished to determine the enthalpy change resulting from the complete oxidation of 1 g of glucose to CO_2 and H_2O by muscle tissue. To make such a measurement directly would present enormous experimental difficulties. For one thing, the enthalpy changes resulting from the numerous metabolic reactions not involving glucose oxidation that normally occur in living muscle tissue would greatly interfere with our enthalpy measurement. Since enthalpy is a state function, however, we can measure glucose's enthalpy of combustion in any apparatus of our choosing, say, a constant pressure calorimeter rather than a muscle, and still obtain the same value. This, of course, is true whether or not we know the mechanism through which muscle converts glucose to CO_2 and H_2O as long as we can establish that these substances actually are the final metabolic products. *In general, the change of enthalpy in any hypothetical reaction pathway can be determined from the enthalpy change in any other reaction pathway between the same reactants and products.*

We stated earlier in the chapter that thermodynamics serves to indicate whether a particular process occurs spontaneously. Yet the first law of thermodynamics cannot, by itself, provide the basis for such an indication, as the following example demonstrates. If two objects at different temperatures are brought into contact, we know that heat spontaneously flows from the hotter object to the colder one, never vice versa. Yet either process is consistent with the first law of thermodynamics since the aggregate energy of the two objects is independent of their temperature distribution. Consequently, we must seek a criterion of spontaneity other than only conformity to the first law of thermodynamics.

2 SECOND LAW OF THERMODYNAMICS: THE UNIVERSE TENDS TOWARD MAXIMUM DISORDER

When a swimmer falls into the water (a spontaneous process), the energy of the coherent motion of his body is converted to that of the chaotic thermal motion of the surrounding water molecules. The reverse process, the swimmer being ejected from still water by the sudden coherent motion of the surrounding water molecules, has never been witnessed even though such a phenomenon violates neither the first law of thermodynamics nor Newton's laws of motion. This is because *spontaneous processes are characterized by the conversion of order* (in this case the coherent motion of the swimmer's body) *to chaos* (here the random thermal motion of the water molecules). The **second law of thermodynamics,** which expresses this phenomenon, therefore provides a criterion for determining whether a process is spontaneous. Note that thermodynamics says nothing about the rate of a process; that is the purview of **chemical**

kinetics (Chapter 14). Thus a spontaneous process might proceed at only an infinitesimal rate.

A. Spontaneity and Disorder

The second law of thermodynamics states, in accordance with all experience, that *spontaneous processes occur in directions that increase the overall **disorder** of the universe,* that is, of the system and its surroundings. Disorder, in this context, is defined as the number of equivalent ways, W, of arranging the components of the universe. To illustrate this point, let us consider an isolated system consisting of two bulbs of equal volume containing a total of N identical molecules of ideal gas (Fig. 3-1). When the stopcock connecting the bulbs is open, there is an equal probability that a given molecule will occupy either bulb, so there are a total of 2^N equally probable ways that the N molecules may be distributed among the two bulbs. Since the gas molecules are indistinguishable from one another, there are only $(N + 1)$ different states of the system: those with 0, 1, 2, ..., $(N - 1)$, or N molecules in the left bulb. Probability theory indicates that the number of (indistinguishable) ways, W_L, of placing L of the N molecules in the left bulb is

$$W_L = \frac{N!}{L!(N - L)!}$$

The probability of such a state occurring is its fraction of the total number of possible states: $W_L/2^N$.

For any value of N, the state that is most probable, that is, the one with the highest value of W_L, is the one with half of the molecules in one bulb ($L = N/2$ for N even). As N becomes large, the probability that L is nearly equal to $N/2$ approaches unity: For instance, when $N = 10$ the probability that L is within 20% of $N/2$ (that is, 4, 5, or 6) is 0.66, whereas for $N = 50$ this probability (that L is in the range 20–30) is 0.88. For a chemically significant number of molecules, say

(a)

(b)

Figure 3-1   Two bulbs of equal volumes connected by a stopcock. In (*a*), a gas occupies the left bulb, the right bulb is evacuated, and the stopcock is closed. When the stopcock is opened (*b*), the gas molecules diffuse back and forth between the bulbs and eventually become distributed so that half of them occupy each bulb.

Figure 3-2 **The improbability of even a small amount of order.** Consider a simple "universe" consisting of a square array of 9 positions that collectively contain 4 identical "molecules" (*red dots*). If the 4 molecules are arranged in a square, we call the arrangement a "crystal"; otherwise we call it a "gas." The total number of distinguishable arrangements of our 4 molecules in 9 positions is given by

$$W = \frac{9 \cdot 8 \cdot 7 \cdot 6}{4 \cdot 3 \cdot 2 \cdot 1} = 126$$

Here, the numerator indicates that the first molecule may occupy any of the universe's 9 positions, the second molecule may occupy any of the 8 remaining unoccupied positions, and so on, whereas the denominator corrects for the number of indistinguishable arrangements of the 4 identical molecules. Of the 126 arrangements this universe can have, only 4 are crystals (*black squares*). Thus, even in this simple universe, there is a more than 30-fold greater probability that it will contain a disordered gas, when arranged at random, than an ordered crystal. [Illustration, Irving Geis. Image from the Irving Geis Collection, Howard Hughes Medical Institute. Reprinted with permission.]

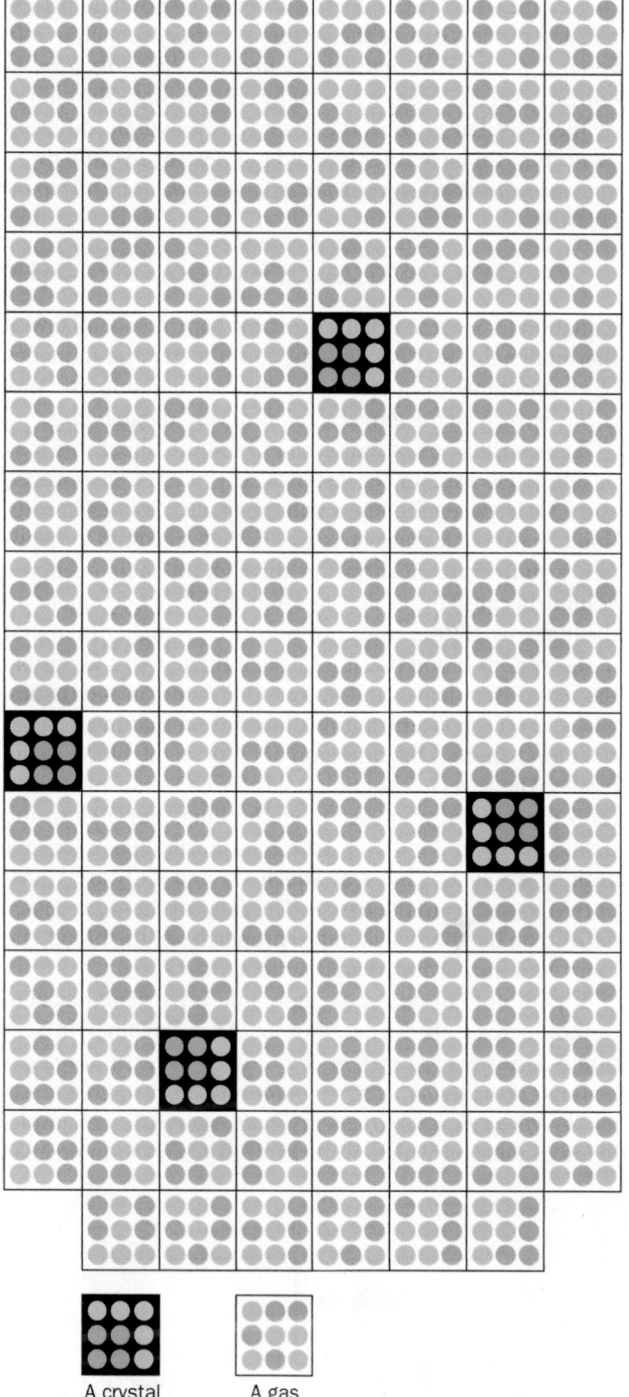

A crystal   A gas

$N = 10^{23}$, the probability that the number of molecules in the left bulb differs from those in the right by as insignificant a ratio as 1 molecule in every 10 billion is 10^{-434}, which, for all intents and purposes, is zero. Therefore, the reason the number of molecules in each bulb of the system in Fig. 3-1*b* is always observed to be equal is not because of any law of motion; the energy of the system is the same for any arrangement of the molecules. *It is because the aggregate probability of all other states is so utterly insignificant* (Fig. 3-2). By the same token, the reason that our swimmer is never thrown out of the water or even noticeably disturbed by the chance coherent motion of the surrounding water molecules is that the probability of such an event is nil.

B. *Entropy*

In chemical systems, *W*, the number of equivalent ways of arranging a system in a particular state, is usually inconveniently immense. For example, when the above twin-bulb system contains N gas molecules, so $W_{N/2} \approx 10^{N \ln 2}$ so that for $N = 10^{23}$, $W_{5 \times 10^{22}} \approx 10^{7 \times 10^{22}}$. In order to be able to deal with W more easily, we define, as did Ludwig Boltzmann in 1877, a quantity known as **entropy** (Greek: *en*, in + *trope*, turning):

$$S = k_B \ln W \qquad [3.5]$$

that increases with W but in a more manageable way. Here k_B is the **Boltzmann constant** (Table 3-1). For our twin-bulb system, $S = k_B N \ln 2$, so the entropy of the system in its most probable state is proportional to the number of gas molecules it contains. Note that *entropy is a state function because it depends only on the parameters that describe a state.*

The laws of random chance cause any system of reasonable size to spontaneously adopt its most probable arrangement, the one in which entropy is a maximum, simply because this state is so overwhelmingly probable. For example, assume that all N molecules of our twin-bulb system are initially placed in the left bulb (Fig. 3-1*a*; $W_N = 1$ and $S = 0$ since there is only one way of doing this). After the stopcock is opened, the molecules will randomly diffuse in and out of the right bulb until eventually they achieve their most probable (maximum entropy) state, that with half of the molecules in each bulb. The gas molecules will subsequently continue to diffuse back and forth between the bulbs, but there will be no further macroscopic (net) change in the system. The system is therefore said to have reached **equilibrium.**

According to Eq. [3.5], the foregoing spontaneous expansion process causes the system's entropy to increase. In general, *for any constant energy process ($\Delta U = 0$), a spontaneous process is characterized by $\Delta S > 0$*. Since the energy of the universe is constant (energy can assume different forms but can be neither created nor destroyed), *any spontaneous process must cause the entropy of the universe to increase:*

$$\Delta S_{system} + \Delta S_{surroundings} = \Delta S_{universe} > 0 \qquad [3.6]$$

Equation [3.6] is the usual expression for the second law of thermodynamics. It is a statement of the general tendency of all spontaneous processes to disorder the universe; that is, *the entropy of the universe tends toward a maximum.*

The conclusions based on our twin-bulb apparatus may be applied to explain, for instance, why blood transports O_2 and CO_2 between the lungs and the tissues. Solutes in solution behave analogously to gases in that they tend to maintain a uniform concentration throughout their occupied volume because this is their most probable arrangement. In the lungs, where the concentration of O_2 is higher than that in the venous blood passing through them, more O_2 enters the blood than leaves it. On the other hand, in the tissues, where the O_2 concentration is lower than that in the arterial blood, there is net diffusion of O_2 from the blood to the tissues. The reverse situation holds for CO_2 transport since the CO_2 concentration is low in the lungs but high in the tissues. Keep in mind, however, that thermodynamics says nothing about the rates at which O_2 and CO_2 are transported to and from the tissues. The rates of these processes depend on the physicochemical properties of the blood, the lungs, and the cardiovascular system.

Equation [3.6] does not imply that a particular system cannot increase its degree of order. As is explained in Section 3-3, however, *a system can only be ordered at the expense of disordering its surroundings to an even greater extent by the application of energy to the system.* For example, living organisms, which are organized from the molecular level upward and are therefore particularly well ordered, achieve this order at the expense of disordering the nutrients they consume. Thus, *eating is as much a way of acquiring order as it is of gaining energy.*

A state of a system may constitute a distribution of more complicated quantities than those of gas molecules in a bulb or simple solute molecules in a solvent. For example, if our system consists of a protein molecule in aqueous solution, its various states differ, as we shall see, in the conformations of the protein's amino acid residues and in the distributions and orientations of its associated water molecules. The second law of thermodynamics applies here because a protein molecule in aqueous solution assumes its native conformation largely in response to the tendency of its surrounding water structure to be maximally disordered (Section 8-4C).

C. Measurement of Entropy

In chemical and biological systems, it is impractical, if not impossible, to determine the entropy of a system by counting the number of ways, W, it can assume its most probable state. An equivalent and more practical definition of entropy was proposed in 1864 by Rudolf Clausius: For spontaneous processes

$$\Delta S \geq \int_{initial}^{final} \frac{dq}{T} \qquad [3.7]$$

where T is the absolute temperature at which the change in heat occurs. The proof of the equivalence of our two definitions of entropy, which requires an elementary knowledge of statistical mechanics, can be found in many physical chemistry textbooks. It is evident, however, that any system becomes progressively disordered (its entropy increases) as its temperature rises (e.g., Fig. 3-3). The equality in Eq. [3.7] holds only for processes in which the system remains in equilibrium throughout the change; these are known as **reversible processes.**

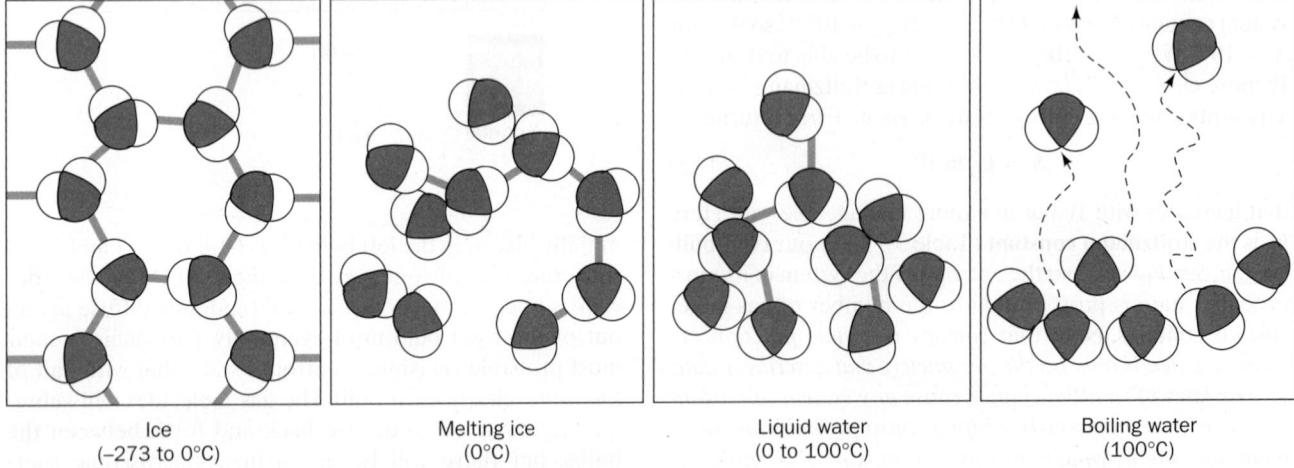

| Ice | Melting ice | Liquid water | Boiling water |
| (−273 to 0°C) | (0°C) | (0 to 100°C) | (100°C) |

Figure 3-3 **Relationship of entropy and temperature.** The structure of water, or any other substance, becomes increasingly disordered, that is, its entropy increases, as its temperature rises.

For the constant temperature conditions typical of biological processes, Eq. [3.7] reduces to

$$\Delta S \geq \frac{q}{T} \qquad [3.8]$$

Thus the entropy change of a reversible process at constant temperature can be determined straightforwardly from measurements of the heat transferred and the temperature at which this occurs. However, since a process at equilibrium can only change at an infinitesimal rate (equilibrium processes are, by definition, unchanging), real processes can approach, but can never quite attain, reversibility. Consequently, *the universe's entropy change in any real process is always greater than its ideal (reversible) value.* This means that when a system departs from and then returns to its initial state via a real process, the entropy of the universe must increase even though the entropy of the system (a state function) does not change.

3 FREE ENERGY: THE INDICATOR OF SPONTANEITY

The disordering of the universe by spontaneous processes is an impractical criterion for spontaneity because it is rarely possible to monitor the entropy of the entire universe. Yet the spontaneity of a process cannot be predicted from a knowledge of the system's entropy change alone. This is because exothermic processes ($\Delta H_{system} < 0$) may be spontaneous even though they are characterized by $\Delta S_{system} < 0$. For example, 2 mol of H_2 and 1 mol of O_2, when sparked, react in a decidedly exothermic reaction to form 2 mol of H_2O. Yet two water molecules, each of whose three atoms are constrained to stay together, are more ordered than are the three diatomic molecules from which they formed. Similarly, under appropriate conditions, many **denatured** (unfolded) proteins will spontaneously fold to assume their highly ordered **native** (normally folded) conformations (Section 9-1A). What we really want, therefore, is a state function that predicts whether or not a given process is spontaneous. In this section, we consider such a function.

A. *Gibbs Free Energy*

The **Gibbs free energy,**

$$G = H - TS \qquad [3.9]$$

which was formulated by J. Willard Gibbs in 1878, is the required indicator of spontaneity for constant temperature and pressure processes. For systems that can only do pressure–volume work ($w' = 0$), combining Eqs. [3.4] and [3.9] while holding T and P constant yields

$$\Delta G = \Delta H - T\,\Delta S = q_P - T\,\Delta S \qquad [3.10]$$

But Eq. [3.8] indicates that $T\,\Delta S \geq q$ for spontaneous processes at constant T. Consequently, $\Delta G \leq 0$ *is the criterion of spontaneity we seek* for the constant T and P conditions that are typical of biochemical processes.

Table 3-2 **Variation of Reaction Spontaneity (Sign of ΔG) with the Signs of ΔH and ΔS**

ΔH	ΔS	$\Delta G = \Delta H - T\,\Delta S$
−	+	The reaction is both enthalpically favored (exothermic) and entropically favored. It is spontaneous (exergonic) at all temperatures.
−	−	The reaction is enthalpically favored but entropically opposed. It is spontaneous only at temperatures *below* $T = \Delta H/\Delta S$.
+	+	The reaction is enthalpically opposed (endothermic) but entropically favored. It is spontaneous only at temperatures *above* $T = \Delta H/\Delta S$.
+	−	The reaction is both enthalpically and entropically opposed. It is *un*spontaneous (endergonic) at all temperatures.

Spontaneous processes, that is, those with negative ΔG values, are said to be **exergonic** (Greek: *ergon,* work); they can be utilized to do work. Processes that are not spontaneous, those with positive ΔG values, are termed **endergonic;** they must be driven by the input of free energy (through mechanisms discussed in Section 3-4C). Processes at equilibrium, those in which the forward and backward reactions are exactly balanced, are characterized by $\Delta G = 0$. Note that the value of ΔG varies directly with temperature. This is why, for instance, the native structure of a protein, whose formation from its denatured form has both $\Delta H < 0$ and $\Delta S < 0$, predominates below the temperature at which $\Delta H = T\,\Delta S$ (the **denaturation temperature**), whereas the denatured protein predominates above this temperature. The variation of the spontaneity of a process with the signs of ΔH and ΔS is summarized in Table 3-2.

B. *Free Energy and Work*

When a system at constant temperature and pressure does non-P–V work, Eq. [3.10] must be expanded to

$$\Delta G = q_P - T\,\Delta S + w' \qquad [3.11]$$

or, because $T\,\Delta S \geq q_P$ (Eq. [3.8]),

$$\Delta G \leq w'$$

so that

$$\Delta G \geq -w' \qquad [3.12]$$

Since P–V work is unimportant in biological systems, ΔG *for a biological process represents its maximum recoverable work.* The ΔG of a process is therefore indicative of the maximum charge separation it can establish, the maximum concentration gradient it can generate (Section 3-4A), the maximum muscular activity it can produce, and so on. In fact, for real processes, which can only approach reversibility, the inequality in Eq. [3.12] holds, so that *the work put into any system can never be fully recovered.* This is indicative of the inherent dissipative character of nature. Indeed, as we have seen, it is precisely this dissipative character that provides the overall driving force for any change.

It is important to reiterate that a large negative value of ΔG does not ensure a chemical reaction will proceed at a measurable rate. This depends on the detailed mechanism of the reaction, which is independent of ΔG. For instance, most biological molecules, including proteins, nucleic acids, carbohydrates, and lipids, are thermodynamically unstable to hydrolysis but, nevertheless, spontaneously hydrolyze at biologically insignificant rates. Only with the introduction of the proper enzymes will the hydrolysis of these molecules proceed at a reasonable pace. Yet a catalyst, which by definition is unchanged by a reaction, cannot affect the ΔG of a reaction. Consequently, *an enzyme can only accelerate the attainment of thermodynamic equilibrium; it cannot, for example, promote a reaction that has a positive ΔG.*

4 CHEMICAL EQUILIBRIA

The entropy (disorder) of a substance increases with its volume. For example, as we have seen for our twin-bulb apparatus (Fig. 3-1), a collection of gas molecules, in occupying all of the volume available to it, maximizes its entropy. Similarly, dissolved molecules become uniformly distributed throughout their solution volume. Entropy is therefore a function of concentration.

If entropy varies with concentration, so must free energy. Thus, as is shown in this section, the free energy change of a chemical reaction depends on the concentrations of both its reactants and its products. This phenomenon is of great biochemical significance because enzymatic reactions can proceed in either direction depending on the relative concentrations of their reactants and products. Indeed, the directions of many enzymatically catalyzed reactions depend on the availability of their **substrates** (reactants) and on the metabolic demand for their products (although most metabolic pathways operate unidirectionally; Section 16-6C).

A. Equilibrium Constants

The relationship between the concentration and the free energy of a substance A, which is derived in the appendix to this chapter, is approximately

$$\overline{G}_A - \overline{G}_A^\circ = RT \ln[A] \qquad [3.13]$$

where $\overline{G}_A$ is known equivalently as the **partial molar free energy** or the **chemical potential** of A (the bar indicates the quantity per mole), $\overline{G}_A^\circ$ is the partial molar free energy of A in its **standard state** (see Section 3-4B), R is the gas constant (Table 3-1), and [A] is the molar concentration of A. Thus for the general reaction,

$$aA + bB \rightleftharpoons cC + dD$$

since free energies are additive and the free energy change

of a reaction is the sum of the free energies of the products less those of the reactants, the free energy change for this reaction is

$$\Delta G = c\overline{G}_C + d\overline{G}_D - a\overline{G}_A - b\overline{G}_B \qquad [3.14]$$

Substituting this relationship into Eq. [3.13] yields

$$\Delta G = \Delta G^\circ + RT \ln\left(\frac{[C]^c[D]^d}{[A]^a[B]^b}\right) \qquad [3.15]$$

where ΔG° is the free energy change of the reaction when all of its reactants and products are in their standard states. Thus the expression for the free energy change of a reaction consists of two parts: (1) a constant term whose value depends only on the reaction taking place, and (2) a variable term that depends on the concentrations of the reactants and the products, the stoichiometry of the reaction, and the temperature.

For a reaction at equilibrium, there is no *net* change because the free energy of the forward reaction exactly balances that of the backward reaction. Consequently, $\Delta G = 0$, so that Eq. [3.15] becomes

$$\Delta G^\circ = -RT \ln K_{eq} \qquad [3.16]$$

where K_{eq} is the familiar **equilibrium constant** of the reaction:

$$K_{eq} = \frac{[C]_{eq}^c [D]_{eq}^d}{[A]_{eq}^a [B]_{eq}^b} = e^{-\Delta G^\circ/RT} \qquad [3.17]$$

and the subscript "eq" in the concentration terms indicates their equilibrium values. (The equilibrium condition is usually clear from the context of the situation, so that equilibrium concentrations are often expressed without this subscript.) *The equilibrium constant of a reaction may therefore be calculated from standard free energy data and vice versa.* Table 3-3 indicates the numerical relationship between ΔG° and K_{eq}. Note that a 10-fold variation of K_{eq} at 25°C corresponds to a 5.7 kJ · mol⁻¹ change in ΔG°, which is less than half of the free energy of even a weak hydrogen bond.

Equations [3.15] through [3.17] indicate that when the reactants in a process are in excess of their equilibrium

Table 3-3 **Variation of K_{eq} with ΔG° at 25°C**

K_{eq}	ΔG° (kJ · mol⁻¹)
10^6	−34.3
10^4	−22.8
10^2	−11.4
10^1	−5.7
10^0	0.0
10^{-1}	5.7
10^{-2}	11.4
10^{-4}	22.8
10^{-6}	34.3

concentrations, the net reaction will proceed in the forward direction until the excess reactants have been converted to products and equilibrium is attained. Conversely, when products are in excess, the net reaction proceeds in the reverse direction so as to convert products to reactants until the equilibrium concentration ratio is likewise achieved. Thus, as **Le Châtelier's principle** states, *any deviation from equilibrium stimulates a process that tends to restore the system to equilibrium. All isolated systems must therefore inevitably reach equilibrium.* Living systems escape this thermodynamic cul-de-sac by being open systems (Section 16-6A).

The manner in which the equilibrium constant varies with temperature is seen by substituting Eq. [3.10] into Eq. [3.16] and rearranging:

$$\ln K_{eq} = \frac{-\Delta H^\circ}{R}\left(\frac{1}{T}\right) + \frac{\Delta S^\circ}{R} \qquad [3.18]$$

where H° and S° represent enthalpy and entropy in the standard state. If ΔH° and ΔS° are independent of temperature, as they often are to a reasonable approximation, a plot of $\ln K_{eq}$ versus $1/T$, known as a **van't Hoff plot,** yields a straight line of slope $-\Delta H^\circ/R$ and intercept $\Delta S^\circ/R$. This relationship permits the values of ΔH° and ΔS° to be determined from measurements of K_{eq} at two (or more) different temperatures. Calorimetric data, which until recent decades were quite difficult to measure for biochemical processes, are therefore not required to obtain the values of ΔH° and ΔS°. Consequently, most biochemical thermodynamic data have been obtained through the application of Eq. [3.18]. However, the development of the **scanning microcalorimeter** has made the direct measurement of ΔH (q_P) for biochemical processes a practical alternative. Indeed, a discrepancy between the values of ΔH° for a reaction as determined calorimetrically and from a van't Hoff plot suggests that the reaction occurs via one or more intermediate states in addition to the initial and final states implicit in the formulation of Eq. [3.18].

B. *Standard Free Energy Changes*

Since only free energy differences, ΔG, can be measured, not free energies themselves, it is necessary to refer these differences to some standard state in order to compare the free energies of different substances (likewise, we refer the elevations of geographic locations to sea level, which is arbitrarily assigned the height of zero). By convention, the free energy of all pure elements in their standard state of 25°C, 1 atm, and in their most stable form (e.g., O_2 not O_3), is defined to be zero. The **free energy of formation** of any nonelemental substance, ΔG_f°, is then defined as the change in free energy accompanying the formation of 1 mol of that substance, in its standard state, from its component elements in their standard states. The standard free energy change for any reaction can be calculated according to

$$\Delta G^\circ = \sum \Delta G_f^\circ(\text{products}) - \sum \Delta G_f^\circ(\text{reactants}) \qquad [3.19]$$

Table 3-4   Free Energies of Formation of Some Compounds of Biochemical Interest

Compound	$-\Delta G_f^\circ$ (kJ · mol^{-1})
Acetaldehyde	139.7
Acetate	369.2
Acetyl-CoA	374.1[a]
cis-Aconitate^{3-}	920.9
$CO_2(g)$	394.4
$CO_2(aq)$	386.2
HCO_3^-	587.1
Citrate^{3-}	1166.6
Dihydroxyacetone^{2-}	1293.2
Ethanol	181.5
Fructose	915.4
Fructose-6-phosphate^{2-}	1758.3
Fructose-1,6-bisphosphate^{4-}	2600.8
Fumarate^{2-}	604.2
α-D-Glucose	917.2
Glucose-6-phosphate^{2-}	1760.3
Glyceraldehyde-3-phosphate^{2-}	1285.6
H^+	0.0
$H_2(g)$	0.0
$H_2O(\ell)$	237.2
Isocitrate^{3-}	1160.0
α-Ketoglutarate^{2-}	798.0
Lactate$^-$	516.6
L-Malate^{2-}	845.1
OH^-	157.3
Oxaloacetate^{2-}	797.2
Phosphoenolpyruvate^{3-}	1269.5
2-Phosphoglycerate^{3-}	1285.6
3-Phosphoglycerate^{3-}	1515.7
Pyruvate$^-$	474.5
Succinate^{2-}	690.2
Succinyl-CoA	686.7[a]

[a]For formation from free elements + free CoA (coenzyme A).

Source: Metzler, D.E., *Biochemistry, The Chemical Reactions of Living Cells* (2nd ed.), *pp.* 290–291, Harcourt/Academic Press (2001).

Table 3-4 provides a list of standard free energies of formation, ΔG_f°, for a selection of substances of biochemical significance.

a. Standard State Conventions in Biochemistry

The standard state convention commonly used in physical chemistry defines the standard state of a solute as that with unit **activity** at 25°C and 1 atm (activity is concentration corrected for nonideal behavior, as is explained in the appendix to this chapter; for the dilute solutions typical of biochemical reactions in the laboratory, such corrections are small, so activities can be replaced by concentrations).

However, because biochemical reactions usually occur in dilute aqueous solutions near neutral pH, a somewhat different standard state convention for biological systems has been adopted:

● Water's standard state is defined as that of the pure liquid, so that the activity of pure water is taken to be unity despite the fact that its concentration is 55.5M. In essence, the [H_2O] term is incorporated into the value of the equilibrium constant. This procedure simplifies the free energy expressions for reactions in dilute aqueous solutions involving water as a reactant or product because the [H_2O] term can then be ignored.

● The hydrogen ion activity is defined as unity at the physiologically relevant pH of 7 rather than at the physical chemical standard state of pH 0, where many biological substances are unstable.

● The standard state of a substance that can undergo an acid–base reaction is defined in terms of the total concentration of its naturally occurring ion mixture at pH 7. In contrast, the physical chemistry convention refers to a pure species whether or not it actually exists at pH 0. The advantage of the biochemistry convention is that the total concentration of a substance with multiple ionization states, such as most biological molecules, is usually easier to measure than the concentration of one of its ionic species. Since the ionic composition of an acid or base varies with pH, however, the standard free energies calculated according to the biochemistry convention are valid only at pH 7.

Under the biochemistry convention, the standard free energy changes of substances are customarily symbolized by $\Delta G^{\circ\prime}$ in order to distinguish them from physical chemistry standard free energy changes, ΔG° (note that the value of ΔG for any process, being experimentally measurable, is independent of the chosen standard state; i.e., $\Delta G = \Delta G'$). Likewise, the biochemical equilibrium constant, which is defined by using $\Delta G^{\circ\prime}$ in place of ΔG° in Eq. [3.17], is represented by K'_{eq}.

The relationship between $\Delta G^{\circ\prime}$ and ΔG° is often a simple one. There are three general situations:

1. If the reacting species include neither H_2O nor H^+, the expressions for $\Delta G^{\circ\prime}$ and ΔG° coincide.

2. For a reaction in dilute aqueous solution that yields n H_2O molecules:

$$A + B \rightleftharpoons C + D + n\,H_2O$$

Eqs. [3.16] and [3.17] indicate that

$$\Delta G^\circ = -RT \ln K_{eq} = -RT \ln\left(\frac{[C][D][H_2O]^n}{[A][B]}\right)$$

Under the biochemistry convention, which defines the activity of pure water as unity,

$$\Delta G^{\circ\prime} = -RT \ln K'_{eq} = -RT \ln\left(\frac{[C][D]}{[A][B]}\right)$$

Therefore

$$\Delta G^{\circ\prime} = \Delta G^\circ + nRT \ln[H_2O] \qquad [3.20]$$

where [H_2O] = 55.5M (the concentration of water in aqueous solution), so that for a reaction at 25°C which yields 1 mol of H_2O, $\Delta G^{\circ\prime} = \Delta G^\circ + 9.96$ kJ · mol^{-1}.

3. For a reaction involving hydrogen ions, such as

$$A + B \rightleftharpoons C + HD$$
$$\Big\Updownarrow K$$
$$D^- + H^+$$

where

$$K = \left(\frac{[H^+][D^-]}{[HD]}\right)$$

manipulations similar to those above lead to the relationship

$$\Delta G^{\circ\prime} = \Delta G^\circ - RT \ln(1 + K/[H^+]_0) + RT \ln[H^+]_0 \qquad [3.21]$$

where [H^+]$_0$ = $10^{-7}\,M$, the only value of [H^+] for which this equation is valid. Of course, if more than one ionizable species participates in the reaction and/or if any of them are polyprotic, Eq. [3.21] is correspondingly more complicated.

C. *Coupled Reactions*

The additivity of free energy changes allows an endergonic reaction to be driven by an exergonic reaction under the proper conditions. This phenomenon is the thermodynamic basis for the operation of metabolic pathways, since most of these reaction sequences comprise endergonic as well as exergonic reactions. Consider the following two-step reaction process:

$$(1) \qquad A + B \rightleftharpoons C + D \qquad \Delta G_1$$
$$(2) \qquad D + E \rightleftharpoons F + G \qquad \Delta G_2$$

If $\Delta G_1 \geq 0$, Reaction (1) will not occur spontaneously. However, if ΔG_2 is sufficiently exergonic so that $\Delta G_1 + \Delta G_2 < 0$, then although the equilibrium concentration of D in Reaction (1) will be relatively small, it will be larger than that in Reaction (2). As Reaction (2) converts D to products, Reaction (1) will operate in the forward direction to replenish the equilibrium concentration of D. The highly exergonic Reaction (2) therefore *drives* the endergonic Reaction (1), and the two reactions are said to be **coupled** through their common intermediate, D. That these coupled reactions proceed spontaneously (although not necessarily at a finite rate) can also be seen by summing Reactions (1) and (2) to yield the overall reaction

$$(1+2) \qquad A + B + E \rightleftharpoons C + F + G \qquad \Delta G_3$$

Where $\Delta G_3 = \Delta G_1 + \Delta G_2 < 0$. *As long as the overall pathway (reaction sequence) is exergonic, it will operate in the forward direction.* Thus, the free energy of ATP hydrolysis, a highly exergonic process, is harnessed to drive many oth-erwise endergonic biological processes to completion (Section 16-4C).

APPENDIX Concentration Dependence of Free Energy

To establish that the free energy of a substance is a function of its concentration, consider the free energy change of an ideal gas during a reversible pressure change at constant temperature ($w' = 0$, since an ideal gas is incapable of doing non-P–V work). Substituting Eqs. [3.1] and [3.2] into Eq. [3.9] and differentiating the result yields

$$dG = dq + dw + P\,dV + V\,dP - T\,dS \quad [3.A1]$$

On substitution of the differentiated forms of Eqs. [3.3] and [3.8] into this expression, it reduces to

$$dG = V\,dP \quad [3.A2]$$

The ideal gas equation is $PV = nRT$, where n is the number of moles of gas. Therefore

$$dG = nRT\frac{dP}{P} = nRT\,d\ln P \quad [3.A3]$$

This gas phase result can be extended to the more biochemically relevant area of solution chemistry by application of **Henry's law** for a solution containing the volatile solute A in equilibrium with the gas phase:

$$P_A = K_A X_A \quad [3.A4]$$

Here P_A is the partial pressure of A when its mole fraction in the solution is X_A, and K_A is the **Henry's law constant** of A in the solvent being used. It is generally more convenient, however, to express the concentrations of the relatively dilute solutions of chemical and biological systems in terms of molarity rather than mole fractions. For a dilute solution

$$X_A \approx \frac{n_A}{n_{\text{solvent}}} = \frac{[A]}{[\text{solvent}]} \quad [3.A5]$$

where the solvent concentration, [solvent], is approximately constant. Thus

$$P_A \approx K'_A[A] \quad [3.A6]$$

where $K'_A = K_A/[\text{solvent}]$. Substituting this expression into Eq. [3.A3] yields

$$dG_A = n_A RT\,d(\ln K'_A + \ln[A]) = n_A RT\,d\ln[A] \quad [3.A7]$$

Free energy, as are energy and enthalpy, is a relative quantity that can only be defined with respect to some arbitrary standard state. The standard state is customarily taken to be 25°C, 1 atm pressure, and, for the sake of mathematical simplicity, $[A] = 1$. The integration of Eq. [3.A7] from the standard state, $[A] = 1$, to the final state, $[A] = [A]$, results in

$$G_A - G_A^\circ = n_A RT\ln[A] \quad [3.A8]$$

where G_A° is the free energy of A in the standard state and [A] really represents the concentration ratio [A]/1. Since Henry's law is valid for real solutions only in the limit of infinite dilution, however, the standard state is defined as the entirely hypothetical state of $1M$ solute with the properties that it has at infinite dilution.

The free energy terms in Eq. [3.A8] may be converted from **extensive quantities** (those dependent on the amount of material) to **intensive quantities** (those independent of the amount of material) by dividing both sides of the equation by n_A. This yields

$$\overline{G}_A - \overline{G}_A^\circ = RT\ln[A] \quad [3.A9]$$

Equation [3.A9] has the limitation that it refers to solutions that exactly follow Henry's law, although real solutions only do so in the limit of infinite dilution if the solute is, in fact, volatile. These difficulties can all be eliminated by replacing [A] in Eq. [3.A9] by a quantity, a_A, known as the **activity** of A. This is defined

$$a_A = \gamma_A[A] \quad [3.A10]$$

where γ_A is the **activity coefficient** of A. Equation [3.A9] thereby takes the form

$$\overline{G}_A - \overline{G}_A^\circ = RT\ln a_A \quad [3.A11]$$

in which all departures from ideal behavior, including the provision that the system may perform non-P–V work, are incorporated into the activity coefficient, which is an experimentally measurable quantity. Ideal behavior is only approached at infinite dilution; that is, $\gamma_A \to 1$ as $[A] \to 0$. The standard state in Eq. [3.A11] is redefined as that of unit activity.

The concentrations of reactants and products in most laboratory biochemical reactions are usually so low (on the order of millimolar or less) that the activity coefficients of these various species are nearly unity. Consequently, the activities of most biochemical species under laboratory conditions can be satisfactorily approximated by their molar concentrations:

$$\overline{G}_A - \overline{G}_A^\circ = RT\ln[A] \quad [3.13]$$

However, the activity coefficient of a particular species varies with the total concentration of all other species present as well as with its own concentration. Thus, despite the low concentrations of most biochemical species in the cell, their extraordinarily high combined concentrations (e.g., see Fig. 1-13) make the activity coefficients of the individual species deviate significantly from unity. Unfortunately, it is difficult to determine the values of these quantities in a cellular compartment (where it is likewise difficult to determine the concentration of any given species).

CHAPTER SUMMARY

1 First Law of Thermodynamics: Energy Is Conserved
The first law of thermodynamics,

$$\Delta U = q + w \qquad [3.1]$$

where q is heat and w is work, is a statement of the law of conservation of energy. Energy is a state function because the energy of a system depends only on the state of the system. Enthalpy,

$$H = U + PV \qquad [3.2]$$

where P is pressure and V is volume, is a closely related state function that represents the heat at constant pressure under conditions where only pressure–volume work is possible.

2 Second Law of Thermodynamics: The Universe Tends Toward Maximum Disorder Entropy, which is also a state function, is defined

$$S = k_B \ln W \qquad [3.5]$$

where W, the disorder, is the number of equivalent ways the system can be arranged under the conditions governing it and k_B is the Boltzmann constant. The second law of thermodynamics states that the universe tends toward maximum disorder and hence $\Delta S_{universe} > 0$ for any real process.

3 Free Energy: The Indicator of Spontaneity The Gibbs free energy of a system

$$G = H - TS \qquad [3.9]$$

decreases in a spontaneous, constant pressure process. In a process at equilibrium, the system suffers no net change, so that $\Delta G = 0$. An ideal process, in which the system is always at equilibrium, is said to be reversible. All real processes are irreversible since processes at equilibrium can only occur at an infinitesimal rate.

4 Chemical Equilibria   For a chemical reaction

$$a\text{A} + b\text{B} \rightleftharpoons c\text{C} + d\text{D}$$

the change in the Gibbs free energy is expressed

$$\Delta G = \Delta G^\circ + RT \ln\left(\frac{[\text{C}]^c[\text{D}]^d}{[\text{A}]^a[\text{B}]^b}\right) \qquad [3.15]$$

where ΔG°, the standard free energy change, is the free energy change at 25°C, 1 atm pressure, and unit activities of reactants and products. The biochemical standard state, $\Delta G^{\circ\prime}$, is similarly defined but in dilute aqueous solution at pH 7 in which the activities of water and H^+ are both defined as unity. At equilibrium

$$\Delta G^{\circ\prime} = -RT \ln K'_{eq} = -RT \ln\left(\frac{[\text{C}]^c_{eq}[\text{D}]^d_{eq}}{[\text{A}]^a_{eq}[\text{B}]^b_{eq}}\right)$$

where K'_{eq} is the equilibrium constant under the biochemical convention. An endergonic reaction ($\Delta G > 0$) may be driven by an exergonic reaction ($\Delta G < 0$) if they are coupled and if the overall reaction is exergonic.

REFERENCES

Allen, J.P., *Biophysical Chemistry* Chapters 1–5, Wiley-Blackwell (2008).

Atkins, P.W. and de Paula, J., *Physical Chemistry for the Life Sciences*, Chapters 1–5, Freeman (2006). [Most physical chemistry texts treat thermodynamics in some detail.]

Edsall, J.T. and Gutfreund, H., *Biothermodynamics*, Wiley (1983).

Hammes, G.G., *Physical Chemistry for the Biological Sciences,* Chapters 1 and 2, Wiley (2007).

Haynie, D.T., *Biological Thermodynamics* (2nd ed.), Cambridge University Press (2008).

Tinoco, I., Jr., Sauer, K., Wang, J.C., and Puglisi, J.C., *Physical Chemistry. Principles and Applications in Biological Sciences* (4th ed.), Chapters 2–5, Prentice Hall (2002).

van Holde, K.E., Johnson, W.C., and Ho, P.S., *Principles of Physical Biochemistry* (2nd ed.), Chapter 2, Prentice Hall (2006). [The equivalence of the Boltzmann and Clausius formulations of the second law of thermodynamics is demonstrated in Section 2.3.]

PROBLEMS

1. A common funeral litany is the Biblical verse: "Ashes to ashes, dust to dust." Why might a bereaved family of thermodynamicists be equally comforted by a recitation of the second law of thermodynamics?

2. How many 4-m-high flights of stairs must an overweight person weighing 75 kg climb to atone for the indiscretion of eating a 500-Cal hamburger? Assume that there is a 20% efficiency in converting nutritional energy to mechanical energy. The gravitational force of an object of mass m kg is $F = mg$, where the gravitational constant g is 9.8 m · s^{-2}.

3. In terms of thermodynamic concepts, why is it more difficult to park a car in a small space than it is to drive it out from such a space?

4. It has been said that an army of dedicated monkeys, typing at random, would eventually produce all of Shakespeare's works. How long, on average, would it take 1 million monkeys, each typing on a 46-key keyboard (space included but no shift key) at the rate of 1 keystroke per second, to type the phrase "to be or not to be"? How long, on average, would it take one monkey to do so at a computer if the computer would only accept the correct letter in the phrase and then would shift to its next letter (i.e., the computer knew what it wanted)? What do these results indicate about the probability of order randomly arising from disorder versus order arising through a process of evolution?

5. Show that the transfer of heat from an object of higher temperature to one of lower temperature, but not the reverse process, obeys the second law of thermodynamics.

6. Carbon monoxide crystallizes with its CO molecules arranged in parallel rows. Since CO is a very nearly ellipsoidal molecule, in the absence of polarity effects, adjacent CO molecules could equally well line up in a head-to-tail or a head-to-head fashion. In a crystal consisting of 10^{23} CO molecules, what is the entropy of all the CO molecules being aligned head to tail?

7. The U.S. Patent Office has received, and continues to receive, numerous applications for perpetual motion machines. Perpetual motion machines have been classified as those of the first kind, which violate the first law of thermodynamics, and those of the second kind, which violate the second law of thermodynamics. The fallacy in a perpetual motion machine of the first kind is generally easy to detect. An example would be a motor-driven electrical generator that produces energy in excess of that input by the motor. The fallacy in a perpetual motion machine of the second type, however, is usually more subtle. Take, for example, a ship that uses heat energy extracted from the sea by a heat pump to boil water so as to power a steam engine that drives the ship as well as the heat pump. Show, in general terms, that such a propulsion system would violate the second law of thermodynamics.

8. Using the data in Table 3-4, calculate the values of $\Delta G°$ at 25°C for the following metabolic reactions:

(a) $C_6H_{12}O_6 + 6\ O_2 \rightleftharpoons 6\ CO_2(aq) + 6\ H_2O(\ell)$
    Glucose

(b) $C_6H_{12}O_6 \rightleftharpoons 2\ CH_3CH_2OH + 2\ CO_2(aq)$
    Glucose   Ethanol

(c) $C_6H_{12}O_6 \rightleftharpoons 2\ CH_3CHOHCOO^- + 2H^+$
    Glucose   Lactate

[These reactions, respectively, constitute oxidative metabolism, alcoholic fermentation in yeast deprived of oxygen, and homolactic fermentation in skeletal muscle requiring energy faster than oxidative metabolism can supply it (Section 17-3B).]

*9. The native and denatured forms of a protein are generally in equilibrium as follows:

$$\text{Protein }(denatured) \rightleftharpoons \text{protein }(native)$$

For a certain solution of the protein **ribonuclease A,** in which the total protein concentration is $2.0 \times 10^{-3}\ M$, the concentrations of the denatured and native proteins at both 50 and 100°C are given in the following table:

Temperature (°C)	[Ribonuclease A (denatured)] (M)	[Ribonuclease A (native)] (M)
50	5.1×10^{-6}	2.0×10^{-3}
100	2.8×10^{-4}	1.7×10^{-3}

(a) Determine $\Delta H°$ and $\Delta S°$ for the folding reaction assuming that these quantities are independent of temperature. (b) Calculate $\Delta G°$ for ribonuclease A folding at 25°C. Is this process spontaneous under standard state conditions at this temperature? (c) What is the denaturation temperature of ribonuclease A under standard state conditions?

*10. Using the data in Table 3-4, calculate $\Delta G_f^{°\prime}$ for the following compounds at 25°C: (a) $H_2O(\ell)$; (b) sucrose (sucrose + $H_2O \rightleftharpoons$ glucose + fructose: $\Delta G^{°\prime} = -29.3\ \text{kJ} \cdot \text{mol}^{-1}$); and (c) ethyl acetate (ethyl acetate + $H_2O \rightleftharpoons$ ethanol + acetate$^-$ + H^+: $\Delta G^{°\prime} = -19.7\ \text{kJ} \cdot \text{mol}^{-1}$; the p$K$ of acetic acid is 4.76).

11. Calculate the equilibrium constants for the hydrolysis of the following compounds at pH 7 and 25°C: (a) phosphoenolpyruvate ($\Delta G^{°\prime} = -61.9\ \text{kJ} \cdot \text{mol}^{-1}$); (b) pyrophosphate ($\Delta G^{°\prime} = -33.5\ \text{kJ} \cdot \text{mol}^{-1}$); and (c) glucose-1-phosphate ($\Delta G^{°\prime} = -20.9\ \text{kJ} \cdot \text{mol}^{-1}$).

12. $\Delta G^{°\prime}$ for the isomerization reaction

$$\text{Glucose-1-phosphate(G1P)} \rightleftharpoons \text{glucose-6-phosphate(G6P)}$$

is $-7.1\ \text{kJ} \cdot \text{mol}^{-1}$. Calculate the equilibrium ratio of [G1P] to [G6P] at 25°C.

13. For the reaction A → B at 298 K, the change in enthalpy is $-7\ \text{kJ} \cdot \text{mol}^{-1}$ and the change in entropy is $-25\ \text{J} \cdot \text{K}^{-1} \cdot \text{mol}^{-1}$. Is the reaction spontaneous? If not, should the temperature be increased or decreased to make the reaction spontaneous?

14. Two biochemical reactions have the same $K_{eq} = 5 \times 10^{-8}$ at temperature $T_1 = 298$ K. However, reaction 1 has $\Delta H° = -28\ \text{kJ} \cdot \text{mol}^{-1}$ and Reaction 2 has $\Delta H° = +28\ \text{kJ} \cdot \text{mol}^{-1}$. The two reactions utilize the same reactants. Your lab partner has proposed that you can get more of the reactants to proceed via Reaction 2 rather than Reaction 1 by lowering the temperature of the reaction. Will this strategy work? Why or why not? How much would the temperature have to be raised or lowered to change the value of K_2/K_1 from 1 to 10?

The digestive enzyme bovine carboxypeptidase A showing its central β sheet.

PART

II

BIOMOLECULES

Amino Acids

CHAPTER 4

It is hardly surprising that much of the early biochemical research was concerned with the study of proteins. Proteins form the class of biological macromolecules that have the most well-defined physicochemical properties, and consequently they were generally easier to isolate and characterize than nucleic acids, polysaccharides, or lipids. Furthermore, proteins, particularly in the form of enzymes, have obvious biochemical functions. The central role that proteins play in biological processes has therefore been recognized since the earliest days of biochemistry. In contrast, the task of nucleic acids in the transmission and expression of genetic information was not realized until the late 1940s and their catalytic function only began to come to light in the 1980s, the role of lipids in biological membranes was not appreciated until the 1960s, and the biological functions of polysaccharides are still somewhat mysterious.

In this chapter we study the structures and properties of the monomeric units of proteins, the **amino acids.** It is from these substances that proteins are synthesized through processes that we discuss in Chapter 32. Amino acids are also energy metabolites and, in animals, many of them are essential nutrients (Chapter 26). In addition, as we shall see, many amino acids and their derivatives are of biochemical importance in their own right (Section 4-3B).

1 THE AMINO ACIDS OF PROTEINS

The analyses of a vast number of proteins from almost every conceivable source have shown that *all proteins are composed of the 20 "standard" amino acids listed in Table 4-1.* These substances are known as **α-amino** acids because, with the exception of **proline,** they have a primary amino group and a carboxylic acid group substituent on the same carbon atom (Fig. 4-1; proline has a secondary amino group).

A. General Properties

The pK values of the 20 "standard" α-amino acids of proteins are tabulated in Table 4-1. Here pK_1 and pK_2, respectively, refer to the α-carboxylic acid and α-amino groups, and pK_R refers to the side groups with acid–base properties. Table 4-1 indicates that the pK values of the α-carboxylic acid groups lie in a small range around 2.2 so that above pH 3.5 these groups are almost entirely in their carboxylate forms. The α-amino groups all have pK values near 9.4 and are therefore almost entirely in their ammonium ion forms below pH 8.0. This leads to an important structural point: *In the physiological pH range, both the carboxylic acid and the amino groups of α-amino acids are completely ionized* (Fig. 4-2). An amino acid can therefore act as either an acid or a base. Substances with this property are said to be **amphoteric** and are referred to as **ampholytes** (*ampho*teric electro*lytes*). In Section 4-1D, we shall delve a bit deeper into the acid–base properties of the amino acids.

$$H_2N - \underset{\underset{H}{|}}{\overset{\overset{R}{|}}{C_\alpha}} - COOH$$

Figure 4-1 **General structural formula for α-amino acids.** There are 20 different R groups in the commonly occurring amino acids (Table 4-1).

$$H_3\overset{+}{N} - \underset{\underset{H}{|}}{\overset{\overset{R}{|}}{C}} - COO^-$$

Figure 4-2 **Zwitterionic form of the α-amino acids that occurs at physiological pH values.**

Table 4-1 **Covalent Structures and Abbreviations of the "Standard" Amino Acids of Proteins, Their Occurrence, and the pK Values of Their Ionizable Groups**

Name, Three-Letter Symbol, and One-Letter Symbol	Structural Formula[a]	Residue Mass (D)[b]	Average Occurrence in Proteins (%)[c]	pK_1 α-COOH[d]	pK_2 α-NH$_3^+$[d]	pK_R Side Chain[d]
Amino acids with nonpolar side chains						
Glycine Gly G		57.0	7.1	2.35	9.78	
Alanine Ala A		71.1	8.3	2.35	9.87	
Valine Val V		99.1	6.9	2.29	9.74	
Leucine Leu L		113.2	9.7	2.33	9.74	
Isoleucine Ile I		113.2	6.0	2.32	9.76	
Methionine Met M		131.2	2.4	2.13	9.28	
Proline Pro P		97.1	4.7	1.95	10.64	
Phenylalanine Phe F		147.2	3.9	2.20	9.31	
Tryptophan Trp W		186.2	1.1	2.46	9.41	

(continued)

[a]The ionic forms shown are those predominating at pH 7.0 (except for that of histidine[e]), although residue mass is given for the neutral compound. The C$_\alpha$ atoms, as well as those atoms marked with an asterisk, are chiral centers with configurations as indicated according to Fischer projection formulas. The standard organic numbering system is provided for heterocycles.

[b]The residue masses are given for the neutral residues. For molecular masses of the parent amino acids, add 18.0 D, the molecular mass of H$_2$O, to the residue masses. For side chain masses, subtract 56.0 D, the formula mass of a peptide group, from the residue masses.

[c]The average amino acid composition in the complete SWISS-PROT database (http://www.expasy.ch/sprot/relnotes/relstat.html), Release 55.11.

[d]From Dawson, R.M.C., Elliott, D.C., Elliott, W.H., and Jones, K.M., *Data for Biochemical Research* (3rd ed.), *pp.* 1–31, Oxford Science Publications (1986).

[e]Both the neutral and protonated forms of histidine are present at pH 7.0 because its pK_R is close to 7.0. The imidazole ring of histidine is numbered here according to the biochemistry convention. In the IUPAC convention, N3 of the biochemistry convention is designated N1 and the numbering increases clockwise around the ring.

[f]The three- and one-letter symbols for asparagine *or* aspartic acid are Asx and B, whereas for glutamine *or* glutamic acid they are Glx and Z. The one-letter symbol for an undetermined *or* "nonstandard" amino acid is X.

Table 4-1 **(Continued)**

Name Three-Letter Symbol, and One-Letter Symbol	Structural Formula[a]	Residue Mass (D)[b]	Average Occurrence in Proteins (%)[c]	pK_1 α-COOH[d]	pK_2 α-^{d}NH$_3^{+d}$	pK_R Side Chain[d]
Amino acids with uncharged polar side chains						
Serine Ser S		87.1	6.5	2.19	9.21	
Threonine Thr T		101.1	5.3	2.09	9.10	
Asparagine[f] Asn N		114.1	4.0	2.14	8.72	
Glutamine[f] Gln Q		128.1	3.9	2.17	9.13	
Tyrosine Tyr Y		163.2	2.9	2.20	9.21	10.46 (phenol)
Cysteine Cys C		103.1	1.4	1.92	10.70	8.37 (sulfhydryl)
Amino acids with charged polar side chains						
Lysine Lys K		128.2	5.9	2.16	9.06	10.54 (ε-NH$_3^+$)
Arginine Arg R		156.2	5.5	1.82	8.99	12.48 (guanidino)
Histidine[e] His H		137.1	2.3	1.80	9.33	6.04 (imidazole)
Aspartic acid[f] Asp D		115.1	5.4	1.99	9.90	3.90 (β-COOH)
Glutamic acid[f] Glu E		129.1	6.8	2.10	9.47	4.07 (γ-COOH)

Molecules that bear charged groups of opposite polarity are known as **zwitterions** (German: *zwitter*, hybrid) or **dipolar ions.** The zwitterionic character of the α-amino acids has been established by several methods including spectroscopic measurements and X-ray crystal structure determinations (in the solid state the α-amino acids are zwitterionic because the basic amine group abstracts a proton from the nearby acidic carboxylic acid group). Because amino acids are zwitterions, their physical properties are characteristic of ionic compounds. For instance, most α-amino acids have melting points near 300°C, whereas their nonionic derivatives usually melt around 100°C. Furthermore, amino acids, like other ionic compounds, are more soluble in polar solvents than in nonpolar solvents. Indeed, most α-amino acids are very soluble in water but are largely insoluble in most organic solvents.

B. *Peptide Bonds*

The α-amino acids polymerize, at least conceptually, through the elimination of a water molecule as is indicated in Fig. 4-3. The resulting CO—NH linkage, which was independently characterized in 1902 by Emil Fischer and Franz Hofmeister, is known as a **peptide bond.** Polymers composed of two, three, a few (3–10), and many **amino acid residues** (alternatively called **peptide units**) are known, respectively, as **dipeptides, tripeptides, oligopeptides,** and **polypeptides.** These substances, however, are often referred to simply as "peptides." *Proteins are molecules that consist of one or more polypeptide chains.* These polypeptides range in length from ~40 to ~34,000 amino acid residues (although few have more than 1500 residues) and, since the average mass of an amino acid residue is ~110 D, have molecular masses that range from ~40 to over ~3700 kD.

Polypeptides are **linear polymers;** that is, each amino acid residue is linked to its neighbors in a head-to-tail fashion rather than forming branched chains. This observation reflects the underlying elegant simplicity of the way living systems construct these macromolecules for, as we shall see, the nucleic acids that encode the amino acid sequences

of polypeptides are also linear polymers. This permits the direct correspondence between the monomer (nucleotide) sequence of a nucleic acid and the monomer (amino acid) sequence of the corresponding polypeptide without the added complication of specifying the positions and sequences of any branching chains.

With 20 different choices available for each amino acid residue in a polypeptide chain, it is easy to see that a huge number of different protein molecules can exist. For example, for dipeptides, each of the 20 different choices for the first amino acid residue can have 20 different choices for the second amino acid residue, for a total of $20^2 = 400$ distinct dipeptides. Similarly, for tripeptides, there are 20 possibilities for each of the 400 choices of dipeptides to yield a total of $20^3 = 8000$ different tripeptides. A relatively small protein molecule consists of a single polypeptide chain of 100 residues. There are $20^{100} = 1.27 \times 10^{130}$ possible unique polypeptide chains of this length, a quantity vastly greater than the estimated number of atoms in the universe (9×10^{78}). Clearly, nature can have made only a tiny fraction of the possible different protein molecules. Nevertheless, *the various organisms on Earth collectively synthesize an enormous number of different protein molecules whose great range of physicochemical characteristics stem largely from the varied properties of the 20 "standard" amino acids.*

C. *Classification and Characteristics*

The most common and perhaps the most useful way of classifying the 20 "standard" amino acids is according to the polarities of their side chains **(R groups).** This is because proteins fold to their native conformations largely in response to the tendency to remove their hydrophobic side chains from contact with water and to solvate their hydrophilic side chains (Chapters 8 and 9). According to this classification scheme, there are three major types of amino acids: (1) those with nonpolar R groups, (2) those with uncharged polar R groups, and (3) those with charged polar R groups.

a. The Nonpolar Amino Acid Side Chains Have a Variety of Shapes and Sizes

Nine amino acids are classified as having nonpolar side chains. **Glycine** (which, when it was found to be a component of gelatin in 1820, was the first amino acid to be identified in protein hydrolyzates) has the smallest possible side chain, an H atom. **Alanine** (Fig. 4-4), **valine, leucine,** and **isoleucine** have aliphatic hydrocarbon side chains ranging in size from a methyl group for alanine to isomeric butyl groups for leucine and isoleucine. **Methionine** has a thiol ether side chain that resembles an *n*-butyl group in many of its physical properties (C and S have nearly equal electronegativities and S is about the size of a methylene group). **Proline,** a cyclic secondary amino acid, has conformational constraints imposed by the cyclic nature of its pyrrolidine side chain, which is unique among the "standard" 20 amino acids. **Phenylalanine,** with its phenyl moiety (Fig. 4-4), and **tryptophan,** with its indole group, contain

Figure 4-3 **Condensation of two α-amino acids to form a dipeptide.** The peptide bond is shown in red.

Alanine   **Glutamine**   **Phenylalanine**

Figure 4-4 **Structures of the α-amino acids alanine, glutamine, and phenylalanine.** The amino acids are shown as ball-and-stick models embedded in their transparent space-filling models. The atoms are colored according to type with C green, H white, N blue, and O red.

aromatic side chains, which are characterized by bulk as well as nonpolarity.

b. Uncharged Polar Side Chains Have Hydroxyl, Amide, or Thiol Groups

Six amino acids are commonly classified as having uncharged polar side chains. **Serine** and **threonine** bear hydroxylic R groups of different sizes. **Asparagine** and **glutamine** (Fig. 4-4) have amide-bearing side chains of different sizes. **Tyrosine** has a phenolic group, which, together with the aromatic groups of phenylalanine and tryptophan, accounts for most of the UV absorbance and fluorescence exhibited by proteins (Section 9-1Cb). **Cysteine** has a thiol group that is unique among the 20 amino acids in that it often forms a disulfide bond to another cysteine residue through the oxidation of their thiol groups (Fig. 4-5). This disulfide bond has great importance in protein structure: *It can join separate polypeptide chains or cross-link two cysteines in the same chain.* Two disulfide-linked cysteines are referred to in the older biochemical literature as the amino acid **cystine** because they were originally thought to form a unique amino acid. However, the discovery that cystine residues arise through the cross-linking of two cysteine residues after polypeptide biosynthesis has occurred has caused the name cystine to become less commonly used.

c. Charged Polar Side Chains May Be Positively or Negatively Charged

Five amino acids have charged side chains. The basic amino acids are positively charged at physiological pH values; they are **lysine,** which has a butylammonium side chain, **arginine,** which bears a guanidino group, and **histidine,** which carries an imidazolium moiety. Of the 20 α-amino acids, only histidine, with $pK_R = 6.0$, ionizes within the physiological pH range. At pH 6.0, its imidazole side group is only 50% charged so that histidine is neutral at the basic end of the physiological pH range. As a consequence, histidine side chains often participate in the catalytic reactions of enzymes. The acidic amino acids, **aspartic acid** and **glutamic acid,** are negatively charged above pH 3; in their ionized state, they are often referred to as **aspartate** and **glutamate.** Asparagine and glutamine are, respectively, the amides of aspartic acid and glutamic acid.

The allocation of the 20 amino acids among the three different groups is, of course, somewhat arbitrary. For example, glycine and alanine, the smallest of the amino acids, and tryptophan, with its heterocyclic ring, might just as well be classified as uncharged polar amino acids. Similarly, tyrosine and cysteine, with their ionizable side chains, might also be thought of as charged polar amino acids, particularly at higher pH's, whereas asparagine and glutamine are nearly as polar as their corresponding carboxylates, aspartate and glutamate.

The 20 amino acids vary considerably in their physicochemical properties such as polarity, acidity, basicity, aromaticity, bulk, conformational flexibility, ability to cross-link, ability to hydrogen bond, and chemical reactivity. These several characteristics, many of which are interrelated, are largely responsible for proteins' great range of properties.

Figure 4-5 **The reaction linking two cysteine residues by a disulfide bond.**

D. *Acid–Base Properties*

Amino acids and proteins have conspicuous acid–base properties. The α-amino acids have two or, for those with ionizable side groups, three acid–base groups. The titration curve of glycine, the simplest amino acid, is shown in Fig. 4-6. At low pH values, both acid–base groups of glycine are fully protonated so that it assumes the cationic form $^+H_3NCH_2COOH$. In the course of the titration with a strong base, such as NaOH, glycine loses two protons in the stepwise fashion characteristic of a polyprotic acid.

The pK values of glycine's two ionizable groups are sufficiently different so that the Henderson–Hasselbalch equation:

$$pH = pK + \log\left(\frac{[A^-]}{[HA]}\right) \qquad [2.6]$$

closely approximates each leg of its titration curve. Consequently, the pK for each ionization step is that of the midpoint of its corresponding leg of the titration curve (Sections 2-2A & 2-2C): At pH 2.35 the concentrations of the cationic form, $^+H_3NCH_2COOH$, and the zwitterionic form, $^+H_3NCH_2COO^-$, are equal; similarly, at pH 9.78 the concentrations of the zwitterionic form and the anionic form, $H_2NCH_2COO^-$, are equal. Note that *amino acids never assume the neutral form in aqueous solution.*

The pH at which a molecule carries no net electric charge is known as its **isoelectric point, p*I*.** For the α-amino acids, the application of the Henderson–Hasselbalch equation indicates that, to a high degree of precision,

$$pI = \frac{1}{2}(pK_i + pK_j) \qquad [4.1]$$

where K_i and K_j are the dissociation constants of the two ionizations involving the neutral species. For monoamino, monocarboxylic acids such as glycine, K_i and K_j represent K_1 and K_2. However, for aspartic and glutamic acids, K_i and K_j are K_1 and K_R, whereas for arginine, histidine, and lysine, these quantities are K_R and K_2.

Acetic acid's pK (4.76), which is typical of aliphatic monocarboxylic acids, is ~2.4 pH units higher than the pK_1 of its α-amino derivative glycine. This large difference in pK values of the same functional group is caused, as is discussed in Section 2-2C, by the electrostatic influence of glycine's positively charged ammonium group; that is, its —NH_3^+ group helps repel the proton from its COOH group. Conversely, glycine's carboxylate group increases the basicity of its amino group ($pK_2 = 9.78$) with respect to that of glycine methyl ester (pK = 7.75). However, the —NH_3^+ groups of glycine and its esters are significantly more acidic than are aliphatic amines ($pK \approx 10.7$) because of the electron-withdrawing character of the carboxyl group.

The electronic influence of one functional group on another is rapidly attenuated as the distance between the groups increases. Hence, the pK values of the α-carboxylate groups of amino acids and the side chain carboxylates of aspartic and glutamic acids form a series that is progressively closer in value to the pK of an aliphatic monocarboxylic acid. Likewise, the ionization constant of lysine's side chain amino group is indistinguishable from that of an aliphatic amine.

a. Proteins Have Complex Titration Curves

The titration curves of the α-amino acids with ionizable side chains, such as that of glutamic acid, exhibit the expected three pK values. However, the titration curves of polypeptides and proteins, an example of which is shown in Fig. 4-7, rarely provide any indication of individual pK values because of the large numbers of ionizable groups they represent (typically 30% of a protein's amino acid side chains are ionizable; Table 4-1). Furthermore, the covalent and three-dimensional structure of a protein may cause the pK of each ionizable group to shift by as much as several pH units from its value in the free α-amino acid as a result of the electrostatic influence of nearby charged groups, medium effects arising from the proximity of groups of low dielectric constant, and the effects of hydrogen bonding associations. The titration curve of a protein is also a function of the salt concentration, as is shown in Fig. 4-7, because the salt ions act electrostatically to shield the side chain charges from one another, thereby attenuating these charge–charge interactions.

E. *A Few Words on Nomenclature*

The three-letter abbreviations for the 20 amino acid residues are given in Table 4-1. It is worthwhile memorizing these symbols because they are widely used throughout the biochemical literature, including this text. These abbreviations are, in most cases, taken from the first three letters of the corresponding amino acid's name; they are conversationally pronounced as read.

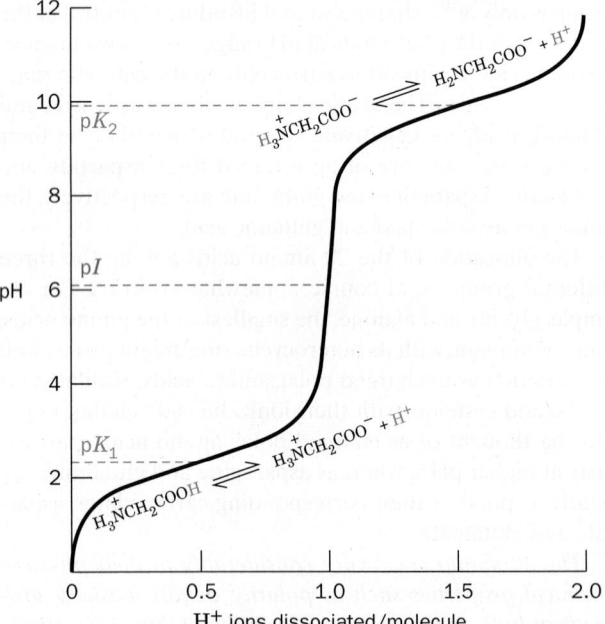

Figure 4-6 **Titration curve of glycine.** Other monoamino, monocarboxylic acids ionize in a similar fashion. [After Meister, A., *Biochemistry of the Amino Acids* (2nd ed.), Vol. 1, *p.* 30, Academic Press (1965).] 🔊**See the Animated Figures**

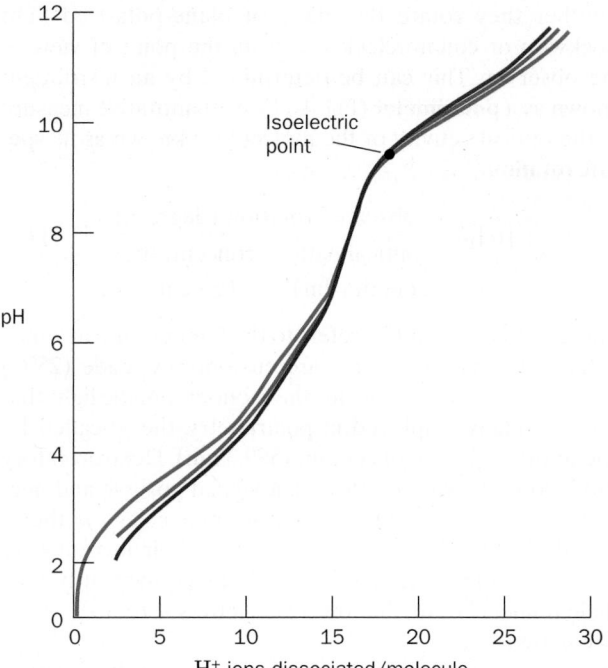

Figure 4-7 **Titration curves of the enzyme ribonuclease A at 25°C.** The concentration of KCl is $0.01M$ for the blue curve, $0.03M$ for the red curve, and $0.15M$ for the green curve. [After Tanford, C. and Hauenstein, J.D., *J. Am. Chem. Soc.* **78**, 5287 (1956).]

Figure 4-8 **The tetrapeptide Ala-Tyr-Asp-Gly.**

The symbol **Glx** means Glu or Gln and, similarly, **Asx** means Asp or Asn. These ambiguous symbols stem from laboratory experience: Asn and Gln are easily hydrolyzed to aspartic acid and glutamic acid, respectively, under the acidic or basic conditions that are usually used to excise them from proteins. Therefore, without special precautions, we cannot determine whether a detected Glu was originally Glu or Gln, and likewise for Asp and Asn.

The one-letter symbols for the amino acids are also given in Table 4-1. This more compact code is often used when comparing the amino acid sequences of several similar proteins and hence should also be memorized. Note that the one-letter symbols are usually the first letter of the amino acid residue's name. However, for those sets of residues that have the same first letter, this is only true of the most abundant residue of the set.

Amino acid residues in polypeptides are named by dropping the suffix **-ine** in the name of the amino acid and replacing it by **-yl.** Polypeptide chains are described by starting at the amino terminus (known as the **N-terminus**) and sequentially naming each residue until the carboxyl terminus (the **C-terminus**) is reached. The amino acid at the C-terminus is given the name of its parent amino acid. Thus the compound shown in Fig. 4-8 is alanyltyrosylaspartylglycine. Of course such names for polypeptide chains of more than a few residues are extremely cumbersome. The use of abbreviations for amino acid residues partially relieves this problem. Thus the foregoing tetrapeptide is Ala-Tyr-Asp-Gly using the three-letter abbreviations and AYDG using the one-letter symbols. Note that these ab-

breviations are always written so that the N-terminus of the polypeptide chain is to the left and the C-terminus is to the right.

The various nonhydrogen atoms of the amino acid side chains are often named in sequence with the Greek alphabet (α, β, γ, δ, ε, ζ, η, …) starting at the carbon atom adjacent to the peptide carbonyl group (the C_α atom). Therefore, as Fig. 4-9 indicates, Glu has a γ-carboxyl group and Lys has a ζ-amino group (alternatively known as an ε-amino group because the N atom is substituent to C_ε). Unfortunately, this labeling system is ambiguous for several amino acids. Consequently, standard numbering schemes for organic molecules are also employed. These are indicated in Table 4-1 for the heterocyclic side chains.

2 OPTICAL ACTIVITY

The amino acids as isolated by the mild hydrolysis of proteins are, with the exception of glycine, all **optically active;** that is, they rotate the plane of plane-polarized light (see below).

Optically active molecules have an asymmetry such that they are not superimposable on their mirror image in the same way that a left hand is not superimposable on its mirror image, a right hand. This situation is characteristic of substances that contain tetrahedral carbon atoms that have four different substituents. The two such molecules

Glu   **Lys**

Figure 4-9 **Greek lettering scheme used to identify the atoms in the glutamyl and lysyl R groups.**

$$
\begin{array}{ccc}
& \text{Cl} & \text{Cl} \\
& \diagup & \diagup \\
\text{H}-\text{C}\cdots\text{F} & & \text{F}\cdots\text{C}-\text{H} \\
& \diagup & \diagup \\
& \text{Br} & \text{Br}
\end{array}
$$

Mirror plane

Figure 4-10 The two enantiomers of fluorochlorobromomethane. The four substituents are tetrahedrally arranged about the central atom with the dotted lines indicating that a substituent lies behind the plane of the paper, a triangular line indicating that it lies above the plane of the paper, and a thin line indicating that it lies in the plane of the paper. The mirror plane relating the enantiomers is represented by a vertical dashed line.

depicted in Fig. 4-10 are not superimposable since they are mirror images. The central atoms in such atomic constellations are known as **asymmetric centers** or **chiral centers** and are said to have the property of **chirality** (Greek: *cheir*, hand). The C_α atoms of all the amino acids, with the exception of glycine, are asymmetric centers. Glycine, which has two H atoms substituent to its C_α atom, is superimposable on its mirror image and is therefore not optically active.

Molecules that are nonsuperimposable mirror images are known as **enantiomers** of one another. Enantiomeric molecules are physically and chemically indistinguishable by most techniques. *Only when probed asymmetrically, for example, by plane-polarized light or by reactants that also contain chiral centers, can they be distinguished and/or differentially manipulated.*

There are three commonly used systems of nomenclature whereby a particular stereoisomer of an optically active molecule can be classified. These are explained in the following sections.

A. *An Operational Classification*

Molecules are classified as **dextrorotatory** (Greek: *dexter*, right) or **levorotatory** (Greek: *laevus*, left) depending on whether they rotate the plane of plane-polarized light clockwise or counterclockwise from the point of view of the observer. This can be determined by an instrument known as a **polarimeter** (Fig. 4-11). A quantitative measure of the optical activity of the molecule is known as its **specific rotation:**

$$
[\alpha]_D^{25} = \frac{\text{observed rotation (degrees)}}{\begin{array}{c}\text{optical path} \\ \text{length (dm)}\end{array} \times \begin{array}{c}\text{concentration} \\ (\text{g}\cdot\text{cm}^{-3})\end{array}} \qquad [4.2]
$$

where the superscript 25 refers to the temperature at which polarimeter measurements are customarily made (25°C) and the subscript D indicates the monochromatic light that is traditionally employed in polarimetry, the so-called D-line in the spectrum of sodium (589.3 nm). Dextrorotatory and levorotatory molecules are assigned positive and negative values of $[\alpha]_D^{25}$. Dextrorotatory molecules are therefore designated by the prefix $(+)$ and their levorotatory enantiomers have the prefix $(-)$. In an equivalent but archaic nomenclature, the lowercase letters *d (dextro)* and *l (levo)* are used.

The sign and magnitude of a molecule's specific rotation depend on the structure of the molecule in a complicated and poorly understood manner. It is not yet possible to predict reliably the magnitude or even the sign of a given molecule's specific rotation. For example, proline, leucine, and arginine, which are isolated from proteins, have specific rotations in pure aqueous solutions of $-86.2°$, $-10.4°$, and $+12.5°$, respectively. Their enantiomers exhibit values of $[\alpha]_D^{25}$ of the same magnitude but of opposite signs. As might be expected from the acid–base nature of the amino acids, these quantities vary with the solution pH.

A problem with this operational classification system for optical isomers is that it provides no presently interpretable indication of the **absolute configuration** (spatial arrangement) of the chemical groups about a chiral center. Furthermore, a molecule with more than one asymmetric center may have an optical rotation that is not obviously

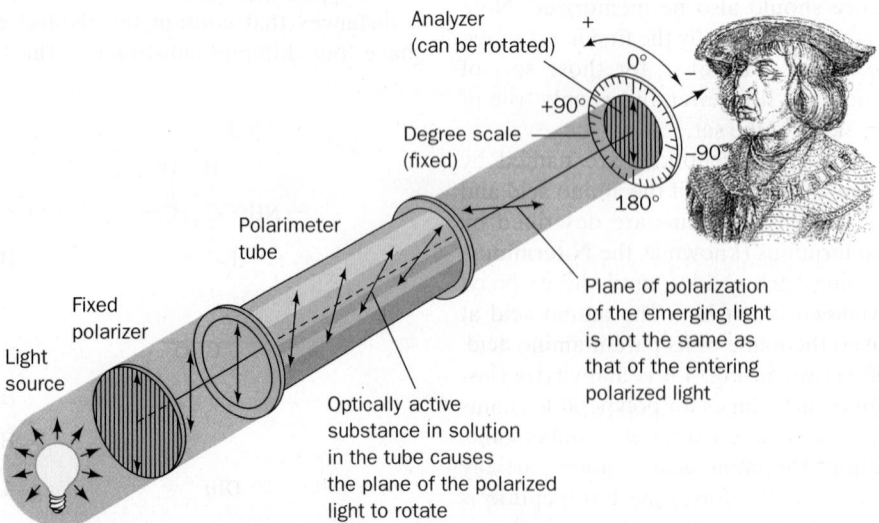

Figure 4-11 Schematic diagram of a polarimeter. This device is used to measure optical rotation.

related to the rotatory powers of the individual asymmetric centers. For this reason, the following relative classification scheme is more useful.

B. *The Fischer Convention*

In this system, the configuration of the groups about an asymmetric center is related to that of **glyceraldehyde,** a molecule with one asymmetric center. By a convention introduced by Fischer in 1891, the $(+)$ and $(-)$ stereoisomers of glyceraldehyde are designated **D-glyceraldehyde** and **L-glyceraldehyde,** respectively (note the use of small uppercase letters). With the realization that there was only a 50% chance that he was correct, Fischer assumed that the configurations of these molecules were those shown in Fig. 4-12. Fischer also proposed a convenient shorthand notation for these molecules, known as **Fischer projections,** which are also given in Fig. 4-12. In the Fischer convention, horizontal bonds extend above the plane of the paper and vertical bonds extend below the plane of the paper as is explicitly indicated by the accompanying geometrical formulas.

The configuration of groups about a chiral center can be related to that of glyceraldehyde by chemically converting these groups to those of glyceraldehyde using reactions of known stereochemistry. For α-amino acids, the arrangement of the amino, carboxyl, R, and H groups about the C_α atom is related to that of the hydroxyl, aldehyde, CH_2OH, and H groups, respectively, of glyceraldehyde. In this way, L-glyceraldehyde and L-α-amino acids are said to have the same relative configurations (Fig. 4-13). Through the use of this method, the configurations of the

Geometric formulas

Fischer projection

Mirror plane

L-Glyceraldehyde   D-Glyceraldehyde

Figure 4-12 Fischer convention configurations for naming the enantiomers of glyceraldehyde. Glyceraldehyde enantiomers are represented by geometric formulas (*top*) and their corresponding Fischer projection formulas (*bottom*). Note that in Fischer projections, all horizontal bonds point above the page and all vertical bonds point below the page. The mirror planes relating the enantiomers are represented by a vertical dashed line. (Fischer projection formulas, as traditionally presented, omit the central C symbolizing the chiral carbon atom. The Fischer projection formulas in this text, however, will generally have a central C.)

L-Glyceraldehyde   L-α-Amino acid

Figure 4-13 Configurations of L-glyceraldehyde and L-α-amino acids.

α-amino acids can be described without reference to their specific rotations.

All α-amino acids derived from proteins have the L stereochemical configuration; that is, they all have the same relative configuration about their C_α atoms. In 1949, it was demonstrated by a then new technique in X-ray crystallography that Fischer's arbitrary choice was correct: The designation of the relative configuration of chiral centers is the same as their absolute configuration. The absolute configuration of L-α-amino acid residues may be easily remembered through the use of the "CORN crib" mnemonic that is diagrammed in Fig. 4-14.

a. Diastereomers Are Chemically and Physically Distinguishable

A molecule may have multiple asymmetric centers. For such molecules, the terms **stereoisomers** and **optical isomers** refer to molecules with different configurations about at least one of their chiral centers, but that are otherwise identical. The term enantiomer still refers to a molecule that is the mirror image of the one under consideration, that is, different in all its chiral centers. Since each asymmetric center in a chiral molecule can have two possible configurations, a molecule with n chiral centers has 2^n different possible stereoisomers and 2^{n-1} enantiomeric pairs. Threonine and isoleucine each have two chiral centers and hence $2^2 = 4$ possible stereoisomers. The forms of threonine and isoleucine that are isolated from proteins, which are by convention called the L forms, are indicated in Table 4-1. The mirror images of the L forms are the D forms. Their other two optical isomers are said to be **diastereomers** (or **allo** forms) of the enantiomeric D and L forms. The relative

Figure 4-14 "CORN crib" mnemonic for the hand of L-amino acids. Looking at the C_α atom from its H atom substituent, its other substituents should read CO—R—N in the clockwise direction as shown. Here CO, R, and N, respectively, represent the carbonyl group, side chain, and main chain nitrogen atom. [After Richardson, J.S., *Adv. Protein Chem.* **34,** 171 (1981).]

COO⁻ COO⁻
H₃N⁺—C—H H—C—NH₃⁺
H—C—OH HO—C—H
CH₃ CH₃

L-Threonine | D-Threonine

Mirror plane

COO⁻ COO⁻
H₃N⁺—C—H H—C—NH₃⁺
HO—C—H H—C—OH
CH₃ CH₃

L-*allo*-Threonine | D-*allo*-Threonine

Figure 4-15 **Fischer projections of threonine's four stereoisomers.** The D and L forms are mirror images as are the D-*allo* and L-*allo* forms. D- and L-threonine are each diastereomers of both D-*allo*- and L-*allo*-threonine.

configurations of all four stereoisomers of threonine are given in Fig. 4-15. Note the following points:

1. The D-*allo* and L-*allo* forms are mirror images of each other, as are the D and L forms. Neither allo form is symmetrically related to either of the D or L forms.

2. In contrast to the case for enantiomeric pairs, diastereomers are physically and chemically distinguishable from one another by ordinary means such as melting points, spectra, and chemical reactivity; that is, they are really different compounds in the usual sense.

A special case of diastereoisomerism occurs when the two asymmetric centers are chemically identical. Two of the four Fischer projections of the sort shown in Fig. 4-15 then represent the same molecule. This is because the two asymmetric centers in this molecule are mirror images of each other. Such a molecule is superimposible on its mirror image and is therefore optically inactive. This so-called **meso** form is said to be **internally compensated.** The three optical isomers of cystine are shown in Fig. 4-16, where it can be seen that the D and L isomers are mirror images of each other as before. Only L-cystine occurs in proteins.

C. *The Cahn–Ingold–Prelog System*

Despite its usefulness, the Fischer scheme is awkward and often ambiguous for molecules with more than one asymmetric center. For this reason, the following absolute nomenclature scheme was formulated in 1956 by Robert Cahn, Christopher Ingold, and Vladimir Prelog. In this system, the four groups surrounding a chiral center are ranked according to a specific although arbitrary priority scheme: *Atoms of higher atomic number bonded to a chiral center are ranked above those of lower atomic number.* For example, the oxygen atom of an OH group takes precedence over the carbon atom of a CH₃ group that is bonded to the same chiral C atom. If any of the first substituent atoms are of the same element, the priority of these groups is established from the atomic numbers of the second, third, etc., atoms outward from the asymmetric center. Hence a CH₂OH group takes precedence over a CH₃ group. There are other rules (given in the references and in many organic chemistry textbooks) for assigning priority ratings to substituents with multiple bonds or differing isotopes. The order of priority of some common functional groups is

$$SH > OH > NH_2 > COOH > CHO$$
$$> CH_2OH > C_6H_5 > CH_3 > {}^2H > {}^1H$$

Note that each of the groups substituent to a chiral center must have a different priority rating; otherwise the center could not be asymmetric.

The prioritized groups are assigned the letters W, X, Y, Z such that their order of priority rating is W > X > Y > Z. To establish the configuration of the chiral center, it is viewed from the asymmetric center toward the Z group (lowest priority). *If the order of the groups* W → X → Y *as seen from this direction is clockwise, then the configuration of the asymmetric center is designated (R)* (Latin: *rectus*, right). *If the order of* W → X → Y *is counterclockwise, the asymmetric center is designated (S)* (Latin: *sinister*, left). L-Glyceraldehyde is therefore designated (*S*)-glyceraldehyde (Fig. 4-17) and, similarly, L-alanine is (*S*)-alanine (Fig. 4-18). In fact, all the L-amino acids from proteins are (*S*)-amino acids, with the exception of L-cysteine, which is (*R*)-cysteine.

A major advantage of this so-called **Cahn–Ingold– Prelog** or **(*RS*) system** is that the chiralities of compounds with multiple asymmetric centers can be unambiguously described. Thus, in the (*RS*) system, L-threonine is (2*S*,3*R*)-threonine, whereas L-isoleucine is (2*S*,3*S*)-isoleucine (Fig. 4-19).

COO⁻ COO⁻
H₃N⁺—C—H H₃N⁺—C—H
CH₂—S—S—CH₂

COO⁻ COO⁻
H—C—NH₃⁺ H—C—NH₃⁺
CH₂—S—S—CH₂

COO⁻ COO⁻
H₃N⁺—C—H H—C—NH₃⁺
CH₂—S—S—CH₂

Mirror plane | Mirror plane

L-Cystine | D-Cystine | *meso*-Cystine

Figure 4-16 **The three stereoisomers of cystine.** The D and L forms are related by mirror symmetry, whereas the meso form has internal mirror symmetry and therefore lacks optical activity.

CHO
HO — C — H
CH₂OH

L-Glyceraldehyde

≡

CHO(X)
(H(Z)) — OH(W)
CH₂OH(Y)

(S)-Glyceraldehyde

Figure 4-17 **The structural formula of L-glyceraldehyde.** Its equivalent (RS) system representation indicates that it is (S)-glyceraldehyde. In the latter drawing, the chiral C atom is represented by the large circle, and the H atom, which is located behind the plane of the paper, is represented by the smaller concentric dashed circle.

COO⁻
H₃N⁺ — C — H
CH₃

L-Alanine

≡

⁻OOC(X)
(H(Z)) — NH₃(W)
H₃C(Y)

(S)-Alanine

Figure 4-18 **The structural formula of L-alanine.** Its equivalent (RS) system representation indicates that it is (S)-alanine.

H
H₃C   OH
⁻OOC   NH₃⁺
H

(2S,3R)-Threonine

H
H₃C   CH₂CH₃
⁻OOC   NH₃⁺
H

(2S,3S)-Isoleucine

Figure 4-19 **Newman projection diagrams of the stereoisomers of threonine and isoleucine derived from proteins.** Here the Cα—Cβ bond is viewed end on. The nearer atom, Cα, is represented by the confluence of the three bonds to its substituents, whereas the more distant atom, Cβ, is represented by a circle from which its three substituents project.

a. Prochiral Centers Have Distinguishable Substituents

Two chemically identical substituents to an otherwise chiral tetrahedral center are geometrically distinct; that is, the center has no rotational symmetry so that it can be unambiguously assigned left and right sides. Consider, for example, the substituents to the C1 atom of ethanol (the CH₂ group; Fig. 4-20a). If one of the H atoms were converted to another group (not CH₃ or OH), C1 would be a chiral center. The two H atoms are therefore said to be **prochiral**. If we arbitrarily assign the H atoms the subscripts *a* and *b* (Fig. 4-20), then Hᵦ is said to be **pro-R** because in sighting from C1 toward Hₐ (as if it were the Z group of a chiral center), the order of priority of the other substituents de-

(a)

OH
Hₐ — C — H_b
CH₃

(b)

H_b (pro-R)
(Hₐ)
H₃C   OH

(c)

H₃C   OH
(H_b)
Hₐ (pro-S)

Figure 4-20 **Views of ethanol.** (*a*) Note that Hₐ and H_b, although chemically identical, are distinguishable: Rotating the molecule by 180° about the vertical axis so as to interchange these two hydrogen atoms does not yield an indistinguishable view of the molecule because the rotation also interchanges the chemically different OH and CH₃ groups. (*b*) Looking from C1 to Hₐ, the *pro-S* hydrogen atom (the dotted circle). (*c*) Looking from C1 to H_b, the *pro-R* hydrogen atom.

creases in a clockwise direction (Fig. 4-20b). Similarly, Hₐ is said to be **pro-S** (Fig. 4-20c).

Planar objects with no rotational symmetry also have the property of prochirality. For example, in many enzymatic reactions, stereospecific addition to a trigonal carbon atom occurs from a particular side of that carbon atom to yield a chiral center (Section 13-2A). If a trigonal carbon is facing the viewer such that the order of priority of its substituents decreases in a clockwise manner (Fig. 4-21a), that face is designated as the **re face** (after *rectus*). The opposite face is designated as the **si face** (after *sinister*) since the priorities of its substituents decrease in the counterclockwise direction (Fig. 4-21b). Comparison of Figs. 4-20b and 4-21a indicates that an H atom adding to the *re* side of acetaldehyde atom C1 occupies the *pro-R* position of the resulting tetrahedral center. Conversely, a *pro-S* H atom is generated by *si* side addition to this trigonal center (Figs. 4-20c and 4-21b).

Closely related compounds that have the same configurational representation under the Fischer DL convention may have different representations under the (RS) system. Consequently, we shall use the Fischer convention in most cases. The (RS) system, however, is indispensable for describing prochirality and stereospecific reactions, so we shall find it invaluable for describing enzymatic reactions.

Figure 4-21 **Views of acetaldehyde.** (*a*) Its *re* face and (*b*) its *si* face.

D. *Chirality and Biochemistry*

The ordinary chemical synthesis of chiral molecules produces **racemic** mixtures of these molecules (equal amounts of each member of an enantiomeric pair) because ordinary chemical and physical processes have no stereochemical bias. Consequently, there are equal probabilities for an asymmetric center of either hand to be produced in any such process. In order to obtain a product with net optical activity, a chiral process must be employed. This usually takes the form of using chiral reagents, although, at least in principle, the use of any asymmetric influence such as light that is plane polarized in one direction can produce a net asymmetry in a reaction product.

One of the most striking characteristics of life is its production of optically active molecules. *The biosynthesis of a substance possessing asymmetric centers almost invariably produces a pure stereoisomer.* The fact that the amino acid residues of proteins all have the L configuration is just one example of this phenomenon. This observation has prompted the suggestion that a simple diagnostic test for the past or present existence of extraterrestrial life, be it on moon rocks or in meteorites that have fallen to Earth, would be the detection of net optical activity in these materials. Any such finding would suggest that the asymmetric molecules thereby detected had been biosynthetically produced. Thus, even though α-amino acids have been extracted from carbonaceous meteorites, the observation that they come in racemic mixtures suggests that they are of chemical rather than biological origin.

One of the enigmas of the origin of life is why terrestrial life is based on certain chiral molecules rather than their enantiomers, that is, on L-amino acids, for example, rather than D-amino acids. Arguments that physical effects such as polarized light might have promoted significant net asymmetry in prebiotically synthesized molecules (Section 1-5B) have not been convincing. Perhaps L-amino acid–based life-forms arose at random and simply "ate" any D-amino acid–based life-forms.

The importance of stereochemistry in living systems is also a concern of the pharmaceutical industry. *Many drugs are chemically synthesized as racemic mixtures, although only one enantiomer has biological activity.* In most cases, the opposite enantiomer is biologically inert and is therefore packaged along with its active counterpart. This is true, for example, of the widely used anti-inflammatory agent **ibuprofen,** only one enantiomer of which is physiologically active (Fig. 4-22). Occasionally, the inactive enan-

Thalidomide

Figure 4-23 **Thalidomide.** This drug was widely used in Europe as a mild sedative in the early 1960s. Its inactive enantiomer (not shown), which was present in equal amounts in the formulations used, causes severe birth defects in humans when taken during the first trimester of pregnancy. Thalidomide was often prescribed to alleviate the nausea (morning sickness) that is common during this period.

tiomer of a useful drug produces harmful effects and must therefore be eliminated from the racemic mixture. The most striking example of this is the drug **thalidomide** (Fig. 4-23), a mild sedative whose "inactive" enantiomer causes severe birth defects. Partly because of the unanticipated problems caused by "inactive" drug enantiomers, **chiral organic synthesis** has become an active area of medicinal chemistry.

3 "NONSTANDARD" AMINO ACIDS

The 20 common amino acids are by no means the only amino acids that occur in biological systems. "Nonstandard" amino acid residues are often important constituents of proteins and biologically active polypeptides. Many amino acids, however, are not constituents of proteins. Together with their derivatives, they play a variety of biologically important roles.

A. *Amino Acid Derivatives in Proteins*

The "universal" genetic code, which is nearly identical in all known life-forms (Section 5-4Bb), specifies only the 20 "standard" amino acids of Table 4-1. Nevertheless, many other amino acids, a selection of which is given in Fig. 4-24, are components of certain proteins. *In all known cases but two (Section 32-2De), however, these unusual amino acids result from the specific modification of an amino acid residue after the polypeptide chain has been synthesized.* Among the most prominent of these modified amino acid residues are **4-hydroxyproline** and **5-hydroxylysine.** Both of these amino acid residues are important structural constituents of the fibrous protein **collagen,** the most abundant protein in mammals (Section 8-2B). Amino acids of proteins that form complexes with nucleic acids are often modified. For example, the chromosomal proteins known as **histones** may be specifically methylated, acetylated, and/or phosphorylated at specific Lys, Arg, and Ser residues (Section 34-3Baa). Several of these derivatized

Ibuprofen

Figure 4-22 **Ibuprofen.** Only the enantiomer shown has anti-inflammatory action. The chiral carbon is red.

Figure 4-24   Some uncommon amino acid residues that are components of certain proteins. All of these residues are modified from one of the 20 "standard" amino acids after polypeptide chain biosynthesis. Those amino acid residues that are derivatized at their N_α position occur at the N-termini of proteins.

amino acid residues are presented in Fig. 4-24. **N-Formyl-methionine** is initially the N-terminal residue of all prokaryotic proteins, but is usually removed as part of the protein maturation process (Section 32-3Ca). **γ-Carboxy-glutamic acid** is a constituent of several proteins involved in blood clotting (Section 35-1Ba). Note that in most cases, these modifications are important, if not essential, for the function of the protein.

D-Amino acid residues are components of many of the relatively short (<20 residues) bacterial polypeptides that are enzymatically rather than ribosomally synthesized. These polypeptides are perhaps most widely distributed as constituents of bacterial cell walls (Section 11-3Ba), which D-amino acids render less susceptible to attack by the **pep-tidases** (enzymes that hydrolyze peptide bonds) that many organisms employ to digest bacterial cell walls. Likewise, D-amino acids are components of many bacterially produced peptide antibiotics including **valinomycin, gramicidin A** (Section 20-2C), and **actinomycin D** (Section 31-2Cc).

D-Amino acid residues are also functionally essential components of several ribosomally synthesized polypeptides of eukaryotic as well as prokaryotic origin. These D-amino acid residues are posttranslationally formed, most probably through the enzymatically mediated inversion of the preexisting L-amino acid residues.

B. Specialized Roles of Amino Acids

Besides their role in proteins, amino acids and their derivatives have many biologically important functions. A few examples of these substances are shown in Fig. 4-25. This alternative use of amino acids is an example of the biological opportunism that we shall repeatedly encounter: *Nature tends to adapt materials and processes that are already present to new functions.*

Amino acids and their derivatives often function as chemical messengers in the communications between cells. For example, glycine, **γ-aminobutyric acid (GABA;** a glutamate

Figure 4-25 Some biologically produced derivatives of "standard" amino acids and amino acids that are not components of proteins.

decarboxylation product), and **dopamine** (a tyrosine derivative) are neurotransmitters (substances released by nerve cells to alter the behavior of their neighbors; Section 20-5C); **histamine** (the decarboxylation product of histidine) is a potent local mediator of allergic reactions; and **thyroxine** (a tyrosine derivative) is an iodine-containing thyroid hormone that generally stimulates vertebrate metabolism (Section 19-1D).

Certain amino acids are important intermediates in various metabolic processes. Among them are **citrulline** and **ornithine,** intermediates in urea biosynthesis (Section 26-2B); **homocysteine,** an intermediate in amino acid metabolism

(Section 26-3Ea); and **S-adenosylmethionine,** a biological methylating reagent (Section 26-3Ea).

Nature's diversity is remarkable. Over 700 different amino acids have been found in various plants, fungi, and bacteria, most of which are α-amino acids. For the most part, their biological roles are obscure although the fact that many are toxic suggests that they have a protective function. Indeed, some of them, such as **azaserine,** are medically useful antibiotics. Many of these amino acids are simple derivatives of the 20 "standard" amino acids although some of them, including azaserine and **β-cyanoalanine** (Fig. 4-25), have unusual structures.

CHAPTER SUMMARY

1 The Amino Acids of Proteins Proteins are linear polymers that are synthesized from the same 20 "standard" α-amino acids through their condensation to form peptide bonds. These amino acids all have a carboxyl group with a pK near 2.2 and an amino substituent with a pK near 9.4 attached to the same carbon atom, the C_α atom. The α-amino acids are zwitterionic compounds, $^+H_3N—CHR—COO^-$, in the physiological pH range. The various amino acids are usually classified according to the polarities of their side chains, R, which are substituent to the C_α atom. Glycine, alanine, valine, leucine, isoleucine, methionine, proline (which is really a secondary amino acid), phenylalanine, and tryptophan are nonpolar amino acids; serine, threonine, asparagine, glutamine, tyrosine, and cysteine are uncharged and polar; and lysine, arginine, histidine, aspartic acid, and glutamic acid are charged and polar. The side chains of many of these amino acids bear acid–base groups, and hence the properties of the proteins containing them are pH dependent.

2 Optical Activity The C_α atoms of all α-amino acids ex-

cept glycine each bear four different substituents and are therefore chiral centers. According to the Fischer convention, which relates the configuration of D- or L-glyceraldehyde to that of the asymmetric center of interest, all the amino acids of proteins have the L configuration; that is, they all have the same absolute configuration about their C_α atom. According to the Cahn–Ingold–Prelog (*RS*) system of chirality nomenclature, they are, with the exception of cysteine, all (*S*)-amino acids. The side chains of threonine and isoleucine also contain chiral centers. A prochiral center has no rotational symmetry, and hence its substituents, in the case of a central atom, or its faces, in the case of a planar molecule, are distinguishable.

3 "Nonstandard" Amino Acids Amino acid residues other than the 20 from which proteins are synthesized also have important biological functions. These "nonstandard" residues result from the specific chemical modifications of amino acid residues in preexisting proteins. Amino acids and their derivatives also have independent biological roles such as neurotransmitters, metabolic intermediates, and poisons.

REFERENCES

History

Vickery, H.B. and Schmidt, C.L.A., The history of the discovery of amino acids, *Chem. Rev.* **9**, 169–318 (1931).

Vickery, H.B., The history of the discovery of the amino acids. A review of amino acids discovered since 1931 as components of native proteins, *Adv. Protein Chem.* **26**, 81–171 (1972).

Properties of Amino Acids

Barrett, G.C. and Elmore, D.T., *Amino Acids and Peptides,* Chapters 1–4, Cambridge University Press (1998).

Cohn, E.J. and Edsall, J.T., *Proteins, Amino Acids and Peptides as Ions and Dipolar Ions,* Academic Press (1943). [A classic work in its field.]

Meister, A., *Biochemistry of the Amino Acids* (2nd ed.), Vol. 1, Academic Press (1965). [A compendium of information on amino acid properties.]

Optical Activity

Cahn, R.S., An introduction to the sequence rule, *J. Chem. Ed.* **41**, 116–125 (1964). [A presentation of the Cahn–Ingold–Prelog system of nomenclature.]

Huheey, J.E., A novel method for assigning *R,S* labels to enantiomers, *J. Chem. Ed.* **63**, 598–600 (1986).

Lamzin, V.S., Dauter, Z., and Wilson, K.S., How nature deals with stereoisomers, *Curr. Opin. Struct. Biol.* **5**, 830–836 (1995). [Discusses proteins synthesized from D-amino acids.]

Mislow, K., *Introduction to Stereochemistry,* Benjamin (1966).

Solomons, T.W.G. and Fryhle, C.B., *Organic Chemistry* (9th ed.), Chapter 5, Wiley (2008). [A discussion of chirality. Most other organic chemistry textbooks contain similar material.]

"Nonstandard" Amino Acids

Fowden, L., Lea, P.J., and Bell, E.A., The non-protein amino acids of plants, *Adv. Enzymol.* **50**, 117–175 (1979).

Fowden, L., Lewis, D., and Tristram, H., Toxic amino acids: their action as antimetabolites, *Adv. Enzymol.* **29**, 89–163 (1968).

Kleinkauf, H. and Döhren, H., Nonribosomal polypeptide formation on multifunctional proteins, *Trends Biochem. Sci.* **8**, 281–283 (1993).

Mor, A., Amiche, M., and Nicholas, P., Enter a new post-transcriptional modification: D-amino acids in gene-encoded peptides, *Trends Biochem. Sci.* **17**, 481–485 (1992).

PROBLEMS

1. Name the 20 standard amino acids without looking them up. Give their three-letter and one-letter symbols. Identify the two standard amino acids that are isomers and the two others that, although not isomeric, have essentially the same molecular mass for the neutral molecules.

2. Draw the following oligopeptides in their predominant ionic forms at pH 7: (a) Phe-Met-Arg, (b) tryptophanyllysylaspartic acid, and (c) Gln-Ile-His-Thr.

3. How many different pentapeptides are there that contain one residue each of Gly, Asp, Tyr, Cys, and Leu?

4. Draw the structures of the following two oligopeptides with their cysteine residues cross-linked by a disulfide bond: Val-Cys, Ser-Cys-Pro.

*5. What are the concentrations of the various ionic species in a 0.1*M* solution of lysine at pH 4, 7, and 10?

6. Derive Eq. [4.1] for a monoamino, monocarboxylic acid (use the Henderson–Hasselbalch equation).

*7. The **isoionic point** of a compound is defined as the pH of a pure water solution of the compound. What is the isoionic point of a 0.1*M* solution of glycine?

8. Normal human hemoglobin has an isoelectric point of 6.87. A mutant variety of hemoglobin, known as **sickle-cell hemoglobin,** has an isoelectric point of 7.09. The titration curve of hemoglobin indicates that, in this pH range, 13 groups change ionization states per unit change in pH. Calculate the difference in ionic charge between molecules of normal and sickle-cell hemoglobin.

9. Indicate whether the following familiar objects are chiral, prochiral, or nonchiral.

(a) A glove
(b) A tennis ball
(c) A good pair of scissors
(d) A screw
(e) This page
(f) A toilet paper roll
(g) A snowflake
(h) A spiral staircase
(i) A flight of normal stairs
(j) A paper clip
(k) A shoe
(l) A pair of glasses

10. Draw four equivalent Fischer projection formulas for L-alanine (see Figs. 4-12 and 4-13).

*11. (a) Draw the structural formula and the Fischer projection formula of (*S*)-3-methylhexane. (b) Draw all the stereoisomers of 2,3-dichlorobutane. Name them according to the (*RS*) system and indicate which of them has the meso form.

12. Identify and name the prochiral centers or faces of the following molecules:

(a) Acetone
(b) Propene
(c) Glycine
(d) Alanine
(e) Lysine
(f) 3-Methylpyridine

13. Write out the dominant structural formula, at pH 12.0, of the pentapeptide Thr-Tyr-His-Cys-Lys. Indicate the positions of its chiral centers and its prochiral centers. Assume that the p*K*'s of its ionizable groups are the same as those in the corresponding free amino acid.

Nucleic Acids, Gene Expression, and Recombinant DNA Technology

CHAPTER 5

Knowledge of how genes are expressed and how they can be manipulated is becoming increasingly important for understanding nearly every aspect of biochemistry. Consequently, although we do not undertake a detailed discussion of these processes until Part V of this textbook, we outline their general principles in this chapter. We do so by describing the chemical structures of nucleic acids, how we have come to know that DNA is the carrier of genetic information, the structure of the major form of DNA, and the general principles of how the information in genes directs the synthesis of RNA and proteins (how genes are expressed) and how DNA is replicated. The chapter ends with a discussion of how DNA is experimentally manipulated and expressed, processes that are collectively referred to as genetic engineering. These processes have revolutionized the practice of biochemistry.

1 NUCLEOTIDES AND NUCLEIC ACIDS

Nucleotides and their derivatives are biologically ubiquitous substances that participate in nearly all biochemical processes:

1. They form the monomeric units of nucleic acids and thereby play central roles in both the storage and the expression of genetic information.

2. Nucleoside triphosphates, most conspicuously ATP (Section 1-3B), are the "energy-rich" end products of the majority of energy-releasing pathways and the substances whose utilization drives most energy-requiring processes.

3. Most metabolic pathways are regulated, at least in part, by the levels of nucleotides such as ATP and ADP. Moreover, certain nucleotides, as we shall see, function as intracellular signals that regulate the activities of numerous metabolic processes.

4. Nucleotide derivatives, such as **nicotinamide adenine dinucleotide** (Section 13-2A), **flavin adenine dinucleotide** (Section 16-2C), and **coenzyme A** (Section 21-2), are required participants in many enzymatic reactions.

5. As components of the enzymelike nucleic acids known as **ribozymes,** nucleotides have important catalytic activities themselves.

A. Nucleotides, Nucleosides, and Bases

*Nucleotides are phosphate esters of a five-carbon sugar (which is therefore known as a **pentose**; Section 11-1A) in which a nitrogenous base is covalently linked to C1' of the sugar residue.* In **ribonucleotides** (Fig. 5-1a), the monomeric units of RNA, the pentose is **D-ribose,** whereas in **deoxyribonucleotides** (or just **deoxynucleotides;** Fig. 5-1b),

Ribonucleotides   **Deoxyribonucleotides**

Figure 5-1 **Chemical structures of (*a*) ribonucleotides and (*b*) deoxyribonucleotides.**

the monomeric units of DNA, the pentose is **2′-deoxy-D-ribose** (note that the "primed" numbers refer to the atoms of the ribose residue; "unprimed" numbers refer to atoms of the nitrogenous base). The phosphate group may be bonded to C5′ of the pentose to form a **5′-nucleotide** (Fig. 5-1) or to its C3′ to form a **3′-nucleotide.** If the phosphate group is absent, the compound is known as a **nucleoside.** A 5′-nucleotide, for example, may therefore be referred to as a **nucleoside-5′-phosphate.** In all naturally occurring nucleotides and nucleosides, the bond linking the nitrogenous base to the pentose C1′ atom (which is called a glycosidic bond; Section 11-1Ca) extends from the same side of the ribose ring as does the C4′—C5′ bond (the so-called β configuration; Section 11-1Ba) rather than from the opposite side (the α configuration). Note that nucleotide phosphate groups are doubly ionized at physiological pH's; that is, *nucleotides are moderately strong acids.*

The nitrogenous bases are planar, aromatic, heterocyclic molecules which, for the most part, are derivatives of either **purine** *or* **pyrimidine.**

Purine   **Pyrimidine**

The structures, names, and abbreviations of the common bases, nucleosides, and nucleotides are given in Table 5-1. The major purine components of nucleic acids are **adenine** and **guanine** residues; the major pyrimidine residues are those of **cytosine, uracil** (which occurs mainly in RNA), and **thymine** (5-methyluracil, which occurs mainly in DNA). The purines form glycosidic bonds to ribose via

Table 5-1   **Names and Abbreviations of Nucleic Acid Bases, Nucleosides, and Nucleotides**

Base Formula	Base (X = H)	Nucleoside (X = ribose[a])	Nucleotide[b] (X = ribose phosphate[a])
	Adenine Ade A	Adenosine Ado A	Adenylic acid Adenosine monophosphate AMP
	Guanine Gua G	Guanosine Guo G	Guanylic acid Guanosine monophosphate GMP
	Cytosine Cyt C	Cytidine Cyd C	Cytidylic acid Cytidine monophosphate CMP
	Uracil Ura U	Uridine Urd U	Uridylic acid Uridine monophosphate UMP
	Thymine Thy T	Deoxythymidine dThd dT	Deoxythymidylic acid Deoxythymidine monophosphate dTMP

[a]The presence of a 2′-deoxyribose unit in place of ribose, as occurs in DNA, is implied by the prefixes "deoxy" or "d." For example, the deoxynucleoside of adenine is deoxyadenosine or dA. However, for thymine-containing residues, which rarely occur in RNA, the prefix is redundant and may be dropped. The presence of a ribose unit may be explicitly implied by the prefixes "ribo" or "r." Thus the ribonucleotide of thymine is ribothymidine or rT.
[b]The position of the phosphate group in a nucleotide may be explicitly specified as in, for example, 3′-AMP and 5′-GMP.

their N9 atoms, whereas pyrimidines do so through their N1 atoms (note that purines and pyrimidines have dissimilar atom numbering schemes).

B. *The Chemical Structures of DNA and RNA*

The chemical structures of the nucleic acids were elucidated by the early 1950s largely through the efforts of Phoebus Levene, followed by the work of Alexander Todd. *Nucleic acids are, with few exceptions, linear polymers of nucleotides whose phosphate groups bridge the 3' and 5' positions of successive sugar residues (e.g., Fig. 5-2).* The phosphates of these **polynucleotides,** the **phosphodiester** groups, are acidic, so that, *at physiological pH's, nucleic acids are polyanions.* Polynucleotides have directionality, that is, each has a **3' end** (the end whose C3' atom is not linked to a neighboring nucleotide) and a **5' end** (the end whose C5' atom is not linked to a neighboring nucleotide).

a. **DNA's Base Composition Is Governed by Chargaff's Rules**

DNA has equal numbers of adenine and thymine residues (A = T) and equal numbers of guanine and cytosine residues (G = C). These relationships, known as **Chargaff's rules,** were discovered in the late 1940s by Erwin Chargaff, who first devised reliable quantitative methods for the separation and analysis of DNA hydrolysates. Chargaff also found that the base composition of DNA from a given organism is characteristic of that organism; that is, it is independent of the tissue from which the DNA is taken as well as the organism's age, its nutritional state, or any other environmental factor. The structural basis for Chargaff's rules is that in double-stranded DNA, G is always hydrogen bonded (forms a **base pair**) with C, whereas A always forms a base pair with T (Fig. 1-16).

DNA's base composition varies widely among different organisms. It ranges from ~25% to 75% G + C in different

(a)

(b)

Figure 5-2 **Chemical structure of a nucleic acid.**
(*a*) The tetranucleotide adenyl-3',5'-uridyl-3',5'-cytidyl-3',5'-guanylyl-3'-phosphate. The sugar atom numbers are primed to distinguish them from the atomic positions of the bases. By convention, a polynucleotide sequence is written with its 5' end at the left and its 3' end to the right. Thus, reading left to right, the phosphodiester bond links neighboring ribose residues in the 5' → 3' direction. The above sequence may be abbreviated ApUpCpGp or just AUCGp (where a "p" to the left and/or right of a nucleoside symbol indicates a 5' and/or a 3' phosphate group, respectively; see Table 5-1 for other symbol definitions). The corresponding deoxytetranucleotide, in which the 2'-OH groups are each replaced by H atoms and the base uracil (U) is replaced by thymine (5-methyluracil; T), is abbreviated d(ApTpCpGp) or d(ATCGp). (*b*) A schematic representation of AUCGp. Here a vertical line denotes a ribose residue, its attached base is indicated by the corresponding one-letter abbreviation, and a diagonal line flanking an optional "p" represents a phosphodiester bond. The atom numbering of the ribose residues, which is indicated here, is usually omitted. The equivalent representation of deoxypolynucleotides differs only by the absence of the 2'-OH groups and the replacement of U by T.

species of bacteria. It is, however, more or less constant among related species; for example, in mammals G + C ranges from 39% to 46%.

RNA, which usually occurs as single-stranded molecules, has no apparent constraints on its base composition. However, double-stranded RNA, which comprises the genetic material of certain viruses, also obeys Chargaff's rules (here A base pairs with U in the same way it does with T in DNA; Fig. 1-16). Conversely, single-stranded DNA, which occurs in certain viruses, does not obey Chargaff's rules. On entering its host organism, however, such DNA is replicated to form a double-stranded molecule, which then obeys Chargaff's rules.

b. Nucleic Acid Bases May Be Modified

Some DNAs contain bases that are chemical derivatives of the standard set. For example, dA and dC in the DNAs of many organisms are partially replaced by N^6-**methyl-dA** and **5-methyl-dC,** respectively.

N^6-**Methyl-dA**   **5-Methyl-dC**

The altered bases are generated by the sequence-specific enzymatic modification of normal DNA (Sections 5-5A and 30-7). The modified DNAs obey Chargaff's rules if the derivatized bases are taken as equivalent to their parent bases. Likewise, many bases in RNAs and, in particular, those in **transfer RNAs (tRNAs;** Section 32-2Aa) are derivatized.

c. RNA but Not DNA Is Susceptible to Base-Catalyzed Hydrolysis

RNA is highly susceptible to base-catalyzed hydrolysis by the reaction mechanism diagrammed in Fig. 5-3 so as to yield a mixture of 2' and 3' nucleotides. In contrast, DNA, which lacks 2'-OH groups, is resistant to base-catalyzed hydrolysis and is therefore much more chemically stable than RNA. This is probably why DNA rather than RNA evolved to be the cellular genetic archive.

2 DNA IS THE CARRIER OF GENETIC INFORMATION

Nucleic acids were first isolated in 1869 by Friedrich Miescher and so named because he found them in the nuclei of **leukocytes** (pus cells) from discarded surgical bandages. The presence of nucleic acids in other cells was demonstrated within a few years, but it was not until some 75 years after their discovery that their biological function was elucidated. Indeed, in the 1930s and 1940s it was widely held, in what was termed the **tetranucleotide hypothesis,** that nucleic acids have a monotonously repeating sequence of all four bases, so that they were not suspected of having a

Figure 5-3 **Mechanism of base-catalyzed RNA hydrolysis.** The base-induced deprotonation of the 2'-OH group facilitates its nucleophilic attack on the adjacent phosphorus atom, thereby cleaving the RNA backbone. The resultant 2',3'-cyclic phosphate group subsequently hydrolyzes to either the 2' or the 3' phosphate.

genetic function. Rather, it was generally assumed that genes were proteins since proteins were the only biochemical entities that, at that time, seemed capable of the required specificity. In this section, we outline the experiments that established DNA's genetic role.

A. *Transforming Principle Is DNA*

The virulent (capable of causing disease) form of pneumococcus (*Diplococcus pneumoniae*), a bacterium that causes pneumonia, is encapsulated by a gelatinous polysaccharide coating that contains the binding sites (known as **O-antigens;** Section 11-3Bc) through which it recognizes the cells it infects. Mutant pneumococci that lack this coating, because of a defect in an enzyme involved in its formation, are not pathogenic (capable of causing disease). The virulent and

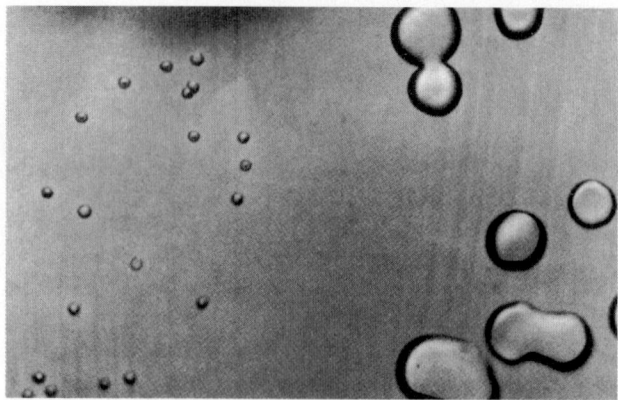

Figure 5-4 **Pneumococci.** The large glistening colonies are virulent S-type pneumococci that resulted from the transformation of nonpathogenic R-type pneumococci (smaller colonies) by DNA from heat-killed S pneumococci. [From Avery, O.T., MacLeod, C.M., and McCarty, M., *J. Exp. Med.* **79,** 153 (1944). Copyright © 1944 by Rockefeller University Press.]

Figure 5-5 **Transgenic mouse.** The gigantic mouse (*left*) grew from a fertilized ovum that had been microinjected with DNA bearing the rat growth hormone gene. His normal-sized littermate (*right*) is shown for comparison. [Courtesy of Ralph Brinster, University of Pennsylvania.]

nonpathogenic pneumococci are known as the S and R forms, respectively, because of the smooth and rough appearances of their colonies in culture (Fig. 5-4).

In 1928, Frederick Griffith made a startling discovery. He injected mice with a mixture of live R and heat-killed S pneumococci. This experiment resulted in the death of most of the mice. More surprising yet was that the blood of the dead mice contained live S pneumococci. The dead S pneumococci initially injected into the mice had somehow **transformed** the otherwise innocuous R pneumococci to the virulent S form. Furthermore, the progeny of the transformed pneumococci were also S; the transformation was permanent. Eventually, it was shown that the transformation could also be made *in vitro* (outside a living organism; literally "in glass") by mixing R cells with a cell-free extract of S cells. The question remained: What is the nature of the **transforming principle?**

In 1944, Oswald Avery, Colin MacLeod, and Maclyn McCarty, after a 10-year investigation, reported that *transforming principle is DNA.* The conclusion was based on the observations that the laboriously purified (few modern fractionation techniques were then available) transforming principle had all the physical and chemical properties of DNA, contained no detectable protein, was unaffected by enzymes that catalyze the hydrolysis of proteins and RNA, and was totally inactivated by treatment with an enzyme that catalyzes the hydrolysis of DNA. *DNA must therefore be the carrier of genetic information.*

Avery's discovery was another idea whose time had not yet come. This seminal advance was initially greeted with skepticism and then largely ignored. Indeed, even Avery did not directly state that DNA is the hereditary material but merely that it has "biological specificity." His work, however, influenced several biochemists, including Erwin Chargaff, whose subsequent accurate determination of DNA base ratios refuted the tetranucleotide hypothesis and thereby indicated that DNA could be a complex molecule.

It was eventually demonstrated that eukaryotes are also subject to transformation by DNA. Thus DNA, which cyto-

logical studies had shown resides in the chromosomes, must also be the hereditary material of eukaryotes. In a spectacular demonstration of eukaryotic transformation, Ralph Brinster, in 1982, microinjected DNA bearing the gene for rat **growth hormone** (a polypeptide) into the nuclei of fertilized mouse eggs (a technique discussed in Section 5-5H) and implanted these eggs into the uteri of foster mothers. The resulting "supermice" (Fig. 5-5), which had high levels of rat growth hormone in their serum, grew to nearly twice the weight of their normal littermates. Such genetically altered animals are said to be **transgenic.**

B. The Hereditary Molecule of Many Bacteriophages Is DNA

Electron micrographs of bacteria infected with bacteriophages show empty-headed phage "ghosts" attached to the bacterial surface (Fig. 5-6). This observation led Roger Herriott to suggest that "the virus may act like a little hypodermic needle full of transforming principle," which it in-

Figure 5-6 **Bacteriophages attached to the surface of a bacterium.** This early electron micrograph shows an *E. coli* cell to which **bacteriophage T5** are adsorbed by their tails. [Courtesy of Thomas F. Anderson, Fox Chase Cancer Center.]

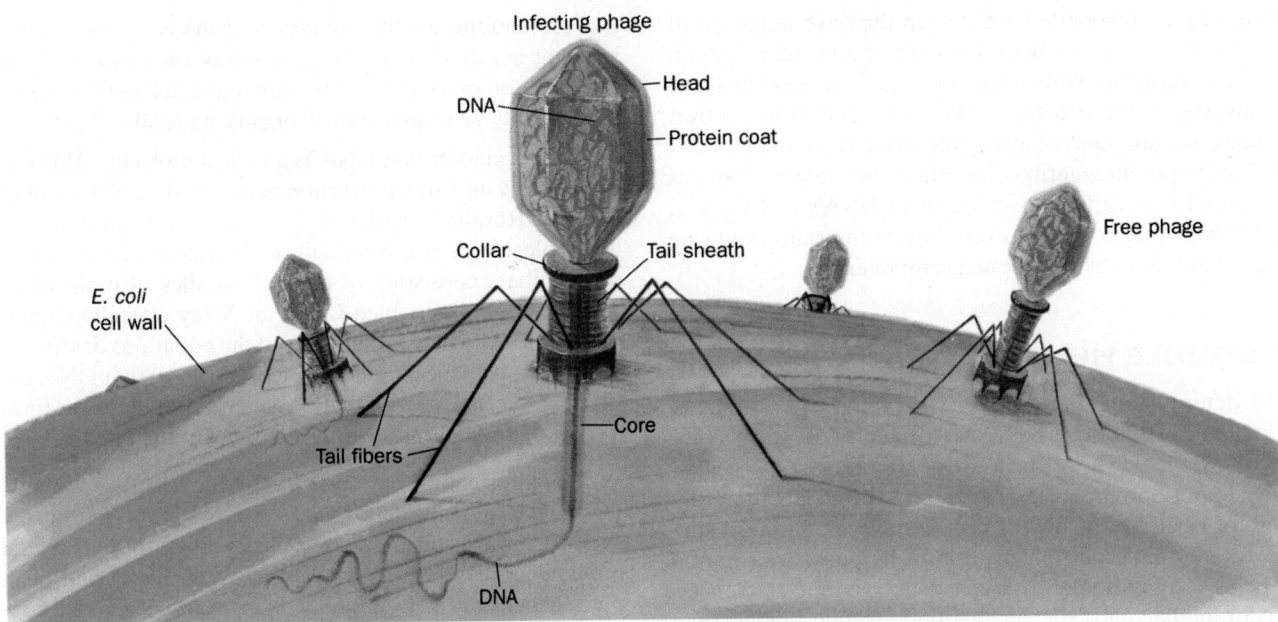

Figure 5-7   Diagram of T2 bacteriophage injecting its DNA into an *E. coli* cell.

jects into the bacterial host (Fig. 5-7). This proposal was tested in 1952 by Alfred Hershey and Martha Chase as is diagrammed in Fig. 5-8. **Bacteriophage T2** was grown on *E. coli* in a medium containing the radioactive isotopes ^{32}P and ^{35}S. This labeled the phage capsid, which contains no P, with ^{35}S, and its DNA, which contains no S, with ^{32}P. These phages were added to an unlabeled culture of *E. coli* and, after sufficient time was allowed for the phages to infect the bacterial cells, the culture was agitated in a kitchen blender so as to shear the phage ghosts from the bacterial cells. This rough treatment neither injured the bacteria nor altered the course of the phage infection. When the phage ghosts were separated from the bacteria (by centrifugation; Section 6-5), the ghosts were found to contain most of the ^{35}S, whereas the bacteria contained most of the ^{32}P. Furthermore, 30% of the ^{32}P appeared in the progeny phages but only 1% of the ^{35}S did so. Hershey and Chase therefore concluded that only the phage DNA was essential for the production of progeny. *DNA therefore must be the hereditary material.* In later years it was shown that, in a process known as **transfection,** purified phage DNA can, by itself, induce a normal phage infection of a properly treated bacterial host (transfection differs from transformation in that the latter results from the recombination of the bacterial chromosome with a fragment of homologous DNA).

In 1952, the state of knowledge of biochemistry was such that Hershey's discovery was much more readily accepted than Avery's identification of the transforming principle had been some 8 years earlier. Within a few months, the first speculations arose as to the nature of the **genetic**

Phage particle with ^{35}S-labeled shell and ^{32}P-labeled DNA

^{35}S

^{32}P

Phage infects *E.coli*; only labeled DNA enters cell

^{35}S phage shells

^{32}P labeled DNA

Parental ^{32}P-labeled DNA replicates. Replica DNA is unlabeled

Unlabeled replica DNA

Phages assemble: only parental DNA is ^{32}P-labeled. Some progeny phages are unlabeled. No ^{35}S shell label remains

Figure 5-8   **The Hershey–Chase experiment.** This experiment demonstrated that only the nucleic acid component of bacteriophages enters the bacterial host during phage infection.

code (the correspondence between the base sequence of a gene and the amino acid sequence of a protein, Section 5-4Bb), and James Watson and Francis Crick were inspired to investigate the structure of DNA. In 1955, it was shown that the somatic cells of eukaryotes have twice the DNA of the corresponding germ cells. When this observation was proposed to be a further indicator of DNA's genetic role, there was little comment even though the same could be said of any other chromosomal component.

3 DOUBLE HELICAL DNA

The determination of the structure of DNA by Watson and Crick in 1953 is often said to mark the birth of modern molecular biology. The **Watson–Crick structure** of DNA is of such importance because, in addition to providing the structure of what is arguably the central molecule of life, it suggested the molecular mechanism of heredity. Watson and Crick's accomplishment, which is ranked as one of science's major intellectual achievements, tied together the less than universally accepted results of several diverse studies:

1. Chargaff's rules. At the time, the relationships A = T and G = C were quite obscure because their significance was not apparent. In fact, even Chargaff did not emphasize them.

2. Correct tautomeric forms of the bases. X-ray, nuclear magnetic resonance (NMR), and spectroscopic investigations have firmly established that the nucleic acid bases are overwhelmingly in the keto tautomeric forms shown in Table 5-1. In 1953, however, this was not generally appreciated. Indeed, guanine and thymine were widely believed to be in their enol forms (Fig. 5-9) because it was thought that the resonance stability of these aromatic molecules would thereby be maximized. Knowledge of the dominant tautomeric forms, which

was prerequisite for the prediction of the correct hydrogen bonding associations of the bases, was provided by Jerry Donohue, an office mate of Watson and Crick and an expert on the X-ray structures of small organic molecules.

3. Information that DNA is a helical molecule. This was provided by an X-ray diffraction photograph of a DNA fiber taken by Rosalind Franklin (Fig. 5-10; DNA, being a thread-like molecule, does not crystallize but, rather, can be drawn out in fibers consisting of parallel bundles of molecules). This photograph enabled Crick, an X-ray crystallographer by training who had earlier derived the equations describing diffraction by helical molecules, to deduce (a) that DNA is a helical molecule and (b) that its planar aromatic bases form a stack of parallel rings which is parallel to the fiber axis.

This information only provided a few crude landmarks that guided the elucidation of the DNA structure. It mostly sprang from Watson and Crick's imaginations through model building studies. Once the Watson–Crick model had been published, however, its basic simplicity combined with its obvious biological relevance led to its rapid acceptance. Later investigations have confirmed the essential correctness of the Watson–Crick model, although its details have been modified.

A. *The Watson–Crick Structure: B-DNA*

Fibers of DNA assume the so-called B conformation, as indicated by their X-ray diffraction patterns, when the counterion is an alkali metal such as Na^+ and the relative humidity is >92%. **B-DNA** *is regarded as the **native** (biologically functional) form of DNA because, for example, its X-ray pattern resembles that of the DNA in intact sperm heads.*

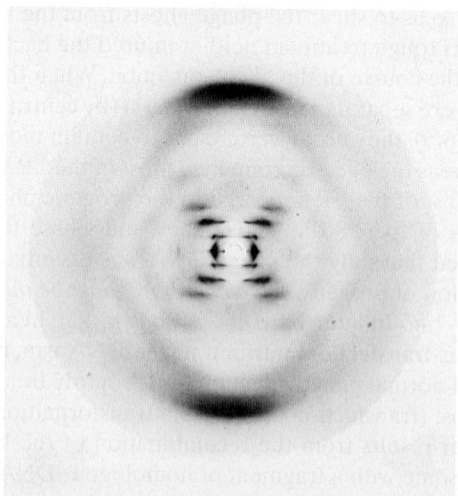

Figure 5-10 **X-ray diffraction photograph of a vertically oriented Na^+ DNA fiber in the B conformation taken by Rosalind Franklin.** This is the photograph that provided key information for the elucidation of the Watson–Crick structure. The central X-shaped pattern of spots is indicative of a helix, whereas the heavy black arcs on the top and bottom of the diffraction pattern correspond to a distance of 3.4 Å and indicate that the DNA structure largely repeats every 3.4 Å along the fiber axis. [Courtesy of Maurice Wilkins, King's College, London.]

(a)

Thymine
(keto *or* lactam form)

Thymine
(enol *or* lactim form)

(b)

Guanine
(keto *or* lactam form)

Guanine
(enol *or* lactim form)

Figure 5-9 **Some possible tautomeric conversions for bases.** (*a*) Thymine and (*b*) guanine residues. Cytosine and adenine residues can undergo similar proton shifts.

The Watson–Crick structure of B-DNA has the following major features:

1. *It consists of two polynucleotide strands that wind about a common axis with a right-handed twist to form an ~20-Å-diameter double helix (Fig. 5-11). The two strands are antiparallel (run in opposite directions) and wrap around each other such that they cannot be separated without unwinding the helix.* The bases occupy the core of the helix and the sugar–phosphate chains are coiled about its periphery, thereby minimizing the repulsions between charged phosphate groups.

2. The planes of the bases are nearly perpendicular to the helix axis. Each base is hydrogen bonded to a base on the opposite strand to form a planar base pair (Fig. 5-11). It is these hydrogen bonding interactions, a phenomenon known as **complementary base pairing,** that result in the specific association of the two chains of the double helix.

3. The "ideal" B-DNA helix has 10 base pairs (**bp**) per turn (a helical twist of 36° per bp) and, since the aromatic bases have van der Waals thicknesses of 3.4 Å and are partially stacked on each other (**base stacking,** Fig. 5-11), the helix has a **pitch** (rise per turn) of 34 Å.

The most remarkable feature of the Watson–Crick structure is that *it can accommodate only two types of base pairs: Each adenine residue must pair with a thymine residue and vice versa, and each guanine residue must pair with a cytosine residue and vice versa.* The geometries of these A · T and G · C base pairs, the so-called **Watson–Crick base pairs,** are shown in Fig. 5-12. It can be seen that *both of these base pairs are interchangeable in that they can replace each other in the double helix without altering the positions of the sugar–phosphate backbone's*

Figure 5-11   Three-dimensional structure of B-DNA. The repeating helix in this ball-and-stick drawing is based on the X-ray structure of the self-complementary dodecamer d(CGCGAATTCGCG) determined by Richard Dickerson and Horace Drew. The view is perpendicular to the helix axis. The sugar–phosphate backbones (*blue with blue-green ribbon outlines*) wind about the periphery of the molecule in opposite directions. The bases (*red*), which occupy its core, form hydrogen bonded base pairs. H atoms have been omitted for clarity. [Illustration, Irving Geis. Image from the Irving Geis Collection, Howard Hughes Medical Institute. Reprinted with permission.] **See Interactive Exercise 1 and Kinemage 2-1**

Figure 5-12   Watson–Crick base pairs. The line joining the C1′ atoms is the same length in both base pairs and makes equal angles with the glycosidic bonds to the bases. This gives DNA a series of pseudo-twofold symmetry axes (often referred to as **dyad axes**) that pass through the center of each base pair (*red line*) and are perpendicular to the helix axis. Note that A · T base pairs associate via two hydrogen bonds, whereas C · G base pairs are joined by three hydrogen bonds. [After Arnott, S., Dover, S.D., and Wonacott, A.J., *Acta Cryst.* **B25,** 2192 (1969).] **See Kinemages 2-2 and 17-2**

C1′ atoms. Likewise, the double helix is undisturbed by exchanging the partners of a Watson–Crick base pair, that is, by changing a G · C to a C · G or an A · T to a T · A. In contrast, any other combination of bases (e.g., A · G or A · C) would significantly distort the double helix since the formation of a non-Watson–Crick base pair would require considerable reorientation of the sugar–phosphate chain.

B-DNA has two deep exterior grooves that wind between its sugar–phosphate chains as a consequence of the helix axis passing through the approximate center of each base pair. However, the grooves are of unequal size (Fig. 5-11) because (1) the top edge of each base pair, as drawn in Fig. 5-12, is structurally distinct from the bottom edge; and (2) the deoxyribose residues are asymmetric. The **minor groove** exposes that edge of a base pair from which its C1′ atoms extend (opening toward the bottom in Fig. 5-12), whereas the **major groove** exposes the opposite edge of each base pair (the top of Fig. 5-12).

Although B-DNA is, by far, the most prevalent form of DNA in the cell, double helical DNAs and RNAs can assume several distinct structures. The structures of these other double helical nucleic acids are discussed in Section 29-1B.

B. DNA Is Semiconservatively Replicated

The Watson–Crick structure can accommodate any sequence of bases on one polynucleotide strand if the opposite strand has the complementary base sequence. This immediately accounts for Chargaff's rules. More importantly, *it suggests that hereditary information is encoded in the sequence of bases on either strand.* Furthermore, each polynucleotide strand can act as a template for the formation of its complementary strand through base pairing interactions (Fig. 1-17). The two strands of the parent molecule must therefore separate so that a complementary daughter strand may be enzymatically synthesized on the surface of each parent strand. This results in two molecules of **duplex** (double-stranded) DNA, each consisting of one polynucleotide strand from the parent molecule and a newly synthesized complementary strand. Such a mode of replication is termed **semiconservative** in contrast with **conservative** replication, which, if it occurred, would result in a newly synthesized duplex copy of the original DNA molecule with the parent DNA molecule remaining intact. The mechanism of DNA replication is the main subject of Chapter 30.

The semiconservative nature of DNA replication was elegantly demonstrated in 1958 by Matthew Meselson and Franklin Stahl. The density of DNA was increased by labeling it with ^{15}N, a heavy isotope of nitrogen (^{14}N is the naturally abundant isotope). This was accomplished by growing *E. coli* for 14 generations in a medium that contained $^{15}NH_4Cl$ as the only nitrogen source. The labeled bacteria were then abruptly transferred to an ^{14}N-containing medium, and the density of their DNA was monitored as a function of bacterial growth by **equilibrium density gradient ultracentrifugation** (a technique for separating macromolecules according to their densities, which Meselson, Stahl, and Jerome Vinograd had developed for the purpose of distinguishing ^{15}N-labeled DNA from unlabeled DNA; Section 6-5Bb).

The results of the Meselson–Stahl experiment are displayed in Fig. 5-13. After one generation (doubling of the cell population), all of the DNA had a density exactly halfway between the densities of fully ^{15}N-labeled DNA and unlabeled DNA. This DNA must therefore contain equal amounts of ^{14}N and ^{15}N, as is expected after one generation of semiconservative replication. Conservative DNA replication, in contrast, would result in the preservation of the parental DNA, so that it maintained its original density, and the generation of an equal amount of unlabeled DNA. After two generations, half of the DNA molecules were unlabeled and the remainder were ^{14}N–^{15}N hybrids. This is also in accord with the predictions of the semiconservative replication model and in disagreement with the conservative replication model. In succeeding generations, the amount of unlabeled DNA increased relative to the amount of hybrid DNA, although the hybrid never totally disappeared. This is again in harmony with semiconservative replication but at odds with conservative replication, which predicts that the fully labeled parental DNA will always be present and that hybrid DNA never forms.

C. Denaturation and Renaturation

When a solution of duplex DNA is heated above a characteristic temperature, its native structure collapses and its two complementary strands separate and assume a flexible and rapidly fluctuating conformational state known as a random coil (Fig. 5-14). This **denaturation** process is accompanied by a qualitative change in the DNA's physical properties. For instance, the characteristic high viscosity of native DNA solutions, which arises from the resistance to deformation of its rigid and rodlike duplex molecules, drastically decreases when the duplex DNA decomposes (denatures) to two relatively freely jointed single strands.

a. DNA Denaturation Is a Cooperative Process
The most convenient way of monitoring the amount of nucleic acid present is by its ultraviolet (UV) absorbance spectrum. A solution containing a solute that absorbs light does so according to the **Beer–Lambert law,**

$$A = -\log\left(\frac{I}{I_0}\right) = \varepsilon cl \qquad [5.1]$$

where A is the solute's **absorbance** (alternatively, its **optical density**), I_0 is the incident intensity of light at a given wavelength λ, I is its transmitted intensity at λ, ε is the **molar extinction coefficient** of the solute at λ, c is its molar concentration, and l is the length of the light path in centimeters. The value of ε varies with λ; a plot of ε versus λ for the

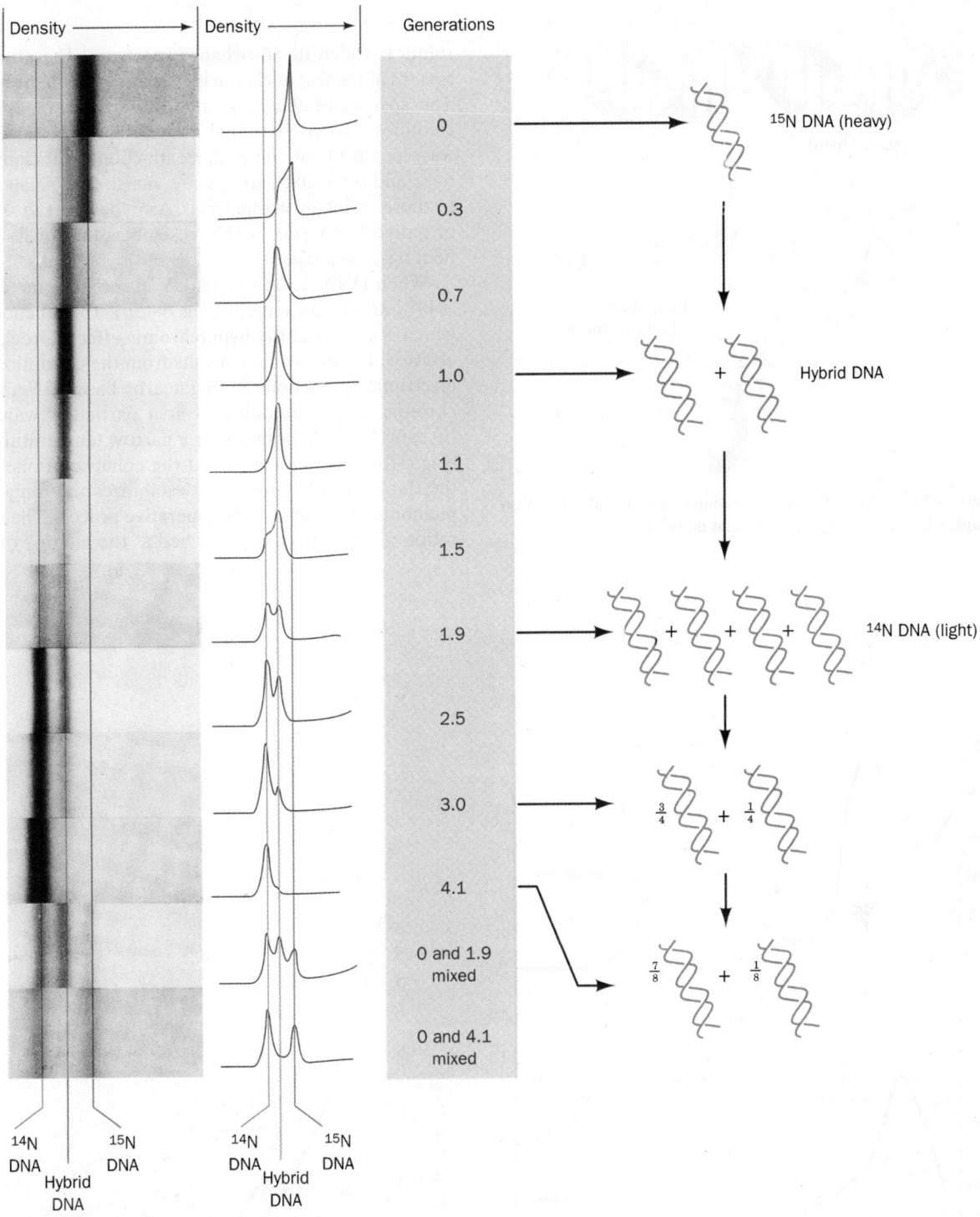

Figure 5-13 Demonstration of the semiconservative nature of DNA replication in *E. coli* by density gradient ultracentrifugation. The DNA was dissolved in an aqueous CsCl solution of density 1.71 g · cm⁻³ and was subjected to an acceleration of 140,000 times that of gravity in an analytical ultracentrifuge (a device in which the rapidly spinning sample can be optically observed). This enormous acceleration induced the CsCl to redistribute in the solution such that its concentration increased with its radius in the ultracentrifuge. Consequently, the DNA migrated within the resulting density gradient to its position of buoyant density. The left panels are ultraviolet absorption photographs of ultracentrifuge cells (DNA strongly absorbs ultraviolet light) and are arranged such that regions of equal density have the same horizontal positions. The middle panels are microdensitometer traces of the corresponding photographs in which the vertical displacement is proportional to the DNA concentration. The buoyant density of DNA increases with its ¹⁵N content. The bands farthest to the right (of greatest radius and density) arise from DNA that is fully ¹⁵N labeled, whereas unlabeled DNA, which is 0.014 g · cm⁻³ less dense, forms the leftmost bands. The bands in the intermediate position result from duplex DNA in which one strand is ¹⁵N-labeled and the other strand is unlabeled. The accompanying interpretive drawings (*right*) indicate the relative numbers of DNA strands at each generation donated by the original parents (*blue*, ¹⁵N labeled) and synthesized by succeeding generations (*red*, unlabeled). [From Meselson, M. and Stahl, F.W., *Proc. Natl. Acad. Sci.* **44**, 671 (1958).] 🐭 **See the Animated Figures**

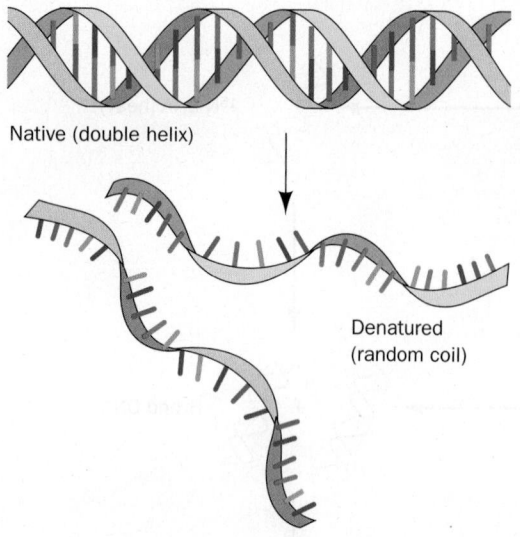

Figure 5-14 **Schematic representation of the strand separation in duplex DNA resulting from its heat denaturation.**

solute is called its **absorbance spectrum.** The absorbance spectra of the five nucleic acid bases are shown in Fig. 5-15*a*. The spectra of the corresponding nucleosides and nucleotides are closely similar above 190 nm because, in this wavelength range, the molar extinction coefficients of ribose and phosphate groups are vanishingly small relative to those of the aromatic bases. As expected, the spectrum of native DNA (Fig. 5-15*b*) resembles that of its component bases in shape.

When DNA denatures, its UV absorbance increases by ~40% at all wavelengths (Fig. 5-15*b*). This phenomenon, which is known as the **hyperchromic effect** (Greek: *hyper,* above + *chroma,* color), results from the disruption of the electronic interactions among nearby bases. DNA's hyperchromic shift, as monitored at a particular wavelength (usually 260 nm), occurs over a narrow temperature range (Fig. 5-16). This indicates that the collapse of one part of the duplex DNA's structure destabilizes the remainder, a phenomenon known as a **cooperative process.** The denaturation of DNA may be described as the melting of a one-

(a)

(b)

Figure 5-15 **UV absorbance spectra of the nucleic acid bases and DNA.** (*a*) Spectra of adenine, guanine, cytosine, thymine, and uracil near pH 7. (*b*) Spectra of native and heat-denatured *E. coli* DNA. Note that denaturation does not change the general shape of the absorbance spectrum but increases its absorbance at all wavelengths. [After Voet, D., Gratzer, W.B., Cox, R.A., and Doty, P., *Biopolymers* **1,** 193 (1963).] 🔊 **See the Animated Figures**

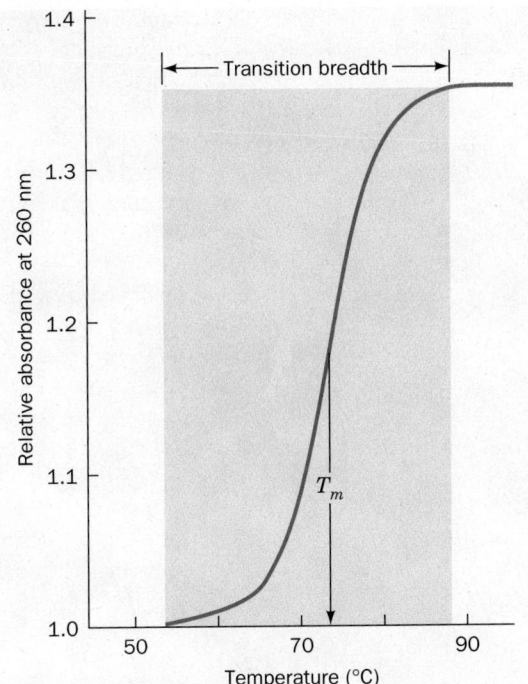

Figure 5-16 **Example of a DNA melting curve.** The relative absorbance is the ratio of the absorbance (customarily measured at 260 nm) at the indicated temperature to that at 25°C. The melting temperature, T_m, is defined as the temperature at which half of the maximum absorbance increase is attained. **See the Animated Figures**

Figure 5-17 **Variation of the melting temperatures, T_m, of DNA with its G + C content.** The DNAs were dissolved in a solution containing 0.15M NaCl and 0.015M sodium citrate. [After Marmur, J. and Doty, P., *J. Mol. Biol.* **5,** 113 (1962).]

completely renatures. Likewise, complementary strands of RNA and DNA, in a process known as **hybridization,** form RNA–DNA hybrid double helices that are only slightly less stable than the corresponding DNA double helices.

dimensional solid, so Fig. 5-16 is referred to as a **melting curve** and the temperature at its midpoint is known as the **melting temperature, T_m.**

The stability of the DNA double helix, and hence its T_m, depends on several factors, including the nature of the solvent, the identities and concentrations of the ions in solution, and the pH. For example, duplex DNA denatures (its T_m decreases) under alkaline conditions that cause some of the bases to ionize and thereby disrupt their base pairing interactions. The T_m increases linearly with the mole fraction of G · C base pairs (Fig. 5-17), which indicates that triply hydrogen bonded G · C base pairs are more stable than doubly hydrogen bonded A · T base pairs.

b. Denatured DNA Can Be Renatured

If a solution of denatured DNA is rapidly cooled to well below its T_m, the resulting DNA will be only partially base paired (Fig. 5-18) because its complementary strands will not have had sufficient time to find each other before the partially base paired structures become effectively "frozen in." If, however, the temperature is maintained ~25°C below the T_m, enough thermal energy is available for short base paired regions to rearrange by melting and reforming but not enough to melt out long complementary stretches. Under such **annealing conditions,** as Julius Marmur discovered in 1960, denatured DNA eventually

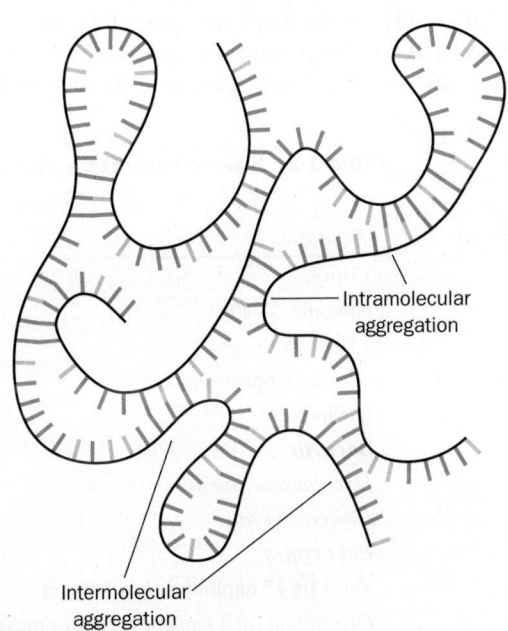

Figure 5-18 **Partially renatured DNA.** A schematic representation showing the imperfectly base paired structures assumed by DNA that has been heat denatured and then rapidly cooled to well below its T_m. Note that both intramolecular and intermolecular aggregation may occur.

D. *The Size of DNA*

DNA molecules are generally enormous (Fig. 5-19). The molecular mass of DNA has been determined by a variety of techniques including ultracentrifugation (Section 6-5A) and through length measurements by electron microscopy [a base pair of Na$^+$ B-DNA has an average molecular mass of 660 D and a length (thickness) of 3.4 Å] and **autoradiography** (a technique in which the position of a radioactive substance in a sample is recorded by the blackening of a photographic emulsion that the sample is laid over or embedded in; Fig. 5-20). The number of base pairs and the **contour lengths** (the end-to-end lengths of the stretched-out native molecules) of the DNAs from a selection of organisms of increasing complexity are listed in Table 5-2. Not surprisingly, an organism's haploid quantity (unique amount) of DNA varies more or less with its complexity (although there are notable exceptions to this generalization, such as the last entry in Table 5-2).

The visualization of DNAs from prokaryotes has demonstrated that their entire **genome** (complement of genetic information) is contained on a single, often circular, length of DNA. Similarly, Bruno Zimm demonstrated that the *largest chromosome of the fruit fly Drosophila melanogaster contains a single molecule of DNA* by comparing the molecular mass of this DNA with the cytologically measured length of DNA contained in the chromosome. Likewise, other eukaryotic chromosomes contain only single molecules of DNA.

The highly elongated shape of duplex DNA (recall B-DNA is only 20 Å in diameter), together with its stiffness, make it extremely susceptible to mechanical damage outside the cell's protective environment (for instance, if the *Drosophila* DNA of Fig. 5-20 were enlarged by a factor of 500,000, it would have the shape and some of the mechanical properties of a 6-km-long strand of uncooked spaghetti). The hydrodynamic shearing forces generated by such ordinary

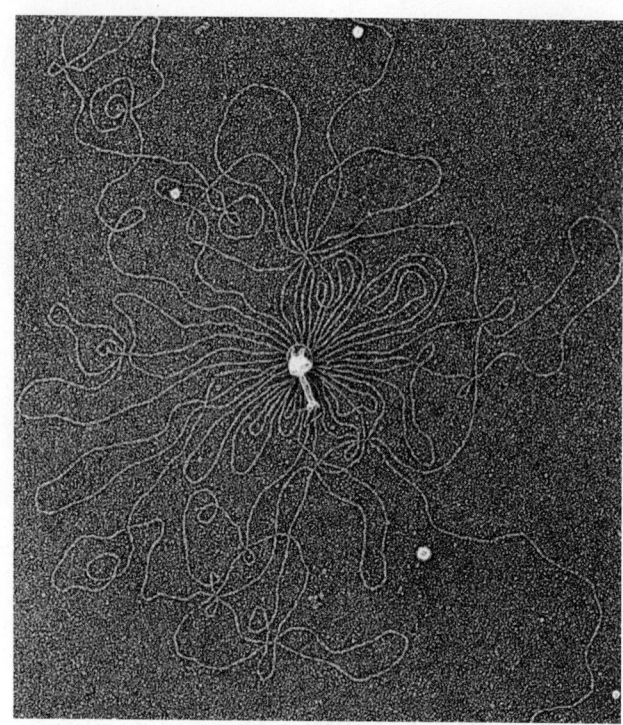

Figure 5-19 **Electron micrograph of a T2 bacteriophage and its DNA.** The phage has been osmotically lysed (broken open) in distilled water so that its DNA spilled out. Without special treatment, duplex DNA, which is only 20 Å in diameter, is difficult to visualize in the electron microscope. In the **Kleinschmidt procedure** used here, DNA is fattened to ~200 Å in diameter by coating it with a denatured basic protein. [From Kleinschmidt, A.K., Lang, D., Jacherts, D., and Zahn, R.K., *Biochim. Biophys. Acta* **61,** 857 (1962).]

laboratory manipulations as stirring, shaking, and pipetting break DNA into relatively small pieces so that the isolation of an intact molecule of DNA requires extremely gentle

Table 5-2 **Sizes of Some DNA Molecules**

Organism	Number of Base Pairs (kb)[a]	Contour Length (μm)
Viruses		
Polyoma, SV40	5.2	1.7
λ Bacteriophage	48.6	17
T2, T4, T6 bacteriophage	166	55
Fowlpox	280	193
Bacteria		
Mycoplasma hominis	760	260
Escherichia coli	4,600	1,600
Eukaryotes		
Yeast (in 17 haploid chromosomes)	12,000	4,100
Drosophila (in 4 haploid chromosomes)	180,000	61,000
Human (in 23 haploid chromosomes)	3,000,000	1,000,000
Lungfish (in 19 haploid chromosomes)	102,000,000	35,000,000

[a]kb = kilobase pair = 1000 base pairs (bp).

Source: Mainly Kornberg, A. and Baker, T.A., *DNA Replication* (2nd ed.), p. 20, Freeman (1992).

Figure 5-20   Autoradiograph of *Drosophila melanogaster* DNA. Lysates of *D. melanogaster* cells that had been cultured with ³H-labeled thymidine were spread on a glass slide and covered with a photographic emulsion that was developed after a 5-month exposure. The white curve, which resulted from the radioactive decay of the ³H, traces the path of the DNA in this photographic positive. The DNA's measured contour length is 1.2 cm. [From Kavenoff, R., Klotz, L.C., and Zimm, B.H., *Cold Spring Harbor Symp. Quant. Biol.* **38**, 4 (1973). Copyright © 1974 by Cold Spring Harbor Laboratory Press.]

handling. Before 1960, when this was first realized, the measured molecular masses of DNA were no higher than ~10 million D (~15 **kb**, where 1 kb = 1 kilobase pair = 1000 bp). DNA fragments of uniform molecular mass and as small as a few hundred base pairs may be generated by **shear degrading** DNA in a controlled manner; for instance, by forcing the DNA solution through a small orifice or by **sonication** (exposure to intense high-frequency sound waves).

4 GENE EXPRESSION AND REPLICATION: AN OVERVIEW

See Guided Exploration 1: Overview of transcription and translation
How do genes function, that is, how do they direct the synthesis of RNA and proteins, and how are they replicated? The answers to these questions form the multifaceted discipline known as **molecular biology.** In 1958, Crick neatly encapsulated the broad outlines of this process in a flow scheme he called the **central dogma of molecular biology:** *DNA directs its own replication and its **transcription** to yield RNA which, in turn, directs its **translation** to form proteins (Fig. 5-21).* Here the term "transcription" indicates that in transferring

Figure 5-21   The central dogma of molecular biology. Solid arrows indicate the types of genetic information transfers that occur in all cells. Special transfers are indicated by the dashed arrows: **RNA-directed RNA polymerase** is expressed both by certain RNA viruses and by some plants; **RNA-directed DNA polymerase (reverse transcriptase)** is expressed by other RNA viruses; and DNA directly specifying a protein is unknown but does not seem beyond the realm of possibility. However, the missing arrows are information transfers the central dogma postulates never occur: protein specifying either DNA, RNA, or protein. In other words, *proteins can only be the recipients of genetic information.* [After Crick, F., *Nature* **227,** 561 (1970).]

information from DNA to RNA, the "language" encoding the information remains the same, that of base sequences, whereas the term "translation" indicates that in transferring information from RNA to proteins, the "language" changes from that of base sequences to that of amino acid sequences (Fig. 5-22). The machinery required to carry out the complex tasks of gene expression and DNA replication in an organized manner and with high fidelity occupies a major portion of every cell. In this section we summarize how gene expression and replication occur to provide the background for understanding the techniques of recombinant DNA technology (Section 5-5). This subject matter is explored in considerably greater detail in Chapters 29 to 34.

A. RNA Synthesis: Transcription

The enzyme that synthesizes RNA is named **RNA polymerase.** It catalyzes the DNA-directed coupling of the **nucleoside triphosphates (NTPs) adenosine triphosphate**

```
DNA    5′ — A–G–A–G–G–T–G–C–T — 3′
       3′ — T–C–T–C–C–A–C–G–A — 5′

              transcription ↓

mRNA   5′ — A–G–A–G–G–U–G–C–U — 3′
tRNAs       U–C–U  C–C–A  C–G–A
            └──┘   └──┘   └──┘
         Arginine Glycine Alanine

              translation ↓

Polypeptide      –Arg–Gly–Ala–
```

Figure 5-22   Gene expression. One strand of DNA directs the synthesis of RNA, a process known as transcription. The base sequence of the transcribed RNA is complementary to that of the DNA strand. The RNAs known as **messenger RNAs (mRNAs)** are translated when molecules of **transfer RNA (tRNA)** align with the mRNA via complementary base pairing between segments of three consecutive nucleotides known as codons. Each type of tRNA carries a specific amino acid. These amino acids are covalently joined by the ribosome to form a polypeptide. Thus, the sequence of bases in DNA specifies the sequence of amino acids in a protein.

Figure 5-23   Action of RNA polymerases.
These enzymes assemble incoming ribonucleoside triphosphates on templates consisting of single-stranded segments of DNA such that the growing strand is elongated in the 5′ to 3′ direction.

(ATP), cytidine triphosphate (CTP), guanosine triphosphate (GTP), and uridine triphosphate (UTP) in a reaction that releases pyrophosphate ion ($P_2O_7^{4-}$):

$$RNA_{n\ \text{residues}} + NTP \longrightarrow RNA_{n+1\ \text{residues}} + P_2O_7^{4-}$$

RNA synthesis proceeds in a stepwise manner in the 5′ → 3′ direction, that is, the incoming nucleotide is appended to the free 3′—OH group of the growing RNA chain (Fig. 5-23). RNA polymerase selects the nucleotide it incorporates into the nascent (growing) RNA chain through the requirement that it form a Watson–Crick base pair with the DNA strand that is being transcribed, the **template strand** (only one of duplex DNA's two strands is transcribed at a time). This is possible because, as the RNA polymerase moves along the duplex DNA it is transcribing, it separates a short (~14 bp) segment of its two strands to form a so-called **transcription bubble,** thereby permitting this portion of the template strand to transiently form a short DNA–RNA hybrid helix with the newly synthesized RNA (Fig. 5-24). Like duplex DNA, a DNA–RNA hybrid helix consists of antiparallel strands, and hence the DNA's template strand is read in its 3′ → 5′ direction.

All cells contain RNA polymerase. In bacteria, one species of this enzyme synthesizes nearly all of the cell's RNA. Certain viruses generate RNA polymerases that syn-

thesize only virus-specific RNAs. Eukaryotic cells contain four or five different types of RNA polymerases that each synthesize a different class of RNA.

a. Transcriptional Initiation Is a Precisely Controlled Process

The DNA template strand contains control sites consisting of specific base sequences that specify both the site at which RNA polymerase initiates transcription (the site on the DNA at which the RNA's first two nucleotides are joined) and the rate at which RNA polymerase initiates transcription at this site. Specific proteins known in prokaryotes as **activators** and **repressors** and in eukaryotes as **transcription factors** bind to these control sites or to other such proteins that do so and thereby stimulate or inhibit transcriptional initiation by RNA polymerase. For the RNAs that encode proteins, which are named **messenger RNAs (mRNAs),** these control sites precede the initiation site (that is, they are "upstream" of the initiation site relative to the RNA polymerase's direction of travel).

The rate at which a cell synthesizes a given protein, or even whether the protein is synthesized at all, is mainly governed by the rate at which the synthesis of the corresponding mRNA is initiated. The way that prokaryotes regulate the rate at which many genes undergo transcriptional initiation

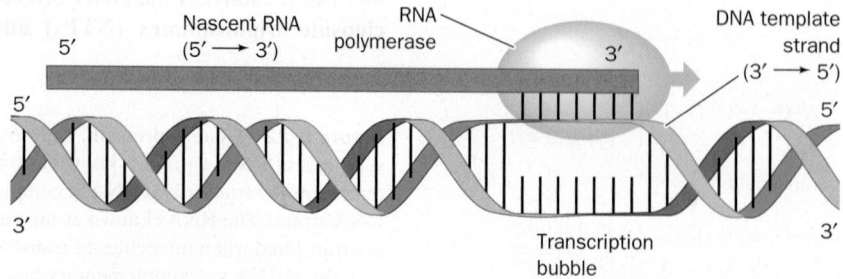

Figure 5-24   Function of the transcription bubble. RNA polymerase unwinds the DNA double helix by about a turn in the region being transcribed, thereby permitting the DNA's template strand to form a short segment of DNA–RNA hybrid double helix with the RNA's newly synthesized 3′ end. As the

RNA polymerase advances along the DNA template (here to the right), the DNA unwinds ahead of the RNA's growing 3′ end and rewinds behind it, thereby stripping the newly synthesized RNA from the template strand.

can be relatively simple. For example, the transcriptional initiation of numerous prokaryotic genes requires only that RNA polymerase bind to a control sequence, known as a **promoter,** that precedes the transcriptional initiation site. However, not all promoters are created equal: RNA polymerase initiates transcription more often at so-called efficient promoters than at those with even slightly different sequences. Thus the rate at which a gene is transcribed is governed by the sequence of its associated promoter.

A more complex way in which prokaryotes control the rate of transcriptional initiation is exemplified by the *E. coli lac* **operon,** a cluster of three consecutive genes (*Z*, *Y*, and *A*) encoding proteins that the bacterium requires to metabolize the sugar **lactose** (Section 11-2B). In the absence of lactose, a protein named the *lac* **repressor** specifically binds to a control site in the *lac* operon known as an **operator** (Section 31-3B). This prevents RNA polymerase from initiating the transcription of *lac* operon genes (Fig. 5-25*a*), thereby halting the synthesis of unneeded proteins. However, when lactose is available, the bacterium meta-

bolically modifies a small amount of it to form the related sugar **allolactose.** This so-called **inducer** specifically binds to the *lac* repressor, thereby causing it to dissociate from the operator DNA so that RNA polymerase can initiate the transcription of the *lac* operon genes (Fig. 5-25*b*).

In eukaryotes, the control sites regulating transcriptional initiation can be quite extensive and surprisingly distant from the transcriptional initiation site (by as much as several tens of thousands of base pairs; Section 34-3). Moreover, the eukaryotic transcriptional machinery that binds to these sites and thereby induces RNA polymerase to commence transcription can be enormously complex (consisting of up to 50 or more proteins; Section 34-3).

b. Transcriptional Termination Is a Relatively Simple Process

The site on the template strand at which RNA polymerase terminates transcription and releases the completed RNA is governed by the base sequence in this region. However, the control of transcriptional termination is rarely involved in the regulation of gene expression. In keeping with this, the cellular machinery that mediates transcriptional termination is relatively simple compared with that involved in transcriptional initiation (Section 31-2D).

c. Eukaryotic RNA Undergoes Post-Transcriptional Modifications

Most prokaryotic mRNA transcripts participate in translation without further alteration. However, most primary transcripts in eukaryotes require extensive **post-transcriptional modifications** to become functional. For mRNAs, these modifications include the addition of a 7-methylguanosine-containing "cap" that is enzymatically appended to the transcript's 5' end and ~250-nucleotide **polyadenylic acid [poly(A)]** "tail" that is enzymatically appended to its 3' end. However, the most striking modification that most eukaryotic transcripts undergo is a process called **gene splicing** in which one or more often lengthy RNA segments known as **introns** (for "intervening sequences") are precisely excised from the RNA and the remaining **exons** (for "expressed sequences") are rejoined in their original order to form the mature mRNA (Fig. 5-26; Section 31-4A). Different mRNAs can be generated from

(*a*) **Absence of inducer**

(*b*) **Presence of inducer**

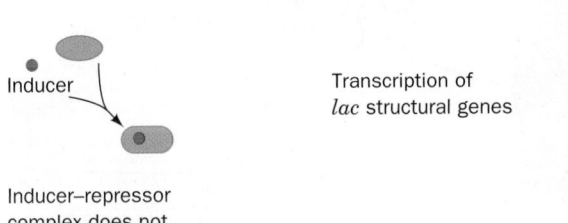

Figure 5-25   Control of transcription of the *E. coli lac* operon. (*a*) In the absence of an inducer such as allolactose, the *lac* repressor binds to the operator (*O*), thereby preventing RNA polymerase from transcribing the *Z*, *Y*, and *A* genes of the *lac* operon. (*b*) On binding inducer, the *lac* repressor dissociates from the operator, which permits RNA polymerase to bind to the promoter (*P*) and transcribe the *Z*, *Y*, and *A* genes. ✑ **See Guided Exploration 2: Regulation of gene expression by the *lac* repressor system**

Figure 5-26   Post-transcriptional processing of eukaryotic mRNAs. Most primary transcripts require further covalent modification to become functional, including the addition of a 5' cap and a 3' poly(A) tail, and splicing to excise its introns from between its exons.

the same gene through the selection of alternate transcriptional initiation sites and/or alternative splice sites, leading to the production of somewhat different proteins, usually in a tissue-specific manner (Section 34-3C).

B. *Protein Synthesis: Translation*

Polypeptides are synthesized under the direction of the corresponding mRNA by **ribosomes,** numerous cytosolic organelles that consist of about two-thirds RNA and one-third protein and have molecular masses of ~2500 kD in prokaryotes and ~4200 kD in eukaryotes. Ribosomal RNAs (**rRNAs**), of which there are several kinds, are transcribed from DNA templates, as are all other kinds of RNA.

a. Transfer RNAs Deliver Amino Acids to the Ribosome

mRNAs are essentially a series of consecutive 3-nucleotide segments known as **codons,** each of which specifies a particular amino acid. However, codons do not bind amino acids. Rather, on the ribosome, *they specifically bind molecules of transfer RNA (tRNA) that are each covalently linked to the corresponding amino acid (Fig. 5-27).* A tRNA typically consists of ~76 nucleotides (which makes it comparable in mass and structural complexity to a medium-sized protein) and contains a trinucleotide sequence, its **anticodon,** which is complementary to the codon(s) specifying its appended amino acid (see below). An amino acid is covalently linked to the 3' end of its corresponding tRNA to form an **aminoacyl-tRNA** (a process called "charging") through the action of an enzyme that specifically recognizes both the tRNA and the amino acid (see below). During

Figure 5-27   Transfer RNA (tRNA) drawn in its "cloverleaf" form. Its covalently linked amino acid residue forms an aminoacyl-tRNA (*top*), and its anticodon (*bottom*), a trinucleotide segment, base pairs with the complementary codon on mRNA during translation.

translation, the mRNA is passed through the ribosome such that each codon, in turn, binds its corresponding aminoacyl-tRNA (Fig. 5-28). As this occurs, the ribosome transfers the amino acid residue on the tRNA to the C-terminal end of

Figure 5-28   Schematic diagram of translation. The ribosome binds an mRNA and two tRNAs and facilitates their specific association through consecutive codon–anticodon interactions. The ribosomal binding site closer to the 5' end of the mRNA binds a **peptidyl-tRNA** (*left,* a tRNA to which the growing polypeptide chain is covalently linked) and is therefore known as the **P site,** whereas the ribosomal site closer to the 3' end of the mRNA binds an aminoacyl-tRNA (*right*) and is hence called the **A site.**

The ribosome catalyzes the transfer of the polypeptide from the peptidyl-tRNA to the aminoacyl-tRNA, thereby forming a new peptidyl-tRNA whose polypeptide chain has increased in length by one residue at its C-terminus. The discharged tRNA in the P site is then ejected, and the peptidyl-tRNA, together with its bound mRNA, is shifted from the A site to the P site, thereby permitting the next codon to bind its corresponding aminoacyl-tRNA in the ribosomal A site.

Figure 5-29 The ribosomal reaction forming a peptide bond. The amino group of the aminoacyl-tRNA in the ribosomal A site nucleophilically displaces the tRNA of the peptidyl-tRNA ester in the ribosomal P site, thereby forming a new peptide bond and transferring the growing polypeptide to the A site tRNA.

the growing polypeptide chain (Fig. 5-29). Hence, *the polypeptide grows from its N-terminus to its C-terminus.*

b. The Genetic Code

The correspondence between the sequence of bases in a codon and the amino acid residue it specifies is known as the **genetic code** (Table 5-3). Its near universality among all forms of life is compelling evidence that life on Earth arose from a common ancestor and makes it possible, for example, to express human genes in *E. coli* (Section 5-5Ga). There are four possible bases (U, C, A, and G) that can occupy each of the three positions in a codon, and hence there are $4^3 = 64$ possible codons. Of these codons, 61 specify amino acids (of which there are only 20) and the remaining three, UAA, UAG, and UGA, are **Stop codons** that instruct the ribosome to cease polypeptide synthesis and release the resulting transcript. All but two amino acids (Met and Trp) are specified by more than one codon and three (Leu, Ser, and Arg) are specified by six codons. Consequently, in a term borrowed from mathematics, the genetic code is said to be **degenerate** (taking on several discrete values).

Note that the arrangement of the genetic code is nonrandom: Most codons that specify a given amino acid, which are known as **synonyms,** occupy the same box in Table 5-3, that is, they differ in sequence only in their third (3′) nucleotide. Moreover, most codons specifying nonpolar amino acid residues have a G in their first position and/or a U in their second position (Table 5-3).

A tRNA may recognize as many as three synonymous codons because the 5′ base of a codon and the 3′ base of a corresponding anticodon do not necessarily interact via a Watson–Crick base pair (Section 32-2D; keep in mind that the codon and the anticodon associate in an antiparallel fashion to form a short segment of an RNA double helix). Thus, cells can have far fewer than the 61 tRNAs that would be required for a 1:1 match with the 61 amino acid–specifying codons, although, in fact, some eukaryotic cells contain over 500 different tRNAs.

c. tRNAs Acquire Amino Acids Through the Actions of Aminoacyl-tRNA Synthetases

In synthesizing a polypeptide, a ribosome does not recognize the amino acid appended to a tRNA but only whether its anticodon binds to the mRNA's codon (the anticodon and the amino acid on a charged tRNA are actually quite distant from one another, as Fig. 5-27 suggests). Thus, *the charging of a tRNA with the proper amino acid is as critical a step for accurate translation as is the proper recognition of a codon by its corresponding anticodon.* The enzymes that catalyze these additions are known as **aminoacyl-tRNA synthetases (aaRSs).** Cells typically contain 20 aaRSs, one for each amino acid, and therefore a given aaRS will charge all the tRNAs that bear codons specifying its corresponding amino acid. Consequently, each aaRS must somehow differentiate its cognate (corresponding) tRNAs from among the many other types of structurally and physically quite similar tRNAs that each cell contains. Although many aaRSs recognize the anticodons of their cognate tRNAs, not all of them do so. Rather, they recognize other sites on their cognate tRNAs.

Table 5-3 **The "Standard" Genetic Code**[a]

$$H_3\overset{+}{N}-\overset{\overset{\displaystyle H}{|}}{\underset{\underset{\displaystyle R}{|}}{C}}-C\overset{\displaystyle O}{\underset{\displaystyle O^-}{<}}$$

First Position (5′ end)	Second Position — U	Second Position — C	Second Position — A	Second Position — G	Third Position (3′ end)
U	UUU Phe	UCU Ser	UAU Tyr	UGU Cys	U
	UUC Phe	UCC Ser	UAC Tyr	UGC Cys	C
	UUA Leu	UCA Ser	UAA STOP	UGA STOP	A
	UUG Leu	UCG Ser	UAG STOP	UGG Trp	G
C	CUU Leu	CCU Pro	CAU His	CGU Arg	U
	CUC Leu	CCC Pro	CAC His	CGC Arg	C
	CUA Leu	CCA Pro	CAA Gln	CGA Arg	A
	CUG Leu	CCG Pro	CAG Gln	CGG Arg	G
A	AUU Ile	ACU Thr	AAU Asn	AGU Ser	U
	AUC Ile	ACC Thr	AAC Asn	AGC Ser	C
	AUA Ile	ACA Thr	AAA Lys	AGA Arg	A
	AUG Met[b]	ACG Thr	AAG Lys	AGG Arg	G
G	GUU Val	GCU Ala	GAU Asp	GGU Gly	U
	GUC Val	GCC Ala	GAC Asp	GGC Gly	C
	GUA Val	GCA Ala	GAA Glu	GGA Gly	A
	GUG Val	GCG Ala	GAG Glu	GGG Gly	G

[a]Nonpolar residues are tan, basic residues are blue, acidic residues are red, and polar uncharged residues are purple.
[b]AUG forms part of the initiation signal as well as coding for internal Met residues.

d. Translation Is Initiated at Specific AUG Codons

Ribosomes read mRNAs in their 5′ to 3′ direction (from "upstream" to "downstream"). The initiating codon is AUG, which specifies a Met residue. However, the tRNA that recognizes this initiation codon differs from the tRNA that delivers a polypeptide's internal Met residues to the ribosome, although both types of tRNA are charged by the same **methionyl-tRNA synthetase (MetRS).**

If a polypeptide is to be synthesized with the correct amino acid sequence, it is essential that the ribosome main-

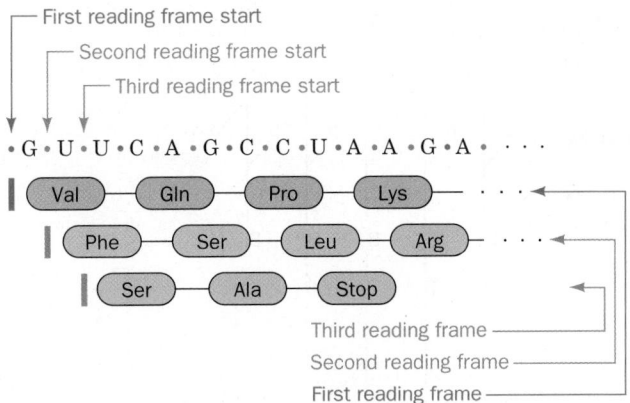

Figure 5-30 **Nucleotide reading frames.** An mRNA might be read in any of three different reading frames, each of which yields a different polypeptide.

tain the proper register between the mRNA and the incoming tRNAs, that is, that the ribosome maintain the correct **reading frame.** As is illustrated in Fig. 5-30, a shift of even one nucleotide along an mRNA will lead to the synthesis of an entirely different polypeptide from the point of the shift onward. Thus, the AUG codon that initiates polypeptide synthesis also sets the polypeptide's reading frame. Yet AUG also specifies a polypeptide's internal Met residues, and an mRNA is likely to contain numerous AUGs in different reading frames. How then does the ribosome select the initiation codon from among the many AUGs in an mRNA? In prokaryotes, the answer is that each mRNA contains a sequence on the upstream (5′) side of the initiating codon (a region that does not encode polypeptide chain) through which the ribosome identifies this codon. In eukaryotes, the answer is simpler; the initiating codon is usually the first AUG that is downstream of the mRNA's 5′ cap.

e. Prokaryotic mRNAs Have Short Lifetimes

In prokaryotes, transcription and translation both take place in the same cellular compartment, the cytosol (Figs. 1-2 and 1-13). Consequently ribosomes often attach to the 5′ end of an mRNA before its synthesis is complete and commence synthesizing the corresponding polypeptide. This is essential because, since the mRNAs in prokaryotes have average lifetimes of only 1 to 3 minutes before being hydrolytically degraded by enzymes known as **nucleases,** the 5′ end of an mRNA may be degraded before its 3′ end is synthesized. This rapid turnover of its mRNAs permits a prokaryote to respond quickly to changes in its environment by synthesizing the proteins appropriate for its new situation within minutes of the change (recall that prokaryotes are adapted to live in environments in which there are rapid fluctuations in the available nutrients; Section 1-2).

Eukaryotic cells, in contrast, mostly lead a more sedentary existence. Their RNAs are transcribed and post-transcriptionally modified in the nucleus, whereas ribosomes occupy the cytosol where translation takes place (Fig. 1-5). Hence, mature mRNAs must be transported from the nucleus to the cytosol in order to participate in translation.

Eukaryotic mRNAs therefore tend to have lifetimes on the order of several days.

f. Proteins Are Subject to Post-Translational Modifications and Degradation

Newly synthesized polypeptides often require post-translational modifications to become functional. In many proteins, the leading (N-terminal) Met residue that was specified by its mRNA's initiating codon is excised by a specific **protease** (an enzyme that hydrolytically cleaves peptide bonds). Proteins are then subject to numerous other chemical modifications at specific residues, including specific proteolytic cleavages, acylation, hydroxylation, methylation, and phosphorylation (Section 4-3A). In addition, eukaryotic proteins, but not prokaryotic proteins, are subject to **glycosylation** (the addition of polysaccharides) at specific sites (Sections 11-3C and 23-3B). Indeed, **glycoproteins** (proteins that have been glycosylated) are the most common type of eukaryotic protein and can consist of up to 90% or more by mass of polysaccharide groups.

All cells have several mechanisms for degrading proteins to their component amino acids. This enables cells to eliminate damaged or abnormal proteins, destroy proteins that are no longer needed, and utilize proteins as nutrients. The lifetime of a protein in a cell can be surprisingly short, as little as a fraction of a minute, although many proteins in eukaryotes have lifetimes of days or weeks. Thus cells are dynamic entities that are constantly turning over most of their components, in particular their RNA and proteins.

C. DNA Replication

The chemical reaction by which DNA is replicated (Fig. 5-31) is nearly identical to that synthesizing RNA (Fig. 5-23), but with two major differences: (1) deoxynucleoside triphosphates **(dNTPs)** rather than nucleoside triphosphates are the reactants and (2) the enzyme that catalyzes the reaction is **DNA polymerase** rather than RNA polymerase. The properties of DNA polymerase result in a third major difference between RNA and DNA synthesis: Whereas RNA polymerase can link together two nucleotides on a DNA template, *DNA polymerase can only extend (in the 5′ to 3′ direction) an existing polynucleotide that is base paired to the DNA's template strand.* Thus, whereas RNA polymerase can initiate RNA synthesis *de novo* (from the beginning), *DNA polymerase requires an oligonucleotide* **primer**, *which it lengthens.*

a. Primers Are RNA

If DNA polymerase cannot synthesize DNA *de novo*, where do primers come from? It turns out that they are not DNA, as might be expected, but rather RNA. In *E. coli*, these RNA primers are synthesized by both RNA polymerase (the same enzyme that synthesizes all other RNAs) and by a special RNA polymerase known as **primase.** DNA polymerase then extends this RNA primer, which is eventually excised and replaced by DNA, as is explained below. This extra complexity in DNA synthesis increases the fidelity of DNA replication. Whereas a cell makes many copies of an RNA and

Figure 5-31 **Action of DNA polymerases.** DNA polymerases assemble incoming deoxynucleoside triphosphates on single-stranded DNA templates such that the growing strand is elongated in the 5′ to 3′ direction.

hence can tolerate an occasional mistake in its synthesis, a mistake (mutation) in the synthesis of DNA, the archive of genetic information, may be passed on to all of the cell's descendants. Since a Watson–Crick base pair is partially stabilized by its neighboring base pairs (a cooperative interaction), the first few base pairs that are formed in a newly synthesized polynucleotide will initially be less stable than the base pairs that are formed later. Consequently, these first few bases are more likely to be erroneously incorporated due to mispairing than those at the end of a longer chain. If a primer were DNA, there would be no way to differentiate it from other DNA so as to selectively replace it with more accurately synthesized DNA. Since the primer is RNA, however, it is readily identified and replaced.

b. DNA's Two Strands Are Replicated in Different Ways

A fourth major difference between RNA and DNA synthesis is that, whereas only one DNA strand at a time is transcribed, in most cases both of its strands are simultaneously replicated. This takes place at a **replication fork,** the junction where the two strands of the parental DNA are pried apart and where the two daughter strands are synthesized (Fig. 1-17), each by a different molecule of DNA polymerase. One of these DNA polymerase molecules continuously copies the parental strand that extends in its 3′ to 5′ direction from the replication fork, thereby synthesizing the resulting daughter strand, which is known as the **leading strand,** in its 5′ to 3′ direction. However, since the second

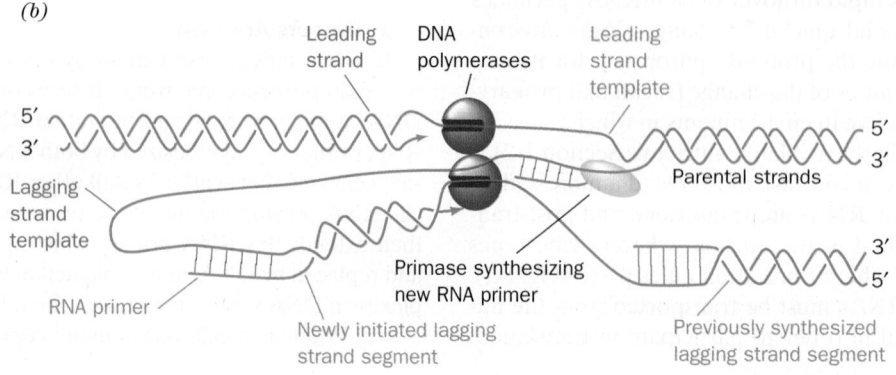

Figure 5-32 **Replication of duplex DNA in *E. coli.*** (*a*) Since the two DNA polymerase molecules at the replication fork are linked together and DNA polymerase can only synthesize DNA in its 5′ to 3′ direction, the leading strand can be synthesized continuously but the lagging strand must be synthesized discontinuously, that is, in segments. (*b*) This is because the lagging strand template can only be copied if it loops around so as to feed through the DNA polymerase in its 3′ to 5′ direction. Consequently, when the DNA polymerase that is synthesizing the lagging strand encounters the previously synthesized lagging strand segment, it releases the lagging strand template and rebinds to it farther upstream so as to extend the next RNA primer to be synthesized.

Figure 5-33 **The 5′ → 3′ exonuclease function of DNA polymerase I.** This enzymatic activity excises up to 10 nucleotides from the 5′ end of a single-strand nick. The nucleotide immediately past the nick (X) may or may not be base paired.

DNA polymerase at the replication fork also synthesizes DNA in the 5′ to 3′ direction and yet must travel with the replication fork, how does it copy the parental strand that extends from the replication fork in its 5′ to 3′ direction? The answer is that *it synthesizes the so-called* **lagging strand** *discontinuously, that is, in pieces (Fig. 5-32a, opposite)*. It does so by binding the looped-around lagging strand template so as to extend its newly synthesized RNA primer in its 5′ to 3′ direction (Fig. 5-32b; in effect, reversing its direction of travel) until it encounters the previously synthesized primer. The DNA polymerase then disengages from the lagging strand template and rebinds to it upstream of its previous position, where it then extends the next RNA primer to be synthesized. Thus the lagging strand is synthesized discontinuously, whereas the leading strand is synthesized continuously. The synthesis of lagging strand primers in *E. coli* is catalyzed by primase, which accompanies the replication fork (Fig. 5-32b), whereas the synthesis of leading strand primers, a much rarer event, occurs most efficiently when both primase and RNA polymerase are present.

c. Lagging Strand Synthesis Requires Several Enzymes

Escherichia coli contains two species of DNA polymerase that are essential for its survival. Of these, **DNA polymerase III (Pol III)** is the DNA replicase, that is, it synthesizes the leading strand and most of the lagging strand. **DNA polymerase I (Pol I)** has a different function, that of removing the RNA primers and replacing them with DNA. Pol I can do so because it has a second enzymatic activity besides that of a DNA polymerase; it is also a **5′ → 3′ exonuclease** (an exonuclease hydrolytically removes one or more nucleotides from the end of a polynucleotide rather than cleaving it at an internal position). The 5′ → 3′ exonuclease function binds to single-strand nicks (places where successive nucleotides are not covalently linked such as on the 5′ side of an RNA primer after the succeeding lagging strand segment has been synthesized). It then excises a 1- to 10-nucleotide segment of the nicked strand in the 5′ to 3′ (5′ → 3′) direction past the nick (Fig. 5-33). Pol I's 5′ → 3′ exonuclease and DNA polymerase activities work in concert, so *as Pol I's 5′ → 3′ exonuclease removes the primer, its DNA polymerase activity replaces this RNA with DNA (Fig. 5-34)*.

The synthesis of the leading strand is completed by the replacement of its single RNA primer with DNA. However, the completion of lagging strand synthesis requires that the nicks between its multiple discontinuously synthesized segments be sealed. This is the job of an independent enzyme named **DNA ligase** that covalently links adjacent 3′-OH and 5′-phosphate groups (Fig. 5-35).

d. Errors in DNA Sequences Are Subject to Correction

In *E. coli*, RNA polymerase has an error rate of ~1 wrong base for every 10^4 nucleotides it transcribes. In contrast, newly replicated DNA contains only ~1 error per 10^8 to 10^{10} base pairs. We have already seen that the use of RNA primers increases the fidelity of lagging strand synthesis. However, the main reason for the enormous fidelity of DNA replication is that both Pol I and Pol III have **3′ → 5′ exonuclease** activities. The 3′ → 5′ exonuclease

Figure 5-34 **Replacement of RNA primers by DNA in lagging strand synthesis.** In *E. coli*, the RNA primer on the 5′ end of a newly synthesized DNA segment is excised through the action of DNA polymerase I's 5′ → 3′ exonuclease activity and is simultaneously replaced by DNA as catalyzed by the enzyme's DNA polymerase activity.

Figure 5-35 **Function of DNA ligase.** DNA ligase seals single-strand nicks in duplex DNA. It does so in a reaction that is powered by the hydrolysis of ATP or a similar compound.

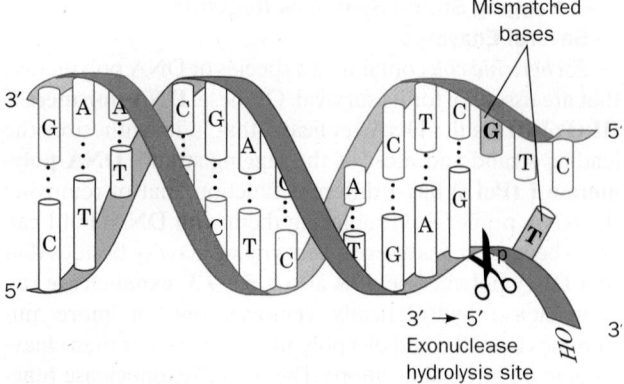

Figure 5-36   **The 3′ → 5′ exonuclease function of DNA polymerase I and DNA polymerase III.** In *E. coli*, this enzymatic activity excises mispaired nucleotides from the 3′ end of a growing DNA strand.

degrades the newly synthesized 3′ end of a daughter strand one nucleotide at a time (Fig. 5-36), thereby annulling the polymerase reaction. This enzymatic function is activated by non-Watson–Crick base pairing and consequently acts to edit out the occasional mistakes made by the polymerase function, thereby greatly increasing the fidelity of replication. However, in addition to this proofreading function on both Pol I and Pol III, all cells contain batteries of enzymes that detect and correct residual errors in replication as well as damage which DNA incurs through the action of such agents as UV radiation and **mutagens** (substances that damage DNA by chemically reacting with it) as well as by spontaneous hydrolysis (Section 30-5). In *E. coli*, Pol I also functions to replace the damaged DNA segments that these enzymes have excised.

5 MOLECULAR CLONING

A major problem in almost every area of biochemical research is obtaining sufficient quantities of the substance of interest. For example, a 10-L culture of *E. coli* grown to its maximum titer of $\sim 10^{10}$ cells $\cdot$ mL^{-1} contains, at most, 7 mg of DNA polymerase I, and many of its proteins are present in far lesser amounts. Yet it is rare that even as much as half of any protein originally present in an organism can be recovered in pure form (Chapter 6). Eukaryotic proteins may be even more difficult to obtain because many eukaryotic tissues, whether acquired from an intact organism or grown in tissue culture, are available only in small quantities. As far as the amount of DNA is concerned, our 10-L *E. coli* culture would contain ~ 0.1 mg of any 1000-bp length of chromosomal DNA (a length sufficient to contain most prokaryotic genes), but its purification in the presence of the rest of the chromosomal DNA (which consists of 4.6 million bp) would be an all but impossible task. These difficulties have been largely eliminated through the development of **molecular cloning** techniques (a **clone** is a collection of identical organisms that are derived from a single ancestor). These methods, which are also referred to as **genetic engineering** and **recombinant DNA** technology, deserve much of the credit for

the enormous progress in biochemistry and the dramatic rise of the biotechnology industry since the late 1970s.

*The main idea of molecular cloning is to insert a DNA segment of interest into an autonomously replicating DNA molecule, a so-called **cloning vector** or **vehicle**, so that the DNA segment is replicated with the vector.* Cloning such a **chimeric vector** (*chimera*: a monster in Greek mythology that has a lion's head, a goat's body, and a serpent's tail) in a suitable **host organism** such as *E. coli* or yeast results in the production of large amounts of the inserted DNA segment. If a cloned gene is flanked by the properly positioned control sequences for transcription and translation, the host may also produce large quantities of the RNA and protein specified by that gene. The techniques of genetic engineering, whose understanding is prerequisite to understanding many of the experiments discussed in this textbook, are outlined in this section.

A. Restriction Endonucleases

In order to effectively carry out molecular cloning, it is necessary to be able to manipulate precisely sequence-defined DNA fragments. This is done through the use of enzymes known as **restriction endonucleases.**

Bacteriophages that propagate efficiently on one bacterial strain, such as *E. coli* K12, have a very low rate of infection ($\sim 0.001\%$) in a related bacterial strain such as *E. coli* B. However, the few viral progeny of this latter infection propagate efficiently in the new host but only poorly in the original host. Evidently, the new host modifies these bacteriophages in some way. What is the molecular basis of this **host-specific modification?** Werner Arber showed that it results from a **restriction–modification system** in the bacterial host, which consists of a restriction endonuclease (alternatively, **restriction enzyme;** endonucleases are enzymes that hydrolytically cleave polynucleotides at internal sites) and a matched **DNA methyltransferase.** *Restriction endonucleases recognize a specific base sequence of four to eight bases in double-stranded DNA and cleave both strands of the duplex.* DNA methyltransferases methylate a specific base (at the amino group of an adenine or either the 5 position or the amino group of a cytosine) in the same base sequence recognized by the matched restriction enzyme.

A restriction enzyme does not cleave its corresponding methylated DNA. A newly replicated strand of bacterial DNA, which is protected from degradation by the methylated parent strand with which it forms a duplex, is methylated before the next cycle of replication. A restriction–modification system therefore protects the bacterium against invasion by foreign (usually viral) DNAs which, once they have been cleaved by a restriction endonuclease, are further degraded by bacterial exonucleases. Invading DNAs are only rarely methylated before being attacked by restriction enzymes. Yet if a viral DNA does become methylated, it is able to reproduce in its new host. Its progeny, however, are no longer methylated in the way that permits them to propagate in the original host (which has different restriction–modification systems).

There are four types of restriction endonucleases, Types I, II, III, and IV. **Type I** and **Type III** restriction enzymes

Table 5-4 **Recognition and Cleavage Sites of Some Type II Restriction Enzymes**

Enzyme	Recognition Sequence[a]	Microorganism
AluI	AG↓C*T	*Arthrobacter luteus*
BamHI	G↓GATC*C	*Bacillus amyloliquefaciens* H
BglI	GCCNNNN↓NGCC	*Bacillus globigii*
BglII	A↓GATCT	*Bacillus globigii*
EcoRI	G↓AA*TTC	*Escherichia coli* RY13
EcoRII	↓CC*(^A_T)GG	*Escherichia coli* R245
EcoRV	GA*T↓ATC	*Escherichia coli* J62 pLG74
HaeII	RGCGC↓Y	*Haemophilus aegyptius*
HaeIII	GG↓C*C	*Haemophilus aegyptius*
HindIII	A*↓AGCTT	*Haemophilus influenzae* R$_d$
HpaII	C↓C*GG	*Haemophilus parainfluenzae*
MspI	C*↓CGG	*Moraxella* species
PstI	CTGCA*↓G	*Providencia stuartii* 164
PvuII	CAG↓C*TG	*Proteus vulgaris*
SalI	G↓TCGAC	*Streptomyces albus* G
TaqI	T↓CGA*	*Thermus aquaticus*
XhoI	C↓TCGAG	*Xanthomonas holcicola*

[a]The recognition sequence is abbreviated so that only one strand, reading 5′ to 3′, is given. The cleavage site is represented by an arrow (↓) and the modified base, where it is known, is indicated by a following asterisk (A* is N^6-methyladenine and C* is 5-methylcytosine). R, Y, and N represent purine nucleotide, pyrimidine nucleotide, and any nucleotide, respectively.

Source: REBASE. The restriction enzyme database (http://rebase.neb.com).

each carry both the endonuclease and the DNA methyl-transferase activity on a single protein molecule. **Type I** restriction enzymes cleave the DNA at a possibly random site located at least 1000 bp from the recognition sequence, **Type III** enzymes do so 24 to 26 bp distant from the recognition sequence, and **Type IV** enzymes cleave methylated DNA. **Type II** restriction enzymes, which were discovered and characterized by Hamilton Smith and Daniel Nathans in the late 1960s, are separate entities from their corresponding DNA methyltransferases. *They cleave DNAs at specific sites within or near the recognition sequence, a property that makes Type II restriction enzymes indispensable biochemical tools for DNA manipulation.* In what follows, we discuss only Type II restriction enzymes.

Nearly 4000 species of Type II restriction enzymes from a variety of bacteria that have over 270 different sequence specificities have been characterized. Several of the more widely used species are listed in Table 5-4. A restriction endonuclease is named by the first letter of the genus of the bacterium that produced it and the first two letters of its species, followed by its serotype or strain designation, if any, and a roman numeral if the bacterium expresses more than one type of restriction enzyme. For example, **EcoRI** is produced by *E. coli* strain RY13.

a. Most Restriction Endonucleases Recognize Palindromic DNA Sequences

Most Type II restriction enzyme recognition sites possess exact twofold rotational symmetry, as is diagrammed in Fig. 5-37. Such sequences are known as **palindromes** (a palindrome is a word, verse, or sentence that reads the

same backward and forward; two examples are "Madam, I'm Adam" and "Sex at noon taxes"). Many restriction enzymes, such as EcoRI (Fig. 5-37*a*), catalyze cleavage of the two DNA strands at positions that are symmetrically staggered about the center of the palindromic recognition sequence. This yields restriction fragments with complementary single-stranded ends that are from one to four nucleotides in length. Restriction fragments with such **cohesive** or **sticky ends** can associate by complementary base pairing with other restriction fragments generated by the same restriction enzyme. Some restriction cuts, such as that of EcoRV (Fig. 5-37*b*), pass through the twofold axis of the

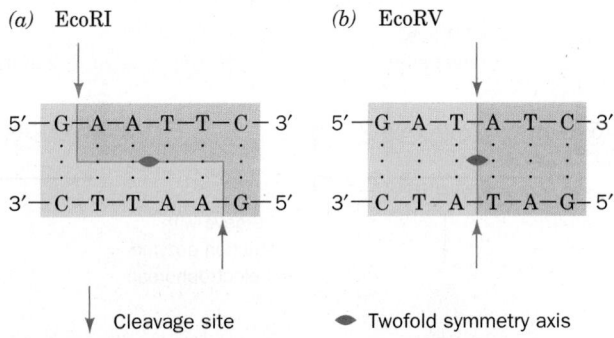

(a) EcoRI   *(b)* EcoRV

↓ Cleavage site   ⬤ Twofold symmetry axis

Figure 5-37 **Restriction sites.** The recognition sequences of the restriction endonucleases (*a*) EcoRI and (*b*) EcoRV have twofold (palindromic) symmetry (*red symbol*). The cleavage sites are indicated (*arrows*). Note that EcoRI generates DNA fragments with sticky ends, whereas EcoRV generates blunt-ended fragments.

palindrome to yield restriction fragments with fully base paired **blunt ends.** Since a given base has a one-fourth probability of occurring at any nucleotide position (assuming the DNA has equal proportions of all bases), a restriction enzyme with an *n*-base pair recognition site produces restriction fragments that are, on average, 4^n base pairs long. Thus AluI (4-bp recognition sequence) and EcoRI (6-bp recognition sequence) restriction fragments should average $4^4 = 256$ and $4^6 = 4096$ bp in length, respectively.

Figure 5-38 **Agarose gel electrophoretogram of restriction digests.** The *Agrobacterium radiobacter* plasmid pAgK84 was digested with (A) BamHI, (B) PstI, (C) BglII, (D) HaeIII, (E) HincII, (F) SacI, (G) XbaI, and (H) HpaI. Lane I contains λ phage DNA digested with HindIII as standards since these fragments have known sizes. The DNA fragments in the electrophoretogram are made visible by fluorescence against a black background. [From Slota, J.E. and Farrand, S.F., *Plasmid* **8**, 180 (1982). Copyright © 1982 by Academic Press.]

b. Restriction-Fragment Length Polymorphisms Provide Markers for Characterizing Genes

The treatment of a DNA molecule with a restriction endonuclease produces a series of precisely defined fragments that can be separated according to size by **gel electrophoresis** (Fig. 5-38). (In gel electrophoresis, charged molecules are applied to one end of a thin slab of polyacrylamide or agarose gel and are separated through the application of an electric field. Under the conditions used to separate DNA fragments, the molecules move according to size, with the smallest fragments moving fastest. Gel electrophoresis is further discussed in Section 6-4B.) Complementary single strands can be separated either by melting the DNA and subjecting it to gel electrophoresis, or by using density gradient ultracentrifugation in alkaline CsCl solution (recall that DNA is denatured under alkaline conditions).

Individuality in humans and other species derives from their genetic polymorphism; homologous human chromosomes differ in sequence, on average, every ~1250 bp. These genetic differences create and eliminate restriction sites (Fig. 5-39). Restriction enzyme digests of the corresponding segments from homologous chromosomes therefore contain fragments with different lengths; that is, these DNAs have **restriction-fragment length polymorphisms (RFLPs;** Fig. 5-40). Since, with the exception of identical twins, each individual has a unique set of RFLPs (its **haplotype**), RFLPs can be used for purposes of identification.

B. *Cloning Vectors*

Plasmids, viruses, and artificial chromosomes are used as cloning vectors in genetic engineering.

a. Plasmid-Based Cloning Vectors

Plasmids are circular DNA duplexes of 1 to 200 kb that contain the requisite genetic machinery, such as a **replication origin** (a site at which DNA replication is initiated; Section 30-3Ca), to permit their autonomous propagation in a bacterial host or in yeast. Plasmids may be considered molecular parasites but in many instances they benefit

Figure 5-39 **Restriction-fragment length polymorphisms.** A mutational change that affects a restriction site in a DNA segment alters the number and sizes of its restriction fragments.

Pedigree and genotypes

Alleles

Figure 5-40   Inheritance of RFLPs according to the rules of Mendelian genetics. Four alleles of a particular gene, each characterized by different restriction markers, can occur in all possible pairwise combinations and segregate independently in each generation (circles in the upper panel represent females and squares represent males). In the *P* (parental) generation, two individuals have heterozygous haplotypes (CD and BD) and the other two have homozygous haplotypes (AA and BB). Their children, the F_1 generation, have the haplotypes AC or BB. Consequently, every individual in the F_2 generation (grandchildren) inherited either an A or a C from their mother and a B from their father. The lower panel shows a gel electrophoretogram of these restriction fragments run in parallel lanes. [Courtesy of Ray White, University of Utah Medical School.]

their host by providing functions, such as resistance to antibiotics, that the host lacks. Indeed, the widespread and alarming appearance, since antibiotics came into use, of an-

tibiotic-resistant pathogens is partially the result of the rapid proliferation among these organisms of plasmids containing genes that confer resistance to antibiotics.

Some types of plasmids, which are present in one or a few copies per cell, replicate once per cell division as does the bacterial chromosome; their replication is said to be under **stringent control.** Most plasmids used in molecular cloning, however, are under **relaxed control;** they are normally present in 10 to as many as 700 copies per cell. Moreover, if protein synthesis in the bacterial host is inhibited, for example, by the antibiotic **chloramphenicol** (Section 32-3Gb), thereby preventing cell division, these plasmids continue to replicate until 2 or 3 thousand copies have accumulated per cell (which represents about half of the cell's total DNA). The plasmids that have been constructed (by genetic engineering techniques; Section 5-5C) for use in molecular cloning are relatively small, replicate under relaxed control, carry genes specifying resistance to one or more antibiotics, and contain a number of conveniently located restriction endonuclease sites into which the DNA to be cloned may be inserted. Indeed, many plasmid vectors contain a strategically located short (<100 bp) segment of DNA known as a **polylinker** that has been synthesized to contain a variety of restriction sites that are not present elsewhere in the plasmid. The *E. coli* plasmid designated **pUC18** (Fig. 5-41) is representative of the cloning vectors presently in use ("pUC" stands for "plasmid-Universal Cloning").

The expression of a chimeric plasmid in a bacterial host was first demonstrated in 1973 by Herbert Boyer and Stanley Cohen. The host bacterium takes up a plasmid when the two are mixed together in a process that is greatly enhanced by the presence of divalent cations such as Ca^{2+}

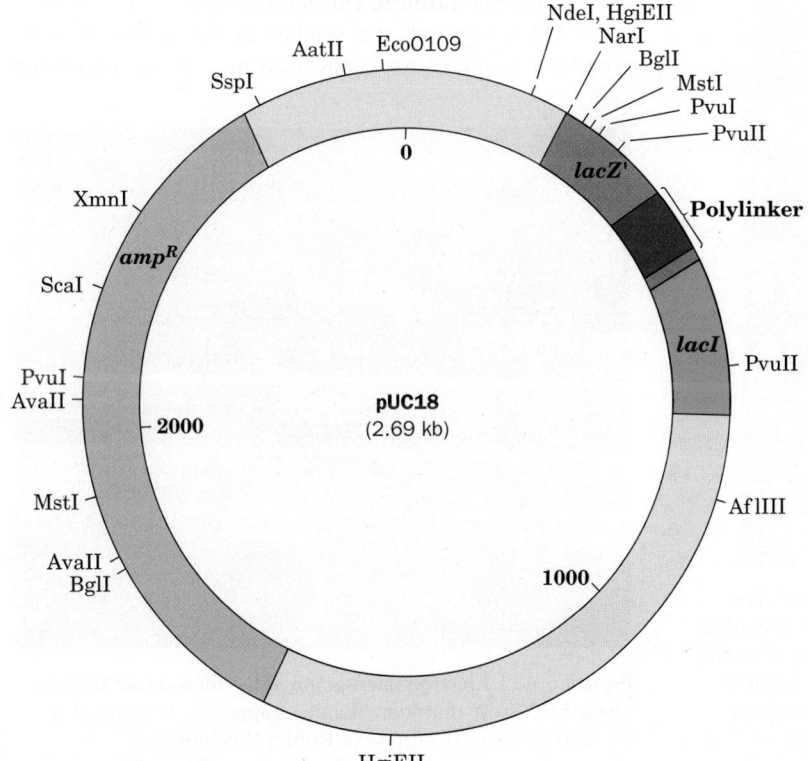

Figure 5-41   The pUC18 cloning vector. A restriction map of the plasmid pUC18 indicates the positions of its ***amp*^{R}, *lacZ'*,** and ***lacI*** genes. The *amp*^{R} gene confers resistance to the antibiotic **ampicillin** (a penicillin derivative; Section 11-3Bb); *lacZ'* is a modified form of the ***lacZ*** gene, which encodes the enzyme **β-galactosidase** (Section 11-2B); and *lacI* encodes the *lac* repressor, a protein that controls the transcription of *lacZ* (Section 5-4Aa). The polylinker, which encodes an 18-residue polypeptide segment inserted near the N-terminus of β-galactosidase, incorporates 13 different restriction sites that do not occur elsewhere in the plasmid.

and brief heating to ~42°C (which increases cell membrane permeability to DNA; such cells are said to be **transformation competent**). Nevertheless, an absorbed plasmid vector becomes permanently established in its bacterial host (transformation) with an efficiency of only ~0.1%.

Plasmid vectors cannot be used to clone DNAs of more than ~10 kb. This is because the time required for plasmid replication increases with plasmid size. Hence intact plasmids with large, unessential (to them) inserts are lost through the faster proliferation of plasmids that have eliminated these inserts by random deletions.

b. Virus-Based Cloning Vectors

Bacteriophage λ (Fig. 5-42) is an alternative cloning vehicle that can be used to clone DNAs of up to 16 kb. The central third of this virus's 48.5-kb genome is not required for phage infection (Section 33-3Aa) and can therefore be replaced by foreign DNAs of up to slightly greater size using techniques discussed in Section 5-5C. The chimeric phage DNA can then be introduced into the host cells by infecting them with phages formed from the DNA by an *in vitro* packaging system (Section 33-3Bc). The use of phages as cloning vectors has the additional advantage that the chimeric DNA is produced in large amounts and in easily purified form.

λ Phages can be used to clone even longer DNA inserts. The viral apparatus that packages DNA into phage heads requires only that the DNA have a specific 16-bp sequence known as a *cos* site located at each end and that these ends be 36 to 51 kb apart (Section 33-3Bc). Placing two *cos* sites the proper distance apart on a plasmid vector yields, via an *in vitro* packaging system, a so-called **cosmid** vector, which can contain foreign DNA of up to ~49 kb. Cosmids have no phage genes and hence, on introduction into a host cell via phage infection, reproduce as plasmids.

The **filamentous bacteriophage M13** (Fig. 5-43) is also a useful cloning vector. It has a single-stranded circular DNA that is contained in a protein tube composed of ~2700 helically arranged identical protein subunits. This number is controlled, however, by the length of the phage DNA being coated; insertion of foreign DNA in a nonessential region of the M13 chromosome results in the production of longer phage particles. Although M13 cloning vectors cannot stably maintain DNA inserts of >1 kb, they are widely used in the production of DNA for sequence analysis (Section 7-2Ba) because these phages directly produce the single-stranded DNA that the technique requires.

Baculoviruses are a large and diverse group of pathogenic viruses that infect mainly insects (but not vertebrates, so that they are safe for laboratory use) and hence can be grown in cultures of insect cells. A segment of the double-stranded DNA that forms the genome of some of these viruses is unnecessary for viral replication in tissue cultures of insect cells and hence can be replaced by a foreign DNA of up to 15 kb.

c. YAC and BAC Vectors

DNA segments larger than those that can be carried by cosmids may be cloned in **yeast artificial chromosomes (YACs)** and in **bacterial artificial chromosomes (BACs).** YACs are linear DNA segments that contain all the molecular paraphernalia required for replication in yeast: a replication

Figure 5-42   Electron micrograph of bacteriophage λ. Bacteriophage λ reproduces in certain strains of *E. coli*. On binding to a susceptible *E. coli*, the DNA contained in the "head" of the phage particle is injected, through its "tail," into the bacterial cell, where it is replicated ~100 times and packaged to form progeny phage (Section 33-3). [Courtesy of A.F. Howatson. From Lewin, B., *Gene Expression*, Vol. 3, Fig. 5.23, Wiley (1977).]

Figure 5-43   Electron micrograph of the filamentous bacteriophage M13. Note that some filaments appear to be pointed at one end (*arrows*). [Courtesy of Robley Williams, Stanford University, and Harold Fisher, University of Rhode Island.]

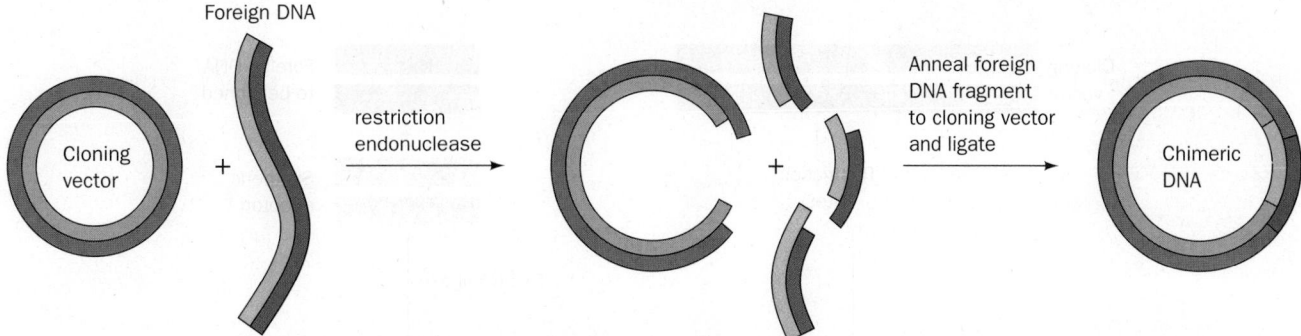

Figure 5-44   Construction of a recombinant DNA molecule. A restriction fragment is inserted in a cloning vector's corresponding restriction cut. The sticky ends of the vector and the foreign DNA anneal and are subsequently covalently joined by DNA ligase to yield a chimeric DNA. **See the Animated Figures**

origin [known as an **autonomously replicating sequence (ARS)**], a **centromere** (the chromosomal segment that attaches to the spindle during mitosis and meiosis), and **telomeres** (the ends of linear chromosomes that permit their replication; Section 30-4D). BACs, which replicate in *E. coli,* are derived from circular plasmids that normally replicate long regions of DNA and are maintained at a level of approximately one copy per cell (properties similar to those of actual chromosomes). These vectors contain the minimal sequences required for autonomous replication, copy-number control, and proper partitioning of the plasmid during cell division. YACs and BACs containing inserts of several hundred kilobase pairs have been successfully cloned.

C. *Gene Manipulation*

A DNA to be cloned is, in many cases, obtained as a sequence-defined fragment through the application of restriction endonucleases (for M13 vectors, the restriction enzymes' requirement of duplex DNA necessitates converting this phage DNA to its double-stranded form through the use of DNA polymerase I). Recall that most restriction endonucleases cleave duplex DNA at specific palindromic sites so as to yield single-stranded ends that are complementary to each other (cohesive or sticky ends; Section 5-5Aa). Therefore, as Janet Mertz and Ron Davis first demonstrated in 1972, *a restriction fragment may be inserted into a cut made in a cloning vector by the same restriction enzyme (Fig. 5-44). The complementary (cohesive) ends of the two DNAs specifically associate under annealing conditions and are covalently joined (spliced) through the action of DNA ligase* (Fig. 5-35; the DNA ligase produced by **bacteriophage T4** must be used for blunt-ended restriction cuts such as those generated by AluI, EcoRV, and HaeIII; Table 5-4). *A great advantage of using a restriction enzyme to construct a chimeric vector is that the DNA insert can be precisely excised from the cloned vector by cleaving it with the same restriction enzyme.*

If the foreign DNA and cloning vector have no common restriction sites at innocuous positions, they may still be spliced, using a procedure pioneered by Dale Kaiser and Paul Berg, through the use of **terminal deoxynucleotidyl trans-**

ferase (terminal transferase). This mammalian enzyme adds nucleotides to the 3′-terminal OH group of a DNA chain; it is the only known DNA polymerase that does not require a template. Terminal transferase and dTTP, for example, can build up poly(dT) tails of ~100 residues on the 3′ ends of the DNA segment to be cloned (Fig. 5-45). The cloning vector is enzymatically cleaved at a specific site and the 3′ ends of the cleavage site are similarly extended with poly(dA) tails. The complementary homopolymer tails are annealed, any gaps resulting from differences in their lengths filled in by DNA polymerase I, and the strands joined by DNA ligase.

A disadvantage of the above technique is that it eliminates the restriction sites that were used to generate the foreign DNA insert and to cleave the vector. It may therefore be difficult to recover the insert from the cloned vector. This difficulty is circumvented by a technique in which a chemically synthesized palindromic "linker" having a restriction site matching that of the cloning vector is appended to both ends of the foreign DNA (the chemical

Figure 5-45   Splicing DNA using terminal transferase. Two DNA fragments may be joined through the generation of complementary homopolymer tails via the action of the enzyme terminal transferase. The poly(dA) and poly(dT) tails shown in this example may be replaced by poly(dC) and poly(dG) tails.

Figure 5-46 **Construction of a recombinant DNA molecule through the use of synthetic oligonucleotide adaptors.** In this ex-ample, the adaptor and the cloning vector have EcoRI restriction sites (*red arrows*).

synthesis of oligonucleotides is discussed in Section 7-6A). The linker is attached to the foreign DNA by blunt end ligation with T4 DNA ligase and then cleaved with the appropriate restriction enzyme to yield the correct cohesive ends for ligation to the vector (Fig. 5-46).

a. Properly Transformed Cells Must Be Selected

Both transformation and the proper construction of chimeric vectors occur with low efficiency. How can one select only those host organisms that have been transformed by the properly constructed vector? In the case of plasmid transformation, this is usually done through a double screen using antibiotics and/or **chromogenic** (color-producing) substrates. For example, the pUC18 plasmid contains the *lacZ'* gene (Fig. 5-41; a modified form of the *lac* operon's *Z* gene; Fig. 5-25). The *lacZ'* gene encodes the enzyme β-galactosidase, which catalyzes the hydrolysis of the bond from O1 of the sugar β-D-galactose to a substituent.

β-D-Galactose

Thus, when grown in the presence of **5-bromo-4-chloro-3-indolyl-β-D-galactoside** (commonly known as **X-gal**), a colorless substance which when hydrolyzed by β-galactosidase yields a blue product,

5-Bromo-4-chloro-3-indolyl-β-D-galactoside (X-gal)
(*colorless*)

H_2O ↓ β-galactosidase

β-D-Galactose + **5-Bromo-4-chloro-3-hydroxyindole**
(*blue*)

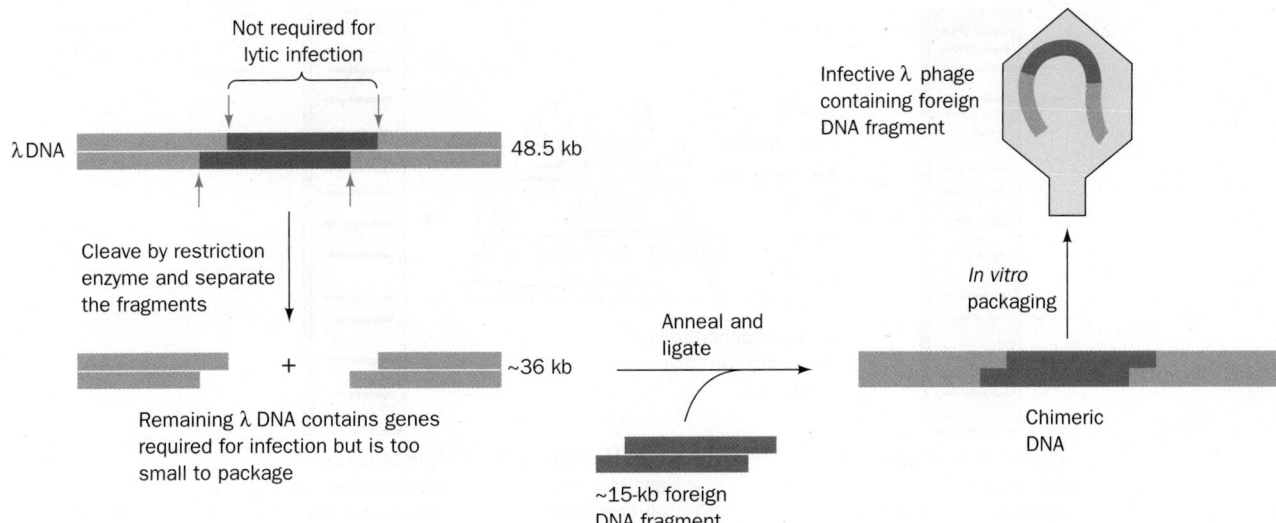

Figure 5-47   Cloning of foreign DNA in λ phages. A nonessential portion of the phage genome can be replaced by a foreign DNA and packaged to form an infectious phage particle only if the foreign DNA is approximately the same size as the DNA segment it replaced. ⚗ **See the Animated Figures**

E. coli transformed by an unmodified pUC18 plasmid form blue colonies. However, *E. coli* transformed by a pUC18 plasmid containing a foreign DNA insert in its polylinker region form colorless colonies because the insert interrupts the protein-encoding sequence of the *lac*Z' gene and hence they lack β-galactosidase activity. Bacteria that have failed to take up any plasmid, which would otherwise also form colorless colonies in the presence of X-gal, are excluded by adding the antibiotic **ampicillin** (Fig. 11-25) to the growth medium. Bacteria that do not contain the plasmid are sensitive to ampicillin, whereas bacteria containing the plasmid will grow, because the plasmid's intact *amp*^{R} gene confers ampicillin resistance. Genes such as *amp*^{R} are therefore known as **selectable markers.**

Genetically engineered λ phage variants contain restriction sites that flank the dispensable central third of the phage genome (Section 5-5Bb). This segment may therefore be replaced, as is described above, by a foreign DNA insert (Fig. 5-47). DNA is only packaged in λ phage heads if its length is from 75 to 105% of the 48.5-kb wild-type λ genome. Consequently, λ phage vectors that have failed to acquire a foreign DNA insert are unable to propagate because they are too short to form infectious phage particles. Cosmid vectors are subject to the same limitation. Moreover, cloned cosmids are harvested by repackaging them into phage particles. Hence, any cosmids that have lost sufficient DNA through random deletion to make them shorter than the above limit are not recovered. This is why cosmids can support the proliferation of large DNA inserts, whereas most other types of plasmids cannot.

D. *The Identification of Specific DNA Sequences: Southern Blotting*

DNA with a specific base sequence may be identified through a procedure developed by Edwin Southern known as the **Southern transfer technique** or, more colloquially, as **Southern blotting** (Fig. 5-48). This procedure takes advantage of the valuable property of nitrocellulose that it tenaciously binds single-stranded (but not duplex) DNA [**nylon** and **polyvinylidene fluoride (PVDF)** membranes also have this property]. Following the gel electrophoresis of double-stranded DNA, the gel is soaked in 0.5*M* NaOH solution, which converts the DNA to its single-stranded form. The gel is then overlaid by a sheet of nitrocellulose paper which, in turn, is covered by a thick layer of paper towels, and the entire assembly is compressed by a heavy plate. The liquid in the gel is thereby forced (blotted) through the nitrocellulose so that the single-stranded DNA binds to it at the same position it had in the gel (the transfer to nitrocellulose can alternatively be accomplished by an electrophoretic process called **electroblotting**). After vacuum drying the nitrocellulose at ~80°C, which permanently fixes the DNA in place, the nitrocellulose sheet is moistened with a minimal quantity of solution containing ^{32}P-labeled single-stranded DNA or RNA (the "probe") that is complementary in sequence to the DNA of interest. The moistened sheet is held at a suitable renaturation temperature for several hours to permit the probe to anneal to its target sequence(s), washed to remove the unbound radioactive probe, dried, and then autoradiographed by placing it for a time (hours to days) over a sheet of X-ray film. The positions of the molecules that are complementary to the radioactive sequences are indicated by a blackening of the developed film. Alternatively, a **phosphorimager,** essentially "electronic film" that detects radioactivity with tenfold greater sensitivity than X-ray film, can be used.

A DNA segment containing a particular base sequence (e.g., an RFLP) may, in this manner, be detected and isolated. The radioactive probe used in this procedure can be the corresponding mRNA if it is produced in sufficient quantity to be isolated [e.g., **reticulocytes** (immature red

Figure 5-48 **Detection of DNAs containing specific base sequences by the Southern transfer technique.**

blood cells), which produce little protein besides **hemoglobin** (the red protein that transports O_2 in the blood), are rich in the mRNAs that specify hemoglobin]. Alternatively, the gene specifying a protein of known amino acid sequence may be found by synthesizing a probe that is a mixture of all oligonucleotides, which, according to the genetic code (Table 5-3), can specify a segment of the gene that has low degeneracy (Fig. 5-49).

Southern blotting may be used for the diagnosis and prenatal detection of genetic diseases. These diseases often result from a specific change in a single gene such as a base substitution, deletion, or insertion. The temperature at which probe hybridization is carried out may be adjusted so that only an oligonucleotide that is perfectly complementary to

a length of DNA will hybridize to it. Even a single base mismatch, under appropriate conditions, will result in a failure to hybridize. For example, the genetic disease **sickle-cell anemia** arises from a single A → T base change in the gene specifying the β subunit of hemoglobin, which causes the amino acid substitution Glu β6 → Val (Section 7-3Aa). A 19-residue oligonucleotide that is complementary to the sickle-cell gene's mutated segment hybridizes, at the proper temperature, to DNA from homozygotes for the sickle-cell gene but not to DNA from normal individuals. An oligonucleotide that is complementary to the gene encoding the β subunit for normal hemoglobin yields opposite results. DNA from sickle-cell heterozygotes (who have one hemoglobin β gene bearing the sickle-cell anemia mu-

— Trp – Lys – Gln – Cys – Met —Peptide segment

^{32}P–UGG–AAA–CAA–UGU–AUG
^{32}P–UGG–AAG–CAA–UGU–AUG
^{32}P–UGG–AAA–CAG–UGU–AUG
^{32}P–UGG–AAG–CAG–UGU–AUG
^{32}P–UGG–AAA–CAA–UGC–AUG
^{32}P–UGG–AAG–CAA–UGC–AUG
^{32}P–UGG–AAA–CAG–UGC–AUG
^{32}P–UGG–AAG–CAG–UGC–AUG

Mixture of all
oligonucleotides
that can encode
the peptide

Figure 5-49 **A degenerate oligonucleotide probe.** Such a probe is a mixture of all oligonucleotides that can encode a polypeptide segment of known sequence. In practice, such a segment is chosen to contain a high proportion of residues specified by low-degeneracy codons. In the pentapeptide segment shown here, Trp and Met are each specified by only one codon and Lys, Gln, and Cys are each specified by two codons that differ in only their third positions (*blue and red*; Table 5-3) for a total of $1 \times 2 \times 2 \times 2 \times 1 = 8$ oligonucleotides. The oligonucleotides are ^{32}P-labeled for use in Southern blotting.

tation and one that is normal) hybridizes to both probes but in reduced amounts relative to the DNAs from homozygotes. The oligonucleotide probes may consequently be used in the prenatal diagnosis of sickle-cell anemia. (Note that the availability of fetal genetic testing has actually increased the number of births because many couples who knew they had a high risk of conceiving a genetically defective child previously chose not to have children.) The use of DNA probes is also rapidly replacing the much slower and less accurate culturing techniques for the identification of pathogenic bacteria.

In a variation of the Southern blotting procedure, specific DNAs may be detected by linking the probe to an enzyme that generates a colored or fluorescent deposit on the blot when exposed to the proper reagents. Alternatively, a probe that is covalently linked to a dye that fluoresces when stimulated by a laser may be used. Such nonradioactive detection techniques are desirable in a clinical setting because of the health hazards, disposal problems, and more cumbersome nature of radiographic methods. Specific RNA sequences may be detected through a different variation of the Southern transfer, punningly named a **Northern transfer (Northern blot),** in which the RNA is immobilized on nitrocellulose paper and detected through the use of complementary RNA or DNA probes.

E. *Genomic Libraries*

In order to clone a particular DNA fragment, it must first be obtained in relatively pure form. The magnitude of this task may be appreciated when it is realized that, for example, a 1-kb fragment of human DNA represents only 0.000033% of the 3 billion-bp human genome. A DNA fragment may be identified by Southern blotting of a restriction digest of the genomic DNA under investigation. In practice, however, it is usually more difficult to identify a particular gene from an organism and then clone it than

it is to clone the organism's entire genome as DNA fragments and then identify the clone(s) containing the sequence(s) of interest. Such a set of cloned fragments is known as a **genomic library.** A genomic library of a particular organism need only be made once since it can be perpetuated for use whenever a new probe becomes available.

Genomic libraries are generated according to a procedure known as **shotgun cloning.** The chromosomal DNA of the organism of interest is isolated, cleaved to fragments of clonable size, and inserted in a cloning vector by the methods described in Section 5-5B. The DNA is fragmented by partial rather than exhaustive restriction digestion (permitting the restriction enzyme to act for only a short time) so that the genomic library contains intact representatives of all the organism's genes, including those whose sequences contain restriction sites. Shear fragmentation by rapid stirring of a DNA solution or by sonication is also used but requires further treatment of the fragments to insert them into cloning vectors. Genomic libraries have been established for numerous organisms including yeast, *Drosophila*, mice, and humans.

a. Many Clones Must Be Screened to Obtain a Gene of Interest

The number of random cleavage fragments that must be cloned to ensure a high probability that a given sequence is represented at least once in the genomic library is calculated as follows: The probability P that a set of N clones contains a fragment that constitutes a fraction f, in bp, of the organism's genome is

$$P = 1 - (1 - f)^N \qquad [5.2]$$

Consequently,

$$N = \log(1 - P)/\log(1 - f) \qquad [5.3]$$

Thus, to have $P = 0.99$ with fragments averaging 10 kb in length, for the 4639-kb *E. coli* chromosome ($f = 0.00216$), $N = 2134$ clones, whereas for the 180,000-kb *Drosophila* genome ($f = 0.0000566$), $N = 83,000$. The use of YAC- or BAC-based genomic libraries therefore greatly reduces the effort necessary to obtain a given gene segment from a large genome.

Since a genomic library lacks an index, it must be screened for the presence of a particular gene. This is done by a process known as **colony** or ***in situ* hybridization** (Fig. 5-50; Latin: *in situ,* in position). The cloned yeast colonies, bacterial colonies, or phage plaques to be tested are transferred, by replica plating (Fig. 1-30), from a master plate to a nitrocellulose filter. The filter is treated with NaOH, which lyses the cells or phages and denatures the DNA so that it binds to the nitrocellulose (recall that single-stranded DNA is preferentially bound to nitrocellulose). The filter is then dried to fix the DNA in place, treated under annealing conditions with a radioactive probe for the gene of interest, washed, and autoradiographed. *Only those colonies or plaques containing the sought-after gene will bind the probe and thereby blacken the film.* The corresponding clones can then be retrieved from the master plate.

Figure 5-50 **Colony (*in situ*) hybridization.** This technique identifies the clones containing a DNA of interest.

Using this technique, even an ~1 million clone human genomic library can be readily screened for the presence of one particular DNA segment.

Many eukaryotic genes and gene clusters span enormous tracts of DNA (Section 34-2); some consist of >1000 kb. With the use of plasmid-, phage-, or cosmid-based genomic libraries, such long DNAs can only be obtained as a series of overlapping fragments (Fig. 5-51). Each gene frag-

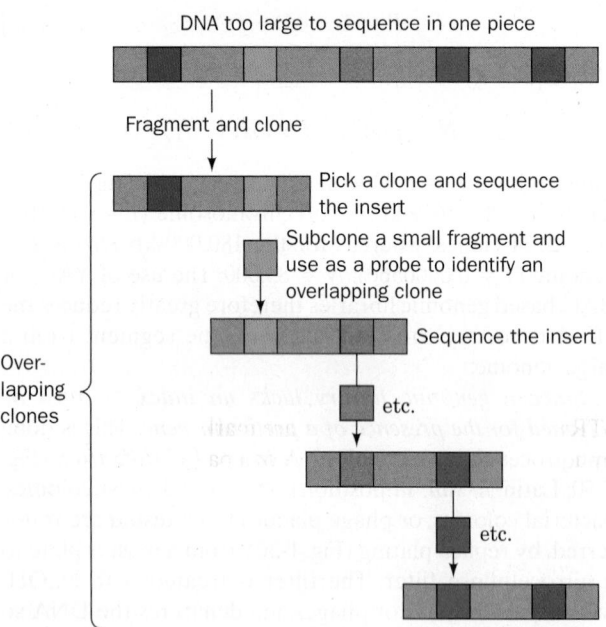

Figure 5-51 **Chromosome walking.** A DNA segment too large to sequence in one piece is fragmented and cloned. A clone is picked and the DNA insert it contains is sequenced. A small fragment of the insert near one end is subcloned (cloned from a clone) and used as a probe to select a clone containing an overlapping insert, which, in turn, is sequenced. The process is repeated so as to "walk" down the chromosome. Chromosome walking can, of course, extend in both directions.

ment that has been isolated is, in turn, used as a probe to identify a successive but partially overlapping fragment of that gene, a process called **chromosome walking.** The use of YACs and BACs, however, greatly reduces the need for this laborious and error-prone process.

F. *The Polymerase Chain Reaction*

See Guided Exploration 3: PCR and site-directed mutagenesis Although molecular cloning techniques are indispensable to modern biochemical research, the use of the **polymerase chain reaction (PCR)** offers a faster and more convenient method of amplifying a specific DNA segment of up to 6 kb. In this technique (Fig. 5-52), which was formulated in 1985 by Kary Mullis, a heat denatured (strand-separated) DNA sample is incubated with DNA polymerase, dNTPs, and two oligonucleotide primers whose sequences flank the DNA segment of interest so that they direct the DNA polymerase to synthesize new complementary strands. Multiple cycles of this process, each doubling the amount of DNA present, geometrically amplify the DNA starting from as little as a single gene copy. In each cycle, the two strands of the duplex DNA are separated by heat denaturation at 95°C, the temperature is then lowered to permit the primers to anneal to their complementary segments on the DNA, and the DNA polymerase directs the synthesis of the complementary strands (Section 5-4C). The use of a heat-stable DNA polymerase, such as those from the thermophilic bacteria *Thermus aquaticus* (**Taq DNA polymerase**) or *Pyroccocus furiosus* (**Pfu DNA polymerase**), both of which are stable at 95°C, eliminates the need to add fresh enzyme after each heat denaturation step. Hence, in the presence of sufficient quantities of primers and dNTPs, the PCR is carried out simply by cyclically varying the temperature in an automatic device called a **thermocycler.**

Twenty cycles of PCR amplification theoretically increase the amount of the target sequence by $2^{20} \approx 10^6$-fold with high specificity (in practice, the number of copies of target sequence largely doubles with each PCR cycle until

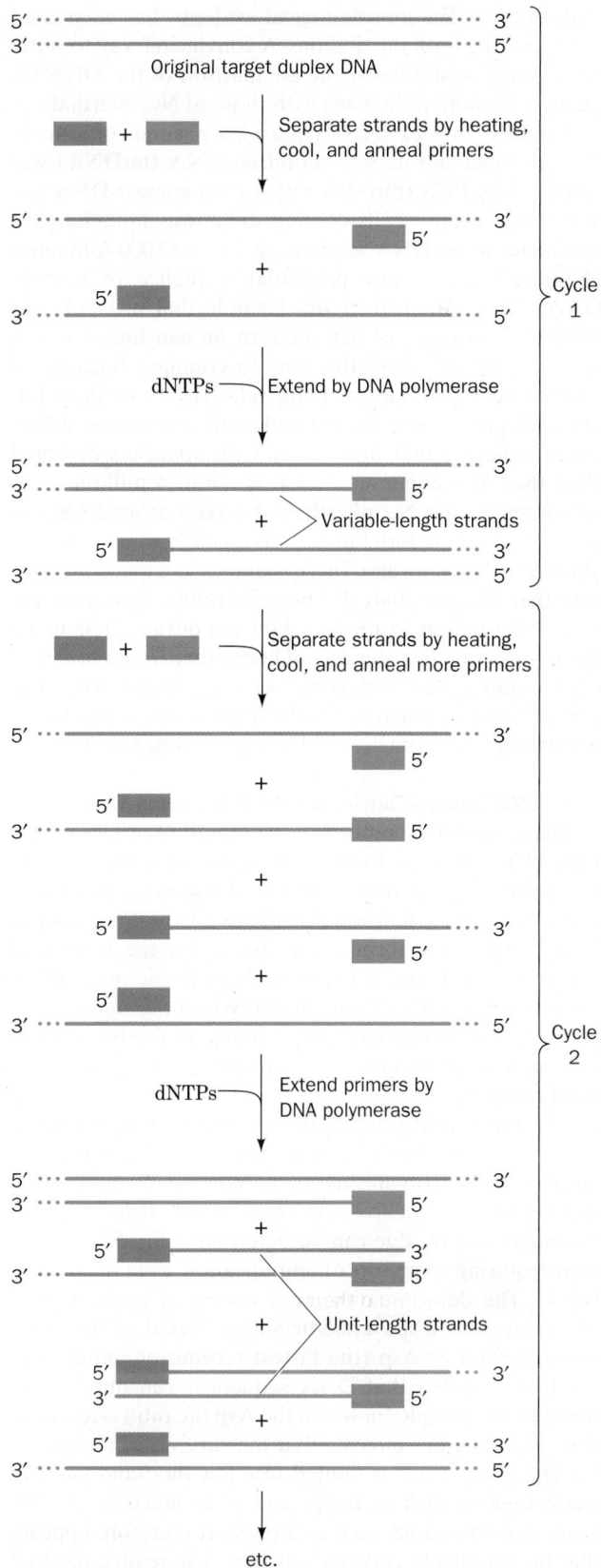

Figure 5-52 **The polymerase chain reaction (PCR).** In each cycle of the reaction, the strands of the duplex DNA are separated by heat denaturation, the preparation is cooled such that synthetic DNA primers anneal to a complementary segment on each strand, and the primers are extended by DNA polymerase. The process is then repeated for numerous cycles. The number of "unit-length" strands doubles with every cycle after the second cycle.

more primer–template complex accumulates than the DNA polymerase can extend during a cycle, whereon the rate of increase of target DNA becomes linear rather than geometric; the actual yield is typically ~20% of the theoretical yield). Indeed, the method has been shown to amplify a target DNA present only once in a sample of 10^5 cells, thereby demonstrating that the method can be used without prior DNA purification (although, as a consequence of this enormous amplification, particular care must be taken that the DNA sample of interest is not contaminated by extraneous DNA that is similar in sequence to that under investigation). The amplified DNA can be characterized by a variety of techniques including RFLP analysis, Southern blotting, and direct sequencing (Section 7-2A). PCR amplification is therefore a form of "cell-free molecular cloning" that can accomplish in an automated *in vitro* procedure requiring as little as 30 minutes what would otherwise take days or weeks via the cloning techniques discussed above.

a. PCR Has Many Uses

PCR amplification has become an indispensable tool in a great variety of applications. Clinically, it is used for the rapid diagnosis of infectious diseases and the detection of rare pathological events such as mutations leading to cancer (Section 19-3Ba). Forensically, the DNA from a single hair, sperm, or drop of blood can be used to identify the donor. This is most commonly done through the analysis of **short tandem repeats (STRs),** segments of DNA that contain repeating sequences of 2 to 7 bp such as $(CA)_n$ and $(ATGC)_n$ and that are scattered throughout the genome [e.g., the human genome contains ~100,000 $(CA)_n$ STRs]. The number of tandem repeats, n, in many STRs is genetically variable [n varies from 1 to 40 for particular $(CA)_n$ STRs] and hence such repeats are markers of individuality (much as are RFLPs). The DNA of a particular STR can be PCR-amplified through the use of primers that are complementary to the unique (nonrepeating) sequences flanking the STR. The number of tandem repeats in that STR from a particular individual can then be determined, either by the measurement of its molecular mass through polyacrylamide gel electrophoresis (Section 6-6C) or by direct sequencing (Section 7-2A). The determination of this number for several well-characterized STRs (those whose numbers of repeats have been determined in numerous individuals of multiple ethnicities), that is, the DNA's haplotype, can unambiguously identify the DNA's donor.

STRs are also widely used to prove or disprove familial relationships. For example, oral tradition suggests that

Thomas Jefferson, the third American president, fathered a son, Eston Hemings (born in 1808), with his slave Sally Hemings (Eston Hemings was said to bear a striking physical resemblance to Jefferson). Only the tips of the Y chromosome undergo recombination (with the X chromosome) and the rest of it is passed unchanged from father to son (except for occasional mutations). The finding that the Y chromosomes of male-line descendants of both Eston Hemings and Jefferson's father's brother (Jefferson had no surviving legitimate sons) had identical STR-based haplotypes indicates that Thomas Jefferson was probably Eston Hemings' father (although this could also be true of any of Jefferson's contemporary male-line relatives).

RNA may also be amplified via PCR by first reverse-transcribing it into a complementary strand of DNA (**cDNA**) through the action of an enzyme named **RNA-directed DNA polymerase** (commonly known as **reverse transcriptase**). This enzyme, which is produced by certain RNA-containing viruses known as **retroviruses** (Section 30-4C), uses an RNA template but is otherwise similar in the reaction it catalyzes to DNA polymerase I.

Variations on the theme of PCR have found numerous applications. For instance, single-stranded DNA (which is required for DNA sequencing; Section 7-2A) can be rapidly generated via **asymmetric PCR,** in which such a small amount of one primer is used that it is exhausted after several PCR cycles. In subsequent cycles, only the strand extended from the other primer, which is present in excess, is synthesized (note that PCR amplification becomes linear rather than geometric after one primer is used up). In cases that primers may anneal to more than one site in the target DNA, **nested primers** can be used to ensure that only the target sequence is amplified. In this technique, PCR amplification is normally carried out using one pair of primers. The products of this amplification are then further amplified through the use of a second pair of primers that anneal to the target DNA within its amplified region. It is highly unlikely that both pairs of primers will incorrectly anneal in a nested fashion to a nontarget DNA, and hence only the target DNA will be amplified.

b. Neanderthals Are Not Ancestors of Modern Humans

PCR is also largely responsible for the budding science of molecular archeology. For example, PCR-based techniques have been used by Svante Pääbo to determine whether or not Neanderthals form a different species from modern human beings. Neanderthals (*Homo neandertalensis*; also called Neandertals) are extinct hominids that were about 30% larger than are modern humans, apparently had great muscular strength, and had low foreheads and protruding brows. According to the radiodated fossil record, they became extinct ~28,000 years ago after having inhabited Europe and western Asia for over 300,000 years. During the latter part of this period they coexisted with our direct ancestors (who might well have been responsible for their demise). Thus, an important anthropological issue is whether Neanderthals constituted an ancient race of *Homo sapiens* ancestral to modern humans or were a separate species. The morphological evidence has been cited as supporting both possibilities. A convincing way to settle this dispute would be by the comparison of the DNA sequences of modern humans with those of Neanderthals.

The DNA was extracted from a 0.4-g sample of a Neanderthal bone, and its mitochondrial DNA (**mtDNA**) was amplified by PCR (mtDNA rather than nuclear DNA was amplified because cells contain numerous mitochondria and hence an mtDNA sequence is 100- to 1000-fold more abundant than is any particular sequence of nuclear DNA). The sequence of the Neanderthal mtDNA was compared to those of 986 modern human lineages of a wide variety of ethnicities and 16 common lineages of chimpanzees (the closet living relatives of modern humans). A phylogenetic tree based on their sequence differences indicates that humans and chimpanzees diverged (had their last common ancestor) about 4 million years ago, humans and Neanderthals diverged around 660,000 years ago, and modern humans diverged from one another about 150,000 years ago. These sequence comparisons indicate that Neanderthals did not contribute significant genetic information to modern humans during their many thousand–year coexistence and hence that *Homo neandertalensis* and *Homo sapiens* are separate species. This conclusion was confirmed by similar analyses of eleven Neanderthal samples from diverse locations throughout Europe.

c. DNA Decays Quickly on the Geological Time Scale

There have been reports in the literature of DNAs that were PCR-amplified from fossils that were several million years old and from amber-entombed insects (amber is fossilized tree resin) that were as old as 135 million years (a phenomenon that formed the "basis" for the novel and movie *Jurassic Park*). Yet, over geological time spans, DNA decomposes, mostly through hydrolysis of the sugar–phosphate backbone and oxidative damage to the bases. How old can a fossil become before its DNA has decayed beyond recognition?

The amino acid residues in hydrated proteins racemize at a rate similar to the rate at which DNA decomposes. Since proteins in organisms are far more abundant than are specific DNA sequences, the enantiomeric (D/L) ratios of an amino acid residue can be determined directly (rather than requiring some sort of amplification, as in the case of DNA). The determination, in a variety of archeological specimens whose age could be authenticated, of the enantiomeric ratio of Asp (the fastest racemizing amino acid residue) revealed that DNA sequences can only be retrieved from samples in which the Asp D/L ratio is less than 0.08. These studies indicate that the survival of recognizable DNA sequences is limited to a few thousand years in warm regions such as Egypt and to as much as 100,000 years in cold regions such as Siberia. It therefore appears that the putatively very ancient DNAs, in reality, resulted from the artifactual amplification of contaminating modern DNAs, particularly those from the human operators carrying out the PCR amplifications. Indeed, the DNA in the above Neanderthal fossil had decomposed to the point that it appeared unlikely that its nuclear DNA could have been successfully sequenced, which is why its mtDNA was amplified instead [although with contemporary methods of DNA sequencing (Section 7-2C) this has now become possible].

Despite the foregoing, there is credible evidence that certain bacterial spores can remain viable almost indefinitely. Bacterial spores, which several bacterial groups including bacilli form under adverse conditions, function to permit the bacterium's survival until conditions become favorable for growth. Bacterial spores have thick protective protein coats, their cytoplasm is partially dehydrated and mineralized, and their DNA is specifically stabilized by specialized proteins (Section 29-1Ba). Thus, a bacillus was cultured from a 25- to 40-million-year-old (Myr) amber-entombed bee after the surface of the amber had been chemically sterilized. Similarly, a halophilic (salt-loving) bacillus was cultured from a tiny (~9 μL) brine-filled inclusion in a surface-sterilized salt crystal from a 250-Myr salt deposit.

G. *Production of Proteins*

One of the most important uses of recombinant DNA technology is the production of large quantities of scarce and/or novel proteins. This is a relatively straightforward procedure for bacterial proteins: A cloned **structural gene** (a gene that encodes a protein) is inserted into an **expression vector,** a plasmid or virus that contains the properly positioned transcriptional and translational control sequences for the protein's expression. With the use of a relaxed control plasmid and an efficient promoter, the production of a protein of interest may reach 30% of the host bacterium's total cellular protein. Such genetically engineered organisms are called **overproducers.**

Bacterial cells often sequester such large amounts of useless (to the bacterium) protein as insoluble and denatured **inclusion bodies** (Fig. 5-53). A protein extracted from such inclusion bodies must therefore be renatured, usually by dissolving it in a solution of **urea** or **guanidinium ion** (substances that induce proteins to denature)

$$H_2N-\overset{\overset{\displaystyle O}{\|}}{C}-NH_2 \qquad\qquad H_2N-\overset{\overset{\displaystyle NH_2^+}{\|}}{C}-NH_2$$

Urea **Guanidinium ion**

and then slowly removing the denaturant via a membrane through which the denaturant but not the protein can pass [**dialysis** or **ultrafiltration** (Section 6-3Bc); protein denaturation and renaturation are discussed in Section 9-1A].

A strategy for avoiding the foregoing difficulty is to engineer the gene for the protein of interest so that is preceded with a bacterial **signal sequence** that directs the protein synthesizing machinery of gram-negative bacteria such as *E. coli* to secrete the protein to their **periplasmic space** (the compartment between their plasma membrane and cell wall; signal sequences are discussed in Section 12-4Ba). The signal sequence is then removed by a specific bacterial protease. Secreted proteins, which are relatively few in number, can be released into the medium by the osmotic disruption (Section 6-1B) of the bacterial outer membrane

Figure 5-53 **Electron micrograph of an inclusion body of the protein prochymosin in an** *E. coli* **cell.** [Courtesy of Teruhiko Beppu, Nikon University, Japan.]

(Section 1-1B; the bacterial cell wall is porous), so their purification is greatly simplified relative to that of intracellular proteins.

Another problem encountered when producing a foreign protein is that the protein may be toxic to the host cell (e.g., producing a protease may destroy the cell's proteins), thus killing the bacterial culture before sufficient amounts of the protein can be generated. One way to circumvent this problem is to place the gene encoding the toxic protein under the control of an inducible promoter, for example, the *lac* promoter in a plasmid that also includes the gene for the *lac* repressor (Section 5-4Aa). Then, the binding of the *lac* repressor to the *lac* promoter will prevent the expression of the foreign protein in the same way that it prevents the expression of the *lac* operon genes (Fig. 5-25*a*). However, after the cells have grown to a high concentration, an inducer is added that releases the repressor from the promoter and permits the expression of the foreign protein (Fig. 5-25*b*). The cells are thereby killed but not before they have produced large amounts of the foreign protein. For the *lac* repressor, the inducer of choice is **isopropylthiogalactoside** (**IPTG;** Section 31-1Aa), a synthetic, nonmetabolizable analog of the *lac* repressor's natural inducer, allolactose.

A problem associated with inserting a DNA segment into a vector, as is indicated in Fig. 5-44, is that any pair of sticky ends that have been made by a given restriction enzyme can be ligated together. Consequently, the products of a ligation reaction will include tandemly (one-after-the-other) linked vectors, inserts, and their various combinations in both linear and circular arrangements. Moreover, in the case of expression systems, 50% of the structural genes that are inserted into circular expression vectors will be installed backward with respect to the expression vector's transcriptional and translational control sequences and hence will not be properly expressed. The efficiency of the ligation process can be greatly enhanced through the use of **directional cloning** (Fig. 5-54). In this process, two different restriction enzymes are employed to generate two different types of sticky ends on both the insert and the vector. In expression systems, these are arranged such that the structural gene can only be inserted into the expression vector in the correct orientation for expression.

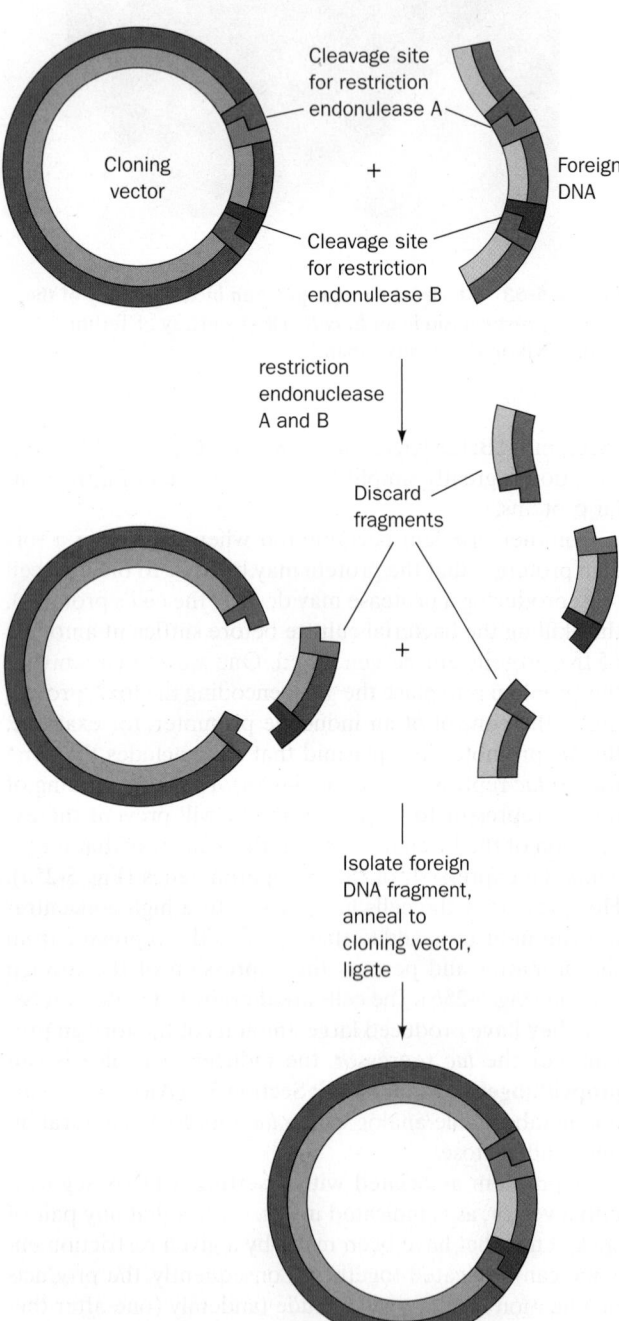

Figure 5-54 Construction of a recombinant DNA molecule by directional cloning. Two restriction enzymes, which yield different sticky ends, are used so that the foreign DNA fragment can only be inserted into the cloning vector in one orientation.

a. Eukaryotic Proteins Can Be Produced in Bacteria and in Eukaryotic Cells

The synthesis of a eukaryotic protein in a prokaryotic host presents several problems not encountered with prokaryotic proteins:

1. The eukaryotic control elements for RNA and protein synthesis are not recognized by bacterial hosts.

2. Bacteria lack the cellular machinery to excise the introns that are present in most eukaryotic transcripts, that is, bacteria cannot carry out gene splicing (Section 5-4Ac).

3. Bacteria lack the enzyme systems to carry out the specific post-translational processing that many eukaryotic proteins require for biological activity (Section 32-5). Most conspicuously, bacteria do not glycosylate proteins (although, in many cases, glycosylation does not seem to affect protein function).

4. Eukaryotic proteins may be preferentially degraded by bacterial proteases (Section 32-6A).

The problem of nonrecognition of eukaryotic control elements can be eliminated by inserting the protein-encoding portion of a eukaryotic gene into a vector containing correctly placed bacterial control elements. The need to excise introns can be circumvented by cloning the cDNA of the protein's mature mRNA. Alternatively, genes encoding small proteins of known sequence can be chemically synthesized (Section 7-6A). Neither of these strategies is universally applicable, however, because few mRNAs are sufficiently abundant to be isolated and the genes encoding many eukaryotic proteins are much larger than can presently be reliably synthesized. Likewise, no general approach has been developed for the post-translational modification of eukaryotic proteins.

The preferential bacterial proteolysis of certain eukaryotic proteins may be prevented by inserting the eukaryotic gene after a bacterial gene such that both have the same reading frame. The resulting **hybrid** or **fusion protein** has an N-terminal polypeptide of bacterial origin that, in some cases, prevents bacterial proteases from recognizing the eukaryotic segment as being foreign. The purification of a fusion protein may be greatly facilitated by the specific binding properties of its N-terminal portion via a process known as **affinity chromatography** (Section 6-3C). Moreover, the formation of a fusion protein may render soluble its otherwise insoluble C-terminal portion. The two polypeptide segments can later be separated by treatment with a protease that specifically cleaves a susceptible site that had been engineered into the boundary between the segments (see below).

The development of cloning vectors that propagate in eukaryotic hosts, such as yeast or cultured animal cells, has led to the elimination of many of the above problems (although post-translational processing, and in particular glycosylation, may vary among different eukaryotes). Baculovirus-based vectors, which replicate in cultured insect cells, have been particularly successful in this regard. Moreover, **shuttle vectors** are available that can propagate in both yeast and *E. coli* and thus transfer (shuttle) genes between these two types of cells.

b. Recombinant Protein Production Has Important Practical Consequences

The ability to synthesize a given protein in large quantities has had an enormous medical, agricultural, and industrial impact. Those that are in routine clinical use include **human insulin** (a polypeptide hormone that regulates fuel metabolism and whose administration is required for survival by certain types of diabetics; Section 27-3B), human growth hormone [**somatotropin,** which induces the proliferation of muscle, bone, and cartilage and is used to stimulate growth in children of abnormally short stature (Section 19-1J); before the advent of recombinant DNA techniques, somatotropin was only available in small quantities from the pituitaries of cadavers], **erythropoietin** (a protein growth factor secreted by the kidney that stimulates the production of red blood cells and is used in the treatment of anemia arising from kidney disease), several types of **colony-stimulating factors** (which stimulate the production and activation of white blood cells and are used clinically to counter the white cell–killing effects of chemotherapy and to facilitate bone marrow transplantation), and **tissue-type plasminogen activator** (**t-PA,** which is used to promote the dissolution of the blood clots responsible for heart attacks and stroke; Section 35-1F). Synthetic vaccines consisting of harmless but immunogenic components of pathogens, for example, **hepatitis B** vaccine, are eliminating the risks attendant in using killed or attenuated viruses or bacteria in vaccines as well as making possible new strategies of vaccine development. The use of recombinant **blood clotting factors** in treating individuals with the inherited disease **hemophilia** (in which these factors are defective; Section 35-1Da) has replaced the need to extract these scarce proteins from large quantities of human blood and has thereby eliminated the high risk that hemophiliacs previously faced of contracting such bloodborne diseases as hepatitis and AIDS. Bovine somatotropin (**bST**) has long been known to stimulate milk production in dairy cows by ~15%. Its use has been made cost-effective, however, by the advent of recombinant DNA technology since bST could previously only be obtained in small quantities from cow pituitaries. Recombinant porcine somatotropin (**pST),** which is administered to growing pigs, induces ~15% greater growth on ~20% less feed while producing leaner meat.

c. Site-Directed and Cassette Mutagenesis Generate Proteins with Specific Sequence Changes

Of equal importance to protein production is the ability to tailor proteins to specific applications by altering their amino acid sequences at specific sites. This is frequently done via a method pioneered by Michael Smith known as **site-directed mutagenesis.** In this technique, an oligonucleotide containing a short gene segment with the desired altered base sequence corresponding to the new amino acid sequence (and synthesized by techniques discussed in Section 7-6Aa) is used as a primer in the DNA polymerase I–smediated replication of the gene of interest. Such a primer can be made to hybridize to the corresponding wild-type sequence if there are only a few mismatched base

Figure 5-55   Site-directed mutagenesis. A chemically synthesized oligonucleotide incorporating the desired base changes is hybridized to the DNA encoding the gene to be altered (*green strand*). The mismatched primer is then extended by DNA polymerase I, thereby generating the mutated gene (*blue strand*). The mutated gene can subsequently be inserted into a suitable host organism so as to yield the mutant DNA, or its corresponding RNA, in quantity, produce a specifically altered protein, and/or generate a mutant organism. **See the Animated Figures**

pairs, and its extension, by DNA polymerase I, yields the desired altered gene (Fig. 5-55). The altered gene can then be inserted in a suitable organism via techniques discussed in Section 5-5C and grown (cloned) in quantity. Similarly, PCR may be used as a vehicle for site-directed mutagenesis simply by using a mutagenized primer in amplifying a gene of interest so that the resulting DNA contains the altered sequence.

Using site-directed mutagenesis, the development of a variant form of the bacterial protease **subtilisin** (Section 15-3Bb) in which Met 222 has been changed to Ala (Met 222 → Ala or M222A) has permitted its use in laundry detergent that contains bleach (which largely inactivates wild-type subtilisin by oxidizing Met 222). **Monoclonal antibodies** (a single species of antibody produced by a clone of an antibody-producing cell; Sections 6-1Da and 35-2Bd) can be targeted against specific proteins and hence are used as antitumor agents. However, since monoclonal antibodies, as usually made, are mouse proteins, they are ineffective as therapeutic agents in humans because humans mount an immune response against mouse proteins. This difficulty has been rectified by "humanizing" monoclonal antibodies by replacing their mouse-specific sequences with those of humans (which the human immune system ignores) through site-directed mutagenesis. Thus the monoclonal antibody known as **trastuzumab** (trade name **Herceptin**), which binds specifically to the growth factor receptor **HER2** that is overexpressed in ~25% of breast cancers, blocks HER2's growth-signaling activity, thereby causing the tumor to stop growing or even regress.

In an alternative mutagenesis technique called **cassette mutagenesis,** complementary oligonucleotides containing the mutation(s) of interest are chemically synthesized (Section 7-6Aa) and annealed to create a duplex "cassette." The cassette is then ligated into the target gene, which must therefore contain an appropriately placed unique restriction site (which can be introduced through site-directed mutagenesis; the cassette must, of course, have the corresponding sticky ends) or, if the cassette is to replace an existing segment of the target gene, two possibly different restriction sites flanking the replaceable segment. Cassette mutagenesis is particularly useful for the insertion of short peptide sequences into a protein of interest (e.g., for the introduction of a proteolytic target site for the cleavage of a fusion protein), when a specific region of the protein is to be subjected to extensive and/or repeated mutagenesis, and for the generation of proteins containing all possible sequences in a short segment (by synthesizing a mixture of cassettes containing all possible variants of the corresponding codons; Section 7-6C).

We will see numerous instances throughout this textbook of protein function being mutagenically character-

ized through the replacement of a specific residue(s) or a polypeptide segment suspected of having an important mechanistic or structural role. Indeed, mutagenesis has become an indispensable tool in the practice of enzymology.

d. Reporter Genes Can Be Used to Monitor Transcriptional Activity

The rate at which a structural gene is expressed depends on its upstream control sequences. Consequently, the rate of expression of a gene can be monitored by replacing its protein-encoding portion with or fusing it in frame to a gene expressing a protein whose presence can be easily determined. An already familiar example of such a **reporter gene** is the *lacZ* gene in the presence of X-gal (Section 5-5Ca) because its level of expression is readily quantitated by the intensity of the blue color that is generated. Although numerous reporter genes have been developed, that which has gained the greatest use encodes **green fluorescent protein (GFP).** GFP, a product of the bioluminescent jellyfish *Aequorea victoria,* fluoresces with a peak wavelength of 508 nm (green light) when irradiated by UV or blue light (optimally 400 nm). This nontoxic protein, whose use was pioneered by Osamu Shimomura and Martin Chalfie, is intrinsically fluorescent; its light-emitting group is the product of the spontaneous cyclization and oxidation by O_2 of three consecutive residues, Ser-Tyr-Gly, to yield a conjugated system of double bonds that gives the protein its fluorescent properties.

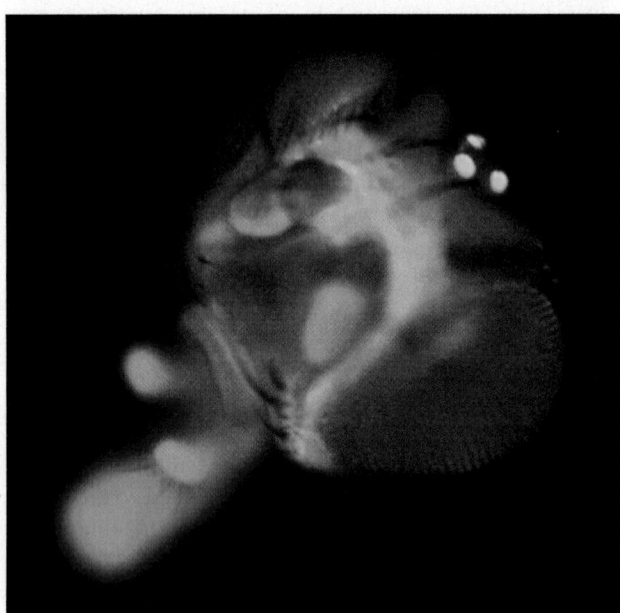

Figure 5-56 **Use of green fluorescent protein (GFP) as a reporter gene.** The gene for GFP was placed under the control of the *Drosophila per* gene promoter and transformed into *Drosophila.* The *per* gene encodes a so-called clock protein that is involved in controlling the fruit fly's circadian (daily) rhythm. The intensity of the green fluorescence of the isolated fly head seen here, which also occurs in other body parts, follows a daily oscillation that can be reset by light. These observations indicate that individual cells in *Drosophila* have photoreceptors and suggest that each of these cells possesses an independent clock. Evidently the head, which was previously thought to be the fly's master oscillator, does not coordinate all of its rhythms. [Courtesy of Steve A. Kay, The Scripps Research Institute, La Jolla, California.]

Fluorophore of green fluorescent protein

Hence GFP requires no substrate or small molecule cofactor to fluoresce as do other highly fluorescent proteins. Its presence can therefore be monitored through the use of UV light or a fluorometer, and its cellular location can be determined through fluorescence microscopy (Fig. 5-56). Consequently, when the GFP gene is placed under control of the gene expressing a particular protein (GFP's fluorescence

Figure 5-57 **Fluorescence of solutions of GFP and ten of its variants.** Note that each of these proteins fluoresces with a different color. [Courtesy of Roger Tsien, University of California at San Diego.]

is unaffected by the formation of a fusion protein), the protein's expressional activity can be readily determined. In fact, a number of GFP variants with distinct sets of excitation and emission wavelengths were developed by Roger Tsien through genetic engineering (Fig. 5-57) and hence the expressional activities of several different genes can be monitored simultaneously. Moreover, the development of pH-sensitive GFP variants permits the monitoring of the pH in subcellular compartments.

H. *Transgenic Organisms and Gene Therapy*

For many purposes it is preferable to tailor an intact organism rather than just a protein—true genetic engineering. Multicellular organisms expressing a foreign (from another organism) gene are said to be **transgenic** and their transplanted foreign genes are often referred to as **transgenes.** For the change to be permanent, that is, heritable, a transgene must be stably integrated into the organism's germ cells. For mice, in techniques pioneered by Mario Capecchi, Martin Evans, and Oliver Smithies, this is accomplished by microinjecting (transfecting) DNA encoding the desired altered characteristics into a **pronucleus** of a fertilized ovum (Fig. 5-58; a fertilized ovum contains two pronuclei,

Figure 5-58 **Microinjection of DNA into the pronucleus of a fertilized mouse ovum.** The ovum is being held in place by gentle suction from the pipette on the left. [Science Vu/Visuals Unlimited.]

one from the sperm and the other from the ovum, which eventually fuse to form the nucleus of the one-celled embryo) and implanting it into the uterus of a foster mother. The DNA integrates at a random position in the genome of the pronucleus through a poorly understood mechanism. Alternatively, an **embryonic stem cell** (an undifferentiated embryonic cell that can give rise, *in utero,* to an entire organism) may be transfected with an altered gene, which occasionally replaces its normal counterpart via recombination. A normal gene may, in this manner, be "knocked out" (permanently inactivated) by transfection with a defective version of the gene. With either method, mating heterozygotes for the altered gene yields progeny that are homozygotes for the altered gene. The use of transgenic mice, and in particular **knockout mice,** has greatly enhanced our understanding of vertebrate gene expression (Section 34-3).

a. Transgenic Organisms Have Many Uses

Procedures are being developed to generate transgenic farm animals such as cows, goats, pigs, and sheep. Animals may thus be induced to grow larger on lesser amounts of feed and/or to be resistant to particular diseases, although this will require a greater understanding of the genes involved than is presently available. An intriguing application of transgenic farm animals is for them to secrete pharmaceutically useful proteins, such as human growth hormone and blood clotting factors, into their milk. Such a transgenic cow, it is expected, will yield several grams of a foreign protein per liter of milk (tens of kilograms per year), which can thereby be produced far more economically than it can by bacteria. A small herd of such "pharm animals" could satisfy the world's need for a particular medicinally useful protein.

The transplantation between humans of organs such as hearts, lungs, livers, and kidneys (a process known as **allotransplantation;** Greek: *allos*, other) has saved tens of thousands of lives since the late 1960s. However, the demand for transplantable organs has so outstripped the supply that as little as 5% of the organs needed in the United States become available. This organ shortage could be entirely eliminated if organs from human-sized animals such as pigs could be transplanted into humans (a process known as **xenotransplantation;** Greek: *xenos*, strange or foreign). However, the xenotransplantation of a pig organ into a human results in the destruction of that organ in as little as a six minutes through a series of **complement system**–mediated reactions that are triggered by the foreign antigens lining the blood vessels of the xenograft (the complement system constitutes the body's first line of immunological defenses; Section 35-2F). This **hyperacute rejection** occurs because the porcine tissue lacks the human proteins that inhibit the human complement system. However, when the organs from pigs that were made transgenic in these human inhibitory proteins were transplanted into baboons, the hyperacute rejection of these organs did not occur. Thus, although not all the problems of xenotransplantation have been eliminated (baboons with transplanted pig hearts have survived as long as six months), it now seems likely that genetic engineering techniques will eventually make xenotransplantation a practical alternative to allo-transplantation.

Transgenic plants are becoming increasingly available, promising a significant extension of the "green revolution" that has changed the face of agriculture throughout the world since the 1950s. For example, during sporulation, various strains of the soil microbe *Bacillus thuringiensis* **(Bt)** express proteins that specifically bind to the intestinal cells of certain insects in a manner that lyses these cells, thereby killing the insect through starvation and infection. These so-called **δ-endotoxins** (also known as **crystal proteins** because Bt spores contain them in microcrystalline form) are innocuous to vertebrates and, hence, Bt spores have been used to control such pests as the **gypsy moth.** Unfortunately, Bt decays after a short time. However, the gene for a δ-endotoxin has been cloned into corn, where, for example, it confers protection against the **European corn borer** (a commercially significant pest that, for much of its life cycle, lives inside the corn plant, where it is largely inaccessible to chemical insecticides). The use of such **Bt corn,** which is now widely planted in the United States, has greatly reduced the need for chemical insecticides. δ-Endotoxin genes have likewise been successfully cloned into a variety of agriculturally significant plants including potatoes, soybeans, and cotton.

Among the properties of crop plants that have been generated through genetic engineering are increased herbicide resistance (which permits the more selective use of herbicides to control weeds); resistance to viruses, bacteria, and fungi; control of ripening (to permit a crop to be brought to market at the optimum time); altered plant architecture such as height (which can improve crop productivity); increased tolerance to environmental stresses such as cold, heat, drought and salinity; and modified or additional vitamins, starch, proteins, and oils (for improved nutritional properties and the production of sustainable supplies of raw materials). For example, a strain of rice has been developed that contains the foreign genes encoding the enzymes that synthesize **β-carotene** (Section 24-2Ad), an orange pigment that is a precursor of **vitamin A.** This genetically modified rice, which is named golden rice (Fig. 5-59), could help alleviate vitamin A deficiency, which afflicts ~400 million people and causes an estimated 1.5 million deaths and 400,000 cases of irreversible blindness per year, mainly in children. Unfortunately, the irrational fear of genetically modified organisms fostered by certain environmental and anti-globalization organizations has led to a thicket of regulations on transgenic crops that, as yet, has prevented golden rice from being made available to farmers in the poor countries where vitamin A deficiency is widespread.

b. Gene Therapy Has Enormous Medical Potential

Gene therapy, the transfer of new genetic material to the cells of an individual resulting in therapeutic benefit to that individual, has been under clinical investigation since 1990, when W. French Anderson and Michael Blaese employed this technology with two children (Section 28-4A) in an effort to alleviate their **severe combined immunodeficiency disease** (**SCID;** any of several genetic diseases that

Figure 5-59   Golden rice. The rice grains on the left are the wild type. The grains on the right have been engineered to synthesize β-carotene, which gives them their yellow color. [Courtesy of Golden Rice Humanitarian Board, http://www.goldenrice.org.]

so impair the immune system that a victim must be kept in a sterile environment in order to survive). Around 4000 genetic diseases are presently known and are therefore potential targets of gene therapy.

Hundreds of gene-transfer protocols are currently under development for use in gene therapy. Ideally, these would deliver the gene(s) of interest to only their target cells such that the gene is permanently installed in these cells in a way that it undergoes sustained expression at the proper level and without causing any adverse side effects. Both nonviral and viral vectors are under development. Nonviral vectors deliver naked DNA to target cells, for example, by direct injection and via **liposomes** (membrane enveloped vesicles that are designed to be taken up by specific cells; Section 12-2B). A large variety of viral vectors are under investigation including those whose genetic material is DNA and those in which it is RNA. These viral vectors have been engineered so as to replace the genes encoding essential viral proteins with therapeutic genes. Hence, cells that have been "infected" by these "viruses" contain the therapeutic genes in their chromosomes but they lack the genetic information to replicate the virus.

There are three categories of gene therapy:

1. In the *ex vivo* (out of the body) approach, cells are removed from the body, incubated with a vector, and then returned to the body. This procedure is usually done with bone marrow cells, which are blood cell precursors.

2. In the *in situ* approach, the vector is applied directly to affected tissues. Such methods are being developed, for example, to treat tumors by injecting into the tumor a vector bearing the gene for a toxin or a gene that would make the tumor susceptible to a chemotherapeutic agent or to attack by the immune system; and to treat **cystic fibrosis,** by inhaling an aerosol containing a vector encoding the normal protein. (Cystic fibrosis, one of the most common genetic diseases, is caused by a defect in a protein involved in the secretion of chloride ion in the lungs and other tissues. This causes the secretion of abnormally thick mucus, which results in recurrent and often damaging lung infections leading to early death.)

3. In the *in vivo* (in the body) approach, the vector would be injected directly into the bloodstream. There are, as yet, no clinical examples of this approach, although vectors to do so must ultimately be developed if gene therapy is to fulfill its promise.

In the first well-documented clinical success of gene therapy, Alain Fischer reported that several infants appear to have been cured, via the *ex vivo* treatment of their bone marrow cells, of a form of SCID called **SCID-X1** (which is caused by a mutation in the gene encoding the **γc cytokine receptor,** a receptor for certain protein growth factors, whose proper function is essential for the differentiation, growth, and survival of the white blood cells known as **T cells;** Section 35-2A). However, because the viral vector containing the γc cytokine receptor gene integrates into the genome at random, the location of the transgene may affect the expression of other genes, triggering cancer (Section 34-4Ca). In fact, two children have developed leukemia (a white blood cell cancer) as a result of gene therapy for SCID-X1.

Recently, several patients with the rare inherited blinding disease **Leber's congenital amaurosis 2 (LCA2;** which results in degeneration of the retina) have shown improved vision without significant adverse side effects, after the subretinal injection of a viral vector bearing a normal copy of the defective gene responsible for LCA2. This gene, named **RPE65,** encodes an enzyme that helps convert vitamin A to **retinal** (Section 12-3Ab), the light-sensing pigment of the eye. Evidently, steady progress is being made in our understanding of the requirements for the construction and administration of effective gene therapy vectors. It therefore seems likely that, over the next few decades, gene therapy will revolutionize the practice of medicine.

I. *Social, Ethical, and Legal Considerations*

In the early 1970s, when strategies for genetic engineering were first being discussed, it was realized that little was known about the safety of the proposed experiments. Certainly it would be foolhardy to attempt experiments such as introducing the gene for **diphtheria toxin** (Section 32-3Ge) into *E. coli* so as to convert this human symbiont into a deadly pathogen. But what biological hazards would result, for example, from cloning tumor virus genes in *E. coli* (a useful technique for analyzing these viruses)? Consequently, in 1975, molecular biologists declared a voluntary moratorium on molecular cloning experiments until these risks could be assessed. There ensued a spirited debate, at first among molecular biologists and later in the public arena, between two camps: those who thought that the enormous potential benefits of recombinant DNA research warranted its continuation once adequate safety precautions had been instituted, and those who felt that its potential dangers were so great that it should not be pursued under any circumstances.

The former viewpoint eventually prevailed with the promulgation, in 1976, of a set of U.S. government regula-

tions for recombinant DNA research. Experiments that are obviously dangerous were forbidden. In other experiments, the escape of laboratory organisms was to be prevented by both physical and biological containment. By biological containment it is meant that vectors will only be cloned in host organisms with biological defects that prevent their survival outside the laboratory. For example, χ1776, the first approved "safe" strain of *E. coli*, has among its several defects the requirement for diaminopimelic acid, an intermediate in lysine biosynthesis (Section 26-5Ba), which is neither present in human intestines nor commonly available in the environment.

As experience with recombinant DNA research accumulated, it became evident that the foregoing reservations were largely groundless. *No genetically altered organism yet reported has caused an unexpected health hazard.* Indeed, recombinant DNA techniques have, in many cases, eliminated the health hazards of studying dangerous pathogens such as the virus causing **acquired immune deficiency syndrome (AIDS)**. Consequently, since 1979, the regulations governing recombinant DNA research have been gradually relaxed.

There are other social, ethical, and legal considerations that will have to be faced as new genetic engineering techniques become available (Fig. 5-60). Recombinant erythropoietin is now routinely prescribed to treat the effects of certain types of kidney disease. However, should athletes be permitted to use this protein, as many reportedly have, to increase the number of red cells in their blood and hence its oxygen-carrying capacity (a dangerous procedure if uncontrolled since the increased number of cells in the blood can put a great strain on the heart)? Few would dispute the use of gene therapy, if it can be developed, to cure such devastating genetic defects as **sickle-cell anemia** (a painful and debilitating condition caused by deformed red blood cells that often results in early death; Section 10-3B) and **Tay-Sachs disease** (which is caused by the absence of the lysosomal enzyme **hexosaminidase A** and results in progressive neuronal dysfunction that is invariably fatal by around age 3; Section 25-8Ce). If, however, it becomes possible to alter complex (i.e., multigene) traits such as athletic ability or intelligence, which changes would be considered desirable, under what circumstances would they be made, and who would decide whether to make them? Should gene therapy be used on individuals with inherited diseases only to correct defects in their somatic cells or should it also be used to alter genes in their germ cells, which could then be transmitted to succeeding generations? Animals such as sheep, cows, dogs, and mice have already been cloned. Should humans with particularly desirable traits, either naturally occurring or generated through genetic engineering, be cloned? When it becomes easy to determine an individual's genetic makeup, should this information be used, for example, in evaluating applications for educational and employment opportunities, or in assessing a person's eligibility for health insurance (which has recently been made illegal in the United States)? These conundrums have led to the advent of a branch of philosophy named **bioethics** designed to deal with them.

Figure 5-60 [Drawing by T.A. Bramley, *in* Andersen, K., Shanmugam, K.T., Lim, S.T., Csonka, L.N., Tait, R., Hennecke, H., Scott, D.B., Hom, S.S.M., Haury, J.F., Valentine, A., and Valentine, R.C., *Trends Biochem. Sci.* **5**, 35 (1980). Copyright © Elsevier Biomedical Press, 1980. Used by permission.]

CHAPTER SUMMARY

1 Nucleotides and Nucleic Acids A nucleotide consists of either a ribose or a 2′-deoxyribose residue whose C1′ atom forms a glycosidic bond with a nitrogenous base and whose 3′ or 5′ position is esterified to a phosphate group. Nucleosides lack the phosphate groups of nucleotides. The nitrogenous bases in the great majority of nucleotides are the purines adenine and guanine and the pyrimidines cytosine and either thymine in DNA or uracil in RNA. Nucleic acids are linear polymers of nucleotides containing either ribose residues in RNA or deoxyribose residues in DNA and whose 3′ and 5′ positions are linked by phosphate groups. In double helical DNAs and RNAs, the base compositions obey Chargaff's rules: A = T(U) and G = C. RNA, but not DNA, is susceptible to base-catalyzed hydrolysis.

2 DNA Is the Carrier of Genetic Information Extracts of virulent S-type pneumococci transform nonpathogenic R-type pneumococci to the S form. The transforming principle is DNA. Similarly, radioactive labeling has demonstrated that the genetically active substance of bacteriophage T2 is its DNA. The viral capsid serves only to protect its enclosed DNA and to inject it into the bacterial host. This establishes that DNA is the hereditary molecule.

3 Double Helical DNA B-DNA consists of a right-handed double helix of antiparallel sugar–phosphate chains with ~10 bp per turn of 34 Å and with its bases nearly perpendicular to the helix axis. Bases on opposite strands hydrogen bond in a geometrically complementary manner to form A · T and G · C Watson–Crick base pairs. DNA replicates in a semiconservative manner, as has been demonstrated by the Meselson–Stahl experiment. When heated past its melting temperature, T_m, DNA denatures and undergoes strand separation. This process may be monitored by the hyperchromism of the DNA's UV spectrum. Denatured DNA can be renatured by maintaining it at ~25°C below its T_m. DNA occurs in nature as molecules of enormous lengths which, because they are also quite stiff, are easily mechanically cleaved by laboratory manipulations.

4 Gene Expression and Replication: An Overview Genes are expressed according to the central dogma of molecular biology: DNA directs its own replication and its transcription to yield RNA, which, in turn, directs its translation to form proteins. RNA is synthesized from ribonucleoside triphosphates on DNA templates by RNA polymerase, a process in which the DNA template strand is read in its 3′ to 5′ direction and the

RNA is synthesized in its 5′ to 3′ direction. The rate at which a particular gene is transcribed is governed by control sites, which, for mRNAs, are located upstream of the transcriptional initiation site and can be quite extensive, particularly in eukaryotes. Eukaryotic mRNAs often require substantial post-transcriptional modifications, including gene splicing (the excision of introns and the rejoining of their flanking exons), to become functional.

mRNAs direct the ribosomal synthesis of polypeptides. In this process, ribosomes facilitate the binding of the mRNA's codons to the anticodons of tRNAs bearing their cognate amino acids, and the ribosomes then catalyze the formation of peptide bonds between successive amino acids. The correspondence between codons and the amino acid carried by the tRNAs that bind to them is called the genetic code. Enzymes known as aminoacyl-tRNA synthetases covalently link their corresponding tRNAs to their cognate amino acids. The selection of the correct initiation site on the mRNA also sets the reading frame for the polypeptide being synthesized. Newly synthesized proteins often require post-translational modifications to be functional, including specific proteolytic cleavages and, in eukaryotes only, glycosylation. The lifetime of a protein in a cell varies from fractions of a minute to days or weeks.

DNA is synthesized from deoxynucleoside triphosphates by DNA polymerase, an enzyme that can only extend existing polynucleotides bound to the template DNA and hence requires a primer. In cells, primers are RNA, which are synthesized on DNA templates by an RNA polymerase. The replication of both strands of duplex DNA takes place at a replication fork. In *E. coli*, duplex DNA replication is carried out by two molecules of DNA polymerase III, one of which synthesizes the leading strand and the other of which synthesizes the lagging strand. The leading strand is synthesized continuously. However, since all DNA polymerases can only extend DNA in its 5′ to 3′ direction, the lagging strand template must loop around to be read in its 3′ to 5′ direction, which requires that the lagging strand be synthesized discontinuously. The RNA primers for the lagging strand are synthesized by primase, and once a lagging strand segment has been synthesized, its primer is replaced through the combined actions of DNA polymerase I's 5′ → 3′ exonuclease and DNA polymerase activities. The single-strand nicks between successive lagging strand segments are then sealed by DNA ligase. Both DNA polymerase I and DNA polymerase III also have 3′ → 5′ exonuclease activities that function to proofread the newly synthesized DNA for mispairing errors and excise the mispaired nucleotides.

5 Molecular Cloning Molecular cloning techniques have revolutionized the practice of biochemistry. Defined DNA fragments are generated through the use of Type II restriction endonucleases (restriction enzymes), which cleave DNA at specific and usually palindromic sequences of four to six bases. Restriction-fragment length polymorphisms (RFLPs) provide markers for identifying chromosomal differences and hence are useful in identity tests and in establishing familial relationships. A DNA fragment may be produced in large quantities by inserting it, using recombinant DNA techniques, into a suitable cloning vector. These may be genetically engineered plasmids, viruses, cosmids, yeast artificial chromosomes (YACs), or bacterial artificial chromosomes (BACs). The DNA to be cloned is usually obtained as a restriction fragment so that it can be specifically ligated into a corresponding restriction cut in the cloning vector. Gene splicing may also occur through the generation of complementary homopolymer tails on the DNA fragment and the cloning vector or through the use of synthetic palindromic linkers containing restriction sequences. The introduction of a recombinant cloning vector into a suitable host organism permits the foreign DNA segment to be produced in nearly unlimited quantities. Those cells that have been properly transformed by a vector can be chosen through the use of selectable markers and chromogenic substrates. Specific base sequences may be detected in DNA by Southern blotting and in RNA by the similar Northern blotting. A particular gene may be isolated through the screening of a genomic library of the organism producing the gene. The polymerase chain reaction (PCR) is a particularly fast and convenient method of identifying and obtaining specific sequences of DNA. Genetic engineering techniques may be used to produce otherwise scarce or specifically altered proteins in large quantities and to monitor gene expression through the use of reporter genes such as green fluorescent protein. They are also used to produce transgenic plants and animals and in gene therapy. The development of recombinant DNA techniques has generated numerous social, ethical, and legal issues whose resolution will govern how biotechnology is used.

REFERENCES

Useful Websites

REBASE. The restriction enzyme database. http://rebase.neb.com

The Role of DNA

Avery, O.T., MacLeod, C.M., and McCarty, M., Studies on the chemical nature of the substance inducing transformation of pneumococcal types, *J. Exp. Med.* **79**, 137–158 (1944). [The milestone report identifying transforming principle as DNA.]

Hershey, A.D. and Chase, M., Independent functions of viral proteins and nucleic acid in growth of bacteriophage, *J. Gen. Physiol.* **36**, 39–56 (1952).

McCarty, M., *The Transforming Principle,* Norton (1985). [A chronicle of the discovery that genes are DNA.]

Palmiter, R.D., Brinster, R.L., Hammer, R.E., Trumbauer, M.E., Rosenfeld, M.G., Birmberg, N.C., and Evans, R.M., Dramatic growth of mice that develop from eggs microinjected with metallothionein–growth hormone fusion genes, *Nature* **300**, 611–615 (1982).

Stent, G.S., Prematurity and uniqueness in scientific discovery, *Sci. Am.* **227**(6), 84–93 (1972). [A fascinating philosophical discourse on what it means for discoveries such as Avery's to be "ahead of their time" and on the nature of creativity in science.]

Structure and Properties of B-DNA

Bloomfield, V.A., Crothers, D.M., and Tinoco, I., Jr., *Nucleic Acids. Structures, Properties, and Functions*, University Science Books (2000).

Crick, F., *What Mad Pursuit*, Basic Books (1988). [A scientific autobiography.]

Judson, H.F., *The Eighth Day of Creation*, Part I, Simon & Schuster (1979). [A fascinating narrative on the discovery of the DNA double helix.]

Manchester, K.L., Historical opinion: Erwin Chargaff and his 'rules' for the base composition of DNA: why did he fail to see the possibility of complementarity? *Trends Biochem. Sci.* **33**, 65–70 (2008).

Meselson, M. and Stahl, F.W., The replication of DNA in *Escherichia coli, Proc. Natl. Acad. Sci.* **44**, 671–682 (1958). [The classic paper establishing the semiconservative nature of DNA replication.]

Saenger, W., *Principles of Nucleic Acid Structure*, Springer-Verlag (1984).

Sayre, A., *Rosalind Franklin and DNA*, Norton (1975) [A biography which argues that Rosalind Franklin, who died in 1958, deserves far more credit than is usually accorded her for the discovery of the structure of DNA.]; *and* Piper, A., Light on a dark lady, *Trends Biochem. Sci.* **23**, 151–154 (1998). [A biographical memoir on Rosalind Franklin.]

Schlenk, F., Early nucleic acid chemistry, *Trends Biochem. Sci.* **13**, 67–68 (1988).

Voet, D. and Rich, A., The crystal structures of purines, pyrimidines and their intermolecular structures, *Prog. Nucleic Acid Res. Mol. Biol.* **10**, 183–265 (1970).

Watson, J.D., *The Double Helix. A Personal Account of the Discovery of the Structure of DNA*, Simon & Schuster (1968). [A provocative autobiographical narrative.]

Watson, J.D. and Crick, F.H.C., Molecular structure of nucleic acids, *Nature* **171**, 737–738 (1953); *and* Genetical implications of the structure of deoxyribonucleic acid, *Nature* **171**, 964–967 (1953). [The seminal papers that are widely held to mark the origin of modern molecular biology.]

Wing, R., Drew, H., Takano, T., Broka, C., Tanaka, S., Itakura, K., and Dickerson, R.E., Crystal structure analysis of a complete turn of B-DNA, *Nature* **287**, 755–758 (1980). [The first X-ray crystal structure of a segment of B-DNA, which largely confirmed the less reliable fiber diffraction-based structure of B-DNA proposed by Watson and Crick.]

Zimm, B.H., One chromosome: one DNA molecule, *Trends Biochem. Sci.* **24**, 121–123 (1999). [A scientific reminiscence on how it was established that chromosomes each contain only one piece of DNA.]

Molecular Cloning

Bainbridge, J.W.B., et al., Effect of gene therapy on visual function in Leber's congenital amaurosis; *and* Maguire, A.M., et al., Safety and efficacy of gene transfer for Leber's congenital amaurosis, *New Eng. J. Med.* **358**, 2231–2239; *and* 2240–2248 (2008).

Brown, T.A., *Gene Cloning and DNA Analysis, An Introduction* (5th ed.), Blackwell Publishing (2006).

Cavazzana-Calvo, M., et al., Gene therapy of human severe combined immunodeficiency (SCID)-X1 disease, *Science* **288**, 669–672 (2000).

Clark, D.P. and Pazdernik, N.J., *Biotechnology. Applying the Genetic Revolution,* Academic Press (2009).

Cooper, A. and Wayne, R., New uses for old DNA, *Curr. Opin. Biotech.* **9**, 49–53 (1998). [Reviews the successes and pitfalls of studying ancient DNAs.]

Cooper, D.K.C., Gollackner, B., and Sachs, D.H., Will the pig solve the transplantation backlog? *Annu. Rev. Med.* **53**, 133–147 (2002).

Dale, J.W. and von Schantz, M., *From Genes to Genomes. Concepts and Application of DNA Technology* (2nd ed.), Wiley (2007).

Foster, E.A., Jobling, M.A., Taylor, P.G., Donnelly, P., de Knijff, P., Mieremet, R., Zerjal, T., and Tyler-Smith, C., Jefferson fathered slave's last child, *Nature* **396**, 27–28 (1998).

Gilbert, S.F., Tyler, A.L., and Zackin, E.J., *Bioethics and the New Embryology: Springboards for Debate,* Sinauer (2005). [A balanced discussion of many of the most contentious bioethical issues of our time.]

Green, R.E., et al., Analysis of one million base pairs of Neanderthal DNA, *Nature* **444**, 330–336 (2006); Noonan, J.P., et al., Sequencing and analysis of Neanderthal genomic DNA, *Science* **314**, 1113–1118 (2006); *and* Green, R.E., et al., A complete Neandertal mitochondrial genome sequence determined by high-throughput sequencing, *Cell* **134**, 416–426 (2008).

Krings, M., Stone, A., Schmitz, R.W., Krainitzki, H., Stoneking, M., and Pääbo, S., Neandertal DNA sequences and the origin of modern humans; *and* Lindahl, T., Facts and artifacts of ancient DNA, *Cell* **90**, 19–30; *and* 1–3 (1997); *and* Ovchinnikov, I.V., Götherström, A., Romanova, G.P., Kharitonov, V.M., Lidén, K., and Goodwin, W., Molecular analysis of Neanderthal DNA from the northern Caucasus, *Nature* **404**, 490–493 (2000).

Lodge, J., Lund, P., and Minchlin, S. *Gene Cloning: Principles and Applications,* Taylor and Francis (2007).

Mullis, K.B., The unusual origin of the polymerase chain reaction. *Sci. Am.* **262**(4): 56–65 (1990).

Nicholl, D.S.T., *An Introduction to Genetic Engineering* (2nd ed.), Cambridge University Press (2003).

Primrose, S.B. and Twyman, R.M., *Principles of Gene Manipulation and Genomics* (7th ed.), Blackwell Publishing (2006).

Sambrook, J. and Russel, D.W., *Molecular Cloning* (3rd ed.), Cold Spring Harbor Laboratory (2001). [A three-volume collection of laboratory protocols with accompanying background explanations.]

Tsien, R.Y., The green fluorescent protein, *Annu. Rev. Biochem.* **67**, 509–544 (1998).

Verma, I.M. and Weitzman, M.D., Gene therapy: Twenty-first century medicine, *Annu. Rev. Biochem.* **74**, 711–738 (2005).

Vreeland, R.H., Rosenzweig, W.D., and Powers, D.W., Isolation of a 250 million-year-old halotolerant bacterium from a primary salt crystal, *Nature* **407**, 897–900 (2000).

Watson, J.D., Meyers, R.M., Caudy, A.A., and Witkowski, J.A., *Recombinant DNA. Genes and Genomes—A Short Course* (3rd ed.), Freeman (2007). [An exposition of the methods, findings, and results of recombinant DNA technology and research.]

PROBLEMS

1. The base sequence of one of the strands of a 20-bp duplex DNA is

5′-GTCCGTTCGACGGTACATC-3′

What is the base sequence of its complementary strand?

2. Non-Watson–Crick base pairs are of biological importance. For example: (a) **Hypoxanthine** (6-oxopurine) is often one of the bases of the anticodon of tRNA. With what base on mRNA is hypoxanthine likely to pair? Draw the structure of this base pair. (b) The third position of the codon–anticodon interaction between tRNA and mRNA is often a G · U base pair. Draw a plausible structure for such a base pair. (c) Many species of tRNA contain a hydrogen bonded U · A · U assembly. Draw two plausible structures for this assembly in which each U forms at least two hydrogen bonds with the A. (d) Mutations may arise during DNA replication when mispairing occurs as a result of the transient formation of a rare tautomeric form of a base. Draw the structure of a base pair with proper Watson–Crick geometry that contains a tautomeric form of adenine. What base sequence change would be caused by such mispairing?

3. (a) What is the molecular mass and contour length of a segment of B-DNA that specifies a 40-kD protein? (b) How many helical turns does this DNA have and what is its axial ratio (length to width ratio)?

*4. The antiparallel orientation of complementary strands in duplex DNA was elegantly demonstrated in 1960 by Arthur Kornberg by **nearest-neighbor analysis.** In this technique, DNA is synthesized by DNA polymerase I from one deoxynucleoside triphosphate whose α-phosphate group is radioactively labeled with ^{32}P and three unlabeled deoxynucleoside triphosphates. The resulting product is treated with an enzyme, DNase I, that catalyzes the hydrolysis of the phosphodiester bonds on the 3′ sides of all deoxynucleotides.

$$ppp^*A + pppC + pppG + pppT$$

$$PP_i \leftarrow | \text{ DNA polymerase}$$

$$\cdots pCpTp^*ApCpCp^*ApGp^*Ap^*ApTp\cdots$$

$$H_2O \leftarrow | \text{ DNase I}$$

$$\cdots + Cp + Tp^* + Ap + Cp + Cp^* + Ap + Gp^* + Ap^* + Ap + Tp + \cdots$$

In this example, the relative frequencies of occurrence of ApA, CpA, GpA, and TpA in the DNA can be determined by measuring the relative amounts of Ap*, Cp*, Gp*, and Tp*, respectively, in the product (where p* represents a ^{32}P-labeled phosphate group). The relative frequencies with which the other 12 dinucleotides occur may likewise be determined by labeling, in turn, the other 3 nucleoside triphosphates in the above reactions. There are equivalencies between the amounts of certain pairs of dinucleotides. However, the identities of these equivalencies depend on whether the DNA consists of parallel or antiparallel strands. What are these equivalencies in both cases?

5. What would be the effect of the following treatments on the melting curve of B-DNA dissolved in 0.5M NaCl solution? Explain. (a) Decreasing the NaCl concentration. (b) Squirting the DNA solution, at high pressure, through a narrow orifice. (c) Bringing the solution to 0.1M adenine. (d) Heating the solution to 25ºC above the DNA's melting point and then rapidly cooling it to 35ºC below the DNA's melting point.

6. What is the mechanism of alkaline denaturation of duplex DNA? [Hint: Some of the bases are relatively acidic.]

7. The following duplex DNA is transcribed from right to left as is printed here.

5′-TCTGACTATTCAGCTCTCTGGCACATAGCA-3′
3′-AGACTGATAAGTCGAGAGACCGTGTATCGT-5′

(a) Identify the template strand. (b) What is the amino acid sequence of the polypeptide that this DNA sequence encodes? Assume that translation starts at the first initiation codon. (c) Why doesn't the UGA sequence in the mRNA transcript cause transcription to terminate?

8. After undergoing splicing, a mature mRNA has the following sequence, where the vertical line indicates the position of the splice junction (the nucleotides from between which an intron had been removed).

5′-CUAGAUGGUAG|
GUACGGUUAUGGGAUAACUCUG-3′

(a) What is the sequence of the polypeptide specified by this mRNA? Assume that translation starts at the first initiation codon. (b) What would the polypeptide sequence be if the splicing system had erroneously deleted the GU on the 3′ side of the splice junction? (c) What would the polypeptide sequence be if the splicing system had erroneously failed to excise a G at the splice junction? (d) Is there any relationship between the polypeptides specified in b and c and, if so, why?

9. Explain why the charging of a tRNA with the correct amino acid is equally as important for accurate translation as is the correct recognition of a codon by its corresponding aminoacyl-tRNA.

10. Describe how to select recombinant clones if a foreign DNA is inserted into the polylinker site of pUC18 and then introduced into *E. coli* cells.

11. Describe the possible outcome of a PCR experiment in which (a) one of the primers is inadvertently omitted from the reaction mixture; (b) one of the primers is complementary to several sites in the starting DNA sample; (c) there is a single-stranded break in the target DNA sequence, which is present in only one copy in the starting sample; (d) there is a double-stranded break in the target DNA sequence, which is present in only one copy in the starting sample.

12. In DNA replication, the leading strand and lagging strand are so named because any particular portion of the lagging strand is always synthesized after the corresponding portion of the leading strand has been synthesized. Explain why this must be the case.

13. SV40 DNA is a circular molecule of 5243 bp that is 40% G + C. In the absence of sequence information, how many restriction cuts would TaqI, EcoRII, PstI, and HaeII be expected to make, on average, in SV40 DNA?

14. Which of the restriction endonucleases listed in Table 5-4 produce blunt ends? Which sets of them are **isoschizomers** (enzymes that have the same recognition sequence but do not necessarily cleave at the same sites; Greek: *isos,* equal + *schizein,* to cut); which of them are **isocaudamers** (enzymes that produce identical sticky ends; Latin: *cauda,* tail)?

15. The plasmid pBR322 contains the *amp*^R and *tet*^R genes, which respectively confer resistance to the antibiotics ampicillin and **tetracycline** (Section 32-3Gd). The *tet*^R gene contains a cleavage site for the restriction enzyme SalI, the only such site in the entire plasmid. Describe how one can select for *E. coli* that had been transformed by pBR322 that contains a foreign DNA insert in its SalI site.

16. A blood stain from a crime scene and blood samples from four suspects were analyzed by PCR using fluorescent primers associated with three STR loci (positions): D3S1358, vWA, and FGA. The resulting electrophoretograms are shown below. The numbers beneath each peak identify the allele (*green box*) and the height of the peak in relative fluorescence units (*red box*). (a) Since everyone has two copies of each chromosome and therefore two copies of each gene, what accounts for the appearance of only one allele at some loci? (b) Which suspect, if any, is a possible source of the blood? (c) Could the suspect be identified using just one of the three STR loci? (d) What can you conclude about the amount of DNA obtained from Suspect 1 compared to Suspect 4?

17. How many yeast DNA fragments of average length 5 kb must be cloned in order to be 90%, 99%, and 99.9% certain that a genomic library contains a particular segment? The yeast genome consists of 12,100 kb.

18. Many of the routine operations in genetic engineering are carried out using commercially available "kits." Genbux Inc., a prospective manufacturer of such kits, has asked your advice on the feasibility of supplying a kit of intact λ phage cloning vectors with the nonessential central section of their DNA already removed. Presumably a "gene jockey" could then grow the required amount of phage, isolate its DNA, and restriction cleave it without having to go to the effort of separating out the central section. What advice would you give the company?

19. Indicate the sequences of the two 10-residue primers that could be used to amplify the central 40-nucleotide region of the following 50-nucleotide single-stranded DNA by PCR.

5′-AGCTGGACCACTGATCATTGACTGCTAGCGTCA
GTCCTAGTAGACTGACG-3′

20. A protein segment of sequence -Phe-Cys-Gly-Val-Leu-His-Lys-Met-Glu-Thr- is encoded by the following DNA segment:

5′-TTCTGCGGAGTCCTACACAAGATGGAGACA-3′

Design an 18-base oligonucleotide that could be used to change the protein's Leu-His segment to Ile-Pro via site-directed mutagenesis.

[From Thompson, W.C., Ford, S., Doom, T., Raymer, M., and Krane, D.E., *The Champion* **27**, 16–25 (2003).]

Techniques of Protein and Nucleic Acid Purification

A major portion of most biochemical investigations involves the purification of the materials under consideration because these substances must be relatively free of contaminants if they are to be properly characterized. This can be a formidable task because a typical cell contains thousands of different substances, many of which closely resemble other cellular constituents in their physical and chemical properties. Furthermore, the material of interest may be unstable and exist in vanishingly small amounts. Typically, a substance that comprises <0.1% of a tissue's dry weight must be brought to ~98% purity. Purification problems of this magnitude would be considered unreasonably difficult by most synthetic chemists. It is therefore hardly surprising that our understanding of biochemical processes has by and large paralleled our ability to purify biological materials.

This chapter presents an overview of the most commonly used techniques for the isolation, the purification, and, to some extent, the characterization of proteins and nucleic acids, as well as other types of biological molecules. These methods are the basic tools of biochemistry whose operation dominates the day-to-day efforts of the practicing biochemist. Furthermore, many of these techniques are routinely used in clinical applications. Indeed, *a basic comprehension of the methods described here is necessary for an appreciation of the significance and the limitations of much of the information presented in this text*. This chapter should therefore be taken as reference material to be consulted repeatedly as the need arises while reading other chapters. Many of the techniques used for protein and nucleic acid fractionation are similar. Consequently, we shall first focus on how proteins are purified and only then concentrate on how these techniques are used in nucleic acid fractionation.

1 PROTEIN ISOLATION

Proteins constitute a major fraction of the mass of all organisms. A particular protein, such as **hemoglobin** in red blood cells, may be the dominant substance present in a tissue. Alternatively, a protein such as the *lac* **repressor** of *E. coli* (Section 31-3B) may normally have a population of only a few molecules per cell. Similar techniques are used for the isolation and purification of both proteins, although, in general, the lower the initial concentration of a substance, the more effort is required to isolate it in pure form.

In this section we discuss the care and handling of proteins and outline the general strategy for their purification. For many proteins, the isolation and purification procedure is an exacting task requiring days of effort to obtain only a few milligrams or less of the desired product. However, as we shall see, modern analytical techniques have achieved such a high degree of sensitivity that this small amount of material is often sufficient to characterize a protein extensively. You should note that the techniques described in this chapter are applicable to the separations of most types of biological molecules.

A. Selection of a Protein Source

Proteins with identical functions generally occur in a variety of organisms. For example, most of the enzymes that mediate basic metabolic processes or that are involved in the expression and transmission of genetic information are common to all cellular life. Of course, there is usually considerable variation in the properties of a particular protein from various sources. In fact, different variants of a given protein may occur in different tissues from the same organism or even in different compartments in the same cell. Therefore, if flexibility of choice is possible, the isolation of a protein may be greatly simplified by a judicious choice of the protein source. This choice should be based on such criteria as the ease of obtaining sufficient quantities of the tissue from which the protein is to be isolated, the amount of the chosen protein in that tissue, and any properties peculiar to the specific protein chosen that would aid in its stabilization and isolation. Tissues from domesticated animals such as chickens, cows, pigs, or rats are often chosen. Alternative sources might be easily obtainable microorganisms such as *E. coli* or **baker's yeast (*Saccharomyces cerevisiae*).** We shall see, however, that proteins from a vast variety of organisms have been studied.

Molecular cloning methods (Section 5-5) have rapidly become equally if not more important protein production techniques. Almost any protein-encoding gene can be isolated from its parent organism, specifically altered (genetically engineered) if desired, and expressed at high levels (overproduced) in a conveniently grown organism such as *E. coli* or yeast, where it may constitute up to 30% of the overproducer's total cell protein. This high level of protein production generally renders the cloned protein far easier to isolate than it would be from its parent organism (in which it may normally occur in vanishingly small amounts).

B. Methods of Solubilization

The first step in the isolation of a protein, or any other biological molecule, is to get it into solution. In some cases, such as with blood serum proteins, nature has already done so. However, a protein must usually be liberated from the cells that contain it. The method of choice for this procedure depends on the mechanical characteristics of the source tissue as well as on the location of the required protein in the cell.

If the protein of interest is located in the cytosol of the cell, its liberation requires only the breaking open **(lysis)** of the cell. In the simplest and gentlest method of doing so, which is known as **osmotic lysis,** the cells are suspended in a **hypotonic solution;** that is, a solution in which the total molar concentration of solutes is less than that inside the cell in its normal physiological state. Under the influence of osmotic forces, water diffuses into the more concentrated intracellular solution, thereby causing the cells to swell and burst. This method works well with animal cells, but with cells that have a cell wall, such as bacteria or plant cells, it is usually ineffective. The use of an enzyme, such as **lysozyme,** which chemically degrades bacterial cell walls (Section 15-2),

is sometimes effective with such cells. Detergents or organic solvents such as acetone or toluene are also useful in lysing cells, but care must be exercised in their use as they may denature the protein of interest (Section 8-4E).

Many cells require some sort of mechanical disruption process to break them open. This may include several cycles of freezing and thawing, grinding with sand, alumina, or glass beads, or the use of a high-speed blender (similar to the familiar kitchen appliance), a **homogenizer** (an implement for crushing tissue between a closely fitting piston and sleeve that may be manually or mechanically driven), a **French press** (a device that shears open cells by squirting them at high pressure through a small orifice), or a **sonicator** (which breaks open cells through ultrasonic vibrations). Once the cells have been broken open, the crude **lysate** may be filtered or centrifuged to remove the particulate cell debris, thereby leaving the protein of interest in the supernatant solution.

If the required protein is a component of subcellular assemblies such as membranes or mitochondria, a considerable purification of the protein can be effected by first separating the subcellular assembly from the rest of the cellular material. This is usually accomplished by **differential centrifugation,** a process in which the cell lysate is centrifuged at a speed that removes only the cell components denser than the desired organelle followed by centrifugation at a speed that spins down the component of interest. The required protein is then usually separated from the purified subcellular component by extraction with concentrated salt solutions or, in the case of proteins tightly bound to membranes, with the use of detergent solutions or organic solvents, such as butanol, that solubilize lipids.

C. Stabilization of Proteins

Once a protein has been removed from its natural environment, it becomes exposed to many agents that can irreversibly damage it. These influences must be carefully controlled at all stages of a purification process or the yield of the desired protein may be greatly reduced or even eliminated.

The structural integrity of many proteins is sensitive to pH as a consequence of their numerous acid–base groups. To prevent damage to biological materials due to variations in pH, they are routinely dissolved in buffer solutions effective in the pH range over which the material is stable.

Proteins are easily denatured by high temperatures. Although the thermal stabilities of proteins vary widely, many of them slowly denature above 25°C. Therefore, the purification of proteins is normally carried out at temperatures near 0°C. However, there are numerous proteins that require lower temperatures, some even lower than –100°C, for stability. Conversely, some **cold-labile** proteins become unstable below characteristic temperatures.

The thermal stability characteristics of a protein can sometimes be used to advantage in its purification. A heat-stable protein in a crude mixture can be greatly purified by briefly heating the mixture so as to denature and precipitate most of the contaminating proteins without affecting the desired protein.

Cells contain **proteases** (enzymes that catalyze the hydrolytic cleavage of peptide bonds) and other degradative enzymes that, on cell lysis, are liberated into solution along with the protein of interest. Care must be taken that the protein is not damaged by these enzymes. Degradative enzymes may often be rendered inactive at pH's and temperatures that are not harmful to the protein of interest. Alternatively, these enzymes can often be specifically inhibited by chemical agents without affecting the desired protein. Of course, as the purification of a protein progresses, more and more of these degradative enzymes are eliminated.

Some proteins are more resistant than others to proteolytic degradation. The purification of a protein that is particularly resistant to proteases may be effected by maintaining conditions in a crude protein mixture under which the proteolytic enzymes present are active. This so-called **autolysis** technique simplifies the purification of the resistant protein because it is generally far easier to remove selectively the degradation products of contaminating proteins than it is the intact proteins.

Many proteins are denatured by contact with the air–water interface, and, at low concentrations, a significant fraction of the protein present may be lost by adsorption to surfaces. Hence, a protein solution should be handled so as to minimize frothing and should be kept relatively concentrated. There are, of course, other factors to which a protein may be sensitive, including the oxidation of cysteine residues to form disulfide bonds; heavy metal contaminants, which may irreversibly bind to the protein; and the salt concentration and polarity of the solution, which must be kept within the stability range of the protein. Finally, many microorganisms consider proteins to be delicious, so protein solutions should be stored under conditions that inhibit the growth of microorganisms [e.g., in a refrigerator and/or with small amounts of a toxic substance that does not react with proteins, such as **sodium azide** (NaN_3)].

D. *Assay of Proteins*

To purify any substance, some means must be found for quantitatively detecting its presence. A particular protein rarely comprises more than a few percent by weight of its tissue of origin and is usually present in much smaller amounts. Yet much of the material from which it is being extricated closely resembles the protein of interest. Accordingly, an assay must be specific for the protein being purified and highly sensitive to its presence. Furthermore, the assay must be convenient to use because it may be done repeatedly, often at every stage of the purification process.

Among the most straightforward of protein assays are those for enzymes that catalyze reactions with readily detectable products. Perhaps such a product has a characteristic spectroscopic absorption or fluorescence that can be monitored. Alternatively, the enzymatic reaction may consume or generate acid so that the enzyme can be assayed by acid–base titrations. If an enzymatic reaction product is not easily quantitated, its presence may still be revealed by further chemical treatment to yield a more readily observable product. Often, this takes the form of a **coupled enzymatic reaction,** in which the product of the enzyme being assayed is converted, by an added enzyme, to an observable substance.

Proteins that are not enzymes may be assayed through their ability to bind specific substances or through observation of their biological effects. For example, receptor proteins are often assayed by incubating them with a radioactive molecule that they specifically bind, passing the mixture through a protein-retaining filter, and then measuring the amount of radioactivity bound to the filter. The presence of a hormone may be revealed by its effect on some standard tissue sample or on a whole organism. The latter type of assays are usually rather lengthy procedures because the response elicited by the assay may take days to develop. In addition, their reproducibility is often less than satisfactory because of the complex behavior of living systems. Such assays are therefore used only when no alternative procedure is available.

a. Immunochemical Techniques Can Readily Detect Small Quantities of Specific Proteins

Immunochemical procedures provide protein assay techniques of high sensitivity and discrimination. These methods employ **antibodies,** proteins that are produced by an animal's immune system in response to the introduction of a foreign protein and that specifically bind to the foreign protein (antibodies and the immune system are discussed in Section 35-2).

Antibodies extracted from the blood serum of an animal that has been immunized against a particular protein are the products of many different antibody-producing cells. They therefore form a heterogeneous mixture of molecules, which vary in their exact specificities and binding affinities for their target protein. Antibody-producing cells normally die after a few cell divisions, so one of them cannot be cloned to produce a single species of antibody in useful quantities. Such **monoclonal antibodies** may be obtained, however, by fusing a cell producing the desired antibody with a cell of an immune system cancer known as a **myeloma** (Section 35-2Bd). The resulting **hybridoma** cell has an unlimited capacity to divide and, when raised in cell culture, produces large quantities of the monoclonal antibody.

A protein can be directly detected, or even isolated, through its precipitation by its corresponding antibodies. Alternatively, in a so-called **radioimmunoassay,** a protein can be indirectly detected by determining the degree with which it competes with a radioactively labeled standard for binding to the antibody (Section 19-1Aa). In an **enzyme-linked immunosorbent assay** (**ELISA;** Fig. 6-1):

1. An antibody against the protein of interest is immobilized on an inert solid such as polystyrene.

2. The solution being assayed for the protein is applied to the antibody-coated surface under conditions in which the antibody binds the protein. The unbound protein is then washed away.

3. The resulting protein–antibody complex is further reacted with a second protein-specific antibody to which an easily assayed enzyme has been covalently linked.

1 Immobilize first antibody on solid support

First antibody

Solid support

2 Incubate with protein-containing sample

Protein

3 Add a second antibody that is covalently linked to an assayable enzyme

4 Wash and assay the enzyme

Substrate — Detectable product   Substrate — Detectable product

Enzyme — Second antibody

Figure 6-1   An enzyme-linked immunosorbent assay (ELISA).
See the Animated Figures

4. After washing away any unbound antibody-linked enzyme, the enzyme in the immobilized antibody–protein–antibody–enzyme complex is assayed, thereby indicating the amount of the protein present.

Both radioimmunoassays and ELISAs are widely used to detect small amounts of specific proteins and other biological substances in both laboratory and clinical applications. For example, a commonly available pregnancy test, which is reliably positive within a few days post conception, uses an ELISA to detect the placental hormone **chorionic gonadotropin** (Section 19-1I) in the mother's urine.

E. General Strategy of Protein Purification

The fact that proteins are well-defined substances was not widely accepted until after 1926, when James Sumner first crystallized an enzyme, jack bean **urease.** Before that, it was thought that the high molecular masses of proteins resulted from a colloidal aggregation of rather ill-defined and mysterious substances of lower molecular mass. Once it was realized that it was possible, in principle, to purify proteins, work to do so began in earnest.

In the first half of the twentieth century, the protein purification methods available were extremely crude by today's standards. Protein purification was an arduous task that was as much an art as a science. Usually, the development of a satisfactory purification procedure for a given protein was a matter of years of labor ultimately involving huge quantities of starting material. Nevertheless, by 1940, ~20 enzymes had been obtained in pure form.

Since then, tens of thousands of proteins have been purified and characterized to varying extents. Modern techniques of separation have such a high degree of discrimination that one can now obtain, in quantity, a series of proteins with such similar properties that only a few years ago their mixture was thought to be a pure substance. Nevertheless, the development of an efficient procedure for the purification of a given protein may still be an intellectually challenging and time-consuming task.

Proteins are purified by fractionation procedures. In a series of independent steps, the various physicochemical properties of the protein of interest are utilized to separate it progressively from other substances. The idea here is not necessarily to minimize the loss of the desired protein, but to eliminate selectively the other components of the mixture so that only the required substance remains.

It may not be philosophically possible to prove that a substance is pure. However, *the operational criterion for establishing purity takes the form of the method of exhaustion: the demonstration, by all available methods, that the sample of interest consists of only one component.* Therefore, as new separation techniques are devised, standards of purity may have to be revised. Experience has shown that when a sample of material previously thought to be a pure substance is subjected to a new separation technique, it occasionally proves to be a mixture of several components.

The characteristics of proteins and other biomolecules that are utilized in the various separation procedures are solubility, ionic charge, polarity, molecular size, and binding specificity for other biological molecules. Some of the procedures we shall discuss and the protein characteristics they depend on are as follows:

Characteristic	Procedure
Solubility	1. Salting in 2. Salting out
Ionic Charge	1. Ion exchange chromatography 2. Electrophoresis 3. Isoelectric focusing
Polarity	1. Adsorption chromatography 2. Paper chromatography 3. Reverse-phase chromatography 4. Hydrophobic interaction chromatography
Molecular Size	1. Dialysis and ultrafiltration 2. Gel electrophoresis 3. Gel filtration chromatography 4. Ultracentrifugation
Binding Specificity	1. Affinity chromatography

In the remainder of this chapter, we discuss these separation procedures.

2 SOLUBILITIES OF PROTEINS

A protein's multiple acid–base groups make its solubility properties dependent on the concentrations of dissolved salts, the polarity of the solvent, the pH, and the temperature.

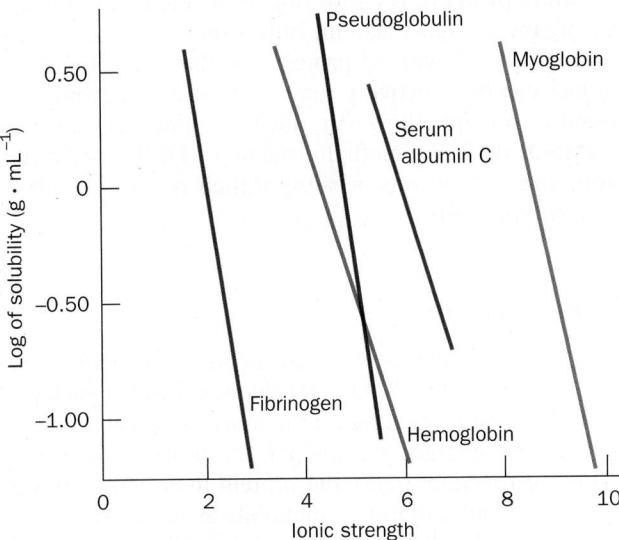

Figure 6-2 Solubilities of several proteins in ammonium sulfate solutions. [After Cohn, E.J. and Edsall, J.T., *Proteins, Amino Acids and Peptides, p.* 602, Academic Press (1943).]

Figure 6-3 Solubility of carboxy-hemoglobin at its isoelectric point as a function of ionic strength and ion type. Here *S* and *S'* are, respectively, the solubilities of the protein in the salt solution and in pure water. The logarithm of their ratios is plotted so that the solubility curves can be placed on a common scale. [After Green, A.A., *J. Biol. Chem.* **95,** 47 (1932).]

Different proteins vary greatly in their solubilities under a given set of conditions: Certain proteins precipitate from solution under conditions in which others remain quite soluble. This effect is routinely used as a basis for protein purification.

A. *Effects of Salt Concentrations*

The solubility of a protein in aqueous solution is a sensitive function of the concentrations of dissolved salts (Figs. 6-2 through 6-4). The salt concentration in Figs. 6-2 and 6-3 is expressed in terms of the **ionic strength,** *I,* which is defined

$$I = \frac{1}{2} \sum c_i Z_i^2 \qquad [6.1]$$

where c_i is the molar concentration of the *i*th ionic species and Z_i is its ionic charge. The use of this parameter to account for the effects of ionic charges results from theoretical considerations of ionic solutions. However, as Fig. 6-3 indicates, a protein's solubility at a given ionic strength varies with the types of ions in solution. The order of effectiveness of these various ions in influencing protein solubility is quite similar for different proteins and is apparently mainly due to the ions' size and hydration.

The solubility of a protein at low ionic strength generally increases with the salt concentration (left side of Fig. 6-3 and the different curves of Fig. 6-4). The explanation of this **salting in** phenomenon is that as the salt concentration of the protein solution increases, the additional counterions more effectively shield the protein molecules' multiple ionic charges and thereby increase the protein's solubility.

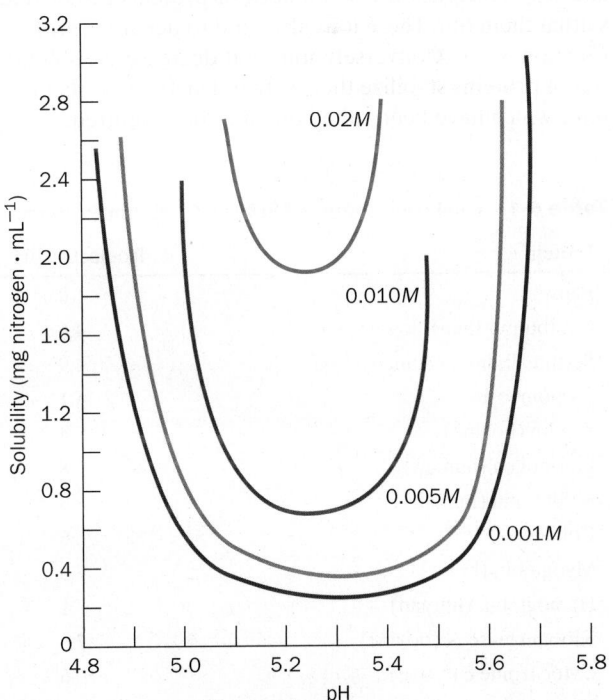

Figure 6-4 Solubility of β-lactoglobulin as a function of pH at several NaCl concentrations. [After Fox, S. and Foster, J.S., *Introduction to Protein Chemistry, p.* 242, Wiley (1975).]

At high ionic strengths, the solubilities of proteins, as well as those of most other substances, decrease. This effect, known as **salting out**, is primarily a result of the competition between the added salt ions and the other dissolved solutes for molecules of solvation. At high salt concentrations, so many of the added ions are solvated that the amount of bulk solvent available becomes insufficient to dissolve other solutes. In thermodynamic terms, the solvent's activity (effective concentration; Appendix to Chapter 3) is decreased. Hence, solute–solute interactions become stronger than solute–solvent interactions and the solute precipitates.

Salting out is the basis of one of the most commonly used protein purification procedures. Figure 6-2 shows that the solubilities of different proteins vary widely as a function of salt concentration. For example, at an ionic strength of 3, fibrinogen is much less soluble than the other proteins in Fig. 6-2. *By adjusting the salt concentration in a solution containing a mixture of proteins to just below the precipitation point of the protein to be purified, many unwanted proteins can be eliminated from the solution. Then, after the precipitate is removed by filtration or centrifugation, the salt concentration of the remaining solution is increased so as to precipitate the desired protein.* In this manner, a significant purification and concentration of large quantities of protein can be conveniently effected. Consequently, salting out is often the initial step in protein purification procedures. Ammonium sulfate is the most commonly used reagent for salting out proteins because its high solubility ($3.9M$ in water at $0°C$) permits the achievement of solutions with high ionic strengths (up to 23.4 in water at $0°C$).

Certain ions, notably I^-, ClO_4^-, SCN^-, Li^+, Mg^{2+}, Ca^{2+}, and Ba^{2+}, increase the solubilities of proteins rather than salting them out. These ions also tend to denature proteins (Section 8-4E). Conversely, ions that decrease the solubilities of proteins stabilize their native structures, so that proteins which have been salted out are not denatured.

Table 6-1 **Isoelectric Points of Several Common Proteins**

Protein	Isoelectric pH
Pepsin	<1.0
Ovalbumin (hen)	4.6
Serum albumin (human)	4.9
Tropomyosin	5.1
Insulin (bovine)	5.4
Fibrinogen (human)	5.8
γ-Globulin (human)	6.6
Collagen	6.6
Myoglobin (horse)	7.0
Hemoglobin (human)	7.1
Ribonuclease A (bovine)	7.8
Cytochrome *c* (horse)	10.6
Histone (bovine)	10.8
Lysozyme (hen)	11.0
Salmine (salmon)	12.1

B. Effects of Organic Solvents

Water-miscible organic solvents, such as acetone and ethanol, are generally good protein precipitants because their low dielectric constants reduce the solvating power of their aqueous solutions for dissolved ions such as proteins. The different solubilities of proteins in these mixed solvents form the basis of a useful fractionation technique. This procedure is normally used near $0°C$ or less because, at higher temperatures, organic solvents tend to denature proteins. The lowering of the dielectric constant by organic solvents also magnifies the differences in the salting out behavior of proteins, so that these two techniques can be effectively combined. Some water-miscible organic solvents, however, such as dimethyl sulfoxide (DMSO) or *N,N*-dimethylformamide (DMF), are rather good protein solvents because of their relatively high dielectric constants.

C. Effects of pH

Proteins generally bear numerous ionizable groups that have a variety of p*K*'s. At a pH characteristic for each protein, the positive charges on the molecule exactly balance its negative charges. At this pH, the protein's isoelectric point, p*I* (Section 4-1D), the protein molecule carries no net charge and is therefore immobile in an electric field.

Figure 6-4 indicates that the solubility of the protein β-lactoglobulin is a minimum near its p*I* of 5.2 in dilute NaCl solutions and increases more or less symmetrically about the p*I* with changes in pH. This solubility behavior, which is shared by most proteins, is easily explained. Physicochemical considerations suggest that the solubility properties of uncharged molecules are insensitive to the salt concentration. To a first approximation, therefore, a protein at its isoelectric point should not be subject to salting in. Conversely, as the pH is varied from a protein's p*I*, that is, as the protein's net charge increases, it should be increasingly subject to salting in because the electrostatic interactions between neighboring molecules that promote aggregation and precipitation should likewise increase. Hence, *in solutions of moderate salt concentrations, the solubility of a protein as a function of pH is expected to be at a minimum at the protein's pI and to increase about this point with respect to pH.*

Proteins vary in their amino acid compositions and therefore, as Table 6-1 indicates, in their p*I*'s. This phenomenon is the basis of a protein purification procedure known as **isoelectric precipitation** in which the pH of a protein mixture is adjusted to the p*I* of the protein to be isolated so as to selectively minimize its solubility. In practice, this technique is combined with salting out so that the protein being purified is usually salted out near its p*I*.

D. Crystallization

Once a protein has been brought to a reasonable state of purity, it may be possible to crystallize it. This is usually

(a)   (b)   (c)

(d)   (e)   (f)

Figure 6-5   Protein crystals. (*a*) Azurin from *Pseudomonas aeruginosa*, (*b*) flavodoxin from *Desulfovibrio vulgaris*, (*c*) rubredoxin from *Clostridium pasteurianum*, (*d*) azidomet myohemerythrin from the marine worm *Siphonosoma funafuti*, (*e*) lamprey hemoglobin, and (*f*) bacteriochlorophyll *a* protein from *Prosthecochloris aestuarii*. These proteins are colored because of their associated chromophores (light-absorbing groups); proteins are colorless in the absence of such bound groups. [Parts *a–c* courtesy of Larry Sieker, University of Washington; Parts *d* and *e* courtesy of Wayne Hendrickson, Columbia University; and Part *f* courtesy of John Olsen, Brookhaven National Laboratories, and Brian Matthews, University of Oregon.]

done by bringing the protein solution just past its saturation point with the types of precipitating agents discussed above. On standing for a time (as little as a few minutes, as much as several months), often while the concentration of the precipitating agent is being slowly increased, the protein may precipitate from the solution in crystalline form. It may be necessary to attempt the crystallization under different solution conditions and with various precipitating agents before crystals are obtained. The crystals may range in size from microscopic to 1 mm or more across. Crystals of the latter size, which generally require great care to grow, may be suitable for X-ray crystallographic analysis (Section 8-3A). Several such crystals are shown in Fig. 6-5.

3 CHROMATOGRAPHIC SEPARATIONS

In 1903, the Russian botanist Mikhail Tswett described the separation of plant leaf pigments in solution through the use of solid adsorbents. He named this process **chromatography** (Greek: *chroma*, color + *graphein*, to write), presumably because of the colored bands that formed in the adsorbents as the components of the pigment mixtures separated from one another (and possibly because Tswett means "color" in Russian).

Modern separation methods rely heavily on chromatographic procedures. In all of them, a mixture of substances

to be fractionated is dissolved in a liquid or gaseous fluid known as the **mobile phase.** The resultant solution is percolated through a column consisting of a porous solid matrix known as the **stationary phase,** which in certain types of chromatography may be associated with a bound liquid. The interactions of the individual solutes with the stationary phase act to retard their progress through the matrix in a manner that varies with the properties of each solute. If the mixture being fractionated starts its journey through the column in a narrow band, the different retarding forces on each component that cause them to migrate at different rates will eventually cause the mixture to separate into bands of pure substances.

The power of chromatography derives from the continuous nature of the separation processes. A single purification step (or "theoretical plate" as it is often termed in analogy with distillation processes) may have very little tendency to separate a mixture into its components. However, since this process is applied in a continuous fashion so that it is, in effect, repeated hundreds or even hundreds of thousands of times, the segregation of the mixture into its components ultimately occurs. The separated components can then be collected into separate fractions for analysis and/or further fractionation.

The various chromatographic methods are classified according to their mobile and stationary phases. For example, in gas–liquid chromatography the mobile and stationary phases are gaseous and liquid, respectively, whereas in liquid–liquid chromatography they are immiscible liquids, one of which is bound to an inert solid support. Chromatographic methods may be further classified according to the nature of the dominant interaction between the stationary phase and the substances being separated. For example, if the retarding force is ionic in character, the separation technique is referred to as **ion exchange chromatography,** whereas if it is a result of the adsorption of the solutes onto a solid stationary phase, it is known as **adsorption chromatography.**

As has been previously mentioned, a cell contains huge numbers of different components, many of which closely resemble one another in their various properties. Therefore, the isolation procedures for most biological substances incorporate a number of independent chromatographic steps in order to purify the substance of interest according to several criteria. In this section, the most commonly used of these chromatographic procedures are described.

A. *Ion Exchange Chromatography*

*In the process of **ion exchange,** ions that are electrostatically bound to an insoluble and chemically inert matrix are reversibly replaced by ions in solution.*

$$R^+A^- + B^- \rightleftharpoons R^+B^- + A^-$$

Here, R^+A^- is an **anion exchanger** in the A^- form and B^- represents anions in solution. **Cation exchangers** similarly bear negatively charged groups that reversibly bind

Fractions sequentially collected

Figure 6-6 **Ion exchange chromatography using stepwise elution.** Here the tan region of the column represents the ion exchanger and the colored bands represent the various proteins. (*a*) The protein mixture is bound to the topmost portion of the ion exchanger in the chromatography column. (*b*) As the elution progresses, the various proteins separate into discrete bands as a consequence of their different affinities for the ion exchanger under the prevailing solution conditions. Here the first band of protein (*red*) has passed through the column and is being isolated as a separate fraction, whereas the other, less mobile, bands remain near the top of the column. (*c*) The salt concentration in the elution buffer is increased to increase the mobility of and thus elute the remaining bands. (*d*) The elution diagram of the protein mixture from the column. See the Animated Figures

cations. Polyanions and polycations therefore bind to anion and cation exchangers, respectively. However, proteins and other **polyelectrolytes** (polyionic polymers) that bear both positive and negative charges can bind to both cation and anion exchangers depending on their net charge. *The affinity with which a particular polyelectrolyte binds to a given ion exchanger depends on the identities and concentrations of the other ions in solution because of the competition among these various ions for the binding sites on the ion exchanger. The binding affinities of polyelectrolytes bearing acid–base groups are also highly pH dependent because of the variation of their net charges with pH.* These principles are used to great advantage in isolating biological molecules by **ion exchange chromatography** (Fig. 6-6), as described below.

In purifying a given protein (or some other polyelectrolyte), the pH and the salt concentration of the buffer solution in which the protein is dissolved are chosen so that the desired protein is strongly bound to the selected ion ex-

changer. A small volume of the impure protein solution is applied to the top of a column in which the ion exchanger has been packed, and the column is washed with this buffer solution.

Various proteins bind to the ion exchanger with different affinities. As the column is washed with the buffer, a process known as **elution,** *those proteins with relatively low affinities for the ion exchanger move through the column faster than the proteins that bind to the ion exchanger with higher affinities.* This occurs because the progress of a given protein through the column is retarded relative to that of the solvent due to interactions between the protein molecules and the ion exchanger.

The greater the binding affinity of a protein for the ion exchanger, the more it will be retarded. Thus, proteins that bind tightly to the ion exchanger can be eluted by changing the elution buffer to one with a higher salt concentration (and/or a different pH), a process called **stepwise elution.**

With the use of a fraction collector, purification of a substance can be effected by selecting only those fractions of the column effluent that contain it. Chromatographically separated materials may be detected in a variety of ways. The contents of the column effluent may be directly monitored through column-mounted detectors according to its UV absorbance at a specific wavelength [often 280 nm for proteins (because the aromatic side chains of Phe, Trp, and Tyr have strong absorbances at this wavelength; Section 9-1Cb) and 260 nm for nucleic acids (their absorption maximum; Fig. 5-15*b*)], its fluorescence, its radioactivity, its refractive index, its pH, or its electrical conductivity. These properties may also be measured for the individual column fractions after the chromatographic run has been completed. In addition, biomolecules may be detected through their enzymatic and biological activities, as is discussed in Section 6-1D.

a. Gradient Elution Improves Chromatographic Separations

The purification process can be further improved by washing the protein-loaded column using the method of **gradient elution.** Here the salt concentration and/or pH is continuously varied as the column is eluted so as to release sequentially the various proteins that are bound to the ion exchanger. This procedure generally leads to a better separation of proteins than does elution of the column by a single solution or stepwise elution.

Many different types of elution gradients have been successfully employed in purifying biological molecules. The most widely used of these is the **linear gradient,** in which the concentration of the eluant solution varies linearly with the volume of solution passed. A simple device for generating such a gradient is illustrated in Fig. 6-7. Here the solute concentration, *c,* in the solution being withdrawn from the mixing chamber, is expressed by

$$c = c_2 - (c_2 - c_1)f \qquad [6.2]$$

where c_1 is the solution's initial concentration in the mixing chamber, c_2 is its concentration in the reservoir chamber, and *f* is the remaining fraction of the combined volumes of the solutions initially present in both reservoirs. Linear gradients of increasing salt concentration are probably more commonly used than all other means of column elution. However, gradients of different shapes can be generated by using two or more chambers of different cross-sectional areas or programmed mixing devices.

Figure 6-7   Device for generating a linear concentration gradient. Two connected open chambers, which have identical cross-sectional areas, are initially filled with equal volumes of solutions of different concentrations. As the solution of concentration c_1 drains out of the mixing chamber, it is partially replaced by a solution of concentration c_2 from the reservoir chamber. The concentration of the solution in the mixing chamber varies linearly from its initial concentration, c_1, to the final concentration, c_2, as is expressed by Eq. [6.2].

b. Several Types of Ion Exchangers Are Available

Ion exchangers consist of charged groups covalently attached to a support matrix. The chemical nature of the charged groups determines the types of ions that bind to the ion exchanger and the strength with which they bind. The chemical and mechanical properties of the support matrix govern the flow characteristics, ion accessibility, and stability of the ion exchanger.

Several classes of materials, colloquially referred to as **resins,** are in general use as support matrices for ion exchangers in protein purification, including cellulose (Fig. 6-8), polystyrene, agarose gels, and cross-linked dextran gels (Section 6-3Bb). Table 6-2 contains descriptions of some commercially available ion exchangers in common use.

Cellulosic ion exchangers are among the materials most commonly employed to separate biological molecules. The cellulose, which is derived from wood or cotton, is lightly derivatized with ionic groups to form the ion exchanger. The most often used cellulosic anion exchanger is **diethylaminoethyl (DEAE)-cellulose,** whereas **carboxymethyl**

DEAE: $R = -CH_2-CH_2-\overset{+}{N}H(CH_2CH_3)_2$
CM: $R = -CH_2-COO^-$

Figure 6-8   Molecular formulas of cellulose-based ion exchangers.

Table 6-2 **Some Biochemically Useful Ion Exchangers**

Name[a]	Type	Ionizable Group	Remarks
DEAE-cellulose	Weakly basic	Diethylaminoethyl $-CH_2CH_2N(C_2H_5)_2$	Used to separate acidic and neutral proteins
CM-cellulose	Weakly acidic	Carboxymethyl $-CH_2COOH$	Used to separate basic and neutral proteins
P-cellulose	Strongly and weakly acidic	Phosphate $-OPO_3H_2$	Dibasic; binds basic proteins strongly
Bio-Rex 70	Weakly acidic, polystyrene-based	Carboxylic acid $-COOH$	Used to separate basic proteins and amines
DEAE-Sephadex	Weakly basic cross-linked dextran gel	Diethylaminoethyl $-CH_2CH_2N(C_2H_5)_2$	Combined chromatography and gel filtration of acidic and neutral proteins
SP-Sepharose	Strongly acidic cross-linked agarose gel	Methyl sulfonate $-CH_2SO_3H$	Combined chromatography and gel filtration of basic proteins

[a]Sephadex and Sepharose are products of GE Healthcare; Bio-Rex resins are products of BioRad Laboratories.

(CM)-cellulose is the most popular cellulosic cation exchanger (Fig. 6-8).

Gel-type ion exchangers can have the same sorts of charged groups as do cellulosic ion exchangers. The advantage of gel-type ion exchangers is that they combine the separation properties of gel filtration (Section 6-3B) with those of ion exchange. Because of their high degree of substitution of charged groups, which results from their porous structures, these gels have a higher loading capacity than do cellulosic ion exchangers.

One disadvantage of cellulosic and gel-type matrices is that they are easily compressed (usually by the high pressures resulting from attempts to increase the eluant flow rate), thereby greatly reducing eluant flow. This problem has been alleviated by the fabrication of noncompressible matrices such as derivatized **silica**

Silica

or coated glass beads. Such materials allow very high flow rates and pressures, even when they are very finely powdered, and hence permit more effective chromatographic separations (see HPLC in Section 6-3Dh).

B. Gel Filtration Chromatography

In **gel filtration chromatography,** *which is also called* **size exclusion** *and* **molecular sieve chromatography,** *molecules are separated according to their size and shape.* The stationary phase in this technique consists of beads of a hydrated, spongelike material containing pores that span a relatively narrow size range of molecular dimensions. If an aqueous

solution containing molecules of various sizes is passed through a column containing such "molecular sieves," the molecules that are too large to pass through the pores are excluded from the solvent volume inside the gel beads. These larger molecules therefore traverse the column more rapidly, that is, in a smaller eluant volume, than the molecules that pass through the pores (Fig. 6-9).

The molecular mass of the smallest molecule unable to penetrate the pores of a given gel is said to be the gel's **exclusion limit.** This quantity is to some extent a function of molecular shape because elongated molecules, as a consequence of their higher radius of hydration, are less likely to penetrate a given gel pore than spherical molecules of the same molecular volume.

The behavior of a molecule on a particular gel column can be quantitatively characterized. If V_x is the volume occupied by the gel beads and V_0, the **void volume,** is the volume of the solvent space surrounding the beads, then V_t, the total **bed volume** of the column, is simply their sum:

$$V_t = V_x + V_0 \qquad [6.3]$$

V_0 is typically ~35% of V_t.

The **elution volume** of a given solute, V_e, is the volume of solvent required to elute the solute from the column after it has first contacted the gel. The void volume of a column is easily measured as the elution volume of a solute whose molecular mass is larger than the exclusion limit of the gel. The behavior of a particular solute on a given gel is therefore characterized by the ratio V_e/V_0, the **relative elution volume,** a quantity that is independent of the size of the particular column used.

Molecules with molecular masses ranging below the exclusion limit of a gel will elute from the gel in the order of their molecular masses, with the largest eluting first. This is because the pore sizes in any gel vary over a limited range, so that larger molecules have less of the gel's interior volume available to them than do smaller molecules. This effect is the basis of gel filtration chromatography.

Figure 6-9   Gel filtration chromatography. (*a*) A gel bead, whose periphery is represented by a dashed line, consists of a gel matrix (*wavy solid lines*) that encloses an internal solvent space. Smaller molecules (*red dots*) can freely enter the internal solvent space of the gel bead from the external solvent space. However, larger molecules (*blue dots*) are too large to penetrate the gel pores. (*b*) The sample solution begins to enter the gel column (in which the gel beads are now represented by brown spheres).

(*c*) The smaller molecules can penetrate the gel and consequently migrate through the column more slowly than the larger molecules that are excluded from the gel. (*d*) The larger molecules emerge from the column to be collected separately from the smaller molecules, which require additional solvent for elution from the column. (*e*) The elution diagram of the chromatogram indicating the complete separation of the two components, with the larger component eluting first. 🐚 **See the Animated Figures**

a. Gel Filtration Chromatography Can Be Used to Estimate Molecular Masses

There is a linear relationship between the relative elution volume of a substance and the logarithm of its molecular mass over a considerable molecular mass range (Fig. 6-10). If a plot such as Fig. 6-10 is made for a particular gel filtration column using macromolecules of known molecular masses, *the molecular mass of an unknown substance can be estimated from its position on the plot. The precision of this technique is limited by the accuracy of the underlying assumption that the known and unknown macromolecules have identical shapes.* Nevertheless, gel filtration chromatography is often used to estimate molecular masses because it can be applied to quite impure samples (providing that the molecule of interest can be identified) and because it can be rapidly carried out using simple equipment.

b. Most Gels Are Made from Dextran, Agarose, or Polyacrylamide

The most commonly used materials for making chromatographic gels are **dextran** (a high molecular mass polymer of

glucose produced by the bacterium *Leuconostoc mesenteroides*), **agarose** (a linear polymer of alternating D-galactose and 3,6-anhydro-L-galactose from red algae), and **polyacrylamide** (Section 6-4B). The properties of several gels that are commonly employed in separating biological molecules are listed in Table 6-3. The porosity of dextran-based gels, sold under the trade name Sephadex, is controlled by the molecular mass of the dextran used and the introduction of glyceryl ether units that cross-link the hydroxyl groups of the polyglucose chains. The several classes of Sephadex that are available have exclusion limits between 0.7 and 600 kD. The pore size in polyacrylamide gels is similarly controlled by the extent of cross-linking of neighboring polyacrylamide molecules (Section 6-4B). They are commercially available under the trade name of Bio-Gel P and have exclusion limits between 0.2 and 400 kD. Very large molecules and supramolecular assemblies can be separated using agarose gels, sold under the trade names Sepharose and Bio-Gel A, which have exclusion limits ranging up to 150,000 kD.

Gel filtration is often used to "desalt" a protein solution. For example, an ammonium sulfate–precipitated protein

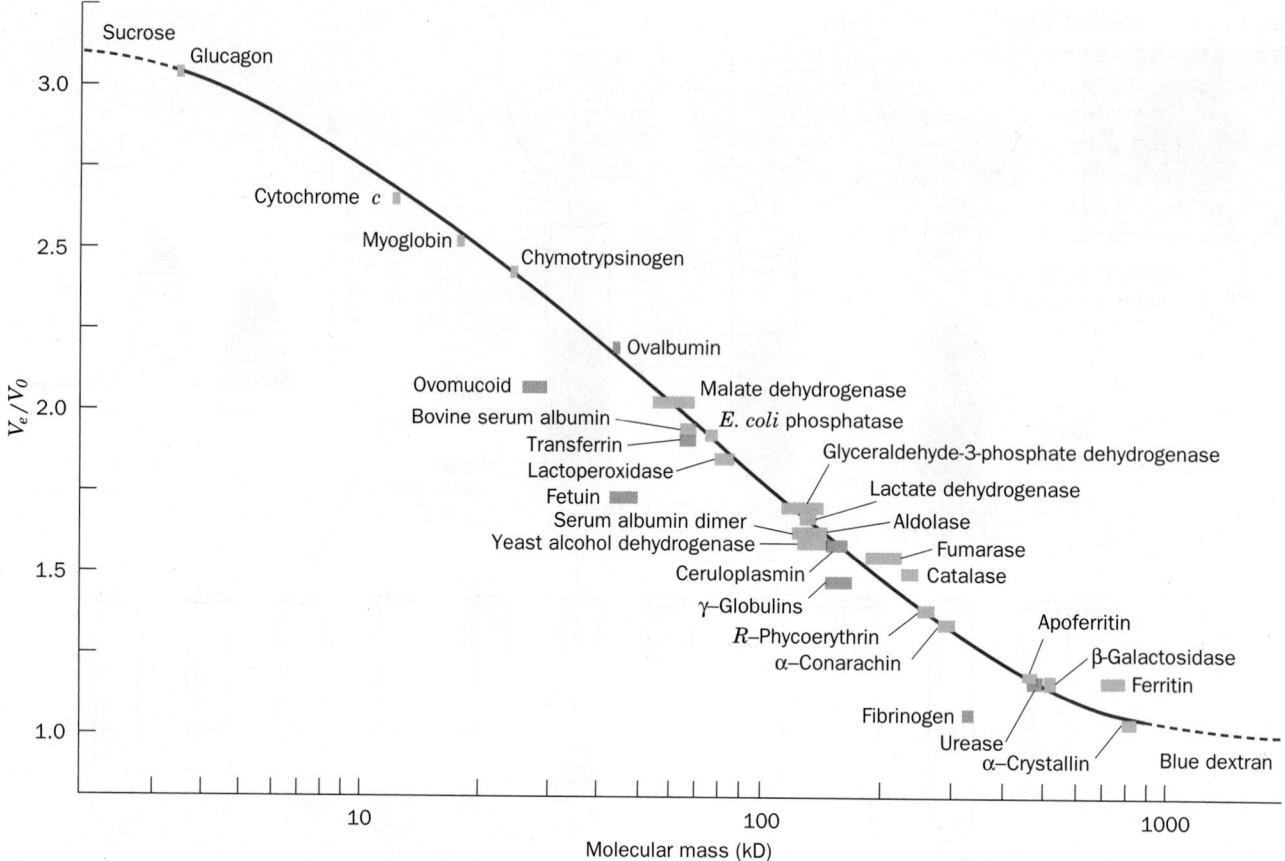

Figure 6-10 **Molecular mass determination by gel filtration chromatography.** The graph shows the relative elution volume versus the logarithm of molecular mass for a variety of proteins on a cross-linked dextran column (Sephadex G-200) at pH 7.5. Orange bars represent glycoproteins (proteins with attached carbohydrate groups). [After Andrews, P., *Biochem. J.* **96,** 597 (1965).]

Table 6-3 **Some Commonly Used Gel Filtration Materials**

Name[a]	Type	Fractionation Range (kD)
Sephadex G-10	Dextran	0.05–0.7
Sephadex G-25	Dextran	1–5
Sephadex G-50	Dextran	1–30
Sephadex G-100	Dextran	4–150
Sephadex G-200	Dextran	5–600
Sephacryl S-100	Dextran, cross-linked	1–100
Sephacryl S-200	Dextran, cross-linked	5–250
Sephacryl S-300	Dextran, cross-linked	4–150
Sephacryl S-400	Dextran, cross-linked	20–8000
Bio-Gel P-2	Polyacrylamide	0.1–1.8
Bio-Gel P-6	Polyacrylamide	1–6
Bio-Gel P-10	Polyacrylamide	1.5–20
Bio-Gel P-30	Polyacrylamide	2.5–40
Bio-Gel P-100	Polyacrylamide	5–100
Sepharose 6B	Agarose	10–4,000
Sepharose 4B	Agarose	60–20,000
Sepharose 2B	Agarose	70–40,000

[a]Sephadex, Sephacryl, and Sepharose are products of GE Healthcare; Bio-Gel gels are products of BioRad Laboratories.

can be easily freed of ammonium sulfate by dissolving the protein precipitate in a minimum volume of suitable buffer and applying this solution to a column of gel with an exclusion limit less than the molecular mass of the protein. On elution of the column with buffer, the protein will precede the ammonium sulfate through the column.

Dextran and agarose gels can be derivatized with ionizable groups such as DEAE and CM to form ion exchange gels (Section 6-3Ab). Substances that are chromatographed on these gels are therefore subject to separation according to their ionic charges as well as their sizes and shapes.

c. Dialysis Is a Form of Molecular Filtration

Dialysis is a process that separates molecules according to size through the use of semipermeable membranes containing pores of less than macromolecular dimensions. These pores allow small molecules, such as those of solvents, salts, and small metabolites, to diffuse across the membrane but block the passage of larger molecules. **Cellophane** (cellulose acetate) is the most commonly used dialysis material, although several other substances such as cellulose and **collodion** are similarly employed. These are available in a wide variety of **molecular weight cutoff** values (the size of the smallest particle that cannot penetrate the membrane) that range from 0.1 to 500 kD.

Dialysis (which is not considered to be a form of chromatography) is used to change the solvent in which macromolecules are dissolved. A macromolecular solution is sealed inside a dialysis bag (usually made by knotting dialysis membrane tubing at both ends), which is immersed in a relatively large volume of the new solvent (Fig. 6-11*a*). After several hours of stirring, the solutions will have equilibrated, but with the macromolecules remaining inside the dialysis bag (Fig. 6-11*b*). The process can be repeated several times to replace one solvent system completely by another.

Dialysis has been largely supplanted by a related technique known as **ultrafiltration** in which a macromolecular solution is forced, under pressure or by centrifugation, through a semipermeable membranous disk, which can be made from a variety of materials including cellulose acetate, nylon, and polyvinylidene fluoride (PVDF). Solvent and small solutes pass through the membrane, leaving behind a more concentrated macromolecular solution. Ultrafiltration can thus be used to desalt a macromolecular solution. Since ultrafiltration membranes with different pore sizes are available, ultrafiltration can also be used to separate different-sized macromolecules.

Solvent may also be removed from a sample solution through **lyophilization** (freeze-drying), a process in which the solution is frozen and the solvent sublimed away under vacuum. Lyophilization is usually used to prepare biological materials for long-term storage or transport.

C. *Affinity Chromatography*

A striking characteristic of many proteins is their ability to bind specific molecules tightly but noncovalently. This property can be used to purify such proteins by **affinity chromatography** (Fig. 6-12). In this technique, a molecule,

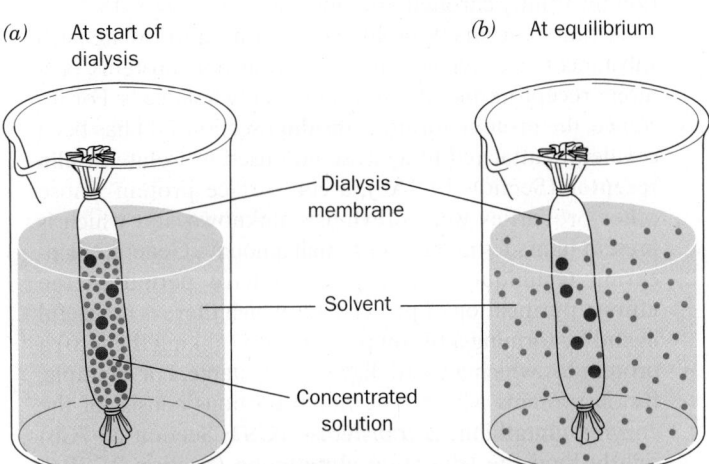

(a) At start of dialysis
(b) At equilibrium

Dialysis membrane

Solvent

Concentrated solution

Figure 6-11 **Use of dialysis to separate small and large molecules.** (*a*) Only small molecules can diffuse through the pores in the bag, which is shown here as a tube knotted at both ends. (*b*) At equilibrium, the concentrations of small molecules are nearly the same inside and outside the bag, whereas the macromolecules remain in the bag.

Solid resin matrix

Specific binding of molecule to matrix ligand

Matrix-anchored ligand

Macromolecules with differing ligand-binding sites

Figure 6-12 **Affinity chromatography.** A ligand (*yellow*) is covalently anchored to a porous matrix. The sample mixture (whose ligand-binding sites are represented by the cutout squares, semicircles, and triangles) is passed through the column. Only certain molecules (represented by orange circles) specifically bind to the ligand; the others are washed through the column.

known as a **ligand** (in analogy with the ligands of coordination compounds), which specifically binds to the protein of interest, is covalently attached to an inert and porous matrix. *When an impure protein solution is passed through this chromatographic material, the desired protein binds to the immobilized ligand, whereas other substances are washed through the column with the buffer. The desired protein can then be recovered in highly purified form by changing the elution conditions such that the protein is released from the chromatographic matrix.* The great advantage of affinity chromatography is its ability to exploit the desired protein's unique biochemical properties rather than the small differences in physicochemical properties between proteins that other chromatographic methods must utilize.

The chromatographic matrix in affinity chromatography must be chemically inert, have high porosity, and have large numbers of functional groups capable of forming covalent linkages to ligands. Of the few materials available that meet these criteria, agarose, which has numerous free hydroxyl groups, is by far the most widely used. If the ligand has a primary amino group that is not essential for its binding to the protein of interest, the ligand can be covalently linked to the agarose in a two-step process (Fig. 6-13):

Figure 6-13   Covalent linking of ligand to agarose. The formation of cyanogen bromide–activated agarose (*top*) and its reaction with a primary amine to form a covalently attached ligand for affinity chromatography (*bottom*).

1. Agarose is reacted with **cyanogen bromide** to form an "activated" but stable intermediate (which is commercially available).

2. Ligand reacts with the activated agarose to form covalently bound product.

Many proteins are unable to bind their cyanogen bromide–coupled ligands due to steric interference with the agarose matrix. This problem is alleviated by attaching the ligand to the agarose by a flexible "spacer" group. This is conveniently done through the use of commercially available activated resins. One such resin is "epoxy-activated" agarose, in which a spacer group (containing, e.g., a chain of 12 atoms) links the resin to a reactive epoxy group. The epoxy group can react with many of the nucleophilic groups on ligands, thereby permitting the ligand of choice to be covalently linked to the agarose via a tether of defined length (Fig. 6-14).

The ligand used in the affinity chromatography isolation of a particular protein must have an affinity for the protein high enough to immobilize it on the agarose gel but not so high as to prevent its subsequent release. If the ligand is a substrate for an enzyme being isolated, the chromatography conditions must be such that the enzyme does not function catalytically or the ligand will be destroyed.

After a protein has been bound to an affinity chromatography column and washed free of impurities, it must be released from the column. One method of doing so is to elute the column with a solution of a compound that has higher affinity for the protein-binding site than the bound ligand. Another is to alter the solution conditions such that the protein–ligand complex is no longer stable, for example, by changes in pH, ionic strength, and/or temperature. However, care must be taken that the solution conditions are not so inhospitable to the protein being isolated that it is irreversibly damaged. An example of protein purification by affinity chromatography is shown in Fig. 6-15.

Affinity chromatography has been used to isolate such substances as enzymes, antibodies, transport proteins, hormone receptors, membranes, and even whole cells. For instance, the protein hormone **insulin** (Section 7-1) has been covalently attached to agarose and used to isolate **insulin receptor** (Section 19-3Ac), a cell-surface protein whose other properties were previously unknown and which is present in tissues in only very small amounts. Genetic engineering techniques (Section 5-5G) have permitted the affinity purification of proteins for which there is no useful ligand by forming a fusion protein with (linking them to) a protein for which a useful ligand is available. For example, fusion proteins whose N-terminal portions consist of the enzyme **glutathione-S-transferase (GST;** Section 25-7Cb) tightly bind the tripeptide **glutathione** (Section 21-2Ba) and hence are readily purified by affinity chromatography on glutathione–agarose.

The separation power of affinity chromatography for a specific protein is often far greater than that of other chromatographic techniques. Indeed, the replacement of many chromatographic steps in a tried-and-true protein isolation

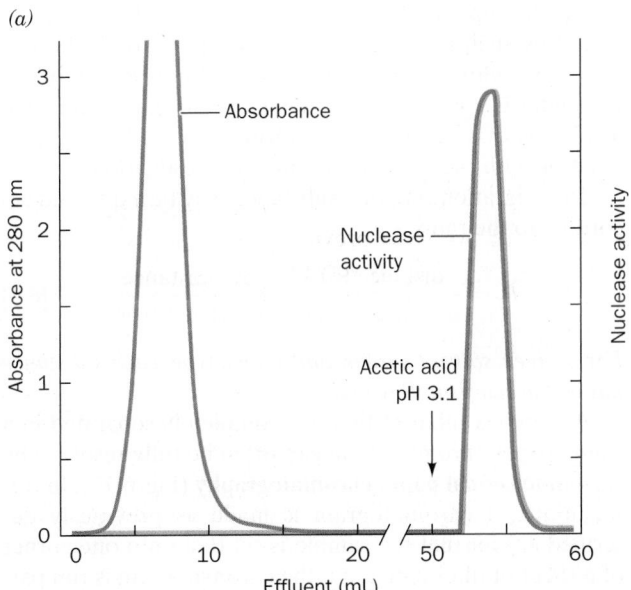

Figure 6-14   Derivatization of epoxy-activated agarose. Various types of nucleophilic groups can be covalently attached to epoxy-activated agarose via reaction with its epoxide groups.

R = Ligand

protocol by a single affinity chromatographic step often results in purer protein in higher yield.

a. Immunoaffinity Chromatography Employs the Binding Specificity of Monoclonal Antibodies

A melding of immunochemistry with affinity chromatography has generated a powerful method for purifying biological molecules. Cross-linking monoclonal antibodies (Section 6-1Da) to a suitable column material yields a substance that will bind only the protein against which the antibody has been raised. Such **immunoaffinity chromatography** can achieve a 10,000-fold purification in a single step. Disadvantages of immunoaffinity chroma-

tography include the technical difficulty of producing monoclonal antibodies and the harsh conditions that are often required to elute the bound protein.

D. *Other Chromatographic Techniques*

A number of other chromatographic techniques are of biochemical value. These are briefly discussed below.

a. Adsorption Chromatography Separates Nonpolar Substances

In **adsorption chromatography** (the original chromatographic method), molecules are physically adsorbed on the

(a)

(b)

Figure 6-15 (*a*) The purification of **staphylococcal nuclease** (a DNA-hydrolyzing enzyme) by affinity chromatography. The compound shown in Part *b*, whose bisphosphothymidine moiety specifically binds to the enzyme, was covalently linked to cyanogen bromide–activated agarose. The column was equilibrated with $0.05M$ borate buffer, pH 8.0, containing $0.01M$ $CaCl_2$, and approximately 40 mg of partially purified material was applied to the column. After 50 mL of buffer had been passed through the column to wash away the unbound material, $0.1M$ acetic acid was added to elute the enzyme. All of the original enzymatic activity, comprising 8.2 mg of pure nuclease, was recovered. [After Cuatrecasas, P., Wilchek, M., and Anfinsen, C.B., *Proc. Natl. Acad. Sci.* **61,** 636 (1968).]

surface of an insoluble substance such as **alumina** (Al_2O_3), charcoal, **diatomaceous earth** (also called **kieselguhr,** the siliceous fossils of unicellular organisms known as diatoms), finely powdered sucrose, or **silica gel** (silicic acid), through van der Waals and hydrogen bonding associations. The molecules are then eluted from the column by a pure solvent such as chloroform, hexane, or ethyl ether or by a mixture of such solvents. The separation process is based on the partition of the various substances between the polar column material and the nonpolar solvent. This procedure is most often used to separate nonpolar molecules rather than proteins.

b. Hydroxyapatite Chromatography Separates Proteins

Proteins are adsorbed by gels of crystalline **hydroxyapatite,** an insoluble form of calcium phosphate with empirical formula $Ca_5(PO_4)_3OH$. The separation of the proteins occurs on gradient elution of the column with phosphate buffer (the presence of other anions is unimportant). The physicochemical basis of this fractionation procedure is not fully understood but apparently involves the adsorption of anions to the Ca^{2+} sites and cations to the PO_4^{3-} sites of the hydroxyapatite crystalline lattice.

c. Paper Chromatography Separates Small Polar Molecules

Paper chromatography, developed in 1941 by Archer Martin and Richard Synge, played an indispensable role in biochemical analysis due to its ability to efficiently separate small molecules such as amino acids, oligopeptides, nucleotides, and oligonucleotides and its requirement for only the simplest of equipment. Although paper chromatography has been supplanted by the more modern techniques discussed in this chapter, we briefly describe it here because of its historical importance and because many of its principles and ancillary techniques are directly applicable to the more modern techniques.

In paper chromatography (Fig. 6-16), a few drops of solution containing a mixture of the components to be separated are applied (spotted) ~2 cm above one end of a strip of filter paper. After drying, that end of the paper is dipped into a solvent mixture consisting of aqueous and organic components; for example, water/butanol/acetic acid in a 4:5:1 ratio, 77% aqueous ethanol, or 6:7:7 water/*t*-amyl alcohol/pyridine. The paper should also be in contact with the equilibrium vapors of the solvent. The solvent soaks into the paper by capillary action because of the fibrous nature of the paper. The aqueous component of the solvent binds to the cellulose of the paper and thereby forms a stationary gel-like phase with it. The organic component of the solvent continues migrating, thus forming the mobile phase.

The rates of migration of the various substances being separated are governed by their relative solubilities in the polar stationary phase and the nonpolar mobile phase. In a single step of the separation process, a given solute is distributed between the mobile and stationary phases accord-

Ascending paper chromatography

Glass jar

Clip

Support rod

Paper

Solvent front

Sample

Solvent for development

Figure 6-16   **Experimental arrangement for paper chromatography.**

ing to its **partition coefficient,** an equilibrium constant defined as

$$K_p = \frac{\text{concentration in stationary phase}}{\text{concentration in mobile phase}} \qquad [6.4]$$

The molecules are therefore separated according to their polarities, with nonpolar molecules moving faster than polar ones.

After the solvent front has migrated an appropriate distance, the **chromatogram** is removed from the solvent and dried. The separated materials, if not colored, may be detected by such means as their radioactivity, their fluorescence or ability to quench the normal fluorescence of paper under UV light, or by spraying the chromatogram with a solution of a substance that forms a colored product on reaction with the substance(s) under investigation.

The migration rate of a substance may be expressed according to the ratio

$$R_f = \frac{\text{distance traveled by substance}}{\text{distance traveled by solvent front}} \qquad [6.5]$$

For a given solvent system and paper type, each substance has a characteristic R_f value.

A complex mixture that is incompletely separated in a single paper chromatogram can often be fully resolved by **two-dimensional paper chromatography** (Fig. 6-17). In this technique, a chromatogram is made as previously described except that the sample is spotted onto one corner of a sheet of filter paper and the chromatogram is run parallel to an edge of the paper. After the chromatography has been completed and the paper dried, the chromatogram is

Origin

Direction of flow of first solvent

Separation if only second solvent is used

Direction of flow of second solvent

Separation if both solvents are used sequentially

Separation if only first solvent is used

Figure 6-17 **Two-dimensional paper chromatography.**

rotated 90° and is chromatographed parallel to the second edge using another solvent system. Since each compound migrates at a characteristic rate in a given solvent system, the second chromatographic step should greatly enhance the separation of the mixture into its components.

d. Thin Layer Chromatography Is Used to Separate Organic Molecules

In **thin layer chromatography (TLC),** a thin (~0.25 mm) coating of a solid material spread on a glass or plastic plate is utilized in a manner similar to that of the paper in paper chromatography. In the case of TLC, however, the chromatographic material can be a variety of substances such as ion exchangers, gel filtration agents, and physical adsorbents. According to the choice of solvent for the mobile phase, the separation may be based on adsorption, partition, gel filtration, ion exchange processes, or some combination of these. The advantages of thin layer chromatography in convenience, rapidity, and high resolution have led to its routine use in the analysis of organic molecules.

e. Reverse-Phase Chromatography Separates Nonpolar Substances Including Denatured Proteins

Reverse-phase chromatography (RPC) is a form of liquid–liquid partition chromatography in which the polar character of the phases is reversed relative to that of paper chromatography: The stationary phase typically consists of a nonpolar liquid immobilized on silica substituted with *n*-alkyl chains such as C_8 and C_{18}, and the mobile phase is a more polar liquid. Reverse-phase chromatography was first developed to separate mixtures of nonpolar substances such as lipids but has also been found to be effective in separating polar substances such as oligonucleotides and proteins, provided that they have exposed nonpolar areas. Although nonpolar side chains tend to inhabit the water-free interiors of native proteins (Section 8-3Bb),

denaturation results in the exposure of these side chains to the solvent. Even when the protein is still in the native state, a significant fraction of these hydrophobic groups are at least partially exposed to the solvent at the protein surface. Consequently, under suitable conditions, proteins hydrophobically interact with the nonpolar groups on an immobilized matrix. The hydrophobic interactions in RPC are strong, so the eluting mobile phase must be highly non-polar (containing high concentrations of organic solvents such as acetonitrile) to dislodge adsorbed substances from the stationary phase. RPC therefore usually denatures proteins.

f. Hydrophobic Interaction Chromatography Separates Native Proteins on the Basis of Surface Hydrophobicity

Hydrophobic interactions form the basis not only of RPC but of **hydrophobic interaction chromatography (HIC).** However, whereas the stationary phase in RPC is strongly hydrophobic in character, often resulting in protein denaturation, in HIC it is a hydrophilic substance, such as an agarose gel, that is only lightly substituted with hydrophobic groups, usually octyl or phenyl residues. The resulting hydrophobic interactions in HIC are therefore relatively weak, so proteins maintain their native structures. The eluants in HIC, whose gradients must progressively reduce these weak hydrophobic interactions, are aqueous buffers with, for example, decreasing salt concentrations (hydrophobic interactions are strengthened by increased ionic strength; Section 6-2A), increasing concentrations of detergents, or increasing pH. Thus, HIC separates native proteins according to their degree of surface hydrophobicity, a criterion that differs from those on which other types of chromatography are based.

g. Metal Chelation Affinity Chromatography Separates Proteins with Metal-Chelating Groups

In **metal chelation affinity chromatography,** a divalent metal ion such as Mn^{2+}, Zn^{2+}, or Ni^{2+} is attached to a chromatographic matrix such as agarose beads covalently linked to metal-chelating groups under conditions that proteins bearing metal-chelating groups (e.g., multiple His or Cys side chains) are retained. Recombinant DNA techniques (Section 5-5G) can be used to append a segment of six consecutive His residues, known as a **His Tag,** to the N- or C-terminus of the polypeptide to be isolated. This creates a metal ion-binding site that allows the recombinant protein to be purified by metal chelation affinity chromatography. After the protein has been eluted, usually by altering the pH, the His-Tag can be removed by the action of a specific protease whose recognition sequence cleaves the $(His)_6$ sequence from the rest of the protein.

h. HPLC Has Permitted Greatly Improved Separations

In **high-performance liquid chromatography (HPLC),** a separation may be based on adsorption, ion exchange, size exclusion, HIC, or RPC as previously described. The separations are greatly improved, however, through the use of high-resolution columns, and the column retention times are much reduced. The narrow and relatively long columns are packed with a noncompressible matrix of fine (1–10 μm in diameter) silica beads, whose available hydroxyl groups can be derivatized with many of the commonly used functional groups of ion exchange chromatography, RPC, HIC, or affinity chromatography. Alternatively, glass or plastic beads may be coated with a thin layer of the stationary phase. The mobile phase is one of the solvent systems previously discussed, including gradient elutions with binary or even ternary mixtures. In the case of HPLC, however, the mobile phase is forced through the tightly packed column at pressures of up to 15,000 psi (pounds per square inch), leading to greatly reduced analysis times. The elutants are detected as they leave the column according to their UV absorption, refractive index, or fluorescence.

The advantages of HPLC are

1. Its high resolution, which permits the routine purification of mixtures that have defied separation by other techniques.

2. Its speed, which permits separations to be accomplished in as little as a few minutes.

3. Its high sensitivity, which, in favorable cases, permits the quantitative estimation of less than picomole quantities of materials.

4. Its capacity for automation.

Thus, few biochemistry laboratories now function without access to at least one HPLC system. HPLC is also often utilized in the clinical analyses of body fluids because it can rapidly, routinely, and automatically yield reliable quantitative estimates of nanogram quantities of biological materials such as vitamins, steroids, lipids, and drug metabolites.

4 ELECTROPHORESIS

Electrophoresis, the migration of ions in an electric field, is widely used for the analytical separation of biological molecules. The laws of electrostatics state that the electrical force, $F_{electric}$, on an ion with charge q in an electric field of strength E is expressed by

$$F_{electric} = qE \qquad [6.6]$$

The resulting electrophoretic migration of the ion through the solution is opposed by a frictional force

$$F_{friction} = vf \qquad [6.7]$$

where v is the rate of migration (velocity) of the ion and f is its **frictional coefficient.** *The frictional coefficient is a measure of the drag that the solution exerts on the moving ion and is dependent on the size, shape, and state of solvation of the ion as well as on the viscosity of the solution (Section 6-5A).* In a constant electric field, the forces on the ion balance each other:

$$qE = vf \qquad [6.8]$$

so that each ion moves with a constant characteristic velocity. An ion's **electrophoretic mobility,** μ, is defined

$$\mu = \frac{v}{E} = \frac{q}{f} \qquad [6.9]$$

The electrophoretic (ionic) mobilities of several common small ions in H_2O at 25°C are listed in Table 2-2.

Equation [6.9] really applies only to ions at infinite dilution in a nonconducting solvent. In aqueous solutions, polyelectrolytes such as proteins are surrounded by a cloud of counterions, which impose an additional electric field of such magnitude that Eq. [6.9] is, at best, a poor approximation of reality. Unfortunately, the complexities of ionic solutions have, so far, precluded the development of a theory that can accurately predict the mobilities of polyelectrolytes. Equation [6.9], however, correctly indicates that molecules at their isoelectric points, p*I*, have zero electrophoretic mobility. Furthermore, for proteins and other polyelectrolytes that have acid–base properties, the ionic charge, and hence the electrophoretic mobility, is a function of pH.

The use of electrophoresis to separate proteins was first reported in 1937 by the Swedish biochemist Arne Tiselius. The technique he introduced, **moving boundary electrophoresis,** was one of the few powerful analytical techniques available in the early years of protein chemistry. However, since this method takes place entirely in solution, preventing the convective mixing of the migrating proteins necessitates a cumbersome apparatus that requires very large samples. Moving boundary electro-phoresis has therefore been supplanted by **zone electrophoresis,** a technique in which the sample is constrained to move in a solid support such as filter paper, cellulose acetate, or, most commonly, a gel. This largely eliminates the convective mixing of the sample that limits the resolution achievable by moving boundary electrophoresis. Moreover, in zone electrophoresis, the various sample components migrate as discrete bands (zones) and hence only small quantities of materials are required.

A. *Paper Electrophoresis*

In **paper electrophoresis,** the sample is applied to a point on a strip of filter paper or cellulose acetate moistened with buffer solution. The ends of the strip are immersed in separate reservoirs of buffer in which the electrodes are placed (Fig. 6-18). On application of a direct current (often of ~20 V · cm^{-1}), the ions of the sample migrate toward the electrodes of opposite polarity at characteristic rates to eventually form discrete bands. An ion's migration rate is influenced, to some extent, by its interaction with the support matrix but is largely a function of its charge. On completion of the electrophoretogram (which usually takes several hours), the strip is dried and the sample components are located using the same detection methods employed in paper chromatography (Section 6-3D).

Paper electrophoresis and paper chromatography are superficially similar. However, *paper electrophoresis separates ions largely on the basis of their ionic charges, whereas paper chromatography separates molecules on the basis of their polarities.* The two methods can be combined in a two-dimensional technique called **fingerprinting** in which a sample is first treated as in two-dimensional paper chromatography (Section 6-3D) but is subjected to electrophoresis in place of the second chromatographic step. Molecules are thereby separated according to both their charge and their polarity.

B. *Gel Electrophoresis*

Gel electrophoresis, which is among the most powerful and yet conveniently used methods of macromolecular separation, has supplanted paper electrophoresis. The gels in common use, polyacrylamide and agarose, have pores of molecular dimensions whose sizes can be specified. *The molecular separations are therefore based on gel filtration as well as the electrophoretic mobilities of the molecules being separated.* The gels in gel electrophoresis, however, retard large molecules relative to smaller ones, the reverse of what occurs in gel filtration chromatography, because there is no solvent space in gel electrophoresis analogous to that between the gel beads in gel filtration chromatography (electrophoretic gels are often directly cast in the electrophoresis device, although precast gels are also widely used). Since the molecules in a sample cannot leave the gel, the electrophoretic movement of larger molecules is impeded relative to that of smaller molecules.

In **polyacrylamide gel electrophoresis (PAGE),** gels are made by the free radical–induced polymerization of **acrylamide** and **N,N'-methylenebisacrylamide** in the buffer of choice (Fig. 6-19). The gel is usually cast as a thin rectangular slab in which several samples can be simultaneously

Figure 6-18 **Paper electrophoresis.** (*a*) A diagram of the apparatus used. The sample is applied to a point on the buffer-moistened paper. The ends of the paper are dipped into reservoirs of buffer in which the electrodes are immersed, and an electric field is applied. (*b*) The completed paper electrophoretogram. Note that positive ions (cations) have migrated toward the cathode and negative ions (anions) have migrated toward the anode. Uncharged molecules remain at the point of sample application.

Figure 6-19 **Polymerization of acrylamide and N,N'-methylenebisacrylamide to form a cross-linked polyacrylamide gel.** The polymerization is induced by free radicals resulting from the chemical decomposition of **ammonium persulfate** ($S_2O_8^{2-} \rightarrow 2SO_4^{-}\cdot$) or the photodecomposition of riboflavin in the presence of traces of O_2. In either case, **N,N,N',N'-tetramethylethylenediamine (TEMED),** a free radical stabilizer, is usually added to the gel mixture. The physical properties of the gel and its pore size are controlled by the proportion of polyacrylamide in the gel and its degree of cross-linking. The most commonly used polyacrylamide concentrations are in the range 3–15%, with the amount of N, N'-methylenebisacrylamide usually fixed at 5% of the total acrylamide present.

Figure 6-20 **Apparatus for slab gel electrophoresis.** Samples, applied in slots that have been cast in the top of the gel, are electrophoresed in parallel lanes.

analyzed in parallel lanes (Fig. 6-20), a good way of comparing similar samples. The buffer, which is the same in both reservoirs and the gel, has a pH (usually ~9 for proteins) such that the macromolecules have net negative charges and hence migrate to the anode in the lower reservoir. Each sample, which can contain as little as 10 μg of macromolecular material, is dissolved in a minimal amount of a relatively dense glycerol or sucrose solution to prevent it from mixing with the buffer in the upper reservoir and is applied in preformed slots at the top of the gel (Fig. 6-20). Alternatively, the sample may be contained in a short length of "sample gel," whose pores are too large to impede macromolecular migration. A direct current of ~300 V is passed through the gel for a time sufficient to separate the macromolecular components into a series of discrete bands (30–90 min), the gel is removed from its holder, and the bands are visualized by an appropriate method (see below). Using this technique, a protein mixture of 0.1 to 0.2 mg can be resolved into as many as 20 discrete bands.

a. Disc Electrophoresis Has Improved Resolution

The narrowness of the bands in the foregoing method, and therefore the resolution of the separations, is limited by the length of the sample column as it enters the gel. The bands are greatly sharpened by an ingenious technique known as **discontinuous pH** or **disc electrophoresis,** which requires a two-gel system and several different buffers (Fig. 6-21). The "running gel," in which the separation takes place, is prepared as described previously and then overlayered by a short (1 cm), large-pored "stacking" or "spacer gel." The buffer in the lower reservoir and in the running gel is as described before, while that in the sample solution (or gel) and in the stacking gel has a pH about two units less than that of the lower reservoir. The pH of the buffer in the upper reservoir, which must contain a weak acid (usu-

ally glycine, pK_2 = 9.78), is adjusted to a pH near that of the lower reservoir.

When the current is switched on, the buffer ions from the upper reservoir migrate into the stacking gel as the stacking gel buffer ions migrate ahead of them. As this occurs, the upper reservoir buffer ions encounter a pH that is much lower than their pK. They therefore assume their uncharged (or, in the case of glycine, zwitterionic) form and become electrophoretically immobile. This causes a deficiency of charge carriers, that is, a high electrical resistance R, in this region which, because of the requirement of a constant current, I, throughout the electrical circuit, results, according to Ohm's law ($E = IR$), in a highly localized increase in the electric field, E. In response to this increased field, the macromolecular anions migrate rapidly until they reach the region containing the stacking gel buffer ions, where they slow down because at that point there is no ion deficiency. *This effect causes the macromolecular ions to approach the running gel as stacks of very narrow (~0.01 mm thick) bands or disks that are ordered according to their mobilities* and lie between the migrating ions of the upper reservoir and those of the stacking gel. As the macromolecular ions enter the running gel, they slow down as a result of gel filtration effects. This permits the upper reservoir buffer ions to overtake the macromolecular bands and, because of the running gel's higher pH, assume their fully charged form as they too enter the gel. The charge carrier deficiency therefore disappears and from this point on the electrophoretic separation proceeds normally. However, *the compactness of the macromolecular bands entering the running gel greatly increases the resolution of the macromolecular separations* (e.g., Fig. 6-22).

b. Agarose Gels Are Used to Separate Large Molecules Electrophoretically

The very large pores needed for the PAGE of large molecular mass compounds (>200 kD) requires gels with such low polyacrylamide concentrations (<2.5%) that they are too soft to be usable. This difficulty is circumvented by using agarose (Fig. 6-13). For example, a 0.8% agarose gel is

Figure 6-21 **Diagram of a disc electrophoresis apparatus.**

which is the most widely used dye for this purpose, is applied by soaking the gel in an acidic, alcoholic solution containing the dye. This fixes the protein in the gel by denaturing it and complexes the dye to the protein. Excess dye is removed by extensively washing the gel with an acidic solution or by electrophoretic destaining. Protein bands containing as little as 0.1 μg can thereby be detected. Gel bands containing less than this amount of protein may be visualized with **silver stain,** which is ~50 times more sensitive but more difficult to apply. The recently developed **SYPRO dyes,** which strongly fluoresce under UV light when bound to protein, are equally sensitive as silver stain but easier to apply. **Fluorescamine,** a widely used protein stain, is a nonfluorescent molecule that reacts with primary amines, such as lysine residues, to yield an addition product that is highly UV-fluorescent.

**Figure 6-22 Disc electrophoresis of human serum in a 0.5 ×
4.0–cm polyacrylamide gel column.** The proteins were visualized
by staining them with **amido black.** [From B. J. Davis, Annals of
the New York Academy of Science 121, 404 (1964), Fig. 8.]

Fluorescamine
(nonfluorescent)

Fluorescamine adduct
(highly fluorescent)

used for the electrophoretic separation of nucleic acids
with molecular masses of up to 50,000 kD.

c. Gel Bands May Be Detected by Staining, Radioactive Counting, or Immunoblotting

Bands resulting from a gel electrophoretic separation
can be located by a variety of techniques. Proteins are often
visualized by staining. **Coomassie brilliant blue,**

R250: R = H
G250: R = CH₃

Coomassie brilliant blue

Proteins, as well as other substances, can be detected
through the UV absorption of a gel along its length. If the
sample is radioactive, the gel may be dried under vacuum
to form a cellophane-like material or, alternatively, covered
with plastic wrap, and then clamped over a sheet of X-ray
film. After a time (from a few minutes to many weeks
depending on the radiation intensity) the film is developed,
and the resulting autoradiograph shows the positions of
the radioactive components by a blackening of the film [alternatively, a phosphorimager (Section 5-5D) can be used
to reveal the locations of the radioactive components
within even a few seconds]. A gel may also be sectioned
widthwise into many slices and the level of radioactivity in
each slice determined using a **scintillation counter.** The latter method yields quantitatively more accurate results than
autoradiography. Sample materials can also be eluted from
gel slices for identification and/or further treatment.

If an antibody to the protein of interest is available, it is
possible to specifically detect this protein on a gel in the
presence of many other proteins by an **immunoblot** (also
known as a **Western blot**). This procedure is a variation of
Southern blotting (Section 5-5D) that uses a technique
similar to ELISA (Section 6-1Da) to detect the protein(s)
of interest (Fig. 6-23):

1. A completed gel electrophoretogram is blotted onto
a sheet of nitrocellulose (much like in Fig. 5-48), which
strongly and nonspecifically binds proteins [nylon or
polyvinylidene fluoride (PVDF) membranes may also be
used].

1. Perform gel electrophoresis on a sample containing the protein of interest

Blot the proteins from the gel onto nitrocellulose

2. Block the unoccupied binding sites on the nitrocellulose with casein

3. Incubate with rabbit antibody to the protein of interest

4. Wash and incubate with an enzyme-linked goat anti-rabbit antibody

5. Assay the linked enzyme with a colorimetric reaction

Weight
Paper towels
Nitrocellulose sheet
Wick
Buffer solution
Gel electrophoretogram containing the protein of interest

Nitrocellulose replica of gel electrophoretogram

Binding of primary antibody

Binding of enzyme-linked secondary antibody

Immunoblot

Figure 6-23   **Detection of proteins by immunoblotting.**

2. The excess adsorption sites on the nitrocellulose are blocked with a nonspecific protein such as **casein** (milk protein; nonfat milk itself is often used) to prevent the nonspecific adsorption of the antibodies (which are also proteins) used in Steps 3 and 4.

3. The blot is treated with antibody to the protein of interest (the primary antibody). This is usually a rabbit antibody.

4. After washing away the unbound primary antibody, the blot is incubated with a goat antibody, directed against all rabbit antibodies, to which an easily assayed enzyme has been covalently linked (the secondary antibody).

5. After washing away the unbound secondary antibody, the enzyme in the bound secondary antibody is assayed with a color-producing reaction, causing colored bands to appear on the nitrocellulose where the protein of interest is bound.

Alternatively, the primary antibody used in Step 3 may be labeled with the radioactive isotope ^{125}I, the unbound antibody washed away, and the position of the bound protein on the blot revealed by autoradiography.

C. *SDS–PAGE*

Soaps and detergents are amphipathic molecules (Section 2-1Ba) that are strong protein denaturing agents for reasons explained in Section 8-4E. **Sodium dodecyl sulfate (SDS),**

$$[CH_3 - (CH_2)_{10} - CH_2 - O - SO_3^-]Na^+$$

Sodium dodecyl sulfate (SDS)

a detergent that is often used in biochemical preparations, binds quite tenaciously to proteins, causing them to assume a rodlike shape. Most proteins bind SDS in the same ratio, 1.4 g of SDS per gram of protein (about one SDS molecule for every two amino acid residues). The large negative charge that the SDS imparts masks the protein's intrinsic charge so that SDS-treated proteins tend to have identical charge-to-mass ratios and similar shapes. Consequently, *the electrophoresis of proteins in an SDS-containing polyacrylamide gel separates them in order of their molecular masses because of gel filtration effects.* Figure 6-24 provides an example of the resolving power and the reproducibility of **SDS–PAGE.**

The molecular masses of "normal" proteins are routinely determined to an accuracy of 5 to 10% through SDS–PAGE. The relative mobilities of proteins on such gels vary linearly with the logarithm of their molecular masses (Fig. 6-25). In practice, a protein's molecular mass is determined by electrophoresing it together with several "marker" proteins of known molecular masses which bracket that of the protein of interest.

Many proteins consist of more than one polypeptide chain (Section 8-5A). SDS treatment disrupts the noncovalent interactions between these subunits. Therefore, SDS–PAGE yields the molecular masses of the protein's subunits rather than that of the intact protein unless the subunits are disulfide linked. However, mercaptoethanol is usually added to SDS–PAGE gels so as to reductively cleave these disulfide bonds (Section 7-1B).

D. *Isoelectric Focusing*

A protein has charged groups of both polarities and therefore has an isoelectric point, pI, the pH at which it is immobile in an electric field (Section 4-1D). *If a mixture of proteins is electrophoresed through a solution having a stable pH gradient in which the pH smoothly increases from anode to cathode, each protein will migrate to the position in the pH gradient corresponding to its isoelectric point.* If a protein molecule diffuses away from this position, its net charge will change as it moves into a region of different pH, and the resulting electrophoretic forces will move it back to its isoelectric position. Each species of protein is

Figure 6-24 **SDS–PAGE.** The SDS–polyacrylamide disc electrophoretogram shows separation of proteins from the supernatant (*left*) and membrane fractions (*right*) of various strains of the bacterium *Salmonella typhimurium*. Samples of 200 μg of protein each were run in parallel lanes on a 35-cm-long, 0.8-mm-thick slab gel containing 10% polyacrylamide. The lane marked MW contains molecular weight standards. [Courtesy of Giovanna F. Ames, University of California at Berkeley.]

Figure 6-25 **Logarithmic relationship between the molecular mass of a protein and its relative electrophoretic mobility in SDS–PAGE.** This relationship is plotted for 37 polypeptides ranging from 11 to 70 kD. [After Weber, K. and Osborn, M., *J. Biol. Chem.* **244,** 4406 (1969).]

thereby "focused" into a narrow band about its isoelectric point that may be as thin as 0.01 pH unit. Hence, this process is called **isoelectric focusing (IEF).**

A pH gradient produced by mixing two different buffers together in continuously varying ratios is unstable in an electric field because the buffer ions migrate to the electrode of opposite polarity. Rather, the pH gradient in IEF is formed by a mixture of low molecular mass (600–900 D) oligomers bearing aliphatic amino and carboxylic acid groups (Fig. 6-26) that have a range of isoelectric points. Under the influence of an electric field in solution, these **ampholytes** (amphoteric electrolytes) will each migrate to their isoelectric points. Consequently, the most acidic ampholytes gather at the anode and the progressively more basic ones position themselves ever closer to the cathode. The pH gradient, which is maintained by the ~1000 V electric field, arises from the buffering action of these ampholytes. The convection of the pH gradient is prevented by carrying out IEF in a lightly cross-linked polyacrylamide gel in the form of a rod or slab. IEF gels often con-

tain ~6*M* urea, a powerful protein denaturing agent that, unlike SDS, is uncharged and hence cannot directly affect the charge of a protein.

An alternative form of IEF utilizes gels containing **immobilized pH gradients.** These are made from acrylamide derivatives that are covalently linked to ampholytes. Through the use of a gradient maker (e.g., Fig. 6-7), a gel containing an immobilized pH gradient is polymerized from a continuously varying mixture of acrylamide derivatives with different p*K*'s so that the gel's pH varies smoothly from one end to the other.

The fact that IEF separates proteins into sharp bands makes it a useful analytical and preparative tool. In fact, many protein preparations once thought to be homogeneous have been resolved into several components by IEF. IEF can be combined with electrophoresis in an extremely powerful two-dimensional separation technique named

$$-CH_2-N-(CH_2)_n-N-CH_2-$$
$$(CH_2)_n \qquad R$$
$$NR_2$$

$$n = 2 \text{ or } 3$$
$$R = H \text{ or } -(CH_2)_n-COOH$$

Figure 6-26 **General formula of the ampholytes used in isoelectric focusing.**

← Isoelectric focusing →

← SDS-PAGE →

Figure 6-27   Two-dimensional (2D) gel electrophoresis. This autoradiogram shows the separation of *E. coli* proteins by 2D gel electrophoresis (isoelectric focusing horizontally and SDS–PAGE vertically). A 10-μg sample of proteins from *E. coli* that had been labeled with ^{14}C-amino acids was subjected to isoelectric focusing in a 2.5 × 130–mm tube of a urea-containing polyacrylamide gel. The gel was then extruded from its tube, placed in contact with one edge of an SDS–polyacrylamide slab gel, and subjected to electrophoresis. Over 1000 spots were counted on the original autoradiogram, which resulted from an 825-h exposure. [Courtesy of Patrick O'Farrell, University of California at San Francisco.]

two-dimensional (2D) gel electrophoresis; Fig. 6-27). Up to 5000 proteins can be observed on a single two-dimensional electrophoretogram. Hence two-dimensional gel electrophoresis is a valuable tool for **proteomics** (the study of the **proteome,** which, in analogy with the term "genome," is defined as the aggregate of all the proteins expressed by a cell or organism, but with emphasis on their quantitation, localization, modifications, interactions, and activities, as well as their identification). Individual protein bands in a stained gel can be cut out from the gel (with a scalpel or by a robot guided by a digitized image of the gel acquired by an optical scanner or a digital camera), destained, and the protein eluted from the gel fragment for identification and/or characterization, often by mass spectrometry (Section 7-1J). Variant proteins can be found by comparing the positions and intensities of the bands in 2D gels of similar preparations. This can be done with the aid of a computer after acquiring digitized images of the stained gels. Numerous reference 2D gels are publicly available for this purpose in the Web-accessible databases listed at http://www.expasy.org/swiss-2dpage?de. These databases contain images of 2D gels of a variety of organisms and tissues and identify many of their component proteins.

E. *Capillary Electrophoresis*

Although gel electrophoresis in its various forms is a common and highly effective method for separating charged molecules, it typically requires an hour or more for a run and is difficult to quantitate and automate. These disadvantages are largely overcome through the use of **capillary electrophoresis (CE),** a technique in which electrophoresis is carried out in very thin (10 to 100 μm inner diameter) capillary tubes made of silica, glass, or plastic. Such narrow capillaries rapidly dissipate heat and hence permit the use of high electric fields (typically 100 to 300 V · cm[−1], about 10 times that of most other electrophoretic techniques), which reduces separation times to a few minutes. These rapid separations, in turn, minimize band broadening caused by diffusion, thereby yielding extremely sharp separations. Capillaries can be filled with buffer (as in moving boundary electrophoresis, but here the capillary's narrow bore all but eliminates convective mixing), SDS–polyacrylamide gel (separation according to molecular mass; Section 6-4C), or ampholytes (isoelectric focusing; Section 6-4D). These CE techniques have extremely high resolution and can be automated in much the same way as is HPLC, that is, with automatic sample loading and on-line sample detection. Since CE can only separate small amounts of material, it is largely limited to use as an analytical technique.

5 ULTRACENTRIFUGATION

If a container of sand and water is shaken and then allowed to stand quietly, the sand will rapidly sediment to the bottom of the container due to the influence of Earth's gravity (an acceleration *g* equal to 9.81 m · s[−2]). Yet macromolecules in solution, which experience the same gravitational field, do not exhibit any perceptible sedimentation because their random thermal (Brownian) motion keeps them uniformly distributed throughout the solution. *Only when they are subjected to enormous accelerations will the sedimentation behavior of macromolecules begin to resemble that of sand grains.*

The ultracentrifuge was developed around 1923 by the Swedish biochemist The Svedberg. Using this instrument, Svedberg first demonstrated that proteins are macromolecules with homogeneous compositions and that many proteins are composed of subunits. Within a few decades, ultracentrifugation became an indispensable tool for the fractionation of proteins, nucleic acids, and subcellular particles. Modern ultracentrifuges can attain rotational speeds as high as 150,000 rpm (revolutions per minute) so as to generate centrifugal fields in excess of 1 million × *g*. In this section we outline the theory and practice of ultracentrifugation.

A. *Sedimentation*

The rate at which a particle sediments in the ultracentrifuge is related to its mass. The force, $F_{sedimentation}$, acting to sediment a particle of mass *m* that is located a distance *r* from a point about which it is revolving with angular velocity ω (in radians · s[−1]) is the centrifugal force ($m\omega^2 r$) on the particle less the buoyant force ($V_p\rho\omega^2 r$) exerted by the solution:

$$F_{sedimentation} = m\omega^2 r - V_p\rho\omega^2 r \qquad [6.10]$$

Here V_p is the particle volume and ρ is the density of the solution. However, the motion of the particle through the solution, as we have seen in our study of electrophoresis, is opposed by the frictional force:

$$F_{friction} = vf \qquad [6.7]$$

where $v = dr/dt$ is the rate of migration of the sedimenting particle and f is its frictional coefficient. The particle's frictional coefficient can be determined from measurements of its rate of diffusion.

Under the influence of gravitational (centrifugal) force, the particle accelerates until the forces on it exactly balance:

$$m\omega^2 r - V_p\rho\omega^2 r = vf \qquad [6.11]$$

The mass of 1 mol of particles, M, is

$$M = mN \qquad [6.12]$$

where N is Avogadro's number (6.022×10^{23}). Thus, a particle's volume, V_p, may be expressed in terms of its molar mass:

$$V_P = \overline{V}m = \frac{\overline{V}M}{N} \qquad [6.13]$$

where $\overline{V}$, the particle's **partial specific volume,** is the volume change when 1 g (dry weight) of particles is dissolved

in an infinite volume of the solute. For most proteins dissolved in pure water at 20°C, $\overline{V}$ is near 0.73 cm³·g⁻¹ (Table 6-4). Indeed, for proteins of known amino acid composition, $\overline{V}$ is closely approximated by the sum of the partial specific volumes of its component amino acid residues, thereby indicating that the atoms in proteins are closely packed (Section 8-3Bc).

a. A Particle May Be Characterized by Its Sedimentation Rate

Substituting Eqs. [6.12] and [6.13] into Eq. [6.11] yields

$$vf = \frac{M(1 - \overline{V}\rho)\omega^2 r}{N} \qquad [6.14]$$

Now define the **sedimentation coefficient,** s, as

$$s = \frac{v}{\omega^2 r} = \frac{1}{\omega^2}\left(\frac{d\ln r}{dt}\right) = \frac{M(1 - \overline{V}\rho)}{Nf} \qquad [6.15]$$

The sedimentation coefficient, a quantity that is analogous to the electrophoretic mobility (Eq. [6.9]) in that it is a velocity per unit force, is usually expressed in units of 10^{-13} s, which are known as **svedbergs (S).** For the sake of uniformity, the sedimentation coefficient is customarily corrected to the value that would be obtained at 20°C in a solvent with the density and viscosity of pure water. This is symbolized $s_{20,w}$. Table 6-4 and Fig. 6-28 indicate the values of $s_{20,w}$ in svedbergs for a variety of biological materials.

Equation [6.15] indicates that a particle's mass, $m = M/N$, can be determined from the measurement of its sedimentation coefficient, s, and the solution density, ρ, if its

Table 6-4 **Physical Constants of Some Proteins**

Protein	Molecular Mass (kD)	Partial Specific Volume, $\overline{V}_{20,w}$ (cm³·g⁻¹)	Sedimentation Coefficient, $s_{20,w}$ (S)	Frictional Ratio, f/f_0
Lipase (milk)	6.7	0.714	1.14	1.190
Ribonuclease A (bovine pancreas)	12.6	0.707	2.00	1.066
Cytochrome c (bovine heart)	13.4	0.728	1.71	1.190
Myoglobin (horse heart)	16.9	0.741	2.04	1.105
α-Chymotrypsin (bovine pancreas)	21.6	0.736	2.40	1.130
Crotoxin (rattlesnake)	29.9	0.704	3.14	1.221
Concanavalin B (jack bean)	42.5	0.730	3.50	1.247
Diphtheria toxin	70.4	0.736	4.60	1.296
Cytochrome oxidase (*P. aeruginosa*)	89.8	0.730	5.80	1.240
Lactate dehydrogenase H (chicken)	150	0.740	7.31	1.330
Catalase (horse liver)	222	0.715	11.20	1.246
Fibrinogen (human)	340	0.725	7.63	2.336
Hemocyanin (squid)	612	0.724	19.50	1.358
Glutamate dehydrogenase (bovine liver)	1015	0.750	26.60	1.250
Turnip yellow mosaic virus protein	3013	0.740	48.80	1.470

Source: Smith, M.H., in Sober, H.A. (Ed.), *Handbook of Biochemistry and Molecular Biology* (2nd ed.), p. C-10, CRC Press (1970).

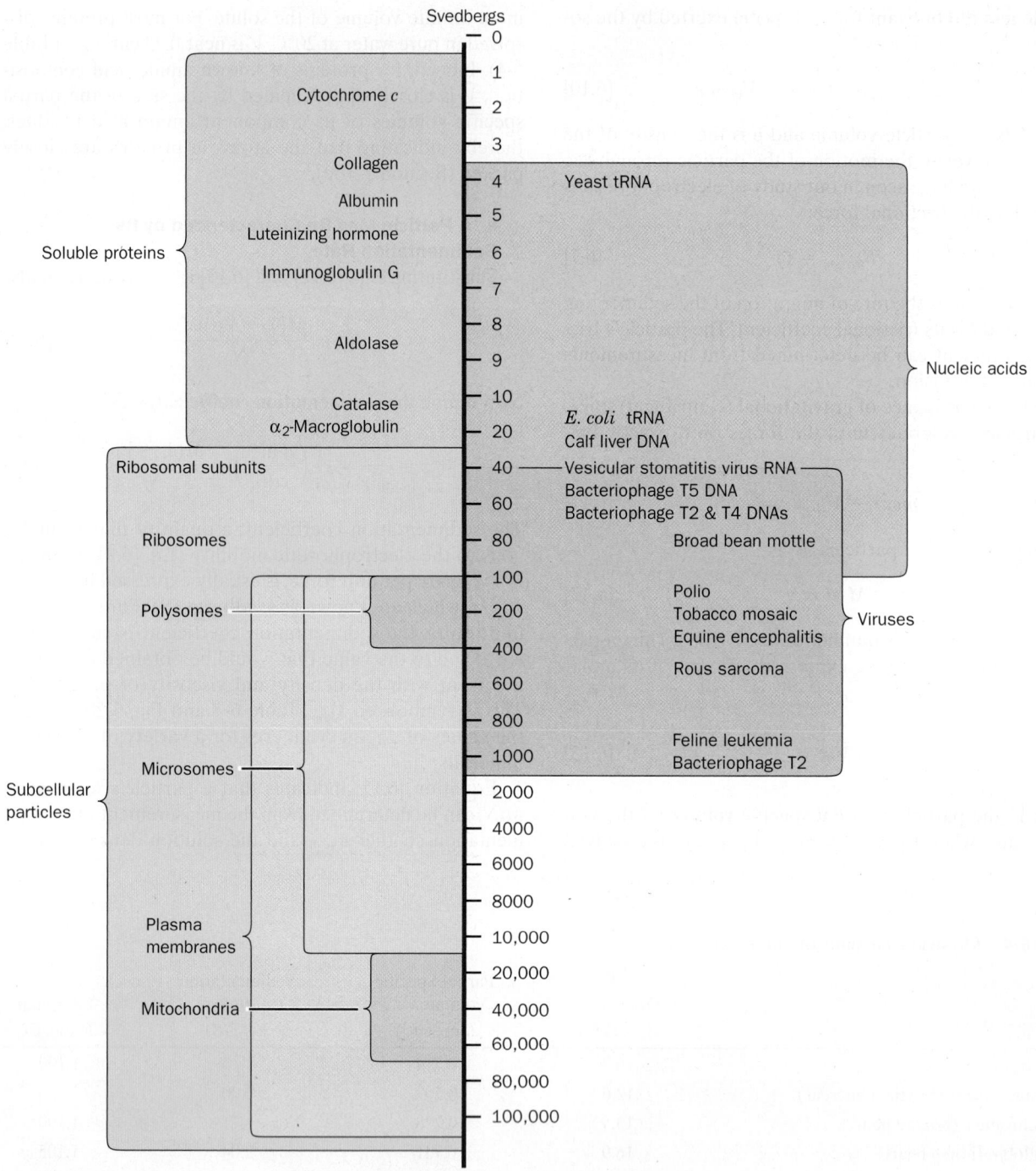

Figure 6-28 **Sedimentation coefficients in svedbergs (S) for some biological materials.**
[After a diagram supplied by Beckman Coulter, Inc.]

frictional coefficient, *f,* and its partial specific volume, $\overline{V}$, are known. Indeed, before about 1970, most macromolecular mass determinations were made using the **analytical ultracentrifuge,** a device in which the sedimentation rates of molecules under centrifugation can be optically measured (the masses of macromolecules are too high to be accurately determined by such classic physical techniques as melting point depression or osmotic pressure measurements).

Although the advent of much simpler molecular mass determination methods, such as gel filtration chromatography (Section 6-3Ba) and SDS–PAGE (Section 6-4C), had caused analytical ultracentrifugation to largely fade from use, recently developed instrumentation has led to a resurgence in the use of analytical ultracentrifugational measurements. They are particularly useful in characterizing systems of associating macromolecules.

b. The Frictional Ratio Is Indicative of Molecular Solvation and Shape

For an unsolvated spherical particle of radius r_p the frictional coefficient is determined according to the **Stokes equation:**

$$f = 6\pi\eta r_p \qquad [6.16]$$

where η is the **viscosity** of the solution. Solvation increases the frictional coefficient of a particle by increasing its effective or **hydrodynamic volume.** Furthermore, f is minimal when the particle is a sphere. This is because a nonspherical particle has a larger surface area than a sphere of equal volume and therefore must, on the average, present a greater surface area toward the direction of movement than a sphere.

The frictional coefficient, f, of a particle of known mass and partial specific volume can be ultracentrifugationally determined using Eq. [6.15]. The effective or **Stokes radius** of a particle in solution can be calculated by solving Eq. [6.16] for r_p, given the experimentally determined values of f and η. Conversely, the minimal frictional coefficient of a particle, f_0, can be calculated from the mass and the partial specific volume of the particle by assuming it to be spherical ($V_p = \frac{4}{3}\pi r_p^3$) and unsolvated:

$$f_0 = 6\pi\eta\left(\frac{3M\overline{V}}{4\pi N}\right)^{1/3} \qquad [6.17]$$

If the **frictional ratio,** f/f_0, of a particle is much greater than unity, it must be concluded that the particle is highly solvated and/or significantly elongated. The frictional ratios of a selection of proteins are presented in Table 6-4. The globular proteins, which are known from structural studies to be relatively compact and spheroidal (Section 8-3B), have frictional ratios ranging up to ~1.5. Fibrous molecules such as DNA and the blood clotting protein **fibrinogen** (Section 35-1Aa) have larger frictional ratios. On denaturation, the frictional coefficients of globular proteins increase by as much as twofold because denatured proteins assume flexible and fluctuating **random coil** conformations in which all parts of the molecule are in contact with solvent (Section 8-1D).

B. *Preparative Ultracentrifugation*

Preparative ultracentrifuges, which as their name implies are designed for sample preparation, differ from analytical ultracentrifuges in that they lack sample observation facilities. Preparative rotors contain cylindrical sample tubes whose axes may be parallel, at an angle, or perpendicular to the rotor's axis of rotation, depending on the particular application (Fig. 6-29).

In the derivation of Eq. [6.15], it was assumed that sedimentation occurred through a homogeneous medium. Sedimentation may be carried out in a solution of an inert substance, however, such as sucrose or CsCl, in which the concentration, and therefore the density, of the solution increases from the top to the bottom of the centrifuge tube.

Figure 6-29   A selection of preparative ultracentrifuge rotors. The sample tubes of the swinging bucket rotors (*rear*) are hinged so that they swing from the vertical to the horizontal position as the rotor starts spinning, whereas the sample tubes of the other rotors have a fixed angle relative to the rotation axis. [Courtesy of Beckman Coulter, Inc.]

The use of such **density gradients** greatly enhances the resolving power of the ultracentrifuge. Two applications of density gradients are widely employed: (1) **zonal ultracentrifugation** and (2) **equilibrium density gradient ultracentrifugation.**

a. Zonal Ultracentrifugation Separates Particles According to Their Sedimentation Coefficients

In zonal ultracentrifugation, a macromolecular solution is carefully layered on top of a density gradient prepared by use of a device resembling that diagrammed in Fig. 6-7. The purpose of the density gradient is to allow smooth passage of the various macromolecular zones by damping out convective mixing of the solution. Sucrose, which forms a syrupy and biochemically benign solution, is commonly used to form a density gradient for zonal ultracentrifugation. The density gradient is normally rather shallow because the maximum density of the solution must be less than that of the least dense macromolecule of interest. Nevertheless, consideration of Eq. [6.15] indicates that the sedimentation rate of a macromolecule is a more sensitive function of molecular mass than density. Consequently, *zonal ultracentrifugation separates similarly shaped macromolecules largely on the basis of their molecular masses.*

During centrifugation, each species of macromolecule moves through the gradient at a rate largely determined by its sedimentation coefficient and therefore travels as a zone that can be separated from other such zones as is diagrammed in Fig. 6-30. After centrifugation, fractionation is commonly effected by puncturing the bottom of the celluloid centrifuge tube with a needle, allowing its contents to drip out, and collecting the individual zones for subsequent analysis.

Figure 6-30 **Zonal ultracentrifugation.** The sample is layered onto a sucrose gradient (*left*). During centrifugation (*middle*), each particle sediments at a rate that depends largely on its mass.

After the end of the run, the centrifugation tube is punctured and the separated particles (zones) are collected (*right*).

b. Equilibrium Density Gradient Ultracentrifugation Separates Particles According to Their Densities

In **equilibrium density gradient ultracentrifugation** [alternatively, **isopycnic ultracentrifugation** (Greek: *isos,* equal + *pyknos,* dense)], the sample is dissolved in a relatively concentrated solution of a dense, fast-diffusing (and therefore low molecular mass) substance, such as CsCl or Cs_2SO_4, and is spun at high speed until the solution achieves equilibrium. *The high centrifugal field causes the low molecular mass solute to form a steep density gradient (Fig. 6-31) in which the sample components band at posi-*

tions where their densities are equal to that of the solution; that is, where $(1 - \overline{V}\rho)$ *in Eq. [6.15] is zero (Fig. 6-32).* These bands are collected as separate fractions when the sample tube is drained as described above. The salt concentration in the fractions and hence the solution density is easily determined with an **Abbé refractometer,** an optical instrument that measures the refractive index of a solution. *The equilibrium density gradient technique is often the method of choice for separating mixtures whose components have a range of densities.* These substances include nucleic acids, viruses, and certain subcellular organelles such as ribosomes. However, isopycnic ultracentrifugation is rather ineffective for the fractionation of protein mixtures because most proteins have similar densities (high salt concentrations also salt out or possibly denature proteins). Density gradient ultracentrifugation in an analytical ultracentrifuge was used to show that DNA is semiconservatively replicated (Section 5-3B).

6 NUCLEIC ACID FRACTIONATION

In the preceding parts of this chapter we considered the most commonly used procedures for isolating and, to some extent, characterizing proteins. Most of these methods, often with some modification, are also regularly used to fractionate nucleic acids according to size, composition, and sequence. There are also many techniques that are applicable only to nucleic acids. In this section we outline some of the most useful of the procedures employed in the separation of nucleic acids.

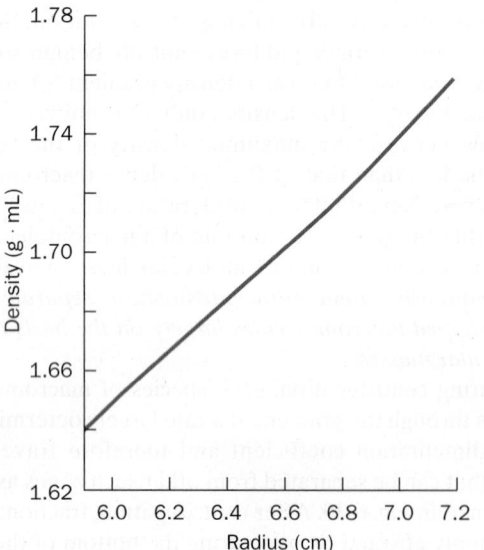

Figure 6-31 **Equilibrium density distribution of a CsCl solution in an ultracentrifuge spinning at 39,460 rpm.** The initial density of the solution was 1.7 g · mL^{-1}. [After Ifft, J.B., Voet, D.H., and Vinograd, J., *J. Phys. Chem.* **65,** 1138 (1961).]

A. *Solution Methods*

Nucleic acids in cells are invariably associated with proteins. Once cells have been broken open (Section 6-1B),

1. Uniform mixture of sample and gradient-forming substance

2. Centrifugation

3. Gradient is formed and samples band at their isopycnic positions

Figure 6-32 **Isopycnic ultracentrifugation.** The centrifugation starts with a uniform mixture of a macromolecular sample dissolved in a solution of a dense, fast-diffusing solute such as CsCl (*left*). At equilibrium in a centrifugal field, the solute forms a density gradient in which the macromolecules migrate to their positions of buoyant density (*right*).

their nucleic acids must be deproteinized. This can be accomplished by shaking (very gently if high molecular mass DNA is being isolated; Section 5-3D) the aqueous solution containing the protein–nucleic acid complex with a 25:24:1 mixture of phenol, chloroform, and isoamyl alcohol. The protein is thereby denatured and extracted into the water-immiscible organic phase, which is separated from the nucleic acid–containing aqueous phase by centrifugation (when a large amount of protein is present, it forms a white precipitate between the organic and aqueous phases). Alternatively (or in addition), protein can be dissociated from the nucleic acids by such denaturing agents as detergents, guanidinium chloride, or high salt concentrations, and/or it can be enzymatically degraded by proteases. In all cases, the nucleic acids, a mixture of RNA and DNA, can then be isolated by precipitation with ethanol. The RNA can be recovered from such a precipitate by treating it with **pancreatic DNase** to eliminate the DNA. Conversely, the DNA can be freed of RNA by treatment with **RNase.** Alternatively, RNA and DNA may be separated by ultracentrifugation (Section 6-6D).

In all these and subsequent manipulations, the nucleic acids must be protected from degradation by nucleases that occur both in the experimental materials and on human hands. Nucleases may be inhibited by the presence of chelating agents such as **ethylenediaminetetraacetic acid (EDTA),**

$$\begin{array}{ccc} \text{HOOC}-\text{H}_2\text{C} & & \text{CH}_2-\text{COOH} \\ & \text{N}-\text{CH}_2-\text{CH}_2-\text{N} & \\ \text{HOOC}-\text{H}_2\text{C} & & \text{CH}_2-\text{COOH} \end{array}$$

Ethylenediaminetetraacetic acid (EDTA)

which sequester the divalent metal ions that nucleases require for activity. In cases where no nuclease activity can be tolerated, all glassware must be autoclaved to heat denature the nucleases and the experimenter should wear plastic gloves. Nevertheless, nucleic acids are generally easier to handle than proteins because their lack, in most cases, of a complex tertiary structure makes them relatively tolerant of extreme conditions.

B. *Chromatography*

Many of the chromatographic techniques that are used to separate proteins (Section 6-3) are also applicable to nucleic acids. Oligonucleotides can be readily separated HPLC, particularly that using reverse-phase chromatography. Larger nucleic acids are often separated by procedures that include ion exchange chromatography and gel filtration chromatography.

a. **Hydroxyapatite Can Be Used to Isolate and Fractionate DNA**

Hydroxyapatite (a form of calcium phosphate; Section 6-3Db) is particularly useful in the chromatographic purification and fractionation of DNA. Double-stranded DNA binds to hydroxyapatite more tightly than do most other molecules. Consequently, DNA can be rapidly isolated by passing a cell lysate through a hydroxyapatite column, washing the column with a phosphate buffer of concentration low enough to release only the RNA and proteins, and then eluting the DNA with a more concentrated phosphate solution. In addition, single-stranded DNA elutes from hydroxyapatite at a lower phosphate concentration than does double-stranded DNA.

b. Messenger RNAs Can Be Isolated by Affinity Chromatography

Affinity chromatography (Section 6-3C) is useful in isolating specific nucleic acids. For example, most eukaryotic messenger RNAs (mRNAs) have a poly(A) sequence at their 3′ ends (Section 5-4Ac). They can be isolated on an agarose or cellulose matrix to which poly(U) has been covalently attached. The poly(A) sequences specifically bind to the complementary poly(U) in high salt and at low temperatures and can later be released by altering these conditions. Moreover, if the (partial) sequence of an mRNA is known (e.g., as inferred from the corresponding protein's amino acid sequence), the complementary DNA strand may be synthesized (via methods discussed in Section 7-6Aa) and used to isolate that particular mRNA.

C. *Electrophoresis*

Nucleic acids of a given type may be separated by polyacrylamide gel electrophoresis (Sections 6-4B and 6-4C) because their electrophoretic mobilities in such gels vary inversely with their molecular masses. However, DNAs of more than a few thousand base pairs are too large to penetrate even a weakly cross-linked polyacrylamide gel. This difficulty is partially overcome through the use of agarose gels. By using gels with an appropriately low agarose content, relatively large DNAs in various size ranges may be fractionated. In this manner, plasmids, for example, may be separated from the larger chromosomal DNA of bacteria.

a. Duplex DNA Is Detected by Selectively Staining It with Intercalation Agents

The various DNA bands in a gel must be detected if they are to be isolated. Double-stranded DNA is readily stained by planar aromatic cations such as **ethidium ion** and **acridine orange.**

Ethidium

Acridine orange

These dyes bind to duplex DNA by **intercalation** (slipping in between the stacked base pairs), where they exhibit a fluorescence under UV light that is far more intense than that of the free dye. As little as 50 ng of DNA may be detected in a gel by staining it with ethidium bromide (Fig. 6-33).

Figure 6-33 **Agarose gel electrophoretogram of double helical DNA.** After electrophoresis, the gel was soaked in a solution of ethidium bromide, washed, and photographed under UV light. The fluorescence of the ethidium cation is strongly enhanced by binding to DNA, so each fluorescent band marks a different sized DNA fragment. [Klaus Guldbrandsen/Photo Researchers, Inc.]

Single-stranded DNA and RNA also stimulate the fluorescence of ethidium ion but to a lesser extent than does duplex DNA. Disadvantages of using these dyes are that they are mutagens and therefore must be handled and disposed of with caution and that exposure to UV light damages DNA (Section 30-5Aa). **SYBR Safe,** an equally sensitive nonmutagenic dye that fluoresces when it is bound to double-stranded DNA and excited by blue light, avoids these difficulties.

b. Very Large DNAs Are Separated by Pulsed-Field Gel Electrophoresis

The sizes of the DNAs that can be separated by conventional gel electrophoresis are limited to ~100,000 bp even when gels containing as little as 0.1% agarose (which makes an extremely fragile gel) are used. However, the development of **pulsed-field gel electrophoresis (PFGE)** by Charles Cantor and Cassandra Smith extended this limit to DNAs with up to 10 million bp (6.6 million kD). The electrophoresis apparatus used in PFGE has two or more pairs of electrodes arrayed around the periphery of an agarose slab gel. The different electrode pairs are sequentially pulsed for times varying from 0.1 to 1000 s depending on the sizes of the DNAs being separated. Gel electrophoresis of DNA requires that these elongated molecules worm their way through the gel's labyrinthine channels more or less in the direction from the cathode to the anode. If the direction of the electric field abruptly changes, these DNAs must reorient their long axes along the new direction of the field before they can continue their passage through the gel. The time required to reorient very long gel-embedded

DNA molecules evidently increases with their size. Consequently, a judicious choice of electrode distribution and pulse lengths causes shorter DNAs to migrate through the gel faster than longer DNAs, thereby effecting their separation (Fig. 6-34).

D. *Ultracentrifugation*

Equilibrium density gradient ultracentrifugation (Section 6-5Bb) in CsCl constitutes one of the most commonly used DNA separation procedures. The buoyant density, ρ, of double-stranded Cs^+ DNA depends on its base composition:

$$\rho = 1.660 + 0.098X_{G+C} \qquad [6.18]$$

where X_{G+C} is the mole fraction of G + C. Hence a CsCl density gradient fractionates DNA according to its base composition. For example, eukaryotic DNAs often contain minor fractions that band separately from the major species. Some of these **satellite bands** consist of mitochondrial and chloroplast DNAs. Another important class of satellite DNA is composed of **repetitive sequences** that are short segments of DNA tandemly repeated hundreds, thousands, and in some cases millions of times in a genome (Section 34-2B). Likewise, plasmids may be separated from bacterial chromosomal DNA by equilibrium density gradient ultracentrifugation.

Single-stranded DNA is ~0.015 g $\cdot$ cm^{-3} denser than the corresponding double-stranded DNA so that the two may be separated by equilibrium density gradient ultracentrifugation. RNA is too dense to band in CsCl but does so in Cs_2SO_4 solutions. RNA–DNA hybrids will band in CsCl but at a higher density than the corresponding duplex DNA.

RNA may be fractionated by zonal ultracentrifugation through a sucrose gradient (Section 6-5Ba). RNAs are separated by this technique largely on the basis of their size. In fact, ribosomal RNA, which constitutes the major portion of cellular RNA, is classified according to its sedimentation rate; for example, the RNA of the *E. coli* small ribosomal subunit is known as **16S RNA** (Section 32-3A).

Figure 6-34 **Pulsed-field gel electrophoresis (PFGE) of a set of yeast chromosomes.** The yeast chromosomes, which were run as identical samples in the 13 inner lanes, have sizes 260, 290, 370, 460, 580/600, 700, 780, 820, and 850 kb. The two outer lanes, which provide molecular mass standards, contained twenty successively larger multimers of a ~43.5-kb bacteriophage λ DNA (top to bottom) to an observed limit of ~850 kb. [Electrophoretogram by Margit Burmeister, University of Michigan, *in* Wilson, K. and Walker, J., *Principles and Techniques of Biochemistry and Molecular Biology* (6th ed.), *p*. 477, Cambridge University Press (2005).]

CHAPTER SUMMARY

1 Protein Isolation Macromolecules in cells are solubilized by disrupting the cells by various chemical or mechanical means such as detergents or blenders. Partial purification by differential centrifugation is used after cell lysis to remove cell debris or to isolate a desired subcellular component. When out of the protective environment of the cell, proteins and other macromolecules must be treated so as to prevent their destruction by such influences as extremes of pH and temperature, enzymatic and chemical degradation, and rough mechanical handling. The state of purity of a substance being isolated must be monitored throughout the purification procedure by a specific assay.

2 Solubilities of Proteins Proteins are conveniently purified on a large scale by a fractional precipitation process called salting out, in which protein solubilities are varied by changing the salt concentration or pH.

3 Chromatographic Separations Ion exchange chromatography employs support materials such as cellulose or cross-linked dextran gels. Separations are based on differential electrostatic interactions between charged groups on the ion exchange materials and those on the substances being separated. Molecules may be located through their UV absorbance, fluorescence, radioactivity, or enzymatic activity. In gel filtration chromatography, molecules are separated according to their size and shape through the use of cross-linked dextran, polyacrylamide, or agarose beads that have pores of molecular dimensions. A calibrated gel filtration column can be used to estimate the molecular masses of macromolecules. Affinity chromatography separates biomolecules according to their unique biochemical abilities to bind other molecules specifically. High-performance liquid chromatography (HPLC) utilizes any of the foregoing separation techniques but uses high-resolution chromatographic materials, high solvent pressures, and automatic solvent mixing and monitoring systems so as to obtain much greater degrees of separation than are achieved with the more conventional chromatographic procedures.

Adsorption chromatography, thin layer chromatography (TLC), reverse-phase chromatography (RPC), hydrophobic interaction chromatography (HIC), and metal chelate affinity chromatography also have valuable biochemical applications.

4 Electrophoresis In electrophoresis, charged molecules are separated according to their rates of migration in an electric field on a solid support such as paper, cellulose acetate, cross-linked polyacrylamide, or agarose. Gel electrophoresis employs a cross-linked polyacrylamide or agarose gel support, so that molecules are separated according to size by gel filtration as well as according to charge. The separated molecules may be visualized by means of stains, autoradiography, or immunoblotting. The anionic detergent sodium dodecyl sulfate (SDS) denatures proteins and uniformly coats them so as to give most proteins a similar charge density and shape. SDS–PAGE may be used to estimate macromolecular masses. In isoelectric focusing (IEF), macromolecules are immersed in a stable pH gradient and subjected to an electric field that causes them to migrate to their isoelectric positions. In capillary electrophoresis, the use of thin capillary tubes and high electric fields permits rapid and highly resolved separations of small amounts of material.

5 Ultracentrifugation In ultracentrifugation, molecules are separated by subjecting them to gravitational fields large enough to counteract diffusional forces. Molecules may be separated and their molecular masses estimated from their rates of sedimentation through a solvent or a preformed gradient of an inert low molecular mass material such as sucrose. Alternately, molecules may be separated according to their buoyant densities in a solution with a density gradient of a dense, fast-diffusing substance such as CsCl. The deviation of a molecule's frictional ratio from unity is indicative of its degrees of solvation and elongation.

6 Nucleic Acid Fractionation Nucleic acids can be fractionated by many of the techniques that are used to separate proteins. Hydroxyapatite chromatography separates single-stranded DNA from double-stranded DNA. Polyacrylamide or agarose gel electrophoresis separates DNA largely on the basis of size. Very large DNAs can be separated by pulsed-field gel electrophoresis (PFGE) on agarose gels. DNAs may be fractionated according to their base composition by CsCl density gradient ultracentrifugation. Different species of RNA can be separated by zonal ultracentrifugation through a sucrose gradient.

REFERENCES

General

Ahmed, H., *Principles and Reactions of Protein Extraction, Purification, and Characterization,* CRC Press (2005).

Bonner, P.L.R., *Protein Purification,* Taylor & Francis (2007).

Boyer, R.F., *Biochemistry Laboratory: Modern Theory and Techniques,* Benjamin-Cummings (2006).

Burgess, R.R. and Deutscher, M.P. (Eds.), *Guide to Protein Purification* (2nd ed.), *Methods Enzymol.* **463,** (2009).

Harding, S.E. and Chowdhry, B.Z. (Eds.), *Protein Ligand Interactions: Structure and Spectroscopy. A Practical Approach,* Oxford University Press (2001). [Contains descriptions of a variety of physical techniques for studying proteins and their interactions with other molecules.]

Meyers, R.A., *Proteins. From Analytics to Structural Genomics,* Vol. 2, Chapters 20–24, Wiley–VCH (2007).

Ninfa, A.J., Ballou, D.P., and Benore, M., *Fundamental Laboratory Approaches for Biochemistry and Biotechnology* (2nd ed.), Wiley (2010).

Pingoud, A., Urbanke, C., Hoggett, J., and Jeltsch, A., *Biochemical Methods. A Concise Guide for Students and Researchers,* Wiley–VCH (2002).

Roe, S. (Ed.), *Protein Purification Techniques. A Practical Approach* (2nd ed.); and *Protein Purification Applications. A Practical Approach* (2nd ed.), Oxford University Press (2001).

Tinoco, I., Sauer, K., Wang, J.C., and Puglisi, J.C., *Physical Chemistry. Principles and Applications in Biological Sciences* (4th ed.), Chapter 6, Prentice-Hall (2002).

Simpson, R.J., Adams, P.D., and Golemis, E.A. (Eds.), *Basic Methods in Protein Purification and Analysis. A Laboratory Manual,* Cold Spring Harbor Laboratory Press (2009).

Structural Genomics Consortium, et al., Protein production and purification, *Nature Methods* **5,** 135–146 (2008). [Discusses methods for most efficiently producing and purifying recombinant proteins.]

Walker, J.M. (Ed.), *The Protein Protocols Handbook* (2nd ed.), Humana Press (2002).

Wilson, K. and Walker, J.M. (Eds.), *Principles and Techniques of Biochemistry and Molecular Biology* (6th ed.), Chapters 10 and 11, Cambridge University Press (2005).

Solubility and Crystallization

Arakawa, T. and Timasheff, S.N., Theory of protein solubility, *Methods Enzymol.* **114,** 49–77 (1985).

Ducruix, A. and Giegé, R. (Eds.), *Crystallization of Nucleic Acids and Proteins. A Practical Approach,* (2nd ed.), Oxford University Press (1999).

McPherson, A., *Crystallization of Biological Macromolecules,* Cold Spring Harbor Laboratory Press (1999).

Chromatography

Dean, P.D.G., Johnson, W.S., and Middle, F.A. (Eds.), *Affinity Chromatography. A Practical Approach,* IRL Press (1985).

Fischer, L., Gel filtration chromatography (2nd ed.), *in* Work, T.S. and Burdon, R.H. (Eds.), *Laboratory Techniques in Biochemistry and Molecular Biology,* Vol. 1, Part II, North-Holland Biomedical Press (1980).

Meyer, V.R., *Practical High-Performance Liquid Chromatography* (2nd ed.), Wiley (1994).

Oliver, R.W.A. (Ed.), *HPLC of Macromolecules. A Practical Approach* (2nd ed.), IRL Press (1998).

Rossomando, E.F., *HPLC in Enzymatic Analysis* (2nd ed.), Wiley (1998).

Weston, A. and Brown, P.R., *HPLC and CE. Principles and Practice,* Academic Press (1997).

Electrophoresis

Altria, K.D., *Capillary Electrophoresis Guidebook,* Humana Press (1996).

Baker, D.R., *Capillary Electrophoresis,* Wiley (1995).

Burmeister, M. and Ulanovsky, L., *Pulsed-Field Gel Electrophoresis,* Humana Press (1992).

Gersten, D.M., *Gel Electrophoresis: Proteins,* Wiley (1996).

Griffin, T.J. and Aebersold, R., Advances in proteome analysis by mass spectrometry, *J. Biol. Chem.* **276,** 45497–45500 (2001).

Hames, B.D. (Ed.), *Gel Electrophoresis of Proteins. A Practical Approach* (3rd ed.), IRL Press (1998).

Jones, P., *Gel Electrophoresis: Essential Techniques,* Wiley (1999).

Karger, B.L., Chu, Y.-H., and Foret, F., Capillary electrophoresis of proteins and nucleic acids, *Annu. Rev. Biophys. Biomol. Struct.* **24,** 579–610 (1995).

Monaco, A.P. (Ed.), *Pulsed Field Gel Electrophoresis. A Practical Approach*, IRL Press (1995).

Righetti, P.G., Immobilized pH gradients: Theory and methodology, *in* Burdon, R.H. and van Knippenberg, P.H. (Eds.), *Laboratory Techniques in Biochemistry and Molecular Biology*, Vol. 20, Elsevier (1990). [Discusses isoelectric focusing.]

Wehr, T., Rodríeguez-Diaz, R., and Zhu, M., *Capillary Electrophoresis of Proteins*, Marcel Dekker (1999).

Ultracentrifugation

Graham, J., *Biological Centrifugation,* Bios Scientific Publishers (2001).

Harding, S.E., Rowe, A.J., and Horton, J.C. (Eds.), *Analytical Ultracentrifugation in Biochemistry and Polymer Science*, Royal Society of Chemistry (1992).

Hesley, P., Defining the structure and stability of macromolecular assemblages in solution: The re-emergence of analytical ultracentrifugation as a practical tool, *Structure* **4,** 367–373 (1996).

Laue, T., Biophysical studies by ultracentrifugation, *Curr. Opin. Struct. Biol.* **11,** 579–583 (2001); *and* Laue, T.M. and Stafford, W.F., III, Modern applications of analytical ultracentrifugation, *Annu. Rev. Biophys. Biomol. Struct.* **28,** 75–100 (1999).

Mächtle, W. and Börger, L., *Analytical Ultracentrifugation,* Springer-Verlag (2006).

Schachman, H.K., *Ultracentrifugation in Biochemistry*, Academic Press (1959). [A classic treatise on ultracentrifugation.]

Schuster, T.M. and Toedt, J.M., New revolutions in the evolution of analytical ultracentrifugation, *Curr. Opin. Struct. Biol.* **6,** 650–658 (1996).

Stafford, W.F., III, Sedimentation velocity spins a new weave for an old fabric, *Curr. Opin. Biotech.* **8,** 14–24 (1997).

BIOINFORMATICS EXERCISES

Bioinformatics Exercises are available at www.wiley.com/college/voet.

Chapter 6

Electrophoresis
1. **One-Dimensional Electrophoresis.** Perform an SDS–PAGE electrophoresis simulation with known and unknown proteins.
2. **Two-Dimensional Electrophoresis.** Explore the predicted and observed electrophoretic parameters (p*I*, molecular mass, and fragmentation pattern) for a known protein.

PROBLEMS

1. What are the ionic strengths of 1.0*M* solutions of NaCl, $(NH_4)_2SO_4$, and K_3PO_4? In which of these solutions would a protein be expected to be most soluble; least soluble?

2. An **isotonic saline solution** (one that has the same salt concentration as blood) is 0.9% NaCl. What is its ionic strength?

3. In what order will the following amino acids be eluted from a column of P-cellulose ion exchange resin by a buffer at pH 6: arginine, aspartic acid, histidine, and leucine?

4. In what order will the following proteins be eluted from a CM-cellulose ion exchange column by an increasing salt gradient at pH 7: fibrinogen, hemoglobin, lysozyme, pepsin, and ribonuclease A (see Table 6-1)?

5. What is the order of elution of the following proteins from a Sephadex G-50 column: catalase, α-chymotrypsin, concanavalin B, lipase, and myoglobin (see Table 6-4)?

6. Estimate the molecular mass of an unknown protein that elutes from a Sephadex G-50 column between cytochrome *c* and ribonuclease A (see Table 6-4).

7. A gel chromatography column of Bio-Gel P-30 with a bed volume of 100 mL is poured. The elution volume of the protein hexokinase (96 kD) on this column is 34 mL. That of an unknown protein is 50 mL. What are the void volume of the column, the volume occupied by the gel, and the relative elution volume of the unknown protein?

8. What chromatographic method would be suitable for separating the following pairs of substances? (a) Ala-Phe-Lys, Ala-Ala-Lys; (b) lysozyme, ribonuclease A (see Table 6-1); and (c) hemoglobin, myoglobin (see Table 6-1).

9. What is the order of the R_f values of the following amino acids in their paper chromatography with a water/butanol/acetic acid solvent system in which the pH of the aqueous phase is 4.5: alanine, aspartic acid, lysine, glutamic acid, phenylalanine, and valine?

10. What fractionation procedure could be used to purify Protein 1 from a mixture of three proteins whose amino acid compositions are as follows?

1. 25% Ala, 20% Gly, 20% Ser, 10% Ile, 10% Val, 5% Asn, 5% Gln, 5% Pro
2. 30% Gln, 25% Glu, 20% Lys, 15% Ser, 10% Cys
3. 25% Asn, 20% Gly, 20% Asp, 20% Ser, 10% Lys, 5% Tyr

All three proteins are similar in size and p*I*, and there is no antibody available for Protein 1.

*11. Purification tables are often used to keep track of the yield and purification of a protein. The specific activity is the ratio of the amount of the protein of interest, here myoglobin (Mb), obtained at given step (in μmol or enzyme units) divided by the amount (mg) of total protein. The yield is the ratio of the amount of the protein of interest obtained at a given purification step (in μmol or enzyme units) divided by the original amount present in the crude extract, often converted to percent yield by multiplying by 100. The fold purification is the ratio of the specific activity of the purified protein to that of the crude extract.

(a) For the purification table below, calculate the specific activity, % yield, and fold purification for the empty cells.
(b) Which step, DEAE or affinity chromatography, causes the greatest loss of Mb?
(c) Which step causes the greater purification of Mb?
(d) If you could use only one purification step, which technique would be best?

Purification Table (for Problem 11)

Purification Step	mg Total Protein	μmol Mb	Specific Activity (μmol protein/mg total protein)	% Yield	Fold Purification
1. Crude extract	1550	0.75	100	1	
2. DEAE-cellulose chromatography	550	0.35			
3. Affinity chromatography	5.0	0.28			

12. The neurotransmitter γ-aminobutyric acid is thought to bind to specific receptor proteins in nerve tissue. Design a procedure for the partial purification of such a receptor protein.

13. A mixture of amino acids consisting of arginine, cysteine, glutamic acid, histidine, leucine, and serine is applied to a strip of paper and subjected to electrophoresis using a buffer at pH 7.5. What are the directions of migration of these amino acids and what are their relative mobilities?

*14. Sketch the appearance of a fingerprint of the following tripeptides: Asn-Arg-Lys, Asn-Leu-Phe, Asn-His-Phe, Asp-Leu-Phe, and Val-Leu-Phe. Assume the paper chromatographic step is carried out using a water/butanol/acetic acid solvent system (pH 4.5) and the electrophoretic step takes place in a buffer at pH 6.5.

15. What is the molecular mass of a protein that has a relative electrophoretic mobility of 0.5 in an SDS–polyacrylamide gel such as that of Fig. 6-25?

16. Explain why the molecular mass of fibrinogen is significantly overestimated when measured using a calibrated gel filtration column (Fig. 6-10) but can be determined with reasonable accuracy from its electrophoretic mobility on an SDS–polyacrylamide gel (see Table 6-4).

17. (a) What would be the relative arrangement of the following proteins after they had been subjected to isoelectric focusing: insulin, cytochrome *c*, histone, myoglobin, and ribonuclease A? (b) Sketch the appearance of a two-dimensional gel electrophoretogram of cytochrome *c*, myoglobin, and ribonuclease A (see Tables 6-1 and 6-4).

18. Calculate the centrifugal acceleration, in gravities (*g*'s), on a particle located 6.5 cm from the axis of rotation of an ultracentrifuge rotating at 60,000 rpm (1 *g* = 9.81 m · s^{-2}).

19. In a dilute buffer solution at 20°C, rabbit muscle aldolase has a frictional coefficient of 8.74×10^{-8} g · s^{-1}, a sedimentation coefficient of 7.35 S, and a partial specific volume of 0.742 cm^3 · g^{-1}. Calculate the molecular mass of aldolase assuming the density of the solution to be 0.998 g · cm^{-3}.

*20. The sedimentation coefficient of a protein was measured by observing its sedimentation at 20°C in an ultracentrifuge spinning at 35,000 rpm.

Time, *t* (min)	Distance of Boundary from Center of Rotation, *r* (cm)
4	5.944
6	5.966
8	5.987
10	6.009
12	6.032

The density of the solution is 1.030 g · cm^{-3}, the partial specific volume of the protein is 0.725 cm^3 · g^{-1}, and its frictional coefficient is 3.72×10^{-8} g · s^{-1}. Calculate the protein's sedimentation coefficient, in svedbergs, and its molecular mass.

Covalent Structures of Proteins and Nucleic Acids

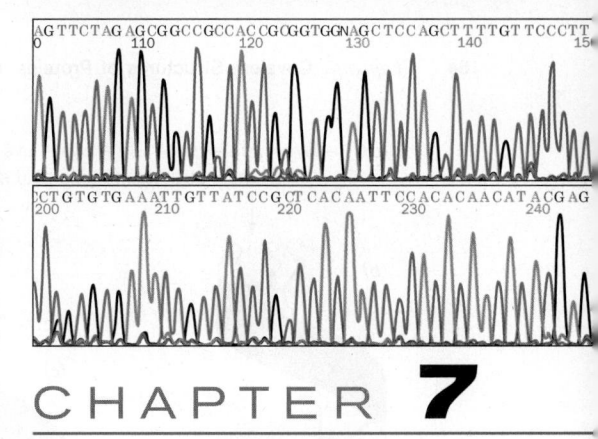

CHAPTER 7

Proteins are at the center of the action in biological processes. They function as enzymes, which catalyze the complex set of chemical reactions that are collectively referred to as life. Proteins serve as regulators of these reactions, both directly as components of enzymes and indirectly in the form of chemical messengers, known as hormones, as well as the receptors for those hormones. They act to transport and store biologically important substances such as metal ions, O_2, glucose, lipids, and many other molecules. In the form of muscle fibers and other contractile assemblies, proteins generate the coordinated mechanical motion of numerous biological processes, such as the separation of chromosomes during cell division and the movement of your eyes as you read this page. Proteins, such as **rhodopsin** in the retina of your eye, acquire sensory information that is processed through the actions of nerve cell proteins. The proteins of the immune system, such as the **immunoglobulins,** form an essential biological defense system in higher animals. Proteins are major active elements in, as well as products of, the expression of genetic information. However, proteins also have important passive roles, such as that of **collagen,** which provides bones, tendons, and ligaments with their characteristic tensile strength. Clearly, there is considerable validity to the old cliché that proteins are the "building blocks" of life.

The function of DNA as the genetic archive and the association of RNA with protein synthesis have been known since the mid-twentieth century. However, it was not until the 1970s that it became clear that RNA can form structures whose complexities rival those of proteins, and it was not until the mid-1980s that it was shown that RNA has biologically important catalytic functions.

Protein and nucleic acid function can best be understood in terms of their structures, that is, the three-dimensional relationships between their component atoms. The structural descriptions of proteins and nucleic acids, as well as those of other polymeric materials, have been traditionally described in terms of four levels of organization (Fig. 7-1):

1. Primary structure (1° structure), which for a protein is the amino acid sequence of its polypeptide chain(s) and for a nucleic acid is its base sequence.

2. Secondary (2°) structure, which is the local spatial arrangement of a polypeptide's or a nucleic acid's backbone atoms without regard to the conformations of their side chains or bases.

3. Tertiary (3°) structure, which refers to the three-dimensional structure of an entire polypeptide or polynucleotide chain. The distinction between secondary and tertiary structures is, of necessity, somewhat vague; in practice, the term "secondary structure" alludes to easily characterized structural entities such as helices.

(a)   – Lys – Ala – His – Gly – Lys – Lys – Val – Leu – Gly - Ala –

Primary structure (amino acid sequence in a polypeptide chain)

(b)

(c)

(d) β₂   β₁

Secondary
structure
(helix)

Tertiary structure:
one complete protein chain
(β chain of hemoglobin)

β

Quaternary structure:
the four separate chains
of hemoglobin assembled
into an oligomeric protein

© IRVING GEIS

α₂   α₁

Figure 7-1   **The structural hierarchy in proteins.** (*a*) Primary structure, (*b*) secondary structure,
(*c*) tertiary structure, and (*d*) quaternary structure. [Illustration, Irving Geis. Image from the
Irving Geis Collection, Howard Hughes Medical Institute. Reprinted with permission.]

4. Most proteins are composed of two or more polypeptide chains, loosely referred to as **subunits,** which associate through noncovalent interactions and, in some cases, disulfide bonds. A protein's **quaternary (4°) structure** refers to the spatial arrangement of its subunits. A nucleic acid's quaternary structure is similarly defined.

In this chapter, we discuss the 1° structures of proteins and nucleic acids: how they are elucidated and their biological and evolutionary significance. We also survey the field of bioinformatics as well as methods of chemically synthesizing polypeptide and oligonucleotide chains. The 2°, 3°, and 4° structures of proteins and nucleic acids, as we shall see, are a consequence of their 1° structures. For proteins, these topics are treated in Chapters 8 and 9, whereas for nucleic acids, they are discussed mainly in Chapters 29, 31, and 32.

1 PRIMARY STRUCTURE DETERMINATION OF PROTEINS

✏️ **See Guided Exploration 4: Protein sequence determination** The first determination of the complete amino acid sequence of a protein, that of the bovine polypeptide hormone **insulin** by Frederick Sanger in 1953, was of enormous biochemical significance in that it definitively established that proteins have unique covalent structures. Since that time, the amino

acid sequences of tens of thousands of proteins have been elucidated. This extensive information has been of central importance in the formulation of modern concepts of biochemistry for several reasons:

1. The knowledge of a protein's amino acid sequence is essential for an understanding of its molecular mechanism of action as well as being prerequisite for the elucidation of its three-dimensional structure by both X-ray crystallography and nuclear magnetic resonance (NMR) spectroscopy (Section 8-3A).

2. Sequence comparisons among analogous proteins from the same individual, from members of the same species, and from members of related species have yielded important insights into how proteins function and have indicated the evolutionary relationships among the proteins and the organisms that produce them. These analyses, as we shall see in Section 7-3, complement and extend corresponding taxonometric studies based on anatomical comparisons.

3. Amino acid sequence analyses have important clinical applications because many inherited diseases are caused by mutations leading to an amino acid change in a protein. Recognition of this fact has led to the development of valuable diagnostic tests for many such diseases and, in many cases, to symptom-relieving therapy.

The elucidation of the 51-residue primary structure of insulin (Fig. 7-2) was the labor of many scientists over a

A chain

Gly—Ile—Val—Glu—Gln—Cys—Cys—Ala—Ser—Val—Cys—Ser—Leu—Tyr—Gln—Leu—Glu—Asn—Tyr—Cys—Asn
        5   10   15   21

B chain

Phe—Val—Asn—Gln—His—Leu—Cys—Gly—Ser—His—Leu—Val—Glu—Ala—Leu—Tyr—Leu—Val—Cys—Gly—Glu—Arg—Gly—Phe—Phe—Tyr—Thr—Pro—Lys—Ala
        5   10   15   20   25   30

Figure 7-2 **Primary structure of bovine insulin.** Note the intrachain and interchain disulfide bond linkages.

period of a decade that altogether utilized ~100 g of protein. Procedures for primary structure determination have since been so refined and automated that proteins of similar size can be sequenced by an experienced technician in a few days using only a few micrograms of protein. The sequencing of the 1021-residue enzyme **β-galactosidase** in 1978 signaled that the sequence analysis of almost any protein could be reasonably attempted. Despite these technical advances, the basic procedure for primary structure determination using the techniques of protein chemistry is that developed by Sanger. The procedure consists of three conceptual parts, each of which requires several laboratory steps:

1. **Prepare the protein for sequencing:**

   a. Determine the number of chemically different polypeptide chains (subunits) in the protein.
   b. Cleave the protein's disulfide bonds.
   c. Separate and purify the unique subunits.

2. **Sequence the polypeptide chains:**

   a. Fragment the individual subunits at specific points to yield peptides small enough to be sequenced directly.
   b. Separate and purify the fragments.
   c. Determine the amino acid sequence of each peptide fragment.
   d. Repeat Step 2a with a fragmentation process of different specificity so that the subunit is cleaved at peptide bonds different from before. Separate these peptide fragments as in Step 2b and determine their amino acid sequences as in Step 2c.

3. **Organize the completed structure:**

   a. Span the cleavage points between one set of peptide fragments by the other. By comparison, the sequences of these sets of polypeptides can be arranged in the order that they occur in the subunit, thereby establishing its amino acid sequence.
   b. Elucidate the positions of the disulfide bonds, if any, between and within the subunits.

We discuss these various steps in the following sections.

A. *End Group Analysis: How Many Different Types of Subunits?*

Each polypeptide chain (if it is not chemically blocked or circular) has an N-terminal residue and a C-terminal residue. By identifying these **end groups,** we can establish the number of chemically distinct polypeptides in a protein. For example, insulin has equal amounts of the N-terminal residues Phe and Gly, which indicates that it has equal numbers of two nonidentical polypeptide chains.

a. N-Terminus Identification

There are several effective methods by which a polypeptide's N-terminal residue may be identified. **1-Dimethyl-aminonaphthalene-5-sulfonyl chloride (dansyl chloride)** reacts with primary amines (including the ε-amino group of Lys) to yield dansylated polypeptides (Fig. 7-3). Acid hydrolysis liberates the N-terminal residue as a **dansylamino acid,** which exhibits such intense yellow fluorescence that it can be chromatographically identified from as little as 100 picomoles of material [1 picomole (pmol) = 10^{-12} mol].

In the most useful method of N-terminal residue identification, the **Edman degradation** (named after its inventor, Pehr Edman), **phenylisothiocyanate (PITC, Edman's reagent)** reacts with the N-terminal amino groups of proteins under mildly alkaline conditions to form their **phenylthiocarbamyl (PTC)** adduct (Fig. 7-4). This product is treated with an anhydrous strong acid such as trifluoroacetic acid, which cleaves the N-terminal residue as its **thiazolinone** derivative but does not hydrolyze other peptide bonds. *The Edman degradation therefore releases the N-terminal amino acid residue but leaves intact the rest of the polypeptide chain.* The thiazolinone-amino acid is selectively extracted into an organic solvent and is converted to the more stable **phenylthiohydantoin (PTH)** derivative by treatment with aqueous acid. This PTH-amino acid is most commonly identified by comparing its retention time on HPLC with those of known PTH-amino acids.

The most important difference between the Edman degradation and other methods of N-terminal residue identification is that *we can determine the amino acid sequence of a polypeptide chain from the N-terminus inward by subjecting the polypeptide to repeated cycles of the Edman degradation and, after every cycle, identifying the newly liberated PTH-amino acid.* This technique has been automated, resulting in great savings of time and materials (Section 7-1F).

b. C-Terminus Identification

There is no reliable chemical procedure comparable to the Edman degradation for the sequential end group analysis from the C-terminus of a polypeptide. This can be done enzymatically, however, using **exopeptidases** (enzymes that

Figure 7-3   **The reaction of dansyl chloride in end group analysis.**

cleave a terminal residue from a polypeptide). One class of exopeptidases, the **carboxypeptidases,** catalyzes the hydrolysis of the C-terminal residues of polypeptides:

Carboxypeptidases, like all enzymes, are highly specific (selective) for the chemical identities of the substances whose reactions they catalyze (Section 13-2). The side chain specificities of the various carboxypeptidases in common use are listed in Table 7-1. The second type of exopeptidase

listed in Table 7-1, the **aminopeptidases,** sequentially cleave amino acids from the N-terminus of a polypeptide and have been similarly used to determine N-terminal sequences.

Why can't carboxypeptidases be used to determine amino acid sequences? If a carboxypeptidase cleaved all C-terminal residues at the same rate, irrespective of their identities, then by following the course of appearance of the various free amino acids in the reaction mixture (Fig. 7-5a), the sequence of several amino acids at the C-terminus could be determined. If, however, the second amino acid residue, for example, were cleaved at a much faster rate than the first, both amino acids would appear to be released simultaneously (Fig. 7-5b). Carboxypeptidases, in fact, exhibit selectivity toward side chains, so their use, either singly or in mixtures, rarely reveals the order of more than the first few C-terminal residues of a polypeptide.

C-Terminal residues with a preceding Pro residue are not subject to cleavage by carboxypeptidases A and B (Table 7-1). Chemical methods are therefore usually employed to identify their C-terminal residue. In the most reliable such chemical method, **hydrazinolysis,** a polypeptide is treated with anhydrous **hydrazine** at 90°C for 20 to 100 h

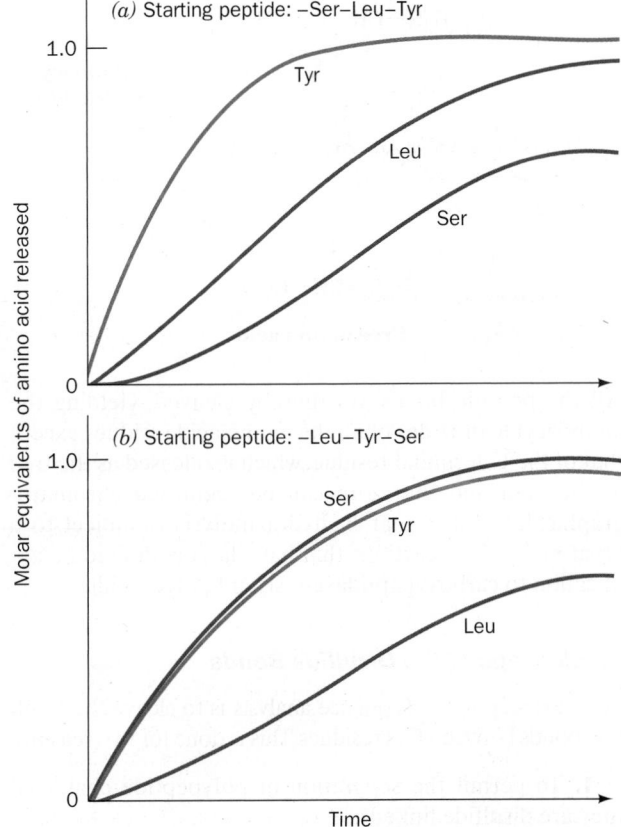

Phenylisothiocyanate (PITC)

Polypeptide

OH⁻

PTC polypeptide

anhydrous F₃CCOOH

Thiazolinone derivative

+

Original polypeptide less its N-terminal residue

H⁺

PTH-amino acid

Figure 7-4 **The Edman degradation.** Note that the reaction occurs in three separate stages that each require quite different conditions. Amino acid residues can therefore be sequentially removed from the N-terminus of a polypeptide in a controlled stepwise fashion. **See the Animated Figures**

(a) Starting peptide: –Ser–Leu–Tyr

1.0

Tyr

Leu

Ser

0

Molar equivalents of amino acid released

(b) Starting peptide: –Leu–Tyr–Ser

1.0

Ser

Tyr

Leu

0

Time

Figure 7-5 **Hypothetical rate of the carboxypeptidase-catalyzed release of amino acids from peptides having the indicated C-terminal sequences.** (*a*) All bonds cleaved at the same rate. (*b*) Ser removed slowly, Tyr cleaved rapidly, and Leu cleaved at an intermediate rate.

Table 7-1   **Specificities of Various Exopeptidases**

Enzyme	Source	Specificity[a]
Carboxypeptidase A	Bovine pancreas	$R_n \neq$ Arg, Lys, Pro; $R_{n-1} \neq$ Pro
Carboxypeptidase B	Bovine pancreas	$R_n \neq$ Arg, Lys, $R_{n-1} \neq$ Pro
Carboxypeptidase C	Citrus leaves	All free C-terminal residues; pH optimum = 3.5
Carboxypeptidase Y	Yeast	All free C-terminal residues, but slowly with $R_n =$ Gly
Leucine aminopeptidase	Porcine kidney	$R_1 \neq$ Pro
Aminopeptidase M	Porcine kidney	All free N-terminal residues

[a] R_1 = the N-terminal residue; R_n = the C-terminal residue.

in the presence of a mildly acidic ion exchange resin (which acts as a catalyst):

Polypeptide

+

NH_2—NH_2

Hydrazine

acidic ion exchange resin catalyst

Free amino acid

All the peptide bonds are thereby cleaved, yielding the aminoacyl hydrazides of all the amino acid residues except that of the C-terminal residue, which is released as the free amino acid and therefore can be identified chromatographically. Unfortunately, hydrazinolysis is subject to a great many side reactions that have largely limited its application to carboxypeptidase-resistant polypeptides.

B. Cleavage of the Disulfide Bonds

The next step in the sequence analysis is to cleave the disulfide bonds between Cys residues. This is done for two reasons:

1. To permit the separation of polypeptide chains (if they are disulfide linked).

2. To prevent the native protein conformation, which is stabilized by disulfide bonds, from obstructing the action of the proteolytic (protein-cleaving) agents used in primary structure determinations (Section 7-1D).

Disulfide bond locations are established in the final step of the sequence analysis (Section 7-1H).

Disulfide bonds are most often cleaved reductively by treatment with sulfhydryl-containing compounds such as **2-mercaptoethanol:**

Cystine   **2-Mercaptoethanol**

Cysteine

In order to expose all disulfide groups to the reducing agent, the reaction is usually carried out under conditions that denature the protein. The resulting free sulfhydryl groups are alkylated, usually by treatment with **iodoacetic acid,** to prevent the reformation of disulfide bonds through oxidation by O_2. *S*-Alkyl derivatives are stable in air and under the conditions used for the subsequent cleavage of peptide bonds.

$$Cys-CH_2-SH \quad + \quad ICH_2COO^-$$

Cysteine   **Iodoacetate**

$$Cys-CH_2-S-CH_2COO^- \quad + \quad HI$$

***S*-Carboxymethylcysteine**

C. *Separation, Purification, and Characterization of the Polypeptide Chains*

A protein's nonidentical polypeptides must be separated and purified in preparation for their amino acid sequence determination. Subunit dissociation, as well as denaturation, occurs under acidic or basic conditions, at low salt concentrations, at elevated temperatures, or through the use of denaturing agents such as urea, guanidinium ion (Section 5-5G), or detergents such as sodium dodecyl sulfate (SDS; Section 6-4C). The dissociated subunits can then be separated by methods described in Chapter 6 that capitalize on small differences in polypeptide size and polarity. Ion exchange and gel filtration chromatography, usually by HPLC (Section 6-3Dg), are most often used.

It is, of course, desirable to know the number of residues in the polypeptide to be sequenced, which can be estimated from its molecular mass (~110 D/residue). Molecular mass can be measured with an accuracy of no better than 5 to 10% by the more traditional laboratory techniques of gel filtration chromatography and SDS–PAGE (Sections 6-3Ba and 6-4C). In recent years, however, mass spectrometry (Section 7-1I) has provided a faster and far more accurate means to determine the molecular masses of macromolecules. Mass spectrometry can determine the molecular masses of picomolar amounts of >100 kD polypeptides with accuracies of ~0.01%.

D. *Specific Peptide Cleavage Reactions*

Polypeptides that are longer than 40 to 100 residues cannot be directly sequenced (Section 7-1F). Polypeptides of greater length must therefore be cleaved, either enzymatically or chemically, to fragments small enough to be sequenced (the polypeptides produced by *E. coli* and humans have broad distributions of lengths that average ~320 and ~470 residues, respectively). In either case, the cleavage process must be complete and highly specific so that the aggregate sequence of a subunit's peptide fragments, when correctly ordered, is that of the intact subunit.

a. Trypsin Specifically Cleaves Peptide Bonds after Positively Charged Residues

Endopeptidases (enzymes that catalyze the hydrolysis of internal peptide bonds), like exopeptidases, have side chain requirements for the residues flanking the **scissile** (to be cleaved) peptide bond. The side chain specificities of the endopeptidases most commonly used to fragment polypeptides are listed in Table 7-2. The digestive enzyme **trypsin** has the greatest specificity and is therefore the

Table 7-2   **Specificities of Various Endopeptidases**

Enzyme	Source	Specificity	Comments
		$$\begin{array}{ccccc} & R_{n-1} & O & & R_n & O \\ & | & \| & & | & \| \\ -NH-CH & - & C & -NH-CH & - & C- \end{array}$$ ↑ **Scissile** **peptide bond**	
Trypsin	Bovine pancreas	R_{n-1} = positively charged residues: Arg, Lys; $R_n \neq$ Pro	Highly specific
Chymotrypsin	Bovine pancreas	R_{n-1} = bulky hydrophobic residues: Phe, Trp, Tyr; $R_n \neq$ Pro	Cleaves more slowly for R_{n-1} = Asn, His, Met, Leu
Elastase	Bovine pancreas	R_{n-1} = small neutral residues: Ala, Gly, Ser, Val; $R_n \neq$ Pro	
Thermolysin	*Bacillus thermoproteolyticus*	R_n = Ile, Met, Phe, Trp, Tyr, Val; $R_{n-1} \neq$ Pro	Occasionally cleaves at R_n = Ala, Asp, His, Thr; heat stable
Pepsin	Bovine gastric mucosa	R_n = Leu, Phe, Trp, Tyr; $R_{n-1} \neq$ Pro	Also others; quite nonspecific; pH optimum 2
Endopeptidase Arg-C	Mouse submaxillary gland	R_{n-1} = Arg	May cleave at R_{n-1} = Lys
Endopeptidase Asp-N	*Pseudomonas fragi*	R_n = Asp	May cleave at R_n = Glu
Endopeptidase Glu-C	*Staphylococcus aureus*	R_{n-1} = Glu	May cleave at R_{n-1} = Gly
Endopeptidase Lys-C	*Lysobacter enzymogenes*	R_{n-1} = Lys	May cleave at R_{n-1} = Asn

most valuable member of the arsenal of peptidases used to fragment polypeptides. It cleaves peptide bonds on the C-side (toward the carboxyl terminus) of the positively charged residues Arg and Lys if the next residue is not Pro:

NH$_3^+$
|
CH$_2$ **Lys**
| **(or Arg)**
CH$_2$
|
CH$_2$
|
CH$_2$ O   R   O
| ‖ | ‖
··· —NH—CH—C—NH—CH—C— ···

Any amino acid
residue but Pro

H$_2$O ——| trypsin

NH$_3^+$
|
CH$_2$
|
CH$_2$
|
CH$_2$
|
CH$_2$ O   R   O
| ‖ | ‖
··· —NH—CH—C—O$^-$ + H$_3$N$^+$—CH—C— ···

The other endopeptidases listed in Table 7-2 exhibit broader side chain specificities than trypsin and often yield a series of peptide fragments with overlapping sequences. However, through **limited proteolysis,** that is, by adjusting reaction conditions and limiting reaction times, these less specific endopeptidases can yield useful peptide fragments. This is because the complex native structure of a protein (subunit) buries many otherwise enzymatically susceptible peptide bonds beneath the surface of the protein molecule. With proper conditions and reaction times, only those peptide bonds in the native protein that are initially accessible to the peptidase will be hydrolyzed. Limited proteolysis is often employed to generate peptide fragments of useful size from subunits that have too many or too few Arg and Lys residues to do so with trypsin (although if too many are present, limited proteolysis with trypsin may also yield useful fragments).

b. Cyanogen Bromide Specifically Cleaves Peptide Bonds after Met Residues

Several chemical reagents promote peptide bond cleavage at specific residues. The most useful of these, **cyanogen bromide** (CNBr), causes specific and quantitative cleavage

on the C-side of Met residues to form a **peptidyl homoserine lactone:**

S: ⟶ C≡N
| ↘Br
CH$_2$
|
CH$_2$ **Cyanogen**
| **bromide**
··· —NH—CH
           _O
           C
           ‖
           O
: NH—CH—C— ···
         | ‖
         R   O

⟶ Br$^-$

CH$_3$
|
$^+$S—C≡N
|
CH$_2$
|
CH$_2$
|
··· —NH—CH
           _O
           C
           ‖
           O
: NH—CH—C— ···
         | ‖
         R   O

CH$_3$
|
S—C≡N   **Methyl**
         thiocyanate
+
        CH$_2$
       /   \
   CH$_2$ O
··· —NH—C   C
         | ‖
         H   N—CH—C— ···
                  | ‖
                  R   O

H$_2$O —|

        CH$_2$
       /   \
   CH$_2$ O
Peptidyl
homoserine ··· —NH—C   C
lactone | _
                  H   O
+
Aminoacyl H$_3$N$^+$—CH—C— ···
peptide | ‖
                    R   O

The reaction is performed in an acidic solvent (0.1*M* HCl or 70% formic acid) that denatures most proteins so that cleavage normally occurs at all Met residues.

A peptide fragment generated by a specific cleavage process may still be too large to sequence. In that case, after its purification, it can be subjected to a second round of fragmentation using a different cleavage process.

E. Separation and Purification of the Peptide Fragments

Once again we must employ separation techniques, this time to isolate the peptide fragments of specific cleavage operations for subsequent sequence determinations. The nonpolar residues of peptide fragments are not excluded from the aqueous environment as they are in native proteins (Chapter 8). Consequently, many peptide fragments aggregate, precipitate, and/or strongly adsorb to chromatographic materials, which can result in unacceptable peptide losses. Until around 1980, the trial-and-error development of methods that could satisfactorily separate a mixture of peptide fragments constituted the major technical challenge of a protein sequence determination, as well as its most time-consuming step. Such methods involved the use of denaturants, such as urea and SDS, to solubilize the peptide fragments, and the selection of chromatographic materials and conditions that would reduce their adsorptive losses. The advent of reverse-phase chromatography by HPLC (Section 6-3Dh), however, has largely reduced the separation of peptide fragments to a routine procedure.

F. Sequence Determination

Once the manageably sized peptide fragments that were formed through specific cleavage reactions have been isolated, their amino acid sequences can be determined. *This is done through repeated cycles of the Edman degradation* (Section 7-1Aa). An automated device for doing so was first developed by Edman and Geoffrey Begg. In modern sequencers, the peptide sample is adsorbed onto a polyvinylidene fluoride (PVDF) membrane or dried onto glass fiber paper which is impregnated with **polybrene** (a polymeric quaternary ammonium salt). In either case, the peptide is immobilized but is readily accessible to Edman reagents. Accurately measured quantities of reagents, either in solution or as vapors in a stream of argon (which minimizes peptide loss), are then delivered to the reaction cell at programmed intervals. The thiazolinone-amino acids are automatically removed, converted to the corresponding PTH-amino acids (Fig. 7-4), and identified via HPLC. Such instruments are capable of processing up to one residue per hour.

A peptide's 40 to 60 N-terminal residues can usually be identified (100 or more with the most advanced systems) before the cumulative effects of incomplete reactions, side reactions, and peptide loss make further amino acid identification unreliable. As little as 0.1 pmol of a PTH-amino acid can be reliably identified by the UV detector–equipped reverse-phase HPLC systems used in advanced sequencers. Consequently, a peptide's N-terminal 5 to 25 residues can, respectively, be determined with as little as 1 to 10 pmol of the peptide—invisibly small amounts.

G. Ordering the Peptide Fragments

With the peptide fragments individually sequenced, what remains is to elucidate the order in which they are connected in the original polypeptide. *We do so by comparing the amino acid sequences of one set of peptide fragments with those of a second set whose specific cleavage sites overlap those of the first set* (Fig. 7-6). The overlapping peptide segments must be of sufficient length to identify each cleavage site uniquely, but as there are 20 possibilities for each amino acid residue, an overlap of only a few residues is usually enough.

Figure 7-6 **The amino acid sequence of a polypeptide chain is determined by comparing the sequences of two sets of mutually overlapping peptide fragments.** In this example, the two sets of peptide fragments are generated by cleaving the polypeptide after all its Arg and Lys residues with trypsin and, in a separate reaction, after all its Met residues by treatment with cyanogen bromide. The order of the first two tryptic peptides is established, for example, by the observation that the Gly-Ala-Lys-Leu-Pro-Met cyanogen bromide peptide has its N- and C-terminal sequences in common with the C- and N-termini, respectively, of the two tryptic peptides. In this manner the order of the peptide fragments in their parent polypeptide chain is established. **See the Animated Figures**

H. *Assignment of Disulfide Bond Positions*

The final step in an amino acid sequence analysis is to determine the positions (if any) of the disulfide bonds. This is done by cleaving a sample of the native protein under conditions that leave its disulfide bonds intact. The resulting peptide fragments are then separated by reverse-phase HPLC. The disulfide-linked peptide fragments are easily identified because, for each such linkage, two of the original peptide fragments will be replaced by a larger one. The disulfide-linked fragments are then subjected to Edman degradation. Although such a fragment yields two PTH-amino acids in each step of this process (at least initially), their locations within the predetermined amino acid sequence of the protein are readily surmised, thereby establishing the positions of the disulfide bonds.

I. *Peptide Characterization and Sequencing by Mass Spectrometry*

Mass spectrometry (MS) has emerged as an important technique for characterizing and sequencing polypeptides. *MS accurately measures the mass-to-charge (m/z) ratio for ions in the gas phase* (where *m* is the ion's mass and *z* is its

charge). Yet, until about 1985, macromolecules such as proteins and nucleic acids could not be analyzed by MS. This was because the method by which mass spectrometers produced gas phase ions destroyed macromolecules: vaporization by heating followed by ionization via bombardment with electrons. However, the development of two techniques has eliminated this roadblock:

1. Electrospray ionization (ESI; Fig. 7-7*a*), a technique pioneered by John Fenn in which a solution of a macromolecule such as a peptide is sprayed from a narrow capillary tube maintained at high voltage (~4000 V), forming fine, highly charged droplets from which the solvent rapidly evaporates. This yields a series of gas phase macromolecular ions that typically have ionic charges in the range +0.5 to +2 per kilodalton. For polypeptides, the ionic charges result from the protonation of basic side chains such as Lys and Arg [$(M + nH)^{n+}$ ions].

2. Matrix-assisted laser desorption/ionization (MALDI; Fig. 7-7*b*), in which the macromolecule is embedded in a crystalline matrix of a low molecular mass organic molecule (usually prepared by drying a droplet of solution containing the macromolecule and a large excess of the organic molecule) and irradiated with intense short

(a) Electrospray ionization (ESI)

(b) Matrix-assisted laser desorption/ionization (MALDI)

Figure 7-7 **Generation of the gas phase ions required for the mass spectrometric analysis of proteins.** (*a*) By electrospray ionization (ESI) and (*b*) by matrix-assisted laser desorption/ionization (MALDI). In ESI, a stream of dry N_2 or

some other gas is used to promote the evaporation of the solvent from the droplets. [After Fitzgerald, M.C. and Siuzdak, G., *Chem. Biol.* **3,** 708 (1996).]

Figure 7-8   The ESI-MS spectrum of the 16,951-D horse heart protein apomyoglobin. The measured *m/z* ratios and the inferred charges for most of the peaks are indicated. Note the bell-shaped distribution of the peaks, which is typical of ESI-MS spectra. The peaks all have shoulders because the polypeptide's component elements contain small admixtures of heavier isotopes (e.g., naturally abundant carbon consists of 98.9% ^{12}C and 1.1% ^{13}C and naturally abundant sulfur consists of 0.8% ^{33}S, 4.2% ^{34}S, and 95.0% ^{35}S). [After Yates, J.R., *Methods Enzymol.* **271,** 353 (1996).]

(nanosecond) pulses of laser light at a wavelength absorbed by the matrix material but not the macromolecule. The energy absorbed by the matrix ejects the intact macromolecules from its surface into the gas phase, usually with a charge of +1 but, with larger molecules, occasionally with charges of +2, +3, etc. For polypeptides, **gentisic acid** (2,5-dihydroxybenzoic acid) is one of the few substances found to have satisfactory properties as a matrix. Through the use of MALDI, polypeptides of >400 kD have been characterized.

In both of these techniques, the gas phase macromolecular ions are directed into the mass spectrometer, which measures their *m/z* values with an accuracy of >0.01%. Consequently, if an ion's *z* value can be determined, its molecular mass can be determined with far greater accuracy than by any other method. For example, Fig. 7-8 shows the ESI-based mass spectrum **(ESI-MS)** of the 16,951-D protein **myoglobin.** Note that successive peaks in this spectrum differ by a single ionic charge with the rightmost peak corresponding to an $(M + 9H)^{9+}$ ion. Consequently, for the mass spectrum of a macromolecule of molecular mass M containing two adjacent peaks with *m/z* values of p_1 and p_2 arising from ions with charges of z_1 and $z_1 - 1$,

$$p_1 = \frac{M + z_1}{z_1} \qquad [7.1]$$

and

$$p_2 = \frac{M + z_1 - 1}{z_1 - 1} \qquad [7.2]$$

These two linear equations can therefore be readily solved for their two unknowns, M and z_1.

Since most mass spectrometers are limited to detecting ions with *m/z* values less than several thousand, the use of ESI-MS has the advantage that the high ionic charges of the ions it produces has permitted the analysis of compounds with molecular masses >100 kD. Another advantage of ESI-MS is that it can be configured to operate in a continuous flow mode with the effluent of an HPLC or CE system. ESI-MS is used in this way, for example, to characterize the tryptic digest of a protein by determining the molecular masses of its component peptides (Section 7-1J).

a. Peptide Sequencing by Mass Spectrometry

Short polypeptides (<25 residues) can be directly sequenced though the use of a **tandem mass spectrometer (MS/MS;** two mass spectrometers coupled in series; Fig. 7-9). The first mass spectrometer functions to select the peptide ion of interest from other peptide ions as well as any contaminants that may be present. The selected peptide ion (P_3 in Fig. 7-9*b*) is then passed into a collision cell, where it

Figure 7-9 **The use of a tandem mass spectrometer (MS/MS) in amino acid sequencing.** (*a*) An MS/MS consists of an ion source (here shown as an ESI system), a first mass spectrometer (MS-1), a collision cell, a second mass spectrometer (MS-2), and a detector. (*b*) The ion source generates gas phase peptide ions, P_1, P_2, etc., from a digest of the protein being analyzed. These peptides are separated by MS-1 according to their *m/z* values and one of them, here P_3, is directed into the collision cell, where it collides with helium atoms. This treatment induces the breakdown of the polypeptide ion to yield the fragments F_1, F_2, etc., which are directed into MS-2, where their *m/z* values are determined. [Part *a* after Yates, J.R., *Methods Enzymol.* **271**, 358 (1996); Part *b* after Biemann, K. and Scoble, H.A., *Science* **237**, 992 (1987).]

collides with chemically inert atoms such as helium. The energy thereby imparted to a peptide ion causes it to fragment predominantly at only one of its several peptide bonds, yielding one or two charged fragments (Fig. 7-10). The molecular masses of the charged fragments are then determined by the second mass spectrometer.

By comparing the molecular masses of successively larger members of a family of fragments, the molecular masses and therefore the identities of the corresponding amino acid residues can be determined. The sequence of an entire polypeptide can thus be elucidated (although MS cannot distinguish the isomeric residues Ile and Leu because they have exactly the same mass and cannot always reliably distinguish Gln and Lys residues because their molecular masses differ by only 0.036 D). Computerization of this comparison process has reduced the time required to sequence a (short) polypeptide to only a few minutes as compared to the 30 to 50 min required per cycle of Edman degradation. The reliability of this process has been increased through the computerized matching of a measured mass spectrum with those of peptides of known sequence as maintained in databases.

The sequences of several polypeptides in a mixture can be determined, even in the presence of contaminants, by sequentially selecting the corresponding polypeptide ions in the first mass spectrometer of the tandem instrument. Hence, in separating and purifying the polypeptide fragments of a protein digest in preparation for their sequencing, less effort need be expended for MS/MS-based as compared to Edman techniques. MS/MS can also be used to sequence peptides with chemically blocked N-termini (a

common eukaryotic post-translational modification that prevents Edman degradation) and to characterize other post-translational modifications such as phosphorylations (Section 4-3A) and glycosylations (Section 11-3C). Finally, MS/MS can be used to locate disulfide bonds by chemically or enzymatically fragmenting both the disulfide-cleaved protein and the intact protein. From the disappearance of peaks from the mass spectrum of the latter relative to that of the former and the appearance of new peaks, the positions of the disulfide bonds in the protein can be deduced. Thus, MS/MS has become an indispensable tool for the characterization of polypeptides.

J. Peptide Mapping

Once the primary structure of a protein has been elucidated, that of a nearly identical protein, such as one arising from a closely related species, a mutation, or a chemical modification, can be more easily determined. This was originally done through the combined paper chromatography and paper electrophoresis (Section 6-4A) of partial protein digests, a technique synonymously known as **fingerprinting** or **peptide mapping.** The peptide fragments incorporating the amino acid variations migrate to different positions on their fingerprint (peptide map) than do the corresponding peptides of the original protein (Fig. 7-11). The variant peptides could then be eluted and sequenced, thereby identifying the changes in the protein without the need to sequence it in its entirety.

In more recent times, peptide mapping has come to mean any method that fragments a protein in a repro-

(a)

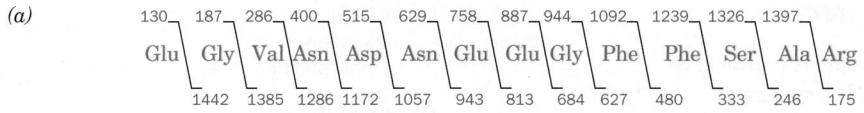

	130	187	286	400	515	629	758	887	944	1092	1239	1326	1397	
Glu	Gly	Val	Asn	Asp	Asn	Glu	Glu	Gly	Phe	Phe	Ser	Ala	Arg
	1442	1385	1286	1172	1057	943	813	684	627	480		333	246	175

(b)

Figure 7-10 **The tandem mass spectrum of the doubly charged ion of the 14-residue human [Glu¹] fibrinopeptide B (*m/z* = 786).** (*a*) The peptide's sequence. The upper and lower rows of numbers are the molecular masses of the charged N-terminal and C-terminal fragments, respectively, that are formed by the cleavage indicated by their connecting diagonal line. (*b*) The mass spectrum of the fragmented peptide with the *m/z* values of the most abundant fragments indicated above the corresponding peaks. The energy of the collisions in the collision cell has been adjusted so that each peptide ion fragments an average of only once. Note that the predominant fragments under these conditions have $z = 1$ and contain the intact peptide's C-terminus. [After Yates, J.R., *Methods Enzymol.* **271**, 354 (1996).]

ducible manner and then separates the resulting peptides to yield a pattern that can be used to distinguish differences between related proteins. Thus, peptide mapping can be carried out by two-dimensional gel electrophoresis or by such high resolution one-dimensional techniques as

HPLC, SDS–PAGE, IEF, or CE (Sections 6-3 and 6-4). With any of these methods, variant peptides can be isolated and sequenced to establish the sequence differences between the related proteins.

(a)

(b)

Figure 7-11 **Peptide mapping.** A comparison of the fingerprints of trypsin-digested (*a*) hemoglobin A (HbA) and (*b*) hemoglobin S (HbS) shows two peptides that differ in these two forms of hemoglobin (*boxes*). These peptides constitute the eight N-terminal residues of the β subunit of hemoglobin. Their amino acid sequences are

Hemoglobin A Val—His—Leu—Thr—Pro—**Glu**—Glu—Lys
Hemoglobin S Val—His—Leu—Thr—Pro—**Val**—Glu—Lys
β1   2   3   4   5   **6**   7   8

[Courtesy of Corrado Baglioni, State University of New York at Albany.]

2 NUCLEIC ACID SEQUENCING

The basic strategy of nucleic acid sequencing is identical to that of protein sequencing (Section 7-1). It involves

1. The specific degradation and fractionation of the polynucleotide of interest to fragments small enough to be fully sequenced.

2. The sequencing of the individual fragments.

3. The ordering of the fragments by repeating the preceding steps using a degradation procedure that yields a set of polynucleotide fragments that overlap the cleavage points in the first such set.

Before about 1975, however, nucleic acid sequencing techniques lagged far behind those of protein sequencing, largely because there were no available endonucleases that were specific for sequences greater than a nucleotide. Rather, RNAs were cleaved into relatively short fragments by partial digestion with enzymes such as **ribonuclease T1** (from *Aspergillus oryzae*), which cleaves RNA after guanine residues, or pancreatic **ribonuclease A,** which does so after pyrimidine residues. Moreover, there is no reliable polynucleotide reaction analogous to the Edman degradation for proteins (Section 7-1A). Consequently, the polynucleotide fragments were sequenced by their partial digestion with either of two exonucleases: **snake venom phosphodiesterase,** which removes residues from the 3′ end of polynucleotides (Fig. 7-12), or **spleen phosphodiesterase,** which does so from the 5′ end. The resulting oligonucleotide fragments were identified from their chromatographic and electrophoretic mobilities. Sequencing RNA in this manner is a lengthy and painstaking procedure.

The first biologically significant nucleic acid to be sequenced was that of yeast **alanine tRNA** (Section 32-2A). The sequencing of this 76-nucleotide molecule by Robert Holley, a labor of 7 years, was completed in 1965, some 12 years after Frederick Sanger had determined the amino acid sequence of insulin. This was followed, at an accelerating pace, by the sequencing of numerous species of tRNAs and the **5S ribosomal RNAs** (Section 32-3A) from several organisms. The art of RNA sequencing by these techniques reached its zenith in 1976 with the sequencing, by Walter Fiers, of the entire 3569-nucleotide genome of the **bacteriophage MS2.** In contrast, DNA sequencing was in a far more primitive state because of the lack of available DNA endonucleases with any sequence specificity.

After 1975, dramatic progress was made in nucleic acid sequencing technology. Three advances made this possible:

1. The discovery of restriction endonucleases to enable the cleavage of DNA at specific sequences (Section 5-5A).

2. The development of molecular cloning techniques to permit the acquisition of almost any identifiable DNA segment in the amounts required for sequencing (Section 5-5).

3. The development of DNA sequencing techniques.

These procedures are largely responsible for the enormous advances in our understanding of molecular biology that

```
G C A C U U G A
          │  snake venom
          │  phosphodiesterase
          ▼
G C A C U U G A
G C A C U U G
G C A C U U
G C A C U
G C A C
G C A
G C   + Mononucleotides
```

Figure 7-12 **Sequence determination of an oligonucleotide by partial digestion with snake venom phosphodiesterase.** This enzyme sequentially cleaves the nucleotides from the 3′ end of a polynucleotide that has a free 3′-OH group. Partial digestion of an oligonucleotide with snake venom phosphodiesterase yields a mixture of fragments of all lengths, as indicated, that may be chromatographically separated. Comparison of the base compositions of pairs of fragments that differ in length by one nucleotide establishes the identity of the 3′-terminal nucleotide of the larger fragment. In this way the base sequence of the oligonucleotide may be elucidated.

have been made over the past three decades and which we discuss in succeeding chapters. DNA sequencing techniques are the subject of this section.

The pace of nucleic acid sequencing has become so rapid that directly determining a protein's amino acid sequence is far more time-consuming than determining the base sequence of its corresponding gene (although amino acid and base sequences provide complementary information; Section 7-2D). There has been such a flood of DNA sequence data—over 300 billion nucleotides in over 200 million sequences as of 2010, and doubling every ~18 months—that only computers can keep track of them. The first complete genome sequence to be determined, that of the gram-negative bacterium *Haemophilus influenzae,* was reported in 1995 by J. Craig Venter. By 2010, the genome sequences of over 1000 prokaryotes had been reported (with many more being determined) as well as sequences of over 120 eukaryotes (with many more in progress), including those of humans and many other vertebrates, insects, worms, plants, and fungi (Table 7-3).

A. *The Sanger Method*

✏ See Guided Exploration 5: **DNA sequence determination by the chain-terminator method** After 1975, several methods were developed for the rapid sequencing of long stretches of DNA. Here we discuss the **Sanger method,** formulated by Frederick Sanger (the same individual who pioneered the amino acid sequencing of proteins), which is mainly responsible for the vast number of DNA sequences that have been elucidated.

The Sanger method (*alternatively called the* **chain-terminator method** *and the* **dideoxy method***) utilizes the E. coli enzyme DNA polymerase I (Section 5-4Cc) to synthesize complementary copies of the single-stranded DNA*

Table 7-3 **Some Sequenced Genomes**

Organism	Haploid Genome Size (kb)	Number of Chromosomes
Carsonella ruddii (insect endosymbiont; smallest known cellular genome)	160	1
Mycoplasma genitalium (human parasite)	580	1
Rickettsia prowazekii (bacterium; cause of typhus; putative mitochondial relative)	1,112	1
Haemophilus influenzae (bacterium; human pathogen)	1,830	1
Archaeoglobus fulgidus (hyperthermophilic, sulfate-reducing archaeon)	2,178	1
Synechocystis sp. (cyanobacterium)	3,573	1
Mycobacterium tuberculosis (cause of tuberculosis)	4,412	1
Escherichia coli (bacterium, human symbiont)	4,639	1
Saccharomyces cerevisiae (baker's or budding yeast)	12,070	16
Plasmodium falciparum (protozoan; cause of malaria)	23,000	14
Caenorhabditis elegans (nematode worm)	97,000	6
Drosophila melanogaster (fruit fly)	180,000	4
Arabidopsis thaliana (flowering plant)	119,200	5
Oryza sativa (rice)	389,000	12
Danio rerio (zebra fish)	1,700,000	25
Gallus gallus (chicken)	1,200,000	40
Ornithorhynchus anatinus (platypus)	1,840,000	31
Mus musculus (mouse)	2,500,000	20
Canis familiaris (dog)	2,400,000	40
Pan troglodytes (chimpanzee)	3,100,000	24
Homo sapiens (human)	3,038,000	23

Source: http://www.ncbi.nlm.nih.gov/sites/entrez?db=genome.

being sequenced. As we have previously seen, under the direction of the strand being replicated (the template strand), DNA polymerase I assembles the four deoxynucleoside triphosphates (dNTPs), dATP, dCTP, dGTP, and dTTP, into a complementary polynucleotide chain that it elongates in the 5′ to 3′ direction (Fig. 5-31). To initiate DNA synthesis, DNA polymerase requires a primer in a stable base paired complex with the template DNA. If the DNA being sequenced is a restriction fragment, as it usually is, it begins and ends with a restriction site. The primer can therefore be a short DNA segment containing this restriction site annealed to the strand being replicated. The template DNAs are obtained in sufficient quantity to sequence by cloning them in M13-based vectors (Section 5-5Bb) or by PCR (Section 5-5F), both of which yield the required single-stranded DNAs.

DNA polymerase I's 5′ → 3′ exonuclease activity (Fig. 5-33) is catalyzed by a separate active site from those which mediate its polymerase and 3′ → 5′ exonuclease (Fig. 5-36) functions. This is demonstrated by the observation that on proteolytic cleavage of the enzyme into two fragments, the larger C-terminal fragment, which is known as the **Klenow fragment,** possesses the full polymerase and 3′ → 5′ exonuclease activities of the intact enzyme, whereas the smaller N-terminal fragment has its 5′ → 3′ exonuclease activity. Only the Klenow fragment is used in DNA sequencing to ensure that all replicated chains have the same 5′ terminus.

a. The Synthesis of Labeled DNA by DNA Polymerase Is Terminated after Specific Bases

In the Sanger method (Fig. 7-13), *the DNA to be sequenced is incubated with the Klenow fragment of DNA polymerase I, a suitable primer, and the four deoxynucleoside triphosphates (dNTPs). Either at least one dNTP (usually dATP) or the primer is α-^{32}P-labeled. In addition, a small amount of the* **2′,3′-dideoxynucleoside triphosphate (ddNTP)**

2′,3′-Dideoxynucleoside triphosphate

of one of the bases is added to the reaction mixture. When the dideoxy analog is incorporated in the growing polynucleotide in place of the corresponding normal nucleotide, chain growth is terminated because of the absence of a 3′-OH group. By using only a small amount of the ddNTP, *a series of truncated chains is generated, each of which is terminated by the dideoxy analog at one of the positions occupied by the corresponding base.* Each of the four ddNTPs is reacted in a separate vessel.

The four reaction mixtures are simultaneously electrophoresed in parallel lanes on a **sequencing gel.** This is a

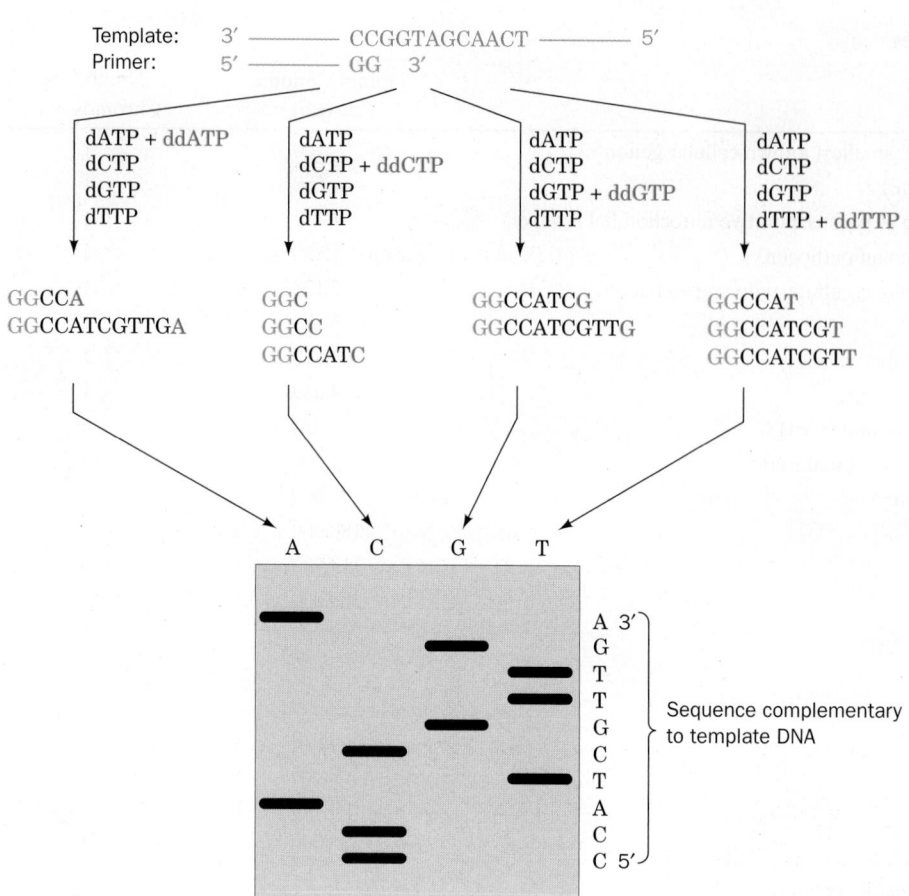

Template: 3' ———— CCGGTAGCAACT ———— 5'
Primer: 5' ———— GG 3'

dATP + ddATP	dATP	dATP	dATP
dCTP	dCTP + ddCTP	dCTP	dCTP
dGTP	dGTP	dGTP + ddGTP	dGTP
dTTP	dTTP	dTTP	dTTP + ddTTP

GGCCA   GGC   GGCCATCG   GGCCAT
GGCCATCGTTGA   GGCC   GGCCATCGTTG   GGCCATCGT
               GGCCATC   GGCCATCGTT

A   C   G   T

A 3'
G
T
T
G
C
T
A
C
C 5'

Sequence complementary to template DNA

Figure 7-13 **Flow diagram of the Sanger (chain-terminator or dideoxy) method of DNA sequencing.** The symbol ddATP represents dideoxyadenosine triphosphate, etc. The sequence that is determined by reading the gel from bottom to top (from the smallest to the largest fragment) is complementary to the sequence of the template DNA.

long, thin (as little as 0.1 mm by up to 100 cm) polyacrylamide slab. It contains ~7M urea and is run at ~70°C so as to eliminate all hydrogen bonding associations. *These conditions ensure that the DNA fragments separate only according to their size. The sequence of the DNA that is complementary to the template DNA can then be directly read off an autoradiogram of the sequencing gel, from bottom to top,* as is indicated in Fig. 7-14. Indeed, computerized devices are available to aid in doing so. However, a single gel is incapable of resolving much more than 300 to 400 consecutive fragments. This limitation is circumvented by generating two sets of gels, one run for a longer time and perhaps at a higher voltage than the other, to obtain the sequence of up to 800-bp DNA fragments.

Improved gels can be obtained through the use of dNTPs whose α-phosphate groups are radioactively labeled with ^{35}S rather than ^{32}P.

$$\overset{O}{\underset{O^-}{\underset{\|}{-O-P}}}-\overset{O}{\underset{O^-}{\underset{\|}{P}}}-\overset{^{35}S}{\underset{O^-}{\underset{\|}{P}}}-O-CH_2$$

α-Thio-[^{35}S]dNTP

This is because the β particles emitted by ^{35}S nuclei have

Figure 7-14 **Autoradiograph of a sequencing gel.** DNA fragments were produced by the Sanger method of DNA sequencing. A second loading of the gel (*right*) was made 90 min after the initial loading (*left*). The deduced sequence of 140 nucleotides is written alongside. [From Hindley, J., DNA sequencing, *in* Work, T.S. and Burdon, R.H. (Eds.), *Laboratory Techniques in Biochemistry and Molecular Biology,* Vol. 10, *p.* 82, Elsevier (1983). Used by permission.]

less energy and hence shorter path lengths than those of ^{32}P, thereby yielding sharper gel bands. More readily interpretable gels may also be obtained by replacement of the Klenow fragment with DNA polymerases either from **bacteriophage T7 (T7 DNA polymerase,** which is less sensitive to the presence of ddNTPs than is Klenow fragment and hence yields gel bands of more even intensities) or from thermophilic bacteria such as *Thermus aquaticus* (*Taq* polymerase; Section 5-5F) that are stable above 90°C and hence can be used at the temperatures required to denature particularly stable segments of DNA.

b. RNA May Be Sequenced through Its Transcription to cDNA

RNA can be readily sequenced by only a slight modification of the above DNA sequencing procedures. The RNA to be sequenced is transcribed into a complementary strand of DNA (cDNA) through the action of reverse transcriptase (Section 5-5Fa). The resulting cDNA may then be sequenced normally.

c. The Sanger Method Has Been Automated

In order to sequence large tracts of DNA such as entire chromosomes, the Sanger method has been greatly accelerated through automation. This required that the above-described radiolabeling techniques, which are not readily automated, be replaced by fluorescence labeling techniques (with the added benefit of eliminating the health hazards and storage problems of using radiolabeled nucleotides). In the most widely used such technique, each of the four ddNTPs used to terminate chain extension is covalently linked to a differently fluorescing dye, the chain-extension reactions are carried out in a single vessel containing all four of these labeled ddNTPs, and the resulting fragment mixture is subjected to sequencing gel electrophoresis in a single lane (Fig. 7-15*a*). As each fragment exits the gel, its terminal base is identified according to its characteristic fluorescence spectrum by a laser-activated fluorescence detection system (Fig. 7-15*b*).

(*a*)

Template: 3′ ——— CCGGTAGCAACT ——— 5′
Primer: 5′ ——— GG 3′

DNA polymerase | dATP + dCTP + dGTP + dTTP + ddATP—⬤ + ddCTP—⬤ + ddGTP—◐ + ddTTP—⬤

GGCCATCGTTGA—⬤
GGCCATCGTTG—◯
GGCCATCGTT—⬤
GGCCATCGT—⬤
GGCCATCG—◯   Dye-labeled
GGCCATC—⬤   DNA segments
GGCCAT—⬤
GGCCA—⬤
GGCC—⬤
GGC—⬤

DNA migration

Dye-labeled DNA segments are subjected to electrophoresis in a capillary gel

Laser   Detector

(*b*)

Figure 7-15 **Automated DNA sequencing using the Sanger method.** (*a*) The primer that is base paired to the template strand being sequenced is extended in the presence of four differently fluorescently labeled dideoxynucleoside triphosphates (ddNTPs). The resulting mixture of dye-labeled DNA strands of all lengths is subject to gel electrophoresis in a capillary tube, thereby separating them according to size with smaller polynucleotides migrating faster than larger ones. As each polynucleotide passes the detector, it's 3′-terminal nucleotide is identified according to the spectrum of its laser-stimulated fluorescence. (*b*) A portion of the output of an automated DNA sequencing system. Each of the four differently colored curves indicates the fluorescence intensity of a particular dye that is linked to a specific ddNTP in terminating the primer extension reaction (green, blue, black, and red for ddATP, ddCTP, ddGTP, and ddTTP, respectively; the letters above the bands identify the bases and the numbers indicate their positions in the DNA segment being sequenced). [Courtesy of Mark Adams, The Institute for Genomic Research, Gaithersburg, Maryland.]

The fluorescence detectors used in these devices, which have error rates of ~1%, are computer-controlled and hence data acquisition is automated. In the most advanced such systems, the sequencing gel is contained in an array of up to 96 capillary tubes rather than in a slab-shaped apparatus, sample preparation and loading are performed by robotic systems, and electrophoresis and data analysis are fully automated. Such systems can simultaneously sequence 96 DNA samples averaging ~650 bases each with a turn-around time of ~1 h and hence can identify up to 1.6 million bases per day—all with only ~15 min of human attention (vs the ~25,000 bases per year that a skilled operator can identify using the above-described manual methods). Nevertheless, one such system would require ~10 years of uninterrupted operation to sequence the 3 billion-bp human genome with only two sets of overlapping fragments. However, to ensure the complete coverage of a large tract of DNA (Section 5-5Ea) and to reduce its error rate to <0.01%, at least 10 sets of overlapping segments must be sequenced (Section 7-2B). Hence, the major sequencing centers, where most genome sequencing is carried out, each have over 100 such sequencing systems in factory-like settings.

B. *Genome Sequencing*

The major technical challenge in sequencing a genome is not the DNA sequencing itself but, rather, assembling the tens of thousands to tens of millions of sequenced segments (depending on the size of the genome) into contiguous blocks (called **contigs**) and assigning them to their proper positions in the genome. One way that contigs might be ordered is through chromosome walking (Section 5-5Ea). However, to do so for a eukaryotic genome would be prohibitively time-consuming and expensive (e.g., to "walk" the 125 million-bp length of an average length human chromosome using ~10-kb inserts from a plasmid library would require a minimum of $1.25 \times 10^{8}/10,000 = 12,500$ labor-intensive "steps").

a. Map-Based Genome Sequencing

A more efficient technique of genome sequencing, the **map-based sequencing** strategy (Fig. 7-16*a*), was developed in the late 1980s. In this approach, low resolution physical maps of each chromosome are prepared by identifying shared landmarks on overlapping ~250-kb inserts that are cloned in yeast artificial chromosomes (YACs). These land-

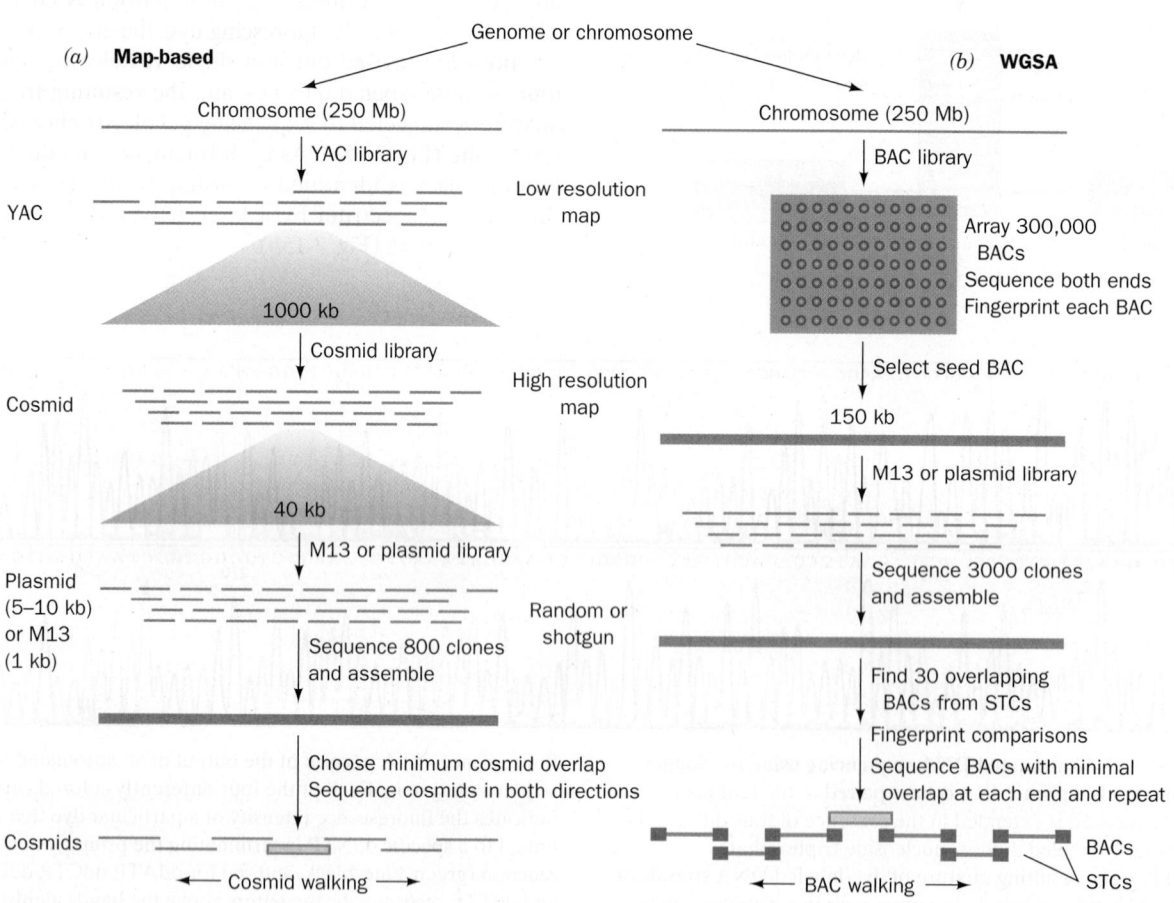

Figure 7-16 **Genome sequencing strategies.** (*a*) The map-based strategy uses three sets of progressively smaller cloned inserts and assembles the sequenced inserts through cosmid walking and the use of landmarks such as STSs and ESTs (see text). (*b*) The WGSA strategy uses only two levels of cloning and employs sophisticated computer algorithms as well as STCs to assemble the sequenced inserts into finished chromosomes. [After Venter, J.C., Smith, H.O., and Hood, L., *Science* **381,** 365 (1996).]

marks often take the form of 200- to 300-bp segments known as **sequence-tagged sites (STSs),** whose exact sequence occurs nowhere else in the genome. Hence, two clones that contain the same STS must overlap. The STS-containing inserts are then randomly fragmented (usually by sonication; Section 5-3D) into ~40-kb segments that are subcloned into cosmid vectors so that a high resolution map can be constructed by identifying their landmark overlaps. The cosmid inserts are then randomly fragmented into overlapping 5- to 10-kb or 1-kb segments for insertion into plasmid or M13 vectors (shotgun cloning; Section 5-5E). These inserts (~800 M13 clones per cosmid) are then sequenced (~400 bp per clone) and the resulting so-called **reads** are assembled computationally into contigs to yield the sequence of their parent cosmid insert (with a redundancy of 400 bp per clone × 800 clones per cosmid/40,000 bp per cosmid = 8). Finally, the cosmid inserts are assembled, through **cosmid walking** (the computational analog of chromosome walking; Fig. 5-51), using their landmark overlaps (with landmarks ideally spaced at intervals of 100 kb or less), to yield the sequences of the YAC inserts which are then assembled, using their STSs, to yield the chromosome's sequence.

The genomes of most complex eukaryotes contain numerous tracts of **repetitive sequences,** that is, segments of DNA that are tandemly repeated hundreds, thousands, and in some cases millions of times (Section 34-2B). Lengthy tracts of repetitive sequences easily confound the foregoing assembly process, leading to gaps in the sequence. Moreover, such repetitive sequences greatly exacerbate the difficulty of finding properly spaced STSs. To partially circumvent the latter difficulty, STS-like sequences of cDNAs, known as **expressed sequence tags (ESTs),** are used in place of STSs. Since the mRNAs from which cDNAs are reverse transcribed encode proteins, they are unlikely to contain repetitive sequences.

b. The Whole Genome Shotgun Assembly Strategy

Although the initial goal of the **human genome project** of identifying STSs and ESTs every ~100 kb in the human genome was achieved, advances in computational and cloning technology permitted a more straightforward sequencing procedure that eliminates the need for both the low resolution (YAC) and high resolution (cosmid) mapping steps. In this so-called **whole genome shotgun assembly (WGSA)** strategy, which was formulated by Venter, Hamilton Smith, and Leroy Hood, a genome is randomly fragmented, a large number of cloned fragments are sequenced, and the genome is assembled by identifying overlaps between pairs of fragments. Statistical considerations indicate that, using this strategy, the probability that a given base is not sequenced is ideally e^{-c}, where c is the redundancy of coverage [$c = LN/G$, where L is the average length of the reads in nucleotides (**nt**), N is the number of reads, and G is the length of the genome in nt], the aggregate length of the gaps between contigs is Ge^{-c}, and the average gap size is G/N. Moreover, without a long-range physical and/or genetic map of the genome being sequenced, the order of the contigs and their relative orientations would be unknown.

For bacterial genomes, the WGSA strategy is carried out straightforwardly by sequencing tens of thousands of fragments and assembling them (a task that required the development of computer algorithms capable of assembling contigs from very large numbers of reads). Then, in a task known as **finishing,** the gaps between contigs are filled in by several techniques including synthesizing PCR primers complementary to the ends of the contigs and using them to isolate the missing segments (chromosome walking; bacterial genomes have few if any repetitive sequences).

For eukaryotic genomes, their much greater sizes require that the WGSA strategy be carried out in stages as follows (Fig. 7-16*b*). A bacterial artificial chromosome (BAC) library of ~150-kb inserts is generated (for the human genome, an ~15-fold redundancy, which would still leave ~900 bases unsequenced, would require ~300,000 such clones; BACs are used because they are subject to fewer technical difficulties than are YACs). The insert in each of these BAC clones is identified by sequencing ~500 bp in from each end to yield segments known as **sequence-tagged connectors** (**STCs** or **BAC-ends;** which for the above 300,000 clones would collectively comprise ~300,000 kb, that is, 10% of the entire human genome). One BAC insert is then fragmented and shotgun cloned into plasmid or M13 vectors (so as to yield ~3000 overlapping clones), and the fragments are sequenced and assembled into contigs. The sequence of this "seed" BAC is then compared with the database of STCs to identify the ~30 overlapping BAC clones. The two with minimal overlap at either end are then selected, sequenced, and the operation repeated until the entire chromosome is sequenced (BAC walking), which for the human genome required 27 million sequencing reads. This process is also confounded by repetitive sequences.

The WGSA strategy is readily automated through robotics and hence is faster and less expensive than the map-based strategy. Indeed, most known genome sequences have been determined using the WGSA strategy, many in a matter of a few months, and its advent reduced the time to sequence the human genome by several years. Nevertheless, it appears that for eukaryotic genomes, most of the residual errors in a WGSA-based genome sequence [mainly the failure to recognize long (>15 kb) segments that have nearly (>97%) identical sequences] can be eliminated by finishing it through the use of some of the techniques of the map-based strategy.

c. The Human Genome Has Been Sequenced

The "rough draft" of the human genome was reported in 2001 by two independent groups: the publicly funded International Human Genome Sequencing Consortium (IHGSC; a collaboration involving 20 sequencing centers in six countries), led by Francis Collins, Eric Lander, and John Sulston, which used the map-based strategy; and a privately funded group, mainly from Celera Genomics, led by Venter, which used the WGSA strategy. The IHGSC-determined genome sequence was a conglomerate from numerous anonymous individuals, whereas that from Celera Genomics was derived from five individuals but mainly from Venter. These draft sequences lacked ~10% of the gene-rich chromosomal regions known as **euchromatin** (Section 34-1) and much of the largely if not entirely unexpressed chromosomal regions known as **constitutive heterochromatin** (Section 34-1; which consists of highly repetitive sequences that are mainly associated with the chromosomal centromeres; Section 34-3A). Moreover, both draft assemblies had sequencing error rates of ~1% and contained ~160,000 gaps so that the order and orientations of many contigs within local regions had not been established. But even this imperfect data greatly accelerated the pace of genetic research such that, for example, the genes for hundreds of inherited diseases were identified and cloned far more rapidly than had previously been possible.

In 2004, the IHGSC reported the finished sequence of the human genome. It covered ~99% of the euchromatic genome (2.851 billion nt of the entire 3.038-billion nt genome) with an error rate of <0.001% and had only 281 gaps, all of which were in regions of repetitive sequence. In 2007, Venter reported the finished sequence of his own **diploid genome** (that of all 46 chromosomes; previously reported genome sequences were those of **haploid genomes**, that is, of one member of each homologous chromosomal pair). These stunning achievements, the culmination of over a decade of intense effort by hundreds of scientists, is revolutionizing the way both biochemistry and medicine are viewed and practiced. A few of the major observations that have been made are as follows:

1. About 45% of the human genome consists of repeating sequences of various lengths.

2. Only ~28% of the genome is transcribed to RNA.

3. Only 1.2% of the genome (~4% of the transcribed RNA) encodes protein.

4. The human genome appears to contain only ~23,000 protein-encoding genes [also known as **open reading frames (ORFs)**] rather than the 50,000 to 140,000 ORFs that had previously been predicted based mainly on extrapolations (and the ~30,000 ORFs predicted from the rough draft). This compares with the ~6600 ORFs in yeast, ~14,000 in *Drosophila*, ~19,000 in *C. elegans*, and ~25,500 in *Arabadopsis*. Note that these numbers will almost certainly change as our presently imperfect ability to recognize ORFs improves.

5. Only a small fraction of human protein families is unique to vertebrates; most occur in other if not all lifeforms.

6. Two randomly selected human genomes differ, on average, by only 1 nucleotide per 1000; that is, any two people are likely to be ~99.9% genetically identical.

The obviously greater complexity of humans (vertebrates) relative to "lower" (nonvertebrate) forms of life is unlikely to be due to the not much larger numbers of ORFs that vertebrates encode. Rather, it appears that vertebrate proteins themselves are more complex than those of nonver-

tebrates; that is, vertebrate proteins tend to have more domains (modules) than invertebrate proteins and these modules are more often selectively expressed through alternative gene splicing (Section 5-4Ac). Thus, many vertebrate genes encode several different although similar proteins. Moreover, mounting evidence indicates that vertebrate genomes encode large numbers of short RNA segments that participate in controlling gene expression (Section 31-4At).

The genomes of eukaryotes, including that of *Homo sapiens,* can be explored at http://www.ncbi.nlm.nih.gov/projects/mapview/.

C. *Next Generation DNA Sequencing Technologies*

One of the goals of the human genome project is to sequence an individual's genome at an affordable price (US$1000 is the figure that is often quoted). This would permit the comparison of many thousands of human genome sequences and hence the correlation of specific sequences with susceptibility to particular diseases. This, in turn, would usher in an age of personalized medicine when the treatment of active disease and the prevention of anticipated disease would be tailored to an individual's genetic makeup.

Like most aspects of science, genome sequencing is technology-driven. Thus the IHGSC-determined sequence cost ~US$300 million and took over a decade to complete. In contrast, using "next generation" sequencing technology (see below), the diploid genome of James Watson (of Watson–Crick fame) was sequenced in 2 months at a cost of <US$1 million—the third human genome to be sequenced. As even newer technologies are developed, the price and time to sequence a human genome is expected to drop even more precipitously.

All of the several available next generation sequencing technologies eliminate the time-consuming cloning steps used by the Sanger sequencing-based methods (Fig. 7-16). They do so by amplifying single isolated molecules of DNA and then sequencing them in a massively parallel way.

a. The 454 Sequencing System

The Watson genome was sequenced using a system developed by 454 Life Sciences that employs the following methodology (Fig. 7-17). The genomic DNA is randomly sheared to small (300–500 bp) fragments and ligated to adaptors, which in turn are specifically bound by ~30-μm-in-diameter "DNA capture" beads under dilution conditions such that, at most, one DNA fragment is bound to each bead. The beads are suspended in a PCR mixture containing dNTPs, primers complementary to the adaptors, and *Taq* DNA polymerase. The suspension is emulsified with oil such that each aqueous droplet contains only one bead, that is, each bead is contained in its own microreactor, thus preventing the introduction of competing or contaminating sequences. The PCR (Section 5-5F) is carried out by thermocycling until ~10 million identical DNA fragments are bound to each DNA capture bead. The emulsion is broken by the addition of isopropanol, the DNA is denatured, and the resulting single-stranded DNA-

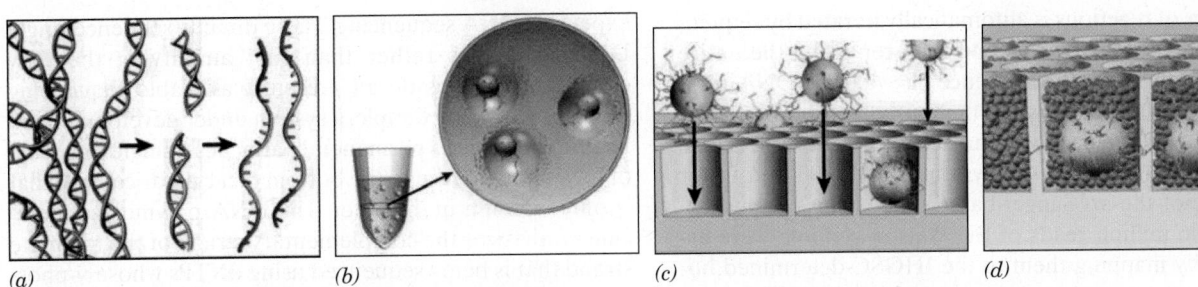

Figure 7-17 Sample preparation for the 454 sequencing system. (*a*) Genomic DNA is isolated, fragmented, linked to adaptors, and denatured to yield single strands. (*b*) The single strands are bound to DNA capture beads under dilution conditions in which, at most, one single strand binds per bead, the beads are captured in a PCR-reaction-mixture-in-oil emulsion, and PCR amplification of the DNA occurs within each bead-containing droplet. (*c*) The emulsion is broken by the addition of isopropanol, the now double-stranded DNA is denatured, and the resulting beads carrying single-stranded DNA clones are deposited in the wells of a fiber-optic slide. (*d*) Smaller beads linked to the enzymes ATP sulfurylase and luciferase are deposited into each well. [Courtesy of Jonathan Rothberg, 454 Life Sciences Corporation, Branford, Connecticut.]

bearing beads are deposited into 75-picoliter wells (1 picoliter = 10^{-12} L) on a fiber-optic slide with one bead per well. The slide contains ~1.6 million wells.

The DNA on each of the beads is sequenced using a series of coupled enzymatic reactions that are collectively known as **pyrosequencing** (Fig. 7-18):

1. A solution containing only one of the four dNTPs is flowed over the bead-containing slide. If that dNTP is complementary to the first unpaired base on a template strand, DNA polymerase catalyzes its addition to the primer strand and releases pyrophosphate ion (Fig. 5-31).

2. In a reaction catalyzed by the enzyme **ATP sulfurylase,** the pyrophosphate ion reacts with **adenosine-5'-phosphosulfate** to yield ATP.

3. In a reaction catalyzed by the firefly enzyme **luciferase,** the ATP reacts with **luciferin** and O_2 to yield **oxyluciferin** and a flash of visible light (a phenomenon named **chemiluminescence**). The well from which the light flash emanated together with its intensity is recorded by an imaging system, thus identifying those wells in which the foregoing nucleotide was added to the primer strand. The intensity of the light is proportional to the number of nucleotides reacted so that when two or more consecutive nucleotides of the same type are added to the primer strand, their number is determined.

4. In preparation for the next reaction cycle, any unreacted dNTPs and ATP are hydrolyzed to mononucleotides (NMPs) and phosphate ion by a wash containing the enzyme **apyrase.**

$$DNA_{n\ residues} + dNTP \xrightarrow[\text{1}]{\text{DNA polymerase}} DNA_{n+1\ residues} + P_2O_7^{4-}$$
$$\textbf{Pyrophosphate}$$

$$P_2O_7^{4-} + Adenosine-\overset{\overset{\displaystyle O}{\|}}{\underset{\underset{\displaystyle O^-}{|}}{P}}-OSO_3^{2-} \xrightarrow[\text{2}]{\text{ATP sulfurylase}} ATP + SO_4^{2-}$$
$$\textbf{Adenosine-5'-phosphosulfate}$$

$$ATP + \textbf{Luciferin} + O_2 \xrightarrow[\text{3}]{\text{luciferase}} light + \textbf{Oxyluciferin}$$
$$+ CO_2 + AMP + P_2O_7^{4-}$$

$$(d)NTP + 2\,H_2O \xrightarrow[\text{4}]{\text{apyrase}} (d)NMP + 2\,PO_4^{3-}$$

Figure 7-18 The reactions of pyrosequencing.

This series of reactions is automatically iterated by sequentially using all four dNTPs and then repeating the entire process. In this way, the sequence of ~400,000 DNA fragments (reads) can be simultaneously determined to a length of >400 nt, each with an accuracy of ~99%, in one 4-h run (and hence the 454 system is over 300-fold faster than state-of-the-art Sanger-based sequencing systems).

The 106 million reads of the Watson genome were assembled by mapping them to the IHGSC-determined human genome sequence (using programs similar to those described in Section 7-4Bg), thus bypassing the computationally difficult process of assembling the genome based on overlaps as well as eliminating the need for finishing. In fact, the relatively short read lengths of the 454 system (maximally ~400 nt vs ~800 nt for Sanger sequencing) makes it difficult to shotgun assemble a eukaryotic genome *de novo* (anew) so that Sanger sequencing is still useful. However, the ~4 million reads of the Watson genome with no or poor alignment to the IHGSC sequence were placed according to their overlaps [which identified 3.3 million single-base differences, which are known as **single nucleotide polymorphisms (SNPs;** pronounced "snips"), relative to the IHGSC sequence and the gain or loss of numerous chromosomal segments ranging in length up to 1.5 million bp].

The read lengths of the 454 system are sufficient to assemble bacterial genomes *de novo,* although a 15-fold redundancy is necessary to do so accurately versus a 6- to 8-fold redundancy for Sanger sequencing. Nevertheless, with the use of a 454 system, the sequence of a typical bacterial genome can be determined in less than a week at a cost of several thousand U.S. dollars.

b. Other DNA Sequencing Technologies

Several other next generation DNA sequencing platforms, each using a different although massively parallel sequencing technology, are available. For example, the SOLiD system from Applied Biosystems can simultaneously sequence ~180 million DNA fragments with read lengths of up to 50 nt each for a total of up to ~9 billion nt in a single run, whereas the Genome Analyzer from Illumina (originally Solexa) can simultaneously sequence ~50 million DNA fragments with read lengths of up to 50 nt each for a total of up to 2.5 billion nt in a single run. The short read of both these systems makes them unsuitable for use in *de novo* genome sequencing. Nevertheless, by mid-2010, several hundred human genome sequences had been determined using the Illumina system and identifying the sequence changes in its reads relative to previously determined human genome sequences (a process called **resequencing**). Moreover, there are many applications for which systems generating short reads are well suited, such as selectively resequencing portions of the genomes of numerous individuals in order to discover disease-related SNPs, identifying the genetic changes in cancerous tumors, and screening for mutations among related populations of bacteria.

The so-called third generation sequencing technologies that are now on the horizon promise even faster and less expensive DNA sequencing. These directly sequence single DNA molecules rather than first amplifying them by cloning or PCR as do all presently available sequencing technologies. For example, a system under development by Pacific Biosciences permanently attaches a single molecule of DNA polymerase to the bottom of a cylindrical well that is only ~50 nm in diameter. The DNA polymerase molecule synthesizes the complementary strand of the template strand that is being sequenced using dNTPs whose γ-phosphate groups are covalently linked to a fluorescent dye, with a differently fluorescing dye for each of the four bases. The volume of the wells is so small (2×10^{-20} L) that freely diffusing dye-linked dNTP molecules in solution rarely enter a well and then for only a few microseconds before diffusing away. In contrast, a DNA polymerase that is incorporating a dNTP into a growing DNA chain holds it for tens of milliseconds before releasing the now dye-linked pyrophosphate ion into the solution and commencing another round of synthesis. During this time the laser-excited fluorescent dNTP emits light that is detected by a sophisticated optical system that only observes the light emanating from the bottom of the well, thus identifying the dNTP that is being incorporated. This system has been shown to produce reads of tens of thousands of nucleotides in each of thousands of wells. Even farther in the future are systems that pass single molecules of DNA through tiny holes (nanopores) and identify their bases by measuring the subtle electrical changes as each base passes through the nanopore.

D. Nucleic Acid Sequencing versus Amino Acid Sequencing

The amino acid sequences of proteins are specified by the base sequences of nucleic acids (Section 5-4Bb). Consequently, with a knowledge of the genetic code (Table 5-3) and the nature of transcriptional and translational initiation sequences (Sections 31-3 and 32-3C), a protein's primary structure can be inferred from that of a corresponding nucleic acid. Techniques for sequencing nucleic acids initially lagged far behind those for proteins, but by the late 1970s, DNA sequencing methods had advanced to the point that it became far easier to sequence a DNA segment than the protein it specified. Although the great majority of known protein primary structures have been inferred from DNA sequences, direct protein sequencing remains an important biochemical tool for several reasons:

1. Disulfide bonds can be located only by protein sequencing.

2. Many proteins are modified after their biosynthesis by the excision of certain residues and by the specific derivatization of others (Section 32-5). The identities of these modifications, which are often essential for the protein's biological function, can be determined only by directly sequencing the protein.

3. One of the most effective ways of identifying the gene that encodes a protein of interest is to determine the amino acid sequence of at least a portion of the protein, infer the base sequence of the DNA segment that encodes this polypeptide segment, chemically synthesize this DNA, and use it to identify and isolate the gene(s) containing its base sequence through Southern blotting or PCR (Sections 5-5D and 5-5F). This process is known as **reverse genetics** because, in prokaryotes, genetics has been traditionally used to characterize proteins rather than vice versa. Of course, for organisms whose genomes have been sequenced, this process can be carried out *in silico* (by computer).

4. The "standard" genetic code is not universal: Those of mitochondria and certain protozoa are slightly different (Section 32-1Db). In addition, in certain species of protozoa, the RNA transcripts are "edited"; that is, their sequences are altered before they are translated (Sections 31-4Ar and 31-4As). These genetic code anomalies were discovered by comparing the amino acid sequences of proteins and the base sequences of their corresponding genes. If there are other genetic code anomalies, they will no doubt be discovered in a like manner.

3 CHEMICAL EVOLUTION

Individuals, as well as whole species, are characterized by their genomes. An organism's genome specifies the amino acid sequences of all members of its **proteome** (all of the proteins encoded by its genome) together with their quantity and schedule of appearance in each cell. An organism's proteomic composition is therefore the direct consequence of its genomic composition.

In this section, we concentrate on the evolutionary aspects of amino acid sequences, the study of the **chemical evolution** of proteins. Evolutionary changes, which stem from random mutational events, often alter a protein's primary structure. A mutational change in a protein, if it is to be propagated, must somehow increase, or at least not decrease, the probability that its owner will survive to reproduce. Many mutations are deleterious and often lethal in their effects and therefore rapidly die out. On rare occasions, however, a mutation arises that, as we shall see below, improves the fitness of its host in its natural environment.

A. *Sickle-Cell Anemia: The Influence of Natural Selection*

Hemoglobin, the red blood pigment, is a protein whose major function is to transport oxygen throughout the body. A molecule of hemoglobin is an $\alpha_2\beta_2$ tetramer; that is, it consists of two identical α chains and two identical β chains (Fig. 7-1*d*). Hemoglobin is contained in the **erythrocytes** (red blood cells; Greek: *erythros*, red + *kytos,* a hollow vessel) of which it forms ~33% by weight in normal individuals, a concentration that is nearly the same as it has in the crystalline state. In every cycle of their voyage through the circulatory system, the erythrocytes, which are normally

(a)

(b)

Figure 7-19 **Scanning electron micrographs of human erythrocytes.** (*a*) Normal human erythrocytes revealing their biconcave disklike shape. [David M. Phillips/Visuals Unlimited.] (*b*) Sickled erythrocytes from an individual with sickle-cell anemia. [Bill Longcore/Photo Researchers, Inc.]

flexible biconcave disks (Fig. 7-19*a*), must squeeze through capillary blood vessels smaller in diameter than they are.

In individuals with the inherited disease **sickle-cell anemia,** many erythrocytes assume an irregular crescentlike shape under conditions of low oxygen concentration typical of the capillaries (Fig. 7-19*b*). This "sickling" increases the erythrocytes' rigidity, which hinders their free passage through the capillaries. The sickled cells therefore impede the flow of blood in the capillaries such that, in a sickle-cell "crisis," the blood flow in some areas may be completely blocked, thereby giving rise to extensive tissue damage and excruciating pain. Moreover, individuals with sickle-cell anemia suffer from severe **hemolytic anemia** (a condition characterized by red cell destruction) because the increased mechanical fragility of their erythrocytes halves the normal 120-day lifetime of these cells. The debilitating effects of this disease are such that, before the latter half of the twentieth century, individuals with sickle-cell anemia rarely survived to maturity (although modern treatments by no means constitute a cure).

a. Sickle-Cell Anemia Is a Molecular Disease

In 1945, Linus Pauling correctly hypothesized that *sickle-cell anemia, which he termed a **molecular disease,** is a result of the presence of a mutant hemoglobin.* Pauling and

Table 7-4   Amino Acid Sequences of Cytochromes c from 38 Species

Group	Species	Sequence (positions −9 … 40)
Mammals	Human, chimpanzee	a GDVEKGKKIFIMKCSQCHTVEKGGKHKTGPNLHGLFGRKTGQA
	Rhesus monkey	a GDVEKGKKIFIMKCSQCHTVEKGGKHKTGPNLHGLFGRKTGQA
	Horse	a GDVEKGKKIFVQKCAQCHTVEKGGKHKTGPNLHGLFGRKTGQA
	Donkey	a GDVEKGKKIFVQKCAQCHTVEKGGKHKTGPNLHGLFGRKTGQA
	Cow, pig, sheep	a GDVEKGKKIFVQKCAQCHTVEKGGKHKTGPNLHGLFGRKTGQA
	Dog	a GDVEKGKKIFVQKCAQCHTVEKGGKHKTGPNLHGLFGRKTGQA
	Rabbit	a GDVEKGKKIFVQKCAQCHTVEKGGKHKTGPNLHGLFGRKTGQA
	California gray whale	a GDVEKGKKIFVQKCAQCHTVEKGGKHKTGPNLHGLFGRKTGQA
	Great gray kangaroo	a GDVEKGKKIFVQKCAQCHTVEKGGKHKTGPNINGIFGRKTGQA
Other vertebrates	Chicken, turkey	a GDIEKGKKIFVQKCSQCHTVEKGGKHKTGPNLHGLFGRKTGQA
	Pigeon	a GDIEKGKKIFVQKCSQCHTVEKGGKHKTGPNLHGLFGRKTGQA
	Pekin duck	a GDVEKGKKIFVQKCSQCHTVEKGGKHKTGPNLHGLFGRKTGQA
	Snapping turtle	a GDVEKGKKIFVQKCAQCHTVEKGGKHKTGPNLNGLIGRKTGQA
	Rattlesnake	a GDVEKGKKIFTMKCSQCHTVEKGGKHKTGPNLHGLFGRKTGQA
	Bullfrog	a GDVEKGKKIFVQKCAQCHTCEKGGKHKVGPNLYGLIGRKTGQA
	Tuna	a GDVAKGKKTFVQKCAQCHTVENGGKHKVGPNLWGLFGRKTGQA
	Dogfish	a GDVEKGKKVFVQKCAQCHTVENGGKHKTGPNLSGLFGRKTGQA
Insects	Samia cynthia (a moth)	h GVPAGNAENGKKIFVQRCAQCHTVEAGGKHKVGPNLHGFYGRKTGQA
	Tobacco hornworm moth	h GVpAGNADNGKKIFVQRCAQCHTVEAGGKHKVGPNLHGFFGRKTGQA
	Screwworm fly	h GVPAGDVEKGKKIFVQRCAQCHTVEAGGKHKVGPNLHGLFGRKTGQA
	Drosophila (fruit fly)	h GVPAGDVEKGKKLFVQRCAQCHTVEAGGKHKVGPNLHGLIGRKTGQA
Fungi	Baker's yeast	h TEFKAGSAKKGATLFKTRCLQCHTVEKGGPHKVGPNLHGIFGRHSGQA
	Candida krusei (a yeast)	h PAPFEQGSAKKGATLFKTRCAQCHTIEAGGPHKVGPNLHGIFSRHSGQA
	Neurospora crassa (a mold)	h GFSAGDSKKGANLFKTRCAQCHTLEEGGGNKIGPALHGLFGRKTGSV
Higher plants	Wheat germ	a ASFSEAPPGNPDAGAKIFKTKCAQCHTVDAGAGHKQGPNLHGLFGRQSGTT
	Buckwheat seed	a ATFSEAPPGNIKSGEKIFKTKCAQCHTVEKGAGHKQGPNLNGLFGRQSGTT
	Sunflower seed	a ASFAEAPPGDPTTGAKIFKTKCAQCHTVEKGAGHKQGPNLNGLFGRQSGTT
	Mung bean	a ASFBEAPPGBSKSGEKIFKTKCAQCHTVDKGAGHKQGPNLNGLFGRQSGTT
	Cauliflower	a ASFBEAPPGBSKAGEKIFKTKCAQCHTVDKGAGHKQGPNLNGLFGRQSGTT
	Pumpkin	a ASFBEAPPGBSKAGEKIFKTKCAQCHTVDKGAGHKQGPNLNGLFGRQSGTT
	Sesame seed	a ASFBEAPPGBVKSGEKIFKTKCAQCHTVDKGAGHKQGPNLNGLFGRQSGTT
	Castor bean	a ASFBEAPPGBVKAGEKIFKTKCAQCHTVEKGAGHKQGPNLNGLFGRQSGTT
	Cottonseed	a ASFZEAPPGBAKAGEKIFKTKCAQCHTVDKGAGHKQGPNLNGLFGRQSGTT
	Abutilon seed	a ASFZEAPPGBAKAGEKIFKTKCAQCHTVEKGAGHKQGPNLNGLFGRQSGTT

Number of different amino acids: 1 3 5 5 5 1 3 3 4 1 4 3 2 1 3 1 1 1 1 4 2 4 1 2 3 2 1 4 1 1 2 1 5 1 3 3 2 1 3 2 1 3 3

(continued)

[a] The amino acid side chains have been shaded according to their polarity characteristics so that an invariant or conservatively substituted residue is identified by a vertical band of a single color. The letter a at the beginning of the chain indicates that the N-terminal amino group is acetylated; an h indicates that the acetyl group is absent.

Source: Dickerson, R.E., *Sci. Am.* **226**(4), 58–72 (1972), with corrections from Dickerson, R.E. and Timkovich, R., *in* Boyer, P.D. (Ed.), *The Enzymes* (3rd ed.), Vol. 11, *pp.* 421–422, Academic Press (1975). [Illustration, Irving Geis. Image from the Irving Geis Collection, Howard Hughes Medical Institute. Reprinted with permission.]

his co-workers subsequently demonstrated, through electrophoretic studies, that normal human hemoglobin **(HbA)** has an anionic charge that is around two units more negative than that of sickle-cell hemoglobin **(HbS;** Fig. 7-20).

In 1956, Vernon Ingram developed the technique of fingerprinting peptides (Section 7-1J) in order to pinpoint the difference between HbA and HbS. Ingram's fingerprints of tryptic digests of HbA and HbS revealed that their α subunits are identical but that their β subunits differ by a variation in one tryptic peptide (Fig. 7-11). Sequencing studies eventually indicated that this difference arises from the replacement of the Glu β6 of HbA (the Glu in the sixth position of each β chain) with Val in HbS (Glu β6 → Val), thus accounting for the charge difference observed by Pauling. This was the first time an inherited disease was shown to arise from a specific amino acid change in a protein. *This*

Figure 7-20 The electrophoretic pattern of hemoglobins from normal individuals and from those with the sickle-cell trait and sickle-cell anemia. [From Montgomery, R., Dryer, R.L., Conway, T.W., and Spector, A.A., *Biochemistry, A Case Oriented Approach* (4th ed.), p. 87. Copyright © 1983 C.V. Mosby Company, Inc.]

Table 7-4   (*Continued*)

| | 45 | 50 | 55 | 60 | 65 | 70 | 75 | 80 | 85 | 90 | 95 | 100 | 104 |

```
PGYSYTAAANKNKGIIWGEDTLMEYLENPKKYIPGTKMIFVGIKKKEERADLIAYLKKATNE
PGYSYTAAANKNKGIIWGEDTLMEYLENPKKYIPGTKMIFVGIKKKEERADLIAYLKKATNE
PGFTYTDANKNKGITWKEETLMEYLENPKKYIPGTKMIFAGIKKKTEREDLIAYLKKATNE
PGFSYTDANKNKGITWKEETLMEYLENPKKYIPGTKMIFAGIKKKTEREDLIAYLKKATNE
PGFSYTDANKNKGITWGEETLMEYLENPKKYIPGTKMIFAGIKKKGEREDLIAYLKKATNE
PGFSYTDANKNKGITWGEETLMEYLENPKKYIPGTKMIFAGIKKTGERADLIAYLKKATKE
VGFSYTDANKNKGITWGEDTLMEYLENPKKYIPGTKMIFAGIKKKDERADLIAYLKKATNE
VGFSYTDANKNKGITWGEETLMEYLENPKKYIPGTKMIFAGIKKKGERADLIAYLKKATNE
PGFTYTDANKNKGIIWGEDTLMEYLENPKKYIPGTKMIFAGIKKKGERADLIAYLKKATNE

EGFSYTDANKNKGITWGEDTLMEYLENPKKYIPGTKMIFAGIKKKSERVDLIAYLKDATSK
EGFSYTDANKNKGITWGEDTLMEYLENPKKYIPGTKMIFAGIKKKAERADLIAYLKQATAK
EGFSYTDANKNKGITWGEDTLMEYLENPKKYIPGTKMIFAGIKKKSERADLIAYLKDATSK
EGFSYTEANKNKGITWGEETLMEYLENPKKYIPGTKMVFTGLSKKKERTNLIAYLKEKTAA
VGYSYTAANKNKGIIWGDDTLMEYLENPKKYIPGTKMIFAGIKKKGERQDLIAYLKSACSK
AGFSYTDANKNKGITWGEDTLMEYLENPKKYIPGTKMIFAGIKKKGERQDLIAYLKSACSK
EGYSYTDANKSKGIVWNNDTLMEYLENPKKYIPGTKMIFAGIKKKGERQDLVAYLKSATS-
QGFSYTDANKSKGITWQQETLRIYLENPKKYIPGTKMIFAGLKKKSERQDLIAYLKKTAAS

PGFSYSNANKAKGITWGDDTLFEYLENPKKYIPGTKMVFAGLKKKANERADLIAYLKESTK-
PGFSYSNANKAKGITWQDDTLFEYLENPKKYIPGTKMVFAGLKKKANERADLIAYLKQATK-
AGFAYTNANKAKGITWQDDTLFEYLENPKKYIPGTKMIFAGLKKPNERGDLIAYLKSATK-
AGFAYTNANKAKGITWQDDTLFEYLENPKKYIPGTKMIFAGLKKPNERGDLIAYLKSATK-

QGYSYTDANIKKNVLWDENNMSEYLTNPXKYIPGTKMAFGGLKKEKDRNDLITYLKKACE-
QGYSYTDANKRAGVEWAEPTMSDYLENPXKYIPGTKMAFGGLKKAKDRNDLVTYMLEASK-
DGYAYTDANKQKGITWDENTLFEYLENPXKYIPGTKMAFGGLKKDKDRNDIITFMKEATA-

AGYSYSAANKNKAVEWEEENTLYDYLLNPXKYIPGTKMVFPGLXKPQDRADLIAYLKKATSS
AGYSYSAANKNKAVTWGEDTLYEYLLNPXKYIPGTKMVFPGLXKPQERADLIAYLKDSTE-
AGYSYSAANKNMAVIWEENTLYDYLENPXKYIPGTKMVFPGLXKPQERADLIAYLKTSTA-
AGYSYSTANKNMAVIWEEKTLYDYLENPXKYIPGTKMVFPGLXKPQDRADLIAYLKESTA-
AGYSYSAANKNKAVEWEEKTLYDYLENPXKYIPGTKMVFPGLXKPQDRADLIAYLKEATA-
PGYSYSAANKNRAVIWEEKTLYDYLENPXKYIPGTKMVFPGLXKPQDRADLIAYLKEATA-
AGYSYSAANKNMAVQWGENTLYAYLENPXKYIPGTKMVFPGLXKPQERADLIAYLKEATA-
AGYSYSAANKNMAVQWGENTLYDYLENPXKYIPGTKMVFPGLXKPQDRADLIAYLKESTA-
PGYSYSAANKNMAVNWGENTLYDYLENPXKYIPGTKMVFPGLXKPQDRADLIAYLKESTA-
```

6 1 2 3 1 2 5 1 1 2 6 4 3 2 7 1 7 4 5 2 2 5 4 1 1 3 1 1 1 1 1 1 1 1 1 1 1 3 1 5 1 2 2 1 6 9 2 1 7 2 2 2 2 2 2 2 6 4 4 5 4

Hydrophilic, acidic: [D] Asp [E] Glu

Hydrophilic, basic: [H] His [K] Lys [R] Arg [X] TrimethylLys

Polar, uncharged: [B] Asn or Asp [G] Gly [N] Asn [Q] Gln [S] Ser [T] Thr [W] Trp [Y] Tyr [Z] Gln or Glu

Hydrophobic: [A] Ala [C] Cys [F] Phe [I] Ile [L] Leu [M] Met [P] Pro [V] Val

---

mutation causes deoxygenated HbS to aggregate into filaments of sufficient size and stiffness to deform erythrocytes—a remarkable example of the influence of primary structure on quaternary structure. The structure of these filaments is further discussed in Section 10-3B.

b. The Sickle-Cell Trait Confers Resistance to Malaria

Sickle-cell anemia is inherited according to the laws of Mendelian genetics (Section 1-4B). The hemoglobin of individuals who are homozygous for sickle-cell anemia is almost entirely HbS. In contrast, individuals heterozygous for sickle-cell anemia have hemoglobin that is ~40% HbS

(Fig. 7-20). Such persons, who are said to have the **sickle-cell trait,** lead a normal life even though their erythrocytes have a shorter lifetime than those of normal individuals.

The sickle-cell trait and disease occur mainly in persons of equatorial African descent. The regions of equatorial Africa where **malaria** is a major cause of death (contributing to childhood mortality rates as high as 50%), as Fig. 7-21 indicates, coincide closely with those areas where the sickle-cell gene is prevalent (possessed by as much as 40% of the population in some places). This observation led Anthony Allison to the discovery that *individuals heterozygous for HbS are resistant to malaria, that is, they are less likely to die of a malarial infection.*

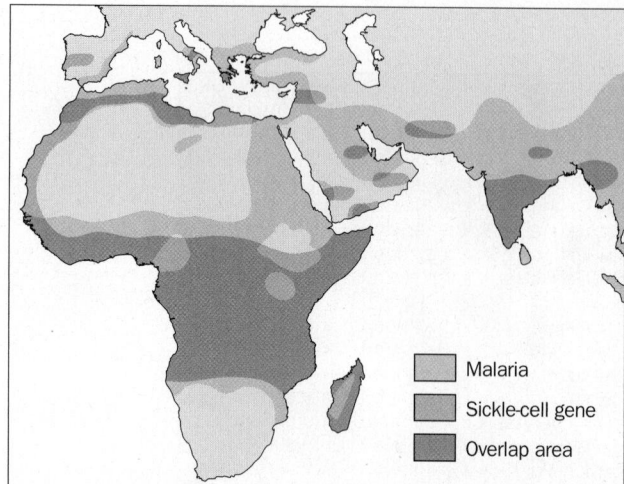

Figure 7-21 A map indicating the regions of the world where malaria caused by *P. falciparum* was prevalent before 1930, together with the distribution of the sickle-cell gene.

Malaria is one of the most lethal infectious diseases that presently afflict humanity: Of the 2.5 billion people living within malaria-endemic areas, 100 million are clinically ill with the disease at any given time and at least 1 million, mostly very young children, die from it each year. In Africa, malaria is caused by the mosquito-borne protozoan *Plasmodium falciparum*, which resides within an erythrocyte during much of its 48-h life cycle. Plasmodia increase the acidity of the erythrocytes they infect by ~0.4 pH units and cause them to adhere to a specific protein lining capillary walls by protein knobs that develop on the erythrocyte surfaces (the spleen would otherwise remove the infected erythrocytes from the circulation, thereby killing the parasites). Death often results when so many erythrocytes are lodged in a vital organ (such as the brain in cerebral malaria) that its blood flow is significantly impeded.

How does the sickle-cell trait confer malarial resistance? Normally, ~2% of the erythrocytes of individuals with the sickle-cell trait are observed to sickle under the low oxygen concentration conditions found in the capillaries. However, the lowered pH of infected erythrocytes increases their proportion of sickling in the capillaries to ~40%. Thus, during the early stages of a malarial infection, parasite-enhanced sickling probably causes the preferential removal of infected erythrocytes from the circulation. In the later stages of infection, when the parasitized erythrocytes are attached to the capillary walls, the sickling induced by this low oxygen environment may mechanically and/or metabolically disrupt the parasite. Consequently, bearers of the sickle-cell trait in a malarial region have an adaptive advantage: The fractional population of heterozygotes (sickle-cell trait carriers) in such areas increases until their reproductive advantage becomes balanced by the inviability of the correspondingly increasing proportion of homozygotes (those with sickle-cell disease). Thus *sickle-cell anemia provides a classic Darwinian*

example of a single mutation's adaptive consequences in the ongoing biological competition among organisms for the same resources.

B. Species Variations in Homologous Proteins: The Effects of Neutral Drift

The primary structures of a given protein from related species closely resemble one another. If one assumes, according to evolutionary theory, that related species have evolved from a common ancestor, then it follows that each of their proteins must have likewise evolved from the corresponding protein in that ancestor.

A protein that is well adapted to its function, that is, one that is not subject to significant physiological improvement, nevertheless continues evolving. The random nature of mutational processes will, in time, change such a protein in ways that do not significantly affect its function, a process called **neutral drift** (deleterious mutations are, of course, rapidly rejected through natural selection). *Comparisons of the primary structures of* **homologous proteins** *(evolutionarily related proteins) therefore indicate which of the proteins' residues are essential to its function, which are of lesser significance, and which have little specific function.* If, for example, we find the same side chain at a particular position in the amino acid sequence of a series of related proteins, we can reasonably conclude that the chemical and/or structural properties of that so-called **invariant residue** uniquely suit it to some essential function of the protein. Other amino acid positions may have less stringent side chain requirements so that only residues with similar characteristics (e.g., those with acidic properties: Asp and Glu) are required; such positions are said to be **conservatively substituted.** On the other hand, many different amino acid residues may be tolerated at a particular amino acid position, which indicates that the functional requirements of that position are rather nonspecific. Such a position is called **hypervariable.**

a. Cytochrome c Is a Well-Adapted Protein

To illustrate these points, let us consider the primary structure of a nearly universal eukaryotic protein, **cytochrome c.** Cytochrome *c* has a single polypeptide chain that, in vertebrates, consists of 103 or 104 residues, but in other phyla has up to 8 additional residues at its N-terminus. It occurs in the mitochondrion as part of the **electron-transport chain,** a complex metabolic system that functions in the terminal oxidation of nutrients to produce adenosine triphosphate (ATP) (Section 22-2). The role of cytochrome *c* is to transfer electrons from a large enzyme complex known as **cytochrome *c* reductase** to one called **cytochrome *c* oxidase.**

It appears that the electron-transport chain took its present form between 1.5 and 2 billion years ago as organisms evolved the ability to respire (Section 1-5Cb). Since that time, the components of this multienzyme system have changed very little, as is evidenced by the observation that the cytochrome *c* from any eukaryotic organism, say a pigeon, will react *in vitro* with the cytochrome oxidase from any other eukaryote, for instance, wheat. Indeed, hybrid cytochromes *c* consisting of covalently linked fragments from such distantly related species as horse and yeast (prepared via techniques of genetic engineering) exhibit biological activity.

b. Protein Sequence Comparisons Yield Taxonometric Insights

Emanuel Margoliash, Emil Smith, and others elucidated the amino acid sequences of the cytochromes *c* from over 100 widely diverse eukaryotic species ranging in complexity from yeast to humans. The sequences from 38 of these organisms are arranged in Table 7-4 (page 186) so as to maximize the similarities between vertically aligned residues (methods of sequence alignment are discussed in Section 7-4B). The various residues in the table have been colored according to their physical properties in order to illuminate the conservative character of the amino acid substitutions. Inspection of Table 7-4 indicates that cytochrome *c* is an evolutionarily conservative protein. A total of 38 of its 105 residues (23 in all that have been sequenced) are invariant and most of the remaining residues are conservatively substituted (see the bottom row of Table 7-4). In contrast, there are 8 positions that each accommodate six or more different residues and, accordingly, are described as being hypervariable.

The clear biochemical role of certain residues makes it easy to surmise why they are invariant. For instance, His 18 and Met 80 form ligands to the redox-active Fe atom of cytochrome *c*; the substitution of any other residues in these positions inactivates the protein. However, the biochemical significance of most of the invariant and conservatively substituted residues of cytochrome *c* can only be profitably assessed in terms of the protein's three-dimensional structure and is therefore deferred until Section 9-6A. In what follows, we consider what insights can be gleaned solely from the comparisons of the amino acid sequences of related proteins. The conclusions we draw are surprisingly far reaching.

The easiest way to compare the evolutionary differences between two homologous proteins is simply to count the amino acid differences between them (more realistically, we should infer the minimum number of DNA base changes to convert one protein to the other but, because of the infrequency with which mutations are accepted, counting amino acid differences yields similar information). Table 7-5 is a tabulation of the amino acid sequence differences among 22 of the cytochromes *c* listed in Table 7-4. It has been boxed off to emphasize the relationships among groups of similar species. The order of these differences largely parallels that expected from classical taxonomy. Thus primate cytochromes *c* more nearly resemble those of other mammals than they do, for example, those of insects (8–12 differences for mammals vs 26–31 for insects). Similarly, the cytochromes *c* of fungi differ as much from those of mammals (45–51 differences) as they do from those of insects (41–47) or higher plants (47–54).

Table 7-5 **Amino Acid Difference Matrix for 26 Species of Cytochrome *c*[a]**

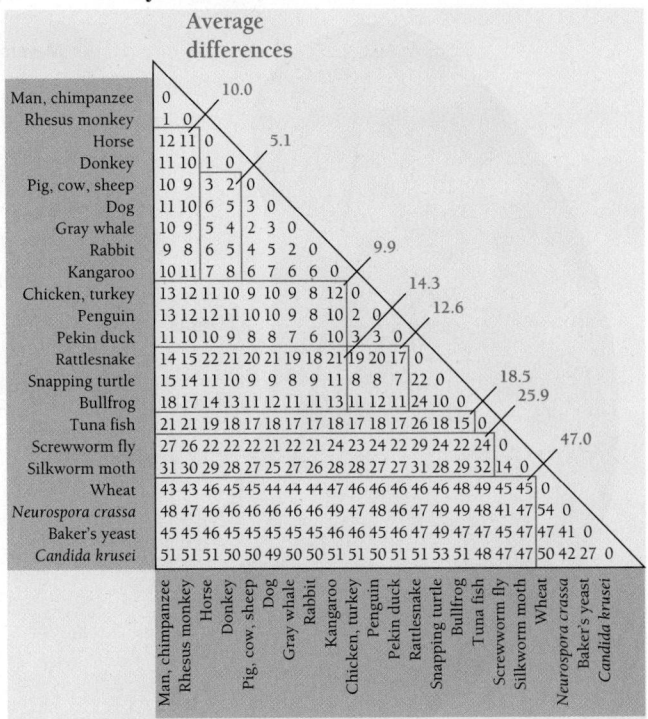

	1	2	3	4	5	6	7	8	9	10	11	12	13	14	15	16	17	18	19	20	21	22
Man, chimpanzee	0																					
Rhesus monkey	1	0																				
Horse	12	11	0																			
Donkey	11	10	1	0																		
Pig, cow, sheep	10	9	3	2	0																	
Dog	11	10	6	5	3	0																
Gray whale	10	9	5	4	2	3	0															
Rabbit	9	8	6	5	4	5	2	0														
Kangaroo	10	11	7	8	6	7	6	6	0													
Chicken, turkey	13	12	11	10	9	10	9	8	12	0												
Penguin	13	12	11	12	11	10	9	8	10	2	0											
Pekin duck	11	10	10	9	8	8	7	6	10	3	3	0										
Rattlesnake	14	15	22	21	20	21	19	18	21	19	20	17	0									
Snapping turtle	15	14	11	10	9	9	8	9	11	8	8	7	22	0								
Bullfrog	18	17	14	13	11	12	11	11	13	11	12	11	24	10	0							
Tuna fish	21	21	19	18	17	18	17	18	17	18	16	18	26	18	15	0						
Screwworm fly	27	26	22	22	22	21	22	21	24	23	24	22	29	24	22	24	0					
Silkworm moth	31	30	29	28	27	25	27	26	28	28	27	27	31	28	29	32	14	0				
Wheat	43	43	46	45	45	44	44	44	47	46	46	46	46	46	48	49	45	45	0			
Neurospora crassa	48	47	46	46	46	46	46	46	49	47	48	46	47	49	49	48	41	47	54	0		
Baker's yeast	45	45	46	45	45	45	45	45	46	45	46	47	49	47	47	45	47	47	47	41	0	
Candida krusei	51	51	51	50	50	49	50	50	51	51	50	51	51	53	51	48	47	47	50	42	27	0

Average differences (diagonal bracket labels): 10.0, 5.1, 9.9, 14.3, 12.6, 18.5, 25.9, 47.0

Column key: 1 Man, chimpanzee; 2 Rhesus monkey; 3 Horse; 4 Donkey; 5 Pig, cow, sheep; 6 Dog; 7 Gray whale; 8 Rabbit; 9 Kangaroo; 10 Chicken, turkey; 11 Penguin; 12 Pekin duck; 13 Rattlesnake; 14 Snapping turtle; 15 Bullfrog; 16 Tuna fish; 17 Screwworm fly; 18 Silkworm moth; 19 Wheat; 20 Neurospora crassa; 21 Baker's yeast; 22 Candida krusei.

[a]Each table entry indicates the number of amino acid differences between the cytochromes *c* of the species noted to the left of and below that entry.

[Table copyrighted © by Irving Geis.]

Through the analysis of data such as those in Table 7-5, *a phylogenetic tree (Section 1-1B) can be constructed that indicates the ancestral relationships among the organisms which produced the proteins* (the methods used to construct phylogenetic trees are discussed in Section 7-4C). That for cytochrome *c* is sketched in Fig. 7-22. Similar trees have been derived for other proteins. Each branch point of a tree indicates the probable existence of a common ancestor for all the organisms above it. The relative evolutionary distances between neighboring branch points are expressed as the number of amino acid differences per 100 residues of the protein (*percentage of accepted point mutations*, or **PAM units**). This furnishes a quantitative measure of the degree of relatedness of the various species that macroscopic taxonomy cannot provide. Note that the evolutionary distances of modern cytochromes *c* from the lowest branch point on their tree are all approximately equal. Evidently, the cytochromes *c* of the so-called lower forms of life have evolved to the same extent as those of the higher forms.

c. Proteins Evolve at Characteristic Rates

The evolutionary distances between various species can be plotted against the time when, according to radio-dated fossil records, the species diverged. For cytochrome

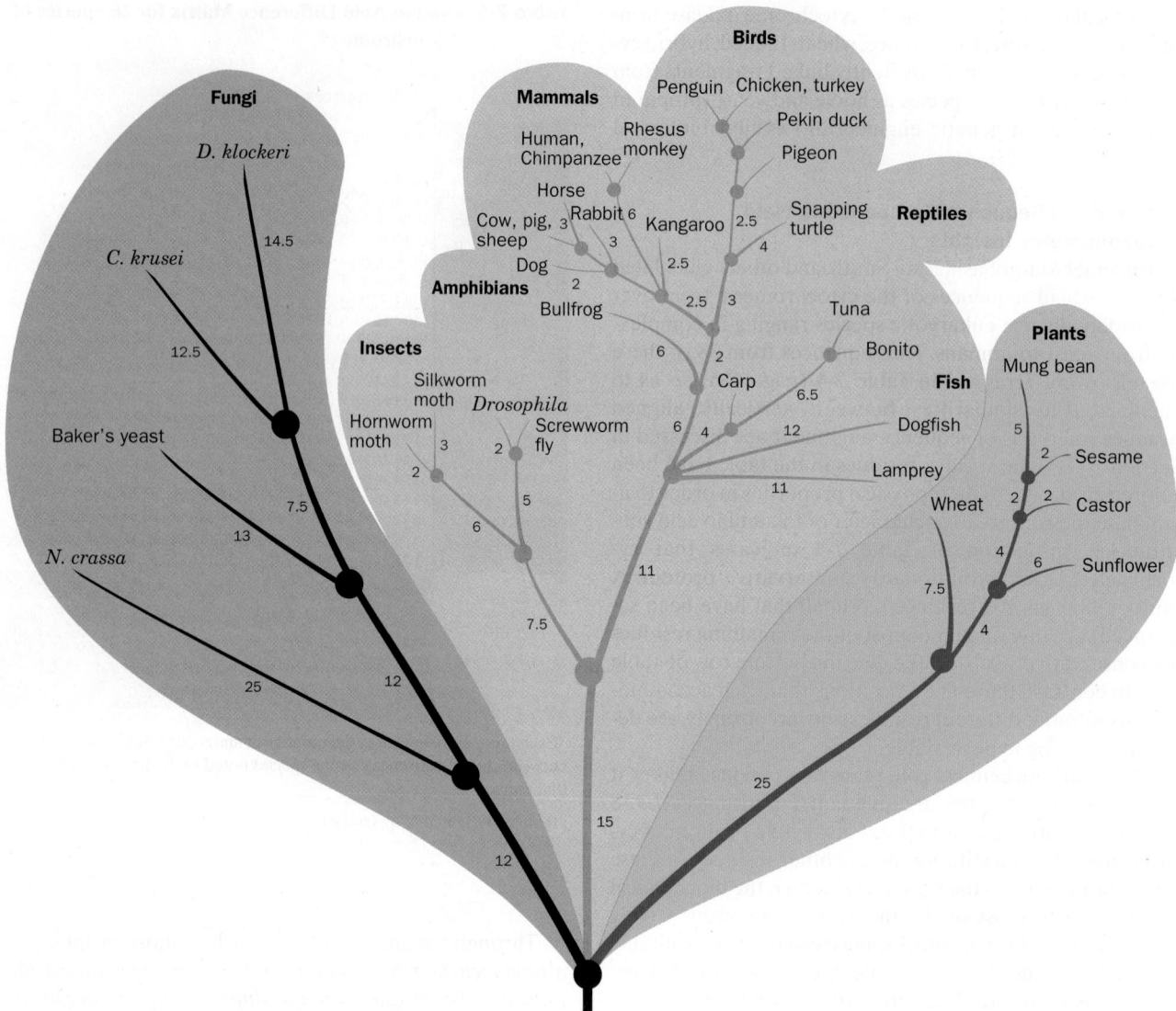

Figure 7-22 **Phylogenetic tree of cytochrome *c*.** The tree was generated by the computer-aided analysis of difference data such as that in Table 7-5 (Section 7-4C). Each branch point indicates the existence of an organism deduced to be ancestral to the species connected above it. The numbers beside each branch indicate the inferred differences, in PAM units, between the cytochromes *c* of its flanking branch points or species. [After Dayhoff, M.O., Park, C.M., and McLaughlin, P.J., *in* Dayhoff, M.O. (Ed.), *Atlas of Protein Sequence and Structure, p. 8,* National Biomedical Research Foundation (1972).]

c, this plot is essentially linear, thereby indicating that cytochrome *c* has accumulated mutations at a constant rate over the geological time scale (Fig. 7-23). This is also true for the other three proteins whose rates of evolution are plotted in Fig. 7-23. Each has its characteristic rate of change, known as a **unit evolutionary period,** which is defined as the time required for the amino acid sequence of a protein to change by 1% after two species have diverged. For cytochrome *c,* the unit evolutionary period is 20.0 million years. Compare this with the much less variant **histone H4** (600 million years) and the more variant hemoglobin (5.8 million years) and **fibrinopeptides** (1.1 million years).

The foregoing information does not imply that the rates of mutation of the DNAs specifying these proteins differ, but rather that *the rate that mutations are accepted into a protein depends on the extent that amino acid changes affect its function.* Cytochrome *c,* for example, is a rather small protein that, in carrying out its biological function, must interact with large protein complexes over much of its surface area. Any mutational change to cytochrome *c* will, most likely, affect these interactions unless, of course, the complexes simultaneously mutate to accommodate the change, a very unlikely occurrence. This accounts for the evolutionary stability of cytochrome *c.* Histone H4 is a protein that binds to DNA in eukaryotic chromosomes (Section 34-1A). Its central role in packaging the genetic archives evidently makes it extremely intolerant of any mutational changes. Indeed, histone H4 is so well adapted to its function that the histones H4 from peas and cows,

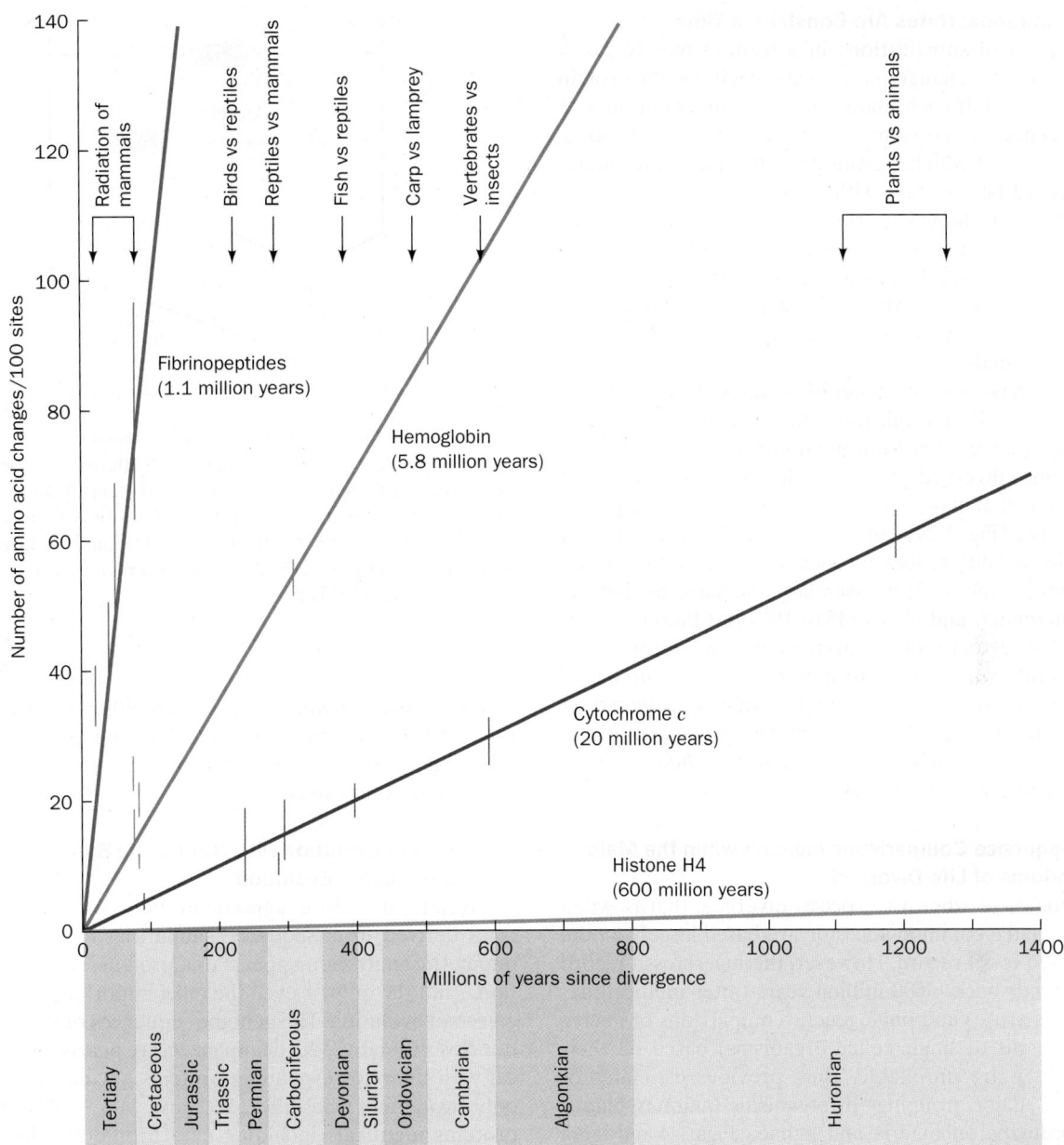

Figure 7-23 **Rates of evolution of four unrelated proteins.** The graph was constructed by plotting the average differences, in PAM units, of the amino acid sequences on two sides of a branch point of a phylogenetic tree (corrected to allow for more than one mutation at a given site) versus the time, according to the fossil record, since the corresponding species diverged from their common ancestor. The error bars indicate the experimental scatter of the sequence data. Each protein's rate of evolution, which is inversely proportional to the slope of its line, is indicated beside the line as its unit evolutionary period. [Illustration, Irving Geis. Image from the Irving Geis Collection, Howard Hughes Medical Institute. Reprinted with permission.]

species that diverged ~1.2 billion years ago, differ by only two conservative changes in their 102 amino acids. Hemoglobin, like cytochrome *c*, is an intricate molecular machine (Section 10-2). It functions as a free floating molecule, however, so that its surface groups are usually more tolerant of change than are those of cytochrome *c* (although not in the case of HbS; Section 10-3B). This accounts for hemoglobin's greater rate of evolution. The fibrinopeptides are polypeptides of ~20 residues that are proteolytically cleaved from the vertebrate protein **fibrinogen** when it is

converted to **fibrin** in the blood clotting process (Section 35-1A). Once they have been excised, the fibrinopeptides are discarded, so there is relatively little selective pressure on them to maintain their amino acid sequence and thus their rate of variation is high. If it is assumed that the fibrinopeptides are evolving at random, then the foregoing unit evolutionary periods indicate that in hemoglobin only 1.1/5.8 = 1/5 of the random amino acid changes are acceptable, that is, innocuous, whereas this quantity is 1/18 for cytochrome *c* and 1/550 for histone H4.

d. Mutational Rates Are Constant in Time

Amino acid substitutions in a protein mostly result from single base changes in the gene specifying the protein (Section 5-4B). If such **point mutations** mainly occur as a consequence of errors in the DNA replication process, then the rate at which a given protein accumulates mutations would be constant with respect to numbers of cell generations. If, however, the mutational process results from the random chemical degradation of DNA, then the mutation rate would be constant with absolute time. To choose between these alternative hypotheses, let us compare the rate of cytochrome *c* divergence in insects with that in mammals.

Insects have shorter generation times than mammals. Therefore, if DNA replication were the major source of mutational error, then from the time the insect and mammalian lines diverged, insects would have evolved further from plants than have mammals. However, a simple phylogenetic tree (Fig. 7-24) indicates that the average number of amino acid differences between the cytochromes *c* of insects and plants (45.2) is essentially the same as that between mammals and plants (45.0). We must therefore conclude that cytochrome *c* accumulates mutations at a uniform rate with respect to time rather than number of cell generations. This, in turn, implies that *point mutations in DNA accumulate at a constant rate with time, that is, through random chemical change, rather than resulting mainly from errors in the replication process.*

e. Sequence Comparisons Indicate when the Major Kingdoms of Life Diverged

Estimates of when two species diverged, that is, when they last had a common ancestor, are based largely on the radiodated fossil record. However, the macrofossil record only extends back ~600 million years (after multicellular organisms arose) and phylogenetic comparisons of microfossils (fossils of single-celled organisms) based on their morphology are unreliable. Thus, previous estimates of when the major groupings of organisms (animals, plants, fungi, protozoa, eubacteria, and archaea; Figs. 1-4 and 1-11) diverged from one another (e.g., the right side of Fig. 7-23) are only approximations based mainly on considerations of shared characteristics.

The burgeoning databases of amino acid sequences (Section 7-4A) permitted Russell Doolittle to compare the sequences of a large variety of enzymes that each have homologous representatives in many of the above major groupings (531 sequences in 57 different enzymes). This analysis is consistent with the existence of a **molecular clock** that can provide reliable estimates of when these groupings diverged. This molecular clock, which is based on the supposition that homologous sequences diverge at a uniform rate, was calibrated using sequences from vertebrates for which there is a reasonably reliable fossil record. This analysis indicates that animals, plants, and fungi last had a common ancestor ~1 billion years ago, with plants having diverged from animals slightly before fungi; that the major protozoan lineages separated from those of other eukarya ~1.2 billion years ago; that eukarya last shared a

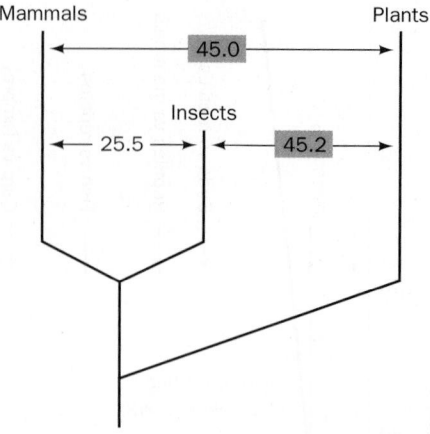

Figure 7-24 **Phylogenetic tree for cytochrome *c*.** The tree shows the average number of amino acid differences between cytochromes *c* from mammals, insects, and plants. Mammals and insects have diverged equally far from plants since their common branch point. [Adapted from Dickerson, R.E. and Timkovitch, R., *in* Boyer, P.D. (Ed.), *The Enzymes* (3rd ed.), Vol. 11, *p.* 447, Academic Press (1975).]

common ancestor with archaea ~1.8 billion years ago and with bacteria slightly more than 2 billion years ago; and that gram-positive and gram-negative bacteria diverged ~1.4 billion years ago.

f. Protein Evolution May Not Be the Basis of Organismal Evolution

Despite the close agreement between phylogenetic trees derived from sequence similarities and classic taxonometric analyses, it appears that protein sequence evolution is not the only or even the most important basis of organismal evolution. The genome sequences of humans and our closest relative, the chimpanzee, are nearly 99% identical with their corresponding proteins having, on average, only two amino acid differences and with ~29% of these proteins identical (including cytochrome *c*). This is the level of homology observed among sibling species of fruit flies and mammals. Yet the anatomical and behavioral differences between human and chimpanzee are so great that these species have been classified in separate families. *This suggests that the rapid divergence of human and chimpanzee stems from relatively few mutational changes in the segments of DNA that control gene expression, that is, how much of each protein will be made, where, and when.* Such mutations do not necessarily change protein sequences but can result in major organismal alterations.

C. *Evolution through Gene Duplication*

Most proteins have extensive sequence similarities with other proteins from the same organism. Such proteins arose through **gene duplication,** a result of an aberrant genetic recombination event in which a single chromosome acquired both copies of the primordial gene in question (the mechanism of genetic recombination is discussed in

Section 30-6A). *Gene duplication is a particularly efficient mode of evolution because one of the duplicated genes can evolve a new functionality through natural selection while its counterpart continues to direct the synthesis of the presumably essential ancestral protein.*

The **globin** family of proteins, which includes hemoglobin and **myoglobin,** provides an excellent example of evolution through gene duplication. Hemoglobin transports oxygen from the lungs (or gills or skin) to the tissues. Myoglobin, which occurs in muscles, facilitates rapid oxygen diffusion through these tissues and also functions as an oxygen storage protein. *The sequences of hemoglobin's α and β subunits (recall that hemoglobin is an $\alpha_2\beta_2$ tetramer) and myoglobin (a monomer) are quite similar.*

The globin family's phylogenetic tree indicates that its members, in humans, arose through the following chain of events (Fig. 7-25):

1. The primordial globin probably functioned simply as an oxygen-storage protein. Indeed, the globins in certain modern invertebrates still have this function. For example, treating a *Planorbis* snail with CO (the binding of which prevents globins from binding O_2; Section 10-1A) does not affect its behavior in well-aerated water, but if the oxygen concentration is reduced, a poisoned *Planorbis* becomes even more sluggish than a normal one.

2. Duplication of a primordial globin gene, ~1.1 billion years ago, permitted the resulting two genes to evolve sep-

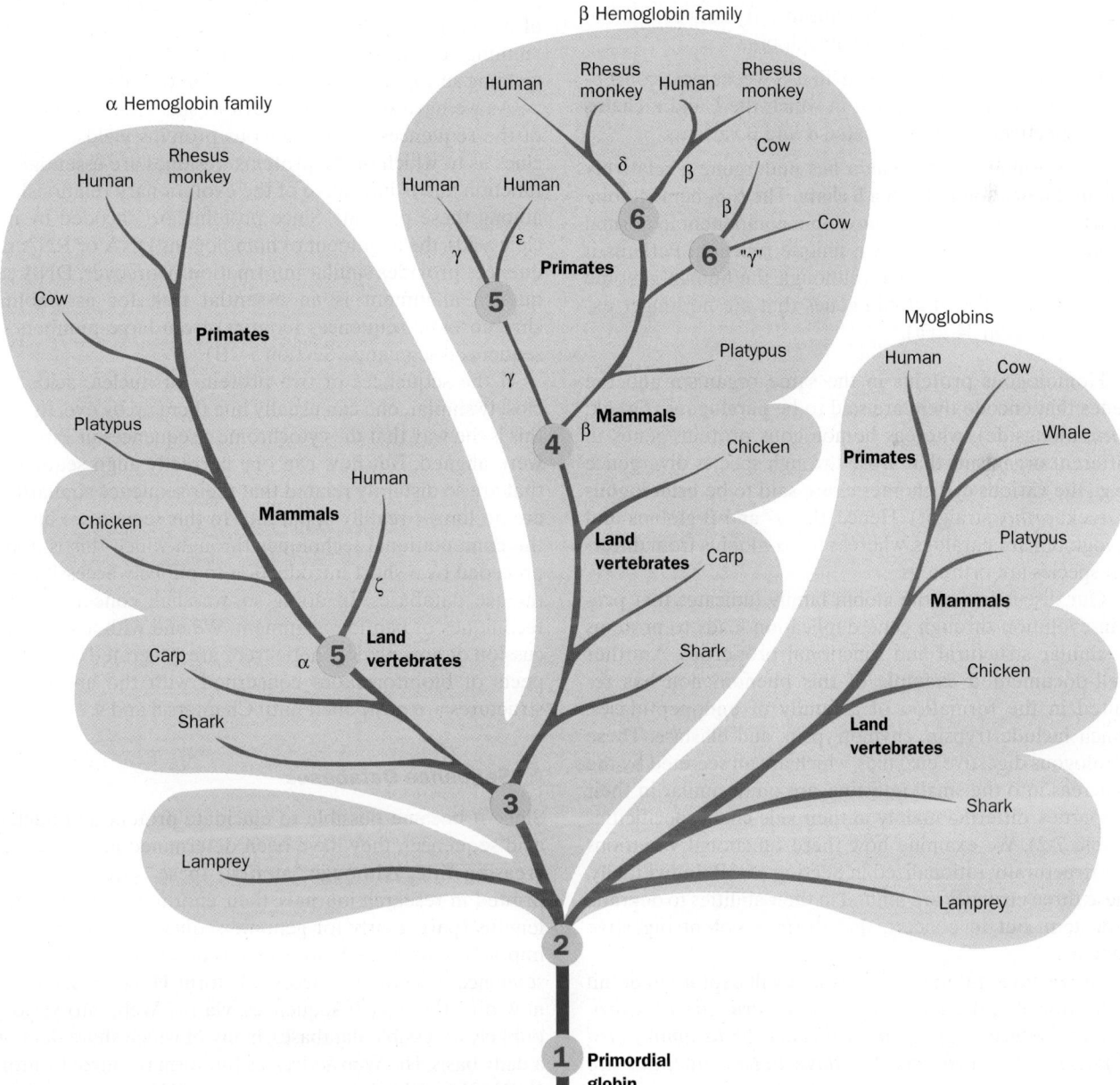

Figure 7-25 **Phylogenetic tree of the globin family.** The circled branch points represent gene duplications and unmarked branch points are species divergences. [After Dickerson, R.E. and Geis, I., *Hemoglobin, p. 82,* Benjamin/Cummings (1983).]

arately so that, largely by a series of point mutations, a monomeric hemoglobin arose that had the lower oxygen-binding affinity required for it to transfer oxygen to the developing myoglobin. Such a monomeric hemoglobin can still be found in the blood of the **lamprey,** a primitive vertebrate that, according to the fossil record, has maintained its eel-like morphology for over 425 million years.

3. Hemoglobin's tetrameric character is a structural feature that greatly increases its ability to transport oxygen efficiently (Section 10-2C). This provided the adaptive advantage that gave rise to the evolution of the β chain from a duplicated α chain.

4. In fetal mammals, oxygen is obtained from the maternal circulation. **Fetal hemoglobin,** an $\alpha_2\gamma_2$ tetramer in which the **γ chain** is a gene-duplicated β chain variant, evolved to have an oxygen-binding affinity between that of normal adult hemoglobin and myoglobin.

5. Human embryos, in their first 8 weeks post conception, make a $\zeta_2\varepsilon_2$ hemoglobin in which the ζ and ε **chains** are, respectively, gene-duplicated α and β variants.

6. In primates, the β chain has undergone a relatively recent duplication to form a **δ chain.** The $\alpha_2\delta_2$ hemoglobin, which occurs as a minor hemoglobin component in normal adults (~1%), has no known unique function. Perhaps it may eventually evolve one (although the human genome contains the relics of globin genes that are no longer expressed; Section 34-2Fa).

Homologous proteins in the same organism and the genes that encode them are said to be **paralogous** (Greek: *para,* alongside), whereas homologous proteins/genes in different organisms that arose through species divergence (e.g., the various cytochomes *c*) are said to be **orthologous** (Greek: *ortho,* straight). Hence, the α- and β-globins and myoglobin are paralogs, whereas the α-globins from different species are orthologs.

Our discussion of the globin family indicates that protein evolution through gene duplication leads to proteins of similar structural and functional properties. Another well-documented example of this phenomenon has resulted in the formation of a family of endopeptidases, which include trypsin, chymotrypsin, and elastase. These paralogous digestive enzymes, which are all secreted by the pancreas into the small intestine, are quite similar in their properties, differing mainly in their side chain specificities (Table 7-2). We examine how these functional variations are structurally rationalized in Section 15-3B. Individually, these three enzymes are limited in their abilities to degrade a protein, but in concert, they form a potent digestive system.

As we have stated previously and will explore in detail in Section 9-1, the *three-dimensional structure of a protein, and hence its function, is dictated by its amino acid sequence.* Most proteins that have been sequenced are more or less similar to several other known proteins. In fact, *many proteins are mosaics of sequence motifs that occur in a variety of other proteins.* It therefore seems likely that most of the myriads of proteins in any given organism have arisen through gene duplications. This suggests that the appearance of a protein with a novel sequence and function is an extremely rare event in biology—one that may not have occurred since early in the history of life.

4 BIOINFORMATICS: AN INTRODUCTION

See Guided Exploration 6: Bioinformatics The enormous profusion of sequence and structural data that have been generated over the last decades has led to the creation of a new field of inquiry, **bioinformatics,** which is loosely defined as being at the intersection of biotechnology and computer science. It is the computational tools that the practitioners of bioinformatics have produced that have permitted the "mining" of this treasure trove of biological data, thereby yielding surprisingly far-reaching biomolecular insights.

As we have seen in the previous section, the alignment of the sequences of homologous proteins yields valuable clues as to which of the proteins' residues are essential for function and is indicative of the evolutionary relationships among these proteins. Since proteins are encoded by nucleic acids, the alignment of homologous DNA or RNA sequences provides similar information. Moreover, DNA sequence alignment is an essential task for assembling chromosomal sequences (contigs) from large numbers of sequenced segments (Section 7-2B).

If the sequences of two proteins or nucleic acids are closely similar, one can usually line them up by eye. In fact, this is the way that the cytochrome *c* sequences in Table 7-4 were aligned. But how can one correctly align sequences that are so distantly related that their sequence similarities are no longer readily apparent? In this section we discuss the computational techniques through which this is done, preceded by a short introduction to publicly accessible sequence databases. In doing so, we shall concentrate on techniques of peptide alignment. We end with a short discussion of how phylogenetic trees are generated. Those aspects of bioinformatics concerned with the analysis of structures are postponed until Chapters 8 and 9.

A. *Sequence Databases*

Since it became possible to elucidate protein and nucleic acid sequences, they have been determined at an ever increasing rate. Although, at first, these sequences were printed in research journals, their enormous numbers and lengths (particularly for genome sequences) now make it impractical to do so. Moreover, it is far more useful to have sequences in computer-accessible form. Hence, researchers now directly deposit sequences, via the Web, into various publicly accessible databases, many of which share data on a daily basis. The Web addresses [uniform resource locators (URLs)] for the major protein and DNA sequence databases are listed in Table 7-6. The URLs for a variety of specialized sequence databases (e.g., those of specific organisms or organelles) can be found at the Life Science

Table 7-6   Websites of the Major Protein and DNA Sequence Data Banks

Data Banks Containing Protein Sequences

Swiss-Prot Protein Knowlegebase:
http://www.expasy.org/sprot/

Protein Information Resource (PIR):
http://pir.georgetown.edu/

Protein Research Foundation (PRF):
http://www.prf.or.jp/

Data Banks Containing Gene and Genome Sequences

GenBank:
http://www.ncbi.nlm.nih.gov/Genbank/

EMBL Nucleotide Sequence Database:
http://www.ebi.ac.uk/embl/

DNA Data Bank of Japan (DDBJ):
http://www.ddbj.nig.ac.jp/

GenomeNet:
http://www.genome.jp/

Genomes OnLine Database (GOLD):
http://genomesonline.org/

Directory (http://www.expasy.ch/links.html). That website also contains links to numerous other biochemically useful databases as well a great variety of computer tools for biomolecular analyses, bibliographic references, tutorials, and many other websites of biomedical interest. (Note that websites evolve far faster than organisms: Even well-established websites change addresses or even disappear with little warning and useful new websites appear on an almost daily basis.)

As an example of a sequence database, let us describe (in cursory detail) the annotated protein sequence database named Swiss-Prot. A sequence record in Swiss-Prot begins with the proteins' ID code of the form X_Y where X is an up-to-four-character mnemonic indicating the protein's name (e.g., CYC for cytochrome *c* and HBA for hemoglobin α chain) and Y is an up-to-five-character identification code indicating the protein's biological source that usually consists of the first three letters of the genus and the first two letters of the species [e.g., CANFA for *Canis familiaris* (dog)]. However, for the most commonly encountered organisms, Y is instead a self-explanatory code (e.g., BOVIN or ECOLI). This is followed by an accession number such as P04567, which is assigned by the database to ensure a stable way of identifying an entry from release to release, even if it becomes necessary to change its ID code. The entry continues with the date the entry was entered into Swiss-Prot and when it was last modified and annotated, a list of pertinent references (which are linked to MedLine), a description of the protein, and its links to other databases. A Feature Table describes regions or sites of interest in the protein such as disulfide bonds, post-translational modifications, elements of local secondary structure, binding sites, and conflicts between different

references. The entry ends with the length of the peptide in residues, its molecular weight, and finally its sequence using the one-letter code (Table 4-1). Other sequence databases are similarly constructed.

B. *Sequence Alignment*

One can quantitate the sequence similarity of two polypeptides or two DNAs by determining their number of aligned residues that are identical. For example, human and dog cytochromes *c*, which differ in 11 of their 104 residues (Table 7-5) are $[(104 - 11)/104] \times 100 = 89\%$ identical, whereas human and baker's yeast cytochromes *c* are $[(104 - 45)/104] \times 100 = 57\%$ identical. Table 7-4 indicates that baker's yeast cytochrome *c* has 5 residues at its N-terminus that human cytochrome *c* lacks but lacks the human protein's C-terminal residue. When determining percent identity, the length of the shorter peptide/DNA is, by convention, used in the denominator. One can likewise calculate the percent similarity between two peptides, once it is decided which amino acid residues are to be considered similar (e.g., Asp and Glu).

a. The Homology of Distantly Related Proteins May Be Difficult to Recognize

Let us examine how proteins evolve by considering a simple model. Assume that we have a 100-residue protein in which all point mutations have an equal probability of being accepted and occur at a constant rate. Thus at an evolutionary distance of one PAM unit (Section 7-3Bb), the original and evolved proteins are 99% identical. At an evolutionary distance of two PAM units, they are $(0.99)^2 \times 100 = 98\%$ identical, whereas at 50 PAM units they are $(0.99)^{50} \times 100 = 61\%$ identical. Note that the latter quantity is not 50%, as one might naively expect. This is because *mutation is a stochastic (probablistic or random) process: At every stage of evolution, each residue has an equal chance of mutating*. Hence some residues may change twice or more before others change even once. Consequently, a plot of percent identity versus evolutionary distance (Fig. 7-26*a*) is an exponential curve that approaches but never equals zero. Even at quite large evolutionary distances, original and evolved proteins still have significant sequence identities.

Real proteins evolve in a more complex manner than our simple model predicts. This is in part because certain amino acid residues are more likely to form accepted mutations than others and in part because the distribution of amino acids in proteins is not uniform (e.g., 9.7% of the residues in proteins, on average, are Leu but only 1.1% are Trp; Table 4-1). Consequently real proteins evolve even more slowly than in our simple model (Fig. 7-26*b*).

At what point in the evolutionary process does homology become unrecognizable? If identical length polypeptides of random sequences were of uniform amino acid composition, that is, if they consisted of 5% of each of the 20 amino acids, then they would exhibit, on average, 5% identity. However, since mutations occur at random, there is considerable variation in such numbers. Thus, statistical

(a)

(b)

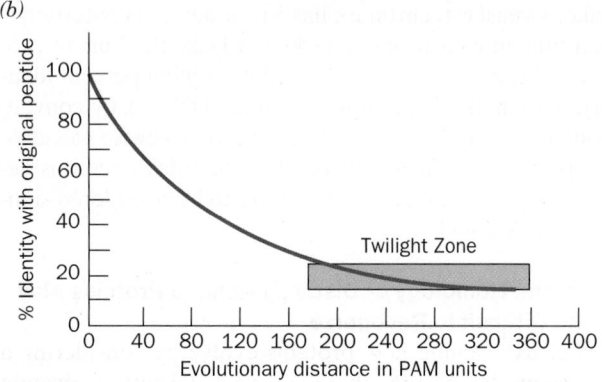

Figure 7-26 **Rate of sequence change in evolving proteins.** (*a*) For a protein that is evolving at random and that initially consists of 5% of each of the 20 "standard" amino acid residues. (*b*) For a protein of average amino acid composition evolving as is observed in nature, that is, with certain residue changes more likely to be accepted than others and with occasional insertions and deletions. [Part *b* after Doolittle, R.F., *Methods Enzymol.* **183,** 103 (1990).]

considerations reveal that there is a 95% probability that such 100-residue peptides are between 0 and 10% identical. But, as we have seen for cytochrome *c,* homologous peptides may have different lengths because one may have more or fewer residues at its N- or C-termini. If we therefore permit our random 100-residue peptides to shift in their alignment by up to 5 residues, the average expected identity for the best alignment increases to 8%, with 95% of such comparisons falling between 4% and 12%. Consequently, one in 20 of such comparisons will be out of this range (>12% or <4%) and one in 40 will exhibit >12% sequence identity.

But this is not the whole story because mutational events may result in the insertion or deletion of one or more residues within a chain. Thus, one chain may have gaps relative to another. Yet, if we permit an unlimited number of gaps, we can always get a perfect match between any two chains. For example, two 15-residue peptides that have only one match (using the one-letter code; Table 4-1)

SQMC I L F K A Q M N Y G H
M F Y A C R L P M G A H Y W L

would have a perfect match over their aligned portions if we allowed unlimited gapping:

SQMC I L F K A Q M N Y G H
– – M – – – F – – – – – Y – – A C R L P M G A H Y W L

Thus we cannot allow unlimited gapping to maximize a match between two peptides, but neither can we forbid all gapping because insertions and deletions (collectively called **indels**) really do occur. Consequently, for each allowed gap, we must impose some sort of penalty in our alignment algorithm that strikes a balance between finding the best alignment between distantly related peptides and rejecting improper alignments. But if we do so (using methods discussed below), *unrelated proteins will exhibit sequence identities in the range 15% to 25%. Yet distantly related proteins may have similar levels of sequence identity.* This is the origin of the **twilight zone** in Fig. 7-26*b.* It requires the sophisticated alignment algorithms we discuss below to differentiate homologous proteins in the twilight zone from those that are unrelated.

b. Sequence Alignment Using Dot Matrices

How does one perform a sequence alignment between two polypeptides (a **pairwise alignment**)? The simplest way is to construct a **dot matrix** (alternatively, a **dot plot** or **diagonal plot**): Lay the sequence of one polypeptide horizontally and that of the other vertically and place a dot in the resulting matrix wherever the residues are identical. A dot plot of a peptide against itself results in a square matrix with a row of dots along the diagonal and a scattering of dots at points where there are chance identities. If the peptides are closely similar, there are only a few absences along the diagonal (e.g., Fig. 7-27*a*), whereas distantly related peptides have a large number of absences along the diagonal and a shift in its position wherever one peptide has a gap relative to the other (e.g., Fig. 7-27*b*).

Once an alignment has been established, it should be scored in some way to determine if it has any relationship to reality. A simple but effective way to calculate an **alignment score (AS)** is to add 10 for every identity but those of Cys, which count 20 (because Cys residues often have indispensable functions), and then subtract 25 for every gap. Furthermore, we can calculate the **normalized alignment score (NAS)** by dividing the AS by the number of residues in the shortest of the two polypeptides and multiplying by 100. Thus, for the alignment of the human hemoglobin α chain (141 residues) and human myoglobin (153 residues; Fig. 7-28), AS = 37 × 10 + 1 × 20 − 1 × 25 = 365 and NAS = (365/141) × 100 = 259. Statistical analysis (Fig. 7-29) indicates that this NAS is indicative of homology. Note that a perfect match would result in NAS = 1000 in the absence of Cys residues or gaps. An acceptable NAS decreases with peptide length because a high proportion of matches is more likely to occur between short peptides than between long ones (e.g., 2 matches in 10 residues is more likely to occur at random than 20 matches in 100 residues, although both have NAS = 200).

(a) Tuna fish cytochrome *c*

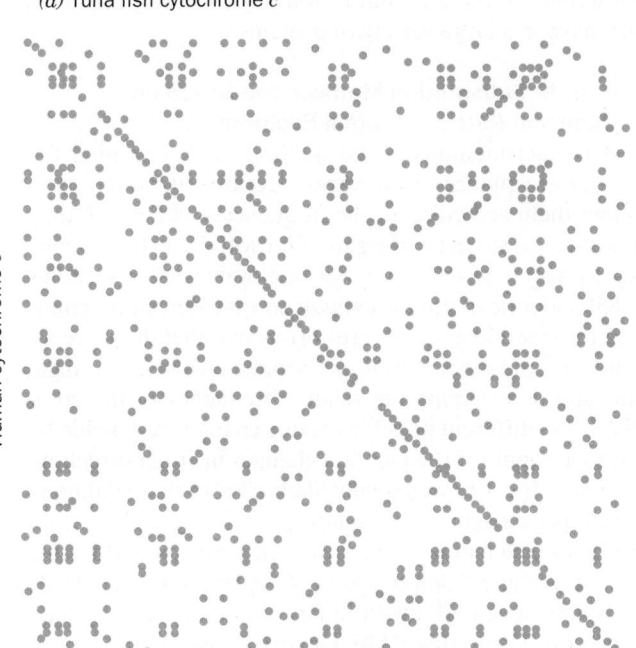

(b) Rhodospirillum rubrum cytochrome *c*₂

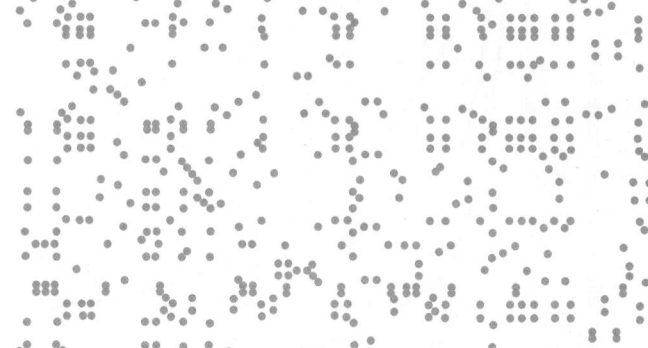

Figure 7-27 **Sequence alignment with dot matrices.** Dot plots show alignments of (*a*) human cytochrome *c* (104 residues) vs tuna fish cytochrome *c* (103 residues) and (*b*) human cytochrome *c* vs *Rhodospirillum rubrum* cytochrome *c*₂ (a bacterial *c*-type cytochrome consisting of 112 residues). The N-termini of these peptides are at the top and left of the diagrams. The two proteins in Part *a* have 82 identities, whereas those in Part *b* have 40

identities. The diagonal in Part *b* is more clearly seen if the diagram is viewed edgewise from its lower right corner. Note that there are two horizontal displacements of this diagonal, one near its center and other toward the C-terminus. This is indicative of inserts in the *Rhodospirillum* protein relative to the human protein. [After Gibbs, A.J. and McIntyre, G.A., *Eur. J. Biochem.* **16,** 2 (1970).]

c. Alignments Should Be Weighted According to the Likelihood of Residue Substitutions

The foregoing techniques can easily (although tediously) be carried out by hand, particularly when there is an obvious alignment. But what if we have numerous polypeptides with which we want to align a new sequence (and, typically, one checks newly determined sequences against all other known sequences). Moreover, alignments in the twilight zone are difficult to discern. We must therefore turn to computerized statistical analyses so as to be

able to distinguish distant evolutionary relationships from chance similarities with maximum sensitivity.

A dot matrix can be cast in a more easily mathematized form by replacing each dot (match) with a 1 and each nonmatch with a 0. Then a self-dot matrix would become a square diagonal matrix (having all 1's along its diagonal) with a few off-diagonal 1's, and two closely related peptides would have several diagonal positions with 0's. But this is a particularly rigid system: It does not differentiate between conservative substitutions and those that are likely to be

Figure 7-28 **Optimal alignments of human myoglobin (Mb, 153 residues) and the human hemoglobin α chain (Hbα, 141 residues).** Identical residues are shaded in blue and gaps are

indicated by dashes. [After Doolittle, R.F., *Of URFs and ORFs,* University Science Books (1986).]

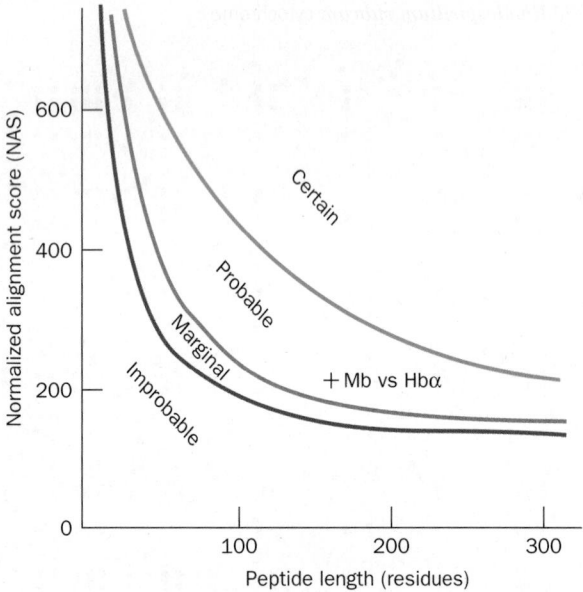

Figure 7-29   Guide to the significance of normalized alignment scores (NAS) in the comparison of peptide sequences. Note how the significance of an NAS varies with peptide length. The position of the Mb vs Hbα alignment (Fig. 7-28) is indicated. [After Doolittle, R.F., *Methods Enzymol.* **183,** 102 (1990).]

hypervariable. Yet it is clear that certain substitutions are more readily accepted than others. What are these favored substitutions, how can we obtain a quantitative measure of them, and how can we use this information to increase the confidence with which we can align distantly related peptides?

One way we might assign a weight (a quantity that increases with the probability of occurrence) to a residue change is according to the genetic code (Table 5-3). Thus residue changes that require only a single base change [e.g., Leu (CUX) → Pro (CCX)] are likely to occur more often and hence would be assigned a greater weight than a residue change that requires two base changes [e.g., Leu (CUX) → Thr (ACX)], which likewise would be assigned a greater weight than a residue change that requires three base changes [e.g., His (CA$_C^U$) → Trp (UGG)]. Of course, no change (the most probable event) would be assigned the greatest weight of all. However, such a scheme only weights the probability of a mutation occurring, not that of a mutation being accepted, which depends on its Darwinian fitness. In fact, over half of the possible single-base residue changes are between physically dissimilar residues, which are therefore less likely to be accepted.

A more realistic weighting scheme would be to assign some sort of relative probability to two residues being exchanged according to their physical similarity. Thus it would seem that a Lys → Arg mutation is more likely to be accepted than, say, a Lys → Phe mutation. However, it is by no means obvious how to formulate such a weighting scheme based on theoretical considerations because it is unclear how to evaluate the various different sorts of prop-

erties that suit the different residues to the many functions they have in a large variety of proteins.

d. PAM Substitution Matrices Are Based on Observed Rates of Protein Evolution

An experimentally based method of determining the rates of acceptance of the various residue exchanges is to weight them according to the frequencies with which they are observed to occur. Margaret Dayhoff did so by comparing the sequences of a number of closely related proteins (>85% identical; similar enough so that we can be confident that their alignments are correct and that there are insignificant numbers of multiple residue changes at a single site) and determining the relative frequency of the $20 \times 19/2 = 190$ different possible residue changes (we divide by two to account for the fact that changes in either direction, A → B or B → A, are equally likely). From these data one can prepare a symmetric square matrix, 20 elements on a side, whose elements, M_{ij}, indicate the probability that, in a related sequence, amino acid i will replace amino acid j after some specified evolutionary interval—usually one PAM unit. Using this **PAM-1** matrix, one can generate a mutation probability matrix for other evolutionary distances, say N PAM units, by multiplying the matrix by itself N times ($[M]^N$), thereby generating a PAM-N matrix. Then an element of the **relatedness odds matrix,** R, is

$$R_{ij} - M_{ij}/f_i = q_{ij}/f_if_j \qquad [7.3]$$

where M_{ij} is now an element of the PAM-N matrix, f_i is the probability that the amino acid i will occur in the second sequence by chance, and q_{ij} is the observed frequency with which residues of types i and j replace each other in a set of aligned polypeptides. Thus, R_{ij} is the probability that amino acid i will replace amino acid j (or vice versa) per occurrence of i per occurrence of j. When two polypeptides are compared with each other, residue by residue, the R_{ij}'s for each position are multiplied to obtain the **relatedness odds** for the entire polypeptide. For example, when the hexapeptide A-B-C-D-E-F evolves to the hexapeptide P-Q-R-S-T-U,

$$\begin{aligned} \text{Relatedness odds} = R_{AP} \times R_{BQ} \times R_{CR} \times R_{DS} \\ \times R_{ET} \times R_{FU} \qquad [7.4] \end{aligned}$$

A more convenient way of making this calculation is to take the logarithm of each R_{ij}, thereby yielding the **log odds substitution matrix.** Then we add the resulting matrix elements rather than multiplying them to obtain the **log odds.** Thus for our hexapeptide pair:

$$\begin{aligned} \text{log odds} = \log R_{AP} + \log R_{BQ} + \log R_{CR} \\ + \log R_{DS} + \log R_{ET} + \log R_{FU} \qquad [7.5] \end{aligned}$$

It is a peptide pair's log odds that we wish to maximize in obtaining their best alignment, that is, we will use log odds values as our alignment scores.

Table 7-7 is the **PAM-250** log odds substitution matrix with all of its elements multiplied by 10 for readability (which only adds a scale factor). Each diagonal element in

Table 7-7 **The PAM-250 Log Odds Substitution Matrix**

	C	S	T	P	A	G	N	D	E	Q	H	R	K	M	I	L	V	F	Y	W
C Cys	12																			
S Ser	0	2																		
T Thr	-2	1	3																	
P Pro	-3	1	0	6																
A Ala	-2	1	1	1	2															
G Gly	-3	1	0	-1	1	5														
N Asn	-4	1	0	-1	0	0	2													
D Asp	-5	0	0	-1	0	1	2	4												
E Glu	-5	0	0	-1	0	0	1	3	4											
Q Gln	-5	-1	-1	0	0	-1	1	2	2	4										
H His	-3	-1	-1	0	-1	-2	2	1	1	3	6									
R Arg	-4	0	-1	0	-2	-3	0	-1	-1	1	2	6								
K Lys	-5	0	0	-1	-1	-2	1	0	0	1	0	3	5							
M Met	-5	-2	-1	-2	-1	-3	-2	-3	-2	-1	-2	0	0	6						
I Ile	-2	-1	0	-2	-1	-3	-2	-2	-2	-2	-2	-2	-2	2	5					
L Leu	-6	-3	-2	-3	-2	-4	-3	-4	-3	-2	-2	-3	-3	4	2	6				
V Val	-2	-1	0	-1	0	-1	-2	-2	-2	-2	-2	-2	-2	2	4	2	4			
F Phe	-4	-3	-3	-5	-4	-5	-4	-6	-5	-5	-2	-4	-5	0	1	2	-1	9		
Y Tyr	0	-3	-3	-5	-3	-5	-2	-4	-4	-4	0	-4	-4	-2	-1	-1	-2	7	10	
W Trp	-8	-2	-5	-6	-6	-7	-4	-7	-7	-5	-3	2	-3	-4	-5	-2	-6	0	0	17
	C	S	T	P	A	G	N	D	E	Q	H	R	K	M	I	L	V	F	Y	W
	Cys	Ser	Thr	Pro	Ala	Gly	Asn	Asp	Glu	Gln	His	Arg	Lys	Met	Ile	Leu	Val	Phe	Tyr	Trp

Source: Dayhoff, M.O. (Ed.), *Atlas of Protein Sequence and Structure,* Vol. 5, Supplement 3, *p.* 352, National Biomedical Research Foundation (1978).

the matrix indicates the mutability of the corresponding amino acid, whereas the off-diagonal elements indicate their exchange probabilities. A neutral (random) score is 0, whereas an amino acid pair with a score of –3 exchanges with only $10^{-3/10} = 0.50$ of the frequency expected at random. This substitution matrix has been arranged such that the amino acid residues most likely to replace each other in related proteins (the pairs that have the highest log R_{ij} values) are grouped together. Note that this grouping is more or less what is expected from their physical properties.

Identities (no replacement) tend to have the highest values in Table 7-7. Trp and Cys (diagonal values 17 and 12) are the residues least likely to be replaced, whereas Ser, Ala, and Asn (all 2) are the most readily mutated. The residue pair least likely to exchange is Cys and Trp (-8), whereas the pair most likely to exchange is Tyr and Phe (7), although these latter residues are among the least likely to exchange with other residues (mostly negative entries). Similarly, charged and polar residues are unlikely to exchange with nonpolar residues (entries nearly always negative).

The confidence that one can align sequences known to be distantly related has been investigated as a function of PAM values (N). The PAM-250 log odds substitution matrix tends to yield the best alignments, that is, the highest alignment scores relative to those derived using substitution matrices based on larger or smaller PAM values. Note that Fig. 7-26*b* indicates that at 250 PAMs, 80% of the residues in the original polypeptide have been replaced.

e. Sequence Alignment Using the Needleman–Wunsch Algorithm

The use of a log odds substitution matrix to find an alignment is straightforward (although tedious). When comparing two sequences, rather than just making a matrix with 1's at all matching positions, one enters the appropriate value in the log odds substitution matrix at every position. Such a matrix represents all possible pair combinations of the two sequences. In Fig. 7-30*a*, we use the PAM-250 log odds matrix with a 10-residue peptide horizontal and an 11-residue peptide vertical. Thus, the alignment of these two peptides must have at least one gap or overhang, assuming a significant alignment can be found at all.

An algorithm for finding the best alignment between two polypeptides (that with the highest log odds value) was formulated by Saul Needleman and Christian Wunsch. One starts at the lower right corner (C-termini) of the matrix, position (M, N) (where in Fig. 7-30*a*, $M = 11$ and $N = 10$), and adds its value (here 2) to the value at position ($M - 1, N - 1$) [here 12, so that the value at position ($M - 1, N - 1$), that is, (10, 9), becomes 14 in the transformed matrix]. Continuing this process in an iterative manner, add to the value of the element at position (i, j) the maximum value of the elements ($p, j + 1$), where $p = i + 1, i + 2, \ldots, M$, and those of ($i + 1, q$), where $q = j + 1, j + 2, \ldots, N$. Figure 7-30*b* shows this process at an intermediate stage with the original value of the (6, 5) position in a small box and the transformed values of the positions ($p, 6$), where $p = 7, 8, \ldots, 11$, together with positions ($7, q$), where $q = 6, 7, \ldots, 10$, in the L-shaped box. The maximum value of the matrix elements in this L-shaped box is 19 and hence this is the value to add to the value (0) at position (6, 5) to yield the value 19 in the transformed matrix. This process is iterated, from the lower right toward the upper left of the matrix, until all its elements have been so treated, yielding the fully transformed matrix shown in Fig. 7-30*c*. The **Needleman–Wunsch algorithm** thereby yields the log odds values for all possible alignments of the two sequences.

The best alignment (that with the highest log odds value) is found by tracing the ridgeline of the transformed matrix (Fig. 7-30*c*) from its maximum value at or near the upper left (N-terminus) to that at or near the lower right (C-terminus). This is because the alignment of a particular residue pair is independent of the alignment of any other residue pair, and hence the best score up to any point in an alignment is the best score through the previous step plus the incremental score of the new step. This additive scoring scheme is based on the assumption that mutations at different sites are independently accepted, which appears to be an adequate characterization of protein evolution even though specific interactions between residues are known to have critical structural and functional roles in proteins.

The line connecting the aligned residue pairs (those circled in Fig. 7-29*c*) must always extend down and to the right. This is because a move up or to the left, or even straight down or straight to the right, would imply that a residue in one peptide aligned with more than one residue in the other peptide. Any allowed deviation from a move of ($+1, +1$) implies the presence of a gap. The best alignment of the two polypeptides, that connected by the lines in Fig. 7-30*c*, is indicated in Fig. 7-30*d*. Note that this alignment is ambiguous; the alignment of S in the 10-mer with either T *(a)* Comparison matrix

	V	E	D	Q	K	L	S	K	C	N
V	4	-2	-2	-2	-2	2	-1	-2	-2	-2
E	-2	4	3	2	0	-3	0	0	-5	1
N	-2	1	2	1	1	-3	1	1	-4	2
K	-2	0	0	1	5	-3	0	5	-5	1
L	2	-3	-4	-2	-3	6	-3	-3	-6	-3
T	0	0	0	-1	0	-2	1	0	-2	0
R	-2	-1	-1	1	3	-3	0	3	-4	0
P	-1	-1	-1	0	-1	-3	1	0	-3	0
K	-2	0	0	1	5	-3	0	5	-5	0
C	-2	-5	-5	-5	-5	-6	0	-5	12	-4
D	-2	3	4	2	0	-4	0	0	-5	2

(c) Transformed matrix

	V	E	D	Q	K	L	S	K	C	N
V	(41)	33	31	29	24	22	18	12	0	-2
E	31	(37)	35	33	26	17	19	14	-3	1
N	29	32	(33)	32	27	17	20	15	-2	2
K	24	26	26	27	(31)	17	19	19	-3	1
L	25	20	18	21	17	(26)	16	11	-4	-3
T	23	23	23	22	19	18	(20)	14	0	0
R	18	19	19	21	23	17	19	17	-2	0
P	18	18	18	19	18	16	(20)	14	-1	0
K	12	14	14	15	19	11	14	(19)	-3	0
C	2	-1	-3	-3	-3	-4	2	-3	(14)	-4
D	-2	3	4	2	0	-4	0	0	-5	(2)

(b) Transforming the matrix by the
Needleman–Wunsch alignment scheme

	V	E	D	Q	K	L	S	K	C	N
V	4	-2	-2	-2	-2	2	-1	-2	-2	-2
E	-2	4	3	2	0	-3	0	0	-5	1
N	-2	1	2	1	1	-3	1	1	-4	2
K	-2	0	0	1	5	-3	0	5	-5	1
L	2	-3	-4	-2	3	6	-3	-3	-6	-3
T	0	0	0	-1	0	-2	1	0	-2	0
R	-2	-1	-1	1	-3	17	19	17	-2	0
P	-1	-1	-1	0	-1	16	20	14	-1	0
K	-2	0	0	1	5	11	14	19	-3	0
C	-2	-5	-5	-5	-5	-4	2	-3	14	-4
D	-2	3	4	2	0	-4	0	0	-5	2

(d) Alignment

VEDQKLS - - KCN
VEN - KLTRPKCD

or

VEDQKL - - SKCN
VEN - KLTRPKCD

Figure 7-30 **Use of the Needleman–Wunsch alignment algorithm in the alignment of a 10-residue peptide (*horizontal*) with an 11-residue peptide (*vertical*).** (*a*) The comparison matrix, whose elements are the corresponding entries in the PAM-250 log odds substitution matrix (Table 7-7). (*b*) The Needleman–Wunsch transformation after several steps starting from the lower right. The numbers in red have already been transformed. The Needleman–Wunsch score of the T–K alignment (*small box*) is the sum of its PAM-250 value (0) plus the maximum of the quantities in the L-shaped box (19). The text explains the mechanics of the transformation process. (*c*) The completed Needleman–Wunsch matrix. The best alignment follows the ridgeline of the matrix as is described in the text. The aligned residues are those whose corresponding elements are circled. Note the ambiguity in this alignment. (*d*) The resulting two equivalent peptide alignments, with the aligned identical residues colored green.

or P in the 11-mer yields the same log odds value, and hence we have insufficient information to choose between them. The overall alignment score is the maximum value of the transformed matrix, here 41, which occurs at the upper left of the alignment (Fig. 7-30*c*).

The Needleman–Wunsch algorithm optimizes the global alignment of two peptides, that is, it maximizes the alignment score over the whole of the two sequences (and

does so even if it has no biological significance). However, since many proteins are modularly constructed from sequence motifs that occur in a variety of other proteins, a better approach would be to optimize the local alignment of two peptides, that is, maximize the alignment score only over their homologous regions. A variant of the Needleman–Wunsch algorithm, formulated by Temple Smith and Michael Waterman, was widely used to do so. This

Smith–Waterman algorithm exploits the property of the substitution matrix-based scoring system that the cumulative score for an alignment path decreases in regions in which the sequences are poorly matched. Where the cumulative score drops to zero, the Smith–Waterman algorithm terminates the extension of an alignment path. Two peptides may have several such local alignments.

f. Gap Penalties

If there are gaps in the alignment, one should now subtract the gap penalty from the overall alignment score to obtain the final alignment score. Since a single mutational event can insert or delete more than one residue, a long gap should be penalized only slightly more than a short gap. Consequently, gap penalties have the form $a + bk$, where a is the penalty for opening the gap, k is the length of the gap in residues, and b is the penalty for extending the gap by one residue. Current statistical theory provides little guidance for optimizing a and b, but empirical studies suggest that $a = -8$ and $b = -2$ are appropriate values for use with the PAM-250 matrix. Thus the final alignment score for both alignments in Fig. 7-29*d* (which have both a 1-residue gap and a 2-residue gap) is $41 - (8 + 2 \times 1) - (8 + 2 \times 2) = 19$.

g. Pairwise Alignments Using BLAST

The Needleman–Wunsch algorithm and later the Smith–Waterman algorithm (in their computerized forms) were widely used in the 1970s and 1980s to find relationships between proteins. However, the need to compare every newly determined sequence with the huge and rapidly growing number of sequences in publicly available databases requires that this process be greatly accelerated. Modern computers can do so using sequence alignment programs that employ sophisticated **heuristic algorithms** (algorithms that make educated "guesses") but at the risk of obtaining suboptimal results (in the case of sequence alignments, the heuristic algorithms are based on knowledge of how sequences evolve). Consequently, in what follows, we shall describe how these programs are used rather than how they work.

The PAM-250 substitution matrix is based on an extrapolation: Its calculation assumes that the rate of mutation over one PAM unit of evolutionary distance is the same as that over 250 PAM units. This may not be the case. After all, homologous proteins that are separated by large evolutionary distances may diverge in their function and hence their rates of evolution may change (recall that different proteins have different rates of evolution; Fig. 7-23). To account for this possibility, and because of the huge amount of sequence data that had become available since the PAM matrices were calculated in the mid-1970s, a log odds substitution matrix based on ~2000 blocks of aligned sequences that lacked gaps taken from ~500 groups of related proteins was calculated. The substitution matrix that gives the most sensitive performance for ungapped alignments is called **BLOSUM62** [blosum (pronounced "blossom") for *blo*ck *su*bstitution *m*atrix; the 62 indicates that all blocks of aligned polypeptides in which there is ≥62% identity are weighted as a single sequence in order to reduce con-

tributions from closely related sequences], whereas **BLOSUM45** appears to perform better for alignments with gaps. Sequence alignments based on the BLOSUM62 or BLOSUM45 matrices are more sensitive than are those based on the PAM-250 matrix.

BLAST (for *b*asic *l*ocal *a*lignment *s*earch *t*ool) is the most widely used, publicly available software package for making pairwise sequence alignments—both for polypeptides and for polynucleotides. This program uses a heuristic approach that approximates the Smith–Waterman algorithm so as to obtain the optimum mix of sensitivity (the ability to identify distantly related sequences) and selectivity (the avoidance of unrelated sequences with spuriously high alignment scores). It pairwise aligns up to a user-selected number of subject sequences (default 100) in the chosen database(s) that are the most similar to the query sequence. BLAST, which was originated by Stephen Altschul, is publicly available, free of charge, for interactive use over the Web (http://www.ncbi.nlm.nih.gov/BLAST/ Blast.cgi) on a server at the National Center for Biotechnology Information (NCBI). Let us discuss the BLAST system in its comparison of proteins (protein blast or blastp).

Protein databases presently contain ~900,000 nonredundant peptide sequences. BLAST therefore minimizes the time it spends on a sequence region whose similarity with the query sequence has little chance of exceeding some minimal alignment score. Pairwise alignments (e.g., Fig. 7-31*a*), which by default are found using BLOSUM62 (substitution matrices and gap penalties can be optionally selected under "Algorithm parameters"), are listed in order of decreasing statistical significance and are presented in a manner that indicates the positions of both the identical residues and similar residues in the query and subject sequences. The number of identical residues, positives (those residue pairs whose exchange has a positive value in the substitution matrix used), and gaps over the length of the alignment are indicated. BLAST assesses the statistical significance of an alignment in terms of its "E value" (E for Expect), which is the number of alignments with at least the same score that would have been expected to occur in the database by chance. For example, an alignment with an E value of 5 is statistically insignificant, whereas one with an E value of 0.01 is significant, and one with an E value of 1×10^{-20} offers extremely high confidence that the query and subject sequences are homologous. BLAST also reports a "bit score" for each alignment, which is a type of normalized alignment score.

h. Multiple Sequence Alignments with CLUSTAL

BLAST makes only pairwise alignments. To simultaneously align more than two sequences, that is, to obtain a **multiple sequence alignment** such as Table 7-4, a different program must used. Perhaps the most widely used such program, **ClustalW2**, is publicly available for interactive use over the Web at http://www.ebi.ac.uk/Tools/clustalw2/. The input for the program is a file containing all the sequences (either peptides or DNA) to be aligned. As with BLAST, the user can select the substitution matrix and the gap penalty parameters that ClustalW2 uses. ClustalW2 begins by finding all possible pairwise alignments of the input sequences. This permits the

(a) **BLAST pairwise alignment**

>sp|P38524|HIP2_ECTVA HIGH POTENTIAL IRON-SULFUR PROTEIN, ISOZYME 2 (HIPIP 2)
Length = 71

Score = 44.3 bits (103), Expect = 1e-04 Method: Compositional matrix adjust.
Identities = 27/70 (38%), Positives = 35/70 (50%), Gaps = 4/70 (5%)

```
Query: 1   EPRAEDGHAHDYVNEAADASG--HPRYQEGQLCENCAFWGEAVQDGWGRCTHPDFDEVLVKAEGWCSVY   68
           E  +ED  A    +   DAS   HP Y+EGQ C NC   + +A    WG C+    F   LV  A GWC+ +
Sbjct: 2   ERLSEDDPAAQALEYRHDASSVQHPAYEEGQTCLNCLLYTDASAQDWGPCS--VFPGKLVSANGWCTAW   69
```

(b) **Clustal W2 Multiple Sequence Alignment**

Figure 7-31   Examples of peptide sequence alignments. *(a)* A BLAST pairwise alignment. The aligned proteins are **high potential iron–sulfur proteins (HiPIPs),** small bacterial proteins whose sequences are archived in the SWISS-PROT database. The amino acid residues are indicated by their one-letter codes (Table 4-1) and gaps are indicated by dashes. The query sequence is HiPIP Isozyme 1 from *Halorhodospira halophila* (HIP1_HALHA, SWISS-PROT accession number P04168; **isozymes** are catalytically and structurally similar but genetically distinct enzymes from the same organism) and the subject sequence is HiPIP Isozyme 2 from *Ectothiorhodospira vacuolata* (HPI2_ECTVA, SWISS-PROT accession number P38524). The first two lines (*green*) identify the subject sequence and indicates its length in residues. This is followed by an assortment of alignment statistics (*black*). The query and subject sequences are then shown vertically aligned (*blue*) with the line between them (*black*) indicating residues that are identical (by their one-letter codes) and similar [by a plus (+)]. The output of BLAST consists of a series of such pairwise alignments. *(b)* A ClustalW2 multiple sequence alignment of five HiPIP sequences: the two foregoing sequences, their corresponding Isozymes 1 and 2 (P38941 and P04169), and HiPIP from *Rhodocyclus gelatinosus* (P00265). Confidently aligned residues are shaded according to residue type. The bar graph below the alignment is indicative of the alignment quality based on the BLOSUM62 score of the observed substitutions. Note that the alignment of the residues numbered 9 through 29 of P04168 (the query sequence in Part *a*) with P38524 differs from that in Part *a*.

program to determine the relationships of the input sequences with one another based on their similarity scores and hence generate a crude phylogenetic tree called a dendrogram. Then, starting with the highest scoring pairwise alignment, it sequentially carries out realignments on the basis of the remaining sequences, which it adds in order of their decreasing relationships with the previously added sequences, while opening up gaps as necessary. The output of ClustalW2 is the aligned sequences (e.g., Fig. 7-31*b*).

Since multiple sequence alignment programs are easily confounded by such anomalies as sequences that are not homologous or that contain homologous segments in different orders, multiple sequence alignments should be carefully inspected to determine if they are sensible and, if necessary, corrected and trimmed by hand. Indeed, in Fig. 7-31*a*, the alignment of the first 21 residues of the query sequence (P04168) with the subject sequence (P38524) differs from that in Fig. 7-30*b*.

i. The Use of Profiles Extends the Sensitivities of Sequence Alignments

Multiple sequence alignments can be used to improve the sensitivity of similarity searches, that is, to detect weak but significant sequence similarities. For example, in pair-

wise alignments, peptide A may appear to be similar to peptide B and peptide B to be similar to peptide C but peptides A and C may not appear to be similar. However, a multiple sequence alignment of peptides A, B, and C will reveal the similarities between peptides A and C. This idea has been extended through the construction of **profiles** (also called **position-specific score matrices**) that take into account the fact that some residues in a given protein are structurally and/or functionally more important than others and therefore are less subject to evolutionary change. Hence, at each residue position in a multiple sequence alignment, highly conserved residues are assigned a large positive score, weakly conserved positions are assigned a score near zero, and unconserved residues are assigned a large negative score. Many profile-generating algorithms are based on statistical models called **hidden Markov models (HMMs).** The use of such conservation patterns has been successful in finding sequences that are so distantly related to a query sequence (so far into the twilight zone) that BLAST would not consider them to have significant sequence similarity.

The program **PSI-BLAST** (for *P*osition-*S*pecific *I*terated *BLAST*), which is also available for use at http://www.ncbi.nlm.nih.gov/blast/Blast.cgi, uses the results of a BLAST search with a query sequence to generate a profile and then employs the profile to search for new alignments. This is an iterative process in which the profile generated after each alignment search is used to make a new alignment search, etc., until no further significant alignments are found. For instance, for the query sequence used in Fig. 7-31*a* (HIP1_HALHA; SWISS-PROT accession number P04168), BLAST finds only five sequences (called hits) that have E values less than 0.001 in the SWISS-PROT database (which includes a self-alignment). In contrast, PSI-BLAST finds 16 such hits after three iterations (and no additional hits in the fourth iteration, whereupon the search is said to have converged). Thus the use of profile analysis makes possible the detection of subtle but significant sequence relationships that, as we shall see in succeeding chapters, lead to considerable evolutionary and functional insights.

j. Structural Genes Should Be Aligned as Polypeptides

In many cases, only the base sequence of the DNA encoding a protein is known. Indeed, most of the known protein sequences have been inferred from DNA sequences. Although BLAST and ClustalW2 are both capable of aligning nucleic acid sequences (and are routinely used to do so), one should compare the inferred amino acid sequences of structural genes rather than just their base sequences. This is because amino acid sequence comparisons can routinely identify sequences that shared a common ancestor over 1 billion years ago (e.g., those of cytochrome *c* and histone H4; Fig. 7-23), whereas it is rare to detect homologies in noncoding DNA sequences that diverged >200 million years (Myr) ago and in coding sequences that diverged >600 Myr ago. The reasons for this are threefold:

1. DNA has only four different bases, whereas peptides consist of 20 different amino acid residues. Consequently, it is much easier to find spurious alignments with DNA, at least for short segments, than with peptides (a dot plot of two unrelated DNAs has, on average, 25% of its spaces filled vs 5% for unrelated polypeptides).

2. DNA evolves much more quickly than proteins. In the coding regions of structural genes, 24% of single base changes encode the same amino acid. There are few evolutionary constraints to maintain the sequence identity of these bases or of the gene's noncoding regions (e.g., those containing introns). Consequently, the evolutionary constraints on proteins are more stringent than those on DNA.

3. Direct DNA sequence alignments do not use amino acid substitution matrices such as PAM-250 and BLOSUM62 and, hence, are not constrained by the evolutionary information implicit in these matrices (although there are analogous 4 × 4 matrices for base substitutions).

If the base sequence of a structural gene is known, its putative control regions, particularly its start and stop codons, can usually be identified. This, in turn, reveals which of two complementary DNA strands is the so-called **sense strand** (which has the same sequence as the mRNA transcribed from the DNA) and indicates its correct reading frame. Even if it is unclear that a DNA segment that is flanked by what appear to be start and stop codons actually encodes a protein, one can compare the amino acid sequences from all six possible reading frames (three each from the DNA's two complementary strands). In fact, BLAST does so automatically when aligning peptide sequences based on DNA sequences.

C. Construction of Phylogenetic Trees

Phylogenetic trees were first made by Linnaeus, the eighteenth century biologist who originated the system of taxonomy (biological classification) still in use today. These trees (e.g., Fig. 1-4) were originally based on morphological characteristics, whose measurements were largely subjective. It was not until the advent of sequence analysis, however, that the generation of phylogenetic trees was put on a firm quantitative basis (e.g., Fig. 7-22). In the following paragraphs we discuss the characteristics of phylogenetic trees and outline how they are generated.

Figure 7-32*a* is a phylogenetic tree diagramming the evolutionary relationships between four homologous genes, A, B, C, and D. The tree consists of four **leaves** or **external nodes,** each representing one of these genes, and two **branch points** or **internal nodes,** which represent ancestral genes, with the length of each **branch** indicating the degree of difference between the two nodes they connect. All branch points are binary, that is, one gene is considered to

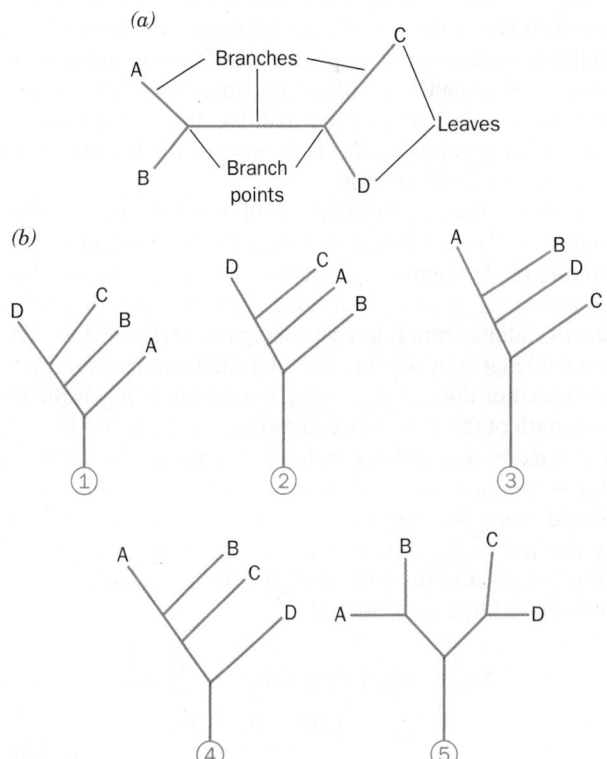

Figure 7-32 **Phylogenetic trees.** (*a*) An unrooted tree with four leaves (A, B, C, and D) and two branch points. (*b*) The five rooted trees that can be generated from the unrooted tree in Part *a*. The roots are drawn and numbered in red.

give rise to only two descendents at a time so that branches can only bifurcate (although branch points may turn out to be so close together that their order cannot be determined; e.g., the root of Fig. 7-22). Note that this is an **unrooted tree,** that is, it indicates the relationships between the four genes but provides no information as to the evolutionary events through which they arose. The five different evolutionary pathways that are possible for our unrooted tree are diagrammed in Fig. 7-32*b* as different **rooted trees,** where the node at which the root joins the tree represents the four genes' last common ancestor. With knowledge of only the A, B, C, and D genes, phylogenetic analysis cannot distinguish between these rooted trees. In order to find a tree's root, it is necessary to obtain the sequence of an **outgroup,** a homologous gene that is less closely related to the genes in the tree than they are to each other. This permits the root of the tree to be identified and hence the pathway through which the genes evolved to be elucidated.

The number of different bifurcating trees with the same *n* leaves increases extremely rapidly with *n* (e.g., for *n* = 10, it is over 2 million). Unfortunately, *there is no exact method for generating an optimal phylogenetic tree.* Indeed, there is no general agreement on what constitutes an optimal tree. Consequently, numerous methods have been formulated for constructing phylogenetic trees based on sequence alignments.

In one class of methods for constructing phylogenetic trees, the sequence data are converted into a **distance matrix,** which is a table showing the evolutionary distances between all pairs of genes in the data set (e.g., Table 7-5). Evolutionary distances are the number of sequence differences between two genes (ideally corrected for the possibility of multiple mutations at a given site). These quantities are used to calculate the lengths of the branches in a tree under the assumption that they are additive, that is, the distance between any pair of leaves is the sum of the lengths of the branches connecting them.

Perhaps the conceptually simplest (if it can be called that) way of generating a phylogenetic tree is the **neighbor-joining (N-J) method,** in which it is initially assumed that there is only one internal node, Y, and hence that all *N* leaves radiate from it in a starlike pattern (Fig. 7-33*a*). The branch lengths in the star are then calculated according to such relationships as $d_{AB} = d_{AY} + d_{BY}$ (where d_{AB} is the total length of the branches connecting leaves A and B, etc.), $d_{AC} = d_{AY} + d_{CY}$, and $d_{BC} = d_{BY} + d_{CY}$, so that, for instance, $d_{AY} = \frac{1}{2}(d_{AB} + d_{AC} - d_{BC})$. A pair of leaves is then transferred from the star to a new internal node, X, that is connected to the center of the star by a new branch, XY (Fig. 7-33*b*), and the sum of all the branch lengths, S_{AB}, in this revised tree is calculated:

$$S_{AB} = d_{AX} + d_{BX} + d_{XY} + \sum_{k \neq AB}^{N} d_{kY}$$

$$= \frac{d_{AB}}{2} + \frac{[2Q - R_A - R_B]}{2(N-2)} \qquad [7.6]$$

where

$$Q = \sum_{i=1}^{N} \sum_{j=1}^{i-1} d_{ij} \qquad [7.7]$$

(*a*)

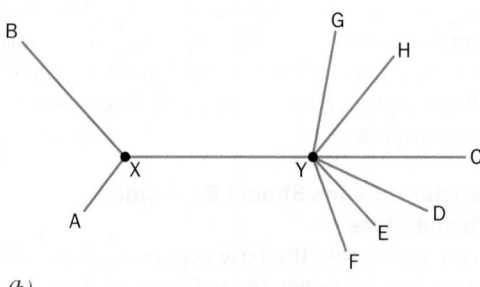

(*b*)

Figure 7-33 **Manipulations employed in the neighbor-joining method for the construction of a phylogenetic tree.** (*a*) The starting configuration. (*b*) The transfer of leaves A and B to a new branch point that is connected to the central star (*red*).

(that is, the sum of all the off-diagonal elements in the unique half of the distance matrix),

$$R_A = \sum_{i=1}^{N} d_{Ai} \qquad [7.8]$$

(that is, the sum of the elements in the Ath row of the distance matrix), and

$$R_B = \sum_{i=1}^{N} d_{Bi} \qquad [7.9]$$

The two leaves are then returned to their initial positions, replaced by a second pair of leaves, and the total branch length again calculated. The process is repeated until all possible $N(N-1)/2$ pairs of leaves have been so treated. The pair yielding the smallest value of S_{ij} (the shortest total branch length) in this process, which will be nearest neighbors in the final tree, are combined into a single unit of their average length, yielding a star with one less branch. If leaves A and B are chosen as neighbors, then the lengths of the branches connecting them are estimated to be

$$d_{AX} = \frac{d_{AB}}{2} + \frac{R_A - R_B}{2(N-2)} \qquad [7.10]$$

$$d_{BX} = d_{AB} - d_{AX} \qquad [7.11]$$

and

$$d_{XY} = \frac{(N-1)(R_i + R_j) - 2Q - (N^2 - 3N + 2)d_{AB}}{2(N-2)(N-3)} \qquad [7.12]$$

Figure 7-34 **An unrooted phylogenetic tree of the five HiPIP sequences that are aligned in Fig. 7-31b.** The tree was generated by Clustal using the neighbor-joining method. The numbers indicate the relative lengths of the associated branches.

Assuming that S_{AB} has the lowest value of all the S_{ij}, a new distance matrix is calculated whose elements, d'_{ij}, are the same as the d_{ij} with the exception that $d'_{A-B,i} = d'_{i,A-B} = (d_{Ai} + d_{Bi})/2$, where $d'_{A-B,i}$ is the distance between the averaged leaves A and B and leaf i. The entire process is then iterated so as to find all pairs of nearest-neighbor sequences, thereby generating a phylogenetic tree. Figure 7-34 is an unrooted phylogenetic tree that Clustal generated from the multiple sequence alignment shown in Fig. 7-31b using the N-J method.

The N-J method is representative of **distance-based** tree-building procedures. There are two other types of tree-building criteria that are in widespread use:

1. Maximum parsimony (MP), which is based on the principle of Occam's razor: The best explanation of the data is the simplest. Thus, MP-based methods assume (perhaps inaccurately) that evolution occurs via the fewest possible genetic changes and hence that the best phylogenetic tree requires the smallest number of sequence changes to account for a multiple sequence alignment.

2. Maximum likelihood (ML), which finds the tree and branch lengths that have the highest probability of yielding the observed multiple sequence alignment. This, in turn, requires an evolutionary model that indicates the probability of occurrence for each type of residue change (e.g., the PAM substitution matrices).

Since the number of possible trees increases very rapidly with the number of leaves, the construction of a phylogenetic tree is a computer-intensive task for even relatively small sets of aligned sequences (e.g., $N = 20$, although distance-based methods require far fewer computations than do MP- or ML-based methods). And, because of the ambiguities inherent in all known tree-building procedures, statistical tests have been developed to check the validity of any particular tree.

5 CHEMICAL SYNTHESIS OF POLYPEPTIDES

In this section we describe methods for the chemical synthesis of polypeptides from amino acids. The ability to manufacture polypeptides not available in nature has considerable biomedical potential:

1. To investigate the properties of polypeptides by systematically varying their side chains.

2. To obtain polypeptides with unique properties, particularly those with nonstandard side chains or with isotopic labels incorporated in specific residues (neither of which is easily accomplished using biological methods).

3. To manufacture pharmacologically active polypeptides that are biologically scarce or nonexistent.

One of the most promising applications of polypeptide synthesis is the production of synthetic vaccines. Vaccines, which have consisted of viruses that have been "killed" (inactivated) or attenuated ("live" but mutated so as not to cause disease in humans), stimulate the immune system to synthesize antibodies specifically directed against these viruses, thereby conferring immunity to them (the immune response is discussed in Section 35-2A). The use of such vaccines, however, is not without risk; attenuated viruses, for example, may mutate to a virulent form and "killed" virus vaccines have, on several occasions, caused disease because they contained "live" viruses. Moreover, it is difficult to culture many viruses and therefore to obtain sufficient material for vaccine production. Such problems would be eliminated by preparing vaccines from synthetic polypeptides that have the amino acid sequences of viral **epitopes** (antigenic determinants; molecular groupings that stimulate the immune system to manufacture antibodies against them). Indeed, several such synthetic vaccines are already in general use.

The first polypeptides to be chemically synthesized were composed of only one type of amino acid and are therefore known as **homopolypeptides.** Such compounds as **polyglycine, polyserine,** and **polylysine** are easily synthesized according to classic methods of polymer chemistry. They have served as valuable model compounds in studying the physicochemical properties of polypeptides, such as conformational behavior and interactions with the aqueous environment.

The first chemical synthesis of a biologically active polypeptide was that of the nonapeptide (9-amino acid residue) hormone **oxytocin** by Vincent du Vigneaud in 1953:

$$
\begin{array}{c}
\overbrace{\qquad\quad S-S\qquad\quad} \\
\underset{1}{Cys}-Tyr-Ile-Gln-Asn-Cys-Pro-Leu-\underset{9}{Gly}
\end{array}
$$

Oxytocin

Improvements in polypeptide synthesis methodology since then have led to the synthesis of numerous biologically active polypeptides and several proteins.

A. *Synthetic Procedures*

Polypeptides are chemically synthesized by covalently linking (coupling) amino acids, one at a time, to the terminus of a growing polypeptide chain. Imagine that a polypeptide is being synthesized from its C-terminus toward its N-terminus; that is, the growing chain ends with a free amino group. Then each amino acid being added to the chain must already have its own α-amino group chemically protected (blocked) or it would react with other like molecules as well as with the N-terminal amino group of the chain. Once the new amino acid is coupled, its now N-terminal amino group must be deprotected (deblocked) so that the next peptide bond can be formed. *Every cycle of amino acid addition therefore requires a coupling step and a deblocking step.* Furthermore, reactive side chains must be blocked to prevent their participation in the coupling reactions, and then deblocked in the final step of the synthesis.

The reactions that were originally developed for synthesizing polypeptides such as oxytocin take place entirely in solution. The losses that are incurred on isolation and purification of the reaction product in each of the many steps, however, contribute significantly to the low yields of final polypeptide. This difficulty was ingeniously circumvented in 1962 by Bruce Merrifield, through his development of **solid phase peptide synthesis (SPPS).** In SPPS, a growing polypeptide chain is covalently anchored, usually by its C-terminus, to an insoluble solid support such as beads of polystyrene resin, and the appropriately blocked amino acids and reagents are added in the proper sequence (Fig. 7-35). This permits the quantitative recovery and purification of intermediate products by simply filtering and washing the beads.

When polypeptide chains are synthesized by amino acid addition to their N-terminus (the opposite direction to that in protein biosynthesis; Section 5-4Ba), the α-amino group of each sequentially added amino acid must be chemically protected during the coupling reaction. The *tert*-**butyloxycarbonyl (Boc)** group is frequently used for this purpose,

$$(CH_3)_3C-O-\overset{\overset{\displaystyle O}{\|}}{C}-Cl + H_2N-\overset{\overset{\displaystyle R}{|}}{C}H-\overset{\overset{\displaystyle O}{\|}}{C}\diagdown_{O^-}$$

t-Butyloxycarbonyl   **α-Amino acid**
chloride

$\searrow$ HCl

$$(CH_3)_3C-O-\overset{\overset{\displaystyle O}{\|}}{C}-NH-\overset{\overset{\displaystyle R}{|}}{C}H-\overset{\overset{\displaystyle O}{\|}}{C}\diagdown_{O^-}$$

Boc-amino acid

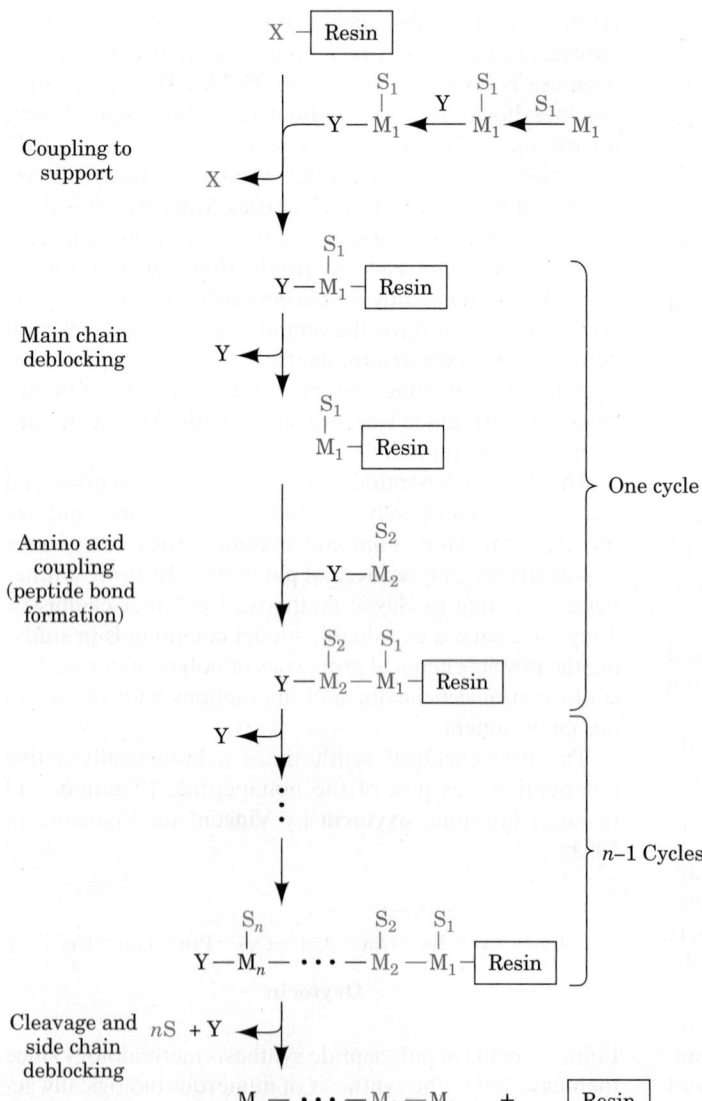

Coupling to support

Main chain deblocking

Amino acid coupling (peptide bond formation)

One cycle

n–1 Cycles

Cleavage and side chain deblocking

Figure 7-35   Flow diagram for polypeptide synthesis by the solid phase method. The symbol M_i represents the *i*th amino acid residue to be added to the polypeptide, S_i is its side chain protecting group, and Y represents the main chain protecting group. The specific reactions are discussed in the text. [After Erikson, B.W. and Merrifield, R.B., *in* Neurath, H. and Hill, R.L. (Eds.), *The Proteins* (3rd ed.), Vol. 2, *p.* 259, Academic Press (1979).]

as is the **9-fluorenylmethoxycarbonyl (Fmoc)** group:

**9-Fluorenylmethoxycarbonyl
(Fmoc) group**

Both groups undergo analogous reactions, although, in what follows, we shall only discuss the Boc group.

a. Anchoring the Chain to the Inert Support

The first step in SPPS is the coupling of the C-terminal amino acid to a solid support. The most commonly used support is a cross-linked polystyrene resin with pendant chloromethyl groups. Resin coupling occurs through the following reaction:

Boc-amino acid   **Inert
support**

$$\downarrow Et_3N$$

$$-\text{Resin}- + Et_3\overset{+}{N}HCl^-$$

and the resulting α-amino acid–derivatized resin is filtered and washed. The amino group is then deblocked by treatment with an anhydrous acid such as trifluoroacetic acid, which leaves intact the alkylbenzyl ester bond to the support resin:

Boc-amino acid

$$\downarrow CF_3COOH \text{ in } CH_2Cl_2$$

Isobutylene

b. Coupling the Amino Acids

The reaction coupling two amino acids through a peptide bond is endergonic and therefore must be activated to obtain significant yields. **Carbodiimides** $(R-N=C=N-R')$ such as **dicyclohexylcarbodiimide (DCCD)** are commonly used coupling agents:

Boc-amino acid

**Dicyclohexyl-
carbodiimide
(DCCD)**

O-**Acylurea intermediate**

**Resin-bound
amino acid**

Dipeptidyl-resin

+

N,N'-**Dicyclohexylurea**

The *O*-acylurea intermediate that results from the reaction of DCCD with the carboxyl group of a Boc-protected α-amino acid readily reacts with the resin-bound α-amino acid to form the desired peptide bond in high yield. By subsequently alternating the deblocking and coupling reactions, a polypeptide with the desired amino acid sequence can be synthesized. *The repetitive nature of these operations allows the SPPS method to be easily automated.*

During the course of peptide synthesis, many of the side chains also require protection to prevent their reaction with the coupling agent. Although there are many different blocking groups, the benzyl group is the most widely used (Fig. 7-36).

c. Releasing the Polypeptide from the Resin

The final step in SPPS is the cleavage of the polypeptide from the solid support. The benzyl ester link from the polypeptide's C-terminus to the support resin may be cleaved by treatment with liquid HF:

The Boc group linked to the polypeptide's N-terminus as well as the benzyl groups protecting its side chains are also removed by this treatment.

B. *Problems and Prospects*

The steps just outlined seem simple enough, but they are not as straightforward as we have implied. A major difficulty with the entire procedure is its low cumulative yield. Let us examine the reasons for this. To synthesize a polypeptide chain with n peptide bonds requires at least $2n$ reaction steps—one for coupling and one for deblocking each residue. If a protein-sized polypeptide is to be synthesized in reasonable yield, then each reaction step must be essentially quantitative; anything less greatly reduces the yield of final product. For example, in the synthesis of a 101-residue polypeptide chain, in which each reaction step occurs with an admirable 98% yield through 200 reaction steps, the overall yield is only $0.98^{200} \times 100 = 2\%$. Therefore, although oligopeptides can be routinely made, the synthesis of large polypeptides requires almost fanatical attention to chemical detail.

An ancillary problem is that the newly liberated synthetic polypeptide must be purified. This may be a difficult

Figure 7-36 **A selection of amino acids with benzyl-protected side chains and a Boc-protected α-amino group.** These substances can be used directly in the coupling reactions forming peptide bonds.

task because a significant level of incomplete reactions and/or side reactions at every stage of SPPS will result in almost a continuum of closely related products for large polypeptides. The use of reverse-phase HPLC techniques (Section 6-3Dh), however, greatly facilitates this purification process, and the quality of both intermediate and final products can be readily assessed through mass spectrometric techniques (Section 7-1I).

Using automated SPPS, Merrifield synthesized the nonapeptide hormone **bradykinin** in 85% yield:

Arg—Pro—Pro—Gly—Phe—Ser—Pro—Phe—Arg

Bradykinin

However, it was only in 1988, through steady progress in improving reaction yields (to >99.5% on average) and eliminating side reactions, that it became possible to synthesize ~100-residue polypeptides of reasonable quality. Thus, Stephen Kent synthesized the 99-residue **HIV-1 protease** [an enzyme that is essential for the maturation of

Figure 7-37 **The native chemical ligation reaction.** Peptide-1 has a C-terminal thioester group (R is an alkyl group) and Peptide-2 has a N-terminal Cys residue. The reaction, which occurs in aqueous solution at pH 7, is initiated by the nucleophilic attack of Peptide-2's Cys thiol group on Peptide-1's thioester group to yield, in a thiol exchange reaction, a new thioester group. This intermediate (as is indicated by the square brackets), undergoes a rapid intramolecular nucleophilic attack to yield a native peptide bond at the ligation site. [After Dawson, P.E., Muir, T.W., Clark-Lewis, I., and Kent, S.B.H., *Science* **266,** 777 (1994).]

human immunodeficiency virus-1 (**HIV-1,** an **AIDS virus;** Section 15-4C)] in such high yield and purity that, after being renatured (folded to its native conformation; Section 9-1A), it exhibited full biological activity. Indeed, this synthetic protein was crystallized and its X-ray structure was shown to be identical to that of biologically synthesized HIV-1 protease. Kent also synthesized HIV-1 protease from D-amino acids and experimentally verified, for the first time, that such a protein has the opposite chirality of its biologically produced counterpart. Moreover, this D-amino acid protease catalyzes the cleavage of its target polypeptide made from D-amino acids but not that made from L-amino acids as does naturally occurring HIV-1 protease.

Despite the foregoing, the accumulation of resin-bound side products limits the sizes of the polypeptides that can routinely be synthesized by SPPS to ~60 residues. Kent has partially circumvented this limitation through the development of the so-called **native chemical ligation** reaction, which links together two polypeptides in a peptide bond to routinely yield polypeptides as large as ~120 residues (Fig. 7-37). Moreover, several peptide segments can be consecutively linked by native chemical ligation, so that the chemical synthesis of polypeptides consisting of several hundred residues can be anticipated. In fact, using this technique, Kent synthesized the 203-residue "covalent dimer" of HIV-1 protease (Section 15-4C) from four synthetic peptides—the largest linear polypeptide yet synthesized—and showed that it was fully enzymatically active.

6 CHEMICAL SYNTHESIS OF OLIGONUCLEOTIDES

Molecular cloning techniques (Section 5-5) have permitted the genetic manipulation of organisms in order to investigate their cellular machinery, change their characteristics, and produce scarce or specifically altered proteins in large quantities. *The ability to chemically synthesize DNA oligonucleotides of specified base sequences is an indispensable part of this powerful technology.* Thus, as we have seen, specific oligonucleotides are required as probes in Southern blotting (Section 5-5D) and *in situ* hybridization (Section 5-5Ea), as primers in PCR (Section 5-5F), and to carry out site-directed mutagenesis (Section 5-5Gc).

A. Synthetic Procedures

The basic strategy of oligonucleotide synthesis is analogous to that of polypeptide synthesis (Section 7-5A): *A suitably protected nucleotide is coupled to the growing end of the oligonucleotide chain, the protecting group is removed, and the process is repeated until the desired oligonucleotide has been synthesized.* The first practical technique for DNA synthesis, the **phosphodiester method,** which was developed by H. Gobind Khorana in the 1960s, is a laborious process in which all reactions are carried out in solution and the products must be isolated at each stage of the multistep synthesis. Khorana, nevertheless,

Dimethoxytrityl

R : N≡C—CH₂—CH₂—
β-Cyanoethyl

1. **Detritylation**

2. **Coupling**

Tetrazole

3. **Capping of unreacted 5' end**

Capped failure sequence (no further extension)

4. **Oxidation**

Figure 7-38 **Reaction cycle in the phosphoramidite method of oligonucleotide synthesis.** Here B₁, B₂, and B₃ represent protected bases, and S represents an inert solid phase support such as controlled-pore glass.

used this method, in combination with enzymatic techniques, to synthesize a 126-nucleotide tRNA gene, a project that required several years of intense effort by numerous skilled chemists.

a. The Phosphoramidite Method

By the early 1980s, these difficult and time-consuming processes had been supplanted by much faster solid phase methodologies that permitted oligonucleotide synthesis to be automated. The presently most widely used chemistry, which was formulated by Robert Letsinger and further developed by Marvin Caruthers, is known as the **phosphoramidite method.** This series of nonaqueous reactions adds a single nucleotide to a growing oligonucleotide chain as follows (Fig. 7-38):

1. The **dimethoxytrityl (DMTr)** protecting group at the 5′ end of the growing oligonucleotide chain (which is anchored via a linking group at its 3′ end to a solid support, S) is removed by treatment with an acid such as **trichloroacetic acid** (Cl_3CCOOH).

2. The newly liberated 5′ end of the oligonucleotide is coupled to the 3′-phosphoramidite derivative of the next deoxynucleoside to be added to the chain. The coupling agent in this reaction is **tetrazole,** which protonates the incoming nucleotide's diisopropylamine moiety so that it becomes a good leaving group. Modified nucleosides (e.g., containing a fluorescent label) can be incorporated into the growing oligonucleotide at this stage. Likewise, a mixture of oligonucleotides containing different bases at this position can be synthesized by adding the corresponding mixture of nucleosides.

3. Any unreacted 5′ end (the coupling reaction has a yield of over 99%) is capped by acetylation so as to block its extension in subsequent coupling reactions. This prevents the extension of erroneous oligonucleotides (failure sequences).

4. The phosphite triester group resulting from the coupling step is oxidized with to the more stable phosphotriester, thereby yielding a chain that has been lengthened by one nucleotide.

This reaction sequence, in commercially available automated synthesizers, can be repeated up to ~250 times with cycle times of 20 to 30 min. Once an oligonucleotide of desired sequence has been synthesized, it is treated with concentrated NH_4OH to release it from its support and remove its various blocking groups, including those protecting the exocyclic amines on the bases. The product can then be separated from the failure sequences and protecting groups by reverse-phase HPLC and/or gel electrophoresis.

The largest DNA molecule yet synthesized is the entire 582,970-bp genome of *Mycoplasma genitalium* (among the smallest bacterial genomes). Initially, overlapping 5- to 7-kb "cassettes" were constructed by sequentially ligating chemically synthesized double-stranded oligonucleotides of ~50 bp. The resulting 101 cassettes were then joined in

stages by utilizing their overlapping ends. Sequencing the final product confirmed that it had the correct sequence.

B. *DNA Microarrays*

The sequencing of the human genome (Section 7-2B) is really only the means to a highly complex end. The questions of real biochemical significance are these: What are the functions of its ~23,000 genes? In which cells, under what circumstances, and to what extent are each of them expressed? How do their gene products interact to yield a functional organism? And what are the medical consequences of variant genes? The traditional method of addressing such questions, the one-gene-at-a-time approach, is simply incapable of acquiring the vast amounts of data necessary to answer these questions. What is required are methods that can globally analyze biological processes, that is, techniques that can simultaneously monitor all the components of a biological system.

A technology that is capable of making such global assessments involves the use of **DNA microarrays** (also called **DNA chips;** Fig. 7-39). These are arrays of different

Figure 7-39 **DNA microarrays.** This ~6000-gene microarray contains most of the genes from baker's yeast, one per spot. The microarray had been hybridized to the cDNAs derived from mRNAs extracted from yeast. The cDNAs derived from cells that were grown in glucose were labeled with a red-fluorescing dye, whereas the cDNAs derived from cells grown in the absence of glucose were labeled with a green-fluorescing dye. Thus the red and green spots, respectively, reveal those genes that are transcriptionally activated by the presence or absence of glucose, whereas the yellow spots (*red plus green*) indicate genes whose expression is unaffected by the level of glucose. [Courtesy of Patrick Brown, Stanford University School of Medicine.]

Figure 7-40   The photolithographic synthesis of a DNA microarray. In Step 1 of the process, oligonucleotides that are anchored to a glass or silicon surface, and each having a photosensitive protecting group (*filled red square*) at their 5′ ends, are exposed to light through a mask that only permits the illumination of the oligonucleotides destined to be coupled, for example, to a T residue. The light deprotects these oligonucleotides so that only they react with the activated T nucleotide that is incubated with the chip in Step 2. The entire process is repeated in Steps 3 and 4 with a different mask for G residues, and in subsequent reaction cycles for A and C residues, thereby extending all of the oligonucleotides by one residue. This quadruple cycle is then repeated for as many nucleotides as are to be added to form the final set of oligonucleotides. Each "feature" (position) on the microarray contains at least 1 million identical oligonucleotides. [After Pease, A.C., Solas, D., Sullivan, E.J., Cronin, M.T., Holmes, C.P., and Fodor, S.P.A., *Proc. Natl. Acad. Sci.* **91**, 5023 (1994).]

DNA molecules anchored to the surface of a glass, silicon, or nylon wafer in an ~1-cm-wide square grid. In one of several methodologies that are presently used to manufacture DNA microarrays, large numbers (up to ~1 million) of different oligonucleotides are simultaneously synthesized via a combination of photolithography (the process used to fabricate electronic chips) and solid phase DNA synthesis. In this process (Fig. 7-40), which was developed by Stephen Fodor, the nucleotides from which the oligonucleotides are synthesized each have a photochemically removable protective group at their 5′ end that has the same function as the DMTr group in conventional solid phase DNA synthesis (Fig. 7-38). At a given stage in the synthetic procedure, oligonucleotides that, for example, require a T at their next position are deprotected by shining light on them through a mask that blocks the light from hitting those grid posi-

tions that require a different base at this next position (in an alternative methodology called maskless array synthesis, an array of individually programmable micromirrors directs light to the desired locations). The chip is then incubated with a solution of activated thymidine nucleotide, which couples only to the deprotected oligonucleotides. After washing away the unreacted thymidine nucleotide, the process is repeated with different masks (light patterns for maskless array synthesis) for each of the remaining three bases. By repeating these four steps N times, an array of all 4^N possible N-residue sequences can be synthesized simultaneously in $4N$ coupling cycles, where $N \leq 30$ (up to 100 for maskless array synthesis). A DNA microarray is shown in Fig. 7-41.

In one application of DNA microrrays, *L*-residue oligonucleotides (the probes) are arranged in an array of *L*

Figure 7-41   A DNA microarray assembly. This so-called GeneChip protects its enclosed DNA microarray and provides a convenient hybridization chamber. To interrogate it requires a specialized fluorescence measurement device. [Courtesy of Affymetrix, Inc., Santa Clara, California.]

columns by 4 rows for a total of 4*L* sequences. The probe in the array's *M*th column has the "standard" sequence with the exception of the probe's *M*th position, where it has a different base, A, C, G, or T, in each row. Thus, one of the four probe DNAs in every column will have the standard sequence, whereas the other three will differ from the standard DNA by only one base change. The probe array is then hybridized with the complementary DNA or RNA, whose variation relative to the standard DNA is to be determined, and the unhybridized DNA or RNA is washed away. This "target" DNA or RNA is fluorescently labeled so that, when interrogated by a laser, the positions on the probe array to which it binds are revealed by fluorescent spots. Since hybridization conditions can be adjusted so that a single base mismatch will significantly reduce the level of binding, a target DNA or RNA that varies from the complement of the standard DNA by a single base change at its *M*th position, say C to A, would be readily detected by an increased fluorescence at the row corresponding to A in the *M*th column relative to that at other positions [a target DNA or RNA that was exactly complementary to the standard DNA would exhibit high fluorescence in each of its columns at the base position (row) corresponding to the standard sequence]. The intensity of the fluorescence at every position in the array, and hence the sequence variation from the standard DNA, is rapidly determined with a computerized fluorescence scanning device. In this manner, single nucleotide polymorphisms (SNPs) can be automatically detected. It is becoming increasingly apparent that genetic variations, and in particular SNPs, are largely responsible for an individual's susceptibility to many diseases as well as to adverse reactions to drugs (side effects; Section 15-4B).

In an alternative DNA microarray methodology, different DNAs are robotically spotted (deposited) to precise locations on a coated glass surface. These DNAs most often consist of PCR-amplified inserts from cDNA clones or expressed sequence tags (ESTs), which are usually robotically synthesized. The DNAs are spotted in nanoliter-sized droplets that rapidly evaporate leaving the DNA attached to the glass substrate. Up to 30,000 DNAs representing all of the genes in an organism can be spotted onto a single glass "chip." Such DNA microarrays, many of which are commercially available, are used to monitor the level of expression of the genes in a tissue of interest by the degree of hybridization of its fluorescently labeled mRNA or cDNA population. They can therefore be used to determine the pattern of gene expression (the **expression profile**) in different tissues of the same organism (e.g., Fig. 7-42) and how specific diseases and drugs (or drug candidates) affect gene expression. Thus, DNA microarrays are the major tool for the study of a cell's **transcriptome** (which, in analogy with the word "genome," is the entire collection of RNAs that the cell transcribes). Moreover, DNA microarrays are used to specifically identify infectious agents by detecting unique segments of their DNA. Consequently, DNA microarrays hold enormous promise for understanding the interplay of genes during cell growth and changes in the environment, for the characterization and diagnosis of both infectious and noninfectious diseases (e.g., cancers), for identifying genetic risk factors and designing specific treatments, and as tools in drug development.

C. *SELEX*

Although nucleic acids such as DNA and mRNA are widely thought of as being passive molecular "tapes" whose structures have only incidental significance, it is now abundantly clear that certain nucleic acids, for example, tRNAs and ribosomal RNAs (Sections 32-2B and 32-3A), have functions that require that these molecules maintain well-defined three-dimensional structures. This has led to the development of a process called **SELEX** (for *S*ystematic *E*volution of *L*igands by *EX*ponential enrichment) for generating single-stranded oligonucleotides that specifically bind target molecules with high affinity and specificity.

SELEX, a technique pioneered by Larry Gold, starts with a library of polynucleotides that have fixed known sequences at their 5′ and 3′ ends and a central region of randomized sequences. Such sequences are synthesized via solid phase methods by adding a mixture of all four nucleoside triphosphates rather than a single nucleoside triphosphate at the positions to be randomized. For a variable region that is 30 nucleotides long, this will yield a mixture of $4^{30} \approx 10^{18}$ different oligonucleotides. The next step in the SELEX process is to select those sequences in the mixture that selectively bind to a target molecule, X. Those oligonucleotides that bind to X with higher than average affinity can be separated from the other oligonucleotides in the mixture by a variety of techniques including affinity

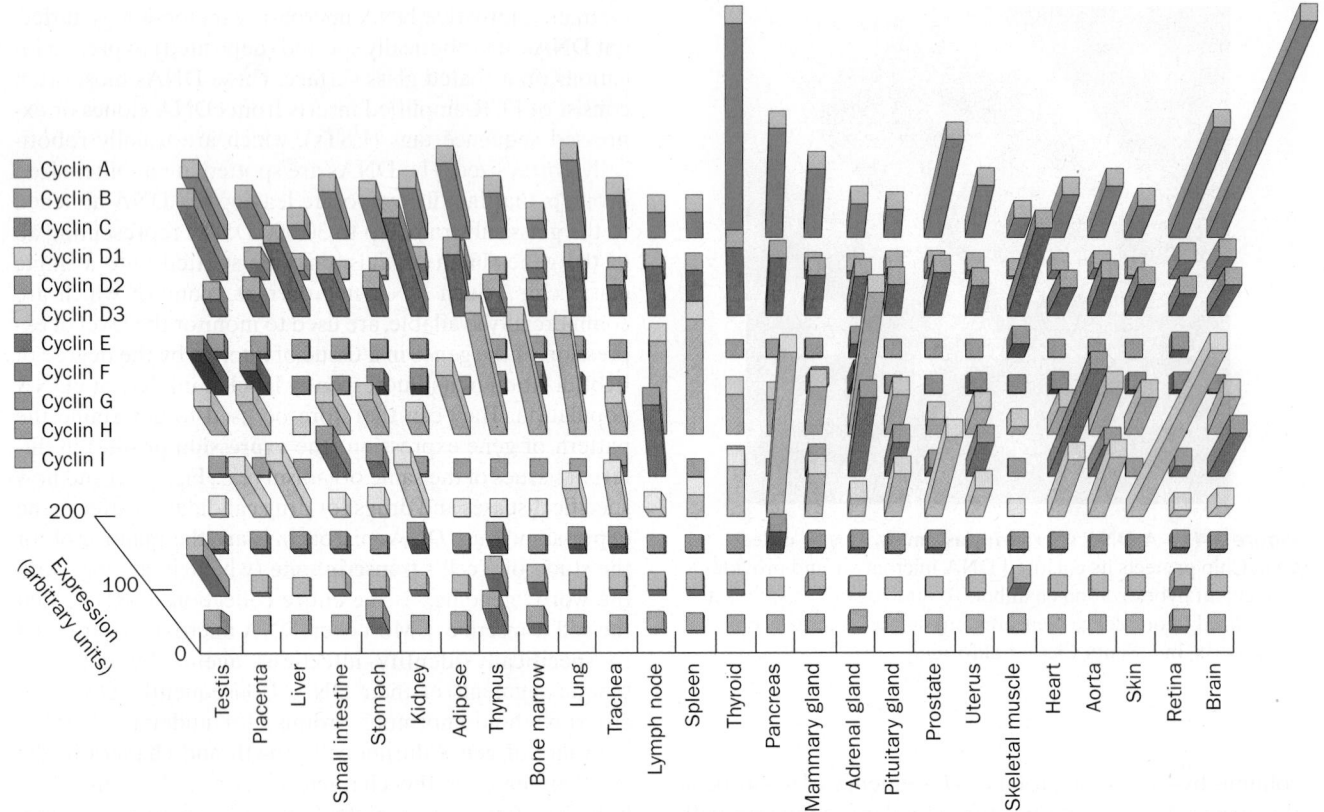

Cyclin A
Cyclin B
Cyclin C
Cyclin D1
Cyclin D2
Cyclin D3
Cyclin E
Cyclin F
Cyclin G
Cyclin H
Cyclin I

Expression (arbitrary units)

200
100
0

Testis, Placenta, Liver, Small intestine, Stomach, Kidney, Adipose, Thymus, Bone marrow, Lung, Trachea, Lymph node, Spleen, Thyroid, Pancreas, Mammary gland, Adrenal gland, Pituitary gland, Prostate, Uterus, Skeletal muscle, Heart, Aorta, Skin, Retina, Brain

Figure 7-42 Variation in the expression of genes that encode proteins known as cyclins (Section 34-4D) in human tissues. The levels of hybridization of the various cyclin mRNAs produced in each tissue were measured through the use of DNA microarrays. [After Gerhold, D., Rushmore, T., and Caskey, C.T., *Trends Biochem. Sci.* **24,** 172 (1999).]

chromatography (Section 6-3C) with X linked to a suitable matrix and the precipitation of X–oligonucleotide complexes by anti-X antibodies. The oligonucleotides are then separated from X and amplified by cloning or PCR. This selection and amplification process is then repeated for several cycles (usually 10–15), thereby yielding oligonucleotides that tenaciously bind X. Several such enrichment cycles are necessary because those few oligonucleotides in a library with the highest affinity for X must compete for binding X with the far more abundant weak binders.

The results of SELEX are remarkable. Its oligonucleotide products, which are known as **aptamers,** bind to their target molecule, X, with dissociation constants that are typically around 10^{-9} M if X is a protein and between 10^{-3} M and 10^{-6} M if X is a small molecule (for the reaction $A + B \rightleftharpoons A \cdot B$, the dissociation constant, $K_D = [A][B]/[A \cdot B]$). Moreover, aptamers bind to their target molecules with great specificity. Thus, aptamers selected for their affinity toward a particular member of a protein family do not bind other members of that family. Aptamers bind a wide variety of target molecules including simple ions, small molecules, proteins, organelles, and even whole cells. These various properties give aptamers great promise both as diagnostic tools and as therapeutic agents.

CHAPTER SUMMARY

1 Primary Structure Determination of Proteins The initial step in the amino acid sequence determination of a protein is to ascertain its content of chemically different polypeptides by end group analysis. The protein's disulfide bonds are then chemically cleaved, the different polypeptides are separated and purified, and their amino acid compositions are determined. Next, the purified polypeptides are specifically cleaved, by enzymatic or chemical means, to smaller peptides that are separated, purified, and then sequenced by (automated) Edman degradation. Repetition of this process, using a cleavage method of different specificity, generates overlapping peptides whose amino acid sequences, when compared to those of the first group of peptide fragments, indicate their order in the parent polypeptide. The primary structure determination is completed by establishing the positions of the disulfide bonds, which requires the degradation of the protein with its disulfide bonds intact. Mass spectrometry using electrospray ionization (ESI) or matrix-assisted laser desorption/ionization (MALDI) to vaporize and ionize peptides can be used to determine their molecular masses. By using these techniques in a tandem mass spectrometer equipped with a collision cell, the sequences of short peptides can be rapidly determined. Once the primary structure of a protein is known, its minor variants, which may arise from mutations or chemical modifications, can be readily identified by peptide mapping.

2 Nucleic Acid Sequencing Nucleic acids may be sequenced by the same basic strategy used to sequence proteins. In the Sanger method, the DNA fragment to be sequenced is replicated by a DNA polymerase in the presence of all four deoxynucleoside triphosphates, one of which is α-^{32}P-labeled, and a small amount of the dideoxy analog of one of the deoxynucleoside triphosphates. This results in a series of ^{32}P-labeled chains that are terminated after the various positions occupied by the corresponding base. The electrophoresis of the four differently terminated DNA samples in parallel lanes of a sequencing gel resolves fragments that differ in size by one nucleotide. An autoradiograph of the sequencing gel containing the four sets of fragments reveals the base sequence of the complementary DNA. RNA may be sequenced by determining the sequence of its corresponding cDNA. Automated methods, which use fluorescently labeled DNA fragments, have greatly accelerated DNA sequence determinations. They have been used to sequence very large tracts of DNA such as the chromosomes of the human genome.

In the conventional strategy for genome sequencing, the genome is mapped by sequencing large numbers of landmark sequences such as sequence-tagged sites (STSs) and expressed sequence tags (ESTs). Landmark-containing YAC clones are then fragmented, the fragments subcloned and sequenced, and the sequences are computationally assembled into contigs, which are further assembled into chromosomes based on the landmark sequences. Advances in cloning and computational techniques permitted genome sequencing via a shotgun strategy. For bacterial genomes this is carried out straightforwardly. For eukaryotic genomes, large segments are cloned in bacterial artificial chromosomes (BACs), whose ends are sequenced to yield sequence-tagged connectors (STCs). The BAC inserts are fragmented, shotgun cloned, and the clones sequenced and assembled into contigs. The STCs are used to further assemble the sequenced BAC inserts into chromosomes via BAC walking. Next generation DNA sequencing technologies, such as those employing pyrosequencing, have enormously accelerated the rate that DNA is being sequenced relative to Sanger-based techniques. Even faster single molecule sequencing techniques are under development.

3 Chemical Evolution Sickle-cell anemia is a molecular disease of individuals who are homozygous for a gene specifying an altered β chain of hemoglobin. Fingerprinting and sequencing studies have identified this alteration as arising from a point mutation that changed Glu β6 to Val. In the heterozygous state, the sickle-cell trait confers resistance to malaria without causing deleterious effects. This accounts for its high incidence in populations living in malarial regions. The cytochromes c from many eukaryotic species contain many

amino acid residues that are invariant or conservatively substituted. Hence, this protein is well adapted to its function. The amino acid differences between the various cytochromes c have permitted the generation of their phylogenetic tree, which closely parallels that determined by classical taxonometry. The number of sequence differences between homologous proteins from related species plotted against the time when, according to the fossil record, these species diverged from a common ancestor reveals that point mutations are accepted into proteins at a constant rate. Proteins whose functions are relatively intolerant to sequence changes evolve more slowly than those that are more tolerant to such changes. Phylogenetic analysis of the globin family—myoglobin and the α and β chains of hemoglobin—reveals that these proteins arose through gene duplication. In this process, the original function of the protein is maintained, while the duplicated copy evolves a new function. Many, if not most, proteins have evolved through gene duplication.

4 Bioinformatics: An Introduction The sequences of nucleic acids and proteins are archived in large and rapidly growing databases that are publicly available over the World Wide Web. The alignment of homologous sequences provides insights into their function and evolution. The alignment of distantly related sequences requires statistically and computationally sophisticated algorithms such as BLAST and ClustalW2. The construction of an optimal phylogenetic tree based on aligned sequences is a computationally intensive process for which there is no known exact solution.

5 Chemical Synthesis of Polypeptides The strategy of polypeptide chemical synthesis involves coupling amino acids, one at a time, to the N-terminus of a growing polypeptide chain. The α-amino group of each amino acid must be chemically protected during the coupling reaction and then unblocked before the next coupling step. Reactive side chains must also be chemically protected but then unblocked at the conclusion of the synthesis. The difficulties in recovering the intermediate product of each of the many steps of such a synthesis has been eliminated by the development of solid phase synthesis techniques. These methods have led to the synthesis of numerous biologically active polypeptides and are capable of synthesizing small proteins in useful amounts. The use of the native chemical ligation technique significantly increases the sizes of the polypeptides that can be chemically synthesized.

6 Chemical Synthesis of Oligonucleotides Oligonucleotides are indispensable to recombinant DNA technology. They are used to identify normal and mutated genes and to alter specific genes through site-directed mutagenesis. Oligonucleotides of defined sequence are efficiently synthesized by the phosphoramidite method, a cyclic, nonaqueous, solid phase process that has been automated. DNA chips, which can be manufactured through photolithographically based DNA synthesis or by the robotic deposition of DNA on a glass substrate, are finding increasing numbers of applications. These include the detection of single nucleotide polymorphisms, monitoring patterns of gene expression, and identifying infectious agents. SELEX is a technique for selecting and amplifying those members of a large polynucleotide library that specifically bind to a target molecule. The resulting aptamers, which have remarkably high affinity and specificity for their target molecules, may be useful diagnostic and therapeutic agents.

REFERENCES

Protein Sequencing

Burlingame, A.L. (Ed.), Biological Mass Spectrometry, *Methods Enzymol.* **402** (2005).

Domon, B. and Aebersold, R., Mass spectrometry and protein analysis, *Science* **312**, 212–217 (2006).

Findlay, J.B.C. and Geisow, M.J. (Eds.), *Protein Sequencing. A Practical Approach*, IRL Press (1989).

James, P. (Ed.), *Proteome Research: Mass Spectrometry*, Springer-Verlag (2001).

Mann, M., Hendrickson, R.C., and Pandey, A., Analysis of proteins and proteomes by mass spectrometry, *Annu. Rev. Biochem.* **70**, 437–473 (2001).

Rappsilber, J. and Mann, M., What does it mean to identify a protein in proteomics? *Trends Biochem. Sci.* **27**, 74–78 (2002).

Sanger, F., Sequences, sequences, and sequences, *Annu. Rev. Biochem.* **57**, 1–28 (1988). [A scientific autobiography that provides a glimpse of the early difficulties in sequencing proteins.]

Simpson, R.J. and Reid, G.E., Sequence analysis of gel-resolved proteins, *in* Hames, B.D. (Ed.), *Gel Electrophoresis of Proteins. A Practical Approach* (3rd ed.) pp. 237–267, Oxford University Press (1998).

Wilson, K. and Walker, J.M. (Eds.), *Principles and Techniques of Biochemistry and Molecular Biology*, 6th ed., Chapter 9, Cambridge University Press (2005). [Discusses mass spectroscopic techniques.]

Wilm, M., Mass spectrometric analysis of proteins, *Adv. Prot. Chem.* **54**, 1–30 (2000).

Nucleic Acid Sequencing

Brown, T.A., *Genomes* (3rd ed.), Chapters 3 and 4, Garland Science (2007).

Graham, C.A. and Hill, A.J.M. (Eds.), *DNA Sequencing Protocols* (2nd ed.), Humana Press (2001).

Primrose, S.B. and Twyman, R.M., *Principles of Gene Manipulations and Genomics* (7th ed.), Chapter 7, Blackwell Publishing (2006).

Turner, S., Real-time DNA sequencing from single polymerase molecules, *Science* **323**, 133–138 (2009). [A proof-of-concept report for the Pacific Biosciences method of DNA sequencing.]

Watson, J.D., Myers, R.M., Caudy, A.A., and Witkowski, J.A., *Recombinant DNA. Genes and Genomes—A Short Course* (3rd ed.), Chapters 10 and 11, Cold Spring Harbor Laboratory Press (2007).

Venter, J.C., Smith, H.O., and Hood, L., A new strategy for genome sequencing, *Science* **381**, 364–366 (1996). [Describes the shotgun strategy for genome sequencing.]

Sequence of The Human Genome

Gregory, S.G., The DNA sequence and the biological annotation of human chromosome 1, *Nature* **441**, 315–321 (2006). [References to the finished sequences of the other 23 human chromosomes are given on pp. 306–307 of the preceding reference by Watson, J.D., et al.]

International Human Genome Sequencing Consortium, Initial sequencing and analysis of the human genome, *Nature* **409**, 860–921 (2001); *and* Venter, J.C., et al., The sequence of the human genome, *Science* **291**, 1304–1351 (2001). [Draft sequences of the human genome.]

International Human Genome Sequencing Consortium, Finishing the euchromatic sequence of the human genome, *Nature* **431**, 931–945 (2004).

Levy, S., et al., The diploid sequence of an individual human, *PLoS Biol.* **5**(10): e254–e286 (2007). [J. Craig Venter's diploid genome sequence.]

Next Generation Sequencing Technologies

Bentley, D.R., et al., Accurate whole human genome sequencing using reversible terminator chemistry; Wang, J., et al., The diploid genome sequence of an Asian individual; *and* Ley, T.J., et al., DNA sequencing of a cytogenetically normal acute myeloid leukaemia genome, *Nature* **456**, 53–59; 60–65; *and* 66–72 (2008). [Describes the sequencing of three human genomes using the Illumina system.]

Margulies, M., et al., Genome sequencing in microfabricated high-density picolitre reactors, *Nature* **437**, 376–380 (2005). [Describes the 454 sequencing system.]

MacLean, D., Jones, J.D.G., and Studholme, D.J., Application of 'next-generation' sequencing technologies to microbial genetics, *Nature. Rev. Microbiol.* **7**, 287–296 (2009).

Mitchelson, K.R. (Ed.), *New High Throughput Technologies for DNA Sequencing and Genomics,* Elsevier (2007).

von Bubnoff, A., Next-generation sequencing: The race is on, *Cell* **132**, 721–723 (2008).

Wheeler, D.A., et al., The complete genome of an individual by massively parallel DNA sequencing, *Nature* **452**, 872–881 (2008). [James Watson's diploid DNA sequence.]

Chemical Evolution

Allison, A.C., The discovery of resistance to malaria of sickle-cell heterozygotes, *Biochem. Mol. Biol. Educ.* **30**, 279–287 (2002).

Benton, M.J. and Ayala, F.J., Dating the tree of life, *Science* **300**, 1698–1700 (2003).

Dickerson, R.E., The structure and history of an ancient protein, *Sci. Am.* **226**(4), 58–72 (1972). [Discusses the evolution of cytochrome *c*.]

Dickerson, R.E. and Geis, I. *Hemoglobin*, Chapter 3, Benjamin/Cummings (1983). [A detailed discussion of globin evolution.]

Dickerson, R.E. and Timkovich, R., Cytochromes *c*, *in* Boyer, P.D. (Ed.), *The Enzymes* (3rd ed.), Vol. 11, *pp.* 397–547, Academic Press (1975). [Contains a detailed analysis of cytochrome *c* sequence studies.]

Doolittle, R.F., Feng, D.F., Tsang, S., Cho, G., and Little, E., Determining divergence times of the major kingdoms of living organisms with a protein clock, *Science* **271**, 470–477 (1996).

Ingram, V.M., A case of sickle-cell anaemia: A commentary, *Biochim. Biophys. Acta* **1000**, 147–150 (1989). [A scientific memoir on the development of fingerprinting to characterize sickle-cell hemoglobin.]

Moore, G.R. and Pettigrew, G.W., *Cytochromes c. Evolutionary, Structural and Physicochemical Aspects,* Springer-Verlag (1990).

Nagel, R.L. and Roth, E.F., Jr., Malaria and red cell genetic defects, *Blood* **74**, 1213–1221 (1989). [An informative review of the various genetic "defects," including sickle-cell anemia, that inhibit malaria and how they might do so.]

Strasser, B.J., Sickle cell anemia, a molecular disease, *Science* **286**, 1488–1490 (1999). [A historical account of Pauling's work on sickle-cell anemia.]

Bioinformatics

Altschul, S.F. and Koonin, E.V., Iterated profile searches with PSI-BLAST—a tool for discovery in protein data bases, *Trends Biochem. Sci.* **23**, 444–447 (1998).

Baxevanis, A.D. and Ouellette, B.F.F. (Eds.), *Bioinformatics. A Practical Guide to the Analysis of Genes and Proteins* (3rd ed.), Wiley-Interscience (2005).

Bork, P. (Ed.), *Analysis of Amino Acid Sequences, Adv. Prot. Chem.* **54** (2000). [Contains informative articles on sequence alignment and phylogenetic tree generation.]

Database Issue, *Nucleic Acids Res.* **38** (Database Issue) (2010). [Annually updated descriptions of numerous databases of biomolecular interest. The database issue has always been published in January.]

Doolittle, R.F. (Ed.), Molecular Evolution: Computer Analysis of Proteins and Nucleic Acids; and Computing Methods for Macromolecular Sequence Analysis, *Methods Enzymol.* **183** (1990); *and* **266** (1996).

Doolittle, R.F., *Of URFs and ORFs. A Primer of How to Analyze Derived Amino Acid Sequences,* University Science Books (1986).

Edgar, R.C. and Batzoglou, S., Multiple sequence alignment, *Curr. Opin. Struct. Biol.* **16,** 368–373 (2006).

Gibson, G. and Muse, S.V., *A Primer of Genomic Science,* Sinauer Associates (2002).

Henikoff, S., Scores for sequence searches and alignments, *Curr. Opin. Struct. Biol.* **6,** 353–360 (1996). [Describes BLOSUM matrices.]

Jeanmougin, F. and Thompson, J.D., Multiple sequence alignment with Clustal X, *Trends Biochem. Sci.* **23,** 403–405 (1998).

Jones, D.T. and Swindells, M.B., Getting the most from PSI-BLAST, *Trends Biochem. Sci.* **27,** 161–164 (2002).

Lesk, A.M., *Introduction to Bioinformatics,* (3rd ed.), Oxford University Press (2008).

Mann, M. and Pandey, A., Use of mass spectrometry–derived data to annotate nucleotide and protein sequence databases, *Trends Biochem. Sci.* **26,** 54–61 (2001).

Mount, D.W., *Bioinformatics: Sequence and Genome Analysis* (2nd ed.), Cold Spring Harbor Laboratory Press (2004).

Needleman, S.B. and Wunsch, C.D., A general method applicable to the search for similarities in the amino acid sequence of two proteins, *J. Mol. Biol.* **48,** 443–453 (1970). [The formulation of the Needleman–Wunsch algorithm.]

Pagel, M., Inferring the historical patterns of biological evolution, *Nature* **401,** 877–884 (1999). [A review.]

Pei, J., Multiple sequence alignment, *Curr. Opin. Struct. Biol.* **18,** 382–386 (2008).

Xiong, J., *Essential Bioinformatics,* Chapters 1–6, Cambridge University Press (2006).

Zvelebil, M. and Baum, J.O., *Understanding Bioinformatics,* Chapters 3–8, Garland Science (2008).

Polypeptide Synthesis

Atherton, E. and Sheppard, R.C., *Solid Phase Peptide Synthesis. A Practical Approach,* IRL Press (1989).

Barrett, G.C. and Elmore, D.T., *Amino Acids and Peptides,* Chapter 7, Cambridge University Press (1998).

Bodanszky, M., *Principles of Peptide Synthesis,* Springer-Verlag (1993).

Dawson, P.E. and Kent, S.B.H., Synthesis of native proteins by chemical ligation, *Annu. Rev. Biochem.* **69,** 923–960 (2000).

Fields, G.B. (Ed.), Solid-Phase Peptide Synthesis, *Methods. Enzymol.* **289** (1997).

Kent, S.B.H., Alewood, D., Alewood, P., Baca, M., Jones, A., and Schnölzer, M., Total chemical synthesis of proteins: Evolution of solid phase synthetic methods illustrated by total chemical synthesis of the HIV-1 protease, *in* Epton, R. (Ed.), *Innovation & Perspectives in Solid Phase Synthesis,* SPPC Ltd. (1992); *and* Milton, R.C. deL., Milton, S.C.F., and Kent, S.B.H., Total chemical synthesis of a D-enzyme: The enantiomers of HIV-1 protease show demonstration of reciprocal chiral substrate specificity, *Science* **256,** 1445–1448 (1992).

Merrifield, B., Solid phase synthesis, *Science* **232,** 342–347 (1986).

Nilsson, B.L., Soellner, M.B., and Raines, R.T., Chemical synthesis of proteins, *Annu. Rev. Biophys. Biomol. Struct.* **34,** 91–118 (2005).

Torbeev, V.Yu. and Kent, S.B.H., Convergent chemical synthesis and crystal structure of a 203 amino acid "covalent dimer" HIV-1 protease enzyme molecule, *Angew. Chem. Int. Ed.* **46,** 1667–1670 (2007).

Wilken, J. and Kent, S.B.H., Chemical protein synthesis, *Curr. Opin. Biotech.* **9,** 412–426 (1998).

Chemical Synthesis of Oligonucleotides

Brewster, J.L., Beason, K.B., Eckdahl, T.T., and Evans, I.M., The microarray revolution, *Biochem. Mol. Biol. Educ.* **32,** 217–227 (2004).

Caruthers, M.H., Beaton, G., Wu, J.V., and Wiesler, W., Chemical synthesis of deoxynucleotides and deoxynucleotide analogs, *Methods Enzymol.* **211,** 3–20 (1992); *and* Caruthers, M.H., Chemical synthesis of DNA and DNA analogues, *Acc. Chem. Res.* **24,** 278–284 (1991).

Gibson, D.G., et al., Complete chemical synthesis, assembly, and cloning of a *Mycoplasma genitalium* genome, *Science* **319,** 1215–1220 (2008).

Gold, L., The SELEX process: A surprising source of therapeutic and diagnostic compounds, *Harvey Lectures* **91,** 47–57 (1997).

Hermann, T. and Patel, D.J., Adaptive recognition by nucleic acid aptamers, *Science* **287,** 820–825 (2000). [A review.]

Schena, M., *Microarray Analysis,* Wiley-Liss (2003).

Staughton,, R.B., Applications of DNA microarrays in biology, *Annu. Rev. Biochem.* **74,** 53–82 (2005).

Watson, J.D., Meyers, R.M., Caudy, A.A., and Witkowski, J.A., *Recombinant DNA. Genes and Genomes—A Short Course* (3rd ed.), Chapter 13, Freeman (2007).

Wilson, D.S. and Szostak, J.W., *In vitro* selection of functional nucleic acids, *Annu. Rev. Biochem.* **68,** 611–647 (1999). [Discusses SELEX].

Young, R., Biomedical discovery with DNA arrays, *Cell* **102,** 9–15 (2000).

BIOINFORMATICS EXERCISES

Bioinformatics Exercises are available at www.wiley.com/college/voet

Chapter 7

Databases for the Storage and "Mining" of Genome Sequences and to Compare and Identify Related Protein Sequences
1. **Finding Databases.** Locate databases for genome sequences and explore the meaning of terms related to them.
2. **The Institute for Genomic Research.** Explore a prokaryotic genome and find listings for eukaryotic genomes.
3. **Analyzing a DNA Sequence.** Given a DNA sequence, identify its open reading frame and translate it into a protein sequence.
4. **Sequence Homology.** Perform a BLAST search for homologs of a protein sequence.

5. Plasmids and Cloning. Predict the sizes of the fragments produced by the action of various restriction enzymes on plasmids.

6. Obtaining Protein Sequences from BLAST. Using a known protein sequence, find and retrieve the sequences of related proteins from other organisms.

7. Multiple Sequence Alignment. Examine the various sequences for similarities, and view their relatedness as a tree diagram.

8. Phylogenetic Trees. Create a tree based on cytochrome *c*.

9. Proteomics and 2D Gel Electrophoresis. The use of 2D gel electrophoresis to examine dihydrofolate reductase.

10. Proteomics. Gain a broader understanding of proteomics.

11. Teaching and Learning Resources. Some proteomics web sites.

12. Microarrays. Learn how microarrays are made and used.

PROBLEMS

Note: Amino acid compositions of polypeptides with unknown sequences are written in parentheses with commas separating amino acid abbreviations such as in (Gly, Tyr, Val). Known amino acid sequences are written with residue names in order and separated by hyphens; for example, Tyr-Val-Gly.

1. State the cleavage pattern of the following polypeptides by the indicated agents.

a. Ser-Ala-Phe-Lys-Pro by chymotrypsin
b. Thr-Cys-Gly-Met-Asn by NCBr
c. Leu-Arg-Gly-Asp by carboxypeptidase A
d. Gly-Phe-Trp-Asp-Phe-Arg by endopeptidase Asp-N
e. Val-Trp-Lys-Pro-Arg-Glu by trypsin

2. A protein is subjected to end group analysis by dansyl chloride. The liberated dansylamino acids are found to be present with a molar ratio of two parts Ser to one part Ala. What conclusions can be drawn about the nature of the protein?

3. A protein is subjected to degradation by carboxypeptidase B. Within a short time, Arg and Lys are liberated, following which no further change is observed. What does this information indicate concerning the primary structure of the protein?

4. Before the advent of the Edman degradation, the primary structures of proteins were elucidated through the use of partial acid hydrolysis. The resulting oligopeptides were separated and their amino acid compositions were determined. Consider a polypeptide with amino acid composition (Ala$_2$, Asp, Cys, Leu, Lys, Phe, Pro, Ser$_2$, Trp$_2$). Treatment with carboxypeptidase A released only Leu. Oligopeptides with the following compositions were obtained by partial acid hydrolysis:

(Ala, Lys)	(Ala, Ser$_2$)	(Cys, Leu)
(Ala, Lys, Trp)	(Ala, Trp)	(Cys, Leu, Pro)
(Ala, Pro)	(Asp, Lys, Phe)	(Phe, Ser, Trp)
(Ala, Pro, Ser)	(Asp, Phe)	(Ser, Trp)
(Ser$_2$, Trp)		

Determine the amino acid sequence of the polypeptide.

*5. A polypeptide is subjected to the following degradative techniques resulting in polypeptide fragments with the indicated amino acid sequences. What is the amino acid sequence of the entire polypeptide?

I. Cyanogen bromide treatment:

1. Asp-Ile-Lys-Gln-Met
2. Lys
3. Lys-Phe-Ala-Met
4. Tyr-Arg-Gly-Met

II. Trypsin hydrolysis:

5. Gln-Met-Lys
6. Gly-Met-Asp-Ile-Lys
7. Phe-Ala-Met-Lys
8. Tyr-Arg

6. Treatment of a polypeptide by 2-mercaptoethanol yields two polypeptides that have the following amino acid sequences:

1. Ala-Phe-Cys-Met-Tyr-Cys-Leu-Trp-Cys-Asn
2. Val-Cys-Trp-Val-Ile-Phe-Gly-Cys-Lys

Chymotrypsin-catalyzed hydrolysis of the intact polypeptide yields polypeptide fragments with the following amino acid compositions:

3. (Ala, Phe)
4. (Asn, Cys$_2$, Met, Tyr)
5. (Cys, Gly, Lys)
6. (Cys$_2$, Leu, Trp$_2$, Val)
7. (Ile, Phe, Val)

Indicate the positions of the disulfide bonds in the original polypeptide.

7. A polypeptide was subjected to the following treatments with the indicated results. What is its primary structure?

I. Acid hydrolysis:

1. (Ala, Arg, Cys, Glx, Gly, Lys, Leu, Met, Phe, Thr)

II. Aminopeptidase M:

2. No fragments

III. Carboxypeptidase A + carboxypeptidase B:

3. No fragments

IV. Trypsin followed by Edman degradation of the separated products:

4. Cys-Gly-Leu-Phe-Arg
5. Thr-Ala-Met-Glu-Lys

*8. While on an expedition to the Amazon jungle, you isolate a polypeptide you suspect of being the growth hormone of a newly discovered species of giant spider. Unfortunately, your portable sequencer was so roughly treated by the airport baggage handlers that it refuses to provide the sequence of more than four consecutive amino acid residues. Nevertheless, you persevere and obtain the following data:

I. Hydrazinolysis:

1. (Val)

II. Dansyl chloride treatment followed by acid hydrolysis:

2. (Dansyl-Pro)

III. Trypsin followed by Edman degradation of the separated fragments:

3. Gly-Lys
4. Phe-Ile-Val
5. Pro-Gly-Ala-Arg
6. Ser-Arg

Provide as much information as you can concerning the amino acid sequence of the polypeptide.

9. In taking the electrospray ionization mass spectrum of an unknown protein, you find that four successive peaks have m/z values of 953.9, 894.4, 841.8, and 795.1. What is the molecular mass of the protein and what are the ionic charges of the ions responsible for the four peaks?

10. Figure 7-43 pictures an autoradiograph of the sequencing gel of a DNA that was treated according to the Sanger method of DNA sequencing. What is the sequence of the template strand corresponding to bases 50 to 100? If there are any positions on the gel where a band seems to be absent, leave a question mark in the sequence for the indeterminate base.

11. Using Table 7-5, compare the relatedness of fungi to higher plants and to animals. Fungi are sometimes said to be non-green plants. In light of your analysis, is this a reasonable classification?

12. Below is a list of the first 10 residues of the B helix in myoglobin from different organisms.

Position	1	2	3	4	5	6	7	8	9	10
Human	D	I	P	G	H	G	Q	E	V	L
Chicken	D	I	A	G	H	G	H	E	V	L
Alligator	K	L	P	E	H	G	H	E	V	I
Turtle	D	L	S	A	H	G	Q	E	V	I
Tuna	D	Y	T	T	M	G	G	L	V	L
Carp	D	F	E	G	T	G	G	E	V	L

Based on this information, which positions (a) appear unable to tolerate substitutions, (b) can tolerate conservative substitution, and (c) are highly variable?

← 400
← 350
← 300
← 250
← 200
← 150
← 100
← 75
← 50

GCAT

Figure 7-43 [Courtesy of Barton Slatko, New England Biolabs Inc., Beverly, Massachusetts.]

13. The inherited hemoglobin disease **β-thalassemia** is common to people from around the Mediterranean Sea and areas of Asia where malaria is prevalent (Fig. 7-21). The disease is characterized by a reduction in the amount of synthesis of the β chain of hemoglobin. Heterozygotes for the β-thalassemia gene, who are said to have **thalassemia minor,** are only mildly affected with adverse symptoms. Homozygotes for this gene, however, suffer from **Cooley's anemia** or **thalassemia major;** they are so severely afflicted that they do not survive their childhood. About 1% of the children born in the malarial regions around the Mediterranean Sea have Cooley's anemia. Why do you suppose the β-thalassemia gene is so prevalent in this area? Justify your answer.

14. Leguminous plants synthesize a monomeric oxygen-binding globin known as **leghemoglobin.** (Section 26-6). From your knowledge of biology, sketch the evolutionary tree of the globins (Fig. 7-25) with leghemoglobin included in its most likely position.

15. Since point mutations mostly arise through random chemical change, it would seem that the rate at which mutations appear in a gene expressing a protein should vary with the size of the gene (number of amino acid residues the gene expresses). Yet, even though the rates at which proteins evolve vary quite widely, these rates seem to be independent of protein size. Explain.

16. Sketch the self-dot matrices of the following: (a) A 100-residue peptide with nearly identical segments spanning its residues 20 to 40 and 60 to 80. (b) A 100-nucleotide DNA that is palindromic.

*17. (a) Using the PAM-250 log odds substitution matrix and the Needleman–Wunsch algorithm, find the best alignment of the two peptides PQRSTV and PDLRSCSV. (b) What is its alignment score using a gap penalty of –8 for opening a gap and –2 for every "residue" in the gap? (c) What is its normalized alignment score (NAS) using the scoring system of 10 for each identity, 20 for each identity involving Cys, and –25 for every gap? (d) Is this NAS indicative of a homology? Explain.

18. You have been given an unknown protein to identify. You digest it with trypsin and, by using Edman degradation, find that one of the resulting peptide fragments has the sequence GIIWGEDTLMEYLENPK. Using BLAST, find the most probable identity (or identities) of the unknown protein. [To carry out a search on the BLAST server, go to http://www.ncbi.nlm.nih.gov/ BLAST/Blast.cgi and, under "Basic BLAST," click on "protein blast" (which is BLAST for comparing an amino acid query sequence against a protein sequence database). In the window that comes up, enter the above sequence (without spaces or punctuation) into the "Enter Query Sequence" box, from the "Database" pulldown menu select "Non-redundant protein sequences (nr)," under "Program Selection" select the "blastp (protein-protein BLAST)" button, and click on the "BLAST" button at the bottom of the window.]

*19. A desiccated dodo bird in a reasonable state of preservation has been found in a cave on Mauritius Island. You have been given a tissue sample in order to perform biochemical analyses and have managed to sequence its cytochrome *c*. The amino acid difference matrix for a number of birds including the dodo is shown here.

Chicken, turkey	0				
Penguin	2	0			
Pigeon	4	4	0		
Pekin duck	3	3	3	0	
Dodo	4	4	2	3	0

(a) Determine the phylogenetic tree for these species using the neighbor-joining method. (b) To which of the other birds does the dodo appear most closely related? (c) What additional information do you require to find the root of this tree? Without making further computations, indicate the most likely possibilities.

20. In a tragic accident, the desiccated dodo discussed in Problem 19 has been eaten by a deranged cat. The sequence of the dodo cytochrome *c* that you had determined before the accident led you to suspect that this cytochrome *c* has some unique biochemical properties. To test this hypothesis, you are forced to synthesize dodo cytochrome *c* by chemical means. As with other known avian cytochromes *c,* that of the dodo consists of 104 amino acid residues. In planning a solid phase synthesis, you expect a 99.7% yield for each coupling step and a 99.3% yield for each deblocking step. The cleavage of the completed polypeptide from the resin and the side chain deblocking step should yield an 80% recovery of product. (a) What percentage of the original resin-bound C-terminal amino acid can you expect to form unmodified dodo cytochrome *c* if you synthesize the entire polypeptide in a single run? (b) You have found that dodo cytochrome *c* has a Cys residue at its position 50. What would be your overall yield if you used the native chemical ligation reaction (assume it has a 75% yield) to synthesize dodo cytochrome *c*? Compare this yield with that from Part a and discuss the implications of this comparison for synthesizing long polypeptides.

21. You have manufactured a DNA chip consisting of 4 rows and 10 columns in which the *M*th column contains DNAs of the sequence 5'-GACCTGACGT-3' but with a different base in the *M*th position for each of the 4 rows (from top to bottom, G, A, T, and C). Draw the appearance of the chip after it is hybridized to fluorescently labeled RNA of sequence (a) 5'-ACGUCAGGUC-3' and (b) 5'-ACGUCUGGUC-3'.

Three-Dimensional Structures of Proteins

CHAPTER **8**

The properties of a protein are largely determined by its three-dimensional structure. One might naively suppose that since proteins are all composed of the same 20 types of amino acid residues, they would be more or less alike in their properties. Indeed, **denatured** (unfolded) proteins have rather similar characteristics, a kind of homogeneous "average" of their randomly dangling side chains. However, the three-dimensional structure of a **native** (physiologically folded) protein is specified by its primary structure, so that it has a unique set of characteristics.

In this chapter, we shall discuss the structural features of proteins, the forces that hold them together, and their hierarchical organization to form complex structures. This will form the basis for understanding the structure–function relationships necessary to comprehend the biochemical roles of proteins. Detailed consideration of the dynamic behavior of proteins and how they fold to their native structures is deferred until Chapter 9.

1 SECONDARY STRUCTURE

A polymer's **secondary structure (2° structure)** is defined as the local conformation of its backbone. For proteins, this has come to mean the specification of regular polypeptide backbone folding patterns: helices, pleated sheets, and turns. However, before we begin our discussion of these basic structural motifs, let us consider the geometrical properties of the peptide group because its understanding is prerequisite to that of any structure containing it.

A. *The Peptide Group*

In the 1930s and 1940s, Linus Pauling and Robert Corey determined the X-ray structures of several amino acids and dipeptides in an effort to elucidate the structural constraints on the conformations of a polypeptide chain. These studies indicated that *the peptide group has a rigid, planar structure (Fig. 8-1), which,* Pauling pointed out, *is a*

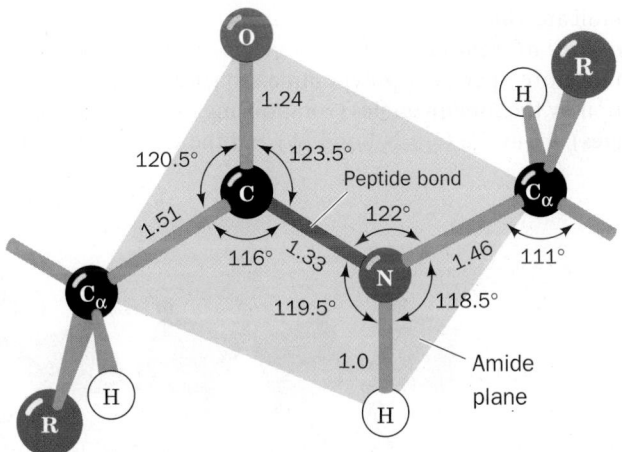

Figure 8-1 **The trans-peptide group.** The standard dimensions (in angstroms, Å, and degrees, °) of this planar group were derived by averaging the corresponding quantities in the X-ray crystal structures of amino acids and peptides. [After Marsh, R.E. and Donohue, J., *Adv. Protein Chem.* **22,** 249 (1967).]

See Kinemage Exercise 3-1

consequence of resonance interactions that give the peptide bond an ~40% double-bond character:

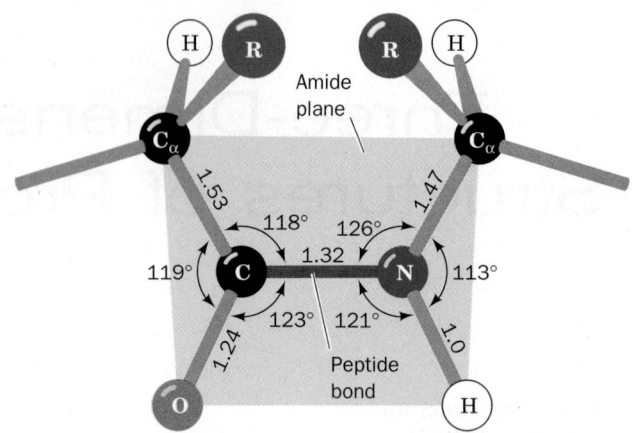

Figure 8-2   **The cis-peptide group.** See Kinemage Exercise 3-1

This explanation is supported by the observations that a peptide's C—N bond is 0.13 Å shorter than its N—C_α single bond and that its C=O bond is 0.02 Å longer than that of aldehydes and ketones. The peptide bond's resonance energy has its maximum value, ~85 kJ · mol^{-1}, when the peptide group is planar because its π-bonding overlap is maximized in this conformation. This overlap, and thus the resonance energy, falls to zero as the peptide bond is twisted to 90° out of planarity, thereby accounting for the planar peptide group's rigidity. (The positive charge on the above resonance structure should be taken as a formal charge; quantum mechanical calculations indicate that the peptide N atom, in fact, has a partial negative charge arising from the polarization of the C—N σ bond.)

Peptide groups, with few exceptions, assume the trans conformation: that in which successive C_α atoms are on opposite sides of the peptide bond joining them (Fig. 8-1). This is partly a result of steric interference, which causes the cis conformation (Fig. 8-2) to be ~8 kJ · mol^{-1} less stable than the trans conformation (this energy difference is somewhat less in peptide bonds followed by a Pro residue and, in fact, ~10% of the Pro residues in proteins follow a cis peptide bond, whereas cis peptides are otherwise extremely rare).

a. Polypeptide Backbone Conformations May Be Described by Their Torsion Angles

The above considerations are important because they indicate that *the backbone of a protein is a linked sequence of rigid planar peptide groups* (Fig. 8-3). We can therefore specify a polypeptide's backbone conformation by the **torsion angles** (rotation angles or **dihedral angles**) about the C_α—N bond (φ) and the C_α—C bond (ψ)

of each of its amino acid residues. These angles, φ and ψ, are both defined as 180° when the polypeptide chain is in its planar, fully extended (all-trans) conformation and increase for a clockwise rotation when viewed from C_α (Fig. 8-4).

There are several steric constraints on the torsion angles, φ and ψ, of a polypeptide backbone that limit its conformational range. The electronic structure of a single (σ) bond, such as a C—C bond, is cylindrically symmetrical about its bond axis, so that we might expect such a bond to exhibit free rotation. If this were the case, then in ethane, for example, all torsion angles about the C—C bond would be equally likely. Yet certain conformations in ethane are favored due to quantum mechanical effects arising from the interactions of its molecular orbitals. The **staggered conformation** (Fig. 8-5a; torsion angle = 180°) is ethane's most stable arrangement, whereas the **eclipsed conformation** (Fig. 8-5b; torsion angle = 0°) is least stable. The energy difference between the staggered and eclipsed conformations in ethane is ~12 kJ · mol^{-1}, a quantity that represents an **energy barrier** to free rotation about the C—C single bond. Substituents other than hydrogen exhibit greater steric interference; that is, they increase the size of this energy barrier due to their greater bulk.

Figure 8-3   **A polypeptide chain in its fully extended conformation showing the planarity of each of its peptide groups.** [Illustration, Irving Geis. Image from the Irving Geis Collection, Howard Hughes Medical Institute. Reprinted with permission.]

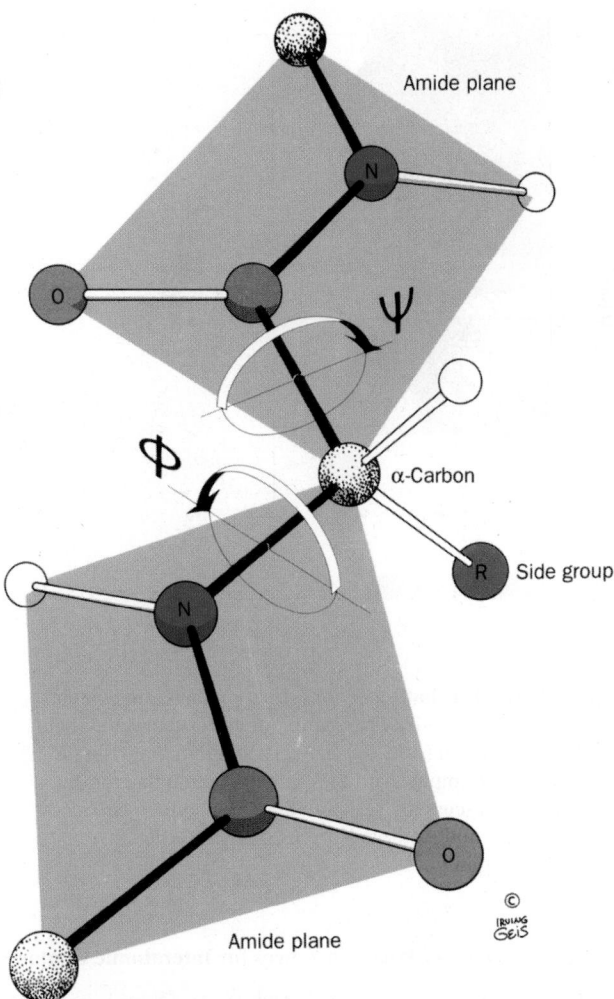

Figure 8-4   The torsional degrees of freedom in a peptide unit. The only reasonably free movements are rotations about the C_α—N bond (ϕ) and the C_α—C bond (ψ). The torsion angles are both 180° in the conformation shown and increase, as is indicated, in a clockwise manner when viewed from C_α. [Illustration, Irving Geis. Image from the Irving Geis Collection, Howard Hughes Medical Institute. Reprinted with permission.] **See Kinemage Exercise 3-1**

(a) Staggered   **(b) Eclipsed**

Figure 8-5   Conformations of ethane. Newman projections indicating the (*a*) staggered conformation and (*b*) eclipsed conformation of ethane.

Indeed, with large substituents, some conformations may be sterically forbidden.

b. Allowed Conformations of Polypeptides Are Indicated by the Ramachandran Diagram

The sterically allowed values of ϕ and ψ can be determined by calculating the distances between the atoms of a tripeptide at all values of ϕ and ψ for the central peptide unit. Sterically forbidden conformations, such as that shown in Fig. 8-6, are those in which any nonbonding interatomic distance is less than its corresponding van der Waals distance. Such information is summarized in a **conformation map** or **Ramachandran diagram** (Fig. 8-7), which was invented by G.N. Ramachandran.

Figure 8-7 indicates that 77% of the Ramachandran diagram (most combinations of ϕ and ψ) is conformationally inaccessible to a polypeptide chain. The particular regions of the Ramachandran diagram that represent allowed conformations depend on the van der Waals radii chosen to calculate it. But with any realistic set of values, such as that in Table 8-1, *only three small regions of the conformational map are physically accessible to a polypeptide chain*. Nevertheless, as we shall see, all of the common types of regular secondary structures found in proteins fall within allowed regions of the Ramachandran diagram. Indeed, the

Figure 8-6   Steric interference between adjacent residues. The collision between a carbonyl oxygen and the following amide hydrogen prevents the conformation $\phi = -60°, \psi = 30°$. [Illustration, Irving Geis. Image from the Irving Geis Collection, Howard Hughes Medical Institute. Reprinted with permission.] **See Kinemage Exercise 3-1.**

Figure 8-7 **The Ramachandran diagram.** It shows the sterically allowed ϕ and ψ angles for poly-L-alanine and was calculated using the van der Waals distances in Table 8-1. Regions of "normally allowed" ϕ and ψ angles are shaded in blue, whereas green-shaded regions correspond to conformations having "outer limit" van der Waals distances. The conformation angles, ϕ and ψ, of several secondary structures are indicated below:

Secondary Structure	ϕ (deg)	ψ (deg)
Right-handed α helix (α)	−57	−47
Parallel β pleated sheet (↑↑)	−119	113
Antiparallel β pleated sheet (↑↓)	−139	135
Right-handed 3_{10} helix (3)	−49	−26
Right-handed π helix (π)	−57	−70
2.2_7 ribbon (2)	−78	59
Left-handed polyglycine II and polyproline II helices (II)	−79	150
Collagen (C)	−51	153
Left-handed α helix (α_L)	57	47

[After Flory, P.J., *Statistical Mechanics of Chain Molecules, p.* 253, Interscience (1969); *and* IUPAC-IUB Commission on Biochemical Nomenclature, *Biochemistry* **9**, 3475 (1970).]

Figure 8-8 **Distribution of conformation angles in proteins.** The conformation angle distribution of all residues but Gly in 207 high-resolution ($\leq$1.2 Å) X-ray structures comprising 25,327 residues is superimposed on the Ramachandran diagram (resolution is discussed in Section 8-3Aa). [Courtesy of Scott Hollingsworth and Andrew Karplus, Oregon State University, Corvallis, Oregon.]

Table 8-1 **van der Waals Distances for Interatomic Contacts**

Contact Type	Normally Allowed (Å)	Outer Limit (Å)
H $\cdots$ H	2.0	1.9
H $\cdots$ O	2.4	2.2
H $\cdots$ N	2.4	2.2
H $\cdots$ C	2.4	2.2
O $\cdots$ O	2.7	2.6
O $\cdots$ N	2.7	2.6
O $\cdots$ C	2.8	2.7
N $\cdots$ N	2.7	2.6
N $\cdots$ C	2.9	2.8
C $\cdots$ C	3.0	2.9
C $\cdots$ CH$_2$	3.2	3.0
CH$_2$ $\cdots$ CH$_2$	3.2	3.0

Source: Ramachandran, G.N. and Sasisekharan, V., *Adv. Protein Chem.* **23**, 326 (1968).

observed conformational angles of most non-Gly residues in proteins whose X-ray structures have been determined lie in these allowed regions (Fig. 8-8).

Most points that fall in forbidden regions of Fig. 8-8 lie between its two fully allowed areas near $\psi = 0$. However, these "forbidden" conformations, which arise from the collision of successive amide groups, are allowed if twists of only a few degrees about the peptide bond are permitted. This is not unreasonable since the peptide bond offers little resistance to small deformations from planarity.

Gly, the only residue without a C$_\beta$ atom, is much less sterically hindered than the other amino acid residues. This is clearly apparent in comparing the Ramachandran diagram for Gly in a polypeptide chain (Fig. 8-9) with that of other residues (Fig. 8-7). In fact, Gly often occupies positions where a polypeptide backbone makes a sharp turn

Figure 8-9 **The Ramachandran diagram of Gly residues in a polypeptide chain.** "Normally allowed" regions are shaded in blue, whereas green-shaded regions correspond to "outer limit" van der Waals distances. Gly residues have far greater conformational freedom than do other (bulkier) amino acid residues, as the comparison of this figure with Fig. 8-7 indicates. [After Ramachandran, G.N. and Sasisekharan, V., *Adv. Protein Chem.* **23,** 332 (1968).]

which, with any other residue, would be subject to steric interference.

Figure 8-7 was calculated for three consecutive Ala residues. Similar plots for larger residues that are unbranched at C_β, such as Phe, are nearly identical. In Ramachandran diagrams of residues that are branched at C_β, such as Thr, the allowed regions are somewhat smaller than for Ala. The cyclic side chain of Pro limits its ϕ to the range $-60° \pm 25°$, making it, not surprisingly, the most conformationally restricted amino acid residue. The conformations of residues in chains longer than tripeptides are even more restricted than the Ramachandran diagram indicates because a polypeptide chain with all its ϕ and ψ angles allowed nevertheless cannot assume a conformation in which it passes through itself. We shall see, however, that despite the great restrictions that peptide bond planarity and side chain bulk place on the conformations of a polypeptide chain, different unique primary structures have correspondingly unique three-dimensional structures.

B. *Helical Structures*

🔖 **See Guided Exploration 7: Stable helices in proteins: The α helix**

Helices are the most striking elements of protein 2° structure. If a polypeptide chain is twisted by the same amount about each of its C_α atoms, it assumes a helical conformation. As an alternative to specifying its ϕ and ψ angles, a helix may be characterized by the number, n, of peptide units per helical turn and by its **pitch,** p, the distance the helix rises along its axis per turn. Several examples of helices are diagrammed in Fig. 8-10. Note that a helix has chirality; that

$n = 5$ $n = 4$ $n = 3$ $n = 2$ $n = -3$

Figure 8-10 **Examples of helices.** These provide definitions of the helical pitch, p, the number of repeating units per turn, n, and the helical rise per repeating unit, $d = p/n$. Right- and left-handed helices are defined, respectively, as having positive and negative

values of n. For $n = 2$, the helix degenerates to a nonchiral ribbon. For $p = 0$, the helix degenerates to a closed ring. [Illustration, Irving Geis. Image from the Irving Geis Collection, Howard Hughes Medical Institute. Reprinted with permission.]

is, it may be either right handed or left handed (a right-handed helix turns in the direction that the fingers of a right hand curl when its thumb points along the helix axis in the direction that the helix rises). In proteins, moreover, *n* need not be an integer and, in fact, rarely is.

A polypeptide helix must, of course, have conformation angles that fall within the allowed regions of the Ramachandran diagram. As we have seen, this greatly limits the possibilities. Furthermore, if a particular conformation is to have more than a transient existence, it must be more than just allowed, it must be stabilized. The "glue" that holds polypeptide helices and other 2° structures together is, in part, hydrogen bonds.

a. The α Helix

*Only one helical polypeptide conformation has simultaneously allowed conformation angles and a favorable hydrogen bonding pattern: the **α helix** (Fig. 8-11), a particularly rigid arrangement of the polypeptide chain.* Its discovery through model building, by Pauling in 1951, ranks as one of the landmarks of structural biochemistry.

For a polypeptide made from L-α-amino acid residues, the α helix is right handed with torsion angles $\phi = -57°$ and $\psi = -47°$, $n = 3.6$ residues per turn, and a pitch of 5.4 Å. (An α helix of D-α-amino acid residues is the mirror image of that made from L-amino acid residues: It is left handed with conformation angles $\phi = +57°$, $\psi = +47°$, and $n = -3.6$ but with the same value of *p*.)

Figure 8-11 indicates that the hydrogen bonds of an α helix are arranged such that the peptide N—H bond of the *n*th residue points along the helix toward the peptide C=O group of the $(n - 4)$th residue. This results in a strong hydrogen bond that has the nearly optimum $N \cdots O$ distance of 2.8 Å. In addition, the core of the α helix is tightly packed; that is, its atoms are in van der Waals contact across the helix, thereby maximizing their association energies (Section 8-4Ab). The R groups, whose positions, as we saw, are not fully dealt with by the Ramachandran diagram, all project backward (downward in Fig. 8-11) and outward from the helix so as to avoid steric interference with the polypeptide backbone and with each other. Such an arrangement can also be seen in Fig. 8-12. Indeed, a major reason why the left-handed α helix has never been observed (its helical parameters are but mildly forbidden; Fig. 8-7) is that its side chains contact its polypeptide backbone too closely. Note, however, that 1 to 2% of the individual non-Gly residues in proteins assume this conformation (Fig. 8-8).

The α helix is a common secondary structural element of both fibrous and globular proteins. In globular proteins, α helices have an average span of ~12 residues, which corresponds to over three helical turns and a length of 18 Å. However, α helices with over 140 residues are known.

b. Other Polypeptide Helices

Figure 8-13 indicates how hydrogen bonded polypeptide helices may be constructed. The first two, the **2.2₇ ribbon** and the **3₁₀ helix**, are described by the notation, n_m,

Figure 8-11   The right-handed α helix. Hydrogen bonds between the N—H groups and the C=O groups that are four residues back along the polypeptide chain are indicated by dashed lines. [Illustration, Irving Geis. Image from the Irving Geis Collection, Howard Hughes Medical Institute. Reprinted with permission.]

See Kinemage Exercise 3-2 and the Animated Figures

where *n*, as before, is the number of residues per helical turn and *m* is the number of atoms, including H, in the ring that is closed by the hydrogen bond. With this notation, an α helix is a 3.6₁₃ helix.

Figure 8-12 **Stereo, space-filling representation of an α helical segment of sperm whale myoglobin (its E helix) as determined by X-ray crystal structure analysis.** Backbone atoms are colored according to type (C green, N blue, O red, and H white) and the side chain atoms are gold. Instructions for viewing stereo diagrams are given in the appendix to this chapter. [Based on an X-ray structure by Ilme Schlichting, Max Planck Institut für Molekulare Physiologie, Dortmund, Germany. PDBid 1A6M (for the definition of "PDBid," see Section 8-3Ca).] **See Kinemage Exercise 3-2**

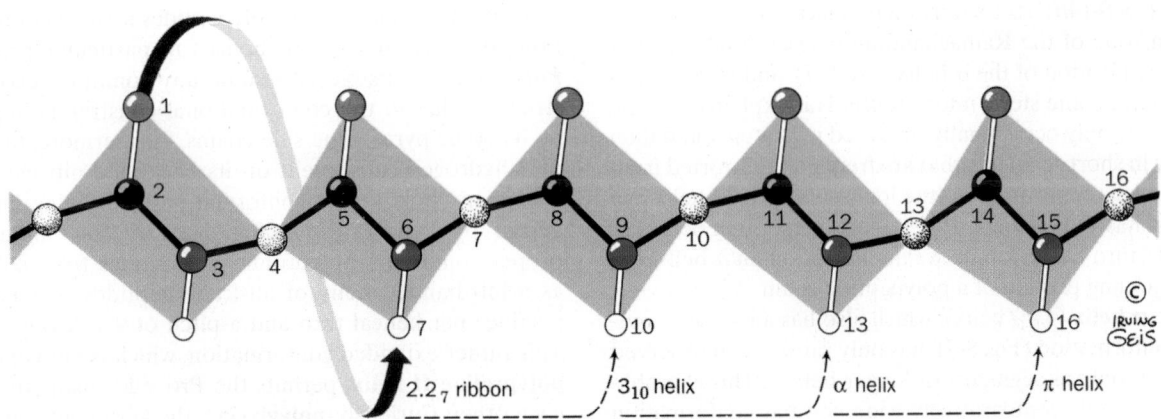

Figure 8-13 **The hydrogen bonding pattern of several polypeptide helices.** In the cases shown, the polypeptide chain is helically wound such that the N—H group on residue *n* forms a hydrogen bond with the C=O groups on residues *n* − 2, *n* − 3, *n* − 4, or *n* − 5. [Illustration, Irving Geis. Image from the Irving Geis Collection, Howard Hughes Medical Institute. Reprinted with permission.]

(a)   *(b)*   *(c)*

3₁₀ helix   α helix   π helix

Figure 8-14 **Comparison of the two polypeptide helices that occasionally occur in proteins with the commonly occurring α helix.** (*a*) The 3_{10} helix, which has 3.0 peptide units per turn and a pitch of 6.0 Å, making it thinner and more elongated than the α helix. (*b*) The α helix, which has 3.6 peptide units per turn and a pitch of 5.4 Å (also see Fig. 8-11). (*c*) The π helix, which has 4.4 peptide units per turn and a pitch of 5.2 Å, making it wider and shorter than the α helix. The peptide planes are indicated. [Illustration, Irving Geis. Image from the Irving Geis Collection, Howard Hughes Medical Institute. Reprinted with permission.]

The right-handed 3_{10} helix (Fig. 8-14*a*), which has a pitch of 6.0 Å, is thinner and rises more steeply than does the α helix (Fig. 8-14*b*). Its torsion angles place it in a mildly forbidden zone of the Ramachandran diagram that is rather near the position of the α helix (Fig. 8-7), and its R groups experience some steric interference. This explains why the 3_{10} helix is only occasionally observed in proteins, and then mostly in short segments that are frequently distorted from the ideal 3_{10} conformation (the longest known 3_{10} helix in a protein has 15 residues). The 3_{10} helix most often occurs as a single-turn transition between one end of an α helix and the adjoining portion of a polypeptide chain.

The **π helix** (4.4_{16} helix), which also has a mildly forbidden conformation (Fig. 8-7), has only rarely been observed and then only as segments of longer helices. This is probably because its comparatively wide and flat conformation (Fig. 8-14*c*) results in an axial hole that is too small to admit water molecules but too wide to allow van der Waals associations across the helix axis; this greatly reduces its stability relative to more closely packed conformations. The 2.2_7 ribbon, which, as Fig. 8-7 indicates, has strongly forbidden conformation angles, has never been observed.

Certain synthetic homopolypeptides assume conformations that are models for helices in particular proteins. **Polyproline** is unable to assume any common secondary structure due to the conformational constraints imposed by its cyclic pyrrolidine side chains. Furthermore, the lack of a hydrogen substituent on its backbone nitrogen precludes any polyproline conformation from being knit together by hydrogen bonding. Nevertheless, under the proper conditions, polyproline precipitates from solution as a left-handed helix of all-trans peptides that has 3.0 residues per helical turn and a pitch of 9.4 Å (Fig. 8-15). This rather extended conformation, which is known as the **polyproline II helix,** permits the Pro side chains to avoid each other. Curiously, **polyglycine,** the least conformationally constrained polypeptide, precipitates from solution as a helix whose parameters are essentially identical to those of polyproline, the most conformationally constrained polypeptide (although the polyglycine helix may be either right or left handed because Gly is nonchiral). The structures of the polyglycine and polyproline helices are of biological significance because they form the basic structural motif of collagen, a structural protein that contains a remarkably high proportion of both Gly and Pro (Section 8-2B). In addition, the polyproline II helical conformation is commonly assumed by polypeptide segments of up to 12 residues, even though it lacks intrahelical hydrogen bonds.

C. Beta Structures

See **Guided Exploration 8: Hydrogren bonding in β sheets and Guided Exploration 9: Secondary structures in proteins** In 1951, the year that they proposed the α helix, Pauling and Corey also postulated the existence of a different polypeptide secondary structure, the **β pleated sheet.** As with the α helix, the β pleated sheet's conformation has repeating ϕ and ψ angles that fall in the allowed region of the Ramachandran diagram (Fig. 8-7) and utilizes the full hydrogen bonding capacity of the polypeptide backbone. *In β pleated sheets, however, hydrogen bonding occurs between neighboring polypeptide chains* rather than within one as in α helices.

β Pleated sheets come in two varieties:

1. The antiparallel β pleated sheet, in which neighboring hydrogen bonded polypeptide chains run in opposite directions (Fig. 8-16a).

2. The parallel β pleated sheet, in which the hydrogen bonded chains extend in the same direction (Fig. 8-16b).

The conformations in which these β structures are optimally hydrogen bonded vary somewhat from that of a fully extended polypeptide (ϕ = ψ = ±180°), as indicated in

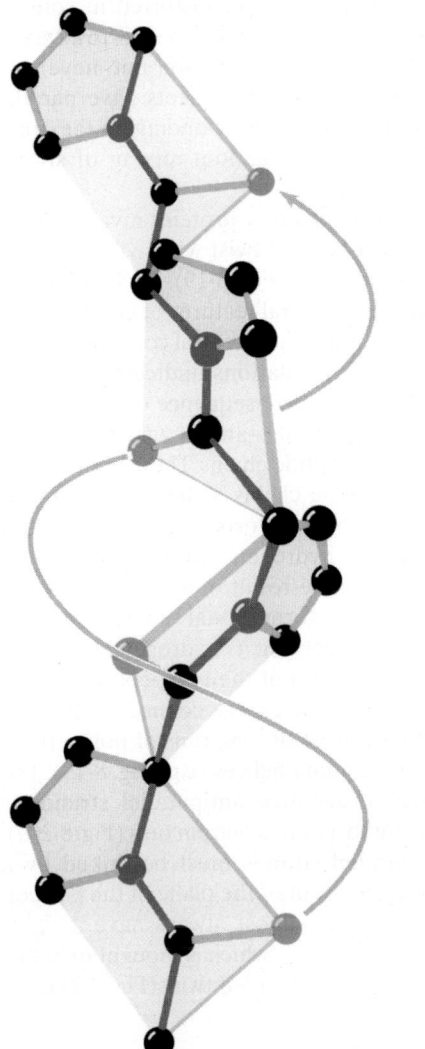

Figure 8-15 **The polyproline II helix.** Polyglycine forms a nearly identical helix **(polyglycine II)**. [Illustration, Irving Geis. Image from the Irving Geis Collection, Howard Hughes Medical Institute. Reprinted with permission.]

(a) **Antiparallel**

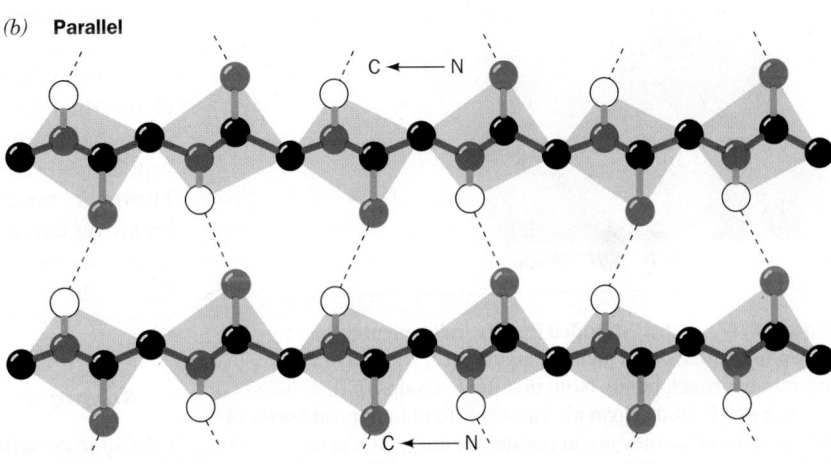

(b) **Parallel**

Figure 8-16 **β Pleated sheets.** Hydrogen bonds are indicated by dashed lines and side chains are omitted for clarity. (*a*) The antiparallel β pleated sheet. (*b*) The parallel β pleated sheet. [Illustration, Irving Geis. Image from the Irving Geis Collection, Howard Hughes Medical Institute. Reprinted with permission.] See **Kinemage Exercise 3-3 and the Animated Figures**

7.0 Å

Figure 8-17 **A two-stranded β antiparallel pleated sheet drawn to emphasize its pleated appearance.** Dashed lines indicate hydrogen bonds. Note that the R groups (*purple balls*) on each polypeptide chain alternately extend to opposite sides of the sheet and that they are in register on adjacent chains. [Illustration, Irving Geis. Image from the Irving Geis Collection, Howard Hughes Medical Institute. Reprinted with permission.]

See Kinemage Exercise 3-3

Fig. 8-7. They therefore have a rippled or pleated edge-on appearance (Fig. 8-17), which accounts for the appellation "pleated sheet." In this conformation, successive side chains of a polypeptide chain extend to opposite sides of the pleated sheet with a two-residue repeat distance of 7.0 Å.

β Sheets are common structural motifs in proteins. In globular proteins, they consist of from 2 to as many as 22 polypeptide strands, the average being 6 strands, which have an aggregate width of ~25 Å. The polypeptide chains in a β sheet are known to be up to 15 residues long, with the average being 6 residues that have a length of ~21 Å. A 7-stranded antiparallel β sheet, for example, occurs in the jack bean protein **concanavalin A** (Fig. 8-18).

Parallel β sheets of less than five strands are rare. This observation suggests that parallel β sheets are less stable than antiparallel β sheets, possibly because the hydrogen bonds of parallel sheets are distorted in comparison to those of the antiparallel sheets (Fig. 8-16). Mixed parallel–antiparallel β sheets are common but, nevertheless, only ~20% of the strands in β sheets have parallel bonding on one side and antiparallel bonding on the other (vs an expected 50% for the random mixing of strand directions).

The β pleated sheets in globular proteins invariably exhibit a pronounced right-handed twist when viewed along their polypeptide strands (e.g., Fig. 8-19). Such twisted β sheets are important architectural features of globular proteins since β sheets often form their central cores (Fig. 8-19). Conformational energy calculations indicate that a β sheet's right-handed twist is a consequence of nonbonded interactions between the chiral L-amino acid residues in the sheet's extended polypeptide chains. These interactions tend to give the polypeptide chains a slight right-handed helical twist (Fig. 8-19) which distorts and hence weakens the β sheet's interchain hydrogen bonds. A particular β sheet's geometry is thus the result of a compromise between optimizing the conformational energies of its polypeptide chains and preserving its hydrogen bonds.

The **topology** (connectivity) of the polypeptide strands in a β sheet can be quite complex; the connecting links of these assemblies often consist of long runs of polypeptide chain which frequently contain helices (e.g., Fig. 8-19). The link connecting two consecutive antiparallel strands is topologically equivalent to a simple hairpin turn (Fig. 8-20a). However, tandem parallel strands must be linked by a crossover connection that is out of the plane of the β sheet. Such crossover connections almost always have a right-handed helical sense (Fig. 8-20b), which is thought to better fit the β sheets' inherent right-handed twist (Fig. 8-21).

D. Nonrepetitive Structures

Globular proteins consist of, on average, ~31% α helix and ~28% β sheet. The protein's remaining polypeptide segments are said to have a **coil** or **loop conformation.** That is not to say that these nonrepetitive secondary structures are any less ordered than are helices or β sheets; they are simply irregular and hence more difficult to describe. You should therefore not confuse the term coil conformation with the

Figure 8-18   Stereo, space-filling representation of the 7-stranded antiparallel β pleated sheet in jack bean concanavalin A as determined by X-ray crystal structure analysis. The β strands are approximately horizontal with their backbone atoms colored according to type (C green, N blue, O red, and H white) and their side chains represented by magenta spheres. Instructions for viewing stereo drawings are given in the appendix to this chapter. [Based on an X-ray structure by Gerald Edelman, The Rockefeller University. PDBid 2CNA.] **See Kinemage Exercise 3-3.**

(a)

(b)

(c)

Figure 8-19   Polypeptide chain folding in proteins illustrating the right-handed twist of β sheets. In these ribbon drawings, the α helices shown as cyan helices, the strands of β sheets are represented by green arrows pointing toward the C-terminus, and the remaining portions of the backbone are drawn as orange worms. Side chains are not shown. (*a*) Bovine **carboxypeptidase A,** a 307-residue protein, contains an 8-stranded mixed β sheet that forms a saddle-shaped curved surface with a right-handed twist. (*b*) Chicken muscle **triose phosphate isomerase,** a 247-residue enzyme, contains an 8-stranded parallel β sheet that forms a cylindrical structure known as a **β barrel,** here viewed from the top. Note that the crossover connections between successive strands of the β barrel, which each consist predominantly of an α helix, are outside the β barrel and have a right-handed helical sense. (*c*) Side view of triose phosphate isomerase. Its N-terminus (N) and C-terminus (C) are indicated. [Part *a* based on an X-ray structure by William Lipscomb, Harvard University. PDBid 3CPA. Parts *b* and *c* based on an X-ray structure by David Phillips, Oxford University, U.K. PDBid 1TIM.] **See Interactive Exercise 2**

(a) *(b)* *(c)*

Figure 8-20 Connections between adjacent polypeptide strands in β pleated sheets. (*a*) The hairpin connection between antiparallel strands is topologically in the plane of the sheet. (*b*) A right-handed crossover connection between successive strands of a parallel β sheet. Nearly all such crossover connections in proteins have this chirality (see, e.g., Fig. 8-19*b*). (*c*) A left-handed crossover connection between parallel β sheet strands. Connections with this chirality are rare. [After Richardson, J.S., *Adv. Protein Chem.* **34,** 290, 295 (1981).]

term **random coil,** which refers to the totally disordered and rapidly fluctuating set of conformations assumed by denatured proteins and other polymers in solution.

Globular proteins consist largely of approximately straight runs of secondary structure joined by stretches of polypeptide that abruptly change direction. Such **reverse turns** or **β bends** (so named because they often connect successive strands of antiparallel β sheets) almost always occur at protein surfaces; indeed, they partially define these surfaces. Most reverse turns involve four successive amino acid residues more or less arranged in one of two ways, Type I and Type II, that differ by a 180° flip of the peptide unit linking residues 2 and 3 (Fig. 8-22). Both types of β bends contain a hydrogen bond, although deviations from these ideal conformations often disrupt this hydrogen bond. Type I β bends may be considered to be distorted sections of 3_{10} helix. In Type II β bends, the oxygen atom of residue 2 crowds the C_β atom of residue 3, which is therefore usually Gly. Residue 2 of either type of β bend is often Pro since it can facilely assume the required conformation.

Many proteins have regions that are truly disordered. Extended, charged surface groups such as Lys side chains or the N- or C-termini of polypeptide chains are good examples: They often wave around in solution because there are few forces to hold them in place (Section 8-4). Often entire peptide chain segments are disordered. Such segments may have functional roles, such as the binding of a specific molecule, so they may be disordered in one state of the protein (molecule absent) and ordered in another (molecule bound). This is one mechanism whereby a protein can interact flexibly with another molecule in the performance of its biological function.

2 FIBROUS PROTEINS

Fibrous proteins are highly elongated molecules whose secondary structures are their dominant structural motifs. Many fibrous proteins, such as those of skin, tendon, and bone, function as structural materials that have a protective, connective, or supportive role in living organisms. Others, such as muscle and ciliary proteins, have motive functions. In this section, we shall discuss structure–function relationships in two common and well-characterized fibrous proteins: keratin and collagen (muscle and ciliary proteins are considered in Section 35-3). The structural simplicity of these proteins relative to those of globular proteins (Section 8-3) makes them particularly amenable to understanding how their structures suit them to their biological roles.

Fibrous molecules rarely crystallize and hence are usually not subject to structural determination by single-crystal X-ray structure analysis (Section 8-3A). Rather than crystallizing, they associate as fibers in which their long molecular axes are more or less parallel to the fiber axis but in which they lack specific orientation in other directions. The X-ray diffraction pattern of such a fiber, Fig. 8-23, for

Figure 8-21 Origin of a right-handed crossover connection. A possible folding scheme illustrates how right-handed polypeptide chain twisting favors the formation of right-handed crossover connections between successive strands of a parallel β sheet.

(a) **Type I β bend**

(b) **Type II β bend**

Figure 8-22 **Reverse turns in polypeptide chains.** (a) A Type I β bend, which has the following torsion angles:

$$\phi_2 = -60°/\psi_2 = -30°$$

$$\phi_3 = -90°/\psi_3 = 0°$$

(b) A Type II β bend, which has the following torsion angles:

$$\phi_2 = -60°/\psi_2 = 120°$$

$$\phi_3 = -90°/\psi_3 = 0°$$

Variations from these ideal conformation angles by as much as 30° are common. Hydrogen bonds are represented by dashed lines. [Illustration, Irving Geis. Image from the Irving Geis Collection, Howard Hughes Medical Institute. Reprinted with permission.] **See Kinemage Exercise 3-4**

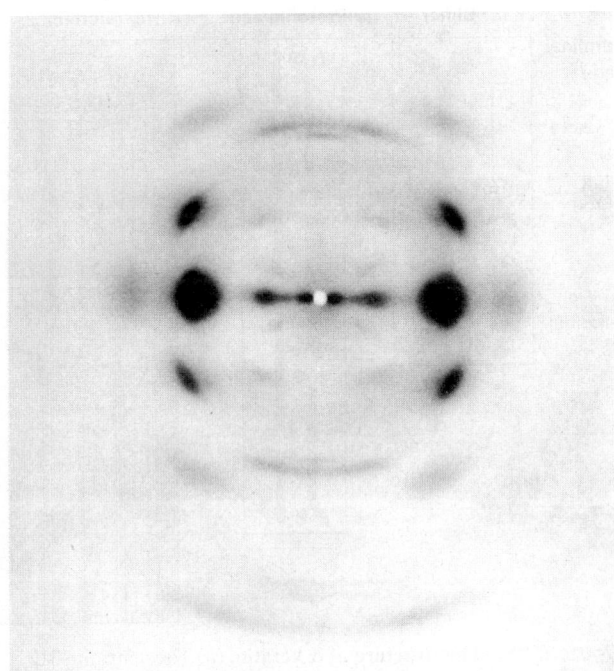

Figure 8-23 **X-ray diffraction photograph of a fiber of *Bombyx mori* silk.** The photograph was obtained by shining a collimated beam of monochromatic X-rays through the silk fiber and recording the diffracted X-rays on a sheet of photographic film placed behind the fiber. The photograph has only a few spots and thus contains little structural information. [From March, R.E., Corey, R.B., and Pauling, L., *Biochim. Biophys. Acta* **16,** 5 (1955).]

example, contains little information, far less than would be obtained if the fibrous protein could be made to crystallize. Consequently, the structures of fibrous proteins are not known in great detail. Nevertheless, the original X-ray studies of proteins were carried out in the early 1930s by William Astbury on such easily available protein fibers as wool and tendon. Since the first X-ray crystal structure of a protein was not determined until the late 1950s, these fiber studies constituted the first tentative steps in the elucidation of the structural principles governing proteins and formed much of the experimental basis for Pauling's formulation of the α helix and β pleated sheet.

A. α Keratin—A Helix of Helices

Keratin is a mechanically durable and chemically unreactive protein that occurs in all higher vertebrates. It is the principal component of their horny outer epidermal layer, comprising up to 85% of the cellular protein, and its related appendages such as hair, horns, nails, and feathers. Keratins have been classified as either **α keratins,** which occur in mammals, or **β keratins,** which occur in birds and reptiles. Mammals have over 50 keratin genes, which are expressed in a tissue-specific manner and whose products are classified as belonging to families of relatively acidic (Type I) and relatively basic (Type II) polypeptides. Keratin filaments, which form the intermediate filaments of skin cells (Section 1-2Ae), must contain at least one member of each type.

Electron microscopic studies indicate that hair, which is composed mainly of α keratin, consists of a hierarchy of structures (Figs. 8-24 and 8-25). A typical hair is ~20 μm in diameter and is constructed from dead cells, each of which contains packed **macrofibrils** (~2000 Å in diameter) that are oriented parallel to the hair fiber (Fig. 8-24). The macrofibrils are constructed from **microfibrils** (~80 Å wide) that are cemented together by an amorphous protein matrix of high sulfur content.

Moving to the molecular level, the X-ray diffraction pattern of α keratin resembles that expected for an α helix (hence the name α keratin). Yet α keratin exhibits a 5.1-Å spacing rather than the 5.4-Å distance corresponding to the pitch of the α helix. This observation, together with a variety of physical and chemical evidence, suggests that *α keratin polypeptides form closely associated pairs of α helices in which each pair is composed of a Type I and a Type II keratin chain twisted in parallel into a left-handed coil* (Fig. 8-25a). The normal 5.4-Å repeat distance of each α helix in the pair is thereby tilted with respect to the axis of this assembly, yielding the observed 5.1-Å spacing. This assembly is said to have a **coiled coil** structure because each α helix axis itself follows a helical path.

The conformation of α keratin's coiled coil is a consequence of its primary structure: The central ~310-residue segment of each polypeptide chain has a heptad (7-residue) pseudorepeat, *a-b-c-d-e-f-g*, with nonpolar residues predominating at positions *a* and *d*. Since an α helix has 3.6 residues per turn, α keratin's *a* and *d* residues line up on one side of the α helix to form a hydrophobic strip that promotes its lengthwise associ-

ation with a similar strip on another such α helix (Fig. 8-26; hydrophobic residues, as we shall see in Section 8-4C, have a strong tendency to associate). Indeed, the slight discrepancy between the 3.6 residues per turn of a normal α helix and the ~3.5-residue repeat of α keratin's hydrophobic strip is responsible for the coiled coil's coil. The resulting 18° inclination of the α helices relative to one another permits the helical ridges formed by the side chains on one helix to fit into the grooves between these ridges on the other helix, thereby greatly increasing their favorable interactions. Coiled coils, as we shall see, are common components of globular proteins as well as of other fibrous proteins.

The higher order substructure of α keratin is poorly understood. The N- and C-terminal portions of each polypeptide probably have a flexible conformation and facilitate the assembly of the coiled coils into ~30-Å-wide protofilaments. These are thought to consist of two staggered antiparallel rows of head-to-tail aligned coiled coils (Fig. 8-25b). Two such protofilaments are thought to comprise an ~50-Å-wide protofibril, four of which, in turn, coil around each other to form a microfibril (Fig. 8-25c).

α Keratin is rich in Cys residues, which form disulfide bonds that cross-link adjacent polypeptide chains. This accounts for α keratin's insolubility and resistance to stretching, two of its most important biological properties. The α keratins are classified as "hard" or "soft" according to whether they have a high or low sulfur content. Hard keratins, such as those of hair, horn, and nail, are less pliable

Figure 8-25 **The structure of α keratin.** (*a*) The central ~310 residues of one polypeptide chain each of Types I and II α keratins associate in a dimeric coiled coil. The conformations of the polypeptides' globular N- and C-terminal domains are unknown. (*b*) Protofilaments are formed from two staggered and antiparallel rows of associated head-to-tail coiled coils. (*c*) The protofilaments dimerize to form a protofibril, four of which form a microfibril. The structures of these latter assemblies are poorly characterized but are thought to form helical arrays.

Figure 8-24 **The macroscopic organization of hair.** [Illustration, Irving Geis. Image from the Irving Geis Collection, Howard Hughes Medical Institute. Reprinted with permission.]

(a)

(b)

Figure 8-26 **The two-stranded coiled coil.** (*a*) View down the coil axis showing the interactions between the nonpolar edges of the α helices. The α helices have the pseudorepeating heptameric sequence *a-b-c-d-e-f-g* in which residues *a* and *d* are predominantly nonpolar. [After McLachlan, A.D. and Stewart, M., *J. Mol. Biol.* **98,** 295 (1975).] (*b*) Side view of the polypeptide backbones drawn in stick form (*left*) and of the entire polypeptides drawn in space-filling form (*right*). The atoms are colored according to type with C green in one chain and cyan in the other, N blue, O red, and S yellow. The 81-residue chains are parallel with their N-terminal ends above. Note that in the space-filling model the side chains contact each other. This coiled coil is a portion of the muscle protein tropomyosin (Section 35-3Ac). [Based on an X-ray structure by Carolyn Cohen, Brandeis University. PDBid 1IC2.] **See Kinemage Exercises 4-1 and 4-2**

than soft keratins, such as those of skin and callus, because the disulfide bonds resist any forces tending to deform them. The disulfide bonds can be reductively cleaved with mercaptans (Section 7-1B). Hair so treated can be curled and set in a "permanent wave" by application of an oxidizing agent which reestablishes the disulfide bonds in the new "curled" conformation. Although the insolubility of α keratin prevents most animals from digesting it, the clothes moth larva,

which has a high concentration of mercaptans in its digestive tract, can do so to the chagrin of owners of woolen clothing.

The springiness of hair and wool fibers is a consequence of the coiled coil's tendency to untwist when stretched and to recover its original conformation when the external force is relaxed. After some of its disulfide bonds have been cleaved, however, an α keratin fiber can be stretched to over twice its original length by the application of moist heat. In this process, as X-ray analysis indicates, the α helical structure extends with concomitant rearrangement of its hydrogen bonds to form a β pleated sheet. β Keratin, such as that of feathers, exhibits a similar X-ray pattern in its native state (hence the name β sheet).

a. Keratin Defects Result in a Loss of Skin Integrity

The inherited skin diseases **epidermolysis bullosa simplex (EBS)** and **epidermolytic hyperkeratosis (EHK)** are characterized by skin blistering arising from the rupture of epidermal basal cells (Fig. 1-14*d*) and suprabasal cells, respectively, as caused by mechanical stresses that normally would be harmless. Symptomatic variations in these conditions range from severely incapacitating, particularly in early childhood, to barely noticeable. In families afflicted with EBS, sequence abnormalities may be present in either keratin 14 or keratin 5, the dominant Types I and II keratins in basal skin cells. EHK is similarly caused by defects in keratins 1 or 10, the dominant Types I and II keratins in suprabasal cells (which arise through the differentiation of basal cells, a process in which the synthesis of keratins 14 and 5 is switched off and that of keratins 1 and 10 is turned on). These defects evidently interfere with normal filament formation, thereby demonstrating the function of the keratin cytoskeleton in maintaining the mechanical integrity of the skin.

B. *Collagen—A Triple Helical Cable*

Collagen (Greek: *kolla,* glue) occurs in all multicellular animals and is the most abundant protein of vertebrates, comprising ~30% of their protein mass. It is an extracellular protein that is organized into insoluble fibers of great tensile strength. This suits collagen to its role as the major stress-bearing component of **connective tissues** such as bone, teeth, cartilage, tendon, ligament, and the fibrous matrices of skin and blood vessels. Collagen occurs in virtually every tissue.

Vertebrates have 46 genetically distinct polypeptide chains comprising 28 distinct collagen types that occur in different tissues of the same individual. The most prominent of these are listed in Table 8-2. A single molecule of

Table 8-2 **The Most Abundant Types of Collagen**

Type	Chain Composition	Distribution
I	[α1(I)]₂α2(I)	Skin, bone, tendon, blood vessels, cornea
II	[α1(II)]₃	Cartilage, intervertebral disk
III	[α1(III)]₃	Blood vessels, fetal skin

Source: Eyre, D.R., *Science* **207,** 1316 (1980).

Type I collagen is composed of three polypeptide chains with an aggregate molecular mass of ~285 kD. It has a rod-like shape with a length of ~3000 Å and a width of ~14 Å.

Collagen has a distinctive amino acid composition: Nearly one-third of its residues are Gly; another 15 to 30% of them are Pro and **4-hydroxyprolyl (Hyp)** *residues:*

4-Hydroxyprolyl residue (Hyp)   **3-Hydroxyprolyl residue**

5-Hydroxylysyl residue (Hyl)

3-Hydroxyprolyl and **5-hydroxylysyl (Hyl)** residues also occur in collagen but in smaller amounts. Radioactive labeling experiments have established that these nonstandard hydroxylated amino acids are not incorporated into collagen during polypeptide synthesis: If ^{14}C-labeled 4-hydroxyproline is administered to a rat, the collagen synthesized is not radioactive, whereas radioactive collagen is produced if the rat is fed ^{14}C-labeled proline. The hydroxylated residues appear after the collagen polypeptides are synthesized, when certain Pro residues are converted to Hyp in a reaction catalyzed by the enzyme **prolyl hydroxylase.**

Hyp confers stability on collagen, possibly through intramolecular hydrogen bonds that involve bridging water molecules. If, for example, collagen is synthesized under conditions that inactivate prolyl hydroxylase, it loses its native conformation (denatures) at 24°C, whereas normal collagen denatures at 39°C (heat-denatured collagen is known as **gelatin**). Prolyl hydroxylase requires **ascorbic acid (vitamin C)**

Ascorbic acid (vitamin C)

to maintain its enzymatic activity. In the vitamin C deficiency disease **scurvy,** the collagen synthesized cannot form fibers properly. This results in the skin lesions, blood vessel fragility, and poor wound healing that are symptomatic of this ultimately fatal vitamin deficiency disease.

a. Collagen Has a Triple Helical Structure

The amino acid sequence of bovine collagen α1(I), *which is similar to that of other collagens, consists of monotonously repeating triplets of sequence Gly-X-Y over a continuous 1011-residue stretch of its 1042-residue polypeptide chain* (Fig. 8-27). Here X is often Pro (~28%) and Y is often Hyp (~38%). The restriction of Hyp to the Y position stems from the specificity of prolyl hydroxylase. Hyl is similarly restricted to the Y position.

The high Gly, Pro, and Hyp content of collagen suggests that its polypeptide backbone conformation resembles those of the polyglycine II and polyproline II helices (Fig. 8-15). X-ray fiber diffraction and model building studies by Alexander Rich and Francis Crick and by Ramachandran led them to independently propose, in 1955, that *collagen's three polypeptide chains, which individually resemble polyproline II helices, are parallel and wind around each other with a gentle, right-handed, ropelike twist to form a triple helical structure* (Fig. 8-28). It was not until 1994, however, that Helen Berman and Barbara Brodsky confirmed this model through their X-ray crystal structure determination of the collagenlike polypeptide (Pro-Hyp-Gly)$_{10}$ in which the fifth Gly is replaced by Ala (Fig. 8-29a). In this structure, every third residue of each polypeptide chain passes through the center of the triple helix, which is so crowded that only a Gly side chain can fit there (Fig. 8-29b). This crowding explains the absolute requirement for a Gly at every third position of a collagen polypeptide chain (Fig. 8-27). It also requires that the three polypeptide chains be staggered so that the Gly, X, and Y residues from the three chains occur at similar levels (Fig. 8-30). The staggered peptide groups are oriented such that the N—H of each Gly makes a strong hydrogen bond with the carbonyl oxygen of

Figure 8-27   **The amino acid sequence at the C-terminal end of the triple helical region of the bovine α1(I) collagen chain.** Note the repeating triplets Gly-X-Y, where X is often Pro and Y is often Hyp. Here Gly is shaded in purple, Pro in tan, and Hyp and Hyp* (3-hydroxyPro) in brown. [From Bornstein, P. and Traub, W., *in* Neurath, H. and Hill, R.L. (Eds.), *The Proteins* (3rd ed.), Vol. 4, *p.* 483, Academic Press (1979).]

(a)

(b)

Figure 8-28   The triple helix of collagen. This diagram indicates how the left-handed polypeptide helices are twisted together to form a right-handed superhelical structure. Ropes and cables are similarly constructed from hierarchies of fiber bundles that are alternately twisted in opposite directions. An individual collagen polypeptide helix has 3.3 residues per turn and a pitch of 10.0 Å (in contrast to polyproline II's 3.0 residues per turn and pitch of 9.4 Å; Fig. 8-15). The collagen triple helix has 10 Gly-X-Y units per turn and a pitch of 86.1 Å. [Illustration, Irving Geis. Image from the Irving Geis Collection, Howard Hughes Medical Institute. Reprinted with permission.]

Figure 8-29   Structure of a collagen model peptide. In this X-ray structure of (Pro-Hyp-Gly)$_{10}$, the fifth Gly of each peptide has been replaced by Ala. (*a*) A stick model of the central portion of the triple helix oriented with its N-termini at the bottom. The C atoms of the three chains are colored orange, magenta, and gray. The N and O atoms on all chains are blue and red. Note how the replacement of Gly with the bulkier Ala (C atoms green) distorts the triple helix. (*b*) This view from the N-terminus down the helix axis shows the interchain hydrogen bonding associations. Three consecutive residues from each chain are shown in stick form (C atoms green). Hydrogen bonds are represented by dashed lines from Gly N atoms to Pro O atoms in adjacent chains. Dots represent the van der Waals surfaces of the backbone atoms of the central residue in each chain. Note the close packing of the atoms along the triple helix axis. The substitution of a centrally located Gly C$_\alpha$ atom (CH$_2$ group) by any other residue would distort the triple helix. [Based on an X-ray structure by Helen Berman, Rutgers University, and Barbara Brodsky, UMDNJ—Robert Wood Johnson Medical School. PDBid 1CAG.] **See Kinemage Exercises 4-3 and 4-4**

Figure 8-30 A schematic diagram showing the interchain hydrogen bonding (*dashed lines*) in the Gly-containing regions of the triple helix. This is a cylindrical projection with Chain 1 repeated on the right for clarity. Note that the three chains are each vertically staggered by one residue so that a Gly, a Pro, and a Hyp from the three different chains occur on the same level. [After Bella, J., Eaton, M., Brodsky, B., and Berman, H.M., *Science* **266**, 78 (1994).] **See Kinemage Exercises 4-3 and 4-4**

Chain 1   Chain 2   Chain 3   Chain 1

an X (Pro) residue on a neighboring chain. The bulky and relatively inflexible Pro and Hyp residues confer rigidity on the entire assembly.

Collagen's well-packed, rigid, triple helical structure is responsible for its characteristic tensile strength. As with the twisted fibers of a rope, the extended and twisted polypeptide chains of collagen convert a longitudinal tensional force to a more easily supported lateral compressional force on the almost incompressible triple helix. This occurs because the oppositely twisted directions of collagen's polypeptide chains and triple helix (Fig. 8-28) prevent the twists from being pulled out under tension (note that successive levels of fiber bundles in ropes and cables are likewise oppositely twisted). The successive helical hierarchies in other fibrous proteins exhibit similar alternations of twist directions, for example, keratin (Section 8-2A) and muscle (Section 35-3Aa).

b. Collagen Is Organized into Fibrils

Types I, II, III, V, and XI collagens form distinctive banded fibrils (Fig. 8-31) that are mostly, if not entirely, composed of several different types of collagens. These fibrils have a periodicity of ~670 Å and a diameter of 100 to 2000 Å depending on the types of collagen they contain and their tissue of origin (the other collagen types form different sorts of aggregates such as networks; we will not discuss them further). X-ray fiber diffraction studies reveal that the molecules in fibrils of Type I collagen are packed in a hexagonal array. Computerized model building studies further indicate that these collagen molecules are precisely staggered parallel to the fibril axis (Fig. 8-32). The darker

Figure 8-31 **Electron micrograph of collagen fibrils from skin.** [Courtesy of Jerome Gross, Massachusetts General Hospital.]

portions of the banded structures correspond to the 400-Å "holes" on the surface of the fibril between head-to-tail aligned collagen molecules. Structural and energetic considerations suggest that the conformations of individual collagen molecules, much like those of individual α helices and β sheets, are but marginally stable (Section 8-4). The driving force for the assembly of collagen molecules into a fibril is apparently provided by the added hydrophobic interactions within the fibrils in a manner analogous to the packing of secondary structural elements to form a globular protein (Section 8-3B).

Collagen contains covalently attached carbohydrates in amounts that range from ~0.4 to 12% by weight, depending on the collagen's tissue of origin. The carbohydrates, which consist mostly of glucose, galactose, and their disaccharides, are covalently attached to collagen at its Hyl residues by specific enzymes:

Although the function of carbohydrates in collagen is unknown, the observation that they are located in the "hole" regions of the collagen fibrils suggests that they are involved in directing fibril assembly.

c. Collagen Fibrils Are Covalently Cross-Linked

Collagen's insolubility in solvents that disrupt hydrogen bonding and ionic interactions is explained by the observation that it is both intramolecularly and intermolecularly covalently cross-linked. The cross-links cannot be disulfide bonds, as in keratin, because collagen is almost devoid of Cys residues. Rather, they are derived from Lys and His side chains in reactions such as those in Fig. 8-33. **Lysyl oxidase,** a Cu-containing enzyme that converts Lys residues to those of the aldehyde **allysine,** is the only enzyme implicated in this cross-linking process. Up to four side chains can be covalently bonded to each other. The cross-links do not form at random but, instead, tend to occur near the N- and C-termini of the collagen molecules.

The importance of cross-linking to the normal functioning of collagen is demonstrated by the disease **lathyrism,**

Collagen molecule

Packing of molecules

Hole zone ———— Overlap zone
0.6D   0.4D

Figure 8-32 **Banded appearance of collagen fibrils.** The banded appearance in the electron microscope arises from the schematically represented staggered arrangement of collagen molecules (*top*) that results in a periodically indented surface. *D*, the distance between cross striations, is ~670 Å, so the length of a 3000-Å-long collagen molecule is 4.4*D*. [Courtesy of Karl A. Piez, Collagen Corporation.]

which occurs in humans and other animals as a result of the regular ingestion of seeds from the sweet pea *Lathyrus odoratus*. The symptoms of this condition are serious abnormalities of the bones, joints, and large blood vessels, which are caused by an increased fragility of the collagen fibers. The causative agent of lathyrism, **β-aminopropionitrile,**

$$N \equiv C-CH_2-CH_2-NH_3^+$$

β-Aminopropionitrile

inactivates lysyl oxidase by covalently binding to its active site. This results in markedly reduced cross-linking in the collagen of lathrytic animals.

Figure 8-33 **A biosynthetic pathway for cross-linking Lys, Hyl, and His side chains in collagen.** The first step in the reaction is the lysyl oxidase–catalyzed oxidative deamination of Lys to form the aldehyde allysine. Two such aldehydes then undergo an aldol condensation to form **allysine aldol.** This product can react with His to form **aldol histidine,** which in turn, can react with Hyl to form a Schiff base (an imine bond), thereby cross-linking four side chains.

Table 8-3   **The Arrangement of Collagen Fibrils in Various Tissues**

Tissue	Arrangement
Tendon	Parallel bundles
Skin	Sheets of fibrils layered at many angles
Cartilage	No distinct arrangement
Cornea	Planar sheets stacked crossways so as to minimize light scattering

Figure 8-34   Distorted structure in abnormal collagen. This schematic diagram shows hydrogen bonding interactions in the Ala-containing portions of the X-ray structure of (Pro-Hyp-Gly)$_{10}$ in which the fifth Gly is replaced by Ala. This cylindrical projection is in the style of Fig. 8-30. Note how the Ala side chains (*large green balls*) distort the triple helix so as to disrupt the normally occurring Gly NH $\cdots$ Pro O hydrogen bonds and replace them with water-bridged hydrogen bonds. [After Bella, J., Eaton, M., Brodsky, B., and Berman, H.M., *Science* **266**, 78 (1994).]

The degree of cross-linking of the collagen from a particular tissue increases with the age of the animal. This is why meat from older animals is tougher than that from younger animals. In fact, individual molecules of collagen (called **tropocollagen**) can only be extracted from the tissues of very young animals. Collagen cross-linking is not the central cause of aging, however, as is demonstrated by the observation that lathyrogenic agents do not slow the aging process.

The collagen fibrils in various tissues are organized in ways that largely reflect the functions of the tissues (Table 8-3). Thus tendons (the "cables" connecting muscles to bones), skin (a tear-resistant outer fabric), and cartilage (which has a load-bearing function) must support stress in predominantly one, two, and three dimensions, respectively, and their component collagen fibrils are arrayed accordingly. How collagen fibrils are laid down in these arrangements is unknown. However, some of the factors guiding collagen molecule assembly are discussed in Sections 32-5Aa and 32-5Ba.

d. Collagen Defects Are Responsible for a Variety of Human Diseases

Numerous heritable disorders of collagen are known. Mutations of Type I collagen, which constitutes the major structural protein in most human tissues, usually result in **osteogenesis imperfecta (brittle bone disease).** The severity of this disease varies with the nature and position of the mutation: Even a single amino acid change can have lethal consequences. Mutations may affect the structure of the collagen molecule or how it forms fibrils. For example, the substitution of Ala for the central Gly in each polypeptide chain of the structure shown in Fig. 8-29a, which reduces the denaturation temperature of this model compound from 62°C to 29°C, locally distorts the collagen triple helix. The need to accommodate the three additional methyl groups in the tightly packed interior of the triple helix pries apart the polypeptide chains in the region of the substitutions so as to rupture the hydrogen bonds that would otherwise link the main chain N—H group of each Ala (normally Gly) to the carbonyl oxygen of the adjacent Pro in a neighboring chain (Fig. 8-34). Rather, these hydrogen bonding groups are bridged by water molecules that insinuate themselves into the distorted part of the structure. Similar distortions almost certainly occur in the Gly → X mutated collagens responsible for such diseases as osteoge-

nesis imperfecta. Such mutations tend to be dominant because they affect either the folding of the triple helix or fibril formation even when normal chains are also members of the triple helix. All known amino acid changes within Type I collagen's triple helical region result in abnormalities, indicating that the structural integrity of this region is essential for proper collagen function.

Many collagen disorders are characterized by deficiencies in the amount of a particular collagen type synthesized or by abnormal activities of collagen-processing enzymes such as lysyl hydroxylase or lysyl oxidase. One group of at least 10 different collagen deficiency diseases, the **Ehlers–Danlos syndromes,** are all characterized by hyperextensibility of the joints (really the ligaments holding them together) and skin. This is because these tissues also contain large amounts of **elastin,** a protein with rubberlike elastic properties. Consequently, the loss of the rigidity normally conferred by collagen coupled with the presence of elastin results in the hyperextensibility of the affected tissues. Several degenerative diseases exhibit collagen abnormalities in certain tissues, including cartilage in **osteoarthritis** and the fibrous **atherosclerotic plaques** in human arteries.

3 GLOBULAR PROTEINS

Globular proteins comprise a highly diverse group of substances that, in their native states, exist as compact spheroidal molecules. Enzymes are globular proteins, as are transport and receptor proteins. In this section we consider the tertiary structures of globular proteins. However, since most of our detailed structural knowledge of proteins, and thus to a large extent their function, has resulted from X-ray crystal structure determinations of globular proteins and, more recently, from their nuclear magnetic resonance (NMR) structure determinations, we begin this section with a discussion of the capabilities and limitations of these powerful techniques.

A. *Interpretation of Protein X-Ray and NMR Structures*

X-ray crystallography is a technique that directly images molecules. X-rays must be used to do so because, according to optical principles, the uncertainty in locating an object is approximately equal to the wavelength of the radiation used to observe it (covalent bond distances and the wavelengths of the X-rays used in structural studies are both ~1.5 Å; individual molecules cannot be seen in a light microscope because visible light has a minimum

Figure 8-35   X-ray diffraction photograph of a single crystal of sperm whale myoglobin. The intensity of each diffraction maximum (the darkness of each spot) is a function of the myoglobin crystal's electron density. The photograph contains only a small fraction of the total diffraction information available from a myoglobin crystal. [Courtesy of John Kendrew, Cambridge University, U.K.]

wavelength of 4000 Å). There is, however, no such thing as an X-ray microscope because there are no X-ray lenses. Rather, a crystal of the molecule to be imaged is exposed to a collimated beam of X-rays and the consequent diffraction pattern is recorded by a radiation detector or, now infrequently, on photographic film (Fig. 8-35). The X-rays used in structural studies are produced by laboratory X-ray generators or, increasingly often, by **synchrotrons,** a type of particle accelerator that produces X-rays of far greater intensity. The intensities of the diffraction maxima (darkness of the spots on a film) are then used to construct mathematically the three-dimensional image of the crystal structure through methods that are beyond the scope of this text. In what follows, we discuss some of the special problems associated with interpreting the X-ray crystal structures of proteins.

X-rays interact almost exclusively with the electrons in matter, not the far more massive nuclei. An X-ray structure is therefore an image of the **electron density** of the object under study. Such **electron density maps** may be presented as a series of parallel sections through the object. On each section, the electron density is represented by contours (Fig. 8-36*a*) in the same way that altitude is represented by the contours on a topographic map. A stack of such sections, drawn on transparencies, yields a three-dimensional electron density map (Fig. 8-36*b*). Modern structural analysis, however, is carried out with the aid of computers that graphically display electron density maps that are contoured in three dimensions (Fig. 8-36*c*).

a. Most Protein Crystal Structures Exhibit Less than Atomic Resolution

The molecules in protein crystals, as in other crystalline substances, are arranged in regularly repeating three-dimensional lattices. Protein crystals, however, differ from those of most small organic and inorganic molecules in being highly hydrated; they are typically 40 to 60% water by volume. The aqueous solvent of crystallization is necessary for the structural integrity of the protein crystals, as J.D. Bernal and Dorothy Crowfoot Hodgkin first noted in 1934 when they carried out the original X-ray studies of protein crystals. This is because water is required for the structural integrity of native proteins themselves (Section 8-4).

The large solvent content of protein crystals gives them a soft, jellylike consistency, so that their molecules often lack the rigid order characteristic of crystals of small molecules such as NaCl or glycine. The molecules in a protein crystal are typically disordered by more than an angstrom, so that the corresponding electron density map lacks information concerning structural details of smaller size. The crystal is therefore said to have a resolution limit of that size. Protein crystals typically have resolution limits in the range 1.5 to 3.0 Å, although some are better ordered (have higher resolution, that is, a lesser resolution limit) and many are less ordered (have lower resolution).

Since an electron density map of a protein must be interpreted in terms of its atomic positions, the accuracy and

(a)   (b)

Figure 8-36 **Electron density maps of proteins.** (*a*) A section through the 2.0-Å-resolution electron density map of sperm whale myoglobin, which contains the heme group (*red*). The large peak at the center of the map represents the electron-dense Fe atom. [After Kendrew, J.C., Dickerson, R.E., Strandberg, B.E., Hart, R.G., Davies, D.R., Phillips, D.C., and Shore, V.C., *Nature* **185,** 434 (1960).] (*b*) A portion of the 2.4-Å-resolution electron density map of myoglobin constructed from a stack of contoured transparencies. Dots have been placed at the positions deduced for the nonhydrogen atoms. The heme group is seen edge-on together with its two associated His residues and a water molecule, W. An α helix, the so-called E helix (Fig. 8-12), extends across the bottom of the map. Another α helix, the C helix, extends into the plane of the paper on the upper right. Note the hole along its axis. [Courtesy of John Kendrew, Cambridge University, U.K.] (*c*) A thin section through the 1.5-Å-resolution electron density map of *E. coli* 6-hydroxymethyl-7,8-dihydropterin pyrophosphokinase (which catalyzes the first reaction in the biosynthesis of folic acid; Section 26-4D) contoured in three dimensions. Only a single contour level (*cyan*) is shown, together with an atomic model of the corresponding polypeptide segments colored according to atom type (C yellow, O red, and N blue, with a water molecule represented by a red sphere). [Courtesy of Xinhua Ji, NCI–Frederick Cancer Research and Development Center, Frederick, Maryland.]

(c)

even the feasibility of a crystal structure analysis depends on the crystal's resolution limit. Indeed, the ability to obtain crystals of sufficiently high resolution is a major limiting factor in determining the X-ray crystal structure of a protein or other macromolecule. Figure 8-37 indicates how the quality (degree of focus) of an electron density map varies with its resolution limit. At 6-Å resolution, the presence of a molecule the size of diketopiperazine is difficult to discern. At 2.0-Å resolution, its individual atoms cannot yet be distinguished, although its molecular shape has be-

come reasonably evident. At 1.5-Å resolution, which roughly corresponds to a bond distance, individual atoms become partially resolved. At 1.1-Å resolution, atoms are clearly visible.

Most protein crystal structures are too poorly resolved for their electron density maps to reveal clearly the positions of individual atoms (e.g., Fig. 8-36). Nevertheless, the distinctive shape of the polypeptide backbone usually permits it to be traced, which, in turn, allows the positions and orientations of its side chains to be deduced (e.g., Fig. 8-37*c*).

(a) 6.0-Å resolution   **(b) 2.0-Å resolution**   **(c) 1.5-Å resolution**   **(d) 1.1-Å resolution**

Figure 8-37 **Sections through the electron density map of diketopiperazine calculated at the indicated resolution levels.** Hydrogen atoms are not apparent in this map because of their low electron density. [After Hodgkin, D.C., *Nature* **188,** 445 (1960).]

Yet side chains of comparable size and shape, such as those of Leu, Ile, and Thr, cannot always be differentiated with a reasonable degree of confidence (hydrogen atoms, having but one electron, are only visible in the few macromolecular X-ray structures with resolution limits less than ~1.2 Å), so that a protein structure cannot be elucidated from its electron density map alone. Rather, the primary structure of the protein must be known, thereby permitting the sequence of amino acid residues to be fitted to its electron density map. Mathematical refinement can then reduce the uncertainty in the crystal structure's atomic positions to as little as 0.1 Å (in contrast, positional errors in the most accurately determined small molecule X-ray structures are as little as 0.001 Å).

b. Most Crystalline Proteins Maintain Their Native Conformations

What is the relationship between the structure of a protein in a crystal and that in solution, where globular proteins normally function? Several lines of evidence indicate that *crystalline proteins assume very nearly the same structures that they have in solution:*

1. A protein molecule in a crystal is essentially in solution because it is bathed by the solvent of crystallization over all of its surface except for the few, generally small patches that contact neighboring protein molecules. In fact, the 40 to 60% water content of typical protein crystals is similar to that of many cells (e.g., see Fig. 1-13).

2. A protein may crystallize in one of several forms or "habits," depending on crystallization conditions, that differ in how the protein molecules are arranged in space relative to each other. In the numerous cases in which different crystal forms of the same protein have been independently analyzed, the molecules have virtually identical conformations. Similarly, in the several cases in which both the X-ray crystal structure and the solution NMR structure of the same protein have been determined, the

two structures are, for the most part, identical to within experimental error (see below). Evidently, crystal packing forces do not greatly perturb the structures of protein molecules.

3. The most compelling evidence that crystalline proteins have biologically relevant structures, however, is the observation that many enzymes are catalytically active in the crystalline state. The catalytic activity of an enzyme is very sensitive to the relative orientations of the groups involved in binding and catalysis (Chapter 15). Active crystalline enzymes must therefore have conformations that closely resemble their solution conformations.

c. Protein Structure Determination by NMR

The determination of the three-dimensional structures of small globular proteins in aqueous solution has become possible, since the mid 1980s, through the development of **two-dimensional (2D) NMR spectroscopy** (and, more recently, of 3D and 4D techniques), in large part by Kurt Wüthrich. Such NMR measurements, whose description is beyond the scope of this text, yield the interatomic distances between specific protons that are <5 Å apart in a protein of known sequence. The interproton distances may be either through space, as determined by nuclear Overhauser effect spectroscopy (NOESY, Fig. 8-38*a*), or through bonds, as determined by correlated spectroscopy (COSY). These distances, together with known geometric constraints such as covalent bond distances and angles, group planarity, chirality, and van der Waals radii, are used to compute the protein's three-dimensional structure. However, since interproton distance measurements are imprecise, they are insufficient to imply a unique structure. Rather, they are consistent with an ensemble of closely related structures. Consequently, an NMR structure of a protein (or any other macromolecule with a well-defined structure) is often presented as a representative sample of structures that are consistent with the constraints (e.g., Figure 8-38*b*). The "tightness" of a bundle of such structures is

(a)

(b)

Figure 8-38   **The 2D proton NMR structures of proteins.** (*a*) A NOESY spectrum of a protein presented as a contour plot with two frequency axes, ω_1 and ω_2. The conventional 1D-NMR spectrum of the protein, which occurs along the diagonal of the plot ($\omega_1 = \omega_2$), is too crowded with peaks to be directly interpretable (even a small protein has hundreds of protons). The off-diagonal peaks, the so-called cross peaks, each arise from the interaction of two protons that are <5 Å apart in space and whose 1D-NMR peaks are located where the horizontal and vertical lines through the cross peak intersect the diagonal [a **nuclear Overhauser effect (NOE)**]. For example, the line to the left of the spectrum represents the extended polypeptide chain with its N- and C-terminal ends identified by the letters N and C and with the positions of four protons, a to d, represented by small circles. The dashed arrows indicate the diagonal NMR peaks to which these protons give rise. Cross peaks, such as i, j, and k, which are each located at the intersections of the horizontal and vertical lines through two diagonal peaks, are indicative of an NOE between the corresponding two protons, indicating that they are <5 Å apart. These distance relationships are schematically indicated by the three looped structures drawn below the spectrum. Note that the assignment of a distance relationship between two protons in a polypeptide requires that the NMR peaks to which they give rise and their positions in the polypeptide be known, which requires that the polypeptide's amino acid sequence has been previously determined. [After Wüthrich, K., *Science* **243**, 45 (1989).] (*b*) The NMR structure of a 64-residue polypeptide comprising the **Src protein SH3 domain** (Section 19-3C). The drawing represents 20 superimposed structures that are consistent with the 2D- and 3D-NMR spectra of the protein (each calculated from a different, randomly generated starting structure). The polypeptide backbone, as represented by its connected C_α atoms, is white and its Phe, Tyr, and Trp side chains are yellow, red, and blue, respectively. It can be seen that the polypeptide backbone folds into two three-stranded antiparallel β sheets that form a sandwich. [Courtesy of Stuart Schreiber, Harvard University.]

indicative both of the accuracy with which the structure is known, which in the most favorable cases is roughly comparable to that of an X-ray crystal structure with a resolution of 2 to 2.5 Å, and of the conformational fluctuations that the protein undergoes (Section 9-4). Although present NMR methods are limited to determining the structures of macromolecules with molecular masses no greater than ~100 kD, advances in NMR technology suggest that this limit may increase to ~1000 kD or more.

In most of the several cases in which both the NMR and X-ray crystal structures of a particular protein have been determined, the two structures are in good agreement. There are, however, a few instances in which there are real differences between the corresponding X-ray and NMR structures. These, for the most part, involve surface residues that, in the crystal, participate in inter-

molecular contacts and are thereby perturbed from their solution conformations. NMR methods, besides providing mutual cross-checks with X-ray techniques, can determine the structures of proteins and other macromolecules that fail to crystallize. Moreover, since NMR can probe motions over time scales spanning 10 orders of magnitude, it can be used to study protein folding and dynamics (Chapter 9).

d. Protein Molecular Structures Are Most Effectively Illustrated in Simplified Form

The several hundred nonhydrogen atoms of even a small protein makes understanding a protein's detailed structure a considerable effort. This complexity makes building a skeletal or ball-and-stick model of a protein such a time-consuming task that such models are rarely available. Moreover, a drawing of a protein showing all its nonhydrogen atoms (e.g., Fig. 8-39*a*) is too complicated to be of much use. In order to be intelligible, drawings of proteins must be selectively simplified. One way of doing so is to represent the polypeptide backbone only by its C_α atoms (its **C_α backbone**) and to display only a few key side chains (e.g., Fig. 8-39*b*). A further level of abstraction may be obtained by representing the protein in a cartoon form that emphasizes its secondary structure (e.g., Figs. 8-39c and 8-19). Computer-generated drawings of space-filling models, such as Figs. 8-12 and 8-18, may also be em-ployed to illustrate certain features of protein structures. However, the most instructive way to examine a macro-molecular structure is through the use of interactive computer graphics programs. The use of such programs is discussed in Section 8-3Cc.

B. *Tertiary Structure*

The **tertiary structure (3° structure)** of a protein is its three-dimensional arrangement; that is, the folding of its 2° structural elements, together with the spatial dispositions of its side chains. The first protein X-ray structure, that of sperm

(a)

Figure 8-39 Representations of the X-ray structure of sperm whale myoglobin. (*a*) The protein and its bound heme are drawn in stick form, with protein C atoms green, heme C atoms red, N atoms blue, and O atoms red. The Fe and its bound water molecule are shown as orange and gray spheres and hydrogen bonds are gray. In this one-of-a-kind painting of the first known protein structure, the artist has employed "creative distortions" to emphasize the protein's structural features, particularly its α helices. (*b*) A diagram in which the protein is represented by its computer-generated Cα backbone, with its Cα atoms, shown as balls, consecutively numbered from the N-terminus. The 153-residue polypeptide chain is folded into eight α helices (highlighted here by hand-drawn envelopes), designated A through H, that are connected by short polypeptide links. The protein's bound heme group (*purple,* with its Fe atom represented by a red sphere), in complex with a water molecule (*orange sphere*), is shown together with its two closely associated His side chains (*blue*). One of the heme group's propionic acid side chains has been displaced for clarity. Hydrogen atoms are not visible in the X-ray structure. (*c*) A computer-generated cartoon drawing in an orientation similar to that of Part *b*, emphasizing the protein's secondary structure. Here helices are green and the intervening coil regions are yellow. The heme group with its bound O₂ molecule and its two associated His side chains are shown in ball-and-stick form with C magenta, N blue, O red, and Fe orange. [Parts *a* and *b* are based on an X-ray structure by John Kendrew, MRC Laboratory of Molecular Biology, Cambridge, U.K. PDBid 1MBN. [Illustration, Irving Geis. Image from the Irving Geis Collection, Howard Hughes Medical Institute. Reprinted with permission.] Part *c* is based on an X-ray structure by Simon Phillips, University of Leeds, Leeds, U.K. PDBid 1MBO.] 🖱️**See Kinemage Exercise 6-1**

(b)

(c)

whale **myoglobin,** was elucidated in the late 1950s by John Kendrew and co-workers. Its polypeptide chain follows such a tortuous, wormlike path (Fig. 8-39) that these investigators were moved to indicate their disappointment at its lack of regularity. In the intervening years, ~70,000 protein structures have been reported. Each of them is a unique, highly complicated entity. Nevertheless, their tertiary structures have several outstanding features in common, as we shall see below.

a. Globular Proteins May Contain Both α Helices and β Sheets

The major types of secondary structural elements, α helices and β pleated sheets, commonly occur in globular proteins but in varying proportions and combinations. Some proteins, such as myoglobin, consist only of α helices spanned by short connecting links that have coil conformations (Fig. 8-39). Others, such as concanavalin A, have a large proportion of β sheets but are devoid of α helices (Fig. 8-40). Most proteins, however, have significant amounts of both types of secondary structure (on average, ~31% α helix and ~28% β sheet, with nearly all of their inner cores so arranged). Human **carbonic anhydrase** (Fig. 8-41) as well as carboxypeptidase A and triose phosphate isomerase (Fig. 8-19) are examples of such proteins.

b. Side Chain Location Varies with Polarity

The primary structures of globular proteins generally lack the repeating or pseudorepeating sequences that are responsible for the regular conformations of fibrous proteins. The amino acid side chains in globular proteins are, nevertheless, spatially distributed according to their polarities:

1. *The nonpolar residues Val, Leu, Ile, Met, and Phe largely occur in the interior of a protein, out of contact with the aqueous solvent.* The hydrophobic interactions that promote this distribution, which are largely responsible for the three-dimensional structures of native proteins, are further discussed in Section 8-4C.

2. *The charged polar residues Arg, His, Lys, Asp, and Glu are largely located on the surface of a protein in contact with the aqueous solvent.* This is because the immersion of an ion in the virtually anhydrous interior of a protein results in the uncompensated loss of much of its hydration energy. In the instances that these groups occur in the interior of a protein, they often have a specific chemical function such as promoting catalysis or participating in metal ion binding (e.g., the metal ion–ligrading His residues in Figs. 8-39 and 8-41).

3. The uncharged polar groups Ser, Thr, Asn, Gln, Tyr, and Trp are usually on the protein surface but frequently

Figure 8-40 **X-ray structure of the jack bean protein concanavalin A.** This protein largely consists of antiparallel β pleated sheets, here represented by flat arrows pointing toward the polypeptide chain's C-terminus. The polypeptide chain is colored in rainbow order from its N-terminus (*blue*) to its C-terminus (*red*). The spheres represent protein-bound Mn^{2+} (*magenta*) and Ca^{2+} (*cyan*) ions. The front sheet is shown in a space-filling representation in Fig. 8-18 but viewed from the opposite side as seen here. [Based on an X-ray structure by George Reeke, Jr., Joseph Becker, and Gerald Edelman, The Rockefeller University. PDBid 2CNA.]

Figure 8-41 **X-ray structure of human carbonic anhydrase.** The polypeptide backbone is drawn in ribbon form colored in rainbow order from its N-terminus (*blue*) to its C-terminus (*red*). The purple sphere in the center represents a Zn^{2+} ion that is coordinated by three His side chains in stick form colored with C yellow and N blue. Note that the C-terminus is tucked through the plane of a surrounding loop of polypeptide chain, so that carbonic anhydrase is one of the rare native proteins in which a polypeptide chain forms a knot. [Based on an X-ray structure by T. Alwyn Jones, Uppsala University, Uppsala, Sweden. PDBid 2CAB.] **See Interactive Exercise 3**

(a)

(b)

Figure 8-42 **X-ray structure of horse heart cytochrome *c*.** The protein (*blue*) is illuminated by the Fe atom of its heme group (*orange*). In Part *a*, the hydrophobic side chains are red, and in Part *b*, the hydrophilic side chains are green. [Based on an X-ray structure by Richard Dickerson, UCLA; [Illustration, Irving Geis. Image from the Irving Geis Collection, Howard Hughes Medical Institute. Reprinted with permission.] **See Kinemage Exercise 5 and Interactive Exercise 4**

occur in the interior of the molecule. In the latter case, these residues are almost always hydrogen bonded to other groups in the protein. In fact, *nearly all buried hydrogen bond donors form hydrogen bonds with buried acceptor groups;* in a sense, the formation of a hydrogen bond "neutralizes" the polarity of a hydrogen bonding group.

Such a side chain distribution is clearly apparent in Fig. 8-42, which displays the X-ray structure of cytochrome *c*. This arrangement is also seen in Fig. 8-43, which shows the surface and interior exposures of the amino acid side chains of myoglobin's H helix, and in Fig. 8-44, which shows one of the antiparallel β pleated sheets of concanavalin A.

c. Globular Protein Cores Are Efficiently Arranged with Their Side Chains in Relaxed Conformations

Globular proteins are quite compact; there is very little space inside them, so that water is largely excluded from their interiors. The micellelike arrangement of their side chains (polar groups outside, nonpolar groups inside) has led to their description as "oil drops with polar coats." This generalization, although picturesque, lacks precision. The **packing density** (ratio of the volume enclosed by the van der Waals envelopes of the atoms in a region to the total volume of the region) of the internal regions of globular proteins averages ~0.75, which is in the same range as that of molecular crystals of small organic molecules. In comparison, equal-sized close-packed spheres have a packing density of 0.74, whereas organic liquids (oil drops) have packing densities that are mostly between 0.60 and 0.70. *The interior of a protein is therefore more like a molecular crystal than an oil drop; that is, it is efficiently packed.*

The bonds of protein side chains, including those occupying protein cores, almost invariably have low-energy staggered torsion angles (Fig. 8-5*a*). Evidently, interior side chains adopt relaxed conformations despite their profusion of intramolecular interactions (Section 8-4).

d. Large Polypeptides Form Domains

Polypeptide chains that consist of more than ~200 residues usually fold into two or more globular clusters known as **domains,** which often give these proteins a bi- or multilobal appearance. Most domains consist of 100 to 200 amino acid residues and have an average diameter of ~25 Å.

(a)

Residues
exposed
to surface

(c)

(b)

Figure 8-43   **The H helix of sperm whale myoglobin.** (*a*) A **helical wheel** representation in which side chain positions about the α helix are projected down the helix axis onto a plane. Here each residue is identified both according to its sequence in the polypeptide chain and according to its order in the H helix. The residues lining the side of the helix facing the protein's interior regions are all nonpolar (*orange*). The other residues, except Leu 137, which contacts the protein segment linking helices E and F (Fig. 8-39*b*), are exposed to the solvent and are all more or less polar (*purple*). (*b*) A stick model, viewed as in Part *a*, in which the polypeptide backbone is gray, nonpolar side chains are orange, and polar side chains are purple. (*c*) A space-filling model, viewed from the bottom of Part *b* such that the helix axis is vertical and colored as in Part *b*. [Parts *b* and *c* based on an X-ray structure by Ilme Schlichting, Max Planck Institut für Molekulare Physiologie, Dortmund, Germany. PDBid 1A6M.]
See Kinemage Exercise 3-2 for another example

Each subunit of the enzyme **glyceraldehyde-3-phosphate dehydrogenase,** for example, has two distinct domains (Fig. 8-45). A polypeptide chain wanders back and forth within a domain, but neighboring domains are usually connected by one, or less commonly two, polypeptide segments. *Domains are therefore structurally independent units that each have the characteristics of a small globular protein.* Indeed, limited proteolysis of a multidomain protein often liberates its domains without greatly altering their structures or enzymatic activities. Nevertheless, the domain structure of a protein is not always obvious since its domains may make such extensive contacts with each other that the protein appears to be a single globular entity.

An inspection of the various protein structures diagrammed in this chapter reveals that domains consist of

two or more layers of secondary structural elements. The reason for this is clear: At least two such layers are required to seal off a domain's hydrophobic core from the aqueous environment.

Domains often have a specific function, such as the binding of a small molecule. In Fig. 8-45, for example, the dinucleotide **nicotinamide adenine dinucleotide (NAD⁺)**

Nicotinamide adenine dinucleotide (NAD⁺)

binds to the first domain of the enzyme glyceraldehyde-3-phosphate dehydrogenase. Small molecule binding sites in multidomain proteins often occur in the clefts between domains; that is, the small molecules are bound by groups from two domains. This arrangement arises, in part, from

Figure 8-44   A space-filling model of an antiparallel β sheet from concanavalin A. The β sheet is shown in side view with the interior of the protein (the surface of a second antiparallel β sheet; Fig. 8-40) to the right and the exterior to the left. The main chain is gray, nonpolar side chains are orange, and polar side chains are purple. See Kinemage Exercise 3-3

Figure 8-45   One subunit of the enzyme glyceraldehyde-3-phosphate dehydrogenase from *Bacillus stearothermophilus.* The polypeptide folds into two distinct domains. The N-terminal domain (*light blue,* residues 1–146) binds NAD^+ (drawn in stick form with C green, N blue, O red, and P magenta) near the C-terminal ends of its parallel β strands, and the C-terminal domain (*orange,* residues 148–333) binds glyceraldehyde-3-phosphate (not shown). [Based on an X-ray structure by Alan Wonacott, Imperial College, London, U.K., PDBid 1GD1.] See Interactive Exercise 5

the need for a flexible interaction between the protein and the small molecule that the relatively pliant covalent connection between the domains can provide.

e. Supersecondary Structures Are the Building Blocks of Proteins

Certain groupings of secondary structural elements, named **supersecondary structures** or **motifs,** occur in many unrelated globular proteins:

1. The most common form of supersecondary structure is the **βαβ motif** (Fig. 8-46*a*), in which the usually right-handed crossover connection between two consecutive parallel strands of a β sheet consists of an α helix.

2. Another common supersecondary structure, the **β hairpin motif** (Fig. 8-46*b*), consists of an antiparallel β sheet formed by sequential segments of polypeptide chain that are connected by relatively tight reverse turns.

3. In an **αα motif** (Fig. 8-46*c*), two successive antiparallel α helices pack against each other with their axes inclined so

Figure 8-46   Schematic diagrams of supersecondary structures. (*a*) A βαβ motif, (*b*) a β hairpin motif, (*c*) an αα motif, and (*d*) a Greek key motif, showing how it is constructed from a folded-over β hairpin.

as to permit their contacting side chains to interdigitate efficiently. Such energetically favorable associations stabilize the coiled coil conformation of α keratin (Section 8-2A).

4. In the **Greek key motif** (Fig. 8-46*d;* named after an ornamental design commonly used in ancient Greece; see inset), a β hairpin is folded over to form a four-stranded antiparallel β sheet. Of the 10 possible ways of connecting the strands of a four-stranded antiparallel β sheet, the two that form Greek key motifs are, by far, the most common in proteins of known structure.

Groups of motifs combine in overlapping and nonoverlapping ways to form the tertiary structure of a domain, which is called a **fold.**

The number of possible different folds would, of course, seem to be unlimited. However, comparisons of the now large number of known protein structures have revealed that *few protein folds are unique; that is, most proteins of known structure have folds that also occur in unrelated proteins.* Indeed, theoretical considerations suggest there are less than 8000 naturally occurring folds. Of

(a)   N   C   *(b)*   N   C

Figure 8-47   **X-ray structures of 4-helix bundle proteins.** (*a*) *E. coli* cytochrome b_{562} and (*b*) human growth hormone. The proteins are represented by their peptide backbones drawn in ribbon form colored in rainbow order from N-terminus (*red*) to C-terminus (*blue*). Cytochrome b_{562}'s bound heme group is shown in ball-and-stick form with C magenta, N blue, O red, and Fe orange. The inset below each ribbon diagram is a topological diagram indicating the connectivity of the α helices in each 4-helix bundle. Cytochrome b_{562} (106 residues) has up-down-up-down topology, whereas human growth hormone (191 residues) has up-up-down-down topology. Note that the N- and C-terminal helices of human growth hormone are longer than its other two helices, so that, at one end, these longer helices associate as an αα motif. [Based on X-ray structures by (*a*) F. Scott Matthews, Washington University School of Medicine, and (*b*) Alexander Wlodawer, National Cancer Institute, Frederick, Maryland. PDBids (*a*) 256B and (*b*) 1HGU.]

these, ~1200 have already been observed, with around half of the proteins of known structure belonging to only 20 fold groups.

Although there are numerous ways in which domain structures might be categorized (see, e.g., Section 8-3C), perhaps the simplest way to do so is to classify them as **α domains** (containing secondary structural elements that are exclusively α helices), **β domains** (containing only β sheets), and **α/β domains** (containing both α helices and β sheets). The **α/β domain** category may be further divided into two main groups: **α/β barrels** and **open β sheets.** In the following paragraphs we describe some of the most common folds in each of these domain categories.

f. α Domains

We are already familiar with a fold that contains only α helices: the **globin fold,** which contains 8 helices in two layers and occurs in myoglobin (Fig. 8-39) and in both the α and β chains of hemoglobin (Section 10-2B). In another common all-α fold, two αα motifs combine to form a **4-helix bundle** such as occurs in **cytochrome b_{562}** (Fig. 8-47a, opposite). The helices in this fold are inclined such that their contacting side chains intermesh and are therefore out of contact with the surrounding aqueous solution. They are consequently largely hydrophobic. The 4-helix bundle is a relatively common fold that occurs in a variety of proteins. However, not all of them have the up-down-up-down topology (connectivity) of cytochrome b_{562}. For example, human **growth hormone** is a 4-helix bundle with up-up-down-down topology (Fig. 8-47b). The successive parallel

helices in this fold are, of necessity, joined by longer loops than those joining successive antiparallel helices.

Different types of α domains occur in **transmembrane proteins.** We study these proteins in Section 12-3A.

g. β Domains

β Domains contain from 4 to >10 predominantly antiparallel β strands that are arranged into two sheets that pack against each other to form a **β sandwich.** For example, the **immunoglobulin fold** (Fig. 8-48), which forms the basic domain structure of most immune system proteins (Section 35-2Be), consists of a 4-stranded antiparallel β sheet in face-to-face contact with a 3-stranded antiparallel β sheet. Note that the strands in the two sheets are not parallel to one another, a characteristic of stacked β sheets. The side chains between the two stacked β sheets are out of contact with the aqueous medium and thereby form the domain's hydrophobic core. Since successive residues in a β strand alternately extend to opposite sides of the β sheet (Fig. 8-17), these residues are alternately hydrophobic and hydrophilic.

The inherent curvature of β sheets (Section 8-1C) often causes sheets of more than 6 strands to roll up into **β barrels.** Indeed, β sandwiches may be considered to be flattened β barrels. Several different β barrel topologies have been observed, most commonly:

1. The **up-and-down β barrel,** which consists of 8 successive antiparallel β strands that are arranged like the staves of a barrel. An example of an up-and-down β barrel

Figure 8-48 **The immunoglobulin fold.** The X-ray structure of the N-terminal domain of the human immunoglobulin fragment **Fab New** shows its immunoglobulin fold. The peptide backbone of this 103-residue domain is drawn in ribbon form colored in rainbow order from N-terminus (*red*) to C-terminus (*blue*) with its β strands represented by flat arrows pointing toward the C-terminus. The inset is the topological diagram of the immunoglobulin fold showing the connectivity of its stacked 4-stranded and 3-stranded antiparallel β sheets. [Based on an X-ray structure by Roberto Poljak, Johns Hopkins School of Medicine. PDBid 7FAB.]

Figure 8-49   **Retinol binding protein.** Its X-ray structure shows its up-and-down β barrel (residues 1–142 of this 182-residue protein). Its peptide backbone is drawn in ribbon form colored in rainbow order from N-terminus (*red*) to C-terminus (*blue*). Note that each β strand is linked via a short loop to its clockwise-adjacent strand as seen from the top. The protein's bound retinol molecule is represented by a gray ball-and-stick model. The inset is the protein's topological diagram. [Based on an X-ray structure by T. Alwyn Jones, Uppsala University, Uppsala, Sweden. PDBid 1RBP.]

occurs in **retinol binding protein** (Fig 8-49), which functions to transport the nonpolar visual pigment precursor **retinol (vitamin A)** in the bloodstream:

Retinol

2. A fold consisting of two Greek key motifs, which thereby constitutes an alternative way of connecting the strands of an 8-stranded antiparallel β barrel. Figure 8-50 indicates how two Greek key motifs in the C-terminal domain of the eye lens protein **γ-B crystallin** are arranged to form a β barrel.

3. The **jelly roll** or **Swiss roll barrel** (which is named for its topological resemblance to these rolled up pastries), in which a 4-segment β hairpin is rolled up into an 8-stranded antiparallel β barrel of yet different topology, as is diagrammed in Fig. 8-51a. The X-ray structure of the enzyme **peptide-N^4-(N-acetyl-β-D-glucosaminyl)asparagine amidase F** contains a domain consisting of a jelly roll barrel (Fig. 8-51b).

h. α/β Barrels

In α/β domains, a central parallel or mixed β sheet is flanked by α helices. The α/β barrel, which is diagrammed in Fig. 8-19b, is a remarkably regular structure that consists of 8 tandem βα units (essentially 8 overlapping βαβ motifs) wound in a right-handed helical sense to form an inner 8-stranded parallel β barrel concentric with an outer barrel of 8 α helices. Each β strand is approximately antiparallel to the succeeding α helix and all are inclined at around the same angle to the barrel axis. Figure 8-52 shows the X-ray structure of chicken **triose phosphate isomerase (TIM),** determined by David Phillips, which consists of an α/β barrel. This is the first known structure of an α/β barrel, which is therefore also called a **TIM barrel.**

The side chains that point inward from the α helices interdigitate with the side chains that point outward from the β strands. A large fraction (~40%) of these side chains are those of the branched aliphatic residues Ile, Leu, and Val. The side chains that point inward from the β strands tend to be bulky and hence fill the core of the β barrel (contrary to the impression that Figs. 8-19b and 8-52 might provide, α/β barrels, with one known exception, do not have hollow cores). Those side chains that fill the ends of the barrel are in contact with solvent and hence tend to be polar, whereas those in its center are out of contact with solvent and are therefore nonpolar. Thus, α/β barrels have four layers of polypeptide backbone interleaved by regions of hydrophobic

(a)   N   (b)

Figure 8-50   X-ray structure of the C-terminal domain of bovine γ-B crystallin. (*a*) A topological diagram showing how its two Greek key motifs are arranged in a β barrel. One Greek key motif is formed by β strands 1 to 4 and the other (*blue*) is formed by β strands 5 to 8. [After Branden, C. and Tooze, J., *Introduction to Protein Structure* (2nd ed.), *p.* 75, Garland (1999).] (*b*) The 83-residue peptide backbone displayed in ribbon form.

Here the members of an antiparallel pair of β strands in a Greek key motif are colored alike with the N-terminal Greek key colored red (strands 1 & 4) and orange (2 & 3) and the C-terminal Greek key colored blue (5 & 8) and cyan (6 & 7). The N-terminal domain of this two-domain protein is nearly superimposable on its C-terminal domain. [Based on an X-ray structure by Tom Blundell, Birkbeck College, London, U.K. PDBid 4GCR.]

Figure 8-51   X-ray structure of the enzyme peptide-N^4-(*N*-acetyl-β-D-glucosaminyl)asparagine amidase F from *Flavobacterium meningosepticum*. (*a*) A diagram indicating how an 8-stranded β barrel is formed by rolling up a 4-segment β hairpin. A topological diagram of the jelly roll barrel is also shown. [After Branden, C. and Tooze, J., *Introduction to Protein Structure* (2nd ed.), *pp.* 77–78, Garland (1999).] (*b*) A ribbon diagram of the domain formed by residues 1 to 140 of this 314-residue enzyme. Here the two β strands in each segment of the β hairpin are colored alike, with strands 1 & 8 (the N- and C-terminal strands) red, strands 2 & 7 orange, strands 3 & 6 cyan, and strands 4 & 5 blue. [Based on an X-ray structure by Patrick Van Roey, New York State Department of Health, Albany, New York. PDBid 1PNG.]

Figure 8-52 **X-ray structure of the 247-residue enzyme triose phosphate isomerase (TIM) from chicken muscle.** The protein is viewed approximately along the axis of its α/β barrel. The peptide backbone is shown as a ribbon diagram with its successive βα units colored, from N- to C-terminus, in rainbow order, red to blue. The inset is its topological diagram (with α helices represented by rectangles). Two other views of this protein are drawn in Figures 8-19*b* and 8-19*c* [Based on an X-ray structure by David Phillips, Oxford University, U.K. PDBid 1TIM.]

side chains. In contrast, both α domains and β domains consist of two layers of polypeptide backbone sandwiching a hydrophobic core.

Around 10% of known enzyme structures contain an α/β barrel, making it the most common fold assumed by enzymes. Moreover, nearly all known α/β barrel proteins are enzymes. Intriguingly, the active sites of α/β barrel enzymes are almost all located in funnel-shaped pockets formed by the loops that link the C-termini of the β strands to the succeeding α helices and hence surround the mouth of the β barrel, an arrangement that has no obvious structural rationale. Thus, despite the observation that few α/β barrel proteins exhibit significant sequence homology, it has been postulated that all of them descended from a common ancestor and are therefore (distantly) related by **divergent evolution.** On the other hand, it has been argued that the α/β barrel is structurally so well suited for its enzymatic roles that it independently arose several times and hence that α/β enzymes are related by **convergent evolution** (i.e., nature has discovered the same fold on several

occasions). Convincing evidence supporting either view has not been forthcoming, so that the nature of the evolutionary relationships among the α/β barrel enzymes remains an open question.

i. Open β Sheets

We have previously encountered examples of an open β sheet in the structures of carboxypeptidase A (Fig. 8-19*a*) and the N-terminal domain of glyceraldehyde-3-phosphate dehydrogenase (Fig. 8-45). Their topological diagrams are drawn in Fig. 8-53. The X-ray structures and topological diagrams of two other such proteins, those of the enzymes **lactate dehydrogenase** (N-terminal domain) and **adenylate kinase,** are shown in Fig. 8-54. Such folds consist of a central parallel or mixed β sheet flanked on both sides by α helices that form the right-handed crossover connections between successive parallel β strands (Fig. 8-20*b*). The strands of such a β sheet do not follow their order in the peptide sequence. Rather, the β sheet has a long crossover that reverses the direction of the succeeding section of sheet and turns it upside down, thereby putting its helical crossovers on the opposite side of the sheet from those of the preceding section (Fig. 8-55). Such assemblies are therefore also known as **doubly wound sheets** (in contrast to singly wound α/β barrels, whose helices are all on the same side of their β sheets). Doubly wound sheets consist of three layers of polypeptide backbone interspersed by regions of hydrophobic side chains (in contrast to four such layers for α/β barrels and two for α domains and β domains). Note that both types of domains containing parallel β sheets are hydrophobic on both sides of the sheet, whereas antiparallel sheets are hydrophobic on only one side. This additional stabilization of parallel β sheets probably compensates for the reduced strength of their nonlinear hydrogen bonds relative to the linear hydrogen bonds of antiparallel β sheets (Fig. 8-16).

Figure 8-53 **Topological diagrams of (*a*) carboxypeptidase A and (*b*) the N-terminal domain of glyceraldehyde-3-phosphate dehydrogenase.** The X-ray structures of these proteins are diagrammed in Figs. 8-19*a* and 8-45. The thin black vertical arrows mark the proteins' topological switch points.

(a)

(b)

Figure 8-54 X-ray structures of open β sheet–containing enzymes. (*a*) Dogfish lactate dehydrogenase, N-terminal domain (residues 20–163 of this 330-residue protein), and (*b*) porcine adenylate kinase (195 residues). The peptide backbones are shown as ribbon diagrams with successive βα units colored, from N- to C-terminus, in rainbow order, red to blue. In Part *b*, structural elements that are not components of the open β sheet are gray. The insets are topological diagrams of these proteins, with the thin black vertical arrows marking their topological switch points. [Based on X-ray structures by (*a*) Michael Rossmann, Purdue University, and (*b*) Georg Schulz, Institut für Organische Chemie und Biochemie, Freiburg, Germany. PDBids (*a*) 6LDH and (*b*) 3ADK.]

There are few geometric constraints on the number of strands in open β sheets; they have been observed to contain from 4 to 10 β strands, with 6 β strands being the most common. Since the position where the chain reverses its winding direction, the so-called **topological switch point,** can occur between any consecutive α helix and β strand, doubly wound sheets can have many different folds. Moreover, some β strands may run in an antiparallel direction to yield mixed sheets (e.g., carboxypeptidase A; Figs. 8-19*a* and 8-53*a*), and in several instances, there is more than one topological switch point (e.g., adenylate kinase; Fig. 8-54*b*).

The open β sheet is the most common domain structure that occurs in globular proteins. Moreover, nearly all such

Figure 8-55 Doubly wound sheets. (*a*) A schematic diagram of a doubly wound sheet indicating how the long crossover (*green*) between its N- and C-terminal segments (*red and blue*) reverses the direction of the sheet's C-terminal segment and places its α helical crossover connections on the opposite side of the sheet. (*b*) The corresponding topological diagram, with the thin black vertical arrow marking the topological switch point. [After Branden, C. and Tooze, J., *Introduction to Protein Structure* (2nd ed.), *p.* 49, Garland (1999).]

proteins are enzymes, many of which bind mono- or dinucleotides. In fact, the fold exemplified by lactate dehydrogenase (**LDH;** Fig. 8-54a) is known as the **dinucleotide-
binding fold** or **Rossmann fold** (after Michael Rossmann,
who first pointed out its significance). This is because
mononucleotide units are commonly bound by βαβαβ
units, of which LDH (which binds the dinucleotide NAD$^+$)
has two. In some proteins, the second α helix in a βαβαβ
unit is replaced by a length of nonhelical polypeptide. This
occurs, for example, between the β5 and β6 strands of
glyceraldehyde-3-phosphate dehydrogenase (Figs. 8-45
and 8-53b), which also binds NAD$^+$.

At the topological switch point of an open β sheet, the
loops emerging from the C-terminal ends of the flanking β
strands go to opposite sides of the sheet and thereby form
a crevice between them. In nearly all of the >100 known
structures of open β sheet–containing enzymes, as Carl-
Ivar Brändén first pointed out, this crevice forms at least
part of the enzyme's active site. Thus the active sites of both
α/β barrels and the open β sheet enzymes are formed by
loops emanating from the C-terminal ends of the β strands.
In contrast, the active sites of enzymes that have other
types of domain structures exhibit no apparent regularities
in the positions of their active sites.

C. *Structural Bioinformatics*

In Section 7-4, we discussed the rapidly developing field of
inquiry known as bioinformatics as it applies to the sequences of proteins and nucleic acids, that is, how sequence
alignments are determined and how phylogenetic trees are
generated. An equally important aspect of bioinformatics,

which we discuss below, is how macromolecular structures
are displayed and compared.

a. The Protein Data Bank

The atomic coordinates of most known macromolecular
structures are archived in the **Protein Data Bank (PDB).**
Indeed, most scientific journals that publish macromolecular structures require that authors deposit their structure's
coordinates in the PDB. The PDB contains the atomic coordinates of ~70,000 macromolecular structures (proteins,
nucleic acids, and carbohydrates as determined by X-ray
and other diffraction-based techniques, NMR, and electron
microscopy) and is growing exponentially at a rate that is
presently ~7500 structures per year. The PDB's Website
(URL), from which these coordinates are publicly available, is listed in Table 8-4.

Each independently determined structure in the PDB is
assigned a unique four-character identifier (its **PDBid**) in
which the first character must be a digit (1–9) and no distinction is made between uppercase and lowercase letters (e.g.,
1 MBO is the PDBid of the myoglobin structure illustrated
in Fig. 8-39c, although PDBids do not necessarily have a relationship to the corresponding macromolecule's name). A
coordinate file begins with information that identifies/de-
scribes the macromolecule, the date the coordinate file was
deposited, its source (the organism from which it was obtained), the author(s) who determined the structure, and key
journal references. The file continues with a synopsis of how
the structure was determined together with indicators of its
accuracy and information that could be helpful in its interpretation, such as a description of its symmetry and which
residues, if any, were not observed. The sequences of the

Table 8-4   **Structural Bioinformatics Websites (URLs)**

Structural Databases
Protein Data Bank (PDB): http://www.rcsb.org/
Nucleic Acid Database: http://ndbserver.rutgers.edu/
Molecular Modeling Database (MMDB): http://www.ncbi.nlm.nih.gov/Structure/MMDB/mmdb.shtml
Most Representative NMR Structure in an Ensemble: http://www.ebi.ac.uk/msd-srv/olderado
PISA (Protein Interfaces, Surfaces and Assemblies): http://www.ebi.ac.uk/msd-srv/prot_int/pistart.html
Proteopedia: http://www.proteopedia.org

Molecular Graphics Programs/Plug-ins
Cn3D: http://www.ncbi.nlm.nih.gov/Structure/CN3D/cn3d.shtml
FirstGlance: http://molvis.sdsc.edu/fgij/
Jmol: http://jmol.sourceforge.net/
KiNG: http://kinemage.biochem.duke.edu/software/king.php
Swiss-Pdb Viewer: http://spdv.vital-it.ch

Structural Classification Algorithms
CATH (class, architecture, topology, and homologous superfamily): http://www.cathdb.info/index.html
CE (combinatorial extension of optimal pathway): http://cl.sdsc.edu/
Pfam (protein families): http://pfam.sanger.ac.uk/ or http://pfam.janelia.org/
SCOP (structural classification of proteins): http://scop.mrc-lmb.cam.ac.uk/scop/
VAST (vector alignment search tool): http://www.ncbi.nlm.nih.gov/Structure/VAST/

structure's various chains are then listed together with the descriptions and formulas of its so-called HET (for heterogen) groups, which are molecular entities that are not among the "standard" amino acid or nucleotide residues (e.g., organic molecules such as the heme group, nonstandard residues such as Hyp, metal ions, and bound water molecules). The positions of the structure's secondary structural elements and its disulfide bonds are then provided.

The bulk of a PDB file consists of a series of ATOM (for "standard" residues) and HETATM (for heterogens) records (lines), each of which provides the coordinates for one atom in the structure. An ATOM or HETATM record identifies its corresponding atom according to its serial number (usually just its sequence in the list), atom name (e.g., C and O for an amino acid residue's carbonyl C and O atoms, CA and CB for C_α and C_β atoms, N1 for atom N1 of a nucleic acid base, C4* for atom C4′ of a ribose or deoxyribose residue), residue name [e.g., PHE, G (for a guanosine residue), HEM (for a heme group), MG (for an Mg^{2+} ion), and HOH (for a water molecule)], chain identifier (e.g., A, B, C, etc., for structures consisting of more than one chain, whether or not the chains are chemically identical), and the residue sequence number in the chain. The record then continues with the atom's Cartesian (orthogonal) coordinates (X, Y, Z), in angstroms relative to an arbitrary origin, the atom's occupancy (which is the fraction of sites that actually contain the atom in question, a quantity that is usually 1.00 but, for groups that have multiple conformations or for molecules/ions that are only partially bound to a protein, may be a positive number less than 1.00), and its isotropic temperature factor (a quantity that is indicative of the atom's thermal motion, with larger numbers denoting a greater degree of motion). The ATOM records are listed in the order of the residues in a chain. For NMR-based structures, the PDB file contains a full set of ATOM and HETATM records for each member of the ensemble of structures that were calculated in solving the structure (Section 8-3A; the most representative member of such a coordinate set can be obtained from http://www.ebi.ac.uk/msd-srv/olderado). PDB files usually end with CONECT (connectivity) records, which denote the nonstandard connectivities between ATOMs such as disulfide bonds and hydrogen bonds as well as connectivities between HETATMs.

A particular PDB file may be located according to its PDBid or, if this is unknown, by searching with a variety of criteria including a protein's name, its source, the author(s), keywords, and/or the experimental technique used to determine the structure. Selecting a particular macromolecule in the PDB initially displays a Structure Summary page with options for interactively viewing the structure, for viewing or downloading the coordinate file, and for classifying or analyzing the structure in terms of its geometric properties and sequence (see below).

b. The Nucleic Acid Database

The **Nucleic Acid Database (NDB)** archives the atomic coordinates of structures that contain nucleic acids. Its coordinate files have substantially the same format as do those of the PDB, where this information is also kept. However, the NDB's organization and search algorithms are specialized for dealing with nucleic acids. This is useful, in part, because many nucleic acids of known structure are identified only by their sequences rather than by names, as are proteins (e.g., myoglobin), and consequently could easily be overlooked in a search of the PDB.

c. Viewing Macromolecular Structures in Three Dimensions

The most informative way to examine a macromolecular structure is through the use of molecular graphics programs that permit the user to interactively rotate a macromolecule and thereby perceive its three-dimensional structure. This impression may be further enhanced by simultaneously viewing the macromolecule in stereo. Most molecular graphics programs use PDB files as input. The programs described here can be downloaded from the Internet addresses listed in Table 8-4, some of which also provide instructions for the program's use.

Jmol, which functions as both a Web browser–based applet or as a standalone program, allows the user to display user-selected macromolecules in a variety of colors and formats (e.g., ball and stick, backbone, wireframe, space-filling, and cartoon). The Interactive Exercises on the Website that accompanies this textbook (http://wiley.com/college/voet/) all use Jmol (this site also contains a Jmol tutorial). **FirstGlance** uses Jmol to display macromolecules via a user-friendly interface. **KiNG,** which also has Web browser–based and standalone versions, displays the so-called **Kinemages** on this textbook's accompanying Website. KiNG provides a generally more author-directed user environment than does Jmol. Macromolecules can be displayed directly from their corresponding PDB page using Jmol, KiNG, and several other viewers. The **Swiss-Pdb Viewer** (also called **DeepView**), in addition to displaying molecular structures, provides tools for basic model building, homology modeling, energy minimization, and multiple sequence alignment. One advantage of the Swiss-PDB Viewer is that it allows users to easily superimpose two or more models. **Proteopedia** is a 3D interactive encyclopedia of proteins and other macromolecules that resembles Wikipedia in that it is user edited. It uses mainly Jmol as a viewer.

d. Structural Classification and Comparison

Most proteins are structurally related to other proteins. Indeed, as we shall see in Section 9-6, *evolution tends to conserve the structures of proteins rather than their sequences.* The computational tools described below facilitate the classification and comparison of protein structures. They can be accessed directly via their websites (Table 8-4) and, in some cases, accessed directly from the PDB. Studies using these programs yield functional insights, reveal distant evolutionary relationships that are not apparent from sequence comparisons (Section 7-4B), generate libraries of unique folds for structure prediction, and provide indications as to why certain types of structures are preferred over others.

1. CATH (for *C*lass, *A*rchitecture, *T*opology, and *H*omologous superfamily), as its name suggests, categorizes proteins in a four-level structural hierarchy. (1) "Class," the highest level, places the selected protein in one of four categories of gross secondary structure: Mainly α, Mainly β, α/β (having both α helices and β sheets), and Few Secondary Structures. (2) "Architecture" is the description of the gross arrangement of secondary structure independent of topology. (3) "Topology" is indicative of both the overall shape and connectivity of the protein's secondary structures. (4) "Homologous superfamily" is those proteins of known structure that are homologous to (share a common ancestor with) the selected protein. For 1 MBO (sperm whale myoglobin), the CATH classification is Class (C): Mainly alpha; Architecture (A): Orthogonal bundle; Topology (T): Globin-like; and Homologous superfamily (H): Globin. CATH permits the user to navigate up and down the various hierarchies and thereby structurally compare them.

2. CE (for *C*ombinatorial *E*xtension of the optimal path) finds all proteins in the PDB that can be structurally aligned with the query structure to within user-specified geometric criteria. The amino acid sequences of these proteins are also aligned on the basis of this structural alignment rather than sequence alignment (Section 7-4B). The structurally aligned proteins can be simultaneously displayed by a Java applet called Compare3D that displays both the aligned C_α backbones and the structurally aligned sequences. CE can likewise align and display two user-selected structures. The atomic coordinates of the aligned structures can also be downloaded in PDB format for use with other programs such as display by Jmol and KiNG.

3. Pfam (for *P*rotein *fam*ilies) is a database of nearly 11,000 multiple sequence alignments of protein domains (called **Pfam families**). Using Pfam, one can analyze a protein for Pfam matches (74% of proteins have at least one match in Pfam), view Pfam family annotations and alignments, determine the domain organization of a protein based on its sequence or its structure, find groups of related Pfam families (called **clans**), examine the phylogenetic tree of a Pfam family, and view the occurrence of a protein's domains across different species.

4. SCOP (*S*tructural *C*lassification *O*f *P*roteins) classifies protein structures based mainly on manually generated topological considerations according to a 6-level hierarchy: Class [All-α, All-β, α/β (having α helices and β strands that are largely interspersed), α + β (having α helices and β strands that are largely segregated), and Multi-domain (having domains of different classes)], Fold (groups that have similar arrangements of secondary structural elements), Superfamily (indicative of distant evolutionary relationships based on structural criteria and functional features), Family (indicative of near evolutionary relationships based on sequence as well as on structure), Protein, and Species. For 1 MBO these are Class: All-α; Fold: Globin-like; Superfamily: Globin-like; Family: Globins; Protein: Myoglobin; and Species: Sperm whale (*Physeter catodon*). SCOP permits the user to navigate through its treelike hierarchical organization and lists the known members of any particular branch. Thus, with 1MBO, SCOP displays a list of the 174 structures in the PDB that contain sperm whale myoglobin (a protein that has received far more structural study than most).

5. VAST (*V*ector *A*lignment *S*earch *T*ool), a component of the National Center for Biotechnology Information (NCBI) Entrez system, reports a precomputed list of proteins of known structure that structurally resemble the query protein. The VAST system uses the **Molecular Modeling Database (MMDB),** an NCBI-generated database that is derived from PDB coordinates but in which molecules are represented by connectivity graphs rather than sets of atomic coordinates. VAST displays the superposition of the query protein in its structural alignment with a user-selected list of the related proteins using **Cn3D** [a molecular graphics program that displays MMDB files and that is publicly available for a variety of computer platforms (Table 8-4)]. VAST also reports the structure-based sequence alignment of these proteins.

In addition, several "Structure Analysis" tools can be invoked from a PDB Structure Summary page. The "Sequence Details" page provides the sequence of each chain in the structure and, for polypeptides, indicates the secondary structure of each of its residues.

e. The Structural Genomics Project

As we shall see in Section 9-3B, proteins with similar sequences are likely to have similar three-dimensional structures. Yet, of the ~7 million polypeptides of known sequence, only <40,000 have known structures comprising ~1200 of the estimated 8000 different protein folds. Consequently, ~40% of the open reading frames (ORFs; DNA sequences that appear to encode proteins) in known genomes specify proteins whose structures and functions are unknown. The need to better characterize such proteins has led to the **structural genomics** project, a loosely organized international consortium of structure determination centers dedicated to elucidating the structures of representative proteins from every protein family and hence making the structures of most proteins readily obtainable from their gene sequences.

Traditionally, the determination of protein structures by X-ray and NMR techniques has been carried out through hypothesis-driven research, that is, the proteins are being studied to solve specific biochemical problems, and hence these proteins tend to be functionally well characterized. In contrast, structural genomics endeavors to determine the structures of large numbers of uncharacterized proteins (except for their sequences) and has a long-range goal of determining the structures of all members of selected **proteomes** (the collections of all proteins encoded by the corresponding genomes). To effectively do so required that the rate of protein structure determination be greatly accelerated. Hence the first phase of the structural genomics project concentrated on developing high-throughput (i.e., robotic) methods of expressing, purifying, crystallizing, and structurally elucidating the proteins to be studied. This has been followed, since 2005, by a production phase in which large numbers of protein structures have been determined.

Indeed, ~14% of the structures in the PDB are products of the structural genomics project (a percentage that is increasing) with many of these proteins having novel folds. Unfortunately, the knowledge of a protein's structure does not necessarily reveal its function. Hence even though the cost of determining a protein's X-ray structure by the structural genomics project appears to be significantly less than that by hypothesis-driven research (although the latter are adopting high-throughput methods), it is still a matter of debate as to whether the structural genomics project is a cost-effective way of obtaining biochemically useful information relative to hypothesis-driven research.

4 PROTEIN STABILITY

Incredible as it may seem, thermodynamic measurements indicate that *native proteins are only marginally stable entities under physiological conditions.* The free energy required to denature them is ~0.4 kJ · mol^{-1} of amino acid residues, so that 100-residue proteins are typically stable by only around 40 kJ · mol^{-1}. In contrast, the energy required to break a typical hydrogen bond is ~20 kJ · mol^{-1}. The various noncovalent influences to which proteins are subject—electrostatic interactions (both attractive and repulsive), hydrogen bonding (both intramolecular and to water), and hydrophobic forces—each have energetic magnitudes that may total thousands of kilojoules per mole over an entire protein molecule. Consequently, *a protein structure arises from a delicate balance among powerful countervailing forces.* In this section we discuss the nature of these forces and end by considering protein denaturation, that is, how these forces can be disrupted.

A. *Electrostatic Forces*

Molecules are collections of electrically charged particles and hence, to a reasonable degree of approximation, their interactions are determined by the laws of classical electrostatics (more exact calculations require the application of quantum mechanics). The energy of association, U, of two electric charges, q_1 and q_2, that are separated by the distance r is found by integrating the expression for Coulomb's law, Eq. [2.1], to determine the work necessary to separate these charges by an infinite distance:

$$U = \frac{kq_1q_2}{Dr} \qquad [8.1]$$

Here $k = 9.0 \times 10^9$ J · m · C^{-2} and D is the dielectric constant of the medium in which the charges are immersed (recall that $D = 1$ for a vacuum and, for the most part, increases with the polarity of the medium; Table 2-1). The dielectric constant of a molecule-sized region is difficult to estimate. For the interior of a protein, it is usually taken to be in the range 3 to 5 in analogy with the measured dielectric constants of substances that have similar polarities, such as benzene and diethyl ether.

Coulomb's law is only valid for point or spherically symmetric charges that are immersed in a medium of constant D. However, proteins are by no means spherical and their internal D values vary with position. Moreover, a protein in solution associates with mobile ions such as Na$^+$ and Cl$^-$, which modulate the protein's electrostatic potential. Consequently, calculating the electrostatic potential of a protein requires mathematically sophisticated and computationally intensive algorithms that are beyond the scope of this text. These methods are widely used to calculate the surface electrostatic potentials of proteins using a program called **GRASP** (for *G*raphical *R*epresentation and *A*nalysis of *S*urface *P*roperties) written by Anthony Nicholls, Kim Sharp, and Barry Honig. Figure 8-56 shows a GRASP diagram of human growth hormone in which the protein's surface is colored according to its electrostatic potential. Such diagrams are useful for assessing how a protein might associate with charged molecules such as other proteins, nucleic acids, and substrates. Similar computations are used to predict the pK's of protein surface groups, which can have significant application in the elucidation of an enzyme's mechanism of action (Section 15-1).

a. Ionic Interactions Are Strong but Do Not Greatly Stabilize Proteins

The association of two ionic protein groups of opposite charge is known as an **ion pair** or **salt bridge**. According to Eq. [8.1], the energy of a typical ion pair, say the carboxyl

Figure 8-56   A GRASP diagram of human growth hormone.
The diagram shows the protein's surface colored according to its electrostatic potential, with its most negative areas dark red, its most positive areas dark blue, and its neutral areas white. The protein's orientation is the same as that in Fig. 8-47*b*. [Based on an X-ray structure by Alexander Wlodawer, National Cancer Institute, Frederick, Maryland. PDBid 1HGU.]

Figure 8-57 **Examples of ion pairs in myoglobin.** In each case, oppositely charged side chain groups from residues far apart in sequence closely approach each other through the formation of ion pairs.

group of Glu and the ammonium group of Lys, whose charge centers are separated by 4.0 Å in a medium of dielectric constant 4, is –86 kJ · mol⁻¹ (one electronic charge = 16.0 × 10⁻¹⁹ C). However, free ions in aqueous solution are highly solvated, and the formation of a salt bridge has the entropic penalty of localizing the salt bridge's charged side chains. Consequently, the free energy of solvation of two separated ions is about equal to the free energy of formation of their unsolvated ion pair. *Ion pairs therefore contribute little stability toward a protein's native structure.* This accounts for the observations that although ~75% of charged residues occur in ion pairs (e.g., Fig. 8-57), very few ion pairs are buried (unsolvated), and ion pairs that are exposed to the aqueous solvent tend to be but poorly conserved among homologous proteins.

b. Dipole–Dipole Interactions Are Weak but Significantly Stabilize Protein Structures

The noncovalent associations between electrically neutral molecules, collectively known as **van der Waals forces,** arise from electrostatic interactions among permanent and/or induced dipoles. These forces are responsible for numerous interactions of varying strengths between nonbonded neighboring atoms. (The hydrogen bond, a special class of dipolar interaction, is considered separately in Section 8-4B.)

Interactions among permanent dipoles are important structural determinants in proteins because many of their groups, such as the carbonyl and amide groups of the peptide backbone, have permanent dipole moments. These interactions are generally much weaker than the charge–charge interactions of ion pairs. Two carbonyl groups, for example, each with dipoles of 4.2×10^{-30} C · m (1.3 debye units) that are oriented in an optimal head-to-tail arrangement (Fig. 8-58a) and separated by 5 Å in a medium of dielectric constant 4, have a calculated attrac-

tive energy of only –9.3 kJ · mol⁻¹. Furthermore, these energies vary with r^{-3}, so they rapidly attenuate with distance. In α helices, however, the negative ends of the dipolar amide and carbonyl groups of the polypeptide backbone all point in the same direction (Fig. 8-11), so that their interactions and bond dipoles are additive (these groups, of course, also form hydrogen bonds, but here we are concerned with their residual electric fields). The α helix therefore has a significant dipole moment that is positive toward the N-terminus and negative toward the C-terminus. Consequently, *in the low dielectric constant core of a protein, dipole–dipole interactions significantly influence protein folding.*

A permanent dipole also induces a dipole moment on a neighboring group so as to form an attractive interaction (Fig. 8-58b). Such dipole–induced dipole interactions are generally much weaker than are dipole–dipole interactions.

Although nonpolar molecules are nearly electrically neutral, at any instant they have a small dipole moment resulting from the rapid fluctuating motions of their electrons. This transient dipole moment polarizes the electrons in a neighboring group (Fig. 8-58c) such that, near their van der Waals contact distances, the groups are attracted to one another (a quantum mechanical effect that cannot be explained in terms of only classical physics). These so-called **London dispersion forces** are extremely weak. The 8.2-kJ · mol⁻¹ heat of vaporization of CH_4, for example, indicates that the attractive interaction of a nonbonded H · · · H contact between neighboring CH_4 molecules is roughly −0.3 kJ · mol⁻¹ (in the liquid, a CH_4 molecule touches its 12 nearest neighbors with ~2 H · · · H contacts each).

London forces are only significant for contacting groups because their association energy is proportional to r^{-6}. Nevertheless, *the great numbers of interatomic contacts in the closely packed interiors of proteins make London forces*

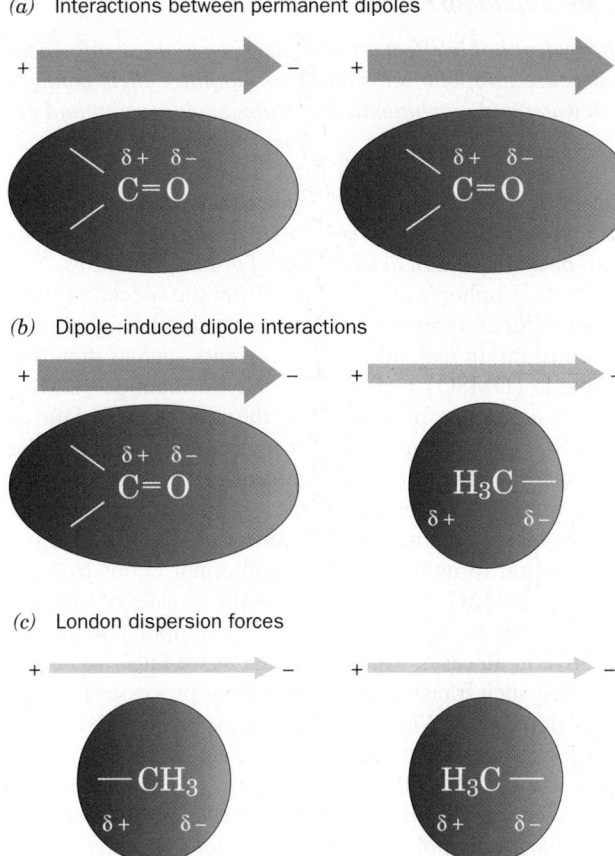

(a) Interactions between permanent dipoles

(b) Dipole–induced dipole interactions

(c) London dispersion forces

Figure 8-58 **Dipole–dipole interactions.** The strength of each dipole is represented by the thickness of the accompanying arrow. (*a*) Interactions between permanent dipoles. These interactions, here represented by carbonyl groups lined up head to tail, may be attractive, as shown, or repulsive, depending on the relative orientations of the dipoles. (*b*) Dipole–induced dipole interactions. A permanent dipole (here shown as a carbonyl group) induces a dipole in a nearby group (here represented by a methyl group) by electrostatically distorting its electron distribution (*shading*). This always results in an attractive interaction. (*c*) London dispersion forces. The instantaneous charge imbalance (*shading*) resulting from the motions of the electrons in a molecule (*left*) induces a dipole in a nearby group (*right*); that is, the motions of the electrons in neighboring groups are correlated. This always results in an attractive interaction.

a major influence in determining their conformations. London forces also provide much of the binding energy in the sterically complementary interactions between proteins and the molecules that they specifically bind.

B. *Hydrogen Bonding Forces*

Hydrogen bonds (D—H ··· A), as we discussed in Section 2-1Aa, are predominantly electrostatic interactions (but with ~10% covalent character) between a weakly acidic donor group (D—H) and an acceptor (A) that bears a lone pair of electrons. In biological systems, D and A can both be the highly electronegative N and O atoms and occasion-

ally S atoms. In addition, a relatively acidic C—H group (e.g., a C_α—H group) can act as a weak hydrogen bond donor, and the polarizable π electron system of an aromatic ring (e.g., that of Trp) can act as a weak acceptor.

Hydrogen bonds have association energies that are normally in the range -12 to -40 kJ · mol^{-1} (but only around -8 to -16 kJ · mol^{-1} for C—H ··· A and D—H ··· π hydrogen bonds and -2 to -4 kJ · mol^{-1} for C—H ··· π hydrogen bonds), values which are between those for covalent bonds and van der Waals forces. Hydrogen bonds (**H bonds**) are much more directional than are van der Waals forces but less so than are covalent bonds. The D ··· A distance is normally in the range 2.7 to 3.1 Å, although since H atoms are unseen in all but the very highest resolution macromolecular X-ray structures, a possible D—H ··· A interaction (where D and A are either N or O) is assumed to be a H bond if its D ··· A distance is significantly less than the 3.7 Å sum of a D—H bond length (~1.0 Å) and the van der Waals contact distance between H and A (~2.7 Å). Keep in mind, however, that there is no rigid cutoff distance beyond which H bonds cease to exist because the energy of an H bond, which is mainly electrostatic in character, varies inversely with the distance between the negative and positive centers (Eq. [8.1]).

H bonds tend to be linear, with the D—H bond pointing along the acceptor's lone pair orbital (or, in D—H ··· π hydrogen bonds, roughly perpendicular to the aromatic ring and pointing at its center with the distance from the D atom to the center of the aromatic ring normally in the range 3.2–3.8 Å). Large deviations from this ideal geometry are not unusual, however. For example, in the H bonds of both α helices (Fig. 8-11) and antiparallel β pleated sheets (Fig. 8-16*a*), the N—H bonds point approximately along the C=O bonds rather than along an O lone pair orbital, and in parallel β pleated sheets (Fig. 8-16*b*), the H bonds depart significantly from linearity. Indeed, many of the H bonds in proteins are members of networks in which each donor is H bonded to two acceptors (**a bifurcated hydrogen bond**) and each acceptor is H bonded to two donors. For example, although the H bonds in ideal α helices form between the N—H group at residue *n* and the C=O group at residue $n - 4$ ($n \rightarrow n - 4$ H bonds; Fig. 8-11), many of the N—H groups in real α helices associate via bifurcated H bonds with two adjacent C=O groups to form both $n \rightarrow n - 4$ and $n \rightarrow n - 3$ H bonds.

a. Hydrogen Bonds Only Weakly Stabilize Proteins

A protein's internal H bonding groups are arranged such that most possible H bonds are formed (Section 8-3B). Clearly, H bonding has a major influence on the structures of proteins. However, an unfolded protein makes most of its H bonds with the water molecules of the aqueous solvent (water, it will be recalled, is a strong H bonding donor and acceptor). The free energy of stabilization that internal H bonds confer on a native protein is therefore equal to the difference in the free energy of H bonding between the native protein and the unfolded protein. Consequently, it might be expected that H bonds do not stabilize (and perhaps even slightly destabilize) the structure of a native protein relative to its unfolded state. However, since H bonding interactions are largely electrostatic in nature, they are likely to be stronger in the low polarity interior of a protein than they are in the high polarity aqueous medium. Moreover, there may be an entropic effect that destabilizes the H bonds between water and an unfolded polypeptide relative to intraprotein H bonds: The water molecules that are H bonded to a polypeptide are likely to be more positionally and orientationally constrained (ordered) than those that are H bonded to only other water molecules, thus favoring the formation of intraprotein H bonds. These effects may very well account for the observation that the mutagenic removal of an H bond from a protein generally reduces the protein's stability by −2 to 8 kJ · mol^{-1}.

Despite their low stability, *a protein's hydrogen bonds provide a structural basis for its native folding pattern:* If a protein folded in a way that prevented some of its internal H bonds from forming, their free energy would be lost and such conformations would be less stable than those that are fully H bonded. In fact, the formation of α helices and β sheets efficiently satisfies the polypeptide backbone's H bonding requirements. This argument also applies to the van der Waals forces discussed in the previous section.

b. Most Hydrogen Bonds in Proteins Are Local

How can as complex a molecule as a protein fold so as to make nearly all of its potential H bonds? The answer to this question was revealed by a survey of the H bonds in high resolution protein X-ray structures by Ken Dill and George Rose: *Most of the H bonds in a protein are local, that is, they involve donors and acceptors that are close together in sequence and hence can readily find their H bonding mates.*

1. On average, 68% of the H bonds in proteins are between backbone atoms. Of these, ~1/3 form $n \rightarrow n - 4$ H bonds (as in ideal α helices), ~1/3 form $n \rightarrow n - 3$ H bonds (as in reverse turns and ideal 3_{10} helices), and ~1/3 are between paired strands in β sheets. In fact, only ~5% of the H bonds between backbone atoms are not wholly within a helix, sheet, or turn.

2. Hydrogen bonds between side chains and backbones are clustered at **helix-capping** positions. In an α helix, the first four N—H groups and the last four C＝O groups cannot form H bonds within the helix (which accounts for half the potential H bonds involving backbone atoms in an α helix of 12 residues, the average length of α helices). These potential H bonds are often made with nearby side chains. In particular, ~1/2 of the N-terminal N—H groups of α helices form H bonds with polar side chains that are 1 to 3 residues distant, and ~1/3 of their C-terminal C＝O groups form H bonds with polar side chains that are 2 to 5 residues distant.

3. Over half the H bonds between side chains are between charged residues (i.e., they form salt bridges) and are therefore located on protein surfaces between and within surface loops (e.g., Fig. 8-57). However, ~85% of the remaining side chain–side chain H bonds are between side chains that are 1 to 5 residues apart. Hence with the exception of those in salt bridges, side chain–side chain H bonds also tend to be local.

C. *Hydrophobic Forces*

*The **hydrophobic effect** is the name given to those influences that cause nonpolar substances to minimize their contacts with water and amphipathic molecules, such as soaps and detergents, to form micelles in aqueous solutions (Section 2-1Ba).* Since native proteins form a sort of intramolecular micelle in which their nonpolar side chains are largely out of contact with the aqueous solvent, *hydrophobic interactions must be an important determinant of protein structures.*

The hydrophobic effect derives from the special properties of water as a solvent, only one of which is its high dielectric constant. In fact, other polar solvents, such as dimethyl sulfoxide (DMSO) and *N,N*-dimethylformamide (DMF), tend to denature proteins. The thermodynamic data of Table 8-5 provide considerable insight as to the origin of the hydrophobic effect because the transfer of a hydrocarbon from water to a nonpolar solvent resembles the transfer of a nonpolar side chain from the exterior of a protein in aqueous solution to its interior. The isothermal Gibbs free energy changes ($\Delta G = \Delta H - T \Delta S$) for the transfer of a hydrocarbon from an aqueous solution to a nonpolar solvent is negative in all cases, which indicates, as we know to be the case, that such transfers are spontaneous processes (oil and water don't mix). What is perhaps unexpected is that these transfer processes are endothermic (positive ΔH) for aliphatic compounds and athermic ($\Delta H = 0$) for aromatic compounds; that is, *it is enthalpically more or equally favorable for nonpolar molecules to dissolve in water than in nonpolar media.* In contrast, the entropy component of the unitary free energy change, $-T \Delta S_u$ (see footnote *a* to Table 8-5), is large and negative in all cases. Evidently, *the transfer of a hydrocarbon from an aqueous medium to a nonpolar medium is entropically driven. The same is true of the transfer of a nonpolar protein group from an aqueous environment to the protein's nonpolar interior.*

What is the physical mechanism whereby nonpolar entities are excluded from aqueous solutions? Recall that entropy is a measure of the order of a system; it decreases with increasing order (Section 3-2). Thus the decrease in entropy when a nonpolar molecule or side chain is solvated by water (the reverse of the foregoing process) must be due to an ordering process. This is an experimental observation, not a theoretical conclusion. The magnitudes of the entropy changes are too large to be attributed only to changes in the conformations of the hydrocarbons; rather, as Henry Frank and Marjorie Evans pointed out in 1945, *these entropy changes mainly arise from some sort of ordering of the water structure.*

Liquid water has a highly ordered and extensively H bonded structure (Section 2-1A). The insinuation of a nonpolar group into this structure disrupts it: A nonpolar group can neither accept nor donate H bonds, so the water molecules at the surface of the cavity occupied by the nonpolar group cannot H bond to other molecules in their usual fashion. In order to recover the lost H bonding energy, these surface waters must orient themselves so as to form an H bonded network enclosing the cavity (Fig. 8-59). This orientation constitutes an ordering of the water structure, since the number of ways that water molecules can form H bonds

Table 8-5 **Thermodynamic Changes for Transferring Hydrocarbons from Water to Nonpolar Solvents at 25°C**[a]

Process	ΔH (kJ · mol^{-1})	$-T\Delta S_u$ (kJ · mol^{-1})	ΔG_u (kJ · mol^{-1})
CH$_4$ in H$_2$O $\rightleftharpoons$ CH$_4$ in C$_6$H$_6$	11.7	−22.6	−10.9
CH$_4$ in H$_2$O $\rightleftharpoons$ CH$_4$ in CCl$_4$	10.5	−22.6	−12.1
C$_2$H$_6$ in H$_2$O $\rightleftharpoons$ C$_2$H$_6$ in benzene	9.2	−25.1	−15.9
C$_2$H$_4$ in H$_2$O $\rightleftharpoons$ C$_2$H$_4$ in benzene	6.7	−18.8	−12.1
C$_2$H$_2$ in H$_2$O $\rightleftharpoons$ C$_2$H$_2$ in benzene	0.8	−8.8	−8.0
Benzene in H$_2$O $\rightleftharpoons$ liquid benzene[b]	0.0	−17.2	−17.2
Toluene in H$_2$O $\rightleftharpoons$ liquid toluene[b]	0.0	−20.0	−20.0

[a]ΔG_u, the **unitary Gibbs free energy change**, is the Gibbs free energy change, ΔG, corrected for its concentration dependence so that it reflects only the inherent properties of the substance in question and its interaction with solvent. This relationship, according to Equation [3.13], is

$$\Delta G_u = \Delta G - nRT \ln \frac{[A_f]}{[A_i]}$$

where $[A_i]$ and $[A_f]$ are the initial and final concentrations of the substance under consideration, respectively, and n is the number of moles of that substance. Since the second term in this equation is a purely entropic term (concentrating a substance increases its order), ΔS_u, the **unitary entropy change**, is expressed

$$\Delta S_u = \Delta S + nR \ln \frac{[A_f]}{[A_i]}$$

[b]Data measured at 18°C.

Source: Kauzmann, W., *Adv. Protein Chem.* **14,** 39 (1959).

about the surface of a nonpolar group is less than the number of ways that they can H bond in bulk water.

Unfortunately, the complexity of liquid water's basic structure (Section 2-1Ac) has not yet allowed a detailed structural description of this ordering process. One model that has been proposed is that water forms quasi-crystalline H bonded cages about the nonpolar groups similar to those of **clathrates** (Fig. 8-60). The magnitudes of the entropy changes that result when nonpolar substances are dissolved in water, however, indicate that the resulting water structures can only be slightly more ordered than bulk water. They also must be quite different from that of ordinary ice,

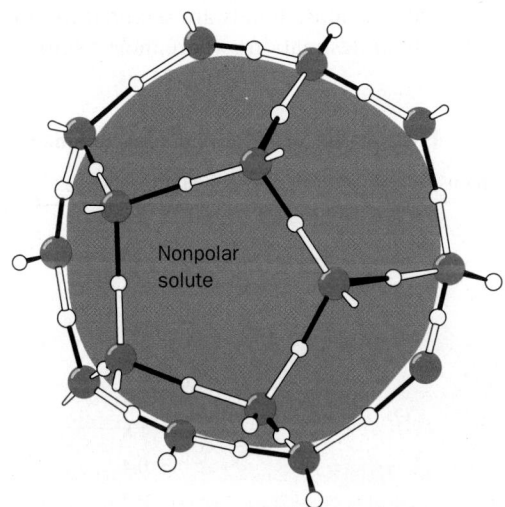

Figure 8-59   The orientational preference of water molecules next to a nonpolar solute. In order to maximize their H bonding energy, these water molecules tend to straddle the inert solute such that, for relatively small solutes, two or three of their tetrahedral directions are tangential to its surface. This permits them to form H bonds (*black*) with neighboring water molecules lining the nonpolar surface. However, for larger (flatter) nonpolar solutes, the adjacent water molecules are each geometrically limited to participating in no more than three H bonds. In either case, the ordering of water molecules extends several layers of water molecules beyond the first hydration shell of the nonpolar solute. [Illustration, Irving Geis. Image from the Irving Geis Collection, Howard Hughes Medical Institute. Reprinted with permission.]

Figure 8-60   Structure of the clathrate (*n*-C$_4$H$_9$)$_3$S$^+$F$^-$ · 23H$_2$O. Clathrates are crystalline complexes of nonpolar compounds with water (usually formed at low temperatures and high pressures) in which the nonpolar molecules are enclosed, as shown, by a polyhedral cage of tetrahedrally H bonded water molecules (here represented by only their oxygen atoms). The H bonding interactions of one such water molecule (*arrow*) are shown in detail. [Illustration, Irving Geis. Image from the Irving Geis Collection, Howard Hughes Medical Institute. Reprinted with permission.]

Figure 8-61 **Hydropathic index plot for bovine chymotrypsinogen.** The sum of the hydropathies of nine consecutive residues (see Table 8-6) are plotted versus the residue sequence number. A large positive hydropathic index is indicative of a hydrophobic region of the polypeptide chain, whereas a large negative value is indicative of a hydrophilic region. The bars above the midpoint line denote the protein's interior regions, as determined by X-ray crystallography, and the bars below the midpoint line indicate the protein's exterior regions. [After Kyte, J. and Doolittle, R.F., *J. Mol. Biol.* **157,** 111 (1982).]

because, for instance, the solvation of nonpolar groups by water causes a large decrease in water volume (e.g., the transfer of CH_4 from hexane to water shrinks the water solution by 22.7 mL · mol^{-1} of CH_4), whereas the freezing of water results in a 1.6-mL · mol^{-1} expansion.

The unfavorable free energy of hydration of a nonpolar substance caused by its ordering of the surrounding water molecules has the net result that *the nonpolar substance is excluded from the aqueous phase.* This is because the surface area of a cavity containing an aggregate of nonpolar molecules is less than the sum of the surface areas of the cavities that each of these molecules would individually occupy. The aggregation of the nonpolar groups thereby minimizes the surface area of the cavity and therefore the entropy loss of the entire system. In a sense, the nonpolar groups are squeezed out of the aqueous phase by the hydrophobic interactions. Thermodynamic measurements indicate that the free energy change of removing a —CH$_2$— group from an aqueous solution is about −3 kJ · mol^{-1}. Although this is a relatively small amount of free energy, *in molecular assemblies involving large numbers of nonpolar contacts, hydrophobic interactions are a potent force.*

Walter Kauzmann pointed out in 1958 that *hydrophobic forces are a major influence in causing proteins to fold into their native conformations.* Figure 8-61 indicates that the amino acid side chain **hydropathies** (indexes of combined hydrophobic and hydrophilic tendencies; Table 8-6) are, in fact, good predictors of which portions of a polypeptide chain are inside a protein, out of contact with the aqueous solvent, and which portions are outside, in contact with the aqueous solvent. In proteins, the effects of hydrophobic forces are often termed **hydrophobic bonding,** presumably to indicate the specific nature of protein folding under the influence of the hydrophobic effect. You should keep in mind, however, that hydrophobic bonding does not generate the directionally specific interactions usually associated with the term "bond."

D. *Disulfide Bonds*

Since disulfide bonds form as a protein folds to its native conformation (Section 9-1A), they function to stabilize its three-dimensional structure. The relatively reducing chemical character of the cytoplasm, however, greatly diminishes the stability of intracellular disulfide bonds. In fact, almost all proteins with disulfide bonds are secreted to more oxidized extracellular destinations, where their disulfide bonds

Table 8-6 **Hydropathy Scale for Amino Acid Side Chains**

Side Chain	Hydropathy
Ile	4.5
Val	4.2
Leu	3.8
Phe	2.8
Cys	2.5
Met	1.9
Ala	1.8
Gly	−0.4
Thr	−0.7
Ser	−0.8
Trp	−0.9
Tyr	−1.3
Pro	−1.6
His	−3.2
Glu	−3.5
Gln	−3.5
Asp	−3.5
Asn	−3.5
Lys	−3.9
Arg	−4.5

Source: Kyte, J. and Doolitle, R.F., *J. Mol. Biol.* **157,** 110 (1982).

are effective in stabilizing protein structures [secreted proteins fold to their native conformations—and hence form their disulfide bonds—in the endoplasmic reticulum (Section 12-4Ba), which, unlike other cellular compartments, has an oxidizing environment]. Apparently, the relative "hostility" of extracellular environments toward proteins (e.g., uncontrolled temperatures and pH's) requires the additional structural stability conferred by disulfide bonds.

E. *Protein Denaturation*

The low conformational stabilities of native proteins make them easily susceptible to denaturation by altering the balance of the weak nonbonding forces that maintain the native conformation. When a protein in solution is heated, its conformationally sensitive properties, such as optical rotation (Section 4-2A), viscosity, and UV absorption, change abruptly over a narrow temperature range (e.g., Fig. 8-62). *Such a nearly discontinuous change indicates that the native protein structure unfolds in a cooperative manner: Any partial unfolding of the structure destabilizes the remaining structure, which must simultaneously collapse to the random coil.* The temperature at the midpoint of this process is known as the protein's **melting temperature,** T_m, in analogy

with the melting of a solid. Most proteins have T_m values well below 100°C. Recall that nucleic acids likewise have characteristic T_m's (Section 5-3Ca).

In addition to high temperatures, proteins are denatured by a variety of other conditions and substances:

1. pH variations alter the ionization states of amino acid side chains (Table 4-1), which changes protein charge distributions and H bonding requirements.

2. Detergents, some of which significantly perturb protein structures at concentrations as low as 10^{-6} *M*, hydrophobically associate with the nonpolar residues of a protein, thereby interfering with the hydrophobic interactions responsible for the protein's native structure.

3. High concentrations of water-soluble organic substances, such as aliphatic alcohols, interfere with the hydrophobic forces stabilizing protein structures through their own hydrophobic interactions with water. Organic substances with several hydroxyl groups, such as ethylene glycol or sucrose,

Ethylene glycol   **Sucrose**

however, are relatively poor denaturants because their H bonding ability renders them less disruptive of water structure.

The influence of salts is more variable. Figure 8-63 shows the effects of a number of salts on the T_m of bovine

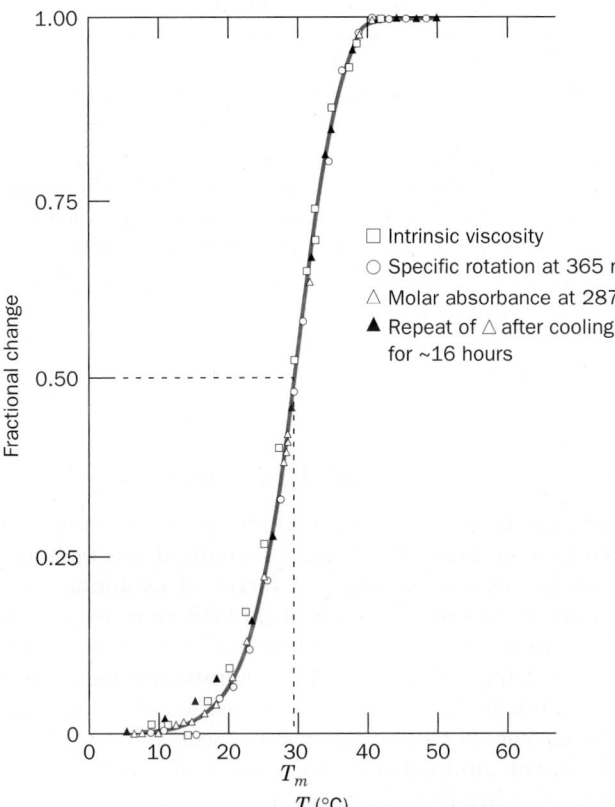

Figure 8-62 **Protein denaturation.** The heat-induced denaturation of bovine pancreatic ribonuclease A (RNase A) in an HCl–KCl solvent at pH 2.1 and 0.019 ionic strength was monitored by several conformationally sensitive techniques. The curve is drawn only through the points Δ. The melting temperature, T_m, is defined as the temperature at the midpoint of the transition. Compare the shape of this melting curve with that of duplex DNA (Fig. 5-16). [After Ginsburg, A. and Carroll, W.R., *Biochemistry* **4,** 2169 (1965).]

Figure 8-63 **Melting temperature of RNase A as a function of the concentrations of various salts.** All solutions also contained 0.15*M* KCl and 0.013*M* sodium cacodylate buffer, pH 7. [After von Hippel, P.J. and Wong, K.Y., *J. Biol. Chem.* **10,** 3913 (1965).]

pancreatic **ribonuclease A (RNase A).** Some salts, such as $(NH_4)_2SO_4$ and KH_2PO_4, stabilize the native protein structure (raise its T_m); others, such as KCl and NaCl, have little effect; and yet others, such as KSCN and LiBr, destabilize it. The order of effectiveness of the various ions in stabilizing a protein, which is largely independent of the identity of the protein, parallels their capacity to salt out proteins (Section 6-2A). This order is known as the **Hofmeister series:**

Anions: $SO_4^{2-} > H_2PO_4^- > CH_3COO^- > Cl^-$
$> Br^- > I^- > ClO_4^- > SCN^-$

Cations: $NH_4^+, Cs^+, K^+, Na^+ > Li^+ > Mg^{2+}$
$> Ca^{2+} > Ba^{2+}$

The ions in the Hofmeister series that tend to denature proteins, I^-, ClO_4^-, SCN^-, Li^+, Mg^{2+}, Ca^{2+}, and Ba^{2+}, are said to be **chaotropic.** This list should also include the guanidinium ion (Gu^+) and the nonionic urea, which, in concentrations in the range 5 to 10 M, are the most commonly used protein denaturants. The effect of the various ions on proteins is largely cumulative: GuSCN is a much more potent denaturant than the often used GuCl, whereas Gu_2SO_4 stabilizes protein structures.

Chaotropic agents increase the solubility of nonpolar substances in water. Consequently, their effectiveness as denaturing agents stems from their ability to disrupt hydrophobic interactions, although the manner in which they do so is not well understood. Conversely, those substances listed that stabilize proteins strengthen hydrophobic forces, thus increasing the tendency of water to expel proteins. This accounts for the correlation between the abilities of an ion to stabilize proteins and to salt them out.

F. *Explaining the Stability of Thermostable Proteins*

Certain species of bacteria known as **hyperthermophiles** grow at temperatures near 100°C (they live in such places as hot springs and submarine hydrothermal vents, with the most extreme, an Fe(III)-reducing archaeon, growing at 121°C and remaining viable as high as 130°C). These organisms have many of the same metabolic pathways as do **mesophiles** (organisms that grow at "normal" temperatures). Yet, most mesophilic proteins denature at the temperatures at which hyperthermophiles thrive. What is the structural basis for the thermostability of hyperthermophilic proteins?

The difference in the thermal stabilities of the corresponding (hyper)thermophilic and mesophilic proteins does not exceed ~100 kJ · mol^{-1}, the equivalent of a few noncovalent interactions. This is probably why comparisons of the X-ray structures of hyperthermophilic enzymes with their mesophilic counterparts have failed to reveal any striking differences between them. These proteins exhibit some variations in secondary structure but no more so than is often the case for homologous proteins from distantly related mesophiles. However, several of these thermostable enzymes have a superabundance of salt bridges on their surfaces, many of which are arranged in extensive

networks. Indeed, one such network from *Pyrococcus furiosus* **glutamate dehydrogenase** consists of 18 side chains.

The idea that salt bridges can stabilize a protein structure appears to contradict the conclusion of Section 8-4Aa that ion pairs are, at best, marginally stable. The key to this apparent paradox is that the salt bridges in thermostable proteins form networks. Thus, the gain in charge–charge free energy on associating a third charged group with an ion pair is comparable to that between the members of this ion pair, whereas the free energy lost on desolvating and immobilizing the third side chain is only about half that lost in bringing together the first two side chains. The same, of course, is true for the addition of a fourth, fifth, etc., side chain to a salt bridge network.

Not all thermostable proteins have such a high incidence of salt bridges. Structural comparisons suggest that these proteins are stabilized by a combination of small effects, the most important of which are an increased size in the protein's hydrophobic core, an increased size in the interface between its domains and/or subunits, and a more tightly packed core as evidenced by a reduced surface-to-volume ratio.

The fact that the proteins of hyperthermophiles and mesophiles are homologous and carry out much the same functions indicates that mesophilic proteins are by no means maximally stable. This, in turn, strongly suggests *that the marginal stability of most proteins under physiological conditions (averaging ~0.4 kJ/mol of amino acid residues) is an essential property that has arisen through evolutionary design.* Perhaps this marginal stability helps confer the structural flexibility that many proteins require to carry out their physiological functions (Section 9-4). Other possibilities are that it may facilitate the elimination of otherwise stable non-native conformations (Section 9-2C), it may promote the unfolding of proteins so as to permit their insertion into or transport through membranes (Section 12-4E), and/or it may expedite their programmed degradation (Section 32-6).

5 QUATERNARY STRUCTURE

Proteins, because of their multiple polar and nonpolar groups, stick to almost anything; anything, that is, but other proteins. This is because the forces of evolution have arranged the surface groups of proteins so as to prevent their association under physiological conditions. If this were not the case, their resulting nonspecific aggregation would render proteins functionally useless (recall, e.g., the consequences of sickle-cell anemia; Section 7-3A). In his pioneering ultracentrifugational studies on proteins, however, The Svedberg discovered that some proteins are composed of more than one polypeptide chain. Subsequent studies established that this is, in fact, true of most proteins, including nearly all those with molecular masses >100 kD. Furthermore, these polypeptide **subunits** associate in a geometrically specific manner. The spatial arrangement of these subunits is known as a protein's **quaternary structure (4° structure).**

There are several reasons why multisubunit proteins are so common. In large assemblies of proteins, such as collagen fibrils, the advantages of subunit construction over the synthesis of one huge polypeptide chain are analogous to those of using prefabricated components in constructing a building: Defects can be repaired by simply replacing the flawed subunit rather than the entire protein, the site of subunit manufacture can be different from the site of assembly into the final product, and the only genetic information necessary to specify the entire edifice is that specifying its few different self-assembling subunits. In the case of enzymes, increasing a protein's size tends to better fix the three-dimensional positions of the groups forming the enzyme's active site. Increasing the size of an enzyme through the association of identical subunits is more efficient, in this regard, than increasing the length of its polypeptide chain since each subunit has an active site. Additionally, in some multimeric enzymes, the active site occurs at the interface between subunits where it is comprised of groups from two or more subunits. More importantly, however, the subunit construction of many enzymes provides the structural basis for the regulation of their activities. Mechanisms for this indispensable function are discussed in Sections 10-4 and 13-4.

In this section we discuss how the subunits of multisubunit proteins associate, what sorts of symmetries they have, and how their stoichiometries may be determined.

A. *Subunit Interactions*

A multisubunit protein may consist of identical or nonidentical polypeptide chains. Recall that hemoglobin, for example, has the subunit composition $\alpha_2\beta_2$. We shall refer to proteins with identical subunits as **oligomers** and to these identical subunits as **protomers**. A protomer may therefore consist of one polypeptide chain or several unlike polypeptide chains. In this sense, hemoglobin is a **dimer** (oligomer of two protomers) of $\alpha\beta$ protomers (Fig. 8-64).

The association of two subunits typically buries 1000 to 2000 Å^2 of surface area (minimally ~600 Å^2) that would otherwise be exposed to solvent. The resulting contact regions superficially resemble the interiors of single subunit proteins: They contain closely packed nonpolar side chains, hydrogen bonds, and in some cases, interchain disulfide bonds. However, protein–protein interfaces differ from subunit interiors in several respects:

1. They tend to have hydrophobicities between those of protein interiors and exteriors. In particular, the subunit interfaces of proteins that dissociate *in vivo* have lesser hydrophobicities than do permanent interfaces.

2. An average of ~77% of intersubunit hydrogen bonds are between side chains. In contrast, an average of ~68% of the hydrogen bonds within subunits are between backbone atoms. This is mainly because secondary structural elements are not continued across subunit boundaries (with the occasional exception of β sheets; see below).

3. Around 56% of protein–protein interfaces contain salt bridges. Th\ese contribute to the specificity as well as to the stability of subunit associations.

In addition, there are negligibly few hydrogen bonds and salt bridges at the edges of the contact regions. Not surprisingly, the residues at protein–protein interfaces are evolutionarily well conserved compared to other surface residues.

B. *Symmetry in Proteins*

In the vast majority of oligomeric proteins, the protomers are symmetrically arranged; that is, the protomers occupy geometrically equivalent positions in the oligomer. This implies that each protomer has exhausted its capacity to bind to other protomers; otherwise, higher oligomers would form. As a result of this limited binding capacity, protomers pack about a single point to form a closed shell, a phenomenon known as **point symmetry.** Proteins cannot have inversion or mirror symmetry, however, because such symmetry operations convert chiral L-residues to D-residues. Thus, *proteins can only have rotational symmetry.*

Figure 8-64 **The quaternary structure of hemoglobin.** The α_1, α_2, β_1, and β_2 subunits in this stereo, space-filling drawing are colored yellow, green, blue, and cyan, respectively. Heme groups are red. The protein is viewed along its molecular 2-fold rotation axis, which relates the $\alpha_1\beta_1$ protomer to the $\alpha_2\beta_2$ protomer.

Instructions for viewing stereo drawings are given in the appendix to this chapter. [Based on an X-ray structure by Max Perutz, MRC Laboratory of Molecular Biology, Cambridge, U.K. PDBid 2DHB.]

Various types of rotational symmetry occur in proteins, as X-ray crystal structure determinations have shown:

1. Cyclic symmetry

In the simplest type of rotational symmetry, **cyclic symmetry,** subunits are related (brought to coincidence) by a single axis of rotation (Fig. 8-65a). Objects with 2-, 3-, . . ., or n-fold rotational axes are said to have C_2, C_3, . . ., or C_n symmetry, respectively. An oligomer with C_n symmetry consists of n protomers that are related by $(360/n)°$ rotations. C_2 symmetry is the most common symmetry in proteins; higher cyclic symmetries are relatively rare.

A common mode of association between protomers related by a twofold rotation axis is the continuation of a β sheet across subunit boundaries. In such cases, the 2-fold axis is perpendicular to the β sheet so that two symmetry equivalent β strands hydrogen bond in an antiparallel fashion. In this manner, the sandwich of two 4-stranded β sheets in a protomer of **transthyretin** (also known as **prealbumin**) is extended across a 2-fold axis to form a sandwich of two 8-stranded β sheets (Fig. 8-66). Hemoglobin's two αβ protomers are also related by C_2 symmetry (Fig. 8-64).

2. Dihedral symmetry

Dihedral symmetry (D_n), a more complicated type of rotational symmetry, is generated when an n-fold rotation axis and a 2-fold rotation axis intersect at right angles (Fig. 8-65b). An oligomer with (D_n) symmetry consists of 2n protomers. D_2 symmetry is, by far, the most common type of dihedral symmetry in proteins.

Hemoglobin's α and β subunits have such similar structures that, in the hemoglobin $\alpha_2\beta_2$ tetramer, they are related by pseudo-2-fold rotational axes that are perpendicular to the tetramer's exact 2-fold axis (lie in the plane of Fig. 8-64; Section 10-2B). Hence the tetramer is said to be **pseudosymmetric** and have pseudo-D_2 symmetry. The X-ray structure of **glutamine synthetase** reveals that this enzyme consists of 12 identical subunits that are related by D_6 symmetry (Fig. 8-67).

Under the proper conditions, many oligomers with D_n symmetry dissociate into two oligomers, each with C_n symmetry (and which are related by the 2-fold rotation axes in the D_n oligomer). These, in turn, dissociate to their component protomers under more stringent denaturing conditions.

3. Other rotational symmetries

The only other types of rotationally symmetric objects are those that have the rotational symmetries of a tetrahedron (T), a cube or octahedron (O), or a dodecahedron or icosahedron (I), and hence have 12, 24, and 60 equivalent positions, respectively (Fig. 8-65c). Certain multienzyme complexes are based on octahedral symmetry (Section 21-2A), whereas others have icosahedral symmetry (Section 21-2Aa). The protein coats of the so-called spherical viruses also have icosahedral symmetry (Section 33-2A).

C_2

(a) Cyclic symmetries

C_3

C_5

D_2

(b) Dihedral symmetries

D_4

D_3

T

(c) Tetrahedral symmetry

O

Octahedral (cubic) symmetry

I

Icosahedral symmetry

Figure 8-65 **Some possible symmetries of proteins with identical protomers.** The lenticular shape, the triangle, the square, and the pentagon at the ends of the dashed lines indicate, respectively, the unique 2-fold, 3-fold, 4-fold, and 5-fold rotational axes of the objects shown. (*a*) Assemblies with the cyclic symmetries C_2, C_3, and C_5. (*b*) Assemblies with the dihedral symmetries D_2, D_4, and D_3. In these objects, a 2-fold axis is perpendicular to the vertical 2-, 4-, and 3-fold axes. (*c*) Assemblies with *T, O,* and *I* symmetry. Note that the tetrahedron has some but not all of the symmetry elements of the cube, and that the cube and the octahedron have the same symmetry. [Illustration, Irving Geis. Image from the Irving Geis Collection, Howard Hughes Medical Institute. Reprinted with permission.] 𝒮 **See the Animated Figures**

Figure 8-66 **A dimer of transthyretin as viewed down its 2-fold axis (*black lenticular symbol*).** Each protomer, which is colored in rainbow order from its N-terminus (*blue*) to its C-terminus (*red*), consists of a β barrel (really a β sandwich) containing two Greek keys (Fig. 8-50*a*). Note how both of its β sheets are continued in an antiparallel fashion in the symmetry-related protomer to form a sandwich of two 8-stranded β sheets. Two of these dimers associate back to back in the native protein to form a tetramer with D_2 symmetry. [Based on an X-ray structure by Colin Blake, Oxford University, U.K. PDBid 2PAB.]

a. Helical Symmetry

Some protein oligomers have **helical symmetry** (Fig. 8-68). The chemically identical subunits in a helix are not strictly equivalent because, for instance, those at the ends of the helix have a different environment than those in the middle. Nevertheless, the surroundings of all subunits in a long helix, except those near its ends, are sufficiently similar that the subunits are said to be **quasi-equivalent.** The subunits of many structural proteins, for example, those of **actin** (Section 35-3Ad) and **tubulin** (Section 35-3Gc), assemble into fibers with helical symmetry.

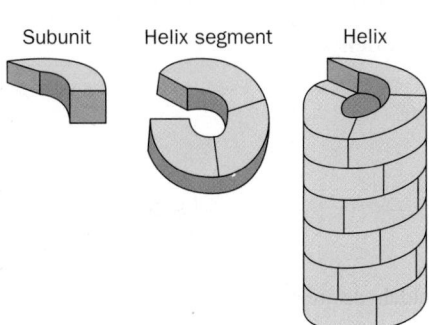

Subunit   Helix segment   Helix

(a)

(b)

Figure 8-67 **X-ray structure of glutamine synthetase from *Salmonella typhimurium*.** The enzyme consists of 12 identical subunits, here drawn in worm form, arranged with D_6 symmetry. (*a*) View down the 6-fold axis of symmetry showing only the six subunits of the upper ring in alternating blue and green. The subunits of the lower ring are roughly directly below those of the upper ring. The protein, including its side chains (*not shown*), has a diameter of 143 Å. The six active sites shown are marked by pairs of bound Mn^{2+} ions (*magenta spheres*). Also drawn in one active site are ADP (*cyan*) and the inhibitor phosphinothricin (*red*). (*b*) Side view along one of the protein's 2-fold axes showing only its eight nearest subunits. The molecule extends 103 Å along the 6-fold axis, which is vertical in this view. [Based on an X-ray structure by David Eisenberg, UCLA. PDBid 1FPY.]

Figure 8-68 **A helical structure composed of a single kind of subunit.**

b. Obtaining the Atomic Coordinates of a Biologically Functional Quaternary Structure

Crystals consist of three-dimensional lattices of identical **unit cells** (the smallest portion of a crystal lattice that is repeated by translation) that usually have internal symmetry. The crystal's **asymmetric unit** is the unique portion of the unit cell from which the entire unit cell can be generated through the operation of its symmetry elements. In a crystal of a symmetrical protein, if one or more of protein's symmetry axes are coincident with unit cell's symmetry axes, the asymmetric unit would contain a subset of the protein's protomers (often only one), which would be related to the other such subsets by **crystallographic symmetry.** Alternatively, the asymmetric units in crystals of many oligomeric proteins contain one or more entire proteins, in which case their protomers are said to be related by **noncrystallographic symmetry.**

A Protein Data Bank (PDB) coordinate file for an X-ray crystal structure contains the atomic coordinates of the protomers occupying an asymmetric unit. The entire crystal structure can then be generated through the application of its crystallographic symmetry. Thus, a symmetric protein's PDB coordinate file might contain the coordinates of only one of its several symmetry-related protomers. Moreover, in some cases, the intermolecular contacts in a crystal may sufficiently resemble the contacts between the protomers in an oligomer so that its quaternary structure may be ambiguous.

To alleviate these difficulties, computerized procedures have been devised to generate the coordinates of the most probable biologically functional molecule based on several criteria, including maximizing the solvent-accessible surface area that is buried on forming the oligomer. The coordinates of the most probable quaternary structures of macromolecules whose structures have been determined by X-ray crystallography are publicly available at http://www.ebi.ac.uk/msd-srv/prot_int/pistart.html. The biological unit may also be viewed directly from the corresponding PDB Structure Summary page, although note that the two algorithms do not always agree.

C. Determination of Subunit Composition

In the absence of an X-ray or NMR structure, the number of different types of subunits in an oligomeric protein may be determined by end group analysis (Section 7-1A). In principle, the subunit composition of a protein may be determined by comparing its molecular mass with those of its component subunits. In practice, however, experimental difficulties, such as the partial dissociation of a supposedly intact protein and uncertainties in molecular mass determinations, often provide erroneous results.

a. Cross-Linking Agents Stabilize Oligomers

A method for 4° structure analysis, which is especially useful for oligomeric proteins that decompose easily, employs **cross-linking agents** such as **dimethylsuberimidate** or **glutaraldehyde** (Fig. 8-69). If carried out at sufficiently low protein concentrations to eliminate intermolecular reactions, cross-linking reactions will covalently join only the

Figure 8-69 **Cross-linking agents.** Dimethylsuberimidate and glutaraldehyde are bifunctional reagents that react to covalently cross-link two Lys residues.

subunits in a molecule that are no farther apart than the length of the cross-link (assuming, of course, that the proper amino acid residues are present). The molecular mass of a cross-linked protein therefore places a lower limit on its number of subunits. Such studies can also provide some indication of the distance between subunits, particularly if a series of cross-linking agents with different lengths is employed.

APPENDIX Viewing Stereo Pictures

Although we live in a three-dimensional world, the images that we see have been projected onto the two-dimensional plane of our retinas. Depth perception therefore involves binocular vision: The slightly different views perceived by each eye are synthesized by the brain into a single three-dimensional impression.

Two-dimensional pictures of complex three-dimensional objects are difficult to interpret because most of the information concerning the third dimension is suppressed. This information can be recovered by presenting each eye with the image only it would see if the three-dimensional object were actually being viewed. A **stereo pair** therefore consists of two images, one for each eye. Corresponding points of stereo pairs are generally separated by ~6 cm, the average distance between human eyes. Stereo drawings are usually computer generated because of the required precision of the geometric relationship between the members of a stereo pair.

In viewing a stereo picture, one must overcome the visual habits of a lifetime because each eye must see its corresponding view independently. Viewers are commercially available to aid in this endeavor. However, with some training and practice, equivalent results can be obtained without their use.

To train yourself to view stereo pictures, you should become aware that each eye sees a separate image. Hold your finger up about a foot (30 cm) before your eyes while fixing your gaze on some object beyond it. You may realize that you are seeing two images of your finger. If, after some concentration, you are still aware of only one image, try blinking your eyes alternately to ascertain which of your eyes is seeing the image you perceive. Perhaps alternately covering and uncovering this dominant eye while staring past your finger will help you become aware of the independent workings of your eyes.

The principle involved in seeing a stereo picture is to visually fuse the left member of the stereo pair seen by the left eye with the right member seen by the right eye. To do this, sit comfortably at a desk, center your eyes about a foot over a stereo drawing such as Fig. 8-70 and stare through it at a point about a foot (30 cm) below the drawing. Try to visually fuse the central members of the four out-of-focus images you see. When you have succeeded, your visual system will "lock onto" it and this fused central image will appear three-dimensional. Ignore the outer images. You may have to slightly turn the book, which should be held perfectly flat, or your head in order to bring the two images to the same level. It may help to place the book near the edge of a desk, center your finger about a foot below the drawing, and fixate on your finger while concentrating on the stereo pair. Another trick is to hold your flattened hand or an index card between your eyes so that the left eye sees only the left half of the stereo pair and the right eye sees only the right half and then fuse the two images you see.

The final step in viewing a stereo picture is to focus on the image while maintaining fusion. This may not be easy because our ingrained tendency is to focus on the point at which our gaze converges. It may help to move your head closer to or farther from the picture. Most people (including the authors) require a fair amount of practice to become proficient at seeing stereo without a viewer. However, the three-dimensional information provided by stereo pictures, not to mention their aesthetic appeal, makes it worth the effort. In any case, the few stereo figures used in this textbook have been selected for their visual clarity without the use of stereo; stereo will simply enhance their impression of depth.

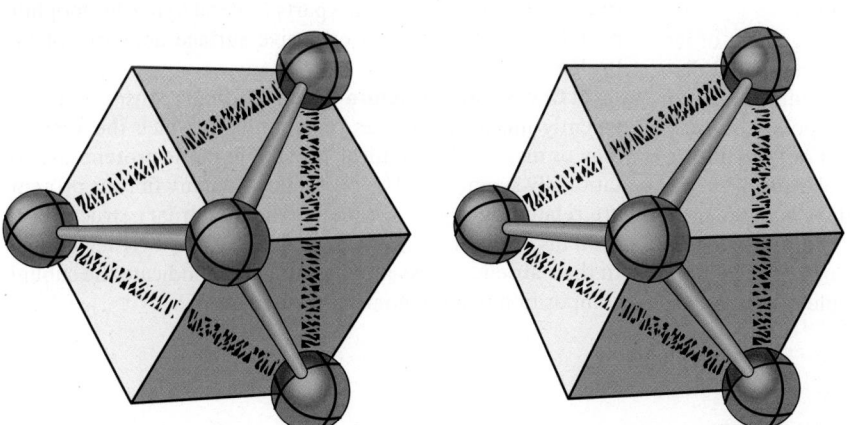

Figure 8-70 **Stereo drawing of a tetrahedron inscribed in a cube.** When properly viewed, the apex of the tetrahedron should appear to be pointing toward the viewer.

CHAPTER SUMMARY

1 Secondary Structure The peptide group is constrained by resonance effects to a planar, trans conformation. Steric interactions further limit the conformations of the polypeptide backbone by restricting the torsion angles, ϕ and ψ, of each peptide group to three small regions of the Ramachandran diagram. The α helix, whose conformation angles fall within the allowed regions of the Ramachandram diagram, is held together by hydrogen bonds. The 3_{10} helix, which is more tightly coiled than the α helix, lies in a mildly forbidden region of the Ramachandran diagram. Its infrequent occurrences are most often as single-turn terminations of α helices. In the parallel and antiparallel β pleated sheets, two or more almost fully extended polypeptide chains associate such that neighboring chains are hydrogen bonded. These β sheets have a right-handed curl when viewed along their polypeptide chains. The polypeptide chain often reverses its direction through a β bend. Other arrangements of the polypeptide chain, which are collectively known as coil conformations, are more difficult to describe but are no less ordered than are α or β structures.

2 Fibrous Proteins The mechanical properties of fibrous proteins can often be correlated with their structures. Keratin, the principal component of hair, horn, and nails, forms protofibrils that consist of two pairs of α helices in which the members of each pair are twisted together into a left-handed coil. The pliability of keratin decreases as the content of disulfide cross-links between the protofibrils increases. Collagen is the major protein component of connective tissue. Its every third residue is Gly, and many of the others are Pro and Hyp. This permits collagen to form a ropelike triple helical structure that has great tensile strength. Collagen molecules aggregate in a staggered array to form fibrils that are covalently cross-linked by groups derived from their His and Lys side chains. Mutations in collagen or the inactivation of the enzymes that process it usually cause a loss in the structural integrity of the affected tissues.

3 Globular Proteins The accuracies of protein X-ray structure determinations are limited by crystalline disorder to resolutions that are mostly in the range 1.5 to 3.0 Å. This requires that a protein's structure be determined by fitting its primary structure to its electron density map. Several lines of evidence indicate that protein crystal structures are nearly identical to their solution structures. The structures of small proteins may also be determined in solution by 2D- or multi-dimensional-NMR techniques which, for the most part, yield results similar to those of X-ray crystal structures.

A globular protein's 3° structure is the arrangement of its various elements of 2° structure together with the spatial dispositions of its side chains. Its amino acid residues tend to segregate according to residue polarity. Nonpolar residues preferentially occupy the interior of a protein out of contact with the aqueous solvent, whereas charged polar residues are located on its surface. Uncharged polar residues may occur at either location but, if they are internal, they form hydrogen bonds with other protein groups. The interior of a protein molecule resembles a crystal of an organic molecule in its packing

efficiency. Larger proteins often fold into two or more domains that may have functionally and structurally independent properties. Certain groupings of secondary structural elements, known as motifs or supersecondary structures, repeatedly occur as components of globular proteins. They combine in numerous ways to form a fold, that is, the tertiary structure of a domain. Among the common folds in globular proteins are the globin fold, 4-helix bundles, various β barrels (up-and-down, Greek key, and jelly roll), the α/β barrel, and the various open β sheets. The atomic coordinates of the ~70,000 proteins and nucleic acids whose structures have been determined are archived in the Protein Data Bank (PDB). These macromolecules may be visually examined through the use of a variety of molecular graphics programs and may be analyzed by programs that compare and classify their structures in various ways.

4 Protein Stability Proteins have marginally stable native structures that form as a result of a fine balance among the various noncovalent forces to which they are subject: ionic and dipolar interactions, hydrogen bonding, and hydrophobic forces. Ionic interactions are relatively weak in aqueous solutions due to the solvating effects of water. The various interactions among permanent and induced dipoles, which are collectively referred to as van der Waals forces, are even weaker and are effective only at short range. Nevertheless, because of their large numbers, they cumulatively have an important influence on protein structures. Hydrogen bonding forces are far more directional than are other noncovalent forces. They add little stability to a protein structure, however, because the H bonds that native proteins form internally are only marginally stronger than those that unfolded proteins form with water. Yet, because most H bonds are local, a protein can only fold stably in ways such that almost all of its possible internal H bonds are formed. Hence H bonding is important in specifying the native structure of a protein. Hydrophobic forces arise from the unfavorable ordering of water structure that results from the hydration of nonpolar groups. By folding such that its nonpolar groups are out of contact with the aqueous solvent, a protein minimizes these unfavorable interactions. The fact that most protein denaturants interfere with the hydrophobic effect demonstrates the importance of hydrophobic forces in stabilizing native protein structures. Disulfide bonds often stabilize the native structures of extracellular proteins. The proteins from (hyper)thermophilic organisms have higher T_m's than their mesophilic counterparts. Several hyperthermophilic proteins are stabilized by extensive surface networks of salt bridges.

5 Quaternary Structure Many proteins consist of noncovalently linked aggregates of subunits in which the subunits may or may not be identical. Most oligomeric proteins are rotationally symmetric. The protomers in many fibrous proteins are related by helical symmetry. The quaternary structures of proteins are best elucidated using X-ray or NMR techniques. In their absence, cross-linking studies can indicate the subunit composition of an oligomeric protein.

REFERENCES

General

Branden, C. and Tooze, J., *Introduction to Protein Structure* (2nd ed.), Garland (1999).

Cox, M.M. and Phillips, G.N., Jr. (Eds.), *Handbook of Proteins. Structure Function and Methods*, Vol. 1, Wiley (2007). [Contains an extensive collection of informative articles on protein structure.]

Dickerson, R.E. and Geis, I., *The Structure and Action of Proteins*, Benjamin/Cummings (1969). [A classic and marvelously illustrated exposition of the fundamentals of protein structure.]

Finkelstein, A.V. and Ptitsyn, O.B., *Protein Physics. A Course of Lectures*, Academic Press (2002).

Kyte, J., *Structure in Protein Chemistry* (2nd ed.), Garland (2007).

Lesk, A.M., *Introduction to Protein Architecture*, Oxford University Press (2004).

Petsko, G.A. and Ringe, D., *Protein Structure and Function*, New Science Press (2004).

Tanford, C. and Reynolds, J., *Nature's Robots*, Oxford University Press (2001). [A history of proteins.]

Secondary Structure

Hollingsworth, S.A., Berkholz, D.S., and Karplus, A., On the occurrence of linear groups in proteins, *Protein Science* **18**, 1321–1325 (2009). [Shows that linear groups (groups of sequential residues with closely similar values of ϕ and ψ) largely occupy three regions of the Ramachandran diagram.]

Milner-White, E.J., The partial charge of the nitrogen atom in peptide bonds, *Protein Sci.* **6**, 2477–2482 (1997). [Discusses the origin of the peptide N atom's partial negative charge.]

Toniolo, C. and Benedetti, E., The polypeptide 3_{10}-helix, *Trends Biochem. Sci.* **16**, 350–353 (1991).

Fibrous Proteins

Brodsky, B. and Persikov, A.V., Molecular structure of the collagen triple helix, *Adv. Prot. Chem.* **70**, 301–339 (2005).

Byers, P.H., Disorders of collagen synthesis and structure, Chapter 205 *in* Valle, D. (Ed.), *The Online Metabolic & Molecular Bases of Inherited Disease*, http://www.ommbid.com/.

Herrmann, H. and Aebi, U., Intermediate filaments: Molecular structure, assembly mechanism, and integration into functionally distinct intracellular scaffolds, *Annu. Rev. Biochem.* **73**, 749–789 (2004).

Myllyharju, J. and Kivirikko, K.I., Collagens, modifying enzymes and their mutations in humans, flies and worms, *Trends Genet.* **20**, 33–43 (2004).

Orgel, J.P.R.O., Miller, A., Irving, T.C., Fischetti, R.F., Hammersley, A.P., and Wess, T.J., The in situ supermolecular structure of type I collagen, *Structure* **9**, 1061–1069 (2001).

Parry, D.A.D., Microdissection of the sequence and structure of intermediate filament chains, *Adv. Prot. Chem.* **70**, 113–142 (2005).

Shoulders, M.D. and Raines, R.T., Collagen structure and stability, *Annu. Rev. Biochem.* **78**, 929–958 (2009).

Wess, T.J., Collagen fibril form and function, *Adv. Prot. Chem.* **70**, 341–374 (2005).

Macromolecular Structure Determination

Blow, D., *Outline of Crystallography for Biologists*, Oxford University Press (2002).

Brünger, A.T., X-Ray crystallography and NMR reveal complementary views of structure and dynamics, *Nature Struct. Biol.* **4**, 862–865 (1997).

Doublié, S. (Ed.), *Macromolecular Crystallography Protocols*, Vols. 1 and 2, Humana Press (2007).

Drenth, J., *Principles of Protein X-Ray Crystallography* (3rd ed.), Springer (2007).

Ferentz, A.E. and Wagner, G., NMR spectroscopy: A multifaceted approach to macromolecular structure, *Q. Rev. Biophys.* **33**, 29–65 (2000).

Fernández, C. and Wider, G., TROSY in NMR studies of the structure and function of large biological macromolecules, *Curr. Opin. Struct. Biol.* **13**, 570–580 (2003).

Joachimiak, A. (Ed.), Structural Genomics, Parts A–C, *Adv. Prot. Chem. Struct. Biol.* **75–77** (2008, 2009, 2010).

Lattman, E.E. and Loll, P.J., *Protein Crystallography. A Concise Guide*, Johns Hopkins University Press (2008).

McPherson, A., *Introduction to Macromolecular Crystallography*, Wiley-Liss (2003).

Messerschmidt, A., *X-Ray Crystallography of Biomacromolecules. A Practical Guide*, Wiley-VCH (2007).

Reik, R., Pervushkin, K., and Wüthrich, K., TROSY and CRINEPT: NMR with large molecular and supramolecular structures in solution, *Trends Biochem. Sci.* **25**, 462–468 (2000).

Rhodes, G., *Crystallography Made Crystal Clear: A Guide for Users of Macromolecular Models* (3rd ed.), Academic Press (2006).

Rule, G.S., and Hitchens, T.K., *Fundamentals of Protein NMR Spectroscopy*, Springer (2006).

Rupp, B., *Biomolecular Crystallography: Principles, Practice, and Applications to Structural Biology*, Garland Science (2010).

Structure **15**, 1342–1356 (2007); **15**, 1517–1529 (2007); and **16**, 1–18 (2008). [A series of Opinion articles on the pros and cons of the structural genomics project and, in particular, on the NIH-sponsored Protein Structure Initiative (PSI).]

Sundström, M., Norin, M., and Edwards, A., *Structural Genomics and High Throughput Structural Biology*, Taylor & Francis (2006).

Terwilliger, T.C., Stuart, D., and Yokoyama, S., Lessons from structural genomics, *Annu. Rev. Biophys.* **38**, 371–383 (2009).

Tugarinov, V., Hwang, P.M., and Kay, L.E., Nuclear magnetic resonance spectroscopy of high-molecular weight proteins, *Annu. Rev. Biochem.* **73**, 107–146 (2004).

Tzakos, A.G., Grace, C.R.R., Lukavsky, P.J., and Reik, R., NMR techniques for very large proteins and RNAs in solution, *Annu. Rev. Biophys.* **35**, 319–341 (2006).

Wüthrich, K., NMR—This other method for protein and nucleic structure determination, *Acta Cryst.* **D51**, 249–270 (1995); *and* Protein structure determination in solution by nuclear magnetic resonance spectroscopy, *Science* **243**, 45–50 (1989).

Globular Proteins

Chothia, C. and Finkelstein, A.V., The classification and origins of protein folding patterns, *Annu. Rev. Biochem.* **59**, 1007–1039 (1990).

Edwards, A., Large-scale structural biology of the human proteome, *Annu. Rev. Biochem.* **78**, 541–568 (2009). [Discusses the impact of structural genomics on determining the structures of human proteins.]

Farber, G.K., An α/β-barrel full of evolutionary trouble, *Curr. Opin. Struct. Biol.* **3**, 409–412 (1993); *and* Farber, G.K. and

Petsko, G.A., The evolution of α/β barrel enzymes, *Trends Biochem. Sci.* **15,** 228–234 (1990).

Garratt, R.C. and Orengo, C.A., *The Protein Chart,* Wiley-VCH (2008). [A beautifully illustrated chart of protein structures that shows their diversity and complexity.]

Goodsell, D.S., Visual methods from atoms to cells, *Structure* **13,** 347–364 (2005).

Gruber, M. and Lupas, A.N., Historical review: Another 50th anniversary—new periodicities in coiled coils, *Trends Biochem. Sci.* **28,** 679–685 (2003).

Gu, J. and Bourne, P. (Eds.), *Structural Bioinformatics* (2nd ed.) Wiley-Blackwell (2009).

Janin, J., Bahadur, R.P., and Chakrabarti, P., Protein–protein interaction and quaternary structure, *Q. Rev. Biophys.* **41,** 133–180 (2008).

Orengo, C.A. and Thornton, J.M., Protein families and their evolution—a structural perspective, *Annu. Rev. Biochem.* **74,** 867–900 (2005).

Richards, F.M., Areas, volumes, packing, and protein structure, *Annu. Rev. Biophys. Bioeng.* **6,** 151–176 (1977).

Richardson, J.S., The anatomy and taxonomy of protein structures, *Adv. Protein Chem.* **34,** 168–339 (1981). [A detailed discussion of the structural principles governing globular proteins accompanied by an extensive collection of their cartoon representations.]

Richardson, J.S. and Richardson, D.C., Principles and patterns of protein conformation, *in* Fasman, G.D. (Ed.), *Prediction of Protein Structure and the Principles of Protein Conformation,* pp. 1–98, Plenum Press (1989). [A comprehensive account of protein conformations based on X-ray structures.]

Protein Stability

Baldwin, R.L., and Baker, D. (Eds.), Peptide Solvation and H-Bonds, *Adv. Protein. Chem.* **72** (2006).

Bolen, D.W. and Rose, G.D., Structure and energetics of the hydrogen-bonded backbone in protein folding, *Annu. Rev. Biochem.* **77,** 339–362 (2008).

Burley, S.K. and Petsko, G.A., Weakly polar interactions in proteins, *Adv. Protein Chem.* **39,** 125–189 (1988).

Chandler, D., Interfaces and the driving force of hydrophobic assembly, *Nature* **437,** 640–647 (2005).

Derewenda, Z.S., Lee, L., and Derewenda, U., The occurrence of C—H···O hydrogen bonds in proteins, *J. Mol. Biol.* **252,** 248–262 (1995).

Fersht, A., *Structure and Mechanism in Protein Science,* Chapter 11, Freeman (1999).

Fersht, A.R. and Serrano, L., Principles of protein stability derived from protein engineering experiments, *Curr. Opin. Struct. Biol.* **3,** 75–83 (1993). [Discusses how the roles of specific side chains in proteins can be quantitatively determined by mutationally changing them and calorimetrically measuring the stabilities of the resulting proteins.]

Goldman, A., How to make my blood boil, *Structure* **3,** 1277–1279 (1995). [Discusses the stabilities of hyperthermophilic proteins.]

Hendsch, Z. and Tidor, B., Do salt bridges stabilize proteins? A continuum electrostatic analysis, *Protein. Sci.* **3,** 211–226 (1994).

Honig, B. and Nichols, A., Classical electrostatics in biology and chemistry, *Science* **268,** 1144–1149 (1995).

Jaenicke, R. and Böhm, G., The stability of proteins in extreme environments, *Curr. Opin. Struct. Biol.* **8,** 738–748 (1998).

Jeffrey, G.A. and Saenger, W., *Hydrogen Bonding in Biological Structures,* Springer-Verlag (1991).

Jones, S. and Thornton, J.M., Principles of protein–protein interactions, *Proc. Natl. Acad. Sci.* **93,** 13–20 (1996).

Karshikoff, A. and Ladenstein, R., Ion pairs and the thermotolerance of proteins from hyperthermophiles: A 'traffic rule' for hot roads, *Trends Biochem. Sci.* **26,** 550–556 (2001).

Kauzmann, W., Some factors in the interpretation of protein denaturation, *Adv. Protein Chem.* **14,** 1–63 (1958). [A classic review that first pointed out the importance of hydrophobic bonding in stabilizing proteins.]

Martin, T.W. and Derewenda, Z.S., The name is bond—H bond, *Nature Struct. Biol.* **6,** 403–406 (1999). [Reviews the history of the concept of hydrogen bonding and discusses X-ray scattering experiments that demonstrated that hydrogen bonds have ~10% covalent character.]

Matthews, B.W., Studies on protein stability with T4 lysozyme, *Adv. Protein Chem.* **46,** 249–278 (1995). [A distillation of the results of stability studies on a large number of mutant varieties of lysozyme from bacteriophage T4, many of which have also been structurally determined by X-ray analysis.]

Mattos, C., Protein–water interactions in a dynamic world, *Trends Biochem. Sci.* **27,** 203–208 (2002).

Ramachandran, G.N. and Sasisekharan, V., Conformation of polypeptides and proteins, *Adv. Protein Chem.* **23,** 283–437 (1968). [A classic paper.]

Rees, D.C. and Adams, M.W.W., Hyperthermophiles: Taking the heat and loving it, *Structure* **3,** 251–254 (1995).

Richards, F.M., Folded and unfolded proteins: An introduction, *in* Creighton, T.E. (Ed.), *Protein Folding,* pp. 1–58, Freeman (1992).

Schellman, J.A., The thermodynamic stability of proteins, *Annu. Rev. Biophys. Biophys. Chem.* **16,** 115–137 (1987).

Steiner, T. and Koellner, G., Hydrogen bonds with π-acceptors in proteins: Frequencies and role in stabilizing local 3D structures, *J. Mol. Biol.* **305,** 535–557 (2001).

Stickle, D.F., Presta, L.G., Dill, K.A., and Rose, G.D., Hydrogen bonding in globular proteins, *J. Mol. Biol.* **226,** 1143–1159 (1992).

Tanford, C., How protein chemists learned about the hydrophobic factor, *Protein Science* **6,** 1358–1366 (1997). [A historical narrative.]

Weiss, M.S., Brandl, M., Sühnel, J., Pal, D., and Hilgenfeld, R., More hydrogen bonds for the (structural) biologist, *Trends Biochem. Sci.* **26,** 521–523 (2001). [Discusses C—H···O and C—H···π hydrogen bonds.]

Yang, A.-S. and Honig, B., Electrostatic effects on protein stability, *Curr. Opin. Struct. Biol.* **2,** 40–45 (1992).

Quaternary Structure

Eisenstein, E. and Schachman, H.K., Determining the roles of subunits in protein function, *in* Creighton, T.E. (Ed.), *Protein Function. A Practical Approach,* pp. 135–176, IRL Press (1989).

Goodsell, D.S. and Olson, J., Structural symmetry and protein function, *Annu. Rev. Biophys. Biomol. Struct.* **29,** 105–153 (2000).

Sheinerman, F.B., Norel, R., and Honig, B., Electrostatic aspects of protein–protein interactions, *Curr. Opin. Struct. Biol.* **10,** 153–159 (2000).

BIOINFORMATICS EXERCISES

Bioinformatics Exercises are available at www.wiley.com/college/voet

Chapter 8

Visualization of Three-Dimensional Protein Structures
1. **3D Structural Alignment of Proteins at the Protein Data Bank.** Align the structures of homologous proteins.
2. **Exercises with Jmol.** Inspect the 3D structure of a protein.
3. **Exercises with PyMOL.** Inspect the 3D structure of a protein.

PROBLEMS

1. What is the length of an α helical section of a polypeptide chain of 20 residues? What is its length when it is fully extended (all trans)?

*2. From an examination of Fig. 8-7, it is apparent that the polypeptide conformation angle ϕ is more constrained than is ψ. By referring to Fig. 8-4, or better yet, by examining a molecular model, indicate the sources of the steric interference that limit the allowed values of ϕ when ψ = 180°.

3. For a polypeptide chain made of γ-amino acids, state the nomenclature of the helix analogous to the 3_{10} helix of α-amino acids. Assume the helix has a pitch of 9.9 Å and a rise per residue of 3.2 Å.

*4. Table 8-7 (page 276) gives the torsion angles, ϕ and ψ, of hen egg white lysozyme for residues 24–73 of this 129-residue protein. (a) What is the secondary structure of residues 26–35? (b) What is the secondary structure of residues 42–53? (c) What is the probable identity of residue 54? (d) What is the secondary structure of residues 56–68? (e) What is the secondary structure of residues 69–71? (f) What additional information besides the torsion angles, ϕ and ψ, of each of its residues is required to define the three-dimensional structure of a protein?

5. Hair splits most easily along its fiber axis, whereas fingernails tend to split across the finger rather than along it. What are the directions of the keratin fibrils in hair and in fingernails? Explain your reasoning.

6. What is the growth rate, in turns per second, of the α helices in a hair that is growing 15 cm · year^{-1}?

7. Can polyproline form a collagenlike triple helix? Explain.

8. As Mother Nature's chief engineer, you have been asked to design a five-turn α helix that is destined to have half its circumference immersed in the interior of a protein. Indicate the helical wheel projection of your prototype α helix and its amino acid sequence (see Fig. 8-43*a*).

9. β-Aminopropionitrile is effective in reducing excessive scar tissue formation after an injury (although its use is contraindicated by side effects). What is the mechanism of action of this lathyrogen?

*10. Using your network browser, visit the Protein Data Bank (PDB) at http://www.rcsb.org/ (for this exercise, you should use some other browser besides Safari because it is not fully compatible with CE). To explore the structure of γ-B crystallin, click the "Enter a PDB id or keyword" button, enter "4GCR" in the field to its right, and click the "Site Search" button. On the left of the "Structure Summary" window that comes up, click on "PDB File" under "Display Files" and inspect the file that appears. (a) How many residues does this protein have and how many water molecules have been found to be associated with it in this crystal structure? (b) Draw skeletal diagrams of an Arg, a Glu, and a Tyr residue and label their atoms using the nomenclature employed in the PDB file. (c) What are the atomic coordinates of the S atom of Cys 32? What is the identity of atom 1556, which is labeled "OXT"? (d) Return to the Structure Summary page and, under "Display Options" on the right, examine the structure of the protein using the viewer of your choice. Can you see that the protein is composed of two well-separated and apparently similar domains? (e) To structurally classify this protein, return to the Structure Summary page and state the different ways that SCOP and CATH (near the bottom of the page) classify this protein. (f) To structurally compare the protein's two domains, point your browser to CE (http://cl.sdsc.edu/) and, in the subsequent window, click on "Two Chains." In the window that comes up, for both Chain 1 and Chain 2, click on the "PDB:" button, enter "4GCR" in the box to its right, and check the "Use Fragment From:" box. Then for Chain 1 enter the range 1 to 83 and for Chain 2 enter the range 84 to 174. Then click on the "Calculate Alignment" button (at the top). The window that appears shows the aligned sequences of the two peptides based on their structure. What is the percent identity of the two peptides (upper right)? Describe the gaps, if any, in the alignment. Now click on the "Press to Start Compare3D" button. The window that comes up shows the superimposed C_α backbones of the two peptide segments in blue and magenta. Rotate this model by clicking on it and dragging your mouse. Describe what you see. What is the "RMSD (A)" (root mean square deviation in Å) of the superimposed backbone segments? What is the significance of the short white segment of chain? What are the two gray segments? Click on specific residues in the sequences below to highlight their positions in the aligned structures.

*11. Using a molecular graphics viewer, inspect the structures of the following proteins as indicated by their PDBids. Draw the corresponding topological diagrams and name the fold, if it is a standard one, for each domain in the protein. (a) 1 RCP, (b) 1 RCB, (c) 1 TNF, (d) 2 CMD, (e) 1 RHD, and (f) 2 TAA. [Note: Although all of these proteins consist of only one type of subunit, some of the PDB files contain the coordinates for more

Table 8-7 **Torsion Angles (φ, ψ) for Residues 24 to 73 of Hen Egg White Lysozyme**

Residue Number	Amino Acid	φ (deg)	ψ (deg)
24	Ser	−60	147
25	Leu	−49	−32
26	Gly	−67	−34
27	Asn	−58	−49
28	Trp	−66	−32
29	Val	−82	−36
30	Cys	−69	−44
31	Ala	−61	−44
32	Ala	−72	−29
33	Lys	−66	−65
34	Phe	−67	−23
35	Glu	−81	−51
36	Ser	−126	−8
37	Asn	68	27
38	Phe	79	6
39	Asn	−100	109
40	Thr	−70	−18
41	Glu	−84	−36
42	Ala	−30	142
43	Thr	−142	150
44	Asn	−154	121
45	Arg	−91	136
46	Asn	−110	174
47	Thr	−66	−20
48	Asp	−96	36
49	Gly	95	−75
50	Ser	−18	138
51	Thr	−131	157
52	Asp	−115	130
53	Tyr	−126	146
54	xxx	67	−179
55	Ile	−42	−37
56	Leu	−107	14
57	Gln	35	54
58	Ile	−72	133
59	Asn	−76	153
60	Ser	−93	−3
61	Arg	−83	−19
62	Trp	−133	−37
63	Trp	−91	−32
64	Cys	−151	143
65	Asn	−85	140
66	Asp	133	8
67	Gly	73	−8
68	Arg	−135	17
69	Thr	−122	83
70	Pro	−39	−43
71	Gly	−61	−11
72	Ser	−45	122
73	Arg	−124	146

Source: Imoto, T., Johnson, L.N., North, A.C.T., Phillips, D.C., and Rupley, J.A., *in* Boyer, P.D. (Ed.), *The Enzymes* (3rd ed.), Vol. 7, *pp.* 693–695, Academic Press (1972).

than one chain. Your task will be simplified if you display only one of these chains (their structures are essentially identical). If you use the stand-alone version of Jmol as your viewer (a Jmol tutorial is on the "Student Companion Site" at http://www.wiley.com/college/voet/), this is easily done by typing, for example, "restrict *a" into the Console window to restrict the display to the A chain. For proteins consisting of more than one domain, inspect each domain individually using the same technique (for example, after typing "restrict *a", typing "restrict 1-20" will display only residues 1–20 of the A chain). A polypeptide chain is most easily followed if it is displayed in Backbone, Cartoons, Ribbons, Strands, or Trace form and assigned Group colors by opening the Console and entering the command "color group all" (which colors the chain, from N- to C-terminus, in rainbow order, blue to red).]

12. It is often stated that proteins are quite large compared to the molecules they bind. However, what constitutes a large molecule depends on your point of view. Calculate the ratio of the volume of a hemoglobin molecule (65 kD) to that of the four O_2 molecules it binds. Also calculate the ratio of the volume of a typical office ($4 \times 4 \times 3$ m) to that of a typical office worker (70 kg) that occupies it. Assume that the molecular volumes of hemoglobin and O_2 are in equal proportion to their molecular masses and that the office worker has a density of 1.0 g $\cdot$ cm^{-3}. Compare these ratios. Is this the answer you expected?

13. Why are London dispersion forces always attractive?

14. Membrane-bound proteins are generally closely associated with the nonpolar groups of lipid molecules (Section 12-3A). Explain how detergents affect the structural integrity of membrane-bound proteins in comparison to their effects on normal globular proteins.

15. The coat protein of **tomato bushy stunt virus** consists of 180 chemically identical subunits, each of which is composed of ~386 amino acid residues. The probability that a wrong amino acid residue will be biosynthetically incorporated in a polypeptide chain is 1 part in 3000 per residue. Calculate the average number of coat protein subunits that would have to be synthesized in order to produce a perfect viral coat. What would this number be if the viral coat were a single polypeptide chain with the same number of residues that it actually has?

16. State the rotational symmetries of each of the following objects: (a) a starfish, (b) a square pyramid, (c) a rectangular box, and (d) a trigonal bipyramid.

*17. Through the use of your favorite molecular graphics viewer, state the rotational symmetries of the proteins with the following PDBids: (a) 1 TIM, (b) 1 TNF, (c) 6 PFK, and (d) 1 AIY. [Note: When using Jmol, individual polypeptide chains are most easily differentiated by displaying the protein in Backbone, Cartoons, Ribbons, Strands, or Trace form and assigning them Chain colors (e.g., Color > Structure > Cartoon > By Scheme > Chain), which makes each polypeptide chain a different color.]

18. Myoglobin and the subunits of hemoglobin are polypeptides of similar size and structure. Compare the expected ratio of nonpolar to polar amino acid residues in myoglobin and in hemoglobin.

19. Sickle-cell hemoglobin (HbS) differs from normal human adult hemoglobin (HbA) by a single mutational change, Glu $\beta 6 \rightarrow$ Val, which causes the HbS molecules to aggregate under proper conditions (Section 7-3A). Under certain conditions, the HbS filaments that form at body temperature disaggregate when the temperature is lowered to $0°C$. Explain.

20. Indicate experimental evidence that is inconsistent with the hypothesis that urea and guanidinium ion act to denature proteins by competing for their internal hydrogen bonds.

21. Proteins in solution are often denatured if the solution is shaken violently enough to cause foaming. Indicate the mechanism of this process. (*Hint:* The nonpolar groups of detergents extend into the air at air–water interfaces.)

22. An oligomeric protein in a dilute buffer at pH 7 dissociates to its component subunits when exposed to the following agents. Which of these observations would not support the contention that the quaternary structure of the protein is stabilized exclusively by hydrophobic interactions? Explain. (a) $6M$ guanidinium chloride, (b) 20% ethanol, (c) $2M$ NaCl, (d) temperatures below $0°C$, (e) 2-mercaptoethanol, (f) pH 3, and (g) $0.01M$ SDS.

*23. The SDS–polyacrylamide gel electrophoresis of a protein yields two bands corresponding to molecular masses of 10 and 17 kD. After cross-linking this protein with dimethylsuberimidate under sufficient dilution to eliminate intermolecular cross-linking, SDS–polyacrylamide gel electrophoresis of the product yields 12 bands with molecular masses 10, 17, 20, 27, 30, 37, 40, 47, 54, 57, 64, and 74 kD. Assuming that dimethylsuberimidate can cross-link only contacting subunits, diagram the quaternary structure of the protein.

*24. Mammals possess two genetically distinct but closely related forms of lactate dehydrogenase (LDH), the M type (which predominates in skeletal muscle) and the H type (which predominates in heart tissue). Before the X-ray structure of LDH was known, the oligomeric state of LDH was determined by dissociating M type and H type LDH to their component subunits and then reconstituting the mixture. This treatment yielded five electrophoretically distinct **isozymes** (catalytically and structurally similar but genetically distinct enzymes from the same organism), M_4, M_3H, M_2H_2, MH_3, and H_4, thereby demonstrating that LDH is a tetramer. What are the relative amounts of each isozyme formed when equimolar amounts of M_4 and H_4 are so hybridized?

Protein Folding, Dynamics, and Structural Evolution

CHAPTER 9

In the preceding chapters, we saw how proteins are constructed from their component parts. This puts us in a similar position to a mechanic who has learned to take apart and put together an automobile engine without any inkling of how the engine works. What we need in order to understand the workings of a protein is knowledge of the types of internal motions it can and must undergo in order to carry out its biological function as well as how it achieves its native structure. Put in terms of our deprived auto mechanic, we wish to understand the operations of the "gears" and "levers" with which proteins carry out their function. This is a problem of enormous complexity whose solution we are only beginning to understand. We shall see in later chapters, for example, that even though the catalytic mechanisms of many enzymes of known structure have been studied in great detail, it cannot be said that we fully understand any of these mechanisms. This is because our comprehension of the ways in which a protein's component groups interact is far from complete. As far as proteins are concerned, we have not greatly surpassed our hypothetical mechanic's level of understanding.

In this third of four chapters on protein structure, we consider the temporal behavior of proteins. Specifically, we first take up the problem of how random coil polypeptides fold to their native structures and how this process is facilitated by other proteins. This is followed by a consideration of the progress that has been made in predicting protein structures based on their sequences and in understanding their dynamic properties, that is, the nature and functional significance of their internal motions. Next we consider the diseases caused by proteins taking up alternate conformations. We end by extending the discussions we began in Section 7-3 on protein evolution but do so in terms of the three-dimensional structures of proteins.

1 PROTEIN FOLDING: THEORY AND EXPERIMENT

Solving the so-called **protein folding problem,** that is, determining how and why proteins fold to their native states, is considered to be one of the grand challenges of biochemistry. Early notions of protein folding postulated the existence of "templates" that somehow caused proteins to assume their native conformations. Such an explanation begs the question of how proteins fold because, even if it were true, one would still have to explain how the template achieved its conformation. In fact, *proteins spontaneously fold into their native conformations under physiological conditions.* This implies that *a protein's primary structure dictates its three-dimensional structure.* In general, under the proper conditions, biological structures are **self-assembling** so that they have no need of external templates to guide their formation.

A. Protein Renaturation

Although evidence had been accumulating since the 1930s that proteins could be reversibly denatured, it was not until 1957 that the elegant experiments of Christian Anfinsen on bovine pancreatic **RNase A** put **protein renaturation** on a quantitative basis. RNase A, a 124-residue single-chain protein, is completely unfolded and its four disulfide bonds reductively cleaved in an $8M$ urea solution containing 2-mercaptoethanol (Fig. 9-2). Yet dialyzing away the urea and exposing the resulting solution to O_2 at pH 8

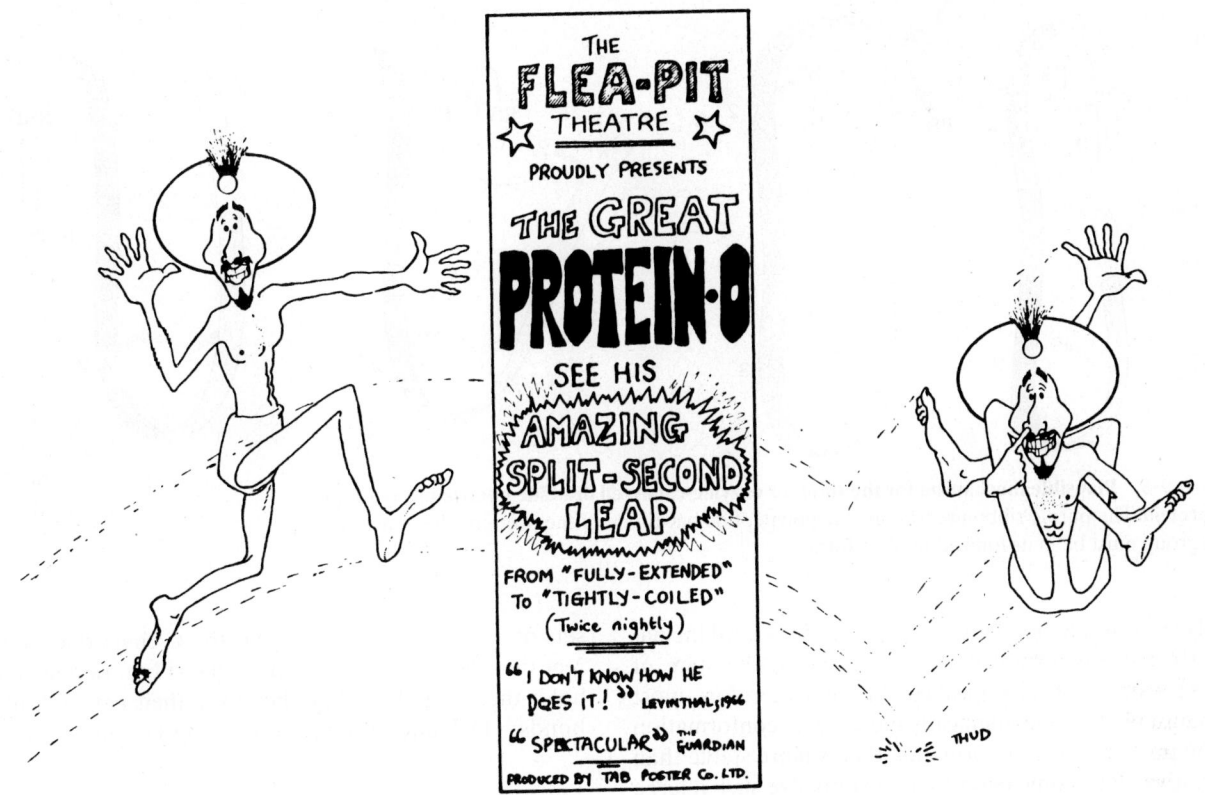

Figure 9-1 [Drawing by T.A. Bramley, *in* Robson, B., *Trends Biochem. Sci.* **1,** 50 (1976). Copyright © Elsevier Biomedical Press, 1976. Used by permission.]

Figure 9-2 Reductive denaturation and oxidative renaturation of RNase A. [Illustration, Irving Geis. Image from the Irving Geis Collection, Howard Hughes Medical Institute. Reprinted with permission.]

yields a protein that is virtually 100% enzymatically active and physically indistinguishable from native RNase A. The protein must therefore have spontaneously renatured. Any reservations that this occurs only because RNase A is really not totally denatured by $8M$ urea have been satisfied by the chemical synthesis of enzymatically active RNase A (Section 7-5).

The renaturation of RNase A demands that its four disulfide bonds reform. The probability of one of the eight Cys residues from RNase A randomly reforming a disulfide bond with its proper (native) mate among the other seven Cys residues is $\frac{1}{7}$; that of one of the remaining six Cys residues then randomly reforming its proper disulfide bond is $\frac{1}{5}$; etc. The overall probability of RNase A reforming its four native disulfide links at random is

$$\frac{1}{7} \times \frac{1}{5} \times \frac{1}{3} \times \frac{1}{1} = \frac{1}{105}$$

Clearly, the disulfide bonds from RNase A do not randomly reform under renaturing conditions.

If the RNase A is reoxidized in $8M$ urea so that its disulfide bonds reform while the polypeptide chain is a random coil, then after removal of the urea, the RNase A is, as expected, only ~1% enzymatically active. This "scrambled" RNase A can be made fully active by exposing it to a trace of 2-mercaptoethanol, which, over about a 10-h period,

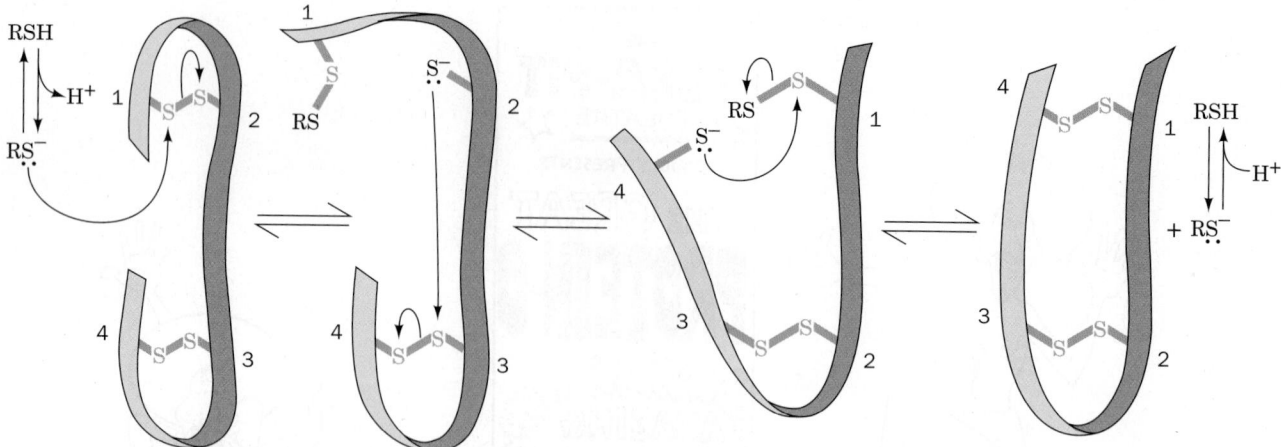

Figure 9-3   **Plausible mechanism for the thiol- or enzyme-catalyzed disulfide interchange reaction in a protein.** The purple ribbon represents the polypeptide backbone of the protein. The attacking thiol group must be in its ionized thiolate form.

catalyzes disulfide bond interchange reactions until the native structure is achieved (Fig. 9-3). The native state of RNase A under physiological conditions is therefore, most probably, its thermodynamically most stable conformation (if the protein has a conformation that is more stable than the native state, conversion to it must involve such a large activation barrier so as to make it kinetically inaccessible; rate processes are discussed in Section 14-1C).

The time for renaturation of "scrambled" RNase A is reduced to ~2 min through the use of an enzyme, **protein disulfide isomerase (PDI),** that catalyzes disulfide interchange reactions. (In fact, the supposition that *in vivo* folding to the native state requires no more than a few minutes prompted the search that led to this enzyme's discovery.) PDI itself contains two active site Cys residues, which must be in the —SH form for the isomerase to be active. The enzyme evidently catalyzes the random cleavage and reformation of a protein's disulfide bonds (Fig. 9-3), thereby interchanging them as the protein progressively attains thermodynamically more favorable conformations. PDI is further discussed in Section 9-2A.

a. Post-Translationally Modified Proteins May Not Readily Renature

Many "scrambled" proteins are renatured through the action of PDI and are unaffected by it in their native state (their PDI-cleaved disulfide bonds rapidly reform because these native proteins are in their most stable local conformations). In post-translationally modified proteins, however, the disulfide bonds may serve to hold the protein in its otherwise unstable native state. For instance, the 51-residue polypeptide hormone **insulin,** which consists of two polypeptide chains joined by two disulfide bonds (Fig. 7-2), is inactivated by PDI. This observation led to the discovery that insulin is derived from a single-chain, 84-residue precursor named **proinsulin** (Fig. 9-4). Only after its disulfide bonds have formed is proinsulin converted to the two-chained active hormone by the specific proteolytic excision of an internal 33-residue segment known as its C chain. Nevertheless, two

sets of observations suggest that the C chain does not direct the folding of the A and B chains but, rather, simply holds them together while they form their native disulfide bonds: (1) Under proper renaturing conditions, native

Figure 9-4   **Primary structure of porcine proinsulin.** Its C chain (*brown*) is proteolytically excised from between its A and B chains to form the mature hormone. [After Chance, R.E., Ellis, R.M., and Brommer, W.W., *Science* **161,** 165 (1968).]

insulin is obtained from scrambled insulin in 25 to 30% yield, which increases to 75% when the A and B chains are chemically cross-linked; and (2) sequence comparisons of proinsulins from a variety of species indicate that mutations are accepted into the C chain at a rate which is eight times that for the A and B chains.

B. *Determinants of Protein Folding*

In Section 8-4, we discussed the various interactions that stabilize native protein structures. In this section we extend the discussion by considering how these interactions are organized in native proteins. Keep in mind that only a small fraction of the myriads of possible polypeptide sequences are likely to have unique stable conformations. Evolution has, of course, selected such sequences for use in biological systems.

a. Helices and Sheets May Predominate in Proteins Simply because They Fill Space Efficiently

Why do proteins contain such a high proportion ($\sim 60\%$, on average) of α helices and β pleated sheets? Hydrophobic interactions, although the dominant influence responsible for the compact nonpolar cores of proteins, lack the specificity to restrict polypeptides to particular conformations. Similarly, the observation that polypeptide segments in the coil conformation are no less hydrogen bonded than helices and sheets suggests that the conformations available to polypeptides are not greatly limited by their hydrogen bonding requirements. Rather, as Ken Dill has shown, it appears that helices and sheets form largely as a consequence of steric constraints in compact polymers. Exhaustive simulations of the conformations which simple flexible chains (such as a string of pearls) can assume indicate that the proportion of helices and sheets increases dramatically with a chain's level of compaction (number of intrachain contacts); that is, helices and sheets are particularly compact entities. Thus, most ways to compact a chain involve the formation of helices and sheets. In native proteins, such elements of secondary structure are fine tuned to form α helices and β sheets by short-range forces such as hydrogen bonding, ion pairing, and van der Waals interactions. It is probably these less dominant but more specific forces that "select" the unique native structure of a protein from among its relatively small number of hydrophobically generated compact conformations (recall that most hydrogen bonds in proteins link residues that are close together in sequence; Section 8-4Bb).

b. Protein Folding Is Directed Mainly by Internal Residues

Numerous protein modification studies have been aimed at determining the role of various classes of amino acid residues in protein folding. In one particularly revealing study, the free primary amino groups of RNase A (Lys residues and the N-terminus) were derivatized with 8-residue chains of poly-DL-alanine. Intriguingly, these large, water-soluble poly-Ala chains could be simultaneously coupled to RNase's 11 free amino groups without significantly altering the protein's native conformation or its ability to refold. Since these free amino groups are all located on the exterior of RNase A, this observation suggests that *it is largely a protein's internal residues that direct its folding to the native conformation.* Similar conclusions have been reached from studies of protein structure and evolution (Section 9-6): Mutations that change surface residues are accepted more frequently and are less likely to affect protein conformations than are changes of internal residues. It is therefore not surprising that the perturbation of protein folding by limited concentrations of denaturing agents indicates that *protein folding is driven by hydrophobic forces.*

c. Protein Structures Are Hierarchically Organized

Large protein subunits consist of domains, that is, of contiguous, compact, and physically separable segments of the polypeptide chain. Furthermore, as George Rose showed, domains consist of subdomains, which in turn consist of sub-subdomains, etc. Conceptually, this means that if a polypeptide segment of any length in a native protein is viewed as a tangle of string, a single plane can be found that divides the string into only two segments rather than many smaller segments (such as would happen if a ball of yarn were cut in this way). This is readily demonstrated by coloring the first $n/2$ residues of an n-residue domain red and the second $n/2$ residues blue. If this process is iterated, as is shown in Fig. 9-5 for high potential iron–sulfur protein (HiPIP), it is clear that at every stage of the process, the red

Figure 9-5 **Hierarchical organization of globular proteins.** Here the X-ray structure of high potential iron–sulfur protein (HiPIP) is represented by its C_α atoms shown as spheres. In the top drawing, the first $n/2$ residues of this n-residue protein (where $n = 71$) are colored red and the remaining $n/2$ residues are colored blue. In the second row, the process is iterated such that, on the right, for example, the first and last halves of the second half of the protein are red and blue, with the remainder of the chain gray. In the third row, the process is again iterated. Note that at each stage of this hierarchy, the red and blue regions do not intermingle. [Courtesy of George Rose, Johns Hopkins University, and Robert Baldwin, Stanford University School of Medicine.]

and blue regions do not interpenetrate. Evidently, *protein structures are organized hierarchically,* that is, polypeptide chains form locally compact structures that associate with similar adjacent (in sequence) structures to form larger compact structures, etc. This structural organization is, of course, consistent with the observation that hydrogen bonding interactions in proteins are mostly local (Section 8-4Bb). It also has important implications for how polypeptides fold to form native proteins (Section 9-1C).

d. Protein Structures Are Highly Adaptable

Globular proteins have packing densities comparable to those of organic crystals (Section 8-3Bc) because the side chains in a protein's interior fit together with exquisite complementarity. To ascertain whether this phenomenon is an important determinant of protein structure, Eaton Lattman and Rose analyzed 67 globular proteins of known structure for the existence of preferred interactions between side chains. They found none, thereby indicating that, at least in globular proteins, *the native fold determines the packing but packing does not determine the native fold.* This view is corroborated by the widespread occurrence of protein families whose members assume the same fold even though they may be so distantly related as to have no recognizable sequence similarity (e.g., the α/β barrel proteins; Section 8-3Bh).

The foregoing study indicates that *there are a large number of ways in which a protein's internal residues can pack together efficiently.* This was perhaps most clearly shown by Brian Matthews in an extensive series of studies on **T4 lysozyme** (a product of bacteriophage T4) in which the X-ray structures of over 300 mutant varieties of this 164-residue monomeric enzyme were compared. Replacements of one or a few residues in T4 lysozyme's hydrophobic core were accommodated mainly by local shifts in the protein backbone rather than by any global structural changes. In many cases, T4 lysozyme could accommodate the insertion of up to four residues without a major structural change or even a loss of enzymatic activity. Moreover, assays of the enzymatic activities of 2015 single-residue substitutions in T4 lysozyme indicated that only 173 of these mutants had significantly decreased enzymatic activity. Clearly protein structures are highly resilient.

e. Secondary Structure Can Be Context-Dependent

The structure of a native protein is determined by its amino acid sequence, but to what extent is the conformation of a given polypeptide segment influenced by the surrounding protein? The NMR structure of **protein GB1** (the B1 domain of streptococcal **protein G,** which helps the bacterium evade the host's immunological defenses by binding to the antibody protein **immunoglobulin G**) reveals that this 56-residue domain, which lacks disulfide bonds, consists of a long α helix lying across a 4-stranded mixed β sheet (Fig. 9-6). In mutagenesis experiments by Peter Kim, the 11-residue "chameleon" sequence AWTVEKAFKTF was made to replace either residues 23 to 33 of GB1's α helix (AATAEKFVFQY in GB1; a 7-residue change) to yield Chm-α, or residues 42 to 52 of its C-terminal β hairpin

Figure 9-6   NMR structure of protein GB1. Residues 23 to 33 are green and residues 42 to 53 are cyan. The 11-residue chameleon sequence AWTVEKAFKTF can occupy either of these positions without significantly altering the native protein's backbone conformation. [NMR structure by Angela Gronenborn and Marius Clore, National Institutes of Health, Bethesda, Maryland. PDBid 1GB1.]

(EWTYDDATKTF in GB1; a 5-residue change) to yield Chm-β. Both Chm-α and Chm-β display reversible thermal unfolding typical of compact single-domain globular proteins, and their 2D NMR spectra indicate that each assumes a structure similar to that of native GB1. Yet NMR measurements also demonstrate that the isolated chameleon peptide (Ac-AWTVEKAFKTF-NH₂, where Ac is acetyl) is unfolded in solution, which indicates that this sequence has no strong preference for either an α helix or a β sheet conformation. This suggests that the information specifying α helix or β sheet secondary structures can be nonlocal; that is, context-dependent effects may be important in protein folding (but see Section 9-1Ci).

f. Changing the Fold of a Protein

Proteins that share as little as ~20% sequence identity may be structurally similar. To what degree must a protein's sequence be changed in order to convert its fold to that of another protein? This question was answered, at least for the protein GB1, by the finding that changing 50% of its 56 residues converted its fold to that of **Rop protein** (Rop for *r*epressor *of p*rimer; a transcriptional regulator). Rop is a homodimer whose 63-residue subunits each form an αα motif (Fig. 8-46c) that dimerizes with its 2-fold axis perpendicular to the helix axes to form a 4-helix bundle

Figure 9-7 **X-ray structure of Rop protein, a homodimer of αα motifs that associate to form a 4-helix bundle.** On a change of 50% of its residues, protein GB1, whose structure is shown in Fig. 9-6, assumes the structure of Rop protein. One of the subunits of the structure shown here is colored according to the sequence of the GB1-derived polypeptide with purple residues identical in both native proteins, magenta residues unchanged from native GB1, cyan residues identical to those in native Rop, and green residues different from those in either native protein. The N-terminus of this subunit is at the lower right. [Based on an X-ray structure by Demetrius Tsernoglou, Università di Roma, Rome, Italy. PDBid 1ROP.]

(Fig. 9-7). Fifty percent of the residues of GB1 were changed based largely on a secondary structure prediction algorithm (Section 9-3Ad), energy minimization, and visual modeling to yield a new polypeptide named Janus (after the two-faced Roman god of new beginnings) that is 41% identical to Rop. In this manner, GB1 residues with high helix-forming propensities were retained, whereas in regions required to be α helical, a number of residues with high β sheet–forming propensities were replaced (helix- and sheet–forming propensities are discussed in Section 9-3Aa); hydrophobic residues were incorporated at the appropriate *a* and *d* positions of a heptad repeat (Fig. 8-26) to form the core of Rop's 4-helix bundle; and residue changes were made to mimic Rop's distribution of surface charges. Fluorescence

and NMR measurements reveal that Janus assumes a stable Rop-like conformation. These studies indicate that not all residues have equally important roles in specifying a particular fold. Indeed, the Janus sequence is more closely related to that of GB1 (50% identity) than to that of Rop (41% identity), even though Janus structurally resembles Rop but not GB1.

g. Many Proteins Are Natively Unfolded

In recent years it has become evident that many entire native proteins and long protein segments (>30 residues) are fully unfolded. Such intrinsically disordered proteins lack specific tertiary structures and are therefore composed of ensembles of conformations. They are characterized by low sequence complexity, a low proportion of the bulky hydrophobic amino acids that form the cores of globular proteins (Val, Leu, Ile, Met, Phe, Trp, and Tyr), and a high proportion of certain polar and charged amino acids (Gln, Ser, Pro, Glu, Lys, Gly, and Ala). Structure prediction techniques based on amino acid sequences (Section 9-3) indicate that an organism's proportion of natively disordered proteins increases with its complexity with ~2% of archeal proteins, ~4% of eubacterial proteins, and ~33% of eukaryotic proteins predicted to contain long disordered regions.

Most natively disordered proteins specifically bind to some other molecule such as a protein, a nucleic acid, or a membrane component, and in doing so fold into stable secondary or tertiary structures. For example, the phosphorylated kinase-inducible domain (pKID) of the transcription factor **cyclic AMP response element-binding protein (CREB)** is disordered when free in solution, but folds to an ordered conformation when it binds to the KID-binding domain of **CREB-binding protein (CBP,** Fig. 9-8). Apparently, the increased flexibility of natively disordered proteins enables them to perform a relatively unhindered conformational

Figure 9-8 **The binding of the pKID domain of rat CREB to the KID-binding domain of mouse CBP.** The pKID, whose backbone is drawn in worm form (*pink*), is unstructured when free in solution (*left*) but forms two perpendicular helices when bound to the KID-binding domain (*right*). The image on the right shows the NMR structure of the pKID–KID-binding domain complex with the side chains of pKID phosphoSer 133 and Leu 141 drawn in ball-and-stick form with C green, O red, and P yellow and with the KID-binding domain (*gray*) represented by its solvent-accessible surface. [Courtesy of Peter Wright, Scripps Research Institute, La Jolla, California. PDBid 1KDX.]

search when binding to their target molecules. It has also been suggested that a structured globular protein would have to be two to three times larger than a disordered protein to provide the same size intermolecular interface and hence the use of disordered proteins provides genetic economy and reduces intracellular crowding. Disordered regions may also aid in the transport of proteins across membranes (Section 12-4Ea) and facilitate selective protein degradation (Section 32-6B).

The functions of natively disordered proteins are quite varied. Their most common function appears to be binding to specific DNA sequences to facilitate such processes as replication, transcription, repair, and transposition (Chapter 30). However, they have also been implicated in a variety of other functions including intracellular signal transduction (Chapter 19), forming phosphorylation sites in proteins whose activities are regulated by phosphorylation (Section 18-3C), and in aiding other proteins and RNAs to fold to their native conformations (Section 9-2C).

C. *Folding Pathways*

How does a protein fold to its native conformation? We, of course, cannot hope to answer this question in detail until we better understand why native protein structures are stable. Moreover, as one might guess, the folding process itself is one of enormous complexity. Nevertheless, as we shall see below, the broad outlines of how proteins fold to their native conformations are beginning to come into focus.

The simplest folding mechanism one might envision is that a protein randomly explores all of the conformations available to it until it eventually "stumbles" onto its native conformation. A back-of-the-envelope calculation first made by Cyrus Levinthal, however, convincingly demonstrates that this cannot possibly be the case: Assume that the $2n$ backbone torsional angles, ϕ and ψ, of an n-residue protein each have three stable conformations. This yields $3^{2n} \approx 10^n$ possible conformations for the protein, which is a gross underestimate, if only because the side chains are ignored. If a protein can explore new conformations at the rate at which single bonds can reorient, it can find $\sim 10^{13}$ conformations per second, which is, no doubt, an overestimate. We can then calculate the time, t, in seconds, required for a protein to explore all the conformations available to it:

$$t = \frac{10^n}{10^{13}\,\text{s}^{-1}} \qquad [9.1]$$

For a small protein of $n = 100$ residues, $t = 10^{87}$ s, which is immensely more than the apparent age of the universe (~ 13.7 billion years $= 4.3 \times 10^{17}$ s).

It would obviously take even the smallest protein an absurdly long time fold to its native conformation by randomly exploring all its possible conformations, an inference known as the **Levinthal paradox.** Yet several proteins fold to their native conformations in microseconds. Therefore, as Levinthal suggested, *proteins must fold by some sort of ordered pathway or set of pathways in which the approach to the native state is accompanied by sharply increasing conformational stability (decreasing free energy).*

a. Rapid Measurements Are Required to Monitor Protein Folding

Folding studies on several small single-domain proteins, including RNase A, cytochrome *c,* and **apomyoglobin** (myoglobin that lacks its heme group), indicate that these proteins fold to a significant degree within one millisecond or less of being brought to native conditions. Hence, if the earliest phases of the folding process are to be observed, denatured proteins must be brought to native conditions in significantly less time. This is most often done using a rapid mixing device such as a **stopped-flow** apparatus (Fig. 9-9) in which a protein solution at a pH that denatures the protein or containing guanidinium chloride or urea at a concentration that does so is rapidly changed in pH or diluted to initiate folding. Most such instruments have "dead times" (the interval between the times when mixing is initiated and meaningful measurements can first be made) of ~ 0.5 ms. However, recently developed ultrarapid mixing devices have dead times of at little as 40 μs.

An alternative technique involves the refolding of **cold denatured proteins.** [For proteins whose folding has both ΔH and ΔS positive, a decrease in temperature is destabilizing (Table 3-2). Since $\Delta G = \Delta H - T\Delta S$, these proteins are unstable, that is denature, when $T < \Delta H/\Delta S$. For many of these proteins, solution conditions can be found for which this temperature is $>0°$C.] The refolding of the cold-denatured protein is initiated by a so-called **temperature-jump** in which the solution is heated with an infrared laser pulse by 10 to 30°C in <100 ns.

With either of the above methods, the folding protein must be monitored by a technique that can report rapid structural changes in a protein. The three such techniques that have been most extensively used are (1) **circular dichroism (CD)** spectroscopy, (2) **pulsed HD exchange** followed by 2D-NMR spectroscopy or mass spectrometry, and (3) **fluorescence resonance energy transfer (FRET).** We discuss these methods below.

b. The Circular Dichroism Spectrum of a Protein Is Indicative of Its Conformation

Polypeptides absorb strongly in the ultraviolet (UV) region of the spectrum ($\lambda = 100$ to 400 nm) largely because their aromatic side chains (those of Phe, Trp, and Tyr) have particularly large molar extinction coefficients (Section 5-3Ca)

Figure 9-9   A stopped-flow device. The reaction is initiated by simultaneously and rapidly discharging the contents of both syringes through the mixer. On hitting the stop switch, the stopping syringe triggers the computer to commence optically monitoring the reaction (via its UV/visible, fluorescence, or CD spectrum).

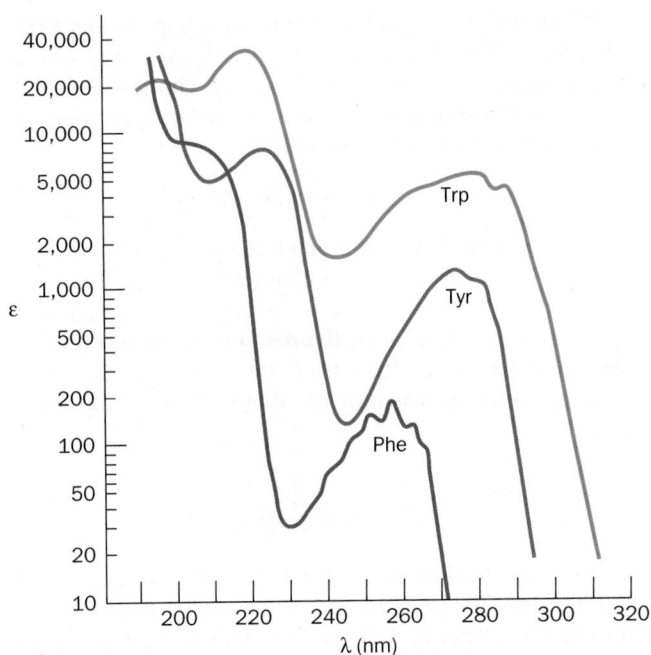

Figure 9-10 **UV absorbance spectra of the three aromatic amino acids, phenylalanine, tryptophan, and tyrosine.** Note that the molar absorbance, ε, is displayed on a log scale. [After Wetlaufer, D.B., *Adv. Prot. Chem.* **7**, 310 (1962).]

Figure 9-11 **Circular dichroism (CD) spectra of polypeptides.** Polypeptides in the α helix, β sheet, and random coil (rc) conformations were determined from the CD spectra of proteins of known X-ray structures. By comparing these spectra with the absorption spectra in Fig. 9-10, it can be seen that $\Delta\varepsilon = \varepsilon_L - \varepsilon_R$ is a small difference of two large numbers. [After Saxena, V.P. and Wetlaufer, D.B., *Proc. Natl. Acad. Sci.* **66**, 971 (1971).]

in this spectral region (ranging into the tens of thousands; Fig. 9-10). However, polypeptides do not absorb visible light (λ = 400 to 800 nm), so that they are colorless.

For chiral molecules such as proteins, ε has different values for left and right circularly polarized light, ε_L and ε_R. The variation with λ of the difference in these quantities, $\Delta\varepsilon = \varepsilon_L - \varepsilon_R$, constitutes the **circular dichroism (CD)** spectrum of the solute of interest (for nonchiral molecules $\varepsilon_L = \varepsilon_R$ and hence they have no CD spectrum). In proteins, α helices, β sheets, and random coils exhibit characteristic CD spectra (Fig. 9-11). Hence the CD spectrum of a polypeptide provides a rough estimate of its secondary structure.

c. Pulsed H/D Exchange Provides Structural Details on How Proteins Fold

Pulsed H/D exchange, a method devised by Walter Englander and Robert Baldwin, is the only known technique that can follow the time course of individual residues in a folding protein. Weakly acidic protons (^{1}H), such as those of amine and hydroxyl groups (X—H), exchange with those of water, a process known as **hydrogen exchange** that can be demonstrated with the use of deuterated water [D_2O; deuterium (D or ^{2}H) is a stable isotope of ^{1}H]:

$$X{-}H + D_2O \rightleftharpoons X{-}D + HOD$$

Since ^{1}H has an NMR spectrum in a different frequency range from that of D, the exchange of ^{1}H for D can be readily followed by NMR spectroscopy. Under physiological conditions, small organic molecules, such as amino acids and dipeptides, completely exchange their weakly acidic protons for D in times ranging from milliseconds to seconds.

Proteins bear numerous exchangeable protons, particularly those of its backbone amide groups. However, protons that are engaged in hydrogen bonding do not exchange with solvent and, moreover, groups in the interior of a native protein are not in contact with solvent.

Through the use of 2D-NMR (Section 8-3Ac), pulsed H/D exchange can be used to follow the time course of protein folding. The protein of interest, usually with its native disulfide bonds intact, is denatured by guanidinium chloride or urea in D_2O solution such that all of the protein's peptide nitrogen atoms become deuterated (N—D). Folding is then initiated in a stopped-flow apparatus by diluting the denaturant solution with ^1H_2O while the pH is simultaneously lowered so as to arrest hydrogen exchange (near neutrality, hydrogen exchange reactions are catalyzed by OH$^-$ and, therefore, their rates are highly pH dependent). After a preset folding time, t_f, the pH is rapidly increased (using a third independently triggered syringe; the so-called labeling pulse) to initiate hydrogen exchange. Peptide nitrogen atoms whose D atoms have not formed hydrogen bonds by time t_f exchange with ^{1}H, whereas those that are hydrogen bonded at t_f, and hence unavailable for hydrogen exchange, remain deuterated. After a short time (10 to 40 ms), the labeling pulse is terminated by rapidly lowering the pH (with a fourth syringe). Folding is then allowed to go to completion and the H/D ratio at each exchangeable site is determined by 2D-NMR (the peaks in the 2D proton NMR spectrum must have been previously assigned). By repeating the analysis for several values of t_f, the time course of hydrogen bond formation at each residue can be determined.

Pulsed H/D exchange–NMR studies do not directly indicate the structures of the folding intermediates. However, if the native structure of the protein under investigation is known (as it almost always is for proteins whose folding is being investigated) and if it is assumed that the protein folds without forming secondary structures not present in the native protein, then the 2D-NMR spectra reveal the time course of the formation of the elements of the native structure together with how fast they are excluded from the bulk solvent.

The time course of protein folding may also be followed by combining pulsed H/D exchange with mass spectrometry. In this method, a partially deuterated protein is fragmented by pepsin (a protease that functions under the acidic conditions necessary to prevent further hydrogen exchange; Table 7-2), the resulting fragments separated by HPLC, and their degree of deuteration determined by mass spectrometry. This method does not yield the residue-level structural information that NMR provides. However, unlike NMR, it can determine if a sample contains subpopulations of protein fragments with different degrees of deuteration and hence that have followed different folding pathways.

d. Fluorescence Resonance Energy Transfer Monitors Distances

Fluorescence is the phenomenon whereby an electronically excited molecule or group decays to its ground state by emitting a photon. The initial excited state rapidly decays, via nonradiative processes (e.g., heating; Section 24-2Aa), to an excited state of lower energy before the photon is emitted. Hence the molecule or group's emission spectrum has a longer wavelength than its absorption spectrum (Fig. 9-12).

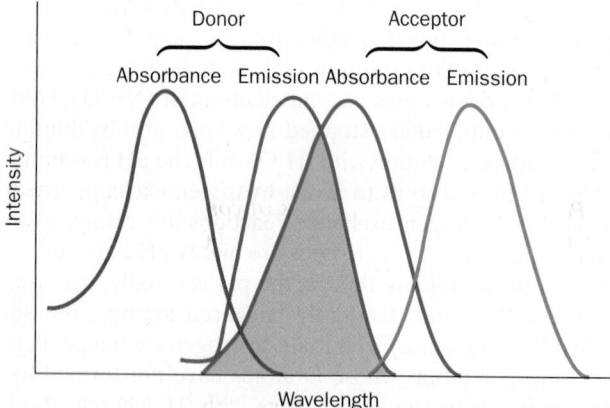

Figure 9-12 **Schematic diagram of the absorption and emission spectra of a donor and an acceptor in fluorescence resonance energy transfer (FRET).** Note that the absorption spectrum occurs at shorter wavelengths than the corresponding emission spectrum and that the donor's emission spectrum must overlap the acceptor's absorption spectrum (*cyan*) for FRET to occur.

When two fluorescent molecules or groups, a donor (D) and an acceptor (A), are within 100 Å of each other and D is electronically excited (say, by a laser with a wavelength within its absorption spectrum), some of the excitation energy will be transferred from D to A,

$$D^* + A \rightarrow D + A^*$$

where the asterisk indicates an electronically excited state. A will then fluoresce with its characteristic emission spectrum (Fig. 9-12), whose intensity can be measured. This phenomenon is known as **fluorescence resonance energy transfer (FRET).** Its efficiency E, the fraction of the energy transferred to the acceptor per donor excitation event, is given by

$$E = \frac{1}{1 + (r/R_0)^6} \qquad [9.2]$$

where r is the distance between D and A, and R_0, their **Förster distance** (named after Theodor Förster, who formulated the theory for the mechanism of long-range nonradiative energy transfer), is the value of r at which the FRET efficiency is 50%. R_0 varies with the degree of overlap between the donor's emission spectrum and the acceptor's absorption spectrum (Fig. 9-12) as well as with the relative orientation of the donor and acceptor. Hence, the intensity of the acceptor's fluoresence is indicative of the distance between D and A as well as their relative orientation.

In proteins, D and A can be the side chains of Trp or Tyr residues. The number and placement of these residues in the protein of interest can be manipulated by site-directed mutagenesis (Section 5-5Gc). Alternatively, fluorescent groups may be covalently linked to reactive side chains such as Cys, which can also be placed via site-directed mutagenasis. FRET measurements can then be used to track how the distances between specific residues vary with time in a folding protein.

e. The Earliest Protein Folding Events Are Initiated by a Hydrophobic Collapse

Stopped-flow–CD measurements indicate that *for many, if not all, small single-domain proteins, much of the secondary structure that is present in native proteins forms within a few milliseconds of when folding is initiated.* This is called the **burst phase** because subsequent folding events occur over much longer time intervals. Pulsed H/D exchange measurements of these small proteins show that some protection against hydrogen exchange in some secondary structural elements develops by ~5 ms after folding initiation.

Since globular proteins contain a compact hydrophobic core, it seems likely that the driving force in protein folding is a so-called **hydrophobic collapse,** in which the protein's hydrophobic groups coalesce so as to expel most of their surrounding water molecules. The polypeptide's radius of gyration is thereby dramatically reduced (from ~30 to ~15 Å for a 100-residue polypeptide), a phenomenon that is generally characteristic of polymers on being transferred from a good to a poor solvent.

This hydrophobic collapse mechanism is consistent with the observation that the hydrophobic dye **8-anilino-1-naphthalenesulfonate (ANS)**

**8-Anilino-1-naphthalene
sulfonate (ANS)**

binds to folding proteins. ANS undergoes a significant enhancement of its fluorescence when it occupies a nonpolar environment, an enhancement that is observed within the burst phase when ANS is present in a solution of a folding protein. Since ANS is expected to preferentially bind to hydrophobic groups, this indicates that the hydrophobic core of a protein rapidly forms once folding has been initiated.

The initial collapsed state of a folding protein is known as a **molten globule.** Such a species has a radius of gyration that is only 5 to 15% greater than that of the native protein and has significant amounts of the native secondary structure and overall fold. However, a molten globule's side chains are extensively disordered, its structure fluctuates far more than that of the native protein, and it has only marginal thermodynamic stability. Nevertheless, to continue folding toward its native state, the polypeptide chain need not undergo large rearrangements in the crowded core of the partially folded protein.

f. Nativelike Tertiary Structure Appears During Intermediate Folding Events

After the burst phase, small proteins exhibit increased ANS binding, further changes in their CD spectrum, and enhanced protection against H/D exchange. These intermediate folding events typically occur over a time interval of 5 to 1000 ms. This is the stage at which the protein's secondary structure becomes stabilized and its tertiary structure begins to form. These nativelike elements are thought to take the form of subdomains that are not yet properly docked to each other. Side chains are probably still mobile, so that, at this stage of folding, the protein can be described as an ensemble of closely related and rapidly interconverting structures.

g. Final Folding Events Often Require Several Seconds

In the final stage of folding, a protein achieves its native structure. To do so, the polypeptide must undergo a series of complex motions that permit the attainment of the relatively rigid native core packing and hydrogen bonding, while expelling the remaining water molecules from its hydrophobic core. For small single-domain proteins, this takes place over a time interval of several seconds or less.

h. Landscape Theory of Protein Folding

The classic view of protein folding was that proteins fold through a series of well-defined intermediates. The folding of a random coil polypeptide was thought to begin with the random formation of short stretches of 2° structure, such as α helices and β turns, that acted as **nuclei** (scaffolding) for the stabilization of additional ordered regions of the protein. Nuclei with the proper nativelike structure then grew by the diffusion, random collision, and adhesion of two or more such nuclei. The stabilities of these ordered regions were thought to increase with size, so, after having randomly reached a certain threshold size, they spontaneously grew in a cooperative fashion until they formed a nativelike domain. Finally, through a series of relatively small conformational adjustments, the domain rearranged to the more compact 3° structure of the native conformation.

The advent of experimental methods that could observe early events in protein folding led to a somewhat different view of how proteins fold. In this so-called **landscape theory,** which was formulated in large part by Peter Wolynes, Baldwin, and Dill, folding is envisioned to occur on an **energy surface** or landscape that represents the conformational energy states available to a polypeptide under the prevailing conditions. The horizontal coordinates of a point on this surface represent a particular conformation of the polypeptide, that is, the values of ϕ and ψ for each of its amino acid residues and the torsion angles for each of its side chains (but here projected onto two dimensions from its multidimensional space). The vertical coordinate of a point on the energy surface represents the polypeptide's internal free energy in this conformation. The above-described measurements indicate that the energy surface of a folding polypeptide is funnel-shaped, with the native state represented by the bottom of the funnel, the global (overall) free energy minimum (Fig. 9-13*a*). The width of the funnel at any particular height (free energy) above the native state is indicative of the number of conformational states with that free energy, that is, the entropy of the polypeptide.

Polypeptides fold via a series of conformational adjustments that reduce their free energy and entropy until the native state is reached. Since a collection of unfolded polypeptides all have different conformations (have different positions on the folding funnel), they cannot follow precisely the same pathway in folding to the native state. If the polypeptide actually folded to its native state via a random conformational search, as Levinthal conjectured, its energy surface would resemble a flat disk with a single small hole, much like the surface of a golf course (Fig. 9-13*b*). Thus, it would take an enormously long time for a polypeptide (a golf ball) to achieve the native state (to fall in the hole) via a random conformational search (by rolling about aimlessly on the surface of the golf course).

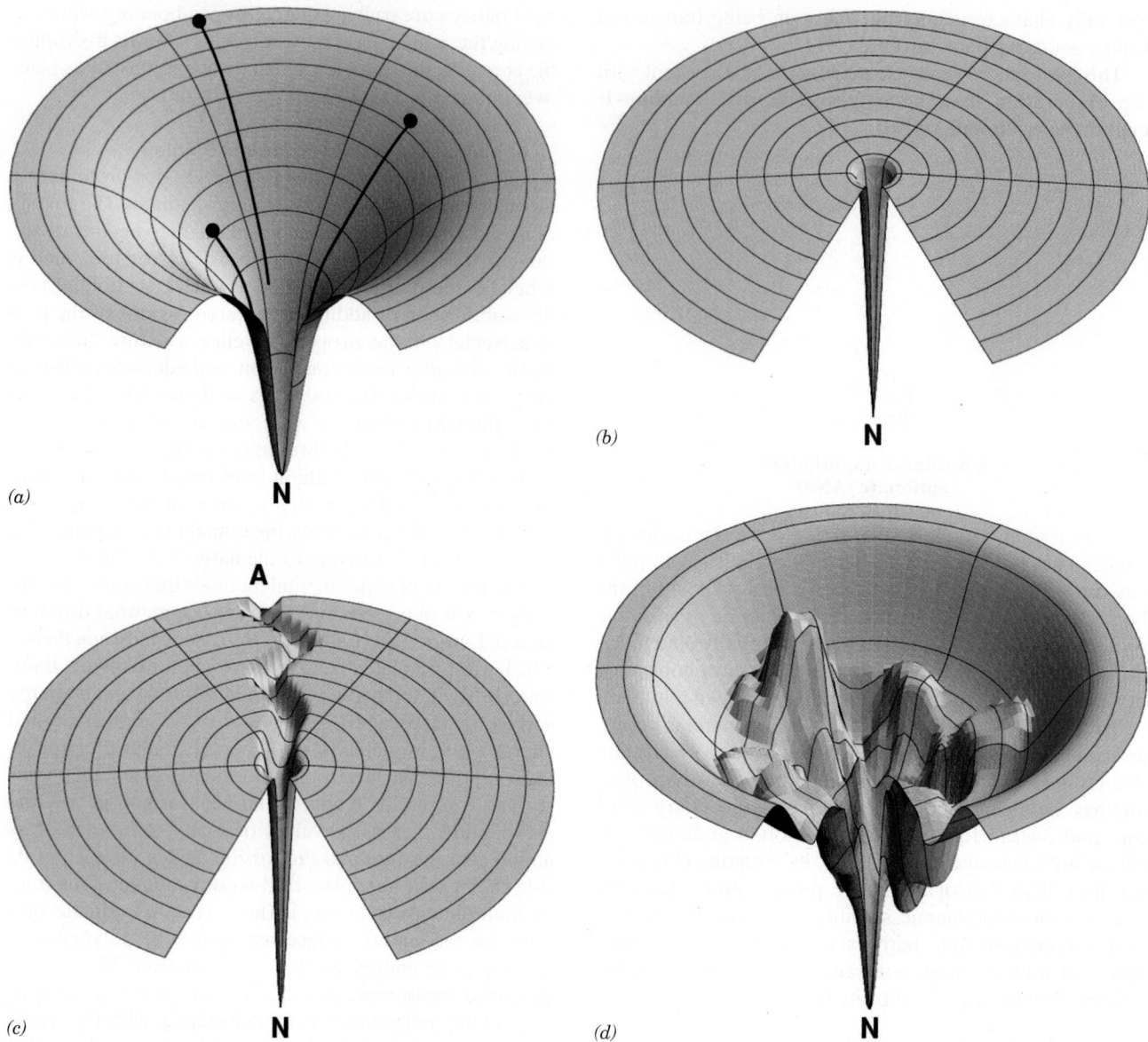

(a)

(b) **N**

(c) **N**

(d) **N**

Figure 9-13 **Folding funnels.** (*a*) An idealized funnel landscape. As the chain forms increasing numbers of intrachain contacts, its internal free energy (its height above the native state, N) decreases together with its conformational freedom (the width of the funnel). Polypeptides with differing conformations (*black dots*) follow different pathways (*black lines*) in achieving the native fold. (*b*) The Levinthal "golf course" landscape in which the chain must search for the native fold (the hole) randomly, that is, on a level energy surface. (*c*) The classic folding landscape in which the chain must search at random on a level energy surface until it encounters the canyon that leads it to the native state. (*d*) A rugged energy surface containing local minima in which a folding polypeptide can become transiently trapped. The folding funnels of real proteins are thought to have such topographies. [Courtesy of Ken Dill, University of California at San Francisco.]

The energy surface of a protein that follows the classic view of protein folding would have a deep radial groove in its disklike surface that slopes toward the hole representing the native state (Fig. 9-13*c*). The extent of the conformational search to randomly find this groove would be much reduced relative to the Levinthal model, so that such a polypeptide would readily fold to its native state. However, the conformational search for the pathway (groove) leading to the native state would still take time, so that the polypeptide would require perhaps several seconds to start down the folding pathway.

The observation that many polypeptides acquire significant nativelike structure within fractions of a millisecond after folding commences indicates that their energy surfaces are, in fact, funnel-shaped; that is, they tend to slope toward the native conformation at all points. Thus, the various pathways followed by initially unfolded polypeptides in folding to their native state are analogous to the various trajectories that could be taken by skiers initially distributed around the top of a bowl-shaped valley to reach the valley's lowest point. Apparently, *there is no single pathway or closely related set of pathways that a polypeptide must follow in folding to its native conformation.*

The foregoing does not imply that the surface of the folding funnel is necessarily smooth, as is drawn in Fig. 9-13*a*. Indeed, landscape theory suggests that this energy surface has a relatively rugged topography, that is, has many local energy minima and maxima (Fig. 9-13*d*). Consequently, in following any particular folding pathway, a polypeptide is likely to become trapped in a local minimum until it randomly acquires sufficient thermal energy to surmount this kinetic barrier and continue the folding process. Thus, in landscape theory, the local energy maxima (transition states; Section 14-1C) that govern the rate of folding are not specific structures as the classic theory of protein folding suggests but, rather, are ensembles of structures.

i. Protein Folding Is Hierarchical

The observation that protein structures are hierarchically organized (Section 9-1B) suggests that they also fold in a hierarchic manner. By this it is meant that folding begins with the formation of marginally stable nativelike microstructures known as **foldons** (e.g., Fig. 9-14) that are local in sequence and that these foldons diffuse and collide with nearby (in sequence) foldons to yield intermediates of increasing complexity and stability that sequentially grow to form the native protein. In contrast, in nonhierarchical folding, a protein's tertiary structure would not only stabilize its local structures but also determine them. Landscape theory is consistent with hierarchical folding, whereas the classic theory of protein folding is more in accord with nonhierarchical folding. Moreover, since a polypeptide *in vivo* begins folding as it is being synthesized, that is, as it is extruded from the ribosome, it would seem that it would most readily achieve its native state if it folded in a hierarchical manner.

Several lines of evidence indicate that proteins, in fact, fold in a hierarchical manner.

1. H/D exchange studies have established the existence of foldons in numerous proteins. Indeed, it appears that foldons rather than individual amino acid residues carry out the unit steps in protein folding pathways.

2. Many peptide fragments excised from proteins either form or exhibit a tendency to form foldons in the absence of long-range (3°) interactions. Moreover, when proteins such as cytochrome *c* and apomyoglobin are brought to a pH sufficiently low to destabilize their native structures, their foldons persist.

3. The boundaries of helices in native proteins are fixed by their flanking sequences (Section 9-3) rather than by 3° interactions.

4. The folding rates of proteins increase, on average, with the degree to which their native contacts are local. Thus fast folders tend to have a high proportion of helices and tight turns, whereas slow folders tend to have a high proportion of β sheets.

In Section 9-1B we saw that in protein GB1 (Fig. 9-6), the 11-residue "chameleon" sequence assumed either an α helix or a β hairpin, depending on its position in the protein.

Figure 9-14 **Ribbon diagram of cytochrome *c*.** Its several foldon units are shown in different colors. Its heme group and several of its functionally important side chains are drawn in ball-and-stick form with C black, N blue, O red, S yellow, and Fe a large red sphere. [Courtesy of Walter Englander, University of Pennsylvania.]

Thus, its conformation appears to be determined by its context rather than by local interactions. However, computer simulations suggest that the conformation of the chameleon sequence is actually determined by local interactions beyond its boundaries.

The folds of native proteins, as we have seen, are highly resistant to sequence changes. Evidently, *the sequence information specifying a particular fold is both distributed throughout the polypeptide chain and highly overdetermined.* It is these characteristics that appear to be responsible for hierarchical folding.

j. Primary Structures Determine Protein Folding Pathways as Well as Structures

The above discussions suggest that *protein primary structures evolved to specify efficient folding pathways as well as stable native conformations.* Evidence corroborating this hypothesis has been obtained by Jonathan King in his study of the renaturation of the **tail spike protein** of bacteriophage P22. The tail spike protein is a trimer of identical 76-kD polypeptides, whose T_m = 88°C. However, certain mutant varieties of the protein fail to renature at 39°C. Nevertheless, at 30°C, these mutant proteins fold to structures whose properties, including their T_m's, are indistinguishable from that of the wild-type tail spike protein. The amino acid changes causing these temperature-sensitive folding mutations apparently act to destabilize intermediate states in the folding process but do not affect the native protein's stability. This observation suggests that *a protein's amino acid sequence dictates its native structure by specifying how it folds to its native conformation.*

This hypothesis is supported by the observation that, in native proteins, a greater number of polar residues than would be randomly expected occupy helix-capping positions (Section 8-4Bb) even though they do not make helix-capping hydrogen bonds. This suggests that they do so as the helix forms so as to facilitate the protein's proper folding.

2 FOLDING ACCESSORY PROTEINS

Most unfolded proteins renature *in vitro* over periods ranging from minutes to days and, quite often, with low efficiency,

that is, with a large fraction of the polypeptide chains assuming quasi-stable non-native conformations and/or forming nonspecific aggregates. *In vivo*, however, polypeptides efficiently fold to their native conformations as they are being synthesized, a process that normally requires a few minutes or less. This is because all cells contain three types of accessory proteins that function to assist polypeptides in folding to their native conformations and in assembling to their 4° structures: **protein disulfide isomerases, peptidyl prolyl cis–trans isomerases,** and **molecular chaperones.** We discuss these essential proteins in this section.

(a)

Non-native S–S bonds   Mixed disulfide   Native S–S bonds

Figure 9-15   Reactions catalyzed by protein disulfide isomerase (PDI). (*a*) Reduced PDI catalyzes the rearrangement of the non-native disulfide bonds in a substrate protein (*purple ribbon*) via disulfide interchange to yield the native disulfide bonds (*horizontal reactions*). If a disulfide bond between PDI and the substrate protein is resistant to disulfide interchange, it is reduced by PDI's second SH group to yield reduced substrate protein and oxidized PDI (*vertical reaction and dashed curved arrow*). (*b*) The oxidized PDI-dependent synthesis of disulfide bonds in proteins. The reaction occurs with the intermediate formation of a mixed disulfide between PDI and the protein. The reduced PDI reaction product reacts with cellular oxidizing agents to regenerate oxidized PDI. See the Animated Figures

(b)

Reduced protein   Mixed disulfide   Oxidized (native) protein

A. Protein Disulfide Isomerase

Protein disulfide isomerase (PDI), which we encountered in Section 9-1A, is an ~510-residue eukaryotic enzyme that inhabits the lumen of the endoplasmic reticulum, where disulfide-containing proteins fold and are post-translationally processed (Section 12-4B). In its reduced form, PDI catalyzes disulfide interchange reactions, thereby facilitating the shuffling of the disulfide bonds in proteins (Fig. 9-15a, *horizontal reactions*) until they achieve their native pairings, which are resistant to further rearrangement. Moreover, PDI must facilitate the correct folding of those proteins that denature in the absence of their native disulfide bonds. Intriguingly, PDI is also the β subunit of the $\alpha_2\beta_2$ heterotetramer prolyl hydroxylase, the enzyme that hydroxylates the Pro residues of collagen (Section 8-2B). The significance of this latter finding is unknown.

Sequence comparisons indicate that PDI contains four ~100-residue domains that are arranged, from N- to C-terminus, as a–b–b′–a′, in which domains a and a′ are homologs that are 30% identical in sequence. They are also homologous to the ubiquitous disulfide-containing redox protein **thioredoxin** (Section 28-3Ae), and hence belong to the thioredoxin superfamily. Prokaryotes have enzymes with functions similar to those of PDI that also assume the thioredoxin fold.

PDI's a and a′ domains each contain the active site sequence motif Cys-Gly-His-Cys, in which the first Cys residue, in its —SH form, participates in the disulfide interchange reaction diagrammed in Fig. 9-15a (the catalytic motif of the thioredoxin superfamily is Cys-X-X-Cys, where X is any amino acid residue). If the second Cys residue is mutated, PDI's isomerization activity drops to

<1% of the wild type and it accumulates in disulfide-linkage to substrate proteins. This suggests that this second Cys residue functions, in its —SH form, to release PDI from the otherwise stable disulfide bonds that its first Cys residue occasionally forms with substrate proteins, thereby yielding reduced substrate proteins and PDI with a disulfide bond linking its two active site Cys residues (Fig. 9-15a, *vertical reaction*).

The X-ray structure of yeast PDI, determined by William Lennarz and Hermann Schindelin, reveals that it adopts a twisted U-shape in which the N-terminal active site Cys S atoms of the a and a′ domains face each other across the top of the U at a distance of 28 Å (Fig. 9-16). As expected, a and a′ have folds that are similar to each other (Fig. 9-17, *top*) and to those of other members of the thioredoxin superfamily. Surprisingly, although b and b′ exhibit no significant sequence similarity with a and a′ or with each other, they also adopt the thioredoxin fold (Fig. 9-17, *bottom*). Nevertheless, both b and b′ lack Cys residues and hence cannot participate directly in the catalytic reaction. The b and b′ domains share an extensive interface (burying ~700 Å²) and hence appear to be rigidly linked together, whereas the a–b and a′–b′ interfaces are negligibly small (burying ~200 Å²). This suggests that the a and a′ domains are flexibly linked to a rigid base formed by the b and b′ domains, thereby enabling PDI to accommodate a diverse set of substrates of up to ~100 residues within the U.

The inner face of the U has a continuous hydrophobic surface that also surrounds the a and a′ active sites. This surface appears to be essential for the binding of PDI to its substrate proteins, which tend to be partially or fully unfolded

Figure 9-16   X-ray structure of yeast protein disulfide isomerase (PDI). The protein is represented by its transparent molecular surface with its polypeptide chain in ribbon form with its a, b, b′, and a′ domains colored magenta, cyan, yellow, and red, respectively. The 16-residue loop, x, linking the b′ and a′ domains is blue and the C-terminal extension, c, is green. The side chains of the N-terminal active site Cys residues of the a and a′ domains are drawn in space-filling form with C green and S yellow. [Based on an X-ray structure by William Lennarz and Hermann Schindelin, State University of New York, Stony Brook, New York. PDBid 2B5E.]

Figure 9-17   Structural comparison of the a, b, b′, and a′ domains of yeast PDI. The domains are shown in similar orientations and drawn in ribbon form colored in rainbow order from their N-terminus (*blue*) to their C-terminus (*red*). The side chains of the N-terminal active site Cys residues of the a and a′ domains are drawn in space-filling form with C green and S yellow. [Based on an X-ray structure by William Lennarz and Hermann Schindelin, State University of New York, Stony Brook, New York. PDBid 2B5E.]

and hence have exposed hydrophobic groups. Moreover, as we shall see in Section 9-2C, PDI's hydrophobic surface facilitates the proper folding of its unfolded substrate proteins. Efficient catalysis of disulfide bond rearrangement requires that reduced PDI be intact, thus suggesting that PDI's two active sites act in concert. The isomerase reaction is driven by the release of conformational strain in the unfolded substrate protein as it folds to its native conformation.

Disulfide bonds in native proteins are usually buried and frequently occur in hydrophobic environments. Indeed, it is probably the burial of the correctly paired Cys residues in a native protein that terminates the action of PDI. However, the N-terminal S atoms in the both the a and a' active sites of PDI are exposed on the protein surface. Although their disulfide bonds almost always stabilize proteins (Section 8-4D) and are usually unreactive, oxidized a and a' are less stable than their reduced forms and therefore have highly reactive, that is, strongly oxidizing, disulfide bonds. This permits oxidized PDI to directly introduce disulfide bonds into newly synthesized and hence reduced polypeptides via a disulfide interchange mechanism (Fig. 9-15b). For this latter process to continue, reduced PDI must be reoxidized (its disulfide bond reformed) by cellular oxidizing agents.

B. *Peptidyl Prolyl Cis–Trans Isomerase*

Although polypeptides are probably biosynthesized with almost all of their X–Pro peptide bonds (where X is any amino acid residue) in the trans conformation, ~10% of these bonds assume the cis conformation in globular proteins because, as we have seen in Section 8-1A, the energy difference between their cis and trans conformations is relatively small. **Peptidyl prolyl cis–trans isomerases (PPIs;** alternatively known as **rotamases)** catalyze the otherwise slow interconversion of X–Pro peptide bonds between their cis and trans conformations, thereby accelerating the folding of Pro-containing polypeptides. Two structurally unrelated families of PPIs, collectively named the **immunophilins,** have been characterized: the **cyclophilins** (so named because they are inhibited by the immunosuppressive drug **cyclosporin A,**

Cyclosporin A

a fungally produced 11-residue cyclic peptide) and the fam-

ily for which the 12-kD **FK506 binding protein (FKBP12)** is prototypic (**FK506**

FK506

is a fungally produced macrocyclic lactone that is also an immunosuppressive drug; medicinal chemists tend to identify the often huge numbers of related drug candidates they deal with by serial numbers rather than by trivial names).

The X-ray structure of human cyclophilin in complex with succinyl-Ala-Ala-Pro-Phe-*p*-nitroanilide reveals that this model substrate binds to the enzyme with its Ala–Pro peptide bond in the cis conformation and that it could not do so if it had the trans conformation. This suggests that the enzyme predominantly catalyzes the trans to cis isomerization of peptidyl-prolyl amide bonds. In addition, the Arg 55 → Ala mutation in cyclophilin reduces its enzymatic activity 100-fold. This, together with the observation that Arg 55 is positioned so that it could hydrogen bond to the N atom of the Ala–Pro peptide bond (although it does not do so in the crystal structure) suggests that the formation of a hydrogen bond from Arg 55 to this N atom facilitates the cis–trans isomerization by deconjugating and hence weakening the peptidyl-prolyl amide bond.

a. Cyclosporin A and FK506 Are Clinically Important Immunosuppressive Agents

Cyclosporin A and FK506 are highly effective agents for the treatment of autoimmune disorders and for preventing organ-transplant rejection. Indeed, until the advent of cyclosporin A in the early 1980s, the long-term survival of a transplanted organ (and its recipient) was a rare occurrence. The more recently discovered FK506 is an even more potent immunosuppressant. The immunosuppressive properties of both cyclosporin A and FK506 stem from the abilities of their respective complexes with cyclophilin and FKBP12 to prevent the expression of genes involved in the activation of **T lymphocytes** (the immune system cells responsible for **cellular immunity;** the immune response is discussed in Section 35-2) by interfering with these cells' intracellular signaling pathways. Enigmatically, there is no obvious relationship between the immunophilins' immunosuppressive properties and rotamase activities: Both cyclosporin A and FK506 are effective immunosuppressants at concentrations far below those of the cyclophilin and FKBP12 in cells; and mutational changes that destroy cyclophilin's rotamase activity do not eliminate its ability to bind cyclosporin A or the ability of the resulting complex to interfere with T lymphocyte signaling. This conundrum is explained in Section 19-3Ff.

C. *Molecular Chaperones: The GroEL/ES System*

Newly synthesized and hence unfolded proteins contain numerous solvent-exposed hydrophobic groups. Moreover, proteins *in vivo* fold in the presence of extremely high concentrations of other macromolecules (~300 g/L, which occupy ~25% of the available volume). Consequently, unfolded proteins *in vivo,* particularly larger proteins (those >15 kD), have a great tendency to form both intramolecular and intermolecular aggregates. **Molecular chaperones,** which are also known as **heat shock proteins** (so named because their rates of synthesis increase at elevated temperatures), are proteins that function to prevent or reverse such improper associations, particularly in multidomain and multisubunit proteins. They do so by binding to an unfolded or aggregated polypeptide's solvent-exposed hydrophobic surfaces and subsequently releasing them, often repeatedly, in a manner that facilitates their proper folding and/or 4° assembly. Most molecular chaperones are **ATPases** (enzymes that catalyze ATP hydrolysis), which bind to unfolded polypeptides and apply the free energy of ATP hydrolysis to effect their release in a favorable manner. Thus it appears, as John Ellis has pointed out, that molecular chaperones function analogously to their human counterparts: *They inhibit inappropriate interactions between potentially complementary surfaces and disrupt unsuitable liasons so as to facilitate more favorable associations.*

The molecular chaperones comprise several unrelated classes of proteins that have somewhat different functions including:

1. The **heat shock proteins 70 (Hsp70),** which are ~70-kD monomeric proteins that are highly conserved in both prokaryotes and eukaryotes (in which different species occur in the cytosol, the endoplasmic reticulum, mitochondria, and chloroplasts; the *E. coli* Hsp70 is called **DnaK** because it was discovered through the isolation of mutants that do not support the growth of bacteriophage λ and hence was initially thought to participate in DNA replication). They function in an ATP-driven process to reverse the denaturation and aggregation of proteins (processes that are accelerated at elevated temperatures), to facilitate the proper folding of newly synthesized polypeptides as they emerge from the ribosome, to unfold proteins in preparation for their transport through membranes (Section 12-4Ea), and to subsequently help them refold. Hsp70 works in association with the **cochaperone** protein **Hsp40**

(**DnaJ** in *E. coli*) to bind and release small hydrophobic regions of misfolded proteins.

2. Trigger factor, which is a ribosome-associated prokaryotic protein. It prevents the intra- and intermolecular aggregation of newly synthesized polypeptides as they emerge from the ribosome by shielding their hydrophobic segments. Unlike most other chaperones, trigger factor does not bind ATP. Trigger factor and the Hsp70/40 system appear to have redundant functions: *E. coli* can tolerate the loss of either one but the loss of both is lethal above 30°C and is accompanied by the massive aggregation of newly synthesized proteins. Trigger factor and Hsp70/40 are the first chaperones that newly synthesized polypeptides encounter. Subsequently, many of the resulting partially folded proteins are handed off to other chaperones, such as those listed below, to complete the folding process. Eukaryotes lack a homolog of trigger factor but contain other small chaperones that may have similar functions.

3. The **chaperonins,** which are heat shock proteins that form large, multisubunit, cagelike assemblies that are universal components of prokaryotes and eukaryotes. They bind improperly folded globular proteins via their exposed hydrophobic surfaces and then, in an ATP-driven process, induce the protein to fold while enveloping it in an internal cavity, thereby protecting the folding protein from nonspecific aggregation with other unfolded proteins (see below). There are two classes of chaperonins: the **Group I chaperonins,** which occur in eubacteria, mitochondria, and chloroplasts, and the **Group II chaperonins,** which occur in archaea and eukaryotes.

4. The **Hsp90** proteins, which are homodimeric, ATP-dependent, eukaryotic proteins of ~730-residue subunits that mainly facilitate the late stage folding of proteins involved in signaling, including **steroid hormone receptors** (Section 34-3Bn) and **receptor tyrosine kinases** (Section 19-3A). Like other chaperones, they do so by binding to exposed hydrophobic surfaces of their substrate proteins so as to prevent nonspecific aggregation. Unlike other chaperones, however, Hsp90 proteins have a regulatory role in that they induce conformational changes in nativelike substrate proteins that result in their activation or stabilization. They do so through their interactions with a large variety of cochaperones. Hsp90 proteins are among the most abundant proteins in eukaryotes, constituting 1 to 2% of their soluble proteins under normal conditions and 4 to 6% under stressful conditions that destabilize proteins such as high temperatures.

5. The **nucleoplasmins,** which are decameric, acidic, nuclear proteins whose presence is required for the proper *in vivo* assembly of **nucleosomes** (particles in which eukaryotic DNA is packaged) from their component DNA and histones (Section 34-1B).

In the following paragraphs we concentrate on the structure and function of the chaperonins, as these are the best characterized molecular chaperones. This discussion also constitutes our introduction to the dynamic functions of proteins, that is, to proteins as molecular machines.

a. The GroEL/ES System Forms a Large Cavity in Which Substrate Protein Folds

Group I chaperonins consist of two families of proteins that work in concert: (1) the **Hsp60** proteins (**GroEL** in *E. coli* and **Cpn60** in chloroplasts), which, as electron microscopic images first revealed, consist of 14 identical ~60-kD subunits arranged in two apposed rings of 7 subunits each (Fig. 9-18); and (2) the **Hsp10 proteins** (**GroES** in *E. coli* and **Cpn10** in chloroplasts), which form single heptameric rings of identical ~10-kD proteins. These proteins, which are essential to the survival of *E. coli* under all conditions tested, facilitate the folding of improperly folded proteins to their native conformations (their discovery in *E. coli* as being necessary for the growth of certain bacteriophages is why they have the designation "Gro").

The X-ray structure of GroEL (Fig. 9-19), determined by Arthur Horwich and Paul Sigler, shows, as expected, that GroEL's 14 identical 547-residue subunits associate to form a porous thick-walled hollow cylinder that consists of two 7-fold symmetric rings of subunits stacked back to back with 2-fold symmetry to yield a complex with D_7 symmetry (Section 8-5B). Each GroEL subunit consists of three domains: a large equatorial domain (residues 1–135 and 410–547) that forms the waist of the protein and holds its subunits together through both intra- and inter-ring interactions, a loosely structured apical domain (residues 191–376) that forms the open ends of the GroEL cylinder, and a small intermediate domain (residues 136–190 and 377–409) that connects the equatorial and apical domains. The X-ray structure suggests that GroEL encloses an ~45-Å-diameter central channel that runs the length of the complex. We shall see below that this channel, in part, forms the chambers in which partially folded proteins fold to their native states. However, both electron microscopy–based images and neutron scattering studies indicate that the channel is obstructed in its equatorial region, so that proteins cannot pass between two GroEL rings. The obstruction is apparently caused by each subunit's N-terminal 5 residues and C-terminal 22 residues, which are not seen in the X-ray structure and hence are almost certainly disordered.

The X-ray structure of GroEL with **ATPγS** bound to each subunit (ATPγS is a poorly hydrolyzable analog of ATP in which S replaces one of the O atoms substituent to P_γ)

$$^-O-\overset{\overset{\displaystyle S}{\|}}{\underset{\underset{\displaystyle O^-}{|}}{P}}-O-\overset{\overset{\displaystyle O}{\|}}{\underset{\underset{\displaystyle O^-}{|}}{P}}-O-\overset{\overset{\displaystyle O}{\|}}{\underset{\underset{\displaystyle O^-}{|}}{P}}-O-CH_2$$

ATPγS

indicates that ATP binds to a pocket in the equatorial domain that opens onto the central channel. The residues forming this pocket are highly conserved among chaperonins. The only significant differences between the structures of the GroEL–ATPγS complex and that of GroEL alone are modest movements of the residues in the vicinity of the ATP pocket.

The X-ray structure of GroES (Fig. 9-20), determined by Lila Gierasch and Johann Deisenhofer, shows that this protein's 7 identical 97-residue subunits form a domelike structure with C_7 symmetry. Each GroES subunit consists of an irregular antiparallel β barrel from which two β hairpins project. One of these β hairpins (residues 47–55) extends from the top of the β barrel toward the protein's 7-fold axis, where it interacts with the other such β hairpins to form the roof of the dome. The second β hairpin (residues 16–33) extends from the opposite side of the β barrel outward from the bottom outer rim of the dome. This so-called mobile loop is seen in only one of GroES's 7 subunits; it is apparently disordered in the other subunits in agreement with the results of NMR studies of uncomplexed GroES in solution. The inner surface of the GroES dome is lined with hydrophilic residues.

Both electron microscopic and neutron scattering studies reveal that partially unfolded proteins bind in the mouth of the GroEL barrel in a manner reminiscent of a cork in a champagne bottle (Fig. 9-18). Mutations that impair polypeptide binding to GroEL all map to a poorly

Figure 9-18 **Electron micrograph–derived 3D image of the Hsp60 chaperonin from the photosynthetic bacterium** ***Rhodobacter sphaeroides.*** Hsp60 consists of 14 identical ~60-kD subunits arranged to form two apposed rings of 7 subunits, each surrounding a central cavity. The image of Hsp60, which is viewed with its 7-fold axis tipped toward the viewer, indicates that each subunit consists of two major domains, one in contact with the opposing heptameric ring, and the other at the end of the cylindrical protein molecule. The spherical density occupying the protein's central cavity is thought to represent a bound polypeptide. The cavity provides a protected microenvironment in which a polypeptide can fold. [Courtesy of Helen Saibil and Steve Wood, Birkbeck College, London, U.K.]

(a)

(b)

Figure 9-19 **X-ray structure of GroEL.** (*a*) Side view perpendicular to the 7-fold axis in which the seven identical subunits of the lower ring are gold and those of the upper ring are silver, with the exception of the two subunits nearest the viewer, whose equatorial, intermediate, and apical domains are colored blue, green, and red on the right subunit and cyan, yellow, and magenta on the left subunit. The two rings of the complex are held together through side chain interactions that are not seen in this drawing. (*b*) Top view along the 7-fold axis in which only the upper ring is shown for the sake of clarity. Note the large central channel that appears to run the length of the protein. [Based on an X-ray structure by Axel Brünger, Arthur Horwich, and Paul Sigler, Yale University. PDBid 1OEL.]

Figure 9-20 **X-ray structure of GroES as viewed along its 7-fold axis.** The mobile loop of only one of the protein's 7 identical subunits (*left*) is visible in the structure. The polypeptide segments that flank the mobile loop are yellow. [Courtesy of Johann Deisenhofer, University of Texas Southwest Medical Center, Dallas.]

resolved (and presumably flexible) segment at the top of the apical domain that, in the structure of GroEL alone, faces the central channel. In fact, changing any of nine highly conserved hydrophobic residues in this region to a hydrophilic residue abolishes polypeptide binding. It therefore seems likely that these residues provide the binding site(s) for non-native polypeptides. Interestingly, mutations of these same residues also abolish the binding of GroES.

The X-ray structure of the GroEL–(ADP)$_7$–GroES complex (Fig. 9-21), also determined by Horwich and Sigler, provides considerable insight into how this chaperonin carries out its function. In this complex, a GroES heptamer and the 7 ADPs are bound to the same GroEL ring (the so-called cis ring; the opposing GroEL ring is known as the trans ring) such that the GroES cap closes over the GroEL cis ring barrel like a lid on a pot, thereby forming a bullet-shaped complex with C_7 symmetry. The trans ring subunits have conformations that closely resemble those in the structure of GroEL alone. In contrast, the apical and intermediate domains of the cis ring have undergone large *en bloc* movements relative to their positions in GroEL alone (Fig. 9-22). This widens and elongates the cis cavity in a way that more than doubles its volume (from 85,000 to 175,000 Å^3; Fig. 9-21*c*), thereby permitting it to enclose a partially folded substrate protein of up to ~70 kD. *These en bloc movements are concerted, that is, they occur simultaneously*

Figure 9-21 **X-ray structure of the GroEL–(ADP)₇–GroES complex.** (*a*) A space-filling drawing as viewed perpendicularly to the complex's 7-fold axis with the GroES ring orange, the cis ring of GroEL green, and the trans ring of GroEL red with one subunit in each ring shaded more brightly. The dimensions of the complex are indicated. Note the different conformations of the two GroEL rings. The ADPs, whose binding sites are in the base of each cis ring GroEL subunit, are not seen because they are surrounded by protein. (*b*) As in Part *a* but viewed along the 7-fold axis. (*c*) As in Part *a* but with the two GroEL subunits closest to the viewer in both the cis and the trans rings removed to expose the interior of the complex. The level of fog increases with the distance from the viewer. Note the much larger size of the cavity formed by the cis ring and GroES in comparison to that of the trans ring. [Based on an X-ray structure by Paul Sigler, Yale University. PDBid 1AON.]

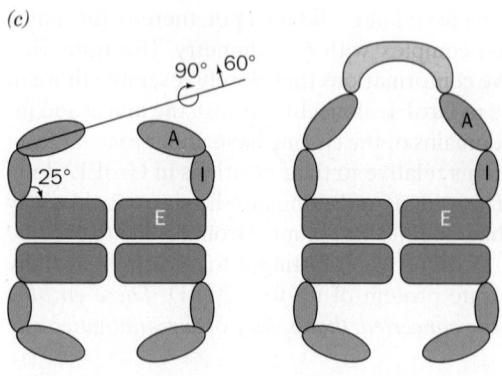

Figure 9-22 **Domain movements in GroEL.** (*a*) Ribbon diagram of a single subunit of GroEL in the X-ray structure of GroEL alone. Its equatorial, intermediate, and apical subunits are colored blue, green, and red. The inset shows a space-filling drawing of GroEL with the colored subunit oriented identically. Circles and arrows indicate the pivot points for domain movements. (*b*) A GroEL subunit in the X-ray structure of GroEL –(ADP)₇–GroES displayed as in Part *a*. The ADP, which is bound in a pocket at the top of the equatorial domain, is shown in space-filling form in pink. (*c*) Schematic diagram indicating the conformational changes in GroEL when it binds GroES. Its equatorial (E), intermediate (I), and apical (A) domains are colored as in Part *a* and GroES is yellow. The arrows indicate the extent of the domain movements in the cis ring of GroEL. [Parts *a* and *b* courtesy of Arthur Horwich, Yale University; Part *c* after Richardson, A., Landry, S.J., and Georgopoulos, C., *Trends Biochem. Sci.* **23,** 138 (1998).]

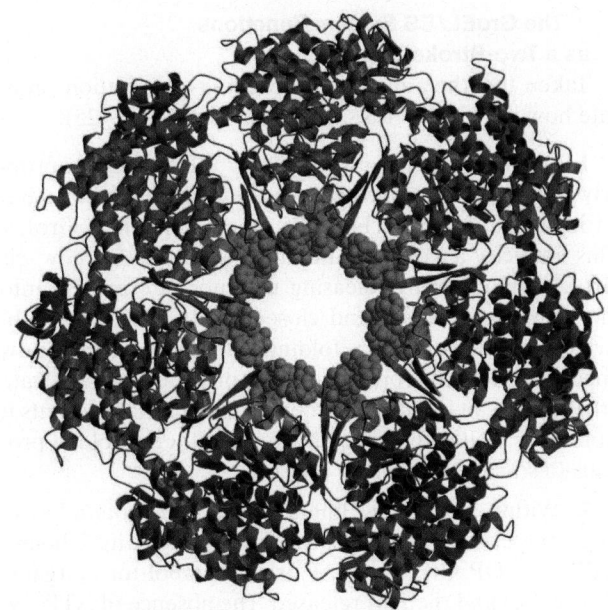

Figure 9-23 **Apical domain of GroEL in complex with a tight-binding 12-residue polypeptide (SWMTTPWGFLHP).** To generate this drawing, the C_α atoms of the apical domain in the X-ray structure of the complex were superimposed on those of the apical domains in the X-ray structure of GroEL alone (Fig. 9-19). Each apical subunit is represented by a ribbon diagram in which the two helices involved in binding the polypeptide (helices H and I in Fig. 9-22a) are red and the remainder of the subunit is green. The polypeptides are shown in space-filling form in red. [Courtesy of Lingling Chen, Yale University. PDBid 1DKD.]

in all seven subunits of a GroEL ring, most probably because if one GroEL subunit did not undergo these conformational shifts, it would mechanically block its adjacent subunits from doing so.

In forming the GroEL–(ADP)$_7$–GroES complex, the ADP becomes completely enclosed by protein through the collapse of the intermediate domain onto the equatorial domain (Fig. 9-22b). This movement activates GroEL's ATPase function by shifting the side chain of its catalytically essential Asp 398, which extends from the L helix of the equatorial domain, into its catalytically active position near the ADP's β phosphate group. Electron microscopy studies at 10-Å resolution by Horwich and Helen Saibil reveal that similar movements occur when ATP binds to GroEL.

The hydrophobic groups lining the inner surface of the trans ring's apical domain, which extend from its and H and I helices and an underlying loop (Fig. 9-22), presumably bind to the improperly exposed hydrophobic groups of substrate proteins. Indeed, an X-ray structure of the apical domain of GroEL in complex with a 12-residue peptide that binds strongly to GroEL reveals that this peptide binds to these exposed hydrophobic groups (Fig. 9-23). However, in the cis ring of the GroEL–(ADP)$_7$–GroES complex, these hydrophobic groups participate either in binding GroES via its flexible loops or in stabilizing the newly formed interface between the rotated and elevated apical domains. Consequently, *these hydrophobic groups are no longer exposed on the inner surface of the cis cavity (Fig. 9-24), thereby depriving a substrate protein of its binding sites.*

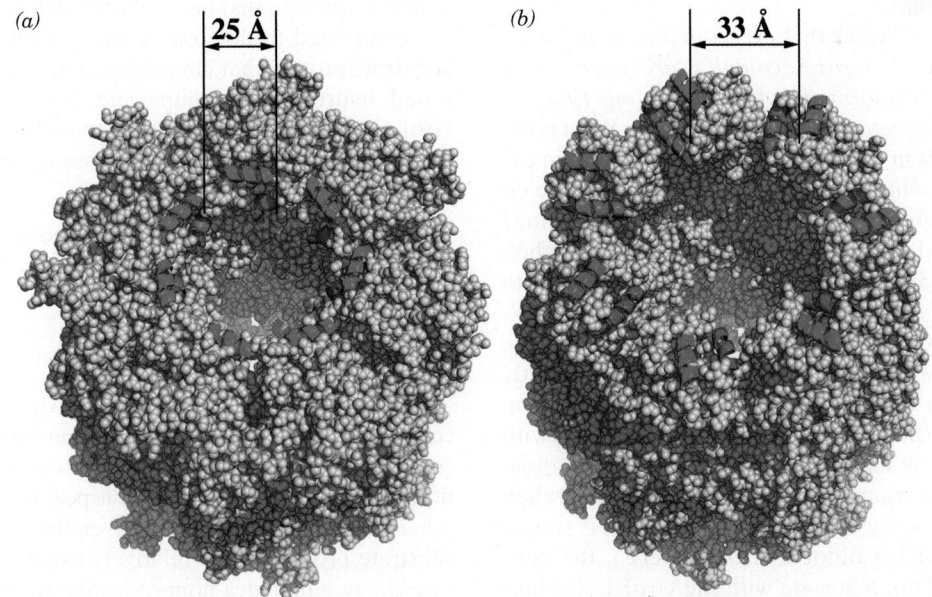

Figure 9-24 **Movements of the polypeptide-binding helices of GroEL.** (a) A space-filling drawing of GroEL in the structure of GroEL alone and (b) in the structure of GroEL–(ADP)$_7$–GroES. The GroEL cis and trans rings are pale cyan and pale yellow and the cis ring's H and I helices (Figs. 9-22a,b), which form most of the hydrophobic binding sites for improperly folded proteins, are drawn in cartoon form and colored green and red, respectively. On the addition of GroES and ATP to GroEL, neighboring binding sites separate by 8 Å and non-neighboring sites separate by up to 20 Å. A substrate protein initially bound to two of these sites will likely be forcibly stretched and hence partially unfolded before being released as the binding sites become occluded. [After drawings by George Lorimer, University of Maryland; and Walter Englander, University of Pennsylvania. PDBids 1OEL and 1AON.]

b. GroEL/ES Undergoes Coordinated Conformational Changes That Are Paced by ATP Binding and Hydrolysis

The binding of ATP and GroES to the cis ring of GroEL strongly inhibits their binding to the trans ring. The X-ray structure of the GroEL–(ADP)$_7$–GroES complex suggests that this occurs through concerted small conformational shifts in the GroEL equatorial domains that apparently prevent the trans ring from assuming the conformation of the cis ring. However, once the cis ring has hydrolyzed its bound ATP (which it is committed to do once its nucleotide binding sites close off and its ATPase active sites form), the trans ring can bind ATP and the resulting conformational shifts release GroES from the cis ring. This explains why a mutant form of GroEL that has only a single ring (and hence is known as SR1) can bind substrate protein and GroES but does not release them after it hydrolyzes its bound ATP. *The proper functioning of GroEL requires two rings, even though their central cavities are unconnected.*

A mutant form of GroEL, D398A (in which Asp 398 has been changed to Ala), binds but cannot hydrolyze ATP. In the presence of ATP, D398A GroEL binds GroES together with substrate protein. However, it does not release GroES or the protein when the trans ring is exposed to ATP, as is the case when the cis ring can hydrolyze ATP. Evidently, *ATP's γ-phosphate group provides strong contacts that stabilize the GroEL–GroES interaction. When the ATP in the cis ring is hydrolyzed, the resulting phosphate group is released and these interactions are lost.*

c. ATP Hydrolysis in the Cis Ring Must Occur before Substrate Protein and GroES Can Bind to the Trans Ring

The foregoing indicates that *events in the cis and trans rings of the GroEL–GroES complex are coordinated through concerted conformational changes in one ring that influence the conformation of the opposing ring.* What is the sequence of events in the trans ring relative to those in the cis ring, that is, at what stage of the folding cycle in the cis ring do substrate protein and GroES bind to the trans ring? Horwich answered this question using fluorescence labeling techniques. D398A GroEL that had been mixed with ADP and GroES so as to form a stable complex [D398A GroEL–(ADP)$_7$–GroES] was then mixed with a substrate protein to which a fluorescent group had been covalently linked. When this mixture was subjected to gel filtration chromatography (Section 6-3B), the label migrated with the GroEL, thereby indicating that the substrate protein had bound to the complex's trans ring. However, when the initial complex was instead made with ATP (recall that D398A GroEL cannot hydrolyze ATP), the substrate protein did not associate with the GroEL. In similar experiments, fluorescently labeled GroES associated with preformed D398A GroEL–(ADP)$_7$–GroES in the presence of ATP but not with preformed D398A GroEL–(ATP)$_7$–GroES. Evidently, *the cis ring of the GroEL–GroES complex must hydrolyze its bound ATP before the trans ring can bind either substrate protein or GroES + ATP.*

d. The GroEL/ES System Functions as a Two-Stroke Engine

Taken together, all of the preceding observations indicate how the GroEL/ES system functions (Fig. 9-25):

1. A GroEL ring that is binding 7 ATP and an improperly folded substrate protein via the hydrophobic patches on its apical domains (Fig. 9-25, *upper left*) binds GroES. This induces a conformational change in the now cis GroEL ring, thereby releasing the substrate protein into the resulting enlarged and closed cavity, where the substrate protein commences folding. The cavity, which is now lined only with hydrophilic groups, provides the substrate protein with an isolated microenvironment that prevents it from nonspecifically aggregating with other unfolded proteins (a so-called **Anfinsen cage**).

2. Within ~10 s (the time the substrate protein has to fold), the cis ring catalyzes the hydrolysis of its 7 bound ATPs to ADP + P$_i$ (where P$_i$ is the symbol for inorganic phosphate) and the P$_i$ is released. The absence of ATP's γ-phosphate group weakens the interactions that bind GroES to GroEL.

3. A second molecule of substrate protein binds to the trans ring followed by 7 ATP.

4. The binding of substrate protein and ATP to the trans ring induces the cis ring to release its bound GroES, 7 ADP, and the now possibly natively folded substrate protein. This leaves only ATP and substrate protein bound to the previous trans ring of GroEL, which becomes the cis ring on binding GroES when the complex again cycles through Step 1.

Substrate protein that has not achieved its native state or is not committed to do so is readily recaptured by GroEL. Substrate protein that has achieved its native fold lacks exposed hydrophobic groups and hence cannot bind to GroEL. It is the irreversible hydrolysis of ATP that drives the folding cycle in only the direction indicated in Fig. 9-25.

e. GroEL Unfolds Its Substrate Proteins before Facilitating Their Refolding

How does the foregoing cycle promote the proper folding of an improperly folded protein? Two models, not mutually exclusive, have received the most consideration:

1. The Anfinsen cage model, in which the GroEL/ES complex provides the substrate protein with a protected microenvironment in which it can fold to its native conformation without interference by nonspecific aggregation with other misfolded proteins. Moreover, the confinement of the substrate protein to the relatively small volume of the cis ring cavity eliminates nonproductive folding pathways involving extended conformations, and the hydrophilic character of the cavity walls promotes productive folding pathways by favoring the burial of hydrophobic residues. In terms of landscape theory (Section 9-1Ch), this would smooth the walls of the folding funnel (Fig. 9-13d) and thus facilitate the folding of the substrate protein toward its global free energy minimum, that is, its native state.

Figure 9-25 **Reaction cycle of the GroEL/ES chaperonin system in protein folding.** See the text for an explanation.

2. The **iterative annealing** model, in which the ATP-driven unfolding of a misfolded and conformationally trapped substrate protein followed by its release permits it to resume folding to its native state. This would occur through the binding of a misfolded protein to the hydrophobic patches on two or more of the GroEL cis rings's seven apical domains, followed by the stretching and ultimate release of the protein as GroEL changes conformation on binding ATP and GroES [recall that these patches are further apart in the GroEL–(ADP)$_7$–GroES complex than they are in GroEL alone; Fig. 9-24]. In terms of landscape theory, this stretching would expel the substrate protein (raise its free energy) from a local energy minimum in which it had become trapped and thereby permit it to continue, but not necessarily complete, its conformational journey down the folding funnel toward its native state.

Fluorescence resonance energy transfer (FRET; Section 9-1Cd) measurements by Hays Rye indicate that the forced unfolding of substrate protein by GroEL enhances its rate of folding. **Ribulose-1,5-bisphosphate carboxylase oxygenase (RuBisCO;** Sections 24-3Ac and 24-3C) from *Rhodospiril-*

lum rubrum, which requires GroEL/ES to fold to its native state, was covalently labeled on its N- and C-terminal domains by acceptor and donor fluorescent probes, respectively (which does not affect RuBisCO's stability or its GroEL-mediated folding rate). FRET measurements indicated that on binding to the trans ring of a GroEL–(ADP)$_7$–GroES complex, the fluorescently labeled RuBisCO's end-to-end distance increases slightly. However, on the subsequent addition of ATP, this distance greatly increases within 0.2 s and then decreases to less than its original value over a ~5 s period. This indicates that the initial binding of ATP to GroEL significantly unfolds its bound substrate protein, which then folds to a more compact state within the now cis cavity of GroEL/ES. Further measurements indicate that the fraction of unfolded RuBisCO that folds to its native state increases with the extent of its unfolding previous to being released into the cis cavity.

As diagrammed in Fig. 9-25, the GroEL/ES system releases its substrate protein after each reaction cycle, whether or not the protein is properly folded. In contrast, SR1, the single-ring mutant form of GroEL, in its complex with GroES, cannot release its bound substrate protein.

Nevertheless, the trapped substrate protein refolds nearly quantitatively to its native state over a period of several minutes, about the same rate as it does so in the cycling system. Evidently, the efficiency with which a substrate protein folds to its native state varies with the length of time that it spends in the cis cavity (Anfinsen cage). Then why hasn't a GroEL/ES system evolved that allows an unfolded protein to complete its folding before it is released? The answer may be that the release of the substrate protein from the cis cavity with each turn of the GroEL/ES cycle is a protective mechanism that prevents irretrievably damaged proteins from permanently clogging GroEL. In a cycling system, a substrate protein spends only a fraction of the time in a cis cavity. Thus, since forced unfolding increases folding efficiency, it is a major contributor to the GroEL/ES system's multilayered protein folding mechanism. Moreover, forced unfolding explains how the GroEL/ES system is able to facilitate the folding of several proteins that are too large to completely fit inside the GroEL cavity.

A variety of experiments indicate that substrate proteins bound to the open ring of GroEL alone are largely unstructured. For example, NMR measurements indicate that the 21-kD enzyme **dihydrofolate reductase (DHFR;** Section 28-3Bd) bound to GroEL or SR1 has no stable structure, and hydrogen exchange measurements (Section 9-1Cc) on several substrate proteins bound to GroEL indicate that they exhibit little or no secondary structure. Moreover, FRET measurements on the 41-kD **maltose binding protein** bound to the trans ring of the GroEL–GroES complex reveal that it undergoes a rapid conformational expansion on ATP addition (as does RuBisCO), and NMR measurements indicate that DHFR inside the SR1–GroES cavity follows the same folding trajectory as does DHFR free in solution. Thus GroEL/ES-mediated folding appears to be an all-or-none process rather than an iterative one in which the substrate protein progressively acquires more native-like structure with each round of folding. This suggests that each time a substrate protein binds to the trans ring of GroEL–GroES, it is raised to the top of its folding funnel in an ATP-driven process from which it commences folding via a different trajectory.

Typically, only ~5% of substrate proteins fold to their native state in each reaction cycle. Thus, to fold half the substrate protein present would require $\log(1 - 0.5)/\log(1 - 0.05) \approx 14$ reaction cycles and hence $7 \times 14 = 98$ ATPs. This may seem like a profligate use of ATP, but it is only a fraction of the 1200 ATPs expended in ribosomally synthesizing a 300-residue protein from its component amino acids (4 ATPs per residue; Sections 32-2C and 32-3D), not to mention the far greater number of ATPs required to synthesize these amino acids (Section 26-5).

f. GroEL/ES Is Required for the Folding of ~85 *E. coli* Proteins *in vivo*

The GroEL/ES system only interacts *in vivo* with a subset of *E. coli* proteins. Ulrich Hartl identified these proteins by modifying GroES to have a C-terminal His$_6$ segment (a His-Tag) and isolating the resulting GroEL–GroES–

substrate protein complexes from *E. coli* lysates by metal chelation affinity chromatography (Section 6-3Dg). These complexes were separated by SDS–PAGE (Section 6-4C) and the substrate proteins identified by mass spectrometry (Section 7-1I).

Approximately 250 of *E. coli's* ~2400 cytosolic proteins were found to be associated with GroEL/ES. Of these, ~165 proteins either show little tendency to aggregate during folding or can utilize other chaperone proteins such as trigger factor or DnaK/J to fold to their native states. However, the remaining ~85 proteins have an absolute dependence on the GroEL/ES system for folding, that is, they invariably aggregate in the absence of GroEL/ES. Thirteen of these proteins are indispensible for *E. coli* viability, thereby explaining why GroEL/ES is also essential for *E. coli* viability. About 75% to 80% of the GroEL/ES binding sites are occupied by the ~85 GroEL/ES-dependent proteins, even though they have only low to intermediate abundance in the *E. coli* cytosol.

What are the characteristics of proteins that are obligate substrates of GroEL/ES? Analysis, using the SCOP database (Section 8-3Cd), of those proteins of known structure or with homologs of known structure revealed that many of them contain α/β domains (Section 8-3Bh). In particular, ~35% by mass of all GroEL/ES substrate proteins contain α/β barrels (also called TIM barrels; Section 8-3Bh), even though they comprise only ~6% of the cytosol's total protein mass. These proteins, whose molecular masses range from 23 to 54 kD, are stabilized by numerous long range (in sequence) interactions and hence would be expected to have particularly rugged folding funnels with many local free energy minima that could trap the unaided protein.

What are the substrate protein sequence motifs that bind to GroEL? During the GroEL/ES cycle, the GroES mobile loops (sequence GGIVLTGSA) displace these motifs (Section 9-2Ca), thus suggesting that they have similar sequences. Moreover, to be stretched by GroEL, a substrate protein must have at least two such motifs separated by at least 10 residues. By searching GroEL's ~250 substrate proteins for motifs with these characteristics, George Lorimer and Devarajan Thirumalai found that they have the consensus sequence P_HHH_P_H, where P, H, and _ respectively represent polar, hydrophobic, and any residues and where the core sequence is P_HHH. This is corroborated by the observation that in natively folded substrate proteins of known structure, nearly all of these sequence motifs are buried (<50% of their surface area is solvent-accessible), although since they occur in helices, sheets, and loops, they apparently have little other structural preferences.

The obligate GroEL/ES substrate proteins belong to fold classes that tend to have a greater number of superfamilies than do other *E. coli* proteins. This suggests that GroEL/ES may have facilitated the evolutionary diversification of certain protein folds, perhaps by "buffering" mutations that would otherwise cause severe aggregation. Indeed, the GroEL/ES system likely played an essential role in the evolution of the α/β barrel into the most versatile structural platform for enzymatic functions (Section 8-3Bh).

g. Group II Chaperonins Have Built-In Lids

Group II chaperonins structurally and functionally resemble Group I chaperonins but consist of back-to-back rings of 8 or 9 subunits and have no corresponding cochaperones such as GroES. Archaeal Group II chaperonins, which are called **thermosomes,** consist of 1 to 3 different types of subunits. Eukaryotic Group II chaperonins, which are named **TRiC** (for *T*CP-1 *ri*ng *c*omplex; TCP for *T*-complex *p*olypeptide) or alternatively **CCT** (for *c*haperonin-*c*ontaining *T*CP-1), have dual octameric rings, each consisting of eight genetically distinct but homologous subunits arranged in a specific order. Like GroEL/ES, each of the TRiC subunits couple the hydrolysis of ATP to the folding of substrate proteins. Around 10% of eukaryotic cytosolic proteins transiently interact with TRiC, many of which have an absolute requirement for TRiC-aided folding. These include a variety of essential structural and regulatory proteins including the muscle proteins **actin** and **myosin** (Section 35-3A), the major microtubule components **tubulins α** and **β** (Section 35-3G), and proteins that participate in signal transduction (Chapter 19) and cell cycle regulation (Section 34-4D). The only characteristic that these proteins appear to have in common is that they form homo- or hetero-oligomeric complexes.

Cryo-electron microscopic (cryo-EM) studies of bovine testes TRiC by Wah Chiu and Judith Frydman (Fig. 9-26) revealed that, unlike GroEL, its overall shape is closer to spherical than cylindrical. [In cryo-EM, the sample is cooled to near liquid N_2 temperatures ($-196°C$) so rapidly (in a few milliseconds) that the water in the sample does not have time to crystallize but, rather, assumes a vitreous (glasslike) state. Consequently, the sample remains hydrated and hence retains its native shape to a greater extent than in conventional electron microscopy (in which the sample is vacuum dried).] Like GroEL, TRiC's subunits each consist of equatorial, intermediate, and apical domains. However, the tip of each TRiC apical domain has a helical extension that GroEL lacks. TRiC has two conformational states: an open state (Fig. 9-26a) in which each helical extension is oriented more or less tangentially to the inner portion of the ring, and a closed state (Fig. 9-26b) in which each helical extension has swung to a more radial

(a)

Top view   Side view

Open TRiC model versus EM open state

(b)

Top view   Side view

Closed TRiC model versus EM closed state

Figure 9-26 **Structure of bovine testes TRiC.** (*a*) Its open state and (*b*) its closed state. Surface diagrams of the cryo-EM-based structures at ~16 Å resolution (*transparent gray*) are shown as viewed along the protein's 8-fold axis (*left*) and as viewed from the side with the 8-fold axis tipped toward the viewer (*right*). The X-ray structure of a homologous archaeal chaperonin was modeled into the upper ring of the cryo-EM-based image with each subunit represented by a differently colored ribbon diagram. [Courtesy of Judith Frydman, Stanford University.]

orientation so as to form an iris-like lid that closes off the TRiC cavity in a manner similar to the way that GroES closes off the GroEL cavity. A lidless mutant of TRiC still hydrolyzes ATP and binds unfolded actin with wild-type affinity but is unable to induce its folding. This suggests that the TRiC lid has a GroES-like function in coupling ATP hydrolysis to the productive folding of substrate protein. Interestingly, the putative substrate binding sites in TRiC are not all hydrophobic as some of its subunits are lined with polar residues. Perhaps each different substrate protein interacts with a specific combination of binding sites on the apical domains in the open state of TRiC.

h. The Concept of Self-Assembly Must Take Accessory Proteins into Account

Many proteins can fold/assemble to their native conformations in the absence of accessory proteins, albeit often with low efficiency. Moreover, accessory proteins are not components of the native proteins whose folding/assembly they facilitate. Hence, accessory proteins must mediate the proper folding/assembly of a polypeptide to a conformation/complex governed solely by the polypeptide's amino acid sequence. Nevertheless, the concept that proteins are self-assembling entities must be modified to incorporate the effects of accessory proteins.

3 PROTEIN STRUCTURE PREDICTION AND DESIGN

Since the primary structure of a protein specifies its three-dimensional structure, it should be possible, at least in principle, to predict the native structure of a protein from a knowledge of only its amino acid sequence. This might be done using theoretical methods based on physicochemical principles, or by empirical methods in which predictive schemes are distilled from the analyses of known protein structures. Theoretical methods, which usually attempt to determine the minimum energy conformation of a protein, are mathematically quite sophisticated and require extensive computations. The enormous difficulty in making such calculations sufficiently accurate and yet computationally tractable has, so far, limited their success. Nevertheless, an understanding of how and why proteins fold to their native structures must ultimately be based on such theoretical methods. In this section we outline various methods that have been used to predict the secondary and tertiary structures of proteins and end with a discussion of a related technique, that of designing proteins that will have a particular structure.

A. Secondary Structure Prediction

The most reliable way to determine the secondary structure taken up by a polypeptide is to map its amino acid sequence onto that of a homolog of known structure. If, however, no such structure is available, the above-mentioned predictive methods must be employed. Here we discuss the use of empirical methods for secondary structure prediction. The theoretical methods discussed in the following

section to predict a polypeptide's tertiary structure will, of necessity, also predict its secondary structure.

a. The Chou–Fasman Method

Empirical methods have had reasonable success in secondary structure prediction. Clearly, certain amino acid sequences limit the conformations available to a polypeptide chain in an easily understood manner. For example, a Pro residue cannot fit into the interior portions of a regular α helix or β sheet because its pyrrolidine ring would fill the space normally occupied by part of an abutting segment of chain and because it lacks the backbone N—H group with which to contribute a hydrogen bond. Likewise, steric interactions between several consecutive amino acid residues with side chains branched at C_β (e.g., Ile and Thr) will destabilize an α helix. Furthermore, there are more subtle effects that may not be apparent without a detailed analysis of known protein structures. Here we shall discuss simple empirical methods for predicting the positions of α helices, β sheets, and reverse turns in proteins of known sequence.

The empirical structure prediction scheme developed by Peter Chou and Gerald Fasman can be readily applied by hand and is reasonably reliable. Its use requires two definitions. The frequency, f_α, with which a given residue occurs in an α helix in a set of protein structures is defined as

$$f_\alpha = \frac{n_\alpha}{n} \qquad [9.3]$$

Table 9-1   **Propensities and Classifications of Amino Acid Residues for α Helical and β Sheet Conformations**

Residue	P_α	Helix Classification	P_β	Sheet Classification
Ala	1.42	H_α	0.83	i_β
Arg	0.98	i_α	0.93	i_β
Asn	0.67	b_α	0.89	i_β
Asp	1.01	I_α	0.54	B_β
Cys	0.70	i_α	1.19	h_β
Gln	1.11	h_α	1.10	h_β
Glu	1.51	H_α	0.37	B_β
Gly	0.57	B_α	0.75	b_β
His	1.00	I_α	0.87	h_β
Ile	1.08	h_α	1.60	H_β
Leu	1.21	H_α	1.30	h_β
Lys	1.16	h_α	0.74	b_β
Met	1.45	H_α	1.05	h_β
Phe	1.13	h_α	1.38	h_β
Pro	0.57	B_α	0.55	B_β
Ser	0.77	i_α	0.75	b_β
Thr	0.83	i_α	1.19	h_β
Trp	1.08	h_α	1.37	h_β
Tyr	0.69	b_α	1.47	H_β
Val	1.06	h_α	1.70	H_β

Source: Chou, P.Y. and Fasman, G.D., *Annu. Rev. Biochem.* **47,** 258 (1978).

where n_α is the number of amino acid residues of the given type that occur in α helices and n is the total number of residues of this type in the set. The propensity of a particular amino acid residue to occur in an α helix is defined as

$$P_\alpha = \frac{f_\alpha}{\langle f_\alpha \rangle} \qquad [9.4]$$

where $\langle f_\alpha \rangle$ is the average value of f_α for all 20 residues. Accordingly, a value of $P_\alpha > 1$ indicates that a residue occurs with greater than average frequency in an α helix. The propensity, P_β, of a residue to occur in a β sheet is similarly defined.

Table 9-1 contains a list of α and β propensities based on the analysis of 29 X-ray structures. In accordance with its value of a given propensity, a residue is classified as a strong former (H), former (h), weak former (I), indifferent former (i), breaker (b), or strong breaker (B) of that secondary structure. Using these data, Chou and Fasman formulated the following empirical rules (the **Chou–Fasman method**) to predict the secondary structures of proteins:

1. A cluster of four helix-forming residues (H_α or h_α, with I_α counting as one-half h_α) out of six contiguous residues will nucleate a helix. The helix segment propagates in both directions until the average value of P_α for a tetrapeptide segment falls below 1.00. A Pro residue, however, can occur only at the N-terminus of an α helix.

2. A cluster of three β sheet formers (H_β or h_β) out of five contiguous residues nucleates a sheet. The sheet is propagated in both directions until the average value of P_β for a tetrapeptide segment falls below 1.00.

3. For regions containing both α- and β-forming sequences, the overlapping region is predicted to be helical if its average value of P_α is greater than its average value of P_β; otherwise a sheet conformation is assumed.

These easily applied empirical rules predict the α helix and β sheet strand positions in a protein with an average reliability of ~50% and, in the most favorable cases, ~80% (Fig. 9-27; note, however, that because proteins consist, on average, of ~31% α helix and ~28% β sheet, random predictions of these secondary structures would average ~30% correct).

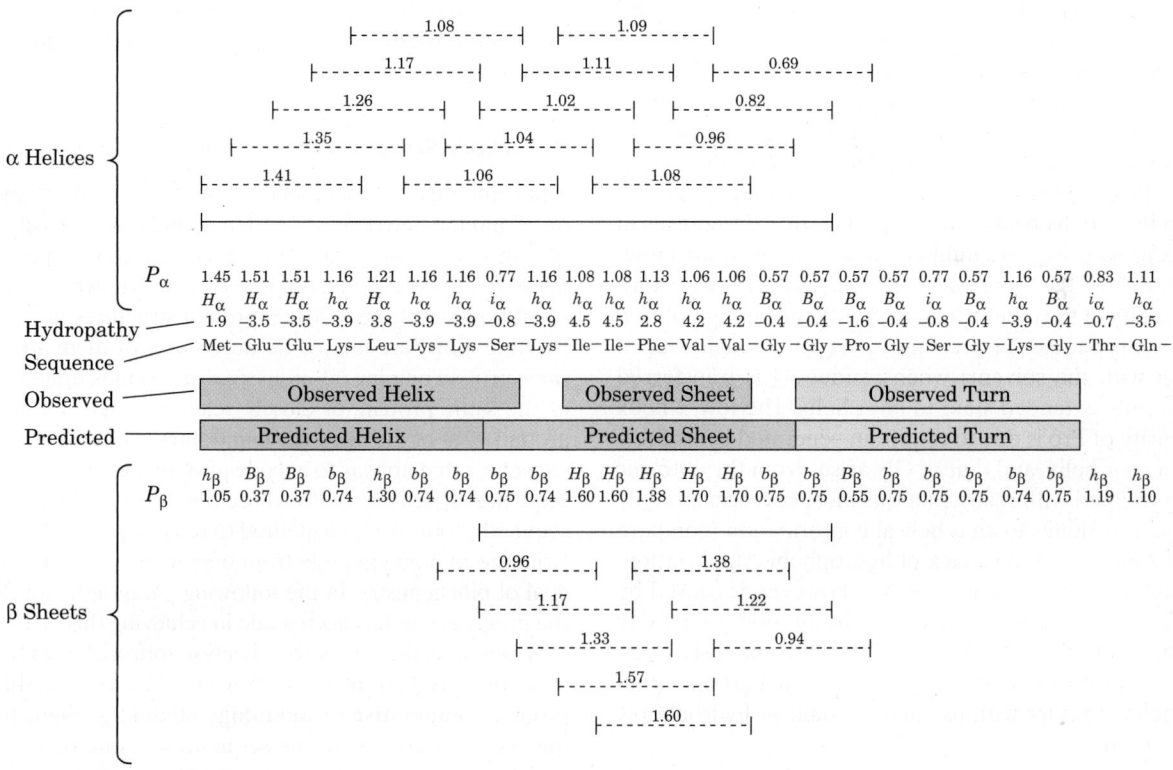

Figure 9-27 **Secondary structure prediction.** The prediction of α helices and β sheets was made by the Chou–Fasman method and the prediction of reverse turns by the method of Rose for the N-terminal 24 residues of adenylate kinase. The helix and sheet propensities and classifications are taken from Table 9-1. The solid lines indicate all hexapeptide sequences that can nucleate an α helix (*top*) and all pentapeptide sequences that can nucleate a β sheet (*bottom*), as is explained in the text. The average helix and sheet propensities for each tetrapeptide segment in the helix and sheet regions are given above the corresponding dashed lines. Twelve of the 15 residues are observed to have their predicted secondary structures (*middle*), so that the prediction accuracy, in this case, is 80%. Reverse turns are predicted to occur in sequences in which the hydropathy (Table 8-6) is a minimum and which do not occur in regions predicted to be helical. The region that matches this criterion is observed to have a reverse turn. [After Schultz, G.E. and Schirmer, R.H., *Principles of Protein Structure*, p. 121, Springer-Verlag (1979).]

b. Reverse Turns Are Characterized by a Minimum in Hydrophobicity Along a Polypeptide Chain

The positions of reverse turns can also be predicted by the Chou–Fasman method. However, since a reverse turn usually consists of four consecutive residues, each with a different conformation (Section 8-1D), their prediction algorithm is necessarily more cumbersome than those for sheets and helices.

Rose has proposed a simpler empirical method for predicting the positions of reverse turns. Reverse turns nearly always occur on the surface of a protein and, in part, define that surface. Since the core of a protein consists of hydrophobic groups and its surface is relatively hydrophilic, reverse turns occur at positions along a polypeptide chain where the hydropathy (Table 8-6) is a minimum. Using these criteria for partitioning a polypeptide chain, we can deduce the positions of most reverse turns by inspection (Fig. 9-27). Since this method often predicts reverse turns to occur in helical regions (helices are all turns), it should be applied only to regions that are not predicted to be helical.

c. Physical Basis of α Helix Propensity

Why do amino acid residues have such different propensities for forming α helices? This question has been answered, in part, by Matthews through the structural and thermodynamic analysis of T4 lysozyme (Section 9-1Bd) in which Ser 44, a solvent-exposed residue in the middle of a 12-residue (3.3-turn) α helix, was mutagenically replaced, in turn, by all 19 other amino acids. The X-ray structures of 13 of these variant proteins revealed that, with the exception of Pro, the substitutions caused no significant distortion to the α helix backbone and, hence, that differences in α helix propensities are unlikely to arise from strain. However, for 17 of the amino acids (all but Pro, Gly, and Ala), the stability of the α helix increases with the amount of side chain hydrophobic surface that is buried (brought out of contact with the solvent) when residue 44 is transferred from a fully extended state to an α helix. The low α helix propensity of Pro is due to the strain generated by its presence in an α helix, and that of Gly arises from the entropic cost associated with restricting this most conformationally flexible of residues to an α helical conformation (compare Figs. 8-7 and 8-9) and its lack of hydrophobic stabilization. The high α helix propensity of Ala, however, is caused by its lack of a γ substituent (possessed by all residues but Gly and Ala) and hence the absence of the entropic cost associated with conformationally restricting such a group within an α helix together with its small amount of hydrophobic stabilization.

d. Computer-Based Secondary Structure Prediction Algorithms

A number of sophisticated computer-based secondary structure prediction algorithms have been developed. Most of them, like the Chou–Fasman method, employ sets of parameters whose values are determined by the analysis of (learning from) a set of nonhomologous proteins with known structures, in some cases coupled with energy mini-

mization techniques. These algorithms are typically ~60% accurate in predicting which of three conformational states, helix, sheet, or coil, a given residue in a protein adopts. However, a significant increase in accuracy has been gained (to over 80%) by employing evolutionary information through the use of multiple sequence alignments. This is because knowledge of the distribution of residue identities at and around each position in a series of homologous and presumably structurally similar proteins provides a better indication of the protein's structural tendencies than does a single sequence.

Several secondary structure prediction algorithms are freely available over the Web. Among them is **Jpred3** (http://www.compbio.dundee.ac.uk/www-jpred/), which classifies residue conformations as being either helical (H), extended/β sheet (E), or coil (−) with 81.5% reliability. It requires as input either the sequence of a single polypeptide or a multiple sequence alignment. However, if Jpred3 is supplied with only a single sequence, it will first use PSI-BLAST (Section 7-4Bi) to construct a multiple sequence alignment.

Although we have seen that secondary structure is mainly dictated by local sequences, we have also seen that tertiary structure can influence secondary structure (Section 9-1Be). The inability of sophisticated secondary structure prediction schemes to surpass ~80% reliability is therefore partially explained by their failure to take tertiary interactions into account.

B. Tertiary Structure Prediction

The sequence databases (Section 7-4A) contain the sequences of ~7 million polypeptides, and the rapid rate at which entire genomes are being sequenced (Section 7-2C) promises that many more such sequences will soon be known. Yet, only a small fraction of the ~70,000 protein structures in the PDB (Section 8-3B) are unique because many of them are of the same protein binding different small molecules, mutant forms of the same protein, or closely related proteins. Moreover, around 40% of the **open reading frames (ORFs;** nucleic acid sequences that appear to encode proteins) in known genome sequences specify proteins whose function is unknown. Consequently, formulating a method to reliably predict the native structure of a polypeptide from only its sequence is a major goal of biochemistry. In the following paragraphs we discuss the progress that has been made in achieving this goal.

There are currently several major approaches to tertiary structure prediction. The simplest and most reliable approach, **comparative** or **homology modeling,** aligns the sequence of interest with the sequences of one or more homologous proteins of known structure, compensating for amino acid substitutions as well as insertions and deletions (indels) through modeling and energy minimization calculations. For proteins with as little as 30% sequence identity, this method can yield a root-mean-square deviation (rmsd) between the predicted and observed positions of corresponding C_α atoms of the "unknown" protein (once its structure has been determined) of as little as ~2.0 Å. However, the accuracy of this method decreases precipitously (the rmsd's rapidly increase) as the degree of sequence identity drops below 30%. Conversely, for polypeptides that are >60% identical, a homology model may have rmsd's of ~1 Å (the accuracy of the atomic positions in an ~2.5-Å-resolution X-ray structure).

There are numerous instances of proteins that are structurally similar even though their sequences have diverged to such an extent that they have no apparent similarity. **Fold recognition** or **threading** is a computational technique that attempts to determine the unknown fold of a protein by ascertaining whether its sequence is compatible with any of the members of a library of known protein structures. It does so by placing the "unknown" protein's residues along the backbone of a known protein structure, determining the stability of the side chains of the unknown protein in that arrangement, and then sliding (threading) the sequence of the unknown protein along that of the known protein by one residue and repeating the calculation, etc., while allowing for the possibility of indels. If the "correct" fold can be found (and there is no guarantee that the fold of the unknown protein will resemble that of any member of the library), the resulting model can be improved via homology modeling. This method has yielded encouraging results, although it cannot yet be considered to be reliable. Of course, as sequence alignment algorithms (Section 7-4B) improve in their ability to recognize distant homologs, sequences that previously would have been candidates for fold recognition can instead be directly treated by comparative modeling.

Since the native structure of a protein depends only on its amino acid sequence, it should be possible, in principle, to predict the structure of a protein based only on its physicochemical properties (e.g., solvent interactions, atomic volume, charge, hydrogen bonding properties, van der Waals interactions, and bond torsion angle potentials for all of its atoms). A major problem faced by such *de novo* (Latin: from the beginning; synonymously *ab initio*) **methods** is that polypeptide chains have astronomical numbers of non-native low-energy conformations, so that it requires extensive and highly detailed calculations to determine a polypeptide's lowest energy conformation.

To assess the effectiveness of the numerous *de novo* algorithms that have been formulated, as well as other structure prediction schemes, a **Critical Assessment of Structure Prediction (CASP)** has been held every 2 years starting in 1994. CASP participants are provided with the sequences of proteins whose structures will soon be determined by X-ray crystallography or NMR spectroscopy and submit their predicted structures of these proteins for comparison with the subsequently experimentally determined structures. Over the years, *de novo* methods have steadily improved from being little better than random guesses to predicting the folding topologies of <200-residue proteins with a success rate of ~20% and occasionally with near-atomic accuracy. **Rosetta,** the most consistently successful *de novo* algorithm in the past few CASPs, which was formulated by David Baker, is enormously computer-intensive. To satisfy its computational needs, Baker has organized a distributed computer network known as Rosetta@home that uses the

Figure 9-28   **Examples of successful modeling predictions by Rosetta.** Each panel shows the superposition of a predicted model (*gray*) with the corresponding experimentally determined X-ray structure colored in rainbow order from its N-terminus (*blue*) to its C-terminus (*red*) with core side chains drawn as sticks. (*a*) A protein of unknown function from *Thermus thermophilus* HB8 (PDBid 1WHZ). The backbones are aligned with an accuracy of 1.6 Å over 70 residues. (*b*) Protein BH3980 (10176605) from *Bacillus halodurans* (PDBid 2HH6). The backbones are aligned with an accuracy of 1.4 Å over 90 residues. [Courtesy of Gautam Dantas, Washington University School of Medicine.]

otherwise idle time of nearly 100,000 volunteered computers so that an average of ~500,000 CPU-hours can be devoted to predicting the structure of each domain. Examples of successful protein structure predictions by Rosetta are shown in Fig. 9-28.

C. *Protein Design*

Although we have not yet fully solved the protein folding problem, considerable progress has been made in solving the inverse problem: generating polypeptide sequences to assume specific three-dimensional structures, that is, **protein design.** This is probably because a polypeptide can be "overengineered" to take up a desired conformation. Consequently, protein design has provided insights into protein folding and stability, and it promises to yield useful proteins that are "made to order." Protein design begins with a target structure such as a 4-helix bundle and attempts to find an amino acid sequence that will form this structure. The designed polypeptide is then synthesized and its structure elucidated.

Successful design requires not only that the desired fold be stable but that other folds be significantly less stable (by ~15–40 kJ · mol^{-1}). Otherwise a sequence that has been found to be the most stable in the desired conformation may actually be more stable in other conformations. Before such **negative design** concepts were implemented, efforts to design proteins typically yielded an ensemble of molten globulelike states rather than the desired folds.

Most successful protein design projects have redesigned naturally occurring proteins so as to enhance their stability or provide them with new functionalities. Because of the strict steric constraints in the cores of globular proteins, this has largely yielded proteins whose internal side chain packings resemble those of the original proteins. Consequently, designing a protein with a novel fold should be a more rigorous test of protein design methods. Indeed, it is unlikely that a polypeptide of arbitrary sequence will have a stable native structure.

Baker designed a topologically novel, 93-residue, α/β protein he named **Top7** as follows. A rough two-dimensional model of the target protein was created and structural constraints that defined its topology (e.g., hydrogen bonds and reverse turns) were identified. Rosetta was then used to generate 172 three-dimensional, backbone-only models with the required topology by assembling 3- and 9-residue fragments with the required secondary structures from the Protein Data Bank (PDB). Side chains were initially placed by considering all sets of energetically allowed torsion angles (which are known as **rotamers**) for each type of side chain but Cys at the 71 core positions and for only polar residues at the remaining 22 surface positions and using Rosetta to identify the lowest energy structures. These models were then structurally optimized through 15 cycles of using Rosetta to calculate the lowest energy backbone conformation for a fixed amino acid sequence followed by sequence redesign as previously described, ultimately yielding Top7. Although the structural differences between Top7 and its initial model were small (their backbones have an rmsd of 1.1 Å), they had dramatic changes in sequence (with only 31% of the residues in Top7 being identical to those in its initial model).

A gene for Top7 with a C-terminal His tag was synthesized (Section 7-6A) and expressed and the protein was purified by metal chelation affinity chromatography (Section 6-3Dg) followed by anion exchange chromatography (Section 6-3A). Top7 is highly soluble in aqueous solution and is monomeric as indicated by gel filtration chromatography (Section 6-3B). It is remarkably stable: Its circular dichroism (CD; Section 9-1Ca) spectrum at 98°C closely resembles that at 25°C. The X-ray structure of Top7 is all but identical, within experimental error, to the structure of the designed model: Their rmsd over all backbone atoms is 1.17 Å and many of their core side chains are effectively superimposable (Fig. 9-29). Evidently, protein folds that have not been observed in nature are not only physically possible but can be highly stable.

Baker also used the principles of protein design to generate enzymes that catalyze nonbiological reactions. He did so by grafting designed constellations of side chains onto the surfaces of naturally occurring proteins so as to form the desired active sites. As we shall see in Chapter 15, the catalytic activities of enzymes are extraordinarily sensitive to the positions of their catalytic groups.

4 PROTEIN DYNAMICS

The fact that X-ray studies yield time-averaged "snapshots" of proteins may leave the false impression that proteins have fixed and rigid structures. In fact, as is becoming increasing

Figure 9-29   **Superposition of the designed model of Top7 (*blue*) with its X-ray structure (*red*).** Core side chains are drawn as sticks. [Courtesy of Gautam Dantas, Washington University School of Medicine. PDBid 1QYS.]

clear, *proteins are flexible and rapidly fluctuating molecules whose structural mobilities have considerable functional significance.* For example, X-ray studies indicate that the heme groups of myoglobin and hemoglobin are so surrounded by protein that there is no clear path for O_2 to approach or escape from its binding pocket. Yet we know that myoglobin and hemoglobin readily bind and release O_2. These proteins must therefore undergo conformational fluctuations, **breathing motions,** that permit O_2 reasonably free access to their heme groups (Fig. 9-30). The three-dimensional structures of myoglobin and hemoglobin undoubtedly evolved the flexibility to facilitate the diffusion of O_2 to its binding pocket.

Figure 9-30   **Conformational fluctuations in myoglobin.** An artist's conception of the "breathing" motions in myoglobin that permit the escape of its bound O_2 molecule (*double red spheres*). The dotted lines trace a trajectory an O_2 molecule might take in worming its way through the rapidly fluctuating protein before finally escaping. O_2 binding presumably resembles the reverse of this process. [Illustration, Irving Geis. Image from the Irving Geis Collection, Howard Hughes Medical Institute. Reprinted with permission.]

The intramolecular motions of proteins have been classified into three broad categories according to their coherence:

1. Atomic fluctuations, such as the vibrations of individual bonds, which have time periods ranging from 10^{-15} to 10^{-11} s and spatial displacements between 0.01 and 1 Å.

2. Collective motions, in which groups of covalently linked atoms, which vary in size from amino acid side chains to entire domains, move as units with time periods ranging from 10^{-12} to 10^{-3} s and spatial displacements between 0.01 and >5 Å. Such motions may occur frequently or infrequently compared with their characteristic time period.

3. Triggered conformational changes, in which groups of atoms varying in size from individual side chains to complete subunits move in response to specific stimuli such as the binding of a small molecule, for example, the binding of ATP to GroEL (Section 9-2Ca). Triggered conformational changes occur over time spans ranging from 10^{-9} to 10^{3} s and result in atomic displacements between 0.5 and >10 Å.

In this section, we discuss how these various motions are characterized and their structural and functional significance. We shall mainly be concerned with atomic fluctuations and collective motions; triggered conformational changes are considered in later chapters in connection with specific proteins.

a. Proteins Have Mobile Structures

X-ray crystallographic analysis is a powerful technique for the analysis of motion in proteins; it reveals not only the average positions of the atoms in a crystal, but also their mean-square displacements from those positions. X-ray analysis indicates, for example, that myoglobin has a rigid core surrounding its heme group and that the regions toward the periphery of the molecule have a more mobile character. Similarly, the apical domain of GroEL and the mobile loop of GroES are both highly flexible in the individual proteins, but when they interact in the GroEL–GroES–(ADP)$_7$ complex, they become significantly more rigid (Fig. 9-31; Section 9-2Ca). Indeed, as we have seen (Section 9-1Bg), portions of the binding sites of many proteins rigidify on binding their target molecules.

Molecular dynamics simulations, a theoretical technique pioneered by Martin Karplus, has revealed the nature of the atomic motions in proteins. In this technique, the atoms of a protein of known structure and its surrounding solvent are initially assigned random motions with velocities that are collectively characteristic of a chosen temperature. Then, after a time step of ~1 femtosecond (1 fs = 10^{-15} s) the aggregate effects of the various interatomic

(a)

(b)

Figure 9-31 **The mobility of the GroEL subunit.** (*a*) In the X-ray structure of GroEL alone and (*b*) in the X-ray structure of GroEL–GroES–(ADP)$_7$. The polypeptide backbone is colored in rainbow order according to its degree of thermal motion, with blue being the least mobile (cool) and red being the most mobile (hot). The subunits are oriented as in Fig. 9-22*a,b*. Note that the outer end of the apical domain, which functions to bind both substrate protein and the mobile loop of GroES (Section 9-2Ca), is more mobile in GroEL alone (*red and red-orange*) than it is in GroEL–GroES–(ADP)$_7$ (*orange and yellow*). [Based on X-ray structures by Axel Brünger, Arthur Horwich, and Paul Sigler, Yale University. PDBids (*a*) 1OEL and (*b*) 1AON.]

forces in the system (those due to departures from ideal co-valent bond lengths, angles, and torsion angles as well as noncovalent interactions) on the velocities of each of its atoms are calculated according to Newton's equations of motion. Since all the atoms in the system will have moved after this time step (by a distance that is only a small fraction of a bond length), the interatomic forces (their potential field) on each atom will likewise have changed (although by only a small amount). Then, using this altered potential field together with the new positions and velocities of the atoms, the calculation is repeated for an additional time step. This computationally intensive process has been iterated for up to ~1 μs for ~100-residue proteins (a time that is increasing with the available computational power), thereby yielding a record of the positions and velocities of all the atoms in the system over this time period.

Molecular dynamics simulations (e.g., Fig. 9-32) have revealed that *a protein's native structure really consists of a large collection of conformational substates that have essentially equal stabilities.* These substates, which each have slightly different atomic arrangements, randomly interconvert at rates that increase with temperature. Consequently, the interior of a protein typically has a fluidlike character for structural displacements of up to ~2 Å, that is, over excursions that are somewhat larger than a bond length.

Gregory Petsko and Dagmar Ringe have demonstrated the functional significance of the internal motions in proteins. Both experimental and theoretical evidence indicates that below ~220 K (−53°C), collective motions in proteins are arrested, leaving atomic fluctuations as the dominant intramolecular motions. For example, X-ray studies have shown that, at 228 K, the enzyme RNase A, in its crystalline form, readily binds an unreactive substrate analog (protein crystals generally contain large solvent-filled channels through which small molecules rapidly diffuse; at low temperatures, the water is prevented from freezing by the addition of an antifreeze such as methanol). Yet, when the same experiment is performed at 212 K, the substrate analog does not bind to the enzyme, even after 6 days of exposure. Likewise, at 228 K, substrate-free solvent washes bound substrate analog out of the crystal within minutes but, if the temperature is first lowered to 212 K, the substrate analog remains bound to the crystalline enzyme for at least 2 days. Evidently, RNase A assumes a glasslike state below 220 K that is too rigid to bind or release substrate. In terms of landscape theory, this is interpreted as the protein being trapped in a single energy well.

b. Protein Core Mobility Is Revealed by Aromatic Ring Flipping

The rate at which an internal Phe or Tyr ring in a protein undergoes 180° "flips" about its C_β—C_γ bonds is indicative of the surrounding protein's rigidity. This is because, in the close packed interior of a protein, these bulky asymmetric groups can move only when the surrounding groups move aside transiently (although note that these rings have the shape of flattened ellipsoids rather than thin disks).

NMR spectroscopy can determine the mobilities of protein groups over a wide range of time scales. Consequently, the rate at which a particular aromatic ring in a protein flips is best inferred from an analysis of its NMR spectrum (infrequent motions such as ring flipping are not detected by X-ray crystallography since this technique only reveals the average structure of a protein). NMR measurements indicate that the ring flipping rate varies from over 10^6 s^{-1} to one of immobility (<1 s^{-1}) depending on both the protein and the location of the aromatic ring within the protein. For example, **bovine pancreatic trypsin inhibitor (BPTI)** is a 58-residue monomeric protein that has eight Phe and Tyr residues. At 4°C, four of these Phe and Tyr rings flip at rates $>5 \times 10^4$ s^{-1}, whereas the remaining four rings flip at rates

(a)

(b)

Figure 9-32 **The internal motions of myoglobin as determined by a molecular dynamics simulation.** Several "snapshots" of the molecule calculated at intervals of 5×10^{-12} s are superimposed. (*a*) The C_α backbone and the heme group. The backbone is shown in blue, the heme in yellow, and the His residue liganding the Fe in orange. (*b*) An α helix. The backbone is shown in blue, the side chains in green, and the helix hydrogen bonds as dashed orange lines. Note that the helices tend to move in a coherent fashion so as to retain their shape. [Courtesy of Martin Karplus, Harvard University.]

ranging between 30 and $<1 \ s^{-1}$. These ring-flipping rates sharply increase with temperature, as expected.

c. Infrequent Motions Can Be Detected through Hydrogen Exchange

Conformational changes occurring over time spans of more than several seconds can be chemically characterized through hydrogen exchange studies (Sections 9-1Cc). These show that the exchangeable protons of native proteins exchange at rates that vary from milliseconds to many years (Fig. 9-33). Protein interiors, as we have seen (Section 8-3B), are largely excluded from contact with their surrounding aqueous solvent, and, moreover, protons cannot exchange with solvent while they are engaged in hydrogen bonding. The observation that the internal protons of proteins do, in fact, exchange with solvent must therefore be a consequence of transient local unfolding or "breathing" that physically and chemically exposes these exchangeable protons to the solvent. Hence, *the rate at which a particular proton undergoes hydrogen exchange is a reflection of the conformational mobility of its surroundings.* This hypothesis

Figure 9-33 **The hydrogen–tritium "exchange-out" curve for hemoglobin that has been pre-equilibrated with tritiated water.** The vertical axis expresses the ratio of exchangeable protons to heme Fe atoms. Exchange-out was initiated by replacing the protein's tritiated water solvent with untritiated water through rapid gel filtration (Section 6-3B). As the exchange-out proceeded, additional gel filtration separations were performed and the amount of tritium remaining bound to the protein was measured. At the arrow, O_2 was added to exchanging deoxyhemoglobin (hemoglobin lacking bound O_2). The changing slopes of these curves indicate that the hydrogen exchange rates of the ~80 exchangeable protons of each hemoglobin subunit vary by factors of many decades and that O_2 binding increases the exchange rates for ~10 of these protons (the structural changes that O_2 binding induces in hemoglobin are discussed in Section 10-2). [After Englander, S.W. and Mauel, C., *J. Biol. Chem.* **247,** 2389 (1972).]

is corroborated by the observation that the hydrogen exchange rates of proteins decrease as their denaturation temperatures increase and that these exchange rates are sensitive to the proteins' conformational states (Fig. 9-33).

5 CONFORMATIONAL DISEASES: AMYLOID AND PRIONS

Most proteins in the body maintain their native conformations or, if they become partially denatured, are either renatured through the auspices of molecular chaperones (Section 9-2C) or are proteolytically degraded (Section 32-6). However, ~35 different, often fatal, human diseases are associated with the extracellular deposition of normally soluble proteins in certain tissues in the form of insoluble aggregates known as **amyloid** (starchlike; a misnomer because it was originally thought that this material resembled starch). These include **Alzheimer's disease** and **Parkinson's disease,** neurodegenerative diseases that mainly strike the elderly; the **transmissible spongiform encephalopathies (TSEs),** a family of infectious neurodegenerative diseases that are propagated in a most unusual way; and the **amyloidoses,** a series of diseases caused by the deposition of an often mutant protein in organs such as the heart, liver, or kidney. The deposition of amyloid interferes with normal cellular function, resulting in cell death and eventual organ failure.

Although the various types of amyloidogenic proteins are unrelated and their native structures have widely different folds, their amyloid forms have remarkably similar core structures: Each consists of an array of ~10-nm-diameter **amyloid fibrils** (Fig. 9-34) in which, as infrared, NMR, and X-ray diffraction methods indicate, certain segments

Figure 9-34 **An electron micrograph of amyloid fibrils of the protein PrP 27–30 (Section 9-5Ce).** They are visually indistinguishable from amyloid fibrils formed by other proteins. The black dots are colloidal gold beads that are coupled to anti-PrP antibodies that are adhering to the PrP 27–30. [Courtesy of Stanley Prusiner, University of California at San Francisco Medical Center.]

of the proteins form extended β sheets whose planes extend parallel to the fibril axis so that their β strands are perpendicular to the fibril axis (see below). Thus, *these proteins each have two radically different stable conformations, their native forms and their amyloid forms.*

We begin this section with a discussion of the amyloidoses as exemplified by **islet amyloid polypeptide (IAPP; also called amylin)** and certain mutants forms of lysozyme. We then consider Alzheimer's disease and finally the TSEs and their bizarre mode of propagation.

A. *Amyloid Diseases*

Many amyloidogenic proteins are mutant forms of normally occurring proteins. These include lysozyme (an enzyme that hydrolyzes bacterial cell walls; Section 15-2) in the disease **lysozyme amyloidosis, transthyretin** [Fig. 8-66; a blood plasma protein that functions as a carrier for the water-insoluble thyroid hormone **thyroxin** (Section 19-1D) as well as retinol through its association with retinol binding protein (Section 8-3Bg)] in **familial amyloidotic polyneuropathy,** and **fibrinogen** (the precursor of **fibrin,** which forms blood clots; Section 35-1A) in **fibrinogen amyloidosis.** Most such diseases do not present (become symptomatic) until the third to seventh decades of life and typically progress over 5 to 15 years ending in death.

a. IAPP Forms a Cross-β Spine Structure

IAPP is a 37-residue peptide that is associated with **type 2 diabetes mellitus** (also called **non-insulin dependent** and **maturity-onset diabetes mellitus;** Section 27-4B), an often fatal disease that afflicts ~100 million mainly older people worldwide. IAPP is expressed and secreted by the pancreatic β islet cells, which also synthesize the polypeptide hormone insulin (Fig. 7-2; whose lack is responsible for **type I** or **juvenile-onset diabetes mellitis).** Although the role of IAPP in the development of type 2 diabetes is unclear, the pancreases of 95% of individuals with type 2 diabetes contain amyloid deposits of IAPP with the extent of this deposition increasing with the severity of the disease. Interestingly, mouse IAPP, which differs from human IAPP in 6 of its 37 residues, does not form amyloid and mice do not develop type 2 diabetes, although transgenic mice that express human IAPP sometimes do so.

Attempts to crystallize IAPP, either in its native conformation or its amyloid form, have been unsuccesful. However, through the use of structure prediction techniques (Section 9-3B), Baker and David Eisenberg identified two of its segments that have high fibril-forming potential: NNFGAIL and SSTNVG, which comprise IAPP's residues 21 to 27 and 28 to 33 (5 of the 6 residue differences between human and mouse IAPP occur in the segment 23–29). Both of these peptides form amyloidlike fibrils as well as very thin needle-shaped crystals.

The X-ray structure of SSTNVG, determined by Eisenberg, reveals that this hexapeptide forms an extended parallel β sheet with two such sheets facing each other such that their protruding side chains interdigtate so tightly that they completely exclude water (Fig. 9-35a). The X-ray structure of NNFGAIL is similar but contains a pronounced bend in its backbone, such that the interface between the two β sheets is formed between main chains (Fig. 9-35b). Such structures, which are known as **cross-β spines,** are assumed by a variety of other amyloid-forming peptides, although many of them contain antiparallel rather than parallel β sheets.

Figures 9-35c–e exhibit a model of the IAPP amyloid fibril based on the forgoing X-ray structures. Its is a 4-sheet,

Figure 9-35 **The IAPP amyloid fibril.** (*a*) The X-ray structure of SSTNVG drawn in ball-and-stick form as viewed along the planes of the β sheets that they form. Atoms are colored according to type with C on one chain white and the other magenta, N blue, O red, and water molecules represented by yellow spheres. (*b*) The X-ray structure of NNFGAIL viewed and colored as in Part *a*. (*c*) Model of the fibril as viewed along its axis (*black dot*). Atoms are colored according to type with the C atoms of the SSTNVG segment green, those of the NNFGAIL segment light blue, those of the N-terminal 20 residues and the C-terminal 4 residues light red or white, N blue, O red, and S atoms, which form disulfide bonds, yellow. (*d*) Schematic view along the fibril axis in which the SSTNVG segments are green, the NNFGAIL segments are light blue, and the modeled residues are white. (*e*) View perpendicular to the fibril axis represented as in Part *d*. The helix has a gentle left-handed twist of 3.4 Å per layer so that it makes a quarter turn every 125 Å. [Courtesy of David Eisenberg, UCLA. PDBids 3DG1 for SSTNVG and 3DGJ for NNFGAIL.]

left-handed helix which has a pitch (rise per turn) of 500 Å. The SSTNVG segment's cross-β spine is centered on the fibril axis. It is extended by NNFGAIL segments that also form hairpin turns. IAPP's final 4 residues were modeled to complete the inner β strands and its initial 20 residues were modeled to to form the outer strands. The calculated X-ray diffraction pattern of this assembly closely resembles the observed diffraction pattern of IAPP fibrils.

Other amyloid-forming proteins have segments that form similar cross-β spines. However, the structures of the loops connecting the β strands must vary with the identity of the protein.

b. Amyloidogenic Lysozyme Variants Have Conformationally Flexible Native Structures

There are two known amyloidogenic variants of the 130-residue human lysozyme, I56T and D67H. These form amyloid fibrils that are deposited in the viscera (internal organs), usually resulting in death by the fifth decade. The amyloid fibrils consist exclusively of the variant lysozymes, thereby explaining why these mutations are dominant. Structural studies on these variant proteins have shed light on how they form amyloid fibrils.

The X-ray structures of both mutant lysozymes resemble that of the wild-type enzyme. However, the replacement of Asp 67 by His interrupts a network of hydrogen bonds that stabilizes the domain containing the structure's only β sheet (its so-called β domain), resulting in the movements of the β sheet and an adjoining loop away from each other by displacements of up to 11 Å (Fig. 9-36). Although the replacement of Ile 56 by Thr causes only subtle changes in the protein structure, it insinuates a hydrophilic residue in a critical hydrophobic interface that links the protein's two domains.

The melting temperatures (T_m's) of both variants are at least 10°C less than those of the wild-type enzyme, and both variants eventually lose all enzymatic activity when incubated at physiological temperature and pH (37°C and 7.4), conditions under which wild-type lysozyme remains fully active. The variants also aggregate on heating *in vitro,* and a variety of physical measurements indicate that, in doing so, they form amyloidlike fibrils. In hydrogen exchange experiments (Section 9-1Cc), wild-type lysozyme strongly protects 55 protons from exchange with D_2O under conditions (37°C and pH 5) in which these protons are essentially unprotected in the amyloidogenic variants, thereby confirming that the native protein's tertiary structure is greatly loosened in the two mutant forms. This suggests that the partially folded, aggregation-prone forms are in dynamic equilibrium with the native conformation, even under conditions in which the native state is thermodynamically stable [keep in mind that the ratio of unfolded (U) to native (N) protein molecules in the reaction N $\rightleftharpoons$ U is governed by Eq. [3.17]: $[U]/[N] = e^{-\Delta G^{\circ\prime}/RT}$, where $\Delta G^{\circ\prime}$ is the standard free energy of unfolding, so that as $\Delta G^{\circ\prime}$ decreases, the proportion of U increases]. It has therefore been proposed that lysozyme fibrillogenesis is initiated by the association of the β domains of two partially unfolded

Figure 9-36 **Superpositions of wild-type human lysozyme and its D67H mutant.** Wild-type lysozyme is gray and its D76H mutant is color-ramped in rainbow order from blue at its N-terminus to red and back to blue at its C-terminus. The white arrows indicate the conformational shifts of residues 45 to 54 and 67 to 75 in the D67H mutant relative to those in the wild-type protein. The four disulfide bonds present in both the wild-type and mutant protein are shown in yellow. The positions of residues 56 and 67 are indicated. [Courtesy of Margaret Sunde, University of Oxford, UK and Colin Blake, University of Oxford, UK.]

lysozyme variants to form a more extensive β sheet. This would provide a template or nucleus for the recruitment of additional polypeptide chains to form the growing fibril in a process that may involve the conformational conversion of α helices to β strands. Such an autocatalytic refolding process may be a general mechanism for amyloid fibrillogenesis. However, the several decades that many hereditary amyloid diseases require to become symptomatic suggest that the spontaneous generation of an amyloid nucleus is a rare event, that is, has a high free energy of activation (activation barriers and their relationship to reaction rates are discussed in Section 14-1C).

B. *Alzheimer's Disease*

Alzheimer's disease (AD), a neurodegenerative condition that afflicts ~20 milion mainly elderly people worldwide (~10% of those over the age of 65 and ~50% of those over 85), causes devastating mental deterioration and eventual death. It is characterized by brain tissue containing abundant amyloid plaques (deposits) surrounded by dead and dying neurons. In addition, many neuronal cell bodies contain abnormal ~20-nm-diameter fibers known as **neurofibrillary tangles.** The amyloid plaques consist mainly of amyloid fibrils of a 40- to 42-residue peptide named **amyloid-β**

peptide (Aβ) [the neurofibrillary tangles, which we shall not further discuss, consist of a hyperphosphorylated form of a protein named **tau** that is normally associated with microtubules (Section 1-2Ae)].

The sequence of the gene encoding Aβ, which was identified via reverse genetics (Section 7-2D) based on the sequence of Aβ, reveals that Aβ is a segment of a 770-residue transmembrane protein named **Aβ precursor protein (βPP;** transmembrane proteins are discussed in Section 12-3A). βPP has a receptorlike sequence (Section 19-2B) although its normal function is unknown. Aβ is excised from βPP in a multistep process through the actions of two membrane-anchored proteolytic enzymes dubbed β- and γ-**secretases.**

It had been hotly debated whether Aβ causes AD or is merely a product of its neurodegenerative processes. This argument was largely put to rest by the observation that microinjecting 200 pg of fibrillar but not soluble Aβ (the approximate quantity in a single Aβ plaque) into the cerebral cortexes of aged but not young rhesus monkeys causes marked neuronal loss and other microscopic changes characteristic of AD as far as 1.5 mm from the injection site. Evidently, *the neurotoxic agents in AD are the Aβ-containing amyloid fibrils before their deposition in amyloid plaques.*

The age-dependence of AD suggests that **β-amyloid** deposition is an ongoing process, at least in the later decades of life. Indeed, there are several rare variants of the βPP gene with mutations in their Aβ regions that result in the onset of AD as early as the fourth decade of life. These mutatations have been shown to affect the proteolytic processing of βPP in a way that increases the rate of Aβ production. A similar phenomenon is seen in **Down's syndrome,** a condition characterized by mental retardation and a distinctive physical appearance caused by the trisomy (3 copies per cell) of chromosome 21 rather than the normal two copies. Individuals with Down's syndrome invariably develop AD by their fortieth year. This is because the gene encoding βPP is located on chromosome 21, and hence individuals with Down's syndrome produce βPP and presumably Aβ at an accelerated rate.

A second gene that has been implicated in the premature onset of AD encodes the cholesterol transport protein **apolipoprotein E (apoE;** Section 12-5Bd). The *apoE* gene has several normally occurring variants (alleles) in the population, one of which, *apoE4,* is a major risk factor for both the development of AD and its earlier onset. Moreover, AD victims with *apoE4* have significantly higher densities of β-amyloid plaques in their brain tissue than AD victims with other apoE variants. These observations motivated experiments showing that ApoE4 induces enhanced aggregation of synthetic Aβ *in vitro.* This suggests that ApoE4 facilitates Aβ aggregation *in vivo* (although another possibility is that ApoE4 inhibits the clearance of Aβ from the extracellular spaces).

There is, at present, no known treatment that arrests the progress of AD. However, the foregoing suggests several strategies for therapeutic intervention, including decreasing the rate of production of Aβ through the administration of substances that inhibit the action of β- or γ-secretase and through the administration of agents that would interfere with the formation of β-amyloid fibrils from soluble Aβ.

C. *Prion Diseases*

Certain infectious diseases that affect the mammalian central nervous system were originally classified as being caused by "slow viruses" because they take months, years, or even decades to develop. Among them are **scrapie,** a neurological disorder of sheep and goats, so named for the tendency of infected sheep to scrape off their wool [they rub against fences in an effort to stay upright due to ataxia (loss of muscle coordination)]; **bovine spongiform encephalopathy (BSE** or **mad cow disease),** which similarly afflicts cattle; and **kuru,** a degenerative brain disease that occurred among the Fore people of Papua New Guinea (kuru means trembling) and that was transmitted by ritual cannibalism. There is also a sporadic (apparently spontaneously arising) human disease with similar symptoms, **Creutzfeldt–Jakob disease (CJD),** a rare, progressive, cerebellar disorder, which resembles and may be identical to kuru. These diseases, all of which are ultimately fatal, have similar symptoms, which suggests that they are closely related. Since, in all of these diseases, neurons develop large vacuoles that gives brain tissue a spongelike microscopic appearance, they are collectively known as **transmissible spongiform encephalopathies (TSEs).** None of the TSEs exhibit any sign of an inflammatory process or fever, which indicates that the immune system, which is not impaired by the disease, is not activated by it.

The classic technique for isolating the agent causing an infectious disease involves the fractionation of diseased tissue as monitored by assays for the disease. The long incubation time for scrapie, the most extensively studied "slow virus" disease, enormously hampered initial efforts to characterize its disease agent. Indeed, in the early work on scrapie in the 1930s, an entire herd of sheep and several years of observation were necessary to evaluate the results of a single fractionation. Assays for scrapie were greatly accelerated, however, by the discovery that Syrian hamsters, after intracerebral inoculation of the scrapie agent, develop the disease in a time, minimally 60 days, that decreases as the dose given is increased. Using a hamster assay, Stanley Prusiner purified the scrapie agent to a high degree and was instrumental in characterizing it.

a. Scrapie Is Caused by Prion Protein

The scrapie agent apparently is a single species of protein. This astonishing conclusion was established by the observations that the scrapie agent is inactivated by substances that modify proteins, such as proteases, detergents, phenol, urea, and reagents that react with specific amino acid side chains, whereas it is unaffected by agents that alter nucleic acids, such as nucleases, UV irradiation, and substances that specifically react with nucleic acids. For example, scrapie agent is inactivated by treatment with **diethylpyrocarbonate,** which carboxyethylates the His residues of proteins (Fig. 9-37a), but is unaltered by the cytosine-specific reagent **hydroxylamine** (Fig. 9-37b). In

(a)

Diethylpyrocarbonate **Ethylcarboxamido-His**

(b)

Cytosine

(c)

Ethylcarboxamido-His **Hydroxylamine**

His

Figure 9-37 Evidence that the scrapie agent is a protein.
(*a*) Scrapie agent is inactivated by treatment with
diethylpyrocarbonate, which specifically reacts with His side
chains. (*b*) Scrapie agent is unaffected by treatment by
hydroxylamine, which reacts with cytosine residues. (*c*) However,
hydroxylamine rescues diethylpyrocarbonate-inactivated scrapie
reagent, presumably by the reaction shown.

fact, the infectivity of diethylpyrocarbonate-inactivated
scrapie agent is restored by treatment with hydroxylamine,
presumably by the reaction shown in Fig. 9-37*c*.

The novel properties of scrapie agent, which distinguish
it from viruses and plasmids, have resulted in its being
termed a **prion** (for *pro*teinaceous *in*fectious particle that
lacks nucleic acid). The scrapie protein, which is named
PrP (for *P*rion *P*rotein), consists of 280 mostly hydrophobic residues. This hydrophobicity, as we shall see below,
causes partially proteolyzed PrP to aggregate as clusters of
rodlike particles. There is a close resemblance between
these clusters and the amyloid fibrils that are seen on electron microscopic examination of prion-infected brain tissue
(Fig. 9-34). In fact, brain tissue from CJD victims contains

protease-resistant protein that cross-reacts with antibodies
raised against scrapie PrP.

b. PrP Is a Widely Expressed Product of a Normal Cellular Gene That Has No Known Function

The bizarre composition of prions immediately raises
the question: How are they synthesized? Three possibilities
have been suggested:

1. Despite all evidence to the contrary, prions contain a
nucleic acid genome that is somehow shielded from detection; that is, prions are conventional viruses. The enormous
and still growing body of information concerning the
nature of prions, however, makes this notion increasingly
untenable.

2. Prions might somehow specify their own amino acid
sequence by "reverse translation" to yield a nucleic acid that
is normally translated by the cellular system. Such a process,
of course, would directly contravene the "central dogma" of
molecular biology (Section 5-4), which states that genetic information flows unidirectionally from nucleic acids to proteins. Alternatively, prions might directly catalyze their own
synthesis. Such protein-directed protein synthesis is likewise
unknown (although many small bacterial polypeptides are
enzymatically rather than ribosomally synthesized).

3. Susceptible cells carry a gene that codes for the corresponding PrP. Infection of such cells by prions activates
this gene and/or alters its protein product in some autocatalytic way.

The latter hypothesis seems to be the most plausible mechanism of prion replication. Indeed, the use of oligonucleotide probes complementary to the PrP gene (which is
named ***Prn-p*** for *prion* *p*rotein), as inferred from the
amino acid sequence of PrP's N-terminus (Section 7-2D),
established that the brains of both scrapie-infected and
normal mice contain *Prn-p*. The most surprising discovery,
however, is that *Prn-p is transcribed at similar levels in both
normal and scrapie-infected brain tissue.* Moreover, the use
of the above probes has revealed that *Prn-p genes occur in
all vertebrates so far tested, including humans, as well as in
invertebrates such as Drosophila.* This evolutionary conservation suggests that PrP, a membrane-anchored protein
(via glycosylphosphatidylinositol groups; Section 12-3Bc)
that occurs mainly on neuron surfaces, has an important
function. Thus it came as a further surprise that knockout
mice (Section 5-5H) in which both *Prn-p* genes have been
disrupted appear to be normal and that mating two such
Prn-p$^{0/0}$ mice gives rise to normal *Prn-p*$^{0/0}$ progeny (although there is some evidence that *Prn-p*$^{0/0}$ mice develop
neurological abnormalities late in life). Nevertheless, evidence is accumulating that PrP is normally a cell-surface
signal receptor, although the identity of its corresponding
signal and its consequences are as yet unknown.

c. Scrapie Disease Requires the Expression of the Corresponding PrP^C Protein

Prn-p$^{0/0}$ mice remain completely free of scrapie symptoms after inoculation with a dose of mouse scrapie PrP

(**PrP^{Sc}**; Sc for scrapie) that causes wild-type (*Prn-p*^{+/+}) mice to die of scrapie within 6 months after inoculation. Evidently, *PrP^{Sc} induces the conversion of normal PrP (**PrP^C**; C for cellular) to PrP^{Sc}*. This unorthodox notion, the so-called **prion hypothesis,** is supported by the observation that when wild-type mice are inoculated with PrP^{Sc} that has been continuously passaged (incubated) in hamsters, the incubation time for developing disease symptoms is, at first, 500 days but then, in all further passages in mice, diminishes to 140 days. Conversely, when PrP^{Sc} that has been passaged in mice is inoculated into hamsters, the incubation time is first 400 days but subsequently shortens to 75 days. This suggests that the conversion of host PrP^C (whose sequence in mice differs from that in hamsters) to PrP^{Sc} by a foreign PrP^{Sc} is a rare event; once it has occurred, however, the newly formed host PrP^{Sc} catalyzes the conversion much more efficiently. Indeed, after inoculation with hamster PrP^{Sc}, transgenic mice expressing hamster PrP have incubation times that are reduced to between 48 and 250 days, depending on the transgenic line.

The foregoing experiments provide indirect support for the prion hypothesis. However, direct support has recently been provided by the demonstration that PrP^{Sc} induces the conversion of PrP^C to PrP^{Sc} in a cell-free system.

d. Mutant *Prn-p* Genes Give Rise to Prion Diseases

Three dominantly inherited neurodegenerative disorders in humans have been traced to mutations in the *Prn-p* gene. These are **familial CJD, Gerstmann–Sträussler–Scheinker syndrome (GSS),** and **fatal familial insomnia (FFI).** All of them are extremely rare. In fact, FFI has been found in only five families. The mutant PrP^{Sc}s causing these diseases are nevertheless infectious.

e. PrP^{Sc} Is a Stable Conformational Variant of PrP^C

The NMR structure of residues 23 to 230 of the 280-residue human PrP^C, determined by Kurt Wüthrich, consists of a flexibly disordered (and hence unobserved) 98-residue N-terminal "tail" and a 110-residue C-terminal globular domain containing three α helices and a short 2-stranded antiparallel β sheet (Fig. 9-38*a*). As expected, this structure closely resembles those of the homologous mouse and hamster PrP^Cs.

How does PrP^{Sc} differ from PrP^C? The direct sequencing of PrP^{Sc} indicates that its amino acid sequence is identical to that deduced from the *Prn-p* gene sequence, thereby eliminating any post-transcriptional sequence variation as a possible cause for the pathogenic properties of PrP^{Sc}. Furthermore, mass spectrometric studies on PrP^{Sc} designed to reveal previously uncharacterized post-translational modifications indicated that, in fact, PrP^{Sc} and PrP^C are chemically identical. Thus, although the possibility that only a small fraction of PrP^{Sc} is chemically modified has not been eliminated, it seems more likely that PrP^{Sc} and PrP^C differ in their secondary and/or tertiary structures. Unfortunately, the insolubility of PrP^{Sc} (see below) has precluded its structural determination. However, CD measurements show that, in fact, the conformations of PrP^{Sc} and PrP^C are quite different: PrP^C has a high (~40%) α helix content but little (~3%) β sheet content (in good agreement with the NMR structure of its globular domain), whereas PrP^{Sc} has a lesser (~30%) α helix content but a high (~45%) β sheet content. In a plausible model of PrP^{Sc} (Fig. 9-38*b*), its N-terminal region has refolded to form a so-called **β helix** in which the polypeptide strand forms a left-handed helix containing three parallel β sheets. Only the two C-terminal helices of PrP^C, which are

(*a*)

(*b*)

Figure 9-38 Prion protein conformations. (*a*) The NMR structure of human prion protein (PrP^C). The protein is drawn in ribbon form colored according to its secondary structure with helices red, β sheets green, and other segments orange. Its disulfide bond is shown in stick form in yellow. Its N-terminal "tail" (residues 23–121) is flexibly disordered (the protein's N-terminal 23 residues had been post-translationally excised). (*b*) A plausible model for the structure of PrP^{Sc} represented as in Part *a*. [Courtesy of Fred Cohen, University of California at San Francisco. Part *a* based on an NMR structure by Kurt Wüthrich, Eidgenössische Technische Hochschule, Zurich, Switzerland. PDBid 1QLX.]

joined by a disulfide bond, maintain their original conformation. The high β sheet content of PrP^{Sc} would, presumably, facilitate the aggregation of PrP^{Sc} as amyloid fibrils. *Evidently, the PrP^C → PrP^{Sc} conformational change is autocatalytic; that is, PrP^{Sc} induces PrP^C to convert to PrP^{Sc}.* In fact, PrP^{Sc} in a cell-free system has been shown to catalyze the conversion of PrP^C from an uninfected source to PrP^{Sc}.

In cells, PrP^{Sc} is deposited in cytosolic vesicles rather than being anchored to the cell-surface membrane as is PrP^C. Both PrP^C and PrP^{Sc} are subject to eventual proteolytic degradation in the cell (Section 32-6). However, although PrP^C is completely degraded, PrP^{Sc} only loses its N-terminal 67 residues to form a 27- to 30-kD protease-resistant core, known as **PrP 27–30,** which still exhibits a high β sheet content. *PrP 27–30 then aggregates to form the amyloid plaques that appear to be directly responsible for the neuronal degeneration characteristic of prion diseases.*

According to the prion hypothesis, sporadically occurring prion diseases such as CJD (which strikes one person per million per year) arise from the spontaneous although infrequent conversion of sufficient quantities of PrP^C to PrP^{Sc} to support the autocatalytic conformational isomerization reaction. This model is corroborated by the observation that transgenic mice that overexpress wild-type *Prn-p* invariably develop scrapie late in life. The prion hypothesis similarly explains inherited prion diseases such as FFI as arising from a lower free energy barrier and hence higher rate for the conversion of the mutant PrP^C to PrP^{Sc} relative to that of normal PrP^C.

f. Prions Have Different Strains

Prions from different sources, when passaged in mice or hamsters, reproducibly exhibit characteristic incubation times, neurological symptoms, and neuropathologies. *Evidently, there are different strains of prions, each of whose corresponding PrP^{Sc}s must have a different stable conformation and induce PrP^C to take up this conformation.* The existence of different prion strains (as many as 30 for scrapie in sheep and at least 4 for CJD in humans) was cited as evidence against the prion hypothesis. However, there is now ample physical evidence that the PrP^{Sc}s in different prion strains have different structures.

BSE or mad cow disease was first reported in the U.K. in late 1985. It soon became an epidemic that, in total, infected ~2 million cattle in the U.K. BSE is surmised to have arisen as a consequence of feeding cattle meat-and-bone meal made from scrapie-infected sheep (and eventually from BSE-infected cattle). BSE, which has an ~5-year incubation period, was unknown before 1985, most likely because the process for manufacturing meat-and-bone meal was changed in the late 1970s from a way that fully inactivates scrapie prions to one that fails to do so. In 1988, the U.K. banned the feeding of ruminants with ruminant-derived protein (other than milk), so that, following its peak in 1993, the BSE epidemic rapidly abated (a process accelerated by the slaughter of large numbers of cattle at risk of having BSE). However, since humans consumed meat from BSE-infected cattle for over a decade, the question remained, had BSE been transmitted to humans? It

should be noted that scrapie-infected sheep have long been consumed worldwide and yet the incidence of CJD in mainly meat-eating countries such as the U.K. (in which sheep are particularly abundant) is no greater than that in largely vegetarian countries such as India. Nevertheless, in 1994, several cases of CJD in teenagers and young adults were reported in the U.K., although heretofore CJD before the age of 40 was extremely rare (its average age of onset is ~64). Individuals with this **new variant CJD (vCJD or nvCJD),** of which there have been ~200 cases yet reported, almost entirely in the U.K., have neurological symptoms and neuropathology that are atypical for sporadic CJD. Moreover, when transmitted to mice expressing bovine PrP^C, vCJD has an incubation time, neurological symptoms, and neuropathology indistinguishable from that caused by BSE. It therefore seems highly likely that vCJD is caused by a prion strain that humans acquired by eating meat products from BSE-infected cattle.

g. Prions Occur in Yeast

Although prions were originally defined to be scrapielike infectious pathogens, it is now evident that this definition must be broadened to include all proteins with stable conformational variants that catalyze their own formation from "wild-type" protein. For example, *Saccharomyces cerevisiae* (baker's yeast) can harbor a genetic element designated **[URE3]** that, in sexual reproduction with cells that lack [URE3], is inherited by all progeny rather than according to the rules of Mendelian genetics (Section 1-4B). Yet [URE3] is a chromosomal gene rather than a plasmid-based or mitochondrial gene (which would account for its non-Mendelian inheritance).

[URE3] is identical to the chromosomal gene ***URE2,*** which specifies a protein, **Ure2,** that in the presence of yeast's preferred nitrogen sources (ammonia or glutamine) represses the expression of the proteins required to metabolize yeast's less preferred nitrogen sources (e.g., proline). Yeast that have the [URE3] phenotype (trait) lack this regulation of nitrogen metabolism (nitrogen metabolism is discussed in Chapter 26). However, [URE3] yeast can be "cured" of this condition by treatment with 5 mM guanidinium chloride; that is, they and their progeny then exhibit normal regulation of nitrogen metabolism. Nevertheless, about one yeast cell per million spontaneously reverts to the [URE3] phenotype. This is because Ure2 has a "wild-type" conformational state, which regulates nitrogen metabolism, and a [URE3] form, which catalyzes its own formation from "wild-type" Ure2 to yield amyloid fibers that do not influence nitrogen metabolism. Thus, *Ure2 is a type of prion.*

The yeast genetic element **[PSI]** encodes a protein, **Sup35,** with similar prionlike properties that participates in transcriptional termination (Section 32-3E). Indeed, the introduction of Sup35 in its [PSI] conformation into the cytoplasm of yeast containing "wild-type" Sup35 induced the formation of the [PSI] phenotype, an experiment that constituted the first direct evidence supporting the prion hypothesis. Moreover, Sup35 can adopt several different fiber conformations *in vitro,* which when introduced into [*psi*$^-$]

cells, produce clearly distinguishable strain variants. Several other fungal proteins that can form prions have also been characterized.

6 STRUCTURAL EVOLUTION

Proteins, as we discussed in Section 7-3, evolve through point mutations and gene duplications. Over eons, through processes of natural selection and/or neutral drift, homologous proteins thereby diverge in character and develop new functions. How these primary structure changes affect function, of course, depends on the protein's three-dimensional structure. In this section, we explore the effects of evolutionary change on protein structures.

A. Structures of Cytochromes c

See Guided Exploration 10: Protein evolution The *c*-type cytochromes are small globular proteins that contain a covalently bound heme group (**iron–protoporphyrin IX;** Fig. 9-39). The X-ray structures of the cytochromes *c* from horse (Fig. 8-42), tuna, bonito, rice, and yeast are closely similar and thus permit the structural significance of cytochrome *c*'s amino acid sequences (Section 7-3B) to be assessed. The internal residues of cytochrome *c*, particularly those lining its heme pocket, tend to be invariant or conservatively substituted, whereas surface positions have greater variability. This observation is, in part, an indication of the more exacting packing requirements of a protein's internal regions compared to those of its surface (Section 8-3Bc).

Certain invariant or highly conserved residues (Table 7-4) have specific structural and/or functional roles in cytochrome *c*:

1. The invariant Cys 14, Cys 17, His 18, and Met 80 residues form covalent bonds with the heme group (Fig. 9-39).

2. The nine invariant or highly conserved Gly residues occupy close-fitting positions in which larger side chains would significantly alter the protein's three-dimensional structure.

3. The highly conserved Lys residues 8, 13, 25, 27, 72, 73, 79, 86, and 87 are distributed in a ring around the exposed edge of the otherwise buried heme group. There is considerable evidence that this unusual constellation of positive charges specifically associates with complementary sets of negative charges on the physiological reaction partners of cytochrome *c*, cytochrome *c* reductase, and cytochrome *c* oxidase.

a. Prokaryotic c-Type Cytochromes Are Structurally Related to Cytochrome c

Although cytochrome *c* occurs only in eukaryotes, similar proteins known as ***c*-type cytochromes** are common in prokaryotes, where they function to transfer electrons at analogous positions in a variety of respiratory and photosynthetic electron-transport chains. Unlike the eukaryotic proteins, however, the prokaryotic *c*-type cytochromes

Figure 9-39   Molecular formula of iron–protoporphyrin IX (heme). In *c*-type cytochromes, the heme is covalently bound to the protein (*red*) by two thioether bonds linking what were the heme vinyl groups to two Cys residues that occur in the sequence Cys-X-Y-Cys-His (residues 14–18 in Table 7-4). Here X and Y symbolize any amino acid residues. A fifth and sixth ligand to the Fe atom, both normal to the heme plane, are formed by a side chain nitrogen of His 18 and the sulfur of Met 80. The iron atom, which is thereby octahedrally liganded, can stably assume either the Fe(II) or the Fe(III) oxidation state. Heme also occurs in myoglobin and hemoglobin but without the thioether bonds or the Met ligand.

exhibit considerable sequence variability among species. For example, the numerous bacterial *c*-type cytochromes whose primary structures are known have from 82 to 134 amino acid residues, whereas eukaryotic cytochromes *c* have a narrower range, from 103 to 112 residues. The primary structures of several representative *c*-type cytochromes have few obvious similarities (Fig. 9-40). Yet their X-ray structures closely resemble each other, particularly in their backbone conformations and side chain packing in the regions surrounding the heme group (Fig. 9-41). Furthermore, most of them have aromatic rings in analogous positions and orientations relative to their heme groups as well as similar distributions of positively charged Lys residues about the perimeters of their heme crevices. The major structural differences among these *c*-type cytochromes stem from various loops of polypeptide chain that are located on their surfaces.

Before the advent of sophisticated sequence alignment algorithms such as BLAST (Section 7-4Bg), the correct alignments of analogous *c*-type cytochrome residues (thin lines in Fig. 9-40) could not have been made on the basis of only their primary structures: These proteins have diverged so far that their three-dimensional structures were essential guides for this task. Three-dimensional structures are evidently more indicative of the similarities among these distantly related proteins than are primary structures. *It is the essential structural and functional elements of proteins, rather than their amino acid residues, that are conserved during evolutionary change.*

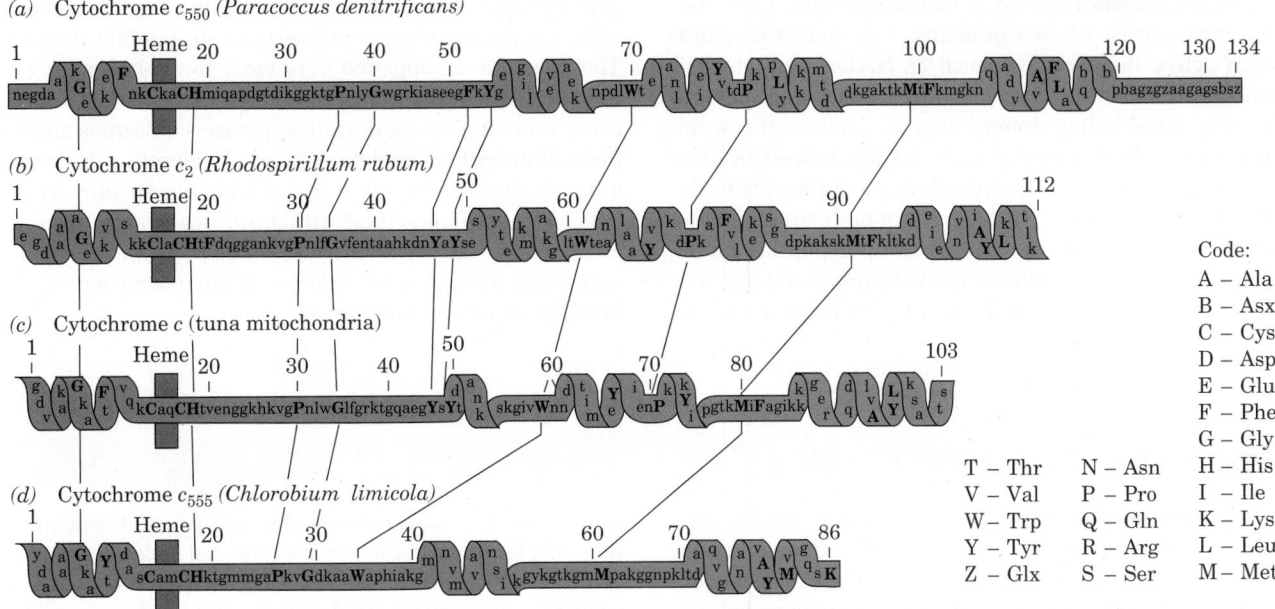

(a) Cytochrome c₅₅₀ (*Paracoccus denitrificans*)

(b) Cytochrome c₂ (*Rhodospirillum rubrum*)

(c) Cytochrome c (tuna mitochondria)

(d) Cytochrome c₅₅₅ (*Chlorobium limicola*)

Code:
A – Ala
B – Asx
C – Cys
D – Asp
E – Glu
F – Phe
G – Gly

T – Thr	N – Asn	H – His
V – Val	P – Pro	I – Ile
W – Trp	Q – Gln	K – Lys
Y – Tyr	R – Arg	L – Leu
Z – Glx	S – Ser	M – Met

Figure 9-40   Primary structures of some representative c-type cytochromes. (*a*) Cytochrome c₅₅₀ (the subscript indicates the protein's peak absorption wavelength in visible light, in nm) from *Paracoccus denitrificans*, a respiring bacterium that can use nitrate as an oxidant. (*b*) Cytochrome c₂ (the subscript has only historical significance) from *Rhodospirillum rubrum*, a purple photosynthetic bacterium. (*c*) Cytochrome c from tuna mitochondria. (*d*) Cytochrome c₅₅₅ from *Chlorobium limicola*, a green photosynthetic bacterium that utilizes H₂S as a hydrogen source. Thin lines connect structurally significant or otherwise invariant residues (*uppercase*). Helical regions are indicated to facilitate structural comparisons with Fig. 9-41. [After Salemme, F.R., *Annu. Rev. Biochem.* **46,** 307 (1977).]

B. Gene Duplication

Gene duplication may promote the evolution of new functions through structural evolution (Section 7-3C). In over half of the multidomain proteins of known structure, two or more of the domains are structurally quite similar. Consider, for example, the four domains of yeast protein disulfide isomerase (PDI; Fig. 9-17). It seems highly unlikely that these complex but topologically similar domains could have independently evolved their present structures—a process known as **convergent evolution**. Almost certainly, they arose through duplications of the gene specifying an ancestral domain accompanied by the fusion of the resulting four genes to yield a single gene specifying a polypeptide that folds into four similar domains. The differences between the four domains are therefore due to their **divergent evolution.**

Structurally similar domains often occur in proteins whose other domains bear no resemblance to one another.

(a) *Paracoccus* c₅₅₀
134 amino acid residues

(b) *Rhodospirillum* c₂
112 amino acid residues

(c) Tuna c
103 amino acid residues

(d) *Chlorobium* c₅₅₅
86 amino acid residues

Figure 9-41   Three-dimensional structures of the c-type cytochromes whose primary structures are displayed in Fig. 9-40. The polypeptide backbones (*blue*) are shown in analogous orientations such that their heme groups (*red*) are viewed edge on. The Cys, Met, and His side chains that covalently link the heme to the protein are also shown. (*a*) Cytochrome c₅₅₀ from *P. denitrificans*. (*b*) Cytochrome c₂ from *Rs. rubrum*. (*c*) Tuna cytochrome c. (*d*) Cytochrome c₅₅₅ from *C. limicola*. [Illustration, Irving Geis. Image from the Irving Geis Collection, Howard Hughes Medical Institute. Reprinted with permission.] **See Kinemage Exercise 5-1**

The redox enzymes known as **dehydrogenases,** for example, each consist of two domains: a domain that binds redox-active dinucleotides such as NAD^+ and which is structurally similar in all the dehydrogenases, and a dissimilar substrate-binding domain that determines the specificity and mode of action of each enzyme. Indeed, in some dehydrogenases, such as glyceraldehyde-3-phosphate dehydrogenase (Fig. 8-45), the dinucleotide-binding domain occurs at the N-terminal end of the polypeptide chain, whereas in others it occurs at the C-terminal end. Each of these dehydrogenases must have arisen by the fusion of the gene specifying an ancestral dinucleotide-binding domain with a gene encoding a proto-substrate-binding domain. This must have happened very early in evolutionary history, perhaps in the precellular stage (Section 1-5Ca), because there are no significant sequence similarities among these dinucleotide-binding domains. Evidently, *a domain is as much a unit of evolution as it is a unit of structure. By genetically combining these structural modules in various ways, nature can develop new functions far more rapidly than it can do so by the evolution of completely new structures through point mutations.*

CHAPTER SUMMARY

1 Protein Folding: Theory and Experiment   Under renaturing conditions, many proteins fold to their native structures in a matter of seconds. Helices and sheets, which together constitute ~60% of the average protein, are so common because they efficiently fill space. Proteins are hierarchically organized, that is, they consist of domains, which consist of subdomains, etc. They are highly tolerant of sequence changes, to which they adapt by local rather than global structural alterations. Some proteins are natively unfolded, although they assume stable structures when binding to their target molecules.

The rapidity with which proteins renature indicates that they fold in an ordered manner rather than via a random search of all their possible conformations. Thus the study of protein folding requires rapid mixing and observational techniques such as stopped-flow devices, circular dichroism (CD), pulsed H/D exchange followed by NMR, and fluorescence resonance energy transfer (FRET). The folding of small single-domain proteins is initiated by a hydrophobic collapse to yield a molten globule, which appears within ~5 ms. This is followed by the stabilization of secondary structure and then the formation of tertiary structure to yield the native protein in a matter of several seconds. Folding is thought to follow landscape theory, which postulates that a polypeptide folds via a folding funnel and hence can take any of a great variety of pathways to reach its native state. This is consistent with the finding that proteins fold in a hierarchical manner. The sequence of a protein appears to specify its folding pathway as well as its native structure.

2 Folding Accessory Proteins   Even though it is clear that a protein's primary structure dictates its three-dimensional structure, many proteins require the assistance of accessory proteins such as protein disulfide isomerase (PDI), peptidyl prolyl cis–trans isomerases, and molecular chaperones to fold/assemble to their native structures. PDI consists of four thioredoxinlike domains, two of which contain exposed Cys residues that form disulfide bonds, either internally or with another protein in a disulfide interchange reaction. Two families of peptidyl prolyl cis–trans isomerases have been characterized, the cyclophilins, which bind cyclosporin A, and FK506 binding protein, which binds FK506.

The chaperonins, such as GroEL and GroES, stimulate the proper folding of certain misfolded proteins through a cyclic sequence of concerted conformational changes that is driven by the binding and hydrolysis of ATP. GroES is a cap-shaped heptamer and GroEL is a 14-mer arranged in two apposed heptameric rings that form two unconnected end-to-end hollow barrels. Together, GroEL and GroES form a bullet-shaped complex that contains a closed cavity (an Anfinsen cage) in which misfolded proteins can fold without interference by aggregation with other misfolded proteins. In doing so, GroEL/ES partially unfolds a misfolded and conformationally trapped protein of up to ~70 kD and releases it so as to enable it to travel down its folding funnel via a new route. Such proteins undergo an average of 14 cycles of binding and release before achieving their native folds. Many of the ~85 *E. coli* proteins that stringently require the GroEL/ES system for proper folding contain α/β domains, whose structural complexity is largely responsible for their misfolding. Eukaryotic Group II chaperonins have built in lids that appear to function analogously to GroES.

3 Protein Structure Prediction and Design   The prediction of protein secondary structures from only amino acid sequences has been reasonably successful using empirical techniques such as the Chou–Fasman method. However, sophisticated computational techniques yield more reliable predictions. Comparative (homology) modeling can provide accurate tertiary structures for polypeptides with >30% identity to a protein of known structure. Fold recognition (threading) techniques have only been marginally successful for determining the structures of proteins that have no apparent homology with proteins of known structure. (Modern) *de novo* structure determination methods yield correct folding topologies with a success rate of ~20% and occasionally reasonably accurate atomic models. The reverse process, computationally based protein design, has been more successful, in part, because one can "overengineer" a protein to take up a desired conformation.

4 Protein Dynamics   Proteins are flexible and fluctuating molecules whose group motions have characteristic periods ranging from 10^{-15} to over 10^3 s. X-ray analysis, which reveals the average atomic mobilities in a protein, indicates that proteins tend to be more mobile at their peripheries than in their interiors. Molecular dynamics simulations indicate that native protein structures each consist of a large number of closely related and rapidly interconverting conformational substates of nearly equal stabilities. Without this flexibility, enzymes would be nonfunctional. The rates of aromatic ring flipping, as revealed by NMR measurements, indicate that internal group mobilities within proteins vary both with the protein and with the position within the protein. The exchange of a protein's internal protons with solvent requires its transient local unfolding. Hydrogen exchange studies therefore demonstrate that proteins have a great variety of infrequently occurring internal motions.

5 Conformational Diseases: Amyloids and Prions   A number of often fatal human diseases are associated with the deposition of amyloid in the brain and other organs. Although the various amyloidogenic proteins are unrelated in both sequence and native structure, all form similar amyloid fibrils that consist mainly of β sheets whose planes extend along the fibril axis. The two known human lysozyme variants that have amyloidogenic properties are conformationally much looser than wild-type lysozyme. In Alzheimer's disease, a neurodegenerative disease of mainly the elderly, the proteolysis of Aβ-precursor protein (βPP) in brain tissue yields the 40- to 42-residue amyloid-β protein (Aβ), which forms the amyloid fibrils that kill neurons.

Humans and other mammals are subject to infectious neurodegenerative diseases such as scrapie, which are caused by prions. Prions appear to consist of only a single species of protein named PrP. PrP exists in two forms: the normal cellular form, PrP^C, a conserved membrane-anchored cell-surface protein on neurons; and PrP^{Sc}, which although chemically identical to PrP^C, has a different conformation. PrP^{Sc} autocatalytically converts PrP^C to PrP^{Sc} thereby accounting for the infectious properties of PrP^{Sc} and the observation that $Prn-p^{0/0}$ mice are resistant to scrapie. PrP^{Sc} is proteolytically degraded in the cell to form a protease-resistant core, PrP 27–30, that aggregates to form the neurotoxic amyloid fibrils thought to be responsible for the symptoms of prion diseases. A given species of PrP may take up several different self-propagating fibrous conformations to yield different prion strains. Fungi, such as yeast, also have proteins with prionlike properties.

6 Structural Evolution   The X-ray structures of eukaryotic cytochromes *c* demonstrate that internal residues and those having specific structural and functional roles tend to be conserved during evolution. Prokaryotic *c*-type cytochromes from a variety of organisms structurally resemble each other and those of eukaryotes even though they have little sequence similarity. This indicates that the three-dimensional structures of proteins rather than their amino acid sequences are conserved during evolutionary change. The structural similarities between the domains in many multidomain proteins indicate that these proteins arose through the duplication of the genes specifying the ancestral domains followed by their fusion. For example, the structural resemblance between the dinucleotide-binding domains of dehydrogenases suggests that these proteins arose by duplication of a primordial dinucleotide-binding domain followed by its fusion with a gene specifying a proto-substrate-binding domain. In this manner, proteins with new functions can evolve much faster than by a series of point mutations.

REFERENCES

Protein Folding

Anfinsen, C.B., Principles that govern the folding of protein chains, *Science* **181**, 223–230 (1973). [A Nobel laureate explains how he got his prize.]

Aurora, R. and Rose, G.D., Helix capping, *Protein Sci.* **7**, 21–38 (1998). [Summarizes the evidence that helix capping interactions stabilize helices.]

Baldwin, R.L., Pulsed H/D-exchange studies of folding intermediates, *Curr. Opin. Struct. Biol.* **3**, 84–91 (1993).

Baldwin, R.L., Protein folding from 1961 to 1982, *Nature Struct. Biol.* **6**, 814–817 (1999). [An intellectual history.]

Baldwin, R.L. and Rose, G.D., Is protein folding hierarchic? I. Local structure and peptide folding; *and* II. Folding intermediates and transition states, *Trends Biochem. Sci.* **24**, 26–33; *and* 77–83 (1999).

Behe, M., Lattman, E.E., and Rose, G.D., The protein folding problem: The native fold determines the packing but does packing determine the native fold? *Proc. Natl. Acad. Sci.* **88**, 4195–4199 (1991).

Betts, S. and King, J., There's a right way and a wrong way: *in vivo* and *in vitro* folding, misfolding and subunit assembly of the P22 tailspike, *Structure* **7**, R131–R139 (1999).

Buchner, J. and Kiefhaber, T., *Protein Folding Handbook,* Wiley-VCH (2005). [An authoritative 5-volume work on most aspects of protein folding.]

Dalal, S., Balasubramanian, S., and Regan, L., Protein alchemy: Changing β-sheet into α-helix, *Nature Struct. Biol.* **4**, 548–552 (1997). [Reports the sequence changes in protein GB1 that cause it to assume the fold of Rop protein.]

Dill, K.A. and Chan, H.S., From Levinthal to pathways to funnels, *Nature Struct. Biol.* **4**, 10–19 (1997). [Reviews the landscape theory of protein folding.]

Dill, K.A., Ozkan, S.B., Shell, M.S., and Weikl, T.R., The protein folding problem, *Annu. Rev. Biophys.* **37**, 289–316 (2008).

Dunker, A.K., Silman, I., Uversky, V.N., and Sussman, J.L., Function and structure of inherently disordered proteins, *Curr. Opin. Struct. Biol.* **18**, 756–764 (2008).

Dyson, H.J. and Wright, P.E., Intrinsically unstructured proteins and their functions, *Nature. Rev. Mol. Cell. Biol.* **6**, 197–208 (2005).

Englander, S.W., Protein folding intermediates and pathways studied by hydrogen exchange, *Annu. Rev. Biophys. Biomol. Struct.* **29**, 213–238 (2000).

Englander, W.S., Mayne, L., and Krishna, M.M.G., Protein folding and misfolding: mechanism and principles, *Q. Rev. Biophys.* **40**, 287–326 (2007).

Fersht, A., *Structure and Mechanism in Protein Science,* Chapters 17–19, Freeman (1999).

Fink, A.L., Natively unfolded proteins, *Curr. Opin. Struct. Biol.* **15**, 35–41 (2005).

Fitzkee, N.C., Fleming, P.J., Gong, H., Panasik, N., Jr., Street, T.O., and Rose, G.D., Are proteins made from a limited parts list? *Trends Biochem. Sci.* **30**, 73–80 (2005).

Gillespie, B. and Plaxco, K.W., Using protein folding rates to test protein folding theories, *Annu. Rev. Biochem.* **73**, 837–859 (2004).

Kubelka, J., Hofrichter, J., and Eaton, W.A., The protein folding 'speed limit,' *Curr. Opin. Struct. Biol.* **14**, 76–88 (2004).

Matthews, B.W., Studies on protein stability with T4 lysozyme, *Adv. Prot. Chem.* **46**, 249–278 (1995).

Meyers, R.A., *Proteins. From Analytics to Structural Genomics,* Vol. 1, Chapters 1, 2, and 4, Wiley-VCH (2007). [Discusses X-ray crystallography, NMR spectroscopy, and circular dichroism of proteins.]

Minor, D.L., Jr. and Kim, P.S., Context-dependent secondary structure formation of a designed protein sequence, *Nature* **380,** 730–734 (1996). [Describes the position-dependent conformation of the chameleon sequence in protein GB1.]

Oliveberg, A. and Wolynes, P.G., The experimental survey of protein-folding energy landscapes, *Q. Rev. Biophys.* **36,** 245–288 (2006).

Onuchic, J.N. and Wolynes, P.G., Theory of protein folding, *Curr. Opin. Struct. Biol.* **14,** 70–75 (2004).

Pain, R.H. (Ed.), *Mechanisms of Protein Folding* (2nd ed.), Oxford University Press (2000).

Piston, D.W. and Kremers, G.-J., Fluorescent protein FRET: the good, the bad and the ugly, *Trends Biochem. Sci.* **32,** 407–414 (2007).

Roder, H. and Shastry, M.C.R., Methods for exploring early events in protein folding, *Curr. Opin. Struct. Biol.* **9,** 620–626 (1999).

Udgaonkar, J.B., Multiple routes and structural heterogeneity in protein folding, *Annu. Rev. Biophys.* **37,** 489–510 (2008).

Wang, C.C. and Tsou, C.L., The insulin A and B chains contain sufficient structural information to form the native molecule, *Trends Biochem. Sci.* **16,** 279–281 (1991).

Folding Accessory Proteins

Booth, C.R., Meyer, A.S., Cong, Y., Topf, M., Sali, A., Ludtke, S.J., Chiu, W., and Frydman, J., Mechanism of lid closure in the eukaryotic chaperonin TRiC/CCT, *Nature Struct. Biol.* **15,** 746–753 (2008).

Chen, L. and Sigler, P.B., The crystal structure of a GroEL/peptide complex: Plasticity as a basis for substrate diversity, *Cell* **99,** 757–768 (1999).

Clark, P.L., Protein folding in the cell: reshaping the folding funnel, *Trends Biochem. Sci.* **29,** 527–534 (2004).

Ellis, R.J., Macromolecular crowding: Obvious but underappreciated. *Trends Biochem. Sci.* **26,** 597–604 (2001).

Ellis, R.J., Molecular chaperones: assisting assembly in addition to folding, *Trends Biochem. Sci.* **31,** 395–401 (2006).

Frydman, J., Folding of newly translated proteins *in vivo*: The role of molecular chaperones, *Annu. Rev. Biochem.* **70,** 603–649 (2001).

Gruber, C.W., Cemazar, M., Heras, B., Martin, J.L., and Craik, D.J., Protein disulfide isomerase: the structure of oxidative folding, *Trends Biochem. Sci.* **31,** 455–464 (2006).

Hartl, F.U., and Hayer-Hartl, M., Molecular chaperones in the cytosol: From nascent chain to unfolded protein, *Science* **295,** 1852–1858 (2002).

Horst, R., Bertelson, E.B., Fiaux, J., Wider, G., Horwich, A.L., and Wüthrich, K., Direct NMR observation of a substrate protein bound to the chaperonin GroEL, *Proc. Natl. Acad. Sci.* **102,** 12748–12753 (2005); *and* Horst, R., Fenton, W.A., Englander, S.W., Wüthrich, K., and Horwich, A.L., Folding trajectories of human dihydrofolate reductase inside the GroEL–GroES chaperonin cavity and free in solution, *Proc. Natl. Acad. Sci.* **104,** 20788–20792 (2007).

Horwich, A.R. (Ed.), Protein Folding in the Cell, *Adv. Prot. Chem.* **59** (2002). [Contains authoritative articles on a variety of folding accessory proteins.]

Horwich, A.R., Farr, G.W., and Fenton, W.A., GroEL–GroES-mediated protein folding, *Chem. Rev.* **106,** 1917–1930 (2006); *and* Horwich, A.R., Fenton, W.A., Chapman, E., and Farr, G.W., Two families of chaperonin: Physiology and mechanism, *Annu. Rev. Cell Dev. Biol.* **23,** 115–145 (2007).

Kerner, M.J., et al., Proteome-wide analysis of chaperone-dependent protein folding in *Eschericia coli, Cell* **122,** 209–220 (2005).

Lin, Z. and Rye, H.S., GroEL-mediated protein folding: Making the impossible, possible, *Crit. Rev. Biochem. Mol. Biol.* **41,** 211–239 (2006).

Mamathambika, B.S., and Bardwell, J.C., Disulfide-linked protein folding pathways, *Annu. Rev. Cell Dev. Biol.* **24,** 211–235 (2008).

Morano, K.A., New tricks for an old dog. The evolving world of Hsp70, *Ann. N.Y. Acad. Sci.* **1113,** 1–14 (2007).

Ransom, N.A., Farr, G.W., Roseman, A.M., Gowen, B., Fenton, W.A., Horwich, A.L., and Saibil, H.R., ATP-bound states of GroEL captured by cryo-electron microscopy, *Cell* **107,** 869–879 (2001).

Saibil, H.R., Chaperone machines in action, *Curr. Opin. Struct. Biol.* **18,** 35–42 (2008).

Schiene, C. and Fischer, G., Enzymes that catalyse the restructuring of proteins, *Curr. Opin. Struct. Biol.* **10,** 40–45 (2000). [Discussess protein disulfide isomerases and peptidyl prolyl cis–trans isomerases.]

Schreiber, S.L., Chemistry and biology of immunophilins and their immunosuppressive ligands, *Science* **251,** 238–287 (1991).

Sharma, S., Chakraborty, K., Müller, B.K., Astola, N., Tang, Y.-C., Lamb, D.C., Hayer-Hartl, M., and Hartl, F.U., Monitoring protein conformation along the pathway of chaperon-assisted folding, *Cell* **133,** 142–153 (2008).

Shtilerman, M., Lorimer, G.H., and Englander, S.W., Chaperonin function: Folding by forced unfolding, *Science* **284,** 822–825 (1999).

Spiess, C., Meyer, A.S., Reissmann, S., and Frydman, J., Mechanism of the eukaryotic chaperonin: protein folding in the chamber of secrets, *Trends Cell Biol.* **14,** 598–604 (2004).

Stan, G., Brooks, B.R., Lorimer, G.H., and Thirumalai, D., Residues in substrate proteins that interact with GroEL in the capture process are buried in the native state, *Proc. Natl. Acad. Sci.* **103,** 4433–4438 (2006).

Thirumalai, D. and Lorimer, G.H., Chaperone-mediated protein folding, *Annu. Rev. Biophys. Biomol. Struct.* **30,** 245–269 (2001).

Tian, G., Xiang, S., Noiva, R., Lennarz, W.J., and Schindelin, H., The crystal structure of yeast protein disulfide isomerase suggests cooperativity between its active sites, *Cell* **124,** 61–73 (2006).

Wandinger, S.K., Richter, K., and Buchner, J., The Hsp90 chaperone machinery, *J. Biol. Chem.* **283,** 18473–18477 (2008); *and* Pearl, L.H. and Prodromou, C., Structure and mechanism of the Hsp90 molecular chaperone machinery, *Annu. Rev. Biochem.* **75,** 271–294 (2006).

Xu, Z., Horwich, A.L., and Sigler, P.B., The crystal structure of the asymmetric GroEL–GroES–(ADP)$_7$ chaperonin complex, *Nature* **388,** 741–750 (1997).

Zhao, Y. and Ke, H., Crystal structure implies that cyclophilin predominantly catalyzes the *trans* to *cis* isomerization, *Biochemistry* **35,** 7356–7361 (1996).

Protein Structure Prediction and Design

Baxevanis, A.D. and Ouellette, B.F.F. (Eds.), *Bioinformatics. A Practical Guide to the Analysis of Genes and Proteins* (3rd ed.), Chapters 8 and 9, Wiley-Interscience (2005).

Blaber, M., Zhang, X., and Matthews, B.W., Structural basis of amino acid α helix propensity, *Science* **260,** 1637–1640 (1993).

Bujnicki, J.M. (Ed.), *Prediction of Protein Structures, Functions, and Interactions,* Wiley (2009).

Chou, P.Y. and Fasman, G.D., Empirical predictions of protein structure, *Annu. Rev. Biochem.* **47,** 251–276 (1978). [Exposition of a particularly simple method of protein secondary structure prediction.]

Cuff, J.A. and Barton, G.J., Evaluation and improvement of multiple sequence methods for protein secondary structure prediction, *Proteins* **34**, 508–519 (1999). [The principles behind Jpred3.]

Das, R. and Baker, D., Macromolecular modeling with Rosetta, *Annu. Rev. Biochem.* **77**, 363–382 (2008).

DeGrado W.F., Summa, S.M., Pavone, V., Nastri, F., and Lombardi, A., De novo design and structural characterization of proteins and metalloproteins, *Annu. Rev. Biochem.* **68**, 779–819 (1999).

Kuhlman, B., Dantas, G., Ireton, G.C., Varani, G., Stoddard, B.L., and Baker, D., Design of a novel protein fold with atomic-level accuracy, *Science* **302**, 1364–1368 (2003). [The design of Top7.]

Lesk, A.M., *Introduction to Bioinformatics* (3rd ed.), pp. 333–358, Oxford University Press (2008).

Mirny, L. and Shakhnovitch, E., Protein folding theory: From lattice to all-atom models, *Annu. Rev. Biophys. Biomol. Struct.* **30**, 361–396 (2001).

Moult, J., Fidelis, K., Kryshtofovych, A., Rost, B., and Tramontano, A., Critical assessment of methods of protein structure prediction—Round VIII, *Proteins* **77** (Issue S9), 1–4 (2009). [The summary article of the issue of *Proteins: Structure, Function, and Bioinformatics* that reports the results of CASP8.]

Rose, G.D., Prediction of chain turns in globular proteins on a hydrophobic basis, *Nature* **272**, 586–590 (1978).

Tramontano, A., *Protein Structure Prediction. Concepts and Applications*, Wiley-VCH (2006).

Zaki, M.J. and Bystroff, C. (Eds.), *Protein Structure Prediction* (2nd ed.), Humana Press (2008).

Protein Dynamics

Henzler-Wildman, K. and Kern, D., Dynamic personalities of proteins, *Nature* **450**, 964–972 (2007).

Karplus, M. and McCammon, A., Molecular dynamics simulations of biomolecules, *Nature Struct. Biol.* **9**, 646–651 (2002).

Palmer, A.G., III, Probing molecular motion by NMR, *Curr. Opin. Struct. Biol.* **7**, 732–737 (1997).

Protein dynamics, *Science* **324**, 197–215 (2009). [A special section containing four articles.]

Rasmussen, B.F., Stock, A.M., Ringe, D., and Petsko, G.A., Crystalline ribonuclease A loses function below the dynamical transition at 220 K, *Nature* **357**, 423–424 (1992).

Ringe, D. and Petsko, G.A., Mapping protein dynamics by X-ray diffraction, *Prog. Biophys. Mol. Biol.* **45**, 197–235 (1985).

Scheraga, H.A., Khalili, M., and Liwo, A., Protein-folding dynamics: Overview of molecular simulation techniques, *Annu. Rev. Phys. Chem.* **58**, 57–83 (2007).

Conformational Diseases

Booth, D.R., et al., Instability, unfolding and aggregation of human lysozyme variants underlying amyloid fibrillogenesis, *Nature* **385**, 787–73 (1997); *and* Funahashi, J., Takano, K., Ogasahara, K., Yamagata, Y., and Yutani, K., The structure, stability, and folding process of amyloidogenic mutant lysozyme, *J. Biochem.* **120**, 1216–1223 (1996).

Büeler, H., Aguzzi, A., Sailer, A., Greiner, R.A., Autenreid, P., Aguet, M., and Weissmann, C., Mice devoid of PrP are resistant to scrapie, *Cell* **73**, 1339–1347 (1993); *and* Büeler, H., Fischer, M., Lang, Y., Bluethmann, H., Lipp, H.-P., DeArmond, S.J., Prusiner, S.B., Aguet, M., and Weissmann, C., Normal development and behaviour of mice lacking the neuronal cell-surface PrP protein, *Nature* **356**, 577–582 (1992).

Buxbaum, J.N. and Tagoe, C.E., The genetics of amyloidoses, *Annu. Rev. Med.* **51**, 543–569 (2000).

Caughey, B., Baron, G.S., Chesebro, B., and Jeffrey, M., Getting a grip on prions: oligomers, amyloids, and pathological membrane interactions, *Annu. Rev. Biochem.* **78**, 177–204 (2009).

Chien, P., Weissman, J.S., and DePace, J.H., Emerging principles of conformation-based inheritance, *Annu. Rev. Biochem.* **73**, 617–656 (2004).

Chiti, F. and Dobson, C.M., Protein misfolding, functional amyloid, and human disease, *Annu. Rev. Biochem.* **75**, 333–366 (2006).

Collinge, J. and Clarke, A.R., A general model of prion strains and their pathogenicity, *Science* **318**, 930–936 (2007).

Deleault, N.R., Harris, B.T., Rees, J.R., and Supattapone, S., Formation of native prions from minimal components *in vitro*, *Proc. Natl. Acad. Sci.* **104**, 9741–9746 (2007).

Geula, C., Wu, C.-K., Saroff, D., Lorenzo, A., Yuan, M., and Yankner, B.A., Aging renders the brain vulnerable to amyloid β-protein neurotoxicity, *Nature Med.* **4**, 827–831 (1998).

Goedert, M. and Spillantini, M.G., A century of Alzheimer's disease, *Science* **314**, 777–781 (2006).

Gregersen, N., Bross, P., Vang, S., and Christiensen, J.H., Protein misfolding and human disease, *Annu. Rev. Genomics Hum. Genet.* **7**, 103–124 (2006).

Hardy, J. and Selkoe, D.J., The amyloid hypothesis of Alzheimer's disease: Progress and problems on the road to therapeutics, *Science* **297**, 353–356 (2002).

Jackson, G.S. and Clarke, A.R., Mammalian prion proteins, *Curr. Opin. Struct. Biol.* **10**, 69–74 (2000).

Kajava, A., Squire, J.M., and Parry, D.A.D. (Eds.), Fibrous Proteins: Amyloids, Prions and Beta Proteins, *Adv. Prot. Chem.* **73** (2006). [The last four chapters are on various aspects of amyloids and prions.]

Moore, R.A., Taubner, L.M., and Priola, S.A., Prion misfolding and disease, *Curr. Opin. Struct. Biol.* **19**, 14–22 (2009).

Pan, K.M., Baldwin, M., Nguyen, J., Gasset, M., Serban, A., Groth, D., Mehlhorn, I., Huang, Z., Fletterick, R.J., Cohen, F.E., and Prusiner, S.B., Conversion of α-helices into β-sheet features in the formation of the scrapie prion proteins, *Proc. Natl. Acad. Sci.* **90**, 10962–10966 (1993).

Prusiner, S.B. (Ed.), *Prion Biology and Diseases* (2nd ed.), Cold Spring Harbor Laboratory Press (2004); *and* Prion diseases, *in* Valle, D. (Ed.), *The Online Metabolic & Molecular Bases of Inherited Disease,* http://www.ommbid.com/, Chap. 224.

Rochet, J.C. and Lansbury, P.T., Jr., Amyloid fibrillogenesis: Themes and variations, *Curr. Opin. Struct. Biol.* **10**, 60–68 (2000).

Sawaya, M.R., et al., Atomic structures of amyloid cross-β spines reveal varied steric zippers, *Nature* **447**, 453–457 (2007); *and* Nelson, R., Sawaya, M.R., Balbirnie, M., Madsen, A.Ø., Riekel, C., Grothe, R., and Eisenberg, D., Structure of the cross-β spine of amyloid-like fibrils, *Nature* **435**, 773–778 (2005).

Selkoe, D.J., Cell biology of protein misfolding: the examples of Alzheimer's and Parkinson's diseases, *Nature Cell Biol.* **6**, 1054–1061 (2004).

Soto, C., Estrada, L., and Castilla, J., Amyloids, prions and the inherent infectious nature of misfolded proteins, *Trends Biochem. Sci.* **31**, 150–155 (2006).

Sparrer, H.E., Santoso, A., Szoka, F.C., Jr., and Weissman, J.S., Evidence for the prion hypothesis: Induction of the yeast [*PSI*+] factor by *in vitro*-converted Sup35 protein, *Science* **289**, 595–599 (2000).

Tuite, M.F., Yeast prions and their prion-forming domain, *Cell* **100**, 289–292 (2000).

Weissmann, C., The state of the prion, *Nature Rev. Microbiol.* **2**, 861–862 (2004).

Wiltzius, J.J.W., Sievers, S.A., Sawaya, M.R., Cascio, D., Popov, D., Riekel, C., and Eisenberg, D., Atomic structure of the cross-β spine of islet amyloid polypeptide (amylin), *Prot. Sci.* **17**, 1467–1474 (2008).

Zahn, R., Liu, A., Lührs, T., Riek, R., von Schroetter, C., Garcia, F.L., Billeter, M., Calzolai, L., Wider, G., and Wüthrich, K., NMR solution structure of the human prion protein, *Proc. Natl. Acad. Sci.* **97**, 145–150 (2000); *and* Liu, H., Farr-Jones, S., Ulyanov, N.B., Llinas, M., Marquee, S., Groth, D., Cohen, F.E., Prusiner, S.B., and James, T.L., Solution structure of Syrian hamster prion protein rPrP(90–231), *Biochemistry* **38**, 5362–5377 (1999).

Structural Evolution

Bajaj, M. and Blundell, T., Evolution and the tertiary structure of proteins, *Annu. Rev. Biophys. Bioeng.* **13**, 453–492 (1983).

Dickerson, R.E., Timkovitch, R., and Almassy, R.J., The cytochrome fold and the evolution of bacterial energy metabolism, *J. Mol. Biol.* **100**, 473–491 (1976).

Eventhoff, W. and Rossmann, M., The structures of dehydrogenases, *Trends Biochem. Sci.* **1**, 227–230 (1976).

Lesk, A.M., NAD-binding domains of dehydrogenases, *Curr. Opin. Struct. Biol.* **5**, 775–783 (1995).

Moore, A.D., Björklund, Å.K., Ekman, D., Bornberg-Baur, E., and Elofsson, A., Arrangements in the modular evolution of proteins, *Trends Biochem. Sci.* **33**, 444–451 (2008).

Scott, R.A. and Mauk, A.G. (Eds.), *Cytochrome c. A Multidisciplinary Approach,* University Science Books (1996).

PROBLEMS

1. How long will it take the polypeptide backbone of a 6-residue folding nucleus to explore all its possible conformations? Repeat the calculation for 10-, 15-, and 20-residue folding nuclei. Why, in the classic view of protein folding, are folding nuclei thought to be no larger than 15 residues?

*2. Consider a protein with 10 Cys residues. On air oxidation, what fraction of the denatured and reduced protein will randomly reform the native set of disulfide bonds if: (a) The native protein has five disulfide bonds? (b) The native protein has three disulfide bonds?

3. Why are β sheets more commonly found in the hydrophobic interiors of proteins than on their surfaces?

4. Under physiological conditions, polylysine assumes a random coil conformation. Under what conditions might it form an α helix?

5. Explain how landscape theory is consistent with the observation that many small proteins appear to fold to their native conformations without detectable intermediates, that is, via two-state mechanisms.

6. Explain why Pro residues can occupy the N-terminal turn of an α helix.

7. Explain why β sheets are less likely to form than α helices during the earliest stages of protein folding.

8. Molten globules are thought to be predominantly stabilized by hydrophobic forces. Why aren't hydrogen bonding forces implicated in doing so?

*9. The GroEL/ES cycle diagrammed in Fig. 9-25 only circulates in the clockwise direction. Explain the basis for this irreversibility in terms of the sequence of structural and binding changes in the GroEL/ES system.

*10. Predict the secondary structure of the C peptide of proinsulin (Fig. 9-4) using the methods of Chou–Fasman and Rose.

11. As Mother Nature's chief engineer, now certified as a master helix builder, you are asked to repeat Problem 8-8 with the stipulation that the α helix really be helical. Use Table 9-1.

*12. Predict the secondary structure of the N-terminal domain of yeast protein disulfide isomerase using Jpred3 (http://www.compbio.dundee.ac.uk/www-jpred/). How does this prediction compare with the observed structure of this domain (PDBid 2B5E; the a domain in Fig. 9-17)? [To enter the sequence of this domain into Jpred3, first point your browser at the PDB (http://www.rcsb.org/pdb), enter the PDBid 2B5E, click on the "Sequence Details" tab at the top of the resulting page, and determine the sequence range of the N-terminal domain. Then click on the UniProt reference (P17967), scroll down to the sequence, copy the relevant portion to the Jpred3 input box, edit out everything but the sequence, and click on the "Make Prediction" button. On the page that comes up indicating that sequence matches were found in the PDB, click on the "continue" button. When the Results page appears (you may have to wait some time for it), click on "View Simple" to see the Jpred3 prediction. The observed secondary structure of the N-terminal domain is diagrammed on the foregoing "Sequence Details" page.]

13. Indicate the probable effects of the following mutational changes on the structure of a protein. Explain your reasoning. (a) Changing a Leu to a Phe, (b) changing a Lys to a Glu, (c) changing a Val to a Thr, (d) changing a Gly to an Ala, and (e) changing a Met to a Pro.

14. Explain why Trp rings are usually completely immobile in proteins that have rapidly flipping Phe and Tyr rings.

15. Explain why *Prn-p*$^{0/0}$ mice are resistant to scrapie. What might be the susceptibility of heterozygous *Prn-p*$^{+/0}$ mice to scrapie?

*16. Discuss the merits of the hypothesis that the dinucleotide-binding domains of the dehydrogenases arose by convergent evolution.

Hemoglobin:
Protein Function
in Microcosm

CHAPTER 10

The existence of hemoglobin (Greek: *haimato,* blood), the red blood pigment, is evident to every child who scrapes a knee. Its brilliant red color, widespread occurrence, and ease of isolation have made it an object of inquiry since ancient times. Indeed, the early history of protein chemistry is essentially that of hemoglobin. The observation of crystalline hemoglobin was first reported by Friedrich Hünefeld in 1840, and by 1909 Edward Reichert and Amos Brown had published a photographic atlas of hemoglobin crystals from several hundred species. In contrast, it was not until 1926 that crystals of an enzyme, those of jack bean **urease,** were first reported. Hemoglobin was one of the first proteins to have its molecular mass accurately determined, the first protein to be characterized by ultracentrifugation, the first to be associated with a specific physiological function (that of oxygen transport), and, in sickle-cell anemia, the first in which a point mutation was demonstrated to cause a single amino acid change (Section 7-3Aa). Theories formulated to account for the cooperative binding of oxygen to hemoglobin (Section 10-4) have also been successful in explaining the control of enzyme activity. The first protein X-ray structures to be elucidated were those of hemoglobin and myoglobin (Greek: *myos,* muscle). This central role in the development of protein chemistry together with its enzymelike O_2-binding properties have caused hemoglobin to be dubbed an "honorary enzyme."

Hemoglobin is not just a simple oxygen tank. Rather, it is a sophisticated oxygen delivery system that provides the proper amount of oxygen to the tissues under a wide variety of circumstances. In this chapter, we discuss hemoglobin's properties, structure, and mechanism of action, both to understand the workings of this physiologically essential molecule and to illustrate the principles of protein structure that we have developed in the preceding chapters. We also consider the properties of abnormal hemoglobins and their relationship to human disease. Finally, we discuss theories of cooperative interactions among proteins, both to better understand the properties of hemoglobin and to set the stage for our later consideration of how enzyme action is regulated.

1 HEMOGLOBIN AND MYOGLOBIN FUNCTION

Hemoglobin **(Hb),** as we have seen in Chapters 7 and 8, is a 65-kD heterotetramer, $\alpha_2\beta_2$ (alternatively, a dimer of $\alpha\beta$ protomers). The α and β subunits are structurally and evolutionarily related to each other and to myoglobin **(Mb),** the 18-kD monomeric oxygen-binding protein of skeletal and heart muscle (Section 7-3C).

Hemoglobin transports oxygen from the lungs, gills, or skin of an animal to its capillaries for use in respiration. Very small organisms do not require such a protein because their respiratory needs are satisfied by the simple passive diffusion of O_2 through their bodies. However, since the transport rate of a diffusing substance varies inversely with the square of the distance it must diffuse, the O_2 diffusion rate through tissue thicker than ~1 mm is too slow to support life. The evolution of organisms as large and complex as annelids (e.g., earthworms) therefore required the development of circulatory systems that actively transport O_2 and nutrients to the tissues. The blood of such organisms must contain an oxygen transporter such as Hb because the solubility of O_2 in **blood plasma** (the fluid component of blood) is too low ($\sim 10^{-4}\ M$ under physiological conditions) to carry sufficient amounts of O_2 for metabolic needs. In contrast, whole blood, which normally contains ~ 150 g of $Hb \cdot L^{-1}$, can carry O_2 at concentrations as high as $0.01 M$, about the same as in air.

Although Mb was originally assumed to store oxygen, it is now clear that this function is significant only in aquatic mammals such as seals and whales, which have Mb concentrations in their muscles 10- to 30-fold greater than those in terrestrial mammals. It would seem more likely that Mb's major physiological role in terrestrial mammals is to facilitate oxygen transport in rapidly respiring muscle. The rate at which O_2 can diffuse from the capillaries to the tissues, and thus the level of respiration, is limited by oxygen's low solubility in aqueous solution. Mb increases the effective solubility of O_2 in muscle, the most rapidly respiring tissue under conditions of high exertion. Hence, in rapidly respiring muscle, Mb functions as a kind of molecular bucket brigade to facilitate O_2 diffusion. It therefore came as a surprise when knockout mice for the gene encoding Mb exhibited no obvious abnormalities (except for the pale color of their muscles) and exhibited normal exercise capacity and response to low oxygen levels. However, further investigations revealed that these knockout mice had compensatory adaptations including increases in their **hematocrit** (the fraction of blood volume occupied by red blood cells; normally $\sim 45\%$) and the capillary density in their muscles. Moreover, many of the mutant embryos died *in utero* due to cardiovascular defects. Additional physiological functions for Mb have recently been recognized: the detoxification of the highly reactive biological signaling molecule **nitric oxide (NO)** through its conversion to **nitrate ion** (NO_3^-) under normal conditions and its synthesis from **nitrite ion** (NO_2^-) under **hypoxic** (having an inadequate supply of O_2) conditions (see below).

In this section, we begin our discussions of hemoglobin by considering its chemical and physical properties and how they relate to its physiological function. Hemoglobin structure and the mechanisms by which it carries out these physiological functions are discussed in Section 10-2.

A. Heme

Myoglobin and each of the four subunits of hemoglobin noncovalently bind a single heme group (Fig. 10-1; spelled "haem" in British English). This is the same group that occurs in the cytochromes (Section 9-6A) and in certain redox enzymes such as **catalase.** Heme is responsible for the characteristic red color of blood and is the site at which each **globin** monomer binds one molecule of O_2 (globins are the heme-free proteins of Hb and Mb). The heterocyclic ring system of heme is a **porphyrin** derivative; it consists of four **pyrrole** rings (labeled A–D in Fig. 10-1) linked by methene bridges. The porphyrin in heme, with its particular arrangement of four methyl, two propionate, and two vinyl substituents, is known as **protoporphyrin IX.** Heme, then, is protoporphyrn IX with a centrally bound iron atom. *In Hb and Mb, the iron atom normally remains in the*

Figure 10-1   **The heme group.** Fe(II)–heme (ferroprotoporphyrin IX) is shown liganded to His and O_2 as it is in oxygenated myoglobin and oxygenated hemoglobin. Note that the heme is a conjugated system so that, although two of its Fe—N bonds are coordinate covalent bonds (bonds in which the bonding electron pair is formally contributed by only one of the atoms forming the bond), all of the Fe—N bonds are equivalent. The pyrrole ring lettering scheme is shown.

Fe(II) (ferrous) oxidation state whether or not the heme is oxygenated (binds O_2).

The Fe atom in deoxygenated Hb and Mb is 5-coordinated by a square pyramid of N atoms: four from the porphyrin and one from a His side chain of the protein. On oxygenation, the O_2 binds to the Fe(II) on the opposite side of the porphyrin ring from the His ligand, so that the Fe(II) is octahedrally coordinated; that is, the ligands occupy the six corners of an octahedron centered on the Fe atom (Fig. 10-1). *Oxygenation changes the electronic state of the Fe(II)–heme, as is indicated by the color change of blood from the dark purplish hue characteristic of venous blood to the brilliant scarlet color of arterial blood and blood from a cut finger* (Fig. 10-2).

Certain small molecules, such as CO, NO, CN^-, and H_2S, coordinate to the sixth liganding position of the Fe(II) in Hb and Mb with much greater affinity than does O_2. This, together with their similar binding to the hemes of cytochromes, accounts for the highly toxic properties of these substances.

The Fe(II) of Hb and Mb can be oxidized to Fe(III) to form **methemoglobin (metHb)** and **metmyoglobin (metMb).** MetHb and metMb do not bind O_2; their Fe(III) is already octahedrally coordinated with an H_2O molecule in the sixth liganding position. The brown color of dried blood and old meat is that of metHb and metMb. Erythrocytes contain the enzyme **methemoglobin reductase,** which converts the

Figure 10-2 **The visible absorption spectra of oxygenated and deoxygenated hemoglobins.**

small amount of metHb that spontaneously forms back to the Fe(II) form.

a. Myoglobin Both Detoxifies and Synthesizes Nitric Oxide

NO, which is synthesized in many tissues, functions as a locally active signaling molecule, most notably to induce vasodilation (Section 19-1L). Once NO has delivered its message, it is important that it be rapidly eliminated to prevent its interference with subsequent NO signals (or lack of them). Moreover, NO is a highly reactive and hence toxic substance. In muscle, under normal O_2 concentrations, NO is detoxified through its reaction with oxygenated myoglobin **(oxyMb)** to yield nitrate ion and metmyoglobin:

$$NO + MbO_2 \rightarrow NO_3^- + metMb$$

Since the metMb is subsequently reduced to Mb through the action of an intracellular **metmyoglobin reductase,** myoglobin functions as an enzyme in this process. Oxygenated hemoglobin **(oxyHb)** likewise detoxifies the NO that is present in blood.

Paradoxically, the tissue damage caused by **ischemia** (inadequate blood flow, such as in a heart attack or stroke) is often exacerbated when blood flow to the deprived tissues is reestablished. This so-called **reperfusion injury** is caused, in part, by the highly destructive **reactive oxygen species (ROS),** such as the **superoxide ion** ($O_2^{-\cdot}$) and the **hydroxyl radical** ($\cdot OH$), initially produced by hypoxic mitochondria when the blood supply is restored. Nitrite ion is normally present in blood and tissues at low (micromolar) concentrations. Under hypoxic conditions, deoxyMb catalyzes the formation of NO from nitrite ion:

$$NO_2^- + H^+ + Mb \rightarrow NO + OH^- + metMb$$

NO suppresses the ROS-producing mitochondrial electron-transport chain (Chapter 22) and thereby protects against reperfusion injury. Indeed, the administration of nitrite ion

to the hearts of normal mice immediately before their blood supply was cut off for 30 minutes provided a significant protective effect against tissue damage, but afforded no such protection to the hearts of Mb-knockout mice.

b. Other Globins and Oxygen-Transport Proteins

Hb and Mb belong to the globin superfamily, whose members occur widely in all three kingdoms of life, where most participate in a variety of enzymatic and O_2-sensing functions. Evidently, Hb's O_2-transport function is a relatively recent evolutionary adaptation.

Invertebrate hemoglobins vary in quaternary structure from dimers to as many as 180 subunits. The larger of these, which are known as **erythrocruorins** (Latin: *cruor,* blood), are extracellular proteins rather than being packaged within cells; their large sizes (up to 3.5 million kD) permit them to be retained within their circulatory systems. The related **chlorocruorins** (Greek, *chloros,* pale green), which occur in the blood of certain annelids, contain a porphyrin that differs from protoporphyrin IX by the replacement of its A-ring vinyl group with a formyl group. Consequently, chlorocruorins are green when deoxygenated and light red when oxygenated.

Antarctic icefish, the only adult vertebrates that lack hemoglobin—their blood is colorless—are viable because of their reduced need for O_2 at low temperatures combined with the relatively high aqueous solubility of O_2 at the $-1.9°C$ temperature of their environment (recall that the solubilities of gases increase with decreasing temperature).

Many invertebrates lack Hb but instead produce one of two alternative types of O_2-transport proteins:

1. Hemocyanins, which occur in many species of mollusks and arthropods, form a family of large, multimeric, extracellular proteins that differ in their primary through quaternary structures. However, they have highly similar O_2-binding sites that each contain two Cu ions. Hemocyanins are blue in complex with O_2 and colorless otherwise.

2. Hemerythrins, which are intracellular proteins, occur in only a few species of marine worms. Their O_2-binding sites contain two non-heme Fe ions. Hemerythrins are violet-pink in complex with O_2 and colorless otherwise.

Vertebrates also express two recently discovered globins: **neuroglobin,** which is present mainly in brain, retina, and endocrine tissues, and **cytoglobin,** which occurs in most tissues. Neuroglobin protects neurons (nerve cells) from damage under conditions of hypoxia, most likely by preventing reperfusion injury in much the same way as does myoglobin in muscle. Cytoglobin may have similar functions.

B. *Oxygen Binding*

The binding of O_2 to myoglobin is described by a simple equilibrium reaction

$$Mb + O_2 \rightleftharpoons MbO_2$$

with dissociation constant

$$K = \frac{[Mb][O_2]}{[MbO_2]} \qquad [10.1]$$

(biochemists usually express equilibria in terms of dissociation constants, the reciprocals of the more chemically traditional association constants). The O_2 dissociation of Mb may be characterized by its **fractional saturation,** Y_{O_2}, defined as the fraction of O_2 binding sites occupied by O_2.

$$Y_{O_2} = \frac{[MbO_2]}{[Mb] + [MbO_2]} = \frac{[O_2]}{K + [O_2]} \qquad [10.2]$$

Since O_2 is a gas, its concentration is conveniently expressed by its partial pressure, pO_2 (also called the **oxygen tension**). Equation [10.2] may therefore be expressed:

$$Y_{O_2} = \frac{pO_2}{K + pO_2} \qquad [10.3]$$

Now define p_{50} as the value of pO_2 when $Y_{O_2} = 0.50$, that is, when half of myoglobin's O_2-binding sites are occupied. Substituting this value into Eq. [10.3] and solving for K yields $K = p_{50}$. Hence our expression for the fractional saturation of Mb finally becomes

$$Y_{O_2} = \frac{pO_2}{p_{50} + pO_2} \qquad [10.4]$$

a. Hemoglobin Cooperatively Binds O_2

Myoglobin's O_2-dissociation curve (Fig. 10-3) closely follows the hyperbolic curve described by Eq. [10.4]; its p_{50} is 2.8 torr (1 torr = 1 mm Hg at 0°C = 0.133 kPa; 760 torr = 1 atm). Mb therefore gives up little of its bound O_2

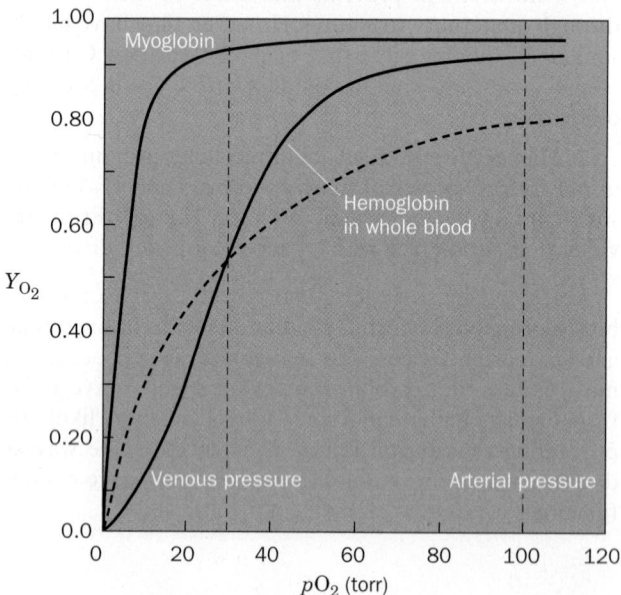

1.00 — Myoglobin

0.80

0.60 — Hemoglobin in whole blood

Y_{O_2}

0.40

0.20

Venous pressure   Arterial pressure

0.0

0   20   40   60   80   100   120

pO_2 (torr)

Figure 10-3   Oxygen-dissociation curves of Mb and of Hb in whole blood. The normal sea level values of human arterial and venous pO_2 values are indicated. The dashed line is a hyperbolic O_2-dissociation curve with the same p_{50} as Hb (26 torr). **See the Animated Figures**

over the normal physiological range of pO_2 in blood (100 torr in arterial blood and 30 torr in venous blood); for example, $Y_{O_2} = 0.97$ at $pO_2 = 100$ torr and 0.91 at 30 torr. In contrast, hemoglobin's O_2-dissociation curve (Fig. 10-3), which has a **sigmoidal** shape (S shape) that Eq. [10.4] does not describe, indicates that the amount of O_2 bound by Hb changes significantly over the normal physiological range of pO_2 in blood. For example, $Y_{O_2} = 0.95$ at 100 torr and 0.55 at 30 torr in whole blood for a difference in Y_{O_2} of 0.40. Mb therefore binds O_2 under conditions in which Hb releases it. Thus, the two proteins form a sophisticated O_2 transport system that delivers O_2 from lung to muscle (where pO_2 may be <20 torr). Hemoglobin's sigmoidal O_2-dissociation curve is of great physiological importance; *it permits the blood to deliver much more O_2 to the tissues than it could if Hb had a hyperbolic O_2-dissociation curve with the same p_{50} (26 torr; dashed curve in Fig. 10-3).* Such a hyperbolic curve has $Y_{O_2} = 0.79$ at 100 torr and 0.54 at 30 torr for a difference in Y_{O_2} of only 0.25.

A sigmoidal dissociation curve is diagnostic of a **cooperative interaction** between a protein's small molecule binding sites; that is, the binding of one small molecule affects the binding of others. In this case, the binding of O_2 increases the affinity of Hb for binding additional O_2. The structural mechanism of hemoglobin cooperativity is discussed in Section 10-2C.

b. The Hill Equation Phenomenologically Describes Hemoglobin's O_2-Binding Curve

The earliest attempt to analyze hemoglobin's sigmoidal O_2-dissociation curve was formulated by Archibald Hill in 1910. We shall follow his analysis in general form because it is useful for characterizing the cooperative behavior of oligomeric enzymes as well as that of hemoglobin.

Consider a protein E consisting of n subunits that can each bind a molecule S, which, in analogy with the substituents of metal ion complexes, is known as a **ligand.** Assume that the ligand binds with infinite cooperativity,

$$E + nS \rightleftharpoons ES_n$$

that is, the protein either has all or none of its ligand-binding sites occupied, so that there are no observable intermediates ES_1, ES_2, etc. The dissociation constant for this reaction is

$$K = \frac{[E][S]^n}{[ES_n]} \qquad [10.5]$$

and, as before, its fractional saturation is expressed:

$$Y_S = \frac{n[ES_n]}{n([E] + [ES_n])} \qquad [10.6]$$

Combining Eqs. [10.5] and [10.6] yields

$$Y_S = \frac{[E][S]^n/K}{[E](1 + [S]^n/K)}$$

which on algebraic rearrangement and cancellation of terms becomes the **Hill equation:**

$$Y_S = \frac{[S]^n}{K + [S]^n} \qquad [10.7]$$

which, in a manner analogous to Eq. [10.4], describes the degree of saturation of a multisubunit protein as a function of ligand concentration.

Infinite ligand-binding cooperativity (n equal to the number of protein subunits), as assumed in deriving the Hill equation, is a physical impossibility. Nevertheless, n may be taken to be a noninteger parameter related to the degree of cooperativity among interacting ligand-binding sites rather than the number of subunits per protein. The Hill equation then becomes a useful empirical curve-fitting relationship rather than an indicator of a particular model of ligand binding. *The quantity n, the* **Hill constant,** *increases with the degree of cooperativity of a reaction and thereby provides a convenient, although simplistic, characterization of a ligand-binding reaction.* If $n = 1$, Eq. [10.7] describes a hyperbola, as do Eqs. [10.3] and [10.4] for Mb, and the ligand-binding reaction is said to be **noncooperative.** A reaction with $n > 1$ is described as **positively cooperative:** Ligand binding increases the affinity of E for further ligand binding (cooperativity is infinite in the limit that n is equal to the number of ligand-binding sites in E). Conversely, if $n < 1$, the reaction is termed **negatively cooperative:** Ligand binding reduces the affinity of E for subsequent ligand binding.

c. Hill Equation Parameters May Be Graphically Evaluated

The Hill constant, n, and the dissociation constant, K, that best describe a saturation curve can be graphically determined by rearranging Eq. [10.7] as follows:

$$\frac{Y_S}{1 - Y_S} = \frac{\dfrac{[S]^n}{K + [S]^n}}{1 - \dfrac{[S]^n}{K + [S]^n}} = \frac{[S]^n}{K}$$

and then taking the log of both sides to yield a linear equation:

$$\log\left(\frac{Y_S}{1 - Y_S}\right) = n \log[S] - \log K \qquad [10.8]$$

The linear plot of $\log[Y_S/(1 - Y_S)]$ versus $\log[S]$, the **Hill plot,** has a slope of n and an intercept on the $\log[S]$ axis of $(\log K)/n$ (recall that the linear equation $y = mx + b$ describes a line with a slope of m and an x intercept of $-b/m$).

For Hb, if we substitute pO_2 for $[S]$ as was done for Mb, the Hill equation becomes:

$$Y_{O_2} = \frac{(pO_2)^n}{K + (pO_2)^n} \qquad [10.9]$$

As in Eq. [10.4], let us define p_{50} as the value of pO_2 at $Y_{O_2} = 0.50$. Then, substituting this value into Eq. [10.9],

$$0.50 = \frac{(p_{50})^n}{K + (p_{50})^n}$$

so that

$$K = (p_{50})^n \qquad [10.10]$$

Substituting this result back into Eq. [10.9] yields

$$Y_{O_2} = \frac{(pO_2)^n}{(p_{50})^n + (pO_2)^n} \qquad [10.11]$$

(*Note:* Equation [10.4] is a special case of Eq. [10.11] with $n = 1$). Equation [10.8] for the Hill plot of Hb therefore takes the form

$$\log\left(\frac{Y_{O_2}}{1 - Y_{O_2}}\right) = n \log pO_2 - n \log p_{50} \qquad [10.12]$$

so that this plot has a slope of n and an intercept on the log pO_2 axis of log p_{50}.

Figure 10-4 shows the Hill plots for Mb and Hb. For Mb it is linear with a slope of 1, as expected. Although Hb does

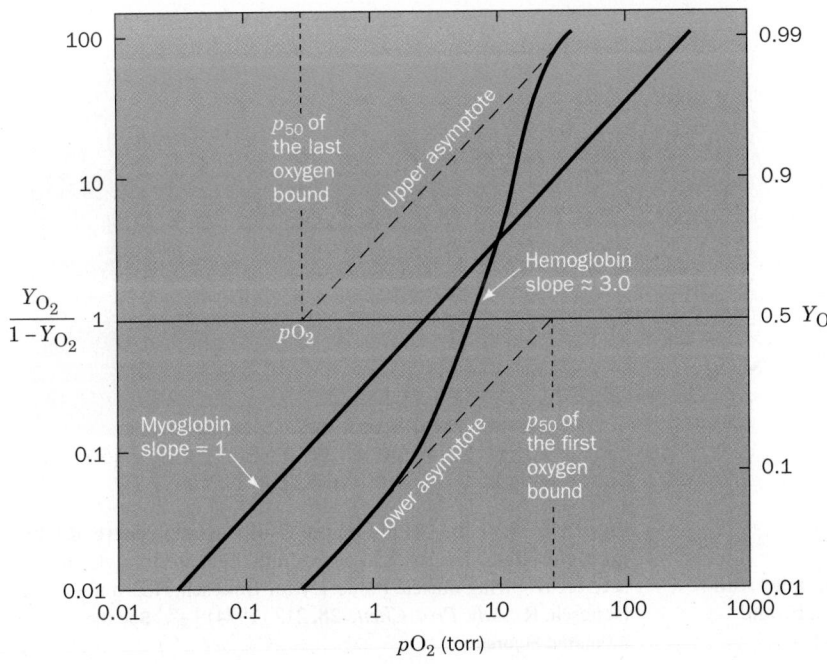

Figure 10-4 **Hill plots for Mb and purified ("stripped") Hb.** Note that this is a log–log plot. Hence the horizontal axis, $\log[Y_{O_2}/(1 - Y_{O_2})] = 0$, occurs where $Y_{O_2}/(1 - Y_{O_2}) = 1$ (and $pO_2 = p_{50}$).

not bind O_2 in a single step as is assumed in deriving the Hill equation, its Hill plot is essentially linear for values of Y_{O_2} between 0.1 and 0.9. Its maximum slope, which occurs near $pO_2 = p_{50}$ $[(Y_{O_2} = 0.5; Y_{O_2}/(1 - Y_{O_2}) = 1]$, is normally taken to be the Hill constant. For normal human Hb, the Hill constant is between 2.8 and 3.0; that is, hemoglobin oxygen binding is highly, but not infinitely, cooperative. Many abnormal hemoglobins exhibit smaller Hill constants (Section 10-3A), indicating that they have a less than normal degree of cooperativity. At Y_{O_2} values near 0, when few Hb molecules have bound even one O_2 molecule, the Hill plot of Hb assumes a slope of 1 (Fig. 10-4, lower asymptote) because the Hb subunits independently compete for O_2 as do molecules of Mb. At Y_{O_2} values near 1, when at least three of each of hemoglobin's four O_2-binding sites are occupied, the Hill plot also assumes a slope of 1 (Fig. 10-4, upper asymptote) because the few remaining unoccupied sites are on different molecules and therefore bind O_2 independently.

Extrapolating the lower asymptote in Fig. 10-4 to the horizontal axis indicates, according to Eq. [10.11], that $p_{50} = 30$ torr for binding the first O_2 to Hb. Likewise, extrapolating the upper asymptote yields $p_{50} = 0.3$ torr for binding hemoglobin's fourth O_2. Thus *the fourth O_2 to bind to Hb does so with 100-fold greater affinity than the first.* This difference, as we shall see in Section 10-2C, is entirely due to the influence of the globin on the O_2 affinity of heme. It corresponds to a free energy difference of 11.4 kJ · mol^{-1} between binding the first and binding the last O_2 to Hb (Section 3-4A).

More sophisticated mathematical models than the Hill equation have been developed for analyzing the cooperative binding of ligands to proteins. We examine some of them in Section 10-4.

d. Globin Prevents Oxyheme from Autooxidizing

Globin not only modulates the O_2-binding affinity of heme, but makes reversible O_2 binding possible. Fe(II)–heme by itself is incapable of binding O_2 reversibly. Rather, in the presence of O_2 it autooxidizes irreversibly to the Fe(III) form through the intermediate formation of a complex consisting of an O_2 bridging the Fe atoms of two hemes. This reaction can be inhibited by derivatizing the heme with bulky groups that sterically prevent the close face-to-face approach of two hemes. Such **picket-fence** Fe(II)–porphyrin complexes (Fig. 10-5), which James Collman first synthesized, bind O_2 reversibly. The back side of this porphyrin is unhindered and is complexed with a substituted imidazole in a manner similar to that in Mb and Hb. In fact, the O_2 affinity of the picket-fence complex is similar to that of Mb. Thus, the globins of Mb and Hb function to prevent the autooxidation of oxyheme by surrounding it, rather like a hamburger bun surrounds a hamburger, so that only its propionate side chains are exposed to the aqueous solvent (Section 10-2B).

C. *Carbon Dioxide Transport and the Bohr Effect*

In addition to being an O_2 carrier, *Hb plays an important role in the transport of CO_2 by the blood.* When Hb (but not Mb) binds O_2 at physiological pH's, it undergoes a conformational change (Section 10-2Ba) that makes it a slightly stronger acid. It therefore releases protons on binding O_2:

$$Hb(O_2)_n + O_2 \rightleftharpoons Hb(O_2)_{n+1} + xH^+$$

where $n = 0, 1, 2,$ or 3 and $x \approx 0.6$ under physiological conditions. Conversely, *increasing the pH, that is, removing protons, stimulates Hb to bind O_2* (Fig. 10-6). This phenomenon, whose molecular basis is discussed in Section 10-2E, is known as the **Bohr effect** after Christian Bohr (the father

Figure 10-5 A picket-fence Fe(II)–porphyrin complex with bound O₂. [After Collman, J.P., Brauman, J.I., Rose, E., and Suslick, K.S., *Proc. Natl. Acad. Sci.* **75**, 1053 (1978).]

Figure 10-6 Effect of pH on the O₂-dissociation curve of Hb: the Bohr effect. The vertical dashed line indicates the pO_2 in actively respiring muscle tissue. [After Benesch, R.E. and Benesch, R., *Adv. Prot. Chem.* **28**, 212 (1974).] See the **Animated Figures**

of the pioneering atomic physicist Niels Bohr), who first reported it in 1904.

a. The Bohr Effect Facilitates O₂ Transport

The ~0.8 molecules of CO_2 formed per molecule of O_2 consumed by respiration diffuse from the tissues to the capillaries largely as dissolved CO_2 as a result of the slowness of the reaction forming bicarbonate:

$$CO_2 + H_2O \rightleftharpoons H^+ + HCO_3^-$$

This reaction, however, is catalyzed in the erythrocyte by carbonic anhydrase (Fig. 8-41). Accordingly, most of the CO_2 in the blood is carried in the form of bicarbonate (in the absence of carbonic anhydrase, the hydration of CO_2 would equilibrate 100-fold more slowly, so bubbles of the only slightly soluble CO_2 would form in the blood and tissues).

In the capillaries, where pO_2 is low, the H^+ generated by bicarbonate formation is taken up by Hb, which is thereby induced to unload its bound O_2. This H^+ uptake, moreover, facilitates CO_2 transport by stimulating bicarbonate formation. Conversely, in the lungs, where pO_2 is high, O_2 binding by Hb releases the Bohr protons, which drive off the CO_2. These reactions are closely matched, so they cause very little change in blood pH.

The Bohr effect provides a mechanism whereby additional O_2 can be supplied to highly active muscles. Such muscles generate acid (Section 17-3A) so fast that they lower the pH of the blood passing through them from 7.4 to 7.2. At pH 7.2, Hb releases ~10% more O_2 at the <20 torr pO_2 in these muscles than it does at pH 7.4 (Fig. 10-6).

b. CO₂ and Cl⁻ Modulate Hemoglobin's O₂ Affinity

CO_2 modulates O_2 binding directly and by combining reversibly with the N-terminal amino groups of blood proteins to form **carbamates:**

$$R-NH_2 + CO_2 \rightleftharpoons R-NH-COO^- + H^+$$

The conformation of deoxygenated Hb **(deoxyHb),** as we shall see in Section 10-2Ba, is significantly different from that of oxygenated Hb **(oxyHb).** Consequently, deoxyHb binds more CO_2 as carbamate than does oxyHb. CO_2 like H^+, is therefore a modulator of hemoglobin's O_2 affinity: A high CO_2 concentration, as occurs in the capillaries, stimulates Hb to release its bound O_2. Note the complexity of this $Hb-O_2-CO_2-H^+$ equilibrium: The protons released by carbamate formation are, in part, taken up through the Bohr effect, thereby increasing the amount of O_2 that Hb would otherwise release. Although the difference in CO_2 binding between the oxy and deoxy states of hemoglobin accounts for only ~5% of the total blood CO_2, it is nevertheless responsible for around half of the CO_2 transported by blood. This is because only ~10% of the total blood CO_2 turns over in each circulatory cycle.

Cl^- is also bound more tightly to deoxyHb than to oxyHb (Section 10-2E). Accordingly, hemoglobin's O_2 affinity also varies with $[Cl^-]$. HCO_3^- freely permeates the

erythrocyte membrane (Section 12-3D), so that once formed, it equilibrates with the surrounding plasma. The need for charge neutrality on both sides of the membrane, however, requires that Cl^-, which also freely permeates the membrane, replace the HCO_3^- that leaves the erythrocyte (the erythrocyte membrane is impermeable to cations). Consequently, $[Cl^-]$ in the erythrocyte is greater in the venous blood than it is in the arterial blood. *Cl^- is therefore also a modulator of hemoglobin's O_2 affinity.*

D. Effect of BPG on O₂ Binding

Purified (stripped) hemoglobin has a much greater O_2 affinity than does hemoglobin in whole blood (Fig. 10-7). This observation led Joseph Barcroft, in 1921, to speculate that blood contains some other substance that complexes with Hb so as to reduce its O_2 affinity. In 1967, Reinhold and Ruth Benesch demonstrated that this substance is **D-2,3-bisphosphoglycerate (BPG)**

$$\begin{array}{c} O^- \quad O \\ \diagdown \! \diagup \\ C \\ | \\ H-C-OPO_3^{2-} \\ | \\ H-C-OPO_3^{2-} \\ | \\ H \end{array}$$

D-2,3-Bisphosphoglycerate (BPG)

[previously known as **2,3-diphosphoglycerate (DPG)].** BPG binds tightly to deoxyHb in a 1:1 mole ratio ($K = 1.5 \times 10^{-5}$ M) but only weakly to oxyHb. The presence of BPG therefore decreases hemoglobin's oxygen affinity by keeping it in the deoxy conformation; for example, the p_{50} of stripped hemoglobin is increased from 12 to 22 torr by 4.7 mM BPG, its normal concentration in erythrocytes

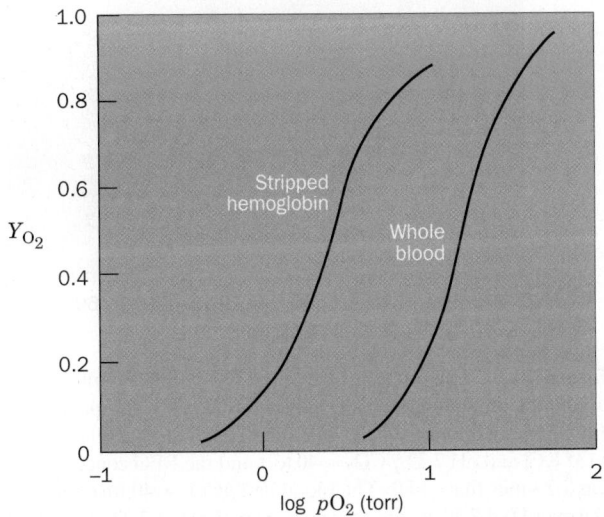

Figure 10-7 Comparison of the O₂-dissociation curves of "stripped" Hb and whole blood in 0.01M NaCl at pH 7.0. [After Benesch, R.E. and Benesch, R., *Adv. Prot. Chem.* **28,** 217 (1974).]

(similar to that of Hb). Organic polyphosphates, such as **inositol hexaphosphate (IHP)**

Inositol hexaphosphate (IHP)

and ATP, also have this effect on Hb. In fact, in birds, IHP functionally replaces BPG and ATP does so in fish and in most amphibians. The ~2 mM ATP normally present in mammalian erythrocytes is prevented from binding to Hb by its complexation with Mg^{2+}.

BPG has an indispensable physiological function: In arterial blood, where pO_2 is ~100 torr, Hb is ~95% saturated with O_2, but in venous blood, where pO_2 is ~30 torr, it is only ~55% saturated (Fig. 10-3). Consequently, in passing through the capillaries, Hb unloads ~40% of its O_2. *In the absence of BPG, little of this O_2 is released since hemoglobin's O_2 affinity is increased, thus shifting the O_2-dissociation curve significantly toward lower pO_2* (Fig. 10-8, *left*).

CO_2 and BPG independently modulate hemoglobin's O_2 affinity. Figure 10-8 indicates that stripped Hb can be made to have the same oxygen-dissociation curve as the Hb in whole blood by adding CO_2 and BPG in the concentrations found in erythrocytes (the pH and [Cl^-] are also the same). Hence, *the presence of these four substances in whole blood—BPG, CO_2, H^+, and Cl^-—accounts for the O_2-binding properties of Hb.*

a. Increased BPG Levels Are Partially Responsible for High-Altitude Adaptation

High-altitude adaptation is a complex physiological process that involves an increase in the amount of hemoglobin per erythrocyte and in the number of erythrocytes. It normally requires several weeks to complete. Yet, as is clear to anyone who has quickly climbed to high altitude, even a 1-day stay there results in a noticeable degree of adaptation. This effect results from a rapid increase in the erythrocyte BPG concentration (Fig. 10-9; BPG, which cannot pass through the erythrocyte membrane, is synthesized in the erythrocyte; Section 17-2Hb). The consequent decrease in O_2-binding affinity, as indicated by its elevated

Figure 10-8   The effects of BPG and CO_2, both separately and combined, on hemoglobin's O_2-dissociation curve compared with that of whole blood (*red curve*). In the Hb solutions, which were 0.1M KCl and pH 7.22, pCO_2 = 40 torr and the BPG concentration was 1.2 times that of Hb. The blood had pCO_2 = 40 torr and a plasma pH of 7.40, which corresponds to a pH of 7.22 inside the erythrocyte. [After Kilmartin, J.V. and Rossi-Bernardi, L., *Physiol. Rev.* **53**, 884 (1973).] 👁🗨 **See the Animated Figures**

Figure 10-9   The effect of high-altitude exposure on the p_{50} and the BPG concentration of blood in sea level–adapted individuals. The region on the right marked "Sea level" indicates the effects of exposure to sea level on high altitude–adapted individuals. [After Lenfant, C., Torrance, J.D., English, E., Finch, C.A., Reynafarje, C., Ramos, J., and Faura, J., *J. Clin. Invest.* **47**, 2653 (1968).]

p_{50}, increases the amount of O_2 that hemoglobin unloads in the capillaries (Fig. 10-10). Similar increases in BPG concentration occur in individuals suffering from disorders that limit the oxygenation of the blood (hypoxia), such as various anemias and cardiopulmonary insufficiency.

b. Fetal Hemoglobin Has a Low BPG Affinity

The effects of BPG also help supply the fetus with oxygen. A fetus obtains its O_2 from the maternal circulation via the placenta. This process is facilitated because fetal hemoglobin (HbF) has a higher O_2 affinity than does maternal hemoglobin (HbA; recall that HbF has the subunit composition $\alpha_2\gamma_2$, in which the γ subunit is a variant of HbA's β subunit; Section 7-3C). BPG occurs in about the same concentrations in adult and fetal erythrocytes but binds more tightly to deoxyHbA than to deoxyHbF; this accounts for HbF's greater O_2 affinity. In the next section we shall develop the structural rationale for the effect of BPG and for the other aspects of O_2 binding.

Figure 10-10 **The O_2-dissociation curves of blood adapted to sea level (*black curve*) and to high altitude (*red curve*).** Between the sea level arterial and venous pO_2 values of 100 and 30 torr, respectively, Hb normally unloads 38% of the O_2 it can maximally carry. However, when the arterial pO_2 drops to 55 torr, as it does at an altitude of 4500 m, this difference is reduced to 30% in nonadapted blood. High-altitude adaptation increases the BPG concentration in erythrocytes, which shifts the O_2-dissociation curve of Hb to the right. The amount of O_2 that Hb delivers to the tissues is thereby restored to 37% of its maximum load.

2 STRUCTURE AND MECHANISM

The determination of the first protein X-ray structures, those of sperm whale myoglobin by John Kendrew in 1959 and of human deoxyhemoglobin and horse methemoglobin by Max Perutz shortly thereafter, ushered in a revolution in biochemical thinking that has reshaped our understanding of the chemistry of life. Before the advent of protein crystallography, macromolecular structures, if they were considered at all, were thought of as having a rather hazy existence of uncertain biological significance. However, as the elucidation of macromolecular structures has continued at an ever quickening pace, it has become clear that *life is based on the interactions of complex, structurally well-defined macromolecules.*

The story of hemoglobin's structural determination is a tale of enormous optimism and tenacity. Perutz began this study in 1937 at Cambridge University as a graduate student of J. D. Bernal (who, with Dorothy Crowfoot Hodgkin, had taken the first X-ray diffraction photographs of hydrated protein crystals in 1934). In 1937, the X-ray crystal structure determination of even the smallest molecule required many months of hand computation, and the largest structure yet determined was that of the dye phthalocyanin, which has 40 nonhydrogen atoms. Since hemoglobin has ~4500 nonhydrogen atoms, it must have seemed to Perutz's colleagues that he was pursuing an impossible goal. Nevertheless, the laboratory director, Lawrence Bragg (who in 1912, with his father William Bragg, had determined the first X-ray structure, that of NaCl), realized the tremendous biological significance of determining a protein structure and supported the project.

It was not until 1953 that Perutz finally hit on the method that would permit him to solve the X-ray structure of hemoglobin, that of isomorphous replacement. Kendrew, a colleague of Perutz, used this technique to solve the X-ray structure of sperm whale myoglobin, first at low resolution in 1957, and then at high resolution in 1959. Hemoglobin's greater complexity delayed its low-resolution structural determination until 1959, and it was not until 1968, over 30 years after he had begun the project, that Perutz and his associates obtained the high-resolution X-ray structure of horse methemoglobin. Those of human and horse deoxyhemoglobins followed shortly thereafter. Since then, the X-ray structures of hemoglobins from numerous different species, from mutational variants, and with different bound ligands have been elucidated. This, together with many often ingenious physicochemical investigations, has made hemoglobin the most intensively studied, and perhaps the best understood, of proteins.

In this section, we examine the molecular structures of myoglobin and hemoglobin and consider the structural basis of hemoglobin's oxygen-binding cooperativity, the Bohr effect, and BPG binding.

A. *Structure of Myoglobin*

Myoglobin consists of eight helices (labeled A–H) that are linked by short polypeptide segments to form an ellipsoidal

Table 10-1   **The Amino Acid Sequences of the α and β Chains of Human Hemoglobin and of Human Myoglobin**[a,b]

Helix Boundaries	A1	A16 B1	B16 / C1	C7	D1	D7 / E1

Hb α ....
V—L S P A D K T N V K A A W G K V G A H A G E Y G A E A L E R M F L S F P T T K T Y F P H F — D L S H — — — — — G S A Q V K G H G K K V A D A L T
(numbering: 1 ... 5 ... 10 ... 15 ... 20 ... 25 ... 30 ... 35 ... 40 ... 45 ... 50 ... 55 ... 60 ... 65)

Hb β ....
V H L T P E E K S A V T A L W G K V — — N V D E V G G E A L G R L L V V Y P W T Q R F F E S F G D L S T P D A V M G N P K V K A H G K K V L G A F S
(numbering: 1 ... 5 ... 10 ... 15 ... 20 ... 25 ... 30 ... 35 ... 40 ... 45 ... 50 ... 55 ... 60 ... 65 ... 70)

Mb ....
G—L S D G E W Q L V L N V W G K V E A D I P G H G Q E V L I R L F K G H P E T L E K F D K F K H L K S E D E M K A S E D L K K H G A T V L T A L G
(numbering: 1 ... 5 ... 10 ... 15 ... 20 ... 25 ... 30 ... 35 ... 40 ... 45 ... 50 ... 55 ... 60 ... 65 ... 70)

molecule of approximate dimensions 44 × 44 × 25 Å (Fig. 10-11; see also Fig. 8-39). The helices range in length from 7 to 26 residues and incorporate 121 of myoglobin's 153 residues (Table 10-1). They are largely α helical but with some distortions from this geometry such as a tightening of the final turns of helices A, C, E, and G to form segments of 3_{10} helix.

In a helix numbering convention peculiar to globins, residues are designated according to their position in a helix or interhelical segment. For example, residue B5 is the fifth residue from the N-terminus of the B helix and residue FG3 is the third residue from the N-terminus in the nonhelical segment connecting helices F and G. The nonhelical N- and C-terminal segments are designated NA and HC, respectively. The usual convention of sequentially numbering all amino acid residues from the N-terminal residue of the polypeptide is also used, and often both conventions are used together. For example, Glu EF7(83) of human Mb is the 83rd residue from its N-terminus and the 7th residue in the nonhelical segment connecting its E and F helices.

The heme is tightly wedged in a hydrophobic pocket formed mainly by helices E and F but which includes contacts with helices B, C, G, and H as well as the CD and FG segments. The fifth ligand of the heme Fe(II) is His F8, the **proximal** (near) **histidine.** In oxyMb, the Fe(II) is positioned 0.22 Å out of the heme plane on the side of the proximal His and is coordinated by O_2 with the bent geometry shown in Fig. 10-12. His E7, the **distal** (distant) **histidine,** hydrogen bonds to the O_2. In deoxyMb, the sixth liganding position of the Fe(II) is unoccupied because the distal His is too far away from the Fe(II) to coordinate with it. Furthermore, the Fe(II) has moved to a point 0.55 Å out of the heme plane. Other structural changes in Mb on changing oxygenation states consist of small motions of various chain segments and slight readjustments of side chain conformations. *By and large, however, the structures of oxy- and deoxyMb are nearly superimposable.*

B. *Structure of Hemoglobin*

The hemoglobin tetramer is a spheroidal molecule of dimensions 64 × 55 × 50 Å. Its two αβ protomers are symmetrically related by a twofold rotation (Fig. 10-13; see also Fig. 8-64). *The tertiary structures of the α and β subunits are remarkably similar, both to each other and to that of Mb (Figs. 10-11 and 10-13), even though only 18% of the corresponding residues are identical among these three polypeptides (Table 10-1) and there is no D helix in hemoglobin's α subunit. Indeed, the α and β subunits in the tetramer are related by pseudo- (inexact) 2-fold rotations so that the subunits occupy the vertices of a tetrahedron (pseudo-D_2 symmetry; Section 8-5B).*

The polypeptide chains of Hb are arranged such that there are extensive interactions between unlike subunits. The $α_1$–$β_1$ interface (and its $α_2$–$β_2$ symmetry equivalent) involves 35 residues, whereas the $α_1$–$β_2$ (and $α_2$–$β_1$) interface involves 19 residues. These associations are predominantly

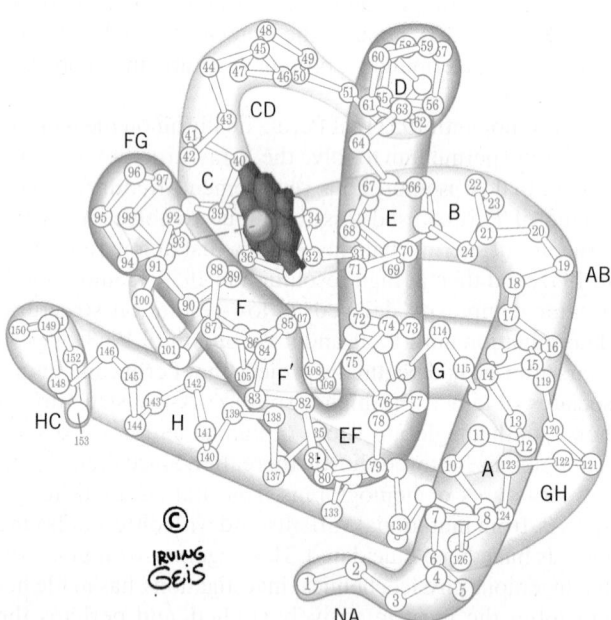

Figure 10-11   Structure of sperm whale myoglobin. Its 153 $C_α$ positions are numbered from the N-terminus and its eight helices are sequentially labeled A through H. The last half of the EF corner is now regarded as a turn of helix and is therefore designated the F′ helix. The heme group is shown in red. Also see Fig. 8-39. [Illustration, Irving Geis. Image from the Irving Geis Collection, Howard Hughes Medical Institute. Reprinted with permission.] Based on an X-ray structure by John Kendrew, MRC Laboratory of Molecular Biology, Cambridge, U.K. PDBid 1MBN.] See Kinemage Exercise 6-1

	E19		F1	F9	G1			G19	H1				H19	H21	H26

Hb α...NAVAHVDDMPNALSALSDLHAHKLRVDPVNFKLLSHCLLVTLAAHLPAEFTPAVHASLDKFLASVSTVLTSKYR

Hb β...DGLAHLDNLKGTFATLSELHCDKLHVDPENFRLLGNVLVCVLAHHFGKEFTPPVQAAYQKVVAGVANALAHKYH

Mb.......GILKKKGHHEAEIKPLAQSHATKHKIPVKYLEFISECIIQVLQSKHPGDFGADAQGAMNKALELFRKDMASNYKELGFQG

[a]The residues have been aligned in structurally analogous positions. The blue boxes shade the residues that are identical in both Hb chains, the purple boxes shade the residues that are identical in both Hb chains and in Mb, and the dark purple boxes shade residues that are invariant in all vertebrate Hb and Mb chains (Thr C4, Phe CD1, Leu F4, His F8, and Tyr HC2).

[b]The first and last residues in helices A–H are indicated, whereas the residues between helices constitute the intervening "segments." The refined Hb structure reveals that much of what is designated the EF segment is really helical in both chains: It encompasses residues EF4–F2 and is designated the F' helix.

Source: Dickerson, R.E. and Geis, I., *Hemoglobin,* pp. 68–69, Benjamin/Cummings (1983).

Figure 10-12   The heme complex in oxyMb. In the upper drawing, atoms are represented as spheres of van der Waals radii. The lower drawing shows the corresponding skeletal model with a dashed line representing the hydrogen bond between the distal His and the bound O_2. [After Phillips, S.E.V., *J. Mol. Biol.* **142,** 544 (1980). PDBid 1MBO.] 🐭 **See Kinemage Exercise 6-1**

hydrophobic in character, although numerous hydrogen bonds and several ion pairs are also involved (Section 10-2C). In contrast, contacts between like subunits, α_1–α_2 and β_1–β_2, are few and largely polar in character. This is because like subunits face each other across an ~20-Å-diameter solvent-filled channel that parallels the 50-Å length of the exact 2-fold axis (Figs. 8-64 and 10-13).

a. Oxy- and Deoxyhemoglobins Have Different Quaternary Structures

Oxygenation causes such extensive quaternary structural changes to Hb that oxy- and deoxyHb have different crystalline forms; indeed, crystals of deoxyHb shatter on exposure to O_2. The crystal structures of hemoglobin's oxy and deoxy forms therefore had to be determined independently. *The quaternary structural change preserves hemoglobin's exact 2-fold symmetry and takes place entirely across its α_1–β_2 (and α_2–β_1) interface.* The α_1–β_1 (and α_2–β_2) contact is unchanged, presumably as a result of its more extensive close associations. This contact provides a convenient frame of reference from which the oxy and deoxy conformations may be compared. Viewed in this way, oxygenation rotates the $\alpha_1\beta_1$ dimer ~15° with respect to the $\alpha_2\beta_2$ dimer (Fig. 10-14), so that some atoms at the α_1–β_2 interface shift by as much as 6 Å relative to each other (compare Figs. 10-13*a* and 10-13*b*).

The quaternary conformation of deoxyHb is named the **T state** (T for "tense"). That of oxyHb, which is essentially independent of the ligand used to induce it (e.g., O_2, met, CO, CN^-, and NO hemoglobins all have the same quaternary structure), is called the **R state** (R for "relaxed"). Similarly, the tertiary conformational states for the deoxy and liganded subunits are designated as the **t** and **r states,** respectively. The structural differences between the quaternary and tertiary conformations are described in the following subsection in terms of hemoglobin's O_2-binding mechanism.

(a)

Figure 10-13 **The X-ray structures of (*a*) deoxyHb and (*b*) oxyHb as viewed down their exact 2-fold axes.** The C_α atoms, numbered from each N-terminus, and the heme groups are shown. The Hb tetramer contains a solvent-filled central channel paralleling its 2-fold axis, whose flanking β chains draw closer together on oxygenation (compare the lengths of the double-headed arrows). In the deoxy state, His FG4(97)β (*small single-headed arrow*) fits between Thr C6(41)α and Pro CD2(44)α (*lower right and upper left*). The relative movements of the two

αβ protomers on oxygenation (*large gray arrows in Part b*) shift His FG4(97)β to a new position between Thr C3(38)α and Thr C6(41)α. See Fig. 8-64 for a similarly viewed space-filling model of deoxyHb. [Illustration, Irving Geis. Image from the Irving Geis Collection, Howard Hughes Medical Institute. Reprinted with permission.] Based on X-ray structures by Max Perutz, MRC Laboratory of Molecular Biology, Cambridge, U.K. PDBids (*a*) 2DHB and (*b*) 2MHB.] **See Kinemage Exercises 6-2 and 6-3**

C. *Mechanism of Oxygen-Binding Cooperativity*

The positive cooperativity of O_2 binding to Hb arises from the effect of the ligand-binding state of one heme on the ligand-binding affinity of another. Yet the distances of 25 to 37 Å between the hemes in an Hb molecule are too large for these heme–heme interactions to be electronic in character. Rather, *they are mechanically transmitted by the protein.* The elucidation of how this occurs has motivated much of the structural research on Hb for the past four decades.

X-ray crystal structure analysis has provided "snapshots" of the R and T states of Hb in various states of ligation but

does not indicate how the protein changes states. It is difficult to determine the sequence of events that result in such transformations because to do so requires an understanding of the inner workings of proteins that is presently lacking. It is as if you were asked to explain the mechanism of a complicated mechanical clock from its out-of-focus photographs when you had only a hazy notion of how gears, levers, and springs might function. Nevertheless, largely on the basis of the X-ray structures of Hb, Perutz formulated the following mechanism of Hb oxygenation, the **Perutz mechanism.**

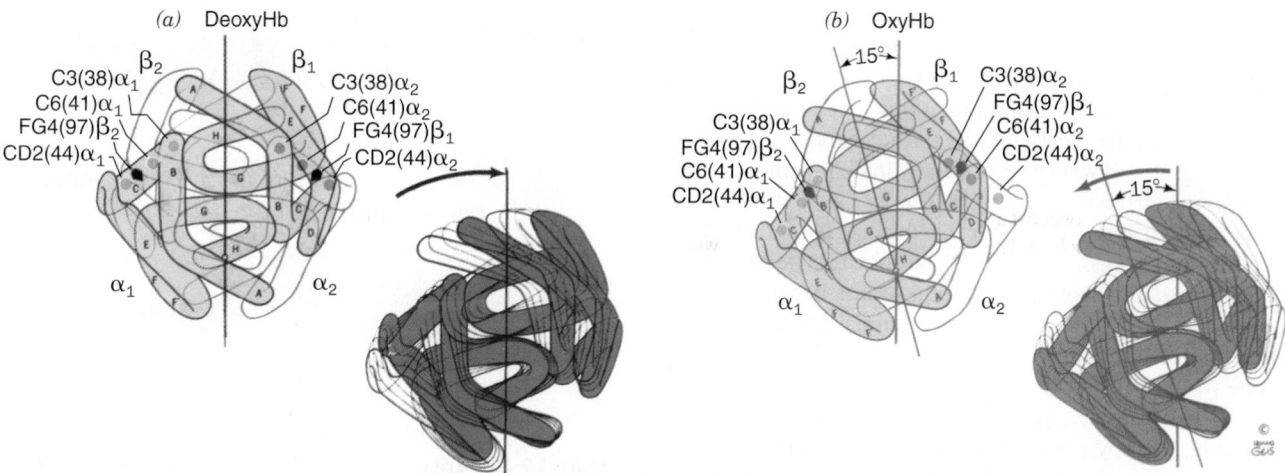

(b)

Figure 10-13 (*continued*)

Figure 10-14 **The major structural differences between the quaternary conformations of (*a*) deoxyHb and (*b*) oxyHb.** On oxygenation, the $\alpha_1\beta_1$ (*shaded*) and $\alpha_2\beta_2$ (*outline*) dimers move, as indicated on the right, as rigid units such that there is an $\sim15°$ off-center rotation of one protomer relative to the other that preserves the molecule's exact 2-fold symmetry. Note how the position of His FG4β (*pentagons*) changes with respect to Thr C3α, Thr C6α, and Pro CD2α (*yellow dots*) at the $\alpha_1-\beta_2$ and $\alpha_2-\beta_1$ interfaces. The view is from the right side relative to that in Fig. 10-13. [Illustration, Irving Geis. Image from the Irving Geis Collection, Howard Hughes Medical Institute. Reprinted with permission.]

Figure 10-15 **The heme group and its environment in the unliganded α chain of human Hb.** Only selected side chains are shown and the heme D propionate group is omitted for clarity. The F helix runs along the left side of the drawing. The close contact between the proximal His and the heme group that inhibits oxygenation of t-state hemes is indicated by a dashed red line. [After Gelin, B.R., Lee, A.W.N., and Karplus, M., *J. Mol. Biol.* **171,** 542 (1983). PDBid 2HHB.] 🖰 **See Kinemage Exercise 6-4**

Leu F4 (83)

Leu H19 (136)

Close contact

Leu F7 (86)

His F8 (87)

A

Lys E10 (61)

B

Val E11 (62)

Leu FG3 (91)

D

C

His E7 (58)

Val FG5 (93)

Leu G8 (101)

Phe CD4 (46)

Tyr C7 (42)

a. The Movement of Fe(II) Into the Heme Plane Triggers the T → R Conformational Shift

In the t state, the Fe(II) is situated ~0.6 Å out of the heme plane on the side of the proximal His because of a pyramidal doming of the porphyrin skeleton and because the Fe—$N_{porphyrin}$ bonds are too long to allow the Fe to lie in the porphyrin plane (Figs. 10-15 and 10-16). The change in the heme's electronic state on binding O_2, however, causes the doming to subside and the Fe—$N_{porphyrin}$ bonds to contract by ~0.1 Å. Consequently, on changing from the t to the r state, the Fe(II) moves to the center of the heme plane (Fig. 10-16) where O_2 can coordinate it without steric interference from the porphyrin. The Fe's movement drags the proximal His along with it, which tilts the attached F helix and translates it ~1 Å across the heme plane (Fig. 10-16). This lateral translation occurs because, in the t state, the imidazole ring of the proximal His is oriented such that its direct movement of ~0.6 Å toward the heme plane would cause it to collide with the heme (Figs. 10-15 and 10-16); however, the F helix shift reorients the imidazole ring, thereby permitting the Fe(II) to move into the heme plane. In addition, in the t state of the β but not the α subunits, Val E11 partially occludes the O_2-binding pocket so that it must be moved aside before O_2 binding can occur.

b. The α₁–β₂ and α₂–β₁ Contacts Have Two Stable Positions

As we saw above, the difference between hemoglobin's R and T conformations occurs mainly in the α₁–β₂ (and the

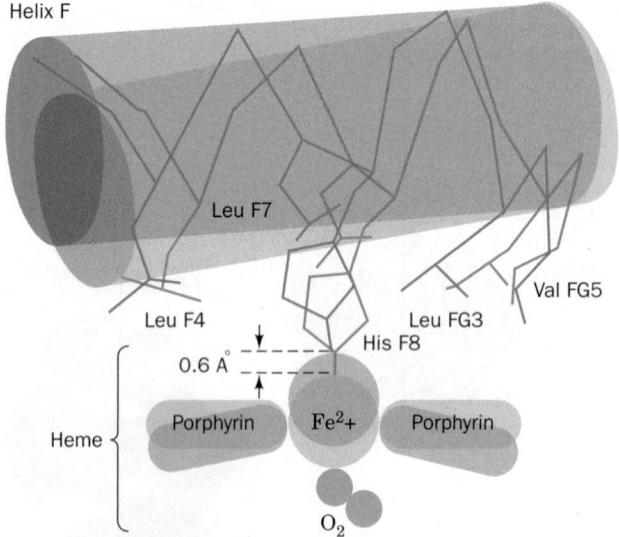

Figure 10-16 **Triggering mechanism for the T → R transition in Hb.** In the T form (*blue*), the Fe is ~0.6 Å above the mean plane of the domed porphyrin ring. On assuming the R form (*red*), the Fe moves into the plane of the now undomed porphyrin, where it can readily bind O_2, and, in doing so, pulls the proximal His F8 and its attached F helix with it. The Fe—O_2 bond is thereby strengthened because of the relaxation of the steric interference between the O_2 and the heme. 🖰 **See Kinemage Exercise 6-4 and the Animated Figures**

(a) **T State (deoxy)**   (b) **R State (oxy)**

Figure 10-17 **The $\alpha_1 C–\beta_2 FG$ interface of Hb in (a) the T state and (b) the R state.** The upper drawings show the C helix in ribbon form (*purple*) and its contacting portion of the FG region in ball-and-stick form colored according to atom type (C green, N blue, and O red). The dots outline the contacting van der Waals surfaces and are also colored according to atom type. The lower drawings are the corresponding schematic diagrams of the $\alpha_1 C–\beta_2 FG$ contact. On a T → R transformation, this contact snaps from one position to the other with no stable intermediate

(note how, in both conformations, the knobs formed by the side chains of His 97β and Asp 99β fit between the grooves on the C helix formed by the side chains of Thr 38α, Thr 41α, and Pro 44α). The subunits are joined by different hydrogen bonds in the two quarternary states. Figures 10-13 and 10-14 provide additional structural views of these interactions. [Based on X-ray structures by Giulio Fermi, Max Perutz, and Boaz Shaanan, MRC Laboratory of Molecular Biology, Cambridge, U.K. PDBids (a) 2HHB and (b) 1HHO.] **See Kinemage Exercise 6-5**

symmetrically related $\alpha_2–\beta_1$) interface, which consists of the C helix and FG segment of α_1, respectively, contacting the FG segment and C helix of β_2. The quaternary change results in a 6-Å relative shift at the $\alpha_1 C–\beta_2 FG$ interface (Fig. 10-14). In the T state, His FG4(97)β is in contact with

Thr C6(41)α (Figs. 10-13*a* and 10-17*a*), whereas in the R state it is in contact with Thr C3(38)α, one turn back along the C helix (Figs. 10-13*b* and 10-17*b*). In both conformations, the "knobs" on one subunit mesh nicely with the "grooves" on the other (Fig. 10-17). An intermediate position, however, would be severely strained because it would bring His FG4(97)β and Thr C6(41)α too close together (knobs on knobs). Hence *these contacts, which are joined by different but equivalent sets of hydrogen bonds in the two states (Fig. 10-17), act as a binary switch that permits only two stable positions of the subunits relative to each other.* In contrast, the quaternary change causes only a 1-Å shift at the α_1FG–β_2C contact, so its side chains maintain the same associations throughout the change. *These side chains therefore act as flexible joints or hinges about which the α_1 and β_2 subunits pivot during the quaternary change.*

c. The T State Is Stabilized by a Network of Salt Bridges That Must Break to Form the R State

The R state is stabilized by ligand binding. But in the absence of ligand, why is the T state more stable than the R state? In the electron density maps of R-state Hb, the C-terminal residues of each subunit (Arg 141α and His 146β) appear as a blur, which suggests that these residues are free to wave about in solution. Maps of the T form, however, show these residues firmly anchored in place via several intersubunit and intrasubunit salt bridges, which evidently help stabilize the T state (Fig. 10-18). *The structural changes accompanying the T → R transition tear away these salt bridges in a process driven by the Fe—O_2 bonds' energy of formation.*

d. Hemoglobin's O_2-Binding Cooperativity Derives from the T → R Conformational Shift

The hemoglobin molecule resembles a finely tooled mechanism that has very little slop. The binding of O_2 requires a series of tightly coordinated movements:

1. The Fe(II) of any subunit cannot move into its heme plane without the reorientation of its proximal His

so as to prevent this residue from bumping into the porphyrin ring.

2. The proximal His is so tightly packed by its surrounding groups that it cannot reorient unless this movement is accompanied by the previously described translation of the F helix across the heme plane.

3. The F helix translation is only possible in concert with the quaternary shift that steps the α_1C–β_2FG contact one turn along the α_1C helix.

4. The inflexibility of the α_1–β_1 and α_2–β_2 interfaces requires that this shift simultaneously occur at both the α_1–β_2 and the α_2–β_1 interfaces.

Consequently, *no one subunit or dimer can greatly change its conformation independently of the others.* Indeed, *the two stable positions of the α_1C–β_2FG contact limit the Hb molecule to only two quaternary forms, R and T.*

We are now in a position to structurally rationalize hemoglobin's O_2-binding cooperativity. Any deoxyHb subunit binding O_2 is constrained to remain in the t state by the T conformation of the tetramer. However, *the t state has reduced O_2 affinity,* most probably because its Fe—O_2 bond is stretched beyond its normal length by the steric repulsions between the heme and the O_2 and in the β subunits, by the need to move Val E11 out of the O_2-binding site. As more O_2 is bound to the Hb tetramer, this strain, which derives from the Fe—O_2 bond energy, accumulates in the liganded subunits until it is of sufficient strength to snap the molecule into the R conformation. *All the subunits are thereby converted to the r state whether or not they are liganded. Unliganded subunits in the r state have an increased O_2 affinity because they are already in the O_2-binding conformation.* This accounts for the high O_2 affinity of nearly saturated Hb.

(a) α Chains

(b) β Chains

Figure 10-18 Networks of salt bridges and hydrogen bonds in deoxyHb. These bonds, which involve the last two residues of *(a)* the α chains and *(b)* the β chains, are all ruptured in the T → R transition. The two groups that participate in the Bohr effect by

becoming partially deprotonated in the R state are indicated by white plus signs. [Illustration, Irving Geis. Image from the Irving Geis Collection, Howard Hughes Medical Institute. Reprinted with permission.]

e. Hemoglobin's Sigmoidal O₂-Binding Curve Is a Composite of Its Hyperbolic R- and T-State Curves

The relative stabilities of the T and R states, as indicated by their free energies, vary with fractional saturation (Fig. 10-19a). In the absence of ligand, the T state is more stable than the R state, and vice versa when all ligand-binding sites are occupied. The formation of Fe—O₂ bonds causes the free energy of both the T and the R states to decrease (become more stable) with oxygenation, although the rate of this decrease is smaller for the T state as a result of the strain that liganding imposes on t-state subunits. The R $\rightleftharpoons$ T transformation is, of course, an equilibrium process, so that Hb molecules, at intermediate levels of fractional saturation (1, 2, or 3 bound O₂ molecules), continually interconvert between the R and the T states.

The O₂-binding curve of Hb can be understood as a composite of those of its R and T states (Fig. 10-19b). For pure states, such as R or T, these curves are hyperbolic because ligand binding at one protomer is unaffected by the state of other protomers in the absence of a quaternary structural change. At low pO_2's, Hb follows the low-affinity T-state curve and at high pO_2's, it follows the high-affinity R-state curve. At intermediate pO_2's, Hb exhibits an O₂ affinity that changes from T-like to R-like as pO_2 increases. The switchover results in the sigmoidal shape of hemoglobin's O₂-binding curve.

D. *Testing the Perutz Mechanism*

The Perutz mechanism is a description of the dynamic behavior of Hb that is largely based on the static structures of its R and T end states. Accordingly, without the direct demonstration that Hb actually follows the postulated

pathway in changing conformational states, the Perutz mechanism must be taken as being at least partially conjectural. Unfortunately, the physical methods that can follow dynamic changes in proteins are, as yet, incapable of providing detailed descriptions of these changes. Nevertheless, certain aspects of the Perutz mechanism are supported by static measurements, as is described below and in Section 10-3.

a. C-Terminal Salt Bridges Are Required to Maintain the T State

The proposed function of the C-terminal salt bridges in stabilizing the T state has been corroborated by chemically modifying human Hb. Removal of the C-terminal Arg 141α (by treating isolated α chains with carboxypeptidase B followed by reconstitution) drastically reduces the cooperativity of O₂ binding (Hill constant of 1.7, reduced from its normal value of 2.8). Cooperativity is abolished by the further removal of the other C-terminal residue, His 146β (Hill constant of ~1.0). Apparently, in the absence of its C-terminal salt bridges, the T form of Hb is unstable. Indeed, human deoxy-Hb, with its C-terminal residues removed, crystallizes in a form very similar to that of normal human oxyHb.

b. Fe—O₂ Bond Tension Has Been Spectroscopically Demonstrated

If movement of the Fe into the heme plane on oxygenation is mechanically coupled via the proximal His to the T → R transformation, then conversely, forcing oxyHb into the T form must exert a tension on the Fe, through the proximal His, that tends to pull the Fe out of the heme plane. Perutz demonstrated the existence of this tension as follows.

Figure 10-19   Free energy and saturation curves for O₂ binding to hemoglobin. *(a)* The variation of the free energies of hemoglobin's T and R states with their fractional saturation, Y_{O_2}. In the absence of O₂ the T state is more stable, and when saturated with O₂ the R state is more stable. The free energy of both states is reduced with increasing oxygenation as a consequence of O₂ liganding. The Fe(II)—O₂ bonding is more

exergonic in the R state than it is in the T state, however, so that the relative stabilities of these two states reverse order at intermediate levels of oxygenation. *(b)* The sigmoid O₂-binding curve of Hb *(purple)* is a composite of its hyperbolic R-state *(red)* and T-state *(blue)* binding curves: It is more T-like at lower pO_2 values and more R-like at higher pO_2 values.

IHP's six phosphate groups cause it to bind to deoxyHb with much greater affinity than does BPG (the structural basis of BPG binding to Hb is discussed in Section 10-2F); the presence of IHP therefore tends to force Hb into the T state. Conversely, nitric oxide (NO) binds to Hb far more strongly than does O_2 and thereby tends to force Hb into the R state. Spectroscopic analysis indicates the consequences of simultaneously binding both NO and IHP to Hb:

1. The NO, as expected, pulls the Fe into the plane of the heme.

2. The IHP forces the Hb molecule into the T state, which through the "gears and levers" coupling the 4° and 3° conformational changes, pulls the proximal His in the opposite direction, away from the Fe.

The bond between the proximal His and the Fe lacks the strength to withstand these two opposing "irresistible" forces; it simply breaks. The spectroscopic observation of this phenomenon therefore confirms the existence of the heme–protein tension predicted by the Perutz mechanism.

c. Detaching the Proximal His from the F Helix Eliminates Most Cooperativity

In a further experimental investigation of the origin of cooperativity in hemoglobin, Chien Ho mutagenically changed the proximal His residue to Gly on only the α subunits, on only the β subunits, and on both the α and β subunits. The missing imidazole ring of the proximal His was then replaced by imidazole (which a variety of evidence indicates ligands the heme Fe as does the proximal His). This, in effect, detaches the proximal His from the protein, thereby cutting the covalent bond that, according to the Perutz model, links the ligand-induced movement of the Fe into the heme plane to the accompanying movement of helix F. In all three cases, this proximal detachment, in agreement with the Perutz model, significantly increases hemoglobin's ligand-binding affinity, reduces its cooperativity, and prevents its T → R quaternary switch. However, these mutant hemoglobins exhibit a small amount of residual cooperativity, suggesting that the heme groups also communicate via pathways that do not require covalent coupling between the F helix and the proximal His. These pathways may involve movements of protein groups in contact with the heme (see Figs. 10-12 and 10-15) in response to the subsidence of heme doming on ligand binding. They may also involve movements of the distal His residues of the α and β subunits, and/or the movement of Val E11 of the β subunits, all of whose side chains must move aside when ligand binds to Hb.

E. *Origin of the Bohr Effect*

The Bohr effect, hemoglobin's release of H^+ on binding O_2, is also observed when Hb binds other ligands. *It arises from pK changes of several groups caused by changes in their local environments that accompany hemoglobin's T → R transition.* The groups involved include the N-terminal amino groups of the α subunits and the C-terminal His of the β subunits. These have been identified through chemi-

Terminal amino group   **Cyanate**   **Carbamoylated terminal amino group**

Figure 10-20 **Reaction of cyanate with the unprotonated (nucleophilic) forms of primary amino groups.** At physiological pH's, N-terminal amino groups, which have pK's near 8.0, readily react with cyanate. Lys ε-amino groups (pK ≈ 10.8), however, are fully protonated under these conditions and are therefore unreactive.

cal and structural studies, and their quantitative contributions to the Bohr effect have been estimated.

Reaction of the α subunits of Hb with **cyanate** results in the specific **carbamoylation** of the N-terminal amino groups (Fig. 10-20). When such carbamoylated α subunits are mixed with normal β subunits, the resulting reconstituted Hb lacks 20 to 30% of the normal Bohr effect. The reason for this is seen on comparing the X-ray structure of deoxyHb with that of carbamoylated deoxyHb. In deoxyHb, a Cl^- ion binds between the N-terminal amino group of Val 1α₂ and the guanidino group of Arg 141α₁ (the C-terminal residue; Fig. 10-18a). This Cl^- is absent in carbamoylated deoxyHb. It is also absent in normal R-state Hb because its C-terminal residues are not held in place by salt bridges (which partially accounts for the preferential binding of Cl^- to deoxyHb; Section 10-1C). N-Terminal amino groups of polypeptides normally have pK's near 8.0. On deoxyHb α subunits, however, the N-terminal amino group is electrostatically influenced by its closely associated Cl^- to increase its positive charge by binding protons more tightly, that is, to increase its pK. Since at the pH of blood (7.4) N-terminal amino groups are normally only partially charged, this pK shift causes them to bind significantly more protons in the T state than in the R state.

The Hb β chain also contributes to the Bohr effect. Removal of its C-terminal residue, His 146β, reduces the Bohr effect by 40%. In normal deoxyHb, the imidazole ring of His 146β associates with the carboxylate of Asp 94β on the same subunit (Fig. 10-18b) to form a salt bridge that is absent in the R state. Proton NMR measurements indicate that formation of this salt bridge increases the pK of the imidazole group from 7.1 to 8.0. This effect more than accounts for His 146β's share of the Bohr effect.

We have not yet accounted for about 30 to 40% of the Bohr effect. It is largely due to small contributions from the numerous surface-exposed His residues whose environments are altered on hemoglobin's T → R transition [since His is the only residue with an intrinsic pK (6.04) in the physiological range, small changes in its pK will significantly alter the number of protons it binds]. Indeed, NMR measurements by Ho indicate that the T → R transition induces small shifts in the pK's of these various His residues, although, interestingly, some of these shifts are in the direction that diminishes the magnitude of the Bohr effect.

F. *Structural Basis of BPG Binding*

BPG decreases the oxygen-binding affinity of Hb by preferentially binding to its deoxy state (Section 10-1D). The binding of the physiologically quadruply charged BPG to deoxyHb is weakened by high salt concentrations, which suggests that this association is ionic in character. This explanation is corroborated by the X-ray structure of a BPG–deoxyHb complex, which indicates that BPG binds in the central cavity of deoxyHb on its 2-fold axis (Fig. 10-21). The anionic groups of BPG are within hydrogen bonding and salt bridging distances of the cationic Lys EF6(82), His H21(143), His NA2(2), and N-terminal amino groups of both β subunits (Fig. 10-21). The T → R transformation brings the two β H helices together, which narrows the central cavity (compare Figs. 10-13a and 10-13b) and expels the BPG. It also widens the distance between the β N-terminal amino groups from 16 to 20 Å, which prevents their simultaneous hydrogen bonding with BPG's phosphate groups. BPG therefore stabilizes the T conformation of Hb by crosslinking its β subunits. This shifts the T ⇌ R equilibrium toward the T state, which lowers hemoglobin's O_2 affinity.

The structure of the BPG–deoxyHb complex also indicates why fetal hemoglobin (HbF) has a reduced affinity for BPG relative to HbA (Section 10-1D). The cationic His H21(143)β of HbA is changed to an uncharged Ser residue in HbF's β-like γ subunit, thereby eliminating a pair of ionic interactions stabilizing the BPG–deoxyHb complex (Fig. 10-21).

The excess positive charge lining Hb's central cavity is also partially responsible for the allosteric effect of Cl^- ions in stabilizing the T state relative to the R state (the remainder being due to the participation of Cl^- in the T-state salt bridge networks; Fig. 10-18a). The central cavity is larger in the T state than in the R state (Fig. 10-13), so that more Cl^- ions occupy this channel in the T state than in the R state. The additional Cl^- ions, through electrostatic shielding, reduce the mutual repulsions of the positive charges, thereby stabilizing the T state.

G. *Role of the Distal Histidine Residue*

O_2 binding paradoxically protects the heme iron from autooxidation: The rate of Mb oxidation decreases as the partial pressure of O_2 increases. This is because heme iron oxidation is catalyzed by protons that are reduced by the heme iron and that in turn reduce O_2 in the solvent to **superoxide ion** (O_2^-). Bound O_2 evidently shields the Fe from the attacking protons.

The mutagenic replacement of the distal His residue in Mb by any other residue reduces Mb's oxygen affinity and increases its rate of autooxidation. Asp, a proton source, at this position increases the rate of Mb autooxidation by 350-fold, the largest increase of all residue replacements, whereas Phe, Met, and Arg provide only 50-fold accelerations, the smallest observed increases. The imidazole ring of the distal His, which has a pK of 5.5 and is therefore neutral at neutral pH and whose unprotonated N_ε atom faces the heme pocket (Fig. 10-12), acts as a proton trap, thereby

Figure 10-21 **Binding of BPG to deoxyHb.** The view is down the molecule's exact twofold axis (the same view as in Fig. 10-13a). BPG (*red*), with its five anionic groups, binds in the central cavity of deoxyHb, where it is surrounded by a ring of eight cationic side chains (*blue*) extending from the two β subunits. In the R state, the central cavity is too narrow to admit BPG (Fig. 10-13b). The arrangement of salt bridges and hydrogen bonds between the α_1 and β_2 subunits that partially stabilizes the T state (Fig. 10-18b) is indicated at the lower right. [Illustration, Irving Geis. Image from the Irving Geis Collection, Howard Hughes Medical Institute. Reprinted with permission.] **See Kinemage Exercise 6-3**

protecting the Fe from protons. Thus, to quote Perutz, "Evolution is a brilliant chemist."

3 ABNORMAL HEMOGLOBINS

Mutant hemoglobins provided the original opportunity to study structure–function relationships in proteins because Hb is a readily isolated protein of known structure that has a large number of well-characterized naturally occurring variants. The examination of individuals with physiological disabilities, together with the routine electrophoretic screening of human blood samples, has led to the discovery of over 1000 variant hemoglobins, >90% of which result from single amino acid substitutions in a globin polypeptide chain (a compendium of variant human hemoglobins is located at http://globin.cse.psu.edu/). In this section, we consider the nature of these **hemoglobinopathies.** Hemoglobin diseases characterized by defective globin synthesis, the **thalassemias,** are the subject of Section 34-2G. It should be noted that ~300,000 individuals with serious hemoglobin disorders are born every year and that ~5% of the world's population are carriers of an inherited variant hemoglobin.

A. *Molecular Pathology of Hemoglobin*

The physiological effect of an amino acid substitution on Hb can, in most cases, be understood in terms of its molecular location:

1. Changes in surface residues

Changes of surface residues are usually innocuous because most of these residues have no specific functional role [although sickle-cell Hb (HbS) is a glaring exception to this generalization; Section 10-3Ba]. For example, **HbE** [Glu B8(26)$\beta \rightarrow$ Lys], the most common human Hb mutant after HbS (possessed by up to 10% of the population in parts of Southeast Asia), has no clinical manifestations in either heterozygotes or homozygotes. About half of the known Hb mutations are of this type and have been discovered only accidentally or through surveys of large populations.

2. Changes in internally located residues

Changing an internal residue often destabilizes the Hb molecule. The degradation products of these hemoglobins, particularly those of heme, form granular precipitates (known as **Heinz bodies**) that hydrophobically adhere to the erythrocyte cell membrane. The membrane's permeability is thereby increased, causing premature cell lysis. Carriers of unstable hemoglobins therefore suffer from **hemolytic anemia** of varying degrees of severity.

The structure of Hb is so delicately balanced that small structural changes may render it nonfunctional. This can occur through the weakening of the heme–globin association or as a consequence of other conformational changes. For instance, the heme group is easily dislodged from its closely fitting hydrophobic binding pocket. This occurs in **Hb Hammersmith** (Hb variants are often named after the locality of their discovery), in which Phe CD1(42)β, an invariant residue that wedges the heme into its pocket (see Fig. 10-12), is replaced by Ser. The resulting gap permits water to enter the heme pocket, which causes the hydrophobic heme to drop out easily (Phe CD1 and the proximal His F8 are the only invariant residues among all known hemoglobins). Similarly, in **Hb Bristol,** the substitution of Asp for Val E11(67)β, which partially occludes the O_2 pocket, places a polar group in contact with the heme. This weakens the binding of the heme to the protein, probably by facilitating the access of water to the subunit's otherwise hydrophobic interior.

Hb may also be destabilized by the disruption of elements of its 2°, 3°, and/or 4° structures. The instability of **Hb Bibba** results from the substitution of a helix-breaking Pro for Leu H19(136)α. Likewise, the instability of **Hb Savannah** is caused by the substitution of Val for the highly conserved Gly B6(24)β, which is located on the B helix where it crosses the E helix with insufficient clearance for side chains larger than an H atom (Fig. 10-13). The α_1–β_1 contact, which does not significantly dissociate under physiological conditions, may do so on structural alteration. This occurs in **Hb Philly,** in which Tyr C1(35)β, which participates in the hydrogen bonded network that helps knit together the α_1–β_1 interface, is replaced by Phe.

3. Changes stabilizing methemoglobin

Changes at the O_2-binding site that stabilize the heme in the Fe(III) oxidation state eliminate the binding of O_2 to the defective subunits. Such methemoglobins are designated **HbM** and individuals carrying them are said to have **methemoglobinemia.** These individuals usually have bluish skin, a condition known as **cyanosis,** which results from the presence of deoxyHb in their arterial blood.

All known methemoglobins arise from substitutions that provide the Fe atom with an anionic oxygen atom ligand. In **Hb Boston,** the substitution of Tyr for His E7(58)α (the distal His, which protects the heme from oxidation; Section 10-2G) results in the formation of a 5-coordinate Fe(III) complex, with the phenolate ion of the mutant Tyr E7 displacing the imidazole ring of His F8(87) as the apical ligand (Fig. 10-22*a*). In **Hb Milwaukee,** the γ-carboxyl group of the Glu that replaces Val E11(67)β forms an ion pair with a 5-coordinate Fe(III) complex (Fig. 10-22*b*). Both the phenolate and glutamate ions in these methemoglobins so stabilize the Fe(III) oxidation state that methemoglobin reductase is ineffective in converting them to the Fe(II) form.

Individuals with HbM are alarmingly cyanotic and have blood that is chocolate brown, even when their normal subunits are oxygenated. In northern Japan, this condition is named "black mouth" and has been known for centuries; it is caused by the presence of **HbM Iwate** [His F8(87)$\alpha \rightarrow$ Tyr]. Methemoglobins have Hill constants of ~1.2. This indicates a reduced cooperativity in comparison with HbA even though HbM, which can bind only two oxygen molecules, can have a maximum Hill constant of 2 (the unmutated α or β chains remain functional). Surprisingly, heterozygotes with HbM, which have an average of one nonfunctional α or β subunit per Hb molecule, have no apparent physical disabilities. Evidently, the amount of O_2 released in their capillaries is within normal limits. Homozygotes of HbM, however, are unknown; this condition is, no doubt, lethal.

4. Changes at the α_1–β_2 contact

Changes at the α_1–β_2 contact often interfere with hemoglobin's quaternary structural changes. Most such hemoglobins have an increased O_2 affinity so that they release less than normal amounts of O_2 in the tissues. Individuals with such defects compensate for it by increasing their hematocrit (concentration of erythrocytes in their blood). This condition, which is named **polycythemia,** often gives them a ruddy complexion. Some amino acid substitutions at the α_1–β_2 interface instead result in a reduced O_2 affinity. Individuals carrying such hemoglobins are cyanotic.

Amino acid substitutions at the α_1–β_2 contact may change the relative stabilities of hemoglobin's R and T forms,

Figure 10-22 **Mutations stabilizing the Fe(III) oxidation state of heme.** (*a*) Alterations in the heme pocket of the α subunit on changing from deoxyHbA to Hb Boston [His E7(58)α → Tyr]. The phenolate ion of the mutant Tyr becomes the fifth ligand of the Fe atom, thereby displacing the proximal His [F8(87)*a*]. [After Pulsinelli, P.D., Perutz, M.F., and Nagel, R.L., *Proc. Natl. Acad. Sci.* **70,** 3870 (1973).] (*b*) The structure of the heme pocket of the β subunit in Hb Milwaukee [Val E11(67)β → Glu]. Here the mutant Glu residue's carboxyl group forms an ion pair with the heme iron atom so as to stabilize its Fe(III) state. [From Perutz, M.F., Pulsinelli, P.D., and Ranney, H.M., *Nature New Biol.* **237,** 259 (1972).]

thereby altering its O_2 affinity. For example, the replacement of Asp G1(99)β by His in **Hb Yakima** eliminates the hydrogen bond at the α_1–β_2 contact that stabilizes the T form of Hb (Fig. 10-17*a*). The interloping imidazole ring also acts as a wedge that pushes the subunits apart and displaces them toward the R state. This change shifts the T → R equilibrium almost entirely to the R state, which results in Hb Yakima having an increased O_2 affinity (p_{50} = 12 torr under physiological conditions vs 26 torr for HbA) and a total lack of cooperativity (Hill constant = 1.0). In contrast, the replacement of Asn G4(102)β by Thr in **Hb Kansas** eliminates the hydrogen bond in the α_1–β_2 contact that stabilizes the R state (Fig. 10-17*b*), so that this Hb variant remains in the T state on binding O_2. Hb Kansas therefore has a low O_2 affinity (p_{50} = 70 torr) and a low cooperativity (Hill constant = 1.3).

B. *Molecular Basis of Sickle-Cell Anemia*

Most harmful Hb variants occur in only a few individuals, in many of whom the mutation apparently originated. However, ~10% of American blacks and as many as 25% of African blacks are heterozygotes for **sickle-cell hemoglobin (HbS).** HbS arises, as we have seen (Section 7-3Aa), from the substitution of a hydrophobic Val residue for the hydrophilic surface residue Glu A3(6)β (Fig. 10-13). The prevalence of HbS results from the protection it affords heterozygotes against malaria. However, homozygotes for HbS, of which there are ~ 50,000 in the United States, are severely afflicted by hemolytic anemia together with painful, debilitating, and sometimes fatal blood flow blockages

caused by the irregularly shaped and inflexible erythrocytes characteristic of the disease (Fig. 7-19*b*).

a. HbS Fibers Are Stabilized by Intermolecular Contacts Involving Val β6 and Other Residues

The sickling of HbS-containing erythrocytes results from the aggregation (polymerization) of deoxyHbS into rigid fibers that extend throughout the length of the cell (Fig. 10-23).

Figure 10-23 **Electron micrograph of deoxyHbS fibers spilling out of a ruptured erythrocyte.** [Courtesy of Robert Josephs, University of Chicago.]

Figure 10-24 **The 220-Å-diameter fibers of deoxyHbS.** (*a*) An electron micrograph of a negatively stained fiber. The accompanying cutaway interpretive drawing indicates the relationship between the inner and outer strands; each sphere represents an individual HbS molecule. The fiber has a layer repeat distance of 64 Å and a moderate twist such that it repeats every 350 Å along the fiber axis. [Courtesy of Stuart Edelstein, University of Geneva.] (*b*) A model, viewed in cross section, of the HbS fiber based on the crystal structure of HbS and three-dimensional reconstructions of electron micrographs of HbS fibers. The residues in the 14 HbS molecules are represented by spheres centered on their C_α positions. The residues making inter-double strand, intra-double strand lateral, and intra-double strand axial contacts are colored red, green, and blue, respectively, with lighter and darker toned residues making intermolecular contacts of <8 Å and <5 Å, respectively. The α and β chain residues outside the contact regions are colored white. [Courtesy of Stanley Watowich, Leon Gross, and Robert Josephs, University of Chicago.]

Electron microscopy indicates that these fibers are ~220-Å-diameter elliptical rods consisting of 14 hexagonally packed and helically twisting strands of deoxyHbS molecules that associate in parallel pairs (Figs. 10-24 and 10-25*a*).

The structural relationship among the HbS molecules in the pairs of parallel HbS strands has been established by the X-ray structure analysis of deoxyHbS crystals. When this crystal structure was first determined, it was unclear whether the intermolecular contacts in the crystal resembled those in the fiber. However, the subsequent observation that HbS fibers slowly convert to these crystals with little change in their overall X-ray diffraction pattern indicates that the fibers structurally resemble the crystals. The crystal structure of deoxyHbS consists of double filaments of HbS molecules whose several different intermolecular contacts are diagrammed in Fig. 10-25*b*. Only one of the two Val 6β's per Hb molecule contacts a neighboring molecule. In this contact, the mutant Val side chain occupies a hydrophobic surface pocket on the β subunit of an adja-

cent molecule whose Val 6β does not make an intermolecular contact (Fig. 10-25*c*). This pocket is absent in oxyHb. Other contacts involve residues that also occur in HbA, including Asp 73β and Glu 23α (Fig. 10-25*b*). The observation that deoxyHbA does not aggregate into fibers, however, even at very high concentrations, indicates that *the contact involving Val 6β is essential for fiber formation.* This conclusion is corroborated by the observation that a genetically engineed human Hb in which Glu 6β is replaced by Ile (which differs from Val by an additional CH_2 group and is therefore even more hydrophobic) has half the solubility of HbS in 1.8*M* phosphate.

The importance of the other intermolecular contacts to the structural integrity of HbS fibers has been demonstrated by studying the effects of other mutant hemoglobins on HbS gelation (polymerization). For example, the doubly mutated **Hb Harlem** (Glu 6β → Val + Asp 73β → Asn) requires a higher concentration to gel than does HbS (Glu 6β → Val); similarly, mixtures of HbS and **Hb Korle-Bu** (Asp 73β → Asn) gel less readily than equivalent

Figure 10-25 **Structure of the deoxyHbS fiber.** (*a*) The arrangement of the deoxyHbS molecules in the fiber. The yellow dots represent the side chains of Glu $6\beta_2$. [Illustration, Irving Geis/Geis Archives Trust. Copyright Howard Hughes Medical Institute. Reproduced with permission.] (*b*) A schematic diagram indicating the intermolecular contacts in the crystal structure of deoxyHbS. The white-lettered residues are implicated in forming these contacts. Note that the only intermolecular association in which the mutant residue Val 6β participates involves subunit β_2;

Val 6 of subunit β_1 is free. [After Wishner, B.C., Ward, K.B., Lattman, E.E., and Love, W.E., *J. Mol. Biol.* **98,** 192 (1975).] (*c*) The mutant Val $6\beta_2$ fits neatly into a hydrophobic pocket formed mainly by Phe 85 and Leu 88 of an adjacent β_1 subunit. This pocket, which is located between helices E and F at the periphery of the heme pocket, is absent in oxyHb and is too hydrophobic to contain the normally occurring Glu 6β side chain. [Illustration, Irving Geis. Image from the Irving Geis Collection, Howard Hughes Medical Institute. Reprinted with permission.]

mixtures of HbS and HbA. These observations suggest that Asp 73β occupies an important intermolecular contact site in HbS fibers (Fig. 10-25*b*). Likewise, the observation that hybrid tetramers consisting of α subunits from **Hb Memphis** (Glu $23\alpha \rightarrow$ Gln) and β subunits from HbS gel less readily than does HbS indicates that Glu 23α also participates in the polymerization of HbS fibers (Fig. 10-25*b*). The other white-lettered residues in Fig. 10-25*b* have been similarly implicated in sickling interactions.

b. The Initiation of HbS Gelation Is a Complex Process

The gelation of HbS, both in solution and within the red cell, follows an unusual time course. A solution of HbS can be brought to conditions under which it will gel by lowering the pO$_2$, raising the HbS concentration, and/or raising the temperature. *On achieving gelation conditions, there is a reproducible delay that varies according to conditions from milliseconds to days: During this time, no HbS fibers can be*

(a)

(b)

Figure 10-26 **Time course of deoxyHbS gelation.** (*a*) The extent of gelation as monitored calorimetrically (*yellow*) and optically (*purple*). Gelation of the 0.233 g · mL^{-1} deoxyHbS solution was initiated by rapidly increasing the temperature from 0°C, where HbS is soluble, to 20°C; t_d is the delay time. (*b*) A

log–log plot showing the concentration dependence of $1/t_d$ for the gelation of deoxyHbS at 30°C. The slope of this line is ~30. [After Hofrichter, J., Ross, P.D., and Eaton, W.A., *Proc. Natl. Acad. Sci.* **71**, 4865, 4867 (1974).]

detected. Only after the delay do fibers first appear, and gelation is then completed in about half the delay time (Fig. 10-26*a*).

William Eaton and James Hofrichter discovered that the delay time, t_d, has a concentration dependence described by

$$\frac{1}{t_d} = k \left(\frac{c_t}{c_s} \right)^n \qquad [10.13]$$

where c_t is the total deoxyHbS concentration prior to gelation, c_s is the solubility of deoxyHbS measured after gelation is complete, and k and n are constants. Graphical analysis of the data indicates that $k \approx 10^{-7}$ s^{-1} and that n is between 30 and 50 (Fig. 10-26*b*). This is a remarkable result: *No other known solution process even approaches a 30th power concentration dependence.*

A two-stage process accounts for Eq. [10.13]:

1. At first, HbS molecules sequentially aggregate to form a **nucleus** consisting of *m* HbS molecules (Fig. 10-27*a*):

$$\text{HbS} \rightleftharpoons (\text{HbS})_2 \rightleftharpoons (\text{HbS})_3 \rightleftharpoons \cdots$$
$$\rightleftharpoons (\text{HbS})_m \longrightarrow \text{Growth}$$

Prenuclear aggregates are unstable and easily decompose, but once a nucleus has formed it assumes a stable structure that rapidly elongates to form an HbS fiber.

2. Once a fiber has formed, it can nucleate the growth of other fibers (Fig. 10-27*b*). These newly formed fibers, in turn, nucleate the growth of yet other fibers, etc., so that this latter process is autocatalytic.

The initial **homogeneous nucleation** process (taking place in solution) accounts for the very high concentration dependence in Eq. [10.13], whereas the secondary **heterogeneous nucleation** process (taking place on a surface—that of a fiber in this case) is responsible for the rapid onset of gelation (Fig. 10-26*a*).

The foregoing kinetic hypothesis suggests why sickle-cell anemia is characterized by episodic "crises" caused by blood flow blockages. HbS fibers dissolve essentially instantaneously on oxygenation, so that none are present in arterial blood. Erythrocytes take from 0.5 to 2 s to pass through the capillaries, where deoxygenation renders HbS insoluble. If the delay time, t_d, for sickling is greater than this transit time, no blood flow blockage occurs (although sickling that occurs in the veins damages the erythrocyte membrane). However, Eq. [10.13] indicates that small increases in HbS concentration, c_t, and/or small decreases in HbS solubility, c_s, caused by conditions known to trigger sickle-cell crises, such as dehydration, O$_2$ deprivation, and fever, result in significant decreases of t_d. Once a blockage occurs, the resulting lack of O$_2$ and slowdown of blood flow in the area compound the situation.

The kinetic hypothesis of sickling has profound clinical implications for the treatment of sickle-cell anemia. Heterozygotes of HbS, whose blood usually contains ~60% HbA and 40% HbS, rarely show any symptoms of sickling. The t_d for the gelation of their Hb is ~10^6-fold greater than that of homozygotes. Accordingly, a treatment of sickle-cell anemia that increases t_d by this amount, which corresponds to decreasing the ratio c_t/c_s by a factor of ~1.6, would relieve the symptoms of this disease. This has suggested three different therapeutic strategies (besides gene therapy; Section 5-5Hb) to increase t_d, and thus inhibit HbS gelation:

1. The disruption of intermolecular interactions, thus increasing c_s. Of particular interest are compounds that have been designed with the aid of the X-ray structure of HbS to bind stereospecifically to its intermolecular contact regions. However, a large amount of any such compound would be necessary to bind to the ~400 g of hemoglobin in the human body. Consequently, no antisickling drug yet tested has had a sufficiently high ratio of efficacy to toxicity to merit clinical use.

(a) Homogeneous nucleation

(b) Heterogeneous nucleation

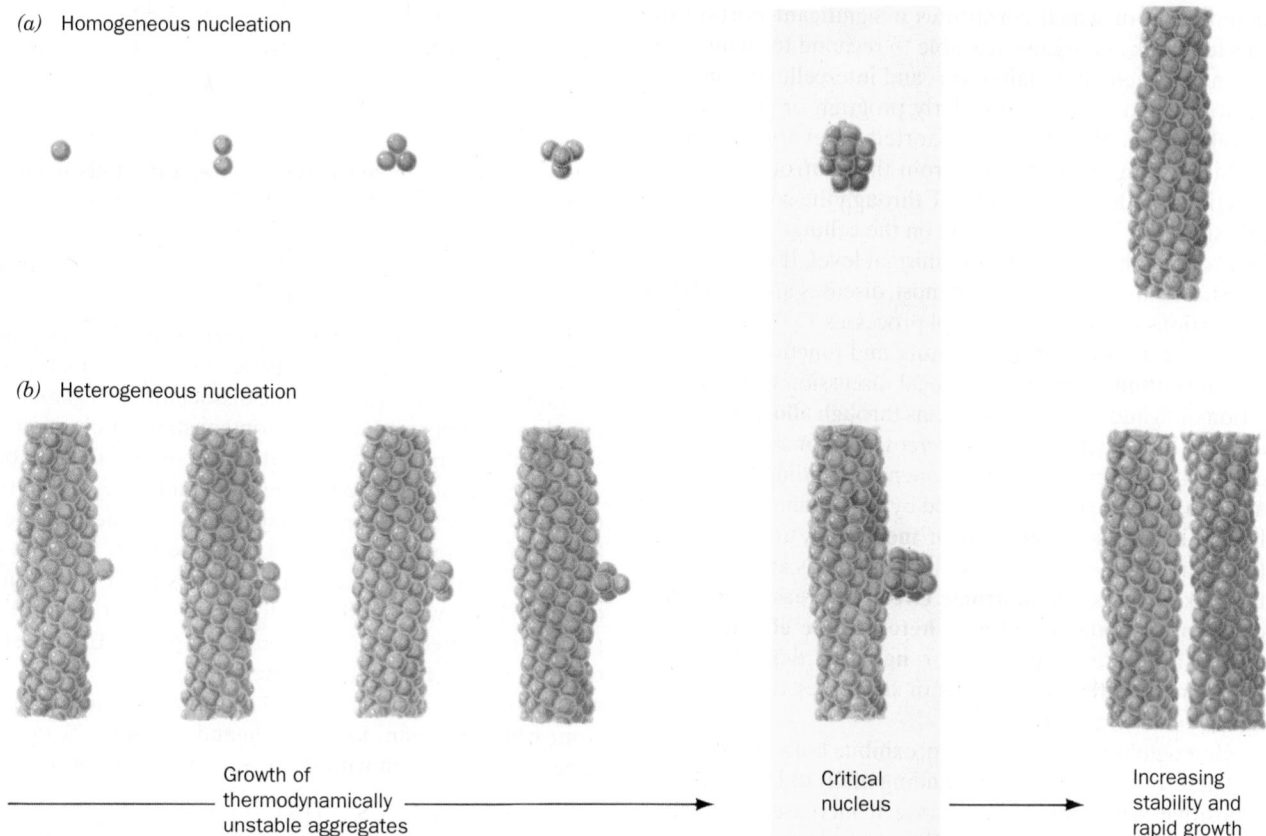

Growth of
thermodynamically
unstable aggregates

Critical
nucleus

Increasing
stability and
rapid growth

Figure 10-27 **Double nucleation mechanism for deoxyHbS gelation.** (*a*) The initial aggregation of HbS molecules (*spheres*) occurs very slowly because this process is thermodynamically unfavorable and hence the intermediates tend to decompose rather than grow. However, once an aggregate reaches a certain size, the **critical nucleus,** its further growth becomes thermodynamically favorable, leading to rapid fiber formation. (*b*) Each fiber, in turn, can nucleate the growth of other fibers, leading to the explosive appearance of polymer. [After Ferrone, F.A., Hofrichter, J., and Eaton, W.A., *J. Mol. Biol.* **183,** 614 (1985).]

2. The use of agents that increase hemoglobin's O_2 affinity, thus decreasing c_t. For example, the administration of cyanate carbamoylates the N-terminal amino groups of Hb (Fig. 10-20). This treatment eliminates some of the salt bridges that stabilize the T state (Section 10-2E) and thereby increases the O_2 affinity of Hb. Although cyanate is an effective *in vitro* antisickling agent, its clinical use has been discontinued because of toxic side effects, cataract formation and peripheral nervous system damage, that probably result from the carbamoylation of proteins other than Hb.

3. Lowering the HbS concentration (c_t) in erythrocytes. Agents that alter erythrocyte membrane permeability so as to permit the influx of water have promise in this regard.

The first, and as yet the only, effective treatment for sickle-cell anemia is a variation of the latter strategy through the administration of **hydroxyurea.**

$$H_2N-\overset{\overset{\displaystyle O}{\|}}{C}-NH-OH$$

Hydroxyurea

Adults with sickle-cell anemia have two types of red blood cells: S cells, which contain only HbS; and F cells, which contain ~20% HbF and the remainder HbS. In most adults, the fraction of F cells is ~30%. However, in those treated with hydroxyurea, this fraction increases to ~50%. Although the mechanism by which hydroxyurea stimulates the production of F cells is unknown, the mechanism by which increased levels of F cells prevent sickling seems clear. F cells contain three species of hemoglobin: HbS ($\alpha_2\beta_2^S$), HbF ($\alpha_2\gamma_2$), and their hybrid ($\alpha_2\beta^S\gamma$), where β^S subunits are the sickle-cell variants of the normal β subunits. Since neither HbF nor the $\alpha_2\beta^S\gamma$ hybrid Hb can form sickle-cell fibers, they act to dilute the HbS in a cell. This, in turn, increases the time it takes the F cells to sickle by a factor of ~1000, so that F cells do not significantly sickle in the period (10–20 s) it takes them pass from the tissues to the lungs, where they are oxygenated. Thus, the greater the proportion of F cells in the blood, the smaller the proportion of S cells that can sickle.

4 ALLOSTERIC REGULATION

One of the outstanding characteristics of life is the high degree of control exercised in almost all of its processes. Through a great variety of regulatory mechanisms, the exploration of which constitutes a significant portion of this textbook, an organism is able to respond to changes in its environment, maintain intra- and intercellular communications, and execute an orderly program of growth and development. Regulation is exerted at every organizational level in living systems, from the control of rates of reactions on the molecular level, through the control of expression of genetic information on the cellular level, to the control of behavior on the organismal level. It is therefore not surprising that many, if not most, diseases are caused by aberrations in biological control processes.

Our exploration of the structure and function of hemoglobin continues with a theoretical discussion of the regulation of ligand binding to proteins through **allosteric interactions** (Greek: *allos,* other + *stereos,* solid or space). These cooperative interactions occur when the binding of one ligand at a specific site is influenced by the binding of another ligand, known as an **effector** or **modulator,** at a different (allosteric) site on the protein. If the ligands are identical, this is known as a **homotropic effect,** whereas if they are different, it is described as a **heterotropic effect.** These effects are termed **positive** or **negative** depending on whether the effector increases or decreases the protein's ligand-binding affinity.

Hemoglobin, as we have seen, exhibits both homotropic and heterotropic effects. The binding of O_2 to Hb results in a positive homotropic effect since it increases hemoglobin's O_2 affinity. In contrast, BPG, CO_2, H^+, and Cl^- are negative heterotropic effectors of O_2 binding to Hb because they decrease its affinity for O_2 (negative) and are chemically different from O_2 (heterotropic). The O_2 affinity of Hb, as we have seen, depends on its quaternary structure. *In general, allosteric effects result from interactions among subunits of oligomeric proteins.*

Even though hemoglobin catalyzes no chemical reaction, it binds ligands in the same manner as do enzymes. Since an enzyme cannot catalyze a reaction until after it has bound its **substrate(s)** [the molecule(s) undergoing reaction], the enzyme's catalytic rate varies with its substrate-binding affinity. Consequently, the cooperative binding of O_2 to Hb is taken as a model for the allosteric regulation of enzyme activity. Indeed, in this section, we shall consider several models of allosteric regulation that, for the most part, were formulated to explain the O_2-binding properties of Hb. Following this, we shall compare these models with the realities of Hb behavior.

A. *The Adair Equation*

The derivation of the Hill equation (Section 10-1B) is predicated on the assumption of all-or-none O_2 binding. The observation of partially oxygenated Hb molecules, however, led Gilbert Adair, in 1924, to propose that the binding of ligands to proteins occurs sequentially with dissociation constants that are not necessarily equal. The expression for the saturation function under this model is straightforwardly derived.

For a protein such as Hb with four ligand-binding sites, the reaction sequence is

$$
\begin{aligned}
E + S &\rightleftharpoons ES & k_1 &= 4K_1 \\
ES + S &\rightleftharpoons ES_2 & k_2 &= \tfrac{3}{2}K_2 \\
ES_2 + S &\rightleftharpoons ES_3 & k_3 &= \tfrac{2}{3}K_3 \\
ES_3 + S &\rightleftharpoons ES_4 & k_4 &= \tfrac{1}{4}K_4
\end{aligned}
$$

where the K_i are the **macroscopic** or **apparent dissociation constants** for binding the ith ligand to the protein,

$$
K_i = \frac{[ES_{i-1}][S]}{[ES_i]} \qquad [10.14]
$$

and the k_i are the **microscopic** or **intrinsic dissociation constants,** that is, the individual dissociation constants for the ligand-binding sites. The intrinsic dissociation constants are equal to the apparent dissociation constants multiplied by **statistical factors,** $4, \tfrac{3}{2}, \tfrac{2}{3}$, and $\tfrac{1}{4}$, that account for the number of ligand-binding sites on the protein molecule. The statistical factor 4 derives from the fact that a tetrameric protein E bears four sites that can bind ligand to form ES (that is, the concentration of ligand-binding sites is 4[E]) but only one site from which ES can dissociate ligand to form E (that is, the concentration of bound ligand is 1[E]); the statistical factor $\tfrac{3}{2}$ is a result of there being three remaining sites on ES that can bind ligand to form ES_2 and two sites from which ES_2 can dissociate ligand to form ES; etc. In general, for a protein with n equivalent binding sites:

$$
k_i = \frac{(n-i+1)[ES_{i-1}][S]}{i[ES_i]} = \left(\frac{n-i+1}{i}\right)K_i \quad [10.15]
$$

since $(n-i+1)[ES_{i-1}]$ is the concentration of free ligand-binding sites in ES_{i-1} and $i[ES_i]$ is the concentration of bound ligand on ES_i. Therefore, solving sequentially for the concentration of each protein–ligand species in a tetrameric protein, we obtain:

$$
\begin{aligned}
[ES] &= [E][S]/K & &= 4[E][S]/k_1 \\
[ES_2] &= [ES][S]/K_2 = \tfrac{3}{2}[ES][S]/k_2 &&= 6[E][S]^2/k_1k_2 \\
[ES_3] &= [ES_2][S]/K_3 = \tfrac{2}{3}[ES_2][S]/k_3 &&= 4[E][S]^3/k_1k_2k_3 \\
[ES_4] &= [ES_3][S]/K_4 = \tfrac{1}{4}[ES_3][S]/k_4 &&= [E][S]^4/k_1k_2k_3k_4
\end{aligned}
$$

The fractional saturation of ligand binding, the fraction of occupied ligand-binding sites divided by the total concentration of ligand-binding sites, is expressed

$$
Y_S = \frac{[ES] + 2[ES_2] + 3[ES_3] + 4[ES_4]}{4([E] + [ES] + [ES_2] + [ES_3] + [ES_4])} \quad [10.16]
$$

so that, substituting in the above relationships and canceling terms, we obtain

$$
Y_S = \frac{\dfrac{[S]}{k_1} + \dfrac{3[S]^2}{k_1k_2} + \dfrac{3[S]^3}{k_1k_2k_3} + \dfrac{[S]^4}{k_1k_2k_3k_4}}{1 + \dfrac{4[S]}{k_1} + \dfrac{6[S]^2}{k_1k_2} + \dfrac{4[S]^3}{k_1k_2k_3} + \dfrac{[S]^4}{k_1k_2k_3k_4}} \quad [10.17]
$$

This is the **Adair equation** for four ligand-binding sites. Equations describing ligand binding to proteins with different numbers of binding sites are similarly derived.

If the microscopic dissociation constants of the Adair equation are not equal, the fractional saturation curve will

Table 10-2 **Adair Constants for Hemoglobin A at pH 7.40**

Solution	k_1 (torr)	k_2 (torr)	k_3 (torr)	k_4 (torr)
Stripped	8.8	6.1	0.85	0.25
0.1*M* NaCl	41.	13.	12.	0.14
2 m*M* BPG	74.	112.	23.	0.24
0.1*M* NaCl + 2 m*M* BPG	97.	43.	119.	0.09

Source: Tyuma, I., Imai, K., and Shimizu, K., *Biochemistry* **12**, 1493, 1495 (1973).

describe cooperative ligand binding. Decreasing and increasing values of these constants lead to positive and negative cooperativity, respectively. Of course, the values of the microscopic dissociation constants may also alternate so that, for example, $k_1 < k_2 > k_3 < k_4$.

In our discussion of the O_2-dissociation curve of Hb (Section 10-1B), we have seen how its values of k_1 and k_4 may be obtained by extrapolating the lower and upper asymptotes of the Hill plot to the log pO_2 axis. The remaining microscopic dissociation constants can be evaluated by fitting Eq. [10.17] to the Hill plot. The values of these **Adair constants** for Hb are given in Table 10-2. Note that k_4 is relatively insensitive to the presence of BPG. Hb therefore binds and releases its last O_2 almost independently of the BPG concentration.

Although the Adair equation is the most general relationship describing ligand binding to a protein and is widely used to do so, it provides no physical insight as to why the various microscopic dissociation constants differ from each other. Yet, if the protein consists, as so many do, of identical subunits that are symmetrically related, it is desirable to understand how ligand binding at one site influences the ligand-binding affinity at a seemingly identical site. This need led to the development of models for ligand binding that rationalize how the binding sites of oligomeric proteins can exhibit different affinities. Two of these models are described in the following sections.

B. *The Symmetry Model*

Perhaps the most elegant model for describing cooperative ligand binding to a protein is the **symmetry model** of allosterism, which was formulated in 1965 by Jacques Monod, Jeffries Wyman, and Jean-Pierre Changeux. This model, alternatively termed the **MWC model,** is defined by the following rules:

1. An allosteric protein is an oligomer of protomers that are symmetrically related (for hemoglobin, we shall assume, for the sake of algebraic simplicity, that all four subunits are functionally identical).

2. Each protomer can exist in (at least) two conformational states, designated T and R; these states are in equilibrium whether or not ligand is bound to the oligomer.

3. The ligand can bind to a protomer in either conformation. *Only the conformational change alters the affinity of a protomer for the ligand.*

4. *The molecular symmetry of the protein is conserved during conformational change.* Protomers must therefore change conformation in a concerted manner, which implies that the conformation of each protomer is constrained by its association with the other protomers; in other words, there are no oligomers that simultaneously contain R- and T-state protomers.

For a ligand S and an allosteric protein consisting of n protomers, these rules imply the following equilibria for conformational conversion and ligand-binding reactions (for the sake of brevity, $T_i \equiv TS_i$ and $R_i \equiv RS_i$).

$$T_0 \rightleftharpoons R_0$$

$$\begin{aligned} T_0 + S &\rightleftharpoons T_1 & R_0 + S &\rightleftharpoons R_1 \\ T_1 + S &\rightleftharpoons T_2 & R_1 + S &\rightleftharpoons R_2 \quad [10.18] \\ &\vdots & &\vdots \\ T_{n-1} + S &\rightleftharpoons T_n & R_{n-1} + S &\rightleftharpoons R_n \end{aligned}$$

This is illustrated in Fig. 10-28 for a tetramer.

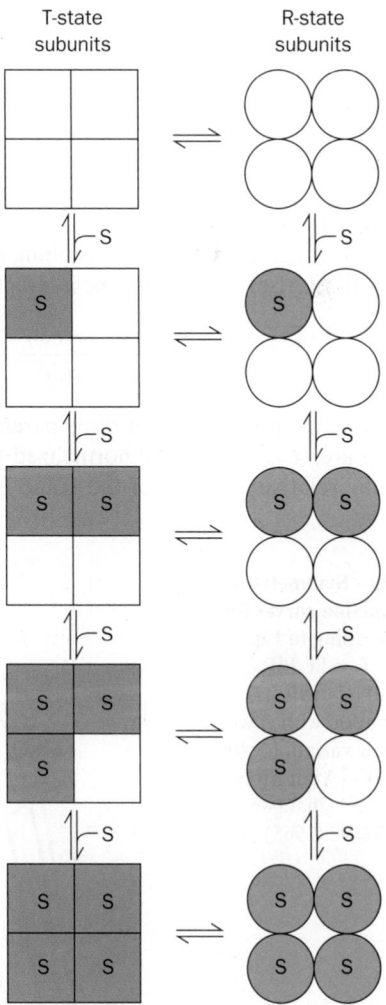

Figure 10-28 **The species and reactions permitted under the symmetry model of allosterism.** Squares and circles represent T- and R-state protomers, respectively.

The equilibrium constant L for the conformational interconversion of the oligomeric protein in the absence of ligand is expressed

$$L = \frac{[T_0]}{[R_0]} \tag{10.19}$$

The microscopic dissociation constant for the R state, k_R, which according to Rule 3 is independent of the number of ligands bound to R, is expressed according to Eq. [10.15]:

$$k_R = \left(\frac{n - i + 1}{i}\right)\frac{[R_{i-1}][S]}{[R_i]} \quad (i = 1, 2, 3, \ldots, n) \tag{10.20}$$

The microscopic dissociation constant for ligand binding to the T state, k_T, is similarly expressed. The fractional saturation, Y_s, for ligand binding is

$$Y_S = \frac{([R_1] + 2[R_2] + \cdots + n[R_n]) + ([T_1] + 2[T_2] + \cdots + n[T_n])}{n\{([R_0] + [R_1] + \cdots + [R_n]) + ([T_0] + [T_1] + \cdots [T_n])\}} \tag{10.21}$$

We shall make two definitions:

$$\alpha = [S]/k_R \qquad c = k_R/k_T$$

α may be considered a normalized ligand concentration. c is the ratio of the ligand-binding dissociation constants; c increases with the ligand-binding affinity of the T state relative to that of the R state. Then, combining the foregoing relationships as is shown in Section A of the Appendix to this chapter, we obtain the equation describing the symmetry model of allosterism for homotropic interactions:

$$Y_S = \frac{\alpha(1 + \alpha)^{n-1} + Lc\alpha(1 + c\alpha)^{n-1}}{(1 + \alpha)^n + L(1 + c\alpha)^n} \tag{10.22}$$

Note that this equation depends on three parameters, α, c, and L, which are, respectively, the normalized ligand concentration, the relative affinities of the T and R states for ligand, and the relative stabilities of the T and R states. In contrast, the Hill equation (Section 10-1B) has but two parameters, K and n, whereas the number of parameters in the Adair equation is equal to the number of ligand-binding sites on the protein.

a. Homotropic Interactions

Let us examine the nature of the symmetry model by plotting Eq. [10.22] for a tetramer ($n = 4$) as a function of α for different values of the parameters L and c (Fig. 10-29). Three major points are evident from an inspection of these plots:

1. The degree of upward curvature exhibited by the initial sections of these sigmoid curves is indicative of their level of cooperativity.

2. When only the R state binds ligand ($c = 0$), the ligand-binding cooperativity increases as the oligomer's conformational preference for the non–ligand-binding T state increases (L increases; Fig. 10-29a). *For high L values, if a single ligand is to bind, it must "force" the protein into its less preferred R state. The requirement that all protomers change their conformational states in a concerted manner causes the remaining three ligand-binding sites to become available.* The binding of the first ligand therefore promotes the binding of subsequent ligands, which is the essence of a positive homotropic effect. Note that cooperativity and ligand-binding affinity are different quantities; in fact, for $c = 0$, curves indicative of high ligand-binding affinity (those with low L) exhibit low cooperativity and vice versa.

3. When the T state is highly preferred (L is large), ligand-binding cooperativity increases with the R state's ligand-binding affinity relative to that of the T state (decreasing c; Fig. 10-29b). At low ligand concentrations (low α) the amount of ligand bound (Y_S) increases with the ligand-binding affinity of the T state (increasing c) since the protein is largely in the T state. As α increases, however, the amount of ligand bound to the intrinsically less stable R state eventually surpasses that of the T state, thereby resulting in a cooperative effect. This is because *the free*

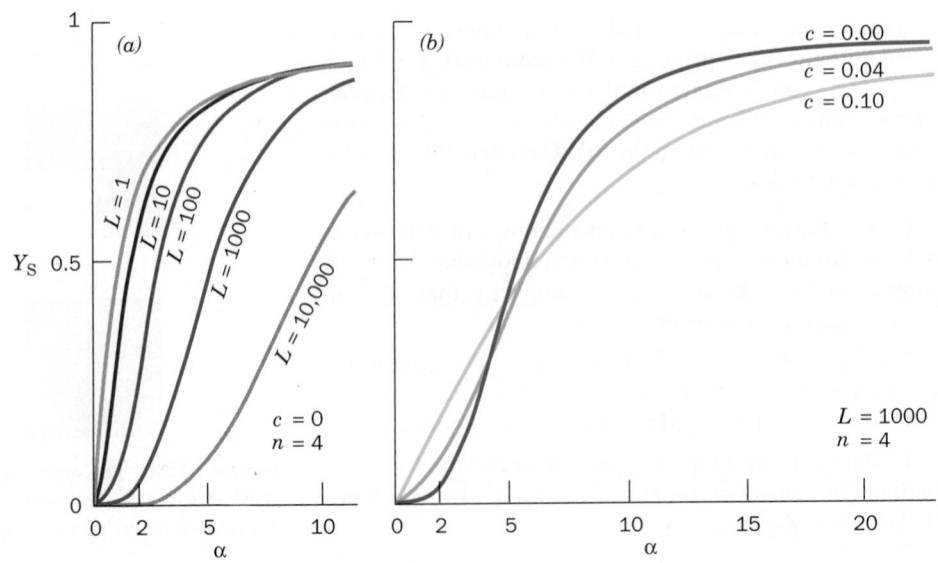

Figure 10-29 **Symmetry model saturation function curves for tetramers according to Eq. [10.22].** Here $L = [T_0]/[R_0]$, $c = k_R/k_T$, and $\alpha = [S]/k_R$. (a) Their variation with L when $c = 0$. (b) Their variation with c when $L = 1000$. [After Monod, J., Wyman, J., and Changeux, J.P., *J. Mol. Biol.* **12**, 92 (1965).]

energy of ligand binding stabilizes the R state with respect to the T state.

b. Heterotropic Interactions

The symmetry model of allosterism is also capable of accounting for heterotropic effects. This comes about by assuming that each protomer has specific and independent binding sites for the three types of ligands: a substrate, S, that for simplicity let us assume binds only to the R state ($c = 0$); an **activator,** A, that also binds only to the R state; and an **inhibitor, I,** that binds only to the T state (Fig. 10-30). Then, through the derivation in Section B of the Appendix to this chapter, we obtain a more general equation for the symmetry model that describes heterotropic interactions as well as homotropic interactions:

$$Y_S = \frac{\alpha(1 + \alpha)^{n-1}}{(1 + \alpha)^n + \dfrac{L(1 + \beta)^n}{(1 + \gamma)^n}} \quad [10.23]$$

where $\alpha = [S]/k_R$ as before and, analogously, $\beta = [I]/k_I$ and $\gamma = [A]/k_A$.

Note that this equation differs from Eq. [10.22] for $c = 0$ only in that the second term in the denominator is modulated by terms related to the amounts of activator and inhibitor bound to the oligomer.

Figure 10-31 indicates the consequences of effector binding to a tetramer that follows this model:

1. Activator binding ($\gamma > 0$) increases the concentration of the substrate-binding R state (the second term in the denominator of Eq. [10.23] decreases) because it is the only state capable of binding activator. *The presence of activator therefore increases the protein's substrate-binding affinity* (a positive heterotropic effect), although it decreases the protein's degree of substrate-binding cooperativity (compare Curves 1 and 2 in Fig. 10-31). (*Note:* There is nothing in the derivation of Eq. [10.23] that differentiates the roles of substrate and activator; consequently, the substrate and the activator each bind to the protein with a positive homotropic effect as well as being positive heterotropic effectors of each other.)

2. *The presence of inhibitor ($\beta > 0$), which only binds to the T state, reduces the binding affinity for substrate* (a negative heterotropic effect) by increasing the concentration of the T state (the second term in the denominator of Eq. [10.23] increases). Therefore, since substrate must "work harder" to convert the oligomer to the substrate-binding R state, inhibitor increases the cooperativity of substrate binding (compare Curves 2 and 3 of Fig. 10-31), as well as that for activator binding.

The model derived here is a rather simple one. In a more realistic but algebraically much more complicated symmetry model, all types of ligands would bind to both conformational states of the oligomer. Nevertheless, this model demonstrates that *both homotropic and heterotropic effects can be explained solely by the requirement that the molecular symmetry of the oligomer be conserved rather than by the existence of any direct interactions between ligands.* In Section 10-4D, we compare the theoretical predictions of

Figure 10-30 **Heterotropic interactions in the symmetry model of allosterism.** Heterotropic effects arise when substrates and activators bind exclusively (or at least preferentially) to the R state (*right*), and inhibitors bind exclusively (or at least preferentially) to the T state (*left*). The binding of substrate and/or activator to the oligomer therefore facilitates the further binding of substrate and activator. Conversely, the binding of inhibitor prevents (or at least inhibits) the oligomer from binding substrate or activator.

the symmetry model with our experimentally based model of hemoglobin oxygen binding.

C. *The Sequential Model*

The symmetry model provides a reasonable rationalization for the ligand-binding properties of many proteins. There are, however, several valid objections to it. Foremost of these is that it is difficult to believe that oligomeric symmetry is invariably preserved in all proteins so that there are never any hybrid conformations such as $R_{n-2}T_2$. Furthermore,

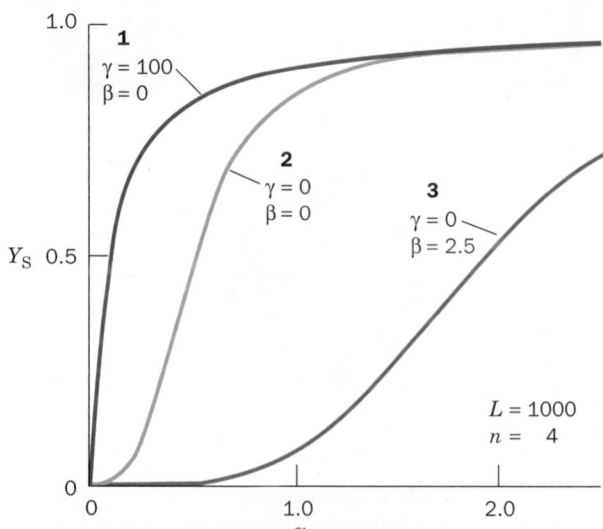

Figure 10-31 **The effects of allosteric activator ($\gamma = [A]/k_A$) and inhibitor ($\beta = [I]/k_I$) on the shape of the fractional saturation curve for substrate ($\alpha = [S]/k_R$) according to Eq. [10.23] for tetramers.** [After Monod, J., Wyman, J., and Changeux, J.-P., *J. Mol. Biol.* **12**, 94 (1965).]

there are well-established instances of negative homotropic effects (e.g., in the GroEL–GroES complex, the binding of ATP to the cis ring of GroEL prevents ATP from binding to the trans ring; Section 9-2Cd), although the symmetry model, which permits only positive homotropic effects, is unable to account for them.

The symmetry model implicitly assumes Emil Fischer's "lock-and-key" model of ligand binding in which ligand-binding sites of proteins are rigid and complementary in shape to their ligand (Fig. 10-32, *left*). A more sophisticated extension of the lock-and-key model, known as the **induced-fit hypothesis,** postulates that *a flexible interaction between ligand and protein induces a conformational change in the protein, which results in its increased ligand-binding affinity* (Fig. 10-32, *right*). The observation, through X-ray crystal structure analysis, that such conformational changes occur in numerous proteins has established the validity of the induced-fit hypothesis.

Daniel Koshland, George Némethy, and David Filmer adapted the induced-fit hypothesis to explain allosteric effects. *In the resulting **sequential model** (alternatively, the **induced-fit** or **KNF model**), ligand binding induces a conformational change in a subunit; cooperative interactions arise through the influence that these conformational changes have on neighboring subunits* (Fig. 10-33). If, for example, they increase the neighbor's ligand-binding affinity, then ligand binding is positively cooperative. *The strengths of these interactions depend on the degree of mechanical coupling*

between subunits. In the limit of very strong coupling, conformational changes become concerted, so that the oligomer maintains its symmetry (the symmetry model). With looser coupling, however, conformational changes occur sequentially as more and more ligand is bound (Fig. 10-34). Thus, *the essence of the sequential model is that a protein's ligand-binding affinity varies with its number of bound ligands, whereas in the symmetry model this affinity depends only on the protein's quaternary state.*

The degree of coupling between oligomer subunits depends on how these subunits are arranged, that is, on the protein's symmetry. Consequently, in the sequential model, the fractional saturation has a different algebraic form for each oligomeric symmetry. The form of the Adair equation (Eq. [10.17] for a tetramer) similarly depends on the number of subunits in the protein. In fact, the sequential model of allosterism may be considered an extension of the Adair model that provides a physical rationalization for the values of its microscopic dissociation constants, k_i.

D. Hemoglobin Cooperativity

Hemoglobin's fractional saturation curve is closely approximated by both the symmetry model and the sequential model (Fig. 10-35). Clearly such curves cannot by themselves be used to differentiate between these two models, if, in fact, either is correct. It is of interest, however, to compare these models with the mechanistic model of Hb we developed in Section 10-2C.

Hb, of course, is not composed of identical subunits, as the symmetry model demands. At least to a first approximation, however, the functional differences of Hb's closely related α and β subunits may be ignored (although their structural differences are essential to the molecular mechanism of Hb cooperativity). *To this approximation, Hb largely follows the symmetry model, although it also exhibits some features of the sequential model.* The quaternary T → R conformation change is concerted as the symmetry model requires. Yet ligand binding to the T state does cause small tertiary structural changes as the sequential model predicts. This phenomenon is evident in the X-ray structure of crystals of human Hb whose α subunits are fully oxygenated and whose β subunits are unliganded. This particular crystal form constrains

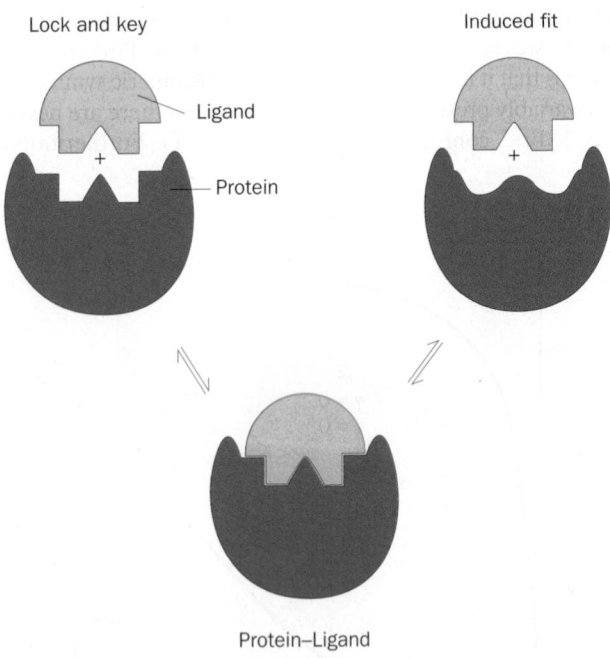

Figure 10-32   Models of ligand binding. In the lock-and-key mechanism of ligand binding (*left*), proteins are postulated to have preformed ligand-binding sites that are complementary in shape to their ligand. Under the induced-fit mechanism, a protein does not have this complementary binding site in the absence of ligand (*right*). Rather, the ligand induces a conformational change at the binding site that results in the complementary interaction.

Figure 10-33   The sequential model of allosterism. Substrate binding to the low-affinity T state induces conformational changes in unliganded subunits that give them ligand-binding affinities between those of the low-affinity T state and the high-affinity R state.

Figure 10-34 **Sequential binding of ligand in the sequential model of allosterism.** Ligand binding progressively induces conformational changes in the subunits, with the greatest changes occurring in those subunits that have bound ligand. The coupling between subunits is not necessarily of sufficient strength to maintain the symmetry of the oligomer as it is in the symmetry model.

the partially liganded Hb to remain in the T state. Nevertheless, its α subunit Fe's are 0.15 Å closer to the still domed porphyrins than they are in deoxyHb (25% of the total distance moved in the T → R transition). *Such tertiary structural changes are undoubtedly responsible for the buildup of strain that eventually triggers the T → R transition.*

A more telling but more difficult question to answer is does ligand-binding cooperativity in Hb arise solely from the T → R transition, in agreement with the symmetry model, or do the T and R states themselves exhibit at least some degree of cooperativity, in accord with the sequential model? Put another way, does the ligand-binding affinity of Hb subunits depend only on Hb's quaternary state (symmetry model) or does this affinity vary with the number of ligands bound to Hb (sequential model)? Although this matter has been under experimental scrutiny and a subject of spirited debate for over four decades, it is still not fully resolved. For example, the determination, by Andrea Mozzarelli and Eaton, of the Hill plot of the above-described single crystals of T-state Hb via sophisticated optical spectroscopic techniques is indicative of noncooperative ligand binding, in accord with the symmetry model. Likewise, time-resolved spectroscopic measurements covering the picosecond to microsecond time scales after ligand binding was initiated indicate that the conformational changes that Hb initially undergoes on binding ligand are consistent with the symmetry model. However, detailed thermodynamic analysis, by Gary Ackers, of the interactions associated with the formation of each of Hb's 10 different ligation microstates (Fig. 10-36) indicates that the Hb tetramer undergoes a T → R transition

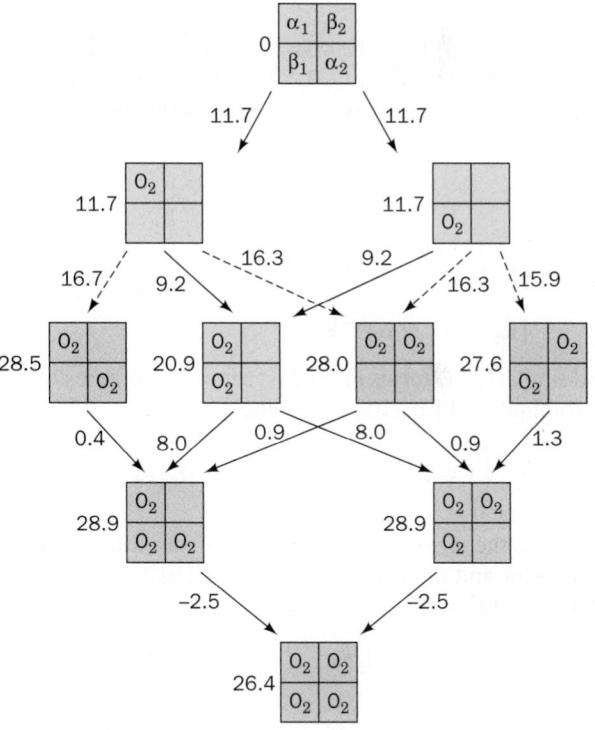

Figure 10-36 **Free energy penalties for binding O_2 to various ligation states of Hb tetramers relative to O_2-binding to noncooperative Hb αβ dimers.** Only the ten unique liganding states are diagrammed (an additional six states are related to those shown by Hb's 2-fold symmetry). Penalties (in kJ · mol^{-1}) for individual binding steps are shown beside the arrows. Cumulative free energy penalties are shown to the left of each Hb tetramer state. Ligation states that predominantly assume the T state are blue and those that predominantly assume the R state are red. The preferred pathways, those in which the free energy penalty of ligand binding progressively decreases with each successively bound ligand (those with all solid arrows), pass through the T state in which O_2 is bound to both sites on one αβ dimer before its conversion to the R state. Note that the T → R transition predominantly occurs via pathways in which at least one subunit in each αβ protomer is liganded. [Based on data from Ackers, G.K., *Adv. Prot. Chem.* **51,** 193 (1998).]

Figure 10-35 **The sequential and the symmetry models of allosterism can provide equally good fits to the measured O_2-dissociation curve of Hb.** [After Koshland, D.E., Jr., Némethy, G., and Filmer, D., *Biochemistry* **5,** 382 (1966).]

only when at least one liganding site on each of its αβ protomers is occupied. This heretofore unrecognized symmetry is inconsistent with the symmetry model. Moreover, as Takashi Yonetani has shown, heterotropic effectors such as BPG can decrease Hb's oxygen affinity without changing its quaternary state. Evidently, cooperativity arises both from concerted quaternary switching (as called for by the symmetry model) and from sequential modulation of ligand binding within each quaternary state through ligand-induced alterations in tertiary structure (in accord with the sequential model).

APPENDIX Derivations of Symmetry Model Equations

A. *Homotropic Interactions–Equation [10.22]*

The fractional saturation Y_S for ligand binding is expressed:

$$Y_S = \frac{([R_1] + 2[R_2] + \cdots + n[R_n]) + ([T_1] + 2[T_2] + \cdots + n[T_n])}{n\{([R_0] + [R_1] + \cdots + [R_n]) + ([T_0] + [T_1] + \cdots + [T_n])\}} \quad [10.21]$$

Defining $\alpha = [S]/k_R$ and $c = k_R/k_T$, and using Eq. [10.20] to substitute $[R_{n-1}]$ for $[R_n]$, $[R_{n-2}]$ for $[R_{n-1}]$ etc., the terms enclosed by the first set of parentheses in the numerator of Eq. [10.21] are reduced to

$$[R_0]\left\{ n\alpha + \frac{2n(n-1)\alpha^2}{2} + \cdots + \frac{n\,n!\alpha^n}{n!} \right\}$$

$$= [R_0]\alpha n\left\{ 1 + \frac{2(n-1)\alpha}{2} + \cdots + \frac{n(n-1)!\alpha^{n-1}}{n(n-1)!} \right\}$$

$$= [R_0]\alpha n(1 + \alpha)^{n-1}$$

and similarly, the terms in the first set of parentheses of the denominator of Eq. [10.21] become

$$[R_0]\left\{ 1 + n\alpha + \cdots + \frac{n!\alpha^n}{n!} \right\} = [R_0](1 + \alpha)^n$$

Likewise, the terms in the second set of parentheses of the numerator and the denominator of Eq. [10.21] assume the respective forms

$$[T_0]([S]/k_T)n(1 + [S]/k_T)^{n-1} = L[R_0]c\alpha n(1 + c\alpha)^{n-1}$$

and

$$[T_0](1 + [S]/k_T)^n = L[R_0](1 + c\alpha)^n$$

Accordingly,

$$Y_S = \frac{[R_0]\alpha n(1 + \alpha)^{n-1} + L[R_0]c\alpha n(1 + c\alpha)^{n-1}}{n\{[R_0](1 + \alpha)^n + L[R_0](1 + c\alpha)^n\}}$$

which, on cancellation of terms, yields the equation describing the symmetry model for homotropic interactions:

$$Y_S = \frac{\alpha(1 + \alpha)^{n-1} + Lc\alpha(1 + c\alpha)^{n-1}}{(1 + \alpha)^n + L(1 + c\alpha)^n} \quad [10.22]$$

B. *Heterotropic Interactions–Equation [10.23]*

For an oligomer that binds activator A and substrate S to only its R state, and inhibitor I to only its T state, the fractional saturation for substrate, Y_S, the fraction of substrate-binding sites occupied by substrate, is expressed:

$$Y_S = \frac{\displaystyle\sum_{i=1}^{n}\sum_{j=0}^{n} i[R_{i,j}]}{n\left(\displaystyle\sum_{i=0}^{n}\sum_{j=0}^{n}[R_{i,j}] + \sum_{k=0}^{n}[T_k] \right)}$$

Here the subscripts i, j, and k indicate the respective numbers of S, A, and I molecules that are bound to one oligomer; that is, $R_{i,j} \equiv RS_iA_j$ and $T_k \equiv TI_k$. Then defining $\alpha = [S]/k_R$ and following the foregoing derivation of Eq. [10.22]:

$$Y_S = \frac{\left(\displaystyle\sum_{j=0}^{n}[R_{0,j}]\right)\alpha n(1 + \alpha)^{n-1}}{n\left\{\left(\displaystyle\sum_{j=0}^{n}[R_{0,j}]\right)(1 + \alpha)^n + \displaystyle\sum_{k=0}^{n}[T_k]\right\}} = \frac{\alpha(1 + \alpha)^{n-1}}{(1 + \alpha)^n + L'}$$

where

$$L' = \sum_{k=0}^{n}[T_k] / \sum_{j=0}^{n}[R_{0,j}]$$

In analogy with the definition of α, we define $\beta = [I]/k_I$ and $\gamma = [A]/k_A$, and again follow the derivation of Eq. [10.22] to obtain

$$\sum_{k=0}^{n}[T_k] = [T_0](1 + \beta)^n$$

and

$$\sum_{j=0}^{n}[R_{0,j}] = [R_{0,0}](1 + \gamma)^n$$

so that

$$L' = \frac{L(1 + \beta)^n}{(1 + \gamma)^n}$$

The symmetry model equation extended to include heterotropic effects is therefore expressed

$$Y_S = \frac{\alpha(1 + \alpha)^{n-1}}{(1 + \alpha)^n + \dfrac{L(1 + \beta)^n}{(1 + \gamma)^n}} \quad [10.23]$$

CHAPTER SUMMARY

1 Hemoglobin Function The heme group in myoglobin and in each subunit of hemoglobin reversibly binds O_2. In deoxyHb, the Fe(II) is five-coordinated to the four pyrrole nitrogen atoms of the protoporphyrin IX and to the protein's proximal His. On oxygenation, O_2 becomes the sixth ligand of Fe(II). Mb has a hyperbolic fractional saturation curve (Hill constant, $n = 1$). However, that of Hb is sigmoidal ($n \approx 2.8$) as a consequence of its cooperative O_2 binding: Hb binds its fourth O_2 with 100-fold greater affinity than its first O_2. The variation of O_2 affinity with pH, the Bohr effect, causes Hb to release O_2 in the tissues in response to the binding of protons liberated by the hydration of CO_2 to HCO_3^-. Hb facilitates the transport of CO_2, both directly, by binding CO_2 as N-terminal carbamate, and indirectly, by increasing the concentration of HCO_3^- through the Bohr effect. The presence of BPG in erythrocytes, which only binds to deoxyHb, further modulates the O_2 affinity of Hb. Short-term high-altitude adaptation results from an increase of BPG concentration in the erythrocytes, which increases the amount of O_2 delivered to the tissues by decreasing hemoglobin's O_2 affinity.

2 Structure and Mechanism The α and β subunits of Hb consist mostly of seven or eight consecutive helices arranged to form a hydrophobic pocket that almost completely envelops the heme. Oxygen binding moves the Fe(II) from a position ~0.6 Å out of the heme plane on the side of the proximal His to the center of the heme, thereby relieving the steric interference that would otherwise occur between the bound O_2 and the porphyrin. The Fe(II) pulls the attached proximal His after it in a motion that can only occur if its imidazole ring reorients so as to avoid collision with the heme. In the T → R conformational transition, the symmetry equivalent $\alpha_1 C$–$\beta_2 FG$ and $\alpha_2 C$–$\beta_1 FG$ contacts simultaneously shift between two stable positions. Intermediate positions are sterically prevented, so that these contacts act as a two-position conformational switch.

The Perutz mechanism of O_2 binding proposes that the low O_2 affinity of the T state arises from strain that prevents the Fe(II) from moving into the heme plane to form a strong Fe—O_2 bond. This strain is relieved by the concerted 4° shift of the Hb molecule to the high O_2 affinity R state. The quaternary shift is opposed by a network of salt bridges in the T state that involve the C-terminal carboxyl groups and that are ruptured in the R state. The stability of the R state relative to the T state increases with the degree of oxygenation as a result of the strain of binding O_2 in the T state. The existence of this strain has been demonstrated through the breakage of the Fe(II)–proximal His bond on hemoglobin's simultaneous binding of IHP, a tight-binding BPG analog that forces Hb into the T state, and NO, a strong ligand that forces Hb into the R state. Conversely, mutagenically detaching the proximal His from the protein eliminates most of hemoglobin's cooperativity.

The Bohr effect results from increases in the pK's of the α N-terminal amino group and His 146β on forming the T-state salt bridges. Surface-exposed His residues also participate in the Bohr effect. BPG binding occurs in the central cavity of T-state Hb through several salt bridges. The distal His residue protects deoxyHb from autooxidation by taking up the protons that would otherwise catalyze the oxidation of the heme Fe.

3 Abnormal Hemoglobins Over 1000 mutant varieties of Hb are known. About half of them are innocuous because they result in surface residue changes. However, alterations of internal residues often disrupt the structure of Hb, which causes hemolytic anemia. Changes at the O_2-binding site that stabilize the Fe(III) state eliminate O_2 binding to these subunits, which results in cyanosis. Mutations affecting subunit interfaces may stabilize either the R state or the T state, which, respectively, increase and decrease hemoglobin's O_2 affinity. Sickle-cell anemia is caused by the homozygous Hb mutant Glu 6β → Val, which promotes the gelation of the resulting deoxyHbS to form rigid 14-strand fibers that deform erythrocytes. Under gelation conditions, fiber growth occurs via a two-stage nucleation mechanism, resulting in a delay time that varies with the 30th to 50th power of the initial HbS concentration. Agents that increase this delay time to longer than the transit times of erythrocytes through the capillaries should therefore prevent sickle-cell crises and thus relieve the symptoms of sickle-cell anemia.

4 Allosteric Regulation The Adair equation rationalizes the O_2-binding cooperativity of Hb by assigning a separate dissociation constant to each O_2 bound. Positive cooperativity results if these constants decrease sequentially. However, the Adair equation offers no physical insight as to why this occurs. The symmetry model proposes that symmetrical oligomers can exist in one of two conformational states, R and T, that differ in ligand-binding affinity. Ligand binding to the high-affinity state forces the oligomer to assume this conformation, which facilitates the binding of additional ligand. This homotropic model is extended to heterotropic effects by postulating that activator and substrate can bind only to the R state and inhibitor can bind only to the T state. The binding of activator forces the oligomer into the R state, which facilitates the binding of substrate and additional activator. The binding of inhibitor, however, forces the oligomer into the T state, which prevents substrate and activator binding. The sequential model postulates that an induced fit between ligand and substrate confers conformational strain on the protein that alters its affinity for binding other ligands without requiring the oligomer to maintain its symmetry. The Perutz mechanism for O_2 binding to Hb is structurally largely consistent with the symmetry model but exhibits some elements of the sequential model. However, Hb's ligand-binding cooperativity is in full accord with the symmetry model.

REFERENCES

General

Bunn, F.H. and Forget, B.G., *Hemoglobin: Molecular, Genetic and Clinical Aspects,* Saunders (1986). [A valuable compendium on normal and abnormal hemoglobins.]

Dickerson, R.E. and Geis, I., *Hemoglobin,* Benjamin/Cummings (1983). [A beautifully written and lavishly illustrated treatise on the structure, function, and evolution of hemoglobin.]

Ferry, G., *Max Perutz and the Secret of Life,* Cold Spring Harbor Laboratory Press (2007). [A definitive biography.]

Hendgen-Cotta, U.B., et al., Nitrite reductase activity of myoglobin regulates respiration and cellular viability in myocardial ischemia–reperfusion injury, *Proc. Natl. Acad. Sci.* **105,** 10256–10261 (2008).

Judson, H.F., *The Eighth Day of Creation* (expanded edition), Chapters 9 and 10, Cold Spring Harbor Laboratory Press (1996). [Includes a fascinating historical account of how our present perception of hemoglobin structure and function came about.]

Lukin, J.A. and Ho, C., The structure–function relationship of hemoglobin in solution at atomic resolution, *Chem. Rev.* **104,** 1219–1230 (2004).

Royer, W.E., Jr., Zhu, H., Gorr, T.A., Flores, J.F., and Knapp, J.E., Allosteric hemoglobin assembly: diversity and similarity, *J. Biol. Chem.* **280,** 27477–27480 (2005). [Reviews the different types of hemoglobins.]

Vinogradov, S.N. and Moens, L., Diversity of globin function: enzymatic, transport, storage, and sensing, *J. Biol. Chem.* **283,** 8773–8777 (2008). [Discusses globin-mediated reactions with NO as well as neuroglobin and cytoglobin.]

Structures of Myoglobin, Hemoglobin, and Model Compounds

Brunori, M., Nitric oxide moves myoglobin to center stage, *Trends Biochem. Sci.* **26,** 209–210 (2001).

Fermi, G., Perutz, M.F., Shaanan, B., and Fourme, R., The crystal structure of human deoxyhaemoglobin at 1.74 Å, *J. Mol. Biol.* **175,** 159–174 (1984).

Jameson, G.B., Molinaro, F.S., Ibers, J.A., Collman, J.P., Brauman, J.I., Rose, E., and Suslick, K.S., Models for the active site of oxygen-binding hemoproteins. Dioxygen binding properties and the structures of (2-methylimidazole)-*meso*-tetra(α,α,α,α-*o*-pivalamidophenyl)porphinato iron(II)-ethanol and its dioxygen adduct, *J. Am. Chem. Soc.* **102,** 3224–3237 (1980). [The picket-fence complex.]

Liddington, R., Derewenda, Z., Dodson, G., and Harris, D., Structure of the liganded T state of haemoglobin identifies the origin of cooperative oxygen binding, *Nature* **331,** 725–728 (1988).

Ordway, G.A. and Garry, D.J., Myoglobin: an essential hemoprotein in striated muscle, *J. Exp. Biol.* **207,** 341–3446 (2004).

Phillips, S.E.V., Structure and refinement of oxymyoglobin at 1.6 Å resolution, *J. Mol. Biol.* **142,** 531–554 (1980).

Shaanan, B., Structure of human oxyhaemoglobin at 2.1 Å resolution, *J. Mol. Biol.* **171,** 31–59 (1983).

Takano, T., Structure of myoglobin refined at 2.0 Å resolution, *J. Mol. Biol.* **110,** 537–568, 569–584 (1977).

Mechanism of Hemoglobin Oxygen Binding

Baldwin, J. and Chothia, C., Haemoglobin: The structural changes related to ligand binding and its allosteric mechanism, *J. Mol. Biol.* **129,** 175–220 (1979). [The exposition of a detailed mechanism of O₂ binding to Hb based on the structures of oxyHb and deoxyHb.]

Barrick, D., Ho, N.T., Simplaceanu, V., Dahlquist, F.W., and Ho, C., A test of the role of the proximal histidines in the Perutz model for cooperativity in haemoglobin, *Nature Struct. Biol.* **4,** 78–83 (1997). [Describes the experiments in which the proximal His is detached from the F helix.]

Gelin, B.R., Lee, A.W.-N., and Karplus, M., Haemoglobin tertiary structural change on ligand binding, *J. Mol. Biol.* **171,** 489–559 (1983). [A theoretical study of the dynamics of O₂ binding to Hb.]

Perutz, M.F., Stereochemistry of cooperative effects in haemoglo-bin, *Nature* **228,** 726–734 (1970). [The landmark paper in which the Perutz mechanism was first proposed. Although many of its details have since been modified, the basic model remains intact.]

Perutz, M.F., Regulation of oxygen affinity of hemoglobin, *Annu. Rev. Biochem.* **48,** 327–386 (1979). [An examination of the Perutz mechanism in light of structural and spectroscopic data.]

Perutz, M.F., Mechanisms of cooperativity and allosteric regulation in proteins, *Q. Rev. Biophys.* **22,** 139–236 (1989). [Contains a detailed structural description of allosterism in hemoglobin.]

Perutz, M.F., Wilkinson, A.J., Paoli, M., and Dodson, G.G., The stereochemical mechanism of the cooperative effects in hemoglobin revisited, *Annu. Rev. Biophys. Biomol. Struct.* **27,** 1–34 (1998).

Bohr Effect and BPG Binding

Arnone, A., X-ray studies of the interaction of CO₂ with human deoxyhaemoglobin, *Nature* **247,** 143–145 (1974).

Benesch, R.E. and Benesch, R., The mechanism of interaction of red cell organic phosphates with hemoglobin, *Adv. Prot. Chem.* **28,** 211–237 (1974).

Kilmartin, J.V. and Rossi-Bernardi, L., Interactions of hemoglobin with hydrogen ion, carbon dioxide and organic phosphates, *Physiol. Rev.* **53,** 836–890 (1973).

Lenfant, C., Torrance, J., English, E., Finch, C.A., Reynafarje, C., Ramos, J., and Faura, J., Effect of altitude on oxygen binding by hemoglobin and on organic phosphate levels, *J. Clin. Invest.* **47,** 2652–2656 (1968).

Perutz, M.F., Kilmartin, J.V., Nishikura, K., Fogg, J.H., and Butler, P.J.G., Identification of residues contributing to the Bohr effect of human haemoglobin, *J. Mol. Biol.* **138,** 649–670 (1980).

Richard, V., Dodson, G.G., and Mauguen, Y., Human deoxy-haemoglobin-2,3-diphosphoglycerate complex low-salt structure at 2.5 Å resolution, *J. Mol. Biol.* **233,** 270–274 (1993).

Sun, D.P., Zou, M., Ho, N.T., and Ho, C., Contribution of surface histidyl residues in the α-chain of the Bohr effect of human normal adult hemoglobin: Roles of global electrostatic effects, *Biochemistry* **36,** 6663–6673 (1997).

Abnormal Hemoglobins

Baudin-Chich, V., Pagnier, J., Marden, M., Bohn, B., Lacaze, N., Kister, J., Schaad, O., Edelstein, S.J., and Poyart, C., Enhanced polymerization of recombinant human deoxyhemoglobin β6 Glu → Ile, *Proc. Natl. Acad. Sci.* **87,** 1845–1849 (1990).

Bunn, F.H., Pathogenesis and treatment of sickle cell disease, *New Engl. J. Med.* **337,** 762–769 (1997).

Bunn, F.H., Human hemoglobins: sickle hemoglobin and other mutants, *in* Stamatoyannopoulos, G., Majerus, P.W., Perlmutter, R.M., and Varmus, H. (Eds.), *The Molecular Basis of Blood Diseases* (3rd ed.), Chapter 7, Elsevier (2001).

Eaton, W.A. and Hofrichter, J., Sickle cell hemoglobin polymerization, *Adv. Prot. Chem.* **40,** 63–279 (1990). [An exhaustive review of HbS polymerization.]

Eaton, W.A. and Hofrichter, J., The biophysics of sickle cell hydroxyurea therapy, *Science* **268,** 1142–1143 (1995).

Harrington, D.J., Adachi, K., and Royer, W.E., Jr., The high resolution crystal structure of deoxyhemoglobin S, *J. Mol. Biol.* **272,** 398–407 (1997).

Nagel, R.L., Haemoglobinopathies due to structural mutations, *in* Provan, D. and Gribben, J. (Eds.), *Molecular Haematology, pp.* 121–133, Blackwell Science (2000).

Perutz, M., *Protein Structure. New Approaches to Disease and Therapy,* Chapter 6, Freeman (1992).

Perutz, M.F. and Lehmann, H., Molecular pathology of human

haemoglobin, *Nature* **219**, 902–909 (1968). [A ground-breaking study correlating the clinical symptoms and inferred structural alterations of numerous mutant hemoglobins.]

Steinberg, M.H., Management of sickle cell disease, *New Engl. J. Med.* **340**, 1021–1030 (1999).

Strasser, B.J., Sickle-cell anemia, a molecular disease, *Science* **286**, 1488–1490 (1999). [A short history of Pauling's characterization of sickle-cell anemia.]

Watowich, S.J., Gross, L.J., and Josephs, R., Intermolecular contacts within sickle hemoglobin fibers, *J. Mol. Biol.* **209**, 821–828 (1989).

Weatherall, D.J., Clegg, J.B., Higgs, D.R., and Wood, W.G., The hemoglobinopathies, *in* Valle, D. (Ed.), *The Online Metabolic & Molecular Bases of Inherited Disease* http://ommbid.com/. [A detailed review of abnormal hemoglobins.]

Allosteric Regulation

Ackers, G.A. and Holt, J.M., Asymmetric cooperativity in a symmetric tetramer: human hemoglobin, *J. Biol. Chem.* **281**, 11441–11443 (2006); *and* Ackers, G.A., Deciphering the molecular code of hemoglobin allostery, *Adv. Prot. Chem.* **51**, 185–253 (1998).

Eaton, W.A., Henry, E.R., Hofrichter, J., Bettati, S., Viappiani, C., and Mozzarelli, A., Evolution of allosteric models for hemoglobin, *IUBMB Life* **59**, 586–589 (2007). [A critical review.]

Fersht, A., *Structure and Mechanism in Protein Science,* Chapter 10, Freeman (1999).

Koshland, D.E., Jr., Némethy, G., and Filmer, D., Comparison of experimental binding data and theoretical models in proteins containing subunits, *Biochemistry* **5**, 365–385 (1966). [The formulation of the sequential model of allosteric regulation.]

Monod, J., Wyman, J., and Changeux, J.P., On the nature of allosteric transitions: A plausible model, *J. Mol. Biol.* **12**, 88–118 (1965). [The exposition of the symmetry model of allosteric regulation.]

Yonetani, T. and Laberge, M., Protein dynamics explain the allosteric behaviours of hemoglobin, *Biochim. Biophys. Acta* **1784**, 1146–1158 (2008).

PROBLEMS

1. The urge to breathe in humans results from a high blood CO_2 content; there are no direct physiological sensors of blood pO_2. Skindivers often **hyperventilate** (breathe rapidly and deeply for several minutes) just before making a protracted dive in the belief that they will thereby increase the O_2 content of their blood. This belief results from the fact that hyperventilation represses the breathing urge by expelling significant quantities of CO_2 from the blood. In light of what you know about the properties of hemoglobin, is hyperventilation a useful procedure? Is it safe? Explain.

2. Explain why the Hill constant, n, can never be larger than the number of ligand-binding sites on the protein.

*3. In the Bohr effect, protonation of the N-terminal amino groups of hemoglobin's α chains is responsible for ~30% of the 0.6 mol of H^+ that combine with Hb on the release of 1 mol of O_2 at pH 7.4. Assuming that this group has $pK = 7.0$ in oxyHb, what is its pK in deoxyHb?

4. As one of the favorites to win the La Paz, Bolivia, marathon, you have trained there for the several weeks it requires to become adapted to its 3700-m altitude. A manufacturer of running equipment who sponsors an opponent has invited you for the weekend to a prerace party at a beach house near Lima, Peru, with the assurance that you will be flown back to La Paz at least a day before the race. Is this a token of his respect for you or an underhanded attempt to handicap you in the race? Explain (see Fig. 10-9).

5. In active muscles, the pO_2 may be 10 torr at the cell surface and 1 torr at the mitochondia (the organelles where oxidative metabolism occurs). How might myoglobin ($p_{50} = 2.8$ torr) facilitate the diffusion of O_2 through these cells? Active muscles consume O_2 much faster than do other tissues. Could myoglobin also be an effective O_2-transport protein in other tissues? Explain.

6. Erythrocytes that have been stored for over a week in standard acid–citrate–dextrose medium become depleted in BPG. Discuss the merits of using fresh versus week-old blood in blood transfusions.

7. The following fractional saturation data have been measured for a certain blood sample:

pO_2	Y_{O_2}	pO_2	Y_{O_2}
20	0.14	60	0.59
30	0.26	70	0.66
40	0.39	80	0.72
50	0.50	90	0.76

What are the Hill constant and the p_{50} of this blood sample? Are they normal?

8. An anemic individual, whose blood has only half the normal Hb content, may appear to be in good health. Yet a normal individual is incapacitated by exposure to sufficient carbon monoxide to occupy half of his/her heme sites (pCO of 1 torr for ~1 h; CO binds to Hb with 200-fold greater affinity than does O_2). Explain.

*9. The X-ray structure of Hb Rainier (Tyr 145β → Cys) indicates that the mutant Cys residue forms a disulfide bond with Cys 93β of the same subunit. This holds the β subunit's C-terminal residue in a quite different orientation than it assumes in HbA. How would the following quantities for Hb Rainier compare with those of HbA? Explain. (a) The oxygen affinity, (b) the Bohr effect, (c) the Hill constant, and (d) the BPG affinity.

10. The crocodile, which can remain under water without breathing for up to 1 h, drowns its air-breathing prey and then dines at its leisure. An adaptation that aids the crocodile in doing so is that it can utilize virtually 100% of the O_2 in its blood, whereas humans, for example, can extract only ~65% of the O_2 in their blood. Crocodile Hb does not bind BPG. However, crocodile deoxyHb preferentially binds HCO_3^-. How does this help the crocodile obtain its dinner?

11. The gelation time of an equimolar mixture of HbA and HbS is less than that of a solution of only HbS in the same concentration that it has in the mixture. What does this observation imply about the participation of HbA in the gelation of HbS?

12. The severely anemic condition of homozygotes for HbS results in an elevated BPG content in their erythrocytes. Discuss whether or not this is a beneficial effect.

13. As organizer of an expedition that plans to climb several very high mountains, it is your responsibility to choose its members. Each of the applicants for one of the positions on the team is a heterozygote for one of the following variant hemoglobins: (1) HbS, (2) **Hb Hyde Park** [His F8(92)$\beta \rightarrow$ Tyr], (3) **Hb Riverdale–Bronx** [Gly B6(24)$\beta \rightarrow$ Arg], (4) **Hb Memphis** [Glu B4(23)$\alpha \rightarrow$ Gln], and (5) **Hb Cowtown** [His HC3(146)$\beta \rightarrow$ Leu]. Assuming that all of these candidates are equal in ability at low altitudes, which one would you choose for the position? Explain your reasoning.

14. Show that the Adair equation for a tetramer reduces to the Hill equation for $k_1 \approx k_2 \approx k_3 \gg k_4$ and to a hyperbolic relationship for $k_1 = k_2 = k_3 = k_4$.

15. Derive the equilibrium constant for the reaction $R_2 \rightleftharpoons T_2$ for a symmetry model n-mer in terms of the parameters $L, c,$ and α.

*16. Derive the equation for the fraction of protein molecules in the R state, $\overline{R}$, for the homotropic symmetry model in terms of the parameters $n, L, c,$ and α. Plot this function versus α for $n = 4$, $L = 1000$, and $c = 0$ and discuss its physical significance.

17. In the symmetry model of allosterism, why must an inhibitor (which causes a negative heterotropic effect with the substrate) undergo a positive homotropic effect?

18. At low concentrations, the hemoglobin tetramer reversibly dissociates into two $\alpha_1\beta_1$ dimers. What is the Hill constant for O_2 binding to these dimers? Explain.

19. Describe the nature of the allosteric changes (homotropic or heterotropic, positive or negative) that take place in the GroEL/ES system during the various stages of its catalytic cycle (Fig. 9-25).

Sugars and Polysaccharides

Carbohydrates or **saccharides** (Greek: *sakcharon,* sugar) are essential components of all living organisms and are, in fact, the most abundant class of biological molecules. The name carbohydrate, which literally means "carbon hydrate," stems from their chemical composition, which is roughly $(C \cdot H_2O)_n$, where $n \geq 3$. The basic units of carbohydrates are known as **monosaccharides.** Many of these compounds are synthesized from simpler substances in a process named **gluconeogenesis** (Section 23-1). Others (and ultimately nearly all biological molecules) are the products of **photosynthesis** (Section 24-3), the light-powered combination of CO_2 and H_2O through which plants and certain bacteria form "carbon hydrates." The metabolic breakdown of monosaccharides (Chapters 17 and 21) provides much of the energy used to power biological processes. Monosaccharides are also principal components of nucleic acids (Section 5-1A), as well as important elements of complex lipids (Section 12-1D).

Oligosaccharides consist of a few covalently linked monosaccharide units. They are often associated with proteins **(glycoproteins)** and lipids **(glycolipids)** in which they have both structural and regulatory functions (glycoproteins and glycolipids are collectively called **glycoconjugates**). **Polysaccharides** consist of many covalently linked monosaccharide units and have molecular masses ranging well into the millions of daltons. They have indispensable structural functions in all types of organisms but are most conspicuous in plants because **cellulose,** their principal structural material, comprises up to 80% of their dry weight. Polysaccharides such as **starch** in plants and **glycogen** in animals serve as important nutritional reservoirs.

The elucidation of the structures and functions of carbohydrates has lagged well behind those of proteins and nucleic acids. This can be attributed to several factors. Carbohydrate compounds are often heterogeneous, both in size and in composition, which greatly complicates their physical and chemical characterization. They are not subject to the types of genetic analysis that have been invaluable in the study of proteins and nucleic acids because saccharide sequences are not genetically specified but are built up through the sequential actions of specific enzymes (Section 23-3B). Furthermore, it has been difficult to establish assays for the biological activities of polysaccharides because of their largely passive roles. Nevertheless, it is abundantly clear that carbohydrates are essential elements in many, if not most, biological processes.

In this chapter, we explore the structures, chemistry, and, to a limited extent, the functions of carbohydrates, alone and in association with proteins. Glycolipid structures are considered in Section 12-1D. The biosynthesis of complex carbohydrates is discussed in Section 23-3.

1 MONOSACCHARIDES

Monosaccharides or **simple sugars** are aldehyde or ketone derivatives of straight-chain polyhydroxy alcohols containing at least three carbon atoms. Such substances, for example, **D-glucose** and **D-ribulose,** cannot be hydrolyzed to form simpler saccharides.

D-Glucose   **D-Ribulose**

In this section, the structures of the monosaccharides and some of their biologically important derivatives are discussed.

A. *Classification*

Monosaccharides are classified according to the chemical nature of their carbonyl group and the number of their C atoms. If the carbonyl group is an aldehyde, as in glucose, the sugar is an **aldose.** If the carbonyl group is a ketone, as in ribulose, the sugar is a **ketose.** The smallest monosaccharides, those with three carbon atoms, are **trioses.** Those with four, five, six, seven, etc., C atoms are, respectively, **tetroses, pentoses, hexoses, heptoses,** etc. These terms may be combined so that, for example, glucose is an **aldohexose,** whereas ribulose is a **ketopentose.**

Examination of D-glucose's molecular formula indicates that all but two of its six C atoms—C1 and C6—are chiral centers, so that D-glucose is one of $2^4 = 16$ stereoisomers that comprise all possible aldohexoses. In general, *n*-carbon aldoses have 2^{n-2} stereoisomers. The stereochemistry and names of the D-aldoses are presented in Fig. 11-1. Emil

Fischer elucidated these configurations for the aldohexoses in 1896. According to the Fischer convention (Section 4-2B), D sugars have the same absolute configuration at the asymmetric center farthest removed from their carbonyl group as does D-*glyceraldehyde.* The L sugars, in accordance with this convention, are mirror images of their D counterparts, as is shown below in Fischer projection for glucose.

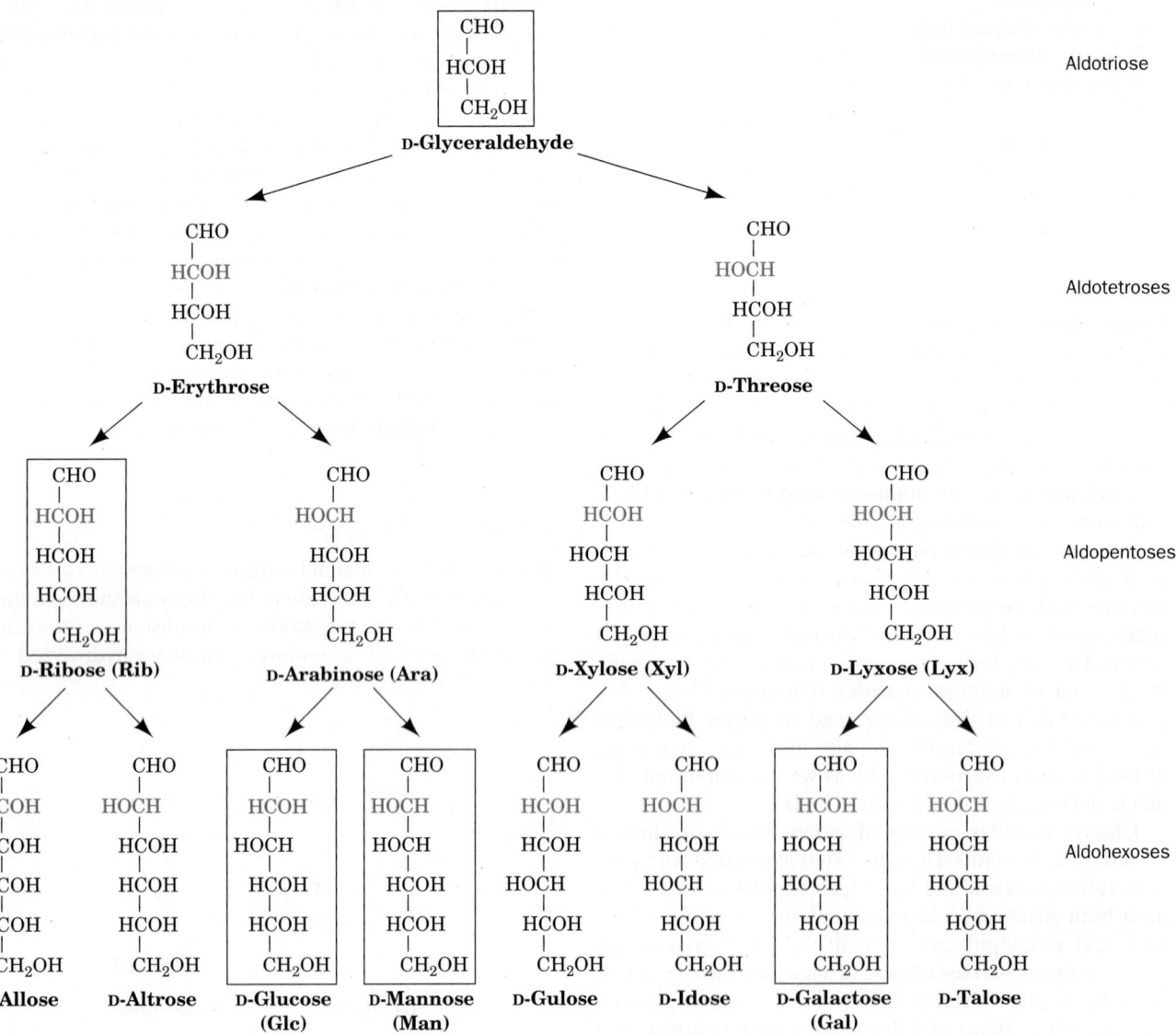

Figure 11-1 **The stereochemical relationships, shown in Fischer projection, among the D-aldoses with three to six carbon atoms.** The arrows indicate stereochemical relationships (not biosynthetic reactions). The configuration about C2 (*red*) distinguishes the members of each pair. The L- counterparts of these 15 sugars are their mirror images. The biologically most common aldoses are boxed.

Sugars that differ only by the configuration about one C atom are known as **epimers** of one another. Thus D-glucose and **D-mannose** are epimers with respect to C2, whereas D-glucose and **D-galactose** are epimers with respect to C4 (Fig. 11-1). However, D-mannose and D-galactose are not epimers of each other because they differ in configuration about two of their C atoms.

D-Glucose is the only aldose that commonly occurs in nature as a monosaccharide. However, it and several other monosaccharides including D-glyceraldehyde, D-ribose, D-mannose, and D-galactose are important components of larger biological molecules. L Sugars are biologically much less abundant than D sugars.

The position of their carbonyl group gives ketoses one less asymmetric center than their isomeric aldoses (e.g., compare D-fructose and D-glucose). *n*-Carbon ketoses therefore have 2^{n-3} stereoisomers. Those with their ketone

function at C2 are the most common form (Fig. 11-2). Note that some of these ketoses are named by the insertion of *-ul-* before the suffix *-ose* in the name of the corresponding aldose; thus **D-xylulose** is the ketose corresponding to the aldose **D-xylose**. **Dihydroxyacetone, D-fructose,** D-ribulose, and D-xylulose are the biologically most prominent ketoses.

B. *Configurations and Conformations*

Alcohols react with the carbonyl groups of aldehydes and ketones to form **hemiacetals** and **hemiketals,** respectively (Fig. 11-3). The hydroxyl and either the aldehyde or the ketone functions of monosaccharides can likewise react intramolecularly to form cyclic hemiacetals and hemiketals (Fig. 11-4). The configurations of the substituents to each carbon atom of these sugar rings are conveniently represented by their **Haworth projection formulas.**

A sugar with a six-membered ring is known as a **pyranose** in analogy with **pyran,** the simplest compound containing such a ring. Similarly, sugars with five-membered rings are designated **furanoses** in analogy with **furan.**

Pyran   **Furan**

The cyclic forms of glucose and fructose with six- and five-membered rings are therefore known as **glucopyranose** and **fructofuranose,** respectively.

a. Cyclic Sugars Have Two Anomeric Forms

The Greek letters preceding the names in Fig. 11-4 still need to be explained. The cyclization of a monosaccharide renders the former carbonyl carbon asymmetric. The resulting pair of diastereomers are known as **anomers** and the hemiacetal or hemiketal carbon is referred to as the **anomeric** carbon. In the α anomer, the OH substituent to the anomeric carbon is on the opposite side of the sugar ring from the CH_2OH group at the chiral center that

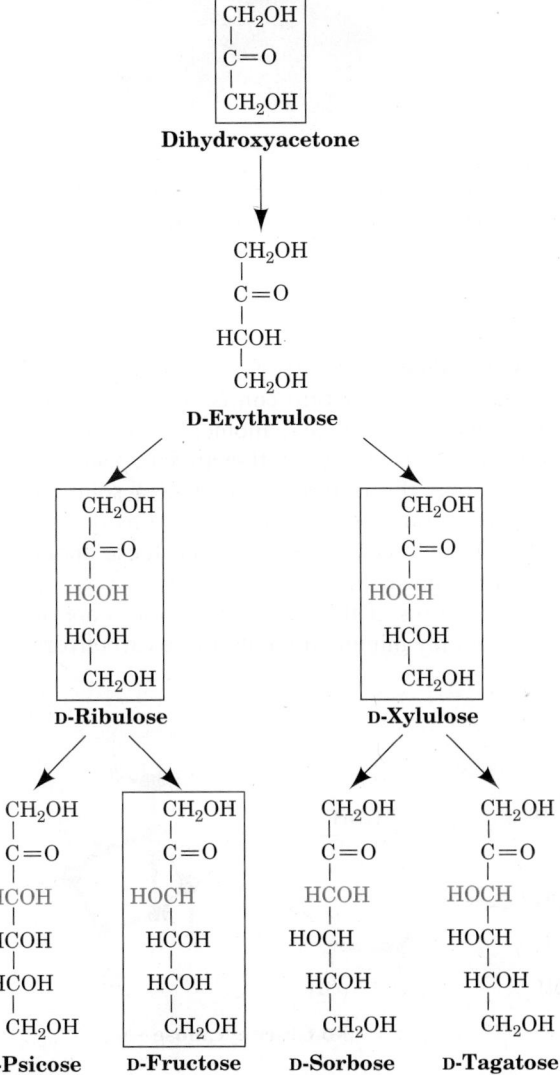

Figure 11-2   The stereochemical relationships among the D-ketoses with three to six carbon atoms. The configuration about C3 (*red*) distinguishes the members of each pair. The biologically most common ketoses are boxed.

Figure 11-3   The reactions of alcohols with (*a*) aldehydes to form hemiacetals and (*b*) ketones to form hemiketals.

(a)

D-Glucose
(linear form)

β-D-Glucopyranose
(Haworth projection)

(b)

D-Fructose
(linear form)

β-D-Fructofuranose
(Haworth projection)

Figure 11-4   Cyclization reactions for hexoses. *(a)* D-Glucose in its linear form reacting to yield the cyclic hemiacetal β-D-glucopyranose and *(b)* D-fructose in its linear form reacting to yield the hemiketal β-D-fructofuranose. The cyclic sugars are shown both as Haworth projections and in stick form embedded in their semitransparent space-filling models with C green, H white, and O red.

designates the D or L configuration (C5 in hexoses). The other anomer is known as the β form (Fig. 11-5).

The two anomers of D-glucose, as any pair of diastereomers, have different physical and chemical properties. For example, the values of the specific optical rotation, $[\alpha]_D^{20}$, for α-D-glucose and β-D-glucose are, respectively, +112.2° and +18.7°. When either of these pure substances is dissolved in water, however, the specific optical rotation of the solution slowly changes until it reaches an equilibrium value of $[\alpha]_D^{20} = +52.7°$. This phenomenon is known

as **mutarotation;** in glucose, it results from the formation of an equilibrium mixture consisting of 63.6% of the β anomer and 36.4% of the α anomer (the optical rotations of separate molecules in solution are independent of each other so that the optical rotation of a solution is the weighted average of the optical rotations of its components). The interconversion between these anomers occurs via the linear form of glucose (Fig. 11-5). Yet, since the linear forms of these monosaccharides are normally present in only minute amounts, these carbohydrates are

α-D-Glucopyranose

D-Glucose
(linear form)

β-D-Glucopyranose

Figure 11-5   The anomeric monosaccharides α-D-glucopyranose and β-D-glucopyranose, drawn as both Haworth projections and ball-and-stick models. These pyranose sugars interconvert through the linear form of D-glucose and differ only by the configurations about their anomeric carbon atoms, C1.

🔖 **See Kinemage Exercise 7-1**

Figure 11-6 **Conformations of the cyclohexane ring.** (*a*) In the boat conformation, substituents at the "bow" and "stern" (*red*) are sterically crowded, whereas those along its sides (*green*) are eclipsed. (*b*) In the chair conformation, the substituents that extend parallel to the ring's threefold rotation axis are designated axial [*a*] and those that extend roughly outward from this symmetry axis are designated equatorial [*e*]. The equatorial substituents about the ring are staggered so that they alternately extend above and below the mean plane of the ring.

Figure 11-7 **The two alternative chair conformations of β-D-glucopyranose.** In the conformation on the left, which predominates, the relatively bulky OH and CH_2OH substituents all occupy equatorial positions, whereas in that on the right (drawn in ball-and-stick form in Fig. 11-5, *right*) they occupy the more crowded axial positions. **See Kinemage Exercise 7-1**

accurately described as cyclic polyhydroxy hemiacetals or hemiketals.

b. Sugars Are Conformationally Variable

Hexoses and pentoses may each assume pyranose or furanose forms. The equilibrium composition of a particular monosaccharide depends somewhat on conditions but mostly on the identity of the monosaccharide. For instance, NMR measurements indicate that whereas glucose almost exclusively assumes its pyranose form in aqueous solutions, fructose is 67% pyranose and 33% furanose, and ribose is 75% pyranose and 25% furanose (although in polysaccharides, glucose, fructose, and ribose residues are exclusively in their respective pyranose, furanose, and furanose forms). Although, in principle, hexoses and larger sugars can form rings of seven or more atoms, such rings are rarely observed because of the greater stabilities of the five- and six-membered rings that these sugars can also form. The internal strain of three- and four-membered sugar rings makes them unstable with respect to linear forms.

The use of Haworth formulas may lead to the erroneous impression that furanose and pyranose rings are planar. This cannot be the case, however, because all of the atoms in these rings are tetrahedrally (sp^3) hybridized. The pyranose ring, like the cyclohexane ring, may assume a **boat** or a **chair** conformation (Fig. 11-6). The relative stabilities of these various conformations depend on the stereochemical interactions between the substituents on the ring. The boat con-

former crowds the substituents on its "bow" and "stern" and eclipses those along its sides, so that in cyclohexane it is ~25 $kJ \cdot mol^{-1}$ less stable than the chair conformer. The ring substituents on the chair conformer (Fig. 11-6*b*) fall into two geometrical classes: the rather close-fitting **axial** groups that extend parallel to the ring's threefold rotational axis and the staggered, and therefore minimally encumbered, **equatorial** groups. Since the axial and equatorial groups on a cyclohexane ring are conformationally interconvertible, a given ring has two alternative chair forms (Fig. 11-7); the one that predominates usually has the lesser crowding among its axial substituents. The conformational situation of a group directly affects its chemical reactivity. For example, equatorial OH groups on pyranoses esterify more readily than do axial OH groups. Note that β-D-glucose is the only D-aldohexose that can simultaneously have all five non-H substituents in the equatorial position (left side of Fig. 11-7). Perhaps this is why glucose is the most abundant naturally occurring monosaccharide. The conformational properties of furanose rings are discussed in Section 29-2Ab in relation to their effects on the conformations of nucleic acids.

C. Sugar Derivatives

a. Polysaccharides Are Held Together by Glycosidic Bonds

The chemistry of monosaccharides is largely that of their hydroxy and carbonyl groups. For example, in an acid-catalyzed reaction, the anomeric hydroxyl of a sugar reversibly condenses with alcohols to form α- and β-**glycosides** (Greek: *glykys,* sweet) (Fig. 11-8). The bond connecting the anomeric carbon to the acetal oxygen is termed a **glycosidic**

Figure 11-8 **The acid-catalyzed condensation of α-D-glucose with methanol to form an anomeric pair of methyl-D-glucosides.**

bond. *Polysaccharides are held together by glycosidic bonds between neighboring monosaccharide units.* The glycosidic bond is therefore the carbohydrate analog of the peptide bond in proteins. The bond in a nucleoside linking its ribose residue to its base is also a glycosidic bond (Section 5-1A).

The hydrolysis of glycosidic bonds is catalyzed by enzymes known as **glycosidases** that differ in specificity according to the identity and anomeric configuration of the glycoside but are often rather insensitive to the identity of the alcohol residue. Under basic or neutral conditions and in the absence of glycosidases, however, the glycosidic bond is stable, so glycosides do not undergo mutarotation as do monosaccharides. The methylation of the non-anomeric OH groups of monosaccharides requires more drastic conditions than is required for the formation of methyl glycosides, such as treatment with dimethyl sulfate.

b. Oxidation–Reduction Reactions

Because the cyclic and linear forms of aldoses and ketoses interconvert so readily, these sugars undergo reactions typical of aldehydes and ketones. Mild oxidation of an aldose, either chemically or enzymatically, results in the conversion of its aldehyde group to a carboxylic acid function, thereby yielding an **aldonic acid** such as **gluconic acid.** Aldonic acids are named by appending the suffix *-onic acid* to the root name of the parent aldose.

$$\begin{array}{c} ^1\text{COOH} \\ | \\ \text{H} - ^2\text{C} - \text{OH} \\ | \\ \text{HO} - ^3\text{C} - \text{H} \\ | \\ \text{H} - ^4\text{C} - \text{OH} \\ | \\ \text{H} - ^5\text{C} - \text{OH} \\ | \\ ^6\text{CH}_2\text{OH} \end{array}$$

D-Gluconic acid

Saccharides bearing anomeric carbon atoms that have not formed glycosides are termed **reducing sugars** because of the facility with which the aldehyde group reduces mild oxidizing agents. A classic test for the presence of a reducing sugar is the reduction of Ag^+ in an ammonia solution

(**Tollens' reagent**) to yield a metallic silver mirror lining on the inside of the reaction vessel.

The specific oxidation of the primary alcohol group of aldoses yields **uronic acids,** which are named by appending *-uronic acid* to the root name of the parent aldose. **D-Glucuronic acid, D-galacturonic acid,** and **D-mannuronic acid** are important components of many polysaccharides.

D-Glucuronic acid   **D-Galacturonic acid**   **D-Mannuronic acid**

Uronic acids can assume the pyranose, furanose, and linear forms.

Both aldonic and uronic acids have a strong tendency to internally esterify so as to form five- and six-membered lactones (Fig. 11-9). **Ascorbic acid (vitamin C,** Fig. 11-10) is a γ-lactone that is synthesized by plants and almost all animals except primates and guinea pigs. Its prolonged deficiency in the diet of humans results in the disease known as **scurvy,** which is caused by the impairment of collagen formation (Section 8-2B). Scurvy generally results from a lack of fresh food. This is because, under physiological conditions, ascorbic acid is reversibly oxidized to **dehydroascorbic acid,** which, in turn, is irreversibly hydrolyzed to the vitamin-inactive **diketogulonic acid** (Fig. 11-10).

Aldoses and ketoses may be reduced under mild conditions, for example, by treatment with $NaBH_4$ to yield acyclic polyhydroxy alcohols known as **alditols,** which are named by appending the suffix *-itol* to the root name of the parent aldose. **Ribitol** is a component of flavin coenzymes (Section 16-2C), and **glycerol** and the cyclic polyhydroxy alcohol *myo*-**inositol** are important lipid components

D-Glucono-δ-lactone   **D-Glucurono-δ-lactone**

Figure 11-9   **D-Glucono-δ-lactone and D-glucurono-δ-lactone are, respectively, the lactones of D-gluconic acid and D-glucuronic acid.** The δ indicates that the O atom closing the lactone ring is also substituent to C_δ.

L-Ascorbic acid   **L-Dehydroascorbic acid**   **L-Diketogulonic acid**

Figure 11-10   **The reversible oxidation of L-ascorbic acid to L-dehydroascorbic acid.** This is followed by the physiologically irreversible hydrolysis of its lactone ring to form L-diketogulonic acid.

(Section 12-1). **Xylitol** is a sweetener that is used in "sugarless" gum and candies.

Ribitol

Xylitol

Glycerol

myo-**Inositol**

c. Other Biologically Important Sugar Derivatives

Monosaccharide units in which an OH group is replaced by H are known as **deoxy sugars.** The biologically most important of these is **β-D-2-deoxyribose,** the sugar compo-

nent of DNA's sugar–phosphate backbone (Section 5-1A). **L-Rhamnose** and **L-fucose** are widely occurring polysaccharide components.

β-D-2-Deoxyribose

α-L-Rhamnose
(6-deoxy-L-mannose)

α-L-Fucose
(6-deoxy-L-galactose)

In **amino sugars,** one or more OH groups are replaced by an often acetylated amino group. **D-Glucosamine** and **D-galactosamine** are components of numerous biologically important polysaccharides.

α-D-Glucosamine
(2-amino-2-deoxy-
α-D-glucopyranose)

α-D-Galactosamine
(2-amino-2-deoxy-
α-D-galactopyranose)

N-Acetylmuramic acid (NAM)

The amino sugar derivative **N-acetylmuramic acid,** which consists of **N-acetyl-D-glucosamine** in an ether linkage with **D-lactic acid,** is a prominent component of bacterial cell walls (Section 11-3Ba). *N*-Acetylneuraminic acid, which is derived from *N*-acetylmannosamine and pyruvic acid (Fig. 11-11), is an important constituent of glycoproteins (Section 11-3C) and glycolipids (Section 12-1D). *N*-Acetylneuraminic acid and its derivatives are often referred to as **sialic acids.**

Pyruvic acid residue

N-Acetylmannosamine

N-Acetylneuraminic acid (linear form)

N-Acetylneuraminic acid (pyranose form)

Figure 11-11   N-Acetylneuraminic acid in its linear and pyranose forms. Note that its pyranose ring incorporates the pyruvic acid residue (*blue*) and part of the mannose moiety.

2 POLYSACCHARIDES

Polysaccharides, which are also known as **glycans,** consist of monosaccharides linked together by glycosidic bonds. They are classified as **homopolysaccharides** or **heteropolysaccharides** if they consist of one type or more than one type of monosaccharide residue. Homopolysaccharides may be further classified according to the identity of their monomeric unit. For example, **glucans** are polymers of glucose, whereas **galactans** are polymers of galactose. Although monosaccharide sequences of heteropolysaccharides can, in principle, be as varied as those of proteins, they are usually composed of only a few types of monosaccharides that alternate in a repetitive sequence.

Polysaccharides, in contrast to proteins and nucleic acids, form branched as well as linear polymers. This is because glycosidic linkages can be made to any of the hydroxyls of a monosaccharide. Fortunately for structural biochemists, many polysaccharides are linear and those that branch tend to do so in only a few well-defined ways.

In this section, we discuss the structures of the simplest polysaccharides, the disaccharides, and then consider the structures and properties of the most abundant classes of polysaccharides. We begin by outlining how polysaccharide structures are elucidated.

A. Carbohydrate Analysis

The purification of carbohydrates can, by and large, be effected by chromatographic and electrophoretic procedures similar to those used in protein purification (Sections 6-3 and 6-4), although thin layer chromatography (TLC; Section 6-3Dd) is also widely used. Affinity chromatography (Section 6-3C), using immobilized proteins known as **lectins** (Latin: *legere,* to pick or choose), is a particularly powerful technique in this regard. Lectins are sugar-binding proteins that were discovered in plants but are now known to occur in all organisms, where they participate in a wide variety of signaling, cell–cell recognition, and adhesion processes, as well as in targeting newly synthethesized proteins to specific cellular locations (Section 12-4Cg). Lectins recognize one or more specific monosaccharides with particular linkages to other sugars in oligosaccharides, usually with exquisite specificity. Their protein–carbohydrate interactions typically include multiple hydrogen bonds, which often include bridging water molecules, and the packing of hydrophobic sugar faces against aromatic side chains (Fig. 11-12). Among the best characterized lectins are jack bean **concanavalin A** (Fig. 8-40), which specifically binds α-D-glucose and α-D-mannose residues, and wheat germ **agglutinin** (so named because it causes cells to agglutinate or clump together), which specifically binds β-*N*-acetylmuramic acid and α-*N*-acetylneuraminic acid.

Characterization of an oligosaccharide requires that the identities, anomers, linkages, and order of its component monosaccharides be elucidated. The linkages of the monosaccharides may be determined by **methylation analysis** (also called **permethylation analysis),** a technique pioneered by Norman Haworth in the 1930s. *Methyl ethers not at the anomeric C atom are resistant to acid hydrolysis but glycosidic bonds are not. Consequently, if an oligosaccharide is exhaustively methylated and then hydrolyzed, the free OH groups on the resulting methylated monosaccharides mark the former positions of the glycosidic bonds.* Methylated monosaccharides are often identified by **gas–liquid chromatography (GLC;** a technique in which the stationary phase is an inert solid, such as diatomaceous earth,

Figure 11-12 **Carbohydrate binding by a lectin in the X-ray structure of human galectin-2 in complex with the disaccharide lactose.** This lectin primarily binds β-D-galactose residues. The structure is drawn in stick form with the C and O atoms of lactose's galactose (Gal) and glucose (Glc) residues green and red, and the galectin-2 amino acid side chains violet. Hydrogen bonds between the protein side chains and the sugar residues are represented by dashed yellow lines. [Courtesy of Hakon Leffler, Lund University, Sweden. PDBid 1HLC.]

impregnated with a low-volatility liquid, such as silicone oil, and the mobile phase is an inert gas, such as He, into which the sample has been flash evaporated) combined with mass spectrometry (GLC/MS). HPLC techniques may similarly be used. Other mass spectrometric techniques for analyzing nonvolatile substances are discussed in Section 7-1I. Although all aldoses and ketoses with the same number of C atoms are isomers (Figs. 11-1 and 11-2) and hence have identical molecular masses, they have characteristic fragmentation patterns.

The sequence and anomeric configurations of the monosaccharides in an oligosaccharide can be determined through the use of specific **exoglycosidases.** These enzymes specifically hydrolyze their corresponding monosaccharides from the nonreducing ends of oligosaccharides (the ends lacking a free anomeric carbon atom) in a manner analogous to the actions of exopeptidases on proteins (Section 7-1Ab). For example, **β-galactosidase** excises the terminal β anomers of galactose, whereas **α-mannosidase** does so with the α anomers of mannose. Some of these exoglycosidases also exhibit specificity for the **aglycone,** the sugar chains to which the monosaccharide to be excised (the **glycone**) is linked. Through the use of mass spectrometry, the sequence of a polysaccharide may be deduced from the mass decrements generated by exoglycosidases. The use of **endoglycosidases** (hydrolases that cleave glycosidic bonds between nonterminal sugar residues) of varying specificities can also supply useful sequence information. The proton and ^{13}C NMR spectra of oligosaccharides can provide the complete sequence of an oligosaccharide if sufficient material is available. Moreover, two-dimensional

NMR techniques (Section 8-3Ac) can reveal oligosaccharide structures (e.g., see Section 11-2Eb).

B. *Disaccharides*

We begin our studies of polysaccharides by considering disaccharides (Fig. 11-13). **Sucrose,** the most abundant disaccharide, occurs throughout the plant kingdom and is familiar to us as common table sugar. Its structure (Fig. 11-13) was established by methylation analysis as described above and was later confirmed by its X-ray structure. To name a polysaccharide systematically, one must specify its component monosaccharides, their ring types, their anomeric forms, and how they are linked together. Sucrose is therefore *O*-α-D-glucopyranosyl-(1 → 2)-β-D-fructofuranoside, where the symbol (1 → 2) indicates that the glycosidic bond links C1 of the glucose residue to C2 of the fructose residue. Note that since these two positions are the anomeric carbon atoms of their respective monosaccharides, sucrose is not a reducing sugar (as the suffix -*ide* implies).

The hydrolysis of sucrose to D-glucose and D-fructose is accompanied by a change in optical rotation from *dextro* to *levo.* Consequently, hydrolyzed sucrose is sometimes called **invert sugar** and the enzyme that catalyzes this process, **β-D-fructofuranosidase,** is archaically named **invertase.**

Lactose [*O*-β-D-galactopyranosyl-(1 → 4)-D-glucopyranose] or milk sugar (Fig. 11-13) occurs naturally only in milk, where its concentration ranges from 0 to 7% depending on the species. The free anomeric carbon of its glucose residue makes lactose a reducing sugar.

Infants normally express the intestinal enzyme **β-D-galactosidase** or **lactase** that catalyzes the hydrolysis of lactose to its component monosaccharides for absorption into the bloodstream. Many adults, however, including most Africans and almost all Asians, have a low level of this enzyme (as do most adult mammals, since they normally do not encounter milk). Consequently, much of the lactose in any milk they drink moves through their digestive tract to the colon, where its bacterial fermentation produces large quantities of CO_2, H_2, and irritating organic acids. This results in an embarrassing and often painful digestive upset termed **lactose intolerance.** Perhaps this is why Chinese cuisine, which is noted for the wide variety of foodstuffs it employs, is devoid of milk products. However, adult members of populations with a tradition of herding cattle, mainly northern Europeans and certain African groups, continue expressing the lactase gene and hence can drink milk without a problem. Modern food technology has come to the aid of milk lovers who develop lactose intolerance: Milk products in which the lactose has been hydrolyzed enzymatically and lactase-containing pills are now widely available.

There are several common glucosyl–glucose disaccharides. These include **maltose** [*O*-α-D-glucopyranosyl-(1 → 4)-D-glucopyranose], an enzymatic hydrolysis product of starch; **isomaltose,** its α(1 → 6) isomer; and **cellobiose,** its β(1 → 4) isomer, the repeating disaccharide of cellulose. Only a few tri- or higher oligosaccharides occur in significant natural abundance. Not surprisingly, they all occur in plants.

Figure 11-13 **Several common disaccharides.** See Kinemage Exercise 7-2

C. *Structural Polysaccharides: Cellulose and Chitin*

Plants have rigid cell walls (Fig. 1-9) that, in order to maintain their shapes, must be able to withstand osmotic pressure differences between the extracellular and intracellular spaces of up to 20 atm. In large plants, such as trees, the cell

Figure 11-14 **Electron micrograph of the cellulose fibers in the cell wall of the alga *Chaetomorpha melagonium.*** Note that the cell wall consists of layers of parallel fibers. [Biophoto Associates/Photo Researchers.]

Cellulose

Figure 11-15 **The primary structure of cellulose.** Here *n* may be several thousand.

walls also have a load-bearing function. Cellulose, the primary structural component of plant cell walls (Fig. 11-14), accounts for over half of the carbon in the biosphere: $\sim 10^{15}$ kg of cellulose are estimated to be synthesized and degraded annually. Although cellulose is predominantly of vegetable origin, it also occurs in the stiff outer mantles of marine invertebrates known as **tunicates** (urochordates; Fig. 1-11).

The primary structure of cellulose was determined through methylation analysis. Cellulose is a linear polymer of up to 15,000 D-glucose residues (a glucan) linked by $\beta(1 \rightarrow 4)$ glycosidic bonds (Fig. 11-15). As is generally true of large polysaccharides, it has no defined size since, in contrast to proteins and nucleic acids, there is no genetically determined template that directs its synthesis.

X-ray studies of cellulose fibers led Anatole Sarko to tentatively propose the model diagrammed in Fig. 11-16. This highly cohesive, hydrogen bonded structure gives cellulose fibers exceptional strength and makes them water insoluble despite their hydrophilicity.

Figure 11-16 **Proposed structural model of cellulose.** Cellulose fibers consist of ~40 parallel glucan chains arranged in an extended fashion. Each of the $\beta(1 \rightarrow 4)$-linked glucose units in a chain is turned 180° with respect to its preceding residue and is held in this position by intrachain hydrogen bonds (*dashed lines*). The glucan chains line up laterally to form sheets, and these sheets stack vertically such that they are staggered by half the length of a glucose unit. The entire assembly is stabilized by intermolecular hydrogen bonds between glucose units of neighboring chains. Hydrogen atoms not participating in hydrogen bonds have been omitted for clarity. [Illustration, Irving Geis. Image from the Irving Geis Collection, Howard Hughes Medical Institute. Reprinted with permission.]

In plant cell walls, the cellulose fibers are embedded in and cross-linked by a matrix of several polysaccharides that are composed of glucose as well as other monosaccharides. In wood, this cementing matrix also contains a large proportion of **lignin,** a plasticlike phenolic polymer. One has only to watch a tall tree in a high wind to realize the enormous strength of plant cell walls. In engineering terms, they are "composite materials," as is concrete reinforced by steel rods. Composite materials can withstand large stresses because the matrix evenly distributes the stresses among the reinforcing elements.

Although vertebrates themselves do not possess an enzyme capable of hydrolyzing the β(1 → 4) linkages of cellulose, the digestive tracts of herbivores contain symbiotic microorganisms that secrete a series of enzymes, collectively known as **cellulase,** that do so. The same is true of termites. Nevertheless, the degradation of cellulose is a slow process because its tightly packed and hydrogen bonded glucan chains are not easily accessible to cellulase and do not separate readily even after many of their glycosidic bonds have been hydrolyzed. The digestion of fibrous plants such as grass by herbivores is therefore a more complex and time-consuming process than is the digestion of meat by carnivores (cows, e.g., have multichambered stomachs and must chew their cud). Similarly, the decay of dead plants by fungi, bacteria, and other organisms, and the consumption of wooden houses by termites, often takes years.

Chitin is the principal structural component of the exoskeletons of invertebrates such as crustaceans, insects, and spiders and is also a major cell wall constituent of most fungi and many algae. It is estimated that ~10^{14} kg of chitin are produced annually, most of it in the oceans, and therefore that it is almost as abundant as is cellulose. Chitin is a homopolymer of β(1 → 4)-linked *N*-acetyl-D-glucosamine residue (Fig. 11-17). It differs chemically from cellulose only in that each C2-OH group is replaced by an acetamido function. X-ray analysis indicates that chitin and cellulose have similar structures.

D. Storage Polysaccharides: Starch and Glycogen

a. Starch Is a Food Reserve in Plants and a Major Nutrient for Animals

Starch is a mixture of glucans that plants synthesize as their principal food reserve. It is deposited in the cytoplasm of plant cells as insoluble granules composed of **α-amylose** and **amylopectin.** α-Amylose is a linear polymer of several thousand glucose residues linked by α(1 → 4) bonds (Fig. 11-18a). Note that although α-amylose is an isomer of cellulose, it has very different structural properties. This is because cellulose's β-glycosidic linkages cause each successive glucose residue to flip 180° with respect to the preceding residue, so that the polymer assumes an easily packed, fully extended conformation (Fig. 11-16). In contrast, α-amylose's α-glycosidic bonds cause it to adopt an irregularly aggregating helically coiled conformation (Fig. 11-18b).

Amylopectin consists mainly of α(1 → 4)-linked glucose residues but is a branched molecule with α(1 → 6) branch

Figure 11-17 **The primary structure of chitin.** Chitin is a β(1 → 4)-linked homopolymer of *N*-acetyl-D-glucosamine.

(a)

(b)

Figure 11-18 **α-Amylose.** (*a*) The D-glucose residues of α-amylose are linked by α(1 → 4) bonds (*red*). Here *n* is several thousand. (*b*) This regularly repeating polymer forms a left-handed helix with ~6 glucose residues per turn. Note the great differences in structure and properties that result from changing α-amylose's α(1 → 4) linkages to the β(1 → 4) linkages of cellulose (Fig. 11-16). [Illustration, Irving Geis. Image from the Irving Geis Collection, Howard Hughes Medical Institute. Reprinted with permission.]

(a)

Branch

Main chain

$\alpha(1 \rightarrow 6)$ branch point

Amylopectin

Figure 11-19 **Amylopectin.** (*a*) Its primary structure near one of its $\alpha(1 \rightarrow 6)$ branch points (*red*). (*b*) Its bushlike structure with glucose residues at branch points indicated in red. The actual distance between branch points averages 24 to 30 glucose residues. Glycogen has a similar structure but is branched every 8 to 14 residues.

(b)

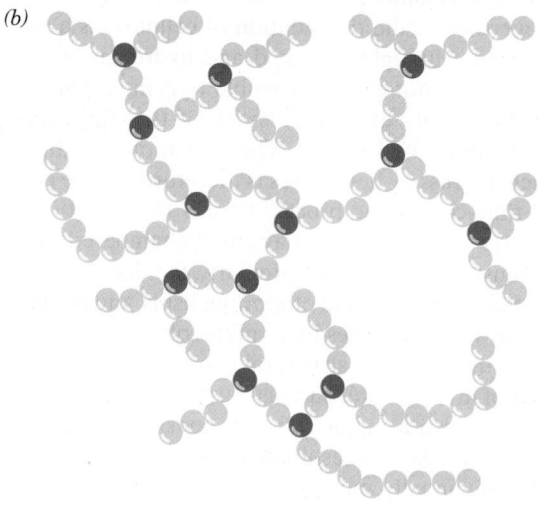

points every 24 to 30 glucose residues on average (Fig. 11-19). Amylopectin molecules contain up to 10^6 glucose residues, which makes them among the largest molecules occurring in nature. The storage of glucose as starch greatly reduces the large intracellular osmotic pressures that would result from its storage in monomeric form because osmotic pressure is proportional to the number of solute molecules in a given volume.

b. Starch Digestion Occurs in Stages

The digestion of starch, the main carbohydrate source in the human diet, begins in the mouth. Saliva contains **α-amylase,** which randomly hydrolyzes all the $\alpha(1 \rightarrow 4)$ glucosidic bonds of starch except its outermost bonds and those next to branches. By the time thoroughly chewed food reaches the stomach, where the acidity inactivates α-amylase, the average chain length of starch has been reduced from several thousand to fewer than eight glucose units. Starch digestion continues in the small intestine under the influence of pancreatic α-amylase, which is similar to the salivary enzyme. This enzyme degrades starch to a mixture of the disaccharide maltose, the trisaccharide **maltotriose,** which contains three $\alpha(1 \rightarrow 4)$-linked glucose residues, and oligosaccharides known as **dextrins** that contain the $\alpha(1 \rightarrow 6)$ branches. These oligosaccharides are hydrolyzed to their component monosaccharides by specific enzymes contained in the brush border membranes of the intestinal mucosa: an **α-glucosidase,** which removes one glucose residue at a time from oligosaccharides, an **α-dextrinase** or **debranching enzyme,** which hydrolyzes $\alpha(1 \rightarrow 6)$ and $\alpha(1 \rightarrow 4)$ bonds, a **sucrase,** and, at least in infants, a lactase. The resulting monosaccharides are absorbed by the intestine and transported to the bloodstream (Section 20-4A).

c. Glycogen Is "Animal Starch"

Glycogen, the storage polysaccharide of animals, is present in all cells but is most prevalent in skeletal muscle and liver, where it occurs as cytoplasmic granules (Fig. 11-20). The primary structure of glycogen resembles that of amylopectin, but glycogen is more highly branched, with branch points occurring every 8 to 14 glucose residues. Glycogen's degree of polymerization is nevertheless similar to that of amylopectin. In the cell, glycogen is degraded for metabolic use by **glycogen phosphorylase,** which phosphorolytically cleaves glycogen's $\alpha(1 \rightarrow 4)$ bonds sequentially inward from its nonreducing ends to yield **glucose-1-phosphate.** Glycogen's highly branched structure, which has many nonreducing ends, permits the rapid mobilization of glucose in times of metabolic need. The $\alpha(1 \rightarrow 6)$ branches of glycogen are cleaved by a debranching enzyme. These enzymes play an important role in glucose metabolism and are discussed further in Section 18-1.

E. *Glycosaminoglycans*

The extracellular spaces, particularly those of connective tissues such as cartilage, tendon, skin, and blood vessel walls, consist of collagen and elastin fibers (Section 8-2B)

Figure 11-20 **Photomicrograph showing the glycogen granules** (*pink*) **in the cytoplasm of a liver cell.** The greenish objects are mitochondria and the yellow object is a fat globule. Note that the glycogen granules tend to aggregate. The glycogen content of liver may reach as high as 10% of its net weight. [CNRI/Science Photo Library/Photo Researchers, Inc.]

embedded in a gel-like matrix known as **ground substance.** Ground substance is composed largely of **glycosaminoglycans (GAGs;** alternatively, **mucopolysaccharides**), unbranched polysaccharides of alternating uronic acid and hexosamine residues. Solutions of GAGs have a slimy, mucuslike consistency that results from their high viscosity and elasticity. In the following paragraphs, we discuss the structural origin of these important mechanical properties.

a. Hyaluronic Acid

Hyaluronic acid (also called **hyaluronan**) is an important GAG component of ground substance, synovial fluid (the fluid that lubricates the joints), and the vitreous humor of the eye. It also occurs in the capsules surrounding certain, usually pathogenic, bacteria. Hyaluronic acid molecules are composed of 250 to 25,000 $\beta(1 \rightarrow 4)$-linked disaccharide units that consist of D-glucuronic acid and *N*-acetyl-D-glucosamine linked by a $\beta(1 \rightarrow 3)$ bond (Fig. 11-21). The anionic character of its glucuronic acid residues causes hyaluronic acid to bind cations such as K^+, Na^+, and Ca^{2+} tightly. X-ray fiber analysis indicates that Ca^{2+} hyaluronate

Figure 11-21 **The disaccharide repeating units of the common glycosaminoglycans.** The anionic groups are drawn in red and the *N*-acetylamido groups are drawn in blue. See Kinemage Exercise 7-3

Figure 11-22   X-ray fiber structure of Ca²⁺ hyaluronate. Three consecutive disaccharide units of the hyaluronate fiber are drawn in stick form with atoms colored according to type with glucuronate C green, N-acetyl-D-glucosamine C cyan, H white, N blue, and O red. Ca²⁺ ions are represented by blue spheres. The hyaluronate polyanion forms an extended, left-handed, single-stranded helix with a pitch of 28.3 Å and ~3 disaccharide units per turn that is stabilized by intramolecular hydrogen bonds (*dashed lines*). The positions of the H atoms are inferred and hence the H atoms of the OH groups are not shown. [Based on a fiber X-ray structure by Struther Arnott, Purdue University. PDBid 4HYA.]

forms an extended, left-handed, single-stranded helix with ~3 disaccharide units per turn (Fig. 11-22).

Hyaluronate's structural features suit it to its biological function. Its high molecular mass and numerous mutually repelling anionic groups make hyaluronate an extended, rigid, and highly hydrated molecule which, in solution, occupies a volume ~1000 times that in its dry state. Hyaluronate solutions therefore have a viscosity that is shear dependent (an object under shear stress has equal and opposite forces applied across its opposite faces). At low shear rates, the hyaluronate molecules form tangled masses that greatly impede flow; that is, the solution is quite viscous. As the shear rate increases, the stiff rodlike hyaluronate molecules tend to line up with the flow and thus offer less resistance to it. This viscoelastic behavior makes hyaluronate solutions excellent biological shock absorbers and lubricants.

Hyaluronic acid and other GAGs (see below) are degraded by **hyaluronidase,** which hydrolyzes their β(1 → 4) linkages. Hyaluronidase occurs in a variety of animal tissues, in bacteria (where it presumably expedites their invasion of animal tissue), and in snake and insect toxins.

b. Other Glycosaminoglycans

Other GAG components of ground substance consist of 50 to 1000 sulfated disaccharide units which occur in proportions that are both tissue and species dependent. The most prevalent structures of these generally heterogeneous substances are (Fig. 11-21)

1. Chondroitin-4-sulfate (Greek: *chondros,* cartilage), a major component of cartilage and other connective tissue,

has N-acetyl-D-galactosamine-4-sulfate residues in place of hyaluronate's N-acetyl-D-glucosamine residues.

2. Chondroitin-6-sulfate is instead sulfated at the C6 position of its N-acetyl-D-galactosamine residues. The two chondroitin sulfates occur separately or in mixtures depending on the tissue.

3. Dermatan sulfate (Greek: *derma,* skin), which is so named because of its prevalence in skin, differs from chondroitin-4-sulfate only by an inversion of configuration about C5 of the β-D-glucuronate residues to form α-L-iduronate. This results from the enzymatic epimerization of these residues after the formation of chondroitin. The epimerization is usually incomplete, so dermatan sulfate also contains glucuronate residues.

4. Keratan sulfate (Greek: *keras,* horn; not to be confused with the protein keratin) consists mainly of alternating β(1 → 4)-linked D-galactose and N-acetyl-D-glucosamine-6-sulfate residues (and hence lacks uronic acid residues). It is a component of cartilage, bone, cornea, as well as hair, nails, and horn. Keratan sulfate is the most heterogeneous of the major GAGs in that its sulfate content is variable and it contains small amounts of fucose, mannose, N-acetylglucosamine, and sialic acid.

5. Heparin is a variably sulfated GAG that consists predominantly of alternating α(1 → 4)-linked residues of L-iduronate-2-sulfate and N-sulfo-D-glucosamine-6-sulfate. It has an average of 2.5 sulfate residues per disaccharide unit, which makes it the most negatively charged polyelectrolyte in mammalian tissues (Fig. 11-23). Heparin, in contrast to the above GAGs, is not a constituent of connective tissue, but occurs almost exclusively in the intracellular granules of the **mast cells** that line arterial walls, especially in the liver, lungs, and skin. It inhibits the clotting of blood, and its release, through injury, is thought to prevent runaway clot formation (Section 35-1Ea). Heparin is therefore in wide clinical use to inhibit blood clotting, for example, in

Figure 11-23   NMR structure of heparin. Three consecutive disaccharide units of this helical polymer are shown in stick form. Atoms are colored according to type with glucosamine C green, iduronate C cyan, H white, N blue, O red, and S yellow. The helical repeat unit is two disaccharides with a pitch of 17.5 Å. Note the high density of anionic sulfate groups. [Based on an NMR structure by Barbara Mulloy and Mark Forster, National Institute for Biological Standards and Control, Herts, U.K. PDBid 1HPN.]

postsurgical patients. **Heparan sulfate,** a ubiquitous cell-surface component as well as an extracellular substance in blood vessel walls and brain, resembles heparin but has a far more variable composition with fewer *N*- and *O*-sulfate groups and more *N*-acetyl groups.

3 GLYCOPROTEINS

Until about 1960, carbohydrates were thought to be rather dull compounds that were probably some sort of inert filler. Protein chemists therefore considered them to be a nuisance that complicated protein "purification." In fact, most eukaryotic proteins are **glycoproteins,** that is, they are covalently associated with carbohydrates. Glycoproteins vary in carbohydrate content from <1% to >90% by weight. They occur in all forms of life and have functions that span the entire spectrum of protein activities, including those of enzymes, transport proteins, receptors, hormones, and structural proteins. Their carbohydrate moieties, as we shall see, have several important biological roles, but in many cases their functions remain enigmatic.

The polypeptide chains of glycoproteins, like those of all proteins, are synthesized under genetic control. Their carbohydrate chains, in contrast, are enzymatically generated and covalently linked to the polypeptide without the rigid guidance of nucleic acid templates. The processing enzymes are generally not available in sufficient quantities to ensure the synthesis of uniform products. Glycoproteins therefore have variable carbohydrate compositions, a phenomenon known as **microheterogeneity,** that compounds the difficulties in their purification and characterization.

In this section we consider the structures and properties of glycoproteins. In particular, we shall study the glycoproteins of connective tissues, those of bacterial cell walls, and several soluble glycoproteins. We end by discussing the general principles of glycoprotein structure and function.

A. *Proteoglycans*

Proteins and glycosaminoglycans in ground substance, in **basal laminae [basement membranes;** the thin matlike extracellular matrix separating **epithelial cells** (the cells lining body cavities and free surfaces) from underlying cells], and in cell-surface membranes aggregate covalently and noncovalently to form a diverse group of macromolecules known as **proteoglycans.** *Proteoglycans consist of a **core protein** to which at least one glycosaminoglycan chain, most often keratan sulfate and/or chondroitin sulfate, is covalently linked.* Numerous types of core proteins have been characterized (Table 11-1). Proteoglycans appear to have multiple roles, most notably as organizers of tissue morphology via their interactions with molecules such as collagen; as selective filters that regulate the traffic of molecules according to their size and/or charge; and as regulators of the activities of other proteins, particularly those involved in signaling (see below).

Electron micrographs such as Fig. 11-24*a* together with reconstitution experiments indicate that proteoglycans can form huge complexes. For example, **aggrecan,** the main proteoglycan component of cartilage, has a bottlebrush-like molecular architecture (Fig. 11-24*b*), whose **proteoglycan subunit** "bristles" are noncovalently attached to a filamentous hyaluronic acid "backbone" at intervals of 200 to 300 Å. Aggrecan has three domains. Its N-terminal domain forms a globular region of 60 to 70 kD that binds noncovalently

Table 11-1 **Properties of Some Proteoglycans**

Proteoglycan	Approximate Core Protein Molecular Mass (kD)	Glycosaminoglycan Type (Number)[a]
Proteoglycans interacting with hyaluronic acid		
Aggrecan	220	CS (~100), KS (~30)
Versican	265–370	CS/DS (10–30)
Neurocan	136	CS (3–7)
Proteoglycans of the basal laminae		
Perlecan	400–467	HS/CS (3)
Agrin	250	HS (3)
Bamacan	138	CS (3)
Small leucine-rich proteoglycans		
Decorin	40	DS/CS (1)
Fibromodulin	42	KS (2–3)
Osteoglycin	35	KS (2–3)

[a]Abbreviations: CS, chondroitin sulfate; DS, dermatan sulfate; HS, heparan sulfate; KS, keratan sulfate.

Source: Iozzo, R.V., *Annu. Rev. Biochem.* **67,** 611, 626, and 624 (1998).

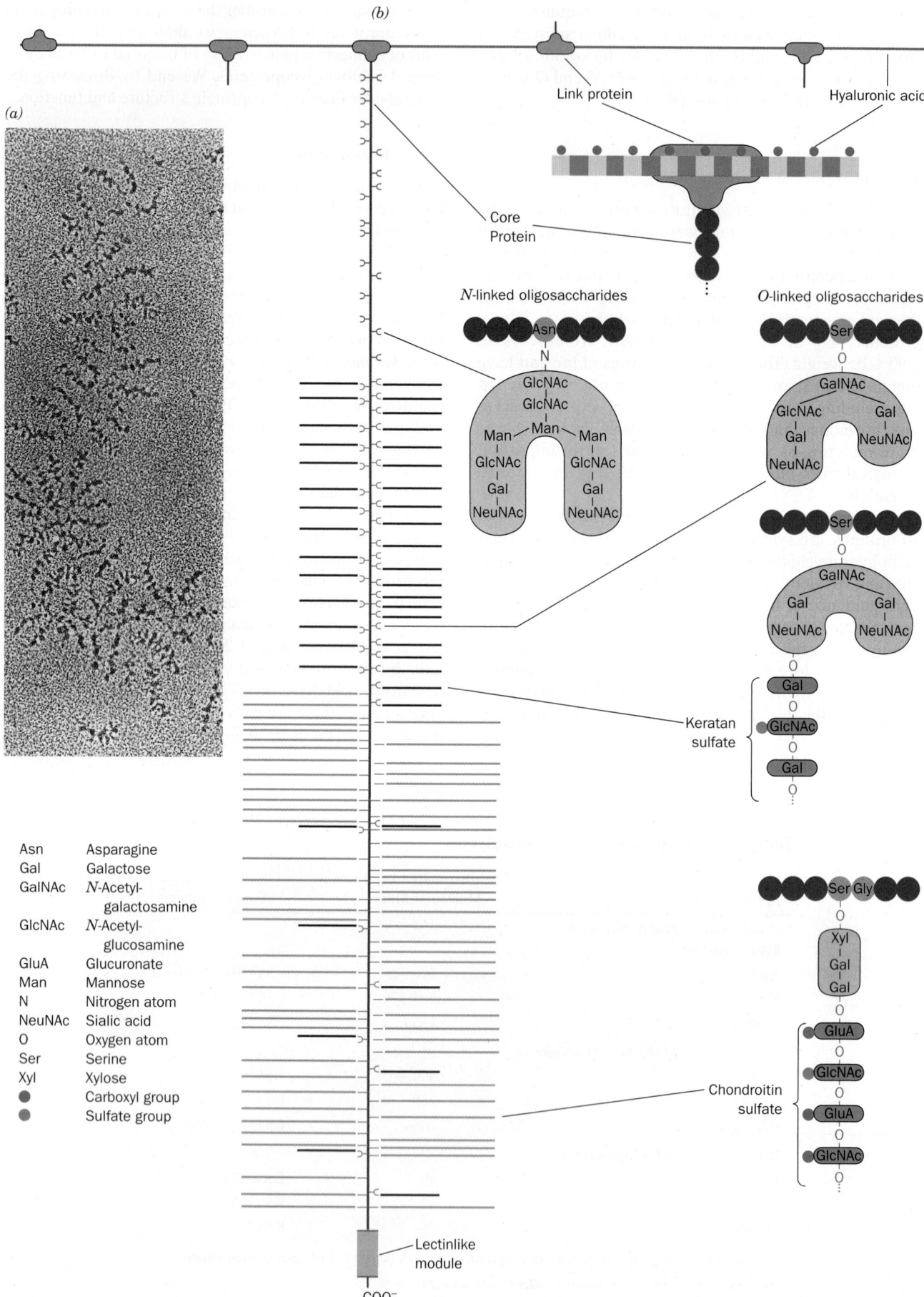

(a)

(b)

Link protein

Hyaluronic acid

Core Protein

N-linked oligosaccharides

O-linked oligosaccharides

Asn

N

GlcNAc
GlcNAc

Man — Man — Man

GlcNAc   GlcNAc

Gal   Gal

NeuNAc   NeuNAc

Ser

O

GalNAc

GlcNAc   Gal

Gal   NeuNAc

NeuNAc

Ser

O

GalNAc

Gal   Gal

NeuNAc   NeuNAc

O

Gal

O

GlcNAc

O

Gal

⋮

Keratan sulfate

Ser Gly

O

Xyl

Gal

Gal

O

GluA

O

GlcNAc

O

GluA

O

GlcNAc

O

⋮

Chondroitin sulfate

Asn	Asparagine
Gal	Galactose
GalNAc	*N*-Acetyl-galactosamine
GlcNAc	*N*-Acetyl-glucosamine
GluA	Glucuronate
Man	Mannose
N	Nitrogen atom
NeuNAc	Sialic acid
O	Oxygen atom
Ser	Serine
Xyl	Xylose
●	Carboxyl group
●	Sulfate group

Lectinlike module

COO⁻

Figure 11-24 (*Opposite*) **Proteoglycans.** (*a*) An electron micrograph showing a central strand of hyaluronic acid, which runs down the field of view, supporting numerous projections, each of which consists of a core protein to which many bushy polysaccharide protrusions are linked. [From Caplan, A.I., *Sci. Am.* **251**(4); 87 (1984). Copyright © 1984 Scientific American, Inc. Used by permission.] (*b*) The bottlebrush model of the proteoglycan aggrecan. The core proteins, one of which is shown extending down through the middle of the diagram, project from the central hyaluronic acid strand. The core is noncovalently anchored to the hyaluronic acid via its globular N-terminal end in an association that is stabilized by link protein. The core has three saccharide-binding regions: (1) the inner region predominantly binds oligosaccharides via the side chain N atoms of Asn residues; (2) the central region binds oligosaccharides, many of which bear keratan sulfate chains, via the side chain O atoms of Ser and Thr residues; and (3) the outer region mainly binds chondroitin sulfate chains that are linked to the core protein via a galactose–galactose–xylose trisaccharide that is bonded to side chain O atoms of Ser residues in the sequence Ser-Gly. The C-terminal end of the aggrecan core protein consists of a lectinlike sequence.

to hyaluronic acid. This attachment is stabilized by the 40- to 60-kD **link protein,** which is similar in sequence to aggrecan's N-terminal domain. Aggrecan's highly extended central domain is covalently linked to a series of polysaccharides, which comprise nearly 90% of this glycoprotein's mass. They divide the central domain into three regions:

1. An N-terminal region, which overlaps the globular hyaluronic acid–binding domain, binds a relatively few carbohydrate chains. These tend to be oligosaccharides that are covalently bonded to the protein via the amide N atoms of specific Asn residues (Section 11-3Ca).

2. A region rich in oligosaccharides, many of which serve as anchor points for keratan sulfate chains. These oligosaccharides are covalently bonded to side chain O atoms of Ser and Thr residues.

3. A C-terminal region rich in chondroitin sulfate chains, which are covalently linked to the side chain O atoms of Ser residues in Ser-Gly dipeptides via galactose–galactose–xylose trisaccharides.

Aggrecan's C-terminal domain contains a lectinlike module, which binds certain monosaccharide units. Thus, aggrecan probably functions to bind together various constituents of the cell surface and the extracellular matrix (see below).

Altogether, a central strand of hyaluronic acid, which varies in length from 4000 to 40,000 Å, noncovalently binds up to 100 associated aggrecan chains, each of which covalently binds ~30 keratan sulfate chains of up to 250 disaccharide units each and ~100 chondroitin sulfate chains of up to 1000 disaccharide units each. This accounts for the enormous molecular masses of the aggrecans, which range up to 220,000 kD, and for their high degree of **polydispersity** (range of molecular masses). Note, however, that many

proteoglycans do not bind to hyaluronic acid (Table 11-1) and hence function as monomers.

a. Cartilage's Mechanical Properties Are Explained by Its Molecular Structure

Cartilage consists largely of a meshwork of collagen fibrils that is filled in by proteoglycans whose chondroitin sulfate and core protein components specifically interact with the collagen. The tensile strength of cartilage and other connective tissues is, as we have seen (Section 8-2Ba), a consequence of their collagen content. Cartilage's characteristic resilience, however, results from its high proteoglycan content. The extended brushlike structure of proteoglycans, together with the polyanionic character of keratan sulfate and chondroitin sulfate, cause this complex to be highly hydrated. The application of pressure on cartilage squeezes water away from these charged regions until charge–charge repulsions prevent further compression. When the pressure is released, the water returns. Indeed, the cartilage in the joints, which lack blood vessels, is nourished by this flow of liquid brought about by body movements. This explains why long periods of inactivity cause joint cartilage to become thin and fragile.

b. Proteoglycans Modulate the Effects of Protein Growth Factors

Proteoglycans have been implicated in a great variety of cellular processes. For example, **fibroblast growth factor** (**FGF;** growth factors are proteins that function to induce their specific target cells to grow and/or differentiate; Section 19-3Aa) binds to heparin or to the heparan sulfate chains of proteoglycans and is only bound to its cell-surface receptor in complex with these glycosaminoglycans. Since the binding of FGF to heparin or heparan sulfate protects FGF from degradation, the release of this growth factor from the extracellular matrix by the proteolysis of proteoglycan core proteins or by the partial degradation of heparan sulfate probably provides an important source of active FGF–glycosaminoglycan complexes. Several other growth factors interact similarly with proteoglycans. Apparently, the abundant and ubiquitous distribution of proteoglycans limits the action of these growth factors on their target cells to short distances from the cells secreting the growth factors, a phenomenon that probably greatly influences the formation and maintenance of tissue architecture.

B. *Bacterial Cell Walls*

Bacteria are surrounded by rigid cell walls (Fig. 1-13) that give them their characteristic shapes (Fig. 1-1) and permit them to live in hypotonic (less than intracellular salt concentration) environments that would otherwise cause them to swell osmotically until their plasma (cell) membranes lysed (burst). Bacterial cell walls are of considerable medical significance because they are responsible for bacterial **virulence** (disease-evoking power). In fact, the symptoms of many bacterial diseases can be elicited in animals merely by the injection of bacterial cell walls. Furthermore, the

(a) Gram-positive bacteria

— Peptidoglycan (cell wall)
— Plasma membrane
— Cytoplasm

(b) Gram-negative bacteria

— Outer membrane
— Peptidoglycan (cell wall)
— Periplasmic space
— Plasma membrane
— Cytoplasm

Figure 11-25   Schematic diagram comparing the cell envelopes of *(a)* gram-positive bacteria and *(b)* gram-negative bacteria.

characteristic **antigens** (immunological markers; Section 35-2) of bacteria are components of their cell walls and capsules, so that injection of preparations of these substances into an animal often invokes its immunity against these bacteria. Consequently, several vaccines that are based on purified bacterial polysaccharides have recently become available, including those against *Streptococcus pneumoniae*, a major cause of pneumonia, and *Neisseria meningitidis*, a major cause of meningitis.

Bacteria are classified as **gram-positive** or **gram-negative** depending on whether or not they take up gram stain (Section 1-1B). Gram-positive bacteria (Fig. 11-25*a*) have a thick (~250 Å) cell wall surrounding their plasma membrane, whereas gram-negative bacteria (Fig. 11-25*b*) have a thin (~30 Å) cell wall covered by a complex outer membrane.

a. Bacterial Cell Walls Have a Peptidoglycan Framework

The cell walls of both gram-positive and gram-negative bacteria consist of covalently linked polysaccharide and polypeptide chains that form a framework that completely encases the cell. This substance, whose molecular structure was elucidated in large part by Jack Strominger, is known as a **peptidoglycan** or **murein** (Latin: *murus*, wall). Its polysaccharide component consists of linear chains of alternating $\beta(1 \rightarrow 4)$-linked *N*-acetylglucosamine **(NAG)** and *N*-acetylmuramic acid **(NAM).** The NAM's lactic acid residue forms an amide bond with a D-amino acid–containing tetrapeptide to form the peptidoglycan repeating unit (Fig. 11-26). Neighboring parallel peptidoglycan chains are covalently cross-linked through their tetrapeptide side chains. In the gram-positive bacterium *Staphylococcus aureus*, whose tetrapeptide has the sequence L-Ala-D-isoglutamyl-L-Lys-D-Ala, this cross-link consists of a pentaglycine chain that extends from the terminal carboxyl group of one tetrapeptide to the ε-amino group of the Lys in a neighboring tetrapeptide.

Atomic force microscopy (AFM; an imaging technique that reports the variation in the force between a probe that is several nanometers in diameter and a surface of interest as the probe is scanned over the surface; its resolution is as little as several Ångstroms) was used by Simon Foster to

image the cell wall of the gram-negative bacterium *Bacillus subtilis* leading to the following model (Fig. 11-27). Several glycan chains are cross-linked much as described above to form a peptidoglycan "rope," which due to its natural twist, forms an ~50-nm-diameter helical cable of up to 50 μm in length that coils around the long axis of the bacterium to form its cell wall. This structure is presumably stabilized by the formation of covalent cross-links between neighboring segments of the coil. The cell walls of gram-negative bacteria appear to be only one layer thick, whereas as those of gram-positive bacteria are postulated to consist of several such layers. How the peptidoglycan imposes cell shape is unknown.

The D-amino acids of peptidoglycans render them resistant to proteases. However, **lysozyme,** an enzyme which is present in tears, mucus, and other vertebrate body secretions, as well as in egg whites, catalyzes the hydrolysis of the $\beta(1 \rightarrow 4)$ glycosidic linkage between NAM and NAG. Consequently, treatment of gram-positive bacteria with lysozyme degrades their cell walls, which results in their lysis (gram-negative bacteria are resistant to lysozyme degradation). Lysozyme was discovered in 1922 by the British bacteriologist Alexander Fleming after he noticed that a bacterial culture had dissolved where mucus from a sneeze had landed. It was Fleming's hope that lysozyme would be a universal antibiotic but, unfortunately, it is clinically ineffective against pathogenic bacteria. The structure and mechanism of lysozyme are examined in detail in Section 15-2.

b. Penicillin Kills Bacteria by Inhibiting Cell Wall Biosynthesis

In 1928, Fleming noticed that the chance contamination of a bacterial culture plate with the mold *Penicillium notatum* lysed nearby bacteria (a clear demonstration of Pasteur's maxim that chance favors a prepared mind). This was caused by the presence of **penicillin** (Fig. 11-28), an antibiotic secreted by the mold. Yet the difficulties of isolating and characterizing penicillin, owing to its instability, led to the passage of over 15 years before penicillin was available for routine clinical use. Penicillin specifically binds to and inactivates enzymes that function to cross-link the peptidoglycan strands of bacterial cell walls. Since cell *(a)*   *N*-Acetylglucosamine   *N*-Acetylmuramic acid

L-Ala

Isoglutamate

L-Lys

D-Ala

(b)

N-Acetylmuramic acid

N-Acetylglucosamine

Peptide chain

Pentaglycine bridge

Figure 11-26   **Chemical structure of peptidoglycan.** *(a)* The repeating unit of peptidoglycan is an NAG–NAM disaccharide whose lactyl side chain forms an amide bond with a tetrapeptide. The tetrapeptide of *S. aureus* is shown. The isoglutamate is so designated because it forms an amide link via its γ-carboxyl group. In some species, its α-carboxylate group is replaced by an amide group to form D-isoglutamine and/or the L-Lys residue may have a carboxyl group appended to its C_ϵ to form **diaminopimelic acid.** *(b)* The *S. aureus* bacterial cell wall peptidoglycan. In other gram-positive bacteria, the Gly_5 connecting bridges shown here may contain different amino acid residues such as Ala or Ser. In gram-negative bacteria, the peptide chains are directly linked via peptide bonds.

wall expansion also requires the action of enzymes that degrade cell walls, *exposure of growing bacteria to penicillin results in their lysis;* that is, penicillin disrupts the normal balance between cell wall biosynthesis and degradation.

However, since no human enzyme binds penicillin, it is of low human toxicity, a therapeutic necessity.

Penicillin-treated bacteria that are kept in a hypertonic medium remain intact, even though they have no cell wall.

Penicillin

Figure 11-28   **Structure of penicillin.** Penicillin contains a thiazolidine ring *(red)* fused to a β-lactam ring *(blue)*. A variable R group is bonded to the β-lactam ring via a peptide linkage. In benzyl penicillin (penicillin G), one of several naturally occurring derivatives that are clinically effective, R is the benzyl group (—CH$_2$φ). In **ampicillin,** a semisynthetic derivative, R is the aminobenzyl group [—CH(NH$_2$)φ].

Figure 11-27   **Model of the *B. subtilis* cell wall.** The cell wall consists of a right-handed helical cable composed of several peptidoglycan strands that wraps about the bacterium's plasma membrane. The cell is ~3 μm long. [Courtesy of Simon Foster, University of Sheffield, U.K.]

Figure 11-29 Enzymatic inactivation of penicillin. Penicillinase inactivates penicillin by catalyzing the hydrolysis of its β-lactam ring to form **penicillinoic acid.**

Such bacteria, which are called **protoplasts** or **spheroplasts,** are spherical and extremely fragile because they are encased by only their plasma membranes. Protoplasts immediately lyse on transfer to a normal medium.

Most bacteria that are resistant to penicillin secrete a **β-lactamase** (also known as **penicillinase**), which inactivates penicillin by hydrolytically cleaving the amide bond of its β-lactam ring (Fig. 11-29). However, the observation that penicillinase activity varies with the nature of penicillin's R group has prompted the semisynthesis of penicillins, such as **ampicillin** (Fig. 11-28), which are clinically effective against penicillin-resistant strains of bacteria. In addition, penicillins are often administered in combination with β-lactamase inhibitors such as **sulbactam.**

Sulbactam

c. Bacterial Cell Walls Are Studded with Antigenic Groups

The surfaces of gram-positive bacteria are covered by **teichoic acids** (Greek: *teichos,* city walls), which account for up to 50% of the dry weight of their cell walls. Teichoic acids are polymers of glycerol or ribitol linked by phosphodiester bridges (Fig. 11-30). The hydroxyl groups of this sugar–phosphate chain are substituted by D-Ala residues and saccharides such as glucose or NAG. Teichoic acids are anchored to the peptidoglycans via phosphodiester bonds to the C6-OH groups of their NAG residues. They often terminate in **lipopolysaccharides** (lipids that contain polysaccharides; Section 12-1).

The outer membranes of gram-negative bacteria (Fig. 11-25b) are composed of complex lipopolysaccharides, proteins, and phospholipids that are organized in a complicated manner. The **periplasmic space,** an aqueous compartment that lies between the plasma membrane and the peptidoglycan cell wall, contains proteins that transport sugars and other nutrients. The outer membrane functions as a barrier to exclude harmful substances (such as gram stain). This accounts for the observation that gram-negative bacteria are less affected by lysozyme and penicillin, as well as by other antibiotics, than are gram-positive bacteria.

The outer surfaces of gram-negative bacteria are coated with complex and often unusual polysaccharides known as **O-antigens** that uniquely mark each bacterial strain (Fig. 11-31). The observation that mutant strains of pathogenic bacteria lacking O-antigens are nonpathogenic suggests

Figure 11-30 Structure of teichoic acid. A segment of a teichoic acid molecule with a glycerol phosphate backbone that bears alternating residues of D-Ala and NAG.

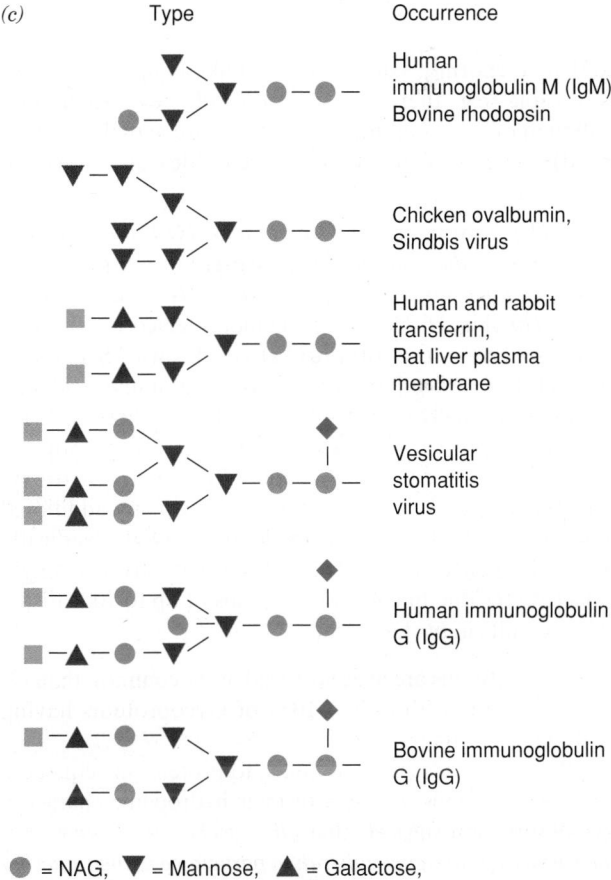

Figure 11-31 **Some of the unusual monosaccharides that occur in the O-antigens of gram-negative bacteria.** These sugars rarely occur in other organisms.

Left column figures: 2-Keto-3-deoxyoctanoate (KDO), L-Glycero-D-mannoheptose, Abequose (Abe), Tyvelose.

that O-antigens participate in the recognition of host cells. O-Antigens, as their name implies, are also the means by which a host's immunological defense system recognizes invading bacteria as foreign (Section 35-2A). As part of the ongoing biological warfare between pathogen and host, O-antigens are subject to rapid mutational alteration so as to generate new bacterial strains that the host does not initially recognize (the mutations are in the genes specifying the enzymes that synthesize the O-antigens).

C. *Glycoprotein Structure and Function*

a. Glycoprotein Carbohydrate Chains Are Highly Diverse

Almost all the secreted and membrane-associated proteins of eukaryotic cells are glycosylated. Indeed, protein glycosylation is more abundant than all other types of posttranslational modifications combined. Oligosaccharides form two types of direct attachments to these proteins: **N-linked** and **O-linked.** Sequence analyses of glycoproteins have led to the following generalizations about these attachments.

1. *In the vast majority of N-glycosidic (N-linked) attachments, an NAG is β-linked to the amide nitrogen of an Asn in the sequence Asn-X-Ser or Asn-X-Thr, where X is any amino acid residue except Pro and only rarely Asp, Glu, Leu, or Trp* (Fig. 11-32a). The oligosaccharides in these linkages usually have a distinctive **core** (innermost sequence; Fig. 11-32b) whose peripheral mannose residues are linked to either mannose or NAG residues. These latter residues may, in turn, be linked to yet other sugar residues,

Figure 11-32 **N-Linked oligosaccharides.** (*a*) All *N*-glycosidic protein attachments occur through a β-*N*-acetylglucosamino–Asn bond in which the Asn occurs in the sequence Asn-X-Ser/Thr (*red*) where X is any amino acid. (*b*) *N*-Linked oligosaccharides usually have the branched (mannose)$_3$(NAG)$_2$ core shown. (*c*) Some examples of *N*-linked oligosaccharides. [After Sharon, N. and Lis, H., *Chem. Eng. News* **59**(13), 28 (1981).] See **Kinemage Exercise 7-4**

(a)

R = H or CH₃

β-Galactosyl-(1→3)-α-*N*-acetylgalactosaminyl-Ser/Thr

(b)

R = H or CH₃

α-Mannosyl-Ser/Thr

Figure 11-33 Some common *O*-glycosidic attachments of oligosaccharides to glycoproteins (*red*).

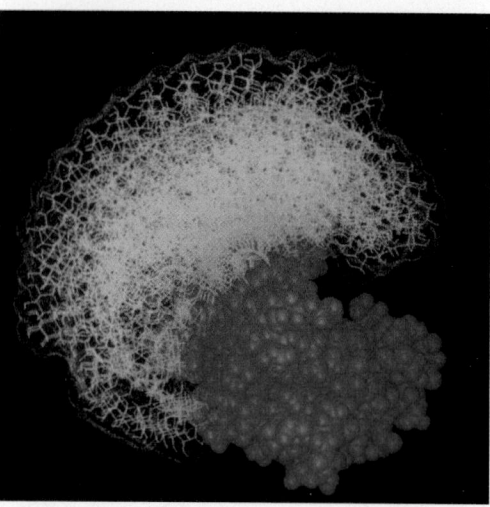

Figure 11-34 Model of oligosaccharide dynamics in bovine pancreatic ribonuclease B (RNase B). The allowed conformations of the (mannose)$_5$(NAG)$_2$ oligosaccharide (*yellow*) that is linked to a single site on the protein (*purple*) are shown in superimposed snapshots. [Courtesy of Raymond Dwek, Oxford University, U.K.]

so that an enormous diversity of *N*-linked oligosaccharides is possible (e.g., there are ~10^{12} possible hexasaccharides, although only a small fraction of them are actually synthesized). Several *N*-linked oligosaccharides are shown in Fig. 11-32*c*.

2. *The most common O-glycosidic (O-linked) attachment involves the disaccharide core β-galactosyl-(1 → 3)-α-N-acetylgalactosamine α-linked to the OH group of either Ser or Thr* (Fig. 11-33*a*). Less commonly, glucose, galactose, mannose, and xylose form α-*O*-glycosides with Ser or Thr (Fig. 11-33*b*). All other hydroxyl-bearing amino acid side chains occasionally form *O*-glycosidic bonds: those with Tyr (e.g., in the protein **glycogenin;** Section 18-2B), 5-hydroxy-Lys (Hyl; e.g., in collagen; Section 8-2Bb), and 4-hydroxy-Pro (Hyp). However, there seem to be few, if any, additional generalizations that can be made about *O*-glycosidically linked oligosaccharides. They vary in size from a single galactose residue in collagen to chains of up to 1000 disaccharide units in proteoglycans.

N-Linked glycans are around 5-fold more common than *O*-linked glycans with only ~10% of glycoproteins having both types of attachments.

Oligosaccharides tend to attach to proteins at sequences that form β bends. Taken with their hydrophilic character, this observation suggests that *oligosaccharides extend from the surfaces of proteins rather than participate in their internal structures.* Indeed, the relatively few glycoprotein X-ray structures that have yet been reported, for example, those of **immunoglobulin G** (Section 35-2Ba) and the influenza virus **hemagglutinin** (Section 33-4Bb), are consistent with this hypothesis. This accounts for the observation that the protein structures of most glycoproteins are unaffected by the removal of their associated oligosaccharides. Both experimental and theoretical studies indicate that oligosaccharides have

mobile and rapidly fluctuating conformations (Fig. 11-34; which accounts for the difficulty in crystallizing them). Thus, representations in which oligosaccharides are shown as having fixed three-dimensional structures do not tell the whole story.

b. Glycoprotein Carbohydrates Have a Variety of Functions

Cells tend to synthesize a large repertoire of a given N-linked glycoprotein, in which each variant species (glycoform) differs somewhat in the sequences, locations, and numbers of its covalently attached oligosaccharides. For example, one of the simplest glycoproteins, bovine pancreatic **ribonuclease B (RNase B),** differs from the well-characterized and carbohydrate-free enzyme RNase A (Section 9-1A) only by the attachment of a single *N*-glycosidically linked oligosaccharide chain. The oligosaccharide has the core sequence diagrammed in Fig. 11-35 with considerable microheterogeneity in the position of a sixth mannose residue. The oligosaccharide does not affect the native enzyme's

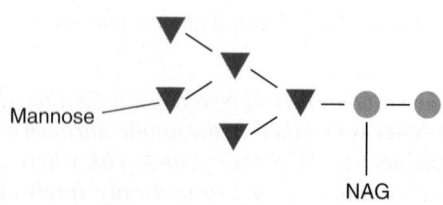

Mannose

NAG

Figure 11-35 The microheterogeneous *N*-linked oligosaccharide of RNase B has the (mannose)$_5$(NAG)$_2$ core shown. A sixth mannose residue occurs at various positions on this core.

conformation, substrate specificity, or catalytic properties. However, RNase A folds to its native state more slowly than does RNase B and tends to aggregate. This suggests that the oligosaccharide functions similarly to a molecular chaperone (Section 9-2C), most likely by shielding a hydrophobic patch on the protein surface.

Human **granulocyte–macrophage colony-stimulating factor (GM-CSF),** a 127-residue protein growth factor that promotes the development, activation, and survival of the white blood cells known as **granulocytes** and **macrophages,** is variably glycosylated at two *N*-linked sites and five *O*-linked sites. Through the generation of mutant varieties of GM-CSF that lack one or both of the *N*-glycosylation sites, it was found that the lifetime of GM-CSF in the bloodstream increases with its level of glycosylation. However, GM-CSF that is produced by *E. coli* and hence is unglycosylated (bacteria rarely glycosylate the proteins they synthesize) has a 20-fold higher specific biological activity than does the naturally occurring glycoprotein.

As the foregoing examples suggest, no generalization can be made about the effects of glycosylation on protein properties; they must be experimentally determined on a case-by-case basis. Nevertheless, it is becoming increasingly evident that glycosylation can affect protein properties in many ways, including protein folding, oligomerization, physical stability, specific bioactivity, rate of clearance from the bloodstream, and protease resistance. Thus, *the species-specific and tissue-specific distribution of glycoforms that each cell synthesizes endows it with a characteristic spectrum of biological properties.*

c. *O*-Linked Glycoproteins Often Have Protective Functions

O-Linked polysaccharides tend not to be uniformly distributed along polypeptide chains. Rather, they are clustered into heavily glycosylated (65–85% carbohydrate by weight) segments in which glycosylated Ser and Thr residues comprise 25 to 40% of the sequence. The carbohydrates' hydrophilic and steric interactions cause these heavily glycosylated regions, which are also rich in Pro and other helix-breaking residues, to assume extended conformations. For example, **mucins,** the protein components of **mucus,** are *O*-linked glycoproteins that can be exceedingly large (up to ~10^7 D) and whose carbohydrate chains are often sulfated and hence mutually repelling. Mucins, which may be membrane-bound or secreted, therefore consist of stiff chains that are devoid of secondary structure and which occupy time-averaged volumes approximating those of small bacteria. Consequently, mucins, at their physiological concentrations, form intertangled networks that comprise the viscoelastic gels that protect and lubricate the mucous membranes that produced them.

Eukaryotic cells, as we shall see in Section 12-3E, have a thick and fuzzy coating of glycoproteins and **glycolipids** named the **glycocalyx** that prevents the close approach of macromolecules and other cells. How, then, can cells interact? Many cell-surface proteins, such as the receptors for various macromolecules, have relatively short and presumably stiff *O*-glycosylated regions that link these glycoproteins' membrane-bound domains to their functional domains. This arrangement is thought to extend the functional domains in a lollipop-like manner above the cell's densely packed glycocalyx, thereby permitting the functional domain to interact with extracellular macromolecules that cannot penetrate the glycocalyx.

d. Oligosaccharide Markers Mediate a Variety of Intercellular Interactions

Glycoproteins are important constituents of plasma membranes (Section 12-3). The location of their carbohydrate moieties can be determined by electron microscopy. The glycoproteins are labeled with lectins that have been conjugated (covalently cross-linked) to **ferritin,** an iron-transporting protein that is readily visible in the electron microscope because of its electron-dense iron hydroxide core. Such experiments, with lectins of different specificities and with a variety of cell types, have demonstrated that *the carbohydrate groups of membrane-bound glycoproteins are, for the most part, located on the external surfaces of cell membranes.* Thus, the viability of cultured cells from multicellular organisms that have any of a large number of glycosylation mutations and the infrequent viability of whole organisms that bear such mutations indicate that oligosaccharides are important for intercellular communications but not for intracellular housekeeping functions.

A further indication that oligosaccharides function as biological markers is the observation that the carbohydrate content of a glycoprotein often governs its metabolic fate. For example, the excision of sialic acid residues from certain radioactively labeled blood plasma glycoproteins by treatment with **sialidase** greatly increases the rate at which these glycoproteins are removed from the circulation. The glycoproteins are taken up and degraded by the liver in a process that depends on the recognition by liver cell receptors of sugar residues such as galactose and mannose, which are exposed by the sialic acid excision. A diverse series of receptors, each specific for a particular type of sugar residue, participates in removing any particular glycoprotein from the blood. A variety of glycoforms for a given glycoprotein therefore probably ensures that it has a range of lifetimes in the blood. *Similar "ticketing" mechanisms probably govern the compartmentation and degradation of glycoproteins within cells.*

The observation that cancerous cells are more susceptible to agglutination by lectins than are normal cells led to the discovery that *there are significant differences between the cell-surface carbohydrate distributions of cancerous and noncancerous cells* (Fig. 11-36). Normal cells stop growing when they touch each other, a phenomenon known as **contact inhibition.** Cancer cells, however, are under no such control and therefore form **malignant tumors** (Section 19-3B).

Carbohydrates are important mediators of cell–cell recognition and have been implicated in related processes such as fertilization, cellular differentiation, the aggregation of cells to form organs, and the infection of cells by

(a)

(b)

Figure 11-36 **The surfaces of (*a*) a normal mouse cell and (*b*) a cancerous cell as seen in the electron microscope.** Both cells were incubated with the ferritin-labeled lectin concanavalin A. The lectin is evenly dispersed on the normal cell but is

aggregated into clusters on the cancerous cell. [Courtesy of Garth Nicolson, The Institute for Molecular Medicine, Huntington Beach, California.]

bacteria and viruses. For example, bacteria initiate infections by attaching to host cells (Fig. 11-37) via bacterial proteins known as **adhesins,** which each specifically bind certain host cell molecules (the adhesins' receptors). In gram-negative bacteria such as *E. coli,* adhesins are often minor components of the heteropolymeric rodlike or-

Figure 11-37 **Scanning electron micrograph of tissue from the inside of a human cheek.** The white cylindrical objects are *E. coli.* The bacteria adhere to mannose residues that are incorporated in the plasma membrane of cheek cells. This is the first step of a bacterial infection. [Courtesy of Fredric Silverblatt and Craig Kuehn, Veterans Administration Hospital, Sepulveda, California.]

ganelles called pili (Fig. 1-3*b*). The so-called P pili that mediate the attachment of the *E. coli* strain that causes urinary tract infections in humans do so via an adhesin named **PapG** protein. This protein specifically binds to the α-D-galactopyranosyl-(1 → 4)-β-D-galactopyranose groups that are present on the surfaces of urinary tract epithelial cells. Electron microscopy studies revealed that the PapG adhesin is located at the end of the P pili's flexible tip, thereby providing this adhesin with considerable steric freedom in binding to its digalactoside receptor.

In the never-ending evolutionary struggle between pathogens and their hosts, mucins have evolved to contain the target oligosaccharides of certain pathogens. These act as decoys that divert these pathogens from their target cells. This, of course, puts selective pressure on the pathogen to evolve a receptor that binds to a different cell-surface oligosaccharide.

D. *Glycomics*

Glycomics, the field of study that structurally and functionally characterizes all the carbohydrates in a given cell type, complements genomics (for DNA) and proteomics (for proteins). It is clear that the **glycome** varies with the species, cell type, developmental stage, and even environmental conditions. However, glycomics is far less developed than genomics or proteomics. There are several reasons for this:

1. The branched structures of oligosaccharides greatly increases their complexity and hence the difficulty in determining their sequences relative to those of polynucleotides and polypeptides, which are invariably linear.

2. The microheterogeneity of oligosaccharides, which often has biological significance (Section 11-3Cb), complicates their characterization relative to polynucleotides and polypeptides, which each have unique primary structures.

3. Because the biosynthesis of carbohydrates is not under direct genetic control, there is no method for amplifying them such as the polymerase chain reaction (PCR; Section 5-5F) for nucleic acids and expression systems for proteins (Section 5-5G). Thus, until recently, the only way of obtaining sufficient quantities of a particular polysaccharide was to isolate it from natural sources.

4. Methods for synthesizing specific oligosaccharides have lagged far behind methods for synthesizing polynucleotides and polypeptides (Sections 7-5 and 7-6). This is due to the branching of oligosaccharides, their large number of functional groups that must be differentially protected during elongation reactions, and the chiral nature of glycosidic bonds. However, in recent years, Peter Seeberger has developed automated, solid-phase methods for synthesizing small oligosaccharides, although these methods are, as yet, incapable of synthesizing all desired oligosaccharides, are time-consuming, and still require considerable expertise.

5. The complexity of an organism's glycome greatly exceeds that of its proteome due to the diversity the glycome's constituent carbohydrates and the number of ways they can interact with one another and with proteins.

A recent advance that has greatly accelerated glycomic research is the development of **carbohydrate microarrays** to identify the carbohydrates that specifically bind to a particular protein, RNA, or even whole cells. In this methodology, which is analogous to the use of DNA microarrays (Section 7-6B), up to several thousand different oligosaccharides are covalently or physically immobilized at specific sites on a solid surface such as a glass slide. A fluorescently labeled protein, RNA, or cell type is then incubated with the microarray, which is subsequently rinsed, and the oligosaccharides to which the protein/RNA/cell binds are identified by the fluorescence at their corresponding positions. In addition to their use in basic research, carbohydrate microarrays have been employed in such diverse applications as the identification of pathogens, the diagnosis of human diseases that are characterized by the presence of certain oligosaccharides, and the development of carbohydrate-based drugs and vaccines.

CHAPTER SUMMARY

Carbohydrates are polyhydroxy aldehydes or ketones of approximate composition $(C \cdot H_2O)_n$ that are important components of biological systems.

1 Monosaccharides The various monosaccharides, such as ribose, fructose, glucose, and mannose, differ in their number of carbon atoms, the positions of their carbonyl groups, and their diastereomeric configurations. These sugars exist almost entirely as cyclic hemiacetals and hemiketals, which, for five- and six-membered rings, are respectively known as furanoses and pyranoses. The two anomeric forms of these cyclic sugars may interconvert by mutarotation. Pyranose sugars have nonplanar rings with boat and chair conformations similar to those of substituted cyclohexanes. Polysaccharides are held together by glycosidic bonds between neighboring monosaccharide units. Glycosidic bonds do not undergo mutarotation. Monosaccharides can be oxidized to aldonic and glycuronic acids or reduced to alditols. An OH group is replaced by H in deoxy sugars and by an amino group in amino sugars.

2 Polysaccharides Carbohydrates can be purified by electrophoretic and chromatographic procedures. Affinity chromatography using lectins has been particularly useful in this regard. The sequences and linkages of polysaccharides may be determined by methylation analysis and by the use of specific exoglycosidases. Similar information may be obtained through NMR spectroscopy and/or mass spectrometric techniques. Cellulose, the structural polysaccharide of plant cell walls, is a linear polymer of $\beta(1 \rightarrow 4)$-linked D-glucose residues. It forms a fibrous hydrogen bonded structure of exceptional strength that in plant cells is embedded in an amorphous matrix. Starch, the food storage polysaccharide of plants, consists of a mixture of the linear $\alpha(1 \rightarrow 4)$-linked glucan α-amylose and the $\alpha(1 \rightarrow 6)$-branched and $\alpha(1 \rightarrow 4)$-linked glucan amylopectin. Glycogen, the animal storage polysaccharide, resembles amylopectin but is more highly branched. Digestion of starch and glycogen is initiated by α-amylase and is completed by specific membrane-bound intestinal enzymes.

3 Glycoproteins Proteoglycans of ground substance are mostly high molecular mass aggregates, many of which structurally resemble a bottlebrush. Their proteoglycan subunits consist of a core protein to which glycosaminoglycans, usually chondroitin sulfate and keratan sulfate, are covalently linked. The rigid framework of a bacterial cell wall consists of chains of alternating $\beta(1 \rightarrow 4)$-linked NAG and NAM that are cross-linked by short polypeptides to form a helical peptidoglycan cable that wraps around the bacterium. Lysozyme cleaves the glycosidic linkages between NAM and NAG of peptidoglycan. Penicillin specifically inactivates enzymes involved in the cross-linking of peptidoglycans. Both of these substances cause the lysis of susceptible bacteria. Gram-positive bacteria have teichoic acids that are linked covalently to their peptidoglycans. Gram-negative bacteria have outer membranes that bear complex and unusual polysaccharides known as O-antigens. These participate in the recognition of host cells and are important in the immunological recognition of bacteria by the host. Oligosaccharides attach to eukaryotic proteins in only a few ways. In *N*-glycosidic attachments, an NAG is invariably bound to the amide nitrogen of Asn in the sequence Asn-X-Ser(Thr). *O*-Glycosidic attachments are made to Ser or Thr in most proteins and to 5-hydroxylysine in collagen.

Oligosaccharides are located on the surfaces of glycoproteins. Glycoproteins have functions that span the entire range of protein activities, although the roles of their carbohydrate moieties are only poorly understood. For example, ribonuclease B differs from the functionally indistinguishable and carbohydrate-free ribonuclease A only by the attachment of a single oligosaccharide of somewhat variable sequence which increases the protein's rate of folding, whereas the biological properties of granulocyte–macrophage colony-stimulating factor are significantly affected by its multiple oligosaccharide chains. The viscoelastic and hence protective properties of mucus largely result from the numerous negatively charged oligosaccharide groups of its component mucins. The carbohydrate moieties of glycoproteins in plasma membranes are invariably located on the external surfaces of the membranes. A glycoprotein's carbohydrate moieties may direct its metabolic fate by governing its uptake by certain cells or cell compartments. Glycoproteins are also important mediators of cell–cell recognition and, in many cases, are the receptors for bacterial attachment, via adhesins, in the initial stages of infection. Glycomics, the carbohydrate analog of genomics and proteomics, seeks to characterize all the carbohydrates in a particular cell type.

REFERENCES

General

Garg, H.G., Cowman, M.K., and Hales, C.A. (Eds.), *Carbohydrate Chemistry, Biology and Medical Applications,* Elsevier (2008).

Lindhorst, T.K., *Essentials of Carbohydrate Chemistry and Biochemistry* (3rd ed.), Wiley-VCH (2007).

Solomons, T.W.G. and Fryhle, C., *Organic Chemistry* (9th ed.), Chapter 22, Wiley (2008). [A general discussion of carbohydrate nomenclature and chemistry. Other comprehensive organic chemistry textbooks have similar material.]

Taylor, M.E. and Drickamer, K., *Introduction to Glycobiology* (2nd ed.), Oxford University Press (2006).

Varki, A., Cummings, R.D., Esko, J.D., Freeze, H.H., Stanley, S., Bertozzi, C.R., Hart, G.W., and Etzler, M.E. (Eds.), *Essentials of Glycobiology* (2nd ed.), Cold Spring Harbor Laboratory Press (2009).

Oligosaccharides and Polysaccharides

Bayer, E.A., Chanzy, H., Lamed, R., and Shoham, Y., Cellulose, cellulases, and cellulosomes, *Curr. Opin. Struct. Biol.* **8,** 548 (1998).

Check, E., How Africa learned to love the cow, *Nature* **444,** 994–996 (2006). [Discusses the evolution of lactose tolerance in African cattle-herding populations.]

Haslam, S.M., North, S.J., and Dell, A., Mass spectrometric analysis of *N*- and *O*-glycosylation of tissues, *Curr. Opin. Struct. Biol.* **16,** 584–591 (2006).

Seeberger, P.H. and Werz, D.B., Synthesis and medical applications of oligosaccharides, *Nature* **446,** 1046–1055 (2007).

Sharon, N. and Lis, H., History of lectins: from hemagglutinins to biological recognition molecules, *Glycobiology* **14,** 53R–62R (2004). [An historical account of lectin research and applications.]

Weis, W.I. and Drickamer, K., Structural basis of lectin–carbohydrate recognition, *Annu. Rev. Biochem.* **65,** 441–473 (1996).

Glycoproteins

Bernfield, M., Götte, M., Park, P.W., Reizes, O., Fitzgerald, M.L., Linecum, J., and Zako, M., Functions of cell surface heparan sulfate proteoglycans, *Annu. Rev. Biochem.* **68,** 729–777 (1999).

Bishop, J.R., Schuksz, M., and Esko, J.D., Heparan sulphate proteoglycans fine-tune mammalian physiology, *Nature* **446,** 1030–1037 (2007). [Reviews the multiple roles of heparan sulfate proteoglycans.]

Bülow, H.E. and Hobert, O., The molecular diversity of glycosaminoglycans shapes animal development, *Annu. Rev. Cell Dev. Biol.* **22,** 375–407 (2006).

Bush, C.A., Martin-Pastor, M., and Imberty, A., Structure and conformation of complex carbohydrates of glycoproteins, glyco-lipids, and bacterial polysaccharides, *Annu. Rev. Biophys. Biomol. Struct.* **28,** 269–293 (1999).

Chain, E., Fleming's contribution to the discovery of penicillin, *Trends Biochem. Sci.* **4,** 143–146 (1979). [An historical account by one of the biochemists who characterized penicillin.]

Drickamer, K. and Taylor, M.E., Evolving views of protein glycosylation, *Trends Biochem. Sci.* **23,** 321–324 (1998).

Esko, J.D. and Lindahl, U., Molecular diversity of heparan sulfate, *J. Clin. Invest.* **108,** 169–173 (2001).

Fukuda, M. (Ed.), Glycobiology; Glycomics; and Functional Genomics, *Methods Enzymol.* **415–417** (2006).

Handel, T.M., Johnson, Z., Crown, S.E., Lau, E.K., Sweeney, M., and Proudfoot, A.E., Regulation of protein function by glycosaminoglycans—as exemplified by chemokines, *Annu. Rev. Biochem.* **74,** 385–410 (2005).

Hayhurst, E.J., Kailas, L., Hobbs, J.K., and Foster, S.J., Cell wall peptidoglycan architecture in *Bacillus subtilis, Proc. Natl. Acad. Sci.* **105,** 14603–14608 (2008).

Iozzo, R.V., Matrix proteoglycans: From molecular design to cellular function, *Annu. Rev. Biochem.* **67,** 609–652 (1998); *and* The biology of the small leucine-rich proteoglycans, *J. Biol. Chem.* **274,** 18843–18846 (1999).

Mitra, N., Sinha, S., Ramya, T.N.C., and Surolia, A., *N*-linked oligosaccharides as outfitters for glycoprotein folding, form and function, *Trends Biochem. Sci.* **31,** 156–163 and 251 (2006). [Summarizes the ways in which oligosaccharides can influence glycoprotein structure.]

Ohtsubo, K. and Marth, J.D., Glycosylation in cellular mechanisms of health and disease, *Cell* **126,** 855–867 (2006).

Perez-Vilar, J. and Hill, R.L., The structure and assembly of secreted mucins, *J. Biol. Chem.* **274,** 31751–31754 (1999).

Rudd, P.M. and Dwek, R.A., Rapid, sensitive sequencing of oligosaccharides from glycoproteins, *Curr. Opin. Biotechnol.* **8,** 488–497 (1997).

Sasisekharan, R., Raman, R., and Prabhakar, V., Glycomics approach to structure–function relationships of glycosaminoglycans, *Annu. Rev. Biomed. Eng.* **8,** 181–231 (2006).

Spiro, R.G., Protein glycosylation: nature, distribution, enzymatic formation, and disease implications of glycopeptide bonds, *Glycobiology* **12,** 43R–56R (2002). [Catalogs the various ways in which saccharides are linked to proteins, and describes the enzymes involved in glycoprotein synthesis.]

Varki, A., Nothing in glycobiology makes sense except in the light of evolution, *Cell* **126,** 841–845 (2006).

Wormald, M.R. and Dwek, R.A., Glycoproteins: Glycan presentation and protein-fold stability, *Structure* **7,** R155–R160 (1999).

PROBLEMS

1. The trisaccharide drawn below is named **raffinose.** What is its systematic name? Is it a reducing sugar?

Raffinose

2. The systematic name of **melezitose** is O-α-D-glucopyranosyl-$(1 \rightarrow 3)$-O-β-D-fructofuranosyl-$(2 \rightarrow 1)$-α-D-glucopyranoside. Draw its molecular formula. Is it a reducing sugar?

3. Name the linear form of D-glucose using the (RS) chirality nomenclature system. [See Section 4-2C. *Hint:* The branch toward C1 has higher priority than the branch toward C6.]

*4. Draw the α-furanose form of D-talose and the β-pyranose form of L-sorbose.

5. The $NaBH_4$ reduction product of D-glucose may be named L-sorbitol or D-glucitol. Explain.

6. How many different disaccharides of D-glucopyranose are possible? How many trisaccharides?

7. A molecule of amylopectin consists of 1000 glucose residues and is branched every 25 residues. How many reducing ends does it have?

8. Most paper is made by removing the lignin from wood pulp and forming the resulting mass of largely unoriented cellulose fibers into a sheet. Untreated paper loses most of its strength when wet with water but maintains its strength when wet with oil. Explain.

*9. Write a chemical mechanism for the acid-catalyzed mutarotation of glucose.

10. The values of the specific rotation, $[\alpha]_D^{20}$, for the α and β anomers of D-galactose are 150.7° and 52.8°, respectively. A mixture that is 20% α-D-galactose and 80% β-D-galactose is dissolved in water at 20° C. What is its initial specific rotation? After several hours, the specific rotation of this mixture reached an equilibrium value of 80.2°. What is its anomeric composition?

11. Name the epimers of D-gulose.

12. Exhaustive methylation of a trisaccharide followed by acid hydrolysis yields equimolar quantities of 2,3,4,6-tetra-O-methyl-D-galactose, 2,3,4-tri-O-methyl-D-mannose, and 2,4,6-tri-O-methyl-D-glucose. Treatment of the trisaccharide with β-galactosidase yields D-galactose and a disaccharide. Treatment of this disaccharide with α-mannosidase yields D-mannose and D-glucose. Draw the structure of the trisaccharide and state its systematic name.

13. The enzyme β-amylase cleaves successive maltose units from the nonreducing end of $\alpha(1 \rightarrow 4)$ glucans. It will not cleave at glucose residues that have an $\alpha(1 \rightarrow 6)$ bond. The end products of the exhaustive digestion of amylopectin by β-amylase are known as **limit dextrins.** Draw a schematic diagram of an amylopectin molecule and indicate what part(s) of it constitutes limit dextrins.

14. One demonstration of P.T. Barnum's maxim that there's a sucker born every minute is that new "reducing aids" regularly appear on the market. An eat-all-you-want nostrum, which was touted as a "starch blocker" [and which the Food and Drug Administration (FDA) eventually banned], contained an α-amylase-inhibiting protein extracted from beans. If this substance had really worked as advertised, which it did not, what unpleasant side effects would have resulted from its ingestion with a starch-containing meal? Discuss why this substance, which inhibits α-amylase *in vitro*, would not do so in the intestines after oral ingestion.

*15. Treatment of a 6.0-g sample of glycogen with Tollens' reagent followed by exhaustive methylation and then hydrolysis yields 3.1 mmol of 2,3-di-O-methylglucose and 0.0031 mmol of 1,2,3-tri-O-methylgluconic acid as well as other products. (a) What fraction of glucose residues occur at $(1 \rightarrow 6)$ branch points, and what is the average number of glucose residues per branch? (b) What are the other products of the methylation–hydrolysis treatment and in what amounts are they formed? (c) What is the average molecular mass of the glycogen?

16. The lysis of a culture of *E. coli* yields a solution with mucuslike viscosity. Adding DNase to the solution greatly reduces this viscosity. What is the physical basis of the viscosity?

17. Instilling methyl-α-D-mannoside into the bladder of a mouse prevents the colonization of its urinary tract by *E. coli*. What is the reason for this effect?

Lipids and Membranes

CHAPTER 12

Membranes function to organize biological processes by compartmentalizing them. Indeed, the cell, the basic unit of life, is essentially defined by its enveloping plasma membrane. Moreover, in eukaryotes, many subcellular organelles, such as nuclei, mitochondria, chloroplasts, the endoplasmic reticulum, and the Golgi apparatus (Fig. 1-5), are likewise membrane bounded.

Biological membranes are organized assemblies of lipids and proteins with small amounts of carbohydrate. Yet they are not impermeable barriers to the passage of materials. Rather, they regulate the composition of the intracellular medium by controlling the flow of nutrients, waste products, ions, etc., into and out of the cell. They do this through membrane-embedded "pumps" and "gates" that transport specific substances against an electrochemical gradient or permit their passage with such a gradient (Chapter 20).

Many fundamental biochemical processes occur on or in a membranous scaffolding. For example, electron transport and oxidative phosphorylation (Chapter 22), processes that oxidize nutrients with the concomitant generation of ATP, are mediated by an organized battery of enzymes that are components of the inner mitochondrial membrane. Likewise, photosynthesis, in which light energy powers the chemical combination of H_2O and CO_2 to form carbohydrates (Chapter 24), occurs in the inner membranes of chloroplasts. The processing of information, such as sensory stimuli or intercellular communications, is generally a membrane-based phenomenon. Thus nerve impulses are mediated by nerve cell membranes (Section 20-5) and the presence of certain substances such as hormones and nutrients is detected by specific membrane-bound receptors (Chapter 19).

In this chapter, we examine the compositions, structures, and formation of biological membranes and related substances. Specific membrane-based biochemical processes, such as those mentioned above, are dealt with in later chapters.

1 LIPID CLASSIFICATION

Lipids (Greek: lipos, fat) are substances of biological origin that are soluble in organic solvents such as chloroform and methanol but are only sparingly soluble, if at all, in water. Hence, they are easily separated from other biological materials by extraction into organic solvents and may be further fractionated by such techniques as adsorption chromatography, thin layer chromatography, and reverse-phase chromatography (Section 6-3D). Fats, oils, certain vitamins and hormones, and most nonprotein membrane components are lipids. In this section, we discuss the structures and physical properties of the major classes of lipids.

A. Fatty Acids

Fatty acids are carboxylic acids with long-chain hydrocarbon side groups (Fig. 12-1). They are rarely free in nature but, rather, occur in esterified form as the major components of the various lipids described in this chapter. The more common biological fatty acids are listed in Table 12-1. In higher plants and animals, the predominant fatty acid residues are those of the C_{16} and C_{18} species **palmitic, oleic, linoleic,** and **stearic acids.** Fatty acids with <14 or >20 carbon atoms are uncommon. *Most fatty acids have an even number of carbon*

Figure 12-1 **Structural formulas of some C_{18} fatty acids.** The double bonds all have the cis configuration.

atoms because they are usually biosynthesized by the concatenation of C_2 units (Section 25-4C). Over half of the fatty acid residues of plant and animal lipids are unsaturated (contain double bonds) and are often polyunsaturated (contain two or more double bonds). Bacterial fatty acids are rarely polyunsaturated but are commonly branched, hydroxylated, or contain cyclopropane rings. Unusual fatty acids also occur as components of the oils and **waxes** (esters of fatty acids and long-chain alcohols) produced by certain plants.

a. The Physical Properties of Fatty Acids Vary with Their Degree of Unsaturation

Table 12-1 indicates that the first double bond of an unsaturated fatty acid commonly occurs between its C9 and C10 atoms counting from the carboxyl C atom (a Δ^9- or 9-double bond). In polyunsaturated fatty acids, the double bonds tend to occur at every third carbon atom toward the methyl terminus of the molecule (such as —CH=CH—CH$_2$—CH=CH—). Double bonds in polyunsaturated fatty acids are almost never conjugated (as in —CH=CH—CH=CH—). Triple bonds rarely occur in fatty acids or any other compound of biological origin. Two important classes of polyunsaturated fatty acids are denoted $n - 3$ (or $\omega - 3$) and $n - 6$ (or $\omega - 6$) fatty acids. This nomenclature identifies the last double-bonded carbon atom as counted from the methyl terminal (ω) end of the chain.

Saturated fatty acids are highly flexible molecules that can assume a wide range of conformations because there is relatively free rotation about each of their C—C bonds. Nevertheless, their fully extended conformation is that of minimum energy because this conformation has the least amount of

Table 12-1 **The Common Biological Fatty Acids**

Symbol[a]	Common Name	Systematic Name	Structure	mp (°C)
Saturated fatty acids				
12:0	Lauric acid	Dodecanoic acid	$CH_3(CH_2)_{10}COOH$	44.2
14:0	Myristic acid	Tetradecanoic acid	$CH_3(CH_2)_{12}COOH$	52
16:0	Palmitic acid	Hexadecanoic acid	$CH_3(CH_2)_{14}COOH$	63.1
18:0	Stearic acid	Octadecanoic acid	$CH_3(CH_2)_{16}COOH$	69.6
20:0	Arachidic acid	Eicosanoic acid	$CH_3(CH_2)_{18}COOH$	75.4
22:0	Behenic acid	Docosanoic acid	$CH_3(CH_2)_{20}COOH$	81
24:0	Lignoceric acid	Tetracosanoic acid	$CH_3(CH_2)_{22}COOH$	84.2
Unsaturated fatty acids (all double bonds are cis)				
16:1n–7	Palmitoleic acid	9-Hexadecenoic acid	$CH_3(CH_2)_5CH{=}CH(CH_2)_7COOH$	−0.5
18:1n–9	Oleic acid	9-Octadecenoic acid	$CH_3(CH_2)_7CH{=}CH(CH_2)_7COOH$	13.4
18:2n–6	Linoleic acid	9,12-Octadecadienoic acid	$CH_3(CH_2)_4(CH{=}CHCH_2)_2(CH_2)_6COOH$	−9
18:3n–3	α-Linolenic acid	9,12,15-Octadecatrienoic acid	$CH_3CH_2(CH{=}CHCH_2)_3(CH_2)_6COOH$	−17
18:3n–6	γ-Linolenic acid	6,9,12-Octadecatrienoic acid	$CH_3(CH_2)_4(CH{=}CHCH_2)_3(CH_2)_3COOH$	
20:4n–4	Arachidonic acid	5,8,11,14-Eicosatetraenoic acid	$CH_3(CH_2)_4(CH{=}CHCH_2)_4(CH_2)_2COOH$	−49.5
20:5n–3	EPA	5,8,11,14,17-Eicosapentaenoic acid	$CH_3CH_2(CH{=}CHCH_2)_5(CH_2)_2COOH$	−54
22:6n–3	DHA	4,7,10,13,16,19-Docosahexenoic acid	$CH_3CH_2(CH{=}CHCH_2)_6CH_2COOH$	
24:1n–9	Nervonic acid	15-Tetracosenoic acid	$CH_3(CH_2)_7CH{=}CH(CH_2)_{13}COOH$	39

[a]Number of carbon atoms: number of double bonds. For unsaturated fatty acids, n is the number of carbon atoms, $n - x$ is the double-bonded carbon atom, and x is the number of that carbon atom counting from the methyl terminal (ω) end of the chain.

Source: Dawson, R.M.C., Elliott, D.C., Elliott, W.H., and Jones, K.M., *Data for Biochemical Research* (3rd ed.), Chapter 8, Clarendon Press (1986).

steric interference between neighboring methylene groups. The melting points (mp) of saturated fatty acids, like those of most substances, increase with molecular mass (Table 12-1).

Fatty acid double bonds almost always have the cis configuration (Fig. 12-1). This puts a rigid 30° bend in the hydrocarbon chain of unsaturated fatty acids that interferes with their efficient packing to fill space. The consequent reduced van der Waals interactions cause fatty acid melting points to decrease with their degree of unsaturation (Table 12-1). Lipid fluidity likewise increases with the degree of unsaturation of their component fatty acid residues. This phenomenon, as we shall see in Section 12-2Cb, has important consequences for membrane properties.

B. Triacylglycerols

The fats and oils that occur in plants and animals consist largely of mixtures of **triacylglycerols** (also referred to as **triglycerides** or **neutral fats**). *These nonpolar, water-insoluble substances are fatty acid triesters of* **glycerol:**

$$
^1CH_2-OH \qquad ^1CH_2-O-\overset{\overset{O}{\|}}{C}-R_1
$$
$$
^2CH-OH \qquad ^2CH\ -O-\overset{\overset{O}{\|}}{C}-R_2
$$
$$
^3CH_2-OH \qquad ^3CH_2-O-\overset{\overset{O}{\|}}{C}-R_3
$$

Glycerol   **Triacylglycerol**

Triacylglycerols function as energy reservoirs in animals and are therefore their most abundant class of lipids even though they are not components of biological membranes.

Figure 12-2 **Scanning electron micrograph of adipocytes.** Each contains a fat globule that occupies nearly the entire cell. [Fred E. Hossler/Visuals Unlimited.]

Triacylglycerols differ according to the identity and placement of their three fatty acid residues. The so-called **simple triacylglycerols** contain one type of fatty acid residue and are named accordingly. For example, **tristearoylglycerol** or **tristearin** contains three stearic acid residues, whereas **trioleoylglycerol** or **triolein** has three oleic acid residues. The more common **mixed triacylglycerols** contain two or three different types of fatty acid residues and are named according to their placement on the glycerol moiety.

$$^1CH_2-{}^2CH-{}^3CH_2$$

1-Palmitoleoyl-2-linoleoyl-3-stearoylglycerol

Fats and oils (which differ only in that fats are solid and oils are liquid at room temperature) are complex mixtures of simple and mixed triacylglycerols whose fatty acid compositions vary with the organism that produced them. Plant oils are usually richer in unsaturated fatty acid residues than are animal fats, as the lower melting points of oils imply.

a. Triacylglycerols Are Efficient Energy Reserves
Fats are a highly efficient form in which to store metabolic energy. This is because fats are less oxidized than are carbohydrates or proteins and hence yield significantly more energy on oxidation. Furthermore, fats, being nonpolar substances, are stored in anhydrous form, whereas glycogen, for example, binds about twice its weight of water under physiological conditions. Fats therefore provide about six times the metabolic energy of an equal weight of hydrated glycogen.

In animals, **adipocytes** (fat cells; Fig. 12-2) are specialized for the synthesis and storage of triacylglycerols. Whereas

(a)

$$
\begin{array}{c}
CH_3 \\
| \\
H_3C - \overset{+}{N} - CH_3 \\
| \\
CH_2 \\
| \\
CH_2 \\
| \\
O \\
| \\
{}^-O - P = O \\
| \\
O
\end{array}
$$

$$
{}^3CH_2 - {}^2\overset{\displaystyle H}{\underset{\displaystyle |}{C}} - {}^1CH_2
$$

1-Stearoyl-2-oleoyl-3-phosphatidylcholine

(b)

Figure 12-4   **The glycerophospholipid 1-stearoyl-2-oleoyl-3-phosphatidylcholine.** (*a*) Molecular formula in Fischer projection. (*b*) Energy-minimized space-filling model with C green, H white, N blue, O red, and P orange. Note how the unsaturated oleyl chain (*left*) is bent compared to the saturated stearoyl chain. [Based on coordinates provided by Richard Venable and Richard Pastor, NIH, Bethesda, Maryland.]

occur at the C1 position of glycerophospholipids, and the C2 position is often occupied by an unsaturated C_{16} to C_{20} fatty acid. Glycerophospholipids are, of course, also named according to the identities of these fatty acid residues (Fig. 12-4). Some glycerophospholipids have common names. For example, phosphatidylcholines are known as **lecithins;** diphosphatidylglycerols, the "double" glycerol phospholipids, are known as **cardiolipins** (because they were first isolated from heart muscle).

Plasmalogens

A plasmalogen

are glycerophospholipids in which the C1 substituent to the glycerol moiety is bonded to it via an α,β-unsaturated ether linkage in the cis configuration rather than through

an ester linkage. **Ethanolamine, choline,** and serine form the most common plasmalogen head groups.

D. *Sphingolipids*

Sphingolipids, which are also major membrane components, are derivatives of the C_{18} amino alcohols **sphingosine, dihydrosphingosine** (Fig. 12-5), and their C_{16}, C_{17}, C_{19}, and C_{20} homologs. Their *N*-acyl fatty acid derivatives, **ceramides,**

A ceramide

occur only in small amounts in plant and animal tissues but form the parent compounds of more abundant sphingolipids:

1. Sphingomyelins, the most common sphingolipids, are ceramides bearing either a phosphocholine (Fig. 12-6) or a phosphoethanolamine moiety, so that they can also be classified as **sphingophospholipids.** *Although sphingomyelins*

OH H   OH
| | |
H₂C ─ C ─ C ─ H
| |
H₃N⁺ CH
‖
HC
|
(CH₂)₁₂
|
CH₃

Sphingosine

OH H   OH
| | |
H₂C ─ C ─ C ─ H
| |
H₃N⁺ CH₂
|
CH₂
|
(CH₂)₁₂
|
CH₃

Dihydrosphingosine

Figure 12-5 **Molecular formulas of sphingosine and dihydrosphingosine.** The chiral centers at C2 and C3 of sphingosine and dihydrosphingosine have the configurations shown in Fischer projection. The double bond in sphingosine has the trans configuration.

differ chemically from phosphatidylcholine and phosphatidylethanolamine, their conformations and charge distributions are quite similar. The membranous myelin sheath that surrounds and electrically insulates many nerve cell axons (Section 20-5Bc) is particularly rich in sphingomyelin.

2. Cerebrosides, the simplest **sphingoglycolipids** (alternatively **glycosphingolipids**), are ceramides with head groups that consist of a single sugar residue. **Galactocerebrosides,** which are most prevalent in the neuronal cell membranes of the brain, have a β-D-galactose head group.

A galactocerebroside

Glucocerebrosides, which instead have a β-D-glucose residue, occur in the membranes of other tissues. *Cerebrosides, in contrast to phospholipids, lack phosphate groups and hence are most frequently nonionic compounds.* The galactose residues of some galactocerebrosides, however, are sulfated at their C3 positions to form ionic compounds known as **sulfatides.** More complex sphingoglycolipids have unbranched oligosaccharide head groups of up to four sugar residues.

(*a*)

Phosphocholine head group

Palmitate residue

A sphingomyelin

(*b*)

Figure 12-6 **A sphingomyelin.** (*a*) Molecular formula in Fischer projection. (*b*) Energy-minimized space-filling model with C green, H white, N blue, O red, and P orange. Note its conformational resemblance to glycerophospholipids (Fig. 12-4). [Based on coordinates provided by Richard Venable and Richard Pastor, NIH, Bethesda, Maryland.]

(a)

(b)

Figure 12-7 **Ganglioside G$_{M1}$.** (*a*) Structural formula with its sphingosine residue in Fischer projection. (*b*) Energy-minimized space-filling model with C green, H white, N blue, O red, and P orange. Gangliosides G$_{M2}$ and G$_{M3}$ differ from G$_{M1}$ only by the sequential absences of the terminal D-galactose and *N*-acetyl-D-galactosamine residues. Other gangliosides have different oligosaccharide head groups. [Based on coordinates provided by Richard Venable and Richard Pastor, NIH, Bethesda, Maryland.]

3. Gangliosides form the most complex group of sphingoglycolipids. They are ceramide oligosaccharides that include among their sugar groups at least one sialic acid residue (*N*-acetylneuraminic acid and its derivatives; Section 11-1Cc). The structures of gangliosides G$_{M1}$, G$_{M2}$, and G$_{M3}$, three of the hundreds that are known, are shown in Fig. 12-7. Gangliosides are primarily components of cell-surface membranes and constitute a significant fraction (6%) of brain lipids. Other tissues also contain gangliosides but in lesser amounts.

Gangliosides have considerable physiological and medical significance. Their complex carbohydrate head groups, which extend beyond the surfaces of cell membranes, act as specific receptors for certain pituitary glycoprotein hormones that regulate a number of important physiological functions (Section 19-1). Gangliosides are also receptors for bacterial protein toxins such as cholera toxin (Section 19-2Cd). There is considerable evidence that gangliosides are specific determinants of cell–cell recognition, so they probably have an important role in the growth and differentiation of tissues as well as in carcinogenesis (cancer generation). Disorders of ganglioside breakdown are responsible for several hereditary **sphingolipid storage diseases,** such as **Tay-Sachs disease,** which are characterized by an invariably fatal neurological deterioration (Section 25-8Ce).

E. *Cholesterol*

Steroids, which are mostly of eukaryotic origin, are derivatives of **cyclopentanoperhydrophenanthrene** (Fig. 12-8). The much maligned **cholesterol** (Fig. 12-9), the most abundant steroid in animals, is further classified as a **sterol** because of its C3-OH group and its branched aliphatic side chain of 8 to 10 carbon atoms at C17.

Cholesterol is a major component of animal plasma membranes, where it is typically present at 30 to 40 mol %, and occurs in lesser amounts in the membranes of their subcellular organelles. Its polar OH group gives it a weak amphiphilic character, whereas its fused ring system provides it with greater rigidity than other membrane lipids. Cholesterol is therefore an important determinant of membrane

Cyclopentanoperhydrophenanthrene

Figure 12-8 **Cyclopentanoperhydrophenanthrene, the parent compound of steroids.** It consists of four fused saturated rings. The standard ring labeling system is indicated.

(a)

Cholesterol

(b)

Figure 12-9 **Cholesterol.** (*a*) Structural formula with the standard numbering system. (*b*) Energy-minimized space-filling model with C green, H white, O red, and P orange. Cholesterol's rigid ring system makes it far less conformationally flexible than are membrane lipids: Its cyclohexane rings can adopt either the boat or the chair conformations (Fig. 11-6) but the chair conformation is highly preferred. [Based on coordinates provided by Richard Venable and Richard Pastor, NIH, Bethesda, Maryland.]

properties. It is also abundant in blood plasma lipoproteins (Section 12-5), where ~70% of it is esterified to long-chain fatty acids to form **cholesteryl esters.**

Cholesteryl stearate

Cholesterol is the metabolic precursor of **steroid hormones,** substances that regulate a great variety of physiological functions including sexual development and carbohydrate metabolism (Section 19-1G). The much-debated role of cholesterol in heart disease is examined in Section 12-5C. Cholesterol metabolism and the biosynthesis of steroid hormones are discussed in Section 25-6.

Plants contain little cholesterol. Rather, the most common sterol components of their membranes are **stigmasterol** and **β-sitosterol**

Stigmasterol

β-Sitosterol

Ergosterol

which differ from cholesterol only in their aliphatic side chains. Yeast and fungi have yet other membrane sterols such as **ergosterol,** which has a C7 to C8 double bond. Prokaryotes, with the exception of mycoplasmas (Section 1-1B), contain little, if any, sterol.

2 PROPERTIES OF LIPID AGGREGATES

The first recorded experiments on the physical properties of lipids were made in 1774 by the American statesman and scientist Benjamin Franklin. In investigating the well-known (at least among sailors) action of oil in calming waves, Franklin wrote:

At length being at Clapham [in London] where there is, on the common, a large pond, which I observed to be one day very

Figure 12-10   An oil monolayer at the air–water interface. The hydrophobic tails of the lipids avoid association with water by extending into the air.

rough with the wind, I fetched out a cruet of oil [probably olive oil] and dropt a little of it in the water. I saw it spread itself with surprising swiftness upon the surface. . . . I then went to the windward side, where [the waves] began to form; and there the oil, though not more than a teaspoonful, produced an instant calm over a space several yards square, which spread amazingly, and extended itself gradually till it reached the lee side, making all that quarter of the pond, perhaps half an acre, as smooth as a looking glass.

This is sufficient information to permit the calculation of the oil layer's thickness (although there is no indication that Franklin made this calculation, we can; see Problem 4). We now know that oil forms a monomolecular layer on the surface of water in which the polar heads of the amphiphilic oil molecules are immersed in the water and their hydrophobic tails extend into the air (Fig. 12-10).

The calming effect of oil on rough water is a consequence of a large reduction in the water's surface tension. An oily surface film has the weak intermolecular cohesion characteristic of hydrocarbons rather than the strong intermolecular attractions of water responsible for its normally large surface tension. Oil, nevertheless, calms only smaller waves; it does not, as Franklin later observed, affect the larger swells.

In this section, we discuss how lipids aggregate to form micelles and bilayers. We shall also be concerned with the physical properties of lipids in bilayers because these aggregates form the structural basis for biological membranes.

A. *Micelles and Bilayers*

In aqueous solutions, amphiphilic molecules, such as soaps and detergents, form micelles (globular aggregates whose hydrocarbon groups are out of contact with water; Section 2-1Ba). This molecular arrangement eliminates unfavorable contacts between water and the hydrophobic tails of the amphiphiles and yet permits the solvation of the polar head groups. Micelle formation is a cooperative process: An assembly of just a few amphiphiles cannot shield its tails from contact with water. Consequently, dilute aqueous solutions of amphiphiles do not form micelles until their concentration surpasses a certain **critical micelle concentration (cmc).** Above the cmc, almost all the added amphiphile aggregates to form micelles. The value of the cmc depends on the identity of the amphiphile and the solution conditions. For amphiphiles with relatively small single tails, such as dodecyl sulfate ion, $CH_3(CH_2)_{11}OSO_3^-$, the cmc is ~1 mM. Those of biological lipids, most of which have two large hydrophobic tails, are generally $<10^{-6}$ M.

a. Single-Tailed Lipids Tend to Form Micelles

The approximate size and shape of a micelle can be predicted from geometrical considerations. Single-tailed amphiphiles, such as soap anions, form spheroidal or ellipsoidal micelles because of their conical shapes (their hydrated head groups are wider than their tails; Fig. 12-11a,b). The number of molecules in such micelles depends on the amphiphile, but for many substances, it is on the order of several hundred. For a given amphiphile, these numbers span a narrow range: Less would expose the hydrophobic core of the micelle to water, whereas more would give the micelle an energetically unfavorable hollow center (Fig. 12-11c). Of course, a large micelle could flatten out to eliminate this hollow center, but the resulting decrease of curvature at the flattened surfaces would also generate empty spaces (Fig. 12-11d).

b. Glycerophospholipids and Sphingolipids Tend to Form Bilayers

The two hydrocarbon tails of glycerophospholipids and sphingolipids give these amphiphiles a more or less cylindrical

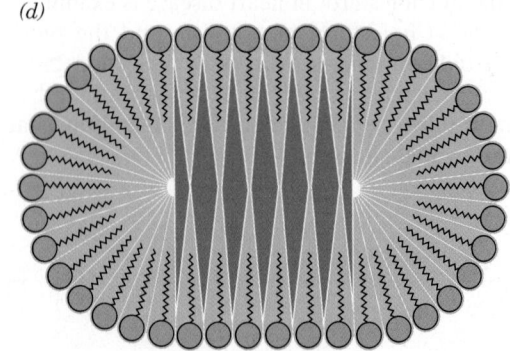

Figure 12-11   Aggregates of single-tailed lipids. The conical van der Waals envelope of single-tailed lipids (*a*) permits them to pack efficiently in forming a spheroidal micelle (*b*). The diameters of these micelles and hence their lipid population largely depend on the length of the tails. Spheroidal micelles composed of many more lipid molecules than the optimal number would have an unfavorable water-filled center (*blue*) (*c*). Such micelles could flatten out to collapse the hollow center, but as such ellipsoidal micelles become elongated they also develop water-filled spaces (*d*).

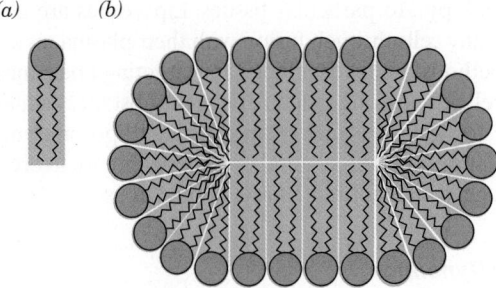

(a) (b)

Figure 12-12 **Bilayer formation by phospholipids.** The cylindrical van der Waals envelope of phospholipids (*a*) causes them to form extended disklike micelles (*b*) that are better described as lipid bilayers.

shape (Fig. 12-12*a*). The steric requirements of packing such molecules together yields large disklike micelles (Fig. 12-12*b*) that are really extended bimolecular leaflets. The existence of such **lipid bilayers** was first proposed in 1925 by Evert Gorter and François Grendel, on the basis of their observation that lipids extracted from erythrocytes covered twice the area when spread as a monolayer at the air–water interface (Fig. 12-10) than in the erythrocyte plasma membrane (the erythrocyte's only membrane). Lipid bilayers typically have thicknesses of ~60 Å, as measured by electron microscopy and X-ray diffraction techniques. Since their two head group layers are each expected to be ~15 Å thick, their ~15-Å-long hydrocarbon tails must be nearly fully extended. We shall see below that *lipid bilayers form the structural basis of biological membranes.*

B. *Liposomes*

A suspension of phospholipids in water forms multilamellar vesicles that have an onionlike arrangement of lipid bilayers (Fig. 12-13*a*). On **sonication** (agitation by ultrasonic vibrations), these structures rearrange to form **liposomes**—closed, self-sealing, solvent-filled vesicles that are bounded by only a single bilayer (Fig. 12-13*b*). They usually have diameters of several hundred Ångstroms and, in a given preparation, are rather uniform in size. Liposomes with diameters of ~1000 Å can be made by injecting an ethanolic solution of phospholipid into water or by dissolving phospholipid in a detergent solution and then dialyzing out the detergent. Once formed, liposomes are quite stable and, in fact, may be separated from the solution in which they reside by dialysis, gel filtration chromatography, or centrifugation. Liposomes with differing internal and external environments can therefore be readily prepared. *Biological membranes consist of lipid bilayers with which proteins are associated (Section 12-3A).* Liposomes composed of synthetic lipids and/or lipids extracted from biological sources (e.g., lecithin from egg yolks) have therefore been extensively studied as models for biological membranes.

a. Lipid Bilayers Are Impermeable to Most Polar Substances

Since biological membranes form cell and organelle boundaries, it is important to determine their ability to partition two aqueous compartments. The permeability of a lipid bilayer to a given substance may be determined by forming liposomes in a solution containing the substance of interest, changing the external aqueous solution, and then measuring the rate at which the substance appears in the new external solution. It has been found in this way that *lipid bilayers are extraordinarily impermeable to ionic and polar substances and that the permeabilities of such substances increase with their solubilities in nonpolar solvents.* This suggests that to penetrate a lipid bilayer, a solute molecule must shed its hydration shell and become solvated by

(a)

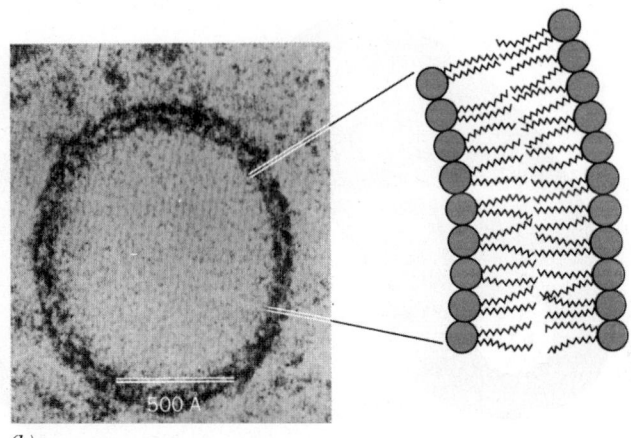

(b)

Figure 12-13 **Lipid bilayers.** (*a*) An electron micrograph of a multilamellar phospholipid vesicle in which each layer is a lipid bilayer. [Courtesy of Alec D. Bangham, Institute of Animal Physiology, Cambridge, U.K.] (*b*) An electron micrograph of a liposome. Its wall, as the accompanying diagram indicates, consists of a bilayer. [Courtesy of Walter Stoeckenius, University of California at San Francisco.]

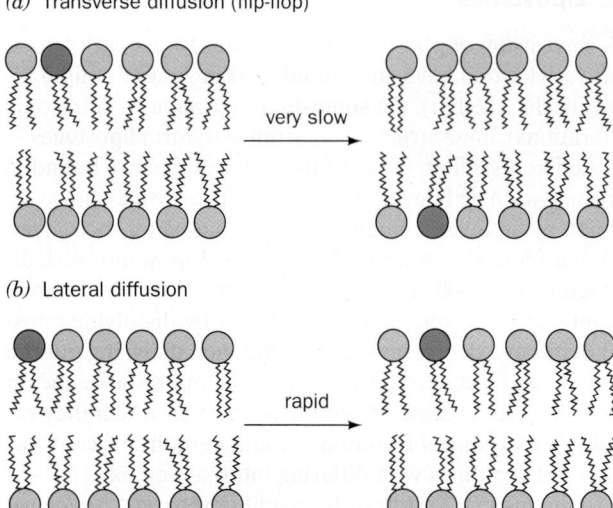

(a) Transverse diffusion (flip-flop)

very slow

(b) Lateral diffusion

rapid

Figure 12-14   Phospholipid diffusion in a lipid bilayer.
(*a*) Transverse diffusion (flip-flop) is defined as the transfer of a phospholipid molecule from one bilayer leaflet to the other. (*b*) Lateral diffusion is defined as the pairwise exchange of neighboring phospholipid molecules in the same bilayer leaflet.

the bilayer's hydrocarbon core. Such a process is highly unfavorable for polar molecules, so that even the ~30-Å thickness of a lipid bilayer's hydrocarbon core forms an effective barrier for polar substances. However, measurements using tritiated water indicate that lipid bilayers are appreciably permeable to water. Despite the polarity of water, its small molecular size makes it significantly soluble in the hydrocarbon core of lipid bilayers and therefore able to permeate them.

The stability of liposomes and their impermeability to many substances make them promising vehicles for the delivery of therapeutic agents, such as drugs, enzymes, and genes

(for gene therapy), to particular tissues. Liposomes are absorbed by many cells through fusion with their plasma membranes. If methods can be developed for targeting liposomes to specific cell populations, then the desired substances could be directed toward particular tissues through liposome microencapsulation. Indeed, a number of liposome-delivered anticancer agents and antibiotics are already in use.

C. *Bilayer Dynamics*

a. Lipid Bilayers Are Two-Dimensional Fluids

The transfer of a lipid molecule across a bilayer (Fig. 12-14*a*), a process termed **transverse diffusion** or a **flip-flop,** is an extremely rare event. This is because a flip-flop requires the polar head group of the lipid to pass through the hydrocarbon core of the bilayer. The flip-flop rates of phospholipids, as measured by several techniques, are characterized by half-times that are minimally several days.

In contrast to their low flip-flop rates, *lipids are highly mobile in the plane of the bilayer* (**lateral diffusion,** Fig. 12-14*b*). The X-ray diffraction patterns of bilayers at physiological temperatures have a diffuse band, centered at a spacing of 4.6 Å, whose width is a measure of the distribution of lateral spacings between the hydrocarbon chains in the bilayer plane. This band, which resembles one in the X-ray diffraction patterns of liquid paraffins, is indicative that *the bilayer is a two-dimensional fluid in which the hydrocarbon chains undergo rapid fluxional (continuously changing) motions involving rotations about their C—C bonds.*

The lateral diffusion rate of lipid molecules can be quantitatively determined from the rate of **fluorescence recovery after photobleaching (FRAP;** Fig. 12-15). A fluorescent group **(fluorophore)** is specifically attached to a bilayer component, and an intense laser pulse focused on a very small area (~3 μm²) is used to destroy (bleach) the fluorophore there. The rate at which the bleached area recovers its fluorescence, as monitored by fluorescence microscopy,

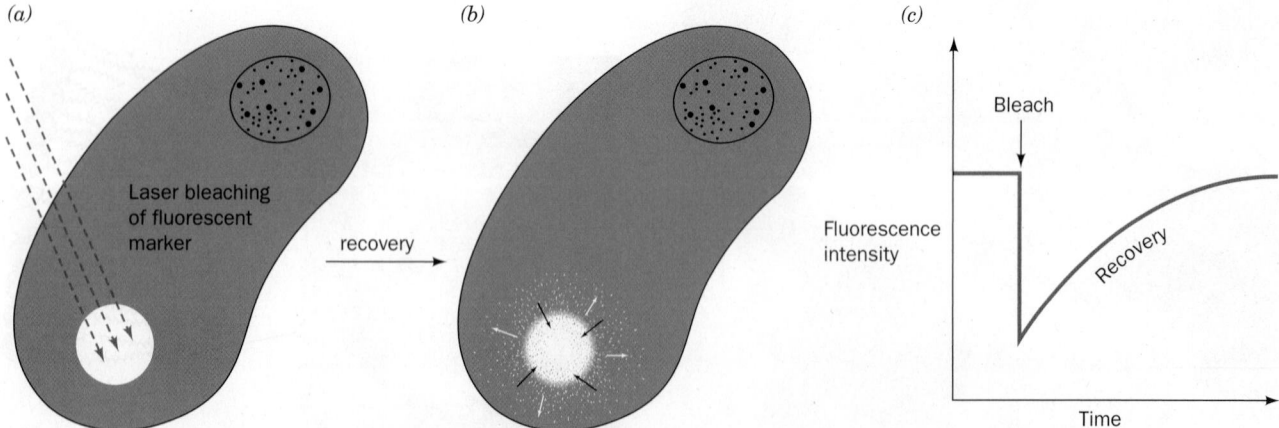

(a)   Laser bleaching of fluorescent marker

recovery

(b)

(c)   Bleach

Fluorescence intensity

Recovery

Time

Figure 12-15   The fluorescence recovery after photobleaching (FRAP) technique. (*a*) An intense laser light pulse bleaches the fluorescent markers (*green*) from a small region of an immobilized cell that has a fluorescence-labeled membrane component. (*b*) The fluorescence of the bleached area, as monitored by

fluorescence microscopy, recovers as the bleached molecules laterally diffuse out of it and intact fluorescence-labeled molecules diffuse into it. (*c*) The fluorescence recovery rate depends on the diffusion rate of the labeled molecule. 🔎 **See Guided Exploration 11: Membrane structure and the fluid mosaic model**

Figure 12-16 Model (snapshot) of a lipid bilayer at an instant in time. The conformations of dipalmitoylphosphatidylcholine molecules in a bilayer surrounded by water were modeled by computer. Atom colors are chain and glycerol C gray except terminal methyl C yellow, ester O red, phosphate P and O green, and choline C and N magenta. Water molecules are represented by translucent blue spheres (those near the bilayer appear dark because they overlap head group atoms). [Courtesy of Richard Pastor and Richard Venable, NIH, Bethesda, Maryland.] **See Guided Exploration 11: Membrane structure and the fluid mosaic model**

indicates the rate at which unbleached and bleached fluorescence-labeled molecules laterally diffuse into and out of the bleached area, respectively. Such observations indicate, as do magnetic resonance measurements, that lipids in bilayers have lateral mobilities similar to those of the molecules in a light machine oil. Lipids in bilayers can therefore diffuse the 1-μm length of a bacterial cell in ~1 s. Methods for tracking the motions of single molecules in membranes have also been developed that link the molecule of interest to a small latex bead, colloidal gold particle, or fluorescent group, and then observe the movement of the label using high speed video techniques.

Molecular dynamics simulations (Section 9-4a) of lipid bilayers (Fig. 12-16) indicate that their lipid tails are highly conformationally mobile due to rotations about their C—C bonds. However, the viscosity of these tails sharply increases closer to the lipid head groups because their lateral mobilities are more constrained by interactions with the more rigid head groups. Note that the methyl ends of the tails from opposite leaflets of the bilayer are frequently interdigitated rather than forming entirely separate layers, as Fig. 12-14 might be taken to suggest. This is particularly true in biological membranes because their various lipid molecules have tails of different lengths and/or are kinked due to the presence of double bonds. Molecular dynamics simulations also indicate that a lipid bilayer is flanked by several layers of ordered water molecules. Moreover, as Fig. 12-16 indicates, water molecules commonly penetrate well below the level of the head groups and glycerol residues. Hence, *a lipid bilayer typically consists of an ~30-Å-thick hydrocarbon core bounded on both sides by ~15-Å-thick interface regions containing rapidly fluctuating conglomerations of head groups, water, glycerol, carbonyl, and methylene groups.*

b. Bilayer Fluidity Varies with Temperature

*As a lipid bilayer cools below a characteristic **transition temperature,** it undergoes a sort of phase change, termed an **order–disorder transition,** in which it becomes a gel-like solid* (Fig. 12-17); that is, it loses its fluidity. Below the transition temperature, the diffuse 4.6-Å X-ray diffraction band characteristic of the lateral spacing between hydrocarbon chains in a liquid-crystalline bilayer is replaced by a sharp 4.2-Å band similar to that exhibited by crystalline paraffins. This indicates that the hydrocarbon chains in a bilayer become fully extended and packed in a hexagonal array as in crystalline paraffins.

The transition temperature of a bilayer increases with the chain length and the degree of saturation of its component fatty acid residues for the same reasons that the melting temperatures of fatty acids increase with these quantities. The transition temperatures of most biological membranes are

Figure 12-17 Order–disorder transition in a lipid bilayer. (*a*) Above the transition temperature, both the lipid molecules as a whole and their nonpolar tails are highly mobile in the plane of the bilayer. Such a state of matter, which is ordered in some directions but not in others, is called a liquid crystal. (*b*) Below the transition temperature, the lipid molecules form a much more orderly array to yield a gel-like solid. [After Robertson, R.N., *The Lively Membranes, pp.* 69–70, Cambridge University Press (1983).]

(a) Above transition temperature

(b) Below transition temperature

Table 12-3   **Lipid Compositions of Some Biological Membranes**[a]

Lipid	Human Erythrocyte	Human Myelin	Beef Heart Mitochondria	*E. coli*
Phosphatidic acid	1.5	0.5	0	0
Phosphatidylcholine	19	10	39	0
Phosphatidylethanolamine	18	20	27	65
Phosphatidylglycerol	0	0	0	18
Phosphatidylinositol	1	1	7	0
Phosphatidylserine	8.5	8.5	0.5	0
Cardiolipin	0	0	22.5	12
Sphingomyelin	17.5	8.5	0	0
Glycolipids	10	26	0	0
Cholesterol	25	26	3	0

[a]The values given are weight percent of total lipid.

Source: Tanford, C., *The Hydrophobic Effect, p.* 109, Wiley (1980).

in the range 10 to 40°C. *Cholesterol, which by itself does not form a bilayer, decreases membrane fluidity near the membrane surface because cholesterol's rigid steroid ring system interferes with the motions of the fatty acid tails causing them to become more ordered.* However, because cholesterol does not extend into the membrane as far as most lipids, it also acts as a spacer that facilitates the increased mobility of the fatty acid tails near their methyl ends. Cholesterol also broadens the temperature range of the order–disorder transition and in high concentrations totally abolishes it. This behavior occurs because cholesterol inhibits the crystallization (cooperative aggregation into ordered arrays) of fatty acid tails by fitting in between them. Thus cholesterol functions as a kind of membrane plasticizer.

The fluidity of biological membranes is one of their important physiological attributes since it permits their embedded proteins to interact (Section 12-3C). The transition temperatures of mammalian membranes are well below body temperatures and hence these membranes all have a fluidlike character. Bacteria and poikilothermic (cold-blooded) animals such as fish modify (through lipid biosynthesis and degradation) the fatty acid compositions of their membrane lipids with the ambient temperature so as to maintain membrane fluidity. For example, the membrane viscosity of *E. coli* at its growth temperature remains constant as the growth temperature is varied from 15 to 43°C.

c. Gaseous Anesthetics Alter Neuronal Membrane Structures

Gaseous anesthetics, such as diethyl ether, cyclopropane, **isoflurane** (CF_3—CHCl—O—CHF_2), and the noble gas Xe, act by interfering with the transmission of nerve impulses in the central nervous system. Since the body excretes these general anesthetics unchanged, it appears that they do not act by chemical means. Rather, experimental evidence, such as the linear correlation of their anesthetic effectiveness with their lipid solubilities, suggests that these nonpolar substances alter the structures of membranes by

Table 12-4   **Compositions of Some Biological Membranes**

Membrane	Protein (%)	Lipid (%)	Carbohydrate (%)	Protein to Lipid Ratio
Plasma membranes:				
Mouse liver cells	46	54	2–4	0.85
Human erythrocyte	49	43	8	1.1
Amoeba	52	42	4	1.3
Rat liver nuclear membrane	59	35	2.0	1.6
Mitochondrial outer membrane	52	48	(2–4)[a]	1.1
Mitochondrial inner membrane	76	24	(1–2)[a]	3.2
Myelin	18	79	3	0.23
Gram-positive bacteria	75	25	(10)[a]	3.0
Halobacterium purple membrane	75	25		3.0

[a]Deduced from the analyses.

Source: Guidotti, G., *Annu. Rev. Biochem.* **41,** 732 (1972).

Figure 12-18 X-ray structure of the integral membrane protein aquaporin-0 (AQP0) in association with lipids. The protein is represented by its surface diagram, which is colored according to charge (red negative, blue positive, and white uncharged). Tightly bound molecules of dimyristoylphosphatidylcholine are drawn in space-filling form with C green, O red, and P orange. Note how the lipid tails closely conform to the nonpolar surface of the protein, thereby solvating it. The arrangement of the two rows of lipid molecules, with phosphorus–phosphorus distances of ~35 Å, matches the dimensions of a lipid bilayer. [Based on an electron crystallographic structure by Stephen Harrison and Thomas Walz, Harvard Medical School. PDBid 2B6O.]

A. *Membrane Proteins*

Membrane proteins are operationally classified according to how tightly they are associated with membranes:

1. *Integral* or *intrinsic proteins are tightly bound to membranes by hydrophobic forces* (Fig. 12-18) *and can only be separated from them by treatment with agents that disrupt membranes.* These include organic solvents, detergents (e.g., those in Fig. 12-19), and chaotropic agents (ions that disrupt water structure; Section 8-4E). Integral proteins

$$CH_3-(CH_2)_{11}-OSO_3^- \quad Na^+$$

Sodium dodecyl sulfate (SDS)

X = H, Y = COO$^-$ Na$^+$ **Sodium deoxycholate**
X = OH, Y = COO$^-$ Na$^+$ **Sodium cholate**
X = OH, Y = CO—NH—(CH$_2$)$_3$—N$^+$(CH$_3$)$_2$—(CH$_2$)$_3$—SO$_3^-$ **CHAPS**

$$CH_3-(CH_2)_n-CH_2-\overset{\overset{\displaystyle CH_3}{|}}{\underset{\underset{\displaystyle CH_3}{|}}{N^+}}-CH_3 \quad Br^-$$

$n = 10$ **Dodecyltriethylammonium bromide (DTAB)**
$n = 15$ **Cetyltrimethylammonium bromide (CTAB)**

$$CH_3-(CH_2)_{11}-(O-CH_2-CH)_n-OH$$

Polyoxyethylenelauryl ether

$n = \ \ 4$ **Brij 30**
$n = 25$ **Brij 35**

$$CH_3-\overset{\overset{\displaystyle CH_3}{|}}{\underset{\underset{\displaystyle CH_3}{|}}{C}}-CH_2-\overset{\overset{\displaystyle CH_3}{|}}{\underset{\underset{\displaystyle CH_3}{|}}{C}}-\text{⟨phenyl⟩}-(O-CH_2-CH_2)_n-OH$$

Polyoxyethylene-*p*-isooctylphenyl ether

$n = \ \ 5$ **Triton X-20**
$n = 10$ **Triton X-100**

Figure 12-19 A selection of the detergents used in biochemical manipulations. Note that they may be anionic, cationic, zwitterionic, or uncharged. Ionic detergents are strongly amphiphilic and therefore tend to denature proteins, whereas neutral detergents are unlikely to do so.

dissolving in their hydrocarbon cores. Nerve impulse transmission, which is a membrane-based phenomenon (Section 20-5), is disrupted by these structural changes to which neuronal membranes seem particularly sensitive.

3 BIOLOGICAL MEMBRANES

Biological membranes are composed of proteins associated with a lipid bilayer matrix. Their lipid fractions consist of complex mixtures that vary according to the membrane source (Table 12-3) and, to some extent, with the diet and environment of the organism that produced the membrane. *Membrane proteins carry out the dynamic processes associated with membranes, and therefore specific proteins occur only in particular membranes.* Protein-to-lipid ratios in membranes vary considerably with membrane function, as is indicated by Table 12-4, although most membranes are at least one-half protein. The myelin membrane, which functions passively as an insulator around certain nerve fibers (Section 20-5Bc), is a prominent exception to this generalization in that it contains only 18% protein.

In this section, we discuss the properties of membrane proteins and their behavior in biological membranes. Following this, we examine specific aspects of biological membranes, namely, the erythrocyte cytoskeleton, the nature of blood groups, gap junctions, and channel-forming proteins. We consider how membranes are assembled and how their component proteins are directed to them in Section 12-4.

tend to aggregate and precipitate in aqueous solutions unless they are solubilized by detergents or water-miscible organic solvents such as butanol or glycerol. Some integral proteins bind lipids so tenaciously that they can be freed from them only under denaturing conditions. Solubilized integral proteins can be purified by many of the protein fractionation methods discussed in Chapter 6.

2. *Peripheral* or *extrinsic proteins are dissociated from membranes by relatively mild procedures that leave the membrane intact,* such as exposure to high ionic strength salt solutions (e.g., $1M$ NaCl), metal chelating agents, or pH changes. Peripheral proteins, for example, cytochrome *c*, are stable in aqueous solution and do not bind lipid. They associate with a membrane by binding at its surface to its lipid head groups and/or its integral proteins through electrostatic and hydrogen bonding interactions. Membrane-free peripheral proteins behave as water-soluble globular proteins and can be purified as such (Chapter 6).

In this subsection we concentrate on integral proteins.

a. Integral Proteins Are Asymmetrically Oriented Amphiphiles

All biological membranes contain integral proteins, which typically comprise ~25% of the proteins encoded by a genome. Their locations on a membrane may be determined through **surface labeling,** a technique employing agents that react with proteins but cannot penetrate membranes. For example, an integral protein on the outer surface of an intact cell membrane binds antibodies elicited against it, but a protein on the membrane's inner surface can do so only if the membrane has been ruptured. Membrane-impermeable protein-specific reagents that are fluorescent or radioactively labeled may be similarly employed. Using such surface-labeling reagents, it has been shown that *some integral proteins are exposed only to a specific surface of a membrane, whereas others, known as* **transmembrane proteins,** *span the membrane.* However, no protein is known to be completely buried in a membrane; that is, all have some exposure to the aqueous environment. Such studies have also established that *biological membranes are asymmetric in that a particular membrane protein is invariably located on only one particular face of a membrane or, in the case of a transmembrane protein, oriented in only one direction with respect to the membrane* (Fig. 12-20).

Integral proteins are amphiphilic; the protein segments immersed in a membrane's nonpolar interior have predominantly hydrophobic surface residues, whereas those portions that extend into the aqueous environment are by and large sheathed with polar residues. For example, proteolytic digestion and chemical modification studies indicate that the erythrocyte transmembrane protein **glycophorin A** (Fig. 12-21) has three domains: (1) a 72-residue externally located *N*-terminal domain that bears 16 carbohydrate chains; (2) a 19-residue sequence, consisting almost entirely of hydrophobic residues, that spans the erythrocyte cell membrane; and (3) a 40-residue cytoplasmic C-terminal

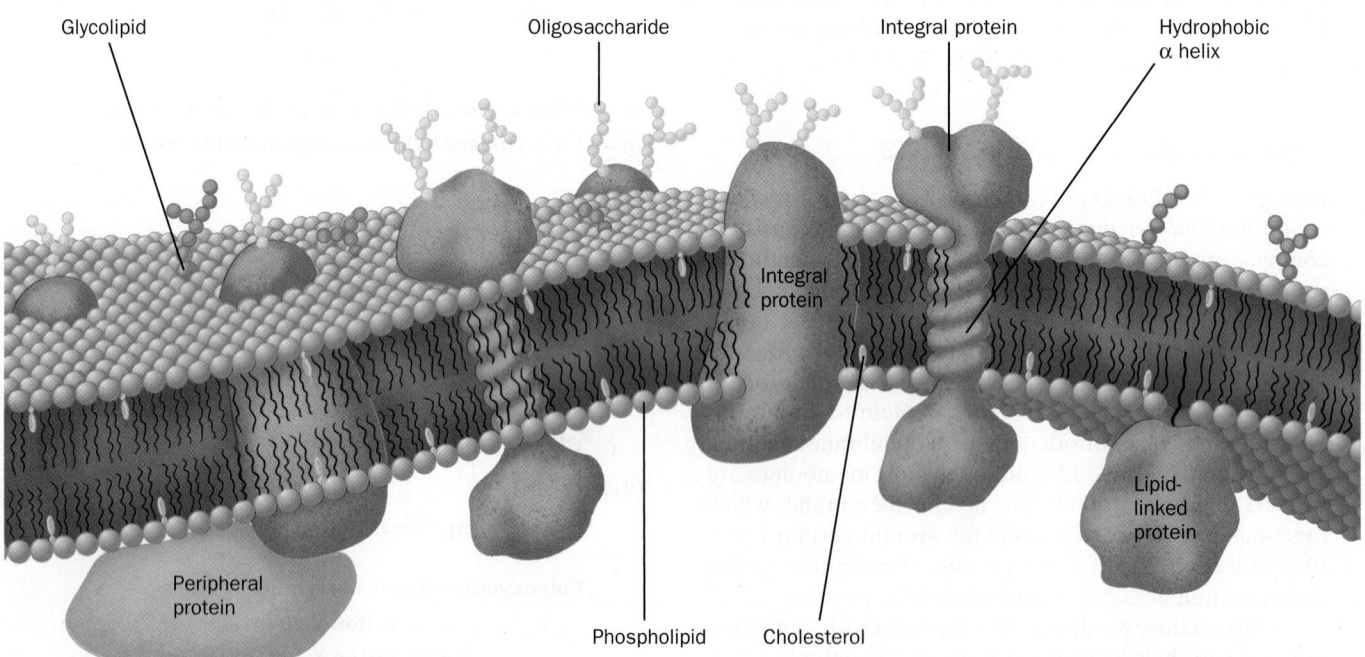

Glycolipid • Oligosaccharide • Integral protein • Hydrophobic α helix • Integral protein • Lipid-linked protein • Peripheral protein • Phospholipid • Cholesterol

Figure 12-20 Diagram of a plasma membrane. Integral proteins (*orange*) are embedded in a bilayer composed of phospholipids (*blue spheres with two wiggly tails*) and cholesterol (*yellow*). The carbohydrate components of glycoproteins (*yellow beaded chains*) and glycolipids (*green beaded chains*) occur only on the external face of the membrane. Most biological membranes have a higher prortion of protein than is drawn here. See Guided Exploration 11: Membrane structure and the fluid mosaic model

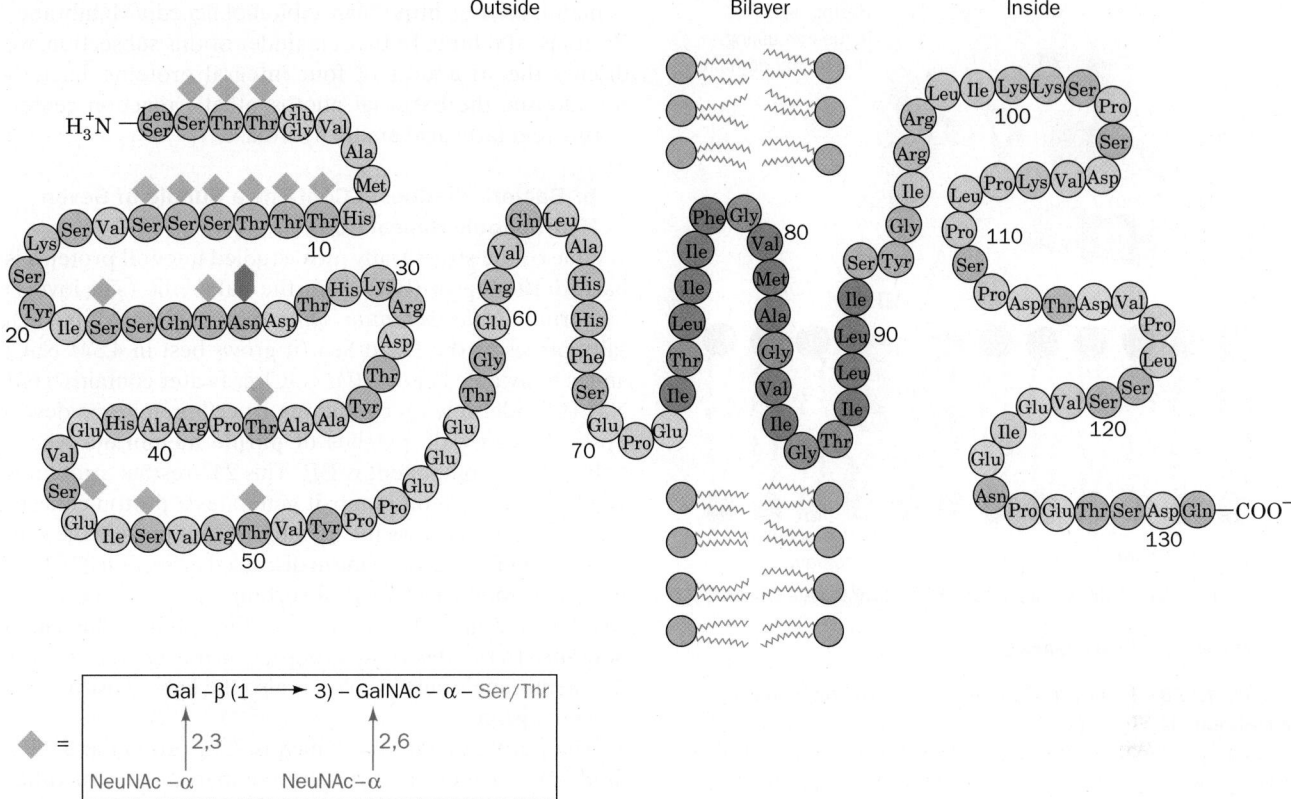

Figure 12-21   The amino acid sequence and membrane location of human erythrocyte glycophorin A. The protein, which is ~60% carbohydrate by weight, bears 15 *O*-linked oligosaccharides (*green diamonds*) and one that is *N*-linked (*dark green hexagon*). The predominant sequence of the *O*-linked oligosaccharides is given below. The protein's transmembrane portion (*brown and purple*) consists of 19 sequential predominantly hydrophobic residues. Its C-terminal portion, which is located on the membrane's cytoplasmic face, is rich in anionic (*pink*) and cationic (*blue*) amino acid residues. There are two common genetic variants of glycophorin A: glycophorin A[M] has Ser and Gly at positions 1 and 5, respectively, whereas they are Leu and Glu in glycophorin A[N]. [Abbreviations: Gal = galactose, GalNAc = *N*-acetylgalactosamine, NeuNAc = *N*-acetylneuraminic acid (sialic acid)]. [After Marchesi, V.T., *Semin. Hematol.* **16,** 8 (1979).]

domain that has a high proportion of charged and polar residues. The transmembrane domain, as is common in many integral proteins, forms an α helix, thereby satisfying the hydrogen bonding requirements of its polypeptide backbone. Indeed, the existence of glycophorin A's single transmembrane helix is predicted by computing the free energy change in transferring α helically folded polypeptide segments from the nonpolar interior of a membrane to water (Fig. 12-22). Similar computations on other integral proteins have also identified their transmembrane helices.

In many integral proteins, the hydrophobic segment(s) anchors the active region of the protein to the membrane. For instance, trypsin cleaves the membrane-bound enzyme

cytochrome b_5 into a polar, enzymatically active ~85-residue N-terminal fragment and an ~50-residue C-terminal fragment that remains embedded in the membrane

Figure 12-22   A plot, for glycophorin A, of the calculated free energy change in transferring 20-residue-long α helical segments from the interior of a membrane to water versus the position of the segment's first residue. Peaks higher than +85 kJ · mol^{-1} are indicative of a transmembrane helix. [After Engleman, D.M., Steitz, T.A., and Goldman, A., *Annu. Rev. Biophys. Biophys. Chem.* **15,** 343 (1986).]

Figure 12-23 **Liver cytochrome b_5 in association with a membrane.** The protein's enzymatically active N-terminal domain (*purple*), whose X-ray structure has been determined, is anchored in the membrane by a hydrophobic and presumably α helical C-terminal segment (*brown*) that begins and ends with hydrophilic segments (*purple*). The amino acid sequence of the horse enzyme indicates that this hydrophobic anchor consists of a 13-residue segment ending 9 residues from the polypeptide's C-terminus (*below*). [Ribbon diagram of the N-terminal domain after a drawing by Jane Richardson, Duke University. Amino acid sequence from Ozols, J. and Gerard, C., *J. Biol. Chem.* **253,** 8549 (1977).]

(Fig. 12-23). *The asymmetric orientation of integral proteins in the membrane is maintained by their infinitesimal flip-flop rates (even slower than those of lipids), which result from the greater sizes of the membrane protein "head groups" in comparison to those of lipids.* The origin of this asymmetry is discussed in Section 12-4.

Relatively few integral proteins have yet been crystallized—and then usually in the presence of detergents, which are but poor substitutes for lipid bilayers. Thus, despite their biological abundance, only ~0.8% of the proteins of known structure are integral proteins (>80% of which are bacterial proteins). A database of these proteins

is maintained at http://blanco.biomol.uci.edu/Membrane_Proteins_xtal.html. In the remainder of this subsection, we discuss the structures of four integral proteins: bacteriorhodopsin, the bacterial photosynthetic reaction center, porins, and fatty acid amide hydrolase.

b. Bacteriorhodopsin Contains a Bundle of Seven Hydrophobic Helical Rods

One of the structurally most studied integral proteins is **bacteriorhodopsin (BR)** from the halophilic (salt loving) bacterium *Halobacterium salinarium* that inhabits such salty places as the Dead Sea (it grows best in 4.3*M* NaCl and is nonviable below 2.0*M* NaCl; seawater contains 0.6*M* NaCl). Under low O_2 conditions, its cell membrane develops ~0.5-μm-wide patches of **purple membrane** whose only protein component is BR. This 247-residue protein is a light-driven proton pump; it generates a proton concentration gradient across the membrane that powers the synthesis of ATP (by a mechanism discussed in Section 22-3Bh). Bacteriorhodopsin's light-absorbing element, **retinal,** is covalently bound to its Lys 216 (Fig. 12-24). This **chromophore** (light-absorbing group), which is responsible for the membrane's purple color, is also the light-sensitive element in vision.

The purple membrane, which is 75% protein and 25% lipid, has an unusual structure compared to most other membranes (Section 12-3C): Its BR molecules are arranged in a highly ordered two-dimensional array (a two-dimensional crystal). This permitted Richard Henderson and Nigel Unwin, through **electron crystallography** (a technique they devised, resembling X-ray crystallography, in which the electron beam of an electron microscope is used to elicit diffraction from two-dimensional crystals), to determine the structure of BR to near-atomic resolution (3.0 Å). The more recently determined 1.9-Å-resolution X-ray structure of BR, based on single crystals of BR dissolved in lipidic cubic phases (mixtures of lipids and water that form a highly convoluted but continuous bilayer that is interpenetrated by aqueous channels), closely resembles that determined by electron crystallography.

Bacteriorhodopsin forms a homotrimer. Each of its subunits consists mainly of a bundle of seven ~25-residue α helical rods that each span the lipid bilayer in directions almost perpendicular to the bilayer plane (Fig. 12-25). BR is therefore said to be **polytopic** (multispanning; Greek: *topos,* place). The ~20-Å spaces between the protein molecules in the purple membrane are occupied by this bilayer (Fig. 12-25*b*). Adjacent α helices, which are largely

Figure 12-24 **Molecular formula of retinal.** Retinal, the prosthetic group of bacteriorhodopsin, forms a Schiff base with Lys 216 of the protein. A similar linkage occurs in **rhodopsin,** the photoreceptor of the eye (Section 19-2B).

(a)   *(b)*

Figure 12-25 **Structure of bacteriorhodopsin.** (*a*) The protein
is shown in ribbon form as viewed from within the membrane
plane and colored in rainbow order from its N-terminus (*blue*) to
its C-terminus (*red*). Its covalently bound retinal is drawn in stick
form (*magenta*). [Based on an X-ray structure by Nikolaus
Grigorieff and Richard Henderson, MRC Laboratory of
Molecular Biology, Cambridge, U.K. PDBid 2BRD.] (*b*) The
X-ray structure of a bacteriorhodopsin trimer with portions of its
surrounding trimers as viewed from the extracellular side of the
membrane. The protein molecules are shown in ribbon form
(*gray*) and their associated lipid tails are shown in ball-and-stick

form in different colors with symmetry-related lipid tails the
same color (the lipid head groups are disordered and hence not
seen). Only the lipids in the extracellular leaflet are shown; those
in the cytoplasmic leaflet have a similar distribution. Note how
the seven antiparallel α helices in each BR monomer are
cyclically arranged in two layers of four and three helices with
helices adjacent in sequence also adjacent in space (the N to C
direction circulates clockwise in this view). [Courtesy of Eva
Pebay-Peyroula, Université Joseph Fourier, Grenoble, France.
PDBid 1AP9.] **See Kinemage Exercise 8-1**

hydrophobic in character, are connected in a head-to-tail
fashion by short polypeptide loops. This arrangement
places the protein's charged residues near the surfaces of
the membrane in contact with the aqueous solvent. The
internal charged residues line the center of the helix
bundle of each monomer so as to form a hydrophilic
channel that facilitates the passage of protons. Other
membrane pumps and channels (Chapter 20) have similar
structures.

c. The Photosynthetic Reaction Center Contains Eleven Transmembrane Helices

The primary photochemical process of photosynthesis
in purple photosynthetic bacteria is mediated by the so-
called **photosynthetic reaction center** (**PRC;** Section 24-2B),
a transmembrane (**TM**) protein consisting of at least
three nonidentical ~300-residue subunits that collectively
bind four **chlorophyll** molecules, four other chromophores,
and a nonheme Fe(II) ion. The 1187-residue photosynthetic reaction center of *Rhodopseudomonas (Rps.)
viridis,* whose X-ray structure was determined in 1984 by
Hartmut Michel, Johann Deisenhofer, and Robert Huber,
was the first TM protein to be described in atomic detail
(Fig. 12-26). The polytopic protein's TM portion consists
of 11 α helices that form a 45-Å-long flattened cylinder
with the expected hydrophobic surface. In later chapters

we shall see that the transmembrane portions of most TM
proteins consist of bundles of one to >20 helices, most of
which are closely perpendicular to the membrane although some may be obliquely oriented and/or not fully
traverse the membrane.

d. Porins Are Channel-Forming Proteins That Contain Transmembrane β Barrels

The outer membranes of gram-negative bacteria (Section 11-3B) protect them from hostile environments but
must nevertheless be permeable to small polar solutes such
as nutrients and waste products. These outer membranes
consequently contain embedded channel-forming proteins
called **porins,** which are usually trimers of identical 30- to
50-kD subunits that permit the passage of solutes of less
than ~600 D. Porins also occur in eukaryotes in the outer
membranes of mitochondria and chloroplasts (thereby
providing a further indication that these organelles are descended from bacteria; Section 1-2A).

The X-ray structures of several different porins have
been elucidated, among them the *Rhodobacter (Rb.) capsulatus* porin, determined by Georg Schulz, and the *E. coli*
OmpF and **PhoE** porins, determined by Johan Jansonius.
The 340- and 330-residue OmpF and PhoE porins share
63% sequence identity but have little sequence similarity
with the 301-residue *Rb. capsulatus* porin. Nevertheless, all

(a)

(b)

Figure 12-26   X-ray structure of the photosynthetic reaction center of *Rps. viridis*. (*a*) The H, M, and L subunits, which are respectively shown as pink, blue, and orange ribbons, collectively have 11 transmembrane helices. The four-heme *c*-type cytochrome (*green*), which does not occur in all species of photosynthetic bacteria, is bound to the external face of the complex. The prosthetic groups are drawn in stick form with C yellow, N blue, and O red with a bound Fe(II) ion represented by a red sphere. The position that the transmembrane protein is thought to occupy in the lipid bilayer is indicated schematically. (*b*) A surface diagram, viewed as in Part *a*, in which hydrophobic residues are tan and polar residues are purple. Note how few polar groups are externally exposed in the portion of the protein that is immersed in the nonpolar region of the lipid bilayer. [Based on an X-ray structure by Johann Deisenhofer, Robert Huber, and Hartmut Michel, Max-Planck-Institut für Biochemie, Martinsreid, Germany. PDBid 1PRC.] 🔎 **See Kinemage Exercise 8-2**

three porins have closely similar structures. Each monomer of these homotrimeric proteins predominantly consists of a 16-stranded antiparallel β barrel which forms a solvent-accessible pore along the barrel axis that has a length of ~55 Å and a minimum diameter of ~7 Å (Fig. 12-27; although note that β barrel membrane proteins with 8, 10, 12, 14, 18, 19, 22, and 24 strands are also known). In the OmpF and PhoE porins, the N- and C-termini associate via a salt bridge in the 16th β strand, thereby forming a pseudocyclic structure (Fig. 12-27*a*). Note that a β barrel fully satisfies the polypeptide backbone's hydrogen bonding potential, as does an α helix. As expected, the side chains at the protein's membrane-exposed surface are nonpolar, thereby forming an ~25-Å-high hydrophobic band encircling the trimer (Fig. 12-27*c*). In contrast, the side chains at the solvent-exposed surface of the protein, including those lining the walls of the aqueous channel, are polar. Possible mechanisms for solute selectivity by these porins are discussed in Section 20-2D.

e. Fatty Acid Amide Hydrolase Binds to Only One Bilayer Leaflet

Not all integral proteins are TM proteins. For example, the enzyme **fatty acid amide hydrolase (FAAH)** is an integral protein that binds to the cytoplasmic leaflet of the plasma membrane. It is therefore said to be **monotopic** as is cytochrome b_5 (Fig. 12-23). FAAH's X-ray structure, determined by Raymond Stevens and Benjamin Cravatt, reveals that each 537-residue subunit of this homodimer consists of an 11-stranded mixed β sheet surrounded by 28 α helices of various lengths (Fig. 12-28). Its membrane-binding segment consists of a helix–turn–helix motif, whose surface faces outward from the body of the protein to form a hydrophobic plateau. Its component nonpolar residues, many of which are aromatic, are interspersed with several basic residues, which presumably interact electrostatically with the membrane's phospholipid head groups. The structure of this hydrophobic motif resembles that observed in the X-ray structures of two other monotopic integral proteins,

(a)

(b)

20 Å

25 Å

8 Å

(c)

cytosol

membrane

Figure 12-27   X-ray crystal structure of the *E. coli* OmpF porin. (*a*) A ribbon diagram of the monomer colored in rainbow order from its N-terminus (*blue*) to its C-terminus (*red*). Each strand of this 16-stranded antiparallel β barrel is inclined by ~45° to the barrel axis. Its C-terminal strand is continued by the N-terminal segment (*bottom right*), thereby forming a pseudo-continuous strand. All porins of known structure have similar structural properties. (*b*) Ribbon diagram of the trimer embedded in its semitransparent surface and viewed along its threefold axis of symmetry from the cell's exterior showing the pore through each subunit. Each subunit is differently colored. Adjacent β strands in adjoining subunits extend essentially perpendicularly to each other. (*c*) A space-filling model of the trimer viewed perpendicular to its 3-fold axis. N atoms are blue, O atoms are red, and C atoms are green, except those of Trp and Tyr side chains, which are white. These latter groups delimit an ~25-Å-high hydrophobic band (*scale at right*) that is immersed in the nonpolar portion of the bacterial outer membrane (with the cell's exterior at the tops of Parts *a* and *c*). Compare the hydrophobic band in this figure with that in Fig. 12-26*b*. [Based on an X-ray structure by Johan Jansonius, University of Basel, Switzerland. PDBid 1OPF.] **See Kinemage Exercise 8-3**

Figure 12-28   X-ray structure of rat liver fatty acid amide hydrolase indicating its proposed disposition in the cytoplasmic leaflet of the plasma membrane. This homodimeric enzyme is viewed along the plane of the membrane with its 2-fold axis of symmetry vertical. It is drawn in ribbon form with each subunit colored in rainbow order from its N-terminus (*blue*) to its C-terminus (*red*) except for its putative membrane-binding motif, which is dark green. The enzyme's bound inhibitor, **methyl arachidonyl fluorophosphonate,** is drawn in space-filling form with C green, O red, and P orange. The membrane model was generated by a molecular dynamics simulation of a palmitoyloleoylphosphatidylethanolamine bilayer. [Based on an X-ray structure by Raymond Stevens and Benjamn Cravatt, Scripps Research Institute, La Jolla, California. PDBid 1MT5.]

prostaglandin H synthase (Section 25-7B) and **squalene-hopene cyclase** (Section 25-6Ad). Nevertheless, these enzymes have no apparent sequence or structural homology, which suggests that they have independently evolved similar modes of membrane integration.

f. Integral Proteins Have Common Structural Features

Hydrophobic forces, as we have seen in Section 8-4, are the dominant interactions stabilizing the three-dimensional structures of water-soluble globular proteins. However, since the membrane-exposed regions of integral proteins are immersed in nonpolar environments, what stabilizes their structures? Analysis of a variety of integral proteins indicates that their membrane-exposed regions have a hydrophobic organization opposite to that of water-soluble proteins: Their membrane-exposed residues are more hydrophobic, on average, than their interior residues, even though these interior residues have average hydrophobicities and packing densities comparable to those of water-soluble proteins. Evidently, *the structures of integral and water-soluble proteins are both stabilized by the exclusion of their interior residues from the surrounding solvent, although in the case of integral proteins, the solvent is the lipid bilayer.* In addition, the low polarity and anhydrous environments of transmembrane proteins are likely to strengthen their hydrogen bonds relative to those of soluble proteins.

In the foregoing TM proteins, those portions of the transmembrane secondary structural elements (helices in BR and the PRC, and β strands in the porins) that contact the bilayer's hydrocarbon core consist mainly of the hydrophobic residues Ala, Ile, Leu, Val, and Phe. The flanking residues, which penetrate the bilayer's interface region, are enriched with Trp, and Tyr. Hence, *TM proteins' hydrophobic transmembrane bands are bordered by rings of Trp and Tyr side chains* (e.g., Fig. 12-27c) *that delineate the water–bilayer interface.* Note that these side chains are oriented such that their polar portions (N and O atoms) extend into the polar regions of the membrane, a phenomenon named **snorkeling.** Lys and Arg side chains near the interface tend to be similarly oriented. In contrast, Phe, Leu, and Ile side chains tend to point toward the membrane core, a phenomenon dubbed **antisnorkeling.**

In each of the foregoing TM proteins, the secondary structural elements that are adjacent in sequence are also adjacent in structure and hence tend to be antiparallel. This relatively simple up–down **topology** results from the constraints associated with the insertion of a folding polypeptide chain into the lipid bilayer (Section 12-4Be).

B. Lipid-Linked Proteins

Lipids and proteins associate covalently to form **lipid-linked** *proteins, whose lipid portions anchor their attached proteins to membranes and mediate protein–protein interactions.* Proteins form covalent attachments with three classes of lipids: (1) isoprenoid groups such as farnesyl and geranylgeranyl residues, (2) fatty acyl groups such as myristoyl and palmitoyl residues, and (3) glycoinositol phospholipids (GPIs). In this subsection, we discuss the properties of these lipid-linked proteins.

a. Prenylated Proteins

A variety of proteins have covalently attached **isoprenoid groups,** mainly the C_{15} **farnesyl** and C_{20} **geranylgeranyl** residues (**isoprene,** a C_5 hydrocarbon, is the chemical unit from which many lipids, including cholesterol and other steroids, are constructed; Section 25-6A).

$$H_2C = \underset{\underset{CH}{|}}{\overset{\overset{CH_3}{|}}{C}} = CH_2$$

Isoprene

Farnesyl residue

Geranylgeranyl residue

The most common isoprenylation (or just **prenylation**) site in proteins is the C-terminal tetrapeptide CaaX, where C is Cys, "a" is often an aliphatic amino acid residue, and X is any amino acid. However, the identity of X is a major prenylation determinant: Proteins are farnesylated when X is Gln, Met, or Ser and geranylgeranylated when X is Leu. In both cases, the prenyl group is enzymatically linked to the Cys sulfur atom via a thioether linkage. The aaX tripeptide is then proteolytically excised and the newly exposed terminal carboxyl group is esterified with a methyl group (Fig. 12-29).

Two other types of prenylation sites have also been characterized: (1) the C-terminal sequence CXC, in which both Cys residues are geranylgeranylated and the terminal carboxyl group is methyl esterified; and (2) the C-terminal sequence CC in which one or both Cys residues are geranylgeranylated but the carboxyl group is not methylated. Proteins that are so prenylated are almost exclusively members of the **Rab** family of small GTP-binding proteins that participate in intracellular membrane trafficking (Section 12-4Db).

What functions are served by protein prenylation? Many prenylated proteins are associated with intracellular membranes, and mutating their Cys prenylation sites blocks their membrane localization. Evidently, *the hydrophobic prenyl group can act to anchor its attached protein to a membrane.* However, this can only be part of the story since proteins with the same prenyl groups may be localized to different intracellular membranes. Moreover, fusing the CaaX motif from a normally prenylated protein to the C-terminus of a normally unprenylated protein yields a hybrid protein that is correctly prenylated and carboxyl methylated but which remains cytosolic. These

(a)

S–Farnesyl cysteine methyl ester

(b)

S–Geranylgeranyl cysteine methyl ester

Figure 12-29   Prenylated proteins. (*a*) A farnesylated protein and (*b*) a geranylgeranylated protein. In both cases, the protein is synthesized with the C-terminal sequence CaaX, where C is Cys, "a" is often an aliphatic amino acid, and X is any amino acid. After the prenyl group is appended to the protein in thioether linkage with the Cys residue, the aaX tripeptide is hydrolytically cleaved away and the new carboxyl terminus is methyl esterified. When X is Ala, Met, or Ser, the protein is farnesylated and when X is Leu, it is geranylgeranylated.

observations suggest that prenylated proteins may interact with specific membrane-bound receptor proteins and hence that *prenylation also facilitates protein–protein interactions.* This idea is corroborated by the observation that, in certain proteins involved in intracellular signaling [for example, **Ras** (Section 19-3Cf) and the so-called **G proteins** (Section 19-2)], prenylation and carboxyl methylation enhance the intersubunit associations that mediate signal transmission.

b. Fatty Acylated Proteins

Two fatty acids are known to be covalently linked to eukaryotic proteins:

1. Myristic acid, a biologically rare saturated C_{14} fatty acid (Table 12-1), which is appended to a protein in amide linkage to the α-amino group of an N-terminal Gly residue. Myristoylation almost always occurs cotranslationally (as the protein is being synthesized), and this attachment is stable, that is, the myristoyl group has a half-life similar to that of the protein to which it is appended.

2. Palmitic acid, a biologically common saturated C_{16} fatty acid, which is joined to a protein in thioester linkage to a specific Cys residue. In some cases, the palmitoylated protein is also prenylated. For example, Ras must be farnesylated and carboxyl methylated as described above before it is palmitoylated at a Cys residue that precedes the protein's C-terminus by several residues. Palmitoylation occurs post-translationally in the cytosol and is reversible.

Fatty acyl groups are thought to function as membrane anchors for proteins, much as do prenyl groups. However, the requirement of many proteins for specific fatty acyl residues suggests that these groups also participate in targeting their attached proteins to specific cellular locations. Indeed, palmitoylated proteins occur almost exclusively on the cytoplasmic face of the plasma membrane, whereas myristoylated proteins are found in a number of subcellular compartments including the cytosol, endoplasmic reticulum, Golgi apparatus, plasma membrane, and nucleus. Many fatty acylated proteins participate in intracellular signaling processes through protein–protein interactions in a manner similar to prenylated proteins. Since the membrane affinities and biological activities of many proteins are enhanced by palmitoylation, the reversibility of palmitoylation appears to be involved in controlling intracellular signaling processes.

c. GPI-Linked Proteins

Glycosylphosphatidylinositol (GPI) groups function to anchor a wide variety of proteins to the exterior surface of the eukaryotic plasma membrane. There is no obvious relationship among the numerous proteins that have GPI anchors, which include enzymes, receptors, immune system proteins, and recognition antigens. Evidently, *GPI groups simply provide an alternative to transmembrane polypeptide domains in binding proteins to the plasma membrane.*

The core structure of GPI anchors consists of phosphatidylinositol (Table 12-2) glycosidically linked to a linear tetrasaccharide composed of three mannose residues

Figure 12-30 Core structure of the GPI anchors of proteins. R_1 and R_2 represent fatty acid residues whose identities vary with the protein. The tetrasaccharide may have a variety of attached sugar residues whose identities also vary with the protein. In addition, the phosphoethanolamine residue may be linked to either of the other two mannose residues in the tetrasaccharide.

and one glucosaminyl residue (Fig. 12-30). The mannose at the nonreducing end of this assembly forms a phospho-ester bond with a phosphoethanolamine residue, which in turn, is amide-linked to the protein's C-terminal carboxyl group. The core tetrasaccharide is generally substituted with a variety of sugar residues that vary with the identity of the protein. There is likewise considerable diversity in the fatty acid residues. The synthesis of GPI anchors is discussed in Section 23-3Bk.

GPI-anchored proteins occur on the exterior surface of the plasma membrane for the same reason as do the carbohydrate residues of glycoproteins (which we discuss in Section 12-4Ca). Proteins destined to be GPI-anchored are synthesized with membrane-spanning C-terminal sequences of 20 to 30 hydrophobic residues (as described in Section 12-4Ba) that are removed during GPI addition. This is corroborated by the observation that GPI-anchored proteins are released from the plasma membrane by treatment with phosphatidylinositol-specific **phospholipases** (Section 19-4B), thereby demonstrating that the mature polypeptides are not embedded in the lipid bilayer.

C. Fluid Mosaic Model of Membrane Structure

See Guided Exploration 11: Membrane structure and the fluid mosaic model The demonstrated fluidity of artificial lipid bilayers suggests that biological membranes have similar properties. This seminal idea was proposed in 1972 by S. Jonathan Singer and Garth Nicolson in their unifying theory of membrane structure known as the **fluid mosaic model.** The theory postulates that integral proteins resemble "icebergs" floating in a two-dimensional lipid "sea" (Fig. 12-20) and that these proteins freely diffuse laterally in the lipid matrix unless their movements are restricted by associations with other cell components.

a. The Fluid Mosaic Model Has Been Verified Experimentally

The validity of the fluid mosaic model has been established in several ways. Perhaps the most vivid is an experiment by Michael Edidin (Fig. 12-31). Cultured mouse cells were fused with human cells by treatment with **Sendai virus** to yield a hybrid cell known as a **heterokaryon.** The mouse cells were labeled with mouse protein–specific

antibodies to which a green-fluorescing dye had been covalently linked (**immunofluorescence**). The proteins on the human cells were similarly labeled with a red-fluorescing marker. On cell fusion, the mouse and human proteins, as seen under the fluorescence microscope, were segregated on the two halves of the heterokaryon. After 40 min at 37°C, however, these proteins had thoroughly intermingled. The addition of substances that inhibit metabolism or protein synthesis did not slow this process, but lowering the temperature below 15°C did. These observations indicate that the mixing process is independent of both metabolic energy and the insertion into the membrane of newly synthesized proteins. Rather, it is a result of the diffusion of existing proteins throughout the fluid membrane, a process that slows as the temperature is lowered.

Fluorescence photobleaching recovery measurements (Fig. 12-15) indicate that membrane proteins vary in their lateral diffusion rates. Some 30 to 90% of these proteins are freely mobile; they diffuse at rates only an order of magnitude or so slower than those of the much smaller lipids, so that they typically take from 10 to 60 min to diffuse the 20-μm length of a eukaryotic cell. Other proteins diffuse more slowly, and some, because of submembrane attachments, are essentially immobile.

The distribution of proteins in membranes may be visualized through electron microscopy using the **freeze-fracture** and **freeze-etch** techniques. In the freeze-fracture procedure, which was devised by Daniel Branton, a membrane specimen is rapidly frozen to near liquid nitrogen temperatures (−196°C). This immobilizes the sample and thereby minimizes its disturbance by subsequent manipulations. The specimen is then fractured with a cold microtome knife, which often splits the bilayer into monolayers (Fig. 12-32). Since the exposed membrane itself would be

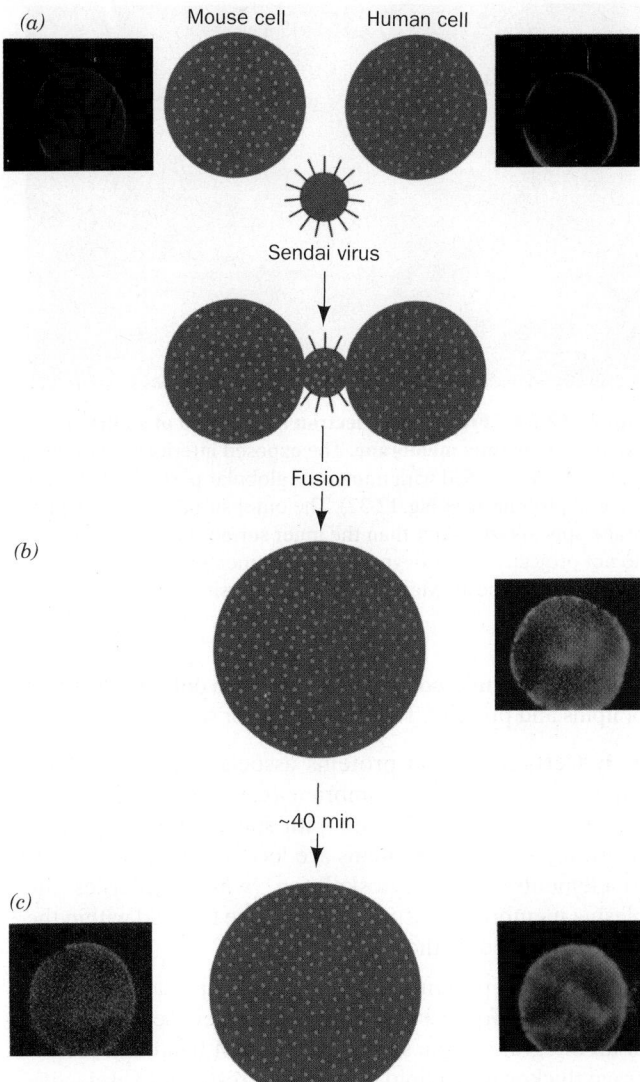

(a) Mouse cell   Human cell

Sendai virus

Fusion

(b)

~40 min

(c)

Figure 12-31 Sendai virus–induced fusion of a mouse cell with a human cell and the subsequent intermingling of their cell-surface components as visualized by immunofluorescence. Human and mouse antigens are labeled with red and green fluorescent markers, respectively. (*a*) The membrane-encapsulated Sendai virus specifically binds to cell-surface receptors on both types of cells and subsequently fuses to their cell membranes. (*b*) This results in the formation of a cytoplasmic bridge between the cells that expands so as to form a heterokaryon. (*c*) After 40 min, the red and green markers are fully intermingled. The photomicrographs were taken through filters that allowed only red or green light to reach the camera; that in Part *b* is a double exposure and those in Part *c* are of the same cell. [Immunofluorescence photomicrographs courtesy of Michael Edidin, Johns Hopkins University.]

Embedded proteins

Split lipid bilayer

Figure 12-32 The freeze-fracture technique. A membrane that has been split by freeze-fracture, as is schematically diagrammed, exposes the interior of the lipid bilayer and its embedded proteins.

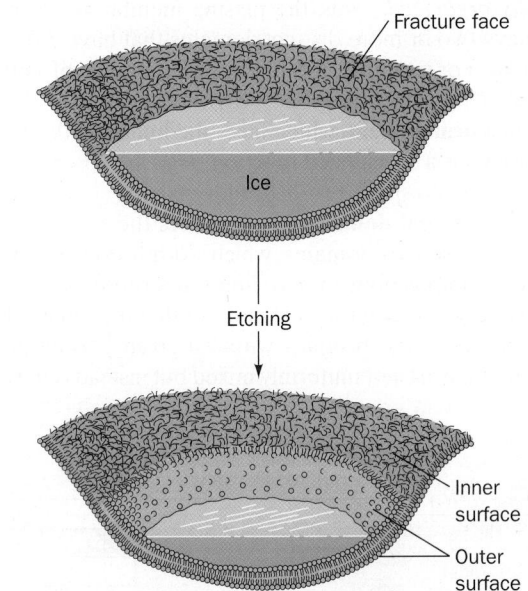

Fracture face

Ice

Etching

Inner surface

Outer surface

Figure 12-33 The freeze-etch procedure. The ice that encases a freeze-fractured membrane (*top*) is partially sublimed away so as to expose the outer membrane surface (*bottom*) for electron microscopy.

destroyed by an electron beam, its metallic replica is made by coating the membrane with a thin layer of carbon, shadowing it (covering it by evaporative deposition under high vacuum) with platinum, and removing the organic matter by treatment with acid. Such a metallic replica can be examined by electron microscopy. In the freeze-etch procedure, the external surface of the membrane adjacent to the

cleaved area revealed by freeze fracture may also be visualized by first subliming (etching) away, at −100°C, some of the ice in which it is encased (Fig. 12-33).

Freeze-etch electron micrographs of most biological membranes show an inner fracture face that is studded with embedded 50- to 85-Å-diameter globular particles (Fig. 12-34) that appear to be distributed randomly. These particles correspond to membrane proteins, as is demonstrated by their disappearance when the membrane is treated with proteases before its freeze fracture. This is further corroborated by the observation that the myelin membrane, which has a low protein content, as well as liposomes composed of only lipids, have smooth inner fracture faces. Outer membrane surfaces also have a relatively smooth appearance (Fig. 12-34) because integral proteins tend not to protrude very far beyond them. The distributions of individual external proteins may be visualized by staining procedures, such as the use of ferritin-labeled antibodies, to yield electron micrographs similar in appearance to Fig. 11-36.

b. Membrane Lipids and Proteins Are Unevenly Distributed

The distribution of lipids between the different sides of biological membranes has been established through the use of phospholipid-hydrolyzing enzymes known as **phospholipases.** Phospholipases cannot pass through membranes, so that only phospholipids on the external surfaces of intact cells are susceptible to their action. Such studies indicate that *the lipids in biological membranes, like the proteins, are asymmetrically distributed between the leaflets of a bilayer* (e.g., Fig. 12-35). Carbohydrates, as we have seen (Section 11-3Cd), are located almost exclusively on the external surfaces of plasma membranes.

Lipids and proteins in plasma membranes may also be laterally organized. Thus, the plasma membranes of most cells have two or more distinct domains that have different functions. For example, the plasma membranes of **epithelial cells** (the cells lining body cavities and free surfaces) have an **apical domain,** which faces the lumen of the cavity and often has a specialized function (e.g., the absorption of nutrients in intestinal brush border cells; Section 20-4A), and a **basolateral domain,** which covers the remainder of the cell. These two domains, which do not intermix, have different compositions of both lipids and proteins.

A variety of measurements indicate that the hundreds of different lipids and proteins within a given plasma membrane domain are not uniformly mixed but instead often seg-

Figure 12-34   Freeze-etch electron micrograph of a human erythrocyte plasma membrane. The exposed interior face of the membrane is studded with numerous globular particles that are integral proteins (see Fig. 12-32). The outer surface of the membrane appears smoother than the inner surface because proteins do not project very far beyond the outer membrane surface. [Courtesy of Vincent Marchesi, Yale University.]

regate to form **microdomains** that contain only certain types of lipids and proteins. This may occur for several reasons:

1. Certain integral proteins associate to form aggregates or patches in the membrane (e.g., BR), which in turn may preferentially associate with specific lipids. Alternatively, some integral proteins are localized by attachments to elements of the cytoskeleton (which underlies the plasma membrane; Section 1-2A) or are trapped within the spaces enclosed by the resulting "fences."

2. Integral proteins may specifically interact with particular lipids. For example, mismatches between the length of an integral protein's hydrophobic TM band and the average thickness of a lipid bilayer may result in the selective accumulation of certain phospholipids around the protein in an annulus of 10 to 20 layers.

3. Divalent metal ions, notably Ca^{2+}, selectively ligate negatively charged head groups such as those of phosphatidylserine, thereby causing these phospholipids to aggregate in the membrane. Such metal ion–induced phase separations are known to regulate the activities of certain membrane-bound enzymes.

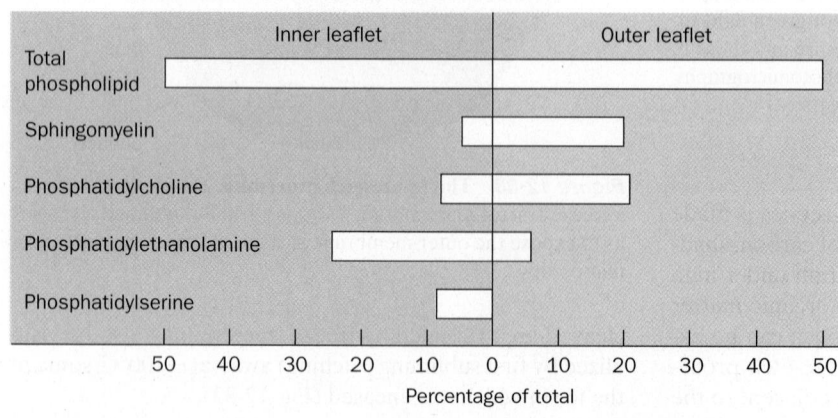

Figure 12-35   Asymmetric distribution of phospholipids in the human erythrocyte membrane. The phospholipid content is expressed as mole percent. [After Rothman, J.E. and Lenard, J., *Science* **194,** 1744 (1977).]

4. Glycosphingolipids (which occur only in the outer leaflet of the plasma membrane) and cholesterol pack together to form mobile **rafts** and ~75-nm-diameter flask-shaped indentations named **caveolae** (Latin for small caves) with which specific proteins preferentially associate. Glycosphingolipids, by themselves, do not form bilayers because their large head groups prevent the requisite close packing of their predominantly saturated hydrophobic tails. Conversely, cholesterol by itself does not form a bilayer due to its small head group. It therefore appears that the glycosphingolipids in these microdomains associate laterally via weak interactions between their carbohydrate head groups with the voids between their tails filled in by cholesterol. The sphingolipid–cholesterol rafts and caveolae are not solublized at 4°C by uncharged detergents such as Triton X-100 (Fig. 12-19). The low density of the resulting **detergent-resistant membranes (DRMs)** allows their isolation by sucrose density gradient ultracentrifugation (Section 6-5Ba), thereby permitting their associated proteins to be identified. Many of the proteins that participate in transmembrane signaling processes (Chapter 19), including GPI-linked proteins, preferentially associate with DRMs. Caveolae, which appear to be rafts with which one or more homologous proteins named **caveolins** are associated, are likewise enriched with proteins that participate in signaling.

It should be noted that all of these aggregates are highly dynamic structures that rapidly exchange both proteins and lipids with their surrounding membrane as a consequence of the weak and transient interactions between membrane components and their interactions with the underlying cytoskeleton. In fact, single molecule tracking techniques (Section 12-2Ca) have demonstrated that lipid molecules in biological membranes undergo a series of short random motions over short time periods (~10 ms) interspersed by large hops (Fig. 12-36), a process called **hop diffusion.** Evidently, *biological membranes are partitioned rather than continuous two-dimensional fluids.*

Figure 12-36 **Hop diffusion of individual colloidal gold–tagged dioleoylphosphatidylethanolamine molecules in a plasma membrane.** The position of each particle was determined at 25-μs intervals (a video frame rate of 40,500 frames · s^{-1}) over a period of 62 ms (2500 steps). Colored lines connect successive positions of the particle, with differently colored segments representing the various plausible regions in the plasma membrane to which the particle appears to have been transiently confined (in the order purple, blue, green, orange, and red). [Courtesy of Akihiro Kasumi, Nagoya University, Japan.]

D. *The Erythrocyte Membrane*

The erythrocyte membrane's relative simplicity, availability, and ease of isolation have made it the most extensively studied and best understood biological membrane. It is therefore a model for the more complex membranes of other cell types. A mature mammalian erythrocyte is devoid of organelles and carries out few metabolic processes; it is essentially a membranous bag of hemoglobin. Erythrocyte membranes can therefore be obtained by osmotic lysis, which causes the cell contents to leak out. The resultant membranous particles are known as erythrocyte **ghosts** because, on return to physiological conditions, they reseal to form colorless particles that retain their original shape. Indeed, by transferring sealed ghosts to another medium, their contents can be made to differ from the external solution.

a. Erythrocyte Membranes Contain a Variety of Proteins

The erythrocyte membrane has a more or less typical plasma membrane composition of about half protein, somewhat less lipid, and the remainder carbohydrate (Table 12-4). Its proteins may be separated by SDS–polyacrylamide gel electrophoresis (Section 6-4C) after first solubilizing the membrane in a 1% SDS solution. The resulting electrophoretogram for a human erythrocyte membrane exhibits seven major and many minor bands when stained with Coomassie brilliant blue (Fig. 12-37). If the electrophoretogram is instead treated with **periodic acid–Schiff's reagent (PAS),** which stains carbohydrates, four so-called PAS bands become evident. The polypeptides corresponding to bands 1, 2, 4.1, 4.2, 5, and 6 are readily extracted from the

Figure 12-37 **SDS–PAGE electrophoretogram of human erythrocyte membrane proteins as stained by Coomassie brilliant blue.** The bands designated 4.1 and 4.2 are not separated with the 1% SDS concentration used. The minor bands are not labeled for the sake of clarity. The positions of the four sialoglycoproteins that would be revealed by PAS staining are indicated. [Courtesy of Vincent Marchesi, Yale University.]

membrane by changes in ionic strength or pH and hence are peripheral proteins. These proteins are located on the inner side of the membrane, as is indicated by the observation that they are not altered by the incubation of intact erythrocytes or sealed ghosts with proteolytic enzymes or membrane-impermeable protein labeling reagents. These proteins are altered, however, if "leaky" ghosts are so treated.

In contrast, bands 3, 7, and all four PAS bands correspond to integral proteins; they can be released from the membrane only by extraction with detergents or organic solvents. Of these, band 3 and PAS bands 1 and 2 correspond to TM proteins, as indicated by their different labeling patterns when intact cells are treated with membrane-impermeable protein-labeling reagents and when these reagents are introduced inside sealed ghosts. The PAS band 1 is a dimer of glycophorin A, which is formed through an SDS-resistant association between the TM helices of the polypeptide chains (Fig. 12-21); this dimer is the protein's native form. The PAS band 2 protein is the monomeric form of glycophorin A.

The transport of CO_2 in blood (Section 10-1C) requires that the erythrocyte membrane be permeable to HCO_3^- and Cl^- (the maintenance of electroneutrality requires that for every HCO_3^- that enters a cell, a Cl^- or some other anion must leave the cell; Section 10-1Cb). The rapid transport of these and other anions across the erythrocyte membrane is mediated by a specific **anion channel** of which there are ~1 million/cell (comprising >30% of the membrane protein). Band 3 protein (929 residues and 5–8% carbohydrate) specifically reacts with anionic protein-labeling reagents that block the anion channel, thereby indicating that the anion channel is composed of band 3 protein. Furthermore, cross-linking studies with bifunctional reagents (Section 8-5Ca) demonstrate that the anion channel is at least a dimer. Hemoglobin and the glycolytic (glucose metabolizing) enzymes **aldolase, phosphofructokinase (PFK),** and the band 6 protein **glyceraldehyde-3-phosphate dehydrogenase (GAPDH;** Section 17-2F) all specifically and reversibly bind to band 3 protein on the cytoplasmic side of the membrane. The functional significance of this observation is unknown.

b. The Erythrocyte's Cytoskeleton Is Responsible for Its Shape and Flexibility

A normal erythrocyte's biconcave disklike shape (Fig. 7-19a) assures the rapid diffusion of O_2 to its hemoglobin molecules by placing them no farther than 1 μm from the cell surface. However, the rim and the dimple regions of an erythrocyte do not occupy fixed positions on the cell membrane. This can be demonstrated by anchoring an erythrocyte to a microscope slide by a small portion of its surface and inducing the cell to move laterally with a gentle flow of isotonic buffer. A point originally on the rim of the erythrocyte will move across the dimple to the rim on the opposite side of the cell from where it began. Evidently, the membrane rolls across the cell while maintaining its shape, much like the tread of a tractor. This remarkable mechanical property of the erythrocyte membrane results from the presence of a submembranous network of proteins that function as a membrane "skeleton"—the cell's cytoskeleton. Indeed, this property is partially duplicated by a

mechanical model consisting of a geodesic sphere (a spheroidal cage) that is freely jointed at the intersections of its struts but constrained from collapsing much beyond a flat surface. When placed inside an evacuated plastic bag, this cage also assumes a biconcave disklike shape.

The fluidity and flexibility imparted to an erythrocyte by its cytoskeleton has important physiological consequences. A slurry of solid particles of a size and concentration equal to that of red cells in blood has the flow characteristics approximating that of sand. Consequently, in order for blood to flow at all, much less for its erythrocytes to squeeze through capillary blood vessels smaller in diameter than they are, erythrocyte membranes, together with their cytoskeletons, must be fluidlike and easily deformable.

The protein **spectrin,** so called because it was discovered in erythrocyte ghosts, accounts for ~75% of the erythrocyte cytoskeleton. It is composed of two similar polypeptide chains, band 1 (α subunit; 2418 residues) and band 2 (β subunit; 2137 residues), which sequence analysis indicates each consist of repeating 106-residue segments that are predicted to fold into triple-stranded α helical coiled coils (Fig. 12-38a,b). Electron microscopy indicates that these large polypeptides are loosely intertwined to form a flexible wormlike αβ dimer that is ~1000 Å long (Fig. 12-38c). Two such heterodimers further associate in a

Figure 12-38 (*Opposite*) **The human erythrocyte cytoskeleton.** (*a*) Structure of an αβ dimer of spectrin. Both of these antiparallel polypeptides contain multiple 106-residue repeats, which are thought to form flexibly connected triple helical bundles. Two of these heterodimers join, head to head, to form an (αβ)₂ heterotetramer. [After Speicher, D.W. and Marchesi, V., *Nature* **311,** 177 (1984).] (*b*) X-ray structure of two consecutive repeats of chicken brain α-spectrin. Each of these 106-residue repeats consists of a down–up–down triple helical bundle in which the C-terminal helix of first repeat (R16; *red*) is continuous, via a 5-residue helical linker (*green*), with the N-terminal helix of the second repeat (R17; *blue*). The helices within each triple helical bundle wrap around each other in a gentle left-handed supercoil that is hydrophobically stabilized by the presence of nonpolar residues at the *a* and *d* positions of heptad repeats on all three of its component α helices (Fig. 8-26). Despite the expected rigidity of α helices, there is considerable evidence that spectrin is a flexible wormlike molecule. [Courtesy of Alfonso Mondragón, Northwestern University. PDBid 1CUN.] (*c*) Electron micrograph of an erythrocyte cytoskeleton that has been stretched to an area 9 to 10 times greater than that of the native membrane. Stretching makes it possible to obtain clear images of the cytoskeleton, which in its native state is so densely packed and irregularly flexed that it is difficult to pick out individual molecules and to ascertain how they are interconnected. Note the predominantly hexagonal network composed of spectrin tetramers cross-linked by junctions containing actin and band 4.1 protein. [Courtesy of Daniel Branton, Harvard University.] (*d*) Model of the erythrocyte cytoskeleton. The so-called junctional complex, which is magnified in this drawing, contains actin, **tropomyosin** (which, in muscle, also associates with actin; Section 35-3Ac), and band 4.1 protein, as well as **adducin, dematin,** and **tropomodulin** (not shown). [After Goodman, S.R., Krebs, K.E., Whitfield, C.F., Riederer, B.M., and Zagen, I.S., *CRC Crit. Rev. Biochem.* 23, 196 (1988).]

(a)

α chain

β chain

N

C

C

N

(b)

C−terminus

Helix B'

R17

Helix A'

Helix C'

Helix B

Linker region

Helix C

Helix A

R16

N−terminus

(c)

(d)

Actin

Tropomyosin

Band 4.1

Spectrin

α

β

Ankyrin

Band 4.2

Anion channel

Glycophorin A

Figure 12-39   **X-ray structure of human ankyrin repeats 13 to 24.** The polypeptide is shown in ribbon form colored in rainbow order from its N-terminus (*blue,* repeat 13) to its C-terminus (*red,* repeat 24). [Based on an X-ray structure by Peter Michaely, University of Texas Southwestern Medical Center, Dallas, Texas. PDBid 1N11.]

head-to-head manner to form an $(\alpha\beta)_2$ heterotetramer. These tetramers, of which there are ~100,000/cell, are cross-linked at both ends by attachments to bands 4.1 and 5 to form a dense and irregular protein meshwork that underlies the erythrocyte plasma membrane (Fig. 12-38*c,d*). Band 5, a globular protein that forms filamentous oligomers, has been identified as **actin,** a common cytoskeletal element in other cells (Section 1-2Ae) and a major component of muscle (Section 35-3Ac). Spectrin also associates with band 2.1, an 1880-residue monomer known as **ankyrin,** which, in turn, binds to band 3, the anion channel protein. This attachment anchors the cytoskeleton to the membrane. Indeed, on solubilization of spectrin and actin by low ionic strength solutions, the erythrocyte ghosts' biconcave shape is lost and their integral proteins, which normally occupy fixed positions in the membrane plane, become laterally mobile.

Ankyrin's N-terminal 798-residue segment consists almost entirely of 24 tandem ~33-residue repeats known as **ankyrin repeats** (Fig. 12-39), which also occur in a variety of other proteins. Each ankyrin repeat consists of two short (8- or 9-residue) antiparallel α helices followed by a long loop. These structures are arranged in a right-handed superhelical stack. The entire assembly forms an elongated concave surface that is postulated to bind various integral proteins as well as spectrin. Immunochemical studies have revealed spectrinlike, ankyrinlike, and band 4.1–like proteins in the cytoskeletons of a variety of tissues.

c. Hereditary Spherocytosis and Elliptocytosis Arise from Defects in the Erythrocyte Cytoskeleton

Individuals with **hereditary spherocytosis** have spheroidal erythrocytes that are relatively fragile and inflexible. These individuals suffer from hemolytic anemia because the spleen, a labyrinthine organ with narrow passages that normally filters out aged erythrocytes (which lose flexibility toward the end of their ~120-day lifetime), prematurely removes spherocytotic erythrocytes. The hemolytic anemia may be alleviated by the spleen's surgical removal. However, the primary defects in spherocytotic cells are reduced

synthesis of spectrin, the production of an abnormal spectrin that binds band 4.1 protein with reduced affinity, or the absence of band 4.1 protein.

Hereditary elliptocytosis (having elongated or elliptical red cells; also known as **hereditary ovalcytosis**), a condition that is common in certain areas of Southeast Asia and Melanesia, confers resistance to malaria in heterozygotes (but apparently is lethal in homozygotes). This condition arises from defects in the erythrocyte anion channel. A common such defect consists of a 9-residue deletion that inactivates this TM protein. The consequent reduced capacity of red cells to import phosphate or sulfate ions may inhibit the intraerythrocytotic growth of rapidly developing malarial parasites.

The camel, the renowned "ship of the desert," provides a striking example of adaptation involving the erythrocyte membrane. This remarkable animal is still active after a loss of water constituting 30% of its body weight and, when thus dehydrated, can drink sufficient water in a few minutes to become fully rehydrated. The rapid uptake of such a large amount of water by the blood, which must deliver it to the cells, would lyse the erythrocytes of most animals. Yet camel erythrocytes, which have the shape of flattened ellipsoids rather than biconcave disks, are resistant to osmotic lysis. Camel spectrin binds to its membrane with particular tenacity, but on spectrin removal, which requires a strong denaturing agent such as guanidinium chloride, camel erythrocytes assume a spherical shape.

E. *Blood Groups*

The outer surfaces of erythrocytes and other eukaryotic cells are covered with complex carbohydrates that are components of plasma membrane glycoproteins and glycolipids. They form a thick, fuzzy cell coating, the **glycocalyx** (Fig. 12-40), which contains numerous identity markers that function in various recognition processes. Human erythrocytes have 30 genetically distinct blood group systems comprised of >600 known **blood group determinants,** although many of these determinants are rare or occur only in certain ethnic groups. Of these systems, only two—the **ABO blood group system** (discovered in 1900 by Karl

Figure 12-40   **The erythrocyte glycocalyx as revealed by electron microscopy using special staining techniques.** It is up to 1400 Å thick and composed of closely packed, 12- to 25-Å-diameter oligosaccharide filaments linked to plasma membrane–associated proteins and lipids. [Courtesy of Harrison Latta, UCLA.]

Landsteiner) and the **rhesus (Rh) blood group system**—have major clinical importance. The various blood groups are identified by means of suitable antibodies or by specific plant lectins.

a. ABO Blood Group Substances Are Carbohydrates

The ABO system consists of three blood group substances, the A, B, and H antigens, which are components of erythrocyte surface sphingoglycolipids. [Antigens are characteristic constellations of chemical groups on macromolecules that elicit the production of specific antibodies when introduced into an animal (Section 35-2Aa). Each antibody molecule can specifically bind to at least two of its corresponding antigen molecules, thereby cross-linking them.] Individuals with type A cells have A antigens on their cell surfaces and carry anti-B antibodies in their serum; those with type B cells, which bear B antigens, carry anti-A antibodies; those with type AB cells, which bear both A and B antigens, carry neither anti-A nor anti-B antibodies; and type O individuals, whose cells bear neither antigen, carry both anti-A and anti-B antibodies. Consequently, the transfusion of type A blood into a type B individual, for example, causes an anti-A antibody–A antigen reaction, which agglutinates (clumps together) the transfused erythrocytes, resulting in an often fatal blockage of blood vessels. The H antigen is discussed below. Anti-A and anti-B antibodies, which are not present at birth, appear to arise through an immune response to A-like and B-like antigens in food and/or to the colonization of the infant gut by bacteria that produce such antigens [the immune system normally suppresses the production of antibodies directed against the body's own antigens (Section 35-2Ac) so, for example, a type A individual does not produce anti-A antibodies].

The ABO blood group substances are not confined to erythrocytes but also occur in the plasma membranes of many tissues as glycolipids of considerable diversity. In fact, in the ~80% of the population known as secretors, these antigens are secreted as *O*-linked components of glycoproteins into various body fluids, including saliva, milk, seminal fluid, gastric juice, and urine. These diverse molecules, which are 85% carbohydrate by weight and have molecular masses ranging up to thousands of kilodaltons, consist of multiple oligosaccharides attached to a polypeptide chain.

The A, B, and H antigens differ only in the sugar residues at their nonreducing ends (Table 12-5). The H antigen occurs in type O individuals; it is also the precursor oligosaccharide of A and B antigens. Type A individuals have a 303-residue glycosyltransferase that specifically adds an *N*-acetylgalactosamine residue to the terminal position of the H antigen, whereas in type B individuals, this enzyme, which differs by four amino acid residues from that of type A individuals, instead adds a galactose residue. In type O individuals, the enzyme is inactive because its synthesis terminates after its 115th residue.

Do the different blood groups confer any biological advantages or disadvantages? Epidemiological studies indicate

Table 12-5 Structures of the A, B, and H Antigenic Determinants in Erythrocytes

Type	Antigen
H	Galβ(1 → 4)GlcNAc··· ↑1,2 L-Fucα
A	GalNAcα(1 → 3)Galβ(1 → 4)GlcNAc··· ↑1,2 L-Fucα
B	Galα(1 → 3)Galβ(1 → 4)GlcNAc··· ↑1,2 L-Fucα

Abbreviations: Gal = galactose, GalNAc = *N*-acetylgalactosamine, GlcNAc = *N*-acetylglucosamine, L-Fuc = L-fucose.

that type A and B individuals are less susceptible to cholera infections than are type O individuals, with the relatively rare type AB individuals being highly resistant to this deadly disease. Apparently, type A and B oligosaccharides block a receptor for the bacterium causing cholera, *Vibrio cholera* (Section 19-2Cd). In addition, type O individuals, particularly nonsecretors, have a higher incidence of peptic (stomach) ulcers. However, type A individuals have a higher incidence of stomach cancer, heart disease, and pernicious anemia (Section 25-2Ee).

F. Gap Junctions

Most cells in multicellular organisms are in metabolic and electrical as well as physical contact with neighboring cells. This contact is brought about by tubular particles, named **gap junctions,** that join discrete regions of neighboring plasma membranes much like hollow rivets (Fig. 12-41). Indeed, these intercellular channels are so widespread that many whole organs are continuous from within. Thus gap junctions are important intercellular communication channels. For example, the synchronized contraction of heart muscle is brought about by flows of ions through gap junctions (heart muscle is not innervated as is skeletal muscle). Likewise, gap junctions serve as conduits for some of the substances that mediate embryonic development; blocking gap junctions with antibodies that bind to them causes developmental abnormalities in species as diverse as hydras, frogs, and mice. Gap junctions also function to nourish cells that are distant from the blood supply, such as bone and lens cells. Thus, it is not surprising that, in humans, defects in gap junctions are associated with certain neurodegenerative diseases, cataracts, deafness, several skin diseases, and developmental anomalies.

Gap junctions consist of a single sort of protein subunit known as a **connexin.** A single gap junction consists of two apposed hexagonal rings of connexins, called **connexons,** one from each of the adjoining plasma membranes

Figure 12-41 **Model of a gap junction.** Gap junctions between adjacent cells consist of two apposed plasma membrane–embedded hexagonal studs that bridge the gap between the cells. Gap junctions hold cells a fixed distance apart—the gap. Small molecules and ions, but not macromolecules, can pass between cells via the gap junction's central channel.

(Fig. 12-41). Cells are normally connected by clusters of hundreds to thousands of connexons. A given animal expresses numerous genetically distinct connexins, with humans, for example, expressing 21 different connexins ranging in molecular mass from 25 to 50 kD. Many types of cells simultaneously express several different species of connexins, and in cells that do so, there is considerable evidence that at least some connexons may be formed from two or more species of connexins. Moreover, the gap junctions joining two cells may consist of two different types of connexons. These various types of gap junctions presumably differ in their selectivities for the substances they transmit.

Mammalian gap junction channels are minimally 16 to 20 Å in diameter, which Werner Loewenstein established by microinjecting single cells with fluorescent molecules of various sizes and observing with a fluorescence microscope whether the fluorescent probe passed into neighboring cells. The molecules and ions that can pass freely between neighboring cells are limited in molecular mass to a maximum of ~1000 D; macromolecules such as proteins and nucleic acids cannot leave a cell via this route.

The diameter of a gap junction channel varies with Ca^{2+} concentration: The channels are fully open when the Ca^{2+} level is $<10^{-7}\ M$ and narrow as the Ca^{2+} concentration increases until, above $5 \times 10^{-5}\ M$, they close. This shutter system is thought to protect communities of interconnected cells from the otherwise catastrophic damage that would result from the death of even one of their members. Cells generally maintain very low cytosolic Ca^{2+} concentrations ($<10^{-7}\ M$) by actively pumping Ca^{2+} out of the cell as well as into their mitochondria and endoplasmic reticulum (Section 20-3B; Ca^{2+} is an important intracellular messenger whose cytosolic concentration is precisely regulated). Ca^{2+} floods back into leaky or metabolically depressed

cells, thereby inducing closure of their gap junctions and sealing them off from their neighbors.

a. Connexins Contain Transmembrane Four-Helix Bundles

The X-ray structure of the gap junction formed by the 226-residue human **connexin 26 (Cx26),** determined by Tomitake Tsukihara, reveals a 12-mer with D_6 symmetry, a height of 155 Å, and a maximal diameter of 92 Å that encloses a central channel (Fig. 12-42*a*). The extracellular portion of each connexon extends 23 Å from the extracellular surface and interdigitates with the opposite connexon by 6 Å to span an intercellular gap of 40 Å. Each Cx26 subunit forms an up–down–up–down four-helix bundle in which both the N- and C-termini occupy the cytosol. The central channel has a diameter of ~40 Å at its cytosolic entrance that funnels down to 14 Å near the extracellular surface of the membrane and then widens to 25 Å in the extracellular space (Fig. 12-42*b*). The positively charged funnel entrance would attract negatively charged molecules. However, the region of maximal channel constriction is negatively charged, which should also affect the channel's charge selectivity.

G. *Channel-Forming Proteins*

A number of bacterial toxins are synthesized as water-soluble monomers that, on interacting with their target membrane via a specific receptor protein, spontaneously insert into the membrane as a TM pore. This process, which for many such **channel-forming toxins (CFTs)** requires their oligomerization, causes the leakage of small ions and molecules from the target cell, thereby killing it through loss of osmotic balance. The formation of only one CFT-based pore is often sufficient to kill a cell.

(a)

(b)

Figure 12-42 X-ray structure of the connexin 26 gap junction.
(*a*) View perpendicular to the protein's 6-fold axis (parallel to
the planes of the membranes) in which the protein is drawn in
ribbon form embedded in its semitransparent molecular surface.
Each connexin of the upper connexon has a different color,
whereas one connexin in the lower connexon is colored in
rainbow order from its N-terminus (*blue*) to its C-terminus (*red*)
with the remaining connexins purple. The extent of the

transmembrane region was deduced from the distribution of
hydrophobic and aromatic residues (Section 12-3Af).
(*b*) Cutaway drawing through the surface diagram of the gap
junction channel. The channel surface is colored according to its
electrostatic potential with red positive, blue negative, and white
neutral. [Part *a* based on an X-ray structure by, and Part *b*
courtesy of Tomitake Tsukihara, University of Osaka, Japan.
PDBid 2ZW3.]

One of the best characterized CFTs is **α-hemolysin,**
which the human pathogen *Staphylococcus aureus* secretes
as a water-soluble 293-residue monomer and which spontaneously inserts into the membranes of erythrocytes and
several other types of cells in the form of heptameric pores.
Even though the α-hemolysin monomer is water-soluble
and lacks clearly hydrophobic segments, the heptamer acts
as a typical TM protein in that it is not released from the
membrane by treatment with high salt, low pH, or
chaotropic agents but, instead, requires treatment with detergents for this to occur.

The X-ray structure of detergent-solublized α-hemolysin,
determined by Eric Gouaux, reveals a striking mushroom-
shaped heptameric complex that is 100 Å in height and
100 Å in diameter (Fig. 12-43*a,b*). A 14- to 46-Å-diameter
solvent-filled channel, which runs along the protein's 7-fold
axis, forms a TM pore. The stem of the mushroom, the
protein's TM segment, consists of a 52-Å-high and 26-Å-
diameter, porinlike, 14-stranded, antiparallel β barrel composed of seven 2-stranded antiparallel β sheets, one from
each subunit (Fig. 12-43*b*). The remainder of each subunit

consists of a β sandwich domain and a rim domain, which
together form a 70-Å-long ellipsoid (Fig. 12-43*c*). Seven of
these ellipsoids are distributed in a ring, thereby forming
the mushroom's cap and rim. The rim domain projects
toward and probably interacts with the membrane's phospholipid head groups via the basic and aromatic residues
that extend from the crevice between the top of the stem
and rim.

A variety of experimental evidence indicates that the
spontaneous formation of the heptameric TM pore occurs
via several discrete steps: (1) the binding of the aqueous
monomer to the membrane surface, probably through the
interaction of the protein's polypeptide loops with the surface groups of the lipid bilayer; (2) the formation of the
heptamer on the surface of the membrane; and (3) the insertion of the 14-stranded β barrel through the membrane
to form the TM pore. The structural details of this process
are as yet unknown, although it seems clear that there is little change in the monomers' secondary structure on their
assembly to form the heptameric TM pore. The reason why
monomers do not form heptamers in aqueous solution is

(a)   *(b)*

Figure 12-43 **X-ray structure of α-hemolysin.** Views (*a*) along and (*b*) perpendicular to the heptameric transmembrane pore's 7-fold axis. Each subunit is drawn with a different color. (*c*) The monomer unit with its three domains drawn in different colors. [Courtesy of Eric Gouaux, Columbia University. PDBid 7AHL.]

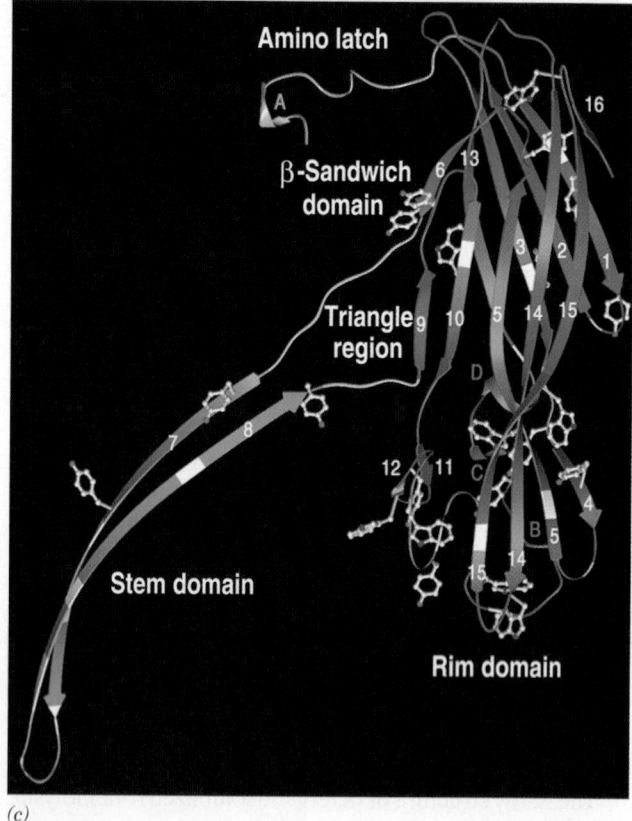

(c)

probably due to differences between the strengths of the intrasubunit interactions in the monomer in aqueous solution and the intersubunit interactions in the heptamer in the membrane.

Not all CFTs form pores using β barrels. Rather, a variety of CFTs, notably several *E. coli* proteins known as **colicins,** form pores that are lined with α helices. Most such pores consist of monomers.

4 MEMBRANE ASSEMBLY AND PROTEIN TARGETING

As cells grow and divide, they synthesize new membranes. How are such asymmetric membranes generated? One way in which this might occur is through self-assembly. Indeed, when the detergent used to disperse a biological membrane is removed, liposomes form in which functional integral proteins are embedded. In most cases, however, these model membranes are symmetrical, both in their lipid distribution between the inner and outer leaflets of the bilayer and in the orientations of their embedded proteins. An alternative hypothesis of membrane assembly is that *it occurs on the scaffolding of preexisting membranes; that is, membranes are generated by the expansion of old ones rather than by the creation of new ones.* In this section we shall see that this is, in fact, how biological membranes are generated. In doing so, we shall consider how proteins

are inserted into and passed through membranes as well as how portions of membranes in the form of vesicles pinch off from one membrane and fuse with another, thereby transporting proteins and lipids between these membranes. These highly complex processes are indicative of the intricacies of biological processes in general.

NO$_2$

O$_2$N   NO$_2$

SO$_3^-$

Trinitrobenzenesulfonic acid (TNBS)

+

$$O$$
$$\parallel$$
$$CH_2-O-C-R_1$$
$$O$$
$$R_2-C-O-CH \qquad O$$
$$CH_2-O-\overset{\parallel}{P}-O-CH_2-CH_2-NH_2$$
$$O^-$$

Phosphatidylethanolamine (PE)

→ H$_2$SO$_3$

$$O$$
$$\parallel$$
$$CH_2-O-C-R_1$$
$$O$$
$$R_2-C-O-CH \qquad O$$
$$CH_2-O-\overset{\parallel}{P}-O-CH_2-CH_2-NH$$
$$O^-$$

NO$_2$

O$_2$N   NO$_2$

Figure 12-44 **Reaction of TNBS with PE.**

A. Lipid Distributions in Membranes

The enzymes involved in the biosynthesis of membrane lipids are mostly integral proteins (Section 25-8). Their substrates and products are themselves membrane components, so that membrane lipids are fabricated on site. Eugene Kennedy and James Rothman demonstrated this to be the case in bacteria through the use of selective labeling. They gave growing bacteria a 1-min pulse of $^{32}PO_4^{3-}$ so as to label radioactively the phosphoryl groups of only the newly synthesized phospholipids. **Trinitrobenzenesulfonic acid (TNBS),** a membrane-impermeable reagent that combines with phosphatidylethanolamine (**PE;** Fig. 12-44), was then immediately added to the cell suspension. Analysis of the resulting doubly labeled membrane showed that none of the TNBS-labeled PE was radioactively labeled. This observation indicates that *newly made PE is synthesized on the cytoplasmic face of the membrane* (Fig. 12-45, *top right*).

a. Membrane Proteins Catalyze Phospholipid Flip-Flops

If an interval of only 3 min is allowed to elapse between the $^{32}PO_4^{3-}$ pulse and the TNBS addition, about half of the ^{32}P-labeled PE is also TNBS labeled (Fig. 12-45, *bottom*). This observation indicates that the flip-flop rate of PE in the bacterial membrane is ~100,000-fold greater than it is in bilayers consisting of only phospholipids (where, it will be recalled, the flip-flop rates have half-times of many days).

How do phospholipids synthesized on one side of the membrane reach its other side so quickly? Phospholipid flip-flops appear to be facilitated in two ways:

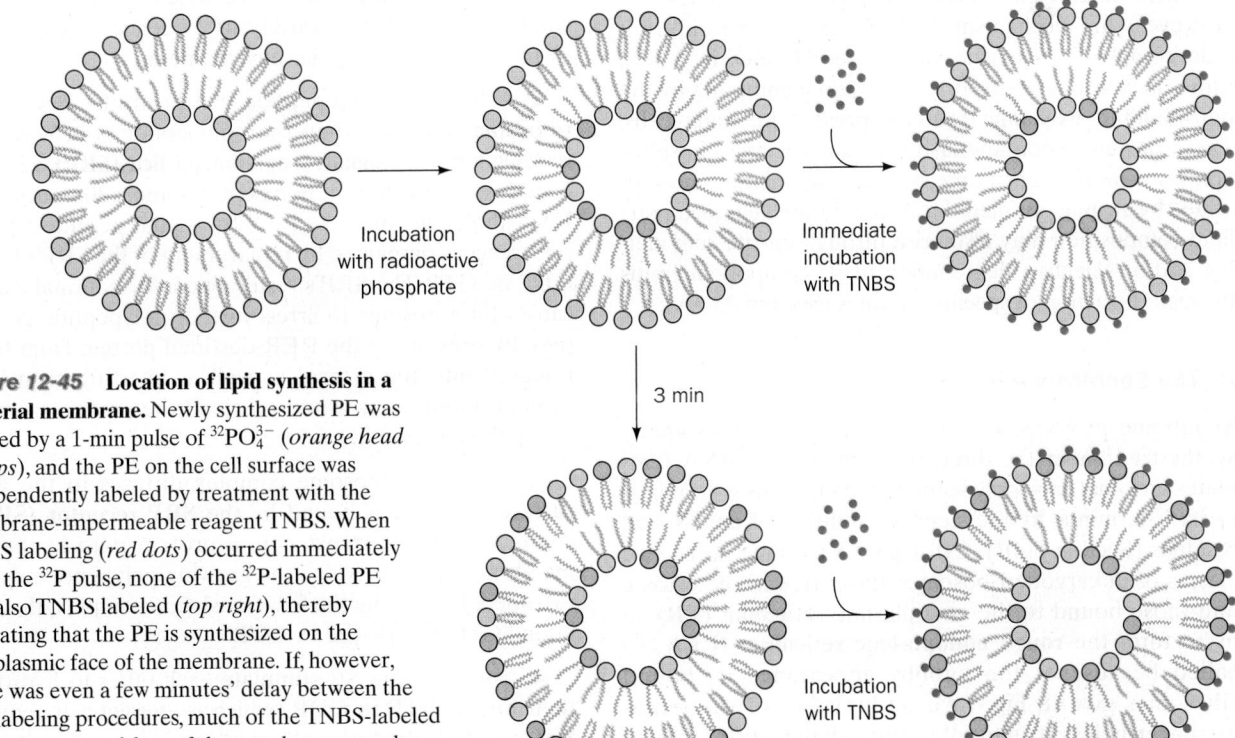

Incubation with radioactive phosphate

Immediate incubation with TNBS

3 min

Incubation with TNBS

Figure 12-45 **Location of lipid synthesis in a bacterial membrane.** Newly synthesized PE was labeled by a 1-min pulse of $^{32}PO_4^{3-}$ (*orange head groups*), and the PE on the cell surface was independently labeled by treatment with the membrane-impermeable reagent TNBS. When TNBS labeling (*red dots*) occurred immediately after the ^{32}P pulse, none of the ^{32}P-labeled PE was also TNBS labeled (*top right*), thereby indicating that the PE is synthesized on the cytoplasmic face of the membrane. If, however, there was even a few minutes' delay between the two labeling procedures, much of the TNBS-labeled PE in the external face of the membrane was also ^{32}P labeled (*bottom*).

1. Membranes contain proteins known as **flippases** that catalyze the flip-flops of specific phospholipids. These proteins tend to equilibrate the distribution of their corresponding phospholipids across a bilayer; that is, the net transport of a phospholipid is from the side of the bilayer with the higher concentration of the phospholipid to the opposite side. Such a process, as we shall see in Section 20-2, is a form of **facilitated diffusion.**

2. Membranes contain proteins known as **phospholipid translocases** that transport specific phospholipids across a bilayer in a process that is driven by ATP hydrolysis. These proteins can transport certain phospholipids from the side of a bilayer that has the lower concentration of the phospholipids being translocated to the opposite side, thereby establishing a nonequilibrium distribution of the phospholipids. Such a process, as we shall see in Section 20-2, is a form of **active transport.**

The observed distribution of phospholipids across membranes (e.g., Fig. 12-35) therefore appears to arise from the membrane orientations of the enzymes that synthesize phospholipids combined with the countervailing tendencies of ATP-dependent phospholipid translocases to generate asymmetric phospholipid distributions and those of flippases to randomize these distributions.

b. A Membrane's Characteristic Lipid Composition Can Arise in Several Ways

In eukaryotic cells, lipids are synthesized on the cytoplasmic face of the endoplasmic reticulum, from where they are transported to other membranes. Perhaps the most important mechanism of lipid transport is the budding off of membranous vesicles from the ER and their subsequent fusion with other membranes (Sections 12-4C and 12-4D). However, this mechanism does not explain the different lipid compositions of the various membranes in a cell. Lipids may also be transported between membranes by the **phospholipid exchange proteins** present in many cells. These proteins spontaneously transfer specific phospholipids, one molecule at a time, between two membranes separated by an aqueous medium. A membrane's characteristic lipid composition may also arise through on-site remodeling and/or selective degradation of its component lipids through the action of specific enzymes (Section 25-8A).

B. The Secretory Pathway

Membrane proteins, as are all proteins, are ribosomally synthesized under the direction of messenger RNA templates such that each polypeptide grows from its N-terminus to its C-terminus by the stepwise addition of amino acid residues (Section 5-4B). Cytologists have long noted two classes of eukaryotic ribosomes, those free in the cytosol and those bound to the endoplasmic reticulum **(ER)** so as to form the **rough endoplasmic reticulum (RER,** so called because of the knobby appearance its bound ribosomes give it; Fig. 1-5). Both classes of ribosomes are nevertheless structurally identical; they differ only in the nature of the polypeptide they are synthesizing. *Free ribosomes synthesize mostly soluble and mitochondrial proteins, whereas membrane-bound ribosomes manufacture TM proteins and proteins destined for secretion, operation within the ER, or incorporation into* **lysosomes** (membranous organelles containing a battery of hydrolytic enzymes that function to degrade and recycle cell components; Section 1-2Ad). These latter proteins initially appear in the RER.

a. The Secretory Pathway Accounts for the Targeting of Many Secreted and Membrane Proteins

How are RER-destined proteins differentiated from other proteins? And how do these large, relatively polar molecules pass through the RER membrane? These processes occur via the **secretory pathway,** which was first described by Günter Blobel, César Milstein, and David Sabatini around 1975. Since ~25% of the different species of proteins synthesized by all types of cells are integral proteins and many others are secreted, *~40% of the various types of proteins that a cell synthesizes must be processed via the secretory pathway or some other protein targeting pathway* (e.g., that which directs proteins to the mitochondrion; Section 12-4E). In this subsection, we first present an overview of the secretory pathway and then discuss its various aspects in detail. The secretory pathway is outlined in Fig. 12-46:

1. *All secreted, ER-resident, and lysosomal proteins, as well as many TM proteins, are synthesized with leading (N-terminal) 13- to 36-residue* **signal peptides.** These signal peptides consist of a 6- to 15-residue hydrophobic core flanked by several relatively hydrophilic residues that usually include one or more basic residues near the N-terminus (Fig. 12-47). Signal peptides otherwise have little sequence similarity. However, a variety of evidence indicates they form α helices in nonpolar environments.

2. When the signal peptide first protrudes beyond the ribosomal surface (when the polypeptide is at least ~40 residues long), the **signal recognition particle (SRP),** a 325-kD complex of six different polypeptides and a 300-nucleotide RNA molecule, binds to both the signal peptide and the ribosome accompanied by replacement of the SRP's bound GDP by GTP. The SRP's resulting conformational change causes the ribosome to arrest further polypeptide growth, thereby preventing the RER-destined protein from being released into the cytosol as well as averting premature protein folding that would preclude the protein from entering the ER (see below).

3. The SRP–ribosome complex diffuses to the RER surface, where it is bound by the **SRP receptor (SR)** in complex with the **translocon,** a protein pore in the ER membrane through which the growing polypeptide will be extruded. In forming the SR–translocon complex, the SR's bound GDP is replaced by GTP.

4. The SRP and SR stimulate each other to hydrolyze their bound GTP to GDP (which is energetically equivalent to ATP hydrolysis), resulting in conformational changes that cause them to dissociate from each other and

Figure 12-46 **The ribosomal synthesis, membrane insertion, and initial glycosylation of an integral protein via the secretory pathway.** **(1)** Protein synthesis is initiated at the N-terminus of the polypeptide, which consists of a 13- to 36-residue signal sequence. **(2)** A signal recognition particle (SRP) binds to the ribosome and the signal sequence emerging from it, thereby arresting polypeptide synthesis. **(3)** The SRP is bound by the transmembrane SRP receptor (SR) in complex with the translocon, thereby bringing together the ribosome and the translocon. **(4)** The SRP and SR hydrolyze their bound GTPs, causing them to dissociate from the ribosome–translocon complex. The ribosome then resumes the synthesis of the polypeptide, which passes through the translocon into the lumen

of the ER. **(5)** Shortly after the entrance of the signal sequence into the lumen of the endoplasmic reticulum, it is proteolytically excised. **(6)** As the growing polypeptide chain passes into the lumen, it commences folding into its native conformation, a process that is facilitated by its interaction with the chaperone protein BiP (not shown). Simultaneously, enzymes initiate the polypeptide's specific glycosylation. Once the protein has folded, it cannot be pulled out of the membrane. At points determined by its sequence, the polypeptide becomes anchored in the membrane (proteins destined for secretion pass completely into the ER lumen). **(7)** Once polypeptide synthesis is completed, the ribosome dissociates into its two subunits. **See the Animated Figures**

from the ribosome–translocon complex. This permits the bound ribosome to resume polypeptide synthesis such that the growing polypeptide's N-terminus passes through the

translocon into the lumen of the ER. Most ribosomal processes, as we shall see in Section 32-3, are driven by GTP hydrolysis.

	Signal peptidase cleavage site
Bovine growth hormone	M M A A G P R T S L L L A F A L L C L P W T Q V V G | A F P
Bovine proalbumin	M K W V T F I S L L L L F S S A Y S | R G V
Human proinsulin	M A L W M R L L P L L A L L A L W G P D P A A A | F V N
Human interferon γ	M K Y T S Y I L A F Q L C I Y L G S L G | C Y C
Human α-fibrinogen	M F S M R I V C L V L S V V G T A W T | A D S
Human IgG heavy chain	M E F G L S W L F L V A I L K G V Q C | E V Q
Rat amylase	M K F V L L L S L I G F C W A | Q Y D
Murine α-fetoprotein	M K W I T P A S L I L L L L H F A A S K | A L H
Chicken lysozyme	M R S L L I L V L C F L P L A A L G | K V F
Zea mays rein protein 22.1	M A T K I L A L L A L L A L L V S A T N A | F I I

Figure 12-47 **N-Terminal sequences of some eukaryotic secretory preproteins.** The hydrophobic cores (*brown*) of most signal peptides are preceded by basic residues (*blue*). [After Watson, M.E.E., *Nucleic Acids Res.* **12,** 5147–5156 (1984).]

5. Shortly after the signal peptide enters the ER lumen, it is specifically cleaved from the growing polypeptide by a membrane-bound **signal peptidase** (polypeptide chains with their signal peptide still attached are known as **preproteins;** signal peptides are alternatively called **presequences**).

6. The nascent (growing) polypeptide starts to fold to its native conformation, a process that is facilitated by its interaction with an ER-resident chaperone protein Hsp70 (Section 9-2C). Enzymes in the ER lumen then initiate **post-translational modification** of the polypeptide, such as the specific attachments of "core" carbohydrates to form glycoproteins (Section 23-3B); the formation of disulfide bonds as facilitated by protein disulfide isomerase (Section 9-2A), an ER-resident protein; and the attachment of GPI anchors (Section 23-3Bk).

7. When polypeptide synthesis is completed, the protein is released from both the ribosome and the translocon, and the ribosome dissociates from the RER. Secretory, ER-resident, and lysosomal proteins pass completely through the RER membrane into the lumen. TM proteins, in contrast, contain one or more hydrophobic ~22-residue TM sequences that remain embedded in the membrane.

The secretory pathway also occurs in prokaryotes for the insertion of certain proteins into the cell membrane (whose exterior is equivalent to the ER lumen). Indeed, all forms of life yet tested have homologous SRPs and SRs. However, in bacteria, the binding of the SRP to the ribosome does not arrest translation.

b. The Cryo-Electron Microscopy Structure of the SRP in Complex with a Translating Ribosome Reveals How the SRP Binds Signal Peptide and Arrests Translation

Mammalian SRPs consist of six polypeptides known as **SRP9, SRP14, SRP19, SRP54, SRP68,** and **SRP72** (where the numbers are their molecular masses in kilodaltons) and an ~300-nucleotide (**nt**) 7S RNA [Fig. 12-48; RNAs are often classified according to their sedimentation rate in Svedberg units (S), which increases with their molecular mass (Section 6-5Aa)]. Many prokaryotic SRPs are much simpler; that in *E. coli* consists of a single polypeptide named **Ffh** that is homologous to SRP54 (Ffh for *Fifty-four homolog*) and a 4.5S RNA (114 nt) that, in part, is predicted to have a secondary structure similar to that portion of the 7S RNA to which SRP54 binds. Indeed, replacing SRP54 with Ffh or vice versa yields functional SRPs, at least *in vitro,* thereby suggesting that the Ffh–4.5S RNA complex is a structurally minimized version of the eukaryotic SRP.

The 12-Å resolution cryo-electron microscopy (cryo-EM)–based structure of canine SRP in complex with a wheat germ ribosome containing a nascent (growing) polypeptide chain was determined by Joachim Frank and Roland Beckmann. The structure reveals that the so-called S domain of the ~270-Å-long SRP binds at the base of the large (60S) ribosomal subunit next to the exit of the tunnel through which newly synthesized polypeptide emerges, whereas the Alu domain bends around the large subunit to contact the ribosome at the interface between its large and small (40S) subunits (Fig. 12-49).

Figure 12-48 **Sequence and secondary structure of canine 7S RNA.** Its various double helical segments (denoted H1 through H8) and loops (denoted L2 and L1.2), are drawn in red and yellow with Watson–Crick base pairs represented by connecting lines and non-Watson–Crick base pairs indicated by dots. The positions at which the various SRP proteins bind to the 7S RNA are indicated in cyan, purple, and gray. [Courtesy of Roland Beckmann, Humboldt University of Berlin, Germany.]

The ribosome–SRP structure was modeled by fitting the much higher resolution X-ray structures of the yeast ribosome (Section 32-3Af) and various SRP fragments to the cryo-EM–derived electron density (Figs. 12-49*d* and 12-50). The model indicates that the 7S RNA consists mainly of a long double helical rod that is bent at two positions named hinge 1 and hinge 2 (RNA, as does DNA, can form a base-paired double helix, although its conformation is distinctly different from that of B-DNA; Section 29-1Bc). The signal sequence exiting the ribosome, which was modeled as an α helix, binds to SRP54, which contacts the ribosome near the mouth of its peptide exit tunnel.

SRP54 consists of three domains: the N-terminal N domain; the central G domain, which contains the SRP's GTPase function and together with the N domain mediates the SRP's interaction with the SRP receptor; and the C-terminal M domain, which is rich in methionine (25 of its 209 residues in humans). The N domain forms a bundle of four antiparallel α helices that closely associates with the G domain, which consists of an open β sheet (Section 8-3Bi) that structurally resembles those of other GTPases. The M domain contains a deep groove that binds the helical signal sequence. The groove is lined almost entirely with hydrophobic residues including many of SRP54's Met residues (the Met side chain has physical properties similar to that of an *n*-butyl group). Its flexible unbranched Met side chain "bristles" presumably provide the groove with the plasticity to bind a variety of different signal sequences so long as they are hydrophobic and form an α helix.

Ribosomes, as we shall see in Section 32-3Dk, employ protein **elongation factors** to deliver aminoacyl-tRNAs and to motivate the sequence of reactions that appends an amino acid residue to the growing polypeptide chain. The eukaryotic SRP's Alu domain, which is required for translational arrest, contacts the ribosome's intersubunit region at the same positions to which the ribosomal elongation factors bind. This suggests that the Alu domain arrests translation by binding to the ribosome with sufficient affinity to block the binding of the ribosome's required elongation factors. This is corroborated by the observation that

Figure 12-49   Cryo-EM structure of a translating wheat germ ribosome in complex with canine SRP at 12 Å resolution. (*a*) Surface diagram showing the small (40S) ribosomal subunit in yellow, the large (60S) ribosomal subunit in blue, the SRP in red, and the tRNA occupying the ribosomal P-site (to which the growing polypeptide chain is covalently linked; Fig. 5-28) in green. C1 to C6 indicate the six positions at which the SRP contacts the ribosome and h1 and h2 indicate the 7S RNA's hinge positions. St and SB stand for stalk and stalk base. (*b*) As in Part *a* but rotated 70° about the vertical axis. (c) As in Part *a* but rotated 90° about the horizontal axis. (*d*) As in Part *b* but with transparent surfaces showing the molecular models of the ribosome and SRP. [Courtesy of Roland Beckmann, Humboldt University of Berlin, Germany.]

bacterial SRP's, which do not arrest translation on binding to a ribosome, lack Alu domains.

c. Secretory Pathway Initiation Is Driven by GTP Hydrolysis

In eukaryotes, the SRP receptor is a heterodimer of subunits named **SRα** and **SRβ.** SRβ is a 271-residue integral protein that has an N-terminal TM segment, whereas SRα is a 638-residue peripheral protein that is apparently membrane-bound through the association of its N-terminal segment with SRβ. Both SRα and SRβ are GTPases.

In *E. coli,* the SR consists of a single 497-residue subunit named **FtsY,** whose C-terminal portion is homologous to that of SRα, although their N-terminal portions have no sequence similarity. Curiously, the X-ray structure of the C-terminal portion of FtsY closely resembles that of the N and G domains of SRP54, with which it shares ~34% identity.

The targeting of the SRP–ribosome complex to the ER membrane is mediated by the GTPase functions of SRP54, SRα, and SRβ. In numerous biological systems, mainly those mediating translation (Section 32-3), vesicle transport (Sections 12-4C and 12-4D), and signal transduction (Section 19-2), *GTPases function as molecular switches that endow the system with unidirectionality and specificity.* These so-called **G proteins** have at least two stable conformations: GDP-bound and GTP-bound. Interconversion between these states only occurs in a unidirectional cycle due to the irreversibility of GTP hydrolysis. In most cases, a G protein must interact with other proteins in order to change conformational states. Thus, GTP hydrolysis often requires stimulation by a specific **GTPase activating protein (GAP),** and the exchange of bound GDP for GTP may require the assistance of a specific **guanine nucleotide exchange factor (GEF;** Section 19-2Ca). The need for these particular factors confers specificity on the system.

The GEF for the SRP is the complex of the newly emerged signal sequence with the M domain of SRP54, which induces the adjoining G domain to exchange its bound GDP for GTP (Fig. 12-46, Stage 2). The formation of the resulting SRP · GTP complex results in a conformational change that locks the SRP to the ribosome, which, in turn, induces translational arrest. The GEF for the SR appears to be an empty translocon, which thereby associates with the resulting SR · GTP complex to which the SRP · GTP–ribosome complex then binds (Fig. 12-46, Stage 3). Evidently, the SRP and SR, both in their GTP forms, act as "molecular matchmakers" to bring together an empty translocon with a ribosome synthesizing a signal sequence–bearing polypeptide. The SRP and the SR then reciprocally stimulate each other's GTPase functions (act as mutual GAPs; neither protein alone has significant GTPase activity) followed by their dissociation, yielding free SRP · GDP and SR · GDP complexes ready to participate in a new

Figure 12-50   Molecular model of the SRP. The transparent cryo-EM–based electron density is shown in white and the ribbon diagrams of the X-ray structures of SRP proteins and RNA fragments that have been docked into it are colored as is indicated in Fig. 12-48. The signal sequence, modeled as an α helix, is green. Note that no atomic resolution structure of the SRP68/72 heterodimer is available. [Courtesy of Roland Beckmann, Humboldt University of Berlin, Germany. PDBid 2G05.]

Figure 12-51   Cryo-EM structure of the eukaryotic SR–SRP–ribosome complex at 8 Å resolution. The complex is oriented and colored as in Fig. 12-49*b* with the SR colored green. [Courtesy of Roland Beckmann, Humboldt University of Berlin, Germany.]

round of the secretory pathway (Fig. 12-46, Stage 4). The release of the SRP and SR permits the now translocon-associated ribosome to recommence translation, thereby extruding the polypeptide it is synthesizing into or through the ER membrane as described below.

In the structure of the SRP–ribosome complex (Fig. 12-49), the S domain of the SRP blocks the binding of the translocon at the mouth of the peptide exit tunnel. However, the cryo-EM–based structure of the SR–SRP–ribosome complex (but lacking SRβ's TM segment), determined by Irmgard Sinning and Beckmann, reveals that the SR contacts both the large ribosomal subunit and the S domain of the SRP (Fig. 12-51) in a way that pushes the SRP S domain in the SRP–ribosome complex away from the peptide exit site. This allows the translocon access to its ribosomal binding site and presumably positions it such that the M domain–bound signal sequence can readily be transferred to it. The position of the eukaryotic SRP's Alu domain on the ribosome is unaffected by the binding of the SR.

d. The Translocon Forms a Transmembrane Pore

How are preproteins transported across or inserted into the RER membrane? In 1975, Blobel postulated that these processes are mediated by a TM channel. However, it was not until 1991 that he was able to show that these channels actually exist through electrophysiological measurements indicating that the RER membrane contains ion-conducting channels. These increase in number when the ribosome-bearing side of the RER is treated with **puromycin** (an antibiotic that causes the ribosome to prematurely release the growing polypeptide; Section 32-3Df), thereby suggesting that the channels are usually plugged by the presence of the polypeptides. By linking fluorescent dyes whose fluorescence is sensitive to the polarity of their environment to a nascent polypeptide, Arthur Johnson demonstrated

that these channels, now called translocons, enclose aqueous pores that completely span the ER membrane.

The various ER transmembrane proteins that comprise the translocon were identified through the use of photoactivatable groups that were attached to signal sequences and mature regions of preproteins. On exposure to light of the proper wavelength, the photoactivatable groups react with nearby proteins to form covalent cross-links, thereby permitting the identification of these proteins. The central component of the translocon, named **Sec61** (Sec for *secretion*) in eukaryotes and the **SecY** complex in prokaryotes, is a heterotrimeric protein. Its α and γ subunits, but not its β subunit, are essential for channel function and are conserved across all kingdoms of life (these subunits are respectively named Sec61α, Sec61β, and Sec61γ in eukaryotes and SecY, SecE, and SecG in bacteria).

The X-ray structure of the SecY complex from the archaeon *Methanococcus jannaschii*, determined by Stephen Harrison and Tom Rapoport, reveals that its α, β, and γ subunits, respectively, have 10, 1, and 1 TM α helices (Fig. 12-52*a,b*). The α subunit's TM helices are wrapped around an hourglass-shaped channel whose minimum diameter is ∼3 Å (Fig. 12-52*c*). The channel is blocked at its extracellular end by a short, relatively hydrophilic helix (blue unnumbered helix in Fig. 12-52*a,b* and yellow helix in Fig. 12-52*c*). A variety of evidence indicates that this helix functions as a plug to prevent small molecules and ions from leaking across the membrane in the absence of a translocating polypeptide and that an incoming signal peptide pushes this helix aside. The γ subunit extends diagonally across the back of the α subunit so as to buttress it. The β subunit makes relatively tenuous contacts with the α subunit, which likely explains why the β subunit is dispensable for translocon function.

A cryo-EM–based structure of a mammalian ribosome–Sec61 complex (Fig. 12-53*a*), determined by Rapoport and Christopher Akey at 11 Å resolution, reveals that a single Sec61 channel is positioned over the ribosome's peptide exit tunnel with Sec61's 6/7 and 8/9 loops extending into the peptide exit tunnel (Fig. 12-53*b*). Indeed, mutating the positively charged residues of the 6/7 and 8/9 loops, which presumably bind to negatively charged ribosomal RNA, abolishes ribosome binding.

How wide is the SecY complex's **protein-conducting channel (PCC)** when it is translocating a polypeptide? At minimum, it would have to be ∼7 Å across (the diameter of an extended anhydrous polypeptide), although if a TM sequence assumed its helical conformation while still in the PCC, the PCC would have to be at least ∼12 Å wide. Such widening could be accomplished by movements of the helices from which the pore ring side chains emanate. The maximum dimensions of the PCC, as estimated from a consideration of the SecY structure (Fig. 12-52), are 15 × 20 Å. Such movements are supported by molecular dynamics calculations (Section 9-4a). Despite the large pore size of an active translocon, the ER membrane's permeability barrier is largely maintained. Evidently, the pore ring fits around the translocating polypeptide chain like a gasket, thereby preventing the passage of small molecules and ions during polypeptide translocation.

(a)   *(b)*   *(c)*

Figure 12-52   X-ray structure of the *M. jannaschii* SecY complex. (*a*) X-ray structure of the complex, with shading indicating the positions of membrane phospholipid head groups (*violet*) and hydrocarbon tails (*pink*). The α subunit of SecY (436 residues) is colored in rainbow order and its helices are numbered from its N-terminus (*dark blue*) to its C-terminus (*red*), the β subunit (74 residues) is magenta, and the γ subunit (53 residues) is purple. (*b*) View of SecY from the cytosol. The translocon's putative lateral gate is on the left between helices 2 and 7. (*c*) Cross-section of the protein-conducting channel as viewed from the bottom of Part *b*. The helix that plugs the channel is yellow and the six hydrophobic side chains that form the narrowest part of the channel, the so-called pore ring, are green. [Courtesy of Stephen Harrison and Tom Rapoport, Harvard Medical School. PDBid 1RH5.]

(a)

(b)

Figure 12-53   Cryo-EM structure of a canine ribosome–Sec61–TRAP complex at 11 Å resolution. (*a*) A surface diagram viewed parallel to the ER membrane. The ribosome's small (S) and large (L) subunits are yellow and blue, a tRNA occupying the ribosome's exit site (Section 32-3Bd) is red, and the Sec61–TRAP complex is magenta. LD is TRAP's lumenal domain. (*b*) A thin slab showing the interface between the ribosome and Sec61. The modeled structures of the ribosome and Sec61, shown as ribbons, are embedded in their transparent surface diagrams, which are colored as in Part *a*. Note how loops 6/7 and 8/9 of Sec61 are inserted into the ribosome's peptide exit tunnel (T), where they interact with RNA helices H7 and H50. L26e is a protein subunit. [Courtesy of Tom Rappoport, Harvard Medical School; and Christopher Akey, Boston University School of Medicine. PDBid 3DKN.]

Figure 12-54 **The stages of polypeptide translocation of a secretory protein.** The red line represents the hydrophobic portion of the signal sequence. The process begins with the insertion of the nascent peptide as a loop into the PCC (*left*). The signal sequence then binds as an ~2-turn helix between the SecY/Sec61 α subunit's helices 2 and 7, which helps displace the plug helix (*middle*). Finally, the signal sequence is excised by the signal protease (*not shown*) and the nascent peptide enters the ER through the PCC (*right*). [Courtesy of Tom Rapoport, Harvard Medical School.]

The translocation of a secretory protein begins with its insertion as a loop into the PCC (Fig. 12-54, *left*). This was established by using a mutant protein whose signal sequence is not excised by the signal protease and showing, through proteolysis experiments, that the protein's N- and C-termini both remained on the ER membrane's cytoplasmic side. Subsequently, as was shown by photo-cross-linking experiments, the signal sequence forms an ~2-turn helix that inserts between TM helices 2 and 7 of Sec61α (Fig. 12-54, *middle*). The separation of helices 2 and 7 (the lateral gate; Fig. 12-52b) helps displace the plug helix, which following signal sequence excision, allows the nascent polypeptide pass through the PCC into the ER (Fig. 12-54, *right*).

Additional components of the mammalian translocon are named **translocating chain-associated membrane protein (TRAM,** ~375 residues; predicted to have 8 TM helices with both its N- and C-termini in the cytosol) and **translocon-associated membrane protein (TRAP;** an ~800-residue heterotetramer with its α, β, and δ subunits each having one TM helix and its γ subunit having four TM helices). Through the use of Sec61-containing liposomes that either did or did not also contain TRAM, Rapoport demonstrated that TRAM is required for the translocation and membrane integration of most but not all preproteins into the liposome. Whether or not a given preprotein requires TRAM for translocation depends on its signal sequence, although no particular characteristic of this sequence appears to be critical for TRAM dependence. TRAP, which is seen in Fig. 12-53a, functions similarly to increase the translocational efficiency of proteins with certain signal sequences.

e. The Translocon Laterally Inserts Transmembrane Helices Into the ER Membrane

In addition to forming a conduit for soluble proteins to enter the ER, *the translocon must insert an integral protein's TM segments into the ER membrane.* The translocon, in concert with the ribosome, recognizes these TM segments and installs them into the lipid bilayer via a lateral gate between helices 2 and 7 in the SecY/Sec61 α subunit (Fig. 12-55).

Monotopic (alternatively, **single-pass**) TM proteins fall into one of three classes:

1. Type I proteins have cleavable N-terminal signal sequences. They are inserted into the membrane much like secretory proteins (Fig. 12-54) but have an ~22-residue hydrophobic **stop-transfer anchor sequence** that the translocon laterally inserts into the membrane as a helix. Hence a type I protein has its N-terminus in the ER.

2. Type II proteins lack a cleavable N-terminal signal sequence. However, they have an ~22-residue hydrophobic **signal-anchor sequence,** not necessarily near the protein's N-terminus, that is recognized by the SRP. The SRP then passes the nascent polypeptide to the translocon, which laterally inserts the signal-anchor sequence into the membrane oriented such that the protein's N-terminus is in the cytoplasm. This requires the polypeptide to loop around inside the translocon before being inserted into the membrane (Fig. 12-56, *lower portion*). The C-terminal segment of a type II protein is presumably extruded into the gap between the ribosome and the translocon (Fig. 12-53a) before being passed through the translocon.

3. Type III proteins, like type II proteins, lack a cleavable N-terminal signal sequence and have a signal-anchor sequence that is not necessarily near the protein's N-terminus. However, the orientation of these signal-anchor sequences in the membrane is opposite to that of type II proteins. Hence the way that type III proteins are inserted into the membrane resembles that of type I proteins (Fig. 12-56, *upper portion*). They differ, however, in that the transfer of a type III protein's N-terminal segment across the membrane cannot be initiated until after its signal-anchor sequence has been synthesized.

For polytopic (alternatively, **multipass** or **type IV**) TM proteins, looping must occur each time an additional helix is to be installed in the membrane. Evidently, the PCC of the

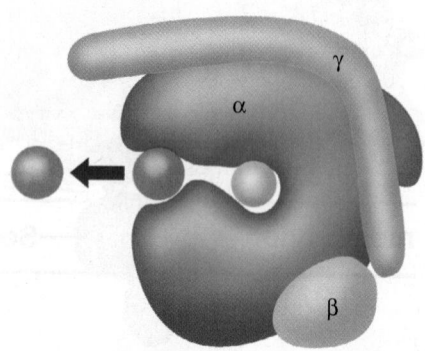

Figure 12-55 **Model for the insertion of a TM helix into a membrane.** The translocon (*blue*) is viewed as in Fig. 12-52b. A polypeptide chain (*yellow*) is shown bound in the translocon's pore during its translocation through the membrane, and a TM helix (*red*) is shown passing through the translocon's lateral gate and being released into the membrane (*arrow*). [Based on a drawing by Dobberstein, B. and Sinning, I., *Science* **303**, 320 (2004).]

Figure 12-56   The generation of types II and III proteins. Here, the red line represents a signal-anchor sequence. The N-terminus of a type III protein must pass through the translocon before its succeeding TM helix is laterally installed in the membrane (*top*). However, for a type II protein, whose N-terminus is retained in the cytoplasm (*bottom*), the subsequently synthesized polypeptide (represented by the loop between the ribosome and the translocon) must pass through the PCC. For polytopic TM proteins, these two processes alternate. [Based on a drawing by Tom Rapoport, Harvard Medical School.]

active translocon has sufficient room for successive TM segments to reverse their direction prior to being inserted into the ER membrane. Helices may be inserted into the membrane either singly or in pairs, depending on their hydrophobicity and their ability to form stable helix–helix interactions.

What controls the orientations of the helices in a TM protein, that is, its **topogenesis?** Most TM proteins, as Gunnar von Heijne pointed out, adopt an orientation such that their cytoplasmically exposed ends, those that are not translocated across the membrane, are more positively charged (have more Arg and Lys residues) than their lumenally exposed ends—the **positive-inside rule.** This appears mainly due to the charge distribution within the translocon, which is oriented with its more positive face on the cytoplasmic side of the membrane (in accordance with the positive-inside rule). In fact, mutating certain charged residues of Sec61α so as to reverse their charge (e.g., changing an Arg to Glu), inverts the orientation of the TM helices it installs in the membrane. Another important influence on the orientation of a TM helix is its hydrophobicity gradient: The more hydrophobic end of a TM helix is preferentially translocated across the membrane.

Despite the foregoing, one might reasonably expect that the membrane orientation of the N-terminal TM helix of a polytopic TM protein dictates the orientations of the succeeding TM helices (many, if not all, of which have yet to be synthesized at the time the N-terminal helix is inserted into the membrane). However, the deletion or insertion of a TM helix from/into a polypeptide does not necessarily change the membrane orientations of the succeeding TM helices: When two successive TM helices have the same preferred orientation, one of them may be forced out of the membrane. Moreover, the topological organization of TM proteins is influenced by the membrane lipid composition.

This suggests that the translocon's lateral gate frequently opens and closes so as to allow its transiting peptide to sample the outside lipid environment and only inserts a peptide segment into the lipid bilayer if it is thermodynamically favorable to do so; that is, helix insertion may be considered as a partitioning between the aqueous environment in the translocon and that of the membrane.

Polytopic TM proteins can fold to their native conformations only after all their TM helices have been inserted into the membrane. This process is guided by packing interactions between helices as well as specific interactions with membrane lipids. Thus, although a TM protein's sequence determines its topology, it does so for a specific membrane lipid environment. Evidently, the lipid composition of a membrane and the topologies of its embedded TM proteins have coevolved.

f. Protein Folding in the ER Is Monitored by Molecular Chaperones

The ER, as does the cytosol, contains a battery of molecular chaperones that assist in protein folding and act as agents of quality control. The best characterized of these is the Hsp70 homolog (Section 9-2C) **BiP** (for *b*inding *p*rotein). BiP associates with many secretory and TM proteins although, if folding proceeds normally, these interactions are weak and short-lived. However, proteins that are improperly folded, incorrectly glycosylated, or improperly assembled form stable complexes with BiP that are often exported, via a poorly understood process involving the translocon called **retrotranslocation,** to the cytosol where they are proteolytically degraded (Section 32-6). The entire process is named **ERAD** (for *ER*-associated *d*egradation). Two other notable ER-resident chaperones are **calreticulin** and **calnexin,** homologous proteins that participate in facilitating and monitoring the folding and assembly of glycoproteins (Section 23-3Bf). The ER also contains protein disulfide isomerases (PDIs; Section 9-2A) and peptidyl prolyl cis–trans isomerases (PPIs; Section 9-2B).

Abnormalities of protein folding and assembly are important mechanisms of disease (e.g., Section 9-5). For instance, **cystic fibrosis** is the most common life-threatening recessive genetic disease in the Caucasian population (affecting one in ~2000 individuals). It occurs in homozygotes for a defective **cystic fibrosis transmembrane regulator (CFTR) protein,** a 1480-residue glycoprotein with 12 TM helices that functions as a Cl⁻ transporter in the plasma membrane of epithelial cells. Individuals with cystic fibrosis produce highly viscous mucus that, in its most damaging effects, blocks the small airways in the lungs. This leads to persistent infections, which cause severe progressive lung degeneration that is usually fatal by around age 30. Although cystic fibrosis is caused by any of more than 1000 known mutations in the CFTR gene, 70% of the cases arise from the deletion of Phe 508 (ΔF508), which is located in a cytoplasmic domain of the CFTR protein (which is initially inserted into the ER membrane). Although this mutant domain in ΔF508 folds to nearly its native conformation (ΔF508 retains almost full biological activity), it does so much more slowly than in the wild-type protein. This results in its retrotranslocation and degradation by an, in this case, overly zealous prote-olytic surveillance system (Section 32-6B).

g. Some Proteins Are Post-Translationally Transported through Membranes

The secretory proteins we discussed pass through the membrane as they are being synthesized by the ribosome, that is, their membrane translocation occurs cotranslation-ally. However, some secretory proteins are translocated only after they have been fully synthesized in the cytoplasm, that is, post-translationally. Nevertheless, both co- and post-translational translocation is mediated by the translocon. Yet, the translocon is a passive pore, that is, it does not pro-vide the free energy that drives translocation. In cotransla-tional translocation, it is the ribosomally mediated extension of the polypeptide that pushes it through the translocon.

In eukaryotes, post-translationally translocated secre-tory proteins have signal sequences that are only moder-ately hydrophobic; they are not recognized by the SRP but still bind to Sec61. These proteins are bound by cytoplas-mic chaperones, which keep them in a loosely folded or un-folded state that permits them to pass through the translo-con. Their translocation is driven by a so-called **Brownian ratchet** mechanism that is mediated by the partnering of Sec61 with the tetrameric TM complex **Sec62/Sec63** and the lumenal Hsp70-like chaperone BiP (Fig. 12-57). The polypeptide in the translocon randomly slides back and forth via Brownian motion. However, in the ER, the so-called J-domain (a homolog of the *E. coli* cochaperone DnaJ; Section 9-2C) on the lumenal surface of Sec63 induces BiP to hydrolyze its bound ATP to ADP. The resulting BiP–ADP complex then binds the polypeptide emerging from the translocon, which prevents it from sliding back to-ward the cytoplasm. When the peptide again slides forward, another BiP–ADP complex binds to it, etc., until the entire polypeptide has entered the ER. BiP eventually exchanges

Figure 12-57 **Scheme for post-translational translocation in eukaryotes.** As the translocating polypeptide enters the ER through Sec61, the BiP–ATP complex binds to Sec63, whose J-domain induces BiP to hydrolyze its bound ATP to ADP. The resulting BiP–ADP complex binds to the emerging peptide so as to prevent its backsliding. As additional peptide segments emerge from Sec61, the process repeats until the entire protein has entered the ER. BiP eventually exchanges its bound ADP for ATP causing it to release the peptide, which then folds to its native conformation. [Courtesy of Tom Rapoport, Harvard Medical School.]

its ADP for ATP, which causes it to release the polypeptide, which then folds to its native conformation.

In bacteria, the motor that drives post-translational translocation is **SecA,** which binds to the cytoplasmic face of the SecY complex and pushes the polypeptide through the translocon via repeated cycles of ATP hydrolysis. SecA is aided in doing so by the cytosolic chaperone **SecB,** which prevents the polypeptide from folding in the cytoplasm.

C. *Vesicle Formation*

Shortly after their polypeptide synthesis is completed, the partially processed transmembrane, secretory, and lysoso-mal proteins appear in the Golgi apparatus (Fig. 1-5), a 0.5- to 1.0-μm-diameter organelle consisting of a stack of 3 to 6 or more (depending on the species) flattened and function-ally distinct membranous sacs known as **cisternae,** where further post-translational processing, mainly glycosylation, occurs (Section 23-3Bg). The Golgi stack (Fig. 12-58) has two distinct faces, each comprised of a network of intercon-nected membranous tubules: the **cis Golgi network (CGN),** which is opposite the ER and is the port through which proteins enter the Golgi apparatus; and the **trans Golgi network (TGN),** through which processed proteins exit to their final destinations. The intervening Golgi stack con-tains at least three different types of sacs, the **cis, medial,** and **trans cisternae,** each of which contains different sets of glycoprotein processing enzymes.

Proteins transit from one end of the Golgi stack to the other while being modified in a stepwise manner, a process that is described in Section 23-3Bg. These proteins are transported via two mechanisms:

1. They are conveyed between successive Golgi com-partments in the cis to trans direction as cargo within mem-branous vesicles that bud off of one compartment and fuse with a successive compartment, a process known as for-ward or **anterograde transport.**

2. They are carried as passengers in Golgi compart-ments that transit the Golgi stack, that is, the cis cisternae eventually become trans cisternae, a process called **cister-nal progression** or **maturation.** This process is mediated through the backward or **retrograde transport** of Golgi-resident proteins from one compartment to the preceeding one via membranous vesicles.

The cisternal progression mechanism has been clearly shown to occur but the significance of the anterograde transport mechanism is as yet unclear. In any case, on reaching the trans Golgi network, the now mature proteins are sorted and sent to their final cellular destinations.

a. Membrane, Secretory, and Lysosomal Proteins Are Transported in Coated Vesicles

The vehicles in which proteins are transported be-tween the RER, the Golgi apparatus, and their final desti-nations, as well as between the different compartments of the Golgi apparatus, are known as **coated vesicles** (Fig. 12-59). This is because these 60- to 150-nm-diameter

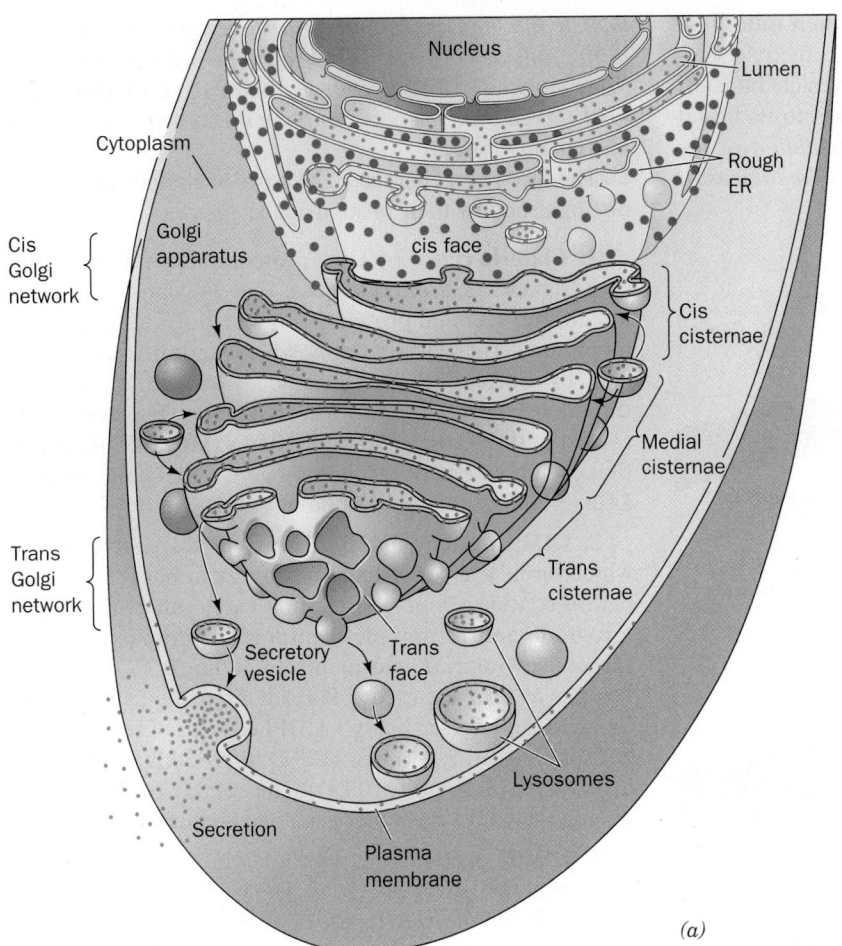

Figure 12-58 **Post-translational processing of proteins.** Proteins destined for secretion, insertion into the plasma membrane, or transport to lysosomes are synthesized by RER-associated ribosomes (*blue dots; top*). As they are synthesized, the proteins (*red dots*) are either translocated into the lumen of the ER or inserted into its membrane. After initial processing in the ER, the proteins are encapsulated in vesicles that bud off from the ER membrane and subsequently fuse with the cis Golgi network. The proteins are progressively processed in the cis, medial, and trans cisternae of the Golgi. Finally, in the trans Golgi network (*bottom*), the completed glycoproteins are sorted for delivery to their final destinations, the plasma membrane, **secretory vesicles,** or lysosomes, to which they are transported by yet other vesicles.

Figure 12-59 **Electron micrographs of coated vesicles.** (*a*) Clathrin-coated vesicles. Note their polyhedral character. [Courtesy of Barbara Pearse, Medical Research Council, Cambridge, U.K.] (*b*) COPI-coated vesicles. (*c*) COPII-coated vesicles. The insets in Parts *b* and *c* show the respective vesicles at higher magnification. [Courtesy of Lelio Orci, University of Geneva, Switzerland.]

(a)

(b)

(c)

membranous sacs are initially encased on their outer (cytosolic) faces by specific proteins that act as flexible scaffolding in promoting vesicle formation. A vesicle buds off from its membrane of origin and later fuses to its target membrane. *This process preserves the orientation of the transmembrane protein* (Fig. 12-60), *so that the lumens of*

the ER and the Golgi cisternae are topologically equivalent to the outside of the cell. This explains why the carbohydrate moieties of TM glycoproteins and the GPI anchors of GPI-linked proteins occur only on the external surfaces of plasma membranes.

The three best characterized types of coated vesicles are distinguished by their protein coats:

1. Clathrin (Fig. 12-59a), a protein that forms a polyhedral framework around vesicles that transport TM, GPI-linked, and secreted proteins from the Golgi to the plasma membrane. The clathrin cages, which were first characterized by Barbara Pearse, can be dissociated to flexible three-legged proteins known as **triskelions** (Fig. 12-61) that consist of three so-called heavy chains (**HC,** 1675 residues) that each bind one of two homologous light chains, **LCa** or **LCb** (~240 residues), at random.

2. COPI protein (Fig. 12-59b; COP for *co*at *p*rotein), which forms what appears to be a fuzzy rather than a polyhedral coating about vesicles that carry out both the anterograde and retrograde transport of proteins between successive Golgi compartments. In addition, COPI-coated vesicles return escaped ER-resident proteins from the Golgi to the ER (see below). COPI consists of seven different subunits (α, 160 kD; β, 110 kD; β', 102 kD; γ, 98 kD; δ, 61 kD; ε, 31 kD; and ζ, 20 kD). The soluble complex comprising the COPI protomer is named **coatomer.**

3. COPII protein (Fig. 12-59c), which transports proteins from the ER to the Golgi. The COPII vesicle components are then recycled by COPI-coated vesicles for participation in another round of vesicle formation (the COPI vesicle components entering the ER are presumably recycled by COPII-coated vesicles). The COPII coat consists of the GTPase **Sar1,** the heterodimer **Sec23/24** in which Sec23 is a Sar1-specific GAP and Sec24 functions in cargo selection,

Figure 12-60   The fusion of a vesicle with the plasma membrane preserves the orientation of the integral proteins embedded in the vesicle bilayer. The inside of the vesicle and the exterior of the cell are topologically equivalent because the same side of the protein is always immersed in the cytosol. Note that any soluble proteins contained within the vesicle would be secreted. In fact, proteins destined for secretion are packaged in membranous secretory vesicles that subsequently fuse with the plasma membrane as shown.

Figure 12-61   Electron micrograph of triskelions. The variable orientations of their legs are indicative of their flexibility. [Courtesy of Daniel Branton, Harvard University.]

and the heterodimer **Sec13/31,** which forms polyhedral cages (see below).

All of the above coated vesicles also carry receptors, which bind the proteins being transported, as well as **fusion proteins,** which mediate the fusion of these vesicles with their target membranes. We discuss these processes below and in Section 12-4D.

b. Clathrin Cages Are Formed by Overlapping Heavy Chains

Clathrin-coated vesicles (**CCVs**) are structurally and functionally better characterized than those coated with COPI or COPII. Clathrin forms polyhedral cages in which, as a cryo-EM study by Harrison, Tomas Kirchhausen, and Thomas Walz has shown most clearly (Fig. 12-62*a*), each vertex is the center (hub) of a triskelion, and its edges,

which are ~225 Å long, are each formed by the interdigitated legs of four triskelions—two antiparallel proximal segments and two distal segments (Fig. 12-62*b*). Such polyhedra (Fig. 12-62*c*), which have 12 pentagonal faces and a variable number of hexagonal faces (for geometric reasons explained in Section 33-2A), are the most parsimonious way of enclosing spheroidal objects in polyhedral cages. The volume enclosed by a clathrin polyhedron, of course, increases with its number of hexagonal faces (a "minicoat" is too small to contain a transport vesicle).

The triskelion's ~475-Å-long legs are each formed by the 1675-residue heavy chains (HCs), which trimerize via their C-terminal domains (Fig. 12-62*b*). In addition to projecting outward from its hub (vertex), each leg curls toward the center of the particle such that three ankles meet and interact ~75 Å below a hub that is two vertices away from each of their hubs.

(a)

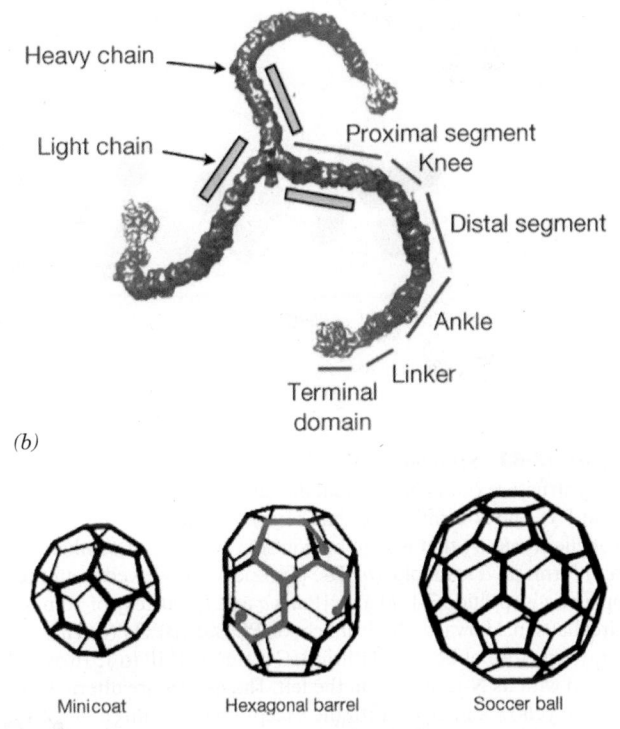

(b)

(c)

Figure 12-62 **Anatomy of clathrin-coated vesicles.** (*a*) A cryo-EM–based image of a light chain–free clathrin cage from bovine brain at 7.9 Å resolution. The particle shown, a so-called hexagonal barrel, which has D_6 symmetry, consists of 36 triskelions. Three of its interdigitated but symmetry unrelated triskelions are are drawn in red, yellow, and green. (*b*) A cryo-EM–based image of a triskelion labeled with the names of its various segments. The N-terminus of each heavy chain occupies the terminal domain and its C-terminus is located in the vertex joining the three heavy chains to form the triskelion. (*c*) Diagrams of the three polyhedral stuctures that are formed when triskelions assemble into clathrin cages *in vitro*. The minicoat

has tetrahedral (*T*) symmetry, the hexagonal barrel has D_6 symmetry, and the soccer ball has icosahedral (*I*) symmetry (symmetry is discussed in Section 8-5B). These polyhedra consist of 28, 36, and 60 triskelions, respectively. The arrangement of one triskelion within the hexagonal barrel is indicated in blue. *In vivo*, clathrin forms membrane-enclosing polyhedral cages with a large range of different sizes (number of hexagons). The hexagonal barrel seen in Part *a* is only ~700 Å in diameter, whereas clathrin-coated membranous vesicles are typically ~1200 Å in diameter or larger. [Courtesy of Stephen Harrison, Tomas Kirchhausen, and Thomas Walz, Harvard Medical School.]

Figure 12-63   Structure of the clathrin heavy chain. (*a*) The X-ray structure of the N-terminal domain and part of the linker of rat HC. The N-terminal domain forms a seven-bladed β propeller (*yellow*) that is seen here in side view, and the linker (*red*) forms an α solenoid. (*b*) The β propeller as viewed from the top along its pseudo-7-fold axis. [Parts *a* and *b* courtesy of Tomas Kirchhausen, Harvard Medical School. PDBid 1BPO.] (*c*) The X-ray structure of bovine clathrin HC residues 1210 to 1516 as viewed with its N-terminus on the left. The helices are alternately colored yellow and green with the exception of the three N-terminal helices, which are colored gray to indicate that they are poorly resolved. The orange, green, and blue bars denote the regions of CHCR5, CHCR6, and CHCR7, respectively. [Courtesy of Peter Hwang, University of California at San Francisco. PDBid 1B89.] (*d*) A backbone model of a triskelion (residues 1–1597) generated by docking the foregoing X-ray structures together with homology models of the remaining CHCRs into the cryo-EM–determined electron density of a heavy chain (Fig. 12-62*a*). [Courtesy of Stephen Harrison, Tomas Kirchhausen, and Thomas Walz, Harvard Medical School. PDBid 1XI4.]

Although the X-ray structure of an entire HC has not been determined, those of its N-terminal portion and a part of its proximal segment have been elucidated:

1. The N-terminal segment (residues 1–494; Fig. 12-63a,b), whose structure was determined by Harrison and Kirchhausen, consists of two domains: (i) an N-terminal seven-bladed **β propeller** in which each structurally similar propeller blade is formed by a four-stranded antiparallel β sheet (Fig. 12-63b; the terminal domain) named the **WD40** sequence motif because it often contains the dipeptide WD and is ~40 residues long; and (ii) a C-terminal linker that consists of 10 α helices of variable lengths (2–4 turns) connected by short loops and arranged in an irregular right-handed helix (a helix of helices, that is, a **superhelix**) named an **α solenoid** (alternatively, an **α-zigzag**).

2. The proximal segment (residues 1210–1516; Fig. 12-63c), whose structure was determined by Peter Hwang and Robert Fletterick, consists of 24 linked α helices that are arranged similarly but more regularly than the above α solenoid to form a rod-shaped right-handed superhelix. The rigidity of this motif is attributed to its continuous hydrophobic core together with the efficient interdigitation of its side chains where its crossing antiparallel α helices come into contact (Section 8-3B).

Sequence and structural alignments indicate that HC residues 537 to 1566 consist of seven homologous ~145-residue **clathrin heavy chain repeats (CHCRs)** that are arranged in tandem and which each contain 10 helices (the proximal segment consists of all of CHRC6 together with the C- and N-terminal portions of CHRC5 and CHRC7; Fig. 12-63c). This has permitted the generation of a backbone model of a triskelion by docking the foregoing X-ray structures and homology models of the CHCRs whose structures have not been experimentally determined in the cryo-EM–determined electron density (Fig. 12-63d; homology modeling is discussed in Section 9-3B). Each HC leg consists of an extended superhelix of linked α helices. Nevertheless, triskelion legs exhibit considerable flexibility (Fig. 12-61), a functional necessity for the formation of different sized vesicles as well as for the budding of a vesicle from a membrane surface, which is accompanied by a change in its curvature. The HC appears to flex mainly along its knee and ankle segments (Fig. 12-62b).

The proximal segment bears extensive hydrophobic surface patches that follow the grooves between adjacent helices. Apparently, the lengthwise association of two proximal segments in a clathrin cage (Fig. 12-62a) is stabilized by the burial of these hydrophobic patches through the complementary packing of the helices of one proximal leg in the grooves on another.

Light chains (LCs) are not required for clathrin cage assembly. Indeed, LCs inhibit heavy chain polymerization *in vitro,* which suggests that they have a regulatory role in preventing inappropriate clathrin cage assembly in the

Figure 12-64   Arrangement of light chains on a clathrin cage. The differences between the cryo-EM–determined electron densities of a hexagonal barrel with and without light chains are shown in yellow with the light chain–free electron density shown in blue. [Courtesy of Stephen Harrison, Tomas Kirchhausen, and Thomas Walz, Harvard Medical School.]

cytosol. Comparison of the cryo-EM structures of intact and LC-free hexagonal barrels reveals that the central portion of an LC consists of a 71-residue helix that binds to a surface formed by the interhelical loops along the HC proximal segment with the C-terminus of the LC closest to the triskelion hub (Fig. 12-64). The segments of the 60% identical LCa and LCb that differ in sequence are largely confined to their N- and C-terminal regions, which do not participate in HC binding and hence are likely to contain sites for the attachment of cytosolic factors that regulate vesicle uncoating.

c. Clathrin-Coated Vesicles Also Participate in Endocytosis

CCVs, as we have seen, transport TM and secretory proteins from the trans Golgi network (TGN) to the plasma membrane (Fig. 12-58). In addition, through a process known as **endocytosis** (discussed in Section 12-5Bc), they act to engulf specific proteins from the extracellular medium by the invagination of a portion of the plasma membrane and to transport them to intracellular destinations.

1. Priming

2. Assembly

3. Release

4. Uncoating

Figure 12-65 **Formation of clathrin-coated vesicles.**
(1) The ARNO-stimulated exchange of ARF1's bound GDP for
GTP frees ARF1 · GDP's protein-bound N-terminal myristoyl
group for insertion into the membrane. **(2)** Membrane-bound
ARF1 · GTP recruits adapter proteins (APs). These, in turn, bind
triskelions, thereby promoting the formation of a clathrin coat,
which causes the vesicle to bud out from the membrane. In
addition, APs bind the transmembrane receptors of cargo
proteins as well as transmembrane cargo proteins. **(3)** The vesicle
is released from the membrane through the action of the GTPase
dynamin. **(4)** Shortly after the vesicle is released from the
membrane, the clathrin coat and the APs dissociate from the
vesicle.

d. The Formation of CCVs Is a Complex Process

The formation of CCVs involves four stages (Fig. 12-65):
(1) priming, (2) assembly, (3) release, and (4) uncoating. We
outline these processes below.

1. Priming: The Activation of ARF1. Vesicle formation
begins with the binding to the membrane of the myristoylated small (181-residue) GTPase named **ARF1** (ARF for
ADP-ribosylation factor, because it was first described as a
cofactor in the cholera toxin–catalyzed ADP-ribosylation of
the GTPases known as heterotrimeric G proteins; Section
19-2). ARFs, which are members of the **Ras** superfamily
(Ras is a small GTPase that participates in intracellular signaling; Section 19-3C), are water-soluble cytosolic proteins
when binding GDP, but when binding GTP they associate
with membranes through the insertion of their N-terminal
myristoyl groups into the bilayer (Section 12-3Bb). The comparison of X-ray structures of ARF1 · GDP and ARF1 ·
GTP, determined by Dagmar Ringe and by Jonathan Goldberg, indicate that this occurs because the N-terminal helix
of ARF1 · GDP together with its appended myristoyl group
are bound in a shallow groove in the protein (Fig. 12-66*a*)
that is absent in ARF1 · GTP (Fig. 12-66*b*).

The guanine nucleotide exchange factor (GEF) for
ARF1, which in humans is called **ARNO** (for **ARF nucleotide-binding site opener;** 399 residues), contains an

(a)　　　　　　　　　　　　　　　　　　　(b)

Figure 12-66 **X-ray structures of (*a*) ARF1 · GDP and (*b*) ARF1 · GMPPNP.** (GMPPNP is a nonhydrolyzable GTP analog in which the O atom linking GTP's β- and γ-phosphorus atoms is replaced by an NH group.) The bound nucleotides are drawn in stick form in white with their phosphorus atoms magenta and their bound Mg^{2+} ions shown as lavender spheres. In ARF1 · GDP, the protein's N-terminal helix (*red*) together with its covalently linked myristoyl group (not present in the X-ray structures) are bound in a shallow hydrophobic groove on the surface of the protein formed in part by the residues of loop λ3.

However, the replacement of GDP by GMPPNP (and presumably GTP) induces a conformational change in residues 37 to 53 (*yellow*) that displaces strand β2 by two residues along strand β3, a shift of 7 Å. The resulting movement of loop λ3 eliminates the binding site for the N-terminus, thereby making the myristoyl group available for membrane insertion (residues 1–17 of the GMPPNP complex are disordered). [Courtesy of Jonathan Goldberg, Memorial Sloan-Kettering Cancer Center, New York. The X-ray structure of ARF1 · GDP was determined by Dagmar Ringe, Brandeis University. PDBid 1HUR.]

~200-residue domain similar to the highly conserved yeast protein **Sec7.** When ARNO or its isolated Sec7 domain is incubated with myristoylated ARF1 · GDP, it fails to catalyze nucleotide exchange unless lipid micelles are also present, thereby suggesting that ARNO is activated only when localized to a membrane surface. Indeed, ARNO contains a **pleckstrin homology (PH) domain,** an ~100-residue module occurring in numerous proteins (Section 19-3Ce) that binds the minor membrane phospholipid **phosphatidylinositol-4,5-bisphosphate (PIP₂),**

OR₁　OR₂

CH₂—CH—CH₂

O—P—O

O⁻

HO　H　HO　OPO_3^{2-}

OH　H　H

H　H

H　OPO_3^{2-}

Phosphatidylinositol-4,5-bisphosphate (PIP₂)

which is also a precursor of compounds that participate in intracellular signaling (Section 19-4A).

2. Assembly: Adaptor Proteins Link Cargo Proteins to the Clathrin Coat. Membrane-bound ARF1 · GTP acts to recruit **adapter proteins (APs)** to the membrane surface. APs bind clathrin HC together with TM proteins that are either receptors that selectively bind soluble cargo proteins inside the budding vesicle or are cargo proteins themselves. APs comprise the central cores of CCVs and, in fact, are the scaffolding on which clathrin cages form. The APs bind clathrin via its N-terminal β propeller domain (Fig. 12-63*a*), which forms the knobs that project inward from clathrin cages (Fig. 12-62*a*). The grooves between the propeller blades on the top face of the β propeller (Fig. 12-63*b*) probably form the AP binding sites.

AP1 is the most common AP contained in the coated vesicles originating from the TGN, whereas the homologous **AP2** predominates in endocytotic vesicles. Both APs are heterotetramers: AP1 consists of the subunits γ, β1 (~110 kD each), μ1 (~50 kD), and σ1 (~17 kD), whereas the corresponding subunits of the better characterized AP2 are named α, β2, μ2, and σ2 (Fig. 12-67). Electron microscopy and X-ray studies indicate that the large subunits each consist of a trunk and an appendage domain joined by

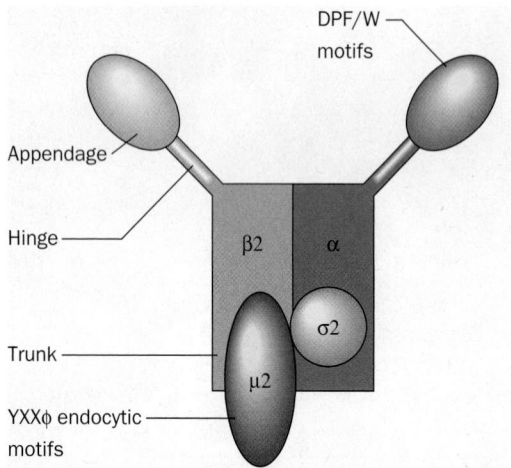

Figure 12-67   Schematic diagram of the AP2 heterotetramer. AP1 has a similar structure. [After Pearse, B.M., Smith, C.J., and Owen, D.J., *Curr. Opin. Struct. Biol.* **10,** 223 (2000).]

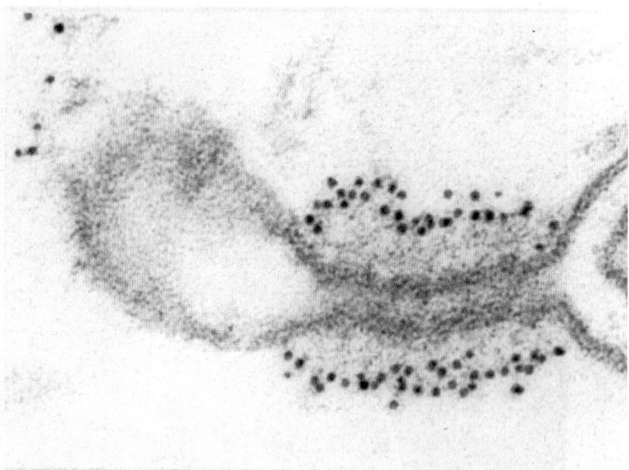

Figure 12-68   Electron micrograph of a budding coated vesicle. The vesicle was incubated with the nonhydrolyzable GTP analog **GTPγS** (in which a terminal O atom on the γ-phosphorus of GTP is replaced by S) and then treated with gold-tagged anti-dynamin antibodies (*black dots*). Note that the dynamin surrounds a long narrow tube at the base of the budding vesicle that has not pinched off from the membrane. [Courtesy of Pietro De Camilli, Yale University School of Medicine.]

a flexible and proteolytically sensitive hinge region (Fig. 12-67). The AP2 hinge region of β2 binds to the clathrin β propeller, whereas the cytoplasmic domains of target proteins bind most commonly to μ2 via YXXϕ sequences (where ϕ is a bulky hydrophobic residue), but in some cases to its α and σ2 subunits via [D/E]XXXL[L/I] sequences, which are known as **dileucine motifs.** This explains why the proteolytic excision of AP2's appendage domain prevents the assembly of clathrin coats, although the remaining AP2 trunk can still bind to membranes that contain proteins bearing a YXXϕ internalization signal. In addition, both AP1 and AP2 bind PIP_2 and mutating their PIP_2-binding sites prevents them from localizing to their target membranes.

Mammals have two additional heterotetrameric APs, **AP3** and **AP4,** both of which function in the TGN. Moreover, database searches for AP homologs have identified a family of monomeric clathrin adapters named **GGAs** (for *G*olgi-localized *γ*-ear-containing *A*RF-binding proteins), whose C-terminal domain is homologous to the appendage or "ear" domain of AP1's γ subunit (and AP2's α subunit; Fig. 12-67). These various adapter proteins participate in the transport of their target proteins between different pairs of membranes so that CCVs are multifunctional entities.

3. Release: Vesicle Scission Is Mediated by Dynamin. The budding of a CCV from its parent membrane appears to be mechanically driven by the formation of the clathrin cage. However, the actual scission of the coated bud from its parent membrane to form a coated vesicle requires the participation of **dynamin,** an ~870-residue GTPase. Dynamin contains a PIP_2-binding PH domain, which recruits dynamin to the membrane. On binding GTP, dynamin forms a helical oligomer that wraps about the base of the budding vesicle so as to squeeze this region down to a thin tube (Fig. 12-68). The oligomerization together with the presence of PIP_2 stimulates dynamin to hydrolyze its bound GTP (dynamin also contains a GAP domain), causing the helical oligomer to lengthen its pitch. However, the

way in which this process releases the vesicle from the membrane is not well understood.

4. Uncoating: The Recycling of Clathrin and Adapter Proteins. Shortly after the formation of a CCV, the clathrin is released as triskelions, thereby recycling them for participation in the formation of additional coated vesicles. This process is mediated by the ATPase **Hsc70** (Hsc for *h*eat *s*hock *c*ognate), an ~650-residue homolog of the chaperone Hsp70 (Section 9-2C) present in all eukaryotic cells, which on ATP hydroysis forms a complex with clathrin. Hsc70 is recruited to the appropriate sites on the clathrin lattice by the ~910-residue cochaperone **auxilin,** which binds to specific sites on the clathrin heavy chains. Auxilin contains a J-domain that induces Hsc70 to hydrolyze its bound ATP to ADP, thereby causing Hsc70 to bind to and dismantle the clathrin lattice. The cryo-EM structure of the clathrin "hexagonal barrel" in complex with Hsc70 and a J-domain-containing fragment of auxilin at 28 Å resolution, determined by Alasdair Steven, indicates that Hsc70 is located within diffuse rings inside the clathrin cage's pentagonal and hexagonal rings (Fig. 12-69). This suggests that triskelions are pried out of the clathrin lattice by the concerted action of up to six Hsc70 molecules. This may occur by a simple clockwise rotation of a triskelion as viewed in Fig. 12-62*a*. On the subsequent exchange of its bound ADP for ATP, the Hsc70 releases its bound triskelions.

Following clathrin release from newly formed vesicles, the APs are also released. This process may be initiated by the hydrolysis of ARF1's bound GTP to GDP, which would release ARF1 from the membrane and, presumably, from binding an AP. In any case, the coating and uncoating of vesicles by clathrin must be closely regulated processes since both occur simultaneously.

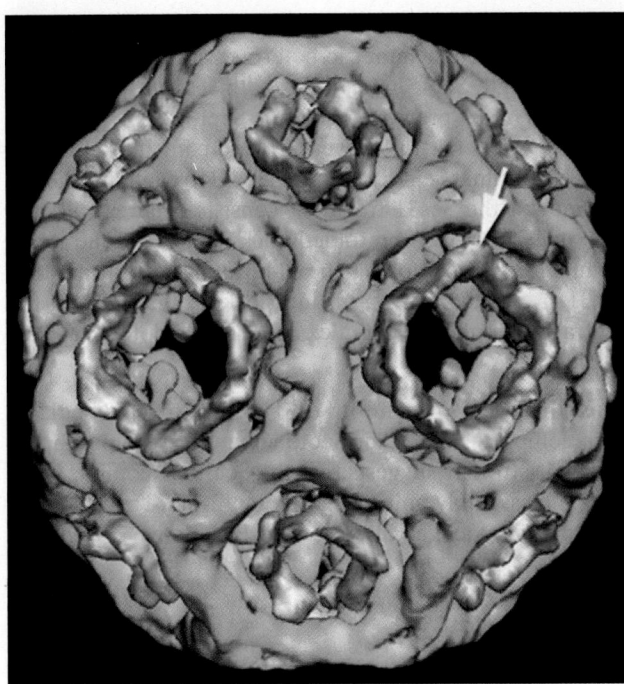

Figure 12-69   Cryo-EM–based image of a clathrin hexagonal barrel in complex with Hsc70 and a J-domain-containing fragment of auxilin at 28 Å resolution. The clathrin cage is gold and electron density attributable to the Hsc70 is blue. The white arrow indicates the position at which the Hsc70 most closely approaches the clathrin lattice. [Courtesy of Alasdair Steven, NIH, Bethesda, Maryland.]

A variety of regulatory and accessory proteins of largely unknown function have also been implicated in CCV formation. Moreover, many of the proteins described above are each present in several isoforms. Hence it is clear that our understanding of this process is far from complete.

e. The Assembly of COPI- and COPII-Coated Vesicles Resembles That of Clathrin-Coated Vesicles

COPI- and COPII-coated vesicles are both assembled in processes, elucidated in large part by Randy Schekman, that resemble CCV assembly:

1. Priming: COPI-coated vesicles are primed identically to CCVs: ARF1 is recruited to the membrane by the ARNO-promoted exchange of its bound GDP for GTP (Fig. 12-65, Step 1). COPII-coated vesicle assembly is similarly primed but by different proteins: **Sar1** (for *secretion-associated* and *Ras*-related protein-*1*) is the small ARF family GTPase that carries out this process, and the exchange of its GDP for GTP is mediated by the transmembrane GEF **Sec12.**

2. Assembly: ARF1 · GTP stoichiometrically recruits intact coatomers to form COPI-coated vesicles. Most of the seven COPI coatomer subunits have homologs in the clathrin system and function accordingly: The β-, γ-, δ-, and ζ-COPs correspond to the β2, α, μ2, and σ2 subunits of AP2, respectively (Fig. 12-67), and the α- and ε-COPs correspond to the clathrin heavy and light chains. In COPII coat formation, Sar1 · GTP recruits the TM complex **Sec23/24,** which in turn recruits cargo proteins and **Sec13/31,** which forms the budding vesicle's polyhedral outer shell (see below).

3. Release: Both COPI- and COPII-coated vesicles spontaneously bud off from their parent membranes; these processes appear to have no requirement for an analog of dynamin as does CCV release.

4. Uncoating: As is the case for CCVs, COPI- and COPII-coated vesicles uncoat shortly after being released from their parent membranes. These processes appear to be initiated by the hydrolysis of the GTPs bound to ARF1 and Sar1, which thereby weaken the attachment of COPI and COPII to their respective vesicles. The GTPase activating protein (GAP) for COPI vesicles, a 415-residue protein named **ARF GAP,** appears to be a component of the COPI coat. In COPII vesicles, Sec23 is the GAP for Sar1.

f. The Components of COPII and Clathrin Cages Are Structurally Similar but Functionally Different

Cryo-EM studies of the COPII component Sec13/31 by Bridget Carragher and William Balch reveal that, *in vitro,* this heterodimer forms a 600-Å-diameter cuboctahedral cage (Fig. 12-70). A cuboctahedron has *O* symmetry (the symmetry of a cube; Section 8-5B) and has 24 edges of equal length; 12 vertices, each of which is formed by the

Figure 12-70   Cryo-EM structure of the human Sec13/31 COPII cage at 30 Å resolution. The views are along the cuboctahedral cage's 2-fold axis (*left*), it 3-fold axis (*middle*), and its 4-fold axis (*right*). The surfaces of the cage elements are colored according to their distance from the center of the cage with blue nearest and yellow farthest. The scale bar is 500 Å long. [Courtesy of Bridget Carragher and William Balch, The Scripps Research Institute, La Jolla, California.]

interection of four edges (in contrast to clathrin cages whose vertices are each formed by the insection of three edges; Fig. 12-62c); and 14 faces, of which 8 are equilateral triangles and 6 are squares. *In vivo,* COPII vesicles are often larger than 600 Å in diameter. However, several larger polyhedra are known whose vertices are each formed by the intersection of four equal-length edges.

Although the full-length Sec13/31 complex has not been crystallized, its limited proteolysis led to two X-ray structures determined by Jonathan Goldberg:

1. That of the 297-residue Sec13 in complex with residues 1 to 411 of the 1297-residue Sec31 (Fig. 12-71a). Sec13 forms six blades of a β propeller and the Sec31 fragment forms a seven-bladed β propeller with its C-terminal segment contributing a seventh blade to the Sec13 β propeller. Each blade of these propellers consists of a WD40 repeat as do the blades of the clathrin β propeller.

2. That of the Sec13/31 edge element (Fig. 12-71b), which is a 2-fold symmetric heterotetramer that contains the full length Sec13 in complex with residues 370–763 of Sec31. As in the previous structure, Sec13 forms six blades of a β propeller with a seventh blade contributed by the here N-terminal segment of the Sec31 fragment. The remainder of the Sec31 fragment consists of an α solenoid with its N-terminal end folded back over itself and its C-terminal end overlapping the C-terminal end of another Sec31 fragment to form an interlocked dimer. Thus, the central portion of the complex consists of a double layer of α solenoids.

Since the same segment of Sec31 passes through Sec13 in both complexes and their Sec13 subunits are superimposable, this strongly suggests that the Sec13/31 complex contains the assembly unit shown in Fig. 12-71c. This assembly unit has been docked into the cryo-EM–determined struc-

β-propeller β-propeller α-solenoid   α-solenoid β-propeller β-propeller

(c)

Figure 12-71 **X-ray structures of portions of the Sec13/31 complex from yeast.** (a) The Sec13/31 vertex element, which consists of Sec13 (*orange*) in complex with residues 1 to 411 of Sec31 (*green*). The complex forms two seven-bladed β propellers with one blade of the mainly Sec13 β propeller contributed by the C-terminal portion of the Sec31 fragment. (b) The Sec13/31 edge element, which is a heterotetramer composed of two molecules each of Sec13 (*red and orange*) and residues 370 to 763 of Sec31 (*light and dark green*). The complex is viewed along its 2-fold axis and oriented as in Part a. Here Sec13 forms β propellers as in Part a and the Sec31 fragment forms a 215-Å-long double layered α solenoid. (c) Molecular model of the Sec13/31 assembly unit drawn as a surface diagram that is colored and oriented as in Parts a and b. [Courtesy of Jonathan Goldberg, Memorial Sloan-Kettering Cancer Center, New York, New York. PDBids 2PM6 and 2PM9.]

Figure 12-72 **Molecular model of the COPII cage viewed approximately along its 3-fold axis.** Its 48 Sec13/31 subunits are drawn in worm form colored as in Fig. 12-71. Four Sec31 β propellers associate to form each vertex of the cuboctahedral cage with the remaining portions of the heterotetrameric Sec13/31 assembly units forming its edges. The inner diameter of the cage is ~520 Å. [Courtesy of Jonathan Goldberg, Memorial Sloan-Kettering Cancer Center, New York, New York.]

ture of the Sec13/31 cage (Fig. 12-70) to yield the model for the COPII cage drawn in Fig. 12-72.

It is instructive to consider the similarities and differences between COPII and clathrin cages. Both consist of seven-bladed β propellers and α solenoids. In COPII cages, all such motifs participate in forming its edges with four Sec31 β propellers associating to form each of its vertices. In contrast, clathrin cages are constructed entirely from their α solenoidal segments with three such segments associating to form each of its vertices and with their β propeller motifs located in the interior of the cage, where they interact with adapter proteins. Moreover, the ~40-Å-diameter edges of COPII cages each consist of a double layer of α solenoids, whereas the ~120-Å-diameter edges of clathrin cages each consist of the interdigitated α solenoidal segments from four triskelions. Evidently, evolution has molded the similar components of these cages to different functions. Sequence analysis of COPI coat proteins have identified α solenoid and β propeller motifs, which suggests that the clathrin-, COPI-, and COPII-coated vesicles arose from the same proto-coatamer.

The C-terminal segment of Sec31, which is not present in the forgoing X-ray structures, contains an apparently unstructured Pro-rich segment (residues 770–1110 are 20% Pro) that has been implicated in binding the Sar1–Sec23/24 complex (which initiates vesicle budding by binding to the cytoplasmic regions of cargo TM proteins). Based on the X-ray structure of the Sar1–Sec23/24 complex and the fact that a cuboctahedral Sec13/31 cage

has has 48 binding sites for this complex, it would appear that the Sar1–Sec23/24 complex forms a 50-Å-thick layer beneath the surface of the COPII cage. Indeed, cryo-EM studies on COPII vesicles assembled from purified Sec13/31 and Sec23/24 complexes reveal that the Sec23/24 complexes form a cage that is concentric to and inside the Sec13/31 cage.

g. Proteins Are Directed to the Lysosome by Carbohydrate Recognition Markers

How are proteins in the ER selected for transport to the Golgi apparatus and from there to their respective membranous destinations? A clue as to the nature of this process is provided by the human hereditary defect known as **I-cell disease** (alternatively, **mucolipidosis II**), which in homozygotes is characterized by severe progressive psychomotor retardation, skeletal deformities, and death by age 10. The lysosomes in the connective tissue of I-cell disease victims contain large inclusions (after which the disease is named) of glycosaminoglycans and glycolipids as a result of the absence of several lysosomal hydrolases. These enzymes are synthesized on the RER with their correct amino acid sequences but, rather than being dispatched to the lysosomes, are secreted into the extracellular medium. This misdirection results from the absence of a mannose-6-phosphate recognition marker on the carbohydrate moieties of these hydrolases because an enzyme required for mannose phosphorylation fails to recognize the lysosomal proteins. The mannose-6-phosphate residues are normally bound by a receptor in the coated vesicles that transport lysosomal hydrolases from the Golgi apparatus to the lysosomes (Section 23-3Bj). Other glycoproteins are directed to their intracellular destinations by similar carbohydrate markers.

h. ER-Resident Proteins Have the C-Terminal Sequence KDEL

Most soluble ER-resident proteins in mammals have the C-terminal sequences KDEL (HDEL in yeast), KKXX, or KXKXXX (where X represents any amino acid residue), whose alteration results in the secretion of the resulting protein. By what means are these proteins selectively retained in the ER? Since many ER-resident proteins freely diffuse within the ER, it seems unlikely that they are immobilized by membrane-bound receptors within the ER. Rather, it has been shown that ER-resident proteins, as do secretory and lysosomal proteins, readily leave the ER via COPII-coated vesicles but that ER-resident proteins are promptly retrieved from the Golgi and returned to the ER in COPI-coated vesicles. Indeed, coatomer binds the Lys residues in the C-terminal KKXX motif of transmembrane proteins, which presumably permits it to gather these proteins into COPI-coated vesicles. Furthermore, genetically appending KDEL to the lysosomal protease **cathepsin D** causes it to accumulate in the ER, but it nevertheless acquires an *N*-acetylglucosaminyl-1-phosphate group, a modification that is made in an early Golgi compartment. Presumably, a membrane-bound receptor in a post-ER compartment binds the KDEL signal and the resulting complex is returned to the ER in a COPI-coated vesicle. **KDEL receptors** have, in fact, been identified in yeast and humans. However, the observation that former KDEL proteins whose KDEL sequences have been deleted are, nevertheless, secreted relatively slowly suggests that there are mechanisms for retaining these proteins in the ER by actively withholding them from the bulk flow of proteins through the secretory pathway.

D. Vesicle Fusion

Vesicles that travel only short distances (<1 μm) between their parent and target membranes (e.g., between neighboring Golgi cisternae) do so via simple diffusion, a process that typically takes from one to several minutes. However, vesicles that have longer distances to commute (e.g., from the TGN to the plasma membrane) are actively transported along cytoskeletal microtubules (Section 1-2A) by the motor proteins **dynein** and **kinesin,** which unidirectionally crawl along microtubule "tracks" in an ATP-driven process (Section 35-3H).

a. Vesicle Fusion Is Most Easily Studied in Yeast and in Synapses

On arriving at its target membrane, a vesicle fuses with it, thereby releasing its contents on the opposite side of the target membrane (Fig. 12-60). How do vesicles fuse and why do they fuse only with their target membranes and not other membranes? Progress in answering these questions has been made mainly by using two experimental approaches, the genetic dissection of this process in yeast and its biochemical analysis in **synapses,** the junctions between neurons (nerve cells) and between neurons and muscles (Fig. 12-73).

When a nerve impulse in the presynaptic cell reaches a synapse, it triggers the fusion of **neurotransmitter-**containing **synaptic vesicles** with the **presynaptic membrane** (a specialized section of the neuron's plasma membrane), thereby releasing the neurotransmitter (a small molecule) into the ~200-Å-wide **synaptic cleft** (the process whereby membranous vesicles fuse with the plasma membrane to release their contents outside the cell is called **exocytosis**). The neurotransmitter rapidly diffuses across the synaptic cleft to the postsynaptic membrane, where it binds to specific receptors that then trigger the continuation of the nerve impulse in the postsynaptic cell (Section 20-5C). The homogenization of nerve tissue causes its presynaptic endings to pinch off and reseal to form **synaptosomes,** which can be readily isolated by density gradient ultracentrifugation for subsequent study.

b. Vesicle Fusion Requires the Coordinated Actions of Many Proteins

Biological membranes do not spontaneously fuse. Indeed, being negatively charged, they strongly repel one another at short distances. These repulsive forces must be overcome if biological membranes are to fuse. As we shall see below, we are just beginning to understand how this complicated process occurs.

(a)

(b)

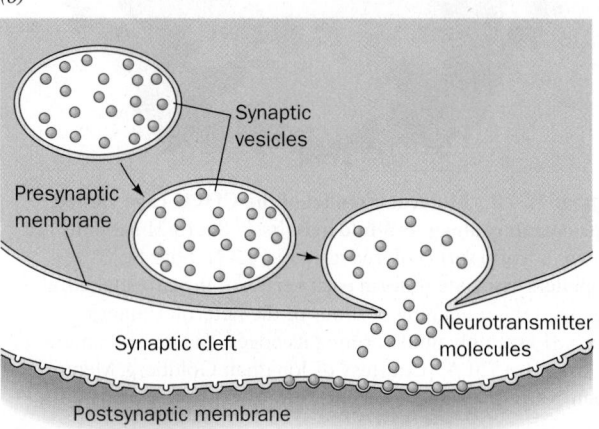

Figure 12-73   **Transmission of nerve impulses across a synaptic cleft.** (*a*) Electron micrograph of a frog neuromuscular junction in which the synaptic vesicles are undergoing exocytosis (*arrows*) with the presynaptic membrane (*top*). [Courtesy of John Heuser, Washington University School of Medicine, St. Louis, Missouri.] (*b*) The neurotransmitter, which is thereby discharged into the synaptic cleft, rapidly (in <0.1 ms) diffuses to the postsynaptic membrane, where it binds to transmembrane receptors, triggering a new nerve impulse.

Studies of the mechanism of vesicle fusion were pioneered by Rothman, who demonstrated that the fusion process is blocked by low concentrations of the cysteine-alkylating agent *N*-ethylmaleimide (NEM),

N-Ethylmaleimide (NEM)   **Cys**

indicating the presence of an **NEM-sensitive fusion (NSF) protein.** NSF is a cytosolic ATPase that does not bind to membranes unless a **soluble NSF attachment protein (SNAP)** is also present. SNAPs bind to membranes in the absence of NSF, demonstrating that SNAPs bind before NSF. SNAPs bind to alkali-extracted membranes, which indicates that **SNAP receptors (SNAREs)** are integral or lipid-linked proteins.

Three classes of proteins appear to participate in all vesicle fusion reactions:

1. Rab proteins, which are small (20–29 kD) GTPases of the Ras superfamily that play a central role in directing vesicle transport. Cells express numerous Rab isoforms, 11 in yeast and 63 in humans, each localized to a specific membrane compartment. Rab proteins have two tandem Cys residues at their C-termini, both of which are geranylgeranylated (Section 12-3Ba). A soluble protein named **GDP dissociation inhibitor (GDI)** binds to Rab · GDP so as to mask its geranylgeranyl groups and thus maintain it in the cytoplasm. However, when Rab · GDP interacts with a cognate **Rab-GEF** on the surface of its target vesicle, the geranylgeranyl groups on the resulting Rab · GTP are unmasked and insert into the vesicle membrane—much like the anchoring of ARF1 · GTP to the Golgi membrane (Fig. 12-65). Rab · GTP then binds to rodlike proteins emanating from the vesicle's target membrane known as **tethering factors** to form a relatively loose association between the two membranes. After vesicle fusion, Rab hydrolyzes its bound GTP to GDP in a process induced by a specific **Rab-GAP** and the resulting Rab · GDP is extracted from the membrane by GDI, thereby recycling the system. Rab proteins are also implicated in initiating the actual membrane fusion step (see below) as well as in the vesicle interactions with the cytoskeleton that function in transporting vesicles to their proper destinations.

2. SNAREs, which form cognate combinations of membrane-associated proteins known as **R-SNAREs** and **Q-SNAREs** (because they contain conserved Arg and Gln residues in their cytoplasmic domains; they were originally named **v-SNAREs** and **t-SNAREs,** respectively, because they are mainly associated with the *v*esicle and *t*arget membranes). The best characterized SNAREs are those functioning at neuronal synapses: **Synaptobrevin** (alteratively, **VAMP** for *v*esicle *a*ssociated *m*embrane *p*rotein) is an R-SNARE, whereas **syntaxin** and **SNAP-25** (for *sy*naptosome *a*ssociated *p*rotein of 25 kD) are Q-SNAREs. *R-SNAREs and Q-SNAREs associate to firmly anchor the vesicle to its previously loosely tethered target membrane,* a process called "docking." The docked complexes, which are described below, are eventually disassembled by NSF in association with a SNAP protein. (Note that SNAP-25 is not a SNAP protein; by curious coincidence, the two independently characterized proteins were assigned the same acronym before it was realized that they are functionally associated.)

3. The **SM proteins** (so called because they are named **Sec1** in yeast and **Munc18** in mammals), which in synapses bind to syntaxin so as to prevent synaptobrevin and SNAP-

25 from binding to it. Mutational studies indicate that these 65- to 70-kD hydrophilic proteins are essential for vesicle fusion.

c. SNAREs Form a Stable Four-Helix Bundle

The R-SNARE synaptobrevin and the Q-SNAREs syntaxin and SNAP-25 form a highly stable complex; boiling SDS solution is required to dissociate it. Synaptobrevin and syntaxin each have a C-terminal TM helix, and SNAP-25 is anchored to the membrane via palmitoyl groups that are linked to Cys residues in its central region. The X-ray structure of the associating portions of this complex (Fig. 12-74a), determined by Reinhard Jahn and Axel Brünger, reveals it to be a bundle of four parallel ~65-residue α helices with two of the helices formed by the N- and C-terminal segments of SNAP-25. Since synaptobrevin is anchored in the vesicle membrane and syntaxin and SNAP-25 are anchored in the target membrane, this so-called core complex firmly ties together the two membranes (Fig. 12-74b).

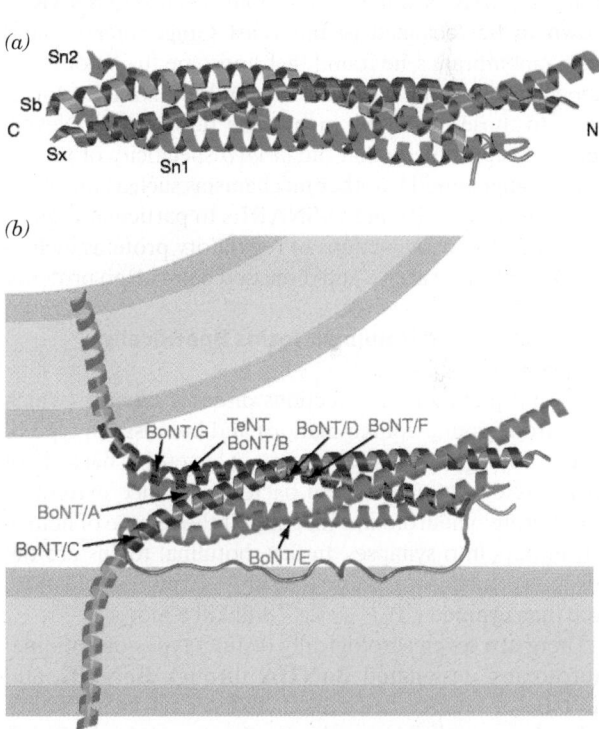

Figure 12-74 **X-ray structure of the syntaxin–synaptobrevin– SNAP-25 core complex.** (*a*) Ribbon diagram showing the syntaxin helix (Sx) in red, the synaptobrevin helix (Sb) in blue, and the N- and C-terminal helices of SNAP-25 (Sn1 and Sn2) in green. (*b*) Model of the synaptic fusion complex linking two membranes (*gray*). The helices of the core complex are colored as in Part *a*. The transmembrane C-terminal extensions of syntaxin and synaptobrevin are modeled as helices (*light green*). The loop connecting the N- and C-terminal helices of SNAP-25 is speculatively represented as an unstructured loop (*brown*). Recall that this loop is anchored to the membrane via Cys-linked palmitoyl groups (*not shown*). The cleavage sites for the various clostridial neurotoxins are indicated by the arrows. [Courtesy of Axel Brünger, Yale University. PDBid 1SFC.]

The four helices of the core complex wrap around each other with a gentle left-handed twist. For the most part, the sequence of each helix has the expected 7-residue repeat, $(a\text{-}b\text{-}c\text{-}d\text{-}e\text{-}f\text{-}g)_n$, with residues a and d hydrophobic (Section 8-2A; note that this property is characteristic of 4- and 3-helix bundles as well as of coiled coils). However, the central layer of side chains along the length of the 4-helix bundle consists of an Arg residue from synaptobrevin that is hydrogen bonded to three Gln side chains, one from syntaxin and one from each of the SNAP-25 helices. These highly conserved polar residues are sealed off from the aqueous environment such that their interactions are enhanced by the low dielectric constant of their environment. It therefore appears that these interactions serve to bring the four helices into proper register.

Since cells contain large numbers of different R-SNAREs and Q-SNAREs (25 in yeast and 36 in humans), it would seem likely that their interactions are at least partially responsible for the specificity that vesicles exhibit in fusing with their target membranes. Indeed, Rothman has shown this to be the case by determining, *in vitro*, the rate of fusion of liposomes bearing different SNAREs. In testing all the R-SNAREs in the yeast genome against Q-SNAREs known to be localized to the yeast Golgi, vacuole, and plasma membranes, he found that liposome fusion only occurs when the combinations of R- and Q-SNAREs correspond to those mediating membrane flow *in vivo*. Nevertheless, it seems likely that the *in vivo* specificity of vesicle fusion is augmented by other mechanisms such as the localization of cognate R- and Q-SNAREs to particular regions in the cell and by the actions of regulatory proteins including, as is indicated above and discussed below, Rab proteins.

d. Tetanus and Botulinus Toxins Specifically Cleave SNAREs

The frequently fatal infectious diseases **tetanus** (which arises from wound contamination) and **botulism** (a type of food poisoning) are caused by certain anaerobic bacteria of the genus *Clostridium*. These bacteria produce extremely potent protein neurotoxins that inhibit the release of neurotransmitters into synapses. In fact, botulinal toxins are the most powerful known toxins; they are ~10 millionfold more toxic than cyanide (10^{-10} g · kg^{-1} will kill a mouse).

There are seven serologically distinct types of botulinal neurotoxins, designated **BoNT/A** through **BoNT/G,** and one type of tetanus neurotoxin, **TeTx.** Each of these homologous proteins is synthesized as a single ~150-kD polypeptide chain that is cleaved by host proteases to yield an ~50-kD L chain that remains disulfide-linked to the ~100-kD H chain (Fig. 12-75). The H chains bind to specific types of neurons (via gangliosides and protein receptors), where they facilitate the uptake of the L chain by endocytosis. *The L chains are proteases, and each cleaves its target SNARE at a specific site* (Fig. 12-74*b*). This prevents the formation of the core complex and thereby halts the exocytosis of synaptic vesicles. The H chain of TeTx specifically binds to inhibitory neurons (which function to moderate excitory nerve impulses) and is thereby responsible for the spastic paralysis characteristic of tetanus. The H chains of the BoNTs instead bind to motor neurons (which

Figure 12-75   **Model of clostridial neurotoxins and their activation by host proteases.** The disulfide bond linking the L and H segments is cleaved after the neurotoxin is taken up by its target neuron.

innervate muscles) and thus cause the flaccid paralysis characteristic of botulism.

The administration of carefully controlled quantities of botulinal toxin (trade name Botox) is medically useful in relieving the symptoms of certain types of chronic muscle spasms. Moreover, this toxin is used cosmetically: Its injection into the skin relaxes the small muscles causing wrinkles and hence these wrinkles disappear for ~3 months.

e. Bilayer Fusion Is Mechanically Induced

The association of Q-SNAREs on a vesicle with an R-SNARE on its target membrane brings the two bilayers into close proximity, yielding a so-called **trans-SNARE complex.** But what induces the fusion of the juxtaposed bilayers? The answer, which is diagrammed in Fig. 12-76, is that the mechanical forces arising from the formation of a ring of several (estimated to be 5–10) trans-SNARE complexes pulls together apposing bilayers. This expels the contacting lipids between them so as to join their outer leaflets, a process known as **hemifusion.** Indeed, the pressure (force/area) within the ring of trans-SNARE complexes is estimated to be 100 to 1000 atm. In the resulting transient structure, no aqueous contact between the two membrane systems has yet been established. However, as the fusion process proceeds (the trans-SNAREs continue zipping up), the two inner leaflets of the now partially joined membranes come together to form a new bilayer, whose component lipids are subsequently similarly expelled to yield a **fusion pore.** The fusion pore then rapidly expands, thereby fully joining the two membranes as well as their contents. Thus, vesicle fusion is driven by the protein folding forming the trans-SNARE complexes.

As we discussed above, liposomes containing the corresponding Q- and R-SNAREs spontaneously fuse. However, this *in vitro* process takes 30 to 40 minutes whereas, for example, the *in vivo* fusion of a synaptic vesicle with the presynaptic membrane takes <0.3 ms (Section 20-5C). This suggests that other proteins such as Rab proteins and/or their **effectors** (proteins with which they interact) participate in mediating the bilayer fusion process.

f. The Structure of the nSec1–Syntaxin Complex Suggests a Function for Rab Protein

The neuronal SM protein, which is named **nSec1,** binds to syntaxin with high affinity to form a complex that is mutually exclusive with the formation of the syntaxin–

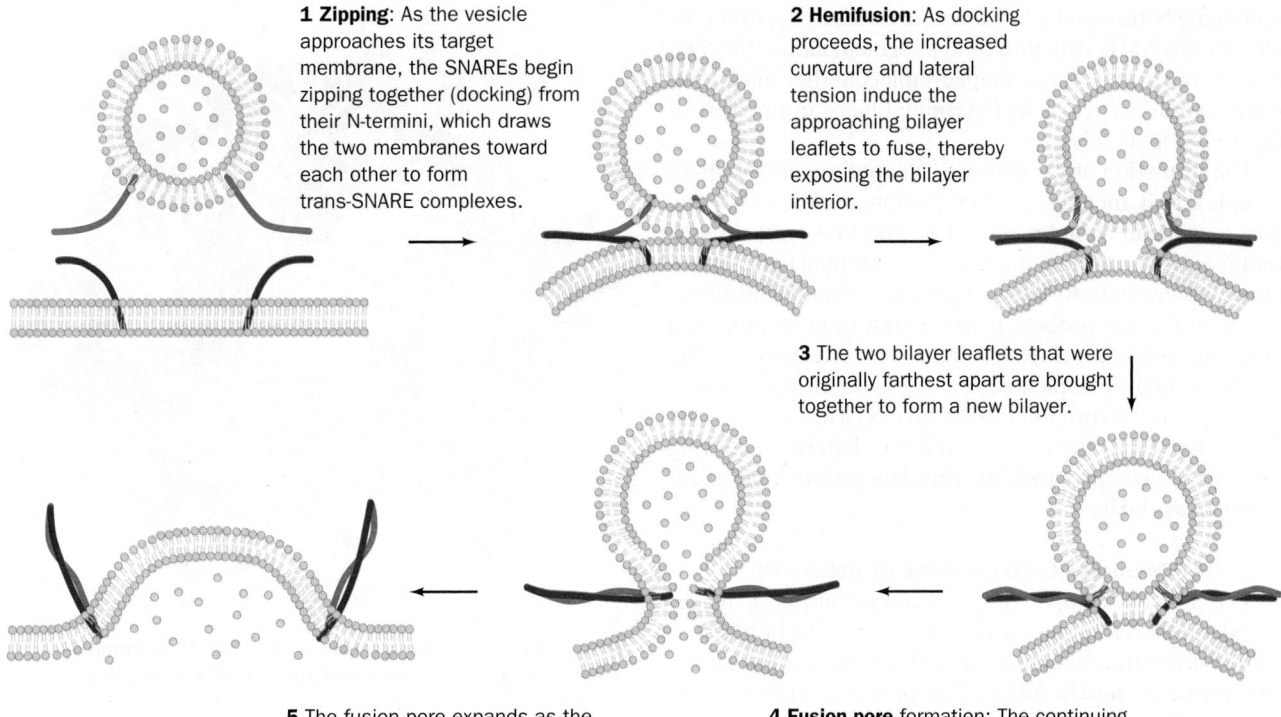

1 Zipping: As the vesicle approaches its target membrane, the SNAREs begin zipping together (docking) from their N-termini, which draws the two membranes toward each other to form trans-SNARE complexes.

2 Hemifusion: As docking proceeds, the increased curvature and lateral tension induce the approaching bilayer leaflets to fuse, thereby exposing the bilayer interior.

3 The two bilayer leaflets that were originally farthest apart are brought together to form a new bilayer.

5 The fusion pore expands as the now fused membrane relaxes yielding cis-SNARE complexes.

4 Fusion pore formation: The continuing SNARE-induced lateral tension causes membrane breakdown, resulting in the formation of a fusion pore.

Figure 12-76 **Model for SNARE-mediated vesicle fusion.** Here the R-SNARE and the Q-SNAREs are schematically represented by red and blue worms. [After a drawing by Chen, Y.A. and Scheller, R.H., *Nature Rev. Mol. Cell Biol.* **2,** 98 (2001).]

synaptobrevin–SNAP-25 complex. The X-ray structure of nSec1 in complex with the cytoplasmic domain of syntaxin (Fig. 12-77), determined by William Weis, reveals that this portion of the 288-residue syntaxin forms an N-terminal up–down–up–down four-helix bundle. Syntaxin's C-terminal helix (but lacking its TM portion) adopts a bent and somewhat irregular conformation, which differs from that in the core complex displayed in Fig. 12-74. In contrast, the

Figure 12-77 **X-ray structure of the complex between nSec1 and syntaxin.** (*a*) Ribbon diagram of syntaxin with its N-terminal 3-helix bundle (Habc) red and the cytoplasmic portion, its C-terminal helix (H3; the segment that forms a component of the core complex), purple. (*b*) Ribbon diagram of nSec1 with its three domains differently colored. (*c*) The nSec1–syntaxin complex colored as in Parts *a* and *b* and viewed such that the nSec1 is rotated by 90° about the vertical axis relative to Part *b*. [Courtesy of William Weis, Stanford University School of Medicine. PDBid 1DN1.]

remaining N-terminal 3-helix bundle is closely superimposable on the NMR structure of this segment alone. The 594-residue nSec1 is an arch-shaped molecule that binds syntaxin, and in particular its C-terminal helix, in the cleft of the arch (Fig. 12-77c).

The formation of the syntaxin–synaptobrevin–SNAP-25 complex that mediates vesicle fusion requires that the nSec1–syntaxin complex dissociate and that syntaxin's N-terminal 3-helix bundle release the C-terminal helix. Mutational studies indicate that a Rab protein and/or its effectors mediate this process. It has therefore been proposed that the binding of Rab and/or its effectors to the nSec1–syntaxin complex causes nSec1 to change conformation, which in turn induces syntaxin's N-terminal 3-helix bundle to release the C-terminal helix, thereby permitting the SNARE complex to form. Thus Rab controls the availability of syntaxin.

g. NSF Mediates Core Complex Disassembly

The SNARE complex in the fused membranes, the so-called **cis-SNARE complex,** must eventually be dissociated in order for its component proteins to participate in a new round of vesicle fusion. This process is mediated by NSF, an ATP-dependent cytosolic protein that binds to SNAREs (SNAP receptors) through the intermediacy of adaptor proteins called SNAPs (soluble NSF attachment proteins). Although it was initially proposed that the NSF-mediated disassembly of the cis-SNARE complex somehow directly drove membrane fusion, it is now clear that NSF functions to recycle SNAREs after their participation in membrane fusion, that is, *NSF functions as an ATP-driven molecular chaperone.* However, since trans-SNARE complexes form spontaneously, *membrane fusion is indirectly driven by NSF-mediated ATP hydrolysis.*

NSF is a hexamer of identical 752-residue subunits. Sequence analysis and limited proteolysis studies indicate that each subunit consists of three domains:

1. An N-terminal so-called N domain (residues 1–205), which mediates NSF's interactions with SNAPs and SNAREs.

2. A D1 domain (206–487), which binds ATP and catalyzes its hydrolysis in a process that drives the disassembly of the cis-SNARE complex.

3. A C-terminal D2 domain (488–752), which is homologous to D1. D2 binds ATP with a much higher affinity than does D1 but hydrolyzes it very slowly, if at all. D2 · ATP mediates the hexamerization of NSF, which is required for NSF activity.

The X-ray structure of the D2 domain of NSF was independently determined by Weis and by Jahn and Brünger. Its wedge-shaped subunits associate to form a 116-Å-diameter and 40-Å-high disk-shaped hexamer that has an ~18-Å-diameter central pore (Fig. 12-78). The ATP is bound near the interface between two subunits, where it presumably helps stabilize their association.

Figure 12-78 **X-ray structure of the NSF D2 hexamer as viewed from its N-terminal end along its 6-fold axis.** Each subunit is differently colored. The bound ATPs are drawn in ball-and-stick form. [Courtesy of Axel Brünger, Yale University. PDBid 1NSF.]

Electron micrographs by Jahn and John Heuser of intact NSF in the presence of ATP have the appearance of an ~120-Å-diameter hexagonal ring with a 30- to 50-Å central opening when seen in top view (Fig. 12-79a) and of a 120-Å by 150-Å rectangle when seen in side view (Fig. 12-79b). The length of the rectangle is about twice the height of the D2 disk, which suggests that D1 forms a D2-like hexagonal disk that stacks on D2. In the presence of ADP, NSF has an identical appearance, which suggests that D1 rapidly hydrolyzes its bound ATP to form ADP. However, in the presence of the nonhydrolyzable ATP analog ATPγS (in which a terminal O atom on the γ-phosphorus atom of ATP is replaced by S), NSF displays six globular feet that are tightly packed around the somewhat smaller hexagonal ring (Fig. 12-79c). Since the hexagonal rings but not the globules are seen when D1–D2 constructs are imaged in the presence of ATPγS, the globules must be the N domains. Evidently, the N domains are held tightly around the central disk of stacked D1 and D2 hexamers when D1 binds ADP but are released when D1 binds ATP.

The mechanism whereby NSF disassembles the cis-SNARE complex is largely unknown. The rod-shaped SNARE core complex (Fig. 12-74a), which is 20 to 25 Å in diameter, is too wide to fit inside the 18-Å-diameter central pore of the D2 hexamer (and presumably the similarly shaped D1 hexamer) without significant structural changes. It is therefore unlikely that the core complex binds inside NSF's central cavity in a manner similar to the way that the GroEL–GroES chaperonin system binds its substrate proteins (Section 9-2Ca). Moreover, electron micrographs indicate that the complex of SNAP and the

Figure 12-79 **Quick-freeze/deep-etch electron micrographs of NSF hexamers.** (*a*) Top and (*b*) side views in the presence of ATP. (*c*) Top view in the presence of ATPγS. [Courtesy of John Heuser, Washington University School of Medicine, St. Louis, Missouri.]

three SNARE proteins binds to one end of NSF in the presence of ATPγS (but not at all in the presence of ADP). Since NSF oligomers containing mixtures of active and inactive D1 domains are unable to disassemble SNARE complexes, it appears that the NSF subunits function in a cooperative manner.

E. *Protein Targeting to Mitochondria*

Although mitochondria contain functioning genetic and protein synthesizing systems, their genomes encode only a handful of inner membrane proteins (13 in humans; 8 in yeast). The vast majority of mitochondrial proteins (~99%), which comprise 10 to 20% of intracellular proteins, are encoded by nuclear genes and are synthesized by cytosolic ribosomes. They must therefore traverse one or both mitochondrial membranes (Section 1-2Ac) to reach their final destinations. In this subsection, we discuss how proteins are imported into mitochondria and are directed to their correct destinations [outer membrane, inner membrane, intermembrane space, or **matrix** (the space enclosed by the inner membrane)]. Our rapidly developing knowledge of this process was elucidated in large part through investigations in yeast and in the pink bread mold *Neurospora crassa* by Walter Neupert, Nikolaus Pfanner, Trevor Lithgow, and Gottfried Schatz. However, there is considerable evidence that this process is well conserved among all eukaryotes. The transport systems we describe

here and in Section 12-4B resemble those that mediate the import of proteins into chloroplasts (in which proteins must cross up to three membranes; Section 1-2Ag) and peroxisomes (Section 1-2Ad).

a. Proteins Must Be Unfolded for Import Into Mitochondria

Most nuclear-encoded mitochondrial proteins are fully synthesized by cytosolic ribosomes before they are imported into mitochondria; that is, they are post-translationally imported. One might expect, therefore, that mitochondrial proteins, many of which are integral proteins, would at least partially fold and/or nonspecifically aggregate in the cytosol before encountering the mitochondrial import system. Yet a variety of evidence indicates that *only unfolded proteins can pass through mitochondrial membranes.* For example, **dihydrofolate reductase (DHFR),** a normally cytosolic enzyme, is imported into yeast mitochondria when it is preceded by the targeting sequence (see below) of a cytosolically synthesized mitochondrial protein. However, the importation of this chimeric protein is arrested by the presence of **methotrexate,** an analog of DHFR's normal substrate **dihydrofolate** (Section 28-3Be), which binds to DHFR with such high affinity that it stabilizes the protein's native conformation.

The import competence of mitochondrially destined proteins is maintained in the cytosol by a variety of ATP-dependent molecular chaperones. These include members of the Hsp70 family (Section 9-2C) and, in mammals, a protein named **mitochondrial import stimulation factor (MSF).** Consequently, the genetically engineered shutdown of Hsp70 production in yeast causes the cells to cytosolically accumulate proteins that would otherwise be imported into the mitochondria. Moreover, the rate of the Hsp70-facilitated mitochondrial import of a protein is enhanced by its prior denaturation by urea. Evidently, Hsp70 functions in this process as an ATP-driven "protein unfoldase."

b. Translocation of Proteins Across the Outer Mitochondrial Membrane

Most cytosolically synthesized matrix proteins have cleavable N-terminal signal sequences that do not interact with the SRP. These presequences consist of 10 to 80 residues that form amphipathic helices with one face positively charged. However, many mitochondrial proteins, including most metabolite carrier proteins of the inner membrane (see below), have poorly characterized internal targeting sequences.

The protein subunits that participate in importing proteins across the outer mitochondrial membrane are called **TOM proteins** (for *t*ranslocase of the *o*uter mitochondrial *m*embrane) and are named Tomxx, where xx is the molecular mass of the subunit in kilodaltons. Likewise, many of the proteins involved in translocating proteins across the inner mitochondrial membrane are called **TIM proteins** (for *t*ranslocase of the *i*nner mitochondrial *m*embrane) and are named Timxx.

Figure 12-80 **Schematic diagram of the mitochondrial protein import machinery in yeast.** See the text for a description. The subunit compositions of these complexes in mitochondria from other organisms are similar. [After Bolender, N., Sickmann, A., Wagner, R., Meisinger, C., and Pfanner, N., *EMBO Rep.* **9**, 42–49 (2008).]

The **TOM complex,** the machinery that imports all mitochondrial proteins through the outer mitochondrial membrane, does so as follows (Fig. 12-80, *top left—green*):

1. The signal sequences of unfolded preproteins associate with the cytoplasmic domains of mitochondrial receptor proteins: N-terminal signal sequences interact mainly with **Tom20** in complex with **Tom22,** whereas internal signal sequences interact mainly with **Tom70.** The NMR structure of a portion of Tom20's cytosolic domain in complex with an 11-residue segment of a presequence peptide (Fig. 12-81), determined by Toshiya Endo and Daisuke Kohda, reveals that the Tom20 domain consists mainly of five helices. Its two N-terminal helices form a nonpolar surface groove in which the helical presequence binds, mainly via hydrophobic interactions rather than ionic interactions. Evidently, Tom20 recognizes the presequence's amphipathic helix but not its positive charges. These positive charges, which are required for mitochondrial import, interact with Tom22.

2. Tom20 and Tom70 deliver preproteins to the **general import pore (GIP),** so called because all nuclear-encoded mitochondrial proteins must pass through it. The GIP is formed by **Tom40,** a polytopic TM protein, which CD measurements indicate consists mainly of β sheets and hence has a TM β barrel structure that presumably resembles that of bacterial porins (Fig. 12-27). Electrophysiological measurements demonstrate that Tom40 contains a cation-selective hydrophilic channel through which precursor proteins are transported. Tom40 is closely associated with three small single pass TM subunits, **Tom5, Tom6,** and **Tom7,** to form

the **TOM core complex.** The deletion of any one of these small subunits has only minor effects but the deletion of all three is lethal. They appear to stabilize the TOM complex but their individual functions are largely unknown. Electron micrographs of the *Neurospora* TOM core complex (Fig. 12-82) reveal an ~70-Å-high (~20 Å larger than the thickness of the lipid bilayer) and ~120-Å-wide particle containing two ~21-Å-diameter pores that presumably are the protein-conducting channels. This agrees with permeability experiments using cations of various sizes, which indicate that the Tom40 pore is ~22 Å in diameter.

3. The forces driving the translocation of polypeptides through the TOM complex remain largely enigmatic. A proposed mechanism, the **acid chain hypothesis,** is that the positively charged presequence is sequentially transferred between acidic (negatively charged) patches to which it binds with successively higher affinities. Such patches are present on the cytoplasmic faces of Tom20, Tom22, and Tom5, as well as on the intermembrane faces of Tom40 and Tom22.

At this stage, the import pathway for mitochondrial proteins splits several ways. We discuss these various pathways below.

c. Translocation of Proteins Into the Matrix

Polypeptides with N-terminal signal sequences, which include the precursors of all matrix-destined proteins, most inner membrane proteins, and many proteins that occupy the intermembrane space (**IMS**), are translocated across the inner mitochondrial membrane by the **TIM23 complex**

Figure 12-81 **NMR structure of the cytoplasmic domain of rat Tom20 in complex with the C-terminal 11-residue segment (GPRLSRLLSYA) of the 22-residue presequence of the rat mitochondrial enzyme aldehyde dehydrogenase.** The diagram is a superposition of the 20 final structures in the NMR analysis (Section 8-3A) in which the residues used to make the superposition are blue (Tom20) and red (presequence) and the remaining residues are gray (Tom20) and orange (presequence). [Courtesy of Toshiya Endo, Nagoya University, Nagoya, Japan, and Daisuke Kohda, Biomolecular Engineering Research Institute, Osaka, Japan. PDBid 1OM2.]

Figure 12-82 **Electron microscopy–based image of the TOM core complex particles from** *Neurospora.* The particles, which are shown in top view (*left*) and side view (*right*), contain two openings that presumably represent the mitochondrial outer membrane's protein-conducting channels. [Courtesy of Stephan Nussberger and Walter Neupert, Universität München, Germany.]

(Fig. 12-80, *bottom left—yellow*). This complex contains a protein channel formed by **Tim23,** which is closely associated with **Tim17.** The peripheral protein **Tim50** binds the polypeptide emerging from the Tom40 channel and passes it to Tim23. Electron microscopy studies indicate that the TOM and TIM23 complexes are in apposition at sites where the inner and outer mitochondrial membranes approach each other most closely. Indeed, **Tim21** transiently associates with Tom22 across this contact site by displacing the emerging signal sequence.

In the presence of methotrexate, the above DHFR chimera becomes stuck in the membrane with the spacer that linked the enzyme to its N-terminal presequence simultaneously spanning the TOM and TIM23 complexes. The N-terminal end of the spacer is presumably trapped in the matrix through its association with **mtHsp70** (see below). Consequently, if the spacer is so short that it cannot span both membranes (less than ~40 residues), no stable translocation intermediate is formed. Thus, it appears that presequences make their way between the TOM and TIM23 complexes without the aid of chaperones.

The translocation of a protein across the inner mitochondrial membrane requires energy in the form of both ATP and an electrostatic potential across the inner mitochondrial membrane. This so-called **membrane potential** (Section 20-1), $\Delta\Psi$, which is metabolically generated (Section 22-3Ba), apparently functions to electrophoretically transport the positively charged N-terminal signal sequence into the matrix (the matrix is negatively charged with respect to the cytosol).

The ATP is utilized by matrix Hsp70 (mtHsp70; alternatively, **mHsp70**), the central component of the **presequence translocase-associated motor (PAM;** Fig. 12-80, *bottom left—orange*). This molecular chaperone binds to **Tim44** on the inner face of the inner mitochondrial membrane, where it is thought to mechanically pull the protein through the Tim23 pore via a Brownian ratchet mechanism (Section 12-4Bg). **Pam 18** (alternatively, **Tim14**), which associates with Tim44, has a J-domain that presumably recruits mtHsp70 and induces it to hydrolyze its bound ATP to ADP, thus activating it to bind the incoming polypeptide. **Pam16** (alternatively **Tim16**), which binds Tim14, is thought to act as a negative regulator of Tim14 by physically blocking its access to mtHsp70. **Pam17** is required for the assembly of the Pam18–Pam16 module. The matrix protein **Mge1** stimulates mtHsp70 to exchange its bound ADP for ATP, thus permitting it to participate in another cycle of the Brownian ratchet.

Once a preprotein, or at least its N-terminal segment, has entered the matrix, its N-terminal signal sequence is excised by **matrix processing peptidase (MPP),** an essential protein. The imported protein then folds/assembles to its native state, a process that is facilitated by a battery of ATP-dependent chaperone proteins including mtHsp70 (only about 10% of which is associated with Tim44) and Hsp60/Hsp10 (homologs of the GroEL/ES system; Section 9-2C).

Some of the polypeptides that are translocated by the TIM23 complex have a stop-transfer anchor sequence. The TIM23 complex laterally inserts the resulting TM helix into the inner mitochondrial membrane (Fig. 12-80, *bottom, far left*) such that its N-terminal portion occupies the matrix, where MPP excises its N-terminal signal sequence.

d. Insertion of Metabolite Carrier Proteins Into the Inner Mitochondrial Membrane

The mitochondrial inner membrane is impermeable to nearly all polar substances and hence contains numerous (35 in yeast) **metabolite carrier proteins** to permit the acquisition of reactants and the delivery of products. The most abundant members of this family are the **ATP–ADP translocator** (which exchanges the ATP synthesized in the matrix for the ADP product of cytosolic ATP hydrolysis; Section 20-4C) and the **phosphate carrier** (which returns the phosphate product of cytosolic ATP hydrolysis to the matrix; Section 22-1Ba). All metabolite carrier proteins have six TM helices with both their N- and C-termini in the IMS.

Most members of the metabolite carrier family lack N-terminal signal sequences and are therefore translocated through the TOM complex via interactions with its Tom70 receptor. Curiously, however, the Tom20–Tom22 complex is the receptor for most other outer membrane proteins that have internal signal sequences. Metabolite carrier proteins are escorted across the IMS by a hexameric complex of the homologous proteins **Tim9** and **Tim10**, (Tim9)$_3$(Tim10)$_3$, which is thought to shield the hydrophobic domains of the metabolite carrier proteins (Fig. 12-80, *middle—blue*). Metabolite carrier proteins in a preparation of mitochondria depleted of Tim9 and Tim10 are not inserted into the GIP, as indicated by their failure to reach a protease-resistant state. This suggests that it is the binding of the Tim9–Tim10 complex to an unfolded metabolite carrier protein that drives its translocation across the outer mitochondrial membrane.

The Tim9–Tim10 complex delivers the metabolite carrier protein to the peripheral protein **Tim12** (a homolog of Tim9 and Tim10), which is associated with the integral proteins **Tim22** (which is homologous to Tim 23), **Tim54,** and **Tim18** to form the **TIM22 complex** (Fig. 12-80, *bottom middle—gold*). *Tim22 then mediates the lateral insertion of the metabolite carrier protein into the inner mitochondrial membrane, where it assembles to form homodimers.* This process occurs via an unknown but membrane potential–dependent mechanism. The functions of Tim54 and Tim18 are unknown.

e. Soluble Proteins Occupying the Intermembrane Space Are Imported via Three Mechanisms

Despite the fact that its width is around that of a membrane bilayer, the IMS contains a large collection of essential proteins. The precursors of some of these proteins are imported, as described above, such that they become anchored to the IMS by a single TM helix that has its N-terminal end in the matrix (Fig. 12-80, *bottom, far left*). Such a protein is then cleaved by an inner membrane protease on the C-terminal side of its TM helix, thereby releasing it into the IMS, where it folds to its native conformation. Since the mature protein lacks a signal sequence, it is no longer subject to importation into the matrix and hence remains in the IMS. **Coproporphyrinogen oxidase,** which participates in heme biosynthesis (Section 26-4Ae), is such a protein.

Many small proteins that lack N-terminal signal sequences are imported, via the TOM complex, into the IMS.

There they assume their native fold, thus trapping them in the IMS—the so-called **folding-trap mechanism.** Such proteins have conserved patterns of Cys and/or His residues that enable them to bind metal ion–containing cofactors in the IMS or to form disulfide bonds, both of which stabilize their native structures. [Note that the latter proteins are among the few intracellular proteins that have disulfide bonds (Section 8-4D). Evidently, the IMS has an oxidative environment.] For example, **apocytochrome c** (cytochrome c without its covalently attached heme group; Fig. 9-39) folds when the IMS-resident enzyme **cytochrome c heme lyase (CCHL)** catalyzes the attachment of its heme group, whereas Tim9, Tim10, and Tim12 each contain twin CX$_3$C motifs that form disulfide bonds.

A third class of IMS-resident proteins remain in the IMS through their association with the inner membrane, that is, they are peripheral proteins. CCHL is a member of this class of proteins.

f. Many Polytopic Inner Membrane Proteins Are First Imported to the Matrix

Many cytosolically synthesized polytopic proteins destined for insertion into the mitochondrion's inner membrane are first imported into the matrix as described above and then inserted into the inner membrane, an indirect routing that reflects the mitochondrion's bacterial origin [the primordial mitochondrion, being a gram-negative bacterium, synthesized all of its proteins in its cytoplasm (the primordial matrix) so that membrane-bound or intermembrane proteins had to be exported to these destinations]. These proteins, for the most part, are synthesized with bipartite N-terminal targeting sequences whose inner (more C-terminal) segments, once exposed by the removal of the above-described N-terminal presequence, direct the proteins to the inner membrane. The insertion of several such proteins into the inner mitochondrial membrane is mediated by the TM protein **Oxa1,** which also occupies the inner mitochondrial membrane (Fig. 12-80, *bottom right, pink*). Oxa1, which binds mitochondrial ribosomes on its matrix side, also inserts mitochondrially synthesized proteins into the inner mitochondrial membrane. As might be expected, Oxa1 is related to a protein that inserts proteins into the inner membrane of gram-negative bacteria.

g. Insertion of β Barrel Proteins Into the Outer Mitochondrial Membrane

The outer membranes of mitochondria and chloroplasts contain proteins, such as porins (Section 12-3Ad) and Tom40, that have TM β barrels. These are the only places in eukaryotic cells that TM β barrels occur, which also reflects the bacterial origins of these organelles (Sections 1-2Ac and 1-2Ag).

β barrel proteins are imported into the IMS by the TOM complex. There they are bound by the Tim9–Tim10 complex, which escorts them to the **SAM complex** (for *s*orting and *a*ssembly *m*achinery; alternatively **TOB complex** for *t*opogenesis of mitochondrial *o*uter membrane *β b*arrel), which in turn inserts them into the outer mitochondrial membrane (Fig. 12-80, *top right—purple*). The SAM complex

Table 12-6   **Characteristics of the Major Classes of Lipoproteins in Human Plasma**

	Chylomicrons	VLDL	IDL	LDL	HDL
Density (g · cm^{-3})	<0.95	<1.006	1.006–1.019	1.019–1.063	1.063–1.210
Particle diameter (Å)	750–12,000	300–800	250–350	180–250	50–120
Particle mass (kD)	400,000	10,000–80,000	5000–10,000	2300	175–360
% Protein[a]	1.5–2.5	5–10	15–20	20–25	40–55
% Phospholipids[a]	7–9	15–20	22	15–20	20–35
% Free cholesterol[a]	1–3	5–10	8	7–10	3–4
% Triacylglycerols[b]	84–89	50–65	22	7–10	3–5
% Cholesteryl esters[b]	3–5	10–15	30	35–40	12
Major apolipoproteins	A-I, A-II, B-48, C-I, C-II, C-III, E	B-100, C-I, C-II, C-III, E	B-100, C-I, C-II, C-III, E	B-100	A-I, A-II, C-I, C-II, C-III, D, E

[a]Surface components.
[b]Core lipids.

is formed by the TM β barrel–containing protein **Sam50** (alternatively, **Tob55** or **Tom55**) in association with **Sam37** (**Mas37/Tom37**), **Sam35** (**Tob35/Tom38**), and **Mdm10** (*m*itochondrial *d*istribution and *m*orphology 10). β barrel proteins are inserted into the outer membrane from its inner side, which presumably is also an evolutionary consequence of the mitochondrion's bacterial origin. Nevertheless, the TOM and SAM complexes are functionally coupled as indicated by the observation that when β barrel proteins are imported into mitochondria lacking Sam50, they accumulate in the TOM complex rather than in the IMS. Sam50 is homologous to the bacterial outer membrane protein **Omp85,** which participates in inserting β barrel proteins into the bacterial outer membrane.

5 LIPOPROTEINS

Lipids and proteins associate noncovalently to form **lipoproteins,** *which function in the blood plasma as transport vehicles for triacylglycerols and cholesterol.* In this section, we discuss the structure, function, and dysfunction of lipoproteins, and how eukaryotic cells take up lipoproteins and other specific proteins from their external medium through receptor-mediated endocytosis.

A. Lipoprotein Structure

Lipids, such as phospholipids, triacylglycerols, and cholesterol, are but sparingly soluble in aqueous solution. Hence, *they are transported by the circulation as components of lipoproteins, globular micellelike particles that consist of a nonpolar core of triacylglycerols and cholesteryl esters surrounded by an amphiphilic coating of protein, phospholipid, and cholesterol.* Lipoproteins have been classified into five broad categories on the basis of their functional and physical properties (Table 12-6):

1. Chylomicrons, which transport exogenous (externally supplied; in this case, dietary) triacylglycerols and cholesterol from the intestines to the tissues.

2–4. Very low density lipoproteins (VLDL), intermediate density lipoproteins (IDL), and low density lipoproteins (LDL), a group of related particles that transport endogenous (internally produced) triacylglycerols and cholesterol from the liver to the tissues (the liver synthesizes triacylglycerols from excess carbohydrates; Section 25-4).

5. High density lipoproteins (HDL), which transport endogenous cholesterol from the tissues to the liver.

Lipoprotein particles undergo continuous metabolic processing, so that they have variable properties and compositions (Table 12-6). Each contains just enough protein, phospholipid, and cholesterol to form an ~20-Å-thick monolayer of these substances on the particle surface (Fig. 12-83). Lipoprotein densities increase with decreasing particle diameter because the density of their outer coating is greater than that of their inner core.

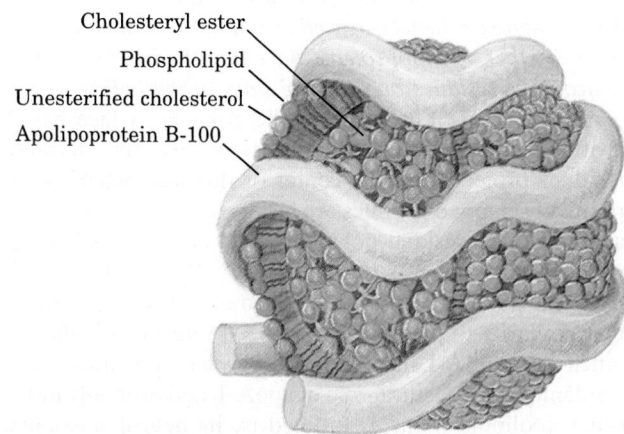

Figure 12-83   **LDL, the major cholesterol carrier of the bloodstream.** This spheroidal particle consists of some 1500 cholesteryl ester molecules surrounded by an amphiphilic coat of 800 phospholipid molecules, 500 cholesterol molecules, and a single 4536-residue molecule of apolipoprotein B-100.

Table 12-7 **Properties of the Major Species of Human Apolipoproteins**

Apolipoprotein	Number of Residues	Molecular Mass[a] (kD)	Function
A-I	243	29	Activates LCAT[b]
A-II	77	17	Inhibits LCAT, activates hepatic lipase
B-48	2152	241	Cholesterol clearance
B-100	4536	513	Cholesterol clearance
C-I	56	6.6	Activates LCAT?
C-II	79	8.9	Activates LPL[c]
C-III	79	8.8	Inhibits LPL, activates LCAT?
D	169	19	Unknown
E	299	34	Cholesterol clearance

[a]All apolipoproteins are monomers but apoA-II, which is a disulfide-linked dimer.
[b]LCAT = lecithin–cholesterol acyltransferase.
[c]LPL = lipoprotein lipase.

a. Apolipoproteins Have Amphipathic Helices That Coat Lipoprotein Surfaces

The protein components of lipoproteins are known as **apolipoproteins** or just **apoproteins.** At least nine apolipoproteins are distributed in significant amounts in the different human lipoproteins (Tables 12-6 and 12-7). Most of them are water-soluble and associate rather weakly with lipoproteins. Hence, they readily transfer between lipoprotein particles via the aqueous phase. CD measurements indicate that *apolipoproteins have a high helix content, which increases when they are incorporated in lipoproteins.* Apparently, the helices are stabilized by a lipid environment, presumably because helices fully satisfy the polypeptide backbone's hydrogen bonding potential in a lipoprotein's water-free interior.

b. The X-ray Structure of ApoA-I Mimics That in HDL

Apolipoprotein A-I (apoA-I) is HDL's major apoprotein. Sequence analysis indicates that apoA-I consists mainly of repeated amphipathic α helices of 11 or 22 residues that provide the protein's lipid-binding regions. *These putative α helices, as well as similar helices that occur in most other apolipoproteins, have their hydrophobic and hydrophilic residues on opposite sides of the helical cylinders* (Fig. 12-84). Furthermore, the polar helix face has a zwitterionic character in that its negatively charged residues project from the center of this face, whereas its positively charged residues are located at its edges. Indeed, a synthetic 22-residue polypeptide of high helix-forming propensity, which was designed by E. Thomas Kaiser to have this polarity distribution but to otherwise have minimal similarity to the repeating apoA-I sequences, behaves much like apoA-I in binding to egg lecithin liposomes. Evidently, the structural role of apoA-I, and probably most other apolipoproteins, is fulfilled by its helical segments rather than by any organized tertiary structure. This suggests that *lipoprotein α helices float on phospholipid surfaces, much like logs on water.* The phospholipids are presumably arrayed with their charged groups bound to oppositely charged residues on the polar face of the helix and with the first few methylene groups of their fatty acid residues in hydrophobic association with the nonpolar face of the helix.

A variety of criteria indicate that apoA-I undergoes significant secondary structural changes on binding lipid. However, apo Δ(1–43)A-I, a truncation mutant that lacks residues 1 to 43 of the 243-residue human apoA-I, has a

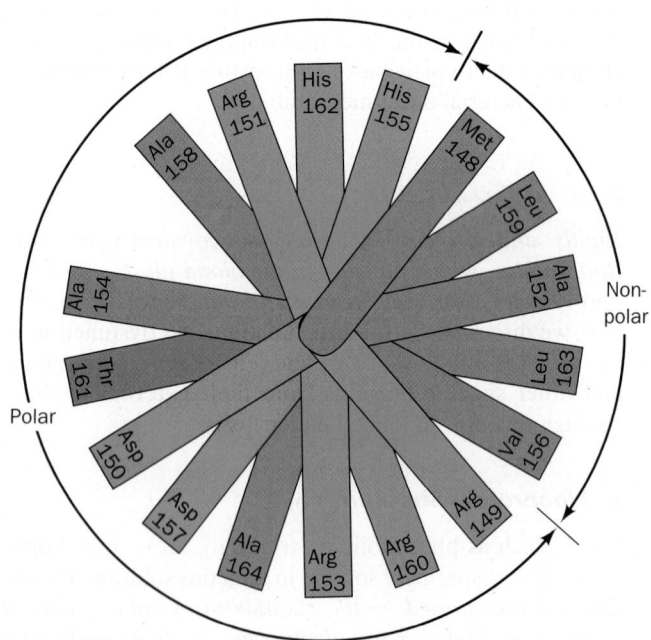

Figure 12-84 **A helical wheel projection of the amphipathic α helix constituting residues 148 to 164 of apolipoprotein A-I.** (In a helical wheel representation, the side chain positions are projected down the helix axis onto a plane.) Note the segregation of nonpolar, acidic, and basic residues to different sides of the helix. Other apolipoprotein helices have similar polarity distributions. [After Kaiser, E.T., *in* Oxender, D.L. and Fox, C.F. (Eds.), *Protein Engineering*, p. 194, Liss (1987).]

(a)

(b)

Figure 12-85 **X-ray structure of human apo Δ(1–43)A-I.** The four monomers of the D_2-symmetric tetramer it forms are drawn in different colors. (*a*) View along the 2-fold axis relating the cyan and yellow subunits and the green and magenta subunits. (*b*) View from the top of Part *a* along the 2-fold axis relating the cyan and magenta subunits and the green and yellow subunits. The third pairings, cyan with green and magenta with yellow, which interact along most of their lengths, probably maintain their identities in HDL particles, whereas the other pairings, whose interactions are less extensive, are unlikely to do so. [Based on an X-ray structure by David Borhani, Southern Research Institute, Birmingham, Alabama, and Christie Brouillette, University of Alabama Medical Center, Birmingham. PDBid 1AV1.]

conformation that closely resembles that of lipid-bound apoA-I, whether or not lipid is present. Lipid-free apo Δ(1–43)A-I is therefore likely to provide a valid structural model for lipid-bound apoA-I.

The X-ray structure of apo Δ(1–43)A-I (Fig. 12-85) was determined by David Borhani and Christie Brouillette. It revealed that, over most of its length, each polypeptide chain forms a pseudocontinuous amphipathic α helix that is punctuated by kinks at Pro residues that are spaced mainly at intervals of 22 residues to form 10 helical segments arranged in the shape of a twisted horseshoe. Two such monomers (e.g., the cyan and magenta monomers in Fig. 12-85) associate in an antiparallel fashion along most of their lengths to form a dimer that has the shape of a twisted elliptical ring. Two such dimers, in turn, associate via their hydrophobic surfaces to form an elliptical tetramer with D_2 symmetry that has outer dimensions of 135×90 Å and an inner hole of 95×50 Å. The surface of this tetrameric ring, which consists of up–down–up–down 4-helix bundles over about three-fourths of its circumference, is hydrophilic with a uniform electrostatic potential, whereas the interior of each 4-helix bundle contains mainly Val and Leu side chains. Since, in this conformation, these hydrophobic residues are unavailable for binding to lipid, it is postulated that they associate in the lipid-free crystal so as to shelter the lipid-binding face of apo Δ(1–43)A-I dimers from contact with water (which fills the spaces in the crystal).

The sizes and shapes of the apo Δ(1–43)A-I dimer and tetramer seem ideal for wrapping around the 50- to 120-Å-diameter HDL particles. Since HDL particles often contain two or four apoA-I monomers, it is proposed that when pairs of apoA-I monomers bind to HDL, they do so as the above-described antiparallel dimer. Its exposed nonpolar side chains could then hydrophobically interact with the HDL particle's buried nonpolar groups. Two such dimers could associate on the surface of an HDL particle to form a tetramer, although, most probably, in a different manner than is seen in the structure of apo Δ(1–43)A-I.

B. *Lipoprotein Function*

The various lipoproteins have different physiological functions, as we discuss below.

a. Chylomicrons Are Delipidated in the Capillaries of Peripheral Tissues

Chylomicrons, which are assembled by the intestinal mucosa, function to keep exogenous triacylglycerols and cholesterol suspended in aqueous solution. These lipoproteins are released into the intestinal lymph (known as **chyle**), which is transported through the lymphatic vessels before draining into the large body veins via the thoracic duct. After a fatty meal, the otherwise clear chyle takes on a milky appearance.

Chylomicrons adhere to binding sites on the inner surface (endothelium) of the capillaries in skeletal muscle and adipose tissue. There, within minutes after entering the bloodstream, the chylomicron's component triacylglycerols are hydrolyzed through the action of **lipoprotein lipase (LPL),** an extracellular enzyme that is activated by **apoC-II.** The tissues then take up the liberated monoacylglycerol and fatty acid hydrolysis products. The chylomicrons shrink as their triacylglycerols are progressively hydrolyzed until they are reduced to cholesterol-enriched **chylomicron remnants.** The chylomicron remnants reenter the circulation by dissociating from the capillary endothelium and are subsequently taken up by the liver, as is

Figure 12-86 **Model for plasma triacylglycerol and cholesterol transport in humans.** [After Brown, M.S. and Goldstein, J.L., *in* Brunwald, E., Isselbacher, K.J., Petersdorf, R.G., Wilson, J.D., Martin, J.B., and Fauci, A.S. (Eds.), *Harrison's Principles of Internal Medicine* (11th ed.), *p.* 1652, McGraw-Hill (1987).]

🔊 **See the Animated Figures**

explained below. *Chylomicrons therefore function to deliver dietary triacylglycerols to muscle and adipose tissue and dietary cholesterol to the liver* (Fig. 12-86, *left*).

b. VLDL Are Degraded Much Like Chylomicrons

VLDL, which are synthesized in the liver as lipid transport vehicles, are also degraded by lipoprotein lipase (Fig. 12-86, *right*). The VLDL remnants appear in the circulation, first as IDL and then as LDL. In the transformation of VLDL to LDL, all their proteins but **apoB-100** are removed and much of their cholesterol is esterified by the HDL-associated enzyme **lecithin–cholesterol acyltransferase (LCAT),** as is discussed below. This enzyme transfers a fatty acid residue from atom C2 of lecithin to cholesterol with the concomitant formation of **lysolecithin** (Fig. 12-87).

ApoB-100, a 4536-residue monomeric glycoprotein (and thus one of the largest monomeric proteins known), has a hydrophobicity approaching that of integral proteins and contains relatively few amphipathic helices. Hence, in contrast to the other, less hydrophobic plasma apolipoproteins, apoB-100 is neither water-soluble nor transferred between lipoprotein particles. Each LDL particle contains but one molecule of apoB-100, which immunoelectron microscopy indicates has an extended form that covers at least half of the particle surface (Fig. 12-83). Chylomicrons, however, contain **apoB-48,** a 2152-residue protein that is identical in sequence to the N-terminal 48% of apoB-100. Indeed, both proteins are encoded by the same gene. The remarkable mechanism by which this gene expresses different length proteins in liver and intestines is discussed in Section 31-4Ar.

c. Cells Take Up Cholesterol through Receptor-Mediated Endocytosis of LDL

Cholesterol, as we have seen, is an essential component of animal cell membranes. The cholesterol may be externally supplied or, if this source is insufficient, internally synthesized (Section 25-6A). Michael Brown and Joseph Goldstein have demonstrated that *cells obtain exogenous cholesterol*

Figure 12-87 Reaction catalyzed by lecithin–cholesterol acyltransferase (LCAT). The transferred acyl group is most often a linoleic acid residue.

*mainly through the endocytosis (engulfment) of LDL in complex with **LDL receptor (LDLR),** a cell-surface transmembrane glycoprotein that specifically binds apoB-100. LDLR also binds chylomicron remnants via their **apoE** components.*

LDLR is an 839-residue glycoprotein that has a 767-residue N-terminal ligand-binding **ectodomain** (extracellular domain; Greek: *ectos,* outside), a 22-residue TM segment that presumably forms an α helix, and a 50-residue C-terminal cytoplasmic domain. The X-ray structure of

LDLR's ectodomain, determined by Brown, Goldstein, and Deisenhofer at pH 5.3 (for reasons explained below), confirms the results of sequence studies indicating that this protein consists of, from N- to C-terminus, seven tandemly repeated ~40-residue Cys-rich modules, two ~40-residue EGF-like domains (EGF is the acronym for **epidermal growth factor,** a hormonally active polypeptide that stimulates cell proliferation; Section 19-3), a six-bladed β propeller domain, and an EGF-like domain (Fig. 12-88). The Cys-rich modules, designated R2 to R7 (R1 is disordered and hence unseen), are arranged in an ~140-Å-long arc that loops around to the EGF-like domains

Figure 12-88 X-ray structure of the extracellular domain of the human LDL receptor at 3.7 Å resolution. The protein is drawn in ribbon form with each of its observed modules given a different color. Its eight bound Ca²⁺ ions are represented by cyan spheres, and two N-linked carbohydrates (a tetrasaccharide and a pentasaccharide) are shown in stick form with C green, N blue, and O red. Disulfide linkages are drawn in yellow. The Cys-rich modules are labeled R2 through R7 and the EGF-like domains are labeled EGF-A through EGF-C. The absence of the R1 module and the fragmented appearance of the R2 and R3 modules are due to the disorder of the missing segments. [After an X-ray structure by Michael Brown, Joseph Goldstein, and Johann Deisenhofer, University of Texas Southwest Medical Center, Dallas, Texas. PDBid 1N7D.]

Figure 12-89   **Freeze-etch electron micrograph of coated pits on the inner surface of a cultured fibroblast's plasma membrane.** Compare this figure with that of clathrin-coated vesicles (Fig. 12-59*a*). [Courtesy of John Heuser, Washington University School of Medicine, St. Louis, Missouri.]

and the β propeller. Each Cys-rich module binds a Ca^{2+} ion and consists of two loops connected by three disulfide bonds. Note that they are devoid of regular secondary structure (helices and sheets). The EGF-like domains likewise each bind a Ca^{2+} ion (except for EGF-C), have three disulfide bonds, and lack regular secondary structure. Modules R4 and R5, which are critical for ligand binding, bind to one face of the β propeller via extensive and conserved side chain interactions. The Cys-rich modules each have a somewhat different conformation, which suggests that they are pliable. Moreover, modules R2, R3,

R6, and R7, and presumably R1, which are also implicated in LDL binding, appear to be unconstrained by interactions with the rest of the protein. This presumably explains why LDLR can bind lipoproteins of varying sizes and compositions.

LDLRs cluster into **coated pits,** which serve to gather the cell-surface receptors that are destined for endocytosis while excluding other cell-surface proteins. The coated pits, which have a clathrin backing (Fig. 12-89), invaginate into the plasma membrane to form clathrin-coated vesicles (Fig. 12-90; Section 12-4C) that subsequently fuse with lysosomes. *Such **receptor-mediated endocytosis** (Fig. 12-91) is a general mechanism whereby cells take up large molecules, each through a corresponding specific receptor.* Indeed, the liver takes up chylomicron remnants in this manner through the mediation of a separate **remnant receptor** that specifically binds apoE.

At neutral pH, LDLR binds LDL via its Cys-rich modules, most importantly R4 and R5. However, in the acidic environment of the endosome, LDLR releases its bound LDL (Fig. 12-91). The X-ray structure of LDLR at pH 5.3 (Fig. 12-88), the pH of the endosome, suggests that this occurs via LDL's displacement from modules R4 and R5 by LDLR's β barrel domain. This model is bolstered by the observation that, at pH 5.3, the interface between modules R4 and R5 and the β barrel domain contains several conserved His-containing salt bridges that presumably form only when these His residues are protonated. Moreover, an LDLR construct in which the EGF-like domains and the β propeller domain have been deleted readily binds LDL but does not release it at acidic pH. Thus it is likely that at neutral pH, LDLR assumes an open and flexible conformation in which the R4 and R5 modules do not associate with the β barrel domain but, instead, are available to bind LDL.

(a)

(b)

Figure 12-90   **Electron micrographs showing the endocytosis of LDL by cultured human fibroblasts.** LDL was conjugated to ferritin so that it appears as dark dots. (*a*) LDL bound to a coated pit on the cell surface. (*b*) The coated pit invaginates and begins

to pinch off from the cell membrane to form a coated vesicle enclosing the bound LDL. [From Anderson, R.G.W., Brown, M.S., and Goldstein, J.L., *Cell* **10,** 356 (1977). Copyright © 1977 by Cell Press.]

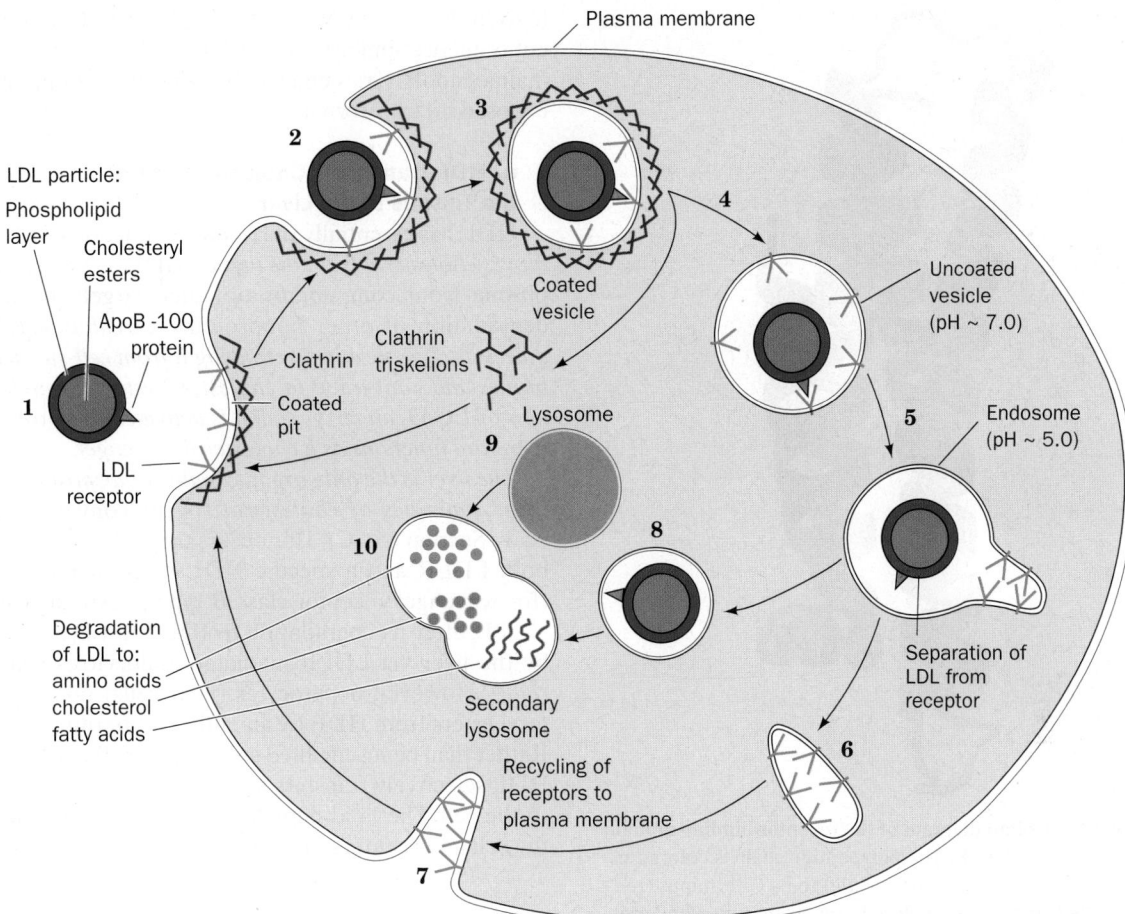

Figure 12-91 **Sequence of events in the receptor-mediated endocytosis of LDL.** LDL specifically binds to LDL receptors (LDLRs) on clathrin-coated pits **(1).** These bud into the cell **(2)** to form coated vesicles **(3),** whose clathrin coats depolymerize as triskelions, resulting in the formation of uncoated vesicles **(4).** These vesicles then fuse with vesicles called **endosomes (5),** which have an internal pH of ~5.0. The acidity induces LDL to dissociate from LDLR. LDL accumulates in the vesicular portion of the endosome, whereas LDLR concentrates in the membrane of an attached tubular structure, which then separates from the endosome **(6)** and subsequently recycles LDLR to the plasma membrane **(7).** The vesicular portion of the endosome **(8)** fuses with a lysosome **(9),** yielding a **secondary lysosome (10),** wherein the apoB-100 component of LDL is degraded to its component amino acids and the cholesteryl esters are hydrolyzed to yield cholesterol and fatty acids. An LDLR molecule cycles in and out of the cell every 10 to 20 minutes during its ~20-hour lifetime.

In the lysosome, as demonstrated by radioactive labeling studies, LDL's apoB-100 is rapidly degraded to its component amino acids (Fig. 12-91). The cholesteryl esters are hydrolyzed by a lysosomal lipase to yield cholesterol, which is subsequently incorporated into the cell membranes. Any excess intracellular cholesterol is reesterified for storage within the cell through the action of **acyl-CoA:cholesterol acyltransferase (ACAT).**

The overaccumulation of cellular cholesteryl esters is prevented by two feedback mechanisms:

1. High intracellular levels of cholesterol suppress the synthesis of LDLR, thus decreasing the rate of LDL accumulation by endocytosis (although LDLR cycles in and out of the cell every 10 to 20 min, it is slowly degraded by the cell such that its half-life is ~20 h).

2. Excess intracellular cholesterol inhibits the biosynthesis of cholesterol (Section 25-6Bb).

d. ApoE's Receptor Binding Domain Contains a Four-Helix Bundle

ApoE is a 299-residue monomeric protein that consists of two independently folded domains: an N-terminal domain that binds strongly to LDLR but only weakly to lipid and a C-terminal domain that binds to the lipoprotein surface but lacks affinity for LDLR. Proteolysis of apoE yields fragments corresponding to apoE's N-terminal domain

Figure 12-92 Ribbon diagram of the receptor-binding domain of human apolipoprotein E. The polypeptide chain is colored in rainbow order from its N-terminus (*blue*) to its C-terminus (*red*). Residues 61, 112, and 158 are colored orange, magenta, and gray, respectively. [Based on an X-ray structure by David Agard, University of California at San Francisco. PDBid 1LPE.]

(residues 1–191) and C-terminal domain (residues 216–299). Sequence analysis suggests that the C-terminal domain is largely composed of helices. The X-ray structure of the N-terminal domain (Fig. 12-92), determined by David Agard, reveals that it consists mainly of five α helices, four of which form an elongated (65 Å) up–down–up–down four-helix bundle. The helices of the four-helix bundle, as expected, are strongly amphipathic, with their hydrophobic residues sequestered inside the protein, out of contact with solvent, whereas their hydrophilic residues are solvent-exposed. The structure appears to be further stabilized by numerous salt bridges on the protein's highly charged surface.

The C-terminal helix of the apoE N-terminal fragment contains nine closely spaced basic residues that are not involved in salt bridges, thereby producing a large positively charged patch on the protein surface. ApoE variants in which one of these basic residues is replaced by a neutral or acidic residue all display reduced affinity for LDLR, thereby suggesting that the patch forms apoE's binding site for LDLR. Hence this C-terminal helix has been dubbed apoE's receptor-binding helix.

LDLR binds apoB-100 and apoE with comparable affinities. ApoB-100 (but not apoB-48) contains a conserved segment that is similar to apoE's receptor-binding

helix, although the two proteins otherwise have no apparent sequence similarity. In VLDL, the receptor-binding domain of apoB-100 is unavailable for receptor binding but is exposed on transformation of the VLDL to LDL.

e. HDL Transports Cholesterol from the Tissues to the Liver

HDL has essentially the opposite function of LDL: *It removes cholesterol from the tissues.* HDL is assembled in the plasma from components obtained largely through the degradation of other lipoproteins. *Circulating HDL acquires its cholesterol by extracting it from cell-surface membranes and converts it to cholesteryl esters through the action of LCAT, an enzyme that is activated by apoA-I. HDL therefore functions as a cholesterol scavenger.*

The liver is the only organ capable of disposing of significant quantities of cholesterol (by its conversion to bile acids; Section 25-6C). This occurs through the mediation of both LDLR and a specific HDL receptor named **SR-BI** (for *scavenger receptor class B type I*). About half of the VLDL, after its degradation to IDL and LDL, is taken up by the liver via LDLR-mediated endocytosis (Fig. 12-86, *right*). However, hepatocytes (liver cells) take up cholesteryl esters from HDL by an entirely different mechanism: Rather than being engulfed and degraded, the SR-BI–bound HDL selectively transfers its component cholesteryl esters to the cell. The lipid-depleted HDL then dissociates from the cell and reenters the circulation.

C. *Lipoprotein Dysfunction in Atherosclerosis and Alzheimer's Disease*

Atherosclerosis, the most common form of **arteriosclerosis** (hardening of the arteries), is characterized by the presence of **atheromas** (Greek: *athera,* mush), arterial thickenings that, on sectioning, exude a pasty yellow deposit of almost pure cholesteryl esters (Fig. 12-93).

Atherosclerosis is a progressive disease that begins as intracellular lipid deposits in the smooth muscle cells of the inner arterial wall. These lesions eventually become fibrous, calcified plaques that narrow and even block the arteries. The resultant roughening of the arterial wall promotes the formation of blood clots, which may also occlude the artery. A blood flow stoppage, known as an **infarction,** causes the death of the deprived tissues. Although atheromas can occur in many different arteries, they are most common in the coronary arteries, the arteries supplying the heart. This results in **myocardial infarctions** or "heart attacks," the most common cause of death in Western industrialized countries.

a. Deficient LDL Receptors Result in Atherosclerosis

The development of atherosclerosis is strongly correlated with the level of plasma cholesterol. This is particularly evident in individuals with **familial hypercholesterolemia (FH).** Homozygotes with this inherited disorder have such high levels of the cholesterol-rich LDL (which is often referred to as "bad cholesterol") in their plasma that their plasma cholesterol levels are three- to fivefold greater than the average level of $\sim$175 mg $\cdot$ 100 mL^{-1}. This situation results in the

Figure 12-93   An atherosclerotic plaque in a coronary artery.
The vessel wall is dramatically thickened as a result of lipid
accumulation and activation of inflammatory processes. [© Eye
of Science/Photo Researchers.]

deposition of cholesterol in their skin and tendons as yellow
nodules known as **xanthomas.** However, far greater damage
is caused by the rapid formation of atheromas that, in homozygotes, cause death from myocardial infarction as early
as the age of 5. Heterozygotes, which comprise ~1 person in
500, are less severely afflicted; they develop symptoms of
coronary artery disease after the age of 30.

*Cells taken from FH homozygotes completely lack func-
tional LDLR, whereas those taken from heterozygotes have
about half of the normal complement. Homozygotes and, to
a lesser extent, heterozygotes are therefore unable to utilize
the cholesterol in LDL.* Rather, their cells must synthesize
most of the cholesterol for their needs. The high level of
plasma LDL in these individuals results from two related
causes:

1. Its decreased rate of degradation because of the lack
of LDLR.

2. Its increased rate of synthesis from IDL due to the
failure of LDLR to take up IDL.

Over 1000 mutations of LDLR that cause FH have been
discovered. These mutants have been grouped into five
classes depending on the nature of the defect in LDLR functioning they cause (Fig. 12-91): (1) failure to produce detectable amounts of protein; (2) partial or complete failure
to be transported to the plasma membrane; (3) impaired lig-
and binding; (4) failure to localize to clathrin-coated pits and
in internalization; and (5) defects in ligand release and recycling. Class 4 mutants are caused by changes in LDLR's so-
called **sorting sequence,** NPXR, in its cytoplasmic domain,
which binds to AP2 in clathrin-coated pits.

*The long-term ingestion of a high-fat/high-cholesterol diet
has an effect similar to although less severe than FH.* A high
intracellular level of cholesterol supresses the synthesis of
LDLR (Section 25-6Bb), thereby reducing the amount of
LDL that a cell takes up from the circulation. Excessive dietary cholesterol, delivered to the tissues via chylomicrons,
therefore contributes to high plasma LDL levels.

b. Scavenger Receptors Take Up Oxidized LDL

Atherosclerotic plaques in individuals with FH contain
macrophages (a type of white blood cell that ingests and, if
possible, destroys a variety of foreign and endogenous substances) that contain so much cholesterol that they are
known as **foam cells.** How do macrophages take up cholesterol? Macrophages from both normal individuals and
those with FH have few LDLRs and therefore take up little native LDL. However, they avidly take up, by endocytosis, LDL that has been chemically modified by the acetylation of its Lys residues (which eliminates this side chain's
positive charge, thereby increasing LDL's negative
charge). The macrophage cell-surface receptors that bind
acetylated LDL are known as **scavenger receptors** because
they also bind certain other polyanionic molecules.

Scavenger receptors avidly take up oxidized LDL. The
unsaturated fatty acids of LDL are highly susceptible to
chemical oxidation but, in the blood, are protected from oxidation by antioxidants. However, these antioxidants are
thought to become depleted when LDL is trapped for an
extended time within the artery walls (where it is thought to
gain access by an injury to the arterial lining through which
plasma leaks into the arterial wall). As a consequence, oxygen radicals convert LDL's unsaturated fatty acids to aldehydes and oxides that react with its Lys residues, thereby
mimicking acetylation. The ingested LDL is degraded as described above and its cholesterol converted to cholesteryl
esters, which accumulate as insoluble residues.

The physiological significance of this scenario has been
demonstrated by the observations that antibodies against
aldehyde-conjugated Lys residues stain atherosclerotic
plaques, that LDL from atherosclerotic plaques binds to
scavenger receptors and produces foam cells *in vitro,* and that
antioxidants inhibit atherosclerosis in rabbits that have an
animal counterpart of FH. It should be noted that tobacco
smoke oxidizes LDL, which may explain why smoking leads
to an increased incidence of atherosclerosis. High plasma levels of LDL, of course, also accelerate LDL uptake.

If this model of atheroma formation is correct, then *the
optimal level of plasma LDL is the lowest concentration
that can adequately supply cholesterol to the cells.* Such a
level, which is thought to be ~25 mg of cholesterol · 100
mL^{-1}, occurs in various mammalian species that are not
naturally susceptible to atherosclerosis as well as in newborn humans. Yet the plasma LDL level in adult Western
men averages ~7-fold higher than this supposed optimal
level. The reasons for such a high plasma cholesterol level
are poorly understood (but see below), although it is clear
that it is affected by diet and by environmental stress. Medical strategies for reducing the level of plasma cholesterol
are considered in Section 25-6Bd.

c. Atherosclerosis Is a Multifactorial Disease

*Epidemiological studies indicate that high plasma levels
of HDL (which is often referred to as "good cholesterol")
are strongly correlated with a low incidence of cardiovascular disease.* Women have HDL levels higher than men and
also less heart disease. Many of the factors that decrease
the incidence of heart disease also tend to increase HDL

levels. These include strenuous exercise, weight loss, certain drugs such as alcohol, and female sex hormones known as **estrogens** (Section 19-1Gb). Conversely, cigarette smoking is inversely related to HDL concentration. Curiously, in communities that have a very low incidence of coronary artery disease, both the mean HDL and LDL concentrations are low. The reasons for these various effects are unknown.

There is also a strong inverse correlation in humans between the risk for atherosclerosis and the plasma level of apoA-I, HDL's major protein component, which is required for its assembly. To investigate whether apoA-I has a direct antiatherogenic effect, mice of a strain that develops diet-induced **fatty streak lesions** in their large blood vessels were genetically modified to express high plasma levels of human apoA-I (fatty streak lesions are the precursors of atherosclerotic plaques, which these mice have too short a lifetime to develop). These transgenic mice are significantly protected from developing fatty streak lesions. Yet transgenic mice that overexpress mouse **apoA-II,** another major HDL protein, develop more and larger fatty streak lesions then their nontransgenic counterparts. Since the plasma HDL–cholesterol levels in the latter transgenic mice are significantly elevated, it appears that both the composition and the level of plasma HDL are important atherosclerotic mediators. Similarly, transgenic mice that express high levels of human apoE or human LDLR resist the elevation in plasma LDL levels that would otherwise be brought on by a cholesterol-rich diet, whereas mice in which the gene encoding apoE has been knocked out rapidly develop atherosclerotic lesions.

Cholesteryl ester transfer protein (CETP) is a plasma protein that exchanges neutral lipids (e.g., cholesteryl esters and triacylglycerols) among lipoproteins and hence functions analogously to phospholipid exchange proteins (Section 12-4Ab). Since VLDL and LDL are triacylglycerol-rich whereas HDL are cholesteryl ester–rich (Table 12-6), CETP mediates the net transport of cholesteryl esters from HDL to VLDL and LDL (and of triacylglycerols in the opposite direction). Consequently, animals that express CETP have higher cholesterol levels in their VLDL and LDL and lower cholesterol levels in their HDL than animals that do not express CETP. Mice of a strain that normally have little or no CETP activity were made transgenic for CETP and fed an atherogenic (high-fat, high-cholesterol) diet. These transgenic mice developed atherosclerotic lesions far more rapidly than their similarly fed nontransgenic counterparts. Since the two types of mice had similar total plasma cholesterol levels, these results suggest that *the progression of atherosclerotic lesions is more a function of how cholesterol is partitioned between lipoproteins than it is of the total plasma cholesterol level.*

Increased risk of atherosclerosis in humans is also associated with elevated plasma levels of lipoprotein **Lp(a),** a variant of LDL in which apoB-100 is tightly associated with the 4259-residue plasma protein **apo(a).** Rodents and most other nonprimate mammals lack the gene for apo(a). However, mice transgenic for human apo(a) rapidly develop fatty streak lesions when given a high-fat diet (approximating human diets in industrialized Western countries).

Apo(a) mainly consists of repeated segments that are homologous to **plasminogen,** a plasma protein that, when activated, functions to proteolytically dismantle blood clots (Section 35-1Fa). The normal function of apo(a) in humans is unknown, although it has been hypothesized that it participates in healing blood vessel wounds.

d. Tangier Disease Eliminates HDL Synthesis

Most cells do not consume cholesterol by converting it to steroid hormones or bile acids, for example, but all cells require cholesterol to maintain membrane fluidity. Cholesterol in excess of this requirement can be esterified by the action of ACAT and stored as cholesteryl esters in intracellular deposits. Cholesterol can also be eliminated from cells by a mechanism illuminated through studies of individuals with **Tangier disease.** In this recessive inherited disorder, almost no HDL is produced, because cells have a defective transport protein, known as **ATP-cassette binding protein A1 (ABCA1).** In normal individuals, ABCA1 functions as a flippase (Section 12-4Aa) that transfers cholesterol, cholesteryl esters, and other lipids from the inner to the outer leaflet of the plasma membrane, from where they are captured by apolA-I to form HDL. Cells lacking ABCA1 cannot dispose of their excess cholesterol and therefore accumulate cholesteryl esters in their cytoplasm. Macrophages thus engorged with lipids contribute to the development of atherosclerosis and consequently individuals with Tangier disease exhibit symptoms similar to those with FH.

e. ApoE4 Is Implicated in Both Cardiovascular Disease and Alzheimer's Disease

There are three common allelic variants of apoE in humans: **apoE2** (occurring in 15% of the population), which has Cys at positions 112 and 158; **apoE3** (78% occurrence), in which these residues are Cys and Arg, respectively (Fig. 12-92 shows the structure of apoE3 with residues 112 and 158 magenta and white); and **apoE4** (7% occurrence), in which these residues are both Arg. These differences have medical significance: ApoE3 has a preference for binding to HDL, whereas apoE4 has a preference for binding to VLDL, which is probably why apoE4 is associated with elevated plasma concentrations of LDL and thus an increased risk of cardiovascular disease. Evidently, changes in apoE's N-terminal domain can affect the function of its C-terminal lipoprotein-binding domain.

ApoE4, as we have seen in Section 9-5B, is also associated with a greatly (16-fold) increased incidence of Alzheimer's disease (AD). This observation is perhaps less surprising when it is realized that apoE is expressed by certain nerve cells and is present in the cerebrospinal fluid, where it functions in mediating cholesterol transport, much as it does in blood plasma (cholesterol is abundant in nerve cell plasma membranes, which mediate neurotransmission; Section 20-5B).

Brain tissue from AD victims reveals numerous extracellular amyloid plaques, which consist of fibrillar deposits of amyloid β (Aβ) peptide that arises through proteolysis of the normally occurring amyloid precursor protein (Section 9-5B). Amyloid plaques appear to be AD's pathogenic agent. Immunochemical staining indicates that apoE is associated with amyloid plaques. *In vitro* experiments demonstrate that both apoE3 and apoE4 form SDS-stable complexes with Aβ peptide that, after long incubation times, aggregate and precipitate from solution as a matrix of fibrils that closely resemble those in amyloid plaques. ApoE4 forms this complex more readily than apoE3 and yields a denser, more extensive matrix.

Comparison of the X-ray structures of apoE4 and apoE3 reveals that there are only minor differences in their backbone conformations, which are restricted to the immediate vicinity of their site of difference (residue 112: Cys in ApoE3 and Arg in ApoE4). The only two side chains in apoE4 that undergo changes in conformation relative to those in apoE3 are Glu 109, which swings around in apoE4 to form a salt bridge with Arg 112, and Arg 61 (orange in Fig. 12-92), which contacts Cys 112 in apoE3 but swings away to accommodate the new salt bridge in apoE4. Thus, both Glu 109 and Arg 61 are candidates for mediating the functional differences between apoE3 and apoE4. However, the mutagenic substitution of Ala for Glu 109 in apoE3 does not significantly alter its preference for binding to HDL over VLDL. In contrast, the substitution of Thr for Arg 61 in apoE4 gives this protein an apoE3-like preference for HDL over VLDL. Evidently, the position of Arg 61 is critical in determining the HDL/VLDL preference of apoE. This hypothesis is supported by the observation that residue 61 is invariably Thr in the 10 apoEs of known sequences from other species. None of these species exhibits the complete pathology of AD, although it remains to be demonstrated that Arg 61 actually contributes to the differential binding of apoE3 and apoE4 to Aβ peptide.

f. ApoE2 Has a Low Affinity for LDL Receptor

ApoE2 binds to LDLR with only 0.1% of the affinity of apoE3 or apoE4. Thus the presence of apoE2 is the underlying cause of **familial type III hyperlipoproteinemia,** which is characterized by elevated plasma cholesterol and triglyceride levels and hence accelerated coronary artery disease.

The defective binding of apoE2 to LDLR is caused by the substitution of Cys for Arg 158, a position (gray in Fig. 12-92) that lies outside the previously identified receptor-binding region, residues 136 to 150 (located in the bottom half of the C-terminal helix in Fig. 12-92). In apoE3, Asp 154 forms a salt bridge with Arg 158 (which is situated one turn farther along the α helix). In apoE2, since Arg 158 is replaced with Cys, this salt bridge cannot form. Rather, as the X-ray structure of apoE2 reveals, Asp 154 forms a salt bridge with Arg 150 (which is situated one turn earlier on the α helix), thereby altering the side chain conformation of this LDLR-binding residue. In fact, the disruption of this abnormal salt bridge by the mutagenic replacement of Asp 154 in ApoE2 with Ala restores LDLR binding affinity to a nearly normal level.

Individuals with type III hyperlipoproteinemia are particularly responsive to a low-fat, low-calorie diet and to a reduction in body weight. It is therefore postulated that the altered lipid composition of lipoproteins under such a regimen causes significant amounts of apoE2 to assume a receptor-active conformation, leading to normal or near-normal rates of lipoprotein clearance from the circulation.

CHAPTER SUMMARY

1 Lipid Classification Fatty acids are long-chain carboxylic acids that may have one or more double bonds that are usually cis. Their anions are amphiphilic molecules that form micelles in water. Fatty acids rarely occur free in nature but rather are components of lipids. The most abundant class of lipids, the triacylglycerols or neutral fats, are nonpolar molecules that constitute the major nutritional store of animals. The lipids that occur in membranes are the phospholipids, the sphingolipids, and, in eukaryotes, cholesterol or similar sterols. Sphingolipids such as cerebrosides and gangliosides have complex carbohydrate head groups that act as specific recognition markers in various biological processes.

2 Properties of Lipid Aggregates The molecular shapes of membrane lipids cause them to aggregate in aqueous solution as bilayers. These form closed vesicles known as liposomes that are useful model membranes and drug delivery systems. Bilayers are essentially impermeable to polar molecules, except for water. Likewise, the flip-flop of a lipid in a bilayer is an extremely rare event. In contrast, bilayers above their transition temperatures behave as two-dimensional fluids in which the individual lipid molecules freely diffuse in the bilayer plane. Cholesterol decreases membrane fluidity and broadens the temperature range of its order–disorder transition by interfering with the orderly packing of the lipids' fatty acid side chains.

3 Biological Membranes Biological membranes contain a high proportion of proteins. Integral proteins, for example, bacteriorhodopsin, the photosynthetic reaction center, porins, and fatty acid amide hydrolase, have nonpolar surface regions that hydrophobically associate with the bilayer core. Peripheral proteins, for example, cytochrome *c,* bind to integral proteins on the membrane surface or to phospholipid head groups via polar interactions. Specific integral proteins are invariably associated with a particular side of the membrane or, if they are transmembrane proteins, have only one orientation. Lipid-linked proteins contain covalently attached isoprenoid, fatty acyl, and/or glycosylphosphatidylinositol (GPI) groups that serve to anchor these proteins to membranes and to mediate protein–protein interactions. According to the fluid mosaic model of membrane structure, integral proteins resemble icebergs floating on a two-dimensional lipid sea. These proteins, as observed by the freeze-fracture and freeze-etch techniques, are randomly distributed in the membrane. Certain lipids and/or proteins may form specific aggregates on one leaflet of a membrane.

The erythrocyte cytoskeleton is responsible for the shape, flexibility, and fluidity of the red cell. Spectrin, the major constituent of the cytoskeleton, is a wormlike $(\alpha\beta)_2$ heterotetramer that is cross-linked by actin oligomers and band 4.1 protein. The resulting protein meshwork is anchored to the membrane by the association of spectrin with ankyrin, which, in turn, binds to band 3 protein, a transmembrane protein that forms an anion channel.

The erythrocyte surface bears the various blood group antigens. The antigens of the ABO system differ in the sugar at a nonreducing end. The ABO blood group substances occur in the plasma membranes of many cells and in the secretions of many individuals.

Gap junctions are hexagonal transmembrane protein tubes that link adjoining cells. The gap junction's central channel, which closes at high intracellular levels of Ca^{2+}, allows small molecules and ions but not macromolecules to pass between cells. The connexin subunits of a gap junction's two face-to-face hexameric connexons each contain four transmembrane helices.

Bacterial channel-forming toxins such as α-hemolysin form oligomers on the outer surface of a target cell's plasma membrane. These insert themselves into the membrane to form pores through which small molecules and ions leak, thereby killing the cell.

4 Membrane Assembly and Protein Targeting New membranes are generated by the expansion of old ones. Lipids are synthesized by membrane-bound enzymes and are deposited on one side of the membrane. They migrate to the other side by flip-flops that are catalyzed by membrane-bound flippases and phospholipid translocases. In eukaryotes, lipids are transported between different membranes by lipid vesicles or by phospholipid-exchange proteins.

In the secretory pathway, transmembrane proteins and proteins destined for secretion are ribosomally synthesized with an N-terminal signal sequence. A signal peptide is bound by an RNA-containing signal recognition particle (SRP), which then arrests polypeptide synthesis. The SRP–ribosome complex binds to the SRP receptor (SR) in complex with the translocon on the endoplasmic reticulum (ER) membrane and, on GTP hydrolysis by both the SRP and SR, resumes polypeptide synthesis. As a protein destined for secretion passes through the translocon into the ER lumen, its signal peptide is removed by an ER-resident signal peptidase, its folding is facilitated through interactions with ER-resident chaperone proteins such as BiP, and its post-translational processing, mainly signal peptide excision and glycosylation, is initiated. Integral proteins, whose transmembrane (TM) segments each contain signal-anchor sequences, also enter the translocon, which laterally installs these TM segments into the ER membrane. The orientation of TM helices in membranes usually obeys the positive-inside rule. Some proteins are wholly synthesized in the cytoplasm before being translocated into the ER.

Proteins are transferred between the ER, the Golgi apparatus (where further post-translational processing takes place), and their final destinations via membranous vesicles that are coated with clathrin, COPI, or COPII. Clathrin-coated vesicles also participate in endocytosis. Polyhedral clathrin cages are formed by triskelions, which are trimers of heavy chains, each of which binds a light chain. Clathrin-coated vesicle formation is primed by the action of ARNO, a guanine nucleotide exchange factor (GEF) that induces the small GTPase ARF1 to exchange its bound GDP for GTP and then insert its myristoyl group into the membrane. ARF1 · GTP then recruits adapter proteins such as AP1 and AP2, which simultaneously bind clathrin heavy chains and TM proteins that are cargo proteins or are receptors for soluble cargo proteins inside the vesicle. The formation of the clathrin cage drives vesicle budding, but its actual release from its parent membrane requires the action of the GTPase dynamin. Shortly after its release, the vesicle uncoats in a process mediated by the chaperone protein Hsc70. COPI- and COPII-coated vesicles undergo similar processes, although they do not require a dynamin-like protein to bud off from their parent membranes. The Sec13/31 components of COPII-coated vesicles form cuboctahedral cages. The receptors in coated vesicles bind their target cargo proteins through specific signals such as the mannose-6-phosphate group that directs proteins to the lysosome or the C-terminal KDEL sequence that retrieves normally ER-resident proteins from the Golgi to the ER.

The fusion of a vesicle with its target membrane is initiated when a Rab protein, a small GTPase, induces the loose tethering of the two membranes. The vesicle is then more firmly anchored (docked) to the membrane through interactions between cognate R-SNAREs on the vesicle and Q-SNAREs on the target membrane. In neurons, synaptic vesicles dock with the presynaptic membrane through the association of the R-SNARE synaptobrevin (VAMP) with the Q-SNAREs syntaxin and SNAP-25 to form a 4-helix bundle. These neuronal SNAREs are specifically cleaved by tetanus and botulinal neurotoxins. The bilayer fusion step probably occurs as a result of the mechanical stresses generated by the formation of the several SNARE complexes at the fusion site. The neuronal SM protein nSec1 binds to syntaxin with high affinity so as to prevent the formation of the SNARE complex. A Rab protein and/or its effectors apparently induce nSec1 to release syntaxin and thereby permit the formation of the SNARE complex. After vesicle fusion, the SNARE complex must be dissociated in order to be recycled. This occurs through the auspices of the ATP-driven molecular chaperone NSF, which binds to the SNARE complex through the intermediacy of a SNAP protein.

Nuclear-encoded mitochondrial proteins are synthesized by cytosolic ribosomes and enter the mitochondrion post-translationally. A protein can only pass through a membrane in its unfolded form and hence must first be unfolded by ATP-driven molecular chaperones such as Hsp70 and MSF. A matrix-directed protein is passed through the mitochondrial outer membrane via the TOM complex, which recognizes the protein's amphipathic and positively charged N-terminal signal sequence. The N-terminal presequence then crosses the intermembrane space to encounter the TIM23 complex, which translocates it through the inner membrane into the matrix. This latter process is driven both by the mitochondrial membrane potential, which electrophoretically draws the positively charged presequence into the matrix, and by the ATP-driven chaperone mtHsp70, which binds to Tim44 in the matrix and pulls the unfolded protein into the matrix via a Brownian ratchet mechanism. MPP then excises the N-terminal signal sequence from the protein, which subsequently folds to its native state as facilitated by a battery of resident chaperones including mtHsp70 and Hsp60/Hsp10. Metabolite carrier proteins, which lack N-terminal presequences but have internal targeting sequences, also enter the intermembrane space via the TOM complex. However, they are then escorted by the

Tim9–Tim10 complex to the TIM22 complex, which laterally inserts them into the inner mitochondrial membrane. Many TM proteins that occupy the inner mitochondrial membrane are inserted there from the matrix by the Oxa1 protein, an indirect pathway that is a relic of the mitochondrion's descent from a gram-negative bacterium. Likewise, proteins that form β barrels are translocated by the TOM complex into the intermembrane space, from which they are inserted into the outer membrane by the SAM complex.

5 Lipoproteins Lipids are transported in the circulation by plasma lipoproteins. These are essentially droplets of triacylglycerols and cholesteryl esters coated with a monolayer of phospholipids, cholesterol, and apolipoproteins. The amphiphilic apolipoprotein helices float on the lipoprotein surface in hydrophobic contact with its lipid interior. Chylomicrons and VLDL function to respectively transport triacylglycerols and cholesterol from the intestines and the liver to the tissues. HDL transports mainly cholesterol from the tissues to the liver, the only organ capable of disposing of significant quantities of cholesterol. The triacylglycerols of chylomicrons and VLDL are degraded by lipoprotein lipase that lines the capillaries. LDL, the cholesterol-containing degradation product of VLDL, binds to cell-surface LDL receptors (LDLRs) and is taken into cells by receptor-mediated endocytosis. The presence of excess intracellular cholesterol inhibits the synthesis of both LDLR and cholesterol. A major cause of atherosclerosis is an excess of plasma LDL, a phenomenon that is particularly evident in individuals with familial hypercholesterolemia, who lack functional LDLRs. The excess LDL becomes oxidized and is taken up by the macrophages that inhabit atherosclerotic plaques via their scavenger receptors. Atherosclerosis, a multifactorial disease, is also correlated with a low concentration of HDL, which functions as a cholesterol scavenger. The apoE2 and apoE4 variants of apoE are implicated in cardiovascular disease, whereas apoE4 is also implicated in Alzheimer's disease.

REFERENCES

General

Edidin, M., Lipids on the frontier: a century of cell-membrane bilayers, *Nature Rev. Mol. Cell Biol.* **4,** 414–418 (2003). [A short account of the history of the study of membranes.]

Luckey, M., *Membrane Structural Biology,* Cambridge University Press (2008).

Mellman, I. and Warren, G., The road taken: Past and future foundations of membrane traffic, *Cell* **100,** 99–112 (2000). [An intellectual history and review of membrane trafficking.]

Tanford, C., *The Hydrophobic Effect: Formation of Micelles and Biological Membranes* (2nd ed.), Wiley–Interscience (1980). [An exposition of the thermodynamic properties of micelles and membranes.]

Vance, D.E. and Vance J.E. (Eds.), *Biochemistry of Lipids, Lipoproteins, and Membranes* (5th ed.), Elsevier (2008).

Lipids and Bilayers

Giles, C.H., Franklin's teaspoon of oil, *Chem. Ind.,* 1616–1624 (1969). [A historical account of Benjamin Franklin's investigations into the effect of oil on waves.]

Gurr, M.I., Harwood, J.L., and Frayn, K.N., *Lipid Biochemistry: An Introduction* (5th ed.), Blackwell Science (2002).

Lasic, D.D., Novel applications of liposomes, *Trends Biotech.* **16,** 307–321 (1998). [Reviews the uses of liposomes as delivery vehicles for drugs, vaccines, gene therapy agents, and cosmetic products.]

Lasic, D.D. and Papahadjopoulos, D. (Eds.), *Medical Applications of Liposomes,* Elsevier (1998).

Lopez, P.H.H. and Schnaar, R.L., Gangliosides in cell recognition and membrane protein regulation, *Curr. Opin. Struct. Biol.* **19,** 549–557 (2009).

Nagle, J.F. and Tristram-Nagle, S., Lipid bilayer structure, *Curr. Opin. Struct. Biol.* **10,** 474–480 (2000). [Explains why it is difficult to quantitatively describe the structure of the lipid bilayer.]

Munro, S., Lipid rafts: elusive or illusive, *Cell* **115,** 377–388 (2003).

Scott, L.H., Modeling the lipid component of membranes, *Curr. Opin. Struct. Biol.* **12,** 495–502 (2002).

van Meer, G., Voelker, D.R., and Feigenson, G.W., Membrane lipids: where they are and how they behave, *Nature Rev. Mol. Cell Biol.* 9, 112–124 (2008).

Membrane Proteins

Bracey, M.H., Hanson, M.A., Masuda, K.R., Stevens, R.C., and Cravatt, B.F., Structural adaptations in a membrane enzyme that terminates endocannabinoid signaling, *Science* **298,** 1793–1796 (2002). [The X-ray structure of fatty acid amide hydrolase.]

Deisenhofer, J. and Michel, H., High-resolution structures of photosynthetic reaction centers, *Annu. Rev. Biophys. Biophys. Chem.* **20,** 247–266 (1991); *and* Deisenhofer, J., Epp, O., Miki, K., Huber, R., and Michel, H., Structure of the protein subunits in the photosynthetic reaction centre of *Rhodopseudomonas viridis* at 3 Å resolution, *Nature* **318,** 618–624 (1985).

Elofsson, A. and von Heijne, G., Membrane protein structure: prediction versus reality, *Annu. Rev. Biochem.* **76,** 125–140 (2007).

Engelman, D.M., Membranes are more mosaic than fluid, *Nature* **438,** 578–580 (2005). [A brief review updating the classic model with more proteins and variable bilayer thickness.]

Ford, R.C. and Holzenburg, A., Electron crystallography of biomolecules: mysterious membranes and missing cones, *Trends Biochem. Sci.* **33,** 38–43 (2008).

Fujiyoshi, Y. and Unwin, N., Electron crystallography of proteins in membranes, *Curr. Opin. Struct. Biol.* **18,** 587–592 (2008).

Grigorieff, N., Ceska, T.A., Downing, K.H., Baldwin, J.M., and Henderson, R., Electron-crystallographic refinement of the structure of bacteriorhodopsin, *J. Mol. Biol.* **259,** 393–421 (1996); *and* Belrhali, H., Nollert, P., Royant, A., Menzel, C., Rosenbusch, J.P., Landau, E.M., and Pebay-Peyroula, E., Protein, lipid and water organization in bacteriorhodopsin crystal: A molecular view of the purple membrane at 1.9 Å resolution, *Structure* **7,** 909–917 (1999).

Grum, V.L., Li, D., MacDonald, R.I., and Mondragón, A., Structures of two repeats of spectrin suggest models of flexibility, *Cell* **98,** 523–535 (1999).

Haupts, U., Tittor, J., and Oesterhelt, D., Closing in on bacteriorhodopsin: Progress in understanding the molecule, *Annu. Rev. Biophys. Biomol. Struct.* **28,** 67–99 (1999).

Killian, J.A. and von Heijne, G., How proteins adapt to a membrane–water interface, *Trends Biochem. Sci.* **25**, 429–434 (2000).

Lacapère, J.-J., Pebay-Peyroula, E., Neumann, J.M., and Etchebest, C., Determining membrane protein structures: still a challenge! *Trends Biochem. Sci.* **32**, 259–270 (2007).

Liang, J., Adamian, L., and Jackups, R., Jr., The membrane–water interface region of membrane proteins: structural bias and the anti-snorkeling effect, *Trends Biochem. Sci.* **30**, 355–359 (2005).

MacKenzie, K.R., Folding and stability of α-helical integral membrane proteins, *Chem. Rev.* **106**, 1931–1977 (2006).

Palsdottir, H. and Hunte, C., Lipids in membrane protein structures, *Biochim. Biophys. Acta* **1666**, 2–18 (2004).

Pebay-Peyroula, E., Rummel, G., Rosenbusch, J.P., and Landau, E.M., X-ray structure of bacteriorhodopsin at 2.5 angstroms from microcrystals grown in lipidic cubic phases, *Science* **277**, 1676–1681 (1997); *and* Gouaux, E., It's not just a phase: Crystallization and X-ray structure of bacteriorhodopsin in lipidic cubic phases, *Structure* **6**, 5–10 (1998).

Popot, J.-L. and Engelman, D.M., Helical membrane protein folding, stability, and evolution, *Annu. Rev. Biochem.* **69**, 881–922 (2000).

Raunser, S. and Walz, T., Electron crystallography as a technique to study structure on membrane proteins in a lipidic environment, *Annu. Rev. Biophys.* **38**, 89–105 (2009).

Rees, D.C., De Antonio, L., and Eisenberg, D., Hydrophobic organization of membrane proteins, *Science* **245**, 510–512 (1989).

Schulz, G.E., β-Barrel membrane proteins, *Curr. Opin. Struct. Biol.* **10**, 443–447 (2000).

Sharom, F.J., and Lehto, M.T., Glycosylphosphatidylinositol-anchored proteins: structure, function, and cleavage by phosphatidylinositol-specific phospholipase C, *Biochem. Cell Biol.* **80**, 535–549 (2002). [Includes a discussion of the GPI anchor and its importance for protein localization and function.]

Subramanium, S., The structure of bacteriorhodopsin: An emerging consensus, *Curr. Opin. Struct. Biol.* **9**, 462–468 (1999). [Compares the six structures of bacteriorhodopsin that have been independently determined by electron or X-ray crystallography and finds them to be remarkably similar.]

Weiss, M.S. and Schulz, G.E., Structure of porin refined at 1.8 Å resolution, *J. Mol. Biol.* **227**, 493–509 (1992).

White, S.H. and Wimley, W.C., Membrane protein folding and stability: Physical principles, *Annu. Rev. Biophys. Biomol. Struct.* **28**, 319–365 (1999).

Wimley, W.C., The versatile β-barrel membrane protein, *Curr. Opin. Struct. Biol.* **13**, 404–411 (2003). [Reviews the basic principles of construction for transmembrane β barrels.]

Lipid-Linked Proteins

Clarke, S., Protein isoprenylation and methylation at carboxyl-terminal cysteine residues, *Annu. Rev. Biochem.* **61**, 355–386 (1992).

Cross, G.A.M., Glycolipid anchoring of plasma membrane proteins, *Annu. Rev. Cell Biol.* **6**, 1–39 (1990).

Englund, P.T., The structure and biosynthesis of glycosyl phosphatidylinositol protein anchors, *Annu. Rev. Biochem.* **62**, 65–100 (1993).

Linder, M.E. and Deschenes, R.J., Palmitoylation: policing protein stability and traffic, *Nature Rev. Mol. Cell Biol.* **8**, 74–84 (2007).

Marshall, C.J., Protein prenylation: A mediator of protein–protein interactions, *Science* **259**, 1865–1866 (1993).

Schafer, W.R. and Rine, J., Protein prenylation: Genes, enzymes, targets, and functions, *Annu. Rev. Genet.* **30**, 209–237 (1992).

Schlesinger, M.J. (Ed.), *Lipid Modification of Proteins*, CRC Press (1993).

Tartakoff, A.M. and Singh, N., How to make a glycoinositol phospholipid anchor, *Trends Biochem. Sci.* **17**, 470–473 (1992).

Zhang, F.L. and Casey, P.J., Protein prenylation: Molecular mechanisms and functional consequences, *Annu. Rev. Biochem.* **65**, 241–269 (1996).

Membrane Structure

Brown, D.A. and London, E., Structure and function of sphingolipid- and cholesterol-rich membrane rafts, *J. Biol. Chem.* **275**, 17221–17224 (2000); *and* Function of lipid rafts in biological membranes, *Annu. Rev. Cell Dev. Biol.* **14**, 111–136 (1998).

Edidin, M., Lipid rafts: From model membranes to cells, *Annu. Rev. Biophys. Biomol. Struct.* **32**, 257–283 (2003).

Engleman, D.M., Membranes are more mosaic than fluid, *Nature* **438**, 578–580 (2005).

Fielding, C.J. (Ed.), *Lipid Rafts and Caveolae*, Wiley-VCH (2006).

Frye, C.D. and Edidin, M., The rapid intermixing of cell surface antigens after formation of mouse–human heterokaryons, *J. Cell Sci.* **7**, 319–335 (1970). [The classic demonstration of membrane fluidity.]

Galbiati, F., Razani, B., and Lisanti, M.P., Emerging themes in rafts and caveolae, *Cell* **106**, 403–411 (2001).

Kusumi, A., Nakada, C., Ritchie, K., Murase, K., Suzuki, K., Murakoshi, H., Kasai, R.S., Kondo, J., and Fujiwara, T., Paradigm shift of the plasma membrane concept from the two-dimensional continuum fluid to the partitioned fluid: High-speed single-molecule tracking of membrane molecules, *Annu. Rev. Biophys. Biomol. Struct.* **34**, 351–378 (2005).

Marguet, D., Lenne, P.-F., Rigneault, H., and He, H.-T., Dynamics in the plasma membrane: how to combine fluidity and order, *EMBO J.* **25**, 3446–3457 (2006).

Singer, S.J. and Nicolson, G.L., The fluid mosaic model of the structure of cell membranes, *Science* **175**, 720–731 (1972). [A landmark paper on membrane structure.]

The Erythrocyte Membrane

Agre, P. and Parker, J.C. (Eds.), *Red Blood Cell Membranes*, Marcel Dekker (1989).

Bennett, V., Ankyrins, *J. Biol. Chem.* **267**, 8703–8706 (1992).

Davies, K.E. and Lux, S.E., Hereditary disorders of the red cell membrane, *Trends Genet.* **5**, 222–227 (1989).

Elgsaeter, A., Stokke, B.T., Mikkelsen, A., and Branton, D., The molecular basis of erythrocyte shape, *Science* **234**, 1217–1223 (1986).

Gallagher, P.G. and Benz, E.J., Jr., The erythrocyte membrane and cytoskeleton: Structure, function, and disorders, *in* Stamatoyannopoulos, G., Majerus, P.W., Perlmutter, R.M., and Varmus, H. (Eds.), *The Molecular Basis of Blood Diseases* (3rd ed.), Chapter 8, Elsevier (2001).

Gilligan, D.M. and Bennett, V., The junctional complex of the membrane skeleton, *Semin. Hematol.* **30**, 74–83 (1993).

Grum, V.L., Li, D., MacDonald, R.I., and Mondragón, A., Structures of two repeats of spectrin suggest models of flexibility, *Cell* **98**, 523–535 (1999); *and* Kusnoki, H., MacDonald, R.I., and Mondragón, A., Structural insights into the stability and flexibility of unusual erythroid repeats, *Structure* **12**, 645–656 (2004).

Jennings, M.L., Structure and function of the red blood cell anion transport protein, *Annu. Rev. Biophys. Biophys. Chem.* **18**, 397–430 (1989).

Liu, S.-C. and Derick, L.H., Molecular anatomy of the red blood cell membrane skeleton: Structure–function relationships, *Semin. Hematol.* **29**, 231–243 (1992).

Luna, E.J. and Hitt, A.L., Cytoskeleton–plasma membrane interactions, *Science* **258**, 955–964 (1992).

Michaely, P., Tomchick, D.R., Machius, M., and Anderson, R.G.W., Crystal structure of a 12 ANK repeat stack from human ankyrinR, *EMBO J.* **21**, 6387–6396 (2002).

Reithmeier, R.A.F., The erythrocyte anion transporter (band 3), *Curr. Opin. Struct. Biol.* **3**, 513–515 (1993).

Schofield, A.E., Reardon, R.M., and Tanner, M.J.A., Defective anion transport activity of the abnormal band 3 in hereditary ovalocytotic red blood cells, *Nature* **355**, 836–838 (1992).

Sedgwick, S.G. and Smerdon, S.J., The ankyrin repeat: a diversity of interactions on a common framework, *Trends Biochem. Sci.* **24**, 311–319 (1999).

Viel, A. and Branton, D., Spectrin: On the path from structure to function, *Curr. Opin. Cell Biol.* **8**, 49–55 (1996).

Yawata, Y., *Cell Membrane. The Red Blood Cell As a Model*, Wiley-VCH (2003).

Blood Groups

Vitala, J. and Järnefelt, J., The red cell surface revisited, *Trends Biochem. Sci.* **10**, 392–395 (1985).

Watkins, H.M., Biochemistry and genetics of the ABO, Lewis and P group systems, *Adv. Human Genet.* **10**, 1–136 (1980).

Yamamoto, F., Clausen, H., White, T., Marken, J., and Hakomori, S., Molecular genetic basis of the histo-blood group ABO system, *Nature* **345**, 229–233 (1990).

Gap Junctions

Goodenough, D.A., Goliger, J.A., and Paul, D.L., Connexins, connexons, and intercellular communication, *Annu. Rev. Biochem.* **65**, 475–502 (1996).

Maeda, S., Nakagawa, S., Suga, M., Yamashita, E., Oshima, A., Fujiyoshi, Y., and Tsikihara, T., Structure of the connexin 26 gap junction channel at 3.5 Å resolution, *Nature* **458**, 597–602 (2009).

Sosinsky, G.E. and Nicholson, B.J., Structural organization of gap junction channels, *Biochim. Biophys. Acta* **1171**, 99–125 (2005). [A detailed review.]

Wei, C.-J., Xu, X., and Lo, C.W., Connexins and cell signaling in development and disease, *Annu. Rev. Cell Dev. Biol.* **20**, 811–838 (2004).

Yeager, M. and Harris, A.L., Gap junction channel structure in the early 21st century: facts and fantasies, *Curr. Opin. Cell Biol.* **19**, 521–528 (2007).

Channel-Forming Proteins

Gouaux, J.E., Channel-forming toxins: Tales of transformation, *Curr. Opin. Struct. Biol.* **7**, 566–573 (1997).

Song, L., Hobaugh, M.R., Shustak, C., Chesley, S., Bayley, H., and Gouaux, J.E., Structure of staphylococcal α-hemolysin, a heptameric transmembrane pore, *Science* **274**, 1859–1866 (1996).

Lipid Asymmetry in Membranes

Devaux, P.E., Protein involvement in transmembrane lipid asymmetry, *Annu. Rev. Biophys. Biomol. Struct.* **21**, 417–439 (1992).

Op den Kamp, J.A.F., Lipid asymmetry in membranes, *Annu. Rev. Biochem.* **48**, 47–71 (1979).

Wirtz, K.W.A., Phospholipid transfer proteins, *Annu. Rev. Biochem.* **60**, 73–99 (1991).

Secretory Pathway

Alberts, B., Johnson, A., Lewis, J., Raff, M. Roberts, K., and Walter, P., *The Molecular Biology of the Cell* (5th ed.), Chapter. 12, Garland Science (2008).

Alder, N.N. and Johnson, A.E., Cotranslational protein biogenesis at the endoplasmic reticulum, *J. Biol. Chem.* **279**, 22787–22790 (2004).

Bowie, J.U., Solving the membrane protein folding problem, *Nature* **438**, 581–589 (2005).

Cross, B.C.S., Sinning, I., Luirink, J., and High, S., Delivering proteins for export from the cytosol, *Nature Rev. Mol. Cell Biol.* **10**, 255–264 (2009).

Doudna, J.A. and Batey, R.T., Structural insights into the signal recognition particle, *Annu. Rev. Biochem.* **73**, 539–557 (2004).

Dowhan, W. and Bogdanov, M., Lipid-dependent membrane protein topogenesis, *Annu. Rev. Biochem.* **78**, 515–540 (2009).

Driessen, A.J.M. and Nouwen, N., Protein translocation across the bacterial cytoplasmic membrane, *Annu. Rev. Biochem.* **77**, 643–667 (2008).

Fewell, S.W., Travers, K.J., Weissman, J.S., and Brodsky, J.L., The action of molecular chaperones in the early secretory pathway, *Annu. Rev. Genet.* **35**, 149–191 (2001).

Halic, M., Becker, T., Pool, M.R., Spahn, C.M.T., Grassucci, R.A., Frank, J., and Beckmann, R., Structure of the signal recognition particle interacting with the elongation-arrested ribosome, *Nature* **427**, 808–814 (2004).

Halic, M. and Beckmann, R., The signal recognition particle and its interaction during protein targeting, *Curr. Opin. Struct. Biol.* **15**, 116–125 (2005).

Halic, M., Gartmann, M., Schlenker, O., Mielke, T., Pool, M.R., Sinning, I., and Beckmann, R. Signal recognition particle receptor exposes the ribosomal translocon binding site, *Science* **312**, 745–747 (2006).

Hegde, R.S. and Bernstein, H.D., The surprising complexity of signal sequences, *Trends Biochem. Sci.* **31**, 563–571 (2006).

Kornfield, S. and Sly, W.S., I-cell disease and pseudo-Hurler polydystrophy disorders of liposomal enzyme phosphorylation and localization, Chapter 138 *in* Valle, D. (Ed.), *The Online Metabolic & Molecular Bases of Inherited Disease*, http://ommbid.com/.

Lippincott-Schwartz, J., Roberts, R.H., and Hirschberg, K., Secretory protein trafficking and organelle dynamics in living cells, *Annu. Rev. Cell Dev. Biol.* **16**, 557–589 (2000).

Lodish, H., Berk, A., Kaiser, K.A., Krieger, M., Scott, M.P., Bretscher, A., Ploegh, H., and Matsudaira, P., *Molecular Cell Biology* (6th ed.), Chapter 13, Freeman (2008).

Ménétret, J.-F., Hegde, R.S., Aguiar, M., Gygi, S.P., Park, E., Rapoport, T.A., and Akey, C.W., Single copies of Sec61 and TRAP associate with a nontranslating mammalian ribosome, *Structure* **16**, 1126–1137 (2008). [A cryo-EM study.]

Osborne, A.R., Rapaport, T.A., and van den Berg, B., Protein translocation by the Sec61/SecY channel, *Annu. Rev. Cell Dev. Biol.* **21**, 529–550 (2005).

Rapoport, T., Protein translocation across the eukaryotic endoplasmic reticulum and bacterial plasma membranes, *Nature* **450**, 663–669 (2007).

Schaffitzel, C., Oswald, M., Berger, I., Ishikawa, T., Abrahams, J.P., Koerten, H.K., Koning, R.I., and Ban, N., Structure of the *E. coli* signal recognition particle bound to a translating ribosome, *Nature* **444**, 503–505 (2006).

van den Berg, B., Clemons, W.M., Jr., Collinson, I., Modis, Y., Hartmann, E., Harrison, S.C., and Rapaport, T.A., X-ray structure of a protein-conducting channel, *Nature* **427**, 36–44 (2004). [The X-ray structure of SecY.]

von Heijne, G., Membrane-protein topology, *Nature Rev. Mol. Cell Biol.* **7**, 909–917 (2006).

Wickner, W. and Schekman, R., Protein translocation across biological membranes, *Science* **310**, 1452–1456 (2005).

Coated Vesicles

Brodsky, F.M., Chen, C.-Y., Knuehl, C., Towler, M.C., and Wakeham, D.E., Biological basket weaving: Formation and function of clathrin-coated vesicles, *Annu. Rev. Cell Dev. Biol.* **17**, 515–568 (2001).

Collins, B.M., McCoy, A.J., Kent, H.M., Evans, P.R., and Owen, D.J., Molecular architecture and functional model of the endocytotic AP2 complex, *Cell* **109**, 523–535 (2002).

Donaldson, J.G. and Lippincott-Schwartz, J., Sorting and signaling at the Golgi complex, *Cell* **101**, 693–696 (2000).

D'Souza-Schorey, C. and Chavrier, P., ARF proteins: roles in membrane traffic and beyond. *Nature Rev. Mol. Cell Biol.* **7**, 347–358 (2006).

Edeling, M.A., Smith, C., and Owen, D., Life of a clathrin coat: insights from clathrin and AP structures, *Nature Rev. Mol. Cell Biol.* **7**, 32–44 (2006).

Evans, P.R. and Owen, D.J., Endocytosis and vesicle trafficking, *Curr. Opin. Struct. Biol.* **12**, 814–821 (2002).

Fath, S., Mancius, J.D., Bi, X., and Goldberg, J., Structure and organization of coat proteins in the COPII cage, *Cell* **129**, 1325–1336 (2007).

Fotin, A., Cheng, Y., Grigorieff, N., Harrison, S.C., Kirchhausen, T., and Walz, T., Molecular model for a complete clathrin lattice from electron cryomicroscopy; *and* Structure of an auxilin-bound clathrin coat and its implications for the mechanism of uncoating, *Nature* **432**, 573–579 *and* 649–643 (2004).

Gürkan, C., Stagg, S.M., LaPointe, and Balch, W.E., The COPII cage: unifying principles of vesicle coat assembly, *Nature Rev. Mol. Cell Biol.* **7**, 727–738 (2006); *and* Stagg, S.M., LaPointe, and Balch, W.E., Structural design of cage and coat scaffolds that direct membrane traffic, *Curr. Opin. Struct. Biol.* **17**, 221–228 (2007).

Heymann, J.B., Iwasaki, K., Yim, Y.-I., Chang, N., Belnap, D.M., Greene, L.E., Eisenberg, E., and Steven, A.C., Visualization of the binding of Hsc70 ATPase to clathrin baskets, *J. Biol. Chem.* **280**, 7156–7161 (2005).

Hinshaw, J.E., Dynamin and its role in membrane fusion, *Annu. Rev. Cell Dev. Biol.* **16**, 483–519 (2000).

Kirchhausen, T., Clathrin, *Annu. Rev. Biochem.* **69**, 699–727 (2000).

Lodish, H., Berk, A., Kaiser, K.A., Krieger, M., Scott, M.P., Bretscher, A., Ploegh, H., and Matsudaira, P., *Molecular Cell Biology* (6th ed.), Chapter 14, Freeman (2008).

McNiven, M.A., Cao, H., Pitts, K.R., and Yoon, Y., The dynamin family of mechanoenzymes: Pinching in new places, *Trends Biochem. Sci.* **25**, 115–120 (2000).

McNiven, M.A., and Thompson, H.M., Vesicle formation at the plasma membrane and trans-Golgi network: the same but different, *Science* **313**, 1591–1594 (2006).

Neufield, E.F., Lysosomal storage diseases, *Annu. Rev. Biochem.* **60**, 257–280 (1991).

Owen, D.J., Collins, B.M., and Evans, P.R., Adaptors for clathrin coats: structure and function, *Annu. Rev. Cell Dev. Biol.* **20**, 151–191 (2004).

Pelham, H.R.B., Maturation of Golgi cisterna directly observed, *Trends Biochem. Sci.* **31**, 601–604 (2006).

Pfeffer, S.R., Unsolved mysteries in membrane traffic, *Anuu. Rev. Biochem.* **76**, 629–645 (2007).

Robinson, M.S., Adaptable adaptors for coated vesicles, *Trends Cell Biol.* **14**, 167–174 (2004).

Roth, M.G., Snapshots of ARF1: Implications for mechanisms of activation and inactivation, *Cell* **97**, 149–152 (1999).

Stagg, S.M., Gürkan, C., Fowler, D.M., LaPointe, P., Foss, T.R., Potter, C.S., Carragher, B., and Balch, W.E., Structure of the Sec13/31 COPII coat cage, *Nature* **439**, 234–238 (2006); and Stagg, S.M., LaPointe, P., Razvi, A., Gürkan, C., Foss, T.R., Potter, C.S., Carragher, B., and Balch, W.E., Structural bases for cargo regulation of COPII coat assembly, *Cell* **134**, 474–484 (2008).

ter Haar, E., Musacchio, A., Harrison, S.C., and Kirchhausen, T., Atomic structure of clathrin: A β propeller terminal domain joins an α zigzag linker, *Cell* **95**, 563–573 (1998).

Traub, L.M., Ticket to ride: selecting cargo for clathrin-regulated internalization, *Nature Rev. Mol. Cell Biol.* **10**, 583–596 (2009).

Ybe, J.A., Brodsky, F.M., Hofmann, K., Lin, K., Liu, S.-H., Chen, L., Earnest, T.N., Fletterick, R.J., and Hwang, P.K., Clathrin self-assembly is mediated by a tandemly repeated superhelix, *Nature* **399**, 371–375 (1999). [The X-ray structure of the clathrin heavy chain proximal leg segment.]

Vesicle Fusion

Alberts, B., Johnson, A., Lewis, J., Raff, M., Roberts, K., and Walter, P., *The Molecular Biology of the Cell* (5th ed.), Chapter 13, Garland Science (2008).

Bonifacino, J.S. and Glick, B.S., The mechanisms of vesicle budding and fusion, *Cell* **116**, 153–166 (2004). [A historical review.]

Brunger, A.T., Weninger, K., Bowen, M., and Chu, S., Single-molecule studies of the neuronal SNARE fusion machinery, *Annu. Rev. Biochem.* **78**, 903–928 (2009); *and* Brünger, A.T., Structure and functions of SNARE and SNARE-interacting proteins, *Q. Rev. Biophys.* **38**, 1–47 (2006).

Chernomordik, L.V. and Kozlov, M.M., Membrane hemifusion: crossing a chasm in two leaps, *Cell* **123**, 375–382 (2005).

Doherty, G.J. and McMahon, H.T., Mechanisms of endocytosis, *Annu. Rev. Biochem.* **78**, 857–902 (2009).

Grosshans, B.L., Ortiz, D., and Novick, P., Rabs and their effectors: Achieving specificity in membrane traffic, *Proc. Natl. Acad. Sci.* **103**, 11821–11827 (2006).

Hanson, P.I., Roth, R., Morisaki, H., Jahn, R., and Heuser, J.E., Structure and conformational changes in NSF and its membrane receptor complexes visualized by quick freeze/deep etch electron microscopy, *Cell* **90**, 523–535 (1997).

Jahn, R. and Scheller, R.H., SNARES—engines for membrane fusion, *Nature Rev. Mol. Cell Biol.* **7**, 631–643 (2006).

Martens, S., and McMahon, H.T., Mechanism of membrane fusion: disparate players and common principles, *Nature Rev. Mol. Cell Biol.* **9**, 543–556 (2008).

May, A.P., Whiteheart, S.W., and Weis, W.I., Unraveling the mechanism of the vesicle transport ATPase NSF, the *N*-ethylmaleimide-sensitive factor, *J. Biol. Chem.* **276**, 21991–21994 (2001).

Mayer, A., Membrane fusion in eukaryotic cells, *Annu. Rev. Cell Dev. Biol.* **18**, 289–314 (2002).

McNew, J.A., Parlati, F., Fukuda, R., Johnston, R.J., Paz, K., Paumet, F., Söllner, T.H., and Rothman, J.E., Compartmental specificity of cellular membrane fusion encoded in SNARE proteins, *Nature* **407**, 153–159 (2000).

Misura, K.M.S., Scheller, R.H., and Weis, W.I., Three-dimensional structure of the neuronal-Sec1–syntaxin 1a complex, *Nature* **404**, 355–362 (2000).

Montecucco, C., Schiavo, G., and Pantano, S., SNARE complexes and neuroexocytosis: how many, how close? *Trends Biochem. Sci.* **30**, 368–372 (2005).

Niemann, H., Blasi, J., and Jahn, R., Clostridial neurotoxins: New tools for dissecting exocytosis, *Trends Cell Biol.* **4**, 179–185 (1994).

Stenmark, H., Rab GTPases as coordinators of vesicle traffic, *Nature Rev. Mol. Cell Biol.* **10**, 513–525 (2009).

Sutton, R.B., Fasshauer, D., Jahn, R., and Brünger, A.T., Crystal structure of a SNARE complex involved in synaptic exocytosis at 2.4 Å resolution, *Nature* **395**, 347–353 (1998).

Ungermann, C. and Langosch, D., Function of SNAREs in intracellular membrane fusion and lipid bilyaer mixing, *J. Cell Sci.* **118**, 3819–3828 (2005).

Yu, R.C., Hanson, P.I., Jahn, R., and Brünger, A.T., Structure of the ATP-dependent oligomerization domain of N-ethylmaleimide sensitive factor complexed with ATP, *Nature Struct. Biol.* **5**, 803–810 (1998); *and* Lenzen, C.U., Steinmann, D., Whiteheart, S.W., and Weis, W.I., Crystal structure of the hexamerization domain of N-ethylmaleimide-sensitive fusion protein, *Cell* **94**, 525– 536 (1998).

Zerial, M. and McBride, H., Rab proteins as membrane organizers, *Nature Rev. Mol. Cell Biol.* **2**, 107–119 (2001).

Mitochondrial and Nuclear Protein Targeting

Abe, Y., Shodai, T., Muto, T., Mihara, K., Torii, H., Nishikawa, S., Endo, T., and Kohda, D., Structural basis of presequence recognition by the mitochondrial protein import receptor Tom20, *Cell* **100**, 551–560 (2000). [An NMR structure.]

Ahting, U., Thun, C., Hegerl, R., Typke, D., Nargang, F.E., Neupert, W., and Nussberger, S., The TOM core complex: The general protein import pore of the outer membrane of mitochondria, *J. Cell Biol.* **147**, 959–968 (1999). [An electron microscopy study of the TOM core complex.]

Bolender, N., Sickmann, A., Wagner, R., Meisinger, C., and Pfanner, N., Multiple pathways for sorting mitochondrial precursor proteins, *EMBO Rep.* **9**, 42–49 (2008).

de Marcos-Lousa, C., Sideris, D.P., and Tokatlidis, K., Translocation of mitochondrial inner-membrane proteins: conformation matters, *Trends Biochem. Sci.* **31**, 259–267 (2006).

Dolezal, P., Likic, V., Tachezy, J., and Lithgow, T., Evolution of the molecular machines for protein import into mitochondria, *Science* **313**, 314–318 (2006).

Neupert, W. and Herrmann, J.M., Translocation of proteins into mitochondria, *Annu. Rev. Biochem.* **76**, 723–749 (2007).

Lipoproteins

Beglova, N. and Blacklow, S.C., The LDL receptor: how acid pulls the trigger, *Trends Biochem. Sci.* **30**, 309–316 (2005).

Berglund, L. and Ramakrishnan, R., Lipoprotein(a). An elusive cardiovascular risk factor, *Arterioscler. Thromb. Vasc. Biol.* **24**, 2219–2226 (2004).

Borhani, D.W., Rogers, D.P., Engler, J.A., and Brouillette, C.G., Crystal structure of truncated human apolipoprotein A-I suggests a lipid-bound conformation, *Proc. Natl. Acad. Sci.* **94**, 12291–12296 (1997).

Brown, M.S. and Goldstein, J.L., A receptor-mediated pathway for cholesterol homeostasis, *Science* **232**, 34–47 (1986). [A Nobel prize address.]

Brown, M.S. and Goldstein, J.L., Koch's postulates for cholesterol, *Cell* **71**, 187–188 (1992).

Gent, J. and Braakman, I., Low-density lipoprotein receptor structure and folding, *Cell Mol. Life Sci.* **61**, 2461–2470 (2004).

Krieger, M., Charting the fate of the "good cholesterol": Identification and characterization of the high-density lipoprotein receptor SR-BI, *Annu. Rev. Biochem.* **68**, 523–558 (1999).

Lawn, R.M., Wade, D.P., Hammer, R.E., Chiesa, G., Verstuyft, J.G., and Rubin, E.M., Atherogenesis in transgenic mice expressing human apolipoprotein(a), *Nature* **360**, 670–672 (1992).

Marotti, K.R., Castle, C.K., Boyle, T.P., Lin, A.H., Murray, R.W., and Melchior, G.W., Severe atherosclerosis in transgenic mice expressing simian cholesteryl ester transfer protein, *Nature* **364**, 73–75 (1993).

Parthasarathy, S., Steinberg, D., and Witzum, J.L., The role of oxidized low-density lipoproteins in the pathogenesis of atherosclerosis, *Annu. Rev. Med.* **43**, 219–225 (1992).

Rudenko, G., Henry, L., Henderson, K., Ichtchenko, K., Brown, M.S., Goldstein, J.L., and Deisenhofer, J., Structure of the LDL receptor extracellular domain at endosomal pH, *Science* **298**, 2353–2358 (2002); *and* Rudenko, G. and Deisenhofer, J., The low-density lipoprotein receptor: ligands, debates, and lore, *Curr. Opin. Struct. Biol.* **13**, 683–689 (2003).

Valle, D. (Ed.), Part 12, Lipids, *The Online Metabolic & Molecular Bases of Inherited Disease*, http://www.ommbid.com/.

Steinberg, D., Low density lipoprotein oxidation and its pathobiological significance, *J. Biol. Chem.* **272**, 20963–20966 (1997).

Weisgraber, K.H. and Mahley, R.W., Human apolipoprotein E: The Alzheimer's disease connection, *FASEB J.* **10**, 1485–1493 (1996).

Zhong, N. and Weisgraber, K. H., Understanding the association of apolipoprotein E4 with Alzheimer disease: clues from its structure, *J. Biol. Chem.* **284**, 6027–6031 (2009).

PROBLEMS

1. Explain the difference in melting points between *trans*-oleic acid (44.5°C) and *cis*-oleic acid (13.4°C).

2. Why do animals that live in cold climates generally have more polyunsaturated fatty acid residues in their fats than do animals that live in warm climates?

*3. How many different isomers of phosphatidylserine, triacylglycerol, and cardiolipin can be made from four types of fatty acids?

4. Estimate the thickness of the surface layer formed by Benjamin Franklin's teaspoon of oil on Clapham pond (1 teaspoon = 5 mL and 1 acre = 4047 m^2).

5. "Hard water" contains a relatively high concentration of Ca^{2+}. Explain why soap is ineffective for washing in hard water.

6. Explain why pure hydrocarbons do not form monolayers on water.

7. Soap bubbles are inside-out bilayers; that is, the polar head groups of the amphiphiles, together with some water, are in apposition, whereas their hydrophobic tails extend into the air. Explain the physical basis of this phenomenon.

8. Describe the action of detergents in extracting integral proteins from membranes. How do they keep these proteins from precipitating? Why do mild detergents such as Triton X-100 bind only to proteins that form lipid complexes?

*9. Is the transmembrane portion of glycophorin A (Fig. 12-21) α helical (use the Chou and Fasman rules; Section 9-3Aa)?

10. The symmetries of oligomeric integral proteins are constrained by the requirement that their subunits must all have the same orientation with respect to the plane of the membrane. What symmetries can these proteins have? Explain. (Protein symmetry is discussed in Section 8-5B.)

11. (a) How many residues must an α helix contain in order to span the 30-Å-thick hydrocarbon core of a lipid bilayer? (b) How many residues in a β sheet are required to span this bilayer core if it is inclined by 30° with respect to the normal to the membrane plane? (c) Why do most transmembrane α helices and β strands have more than these minimum numbers?

12. Explain why antibodies against type A blood group antigens are inhibited by *N*-acetylgalactosamine, whereas anti-B antibodies are inhibited by galactose.

13. (a) Individuals with a certain one of the ABO blood types are said to be "universal donors," whereas those with another type are said to be "universal recipients." What are these blood types? Explain. (b) Antibodies are contained in blood plasma, which is blood with its red and white cells removed. Indicate the various compatiblities of blood plasma from an indiviual with one ABO blood type with an individual with a different ABO blood type. (c) Considering the answers to Parts a and b, why is it possible that there can be a universal donor and a universal recipient for a transfusion of whole blood?

14. Anti-H antibodies are not normally found in human blood. They may, however, be elicited in animals by the injection of human blood. How would such antibodies be expected to react with tissues from individuals with type A, type B, and type O blood groups?

15. *Thermus aquaticus* is a thermophilic bacterium that grows between the temperatures of 50 and 80°C. Although the signal peptide binding groove of its Ffh M domain is lined with hydrophobic groups, only three of them are Met side chains. In contrast, the binding grooves of mesophilic organisms (those that live at normal temperatures) are lined with numerous Met side chains (11 in *E. coli*). In addition, one wall of the binding groove is disordered in the X-ray structure of the *E. coli* M domain but ordered in that of *T. aquaticus* (both proteins were crystallized at room temperature). Suggest a reason for these evolutionary adaptations in *T. aquaticus*.

16. Influenza virus neuraminidase (Section 33-4Bd) is a type II protein in which three Arg residues are located just before the N-terminal end of its signal-anchor sequence. What is the likely effect of mutating all of these Arg residues to Glu?

***17.** In a genetically distinct form of familial hypercholesterolemia, LDL binds to the cell surface but fails to be internalized by endocytosis. Electron microscopy reveals that each mutant cell has its normal complement of coated pits but that ferritin-conjugated LDL does not bind to them. Rather, the bound LDL is uniformly distributed about the noncoated regions of the cell surfaces. Apparently the binding properties of the mutant LDL receptors are normal but they are in the wrong place. What do these data suggest about how LDL receptor is assembled into coated pits?

18. Table 12-6 indicates that the densities of lipoproteins increase as their particle diameters decrease. Explain.

19. Certain types of animal viruses form by budding out from a cell surface much like coated pits bud into the cytoplasm during endocytosis to form coated vesicles. In both cases, the membranous vesicles form on a polyhedral protein scaffolding. Sketch the budding of an animal virus and indicate the location of its membrane relative to its protein shell.

20. Why aren't chylomicrons taken up by LDL receptors?

21. Wolman's disease is a rapidly fatal homozygous defect characterized by a severe deficiency in **cholesteryl ester hydrolase,** the enzyme that catalyzes the hydrolysis of intracellularly located cholesteryl esters. Describe the microscopic appearance of the cells of victims of Wolman's disease.

Bovine pancreatic ribonuclease S in complex with a nonhydrolyzable substrate analog, the dinucleotide phosphonate UpcA.

PART III

MECHANISMS OF ENZYME ACTION

Introduction to Enzymes

CHAPTER **13**

The enormous variety of biochemical reactions that comprise life are nearly all mediated by a series of remarkable biological catalysts known as **enzymes.** Although enzymes are subject to the same laws of nature that govern the behavior of other substances, they differ from ordinary chemical catalysts in several important respects:

1. Higher reaction rates: The rates of enzymatically catalyzed reactions are typically factors of 10^6 to 10^{12} greater than those of the corresponding uncatalyzed reactions and are at least several orders of magnitude greater than those of the corresponding chemically catalyzed reactions.

2. Milder reaction conditions: Enzymatically catalyzed reactions occur under relatively mild conditions: temperatures below 100°C, atmospheric pressure, and nearly neutral pH's. In contrast, efficient chemical catalysis often requires elevated temperatures and pressures as well as extremes of pH.

3. Greater reaction specificity: Enzymes have a vastly greater degree of specificity with respect to the identities of both their **substrates** (reactants) and their products than do chemical catalysts; that is, enzymatic reactions rarely have side products. For example, in the enzymatic synthesis of proteins on ribosomes (Section 32-3), polypeptides consisting of well over 1000 amino acid residues are made all but error free. Yet, in the chemical synthesis of polypeptides, side reactions and incomplete reactions presently limit the lengths of polypeptides that can be accurately produced in reasonable yields to ~200 residues (Section 7-5B).

4. Capacity for control: The catalytic activities of many enzymes vary in response to the concentrations of substances other than their substrates and products. The mechanisms of these control processes include allosteric control, covalent modification of enzymes, and variation of the amounts of enzymes synthesized.

Consideration of these remarkable catalytic properties of enzymes leads to one of the central questions of biochemistry: *How do enzymes work?* We address this issue in this part of the text.

In this chapter, following a historical review, we commence our study of enzymes with a discussion of two clear instances of enzyme action: one that illustrates how enzyme specificity is manifested, and a second that exemplifies the control of enzyme activity. These are by no means exhaustive treatments but are intended to highlight these all-important aspects of enzyme mechanism. We shall encounter numerous other examples of these phenomena in our study of metabolism (Chapters 16–28). These two expositions are interspersed with a consideration of the roles of enzymatic cofactors. The chapter ends with a short synopsis of enzyme nomenclature. In Chapter 14 we take up the formalism of enzyme kinetics because the study of the rates of enzymatically catalyzed reactions provides indispensable mechanistic information. Finally, Chapter 15 is a general discussion of the catalytic mechanisms employed by enzymes, followed by an examination of the mechanisms of several specific enzymes.

1 HISTORICAL PERSPECTIVE

The early history of **enzymology,** the study of enzymes, is largely that of biochemistry itself; these disciplines evolved together from nineteenth century investigations of fermentation and digestion. Research on fermentation is widely considered to have begun in 1810 with Joseph Gay-Lussac's determination that ethanol and CO_2 are the principal products of sugar decomposition by yeast. In 1835, Jacob Berzelius, in the first general theory of chemical catalysis, pointed out that an extract of malt known as **diastase** (now known to contain the enzyme **α-amylase;** Section 11-2Db) catalyzes the hydrolysis of starch more efficiently than does sulfuric acid. Yet, despite the ability of mineral acids to mimic the effect of diastase, it was the inability to reproduce most other biochemical reactions in the laboratory that led Louis Pasteur, in the mid-nineteenth century, to propose that the processes of fermentation could only occur in living cells. Thus, as was common in his era, Pasteur assumed that living systems were endowed with a "vital force" that permitted them to evade the laws of nature governing inanimate matter. Others, however, notably Justus von Liebig, argued that biological processes are caused by the action of chemical substances that were then known as "ferments." Indeed, the name "enzyme" (Greek: *en,* in + *zyme,* yeast) was coined in 1878 by Wilhelm Friedrich Kühne in an effort to emphasize that there is something *in* yeast, as opposed to the yeast itself, that catalyzes the reactions of fermentation. Nevertheless, it was not until 1897 that Eduard Buchner obtained a cell-free yeast extract that could carry out the synthesis of ethanol from glucose (**alcoholic fermentation;** Section 17-3B).

Emil Fischer's discovery, in 1894, that glycolytic enzymes can distinguish between stereoisomeric sugars led to the formulation of his **lock-and-key hypothesis:** *The specificity of an enzyme (the lock) for its substrate (the key) arises from their geometrically complementary shapes.* Yet the chemical composition of enzymes was not firmly established until well into the twentieth century. In 1926, James Sumner, who crystallized the first enzyme, jack bean **urease,** which catalyzes the hydrolysis of urea to NH_3 and CO_2, demonstrated that these crystals consist of protein. Since Sumner's preparations were somewhat impure, however, the protein nature of enzymes was not generally accepted until the mid-1930s, when John Northrop and Moses Kunitz showed that there is a direct correlation between the enzymatic activities of crystalline pepsin, trypsin, and chymotrypsin and the amounts of protein present. Enzymological experience since then has amply demonstrated that enzymes are proteins (although it has more recently been shown that RNA can also have catalytic properties; Section 31-4Ae).

Although the subject of enzymology has a long history, most of our understanding of the nature and functions of enzymes is a product of the last 60 years. Only with the advent of modern techniques for separation and analysis (Chapter 6) has the isolation and characterization of an enzyme become less than a monumental task. It was not until 1963 that the first amino acid sequence of an enzyme, that of **bovine pancreatic ribonuclease A** (Section 15-1Ab), was reported in its entirety, and not until 1965 that the first X-ray structure of an enzyme, that of hen egg white **lysozyme** (Section 15-2A), was elucidated. In the years since then, tens of thousands of enzymes have been purified and characterized to at least some extent, and the pace of this endeavor is rapidly accelerating.

2 SUBSTRATE SPECIFICITY

The noncovalent forces through which substrates and other molecules bind to enzymes are similar in character to the forces that dictate the conformations of the proteins themselves (Section 8-4): Both involve van der Waals, electrostatic, hydrogen bonding, and hydrophobic interactions. In general, a substrate-binding site consists of an indentation or cleft on the surface of an enzyme molecule that is complementary in shape to the substrate (geometric complementarity). Moreover, the amino acid residues that form the binding site are arranged to interact specifically with the substrate in an attractive manner (electronic complementarity; Fig. 13-1). Molecules that differ in shape or functional group distribution from the substrate cannot productively bind to the

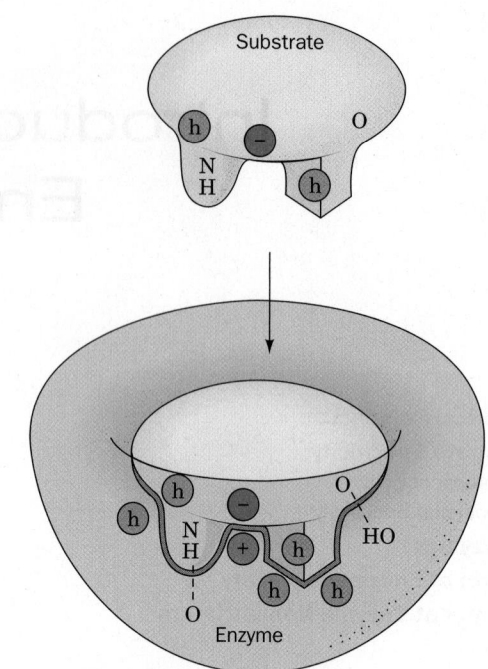

Figure 13-1 **An enzyme–substrate complex illustrating both the geometric and the physical complementarity between enzymes and substrates.** Hydrophobic groups are represented by an h in a brown circle, and dashed lines represent hydrogen bonds.

enzyme; that is, they cannot form enzyme–substrate complexes that lead to the formation of products. The substrate-binding site may, in accordance with the lock-and-key hypothesis, exist in the absence of bound substrate or it may, as suggested by the induced-fit hypothesis (Section 10-4C), form about the substrate as it binds to the enzyme. *X-ray studies indicate that the substrate-binding sites of most enzymes are largely preformed but that most of them exhibit at least some degree of induced fit on binding substrate.*

A. Stereospecificity

Enzymes are highly specific both in binding chiral substrates and in catalyzing their reactions. This **stereospecificity** arises because enzymes, by virtue of their inherent chirality (proteins consist of only L-amino acids), form asymmetric active sites. For example, trypsin readily hydrolyzes polypeptides composed of L-amino acids but not those consisting of D-amino acids. Likewise, the enzymes involved with glucose metabolism (Section 17-2) are specific for D-glucose residues.

Enzymes are absolutely stereospecific in the reactions they catalyze. This was strikingly demonstrated for the case of **yeast alcohol dehydrogenase (YADH)** by Frank Westheimer and Birgit Vennesland. Alcohol dehydrogenase catalyzes the interconversion of ethanol and acetaldehyde according to the reaction

$$CH_3CH_2OH + NAD^+ \underset{YADH}{\overset{}{\rightleftharpoons}} \overset{O}{\overset{\|}{CH_3CH}} + NADH + H^+$$

Ethanol   **Acetaldehyde**

Oxidized form   Reduced form

Nicotinamide

$$+ \ 2\,[\mathrm{H\cdot}] \rightleftharpoons \quad\quad + \ \mathrm{H}^+$$

D-Ribose

Adenosine

X = H   **Nicotinamide adenine dinucleotide (NAD$^+$)**
X = PO$_3^{2-}$ **Nicotinamide adenine dinucleotide phosphate (NADP$^+$)**

Figure 13-2   **The structures and reaction of nicotinamide adenine dinucleotide (NAD$^+$) and nicotinamide adenine dinucleotide phosphate (NADP$^+$).** Their reduced forms are **NADH** and **NADPH.** These substances, which are collectively referred to as the **nicotinamide coenzymes** or **pyridine nucleotides** (nicotinamide is a pyridine derivative), function, as is indicated in later chapters, as intracellular carriers of reducing equivalents (electrons). Note that only the nicotinamide ring is changed in the reaction. Reduction formally involves the transfer of two hydrogen atoms (H·), although the actual reduction may occur via a different mechanism.

The structures of **NAD$^+$** and **NADH** are presented in Fig. 13-2. Ethanol, it will be recalled, is a prochiral molecule (see Section 4-2Ca for a discussion of prochirality):

$$\mathrm{H}_{pro\text{-}S} - \overset{\displaystyle \mathrm{OH}}{\underset{\displaystyle \mathrm{CH_3}}{\mathrm{C}}} - \mathrm{H}_{pro\text{-}R}$$

Ethanol's two methylene H atoms may be distinguished if *the molecule is held in some sort of asymmetric jig* (Fig. 13-3). *The substrate-binding sites of enzymes are, of course, just such jigs because they immobilize the reacting groups of the substrate on the enzyme surface.*

Westheimer and Vennesland elucidated the stereospecific nature of the YADH reaction through the following series of experiments:

1. If the YADH reaction is carried out with deuterated ethanol, the product NADH is deuterated:

$$\mathrm{CH_3CD_2OH}$$
$$+$$

$$\xrightarrow{\text{YADH}}$$

$$\underset{\|}{\overset{O}{\mathrm{CH_3CD}}}$$
$$+$$

$$+ \ \mathrm{H}^+$$

NAD$^+$ **NADD**

Note that the nicotinamide ring of NAD$^+$ is also prochiral.

2. On isolating this NADD and using it in the reverse reaction to reduce normal acetaldehyde, the deuterium is

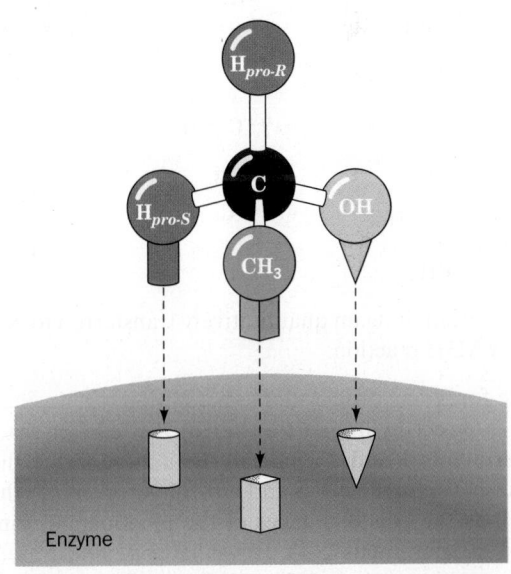

Figure 13-3   **Prochiral differentiation.** The specific attachment of a prochiral center to an enzyme binding site permits the enzyme to differentiate between prochiral groups. Note: If it were possible, the binding of the prochiral molecule's mirror image to the same three sites from the underside of the binding site as pictured here would still result in H$_{pro\text{-}R}$ pointing toward a different position.

quantitatively transferred from the NADD to the acetaldehyde to form the product ethanol:

3. If the enantiomer of the foregoing CH₃CHDOH is made as follows:

none of the deuterium is transferred from the product ethanol to NAD⁺ in the reverse reaction.

4. If, however, this ethanol is converted to its tosylate and then inverted by S_N2 hydrolysis to yield the enantiomeric ethanol,

p-Toluenesulfonyl chloride (tosyl chloride)

the deuterium is again quantitatively transferred to NAD⁺ in the YADH reaction.

The foregoing observations, in addition to showing that there is direct hydrogen transfer in the YADH reaction (Experiments 1 and 2), indicate that the enzyme distinguishes between the *pro-S* and *pro-R* hydrogens of ethanol as well as the *si* and *re* faces of the nicotinamide ring of NAD⁺ (Experiments 2–4). It was later demonstrated, by stereospecific syntheses, that YADH transfers the *pro-R* hydrogen of ethanol to the *re* face of the nicotinamide ring of NAD⁺ as is drawn in the preceding diagrams.

The stereospecificity of YADH is by no means unusual. As we consider biochemical reactions we shall find that nearly all enzymes that participate in chiral reactions are absolutely stereospecific.

a. Stereospecificity in the NADH-Dependent Dehydrogenases May Have Functional Significance

In our exploration of metabolism, we shall encounter numerous species of NADH-dependent dehydrogenases that function to reduce (or oxidize) a great variety of substrates. These various dehydrogenases are more or less equally distributed between those transferring the *pro-R* (*re*-side) and the *pro-S* (*si*-side) hydrogens at C4 of NADH (also known as A-side and B-side transfers).

Yet, despite the fact that *si*- and *re*-side hydrogen transfers to or from the nicotinamide ring yield chemically identical products, a particular specificity of transfer is rigidly maintained within classes of dehydrogenases catalyzing similar reactions in different organisms. Indeed, dehydrogenases that catalyze reactions whose equilibrium constants with their natural substrates in the direction of reduction are $<10^{-12}$ M almost always transfer the nicotinamide's *pro-R* hydrogen, whereas those with equilibrium constants $>10^{-10}$ M generally transfer the *pro-S* hydrogen. Why has evolution so assiduously maintained this stereospecificity? Is it simply the result of a historical accident or does it serve some physiological function?

The NADH hydrogen transferred in a given enzymatic reaction is almost certainly that on the side of the nicotinamide ring facing the substrate. It was therefore widely assumed that the stereospecificity in any given class of dehydrogenases simply arose through a random choice made early in evolutionary history. Once made, this choice became "locked in," because flipping a nicotinamide ring about its glycosidic bond in NADH would result, it was presumed, in its carboxamide group obstructing catalytically essential residues on the enzyme.

In an effort to shed light on this matter, Steven Benner mutated YADH in a manner that the X-ray structure of the closely similar enzyme horse **liver alcohol dehydrogenase (LADH)** suggests permits the *si* face of nicotinamide to bind to the enzyme without interfering with catalysis. The resulting mutant enzyme (Leu 182 → Ala) makes one stereochemical "mistake" every 850,000 turnovers versus one mistake every 7 billion turnovers for wild-type (unmutated) YADH. This 8000-fold decrease in stereospecificity indicates that at least some of the side chains responsible for YADH's stereospecificity are not essential for catalysis and hence strengthens the argument that stereospecificity in the dehydrogenases has functional significance.

B. Geometric Specificity

The stereospecificity of enzymes is not particularly surprising in light of the complementarity of an enzymatic binding site for its substrate. A substrate of the wrong chirality will not fit into an enzymatic binding site for much the same reasons that you cannot fit your right hand into your left glove. *In addition to their stereospecificity, however, most enzymes are quite selective about the identities of the chemical groups on their substrates.* Indeed, such **geometric specificity** is a more stringent requirement than is stereospecificity. After all, your left glove will more or less fit left hands that have somewhat different sizes and shapes than your own.

Enzymes vary considerably in their degree of geometric specificity. A few enzymes are absolutely specific for only one compound. Most enzymes, however, catalyze the reactions of a small range of related compounds. For example, YADH catalyzes the oxidation of small primary and secondary alcohols to their corresponding aldehydes or ketones but none so efficiently as that of ethanol. Even methanol and isopropanol, which differ from ethanol only by the deletion or addition of a CH_2 group, are oxidized by YADH at rates that are, respectively, 25-fold and 2.5-fold slower than that for ethanol. Similarly, $NADP^+$, which differs from NAD^+ only by the addition of a phosphoryl group at the $2'$ position of its adenosine ribose group (Fig. 13-2), does not bind to YADH. On the other hand, there are many enzymes that bind $NADP^+$ but not NAD^+.

Some enzymes, particularly digestive enzymes, are so permissive in their ranges of acceptable substrates that their geometric specificities are more accurately described as preferences. Carboxypeptidase A, for example, catalyzes the hydrolysis of C-terminal peptide bonds to all residues except Arg, Lys, and Pro if the preceding residue is not Pro (Table 7-1). However, the rate of this enzymatic reaction varies with the identities of the residues in the vicinity of the C-terminus of the polypeptide (see Fig. 7-5). Some enzymes are not even very specific in the type of reaction they catalyze. Thus chymotrypsin, in addition to its ability to mediate peptide bond hydrolysis, also catalyzes ester bond hydrolysis.

$$RC\underset{\|}{\overset{O}{}}-NHR' + H_2O \xrightarrow{\text{chymotrypsin}} RC\underset{\|}{\overset{O}{}}-O^- + H_3\overset{+}{N}R'$$
Peptide

$$RC\underset{\|}{\overset{O}{}}-OR' + H_2O \xrightarrow[\substack{\downarrow \\ H^+}]{\text{chymotrypsin}} RC\underset{\|}{\overset{O}{}}-O^- + HOR'$$
Ester

Moreover, the acyl group acceptor in the chymotrypsin reaction need not be water; amino acids, alcohols, or ammonia can also act in this capacity. You should realize, however, that such permissiveness is much more the exception than the rule. Indeed, most intracellular enzymes function *in vivo* (in the cell) to catalyze a particular reaction on a specific substrate.

3 COENZYMES

Enzymes catalyze a wide variety of chemical reactions. Their functional groups can facilely participate in acid–base reactions, form certain types of transient covalent bonds, and take part in charge–charge interactions (Section 15-1). They are, however, less suitable for catalyzing oxidation–reduction reactions and many types of group-transfer processes. Although enzymes catalyze such reactions, they mainly do so in association with small molecule **cofactors,** which essentially act as the enzymes' "chemical teeth."

Cofactors may be metal ions, such as the Zn^{2+} required for the catalytic activity of carboxypeptidase A, or organic molecules known as **coenzymes,** such as the NAD^+ in YADH (Section 13-2A). Some cofactors, for instance NAD^+, are but transiently associated with a given enzyme molecule, so that, in effect, they function as cosubstrates. Other cofactors, known as **prosthetic groups,** are essentially permanently associated with their protein, often by covalent bonds. For example, the heme prosthetic group of hemoglobin is tightly bound to its protein through extensive hydrophobic and hydrogen bonding interactions together with a covalent bond between the heme Fe^{2+} ion and His F8 (Sections 10-1A and 10-2B).

Coenzymes are chemically changed by the enzymatic reactions in which they participate. Thus, in order to complete the catalytic cycle, the coenzyme must be returned to its original state. For prosthetic groups, this can occur only in a separate phase of the enzymatic reaction sequence. For transiently bound coenzymes, such as NAD^+, however, the regeneration reaction may be catalyzed by a different enzyme.

A catalytically active enzyme–cofactor complex is called a **holoenzyme** (Greek: *holos,* whole). The enzymatically inactive protein resulting from the removal of a holoenzyme's cofactor is referred to as an **apoenzyme** (Greek: *apo,* away); that is,

Apoenzyme (*inactive*) + cofactor $\rightleftharpoons$

holoenzyme (*active*)

Table 13-1 lists the most common coenzymes together with the types of reactions in which they participate. We shall

Table 13-1   **The Common Coenzymes**

Coenzyme	Reaction Mediated	Section Discussed
Biotin	Carboxylation	23-1A
Cobalamin (B_{12}) coenzymes	Alkylation	25-2E
Coenzyme A	Acyl transfer	21-2A
Flavin coenzymes	Oxidation–reduction	16-2C
Lipoic acid	Acyl transfer	21-2A
Nicotinamide coenzymes	Oxidation–reduction	13-2A
Pyridoxal phosphate	Amino group transfer	26-1A
Tetrahydrofolate	One-carbon group transfer	26-4D
Thiamine pyrophosphate	Aldehyde transfer	17-3B

Nicotinamide   **Nicotinic acid**
(niacinamide)   **(niacin)**

Figure 13-4 **Structures of nicotinamide and nicotinic acid.**
These vitamins form the redox-active components of the nicotinamide coenzymes NAD$^+$ and NADP$^+$ (compare with Fig. 13-2).

describe the structures of these substances and their reaction mechanisms in the appropriate sections of the textbook.

a. Many Vitamins Are Coenzyme Precursors

Many organisms are unable to synthesize certain portions of essential cofactors and therefore these substances
must be present in the organism's diet; thus they are **vitamins.** In fact, many coenzymes were discovered as growth
factors for microorganisms or substances that cure nutritional deficiency diseases in humans and animals. For example, the NAD$^+$ component **nicotinamide** (alternatively
known as **niacinamide**) or its carboxylic acid analog **nicotinic acid (niacin;** Fig. 13-4), relieves the dietary deficiency
disease in humans known as **pellagra.** Pellagra, which is
characterized by diarrhea, dermatitis, and dementia, was
endemic in the rural southern United States in the early
twentieth century. Most animals, including humans, can
synthesize nicotinamide from the amino acid tryptophan
(Section 28-5A). The corn-rich diet that was prevalent in
the rural South, however, contained little available nicotinamide or tryptophan from which to synthesize it. [Corn
actually contains significant quantities of nicotinamide but
in a form that requires treatment with base before it can
be intestinally absorbed. The Mexican Indians, who are
thought to have domesticated the corn plant, customarily
soak corn meal in lime water—dilute Ca(OH)$_2$ solution—
before using it to make their staple food, tortillas.]

Table 13-2 **Vitamins That Are Coenzyme Precursors**

Vitamin	Coenzyme	Human Deficiency Disease
Biotin	Biocytin	*a*
Cobalamin (B$_{12}$)	Cobalamin (B$_{12}$) coenzymes	Pernicious anemia
Folic acid	Tetrahydrofolate	Megaloblastic anemia
Nicotinamide	Nicotinamide coenzymes	Pellagra
Pantothenate	Coenzyme A	*a*
Pyridoxine (B$_6$)	Pyridoxal phosphate	*a*
Riboflavin (B$_2$)	Flavin coenzymes	*a*
Thiamine (B$_1$)	Thiamine pyrophosphate	Beriberi

^{a}No specific name; deficiency in humans is rare or unobserved.

The vitamins in the human diet that are coenzyme precursors are all **water-soluble vitamins** (Table 13-2). In contrast, the **lipid-soluble vitamins,** such as **vitamins A** and **D,**
are not components of coenzymes, although they are also
required in trace amounts in the diets of many higher animals. The distant ancestors of humans probably had the
ability to synthesize the various vitamins, as do many modern plants and microorganisms. Yet, since vitamins are normally available in the diets of higher animals, which all eat
other organisms, or are synthesized by the bacteria that
normally inhabit their digestive systems, it is believed that
the then superfluous cellular machinery to synthesize them
was lost through evolution.

4 CONTROL OF ENZYMATIC ACTIVITY

An organism must be able to control the catalytic activities
of its component enzymes so that it can coordinate its numerous metabolic processes, respond to changes in its environment, and grow and differentiate, all in an orderly manner. There are two ways that this may occur:

1. *Control of enzyme availability: The amount of a
given enzyme in a cell depends on both its rate of synthesis
and its rate of degradation.* Each of these rates is directly
controlled by the cell. For example, *E. coli* grown in the absence of the disaccharide lactose (Fig. 11-13) lack the enzymes to metabolize this sugar. Within minutes of their exposure to lactose, however, these bacteria commence
synthesizing the enzymes required to utilize this nutrient
(Section 31-1Aa). Similarly, the various tissues of a higher
organism contain different sets of enzymes, although most
of its cells contain identical genetic information. How cells
achieve this control of enzyme synthesis is a major subject
of Part V of this textbook. The degradation of proteins is
discussed in Section 32-6.

2. *Control of enzyme activity: An enzyme's catalytic activity may be directly controlled through conformational or
structural alterations.* The rate of an enzymatically catalyzed reaction is directly proportional to the concentration of its enzyme–substrate complex, which, in turn, varies
with the enzyme and substrate concentrations and with the
enzyme's substrate-binding affinity (Section 14-2A). The
catalytic activity of an enzyme can therefore be controlled
through the variation of its substrate-binding affinity. Recall that Sections 10-1 and 10-4 detail how hemoglobin's
oxygen affinity is allosterically controlled by the binding of
ligands such as O$_2$, CO$_2$, H$^+$, and BPG. These homotropic
and heterotropic effects (ligand binding that, respectively,
alters the binding affinity of the same or different ligands)
result in cooperative (sigmoidal) O$_2$-binding curves such as
those of Figs. 10-6 and 10-8. *An enzyme's substrate-binding
affinity may likewise vary with the binding of small molecule effectors, thereby changing the enzyme's catalytic activity.* In this section we consider the allosteric control of enzymatic activity by examining one particular example:
aspartate transcarbamoylase (ATCase) from *E. coli.* (The
activities of many enzymes are similarly controlled through

Figure 13-5 **The rate of the reaction catalyzed by ATCase as a function of aspartate concentration.** The rates were measured in the absence of allosteric effectors, in the presence of 0.4 mM CTP (inhibition), and in the presence of 2.0 mM ATP (activation). [After Kantrowitz, E.R., Pastra-Landis, S.C., and Lipscomb, W.N., *Trends Biochem. Sci.* **5,** 125 (1980).] 🔗 **See the Animated Figures**

their reversible covalent modification, usually by the phosphorylation of a Ser residue. We study this form of enzymatic control in Section 18-3.)

a. The Feedback Inhibition of ATCase Controls Pyrimidine Biosynthesis

Aspartate transcarbamoylase catalyzes the formation of *N*-carbamoylaspartate from **carbamoyl phosphate** and aspartate:

Carbamoyl phosphate   **Aspartate**

aspartate transcarbamoylase

N-Carbamoylaspartate   + $H_2PO_4^-$

Arthur Pardee demonstrated that this reaction is the first step unique to the biosynthesis of pyrimidines (Section 28-2A), major components of nucleic acids.

The allosteric behavior of *E. coli* ATCase was investigated by John Gerhart and Howard Schachman, who demonstrated that this enzyme exhibits positive homotropic cooperative binding of both its substrates, namely, aspartate and carbamoyl phosphate. Moreover, ATCase is

heterotropically inhibited by **cytidine triphosphate (CTP),** a pyrimidine nucleotide, and is heterotropically activated by **adenosine triphosphate (ATP),** a purine nucleotide. CTP therefore decreases the enzyme's catalytic rate, whereas ATP increases it (Fig. 13-5).

CTP, a product of the pyrimidine biosynthesis pathway (Fig. 13-6), is a nucleic acid precursor (Section 5-4). Consequently, when rapid nucleic acid biosynthesis has depleted a cell's CTP pool, this effector dissociates from ATCase through mass action, thereby deinhibiting the enzyme and increasing the rate of CTP synthesis. Conversely, if the rate of CTP synthesis outstrips its rate of uptake, the resulting excess CTP inhibits ATCase, which, in turn, reduces the rate of CTP synthesis. *This is an example of **feedback inhibition,** a common mode of metabolic control in which the concentration of a biosynthetic pathway product controls the activity of an enzyme near the beginning of that pathway.*

The metabolic significance of the ATP activation of ATCase is that it tends to coordinate the rates of synthesis of purine and pyrimidine nucleotides for nucleic acid biosynthesis. For instance, if the ATP and CTP concentrations are out of balance with ATP in excess, ATCase is activated to synthesize pyrimidines until balance is achieved. (Note: The ATP concentration in cells is normally greater than the CTP concentration because ATP is in greater demand. Hence the ATP concentration required to activate ATCase is higher than the CTP concentration required to inhibit it by an equal amount.) Conversely, if CTP is in excess, the resulting CTP inhibition of ATCase permits purine biosynthesis to attain this balance.

Cytidine triphosphate (CTP)

Figure 13-6 **Schematic representation of the pyrimidine biosynthesis pathway.** CTP, the end product of the pathway, inhibits ATCase, which catalyzes the pathway's first step.

b. Allosteric Changes Alter ATCase's Substrate-Binding Sites

E. coli ATCase (309 kD) has the subunit composition c_6r_6, where c and r represent its catalytic and regulatory subunits (311 and 153 residues). The X-ray structure of ATCase (Fig. 13-7), determined by William Lipscomb, reveals that the catalytic subunits are arranged as two sets of trimers (c_3) in complex with three sets of regulatory dimers (r_2) to form a molecule with the rotational symmetry of a trigonal prism (D_3 symmetry; Section 8-5B). Each regulatory dimer joins two catalytic subunits in different c_3 trimers.

Dissociated catalytic trimers retain their catalytic activity, exhibit a noncooperative (hyperbolic) substrate saturation curve, have a maximum catalytic rate higher than that of intact enzyme, and are unaffected by the presence of either ATP or CTP. The isolated regulatory dimers bind these allosteric effectors but are devoid of enzymatic activity.

Evidently, *the regulatory subunits allosterically reduce the activity of the catalytic subunits in the intact enzyme.*

As allosteric theory predicts (Section 10-4), the activator ATP preferentially binds to ATCase's active (R or high substrate affinity) state, whereas the inhibitor CTP preferentially binds to the enzyme's inactive (T or low substrate affinity) state. Similarly, the unreactive bisubstrate analog **N-(phosphonacetyl)-L-aspartate (PALA)**

N-(Phosphonacetyl)-L-aspartate (PALA)

Carbamoyl phosphate + Aspartate

binds tightly to R-state but not to T-state ATCase (the use of unreactive substrate analogs is common in the study of enzyme mechanisms because they form stable complexes

(a)

(b)

(c)

Figure 13-7   X-ray structure of ATCase from *E. coli*. The T-state enzyme in complex with CTP is viewed (*a*) along the protein's molecular 3-fold axis of symmetry and (*b*) along a molecular 2-fold axis of symmetry perpendicular to the view in Part *a*. The polypeptide chains are drawn in worm form embedded in their transparent molecular surface. The regulatory dimers (*yellow*) join the upper catalytic trimer (*red*) to the lower catalytic trimer (*blue*). CTP is drawn in space-filling form colored according to atom type (C green, O red, N blue, and P orange). (*c*) The R-state enzyme in complex with PALA viewed as in Part *b*. PALA is drawn in space-filling form. Note how the rotation of the regulatory dimers in the T → R transition causes the catalytic trimers to move apart along the 3-fold axis. [Based on X-ray structures by William Lipscomb, Harvard University. PDBids 5AT1 and 8ATC.] See Kinemage Exercise 11-1

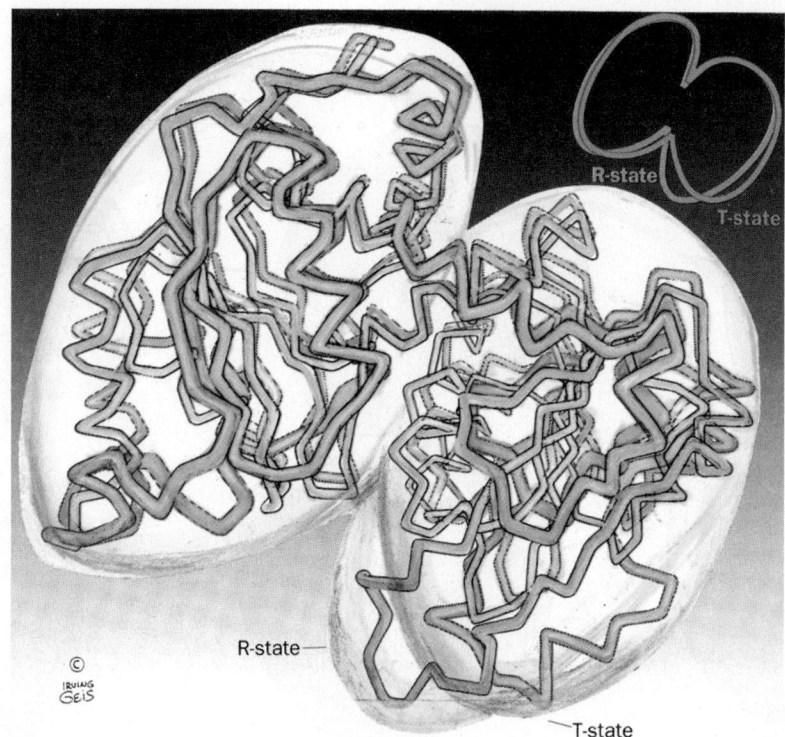

Figure 13-8 **Comparison of the polypeptide backbones of the ATCase catalytic subunit in the T state (*orange*) and the R state (*blue*).** The subunit consists of two domains, with the one on the left containing the carbamoyl phosphate binding site and that on the right forming the aspartic acid binding site. The T → R transition brings the two domains together such that their two bound substrates can react to form product. [Illustration, Irving Geis. Image from the Irving Geis Collection, Howard Hughes Medical Institute. Reprinted with permission.] X-ray structures by William Lipscomb, Harvard University.]

that are amenable to structural study rather than rapidly reacting to form products as do true substrates).

The X-ray structures of the T-state ATCase–CTP complex and the R-state ATCase–PALA complex reveal that the T → R transition maintains the protein's D_3 symmetry. The comparison of these two structures (Fig. 13-7) indicates that in the T → R transition, the enzyme's catalytic trimers separate along the molecular 3-fold axis by ~11 Å and reorient about this axis relative to each other by 12° such that these trimers assume a more nearly eclipsed configuration than is seen in Fig. 13-7*a*. In addition, the regulatory dimers rotate clockwise by 15° about their 2-fold axes and separate by ~4 Å along the 3-fold axis. Such large quaternary shifts are reminiscent of those in hemoglobin (Section 10-2B).

ATCase's substrates, carbamoyl phosphate and aspartate, each bind to a separate domain of the catalytic subunit (Fig. 13-8). The binding of PALA to the enzyme, which presumably mimics the binding of both substrates, induces active site closure in a manner that would bring them together so as to promote their reaction. The resulting atomic shifts, up to 8 Å for some residues (Fig. 13-8), trigger ATCase's T → R quaternary shift. Indeed, *ATCases's tertiary and quaternary shifts are so tightly coupled through extensive* *intersubunit contacts (see below) that they cannot occur independently* (Fig. 13-9). The binding of substrate to one catalytic subunit therefore increases the substrate-binding affinity and catalytic activity of the other catalytic subunits and hence accounts for the enzyme's positively cooperative substrate binding, much as occurs in hemoglobin (Section 10-2C). Thus, low levels of PALA actually activate ATCase by promoting its T → R transition: ATCase has such high affinity for this unreactive bisubstrate analog that the binding of one molecule of PALA converts all six of its catalytic subunits to the R state. Evidently, *ATCase closely follows the symmetry model of allosterism (Section 10-4B).*

c. The Structural Basis of Allosterism in ATCase

What are the interactions that stabilize the T and R states of ATCase and why must their interconversion be concerted? The region of the protein that undergoes the most profound conformational rearrangement with the T → R transition is a flexible loop composed of residues 230 to 250 in the catalytic (*c*) subunit, the so-called 240s loop [the symmetry-related red and blue loops that lie side by side in the T state (center of Fig. 13-7*b*) but are vertically apposed in the R state (center of Fig. 13-7*c*)]. In the T state, each 240s loop forms two intersubunit hydrogen bonds

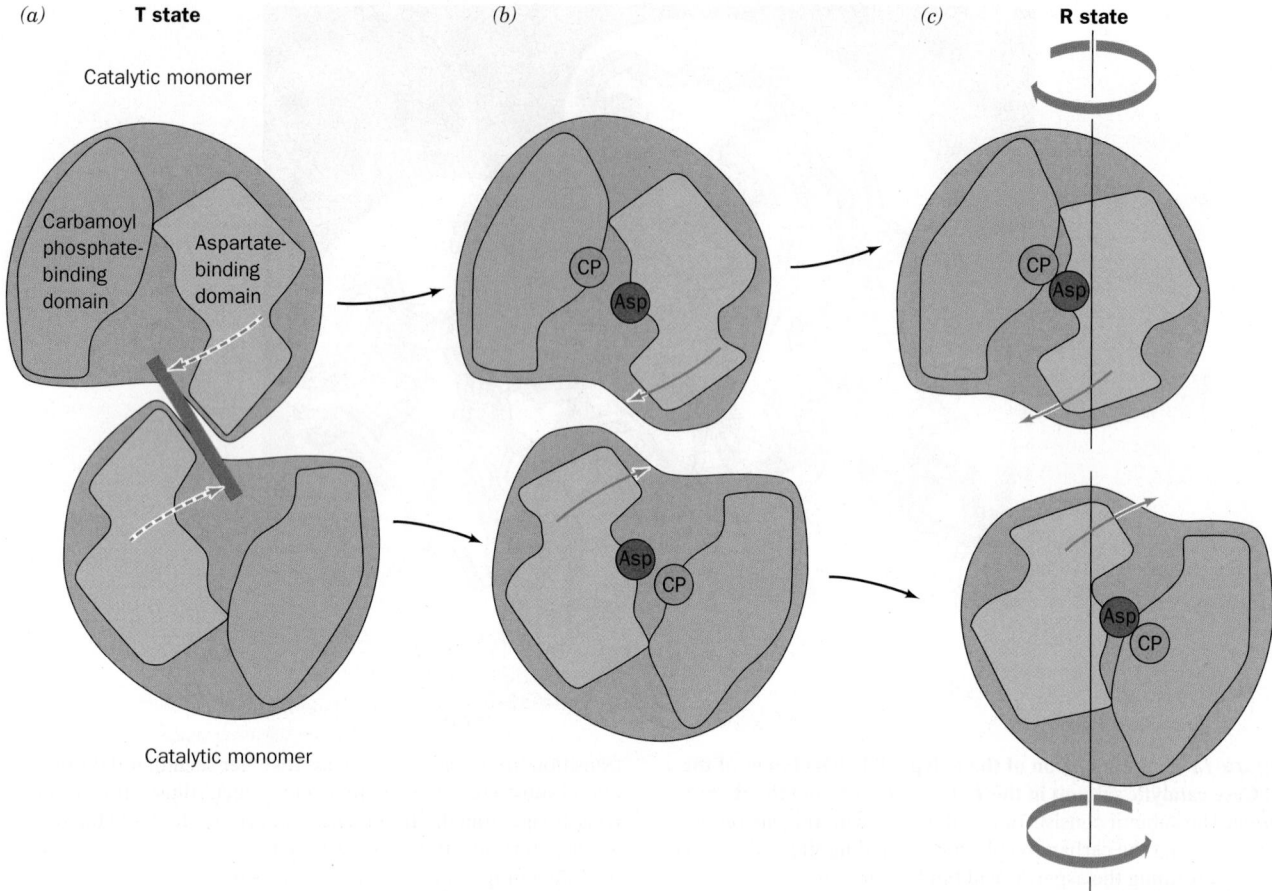

(a) **T state**

Catalytic monomer

Carbamoyl phosphate-binding domain

Aspartate-binding domain

(b)

CP

Asp

Asp

CP

(c) **R state**

CP

Asp

Asp

CP

Catalytic monomer

Figure 13-9 **Schematic diagram indicating the tertiary and quaternary conformational changes in two vertically interacting catalytic ATCase subunits.** (*a*) In the absence of bound substrate the protein is held in the T state because the motions that bring together the two domains of each subunit (*dashed arrows*) are prevented by steric interference (*purple bar*) between the contacting aspartic acid binding domains. (*b*) The binding of carbamoyl phosphate (CP) followed by aspartic acid (Asp) to their respective binding sites causes the subunits to move apart and rotate with respect to each other so as to permit the T → R transition. (*c*) In the R state, the two domains of each subunit come together so as to promote the reaction of their bound substrates to form products. [Illustration, Irving Geis. Image from the Irving Geis Collection, Howard Hughes Medical Institute. Reprinted with permission.] 🐚 **See Kinemage Exercises 11-1 and 11-2**

with the vertically opposite *c* subunit (Fig. 13-7*b*), together with an intrasubunit hydrogen bond. Domain closure as a consequence of substrate binding (Figs. 13-8 and 13-9) ruptures these hydrogen bonds and replaces them, in the R state, with new intrachain hydrogen bonds. The consequent reorientation of the 240s loop is thought to be largely responsible for the quaternary shift to the R state (see below). Since the Glu 239 carboxyl group is the acceptor in all of the above T-state interchain and R-state intrachain hydrogen bonds, this hypothesis is corroborated by the observation that the mutation of Glu 239 to Gln converts ATCase to an enzyme that is devoid of both homotropic and heterotropic effects and that has a quaternary structure midway between those of the R and T states.

What is the structural basis for heterotropic effects in ATCase? Both the inhibitor CTP and the activator ATP bind to the same site on the outer edge of the regulatory (*r*) subunit, about 60 Å away from the nearest catalytic site. CTP binds preferentially to the T state, increasing its stabil-

ity, while ATP binds preferentially to the R state, increasing its stability. The binding of these effectors to their less favored states also has structural consequences. When CTP binds to R-state ATCase, it reorients several residues at the nucleotide binding site, which induces a contraction in the length of the regulatory dimer (r_2). This distortion, through the interactions of residues at the *r–c* interface, causes the catalytic trimers (c_3) to come together by 0.5 Å (become more T-like, that is, less active, which presumably destabilizes the R state). This, in turn, reorients key residues in the enzyme's active sites, thereby decreasing the enzyme's catalytic activity. ATP has essentially opposite effects when binding to the T-state enzyme: It causes the catalytic trimers to move apart by 0.4 Å (become more R-like, that is, more active, which presumably destabilizes the T state), thereby reorienting key residues in the enzyme's active sites so as to increase the enzyme's catalytic activity. The binding of CTP to T-state ATCase does not further compress the catalytic trimers but, nevertheless, perturbs active site residues in a way that further stabilizes the T state. Although the X-ray structure of ATP complexed to R-state ATCase has not yet been reported, it is expected that ATP binding perturbs the R state in a manner analogous but opposite to the binding of CTP to T-state ATCase.

d. Allosteric Transitions in Other Enzymes Resemble Those of Hemoglobin and ATCase

Allosteric enzymes are widely distributed in nature and tend to occupy key regulatory positions in metabolic pathways. Three such enzymes, in addition to hemoglobin and ATCase, have had their X-ray structures determined in both their R and T states: **phosphofructokinase** (Sections 17-2C and 17-4F), **fructose-1,6-bisphosphatase** (Section 23-1Ah), and **glycogen phosphorylase** (Section 18-1A). In all five proteins, quaternary changes, through which binding and catalytic effects are communicated among active sites, are concerted and preserve the symmetry of the protein. This is because each of these proteins has two sets of alternative contacts, which are stabilized largely by hydrogen bonds that mostly involve side chains of opposite charge. In all five proteins, the quaternary shifts are primarily rotations of subunits relative to one another with only small translations. Secondary structures are largely preserved in T → R transitions, which is probably important for mechanically transmitting heterotropic effects over the tens of Ångstroms necessary in these proteins. The ubiquity of these structural features among allosteric proteins of known structures suggests that the control mechanisms of other allosteric enzymes, by and large, follow this model.

5 A PRIMER OF ENZYME NOMENCLATURE

Enzymes, as we have seen throughout the text so far, are commonly named by appending the suffix *-ase* to the name of the enzyme's substrate or to a phrase describing the enzyme's catalytic action. Thus urease catalyzes the hydrolysis of urea and alcohol dehydrogenase catalyzes the oxidation of alcohols to their corresponding aldehydes. Since there were at first no systematic rules for naming enzymes, this practice occasionally resulted in two different names being used for the same enzyme or, conversely, in the same name being used for two different enzymes. Moreover, many enzymes, such as catalase, which mediates the dismutation of H_2O_2 to H_2O and O_2, were given names that provide no clue as to their function; even such atrocities as "old yellow enzyme" had crept into use. In an effort to eliminate this confusion and to provide rules for rationally naming the rapidly growing number of newly discovered

Table 13-3 Enzyme Classification According to Reaction Type

Classification	Type of Reaction Catalyzed
1. Oxidoreductases	Oxidation–reduction reactions
2. Transferases	Transfer of functional groups
3. Hydrolases	Hydrolysis reactions
4. Lyases	Group elimination to form double bonds
5. Isomerases	Isomerization
6. Ligases	Bond formation coupled with ATP hydrolysis

enzymes, a scheme for the systematic functional classification and nomenclature of enzymes was adopted by the International Union of Biochemistry and Molecular Biology (IUBMB).

Enzymes are classified and named according to the nature of the chemical reactions they catalyze. There are six major classes of reactions that enzymes catalyze (Table 13-3), as well as subclasses and sub-subclasses within these classes. Each enzyme is assigned two names and a four-number classification. Its **accepted** or **recommended name** is convenient for everyday use and is often an enzyme's previously used name. Its **systematic name** is used when ambiguity must be minimized; it is the name of its substrate(s) followed by a word ending in *-ase* specifying the type of reaction the enzyme catalyzes according to its major group classification. For example, the Enzyme Nomenclature Database (available from http://www.brenda-enzymes.info/ and from http://www.chem.qmul.ac.uk/iubmb/enzyme/) indicates that the enzyme whose alternative name is lysozyme (Section 11-3Ba) has the systematic name peptidoglycan *N*-acetylmuramoylhydrolase and the Classification Number EC 3.2.1.17. Here "EC" stands for *Enzyme Commission*, the first number (3) indicates the enzyme's major class (hydrolases; Table 13-3), the second number (2) denotes its subclass (glycosylases), the third number (1) designates its sub-subclass (enzymes hydrolyzing *O*- and *S*-glycosyl compounds), and the fourth number (17) is the enzyme's arbitrarily assigned serial number in its sub-subclass. As another example, the enzyme with the recommended name alcohol dehydrogenase has the systematic name alcohol:NAD$^+$ oxidoreductase and the classification number EC 1.1.1.1. In this text, as in general biochemical terminology, we shall most often use the recomended names of enzymes but when ambiguity must be minimized, we shall refer to an enzyme's systematic name.

CHAPTER SUMMARY

2 Substrate Specificity   Enzymes specifically bind their substrates through geometrically and physically complementary interactions. This permits enzymes to be absolutely stereospecific, both in binding substrates and in catalyzing reactions. Enzymes vary in the more stringent requirement of geometric specificity. Some are highly specific for the identity of their substrates, whereas others can bind a wide range of substrates and catalyze a variety of related types of reactions.

3 Coenzymes   Enzymatic reactions involving oxidation–reduction reactions and many types of group-transfer processes are mediated by coenzymes. Many vitamins are coenzyme precursors.

4 Control of Enzyme Activity   Enzymatic activity may be regulated by the allosteric alteration of substrate-binding affinity. For example, the rate of the reaction catalyzed by *E. coli*

ATCase is subject to positive homotropic control by substrates, heterotropic inhibition by CTP, and heterotropic activation by ATP. ATCase has the subunit composition c_6r_6. Its isolated catalytic trimers are catalytically active but not subject to allosteric control. The regulatory dimers bind ATP and CTP. Substrate binding induces a tertiary conformational shift in the catalytic subunits, which increases the subunit's substrate-binding affinity and catalytic efficiency. This tertiary shift is strongly coupled to ATCase's large quaternary T → R conformational shift, thereby accounting for the enzyme's allosteric properties. Other allosteric enzymes appear to operate in a similar manner.

5 A Primer of Enzyme Nomenclature   Enzymes are classified according to their recommended name, their systematic name, and their EC classification number, which is indicative of the type of reaction catalyzed by the enzyme.

REFERENCES

History

Friedmann, H.C. (Ed.), *Enzymes,* Hutchinson Ross (1981). [A compendium of classic enzymological papers published between 1761 and 1974; with commentary.]

Fruton, J.S., *Molecules and Life, pp.* 22–86, Wiley (1972).

Schlenk, F., Early research on fermentation—a story of missed opportunities, *Trends Biochem. Sci.* **10,** 252–254 (1985).

Substrate Specificity

Creighton, D.J. and Murthy, N.S.R.K., Stereochemistry of enzyme-catalyzed reactions at carbon, *in* Sigman, D.S. and Boyer, P.D. (Eds.), *The Enzymes* (3rd ed.), Vol. 19, *pp.* 323–421, Academic Press (1990). [Section II discusses the stereochemistry of reactions catalyzed by nicotinamide-dependent dehydrogenases.]

Fersht, A., *Structure and Mechanism in Protein Science,* Freeman (1999).

Lamzin, V.S., Sauter, Z., and Wilson, K.S., How nature deals with stereoisomers, *Curr. Opin. Struct. Biol.* **5,** 830–836 (1995).

Mesecar, A.D. and Koshland, D.E., Jr., A new model for protein stereospecificity, *Nature* **403,** 614–615 (2000).

Ringe, D., What makes a binding site a binding site? *Curr. Opin. Struct. Biol.* **5,** 825–829 (1995).

Weinhold, E.G., Glasfeld, A., Ellington, A.D., and Benner, S.A., Structural determinants of stereospecificity in yeast alcohol dehydrogenase, *Proc. Natl. Acad. Sci.* **88,** 8420–8424 (1991).

Control of Enzyme Activity

Allewell, N.M., *Escherichia coli* aspartate transcarbamoylase: Structure, energetics, and catalytic and regulatory mechanisms, *Annu. Rev. Biophys. Biophys. Chem.* **18,** 71–92 (1989).

Evans, P.R., Structural aspects of allostery, *Curr. Opin. Struct. Biol.* **1,** 773–779 (1991).

Gouaux, J.E., Stevens, R.C., Ke, H., and Lipscomb, W.N., Crystal structure of the Glu-289 → Gln mutant of aspartate

carbamoyl-transferase at 3.1-Å resolution: An intermediate quaternary structure, *Proc. Natl. Acad. Sci.* **86,** 8212–8216 (1989).

Jin, L., Stec, B., Lipscomb, W.N., and Kantrowitz, E.R., Insights into the mechanisms of catalysis and heterotropic regulation of *Escherichia coli* aspartate transcarbamoylase based upon a structure of the enzyme complexed with the bisubstrate analogue N-phosphonacetyl-L-aspartate at 2.1 Å, *Proteins* **37,** 729–742 (1999).

Kantrowitz, E.R. and Lipscomb, W.N., *Escherichia coli* aspartate transcarbamylase: The molecular basis for a concerted allosteric transition, *Trends Biochem. Sci.* **15,** 53–59 (1990).

Koshland, D.E., Jr., The key–lock theory and the induced fit theory, *Angew. Chem. Int. Ed. Engl.* **33,** 2375–2378 (1994).

Macol, C.P., Tsuruta, H., Stec, B., and Kantrowitz, E.R., Direct structural evidence for a concerted allosteric transition in *Escherichia coli* aspartate transcarbamoylase, *Nature Struct. Biol.* **8,** 423–426 (2001).

Schachman, H.K., Can a simple model account for the allosteric transition of aspartate transcarbamoylase? *J. Biol. Chem.* **263,** 18583–18586 (1988).

Stevens, R.C. and Lipscomb, W.N., A molecular mechanism for pyrimidine and purine nucleotide control of aspartate transcarbamoylase, *Proc. Natl. Acad. Sci.* **89,** 5281–5285 (1992).

Zhang, Y. and Kantrowitz, E.R., Probing the regulatory site of *Escherichia coli* aspartate transcarbamoylase by site specific mutagenesis, *Biochemistry* **31,** 792–798 (1992).

Enzyme Nomenclature

Tipton, K.F., The naming of parts, *Trends Biochem. Sci.* **18,** 113–115 (1993). [A discussion of the advantages of a consistent naming scheme for enzymes and the difficulties of formulating one.]

PROBLEMS

1. Indicate the products of the YADH reaction with normal acetaldehyde and NADH in D_2O solution.

2. Indicate the product(s) of the YADH-catalyzed oxidation of the chiral methanol derivative (R)-TDHCOH (where T is the symbol for tritium).

3. The enzyme fumarase catalyzes the hydration of the double bond of fumarate:

Fumarate   **L-Malate**

Predict the action of fumarase on **maleate,** the cis isomer of fumarate. Explain.

4. Write a balanced equation for the chymotrypsin-catalyzed reaction between an ester and an amino acid.

5. Hominy grits, a regional delicacy of the southern United States, is made from corn that has been soaked in a weak lye (NaOH) solution. What is the function of this unusual treatment?

6. Which of the curves in Fig. 13-5 exhibits the greatest cooperativity? Explain.

7. What are the advantages of having the final product of a multistep metabolic pathway inhibit the enzyme that catalyzes the first step?

8. What are the systematic names and classification numbers for the enzymes whose recommended names are hexokinase, dihydroorotase, and catalase?

9. Which type of enzyme (Table 13-3) catalyzes the following reactions?

(a)

(b)

(c)

(d)

Rates of Enzymatic Reactions

CHAPTER 14

Kinetics is the study of the rates at which chemical reactions occur. A major purpose of such a study is to gain an understanding of a reaction mechanism, that is, a detailed description of the various steps in a reaction process and the sequence with which they occur. Thermodynamics, as we saw in Chapter 3, tells us whether a given process can occur spontaneously but provides little indication as to the nature or even the existence of its component steps. In contrast, *the rate of a reaction and how this rate changes in response to different conditions is intimately related to the path followed by the reaction and is therefore indicative of its reaction mechanism.*

In this chapter, we take up the study of **enzyme kinetics,** a subject that is of enormous practical importance in biochemistry because:

1. It is through kinetic studies that the binding affinities of substrates and inhibitors to an enzyme can be determined and that the maximum catalytic rate of an enzyme can be established.

2. By observing how the rate of an enzymatic reaction varies with the reaction conditions and combining this information with that obtained from chemical and structural studies of the enzyme, the enzyme's catalytic mechanism may be elucidated.

3. Most enzymes, as we shall see in later chapters, function as members of metabolic pathways. The study of the kinetics of an enzymatic reaction leads to an understanding of that enzyme's role in an overall metabolic process.

4. Under the proper conditions, the rate of an enzymatically catalyzed reaction is proportional to the amount of the enzyme present, and therefore most enzyme assays (measurements of the amount of enzyme present) are based on kinetic studies of the enzyme. Measurements of enzymatically catalyzed reaction rates are therefore among the most commonly employed procedures in biochemical and clinical analyses.

We begin our consideration of enzyme kinetics by reviewing chemical kinetics because enzyme kinetics is based on this formalism. Following that, we derive the basic equations of enzyme kinetics, describe the effects of inhibitors on enzymes, and consider how the rates of enzymatic reactions vary with pH. We end by outlining the kinetics of complex enzymatic reactions.

Kinetics is, by and large, a mathematical subject. Although the derivations of kinetic equations are occasionally rather detailed, the level of mathematical skills it requires should not challenge anyone who has studied elementary calculus. Nevertheless, to prevent mathematical detail from obscuring the underlying enzymological principles, the derivations of all but the most important kinetic equations have been collected in the appendix to this chapter. Those who wish to cultivate a deeper understanding of enzyme kinetics are urged to consult this appendix.

1 CHEMICAL KINETICS

Enzyme kinetics is a branch of chemical kinetics and, as such, shares much of the same formalism. In this section we shall therefore review the principles of chemical kinetics so that, in later sections, we can apply them to enzymatically catalyzed reactions.

A. *Elementary Reactions*

A reaction of overall stoichiometry

$$A \longrightarrow P$$

may actually occur through a sequence of **elementary reactions** (simple molecular processes) such as

$$A \longrightarrow I_1 \longrightarrow I_2 \longrightarrow P$$

Here A represents reactants, P products, and I_1 and I_2 symbolize **intermediates** in the reaction. *The characterization of the elementary reactions comprising an overall reaction process constitutes its mechanistic description.*

a. Rate Equations

At constant temperature, elementary reaction rates vary with reactant concentration in a simple manner. Consider the general elementary reaction:

$$aA + bB + \cdots + zZ \longrightarrow P$$

The rate of this process is proportional to the frequency with which the reacting molecules simultaneously come together, that is, to the products of the concentrations of the reactants. This is expressed by the following **rate equation**

$$\text{Rate} = k[A]^a[B]^b \cdots [Z]^z \qquad [14.1]$$

where k is a proportionality constant known as a **rate constant.** The **order** of a reaction is defined as $(a + b + \cdots + z)$, the sum of the exponents in the rate equation. *For an elementary reaction, the order corresponds to the* **molecularity** *of the reaction, the number of molecules that must simultaneously collide in the elementary reaction.* Thus the elementary reaction $A \rightarrow P$ is an example of a **first-order** or **unimolecular** reaction, whereas the elementary reactions $2A \rightarrow P$ and $A + B \rightarrow P$ are examples of **second-order** or **bimolecular** reactions. Unimolecular and bimolecular reactions are common. **Termolecular** reactions are unusual and fourth- and higher order elementary reactions are unknown. This is because the simultaneous collision of three molecules is a rare event; that of four or more molecules essentially never occurs.

B. *Rates of Reactions*

We can experimentally determine the order of a reaction by measuring [A] or [P] as a function of time; that is,

$$v = -\frac{d[A]}{dt} = \frac{d[P]}{dt} \qquad [14.2]$$

where v is the instantaneous rate or **velocity** of the reaction. For the first-order reaction $A \rightarrow P$:

$$v = -\frac{d[A]}{dt} = k[A] \qquad [14.3a]$$

For second-order reactions such as $2A \rightarrow P$:

$$v = -\frac{d[A]}{dt} = k[A]^2 \qquad [14.3b]$$

whereas for $A + B \rightarrow P$, a second-order reaction that is first order in [A] and first order in [B],

$$v = -\frac{d[A]}{dt} = -\frac{d[B]}{dt} = k[A][B] \qquad [14.3c]$$

The rate constants of first- and second-order reactions must have different units. In terms of units, v in Eq. [14.3a] is expressed as $M \cdot s^{-1} = kM$. Therefore, k must have units of reciprocal seconds (s^{-1}) in order for Eq. [14.3a] to balance. Similarly, for second-order reactions, $M \cdot s^{-1} = kM^2$, so that k has the units $M^{-1} s^{-1}$.

The order of a specific reaction can be determined by measuring the reactant or product concentrations as a function of time and comparing the fit of these data to equations describing this behavior for reactions of various orders. To do this we must first derive these equations.

a. First-Order Rate Equation

The equation for [A] as a function of time for a first-order reaction, $A \rightarrow P$, is obtained by rearranging Eq. [14.3a]

$$\frac{d[A]}{[A]} \equiv d \ln [A] = -k \, dt$$

and integrating it from $[A]_o$, the initial concentration of A, to [A], the concentration of A at time t:

$$\int_{[A]_o}^{[A]} d \ln[A] = -k \int_0^t dt$$

This results in

$$\ln[A] = \ln[A]_o - kt \qquad [14.4a]$$

or, by taking the antilogs of both sides,

$$[A] = [A]_o \, e^{-kt} \qquad [14.4b]$$

Equation [14.4a] is a linear equation in terms of the

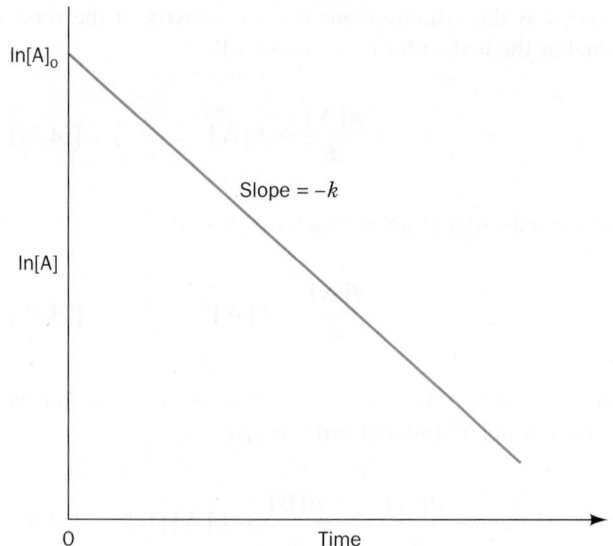

Figure 14-1   **Plot of ln[A] versus time for a first-order reaction.** This illustrates the graphical determination of the rate constant k using Eq. [14.4a].

variables $\ln[A]$ and t as is diagrammed in Fig. 14-1. Therefore, if a reaction is first order, a plot of $\ln[A]$ versus t will yield a straight line whose slope is $-k$, the negative of the first-order rate constant, and whose intercept on the $\ln[A]$ axis is $\ln[A]_o$.

Substances that are inherently unstable, such as radioactive nuclei, decompose through first-order reactions (first-order processes are not just confined to chemical reactions). One of the hallmarks of a first-order reaction is that *the time for half of the reactant initially present to decompose, its **half-time** or **half-life**, $t_{1/2}$, is a constant and hence independent of the initial concentration of the reactant.* This is easily demonstrated by substituting the relationship $[A] = [A]_o/2$ when $t = t_{1/2}$ into Eq. [14.4a] and rearranging:

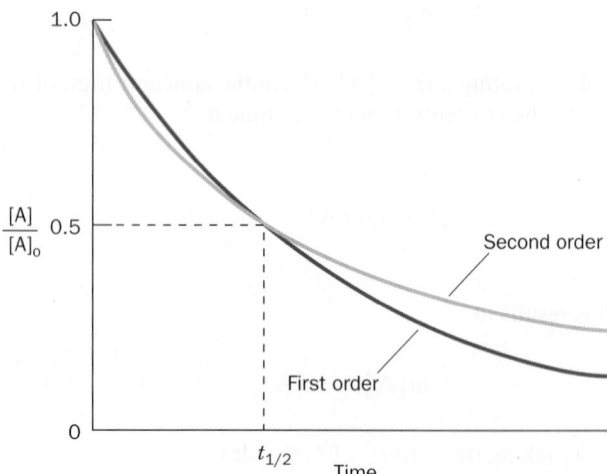

Figure 14-2   **Comparison of the progress curves for first- and second-order reactions that have the same value of $t_{1/2}$.** [After Tinoco, I., Jr., Sauer, K., and Wang, J.C., *Physical Chemistry. Principles and Applications in Biological Sciences* (2nd ed.), p. 291, Prentice-Hall (1985).]

$$\ln\left(\frac{[A]_o/2}{[A]_o}\right) = -kt_{1/2}$$

Thus

$$t_{1/2} = \frac{\ln 2}{k} = \frac{0.693}{k} \qquad [14.5]$$

In order to appreciate the course of a first-order reaction, let us consider the decomposition of ^{32}P, a radioactive isotope that is widely used in biochemical research. It has a half-life of 14 days. Thus, after 2 weeks, one-half of the ^{32}P initially present in a given sample will have decomposed; after another 2 weeks, one-half of the remainder, or three-quarters of the original sample, will have decomposed; etc. The long-term storage of waste ^{32}P therefore presents little problem, since after 1 year (26 half-lives), only 1 part in $2^{26} = 67$ million of the original sample will remain. How much will remain after 2 years? In contrast, ^{14}C, another commonly employed radioactive tracer, has a half-life of 5715 years. Only a small fraction of a given quantity of ^{14}C will decompose over the course of a human lifetime.

b. Second-Order Rate Equation for One Reactant

In a second-order reaction with one type of reactant, $2A \rightarrow P$, the variation of $[A]$ with time is quite different from that in a first-order reaction. Rearranging Eq. [14.3b] and integrating it over the same limits used for the first-order reaction yields

$$\int_{[A]_o}^{[A]} -\frac{d[A]}{[A]^2} = k \int_0^t dt$$

so that

$$\frac{1}{[A]} = \frac{1}{[A]_o} + kt \qquad [14.6]$$

Equation [14.6] is a linear equation in terms of the variables $1/[A]$ and t. Consequently, Eqs. [14.4a] and [14.6] may be used to distinguish a first-order from a second-order reaction by plotting $\ln[A]$ versus t and $1/[A]$ versus t and observing which, if any, of these plots is a straight line.

Figure 14-2 compares the different shapes of the progress curves describing the disappearance of A in first- and second-order reactions having the same half-times. Note that before the first half-time, the second-order progress curve descends more steeply than the first-order curve, but after this time the first-order progress curve is the more rapidly decreasing of the two. The half-time for a second-order reaction is expressed $t_{1/2} = 1/(k[A_o])$ and therefore, in contrast to a first-order reaction, is dependent on the initial reactant concentration.

C. *Transition State Theory*

The goal of kinetic theory is to describe reaction rates in terms of the physical properties of the reacting molecules. A theoretical framework for doing so, which explicitly considers the structures of the reacting molecules and how they collide, was developed in the 1930s, principally by Henry Eyring. This view of reaction processes, known as **transition state theory** or **absolute rate theory,** is the foundation of much of modern kinetics and has provided an

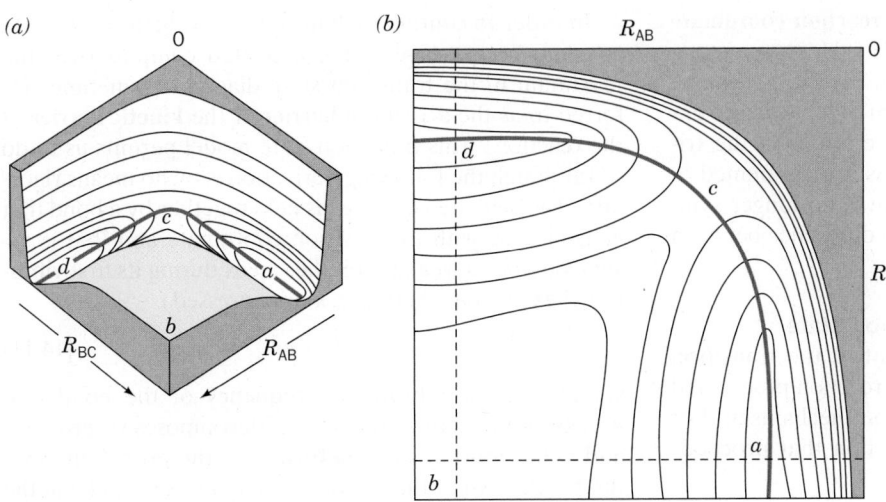

Figure 14-3 **Potential energy of the colinear H + H₂ system as a function of its internuclear distances, R_{AB} and R_{BC}.** The reaction is represented as (*a*) a perspective drawing and (*b*) the corresponding contour diagram. The points *a* and *d* are approaching potential energy minima, *b* is approaching a maximum, and *c* is a saddle point. [After Frost, A.A. and Pearson, R.G., *Kinetics and Mechanism* (2nd ed.), p. 80, Wiley (1961).]

extraordinarily productive framework for understanding how enzymes catalyze reactions.

a. The Transition State

Consider a bimolecular elementary reaction involving three atoms A, B, and C:

$$A—B + C \longrightarrow A + B—C$$

Clearly atom C must approach the diatomic molecule A—B so that, at some point in the reaction, a high-energy (unstable) complex represented as A···B···C exists in which the A—B covalent bond is in the process of breaking while the B—C bond is in the process of forming.

Let us consider the simplest example of this reaction: that of a hydrogen atom with diatomic hydrogen (H_2) to yield a new H_2 molecule and a different hydrogen atom:

$$H_A—H_B + H_C \longrightarrow H_A + H_B—H_C$$

The potential energy of this triatomic system as a function of the relative positions of its component atoms is plotted in Fig. 14-3. Its shape is of two long and deep valleys parallel to the coordinate axes with sheer walls rising toward the axes and less steep ones rising toward a plateau where both coordinates are large (the region of point *b*). The two val-

leys are joined by a pass or saddle near the origin of the diagram (point *c*). The minimum energy configuration is that of an H_2 molecule and an isolated atom, that is, with one coordinate large and the other at the H_2 covalent bond distance [near points *a* (the reactants) and *d* (the products)]. During a collision, the reactants generally approach one another with little deviation from the minimum energy reaction pathway (line *a—c—d*) because other trajectories would require much greater energy. As the atom and molecule come together, they increasingly repel one another (have increasing potential energy) and therefore usually fly apart. *If, however, the system has sufficient kinetic energy to continue its coalescence, it will cause the covalent bond of the H_2 molecule to weaken until ultimately, if the system reaches the saddle point (point c), there is an equal probability that either the reaction will occur or that the system will decompose back to its reactants.* Therefore, at this saddle point, the system is said to be at its **transition state** and hence to be an **activated complex.** Moreover, since the concentration of the activated complex is small, *the decomposition of the activated complex is postulated to be the rate-determining process of this reaction.*

The minimum free energy pathway of a reaction is known as its **reaction coordinate.** Figure 14-4a, which is

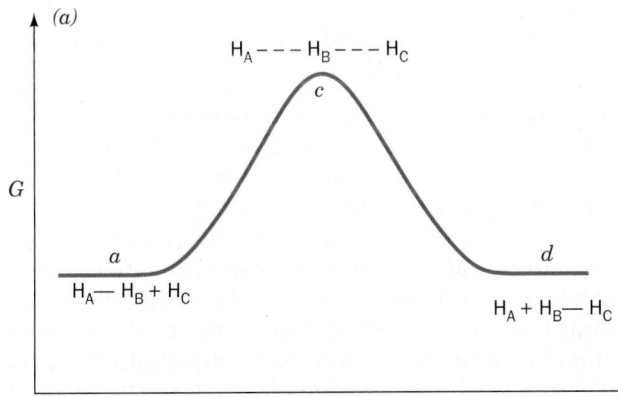

Figure 14-4 **Transition state diagrams.** (*a*) For the H + H₂ reaction. This is a section taken along the *a—c—d* line in

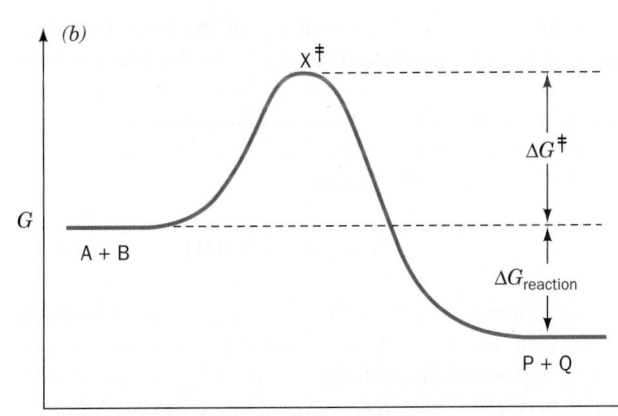

Fig. 14-3. (*b*) For a spontaneous reaction, that is, one in which the free energy decreases.

called a **transition state diagram** or a **reaction coordinate diagram,** shows the free energy of the $H + H_2$ system along the reaction coordinate (line *a—c—d* in Fig. 14-3). It can be seen that the transition state is the point of highest free energy on the reaction coordinate. If the atoms in the triatomic system are of different types, as is diagrammed in Fig. 14-4*b*, the transition state diagram is no longer symmetrical because there is a free energy difference between reactants and products.

b. Thermodynamics of the Transition State

The realization that the attainment of the transition state is the central requirement in any reaction process led to a detailed understanding of reaction mechanisms. For example, consider a bimolecular reaction that proceeds along the following pathway:

$$A + B \xrightleftharpoons{K^\ddagger} X^\ddagger \xrightarrow{k'} P + Q$$

where $X^\ddagger$ represents the activated complex. Therefore, considering the preceding discussion,

$$\frac{d[P]}{dt} = k[A][B] = k'[X^\ddagger] \qquad [14.7]$$

where k is the ordinary rate constant of the elementary reaction and k' is the rate constant for the decomposition of $X^\ddagger$ to products.

In contrast to stable molecules, such as A and P, which occur at energy minima, the activated complex occurs at an energy maximum and is therefore only metastable (like a ball balanced on a pin). Transition state theory nevertheless assumes that $X^\ddagger$ is in rapid equilibrium with the reactants; that is,

$$K^\ddagger = \frac{X^\ddagger}{[A][B]} \qquad [14.8]$$

where $K^\ddagger$ is an equilibrium constant. *This central assumption of transition state theory permits the powerful formalism of thermodynamics to be applied to the theory of reaction rates.*

If $K^\ddagger$ is an equilibrium constant it can be expressed as

$$-RT \ln K^\ddagger = \Delta G^\ddagger \qquad [14.9]$$

where $\Delta G^\ddagger$ is the Gibbs free energy of the activated complex less that of the reactants (Fig. 14-4*b*), T is the absolute temperature, and R ($= 8.3145 \text{ J} \cdot \text{K}^{-1} \text{ mol}^{-1}$) is the gas constant (this relationship between equilibrium constants and free energy is derived in Section 3-4A). Then combining Eqs. [14.7] through [14.9] yields

$$\frac{d[P]}{dt} = k' \, e^{-\Delta G^\ddagger/RT}[A][B] \qquad [14.10]$$

This equation indicates that the rate of a reaction depends not only on the concentrations of its reactants but also decreases exponentially with $\Delta G^\ddagger$. *Thus, the larger the difference between the free energy of the transition state and that of the reactants, that is, the less stable the transition state, the slower the reaction proceeds.*

In order to continue, we must now evaluate k', the rate constant for passage of the activated complex over the maximum in the transition state diagram (sometimes referred to as the **activation barrier** or the **kinetic barrier** of the reaction). This transition state model permits us to do so (although the following derivation is by no means rigorous). The activated complex is held together by a bond that is associated with the reaction coordinate and that is assumed to be so weak that it flies apart during its first vibrational excursion. Therefore, k' is expressed

$$k' = \kappa \nu \qquad [14.11]$$

where ν is the vibrational frequency of the bond that breaks as the activated complex decomposes to products and κ, the **transmission coefficient,** is the probability that the breakdown of the activated complex, $X^\ddagger$, will be in the direction of product formation rather than back to reactants. For most spontaneous reactions in solution, κ is between 0.5 and 1.0; for the colinear $H + H_2$ reaction, we saw that it is 0.5.

We have nearly finished our job of evaluating k'. All that remains is to determine the value of ν. Planck's law states that

$$\nu = \varepsilon/h \qquad [14.12]$$

where, in this case, ε is the average energy of the vibration that leads to the decomposition of $X^\ddagger$, and h ($= 6.6261 \times 10^{-34} \text{ J} \cdot \text{s}$) is Planck's constant. Statistical mechanics tells us that at temperature T, the classical energy of an oscillator is

$$\varepsilon = k_B T \qquad [14.13]$$

where k_B ($= 1.3807 \times 10^{-23} \text{ J} \cdot \text{K}^{-1}$) is a constant of nature known as the **Boltzmann constant** and $k_B T$ is essentially the available thermal energy. Combining Eqs. [14.11] through [14.13]

$$k' = \frac{\kappa k_B T}{h} \qquad [14.14]$$

Then assuming, as is done for most reactions, that $\kappa = 1$ (κ can rarely be calculated with any confidence), the combination of Eqs. [14.7] and [14.10] with [14.14] yields the expression for the rate constant k of our elementary reaction:

$$k = \frac{k_B T}{h} e^{-\Delta G^\ddagger/RT} \qquad [14.15]$$

This equation indicates that *the rate of reaction decreases as its free energy of activation, $\Delta G^\ddagger$, increases.* Conversely, as the temperature rises, so that there is increased thermal energy available to drive the reacting complex over the activation barrier, the reaction speeds up. (Of course, enzymes, being proteins, are subject to thermal denaturation, so that the rate of an enzymatically catalyzed reaction falls precipitously with increasing temperature once the enzyme's denaturation temperature has been surpassed.) Keep in mind, however, that transition state theory is an ideal model; real systems behave in a more complicated, although qualitatively similar, manner.

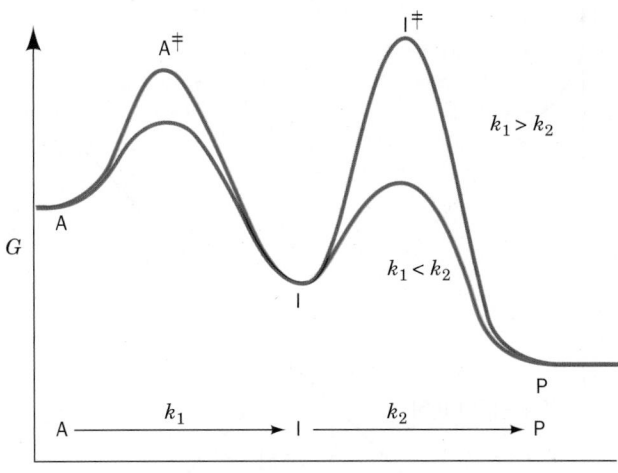

Figure 14-5 **Transition state diagram for the two-step overall reaction A → I → P.** For $k_1 < k_2$ (*green curve*), the first step is rate determining, whereas if $k_1 > k_2$ (*red curve*), the second step is rate determining.

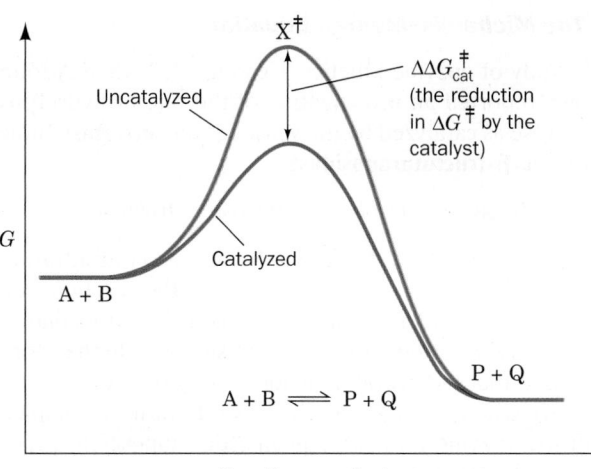

Figure 14-6 **The effect of a catalyst on the transition state diagram of a reaction.** Here $\Delta\Delta G^{\ddagger} = \Delta G_{uncat}^{\ddagger} - \Delta G_{cat}^{\ddagger}$.

c. Multistep Reactions Have Rate-Determining Steps

Since chemical reactions commonly consist of several elementary reaction steps, let us consider how transition state theory treats such reactions. For a multistep reaction such as

$$A \xrightarrow{k_1} I \xrightarrow{k_2} P$$

where I is an intermediate of the reaction, there is an activated complex for each elementary reaction step; the shape of the transition state diagram for such a reaction reflects the relative rates of the elementary reactions involved. For this reaction, if the first reaction step is slower than the second reaction step ($k_1 < k_2$), then the activation barrier of the first step must be higher than that of the second step, and conversely if the second reaction step is the slower (Fig. 14-5). Since the rate of formation of product P can only be as fast as the slowest elementary reaction, *if one reaction step of an overall reaction is much slower than the other, the slow step acts as a "bottleneck" and is therefore said to be the **rate-determining step*** of the reaction.

d. Catalysis Reduces ΔG‡

Biochemistry is, of course, mainly concerned with enzyme-catalyzed reactions. *Catalysts act by lowering the activation barrier for the reaction being catalyzed* (Fig. 14-6). If a catalyst lowers the activation barrier of a reaction by $\Delta\Delta G_{cat}^{\ddagger}$ then, according to Eq. [14.15], the rate of the reaction is enhanced by the factor $e^{\Delta\Delta G_{cat}^{\ddagger}/RT}$. Thus, a 10-fold rate enhancement requires that $\Delta\Delta G_{cat}^{\ddagger} = 5.71$ kJ · mol^{-1}, less than half the energy of a typical hydrogen bond; a millionfold rate acceleration occurs when $\Delta\Delta G_{cat}^{\ddagger} = 34.25$ kJ · mol^{-1}, a small fraction of the energy of most covalent bonds. The rate enhancement is therefore a sensitive function of $\Delta\Delta G_{cat}^{\ddagger}$.

Note that the kinetic barrier is lowered by the same amount for both the forward and the reverse reactions (Fig. 14-6). Consequently, a catalyst equally accelerates the forward and the reverse reactions so that the equilibrium constant for the reaction remains unchanged. The chemical mechanisms through which enzymes lower the activation barriers of reactions are the subject of Section 15-1. There we shall see that the most potent such mechanism often involves the enzymatic binding of the transition state of the catalyzed reaction in preference to the substrate.

2 ENZYME KINETICS

See Guided Exploration 12: Michaelis–Menten kinetics, Lineweaver–Burk plots, and enzyme inhibition The chemical reactions of life are mediated by enzymes. These remarkable catalysts, as we saw in Chapter 13, are individually highly specific for particular reactions. Yet collectively they are extremely versatile in that the many thousand enzymes now known carry out such diverse reactions as hydrolysis, polymerization, functional group transfer, oxidation–reduction, dehydration, and isomerization, to mention only the most common classes of enzymatically mediated reactions. Enzymes are not passive surfaces on which reactions take place but, rather, are complex molecular machines that operate through a great diversity of mechanisms. For instance, some enzymes act on only a single substrate molecule; others act on two or more different substrate molecules whose order of binding may or may not be obligatory. Some enzymes form covalently bound intermediate complexes with their substrates; others do not.

Kinetic measurements of enzymatically catalyzed reactions are among the most powerful techniques for elucidating the catalytic mechanisms of enzymes. The remainder of this chapter is therefore largely concerned with the development of the kinetic tools that are most useful in the determination of enzymatic mechanisms. We begin, in this section, with a presentation of the basic theory of enzyme kinetics.

A. *The Michaelis–Menten Equation*

The study of enzyme kinetics began in 1902 when Adrian Brown reported an investigation of the rate of hydrolysis of sucrose as catalyzed by the yeast enzyme **invertase** (now known as **β-fructofuranosidase**):

$$\text{Sucrose} + H_2O \longrightarrow \text{glucose} + \text{fructose}$$

Brown demonstrated that when the sucrose concentration is much higher than that of the enzyme, the reaction rate becomes independent of the sucrose concentration; that is, the rate is **zero order** with respect to sucrose. He therefore proposed that the overall reaction is composed of two elementary reactions in which the substrate forms a complex with the enzyme that subsequently decomposes to products and enzyme:

$$E + S \underset{k_{-1}}{\overset{k_1}{\rightleftharpoons}} ES \overset{k_2}{\longrightarrow} P + E$$

Here E, S, ES, and P symbolize the enzyme, substrate, **enzyme–substrate complex,** and products, respectively (for enzymes composed of multiple identical subunits, E refers to active sites rather than enzyme molecules). According to this model, *when the substrate concentration becomes high enough to entirely convert the enzyme to the ES form, the second step of the reaction becomes rate limiting and the overall reaction rate becomes insensitive to further increases in substrate concentration.*

The general expression for the **velocity** (rate) of this reaction is

$$v = \frac{d[P]}{dt} = k_2[ES] \qquad [14.16]$$

The overall rate of production of ES is the difference between the rates of the elementary reactions leading to its appearance and those resulting in its disappearance:

$$\frac{d[ES]}{dt} = k_1[E][S] - k_{-1}[ES] - k_2[ES] \qquad [14.17]$$

This equation cannot be explicitly integrated, however, without simplifying assumptions. Two possibilities are

1. Assumption of equilibrium: In 1913, Leonor Michaelis and Maud Menten, building on earlier work by Victor Henri, assumed that $k_{-1} \gg k_2$, so that the first step of the reaction achieves equilibrium.

$$K_S = \frac{k_{-1}}{k_1} = \frac{[E][S]}{[ES]} \qquad [14.18]$$

Here K_S is the dissociation constant of the first step in the enzymatic reaction. With this assumption, Eq. [14.17] can be integrated. Although this assumption is not often correct, in recognition of the importance of this pioneering work, the noncovalently bound enzyme–substrate complex ES is known as the **Michaelis complex.**

2. Assumption of steady state: Figure 14-7 illustrates the progress curves of the various participants in the preceding reaction model under the physiologically common

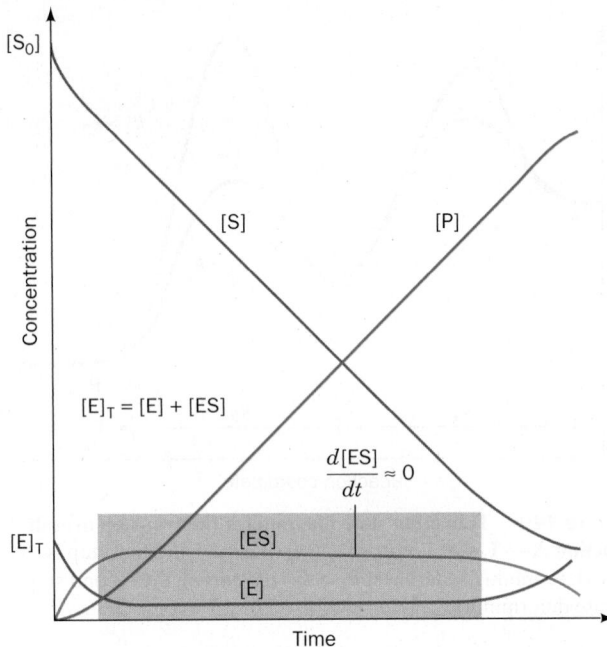

Figure 14-7 **Progress curves for the components of a simple Michaelis–Menten reaction.** Note that with the exception of the transient phase of the reaction, which occurs before the shaded block, the slopes of the progress curves for [E] and [ES] are essentially zero so long as [S] ≫ $[E]_T$ (within the shaded block). [After Segel, I.H., *Enzyme Kinetics, p.* 27, Wiley (1975).] 🖙 **See the Animated Figures**

condition that substrate is in great excess over enzyme. With the exception of the initial stage of the reaction, the so-called **transient phase,** which is usually over within milliseconds of mixing the enzyme and substrate, [ES] remains approximately constant until the substrate is nearly exhausted. Hence, the rate of synthesis of ES must equal its rate of consumption over most of the course of the reaction; that is, [ES] maintains a **steady state.** One can therefore assume with a reasonable degree of accuracy that [ES] is constant; that is,

$$\frac{d[ES]}{dt} = 0 \qquad [14.19]$$

This so-called **steady-state assumption** was first proposed in 1925 by George E. Briggs and John B.S. Haldane.

In order to be of use, kinetic expressions for overall reactions must be formulated in terms of experimentally measurable quantities. The quantities [ES] and [E] are not, in general, directly measurable but the total enzyme concentration

$$[E]_T = [E] + [ES] \qquad [14.20]$$

is usually readily determined. The rate equation for our enzymatic reaction is then derived as follows. Combining Eq. [14.17] with the steady-state assumption, Eq. [14.19], and the conservation condition, Eq. [14.20], yields

$$k_1([E]_T - [ES])[S] = (k_{-1} + k_2)[ES]$$

which on rearrangement becomes

$$[ES](k_{-1} + k_2 + k_1[S]) = k_1[E]_T[S]$$

Dividing both sides by k_1 and solving for [ES],

$$[ES] = \frac{[E]_T[E]}{K_M + [S]}$$

where K_M, which is known as the **Michaelis constant,** is defined

$$K_M = \frac{k_{-1} + k_2}{k_1} \qquad [14.21]$$

The meaning of this important constant is discussed below.

The **initial velocity** of the reaction from Eq. [14.16] can then be expressed in terms of the experimentally measurable quantities $[E]_T$ and $[S]$:

$$v_o = \left(\frac{d[P]}{dt}\right)_{t=t_s} = k_2[ES] = \frac{k_2[E]_T[S]}{K_M + [S]} \qquad [14.22]$$

where t_s is the time when the steady state is first achieved (usually milliseconds after $t = 0$). The use of the initial velocity (operationally taken as the velocity measured before more than ~10% of the substrate has been converted to product) rather than just the velocity minimizes such complicating factors as the effects of reversible reactions, inhibition of the enzyme by product, and progressive inactivation of the enzyme.

The **maximal velocity** of a reaction, V_{max}, occurs at high substrate concentrations when the enzyme is **saturated,** that is, when it is entirely in the ES form:

$$V_{max} = k_2[E]_T \qquad [14.23]$$

Therefore, combining Eqs. [14.22] and [14.23], we obtain

$$v_o = \frac{V_{max}[S]}{K_M + [S]} \qquad [14.24]$$

*This expression, the **Michaelis–Menten equation,** is the basic equation of enzyme kinetics.* It describes a rectangular hyperbola such as is plotted in Fig. 14-8 (although this curve is rotated by 45° and translated to the origin with respect to the examples of hyperbolas seen in most elementary

Figure 14-8 **Plot of the initial velocity v_o of a simple Michaelis–Menten reaction versus the substrate concentration [S].** Points are plotted in 0.5-K_M intervals of substrate concentration between 0.5 K_M and 5 K_M. **See the Animated Figures**

algebra texts). The saturation function for oxygen binding to myoglobin, Eq. [10.4], has the same functional form.

a. Significance of the Michaelis Constant

The Michaelis constant, K_M, has a simple operational definition. At the substrate concentration where $[S] = K_M$, Eq. [14.24] yields $v_o = V_{max}/2$ so that K_M *is the substrate concentration at which the reaction velocity is half-maximal.* Therefore, if an enzyme has a small value of K_M, it achieves maximal catalytic efficiency at low substrate concentrations.

The magnitude of K_M varies widely with the identity of the enzyme and the nature of the substrate (Table 14-1). It is also a function of temperature and pH (see Section 14-4). The Michaelis constant (Eq. [14.21]) can be expressed as

$$K_M = \frac{k_{-1}}{k_1} + \frac{k_2}{k_1} = K_S + \frac{k_2}{k_1} \qquad [14.25]$$

Since K_S is the dissociation constant of the Michaelis complex, as K_S decreases, the enzyme's affinity for substrate

Table 14-1 **Values of K_M, k_{cat}, and k_{cat}/K_M for Some Enzymes and Substrates**

Enzyme	Substrate	K_M (M)	k_{cat} (s^{-1})	k_{cat}/K_M (M$^{-1} \cdot$ s^{-1})
Acetylcholinesterase	Acetylcholine	9.5×10^{-5}	1.4×10^4	1.5×10^8
Carbonic anhydrase	CO_2	1.2×10^{-2}	1.0×10^6	8.3×10^7
	HCO_3^-	2.6×10^{-2}	4.0×10^5	1.5×10^7
Catalase	H_2O_2	2.5×10^{-2}	1.0×10^7	4.0×10^8
Chymotrypsin	*N*-Acetylglycine ethyl ester	4.4×10^{-1}	5.1×10^{-2}	1.2×10^{-1}
	N-Acetylvaline ethyl ester	8.8×10^{-2}	1.7×10^{-1}	1.9
	N-Acetyltyrosine ethyl ester	6.6×10^{-4}	1.9×10^2	2.9×10^5
Fumarase	Fumarate	5.0×10^{-6}	8.0×10^2	1.6×10^8
	Malate	2.5×10^{-5}	9.0×10^2	3.6×10^7
Superoxide dismutase	Superoxide ion ($O_2^{\cdot-}$)	3.6×10^{-4}	1.0×10^6	2.8×10^9
Urease	Urea	2.5×10^{-2}	1.0×10^4	4.0×10^5

increases. K_M is therefore also a measure of the affinity of the enzyme for its substrate providing k_2/k_1 is small compared with K_S, that is, $k_2 < k_{-1}$.

B. *Analysis of Kinetic Data*

There are several methods for determining the values of the parameters of the Michaelis–Menten equation. At very high values of [S], the initial velocity v_o asymptotically approaches V_{max}. In practice, however, it is very difficult to assess V_{max} accurately from direct plots of v_o versus [S] such as Fig. 14-8. Even at such high substrate concentrations as [S] = 10 K_M, Eq. [14.24] indicates that v_o is only 91% of V_{max}, so that the extrapolated value of the asymptote will almost certainly be underestimated.

A better method for determining the values of V_{max} and K_M, which was formulated by Hans Lineweaver and Dean Burk, uses the reciprocal of Eq. [14.24]:

$$\frac{1}{v_o} = \left(\frac{K_M}{V_{max}}\right)\frac{1}{[S]} + \frac{1}{V_{max}} \qquad [14.26]$$

This is a linear equation in $1/v_o$ and $1/[S]$. If these quantities are plotted, in the so-called **Lineweaver–Burk** or **double-reciprocal plot,** the slope of the line is K_M/V_{max}, the $1/v_o$ intercept is $1/V_{max}$, and the extrapolated $1/[S]$ intercept is $-1/K_M$ (Fig. 14-9). A disadvantage of this plot is that most experimental measurements involve relatively high [S] and are therefore crowded onto the left side of the graph. Furthermore, for small values of [S], small errors in v_o lead to large errors in $1/v_o$ and hence to large errors in K_M and V_{max}.

Several other types of plots, each with its advantages and disadvantages, have been formulated for the determination of V_{max} and K_M from kinetic data. With the advent of conveniently available computers, however, kinetic data are commonly analyzed by mathematically sophisticated statistical treatments. Nevertheless, Lineweaver–Burk plots are valuable for the visual presentation of kinetic data as well as being useful in the analysis of kinetic data from enzymes requiring more than one substrate (Section 14-5C).

a. k_{cat}/K_M Is a Measure of Catalytic Efficiency

An enzyme's kinetic parameters provide a measure of its catalytic efficiency. We may define the catalytic constant of an enzyme as

$$k_{cat} = \frac{V_{max}}{[E]_T} \qquad [14.27]$$

This quantity is also known as the **turnover number** of an enzyme because it is the number of reaction processes (turnovers) that each active site catalyzes per unit time. The turnover numbers for a selection of enzymes are given in Table 14-1. Note that these quantities vary by over eight orders of magnitude depending on the identity of the enzyme as well as that of its substrate. Equation [14.23] indicates that for the Michaelis–Menten model, $k_{cat} = k_2$. For enzymes with more complicated mechanisms, k_{cat} may be a function of several rate constants.

When [S] $<<$ K_M, very little ES is formed. Consequently, [E] $\approx$ $[E]_T$, so that Eq. [14.22] reduces to a second-order rate equation:

$$v_o \approx \left(\frac{k_2}{K_M}\right)[E]_T[S] \approx \left(\frac{k_{cat}}{K_M}\right)[E][S] \qquad [14.28]$$

k_{cat}/K_M is the apparent second-order rate constant of the enzymatic reaction; the rate of the reaction varies directly with how often enzyme and substrate encounter one another in solution. *The quantity k_{cat}/K_M is therefore a measure of an enzyme's catalytic efficiency.*

b. Some Enzymes Have Attained Catalytic Perfection

Is there an upper limit on enzymatic catalytic efficiency? From Eq. [14.21] we find

$$\frac{k_{cat}}{K_M} = \frac{k_2}{K_M} = \frac{k_1 k_2}{k_{-1} + k_2} \qquad [14.29]$$

This ratio is maximal when $k_2 >> k_{-1}$, that is, when the formation of product from the Michaelis complex, ES, is fast compared to its decomposition back to substrate and enzyme. Then $k_{cat}/K_M = k_1$, the second-order rate constant for the formation of ES. The term k_1, of course, can be no greater than the frequency with which enzyme and substrate molecules collide with each other in solution. This **diffusion-controlled limit** is in the range of 10^8 to $10^9 \, M^{-1} \cdot s^{-1}$. Thus, enzymes with such values of k_{cat}/K_M must catalyze a reaction almost every time they encounter a substrate molecule. Table 14-1 indicates that several enzymes, namely, catalase, superoxide dismutase, fumarase, acetylcholinesterase, and possibly carbonic anhydrase, have achieved this state of virtual catalytic perfection.

Figure 14-9   **A double-reciprocal (Lineweaver–Burk) plot.** Error bars are $\pm 0.05 \, V_{max}$. The indicated points are the same as those in Fig. 14-8. Note the large effect of small errors at small [S] (large 1/[S]) and the crowding together of points at large [S].

See the Animated Figures

Figure 14-10   Cross section through the active site of human superoxide dismutase (SOD). The enzyme binds both a Cu^{2+} and a Zn^{2+} ion (*orange and cyan spheres*). SOD's molecular surface is represented by a dot surface that is colored according to its electrostatic charge, with red most negative, yellow negative, green neutral, cyan positive, and blue most positive. The electrostatic field vectors are represented by similarly colored arrows. Note how this electrostatic field would draw the negatively charged superoxide ion into its binding site, which is located between the Cu^{2+} ion and Arg 143. [Courtesy of Elizabeth Getzoff, The Scripps Research Institute, La Jolla, California.]

Since the active site of an enzyme generally occupies only a small fraction of its total surface area, how can any enzyme catalyze a reaction every time it encounters a substrate molecule? In the case of **superoxide dismutase (SOD),** it appears that the arrangement of charged groups on the enzyme's surface serves to electrostatically guide the charged substrate to the enzyme's active site (Fig. 14-10). [SOD, which is present in nearly all cells, functions to inactivate the highly reactive and therefore destructive **superoxide radical** $O_2^{\cdot-}$ by catalyzing the reaction $2 O_2^{\cdot-} + 2H^+ \rightarrow H_2O_2 + O_2$; Section 22-4Ch]. Other enzymes, including **acetylcholinesterase** (Section 20-5C), have similar mechanisms to funnel polar substrates to their active sites.

C. *Reversible Reactions*

The Michaelis–Menten model implicitly assumes that enzymatic reverse reactions may be neglected. Yet many enzymatic reactions are highly reversible (have a small free energy of reaction) and therefore have products that back react to form substrates at a significant rate. In this section we therefore relax the Michaelis–Menten restriction of no back reaction and, by doing so, discover some interesting and important kinetic principles.

a. The One-Intermediate Model

Modification of the Michaelis–Menten model to incorporate a back reaction yields the following reaction scheme:

$$E + S \underset{k_{-1}}{\overset{k_1}{\rightleftharpoons}} ES \underset{k_{-2}}{\overset{k_2}{\rightleftharpoons}} P + E$$

(Here ES might just as well be called EP because this model does not specify the nature of the intermediate complex.) The equation describing the kinetic behavior of this model, which is derived in Appendix A of this chapter, is expressed

$$v = \frac{\dfrac{V_{max}^{f}[S]}{K_M^S} - \dfrac{V_{max}^{r}[P]}{K_M^P}}{1 + \dfrac{[S]}{K_M^S} + \dfrac{[P]}{K_M^P}} \qquad [14.30]$$

where

$$V_{max}^{f} = k_2[E]_T \qquad V_{max}^{r} = k_{-1}[E]_T$$

$$K_M^S = \frac{k_{-1} + k_2}{k_1} \qquad K_M^P = \frac{k_{-1} + k_2}{k_{-2}}$$

and

$$[E]_T = [E] + [ES]$$

This is essentially a Michaelis–Menten equation that works backwards as well as forwards. Indeed, at $[P] = 0$, that is, when $v = v_o$, this equation becomes the Michaelis–Menten equation.

b. The Haldane Relationship

At equilibrium (which occurs after the reaction has run its course), $v = 0$, so Eq. [14.30], which holds at equilibrium as well as at steady state, can be solved to yield

$$K_{eq} = \frac{[P]_{eq}}{[S]_{eq}} = \frac{V_{max}^{f} K_M^P}{V_{max}^{r} K_M^S} \qquad [14.31]$$

where $[P]_{eq}$ and $[S]_{eq}$ are the concentrations of P and S at equilibrium. This so-called **Haldane relationship** demonstrates that *the kinetic parameters of a reversible enzymatically catalyzed reaction are not independent of one another. Rather, they are related by the equilibrium constant for the overall reaction, which, of course, is independent of the presence of the enzyme.*

c. Kinetic Data Cannot Unambiguously Establish a Reaction Mechanism

An enzyme that forms a reversible complex with its substrate should likewise form one with its product; that is, it should have a mechanism such as

$$E + S \underset{k_{-1}}{\overset{k_1}{\rightleftharpoons}} ES \underset{k_{-2}}{\overset{k_2}{\rightleftharpoons}} EP \underset{k_{-3}}{\overset{k_3}{\rightleftharpoons}} P + E$$

The equation describing the kinetic behavior of this two-intermediate model, whose derivation is analogous to that described in Appendix A for the one-intermediate model, has a form identical to that of Eq. [14.30]. However, its parameters V_{max}^f, V_{max}^r, K_M^S, and K_M^P are defined in terms of the six kinetic constants of the two-intermediate model rather than the four of the one-intermediate model. In fact, the steady-state rate equations for reversible reactions with three or more intermediates also have this same form but with yet different definitions of the four parameters.

The values of V_{max}^f, V_{max}^r, K_M^S, and K_M^P in Eq. [14.30] can be determined by suitable manipulations of the initial substrate and product concentrations under steady-state conditions. This, however, will not yield the values of the rate constants for our two-intermediate model because there are six such constants and only four equations describing their relationships. Moreover, steady-state kinetic measurements are incapable of distinguishing the number of intermediates in a reversible enzymatic reaction because the form of Eq. [14.30] does not change with the number of intermediates.

The functional identities of the equations describing these reaction schemes may be understood in terms of an analogy between our *n*-intermediate reversible reaction model and a "black box" containing a system of water pipes with one inlet and one drain:

"Black box"

In → → Out

At steady state, that is, after the pipes have filled with water, one can measure the relationship between input pressure and output flow. However, such measurements yield no information concerning the detailed construction of the plumbing connecting the inlet to the drain. This would require additional observations such as opening the black box and tracing the pipes. *Likewise, steady-state kinetic measurements can provide a phenomenological description of enzymatic behavior, but the nature of the intermediates remains indeterminate. Rather, these intermediates must be detected and characterized by independent means such as by spectroscopic analysis.*

The foregoing discussion brings to light a central principle of kinetic analysis: *The steady-state kinetic analysis of a reaction cannot unambiguously establish its mechanism.* This is because no matter how simple, elegant, or rational a mechanism one postulates that fully accounts for kinetic data, there are an infinite number of alternative mechanisms, perhaps complicated, awkward, and seemingly irrational, that can account for these kinetic data equally well. Usually it is the simpler and more elegant mechanism that turns out to be correct, but this is not always the case. *If,*

however, kinetic data are not compatible with a given mechanism, then the mechanism must be rejected. Therefore, although kinetics cannot be used to establish a mechanism unambiguously without confirming data, such as the physical demonstration of an intermediate's existence, the steady-state kinetic analysis of a reaction is of great value because it can be used to eliminate proposed mechanisms.

3 INHIBITION

Many substances alter the activity of an enzyme by combining with it in a way that influences the binding of substrate and/or its turnover number. Substances that reduce an enzyme's activity in this way are known as **inhibitors.**

Many inhibitors are substances that structurally resemble their enzyme's substrate but either do not react or react very slowly compared to substrate. Such inhibitors are commonly used to probe the chemical and conformational nature of a substrate-binding site as part of an effort to elucidate the enzyme's catalytic mechanism. In addition, many enzyme inhibitors are effective chemotherapeutic agents, since an "unnatural" substrate analog can block the action of a specific enzyme. For example, **methotrexate** (also called **amethopterin**) chemically resembles **dihydrofolate.** Methotrexate binds tightly to the enzyme **dihydrofolate reductase,** thereby preventing it from carrying out its normal function, the reduction of dihydrofolate to **tetrahydrofolate,** an essential cofactor in the biosynthesis of the DNA precursor dTMP (Section 28-3Bd):

Dihydrofolate

dihydrofolate reductase

Tetrahydrofolate

$$\overset{O}{\underset{}{C}} - NHCHCH_2CH_2COO^-$$

$$COO^-$$

$$H_2N$$

Methotrexate

↓ dihydrofolate reductase

NO REACTION

Rapidly dividing cells, such as cancer cells, which are actively engaged in DNA synthesis, are far more susceptible to methotrexate than are slower growing cells such as those of most normal mammalian tissues. Hence, methotrexate, when administered in proper dosage, kills cancer cells without fatally poisoning the host.

There are various mechanisms through which enzyme inhibitors can act. In this section, we discuss several of the simplest such mechanisms and their effects on the kinetic behavior of enzymes that follow the Michaelis–Menten model.

A. Competitive Inhibition

A substance that competes directly with a normal substrate for an enzymatic binding site is known as a **competitive inhibitor.** Such an inhibitor usually resembles the substrate to the extent that it specifically binds to the active site but differs from it so as to be unreactive. Thus methotrexate is a competitive inhibitor of dihydrofolate reductase. Similarly, **succinate dehydrogenase,** a citric acid cycle enzyme that functions to convert **succinate** to **fumarate** (Section 21-3F), is competitively inhibited by **malonate,** which structurally resembles succinate but cannot be dehydrogenated:

$$COO^-$$
$$CH_2$$
$$CH_2$$
$$COO^-$$

Succinate

$$\xrightarrow{\text{succinate dehydrogenase}}$$

Fumarate

$$COO^-$$
$$CH_2$$
$$COO^-$$

Malonate

$$\xrightarrow{\text{succinate dehydrogenase}} \text{NO REACTION}$$

The effectiveness of malonate in competitively inhibiting succinate dehydrogenase strongly suggests that the enzyme's substrate-binding site is designed to bind both of the substrate's carboxylate groups, presumably through the influence of two appropriately placed positively charged residues.

The general model for competitive inhibition is given by the following reaction scheme:

$$E + S \underset{k_{-1}}{\overset{k_1}{\rightleftharpoons}} ES \xrightarrow{k_2} P + E$$
$$+$$
$$I$$
$$K_I \Updownarrow$$
$$EI + S \longrightarrow NO\ REACTION$$

Here it is assumed that I, the inhibitor, binds reversibly to the enzyme and is in rapid equilibrium with it so that

$$K_I = \frac{[E][I]}{[EI]} \qquad [14.32]$$

and EI, the enzyme–inhibitor complex, is catalytically inactive. *A competitive inhibitor therefore acts by reducing the concentration of free enzyme available for substrate binding.*

Our goal, as before, is to express v_o in terms of measurable quantities, in this case $[E]_T$, $[S]$, and $[I]$. We begin, as in the derivation of the Michaelis–Menten equation, with the expression for the conservation condition, which must now take into account the existence of EI.

$$[E]_T = [E] + [EI] + [ES] \qquad [14.33]$$

The enzyme concentration can be expressed in terms of [ES] by rearranging Eq. [14.17] under the steady-state condition:

$$[E] = \frac{K_M[ES]}{[S]} \qquad [14.34]$$

That of the enzyme–inhibitor complex is found by rearranging Eq. [14.32] and substituting Eq. [14.34] into it:

$$[EI] = \frac{[E][I]}{K_I} = \frac{K_M[ES][I]}{[S]K_I} \qquad [14.35]$$

Substituting the latter two results into Eq. [14.33] yields

$$[E]_T = [ES]\left\{\frac{K_M}{[S]}\left(1 + \frac{I}{K_I}\right) + 1\right\}$$

which can be solved for [ES] by rearranging it to

$$[ES] = \frac{[E]_T[S]}{K_M\left(1 + \frac{[I]}{K_I}\right) + [S]}$$

so that, according to Eq. [14.22], the initial velocity is expressed

$$v_o = k_2[ES] = \frac{k_2[E]_T[S]}{K_M\left(1 + \frac{[I]}{K_I}\right) + [S]} \qquad [14.36]$$

Then defining

$$\alpha = \left(1 + \frac{[I]}{K_I}\right) \qquad [14.37]$$

and $V_{max} = k_2[E]_T$ as in Eq. [14.23],

$$v_o = \frac{V_{max}[S]}{\alpha K_M + [S]} \qquad [14.38]$$

This is the Michaelis–Menten equation with K_M modulated by α, a function of the inhibitor concentration (which, according to Eq. [14.37], must always be ≥ 1). The value of [S] at $v_o = V_{max}/2$ is therefore αK_M.

Figure 14-11 shows the hyperbolic plot of Eq. [14.38] for various values of α. Note that as $[S] \to \infty$, $v_o \to V_{max}$ for any value of α. The larger the value of α, however, the greater [S] must be to approach V_{max}. Thus, the inhibitor does not affect the turnover number of the enzyme. Rather, the presence of I has the effect of making [S] appear more dilute than it actually is, or alternatively, making K_M appear larger than it really is. Conversely, increasing [S] shifts the substrate-binding equilibrium toward ES. Hence, there is true competition between I and S for the enzyme's substrate-binding site; their binding is mutually exclusive.

Recasting Eq. [14.38] in the double-reciprocal form yields

$$\frac{1}{v_o} = \left(\frac{\alpha K_M}{V_{max}}\right)\frac{1}{[S]} + \frac{1}{V_{max}} \qquad [14.39]$$

A plot of this equation is linear and has a slope of $\alpha K_M/V_{max}$, a 1/[S] intercept of $-1/\alpha K_M$, and a $1/v_o$ intercept of $1/V_{max}$ (Fig. 14-12). *The double-reciprocal plots for a competitive inhibitor at various concentrations of I intersect at $1/V_{max}$ on the $1/v_o$ axis; this is diagnostic for competitive inhibition as compared with other types of inhibition (Sections 14-3B and 14-3C).*

By determining the values of α at different inhibitor concentrations, the value of K_I can be found from Eq. [14.37]. In this way, competitive inhibitors can be used to probe the structural nature of an active site. For example, to ascertain the importance of the various segments of an ATP molecule

for binding to the active site of an ATP-requiring enzyme, one might determine the K_I, say, for ADP, AMP (adenosine monophosphate), ribose, triphosphate ion, etc. Since many of these ATP components are catalytically inactive, inhibition studies are the most convenient means of monitoring their binding to the enzyme.

If the inhibitor binds irreversibly to the enzyme, the inhibitor is classified as an **inactivator,** as is any agent that somehow inactivates the enzyme. Inactivators truly reduce the effective level of $[E]_T$ at all values of [S]. Reagents that modify specific amino acid residues can act in this manner.

Figure 14-11 Competitive inhibition. Plot of the initial velocity v_o of a simple Michaelis–Menten reaction versus the substrate concentration [S] in the presence of different concentrations of a competitive inhibitor.

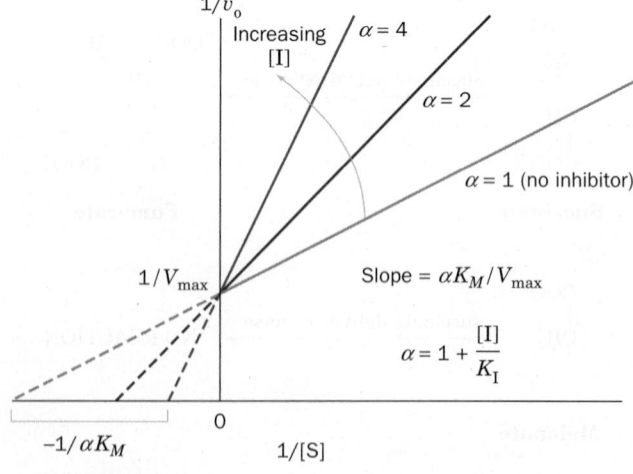

***Figure 14-12* Lineweaver–Burk plot of the competitively inhibited Michaelis–Menten enzyme described by Fig. 14-11.** Note that all lines intersect on the $1/v_o$ axis at $1/V_{max}$. **See the Animated Figures**

B. *Uncompetitive Inhibition*

In **uncompetitive inhibition,** the inhibitor binds directly to the enzyme–substrate complex but not to the free enzyme:

$$E + S \underset{k_{-1}}{\overset{k_1}{\rightleftharpoons}} ES \xrightarrow{k_2} P + E$$
$$+$$
$$I$$
$$K'_I \,\Updownarrow$$
$$ESI \longrightarrow NO\ REACTION$$

The inhibitor-binding step, which has the dissociation constant

$$K'_I = \frac{[ES][I]}{[ESI]} \qquad [14.40]$$

is assumed to be at equilibrium. The binding of the uncompetitive inhibitor, which need not resemble the substrate, is envisioned to cause structural distortion of the active site, thereby rendering the enzyme catalytically inactive. (If the inhibitor binds to enzyme alone, it does so without affecting its affinity for substrate.)

The Michaelis–Menten equation for uncompetitive inhibition, which is derived in Appendix B of this chapter, is

$$v_o = \frac{V_{max}[S]}{K_M + \alpha'[S]} \qquad [14.41]$$

where

$$\alpha' = 1 + \frac{[I]}{K'_I} \qquad [14.42]$$

Inspection of this equation indicates that *at high values of [S], v_o asymptotically approaches V_{max}/α', so that, in contrast to competitive inhibition, the effects of uncompetitive inhibition on V_{max} are not reversed by increasing the substrate concentration.* However, at low substrate concentrations, that is, when $[S] \ll K_M$, the effect of an uncompetitive inhibitor becomes negligible, again the opposite behavior of a competitive inhibitor.

When cast in the double-reciprocal form, Eq. [14.41] becomes

$$\frac{1}{v_o} = \left(\frac{K_M}{V_{max}}\right)\frac{1}{[S]} + \frac{\alpha'}{V_{max}} \qquad [14.43]$$

The Lineweaver–Burk plot for uncompetitive inhibition is linear with slope K_M/V_{max}, as in the uninhibited reaction, and with $1/v_o$ and $1/[S]$ intercepts of α'/V_{max} and $-\alpha'/K_M$, respectively. *A series of Lineweaver–Burk plots at various uncompetitive inhibitor concentrations consists of a family of parallel lines* (Fig. 14-13). *This is diagnostic for uncompetitive inhibition.*

Uncompetitive inhibition requires that the inhibitor affect the catalytic function of the enzyme but not its substrate binding. For single-substrate enzymes it is difficult to conceive of how this could happen with the

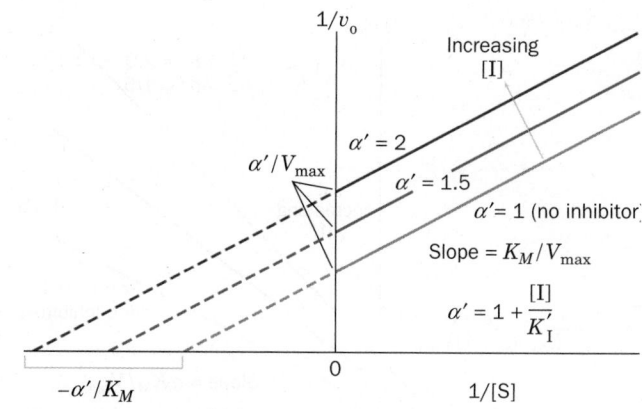

Figure 14-13   Lineweaver–Burk plot of a simple Michaelis–Menten enzyme in the presence of uncompetitive inhibitor. Note that all lines have identical slopes of K_M/V_{max}. 🔊 **See the Animated Figures**

exception of small inhibitors such as protons (see Section 14-4) or metal ions. As we discuss in Section 14-5C, however, uncompetitive inhibition is important for multisubstrate enzymes.

C. *Mixed Inhibition*

If both the enzyme and the enzyme–substrate complex bind inhibitor, the following model results:

$$E + S \underset{k_{-1}}{\overset{k_1}{\rightleftharpoons}} ES \xrightarrow{k_2} P + E$$
$$+ \qquad\qquad +$$
$$I \qquad\qquad I$$
$$K_I \Updownarrow \qquad K'_I \Updownarrow$$
$$EI \qquad\qquad ESI \longrightarrow NO\ REACTION$$

Both of the inhibitor-binding steps are assumed to be at equilibrium but with different dissociation constants:

$$K_I = \frac{[E][I]}{[EI]} \quad\text{and}\quad K'_I = \frac{[ES][I]}{[ESI]} \qquad [14.44]$$

This phenomenon is alternatively known as **mixed inhibition** or **noncompetitive inhibition.** Presumably a mixed inhibitor binds to enzyme sites that participate in both substrate binding and catalysis.

The Michaelis–Menten equation for mixed inhibition, which is derived in Appendix C of this chapter, is

$$v_o = \frac{V_{max}[S]}{\alpha K_M + \alpha'[S]} \qquad [14.45]$$

where α and α' are defined in Eqs. [14.37] and [14.42], respectively. It can be seen from Eq. [14.45] that the name "mixed inhibition" arises from the fact that the denominator has the factor α multiplying K_M as in competitive inhibition (Eq. [14.38]) and the factor α' multiplying [S] as in uncompetitive inhibition (Eq. [14.41]). Mixed inhibitors

Figure 14-14 Lineweaver–Burk plot of a simple Michaelis–Menten enzyme in the presence of a mixed inhibitor. Note that the lines all intersect to the left of the $1/v_o$ axis. The coordinates of this intersection point are given in brackets. When $K_I = K_I'$, $\alpha = \alpha'$ and the lines intersect on the $1/[S]$ axis at $-1/K_M$.

🐭 **See the Animated Figures**

are therefore effective at both high and low substrate concentrations.

The Lineweaver–Burk equation for mixed inhibition is

$$\frac{1}{v_o} = \left(\frac{\alpha K_M}{V_{max}}\right)\frac{1}{[S]} + \frac{\alpha'}{V_{max}} \qquad [14.46]$$

The plot of this equation consists of lines that have slope $\alpha K_M/V_{max}$ with a $1/v_o$ intercept of α'/V_{max} and a $1/[S]$ intercept of $-\alpha'/\alpha K_M$ (Fig. 14-14). Algebraic manipulation of Eq. [14.46] for different values of [I] reveals that this equation describes a family of lines that intersect to the left of the $1/v_o$ axis (Fig. 14-14). For the special case in which $K_I = K_I'$ ($\alpha = \alpha'$), the intersection is, in addition, on the $1/[S]$ axis, a situation which, in an ambiguity of nomenclature, is sometimes described as noncompetitive inhibition.

Table 14-2 provides a summary of the preceding results concerning the inhibition of simple Michaelis–Menten

enzymes. The quantities K_M^{app} and V_{max}^{app} are the "apparent" values of K_M and V_{max} that would actually be observed in the presence of inhibitor for the Michaelis–Menten equation describing the inhibited enzymes.

4 EFFECTS OF pH

Enzymes, being proteins, have properties that are quite pH sensitive. Most proteins, in fact, are active only within a narrow pH range, typically 5 to 9. This is a result of the effects of pH on a combination of factors: (1) the binding of substrate to enzyme, (2) the catalytic activity of the enzyme, (3) the ionization of substrate, and (4) the variation of protein structure (usually significant only at extremes of pH).

a. pH Dependence of Simple Michaelis–Menten Enzymes

The initial rates for many enzymatic reactions exhibit bell-shaped curves as a function of pH (e.g., Fig. 14-15). These curves reflect the ionizations of certain amino acid residues that must be in a specific ionization state for enzyme activity. The following model can account for such pH effects.

$$
\begin{array}{ccc}
\text{E}^- & & \text{ES}^- \\
K_{E2} \big\Vert \text{H}^+ & & K_{ES2} \big\Vert \text{H}^+ \\
\text{EH} + \text{S} \underset{k_{-1}}{\overset{k_1}{\rightleftharpoons}} \text{ESH} \xrightarrow{k_2} \text{P} + \text{EH} \\
K_{E1} \big\Vert \text{H}^+ & & K_{ES1} \big\Vert \text{H}^+ \\
\text{EH}_2^+ & & \text{ESH}_2^+
\end{array}
$$

In this expansion of the simple one substrate–no back reaction mechanism, it is assumed that only EH and ESH are catalytically active.

The Michaelis–Menten equation for this model, which is derived in Appendix D, is

$$v_o = \frac{V'_{max}[S]}{K'_M + [S]} \qquad [14.47]$$

Table 14-2   Effects of Inhibitors on the Parameters of the Michaelis–Menten Equation[a]

Type of Inhibition	V_{max}^{app}	K_M^{app}
None	V_{max}	K_M
Competitive	V_{max}	αK_M
Uncompetitive	V_{max}/α'	K_M/α'
Mixed	V_{max}/α'	$\alpha K_M/\alpha'$

[a] $\alpha = 1 + \dfrac{[I]}{K_I}$ and $\alpha' = 1 + \dfrac{[I]}{K_I'}$.

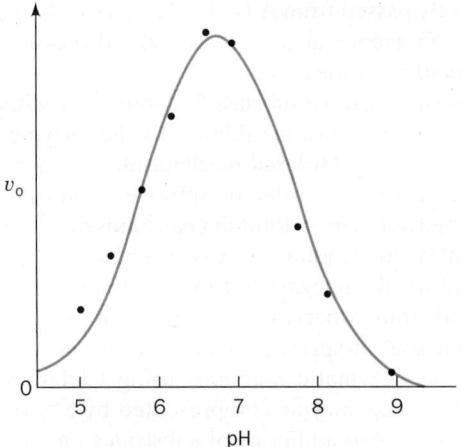

Figure 14-15 **Effect of pH on the initial rate of the reaction catalyzed by the enzyme fumarase.** [After Tanford, C., *Physical Chemistry of Macromolecules,* p. 647, Wiley (1961).]

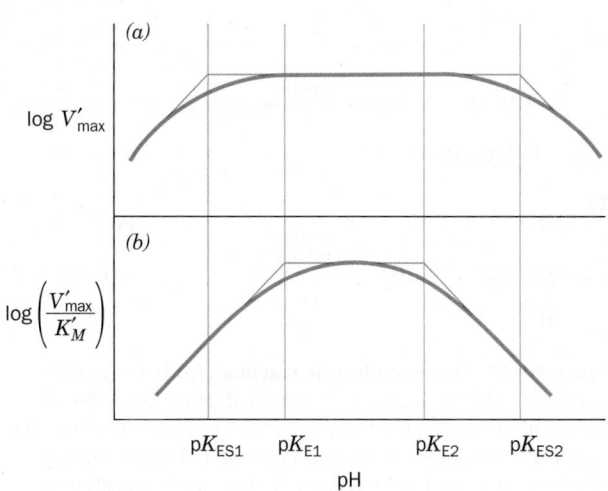

Figure 14-16 **The pH dependence of (*a*) log V'_{max} and (*b*) log (V'_{max}/K'_{M}).** The light blue lines indicate how the values of the molecular ionization constants can be determined by graphical extrapolation.

Here the apparent Michaelis–Menten parameters are defined

$$V'_{max} = V_{max}/f_2 \quad \text{and} \quad K'_M = K_M(f_1/f_2)$$

where

$$f_1 = \frac{[\text{H}^+]}{K_{E1}} + 1 + \frac{K_{E2}}{[\text{H}^+]}$$

$$f_2 = \frac{[\text{H}^+]}{K_{ES1}} + 1 + \frac{K_{ES2}}{[\text{H}^+]}$$

and V_{max} and K_M refer to the active forms of the enzyme, EH and ESH. Note that at any given pH, Eq. [14.47] behaves as a simple Michaelis–Menten equation, but because of the pH dependence of f_1 and f_2, v_o varies with pH in a bell-shaped manner (e.g., Fig. 14-15).

b. Evaluation of Ionization Constants

The ionization constants of enzymes that obey Eq. [14.47] can be evaluated by the analysis of the curves of log V'_{max} versus pH, which provides values of K_{ES1} and K_{ES2} (Fig. 14-16*a*), and of log (V'_{max}/K'_M) versus pH, which yields K_{E1} and K_{E2} (Fig. 14-16*b*). This, of course, entails the determination of the enzyme's Michaelis–Menten parameters at each of a series of different pH's.

The measured pK's often provide valuable clues as to the identities of the amino acid residues essential for enzymatic activity. For example, a measured pK of ~4 suggests that an Asp or Glu residue is essential to the enzyme. Similarly, pK's of ~6 or ~10 suggest the participation of a His or a Lys residue, respectively. However, a given acid–base group may vary by as much as several pH units from its expected value as a consequence of the electrostatic influence of nearby charged groups, as well as of the prox-

imity of regions of low polarity. For example, the carboxylate group of a Glu residue forming a salt bridge with a Lys residue is stabilized by the nearby positive charge and therefore has a lower pK than it would otherwise have; that is, it is more difficult to protonate. Conversely, a carboxylate group immersed in a region of low polarity is less acidic than normal because it attracts protons more strongly than if it were in a region of higher polarity. The identification of a kinetically characterized pK with a particular amino acid residue must therefore be verified by other types of measurements such as the use of group-specific reagents to inactivate a putative essential residue.

5 BISUBSTRATE REACTIONS

We have heretofore been concerned with reactions involving enzymes that require only a single substrate. Yet enzymatic reactions involving two substrates and yielding two products

$$\text{A} + \text{B} \xrightleftharpoons{\text{E}} \text{P} + \text{Q}$$

account for ~60% of known biochemical reactions. Almost all of these so-called **bisubstrate reactions** are either **transferase** reactions in which the enzyme catalyzes the transfer of a specific functional group, X, from one of the substrates to the other:

$$\text{P}-\text{X} + \text{B} \xrightleftharpoons{\text{E}} \text{P} + \text{B}-\text{X}$$

or oxidation–reduction reactions in which reducing equivalents are transferred between the two substrates. For example, the hydrolysis of a peptide bond by trypsin (Section 7-1Da) is the transfer of the peptide carbonyl group from

(a)

$$R_1-\overset{\overset{\displaystyle O}{\|}}{C}-NH-R_2 \;+\; H_2O \;\xrightarrow{\text{trypsin}}\; R_1-\overset{\overset{\displaystyle O}{\|}}{C}-O^- \;+\; H_3\overset{+}{N}-R_2$$

Polypeptide

(b)

$$CH_3-\overset{\overset{\displaystyle H}{|}}{\underset{\underset{\displaystyle H}{|}}{C}}-OH \;+\; NAD^+ \;\xrightarrow[\displaystyle H^+]{\text{alcohol}\atop\text{dehydrogenase}}\; CH_3-\overset{\overset{\displaystyle O}{\|}}{C}H \;+\; NADH$$

Figure 14-17 **Some bisubstrate reactions.** (*a*) In the peptide hydrolysis reaction catalyzed by trypsin, the peptide carbonyl group, with its pendent polypeptide chain, is transferred from the peptide nitrogen atom to a water molecule. (*b*) In the alcohol dehydrogenase reaction, a hydride ion is formally transferred from ethanol to NAD$^+$.

the peptide nitrogen atom to water (Fig. 14-17*a*). Similarly, in the alcohol dehydrogenase reaction (Section 13-2A), a hydride ion is formally transferred from ethanol to NAD$^+$ (Fig. 14-17*b*). Although such bisubstrate reactions could, in principle, occur through a vast variety of mechanisms, only a few types are commonly observed.

A. *Terminology*

We shall follow the nomenclature system introduced by W.W. Cleland for representing enzymatic reactions:

1. Substrates are designated by the letters A, B, C, and D *in the order that they add to the enzyme.*

2. Products are designated P, Q, R, and S *in the order that they leave the enzyme.*

3. Stable enzyme forms are designated E, F, and G with E being the free enzyme, if such distinctions can be made. A stable enzyme form is defined as one that by itself is incapable of converting to another stable enzyme form (see below).

4. The numbers of reactants and products in a given reaction are specified, in order, by the terms **Uni** (one), **Bi** (two), **Ter** (three), and **Quad** (four). A reaction requiring one substrate and yielding three products is designated a Uni Ter reaction. In this section, we shall be concerned with reactions that require two substrates and yield two products, that is, Bi Bi reactions. Keep in mind, however, that there are numerous examples of even more complex reactions.

a. Types of Bi Bi Reactions
Enzyme-catalyzed group-transfer reactions fall under two major mechanistic classifications:

1. Sequential Reactions: *Reactions in which all substrates must combine with the enzyme before a reaction can occur and products can be released are known as **Sequential reactions.*** In such reactions, the group being trans-

ferred, X, is directly passed from A (= P—X) to B, yielding P and Q (= B—X). Hence, such reactions are also called **single-displacement reactions.**

Sequential reactions can be subclassified into those with a compulsory order of substrate addition to the enzyme, which are said to have an **Ordered mechanism,** and those with no preference for the order of substrate addition, which are described as having a **Random mechanism.** In the Ordered mechanism, the binding of the first substrate is apparently required for the enzyme to form the binding site for the second substrate, whereas for the Random mechanism, both binding sites are present on the free enzyme.

Let us describe enzymatic reactions using Cleland's shorthand notation. The enzyme is represented by a horizontal line and successive additions of substrates and release of products are denoted by vertical arrows. Enzyme forms are placed under the line and rate constants, if given, are to the left of the arrow or on top of the line for forward reactions. An **Ordered Bi Bi** reaction is represented:

where A and B are said to be the **leading** and **following** substrates, respectively. Here, only minimal details are given concerning the interconversions of intermediate enzyme forms because, as we have seen for reversible single-substrate enzymes, steady-state kinetic measurements provide no information concerning the number of intermediates in a given reaction step. Many NAD$^+$- and NADP$^+$-requiring dehydrogenases follow an Ordered Bi Bi mechanism in which the coenzyme is the leading reactant.

A **Random Bi Bi** reaction is diagrammed:

Some dehydrogenases and kinases operate through Random Bi Bi mechanisms.

2. Ping Pong Reactions: *Mechanisms in which one or more products are released before all substrates have been added are known as **Ping Pong reactions.*** The **Ping Pong Bi Bi** reaction is represented by

In it, a functional group X of the first substrate A (= P—X) is displaced from the substrate by the enzyme E to yield the first product P and a stable enzyme form F (= E—X) in which X is tightly (often covalently) bound to the enzyme (Ping). In the second stage of the reaction, X is displaced from the enzyme by the second substrate B to yield the second product Q (= B—X), thereby regenerating the original form of the enzyme, E (Pong). Such reactions are therefore also known as **double-displacement reactions.** *Note that in Ping Pong Bi Bi reactions, the substrates A and B do not encounter one another on the surface of the enzyme.* Many enzymes, including chymotrypsin (Section 15-3), transaminases (Section 26-1A), and some flavoenzymes, react with Ping Pong mechanisms.

B. *Rate Equations*

Steady-state kinetic measurements can be used to distinguish among the foregoing bisubstrate mechanisms. In order to do so, one must first derive their rate equations. This can be done in much the same manner as for single-substrate enzymes, that is, solving a set of simultaneous linear equations consisting of an equation expressing the steady-state condition for each kinetically distinct enzyme complex plus one equation representing the conservation condition for the enzyme. This, of course, is a more complex undertaking for bisubstrate enzymes than it is for single-substrate enzymes.

The rate equations for the above described bisubstrate mechanisms in the absence of products are given below in double-reciprocal form.

a. Ordered Bi Bi

$$\frac{1}{v_o} = \frac{1}{V_{max}} + \frac{K_M^A}{V_{max}[A]} + \frac{K_M^B}{V_{max}[B]} + \frac{K_S^A K_M^B}{V_{max}[A][B]} \quad [14.48]$$

b. Rapid Equilibrium Random Bi Bi

The rate equation for the general Random Bi Bi reaction is quite complicated. However, in the special case that both substrates are in rapid and independent equilibrium with the enzyme, that is, when the EAB–EPQ interconversion is rate determining, the initial rate equation reduces to the

following relatively simple form. This mechanism is known as the **Rapid Equilibrium Random Bi Bi** mechanism:

$$\frac{1}{v_o} = \frac{1}{V_{max}} + \frac{K_S^A K_M^B}{V_{max}K_S^B[A]} + \frac{K_M^B}{V_{max}[B]} + \frac{K_S^A K_M^B}{V_{max}[A][B]} \quad [14.49]$$

c. Ping Pong Bi Bi

$$\frac{1}{v_o} = \frac{K_M^A}{V_{max}[A]} + \frac{K_M^B}{V_{max}[B]} + \frac{1}{V_{max}} \quad [14.50]$$

d. Physical Significance of the Bisubstrate Kinetic Parameters

The kinetic parameters in the equations describing bisubstrate reactions have meanings similar to those for single-substrate reactions. V_{max} is the maximal velocity of the enzyme obtained when both A and B are present at saturating concentrations, K_M^A and K_M^B are the respective concentrations of A and B necessary to achieve $\frac{1}{2}V_{max}$ in the presence of a saturating concentration of the other, and K_S^A and K_S^B are the respective dissociation constants of A and B from the enzyme, E.

C. *Differentiating Bisubstrate Mechanisms*

One can discriminate between Ping Pong and Sequential mechanisms from their contrasting properties in linear plots such as those of the Lineweaver–Burk type.

a. Diagnostic Plot for Ping Pong Bi Bi Reactions

A plot of $1/v_o$ versus $1/[A]$ at constant [B] for Eq. [14.50] yields a straight line of slope K_M^A/V_{max} and an intercept on the $1/v_o$ axis equal to the last two terms in Eq. [14.50]. Since the slope is independent of [B], such plots for different values of [B] yield a family of parallel lines (Fig. 14-18). A plot of $1/v_o$ versus $1/[B]$ for different values of [A] likewise yields a family of parallel lines. *Such parallel lines are diagnostic for a Ping Pong mechanism.*

b. Diagnostic Plot for Sequential Bi Bi Reactions

The equations representing the Ordered Bi Bi mechanism (Eq. [14.48]) and the Rapid Equilibrium Random Bi

Figure 14-18 **Double-reciprocal plots for an enzymatic reaction with a Ping Pong Bi Bi mechanism.** (*a*) Plots of $1/v_o$ versus $1/[A]$ at various constant concentrations of B. (*b*) Plots of $1/v_o$ versus $1/[B]$ at various constant concentrations of A.

(a)

(b)

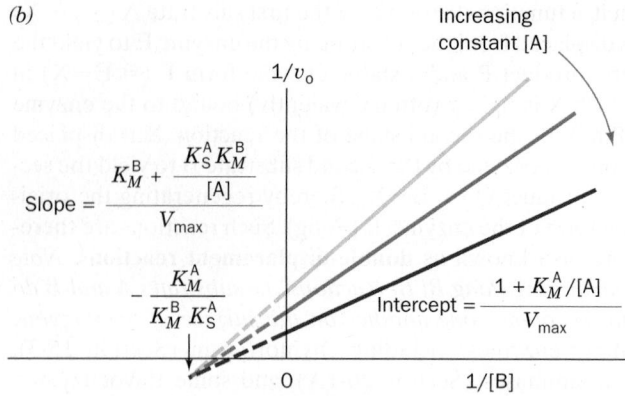

Figure 14-19 **Double-reciprocal plots of an enzymatic reaction with a Sequential Bi Bi mechanism.** (*a*) Plots of $1/v_o$ versus $1/[A]$ at various constant concentrations of B. (*b*) Plots of $1/v_o$ versus $1/[B]$ at various constant concentrations of A. The corresponding plots for Rapid Equilibrium Random Bi Bi reactions have identical appearances; their lines all intersect to the left of the $1/v_o$ axis.

Bi mechanism (Eq. [14.49]) have identical functional dependence on [A] and [B].

Equation [14.48] can be rearranged to

$$\frac{1}{v_o} = \frac{K_M^A}{V_{max}}\left(1 + \frac{K_S^A K_M^B}{K_M^A [B]}\right)\frac{1}{[A]} + \frac{1}{V_{max}}\left(1 + \frac{K_M^B}{[B]}\right) \quad [14.51]$$

Thus plotting $1/v_o$ versus $1/[A]$ for constant [B] yields a linear plot with a slope equal to the coefficient of $1/[A]$ and an intercept on the $1/v_o$ axis equal to the second term of Eq. [14.51] (Fig. 14-19*a*). Alternatively, Eq. [14.48] can be rearranged to

$$\frac{1}{v_o} = \frac{K_M^B}{V_{max}}\left(1 + \frac{K_S^A}{[A]}\right)\frac{1}{[B]} + \frac{1}{V_{max}}\left(1 + \frac{K_M^A}{[A]}\right) \quad [14.52]$$

which yields a linear plot of $1/v_o$ versus $1/[B]$ for constant [A] with a slope equal to the coefficient of $1/[B]$ and an intercept on the $1/v_o$ axis equal to the second term of Eq. [14.52] (Fig. 14-19*b*). *The characteristic feature of these plots, which is indicative of a Sequential mechanism, is that the lines intersect to the left of the $1/v_o$ axis.*

c. Differentiating Random and Ordered Sequential Mechanisms

The Ordered Bi Bi mechanism may be experimentally distinguished from the Random Bi Bi mechanism through **product inhibition studies.** If only one product of the reaction, P or Q, is added to the reaction mixture, the reverse reaction still cannot occur. Nevertheless, by binding to the enzyme, this product will inhibit the forward reaction. For an Ordered Bi Bi reaction, Q (= B—X, the second product to be released) directly competes with A (= P—X, the leading substrate) for binding to E and hence is a competitive inhibitor of A when [B] is fixed (the presence of X in Q = B—X interferes with the binding of A = P—X). However, since B combines with EA, not E, Q is a mixed inhibitor of B when [A] is fixed (Q interferes with both the binding of B to enzyme and with the catalysis of the reaction). Similarly, P, which combines only with EQ, is a mixed inhibitor of A when [B] is held constant and of B when [A] is held constant. In contrast, in a Rapid Equilibrium Bi Bi reaction, since both products as well as both substrates can combine directly with E, both P and Q are competitive inhibitors of A when [B] is constant and of B when [A] is constant. These product inhibition patterns are summarized in Table 14-3.

D. *Isotope Exchange*

Mechanistic conclusions based on kinetic analyses alone are fraught with uncertainties and are easily confounded by inaccurate experimental data. A particular mechanism for an enzyme is therefore greatly corroborated if the mechanism can be shown to conform to experimental criteria other than kinetic analysis.

*Sequential (single-displacement) and Ping Pong (double-displacement) bisubstrate mechanisms may be differentiated through the use of **isotope exchange** studies.* Double-displacement reactions are capable of exchanging an isotope from the first product P back to the first substrate A in

Table 14-3 **Patterns of Product Inhibition for Sequential Bisubstrate Mechanisms**

Mechanism	Product Inhibitor	[A] Variable	[B] Variable
Ordered Bi Bi	P	Mixed	Mixed
	Q	Competitive	Mixed
Rapid Equilibrium Random Bi Bi	P	Competitive	Competitive
	Q	Competitive	Competitive

the absence of the second substrate. Consider an overall Ping Pong reaction catalyzed by the bisubstrate enzyme E

$$P-X + B \overset{E}{\rightleftharpoons} P + B-X$$

in which, as usual, A = P—X, Q = B—X, and X is the group that is transferred from one substrate to the other in the course of the reaction. Only the first step of the reaction can take place in the absence of B. If a small amount of isotopically labeled P, denoted P*, is added to this reaction mixture then, in the reverse reaction, P*—X will form:

Forward reaction   E + P—X ⟶ E—X + P
Reverse reaction   E—X + P* ⟶ E + P*—X

that is, isotopic exchange will occur.

In contrast, let us consider the first step of a Sequential reaction. Here a noncovalent enzyme–substrate complex forms:

$$E + P-X \rightleftharpoons E \cdot P-X$$

Addition of P* cannot result in an exchange reaction because no covalent bonds are broken in the formation of E · P—X; that is, there is no P released from the enzyme to exchange with P*. The demonstration of isotopic exchange for a bisubstrate enzyme is therefore convincing evidence favoring a Ping Pong mechanism.

a. Isotope Exchange in Sucrose Phosphorylase and Maltose Phosphorylase

The enzymes **sucrose phosphorylase** and **maltose phosphorylase** provide two clear-cut examples of how enzymatically catalyzed isotopic exchange reactions are used to differentiate kinetic mechanisms. Sucrose phosphorylase catalyzes the overall reaction

Glucose—fructose + phosphate
Sucrose
⇅ E
Glucose-1-phosphate + fructose

If the enzyme is incubated with sucrose and isotopically labeled fructose in the absence of phosphate, it is observed that the label passes into the sucrose:

Glucose—fructose + fructose*
Sucrose
⇅ E
Glucose—fructose* + fructose

For the reverse reaction, if the enzyme is incubated with glucose-1-phosphate and ^{32}P-labeled phosphate, this label exchanges into the glucose-1-phosphate:

Glucose-1-phosphate + phosphate*
⇅ E
Glucose-1-phosphate* + phosphate

These observations indicate that a tight glucosyl–enzyme complex is formed with the release of fructose, thereby establishing that the sucrose phosphorylase reaction occurs via a Ping Pong mechanism. This finding has been conclusively corroborated by the isolation and characterization of the glucosyl–enzyme complex.

The enzyme **maltose phosphorylase** catalyzes a similar overall reaction:

Glucose—glucose + phosphate
Maltose
⇅ E
Glucose-1-phosphate + glucose

In contrast to sucrose phosphorylase, however, it does not catalyze isotopic exchange between glucose-1-phosphate and [^{32}P]phosphate or between maltose and [^{14}C]glucose. Likewise, a glucosyl–enzyme complex has not been detected. This evidence is consistent with maltose phosphorylase having a sequential mechanism.

APPENDIX Derivations of Michaelis–Menten Equation Variants

A. The Michaelis–Menten Equation for Reversible Reactions–Equation [14.30]

The conservation condition for the reversible reaction with one intermediate (Section 14-2Ca) is

$$[E]_T = [E] + [ES] \qquad [14.A1]$$

The steady-state condition (as well as the equilibrium condition) is

$$\frac{d[ES]}{dt} = k_1[E][S] + k_{-2}[E][P] - (k_{-1} + k_2)[ES] = 0$$
$$[14.A2]$$

so that

$$[E] = \left(\frac{k_{-1} + k_2}{k_1[S] + k_{-2}[P]}\right)[ES] \qquad [14.A3]$$

Substituting this result into Eq. [14.A1] yields

$$[E]_T = \left(\frac{k_{-1} + k_2}{k_1[S] + k_{-2}[P]} + 1\right)[ES] \qquad [14.A4]$$

The velocity of the reaction is expressed

$$v = -\frac{d[S]}{dt} = k_1[E][S] - k_{-1}[ES] \qquad [14.A5]$$

which can be combined with Eq. [14.A3] to give

$$v = \left(\frac{k_1 [S](k_{-1} + k_2)}{k_1 [S] + k_{-2}[P]} - k_{-1} \right)[ES] \quad [14.A6]$$

which, in turn, is combined with Eq. [14.A4] to yield

$$v = \left(\frac{k_1 k_2 [S] - k_{-1} k_{-2}[P]}{k_{-1} + k_2 + k_1 [S] + k_{-2}[P]} \right)[E]_T \quad [14.A7]$$

Dividing the numerator and denominator of this equation by $(k_{-1} + k_2)$ results in

$$v = \left(\frac{k_2 \left(\dfrac{k_1}{k_{-1} + k_2} \right)[S] - k_{-1} \left(\dfrac{k_{-2}}{k_{-1} + k_2} \right)[P]}{1 + \left(\dfrac{k_1}{k_{-1} + k_2} \right)[S] + \left(\dfrac{k_{-2}}{k_{-1} + k_2} \right)[P]} \right)[E]_T \quad [14.A8]$$

Then, if we define the following parameters analogously with the constants of the Michaelis–Menten equation (Eqs. [14.23] and [14.21]),

$$V_{max}^f = k_2 [E]_T \qquad V_{max}^r = k_{-1}[E]_T$$

$$K_M^S = \frac{k_{-1} + k_2}{k_1} \qquad K_M^P = \frac{k_{-1} + k_2}{k_{-2}}$$

we obtain the Michaelis–Menten equation for a reversible one-intermediate reaction:

$$v = \frac{\dfrac{V_{max}^f [S]}{K_M^S} - \dfrac{V_{max}^r [P]}{K_M^P}}{1 + \dfrac{[S]}{K_M^S} + \dfrac{[P]}{K_M^P}} \quad [14.30]$$

B. *Michaelis–Menten Equation for Uncompetitive Inhibition–Equation [14.41]*

For uncompetitive inhibition (Section 14-3B), the inhibitor binds to the Michaelis complex with dissociation constant

$$K_I' = \frac{[ES][I]}{[ESI]} \quad [14.A9]$$

The conservation condition is

$$[E]_T = [E] + [ES] + [ESI] \quad [14.A10]$$

Substituting in Eqs. [14.34] and [14.A9] yields

$$[E]_T = [ES] \left(\frac{K_M}{[S]} + 1 + \frac{[I]}{K_I'} \right) \quad [14.A11]$$

Defining α' analogously to Eq. [14.37] as

$$\alpha' = 1 + \frac{[I]}{K_I'} \quad [14.A12]$$

and v_o and V_{max} as in Eqs. [14.22] and [14.23], respectively,

$$v_o = k_2 [ES] = \frac{V_{max}}{\dfrac{K_M}{[S]} + \alpha'} \quad [14.A13]$$

which on rearrangement yields the Michaelis–Menten equation for uncompetitive inhibition:

$$v_o = \frac{V_{max}[S]}{K_M + \alpha'[S]} \quad [14.41]$$

C. *The Michaelis–Menten Equation for Mixed Inhibition–Equation [14.45]*

In mixed inhibition (Section 14-3C), the inhibitor-binding steps have different dissociation constants:

$$K_I = \frac{[E][I]}{[EI]} \quad \text{and} \quad K_I' = \frac{[ES][I]}{[ESI]} \quad [14.A14]$$

(Here, for the sake of mathematical simplicity, we are making the thermodynamically unsupportable assumption that EI does not react with S to form ESI. Inclusion of this reaction requires a more complex derivation than that given here but leads to results that are substantially the same.) The conservation condition for this reaction scheme is

$$[E]_T = [E] + [EI] + [ES] + [ESI] \quad [14.A15]$$

so that substituting in Eqs. [14.A14]

$$[E]_T = [E]\left(1 + \frac{[I]}{K_I}\right) + [ES]\left(1 + \frac{[I]}{K_I'}\right) \quad [14.A16]$$

Defining α and α' as in Eqs. [14.37] and [14.A12], respectively, Eq. [14.A16] becomes

$$[E]_T = [E]\alpha + [ES]\alpha' \quad [14.A17]$$

Then substituting in Eq. [14.34]

$$[E]_T = [ES]\left(\frac{\alpha K_M}{[S]} + \alpha' \right) \quad [14.A18]$$

Defining v_o and V_{max} as in Eqs. [14.22] and [14.23] results in the Michaelis–Menten equation for mixed inhibition:

$$v_o = \frac{V_{max}[S]}{\alpha K_M + \alpha'[S]} \quad [14.45]$$

D. *The Michaelis–Menten Equation for Ionizable Enzymes–Equation [14.47]*

In the model presented in Section 14-4a to account for the effect of pH on enzymes, the dissociation constants for the ionizations are

$$K_{E2} = \frac{[H^+][E^-]}{[EH]} \qquad K_{ES2} = \frac{[H^+][ES^-]}{[ESH]}$$

$$K_{E1} = \frac{[H^+][EH]}{[EH_2^+]} \qquad K_{ES1} = \frac{[H^+][ESH]}{[ESH_2^+]} \quad [14.A19]$$

Protonation and deprotonation are among the fastest known reactions, so that, with the exception of the few enzymes with extremely high turnover numbers, it can be reasonably assumed that all acid–base reactions are at equilibrium. The conservation condition is

$$[E]_T = [EH]_T + [ESH]_T \qquad [14.A20]$$

where $[E]_T$ is the total enzyme present in any form,

$$\begin{aligned}
[EH]_T &= [EH_2^+] + [EH] + [E^-] \\
&= [EH]\left(\frac{[H^+]}{K_{E1}} + 1 + \frac{K_{E2}}{[H^+]}\right) \\
&= [EH]f_1 \qquad [14.A21]
\end{aligned}$$

and

$$\begin{aligned}
[ESH]_T &= [ESH_2^+] + [ESH] + [ES^-] \\
&= [ESH]\left(\frac{[H^+]}{K_{ES1}} + 1 + \frac{K_{ES2}}{[H^+]}\right) \\
&= [ESH]f_2 \qquad [14.A22]
\end{aligned}$$

Then making the steady-state assumption

$$\frac{d[ESH]}{dt} = k_1[EH][S] - (k_{-1} + k_2)[ESH] = 0 \qquad [14.A23]$$

and solving for [EH]

$$[EH] = \frac{(k_{-1} + k_2)[ESH]}{k_1[S]} = \frac{K_M[ESH]}{[S]} \qquad [14.A24]$$

Therefore, from Eq. [14.A21],

$$[EH]_T = \frac{K_M[ESH]f_1}{[S]} \qquad [14.A25]$$

which, together with Eqs. [14.A20] and [14.A22], yields

$$[E]_T = [ESH]\left(\frac{K_Mf_1}{[S]} + f_2\right) \qquad [14.A26]$$

As in the simple Michaelis–Menten derivation, the initial rate is

$$v_o = k_2[ESH] = \frac{k_2[E]_T}{\left(\frac{K_Mf_1}{[S]}\right) + f_2} = \frac{(k_2/f_2)[E]_T[S]}{K_M(f_1/f_2) + [S]}$$

$$[14.A27]$$

Then defining the "apparent" values of K_M and $V_{max} = k_2[E]_T$ at a given pH:

$$K'_M = K_M(f_1/f_2) \qquad [14.A28]$$

and

$$V'_{max} = V_{max}/f_2 \qquad [14.A29]$$

the Michaelis–Menten equation modified to account for pH effects is

$$v_o = \frac{V'_{max}[S]}{K'_M + [S]} \qquad [14.47]$$

CHAPTER SUMMARY

1 Chemical Kinetics Complicated reaction processes occur through a series of elementary reaction steps defined as having a molecularity equal to the number of molecules that simultaneously collide to form products. The order of a reaction can be determined from the characteristic functional form of its progress curve. Transition state theory postulates that the rate of a reaction depends on the free energy of formation of its activated complex. This complex, which occurs at the free energy maximum of the reaction coordinate, is poised between reactants and products and is therefore also known as the transition state. Transition state theory explains that catalysis results from the reduction of the free energy difference between the reactants and the transition state.

2 Enzyme Kinetics In the simplest enzymatic mechanism, the enzyme and substrate reversibly combine to form an enzyme–substrate complex known as the Michaelis complex, which may irreversibly decompose to form product and the regenerated enzyme. The rate of product formation is expressed by the Michaelis–Menten equation, which is derived under the assumption that the concentration of the Michaelis complex is constant, that is, at a steady state. The Michaelis–Menten equation, which has the functional form of a rectangular hyperbola, has two parameters: V_{max}, the maximal rate of the reaction, which occurs when the substrate concentration is saturating, and K_M, the Michaelis constant, which has the value of the substrate concentration at the half-maximal reaction rate. These parameters may be graphically determined using the

Lineweaver–Burk plot. Physically more realistic models of enzyme mechanisms than the Michaelis–Menten model assume the enzymatic reaction to be reversible and to have one or more intermediates. The functional form of the equations describing the reaction rates for these models is independent of their number of intermediates, so that the models cannot be differentiated using only steady-state kinetic measurements.

3 Inhibition Enzymes may be inhibited by competitive inhibitors, which compete with the substrate for the enzymatic binding site. The effect of a competitive inhibitor may be reversed by increasing the substrate concentration. An uncompetitive inhibitor inactivates a Michaelis complex on binding to it. The maximal rate of an uncompetitively inhibited enzyme is a function of inhibitor concentration, and therefore the effect of an uncompetitive inhibitor cannot be reversed by increasing substrate concentration. In mixed inhibition, the inhibitor binds to both the enzyme and the enzyme–substrate complex to form a complex that is catalytically inactive. The rate equation describing this situation has characteristics of both competitive and uncompetitive reactions.

4 Effects of pH The rate of an enzymatic reaction is a function of hydrogen ion concentration. At any pH, the rate of a simple enzymatic reaction can be described by the Michaelis–Menten equation. However, its parameters V_{max} and K_M vary with pH. By the evaluation of kinetic rate curves as a function of pH, the pK's of an enzyme's ionizable binding and catalytic groups can be determined, which may help identify these groups.

5 Bisubstrate Reactions   The majority of enzymatic reactions are bisubstrate reactions in which two substrates react to form two products. Bisubstrate reactions may have Ordered or Random Sequential mechanisms or Ping Pong Bi Bi mechanisms, among others. The initial rate equations for any of these mechanisms involve five parameters, which are analogous to either Michaelis–Menten equation parameters or equilibrium constants. The various bisubstrate mechanisms may be experimentally differentiated according to the forms of their double-reciprocal plots and from the nature of their product inhibition patterns. Isotope exchange reactions provide an additional, nonkinetic method of differentiating bisubstrate mechanisms.

REFERENCES

Chemical Kinetics

Atkins, P.W. and de Paula, J., *Physical Chemistry for the Life Sciences,* Chapters 6–8, Freeman (2006). [Most physical chemistry textbooks have similar coverage.]

Hammes, G.G., *Principles of Chemical Kinetics,* Academic Press (1978).

Laidler, K.J., *Chemical Kinetics* (3rd ed.), Harper & Row (1987).

Enzyme Kinetics

Biswanger, H., *Enzyme Kinetics: Principles and Methods* (2nd ed.), Wiley–VCH (2008).

Cleland, W.W., Steady state kinetics, *in* Boyer, P.D. (Ed.), *The Enzymes* (3rd ed.), Vol. 2, *pp.* 1–65, Academic Press (1970); *and* Steady-state kinetics, *in* Sigman, D.S. and Boyer, P.D. (Eds.), *The Enzymes* (3rd ed.), Vol. 19, *pp.* 99–158, Academic Press (1990).

Cleland, W.W., Determining the mechanism of enzyme-catalyzed reactions by kinetic studies, *Adv. Enzymol.* **45,** 273 (1977).

Cornish-Bowden, A., *Fundamentals of Enzyme Kinetics* (Revised ed.), Portland Press (1995). [A lucid and detailed account of enzyme kinetics.]

Copeland, R.A., *Enzymes,* VCH (1996).

Dixon, M. and Webb, E.C., *Enzymes* (3rd ed.), Chapter IV, Academic Press (1979). [An almost exhaustive treatment of enzyme kinetics.]

Fersht, A., *Structure and Mechanism in Protein Science,* Chapters 3–7, Freeman (1999).

Gutfreund, H., *Kinetics for the Life Sciences: Receptors, Transmitters, and Catalysts,* Cambridge University Press (1995).

Knowles, J.R., The intrinsic pK_a-values of functional groups in enzymes: Improper deductions from the pH-dependence of steady state parameters, *CRC Crit. Rev. Biochem.* **4,** 165 (1976).

Marangoni, A.G., *Enzyme Kinetics. A Modern Approach,* Wiley (2002).

Piszkiewicz, D., *Kinetics of Chemical and Enzyme Catalyzed Reactions,* Oxford University Press (1977). [A highly readable discussion of enzyme kinetics.]

Purich, D.L. (Ed.), *Contemporary Enzyme Kinetics and Mechanism* (2nd ed.), Academic Press (1996) [A collection of articles on advanced topics.]

Schulz, A.R., *Enzyme Kinetics,* Cambridge (1994).

Segel, I.H., *Enzyme Kinetics,* Wiley–Interscience (1993). [A detailed and understandable treatise providing full explanations of many aspects of enzyme kinetics.]

Tinoco, I., Jr., Sauer, K., Wang, J.C., and Puglisi, J.D., *Physical Chemistry. Principles and Applications for Biological Sciences* (4th ed.), Chapters 7 and 8, Prentice-Hall (2002).

BIOINFORMATICS EXERCISES

Bioinformatics Exercises are available at www.wiley.com/college/voet.

Chapter 13/14

BRENDA Exercise
1. **Enzyme Classes.** Explore enzyme nomenclature.
2. **Pathways.** Investigate metabolic pathways.
3. **Comparing Isoenzymes.** Compare the characteristics of enzymes catalyzing the same reaction.
4. **Functional Enzyme Parameters.** Survey the ranges and distributions of pH values, pH optima, and turnover numbers.
5. **Search by Substructure.** Search the database for enzymes that act on a particular molecule.
6. **Other Activities with the BRENDA Database.**

PROBLEMS

1. The hydrolysis of sucrose:

$$\text{Sucrose} + \text{H}_2\text{O} \longrightarrow \text{glucose} + \text{frutose}$$

takes the following time course.

Time (min)	[Sucrose] (M)
0	0.5011
30	0.4511
60	0.4038
90	0.3626
130	0.3148
180	0.2674

Determine the first-order rate constant and the half-life of the reaction. Why does this bimolecular reaction follow a first-order rate law? How long will it take to hydrolyze 99% of the sucrose initially present? How long will it take if the amount of sucrose initially present is twice that given in the table?

2. By what factor will a reaction at 25°C be accelerated if a catalyst reduces the free energy of its activated complex by 1 kJ · mol^{-1}; by 10 kJ · mol^{-1}?

3. For a Michaelis–Menten reaction, $k_1 = 5 \times 10^7\ M^{-1} \cdot \text{s}^{-1}$, $k_{-1} = 2 \times 10^4\ \text{s}^{-1}$, and $k_2 = 4 \times 10^2\ \text{s}^{-1}$. Calculate K_S and K_M for this reaction. Does substrate binding achieve equilibrium or the steady state?

*4. The following table indicates the rates at which a substrate reacts as catalyzed by an enzyme that follows the Michaelis–Menten mechanism: (1) in the absence of inhibitor; (2) and (3) in the presence of 10 mM concentration, respectively, of either of two inhibitors. Assume $[E]_T$ is the same for all reactions.

[S] (mM)	(1) v_o ($\mu M \cdot s^{-1}$)	(2) v_o ($\mu M \cdot s^{-1}$)	(3) v_o ($\mu M \cdot s^{-1}$)
1	2.5	1.17	0.77
2	4.0	2.10	1.25
5	6.3	4.00	2.00
10	7.6	5.7	2.50
20	9.0	7.2	2.86

(a) Determine K_M and V_{max} for the enzyme. For each inhibitor determine the type of inhibition and K_I and/or K'_I. What additional information would be required to calculate the turnover number of the enzyme? (b) For [S] = 5 mM, what fraction of the enzyme molecules have a bound substrate in the absence of inhibitor, in the presence of 10 mM inhibitor of type (2), and in the presence of 10 mM inhibitor of type (3)?

*5. Ethanol in the body is oxidized to acetaldehyde (CH_3CHO) by liver alcohol dehydrogenase (LADH). Other alcohols are also oxidized by LADH. For example, methanol, which is mildly intoxicating, is oxidized by LADH to the quite toxic product formaldehyde (CH_2O). The toxic effects of ingesting methanol (a component of many commercial solvents) can be reduced by administering ethanol. The ethanol acts as a competitive inhibitor of the methanol by displacing it from LADH. This provides sufficient time for the methanol to be harmlessly excreted by the kidneys. If an individual has ingested 100 mL of methanol (a lethal dose), how much 100 proof whiskey (50% ethanol by volume) must he imbibe to reduce the activity of his LADH toward methanol to 5% of its original value? The adult human body contains ~40 L of aqueous fluids throughout which ingested alcohols are rapidly and uniformly mixed. The densities of ethanol and methanol are both 0.79 g $\cdot$ cm^{-3}. Assume the K_M values of LADH for ethanol and methanol to be 1.0×10^{-3} M and 1.0×10^{-2} M, respectively, and that $K_I = K_M$ for ethanol.

6. The K_M of a Michaelis–Menten enzyme for a substrate is 1.0×10^{-4} M. At a substrate concentration of 0.2M, $v_o = 43$ $\mu M \cdot$ min^{-1} for a certain enzyme concentration. However, with a substrate concentration of 0.02M, v_o has the same value. (a) Using numerical calculations, show that this observation is accurate. (b) What is the best range of [S] for measuring K_M?

7. Why are uncompetitive and mixed inhibitors generally considered to be more effective *in vivo* than competitive inhibitors?

8. Explain why an exact fit to a kinetic model of the experimental parameters describing a reaction does not prove that the reaction follows the model.

9. An enzyme that follows the model for pH effects presented in Section 14-4a has p$K_{ES1} = 4$ and p$K_{ES2} = 8$. What is the pH at which V'_{max} is a maximum for this enzyme? What fraction of V_{max} does V'_{max} achieve at this pH?

10. Derive the initial rate equation for a Rapid Equilibrium Random Bi Bi reaction. Assume the equilibrium constants K_S^A and K_S^B for binding A and B to the enzyme are independent of whether the other substrate is bound (an assumption that constrains $K_M^B = K_S^B$ in Eq. [14.49]).

*11. Consider the following variation of a Ping Pong Bi Bi mechanism.

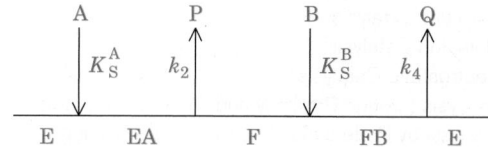

Assume that the substrate-binding reactions are in rapid equilibrium,

$$K_S^A = \frac{[E][A]}{[EA]} \quad \text{and} \quad K_S^B = \frac{[F][B]}{[FB]}$$

that both $[A] \gg [E]_T$ and $[B] \gg [E]_T$, that neither product release reaction is reversible, and that the steady-state approximation is valid. (a) Derive an expression for v_o in terms of K_S^A, K_S^B, k_2, and k_4. (b) Indicate the form of the double-reciprocal plots for $1/v_o$ versus $1/[A]$ for various values of [B]. (c) Indicate the form of the double-reciprocal plots for $1/v_o$ versus $1/[B]$ for various values of [A].

12. Creatine kinase catalyzes the reaction

$$MgADP^- + \text{phosphocreatine} \rightleftharpoons MgATP^{2-} + \text{creatine}$$

which functions to regenerate ATP in muscle. Rabbit muscle creatine kinase exhibits the following kinetic behavior. In the absence of both products, plots of $1/v_o$ versus $1/[MgADP^-]$ at different fixed concentrations of phosphocreatine yield lines that intersect to the left of the $1/v_o$ axis. Similarly, plots of $1/v_o$ versus $1/[\text{phosphocreatine}]$ in the absence of product at different fixed concentrations of $MgADP^-$ yield lines that intersect to the left of the $1/v_o$ axis. In the absence of one of the reaction products, $MgATP^{2-}$ or creatine, plots of $1/v_o$ versus $1/[MgADP^-]$ at different concentrations of the other product intersect on the $1/v_o$ axis. The same is true of the plots of $1/v_o$ versus $1/[\text{phosphocreatine}]$. Indicate a kinetic mechanism that is consistent with this information.

Enzymatic Catalysis

CHAPTER 15

1 Catalytic Mechanisms

Enzymes, as we have seen, cause rate enhancements that are orders of magnitude greater than those of the best chemical catalysts. Yet they operate under mild conditions and are highly specific as to the identities of both their substrates and their products. These catalytic properties are so remarkable that many nineteenth century scientists concluded that enzymes have characteristics that are not shared by substances of nonliving origin. To this day, there are few enzymes for which we understand in more than cursory detail how they achieve their enormous rate accelerations. Nevertheless, it is now abundantly clear that the catalytic mechanisms employed by enzymes are identical to those used by chemical catalysts. Enzymes are simply better designed.

In this chapter we consider the nature of enzymatic catalysis. We begin by discussing the underlying principles of chemical catalysis as elucidated through the study of organic reaction mechanisms. We then embark on a detailed examination of the catalytic mechanisms of several of the best characterized enzymes: **lysozyme** and the **serine proteases.** Their study should lead to an appreciation of the intricacies of these remarkably efficient catalysts as well as of the experimental methods used to elucidate their proper-

ties. We end with a discussion of how drugs are discovered and tested, a process that depends heavily on the principles of enzymology since many drug targets are enzymes. In doing so, we consider how therapeutically effective inhibitors of **HIV-1 protease** were discovered.

1 CATALYTIC MECHANISMS

Catalysis is a process that increases the rate at which a reaction approaches equilibrium. Since, as we discussed in Section 14-1Cb, the rate of a reaction is a function of its free energy of activation ($\Delta G^{\ddagger}$), a catalyst acts by lowering the height of this kinetic barrier; that is, a catalyst stabilizes the transition state with respect to the uncatalyzed reaction. There is, in most cases, nothing unique about enzymatic mechanisms of catalysis in comparison to nonenzymatic mechanisms. *What apparently make enzymes such powerful catalysts are two related properties: their specificity of substrate binding combined with their optimal arrangement of catalytic groups.* An enzyme's arrangement of binding and catalytic groups is, of course, the product of eons of evolution: Nature has had ample opportunity to fine-tune the performances of most enzymes.

The types of catalytic mechanisms that enzymes employ have been classified as:

1. Acid–base catalysis.

2. Covalent catalysis.

3. Metal ion catalysis.

4. Electrostatic catalysis.

5. Proximity and orientation effects.

6. Preferential binding of the transition state complex.

In this section, we examine these various phenomena. In doing so we shall frequently refer to the organic model compounds that have been used to characterize these catalytic mechanisms.

A. Acid–Base Catalysis

General acid catalysis *is a process in which partial proton transfer from a Brønsted acid (a species that can donate protons; Section 2-2A) lowers the free energy of a reaction's transition state.* For example, an uncatalyzed keto–enol tautomerization reaction occurs quite slowly as a result of

Figure 15-1 **Mechanisms of keto–enol tautomerization.** (*a*) Uncatalyzed, (*b*) general acid catalyzed, and (*c*) general base catalyzed.

the high energy of its carbanionlike transition state (Fig. 15-1*a*). Proton donation to the oxygen atom (Fig. 15-1*b*), however, reduces the carbanion character of the transition state, thereby catalyzing the reaction. *A reaction may also be stimulated by* **general base catalysis** *if its rate is increased by partial proton abstraction by a Brønsted base (a species that can combine with a proton; Fig. 15-1c). Some reactions may be simultaneously subject to both processes: a* **concerted general acid–base catalyzed reaction.**

a. Mutarotation Is Catalyzed by Acids and by Bases

The mutarotation of glucose provides an instructive example of acid–base catalysis. Recall that a glucose molecule can assume either of two anomeric cyclic forms through the intermediacy of its linear form (Section 11-1B):

In aqueous solvents, the initial rate of mutarotation of α-D-glucose, as monitored by polarimetry (Section 4-2A), is observed to follow the relationship:

$$v = -\frac{d[\alpha\text{-D-glucose}]}{dt} = k_{obs}[\alpha\text{-D-glucose}] \quad [15.1]$$

where k_{obs} is the reaction's apparent first-order rate constant. The mutarotation rate increases with the concentrations of general acids and general bases; they are thought to catalyze mutarotation according to the mechanism:

This model is consistent with the observation that in aprotic solvents such as benzene, **2,3,4,6-*O*-tetramethyl-α-D-glucose** (a less polar benzene-soluble analog)

2,3,4,6-*O*-Tetramethyl-α-D-glucose

does not undergo mutarotation. Yet, the reaction is catalyzed by the addition of phenol, a weak benzene-soluble acid, together with pyridine, a weak benzene-soluble base, according to the rate equation:

$$v = k[\text{phenol}][\text{pyridine}][\text{tetramethyl-α-D-glucose}]$$
[15.2]

Moreover, in the presence of **α-pyridone,** whose acid and base groups can rapidly interconvert between two tautomeric forms and are situated so that they can simultaneously catalyze mutarotation,

α-Pyridone

Glucose

the reaction follows the rate law

$$v = k'[\text{α-pyridone}][\text{tetramethyl-α-D-glucose}] \quad [15.3]$$

where $k' = 7000M \times k$. This increased rate constant indicates that α-pyridone does, in fact, catalyze mutarotation in a concerted fashion since $1M$ α-pyridone has the same catalytic effect as impossibly high concentrations of phenol and pyridine (e.g., $70M$ phenol and $100M$ pyridine).

Many types of biochemically significant reactions are susceptible to acid and/or base catalysis. These include the hydrolysis of peptides and esters, the reactions of phosphate groups, tautomerizations, and additions to carbonyl groups. The side chains of the amino acid residues Asp, Glu, His, Cys, Tyr, and Lys have pK's in or near the physiological pH range (Table 4-1), which we shall see permits them to act in the enzymatic capacity of general acid and/or base catalysts in analogy with known organic mechanisms. Indeed, the ability of enzymes to arrange several catalytic groups about their substrates makes concerted acid–base catalysis a common enzymatic mechanism.

b. The RNase A Reaction Incorporates General Acid–Base Catalysis

Bovine pancreatic ribonuclease A (RNase A) provides an illuminating example of enzymatically mediated general acid–base catalysis. This digestive enzyme functions to hydrolyze RNA to its component nucleotides. The isolation of **2′,3′-cyclic nucleotides** from RNase A digests of RNA indicates that the enzyme mediates the following reaction sequence:

RNA

2′,3′-Cyclic nucleotide

The RNase A reaction exhibits a pH rate profile that peaks near pH 6 (Fig. 15-2). Analysis of this curve (Section 14-4b), together with chemical derivatization and X-ray studies, indicates that RNase A has two essential His residues, His 12 and His 119, which act in a concerted manner as general acid and base catalysts (the structure of RNase A is sketched in Fig. 9-2). Evidently, the RNase A reaction is a two-step process (Fig. 15-3):

1. His 12, acting as a general base, abstracts a proton from an RNA 2′-OH group, thereby promoting its nucleophilic

Figure 15-2   The pH dependence of V'_{max}/K'_M in the RNase A–catalyzed hydrolysis of cytidine-2′,3′-cyclic phosphate. V'_{max}/K'_M is given in units of $M^{-1} \cdot s^{-1}$. Analysis of this curve (Section 14-4b) suggests the catalytic participation of groups with pK's of 5.4 and 6.4. [After del Rosario, E.J. and Hammes, G.G., *Biochemistry* **8,** 1887 (1969).]

Figure 15-3   The bovine pancreatic RNase A–catalyzed hydrolysis of RNA is a two-step process with the intermediate formation of a 2′,3′-cyclic nucleotide.

attack on the adjacent phosphorus atom while His 119, acting as a general acid, promotes bond scission by protonating the leaving group.

2. The 2′,3′-cyclic intermediate is hydrolyzed through what is essentially the reverse of the first step in which water replaces the leaving group. Thus His 12 acts as a general acid and His 119 as a general base to yield the hydrolyzed RNA and the enzyme in its original state.

B. *Covalent Catalysis*

Covalent catalysis *involves rate acceleration through the transient formation of a catalyst–substrate covalent bond.* The decarboxylation of **acetoacetate,** as chemically catalyzed by primary amines, is an example of such a process (Fig. 15-4). In the first stage of this reaction, the amine nucleophilically attacks the carbonyl group of acetoacetate to form a **Schiff base** (imine bond).

The protonated nitrogen atom of the covalent intermediate then acts as an electron sink (Fig. 15-4, bottom) so as to reduce the otherwise high-energy enolate character of the transition state. The formation and decomposition of the Schiff base occur quite rapidly, so that these steps are not rate determining in this reaction sequence.

a. Covalent Catalysis Has Both Nucelophilic and Electrophilic Stages

As the preceding example indicates, covalent catalysis may be conceptually decomposed into three stages:

1. The nucleophilic reaction between the catalyst and the substrate to form a covalent bond.

2. The withdrawal of electrons from the reaction center by the now electrophilic catalyst.

3. The elimination of the catalyst, a reaction that is essentially the reverse of stage 1.

Reaction mechanisms are somewhat arbitrarily classified as occurring with either **nucleophilic catalysis** or **electrophilic catalysis** depending on which of these effects provides the greater driving force for the reaction, that is, which catalyzes its rate-determining step. The primary amine–catalyzed decarboxylation of acetoacetate is clearly an electrophilically catalyzed reaction since its nucleophilic phase, Schiff base formation, is not its rate-determining step. In other covalently catalyzed reactions, however, the nucleophilic phase may be rate determining.

The nucleophilicity of a substance is closely related to its basicity. Indeed, the mechanism of nucleophilic catalysis resembles that of general base catalysis except that, instead of abstracting a proton from the substrate, the catalyst nucleophilically attacks it so as to form a covalent bond. Consequently, if covalent bond formation is the rate-determining step of a covalently catalyzed reaction, the reaction rate tends to increase with the covalent catalyst's basicity (pK).

An important aspect of covalent catalysis is that the more stable the covalent bond formed, the less facilely it will decompose in the final steps of a reaction. A good covalent catalyst must therefore combine the seemingly contradictory properties of high nucleophilicity and the ability to form a good leaving group, that is, to easily reverse the bond formation step. Groups with high polarizabilities (highly mobile electrons), such as imidazole and thiol functions, have these properties and hence make good covalent catalysts.

Acetoacetate   **Enolate**   **Acetone**

Schiff base
(imine)

Figure 15-4 **The decarboxylation of acetoacetate.** The uncatalyzed reaction mechanism is shown at the top and the reaction mechanism as catalyzed by primary amines is shown at the bottom.

b. Certain Amino Acid Side Chains and Coenzymes Can Serve as Covalent Catalysts

Enzymes commonly employ covalent catalytic mechanisms as is indicated by the large variety of covalently linked enzyme–substrate reaction intermediates that have been isolated. For example, the enzymatic decarboxylation of acetoacetate proceeds, much as described above, through Schiff base formation with an enzyme Lys residue's ε-amino group. The covalent intermediate, in this case, has been isolated through $NaBH_4$ reduction of its imine bond to an amine, thereby irreversibly inhibiting the enzyme. Other enzyme functional groups that participate in covalent catalysis include the imidazole moiety of His, the thiol group of Cys, the carboxyl function of Asp, and the hydroxyl group of Ser. In addition, several coenzymes, most notably **thiamine pyrophosphate** (Section 17-3Ba) and **pyridoxal phosphate** (Section 26-1Aa), function in association with their apoenzymes mainly as covalent catalysts.

C. Metal Ion Catalysis

Nearly one-third of all known enzymes require the presence of metal ions for catalytic activity. There are two classes of metal ion–requiring enzymes that are distinguished by the strengths of their ion–protein interactions:

1. *Metalloenzymes* contain tightly bound metal ions, most commonly transition metal ions such as Fe^{2+}, Fe^{3+}, Cu^{2+}, Zn^{2+}, Mn^{2+}, or Co^{3+}.

2. *Metal-activated enzymes* loosely bind metal ions from solution, usually the alkali and alkaline earth metal ions Na^+, K^+, Mg^{2+}, or Ca^{2+}.

Metal ions participate in the catalytic process in three major ways:

1. By binding to substrates so as to orient them properly for reaction.

2. By mediating oxidation–reduction reactions through reversible changes in the metal ion's oxidation state.

3. By electrostatically stabilizing or shielding negative charges.

In this section we shall be mainly concerned with the third aspect of metal ion catalysis. The other forms of enzyme-mediated metal ion catalysis are considered in later chapters in conjunction with discussions of specific enzyme mechanisms.

a. Metal Ions Promote Catalysis through Charge Stabilization

In many metal ion–catalyzed reactions, the metal ion acts in much the same way as a proton to neutralize negative charge, that is, it acts as a Lewis acid. Yet *metal ions are often much more effective catalysts than protons because metal ions can be present in high concentrations at neutral pH's and can have charges greater than +1.* Metal ions have therefore been dubbed "superacids."

The decarboxylation of **dimethyloxaloacetate,** as catalyzed by metal ions such as Cu^{2+} and Ni^{2+}, is a nonenzymatic example of catalysis by a metal ion:

Dimethyloxaloacetate

Here the metal ion (M^{n+}), which is chelated by the dimethyloxaloacetate, electrostatically stabilizes the developing enolate ion of the transition state. This mechanism is supported by the observation that acetoacetate, which cannot form such a chelate, is not subject to metal ion–catalyzed decarboxylation. Most enzymes that decarboxylate oxaloacetate require a metal ion for activity.

b. Metal Ions Promote Nucleophilic Catalysis via Water Ionization

A metal ion's charge makes its bound water molecules more acidic than free H_2O and therefore a source of OH^- ions even below neutral pH's. For example, the water molecule of $(NH_3)_5Co^{3+}(H_2O)$ ionizes according to the reaction:

$$(NH_3)_5Co^{3+}(H_2O) \rightleftharpoons (NH_3)_5Co^{3+}(OH^-) + H^+$$

with a pK of 6.6, which is ~9 pH units below the pK of free H_2O. *The resulting metal ion–bound hydroxyl group is a potent nucleophile.*

An instructive example of this phenomenon occurs in the catalytic mechanism of **carbonic anhydrase** (Section 10-1C), a widely occurring enzyme that catalyzes the reaction:

$$CO_2 + H_2O \rightleftharpoons HCO_3^- + H^+$$

Carbonic anhydrase contains an essential Zn^{2+} ion that lies at the bottom of an ~15-Å-deep active site cleft (Fig. 8-41), where it is tetrahedrally coordinated by three evolutionarily invariant His side chains and an O atom of either an

(a)

(b)

Figure 15-5 **X-ray structures of human carbonic anhydrase.**
(*a*) Its active site in complex with bicarbonate ion. The
polypeptide is shown in ribbon form (*gold*) with its side chains
shown in stick form colored according to atom type (C green,
N blue, and O red). The protein-bound Zn^{2+} ion (*cyan sphere*) is
tetrahedrally liganded (*gray bonds*) by three invariant His side
chains and the HCO_3^- ion, which is shown in ball-and-stick form.
The HCO_3^- ion also interacts with the protein via van der Waals
contacts (*dot surface colored according to atom type*) and a
hydrogen bonded network (*dashed gray lines*) involving Thr 199
and Glu 106. [Based on an X-ray structure by K.K. Kannan,
Bhabha Atomic Research Center, Bombay, India. PDBid 1HCB.]
(*b*) The active site showing the proton shuttle through which His
64, acting as a general base, abstracts a proton from the
Zn^{2+}-bound H_2O to form an OH^- ion. The polypeptide
backbone is shown in ribbon form (*cyan*), and its side chains and
several bound solvent molecules are shown in ball-and-stick
form with C black, N blue, and O red. The proton shuttle consists
of two water molecules that form a hydrogen bonded network
(*dotted white lines*) that bridges the Zn^{2+}-bound OH^- ion and
His 64 in its "in" conformation. On protonation, His 64 swings to
the "out" conformation. [Courtesy of David Christianson,
University of Pennsylvania.] **See Interactive Exercise 3**

HCO_3^- ion (Fig. 15-5*a*) or a water molecule (Fig. 15-5*b*). The
enzyme has the following catalytic mechanism:

1. We begin with a water molecule bound to the protein
in the Zn^{2+} ion's fourth liganding position (Fig. 15-5*b*). This
Zn^{2+}-polarized H_2O ionizes in a process facilitated through
general base catalysis by His 64 in its "in" conformation.
Although His 64 is too far away from the Zn^{2+}-bound water to directly abstract its proton, these entities are linked
by two intervening water molecules to form a hydrogen
bonded network that is thought to act as a proton shuttle.

Im = imidazole

2. The resulting Zn^{2+}-bound OH^- ion nucleophilically
attacks the nearby enzymatically bound CO_2, thereby converting it to HCO_3^-.

Im = imidazole

In doing so, the Zn^{2+}-bound OH^- group donates a hydrogen bond to Thr 199, which in turn donates a hydrogen
bond to Glu 106 (Fig. 15-5*a*). These interactions orient the
OH^- group with the optimal geometry (see below) for nucleophilic attack on the substrate CO_2.

3. The catalytic site is regenerated by the exchange of
the Zn^{2+}-bound HCO_3^- reaction product for H_2O together

with the deprotonation of His 64. In the latter process, His 64 swings to its "out" conformation (Fig. 15-5b), which may facilitate proton transfer to the bulk solvent.

c. Metal Ions Promote Reactions through Charge Shielding

Another important enzymatic function of metal ions is **charge shielding.** For example, the actual substrates of **kinases** (phosphoryl-transfer enzymes utilizing ATP) are Mg^{2+}–ATP complexes such as

rather than just ATP. Here, the Mg^{2+} ion's role, in addition to its orienting effect, is to shield electrostatically the negative charges of the phosphate groups. Otherwise, these charges would tend to repel the electron pairs of attacking nucleophiles, especially those with anionic character.

D. Electrostatic Catalysis

The binding of substrate generally excludes water from an enzyme's active site. The local dielectric constant of the active site therefore resembles that in an organic solvent, where electrostatic interactions are much stronger than they are in aqueous solutions (Section 8-4A). The charge distribution in a medium of low dielectric constant can greatly influence chemical reactivity. Thus, as we have seen, the pK's of amino acid side chains in proteins may vary by several units from their nominal values (Table 4-1) because of the proximity of charged groups.

Although experimental evidence and theoretical analyses on the subject are still sparse, *there are mounting indications that the charge distributions about the active sites of enzymes are arranged so as to stabilize the transition states of the catalyzed reactions.* Such a mode of rate enhancement, which resembles the form of metal ion catalysis discussed above, is termed **electrostatic catalysis.** Moreover, in several enzymes, *these charge distributions apparently serve to guide polar substrates toward their binding sites so that the rates of these enzymatic reactions are greater than their apparent diffusion-controlled limits (Section 14-2Bb).*

E. Catalysis through Proximity and Orientation Effects

Although enzymes employ catalytic mechanisms that resemble those of organic model reactions, they are far more catalytically efficient than these models. Such efficiency must arise from the specific physical conditions at enzyme catalytic sites that promote the corresponding chemical reactions. The most obvious effects are **proximity** and **orientation:** Reactants must come together with the proper spatial relationship for a reaction to occur. For example, in the

bimolecular reaction of imidazole with *p*-nitrophenylacetate,

p-Nitrophenylacetate
(p-NO$_2$φAc)

Imidazole

k_1

N-Acetylimidazolium

p-Nitrophenolate
(p-NO$_2$φO$^-$)

the progress of the reaction is conveniently monitored by the appearance of the intensely yellow **p-nitrophenolate ion:**

$$\frac{d[p\text{-}NO_2\phi O^-]}{dt} = k_1[\text{imidazole}][p\text{-}NO_2\phi Ac]$$
$$= k_1'[p\text{-}NO_2\phi Ac] \qquad [15.4]$$

where ϕ = phenyl. Here k_1', the pseudo-first-order rate constant, is $0.0018\ s^{-1}$ when [imidazole] = $1M$. However, for the intramolecular reaction

the first-order rate constant $k_2 = 0.043\ s^{-1}$; that is, $k_2 = 24\ k_1'$. Thus, when the $1M$ imidazole catalyst is covalently attached to the reactant, it is 24-fold more effective than when it is free in solution; that is, *the imidazole group in the intramolecular reaction behaves as if its concentration is 24M.* This rate enhancement has contributions from both proximity and orientation.

a. Proximity Alone Contributes Relatively Little to Catalysis

Let us make a rough calculation as to how the rate of a reaction is affected purely by the proximity of its reacting groups. Following Daniel Koshland's treatment, we shall make several reasonable assumptions:

1. Reactant species, that is, functional groups, are about the size of water molecules.

2. Each reactant species in solution has 12 nearest-neighbor molecules, as do packed spheres of identical size.

3. Chemical reactions occur only between reactants that are in contact.

4. The reactant concentration in solution is low enough so that the probability of any reactant species being in simultaneous contact with more than one other reactant molecule is negligible.

Then the reaction:

$$A + B \xrightarrow{k_1} A\!-\!B$$

obeys the second-order rate equation

$$v = \frac{d[A\!-\!B]}{dt} = k_1[A][B] = k_2[A, B]_{pairs} \quad [15.5]$$

where $[A, B]_{pairs}$ is the concentration of contacting molecules of A and B. The value of this quantity is

$$[A, B]_{pairs} = \frac{12[A][B]}{55.5M} \quad [15.6]$$

since there are 12 ways that A can be in contact with B, and $[A]/55.5M$ is the fraction of sites occupied by A in water solution ($[H_2O] = 55.5M$ in dilute aqueous solutions) and hence the probability that a molecule of B will be next to one of A. Combining Eqs. [15.5] and [15.6] yields

$$v = k_1\left(\frac{55.5}{12}\right)[A, B]_{pairs} = 4.6k_1[A, B]_{pairs} \quad [15.7]$$

Thus, in the absence of other effects, this model predicts that for the intramolecular reaction,

$k_2 = 4.6k_1$, which is a rather small rate enhancement. Factors that will increase this value other than proximity alone clearly must be considered.

b. Properly Orienting Reactants and Arresting Their Relative Motions Can Result in Large Catalytic Rate Enhancements

The foregoing theory is, of course, quite simple. For example, it does not take into account the relative orientations of the reacting molecules. Yet molecules are not equally reactive in all directions as Koshland's simple theory assumes. Rather, *they react most readily only if they have the proper relative orientation.* For example, in an S_N2 (bimolecular nucleophilic substitution) reaction, the incoming nucleophile optimally attacks its target C atom along the direction opposite to that of the bond to the leaving group (Fig. 15-6). The approaches of reacting atoms along a trajectory that deviates by as little as 10° from this optimum direction can reduce the reaction rate by as much as a factor of ~100. In a related phenomenon, a molecule may be maximally reactive only when it assumes a conformation that aligns its various orbitals in a way that minimizes the electronic energy of its transition state, an effect termed **stereoelectronic assistance** or **stereoelectronic control.**

Another effect that we have neglected in our treatment of proximity is that of motions of the reacting groups with respect to one another. Yet, in the transition state complex, the reacting groups have little relative motion. In fact, as Thomas Bruice demonstrated, the rates of intramolecular reactions are greatly increased by arresting a molecule's internal motions in a way that increases the mole fraction of the reacting groups that are in a conformation which can

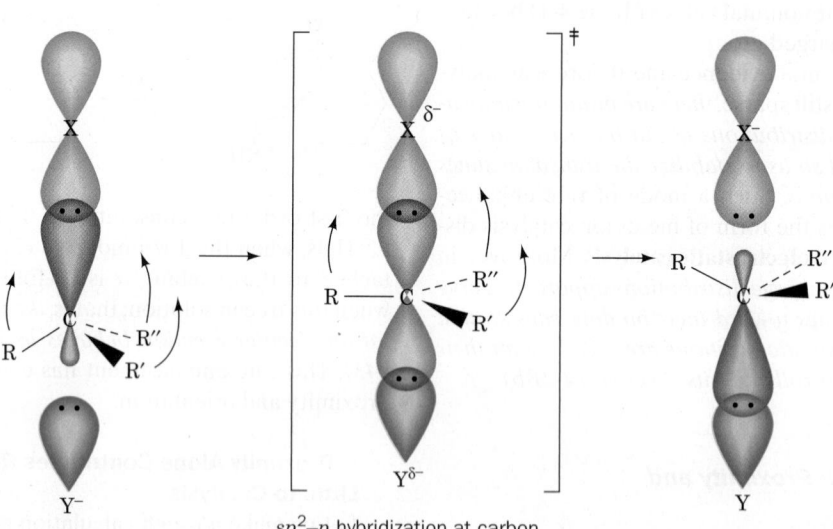

sp^2–p hybridization at carbon

Figure 15-6 **The geometry of an S_N2 reaction.** The attacking nucleophile, Y^-, must approach the tetrahedrally coordinated and hence sp^3-hybridized C atom along the direction opposite that of its bond to the leaving group, X, a process called **backside attack.** In the transition state of the reaction, the C atom becomes trigonal bipyramidally coordinated and hence sp^2–p hybridized, with the p orbital (*blue*) forming partial bonds to X and Y. The three sp^2 orbitals form bonds to the C atom's three other substituents (R, R′, and R″), which have shifted their positions into the plane perpendicular to the X—C—Y axis (*curved arrows*). Any deviation from this optimal geometry would increase the free energy of the transition state, $\Delta G^{\ddagger}$, and hence reduce the rate of the reaction (Eq. [14.15]). The transition state then decomposes to products in which the R, R′, and R″ have inverted their positions about the C atom, which has rehybridized to sp^3, and X^- has been released.

enter the transition state (Table 15-1). Similarly, when an enzyme brings two molecules together in a bimolecular reaction, as William Jencks pointed out, not only does it increase their proximity, but it freezes out their relative translational and rotational motions (decreases their entropy), thereby enhancing their reactivity. Theoretical studies by Bruice indicate that much of this rate enhancement can arise from the enzymatic binding of substrates in a conformation that readily enters the transition state.

Enzymes, as we shall see in Sections 16-2 and 16-3, bind substrates in a manner that both aligns and immobilizes them so as to optimize their reactivities. The free energy required to do so is derived from the specific binding free energy of substrate to enzyme.

F. Catalysis by Preferential Transition State Binding

The rate enhancements effected by enzymes are often greater than can be reasonably accounted for by the catalytic mechanisms so far discussed. However, we have not yet considered one of the most important mechanisms of enzymatic catalysis: *the binding of the transition state to an enzyme with greater affinity than the corresponding substrates or products.* When taken together with the previously described catalytic mechanisms, preferential transition state binding rationalizes the observed rates of enzymatic reactions.

The original concept of transition state binding proposed that enzymes mechanically strained their substrates toward the transition state geometry through binding sites into which undistorted substrates did not properly fit. This so-called **rack mechanism** (in analogy with the medieval torture device) was based on the extensive evidence for the role of strain in promoting organic reactions. For example, the rate of the reaction,

is 315 times faster when R is CH_3 rather than when it is H because of the greater steric repulsions between the CH_3 groups and the reacting groups. Similarly, ring opening reactions are considerably more facile for strained rings such as cyclopropane than for unstrained rings such as cyclohexane. In either process, *the strained reactant more closely resembles the transition state of the reaction than does the corresponding unstrained reactant.* Thus, as was first suggested by Linus Pauling and further amplified by Richard Wolfenden and Gustav Lienhard, *interactions that preferentially bind the transition state increase its concentration and therefore proportionally increase the reaction rate.*

Let us quantitate this statement by considering the kinetic consequences of preferentially binding the transition state of an enzymatically catalyzed reaction involving a single substrate. The substrate S may react to form product P

Table 15-1   Relative Rates of Anhydride Formation for Esters Possessing Different Degrees of Motional Freedom in the Reaction:

Reactants[a]	Relative Rate Constant
$CH_3COO\phi Br$ + CH_3COO^-	1.0
COOϕBr / COO$^-$	~1 × 10³
COOϕBr / COO$^-$	~2.3 × 10⁵
COOϕBr / COO$^-$	~8 × 10⁷

[a]Curved arrows indicate rotational degrees of freedom.

Source: Bruice, T.C. and Lightstone, F.C., *Acc. Chem. Res.* **32,** 127 (1999).

either spontaneously or through enzymatic catalysis:

$$S \xrightarrow{k_N} P$$
$$ES \xrightarrow{k_E} EP$$

The relationships between the various states of these two reaction pathways are indicated in the following scheme:

$$E + S \xrightleftharpoons{K_N^\ddagger} S^\ddagger + E \longrightarrow P + E$$

$$\Updownarrow K_R \qquad \Updownarrow K_T \qquad \Updownarrow$$

$$ES \xrightleftharpoons{K_E^\ddagger} ES^\ddagger \longrightarrow EP$$

where

$$K_R = \frac{[ES]}{[E][S]} \qquad K_T = \frac{[ES^\ddagger]}{[E][S^\ddagger]}$$

$$K_N^\ddagger = \frac{[E][S^\ddagger]}{[E][S]} \qquad K_E^\ddagger = \frac{[ES^\ddagger]}{[ES]}$$

are all association constants. Consequently,

$$\frac{K_T}{K_R} = \frac{[S][ES^\ddagger]}{[S^\ddagger][ES]} = \frac{K_E^\ddagger}{K_N^\ddagger} \qquad [15.8]$$

According to transition state theory, Eqs. [14.7] and [14.14], the rate of the uncatalyzed reaction can be expressed

$$v_N = k_N[S] = \left(\frac{\kappa k_B T}{h}\right)[S^\ddagger] = \left(\frac{\kappa k_B T}{h}\right)K_N^\ddagger[S] \qquad [15.9]$$

Similarly, the rate of the enzymatically catalyzed reaction is

$$v_E = k_E[ES] = \left(\frac{\kappa k_B T}{h}\right)[ES^\ddagger] = \left(\frac{\kappa k_B T}{h}\right)K_E^\ddagger[ES] \quad [15.10]$$

Therefore, combining Eqs. [15.8] to [15.10],

$$\frac{k_E}{k_N} = \frac{K_E^\ddagger}{K_N^\ddagger} = \frac{K_T}{K_R} \quad [15.11]$$

This equation indicates that *the more tightly an enzyme binds its reaction's transition state (K_T) relative to the substrate (K_R), the greater the rate of the catalyzed reaction (k_E) relative to that of the uncatalyzed reaction (k_N); that is, catalysis results from the preferential binding and therefore the stabilization of the transition state ($S^\ddagger$) relative to that of the substrate (S)* (Fig. 15-7).

According to Eq. [14.15], the ratio of the rates of the catalyzed versus the uncatalyzed reaction is expressed

$$\frac{k_E}{k_N} = \exp[(\Delta G_N^\ddagger - \Delta G_E^\ddagger)/RT] \quad [15.12]$$

A rate enhancement factor of 10^6 therefore requires that an enzyme bind its transition state complex with 10^6-fold higher affinity than its substrate, which corresponds to a 34.2 kJ · mol^{-1} stabilization at $25°C$. This is roughly the free energy of two hydrogen bonds. Consequently, *the enzymatic binding of a transition state ($ES^\ddagger$) by two hydrogen bonds that cannot form in the Michaelis complex (ES) should result in a rate enhancement of ~10^6 based on this effect alone.*

It is commonly observed that the specificity of an enzyme is manifested by its turnover number (k_{cat}) rather than by its substrate-binding affinity. In other words, an enzyme binds poor substrates, which have a low reaction rate, as well as or even better than good ones, which have a high reaction rate. Such enzymes apparently use a good substrate's intrinsic binding energy to stabilize the corresponding transition state; that is, *a good substrate does not necessarily bind to its enzyme with high affinity, but does so on activation to the transition state.*

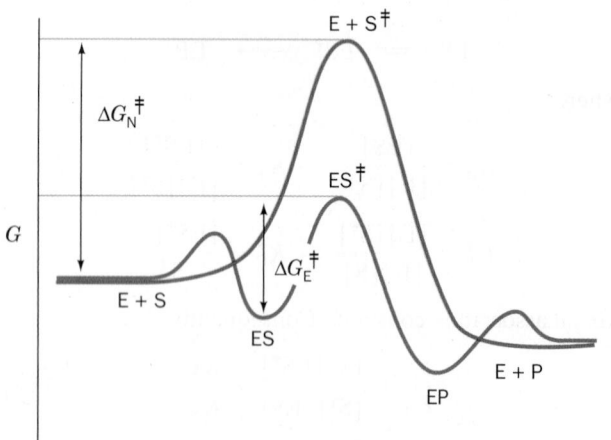

Figure 15-7 **Reaction coordinate diagrams for a hypothetical enzymatically catalyzed reaction involving a single substrate (*blue*) and the corresponding uncatalyzed reaction (*red*).**
🔷 **See the Animated Figures**

a. Transition State Analogs Are Potent Competitive Inhibitors

*If an enzyme preferentially binds its transition state, then it can be expected that **transition state analogs**, stable molecules that resemble $S^\ddagger$ or one of its components, are potent competitive inhibitors of the enzyme.* For example, the reaction catalyzed by **proline racemase** from *Clostridium sticklandii* is thought to occur via a planar transition state:

L-Proline

D-Proline

Planar transition state

Proline racemase is competitively inhibited by the planar analogs of proline, **pyrrole-2-carboxylate** and **Δ-1-pyrroline-2-carboxylate**,

Pyrrole-2-carboxylate   **Δ-1-Pyrroline-2-carboxylate**

both of which bind to the enzyme with 160-fold greater affinity than does proline. These compounds are therefore thought to be analogs of the transition state in the proline racemase reaction. In contrast, **tetrahydrofuran-2-carboxylate**,

Tetrahydrofuran-2-carboxylate

which more closely resembles the tetrahedral structure of proline, is not nearly as good an inhibitor as these compounds. A 160-fold increase in binding affinity corresponds, according to Eq. [15.12], to a 12.6 kJ · mol^{-1} increase in the free energy of binding. This quantity presumably reflects the additional binding affinity that proline racemase has for proline's planar transition state over that of the undistorted molecule.

Hundreds of transition state analogs for various enzymatic reactions have been reported. Some are naturally occurring antibiotics. Others were designed to investigate the mechanisms of particular enzymes and/or to act as specific enzymatic inhibitors for therapeutic or agricultural use. Indeed, as we discuss in Section 15-4C, *the theory that enzymes bind transition states with higher affinity than substrates has led to a rational basis for drug design based on the understanding of specific enzyme reaction mechanisms.*

Figure 15-8 **The alternating NAG–NAM polysaccharide component of bacterial cell walls.** The position of the lysozyme cleavage site is shown.

2 LYSOZYME

In the following two sections, we shall investigate the catalytic mechanisms of several well-characterized enzymes. In doing so, we shall see how enzymes apply the catalytic principles described in Section 15-1. You should note that *the great catalytic efficiency of enzymes arises from their simultaneous use of several of these catalytic mechanisms.*

Lysozyme is an enzyme that destroys bacterial cell walls. It does so, as we saw in Section 11-3Ba, by hydrolyzing the $\beta(1 \rightarrow 4)$ glycosidic linkages from **N-acetylmuramic acid (NAM)** to **N-acetylglucosamine (NAG)** in the alternating NAM–NAG polysaccharide component of cell wall peptidoglycans (Fig. 15-8). It likewise hydrolyzes $\beta(1 \rightarrow 4)$-linked poly(NAG) (chitin), a cell wall component of most fungi. Lysozyme occurs widely in the cells and secretions of vertebrates, where it may function as a bactericidal agent. However, the observation that few pathogenic bacteria are susceptible to lysozyme alone has prompted the suggestion that this enzyme mainly helps dispose of bacteria after they have been killed by other means.

Hen egg white (HEW) lysozyme is the most widely studied species of lysozyme and is one of the mechanistically best understood enzymes. It is a rather small (14.7 kD) and readily available protein (an egg contains ~5 g of it), whose single polypeptide chain consists of 129 amino acid residues and is internally cross-linked by four disulfide bonds (Fig. 15-9). HEW lysozyme catalyzes the hydrolysis of its substrate at a rate that is ~10^8-fold greater than that of the uncatalyzed reaction.

A. Enzyme Structure

The elucidation of an enzyme's mechanism of action requires a knowledge of the structure of its enzyme–substrate complex. This is because, even if the active site residues have been identified through chemical, physical, and genetic means, their three-dimensional arrangements relative to the substrate as well as to each other must be known for an understanding of how the enzyme works. However, an enzyme binds its good substrates only transiently before it catalyzes a reaction and releases the products. Consequently, *most of our knowledge of enzyme–substrate complexes derives from X-ray studies of enzymes in complex with inhibitors or poor substrates* that remain stably bound

to the enzyme for the several hours that are usually required to measure a protein crystal's X-ray diffraction intensities (although techniques for measuring X-ray intensities in less than 1 s have been developed). The large solvent-filled channels that occupy much of the volume of most protein crystals (Section 8-3Aa) often permit the formation of enzyme–inhibitor complexes by the diffusion of inhibitor molecules into crystals of the native protein.

The X-ray structure of HEW lysozyme, which was elucidated by David Phillips in 1965, was the second structure of a protein and the first of an enzyme to be determined at

Figure 15-9 **Primary structure of HEW lysozyme.** The amino acid residues that line the substrate-binding pocket are shown in dark purple.

high resolution. The protein molecule is roughly ellipsoidal in shape with dimensions $30 \times 30 \times 45$ Å (Fig. 15-10). *Its most striking feature is a prominent cleft, the substrate-binding site, that traverses one face of the molecule.* The polypeptide chain contains five helical segments as well as a three-stranded antiparallel β sheet that comprises much of one wall of the binding cleft. As expected, most of the nonpolar side chains are in the interior of the molecule, out of contact with the aqueous solvent.

a. The Nature of the Binding Site

NAG oligosaccharides of less than five residues are but very slowly hydrolyzed by HEW lysozyme (Table 15-2), although these substrate analogs bind to the enzyme's active site and are thus its competitive inhibitors. The X-ray structure of the (NAG)$_3$–lysozyme complex reveals that (NAG)$_3$ is bound at substrate subsites A, B, and C in Fig. 15-10. This inhibitor associates with the enzyme through strong hydrogen

bonding interactions, some of which involve the acetamido groups of residues A and C, as well as through close-fitting hydrophobic contacts. In an example of induced-fit ligand binding (Section 10-4C), there is a slight (~1 Å) closure of lysozyme's binding cleft on binding (NAG)$_3$.

b. Lysozyme's Catalytic Site Was Identified through Model Building

(NAG)$_3$ takes several weeks to hydrolyze under the influence of lysozyme. It was therefore assumed that the complex revealed by X-ray analysis is unproductive; that is, the enzyme's catalytic site occurs at neither the A—B nor the B—C glycosidic bonds. [Presumably, the rare occasions when (NAG)$_3$ is hydrolyzed occur when it binds productively at the catalytic site.]

In order to locate lysozyme's catalytic site, Phillips used model building to investigate how a larger substrate could bind to the enzyme. Lysozyme's active site cleft is long enough to

Figure 15-10 **X-ray structure of HEW lysozyme in complex with (NAG)$_6$.** The protein is represented by its transparent molecular surface with its polypeptide chain in ribbon form colored in rainbow order from its N-terminus (*blue*) to its C-terminus (*red*). The (NAG)$_6$, which is drawn in stick form with its sugar residues designated A, at its nonreducing end, through F, at its reducing end, binds in a deep cleft in the enzyme surface. Rings A, B, and C (colored according to atom type with C green, N blue, and O red) are observed in the X-ray structure of the complex of (NAG)$_3$ with lysozyme; the positions of rings D, E, and F (C cyan, N blue, O red) were inferred by model building. The side chains of lysozyme's active site residues, Glu 35 and Asp 52, which are drawn in space-filling form (C yellow, O red), catalyze the hydrolysis of the glycosidic bond between rings D and E. [Based on an X-ray structure by David Phillips, Oxford University. PDBid 1HEW.] *See* **Interactive Exercise 6 and Kinemage Exercise 9.**

Table 15-2 Rates of HEW Lysozyme-Catalyzed Hydrolysis of Selected Oligosaccharide Substrate Analogs

Compound	$k_{cat} (s^{-1})$
(NAG)$_2$	2.5×10^{-8}
(NAG)$_3$	8.3×10^{-6}
(NAG)$_4$	6.6×10^{-5}
(NAG)$_5$	0.033
(NAG)$_6$	0.25
(NAG–NAM)$_3$	0.5

Source: Imoto, T., Johnson, L.N., North, A.C.T., Phillips, D.C., and Rupley, J.A., *in* Boyer, P.D. (Ed.), *The Enzymes* (3rd ed.), Vol. 7, *p.* 842, Academic Press (1972).

accommodate (NAG)$_6$, which the enzyme rapidly hydrolyzes (Table 15-2). However, the fourth NAG residue (the D-ring in Fig. 15-10) appeared unable to bind to the enzyme because its C6 and O6 atoms too closely contact Glu 35, Trp 108, and the acetamido group of the C-ring. This steric interference could be relieved by distorting the glucose ring from its normal chair conformation to that of a half-chair (Fig. 15-11). *This distortion, which renders atoms C1, C2, C5, and O5 of residue D coplanar, moves the —C6H$_2$OH group from its normal equatorial position to an axial position where it makes no close contacts and can hydrogen bond to the backbone carbonyl group of Gln 57 and the amido group of Val 109* (Fig. 15-12). Continuing the model building, Phillips found that the E- and F-rings apparently bind to the enzyme without distortion and with a number of favorable hydrogen bonding and van der Waals contacts.

We are almost in a position to identify lysozyme's catalytic site. In the enzyme's natural substrate, every second

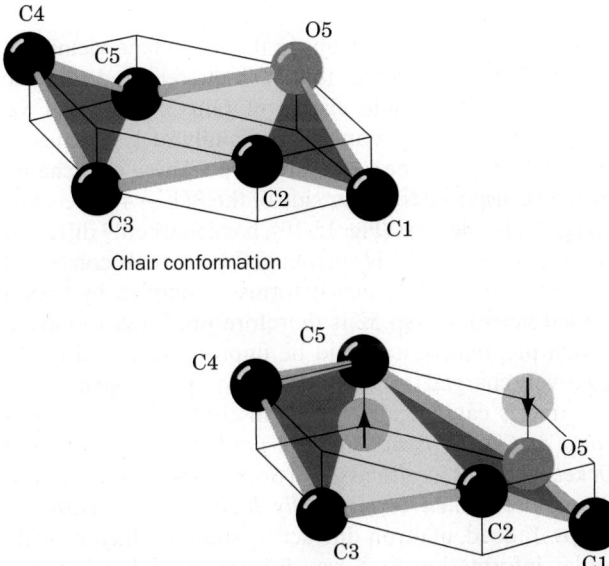

Figure 15-11 **Chair and half-chair conformations.** Hexose rings normally assume the chair conformation. It is postulated, however, that binding by lysozyme distorts the D-ring into the half-chair conformation such that atoms C1, C2, C5, and O5 are coplanar. ⌘ **See the Animated Figures**

Figure 15-12 **Interactions of lysozyme with its substrate.** The view is into the binding cleft with the heavier edges of the rings facing the outside of the enzyme and the lighter ones against the bottom of the cleft. [Illustration, Irving Geis. Image from the Irving Geis Collection, Howard Hughes Medical Institute. Reprinted with permission.] Based on an X-ray structure by David Phillips, Oxford University, U.K. PDBid 4LYZ.] ⌘ **See Kinemage Exercise 9**

residue is an NAM. Model building, however, indicated that its lactyl side chain cannot be accommodated in the binding subsites of either residues C or E. Hence, the NAM residues must bind to the enzyme in subsites B, D, and F.

$$\cdots-NAG-NAM-NAG-NAM-NAG-NAM-\cdots \rightarrow \left(\begin{array}{c}\text{reducing}\\\text{end}\end{array}\right)$$
$$\quad\; A \quad\;\; B \quad\;\;\; C \quad\;\; D \quad\;\; E \quad\;\;\; F$$

The observation that lysozyme hydrolyzes $\beta(1 \rightarrow 4)$ linkages from NAM to NAG indicates that bond cleavage occurs either between rings B and C or rings D and E. Since $(NAG)_3$ is stably bound to but not cleaved by the enzyme while spanning subsites B and C, the probable cleavage site is between rings D and E. This conclusion is supported by John Rupley's observation that lysozyme nearly quantitatively hydrolyzes $(NAG)_6$ between the second and third residues from its reducing terminus (the end with a free C1—OH), just as is expected if the enzyme has six saccharide-binding subsites and cleaves its bound substrate between rings D and E.

The bond that lysozyme cleaves was identified by carrying out the lysozyme-catalyzed hydrolysis of $(NAG)_3$ in $H_2{}^{18}O$. The resulting product had ^{18}O bonded to the C1 atom of its newly liberated reducing terminus, thereby demonstrating that bond cleavage occurs between C1 and the bridge oxygen O1:

Thus, lysozyme catalyzes the hydrolysis of the C1—O1 bond of a bound substrate's D residue. Moreover, *this reaction occurs with retention of configuration, so that the D-ring product remains the β anomer.*

B. Catalytic Mechanism

It remains to identify lysozyme's catalytic groups. The reaction catalyzed by lysozyme, the hydrolysis of a glycoside, is the conversion of an acetal to a hemiacetal. Nonenzymatic acetal hydrolysis is an acid-catalyzed reaction that involves the protonation of a reactant oxygen atom followed by cleavage of its C—O bond (Fig. 15-13). This results in the formation of a resonance-stabilized carbocation that is called an **oxonium ion.** To attain maximum orbital overlap, and thus resonance stabilization, the oxonium ion's R and R' groups must be coplanar with its C, O, and H atoms (stereoelectronic assistance). The oxonium ion then adds water to yield the hemiacetal and regenerate the acid catalyst. In searching for catalytic groups on an enzyme that mediates acetal hydrolysis, we should therefore seek a

Figure 15-13   Mechanism of the nonenzymatic acid-catalyzed hydrolysis of an acetal to a hemiacetal. The reaction involves the protonation of one of the acetal's oxygen atoms followed by cleavage of its C—O bond to form an alcohol (R″OH) and a resonance-stabilized carbocation (oxonium ion). The addition of water to the oxonium ion forms the hemiacetal and regenerates the H^+ catalyst. Note that the oxonium ion's C, O, H, R, and R' atoms all lie in the same plane.

potential acid catalyst and possibly a group that could further stabilize an oxonium ion intermediate.

a. Glu 35 and Asp 52 Are Lysozyme's Catalytic Residues

The only functional groups in the immediate vicinity of lysozyme's reaction center that have the required catalytic properties are the side chains of Glu 35 and Asp 52, residues that are invariant in the family of lysozymes of which HEW lysozyme is the prototype. These side chains, which are disposed to either side of the $\beta(1 \rightarrow 4)$ glycosidic linkage to be cleaved (Fig. 15-10), have markedly different environments. Asp 52 is surrounded by several conserved polar residues with which it forms a complex hydrogen bonded network. Asp 52 is therefore predicted to have a normal pK; that is, it should be unprotonated and hence negatively charged throughout the 3 to 8 pH range in which lysozyme is catalytically active. In contrast, *the carboxyl group of Glu 35 is nestled in a predominantly nonpolar pocket, where, as we discussed in Section 15-1D, it is likely to remain protonated at unusually high pH's for carboxyl groups.* Indeed, neutron diffraction studies, which provide similar information to X-ray diffraction studies but also reveal the positions of hydrogen atoms, indicate that Glu 35 is protonated at physiological pH. The closest approaches in the X-ray structures between the carboxyl O atoms of both Asp 52 and Glu 35 and the C1—O1 bond of NAG D-ring are ~3 Å, which makes them the prime candidates for electrostatic and acid catalysts, respectively.

b. The Lysozyme Catalytic Mechanism Proceeds via a Covalent Intermediate

Lysozyme's catalytic mechanism was largely formulated by Phillips based on the foregoing information. However, as we shall see, further investigations have required that important changes be made to the original version of this mechanism. The presently accepted mechanism occurs as follows (Fig. 15-14):

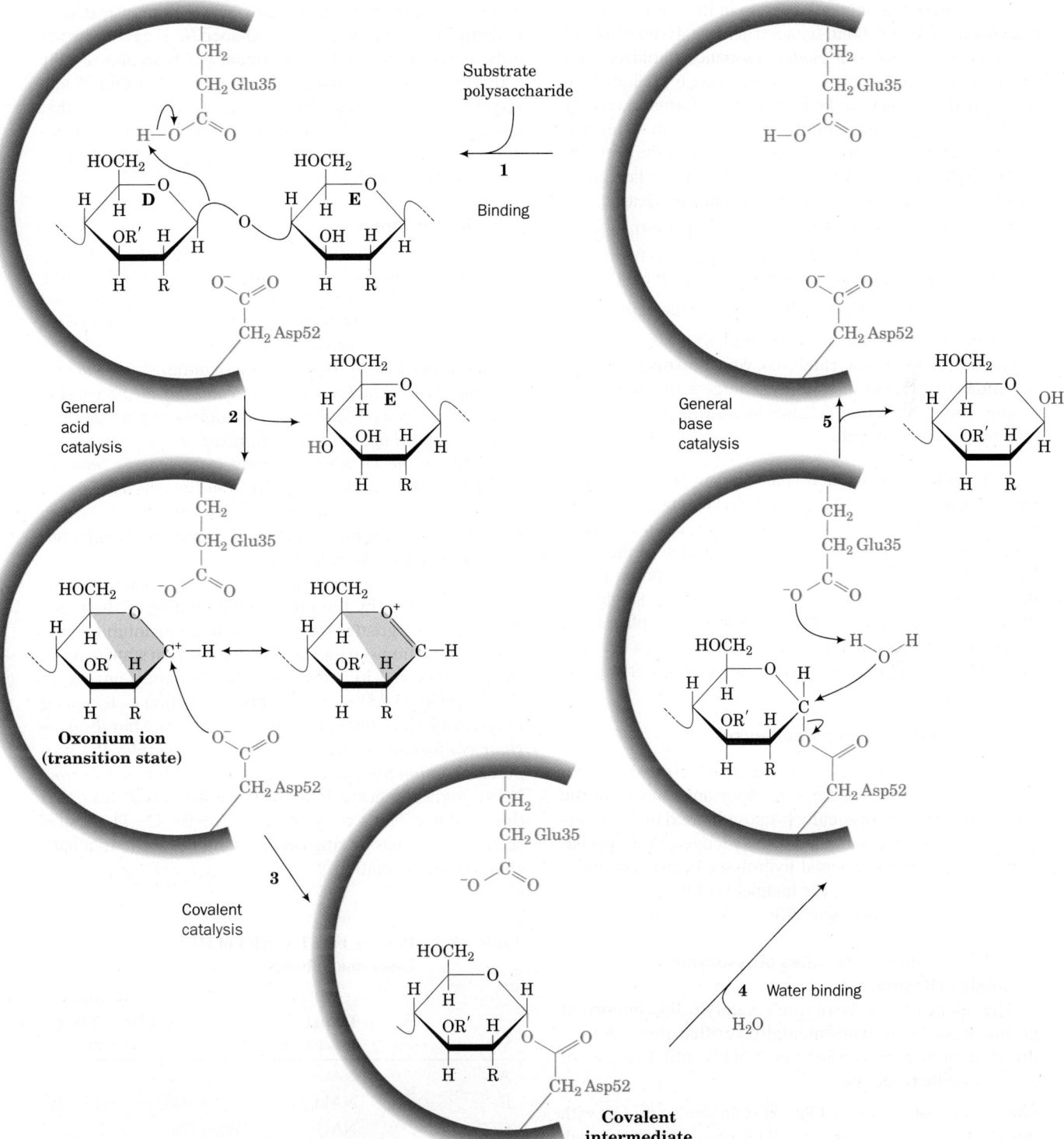

Figure 15-14 **The lysozyme reaction mechanism.** Glu 35 acts as an acid catalyst, and Asp 52 acts as a covalent catalyst. Only the substrate D- and E-rings are shown. R represents the *N*-acetyl group at C2, and R′ represents the CH₃CHCOO⁻ group at C3. The resonance-stabilized oxonium ion transition state requires that C1, C2, C5, and O5 be coplanar (*orange shading*) creating a half-chair conformation. Step 5 includes the participation of an oxonium ion transition state that is not shown. **See Kinemage Exercise 9**

1. Lysozyme attaches to a bacterial cell wall by binding to a hexasaccharide unit. In the process, the D-ring is distorted toward the half-chair conformation in response to the unfavorable contacts that its —C6H2OH group would otherwise make with the protein.

2. Glu 35 transfers its proton to the O1 atom linking the D- and E-rings, the only polar group in its vicinity, thereby cleaving the C1—O1 bond (general acid catalysis). This step converts the D-ring to a planar resonance-stabilized oxonium ion transition state, whose formation is facilitated by the strain distorting it to the half-chair conformation (catalysis by the preferential binding of the transition state). The positively charged oxonium ion is stabilized by the presence of the nearby negatively charged Asp 52 carboxylate group (electrostatic catalysis). The E-ring product is released.

3. The Asp 52 carboxylate group nucleophilically attacks the now electron-poor C1 of the D ring to form a co-valent **glycosyl–enzyme intermediate** (covalent catalysis).

4. Water replaces the E-ring product in the active site.

5. Hydrolysis of the covalent bond with the assistance of Glu 35 (general base catalysis), which involves another oxonium ion transition state, regenerates the active site groups. The enzyme then releases the D-ring product, completing the catalytic cycle.

The double-displacement mechanism diagrammed in Fig. 15-14 (in which the Asp 52 carboxylate group displaces O1 from C1 of the D-ring and is, in turn, displaced by water) allows the incoming water molecule to attach to the same face of the D-ring as the E-ring it replaces. Consequently, the configuration of the D-ring is retained. A single-displacement reaction, in which water directly displaces O1, would invert the configuration at C1 of the D-ring between the substrate and product, a result that is not observed.

C. Testing the Catalytic Mechanism

The forgoing mechanism is the product of over 40 years of enzymatic research. It differs in important ways from the mechanism Phillips originally proposed based on his structural studies of lysozyme and a knowledge of the mechanism of nonenzymatic acetal hydrolysis. In the remainder of this section, we discuss the highlights of these enzymatic studies to illustrate how scientific models evolve.

a. Confirming the Identities of Lysozyme's Catalytic Residues

The identities of lysozyme's catalytically important groups have been experimentally verified through site-directed mutagenesis (Section 5-5Gc) and the use of group-specific reagents:

Glu 35. The mutagenesis of Glu 35 to Gln yields a protein with no detectable catalytic activity (<0.1% of wild type), although it has only a ~1.5-fold decrease in substrate affinity. *Glu 35 must therefore be essential for lysozyme's catalytic activity.*

Asp 52. The mutagenesis of Asp 52 to Asn, which has a polarity comparable to that of Asp but lacks its negative

charge, yields an enzyme with no more than 5% of wild-type lysozyme's catalytic activity even though this mutation causes an ~2-fold increase in the enzyme's affinity for substrate. *Asp 52 is therefore important for enzymatic activity.*

Noninvolvement of Other Amino Acid Residues. Lysozyme's other carboxyl groups besides Glu 35 and Asp 52 do not participate in the catalytic process, as was demonstrated by reacting lysozyme with carboxyl-specific reagents in the presence of substrate. This treatment yields an almost fully active enzyme in which all carboxyl groups but Glu 35 and Asp 52 are derivatized. Other group-specific reagents that modify, for instance, His, Lys, Met, or Tyr residues but induce no major protein structure disruptions, cause little change in lysozyme's catalytic efficiency.

b. Role of Strain

Many of the mechanistic investigations of lysozyme have had the elusive goal of establishing the catalytic role of strain. Not all of these studies, as we shall see, supported the Phillips mechanism, thereby stimulating a series of investigations that eventually settled this issue.

Measurements of the binding equilibria of various oligosaccharides to lysozyme indicate that all saccharide residues except that binding to the D subsite contribute energetically toward the binding of substrate to lysozyme; binding NAM in the D subsite requires a free energy input of 12.1 kJ · mol⁻¹ (Table 15-3). *The Phillips mechanism explains this observation as being indicative of the energy penalty of straining the D-ring from its preferred chair conformation toward the half-chair form.*

As we discussed in Section 15-1Fa, an enzyme that catalyzes a reaction by the preferential binding of its transition state has a greater binding affinity for an inhibitor that has the transition state geometry (a transition state analog) than it does for its substrate. The δ-lactone analog of (NAG)₄ (Fig. 15-15) is presumably a transition state analog of lysozyme since *this compound's lactone ring has the half-chair conformation that geometrically resembles the proposed oxonium ion transition state of the substrate's D-ring.* X-ray studies indicate, in accordance with prediction, that this inhibitor binds to lysozyme's A—B—C—D subsites such that the lactone ring occupies the D subsite in a half-chairlike conformation.

Table 15-3 **Binding Free Energies of HEW Lysozyme Subsites**

Site	Bound Saccharide	Binding Free Energy (kJ · mol⁻¹)
A	NAG	−7.5
B	NAM	−12.3
C	NAG	−23.8
D	**NAM**	**+12.1**
E	NAG	−7.1
F	NAM	−7.1

Source: Chipman, D.M. and Sharon, N., *Science* **165,** 459 (1969).

δ-Lactone analog of (NAG)₄ (left)

Lysozyme transition state

Figure 15-15 Transition state analog inhibition of lysozyme. The δ-lactone analog of (NAG)₄ (*left*) resembles the transition state of the lysozyme reaction (*right*). Note that atoms C1, C2, C5, and O5 in each structure are coplanar (as indicated by orange shading), consistent with the half-chair conformation of the hexose ring.

Despite the foregoing, *the role of substrate distortion in lysozyme catalysis had been questioned.* Theoretical studies by Michael Levitt and Arieh Warshel on substrate binding by lysozyme suggested that the protein is too flexible to mechanically distort the D-ring of a bound substrate. Rather, these calculations implied that transition state stabilization occurs through the displacement by substrate of several tightly bound water molecules from the D subsite. The resulting desolvation of the Asp 52 carboxylate group would significantly enhance its capacity to electrostatically stabilize the transition state oxonium ion. This study therefore concluded that "electrostatic strain" rather than steric strain is the more important factor in stabilizing lysozyme's transition state.

In an effort to obtain further experimental information bearing on the Phillips strain mechanism, Nathan Sharon and David Chipman determined the D subsite–binding affinities of several saccharides by comparing the lysozyme-binding affinities of various substrate analogs. The NAG lactone inhibitor binds to the D subsite with 9.2 $kJ \cdot mol^{-1}$ greater affinity than does NAG. This quantity corresponds, according to Eq. [14.15], to no more than an ~40-fold rate enhancement of the lysozyme reaction as a result of strain (recall that the difference in binding energy between a transition state analog and a substrate is indicative of the enzyme's rate enhancement arising from the preferential binding of the transition state complex). Such an enhancement is hardly a major portion of lysozyme's ~10^8-fold rate enhancement (accounting for only ~20% of the reaction's $\Delta\Delta G_{cat}^{\ddagger}$; Section 14-1Cd). Moreover, an ***N*-acetylxylosamine (XylNAc)** residue,

***N*-Acetylxylosamine residue**

which lacks the sterically hindered —$C6H_2OH$ group of NAM and NAG, has only marginally greater binding affinity for the D subsite ($-3.8 \ kJ \cdot mol^{-1}$) than does NAG ($-2.5 \ kJ \cdot mol^{-1}$). Yet recall that the Phillips mechanism postulates that it is the unfavorable contacts made by this —$C6H_2OH$ group that promotes D-ring distortion. Nevertheless, lysozyme does not hydrolyze saccharides with XylNAc in the D subsite.

The apparent inconsistencies among the foregoing experimental observations were largely rationalized by Michael James' highly accurate (1.5-Å resolution) X-ray crystal structure determination of lysozyme in complex with NAM–NAG–NAM. This trisaccharide binds, as expected, to the B, C, and D subsites of lysozyme. *The NAM in the D subsite, in agreement with the Phillips mechanism, is distorted to the half-chair conformation with its —$C6H_2OH$ group in a nearly axial position due to steric clashes that would otherwise occur with the acetamido group of the C subsite NAG* (although, contrary to the original Phillips mechanism, Glu 35 and Trp 108 are too far away from the —$C6H_2OH$ group to contribute to this distortion). This strained conformation is stabilized by a strong hydrogen bond between the D-ring O6 and the backbone NH of Val 109 (Fig. 15-12; transition state stabilization). Indeed, the mutation of Val 109 to Pro, which lacks the NH group to make such a hydrogen bond, inactivates the enzyme. Lysozyme's lack of hydrolytic activity when XylNAc occupies its D subsite is likewise explained by the absence of this hydrogen bond and the consequent lesser stability of the XylNAc ring's half-chair transition state.

The unexpectedly small free energy differences in binding NAG, NAG lactone, and XylNAc to the D subsite are explained by the observation that undistorted NAG and XylNAc can be modeled into the D subsite as it occurs in the X-ray structure of the lysozyme · NAM–NAG–NAM complex. NAM's bulky lactyl side chain prevents it from binding to the D subsite in this manner.

c. Evidence for a Covalent Intermediate

The original Phillips mechanism proposed that the lysozyme reaction proceeds via the direct attack of a water molecule on C1 of the half-chair-distorted D-ring so as to directly displace the E-ring (a single-displacement reaction) and therefore did not involve the intermediate formation of a covalent bond (much like the mechanism diagrammed in Fig. 15-14 but skipping Step 3). This was based on Phillip's observation that the distance between C1 of the D-ring and a carboxyl O of Asp 52 (which participates in a network of hydrogen bonds that appear to hold this side chain in place) are too long to form a covalent bond (minimally 2.3 Å in the NAM–NAG–NAM complex without significantly disrupting the protein structure vs ~1.4 Å for a C—O single bond). The fact that the lysozyme reaction proceeds with retention of configuration was attributed to the shielding of the reaction intermediate by the enzyme, thereby preventing a water molecule from approaching the bond being cleaved from its back side as normally occurs in a single-displacement reaction. This aspect of the original Phillips mechanism was widely accepted for over 35 years because no enzyme–substrate covalent bond had been detected in any of the numerous experimental studies of hen egg white (HEW) lysozyme.

Despite the foregoing, all other β-glycosidases of known structure that cleave glycosidic linkages with net retention of configuration at the anomeric carbon (as does HEW lysozyme) have been shown to do so via a covalent glycosyl–enzyme intermediate. The active sites of these so-called **retaining β-glycosidases** structurally resemble that of HEW lysozyme. Moreover, there is no direct evidence indicative of the existence of a long-lived oxonium ion at the active site of any retaining β-glycosidase, including HEW lysozyme (the lifetime of a glucosyl oxonium ion in water is ~10^{-12} s, a time only slightly larger than that of a bond vibration). Consequently, there had been a growing suspicion that the HEW lysozyme reaction also proceeds via a covalent intermediate, one between the D-ring's anomeric carbon (C1) and the side chain carboxyl group of Asp 52 to form an ester linkage (as generated by Step 3 in Fig. 15-14). This intermediate presumably reacts with H_2O in what is essentially the reverse of the reaction leading to its formation, thereby yielding the reaction's second product (Steps 4 and 5 of Fig. 15-14). In this double-displacement mechanism, the oxonium ion is proposed to be the transition state on the way to forming the covalent intermediate, rather than being an intermediate itself as it was in the original Phillips mechanism.

If, in fact, HEW lysozyme follows the mechanism in Fig. 15-14, the reason that its covalent intermediate had not been observed is that its rate of breakdown must be much faster than its rate of formation. Hence, if this intermediate is to be experimentally observed, its rate of formation must be made significantly greater than its rate of breakdown. To do so, Stephen Withers capitalized on three phenomena. First, if, as postulated, the reaction goes through an oxonium ion transition state, all steps involving its formation should be slowed by the electron withdrawing effects of substituting F (the most electronegative element) at C2 of the D-ring. Second, mutating Glu 35 to Gln (E35Q) removes the general acid–base that catalyzes the reaction, further slowing all steps involving the oxonium ion transition state. Third, substituting an additional F atom at C1 of the D-ring accelerates the formation of the intermediate because this F is a good leaving group. Making all three of these changes should increase the rate of formation of the proposed covalent intermediate relative to its breakdown and hence should result in its accumulation. Withers therefore incubated E35Q HEW lysozyme with NAG-β(1 → 4)-2-deoxy-2-fluoro-β-D-glucopyranosyl fluoride (**NAG2FGlcF**):

NAG2FGlcF

Electrospray ionization mass spectrometry (ESI-MS; Sec-

tion 7-1I) of this reaction mixture revealed a sharp peak at 14,683 D, consistent with the formation of the proposed covalent intermediate, but no significant peak at or near the 14,314-D molecular mass of the mutant enzyme alone.

The X-ray structure of this covalent complex unambiguously reveals the expected ~1.4-Å-long covalent bond between C1 of the D-ring NAG and a side chain carboxyl O of Asp 52 (Fig. 15-16). This D-ring NAG adopts an undistorted chair conformation, thus indicating that it is a reaction intermediate rather than an approximation of the transition state. The superposition of this covalent complex with that of the above described complex of NAM–NAG–NAM with wild-type HEW lysozyme reveals how this covalent bond forms (Fig. 15-16). The shortening of the 3.2-Å distance between the D-ring NAG C1 and the Asp 52 O in the NAM–NAG–NAM

Figure 15-16 The HEW lysozyme covalent intermediate. The substrate C- and D-rings and Asp 52 are shown in the superposition of the X-ray structures of the covalent complex formed by reacting E35Q lysozyme with NAG2FGlcF (C green, N blue, O red, and F magenta) and the noncovalent complex of wild-type lysozyme with NAM–NAG–NAM (C yellow, N blue, and O red). Note that the covalent bond between Asp 52 and C1 of the D-ring forms when the D-ring in the noncovalent complex relaxes from its distorted half-chair conformation to an undistorted chair conformation and that the side chain of Asp 52 undergoes an ~45° rotation about its C_α—C_β bond. [Based on X-ray structures by David Vocadlo and Stephen Withers, University of British Columbia, Vancouver, Canada; *and* Michael James, University of Alberta, Edmonton, Canada. PDBids 1H6M and 9LYZ.]

complex to ~1.4 Å in the covalent complex is almost entirely a consequence of the relaxation of the D-ring from the half-chair to the chair conformation combined with an ~45°rotation of the Asp 52 side chain about its C_α—C_β bond; the positions of the D-ring O4 and O6 atoms are essentially unchanged. Hence, over 35 years after its formulation, it was shown that *the Phillips mechanism must be altered to take into account the transient formation of this covalent glycosyl–enzyme ester intermediate (covalent catalysis)*. Keep in mind, however, that in order to form this covalent linkage, the D-ring must pass through an oxonium-like transition state, which requires it to transiently assume the half-chair conformation.

3 SERINE PROTEASES

Our next example of enzymatic mechanisms is a diverse group of proteolytic enzymes known as the **serine proteases** (Table 15-4). These enzymes are so named because they have a common catalytic mechanism characterized by the possession of a peculiarly reactive Ser residue that is essential for their enzymatic activity. The serine proteases are the most thoroughly understood family of enzymes as a result of their extensive examination over a nearly 60-year period by kinetic, chemical, physical, and genetic techniques. In this section, we mainly study the best characterized serine proteases, **chymotrypsin, trypsin,** and **elastase.** We also consider how these three enzymes, which are synthesized in inactive forms, are physiologically activated.

A. *Kinetics and Catalytic Groups*

Chymotrypsin, trypsin, and elastase are digestive enzymes that are synthesized by the pancreatic acinar cells (Fig. 1-10c) and secreted, via the pancreatic duct, into the duodenum (the small intestine's upper loop). All of these enzymes catalyze the hydrolysis of peptide (amide) bonds but with different specificities for the side chains flanking the scissile (to be cleaved) peptide bond (recall that chymotrypsin is specific for a bulky hydrophobic residue preceding the scissile peptide bond, trypsin is specific for a positively charged residue, and elastase is specific for a small neutral residue; Table 7-2). Together, they form a potent digestive team.

a. Ester Hydrolysis as a Kinetic Model

That chymotrypsin can act as an esterase as well as a protease is not particularly surprising because the chemical mechanisms of ester and amide hydrolysis are almost identical. The study of chymotrypsin's esterase activity has led to important insights concerning this enzyme's catalytic mechanism. Kinetic measurements by Brian Hartley of the chymotrypsin-catalyzed hydrolysis of *p*-nitrophenylacetate

p-Nitrophenylacetate

— H_2O
chymotrypsin
→ $2H^+$

Acetate   **p-Nitrophenolate**

indicated that the reaction occurs in two phases (Fig. 15-17):

1. The "burst phase," in which the highly colored *p*-nitrophenolate ion is rapidly formed in amounts stoichiometric with the quantity of active enzyme present.

2. The "steady-state phase," in which *p*-nitrophenolate is generated at a reduced but constant rate that is independent of substrate concentration.

Table 15-4   **A Selection of Serine Proteases**

Enzyme	Source	Function
Trypsin	Pancreas	Digestion of proteins
Chymotrypsin	Pancreas	Digestion of proteins
Elastase	Pancreas	Digestion of proteins
Thrombin	Vertebrate serum	Blood clotting
Plasmin	Vertebrate serum	Dissolution of blood clots
Kallikrein	Blood and tissues	Control of blood flow
Complement C1	Serum	Cell lysis in the immune response
Acrosomal protease	Sperm acrosome	Penetration of ovum
Lysosomal protease	Animal cells	Cell protein turnover
Cocoonase	Moth larvae	Dissolution of cocoon after metamorphosis
α-Lytic protease	*Lysobacter enzymogenes*	Possibly digestion
Proteases A and B	*Streptomyces griseus*	Possibly digestion
Subtilisin	*Bacillus subtilis*	Possibly digestion

Source: Stroud, R.M., *Sci. Am.* **231**(1), 86 (1974).

Figure 15-17   Time course of *p*-nitrophenylacetate hydrolysis as catalyzed by two different concentrations of chymotrypsin. The enzyme rapidly binds substrate and releases the first product, *p*-nitrophenolate ion, but the second product, acetate ion, is released more slowly. Consequently, the rate of *p*-nitrophenolate generation begins rapidly (burst phase) but slows as acyl–enzyme complex accumulates until the rate of *p*-nitrophenolate generation approaches that of acetate release (steady state). The extrapolation of the steady state curve to zero time (*dashed lines*) indicates the initial concentration of active enzyme. [After Hartley, B.S. and Kilby, B.A., *Biochem. J.* **56,** 294 (1954).]

These observations have been interpreted in terms of a two-stage reaction sequence in which the enzyme **(1)** rapidly reacts with the *p*-nitrophenylacetate to release *p*-nitrophenolate ion forming a covalent acyl–enzyme intermediate that **(2)** is slowly hydrolyzed to release acetate:

$$CH_3-\overset{\overset{\textstyle O}{\|}}{C}-O-\!\!\!\bigcirc\!\!\!-NO_2 \;+\; Enzyme$$

***p*-Nitrophenylacetate**   **Chymotrypsin**

fast

$$^-O-\!\!\!\bigcirc\!\!\!-NO_2$$

***p*-Nitrophenolate**

$$CH_3-\overset{\overset{\textstyle O}{\|}}{C}-Enzyme$$

Acyl–enzyme intermediate

slow $-H_2O$ $\rightarrow H^+$

$$CH_3-\overset{\overset{\textstyle O}{\|}}{C}-O^- \;+\; Enzyme$$

Acetate

Chymotrypsin evidently follows a Ping Pong Bi Bi mechanism (Section 14-5A). Chymotrypsin-catalyzed amide hydrolysis has been shown to follow a reaction pathway similar to that of ester hydrolysis but with the first step of the reaction, enzyme acylation, being rate determining rather than the deacylation step.

b. Identification of the Catalytic Residues

Chymotrypsin's catalytically important groups were identified by chemical labeling studies. These are described below.

Ser 195. A diagnostic test for the presence of the **active Ser** of serine proteases is its reaction with **diisopropylphosphofluoridate (DIPF):**

$$(Active\ Ser)-CH_2OH \;+\; F-\overset{\overset{\textstyle CH(CH_3)_2}{|}\ \overset{\textstyle O}{|}}{\underset{\underset{\textstyle CH(CH_3)_2}{|}\ \underset{\textstyle O}{|}}{P}}\!\!=\!\!O$$

Diisopropylphospho-fluoridate (DIPF)

$$(Active\ Ser)-CH_2-O-\overset{\overset{\textstyle CH(CH_3)_2}{|}\ \overset{\textstyle O}{|}}{\underset{\underset{\textstyle CH(CH_3)_2}{|}\ \underset{\textstyle O}{|}}{P}}\!\!=\!\!O \;+\; HF$$

DIP–Enzyme

which irreversibly inactivates the enzyme. Other Ser residues, including those on the same protein, do not react with DIPF. *DIPF reacts only with Ser 195 of chymotrypsin, thereby demonstrating that this residue is the enzyme's active Ser.*

The use of DIPF as an enzyme inactivating agent came about through the discovery that organophosphorus compounds such as DIPF are potent nerve poisons. The neurotoxicity of DIPF arises from its ability to inactivate **acetylcholinesterase,** a serine esterase that catalyzes the hydrolysis of **acetylcholine:**

$$(CH_3)_3\overset{+}{N}-CH_2-CH_2-O-\overset{\displaystyle O}{\overset{\|}{C}}-CH_3 \ + \ H_2O$$

Acetylcholine

|acetylcholinesterase

$$(CH_3)_3\overset{+}{N}-CH_2-CH_2-OH \ + \ \overset{\displaystyle O}{\underset{-O}{\overset{\|}{C}}}-CH_3$$

Choline

Acetylcholine is a **neurotransmitter:** It transmits nerve impulses across the **synapses** (junctions) between certain types of nerve cells (Sections 12-4Da and 20-5Cb). The inactivation of acetylcholinesterase prevents the otherwise rapid hydrolysis of the acetylcholine released by a nerve impulse and thereby interferes with the regular sequence of nerve impulses. DIPF is of such great toxicity to humans that it has been used militarily as a nerve gas. Related compounds, such as **parathion** and **malathion,**

Parathion

Malathion

are useful insecticides because they are far more toxic to insects than to mammals.

His 57. A second catalytically important residue was discovered through **affinity labeling.** In this technique, a substrate analog bearing a reactive group specifically binds at the enzyme's active site, where it reacts to form a stable covalent bond with a nearby susceptible group (these reactive substrate analogs have therefore been described as the "Trojan horses" of biochemistry). The affinity labeled groups can subsequently be identified by peptide mapping (Section 7-1J). Chymotrypsin specifically binds **tosyl-L-phenylalanine chloromethyl ketone (TPCK),**

Figure 15-18   Reaction of TPCK with chymotrypsin to alkylate His 57.

because of its resemblance to a Phe residue (one of chymotrypsin's preferred residues; Table 7-2). Active site–bound TPCK's chloromethyl ketone group is a strong alkylating agent; it reacts with His 57 (Fig. 15-18), thereby inactivating the enzyme. The TPCK reaction is inhibited by **β-phenylpropionate,**

β-Phenylpropionate

a competitive inhibitor of chymotrypsin that presumably competes with TPCK for its enzymatic binding site. Moreover, the TPCK reaction does not occur in 8M urea, a denaturing reagent, or with DIP–chymotrypsin, in which the active site is blocked. These observations establish that *His 57 is an essential active site residue of chymotrypsin.*

B. X-Ray Structures

Bovine chymotrypsin, bovine trypsin, and porcine elastase are strikingly homologous: The primary structures of these ~240-residue monomeric enzymes are ~40% identical and their internal sequences are even more alike (in comparison, the α and β chains of human hemoglobin have a 44% sequence identity). Furthermore, *all of these enzymes have an active Ser and a catalytically essential His as well as similar kinetic mechanisms.* It therefore came as no surprise when their X-ray structures all proved to be closely related.

To most conveniently compare the structures of these three digestive enzymes, they have been assigned the same amino acid residue numbering scheme. Bovine chymotrypsin is synthesized as an inactive 245-residue precursor named **chymotrypsinogen** that is proteolytically converted to chymotrypsin (Section 15-3Ea). In what follows, the numbering of the amino acid residues in chymotrypsin, trypsin, and elastase will be that of the corresponding residues in bovine chymotrypsinogen.

The X-ray structure of bovine chymotrypsin was elucidated in 1967 by David Blow. This was followed by the

Figure 15-19 **X-ray structure of bovine trypsin in covalent complex with its inhibitor leupeptin.** The protein, viewed looking into its upper β barrel, is represented by its transparent molecular surface with its polypeptide chain in ribbon form colored in rainbow order from its N-terminus (*blue*) to its C-terminus (*red*), but with loop L1 (residues 185–188) blue-green, loop L2 (residues 221–225) violet, and Tyr 172 and Gly 216 purple. The side chains of the catalytic triad (Ser 195, His 57, and Asp 102) and Asp 189 are drawn in ball-and-stick form colored according to atom type (C green, N blue, O red) with hydrogen bonds represented by dashed black lines. **Leupeptin** (acetyl-Leu-Leu-Arg in which the terminal carboxyl group is replaced by —CHO) is drawn in stick form (C cyan, N blue, O red) with its Arg side chain occupying the enzyme's specificity pocket (*magenta mesh*). [Based on an X-ray structure by Daniel Koshland, Jr., University of California at Berkeley. PDBid 2AGI.] 🖱️ **See Kinemage Exercise 10-1**

Catalytic triad

Gly 193

His 57   Ser 195   Asp 194

Asp 102   Ile 16

Figure 15-20 **The active site residues of trypsin.** Residues are drawn in ball-and-stick form with C green, N blue, and O red and viewed similarly to Fig. 15-19. The catalytic triad consists of Ser 195, His 57, and Asp 102. [Based on an X-ray structure by Daniel Koshland, Jr., University of California at Berkeley. PDBid 2AGI.]

determination of the structures of bovine trypsin (Fig. 15-19) by Robert Stroud and Richard Dickerson, and porcine elastase by David Shotton and Herman Watson. Each of these proteins is folded into two structurally similar domains, each of which have extensive regions of antiparallel β sheets arranged in a 6-stranded β barrel, but contain little helix. *The catalytically essential His 57 and Ser 195 are located in a cleft between the β barrels, as is the invariant (in all serine proteases) Asp 102, which is buried in a solvent-inaccessible pocket. These three residues form a hydrogen bonded constellation referred to as the* **catalytic triad** (Figs. 15-19 and 15-20).

a. The Structural Basis of Substrate Specificity Can Be Quite Complex

The X-ray structures of the above three enzymes suggest the basis for their differing substrate specificities (Fig. 15-21):

1. In chymotrypsin, the bulky aromatic side chain of the preferred Phe, Trp, or Tyr residue (Table 7-2) that con-

tributes the carbonyl group of the scissile peptide fits snugly into a slitlike hydrophobic pocket, the specificity pocket, that is located near the catalytic groups (Fig. 15-19).

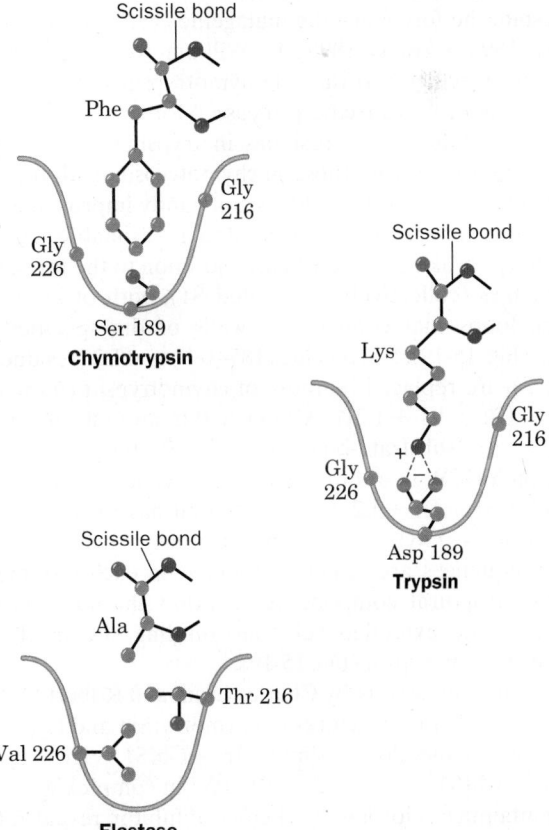

Figure 15-21 **Specificity pockets of three serine proteases.** The side chains of key residues that determine the size and nature of the specificity pocket are shown along with a representative substrate for each enzyme. Chymotrypsin prefers to cleave peptide bonds following large hydrophobic side chains; trypsin prefers Lys or Arg; and elastase prefers Ala, Gly, or Val. [After a drawing in Branden, C. and Tooze, J., *Introduction to Protein Structure* (2nd ed.), Garland Publishing, *p.* 213 (1999).]

2. In trypsin, the residue corresponding to chymotrypsin Ser 189, which lies at the back of the specificity pocket, is the anionic residue Asp (Fig. 15-19). The cationic side chains of trypsin's preferred residues, Arg or Lys, can therefore form ion pairs with this Asp residue. The rest of chymotrypsin's specificity pocket is preserved in trypsin so that it can accommodate the bulky side chains of Arg and Lys.

3. Elastase is so named because it rapidly hydrolyzes the otherwise nearly indigestible Ala, Gly, and Val-rich protein **elastin** (a connective tissue protein with rubber-like elastic properties). Elastase's specificity pocket is largely occluded by the side chains of a Val and a Thr residue that replace two Gly's lining this pocket in both chymotrypsin and trypsin. Consequently elastase, whose specificity pocket is better described as a depression, specifically cleaves peptide bonds after small neutral residues, particularly Ala. In contrast, chymotrypsin and trypsin hydrolyze such peptide bonds extremely slowly because these small substrates cannot be sufficiently immobilized on the enzyme surface for efficient catalysis to occur (Section 15-1E).

Thus, for example, trypsin catalyzes the hydrolysis of peptidyl amide substrates with an Arg or Lys residue preceding the scissile bond with an efficiency, as measured by k_{cat}/K_M (Section 14-2Ba), that is 10^6-fold greater than that for the corresponding Phe-containing substrates. Conversely, chymotrypsin catalyzes the hydrolysis of substrates after Phe, Trp, and Tyr residues 10^4-fold more efficiently than after the corresponding Lys-containing substrates.

Despite the foregoing, the mutagenic change in trypsin of Asp 189 → Ser (D189S) by William Rutter did not switch its specificity to that of chymotrypsin but instead yielded a poor, nonspecific protease. Moreover, even replacing the other three residues in trypsin's specificity pocket that differ from those in chymotrypsin, with those of chymotrypsin, fails to yield a significantly improved enzyme. However, trypsin is converted to a reasonably active chymotrypsin-like enzyme when, in addition to the foregoing changes (collectively designated S1), both of its two surface loops that connect the walls of the specificity pocket (Fig. 15-19), L1 (residues 185–188) and L2 (residues 221–225), are replaced by those of chymotrypsin (termed Tr → Ch[S1 + L1 + L2]). Although this mutant enzyme still has a low substrate-binding affinity, K_S, the additional mutation Y172W in a third surface loop yields an enzyme (Tr → Ch[S1 + L1 + L2 + Y172W]) that has 15% of chymotrypsin's catalytic efficiency. Curiously, these loops, whose sequences are largely conserved in each enzyme, are not structural components of either the specificity pocket or the extended substrate binding site in chymotrypsin or in trypsin (Fig. 15-19).

Careful comparisons, by Charles Craik and Robert Fletterick, of the X-ray structures of chymotrypsin and trypsin with those of the closely similar Tr → Ch[S1 + L1 + L2] and Tr → Ch[S1 + L1 + L2 + Y172W] in complex with a Phe-containing chloromethyl ketone inhibitor reveal the structural basis of substrate specificity in trypsin and chymotrypsin. Efficient catalysis in the serine proteases requires that the enzyme's active site be structurally intact and that the substrate's scissile bond be properly positioned relative to the catalytic triad and other components of the active site (see below). The above mutagenic changes do not affect the structure of the catalytic triad or those portions of the active site that bind the substrate's leaving group (that segment on the C-terminal side of the scissile bond). However, the main chain conformation of the conserved Gly 216 (which forms two hydrogen bonds to the backbone of the third residue before the substrate's scissile bond in an antiparallel β pleated sheet–like arrangement) differs in trypsin and chymotrypsin and adopts a chymotrypsin-like structure in both hybrid proteins. Evidently, if Gly 216 adopts a trypsin-like conformation, the scissile bond in Phe-containing substrates is misoriented for efficient catalysis. Thus, despite the fact that Gly 216 is conserved in trypsin and chymotrypsin, the differing structures of loop L2 in the two enzymes maintain it in distinct conformations.

Loop L1, which interacts with L2 in both trypsin and chymotrypsin, is largely disordered in the X-ray structure of Tr → Ch[S1 + L1 + L2]. Modeling a trypsin-like L1 into Tr → Ch[S1 + L1 + L2] results in severe steric clashes with the chymotrypsin-like L2. Thus, the requirement of a chymotrypsin-like L1 for the efficient catalysis by Tr → Ch[S1 + L1 + L2] appears to arise from the need to permit L2 to adopt a chymotrypsin-like conformation.

The side chain of Tyr 172 extends toward the base of the specificity pocket. The improvement in substrate binding affinity of Tr → Ch[S1 + L1 + L2 + Y172W] over Tr → Ch[S1 + L1 + L2] arises from structural rearrangements in this region of the enzyme caused by the increased bulk and different hydrogen bonding requirements of Trp versus Tyr. These changes appear to improve both the structural stability of residues forming the specificity pocket and their specificity for chymotrypsin-like substrates. These results therefore highlight an important caveat for genetic engineers: *Enzymes are so exquisitely tailored to their functions that they often respond to mutagenic tinkering in unexpected ways.*

b. Evolutionary Relationships Among Serine Proteases

We have seen that sequence and structural homologies among proteins reveal their evolutionary relationships (Sections 7-3 and 9-6). *The great similarities among chymotrypsin, trypsin, and elastase indicate that these proteins evolved through gene duplications of an ancestral serine protease followed by the divergent evolution of the resulting enzymes (Section 7-3C).*

Several serine proteases from various sources provide further insights into the evolutionary relationships among the serine proteases. **Streptomyces griseus protease A (SGPA)** is a bacterial serine protease of chymotryptic specificity that exhibits extensive structural similarity, although only ~20% sequence identity, with the pancreatic serine proteases. The primordial trypsin gene evidently arose before the divergence of prokaryotes and eukaryotes.

There are three known serine proteases whose primary and tertiary structures bear no discernible relationship to each other or to chymotrypsin but which, nevertheless, contain catalytic triads at their active sites whose structures closely resemble that of chymotrypsin:

1. Subtilisin, an endopeptidase that was originally isolated from *Bacillus subtilis.*

2. Wheat germ **serine carboxypeptidase II,** an exopeptidase whose structure is surprisingly similar to that of carboxypeptidase A (Fig. 8-19*a*) even though the latter protease has an entirely different catalytic mechanism from that of the serine proteases (see Problem 3).

3. *E. coli* **ClpP,** which functions in the degradation of cellular proteins (Section 32-6B).

Since the orders of the corresponding active site residues in the amino acid sequences of the four types of serine proteases are quite different (Fig. 15-22), *it seems highly improbable that they could have evolved from a common ancestor serine protease. These proteins apparently constitute a remarkable example of* **convergent evolution:** *Nature seems to have independently discovered the same catalytic mechanism at least four times.* (In addition, **human cytomegalovirus protease,** an essential protein for virus replication that bears no resemblance to the above

proteases, has active site Ser and His residues whose relative positions are similar to those in other serine proteases but lacks an active site Asp residue; it appears to have a catalytic dyad.)

C. *Catalytic Mechanism*

See Guided Exploration 12: The catalytic mechanism of serine proteases The extensive active site homologies among the various serine proteases indicate that they all have the same catalytic mechanism. On the basis of considerable chemical and structural data gathered in many laboratories, the following catalytic mechanism has been formulated for the serine proteases, here given in terms of chymotrypsin (Fig. 15-23):

1. After chymotrypsin has bound substrate to form the Michaelis complex, *Ser 195, in the reaction's rate-determining step, nucleophilically attacks the scissile peptide's carbonyl group to form a complex known as the* **tetrahedral intermediate** *(covalent catalysis).* X-ray studies indicate that Ser 195 is ideally positioned to carry out this nucleophilic attack (proximity and orientation effects). The imidazole ring of His 57 takes up the liberated proton, thereby forming an imidazolium ion (general base catalysis). This process is aided by the polarizing effect of the unsolvated carboxylate ion of Asp 102, which is hydrogen bonded to His 57 (electrostatic catalysis; see Section 15-3Dd). Indeed, the mutagenic replacement of trypsin's Asp

Figure 15-22   Relative positions of the active site residues in subtilisin, chymotrypsin, serine carboxypeptidase II, and ClpP protease. The peptide backbones of Ser 214, Trp 215, and Gly 216 in chymotrypsin, and their counterparts in subtilisin, participate in substrate-binding interactions. [After Robertus, J.D., Alden, R.A., Birktoft, J.J., Kraut, J., Powers, J.C., and Wilcox, P.E., *Biochemistry* **11,** 2449 (1972).] **See Kinemage Exercise 10-2**

Figure 15-23 **Catalytic mechanism of the serine proteases.** The reaction involves **(1)** the nucleophilic attack of the active site Ser on the carbonyl carbon atom of the scissile peptide bond to form the tetrahedral intermediate; **(2)** the decomposition of the tetrahedral intermediate to the acyl–enzyme intermediate through general acid catalysis by the active site Asp-polarized His, followed by loss of the amine product and its replacement by a water molecule; **(3)** the reversal of Step 2 to form a second tetrahedral intermediate; and **(4)** the reversal of Step 1 to yield the reaction's carboxyl product and the active enzyme.

102 by Asn leaves the enzyme's K_M substantially unchanged at neutral pH but reduces its k_{cat} to ~0.05% of its wild-type value. Neutron diffraction studies have demonstrated that *Asp 102 remains a carboxylate ion rather than abstracting a proton from the imidazolium ion to form an uncharged carboxylic acid group.* The tetrahedral

intermediate has a well-defined, although transient, existence. We shall see that *much of chymotrypsin's catalytic power derives from its preferential binding of the transition state leading to this intermediate (transition state binding catalysis).*

2. The tetrahedral intermediate decomposes to the **acyl–enzyme intermediate** under the driving force of proton donation from N3 of His 57 (general acid catalysis). The amine leaving group (R′NH$_2$, the new N-terminal portion of the cleaved polypeptide chain) is released from the enzyme and replaced by water from the solvent.

3 & 4. The acyl–enzyme intermediate (which, in the absence of enzyme, would be a stable compound) is rapidly deacylated by what is essentially the reverse of the previous steps followed by the release of the resulting carboxylate product (the new C-terminal portion of the cleaved polypeptide chain), thereby regenerating the active enzyme. In this process, water is the attacking nucleophile and Ser 195 is the leaving group.

D. *Testing the Catalytic Mechanism*

The formulation of the foregoing model for catalysis by serine proteases has prompted numerous investigations of its validity. In this section we discuss several of the most revealing of these studies.

a. The Tetrahedral Intermediate Is Mimicked in a Complex of Trypsin with Trypsin Inhibitor

Convincing structural evidence for the existence of the tetrahedral intermediate was provided by Robert Huber in an X-ray study of the complex between the 58-residue protein **bovine pancreatic trypsin inhibitor (BPTI)** and trypsin. BPTI binds to and inactivates trypsin, thereby preventing any trypsin that is prematurely activated in the pancreas from digesting that organ (Section 15-3E). BPTI binds to the active site region of trypsin across a tightly packed interface that is cross-linked by a complex network of hydrogen bonds. This complex's 10^{13} M^{-1} association constant, among the largest of any known protein–protein interaction, emphasizes BPTI's physiological importance.

The portion of BPTI in contact with the trypsin active site resembles bound substrate. The side chain of BPTI Lys 15I (here "I" differentiates BPTI residues from trypsin residues) occupies the trypsin specificity pocket (Fig. 15-24a) and the peptide bond between Lys 15I and Ala 16I is positioned as if it were the scissile peptide bond (Fig. 15-24b). What is most remarkable about this structure is that *its active site complex assumes a conformation well along the reaction coordinate toward the tetrahedral intermediate: The side chain oxygen of trypsin Ser 195, the active Ser, is in closer-than-van der Waals contact (2.6 Å) with the pyramidally distorted carbonyl carbon of BPTI's "scissile" peptide.* Despite this close contact, the proteolytic reaction cannot proceed past this point along the reaction coordinate because of the rigidity of the active site complex and because it is so tightly sealed that the leaving group cannot leave and water cannot enter the reaction site.

(a)

(b)

Figure 15-24   Trypsin–BPTI complex. (*a*) The X-ray structure shown as a cutaway surface drawing indicating how trypsin (*red*) binds BPTI (*green*). The green protrusion extending into the red cavity near the center of the figure represents the Lys 15I side chain occupying trypsin's specificity pocket. Note the close complementary fit of these two proteins. [Courtesy of Michael Connolly, New York University.] (*b*) Trypsin Ser 195, the active Ser, is in closer-than-van der Waals contact with the carbonyl carbon of BPTI's scissile peptide, which is pyramidally distorted toward Ser 195. The normal proteolytic reaction is apparently arrested somewhere along the reaction coordinate between the Michaelis complex and the tetrahedral intermediate.

Protease inhibitors are common in nature, where they have protective and regulatory functions. For example, certain plants release protease inhibitors in response to insect bites, thereby causing the offending insect to starve by inactivating its digestive enzymes. Protease inhibitors constitute ~10% of the nearly 200 proteins of blood serum. For instance, **α$_1$-proteinase inhibitor,** which is secreted by the liver, inhibits **leukocyte elastase** (leukocytes are a type of white blood cell; the action of leukocyte elastase is thought to be part of the inflammatory process). Pathological variants of α$_1$-proteinase inhibitor with reduced activity are associated with **pulmonary emphysema,** a degenerative disease of the lungs resulting from the hydrolysis of its elastic fibers. Smokers also suffer from reduced activity of their α$_1$-proteinase inhibitor because of the oxidation of its active site Met residue. Full activity of this inhibitor is not regained until several hours after smoking.

b. Serine Proteases Preferentially Bind the Transition State

Detailed comparisons of the X-ray structures of several serine protease–inhibitor complexes have revealed a further structural basis for catalysis in these enzymes (Fig. 15-25):

1. The conformational distortion that occurs with the formation of the tetrahedral intermediate causes the carbonyl oxygen of the scissile peptide to move deeper into the active site so as to occupy a previously unoccupied position, the **oxyanion hole.**

2. *There it forms two hydrogen bonds with the enzyme that cannot form when the carbonyl group is in its normal trigonal conformation.* These two enzymatic hydrogen bond donors were first noted by Joseph Kraut to occupy corresponding positions in chymotrypsin and subtilisin. He proposed the existence of the oxyanion hole based on the premise that convergent evolution had made the active sites of these unrelated enzymes functionally identical.

3. The tetrahedral distortion, moreover, permits the formation of an otherwise unsatisfied hydrogen bond between the enzyme and the backbone NH group of the residue preceding the scissile peptide. Consequently, *the enzyme binds the tetrahedral intermediate in preference to either the Michaelis complex or the acyl–enzyme intermediate.*

It is this phenomenon that is responsible for much of the catalytic efficiency of serine proteases (see below). In fact, the reason that DIPF is such an effective inhibitor of serine proteases is because its tetrahedral phosphate group makes this compound a transition state analog of the enzyme.

c. The Tetrahedral Intermediate and the Water Molecule Attacking the Acyl–Enzyme Intermediate Have Been Directly Observed

Most enzymatic reactions turn over far too rapidly for their intermediate states to be studied by X-ray or NMR techniques. Consequently, much of our structural knowledge of these intermediate states derives from the study of enzyme–inhibitor complexes or complexes of substrates with inactivated enzymes. Yet the structural relevance of these complexes is subject to doubt precisely because they are catalytically unproductive.

In an effort to rectify this situation for serine proteases, Janos Hajdu and Christopher Schofield searched for peptide–protease complexes that are stable at a pH at which the protease is inactive but that could be rendered active by changing the pH. To do so, they screened libraries of peptides for their ability to bind to porcine pancreatic elastase at pH 3.5 (at which His 57 is protonated and hence unable to act as a general base) through the use of ESI-MS (Section 7-1I). They thereby discovered that YPFVEPI, a heptapeptide segment of the human milk protein **β-casein** that is named **BCM7,** forms a complex with elastase, whose mass is consistent with the formation of an ester linkage between BCM7 and the enzyme. In the presence of $^{18}OH_2$ at pH 7.5 (where elastase is active), the ^{18}O label was incorporated into both BCM7 and the elastase–BCM7 complex, thereby demonstrating that the reaction of BCM7 with elastase is reversible at this pH. Fragmentation studies by fast atom bombardment–tandem mass spectrometry (FAB–MS/MS; Section 7-1I) further revealed that BCM7 that had been incubated with elastase in the presence of $^{18}OH_2$ at pH 7.5 incorporated the ^{18}O label into only its C-terminal Ile residue.

(a)

Oxyanion hole

(b)

Figure 15-25 **Transition state stabilization in the serine proteases.** (*a*) In the Michaelis complex, the trigonal carbonyl carbon of the scissile peptide is conformationally constrained from binding in the oxyanion hole (*upper left*). (*b*) In the tetrahedral intermediate, the now charged carbonyl oxygen of the scissile peptide (the oxyanion) has entered the oxyanion hole, thereby hydrogen bonding to the backbone NH groups of Gly 193 and Ser 195. The consequent conformational distortion permits the NH group of the residue preceding the scissile peptide bond to form an otherwise unsatisfied hydrogen bond to Gly 193. Serine proteases therefore preferentially bind the tetrahedral intermediate. [After Robertus, J.D., Kraut, J., Alden, R.A., and Birktoft, J.J., *Biochemistry* **11,** 4302 (1972).] **See Kinemage Exercise 10-3**

The X-ray structure of the BCM7–elastase complex at pH 5 (Fig. 15-26*a*) revealed that BCM7's C-terminal carboxyl group, in fact, forms an ester linkage with elastase's Ser 195 side chain hydroxyl group to form the expected acyl–enzyme intermediate. Moreover, this X-ray structure reveals the presence of a bound water molecule that appears poised to nucleophilically attack the ester linkage (the distance from this water molecule to BCM7's C-terminal C atom is 3.1 Å and the line between them is nearly perpendicular to the plane of the acyl group). His 57, which is hydrogen bonded to this water molecule, is properly positioned to abstract one of its protons, thereby activating it for the nucleophilic attack (general base catalysis). The carbonyl O atom of the acyl group occupies the enzyme's oxyanion hole such that it is hydrogen bonded to the main chain N atoms of both Ser 195 and Gly 193. This is in agreement with spectroscopic measurements indicating that the acyl–enzyme intermediate's carbonyl group is, in fact, hydrogen bonded to the oxyanion hole. It was initially assumed that the oxyanion hole acts only to stabilize the tetrahedral oxyanion transition state that resides near the tetrahedral intermediate on the catalytic reaction coordinate. However, it now appears that the oxyanion hole also functions to polarize the carbonyl group of the acyl–enzyme intermediate toward an oxyanion (electrostatic catalysis).

The catalytic reaction was initiated in crystals of the BCM7–elastase complex by transferring them to a buffer at pH 9. After soaking in this buffer for 1 min, the crystals were rapidly frozen in liquid N_2 ($-196°C$), thereby arresting the enzymatic reaction (recall that the catalytically essential collective motions of proteins cease at such low temperatures; Section 9-4a). The X-ray structure of such a frozen crystal (Fig. 15-26*b*) revealed that the above acyl–enzyme intermediate had converted to the tetrahedral intermediate, whose oxyanion, as expected, remained hydrogen bonded to the N atoms of Ser 195 and Gly 193. Comparison of this crystal structure with that of the acyl–enzyme intermediate reveals that the enzyme's active site residues do not significantly change their positions in the conversion from the acyl–enzyme intermediate to the tetrahedral intermediate. However, the peptide substrate must do so out of steric necessity when the trigonal planar acyl group converts to the tetrahedral oxyanion (compare Figs. 15-26*a* and 15-26*b*). In response, several enzyme residues that contact the peptide but which are distant from the active site also shift their positions (not shown in Fig. 15-26).

d. The Role of the Catalytic Triad: Low-Barrier Hydrogen Bonds

The earlier literature postulated that the Asp 102-polarized His 57 side chain directly abstracts a proton

(a)

(b)

Figure 15-26 **X-ray structures of porcine pancreatic elastase in complex with the heptapeptide BCM7 (YPFVEPI).** The residues of elastase are specified by the three-letter code and those of BCM7 are specified by the one-letter code. (*a*) The complex at pH 5. The enzyme's active site residues and the heptapeptide (whose N-terminal three residues are disordered) are shown in ball-and-stick form with elastase C green, BCM7 C cyan, N blue, O red, S yellow, and the bond between the Ser 195 O atom and the C-terminal C atom of BCM7 lavender. The enzyme-bound water molecule, which appears poised to nucleophilically attack the acyl–enzyme's carbonyl C atom, is represented by an orange sphere. The dashed gray lines represent the catalytically

important hydrogen bonds and the dotted gray line indicates the trajectory that the bound water molecule presumably follows in nucleophilically attacking the acyl group's carbonyl C atom. (*b*) The complex after being brought to pH 9 for 1 min and then rapidly frozen in liquid nitrogen. The various groups in the structure are represented and colored as in Part *a*. Note that the water molecule in Part *a* has become a hydroxyl substituent (*orange*) to the carbonyl C atom, thereby yielding the tetrahedral intermediate. [Based on X-ray structures by Christopher Schofield and Janos Hajdu, University of Oxford, U.K. PDBids (*a*) 1HAX and (*b*) 1HAZ.]

from Ser 195, thereby converting its weakly nucleophilic —CH_2OH group to a highly nucleophilic alkoxide ion, —CH_2O^-:

"Charge relay system"

In the process, the anionic charge of Asp 102 was thought to be transferred, via a tautomeric shift of His 57, to Ser 195. The catalytic triad was therefore originally named the **charge relay system.** It is now realized, however, that such a mechanism is implausible because an alkoxide ion ($pK \geq$ 15) has far greater proton affinity than does His 57 ($pK \approx$ 7, as measured by NMR techniques). How, then, can Asp 102 nucleophilically activate Ser 195?

A possible solution to this conundrum has been pointed out by W. W. Cleland and Maurice Kreevoy and, independently, by John Gerlt and Paul Gassman. Proton transfers between hydrogen bonded groups (D—H···A) only occur at physiologically reasonable rates when the pK of the proton donor is no more than 2 or 3 pH units greater than that of the protonated form of the proton acceptor (the height of the kinetic barrier, $\Delta G^{\ddagger}$, for the protonation of an acceptor by a more basic donor increases with the difference between the pK's of the donor and acceptor). However, when the pK's of the hydrogen bonding donor (D) and acceptor (A) groups are nearly equal, the distinction between them breaks down: *The hydrogen atom becomes more or less equally shared between them* (D···H···A). Such **low-barrier hydrogen bonds (LBHBs)** are unusually short and strong (they are also known as **short, strong hydrogen bonds**): They have, as studies of model compounds in the gas phase indicate, association free energies as high as −40 to −80

$kJ \cdot mol^{-1}$ versus the −12 to −30 $kJ \cdot mol^{-1}$ for normal hydrogen bonds (the energy of the normally covalent D—H bond is subsumed into the low-barrier hydrogen bonding system) and a D···A length of <2.55 Å for O—H···O and <2.65 Å for N—H···O versus 2.8 to 3.1 Å for normal hydrogen bonds.

LBHBs are unlikely to exist in dilute aqueous solution because water molecules, which are excellent hydrogen bonding donors and acceptors, effectively compete with D—H and A for hydrogen bonding sites. However, LBHBs may exist in nonaqueous solution and in the active sites of enzymes that exclude bulk solvent water. If so, an effective enzymatic "strategy" would be to convert a weak hydrogen bond in the Michaelis complex to a strong hydrogen bond in the transition state, thereby facilitating proton transfer while applying the difference in the free energy between the normal and low-barrier hydrogen bonds to preferentially binding the transition state. In fact, as Perry Frey has shown, the NMR spectrum of the proton linking His 57 to Asp 102 in chymotrypsin (which exhibits a particularly large downfield chemical shift indicative of deshielding) is consistent with the formation of an LBHB in the transition state (Fig. 15-25b; the pK's of protonated His 57 and Asp 102 are nearly equal in the anhydrous environment of the active site complex). This presumably promotes proton transfer from Ser 195 to His 57 as in the charge relay mechanism. Moreover, an ultrahigh (0.78 Å) resolution X-ray structure of *Bacillus lentus* subtilisin by Richard Bott reveals that the hydrogen bond between His 64 and Asp 32 of its catalytic triad has an unusually short N···O distance of 2.62 ± 0.01 Å and that its H atom is nearly centered between the N and O atoms (note that this highly accurate protein X-ray structure is one of the very few in which H atoms are observed and in which short D···A distances are confidently measured).

Although several studies, such as the foregoing, have revealed the existence of unusually short hydrogen bonds in enzyme active sites, it is far more difficult to demonstrate experimentally that they are unusually strong, as LBHBs are predicted to be. In fact, several studies of the strengths of unusually short hydrogen bonds in organic model compounds in nonaqueous solutions suggest that these hydrogen bonds are not unusually strong. Consequently, a lively debate has ensued as to the catalytic significance of LBHBs. Yet if enzymes do not form LBHBs, it remains to be explained how, in numerous widely accepted enzymatic mechanisms that we shall encounter, the conjugate base of an acidic group can abstract a proton from a far more basic group.

e. Much of a Serine Protease's Catalytic Activity Arises from Preferential Transition State Binding

Despite the foregoing, blocking the action of the catalytic triad through the specific methylation of His 57 by treating chymotrypsin with **methyl-*p*-nitrobenzene sulfonate**

His 57

CH₂

$H-N^1$ ... $\overset{3}{N}$ + O_2N-⬡$-\overset{O}{\underset{O}{S}}-O-CH_3$

Methyl-*p*-nitrobenzene sulfonate

↓

His 57

CH₂

$H-N^1$ ... $\overset{3}{\overset{+}{N}}-CH_3$ + O_2N-⬡$-SO_3^-$

yields an enzyme that is a reasonably good catalyst: It enhances the rate of proteolysis by as much as a factor of 2×10^6 over the uncatalyzed reaction, whereas the native enzyme has a rate enhancement factor of $\sim 10^{10}$. Similarly, the mutation of Ser 195, His 57, or even all three residues of the catalytic triad yields enzymes that enhance proteolysis rates by $\sim 5 \times 10^4$-fold over that of the uncatalyzed reaction. Evidently, the catalytic triad provides a nucleophile and is an alternate source and sink of protons (general acid–base catalysis). However, *a large portion of chymotrypsin's rate enhancement must be attributed to its preferential binding of the catalyzed reaction's transition state.*

f. Enzymes Have Free Energy Landscapes That Facilitate Catalysis

How does an enzyme-catalyzed reaction reach its transition state? Recall that, under physiological conditions, proteins are highly dynamic entities with structural fluctuations having periods ranging from $\sim 10^{-15}$ s for bond vibrations to 1 s or more for triggered conformational changes (Section 9-4). The turnover times for enzymatic reactions, $1/k_{cat}$, are mostly in the range 1 μs to 1 s (Table 14-1), but yet the lifetime of a transition state is only around that of a bond vibration (Section 14-1Cb). Thus, even for a reaction with a turnover time as little as 1 μs, every atom in the enzyme–substrate complex undergoes approximately $10^{-6}/10^{-15} = 10^9$ vibrational excursions between turnovers. Apparently, *the transition state is an arrangement of substrate and catalytic groups that occurs extremely rarely through the fluctuations of its component atoms.*

Proteins, as we have seen, are designed by evolution to fold to their native states via a series of conformational adjustments that follow funnel-shaped free energy landscapes (Section 9-1Ch). A variety of structural, mutational, and theoretical studies indicate that *enzyme–sub-*

strate complexes are similarly evolutionarily designed to structurally rearrange themselves through a progression of conformational changes that lead to the formation of the transition state. This explains, for example, why the mutation of a residue far from the active site of an enzyme that does not appear to have a structurally important role, may nevertheless significantly reduce the rate of the reaction that the enzyme catalyzes. Such a mutation perturbs extended hydrogen bonding networks and long range electrostatic interactions in a way that alters the entire enzyme's spectrum of thermal motions. This changes the enzyme's free energy landscape so as to reduce the probability that it will achieve the transition state in a given time period.

E. Zymogens

Most proteolytic enzymes are biosynthesized as somewhat larger inactive precursors known as **zymogens** (enzyme precursors, in general, are known as **proenzymes**). In the case of digestive enzymes, the reason for this is clear: If these enzymes were synthesized in their active forms, they would digest the tissues that synthesized them. Indeed, **acute pancreatitis,** a painful and sometimes fatal condition that can be precipitated by pancreatic trauma, is characterized by the premature activation of the digestive enzymes synthesized by this gland.

a. Serine Proteases Are Autocatalytically Activated

Trypsin, chymotrypsin, and elastase are activated according to the following pathways:

Trypsin. The activation of **trypsinogen,** the zymogen of trypsin, occurs as a two-stage process when trypsinogen enters the duodenum from the pancreas. **Enteropeptidase,** a single-pass transmembrane serine protease that is located in the duodenal mucosa, specifically hydrolyzes trypsinogen's Lys 15—Ile 16 peptide bond, thereby excising its N-terminal hexapeptide (Fig. 15-27). This yields the active enzyme, which has Ile 16 at its N-terminus. Since this

$$\overset{+}{H_3N}-\overset{10}{Val}-(Asp)_4-\overset{15}{Lys}-\overset{16}{Ile}-Val-\cdots$$

Trypsinogen

↓ enteropeptidase or trypsin

$$\overset{+}{H_3N}-Val-(Asp)_4-Lys \; + \; Ile-Val-\cdots$$

Trypsin

Figure 15-27 **Activation of trypsinogen to form trypsin.** Proteolytic excision of the N-terminal hexapeptide is catalyzed by either enteropeptidase or trypsin. The chymotrypsinogen residue numbering is used here; that is, Val 10 is actually trypsinogen's N-terminus and Ile 16 is trypsin's N-terminus.

Figure 15-28   **Activation of chymotrypsinogen by proteolytic cleavage.** Both π- and α-chymotrypsin are enzymatically active. **See Kinemage Exercise 10-4**

activating cleavage occurs at a trypsin-sensitive site (recall that trypsin cleaves after Arg and Lys residues), the small amount of trypsin produced by enteropeptidase also catalyzes trypsinogen activation, generating more trypsin, etc.; that is, trypsinogen activation is autocatalytic.

Chymotrypsin. Chymotrypsinogen is activated by the specific tryptic cleavage of its Arg 15—Ile 16 peptide bond to form **π-chymotrypsin** (Fig. 15-28). π-Chymotrypsin subsequently undergoes autolysis (self-digestion) to specifically excise two dipeptides, Ser 14–Arg 15 and Thr 147–Asn 148, thereby yielding the equally active enzyme **α-chymotrypsin** (heretofore and hereafter referred to as chymotrypsin). The biochemical significance of this latter process, if any, is unknown.

Elastase. Proelastase, the zymogen of elastase, is activated similarly to trypsinogen by a single tryptic cleavage that excises a short N-terminal polypeptide.

b. Biochemical "Strategies" That Prevent Premature Zymogen Activation

Trypsin activates pancreatic **procarboxypeptidases A** and **B** and **prophospholipase A₂** (the action of phospholipase A$_2$ is outlined in Section 25-1) as well as the pancreatic serine proteases. Premature trypsin activation can consequently trigger a series of events that lead to pancreatic self-digestion. Nature has therefore evolved an elaborate defense against such inappropriate trypsin activation. We have already seen (Section 15-3Da) that pancreatic trypsin inhibitor binds essentially irreversibly to any trypsin formed in the pancreas so as to inactivate it. Furthermore, the trypsin-catalyzed activation of trypsinogen (Fig. 15-27) occurs quite slowly, presumably because the unusually large negative charge of its highly evolutionarily conserved N-terminal hexapeptide repels the Asp at the back of trypsin's specificity pocket (Fig. 15-21). Finally, pancreatic zymogens are stored in intracellular vesicles called **zymogen granules** whose membranous walls are thought to be resistant to enzymatic degradation.

c. Zymogens Have Distorted Active Sites

Since the zymogens of trypsin, chymotrypsin, and elastase have all their catalytic residues, why aren't they enzymatically active? Comparisons of the X-ray structures of trypsinogen with that of trypsin and of chymotrypsinogen with that of chymotrypsin show that on activation, the newly liberated N-terminal Ile 16 residue moves from the surface of the protein to an internal position, where its free cationic amino group forms an ion pair with the invariant anionic Asp 194 (Fig. 15-20). Aside from this change, however, the structures of these zymogens closely resemble those of their corresponding active enzymes. Surprisingly, this resemblance includes their catalytic triads, an observation which led to the discovery that these zymogens are actually catalytically active, albeit at a very low level. Careful comparisons of the corresponding enzyme and zymogen structures revealed the reason for this low activity: *The zymogens' specificity pockets and oxyanion holes are improperly formed such that, for example, the amide NH of chymotrypsinogen's Gly 193 points in the wrong direction to form a hydrogen bond with the tetrahedral intermediate (see Fig. 15-25).* Hence, the zymogens' very low enzymatic activity arises from their reduced ability to bind substrate productively and to stabilize the tetrahedral intermediate. These observations provide further structural evidence favoring the role of preferred transition state binding in the catalytic mechanism of serine proteases.

4 DRUG DESIGN

The improvements in medical care over the past several decades are, in large measure, attributable to the development of a huge variety of drugs, which have eliminated or greatly relieved numerous human ailments. Such medications include antibiotics (which have enormously reduced the impact of infectious diseases), anti-inflammatory agents (which reduce the effects of inflammatory diseases such as arthritis), analgesics and anesthetics (which make modern surgical techniques possible), agents that reduce the incidence and severity of cardiovascular disease and stroke, antidepressants, antipsychotics, agents that inhibit stomach acid secretion (which prevent stomach ulcers and heartburn), agents to combat allergies and asthma, immunosuppressants (which make organ transplants possible), agents used for cancer chemotherapy, and a great variety of other substances.

Early human cultures almost certainly recognized both the beneficial and toxic effects of indigenous plant and animal products and used many of them as "medications." Unfortunately, most of these substances were useless or even harmful. Although there were sporadic attempts over the 2500 years preceding the modern era to formulate rational systems of drug discovery, they had little success because they were based mainly on unfounded theories and superstition (e.g., the doctrine of signatures stated that if a plant resembles a particular body part, it must be designed by nature to influence that body part) rather than observation and experiment. Consequently, at the beginning of the 20th century, only three known drugs, apart from folk medicines, were effective in treating specific diseases: (1) **Digitalis,** a heart stimulant extracted from the foxglove plant (Section 20-3Af), was used to treat various heart conditions; (2) **quinine** (Section 26-4Ak), obtained from the bark and roots of the *Cinchona* tree, was used to treat malaria; and (3) mercury was used to treat syphilis (a cure that was often worse than the disease). It was not until several decades later that the rise of the scientific method coupled to the rapidly increasing knowledge of physiology, biochemistry, and chemistry led to effective methods of drug discovery. In fact, the vast majority of drugs in use today were discovered and developed in the past four decades.

In this section we discuss the elements of drug discovery and **pharmacology** (Greek: *pharmacon,* drug; the science of drugs, including their composition, uses, and effects). The section ends with a consideration of one of the major successes of modern drug discovery methods, HIV protease inhibitors.

A. *Techniques of Drug Discovery*

Most drugs act by modifying the function of a particular **receptor** in the body or in an invading pathogen. In most cases, the receptor is a protein to which the drug specifically binds. It may be an enzyme, a transmembrane channel that transports a specific substance into or out of a cell (Chapter 20), and/or a protein that participates in an inter- or intracellular signaling pathway (Chapter 19). In all of these cases, a substance that in binding to a receptor modulates its function is known as an **agonist,** whereas a substance that binds to a receptor without affecting its function but blocks the binding of agonists is called an **antagonist.** The biochemical and physiological effects of a drug and its mechanism of action are referred to as its **pharmacodynamics.**

a. Drug Discovery Is a Complex Procedure

How are new drugs discovered? Nearly all drugs that have been in use for over 15 years were discovered by screening large numbers of synthetic compounds and natural products for the desired effect. Drug candidates that are natural products are usually discovered by the fractionation of the organisms in which they occur, which are often plants used in folk remedies of the conditions of interest. Humans having the condition whose treatment is being sought cannot be used as "guinea pigs" in this initial screening process, and even guinea pigs or other laboratory animals such as mice or dogs (if they can be made to be suitable models of the condition under consideration) are too expensive to use on the many thousands of compounds that are usually tested. Thus, *in vitro* **screens** are initially used, such as the degree of binding of a drug candidate to an enzyme that is implicated in a disease of interest, toxicity toward the target bacteria in the search for a new antibiotic, or effects on a line of cultured mammalian cells. However, as the number of drug candidates is winnowed down, more sensitive screens such as testing in animals are employed.

A drug candidate that exhibits a desired effect is called a **lead compound** (or, colloquially, a lead; pronounced leed). A good lead compound binds to its target receptor with a dissociation constant, $K_D < 1~\mu M$. Such a high affinity is necessary to minimize a drug's less specific binding to other macromolecules in the body and to ensure that only low doses of the drug need be taken. For enzyme inhibitors, the dissociation constant is the inhibitor's K_I or K_I' (Section 14-3). Other common measures of the effect of a drug are the **IC$_{50}$,** the *i*nhibitor *c*oncentration at which an enzyme exhibits 50% of its maximal activity; the **ED$_{50}$,** the *e*ffective *d*ose of a drug required to produce a therapeutic effect in 50% of a test sample; the **TD$_{50}$,** the mean *t*oxic *d*ose required to produce a particular toxic effect in animals; and the **LD$_{50}$,** the mean *l*ethal *d*ose required to kill 50% of a test sample.

For an inhibitor of an enzyme that follows Michaelis–Menten kinetics, the IC$_{50}$ is determined by measuring the ratio v_I/v_o for several values of [I] at constant [S], where v_I is the initial velocity of the enzyme when the inhibitor concentration is [I]. By dividing Eq. [14.24] by Eq. [14.38] with α defined according to Eq. [14.37], we see that

$$\frac{v_I}{v_o} = \frac{K_M + [S]}{K_M\alpha + [S]} = \frac{K_M + [S]}{K_M\left(1 + \dfrac{[I]}{K_I}\right) + [S]} \quad [15.13]$$

When $v_I/v_o = 0.5$ (50% inhibition),

$$[I] = [IC_{50}] = K_I\left(1 + \frac{[S]}{K_M}\right) \quad [15.14]$$

Consequently, if the measurements of v_I/v_o are made with $[S] \ll K_M$, then $[IC_{50}] = K_I$.

The ratio TD_{50}/ED_{50} is defined as a drug's **therapeutic index**, the ratio of the dose of the drug that produces toxicity to that which produces the desired effect. It is, of course, preferable that a drug have a high therapeutic index, but this is not always possible.

b. Cathepsin K Is a Drug Target for Osteoporosis

The development of genomic sequencing techniques (Section 7-2B) and hence the characterization of tens of thousands of previously unknown genes is providing an enormous number of potential drug targets. For example, **osteoporosis** (Greek: *osteon,* bone + *poros,* porous), a condition that afflicts mostly postmenopausal women, is characterized by the progressive loss of bone mass leading to a greatly increased frequency of bone fracture, most often of the hip, spine, and wrist. Bones consist of a protein matrix that is >90% type I collagen (Section 8-2B), in which spindle- or plate-shaped crystals of **hydroxyapatite,** $Ca_5(PO_4)_3OH$, are embedded. Bones are by no means static structures. They undergo continuous remodeling through the countervailing action of two types of bone cells: **osteoblasts** (Greek: *blast,* germ cell), which synthesize bone's protein matrix in which its mineral component is laid down; and **osteoclasts** (Greek: *clast,* broken), which solubilize mineralized bone matrix through the secretion of proteolytic enzymes into an extracellular bone resorption pit, which is maintained at pH 4.5. The acidic solution dissolves the bone's mineral component, thereby exposing its protein matrix to proteolytic degradation. Osteoporosis arises when bone resorption outstrips bone formation.

In the search for a drug target for osteoporosis, a cDNA library (Sections 5-5E and 5-5Fa) was prepared from an **osteoclastoma** (a cancer derived from osteoclasts; normally osteoclasts are very rare cells). Around 4% of these cDNAs encode a heretofore unknown protease, which was named **cathepsin K** (cathepsins are proteases that occur in the lysosome). Further studies, both at the cDNA and protein levels, indicated that cathepsin K is only expressed at high levels in osteoclasts. Microscopic examination of osteoclasts that had been stained with antibodies directed against cathepsin K revealed that this enzyme is localized at the contact site between osteoclasts and the bone resorption pit. Subsequently, it was shown that mutations in the gene encoding cathepsin K are the cause of **pycnodysostosis,** a rare hereditary disease which is characterized by hardened and fragile bones, short stature, skull deformities, and osteoclasts that demineralize bone normally but do not degrade its protein matrix. Evidently, cathepsin K functions to degrade the protein matrix of bone and hence is an attractive drug target for the treatment of osteoporosis. Indeed, several cathepsin K inhibitors are in clinical trials (Section 15-4Bb).

c. SARs and QSARs Are Useful Tools for Drug Discovery

A lead compound is used as a point of departure to design more efficacious compounds. Experience has shown that even minor modifications to a drug candidate can result in major changes in its pharmacological properties. Thus, one might place methyl, chloro, hydroxyl, or benzyl groups at various places on a lead compound in an effort to improve its pharmacodynamics. For most drugs in use today, 5 to 10 thousand related compounds were typically synthesized in generating the medicinally useful drug. These were not random procedures but were guided by experience as medicinal chemists tested various derivatives of a lead compound: For those compounds that had improved efficacy, derivatives were made and tested; etc. This process has been systematized through the use of **structure–activity relationships (SARs):** the determination, via synthesis and screening, of which groups on a lead compound are important for its drug function and which are not. For example, if a phenyl group on a lead compound interacts hydrophobically with a flat region of its receptor, then hydrogenating the phenyl ring to form a nonplanar cyclohexane ring will yield a compound with reduced affinity for the receptor.

A logical extension of the SAR concept is to quantify it, that is, to determine a **quantitative structure–activity relationship (QSAR).** This idea is based on the premise that there is a relatively simple mathematical relationship between the biological activity of a drug and its physicochemical properties. For instance, if the hydrophobicity of a drug is important for its biological activity, then changing the substituents on the drug so as to alter its hydrophobicity will affect its activity. A measure of the substance's hydrophobicity is its **partition coefficient,** P, between the two immiscible solvents, octanol and water, at equilibrium:

$$P = \frac{\text{concentration of drug in octanol}}{\text{concentration of drug in water}} \quad [15.15]$$

Biological activity may be expressed as $1/C$, where C is the drug concentration required to achieve a specified level of biological function (e.g., IC_{50}). Then a plot of $\log 1/C$ versus $\log P$ (the use of logarithms keeps the plot on a manageable scale) for a series of derivatives of the lead compound having a relatively small range of $\log P$ values often indicates a linear relationship (Fig. 15-29a), which can therefore be expressed:

$$\log\left(\frac{1}{C}\right) = k_1 \log P + k_2 \quad [15.16]$$

Here k_1 and k_2 are constants, whose optimum values in this QSAR can be determined by computerized curve-fitting methods. For compounds with a larger range of $\log P$ values, it is likely that a plot of $\log 1/C$ versus $\log P$ will have a maximum value (Fig. 15-29b) and hence be better described by a quadratic equation:

$$\log\left(\frac{1}{C}\right) = k_1 (\log P)^2 + k_2 \log P + k_3 \quad [15.17]$$

Of course, the biological activities of few substances depend only on their hydrophobicities. A QSAR can therefore simultaneously take into account several physicochemical properties of substituents such as their pK values, van der Waals radii, hydrogen bonding energy, and conformation. The values of the constants for each of the terms in a QSAR is indicative of the contribution of that term to the drug's activity. The use of QSARs to optimize the biological activity of a lead compound has proven to be a valuable tool in drug discovery.

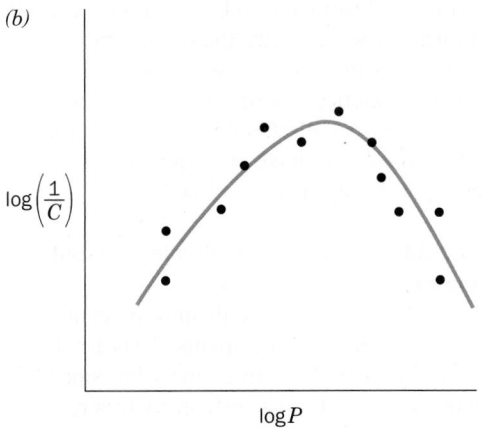

Figure 15-29 **Hypothetical QSAR plots of log(1/C) versus log P for a series of related compounds.** (*a*) A plot that is best described by a linear equation. (*b*) A plot that is best described by a quadratic equation.

d. Structure-Based Drug Design

Since the mid 1980s, dramatic advances in the speed and precision with which a macromolecular structure can be determined by X-ray crystallography and NMR (Section 8-3A) have enabled **structure-based drug design,** a process that greatly reduces the number of compounds that need be synthesized in a drug discovery program. As its name implies, structure-based drug design (also called **rational drug design**) uses the structure of a receptor in complex with a drug candidate to guide the development of more efficacious compounds. Such a structure will reveal, for example, the positions of the hydrogen bonding donors and acceptors in a receptor binding site as well as cavities in the binding site into which substituents might be placed on a drug candidate to increase its binding affinity for the receptor. These direct visualization techniques are usually supplemented with molecular modeling tools such as the computation of the minimum energy conformation of a proposed derivative, quantum mechanical calculations that determine its charge distribution and hence how it would interact electrostatically with the receptor, and docking simulations in which an inhibitor candidate is computationally modeled into the binding site on the receptor to assess potential interactions. Structure-based drug design is an iterative process: The structure of the receptor in complex with a compound with improved properties is determined in an effort to further improve its properties.

e. Combinatorial Chemistry, Fragment-Based Lead Discovery, and High-Throughput Screening

As structure-based methods were developed, it appeared that they would become the dominant mode of drug discovery. However, the advent of **combinatorial chemistry** techniques to rapidly and inexpensively synthesize large numbers of related compounds combined with the development of robotic **high-throughput screening** techniques has caused the drug discovery "pendulum" to again swing toward the "make-many-compounds-and-see-what-they-do" approach. A familiar example of combinatorial chemistry is the parallel synthesis of the large number of different oligonucleotides on a DNA microarray (Section 7-6B). If a lead compound can similarly be synthesized in a stepwise manner from several smaller modules, then the substituents on each of these modules can be varied in parallel to produce a library of related compounds (e.g., Fig. 15-30).

A variety of synthetic techniques have been developed that permit the combinatorial synthesis of thousands of related compounds in a single procedure. Thus, whereas investigations into the importance of a hydrophobic group at a particular position in a lead compound might previously have prompted the individual syntheses of only the ethyl, propyl, and benzyl derivatives of the compound, the use of combinatorial synthesis would permit the generation of perhaps 100 different groups at that position. This would far more effectively map out the potential range of the substituent and possibly identify an unexpectedly active analog. Interestingly, QSAR and computational techniques have been combined in the development of "virtual combinatorial chemistry," a procedure in which libraries of compounds are computationally "synthesized" and "analyzed" to predict their efficacy, thereby again reducing the number of compounds that must actually be synthesized in order to generate an effective drug.

The relatively recently developed approach to drug discovery known as **fragment-based lead discovery (FBLD)** has shown encouraging signs of success. In this method, rather than screening large numbers of potential lead compounds for substances that bind with high affinity to a drug target, only a relatively small number of simple compounds are screened for their ability to bind to the drug target with low affinity. Substances that do so, which likely bind to only a small portion of the drug target's surface area, are then

Figure 15-30 **The combinatorial synthesis of arylidene diamides.** If ten different variants of each R group are used in the synthesis, then 1000 different derivatives will be synthesized.

"grown" by adding chemical groups and/or linking several such fragments together. Thus, FBLD discovers a lead compound one piece at a time rather than all at once.

B. Introduction to Pharmacology

The *in vitro* development of an effective drug candidate is only the first step in the drug development process. *Besides causing the desired response in its isolated target receptor, a useful drug must be delivered in sufficiently high concentration to this receptor where it resides in the human body without causing unacceptable side effects.*

a. Pharmacokinetics Is a Multifaceted Phenomenon

The most convenient form of drug administration is orally (by mouth). In order to reach its target receptor, a drug administered in this way must surmount a series of formidable barriers: (1) It must be chemically stable in the highly acidic (pH 1) environment of the stomach and must not be degraded by the digestive enzymes in the gastrointestinal tract; (2) it must be absorbed from the gastrointestinal tract into the bloodstream, that is, it must pass through several cell membranes; (3) it must not bind too tightly to other substances in the body (e.g., lipophilic substances tend to be absorbed by certain plasma proteins and by fat tissue; anions may be bound by plasma proteins, mainly **albumin;** and cations may be bound by nucleic acids); (4) it must survive derivatization by the battery of enzymes, mainly in the liver, that function to detoxify **xenobiotics** (foreign compounds), as discussed below (note that the intestinal blood flow drains directly into the liver via the portal vein, so that the liver processes all orally ingested substances before they reach the rest of the body); (5) it must avoid rapid excretion by the kidneys; (6) it must pass from the capillaries to its target tissue; (7) if it is targeted to the brain, it must cross the **blood–brain barrier,** which blocks the passage of most polar substances; and (8) if it is targeted to an intracellular receptor, it must pass through the plasma membrane and, possibly, other intracellular membranes. The ways in which a drug interacts with these various barriers is known as its **pharmacokinetics.** Thus, the **bioavailability** of a drug (the extent to which it reaches its site of action, which is usually taken to be the systemic circulation) depends on both the dose given and its pharmacokinetics. Of course, barriers (1) and (2) can be circumvented by injecting the drug [e.g., some forms of penicillin (Fig. 11-28) must be injected because their functionally essential β-lactam rings are highly susceptible to acid hydrolysis], but this mode of drug delivery is undesirable for long-term use.

Since the pharmacokinetics of a drug candidate is as important to its efficacy as is its pharmacodynamics, both must be optimized in producing a medicinally useful drug. The following empirically based rules, formulated by Christopher Lipinski and known as **Lipinski's "rule of five,"** state that an orally administered compound is likely to exhibit poor absorption or permeation if:

1. Its molecular mass is greater than 500 D.

2. It has more than 5 hydrogen bond donors (expressed as the sum of its OH and NH groups).

3. It has more than 10 hydrogen bond acceptors (expressed as the sum of its N and O atoms).

4. Its value of log P is greater than 5.

Drug candidates that disobey Rule 1 are likely to have low solubilities and to only pass through cell membranes with difficulty; those that disobey Rules 2 and/or 3 are likely to be too polar to pass through cell membranes; and those that disobey Rule 4 are likely to be poorly soluble in aqueous solution and hence unable to gain access to membrane surfaces. Thus, *the most effective drugs are usually a compromise; they are neither too lipophilic nor too hydrophilic.* In addition, their pK values are usually in the range 6 to 8 so that they can readily assume both their ionized and unionized forms at physiological pH's. This permits them to cross cell membranes in their unionized form and to bind to their receptor in their ionized form. However, since the concentration of a drug at its receptor depends, as we saw, on many different factors, the pharmacokinetics of a drug candidate may be greatly affected by even small chemical changes. QSARs and other computational tools have been developed to predict these effects but they are, as yet, rather crude.

b. Toxicity and Adverse Reactions Eliminate Most Drug Candidates

The final criteria that a drug candidate must meet are that its use be safe and efficacious in humans. Tests for these properties are initially carried out in animals, but since humans and animals often react quite differently to a particular drug, the drug must ultimately be tested in humans through **clinical trials.** In the United States, clinical trials are monitored by the Food and Drug Administration (FDA) and have three increasingly detailed (and expensive) phases:

Phase I. This phase is primarily designed to test the safety of a drug candidate but is also used to determine its dosage range and the optimal dosage method (e.g., orally vs injection) and frequency. It is usually carried out on a small number (20–100) of normal, healthy volunteers, but in the case of a drug candidate known to be highly toxic (e.g., a cancer chemotherapeutic agent), it is carried out on volunteer patients with the target disease.

Phase II. This phase mainly tests the efficacy of the drug against the target disease in 100 to 500 volunteer patients but also refines the dosage range and checks for side effects. The effects of the drug candidate are usually assessed via **single blind tests,** procedures in which the patient is unaware of whether he/she has received the drug or a control substance. Usually the control substance is a **placebo** (an inert substance with the same physical appearance, taste, etc., as the drug being tested) but, in the case of a life-threatening disease, it is an ethical necessity that the control substance be the best available treatment against the disease.

Phase III. This phase monitors adverse reactions from long-term use as well as confirming efficacy in 1000 to 5000 patients. It pits the drug candidate against control substances through the statistical analysis of carefully designed **double blind tests,** procedures in which neither the patients nor the clinical investigators evaluating the patients' responses to the drug know whether a given patient has received the drug or a control substance. This is done to minimize bias in the subjective judgments the investigators must make.

Currently, only about 5 drug candidates in 5000 that enter preclinical trials reach clinical trials. Of these, only one, on average, is ultimately approved for clinical use, with ~40% of drug candidates passing Phase I trials and ~50% of those passing Phase II trials (most drug candidates that enter Phase III trials are successful). In recent years, the preclinical portion of a drug discovery process has averaged ~3 years to complete, whereas successful clinical trials have usually required an additional 7 to 10 years. These successive stages of the drug discovery process are increasingly expensive, so that to successfully bring a drug to market costs, on average, around $300 million.

The most time-consuming and expensive aspect of a drug development program is identifying a drug candidate's rare adverse reactions. Nevertheless, it is not an uncommon experience for a drug to be brought to market only to be withdrawn some months or years later when it is found to have caused unanticipated life-threatening side effects in as few as 1 in 10,000 individuals (the search for new applications of an approved drug and its postmarketing surveillance are known as its Phase IV clinical trials). For example, in 1997, the FDA withdrew its approval of the drug **fenfluramine (fen),**

Fenfluramine   **Phentermine**

which it had approved in 1973 for use as an appetite suppressant in short-term (a few weeks) weight loss programs. Fenfluramine had become widely prescribed, often for extended periods, together with another appetite suppressant, **phentermine (phen;** approved in 1959), a combination known as **fen-phen** (although the FDA had not approved of the use of the two drugs in combination, once it approves a drug for any purpose, a physician may prescribe it for any other purpose, a practice known as off-label use). The withdrawal of fenfluramine was prompted by over 100 reports of heart valve damage in individuals (mostly women) who had taken fen-phen for an average of 12 months (phentermine was not withdrawn because the evidence indicated that fenfluramine was the responsible agent). This rare side effect had not been observed in the clinical trials of fenfluramine, in part because, being an extremely unusual type of drug reaction, it had not been screened for.

More recently (2004), the widely prescribed analgesic **Vioxx** was withdrawn from use due to its previously undetected cardiac side effects, although the closely related analgesic **Celebrex** remains available (Section 25-7Bb).

c. The Cytochromes P450 Metabolize Most Drugs

Why is it that a drug that is well tolerated by the majority of patients can pose such a danger to others? *Differences in reactions to drugs arise from genetic differences among individuals as well as differences in their disease states, other drugs they are taking, age, sex, and environmental factors.* The **cytochromes P450,** which function in large part to detoxify xenobiotics and participate in the metabolic clearance of the majority of drugs in use, provide instructive examples of these phenomena.

The cytochromes P450 constitute a superfamily of heme-containing enzymes that occur in nearly all living organisms, from bacteria to mammals [their name arises from the characteristic 450-nm peak in their absorption spectra when reacted in their Fe(II) state with CO]. The human genome encodes 57 **isozymes** (catalytically and structurally similar but genetically distinct enzymes from the same organism; also called **isoforms**) of cytochromes P450, about one-third of which occur only in the liver (P450 isozymes are named by the letters "CYP" followed by a number designating its family, an uppercase letter designating its subfamily, and often another number; e.g., CYP2D6). These **monooxygenases** (Fig. 15-31), which in animals are embedded in the endoplasmic reticulum membrane, catalyze reactions of the sort

$$RH + O_2 + 2H^+ + 2e^- \rightleftharpoons ROH + H_2O$$

The electrons (e^-) are supplied by NADPH, which passes them to cytochrome P450's heme prosthetic group via the intermediacy of the enzyme **NADPH–P450 reductase.** Here RH represents a wide variety of usually lipophilic compounds for which the different cytochromes P450 are specific. They include polycyclic aromatic hydrocarbons [PAHs, frequently carcinogenic (cancer-causing) compounds that are present in tobacco smoke, broiled meats, and other pyrolysis products], polycyclic biphenyls (PCBs,

Figure 15-31   **X-ray structure of the human cytochrome P450 CYP2C9 in complex with the blood clotting inhibitor warfarin.** A cutaway diagram of the enzyme's active site region drawn with its surface purple and with its polypeptide backbone in ribbon form colored in rainbow order from N-terminus (*blue*) to C-terminus (*red*). The heme (seen edgewise), the Cys side chain that axially ligands the heme Fe atom, and the warfarin (**coumadin;** Fig. 35-12) are shown in ball-and-stick form with C gray, N blue, O red, S yellow, and Fe orange. [Courtesy of Astex Therapeutics, Limited. PDBid 1OG5.]

which were widely used in electrical insulators and as plasticizers and are also carcinogenic), steroids (in whose syntheses cytochromes P450 participate; Sections 25-6A and 25-6C), and many different types of drugs. The xenobiotics are thereby converted to a more water-soluble form, which aids in their excretion by the kidneys. Moreover, the newly generated hydroxyl groups are often enzymatically conjugated (covalently linked) to polar substances such as glucuronic acid (Section 11-1Cb), glycine, sulfate, and acetate, which further enhances aqueous solubility. The many types of cytochromes P450 in animals, which have different substrate specificities (although these specificities tend to be broad and hence often overlap), are thought to have arisen in response to the numerous toxins which plants produce, presumably to discourage animals from eating them (other P450s function to catalyze specific biosynthetic reactions).

Drug–drug interactions are often mediated by cytochromes P450. For example, if drug A is metabolized by or otherwise inhibits a cytochrome P450 isozyme that metabolizes drug B, then coadministering drugs A and B will cause the bioavailability of drug B to increase above the value it would have had if it alone had been administered. This phenomenon is of particular concern if drug B has a low therapeutic index. Conversely, if, as is often the case, drug A induces the increased expression of the cytochrome P450 isozyme that metabolizes it and drug B, then co-administering drugs A and B may reduce drug B's bioavailability, a

phenomenon that was first noted when certain antibiotics caused oral contraceptives to lose their efficacy. Moreover, if drug B is metabolized to a toxic product, its increased rate of reaction may result in an adverse reaction. Environmental pollutants such as PAHs or PCBs are also known to induce the expression of specific cytochrome P450 isozymes and thereby alter the rates at which certain drugs are metabolized. Finally, some of these same effects may occur in patients with liver disease, as well as arising from age-based, gender-based, and individual differences in liver physiology.

Although the cytochromes P450 presumably evolved to detoxify and/or help eliminate harmful substances, in several cases they have been shown to participate in converting relatively innocuous compounds to toxic agents. For example, **acetaminophen** (Fig. 15-32), a widely used analgesic and antipyretic (fever reducer), is quite safe when taken in therapeutic doses (1.2 g/day for an adult) but in large doses (>10 g) is highly toxic. This is because, in therapeutic amounts, 95% of the acetaminophen present is enzymatically glucuronidated or sulfated at its —OH group to the corresponding conjugates, which are readily excreted. The remaining 5% is converted, through the action of a cytochrome P450 (CYP2E1), to **acetimidoquinone** (Fig. 15-32), which is then conjugated with **glutathione,** a tripeptide with an unusual γ-amide bond that participates in a wide variety of metabolic processes (Section 26-4C). However, when acetaminophen is taken in large amounts, the glucuronida-

Figure 15-32 The metabolic reactions of acetaminophen that convert it to its conjugate with glutathione.

tion and sulfation pathways become saturated and hence the cytochrome P450-mediated pathway becomes increasingly important. If hepatic (liver) glutathione is depleted faster than it can be replaced, acetimidoquinone, a reactive compound, instead conjugates with the sulfhydryl groups of cellular proteins, resulting in often fatal hepatotoxicity.

Many of the cytochromes P450 in humans are unusually **polymorphic,** that is, there are several common alleles (variants) of the genes encoding these enzymes. Alleles that cause diminished, enhanced, and qualitatively altered rates of drug metabolism have been characterized for many of the cytochromes P450. The distributions of these various alleles differ markedly among ethnic groups and hence probably arose to permit each group to cope with the toxins in its particular diet.

Polymorphism in a given cytochrome P450 results in differences between individuals in the rates at which they metabolize certain drugs. For instance, in cases that a cytochrome P450 variant has absent or diminished activity, otherwise standard doses of a drug that the enzyme normally metabolizes may cause the bioavailability of the drug to reach toxic levels. Conversely, if a particular P450 enzyme has enhanced activity (usually because the gene encoding it has been duplicated one or more times), higher than normal doses of a drug that the enzyme metabolizes would have to be administered to obtain the required therapeutic effect. However, if the drug is metabolized to a toxic product, this may result in an adverse reaction. Several known P450 variants have altered substrate specificities and hence produce unusual metabolites, which also may cause harmful side effects.

Experience has amply demonstrated that *there is no such thing as a drug that is entirely free of adverse reactions.* However, as the enzymes and their variants that participate in drug metabolism are characterized and rapid and inexpensive genotyping methods are developed, it may become possible to tailor drug treatment to an individual's genetic makeup rather than to the population as a whole. This rapidly developing area of study is known as **pharmacogenomics** (or, more colloquially, as "personalized genomics" or "personalized medicine").

C. HIV Protease and Its Inhibitors

Acquired immunodeficiency syndrome (AIDS), the only major epidemic attributable to a previously unknown pathogen to appear in the 20th century (it was first described in 1981), is caused by **human immunodeficiency virus type 1 (HIV-1;** the closely related **HIV-2,** which we shall not explicitly discuss here, also causes AIDS and has a similar response to drugs). HIV-1, which was discovered in 1983 by Françoise Barré-Sinoussi, Luc Montagnier, and Robert Gallo, is a **retrovirus,** a family of viruses that were independently characterized in 1970 by David Baltimore and Howard Temin. The retroviral genome is a single-stranded RNA that reproduces inside its host cell by transcribing the RNA to double-stranded DNA in a process mediated by the virally encoded enzyme **reverse transcriptase** (Section 30-4C). The DNA is then inserted into the host cell's chromosomal DNA by a viral enzyme named **integrase** and is passively replicated along with the cell's DNA. However, under activating conditions (which for HIV-1 often is an infection by another pathogen), the retroviral DNA is transcribed, the proteins it encodes are expressed and inserted in or anchored to the host cell plasma membrane, and new **virions** (virus particles) are produced by the budding out of a viral protein–laden segment of plasma membrane so as to enclose viral RNA (Fig. 15-33).

HIV-1 is targeted to and specifically replicates within **helper *T* cells,** essential components of the immune system

Figure 15-33 **The assembly, budding, and maturation of HIV-1.** SU is the surface glycoprotein **gp120** and TM is the transmembrane protein **gp41.** [After Turner, B.G. and Summers, M.F., *J. Mol. Biol.* **285,** 4 (1999).]

(Section 35-2Aa). Unlike most types of retroviruses, HIV-1 eventually kills the cells producing it. Although the helper *T* cells within which HIV-1 are actively replicating are often destroyed by the immune system, those within which the HIV-1 is latent (its DNA is not being transcribed) are not detected by the immune system and hence provide a reservoir of HIV-1 (other types of cells also harbor HIV-1). Consequently, over a several year period after the initial infection (during most of which the host exhibits no obvious symptoms), the host's immune system is steadily depleted until it has deteriorated to the point that the host regularly falls victim to and is eventually killed by opportunistic pathogens that individuals with normally functioning immune systems can readily withstand. It is this latter stage of an HIV infection that is called AIDS. In the absence of effective therapy, AIDS is almost invariably fatal. Through the year 2008, nearly 30 million people had died of AIDS and an estimated 35 million others, largely in sub-Saharan Africa, were HIV-positive, numbers that are increasing at the rate of ~3 million per year. As a consequence of this global catastrophe, HIV has been characterized and effective countermeasures against it have been devised faster than for any other pathogen in history.

a. Reverse Transcriptase Inhibitors Are Only Partially Effective

The first drug to be approved by the FDA (in 1987) to fight AIDS was **3′-azido-3′-deoxythymidine (AZT; zidovudine),**

**3′-Azido-3′-deoxythymidine
(AZT; zidovudine)**

which had first been synthesized in 1964 as a possible anticancer agent (it was ineffective). AZT is a nucleoside analog that, on enzymatic conversion to its triphosphate in the cell (the plasma membrane is impermeable to nucleoside triphosphates), inhibits HIV-1 reverse transcriptase, as do the several other drugs (Section 30-4Ca) that the FDA had approved to treat AIDS prior to 1996. Unfortunately, these agents only slow the progression of an HIV infection but do not stop it. This is in part because they are toxic, mainly to the bone marrow cells that are blood cell precursors, and hence cannot be taken in large doses. More important, however, is that reverse transcriptase, unlike most other DNA polymerases (Section 30-2A), cannot correct its mistakes and hence frequently generates mutations (about one per 10^4 bp and, since the viral genome consists of ~10^4 bp, each viral genome bears, on average, one new mutation). Consequently, *under the selective pressure of an anti-HIV drug such as AZT, the drug's target receptor rapidly evolves to a drug-resistant form.*

b. HIV-1 Polyproteins Are Cleaved by HIV-1 Protease

HIV-1, as do other retroviruses, synthesizes its proteins in the form of **polyproteins,** which each consist of several tandemly linked proteins (Fig. 15-34). HIV-1 encodes two polyproteins, **gag** (55 kD) and **gag–pol** (160 kD), which are both anchored to the plasma membrane via N-terminal myristoylation (Section 12-3Bb). These polyproteins are then cleaved to their component proteins through the action of **HIV-1 protease,** but only after this enzyme has excised itself from gag–pol. This process occurs only after the virion has budded off from the host cell and results in a large structural reorganization of the virion (Fig. 15-33). The virion is thereby converted from its noninfectious immature form to its pathogenic mature form. If HIV-1 protease is inactivated, either mutagenically or by an inhibitor, the virion remains noninfectious. Hence HIV-1 protease is an opportune drug target.

c. Aspartic Proteases and Their Catalytic Mechanism

HIV-1 protease is a member of the **aspartic protease** family (also known as **acid proteases**), so called because these enzymes all contain catalytically essential Asp residues that

Figure 15-34 **HIV-1 polyproteins.** (*a*) The organization of the HIV-1 gag and gag–pol polyproteins. The symbols used are MA, matrix protein; CA, capsid protein; NC, nucleocapsid protein; TF, transmembrane protein; PR, protease; RT, reverse transcriptase; RN, ribonuclease; and IN, integrase. (*b*) The sequences flanking the HIV-1 protease cleavage sites (*red bonds*) indicated in Part *a*.

Cleavage site	Sequence
I	··· Ser - Gln - Asn - Tyr —— Pro - Ile - Val - Gln ···
II	··· Ala - Arg - Val - Leu —— Ala - Glu - Ala - Met ···
III	··· Ala - Thr - Ile - Met —— Met - Gln - Arg - Gly ···
IV	··· Pro - Gly - Asn - Phe —— Leu - Gln - Ser - Arg ···
V	··· Ser - Phe - Asn - Phe —— Pro - Gln - Ile - Thr ···
VI	··· Thr - Leu - Asn - Phe —— Pro - Ile - Ser - Pro ···
VII	··· Ala - Glu - Thr - Phe —— Tyr - Val - Asp - Gly ···
VIII	··· Arg - Lys - Ile - Leu —— Phe - Leu - Asp - Gly ···

¹Asp-Arg-Val-Tyr-Ile-His-Pro-Phe-His-Leu-Val-Ile-His¹³

Angiotensinogen

H_2O ⤵ | renin

¹Asp-Arg-Val-Tyr-Ile-His-Pro-Phe-His-Leu¹⁰ + Val-Ile-His

Angiotensin I

H_2O ⤵ | angiotensin converting enzyme (ACE)

¹Asp-Arg-Val-Tyr-Ile-His-Pro-Phe⁸ + His-Leu

Angiotensin II

Figure 15-35 Renin participation in blood pressure regulation. Renin proteolytically cleaves the 13-residue polypeptide **angiotensinogen** to the 10-residue polypeptide **angiotensin I.** This latter peptide is then cleaved by **angiotensin converting enzyme (ACE)** to the 8-residue polypeptide **angiotensin II,** which, on binding to its receptor, induces vasoconstriction and retention of Na^+ and water by the kidneys, resulting in increased blood pressure. Consequently several inhibitors of renin and of ACE have been developed for the control of **hypertension** (high blood pressure).

occur in the signature sequence Asp–Thr/Ser–Gly. Humans have several known aspartic proteases including **pepsin,** a digestive enzyme secreted by the stomach (its specificity is indicated in Table 7-2) that functions at pH 1 and which was the first enzyme to be recognized (named in 1836 by Theodor Schwann); **chymosin** (formerly **rennin**), a stomach enzyme, occurring mainly in infants, that specifically cleaves a Phe–Met peptide bond in the milk protein **κ-casein,** thereby causing milk to curdle, making it easier to digest

(calf stomach chymosin has been used for millennia to make cheese); **cathepsins D** and **E,** lysosomal proteases that function to degrade cellular proteins; **renin,** which participates in the regulation of blood pressure and electrolyte balance (Fig. 15-35); and **β-secretase** (also known as **memapsin 2**), a transmembrane protein common in brain that participates in cleaving Aβ precursor protein to yield amyloid-β protein (Aβ), which is implicated in Alzheimer's disease (Section 9-5B). In addition, many fungi secrete aspartic proteases, presumably to aid them in invading the tissues they colonize.

Eukaryotic aspartic proteases are ~330-residue monomeric proteins. The X-ray structure of pepsin (Fig. 15-36a), which closely resembles those of other eukaryotic aspartic proteases, reveals that this croissant-shaped protein consists of two homologous domains that are related by approximate 2-fold symmetry (although only about 25 residues in the core β sheets of each domain are closely related by this symmetry). Each domain contains a catalytically essential Asp in an analogous position. The X-ray structures of enzyme–inhibitor complexes of various aspartate proteases indicate that substrates bind in a prominent cleft between the two domains that could accommodate an ~8-residue polypeptide segment in an extended β sheet–like conformation. The active site Asp residues are located at the base of this cleft (Fig. 15-36a).

What is the catalytic mechanism of eukaryotic aspartic proteases? Proteolytic enzymes, in general, have three essential catalytic components:

1. A nucleophile to attack the carbonyl C atom of the scissile peptide to form a tetrahedral intermediate (Ser 195 serves this function in trypsin; Fig. 15-23).

2. An electrophile to stabilize the negative charge that develops on the carbonyl O atom of the tetrahedral intermediate (the H bonding donors lining the oxyanion hole, Gly 193 and Ser 195, do so in trypsin; Fig. 15-25).

(a)

(b)

Figure 15-36 X-ray structure of porcine pepsin. (*a*) Ribbon diagram in which the N-terminal domain (residues 1–172) is gold, the C-terminal domain (residues 173–326) is cyan, the side chains of the active site Asp residues are shown in ball-and-stick form with C green and O red, and the water molecule that is bound by these Asp side chains is represented by a red sphere. The protein is viewed with the pseudo-2-fold axis relating core

portions of the two domains tipped from vertical toward the viewer. (*b*) Enlarged view of the active site Asp residues and their bound water molecule indicating the lengths (in Å) of possible hydrogen bonds (*thin gray bonds*). The X-ray structures of other aspartic proteases exhibit similar interatomic distances. [Based on an X-ray structure by Anita Sielecki and Michael James, University of Alberta, Edmonton, Canada. PDBid 4PEP.]

3. A proton donor so as to make the amide N atom of the scissile peptide a good leaving group (the imidazolium group of His 57 in trypsin; Fig. 15-23).

Pepsin's pH rate profile (Section 14-4) suggests that it has two ionizable essential residues, one with $pK \approx 1.1$ and the other with $pK \approx 4.7$, which are almost certainly the carboxyl groups of its essential Asp residues. At the pH of the stomach, the Asp residue with pK 4.7 is protonated and that with pK 1.1 is partially ionized. This suggests that the ionized carboxyl group acts as a nucleophile to form the putative tetrahedral intermediate. However, no covalent intermediate between an aspartic protease and its substrate has ever been detected.

The two active site Asp residues in eukaryotic aspartic proteases are in close proximity and both appear to form hydrogen bonds to a bridging water molecule that is present in several X-ray structures of eukaryotic aspartic proteases (Fig. 15-36*b*). This, together with a variety of enzymological and kinetic data, led Thomas Meek to propose the following catalytic mechanism for aspartic proteases (Fig. 15-37):

1. An active site Asp carboxylate group, acting as a general base, activates the bound water molecule, the so-called lytic water, to nucleophilically attack the scissile peptide's carbonyl C as an OH^- ion. Proton donation (general acid catalysis) by the second, previously uncharged active site Asp stabilizes the oxyanion that would otherwise form in the resulting tetrahedral intermediate.

2. The N atom of the scissile peptide is protonated by the first Asp (general acid catalysis) resulting, through charge rearrangement and proton transfer to the second Asp (general base catalysis), in amide bond scission.

Aspartic proteases are inhibited by compounds with tetrahedral carbon atoms at a position mimicking a scissile peptide bond (see below). This strongly suggests that these

Figure 15-37 Catalytic mechanism of aspartic proteases. (1) The nucleophilic attack of the enzyme-activated water molecule (*red*) on the carbonyl carbon atom of the scissile peptide bond (*green*) to form the tetrahedral intermediate. This reaction step is promoted by general base catalysis by the Asp on the right and general acid catalysis by the Asp on the left (*blue*). **(2)** The decomposition of the tetrahedral intermediate to form products via general acid catalysis by the Asp on the right and general base catalysis by the Asp on the left.

Figure 15-38 X-ray structure of HIV-1 protease. (*a*) Uncomplexed and (*b*) in complex with its inhibitor saquinavir (structural formula in Fig. 15-41). In each structure, the homodimeric protein is viewed with its 2-fold axis of symmetry vertical and is shown as a ribbon diagram with one subunit gold and the other cyan. The side chains of the active site Asp residues, Asp 25 and Asp 25′, as well as the saquinavir in Part *b*, are shown in ball-and-stick form with C green, N blue, and O red. Note how the β hairpin "flaps" at the top of the uncomplexed enzyme have folded down over the inhibitor in the saquinavir complex. Compare these structures with that of the similarly viewed pepsin in Fig. 15-36*a*. [Part *a* based on an X-ray structure by Tom Blundell, Birkbeck College, London, U.K., and Part *b* based on an X-ray structure by Robert Crowther, Hoffmann-LaRoche Ltd., Nutley, New Jersey. PDBids (*a*) 3PHV and (*b*) 1HXB.] 🔎 **See Interactive Exercise 7**

enzymes preferentially bind their transition states (transition state stabilization), thereby enhancing catalysis.

d. HIV-1 Protease Inhibitors Are Effective Anti-AIDS Agents

HIV-1 protease differs from eukaryotic aspartic proteases in that it is a homodimer of 99-residue subunits. Nevertheless, its X-ray structure (Fig. 15-38*a*), determined independently in 1989 by Alexander Wlodawer, by Manuel Navia and Paula Fitzgerald, and by Tom Blundell, closely resembles those of eukaryotic aspartic proteases. Thus, HIV-1 protease has the enzymatically unusual property that its single active site is formed by two identical symmetrically arranged subunits. Quite possibly HIV-1 protease resembles the putative primordial aspartic protease that, through gene duplication, evolved to form the eukaryotic enzymes (although HIV-1 protease is well suited to the limited amount of genetic information that a virus can carry).

Once the structure of HIV-1 protease became available, intensive efforts were mounted in numerous laboratories to find therapeutically effective inhibitors of this enzyme. In this process, nearly 300 X-ray structures and several

NMR structures have been reported of HIV-1 protease, its mutants, and the proteases of other retroviruses, both alone and in their complexes with a great variety of inhibitors. Hence, HIV-1 protease is perhaps the most exhaustively structurally studied protein.

Comparison of the X-ray structure of HIV-1 protease alone (Fig. 15-38*a*) with that of its complexes with polypeptide-like inhibitors (e.g., Fig. 15-38*b*) reveals that, on binding an inhibitor, the β hairpin "flaps" covering the "top" of the substrate-binding cleft swing down by as much as 7 Å to enclose the inhibitor. Such an inhibitor binds to the 2-fold symmetric enzyme in a 2-fold pseudosymmetric extended conformation such that the inhibitor interacts with the enzyme much like a strand in a β sheet (Fig. 15-39). On the "floor" of the binding cleft, each signature sequence (Asp 25–Thr 26–Gly 27) is located in a loop that is stabilized by a network of hydrogen bonds similar to that observed in eukaryotic aspartic proteases. The inhibitor interacts with the enzyme via a hydrogen bond to the active site residue Asp 25. However, contrary to the case for eukaryotic aspartic proteases (Fig. 15-36*b*), no X-ray structure of an HIV-1 protease contains a water molecule within

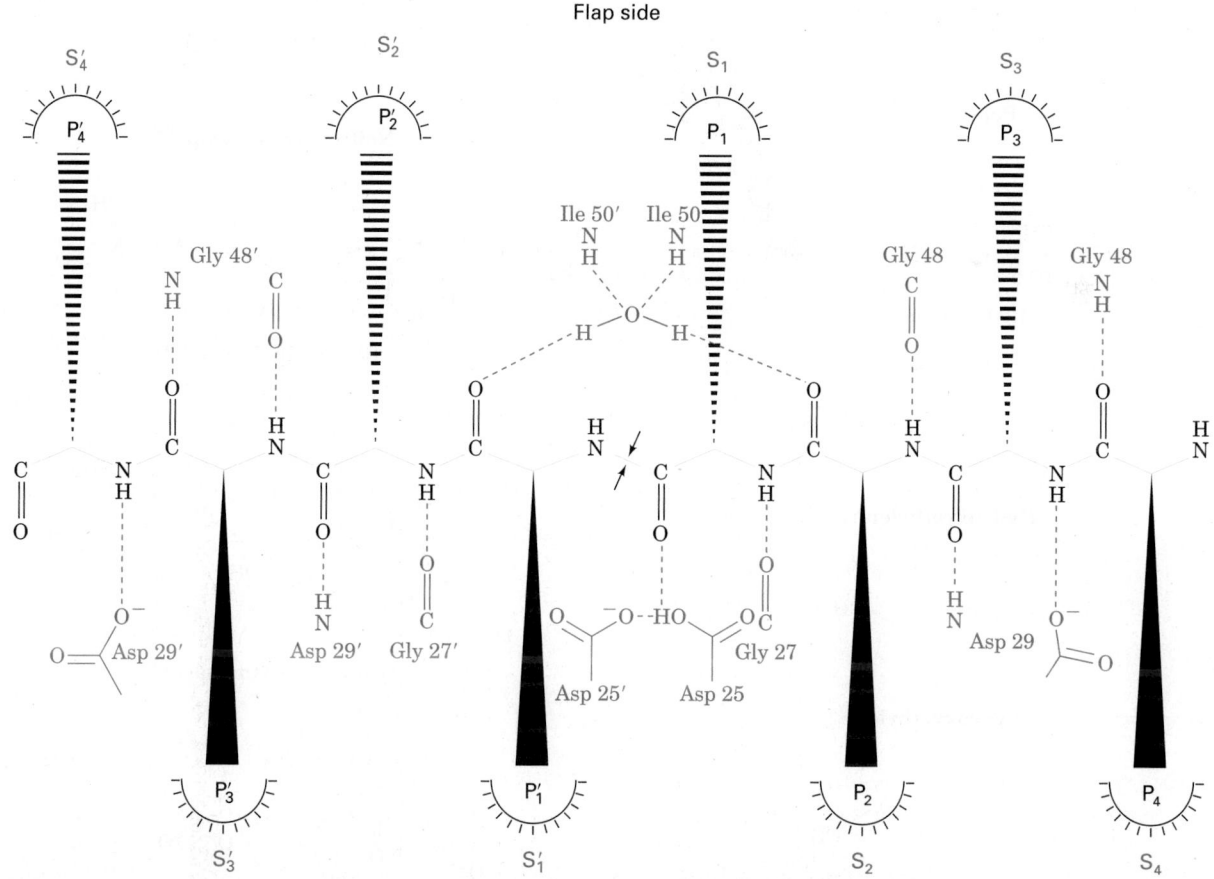

Figure 15-39 **Arrangement of hydrogen bonds between HIV-1 protease and a modeled substrate.** In the nomenclature used here, polypeptide residues in one subunit are assigned primed numbers to differentiate them from the residues of the other subunit; substrate residues on the N-terminal side of the scissile peptide bond are designated $P_1, P_2, P_3, \ldots$, counting toward the

N-terminus; substrate residues on its C-terminal side are designated $P'_1, P'_2, P'_3, \ldots$, counting toward the C-terminus; and the symbols $S_1, S_2, S_3, \ldots$, and $S'_1, S'_2, S'_3, \ldots$, designate the enzyme's corresponding residue-binding subsites. The scissile peptide bond is marked by arrows. [After Wlodawer, A. and Vondrasek, J., *Annu. Rev. Biophys. Biomol. Struct.* **27,** 257 (1998).]

hydrogen bonding distance of Asp 25 or Asp 25'. On the flap side of the binding cleft, the inhibitor interacts with Gly 48 and Gly 48' and with a water molecule that is not the attacking nucleophile but which mediates the contacts between the flaps and the inhibitor backbone.

Although HIV-1 protease specifically cleaves the gag and gag–pol polyproteins at a total of 8 sites (Fig. 15-34*b*), these sites appear to have little in common except that their immediately flanking residues are nonpolar and mostly bulky. Indeed, binding studies indicate that HIV-1 protease's specificity arises from the cumulative effects of the interactions between the enzyme and the amino acids in positions P_4 through P_4'. However, three of the peptides cleaved by HIV-1 have either the sequence Phe-Pro or Tyr-Pro, which are sequences that human aspartic proteases do not cleave. Hence, HIV-1 protease inhibitors containing groups that resemble either of these dipeptides would be unlikely to inhibit essential human aspartic proteases.

An effective HIV-1 protease inhibitor should resemble a substrate with its scissile peptide replaced by a group that the enzyme cannot cleave. Such a group should, preferably, enhance the enzyme's affinity for the inhibitor. Mimics of the tetrahedral intermediate (Fig. 15-37), that is, transition state analogs, are likely to do so. Consequently, a variety of such groups (Fig. 15-40) have been investigated in efforts to synthesize therapeutically effective inhibitors of HIV-1 protease.

Although HIV-1 protease has high *in vitro* affinity for its polypeptide-based inhibitors, these substances have poor oral bioavailability (they are degraded by digestive

Figure 15-40 Comparison of a normal peptide bond (*top*) to a selection of groups (*red*) that are isosteres (stereochemical analogs) of the tetrahedral intermediate in reactions catalyzed by aspartic proteases.

Figure 15-41 Some HIV-1 protease inhibitors that are in clinical use. Note that in addition to its generic (chemical) name, each drug has a proprietary trade name, here in parentheses, under which it is marketed.

proteases) and pharmacokinetics (they do not readily pass through cell membranes.). Consequently, therapeutically effective HIV-1 protease inhibitors must be **peptidomimetics** (peptide mimics), substances that sterically and perhaps physically, but not chemically, resemble polypeptides. The use of peptidomimetics also permits conformational constraints to be imposed on a drug candidate that would not be present in the corresponding polypeptide.

The FDA has approved ten HIV-1 protease inhibitors (Fig. 15-41), the first of which, **saquinavir,** was sanctioned in late 1995. These peptidomimetics have IC_{50}'s against HIV in culture ranging from 2 to 60 nM but have little or no activity against human aspartic proteases (K_I's > 10 μM). They are the first drugs to clearly prolong the lives of AIDS victims. Their development, in each case, was a complex iterative process that required the design, synthesis, and evaluation of numerous related compounds. In several cases, these investigations capitalized on the wealth of experience gained in developing peptidomimetic inhibitors of the aspartic protease renin and in the resulting stockpiles of these compounds.

All the FDA-approved HIV-1 protease inhibitors initially cause a rapid and profound decline in a patient's plasma HIV load, which is often paralleled by immune system recovery. However, as we saw with reverse transcriptase inhibitors, mutant forms of the protease that are resistant to the inhibitor being used arise, usually within 4 to 12 weeks. Moreover, such a mutant protease is likely to be resistant to other HIV-1 protease inhibitors, because all of the HIV-1 protease inhibitors are targeted to the same binding site. This has led to the use of combination therapies in which an HIV-1 protease inhibitor is administered together with one, or more often, two inhibitors of other viral processes (inhibitors of reverse transcriptase, integrase, and viral entry into host cells). This is because any virus that gains resistance to one drug in a regimen will be suppressed by the other drug(s) in that regimen. In addition, the HIV-1 protease inhibitor **ritonavir** has been shown to be a potent inhibitor of the cytochrome P450 isoforms (CYP3A4,5,7) that metabolize other protease inhibitors and hence is usually prescribed in low dosage as an adjunct to another protease inhibitor to improve the latter's pharmacokinetics.

The plasma virus levels in many patients who were placed on combination therapy rapidly became undetectable and have remained so for several years. This, however, does not constitute a cure: If drug therapy is interrupted, the virus will reappear in the plasma because certain tissues in the body harbor latent viruses that are unaffected by and/or inaccessible to drug therapy. Thus, the presently available anti-HIV medications must be taken for a lifetime.

Current anti-HIV therapies are by no means ideal. To maximize their oral bioavailability, some of the different drugs must be taken well before or after a meal but others must be taken with a meal. To minimize the probability of resistant forms of HIV arising, the bioavailability of each drug must be maintained at a certain minimum level and hence each drug must be taken on a rigid schedule. Moreover, these drugs have significant side effects, mainly fatigue, nausea, diarrhea, tingling and numbness with ritonavir, and kidney stones with **indinavir.** Consequently, numerous AIDS patients fail to take their medications properly, which greatly increases the likelihood that they will develop resistance to these drugs and infect others with drug-resistant viruses. Finally, HIV-1 protease inhibitors, being complex molecules, are difficult to synthesize and therefore are relatively expensive, so that in the developing countries in which AIDS is most prevalent, governments and most individuals cannot afford to purchase these drugs, even if they were to be supplied at cost. It is therefore important that anti-HIV therapies be developed that are easy for patients to comply with, are inexpensive, and ideally, will totally eliminate an HIV infection.

CHAPTER SUMMARY

1 Catalytic Mechanisms Most enzymatic mechanisms of catalysis have ample precedent in organic catalytic reactions. Acid- and base-catalyzed reactions occur, respectively, through the donation or abstraction of a proton to or from a reactant so as to stabilize the reaction's transition state complex. Enzymes often employ ionizable amino acid side chains as general acid–base catalysts. Covalent catalysis involves nucleophilic attack of the catalyst on the substrate to transiently form a covalent bond followed by the electrophilic stabilization of a developing negative charge in the reaction's transition state. Various protein side chains as well as certain coenzymes can act as covalent catalysts. Metal ions, which are common enzymatic components, catalyze reactions by stabilizing developing negative charges in a manner resembling general acid catalysis. Metal ion–bound water molecules are potent sources of OH^- ions at neutral pH's. Metal ions also facilitate enzymatic reactions through the charge shielding of bound substrates. The arrangement of charged groups about an enzymatic active site of low dielectric constant in a manner that stabilizes the transition state complex results in the electrostatic catalysis of the enzymatic reaction. Enzymes catalyze reactions by bringing their substrates into close proximity in reactive orientations. The enzymatic binding of the substrates in a bimolecular reaction arrests their relative motions resulting in a rate enhancement. The preferential enzymatic binding of the transition state of a catalyzed reaction over the substrate is an important rate enhancement mechanism. Transition state analogs are potent competitive inhibitors because they bind to the enzyme more tightly than does the corresponding substrate.

2 Lysozyme Lysozyme catalyzes the hydrolysis of $\beta(1 \rightarrow 4)$-linked poly(NAG–NAM), the bacterial cell wall polysaccharide, as well as that of poly(NAG). Lysozyme binds a hexasaccharide so as to distort its D-ring toward the half-chair conformation of the planar oxonium ion transition state. This is followed by cleavage of the C1—O1 bond between the D- and E-rings as promoted by proton donation from Glu 35. The resulting oxonium ion transition state is electrostatically stabilized by the nearby carboxyl group of Asp 52, which then forms a covalent bond with C1. The E-ring leaving group is subsequently replaced by water, which in what is essentially a reversal of the previous reaction sequence, attacks C1 yielding the reaction's second product and regenerating the enzyme. The roles of Glu 35 and Asp 52 in lysozyme catalysis have been verified through mutagenesis studies. Structural and binding studies indicate that strain is of major catalytic importance in the lysozyme mechanism. Mass spectrometry and X-ray studies have shown that the lysozyme reaction proceeds via a covalent glycosyl–enzyme intermediate involving Asp 52 rather than by the noncovalently bound oxonium ion intermediate postulated by the original Phillips mechanism.

3 Serine Proteases   Serine proteases constitute a widespread class of proteolytic enzymes that are characterized by the possession of a reactive Ser residue. The pancreatically synthesized digestive enzymes trypsin, chymotrypsin, and elastase are sequentially and structurally related but have different side chain specificities for their substrates. All have the same catalytic triad, Asp 102, His 57, and Ser 195, at their active sites. The differing side chain specificities of trypsin and chymotrypsin depend in a complex way on the structures of the loops that connect the walls of the specificity pocket, as well as on the charge of the side chain at the base of the specificity pocket. Subtilisin, serine carboxypeptidase II, and ClpP are unrelated serine proteases that have essentially the same active site geometry as do the pancreatic enzymes.

Catalysis in serine proteases is initiated by the nucleophilic attack of the active Ser on the carbonyl carbon atom of the scissile peptide to form the tetrahedral intermediate, a process that may be facilitated by the formation of a low-barrier hydrogen bond between Asp 102 and His 57. The tetrahedral intermediate, which is stabilized by its preferential binding to the enzyme's active site, then decomposes to the acyl–enzyme intermediate under the impetus of proton donation from the Asp 102-polarized His 57. After the replacement of the leaving group by solvent H_2O, the catalytic process is reversed to yield the second product and the regenerated enzyme. The active Ser is not unusually reactive but is ideally situated to nucleophilically attack the activated scissile peptide. The X-ray structure of the trypsin–BPTI complex indicates the existence of the tetrahedral intermediate, whereas X-ray structures of a complex of elastase with the heptapeptide BCM7 have visualized both the acyl–enzyme intermediate and the tetrahedral intermediate. Enzyme–substrate complexes have free energy landscapes that facilitate the formation of the transition state.

The pancreatic serine proteases are synthesized as zymogens to prevent pancreatic self-digestion. Trypsinogen is activated by a single proteolytic cleavage by enteropeptidase. The resulting trypsin similarly activates trypsinogen as well as chymotrypsinogen, proelastase, and other pancreatic digestive enzymes. Trypsinogen's catalytic triad is structurally intact. The

zymogen's low catalytic activity arises from a distortion of its specificity pocket and oxyanion hole, so that it is unable to productively bind substrate or preferentially bind the catalytic reaction's transition state.

4 Drug Design   Drugs act by binding to and thereby modifying the functions of receptors. Many promising drug candidates, which are known as lead compounds, have been found by methods in which a large number of compounds are tested for drug efficacy in an assay that is a suitable surrogate of the disease/condition under consideration. Lead compounds are then chemically manipulated in the search for compounds with improved drug efficacy. Structure–activity relationships (SARs) and quantitative structure–activity relationships (QSARs) are useful tools in this endeavor. Structure-based drug design uses the X-ray and NMR structures of drug candidates in complex with their target proteins, together with a variety of molecular modeling tools, to guide the search for improved drug candidates. However, the advent of combinatorial chemistry, fragment-based lead discovery, and high-throughput screening procedures has extended the "make-many-compounds-and-see-what-they-do" approaches to drug discovery.

In order to reach their target receptors, drugs must have favorable pharmacokinetics, that is, they must readily traverse numerous physical barriers in the body, avoid chemical transformation by enzymes, and not be excreted too rapidly. Most useful drugs are neither too lipophilic nor too hydrophilic so that they can both gain access to the necessary membranes and pass through them. Drug toxicity, dosage, efficacy, and the nature of rare adverse reactions are determined through extensive and carefully designed clinical trials. Most drugs are metabolically cleared through oxidative hydroxylation by one of the numerous cytochrome P450 isozymes. This permits the hydroxylated drugs to be enzymatically conjugated to polar groups such as glucuronic acid and glycine, which increases their rates of excretion by the kidneys. Drug–drug interactions are frequently mediated by cytochromes P450. Polymorphisms among cytochromes P450 are often responsible for the variations among individuals in their response to a given drug, including adverse reactions.

The formulation of HIV-1 protease inhibitors to control HIV infections is one of the major triumphs of modern drug discovery methods. HIV are retroviruses that attack specific immune system cells and thereby degrade the immune system over a period of several years to the point that it is no longer able defend against opportunistic infections. HIV-1 protease functions to cleave the polyproteins in immature HIV-1 virions that have budded out from a host cell, thus generating the mature, infectious form. HIV-1 protease is an aspartic protease that, as do eukaryotic aspartic proteases such as pepsin, uses its two active site Asp residues to activate its bound lytic water molecule as the nucleophile that attacks and thereby cleaves specific peptide bonds in the substrate polyprotein. All of the FDA-approved peptidomimetic inhibitors of HIV-1 protease cause a rapid and profound decrease in plasma HIV levels, although they do not entirely eliminate the virus. They are used in combination with inhibitors of other viral processes to minimize the ability of the rapidly mutating HIV to evolve drug-resistant forms.

REFERENCES

General

Fersht, A., *Structure and Mechanism in Protein Science,* Freeman (1999).

Frey, P.A. and Hegeman, A.D., *Enzymatic Reaction Mechanisms,* Oxford (2007). [A compendium of enzymatic reactions.]

Jencks, W.P., *Catalysis in Chemistry and Enzymology,* Dover (1987). [A classic and, in many ways, still current work.]

Catalytic Mechanisms

Bruice, T.C., Some pertinent aspects of mechanism as determined with small molecules, *Annu. Rev. Biochem.* **45,** 331–373 (1976).

Bruice, T.C. and Benkovic, S.J., Chemical basis for enzyme catalysis, *Biochemistry* **39,** 6267–6274 (2000); *and* Bruice, T.C. and Lightstone, F.C., Ground state and transition state contributions to the rates of intramolecular and enzymatic reactions, *Acc. Chem. Res.* **32,** 127–136 (1999).

Christianson, D.W. and Cox, J.D., Catalysis by metal-activated hydroxide in zinc and manganese metalloenzymes, *Annu. Rev. Biochem.* **68,** 33–57 (1999). [Discusses the enzymatic mechanism of carbonic anhydrase.]

Garcia-Viloca, M., Gao, J., Karplus, M., and Truhlar, D.G., How enzymes work: Analysis by modern rate theory and computer simulations, *Science* **303,** 186–195 (2004).

Hackney, D.D., Binding energy and catalysis, *in* Sigman, D.S. and Boyer, P.D. (Eds.), *The Enzymes* (3rd ed.), Vol. 19, *pp.* 1–36, Academic Press (1990).

Jencks, W.P., Binding energy, specificity, and enzymatic catalysis: The Circe effect, *Adv. Enzymol.* **43,** 219–410 (1975).

Kraut, J., Carroll, K.S., and Herschlag, D., Challenges in enzyme mechanism and energetics, *Annu. Rev. Biochem.* **72,** 517–571 (2003).

Lolis, E. and Petsko, G.A., Transition-state analogues in protein crystallography: Probes of the structural source of enzyme catalysis, *Annu. Rev. Biochem.* **59,** 597–630 (1990).

Schramm, V.L., Enzymatic transition states and transition state analog analogues, *Curr. Opin. Struct. Biol.* **15,** 604–613 (2005).

Wolfenden, R., Analogue approaches to the structure of the transition state in enzyme reactions, *Acc. Chem. Res.* **5,** 10–18 (1972).

Lysozyme

Blake, C.C.F., Johnson, L.N., Mair, G.A., North, A.C.T., Phillips, D.C., and Sarma, V.R., Crystallographic studies of the activity of hen egg-white lysozyme, *Proc. R. Soc. London Ser. B* **167,** 378–388 (1967).

Chipman, D.M. and Sharon, N., Mechanism of lysozyme action, *Science* **165,** 454–465 (1969).

Ford, L.O., Johnson, L.N., Machin, P.A., Phillips, D.C., and Tijan, R., Crystal structure of a lysozyme–tetrasaccharide lactone complex, *J. Mol. Biol.* **88,** 349–371 (1974).

Imoto, T., Johnson, L.N., North, A.C.T., Phillips, D.C., and Rupley, J.A., Vertebrate lysozymes, *in* Boyer, P.D. (Ed.), *The Enzymes* (3rd ed.), Vol. 7, *pp.* 665–868, Academic Press (1972). [An exhaustive review.]

Johnson, L.N., Cheetham, J., McLaughlin, P.J., Acharya, K.R., Barford, D., and Phillips, D.C., Protein–oligosaccharide interactions: Lysozyme, phosphorylase, amylases, *Curr. Top. Microbiol. Immunol.* **139,** 81–134 (1988).

Jollès, P. (Ed.), *Lysozymes: Model Enzymes in Biochemistry and Biology,* Birkhäuser Verlag (1996).

Kirby, A.J., The lysozyme mechanism sorted—after 50 years, *Nature Struct. Biol.* **8,** 737–739 (2001). [Briefly summarizes the theoretical and experimental evidence for a covalent intermediate in the lysozyme mechanism.]

Mooser, G., Glycosidases and glycosyltransferases, *in* Sigman, D.S. (Ed.), *The Enzymes* (3rd ed.), Vol. 20, *pp.* 187–233, Academic Press (1992). [Section II discusses lysozyme.]

Phillips, D.C., The three-dimensional structure of an enzyme molecule, *Sci. Am.* **215**(5), 75–80 (1966).

Schindler, M., Assaf, Y., Sharon, N., and Chipman, D.M., Mechanism of lysozyme catalysis: Role of ground-state strain in subsite D in hen egg-white and human lysozymes, *Biochemistry* **16,** 423–431 (1977).

Secemski, I.I., Lehrer, S.S., and Lienhard, G.E., A transition state analogue for lysozyme, *J. Biol. Chem.* **247,** 4740–4748 (1972). [Binding studies on the lactone derivative of $(NAG)_4$.]

Strynadka, N.C.J. and James, M.N.G., Lysozyme revisited: Crystallographic evidence for distortion of an *N*-acetylmuramic acid residue bound in site D, *J. Mol. Biol.* **220,** 401–424 (1991).

Vocadlo, D.J., Davies, G.J., Laine, R., and Withers, S.G., Catalysis by hen egg-white lysozyme proceeds via a covalent intermediate, *Nature* **412,** 835–838 (2001).

Warshel, A. and Levitt, M., Theoretical studies of enzymatic reactions; dielectric, electrostatic and steric stabilization of the carbonium ion in the reaction of lysozyme, *J. Mol. Biol.* **103,** 227–249 (1976). [Theoretical indications that lysozyme catalysis occurs through electrostatic rather than steric strain.]

Serine Proteases

Blow, D.M., The tortuous story of Asp...His...Ser: Structural analysis of chymotrypsin, *Trends Biochem. Sci.* **22,** 405–408 (1998). [A personal memoir of the structural determination of α-chymotrypsin in the years 1967 through 1969.]

Cleland, W.W., Frey, P.A., and Gerlt, J.A., The low barrier hydrogen bond in enzymatic catalysis, *J. Biol. Chem.* **273,** 25529–25532 (1998).

Corey, D.R. and Craik, C.S., An investigation into the minimum requirements for peptide hydrolysis by mutation of the catalytic triad of trypsin, *J. Am. Chem. Soc.* **114,** 1784–1790 (1992).

Ding, X., Rasmussen, B.F., Petsko, G.A., and Ringe, D., Direct structural observation of an acyl-enzyme intermediate in the hydrolysis of an ester substrate by elastase, *Biochemistry* **33,** 9285–9293 (1994).

Dodson, G. and Wlodawer, A., Catalytic triads and their relatives, *Trends Biochem. Sci.* **23,** 347–352 (1998).

Frey, P.A., Whitt, S.A., and Tobin, J.B., A low-barrier hydrogen bond in the catalytic triad of serine proteases, *Science* **264,** 1927–1930 (1994).

Hammes-Schiffer, S. and Benkovic, S.J., Relating protein motion to catalysis, *Annu. Rev. Biochem.* **75,** 519–541 (2006); *and* Benkovic, S.J., Hammes, G.G., and Hammes-Schiffer, S., Free-energy landscape of enzyme catalysis, *Biochemistry* **47,** 3317–3321 (2008).

Hedstrom, L., Serine protease mechanism and specificity, *Chem. Rev.* **102,** 4501–4523 (2002). [A detailed review.]

Henzler-Wildman, K.A., et al., Intrinsic motions along an enzymatic reaction pathway, *Nature* **450,** 838–844 (2007).

James, M.N.G., Sielecki, A.R., Brayer, G.D., Delbaere, L.T.J., and Bauer, C.A., Structure of product and inhibitor complexes of *Streptomyces griseus* protease A at 1.8 Å resolution, *J. Mol. Biol.* **144,** 45–88 (1980).

Kuhn, P., Knapp, M., Soltis, S.M., Ganshaw, G., Thoene, M., and Bott, R., The 0.78 Å structure of a serine protease: *Bacillus lentus* subtilisin, *Biochemistry* **37,** 13446–13452 (1998).

Liao, D.-I. and Remington, S.J., Structure of wheat serine carboxypeptidase II at 3.5-Å resolution, *J. Biol. Chem.* **265,** 6528–6531 (1990).

Neurath, H., Evolution of proteolytic enzymes, *Science* **224,** 350–357 (1984).

Perona, J.J. and Craik, C.S., Evolutionary divergence of substrate specificity within the chymotrypsin-like serine protease fold, *J. Biol. Chem.* **272,** 29987–29990 (1997); *and* Structural basis of substrate specificity in the serine proteases, *Protein Sci.* **4,** 337–360 (1995).

Perrin, C.L. and Nielson, J.B., "Strong" hydrogen bonds in chemistry and biology, *Annu. Rev. Phys. Chem.* **48,** 511–544 (1997). [A detailed review which concludes that the evidence for the importance of LBHBs in enzymatic reactions is inconclusive.]

Phillips, M.A. and Fletterick, R.J., Proteases, *Curr. Opin. Struct. Biol.* **2,** 713–720 (1992).

Radisky, E.S., Lee, J.M., Lu, C.-J. K., and Koshland, D.E., Jr., Insights into the serine protease mechanism from atomic resolution structures of trypsin reaction intermediates, *Proc. Natl. Acad. Sci.* **103,** 6835–6840 (2006). [Reveals the subtle motions of the catalytic Ser and His residues, the substrate, and the hydrolytic water molecule that favor catalysis during the acylation reaction.]

Roberts, R.M., Mathialagan, N., Duffy, J.Y., and Smith, G.W., Regulation and regulatory role of proteinase inhibitors, *Crit. Rev. Euk. Gene Express.* **5,** 385–435 (1995).

Shan, S., Loh, S., and Herschlag, D., The energetics of hydrogen bonds in model systems: Implications for enzymatic catalysis, *Science* **272,** 97–101 (1996).

Stroud, R.M., Kossiakoff, A.A., and Chambers, J.L., Mechanism of zymogen activation, *Annu. Rev. Biophys. Bioeng.* **6,** 177–193 (1977).

Wang, J., Hartling, J.A., and Flanagan, J.M., The structure of ClpP at 2.3 Å resolution suggests a model for ATP-dependent proteolysis, *Cell* **91,** 447–456 (1997).

Wilmouth, R.C., Edman, K., Neutze, R., Wright, P.A., Clifton, I.J., Schneider, T.R., Schofield, C.J., and Hajdu, J., X-Ray snapshots of serine protease catalysis reveals a tetrahedral intermediate, *Nature Struct. Biol.* **8,** 689–694 (2001); *and* Wilmouth, R.C., Clifton, I.J., Robinson, C.V., Roach, P.L., Aplin, R.T., Westwood, N.J., Hajdu, J., and Schofield, C.J., Structure of a specific acyl-enzyme complex formed between β-casomorphin-7 and porcine pancreatic elastase, *Nature Struct. Biol.* **4,** 456–461 (1997).

Drug Discovery

Ahluwalia, V.K. and Chopra, M., *Medicinal Chemistry,* CRC Press (2008).

Bannwarth, W. and Hinzen, B. (Eds.), *Combinatorial Chemistry. From Theory to Applications* (2nd ed.), Wiley–VCH (2006).

Brunton, L., Lazo, J., and Parker, K. (Eds.), *Goodman & Gilman's The Pharmacologic Basis of Therapeutics* (11th ed.), McGraw-Hill (2006).

Corey, E.J., Czakó, B., and Kürti, L., *Molecules and Medicine,* Wiley (2007). [Discusses the discovery, application, and mode of action of numerous drug molecules.]

Corson, T.W. and Crews, C.M., Molecular understanding and modern application of traditional medicines: triumphs and trials, *Cell* **130,** 769–774 (2007).

Debouck, C. and Metcalf, B., The impact of genomics on drug discovery, *Annu. Rev. Pharmacol. Toxicol.* **40,** 193–208 (2000).

Furge, L.L. and Guengerich, F.P., Cytochrome P450 enzymes in drug metabolism and chemical toxicology, *Biochem. Mol. Biol. Educ.* **34,** 66–74 (2006).

Ingelman-Sundberg, M., Oscarson, M., and McLellan, R.A., Polymorphic human cytochrome P450 enzymes: An opportunity for individualized drug treatment, *Trends Pharmacol. Sci.* **20,** 342–349 (1999).

Jorgenson, W.L., The many roles of computation in drug discovery, *Science* **303,** 1813–1818 (2004).

Katzung, B.G. (Ed.), *Basic & Clinical Pharmacology* (10th ed.), McGraw-Hill (2007).

Lipinski, C. and Hopkins, A., Navigating chemical space for biology and medicine, *Nature* **432,** 855–861 (2004). [Discusses strategies of discovering new drugs.]

MacCoss, M. and Baillie, T.A., Organic chemistry in drug discovery, *Science* **303,** 1810–1813 (2004).

Ohlstein, E.H., Ruffolo, R.R., Jr., and Elliott, J.D., Drug discovery in the next millennium, *Annu. Rev. Pharmacol. Toxicol.* **40,** 177–191 (2000).

Patrick, G.L, *An Introduction to Medicinal Chemistry* (3rd ed.), Oxford University Press (2005).

Smith, D.A. and van der Waterbeemd, H., Pharmacokinetics and metabolism in early drug design, *Curr. Opin. Chem. Biol.* **3,** 373–378 (1999).

Terrett, N.O., *Combinatorial Chemistry,* Oxford University Press (1998).

White, R.E., High-throughput screening in drug metabolism and pharmacokinetic support of drug discovery, *Annu. Rev. Pharmacol. Toxicol.* **40,** 133–157 (2000).

Williams, P.A., Cosme, J., Ward, A., Angove, H.C., Matak, D., and Jhoti, H., Crystal structure of human cytochrome P450 2C9 with bound warfarin, *Nature* **424,** 464–468 (2003).

Wong, L.-L., Cytochrome P450 monooxygenases, *Curr. Opin. Chem. Biol.* **2,** 263–268 (1998).

Zartler, E.R. and Shapiro, M.J., *Fragment-Based Drug Discovery. A Practical Approach,* Wiley (2008).

HIV-1 Protease and Other Aspartic Proteases

Davies, D.R., The structure and function of the aspartic proteases, *Annu. Rev. Biophys. Biophys. Chem.* **19,** 189–215 (1990).

Erickson, J.W. and Burt, S.K., Structural mechanisms of HIV drug resistance, *Annu. Rev. Pharmacol. Toxicol.* **36,** 545–571 (1996).

Flexner, C., Dual protease inhibitor therapy in HIV-infected patients: Pharmacological rationale and clinical benefits, *Annu. Rev. Pharmacol. Toxicol.* **40,** 649–674 (2000).

Kling, J., Blocking HIV's "scissors," *Modern Drug Discovery* **3**(2), 37–45 (2000).

Meek, T.D., Catalytic mechanisms of the aspartic proteases, *in* Sinnott, M. (Ed.), *Comprehensive Biological Catalysis,* Vol. 1, pp. 327–344, Academic Press (1998).

Richman, D.D., HIV chemotherapy, *Nature* **410,** 995–1001 (2001).

Tomesselli, A.G., Thaisrivongs, S., and Heinrikson, R.L., Discovery and design of HIV protease inhibitors as drugs for treatment of AIDS, *Adv. Antiviral Drug Design* **2,** 173–228 (1996).

Turner, B.G. and Summers, M.F., Structural biology of HIV, *J. Mol. Biol.* **285,** 1–32 (1999). [A review.]

Wilk, T. and Fuller, S.D., Towards the structure of human immunodeficiency virus: Divide and conquer? *Curr. Opin. Struct. Biol.* **9,** 231–243 (1999).

Wlodawer, A., Rational approach to AIDS drug design through structural biology, *Annu. Rev. Med.* **53,** 595–614 (2001); *and* Wlodawer, A. and Vondrasek, J., Inhibitors of HIV-1 protease: A major success of structure-assisted drug design, *Annu. Rev. Biophys. Biomol. Struct.* **27,** 249–284 (1998).

BIOINFORMATICS EXERCISES

Bioinformatics Exercises are available at www.wiley.com/college/voet.

Chapter 12
Enzyme Inhibitors and Rational Drug Design.
1. Dihydrofolate Reductase. Examine the structure of an enzyme with an inhibitor bound to it.
2. HIV Protease. Compare the structures of complexes containing HIV protease and an inhibitor.
3. Pharmacogenomics and Single Nucleotide Polymorphisms. Use online databases to find information on cytochrome P450 polymorphisms.

PROBLEMS

1. Explain why γ-pyridone is not nearly as effective a catalyst for glucose mutarotation as is α-pyridone. What about β-pyridone?

2. RNA is rapidly hydrolyzed in alkaline solution to yield a mixture of nucleotides whose phosphate groups are bonded to either the 2′ or the 3′ positions of the ribose residues. DNA, which lacks RNA's 2′ OH groups, is resistant to alkaline degradation. Explain.

3. Carboxypeptidase A, a Zn^{2+}-containing enzyme, hydrolyzes the C-terminal peptide bonds of polypeptides (Fig 8-19a). In the enzyme–substrate complex, the Zn^{2+} ion is coordinated to three enzyme side chains, the carbonyl oxygen of the scissile peptide bond, and a water molecule. A plausible model for the enzyme's reaction mechanism that is consistent with X-ray and enzymological data is diagrammed in Fig. 15-42. What are the roles of the Zn^{2+} ion and Glu 270 in this mechanism?

4. In the following lactonization reaction,

the relative reaction rate when R = CH_3 is 3.4×10^{11} times that when R = H. Explain.

*5. Derive the analog of Eq. [15.11] for an enzyme that catalyzes the reaction:

$$A + B \rightarrow P$$

Assume the enzyme must bind A before it can bind B:

$$E + A + B \rightleftharpoons EA + B \rightleftharpoons EAB \rightarrow EP$$

6. Explain, in thermodynamic terms, why an "enzyme" that stabilizes its Michaelis complex as much as its transition state does not catalyze a reaction.

7. Suggest a transition state analog for proline racemase that differs from those discussed in the text. Justify your suggestion.

8. Wolfenden has stated that it is meaningless to distinguish between the "binding sites" and the "catalytic sites" of enzymes. Explain.

9. Explain why oxalate ($^-OOCCOO^-$) is an inhibitor of oxaloacetate decarboxylase.

10. In light of the information given in this chapter, why are enzymes such large molecules? Why are active sites almost always

Michaelis complex

attack of water

Tetrahedral intermediate

scissle bond scission

Enzyme-product complex

Figure 15-42 **Mechanism of carboxypeptidase A.**

located in clefts or depressions in enzymes rather than on protrusions?

11. Predict the effects on lysozyme catalysis of changing Phe 34, Ser 36, and Trp 108 to Arg, assuming that this change does not significantly alter the structure of the protein.

*12. The incubation of $(NAG)_4$ with lysozyme results in the slow formation of $(NAG)_6$ and $(NAG)_2$. Propose a mechanism for this reaction. What aspect of the lysozyme mechanism is supported by this reaction?

13. How would the lysozyme binding affinity of the following $\beta(1 \rightarrow 4)$-linked tetrasaccharide compare with that of NAG–NAM–NAG–NAM? Explain.

14. A major difficulty in investigating the properties of the pancreatic serine proteases is that these enzymes, being proteins themselves, are self-digesting. This problem is less severe, however, for solutions of chymotrypsin than it is for solutions of trypsin or elastase. Explain.

15. The comparison of the active site geometries of chymotrypsin and subtilisin under the assumption that their similarities have catalytic significance has led to greater mechanistic understanding of both these enzymes. Discuss the validity of this strategy.

16. Benzamidine ($K_I = 1.8 \times 10^{-5} M$) and **leupeptin** ($K_I = 1.8 \times 10^{-7} M$; Fig. 15-19)

are both specific competitive inhibitors of trypsin. Explain their mechanisms of inhibition. Design leupeptin analogs that inhibit chymotrypsin and elastase.

17. Trigonal boronic acid derivatives have a high tendency to form tetrahedral adducts. **2-Phenylethyl boronic acid**

2–Phenylethyl boronic acid

is an inhibitor of subtilisin and chymotrypsin. Indicate the structure of these enzyme–inhibitor complexes.

18. Tofu (bean curd), a high-protein soybean product that is widely consumed in China and Japan, is prepared in such a way as to remove the trypsin inhibitor present in soybeans. Explain the reason(s) for this treatment.

19. Explain why mutating all three residues of trypsin's catalytic triad has essentially no greater effect on the enzyme's catalytic rate enhancement than mutating only Ser 195.

20. Explain why chymotrypsin is not self-activating as is trypsin.

21. Does Lipinski's "rule of five" predict that a hexapeptide would be a therapeutically effective drug? Explain.

22. The preferred antidote for acetaminophen overdose is *N*-acetylcysteine. Explain why the administration of this substance, which must occur within 8 to 16 hours of the overdose, is an effective treatment.

23. Why would the activation of HIV-1 protease before the virus buds from its host cell be disadvantageous to the virus? Explain.

Schematic diagram of the major pathways of energy metabolism.

PART **IV**

METABOLISM

Introduction to Metabolism

CHAPTER 16

Living organisms are not at equilibrium. Rather, they require a continuous influx of free energy to maintain order in a universe bent on maximizing disorder. **Metabolism** is the overall process through which living systems acquire and utilize the free energy they need to carry out their various functions. *They do so by coupling the exergonic reactions of nutrient oxidation to the endergonic processes required to maintain the living state* such as the performance of mechanical work, the active transport of molecules against concentration gradients, and the biosynthesis of complex molecules. How do living things acquire this necessary free energy? And what is the nature of the energy coupling process?

Phototrophs (plants and certain bacteria; Section 1-1A) acquire free energy from the sun through **photosynthesis,** a process in which light energy powers the endergonic reaction of CO_2 and H_2O to form carbohydrates and O_2 (Chapter 24). **Chemotrophs** obtain their free energy by oxidizing organic compounds (carbohydrates, lipids, proteins) obtained from other organisms, ultimately phototrophs. *This free energy is most often coupled to endergonic reactions through the intermediate synthesis of "high-energy" phosphate compounds such as **adenosine triphosphate** (ATP; Section 16-4). In addition to being completely oxidized, nutrients are broken down in a series of metabolic reactions to common intermediates that are used as precursors in the synthesis of other biological molecules.*

A remarkable property of living systems is that, despite the complexity of their internal processes, they maintain a steady state. This is strikingly demonstrated by the observation that, over a 40-year time span, a normal human adult consumes literally tons of nutrients and imbibes over 20,000 L of water, but does so without significant weight change. This steady state is maintained by a sophisticated set of metabolic regulatory systems. In this introductory chapter to metabolism, we outline the general characteristics of metabolic pathways, study the main types of chemical reactions that comprise these pathways, and consider the experimental techniques that have been most useful in their elucidation. We then discuss the free energy changes associated with reactions of phosphate compounds and oxidation–reduction reactions. Finally we consider the thermodynamic nature of biological processes, that is, what properties of life are responsible for its self-sustaining character.

1 METABOLIC PATHWAYS

Metabolic pathways are series of consecutive enzymatic reactions that produce specific products. Their reactants, intermediates, and products are referred to as **metabolites.** Since an organism utilizes many metabolites, it has many metabolic pathways. Figure 16-1 shows a metabolic map for a typical cell with many of its interconnected pathways. Each reaction on the map is catalyzed by a distinct enzyme, of which there are ~4000 known. At first glance, this network seems hopelessly complex. Yet, by focusing on its major areas in the following chapters, for example, the main pathways of glucose oxidation (the shaded areas of Fig. 16-1), we shall become familiar with its most important avenues and their interrelationships. Maps of metabolic pathways in a more readable form can be found on the Web at http://www.expasy.org/cgi-bin/search-biochem-index, http://www.iubmb-nicholson.org/, and http://www.genome. ad.jp/kegg/metabolism.html.

Figure 16-1 **Map of the major metabolic pathways in a typical cell.** The main pathways of glucose metabolism are shaded. [Designed by Donald Nicholson. Published by BDH Ltd., Poole 2, Dorset, England.]

The reaction pathways that comprise metabolism are often divided into two categories:

1. Catabolism, or degradation, in which nutrients and cell constituents are broken down exergonically to salvage their components and/or to generate free energy.

2. Anabolism, or biosynthesis, in which biomolecules are synthesized from simpler components.

The free energy released by catabolic processes is conserved through the synthesis of ATP from ADP and phosphate or through the reduction of the coenzyme NADP$^+$ to NADPH (Fig. 13-2). ATP and NADPH are the major free energy sources for anabolic pathways (Fig. 16-2).

A striking characteristic of degradative metabolism is that *it converts large numbers of diverse substances (carbohydrates, lipids, and proteins) to common intermediates.* These intermediates are then further metabolized in a central oxidative pathway that terminates in a few end products. Figure 16-3 outlines the breakdown of various foodstuffs, first to their monomeric units, and then to the common intermediate, **acetyl-coenzyme A (acetyl-CoA)** (Fig. 21-2).

Biosynthesis carries out the opposite process. *Relatively few metabolites, mainly pyruvate, acetyl-CoA, and the citric acid cycle intermediates, serve as starting materials for a host of varied biosynthetic products.* In the next several chapters we discuss many degradative and biosynthetic pathways in detail. For now, let us consider some general characteristics of these processes.

Five principal characteristics of metabolic pathways stem from their function of generating products for use by the cell:

1. Metabolic pathways are irreversible. A highly exergonic reaction (having a large negative free energy change) is irreversible; that is, it goes to completion. If such a reaction is part of a multistep pathway, it confers directionality on the pathway; that is, it makes the entire pathway irreversible.

2. Catabolic and anabolic pathways must differ. *If two metabolites are metabolically interconvertible, the pathway*

from the first to the second must differ from the pathway from the second back to the first:

This is because if metabolite 1 is converted to metabolite 2 by an exergonic process, the conversion of metabolite 2 to metabolite 1 requires that free energy be supplied in order to bring this otherwise endergonic process "back up the hill." Consequently, the two pathways must differ in at least

Figure 16-3 Overview of catabolism. Complex metabolites such as carbohydrates, proteins, and lipids are degraded first to their monomeric units, chiefly glucose, amino acids, fatty acids, and glycerol, and then to the common intermediate, acetyl-coenzyme A (acetyl-CoA). The acetyl group is then oxidized to CO_2 via the citric acid cycle with the concomitant reduction of NAD$^+$ and FAD. Reoxidation of these latter coenzymes by O_2 via the electron-transport chain and oxidative phosphorylation yields H_2O and ATP.

Figure 16-2 ATP and NADPH are the sources of free energy for biosynthetic reactions. They are generated through the degradation of complex metabolites.

one of their reaction steps. *The existence of independent interconversion routes, as we shall see, is an important property of metabolic pathways because it allows independent control of the two processes.* If metabolite 2 is required by the cell, it is necessary to "turn off" the pathway from 2 to 1 while "turning on" the pathway from 1 to 2. Such independent control would be impossible without different pathways.

3. Every metabolic pathway has a first committed step. Although metabolic pathways are irreversible, most of their component reactions function close to equilibrium. Early in each pathway, however, there is an irreversible (exergonic) reaction that "commits" the intermediate it produces to continue down the pathway.

4. All metabolic pathways are regulated. Metabolic pathways are regulated by laws of supply and demand. In order to exert control on the flux of metabolites through a metabolic pathway, it is necessary to regulate its rate-limiting step. The first committed step, being irreversible, functions too slowly to permit its substrates and products to equilibrate (if the reaction were at equilibrium, it would not be irreversible). Since most of the other reactions in a pathway function close to equilibrium, the first committed step is often one of its rate-limiting steps. Most metabolic pathways are therefore controlled by regulating the enzymes that catalyze their first committed step(s). This is an efficient way to exert control because it prevents the unnecessary synthesis of metabolites further along the pathway when they are not required. Specific aspects of such flux control are discussed in Section 17-4C.

5. Metabolic pathways in eukaryotic cells occur in specific cellular locations. The compartmentation of the eukaryotic cell allows different metabolic pathways to operate in different locations, as is listed in Table 16-1 (these organelles are described in Section 1-2A). For example, ATP is mainly generated in the mitochondrion but much of it is utilized in the cytoplasm. The synthesis of metabolites in specific membrane-bounded subcellular compartments makes their transport between these compartments a vital component of eukaryotic metabolism. Biological membranes are selectively permeable to metabolites because of

the presence in membranes of specific transport proteins. The transport protein that facilitates the passage of ATP through the mitochondrial membrane is discussed in Section 20-4C, along with the characteristics of membrane transport processes in general. The synthesis and utilization of acetyl-CoA are also compartmentalized. This metabolic intermediate is utilized in the cytosolic synthesis of fatty acids but is synthesized in mitochondria. Yet there is no transport protein for acetyl-CoA in the mitochondrial membrane. How cells solve this fundamental problem is discussed in Section 25-4D. In multicellular organisms, compartmentation is carried a step further to the level of tissues and organs. The mammalian liver, for example, is largely responsible for the synthesis of glucose from noncarbohydrate precursors (**gluconeogenesis;** Section 23-1) so as to maintain a relatively constant level of glucose in the circulation, whereas adipose tissue is specialized for the storage and mobilization of triacylglycerols. The metabolic interdependence of the various organs is the subject of Chapter 27.

2 ORGANIC REACTION MECHANISMS

Almost all of the reactions that occur in metabolic pathways are enzymatically catalyzed organic reactions. Section 15-1 details the various mechanisms enzymes have at their disposal for catalyzing reactions: acid–base catalysis, covalent catalysis, metal ion catalysis, electrostatic catalysis, proximity and orientation effects, and transition state binding. Few enzymes alter the chemical mechanisms of these reactions, so *much can be learned about enzymatic mechanisms from the study of nonenzymatic model reactions.* We therefore begin our study of metabolic reactions by outlining the types of reactions we shall encounter and the mechanisms by which they have been observed to proceed in nonenzymatic systems.

Christopher Walsh has classified biochemical reactions into four categories: (1) **group-transfer reactions;** (2) **oxidations and reductions;** (3) **eliminations, isomerizations, and rearrangements;** and (4) **reactions that make or break carbon–carbon bonds.** Much is known about the mechanisms of

Table 16-1   Metabolic Functions of Eukaryotic Organelles

Organelle	Function
Mitochondrion	Citric acid cycle, electron transport and oxidative phosphorylation, fatty acid oxidation, amino acid breakdown
Cytosol	Glycolysis, pentose phosphate pathway, fatty acid biosynthesis, many reactions of gluconeogenesis
Lysosomes	Enzymatic digestion of cell components and ingested matter
Nucleus	DNA replication and transcription, RNA processing
Golgi apparatus	Post-translational processing of membrane and secretory proteins; formation of plasma membrane and secretory vesicles
Rough endoplasmic reticulum	Synthesis of membrane-bound and secretory proteins
Smooth endoplasmic reticulum	Lipid and steroid biosynthesis
Peroxisomes (glyoxisomes in plants)	Oxidative reactions catalyzed by amino acid oxidases and catalase; glyoxylate cycle reactions in plants

these reactions and about the enzymes that catalyze them. The discussions in the next several chapters focus on these mechanisms as they apply to specific metabolic interconversions. In this section we outline the four reaction categories and discuss how our knowledge of their reaction mechanisms derives from the study of model organic reactions. We begin by briefly reviewing the chemical logic used in analyzing these reactions.

A. *Chemical Logic*

A covalent bond consists of an electron pair shared between two atoms. In breaking such a bond, the electron pair can either remain with one of the atoms (**heterolytic bond cleavage**) or separate such that one electron accompanies each of the atoms (**homolytic bond cleavage**) (Fig. 16-4). Homolytic bond cleavage, which usually produces unstable radicals, occurs mostly in oxidation–reduction reactions. Heterolytic C—H bond cleavage involves either carbanion and proton (H^+) formation or carbocation (carbonium ion) and hydride ion (H^-) formation. Since hydride ions are highly reactive species and carbon atoms are slightly more electronegative than hydrogen atoms, bond cleavage in which the electron pair remains with the carbon atom is the predominant mode of C—H bond breaking in biochemical systems. Hydride ion abstraction occurs only if the hydride is transferred directly to an acceptor such as NAD^+ or $NADP^+$.

Compounds participating in reactions involving heterolytic bond cleavage and bond formation are categorized into two broad classes: electron rich and electron deficient. Electron-rich compounds, which are called **nucleophiles** (nucleus lovers), are negatively charged or contain unshared electron pairs that easily form covalent bonds with electron-deficient centers. Biologically important nucleophilic groups include amino, hydroxyl, imidazole, and sulfhydryl functions (Fig. 16-5*a*). The nucleophilic forms of these groups are also their basic forms. Indeed, nucleophilicity and basicity are closely related properties (Sec-

Homolytic:

Radicals

Heterolytic:

Carbanion   Proton

Carbocation   Hydride ion

Figure 16-4   Modes of C—H bond breaking. Homolytic cleavage yields radicals, whereas heterolytic cleavage yields either (*i*) a carbanion and a proton or (*ii*) a carbocation and a hydride ion.

tion 15-1Ba): A compound acts as a base when it forms a covalent bond with H^+, whereas it acts as a nucleophile when it forms a covalent bond with an electron-deficient center other than H^+, usually an electron-deficient carbon atom:

Electron-deficient compounds are called **electrophiles** (electron lovers). They may be positively charged, contain an unfilled valence electron shell, or contain an electronegative atom. The most common electrophiles in biochemical

(*a*) **Nucleophiles**

(*b*) **Electrophiles**

Figure 16-5   Biologically important nucleophilic and electrophilic groups. (*a*) Nucleophiles are the conjugate bases of weak acids such as the hydroxyl, sulfhydryl, amino, and imidazole groups. (*b*) Electrophiles contain an electron-deficient atom (*red*).

systems are H^+, metal ions, the carbon atoms of carbonyl groups, and cationic imines (Fig. 16-5b).

Reactions are best understood if the electron pair rearrangements involved in going from reactants to products can be traced. In illustrating these rearrangements we shall use the **curved arrow convention** in which the movement of an electron pair is symbolized by a curved arrow emanating from the electron pair and pointing to the electron-deficient center attracting the electron pair. For example, imine formation, a biochemically important reaction between an amine and an aldehyde or ketone, is represented:

In the first reaction step, the amine's unshared electron pair adds to the electron-deficient carbonyl carbon atom while one electron pair from its C=O double bond transfers to the oxygen atom. In the second step, the unshared electron pair on the nitrogen atom adds to the electron-deficient carbon atom with the elimination of water. *At all times, the rules of chemical reason prevail:* For example, there are never five bonds to a carbon atom or two bonds to a hydrogen atom.

B. *Group-Transfer Reactions*

The group transfers that occur in biochemical systems involve the transfer of an electrophilic group from one nucleophile to another:

They could equally well be called nucleophilic substitution reactions. The most commonly transferred groups in biochemical reactions are acyl groups, phosphoryl groups, and glycosyl groups (Fig. 16-6):

Figure 16-6 Types of metabolic group-transfer reactions.
(a) Acyl group transfer involves addition of a nucleophile (Y) to the electrophilic carbon atom of an acyl compound to form a tetrahedral intermediate. The original acyl carrier (X) is then expelled to form a new acyl compound. (b) Phosphoryl group transfer involves the in-line (with the leaving group) addition of a nucleophile (Y) to the electrophilic phosphorus atom of a tetrahedral phosphoryl group. This yields a trigonal bipyramidal intermediate whose apical positions are occupied by the leaving group (X) and the attacking group (Y). Elimination of the leaving group to complete the transfer reaction results in the phosphoryl group's inversion of configuration. (c) Glycosyl group transfer involves the substitution of one nucleophilic group for another at C1 of a sugar ring. This reaction usually occurs via a double displacement mechanism in which the elimination of the original glycosyl carrier (X) is accompanied by the intermediate formation of a resonance-stabilized carbocation (oxonuim ion) followed by the addition of the adding nucleophile (Y). The reaction also may occur via a single displacement mechanism in which Y directly displaces X with inversion of configuration.

1. Acyl group transfer from one nucleophile to another almost invariably involves the addition of a nucleophile to the acyl carbonyl carbon atom so as to form a tetrahedral intermediate (Fig. 16-6*a*). Peptide bond hydrolysis, as catalyzed, for example, by chymotrypsin (Section 15-3C), is a familiar example of such a reaction.

2. Phosphoryl group transfer proceeds via the in-line addition of a nucleophile to a phosphoryl phosphorus atom to yield a trigonal bipyramidal intermediate whose apexes are occupied by the adding and leaving groups (Fig. 16-6*b*). The overall reaction results in the tetrahedral phosphoryl group's inversion of configuration. Indeed, chiral phosphoryl compounds have been shown to undergo just such an inversion. For example, Jeremy Knowles has synthesized ATP made chiral at its γ-phosphoryl group by isotopic substitution and demonstrated that this group is inverted on its transfer to glucose in the reaction catalyzed by **hexokinase** (Fig. 16-7).

3. Glycosyl group transfer involves the substitution of one nucleophilic group for another at C1 of a sugar ring (Fig. 16-6*c*). This is the central carbon atom of an acetal. Chemical models of acetal reactions generally proceed via acid-catalyzed cleavage of the first bond to form a

resonance-stabilized carbocation at C1 (an oxonium ion). The lysozyme-catalyzed hydrolysis of bacterial cell wall polysaccharides (Section 15-2Bb) is such a reaction.

C. *Oxidations and Reductions*

Oxidation–reduction (redox) reactions involve the loss or gain of electrons. The thermodynamics of these reactions is discussed in Section 16-5. Many of the redox reactions that occur in metabolic pathways involve C—H bond cleavage with the ultimate loss of two bonding electrons by the carbon atom. These electrons are transferred to an electron acceptor such as NAD^+ (Fig. 13-2). Whether these reactions involve homolytic or heterolytic bond cleavage has not always been rigorously established. In most instances heterolytic cleavage is assumed when radical species are not observed. It is useful, however, to visualize redox C—H bond cleavage reactions as hydride transfers as diagrammed below for the oxidation of an alcohol by NAD^+:

Figure 16-7   The phosphoryl-transfer reaction catalyzed by hexokinase. During its transfer to the 6-OH group of glucose, the γ-phosphoryl group of ATP made chiral by isotopic substitution undergoes inversion of configuration via a trigonal bipyramidal intermediate.

For aerobic organisms, the terminal acceptor for the electron pairs removed from metabolites by their oxidation is molecular oxygen (O_2). Recall that this molecule is a ground state diradical species whose unpaired electrons have parallel spins. The rules of electron pairing (the Pauli exclusion principle) therefore dictate that O_2 can only accept unpaired electrons; that is, electrons must be transferred to O_2 one at a time (in contrast to redox processes in which electrons are transferred in pairs). Electrons that are removed from metabolites as pairs must therefore be passed to O_2 via the electron-transport chain one at a time. This is accomplished through the use of conjugated coenzymes that have stable radical oxidation states and can therefore undergo both $1e^-$ and $2e^-$ redox reactions. One such coenzyme is **flavin adenine dinucleotide (FAD;**

Figure 16-8 The molecular formula and reactions of the coenzyme flavin adenine dinucleotide (FAD). The term "flavin" is synonymous with the isoalloxazine system. The D-ribitol residue is derived from the alcohol of the sugar D-ribose. The FAD may be half-reduced to the stable radical FADH· or fully reduced to $FADH_2$ (*boxes*). Consequently, different FAD-containing enzymes cycle between different oxidation states of FAD. FAD is usually tightly bound to its enzymes, so that this coenzyme is normally a prosthetic group rather than a cosubstrate as is, for example, NAD^+. Consequently, although humans and other higher animals are unable to synthesize the isoalloxazine component of flavins and hence must obtain it in their diets [for example, in the form of **riboflavin (vitamin B_2)**], riboflavin deficiency is quite rare in humans. The symptoms of riboflavin deficiency, which are associated with general malnutrition or bizarre diets, include an inflamed tongue, lesions in the corners of the mouth, and dermatitis.

Fig. 16-8). **Flavins** (substances that contain the **isoalloxazine** ring) can undergo two sequential one-electron transfers or a simultaneous two-electron transfer that bypasses the semiquinone state.

D. *Eliminations, Isomerizations, and Rearrangements*

a. Elimination Reactions Form Carbon–Carbon Double Bonds

Elimination reactions result in the formation of a double bond between two previously single-bonded saturated centers. The substances eliminated may be H_2O, NH_3, an alcohol (ROH), or a primary amine (RNH_2). The dehydration of an alcohol, for example, is an elimination reaction:

Bond breaking and bond making in this reaction may proceed via one of three mechanisms (Fig. 16-9a): (1) concerted; (2) stepwise with the C—O bond breaking first to form a carbocation; or (3) stepwise with the C—H bond breaking first to form a carbanion.

Enzymes catalyze dehydration reactions by either of two simple mechanisms: (1) protonation of the OH group by an acidic group (acid catalysis) or (2) abstraction of the proton by a basic group (base catalysis). Moreover, in a stepwise reaction, the charged intermediate may be stabilized

(a)
Concerted

R—C—C—R' → C=C + H⁺ + OH⁻

Stepwise via a carbocation

Stepwise via a carbanion

(b)

→ trans (anti)

→ cis (syn)

Figure 16-9   Possible elimination reaction mechanisms using dehydration as an example. Reactions may be (a) either concerted, stepwise via a carbocation intermediate, or stepwise via a carbanion intermediate; and may occur with (b) either trans (anti) or cis (syn) stereochemistry.

by an oppositely charged active site group (electrostatic catalysis). The glycolytic enzyme **enolase** (Section 17-2I) and the citric acid cycle enzyme **fumarase** (Section 21-3G) catalyze such dehydration reactions.

Elimination reactions may take one of two possible stereochemical courses (Fig. 16-9b): (1) trans (anti) eliminations, the most prevalent biochemical mechanism, and (2) cis (syn) eliminations, which are biochemically less common.

b. Biochemical Isomerizations Involve Intramolecular Hydrogen Atom Shifts

Biochemical **isomerization reactions** involve the intramolecular shift of a hydrogen atom so as to change the location of a double bond. In such a process, a proton is removed from one carbon atom and added to another. The metabolically most prevalent isomerization reaction is the **aldose–ketose interconversion,** a base-catalyzed reaction that occurs via **enediolate anion** intermediates (Fig. 16-10). The glycolytic enzyme **phosphoglucose isomerase** catalyzes such a reaction (Section 17-2B).

Racemization is an isomerization reaction in which a hydrogen atom shifts its stereochemical position at a molecule's

Figure 16-10   Mechanism of aldose–ketose isomerization. The reaction occurs with acid–base catalysis and proceeds via *cis*-enediolate intermediates.

only chiral center so as to invert that chiral center (e.g., the racemization of proline by proline racemase; Section 15-1Fa). Such an isomerization is called an **epimerization** in a molecule with more than one chiral center.

c. Rearrangements Produce Altered Carbon Skeletons

Rearrangement reactions break and reform C—C bonds so as to rearrange a molecule's carbon skeleton. There are few such metabolic reactions. One is the conversion of **L-methylmalonyl-CoA** to **succinyl-CoA** by **methylmalonyl-CoA mutase,** an enzyme whose prosthetic group is a **vitamin B₁₂** derivative:

L-Methylmalonyl-CoA →(methylmalonyl-CoA mutase)→ Succinyl-CoA

This reaction is involved in the oxidation of fatty acids with an odd number of carbon atoms (Section 25-2Ec) and several amino acids (Section 26-3Ec).

E. Reactions That Make and Break Carbon–Carbon Bonds

Reactions that make and break carbon–carbon bonds form the basis of both degradative and biosynthetic metabolism. The breakdown of glucose to CO_2 involves five such cleavages, whereas its synthesis involves the reverse process. Such reactions, considered from the synthetic direction, involve addition of a nucleophilic carbanion to an electrophilic carbon atom. The most common electrophilic carbon atoms in such reactions are the sp^2-hybridized carbonyl carbon atoms of aldehydes, ketones, esters, and CO_2:

$$-\overset{|}{\underset{|}{C}}{:}^- \;+\; \overset{\diagdown}{\diagup}C{=}O \longrightarrow -\overset{|}{\underset{|}{C}}-\overset{|}{\underset{|}{C}}-OH$$

Stabilized carbanions must be generated to add to these electrophilic centers. Three examples are the **aldol condensation** (catalyzed, e.g., by **aldolase;** Section 17-2D), **Claisen ester condensation (citrate synthase;** Section 21-3A), and the decarboxylation of β-keto acids (**isocitrate dehy-**

drogenase, Section 21-3C; and **fatty acid synthase,** Section 25-4C). In nonenzymatic systems, both the aldol condensation and Claisen ester condensation involve the base-catalyzed generation of a carbanion α to a carbonyl group (Fig. 16-11a,b). The carbonyl group is electron withdrawing and thereby provides resonance stabilization by forming an enolate (Fig. 16-12a). The enolate may be further stabilized by neutralizing its negative charge. Enzymes do so through hydrogen bonding or protonation (Fig. 16-12b), conversion of the carbonyl group to a protonated Schiff base (covalent catalysis; Fig. 16-12c), or by its coordination

Figure 16-11   Examples of C—C bond formation and cleavage reactions. (*a*) Aldol condensation, (*b*) Claisen ester condensation, and (*c*) decarboxylation of a β-keto acid. All three types of reactions involve generation of a resonance-stabilized carbanion followed by addition of this carbanion to an electrophilic center.

(a)

Carbanion   **Enolate**

(b)

Hydrogen-bonded   **Hydrogen-bonded**
carbonyl   **enolate or enol**

(c)

Schiff base   **Schiff base**
carbanion (imine)   **(enamine)**

(d)

Carbanion   **Zn^{2+}–stabilized**
                  enolate

Figure 16-12 **Stabilization of carbanions.** (*a*) Carbanions
adjacent to carbonyl groups are stabilized by the formation of
enolates. (*b*) Carbanions adjacent to carbonyl groups hydrogen
bonded to general acids are stabilized electrostatically or by
charge neutralization. (*c*) Carbanions adjacent to protonated
imines (Schiff bases) are stabilized by the formation of enamines.
(*d*) Metal ions stabilize carbanions adjacent to carbonyl groups
by the electrostatic stabilization of the enolate.

to a metal ion (metal ion catalysis; Fig. 16-12*d*). The decarboxylation of a β-keto acid does not require base catalysis
for the generation of the resonance-stabilized carbanion;
the highly exergonic formation of CO_2 provides its driving
force (Fig. 16-11*c*).

3 EXPERIMENTAL APPROACHES
TO THE STUDY OF METABOLISM

A metabolic pathway can be understood at several levels:

1. In terms of the sequence of reactions by which a specific nutrient is converted to end products, and the energetics of these conversions.

2. In terms of the mechanisms by which each intermediate is converted to its successor. Such an analysis requires
the isolation and characterization of the specific enzymes
that catalyze each reaction.

3. In terms of the control mechanisms that regulate the
flow of metabolites through the pathway. An exquisitely

complex network of regulatory processes renders metabolic pathways remarkably sensitive to the needs of the organism; the output of a pathway is generally only as great
as required.

As you might well imagine, the elucidation of a metabolic pathway on all of these levels is a complex process, involving contributions from a variety of disciplines. Most of
the techniques used to do so involve somehow perturbing
the system and observing the perturbation's effect on
growth or on the production of metabolic intermediates.
One such technique is the use of metabolic inhibitors that
block metabolic pathways at specific enzymatic steps.
Another is the study of genetic abnormalities that interrupt
specific metabolic pathways. Techniques have also been developed for the dissection of organisms into their component organs, tissues, cells, and subcellular organelles, and for
the purification and identification of metabolites as well as
the enzymes that catalyze their interconversions. The use of
isotopic tracers to follow the paths of specific atoms and
molecules through the metabolic maze has become routine.
Techniques utilizing NMR technology are able to trace
metabolites noninvasively as they react *in vivo*. This section
outlines the use of these various techniques.

A. *Metabolic Inhibitors, Growth Studies, and Biochemical Genetics*

a. Pathway Intermediates Accumulate in the Presence of Metabolic Inhibitors

The first metabolic pathway to be completely traced was
the conversion of glucose to ethanol in yeast by a process
known as **glycolysis** (Section 17-1A). In the course of these
studies, certain substances, called **metabolic inhibitors,** were
found to block the pathway at specific points, thereby causing preceding intermediates to build up. For instance,
iodoacetate causes yeast extracts to accumulate fructose-
1,6-bisphosphate, whereas fluoride causes the buildup of
two phosphate esters, 3-phosphoglycerate and 2-phospho-
glycerate. The isolation and characterization of these intermediates was vital to the elucidation of the glycolytic pathway: Chemical intuition combined with this information led
to the prediction of the pathway's intervening steps. Each
of the proposed reactions was eventually shown to occur
in vitro as catalyzed by a purified enzyme.

b. Genetic Defects Also Cause Metabolic Intermediates to Accumulate

Archibald Garrod's realization, in the early 1900s, that
human genetic diseases are the consequence of deficiencies in specific enzymes (Section 1-4Cd) also contributed to
the elucidation of metabolic pathways. For example, on the
ingestion of either phenylalanine or tyrosine, individuals with
the largely harmless inherited condition known as **alcap-
tonuria,** but not normal subjects, excrete **homogentisic acid**
in their urine (Section 26-3Hd). This is because the liver of
alcaptonurics lacks an enzyme that catalyzes the breakdown
of homogentisic acid. Another genetic disease, **phenylketonuria** (Section 26-3Hd), results in the accumulation of

phenylpyruvate in the urine (and which, if untreated, causes severe mental retardation in infants). Ingested phenylalanine and phenylpyruvate appear as phenylpyruvate in the urine of affected subjects, whereas tyrosine is metabolized normally. The effects of these two abnormalities suggested the pathway for phenylalanine metabolism diagrammed in Fig. 16-13. However, the supposition that phenylpyruvate but not tyrosine occurs on the normal pathway of phenylalanine metabolism because phenylpyruvate accumulates in the urine of phenylketonurics has proved incorrect. This indicates the pitfalls of relying solely on metabolic blocks and the consequent buildup of intermediates as indicators of a metabolic pathway. In this case, phenylpyruvate formation was later shown to arise from a normally minor pathway that becomes significant only when the phenylalanine concentration is abnormally high, as it is in phenylketonurics.

c. Metabolic Blocks Can Be Generated by Genetic Manipulation

Early metabolic studies led to the astounding discovery that *the basic metabolic pathways in most organisms are essentially identical.* This metabolic uniformity has greatly facilitated the study of metabolic reactions. A mutation that inactivates or deletes an enzyme in a pathway of interest can be readily generated in rapidly reproducing microorganisms through the use of **mutagens** (chemical agents that induce genetic changes; Section 32-1A), X-rays, or genetic engineering techniques (Section 5-5). Desired mutants are identified by their requirement of the pathway's end product for growth. For example, George Beadle and Edward Tatum proposed a pathway of arginine biosynthesis in the mold *Neurospora crassa* based on their analysis of three arginine-requiring **auxotrophic mutants** (mutants requiring a specific nutrient for growth), which were isolated after X-irradiation (Fig. 16-14). This landmark study also conclusively demonstrated that enzymes are specified by genes (Section 1-4Cd).

d. Genetic Manipulations of Higher Organisms Provide Metabolic Insights

Transgenic organisms (Section 5-5H) constitute valuable resources for the study of metabolism. *They can be used to both create metabolic blocks and to express genes in tissues where they are not normally present.* For example, **creatine kinase** catalyzes the formation of **phosphocreatine** (Section 16-4Cd), a substance that functions to generate ATP rapidly when it is in short supply. This enzyme is normally present in many tissues, including brain and muscle, but not in liver. The introduction of the gene encoding creatine kinase into the liver of a mouse causes the liver to synthesize phosphocreatine when the mouse is fed creatine, as demonstrated by localized *in vivo* NMR techniques (Fig. 16-15; NMR is discussed below). The presence of

Figure 16-13 Pathway for phenylalanine degradation. It was originally hypothesized that phenylpyruvate was a pathway intermediate based on the observation that phenylketonurics excrete ingested phenylalanine and phenylpyruvate as phenylpyruvate. Further studies, however, demonstrated that phenylpyruvate is not a homogentisate precursor; rather, phenylpyruvate production is significant only when the phenylalanine concentration is abnormally high. Instead, tyrosine is the normal product of phenylalanine degradation.

Figure 16-14 Pathway of arginine biosynthesis indicating the positions of genetic blocks. All of these mutants grow in the presence of arginine, but mutant 1 also grows in the presence of the (nonstandard) α-amino acids **citrulline** or **ornithine** and mutant 2 grows in the presence of citrulline. This is because in mutant 1, an enzyme leading to the production of ornithine is absent but enzymes farther along the pathway are normal. In mutant 2, the enzyme catalyzing citrulline production is defective, whereas in mutant 3 an enzyme involved in the conversion of citrulline to arginine is lacking.

phosphocreatine in a transgenic mouse liver protects the animal against the sharp drop in [ATP] ordinarily caused by fructose overload (Section 17-5Aa). This genetic manipulation technique is being used to study mechanisms of metabolic control *in vivo.*

Metabolic pathways are regulated both by controlling the activities of regulatory enzymes (Sections 17-4 and 18-3) and by controlling their concentrations at the level of gene expression (Sections 31-3, 32-4, and 34-3). The important question of how hormones and diet control metabolic processes at the level of gene expression is being addressed through the use of transgenic animals. **Reporter genes** (genes whose products are easily detected; Section 5-5Gd) are placed under the influence of **promoters** (genetic elements that regulate transcriptional initiation; Section 5-4Aa) that control the expression of specific regulatory enzymes, and the resulting composite gene is expressed in animals. The transgenic animals can then be treated with specific hormones and/or diets and the production of the reporter gene product measured. For instance, in an investigation by Richard Hanson, the promoter for the enzyme **phosphoenolpyruvate carboxykinase (PEPCK)** was attached to the structural gene encoding **growth hormone (GH)**. PEPCK, an important regulatory enzyme in **gluconeogenesis** (the synthesis of glucose from noncarbohydrate precursors; Section 23-1), is normally present in liver and kidneys but not in blood. GH, however, is secreted into the blood and its presence there can be readily quantitated by an ELISA (Section 6-1Da). Mice transgenic for PEPCK/GH were fed either a high-carbohydrate/low-protein diet or a high-protein/low-carbohydrate diet, which are known to decrease and increase PEPCK activity, respectively. GH in high concentrations was detected only in the serum of PEPCK/GH mice on a high-protein diet, thereby indicating that the GH was synthesized under the same dietary control as that of the PEPCK expressed by the normal gene. Thus, the activity of PEPCK in PEPCK/GH mice can be continuously monitored, albeit indirectly, through serum GH assays (the direct measurement of PEPCK in mouse liver or kidney requires the sacrifice of the animal and hence can be done only once). Such use of reporter genes has proved to be of great value in the study of the genetic control of metabolism.

Figure 16-15 The expression of creatine kinase in transgenic mouse liver as demonstrated by localized *in vivo* ^{31}P NMR. (*a*) The spectrum of a normal mouse liver after the mouse had been fed a diet supplemented with 2% creatine. The peaks corresponding to inorganic phosphate (P_i), the α, β, and γ phosphoryl groups of ATP, and phosphomonoesters (PME) are labeled. (*b*) The spectrum of the liver of a mouse transgenic for creatine kinase that had been fed a diet supplemented with 2% creatine. The phosphocreatine peak is labeled PCr. [After Koretsky, A.P., Brosnan, M.J., Chen, L., Chen, J., and Van Dyke, T.A., *Proc. Natl. Acad. Sci.* **87,** 3114 (1990)].

Modern techniques also make it possible to insert a mutation that inactivates or deletes an enzyme or control protein in a pathway of interest in higher organisms such as mice (**knockout mice**; Section 5-5H). Knockout mice have proved useful for studying metabolic control mechanisms. For example, PEPCK activity is thought to be controlled exclusively by increasing or decreasing its availability. Diet affects its production, as we have seen. However, this demand-based control is superimposed on the developmental regulation of PEPCK production. The enzyme is not produced at all in early embryos and only appears near birth, when gluconeogenesis is required to supply the glucose that had been previously available *in utero.* One of the proteins thought to be responsible for the developmental regulation of PEPCK production is **CCAAT/enhancer-binding protein α (C/EBPα), a transcription factor** (Section 5-4Aa; transcriptional regulation in eukaryotes is discussed in Section 34-3B). Newborn mice homozygous for the targeted deletion of the *c/ebpα* gene (*c/ebpα* knockout mice) do not produce C/EBPα and therefore do not produce PEPCK. Consequently, their livers cannot synthesize the glucose necessary to maintain adequate blood glucose levels once they are disconnected from the maternal circulation. Indeed, these mice become so hypoglycemic that they die within 8 hours of birth. Clearly C/EBPα has an important role in the developmental regulation of PEPCK.

B. *Isotopes in Biochemistry*

The specific labeling of metabolites such that their interconversions can be traced is an indispensable technique for elucidating metabolic pathways. Franz Knoop formulated this technique in 1904 to study fatty acid oxidation. He fed dogs fatty acids chemically labeled with phenyl groups and isolated the phenyl-substituted end products from their urine. From the differences in these products when the phenyl-substituted starting material contained odd and even numbers of carbon atoms he deduced that fatty acids are degraded in C_2 units (Section 25-2).

a. Isotopes Specifically Label Molecules without Altering Their Chemical Properties

Chemical labeling has the disadvantage that the chemical properties of labeled metabolites differ from those of normal metabolites. This problem is eliminated by labeling molecules of interest with **isotopes** (atoms with the same number of protons but a different number of neutrons in their nuclei). Recall that the chemical properties of an element are a consequence of its electron configuration which, in turn, is determined by its atomic number, not its atomic mass. The metabolic fate of a specific atom in a metabolite can therefore be elucidated by isotopically labeling that position and following its progress through the metabolic pathway of interest. The advent of isotopic labeling and tracing techniques in the 1940s therefore revolutionized the study of metabolism. (**Isotope effects,** which are changes in reaction rates arising from the mass differences between isotopes, are in most instances negligible. Where they are significant, most noticeably between hydrogen and its isotopes deuterium and tritium, they have been used to gain insight into enzymatic reaction mechanisms.)

b. NMR Can Be Used to Study Metabolism in Whole Animals

Nuclear magnetic resonance (NMR) detects specific isotopes due to their characteristic nuclear spins. Among the isotopes that NMR can detect are ^{1}H, ^{13}C, and ^{31}P. Since the NMR spectrum of a particular nucleus varies with its immediate environment, it is possible to identify the peaks corresponding to specific atoms even in relatively complex mixtures.

The development of magnets large enough to accommodate animals and humans, and to localize spectra to specific organs, has made it possible to study metabolic pathways noninvasively by NMR techniques. Thus, ^{31}P NMR can be used to study energy metabolism in muscle by monitoring the levels of ATP, ADP, inorganic phosphate, and phosphocreatine (Figure 16-15). Indeed, a ^{31}P NMR system has been patented to measure the muscular metabolic efficiency and maximum power of race horses while they are walking or running on a motor-driven treadmill in order to identify promising animals and to evaluate the efficacy of their training and nutritional programs.

Isotopically labeling specific atoms of metabolites with ^{13}C (which is only 1.10% naturally abundant) permits the metabolic progress of the labeled atoms to be followed by ^{13}C NMR. Figure 16-16 shows *in vivo* ^{13}C NMR spectra of a rat liver before and after an injection of D-[1-^{13}C]glucose. The ^{13}C can be seen entering the liver and then being converted to glycogen (the storage form of glucose; Chapter 18). ^{1}H NMR techniques are being used to determine the *in vivo* levels of a variety of metabolites in tissues such as brain and muscle.

c. The Detection of Radioactive Isotopes

All elements have isotopes. For example, the atomic mass of naturally occurring Cl is 35.45 D because, at least on Earth, it is a mixture of 55% ^{35}Cl and 45% ^{36}Cl (other isotopes of Cl are present in only trace amounts). Stable isotopes are generally identified and quantitated by mass spectrometry or NMR techniques. Many isotopes, however, are unstable; they undergo **radioactive decay,** a process that involves the emission from the radioactive nuclei of subatomic particles such as helium nuclei (**α particles),** electrons (**β particles),** and/or photons (**γ radiation).** Radioactive nuclei emit radiation with characteristic energies. For example, ^{3}H, ^{14}C, and ^{32}P all emit β particles but with respective energies of 0.018, 0.155, and 1.71 MeV. The radiation from ^{32}P is therefore highly penetrating, whereas that from ^{3}H and ^{14}C is not. (^{3}H and ^{14}C, as all radioactive isotopes, must, nevertheless, be handled with great caution because they can cause genetic damage on ingestion.)

Radiation can be detected by a variety of techniques. Those most commonly used in biochemical investigations are **proportional counting** (known in its simplest form as **Geiger counting), liquid scintillation counting,** and **autoradiography.** Proportional counters electronically detect the ionizations in a gas caused by the passage of radiation. Moreover, they can also discriminate between particles of different energies and thus simultaneously determine the amounts of two or more different isotopes present.

Although proportional counters are quite simple to use, the radiation from two of the most widely used isotopes in

Figure 16-16 **The conversion of [1-^{13}C]glucose to glycogen as observed by localized** *in vivo* **^{13}C NMR.** (*a*) The natural abundance ^{13}C NMR spectrum of the liver of a live rat. Note the resonance corresponding to C1 of glycogen. (*b*) The ^{13}C NMR spectrum of the liver of the same rat ~5 min after it was intravenously injected with 100 mg of [1-^{13}C]glucose (90% enriched). The resonances of the C1 atom of both the α and β anomers of glucose are clearly distinguishable from each other and from the resonance of the C1 atom of glycogen. (*c*) The ^{13}C NMR spectrum of the liver of the same rat ~30 min after the [1-^{13}C]glucose injection. The C1 resonances of both the α and β glucose anomers are much reduced while the C1 resonance of glycogen has increased. [After Reo, N.V., Siegfried, B.A., and Acherman, J.J.H., *J. Biol. Chem.* **259**, 13665 (1984)].

biochemical analysis, ^{3}H and ^{14}C, have insufficient penetrating power to enter a proportional counter's detection chamber with reasonable efficiency. This limitation is circumvented through liquid scintillation counting. In this technique, a radioactive sample is dissolved or suspended in a solution containing fluorescent substances that emit a pulse of light when struck by radiation. The light is detected electronically so that the number of light pulses can be counted. The emitting nucleus can also be identified because the intensity of a light pulse is proportional to the radiation energy (the number of fluorescent molecules excited by a radioactive particle is proportional to the particle's energy).

In autoradiography, radiation is detected by its blackening of photographic film. The radioactive sample is laid on, or in some cases mixed with, the photographic emulsion and, after sufficient exposure time (from minutes to months), the film is developed. Autoradiography is widely used to locate radioactive substances in polyacrylamide gels (e.g., Fig. 6-27). Position-sensitive radiation counters (electronic film) are similarly employed.

d. Radioactive Isotopes Have Characteristic Half-Lives

Radioactive decay is a random process whose rate for a given isotope depends only on the number of radioactive atoms present. It is therefore a simple first-order process whose half-life, $t_{1/2}$, is a function only of the rate constant, k, for the decay process (Section 14-1Ba):

$$t_{1/2} = \frac{\ln 2}{k} = \frac{0.693}{k} \qquad [14.5]$$

Because k is different for each radioactive isotope, each has a characteristic half-life. The properties of some isotopes in common biochemical use are listed in Table 16-2.

Table 16-2 **Some Trace Isotopes of Biochemical Importance**

Stable Isotopes

Nucleus	Natural Abundance (%)
^{2}H	0.012
^{13}C	1.07
^{15}N	0.36
^{18}O	0.20

Radioactive Isotopes

Nucleus	Radiation Type	Half-Life
^{3}H	β	12.31 years
^{14}C	β	5715 years
^{22}Na	β$^+$, γ	2.60 years
^{32}P	β	14.28 days
^{35}S	β	87.2 days
^{45}Ca	β	162.7 days
^{60}Co	β, γ	5.271 years
^{125}I	γ	59.4 days
^{131}I	β, γ	8.02 days

Source: Holden, N.E., *in* Lide, D.R. (Ed.), *Handbook of Chemistry and Physics* (90th ed.), *pp.* 11–57 to 266, CRC Press (2009–2010).

Figure 16-17 **The metabolic origin of the nitrogen atoms in heme.** Only [^{15}N]glycine, of many ^{15}N-labeled metabolites, is an ^{15}N-labeled heme precursor.

e. Isotopes Are Indispensable for Establishing the Metabolic Origins of Complex Metabolites and Precursor–Product Relationships

The metabolic origins of complex molecules such as heme, cholesterol, and phospholipids may be determined by administering isotopically labeled starting materials to animals and isolating the resulting products. One of the early advances in metabolic understanding resulting from the use of isotopic tracers was the demonstration, by David Shemin and David Rittenberg in 1945, that the nitrogen atoms of heme are derived from glycine rather than from ammonia, glutamic acid, proline, or leucine (Section 26-4Aa). They showed this by feeding rats these ^{15}N-labeled nutrients, isolating the heme in their blood, and analyzing it for ^{15}N content. Only when the rats were fed [^{15}N]glycine did the heme contain ^{15}N (Fig. 16-17). This technique was also used to demonstrate that all of cholesterol's carbon atoms are derived from acetyl-CoA (Section 25-6A).

Isotopic tracers are also useful in establishing the order of appearance of metabolic intermediates, their so-called **precursor–product relationships.** An example of such an analysis concerns the biosynthesis of the complex phospholipids called **plasmalogens** and **alkylacylglycerophospholipids** (Section 25-8Ab). Alkylacylglycerophospholipids are ethers, whereas the closely related plasmalogens are vinyl ethers. Their similar structures brings up the interesting

question of their biosynthetic relationship: Which is the precursor and which is the product? Two possible modes of synthesis can be envisioned (Fig. 16-18):

I. The starting material is converted to the vinyl ether (plasmalogen), which is then reduced to yield the ether (alkylacylglycerophospholipid). Accordingly, the vinyl ether would be the precursor and the ether the product.

Figure 16-18 **Two possible pathways for the biosynthesis of ether– and vinyl ether–containing phospholipids. (I)** The vinyl ether is the precursor and the ether is the product. **(II)** The ether is the precursor and the vinyl ether is the product.

II. The ether is formed first and then oxidized to yield the vinyl ether. The ether would then be the precursor and the vinyl ether the product.

Precursor–product relationships can be most easily sorted out through the use of radioactive tracers. A pulse of the labeled starting material is administered to an organism and the specific radioactivities of the resulting metabolic products are followed with time (Fig. 16-19):

$$\text{Starting material*} \longrightarrow A* \longrightarrow B* \longrightarrow \text{later products*}$$

(here the * represents the radioactive label). Metabolic pathways, as we shall see in Section 16-6Ba, normally operate in a steady state; that is, the throughput of metabolites in each of its reaction steps is equal. Moreover, the rates of most metabolic reactions are first order for a given substrate. Making these assumptions, we note that the rate of change of B's radioactivity, [B*], is equal to the rate of passage of label from A* to B* less the rate of passage of label from B* to the pathway's next product:

$$\frac{d[B*]}{dt} = k[A*] - k[B*] = k([A*] - [B*]) \quad [16.1]$$

where k is the pseudo-first-order rate constant for both the conversion of A to B and the conversion of B to its product, and t is time. Inspection of this equation indicates the criteria that must be met to establish that A is the precursor of B (Fig. 16-19):

1. Before the radioactivity of the product [B*] is maximal, $d[B*]/dt > 0$, so $[A*] > [B*]$; that is, *while the radioactivity of a product is rising, it should be less than that of its precursor.*

2. When [B*] is maximal, $d[B*]/dt = 0$, so $[A*] = [B*]$; that is, *when the radioactivity of a product is at its peak, it should be equal to that of its precursor.* This result also implies that *the radioactivity of a product peaks after that of its precursor.*

3. After [B*] begins to decrease, $d[B*]/dt < 0$, so $[A*] < [B*]$; that is, *after the radioactivity of a product has peaked, it should remain greater than that of its precursor.*

Such a determination of the precursor–product relationship between alkylacylglycerophospholipid and plasmalogen, using ^{14}C-labeled starting materials, indicated that the ether is the precursor and the vinyl ether is the product (Fig. 16-18, Scheme II).

C. *Isolated Organs, Cells, and Subcellular Organelles*

In addition to understanding the chemistry and catalytic events that occur at each step of a metabolic pathway, it is important to learn where a given pathway occurs within an organism. Early workers studied metabolism in whole animals. For example, the role of the pancreas in diabetes was established by Frederick Banting and Charles Best in 1921 by surgically removing that organ from dogs and observing that these animals then developed the disease.

The metabolic products produced by a particular organ can be studied by **organ perfusion** or in **tissue slices.** In organ perfusion, a specific organ is surgically removed from an animal and the organ's arteries and veins are connected to an artificial circulatory system. The composition of the material entering the organ can thereby be controlled and its metabolic products monitored. Metabolic processes can be similarly studied in slices of tissue thin enough to be nourished by free diffusion in an appropriate nutrient solution. Otto Warburg pioneered the tissue slice technique in the early twentieth century through his studies of respiration, in which he used a manometer to measure the changes in gas volume above tissue slices as a consequence of their O_2 consumption.

A given organ or tissue generally contains several cell types. **Cell sorters** are devices that can separate cells according to type once they have been treated with the enzymes trypsin and collagenase to destroy the intercellular matrix that binds them into a tissue. This technique allows further localization of metabolic function. A single cell type may also be grown in **tissue culture** for study. Although culturing cells often results in their loss of differentiated function, techniques have been developed for maintaining several cell types that still express their original characteristics.

As discussed in Section 16-1, metabolic pathways in eukaryotes are compartmentalized in various subcellular organelles (Table 16-1). For example, oxidative phosphorylation occurs in the mitochondrion, whereas glycolysis and fatty acid biosynthesis occur in the cytosol. Such observations are made by breaking cells open and fractionating their components by differential centrifugation (Section 6-1B), possibly followed by zonal ultracentrifugation through a sucrose density gradient or by equilibrium density gradient ultracentrifugation in a CsCl density gradient, which, respectively, separate particles according to their size and density (Section 6-5B). The cell fractions are then analyzed for biochemical function.

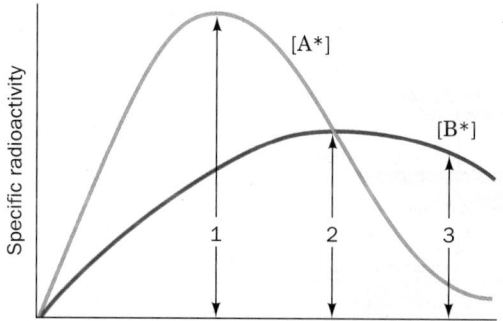

Figure 16-19 **The flow of a pulse of radioactivity from precursor to product.** At point 1, product radioactivity (B*, *purple*) is increasing and is less than that of its precursor (A*, *orange*); at point 2, product radioactivity is maximal and is equal to that of its precursor; and at point 3, product radioactivity is decreasing and is greater than that of its precursor.

D. *Systems Biology*

Metabolism has traditionally been studied by hypothesis-based research: isolating individual enzymes and metabolites and assembling them into metabolic pathways as guided by experimentally testable hypotheses. This is a reductionist approach: the explanation of the workings of a system in terms of its component parts. A different, so-called integrative approach, **systems biology,** has recently emerged with the advent of complete genome sequences, the development of rapid and sensitive techniques for analyzing large numbers of gene transcripts, proteins, and metabolites all at once, and the development of new computational and mathematical tools. Systems biology is discovery-based: collecting and integrating enormous amounts of data in searchable databases so that the properties and dynamics of entire biological networks can be analyzed. As a result, our understanding of the path from genotype to phenotype has expanded. In addition to the central dogma of molecular biology (Section 5-4), that a single gene composed of DNA is transcribed to mRNA, which is translated to a single protein that influences metabolism, we are increasingly taking into account the genome, transcriptome, proteome, and **metabolome** (the complete set of a cell's metabolites) and their interrelationships (Fig. 16-20). The term **bibliome** (Greek: *biblion,* book) has even been coined to denote the systematic incorporation of pre-existing information about reaction mechanisms and metabolic pathways. In the following paragraphs we discuss some of these emerging technologies and new fields of study.

a. Transcriptomics

The overall metabolic capabilities of an organism are encoded by its genome (its entire complement of genes). In principle, it should be possible to reconstruct a cell's metabolic activities from its genomic sequence. However, at present, this can be done only in a general sense. For example, the 4.0-Mb genome of *Vibrio cholerae,* the bacterium that causes cholera, contains a large repertoire of genes encoding transport proteins and enzymes for catabolizing a wide range of nutrients. This is consistent with the complicated lifestyle of *V. cholerae,* which can live on its own, in association with zooplankton, or in the human gastrointestinal tract (where it causes cholera; Section 19-2Cd). However, a simple catalog of an organism's genes does not reveal how these genes function. Thus, some genes are expressed continuously at high levels, whereas others are expressed rarely, for example, only when the organism encounters a particular metabolite.

Creating an accurate picture of gene expression is the goal of **transcriptomics,** the study of a cell's transcriptome (its entire complement of mRNAs). Identifying and quantifying all the transcripts from a single cell type reveals which genes are active. Cells transcribe thousands of genes at once so this study requires the use of DNA microarray technology (Section 7-6B). For example, Fig. 7-39 shows a DNA microarray that indicates the differences in gene expression between yeast grown in the presence and absence of glucose.

Differences in the expression of particular genes have been correlated with many developmental processes or growth patterns. For example, DNA microarrays have been used to profile the patterns of gene expression in tumor cells because different types of tumors express different types and amounts of proteins (Section 34-3B). This information is useful in choosing how best to treat a cancer.

b. Proteomics

The correlation between the amount of a particular mRNA and the amount of its protein product is imperfect. This is because the various mRNAs and their corresponding proteins are synthesized and degraded at different rates. Furthermore, many proteins are post-translationally modified, sometimes in several different ways (e.g., by phosphorylation or glycosylation). Consequently, the number of unique proteins in a cell exceeds the number of unique mRNAs.

A more reliable way than transcriptomics to assess gene expression is to examine a cell's proteome, the complete set of proteins that the cell synthesizes. This **proteomics**

Figure 16-20 **The relationship between genotype and phenotype.** The path from genetic information (genotype) to metabolic function (phenotype) has several steps. Portions of the genome are transcribed to produce the transcriptome, which directs the synthesis of the proteome, whose various activities are responsible for synthesizing and degrading the components of the metabolome.

approach requires that the proteins first be separated, usually by two-dimensional (2D) gel electrophoresis (Section 6-4D). Individual proteins are then identified by using tandem mass spectrometry to obtain amino acid sequence information (Section 7-1Ia) and correlating it with protein sequence databases. All the proteins that are contained in a cell or tissue under a given set of conditions can thereby be catalogued.

One can compare all the proteins synthesized by a cell under two different sets of conditions by using different isotopically labeled reagents that are either contained in the growth medium (e.g., deuterated amino acids) or that are reacted with the cell extract. One technique for labeling cellular proteins uses **isotope-coded affinity tags (ICATs),** which are analogous to the differently fluorescing dyes that are used to label cDNAs.

An ICAT contains three functional elements: an iodoacetyl group to react with Cys residues, a linker that

contains either 8 hydrogen (light) or 8 deuterium (heavy) atoms, and **biotin,** a coenzyme (Section 23-1Ab) that is used as a biotechnology tool because of its extremely tight binding to the protein **avidin** ($K = 10^{-15}$ M; Fig. 16-21a). Avidin is immobilized on a chromatographic resin so that the ICAT-labeled peptides can be isolated by biotin/avidin affinity chromatography (Section 6-3C).

The ICAT procedure is illustrated in Fig. 16-21b. Two protein mixtures representing two different growth conditions are treated with light (d0) or heavy (d8) versions of the ICAT reagent. The labeled protein mixtures are combined and digested with trypsin to form Cys-containing labeled peptides, which are then isolated by biotin/avidin affinity chromatography. Individual peptides are separated by liquid chromatography and detected by mass spectrometry (LC/MS). The ratio of the intensities of the light and heavy peptide signals indicates the relative peptide abundance in the two samples. Tandem mass spectrometry

Figure 16-21   The isotope-coded affinity tag (ICAT) method for quantitative proteome analysis. (*a*) An example of an ICAT reagent that contains an iodoacetyl reactive group, a linker, and a biotin residue. X denotes the position of hydrogen (d0) or deuterium (d8). (*b*) The ICAT strategy for differential labeling of proteins expressed by cells under two different sets of conditions. (**1**) Proteins from states A and B are respectively treated with the light (d0) and heavy (d8) forms of the ICAT reagent. (**2**) The labeled protein mixtures are combined. (**3**) The labeled proteins are digested with trypsin to form Cys-containing labeled peptides. These peptides are then purified by biotin/avidin

affinity chromatography. The purified peptides are analyzed by mass spectrometry in two ways: (**4**) Liquid chromatography followed by mass spectrometry (LC/MS) is used to quantitate the peptides. The ratio of the signal intensities from the corresponding light and heavy peptides indicates the relative peptide abundance in the two mixtures. (**5**) Tandem mass spectrometry (MS/MS) is used to determine the amino acid sequence of each peptide and to thereby identify the protein from which it is derived by comparing the peptide's sequence to those in a database of all known proteins.

(MS/MS) is then used to sequence each peptide and determine its identity. This method was used to identify many of the yeast proteins whose mRNA concentrations increased or decreased when glucose was absent from the growth medium (Fig. 7-39). A hope for the future is that samples from diseased and normal subjects can be compared in this manner to find previously undetected disease markers that would allow early diagnosis of various diseases.

c. Metabolomics

In order to describe a cell's functional state (its phenotype) we need, in addition to the cell's genome, transcriptome, and proteome, a quantitative description of all of the metabolites it contains under a given set of conditions, its metabolome. However, a cell or tissue contains thousands of metabolites with greatly varying properties, so that identifying and quantifying all these substances is a daunting task, requiring many different analytical tools. Consequently, this huge undertaking is often subdivided. For example, **lipidomics** is the subdiscipline of **metabolomics** aimed at identifying and characterizing all lipids in a cell under a particular set of conditions, including how these lipids influence membrane structure, cell signaling, gene expression, cell–cell interactions, etc., whereas **glycomics** similarly identifies and characterizes all the carbohydrates in a cell.

A recently constructed model of the human metabolome—based on 1496 protein-encoding genes, 2004 proteins, 2766 metabolites, and 3311 metabolic and transport reactions—has been used to simulate 288 known metabolic functions in a variety of cell and tissue types. This *in silico* (computerized) model is expected to provide a framework for future advances in human systems biology.

4 THERMODYNAMICS OF PHOSPHATE COMPOUNDS

The endergonic processes that maintain the living state are driven by the exergonic reactions of nutrient oxidation. This coupling is most often mediated through the syntheses of a few types of "high-energy" intermediates whose exergonic consumption drives endergonic processes. These intermediates therefore form a sort of universal free energy "currency" through which free energy–producing reactions "pay for" the free energy–consuming processes in biological systems.

Adenosine triphosphate (ATP; Fig. 16-22), which occurs in all known life-forms, is the "high-energy" intermediate that constitutes the most common cellular energy currency. Its central role in energy metabolism was first recognized in 1941 by Fritz Lipmann and Herman Kalckar. ATP consists of an **adenosine** moiety to which three **phosphoryl groups** ($-PO_3^{2-}$) are sequentially linked via a **phosphoester bond** followed by two **phosphoanhydride bonds**. **Adenosine diphosphate (ADP)** and **5′-adenosine monophosphate (AMP)** are similarly constituted but with only two and one phosphoryl units, respectively.

In this section we consider the nature of phosphoryl-transfer reactions, discuss why some of them are so exergonic, and outline how the cell consumes and regenerates ATP.

A. *Phosphoryl-Transfer Reactions*

Phosphoryl-transfer reactions,

$$R_1-O-PO_3^{2-} + R_2-OH \rightleftharpoons R_1-OH + R_2-O-PO_3^{2-}$$

are of enormous metabolic significance. Some of the most important reactions of this type involve the synthesis and hydrolysis of ATP:

$$ATP + H_2O \rightleftharpoons ADP + P_i$$
$$ATP + H_2O \rightleftharpoons AMP + PP_i$$

where P_i and PP_i, respectively, represent **orthophosphate** (PO_4^{3-}) and **pyrophosphate** ($P_2O_7^{4-}$) in any of their ionization states. *These highly exergonic reactions are coupled to numerous endergonic biochemical processes so as to drive them to completion. Conversely, ATP is regenerated by coupling its formation to a more highly exergonic metabolic process* (the thermodynamics of coupled reactions is discussed in Section 3-4C).

To illustrate these concepts, let us consider two examples of phosphoryl-transfer reactions. The initial step in the metabolism of glucose is its conversion to glucose-6-phosphate (Section 17-2A). Yet the direct reaction of glucose and P_i is thermodynamically unfavorable (Fig. 16-23*a*). In biological systems, however, this reaction is coupled to the exergonic hydrolysis of ATP, so the overall reaction is thermodynamically favorable. ATP can be similarly regenerated by coupling its synthesis from ADP and P_i to the even more exergonic hydrolysis of **phosphoenolpyruvate** (Fig. 16-23*b*; Section 17-2J).

The bioenergetic utility of phosphoryl-transfer reactions stems from their kinetic stability to hydrolysis combined with their capacity to transmit relatively large amounts of free energy. The $\Delta G°′$ values of hydrolysis of several phosphorylated compounds of biochemical importance are tabulated in Table 16-3. The negatives of these values are often referred to as **phosphate group-transfer potentials**; they

Figure 16-22 **The structure of ATP indicating its relationship to ADP, AMP, and adenosine.** The phosphoryl groups, starting with that on AMP, are referred to as the α, β, and γ phosphates. Note the difference between phosphoester and phosphoanhydride bonds.

(a) $\Delta G^{\circ\prime}$ (kJ $\cdot$ mol^{-1})

Endergonic half-reaction 1	P_i + glucose	$\rightleftharpoons$	glucose-6-P + H_2O	+13.8
Exergonic half-reaction 2	ATP + H_2O	$\rightleftharpoons$	ADP + P_i	−30.5
Overall coupled reaction	ATP + glucose	$\rightleftharpoons$	ADP + glucose-6-P	−16.7

(b) $\Delta G^{\circ\prime}$ (kJ $\cdot$ mol^{-1})

Exergonic half-reaction 1

$$CH_2{=}\underset{\displaystyle OPO_3^{2-}}{\overset{\displaystyle COO^-}{C}} + H_2O \rightleftharpoons CH_3{-}\overset{\displaystyle O}{\underset{\displaystyle \|}{C}}{-}COO^- + P_i \qquad -61.9$$

Phosphoenolpyruvate   **Pyruvate**

Endergonic half-reaction 2   ADP + P_i $\rightleftharpoons$ ATP + H_2O +30.5

Overall coupled reaction

$$CH_2{=}\underset{\displaystyle OPO_3^{2-}}{\overset{\displaystyle COO^-}{C}} + ADP \rightleftharpoons CH_3{-}\overset{\displaystyle O}{\underset{\displaystyle \|}{C}}{-}COO^- + ATP \qquad -31.4$$

Figure 16-23   Some overall coupled reactions involving ATP. (*a*) The phosphorylation of glucose to form glucose-6-phosphate and ADP. (*b*) The phosphorylation of ADP by phosphoenolpyruvate to form ATP and pyruvate. Each reaction has been conceptually decomposed into a direct phosphorylation step (half-reaction 1) and a step in which ATP is hydrolyzed (half-reaction 2). Both half-reactions proceed in the direction in which the overall reaction is exergonic ($\Delta G < 0$).

are a measure of the tendency of phosphorylated compounds to transfer their phosphoryl groups to water. Note that ATP has an intermediate phosphate group-transfer potential. Under standard conditions, the compounds above ATP in Table 16-3 can spontaneously transfer a phosphoryl group to ADP to form ATP, which can, in turn, spontaneously transfer a phosphoryl group to the hydrolysis products (ROH form) of the compounds below it.

a. ΔG of ATP Hydrolysis Varies with pH, Divalent Metal Ion Concentration, and Ionic Strength

The ΔG of a reaction varies with the total concentrations of its reactants and products and thus with their ionic states (Eq. [3.15]). The ΔG's of hydrolysis of phosphorylated compounds are therefore highly dependent on pH, divalent metal ion concentration (divalent metal ions such as Mg^{2+} have high phosphate-binding affinities), and ionic strength. Reasonable estimates of the intracellular values of these quantities as well as of [ATP], [ADP], and [P_i] (which are generally on the order of millimolar) indicate that ATP hydrolysis under physiological conditions has $\Delta G \approx -50$ kJ $\cdot$ mol^{-1} rather than the -30.5 kJ $\cdot$ mol^{-1} of its $\Delta G^{\circ\prime}$. Nevertheless, for the sake of consistency in comparing reactions, we shall usually refer to the latter value.

The above situation for ATP is not unique. It is important to keep in mind that *within a given cell, the concentrations of most substances vary both with location and time. Indeed, the concentrations of many ions, coenzymes, and metabolites commonly vary by several orders of magnitude across membranous organelle boundaries.* Unfortunately, it is usually quite difficult to obtain an accurate measurement of the concentration of any particular chemical species in a specific cellular compartment. The ΔG's for most *in vivo* reactions are therefore little more than estimates.

Table 16-3   **Standard Free Energies of Phosphate Hydrolysis of Some Compounds of Biological Interest**

Compound	$\Delta G^{\circ\prime}$ (kJ $\cdot$ mol^{-1})
Phosphoenolpyruvate	−61.9
1,3-Bisphosphoglycerate	−49.4
ATP ($\rightarrow$ AMP + PP$_i$)	**−45.6**
Acetyl phosphate	−43.1
Phosphocreatine	−43.1
ATP ($\rightarrow$ ADP + P$_i$)	**−30.5**
Glucose-1-phosphate	−20.9
PP$_i$	−19.2
Fructose-6-phosphate	−13.8
Glucose-6-phosphate	−13.8
Glycerol-3-phosphate	−9.2

Source: Mostly from Jencks, W.P., *in* Fasman, G.D. (Ed.), *Handbook of Biochemistry and Molecular Biology* (3rd ed.), *Physical and Chemical Data*, Vol. I, *pp.* 296–304, CRC Press (1976).

B. *Rationalizing the "Energy" in "High-Energy" Compounds*

Bonds whose hydrolysis proceeds with large negative values of $\Delta G^{\circ\prime}$ (customarily more negative than -25 kJ $\cdot$ mol^{-1}) are often referred to as **"high-energy" bonds** or **"energy-rich" bonds** and are frequently symbolized by the squiggle ($\sim$). Thus ATP may be represented as AR—P$\sim$P$\sim$P, where A, R, and P symbolize adenyl, ribosyl, and phosphoryl groups, respectively. Yet, the phosphoester bond joining the adenosyl group of ATP to its α-phosphoryl group appears to be not greatly different in electronic character from the so-called "high-energy" bonds bridging its β and γ phosphoryl groups. In fact, none of these bonds have any unusual properties, so the term "high-energy" bond is somewhat of a misnomer. (In any case, it should not be confused with the term "bond energy," which is defined as the energy required to break, not hydrolyze, a covalent bond.) Why then, should the phosphoryl-transfer reactions of ATP be so exergonic? The answer comes from the comparison of the stabilities of the reactants and products of these reactions.

Several different factors appear to be responsible for the "high-energy" character of phosphoanhydride bonds such as those in ATP (Fig. 16-24):

1. The resonance stabilization of a phosphoanhydride bond is less than that of its hydrolysis products. This is because a phosphoanhydride's two strongly electron-withdrawing phosphoryl groups must compete for the lone pair of electrons of its bridging oxygen atom, whereas this competition is absent in the hydrolysis products. In other words, the electronic requirements of the phosphoryl groups are less satisfied in a phosphoanhydride than in its hydrolysis products.

2. Of perhaps greater importance is the destabilizing effect of the electrostatic repulsions between the charged groups of a phosphoanhydride in comparison to that of its hydrolysis products. In the physiological pH range, ATP has three to four negative charges whose mutual electrostatic repulsions are partially relieved by ATP hydrolysis.

3. Another destabilizing influence, which is difficult to assess, is the smaller solvation energy of a phosphoanhydride in comparison to that of its hydrolysis products. Some estimates

suggest that this factor provides the dominant thermodynamic driving force for the hydrolysis of phosphoanhydrides.

A further property of ATP that suits it to its role as an energy intermediate stems from the relative kinetic stability of phosphoanhydride bonds to hydrolysis. Most types of anhydrides are rapidly hydrolyzed in aqueous solution. Phosphoanhydride bonds, however, have unusually large free energies of activation. Consequently, ATP is reasonably stable under physiological conditions but is readily hydrolyzed in enzymatically mediated reactions.

a. Other "High-Energy" Compounds
The compounds in Table 16-3 with phosphate group-transfer potentials significantly greater than that of ATP have additional destabilizing influences:

1. Acyl phosphates. The hydrolysis of **acyl phosphates** (mixed phosphoric–carboxylic anhydrides), such as **acetyl phosphate** and **1,3-bisphosphoglycerate,**

Acetyl phosphate

1,3-Bisphosphoglycerate

is driven by the same competing resonance and differential solvation influences that function in the hydrolysis of phosphoanhydrides. Apparently these effects are more pronounced for acyl phosphates than for phosphoanhydrides.

2. Enol phosphates. The high phosphate group-transfer potential of an **enol phosphate,** such as phosphoenolpyruvate (Fig. 16-23b), derives from its **enol** hydrolysis product being less stable than its **keto** tautomer. Consider the hydrolysis reaction of an enol phosphate as occurring in two steps (Fig. 16-25). The hydrolysis step is subject to the driving forces discussed above. *It is therefore the highly exergonic enol–keto conversion that provides phosphoenolpyruvate with the added thermodynamic impetus to phosphorylate ADP to form ATP.*

3. Phosphoguanidines. The high phosphate group-transfer potentials of **phosphoguanidines,** such as **phosphocreatine** and **phosphoarginine,** largely result from the competing resonances in their **guanidino** group, which are even more pronounced than they are in the phosphate group of phosphoanhydrides (Fig. 16-26). Consequently, phosphocreatine can phosphorylate ADP (see Section 16-4Cd).

Compounds such as **glucose-6-phosphate** or **glycerol-3-phosphate,**

Figure 16-24 Resonance and electrostatic stabilization in a phosphoanhydride and its hydrolysis products. The competing resonances (*curved arrows* from the central O) and charge–charge repulsions (*zigzag line*) between the phosphoryl groups of a phosphoanhydride decrease its stability relative to its hydrolysis products.

α-D-**Glucose-6-phosphate**   L-**Glycerol-3-phosphate**

Hydrolysis

$$\underset{\underset{H}{\overset{|}{C}}{\overset{\overset{COO^-}{|}}{C}}-O{\sim}PO_3^{2-} + H_2O \rightleftharpoons \underset{\underset{H}{\overset{|}{C}}{\overset{\overset{COO^-}{|}}{C}}-O-H + HPO_4^{2-} \qquad \Delta G^{\circ\prime} = -16\ \text{kJ} \cdot \text{mol}^{-1}$$

**Phosphoenol-
pyruvate**

Tautomerization

$$\underset{\underset{H}{\overset{|}{C}}{\overset{\overset{COO^-}{|}}{C}}-O-H \rightleftharpoons \underset{\underset{H}{\overset{|}{\underset{H}{C}}}{\overset{\overset{COO^-}{|}}{C}}=O \qquad \Delta G^{\circ\prime} = -46\ \text{kJ} \cdot \text{mol}^{-1}$$

**Pyruvate
(enol form)**   **Pyruvate
(keto form)**

Overall reaction

$$\underset{\underset{H}{\overset{|}{C}}{\overset{\overset{COO^-}{|}}{C}}-O{\sim}PO_3^{2-} + H_2O \rightleftharpoons \underset{\underset{H}{\overset{|}{\underset{H}{C}}}{\overset{\overset{COO^-}{|}}{C}}=O + HPO_4^{2-} \qquad \Delta G^{\circ\prime} = -61.9\ \text{kJ} \cdot \text{mol}^{-1}$$

Figure 16-25 **Hydrolysis of phosphoenolpyruvate.** The reaction is broken down into two steps, hydrolysis and tautomerization.

which are below ATP in Table 16-3, have no significantly different resonance stabilization or charge separation in comparison with their hydrolysis products. Their free energies of hydrolysis are therefore much less than those of the preceding "high-energy" compounds.

C. The Role of ATP

As Table 16-3 indicates, *in the thermodynamic hierarchy of phosphoryl-transfer agents, ATP occupies the middle rank.* This enables ATP to serve as an energy conduit between "high-energy" phosphate donors and "low-energy" phosphate acceptors (Fig. 16-27). Let us examine the general biochemical scheme of how this occurs.

$$R = CH_2-CO_2^-\ ;\ X = CH_3$$
Phosphocreatine

$$R = CH_2-CH_2-CH_2-\underset{\underset{NH_3^+}{|}}{CH}-CO_2^-\ ;\ X = H$$
Phosphoarginine

Figure 16-26 **Competing resonances in phosphoguanidines.**

In general, the highly exergonic phosphoryl-transfer reactions of nutrient degradation are coupled to the formation of ATP from ADP and P_i through the auspices of various enzymes known as **kinases,** enzymes that catalyze the transfer of phosphoryl groups between ATP and other molecules. Consider the two reactions in Fig. 16-23b. If carried out independently, these reactions would not influence each other. In the cell, however, the enzyme **pyruvate**

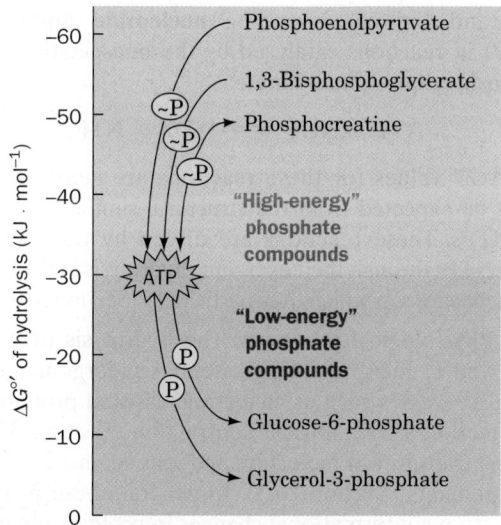

Figure 16-27 **The flow of phosphoryl groups from "high-energy" phosphate donors, via the ATP–ADP system, to "low-energy" phosphate acceptors.**

$$^{-2}O_3P-O-CH_2 \quad O \quad CH_2-OH \quad + \quad ATP \xrightarrow[\Delta G^{\circ\prime} = -14.2 \text{ kJ} \cdot \text{mol}^{-1}]{\text{phosphofructokinase}} \quad ^{-2}O_3P-O-CH_2 \quad O \quad CH_2-O-PO_3^{2-} \quad + \quad ADP$$

Fructose-6-phosphate   **Fructose-1,6-bisphosphate**

Figure 16-28   **The phosphorylation of fructose-6-phosphate by ATP to form fructose-1,6-bisphosphate and ADP.**

kinase couples the two reactions by catalyzing the transfer of the phosphoryl group of phosphoenolpyruvate directly to ADP to result in an overall exergonic reaction (Section 17-2J).

a. Consumption of ATP

In its role as the universal energy currency of living systems, ATP is consumed in a variety of ways:

1. Early stages of nutrient breakdown. The exergonic hydrolysis of ATP to ADP may be enzymatically coupled to an endergonic phosphorylation reaction to form "low-energy" phosphate compounds. We have seen one example of this in the hexokinase-catalyzed formation of glucose-6-phosphate (Fig. 16-23*a*). Another example is the **phosphofructokinase**-catalyzed phosphorylation of **fructose-6-phosphate** to form **fructose-1,6-bisphosphate** (Fig. 16-28). Both of these reactions occur in the first stage of glycolysis (Section 17-2).

2. Interconversion of nucleoside triphosphates. Many biosynthetic processes, such as the synthesis of proteins and nucleic acids, require nucleoside triphosphates other than ATP. These include the ribonucleoside triphosphates CTP, GTP, and UTP, which, together with ATP, are utilized, for example, in the biosynthesis of RNA (Section 31-2) and the deoxyribonucleoside triphosphate DNA precursors dATP, dCTP, dGTP, and dTTP (Section 5-4C). All these **nucleoside triphosphates (NTPs)** are synthesized from ATP and the corresponding **nucleoside diphosphate (NDP)** in reactions catalyzed by the nonspecific enzyme **nucleoside diphosphate kinase:**

$$\text{ATP} + \text{NDP} \rightleftharpoons \text{ADP} + \text{NTP}$$

The $\Delta G^{\circ\prime}$ values for these reactions are nearly zero, as might be expected from the structural similarities among the NTPs. These reactions are driven by the depletion of the NTPs through their exergonic hydrolysis in the biosynthetic reactions in which they participate (Section 3-4C).

3. Physiological processes. The hydrolysis of ATP to ADP and P_i energizes many essential endergonic physiological processes such as chaperone-assisted protein folding (Section 9-2C), muscle contraction (Section 35-3B), and the transport of molecules and ions against concentration gradients (Section 20-3). In general, these processes result from conformational changes in proteins (enzymes) that occur in response to their binding of ATP. This is followed by the exergonic hydrolysis of ATP and release of

ADP and P_i, thereby causing these processes to be unidirectional (irreversible).

4. Additional phosphoanhydride cleavage in highly endergonic reactions. Although many reactions involving ATP yield ADP and P_i **(orthophosphate cleavage)**, others yield AMP and PP_i **(pyrophosphate cleavage)**. In these latter cases, the PP_i is rapidly hydrolyzed to $2P_i$ by **inorganic pyrophosphatase** ($\Delta G^{\circ\prime} = -19.2$ kJ $\cdot$ mol^{-1}) so that *the pyrophosphate cleavage of ATP ultimately results in the hydrolysis of two "high-energy" phosphoanhydride bonds.* The attachment of amino acids to tRNA molecules for protein synthesis is an example of this phenomenon (Fig. 16-29 and Section 32-2C). The two steps of the reaction involving the amino acid are readily reversible because the free energies of hydrolysis of the bonds formed are comparable to that of ATP hydrolysis. The overall reaction is driven to completion by the hydrolysis of PP_i, which is essentially irreversible. Nucleic acid biosynthesis from the appropriate NTPs also releases PP_i (Sections 30-1A and 31-2). The free energy changes of these vital reactions are around zero, so the subsequent hydrolysis of PP_i is essential to drive the synthesis of nucleic acids.

b. Formation of ATP

To complete its intermediary metabolic function, ATP must be replenished. This is accomplished through three types of processes:

1. Substrate-level phosphorylation. ATP may be formed, as is indicated in Fig. 16-23*b*, from phosphoenolpyruvate by direct transfer of a phosphoryl group from a "high-energy" compound to ADP. Such reactions, which are referred to as **substrate-level phosphorylations,** most commonly occur in the early stages of carbohydrate metabolism (Section 17-2).

2. Oxidative phosphorylation and photophosphorylation. Both oxidative metabolism and photosynthesis act to generate a proton (H^+) concentration gradient across a membrane (Sections 22-3 and 24-2D). Discharge of this gradient is enzymatically coupled to the formation of ATP from ADP and P_i (the reverse of ATP hydrolysis). In oxidative metabolism, this process is called **oxidative phosphorylation,** whereas in photosynthesis it is termed **photophosphorylation.** Most of the ATP produced by respiring and photosynthesizing organisms is generated in this manner.

3. Adenylate kinase reaction. The AMP resulting from pyrophosphate cleavage reactions of ATP is converted to

$$R-\underset{\underset{NH_3^+}{|}}{\overset{\overset{H}{|}}{C}}-\overset{\overset{O}{\parallel}}{C}-O^- \ + \ AMP\sim P\sim P \ \rightleftharpoons \ R-\underset{\underset{NH_3^+}{|}}{\overset{\overset{H}{|}}{C}}-\overset{\overset{O}{\parallel}}{C}\sim AMP \ \rightleftharpoons \ R-\underset{\underset{NH_3^+}{|}}{\overset{\overset{H}{|}}{C}}-\overset{\overset{O}{\parallel}}{C}-tRNA$$

Amino acid   ATP   Aminoacyl–adenylate   Aminoacyl–tRNA

tRNA   AMP

$$P\sim P \ \xrightarrow[\text{H}_2\text{O}]{\text{inorganic pyrophosphatase}} \ 2P_i$$

PP$_i$

Figure 16-29 Pyrophosphate cleavage in the synthesis of an aminoacyl–tRNA. Here the squiggle (∼) represents a "high-energy" bond. In the first reaction step, the amino acid is **adenylylated** by ATP. In the second step, a tRNA molecule displaces the AMP moiety to form an aminoacyl–tRNA. The exergonic hydrolysis of pyrophosphate ($\Delta G^{\circ\prime} = -19.2$ kJ · mol^{-1}) drives the reaction forward.

ADP in a reaction catalyzed by the enzyme **adenylate kinase** (Section 17-4Fe):

$$AMP + ATP \rightleftharpoons 2ADP$$

The ADP is subsequently converted to ATP through substrate-level phosphorylation, oxidative phosphorylation, or photophosphorylation.

c. Rate of ATP Turnover

The cellular role of ATP is that of a free energy transmitter rather than a free energy reservoir. The amount of ATP in a cell is typically only enough to supply its free energy needs for a minute or two. Hence, ATP is continually being hydrolyzed and regenerated. Indeed, ^{32}P-labeling experiments indicate that the metabolic half-life of an ATP molecule varies from seconds to minutes depending on the cell type and its metabolic activity. For instance, brain cells have only a few seconds' supply of ATP (which, in part, accounts for the rapid deterioration of brain tissue by oxygen deprivation). *An average person at rest consumes and regenerates ATP at a rate of ~3 mol (1.5 kg) · h^{-1} and as much as an order of magnitude faster during strenuous activity.*

d. Phosphocreatine Provides a "High-Energy" Reservoir for ATP Formation

Muscle and nerve cells, which have a high ATP turnover (a maximally exerting muscle has only a fraction of a second's ATP supply), have a free energy reservoir that functions to regenerate ATP rapidly. In vertebrates, phosphocreatine (Fig. 16-26) functions in this capacity. It is synthesized by the reversible phosphorylation of creatine by ATP as catalyzed by **creatine kinase:**

$$ATP + creatine \rightleftharpoons phosphocreatine + ADP$$
$$\Delta G^{\circ\prime} = +12.6 \text{ kJ} \cdot \text{mol}^{-1}$$

Note that this reaction is endergonic under standard conditions. However, the intracellular concentrations of its reactants and products (typically 4 mM ATP and 0.013 mM ADP) are such that it operates close to equilibrium ($\Delta G \approx 0$). Accordingly, when the cell is in a resting state, so that [ATP] is relatively high, the reaction proceeds with net synthesis of phosphocreatine, whereas at times of high metabolic activity, when [ATP] is low, the equilibrium shifts so as to yield net synthesis of ATP. *Phosphocreatine thereby acts as an ATP "buffer" in cells that contain creatine kinase.* A resting vertebrate skeletal muscle normally has sufficient phosphocreatine to supply its free energy needs for several minutes (but for only a few seconds at maximum exertion). In the muscles of some invertebrates, such as lobsters, phosphoarginine performs the same function. These phosphoguanidines are collectively named **phosphagens.**

5 OXIDATION–REDUCTION REACTIONS

Oxidation–reduction reactions, processes involving the transfer of electrons, are of immense biochemical significance; living things derive most of their free energy from them. In photosynthesis (Chapter 24), CO_2 is **reduced** (gains electrons) and H_2O is **oxidized** (loses electrons) to yield carbohydrates and O_2 in an otherwise endergonic process that is powered by light energy. In aerobic metabolism, which is carried out by all eukaryotes and many prokaryotes, the overall photosynthetic reaction is essentially reversed so as to harvest the free energy of oxidation of carbohydrates and other organic compounds in the form of ATP (Chapter 22). Anaerobic metabolism generates ATP, although in lower yields, through intramolecular oxidation–reductions of various organic molecules, for example, glycolysis (Chapter 17). In certain anaerobic bacteria, ATP is generated through the use of non-O_2 oxidizing agents such as sulfate or nitrate. In this section we outline the thermodynamics of oxidation–reduction reactions in order to understand the quantitative aspects of these crucial biological processes.

A. *The Nernst Equation*

Oxidation–reduction reactions (also known as **redox** or **oxidoreduction reactions**) resemble other types of chemical reactions in that they involve group transfer. For instance, hydrolysis transfers a functional group to water. In oxidation–reduction reactions, the "groups" transferred are electrons, which are passed from an **electron donor (reductant** or **reducing agent)** to an **electron acceptor (oxidant** or **oxidizing agent).** For example, in the reaction

$$Fe^{3+} + Cu^+ \rightleftharpoons Fe^{2+} + Cu^{2+}$$

Cu^+, the reductant, is oxidized to Cu^{2+} while Fe^{3+}, the oxidant, is reduced to Fe^{2+}.

Redox reactions may be divided into two **half-reactions** or **redox couples,** such as

$$Fe^{3+} + e^- \rightleftharpoons Fe^{2+} \text{ (reduction)}$$
$$Cu^+ \rightleftharpoons Cu^{2+} + e^- \text{ (oxidation)}$$

whose sum is the above whole reaction. These half-reactions occur during oxidative metabolism in the vital mitochondrial electron transfer mediated by **cytochrome *c* oxidase** (Section 22-2C5). Note that for electrons to be transferred, both half-reactions must occur simultaneously. In fact, the electrons are the two half-reactions' common intermediate.

a. Electrochemical Cells

A half-reaction consists of an electron donor and its conjugate electron acceptor; in the oxidation half-reaction shown above, Cu^+ is the electron donor and Cu^{2+} is its conjugate electron acceptor. Together these constitute a **conjugate redox pair** analogous to the conjugate acid–base pair (HA and A^-) of a Brønsted acid (Section 2-2A). An important difference between redox pairs and acid–base pairs, however, is that *the two half-reactions of a redox reaction, each consisting of a conjugate redox pair, may be physically separated so as to form an **electrochemical cell*** (Fig. 16-30). In such a device, each half-reaction takes place in its separate **half-cell,** and electrons are passed between half-cells as an electric current in the wire connecting their two electrodes. A salt bridge is necessary to complete the electrical circuit by providing a conduit for ions to migrate in the maintenance of electrical neutrality.

The free energy of an oxidation–reduction reaction is particularly easy to determine through a simple measurement of the voltage difference between its two half-cells. Consider the general redox reaction:

$$A_{ox}^{n+} + B_{red} \rightleftharpoons A_{red} + B_{ox}^{n+}$$

in which *n* electrons per mole of reactants are transferred from reductant (B_{red}) to oxidant (A_{ox}^{n+}). The free energy of this reaction is expressed, according to Eq. [3.15], as

$$\Delta G = \Delta G^\circ + RT \ln\left(\frac{[A_{red}][B_{ox}^{n+}]}{[A_{ox}^{n+}][B_{red}]}\right) \quad [16.2]$$

Equation [3.12] indicates that, under reversible conditions,

$$\Delta G = -w' = -w_{el} \quad [16.3]$$

where w', the non-pressure–volume work, is, in this case, w_{el}, the electrical work required to transfer the *n* moles of electrons through the electric potential difference $\Delta\mathscr{E}$. This, according to the laws of electrostatics, is

$$w_{el} = n\mathscr{F}\,\Delta\mathscr{E} \quad [16.4]$$

where $\mathscr{F}$, the **faraday,** is the electrical charge of 1 mol of electrons (1 $\mathscr{F}$ = 96,485 C · mol^{-1} = 96,485 J · V^{-1} · mol^{-1}, where C and V are the symbols for coulomb and volt). Thus, substituting Eq. [16.4] into Eq. [16.3],

$$\Delta G = -n\mathscr{F}\,\Delta\mathscr{E} \quad [16.5]$$

Combining Eqs. [16.2] and [16.5], and making the analogous substitution for ΔG°, yields the **Nernst equation:**

$$\Delta\mathscr{E} = \Delta\mathscr{E}^\circ - \frac{RT}{n\mathscr{F}}\ln\left(\frac{[A_{red}][B_{ox}^{n+}]}{[A_{ox}^{n+}][B_{red}]}\right) \quad [16.6]$$

which was originally formulated in 1881 by Walther Nernst. Here $\Delta\mathscr{E}$, the **electromotive force (emf)** or **redox potential,** may be described as the "electron pressure" that the electrochemical cell exerts. The quantity $\Delta\mathscr{E}^\circ$, the redox potential when all components are in their standard states, is called the **standard redox potential.** If these standard states refer to biochemical standard states (Section 3-4Ba), then $\Delta\mathscr{E}^\circ$ is replaced by $\Delta\mathscr{E}^{\circ'}$. Note that a positive $\Delta\mathscr{E}$ in Eq. [16.5] results in a negative ΔG; in other words, *a positive $\Delta\mathscr{E}$ is indicative of a spontaneous reaction, one that can do work.*

B. *Measurements of Redox Potentials*

The free energy change of a redox reaction may be determined, as Eq. [16.5] indicates, by simply measuring its redox potential with a voltmeter (Fig. 16-30). Consequently, voltage measurements are commonly employed to characterize the sequence of reactions comprising a metabolic

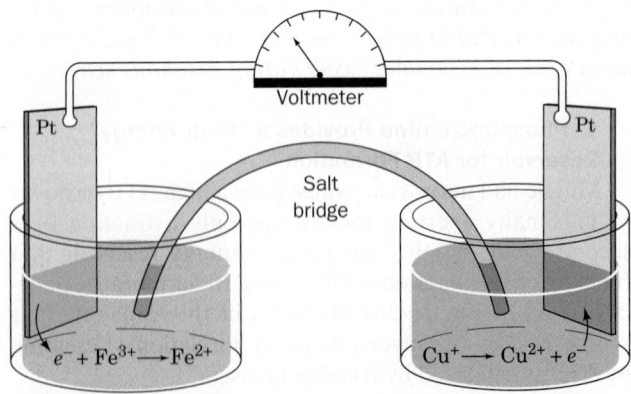

Figure 16-30   Example of an electrochemical cell. The half-cell undergoing oxidation (here $Cu^+ \rightarrow Cu^{2+} + e^-$) passes the liberated electrons through the wire to the half-cell undergoing reduction (here $e^- + Fe^{3+} \rightarrow Fe^{2+}$). Electroneutrality in the two half-cells is maintained by the transfer of ions through the electrolyte-containing salt bridge.

electron-transport pathway (such as mediates, e.g., oxidative metabolism; Chapter 22).

Any redox reaction can be divided into its component half-reactions:

$$A_{ox}^{n+} + ne^- \rightleftharpoons A_{red}$$
$$B_{ox}^{n+} + ne^- \rightleftharpoons B_{red}$$

where, by convention, both half-reactions are written as reductions. These half-reactions can be assigned **reduction potentials,** $\mathscr{E}_A$ and $\mathscr{E}_B$, in accordance with the Nernst equation:

$$\mathscr{E}_A = \mathscr{E}_A^\circ - \frac{RT}{n\mathscr{F}} \ln\left(\frac{[A_{red}]}{[A_{ox}^{n+}]}\right) \quad [16.7a]$$

$$\mathscr{E}_B = \mathscr{E}_B^\circ - \frac{RT}{n\mathscr{F}} \ln\left(\frac{[B_{red}]}{[B_{ox}^{n+}]}\right) \quad [16.7b]$$

For the redox reaction of any two half-reactions:

$$\Delta\mathscr{E}^\circ = \mathscr{E}_{(e^-\text{acceptor})}^\circ - \mathscr{E}_{(e^-\text{donor})}^\circ \quad [16.8]$$

Thus, when the reaction proceeds with A as the electron acceptor and B as the electron donor, $\Delta\mathscr{E}^\circ = \mathscr{E}_A^\circ - \mathscr{E}_B^\circ$ and similarly for $\Delta\mathscr{E}$.

Reduction potentials, like free energies, must be defined with respect to some arbitrary standard. By convention, standard reduction potentials are defined with respect to the standard hydrogen half-reaction

$$2H^+ + 2e^- \rightleftharpoons H_2(g)$$

in which H^+ at pH 0, 25°C, and 1 atm is in equilibrium with $H_2(g)$ that is in contact with a Pt electrode. This half-cell is arbitrarily assigned a standard reduction potential of $\mathscr{E}^\circ = 0$ V $(1 V = 1 J \cdot C^{-1})$. For the biochemical convention, we likewise define the standard (pH = 0) hydrogen half-reaction as having $\mathscr{E}' = 0$ so that the hydrogen half-cell at the biochemical standard state (pH = 7) has $\mathscr{E}^{\circ\prime} = -0.421$ V (Table 16-4). When $\Delta\mathscr{E}$ is positive, ΔG is negative (Eq. [16.5]), indicating a spontaneous process. In combining two half-reactions under standard conditions, the direction of spontaneity therefore involves the reduction of the redox couple with the more positive standard reduction potential. In other words, *the more positive the standard reduction potential, the greater the tendency for the redox couple's oxidized form to accept electrons and thus become reduced.*

Table 16-4 **Standard Reduction Potentials of Some Biochemically Important Half-Reactions**

Half-Reaction	$\mathscr{E}^{\circ\prime}$ (V)
$\frac{1}{2}O_2 + 2H^+ + 2e^- \rightleftharpoons H_2O$	0.815
$NO_3^- + 2H^+ + 2e^- \rightleftharpoons NO_2^- + H_2O$	0.42
Cytochrome $a_3(Fe^{3+}) + e^- \rightleftharpoons$ cytochrome $a_3(Fe^{2+})$	0.385
$O_2(g) + 2H^+ + 2e^- \rightleftharpoons H_2O_2$	0.295
Cytochrome $a(Fe^{3+}) + e^- \rightleftharpoons$ cytochrome $a(Fe^{2+})$	0.29
Cytochrome $c(Fe^{3+}) + e^- \rightleftharpoons$ cytochrome $c(Fe^{2+})$	0.235
Cytochrome $c_1(Fe^{3+}) + e^- \rightleftharpoons$ cytochrome $c_1(Fe^{2+})$	0.22
Cytochrome $b(Fe^{3+}) + e^- \rightleftharpoons$ cytochrome $b(Fe^{2+})$ *(mitochondrial)*	0.077
Ubiquinone $+ 2H^+ + 2e^- \rightleftharpoons$ ubiquinol	0.045
Fumarate$^- + 2H^+ + 2e^- \rightleftharpoons$ succinate$^-$	0.031
FAD $+ 2H^+ + 2e^- \rightleftharpoons FADH_2$ *(in flavoproteins)*	−0.040
Oxaloacetate$^- + 2H^+ + 2e^- \rightleftharpoons$ malate$^-$	−0.166
Pyruvate$^- + 2H^+ + 2e^- \rightleftharpoons$ lactate$^-$	−0.185
Acetaldehyde $+ 2H^+ + 2e^- \rightleftharpoons$ ethanol	−0.197
FAD $+ 2H^+ + 2e^- \rightleftharpoons FADH_2$ *(free coenzyme)*	−0.219
S $+ 2H^+ + 2e^- \rightleftharpoons H_2S$	−0.23
Lipoic acid $+ 2H^+ + 2e^- \rightleftharpoons$ dihydrolipoic acid	−0.29
$NAD^+ + H^+ + 2e^- \rightleftharpoons NADH$	−0.315
$NADP^+ + H^+ + 2e^- \rightleftharpoons NADPH$	−0.320
Cystine $+ 2H^+ + 2e^- \rightleftharpoons$ 2 cysteine	−0.340
Acetoacetate$^- + 2H^+ + 2e^- \rightleftharpoons$ β-hydroxybutyrate$^-$	−0.346
$H^+ + e^- \rightleftharpoons \frac{1}{2}H_2$	−0.421
$SO_4^{2-} + 2H^+ + 2e^- \rightleftharpoons SO_3^{2-} + H_2O$	−0.454
Acetate$^- + 3H^+ + 2e^- \rightleftharpoons$ acetaldehyde $+ H_2O$	−0.581

Source: Mostly from Loach, P.A., *in* Fasman, G.D. (Ed.), *Handbook of Biochemistry and Molecular Biology* (3rd ed.), *Physical and Chemical Data,* Vol. I, pp. 123–130, CRC Press (1976).

a. Biochemical Half-Reactions Are Physiologically Significant

The biochemical standard reduction potentials ($\mathscr{E}°'$) of some biochemically important half-reactions are listed in Table 16-4. The oxidized form of a redox couple with a large positive standard reduction potential has a high affinity for electrons and is a strong electron acceptor (oxidizing agent), whereas its conjugate reductant is a weak electron donor (reducing agent). For example, O_2 is the strongest oxidizing agent in Table 16-4, whereas H_2O, which tightly holds its electrons, is the table's weakest reducing agent. The converse is true of half-reactions with large negative standard reduction potentials. Since electrons spontaneously flow from low to high reduction potentials, they are transferred, under standard conditions, from the reduced products in any half-reaction in Table 16-4 to the oxidized reactants of any half-reaction above it (although this may not occur at a measurable rate in the absence of a suitable enzyme). Thus, in biological systems, the approximate lower limit for a standard reduction potential is -0.421 V because reductants with a lesser value of $\mathscr{E}°'$ would reduce protons to H_2. However, reducing centers in proteins that are protected from water may have lower potentials. Note that the Fe^{3+} ions of the various cytochromes tabulated in Table 16-4 have significantly different redox potentials. This indicates that *the protein components of redox enzymes play active roles in electron-transfer reactions by modulating the redox potentials of their bound redox-active centers.*

Electron-transfer reactions are of great biological importance. For example, in the mitochondrial electron-transport chain (Section 22-2), the primary source of ATP in eukaryotes, electrons are passed from NADH (Fig. 13-2) along a series of electron acceptors of increasing reduction potential (many of which are listed in Table 16-4) to O_2. ATP is generated from ADP and P_i by coupling its synthesis to this free energy cascade. *NADH thereby functions as an energy-rich electron-transfer coenzyme.* In fact, the oxidation of one NADH to NAD^+ supplies sufficient free energy to generate 2.5 ATPs (Section 22-2Bb). The NAD^+/NADH redox couple functions as the electron acceptor in many exergonic metabolite oxidations. In serving as the electron donor in ATP synthesis, it fulfills its cyclic role as a free energy conduit in a manner analogous to ATP. The metabolic roles of redox coenzymes are further discussed in succeeding chapters.

C. Concentration Cells

A concentration gradient has a lower entropy (greater order) than the corresponding uniformly mixed solution and therefore requires the input of free energy for its formation. Consequently, discharge of a concentration gradient is an exergonic process that may be harnessed to drive an endergonic reaction. For example, discharge of a proton concentration gradient (generated by the reactions of the electron-transport chain) across the inner mitochondrial membrane drives the enzymatic synthesis of ATP from ADP and P_i (Section 22-3). Likewise, nerve impulses,

which require electrical energy, are transmitted through the discharge of $[Na^+]$ and $[K^+]$ gradients that nerve cells generate across their cell membranes (Section 20-5B). Quantitation of the free energy contained in a concentration gradient is accomplished by use of the concepts of electrochemical cells.

The reduction potential and free energy of a half-cell vary with the concentrations of its reactants. An electrochemical cell may therefore be constructed from two half-cells that contain the same chemical species but at different concentrations. The overall reaction for such an electrochemical cell may be represented

$$A_{ox}^{n+}(\text{half-cell 1}) + A_{red}(\text{half-cell 2}) \rightleftharpoons$$
$$A_{ox}^{n+}(\text{half-cell 2}) + A_{red}(\text{half-cell 1}) \quad [16.9]$$

and, according to the Nernst equation, since $\Delta\mathscr{E}° = 0$ when the same reaction occurs in both cells,

$$\Delta\mathscr{E} = \frac{RT}{n\mathscr{F}}\ln\left(\frac{[A_{ox}^{n+}(\text{half-cell 2})][A_{red}(\text{half-cell 1})]}{[A_{ox}^{n+}(\text{half-cell 1})][A_{red}(\text{half-cell 2})]}\right)$$

Such **concentration cells** are capable of generating electrical work until they reach equilibrium. This occurs when the concentration ratios in the half-cells become equal ($K_{eq} = 1$). The reaction constitutes a sort of mixing of the two half-cells; the free energy generated is a reflection of the entropy of this mixing. The thermodynamics of concentration gradients as they apply to membrane transport is discussed in Section 20-1.

6 THERMODYNAMICS OF LIFE

One of the last refuges of **vitalism,** the doctrine that biological processes are not bound by the physical laws that govern inanimate objects, was the belief that living things can somehow evade the laws of thermodynamics. This view was partially refuted by elaborate calorimetric measurements on living animals that are entirely consistent with the energy conservation predictions of the first law of thermodynamics. However, the experimental verification of the second law of thermodynamics in living systems is more difficult. It has not been possible to measure the entropy of living matter because the heat, q_p, of a reaction at a constant T and P is only equal to $T\,\Delta S$ if the reaction is carried out reversibly (Eq. [3.8]). Obviously, the dismantling of a living organism to its component molecules for such a measurement would invariably result in its irreversible death. Consequently, the present experimentally verified state of knowledge is that the entropy of living matter is less than that of the products to which it decays.

In this section we consider the special aspects of the thermodynamics of living systems. Knowledge of these matters, which is by no means complete, has enhanced our understanding of how metabolic pathways are regulated, how cells respond to stimuli, and how organisms grow and change with time.

A. *Living Systems Cannot Be at Equilibrium*

Classical or **equilibrium thermodynamics** (Chapter 3) applies largely to reversible processes in closed systems. The fate of any isolated system, as we discussed in Section 3-4A, is that it must inevitably reach equilibrium. For example, if its reactants are in excess, the forward reaction will proceed faster than the reverse reaction until equilibrium is attained ($\Delta G = 0$). In contrast, open systems may remain in a nonequilibrium state as long as they are able to acquire free energy from their surroundings in the form of reactants, heat, or work. While classical thermodynamics provides invaluable information concerning open systems by indicating whether a given process can occur spontaneously, further thermodynamic analysis of open systems requires the application of the more recently elucidated principles of **nonequilibrium** or **irreversible thermodynamics.** In contrast to classical thermodynamics, this theory explicitly takes time into account.

Living organisms are open systems and therefore can never be at equilibrium. As indicated above, they continuously ingest high-enthalpy, low-entropy nutrients, which they convert to low-enthalpy, high-entropy waste products. The free energy resulting from this process is used to do work and to produce the high degree of organization characteristic of life. If this process is interrupted, the organism ultimately reaches equilibrium, which for living things is synonymous with death. For example, one theory of aging holds that senescence results from the random but inevitable accumulation in cells of genetic defects that interfere with and ultimately disrupt the proper functioning of living processes. [The theory does not, however, explain how single-celled organisms or the germ cells of multicellular organisms (sperm and ova), which are in effect immortal, are able to escape this so-called **error catastrophe.**]

Living systems must maintain a nonequilibrium state for several reasons:

1. Only a nonequilibrium process can perform useful work.

2. The intricate regulatory functions characteristic of life require a nonequilibrium state because a process at equilibrium cannot be controlled (similarly, a ship that is dead in the water will not respond to its rudder).

3. The complex cellular and molecular systems that conduct biological processes can be maintained only in the nonequilibrium state. Living systems are inherently unstable because they are degraded by the very biochemical reactions to which they give rise. Their regeneration, which must occur almost simultaneously with their degradation, requires the continuous influx of free energy. For example, the ATP-generating consumption of glucose (Section 17-2), as has been previously mentioned, occurs with the initial consumption of ATP through its reactions with glucose to form glucose-6-phosphate and with fructose-6-phosphate to form fructose-1,6-bisphosphate. Consequently, if metabolism is suspended long enough to exhaust the available ATP supply, glucose metabolism cannot be resumed. Life therefore differs in a fundamental way from a complex machine such as a computer. Both require a throughput of free energy to be active. However, the function of the machine is based on a static structure, so that the machine can be repeatedly switched on and off. Life, in contrast, is based on a self-destructing but self-renewing process, which once interrupted, cannot be reinitiated.

B. *Nonequilibrium Thermodynamics and the Steady State*

In a nonequilibrium process, something (such as matter, electrical charge, or heat) must flow, that is, change its spatial distribution. In classical mechanics, the acceleration of mass occurs in response to force. *Similarly, flow in a thermodynamic system occurs in response to a thermodynamic force* (**driving force**), *which results from the system's nonequilibrium state.* For example, the flow of matter in diffusion is motivated by the thermodynamic force of a concentration gradient; the migration of electrical charge (electric current) occurs in response to a gradient in an electric field (a voltage difference); the transport of heat results from a temperature gradient; and a chemical reaction results from a difference in chemical potential. Such flows are said to be **conjugate** to their thermodynamic force.

A thermodynamic force may also promote a **nonconjugate flow** under the proper conditions. For example, a gradient in the concentration of matter can give rise to an electric current (a concentration cell), heat (such as occurs on mixing H_2O and HCl), or a chemical reaction (the mitochondrial production of ATP through the dissipation of a proton gradient). Similarly, a gradient in electrical potential can motivate a flow of matter (electrophoresis), heat (resistive heating), or a chemical reaction (the charging of a battery). When a thermodynamic force stimulates a nonconjugate flow, the process is called **energy transduction.**

a. Living Things Maintain the Steady State

Living systems are, for the most part, characterized by being in a **steady state.** By this it is meant that all flows in the system are constant, so that the system does not change with time. Some environmental steady-state processes are schematically illustrated in Fig. 16-31. Ilya Prigogine, a pioneer in the development of irreversible thermodynamics, has shown that a steady-state system produces the maximum amount of useful work for a given energy expenditure under the prevailing conditions. *The steady state of an open system is therefore its state of maximum thermodynamic efficiency.* Furthermore, in analogy with Le Châtelier's principle (Section 3-4A), slight perturbations from the steady state give rise to changes in flows that counteract these perturbations so as to return the system to the steady state. *The steady state of an open system is therefore analogous to the equilibrium state of an isolated system; both are stable states.*

In the following chapters we shall see that many biological regulatory mechanisms function to maintain a steady state. For example, the flow of reaction intermediates through a metabolic pathway is often inhibited by an excess of final product and stimulated by an excess of starting material through the allosteric regulation of its key enzymes (Section 13-4).

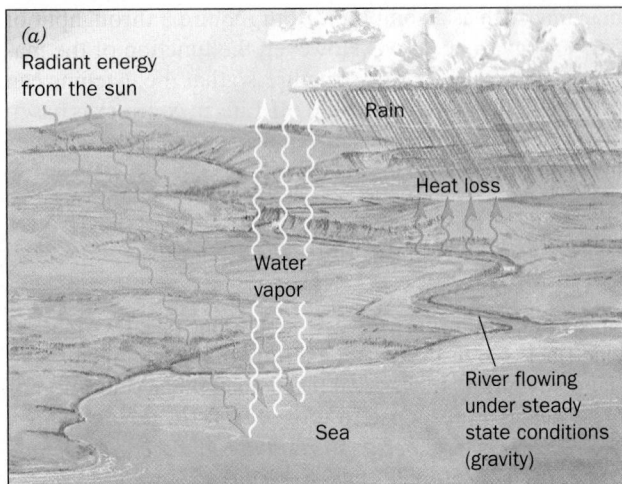

Figure 16-31   Two examples of open systems in a steady state.
(*a*) A constant flow of water in the river occurs under the
influence of the force of gravity. The water level in the reservoir
is maintained by rain, the major source of which is the
evaporation of seawater. Hence the entire cycle is ultimately
powered by the sun. (*b*) The steady state of the biosphere is
similarly maintained by the sun. Plants harness the sun's radiant
energy to synthesize carbohydrates from CO_2 and H_2O. The
eventual metabolism of the carbohydrates by the plants or by the
animals that eat them results in the release of their stored free
energy and the return of the CO_2 and H_2O to the environment to
complete the cycle.

Living things have apparently evolved so as to take maximum
thermodynamic advantage of their environments.

C. *Thermodynamics of Metabolic Control*

a. **Enzymes Selectively Catalyze Required Reactions**
*Biological reactions are highly specific; only reactions that
lie on metabolic pathways take place at significant rates despite the many other thermodynamically favorable reactions
that are also possible.* As an example, let us consider the reactions of ATP, glucose, and water. Two thermodynamically
favorable reactions that ATP can undergo are phosphoryl
transfer to form ADP and glucose-6-phosphate, and hydrolysis to form ADP and P_i (Fig. 16-23*a*). The free energy profiles
of these reactions are diagrammed in Fig. 16-32. ATP hydrolysis is thermodynamically favored over the phosphoryl
transfer to glucose. However, their relative rates are determined by their free energies of activation to their transition

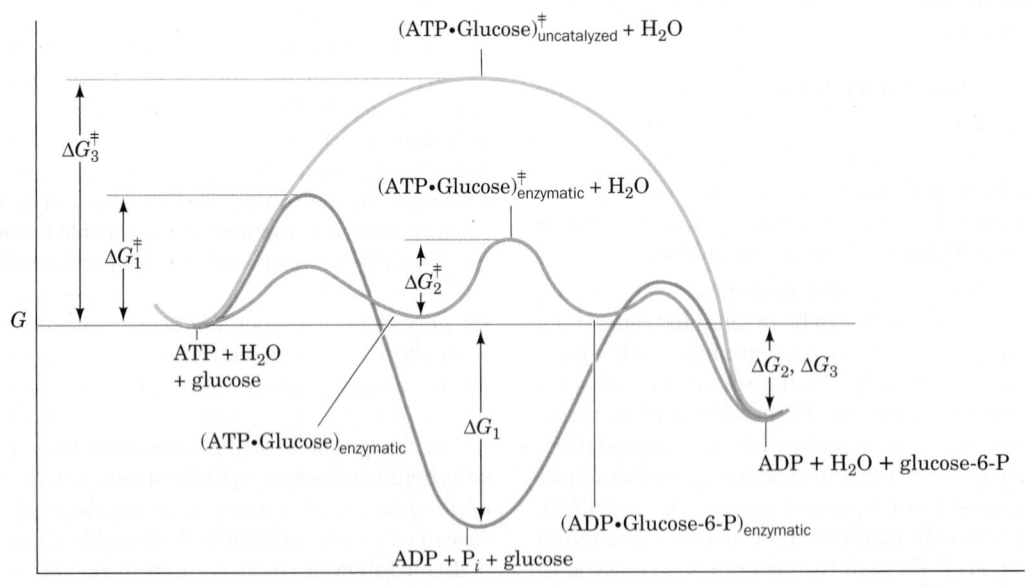

Figure 16-32   Reaction coordinate diagrams. These are (1) the
reaction of ATP and water (*purple curve*), and the reaction of
ATP and glucose (2) in the presence (*orange curve*) and (3) in
the absence (*yellow curve*) of an appropriate enzyme. Although
the hydrolysis of ATP is a more exergonic reaction than the
phosphorylation of glucose (ΔG_1 is more negative than ΔG_2), the
latter reaction is predominant in the presence of a suitable
enzyme because it is kinetically favored ($\Delta G_2^{\ddagger} < \Delta G_1^{\ddagger}$).

states ($\Delta G^{\ddagger}$ values; Section 14-1Cb) and the relative concentrations of glucose and water. The larger $\Delta G^{\ddagger}$, the slower the reaction. In the absence of enzymes, $\Delta G^{\ddagger}$ for the phosphoryl-transfer reaction is greater than that for hydrolysis, so the hydrolysis reaction predominates (although neither reaction occurs at a biologically significant rate).

The free energy barriers of both of the nonenzymatic reactions are far higher than that of the enzyme-catalyzed phosphoryl transfer to glucose. Hence enzymatic formation of glucose-6-phosphate is kinetically favored over the nonenzymatic hydrolysis of ATP. *It is the role of an enzyme, in this case hexokinase, to selectively reduce the free energy of activation of a chemically coupled reaction so that it approaches equilibrium faster than the more thermodynamically favored uncoupled reaction.*

b. Many Enzymatic Reactions Are Near Equilibrium

Although metabolism as a whole is a nonequilibrium process, many of its component reactions function close to equilibrium. The reaction of ATP and creatine to form phosphocreatine (Section 16-4Cd) is an example of such a reaction. The ratio [creatine]/[phosphocreatine] depends on [ATP] because creatine kinase, the enzyme catalyzing this reaction, has sufficient activity to equilibrate the reaction rapidly. The net rate of such an equilibrium reaction is effectively controlled by varying the concentrations of its reactants and/or products.

c. Pathway Throughput Is Regulated by Controlling Enzymes Operating Far from Equilibrium

Other biological reactions function far from equilibrium. For example, the phosphofructokinase reaction (Fig. 16-28)

has an equilibrium constant of $K'_{eq} = 300$ but under physiological conditions in rat heart muscle has the mass action ratio [fructose-1,6-bisphosphate][ADP]/[fructose-6-phosphate][ATP] = 0.03, which corresponds to $\Delta G = -25.7$ kJ · mol^{-1} (Eq. [3.15]). This situation arises from a buildup of reactants because there is insufficient phosphofructokinase activity to equilibrate the reaction. Changes in substrate concentrations therefore have relatively little effect on the rate of the phosphofructokinase reaction; the enzyme is close to saturation. Only changes in the activity of the enzyme, through allosteric interactions, for example, can significantly alter this rate. An enzyme such as phosphofructokinase is therefore analogous to a dam on a river. Substrate **flux** (rate of flow) is controlled by varying its activity (allosterically or by other means), much as a dam controls the flow of a river below the dam by varying the opening of its floodgates (when the water levels on the two sides of the dam are different, that is, when they are not at equilibrium).

Understanding of how reactant flux in a metabolic pathway is controlled requires knowledge of which reactions are functioning near equilibrium and which are far from it. Most enzymes in a metabolic pathway operate near equilibrium and therefore have net rates that are sensitive only to their substrate concentrations. However, as we shall see in the following chapters (particularly in Section 17-4), *certain enzymes, which are strategically located in a metabolic pathway, operate far from equilibrium. These enzymes, which are targets for metabolic regulation by allosteric interactions and other mechanisms, are responsible for the maintenance of a stable steady-state flux of metabolites through the pathway.* This situation, as we have seen, maximizes the pathway's thermodynamic efficiency.

CHAPTER SUMMARY

1 Metabolic Pathways Metabolic pathways are series of consecutive enzymatically catalyzed reactions that produce specific products for use by an organism. The free energy released by degradation (catabolism) is, through the intermediacy of ATP and NADPH, used to drive the endergonic processes of biosynthesis (anabolism). Carbohydrates, lipids, and proteins are all converted to the common intermediate acetyl-CoA, whose acetyl group is then converted to CO_2 and H_2O through the action of the citric acid cycle and oxidative phosphorylation. A relatively few metabolites serve as starting materials for a host of biosynthetic products. Metabolic pathways have five principal characteristics: (1) Metabolic pathways are irreversible; (2) if two metabolites are interconvertible, the synthetic route from the first to the second must differ from the route from the second to the first; (3) every metabolic pathway has an exergonic first committed step; (4) all metabolic pathways are regulated, usually at the first committed step; and (5) metabolic pathways in eukaryotes occur in specific subcellular compartments.

2 Organic Reaction Mechanisms Almost all metabolic reactions fall into four categories: (1) group-transfer reactions; (2) oxidation–reduction reactions; (3) eliminations, isomerizations, and rearrangements; and (4) reactions that make or break

carbon–carbon bonds. Most of these reactions involve heterolytic bond cleavage or formation occurring through the addition of nucleophiles to electrophilic carbon atoms. Group-transfer reactions therefore involve transfer of an electrophilic group from one nucleophile to another. The main electrophilic groups transferred are acyl groups, phosphoryl groups, and glycosyl groups. The most common nucleophiles are amino, hydroxyl, imidazole, and sulfhydryl groups. Electrophiles participating in metabolic reactions are protons, metal ions, carbonyl carbon atoms, and cationic imines. Oxidation–reduction reactions involve loss or gain of electrons. Oxidation at carbon usually involves C—H bond cleavage, with the ultimate loss by C of the two bonding electrons through their transfer to an electron acceptor such as NAD$^+$. The terminal electron acceptor in aerobes is O_2. Elimination reactions are those in which a C=C double bond is created from two saturated carbon centers with the loss of H_2O, NH_3, ROH, or RNH_2. Dehydration reactions are the most common eliminations. Isomerizations involve shifts of double bonds within molecules. Rearrangements are biochemically uncommon reactions in which intramolecular C—C bonds are broken and reformed to produce new carbon skeletons. Reactions that make and break C—C bonds form the basis of both degradative and biosynthetic metabolism. In the synthetic direction, these reactions involve addition of a nucleophilic carbanion to an electrophilic carbon atom. The most common electrophilic carbon atom is the carbonyl carbon, whereas carbanions are usually generated by removal of a proton from a carbon atom adjacent to a carbonyl group or by decarboxylation of a β-keto acid.

3 Experimental Approaches to the Study of Metabolism Experimental approaches employed in elucidating metabolic pathways include the use of metabolic inhibitors, growth studies, and biochemical genetics. Metabolic inhibitors block pathways at specific enzymatic steps. Identification of the resulting intermediates indicates the course of the pathway. Mutations, which occur naturally in genetic diseases or can be induced by mutagens, X-rays, or genetic engineering, may also result in the absence or inactivity of an enzyme. Modern genetic techniques make it possible to express foreign genes in higher organisms (transgenic animals) or inactivate (knock out) a gene and study the effects of these changes on metabolism. When isotopic labels are incorporated into metabolites and allowed to enter a metabolic system, their paths may be traced from the distribution of label in the intermediates. NMR is a noninvasive technique that may be used to detect and study metabolites *in vivo*. Studies on isolated organs, tissue slices, cells, and subcellular organelles have contributed enormously to our knowledge of the localization of metabolic pathways. Systems biology endeavors to quantitatively describe the properties and dynamics of biological networks as a whole through the integration of genomic, transcriptomic, proteomic, and metabolomic information.

4 Thermodynamics of Phosphate Compounds Free energy is supplied to endergonic metabolic processes by the ATP produced via exergonic metabolic processes. ATP's -30.5 kJ · mol^{-1} $\Delta G^{\circ\prime}$ of hydrolysis is intermediate between those of "high-energy" metabolites such as phosphoenolpyruvate and "low-energy" metabolites such as glucose-6-phosphate. The "high-energy" phosphoryl groups are enzymatically transferred to ADP, and the resulting ATP, in a separate reaction, phosphorylates "low-energy" compounds. ATP may also undergo pyrophosphate cleavage to yield PP$_i$, whose subsequent hydrolysis adds further thermodynamic impetus to the reaction. ATP is present in too short a supply to act as an energy reservoir. This function, in vertebrate nerve and muscle cells, is carried out by phosphocreatine, which under low-ATP conditions readily transfers its phosphoryl group to ADP to form ATP.

5 Oxidation–Reduction Reactions The half-reactions of redox reactions may be physically separated to form two electrochemical half-cells. The redox potential for the reduction of A by B,

$$A_{ox}^{n+} + B_{red} \rightleftharpoons A_{red} + B_{ox}^{n+}$$

in which n electrons are transferred, is given by the Nernst equation

$$\Delta\mathscr{E} = \Delta\mathscr{E}^{\circ} - \frac{RT}{n\mathscr{F}}\ln\left(\frac{[A_{red}][B_{ox}^{n+}]}{[A_{ox}^{n+}][B_{red}]}\right)$$

The redox potential of such a reaction is related to the reduction potentials of its component half-reactions, $\mathscr{E}_A$ and $\mathscr{E}_B$, by

$$\Delta\mathscr{E} = \mathscr{E}_A - \mathscr{E}_B$$

If $\mathscr{E}_A > \mathscr{E}_B$, then A_{ox}^{n+} has a greater electron affinity than does B_{ox}^{n+}. The reduction potential scale is defined by arbitrarily setting the reduction potential of the standard hydrogen half-cell to zero. Redox reactions are of great metabolic importance. For example, the oxidation of NADH yields 2.5 ATPs through the mediation of the electron-transport chain.

6 Thermodynamics of Life Living organisms are open systems and therefore cannot be at equilibrium. They must continuously dissipate free energy in order to carry out their various functions and preserve their highly ordered structures. The study of nonequilibrium thermodynamics has indicated that the steady state, which living processes maintain, is the state of maximum efficiency under the constraints governing open systems. Control mechanisms that regulate biological processes preserve the steady state by regulating the activities of enzymes that are strategically located in metabolic pathways.

REFERENCES

Metabolic Studies

Aebersold, R., Quantitative proteome analysis: Methods and applications, *J. Infect. Dis.* **182** (supplement 2), S315–S320 (2003).

Beadle, G.W., Biochemical genetics, *Chem. Rev.* **37**, 15–96 (1945). [A classic review summarizing the "one gene–one enzyme" hypothesis.]

Campbell A.M. and Heyer L.J., *Discovering Genomics, Proteomics and Bioinformatics* (2nd ed.), Pearson Benjamin Cummings, New York (2007). [An interactive introduction to these subjects.]

Cerdan, S. and Seelig, J., NMR studies of metabolism, *Annu. Rev. Biophys. Biophys. Chem.* **19**, 43–67 (1990).

Choi, S. (Ed.), *Introduction to Systems Biology,* Humana Press (2007).

Cooper, T.G., *The Tools of Biochemistry,* Chapter 3, Wiley-Interscience (1977). [A presentation of radiochemical techniques.]

Duarte, N.C., Becker, S.A., Jamshidi, N., Thiele, I., Mo, M.L., Vo, T.D., Srivas, R., and Palsson, B. Ø., Global reconstruction of the human metabolic network based on genomic and bibliomic data, *Proc. Natl. Acad. Sci.* **104**, 1777–1782 (2007).

Freifelder, D., *Biophysical Chemistry* (2nd ed.), Chapters 5 and 6, Freeman (1982). [A discussion of the principles of radioactive counting and autoradiography.]

Go, V.L.W., Nguyen, C.T.H., Harris, D.M., and Lee, W.-N.P., Nutrient-gene interaction: Metabolic genotype-phenotype relationship, *J. Nutr.* **135**, 2016s–3020s (2005).

Hevesy, G., Historical sketch of the biological application of tracer elements, *Cold Spring Harbor Symp. Quant. Biol.* **13**, 129–150 (1948).

Jeffrey, F.M.H., Rajagopal, A., Malloy, C.R., and Sherry, A.D., ^{13}C-NMR: A simple yet comprehensive method for analysis of intermediary metabolism, *Trends Biochem. Sci.* **16**, 5–10 (1991).

Michal, G. (Ed.), *Biochemical Pathways. An Atlas of Biochemistry and Molecular Biology,* Wiley (1999). [An encyclopedic compendium of metabolic pathways.]

Shemin, D. and Rittenberg, D., The biological utilization of glycine for the synthesis of the protoporphyrin of hemoglobin, *J. Biol Chem.* **166**, 621–625 (1946).

Shulman, R.G. and Rothman, D.L., ^{13}C NMR of intermediary metabolism: Implications for systematic physiology, *Annu. Rev. Physiol.* **63**, 15–48 (2001).

Smolin, L.A. and Grosvenor, M.B, *Nutrition: Science and Applications,* Wiley (2008). [A good text for those interested in pursuing nutritional aspects of metabolism].

Suckling, K.E. and Suckling, C.J., *Biological Chemistry,* Cambridge University Press (1980). [Presents the organic chemistry of biochemical reactions.]

Walsh, C., *Enzymatic Reaction Mechanisms,* Chapter 1, Freeman (1979). [A discussion of the types of biochemical reactions.]

Wang, N.-D., Finegold, M.J., Bradley, A., Ou, C.N., Abdelsayed, S.V., Wilde, M.D., Taylor, L.R., Wilson, D.R., and Darlington, G.J., Impaired energy homeostasis in C/EBPα knockout mice, *Science* **269**, 1108–1112 (1995).

Weckwerth, W. (Ed.), *Metabolomics. Methods and Protocols,* Humana Press (2007).

Westheimer, F.H., Why nature chose phosphates, *Science* **235**, 1173–1178 (1987).

Xia, Y., Yu, H., Jansen, R., Seringhaus, M., Baxter, S., Greenbaum, D., Zhao, H., and Gerstein, M., Analyzing cellular biochemistry in terms of molecular networks, *Annu. Rev. Biochem.* **73**, 1051–1087 (2004).

Zhu, H., Bilgin, M., and Snyder, M., Proteomics, *Annu. Rev. Biochem.* **72**, 783–812 (2003).

Bioenergetics

Alberty, R.A., Standard Gibbs free energy, enthalpy and entropy changes as a function of pH and pMg for reactions involving adenosine phosphates, *J. Biol. Chem.* **244**, 3290–3302 (1969).

Alberty, R.A., Calculating apparent equilibrium constants of enzyme-catalyzed reactions at pH 7, *Biochem. Ed.* **28**, 12–17 (2000).

Caplan, S.R., Nonequilibrium thermodynamics and its application to bioenergetics, *Curr. Top. Bioenerg.* **4**, 1–79 (1971).

Crabtree, B. and Taylor, D.J., Thermodynamics and metabolism, *in* Jones, M.N. (Ed.), *Biochemical Thermodynamics*, pp. 333–378, Elsevier (1979).

Dickerson, R.E., *Molecular Thermodynamics,* Chapter 7, Benjamin (1969). [An interesting chapter on the thermodynamics of life.]

Henley, H.J.M., An introduction to nonequilibrium thermodynamics, *J. Chem. Ed.* **41**, 647–655 (1964).

Katchelsky, A. and Curran, P.F., *Nonequilibrium Thermodynamics in Biophysics,* Harvard University Press (1965).

Morowitz, H.J., *Foundations of Bioenergetics,* Academic Press (1978).

BIOINFORMATICS EXERCISES

Bioinformatics Exercises are available at www.wiley.com/college/voet.

Chapter 16

Metabolic Enzymes. Explore databases concerned with metabolism.

PROBLEMS

1. Glycolysis (glucose breakdown) has the overall stoichiometry:

Glucose + 2ADP + 2P$_i$ + 2NAD$^+$ $\longrightarrow$

2 pyruvate$^-$ + 2ATP + 2NADH + 4H$^+$ + 2H$_2$O

whereas that of gluconeogenesis (glucose synthesis) is

2 pyruvate + 6ATP + 2NADH + 4H$^+$ + 6H$_2$O $\longrightarrow$

glucose + 6ADP + 6P$_i$ + 2NAD$^+$

What is the overall stoichiometry of the glycolytic breakdown of 1 mol of glucose followed by its gluconeogenic synthesis? Explain why it is necessary that the pathways of these two processes be independently controlled and why they must differ by at least one reaction.

2. It has been postulated that a trigonal bipyramidal pentacovalent phosphorus intermediate can undergo a vibrational deformation process known as **pseudorotation** in which its apical ligands exchange with two of its equatorial ligands via a tetragonal pyramidal transition state:

**Trigonal bipyramid
[X and Y apical]**   **Trigonal bipyramid
[O2 and O3 apical]**

**Tetragonal pyramidal
transition state**

In a nucleophilic substitution reaction, would two cycles of pseudorotation, so as to place the leaving group (X) in an apical position and the attacking group (Y) in an equatorial position, lead to retention or inversion of configuration on the departure of the leaving group?

3. One **Curie (Ci)** of radioactivity is defined as 3.70×10^{10} disintegrations per second, the number that occurs in 1 g of pure ^{226}Ra. A sample of ^{14}CO$_2$ has a specific radioactivity of 5 μCi · μmol^{-1}. What percentage of its C atoms are ^{14}C?

4. In the hydrolysis of ATP to ADP and P$_i$, the equilibrium concentration of ATP is too small to be measured accurately. A better way of determining K'_{eq}, and hence $\Delta G^{\circ\prime}$ of this reaction, is to break it up into two steps whose values of $\Delta G^{\circ\prime}$ can be accurately determined. This has been done using the following pair of reactions (the first being catalyzed by **glutamine synthetase**):

(1) ATP + glutamate + NH$_4^+$ $\rightleftharpoons$ ADP + P$_i$ + glutamine + H$^+$

$$\Delta G_1^{\circ\prime} = -16.3 \text{ kJ} \cdot \text{mol}^{-1}$$

(2) Glutamate + NH$_4^+$ $\rightleftharpoons$ glutamine + H$_2$O + H$^+$

$$\Delta G_2^{\circ\prime} = -14.2 \text{ kJ} \cdot \text{mol}^{-1}$$

What is the $\Delta G^{\circ\prime}$ of ATP hydrolysis according to these data?

*5. Consider the reaction catalyzed by hexokinase:

ATP + glucose $\rightleftharpoons$ ADP + glucose-6-phosphate

A mixture containing 40 mM ATP and 20 mM glucose was incubated with hexokinase at pH 7 and 25°C. Calculate the equilibrium concentrations of the reactants and products (see Table 16-3).

6. In aerobic metabolism, glucose is completely oxidized in the reaction

Glucose + 6 O$_2$ $\rightleftharpoons$ 6CO$_2$ + 6H$_2$O

with the coupled generation of 32 ATP molecules from 32 ADP and 32 P$_i$. Assuming the ΔG for the hydrolysis of ATP to ADP and

P_i under intracellular conditions is $-50\,\mathrm{kJ \cdot mol^{-1}}$ and that for the combustion of glucose is $-2823.2\,\mathrm{kJ \cdot mol^{-1}}$, what is the efficiency of the glucose oxidation reaction in terms of the free energy sequestered in the form of ATP?

7. Typical intracellular concentrations of ATP, ADP, and P_i in muscles are 5.0, 0.5, and 1.0 m*M*, respectively. At 25°C and pH 7: (a) What is the free energy of hydrolysis of ATP at these concentrations? (b) Calculate the equilibrium concentration ratio of phosphocreatine to creatine in the creatine kinase reaction:

$$\text{Creatine} + \text{ATP} \rightleftharpoons \text{phosphocreatine} + \text{ADP}$$

if ATP and ADP have the above concentrations. (c) What concentration ratio of ATP to ADP would be required under the foregoing conditions to yield an equilibrium concentration ratio of phosphocreatine to creatine of 1? Assuming the concentration of P_i remained 1.0 m*M*, what would the free energy of hydrolysis of ATP be under these latter conditions?

*8. Assuming the intracellular concentrations of ATP, ADP, and P_i, are those given in Problem 7: (a) Calculate the concentration of AMP at pH 7 and 25°C under the condition that the adenylate kinase reaction:

$$2\text{ADP} \rightleftharpoons \text{ATP} + \text{AMP}$$

is at equilibrium. (b) Calculate the equilibrium concentration of AMP when the free energy of hydrolysis of ATP to ADP and P_i is $-55\,\mathrm{kJ \cdot mol^{-1}}$. Assume $[P_i]$ and $([\text{ATP}] + [\text{ADP}])$ remain constant.

9. Using the data in Table 16-4, list the following substances in order of their decreasing oxidizing power: (a) fumarate$^-$, (b) cystine, (c) O_2, (d) NADP$^+$, (e) cytochrome *c* (Fe^{3+}), and (f) lipoic acid.

10. Calculate the equilibrium concentrations of reactants and products for the reaction:

$$\text{Acetoacetate}^- + \text{NADH} + \text{H}^+ \rightleftharpoons$$
$$\beta\text{-hydroxybutyrate}^- + \text{NAD}^+$$

when the initial concentrations of acetoacetate$^-$ and NADH are 0.01 and 0.005*M*, respectively, and β-hydroxybutyrate$^-$ and NAD$^+$ are initially absent. Assume the reaction takes place at 25°C and pH 7.

11. In anaerobic bacteria, the final metabolic electron acceptor is some molecule other than O_2. A major requirement for any redox pair utilized as a metabolic free energy source is that it provides sufficient free energy to generate ATP from ADP and P_i. Indicate which of the following redox pairs are sufficiently exergonic to enable a properly equipped bacterium to utilize them as a major energy source. Assume that redox reactions forming ATP require two electrons and that $\Delta\mathscr{E} = \Delta\mathscr{E}^{\circ\prime}$.

(a) Ethanol + NO$_3^-$ (c) H$_2$ + S
(b) Fumarate$^-$ + SO$_3^{2-}$ (d) Acetaldehyde + acetaldehyde

12. Calculate $\Delta G^{\circ\prime}$ for the following pairs of half-reactions at pH 7 and 25°C. Write a balanced equation for the overall reaction and indicate the direction in which it occurs spontaneously under standard conditions.

(a) (H$^+$/$\frac{1}{2}$H$_2$) and ($\frac{1}{2}$O$_2$ + 2H$^+$/H$_2$O)
(b) (Pyruvate$^-$ + 2H$^+$/lactate$^-$) and (NAD$^+$ + H$^+$/NADH)

*13. The chemiosmotic hypothesis (Section 22-3A) postulates that ATP is generated in the two-electron reaction:

$$\text{ADP} + P_i + 2\text{H}^+ \text{ (low pH)} \rightleftharpoons$$
$$\text{ATP} + \text{H}_2\text{O} + 2\text{H}^+ \text{ (high pH)}$$

which is driven by a metabolically generated pH gradient in the mitochondria. What is the magnitude of the pH gradient required for net synthesis of ATP at 25°C and pH 7, if the steady-state concentrations of ATP, ADP, and P_i are 0.01, 10, and 10 m*M*, respectively?

14. Gastric juice is 0.15*M* HCl. The blood plasma, which is the source of this H$^+$ and Cl$^-$, is 0.10*M* in Cl$^-$ and has a pH of 7.4. Calculate the free energy necessary to produce the HCl in 0.1 L of gastric juice at 37°C.

Glycolysis

CHAPTER 17

At this point we commence our discussions of specific metabolic pathways by considering **glycolysis** (Greek: *glykos*, sweet; *lysis*, loosening), the pathway by which **glucose** is converted via **fructose-1,6-bisphosphate** to **pyruvate** with the generation of 2 mol of ATP per mole of glucose. This sequence of 10 enzymatic reactions, which is probably the most completely understood biochemical pathway, plays a key role in energy metabolism by providing a significant portion of the energy utilized by most organisms and by preparing glucose, as well as other carbohydrates, for oxidative degradation.

In our study of glycolysis, and indeed of all of metabolism, we shall attempt to understand the pathway on four levels:

1. The chemical interconversion steps, that is, the sequence of reactions by which glucose is converted to the pathway's end products.

2. The mechanism of the enzymatic conversion of each pathway intermediate to its successor.

3. The energetics of the conversions.

4. The mechanisms controlling the **flux** (rate of flow) of metabolites through the pathway.

The flux of metabolites through a pathway is remarkably sensitive to the needs of the organism for the products of the pathway. Through an exquisitely complex network of control mechanisms, flux through a pathway is only as great as required.

1 THE GLYCOLYTIC PATHWAY

An overview of glucose metabolism is diagrammed in Fig. 17-1. *Under aerobic conditions, the pyruvate formed by glycolysis is further oxidized by the citric acid cycle (Chapter 21) and oxidative phosphorylation (Chapter 22) to CO_2 and water. Under anaerobic conditions, however, the pyruvate is instead converted to a reduced end product, which is **lactate** in muscle (**homolactic fermentation;** a fermentation is an anaerobic biological reaction process) and ethanol + CO_2 in yeast (**alcoholic fermentation**).*

A. Historical Perspective

The fermentation of glucose to ethanol and CO_2 by yeast (Fig. 17-2) has been a useful process since before the dawn of recorded history. Winemaking and bread baking both exploit this process. Yet the scientific investigation of the mechanism of glycolysis began only in the latter half of the nineteenth century.

In the years 1854 to 1864, Louis Pasteur established that fermentation is caused by microorganisms. It was not until 1897, however, that Eduard Buchner demonstrated that cell-free yeast extracts can also carry out this process.

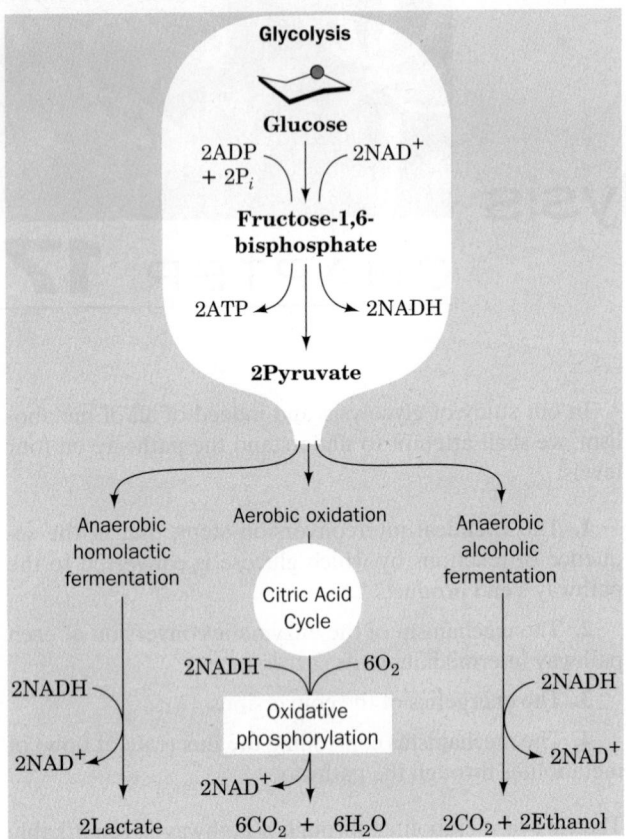

Figure 17-1 Glycolysis. Glycolysis converts glucose to pyruvate while generating two ATPs. Under anaerobic conditions, alcoholic fermentation of pyruvate occurs in yeast, whereas homolactic fermentation occurs in muscle. Under aerobic conditions, pyruvate is oxidized to H_2O and CO_2 via the citric acid cycle (Chapter 21) and oxidative phosphorylation (Chapter 22).

Figure 17-2 Electron micrograph of yeast cells. [Biophoto Associates/Photo Researchers, Inc.]

This discovery refuted the then widely held belief that fermentation, and every other biological process, was mediated by some "vital force" inherent in living matter, and thereby brought glycolysis within the province of chemistry. This was a major step in the development of biochemistry as a science. Although, in principle, the use of cell-free extracts enabled a systematic "dissection" of the reactions involved in the pathway, the complete elucidation of the glycolytic pathway was still a long-range project because analytical techniques for the isolation and identification of intermediates and enzymes had to be developed concurrently.

In the years 1905 to 1910, Arthur Harden and William Young made two important discoveries:

1. Inorganic phosphate is required for fermentation and is incorporated into fructose-1,6-bisphosphate, an intermediate in the process.

2. A cell-free yeast extract can be separated, by dialysis, into two fractions that are both required for fermentation: A nondialyzable heat-labile fraction they named **zymase**; and a dialyzable, heat-stable fraction they called **cozymase.** It was shown later by others that zymase is a mixture of enzymes and that cozymase is a mixture of cofactors: coenzymes such as NAD^+, ATP, and ADP, as well as metal ions.

In their efforts to identify pathway intermediates, the early investigators of glycolysis developed a general technique of metabolic investigation that is in use today: *Reagents are found that inhibit the production of pathway products, thereby causing the buildup of metabolites that can then be identified as pathway intermediates.* Over years of investigation that attempted to identify glycolytic intermediates, various reagents were found that inhibit the production of ethanol from glucose in yeast extracts. The use of different inhibitors results in the accumulation of different intermediates. For example, the addition of iodoacetate to fermenting yeast extracts causes the buildup of fructose-1,6-bisphosphate, whereas addition of fluoride ion induces the accumulation of **3-phosphoglycerate** and **2-phosphoglycerate:**

3-Phosphoglycerate   **2-Phosphoglycerate**

The mechanisms by which these inhibitors act are discussed in Sections 17-2Da and 17-2I, respectively.

One remarkable finding of these studies was that the same intermediates and enzyme activities could be isolated not only from yeast, but from a great variety of other organisms. With few exceptions (see Problem 11 in this chapter), *living things all metabolize glucose by identical pathways. In spite of their enormous diversity, they share a common biochemistry.*

By 1940, the efforts of many investigators had come to fruition with the elucidation of the complete pathway of glycolysis. The work of three of these individuals, Gustav Embden, Otto Meyerhof, and Jacob Parnas, has been commemorated in that glycolysis is alternatively known as the **Embden–Meyerhof–Parnas pathway.** Other major contributors to the elucidation of this pathway were Carl and Gerty Cori, Carl Neuberg, Robert Robison, and Otto Warburg.

B. *Pathway Overview*

Before beginning our detailed discussion of the enzymes of glycolysis, let us first take a moment to survey the overall pathway as it fits in with animal metabolism as a whole. Glucose usually arises in the blood as a result of the breakdown of higher polysaccharides (Sections 11-2B, 11-2Db, and 18-1) or from its synthesis from noncarbohydrate sources (**gluconeogenesis;** Section 23-1). The fate of nonglucose hexoses is discussed in Section 17-5. Glucose enters most cells by specific carriers that transport it from the exterior of the cell into the cytosol (Section 20-2E). *The enzymes of glycolysis are located in the cytosol, where they are only loosely associated, if at all, with cell structures such as membranes.* However, there is considerable circumstantial evidence that successive enzymes in the glycolytic pathway loosely associate, presumably to facilitate the efficient transfer of intermediates between enzymes. Such associations of functionally related enzymes have been referred to as **metabolons.** Nevertheless, no actual complexes of glycolytic enzymes have yet been isolated.

Glycolysis converts glucose to two C_3 units (pyruvate) of lower free energy in a process that harnesses the released free energy to synthesize ATP from ADP and P_i. This process requires a pathway of chemically coupled phosphoryl-transfer reactions (Sections 16-4 and 16-6). Thus the chemical strategy of glycolysis is

1. Add phosphoryl groups to the glucose.

2. Chemically convert phosphorylated intermediates into compounds with high phosphate group-transfer potentials.

3. Chemically couple the subsequent hydrolysis of reactive substances to ATP synthesis.

The 10 enzyme-catalyzed reactions of glycolysis are diagrammed in Fig. 17-3. Note that ATP is used early in the pathway to synthesize phosphoryl compounds (Reactions 1 and 3) but is later resynthesized (Reactions 7 and 10). Glycolysis may therefore be considered to occur in two stages:

Stage I (Reactions 1–5): A preparatory stage in which the hexose glucose is phosphorylated and cleaved to yield two molecules of the triose **glyceraldehyde-3-phosphate.** This process utilizes two ATPs in a kind of energy investment.

Stage II (Reactions 6–10): The two molecules of glyceraldehyde-3-phosphate are converted to pyruvate, with concomitant generation of four ATPs. Glycolysis therefore has a net profit of two ATPs per glucose: Stage I consumes two ATPs; Stage II produces four ATPs.

The overall reaction is

$$\text{Glucose} + 2\text{NAD}^+ + 2\text{ADP} + 2\text{P}_i \longrightarrow$$
$$2\text{NADH} + 2 \text{ pyruvate} + 2\text{ATP} + 2\text{H}_2\text{O} + 4\text{H}^+$$

a. The Oxidizing Power of NAD^+ Must Be Recycled

NAD^+ is the primary oxidizing agent of glycolysis. The NADH produced by this process (Fig. 17-3, Reaction 6) must be continually reoxidized to keep the pathway supplied with NAD^+. There are three common ways that this occurs (Fig. 17-1, *bottom*):

1. Under anaerobic conditions in muscle, NAD^+ is regenerated when NADH reduces pyruvate to lactate (homolactic fermentation; Section 17-3A).

2. Under anaerobic conditions in yeast, pyruvate is decarboxylated to yield CO_2 and acetaldehyde and the latter is reduced by NADH to yield NAD^+ and ethanol (alcoholic fermentation; Section 17-3B).

3. Under aerobic conditions, the mitochondrial oxidation of each NADH to NAD^+ yields 2.5 ATPs (Section 22-2A).

Thus, in aerobic glycolysis, NADH may be thought of as a "high-energy" compound, whereas in anaerobic glycolysis its free energy of oxidation is dissipated as heat.

2 THE REACTIONS OF GLYCOLYSIS

See Guided Exploration 14: Glycolysis overview In this section we examine the reactions of glycolysis more closely, describing the properties of the individual enzymes and their mechanisms. In Section 17-3 we consider the anaerobic fate of pyruvate. Finally, in Section 17-4 we consider the thermodynamics of the entire process and address the problem of how the flux of metabolites through the pathway is controlled. As we study the individual glycolytic enzymes we shall encounter many organic reaction mechanisms (Section 16-2). Indeed, the study of organic reaction mechanisms has been invaluable in understanding the mechanisms by which enzymes catalyze reactions.

Figure 17-3 Degradation of glucose via the glycolytic pathway.
Glycolysis may be considered to occur in two stages. Stage I
(Reactions 1–5): Glucose is phosphorylated and cleaved to form
two molecules of the triose glyceraldehyde-3-phosphate. This
requires the expenditure of two ATPs in an "energy investment"
(Reactions 1 and 3). Stage II (Reactions 6–10): The two
molecules of glyceraldehyde-3-phosphate are converted to
pyruvate with the concomitant generation of four ATPs
(Reactions 7 and 10). 🔊 **See the Animated Figures**

Note that the X-ray structure of each of the 10 glycolytic enzymes has been reported. All of these enzymes are either homodimers or homotetramers with D_2 symmetry (Section 8-5B), whose subunits consist mainly of α/β domains (Section 8-3Bi).

A. *Hexokinase: First ATP Utilization*

Reaction 1 of glycolysis is the transfer of a phosphoryl group from ATP to glucose to form **glucose-6-phosphate (G6P)** in a reaction catalyzed by **hexokinase (HK):**

Glucose

hexokinase
Mg^{2+}

Glucose-6-phosphate (G6P)

A **kinase** is an enzyme that transfers phosphoryl groups between ATP and a metabolite (Section 16-4C). The metabolite that serves as the phosphoryl group acceptor for a specific kinase is identified in the prefix of the kinase name. HK is a relatively nonspecific enzyme contained in all cells that catalyzes the phosphorylation of hexoses such as D-glucose, D-mannose, and D-fructose. Liver cells also contain **glucokinase,** which catalyzes the same reaction but which is primarily involved in the maintenance of blood glucose levels (Section 18-3Fa). The second substrate for HK, as with other kinases, is an Mg^{2+}–ATP complex. In fact, uncomplexed ATP is a potent competitive inhibitor of HK. In what follows, we shall rarely refer to this Mg^{2+} requirement, but keep in mind that it is essential for kinase enzymatic activity (other divalent metal ions such as Mn^{2+} often satisfy the metal ion requirements of kinases *in vitro,* but Mg^{2+} is the normal physiological species).

a. Kinetics and Mechanism of the Hexokinase Reaction

Hexokinase has a Random Bi Bi mechanism in which the enzyme forms a ternary complex with glucose and Mg^{2+}–ATP before the reaction occurs. The Mg^{2+}, by complexing with the phosphate oxygen atoms, is thought to shield their negative charges, making the phosphorus atom

ATP   **Glucose**

Figure 17-4 **The nucleophilic attack of the C6—OH group of glucose on the γ phosphate of an Mg^{2+}–ATP complex.** The position of the Mg^{2+} ion is shown as an example; its actual position(s) has not been conclusively established. In any case, the Mg^{2+} functions to shield the negatively charged groups of ATP and thereby facilitates the nucleophilic attack.

more accessible for the nucleophilic attack of the C6—OH group of glucose (Fig. 17-4).

An important mechanistic question is why does HK catalyze the transfer of a phosphoryl group from ATP to glucose to yield G6P, but not to water to yield ADP + P_i (ATP hydrolysis)? Water is certainly small enough to fit into the phosphoryl acceptor group's enzymatic binding site. Furthermore, phosphoryl transfer from ATP to water is more exergonic than it is to glucose (Table 16-3), particularly since [H$_2$O] = 55.5M and [glucose] = 5 to 10 mM *in vivo*. Yet HK catalyzes phosphoryl transfer to glucose 40,000 times faster than it does to water.

The answer was provided by Thomas Steitz via X-ray structural studies of yeast HK. Comparison of the X-ray structures of HK and the glucose–HK complex indicates that *glucose induces a large conformational change in HK* (Fig. 17-5). The two lobes that form its active site cleft swing together by up to 11.5 Å so as to engulf the glucose in a manner that suggests the closing of jaws. *This movement places the ATP in close proximity to the —C6H$_2$OH group of glucose and excludes water from the active site (catalysis by proximity effects; Section 15-1E).* If the catalytic and reacting groups were in the proper position for reaction while the enzyme was in the open position (Fig. 17-5a), ATP hydrolysis would almost certainly be the dominant reaction. This conclusion is confirmed by the observation that **xylose,** which differs from glucose only by the lack of the —C6H$_2$OH group,

α-D-**Xylose**

greatly enhances the rate of ATP hydrolysis by HK (presumably xylose induces the activating conformational change while water occupies the binding site of the missing

(a)

(b)

Figure 17-5 **Conformational changes in yeast hexokinase on binding glucose.** The enzyme is represented by its transparent molecular surface with its embedded ribbon diagram colored with its large domain purple and its small domain yellow. (*a*) Free hexokinase. (*b*) Hexokinase in complex with glucose drawn in space-filling form with C green and O red. Note the prominent bilobal appearance of the free enzyme. In the enzyme–substrate complex these lobes have swung together by a 17° rotation to engulf the substrate. [Based on an X-ray structure by Igor Polikarpov, Instituto de Física em São Carlos, Brazil. PDBids 1IG8 and 3B8A.] **See Interactive Exercise 8**

hydroxymethyl group). Clearly, *this substrate-induced conformational change in HK is responsible for the enzyme's specificity.* In addition, the active site polarity is reduced by exclusion of water, thereby expediting the nucleophilic reaction process. Other kinases have the same deeply clefted structure as HK (e.g., Section 17-2G) and undergo conformational changes on binding their substrates. This suggests that all kinases have similar mechanisms for maintaining specificity.

B. *Phosphoglucose Isomerase*

Reaction 2 of glycolysis is the conversion of G6P to **fructose-6-phosphate (F6P)** by **phosphoglucose isomerase (PGI; also called glucose-6-phosphate isomerase).** This is the isomerization of an aldose to a ketose:

Glucose-6-phosphate (G6P)

phosphoglucose
isomerase (PGI)

Fructose-6-phosphate (F6P)

Since G6P and F6P both exist predominantly in their cyclic forms (Fig. 11-4 shows these structures for the unphosphorylated sugars), the reaction requires ring opening, followed by isomerization, and subsequent ring closure. The determination of the enzyme's pH dependence led to a hypothesis for amino acid side chain participation in the catalytic mechanism. The catalytic rate exhibits a bell-shaped pH dependence curve with characteristic pK's of 6.7 and 9.3, which suggests the catalytic participation of both a His and a Lys (Section 14-4). Indeed, comparison of the amino acid sequences of PGI from several different organisms reveals that both a His and Lys are conserved. However, a Glu residue is also conserved, and as we have seen for lysozyme (Section 15-2Ba), Glu can have an unusually high pK under certain conditions. In fact, the X-ray structure of PGI reveals that Glu 216 and His 388 form a hydrogen bonded catalytic dyad (resembling the interaction of the Asp and His residues in the catalytic triad of serine proteases; Fig. 15-20), which facilitates the action of His 388 as an acid–base catalyst.

A proposed reaction mechanism for the PGI reaction involves general acid–base catalysis by the enzyme (Fig. 17-6):

Step 1 Substrate binding.

Step 2 An acid, presumably the Lys ε-amino group, catalyzes ring opening.

Step 3 A base, presumably the imidazole portion of the His–Glu dyad, abstracts the acidic proton from C2 to form

Figure 17-6 **Reaction mechanism of phosphoglucose isomerase.** The active site catalytic residues (BH$^+$ and B′) are thought to be Lys and a His–Glu dyad, respectively.

Glucose-6-phosphate (G6P)

cis-**Enediolate intermediate**

Fructose-6-phosphate (F6P)

a *cis*-enediolate intermediate (this proton is acidic because it is α to a carbonyl group).

Step 4 The proton is replaced on C1 in an overall proton transfer. Protons abstracted by bases are labile and exchange rapidly with solvent protons. Nevertheless, Irwin Rose confirmed this step by demonstrating that [2-^{3}H]G6P is occasionally converted to [1-^{3}H]F6P by intramolecular proton transfer before the ^{3}H has had a chance to exchange with the medium.

Step 5 Ring closure to form the product, which is subsequently released to yield free enzyme, thereby completing the catalytic cycle.

PGI, like most enzymes, catalyzes reactions with nearly absolute stereospecificity. To appreciate this, let us compare the proposed enzymatic reaction mechanism with that in the nonenzymatic base-catalyzed isomerization of glucose, fructose, and mannose (Fig. 17-7). Glucose and mannose are epimers of one another because they differ only with

Glucose

cis-**Enediolate intermediate**

Mannose

Fructose

Figure 17-7 **Base-catalyzed isomerization of glucose, mannose, and fructose.** In the absence of enzyme, this reaction is nonstereospecific.

respect to their configuration at one chiral center, C2 (Section 11-1A). In the enediolate intermediate, as well as in the linear form of fructose, C2 lacks chirality. Therefore, in nonenzymatic systems, base-catalyzed isomerization of glucose also results in racemization of C2 with the production of mannose. In the presence of PGI, however, ^{1}H NMR measurements indicate that the rate of the isomerization reaction is several orders of magnitude greater than that of the epimerization reaction. Evidently, PGI shields the face of the enediolate to which H$^+$ must be added to form mannose-6-phosphate.

C. *Phosphofructokinase: Second ATP Utilization*

In Reaction 3 of glycolysis, **phosphofructokinase (PFK)** phosphorylates F6P to yield fructose-1,6-bisphosphate [**FBP** or **F1,6P;** previously known as **fructose-1,6-diphosphate (FDP)**]:

This reaction is similar to the hexokinase reaction (Reaction 1 in Fig. 17-3; Section 17-2A). PFK catalyzes the nucleophilic attack by the C1—OH group of F6P on the electrophilic γ-phosphorus atom of the Mg^{2+}–ATP complex.

PFK plays a central role in the control of glycolysis because it catalyzes one of the pathway's rate-determining reactions. In many organisms the activity of PFK is enhanced allosterically by several substances, including AMP, and inhibited allosterically by several other substances, including ATP and citrate. The control of PFK is exquisitely complex; the mechanism by which it regulates the glycolytic pathway is examined in Section 17-4F.

D. *Aldolase*

Aldolase catalyzes Reaction 4 of glycolysis, the cleavage of FBP to form the two trioses **glyceraldehyde-3-phosphate (GAP)** and **dihydroxyacetone phosphate (DHAP):**

This reaction is an **aldol cleavage (retro aldol condensation)** whose nonenzymatic base-catalyzed mechanism is shown in Fig. 17-8. Note that aldol cleavage between C3 and C4 of FBP requires a carbonyl at C2 and a hydroxyl at C4. Hence, the "logic" of Reaction 2 in the glycolytic pathway, the isomerization of G6P to F6P, is clear. Aldol cleavage of G6P would have resulted in products of unequal carbon chain length, while aldol cleavage of FBP results in two

Figure 17-8 Mechanism for base-catalyzed aldol cleavage. Aldol condensation occurs by the reverse mechanism.

interconvertible C3 compounds that can therefore enter a common degradative pathway. The enolate intermediate in the aldol cleavage reaction is stabilized by resonance, as shown, as a result of the electron-withdrawing character of the carbonyl oxygen atom.

Note that at this point in the pathway the atom numbering system changes. Atoms 1, 2, and 3 of glucose become atoms 3, 2, and 1 of DHAP, thus reversing order. Atoms 4, 5, and 6 become atoms 1, 2, and 3 of GAP (Fig. 17-3).

a. There Are Two Mechanistic Classes of Aldolases

Aldol cleavage is catalyzed by stabilizing its enolate intermediate through increased electron delocalization. There are two types of aldolases that are classified according to the chemistry they employ to stabilize the enolate. In Class I aldolases, which occur in animals and plants, the reaction occurs as follows (Fig. 17-9):

Step 1 Substrate binding.

Step 2 Reaction of the FBP carbonyl group with the ε-amino group of the active site Lys 229 to form an iminium cation, that is, a protonated Schiff base.

Step 3 C3—O4 bond cleavage resulting in enamine formation and the release of GAP. The iminium ion, as we saw in Section 16-2E, is a better electron-withdrawing group than is the oxygen atom of the precursor carbonyl group. Thus, catalysis occurs because the enamine intermediate (Fig. 17-9, Step 3) is more stable than the corresponding enolate intermediate of the base-catalyzed aldol cleavage reaction (Fig. 17-8, Step 2).

Step 4 Protonation of the enamine to an iminium cation.

Step 5 Hydrolysis of this iminium cation to release DHAP, with regeneration of the free enzyme.

Proof for the formation of the Schiff base in Step 2 was provided by "trapping" ^{14}C-labeled DHAP on the enzyme by reacting it with $NaBH_4$, which reduces imines to amines:

$$CH_2OPO_3^{2-}$$
$$^{14}C=\overset{+}{N}H-(CH_2)_4-Enzyme$$
$$CH_2OH$$

$\downarrow$ $NaBH_4$ reduction

$P_i \leftarrow$ hydrolysis

$$CH_2OH \qquad\qquad COO^-$$
$$H-^{14}C-NH-(CH_2)_4-CH$$
$$CH_2OH \qquad\qquad NH_3^+$$

N^6-β-**Glyceryl lysine**

The radioactive product was hydrolyzed and identified as N^6-β-**glyceryl lysine.**

Cys and His residues were initially thought to act as the acid and base catalysts that facilitate the proton transfers in the aldolase reaction because the appropriate group-specific reagents inactivate the enzyme by reacting with these residues. For example, the reaction of a specific Cys residue of aldolase with iodoacetic acid inactivates the enzyme and results in the buildup of the FBP observed in the early glycolysis inhibition studies (Section 17-1A). However, site-directed mutagenesis to Ala of the Cys residue thought to be involved in the catalytic activity results in no loss of enzymatic function. Modification of this Cys residue apparently prevents the conformational changes required for productive substrate binding.

An early X-ray structure of aldolase suggested that a Tyr side chain was positioned to act as the active site acid–base catalyst and that the His was instead necessary for the maintenance of the Tyr's catalytically active orientation. A re-examination of the X-ray data caused yet another modification of the mechanism. The Tyr originally seen at the active site has changed position in this new analysis, so that it is out of reach of the active site. Asp 33 and Lys 229 now appear to be acting as acid–base catalysts. These residues are evolutionarily conserved and their mutagenesis eliminates enzyme activity. This is an excellent example of the caution that must be exercised in the interpretation of chemical modification and structural data, and the power of site-directed mutagenesis in the study of enzyme mechanisms (although see Section 15-3Ba).

Class II aldolases, which occur in fungi, algae, and some bacteria, do not form a Schiff base with the substrate. Rather, a divalent cation, usually Zn^{2+} or Fe^{2+}, polarizes the carbonyl oxygen of the substrate to stabilize the enolate intermediate of the reaction (Fig. 16-12d):

$$CH_2OPO_3^{2-} \qquad\qquad CH_2OPO_3^{2-}$$
$$C=O\cdots Zn^{2+}-Enzyme \longleftrightarrow C-O^-\cdots Zn^{2+}-Enzyme$$
$$HO-C-H \qquad\qquad HO-C-H$$

Are the two classes of aldolases related? Although both classes exhibit the Uni Bi kinetics implicit in their mechanisms, they exhibit only ~15% sequence identity, which places them in the lower end of the twilight zone for establishing homology (Section 7-4Ba). Nevertheless, their X-ray structures reveal that they have the same fold, the α/β barrel. The evolution of this particularly common fold is discussed in Section 8-3Bh.

b. Why Two Classes of Aldolase?

Since glycolysis presumably arose very early in evolutionary history, the existence of two classes of aldolase is unexpected. It had originally been postulated that, since Class I aldolases occur in higher organisms, Class II aldolases must be the more primitive enzyme form, that is, less metabolically capable than are the Class I enzymes. However, the discovery that some organisms simultaneously express both classes of aldolase suggests that both enzyme classes are evolutionarily ancient and equally

Figure 17-9 **Enzymatic mechanism of Class I aldolase.** The reaction involves **(1)** substrate binding; **(2)** Schiff base formation between the enzyme's active site Lys residue and FBP; **(3)** aldol cleavage to form an enamine intermediate of the enzyme and DHAP with release of GAP (shown with its *re* face up); **(4)** tautomerization and protonation to the iminium form of the Schiff base; and **(5)** hydrolysis of the Schiff base with release of DHAP. 🎧 **See the Animated Figures.**

adept at carrying out their metabolic functions. Thus, the expression of both classes of aldolase in some organisms probably represents an ancient metabolic redundancy that evolution has eliminated in most contemporary organisms. Whatever the reason for the occurrence of two classes of aldolase, the fact that Class II aldolases do not occur in mammals makes them an attractive target in the development of antibacterial drugs.

c. Aldolase Is Stereospecific

The aldolase reaction provides another example of the extraordinary stereospecificity of enzymes. In the nonenzymatic aldol condensation to form hexose-1,6-bisphosphate from DHAP and GAP, there are four possible products depending on whether the *pro-R* or *pro-S* hydrogen at C3 of DHAP is removed and whether the resulting carbanion attacks GAP on its *re* or its *si* face:

D-Fructose 1,6-bisphosphate

D-Psicose 1,6-bisphosphate

D-Tagatose 1,6-bisphosphate

D-Sorbose 1,6-bisphosphate

In the enzymatic aldol condensation (Fig. 17-9 in reverse), carbanion formation from the enzyme–DHAP iminium ion (Fig. 17-9, Step 4 in reverse) occurs with removal of only the *pro-S* hydrogen. Attack of this carbanion occurs only on the *si* face of the enzyme-bound GAP carbonyl group, so that only FBP is formed (Fig. 17-9, Step 3 in reverse).

E. *Triose Phosphate Isomerase*

Only one of the products of the aldol cleavage reaction, GAP, continues along the glycolytic pathway (Fig. 17-3). However, DHAP and GAP are ketose–aldose isomers just as are F6P and G6P. Interconversion of GAP and DHAP therefore probably occurs via an enediol or enediolate intermediate in analogy with the phosphoglucose isomerase reaction (Fig. 17-6). **Triose phosphate isomerase (TIM** or

TPI; Figs. 8-19*b* and 8-52) catalyzes this process in Reaction 5 of glycolysis, the final reaction of Stage I:

Glyceraldehyde-3-phosphate (an aldose)

Dihydroxyacetone phosphate (a ketose)

Enediol intermediate

Support for this reaction scheme comes from the use of the transition state analogs **phosphoglycohydroxamate** and **2-phosphoglycolate,** stable compounds whose geometric structures resemble that of the proposed enediol or enediolate intermediate:

Phosphoglyco-hydroxamate

2-Phosphoglycolate

Proposed enediolate intermediate

Since enzymes catalyze reactions by binding the transition state complex more tightly than the substrate (Section 15-1F), phosphoglycohydroxamate and 2-phosphoglycolate should bind more tightly to TIM than does substrate. In fact, phosphoglycohydroxamate and 2-phosphoglycolate bind 155- and 100-fold more tightly to TIM than do either GAP or DHAP.

a. Glu 165 Functions as a General Base

The pH dependence of the TIM reaction is a bell-shaped curve with pK's of 6.5 and 9.5. The similarity of these pK's to the corresponding quantities of the phosphoglucose isomerase reaction suggests the participation of both an acid and a base in the TIM reaction as well. However, pH studies alone, as we have seen, are difficult to interpret in terms of specific amino acid residues since the active site environment may alter the pK of an acidic or basic group.

Affinity labeling reagents have been employed in an effort to identify the base at the active site of TIM. Both **bromohydroxyacetone phosphate** and **glycidol phosphate**

Bromohydroxyacetone   Glycidol phosphate
phosphate

inactivate TIM by forming esters of Glu 165, whose carboxylate group, X-ray studies indicate, is ideally situated to abstract the C2 proton from the substrate (general base catalysis). In fact, the mutagenic replacement of Glu 165 by Asp, which X-ray studies show withdraws the carboxylate group only ~1 Å farther away from the substrate than its position in the wild-type enzyme, reduces TIM's catalytic power ~1000-fold. Note that Glu 165's pK is drastically altered from the 4.1 value of the free amino acid to the observed 6.5 value. This provides yet another striking example of the effect of environment on the properties of amino acid side chains.

b. The TIM Reaction Probably Occurs via Concerted General Acid–Base Catalysis Involving Low-Barrier Hydrogen Bonds

The X-ray structure of yeast TIM in complex with phosphoglycohydroxamate indicates that His 95 is hydrogen bonded to and hence is properly positioned to protonate the carbonyl oxygen atom of GAP (general acid catalysis):

However, NMR studies indicate that His 95 is in its neutral imidazole form rather than its protonated imidazolium form. How can an imidazole N3—H group, which has a highly basic pK of ~14, protonate a carbonyl oxygen atom

that, when protonated, has a very acidic pK of <0? Likewise, how can the Glu 165 carboxylate group (pK 6.5) abstract the C2 proton from GAP (pK ~17)? A plausible answer is that these proton shifts are facilitated by the formation of low-barrier hydrogen bonds (LBHBs). These unusually strong associations (-40 to -80 kJ · mol^{-1} vs -12 to -30 kJ · mol^{-1} for normal hydrogen bonds), as we have seen in the case of the serine protease catalytic triad (Section 15-3Dd), form when the pK's of the hydrogen bonding donor and acceptor groups are nearly equal. They can be important contributors to rate enhancement if they form only in the transition state of an enzymatically catalyzed reaction.

In converting GAP to the enediol (or enediolate) intermediate (Fig. 17-10, *left*), the pK of the protonated form of its carbonyl oxygen, which becomes a hydroxyl group, increases to ~14, which closely matches that of neutral His 95. The resulting LBHB between this hydroxyl group and His 95 permits the neutral imidazole side chain to protonate the oxygen atom. Likewise, as the carbonyl oxygen is protonated, the pK of the C2—H proton decreases to ~7, close to the pK of the Glu 165 carboxylate. It therefore appears that the reaction occurs via simultaneous proton abstraction by Glu 165 and protonation by His 95 (concerted general acid–base catalysis). The LBHBs postulated to form in the transition state, but not in the Michaelis complex, between Glu 165 and C2—H and between His 95 and the carbonyl oxygen atom are thought to provide some of the transition state stabilization necessary to catalyze the reaction. The positively charged side chain of Lys 12, which is probably responsible for the 9.5 pK observed in TIM's pH rate profile, is thought to electrostatically stabilize the negatively charged transition state. The conversion of the enediol(ate) intermediate to DHAP is likewise facilitated by the formation of transition state LBHBs (Fig. 17-10, *right*). In fact, in the high resolution (1.2 Å) X-ray structure of TIM in complex with DHAP, determined by Ann McDermott and Liang Tong, the hydrogen bond between His 95 and O2 of DHAP has an unusually short distance of 2.6 Å. In addition, a carboxylate oxygen atom of Glu 165 forms extraordinarily close contacts of ~3.0 Å with both C1 and C2 of DHAP.

c. A Flexible Loop Both Preferentially Binds and Protects the Enediol Intermediate

The comparison of the X-ray structure of the TIM · phosphoglycohydroxamate complex with that of TIM alone reveals that a 10-residue loop, which is closed over the active site in the enzyme–substrate complex, is flipped up in the unoccupied active site like a hinged lid, a movement that involves main chain shifts of >7 Å (Fig. 17-11). A four-residue segment of this loop makes a hydrogen bond with the phosphate group of the substrate. Mutagenic excision of these four residues neither significantly distorts the protein nor greatly impairs substrate binding. The catalytic power of the mutant enzyme is, nevertheless, reduced 10⁵-fold and it only weakly binds phosphoglycohydroxamate. Evidently, the closed loop preferentially stabilizes the enzymatic reaction's enediol-like transition state.

GAP·TIM Michaelis complex

DHAP·TIM Michaelis complex

Transition state

Transition state

Enediol (or enendiolate) intermediate

Figure 17-10 Proposed enzymatic mechanism of the TIM reaction. The reaction proceeds via the concerted abstraction of the C2—H proton of GAP by the carboxylate group of Glu 165 and the protonation of the GAP carbonyl oxygen atom by the imidazole group of His 95. The p*K*'s of the corresponding donor and acceptor groups participating in each proton transfer process become nearly equal in the transition state and hence are

proposed to form low-barrier hydrogen bonds (*red dashed lines*), which act to stabilize the transition state. The resulting enediol (or possibly the electrostatically stabilized enediolate) intermediate then reacts in a similar fashion with the carboxyl group of Glu 165 protonating C1, while the deprotonated N3 atom of His 95 abstracts the proton on the 2-hydroxyl group to yield DHAP.

Figure 17-11 Ribbon diagram of yeast TIM in complex with its transition state analog 2-phosphoglycolate. A single 248-residue subunit of this homodimeric enzyme is viewed roughly along the axis of its α/β barrel. The enzyme's flexible loop, residues 168 through 177, is cyan and the side chains of Lys 12, His 95, and Glu 165 are purple, magenta, and red, respectively. The 2-phosphoglycolate is shown in space-filling form colored according to atom type (C green, O red, P yellow). [Based on an X-ray structure by Gregory Petsko, Brandeis University. PDBid 2YPI.] **See Interactive Exercise 2 and Kinemage Exercises 12-1 and 12-2**

Figure 17-12 **The spontaneous decomposition of the enediol intermediate in the TIM reaction to form methylglyoxal through the elimination of a phosphate group.** (*a*) This reaction can occur only when the C—O bond to the phosphate group lies in a plane that is nearly perpendicular to that of the enediol so as to permit the formation of a double bond in the intermediate enol product. (*b*) When the C—O bond to the phosphate group lies in a plane that is nearly parallel to that of the enediol, the *p* orbitals on the resulting intermediate product would be perpendicular to each other and hence lack the overlap necessary to form a π bond, that is, a double bond. The resulting unsatisfied bonding capacity greatly increases the energy of the reaction intermediate and hence makes the reaction highly unfavorable.

(*a*)

C—O bond to phosphate in plane perpendicular to plane of molecule

Parallel *p* orbitals maximally overlapped to form a π bond

Enediol intermediate   **Enol**   **Methyl glyoxal**

(*b*)

C—O bond to phosphate in plane of molecule

Perpendicular *p* orbitals have no overlap; a π bond cannot be formed

The closed loop conformation in the TIM reaction provides a striking example of the stereoelectronic control that enzymes can exert on a reaction (Section 15-1Eb). In solution, the enediol intermediate readily breaks down with the elimination of the phosphate at C3 to form the toxic compound **methylglyoxal** (Fig. 17-12*a*). On the enzyme's surface, however, this reaction is prevented because the phosphate group is held by the flexible loop in the plane of the enediol, a position that disfavors phosphate elimination. In order for this elimination to occur, the C—O bond to the phosphate group must lie, as shown in Fig. 17-12*a*, in the plane perpendicular to that of the enediol. This is because, if the phosphate group were to be eliminated while this C—O bond was in the plane of the enediol as diagrammed in Fig. 17-12*b*, the CH₂ group of the resulting enol product would be twisted 90° out of the plane of the rest of the molecule. Such a conformation is energetically prohibitive because it prevents the formation of the enol's double bond by eliminating the overlap between its component *p* orbitals. In the mutant enzyme lacking the flexible loop, the enediol is able to escape: ~85% of the enediol intermediate is released into solution, where it rapidly decomposes to methylglyoxal and P$_i$. Thus, flexible loop closure also ensures that substrate is efficiently transformed to product.

On the basis of the foregoing X-ray structures, it had been widely assumed that the binding of substrate to TIM is **ligand-gated,** that is, it induces loop closure. Yet, if this were the case, the reversibility of the TIM reaction and the chemical resemblance of its reactant and product (GAP and DHAP) make it difficult to rationalize how product could be released. However, NMR measurements by John Williams and McDermott reveal that, in fact, loop motion still occurs when TIM is binding either glycerol-3-phosphate (a substrate analog) or 2-phosphoglycolate (a transition state analog) and is sufficiently fast (on a time scale of 100 μs) to account for the catalytic reaction rate (a turnover time of 230 μs). This is a clear example of how complementary information supplied by X-ray and NMR methods has yielded important insights into an enzymatic mechanism that neither technique alone could have provided.

d. TIM Is a Perfect Enzyme

TIM, as Jeremy Knowles demonstrated, has achieved catalytic perfection in that the rate of bimolecular reaction between enzyme and substrate is diffusion controlled; that is, product formation occurs as rapidly as enzyme and substrate can collide in solution, so that any increase in TIM's catalytic efficiency would not increase the reaction rate (Section 14-2Bb). Because of the high interconversion efficiency of GAP and DHAP, these two metabolites are maintained in equilibrium: $K = [GAP]/[DHAP] = 4.73 \times 10^{-2}$; that is, [DHAP] >> [GAP] at equilibrium. However, *as GAP is utilized in the succeeding reaction of the glycolytic pathway, more DHAP is converted to GAP, so that these compounds maintain their equilibrium ratio.* One common pathway therefore accounts for the metabolism of both products of the aldolase reaction.

Let us now take stock of where we are in our travels down the glycolytic pathway. At this stage, the glucose, which has been transformed into two GAPs, has completed the preparatory stage of glycolysis. This process has required the expenditure of two ATPs. However, this investment has resulted in the conversion of one glucose to two C₃ units, each of which has a phosphoryl group that, with a little chemical artistry, can be converted to a "high-energy" compound (Section 16-4Ba) whose free energy of hydrolysis can be coupled to ATP synthesis. *This energy investment is doubly repaid in the final stage of glycolysis in which the two phosphorylated C₃ units are transformed to two pyruvates with the coupled synthesis of four ATPs per glucose.*

F. *Glyceraldehyde-3-Phosphate Dehydrogenase: First "High-Energy" Intermediate Formation*

Reaction 6 of glycolysis involves the oxidation and phosphorylation of GAP by NAD^+ and P_i as catalyzed by **glyceraldehyde-3-phosphate dehydrogenase (GAPDH;** Figs. 8-45 and 8-53*b*):

Glyceraldehyde-3-phosphate (GAP)

glyceraldehyde-3-phosphate dehydrogenase (GAPDH)

1,3-Bisphosphoglycerate (1,3-BPG)

This is the first instance of the chemical artistry alluded to above. *In this reaction, aldehyde oxidation, an exergonic reaction, drives the synthesis of the acyl phosphate* **1,3-bisphosphoglycerate (1,3-BPG;** *previously called* **1,3-diphosphoglycerate).** Recall that acyl phosphates are compounds with high phosphate group-transfer potential (Section 16-4Ba).

a. Mechanistic Studies

Several key enzymological experiments have contributed to the elucidation of the GAPDH reaction mechanism (Fig. 17-13):

1. GAPDH is inactivated by alkylation with stoichiometric amounts of iodoacetate. The presence of **carboxymethylcysteine** in the hydrolysate of the resulting alkylated enzyme (Fig. 17-13*a*) suggests that GAPDH has an active site Cys sulfhydryl group.

2. GAPDH quantitatively transfers ^3H from C1 of GAP to NAD^+ (Fig. 17-13*b*), thereby establishing that this reaction occurs via direct hydride transfer.

3. GAPDH catalyzes exchange of ^{32}P between $[^{32}P]P_i$ and the product analog **acetyl phosphate** (Fig. 17-13*c*). Such isotope exchange reactions are indicative of an acyl–enzyme intermediate (Section 14-5D).

(*a*)

(*b*)

(*c*)

Acetyl phosphate

Figure 17-13 Some reactions employed in elucidating the enzymatic mechanism of GAPDH. (*a*) The reaction of iodoacetate with an active site Cys residue. (*b*) Quantitative tritium transfer from substrate to NAD^+. (*c*) The enzyme-catalyzed exchange of ^{32}P from phosphate to acetyl phosphate.

Figure 17-14 **Enzymatic mechanism of glyceraldehyde-3-phosphate dehydrogenase.** (1) GAP binds to the enzyme; (2) the active site sulfhydryl group forms a thiohemiacetal with the substrate; (3) NAD$^+$ oxidizes the thiohemiacetal to a thioester; (4) the newly formed NADH is replaced on the enzyme by NAD$^+$; and (5) P_i attacks the thioester, forming the acyl phosphate product, 1,3-BPG, and regenerating the active enzyme.

☜ **See the Animated Figures**

David Trentham has proposed a mechanism for GAPDH based on this information and the results of kinetic studies (Fig. 17-14):

Step 1 GAP binds to the enzyme.

Step 2 The essential sulfhydryl group, acting as a nucleophile, attacks the aldehyde to form a **thiohemiacetal.**

Step 3 The thiohemiacetal undergoes oxidation to an **acyl thioester** by direct transfer of a hydride to NAD$^+$. This intermediate, which has been isolated, has a high group-transfer potential. *The energy of aldehyde oxidation has not been dissipated but has been conserved through the synthesis of the thioester and the reduction of NAD$^+$ to NADH.*

Step 4 Another molecule of NAD$^+$ replaces NADH.

Step 5 The thioester intermediate undergoes nucleophilic attack by P_i to regenerate free enzyme and form 1,3-BPG. This "high-energy" mixed anhydride generates ATP from ADP in the next reaction of glycolysis.

G. *Phosphoglycerate Kinase: First ATP Generation*

Reaction 7 of the glycolytic pathway results in the first formation of ATP together with **3-phosphoglycerate (3PG)** in

a reaction catalyzed by **phosphoglycerate kinase (PGK):**

1,3-Bisphosphoglycerate (1,3-BPG)

3-Phosphoglycerate (3PG)

(*Note:* The name "kinase" is given to any enzyme that transfers a phosphoryl group between ATP and a metabolite. Nothing is implied as to the exergonic direction of transfer.)

(a)

(b)

Figure 17-15   X-ray structures of phosphoglycerate kinase (PGK). (*a*) The yeast enzyme in complex with 3PG and Mg^{2+}–ATP. (*b*) The *T. maritima* enzyme in complex with 3PG and Mg^{2+}–AMPPNP. In both structures, the enzyme is represented by its transparent molecular surface with its embedded ribbon diagram colored with its N-terminal domain yellow and its C-terminal domain purple. The Mg^{2+}–ATP, Mg^{2+}–AMPPNP, and 3PG are drawn in space-filling form

colored according to atom type (ATP and AMPPNP C green, 3PG C cyan, N blue, O red, P orange, and Mg^{2+} pale green). Note the similar overall appearance of PGK and hexokinase (Fig. 17-5), although these proteins unrelated. [Based on X-ray structures by Herman Watson, University of Bristol, U.K.; *and* Günter Auerbach and Robert Huber, Max-Planck-Institut für Biochemie, Martinsreid, Germany. PDBids 3PGK and 1VPE.]

The X-ray structure of yeast PGK in complex with Mg^{2+}–ATP and 3PG, its enzyme–product complex, was determined by Herman Watson (Fig. 17-15*a*). Note its conspicuously bilobal appearance and the close approach between the ATP's γ-phosphate group and one of the 3PG's

carboxylate oxygen atoms. The 3PG and the Mg–ATP are respectively bound to PGK's N- and C-terminal domains.

The X-ray structure of PGK from the thermophile *Thermotoga maritima* in complex with 3PG and the nonhydrolyzable ATP analog **AMPPNP** (ATP with the O atom bridging its β and γ phosphate groups replaced by an NH group) reveals that its two domains have swung together relative to those in the 50% identical yeast PGK (Fig. 17-15*b*). This forms the catalytic site and permits the substrates to react in a water-free environment, much as occurs with hexokinase (Section 17-2Aa).

Figure 17-16 indicates a reaction mechanism for PGK that is consistent with its observed sequential kinetics. The terminal phosphoryl oxygen of ADP nucleophilically attacks the C1 phosphorus atom of 1,3-BPG to form the reaction product.

The energetics of the overall GAPDH–PGK reaction pair are

$$GAP + P_i + NAD^+ \longrightarrow 1,3\text{-BPG} + NADH$$
$$\Delta G^{\circ\prime} = +6.7 \text{ kJ} \cdot \text{mol}^{-1}$$

$$1,3\text{-BPG} + ADP \longrightarrow 3PG + ATP$$
$$\Delta G^{\circ\prime} = -18.8 \text{ kJ} \cdot \text{mol}^{-1}$$

$$GAP + P_i + NAD^+ + ADP \longrightarrow 3PG + NADH + ATP$$
$$\Delta G^{\circ\prime} = -12.1 \text{ kJ} \cdot \text{mol}^{-1}$$

1,3-Bisphosphoglycerate Mg^{2+}–ADP

3-Phosphoglycerate Mg^{2+}–ATP

Figure 17-16   **Mechanism of the PGK reaction.**

Although the GAPDH reaction is endergonic, the strongly exergonic nature of the transfer of a phosphoryl group from 1,3-BPG to ADP makes the overall synthesis of NADH and ATP from GAP, P_i, NAD^+, and ADP favorable.

H. *Phosphoglycerate Mutase*

In Reaction 8 of glycolysis, 3PG is converted to **2-phosphoglycerate (2PG)** by **phosphoglycerate mutase (PGM)**:

A **mutase** catalyzes the transfer of a functional group from one position to another on a molecule. This reaction is a necessary preparation for the next reaction in glycolysis, which generates a "high-energy" phosphoryl compound for use in ATP synthesis.

Figure 17-17 **The active site region of yeast phosphoglycerate mutase (dephospho form).** The substrate, 3PG, which is drawn in ball-and-stick form with C green, O red, and P orange, binds to an ionic pocket whose side chains are drawn in stick form with C green, N blue, and O red. His 8 is phosphorylated in the active enzyme. [Based on an X-ray structure by Jennifer Littlechild, University of Exeter, U.K. PDBid 1QHF.]

a. Reaction Mechanism of PGM

At first sight, the reaction catalyzed by PGM appears to be a simple intramolecular phosphoryl transfer. This is not the case, however. *The active enzyme has a phosphoryl group at its active site, which it transfers to the substrate to form a bisphospho intermediate. This intermediate then rephosphorylates the enzyme to form the product and regenerate the active phosphoenzyme.* The following experimental data permitted the elucidation of PGM's enzymatic mechanism:

1. Catalytic amounts of **2,3-bisphosphoglycerate (2,3-BPG**; previously known as **2,3-diphosphoglycerate)**

2,3-Bisphosphoglycerate (2,3-BPG)

are required for enzymatic activity; that is, 2,3-BPG acts as a reaction primer.

2. Incubation of the enzyme with catalytic amounts of ^{32}P-labeled 2,3-BPG yields a ^{32}P-labeled enzyme. Zelda Rose demonstrated that this was a result of the phosphorylation of a His residue:

Phospho-His residue

3. The enzyme's X-ray structure shows His at the active site (Fig. 17-17). In the active enzyme, His 8 is phosphorylated.

These data are consistent with a mechanism in which the active enzyme contains a phospho-His residue at the active site (Fig. 17-18):

Step 1 3PG binds to the phosphoenzyme in which His 8 is phosphorylated.

Step 2 This phosphoryl group is transferred to the substrate, resulting in an intermediate 2,3-BPG · enzyme complex.

Steps 3 and 4 The complex decomposes to form the product 2PG with regeneration of the phosphoenzyme.

The phosphoryl group on 3PG therefore ends up on the C2 of the next 3PG to undergo reaction.

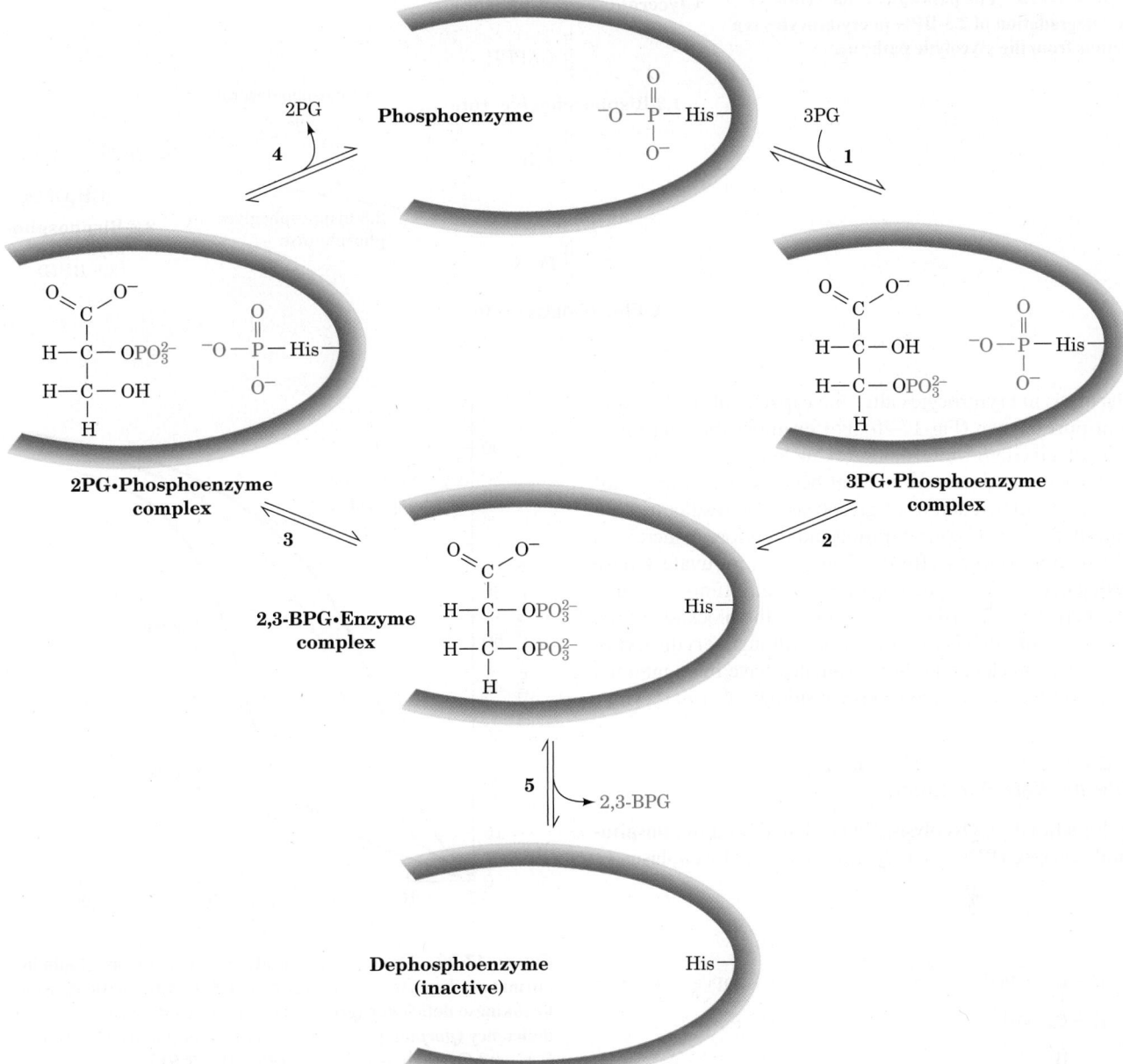

Figure 17-18   **Proposed reaction mechanism for phosphoglycerate mutase.** The active form of the enzyme contains a phospho-His residue at the active site. **(1)** Formation of an enzyme–substrate complex; **(2)** transfer of the enzyme-bound phosphoryl group to the substrate; **(3)** rephosphorylation of the enzyme by the other phosphoryl group of the substrate; and **(4)** release of product regenerating the active phosphoenzyme. **(5)** Occasionally, 2,3-BPG dissociates from the enzyme, leaving it in an inactive, dephospho form that must be rephosphorylated by the reverse reaction.

Occasionally, 2,3-BPG dissociates from the enzyme (Fig. 17-18; Step 5), leaving it in an inactive form. Trace amounts of 2,3-BPG must therefore always be available to regenerate the active phosphoenzyme by the reverse reaction.

b. Glycolysis Influences Oxygen Transport

2,3-BPG specifically binds to deoxyhemoglobin and thereby alters the oxygen affinity of hemoglobin (Section 10-1D). The concentration of 2,3-BPG in erythrocytes is much higher (~5 mM) than the trace amounts required for its use as a primer of PGM. Erythrocytes synthesize and degrade 2,3-BPG by a detour from the glycolytic pathway, diagrammed in Fig. 17-19. **Bisphosphoglycerate mutase** catalyzes the transfer of a phosphoryl group from C1 to C2 of 1,3-BPG. The resulting 2,3-BPG is hydrolyzed to 3PG by **2,3-bisphosphoglycerate phosphatase.** The rate of glycolysis affects the oxygen affinity of hemoglobin through the mediation of 2,3-BPG. Consequently, inherited defects of

Figure 17-19 **The pathway for the synthesis and degradation of 2,3-BPG in erythrocytes is a detour from the glycolytic pathway.**

glycolysis in erythrocytes alter the capacity of the blood to transport oxygen (Fig. 17-20). For example, the concentration of glycolytic intermediates in hexokinase-deficient erythrocytes is less than normal because hexokinase catalyzes the first reaction of glycolysis. This results in a diminished 2,3-BPG concentration and therefore in increased hemoglobin oxygen affinity. Conversely, pyruvate kinase deficiency decreases hemoglobin oxygen affinity through the increase of 2,3-BPG resulting from the blockade of the last reaction in glycolysis. Thus, although erythrocytes, which lack nuclei and other organelles, have but a minimal metabolism, this metabolism is physiologically significant.

I. *Enolase: Second "High-Energy" Intermediate Formation*

In Reaction 9 of glycolysis, 2PG is dehydrated to **phosphoenolpyruvate (PEP)** in a reaction catalyzed by **enolase:**

Figure 17-20 **The oxygen-saturation curves of hemoglobin in normal erythrocytes (*red curve*) and in those from patients with hexokinase deficiency (*green*) and with pyruvate kinase deficiency (*purple*).** [After Delivoria-Papadopoulos, M., Oski, F.A., and Gottlieb, A.J., *Science* **165,** 601 (1969).]

2-Phosphoglycerate   Phosphoenolpyruvate
      (2PG)   (PEP)

The enzyme forms a complex with a divalent cation such as Mg^{2+} before the substrate is bound. A second divalent metal ion then binds to the enzyme. As is mentioned in Section 17-1A, fluoride ion inhibits glycolysis, resulting in the accumulation of 2PG and 3PG. It does so by strongly inhibiting enolase in the presence of P_i. F^- and P_i form a tightly bound complex with the Mg^{2+} at the enzyme's active site, blocking substrate binding and thereby inactivating the enzyme. Enolase's substrate, 2PG, therefore builds up and, as it does so, is equilibrated with 3PG by PGM.

a. Catalytic Mechanism of Enolase

The dehydration (elimination of H_2O) catalyzed by enolase might occur in one of three ways (Fig. 16-9*a*): (1) The —OH group at C3 can leave first, generating a carbocation

at C3; (2) the C2 proton can leave first, generating a carbanion at C2; or (3) the reaction can be concerted. Isotope exchange studies by Paul Boyer demonstrated that the C2 proton of 2PG exchanges with solvent 12 times faster than the rate of PEP formation. However, the C3 oxygen exchanges with solvent at a rate roughly equivalent with the overall reaction rate. This suggests the following mechanism (Fig. 17-21):

Step 1 Rapid carbanion formation at C2 facilitated by a general base on the enzyme. The abstracted proton can readily exchange with the solvent, accounting for its observed rapid exchange rate.

Step 2 Rate-limiting elimination of the —OH group at C3. This is consistent with the slow rate of exchange of this hydroxyl group with solvent.

2-Phosphoglycerate (2 PG)

1 ↓ fast

Figure 17-21 **Proposed reaction mechanism of enolase. (1)** Rapid formation of a carbanion by removal of a proton at C2 by Lys 345 acting as a general base; this proton can rapidly exchange with the solvent. **(2)** Slow elimination of H_2O to form phosphoenolpyruvate with general acid catalysis by Glu 211; the C3 oxygen of the substrate can exchange with solvent only as rapidly as this step occurs.

Delocalized carbanion intermediate

2 ↓ slow

Phosphoenolpyruvate (PEP)

The enolase reaction (Fig. 17-21) is of mechanistic interest because it involves the abstraction of the decidedly nonacidic proton at C2 (pK > 30), followed by the elimination of an OH^- ion, which is a poor leaving group. The X-ray structure of yeast enolase in complex with two Mg^{2+} ions and an equilibrium mixture of 2PG and PEP (enolase's substrate and product), determined by George Reed and Ivan Rayment, reveals that enolase binds 2PG in an intricate complex that involves both Mg^{2+} ions. Mutagenic and enzymological studies indicate that the reaction involves the Lys 345 side chain functioning as a general base

and the Glu 211 side chain functioning as a general acid. Lys 396 and the two Mg^{2+} ions are thought to stabilize the increased negative charge that develops on the carboxylate ion in the delocalized carbanion intermediate.

J. *Pyruvate Kinase: Second ATP Generation*

In Reaction 10 of glycolysis, its final reaction, **pyruvate kinase (PK)** couples the free energy of PEP hydrolysis to the synthesis of ATP to form pyruvate:

Phosphoenolpyruvate (PEP)

| pyruvate kinase (PK)

Pyruvate

$$\Delta G^{\circ\prime} = +14.6 \; \text{kJ·mol}^{-1} \qquad\qquad \Delta G^{\circ\prime} = -46 \; \text{kJ·mol}^{-1}$$

$$\text{Overall } \Delta G^{\circ\prime} = -31.4 \; \text{kJ·mol}^{-1}$$

Figure 17-22   **Mechanism of the reaction catalyzed by pyruvate kinase. (1)** Nucleophilic attack of an ADP β-phosphoryl oxygen atom on the phosphorus atom of PEP to form ATP and enolpyruvate; and **(2)** tautomerization of enolpyruvate to pyruvate.

a. Catalytic Mechanism of PK

The PK reaction, which requires the participation of both monovalent (K^+) and divalent (Mg^{2+}) cations, occurs as follows (Fig. 17-22):

Step 1 A β-phosphoryl oxygen of ADP nucleophilically attacks the PEP phosphorus atom, thereby displacing enolpyruvate and forming ATP. This reaction conserves the free energy of PEP hydrolysis.

Step 2 Enolpyruvate converts to pyruvate. This enol–keto tautomerization is sufficiently exergonic to drive the coupled endergonic synthesis of ATP (Section 16-4Ba).

We can now see the "logic" of the enolase reaction. The standard free energy of hydrolysis of 2PG ($\Delta G^{\circ\prime}$) is only $-17.6 \; \text{kJ·mol}^{-1}$, which is insufficient to drive ATP synthesis ($\Delta G^{\circ\prime} = 30.5 \; \text{kJ·mol}^{-1}$ for ATP synthesis from ADP and P_i). The dehydration of 2PG results in the formation of a "high-energy" compound capable of such synthesis [the standard free energy of hydrolysis of PEP is -61.9 kJ·mol^{-1} (Fig. 16-25)]. In other words, PEP is a "high-energy" compound, 2PG is not.

3 FERMENTATION: THE ANAEROBIC FATE OF PYRUVATE

For glycolysis to continue, NAD^+, which cells have in limited quantities, must be recycled after its reduction to NADH by GAPDH (Fig. 17-3; Reaction 6). In the presence of oxygen, the reducing equivalents of NADH are passed into the mitochondria for reoxidation (Chapter 22). Under anaerobic conditions, on the other hand, the NAD^+ is replenished by the reduction of pyruvate in an extension of the glycolytic pathway. Two processes for the anaerobic replenishment of NAD^+ are homolactic and alcoholic fermentation, which occur in muscle and yeast, respectively.

A. Homolactic Fermentation

In muscle, particularly during vigorous activity when the demand for ATP is high and oxygen has been depleted, **lactate dehydrogenase (LDH)** catalyzes the oxidation of NADH by pyruvate to yield NAD^+ and **lactate**. This reaction is often classified as Reaction 11 of glycolysis:

LDH, as do other NAD^+-requiring enzymes, catalyzes its reaction with absolute stereospecificity: The *pro-R* (A-side) hydrogen at C4 of NADH is stereospecifically transferred to the *re* face of pyruvate at C2 to form L- (or *S*-) lactate. This regenerates NAD^+ for participation in the GAPDH reaction. The hydride transfer to pyruvate is from the same face of the nicotinamide ring as that to acetaldehyde in the alcohol dehydrogenase reaction (Section 13-2A) but from the opposite (*si*) face of the nicotinamide ring as that to GAP in the GAPDH reaction (Section 17-2F).

Mammals have two different types of LDH subunits, the M type and the H type, which together form five tetrameric isozymes: M_4, M_3H, M_2H_2, MH_3, and H_4. Although these hybrid forms occur in most tissues, the H-type subunit predominates in aerobic tissues such as heart muscle, while the M-type subunit predominates in tissues that are subject to anaerobic conditions such as skeletal muscle and liver. H_4

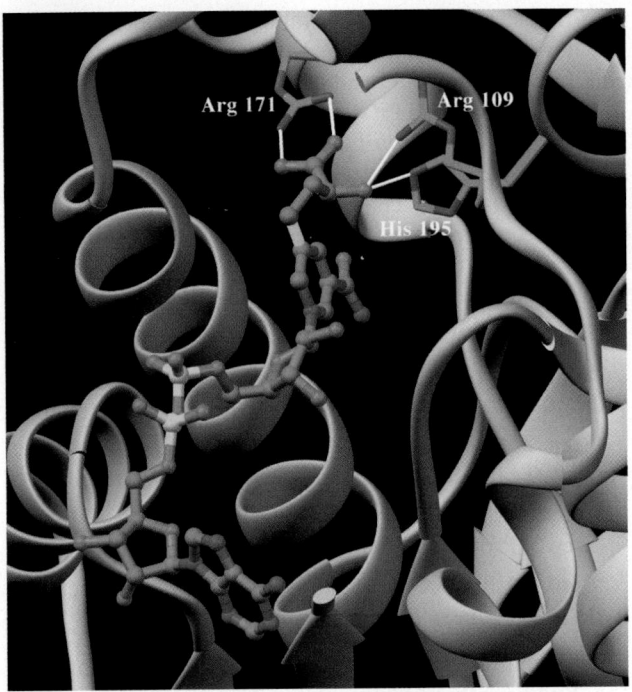

Figure 17-23 **The active site region of porcine H$_4$ LDH in complex with *S-lac*-NAD$^+$, a covalent adduct of lactate and NAD$^+$.** The adduct is shown in ball-and-stick form with C green, N blue, O red, and P yellow, except for the covalent bond between the methylene substituent to lactate atom C3 and nicotinamide atom C5, which is light green. The three LDH side chains that form hydrogen bonds (*white lines*) with the pyruvate residue are shown in stick form with C magenta and N blue. [Based on an X-ray structure by Michael Rossmann, Purdue University. PDBid 5LDH.]

Figure 17-24 **Reaction mechanism of lactate dehydrogenase.** The reaction involves direct hydride transfer from NADH to pyruvate's carbonyl carbon atom accompanied by proton donation from the imidazolium group of His 195 to the pyruvate carbonyl oxygen atom. The latter process is facilitated by the positive charge on the nearby side chain of Arg 109.

LDH has a low K_M for pyruvate and is allosterically inhibited by high levels of this metabolite, whereas the M$_4$ isozyme has a higher K_M for pyruvate and is not inhibited by it. The other isozymes have intermediate properties that vary with the ratio of their two types of subunits. It has therefore been proposed, although not without disagreement, that H-type LDH is better adapted to function in the oxidation of lactate to pyruvate, whereas M-type LDH is more suited to catalyze the reverse reaction.

The X-ray structure of porcine H$_4$ LDH in complex with **S-*lac*-NAD$^+$** (a bisubstrate analog in which atom C3 of lactate is covalently linked to nicotinamide atom C5 of NAD$^+$ via a CH$_2$ group) was determined by Michael Rossmann (Fig. 17-23; he also determined the X-ray structure of dogfish M$_4$ LDH shown in Fig. 8-54*a*). Lactate atom O2, its hydroxyl oxygen, is hydrogen bonded to the side chains of both Arg 109 and His 195, whereas the lactate carboxyl group at C1 is doubly hydrogen bonded to the side chain of Arg 171. On the basis of this structure and extensive enzymological evidence, Rossmann proposed the following mechanism for pyruvate reduction by LDH (Fig. 17-24): *The pro-R hydride is transferred from C4 of NADH's nicotinamide ring to C2 of pyruvate with the concomitant transfer of a proton from the imidazolium moiety of His 195*

to pyruvate O2, thereby yielding NAD$^+$ and lactate. The proton transfer is facilitated by repulsive interactions with the closely associated positively charged side chain of Arg 109. These interactions also serve to properly orient the pyruvate, as does the salt bridge that the pyruvate carboxyl group forms with the side chain of Arg 171.

The overall process of anaerobic glycolysis in muscle can be represented:

$$\text{Glucose} + 2\text{ADP} + 2\text{P}_i \longrightarrow$$
$$2 \text{ lactate} + 2\text{ATP} + 2\text{H}_2\text{O} + 2\text{H}^+$$

Much of the lactate, the end product of anaerobic glycolysis, is exported from the muscle cell via the blood to the liver, where it is reconverted to glucose (Section 23-1C).

Contrary to the widely held belief, it is not lactate buildup in the muscle *per se* that causes muscle fatigue and soreness but the accumulation of glycolytically generated acid (muscles can maintain their workload in the presence of high lactate concentrations if the pH is kept constant; but see Section 27-2B). Indeed, it is well known among hunters that the meat of an animal that has run to exhaustion before being killed has a sour taste. This is a result of lactic acid buildup in the muscles.

B. *Alcoholic Fermentation*

Under anaerobic conditions in yeast, NAD^+ is regenerated in a manner that has been of importance to mankind for thousands of years: the conversion of pyruvate to ethanol and CO_2. Ethanol is, of course, the active ingredient of wine and spirits; CO_2 so produced leavens bread. From the point of view of the yeast, however, alcoholic fermentation has a practical benefit that homolactic fermentation cannot supply: Yeast employ ethanol as a kind of antibiotic to eliminate competing organisms. This is because yeast can grow in ethanol concentrations >12% (2.5*M*), whereas few other organisms can survive in >5% ethanol (recall that ethanol is a widely used antiseptic).

a. TPP Is an Essential Cofactor of Pyruvate Decarboxylase

Yeast produces ethanol and CO_2 via two consecutive reactions (Fig. 17-25). The first reaction is the decarboxylation of pyruvate to form acetaldehyde and CO_2 as catalyzed by **pyruvate decarboxylase (PDC;** an enzyme not present in animals). PDC contains the coenzyme **thiamine pyrophosphate [TPP;** Fig. 17-26; also called **thiamin diphosphate (ThDP)]**, which it binds tightly but noncovalently. The coenzyme is employed because decarboxylation of an α-keto acid such as pyruvate requires the buildup of negative charge on the carbonyl carbon atom in the transition state, an unstable situation:

This transition state may be stabilized by delocalization of the developing negative charge into a suitable "electron sink." The amino acid residues of proteins function poorly in this capacity but TPP does so easily.

*The "business" end of TPP is the **thiazolium ring*** (Fig. 17-26). Its C2—H group is relatively acidic because of the adjacent positively charged quaternary nitrogen atom, which electrostatically stabilizes the carbanion formed on dissociation of the proton. This dipolar carbanion (or **ylid**) is the active form of the coenzyme. The mechanism of PDC catalysis is as follows (Fig. 17-27):

Step 1 Nucleophilic attack by the ylid form of TPP on the carbonyl carbon of pyruvate to form a covalent adduct.

Step 2 Departure of CO_2 to generate a resonance-stabilized carbanion adduct in which the thiazolium ring of the coenzyme acts as an electron sink.

Step 3 Protonation of the carbanion.

Step 4 Elimination of the TPP ylid to form acetaldehyde and regenerate the active enzyme.

This mechanism has been corroborated by the isolation of the **hydroxyethylthiamine pyrophosphate** intermediate (Fig. 17-27).

The X-ray structure of PDC in complex with TPP (Fig. 17-28), which was determined by William Furey and Martin Sax, has suggested a role for TPP's aminopyrimidine ring in the formation of the active ylid. Ylid formation requires a base to remove the C2 proton. Yet PDC has no basic side chain that is properly positioned to do so. The amino group of the enzyme-bound TPP's aminopyrimidine ring is suitably positioned to accept this proton; however, its p*K* is too low to do so efficiently and one of its protons sterically

Figure 17-25 **The two reactions of alcoholic fermentation. (1)** Decarboxylation of pyruvate to form acetaldehyde is followed by **(2)** reduction of the acetaldehyde to ethanol by NADH.

Figure 17-26 **Thiamine pyrophosphate.** The thiazolium ring constitutes its catalytically active functional group.

Figure 17-27 **Reaction mechanism of pyruvate decarboxylase.** **(1)** Nucleophilic attack by the ylid form of TPP on the carbonyl carbon of pyruvate; **(2)** departure of CO_2 to generate a resonance-stabilized carbanion; **(3)** protonation of the carbanion; and **(4)** elimination of the TPP ylid and release of product.

Figure 17-28 **A portion of the X-ray structure of pyruvate decarboxylase from *Saccharomyces uvarum* (brewer's yeast) in complex with its TPP cofactor.** The enzyme's identical 563-residue subunits form a tightly associated dimer, two of which associate loosely to form a tetramer. The TPP and the side chain of Glu 51 are shown in stick form with C green, N blue, O red, S yellow, and P gold. The TPP binds in a cavity situated between the dimer's two subunits (*cyan and magenta*), where it hydrogen bonds to Glu 51. [Based on an X-ray structure by William Furey and Martin Sax, Veterans Administration Medical Center and University of Pittsburgh, Pittsburgh, Pennsylvania. PDBid 1PYD.] 🐟 **See Interactive Exercise 9**

clashes with the C2 proton. It is therefore proposed that the aminopyrimidine is converted to its imino tautomeric form on the enzyme's surface in a reaction involving proton donation by Glu 51 (Fig. 17-29). The imine, in turn, accepts a proton from C2, thereby forming the ylid, with tautomerization back to the amino form. The participation of N1′ and the 4′-amino group of the aminopyrimidine is supported by experiments showing that TPP analogs missing either of these functionalities are catalytically inactive. H/D exchange experiments followed by ^{1}H NMR analysis of the exchange products indicate that when TPP is bound to PDC in complex with the substrate analog **pyruvamide** (CH_3—CO—CO—NH_2), its rate of exchange to form the active species (ylid) is much greater ($>6 \times 10^2$ s^{-1}) than the enzyme's catalytic rate ($k_{cat} = 10$ s^{-1}). Moreover, the mutation of PDC's Glu 51 to Gln reduces the rate of H/D exchange to 1.7 s^{-1}, thereby supporting Glu 51's postulated function of donating a proton to N1′ of TPP's aminopyridine ring.

b. Beriberi Is a Thiamine Deficiency Disease

The ability of TPP's thiazolium ring to add to carbonyl groups and act as an "electron sink" makes it the coenzyme most utilized in α-keto acid decarboxylations. TPP is also

Glu 51

Imine **Predominant form**

↓ fast

Ylid

Figure 17-29   The formation of the active ylid form of TPP in the pyruvate decarboxylase reaction. This reaction requires the participation of TPP's aminopyrimidine ring together with general acid catalysis by Glu 51. The predominant form of the cofactor on the enzyme is the imine, but the rate of formation of the active ylid is fast relative to the enzyme's catalytic rate.

involved in decarboxylation reactions that we shall encounter in other metabolic pathways. Consequently, thiamine (**vitamin B₁**), which is neither synthesized nor stored in significant amounts by the tissues of most vertebrates, is required in their diets. Its deficiency in humans results in an ultimately fatal condition known as **beriberi** (Sinhalese for weakness) that is characterized by neurological disturbances causing pain, paralysis and atrophy (wasting) of the limbs, and/or cardiac failure resulting in edema (the accumulation of fluid in tissues and body cavities). Beriberi was particularly prevalent in the late nineteenth and early twentieth centuries in the rice-consuming areas of Asia after the introduction of steam-powered milling machines that polished the rice

grains to remove their coarse but thiamine-containing outer layers (the previously used milling procedures were less efficient and hence left sufficient thiamine on the grains). Parboiling rice before milling, a process common in India, causes the rice kernels to absorb nutrients from their outer layers, thereby decreasing the incidence of beriberi. Once thiamine deficiency was recognized as the cause of beriberi, enrichment procedures were instituted so that today it has ceased to be a problem except in areas undergoing famine. However, beriberi occasionally develops in chronic alcoholics due to their penchant for drinking but not eating.

c. Reduction of Acetaldehyde and Regeneration of NAD⁺

The acetaldehyde formed by the decarboxylation of pyruvate is reduced to ethanol by NADH in a reaction catalyzed by **alcohol dehydrogenase (ADH)**. Each subunit of the tetrameric yeast ADH (YADH) binds one NADH and one Zn²⁺ ion. The Zn²⁺ ion functions to polarize the carbonyl group of acetaldehyde (Fig. 17-30), so as to stabilize the developing negative charge in the transition state of the reaction (the role of metal ions in enzymes is discussed in Section 15-1C). This facilitates the transfer of NADH's *pro-R* hydrogen (the same atom that LDH transfers) to acetaldehyde's *re* face, forming ethanol with the transferred hydrogen in the *pro-R* position (Section 13-2A).

Both homolactic and alcoholic fermentation have the same function: the anaerobic regeneration of NAD⁺ for continued glycolysis. Their main difference is in their metabolic products.

Mammalian liver ADH (**LADH**) functions to metabolize the alcohols anaerobically produced by intestinal flora as well as those from external sources (the direction of the ADH reaction varies with the relative concentrations of

Figure 17-30   The reaction mechanism of alcohol dehydrogenase involves direct hydride transfer of the *pro-R* hydrogen of NADH to the *re* face of acetaldehyde.

ethanol and acetaldehyde). Each subunit of this dimeric enzyme binds one NAD^+ and two Zn^{2+} ions, although only one of these ions participates directly in catalysis. There is significant amino acid sequence similarity between YADH and LADH, so it is quite likely that both enzymes have the same general mechanism.

C. *Energetics of Fermentation*

Thermodynamics permits us to dissect the process of fermentation into its component parts and to account for the free energy changes that occur. This enables us to calculate the efficiency with which the free energy of degradation of glucose is utilized in the synthesis of ATP. The overall reaction of homolactic fermentation is

$$\text{Glucose} \longrightarrow 2 \text{ lactate} + 2H^+$$
$$\Delta G^{\circ\prime} = -196 \text{ kJ} \cdot \text{mol}^{-1} \text{ of glucose}$$

($\Delta G^{\circ\prime}$ is calculated from the data in Table 3-4 using Eqs. [3.19] and [3.21] adapted for $2H^+$ ions.) For alcoholic fermentation, the overall reaction is

$$\text{Glucose} \longrightarrow 2CO_2 + 2 \text{ ethanol}$$
$$\Delta G^{\circ\prime} = -235 \text{ kJ} \cdot \text{mol}^{-1} \text{ of glucose}$$

Each of these reactions is coupled to the net formation of two ATPs, which requires $\Delta G^{\circ\prime} = +61 \text{ kJ} \cdot \text{mol}^{-1}$ of glucose consumed (Table 16-3). Dividing the $\Delta G^{\circ\prime}$ of ATP formation by that of lactate formation indicates that homolactic fermentation is 31% "efficient"; that is, 31% of the free energy released by this process under standard biochemical conditions is sequestered in the form of ATP. The rest is dissipated as heat, thereby making the process irreversible. Likewise, alcoholic fermentation is 26% efficient under biochemical standard state conditions. Actually, *under physiological conditions, where the concentrations of reactants and products differ from those of the standard state, these reactions have free energy efficiencies of >50%.*

a. **Glycolysis Is Used for Rapid ATP Production**

Anaerobic fermentation utilizes glucose in a profligate manner compared to oxidative phosphorylation: Fermentation results in the production of 2 ATPs per glucose, whereas oxidative phosphorylation yields 32 ATPs per glucose (Chapter 22). This accounts for Pasteur's observation that yeast consumes far more sugar when growing anaerobically than when growing aerobically (the **Pasteur effect;** Section 22-4C). However, *the rate of ATP production by anaerobic glycolysis can be up to 100 times faster than that of oxidative phosphorylation. Consequently, when tissues such as muscle are rapidly consuming ATP, they regenerate it almost entirely by anaerobic glycolysis.* (Homolactic fermentation does not really "waste" glucose since the lactate so produced is aerobically reconverted to glucose by the liver; Section 23-1C).

Skeletal muscles consist of both **slow-twitch** (Type I) and **fast-twitch** (Type II) **fibers.** Fast-twitch fibers, so called because they predominate in muscles capable of short bursts of rapid activity, are nearly devoid of mitochondria, so that they must obtain nearly all of their ATP through anaerobic glycolysis, for which they have a particularly large capacity. Muscles designed to contract slowly and steadily, in contrast, are enriched in slow-twitch fibers that are rich in mitochondria and obtain most of their ATP through oxidative phosphorylation. (Fast- and slow-twitch fibers were originally known as white and red fibers, respectively, because otherwise pale colored muscle tissue, when enriched with mitochondria, takes on the red color characteristic of their heme-containing cytochromes. However, fiber color has been shown to be an imperfect indicator of muscle physiology.)

In a familiar example, the flight muscles of migratory birds such as ducks and geese, which need a continuous energy supply, are rich in slow-twitch fibers and therefore such birds have dark breast meat. In contrast, the flight muscles of less ambitious fliers, such as chickens and turkeys, which are used only for short bursts (often to escape danger), consist mainly of fast-twitch fibers that form white meat. In humans, the muscles of sprinters are relatively rich in fast-twitch fibers, whereas distance runners have a greater proportion of slow-twitch fibers (although their muscles have the same color). World class distance runners have a remarkably high capacity to generate ATP aerobically. This was demonstrated by the noninvasive ^{31}P NMR monitoring of the ATP, P_i, phosphocreatine, and pH levels in their exercising but untrained forearm muscles. These observations suggest that the muscles of these athletes are better endowed genetically for endurance exercise than those of "normal" individuals.

4 METABOLIC REGULATION AND CONTROL

Living organisms, as we saw in Section 16-6, are thermodynamically open systems that tend to maintain a steady state rather than reaching equilibrium (death for living things). *Thus the **flux** (rate of flow) of intermediates through a metabolic pathway is constant; that is, the rates of synthesis and breakdown of each pathway intermediate maintain it at a constant concentration.* Such a state, it will be recalled, is one of maximum thermodynamic efficiency (Section 16-6Ba). *Regulation of the steady state (**homeostasis**) must be maintained in the face of changes in flux through the pathway in response to changes in demand.*

The terms metabolic control and metabolic regulation are often used interchangeably. However, for our purposes we shall give them different definitions: **Metabolic regulation** is the process by which the steady-state flow of metabolites through a pathway is maintained, whereas **metabolic control** is the influence exerted on the enzymes of a pathway in response to an external signal in order to alter the flux of metabolites.

A. *Homeostasis and Metabolic Control*

There are two reasons why metabolic flow must be controlled:

I. To provide products at the rate they are needed, that is, to balance supply with demand.

II. To maintain the steady-state concentrations of the intermediates in a pathway within a narrow range (homeostasis).

Organisms maintain homeostasis for several reasons:

1. In an open system, such as metabolism, the steady state is the state of maximum thermodynamic efficiency (Section 16-6Ba).

2. Many intermediates participate in more than one pathway, so that changing their concentrations may disturb a delicate balance.

3. The rate at which a pathway can respond to a control signal slows if large changes in intermediate concentrations are involved.

4. Large changes in intermediate concentrations may have deleterious effects on cellular osmotic properties.

The concentrations of intermediates and the level of metabolic flux at which a pathway is maintained vary with the needs of the organism through a highly responsive system of precise controls. Such pathways are analogous to rivers that have been dammed to provide a means of generating electricity. Although water is continually flowing in and out of the lake formed by the dam, a relatively constant water level is maintained. The rate of water outflow from the lake is precisely controlled at the dam and is varied in response to the need for electrical power. In this section, we examine the mechanisms by which metabolic pathways in general, and the glycolytic pathway in particular, are controlled in response to biological energy needs.

B. *Metabolic Flux*

Since a metabolic pathway is a series of enzyme-catalyzed reactions, it is easiest to describe the flux of metabolites through the pathway by considering its reaction steps individually. The flux of metabolites, J, through each reaction step is the rate of the forward reaction, v_f, less that of the reverse reaction, v_r:

$$J = v_f - v_r \qquad [17.1]$$

At equilibrium, by definition, there is no flux ($J = 0$), although v_f and v_r may be quite large. At the other extreme, in reactions that are far from equilibrium, $v_f \gg v_r$, so that the flux is essentially equal to the rate of the forward reaction, $J \approx v_f$. *The flux throughout a steady-state pathway is constant and is set (generated) by the pathway's rate-determining step (or steps). Consequently, control of flux through a metabolic pathway requires: (1) that the flux through this **flux-generating step** vary in response to the organism's metabolic requirements and (2) that this change in flux be communicated throughout the pathway to maintain a steady state.*

The classic description of metabolic control and regulation is that every metabolic pathway has a rate-limiting step and is regulated by controlling the rate of this pivotal

enzyme. These so-called regulatory enzymes are almost invariably allosteric enzymes subject to feedback inhibition (Section 13-4) and are often also controlled by covalent modification (which we discuss in Section 18-3).

Several questions arise. Are these regulatory enzymes really rate limiting for the pathway? Is there really only one step in the pathway that is rate limiting, or might there be a number of enzymes contributing to the regulation of the pathway? Does controlling these enzymes really control the flux of metabolites through the pathway or is the function of feedback inhibition really to maintain a steady state? These are complicated questions with complicated answers.

C. *Metabolic Control Analysis*

While it has been common practice to assume that every metabolic pathway has a rate-limiting step, experiments suggest that the situation becomes more complex when these pathways are combined in a living organism. Hence, it is important to develop methods to quantitatively analyze metabolic systems in order to establish mechanisms of control and regulation. **Metabolic control analysis,** developed by Henrik Kacser and Jim Burns and independently by Reinhart Heinrich and Tom Rapoport, provides a framework for considering these problems. It is a way of quantitatively describing the behavior of metabolic systems in response to various perturbations.

a. The Flux Control Coefficient Measures the Sensitivity of the Flux to the Change in Enzyme Concentration

Metabolic control analysis makes no *a priori* assumption that only one step is rate limiting. Instead, it defines a **flux control coefficient, C^J** (where J is an index, not an exponent), to measure the sensitivity of flux to a change in enzyme concentration. The flux control coefficient is defined as the fractional change in flux, J, with respect to the fractional change in enzyme concentration, [E]:

$$C^J = \frac{\partial J/J}{\partial [E]/[E]} = \frac{\partial \ln J}{\partial \ln [E]} \approx \frac{\Delta J/J}{\Delta [E]/[E]} \qquad [17.2]$$

(recall that $\partial x/x = \partial \ln x$).

The flux control coefficient is the analog of the kinetic order of a reaction. If a reaction is first order in substrate concentration, [S], then doubling [S] doubles the rate of the reaction, whereas if the reaction is zero order in [S] (e.g., in a saturated enzymatic reaction), then the reaction rate is insensitive to the value of [S]. Similarly, if the flux control coefficient of an enzyme is 1, then doubling the concentration of the enzyme, [E], doubles the flux through the pathway and if it is zero, the flux is insensitive to the value of [E]. Of course, the flux control coefficient may have some intermediate value between 0 and 1. For example, if a 10% increase in the enzyme concentration increases the flux by only 7.5%, the flux control coefficient would be 0.075/0.10 = 0.75.

The flux through a metabolic system is generally controlled by more than one enzyme. Consequently, the flux control coefficient for each of the participating enzymes is the fraction of the total control on the pathway exerted by that enzyme. Thus, *the sum of all the flux control coefficients involved in controlling a pathway must equal 1.* This is the **additivity theorem** of metabolic control. It indicates that the flux control coefficient of a particular enzyme in a system depends, in part, on the flux control coefficients of the other enzymes in the system, that is, an enzyme's flux control coefficient is a property of the system as a whole. Indeed, *flux control in a metabolic system is shared among all the enzymes in the system,* although most of their flux control coefficients are likely to be small.

b. Recombinant DNA Technology Has Been Used to Measure Flux Control Coefficients

The flux control coefficient is a variable that has been experimentally determined *in vivo* for many enzymes that had previously been assumed to catalyze the rate-determining steps for their pathways. For example, the citric acid cycle enzyme **citrate synthase** (Sections 21-1A and 21-3A) catalyzes an irreversible reaction ($\Delta G^{\circ\prime} = -31.5$ kJ · mol^{-1}) and has therefore long been assumed to be one of the enzymes regulating the flow through the citric acid cycle (Section 21-4). Daniel Koshland determined how the activity of citrate synthase affected the flux through the citric acid cycle via genetic engineering techniques that permitted him to control the concentration of this enzyme *in vivo*. He constructed a plasmid (Fig. 17-31) that contained the gene for citrate synthase under the control of (directly downstream from) a modified ***lac* promoter,** together with the ***lacI*** gene, which encodes the ***lac* repressor** (in the absence of inducer, the *lac* repressor binds to the *lac* promoter and thereby prevents the transcription by RNA polymerase of the genes it controls; Section 5-4Aa), and the ***amp^R*** gene, which confers resistance to the antibiotic **ampicillin.** This plasmid was introduced into a mutant *E. coli* that lacked a gene for citrate synthase and was ampicillin-sensitive. These *E. coli* were grown in the presence of ampicillin (which killed any cells that had not taken up the plasmid) and of varying amounts of **isopropylthiogalactoside (IPTG),** a nonmetabolizable inducer of the *lac* operon (Section 31-1Aa).

Using this system, the citrate synthase concentration was measured as a function of [IPTG] and the growth rate of the *E. coli* was determined as a function of [IPTG] when glucose and/or acetate were the sole carbon sources. When acetate was the sole carbon source, the *E. coli* obtained most of their metabolic energy via the citric acid cycle and their growth varied proportionately with [citrate synthase]. The flux control coefficient of the enzyme in this case approached its maximum value of 1, that is, the flux through the citric acid cycle was almost entirely controlled by the activity of citrate synthase. However, when glucose was also available, the *E. coli* grew rapidly, even when [citrate synthase] was low, and were unaffected by changes in its value. Here, the flux control coefficient was near zero, indicating that the flux through the citric acid cycle was reduced to the point that even low concentrations of citrate synthase were in catalytic excess (evidently, when glucose

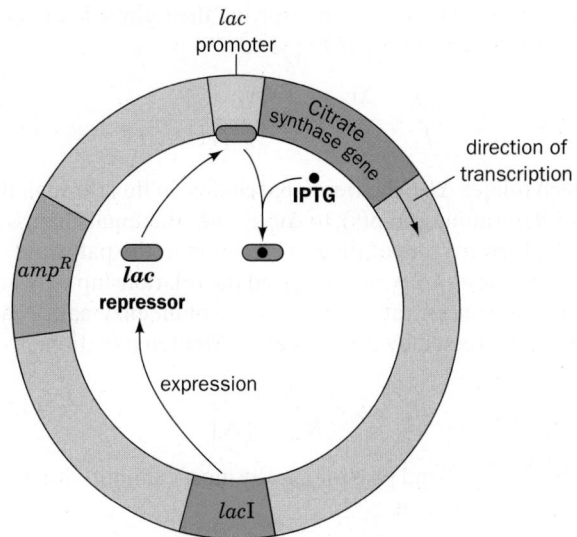

Figure 17-31 Schematic diagram of the plasmid constructed to control the amount of citrate synthase produced by *E. coli*. The *lacI* gene encodes the *lac* repressor, which binds to the *lac* promoter. This prevents the transcription of its immediately downstream gene, which encodes citrate synthase. On binding IPTG, the *lac* repressor releases the *lac* promoter, thereby permitting the expression of citrate synthase. Consequently, the concentration of the nonmetabolizable IPTG controls the level of expression of citrate synthase. The *amp^R* gene encodes a protein that provides resistance to the antibiotic ampicillin. Hence, in the presence of ampicillin, only those otherwise ampicillin-sensitive *E. coli* that have taken up the plasmid will survive.

is present, the citric acid cycle has a secondary role in energy production and biosynthesis in *E. coli*).

c. The Rates of Enzymatic Reactions Respond to Changes in Flux

Let us consider how a constant flux is maintained throughout a metabolic pathway by analyzing the response of an enzyme-catalyzed reaction to a change in the flux of the reaction preceding it. In the following steady-state pathway:

$$S \xrightarrow[\substack{\text{rate-determining} \\ \text{step(s)}}]{J} A \underset{v_r}{\overset{v_f}{\rightleftharpoons}} B \xrightarrow{J} P$$

the flux, J, through the reaction A $\rightleftharpoons$ B, which must be identical to the flux through the rate-determining step(s), is expressed by Eq. [17.1] ($J = v_f - v_r$). If the flux of the rate-determining step increases by the amount ΔJ, the increase must be communicated to the next reaction step in the pathway by an increase in v_f (Δv_f) in order to re-establish the steady state. Qualitatively, we can see that this occurs because an increase in J causes an increase in [A], which in turn causes an increase in v_f. The amount of increase in [A] (Δ[A]) that causes v_f to increase the appropriate amount (Δv_f) is determined as follows:

$$\Delta J = \Delta v_f \qquad [17.3]$$

Dividing Eq. [17.3] by J, multiplying the right side by v_f/v_f, and substituting in Eq. [17.1] yields

$$\frac{\Delta J}{J} = \frac{\Delta v_f}{v_f} \frac{v_f}{J} = \frac{\Delta v_f}{v_f} \frac{v_f}{(v_f - v_r)} \qquad [17.4]$$

which relates $\Delta J/J$, the fractional change in flux through the rate-determining step(s), to $\Delta v_f/v_f$, the fractional change in v_f, the forward rate of the next reaction in the pathway.

In Section 14-2A, we discussed the relationship between substrate concentration and the rate of an enzymatic reaction as expressed by the Michaelis–Menten equation:

$$v_f = \frac{V_{max}^f [A]}{K_M + [A]} \qquad [14.24]$$

In the simplest and physiologically most common situation, $[A] \ll K_M$, so that

$$v_f = \frac{V_{max}^f [A]}{K_M} \qquad [17.5]$$

and

$$\Delta v_f = \frac{V_{max}^f \Delta [A]}{K_M} \qquad [17.6]$$

Hence,

$$\frac{\Delta v_f}{v_f} = \frac{\Delta [A]}{[A]} \qquad [17.7]$$

that is, the fractional change in forward reaction rate is equal to the fractional change in substrate concentration. Then, by substituting Eq. [17.7] into Eq. [17.4], we find that

$$\frac{\Delta J}{J} = \frac{\Delta [A]}{[A]} \frac{v_f}{(v_f - v_r)} \qquad [17.8]$$

This equation relates the fractional change in flux through a metabolic pathway's rate-determining step(s) to the fractional change in substrate concentration necessary to communicate that change to the following reaction steps. *The quantity $v_f/(v_f - v_r)$ is a measure of the sensitivity of a reaction's fractional change in flux to its fractional change in substrate concentration.* This quantity is also a measure of the reversibility of the reaction, that is, how close it is to equilibrium:

1. In an irreversible reaction, v_r approaches 0 (relative to v_f) and therefore $v_f/(v_f - v_r)$ approaches 1. The reaction therefore requires a nearly equal fractional increase in its substrate concentration in order to respond to a fractional increase in flux.

2. As a reaction approaches equilibrium, v_r approaches v_f and hence $v_f/(v_f - v_r)$ approaches infinity. The reaction's response to a fractional increase in flux therefore requires a much smaller fractional increase in its substrate concentration.

Consequently, *the ability of a reaction to communicate a change in flux increases as the reaction approaches equilib-*

rium. A series of sequential reactions that are all near equilibrium therefore have the same flux and maintain concentrations of intermediates in a steady state (homeostasis).

d. The Elasticity Coefficient Measures the Sensitivity of an Enzymatic Reaction to the Change in Substrate Concentration

The ratio $v_f/(v_f - v_r)$, which measures the sensitivity of an enzymatic reaction rate to the change in substrate concentration, is called, in metabolic control analysis, the **elasticity coefficient, ε.** It is the fractional change in the net rate of an enzyme reaction, v, with respect to the fractional change in the substrate concentration, $[A]$:

$$\varepsilon = \frac{\partial v/v}{\partial [A]/[A]} = \frac{\partial \ln v}{\partial \ln [A]} \approx \frac{v_f}{v_f - v_r} \qquad [17.9]$$

(When studying an individual enzymatic reaction, so that the fractional change in the net rate through the reaction, $\Delta v/v$, corresponds to the fractional change in the flux, $\Delta J/J$, and $[A] \ll K_M$, this equation is simply a rearrangement of Eq. [17.8].) The value of the elasticity coefficient depends on the kinetic characteristics of the enzyme and how close to equilibrium the enzyme is functioning. As mentioned above, if an enzyme is functioning far from equilibrium ($v_f \gg v_r$), changing the substrate concentration will have a small effect on the net rate of the enzyme reaction (ε will be close to 1). However, if the enzyme is functioning very close to equilibrium so that both the forward and reverse reaction rates are much faster than the overall net rate, ε approaches infinity and it takes only a tiny change in substrate concentration to adjust to a new flux. Such large elasticity coefficients are therefore associated with maintaining homeostasis.

D. Supply–Demand Analysis

Early studies on control of metabolic pathways focused on individual pathways, ignoring their overall physiological functions. Control was always assumed to reside within the pathway. However, often when enzymes thought to be "rate controlling" in an individual pathway were overproduced in living organisms using genetic engineering techniques, increases in enzyme concentrations of as much as 10-fold had no effect on the flux through the specific pathway studied. The flux control coefficients of the overproduced enzymes were near zero in the *in vivo* system; they were already present in metabolic excess. The flux must somehow have been controlled from outside of the pathway. This is because, as we now realize, it is impossible to separate a pathway from the process(es) that utilizes the product(s) of that pathway (i.e., the living organisms must have reduced the activities of these enzymes in keeping with their metabolic requirements, that is, they maintained homeostasis).

Degradation pathways are inextricably linked to the biosynthetic pathways that utilize their products (Fig. 16-2). This is a **supply–demand process** and both supply and demand are involved in the two metabolic control challenges: flux control and homeostasis. Jan-Hendrik Hofmeyr and Athel Cornish-Bowden have used metabolic control analysis to explore such a system, lumping all of the reactions of the supply pathway(s) together into one block and all of the reactions of the demand pathway(s) into a second block.

Here X is the intermediate that is produced by the supply block for use by the demand block. For the supply block, X is a product and a feedback inhibitor, so that as the concentration of X increases, the rate of flow through the supply block decreases. For the demand block, X is a substrate, so that as the concentration of X increases, the rate of flow through the demand block increases until it becomes saturated. When the flux through the supply block is equal to the flux through the demand block, the concentration of X is in a steady state, the point at which its rate of production is equal to its rate of utilization. This rate defines the actual flux through the supply–demand system and the steady-state concentration of X.

a. The Steady-State Concentration of Intermediates Responds to Changes in Supply and/or Demand

The response of the steady-state concentration of the intermediate, X, to any small change in the rate of the supply or demand block depends entirely on the elasticity coefficients of the two blocks at the steady state. Imagine that the activity of the demand block increases. This would result in a decrease of [X] and a concomitant increase in the flux through the supply block as feedback inhibition is decreased. The shift would continue until the rates through the supply and demand blocks equalize, shifting the system to a new steady state with a lower [X] and higher overall flux. Alternatively, if the activity of the supply block increases, producing a higher value of [X], the demand block would respond by increasing its rate to re-establish a new steady state at this higher [X]. *The higher the elasticity coefficient of the responding block, the smaller the change that [X] must make in order to re-establish a steady state.*

The question remains, where is the control? Is it in the supply block or in the demand block? The answer is that it is in the block for which the elasticity coefficient is lowest. Since it is the change in [X] that causes the readjustment of the steady state and the change in flux, the block for which the largest change in [X] is produced for a given change in rate is the controlling block. The change in flux for a given change in the rate in a particular block is its flux control coefficient, so the *control lies in the block that has the highest flux control coefficient and the lowest elasticity coefficient.* For example, if the supply block has a very high elasticity coefficient and the demand block has a very low elasticity coefficient, increased demand need cause very little decrease in [X] to result in a change in supply rate to reach a

new steady state. However, because of the low elasticity coefficient of the demand block, there will have to be a much larger increase in [X] due to an increase in supply to cause the rate of the demand block to increase enough to reach a new steady state. Consequently, increasing the activity of the demand block would have a much larger effect on the flux than increasing the activity of the supply block. Thus, for this case, the flux is much more sensitive to changes in demand than to changes in supply, that is, the flux control coefficient of the demand block is much greater than that of the supply block.

There is a reciprocal relationship between the flux control coefficient and the elasticity coefficient. The larger the flux control coefficient, the lower the elasticity coefficient, and vice versa. The ratio of the elasticity coefficients of the supply and demand blocks determines the distribution of flux control between supply and demand. *When the ratio of the supply elasticity coefficient to the demand elasticity coefficient is greater than 1, as in our example, flux control lies in the demand portion of the pathway, and vice versa.*

b. The Elasticity Coefficient Describes the Regulation of Steady-State Intermediate Concentrations

In addition to controlling flux through the supply–demand system, the steady-state concentrations of the intermediates are also regulated. We have seen that the larger the elasticity coefficient of a given block, the smaller the change in [X] that is needed to re-establish a steady state and change the flux. Keeping the change in [X] as small as possible while changing the flux and maintaining a steady state is very important. *The larger the elasticity coefficient, the more sensitive the regulation of homeostasis.*

Flux control requires a high flux control coefficient, which requires a low elasticity coefficient. Regulation of homeostasis requires a high elasticity coefficient, which requires a low flux control coefficient. A large difference in the elasticity coefficients of the supply and demand blocks therefore leads to the exclusive control of flux by one or the other of the blocks. *The functions of flux and concentration control are mutually exclusive. If the demand block controls the flux, the function of the supply block is to regulate homeostasis.*

c. Feedback Inhibition Is Required for Homeostasis, Not Flux Control

When the demand block is exerting flux control, an increase in demand results in a decrease in the concentration of X, thereby decreasing feedback inhibition of the supply block. Feedback inhibition might therefore appear to be an essential part of the control process. In fact, this is not the case. Feedback inhibition is not part of the control system but part of the homeostasis system. It determines the range of [X] at which there is a steady state. In the absence of feedback inhibition, the supply block will be insensitive to [X] for most of that concentration range but will become sensitive to [X] near equilibrium, where the demand block could then control the flux. However, this would require such high concentrations of X and the other metabolites in the supply pathway as to be osmotically dangerous. Feedback inhibition maintains homeostasis at physiologically reasonable metabolite concentrations.

E. *Mechanisms of Flux Control*

a. Flux through a Pathway Is Controlled at Its Rate-Determining Step(s)

The metabolic flux through an entire pathway is determined by controlling its rate-determining step(s), which by definition is much slower than the following reaction step(s). The product(s) of the rate-determining step(s) is therefore removed before it can equilibrate with the reactant, so that the rate-determining step(s) functions far from equilibrium and has a large negative free energy change. In an analogous manner, the flow of a river can only be controlled at a dam, which creates a difference in water levels between its upstream and downstream sides; this is a situation that also has a large negative free energy change, in this case resulting from the hydrostatic pressure head. Yet, as we have just seen, the elasticity coefficient, ε, of a nonequilibrium reaction ($v_f >> v_r$) is close to 1; that is, its substrate concentration must double (in the absence of other controlling effects) in order to double the reaction flux rate (Eq. [17.8]). However, some pathway fluxes vary by factors that are much greater than can be explained by changes in substrate concentrations. For example, glycolytic fluxes are known to vary by factors of 100 or more, whereas variations of substrate concentrations over such a large range are unknown. Consequently, although changes in substrate concentration can communicate a change in flux at the rate-determining step(s) to the other (near equilibrium; $v_f \approx v_r$) reaction steps of the pathway, there must be other mechanisms that control the flux of the rate-determining step(s).

The flux through a rate-determining step(s) of a pathway may be altered by several mechanisms:

1. Allosteric control: Many enzymes are allosterically controlled (Section 13-4) by effectors that are often substrates, products, or coenzymes in the pathway but not necessarily of the enzyme in question (feedback regulation). One such enzyme is PFK, an important glycolytic regulatory enzyme (Section 17-4Fb).

2. Covalent modification (enzymatic interconversion): Many enzymes that regulate pathway fluxes have specific sites that may be enzymatically phosphorylated and dephosphorylated at specific Ser, Thr, and/or Tyr residues or covalently modified in some other way. Such enzymatic modification processes, which are themselves subject to control, greatly alter the activities of the modified enzymes. This flux control mechanism is discussed in Section 18-3.

3. Substrate cycles: If v_f and v_r in Eq. [17.8] represent the rates of two opposing nonequilibrium reactions that are catalyzed by different enzymes, v_f and v_r may be independently varied. The flux through such a substrate cycle,

as we shall see in the next section, is more sensitive to the concentrations of allosteric effectors than is the flux through a single unopposed nonequilibrium reaction.

4. Genetic control: Enzyme concentrations, and hence enzyme activities, may be altered by protein synthesis in response to metabolic needs. Genetic control of enzyme concentrations is a major concern of Part V of this textbook.

Mechanisms 1 to 3 can respond rapidly (within seconds or minutes) to external stimuli and are therefore classified as "short-term" control mechanisms. Mechanism 4 responds more slowly to changing conditions (within hours or days in higher organisms) and is therefore referred to as a "long-term" control mechanism.

F. *Regulation of Glycolysis in Muscle*

Elucidation of the flux regulation mechanisms of a given pathway involves the determination of the pathway's regulatory enzymes involved in the rate-determining steps together with the identification of the modulators of these enzymes and their mechanism(s) of modulation. A hypothesis may then be formulated that can be tested *in vivo*. A common procedure for establishing regulatory mechanisms involves three steps.

1. Identification of the rate-determining step(s) of the pathway. One way to do so is to measure the *in vivo* ΔG's of all the reactions in the pathway to determine how close to equilibrium they function. Those that operate far from equilibrium are potential control points; the enzymes catalyzing them may be regulated by one or more of the mechanisms listed above. Another way of establishing the rate-determining step(s) of a pathway is to measure the effect of a known inhibitor on a specific reaction step and on the flux through the pathway as a whole. The ratio of the fractional change in the activity of the inhibited enzyme to the fractional change in the total flux (the flux control coefficient) will vary between 0 and 1. The closer the ratio is to 1, the more involved that enzyme is in the regulation of the total flux through the pathway.

2. *In vitro* identification of allosteric modifiers of the enzymes catalyzing the rate-determining reactions. The mechanisms by which these compounds act are determined from their effects on the enzyme's kinetics. From this information, a model of the allosteric mechanisms for regulating the pathway may be formulated.

3. Measurement of the *in vivo* levels of the proposed regulators under various conditions to establish whether these concentration changes are consistent with the proposed regulation mechanism.

a. Free Energy Changes in the Reactions of Glycolysis

Let us examine the thermodynamics of glycolysis with an eye toward understanding its regulatory mechanisms. This must be done separately for each type of tissue in question because glycolysis is regulated in different tissues

Table 17-1 $\Delta G^{\circ\prime}$ and ΔG of the Reactions of Glycolysis in Heart Muscle[a]

Reaction	Enzyme	$\Delta G^{\circ\prime}$ (kJ $\cdot$ mol^{-1})	ΔG (kJ $\cdot$ mol^{-1})
1	HK	−20.9	−27.2
2	PGI	+2.2	−1.4
3	PFK	−17.2	−25.9
4	Aldolase	+22.8	−5.9
5	TIM	+7.9	~0
6 + 7	GAPDH + PGK	−16.7	−1.1
8	PGM	+4.7	−0.6
9	Enolase	−3.2	−2.4
10	PK	−23.0	−13.9

[a]Calculated from data in Newsholme, E.A. and Start, C., *Regulation in Metabolism, p.* 97, Wiley (1973).

in different ways. We shall confine ourselves to muscle tissue. First we establish the pathway's possible regulation points through the identification of its nonequilibrium reactions. Table 17-1 lists the standard free energy change ($\Delta G^{\circ\prime}$) and the actual physiological free energy change (ΔG) associated with each reaction in the pathway. It is important to realize that the free energy changes associated with the reactions under standard conditions may differ dramatically from those in effect under physiological conditions. For example, the $\Delta G^{\circ\prime}$ for aldolase is +22.8 kJ $\cdot$ mol^{-1}, whereas under physiological conditions in heart muscle it is close to zero, indicating that the *in vivo* activity of aldolase is sufficient to equilibrate its substrates and products. The same is true of the GAPDH + PGK reaction series. Nevertheless, in a steady-state pathway, all the reactions must have $\Delta G < 0$. This is because if $\Delta G > 0$ for any reaction, its flux would be in the reverse direction.

In the glycolytic pathway, only three reactions, those catalyzed by hexokinase (HK), phosphofructokinase (PFK), and pyruvate kinase (PK), function with large negative free energy changes in heart muscle under physiological conditions (Table 17-1). These nonequilibrium reactions of glycolysis are the candidates for the flux-control points. The other glycolytic reactions function near equilibrium: Their forward and reverse rates are much faster than the actual flux through the pathway (although their forward rates must be at least slightly greater than their reverse rates). Consequently, these near-equilibrium reactions are very sensitive to changes in the concentration of pathway intermediates (they have high elasticity coefficients) and

thereby rapidly communicate any changes in flux generated at the rate-determining step(s) throughout the rest of the pathway, ensuring the maintenance of a steady state (homeostasis).

b. Phosphofructokinase Is a Major Target for Regulating the Flux of Glycolysis in Muscle

In vitro kinetic studies of HK, PFK, and PK indicate that each is controlled by a variety of compounds, some of which are listed in Table 17-2. Yet, when the G6P source for glycolysis is glycogen, rather than glucose, as is often the case in skeletal muscle (Section 18-1), the hexokinase reaction is not required. *PFK, an elaborately controlled enzyme functioning far from equilibrium, evidently is the major target for regulating glycolysis in muscle under most conditions.*

PFK (Fig. 17-32a) is a tetrameric enzyme with two conformational states, R and T, that are in equilibrium. ATP is both a substrate and an allosteric inhibitor of PFK. Each subunit has two binding sites for ATP, a substrate site and an inhibitor site. The substrate site binds ATP equally well in either conformation, but the inhibitor site binds ATP almost exclusively in the T state. The other substrate of PFK, F6P, preferentially binds to the R state. Consequently, at high concentrations, ATP acts as a heterotropic allosteric inhibitor of PFK by binding to the T state, thereby shifting the T $\rightleftharpoons$ R equilibrium in favor of the T state and thus decreasing PFK's affinity for F6P (this is similar to the action of 2,3-BPG in decreasing the affinity of hemoglobin for O$_2$; Section 10-2F). In graphical terms, at high concentrations of

Table 17-2   **Some Effectors of the Nonequilibrium Enzymes of Glycolysis**

Enzyme	Inhibitors	Activators[a]
HK	G6P	—
PFK	ATP, citrate, PEP	ADP, AMP, cAMP, FBP, F2,6P, F6P, NH$_4^+$, P$_i$
PK (muscle)	ATP	AMP, PEP, FBP

[a]The activators for PFK are better described as deinhibitors of ATP because they reverse the effect of inhibitory concentrations of ATP.

(a)

Figure 17-32   X-ray structure of PFK. (*a*) Two subunits of the tetrameric enzyme are shown in ribbon form (the other two subunits, which have been omitted for clarity, are related to those shown by a 2-fold rotation about a vertical axis). Each subunit binds its products, FBP (*near the center of each subunit*) and Mg^{2+}–ADP (*lower right and upper left*), together with its activator Mg^{2+}–ADP (*upper right and lower left, in the rear*), all in space-filling form with atoms colored according to type (ADP C green, FBP C cyan, N blue, O red, P orange, and Mg purple). Note the close proximity of the product ADP's β-phosphate group to the phosphoryl group at FBP's 1-position, the group that PFK transfers from ATP to F6P. [Based on an X-ray structure by Phillip Evans, Cambridge University, U.K. PDBid 1PFK.] (*b*) A superposition of those segments of the T-state (*blue*) and R-state (*red*) enzymes that undergo a large conformational rearrangement on the T → R allosteric transition (indicated by the arrows). Residues of the R-state structure are marked by a prime. Also shown are bound ligands: the nonphysiological inhibitor 2-phosphoglycolate (**PGC;** a PEP analog) for the T state, and the cooperative substrate F6P and the activator ADP for the R state. [After Schirmer, T. and Evans, P.R., *Nature* **343,** 142 (1990). PDBids 4PFK and 6PFK.] 🐭 **See Kinemage Exercises 13-1 and 13-2**

(b)

ATP, the hyperbolic (noncooperative) curve of PFK activity versus [F6P] is converted to the sigmoidal (cooperative) curve characteristic of allosteric enzymes (Fig. 17-33; cooperative and noncooperative processes are discussed in Section 10-1Ba). For example, when [F6P] = 0.5 m*M* (the dashed line in Fig. 17-33), the enzyme is nearly maximally active, but in the presence of 1 m*M* ATP, the activity drops to 15% of its original level (a nearly 7-fold decrease).

Figure 17-33 **PFK activity versus F6P concentration.** The various conditions are: blue, no inhibitors (low, noninhibitory [ATP]); green, 1 m*M* ATP (inhibitory); and red, 1 m*M* ATP + 0.1 m*M* AMP. [After data from Mansour, T.E. and Ahlfors, C.E., *J. Biol. Chem.* **243,** 2523–2533 (1968).] 🔗 **See the Animated Figures.**

[Actually, the most potent allosteric effector of PFK is **fructose-2,6-bisphosphate (F2,6P).** We discuss the role of F2,6P in regulating PFK activity when we study the mechanism by which the liver maintains blood glucose concentrations (Section 18-3Fc).]

c. Structural Basis for PFK's Allosteric Change in F6P Affinity

The X-ray structures of PFK from several organisms have been determined for both the R and the T states by Phillip Evans. The R state of PFK is homotropically stabilized by the binding of its substrate fructose-6-phosphate (F6P). In the R state of *Bacillus stearothermophilus* PFK, the side chain of Arg 162 forms a salt bridge with the phosphoryl group of an F6P bound in an active site of another subunit (Fig. 17-32*b*). However, Arg 162 is located at the end of a helical turn that unwinds on transition to the T state. The positively charged side chain of Arg 162 thereby swings away and is replaced by the negatively charged side chain of Glu 161. As a consequence, the doubly negative phosphoryl group of F6P has a greatly diminished affinity for the T-state enzyme. The unwinding of this helical turn, which is obligatory for the R → T transition, is prevented by the binding of the activator ADP to its effector site on PFK in the R state, and facilitated by the binding of ATP to this effector site in the T state. Evidently, the same conformational shift is responsible for both the homotropic and the heterotropic allosteric effects in PFK.

d. AMP Overcomes the ATP Inhibition of PFK

Direct allosteric control of PFK by ATP may superficially appear to be the means by which glycolytic flux is regulated. After all, when [ATP] is high as a result of low metabolic demand, PFK is inhibited and flux through the pathway is low; conversely, when [ATP] is low, flux through

the pathway is high and ATP is synthesized to replenish the pool. Consideration of the physiological variation in ATP concentration, however, indicates that the situation must be more complex. The metabolic flux through glycolysis may vary by 100-fold or more, depending on the metabolic demand for ATP. However, measurements of [ATP] *in vivo* at various levels of metabolic activity indicate that [ATP] varies <10% between rest and vigorous exertion. Yet *there is no known allosteric mechanism that can account for a 100-fold change in flux of a nonequilibrium reaction with only 10% change in effector concentration.* Thus, some other mechanism, or mechanisms, must be responsible for controlling glycolytic flux.

The inhibition of PFK by ATP is relieved by AMP. This results from AMP's preferential binding to the R state of PFK. If a PFK solution containing 1 m*M* ATP and 0.5 m*M* F6P is brought to 0.1 m*M* in AMP, the activity of PFK rises from 10 to 50% of its maximal activity, a 5-fold increase (Fig. 17-33).

[ATP] decreases by only 10% in going from a resting state to one of vigorous activity because it is buffered by the action of two enzymes: creatine kinase (Section 16-4Cd) and, of particular importance to this discussion, **adenylate kinase (AK;** also known as **myokinase).** Adenylate kinase catalyzes the reaction

$$2ADP \rightleftharpoons ATP + AMP \quad K = \frac{[ATP][AMP]}{[ADP]^2} = 0.44$$

which rapidly equilibrates the ADP resulting from ATP hydrolysis in muscle contraction with ATP and AMP.

In muscle, [ATP] is ∼50 times [AMP] and ∼10 times [ADP], so that, *as a result of the adenylate kinase reaction, a 10% decrease in [ATP] will cause over a 4-fold increase in [AMP]* (see Problem 12 in this chapter). Consequently, a metabolic signal consisting of a decrease in [ATP] too small to relieve PFK inhibition is amplified significantly by the adenylate kinase reaction, which increases [AMP] by an amount sufficient to produce a much larger increase in PFK activity.

e. Adenylate Kinase's Internal Motions Act as an Energetic Counterweight to Balance Substrate Binding

Adenylate kinase, like other kinases, must be specific to prevent undesirable phosphoryl-transfer reactions such as hydrolysis. However, once the reaction has occurred, the tightly bound products must be rapidly released to maintain the enzyme's catalytic efficiency. With kinases such as hexokinase and phosphoglycerate kinase, this process is accomplished by the closing of "jaws" on the bound substrates that open when product is formed (Figs. 17-5 and 17-15), a process that is presumably driven by the exergonic free energy change of the reaction the enzyme catalyzes. However, since the AK reaction is energetically neutral (it replaces one phosphodiester bond with another), AK specificity is accomplished by a somewhat different means. Comparison of the X-ray structures, determined

(a)

(b)

Figure 17-34 **Conformational changes in *E. coli* adenylate kinase (AK) on binding substrate.** (*a*) The unliganded enzyme. (*b*) The enzyme with the bound substrate analog Ap₅A. The Ap₅A is drawn in stick form colored according to atom type (C green, N blue, O red, and P yellow). The protein's cyan and blue domains undergo extensive conformational changes on ligand binding, whereas the remainder of the protein (*magenta*), whose orientation is the same in Parts *a* and *b,* largely maintains its conformation. Compare these structures to that of porcine AK (Fig. 8-54*b*). [Based on X-ray structures by Georg Schulz, Institut für Organische Chemie und Biochemie, Freiburg, Germany. PDBids (*a*) 4AKE and (*b*) 1AKE.] 🔖 **See Interactive Exercise 9A**

by Georg Schulz, of unliganded AK with AK in complex with the inhibitory bisubstrate analog **Ap₅A** (two ADPs connected by a fifth phosphate) indicates that two ~30-residue domains of AK close over the Ap₅A, thereby tightly binding it and excluding water (Fig. 17-34). This comparison also suggests how AK avoids falling into the energy well of tight-binding substrates and products. On binding substrate, a portion of the protein remote from the active site increases its chain mobility and thereby "absorbs" some of the free energy of substrate binding (recall that an X-ray structure determination reveals atomic mobilities as well as positions; Section 9-4). This region "resolidifies" on product release. This mechanism, Schulz has hypothesized, acts as an "energetic counterweight" that permits facile product release and hence maintains a high reaction rate.

f. Substrate Cycling Can Increase Flux Sensitivity

Even though a mechanism exists for amplifying the effect of a small change in [ATP] by producing a larger change in [AMP], a 4-fold increase in [AMP] would allosterically increase the activity of PFK by only ~10-fold, an amount insufficient to account for the observed 100-fold increase in glycolytic flux. Small changes in effector concentration (and therefore v_f) can only cause relatively large changes in the flux through a reaction ($v_f - v_r$) if the reaction is functioning close to equilibrium. The reason for this high sensitivity is that for such reactions, the term $v_f/(v_f - v_r)$ in Eq. [17.8] (the elasticity coefficient) is large, that is, the reverse reaction contributes significantly to the value of the net flux. This is not the case for the PFK reaction.

Such equilibrium-like conditions may be imposed on a nonequilibrium reaction if a second enzyme catalyzes the regeneration of substrate from product in a thermodynamically favorable manner. Then v_r is no longer negligible compared to v_f. This situation requires that the forward process (formation of FBP from F6P) and reverse process (breakdown of FBP to F6P) be accomplished by different reactions since the laws of thermodynamics would otherwise be violated. In the following paragraphs, we discuss the nature of such **substrate cycles.**

Under physiological conditions, the reaction catalyzed by PFK:

Fructose-6-phosphate + ATP $\longrightarrow$

fructose-1,6-bisphosphate + ADP

is highly exergonic ($\Delta G = -25.9$ kJ · mol^{-1}, Table 17-1). Consequently, the back reaction has a negligible rate compared to the forward reaction. **Fructose-1,6-bisphosphatase (FBPase),** however, which is present in many mammalian tissues (and which is an essential enzyme in gluconeogenesis; Section 23-1), catalyzes the exergonic hydrolysis of FBP ($\Delta G = -8.6$ kJ · mol^{-1}):

Fructose-1,6-bisphosphate + H$_2$O $\longrightarrow$
$$\text{fructose-6-phosphate} + P_i$$

Note that the combined reactions catalyzed by PFK and FBPase result in net ATP hydrolysis:

$$\text{ATP} + \text{H}_2\text{O} \rightleftharpoons \text{ADP} + P_i$$

Such a set of opposing reactions is known as a substrate cycle because it cycles a substrate to an intermediate and back again. When this set of reactions was discovered, it was referred to as a **futile cycle** since its net result seemed to be the useless consumption of ATP. In fact, when it was found that the PFK activators AMP and F2,6P allosterically inhibit FBPase, it was suggested that only one of these enzymes was functional in a cell under any given set of conditions. It was subsequently demonstrated, however, that both enzymes often function simultaneously at significant rates.

g. Substrate Cycling Can Account for Glycolytic Flux Variation

Eric Newsholme proposed that substrate cycles are not at all "futile" but, rather, have a regulatory function. The *in vivo* activities of enzymes and concentrations of metabolites are extremely difficult to measure, so that their values are rarely known accurately. However, let us make the physiologically reasonable assumption that a 4-fold increase in [AMP], resulting from the adenylate kinase reaction, causes PFK activity (v_f) to increase from 10 to 90% of its maximum and FBPase activity (v_r) to decrease from 90 to 10% of its maximum. The maximum activity of muscle PFK is known from *in vitro* studies to be ~10-fold greater than that of muscle FBPase. Hence, if we assign full activity of PFK to be 100 arbitrary units, then full activity of FBPase is 10 such units. The flux through the PFK reaction in glycolysis under conditions of low [AMP] is

$$J_{low} = v_f(low) - v_r(low) = 10 - 9 = 1$$

where v_f is catalyzed by PFK and v_r by FBPase. The flux under conditions of high [AMP] is

$$J_{high} = v_f(high) - v_r(high) = 90 - 1 = 89$$

Substrate cycling could therefore amplify the effect of changes in [AMP] on the net rate of phosphorylation of F6P. Without the substrate cycle, a 4-fold increase in [AMP] increases the net flux by about 9-fold, whereas with the cycle the same increase in [AMP] causes a $J_{high}/J_{low} =$

$89/1 \approx 90$-fold increase in net flux. Consequently, under the above assumptions, *a 10% change in [ATP] could stimulate a 90-fold change in flux through the glycolytic pathway by a combination of the adenylate kinase reaction and substrate cycles.*

h. Physiological Impact of Substrate Cycling

Substrate cycling, if it has a regulatory function, does not increase the maximum flux through a pathway. On the contrary, it functions to decrease its minimum flux. In a sense, the substrate is put into a "holding pattern." In the case described above, *the cycling of substrate is the energetic "price" that a muscle must pay to be able to change rapidly from a resting state, in which substrate cycling is maximal, to one of sustained high activity.* However, the rate of substrate cycling may itself be under hormonal or nervous control so as to increase the sensitivity of the metabolic system under conditions when high activity (fight or flight) is anticipated (we address the involvement of hormones in metabolic regulation in Sections 18-3E and 18-3F).

In some tissues, substrate cycles function to produce heat. For example, many insects require a thoracic temperature of 30°C to be able to fly. Yet bumblebees are capable of flight at ambient temperatures as low as 10°C. Bumblebee flight muscle FBPase has a maximal activity similar to that of its PFK (10-fold greater than our example for mammalian muscle); furthermore, unlike all other known muscle FBPases, it is not inhibited by AMP. This permits the FBPase and PFK of bumblebee flight muscle to be highly active simultaneously so as to generate heat. Since the maximal rate of FBP cycling possible in bumblebee flight muscle generates only 10 to 15% of the required heat, however, other mechanisms of thermogenesis must also be operative. Nevertheless, FBP cycling is probably significant because, unlike bumblebees, honeybees, which have no FBPase activity in their flight muscles, cannot fly when the temperature is low.

i. Substrate Cycling, Thermogenesis, and Obesity

Many animals, including adult humans, generate some of their body heat, particularly when it is cold, through substrate cycling in muscle and liver, one mechanism of a process known as **nonshivering thermogenesis** (the muscle contractions of shivering or any other movement also produce heat; another mechanism of nonshivering thermogenesis is described in Section 22-3Da). Substrate cycling is stimulated by thyroid hormones (which stimulate metabolism in most tissues; Section 19-1D), as is indicated, for example, by the observation that rats lacking a functioning thyroid gland do not survive at 5°C. Chronically obese individuals tend to have lower than normal metabolic rates, which is probably due, in part, to a reduced rate of nonshivering thermogenesis. Such individuals therefore tend to be cold sensitive. Indeed, whereas normal individuals increase their rate of thyroid hormone activation on exposure to cold, genetically obese animals and obese humans fail to do so.

j. The Overexpression of PFK Does Not Increase the Rate of Glycolysis

PFK has long been thought to be the controlling enzyme of glycolysis. It was therefore expected that increasing the level of expression of PFK in yeast cells via genetic engineering techniques would increase the rate of glycolysis independent of the demand for glycolytic products. It has been amply demonstrated, however, that this is not the case. Although PFK is a major regulatory enzyme of glycolysis, its catalytic activity *in vivo* is controlled by the concentrations of the effectors that reflect the needs of the demand blocks that utilize its products.

Metabolic control analysis, in addition to helping us recognize that control can be shared by several enzymes in a pathway, has also alerted us to the difference between control and regulation. *Although PFK has a major role in regulating the flux through glycolysis, it is controlled, in vivo, by factors outside the pathway.* An increase in the *in vivo* concentration of PFK will therefore not increase the flux through the pathway because these controlling factors adjust the catalytic activity of PFK only to meet the needs of the cell.

5 METABOLISM OF HEXOSES OTHER THAN GLUCOSE

While glucose is the primary end product of the digestion of starch and glycogen (Section 11-2D), three other hexoses are also prominent digestion products: **Fructose,** obtained from fruits and from the hydrolysis of sucrose (table sugar); **galactose,** obtained from the hydrolysis of lactose (milk sugar); and **mannose,** obtained from the digestion of polysaccharides and glycoproteins. After digestion, these monosaccharides enter the bloodstream, which carries them to various tissues. *The metabolism of fructose, galactose, and mannose proceeds by their conversion to glycolytic intermediates, from which point they are broken down in a manner identical to glucose.*

A. Fructose

Fructose is a major fuel source in diets that contain large amounts of sucrose (a disaccharide of fructose and glucose; Fig. 11-13). There are two pathways for the metabolism of fructose; one occurs in muscle and the other occurs in liver. This dichotomy results from the different enzymes present in these various tissues.

Fructose metabolism in muscle differs little from that of glucose. Hexokinase (Section 17-2A), which converts glucose to G6P on entry into muscle cells, also phosphorylates fructose, yielding F6P (Fig. 17-35, *left*). The entry of fructose into glycolysis therefore involves only one reaction step.

Liver contains little hexokinase; rather, it contains glucokinase, which phosphorylates only glucose (Section 17-2A). Fructose metabolism in liver must therefore differ from that in muscle. In fact, liver converts fructose to glycolytic

intermediates through a pathway that involves seven enzymes (Fig. 17-35, *right*):

1. Fructokinase catalyzes the phosphorylation of fructose by ATP at C1 to form **fructose-1-phosphate.** *Neither hexokinase nor phosphofructokinase can phosphorylate fructose-1-phosphate at C6 to form the glycolytic intermediate fructose-1,6-bisphosphate.*

2. Class I aldolase (Section 17-2Da) has several isoenzyme forms. Muscle contains Type A aldolase, which is specific for fructose-1,6-bisphosphate. Liver, however, contains Type B aldolase, which also utilizes fructose-1-phosphate as a substrate (Type B aldolase is also called **fructose-1-phosphate aldolase**). In liver, fructose-1-phosphate therefore undergoes an aldol cleavage (Section 17-2D):

Fructose-1-phosphate $\rightleftharpoons$

dihydroxyacetone phosphate + glyceraldehyde

3. Direct phosphorylation of glyceraldehyde by ATP through the action of **glyceraldehyde kinase** forms the glycolytic intermediate glyceraldehyde-3-phosphate.

4–7. Alternatively, glyceraldehyde is converted to the glycolytic intermediate dihydroxyacetone phosphate by reduction to glycerol by NADH as catalyzed by alcohol dehydrogenase (Reaction 4), phosphorylation to glycerol-3-phosphate by ATP through the action of **glycerol kinase** (Reaction 5), and reoxidation by NAD^+ to dihydroxyacetone phosphate as mediated by glycerol phosphate dehydrogenase (Reaction 6). The DHAP is then converted to GAP by triose phosphate isomerase (Reaction 7).

As this complex series of reactions suggests, the liver has an enormous repertory of enzymes. This is because the liver is involved in the breakdown of a great variety of metabolites. Efficiency in metabolic processing dictates that many of these substances be converted to glycolytic intermediates. The liver, in fact, contains many of the enzymes necessary to do so.

a. Excessive Fructose Depletes Liver P_i

At one time, fructose was thought to have advantages over glucose for intravenous feeding. The liver, however, encounters metabolic problems when the blood concentration of this sugar is too high (higher than can be attained by simply eating fructose-containing foods). When the fructose concentration is high, fructose-1-phosphate may be produced faster than Type B aldolase can cleave it. Intravenous feeding of large amounts of fructose may therefore result in high enough fructose-1-phosphate accumulation to severely deplete the liver's store of P_i. Under these conditions, [ATP] drops, thereby activating glycolysis and lactate production. The lactate concentration in the blood and the consequent low pH under such conditions can reach life-threatening levels.

Fructose intolerance, a genetic disease in which ingestion of fructose causes the same fructose-1-phosphate accumulation as with its intravenous feeding, results from a deficiency of Type B aldolase. This condition appears to be

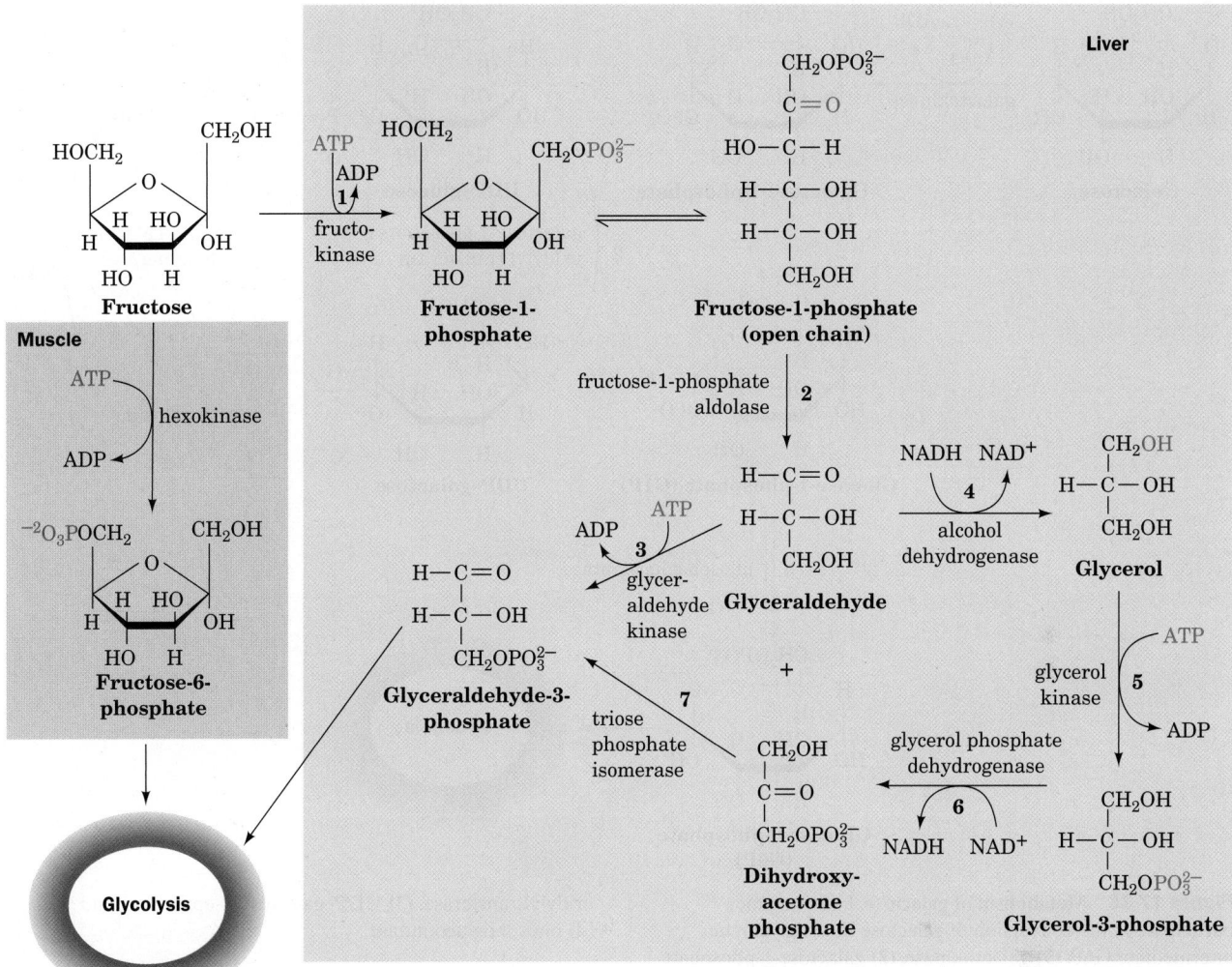

Figure 17-35   Metabolism of fructose. In muscle (*left*), the conversion of fructose to the glycolytic intermediate F6P involves only one enzyme, hexokinase. In liver (*right*), seven enzymes participate in the conversion of fructose to glycolytic intermediates: (**1**) fructokinase, (**2**) fructose-1-phosphate aldolase, (**3**) glyceraldehyde kinase, (**4**) alcohol dehydrogenase, (**5**) glycerol kinase, (**6**) glycerol phosphate dehydrogenase, and (**7**) triose phosphate isomerase.

self-limiting: Individuals with fructose intolerance rapidly develop a strong distaste for anything sweet.

B. Galactose

Galactose comprises half of the milk sugar lactose and is thus a major fuel constituent of dairy products. Galactose and glucose are epimers that differ only in their configuration about C4:

α-D-Glucose α-D-Galactose

The enzymes of glycolysis are specific; they do not recognize the galactose configuration. An epimerization reaction must therefore be carried out before galactose enters the glycolytic pathway. This reaction takes place after the conversion of galactose to its uridine diphosphate derivative. The role of UDP–sugars and other nucleotidyl–sugars is discussed in more detail in Sections 18-2 and 23-3. The entire pathway converting galactose to a glycolytic intermediate, which was elucidated by Luis Leloir and is therefore known as the **Leloir pathway,** involves four reactions (Fig. 17-36):

1. Galactose is phosphorylated at C1 by ATP in a reaction catalyzed by **galactokinase.**

2. Galactose-1-phosphate uridylyltransferase transfers **UDP–glucose's** uridylyl group to galactose-1-phosphate to yield **glucose-1-phosphate (G1P)** and **UDP–galactose** by the reversible cleavage of UDP–glucose's pyrophosphoryl bond.

Figure 17-36   Metabolism of galactose. Four enzymes participate in the conversion of galactose to the glycolytic intermediate G6P: **(1)** galactokinase, **(2)** galactose-1-phosphate uridylyltransferase, **(3)** UDP–galactose-4-epimerase, and **(4)** phosphoglucomutase.

3. UDP–galactose-4-epimerase converts UDP–galactose to UDP–glucose. This enzyme has an associated NAD^+, which suggests that the reaction involves the sequential oxidation and reduction of the hexose C4 atom:

UDP–Galactose   **UDP–Glucose**

4. G1P is converted to the glycolytic intermediate G6P by the action of **phosphoglucomutase** (Section 18-1B).

a. Galactosemia

Galactosemia is a genetic disease characterized by the inability to convert galactose to glucose. Its symptoms include failure to thrive, mental retardation, and in some instances death from liver damage. Most cases of galactosemia involve a mutation in the enzyme catalyzing Reaction 2 of the interconversion, galactose-1-phosphate uridylyltransferase. This reaction is a double displacement in which an enzyme His side chain first nucleophilically attacks the α-phosphoryl group of UDP–glucose, displacing G1P and forming a uridylyl–His intermediate:

UDP–glucose + E–His166 $\rightleftharpoons$
            glucose-1-phosphate + E–His-UMP

Galactose-1-phosphate then displaces the uridylyl group from the enzyme's His to form UDP–galactose:

Galactose-1-phosphate + E–His-UMP $\rightleftharpoons$
            UDP–galactose + E–His

A Gln residue forms hydrogen bonds to the uridylyl group's phosphoryl oxygens to stabilize the intermediate uridylyl–His. Mutation of this Gln to Arg inactivates the enzyme. Formation of UDP–galactose from galactose-1-phosphate

Figure 17-37 Metabolism of mannose. Two enzymes are required to convert mannose to the glycolytic intermediate F6P: **(1)** hexokinase and **(2)** phosphomannose isomerase.

is thus prevented, leading to a buildup of toxic metabolic by-products. For example, the increased galactose concentration in the blood results in a higher galactose concentration in the lens of the eye where this sugar is reduced to **galactitol.**

Galactitol

The presence of this sugar alcohol in the lens eventually causes cataract formation (clouding of the lens).

Galactosemia is treated by a galactose-free diet. Except for the mental retardation, this reverses all symptoms of the disease. The galactosyl units that are essential for the synthesis of glycoproteins (Section 11-3C) and glycolipids (Section 12-1D) may be synthesized from glucose by a reversal of the epimerase reaction. These syntheses therefore do not require dietary galactose.

C. *Mannose*

Mannose, a common component of glycoproteins (Section 11-3C), and glucose are C2 epimers:

α-D-**Glucose**   α-D-**Mannose**

Mannose enters the glycolytic pathway after its conversion to F6P via a two-reaction pathway (Fig. 17-37):

1. Hexokinase (Section 17-2A) converts mannose to mannose-6-phosphate.

2. Phosphomannose isomerase then converts this aldose to the ketose F6P. The mechanism of the phosphomannose isomerase reaction resembles that catalyzed by phosphoglucose isomerase (Section 17-2B); it involves an enediolate intermediate.

CHAPTER SUMMARY

1 The Glycolytic Pathway Glycolysis is the metabolic pathway by which most life-forms degrade glucose to two molecules of pyruvate with the concomitant net generation of two ATPs. The overall reaction:

Glucose + 2NAD$^+$ + 2ADP + 2P$_i$ $\longrightarrow$

$\qquad$ 2NADH + 2pyruvate + 2ATP + 2H$_2$O + 4H$^+$

occurs in 10 enzymatically catalyzed reactions.

2 The Reactions of Glycolysis In the preparatory stage of glycolysis, which encompasses its first five reactions, glucose reacts with two ATPs, in an "energy investment," to form fructose-1,6-bisphosphate, which is subsequently converted to two molecules of glyceraldehyde-3-phosphate. In the second stage of glycolysis, the "payoff" stage, which comprises its last five reactions, glyceraldehyde-3-phosphate reacts with NAD$^+$ and P$_i$ to form the "high-energy" compound 1,3-bisphosphoglycerate. This compound reacts in the last four reactions of

the pathway with two ADPs to form pyruvate and two ATPs per molecule. The mechanisms of the 10 glycolytic enzymes have been elucidated through chemical and kinetic measurements combined with X-ray structural studies. The glycolytic enzymes exhibit stereospecificity in the reactions that they catalyze. In at least two kinases, phosphoryl transfer from substrate to water is prevented by substrate-induced conformational changes that form the active site and exclude water from it.

3 Fermentation: The Anaerobic Fate of Pyruvate The NAD$^+$ consumed in the formation of 1,3-BPG must be regenerated if glycolysis is to continue. In the presence of O$_2$, NAD$^+$ is regenerated by oxidative phosphorylation in the mitochondria. Under anaerobic conditions in muscle, pyruvate is reduced by NADH, yielding lactate and NAD$^+$ in a reaction catalyzed by lactate dehydrogenase. In many muscles, particularly during strenuous activity, the process of homolactic fermentation is a major free energy source. In anaerobic yeast, NAD$^+$ is regenerated by alcoholic fermentation in two reactions. First pyruvate is decarboxylated to acetaldehyde by pyruvate decarboxylase, an enzyme that requires thiamine pyrophosphate as a cofactor. The acetaldehyde is then reduced by NADH to form ethanol and NAD^+ in a reaction catalyzed by alcohol dehydrogenase.

4 Metabolic Regulation and Control Metabolic regulation is the process by which the steady-state flow of metabolites through a pathway is maintained. Metabolic control is the force exerted on the enzymes of the pathway in response to an external signal in order to increase or decrease the flow while maintaining the steady state to the extent possible. Homeostasis is the regulation of the steady state. Metabolic flow must be controlled to balance supply with demand and also to maintain homeostasis. It is possible for more than one enzyme to be rate-limiting in a metabolic pathway. Metabolic control analysis provides a framework for the study of metabolic systems *in vivo* that share control among more than one enzyme, and it quantitatively describes flux control and homeostasis. The flux control coefficient measures the sensitivity of the flux to a change in enzyme concentration. The elasticity coefficient measures the sensitivity of an enzymatic rate to the change in substrate concentration. Both supply and demand are involved in flux control and homeostasis. The response of the steady-state concentration of intermediates to changes in the supply or demand blocks depends entirely on the elasticity coefficients of the two blocks at the steady state. When the supply elasticity coefficient is greater than the demand elasticity coefficient, flux control lies in the demand portion of the pathway, and vice versa. Homeostasis control depends on large elasticity coefficients, whereas flux control requires a low elasticity coefficient and a high flux control coefficient. If the demand block controls the flux, the function of the supply block is to control homeostasis. Feedback

inhibition determines the range of concentration of intermediates at which there is a steady state. It maintains homeostasis at physiologically reasonable metabolite concentrations, sometimes far from their equilibrium values.

The flux through a reaction that is close to equilibrium is very sensitive to changes in substrate concentration. Hence, the steady-state flux through a metabolic pathway can only be regulated by a nonequilibrium reaction. Nonequilibrium reactions are controlled by allosteric interactions, substrate cycles, covalent modification, and genetic (long-term) control mechanisms. In muscle glycolysis, phosphofructokinase (PFK) catalyzes one of the flux-generating steps. Although PFK is inhibited by high concentrations of one of its substrates, ATP, the 10% variation of [ATP] over the range of metabolic activity has insufficient influence on PFK activity to account for the observed 100-fold range in glycolytic flux. [AMP] has a 4-fold variation in response to the 10% variation of [ATP] through the action of adenylate kinase. Although AMP relieves the ATP inhibition of PFK, its concentration variation is also insufficient to account for the observed glycolytic flux range. However, the product of the PFK reaction, fructose-1,6-bisphosphate, is hydrolyzed to F6P by FBPase, which is inhibited by AMP. The substrate cycle catalyzed by these two enzymes confers, at least in principle, the necessary sensitivity of the glycolytic flux to variations in [AMP]. Substrate cycling is an important source of nonshivering thermogenesis.

5 Metabolism of Hexoses Other than Glucose Digestion of carbohydrates yields glucose as the primary product. Other prominent products are fructose, galactose, and mannose. These monosaccharides are metabolized through their conversion to glycolytic intermediates.

REFERENCES

General

Cornish-Bowden, A. (Ed.), *New Beer in an Old Bottle: Eduard Buchner and the Growth of Biochemical Knowledge,* Universitat de València (1997). [The 1897 paper by Eduard Buchner reporting the discovery of cell-free fermentation (in the original German as well as its English and Spanish translations) together with a series of essays discussing the historical context of this discovery and the modern study of multienzyme systems.]

Fersht, A., *Structure and Mechanism in Protein Science,* Freeman (1999).

Frey, P.A. and Hegeman, A.D., *Enzymatic Reaction Mechanisms,* Oxford University Press (2007).

Fruton, J.S., *Molecules and Life: Historical Essays on the Interplay of Chemistry and Biology,* Wiley–Interscience (1974). [Includes a detailed historical account of the elucidation of fermentation.]

Kim, J. and Dang, C.V., Multifaceted roles of glycolytic enzymes, *Trends Biochem. Sci.* **30,** 142–150 (2005). [Discusses the various nonglycolytic functions, mainly regulatory, that glycolytic enzymes have acquired.]

Saier, M.H., Jr., *Enzymes in Metabolic Pathways,* Chapter 5, Harper & Row (1987).

Enzymes of Glycolysis

The Enzymes of Glycolysis: Structure, Activity and Evolution, *Philos. Trans. R. Soc. London Ser. B* **293,** 1–214 (1981). [A

collection of authoritative discussions on the enzymes of glycolysis.]

Allen, S.C. and Muirhead, H., Refined three-dimensional structure of cat-muscle (M1) pyruvate kinase at a resolution of 2.6 Å, *Acta Cryst.* D**52,** 499–504 (1996).

Bennett, W.S., Jr. and Steitz, T.A., Glucose-induced conformational change in yeast hexokinase, *Proc. Natl. Acad. Sci.* **75,** 4848–4852 (1978).

Bernstein, B.E., Michels, P.A.M., and Hol, W.G.J., Synergistic effects of substrate-induced conformational changes in phosphoglycerate activation, *Nature* **385,** 275–278 (1997).

Biesecker, G., Harris, J.I., Thierry, J.C., Walker, J.E., and Wonacott, A.J., Sequence and structure of D-glyceraldehyde-3-phosphate dehydrogenase from *Bacillus stearothermophilus, Nature* **266,** 328–333 (1977).

Cleland, W.W. and Kreevoy, M.M., Low-barrier hydrogen bonds and enzymic catalysis, Science **264,** 1887–1890 (1994); *and* Gerlt, J.A. and Gassman, P.G., Understanding the rates of certain enzyme-catalyzed reactions: Proton abstraction from carbon acids, acyl-transfer reactions, and displacement of phosphodiesters, *Biochemistry* **32,** 11943–11952 (1993).

Dalby, A., Dauter, Z., and Littlechild, J.A., Crystal structure of human muscle aldolase complexed with fructose 1,6-bisphosphate: Mechanistic implications, *Protein Sci.* **8,** 291–297 (1999).

Davenport, R.C., Bash, P.A., Seaton, B.A., Karplus, M., Petsko, G.A., and Ringe, D., Structure of the triosephosphate isomerase–

phosphoglycohydroxamate complex: An analogue of the intermediate on the reaction pathway, *Biochemistry* **30,** 5821–5826 (1991); *and* Lolis, E. and Petsko, G.A., Crystallographic analysis of the complex between triosephosphate isomerase and 2-phosphoglycolate at 2.5 Å resolution: Implications for catalysis, *Biochemistry* **29,** 6619–6625 (1990).

Evans, P.R. and Hudson, P.J., Structure and control of phosphofructokinase from *Bacillus stearothermophilus, Nature* **279,** 500–504 (1979).

Gefflaut, T., Blonski, C., Perie, J., and Willson, M., Class I aldolases: Substrate specificity, mechanism, inhibitors and structural aspects, *Prog. Biophys. Mol. Biol.* **63,** 301–340 (1995).

Hall, D.R., Leonard, G.A., Reed, C.D., Watt, C.I., Berry, A., and Hunter, W.N., The crystal structure of Escherichia coli class II fructose-1,6-bisphosphate aldolase in complex with phosphoglycohydroxamate reveals details of mechanism and specificity, *J. Mol. Biol.* **287,** 383–394 (1999).

Harlos, K., Vas, M., and Blake, C.C.F., Crystal structure of the binary complex of pig muscle phosphoglycerate kinase and its substrate 3-phospho-D-glycerate, *Proteins* **12,** 133–144 (1992).

Jedrzejas, M.J., Structure, function, and evolution of phosphoglycerate mutase: Comparison with fructose-2,6-bisphosphatase, acid phosphatase, and alkaline phosphatase, *Prog. Biophys. Mol. Biol.* **73,** 263–287 (2000).

Jeffrey, C.J., Bahnson, B.J., Chien, W., Ringe, D., and Petsko, G.A., Crystal structure of rabbit phosphoglucose isomerase, a glycolytic enzyme that moonlights as neuroleukin, autocrine motility factor, and differentiation mediator, *Biochemistry* **39,** 955–964 (2000).

Jogl, G., Rozovsky, S., McDermott, A.E., and Tong, L., Optimal alignment for enzymatic proton transfer: Structure of the Michaelis complex of triosphosphate isomerase at 1.2-Å resolution, *Proc. Natl. Acad. Sci.* **100,** 50–55 (2003).

Joseph, D., Petsko, G.A., and Karplus, M., Anatomy of a conformational change: Hinged "lid" motion of the triosephosphate isomerase loop, *Science* **249,** 1425–1428 (1990).

Knowles, J.R., Enzyme catalysis: Not different, just better, *Nature* **350,** 121–124 (1991). [A lucid discussion of TIM's catalytic mechanism.]

Kuby, S.A. (Ed.), *A Study of Enzymes,* Vol. II, CRC Press (1991). [Chapters 17, 18, 19, and 20 discuss the mechanisms of adenylate kinase, PFK, PGI and TIM, and aldolase, respectively. Chapter 4 discusses thiamine-dependent reaction mechanisms.]

Kuser, P., Cupri, F., Bleicher, L., and Polikarpov, I., Crystal structure of yeast hexokinase PI in complex with glucose: A classical "induced fit" example revised, *Proteins* **72,** 731–740 (2008).

Marsh, J.J. and Lebherz, H.G., Fructose-bisphosphate aldolases: An evolutionary history, *Trends Biochem. Sci.* **17,** 110–113 (1992).

Maurer, P.J. and Nowak, T., Fluoride inhibition of yeast enolase. 1. Formation of ligand complexes, *Biochemistry* **20,** 6894–6900 (1981); *and* Nowak, T. and Maurer, P.J., Fluoride inhibition of yeast enolase. 2. Structural and kinetic properties of ligand complexes determined by nuclear relaxation rate studies, *Biochemistry* **20,** 6901–6911 (1981).

Morris, A.J. and Tolan, D.R., Lysine-146 of rabbit muscle aldolase is essential for cleavage and condensation of the C3-C4 bond of fructose 1,6-bis(phosphate), *Biochemistry* **33,** 12291–12297 (1994); *and* Site-directed mutagenesis identifies aspartate 33 as a previously unidentified critical residue in the catalytic mechanism of rabbit aldolase A, *J. Biol. Chem.* **268,** 1095–1100 (1993).

Muirhead, H. and Watson, H. Glycolytic enzymes: From hexose to pyruvate, *Curr. Opin. Struct. Biol.* **2,** 870–876 (1992).

Reed, G.H., Poyner, R.R., Larsen, T.M., Wedekind, J.E., and Rayment, I., Structural and mechanistic studies on enolase, *Curr. Opin. Struct. Biol.* **6,** 736–743 (1996).

Seeholzer, S.H., Phosphoglucose isomerase: A ketol isomerase with aldol C2-epimerase activity, *Proc. Natl. Acad. Sci.* **90,** 1237–1241 (1993).

Williams, J.C. and McDermott, A.E., Dynamics of the flexible loop of triosephosphate isomerase: The loop motion is not ligand-gated, *Biochemistry* **34,** 8309–8319 (1995).

Enzymes of Anaerobic Fermentation

Boyer, P.D. (Ed.), *The Enzymes* (3rd ed.), Vol. 11, Academic Press (1975). [Contains authoritative reviews on alcohol dehydrogenase, lactate dehydrogenase, and the evolutionary and structural relationships among the dehydrogenases.]

Dyda, F., Furey, W., Swaminathan, S., Sax, M., Farrenkopf, B., and Jordan, F., Catalytic centers in the thiamin diphosphate dependent enzyme pyruvate decarboxylase at 2.4-Å resolution, *Biochemistry* **32,** 6165–6170 (1993).

Golbik, R., Neef, H., Hubner, G., Konig, S., Seliger, B., Meshalkina, L., Kochetov, G.A., and Schellenberger, A., Function of the aminopyridine part in thiamine pyrophosphate enzymes, *Bioinorg. Chem.* **19,** 10–17 (1991).

Park, J.H., Brown, R.L., Park, C.R., Cohn, M., and Chance, B., Energy metabolism in the untrained muscle of elite runners as observed by ^{31}P magnetic resonance spectroscopy: Evidence suggesting a genetic endowment for endurance exercise. *Proc. Natl. Acad. Sci.* **85,** 8780–8785 (1988).

Control of Metabolic Flux

Crabtree, B. and Newsholme, E.A., A systematic approach to describing and analyzing metabolic control systems, *Trends Biochem. Sci.* **12,** 4–12 (1987).

Fell, D.A., Metabolic control analysis: A survey of its theoretical and experimental development, *Biochem. J.* **286,** 313–330 (1992).

Fell, D., *Understanding the Control of Metabolism,* Portland Press (1997).

Hofmeyr, J.-H.S. and Cornish-Bowden, A., Regulating the cellular economy of supply and demand, *FEBS Lett.* **476,** 47–51 (2000).

Kacser, H. and Burns, J.A. (with additional comments by Kacser, H. and Fell, D.A.), The control of flux, *Biochem. Soc. Trans.* **23,** 341–366 (1995).

Kacser, H. and Porteous, J.W., Control of metabolism: What do we have to measure? *Trends Biochem. Sci.* **12,** 5–14 (1987).

Lardy, H. and Schrago, E., Biochemical aspects of obesity, *Annu. Rev. Biochem.* **59,** 689–710 (1990).

Newsholme, E.A., Challiss, R.A.J., and Crabtree, B., Substrate cycles: their role in improving sensitivity in metabolic control, *Trends Biochem. Sci.* **9,** 277–280 (1984).

Perutz, M.F., Mechanism of cooperativity and allosteric regulation in proteins, *Q. Rev. Biophys.* **22,** 139–236 (1989). [Section 6 discusses PFK.]

Schaaf, I., Heinisch, J., and Zimmermann, K., Overproduction of glycolytic enzymes in yeast, *Yeast* **5,** 285–290 (1989).

Schirmer, T. and Evans, P.R., Structural basis of the allosteric behaviour of phosphofructokinase, *Nature* **343,** 140–145 (1990).

Walsh, K. and Koshland, D.E., Jr., Characterization of rate-controlling steps *in vivo* by use of an adjustable expression vector, *Proc. Natl. Acad. Sci.* **82,** 3577–3581 (1985).

Metabolism of Hexoses Other Than Glucose

Frey, P.A., The Leloir pathway: A mechanistic imperative for three enzymes to change the stereochemical configuration of a single carbon in galactose, *FASEB J.* **10,** 461–470 (1996).

Valle, D. (Ed.), *The Metabolic & Molecular Bases of Inherited Disease,* http://www.ommid.com. [Chapters 70 and 72 discuss fructose and galactose metabolism and their genetic disorders.]

PROBLEMS

1. Write out the reactions of the glycolytic pathway from glucose to lactate using structural formulas for all intermediates. Learn the names of these intermediates and the enzymes that catalyze the reactions.

2. $\Delta G°'$ for the aldolase reaction is $+22.8$ kJ $\cdot$ mol^{-1}. In the cell, at 37°C, the mass action ratio [DHAP]/[GAP] = 5.5. Calculate the equilibrium ratio of [FBP]/[GAP] when [GAP] is (a) 2×10^{-5} M and (b) 10^{-3} M.

3. The pH dependence of the rate of the triose phosphate isomerase (TIM) reaction has characteristic pK's of 6.5 and 9.5. His 95, a catalytically essential residue, has been shown to have a pK of 4.5. Why doesn't the pH rate curve indicate the existence of this pK?

4. Arsenate (AsO_4^{3-}), a structural analog of phosphate, can act as a substrate for any reaction in which phosphate is a substrate. Arsenate esters, unlike phosphate esters, are kinetically as well as thermodynamically unstable and hydrolyze almost instantaneously. Write a balanced overall equation for conversion of glucose to pyruvate in the presence of ATP, ADP, NAD$^+$, and either (a) phosphate or (b) arsenate. (c) Why is arsenate a poison?

5. When glucose is degraded anaerobically via glycolysis there is no overall oxidation or reduction of the substrate. The fermentation reaction is therefore said to be "balanced." The free energy required for ATP formation is nevertheless obtained from favorable electron-transfer reactions. Which metabolic intermediate is the electron donor and which is the electron acceptor when glucose is degraded by a balanced glycolytic fermentation: (a) in muscle and (b) in yeast?

6. In which carbon atoms of pyruvate would radioactivity be found if glucose metabolized by the glycolytic pathway were labeled with ^{14}C at: (a) C1 and (b) C4? (*Note:* Assume that triose phosphate isomerase is able to equilibrate dihydroxyacetone phosphate and glyceraldehyde-3-phosphate.)

*7. The following reaction is catalyzed by an enzyme very similar to Class I aldolases:

Fructose-6-phosphate **Erythrose-4-phosphate**

transaldolase

Glyceraldehyde-3-phosphate **Sedoheptulose-7-phosphate**

Write a plausible mechanism for this reaction using curved arrows to indicate the electron flow.

8. The half-reactions involved in the LDH reaction and their standard reduction potentials are

Pyruvate + 2H$^+$ + 2e^- $\longrightarrow$ lactate $\mathscr{E}°'$ = -0.185 V

NAD$^+$ + 2H$^+$ + 2e^- $\longrightarrow$ NADH + H$^+$ $\mathscr{E}°'$ = -0.315 V

Calculate ΔG for the reaction at the biochemical standard state under the following conditions:

(a) [lactate]/[pyruvate] = 1;   [NAD$^+$]/[NADH] = 1
(b) [lactate]/[pyruvate] = 160;   [NAD$^+$]/[NADH] = 160
(c) [lactate]/[pyruvate] = 1000;   [NAD$^+$]/[NADH] = 1000
(d) Under what conditions will the reaction spontaneously favor NADH oxidation?
(e) In order for the free energy change of the glyceraldehyde-3-phosphate dehydrogenase reaction to favor glycolysis, the [NAD$^+$]/[NADH] ratio must be maintained close to 10^3. Under anaerobic conditions in mammalian muscle, lactate dehydrogenase performs this function. How high can the [lactate]/[pyruvate] ratio become in muscle cells before the LDH-catalyzed reaction ceases to be favorable in the direction of NAD$^+$ production while maintaining the foregoing [NAD$^+$]/[NADH] ratio constant?

9. Although it is not the primary flux-control point for glycolysis, pyruvate kinase is subject to allosteric regulation. (a) What is the metabolic importance of regulating flux through the pyruvate kinase reaction? (b) What is the advantage of activating pyruvate kinase with fructose-1,6-bisphosphate?

*10. Based on the involvement of thiamine pyrophosphate (TPP) in the pyruvate decarboxylase reaction, which of the following reactions, if any, might be expected to utilize TPP as a cofactor?

(a)

(b)

Write hypothetical mechanisms for each reaction showing where TPP is involved or why it is unnecessary.

11. The glycolytic pathway for glucose breakdown is almost universal. Some bacteria, however, utilize an alternate route called the **Entner–Doudoroff pathway** (Fig. 17-38). Like the glycolytic pathway in yeast, the final product is ethanol. (a) Write balanced equations for the conversion of glucose to ethanol and CO_2 via the Entner–Doudoroff pathway and the yeast alcoholic fermentation. (b) Infer from your stoichiometries why the glycolytic pathway rather than the Entner–Doudoroff pathway is almost universal.

H—C=O   **1**   H—C=O   **2**   O=C—O⁻   **3**   O=C—O⁻

Glucose structure with ATP ADP, Mg²⁺, glucokinase → Glucose-6-phosphate, then NAD⁺ + H₂O, NADH, glucose-6-phosphate dehydrogenase → 6-Phosphogluconate, then H₂O, 6-phosphogluconate dehydrase (analogous to enolase) → 2-Keto-3-deoxy-6-phosphogluconate (KDPG)

$H-\overset{O}{\underset{}{C}}$ ...

Glucose

$(P) \equiv -PO_3^{2-}$

Glucose-6-phosphate (G6P)

6-Phospho-gluconate

2-Keto-3-deoxy-6-phosphogluconate (KDPG)

4 ‖ KDPG-aldolase

NAD⁺   NADH

CO₂
+
CH₃CH₂OH

same reactions as
in glycolysis and
alcoholic fermentation

Ethanol

2ATP + H₂O   2ADP + Pᵢ

O=C—O⁻
|
C=O
|
CH₃

Pyruvate

+

H—C=O
|
H—C—OH
|
CH₂O(P)

Glyceraldehyde-3-phosphate (GAP)

Figure 17-38   **Entner–Doudoroff pathway for glucose breakdown.**

*12. The hydrolysis of ATP to ADP in the cell results in a concomitant change in [AMP] as mediated by adenylate kinase. (a) Assuming that [ATP] >> [AMP] and that the total adenine nucleotide concentration in the cell, A_T = [AMP] + [ADP] + [ATP], is constant, derive an expression for [AMP] in terms of [ATP] and A_T. (b) Assuming an initial [ATP]/[ADP] of 10 and A_T = 5 mM, calculate the ratio of the final to initial values of [AMP] on a 10% decrease of [ATP].

Glycogen Metabolism

CHAPTER 18

Everything should be made as simple as possible, but not simpler.

Albert Einstein

Glucose, a major metabolic fuel source, is degraded via glycolysis to produce ATP (Chapter 17). Higher organisms protect themselves from potential fuel shortage by polymerizing excess glucose for storage as high molecular mass glucans (glucose polysaccharides) that may be readily mobilized in times of metabolic need. In plants, this glucose storage substance is starch, a mixture of the $\alpha(1 \rightarrow 4)$-linked glucan α-amylose (Fig. 11-18) and amylopectin, which differs from α-amylose by the presence of $\alpha(1 \rightarrow 6)$ branches every 24 to 30 residues (Fig. 11-19). In animals, the storage glucan is **glycogen** (Fig. 18-1), which differs from amylopectin only in that its branches occur every 8 to 14 residues. Glycogen occurs in 100- to 400-Å-diameter cytoplasmic granules (Figs. 11-20 and 18-1*c*), which contain up to 120,000 glucose units. They are especially prominent in the cells that make the greatest use of glycogen, muscle (maximally 1–2% glycogen by weight) and liver cells (maximally 10% glycogen by weight, an ~12-h energy supply for the body). Glycogen granules also contain the enzymes

that catalyze glycogen synthesis and degradation as well as some of the enzymes that regulate these processes.

As we shall see in this chapter, glycogen's glucose units are mobilized by their sequential removal from the glucan chains' nonreducing ends (ends lacking a C1—OH group). *Glycogen's highly branched structure is therefore physiologically significant: It permits glycogen's rapid degradation through the simultaneous release of the glucose units at the end of every branch.*

Why does the body go to such metabolic effort to use glycogen for energy storage when fat, which is far more abundant in the body, seemingly serves the same purpose? The answer is threefold:

1. Muscles cannot mobilize fat as rapidly as they can glycogen.

2. The fatty acid residues of fat cannot be metabolized anaerobically (Section 25-2).

3. Animals cannot convert fatty acids to glucose (Section 23-1), so fat metabolism alone cannot adequately maintain essential blood glucose levels (Section 18-3F).

As with all metabolic processes, there are several levels on which glycogen metabolism may be understood. We shall examine this process in order to understand the pathway's thermodynamics and the reaction mechanisms of its individual steps but will emphasize the mechanisms by which glycogen synthesis and breakdown rates are controlled. We began our consideration of metabolic control mechanisms in Section 17-4 with a discussion of the role of allosteric interactions and substrate cycles in the regulation of glycolysis. Glycogen metabolism's more complex control systems provide us with examples of several additional control processes: covalent modification of enzymes and enzyme cascades. In addition, we shall consider glycogen metabolism as a model for the role of hormones in the overall regulatory process. We end the chapter by discussing the consequences of genetic defects in various enzymes of glycogen metabolism.

1 GLYCOGEN BREAKDOWN

Liver and muscle are the two major storage tissues for glycogen. In muscle, the need for ATP results in the conversion of glycogen to glucose-6-phosphate (G6P) for entry into glycolysis. In liver, low blood glucose concentration triggers glycogen breakdown to G6P, which in this case is

(a)

(b)

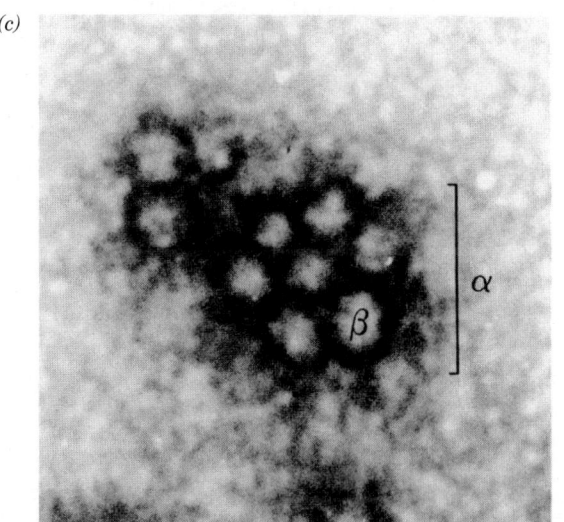

Figure 18-1 **Structure of glycogen.** (*a*) Molecular formula. In the actual molecule the chains are much longer than shown. (*b*) Schematic diagram illustrating its branched structure. Branch points in the actual molecule are separated by 8 to 14 glucosyl units. Note that the molecule, no matter how big, has but one reducing end. (*c*) Electron micrograph of a glycogen granule from rat skeletal muscle. Each granule (α) consists of several spherical glycogen molecules (β) and its associated proteins. [From Calder, P.C., *Int. J. Biochem.* **23**, 1339 (1991). Copyright Elsevier Science. Used with permission.]

hydrolyzed to glucose and released into the bloodstream to reverse this situation.

Glycogen breakdown requires the actions of three enzymes:

1. Glycogen phosphorylase (or simply **phosphorylase**) catalyzes glycogen **phosphorolysis** (bond cleavage by the substitution of a phosphate group) to yield **glucose-1-phosphate (G1P).**

$$\underset{(n \text{ residues})}{\text{Glycogen}} + \text{P}_i \;\rightleftharpoons\; \underset{(n-1 \text{ residues})}{\text{glycogen}} + \text{G1P}$$

This enzyme will only release a glucose unit that is at least five units from a branch point.

2. Glycogen debranching enzyme removes glycogen's branches, thereby permitting the glycogen phosphorylase reaction to go to completion. It also hydrolyzes $\alpha(1 \rightarrow 6)$-linked glucosyl units to yield glucose. Consequently, ~92% of glycogen's glucose residues are converted to G1P. The remaining ~8%, those at the branch points, are converted to glucose.

3. Phosphoglucomutase converts G1P to G6P, which, as we have seen (Section 17-2A), is also formed in the first step of glycolysis through the action of either hexokinase or glucokinase. G6P can either continue along the glycolytic pathway (as in muscle) or be hydrolyzed to glucose (as in liver).

In this section, we discuss the structures and mechanisms of action of these three enzymes.

A. Glycogen Phosphorylase

Glycogen phosphorylase is a dimer of identical 842-residue (97-kD) subunits that catalyzes the controlling step in glycogen breakdown. It is regulated both by allosteric interactions and by covalent modification. *The enzyme-catalyzed modification/demodification process yields two forms of phosphorylase:* **phosphorylase a,** *which has a phosphoryl group esterified to Ser 14 in each of its subunits, and* **phosphorylase b,** *which lacks these phosphoryl groups. Phosphorylase's allosteric inhibitors, ATP, G6P, and glucose, and its allosteric activator, AMP (to name only the enzyme's most prominent* *effectors), interact differently with the phospho- and dephos-phoenzymes, resulting in an extremely sensitive control process.* We study this control process in Section 18-3C.

a. Structural Domains and Binding Sites

The high-resolution X-ray structures of phosphorylase *a* and phosphorylase *b* were determined by Robert Fletterick and Louise Johnson, respectively. The structure of phosphorylase *b,* despite its lack of a Ser-linked phosphate, is very similar to that of phosphorylase *a* (Fig. 18-2). Both structures have two domains, an N-terminal domain (residues 1–484; among the largest known domains), and a C-terminal domain (residues 485–842). The N-terminal domain is further divided into an interface subdomain (residues 1–315), which includes the covalent modification site (Ser 14), the allosteric effector site, and all the intersubunit contacts in the dimer; and a glycogen-binding subdomain (residues 316–484), which contains the "glycogen storage site" (see below). The catalytic site is located at the center of the subunit where these two subdomains come together with the C-terminal domain. In Section 18-3 we discuss the allosteric behavior of glycogen phosphorylase and the conformational differences between phosphorylases *a* and *b*.

Glycogen forms a left-handed helix with 6.5 glucose residues per turn, similar to α-amylose (Fig. 11-18*b*). An ~30-Å-long crevice on the surface of the phosphorylase monomer that has the same radius of curvature as glycogen connects the glycogen storage site, which binds glycogen, to the active site, which phosphorylizes it. *Since this crevice can accommodate four or five sugar residues in a chain but is too narrow to admit branched oligosaccharides, it provides a clear physical rationale for the inability of phosphorylase to cleave glycosyl residues closer than five units from a branch point.* Presumably the glycogen storage site increases the catalytic efficiency of phosphorylase by permitting it to phosphorylize many glucose residues on the same glycogen particle without having to dissociate and reassociate completely between catalytic cycles.

b. Pyridoxal Phosphate Is an Essential Cofactor for Phosphorylase

Phosphorylase contains **pyridoxal-5′-phosphate (PLP)**

Figure 18-2   X-ray structure of rabbit muscle glycogen phosphorylase *a*. The homodimeric enzyme is drawn in ribbon form and viewed along its 2-fold axis. Each subunit consists of an N-terminal domain, which is subdivided into an interface subdomain (residues 1–315) and a glycogen-binding subdomain (residues 316–484), and a C-terminal domain (residues 485–842). The enzyme's several ligands are drawn in stick form colored according to type with N blue, O red, P orange, and C atoms as indicated. The active site is marked by a bound glucose molecule (C yellow). Pyridoxal phosphate (PLP) is covalently linked to the side chain of Lys 678 in the C-terminal domain (C magenta). In addition, the enzyme binds its allosteric effector AMP (C cyan) and **maltoheptaose** (C green), an α(1 → 4)-linked glucose heptamer, which is bound in the enzyme's glycogen storage site. Ser 14-P, the phosphoryl group on Ser 14, is drawn in space-filling form. [X-ray structure coordinates courtesy of Stephen Sprang, University of Texas Southwest Medical Center.] 🖙 See **Kinemage Exercise 14-1**

Pyridoxal-5′-phosphate (PLP)

PLP covalently bound to phosphorylase via a Schiff base to Lys 679

and requires it for activity. This vitamin B$_6$ derivative is covalently linked to phosphorylase via a Schiff base to Lys 679. PLP is similarly linked to a variety of enzymes involved in amino acid metabolism, where it is an essential cofactor in transamination reactions (Section 26-1Aa). The

Figure 18-3 **The reaction mechanism of glycogen phosphorylase.** Here PL is an enzyme-bound pyridoxal group and BH$^+$ is a positively charged amino acid side chain, probably that of Lys 568, necessary for the maintenance of PLP electrical neutrality. **(1)** Formation of an E · P$_i$ · glycogen ternary complex. **(2)** Formation of a shielded oxonium ion intermediate from the α-linked terminal glucosyl residue involving acid catalysis by P$_i$ as facilitated by proton transfer from PLP. The oxonium ion has the half-chair conformation. **(3)** Reaction of P$_i$ with the oxonium ion with overall retention of configuration about C1 to form α-D-glucose-1-phosphate. The glycogen, which has one less residue than before, cycles back to Step 1.

an intriguing example of nature's opportunism in using the same cofactor to perform different chemistries.

Extensive studies on phosphorylase using PLP analogs in which various parts of this molecule are missing or modified indicate that only its phosphate group participates in the catalytic process. Indeed, the X-ray structures of phosphorylase reveal that only PLP's phosphate group is near this enzyme's active site. This phosphoryl group most probably functions as an acid–base catalyst.

c. Kinetics and Reaction Mechanism

The phosphorylase reaction results in the cleavage of the C1—O1 bond from a nonreducing terminal glucosyl unit of glycogen, yielding G1P. This reaction proceeds with retention of configuration, which suggests that the phosphorolysis occurs via a double displacement mechanism (two sequential nucleophilic substitutions, each occurring with inversion of configuration; Fig. 16-6c) involving a covalent glucosyl–enzyme intermediate. Yet, phosphorylase exhibits Rapid Equilibrium Random Bi Bi kinetics (Section 14-5), not Ping Pong kinetics, as would be expected for a double displacement mechanism. Furthermore, all attempts to establish the existence of the putative covalent intermediate have been unsuccessful.

An alternative mechanism (Fig. 18-3), which is compatible with the available kinetic, chemical, and structural data, commences with the formation of a ternary enzyme · P$_i$ · glycogen complex, followed by the generation of an intermediate shielded oxonium ion similar to the transition state in the lysozyme reaction (which also involves

mechanism of PLP participation in the phosphorylase reaction must differ from that in these other enzymes because, for example, reduction of the Schiff base with NaBH$_4$ (—HC=N— → —H$_2$C—NH—) has no effect on the activity of phosphorylase, whereas it inactivates the PLP-requiring enzymes of amino acid metabolism. This is glycosidic bond cleavage in a polysaccharide; Section 15-2B). *Bond cleavage, with its consequent oxonium ion formation, is assisted by protonation of the glycosidic oxygen by the P_i substrate (acid catalysis).* Phosphorylase has no protein nucleophilic or carboxylate groups in the vicinity of the scissile glycosidic bond and hence would be unable to form a covalent intermediate as does lysozyme. However, since the PLP phosphoryl group is within hydrogen bonding distance of the P_i, it appears that bond cleavage is facilitated by the simultaneous protonation of the reacting P_i by the PLP phosphoryl group in a kind of proton relay. The resulting oxonium ion (Fig. 18-3) is stabilized through its formation of an ion pair with the anionic P_i (electrostatic catalysis), which subsequently collapses to yield product, G1P, in a reaction step that is facilitated by the abstraction of a proton from P_i by the PLP phosphoryl group (base catalysis).

Support for the oxonium ion mechanism comes from the observation that **1,5-gluconolactone**

1,5-Gluconolactone

is a potent inhibitor of phosphorylase. 1,5-Gluconolactone has the same half-chair conformation as the proposed oxonium ion, suggesting that it is a transition state analog that mimics the oxonium ion at the active site of phosphorylase (Section 15-1F).

B. *Phosphoglucomutase*

Phosphorylase converts the glucosyl units of glycogen to G1P, which, in turn, is converted by phosphoglucomutase to G6P either for entry into glycolysis in muscle or hydrol-

ysis to glucose in liver. The X-ray structure of rabbit muscle phosphoglucomutase indicates that the active site of this 561-residue monomeric enzyme is largely buried at the bottom of a particularly deep crevice in the protein. The phosphoglucomutase reaction is similar to that catalyzed by phosphoglycerate mutase (Section 17-2Ha). A phosphoryl group is transferred from the active phosphoenzyme to G6P, forming **glucose 1,6-bisphosphate (G1,6P),** which then rephosphorylates the enzyme to yield G1P (Fig. 18-4). An important difference between this enzyme and phosphoglycerate mutase is that the phosphoryl group in phosphoglucomutase is covalently bound to a Ser hydroxyl group rather than to a His imidazole nitrogen.

G1,6P occasionally dissociates from phosphoglucomutase, resulting in the inactivation of this enzyme. The presence of small amounts of G1,6P is therefore necessary to keep phosphoglucomutase fully active. This intermediate is provided by **phosphoglucokinase,** which catalyzes the phosphorylation of the C6—OH group of G1P by ATP.

C. *Glycogen Debranching Enzyme*

Glycogen debranching enzyme, an ~1540-residue monomer, acts as an $\alpha(1 \rightarrow 4)$ transglycosylase (glycosyl transferase) by transferring an $\alpha(1 \rightarrow 4)$-linked trisaccharide unit from a "limit branch" of glycogen to the nonreducing end of another branch (Fig. 18-5). This reaction forms a new $\alpha(1 \rightarrow 4)$ linkage with three more units available for phosphorylase-catalyzed phosphorolysis. The $\alpha(1 \rightarrow 6)$ bond linking the remaining glycosyl residue in the branch to the main chain is hydrolyzed (not phosphorylized) by the same debranching enzyme to yield glucose and debranched glycogen. Thus *debranching enzyme has different active sites for the transferase reaction and the $\alpha(1 \rightarrow 6)$-glucosidase reaction.* The presence of two independent catalytic activities on the same enzyme no doubt improves the efficiency of the debranching process.

Figure 18-4   The mechanism of action of phosphoglucomutase.
(1) The OH group at C1 of G6P attacks the phosphoenzyme to form a dephosphoenzyme–G1,6P intermediate. **(2)** The Ser—OH group on the dephosphoenzyme attacks the phosphoryl group at C6 to regenerate the phosphoenzyme with the formation of G1P.

Figure 18-5 Reactions catalyzed by debranching enzyme. The enzyme transfers the terminal three α(1 → 4)-linked glucose residues from a "limit branch" of glycogen to the nonreducing end of another branch. The α(1 → 6) bond of the residue remaining at the branch point is hydrolyzed by further action of debranching enzyme to yield free glucose. The newly elongated branch is subject to degradation by glycogen phosphorylase.

The maximal rate of the glycogen phosphorylase reaction is much greater than that of the glycogen debranching reaction. Consequently, the outermost branches of glycogen, which comprise nearly half of its residues, are degraded in muscle in a few seconds under conditions of high metabolic demand. Glycogen degradation beyond this point requires debranching and hence occurs more slowly. This, in part, accounts for the fact that a muscle can sustain its maximum exertion for only a few seconds.

D. Thermodynamics of Glycogen Metabolism: The Need for Separate Pathways of Synthesis and Breakdown

The $\Delta G^{\circ\prime}$ (ΔG under standard biochemical conditions) for the phosphorylase reaction is +3.1 kJ · mol^{-1}, so, as Eq. [3.15] indicates, this reaction is at equilibrium ($\Delta G = 0$) at 25°C when $[P_i]/[G1P] = 3.5$. In the cell, however, this concentration ratio varies between 30 and 100, which places ΔG in the range −5 to −8 kJ · mol^{-1}; that is, *under physiological conditions, glycogen breakdown is exergonic.* The synthesis of glycogen from G1P under physio-logical conditions is therefore thermodynamically unfavorable without free energy input. Consequently, *glycogen biosynthesis and breakdown must occur by separate pathways. Thus we encounter a recurrent metabolic strategy: Biosynthetic and degradative pathways of metabolism are almost always different (Section 16-1).* There are two important reasons for this. The first, as we have seen, is that both pathways may be required under similar *in vivo* metabolite concentrations. This situation is thermodynamically impossible if one pathway is just the reverse of the other. The second reason is equally important: Reactions catalyzed by different enzymes can be independently regulated, which permits very fine flux control. We have seen this principle in operation in the glycolytic conversion of fructose-6-phosphate (F6P) to fructose-1,6-bisphosphate (F1,6P) by phosphofructokinase (PFK; Section 17-4F). The reverse process in that case (hydrolysis of F1,6P) is catalyzed by fructose bisphosphatase (FBPase). Independent control of those two enzymes provides precise regulation of glycolytic flux.

Glycogen metabolism, like glycolysis, is exquisitely regulated by the independent control of its synthetic and degradative pathways. In the next section we examine the pathway of glycogen synthesis and, in Section 18-3, we explore the regulatory process.

2 GLYCOGEN SYNTHESIS

Although the thermodynamic arguments presented in Section 18-1D demonstrate that glycogen synthesis and breakdown must occur by separate pathways, it was not thermodynamic arguments that led to the general acceptance of this idea. Rather, it was the elucidation of the cause of **McArdle's disease,** a rare inherited glycogen storage disease that results in painful muscle cramps on strenuous exertion (Section 18-4). The muscle tissue from individuals with McArdle's disease exhibits no glycogen phosphorylase activity and is therefore incapable of glycogen breakdown. Their muscles, nevertheless, contain moderately high quantities of normal glycogen. Clearly, there must be separate pathways for glycogen synthesis and breakdown.

Since the direct conversion of G1P to glycogen and P_i is thermodynamically unfavorable (positive ΔG) under all physiological P_i concentrations, glycogen biosynthesis requires an additional exergonic step. This is accomplished, as Luis Leloir discovered in 1957, by combining G1P with uridine triphosphate (UTP) to form **uridine diphosphate glucose (UDP–glucose or UDPG):**

Uridine diphosphate glucose (UDPG)

UDPG's "high-energy" status permits it to spontaneously donate glucosyl units to the growing glycogen chain.

The enzymes catalyzing the three steps involved in the glycogen synthesis pathway are **UDP–glucose pyrophosphorylase, glycogen synthase,** and **glycogen branching enzyme.** In this section, we examine the reactions catalyzed by these enzymes. Discussion of how these enzymes are controlled is reserved for Section 18-3.

A. *UDP–Glucose Pyrophosphorylase*

UDP–glucose pyrophosphorylase catalyzes the reaction of UTP and G1P (Fig. 18-6). In this reaction, the phosphoryl oxygen of G1P attacks the α phosphorus atom of UTP to form UDPG and release PP_i. The $\Delta G^{\circ\prime}$ of this phosphoanhydride exchange is, as expected, nearly zero. However, the PP_i formed is hydrolyzed in an exergonic reaction by the omnipresent enzyme inorganic pyrophosphatase. The overall reaction for the formation of UDPG is therefore also exergonic:

Figure 18-6   Reaction catalyzed by UDP–glucose pyrophosphorylase. The reaction is a phosphoanhydride exchange in which the phosphoryl oxygen of G1P attacks the α phosphorus atom of UTP to form UDPG and release PP_i. The PP_i is rapidly hydrolyzed by inorganic pyrophosphatase.

	$\Delta G^{\circ\prime}$ (kJ · mol^{-1})
G1P + UTP $\rightleftharpoons$ UDPG + PP_i	~0
H_2O + PP_i $\rightleftharpoons$ $2P_i$	−19.2
Overall: G1P + UTP $\rightleftharpoons$ UDPG + $2P_i$	−19.2

The cleavage of a nucleoside triphosphate to form PP_i is a common biosynthetic strategy. The free energy of PP_i hydrolysis can then be utilized together with the free energy of nucleoside triphosphate hydrolysis to drive an otherwise endergonic reaction to completion (Section 16-4Ca).

B. *Glycogen Synthase*

In the next step of glycogen synthesis, the glycogen synthase reaction, the glucosyl unit of UDPG is transferred to the C4—OH group on one of glycogen's nonreducing ends to form an $\alpha(1 \rightarrow 4)$-glycosidic bond (Fig. 18-7). The glycogen synthase reaction, like those of glycogen phosphorylase and lysozyme, is thought to involve a glucosyl oxonium ion intermediate or transition state since it is also inhibited by 1,5-gluconolactone, an analog that mimics the oxonium ion's half-chair geometry.

The $\Delta G^{\circ\prime}$ for the glycogen synthase reaction is -13.4 $kJ \cdot mol^{-1}$, making the overall reaction spontaneous under the same conditions that glycogen breakdown by glycogen phosphorylase is also spontaneous. The rates of both reactions may then be independently controlled. There is, however, an energetic price for doing so. In this case, *for each molecule of G1P that is converted to glycogen and then regenerated, one molecule of UTP is hydrolyzed to UDP and P_i. The cyclic synthesis and breakdown of glycogen is therefore not a perpetual motion "machine" but, rather, is an "engine" that is powered by UTP hydrolysis.* The UTP is replenished through a phosphate-transfer reaction mediated by **nucleoside diphosphate kinase** (Section 28-1Ba):

$$UDP + ATP \rightleftharpoons UTP + ADP$$

so that UTP hydrolysis is energetically equivalent to ATP hydrolysis. This reaction occurs via a Ping Pong mechanism in which an active site His residue is transiently phosphorylated at its N_δ position much as occurs in the phosphoglycerate mutase reaction of glycolysis (Section 17-2Ha).

Mammals express two ~70% identical isoforms of glycogen synthase, one mainly in muscle, and the other in liver, much as is the case with glycogen phosphorylase. Plants and bacteria express **starch/glycogen synthases** but these employ **ADP–glucose** as glucose donors rather than UDP–glucose and exhibit very little sequence similarity to animal glycogen synthases.

a. Glycogenin Initiates Glycogen Synthesis

Glycogen synthase cannot simply link together two glucose residues; it can only extend an already existing $\alpha(1 \rightarrow 4)$-linked glucan chain. How, then, is glycogen synthesis initiated? The answer is that the first step in glycogen synthesis is the self-catalyzed attachment of a glucose residue to the Tyr 194 OH group of a 332-residue homodimeric protein named **glycogenin.** Glycogenin, which was discovered by William Whelan, further extends the glucan chain by up to ~9 additional UDP–glucose-supplied residues, forming a "primer" for the initiation of glycogen synthesis. Only at this point does glycogen synthase commence glycogen synthesis, which it initiates on the "primer" while tightly complexed to glycogenin. However, these proteins dissociate after the growing glycogen

Figure 18-7 **Reaction catalyzed by glycogen synthase.** The reaction involves a glucosyl oxonium ion intermediate.

Tyr 194

Asp 159

Asp 162

Figure 18-8   **X-ray structure of rabbit muscle glycogenin.** The upper subunit of this homodimeric enzyme is colored in rainbow order from its N-terminus (*blue*) to its C-terminus (*red*) and the lower subunit is pale green. UDPG is drawn in ball-and-stick form with C green, N blue, O red, and P orange. The side chains of Tyr 194, to which the glucose residue at glycogen's reducing end is linked, and Asp 159 and Asp 162, are drawn in stick form with C cyan and O red. Mn^{2+} ions, which are thought to electrostatically stabilize the leaving group, UDP, are represented by lavender spheres. Residues 262 to 332, which are implicated in binding glycogen synthase, are not visible in the X-ray structure and hence disordered. [Based on an X-ray structure by Thomas Hurley and Peter Roach, Indiana University School of Medicine. PDBid 1LL2.]

its catalytic cycle, although the nature of these changes is unknown. Indeed, it is unclear whether a given subunit transfers glucosyl groups to itself or to the opposite subunit.

C. *Glycogen Branching Enzyme*

Glycogen synthase catalyzes only $\alpha(1 \rightarrow 4)$-linkage formation to yield α-amylose. Branching to form glycogen is accomplished by an ~700-residue monomeric enzyme, **amylo-$(1,4 \rightarrow 1,6)$-transglycosylase (glycogen branching enzyme),** which is distinct from glycogen debranching enzyme. Branches are created by the transfer of terminal chain segments consisting of ~7 glucosyl residues to the C6—OH groups of glucose residues on the same or another glycogen chain (Fig. 18-9). Each transferred segment must come from a chain of at least 11 residues, and the new branch point must be at least 4 residues away from other branch points.

Debranching (Section 18-1C) involves breaking and reforming $\alpha(1 \rightarrow 4)$-glycosidic bonds and only the hydrolysis of $\alpha(1 \rightarrow 6)$-glycosidic bonds; branching, on the other hand, involves breaking $\alpha(1 \rightarrow 4)$-glycosidic bonds and reforming $\alpha(1 \rightarrow 6)$ linkages. The need to hydrolyze glycogen's $\alpha(1 \rightarrow 6)$-glycosidic bonds rather than to convert them to $\alpha(1 \rightarrow 4)$ linkages is explained by the energetics of these reactions. The free energy of hydrolysis of an $\alpha(1 \rightarrow 4)$-glycosidic bond is -15.5 kJ · mol^{-1}, whereas that of an $\alpha(1 \rightarrow 6)$-glycosidic bond is only -7.1 kJ · mol^{-1}. Consequently, the hydrolysis of an $\alpha(1 \rightarrow 4)$-glycosidic bond drives the synthesis of an $\alpha(1 \rightarrow 6)$-glycosidic bond, but the reverse reaction is endergonic.

a. Glycogen Particles Are Fabricated to Optimize Glucose Mobilization

The biological function of glycogen is to maximize the density of stored glucose units consistent with the need to rapidly mobilize it under conditions of high metabolic demand. To do so, three related parameters must be optimized: the number of tiers of branches in a glycogen molecule, the number of branches per tier, and the average chain length per tier. For a glycogen molecule with a fixed

granule has reached some minimum size. Analysis of glycogen granules for glycogenin and glycogen synthase shows that they are present in a 1:1 ratio. Evidently, each glycogen molecule (Fig. 18-1c) is associated with one molecule each of glycogenin and glycogen synthase.

The X-ray structure of rabbit muscle glycogenin in complex with UDPG (Fig. 18-8), determined by Thomas Hurley and Peter Roach, reveals that the enzyme's N-terminal portion consists of a Rossmann-like fold (Section 8-3Bi) that is common to the nucleotide-binding domains of most glycosyltransferases. The OH group of its Tyr 194 is 21 and 16 Å distant from C1 of the UDPG's glucosyl group on the same and on the opposite subunit. Moreover, the glucosyltransferase reaction occurs with retention of configuration about the glucosyl C1 atom. This suggests that the glucosyl group is first transferred to an intermediate nucleophile and then transferred to Tyr 194 or to a glucose group that is linked to Tyr 194 (a double displacement reaction). Site-directed mutagenesis has identified Asp 162 as the probable nucleophile, whereas Asp 159 appears to have a role in binding and activating the acceptor molecule. Evidently, glycogenin undergoes large conformational changes during

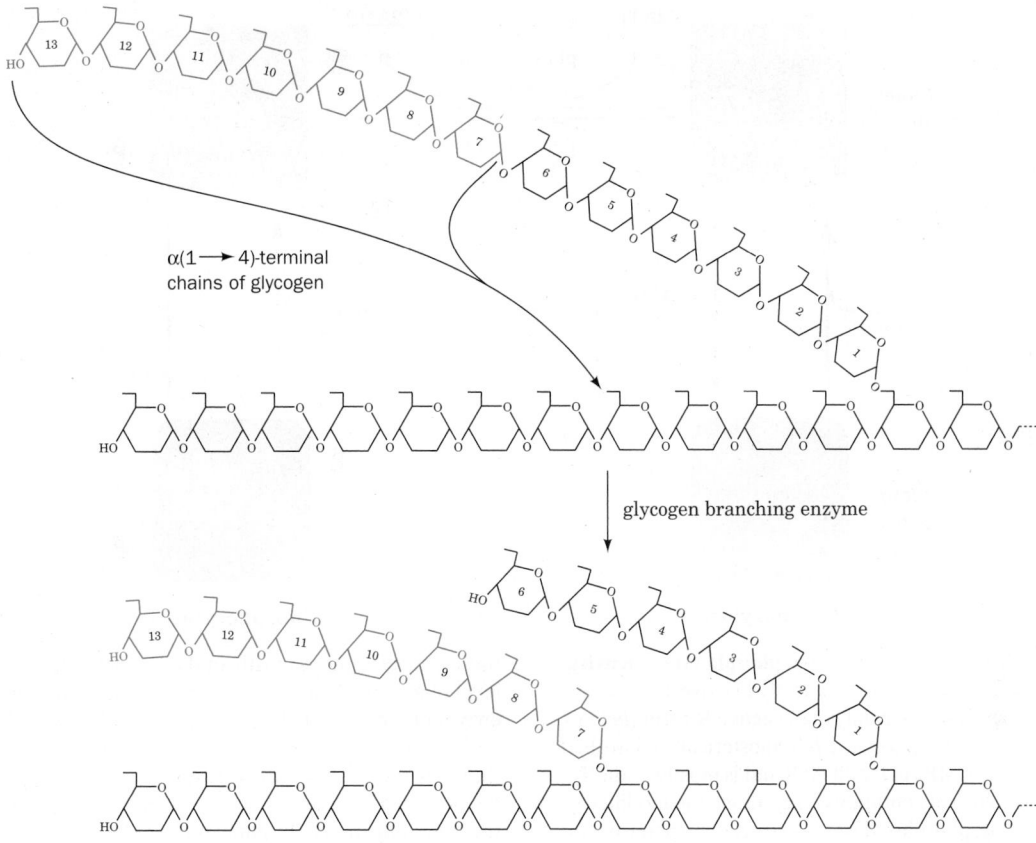

$\alpha(1 \longrightarrow 4)$-terminal
chains of glycogen

glycogen branching enzyme

Figure 18-9   The branching of glycogen. Branches are formed by transferring a 7-residue terminal segment from an $\alpha(1 \rightarrow 4)$-linked glucan chain to the C6—OH group of a glucose residue on the same or another chain.

number of residues, the number of outer branches from which glucose can be mobilized before debranching is required decreases as the average chain length increases (recall that debranching is a slower process than phosphorolysis). However, molecules with longer chains have a greater number of glucose residues that can be phosphorylized between branch points. Since the density of outermost branches is sterically limited, the maximum size of a glycogen molecule decreases as the average number of branches per tier increases. Mature glycogen particles from a variety of animals have ~12 tiers of branches, with ~2 branches per tier and branch lengths averaging ~13 residues. Mathematical analysis suggests that these values are close to optimal for mobilizing the greatest amount of glucose in the shortest possible time.

3 CONTROL OF GLYCOGEN METABOLISM

We have just seen that both glycogen synthesis and breakdown are exergonic under the same physiological conditions. If both pathways operate simultaneously, however,

all that is achieved is wasteful hydrolysis of UTP. This situation is similar to that of the phosphofructokinase–fructose bisphosphatase substrate cycle (Section 17-4Ff). Glycogen phosphorylase and glycogen synthase therefore must be under stringent control such that glycogen is either synthesized or utilized according to cellular needs. The astonishing mechanism of this control is our next topic of discussion. It involves not only allosteric control and substrate cycles but enzyme-catalyzed covalent modification of both glycogen synthase and glycogen phosphorylase. The covalent modification reactions are themselves ultimately under hormonal control through an enzymatic cascade.

A. Direct Allosteric Control of Glycogen Phosphorylase and Glycogen Synthase

As we saw in Section 17-4B, the net flux of reactants, J, through a step in a metabolic pathway is the difference between the forward and reverse reaction velocities, v_f and v_r. The fractional variation in the flux through any step in a pathway with a change in substrate concentration approaches infinity as that reaction step approaches equilibrium ($v_f \approx v_r$; Eq. [17.4]). The flux through a near-equilibrium reaction

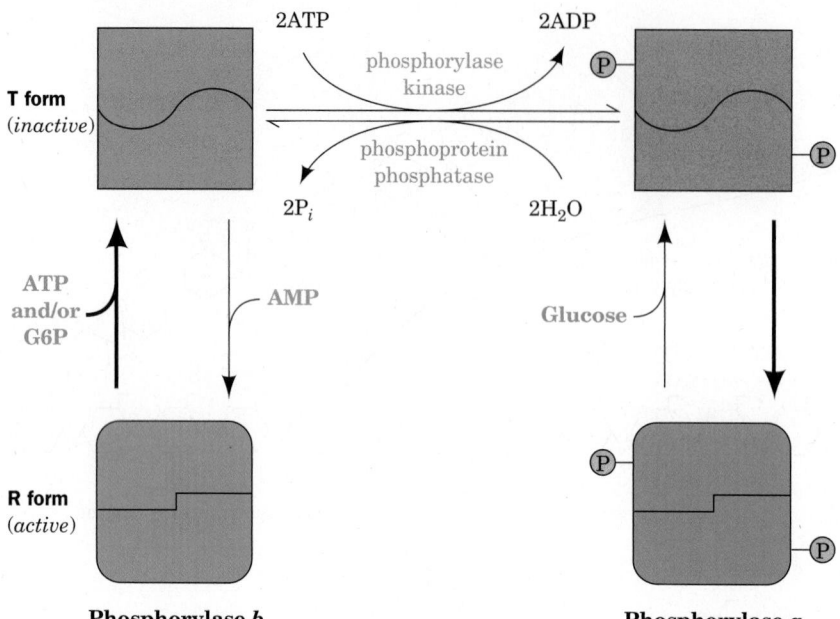

Figure 18-10 **The control of glycogen phosphorylase activity.** The enzyme may assume the enzymatically inactive T conformation (*above*) or the catalytically active R form (*below*). The conformation of phosphorylase *b* is allosterically controlled by effectors such as AMP, ATP, and G6P and is mostly in the T state under physiological conditions. In contrast, the modified form of the enzyme, phosphorylase *a*, is largely unresponsive to these effectors and is mostly in the R state unless there is a high level of glucose. Under usual physiological conditions, the enzymatic activity of glycogen phosphorylase is essentially determined by its rates of modification and demodification. Note that only the T form enzyme is subject to phosphorylation and dephosphorylation, so effector binding influences the rates of these modification/demodification events.

is therefore all but uncontrollable. As we have seen for the case of PFK and FBPase, however, *precise flux control of a pathway is possible when an enzyme functioning far from equilibrium is opposed by a separately controlled enzyme. Then, v_f and v_r vary independently. In fact, under these circumstances, even the flux direction is controllable if v_r can be made larger than v_f.* Exactly this situation occurs in glycogen metabolism through the opposition of the glycogen phosphorylase and glycogen synthase reactions. The rates of both of these reactions are under allosteric control by effectors that include ATP, G6P, and AMP. In muscle, glycogen phosphorylase is activated by AMP and inhibited by ATP and G6P (Fig. 18-10, *left*). Glycogen synthase, on the other hand, is activated by G6P. When there is high demand for ATP (low [ATP], low [G6P], and high [AMP]), glycogen phosphorylase is stimulated and glycogen synthase is inhibited, so flux through this pathway favors glycogen breakdown. When [ATP] and [G6P] are high, the reverse is true and glycogen synthesis is favored.

The structural differences between the active (R) and inactive (T) conformations of glycogen phosphorylase (Figs. 18-11*a* and 18-11*b*) can be understood in terms of the symmetry model of allosterism (Section 10-4B). The T-state enzyme has a buried active site and hence a low affinity for its substrates, whereas the R-state enzyme has an accessible catalytic site and a high-affinity phosphate binding site.

AMP promotes phosphorylase's T (*inactive*) → R (*active*) conformational shift by binding to the R state of the enzyme at its allosteric effector site (Fig. 18-10, *left*). In doing so, AMP's adenine, ribose, and phosphate groups bind to separate segments of the polypeptide chain so as to link the active site, the subunit interface, and the N-terminal region (Fig. 18-11*c*), the latter having undergone a large conformational shift (36 Å for Ser 14) from its position in the T-state enzyme (Figs. 18-11*a* and 18-11*b*). AMP binding also causes glycogen phosphorylase's tower helices (Figs. 18-2 and 18-11) to tilt and pull apart so as to achieve a more favorable packing. These tertiary movements trigger a concerted T → R transition, which largely consists of an ~10° relative rotation of the two subunits about an axis at the subunit interface that is perpendicular to the dimer's 2-fold axis of symmetry. The enzyme's 2-fold symmetry is thereby preserved in accordance with the symmetry model of allosterism (Section 10-4B). The movement of the tower helices also displaces and disorders a loop (the 280s loop, residues 282–286), which covers the T-state active site so as to prevent substrate access. It also causes the Arg 569 side chain, which is located in the active site near the PLP phosphoryl group and the P_i-binding site, to rotate in a way that increases the enzyme's binding affinity for its anionic P_i substrate (Figs. 18-11*a* and 18-11*b*).

Curiously, ATP also binds to the allosteric effector site, but in the T state, so that it inhibits rather than promotes

(a)

T state

(b)

R state

(c)

Figure 18-11 Conformational changes in glycogen phosphorylase. Ribbon diagrams of one subunit of the dimeric enzyme glycogen phosphorylase *b* (*a*) in the T state in the absence of allosteric effectors and (*b*) in the R state with bound AMP. The view is of the lower (*orange*) subunit in Fig. 18-2 rotated by ~45° about a horizontal axis. The tower helix is blue; the N-terminal helix is cyan; the N-terminal residues that change conformation on AMP binding are red; and residues 280 to 286, the 280s loop, which in the R state are disordered and hence not seen, are green. Of the groups that are shown in space-filling representation, the side chain of Ser 14, the phosphorylation site, and AMP, which binds only to the R state enzyme, have C green, N blue, O red, and P orange; the side chain of Arg 569, which reorients in the T → R transition so as to interact with the substrate phosphate, has C cyan and N blue; and sulfate ions bound near Ser 14 and at the active site of the R state enzyme, which mimic the sterically similar phosphate ions, have O red and S orange. The PLP at the active site is drawn in stick form with C magenta, N blue, O red, and P orange. [Based on X-ray structures by Louise Johnson, Oxford University, U.K. PDBids 8GPB and 7GPB.] (*c*) The portion of the glycogen phosphorylase *a* dimer in the vicinity of the dimer interface showing the position of the Ser 14 phosphate group, the AMP bound in the allosteric effector site, and the active site–bound glucose molecule. The view is along the molecular 2-fold axis and hence is similar to that in Fig. 18-2. Residues 5 to 80 and 265 to 310 are, respectively, cyan and blue in one subunit and pink and magenta in the other. The AMP and glucose are shown in space-filling representation with C green, N dark blue, O red, and P orange. The Ser 14 phosphate group is also shown in space-filling representation with O dark red and P yellow. [X-ray structure coordinates courtesy of Stephen Sprang, University of Texas Southwest Medical Center.] See Kinemage Exercises **14-2 and 14-3**

the T → R conformational shift. This is because, as structural analysis indicates, the β and γ phosphate groups of ATP bind to the enzyme such that its ribose and α phosphate groups are displaced relative to those of AMP, thus destabilizing the R state. The inhibitory action of ATP on phosphorylase is therefore simply understood: It competes with

AMP for binding to phosphorylase and, in doing so, prevents the relative motions of the three polypeptide segments required for phosphorylase activation.

The above allosteric interactions are superimposed on an even more sophisticated control system involving covalent modifications (phosphorylation/dephosphorylation) of glycogen phosphorylase and glycogen synthase. These modifications alter the structures of the enzymes so as to change their responses to allosteric regulators. We shall therefore discuss the general concept of covalent modification and how it increases the sensitivity of a metabolic system to effector concentration changes. We subsequently consider the functions of such modifications in glycogen metabolism. Only then will we be in a position to take up the detailed consideration of allosteric control in glycogen metabolism.

B. Covalent Modification of Enzymes by Cyclic Cascades: Effector "Signal" Amplification

Glycogen synthase and glycogen phosphorylase can each be enzymatically interconverted between two forms with different kinetic and allosteric properties through a complex series of reactions known as a **cyclic cascade.** *The interconversion of these different enzyme forms involves distinct, enzyme-catalyzed* **covalent modification** *and* **demodification reactions.**

Compared with other regulatory enzymes, enzymatically interconvertible enzyme systems:

1. Can respond to a greater number of allosteric stimuli.

2. Exhibit greater flexibility in their control patterns.

3. Possess enormous amplification potential in their responses to variations in effector concentrations.

This is because *the enzymes that modify and demodify a target enzyme are themselves under allosteric control. It is therefore possible for a small change in concentration of an allosteric effector of a modifying enzyme to cause a large change in the concentration of an active, modified target enzyme.* Such a cyclic cascade is diagrammed in Fig. 18-12.

a. Description of a General Cyclic Cascade

Figure 18-12*a* shows a general scheme for a cyclic cascade where, by convention, the more active target enzyme form has the subscript *a* and the less active form has the subscript *b*. Here, modification, in this case, phosphorylation, activates the enzyme. Note that the modifying enzymes, F and R, are active only when they have bound their respective allosteric effectors e_1 and e_2. The kinetic mechanisms for the interconversion of the unmodified and modified forms of the target enzyme, E_b and E_a, are indicated in Fig. 18-12*b*.

In the steady state, the fraction of E in the active form, $[E_a]/[E]_T$ (where $[E]_T = [E_a] + [E_b]$ is the total enzyme concentration), determines the rate of the reaction catalyzed by E. This fraction is a function of the total concentrations of the modifying enzymes, $[F]_T$ and $[R]_T$, the concentrations of their allosteric effectors, e_1 and e_2, the dissociation constants of these effectors, K_1 and K_2, and the substrate dissociation constants, K_f and K_r, of the target enzymes, as well as the rate constants, k_f and k_r, for the interconversions themselves (Fig. 18-12). This relationship is obviously quite complex. Nevertheless, it can be shown that, in a cyclic cascade, a relatively small change in the concentration of e_1, the allosteric effector of the modifying enzyme F, can result in a much larger change in $[E_a]/[E]_T$, the fraction of E in the active form. In other words, *the cascade functions to amplify the sensitivity of the system to an allosteric effector.*

We have so far considered the covalent modification of only one enzyme, a **monocyclic cascade.** Imagine a **bicyclic cascade** involving the covalent modification of one of the modifying enzymes (F), as well as the metabolic target enzyme (E) (Fig. 18-13). As you might expect, the amplification potential of a "signal," e_1, as well as the control flexibility of such a system, is enormous.

(a)

$$\text{ATP} + \text{F} \cdot e_1 + \overset{\overset{\text{OH}}{|}}{\boxed{E_b}} \overset{K_f}{\rightleftharpoons} \overset{\overset{\text{OH}}{|}}{\boxed{E_b}} \cdot \text{ATP} \cdot \text{F} \cdot e_1 \xrightarrow{k_f} \text{F} \cdot e_1 + \overset{\overset{\text{P}}{|}}{\boxed{E_a}} + \text{ADP}$$

$$\text{R} \cdot e_2 + \overset{\overset{\text{P}}{|}}{\boxed{E_a}} \overset{K_r}{\rightleftharpoons} \overset{\overset{\text{P}}{|}}{\boxed{E_a}} \cdot \text{R} \cdot e_2 \xrightarrow{k_r} \text{R} \cdot e_2 + \overset{\overset{\text{OH}}{|}}{\boxed{E_b}} + \text{P}_i$$

(b)

Figure 18-12 **A monocyclic enzyme cascade.** (*a*) General scheme, where F and R are, respectively, the modifying and demodifying enzymes. These are allosterically converted from their inactive to their active conformations on binding their respective effectors, e_1 and e_2. The target enzyme, E, is more active in the modified form (E_a) and less active in the unmodified form (E_b). Dashed arrows symbolize catalysis of the indicated reactions. (*b*) Chemical equations for the interconversion of the target enzyme's unmodified and modified forms E_b and E_a.

The activities of both glycogen phosphorylase and glycogen synthase are controlled by bicyclic cascades. Let us now examine the enzymatic interconversions involved in these bicyclic cascades. We shall specifically focus on the covalent modifications of glycogen phosphorylase and glycogen synthase, the structural effects of these covalent modifications, and how these structural changes affect the interactions of their allosteric effectors. We shall then consider the cyclic cascades as a whole, studying the various modification enzymes involved and their "ultimate" allosteric effectors. Finally, we shall see how the various cyclic cascades of glycogen metabolism function in different physiological situations.

C. Glycogen Phosphorylase Bicyclic Cascade

In 1938, Carl and Gerty Cori found that glycogen phosphorylase exists in two forms, the *b* form that requires AMP for activity, and the *a* form that is active without AMP. It nevertheless took 20 years for the development of the protein chemistry techniques through which Edwin Krebs and Edmond Fischer demonstrated, in 1959, that phosphorylases *a* and *b* correspond to forms of the protein in which a specific residue, Ser 14, is enzymatically phosphorylated or dephosphorylated, respectively.

a. Glycogen Phosphorylase: The Cascade's Target Enzyme

The activity of glycogen phosphorylase is allosterically controlled, as we saw, through AMP activation and ATP, G6P, and glucose inhibition (Section 18-3A). Superim-

posed on this allosteric control is control by enzymatic interconversion through a bicyclic cascade involving the actions of three enzymes (Figs. 18-13 and 18-14, *left*):

1. Phosphorylase kinase, which specifically phosphorylates Ser 14 of glycogen phosphorylase *b* (Fig. 18-13, enzyme F_2).

2. Protein kinase A, which phosphorylates and thereby activates phosphorylase kinase (Fig. 18-13, enzyme F_1).

3. Phosphoprotein phosphatase-1, which dephosphorylates and thereby deactivates both glycogen phosphorylase *a* and phosphorylase kinase (Fig. 18-13, enzymes R_1 and R_2).

In an interconvertible enzyme system, the "modified" form of the enzyme bears the prefix *m* and the "original" (unmodified) form bears the prefix *o*, whereas the enzyme's most active and least active forms are identified by the suffixes *a* and *b*, respectively. In this case, *o*-phosphorylase *b* (unmodified, least active) is the form under allosteric control by AMP, ATP, and G6P (Fig. 18-10, *left*). Phosphorylation to yield *m*-phosphorylase *a* (modified, most active) all but removes the effects of these allosteric modulators. In terms of the symmetry model of allosterism (Section 10-4B), *the phosphorylation of Ser 14 shifts the enzyme's T (inactive) $\rightleftharpoons$ R (active) equilibrium in favor of the R state (Fig. 18-10, right).* Indeed, *phosphorylase a's Ser 14-phosphoryl group is analogous to an allosteric activator:* It forms ion pairs with two Arg side chains on the opposite subunit, thereby knitting the subunits together in much the same way as does AMP when it binds tightly to a site between the subunits (Fig. 18-11*b*).

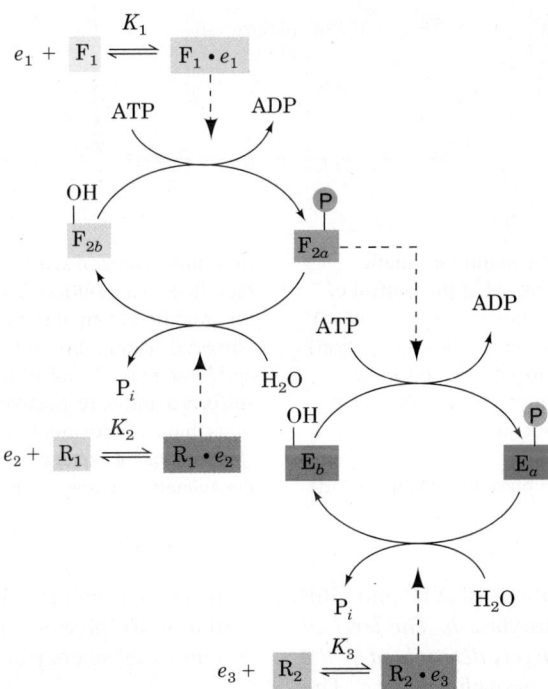

Figure 18-13 **A bicyclic enzyme cascade.** See the legend of Fig. 18-12 for symbol definitions. In a bicyclic cascade, one of the modifying enzymes (F_2) is also subject to covalent

modification. It is active in the modified state (F_{2a}) and inactive in the unmodified state (F_{2b}).

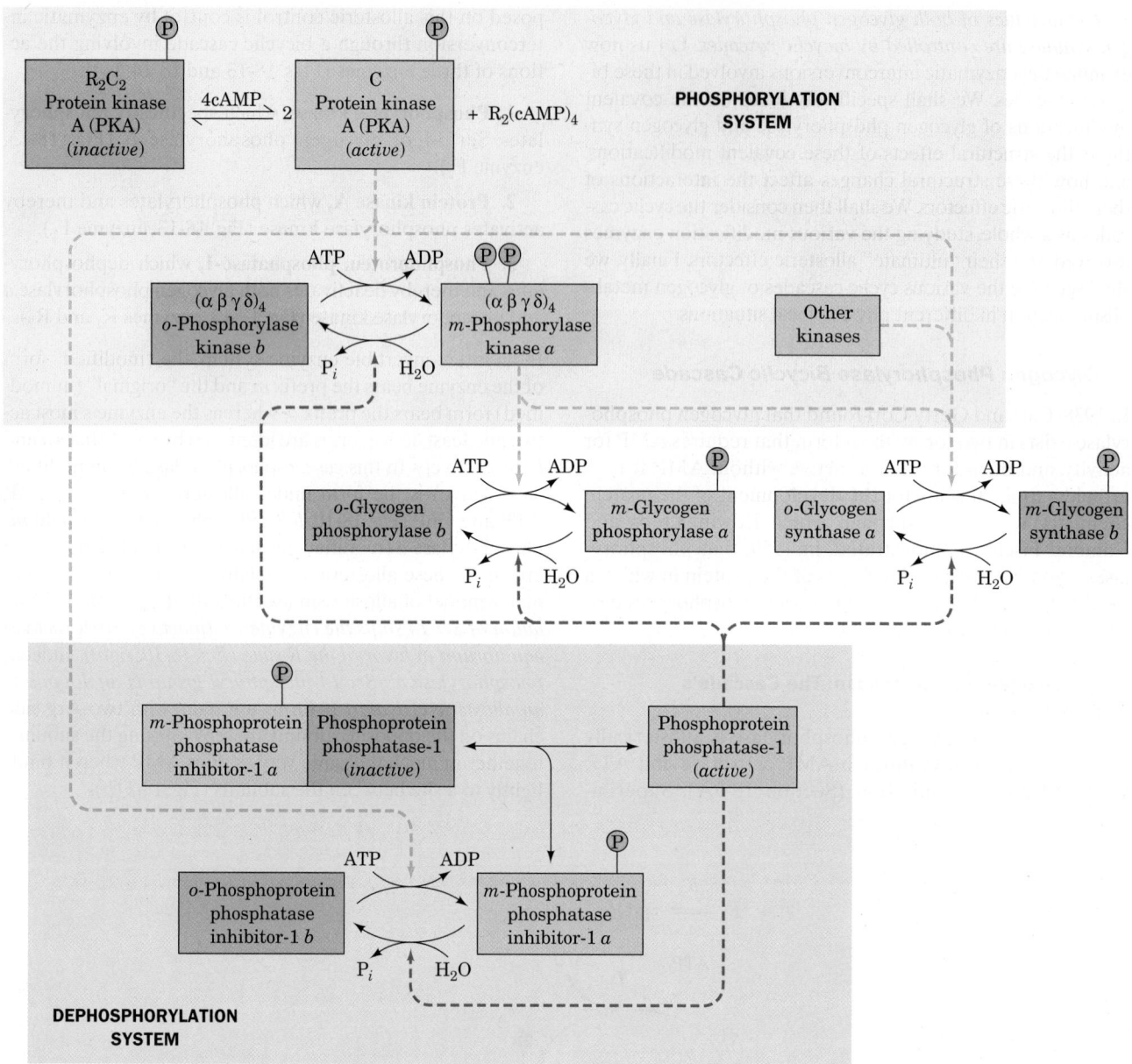

Figure 18-14   Schematic diagram of the major enzymatic modification/demodification systems involved in the control of glycogen metabolism in muscle. Modification (phosphorylation) systems are shaded in yellow, demodification (dephosphorylation) systems are shaded in lavender, active enzymes/inhibitors are shaded in green, and inactive enzymes/inhibitors are shaded in orange. Dashed yellow and purple arrows indicate facilitation of a modification and demodification reaction. Note that glycogen phosphorylase activity is controlled by a bicyclic enzyme cascade (*left*) and glycogen synthase activity is controlled by both a bicyclic and a monocyclic enzyme cascade (*right*). By convention, the modified form of the enzyme bears the prefix *m* and the "original" (unmodified) form bears the prefix *o*. The most active and least active forms of the enzymes are identified by the suffixes *a* and *b*, respectively. Further control of phosphoprotein phosphatase-1 covalent modification is diagrammed in Fig. 18-22.

🔊 **See Guided Exploration 15: Control of Glycogen Breakdown, and the Animated Figures**

In the resting cell, the concentrations of ATP and G6P are high enough to inhibit phosphorylase b. *The level of phosphorylase activity is therefore largely determined by the fraction of the enzyme present as phosphorylase a.* The steady-state fraction of phosphorylated enzyme (E_a) depends on the relative activities of phosphorylase kinase (F_2), protein kinase A (F_1), and phosphoprotein phos-

phatase-1 (R_1 and R_2). This interrelationship is remarkably elaborate for glycogen phosphorylase. Let us consider the actions of these enzymes.

b. Protein Kinase A: A Crucial Regulatory Link

Phosphorylase kinase, which converts phosphorylase b to phosphorylase a, is itself subject to covalent modification

(Fig. 18-14). For phosphorylase kinase to be fully active, Ca^{2+} must be present (see below) and the protein must be phosphorylated.

In both the glycogen phosphorylase and glycogen synthase cascades, the primary intracellular signal, e_1, is **adenosine-3',5'-cyclic monophosphate (3',5'-cyclic AMP or cAMP).** The cAMP concentration in a cell is a function of the ratio of its rate of synthesis from ATP by **adenylate cyclase (AC;** also called **adenylyl cyclase)** and its rate of breakdown to AMP by a enzymes known as **cAMP-phosphodiesterases (cAMP-PDEs;** Section 19-2E):

ATP

PP_i ← adenylate cyclase

**3',5'-Cyclic AMP
(cAMP)**

H_2O ⌐ phosphodiesterase

AMP

AC is, in turn, activated by certain hormones (Sections 18-3Ea and 19-2D).

cAMP is absolutely required for the activity of **protein kinase A [PKA;** *also called* **cAMP-dependent protein kinase (cAPK)],** *an enzyme that phosphorylates specific Ser and/or Thr residues of numerous cellular proteins, including phosphorylase kinase and glycogen synthase.* These proteins all contain PKA's consensus recognition sequence, Arg-Arg-X-Ser/Thr-Y, where Ser/Thr is the phosphorylation site, X is any small residue, and Y is a large hydrophobic residue. In the absence of cAMP, PKA is an inactive heterotetramer consisting of two regulatory (R) and two catalytic (C) subunits, R_2C_2. The cAMP binds to the regulatory subunits so as to cause the dissociation of active catalytic monomers (Fig. 18-14; *top). The intracellular concentration of cAMP therefore determines the fraction of PKA in its active form and thus the rate at which it phosphorylates its substrates.* In fact, in all known eukaryotic cases, the physiological effects of cAMP are exerted through the activation of specific protein kinases.

The X-ray structure of the 350-residue C subunit of mouse PKA in complex with Mg^{2+}–ATP and a 20-residue inhibitor peptide was determined by Susan Taylor and Janusz Sowadski (Fig. 18-15), and that of a similar complex of the porcine heart enzyme was determined by Robert Huber. The C subunit, as are other kinases of known structure (e.g., Figs. 17-5 and 17-15), is bilobal. It has an N-terminal domain that consists of a 5-stranded β sheet and an α helix, and a larger C-terminal domain that is mainly α helical. A deep cleft between the lobes is occupied by the Mg^{2+}–ATP and the segment of the inhibitor peptide that includes the above 5-residue consensus sequence. This cleft therefore contains PKA's catalytic site, with the small domain contributing the nucleotide-binding site and the large subunit supplying the substrate-binding and catalytic residues.

The C subunit of PKA must be phosphorylated at Thr 197 for activity. Thr 197 is part of the so-called activation loop (comprising residues 184–208), which is located at the "mouth" of the cleft between PKA's N- and C-terminal domains. The phosphoryl group at Thr 197 knits together the various components of PKA into its active conformation via extensive interactions with the protein. Most notably, the phosphoryl group interacts with Arg 165, a conserved residue that is adjacent to Asp 166, the catalytic base that activates the substrate protein's target Ser/Thr hydroxyl group for phosphorylation, thereby properly orienting PKA's active site residues.

Protein kinases play key roles in the signaling pathways by which many hormones, growth factors, neurotransmitters, and toxins affect the functions of their target cells (Chapter 19), as well as in controlling metabolic pathways. Indeed, 518 human proteins constituting ~2.3% of human genes are predicted to be protein kinases, which accounts for the observation that ~30% of the proteins in mammalian cells are phosphorylated. The >1000 different protein kinases that have been sequenced share a conserved catalytic core corresponding to residues 40 to 280 of PKA's

Figure 18-15 **X-ray structure of the catalytic (C) subunit of mouse protein kinase A (PKA).** The protein is in complex with ATP and a 20-residue peptide segment of a naturally occurring protein kinase inhibitor. The N-terminal domain is pink and its C-terminal domain is cyan with its activation loop blue. The polypeptide inhibitor is orange and its pseudo–target sequence, Arg-Arg-Asn-Ala-Ile, is purple, with the Ala replacing the Ser to be phosphorylated green (note that the enzyme's true target sequence is Arg-Arg-X-Ser/Thr-Y, where X is a small residue, Y is a large hydrophobic residue, and Ser/Thr, which is replaced by Ala in the polypeptide inhibitor, is the residue that the enzyme phosphorylates). The ATP and the phosphoryl group of phosphoThr 197 are shown in space-filling form and the side chains of the catalytically essential Arg 165, Asp 166, and Thr 197 are shown in stick form, all colored according to atom type (C green, N blue, O red, and P orange). Note that the inhibitor's pseudo–target sequence is in close proximity to the ATP's γ phosphate group, the group that the enzyme transfers. [Based on an X-ray structure by Susan Taylor and Janusz Sowadski, University of California at San Diego. PDBid 1ATP.] 🔖 **See Interactive Exercise 10 and Kinemage Exercise 15-1**

C subunit. In addition to phosphorylating other proteins, many protein kinases are themselves phosphoproteins whose activities are controlled by phosphorylation, often at their activation loops. However, since PKA is normally fully phosphorylated at Thr 197 and resistant to dephosphorylation, it is unclear whether its activity is regulated *in vivo* by phosphorylation/dephosphorylation.

c. PKA's R Subunit Competitively Inhibits Its C Subunit

The R subunit of PKA has a well-defined domain structure that was first characterized by limited proteolysis. It consists of, from N- to C-terminus, a dimerization domain, an autoinhibitor segment, and two tandem homologous cAMP-binding domains, A and B. In the R_2C_2 complex, the autoinhibitor segment, which resembles the C subunit's substrate peptide, binds in the C subunit's active site (as does the inhibitory peptide in Fig. 18-15) so as to block substrate binding. Thus, the R subunit is a competitive inhibitor of PKA's substrate proteins.

Each R subunit cooperatively binds two cAMPs. When the B domain lacks bound cAMP, it masks the A domain so as to prevent it from binding cAMP. However, the binding of cAMP to the B domain triggers a conformational change that permits the A domain to bind cAMP, which, in turn, releases the C subunits from the complex (see below).

Taylor determined the X-ray structure of the R subunit lacking its N-terminal 91 residues and in complex with two cAMPs (Fig. 18-16a). This truncated protein is unable to dimerize but, in the absence of cAMP, forms a tight inactive complex with the C subunit, and on binding cAMP releases active C subunits as do intact R_2 dimers. As previously predicted by sequence alignments, the A and B domains are structurally similar to each other and to the prokaryotic cAMP-binding transcriptional regulator named **catabolite gene activator protein (CAP;** Section 31-3Cb). The autoinhibitory segment, which in the free R subunit is extremely sensitive to proteolysis, has its first 21 residues disordered in the X-ray structure.

The X-ray structure of the truncated R subunit in complex with the C subunit binding AMPPNP (Fig. 18-16b), also determined by Taylor, reveals that the R subunit has undergone a massive conformational reorganization relative to its cAMP-binding structure. The most striking such change is that the 25-residue, bent 2-helix segment linking domains A and B (purple in Fig. 18-16a) has coalesced to form a single straight helix such that the β sandwich in the R subunit's B domain has rotated by ~180° relative to that in the A domain and separated from it. As a consequence of this ~60-Å screwlike shift of the B domain, the elongated globular structure of the cAMP complex changes to a dumbbell-like shape in which the interface between the A and B domains in the cAMP complex is replaced by extensive interactions with the large domain of the C subunit. In addition, the helical regions of the A and B domains undergo extensive conformational changes that eliminate their cAMP binding sites by separating their phosphate-binding pockets from their adenine-binding pockets. All of this positions the R subunit's autoinhibitor segment in the C subunit's active site cleft, thereby inactivating it.

(a)

(b)

Figure 18-16 **X-ray structures of the regulatory (R) subunit of bovine protein kinase A (PKA).** (*a*) The R subunit lacking its N-terminal 91 residues (which form its dimerization domain) in complex with cAMP. The N-terminal region, which includes its autoinhibitor segment, is red, domain A is green, domain B is orange, and the 2-helix segment linking domains A and B is purple. The cAMPs, which are drawn in space-filling form colored according to atom type (C cyan, N blue, O red, and P orange), each bind at the opening of an 8-stranded β sandwich with their phosphate group abutting the N-terminal end of a short helical segment. (*b*) The truncated R subunit in complex with the C subunit that is binding AMPPNP. The R subunit is

colored as in Part *a* and the C subunit is colored as in Fig. 18-15. The C subunit is rotated 180° about the vertical axis relative to Fig. 18-15 and the β sandwich of domain A is oriented similarly to that in Part *a*. Note the dramatic conformational differences between the R subunit in Parts *a* and *b,* which in large part is caused by the coalescence and straightening of the two helices linking domains A and B in the cAMP complex. Also note that the now ordered N-terminal portion of the R subunit's autoinhibitor segment occupies the C subunit's active site cleft. [Based on X-ray structures by Susan Taylor, University of California at San Diego. PDBids 1RGS and 2QCS.]

d. Phosphorylase Kinase: Coordination of Enzyme Activation with [Ca²⁺]

Phosphorylase kinase (PhK) is activated by Ca²⁺ concentrations as low as 10^{-7} M as well as by covalent modification. This 1300-kD enzyme consists of four nonidentical subunits that form the active oligomer $(\alpha\beta\gamma\delta)_4$. The isolated γ subunit is capable of full catalytic activity (ability to convert phosphorylase *b* to phosphorylase *a*), whereas the α, β, and δ subunits are inhibitors of the catalytic reaction.

The δ subunit, which is known as **calmodulin (CaM),** confers Ca²⁺ sensitivity on the complex. When Ca²⁺ binds to calmodulin's four Ca²⁺-binding sites, this ubiquitous eukaryotic regulatory protein undergoes an extensive conformational change (see below) that activates phosphorylase kinase. Glycogen phosphorylase therefore becomes

phosphorylated and the rate of glycogen breakdown increases. The physiological significance of this Ca²⁺ activation process is that nerve impulses trigger muscle contraction through the release of Ca²⁺ from intracellular reservoirs (Section 35-3C). *This transient increase in cytosolic [Ca²⁺] induces both muscle contraction and the increase in glycogen breakdown that supplies glycolysis, which in turn, generates the ATP required for muscle contraction.*

e. Calmodulin: A Ca²⁺-Activated Switch

Calmodulin is a ubiquitous eukaryotic Ca²⁺-binding protein that participates in numerous cellular regulatory processes. In some of these, CaM functions as a monomeric protein, whereas in others (e.g., PhK) it is a subunit of a larger protein. The X-ray structure of this highly conserved

Figure 18-17 **X-ray structure of rat testis calmodulin.** This monomeric 148-residue protein, which is colored in rainbow order from N-terminus (*blue*) to C-terminus (*red*), contains two remarkably similar globular domains separated by a seven-turn α helix. The two Ca^{2+} ions bound to each domain are represented by cyan spheres. The side chains liganding the Ca^{2+} ions are drawn in stick form colored according to atom type (C green, N blue, and O red). [Based on an X-ray structure by Charles Bugg, University of Alabama at Birmingham. PDBid 3CLN.] 🔁 **See Kinemage Exercise 16-1**

148-residue protein, determined by Charles Bugg, has a curious dumbbell-like shape in which CaM's two globular domains are connected by a seven-turn α helix (Fig. 18-17). CaM has two high-affinity Ca^{2+}-binding sites on each of its globular domains, both of which are formed by nearly superimposable helix–loop–helix motifs known as **EF hands** (Fig. 18-18) that also occur in numerous other Ca^{2+}-sensing proteins of known structure. The Ca^{2+} ion in each of these sites is octahedrally coordinated by oxygen atoms from the backbone and side chains of the loop as well as from a protein-associated water molecule.

The binding of Ca^{2+} to either domain of CaM induces a conformational change in that domain, which exposes an otherwise buried Met-rich hydrophobic patch. This patch, in turn, binds with high affinity to the CaM-binding domain of the phosphorylase kinase γ subunit, as well as to the CaM-binding domains of numerous other Ca^{2+}-regulated proteins (many of which interact with CaM that is free in solution), and in doing so modulates the activities of these proteins. These CaM-binding domains have little mutual sequence similarity but are all basic amphiphilic α helices. In fact, ~20-residue segments of these helices, as well as

synthetic amphiphilic helices composed of only Leu, Lys, and Trp residues, are bound by Ca^{2+}–CaM as tightly as the target proteins themselves.

Despite uncomplexed CaM's extended appearance in its X-ray structure (Fig. 18-17), a variety of studies indicate that both of its globular domains can simultaneously bind to a single target helix. Evidently, CaM's central α helix serves as a flexible linker rather than as a rigid spacer, a property that probably further increases the range of target sequences to which CaM can bind. This idea is confirmed by the NMR structure (Fig. 18-19) of $(Ca^{2+})_4$–CaM in complex with its 26-residue CaM-binding target polypeptide of skeletal muscle **myosin light chain kinase** (**MLCK;** a homolog of the PKA C subunit, which phosphorylates and thereby activates the light chains of the muscle protein **myosin;** Section 35-3Da), which was determined by Marius Clore, Angela Gronenborn, and Ad Bax. Indeed, the extended conformation of CaM's central helix in Fig. 18-17 is probably an artifact arising from crystal packing forces, considering that this helix's central two turns contact no other portion of the protein and hence are maximally solvent-exposed (almost all other known α helices are at least partially buried in a protein). Moreover, a polypeptide with the sequence of this helix assumes a random coil conformation in aqueous solution. Nevertheless, the flexible linker is essential to the function of CaM: In the presence of Ca^{2+}, CaM's individual domains (obtained by

Figure 18-18 **EF hand.** The Ca^{2+}-binding sites in many proteins that function to sense the level of Ca^{2+} are formed by helix–loop–helix motifs named EF hands. [After Kretsinger, R.H., *Annu. Rev. Biochem.* **45,** 241 (1976).] 🔁 **See Kinemage Exercise 16-1**

(a)   *(b)*

Figure 18-19   NMR structure of (Ca²⁺)₄–CaM from *Drosophila melanogaster* **in complex with its 26-residue target polypeptide from rabbit skeletal muscle myosin light chain kinase (MLCK).** The N terminal domain of CaM is blue, its C-terminal domain is red, the target polypeptide is green, and the Ca^{2+} ions are represented by cyan spheres. (*a*) A view of the complex in which the N-terminus of the target polypeptide is on the right, and (*b*) the perpendicular view as seen from the right side of Part *a*. In both views, the pseudo-2-fold axis relating the N- and C-terminal domains of CaM is approximately vertical. Note how the middle segment of the long central helix in uncomplexed CaM (Fig. 18-17) has unwound and bent (bottom loop in Part *b*) such that CaM forms a globular protein that largely encloses the helical target polypeptide within a hydrophobic tunnel in a manner resembling two hands holding a rope (the target polypeptide assumes the random coil conformation in solution). However, the conformations of CaM's two globular domains are essentially unchanged by the complexation. Evidently, CaM's bound Ca^{2+} ions serve to organize and stabilize the target binding conformations of its globular domains. [Based on an NMR structure by Marius Clore, Angela Gronenborn, and Ad Bax, National Institutes of Health. PDBid 2BBM.] 🔎 **See Kinemage Exercise 16-2**

tryptic cleavage), when in high concentration, are able to bind their target proteins but fail to even marginally activate them unless present in several hundred-fold excess.

How does Ca^{2+}–CaM activate its target protein kinases? MLCK contains a C-terminal segment whose sequence resembles that of MLCK's target polypeptide on the light chain of myosin but lacks a phosphorylation site. A model of MLCK, based on the X-ray structure of the 30% identical C subunit of PKA, strongly suggests that this autoinhibitor peptide inactivates MLCK by binding in its active site. Indeed, the excision of MLCK's autoinhibitor peptide by limited proteolysis permanently activates this enzyme. MLCK's CaM-binding segment overlaps this autoinhibitor peptide. Evidently, *the binding of Ca²⁺–CaM to this peptide segment extracts the autoinhibitor from MLCK's active site, thereby activating this enzyme* (Fig. 18-20).

Ca^{2+}–CaM's other target proteins, including the phosphorylase kinase γ subunit, are presumably activated in the

Figure 18-20   Schematic diagram of the Ca²⁺–CaM-dependent activation of protein kinases. Autoinhibited kinases have an N- or C-terminal "pseudosubstrate" sequence (*red*) that binds at or near the enzyme's active site (*brown*) so as to inhibit its function. This autoinhibitory segment is in close proximity with or overlaps a Ca^{2+}–CaM-binding sequence. Consequently, Ca^{2+}–CaM (*green*) binds to this sequence so as to extract it from the enzyme's active site, thereby activating the enzyme to phosphorylate other proteins (*purple*). [After Crivici, A. and Ikura, M., *Annu. Rev. Biophys. Biomol. Struct.* **24,** 88 (1995).]

same way. The X-ray structures of two homologous protein kinases support this so-called **intrasteric mechanism,** those of **calmodulin-dependent protein kinase I (CaMKI)** and **twitchin kinase.** Although the details of binding of the autoinhibitory sequence differ for each of these protein kinases, the general mode of autoinhibition and activation by Ca^{2+}–CaM is the same.

PKA's R subunit, as we have seen, contains a similar autoinhibitory sequence adjacent to its two tandem cAMP-binding domains. In this case, however, the autoinhibitory peptide is allosterically ejected from the C subunit's active site by the binding of cAMP to the R subunit (which lacks a Ca^{2+}–CaM binding site).

f. Phosphorylase Kinase's γ Subunit Is Controlled by Multiple Autoinhibitors

Phosphorylase kinase's 386-residue γ subunit consists of an N-terminal kinase domain, which is 36% identical in sequence to the C subunit of PKA, and a C-terminal regulatory domain, which contains a CaM-binding peptide and an overlapping autoinhibitor segment. Evidently, Ca^{2+}–CaM relieves this inhibition, as is diagrammed in Fig. 18-20. This explains why the N-terminal 298-residue segment of the PhK γ subunit, termed **PhKγ$_t$** (t for truncated), displays catalytic activity comparable to that of fully activated PhK but is unaffected by Ca^{2+} or phosphorylation signals.

The X-ray structure of PhKγ$_t$ in complex with ATP and a heptapeptide related to the natural substrate was determined by Johnson (Fig. 18-21). It reveals, as expected, that PhKγ$_t$ structurally resembles PKA (Fig. 18-15) as well as other protein kinases of known structure including CaMKI and twitchin kinase. Comparisons of these various structures shed light on how the catalytic activity of PhK is regulated. Numerous protein kinases, including PKA, are activated by the phosphorylation of Ser, Thr, and/or Tyr residues in their activation loops, which, as we saw in Fig. 18-15, interacts with a conserved Arg residue that thereby correctly positions the adjacent catalytically important Asp residue. However, the PhK γ subunit is not subject to phosphorylation. Rather, its activation loop residue that might otherwise be phosphorylated is Glu 182, whose negative charge mimics the presence of a phosphate group by interacting with Arg 148 so as to correctly position Asp 149 (Fig. 18-21). Thus, the PhK catalytic site maintains an active conformation but, in the absence of Ca^{2+}, is inactivated by the binding of its C-terminal autoinhibitor segment.

Sites on both the α and β subunits of PhK are subject to phosphorylation by PKA (Fig. 18-14). This activates PhK at much lower Ca^{2+} concentrations than otherwise, and full enzyme activity is obtained in the presence of Ca^{2+} only when both these subunits are phosphorylated. The β subunit does, in fact, have an autoinhibitor sequence, suggesting that phosphorylation changes its conformation so as to make it unavailable for inhibiting the γ subunit's active site. This would explain the synergistic effect of phosphorylation and Ca^{2+} on the activity of PhK: Ca^{2+}–CaM sequesters the γ subunit's autoinhibitory segment, whereas phosphorylation of the β subunit removes yet another au-

Figure 18-21 **X-ray structure of rabbit muscle PhKγ$_t$ in complex with ATP and a heptapeptide (RQMSFRL).** This heptapeptide is related in sequence to the enzyme's natural substrate (KQISVRG). The protein is shown in the "standard" protein kinase orientation with its N terminal domain pink, its C-terminal domain cyan, and its activation loop blue. The heptapeptide is orange, with its residue to be phosphorylated (Ser) green. The ATP is shown in space-filling form and the side chains of the catalytically essential Arg 148, Asp 149, and Glu 182 are shown in stick form, all colored according to atom type (C green, N blue, O red, and P orange). Note the structural similarities and differences between this protein and the homologous C subunit of PKA (Fig. 18-15). [After an X-ray structure by Louise Johnson, Oxford University, Oxford, U.K. PDBid 2PHK.]

toinhibitor. The way in which the phosphorylation of the α subunit modulates the activity of PhK is, as yet, unknown.

g. Phosphoprotein Phosphatase-1

The steady-state phosphorylation levels of most enzymes involved in cyclic cascades are maintained by the opposition of kinase-catalyzed phosphorylations and the hydrolytic dephosphorylations catalyzed by phosphoprotein phosphatases. The phosphatase involved in the cyclic cascades controlling glycogen metabolism is phosphoprotein phosphatase-1. This enzyme, as is indicated in Fig. 18-14, hydrolyzes the phosphoryl groups from *m*-glycogen phosphorylase *a*, both the α and β subunits of phosphorylase kinase, and two other proteins involved in glycogen metabolism, as discussed below.

The catalytic subunit of phosphoprotein phosphatase-1 **(PP1),** which is designated **PP1c,** hydrolyzes phosphoryl groups on Ser/Thr residues via a single step mechanism.

The X-ray structure of PP1c indicates that it contains a binuclear metal ion center (both metals are Mn^{2+} in the recombinant enzyme) which, it is proposed, activates a water molecule (promotes its ionization to OH^-, Section 15-1Cb) for nucleophilic attack on the phosphoryl group.

PP1c binds to glycogen through the intermediacy of regulatory proteins in both muscle and liver. In muscle, PP1c is only active when it is bound to glycogen through this glycogen-binding **G_M subunit.** The activity of PP1c and its affinity for the G_M subunit are regulated by phosphorylation of the G_M subunit at two separate sites (Fig. 18-22). Phosphorylation of site 1 by **insulin-stimulated protein kinase** activates phosphoprotein phosphatase-1, whereas phosphorylation of site 2 by PKA (which can also phosphorylate site 1) causes the enzyme to be released into the cytoplasm, where it cannot dephosphorylate the glycogen-bound enzymes of glycogen metabolism.

In the cytosol, PP1c is also inhibited by its binding to the protein **phosphoprotein phosphatase inhibitor 1 (inhibitor-1).** This latter protein provides yet another example of control by enzymatic interconversion: It too is modified by PKA and demodified by PP1c (Fig. 18-14, *bottom*), although, in this case, a Thr, not a Ser, is phosphorylated/dephosphorylated. The protein is a functional inhibitor only when it is phosphorylated. *The concentration of cAMP therefore controls the fraction of an enzyme in its phosphorylated form, not only by increasing the rate at which it is phosphorylated, but also by decreasing the rate at which it is dephosphorylated. In the case of glycogen phosphorylase, an increase in [cAMP] results not only in an* *increase in this enzyme's rate of activation, but also in a decrease in its rate of deactivation.*

The activity of phosphoprotein phosphatase-1 in liver is also controlled by its binding to glycogen through the intermediacy of a glycogen-binding subunit, here named **G_L.** When bound to G_L, PP1c is activated toward dephosphorylation of the glycogen-bound enzymes of glycogen metabolism. However, G_L is not subject to control via phosphorylation as is G_M in the muscle. Rather, the binding of *m*-phosphorylase *a* to G_L strongly inhibits the activity of PP1c by an allosteric mechanism.

Among the major conformational changes that glycogen phosphorylase undergoes in converting from the T to the R state is the movement of the Ser 14-phosphoryl group from the surface of the T-state (inactive) enzyme to a position buried a few angstroms beneath the protein's surface at the dimer interface in the R-state (active) enzyme (Figs. 18-11*b*). Both the R and T forms of phosphorylase *a* strongly bind the $G_L \cdot$ PP1c *complex,* but only in the T-state enzyme is the Ser 14-phosphoryl group accessible for hydrolysis by PP1c. Consequently, under the conditions that phosphorylase *a* converts to the T state (Section 18-3G), PP1c hydrolyzes its now exposed Ser 14-phosphoryl group. This converts *m*-phosphorylase *a* to *o*-phosphorylase *b*, which has only a low affinity for binding the $G_L \cdot$ PP1c complex and hence does not inhibit PP1c. One effect of phosphorylase *a* demodification, therefore, is to relieve the inhibition of PP1c and thus allow it to excise the phosphoryl groups of other susceptible phosphoproteins. Since phosphorylase *a* has a high affinity for the $G_L \cdot$ PP1c complex and is in ~10-fold greater concentration, *relief of PP1c inhibition only occurs when more than*

Figure 18-22 **The antagonistic effects of insulin and epinephrine on glycogen metabolism in muscle.** This occurs through their effects on the phosphoprotein phosphatase-1 catalytic subunit, PP1c, via its glycogen-bound G_M subunit. Green discs and dashed arrows indicate activation.

~90% of the glycogen phosphorylase is in the o-phosphorylase b form. Glycogen synthase is among the proteins that are dephosphorylated by the $G_L \cdot PP1c$ complex when it is no longer inhibited by phosphorylase. However, in contrast to phosphorylase, dephosphorylation activates glycogen synthase. This enzyme is involved in its own bicyclic cascade whose properties we shall now examine.

D. Glycogen Synthase Bicyclic Cascade

Like glycogen phosphorylase, glycogen synthase exists in two enzymatically interconvertible forms:

1. The modified (*m*; phosphorylated) form that is inactive under physiological conditions (the *b* form).

2. The original (*o*; dephosphorylated) form that is active (the *a* form).

m-Glycogen synthase *b* is under allosteric control; it is strongly inhibited by physiological concentrations of ATP, ADP, and P_i and hence the modified enzyme is almost totally inactive *in vivo.* The activity of the unmodified enzyme is essentially independent of these effectors, so a cell's glycogen synthase activity varies with the fraction of the enzyme in its unmodified form.

The mechanistic details of the interconversion of modified and unmodified forms of glycogen synthase are particularly complex and are therefore not as well understood as those of glycogen phosphorylase. It has been clearly established that the fraction of unmodified glycogen synthase is, in part, controlled by a bicyclic cascade involving phosphorylase kinase (PhK) and phosphoprotein phosphatase-1, enzymes that are also involved in the glycogen phosphorylase bicyclic cascade (Fig. 18-14, *right*). This demodification process is facilitated by G6P, whose binding to *m*-glycogen synthase *b* induces it to undergo a conformational change that exposes its phosphoryl groups to the surface of the protein, thereby making them available for dephosphorylation by phosphoprotein phosphatase-1.

Glycogen synthase is phosphorylated at several sites. Several protein kinases are known to at least partially deactivate human muscle glycogen synthase by phosphorylating this homotetramer at 1 or more of 9 Ser residues in the N- and C-terminal segments on its 737-residue subunits. These enzymes include PhK, PKA (so glycogen synthase deactivation may also be considered to occur via a monocyclic cascade), CaMKI (which is activated by the presence of Ca^{2+}), **protein kinase C (PKC;** which responds to the extracellular presence of certain hormones via a mechanism described in Sections 18-3G and 19-4Cb), **AMP-dependent protein kinase (AMPK;** which responds to ATP availability and hence acts as a fuel gauge; Sections 25-5a and 27-1), **glycogen synthase kinase-3β [GSK3β;** which is inhibited by **insulin** (Sections 18-3Ea and 18-3F), whose presence therefore results in the dephosphorylation and hence activation of glycogen synthase], and **casein kinases 1** and **2** (which participate in a variety of cellular control processes). Why glycogen synthase deactivation is so elaborately controlled compared to its activation or the

activation/deactivation of glycogen phosphorylase is unclear, although, whatever the reasons, it closely monitors the organism's metabolic state.

E. Integration of Glycogen Metabolism Control Mechanisms

Whether there is net synthesis or degradation of glycogen and at what rate depends on the relative balance of the active forms of glycogen synthase and glycogen phosphorylase. This, in turn, largely depends on the rates of the phosphorylation and dephosphorylation reactions of the two bicyclic cascades. These cascades, one controlling the rate of glycogen breakdown and the other controlling the rate of glycogen synthesis, are intimately related. They are linked by protein kinase A and phosphorylase kinase, which, through phosphorylation, activate glycogen phosphorylase as they inactivate glycogen synthase (Fig. 18-14). The cascades are also linked by phosphoprotein phosphatase-1, which in liver is inhibited by phosphorylase *a* and therefore is unable to activate (dephosphorylate) glycogen synthase unless it first inactivates (also by dephosphorylation) phosphorylase *a*.

a. Hormones Are Important Regulators of Glycogen Metabolism

Glycogen metabolism is largely regulated by the peptide hormone **insulin** (Fig. 7-2) acting in opposition to **glucagon,** another peptide hormone,

^+H_3N - His - Ser - Glu- Gly - Thr - Phe - Thr - Ser - Asp- Tyr -   10
Ser - Lys - Tyr - Leu - Asp- Ser - Arg - Arg - Ala - Gln-   20
Asp- Phe - Val - Gln - Trp - Leu - Met - Asn - Thr - COO^- 29
Glucagon

together with the adrenal hormones **epinephrine (adrenaline)** and **norepinephrine (noradrenaline):**

X = CH_3 **Epinephrine**
X = H **Norepinephrine**

Hormonal stimulation of cells at their plasma membranes occurs through the mediation of transmembrane proteins called **receptors.** *Different cell types have different complements of receptors and thus respond to different sets of hormones.* For example, both muscle and liver cells have abundant insulin and **adrenergic receptors** (receptors responsive to epinephrine and norepinephrine), whereas glucagon receptors are more prevalent in liver than in skeletal muscle.

b. Second Messengers Mediate Glucagon- and Epinephrine-Stimulated Glycogen Breakdown

*The response to glucagon and epinephrine involves the release inside the cell of molecules known as **second messengers**, that is, intracellular mediators of the externally received hormonal message.* Different receptors act to release different second messengers. Indeed, cAMP was identified by Earl Sutherland as the first known instance of a second messenger through his demonstration that glucagon and epinephrine act at cell surfaces to stimulate adenylyl cyclase (AC) to increase [cAMP] [the mechanism of AC activation, as well as a discussion of other second messengers, including Ca^{2+}, **inositol-1,4,5-trisphosphate (IP$_3$)**, and **diacylglycerol (DAG)**, is elaborated in Sections 19-2D and 19-4A]. Following this discovery, it was realized that cAMP, which is present in all forms of life, is an essential control element in many biological processes.

When hormonal stimulation by glucagon or epinephrine increases the intracellular cAMP concentration, the protein kinase A activity increases, increasing the rates of phosphorylation of many proteins and decreasing their dephosphorylation rates as well. A decrease in dephosphorylation rates, as previously noted, increases the phosphorylation level of phosphoprotein phosphatase inhibitor-1, which in turn inhibits phosphoprotein phosphatase-1. An increase in the concentration of phosphorylase *a* also contributes to the inhibition of phosphoprotein phosphatase-1.

Because of the amplifying properties of the cyclic cascades, a small change in [cAMP] results in a large change in the fraction of enzymes in their phosphorylated forms. When a large fraction of the glycogen metabolism enzymes are present in their phosphorylated forms, the metabolic flux is in the direction of glycogen breakdown, since glycogen phosphorylase is active and glycogen synthase is inactive. When [cAMP] decreases, phosphorylation rates decrease, dephosphorylation rates increase, and the fraction of enzymes in their dephospho forms increases. The resultant activation of glycogen synthase and the inhibition of glycogen phosphorylase cause a change in the flux direction toward net glycogen synthesis.

F. *Maintenance of Blood Glucose Levels*

An important function of the liver is to maintain the blood concentration of glucose, the brain's primary fuel source, at ~5 m*M*. When blood [glucose] decreases beneath this level, usually during exercise or well after meals have been digested, the liver releases glucose into the bloodstream. The process is mediated by the hormone glucagon as follows:

1. Glucose inhibits the pancreatic α cells from secreting glucagon into the bloodstream. When the blood glucose concentration falls, this inhibition is released causing the α cells to secrete glucagon.

2. Glucagon receptors on liver cell surfaces respond to the presence of glucagon by activating adenylate cyclase, thereby increasing the [cAMP] inside these cells.

3. The [cAMP] increase, as described above, triggers an increase in the rate of glycogen breakdown, leading to increased intracellular [G6P].

4. G6P, in contrast to glucose, cannot pass through the cell membrane. However, in liver, which does not employ glucose as a major energy source, the enzyme **glucose-6-phosphatase (G6Pase)** hydrolyzes G6P:

$$G6P + H_2O \longrightarrow glucose + P_i$$

The resulting glucose enters the bloodstream, thereby increasing the blood glucose concentration. Muscle and brain cells, however, lack G6Pase so that they retain their G6P.

G6P hydrolysis requires intracellular G6P transport. G6P is produced in the cytosol, whereas G6Pase is a TM protein that resides in the endoplasmic reticulum (ER) membrane with its active site in the ER lumen. G6P must therefore be imported into the ER by a **G6P translocase** before it can be hydrolyzed. The resulting glucose and P_i are then returned to the cytosol via specific transport proteins (Section 18-3G). A defect in any of the components of this G6P hydrolysis system results in **type I glycogen storage disease** (Section 18-4).

How does this delicately balanced system respond to an increase in blood [glucose]? When blood sugar is high, normally immediately after meals have been digested, glucagon levels decrease and insulin is released from the pancreatic β cells. *The rate of glucose transport across many cell membranes increases in response to insulin (through the insulin-dependent glucose transporter **GLUT4**; Section 20-2E), [cAMP] decreases, and glycogen metabolism therefore shifts from glycogen breakdown to glycogen synthesis.* The mechanism of insulin action is quite complex and is not fully understood (Sections 19-3 and 19-4F), but one of its target enzymes appears to be phosphoprotein phosphatase-1.

In muscle, insulin and epinephrine have antagonistic effects on glycogen metabolism. Epinephrine promotes glycogenolysis by activating the cAMP-dependent phosphorylation cascade, which stimulates glycogen breakdown while inhibiting glycogen synthesis. Insulin, as we saw in Section 18-3Cg, activates insulin-stimulated protein kinase to phosphorylate site 1 on the glycogen-binding G$_M$ subunit of phosphoprotein phosphatase-1 so as to activate this protein to dephosphorylate the enzymes of glycogen metabolism (Fig. 18-22). The storage of glucose as glycogen is thereby stimulated through the inhibition of glycogen breakdown and the stimulation of glycogen synthesis.

In liver, it appears that glucose and glucose-6-phosphate themselves may be the messengers to which the glycogen metabolism system responds. *Glucose inhibits phosphorylase a by binding only to the active site of the enzyme's inactive T state, but in a manner different from that of substrate.* The presence of glucose therefore shifts phosphorylase *a*'s $T \rightleftharpoons R$ equilibrium toward the T state (Fig. 18-10, *right*). This conformational shift, as we saw in Section 18-3Cg, exposes the Ser 14-phosphoryl group to phosphoprotein phosphatase-1, resulting in the demodification of phosphorylase *a*. An increase in glucose concentration therefore

Figure 18-23 **The enzymatic activities of phosphorylase *a* and glycogen synthase in mouse liver in response to an infusion of glucose.** Phosphorylase *a* is rapidly inactivated and, somewhat later, glycogen synthase is activated. [After Stalmans, W., De Wulf, H., Hue, L., and Hers, H.-G., *Eur. J. Biochem.* **41,** 129 (1974).]

promotes inactivation of glycogen phosphorylase *a* through the enzyme's conversion to phosphorylase *b* (Fig. 18-23; i.e., phosphorylase *a* acts as a glucose receptor). The concomitant release of phosphoprotein phosphatase-1 inhibition (recall that it specifically binds to and is thereby inactivated by phosphorylase *a*), moreover, results in the activation (dephosphorylation) of *m*-glycogen synthase *b*. In addition, glucose is converted to G6P by **glucokinase** (see below), which facilitates the dephosphorylation and activation of *m*-glycogen synthase *b* to *o*-glycogen synthase *a*. Above a glucose concentration of 7 m*M*, these processes reverse the flux of glycogen metabolism from breakdown to synthesis. The liver can thereby store the excess glucose as glycogen.

a. Glucokinase Forms G6P at a Rate Proportional to the Glucose Concentration

The liver's function in "buffering" the blood [glucose] is made possible because this organ contains a variant of hexokinase (the first glycolytic enzyme; Section 17-2A) known as glucokinase **(GK;** also called **hexokinase D** and **hexokinase IV).** The hexokinase in most cells obeys Michaelis–Menten kinetics, has a high glucose affinity (K_M < 0.1 m*M*; the value of [glucose] at which the enzyme achieves half-maximal velocity; Section 14-2A), and is inhibited by its reaction product, G6P. GK, in contrast, has a much lower glucose affinity (reaching half of its maximal velocity at ∼5 m*M*) and displays sigmoidal kinetics with a Hill constant (Section 10-1Bb) of 1.5, so *its activity increases rapidly with the blood [glucose] over the normal physiological range* (Fig. 18-24; *see Problem 7 of this chapter*). GK, moreover, is not inhibited by physiological concentrations of G6P. Consequently, the higher the blood [glucose], the faster the liver converts glucose to G6P [liver

cells, unlike most cells, contain ample quantities of the insulin-independent glucose transporter **GLUT2** (Section 20-2Eb) and are therefore freely permeable to glucose; their glucose transport rate is unresponsive to insulin]. Thus at low blood [glucose], the liver does not compete with other tissues for the available glucose supply, whereas at high blood [glucose], when the glucose needs of these tissues are met, glucose in the liver is converted to G6P. The excess glucose in the liver induces the inactivation of glycogen phosphorylase and the release of phosphoprotein phosphatase-1, whereas the resulting G6P allosterically facilitates the activation of glycogen synthase via dephosphorylation. The net result is that the liver converts the excess glucose to glycogen. (Note that GK is a monomeric enzyme, so that its sigmoidal rate increase with [glucose] is a puzzling observation in light of the various allosteric models indicating that monomeric enzymes are incapable of cooperative behavior. Since GK does not exhibit Michaelis–Menten kinetics, the glucose concentration when this enzyme has half its maximal activity is known as its $K_{0.5}$ in analogy with the operational definition of K_M.)

GK is subject to metabolic controls. Emile Van Schaftingen isolated **glucokinase regulatory protein (GKRP;** a 625-residue monomer) from rat liver, which, in the presence of the glycolytic intermediate fructose-6-phosphate (F6P), is a competitive inhibitor of glucokinase. Fructose-1-phosphate (F1P), an intermediate in liver fructose metabolism (Section 17-5A), overcomes this inhibition. Since fructose is normally available only from dietary sources (e.g., sucrose), fructose may be the signal that triggers the uptake of dietary glucose by the liver.

Figure 18-24 **Comparison of the relative enzymatic activities of hexokinase and glucokinase over the physiological blood glucose range.** The affinity of glucokinase for glucose ($K_{0.5}$ = 5 m*M*) is much lower than that of hexokinase (K_M = 0.1 m*M*) and exhibits sigmoidal rather than hyperbolic variation with [glucose]. [The glucokinase curve was generated using the Hill equation (Eq. [10.7]) with K = 10 m*M* and n = 1.5 as determined by Cardenas, M.L., Rabajille, E., and Niemeyer, H., *Eur. J. Biochem.* **145,** 163–171 (1984).]

b. Glucokinase Regulates Glucose Homeostasis via an Intracellular Localization Mechanism

Intracellular localization plays an important role in GK inhibition by GKRP. GK translocates freely between the nucleus and cytoplasm. However, GKRP is located exclusively in the nucleus. When the glucose concentration is low, GK remains bound to GKRP in the nucleus, where it is unavailable to phosphorylate glucose. At increased concentrations of glucose and/or F1P, GK dissociates from GKRP and translocates to the cytoplasm, where it phosphorylates glucose to G6P, starting it on its way toward glycogen synthesis. The antagonism of GK and GKRP provides a major mechanism for controlling glucose phosphorylation and glycogen metabolism in the liver. Their flux control coefficients (Section 17-4Ca) in liver cells are close to +1 for GK and −1 for GKRP (a negative flux control coefficient is indicative of inhibition). This countervailing relationship provides a sensitive mechanism for maintaining glucose homeostasis.

Phosphoglucomutase, which has a high enough activity to equilibrate its substrate and product and therefore functions in either direction, transforms G6P to G1P, which is then converted to glycogen. Some of the G6P is also reconverted to glucose by the action of glucose-6-phosphatase in what amounts to a "futile" cycle. This is apparently the energetic price of effective glucose "buffering" of the blood.

c. Fructose-2,6-Bisphosphate Activates Glycolysis
β-D-Fructose-2,6-bisphosphate (F2,6P)

$$^{-2}O_3P-OH_2C \qquad O \qquad O-PO_3^{2-}$$

**β-D-Fructose-2,6-bisphosphate
(F2,6P)**

is also an important factor in the liver's maintenance of blood [glucose]. *F2,6P, which is not a glycolytic metabolite, is an extremely potent allosteric activator of animal phosphofructokinase (PFK) and an inhibitor of fructose bisphosphatase (FBPase).* F2,6P, which was independently discov-

ered in 1980 by Simon Pilkis, by Emile Van Schaftingen and Henri-Géry Hers, and by Kosaku Uyeda, therefore stimulates glycolytic flux (the F6P–FBP substrate cycle is discussed in Section 17-4Ff).

*The concentration of F2,6P in the cell depends on the balance between its rates of synthesis and degradation by **phosphofructokinase-2 (PFK-2; also called 6PF-2-K)** and **fructose bisphosphatase-2 (FBPase-2; also called F-2,6-Pase)**, respectively* (Fig. 18-25). These enzyme activities are located on different domains of a single ~100-kD homodimeric protein named **PFK-2/FBPase-2.** The X-ray structure of the H256A mutant of rat testis PFK-2/FBPase-2 in complex with F6P, P_i, succinate, and the nonhydrolyzable ATP analog **adenosine-5'-(β,γ-imido)triphosphate (AMPPNP)**

$$^-O-P-NH-P-O-P-CH_2 O$$

**Adenosine-5'-(β,γ-imido)triphosphate
(AMPPNP)**

was determined by Uyeda and Charles Hasemann (Fig. 18-26). It indicates, in agreement with a variety of studies, that the PFK-2 activity resides on each subunit's 246-residue N-terminal domain, whereas the FBPase-2 activity resides on each subunit's 213-residue C-terminal domain. The succinate, which binds in the vicinity of AMPPNP's γ phosphate group, presumably occupies the F6P binding pocket in the PFK-2 active site, whereas the F6P and P_i mark the F2,6P binding site in the FBPase-2 active site. The FBPase-2 domain is structurally related to the glycolytic enzyme phosphoglycerate mutase (PGM; Section 17-2H) and shares a common catalytic mechanism involving a covalent phospho-His intermediate (His 256 in FBPase-2). The PFK-2 domain is structurally related to adenylate kinase (Section 17-4Fe) but not, as had been speculated, to PFK (also called **PFK-1** to distinguish this glycolytic enzyme from PFK-2).

The enzyme activities of PFK-2/FBPase-2 are subject to allosteric regulation by a variety of metabolic intermediates

Figure 18-25   **Formation and degradation of β-D-fructose-2, 6-bisphosphate as catalyzed by PFK-2 and FBPase-2.** These two enzyme activities occur on different domains of the same protein molecule. Dephosphorylation of the liver enzyme activates PFK-2 but deactivates FBPase-2.

Figure 18-26 **X-ray structure of the H256A mutant of rat testis PFK-2/FBPase-2.** The N-terminal PFK-2 domain is blue and the C-terminal FBPase-2 domain is yellow-green. The bound Mg^{2+}–AMPPNP, succinate, F6P, and P_i are shown in space-filling form colored according to atom type (AMPPNP C green, succinate C magenta, F6P C cyan, N blue, O red, Mg^{2+} light green, and P orange). The P_i, which occupies the binding site of the F2,6P's 2-phosphate group, is opposite the site that would be occupied by the side chain of the wild-type enzyme's His 256 (*magenta*), to which it would be transferred in the catalytic reaction. The succinate occupies the presumed F6P-binding pocket of the PFK-2 domain. [Based on an X-ray structure by Kosaku Uyeda and Charles Hasemann, University of Texas Southwestern Medical Center. PDBid 2BIF.]

as well as to phosphorylation/dephosphorylation by protein kinase A and a phosphoprotein phosphatase. Phosphorylation of the liver enzyme at its Ser 32 inhibits its PFK-2 activity and activates its FBPase-2 activity. Thus, the pancreatic α cell's release of glucagon in response to low blood [glucose] results, through an increase in liver [cAMP], in a decreased liver [F2,6P]. This situation, in turn, decreases the PFK-1 activity, thereby inhibiting glycolysis. Hence, the G6P resulting from the concurrent stimulation of glycogen degradation is converted to glucose and secreted as described above rather than being metabolized. Simultaneously, the deinhibition of FBPase (also called **FBPase-1** to distinguish it from FBPase-2) by the decrease of [F2,6P] stimulates **gluconeogenesis,** the formation of glucose from nonglucose precursors such as amino acids by a pathway that effectively reverses glycolytic flux (and in

which FBPase-1 is a key regulatory enzyme; Section 23-1). This process provides a second means of glucose production. Conversely, when the blood [glucose] is high, cAMP levels decrease, the liver PFK-2/FBPase-2 is dephosphorylated by phosphoprotein phosphatase-1 activating PFK-2, which, in turn, causes a rise in [F2,6P]. PFK-1 is therefore activated, FBPase-1 is inhibited, and the net glycolytic flux changes from gluconeogenesis to glycolysis.

The F2,6P control systems in skeletal muscle and in heart muscle function quite differently from that in liver due to the presence in these tissues of different PFK-2/FBPase-2 isozymes. In heart and skeletal muscles, increased glycogen breakdown is coordinated with increased glycolysis rather than increased glucose secretion. This is because phosphorylation of the heart muscle PFK-2/FBPase-2 isozyme occurs at entirely different sites (Ser 406 and Thr 475 of the 530-residue protein) from that of the liver isozyme (Ser 32 of the 470-residue protein) and activates rather than inhibits PFK-2. Consequently, hormones that stimulate glycogen breakdown also increase heart muscle [F2,6BP], thereby stimulating glycolysis as well. The skeletal muscle and testis isozymes lack phosphorylation sites altogether and are therefore not subject to cAMP-dependent phosphorylation control.

G. *Response to Stress*

Epinephrine and norepinephrine, which are often called the "fight or flight" hormones, are released into the bloodstream by the adrenal glands in response to stress. Epinephrine receptors (known as **β-adrenergic receptors;** Section 19-1F) present on the surfaces of liver and muscle cells respond to these hormones just as glucagon receptors respond to the presence of glucagon; they activate adenylate cyclase, thereby increasing intracellular [cAMP]. Indeed, epinephrine also stimulates the pancreatic α cells to release glucagon, which further increases liver [cAMP]. The G6P produced by the consequent glycogen breakdown in muscle enters the glycolytic pathway, thereby generating ATP and helping the muscles cope with the stress that triggered the epinephrine release.

The liver's response to stress, in addition to its response to the glucagon released by pancreatic epinephrine stimulation, involves response to epinephrine stimulation via two types of receptors, β-adrenergic receptors, as discussed above, and **α-adrenergic receptors.** α-Adrenergic receptors act by stimulating **phospholipase C** to release other second messengers, namely inositol-1,4,5-trisphosphate (IP_3), diacylglycerol (DAG), and Ca^{2+} (Fig. 18-27a), which act to reinforce the cells' response to cAMP. As we mentioned in Section 18-3Cf, phosphorylase kinase, which both activates phosphorylase and inactivates glycogen synthase, is fully active only when both phosphorylated and in the presence of increased [Ca^{2+}]. In addition, glycogen synthase is inactivated by phosphorylation by several other Ca^{2+}-dependent protein kinases, including protein kinase C (Section 18-3D). Protein kinase C requires both Ca^{2+} and DAG for activity (Section 19-4C). This dual stimulation of receptors in response to epinephrine causes the liver to produce

Phosphatidylinositol-4,5-bisphosphate (PIP₂)

H_2O

phospholipase C

1,2-Diacylglycerol (DAG)

Inositol-1,4,5-triphosphate (IP₃)

(a)

(b)

Figure 18-27   The liver's response to stress. *(a)* Stimulation of α-adrenergic receptors by epinephrine activates phospholipase C to hydrolyze phosphatidylinositol-4,5-bisphosphate (PIP₂) to inositol-1,4,5-trisphosphate (IP₃) and diacylglycerol (DAG). *(b)* The participation of two second messenger systems: the cAMP-mediated stimulation of glycogenolysis and inhibition of glycogen synthesis triggered by glucagon and β-adrenergic receptor activation; and the IP₃, DAG, and Ca²⁺-mediated stimulation of glycogenolysis and inhibition of glycogen synthesis triggered by α-adrenergic receptor activation. IP₃ stimulates the release of Ca²⁺ from the endoplasmic reticulum, whereas DAG, together with Ca²⁺, activates protein kinase C to phosphorylate and thereby inactivate glycogen synthase. G6Pase occupies the endoplasmic reticulum. Consequently, the cytosolically produced G6P is transported into the endoplasmic reticulum via the **T1 G6P translocase,** where it is hydrolyzed to glucose and Pᵢ. The Pᵢ and glucose are then returned to the cytosol by the **T2** and **T3 transporters,** respectively, and the glucose is exported from the cell via the GLUT2 glucose transporter.

G6P, which is hydrolyzed by G6Pase, resulting in the release of glucose into the bloodstream, thereby further fueling the muscles (Fig. 18-27*b*).

4 GLYCOGEN STORAGE DISEASES

With glycogen metabolism being such a finely controlled system, it is not surprising that genetically determined enzyme deficiencies result in disease states. The study of these disease states and the enzyme deficiencies that cause them has provided insights into the system's balance. In this sense, genetic diseases are valuable research tools. Conversely, the biochemical characterization of the pathways affected by a genetic disease often leads, as we shall see, to useful strategies for its treatment. Many diseases have been characterized that result from inherited deficiencies of one or another of the enzymes of glycogen metabolism. These defects are listed in Table 18-1 and discussed in this section.

Type I: Glucose-6-Phosphatase Deficiency (von Gierke's Disease)

G6Pase catalyzes the final step leading to the release of glucose into the bloodstream by the liver. Deficiency of this enzyme results in an increase of intracellular [G6P], which leads to a large accumulation of glycogen of normal structure in the liver and kidney (recall that G6P inhibits glycogen phosphorylase and activates glycogen synthase) and an inability to increase blood glucose concentration in response to glucagon or epinephrine. Similar difficulties occur when there are defects in the protein that transports glucose across the liver cell plasma membrane (Section 20-2Eb) or in any of the proteins that transport glucose, G6P, or P_i across the endoplasmic reticulum membrane (Section 18-3F; Fig. 18-27*b*). The symptoms of Type I glycogen storage disease include massive liver enlargement, severe **hypoglycemia** (low blood sugar) after a few

hours' fast, and a general failure to thrive. Treatment of the disease has included drug-induced inhibition of glucose uptake by the liver to increase blood [glucose], continuous intragastric feeding overnight, again to increase blood [glucose], oral administration of uncooked corn starch (which is only slowly broken down to glucose), and surgical transposition of the portal vein, which ordinarily feeds the liver directly from the intestines, so as to allow this glucose-rich blood to reach peripheral tissues before it reaches the liver. This latter treatment has the added benefit of allowing the tissues to receive more glucose while decreasing the storage of this glucose as liver glycogen. Liver transplantation has also been successful in the few patients in which this treatment has been tried.

A gene therapy protocol (Section 5-5H) is being developed to correct type I glycogen storage disease. G6Pase-deficient knockout mice (Section 5-5H) have been treated with a viral vector containing the mouse G6Pase gene. This treatment, which delivers G6Pase to the livers of these mice, greatly increases their survival rate and corrects the metabolic abnormalities associated with this glycogen storage disease.

Type II: α-1,4-Glucosidase Deficiency (Pompe's Disease)

This is the most devastating glycogen storage disease. It results in a large accumulation of glycogen of normal structure in the lysosomes of all cells and causes death by cardiorespiratory failure, usually before the age of 1 year. We have not discussed **α-1,4-glucosidase** in the sections on the pathways of glycogen synthesis and breakdown since it is not among those enzymes. It occurs in lysosomes, where it functions to hydrolyze the disaccharide maltose (Section 11-2B) and linear oligosaccharides, as well as the outer branches of glycogen, thereby yielding free glucose. However, this second pathway of glycogen metabolism is not quantitatively important. The reason that lysosomes normally take up and degrade glycogen granules is unknown.

Table 18-1 **Hereditary Glycogen Storage Diseases**

Type	Enzyme Deficiency	Tissue	Common Name	Glycogen Structure
I	Glucose-6-phosphatase	Liver	von Gierke's disease	Normal
II	α-1,4-Glucosidase	All lysosomes	Pompe's disease	Normal
III	Amylo-1,6-glucosidase (glycogen debranching enzyme)	All organs	Cori's disease	Outer chains missing or very short
IV	Amylo-(1,4 → 1,6)-transglycosylase (glycogen branching enzyme)	Liver, probably all organs	Andersen's disease	Very long unbranched chains
V	Glycogen phosphorylase	Muscle	McArdle's disease	Normal
VI	Glycogen phosphorylase	Liver	Hers' disease	Normal
VII	Phosphofructokinase	Muscle	Tarui's disease	Normal
VIII	Phosphorylase kinase	Liver	X-Linked phosphorylase kinase deficiency	Normal
IX	Phosphorylase kinase	All tissues		Normal
0	Glycogen synthase	Liver		Normal, deficient in quantity

Type III: Amylo-1,6-Glucosidase (Glycogen Debranching Enzyme) Deficiency (Cori's Disease)

In this disease, glycogen of abnormal structure containing very short outer chains accumulates in both liver and muscle since, in the absence of debranching enzyme, the glycogen cannot be further degraded. Its hypoglycemic symptoms are similar to, but not as severe as, those of von Gierke's disease (Type I). The low blood sugar, which in this case is a result of the decreased efficiency of glycogen breakdown, is treated with frequent feeding and a high-protein diet [in response to low blood sugar, the liver, through gluconeogenesis (Section 23-1), synthesizes glucose from amino acids]. For unknown reasons, the symptoms of Cori's disease often disappear at puberty.

Type IV: Amylo-(1,4 → 1,6)-Transglycosylase (Glycogen Branching Enzyme) Deficiency (Andersen's Disease)

This is one of the most severe glycogen storage diseases; victims rarely survive past the age of 5 years because of liver dysfunction. Glycogen concentration in liver is not increased but its structure is abnormal, with very long unbranched chains resulting from the lack of glycogen branching enzyme. This decreased branching greatly reduces the solubility of glycogen. It has been suggested that the liver dysfunction may be caused by a "foreign body" immune reaction to the abnormal glycogen.

Type V: Muscle Glycogen Phosphorylase Deficiency (McArdle's Disease)

We have mentioned this condition in connection with the realization that glycogen synthesis and breakdown must occur by different pathways (Section 18-2). Its major symptom, which is most severely manifested in early adulthood, is painful muscle cramps on exertion. This situation is a result of the inability of the glycogen breakdown system to provide sufficient fuel for glycolysis to keep up with the metabolic demand for ATP. Studies by ^{31}P NMR on human forearm muscle have noninvasively corroborated this conclusion by demonstrating that exercise in individuals with McArdle's disease leads to elevated muscle ADP levels compared to those of normal individuals (Fig. 18-28). Curiously, if McArdle's victims continue their exertions after a short rest, their cramps subside. This "second wind" effect has been attributed to the muscles' switch from anaerobic to aerobic metabolism as well as vasodilation, which gives the muscles increased access to the glucose and fatty acids in the blood for use as alternative fuels to glycogen. Liver glycogen phosphorylase is normal in these individuals, implying the presence of different glycogen phosphorylase isozymes in muscle and liver.

Type VI: Liver Glycogen Phosphorylase Deficiency (Hers' Disease)

Patients with a deficiency of liver phosphorylase have symptoms similar to those with mild forms of Type I glycogen storage disease. The hypoglycemia in this case results from the inability of glycogen phosphorylase to respond to the need for glucose production by the liver.

Figure 18-28 **The ADP concentration in human forearm muscles during rest and following exertion in normal individuals and those with McArdle's disease.** The ADP concentration was determined from ^{31}P NMR measurements on intact forearms. [After Radda, G.K., *Biochem. Soc. Trans.* **14,** 522 (1986).]

Type VII: Muscle Phosphofructokinase Deficiency (Tarui's Disease)

The result of a deficiency of the glycolytic enzyme PFK-1 in muscle is an abnormal buildup of the glycolytic metabolites G6P and F6P. High concentrations of G6P increase the activities of glycogen synthase (G6P activates glycogen synthase and inactivates glycogen phosphorylase) and UDP–glucose pyrophosphorylase (G6P is in equilibrium with G1P, a substrate for the enzyme) so that glycogen accumulates in muscle. Other symptoms are similar to those of Type V glycogen storage disease, muscle phosphorylase deficiency, since PFK deficiency prevents glycolysis from keeping up with the ATP demand of muscle contraction.

Type VIII: Liver Phosphorylase Kinase Deficiency (X-Linked Phosphorylase Kinase Deficiency)

Some individuals with symptoms of Type VI glycogen storage disease have liver phosphorylase of normal structure. However, they have a defective phosphorylase kinase, which results in their inability to convert phosphorylase *b* to phosphorylase *a*. The gene encoding the α subunit of phosphorylase kinase resides on the X chromosome and hence Type VIII disease is X-linked rather than autosomal, as are the other glycogen storage diseases.

Type IX: Phosphorylase Kinase Deficiency

Phosphorylase kinase deficiency, when it is autosomal recessive, is caused by a mutation in one of the genes encoding a β, γ, or δ subunit of phosphorylase kinase. Since different organs contain different isozymes of phosphorylase kinase, the symptoms and severity of Type IX disease vary with the affected organs.

Type 0: Liver Glycogen Synthase Deficiency

The activity of liver glycogen synthase is extremely low in individuals with this disease and hence they exhibit hyperglycemia after meals and hypoglycemia at other times. However, the primary lesion may not always be in the synthase itself because other metabolic defects may lead to an imbalance of the glycogen synthase cyclic cascade. The root cause of Type 0 glycogen storage disease is still under investigation. Muscle glycogen synthase deficiency disease has recently been described. Its victims are easily fatigued and develop heart abnormalities that may cause sudden cardiac arrest. Note these are the only known glycogen storage diseases in which there is a deficiency rather than an overabundance of glycogen. This suggests that the complete absence of glycogen is lethal.

Since many of the glycogen storage diseases have similar clinical symptoms, these diseases are best diagnosed via genetic testing.

CHAPTER SUMMARY

1 Glycogen Breakdown   In animals, when glucose is not needed as a source of metabolic energy, it is stored, predominantly in liver and muscle cells, as glycogen, an $\alpha(1 \rightarrow 4)$-linked glucan with $\alpha(1 \rightarrow 6)$ branches every 8 to 14 residues. Glycogen breakdown to glucose-6-phosphate (G6P) is a two-step process. Glycogen phosphorylase catalyzes the phosphorolysis of the glycosidic linkage of a terminal glucosyl residue to form glucose-1-phosphate (G1P). Phosphoglucomutase interconverts G1P and G6P. Glycogen debranching enzyme allows complete degradation of glycogen by catalyzing the transfer of three-residue chains onto the nonreducing ends of other chains and catalyzing the hydrolysis of the remaining $\alpha(1 \rightarrow 6)$-linked glucosyl unit to glucose.

2 Glycogen Synthesis   Glycogen is synthesized from G6P by a pathway different from that of glycogen breakdown. G6P is converted to G1P under the influence of phosphoglucomutase. UDP–glucose pyrophosphorylase utilizes UTP to convert G1P to UDP–glucose, the activated intermediate in glycogen synthesis. The hydrolysis of the PP_i product by inorganic pyrophosphatase drives the reaction to completion. Glucosyl units are transferred from UDP–glucose to the C4—OH group of a terminal residue on a growing glycogen chain by glycogen synthase. The chain is initiated on glycogenin. Branching occurs through the action of glycogen branching enzyme, which transfers ~7-residue segments of $\alpha(1 \rightarrow 4)$-linked chains to the C6—OH group of a glucosyl residue on the same or another glycogen chain.

3 Control of Glycogen Metabolism   The rates at which glycogen is synthesized by glycogen synthase and degraded by glycogen phosphorylase are controlled by the levels of their allosteric effectors such as ATP, AMP, G6P, and glucose. Superimposed on this allosteric control is control by the phosphorylation/dephosphorylation of these enzymes. The kinases and phosphatases that catalyze these modifications are part of amplifying cascades that are ultimately controlled by the hormones glucagon, insulin, and epinephrine, and by Ca^{2+}. Glucagon and epinephrine stimulate glycogen breakdown by stimulating adenylate cyclase to increase the intracellular [cAMP]. cAMP is a "second messenger" that activates protein kinase A (PKA), which, through its activation of phosphorylase kinase, results in the phosphorylation of both glycogen phosphorylase and glycogen synthase. Phosphorylation activates glycogen phosphorylase but inactivates glycogen synthase. In addition, epinephrine causes an increase in the concentrations of other second messengers, namely, inositol-1,4,5-trisphosphate (IP_3), diacylglycerol (DAG), and Ca^{2+}, that reinforce the cAMP-dependent responses. Ca^{2+}, which is also released into muscle cytosol by nerve impulses, binds to calmodulin so as to induce this protein to activate protein kinases via an intrasteric mechanism by which Ca^{2+}–CaM extracts autoinhibitory sequences from the kinases' active sites. A decrease in [cAMP] and/or the presence of insulin leads to the activation of phosphoprotein phosphatase-1 to dephosphorylate glycogen phosphorylase and glycogen synthase.

The phosphoproteins that participate in glycogen metabolism are dephosphorylated through the action of phosphoprotein phosphatase-1, which is only active when it is associated with a glycogen particle through the intermediacy of glycogen-binding G_M or G_L subunits. When [glucose] is high, the liver synthesizes glucose-6-phosphate (G6P) and ultimately glycogen from glucose via the action of glucokinase (GK), which has kinetic properties distinct from those of other hexokinases. When [glucose] is low, GK is inhibited by glucokinase regulatory protein (GKRP) and G6Pase hydrolyzes the G6P product of glycogen breakdown (which is favored at low [glucose]) for export to other tissues. The concentration in liver of F2,6P, an activator of PFK and an inhibitor of FBPase, is also dependent on the rates of cAMP-dependent phosphorylation and dephosphorylation. It is both synthesized and degraded through the action of PFK-2/FBPase-2, whose enzymatic activities are oppositely controlled by both allosteric regulation and phosphorylation/ dephosphorylation.

4 Glycogen Storage Diseases   Glycogen storage diseases are caused by a genetic deficiency of one or another of the enzymes of glycogen metabolism. Ten different deficiencies of varying severity have been reported in humans.

REFERENCES

General

Boyer, P.D. and Krebs, E.G. (Eds.), *The Enzymes* (3rd ed.), Vol. 17, Academic Press (1986). [Contains detailed articles on the enzymes of glycogen metabolism and their control.]

Roach, P.J., Glycogen and its metabolism, *Curr. Mol. Med.* **2,** 101–120 (2002).

Glycogen Metabolism

Browner, M.F. and Fletterick, R.J., Phosphorylase: A biological transducer, *Trends Biochem. Sci.* **17,** 66–71 (1992).

Buchbinder, J.L., Rath, V.L., and Fletterick, R.J., Structural relationships among regulated and unregulated phosphorylases, *Annu. Rev. Biophys. Biomol. Struct.* **30,** 191–209 (2001).

Dai, J.-B., Liu, Y., Ray, W.J., Jr., and Konno, M., The crystal structure of muscle phosphoglucomutase refined at 2.7-angstrom resolution, *J. Biol. Chem.* **267,** 6322–6337 (1992).

Gibbons, B.J., Roach, P.J., and Hurley, T.D., Crystal structure of the autocatalytic initiator of glycogen biosynthesis, glycogenin, *J. Mol. Biol.* **319,** 463–477 (2002); *and* Hurley, T.D., Stout, S., Miner, E., Zhou, J., and Roach, P.J., Requirements for catalysis in mammalian glycogenin, *J. Biol. Chem.* **280,** 23892–23899 (2005).

Johnson, L.N., Glycogen phosphorylase: Control by phosphorylation and allosteric effectors, *FASEB J.* **6,** 2274–2282 (1992); *and* Rabbit muscle glycogen phosphorylase *b*. The structural basis of activation and catalysis, *in* Harding, J.J. and Crabbe, M.J.C. (Eds.), *Post-Translational Modifications of Proteins, pp.* 81–151, CRC Press (1993).

Johnson, L.N. and Barford, D., Glycogen phosphorylase, *J. Biol. Chem.* **265,** 2409–2412 (1990).

Madsen, N.B., Glycogen phosphorylase and glycogen synthetase, *in* Kuby, S.A. (Ed.), *A Study of Enzymes,* Vol. II, *pp.* 139–158, CRC Press (1991).

Meléndez-Hevia, E., Waddell, T.G., and Shelton, E.D., Optimization of molecular design in the evolution of metabolism: The glycogen molecule, *Biochem. J.* **295,** 477–483 (1993).

Palm, D., Klein, H.W., Schinzel, R.S., Bucher, M., and Helmreich, E.J.M., The role of pyridoxal 5′-phosphate in glycogen phosphorylase catalysis, *Biochemistry* **29,** 1099–1107 (1990).

Smythe, C. and Cohen, P., The discovery of glycogenin and the priming mechanism for glycogen biosynthesis, *Eur. J. Biochem.* **200,** 625–631 (1991).

Sprang, S.R., Acharya, K.R., Goldsmith, E.J., Stuart, D.I., Varvill, K., Fletterick, R.J., Madsen, N.B., and Johnson, L.N., Structural changes in glycogen phosphorylase induced by phosphorylation, *Nature* **336,** 215–221 (1988).

Sprang, S.R., Withers, S.G., Goldsmith, E.J., Fletterick, R.J., and Madsen, N.B., Structural basis for the activation of glycogen phosphorylase *b* by adenosine monophosphate. *Science* **254,** 1367–1371 (1991).

Calmodulin and Its Control of Glycogen Metabolism

Babu, Y.S., Sack, J.S., Greenough, T.J., Bugg, C.E., Means, A.R., and Cook, W.J., Three-dimensional structure of calmodulin, *Nature* **315,** 37–40 (1985).

Crivici, A. and Ikura, M., Molecular and structural basis of target recognition by calmodulin, *Annu. Rev. Biophys. Biomol. Struct.* **25,** 85–116 (1995).

Ikura, M., Clore, G.M., Gronenborn, A.M., Zhu, G., and Bax, A., Solution structure of a calmodulin-target peptide complex by multidimensional NMR, *Science* **256,** 632–638 (1992); *and* Meador, W.E., Means, A.R., and Quiocho, F.A., Target enzyme recognition by calmodulin: 2.4 Å structure of a calmodulin-peptide complex, *Science* **257,** 1251–1255 (1992).

James, P., Vorherr, T., and Carafoli, E., Calmodulin-binding domains: Just two faced or multifaceted? *Trends Biochem. Sci.* **20,** 38–42 (1995).

Nakayama, S. and Kretsinger, R.H., Evolution of the EF-hand family of proteins, *Annu. Rev. Biophys. Biomol. Struct.* **23,** 473–507 (1994).

Protein Kinases and Protein Phosphatases

Bollen, M., Keppens, S., and Stalmans, W., Specific features of glycogen metabolism in the liver, *Biochem. J.* **336,** 19–31 (1998).

Bossemeyer, D., Engh, R.A., Kinzel, V., Ponstingl, H., and Huber, R., Phosphotransferase and substrate binding mechanism of the cAMP-dependent protein kinase catalytic subunit from porcine heart as deduced from the 2.0 Å structure of the complex with Mn^{2+} adenyl imidodiphosphate and inhibitor peptide PKI (5–24), *EMBO J.* **12,** 849–859 (1993).

Egloff, M.P., Johnson, D.F., Moorhead, G., Cohen, P.T.W., Cohen, P., and Barford, D., Structural basis for the recognition of regulatory subunits by the catalytic subunit of protein phosphatase 1, *EMBO J.* **16,** 1876–1887 (1997).

Goldberg, J., Huang, H., Kwon, Y., Greengard, P., Nairn, A.C., and Kuriyan, J., Three-dimensional structure of the catalytic subunit of protein serine/threonine phosphatase-1, *Nature* **376,** 745–753 (1995).

Johnson, L.N., Lowe, E.D., Noble, M.E.M., and Owen, D.J., The structural basis for substrate recognition and control by protein kinases, *FEBS Lett.* **430,** 1–11 (1998).

Kim, C., Cheng, C.Y., Saldanha, S.A., and Taylor, S.S., PKA-I holoenzyme structure reveals a mechanism for cAMP-dependent activation, *Cell* **130,** 1032–1043 (2007).

Kobe, B. and Kemp, B.E., Active site-directed protein regulation, *Nature* **402,** 373–376 (1999). [Discusses intrasteric regulation.]

Lowe, E.D., Noble, M.E.M., Skamnaki, V.T., Oikonomakos, N.G., Owen, D.J., and Johnson, L.N., The crystal structure of a phosphorylase kinase peptide substrate complex: Kinase substrate recognition, *EMBO J.* **16,** 6646–6658 (1997).

Manning, G., Whyte, D.B., Martinez, R., Hunter, T., and Sundarsanum, S., The protein kinase complement of the human genome, *Science* **298,** 1912–1934 (2002).

Nordlie, R.C., Foster, J.D., and Lange, A.J., Regulation of glucose production by the liver, *Annu. Rev. Nutr.* **19,** 379–406 (1999).

Smith, C.M., Radzio-Andzelm, E., Akamine, M.P., Madhusudan, and Taylor, S.S., The catalytic subunit of cAMP-dependent protein kinase: Prototype for an extended network of communication, *Prog. Biophys. Mol. Biol.* **71,** 313–341 (1999).

Su, Y., Dostmann, W.R.G., Herberg, F.W., Durick, K., Xuong, N., Ten Eyck, L., Taylor, S.S., and Varughese, K.I., Regulatory subunit of protein kinase A: Structure of deletion mutant with cAMP binding domains, *Science* **269,** 807–813 (1995).

Taylor, S.S., Knighton, D.R., Zheng, J., Sowadski, J.M., Gibbs, C.S., and Zoller, M.J., A template for the protein kinase family, *Trends Biochem. Sci.* **18,** 84–89 (1993); *and* Taylor, S.S., Knighton, D.R., Zheng, J., Ten Eyck, L.F., and Sowadski, J.M., Structural framework for the protein kinase family, *Annu. Rev. Cell Biol.* **8,** 429–462 (1992).

Villafranca, J.E., Kissinger, C.R., and Parge, H.E., Protein serine/threonine phosphatases, *Curr. Opin. Biotech.* **7,** 397–402 (1996).

Glucose-6-Phosphatase, Glucokinase, and PFK-2/FBPase-2

Cornish-Bowden, A. and Cárdenas, M.L., Hexokinase and "glucokinase" in liver metabolism, *Trends Biochem. Sci.* **16,** 281–282 (1991).

de la Iglesia, N., Mukhtar, M., Seoane, J., Guinovart, J.J., and Agius, L., The role of the regulatory protein of glucokinase in the glucose sensory mechanism of the hepatocyte, *J. Biol. Chem.* **275,** 10597–10603 (2000).

Iynedjian, P.B., Mammalian glucokinase and its gene, *Biochem. J.* **293,** 1–13 (1993). [Reviews the function and control of glucokinase.]

Okar, D.A., Manzano, À., Navarro-Sabatè, A., Riera, L., Bartrons, R., and Lange, A.J., PFK-2/FBPase-2: maker and breaker of the essential biofactor fructose-2,6-bisphosphate, *Trends Biochem. Sci.* **26,** 30–35 (2001).

Pilkis, S.J., 6-Phosphofructo-2-kinase/fructose-2,6-bisphosphatase: a metabolic signaling enzyme, *Annu. Rev. Biochem.* **64,** 799–835 (1995).

Rousseau, G.G. and Hue, L., Mammalian 6-phosphofructo-2-kinase/fructose-2,6-bisphosphatase: A bifunctional enzyme that controls glycolysis, *Prog. Nucleic Acid Res. Mol. Biol.* **45**, 99–127 (1993).

Van Schaftingen, E. and Gerin, I., The glucose-6-phosphatase system, *Biochem. J.* **362**, 513–532 (2002).

Van Schaftingen, E., Vandercammen, A., Detheux, M., and Davies, D.R., The regulatory protein of liver glucokinase, *Adv. Enzyme Reg.* **32**, 133–148 (1992).

Yuan, M.H., Mizuguchi, H., Lee, Y.-H., Cook, P.F., Uyeda, K., and Hasemann, C.A., Crystal structure of the H256A mutant of rat testis fructose-6-phosphate-2-kinase/fructose-2,6-bisphosphatase, *J. Biol. Chem.* **274**, 2176–2184 (1999); *and* Haseman, C.A., Istvan, E.S., Uyeda, K., and Deisenhofer, J., The crystal structure of the bifunctional enzyme 6-phosphofructo-2-kinase/fructose-2,6-bisphosphatase reveals distinct homologies, *Structure* **4**, 1017–1029 (1996).

Glycogen Storage Diseases

Bartram, C., Edwards, R.H.T., and Beynon, R.J., McArdle's disease–muscle glycogen phosphorylase deficiency, *Biochim. Biophys. Acta* **1272**, 1–13 (1995). [A review.]

Chen, Y.-T., Glycogen storage diseases, *in* Valle, D. (Ed.), *The Online Metabolic and Molecular Bases of Inherited Disease*, Chapter 71, http://www.ommbid.com/. [Begins with a review of glycogen metabolism.]

Kollberg, G., Tulinius, M., Gilljam, T., Östman-Smith, I., Forsander, G., Jotorp, P., Oldfors, A., and Holme, E., Cardiomyopathy and exercise intolerance in muscle glycogen storage disease 0, *New Engl. J. Med.* **357**, 1507–1514 (2007).

Online Mendelian Inheritance in Man (OMIM), http://www.ncbi. nlm.nih.gov/sites/entrez?db=OMIM [A comprehensive compendium of human genes and genetic phenotypes containing information on all known Mendelian disorders, including the glycogen storage diseases, and involving 12,000 genes.]

Radda, G.K., Control of bioenergetics: From cells to man by phosphorus nuclear-magnetic-resonance spectroscopy, *Biochem. Soc. Trans.* **14**, 517–525 (1986). [Discusses the noninvasive diagnosis of McArdle's disease by ^{31}P NMR.]

Zingone, A., Hiraiwa, H., Pan, C.-J., Lin, B., Chen, H., Ward, J.M., and Chou, J.Y., Correction of glycogen storage disease type 1a in a mouse model by gene therapy, *J. Biol. Chem.* **275**, 828–832 (2000).

PROBLEMS

1. A glycogen molecule consisting of 100,000 glucose residues is branched, on average, every 10 residues with one branch per tier. (a) How many reducing ends does it have? (b) How many tiers of branches does it have, on average?

2. A mature glycogen particle typically has 12 tiers of branches with 2 branches per tier and 13 residues per branch. How many glucose residues are in such a particle?

3. Calculations based on the volume of a glucose residue and the branching pattern of cellular glycogen indicate that a glycogen molecule could have up to 28 branching tiers before becoming impossibly dense. What are the advantages of such a molecule and why is it not found *in vivo*?

4. The free energy of hydrolysis of an $\alpha(1 \rightarrow 4)$ glycosidic bond is -15.5 kJ · mol^{-1}, whereas that of an $\alpha(1 \rightarrow 6)$ glycosidic bond is -7.1 kJ · mol^{-1}. Use these data to explain why glycogen debranching includes three reactions [breaking and reforming $\alpha(1 \rightarrow 4)$ bonds and hydrolyzing $\alpha(1 \rightarrow 6)$ bonds], while glycogen branching requires only two reactions [breaking $\alpha(1 \rightarrow 4)$ bonds and forming $\alpha(1 \rightarrow 6)$ bonds].

5. The complete metabolic oxidation of glucose to CO_2 and O_2 yields 32 ATPs (Section 22-2Bc). What is the fractional energetic cost of storing glucose as glycogen and later metabolizing the glycogen rather than directly metabolizing the glucose? (Recall that glycogen's branched structure results in its degradation to 92% G1P and 8% glucose.)

6. What are the effects of the following on the rates of glycogen synthesis and glycogen degradation: (a) increasing the Ca^{2+} concentration, (b) increasing the ATP concentration, (c) inhibiting adenylate cyclase, (d) increasing the epinephrine concentration, and (e) increasing the AMP concentration?

7. Show that hexokinase activity but not glucokinase activity is insensitive to blood [glucose] over the physiological range. Calculate the ratio of glucokinase to hexokinase activities when [glucose] is 2 mM (hypoglycemic), 5 mM (normal), and 25 mM (diabetic). Assume that $K_M = 0.1$ mM for hexokinase and that both enzymes have the same V_{max}.

8. Compare the properties of a bicyclic cascade with those of a monocyclic cascade.

9. The V_{max} of muscle glycogen phosphorylase is much larger than that of liver. Discuss the functional significance of this phenomenon.

10. How does epinephine act on muscles to prepare them for "fight or flight"?

*11. A complication of glycogen metabolism that we have not discussed is that many protein kinases, including phosphorylase kinase, are autophosphorylating; that is, they can specifically phosphorylate and thereby activate themselves. Discuss how this phenomenon affects glycogen metabolism, taking into consideration the possibilities that phosphorylase kinase autophosphorylation may be an intramolecular or an intermolecular process.

12. Explain the symptoms of von Gierke's disease.

13. A sample of glycogen from a patient with liver disease is incubated with P_i, normal glycogen phosphorylase, and normal debranching enzyme. The ratio of glucose-1-phosphate to glucose formed in this reaction mixture is 100. What is the patient's most likely enzymatic deficiency? What is the probable structure of the patient's glycogen?

Signal Transduction

they control. For example, metabolic pathways, as we have seen, are regulated by the feedback control of allosteric enzymes by metabolites in those pathways or by the covalent modification of these enzymes. In this chapter we consider chemical signaling and how these signals are mediated. We begin by discussing the functions of the major human hormone systems. We then discuss the three major pathways whereby intercellular signals are transduced (converted) to intracellular signals, namely, those that utilize (1) heterotrimeric G proteins, (2) receptor tyrosine kinases, and (3) phosphoinositide cascades. Neurotransmission is discussed in Section 20-5.

1 HORMONES

Hormones are classified according to the distance over which they act (Fig. 19-1):

1. Endocrine hormones act on cells distant from the site of their release. Endocrine hormones, for example, insulin and epinephrine, are synthesized and released in the bloodstream by specialized ductless **endocrine glands.**

2. Paracrine hormones (alternatively, **local mediators**) act only on cells close to the cell that released them. For example, an essential element of the immune response is initiated when a white blood cell known as a **macrophage** that has encountered a specific antigen binds a so-called **T cell** specific for that antigen. The macrophage thereupon releases a **protein growth factor** named **interleukin-1 (IL-1)** that stimulates the bound T cell to proliferate and differentiate (Section 35-2Aa).

3. Autocrine hormones act on the same cell that released them. For example, the response of a T cell to interleukin-1 is enhanced by the T cell's autostimulatory release of the protein growth factor **interleukin-2 (IL-2).**

Living things coordinate their activities at every level of their organization through complex chemical signaling systems. Intercellular signals occur through the mediation of chemical messengers known as **hormones** and, in higher animals, via neuronally transmitted electrochemical impulses. Intracellular communications are maintained by the synthesis or alteration of a great variety of different substances that are often integral components of the processes

We are already familiar with certain aspects of hormonal control. For instance, we have considered how epinephrine, insulin, and glucagon regulate energy metabolism through the intermediacy of cAMP (Sections 18-3E and 18-3G). In this section we extend and systematize this information. Before we do so, you should note that biochemical communications are not limited to intracellular and intercellular signals. Many organisms release substances called **pheromones** that alter the behavior of other organisms of the same species in much the same way as

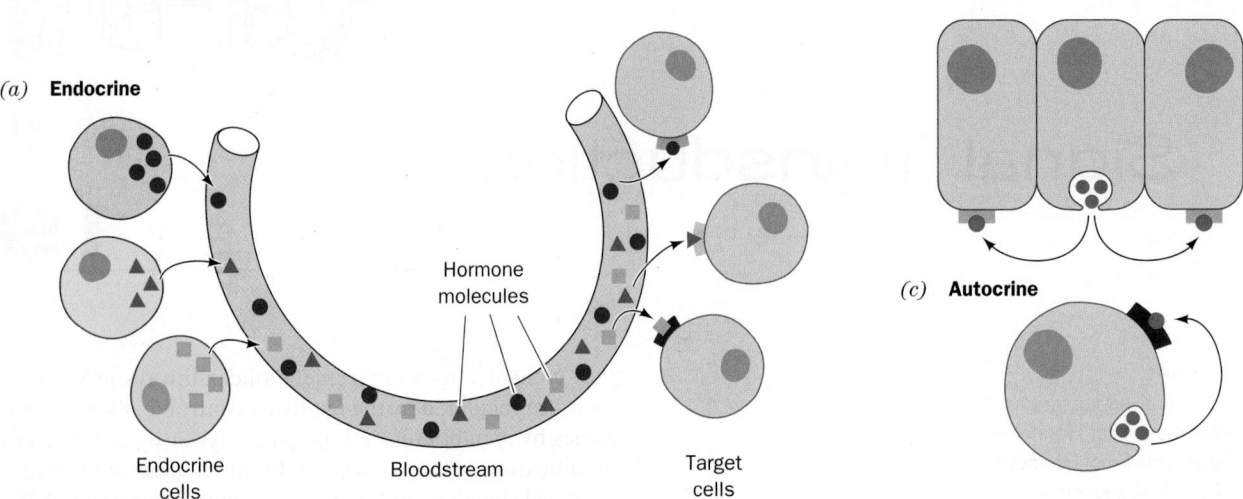

(a) **Endocrine**

(b) **Paracrine**

(c) **Autocrine**

Hormone molecules

Endocrine cells

Bloodstream

Target cells

Figure 19-1 **Classification of hormones.** Hormonal communications are classified according to the distance over which the signal acts: (*a*) endocrine signals are directed at distant cells through the intermediacy of the bloodstream, (*b*) paracrine signals are directed at nearby cells, and (*c*) autocrine signals are directed at the cell that produced them.

hormones. Pheromones are commonly sexual attractants but some have other functions in species, such as ants, that have complex social interactions.

The human endocrine system (Fig. 19-2) secretes a wide variety of hormones (Table 19-1) that enable the body to:

1. Maintain homeostasis (e.g., insulin and glucagon maintain the blood glucose level within rigid limits during feast or famine).

2. Respond to a wide variety of external stimuli (such as the preparation for "fight or flight" engendered by epinephrine and norepinephrine).

3. Follow various cyclic and developmental programs (for instance, sex hormones regulate sexual differentiation, maturation, the menstrual cycle, and pregnancy; Sections 19-1Gb and 19-1I).

Most hormones are either polypeptides, amino acid derivatives, or steroids, although there are important exceptions to this generalization. In any case, *only those cells with a specific receptor for a given hormone will respond to its presence even though nearly all cells in the body may be exposed to the hormone.* Hormonal messages are therefore quite specifically addressed.

In this section, we outline the hormonal functions of the various endocrine glands. Throughout this discussion keep in mind that these glands are not just a collection of independent secretory organs but form a complex and highly interdependent control system. Indeed, as we shall see, the secretion of many hormones is under feedback control through the secretion of other hormones to which the original hormone-secreting gland responds. Much of our

Figure 19-2 **Major glands of the human endocrine system.** Other tissues, the intestines, for example, also secrete endocrine hormones.

Hypothalamus
Pituitary
Parathyroid
Thyroid
Adrenal glands
Pancreas
Kidney
Ovaries
Testes

Table 19-1 **Some Human Hormones**

Hormone	Origin	Major Effects
Polypeptides		
Corticotropin-releasing factor (CRF)	Hypothalamus	Stimulates ACTH release
Gonadotropin-releasing factor (GnRF)	Hypothalamus	Stimulates FSH and LH release
Thyrotropin-releasing factor (TRF)	Hypothalamus	Stimulates TSH release
Growth hormone-releasing factor (GRF)	Hypothalamus	Stimulates growth hormone release
Somatostatin	Hypothalamus	Inhibits growth hormone release
Adrenocorticotropic hormone (ACTH)	Adenohypophysis	Stimulates the release of adrenocorticosteroids
Follicle-stimulating hormone (FSH)	Adenohypophysis	In ovaries, stimulates follicular development, ovulation, and estrogen synthesis; in testes, stimulates spermatogenesis
Luteinizing hormone (LH)	Adenohypophysis	In ovaries, stimulates oocyte maturation and follicular synthesis of estrogens and progesterone; in testes, stimulates androgen synthesis
Chorionic gonadotropin (CG)	Placenta	Stimulates progesterone release from the corpus luteum
Thyrotropin (TSH)	Adenohypophysis	Stimulates T_3 and T_4 release
Somatotropin (growth hormone)	Adenohypophysis	Stimulates growth and synthesis of somatomedins
Met-enkephalin	Adenohypophysis	Opioid effects on central nervous system
Leu-enkephalin	Adenohypophysis	Opioid effects on central nervous system
β-Endorphin	Adenohypophysis	Opioid effects on central nervous system
Vasopressin	Neurohypophysis	Stimulates water resorption by kidney and increases blood pressure
Oxytocin	Neurohypophysis	Stimulates uterine contractions
Glucagon	Pancreas	Stimulates glucose release through glycogenolysis and stimulates lipolysis
Insulin	Pancreas	Stimulates glucose uptake through gluconeogenesis, protein synthesis, and lipogenesis
Gastrin	Stomach	Stimulates gastric acid and pepsinogen secretion
Secretin	Intestine	Stimulates pancreatic secretion of HCO_3^-
Cholecystokinin (CCK)	Intestine	Stimulates gallbladder emptying and pancreatic secretion of digestive enzymes and HCO_3^-
Gastric inhibitory peptide (GIP)	Intestine	Inhibits gastric acid secretion and gastric emptying; stimulates pancreatic insulin release
Parathyroid hormone	Parathyroid	Stimulates Ca^{2+} uptake from bone, kidney, and intestine
Calcitonin	Thyroid	Inhibits Ca^{2+} uptake from bone and kidney
Somatomedins	Liver	Stimulates cartilage growth; have insulinlike activity
Steroids		
Glucocorticoids	Adrenal cortex	Affect metabolism in diverse ways, decrease inflammation, increase resistance to stress
Mineralocorticoids	Adrenal cortex	Maintain salt and water balance
Estrogens	Gonads	Maturation and function of secondary sex organs, particularly in females
Androgens	Gonads	Maturation and function of secondary sex organs, particularly in males; male sexual differentiation
Progestins	Ovaries and placenta	Mediate menstrual cycle and maintain pregnancy
Vitamin D	Diet and sun	Stimulates Ca^{2+} absorption from intestine, kidney, and bone
Amino Acid Derivatives		
Epinephrine	Adrenal medulla	Stimulates contraction of some smooth muscles and relaxes others, increases heart rate and blood pressure, stimulates glycogenolysis in liver and muscle, stimulates lipolysis in adipose tissue
Norepinephrine	Adrenal medulla	Stimulates arteriole contraction, decreases peripheral circulation, stimulates lipolysis in adipose tissue
Triiodothyronine (T_3)	Thyroid	General metabolic stimulation
Thyroxine (T_4)	Thyroid	General metabolic stimulation

understanding of hormonal function has come from careful measurements of hormone concentrations, the effects of changes of these concentrations on physiological functions, and measurements of the affinities with which hormones bind to their receptors. We begin, therefore, with a consideration of how physiological hormone concentrations are measured and how receptor–ligand interactions are quantified.

A. Quantitative Measurements

a. Radioimmunoassays

The serum concentrations of hormones are extremely small, generally between 10^{-12} and 10^{-7} M, so they usually must be measured by indirect means. Biological assays were originally employed for this purpose but they are generally slow, cumbersome, and imprecise. Such assays have therefore been largely supplanted by **radioimmunoassays.** In this technique, which was developed by Rosalyn Yalow, the unknown concentration of a hormone, H, is determined by measuring how much of a known amount of the radioactively labeled hormone, H*, binds to a fixed quantity of anti-H antibody in the presence of H. This competition reaction is easily calibrated by constructing a standard curve indicating how much H* binds to the antibody as a function of [H]. The high ligand affinity and specificity that antibodies possess gives radioimmunoassays the advantages of great sensitivity and specificity.

b. Receptor Binding

Receptors, as do other proteins, bind their corresponding ligands according to the laws of mass action:

$$R + L \rightleftharpoons R \cdot L$$

Here R and L represent receptor and ligand, and the reaction's dissociation constant is expressed:

$$K_L = \frac{[R][L]}{[R \cdot L]} = \frac{([R]_T - [R \cdot L])[L]}{[R \cdot L]} \qquad [19.1]$$

where the total receptor concentration, $[R]_T = [R] + [R \cdot L]$. Equation [19.1] may be rearranged to a form analogous to the Michaelis–Menten equation of enzyme kinetics (Section 14-2A):

$$Y = \frac{[R \cdot L]}{[R]_T} = \frac{[L]}{K_L + [L]} \qquad [19.2]$$

where Y is the fractional occupation of the ligand-binding sites. Equation [19.2] represents a hyperbolic curve (Fig. 19-3a) in which K_L may be operationally defined as the ligand concentration at which the receptor is half-maximally occupied by ligand.

Although K_L and $[R]_T$ may, in principle, be determined from an analysis of a hyperbolic plot such as Fig. 19-3a, the analysis of a linear form of the equation is a simpler procedure. Equation [19.1] may be rearranged to:

$$\frac{[R \cdot L]}{[L]} = \frac{([R]_T - [R \cdot L])}{K_L} \qquad [19.3]$$

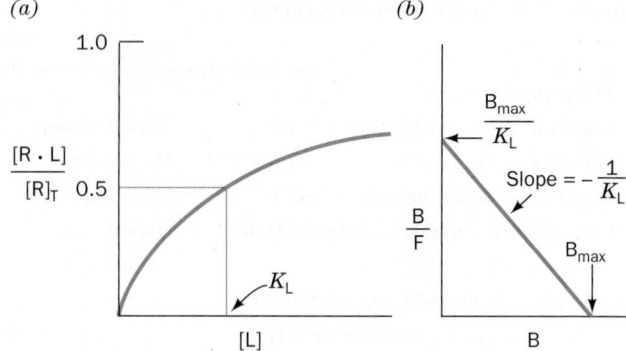

Figure 19-3   **Binding of ligand to receptor.** (*a*) A hyperbolic plot. (*b*) A Scatchard plot. Here $B \equiv [R \cdot L], F \equiv [L]$, and $B_{max} \equiv [R]_T$.

Now, in keeping with customary receptor-binding nomenclature, let us redefine $[R \cdot L]$ as B (for bound ligand), [L] as F (for free ligand), and $[R]_T$ as B_{max}. Then Eq. [19.3] becomes:

$$\frac{B}{F} = \frac{(B_{max} - B)}{K_L} = -\frac{1}{K_L} \cdot B + \frac{B_{max}}{K_L} \qquad [19.4]$$

A plot of B/F versus B, which is known as a **Scatchard plot** (after George Scatchard, its originator), therefore yields a straight line of slope $-1/K_L$ whose intercept on the B axis is B_{max} (Fig. 19-3b). Here, both B and F may be determined by filter-binding assays as follows. Most receptors are insoluble membrane-bound proteins and may therefore be separated from soluble free ligand by filtration (receptors that have been solubilized may be separated from free ligand by filtration, for example, through nitrocellulose since proteins nonspecifically bind to nitrocellulose). Hence, through the use of radioactively labeled ligand, the values of B and F ([R · L] and [L]) may be determined, respectively, from the radioactivity on the filter and that remaining in solution. The rate of R · L dissociation is generally so slow (half-times of minutes to hours) as to cause insignificant errors when the filter is washed to remove residual free ligand.

c. Competitive-Binding Studies

Once the receptor-binding parameters for one ligand have been determined, the dissociation constant of other ligands for the same ligand-binding site may be determined through competitive-binding studies. The model describing this competitive binding is analogous to the competitive inhibition of a Michaelis–Menten enzyme (Section 14-3A):

$$
\begin{array}{c}
R + L \xrightleftharpoons{K_L} R \cdot L \\
+ \\
I \\
K_I \updownarrow \\
R \cdot I + L \longrightarrow \text{No binding}
\end{array}
$$

where I is the competing ligand whose dissociation constant with the receptor is expressed:

$$K_I = \frac{[R][I]}{[R \cdot I]}$$ [19.5]

Thus, in direct analogy with the derivation of the equation describing competitive inhibition:

$$[R \cdot L] = \frac{[R]_T[L]}{K_L\left(1 + \dfrac{[I]}{K_I}\right) + [L]}$$ [19.6]

The relative affinities of a ligand and an inhibitor may therefore be determined by dividing Eq. [19.6] in the presence of inhibitor with that in the absence of inhibitor:

$$\frac{[R \cdot L]_I}{[R \cdot L]_0} = \frac{K_L + [L]}{K_L\left(1 + \dfrac{[I]}{K_I}\right) + [L]}$$ [19.7]

When this ratio is 0.5 (50% inhibition), the competitor concentration is referred to as $[I_{50}]$ in analogy with the $[IC_{50}]$ of drugs that inhibit enzymes (Section 15-4Aa). Thus, solving Eq. [19.7] for K_I at 50% inhibition:

$$K_I = \frac{[I_{50}]}{1 + \dfrac{[L]}{K_L}}$$ [19.8]

B. *Pancreatic Islet Hormones*

The pancreas is a large glandular organ, the bulk of which is an **exocrine gland** dedicated to producing digestive enzymes such as trypsin, RNase A, α-amylase, and phospholipase A_2 that it secretes via the pancreatic duct into the small intestine. However, ~1 to 2% of pancreatic tissue consists of scattered clumps of cells known as **islets of Langerhans,** which comprise an endocrine gland that functions to maintain energy metabolite homeostasis. Pancreatic islets contain three types of cells, each of which secretes a characteristic polypeptide hormone:

1. The α cells secrete glucagon (29 residues; Section 18-3Ea).

2. The β cells secrete insulin (51 residues; Fig. 9-4).

3. The δ cells secrete **somatostatin** (14 residues).

Insulin, which is secreted in response to high blood glucose levels, primarily functions to stimulate muscle, liver, and adipose cells to store glucose for later use by synthesizing glycogen, protein, and fat (Section 27-2). Glucagon, which is secreted in response to low blood glucose, has essentially the opposite effects: It stimulates liver to release glucose through glycogenolysis (Section 18-3E) and gluconeogenesis (Section 23-1) and it stimulates adipose tissue to release fatty acids through lipolysis. Somatostatin, which is also secreted by the hypothalamus (Section 19-1H), inhibits the release of insulin and glucagon from their islet cells and is

therefore thought to have a paracrine function in the pancreas.

Polypeptide hormones, as are other proteins destined for secretion, are ribosomally synthesized as preprohormones, processed in the rough endoplasmic reticulum and Golgi apparatus to form the mature hormone, and then packaged in secretory granules to await the signal for their release by exocytosis (Sections 12-4B, 12-4C, and 12-4D). The most potent physiological stimuli for the release of insulin and glucagon are, respectively, high and low blood glucose concentrations, so that islet cells act as the body's primary glucose sensors. However, the release of these hormones is also influenced by the autonomic (involuntary) nervous system and by hormones secreted by the gastrointestinal tract (Section 19-1C).

C. *Gastrointestinal Hormones*

The digestion and absorption of nutrients are complicated processes that are regulated by the autonomic nervous system in concert with a complex system of polypeptide hormones. Indeed, gastrointestinal peptide hormones are secreted into the bloodstream by a system of specialized cells lining the gastrointestinal tract whose aggregate mass is greater than that of the rest of the endocrine system. Over 20 gastrointestinal hormones have been described. Four of the better characterized gastrointestinal hormones are:

1. Gastrin (17 residues), which is produced by the gastric mucosa, stimulates the gastric secretion of HCl and **pepsinogen** (the zymogen of the digestive protease pepsin). Gastrin release is stimulated by amino acids and partially digested protein as well as by the vagus nerve (which innervates the stomach) in response to stomach distension. Gastrin release is inhibited by HCl and by other gastrointestinal hormones.

2. Secretin (27 residues), which is produced by the mucosa of the duodenum (upper small intestine) in response to acidification by gastric HCl, stimulates the pancreatic secretion of HCO_3^- so as to neutralize this acid.

3. Cholecystokinin (CCK; 8 residues), which is produced by the duodenum, stimulates gallbladder emptying, the pancreatic secretion of digestive enzymes and HCO_3^- (and thus enhances the effect of secretin), and inhibits gastric emptying. CCK is released in response to the products of lipid and protein digestion, that is, fatty acids, monoacylglycerols, amino acids, and peptides.

4. Gastric inhibitory peptide (GIP, also known as **glucose-dependent insulinotropic polypeptide;** 42 residues), which is produced by specialized cells lining the small intestine, is a potent inhibitor of gastric acid secretion, gastric mobility, and gastric emptying. However, GIP's major physiological function is to stimulate pancreatic insulin release. Indeed, the release of GIP is stimulated by the presence of glucose in the gut, which accounts for the observation that, after a meal, the blood insulin level increases before the blood glucose level does.

These gastrointestinal hormones form families of related polypeptides: The C-terminal pentapeptides of gastrin and CCK are identical; secretin, GIP, and glucagon are closely similar. Two other gastrointestinal peptide hormones are discussed in Section 27-3C.

D. *Thyroid Hormones*

The thyroid gland produces two related hormones, triiodothyronine (T₃) and thyroxine (T₄),

X = H   **Triiodothyronine (T₃)**
X = I   **Thyroxine (T₄)**

that stimulate metabolism in most tissues (adult brain is a conspicuous exception). The production of these unusual iodinated amino acids begins with the synthesis of **thyroglobulin,** a 2748-residue protein. Thyroglobulin is posttranslationally modified in a series of biochemically unique reactions (Fig. 19-4):

1. Around 20% of thyroglobulin's 140 Tyr residues are iodinated in a **thyroperoxidase (TPO)-**catalyzed reaction forming **2,5-diiodotyrosyl** residues.

2. Two such residues are oxidatively coupled to yield T_3 and T_4 residues.

3. Mature thyroglobulin itself is hormonally inactive. However, some five or six molecules of the active hormones, T_3 and T_4, are produced by the proteolysis in the lysosome of thyroglobulin on hormonal stimulation of the thyroid (Section 19-1H).

How do thyroid hormones work? T_3 and T_4, being nonpolar substances, are transported by the blood in complex

Figure 19-4 **Biosynthesis of T_3 and T_4 in the thyroid gland.** The pathway involves the iodination, coupling, and hydrolysis (proteolysis) of thyroglobulin Tyr residues. The relatively scarce I^- is actively sequestered by the thyroid gland.

with plasma carrier proteins, primarily **thyroxine-binding globin,** but also **prealbumin** and **albumin.** The hormones then pass through the cell membranes of their target cells into the cytosol, where they bind to a specific protein. Since the resulting hormone–protein complex does not enter the nucleus, it is thought that this complex acts to maintain an intracellular reservoir of thyroid hormones. The true **thyroid hormone receptor** is a chromosomally associated protein and therefore does not leave the nucleus. *The binding of T_3, and to a lesser extent T_4, activates this receptor as a transcription factor (Section 5-4Aa), resulting in increased rates of expression of numerous metabolic enzymes.* High affinity thyroid hormone–binding sites also occur on the inner mitochondrial membrane (the site of electron transport and oxidative phosphorylation; Section 22-1), suggesting that these receptors may directly regulate O_2 consumption and ATP production.

Abnormal levels of thyroid hormones are common human afflictions. **Hypothyroidism** is characterized by lethargy, obesity, and cold dry skin, whereas **hyperthyroidism** has the opposite effects. The inhabitants of areas in which the soil has a low iodine content often develop hypothyroidism accompanied by an enlarged thyroid gland, a condition known as **goiter.** The small amount of NaI usually added to commercially available table salt ("iodized" salt) easily prevents this iodine deficiency disease. Young mammals require thyroid hormone for normal growth and development: Hypothyroidism during the fetal and immediate postnatal periods results in irreversible physical and mental retardation, a syndrome named **cretinism.**

E. Control of Calcium Metabolism

Ca^{2+} forms **hydroxyapatite,** $Ca_5(PO_4)_3OH$, the major mineral constituent of bone, and is an essential element in many biological processes including the mediation of hormonal signals as a second messenger, the triggering of muscle contraction, the transmission of nerve impulses, and blood clotting. The extracellular $[Ca^{2+}]$ must therefore be closely regulated to keep it at its normal level of ~ 1.2 mM. Three hormones have been implicated in maintaining Ca^{2+} homeostasis (Fig. 19-5):

1. Parathyroid hormone (PTH), an 84-residue polypeptide secreted by the parathyroid gland, which increases serum $[Ca^{2+}]$ by stimulating its resorption from bone and kidney and by increasing the dietary absorption of Ca^{2+} from the intestine.

2. Vitamin D, a group of steroidlike substances that act in a synergistic manner with PTH to increase serum $[Ca^{2+}]$.

3. Calcitonin, a 33-residue polypeptide synthesized by specialized thyroid gland cells, which decreases serum $[Ca^{2+}]$ by inhibiting the resorption of Ca^{2+} from bone and kidney.

We shall briefly discuss the functions of these hormones.

a. Parathyroid Hormone

The bones, the body's main Ca^{2+} reservoir, are by no means metabolically inert. They are continually "remodeled" through the action of two types of bone cells: **osteoblasts** which synthesize the collagen fibrils that form the bulk of bone's organic matrix, the scaffolding on which its $Ca_5(PO_4)_3OH$ mineral phase is laid down; and **osteoclasts,** which participate in bone resorption (Section 15-4Ab). *PTH inhibits collagen synthesis by osteoblasts and stimulates bone resorption by osteoclasts. The main effect of PTH, however, is to increase the rate at which the kidneys excrete phosphate, the counterion of Ca^{2+} in bone.* The consequent decreased serum $[P_i]$ causes $Ca_5(PO_4)_3OH$ to leach out of bone through mass action and thus increase serum $[Ca^{2+}]$. In addition, PTH stimulates the production of the active form of vitamin D by the kidney, which, in turn, enhances the transfer of intestinal Ca^{2+} to the blood (see below).

b. Vitamin D

Vitamin D is a group of fat-soluble dietary substances that prevent **rickets,** a disease of children characterized by stunted growth and deformed bones stemming from insufficient bone mineralization (vitamin D deficiency in adults is known as **osteomalacia,** a condition characterized by weakened, demineralized bones). Although rickets was first described in 1645, it was not until the early twentieth century that it was discovered that animal fats, particularly fish liver oils, are effective in preventing this deficiency disease. Moreover, rickets can also be prevented by exposing children to sunlight or just UV light in the wavelength range 230 to 313 nm, regardless of their diets.

The D vitamins, which we shall see are really hormones, are sterol derivatives (Section 12-1E) in which the steroid B ring is disrupted at its 9,10 position. The natural form of the vitamin, **vitamin D_3 (cholecalciferol),** is nonenzymatically

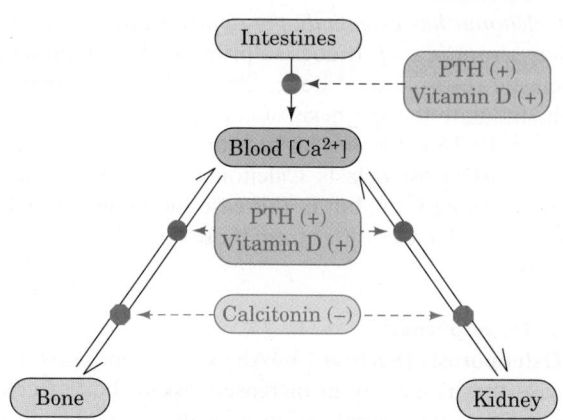

Figure 19-5   **Roles of PTH, vitamin D, and calcitonin in controlling Ca^{2+} metabolism.**

formed in the skin of animals through the photolytic action of UV light on **7-dehydrocholesterol:**

Vitamin D$_2$ (ergocalciferol), which differs from vitamin D$_3$ only by a side chain double bond and methyl group, is formed by the UV irradiation of the plant sterol **ergosterol.** Since vitamins D$_2$ and D$_3$ have essentially identical biological activities, vitamin D$_2$ is commonly used as a vitamin supplement, particularly in milk.

Vitamins D$_2$ and D$_3$ are hormonally inactive as such; they gain biological activity through further metabolic processing, first in the liver and then in the kidney (Fig. 19-6):

1. In the human liver, vitamin D$_3$ is hydroxylated to form **25-hydroxycholecalciferol** in an O$_2$-requiring reaction catalyzed by either of two cytochrome P450's (Section 15-4Bc), **CYP27A1** and **CYP2R1.**

2. The 25-hydroxycholecalciferol is transported to the kidney, where it is further hydroxylated by **CYP27B1** to yield the active hormone **1α,25-dihydroxycholecalciferol [1,25(OH)$_2$D].** *CYP27B1 is activated by PTH, so this reaction is an important control point in Ca^{2+} homeostasis.*

These various nonpolar substances are transported in the bloodstream in complex with **vitamin D-binding protein.**

1,25(OH)$_2$D acts to increase serum [Ca^{2+}] by promoting the intestinal absorption of dietary Ca^{2+} and by stimulating Ca^{2+} release from bone. Intestinal Ca^{2+} absorption is stimulated through increased synthesis of a **Ca^{2+}-binding protein,** which functions to transport Ca^{2+} across the intestinal mucosa. 1,25(OH)$_2$D binds to cytoplasmic receptors in intestinal epithelial cells that, on transport to the nucleus, function as transcription factors for the Ca^{2+}-binding protein. The maintenance of electroneutrality requires that Ca^{2+} transport be accompanied by that of counterions, mostly P$_i$, so that 1,25(OH)$_2$D also stimulates the intestinal absorption of P$_i$. The observation that 1,25(OH)$_2$D, like PTH, stimulates the release of Ca^{2+} and P$_i$ from bone seems paradoxical in view of the fact that low levels of 1,25(OH)$_2$D result in subnormal bone mineralization. Presumably the increased serum [Ca^{2+}] resulting from 1,25(OH)$_2$D-stimulated intestinal uptake of Ca^{2+} causes bone to take up more Ca^{2+} than it loses through direct hormonal stimulation. In addition, vitamin D has been shown to modulate the immune response, provide protection against certain types of cancers, and has been implicated in preventing/reversing heart disease.

Vitamin D, unlike the water-soluble vitamins, is retained by the body, so that excessive intake of vitamin D over long periods causes **vitamin D intoxication** (although note that most individuals, particularly the elderly and those with limited sun exposure, have less than the recommended levels of vitamin D in their blood). The consequent high serum [Ca^{2+}] results in aberrant calcification of a wide variety of soft tissues. The kidneys are particularly prone to calcification, a process that can lead to the formation of kidney stones and ultimately kidney failure. In addition, vitamin D intoxication promotes bone demineralization to the extent that bones are easily fractured. The observation that the level of skin pigmentation in indigenous human populations tends to increase with their proximity to the equator is explained by the hypothesis that skin pigmentation functions to prevent vitamin D intoxication by filtering out excessive solar radiation.

c. Calcitonin

Calcitonin has essentially the opposite effect of PTH; it lowers serum [Ca^{2+}]. It does so primarily by inhibiting osteoclastic resorption of bone. Since PTH and calcitonin both stimulate the synthesis of cAMP in their target cells (Section 19-2A), it is unclear how these hormones can oppositely affect osteoclasts. Calcitonin also inhibits kidney from resorbing Ca^{2+}, but in this case the kidney cells that calcitonin influences differ from those that PTH stimulates to resorb Ca^{2+}.

d. Osteoporosis

Osteoporosis (Section 15-4Ab) is a degenerative bone disease that results in an increased risk of bone fracture due to bone demineralization as well as changes in the amounts and variety of noncollagenous bone proteins. It is most common in postmenopausal women, although it also

Figure 19-6   Activation of vitamin D₃ as a hormone in liver and kidney. Vitamin D₂ (ergocalciferol) is similarly activated.

occurs in elderly men and as a consequence of certain hormonal disorders or the long-term administration of glucocorticoids (Section 19-1G).

Osteoporosis is most commonly treated by the administration of **bisphosphonates** (Fig. 19-7). These drugs bind to bone tissue from which they are absorbed by osteoclasts. There they inhibit **prenyltransferase** (Section 25-6Ac), an enzyme in the pathway forming the prenyl groups that anchor certain proteins to the cell membrane (Section 12-3Ba). This disrupts the contact region between the osteoclast and the bone surface, thus decreasing bone resorption. Bisphosphonate treatment is usually accompanied by calcium and vitamin D supplements and by load-bearing exercise.

F. *Epinephrine and Norepinephrine*

The adrenal glands consist of two distinct types of tissue: the **medulla** (core), which is really an extension of the sympathetic nervous system (a part of the autonomic nervous system), and the more typically glandular **cortex** (outer layer). Here we consider the hormones of the adrenal medulla; those of the cortex are discussed in the following subsection.

Figure 19-7   Some bisphosphonates that are in clinical use against osteoporosis. Each drug has both a generic (chemical) name and proprietary trade name, here in parentheses, under which it is marketed.

The adrenal medulla synthesizes two hormonally active **catecholamines** *(amine-containing derivatives of* **catechol,** *1,2-dihydroxybenzene),* **norepinephrine (noradrenaline)** *and its methyl derivative* **epinephrine (adrenaline):**

R = H   **Norepinephrine (noradrenaline)**
R = CH₃   **Epinephrine (adrenaline)**

These hormones are synthesized from tyrosine, as is described in Section 26-4B, and stored in granules to await their exocytotic release under the control of the sympathetic nervous system.

The biological effects of catecholamines are mediated by two classes of plasma transmembrane receptors, the **α-** and the **β-adrenergic receptors** (also known as **adrenoreceptors**). These TM glycoproteins were originally identified on the basis of their varying responses to certain **agonists** (substances that bind to a hormone receptor so as to evoke a hormonal response) and **antagonists** (substances that bind to a hormone receptor but fail to elicit a hormonal response, thereby blocking agonist action). The β- but not the α-adrenergic receptors, for example, are stimulated by **isoproterenol** but blocked by **propranolol,** whereas α- but not β-adrenergic receptors are blocked by **phentolamine:**

Isoproterenol

Propranolol

Phentolamine

The α- and β-adrenergic receptors, which occur in separate tissues in mammals, generally respond differently and often oppositely to catecholamines. For instance, β-adrenergic receptors, which activate adenylate cyclase, stimulate glycogenolysis and gluconeogenesis in liver (Sections 18-3Ea and 18-3G), glycogenolysis and glycolysis in skeletal muscle,

lipolysis in adipose tissue, the relaxation of smooth (involuntary) muscle in the bronchi and the blood vessels supplying the skeletal (voluntary) muscles, and increased heart action. In contrast, α-adrenergic receptors, whose intracellular effects are mediated either by the inhibition of adenylate cyclase (**α₂ adrenergic receptors;** Section 19-2D) or via the phosphoinositide cascade (**α₁ adrenergic receptors;** Section 19-4A), stimulate smooth muscle contraction in blood vessels supplying peripheral organs such as skin and kidney, smooth muscle relaxation in the lung and gastrointestinal tract, and blood platelet aggregation. *Most of these diverse effects are directed toward a common end: the mobilization of energy resources and their shunting to where they are most needed to prepare the body for sudden action.*

The varying responses and tissue distributions of the α- and β-adrenergic receptors and their subtypes to different agonists and antagonists have important therapeutic consequences. For example, propranolol is used for the treatment of high blood pressure and protects heart attack victims from further heart attacks, whereas epinephrine's bronchodilator effects make it clinically useful in the treatment of **asthma,** a breathing disorder caused by the inappropriate contraction of bronchial smooth muscle.

G. *Steroid Hormones*

a. The Adrenocortical Steroids Mediate a Wide Variety of Metabolic Functions

The adrenal cortex produces at least 50 different **adrenocortical steroids** *(whose synthesis is outlined in Section 25-6C).* These have been classified according to the physiological responses they evoke:

1. The **glucocorticoids** affect carbohydrate, protein, and lipid metabolism in a manner nearly opposite to that of insulin and influence a wide variety of other vital functions, including inflammatory reactions and the capacity to cope with stress.

2. The **mineralocorticoids** largely function to regulate the excretion of salt and water by the kidney.

3. The **androgens** and **estrogens** affect sexual development and function. They are made in larger quantities by the gonads.

Glucocorticoids, the most common of which are **cortisol** (also known as **hydrocortisone**) and **corticosterone,** and the mineralocorticoids, the most common of which is **aldosterone,** are all C_{21} compounds:

Cortisol (hydrocortisone)

Corticosterone

Aldosterone

Steroids, being water insoluble, are transported in the blood in complex with the glycoprotein **transcortin** and, to a lesser extent, by albumin. The steroids (including vitamin D) spontaneously pass through the membranes of their target cells to the cytosol, where they bind to their cognate receptors. The steroid–receptor complexes then migrate to the cell nucleus, where they function as transcription factors to induce, or in some cases repress, the transcription of specific genes (a process that is discussed in Section 34-3Bn). In this way, the glucocorticoids and the mineralocorticoids influence the expression of numerous metabolic enzymes in their respective target tissues. Thyroid hormones, which are also nonpolar, function similarly. However, as we shall see in the following sections, all other hormones act less directly in that they bind to their cognate cell-surface receptors and thereby trigger complex cascades of events within cells that ultimately influence transcription as well as other cellular processes.

Impaired adrenocortical function, either through disease or trauma, results in a condition known as **Addison's disease,** which is characterized by hypoglycemia, muscle weakness, Na^+ loss, K^+ retention, impaired cardiac function, loss of appetite, and a greatly increased susceptibility to stress. The victim, unless treated by the administration of glucocorticoids and mineralocorticoids, slowly languishes and dies without any particular pain or distress. The opposite problem, adrenocortical hyperfunction, which is usually caused by a tumor of the adrenal cortex or the pituitary gland (Section 19-1H), results in **Cushing's syndrome,** which is characterized by fatigue, hyperglycemia, edema (water retention), and a redistribution of body fat to yield a characteristic "moon face." Long-term treatments of various diseases with synthetic glucocorticoids result in similar symptoms.

b. Gonadal Steroids Mediate Sexual Development and Function

The **gonads** *(testes in males, ovaries in females), in addition to producing sperm or ova, secrete steroid hormones (androgens and estrogens) that regulate sexual differentiation,*

the expression of secondary sex characteristics, and sexual behavior patterns. Although testes and ovaries both synthesize androgens and estrogens, the testes predominantly secrete androgens, which are therefore known as **male sex hormones,** whereas ovaries produce mostly estrogens, which are consequently termed **female sex hormones.**

Androgens, of which **testosterone** is prototypic,

Testosterone

β-Estradiol

Progesterone

lack the C_2 substituent at C17 present in glucocorticoids and are therefore C_{19} compounds. Estrogens, such as **β-estradiol,** resemble androgens but lack a C10 methyl group because they have an aromatic A ring and are therefore C_{18} compounds. Interestingly, testosterone is an intermediate in estrogen biosynthesis (Section 25-6C). A second class of ovarian steroids, C_{21} compounds called **progestins,** help mediate the menstrual cycle and pregnancy (Section 19-1I). **Progesterone,** the most abundant progestin, is, in fact, a precursor of glucocorticoids, mineralocorticoids, and testosterone (Section 25-6C).

c. Sexual Differentiation Is Both Hormonally and Genetically Controlled

What factors control sexual differentiation? If the gonads of an embryonic male mammal are surgically removed, that individual will become a phenotypic female. Evidently, *mammals are programmed to develop as females unless embryonically subjected to the influence of testicular hormones.* Indeed, genetic males with absent or nonfunctional cytosolic androgen receptors are phenotypic females, a condition named **testicular feminization.** Curiously, estrogens appear to play no part in embryonic female sexual development, although they are essential for female sexual maturation and function.

Normal individuals have either the XY (male) or the XX (female) genotypes (Section 1-4C). However, those with the abnormal genotypes XXY (**Klinefelter's syndrome**) and X0 (only one sex chromosome; **Turner's syndrome**) are, respectively, phenotypic males and phenotypic females, although both are sterile. Apparently, *the normal Y chromosome confers the male phenotype, whereas its absence results in the female phenotype.* There are, however, rare (1 in 20,000) XX males and XY females. These XX males (who are sterile and have therefore been identified through infertility clinics) have a small segment of a normal Y chromosome translocated onto one of their X chromosomes, whereas XY females are missing this segment.

Early male and female embryos—through the sixth week of development in humans—have identical undifferentiated genitalia. Evidently, the Y chromosome contains a gene, **testes-determining factor (TDF),** that induces the differentiation of testes, whose hormonal secretions, in turn, promote male development. The misplaced chromosomal segments in XY females and XX males have a common 140-kb sequence that contains a structural gene dubbed ***SRY*** (for *s*ex-determining *r*egion of *Y*) that encodes an 80-residue DNA-binding motif. Several sex-reversed XY women have a mutation in the region of their *SRY* gene encoding this DNA-binding domain that eliminates its ability to bind DNA, a mutation that is not present in their father's gene. *SRY* is expressed in embryonic go-

nadal cells previously shown to be responsible for testis determination. Moreover, of eleven XX mice that were made transgenic for *Sry* (the mouse analog of *SRY*), three were males. Thus, *TDF/SRY* is the first clear example of a mammalian gene that controls the development of an entire organ system (development is discussed in Section 34-4B).

H. Control of Endocrine Function: The Hypothalamus and Pituitary Gland

The anterior lobe of the **pituitary gland** (the **adenohypophysis**) and the **hypothalamus,** a nearby portion of the brain, constitute a functional unit that hormonally controls much of the endocrine system. *The neurons of the hypothalamus synthesize a series of polypeptide hormones known as releasing factors and release-inhibiting factors which, on delivery to the adenohypophysis via a direct circulatory connection (their half-lives are on the order of a few minutes), stimulate or inhibit the release of the corresponding trophic hormones into the bloodstream.* Trophic hormones, by definition, stimulate their target endocrine tissues to secrete the hormones they synthesize. Since releasing and release-inhibiting factors, trophic hormones, and endocrine hormones are largely secreted in nanogram, microgram, and milligram quantities per day, respectively, and tend to have progressively longer half-lives, these hormonal systems can be said to form amplifying cascades. Four such systems are prominent in humans (Fig. 19-8; *left*):

1. Corticotropin-releasing factor (**CRF;** 41 residues) causes the adenohypophysis to release **adrenocorticotropic**

Figure 19-8 Hormonal control circuits, indicating the relationships between the hypothalamus, pituitary, and target tissues. Releasing factors and release-inhibiting factors secreted by the hypothalamus signal the adenohypophysis to secrete or stop secreting the corresponding trophic hormones, which, for the most part, stimulate the corresponding endocrine gland(s) to

secrete their respective endocrine hormones. The endocrine hormones, in addition to controlling the growth, differentiation, and metabolism of their corresponding target tissues, influence the secretion of releasing factors and trophic hormones through feedback inhibition. The levels of trophic hormones likewise influence the levels of their corresponding releasing factors.

hormone (**ACTH;** 39 residues), which stimulates the release of adrenocortical steroids. *The entire system is under feedback control: ACTH inhibits the release of CRF and the adrenocortical steroids inhibit the release of both CRF and ACTH. Moreover, the hypothalamus, being part of the brain, is also subject to neuronal control, so the hypothalamus forms the interface between the nervous system and the endocrine system.*

2. Thyrotropin-releasing factor (TRF), a tripeptide with an N-terminal **pyroGlu** residue (a Glu derivative in which the side chain carboxyl group forms an amide bond with its amino group),

$$\text{pyroGlu}$$

Thyrotropin-releasing factor (TRF)

stimulates the adenohypophysis to release the trophic hormone **thyrotropin (thyroid-stimulating hormone; TSH)** which, in turn, stimulates the thyroid to synthesize and release T_3 and T_4. TRF, as are other releasing factors, is present in the hypothalamus in only vanishingly small quantities. It was independently characterized in 1969 by Roger Guillemin and Andrew Schally using extracts of the hypothalami from over 2 million sheep and 1 million pigs.

3. Gonadotropin-releasing factor (GnRF; 10 residues)

$$\overset{1}{\text{pyroGlu}}\text{-His-Trp-Ser-Tyr-Gly-Leu-Arg-Pro-}\overset{10}{\text{Gly}}\text{-NH}_2$$

Gonadotropin-releasing factor (GnRF)

stimulates the adenohypophysis to release **luteinizing hormone (LH)** and **follicle-stimulating hormone (FSH),** which are collectively known as **gonadotropins.** In males, LH stimulates the testes to secrete androgens, whereas FSH promotes spermatogenesis. In females, FSH stimulates the development of ovarian follicles (which contain the immature ova), whereas LH triggers ovulation.

4. Growth hormone-releasing factor (GRF; 44 residues) and **somatostatin** [14 residues; also known as **growth hormone release-inhibiting factor (GRIF)**], stimulate/inhibit the release of **growth hormone (GH)** from the adenohypophysis. GH (also called **somatotropin**), in turn, stimulates generalized growth (see Fig. 5-5 for a striking example of its effect). GH directly accelerates the growth of a variety of tissues (in contrast to TSH, LH, and FSH, which act only indirectly by activating endocrine glands) and induces the liver to synthesize a series of polypeptide growth factors termed **somatomedins** that stimulate cartilage growth and have insulinlike activities.

TSH, LH, and FSH are heterodimeric glycoproteins, which in a given species, all have the same α subunit (92 residues) and a homologous β subunit (114, 114, and 118 residues, respectively, in humans). Human GH consists of a single 191-residue polypeptide chain, which is unrelated to TSH, LH, or FSH.

a. The Neurohypophysis Secretes Oxytocin and Vasopressin

The posterior lobe of the pituitary, the **neurohypophysis,** which is anatomically distinct from the adenohypophysis, secretes two homologous nonapeptide hormones (Fig. 19-8, *right*): **vasopressin** [also known as **antidiuretic hormone (ADH)**], which increases blood pressure and stimulates the kidneys to retain water; and **oxytocin,** which causes contraction of uterine smooth muscle and therefore induces labor:

$$\overset{1}{\text{Cys}}\text{-Tyr-Phe-Gln-Asn-Cys-Pro-}\overset{9}{\text{Arg}}\text{-Gly-NH}_2$$
$$\text{S—S}$$

Human vasopressin

$$\overset{1}{\text{Cys}}\text{-Tyr- Ile -Gln-Asn-Cys-Pro-}\overset{9}{\text{Leu}}\text{-Gly-NH}_2$$
$$\text{S—S}$$

Human oxytocin

The rate of vasopressin release is largely controlled by osmoreceptors, which monitor the osmotic pressure of the blood.

I. *Control of the Menstrual Cycle*

The menstrual cycle and pregnancy are particularly illustrative of the interactions among hormonal systems. The ~28-day human menstrual cycle (Fig. 19-9) begins during menstruation with a slight increase in the FSH level that initiates the development of a new ovarian follicle. As the follicle matures, it secretes estrogens that act to sensitize the adenohypophysis to GnRF. This process culminates in a surge of LH and FSH, which triggers ovulation. The ruptured ovarian follicle, the **corpus luteum,** secretes progesterone and estrogens, which inhibit further gonadotropin

Figure 19-9 **Patterns of hormone secretion during the menstrual cycle in the human female.**

secretion by the adenohypophysis and stimulate the uterine lining to prepare for the implantation of a fertilized ovum. If fertilization does not occur, the corpus luteum regresses, progesterone and estrogen levels fall, and menstruation (the sloughing off of the uterine lining) ensues. The reduced steroid levels also permit a slight increase in the FSH level, which initiates a new menstrual cycle.

A fertilized ovum that has implanted into the hormonally prepared uterine lining soon commences synthesizing **chorionic gonadotropin (CG).** This heterodimeric glycoprotein hormone contains a 145-residue β subunit that has a high degree of sequence identity with those of LH (85%), FSH (45%), and TSH (36%) over their N-terminal 114 residues and the same α subunit. CG stimulates the corpus luteum to continue secreting progesterone rather than regressing and thus prevents menstruation. Pregnancy tests utilize immunoassays that can detect CG in blood or urine within a few days after embryo implantation. Most female oral contraceptives (birth control pills) contain progesterone derivatives, whose ingestion induces a state of pseudopregnancy in that they inhibit the midcycle surge of FSH and LH so as to prevent ovulation.

J. Growth Hormone and Its Receptor

The binding of growth hormone activates its receptor to stimulate growth and metabolism in muscle, bone, and cartilage cells. This 620-residue receptor is a member of a large family of structurally related protein growth factor receptors, which includes those for various interleukins (Section 19-3Eb). All of these receptors consist of an N-terminal extracellular ligand-binding domain, a single transmembrane segment that is almost certainly helical, and a C-terminal cytoplasmic domain that is not homologous within the superfamily but in many cases contains a tyrosine kinase function (Section 19-3A).

The X-ray structure of the 191-residue human growth hormone **(hGH)** in complex with the 238-residue **ectodomain** (extracellular portion; Greek: *ectos,* outside) of its binding protein **(hGHbp),** determined by Abraham de Vos and Anthony Kossiakoff, revealed that this complex consists of two molecules of hGHbp bound to a single hGH molecule (Fig. 19-10). hGH consists largely of an up–up–down–down four-helix bundle, which closely resembles that in the previously determined X-ray structure of porcine GH, although with significant differences that may be caused by the binding of hGH to its receptor. A variety of other protein growth factors with known structures, including many interleukins (Section 19-3Eb), contain similar four-helix bundles. Each hGHbp molecule consists of two structurally similar ~100-residue **fibronectin type III domains,** each of which forms a topologically identical sandwich of a three- and a four-stranded antiparallel β sheet that resembles the immunoglobulin fold (Fig. 8-48). Fibronectin type III domains, so called because they were first observed in the multidomain extracellular matrix glycoprotein **fibronectin,** are among the most common structural modules in receptor ectodomains.

Figure 19-10 **X-ray structure of human growth hormone (hGH) in complex with two molecules of its receptor's extracellular domain (hGHbp).** The proteins are drawn in ribbon form, with the hGH magenta and the two hGHbp molecules, which together bind one molecule of hGH, each colored in rainbow order from N-terminus (*blue*) to C-terminus (*red*). The view is along the axis of the hGH's four-helix bundle with the approximate 2-fold axis relating the two hGHbp molecules vertical. The dotted red lines represent the pathways that the hGHbp chains take in penetrating the membrane. Note that the width of a membrane is actually nearly the ~75-Å height of the hGHbp–hGH–hGHbp complex. [Based on an X-ray structure by Abraham de Vos and Anthony Kossiakoff, Genentech Inc., South San Francisco, California. PDBid 3HHR.] *See Interactive* **Exercise 11**

The two hGHbp molecules bind to hGH with near 2-fold symmetry about an axis that is roughly perpendicular to the helical axes of the hGH four-helix bundle and, presumably, to the plane of the cell membrane to which the intact hGH receptor is anchored (Fig. 19-10). The C-terminal domains of the two hGHbp molecules are almost parallel and in contact with one another. Intriguingly, the two hGHbp molecules use essentially the same residues to bind to sites that are on opposite sides of hGH's four-helix bundle and that have no structural similarity. The X-ray structure is largely consistent with the results of mutational studies designed to identify the hGH and hGHbp residues important for receptor binding.

The ligand-induced dimerization of hGHbp has important implications for the mechanism of signal transduction. The dimerization, which does not occur in the absence of hGH, apparently brings together the intact receptors' intracellular domains in a way that activates an effector protein such as a tyrosine kinase (Section 19-3A). Indeed, hGH mutants that cannot induce receptor dimerization are biologically inactive. Numerous other protein growth factors also induce the dimerization of their receptors.

a. Abnormal GH Production Causes Abnormal Growth

Overproduction of GH, usually a consequence of a pituitary tumor, results in excessive growth. If this condition commences while the skeleton is still growing, that is, before its growth plates have ossified, then this excessive growth is of normal proportions over the entire body, resulting in **gigantism.** Moreover, since excessive GH inhibits the testosterone production necessary for growth plate ossification, such "giants" continue growing throughout their abnormally short lives. If, however, the skeleton has already matured, GH stimulates only the growth of soft tissues, resulting in enlarged hands and feet and thickened facial features, a condition named **acromegaly** (Fig. 19-11). The opposite problem, GH deficiency, which results in insufficient growth **(dwarfism),** can be treated before skeletal maturity by regular injections of hGH (animal GH is ineffective in humans). hGH was, at first, available only from the pituitaries of cadavers and therefore was in extremely short supply. Since the early 1980s,

however, hGH has been synthesized in virtually unlimited amounts via recombinant DNA techniques. Indeed, hGH has been illicitly taken by individuals wishing to increase their athletic prowess (the ban on using hGH could not be enforced until the early 2000s, when blood tests were developed that could distinguish between natural and recombinant hGH).

K. Opioid Peptides

Among the most intriguing hormones secreted by the adenohypophysis are polypeptides that have opiatelike effects on the central nervous system. These include the 31-residue **β-endorphin,** its N-terminal pentapeptide, termed **methionine-enkephalin,** and the closely similar **leucine-enkephalin** (although the enkaphalins are independently expressed).

$$
\begin{array}{l}
\overset{1}{\text{Tyr}} \text{-Gly-Gly-Phe-}\overset{5}{\text{Met}} \text{-Thr-Ser-Glu-Lys-}\overset{10}{\text{Ser}} \\
\overset{11}{\text{Gln}} \text{-Thr-Pro-Leu-}\overset{15}{\text{Val}} \text{-Thr-Leu-Phe-Lys-}\overset{20}{\text{Asn}} \\
\overset{21}{\text{Ala}} \text{-Ile-Val-Lys-}\overset{25}{\text{Asn}} \text{-Ala-His-Lys-Lys-}\overset{30}{\text{Gly}} \\
\overset{31}{\text{Gln}} \text{-}
\end{array}
$$

β-Endorphin

Tyr-Gly-Gly-Phe-Met

Methionine-enkephalin (Met-enkephalin)

Tyr-Gly-Gly-Phe-Leu

Leucine-enkephalin (Leu-enkephalin)

Morphine (an opiate)

These substances bind to **opiate receptors** in the brain and have been shown to be their physiological agonists. The role of these so-called **opioid peptides** has yet to be definitively established, but it appears they are important in the control of pain and emotional states. Pain relief through the use of acupuncture and placebos as well as such phenomena as "runner's high" may be mediated by opioid peptides.

L. The Hormonal Function of Nitric Oxide

Nitric oxide (NO) is a reactive and toxic free radical gas. Thus, it came as a great surprise that *this molecule functions as paracrine signal in regulating blood vessel dilation and serves as a neurotransmitter. It also functions in the immune response.* The role of NO in vasodilation was discovered

Figure 19-11 Acromegaly. The characteristic enlarged features of Akhenaten, the Pharaoh who ruled Egypt in the years 1379–1362 B.C., strongly suggest that he suffered from acromegaly. [Bildarchiv Preussischer Kulturbesitz/Art Resource.]

through the observation that substances such as acetyl-choline (Section 20-5Cb) and bradykinin (Section 7-5B), which act through the phosphoinositide signaling system (Section 19-4) to increase the flow through blood vessels by eliciting smooth muscle relaxation, require an intact **endothelium** overlying the smooth muscle (the endothelium is a layer of cells that lines the inside of certain body cavities such as blood vessels). Evidently, endothelial cells respond to the presence of these vasodilation agents by releasing a diffusible and highly labile substance (half-life ~5 s) that induces the relaxation of smooth muscle cells. This substance was identified as NO, in part, through parallel studies identifying NO as the active metabolite that mediates the well-known vasodilating effects of antianginal organic nitrates such as **nitroglycerin**

$$CH_2-CH-CH_2$$
$$| \quad | \quad |$$
$$O \quad O \quad O$$
$$| \quad | \quad |$$
$$NO_2 \quad NO_2 \quad NO_2$$

Nitroglycerin

(**angina pectoris** is a condition caused by insufficient blood flow to the heart muscle, leading to severe chest pain; nitroglycerin's vasodilating properties were discovered in the nineteenth century through the observation that workers with angina pectoris in a factory that produced nitroglycerin experienced greater pain on weekends).

a. Nitric Oxide Synthase Requires Five Redox-Active Cofactors

NO is synthesized by **NO synthase (NOS),** which catalyzes the NADPH-dependent 5-electron oxidation of L-arginine by O_2 to yield NO and the amino acid **L-citrulline** with the intermediate formation of N^ω-**hydroxy-L-arginine (NOHA;** Fig. 19-12). Three isozymes of NOS have been identified in mammals, **neuronal NOS (nNOS), inducible NOS (iNOS),** and **endothelial NOS (eNOS),** which are also known as **NOS-1, -2,** and **-3,** respectively. These isozymes, which have 50 to 60% sequence identity, are all

homodimeric proteins of 125- to 160-kD subunits that each consist of two domains:

1. An N-terminal, ~500-residue, oxygenase or heme domain that catalyzes both reaction steps of Fig. 19-12 and contains the dimer interface. This domain binds the substrates O_2 and L-arginine and two redox-active prosthetic groups, Fe(III)-heme and **5,6,7,8-tetrahydrobiopterin (H_4B),**

5,6,7,8-Tetrahydrobiopterin

a compound that also functions in the hydroxylation of phenylalanine to tyrosine (Section 26-3Ha). The X-ray structures of the oxygenase domains of nNOS (Fig. 19-13a), iNOS, and eNOS are closely similar.

2. A C-terminal, ~600-residue, reductase domain that supplies the electrons for the NOS reaction. It binds NADPH and two redox-active prosthetic groups, an FAD (Fig. 16-8) and a **flavin mononucleotide (FMN;** FAD lacking its AMP residue; Fig. 22-17a) via three nucleotide binding modules. This domain is homologous to **cytochrome P450 reductase,** an enzyme that participates in detoxification processes (Section 15-4Bc). The X-ray structure of the reductase domain of nNOS, determined by John Tainer and Elizabeth Getzoff, is shown in Fig. 19-13b.

NADPH bound to the reductase domain transmits its electrons via the FAD and then the FMN to the heme in the oxygenase domain. Interestingly, the reductase domain transmits its electrons to the oxygenase domain on the opposite subunit. This was shown by Dennis Stuehr through his construction of an NOS heterodimer in which one subunit was full length and the other consisted of only a oxygenase domain. If a mutation that disrupts the L-arginine

Figure 19-12   The NO synthase (NOS) reaction. The N^ω-hydroxy-L-arginine intermediate is tightly bound to the enzyme.

(a)

(b)

Figure 19-13   X-ray structure of rat nNOS. (*a*) Its N-terminal oxygenase domain in complex with heme, L-arginine, and H_4B. (*b*) Its C-terminal reductase domain in complex with FMN, FAD, and $NADP^+$. In both structures the homodimeric protein is viewed in semitransparent ribbon form with its 2-fold axis vertical and with one subunit colored in rainbow order from N-terminus (*blue*) to C-terminus (*red*) and the other subunit blue-gray. The various bound groups are drawn in space-filling form colored according to atom type (heme and FMN C green, L-arginine and FAD C cyan, H_4B and $NADP^+$ C magenta, N blue, O red, P orange, and Fe red-brown). [Part *a* based on an X-ray structure by Thomas Poulos, University of California at Irvine; and Part *b* based on an X-ray structure by John Tainer and Elizabeth Getzoff, The Scripps Research Institute, La Jolla, California. PDBids 1OM4 and 1TLL.]

binding site was in the full-length subunit, the enzymatic activity of this heterodimer was unaffected, but if it was in the oxygenase domain–only subunit, activity was abolished.

The heme Fe atom is 5-coordinated with its axial ligand supplied by a specific Cys S atom. The L-arginine substrate binds on the opposite side of the heme from this Cys with the N atom to be hydroxylated ~4.0 Å distant from the Fe atom, a distance too large for covalent bond formation. Since O_2 is known to react with the heme Fe atom, it presumably binds between it and this N atom.

NOS requires bound H_4B to produce NO. In the absence of this prosthetic group, NOS efficiently catalyzes the NADPH-mediated oxidation of O_2 to H_2O_2. Investigations by Steuhr have established that H_4B functions as an internal redox agent in that, during the reactions forming both NOHA and NO, it is oxidized to its radical form (H_4B^{+}) and then re-reduced to its reduced form (H_4B). Thus, H_4B does not undergo net oxidation in the NOS reaction (as it does in the reaction hydroxylating phenylalanine to tyrosine; Section 26-3Ha).

NO rapidly diffuses across cell membranes, although its high reactivity prevents it from traveling >1 mm from its site of synthesis (in particular, it efficiently reacts with both oxyhemoglobin and deoxyhemoglobin: $NO + HbO_2 \rightarrow NO_3^- + metHb$; and $NO + Hb \rightarrow HbNO$; Section 10-1A). *The physiological target of NO in smooth muscle cells is* ***guanylate cyclase (GC),*** which catalyzes the reaction of GTP to yield **3′,5′-cyclic GMP (cGMP),**

**3′,5′-Cyclic GMP
(cGMP)**

an intracellular second messenger that resembles 3′,5′-cyclic AMP (cAMP; GC is a homolog of adenylate cyclase; Section 19-2D). cGMP causes smooth muscle relaxation through its stimulation of protein phosphorylation by **cGMP-dependent protein kinase.** NO reacts with GC's heme prosthetic group to yield **nitrosoheme,** whose presence increases GC's activity by up to 200-fold, presumably via a conformation change resembling that in hemoglobin on binding O_2 (Section 10-2Ba; although GC binds O_2 quite poorly).

b. eNOS and nNOS but Not iNOS Are Regulated by [Ca²⁺]

Ca^{2+}–calmodulin activates eNOS and nNOS by binding to the $\sim$30-residue segments linking their oxygenase and reductase domains. Thus, for example, the stimulatory action of vasodilatory agents on the phosphoinositide signaling system (Section 19-4A) in endothelial cells to produce an influx of Ca^{2+} results in the synthesis of NO. Hence, *NO functions to transduce hormonally induced increases in intracellular [Ca²⁺] in endothelial cells to increased rates of production of cGMP in neighboring smooth muscle cells.*

NO produced by nNOS mediates vasodilation through endothelium-independent neural stimulation of smooth muscle. In this signal transduction pathway, which is responsible for the dilation of cerebral and other arteries as well as penile erection (see Section 19-2E), nerve impulses cause an increased $[Ca^{2+}]$ in nerve terminals, thereby stimulating neuronal NOS. The resultant NO diffuses to nearby smooth muscle cells, where it binds to guanylate cyclase and activates it to synthesize cGMP as described above.

Inducible NOS (iNOS) is unresponsive to Ca^{2+} even though it has two tightly bound calmodulin subunits. However, it is transcriptionally induced in macrophages and **neutrophils** (white blood cells that function to ingest and kill bacteria), as well as in endothelial and smooth muscle cells (in contrast, eNOS and nNOS are expressed **constitutively,** that is, at a constant rate). Several hours after exposure to **cytokines** (protein growth factors that regulate the differentiation, proliferation, and activities of many types of cells; Section 19-3Eb) and/or **endotoxins** (bacterial cell wall lipopolysaccharides that elicit inflammatory responses; Section 35-2Fb), these cells begin to produce large quantities of NO and continue to do so for many hours. Activated macrophages and neutrophils also produce superoxide ion (O_2^-), which chemically combines with NO to form the even more toxic **peroxynitrite (OONO⁻,** which rapidly reacts with H_2O to yield the highly reactive **hydroxide radical, OH·,** and NO_2) that they use to kill ingested bacteria. Indeed, NOS inhibitors block the cytotoxic actions of macrophages.

Cytokines and endotoxins induce a long-lasting and profound vasodilation and a poor response to vasoconstrictors such as epinephrine. The sustained release of NO has been implicated in **septic shock** (an often fatal immune system overreaction to bacterial infection that results in a catastrophic reduction in blood pressure), in inflammation-related tissue damage as occurs in autoimmune diseases such as rheumatoid arthritis, and in the damage to neurons in the vicinity of but not directly killed by a stroke (reperfusion injury; Section 10-1Aa). Many of these conditions might be alleviated if drugs can be developed that selectively inhibit iNOS and/or nNOS, while permitting eNOS to carry out its essential function of maintaining vascular tone. Moreover, the administration of NO itself appears to be medically useful. For example, the inhalation of low levels NO has been used to reduce **pulmonary hypertension** (high blood pressure in the lung, an often fatal condition caused by constriction of its arteries) in newborn infants.

2 HETEROTRIMERIC G PROTEINS

We have seen (Section 18-3) that hormones such as glucagon and epinephrine regulate glycogen metabolism by stimulating adenylate cyclase (AC) to synthesize the second messenger cAMP from ATP. The cAMP then binds to protein kinase A (PKA) so as to activate this enzyme to initiate cascades of phosphorylation/dephosphorylation events that ultimately control the activities of glycogen phosphorylase and glycogen synthase. Numerous other extracellular signaling molecules (known agonists, ligands, or **effectors**) also activate the intracellular synthesis of cAMP, thereby eliciting a cellular response. But what is the mechanism through which the binding of an agonist to a receptor induces AC to synthesize cAMP in the cytosol? In answering this question we shall see that the systems that link receptors to AC as well as other effectors have a surprising complexity that endows them with immense capacity for both signal amplification and regulatory flexibility.

A. *Overview*

🔎 See Guided Exploration 16: Mechanisms of hormone signaling involving the adenylate cyclase system Adenylate cyclase, which is located on the plasma membrane's cytosolic surface, and the receptors that activate it, whose agonist-binding sites are exposed to the extracellular space, are separate proteins that do not physically interact. Rather, *they are functionally coupled by* **heterotrimeric G proteins** (Fig. 19-14), so called because they specifically bind the guanine nucleotides GTP and GDP.

AC is activated by a heterotrimeric G protein (often called just a G protein) but only when the G protein is complexed with GTP. However, G protein slowly hydrolyzes GTP to GDP + P_i (at the leisurely rate of 2–3 min^{-1}) and thereby deactivates itself (if G proteins were efficient enzymes, they would be unable to effectively activate AC). G protein is reactivated by the exchange of its bound GDP for GTP, a process that is mediated by the agonist–receptor complex but not by unoccupied receptor. *Heterotrimeric G protein therefore mediates the transduction of an extracellular signal to an intracellular signal (the cAMP). Moreover, the receptor–G protein–AC system amplifies the extracellular signal because each agonist–receptor complex activates many G proteins before it is inactivated by the spontaneous dissociation of the agonist and, during its lifetime, each G protein · GTP–AC complex catalyzes the formation of many cAMP molecules.* In this section, we discuss how this process occurs.

Heterotrimeric G proteins are members of the superfamily of regulatory GTPases that are collectively known as **G proteins** (whether one is referring to a heterotrimeric or some other species of G protein is usually clear from context). G proteins other than heterotrimeric G proteins have a wide variety of essential functions including signal transduction (e.g., **Ras;** Section 19-3Cf), vesicle trafficking (e.g., Arf, dynamin, and Rab; Sections 12-4Cd and 12-4Db), translation (as ribosomal accessory factors; Section 32-3), and targeting [as components of the signal recognition

Figure 19-14   Activation/deactivation cycle for hormonally stimulated AC. (*a*) In the absence of hormone, heterotrimeric G protein binds GDP and AC is catalytically inactive. (*b*) The hormone–receptor complex stimulates the G protein to exchange its bound GDP for GTP. (*c*) The G protein · GTP complex, in turn, binds to and thereby activates AC to produce cAMP. (*d*) The eventual G protein-catalyzed hydrolysis of its bound GTP to GDP causes G protein to dissociate from and hence deactivate AC.

particle (SRP) and the SRP receptor; Section 12-4Ba]. The many G proteins share common structural motifs that bind guanine nucleotides (GDP and GTP) and catalyze the hydrolysis of GTP to GDP + P_i (see below).

B. *G Protein-Coupled Receptors*

The receptors responsible for activating AC and other targets of heterotrimeric G proteins are all integral proteins with 7 transmembrane helices (Fig. 19-15) that have their N-termini in the extracellular space and their C-termini in the cytosol. These **G protein-coupled receptors (GPCRs; also called heptahelical, 7TM,** and **serpentine receptors)** constitute one of the largest known protein families (>800

species in humans, which constitutes 3.5% of the ~23,000 genes in the human genome). They include receptors for nucleosides, nucleotides, Ca^{2+}, catecholamines [epinephrine and norepinephrine as well as **dopamine** (Section 26-4B)] and other biogenic amines (e.g., **histamine** and **serotonin;** Section 26-4B), **eicosanoids (prostaglandins, prostacyclins, thromboxanes, leukotrienes,** and **lipoxins,** derivatives of the C_{20} fatty acid arachidonic acid, which are potent local mediators of numerous important physiological processes; Section 25-7), and for most of the large variety of peptide and protein hormones discussed in Section 19-1. In addition, GPCRs have important sensory functions: They constitute the olfactory (odorant) and gustatory (taste) receptors (of which there are estimated to be 460 different types in humans), as well as the several light-sensing proteins in the retina, which are known as **rhodopsins.** In addition, the GPCRs constitute the most important class of drug targets in the pharmaceutical arsenal (Section 15-4): ~50% of approved drugs elicit their therapeutic effects by selectively interacting with specific GPCRs.

a. GPCRs Have Similar Structures

Despite the foregoing, the structures of only few species of GPCRs have yet been elucidated, mainly due to the apparent flexibilities of their extramembranous segments and the general difficulty of crystallizing transmembrane proteins (Section 12-3Aa). The first GPCR whose X-ray

Figure 19-15   General structure of a G protein-coupled receptor (GPCR).

Cytosol

Plasma
membrane

Extracellular

Figure 19-16   X-ray structure of bovine rhodopsin. The protein
is viewed parallel to the plane of the plasma membrane with its
approximate position therein indicated. The protein is represented
by its transparent molecular surface with its polypeptide chain in
ribbon form colored in rainbow order from its N-terminus (*blue*)
to its C-terminus (*red*). Note its bundle of seven nearly parallel
transmembrane helices. The protein's retinal prosthetic group
(*magenta*) is drawn in space-filling form as are its two Cys-linked
palmitoyl groups and its two *N*-linked oligosaccharide groups
(colored according to atom type with C green, N blue, O red, and
S yellow). [Based on an X-ray structure by Tetsuji Okada, National
Institute of Advanced Industrial Science and Technology, Kyoto,
Japan; and Volker Buss, University of Duisberg-Essen, Duisberg,
Germany. PDBid 1U19.]

structure was reported was bovine rhodopsin (Fig. 19-16).
Rhodopsin consists of the 348-residue protein **opsin** that is
covalently linked to the chromophore **retinal** (Fig. 12-24)
via a Schiff base to Lys 296, much as occurs in the homologous bacteriorhodopsin (Section 12-3Ab), which is a heptahelical light-driven proton pump (Section 22-3Bh). The
absorption of a photon causes the rhodopsin-bound retinal
to isomerize from its ground state 11-*cis* form to its all-*trans*
form. This isomerization is accompanied by a transient conformational change in opsin before the all-*trans*-retinal is
hydrolyzed and dissociated from the opsin (which is
subsequently regenerated by the addition of 11-*cis*-retinal
delivered from adjacent epithelial cells in the retina). It
is this conformational change, which occurs mainly on
rhodopsin's cytosolic surface, that activates its cognate G
protein.

Rhodopsins are unique among GPCRs in that their 11-*cis*
retinal "agonist" is covalently bound to the protein. All

other GPCRs bind diffusible ligands. The X-ray structures
of β_1- and β_2-adrenergic receptors and the human A_{2A}
adenosine receptor, the only other species of GPCRs
whose structures have yet been determined, reveal that
their transmembrane portions closely resemble each other
and that of rhodopsin, and that their bound ligands occupy
positions similar to that of retinal in rhodopsin. Note that the
transmembrane helices of the GPCRs are more or less uniform in size (20–27 residues), but their extramembranous
segments, which largely form their ligand- and G protein-
binding sites, vary widely in length with the identity of the
GPCR (7–595 residues for the N- and C-termini and 5–230
residues for the loops connecting their TM helices).

C. *Heterotrimeric G Proteins: Structure and Function*

Heterotrimeric G proteins, which were first characterized
by Alfred Gilman and Martin Rodbell, are more complex
than Fig. 19-14 implies: They consist, as their name indi-
cates, of three different subunits, α, β, and γ (45, 37, and 9 kD,
respectively), of which it is G_α that binds GDP and GTP
(Fig. 19-17) and hence is a member of the G protein
superfamily. The binding of $G_\alpha \cdot GDP–G_\beta G_\gamma$ to its cognate
ligand–GPCR complex induces the G_α to exchange its
bound GDP for GTP and, in so doing, to dissociate from
$G_\beta G_\gamma$. In contrast, G_β and G_γ bind one another with such
high affinity that they only dissociate under denaturing
conditions. Consequently, we shall henceforth refer to their
complex as $G_{\beta\gamma}$.

Both G_α and $G_{\beta\gamma}$ are membrane-anchored proteins: G_α
through its myristoylation or palmitoylation or both at or
near its N-terminus (Section 12-3Bb), and $G_{\beta\gamma}$ through the
prenylation of G_γ at its C-terminus (Section 12-3Ba). These
lipid modifications stabilize the interactions of G_α with
$G_{\beta\gamma}$, as they localize both to the inner surface of the plasma
membrane.

GTP binding, in addition to decreasing G_α's affinity for
its cognate ligand–GPCR complex, increases its affinity for
its effector, AC. Thus, *it is the binding of $G_\alpha \cdot GTP$ that activates AC* (Fig. 19-17, *left*).

$G_{\beta\gamma}$ can also directly participate in signal transduction:
It activates a wide variety of signaling proteins including
several isoforms of AC (Section 19-2D), certain Na$^+$, K$^+$,
and Ca^{2+}-specific ion channels, various **protein tyrosine kinases** (Section 19-3A), and **phospholipase C-β** (**PLC-β;** a
component of the phosphoinositide signaling system; Sec-
tion 19-4Ba). $G_{\beta\gamma}$ thereby provides an important source of
cross talk between signaling systems.

On the eventual G_α-catalyzed hydrolysis of GTP, the
resulting $G_\alpha \cdot GDP$ complex dissociates from AC and reas-
sociates with $G_{\beta\gamma}$ to reform inactive G protein. *Since G_α
hydrolyzes its bound GTP at a characteristic rate, it func-
tions as a molecular clock that limits the length of time that
both $G_\alpha \cdot GTP$ and $G_{\beta\gamma}$ can interact with their effectors.*

*Several types of ligand–GPCR complexes may activate
the same G protein.* This occurs, for example, in liver cells in
response to the binding of the corresponding hormones to

Figure 19-17 Mechanism of receptor-mediated activation/inhibition of AC. The binding of hormone to a stimulatory receptor, R_s (*left*), induces it to bind G_s protein, which, in turn, stimulates the $G_{s\alpha}$ subunit of this $G_{s\alpha}G_{\beta\gamma}$ heterotrimer to exchange its bound GDP for GTP. The $G_{s\alpha}$ · GTP complex then dissociates from $G_{\beta\gamma}$ and, until it catalyzes the hydrolysis of its bound GTP to GDP, stimulates adenylate cyclase (AC) to convert ATP to cAMP. The binding of hormone to the inhibitory receptor, R_i (*right*), triggers an almost identical chain of events except that the presence of $G_{i\alpha}$ · GTP complex inhibits AC from synthesizing cAMP. R_2C_2 represents protein kinase A (PKA), whose catalytic subunit, C, when activated by the dissociation of the regulatory dimer as R_2 · cAMP$_4$ (Section 18-3Cb), activates its target cellular proteins by catalyzing their phosphorylation. The sites of action of cholera and pertussis toxins are indicated.

glucagon receptors and to β-adrenergic receptors. In such cases, the amount of cAMP produced is the sum of that induced by the individual hormones. G proteins may also act in other ways than by activating AC: They are known, for example, to stimulate the opening of K^+ channels in heart cells and to participate in the phosphoinositide signaling system (Section 19-4A).

Some ligand–GPCR complexes inhibit rather than activate AC (Fig. 19-17, right). These include the α_2-adrenergic receptor and receptors for somatostatin and opioids. The inhibitory effect is mediated by "inhibitory" G protein, **G_i,** which may have the same β and γ subunits as does "stimulatory" G protein, **G_s,** but has a different α subunit, **$G_{i\alpha}$** (41 kD). G_i acts analogously to G_s in that on binding to its corresponding ligand–GPCR complex, its $G_{i\alpha}$ subunit exchanges bound GDP for GTP and dissociates from $G_{\beta\gamma}$. However, $G_{i\alpha}$ inhibits rather than activates AC, through direct interactions and possibly because the liberated $G_{\beta\gamma}$ binds to and sequesters **$G_{s\alpha}$.** The latter mechanism is supported by the observation that liver cell membranes contain far more G_i than G_s. The activation of G_i in such cells would therefore release enough $G_{\beta\gamma}$ to bind the available $G_{s\alpha}$.

$G_{s\alpha}$ and $G_{i\alpha}$ are members of a family of related proteins, many of which have downstream effectors other than AC. This family also includes:

1. $G_{q\alpha}$, which forms a link in the phosphoinositide signaling system (Section 19-4Ba).

2. Transducin ($G_{t\alpha}$), a variant of $G_{i\alpha}$, which transduces visual stimuli by coupling the light-induced conformational change of rhodopsin to the activation of a specific phosphodiesterase, which then hydrolyzes cGMP to GMP. This **cGMP-phosphodiesterase (cGMP-PDE)** is an $\alpha\beta\gamma_2$ heterotetramer that is activated by the displacement of its inhibitory γ subunits **(PDEγ)** by their tighter binding to $G_{t\alpha} \cdot$ GTP. A cation-specific transmembrane channel (Section 20-3A) that is held open by the binding of cGMP closes on the resulting reduction in [cGMP], thereby triggering a nerve impulse (Section 20-5B) indicating that light has been detected.

3. G_{olf}, a variant of $G_{s\alpha}$, which is expressed only in olfactory sensory neurons and participates in odorant signal transduction.

4. $G_{12\alpha}$ and $G_{13\alpha}$, which participate in the regulation of the cytoskeleton.

This heterogeneity in G proteins occurs in the β and γ subunits as well as in the α subunits. In fact, 21 different α subunits, 6 different β subunits, and 12 different γ subunits have been identified in humans, some of which appear to be ubiquitously expressed whereas others are expressed only in specific cells. Thus, a cell may contain several closely related G proteins of a given type that interact with varying specificities with receptors and effectors. This complex signaling system presumably permits cells to respond in a graded manner to a variety of stimuli.

a. G Proteins Often Require Accessory Proteins to Function

The proper physiological functioning of a G protein often requires the participation of several other types of proteins:

1. A GTPase-activating protein (GAP), which as its name implies, stimulates its corresponding G protein to hydrolyze its bound GTP. This rate enhancement can be >2000-fold. The downstream effectors of $G_{t\alpha}$ and $G_{q\alpha}$, cGMP-PDE (Section 19-3E) and PLC-β (Section 19-4Ba), respectively, exhibit GAP activities toward $G_{t\alpha}$ and $G_{q\alpha}$ (which otherwise would hydrolyze GTP at physiologically insignificant rates), but AC does not exhibit GAP activity toward either $G_{s\alpha}$ or $G_{i\alpha}$. However, in humans, a diverse family of 37 **RGS proteins** (for *regulators of G protein signaling*) function as GAPs for G_α subunits by binding most avidly to them when they are in the transition state conformation for hydrolyzing GTP.

2. A guanine nucleotide exchange factor [GEF; alternatively **guanine nucleotide releasing factor (GRF)]**, which induces its corresponding G protein to release its bound GDP. The G protein subsequently binds another guanine nucleotide (GTP or GDP, which most G proteins bind with approximately equal affinities), but since cells maintain a GTP concentration that is 10-fold higher than that of GDP, this, in effect, exchanges the bound GDP for GTP. For heterotrimeric G proteins, the agonist–GPCR complexes function as GEFs.

3. A guanine nucleotide-dissociation inhibitor (GDI). A $G_{\beta\gamma}$ may be regarded as its associated G_α's GDI because

GDP dissociates slowly from isolated G_α subunits but is essentially irreversibly bound by heterotrimers.

b. The X-Ray Structures of G_α Proteins Rationalize Their Functions

The X-ray structures of the C-terminal 325 residues of the 350-residue bovine transducin-α ($G_{t\alpha}$) in its complexes with GDP (Fig. 19-18a,b) and with the poorly hydrolyzable GTP analog **GTPγS**

GTPγS

(Fig. 19-18c,d) were determined by Heidi Hamm and Paul Sigler. $G_{t\alpha}$ consists of two clearly delineated domains connected by two polypeptide linkers: (1) a highly conserved GTPase domain that is structurally similar to those in other G proteins of known structure (and hence is often described as a Ras-like domain), and (2) a helical domain that is unique to heterotrimeric G proteins. Guanine nucleotides bind to $G_{t\alpha}$ in a deep cleft that is flanked by these domains. The X-ray structures of $G_{i\alpha} \cdot$ GTPγS and $G_{s\alpha} \cdot$ GTPγS, both determined by Gilman and Stephen Sprang, closely resemble that of the $G_{t\alpha} \cdot$ GTPγS.

Comparison of the structures of the $G_{t\alpha} \cdot$ GDP and $G_{t\alpha} \cdot$ GTPγS complexes reveals that the presence of GTP's γ phosphate group promotes significant conformational shifts in three loops known as switch regions, all of which are located on the facing side of $G_{t\alpha}$ in Fig. 19-18. The γ phosphate hydrogen bonds to side chains on Switches I and II, thereby pulling these polypeptide segments in toward it and causing Switch II to contact Switch III in a way that pulls it to the right in Fig. 19-18. These concerted conformational shifts cause an extensive cavity over the GDP-binding site to largely fill in the GTPγS complex.

Switches I and II have counterparts in other G proteins of known structure. Portions of these polypeptide segments have been implicated in the interactions of $G_{t\alpha}$ with the cGMP-PDE it activates and in the interactions between the closely related $G_{s\alpha}$ with its target AC (Section 19-2D).

c. The X-Ray Structures of Heterotrimeric G Proteins

The X-ray structures of heterotrimeric G proteins were determined by Gilman and Sprang ($G_i \cdot$ GDP; Fig. 19-19) and by Hamm and Sigler ($G_t \cdot$ GDP). These structures reveal that the G_β subunit (Fig. 19-19b) consists of an N-terminal helical domain followed by a C-terminal domain comprising seven 4-stranded antiparallel β sheets arranged like the blades of a propeller—a β propeller whose blades are each formed by a WD40 sequence motif (Section 12-4Cb)—that surround a water-filled central channel. The WD40 motif occurs in a functionally diverse group of 4- to 8-bladed β-propeller proteins, including the 7-bladed N-terminal domain of the clathrin heavy chain (Section 12-4Cb). The G_γ subunit consists mainly of two helical

(a)

(b)

(c)

(d)

Figure 19-18 **Structural differences between the inactive and active forms of G$_{t\alpha}$ (transducin).** The change in structure is indicated by the comparison of the X-ray structure of G$_{t\alpha}$· GDP in its (*a*) ribbon and (*b*) space-filling forms with that of G$_{t\alpha}$· GTPγS in its (*c*) ribbon and (*d*) space-filling forms, all viewed from the same direction. In the ribbon drawings, helices and sheets are green; the segments linking them are gold; the guanine nucleotides are magenta, except for the γ phosphate of GTPγS, which is yellow; and the bound Mg^{2+} ion is represented by a blue ball. The protein's three switch regions (I, II, and III) are highlighted in cyan. In Part *c*, the two loop regions of the protein that are implicated in its interaction with the

cGMP-phosphodiesterase subunit to which it binds (PDEγ) are pointed out with yellow labels, whereas the three loop regions that are implicated in the interaction of the homologous G$_{s\alpha}$ with adenylate cyclase (AC) are indicated with pink labels. The space-filling models are colored similarly to the ribbon diagrams except for the yellow residues, which here represent those that appear to propagate or stabilize the structural transitions induced by the binding of the γ phosphate group. The box in the space-filling models outlines the cavity in G$_{t\alpha}$· GDP that closes when the GDP is replaced by GTPγS and that has been implicated in modulating the affinity of G$_{t\alpha}$ for G$_{\beta\gamma}$ and for the receptor. [Courtesy of Paul Sigler, Yale University. PDBids 1TAG and 1TND.]

segments joined by a polypeptide link (Fig. 19-19*b*). It is closely associated with G$_\beta$ along its entire extended length through mainly hydrophobic interactions and hence has no tertiary structure. The X-ray structure of isolated G$_{\beta\gamma}$ is essentially identical to that in the G$_\alpha$ · GDP–G$_{\beta\gamma}$ complex, thereby indicating that the structure of G$_{\beta\gamma}$ is unchanged by its association with G$_\alpha$ · GDP.

G$_\alpha$ and G$_{\beta\gamma}$ associate mainly via highly conserved contacts between the Switch I and II regions of G$_\alpha$ and the loops and turns at the bottom of G$_{\beta\gamma}$'s β propeller (Fig. 19-19). In addition, there is a less extensive interaction between the N-terminal helix of G$_\alpha$ (which is disordered in G$_\alpha$ alone) and the first blade of the G$_\beta$ propeller (back side of Fig. 19-19*a*). Comparison of the structure

αN

γN

βN

**GTPase
domain**

**Helical
domain**

(a)

αN

γN

βN

(b)

Figure 19-19 **X-ray structure of the heterotrimeric G protein
G_i.** *(a)* The G_α subunit is violet with its Switch I, II, and III
segments green, blue, and red, respectively, and with its bound
GDP shown in space-filling form with C green, N blue, O red,
and P yellow. The G_β subunit's N-terminal segment is blue and
each blade of its β propeller has a different color. The G_γ subunit
is gold. The view is perpendicular to the axis of the G_β subunit's
β propeller. The plasma membrane would be at the top of the
drawing as inferred from the positions of the N terminus of G_α
and the neighboring C terminus of G_γ, which, *in vivo*, are

lipid-linked to the plasma membrane. However, the orientation
of the protein relative to the plasma membrane is unknown (but
see Fig. 19-20). *(b)* View related to that in Part *a* by a 90° rotation
about its horizontal axis and thus looking from the general
direction of the plasma membrane. The protein is colored as in
Part *a* except that the G_α subunit is mainly gray. [Based on an
X-ray structure by Alfred Gilman and Stephan Sprang,
University of Texas Southwestern Medical Center. PDBid
1GP2.] 🖱 **See Interactive Exercise 12**

of $G_\alpha \cdot GDP-G_{\beta\gamma}$ with that of $G_\alpha \cdot GTP\gamma S$ reveals why
G_α cannot simultaneously bind GTP and $G_{\beta\gamma}$: *In
$G_\alpha \cdot GDP-G_{\beta\gamma}$, the Switch II segment of G_α contacts G_β in a
way that prevents Switch II from assuming the conforma-
tion it requires to bind GTP's γ phosphate.* Moreover, the
conformational changes in Switch II are coordinated with
those in Switch I so that, together, they close over the GDP
bound to $G_\alpha-G_{\beta\gamma}$, thereby accounting for its tight binding
relative to that in $G_\alpha \cdot GDP$.

The question remains, how does a liganded GCPR induce its target G_α subunit to exchange its bound GDP for
GTP? The X-ray structure of opsin in an activated conformation in complex with the 11-residue C-terminal segment
of $G_{t\alpha}$, determined by Oliver Ernst, suggests a partial answer to this question. This helical segment (the right end of
helix α5 in Fig. 19-18*a* and the left end of the nearly horizontal gray helix at the top of Fig. 19-19*b*), which has been
shown to be a major site of interaction with activated
opsin, binds at the cytosolic face of opsin. This prompted
the construction of the model in which the remainder
of $G_{t\alpha}G_{\beta\gamma}$ is appended onto the $G_{t\alpha}$ C-terminal segment
(Fig. 19-20). To avoid a steric clash between $G_{t\alpha}G_{\beta\gamma}$ and the
membrane, $G_{t\alpha}G_{\beta\gamma}$ must be tilted upward by 40°. A simple
mechanism for this process is a reorientation of the loop
connecting helix α5 and strand β6 (Fig. 19-18*a*). Since this

loop participates in binding the guanine ring of GDP, a reasonable assumption is that this conformational adjustment
expels GDP from its binding pocket, which when opsin
subsequently releases $G_{t\alpha}G_{\beta\gamma}$, allows GTP to bind in its
place.

d. Cholera Toxin Stimulates Adenylate Cyclase by Permanently Activating $G_{s\alpha}$

The major symptom of **cholera,** an intestinal disorder
caused by the bacterium *Vibrio cholerae,* is massive diarrhea that, if untreated, frequently results in death from dehydration. This dreaded disease is not an infection in the
usual sense since the vibrio neither invades nor damages
tissues but merely colonizes the intestine, much like *E. coli.*
The catastrophic fluid loss that cholera induces (often over
6 liters per hour!) occurs in response to a bacterial toxin.
Indeed, merely replacing cholera victims' lost water and
salts enables them to survive the few days necessary to immunologically eliminate the bacterial infestation.

Cholera toxin (CT; also known as **choleragen)** is an
87-kD protein of subunit composition AB_5 in which the B
subunits (103 residues each) form a pentagonal ring to which
the A subunit (240 residues) is bound. Previous to CT's secretion, its A subunit is cleaved at a single site by a bacterial protease to yield two fragments, A1 (the N-terminal

Figure 19-20 Model for signal transmission from an activated GPCR to its target heterotrimeric G protein. (*a*) The X-ray structure of activated bovine opsin (*orange,* oriented relative to that in Fig. 19-16 by an ~90° rotation about the vertical axis) in a complex with the 11-residue C-terminal helical segment from $G_{t\alpha}$ (*blue*). The X-ray structure of $G_{t\alpha}G_{\beta\gamma}$ (*gray*) is positioned over the opsin structure such that this portion of the $G_{t\alpha}$'s C-terminal helix (the remainder of which is red) is superimposed on the helical segment bound to opsin. The dashed lines delineate the membrane in which opsin is embedded. Note that this model results in a steric clash between the G protein and the membrane. (*b*) To alleviate this steric clash, the G protein has been rotated by 40° (*lower yellow arrow*) through a conformational change in the loop preceding the C-terminal helix of $G_{t\alpha}$. This loop participates in binding guanine nucleotides (GDP and GTP) and hence it is postulated that its reorganization expels GDP (drawn in space-filling form with C magenta, N blue, and O and P orange) from its binding pocket (*upper yellow arrow*). On the subsequent dissociation of opsin, the $G_{t\alpha}$ binds GTP (which the cell maintains at an ~10-fold higher concentration than GDP), and consequently dissociates from $G_{\beta\gamma}$. [Courtesy of Oliver Ernst, Charité–Universitätsmedizin Berlin, Germany. PDBids 3DQB and 1GOT.]

~195 residues) and A2 (the C-terminal ~45 residues), that remain joined by a disulfide bond. On the binding of CT to its cell surface receptor, ganglioside G_{M1} (Sections 12-1D and 25-8Cd), the nicked A subunit (but not the B subunits) is taken into the cell via receptor-mediated endocytosis and travels backward through the secretory pathway (Section 12-4B) to the Golgi apparatus. From there, it is conducted into the endoplasmic reticulum (ER) via the binding of A2's C-terminal KDEL sequence to a KDEL receptor (which normally functions to retrieve ER-resident proteins that have escaped the ER; Section 12-4Ch). The A1 fragment is then released from A2 and enters the cytoplasm

through the translocon (which normally conducts growing and still unfolded polypeptides into the ER; Section 12-4Bd) via a process in which A1 is unfolded through the chaperonelike action of protein disulfide isomerase (PDI; Section 9-2A).

In the cytoplasm, A1 catalyzes the irreversible transfer of the ADP–ribose unit from NAD^+ to a specific Arg side chain of $G_{s\alpha}$ (Fig. 19-21). This reaction is greatly accelerated by the interaction of A1 with the small Ras-like G protein **ADP-ribosylation factor (ARF)** in complex with GTP, which normally functions to prime the formation of clathrin-coated vesicles (Section 12-4Cd).

Figure 19-21 Mechanism of action of cholera toxin. The cholera toxin's A1 fragment in complex with ARF · GTP catalyzes the ADP-ribosylation of a specific Arg residue on $G_{s\alpha}$ by NAD^+, thereby rendering this subunit incapable of hydrolyzing GTP.

ADP-ribosylated $G_{s\alpha} \cdot GTP$ can activate AC but is incapable of hydrolyzing its bound GTP. As a consequence, the AC remains "locked" in its active state. The epithelial cells of the small intestine normally secrete digestive fluid (an HCO_3^--rich salt solution) in response to small increases in [cAMP] that activate intestinal Na^+ pumps through their phosphorylation by PKA (ion pumps are discussed in Sections 20-3 and 20-4). The ~100-fold rise in intracellular [cAMP] induced by CT causes these epithelial cells to pour out enormous quantities of digestive fluid, thereby producing the symptoms of cholera. CT also affects other tissues *in vitro* but does not do so *in vivo* because CT is not absorbed from the gut into the bloodstream.

The remarkable X-ray structure of CT (Fig. 19-22*a*), determined by Graham Shipley and Edwin Westbrook, reveals that its A2 segment forms an unusual extended helix whose C-terminal end inserts into the B_5 pentamer's solvent-filled central pore, where it is noncovalently anchored. The N-terminal segment of A2 extends beyond the B_5 pentamer so as to tether the wedge-shaped A1 segment to B_5, much like a balloon on a string. The X-ray structure of only B_5 in complex with the pentasaccharide from its G_{M1} receptor (Fig. 19-22*b*), determined by Wim Hol, indicates that this pentasaccharide binds through an extensive hydrogen bonded network to each B subunit on the face of B_5 opposite

that which binds A. The binding of the A subunit or the receptor pentasaccharide to B_5 causes only modest structural changes at their respective binding sites without altering B_5's subunit interfaces. A1 contains an elongated crevice in the vicinity of a catalytically implicated residue, Glu 112, that presumably forms its active site.

Certain strains of *E. coli* cause a diarrheal disease (travelers' diarrhea) similar to, although considerably less severe than, cholera through their production of **heat-labile enterotoxin (LT),** a protein that closely resembles CT (their A and B subunits are >80% identical and form AB_5 toxins that have closely similar X-ray structures) and has the same mechanism of action. The reasons for the difference in severity of these infections are unclear (cholera can be fatal within hours, whereas enterotoxic strains of *E. coli* usually only temporarily incapacitate adults, although they are responsible for the deaths of hundreds of thousands of children annually). It might be due to the modest structural differences between the toxins, differences in the amounts of toxin secreted, and/or variations in microbial ecology.

The foregoing results provide a structural basis for the design of ligands that interfere with the binding of CT and LT to their receptors. Since these receptors occur on the surface of the intestinal epithelium, ligands that compete with them need not pass through any membrane. This

(*a*)

(*b*)

Figure 19-22 **X-ray structure of cholera toxin.** (*a*) The entire AB_5 complex as viewed parallel to the presumed direction of the plane of the plasma membrane to which it binds, extracellular side up. The A1 segment is cyan, the A2 segment is gray, and each B subunit has a different color. Although the A1 and A2 segments in this structure form a continuous polypeptide chain, residues 193–195, which immediately precede the peptide bond that is cleaved on toxin activation, are disordered and hence not visible here (*upper left*). The C-terminal end of the A2 helix binds in the pentamer's central pore. [Based on an X-ray

structure by Graham Shipley, Boston University School of Medicine, and Edwin Westbrook, Northwestern University. PDBid 1XTC.] (*b*) The structure of only the B_5 pentamer in which each subunit is binding CT's G_{M1} receptor pentasaccharide. The structure is viewed as from the bottom of Part *a*. The subunits of the B_5 pentamer are colored as in Part *a* and the pentasaccharides are shown in ball-and-stick form with C green, N blue, and O red. Note the pentamer's large central pore. [Based on an X-ray structure by Wim Hol, University of Washington. PDBid 2CHB.]

greatly increases the usual ~500-D size limit for an effective drug candidate (Section 15-4Ba). Moreover, a large ligand is unlikely to enter the bloodstream and hence would have minimal side effects. Consequently, the synthesis of multivalent ligands that simultaneously bind with high affinity to all five receptor binding sites on an AB_5 molecule has yielded promising lead compounds against CT and LT.

e. Pertussis Toxin ADP-Ribosylates $G_{i\alpha}$

Bordetella pertussis, the bacterium that causes **pertussis** (whooping cough; a disease that is still responsible for ~300,000 infant deaths per year worldwide), produces an AB_5 protein, **pertussis toxin (PT),** that ADP-ribosylates a specific Cys residue in $G_{i\alpha}$. In doing so, it prevents $G_{i\alpha}$ from exchanging its bound GDP for GTP and therefore from inhibiting AC. PT's X-ray structure, determined by Randy Read, reveals that its A and B subunits are structurally homologous to those of CT and LT, although the A subunit of PT extends from the opposite face of its B pentamer relative to that in CT. Moreover, PT's B pentamer consists of four different subunits (one in two copies), each of which is only ~15% identical to the B subunits of CT and LT.

f. Receptors Are Subject to Desensitization

One of the hallmarks of biological signaling systems is that they adapt to long-term stimuli by reducing their response to them, a process named **desensitization.** *These signaling systems therefore respond to changes in stimulation levels rather than to their absolute values.* What is the mechanism of desensitization? In the case of β-adrenergic receptors, an epinephrine–receptor complex, but not receptor alone, is phosphorylated at one or more of the Ser or Thr residues on its C-terminal tail by **β-adrenergic receptor kinase (βARK).** This cytosolic protein is recruited to the plasma membrane through its interaction with $G_{\beta\gamma}$ when it is not bound to $G_{s\alpha}$, which is also a consequence of receptor activation. The phosphorylation of the receptor decreases the ability of epinephrine to influence it, at least in part by reducing the receptor's epinephrine-binding affinity. The phosphorylated receptor, in turn, is bound by either of two 78% identical proteins known as **β-arrestins** in a way that blocks the activated receptor from activating its target G protein. Moreover, the β-arrestin binds to the adaptor protein AP2 in clathrin-coated pits (Section 12-4Cd), whereupon the β-adrenergic receptor–β-arrestin complex is endocytotically sequestered (Section 12-5Bc) in specialized vesicles. The receptor's otherwise extracellular, epinephrine-binding surface then faces the interior of the vesicle and the β-arrestin is located on its outside, from which it is subsequently released. The vesicles are devoid of both heterotrimeric G protein and AC, thus further attenuating the cell's response to epinephrine. If, however, the epinephrine level is reduced, the receptor is slowly dephosphorylated by a phosphatase and exocytotically returned to the cell surface, thus restoring the cell's epinephrine sensitivity. Alternatively, the vesicles may fuse with endosomes for the delivery of their contents to lysosomes, where the receptor is proteolytically degraded (Fig. 12-91). In this latter case, the restoration of epinephrine sensitivity requires the synthesis of additional receptor.

βARK is a member of a family of seven proteins known as **GCPR kinases (GRKs)** that, together with several other types of protein kinases including PKA, phosphorylate the C-terminal tails and/or the cytosolic loops of most agonist-occupied GPCRs. As above, the phosphorylated GPCRs are bound by a β-arrestin and sequestered into endocytotic vesicles, thereby isolating them from their corresponding heterotrimeric G proteins. Moreover, the sites of a given GCPR that are phosphorylated vary in a tissue-specific way that provides further regulatory flexibility.

D. *Adenylate Cyclases*

The heterotrimeric G proteins G_s and G_i function to control the activities of adenylate cyclase (AC). In fact, mammals have 9 known membrane-bound isoforms of AC, AC1 through AC9 (alternatively, AC-I through AC-IX), which are each expressed in a tissue-specific manner and differ in their regulatory properties. These ~120-kD transmembrane glycoproteins each consist of a small N-terminal domain (N), followed by two repeats of a unit consisting of a transmembrane domain (M) followed by two consecutive cytoplasmic domains (C), thus forming the sequence $NM_1C_{1a}C_{1b}M_2C_{2a}C_{2b}$ (Fig. 19-23). The ~30% identical C_{1a} and C_{2a} domains associate to form the AC's catalytic core, whereas C_{1b}, as well as C_{1a} and C_{2a}, bind regulatory molecules. Thus, $G_{i\alpha}$ inhibits AC1, 5, and 6 by binding to C_{1a}; $G_{s\alpha}$ activates all AC isoforms but AC9 by binding to C_{2a}; $G_{\beta\gamma}$ inhibits AC1, 3, and 8 but activates AC2, 4, and 7 by binding

Figure 19-23 **Schematic diagram of a typical mammalian AC.** The M_1 and M_2 domains are each predicted to contain six transmembrane helices. C_{1a} and C_{2a} form the enzyme's pseudosymmetric catalytic core. The domains with which various regulatory proteins are known to interact are indicated. [After Tesmer, J.J.G. and Sprang, S.R., *Curr. Opin. Struct. Biol.* **8,** 713 (1998).]

to C_{2a}; and Ca^{2+}–calmodulin (Ca^{2+}–CaM; Section 18-3Ce) activates AC1, 3, and 8 by binding to C_{1b}. Moreover, the C_{2a} of AC2, 5, and 7 are activated by the phosphorylation of specific Ser/Thr control sites, for example, by **protein kinase C (PKC;** Section 19-4C), whereas the C_{1b} of AC5 and 6 are similarly inhibited by PKA, for example. Clearly, cells can respond to a great variety of stimuli in determining their cAMP levels.

No X-ray structure of an intact AC isoform has yet been reported. However, Sprang has determined the X-ray structure of a hybrid catalytic core consisting of the C_{1a} domain of AC5 (VC_1) and the C_{2a} domain of AC2 (IIC_2) in complex with $G_{s\alpha} \cdot GTP\gamma S$, **ATPαS** (an isomer of ATPγS with the S atom on the α phosphate), and **forskolin**

Forskolin

(a product of the plant *Coleus forskohlii* that activates all ACs but AC9 and functions to lower blood pressure). The $VC_1 \cdot IIC_2$ catalytic core is enzymatically active and is sensitive to both $G_{s\alpha} \cdot GTP$ and forskolin. Its X-ray structure (Fig. 19-24) reveals that VC_1 and IIC_2 form a pseudosymmetric heterodimer that binds ATPγS and forskolin at pseudosymmetrically related sites of their interface. $G_{s\alpha} \cdot GTP\gamma S$ interacts with IIC_2 mainly via its Switch II helix, which binds in a cleft on IIC_2.

The X-ray structure of the catalytically inactive C_{2a} homodimer in complex with two symmetrically arranged forskolin molecules, by James Hurley, provides a perhaps crude model for the inactivated heterodimer. Comparison of these structures (Fig. 19-24) suggests that the binding of $G_{s\alpha} \cdot GTP$ to the $C_{1a} \cdot C_{2a}$ catalytic core pries open the Switch II binding cleft on C_{2a} in a way that mechanically forces C_{1a} to rotate ~10° with respect to C_{2a}. This is postulated to reorient the complex's active site residues such that they can efficiently catalyze the conversion of ATP to cAMP. The conformational change that $G_{s\alpha}$ undergoes on hydrolyzing its bound GTP to GDP (Fig. 19-18) apparently reorients its Switch II region such that it no longer can bind to C_{2a}, thereby causing AC to revert to its inactive conformation.

VC_1 has a cleft that corresponds to the $G_{s\alpha}$-binding cleft on IIC_2. This suggests that this cleft on VC_1 provides the binding site for $G_{i\alpha}$. Indeed, mutagenesis studies on VC_1 are consistent with this hypothesis. However, the cleft on VC_1 is too narrow to accommodate the binding of a Switch II helix. This further suggests that the binding of $G_{i\alpha} \cdot GTP$ to C_{1a} pries open this cleft in a way that reorients

Figure 19-24   X-ray structure of an AC catalytic core. This core consists of dog VC_1 and rat IIC_2 in complex with bovine $G_{s\alpha} \cdot$ GTPγS and forskolin, and is shown with a model of ATP. VC_1 is tan, IIC_2 is violet, and $G_{s\alpha}$, which is only shown in part, is dark gray, with its IIC_2-contacting segments, Switch II and the α3–β5 loop, highlighted in red and blue. The forskolin and the ATP are shown in stick form with C gray, N cyan, O red, and P green. The brown ribbon shows the nonoverlapping portions of the catalytically inactive rat IIC_2 homodimer in which one of its subunits is superimposed on IIC_2 in the VC_1–IIC_2 complex. [Courtesy of Heidi Hamm, Northwestern University Medical School. The X-ray structures of the VC_1–IIC_2–Gs_{α} complex and the IIC_2 homodimer were determined by John Tesmer and Stephen Sprang, University of Texas Southwestern Medical Center, and by James Hurley, NIH. PDBids 1AZS and 1AB8.]

the complex's catalytic residues so as to reduce its catalytic activity.

E. *Phosphodiesterases*

In any chemically based signaling system, the signal molecule must eventually be eliminated in order to control the amplitude and duration of the signal and to prevent interference with the reception of subsequent signals. In the case of cAMP, this second messenger is hydrolyzed to AMP by enzymes known as **cAMP-phosphodiesterases (cAMP-PDEs).**

The PDE superfamily, which includes both cAMP-PDEs and cGMP-PDEs, is encoded in mammals by at least 20 different genes grouped into 12 families (PDE1 through PDE12). Moreover, many of the mRNAs transcribed from these genes have alternative initiation sites and alternative splice sites (Section 34-3C), so that mammals express ~50 PDE isoforms. These are functionally distinguished by their substrate specificities (for cAMP, cGMP, or both) and kinetic properties, their responses (or lack of them) to various activators and inhibitors (see below), and their tissue, cellular, and subcellular distributions. The PDEs have characteristic modular architectures with a conserved ~270-residue catalytic domain near their C-termini and widely divergent regulatory domains or motifs, usually in their N-terminal portions. Some PDEs are membrane-anchored, whereas others are cytosolic.

PDE activity, as might be expected, is elaborately controlled. Depending on its isoform, a PDE may be activated by one or more of a variety of agents including Ca^{2+}–CaM; phosphorylation by PKA, insulin-stimulated protein kinase (Section 18-3Cg), and **calmodulin-dependent protein kinase II;** and the binding of cGMP to a noncatalytic site. However, for some PDEs, cGMP is inhibitory. Phosphorylated PDEs are dephosphorylated by a variety of protein phosphatases including Ca^{2+}–CaM-dependent phosphatase and **protein phosphatase-2A.** Thus, the PDEs provide a means for cross talk between cAMP-based signaling systems and those using other types of signals.

PDEs are inhibited by a variety of drug agents that influence such widely divergent disorders as asthma, congestive heart failure, depression, erectile dysfunction, inflammation, and retinal degeneration. **Sildenafil** (trade name **Viagra**),

Sildenafil (Viagra™)

a compound used to treat erectile dysfunction, specifically inhibits PDE5, which hydrolyzes only cGMP. Sexual stimulation in males causes penile nerves to release NO, which activates guanylate cyclase to produce cGMP. This induces vascular smooth muscle relaxation in the penis, thereby increasing the inflow of blood, which results in an erection. This cGMP is eventually hydrolyzed by PDE5. Sildenafil is therefore an effective treatment in men who produce insufficient NO and hence cGMP to otherwise generate a satisfactory erection.

3 TYROSINE KINASE–BASED SIGNALING

See Guided Exploration 17: Mechanisms of hormone signaling involving the receptor tyrosine kinase system We have seen that glycogen synthesis and breakdown are regulated by the phosphorylation/dephosphorylation of the enzymes that

catalyze these metabolic processes as well as of many of the enzymes that catalyze these modification/demodification processes (Section 18-3). Numerous other processes in multicellular organisms are similarly regulated. In fact, over one-third of the proteins in vertebrates are subject to reversible phosphorylation, and the human genome contains 518 protein kinase genes (the so-called **kinome;** the Protein Kinase Resource at http://pkr.genomics.purdue.edu/ and kinase.com at http://kinase.com/ are databases for the kinome). The vast majority of the phosphorylated amino acid residues are Ser or Thr; only about 1 in 2000 is Tyr. Nevertheless, as we discuss in this section, Tyr phosphorylation is of central importance in regulating a variety of essential cellular processes.

A. *Receptor Tyrosine Kinases*

Many protein growth factors variously control the differentiation, proliferation, migration, metabolic state, and survival of their target cells by binding to their cognate **receptor tyrosine kinases (RTKs).** The RTKs form a diverse family of 58 transmembrane glycoproteins (Fig. 19-25) that each have a C-terminal cytoplasmic **protein tyrosine kinase (PTK)** domain and a single-pass transmembrane segment that is presumably an α helix. As their name indicates, PTKs catalyze the ATP-dependent phosphorylation of their target proteins at specific Tyr residues:

The PTK domains of RTKs are homologous to and, as we shall see, structurally resemble the far more abundant Ser/Thr-specific protein kinases such as PKA (Fig. 18-15).

RTKs are activated by the binding of a cognate protein growth factor to their ectodomains. It seems unlikely that the single transmembrane helix of monomeric RTKs such as **platelet-derived growth factor receptor (PDGFR)** and **epidermal growth factor receptor (EGFR;** Fig. 19-25) has the structural complexity to transmit the ligand-binding state of its ectodomain to its cytoplasmic tyrosine kinase domain. Rather, as we have seen for human growth hormone receptor (which is not an RTK), ligand binding induces receptor dimerization (Fig. 19-10). This, in turn, activates the RTK's PTK activity, as we discuss below. For RTKs that are permanent dimers, such as the **insulin receptor (InsR;** Fig. 19-25), the PTK is thought to be activated by a ligand-induced structural change (probably a counter-rotation of the two protomers that preserves the dimer's 2-fold axis of symmetry) that is transmitted across the membrane.

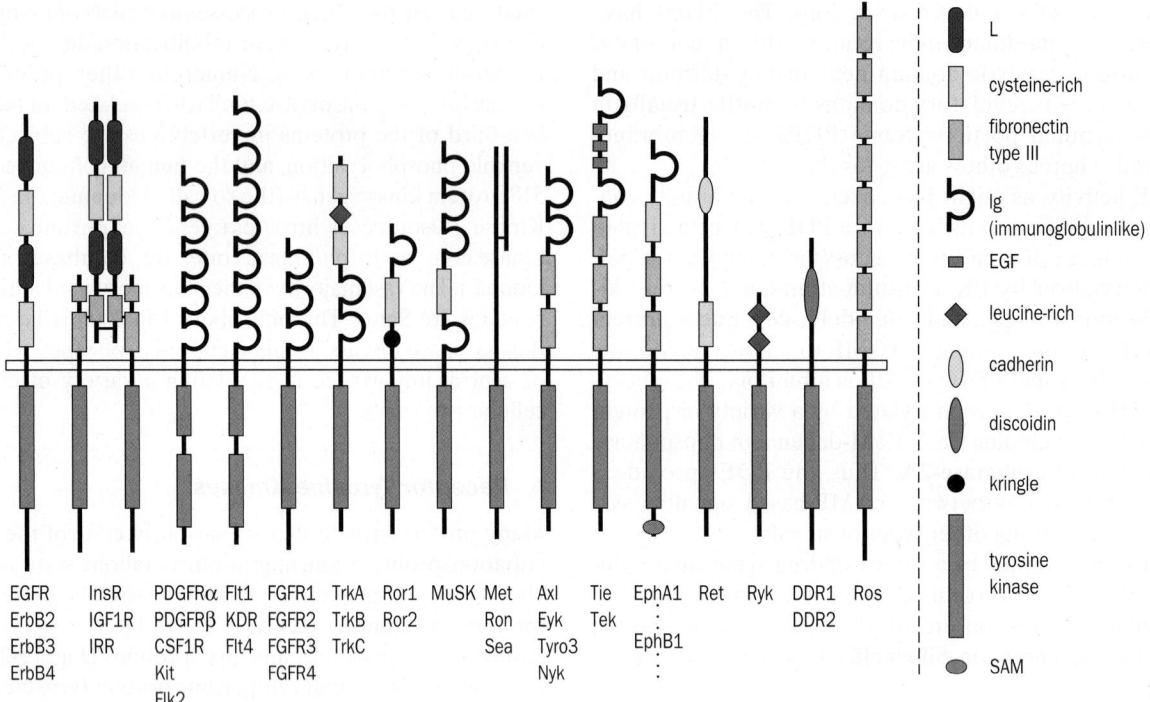

Figure 19-25 **Domain organization in a variety of receptor tyrosine kinase (RTK) subfamilies.** One to five members of each subfamily are depicted. The narrow rectangle that extends horizontally across the diagram represents the plasma membrane, with the extracellular region above it and the cytosol below. The polypeptides are shown only approximately to scale, with their N-termini above. **EGFR, InsR, PDGFR,** and **FGFR** refer to **epidermal growth factor receptor, insulin receptor, platelet-derived growth factor receptor,** and **fibroblast growth factor receptor,** respectively. The RTKs' extracellular portions are modularly constructed from a variety of often repeating domains that are identified at the right of the diagram. Note that the tyrosine kinase domains of PDGFR and **Flt1** subfamilies are interrupted by ~100-residue **kinase inserts** and that the members of the InsR subfamily are $\alpha_2\beta_2$ heterotetramers, whose subunits are disulfide-linked (short horizontal lines). [Courtesy of Stevan Hubbard, New York University School of Medicine.]

a. FGF and Heparin Sulfate Are Required to Activate FGF Receptor

The mammalian **fibroblast growth factors (FGFs)** form a family of at least 21 structurally related proteins (FGF1–21) that regulate a variety of critical biological processes including cell growth, differentiation, and migration and that are expressed in specific spatial and temporal patterns in embryos and adults. FGF-stimulated processes are mediated by four FGF receptors (FGFR1–4), which each bind a unique subset of the FGFs, thereby accounting for the diversity and tight regulation of the foregoing processes. FGFR dimerization in solution requires the presence of heparan sulfate proteoglycans (Section 11-3A) in addition to that of FGF.

FGF receptors each consist of, from N- to C-terminus (Fig. 19-25), three extracellular immunoglobulin-like domains (D1–D3 for domains 1–3), a single transmembrane helix, and a cytoplasmic domain with PTK activity. Of these, only the D2 and D3 domains are involved in FGF binding (in general, only a few of the domains in the extracellular portions of RTKs participate in ligand binding). Moosa Mohammadi determined the X-ray structure of the 2:2:2 complex of FGF2, the D2–D3 segment of FGFR1, and a heparin decasaccharide (Fig. 11-21). It reveals

(Fig. 19-26) that each FGF monomer binds to the D2 and D3 domain on one FGFR subunit and, more tenuously, to the D2 domain on the other subunit, whereas the heparin cross-links each FGF monomer to both D2 domains (whose contacts in the absence of FGF and heparin are insufficient to support appreciable FGFR dimerization).

b. RTK Dimers Are Activated by Autophosphorylation

The dimerization of an RTK (or its conformational change in the case of the insulin receptor subfamily) brings its cytoplasmically located PTK domains into apposition, such that they cross-phosphorylate each other on specific Tyr residues on their activation loops (Fig. 19-27a). This **autophosphorylation** activates the PTK in much the same way as we saw that activation loop phosphorylation induces PKA to phosphorylate its target proteins (Section 18-3Cb). In many cases, the activated PTK further phosphorylates the opposing RTK subunit at specific Tyr residues outside of the PTK domain (Fig. 19-27). This, as we shall see in Sections 19-3Cb and 19-3Cc, provides binding sites for certain cytoplasmic proteins. The activated PTK may also phosphorylate specific Tyr residues on a variety of cytoplasmic proteins. In both cases, as we discuss in Section

Figure 19-26 **X-ray structure of the 2:2:2 complex of FGF2, the D2–D3 portion of FGFR1, and a heparin decasaccharide.** The view is normal to the dimeric complex's 2-fold axis of symmetry, with the plasma membrane below. The polypeptides are represented in ribbon form with the FGF molecules cyan and blue, the D2 and D3 domains of one FGFR monomer lavender and gold, and those of the other magenta and orange. The two heparin decasaccharides are shown in space-filling form with C green, N blue, O red, and S yellow. The D2 and D3 domains of FGFR each have an immunoglobulin fold: a β sandwich comprised of a 3-stranded and a 4-stranded antiparallel β sheet (Section 8-3Bg). [Based on an X-ray structure by Moosa Mohammadi, New York University School of Medicine. PDBid 1FQ9.]

(a)

(b)

Extracellular

Plasma membrane

Cytosol

SH2 domain

Y—P

Phospho-Tyr residue

Src

Tyrosine kinase domain

Tyrosine kinase domain

PI3 kinase

Kinase insert sequence

RasGAP

Tyrosine kinase domain

SHP-2

PLC-γ1

PLC-γ1

Grb2

Shc

PDGF receptor

EGF receptor

Figure 19-27 **Schematic diagrams of RTKs.** (*a*) The PDGF receptor and (*b*) the EGF receptor. Their autophosphorylation sites and the proteins that are activated by binding to these sites via their SH2 domains (all of which are discussed in this chapter) are indicated. Note that almost all of the autophosphoryated Tyr residues that bind to other proteins lie outside the tyrosine kinase domains. [After Pawson, T. and Schlessinger, J., *Curr. Biol.* **3,** 435 (1993).]

19-3D, this causes the activation of the proteins that participate in executing the instructions implied by the extracellular presence of the protein growth factor.

c. The Insulin Receptor PTK Undergoes Major Conformational Changes on Autophosphorylation

How does autophosphorylation activate a PTK? The comparison of the X-ray structures of the PTK domain of the insulin receptor in its inactive unphosphorylated and active triphosphorylated states has done much to answer this question. The insulin receptor is expressed as a single 1382-residue precursor peptide that is proteolytically processed to yield the disulfide-linked α and β subunits (731 and 619 residues) of the mature receptor (Fig. 19-25). The X-ray structure of a 306-residue PTK-containing segment of the β subunit that was phosphorylated at its three

autophosphorylation sites, Tyr residues 1158, 1162, and 1163 (using the precursor numbering system), and in complex with the nonhydrolyzable ATP analog AMPPNP and an 18-residue peptide substrate, was determined by Stevan Hubbard. It reveals (Fig. 19-28*a*) that this PTK structurally resembles other PTKs of known structure as well as protein Ser/Thr kinases such as PKA (Fig. 18-15) and the phosphorylase kinase γ subunit (Fig. 18-21).

Comparison of this structure with that of the unphosphorylated and uncomplexed protein, determined by Hubbard and Wayne Hendrickson, reveals that, on phosphorylation and substrate binding, the PTK's N-terminal lobe undergoes a nearly rigid 21° rotation relative to the C-terminal lobe about the long axis of the protein (Fig. 19-28*b*). This dramatic conformational change closes the active site cleft about the AMPPNP and, presumably, correctly positions

(a)

(b)

Figure 19-28 **X-ray structure of the PTK domain of the insulin receptor.** (*a*) The PTK phosphorylated at Tyr residues 1158, 1162, and 1163 and in complex with AMPPNP and an 18-residue polypeptide substrate. The PTK is shown in the "standard" protein kinase orientation with its N-terminal domain pink, its C-terminal domain cyan, and its activation loop light blue. Its three phosphorylated Tyr side chains are shown in ball-and-stick form with C green, N blue, O red, and P yellow. The AMPPNP, which is identically colored, is shown in space-filling form. The substrate polypeptide is orange (only 6 of its residues are visible) and its phosphorylatable Tyr residue is shown in ball-and-stick form with C magenta and O red. (*b*) The

polypeptide backbones of the phosphorylated and unphosphorylated forms of the insulin receptor PTK domain as superimposed on their C-terminal lobes. The phosphorylated protein is green with its activation loop blue, and the unphosphorylated protein is yellow with its activation loop red. The axis (*black line*) and arrow (*blue*) indicate the rotation required to align the N-terminal domain of the unphosphorylated protein with that of the phosphorylated protein. [Part *a* based on an X-ray structure by and Part *b* courtesy of Stevan Hubbard, New York University Medical School. PDBids 1IR3 for the phosphorylated protein and 1IRK for the unphosphorylated protein.] 🔗 **See Interactive Exercise 13** critical residues for substrate binding and catalysis. All three phosphorylated Tyr residues occur on the PTK's activation loop (residues 1149–1170; recall that many Ser/Thr kinases are also phosphorylated on their activation loops; Section 18-3Cb). The unphosphorylated activation loop threads through the PTK's active site so as to prevent the binding of both ATP and protein substrates. However, on phosphorylation, the activation loop assumes a conformation that does not occlude the active site (Fig. 19-28b) but instead forms part of the substrate recognition site. The phosphate group on Tyr 1163 bridges the activation loop by forming a hydrogen bond with the side chain of the conserved Arg 1155 on one side of the loop and with the main chain N of Gly 1166 on its other side. The phosphate group of Tyr 1162 makes two hydrogen bonds to the side chain of the conserved Arg 1164. However, the phosphate group of Tyr 1168 makes no protein contacts, which suggests that it forms a docking site for downstream signaling proteins (see below). These observations are in agreement with experiments indicating that the tyrosine kinase activity of the insulin receptor increases with the degree of phosphorylation at its three autophosphorylatable Tyr side chains and that full activity is not achieved until Tyr 1163 is phosphorylated. Indeed, nearly all known RTKs have between one and three autophosphorylatable Tyr residues in their activation loops (a major exception being the members of EGFR subfamily; Fig. 19-27b) and, in all phosphorylated protein kinases of known structure, assume similar conformations.

Only the 6 centrally located residues, GDYMNM, of the 18-residue substrate peptide are seen in the foregoing X-ray structure. These include the YMXM sequence found in all efficient substrates of the insulin receptor. This segment associates with the kinase as a strand in a β sheet. Its Met side chains fit into adjacent hydrophobic pockets of the protein, whereas its phosphorylatable Tyr side chain extends toward the γ phosphate group of the AMPPNP (Fig. 19-28a). The specificity of the protein for phosphorylating Tyr rather than Ser or Thr is explained by the observation that the side chain of Tyr, but not those of Ser or Thr, are long enough to reach the active site.

B. *Cancer: The Loss of Control of Growth*

Before we continue our discussion of signaling pathways, let us consider cancer, a group of diseases that is characterized by defects in signal transduction causing uncontrolled growth. Indeed, studies of cancer have greatly increased our understanding of signal transduction and vice versa.

The cells of the body normally remain under strict developmental control. For instance, during embryogenesis, cells must differentiate, proliferate, migrate, and even die in the correct spatial arrangement and temporal sequence to yield a normally functioning organism. In the adult, the cells of certain tissues, such as the intestinal epithelium, the blood-forming tissues of the bone marrow, and those of hair follicles, continue to proliferate. Most adult body cells, however, have permanently ceased doing so.

Cells occasionally lose their developmental controls and commence excessive proliferation. The resulting tumors can be of two types:

1. Benign tumors, such as warts and moles, grow by simple expansion and often remain encapsulated by a layer of connective tissue. Benign tumors are rarely life threatening, although if they occur in an enclosed space such as in the brain or secrete large quantities of certain hormones, they can be lethal.

2. Malignant tumors or **cancers** grow in an invasive manner and shed cells that, in a process known as **metastasis,** colonize new sites in the body. Malignant tumors are almost invariably life threatening; they are responsible for 20% of the mortalities in the United States.

The most obvious and the medically most significant property of cancer cells is that they proliferate uncontrollably. For instance, when grown in a tissue culture dish, normal cells form a monocellular layer on the bottom of the dish and then, through a process termed **contact inhibition,** cease dividing (Fig. 19-29a). In contrast, the growth of malignant cells is unhampered by intercellular contacts; in culture they form multicellular layers (Fig. 19-29b). Moreover, even in the absence of contact inhibition, normal cells are far more limited in their capacity to reproduce than are cancer cells. Normal cells, depending on the species and age of the animal from which they were taken, will only divide in culture 20 to 60 times before they reach **senescence** (a stage at which they cease dividing) and die (a phenomenon that, no doubt, is at the heart of the aging process; Section 30-4Db). *Cancer cells, on the other hand, are immortal; there is no limit to the number of times they can divide.* In fact, some cancer cell lines have been maintained in culture through thousands of divisions spanning 6 decades. Immortal cells, however, are not necessarily malignant: *The hallmark of cancer is immortality combined with uncontrolled growth.*

(*a*) **Normal cells**

Growth medium

Plastic tissue culture dish

(*b*) **Transformed cells**

Figure 19-29 **Growth pattern of vertebrate cells in culture.** (*a*) Normal cells stop growing through contact inhibition once they have formed a confluent monolayer. (*b*) In contrast, transformed cells lack contact inhibition; they pile up to form a multilayer.

Figure 19-30   Variation of the cancer death rate in humans with age. The linearity of this log–log plot can be explained by the hypothesis that several randomly occurring mutations are required to generate a malignancy. The slope of the line suggests that, on the average, five such mutations are required for a malignant transformation.

a. Cancer Is Caused by Carcinogens, Radiation, and Viruses

Most cancers are caused by agents that damage DNA or interfere with its replication or repair. These include a great variety of man-made and naturally occurring substances known as **chemical carcinogens** (Section 30-5F), as well as radiation, both electromagnetic and particulate,

with sufficient energy to break chemical bonds. In addition, *certain viruses induce the formation of malignant tumors in their hosts (see below).*

Almost all malignant tumors result from the **transformation** of a single cell [conversion to the cancerous state; this term should not be confused with the acquisition of genetic information from exogenously supplied DNA (Section 5-2A)], which, then being free of its normal developmental constraints, proliferates. Yet, considering, for example, that the human body consists of around 10^{14} cells, transformation must be a very rare event. One of the major reasons for this, as the age distribution of the cancer death rate indicates (Fig. 19-30), is that *transformation requires a cell or its ancestors to have undergone several independent and presumably improbable carcinogenic changes.* Consequently, exposure to a carcinogen may prime many cells for transformation, but a malignant tumor may not form until decades later when one of these cells suffers a final transforming event.

The viral induction of cancer was first observed in 1911 by Peyton Rous, who demonstrated that cell-free filtrates from certain chicken **sarcomas** (malignant tumors arising from connective tissues) promote new sarcomas in chickens (Fig. 19-31). Although decades were to pass before the significance of this work was appreciated (Rous was awarded the Nobel prize in 1966 at the age of 85), many other such **tumor viruses** have since been characterized. The **Rous sarcoma virus (RSV),** as are all known RNA tumor viruses, is a retrovirus (an RNA virus that replicates its chromosome by copying it to DNA in a process mediated by a virally encoded reverse transcriptase, inserting the DNA into the host's genome, and then transcribing this DNA). It contains a gene, **v-*src*** ("v" for *v*iral, "*src*" for *sarcoma*), which encodes a protein named **v-Src** that mediates

(a)

(b)

Figure 19-31   Transformation of cultured chicken fibroblasts by Rous sarcoma virus. (*a*) Normal cells adhere to the surface of the culture dish, where they assume a flat extended conformation. (*b*) On infection with RVS, these cells become rounded and cluster together in piles. [Courtesy of G. Steven Martin, University of California at Berkeley.]

host cell transformation. v-*src* has therefore been termed an **oncogene** (Greek: *onkos,* mass or tumor).

What is the origin of v-*src* and what is its viral function? Hybridization studies (Section 5-3Cb) by Michael Bishop and Harold Varmus in 1976 led to the remarkable discovery that *uninfected chicken cells contain a gene, c-src ("c" for cellular), that is homologous to v-src.* Moreover, c-*src* is highly conserved in a wide variety of eukaryotes that span the evolutionary scale from *Drosophila* to humans. This observation strongly suggests that c-*src*, which antibodies directed against v-Src indicated is expressed in normal cells, is an essential cellular gene. In fact, *both v-Src and its normal cellular analog, c-Src, function to stimulate cell proliferation* (Section 19-3Ea). Apparently, v-*src* was originally acquired from a cellular source by an initially nontransforming ancestor of RSV. By maintaining the host cell in a proliferative state (cells are usually not killed by RSV infection), v-Src presumably enhances the viral replication rate.

b. Viral Oncogene Products Mimic the Effects of Protein Growth Factors and Hormones

The proteins encoded by many viral oncogenes are analogs of various growth factor and hormone system components. For instance:

1. The **v-*sis*** oncogene of **simian sarcoma virus** encodes a protein secreted by infected cells that is nearly identical to PDGF. Hence, the uncontrolled growth of simian sarcoma virus–infected cells apparently results from the continuous and inappropriate presence of this PDGF homolog.

2. Nearly half of the more than 20 known retroviral oncogenes, including v-*src*, encode PTKs. For example, the **v-*erbB*** oncogene specifies a truncated version of the EGF receptor (Fig. 19-27*b*) that lacks the EGF-binding domain but retains its transmembrane segment and its protein kinase domain. *Evidently, oncogene-encoded PTKs inappropriately phosphorylate the target proteins normally recognized by RTKs, thereby driving the afflicted cells to a state of unrestrained proliferation.*

3. The **v-*ras*** oncogene encodes a protein, **v-Ras,** that functionally resembles the monomeric G-protein **c-Ras** (Section 19-3Cf) in that it is localized on the cytoplasmic side of the mammalian plasma membrane where, when binding GTP, it activates a variety of cellular processes by stimulating the phosphorylation of numerous proteins at specific Ser and Thr residues. Although v-Ras hydrolyzes GTP to GDP, it does so much more slowly than c-Ras. The restraint to protein phosphorylation that GTP hydrolysis would normally impose on c-Ras is thus greatly reduced in v-Ras, thereby transforming the cell.

4. Several viral oncogenes, including **v-*jun*** and **v-*fos*,** encode nuclear proteins whose corresponding normal cellular analogs are synthesized in response to growth factors such as EGF and PDGF that induce mitosis (cell division). Many such proteins, including the v-*jun* and v-*fos* gene products, bind to DNA, strongly suggesting that they influence

its transcription and/or replication. Indeed, **v-*jun*** is 80% identical in sequence to the **proto-oncogene** (normal cellular analog of an oncogene) **c-*jun*,** which encodes a transcription factor named **Jun** (also called **AP-1;** Section 19-3D). Moreover, Jun/AP-1 forms a tight complex with the protein encoded by the proto-oncogene **c-*fos*,** which greatly increases the ability of Jun/AP-1 to stimulate transcription from Jun-responsive genes.

Oncogene products therefore appear to be functionally modified or inappropriately expressed components of elaborate control networks that regulate cell growth and differentiation. The complexity of these networks (as we shall see, cells generally respond to a variety of growth factors, hormones, and transcription factors in partially overlapping ways) is probably why malignant transformation requires several independent carcinogenic events. Note, however, that few human cancers are virally induced; nearly all of them arise from genetic alterations involving proto-oncogenes. We discuss the nature of these alterations in Section 34-4C.

C. Relaying the Signal: Binding Modules, Adaptors, GEFs, and GAPs

Many autophosphorylated RTKs can directly phosphorylate their target proteins. Surprisingly, however, not all RTKs do so. How, then, do they activate their target proteins? The answer, as we shall see, is through a highly diverse and complicated set of interconnected signaling pathways involving cascades of associating proteins.

a. Two-Hybrid Systems Identify Proteins That Interact *in vivo*

Before we consider the interacting proteins that participate in RTK-mediated signal transduction, let us discuss one of the most often used methods to detect their associations *in vivo*, the **two-hybrid system.** This ingenious experimental technique, which was formulated by Stanley Fields, is based on the peculiar bipartite nature of many transcription factors (proteins that bind to the promoters and other upstream control regions of eukaryotic genes and, in doing so, influence the rate at which RNA polymerase initiates the transcription of these genes; Section 5-4Ab). Such transcription factors, as we further discuss in Section 34-3Bi, contain a DNA-binding domain (DBD) that targets the transcription factor to a specific DNA sequence and an activation domain (AD) that recruits RNA polymerase to initiate transcription at a nearby transcriptional initiation site. These two domains function independently, so that a genetically engineered hybrid protein with the DBD of one transcription factor and the AD of another will activate the transcription of the gene for which the DBD is targeted. Moreover, it makes little difference as to whether the DBD is on the N-terminal or the C-terminal side of the AD, regardless of how they are arranged in their parent proteins. Evidently, as long as a DBD and an AD are held in proximity, they function as a transcription factor for the gene(s) to which the DBD is targeted.

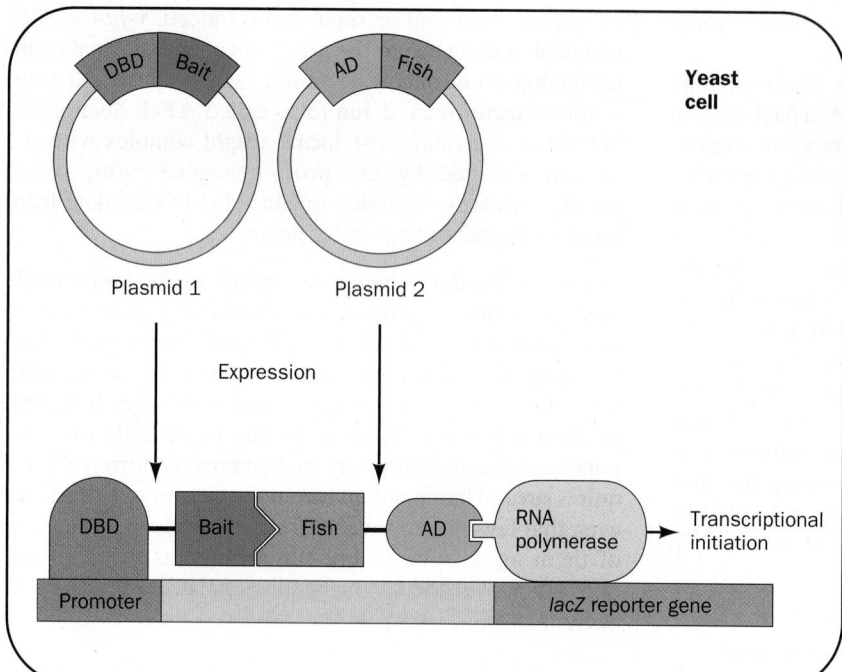

Figure 19-32 **The two-hybrid system.** Two plasmids are expressed in yeast: One encodes a hybrid protein consisting of the DNA-binding domain (DBD) of a transcription factor fused to a "bait" protein, and the other encodes the activation domain (AD) of a transcription factor fused to a "fish" protein. The DBD specifically binds to the promoter of a reporter gene, here *lacZ.* If the "fish" protein associates with the "bait" protein, its attached AD recruits RNA polymerase to initiate the transcription of the reporter gene. The β-galactosidase encoded by the *lacZ* gene is readily detected through the use of X-gal, which turns blue when hydrolyzed by β-galactosidase. Yeast colonies expressing "bait" and "fish" proteins that do not interact remain colorless, as do colonies that do not express both plasmids.

The two-hybrid system employs two different plasmids in yeast (Fig. 19-32): One encodes a hybrid protein consisting of a DBD fused to a so-called bait or probe protein; the other encodes a hybrid protein consisting of an AD fused to a so-called fish or target protein. The DBD is targeted to a reporter gene that has been engineered into the yeast chromosome such as the *E. coli lacZ* gene, which encodes the enzyme β-galactosidase. On culture plates containing X-gal (a colorless compound that turns blue on being hydrolyzed by β-galactosidase; Section 5-5Ca), yeast colonies expressing β-galactosidase turn blue, thereby indicating that the bait and fish proteins they encode associate with one another. Using this technique, cells can be screened for fish proteins that specifically interact with a particular bait protein by properly inserting the various cDNAs (Section 5-5F) derived from the cells into the AD-containing plasmid. A fish protein selected in this way can then be identified by sequencing its cDNA.

b. SH2 Domains Mediate Signal Transduction

Now let us return to our discussion of RTK-mediated signaling. Many (>100) of the diverse cytoplasmic proteins that bind to autophosphorylated receptors, for example, the cytoplasmic PTK **c-Src** (henceforth referred as just Src), certain GTPase activating proteins (GAPs), and **phospholipase C-γ** (Section 19-4Bc), contain one or two conserved ~100-residue modules known as **Src homology domain 2 (SH2;** so named because they were first noticed in tyrosine kinases related to Src; **SH1** refers to their catalytic domains). *SH2 domains specifically bind phosphoTyr residues in their target peptides with high affinity;* they bind their unphosphorylated target peptides weakly if at all. Most of the phosphoTyr residues to which SH2 binds are located in the juxtamembrane (just after the transmem-

brane helix), kinase insert, and C-terminal regions of RTKs; those in the activation loop function mainly to stimulate PTK activity. Indeed, RTK autophosphorylation occurs in two phases: First the activation loops of the RTK are phosphorylated and then the resulting activated PTK phosphorylates the other sites on the opposing subunit of the RTK.

The X-ray and NMR structures of SH2 domains from several proteins, both alone and in complex with phospho-Tyr (pY)-containing polypeptides, have been determined. SH2 is a hemispherically shaped domain, which contains a central 5-stranded antiparallel β sheet that is sandwiched between two nearly parallel α helices. The N- and C-terminal residues of SH2 are in close proximity on the surface opposite the peptide-binding site, which suggests that this domain can be inserted between any two surface residues on a protein without greatly disturbing its fold or function. Indeed, the sequences of a variety of SH2-containing proteins reveal no apparent preference for the location of this domain in a protein.

John Kuriyan determined the X-ray structure of the SH2 domain of Src in complex with an 11-residue polypeptide containing the sequence pYEEI, a tetrapeptide segment that binds to this SH2 domain with high affinity. The 11-residue peptide binds to the SH2 domain in an extended conformation, with contact being made primarily by the pYEEI tetrapeptide (Fig. 19-33a). The phosphoTyr side chain inserts into a small cleft formed, in part, by three highly conserved positively charged residues, including the side chain of an invariant Arg, which contacts the phosphate group. The Ile side chain is similarly inserted into a nearby hydrophobic pocket and the entire tetrapeptide segment interacts very tightly with SH2, although the side chains of the peptide's two central Glu residues do not

(a)

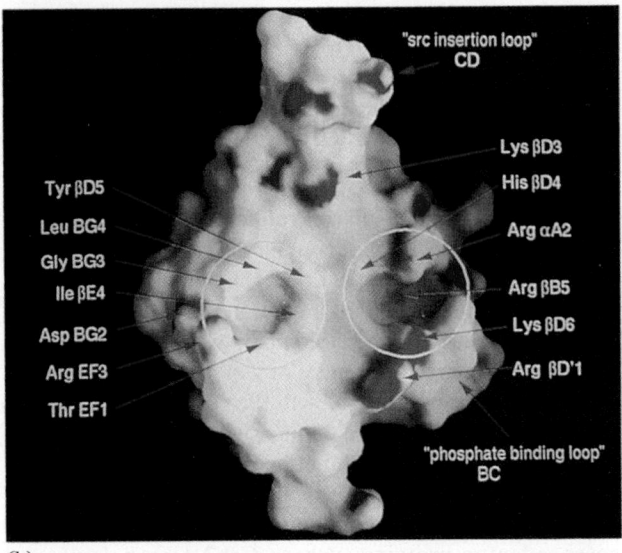

(b)

Figure 19-33 **X-ray structure of the 104-residue Src SH2 domain in complex with an 11-residue polypeptide (EPQpYEEIPIYL) containing the protein's pYEEI target tetrapeptide.** (*a*) A cutaway view of the complex, in which its solvent-accessible surface is represented by red dots, the protein (*pink*) is shown in ribbon form with its side chains in stick form (*light blue*), and the N-terminal 8-residue segment of the bound polypeptide is shown in space-filling form with its backbone yellow, its side chains green, and its phosphate group white [the

N-terminal Pro side chain (*left*) is largely obscured in this view and the C-terminal three residues are disordered]. (*b*) The molecular surface of the protein only, as viewed toward the peptide binding site and colored according to its local electrostatic potential with the most positive regions deep blue and the most negative regions deep red. The binding pockets for the phosphoTyr (*right*) and Ile (*left*) side chains are circled in yellow and important residues are identified by red arrows. [Courtesy of John Kuriyan, The Rockefeller University.]

project toward SH2. Thus, the peptide resembles a two-pronged plug that is inserted into a two-holed socket on SH2 (Fig. 19-33*b*). Comparison of this structure with that of uncomplexed Src SH2 indicates that, on binding peptide, SH2 undergoes only small conformational changes that are localized at its peptide-binding site. These structures provide a simple explanation for why SH2 does not bind the far more abundant phosphoSer- and phosphoThr-containing peptides: The side chains of these residues are too short for their phosphate groups to contact the invariant Arg side chain at the bottom of the phosphoTyr-binding pocket.

c. PTB Domains Also Bind PhosphoTyr-Containing Peptides

A second type of motif that specifically binds to phosphoTyr-containing target peptides is known as the **phosphotyrosine-binding (PTB) domain.** PTB domains specifically bind the consensus sequence NPXpY (where X is any residue) and hence recognize the sequence on the N-terminal side of pY rather than that on its C-terminal side, as do SH2 domains.

The NMR structure of the 195-residue PTB domain of **Shc,** an adaptor protein (see below), in complex with a 12-residue target peptide containing the centrally located sequence NPQpY was determined by Stephen Fesik. The structure consists of a β sandwich comprising two nearly perpendicular antiparallel β sheets flanked by three α helices (Fig. 19-34). The N-terminal segment of the target phosphopeptide assumes an extended conformation that,

Figure 19-34 **The NMR structure of the PTB domain of Shc in complex with a 12-residue polypeptide (HIIENPQpYFSDA) from the Shc binding site of a nerve growth factor (NGF) receptor.** The PTB domain is colored in rainbow order from its N-terminus (*blue*) to its C-terminus (*red*) and the target peptide is magenta. The target peptide's phosphoTyr (pY) side chain and the side chain of Arg 67, which form a salt bridge (*dashed line*), are drawn in stick form with C green, N blue, O red, and P orange. Note that the pY side chain extends from a β turn on the peptide ligand. [Based on an NMR structure by Stephen Fesik, Abbott Laboratories, Abbott Park, Illinois. PDBid 1SHC.]

in effect, forms an additional antiparallel β strand of one of the β sheets. The NPQpY segment forms a β turn in which the phosphate group contacts an Arg side chain that mutational studies indicate is essential for the binding of target peptide.

d. SH3 Domains Bind to Proline-Rich Peptides

Many of the RTKs that contain SH2 domains also have one or more 50- to 75-residue **SH3 domains.** Moreover, SH3 is contained in several membrane-associated proteins that lack an SH2. The SH3 domain, which is unrelated to SH2, binds Pro-rich sequences of 9 or 10 residues containing the motif Pro-X-X-Pro, with the residues surrounding this motif targeting these sequences to specific SH3 domains. The physiological function of SH3 is less apparent than that of SH2 because SH3 occurs in a greater variety of proteins, including receptor and nonreceptor tyrosine kinases, adaptor proteins such as **Grb2** (see below), and structural proteins such as spectrin and myosin. However, the observation that the deletion of the SH3 domain–encoding segments from the proto-oncogenes *Src* and *Abl* (which both encode PTKs) converts them to oncogenes suggests that SH3, much like SH2 and PTB, functions to mediate the interactions between kinases and regulatory proteins. SH2, PTB, and SH3 have therefore been called "molecular velcro."

The X-ray and NMR structures of SH3 domains from several proteins indicate that the SH3 core consists of two 3-stranded, antiparallel β sheets that pack against each other with their strands nearly perpendicular. As with SH2, the close proximity of SH3's N- and C-termini suggests that this domain could be modularly inserted between two residues on the surface of another protein without greatly perturbing either structure. The X-ray structures of the SH3 domains from the tyrosine kinases Abl and **Fyn** in complex with two different 10-residue Pro-rich polypeptides to which they tightly bind were determined by Andrea Musacchio and Matti Saraste. Both decapeptides assume nearly identical conformations with their C-terminal 7 residues in the polyproline II helix conformation (Section 8-1Bb). The peptides bind to SH3 over their entire length in three geometrically complementary cavities (Fig. 19-35), which are mostly occupied by Pro side chains.

e. Other Binding Modules

Several other binding modules have been implicated in mediating signal transduction. These include:

1. The **WW domain** (named after its two highly conserved Trp residues), an ~40-residue module that binds to Pro-rich sequences on its target proteins.

2. The **pleckstrin homology (PH) domain** (so named because was first recognized in **pleckstrin** [for *pl*atelet and *leu*kocyte *C k*inase sub*str*ate prote*in*)], an ~120-residue module that is present in >100 proteins. It structurally resembles the PTB domain (Fig. 19-34) but binds to the inositol head groups of phosphoinositides (Section 19-4). It therefore targets its attached proteins to the inner surface

Figure 19-35 **X-ray structure of the SH3 domain from Abl protein in complex with its 10-residue target Pro-rich polypeptide (APTMPPPLPP).** The protein is represented by its surface diagram (*cyan*) and the peptide is drawn in stick form with Pro C magenta, other C green, N blue, O red, and S yellow. [Based on an X-ray structure by Andrea Musacchio, European Molecular Biology Laboratory, Heidelberg, Germany. PDBid 1ABO.]

of the plasma membrane. In addition to its role in intracellular signaling, the PH domain participates in cytoskeletal organization, regulation of intracellular membrane transport, and modification of membrane lipids. Its structure is discussed in Section 19-4Bb.

3. The **PDZ domain** (named after the three proteins in which it was first described: *PSD*-95, *Dlg*, and *ZO*-1), an ~100-residue module that mainly binds to the C-terminal tripeptide, Ser/Thr-X-Val, of its target proteins.

Many of the proteins that participate in signal transduction consist of several modular units that also occur in several, if not many, other such proteins. These modules may have enzymatic activities (e.g., PTK activity) or bind to specific molecular motifs, such as a phosphoTyr residue within a specific sequence (e.g., SH2 domains) or another protein module (e.g., SH3 domains). Apparently, signaling proteins have arisen through the evolutionary shuffling of these modules to generate different combinations of interactions and activities. Indeed, we shall see that the complex behavior of these signaling proteins is a consequence of the interactions among their various modules.

f. Ras Is Activated by Phosphorylated RTKs via a Grb2–Sos Complex

c-Ras *(or just Ras), a proto-oncogene product, is a monomeric membrane-anchored (by prenylation) G protein* *that lies at the center of an intracellular signaling system: It regulates such essential cellular functions as growth and differentiation through the phosphorylation and hence activation of a variety of proteins.* Ras is the prototypic member of the superfamily of small G proteins, whose 154 members form five principal families, those of (1) Ras; (2) Arf and (3) Rab, which both participate in vesicle trafficking (Sections 12-4Cd and 12-4Db); (4) **Rho,** which is mainly involved in regulating the cytoskeleton (Section 35-3Ed); and (5) **Ran,** which regulates nuclear import and export.

In the signaling pathway described in Section 19-3D, the binding of ligand to RTKs ultimately activates a guanine nucleotide exchange factor (GEF; Section 19-2Ca) to exchange a Ras-bound GDP for GTP. Only Ras · GTP is capable of further relaying the signal. However, as do the homologous and structurally similar α subunits of the heterotrimeric G proteins, Ras eventually hydrolyzes its bound GTP to GDP, thereby halting further signal transduction and limiting the magnitude of the signal generated by the binding of ligand to the receptor. Indeed, the structure of Ras closely resembles those of the GTPase domains of the heterotrimeric G protein G_α subunits, including their Switch I and Switch II regions (Fig. 19-18). Mammalian cells express four Ras homologs: H-Ras, N-Ras, K-Ras 4A, and K-Ras 4B.

Molecular genetic analyses of signaling in a variety of distantly related organisms (notably humans, mice, *Xenopus, Drosophila,* and the nematode worm *Caenorhabditis elegans*) have revealed a remarkably conserved pathway in which *RTKs funnel the signal that they have bound ligand to Ras, which, in turn, relays the signal, via a so-called MAP kinase cascade, to the transcriptional apparatus in the nucleus (Section 19-3D).* Nevertheless, the way in which messages are passed between the RTKs and Ras remained enigmatic for several years until investigations in numerous laboratories revealed their major details. In particular, these studies demonstrated that *two previously characterized proteins, Grb2 and Sos, form a complex that bridges activated RTKs and Ras in a way that induces Ras to exchange its bound GDP for GTP, thereby activating it (i.e., they act as a GEF).*

The mammalian protein Grb2, a 217-residue homolog of **drk** in *Drosophila* and **Sem-5** in *C. elegans,* consists almost entirely of an SH2 domain flanked by two SH3 domains. Sos protein (the 1596-residue product of the *Son of Sevenless* gene, so named because Sos interacts with the *Sevenless* gene product, an RTK that regulates the development of the R7 photoreceptor cell in the *Drosophila* compound eye), which is required for Ras-mediated signaling, contains a central domain homologous to known Ras-GEFs and a Pro-rich sequence in its C-terminal segment similar to known SH3-binding motifs. Moreover, mammalian homologs of Sos **(mSos)** have been shown to specifically stimulate guanine nucleotide exchange in mammalian Ras proteins. Western blotting techniques (Section 6-4Bc) using anti-Grb2 and anti-mSos antibodies indicated that Grb2 binds to the C-terminal segment of mSos but not when one of the Pro residues in Sos' SH3-binding motif has been replaced by Leu or in the presence

of synthetic polypeptides that have these Pro-rich sequences. Similar studies indicated that in the presence of epidermal growth factor (EGF), the EGF receptor (an RTK; Figs. 19-25 and 19-27*b*) specifically binds the Grb2–mSos complex. However, this interaction is blocked by the presence of a phosphopeptide with the sequence of the peptide segment containing one of the activated EGF receptor's phosphoTyr residues. Evidently, Grb2's SH2 domain binds a phosphoTyr-containing peptide segment in an activated RTK while its two SH3 domains bind the Pro-rich sequences of Sos. The GEF function of Sos is thereby stimulated to activate Ras.

g. Grb2, Shc, and IRS Are Adaptors That Recruit Sos to the Vicinity of Ras

The X-ray structure of Grb2, determined by Arnaud Ducruix, reveals that neither of its SH3 domains contacts its SH2 domain (Fig. 19-36). Moreover, although the two SH3 domains are in contact, their interface area is relatively small and hence is more likely to be an artifact of crystallization than a structural feature of Grb2 in solution. It therefore appears that Grb2's SH3 domains are flexibly linked to its SH2 domain. How does the binding of such a pliable **adaptor** (a linker that lacks enzymatic activity) to a phosphorylated RTK stimulate Sos to act as Ras's GEF? Grb2 and Sos bind one another so tightly that they are essentially permanently associated in the cell. Hence, when

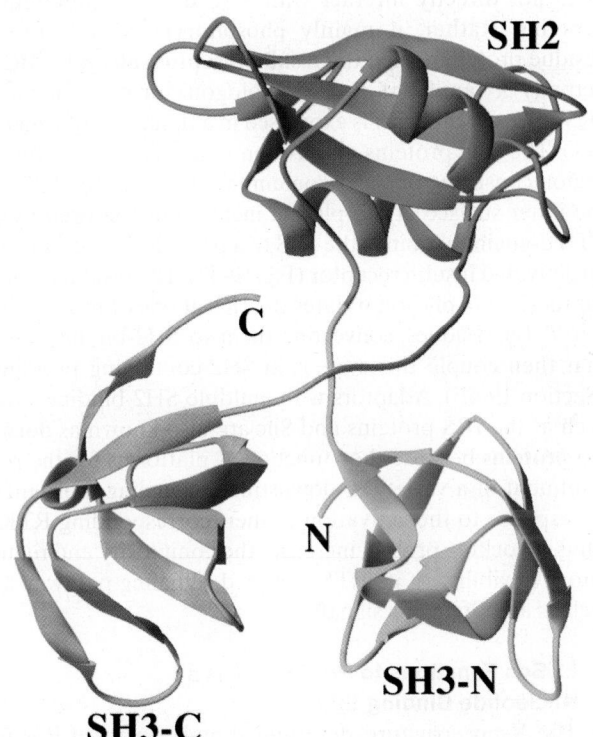

Figure 19-36 **X-ray structure of Grb2.** Its SH2 domain (*green*) is linked to its flanking SH3 domains (*cyan and orange*) via apparently unstructured and hence flexible 4-residue linkers. [Based on an X-ray structure by Arnaud Ducruix, Université de Paris-Sud, Gif sur Yvette Cedex, France. PDBid 1GRI.]

Figure 19-37   Structure of an insulin receptor substrate (IRS).
An IRS contains a PH and a PTB domain at its N-terminus
followed by multiple phosphoTyr-containing binding sites for the
SH2 domains of downstream signaling proteins.

Grb2 binds its target phosphorylated RTK, it recruits Sos
to the inner surface of the plasma membrane, where Sos's
then increased local concentration causes it to more readily bind to the membrane-anchored Ras.

Shc proteins are also adaptors that link activated RTKs
to Ras. Shc proteins consist of an N-terminal PTB domain
(Fig. 19-34), a central effector region (CH1), and a C-terminal
SH2 domain that binds to certain activated RTKs. Moreover, Shc proteins are also major targets of various RTKs,
which phosphorylate them within their CH1 domains at
sequences that then form binding sites for the Grb2 SH2
domain. Activated RTKs may therefore bind Grb2 indirectly via Shc as well as directly. Moreover, in some cases,
an Shc–Grb2–Sos complex that lacks a bound RTK may
activate Ras.

The activated (autophosphorylated) insulin receptor
does not directly interact with SH2 domain–containing
proteins. Rather, it mainly phosphorylates an ~1300-
residue protein named the **insulin receptor substrate** (IRS;
actually a family of four homologous proteins named
IRS1–4, each of which is expressed in a tissue-specific manner). The IRS proteins all have an N-terminal "targeting"
region consisting of a PH domain that localizes the IRS to
the inner surface of the plasma membrane, followed by a
PTB domain that binds the IRS to a phosphoTyr residue of
an activated insulin receptor (Fig. 19-37). The insulin receptor thereupon phosphorylates the IRS at one or more of its
6 to 8 Tyr residues, converting them to SH2-binding sites
that then couple this system to SH2-containing proteins
(Section 19-4F). Adaptors with multiple SH2-binding sites
such as the IRS proteins and Shc are also known as **docking proteins** because they function as platforms for the recruitment of a variety of downstream signaling molecules
in response to the activation of their corresponding RTK.
Thus, a docking protein increases the complexity and regulatory flexibility of its RTK-initiated signaling pathway as
well as amplifying its signal.

h. Sos Functions to Pry Open Ras's Nucleotide Binding Site

The X-ray structure, determined by Kuriyan, of Ras in
complex with a 506-residue GEF-containing segment of
Sos reveals how Sos induces Ras to exchange its normally
tightly bound GDP for GTP. This Sos segment consists
of two α helical domains, of which only the C-terminal

so-called catalytic domain contacts Ras. Ras is packed
against the center of the elongated bowl-shaped catalytic
domain (Fig. 19-38). Those portions of Ras that interact
with the catalytic domain include both its Switch I and
Switch II regions as well as the loop that binds the α and β
phosphates of GDP and GTP, the so-called **P-loop** (although GMP binds to Ras with 10^6-fold less affinity than
does GDP, so that the β phosphate of GDP is largely responsible for its tight binding to Ras). This interaction displaces Switch I relative to its position in the X-ray structure
of Ras in complex with the nonhydrolyzable GTP analog
GMPPNP:

**Guanosine-5′-(β,γ-imido)triphosphate
(GMPPNP)**

Ras's nucleotide binding site is thereby partially opened up
and a Leu and a Glu side chain from Sos are, respectively,
introduced into Ras' Mg^{2+}-binding site and the site that
binds the α phosphate group of GDP/GTP. However, this
interaction does not significantly occlude Ras's guanine
and ribose binding sites. This rationalizes how the Sos–Ras
interaction can be strong enough to displace the normally
tightly bound GDP from Ras but yet weak enough so that

**Figure 19-38   X-ray structure of the complex between Ras and
the GEF-containing region of Sos.** The N-terminal domain of the
Sos region is blue, its catalytic domain is green, and Ras is mainly
gray, with its Switch I and II regions orange and its P loop red.
Conserved regions (the SCRs) among the Ras family of GEFs
are cyan. [Courtesy of John Kuriyan, The Rockefeller University.
PDBid 1BDK.]

GTP (or the 10-fold less abundant GDP) can subsequently displace Sos from Ras.

The different families of small G proteins interact with different classes of GEFs, whose catalytic domains share no sequence similarity and are structurally unrelated [e.g., rhodopsin (Fig. 19-16) is a GEF for $G_{t\alpha}$]. Nevertheless, many of these GEFs share the same general mechanism for promoting GDP–GTP exchange, which suggests that this mechanism arose on several occasions through convergent evolution.

i. GAPs Function to Turn Off Ras-Mediated Signals

Ras hydrolyzes its bound GTP with a rate constant of 0.02 min^{-1} (vs 2–3 min^{-1} for G_{α} subunits), too slowly for effective signal transduction. This led to the discovery of a 120-kD GTPase activating protein, **RasGAP,** that, on binding Ras · GTP, accelerates the rate of GTP hydrolysis by a factor of 10^5. RasGAP's physiological importance as a regulator of Ras-mediated signal transduction is demonstrated by the observation that the relative biological activities of Ras mutants are better correlated with their

Figure 19-39 X-ray structure of the GAP334 · Ras · GDP · AlF₃ complex. The active site regions of the proteins are shown as ribbons with Ras cyan, its Gly 12 magenta, GAP334 yellow, and its finger loop red. The GDP, AlF₃, and the side chains of Ras Asn 61 and GAP334 Arg 789 are drawn in stick form with C green, N blue, O red, F yellow-green, P orange, and Al pink; the nucleophilic water molecule is represented by a red sphere; and hydrogen bonds are shown as dashed lines. [Based on an X-ray structure by Alfred Wittinghofer, Max-Planck-Institut für Molekulare Physiologie, Dortmund, Germany. PDBid 1WQ1.]

resistance to regulation by RasGAP than by their intrinsic GTPase activity.

The mechanism whereby RasGAP activates the GTPase activity of Ras was revealed by the X-ray structure, determined by Alfred Wittinghofer, of the 334-residue GTPase-activating domain of RasGAP (GAP334) bound to Ras in its complex with GDP and AlF₃ (Fig. 19-39). GAP334, which consists of two all-helical domains, interacts with Ras over an extensive surface that includes its Switch I and II regions. The AlF₃, which has trigonal planar symmetry, binds to Ras at the expected position of GTP's γ phosphate group, with the Al atom opposite a bound water molecule that presumably would be the attacking nucleophile in the GTPase reaction. Since Al—F and P—O bonds have similar lengths and phosphoryl-transfer reactions occur via a trigonal bipyramidal transition state (Fig. 16-6*b*), the GDP–AlF₃–H₂O assembly presumably resembles the GTPase reaction's transition state, with the AlF₃ mimicking the planar PO₃ group. Note that Ras · GDP does not by itself bind AlF₃.

GAP334 binds to Ras with GAP334's exposed so-called finger loop inserted into the Ras active site such that the finger loop's Arg 789 side chain interacts with both the Ras-bound GDP's β phosphate and the AlF₃ (Fig. 19-39). In Ras · GTP, this Arg side chain would be in an excellent position to stabilize the developing negative charge in the GTPase reaction's transition state. Indeed, the catalytically more efficient G_{α} subunits contain an Arg residue (Arg 178 in $G_{i\alpha}$), whose guanidinium group occupies a nearly identical position (in $G_{s\alpha}$, this is the Arg side chain that is ADP-ribosylated by cholera toxin; Section 19-2Cd). The main chain carbonyl O of Arg 789 hydrogen bonds to the side chain N of Ras' catalytically important Gln 61. The O of this side chain is thereby positioned to hydrogen bond with the nucleophilic water molecule while its NH₂ group interacts with an F atom of AlF₃ (Fig. 19-37), an arrangement that presumably stabilizes the GTPase reaction's transition state.

j. Oncogenic Mutants of Ras Are GAP-Insensitive

Mutations in Ras of Gly 12 and Gln 61 are its most common oncogenic mutations (an oncogenic form of which is found in ~30% of human cancers). These mutations prevent RasGAP from activating Ras to hydrolyze its bound GTP and hence lock Ras into its active conformation. The foregoing X-ray structure reveals why these mutants are GAP-insensitive. Gly 12 is in such close proximity to the finger loop that even the smallest possible residue change (to Ala) would sterically interfere with the geometry of the transition state through steric clashes with the main chain of Arg 789 (of RasGAP) and the side chain NH₂ of Gln 61. The observation that Gly 12 mutants of Ras bind GTP with nearly wild-type affinity therefore suggests that larger side chains can be tolerated at Ras residue 12 in the Ras–RasGAP Michaelis complex but not in the transition state. The apparent participation of Gln 61 in transition state stabilization confirms that this residue has an essential role in catalysis.

D. MAP Kinase Signaling Cascades

*The signaling pathway downstream of Ras consists of a linear series of Ser/Thr kinases that form a so-called **MAP kinase cascade** (Fig. 19-40). Many of the proteins that par-*

ticipate in MAP kinase cascades are the products of proto-oncogenes:

1. Raf, a Ser/Thr protein kinase, is activated by direct interaction with Ras · GTP (although other signaling

Figure 19-40   The Ras-activated MAP kinase cascade. This signaling cascade begins when an RTK binds its cognate growth factor, thereby inducing the autophosphorylation of this RTK's cytosolic domain. Grb2/Sem-5 binds to the resulting phosphoTyr-containing peptide segment via its SH2 domain and simultaneously binds to Pro-rich segments on Sos via its two SH3 domains. This activates Sos as a guanine nucleotide exchange factor (GEF) to exchange Ras' bound GDP for GTP, which activates Ras to bind to Raf. Then, Raf, a Ser/Thr kinase,

phosphorylates MEK, which in turn phosphorylates MAPK, which then migrates to the nucleus, where it phosphorylates transcription factors such as Fos, Jun, and Myc, thereby modulating gene expression. The MAP kinase cascade eventually returns to its resting state through the actions of protein phosphatases (Section 19-3F) after a GTPase activating protein (GAP) deactivates Ras by inducing it to hydrolyze its bound GTP to GDP. [After Egan, S.E. and Weinberg, R.A., *Nature* **365,** 782 (1993).] 🎞️ **See the Animated Figures**

Figure 19-41   X-ray structure of the Ras binding domain of Raf (RafRBD; *orange*) in complex with Rap1A · GMPPNP (*blue*). The Switch I and II regions of Rap1A are magenta and green, and its bound GMPPNP is shown in space-filling form with C green, N blue, O red, and P yellow. Rap1A · GMPPNP and Ras · GMPPNP have nearly identical structures. [Based on an X-ray structure by Alfred Wittinghofer, Max-Planck-Institut für Molekulare Physiologie, Dortmund, Germany. PDBid 1GUA.]

pathways may activate Raf by phosphorylating it at multiple Ser and Thr residues; see below). The X-ray structure of the Ras homolog **Rap1A** in complex with GMPPNP and the Ras binding domain of Raf (RafRBD), determined by Wittinghofer, reveals that the two proteins associate largely by mutually extending their antiparallel β sheets across a mainly polar interface (Fig. 19-41). Although Ras · GTP has 1300-fold greater affinity for binding to Raf than does Ras · GDP, it is unclear from this structure how GTP hydrolysis by Ras affects the Ras–Raf interface. Quite possibly the conformational change in Ras Switch I perturbs the Ras–Raf interface to the point that it dissociates.

2. Activated Raf phosphorylates a protein alternatively known as **MEK** and **MAP kinase kinase (MKK)** at specific Ser and Thr residues, thereby activating it as a Ser/Thr kinase. [Raf is therefore a **MAP kinase kinase kinase (MKKK)**].

3. Activated MEK phosphorylates a family of proteins named **mitogen-activated protein kinases (MAP kinases** or **MAPKs)** or **extracellular-signal-regulated kinases (ERKs).** For more than marginal activation, a MAPK must be phosphorylated at both its Thr and Tyr residues in the sequence Thr-X-Tyr. MEK (which stands for *MAP* kinase/*ERK-*activating *k*inase) catalyzes both phosphorylations and thus has dual specificity for Ser/Thr and Tyr. The X-ray structure of the unphosphorylated MAP kinase ERK2, determined by Elizabeth Goldsmith, reveals that this protein structurally resembles other protein kinases of known structure and that its Tyr residue that becomes phosphorylated blocks the peptide-binding site in its unphosphorylated form.

4. The activated MAPKs phosphorylate a variety of cytoplasmic and membrane-associated proteins, including Sos and EGFR, at Ser/Thr-Pro motifs. In addition, the MAPKs migrate from the cytosol to the nucleus, where they phosphorylate a large variety of transcription factors including Jun/AP-1, Fos, and **Myc.** These activated transcription factors, in turn, induce the transcription of their target genes (Section 34-4Bd). The effects commissioned by the extracellular presence of the protein growth factor that initiated the signaling cascade are thereby produced.

MAP kinase cascades can be activated in other ways besides by liganded RTKs. For example, Raf may also be activated via its Ser/Thr phosphorylation by **protein kinase C,** which is activated via the phosphoinositide signaling system described in Section 19-4. Alternatively, Ras may be activated by subunits of certain heterotrimeric G proteins. Thus, *the MAP kinase cascade serves to integrate a variety of extracellular signals.*

a. Scaffold and Anchoring Proteins Organize and Position Protein Kinases

Eukaryotic cells contain numerous different MAPK signaling cascades, each with a characteristic set of component kinases, which in mammals comprise at least 14 MKKKs, 7 MKKs, and 12 MAPKs (Fig. 19-42). Although each MAPK is activated by a specific MKK, a given MKK can be activated by more than one MKKK. Moreover, several pathways may be activated by a single type of receptor. How then does a cell prevent inappropriate cross talk between closely related signaling pathways? One way that this occurs is through the use of **scaffold proteins,** proteins that bind some or all of the component protein kinases of a particular signaling cascade so as to ensure that the protein kinases of a given pathway interact only with one another. In addition, a scaffold protein can control the subcellular location of its associated kinases.

The first known scaffold protein was discovered through the genetic analysis of a MAP kinase cascade in yeast, which demonstrated that this protein, **Ste5p,** binds the MKKK, MKK, and MAPK components of the pathway and that, *in vivo,* the scaffold's absence inactivates the pathway. Evidently, the interactions between successive kinase components of this MAP kinase cascade are, by themselves, insufficient for signal transmission.

Figure 19-42 MAP kinase cascades in mammalian cells. Each MAP kinase cascade consists of an MKKK, an MKK, and an MAPK. Various external stimuli may each activate one or more MKKKs, which in turn may activate one or more MKKs. However, the MKKs are relatively specific for their target MAPKs. The activated MAPKs phosphorylate specific transcription factors (e.g., **Elk-1, Ets1, p53, NFAT4, Max**) as well as specific kinases (e.g., **p90rsk, S6 kinase, MAPKAP kinase**). The resulting activated transcription factors and kinases then induce cellular responses such as growth, differentiation, and **apoptosis** (programmed cell death; Section 34-4E). [After Garrington, T.P. and Johnson, G.L., *Curr. Opin. Cell Biol.* **11,** 212 (1999).]

JIP-1 (for *J*NK *i*nteracting *p*rotein-*1*; Fig. 19-43*a*) is a scaffold protein that simultaneously binds **HPK1** (for *h*ematopoietic *p*rogenitor *k*inase-*1*), a Ras analog and

Figure 19-43 Some examples of scaffold proteins that modulate mammalian MAP kinase cascades. (*a*) JIP-1 binds all protein components of the MAP kinase cascade in which HPK1 phosphorylates MKL3 or DLK (MKKKs), which then phosphorylates MKK7, which then phosphorylates JNK (an MAPK). (*b*) MEKK1 (an MKKK) is the kinase for MKK4 and also binds JNK, MKK4's target MAPK. [After Garrington, T.P. and Johnson, G.L., *Curr. Opin. Cell Biol.* **11,** 213 (1999).]

hence an **MKKK kinase (MKKKK);** the MKKKs **MLK3** and **DL3; MKK7;** and the MAPK **JNK** (for *J*un *N*-terminal *k*inase). **MEKK1** is a somewhat different type of scaffold protein (Fig. 19-43*b*); this functional MKKK binds its substrate, **MKK4,** as well as the latter's substrate, JNK.

Protein Ser/Thr kinases may similarly be individually tethered to their sites of action by **anchoring proteins.** For example, protein kinase A (PKA), which participates in numerous parallel signaling pathways including that regulating glycogen metabolism (Section 18-3), associates with several unrelated so-called **A-kinase anchoring proteins (AKAPs).** The different AKAPs, which all bind the regulatory (R) subunits of PKA, target PKA to different subcellular locations (e.g., to vesicle or plasma membranes or to particular receptors) and may also bind other signaling proteins (e.g., PP1, the protein phosphatase that removes phosphate groups installed by PKA; Section 18-3Cg), thus functioning to integrate intracellular signals.

b. Anthrax Lethal Factor Specifically Cleaves MAPKKs

Anthrax, an infectious disease caused by the bacterium *Bacillus anthracis,* affects mainly herbivorous animals such as cattle, sheep, and goats. On rare occasions, however, it may be transmitted to humans (but not between humans), in whom it is often fatal, if untreated, through massive septic shock (Section 19-1Lb). Anthrax spores are significant agents of biological warfare because their inhalation results in inhalational anthrax, a form of the disease that is nearly always fatal. This is because by the time the symptoms of inhalational anthrax become apparent, the bacterial infection has already released so much toxin that eliminating the infection through antibiotic treatment does not reverse the progress of the disease.

Anthrax toxin consists of three proteins that act in concert: **protective antigen (PA), lethal factor (LF),** and **edema factor (EF;** oedema factor in British English). PA, which is named for its use in vaccines, is a 735-residue, 4-domain protein that binds to its host cell-surface receptor (a single pass transmembrane protein) via its C-terminal domain. Most of PA's N-terminal domain is then cleaved away by a cell-surface protease, whereon the remaining membrane-bound portions of PA form cyclic heptamers reminiscent of the cyclic pentamers formed by cholera toxin (Fig. 19-22). The heptameric PA then binds LF and/or EF by their homologous N-terminal domains and mediates their endocytotic uptake into the cell. Indeed, the intravenous administration of only PA and LF rapidly kills animals. EF is a calmodulin-activated adenylate cyclase whose action upsets water homeostasis and hence is probably responsible for the massive **edema** (abnormal buildup of intercellular fluids) seen in cutaneous anthrax infections.

LF is a 776-residue monomeric protease that has only one known cellular target: *It cleaves members of the MAAPKK family of proteins near their N-termini so as to excise the docking sequences for their cognate downstream MAPKs.* It thereby disrupts the signal transduction pathways in which these proteins participate. However, anthrax infection targets mainly macrophages, a type of white blood cell (mice whose blood has been depleted of macrophages are resistant to anthrax). Low levels of LF, which occur in early stages of anthrax infection, cleave MAPKK-3, which inhibits macrophages from releasing but

not producing the inflammatory mediators NO (Section 19-1Lb) and **tumor necrosis factor-α (TNFα;** a cytokine that has opposite effects to most protein growth factors and is largely responsible for the wasting seen in chronic infections). This has the effect of reducing and/or delaying the immune response. In contrast, high levels of LF, which occur in late stages of infection, trigger macrophage lysis, causing the sudden release of NO and TNF-α, which presumably results in the massive septic shock that causes death.

E. *Tyrosine Kinase–Associated Receptors*

Many cell-surface receptors are not members of the receptor families that we have discussed and do not respond to ligand binding by autophosphorylation. These include the receptors for the cytokines (Section 19-3Eb) and **T cell receptors** [which control the proliferation of immune system cells known as T lymphocytes (T cells); Section 35-2D]. *Ligand binding induces these* **tyrosine kinase–associated receptors** *to dimerize (and, in some cases, to trimerize or form even higher oligomers), often with different types of subunits, in a way that activates associated* **nonreceptor tyrosine kinases (NRTKs).** The domain organization for the major subfamilies of NRTKs is diagrammed in Fig. 19-44.

a. The Structure of Src Reveals Its Autoinhibitory Mechanism

Many of the NRTKs that are activated by tyrosine kinase–associated receptors belong to the **Src family,** which contains at least nine members including Src, Fyn, and **Lck.** Most of these ~530-residue membrane-anchored (by myristoylation) proteins have both an SH2 and an SH3 domain and all have a PTK domain. Hence, an Src-related kinase may also be activated by association with an autophosphorylated RTK. Although Src-related kinases are each associated with different receptors, they phosphorylate

Figure 19-44 Domain organization of the major NRTK subfamilies. The N-termini for these polypeptides, which are drawn approximately to scale, are on the left and the domain identification is provided on the right. [Courtesy of Stevan Hubbard, New York University School of Medicine.]

overlapping sets of target proteins. This complex web of interactions explains why different ligands often activate some of the same signaling pathways.

Src, as indicated in Fig. 19-44, consists of, from N- to C-terminus, a myristoylated N-terminal "unique" domain that differs among Src family members, an SH3 domain, an SH2 domain, a PTK domain, and a short C-terminal tail. Phosphorylation of Tyr 416 in the PTK's activation loop activates Src, whereas phosphorylation of Tyr 527 in its C-terminal tail deactivates it. *In vivo,* Src is phosphorylated at either Tyr 416 or Tyr 527, but not at both. The dephosphorylation of Tyr 527 or the binding of external ligands to the SH2 or the SH3 domain activates Src, a state that is then maintained by the autophosphorylation of Tyr 416. When Tyr 527 is phosphorylated and no activating phosphopeptides are available, Src's SH2 and SH3 domains function to deactivate its PTK domain, that is, Src is then autoinhibited.

The X-ray structure of Src · AMPPNP lacking its N-terminal domain and with Tyr 527 phosphorylated, determined by Stephen Harrison and Michael Eck, reveals the structural basis of Src autoinhibition (Fig. 19-45). As biochemical studies had previously shown, the SH2 domain binds phosphoTyr 527, which occurs in the sequence pYNPG rather than the pYEEI sequence characteristic of high affinity Src SH2 target peptides. Although the pYNP segment binds to SH2 as does the pYEE segment in Fig. 19-33*b*, the succeeding residues are poorly ordered in the X-ray structure and, moreover, the SH2 pocket in which the Ile side chain of pYEEI binds is unoccupied. Apparently, the phosphoTyr 527-containing peptide segment binds to the Src SH2 domain with reduced affinity relative to its target peptides.

The SH3 domain binds to the linker connecting the SH2 domain to the N-terminal lobe of the PTK domain. Residues 249 to 253 of this linker form a polyproline II helix that binds to the SH3 domain in much the same way as do SH3's Pro-rich target peptides (Fig. 19-33). However, the only Pro in this segment is residue 250. The polar side chain of Gln 253, which occupies the position of the second Pro in SH3's normal Pro-X-X-Pro target sequence, does not enter the hydrophobic binding pocket that this second Pro would occupy (Fig. 19-35), and hence the path of the peptide deviates from that of Pro-rich target peptides at this point. Apparently, this interaction is also weaker than those with Src's SH3 target peptides.

Src's SH2 and SH3 domains bind on the opposite side of the PTK domain from its active site. How, then, does the conformation shown in Fig. 19-45 inhibit the PTK's activity? The two lobes of Src's PTK domain are, for the most part, closely superimposable on their counterparts in the PTK domain of phosphorylated and hence activated Lck (a Src family member) as well as the C subunit of activated PKA (Fig. 18-15). However, Src helix C (the only helix in the PTK's N-terminal lobe) is displaced from the interface between the N- and C-terminal lobes relative to its counterparts in Lck and PKA. Helix C contains the conserved residue Glu 310 (using Src numbering), which in activated

Figure 19-45   X-ray structure of Src · AMPPNP lacking its N-terminal domain and with Tyr 527 phosphorylated. The SH3 domain is orange, the SH2 domain is magenta, the linker joining the SH2 domain to the PTK domain is green with its 5-residue polyproline II helix gold, the N-terminal lobe of the PTK domain is pink, the C-terminal lobe is cyan with its activation loop blue, and the C-terminal tail is orange. The AMPPNP is shown in space-filling form and Y416 and pY527 are shown in ball-and-stick form, all with C green, N blue, O red, and P yellow. [Based on an X-ray structure by Stephen Harrison and Michael Eck, Harvard Medical School. PDBid 2SRC.]

Lck and PKA projects into the catalytic cleft, where it forms a salt bridge with Lys 295, an important ligand of the substrate ATP's α and β phosphates. However, in inactive Src, Glu 310 forms an alternative salt bridge with Arg 409, whereas Lys 295 instead interacts with Asp 404. In activated Lck, Arg 409 forms a salt bridge with phosphoTyr 416.

The foregoing structural observations suggest the following scenario for Src activation (Fig. 19-46):

1. The dephosphorylation of Tyr 527 and/or the binding of the SH2 and/or SH3 domains to their target peptides (for which SH2 and SH3 have greater affinity than their internal Src binding sites) releases these domains from their PTK-bound positions shown in Fig. 19-45, thus relaxing conformational constraints on the PTK domain. This allows the PTK's active site cleft to open, thereby disrupting the structure of its partially helical activation loop (which occupies a blocking position in the active site cleft; Fig. 19-45) so as to expose Tyr 416 to autophosphorylation.

2. The resulting phosphoTyr 416 forms a salt bridge with Arg 409, which sterically requires the structural reorganization of the activation loop to its active, nonblocking conformation. The consequent rupture of the Glu 310–Arg 409 salt bridge frees helix C to assume its active orientation which, in turn, allows Glu 310 to form its catalytically important salt bridge to Lys 295, thereby activating the Src PTK activity.

Autoinhibited Form   **Active Form**

Figure 19-46 Schematic model of Src activation. See the text for an explanation. The coloring scheme largely matches that in Fig. 19-45, as does the viewpoint. [After Young, M.A.,

Gonfloni, F., Superti-Furga, G., Roux, B., and Kuriyan, J., *Cell* **105,** 115 (2001).]

The above mechanism is, perhaps unexpectedly, critically dependent on the rigidity of the 8-residue linker joining the SH2 and SH3 domains. Thus, replacing three of these linker residues with Gly (whose lack of a C_β atom makes it the least conformationally restricted residue) results in a protein that is no longer deactivated by the phosphorylation of Tyr 527. This is corroborated by molecular dynamics simulations (Section 9-4) indicating that the thermal motions of the SH2 and SH3 domains are highly correlated (move as a unit) when Tyr 527 is phosphorylated but that this correlation is significantly reduced when Tyr 527 is dephosphorylated or when Gly replaces the three linker residues.

b. The JAK-STAT Pathway Relays Cytokine-Based Signals

Cytokines form a diverse group of small soluble proteins that, when secreted by cells, act in an autocrine, paracrine, or endocrine fashion to induce a great variety of responses, including the immune response (Section 35-2Aa), cell proliferation, growth, differentiation, **apoptosis** (programmed cell death; Section 34-4E), and **chemotaxis** (the movement of a motile cell along a concentration gradient of a specific substance). They include the 35 different **interleukins (IL-1** to **IL-35),** the colony stimulating factors [which include **macrophage colony-stimulating factor (M-CSF),** **granulocyte colony-stimulating factor (G-CSF),** and **granulocyte macrophage colony-stimulating factor (GM-CSF);**

macrophages and granulocytes are types of white blood cells], growth hormone (Section 19-1J), **erythropoietin (EPO;** which stimulates the production of erythrocytes), the **interferons** (which protect against viral infection; Section 32-4Ab), the **tumor necrosis factors (TNFs), nerve growth factor (NGF),** and the **chemokines** (which induce chemotaxis in responsive cells). Abnormalities in specific cytokines or their receptors have been implicated in a large variety of diseases and, conversely, several cytokines are therapeutically useful in alleviating certain pathological states [e.g., EPO is used to treat anemia resulting from chronic kidney disease (EPO is produced by the kidneys) and GM-CSF is used to stimulate white cell production following chemotherapy (which kills fast-growing cells such as white cells)].

Most **cytokine receptors,** as we saw for human growth hormone (Section 19-1J), are activated through ligand-induced receptor aggregation of two or more receptor components. For example, the **GM-CSF receptor** consists of two different types of subunits: **GMRα,** which is specific for GM-CSF, and **βc,** which is a common subunit for the GM-CSF, **IL-3,** and **IL-5** receptors. Both of these subunits consist of an N-terminal cytosolic domain, a single transmembrane helix, and a C-terminal ectodomain, which in GMRα consists of two fibronectin type III domains (Section 19-1J) and in βc consists of four such domains.

The X-ray structure of GM-CSF in complex with its receptor's ectodomain, determined by Angel Lopez and

(a)

(b)

Figure 19-47   **X-ray structure of the extracellular portions of the GM-CSF receptor in complex with GM-CSF.** The structure of this 2-fold symmetric 2:2:2 complex of GM-CSF, GMRα, and β$_c$ is viewed (*a*) parallel to the plane of the plasma membrane and (*b*) from the extracellular side of the membrane. The GM-CSF is purple, the GMRα is yellow, one β$_c$ monomer (chain a) is magenta and the other (chain b) is green. The labels indicate the domain names. The observed N-linked carbohydrates are drawn in stick form with Asn C green, carbohydrate C gray, N blue, and O red. The disordered peptides that link the C-termini of the receptor chains to their transmembrane segments are represented by dashed lines. [Modified from a drawing by Angel Lopez, Hanson Institute, Adelaide, Australia, and Michael Parker, St. Vincent's Institute of Medical Research, Fitzroy, Victoria, Australia. PDBid 3CXE.]

Michael Parker, reveals that it consists of a 2:2:2 complex of GM-CSF, GMRα, and β$_c$ (Fig. 19-47). GM-CSF, as had previously been determined, consists of an up–up–down–down four-helix bundle, a topology that occurs only in helical cytokines. Each GM-CSF molecule interacts with both domains of a GMRα subunit and with the N-terminal domain of one β$_c$ subunit (D1) and with the C-terminal domain of the other β$_c$ subunit (D4). The latter phenomenon, in which the domain from one subunit is exchanged with the same domain from an identical subunit to form an intertwined dimer, is known as **domain swapping.** It is a common mechanism for oligomer assembly.

Although the X-ray structures of the ectodomains of several cytokine receptors have been determined, the structures of their cytosolic domains are, as yet, unknown. However, as James Darnell elucidated, the signal that helical cytokines have been bound by their cognate receptors is transmitted within the cell by the **JAK-STAT pathway.** These cytokine receptors form complexes with proteins of the **Janus kinase (JAK)** family of NRTKs, so named because each of its four ~1150-residue members (**JAK1, JAK2, JAK3,** and **Tyk2**) has two PTK domains (Janus is the two-faced Roman god of gates and doorways), although only the C-terminal domain is functional (Fig. 19-44). **STATs** (for *signal transducers* and *activators* of *transcription*) comprise a family of seven ~800-residue proteins that are the only known transcription factors whose activities are regulated by Tyr phosphorylation and that have SH2 domains.

The JAK-STAT pathway functions as is diagrammed in Fig. 19-48:

1. Cytokine binding induces the cytokine receptor to oligomerize.

2. The cytokine receptor's two associated JAKs are thereby brought into apposition (in the case of the GM-CSF receptor, JAK2 binds to the cytosolic domain of β$_c$), whereupon they reciprocally phosphorylate each other and then their associated receptors, a process resembling the autophosphorylation of dimerized RTKs (Section 19-3Ab). Note that unlike most NRTKs, JAKs lack both SH2 and SH3 domains.

3. STATs bind to the phosphoTyr group on their cognate activated receptor via their SH2 domain and are then phosphorylated on a conserved Tyr residue by the associated JAK.

4. Following their dissociation from the receptor, the phosphorylated STATs homo- or heterodimerize via the association of their phosphoTyr residue with the SH2 domain on the opposing subunit.

5. The STAT dimers are translocated to the nucleus, where these now functional transcription factors induce the expression of their target genes in much the same way as do the transcription factors that are phosphorylated by the MAPKs (Fig. 19-40).

c. PTKs Are Targets of Anticancer Drugs

The hallmark of **chronic myelogenous leukemia (CML)** is a specific chromosomal translocation (Section 34-4C) forming the so-called **Philadelphia chromosome** in which the *Abl* gene (which encodes the NRTK Abl) is fused with the *Bcr* gene (which encodes the protein Ser/Thr kinase **Bcr).** The Abl portion of this Bcr–Abl fusion protein is constitutively (continuously, without regulation) activated, probably because its Bcr portion oligomerizes. Hematopoietic stem cells (from which all blood cells are descended) bearing the Philadelphia chromosome are therefore primed to develop CML (malignancy requires several independent genetic alterations; Section 19-3Ba). Without a bone marrow transplant (a high-risk procedure that is unavailable to

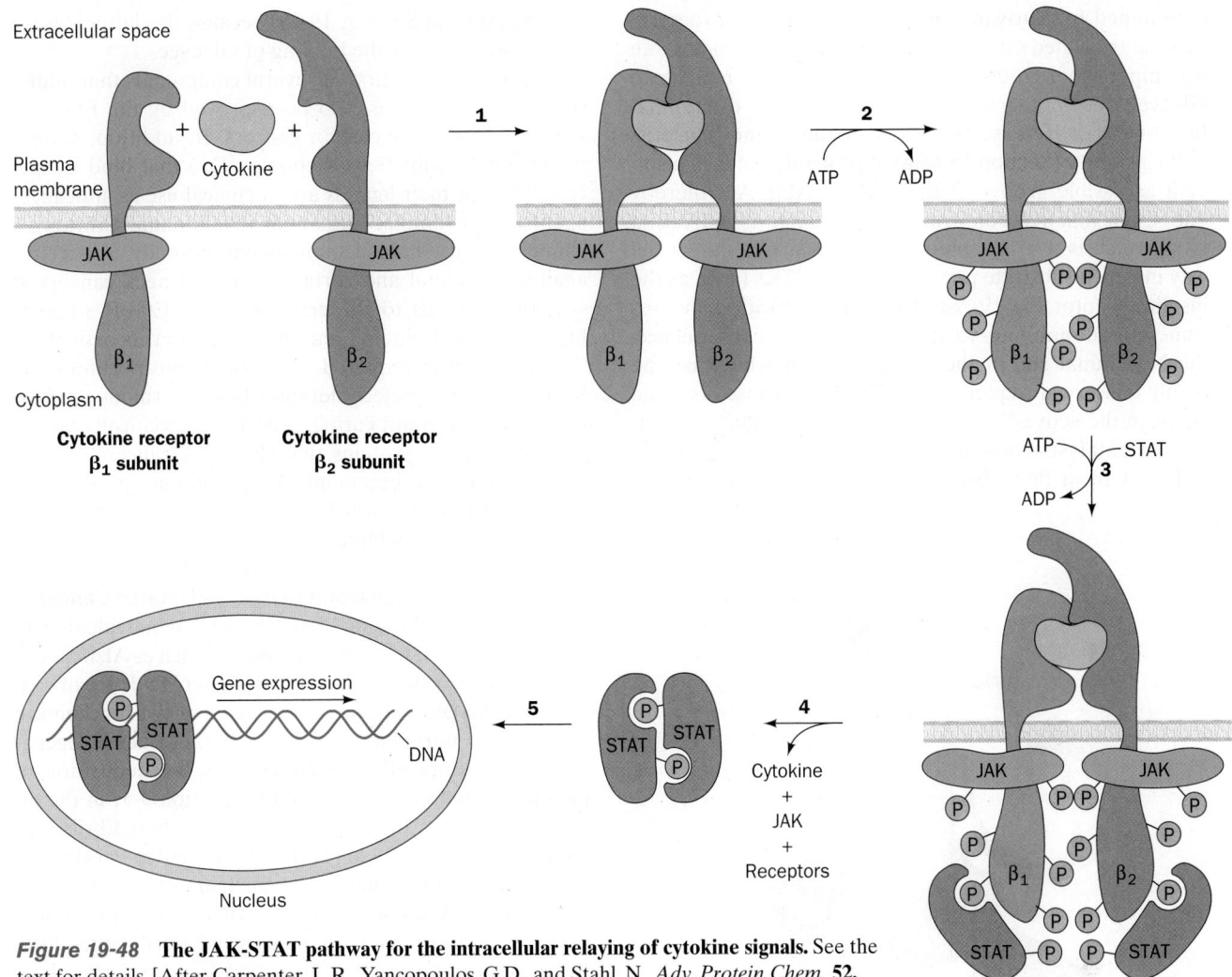

Figure 19-48 **The JAK-STAT pathway for the intracellular relaying of cytokine signals.** See the text for details. [After Carpenter, L.R., Yancopoulos, G.D., and Stahl, N., *Adv. Protein Chem.* **52,** 109 (1999).]

most individuals due to the lack of a suitable donor), CML is invariably fatal, with an average survival time of ~6 years.

An inhibitor of Abl would be expected to prevent the proliferation of, and even kill, CML cells. However, to be an effective anti-CML agent, such a substance must not inhibit other protein kinases because this would almost certainly cause serious side effects. Derivatives of 2-phenylaminopyrimidine bind to Abl with exceptionally high affinity and specificity. One such derivative, **imatinib** (trade name **Gleevec**),

which was developed by Brian Druker and Nicholas Lydon, has caused the remission of symptoms in ~90% of CML patients with almost no serious side effects. This unprecedented performance occurs, in part, because Gleevec is essentially inactive against other PTKs (an exception being the PDGF receptor) as well as protein Ser/Thr kinases.

Abl resembles Src (Fig. 19-44) but lacks Src's C-terminal regulatory phosphorylation site (Figs. 19-45 and 19-46). The X-ray structure of Abl's PTK domain in complex with a truncated form of Gleevec,

Imatinib (Gleevec™)

determined by Kuriyan (Fig. 19-49), reveals, as expected, that the truncated Gleevec binds in Abl's ATP-binding site [the piperazinyl group that this inhibitor lacks relative to Gleevec does not significantly alter its target discrimination but does increase Gleevec's solubility and hence its bioavailability (Section 15-4Ba); it probably binds in a solvent accessible groove at the back of Abl]. Abl thereby adopts an inactive conformation in which its activation loop, which is not phosphorylated, appears to mimic the way in which substrate peptides bind to PTKs (such as the insulin receptor, Fig. 19-28*a*); that is, the activation loop assumes an autoinhibitory conformation. As a consequence, the N-terminal end of the activation loop, which has the highly conserved sequence Asp-Phe-Gly (whose Asp side chain, in the active PTK, ligates an Mg^{2+} ion that is essential for catalysis), assumes a conformation that is quite different from that observed in the X-ray structures of

inactive Abl and Src (Fig. 19-45) because this latter conformation would block the binding of Gleevec.

Gleevec was the first of several compounds that inhibit specific protein kinases to be approved by the FDA for clinical use against certain cancers. In addition, several monoclonal antibodies (Section 6-1Da) that bind to specific PTKs or their ligands are in clinical use as anticancer agents. For example, **cetuximab** (trade name **Erbitux**), a chimeric mouse/human monoclonal antibody, is effective against colorectal and certain head and neck cancers. It specifically binds to the ectodomain of EGFR so as to block its ligand binding and hence prevent its activation, thus resulting in impaired cell growth and proliferation. Such receptor-targeted therapies hold enormous promise for controlling, if not curing, cancers by specifically targeting the aberrant proteins that cause the cancers. In contrast, most chemotherapeutic agents that are presently in use indiscriminately kill fast-growing cells and hence almost always have debilitating side effects.

d. Increased Chaperone Activity Facilitates Cancer

Epidemiological studies have shown that individuals with age-related neurodegenerative diseases such as Alzheimer's and Parkinson's diseases have a far lower incidence of cancer than the general population. What is the biochemical basis of this intriguing observation? These neurodegenerative diseases are all characterized by the deposition of plaques containing amyloid fibrils (Section 9-5) in the affected brain cells. In Alzheimer's disease, these fibrils consist mainly of Aβ protein (Section 9-5B) and in Parkinson's disease they contain mainly **α-synuclein** (a 140-residue soluble protein of unknown function that normally occurs in presynaptic terminals). These amyloid fibrils apparently form because of age-associated reductions in the levels of activity of the chaperone proteins that normally prevent their aggregation.

Many oncogene products, as we have seen, are mutant forms of proteins that participate in signal transduction and hence have altered functionality. Such mutant proteins are usually less stable than their unmutated forms and hence require greater than normal attention from chaperone proteins to maintain their active conformations. In particular, the Hsp90 chaperone proteins (Section 9-2C) facilitate the late stage folding of numerous signaling proteins and, presumably, their oncogenic mutants. In fact, many unmutated signaling proteins are unstable unless they are binding their corresponding ligand or protein and hence are present in cells in complex with Hsp90 (which accounts for Hsp90's high abundance, normally 1–2% of a cell's soluble proteins).

Hsp90 is overexpressed in many types of cancers, a situation that is correlated with resistance to therapy and hence a poor prognosis (and which accounts for the observation that individuals with neurodegenerative diseases rarely get cancer). Consequently, Hsp90 inhibitors are anticancer drugs. Several Hsp90 inhibitors, all of which interfere with its ATPase function, are presently in clinical trials. These substances, unlike most targeted anticancer drugs (e.g., Gleevec), are likely to be effective against a broad range of cancers.

Figure 19-49 **X-ray structure of the Abl PTK domain in complex with a truncated derivative of Gleevec.** The protein is viewed from the right of the "standard" view of protein kinases (e.g., Figs. 19-28*a* and 19-45), with its N-terminal lobe pink, its C-terminal lobe cyan, and its activation loop blue. The truncated Gleevec, which occupies the PTK's ATP-binding site, is shown in space-filling form with C green, N blue, and O red. [Based on an X-ray structure by John Kuriyan, The Rockefeller University. PDBid 1FPU.]

F. *Protein Phosphatases*

As we previously discussed (Section 19-2E), to prevent an intracellular signaling pathway from being stuck in the "on" position, its signals must be rapidly eliminated once the message has been delivered. For proteins with phosphoTyr or phospho-Ser/Thr residues, this task is carried out by a variety of protein phosphatases, of which ~500 are encoded by the human genome (around the same number as species of protein kinases, which suggests that these families have similar levels of complexity). The protein phosphatases, as we shall see, are not just simple housekeeping enzymes but are signal transducers in their own right. Thus, whereas kinases control the amplitude of a signaling response, protein phosphatases control its rate and duration.

a. Protein Tyrosine Phosphatases Also Mediate Signal Transduction

The enzymes that dephosphorylate Tyr residues, the **protein tyrosine phosphatases (PTPs),** which were discovered by Nicholas Tonks, form a large family of diverse proteins that are present in all eukaryotes (humans have 107 PTP genes vs 90 PTK genes). Each PTP contains at least one conserved ~280-residue phosphatase domain that has the 11-residue signature sequence (I/V)HCXAGXGR(S/T)G, the so-called HCX_5R motif, which contains the enzyme's catalytically essential Cys and Arg residues. The reaction proceeds via the nucleophilic attack of the Cys thiolate group on the P atom of the bound phosphoTyr to yield Tyr and a cysteinyl–phosphate intermediate that is subsequently hydrolyzed. The Arg side chain participates in substrate binding and stabilizes the cysteinyl–phosphate intermediate.

The PTPs have been classified into three groups: (1) receptorlike PTPs, (2) intracellular PTPs, and (3) dual-specificity PTPs, which can also dephosphorylate phospho-Ser/Thr residues. The receptorlike PTPs are constructed much like the RTKs (Fig. 19-25); that is, they have, from N- to C-terminus, an ectodomain consisting of often multiple repeating modules that occur in other proteins, a single transmembrane helix, and a cytosolic domain consisting of a catalytically active PTP domain that, in most cases, is followed by a second PTP domain with little or no catalytic activity. Nevertheless, these inactive PTP domains, which are highly conserved, are important for the activity, specificity, and stability of the PTP as a whole.

Biochemical and structural analyses indicate that ligand-induced dimerization of a receptorlike PTP reduces its catalytic activity, probably by blocking its active sites. Intracellular PTPs contain only one PTP domain, which is flanked by regions containing motifs, such as SH2 domains, that participate in protein–protein interactions. Structural studies reveal that the active sites of receptorlike and intracellular PTPs are too deep to bind phospho-Ser/Thr side chains—as we also saw to be the case for both PTK and SH2 domains (Sections 19-3Ac and 19-3Cb). However, the active site pockets of dual-specificity PTPs are sufficiently shallow to bind both phosphoTyr and phospho-Ser/Thr residues.

b. SHP-2 Is Inactivated by Binding Its Unliganded N-Terminal SH2 Domain

The cytoplasmic PTP **SHP-2** (for *SH*2 domain-containing *p*hosphatase 2), which is expressed in all mammalian cells, binds to PTKs that are activated by a variety of ligands, including cytokines, growth factors, and hormones. The 591-residue SHP-2 consists of two tandem SH2 domains, followed by a PTP domain and a 66-residue C-terminal tail that contains Tyr-phosphorylation sites as well as a Pro-rich segment that may bind SH3- or WW-containing proteins. SHP-2's PTP activity is increased ~10-fold on binding peptides with a single phosphoTyr residue and ~100-fold with those having two phosphoTyr residues and at much lower peptide concentrations. SHP-2 binds to both growth factors and certain cytokine receptors via its SH2 domains and, when its C-terminal tail is phosphorylated, also functions as an adaptor to recruit Grb2 so as to activate MAP kinase pathways (Section 19-3D). Mutations in the gene encoding SHP-2 are responsible for ~50% of the cases of **Noonan syndrome,** a relatively common (1 in ~2000 live births) disorder principally characterized by cardiac abnormalities, short stature, learning disabilities, and distinctive facial features.

The X-ray structure of SHP-2 lacking its C-terminal tail (Fig. 19-50), determined by Eck and Steven Shoelson, reveals that the N-terminal SH2 domain (N-SH2) interacts extensively with the PTP domain. N-SH2 inhibits the PTP by inserting its D′E loop far into the PTP's 9-Å-deep catalytic cleft, where the loop interacts with the PTP's catalytic Arg and Cys residues and prevents the active site closure observed in the X-ray structure of a PTP in complex with a phosphopeptide. In contrast, the more C-terminal SH2 domain (C-SH2) does not have a significant interface with either the N-SH2 or the PTP domains.

Figure 19-50 **X-ray structure of the protein tyrosine phosphatase SHP-2.** In this structure, its N-SH2 domain is gold with its D′E loop red, its C-SH2 domain is green, and its PTP domain is cyan, with its 11-residue signature sequence, its CX_5R motif, blue and the side chain of its catalytically essential Cys residue shown in ball-and-stick form with C green and S yellow. [Based on an X-ray structure by Michael Eck and Steven Shoelson, Harvard Medical School. PDBid 2SHP.]

The phosphopeptide-binding sites on both SH2 domains face away from the PTP domain and are therefore fully exposed on the protein surface. However, the comparison of the structure of N-SH2 complexed to a phosphopeptide with that in the above autoinhibited form of SHP-2 indicates that, in the autoinhibited form, N-SH2 adopts a conformation in which it is unable to bind phosphoTyr. Evidently, the conformations of N-SH2's PTP-binding surface and phosphopeptide binding site are allosterically linked such that its binding of PTP and phosphopeptide are mutually exclusive. The C-SH2 domain does not participate in PTP activation, although it almost certainly contributes binding energy and specificity to the binding of a phosphopeptide.

c. Bubonic Plague Virulence Requires a PTP

Bacteria lack PTKs and hence do not synthesize phosphoTyr residues. Nevertheless, PTPs are expressed by bacteria of the genus *Yersinia*, most notably *Yersinia pestis*, the pathogen that causes **bubonic plague** (the flea-transmitted "Black Death," which, since the sixth century, has been responsible for an estimated ~200 million human deaths including about one-third of the European population in the years 1347–1350). The *Y. pestis* PTP, **YopH,** which is required for bacterial virulence, is far more active than other known PTPs. Hence, when *Yersinia* injects YopH into a cell, the cell's phosphoTyr-containing proteins are catastrophically dephosphorylated. Although YopH is only ~15% identical in sequence to mammalian PTPs, it contains all of their invariant residues and their X-ray structures are closely similar. This suggests that an ancestral *Yersinia* acquired a PTP gene from a eukaryote. However, the discovery of a dual-specificity protein phosphatase in a free-living cyanobacterium raises the possibility that PTPs arose before the divergence of eukaryotes and prokaryotes.

d. Cells Contain Several Types of Protein Ser/Thr Phosphatases

The **protein Ser/Thr phosphatases** were first characterized by Earl Sutherland (who also discovered the role of cAMP as a second messenger; Section 18-3Eb) and by Edmond Fischer and Edwin Krebs (who discovered the role of protein phosphorylation in controlling glycogen metabolism; Section 18-3C). The majority of these enzymes are members of two protein families: the **PPP family,** which consists of **PP1, PP2A,** and **PP2B** (PP for *phosphoprotein phosphatase*); and the **PPM family,** which consists of **PP2C.** The PPP and PPM families are unrelated to each other or to the PTKs. We have already considered PP1 in connection with the role of its catalytic subunit, PP1c, in dephosphorylating the proteins that regulate glycogen metabolism as well as the roles of its targeting subunits, G_M and G_L, in binding PP1c to glycogen in muscle and liver (Section 18-3Cg). Indeed, all PP1c's are associated with one or two regulatory (R) subunits that function to modulate the activity of their bound PP1c's, target them to substrates in specific subcellular locations, or modify their substrate specificities. It is the large variety of these mostly

unrelated R subunits that permits the limited number (1–8) of genetically distinct but closely similar (~90% sequence identity) PP1c's in a eukaryotic cell to carry out their diverse functions.

X-ray structures have shown that PPP catalytic centers each contain an Fe^{2+} (or possibly an Fe^{3+}) ion and a Zn^{2+} (or possibly an Mn^{2+}) ion, whereas PPM catalytic centers each contain two Mn^{2+} ions. These binuclear metal ion centers nucleophilically activate water molecules to dephosphorylate substrates in a single reaction step.

e. PP2A Is Structurally Variable and Functionally Diverse

PP2A participates in a wide variety of regulatory processes including those governing metabolism, DNA replication, transcription, and development. It consists of three different subunits:

1. An ~36 kD catalytic subunit (C), whose N-terminal catalytic domain contains the ~280-residue catalytic core common to all PPP family members. Its C-terminal regulatory domain contains an activating binding site for Ca^{2+}–calmodulin, an inactivating Tyr phosphorylation site that is targeted by a variety of PTKs including the EGF and insulin receptors, and a C-terminal autoinhibitory tail. The C subunit is highly conserved from yeast to mammals.

2. An ~65 kD scaffold subunit (A; also called PR65), with which the C subunit is tightly associated in the cell.

3. One of four dissimilar types of regulatory subunits (B, B′, B″, and B9‴) that bind to both the A and C subunits and, to a large extent, control the substrate specificity of PP2A.

All of PP2A's subunits have multiple isoforms and splice variants that are expressed in a tissue-specific and developmentally specific manner, thereby generating an enormous panoply of enzymes that are targeted to different phosphoproteins in distinct subcellular sites. This complexity is a major cause of our limited understanding of how PP2A carries out its diverse cellular functions, even though it comprises between 0.3 and 1% of cellular proteins and, together with PP1, accounts for >90% of the Ser/Thr phosphatase activity in most cells.

The X-ray structure of PP2A's A subunit (Fig. 19-51*a*), determined by David Barford, reveals a remarkable solenoidal protein that consists of 15 imperfect tandem repeats of a 39-residue sequence termed HEAT (because it occurs in proteins named *H*untingtin, *E*F3, *A* subunit of PP2A, and *T*OR1). Successive HEAT repeats, which each consist of two antiparallel helices joined by a short linker, stack on one another with their corresponding helices nearly parallel so as to form an ~100-Å-long right-handed superhelix (helix of helices) with a hooklike shape.

The X-ray structure of a PP2A **holoenzyme** (complete enzyme; Fig. 19-51*b*), determined independently by Yigong Shi and Wenqing Xu, reveals, unexpectedly, that its regulatory subunit consists of 8 tandem HEAT-like repeats arranged like those of the A subunit, despite their lack of sequence similarity. The C subunit binds to the A subunit's concave surface along a ridge of conserved hydrophobic side chains spanning HEAT repeats 11 to 15. The regulatory subunit similarly interacts with the A subunit's HEAT repeats 2 to 8 and also binds to the C subunit via a ridge spanning its own HEAT-like repeats 6 to 8. The highly acidic, convex side of the regulatory subunit (lower part of Fig. 19-51*b*) is thereby left unoccupied, which suggests that it interacts with substrate proteins. The C subunit is structurally similar to the catalytic subunits of PP1 and PP2B.

(a)

Figure 19-51 **X-ray structure of protein phosphatase PP2A.** (*a*) The structure of an isolated scaffold (A) subunit. HEAT repeats, which are drawn here in different colors, each consist of two antiparallel helices joined by a short linker. These stack on one another with their corresponding helices nearly parallel to form an ~100-Å-long right-handed superhelix (helix of helices) with a hooklike shape. Compare this structure to that of a portion of human ankyrin (Fig. 12-39), which also forms a right-handed solenoid, although consisting of ankyrin repeats. [Courtesy of Bostjan Kobe, St. Vincent's Institute of Medical Research, Fitzroy, Victoria, Australia. X-ray structure by David Barford, University of Oxford, U.K. PDBid 1B3U.] (*b*) The structure of a PP2A heterotrimer viewed with the scaffold subunit oriented approximately as in Part *a*. Here the scaffold (A; 589 residues) and regulatory (B′; 449 residues) subunits are drawn in worm form, each colored in rainbow order from its N-terminus (*blue*) to its C-terminus (*red*). In addition, the A subunit is embedded in its transparent molecular surface. The catalytic (C; 309 residues) subunit (*magenta*) is drawn in ribbon form. Note the close structural resemblance of the A and B′ subunits. [Based on an X-ray structure by Yigong Shi, Princeton University. PDBid 2NPP.]

(b)

f. PP2B Is the Target of Immunosuppressant Drugs

PP2B, which is also known **calcineurin (CaN),** is unique among protein Ser/Thr phosphatases in that it is activated by Ca^{2+}. CaN is a heterodimer composed of a catalytic A subunit (CaNA) and a regulatory B subunit (CaNB). CaNA contains an N-terminal catalytic domain followed by a CaNB-binding domain, a calmodulin (CaM)-binding domain, and a C-terminal autoinhibitory segment. CaNB, which has 35% sequence identity with CaM, binds four Ca^{2+} ions via its four EF hand motifs (Section 18-3Ce). CaN is activated by the binding of Ca^{2+} to CaNB and Ca^{2+}–CaM to CaNA.

Calcineurin plays an essential role in the antigen induced proliferation of T cells. As we discuss in Section 35-2D, the binding of an antigenic peptide to a **T cell receptor,** a tyrosine kinase–associated receptor, initiates a complex series of signaling events involving the Src-like PTKs Lck and Fyn, a MAP kinase cascade, and the phosphoinositide cascade (Section 19-4), which, among other things, releases Ca^{2+} into the cytosol. The Ca^{2+}, in turn, activates CaN to dephosphorylate the transcription factor **NFAT$_p$** (for *n*uclear *f*actor of *a*ctivated *T* cells). NFAT$_p$ in complex with CaN is then translocated to the nucleus where, in concert with other transcription factors, it induces some of the early steps in T cell proliferation.

As we discussed Section 9-2B, the fungal products **cyclosporin A (CsA)** and **FK506** are highly effective immunosuppressants that are in clinical use for the prevention of organ-transplant rejection and for the treatment of autoimmune disorders (processes that are mediated by T cells). CsA and FK506, respectively, bind to the peptidyl prolyl cis–trans isomerases (rotamases) **cyclophilin** and **FK506 binding protein (FKBP12),** which are therefore collectively known as **immunophilins.** However, the observation that both CsA and FK506 (also known as **tacrolimus)** are effective immunosuppressants at concentrations far below those of the immunophilins suggests that it is the presence of the cyclophilin · CsA and FKBP12 · FK506 complexes themselves, rather than the inhibition of their rotamase activity, that interferes with T cell proliferation. It is, in fact, the binding of either complex to CaN that prevents it from dephosphorylating NFAT$_p$ and thereby suppresses T cell proliferation.

The X-ray structures of the bovine complex FKBP12 · FK506–CaN, by Manuel Navia, and the corresponding human complex, by Ernest Villafranca, reveal how the FKBP12 · FK506 complex binds to CaN (Fig. 19-52a). The catalytic domain of CaNA, with its binuclear Fe^{2+}–Zn^{2+} center marking its active site, resembles those of other protein Ser/Thr phosphatases of known structure. A 22-residue α helix at the C-terminal end of this phosphatase domain, which extends out from the phosphatase domain by up to 40 Å, provides much of the CaNB-binding site. Beyond this helix, the C-terminal portion of CaNA, which contains the CaM-binding site and the autoinhibitory segment, is not visible due to disorder. However, in the X-ray structure of CaN alone, the autoinhibitory segment is seen to bind in the CaNA active site so as to block the access of substrate phosphoproteins. The structure of CaNB, which

(a)   *(b)*

Figure 19-52 **Calcineurin.** (*a*) The X-ray structure of bovine FKBP12 · FK506–CaN. The CaNA subunit is orange, the CaNB subunit is blue, and the FKBP12 is cyan. The FK506 is drawn in stick form with C green, N blue, and O red; the Fe^{2+} and Zn^{2+} ions in the CaNA active site are, respectively, represented by red and green spheres; and the four Ca^{2+} ions bound to CaNB are represented by pale green spheres. [Based on an X-ray structure by Manual Navia, Vertex Pharmaceuticals, Cambridge, Massachusetts. PDBid 1TCO.] (*b*) The X-ray structure of human cyclophilin · CsA–CaN. The CaN and its bound metal ions are represented and oriented as in Part *a*, the cyclophilin is magenta, and the CsA is drawn in stick form with C green, N blue, and O red. Note the close resemblance between the two structures. [Based on an X-ray structure by Hengming Ke, University of North Carolina, Chapel Hill, North Carolina. PDBid 1M63.]

has four bound Ca^{2+} ions, resembles that of Ca^{2+}–CaM in complex with a helical target peptide (Fig. 18-19) except that CaNB's two globular domains are on the same side of its bound peptide rather than on opposite sides as are those of Ca^{2+}–CaM. CaNB thereby forms a continuous groove in which the CaNA helix binds.

FKBP12 · FK506, a mixed inhibitor of CaN (Section 14-3C), binds to it so as to contact both CaNA and CaNB, with the portion of FK506 that extends out from its FKBP12-binding site forming a significant part of this interface. The structures of FKBP12 and CaN in this complex closely resemble those in the X-ray structures of these proteins alone. It therefore appears that FK506 provides a critical component of this contact. Nevertheless, no part of the FKBP12 · FK506 complex is within 10 Å of CaN's phosphatase site (although the CaN autoinhibitory segment has been displaced). This accounts for the observations that FKBP12 · FK506 strongly inhibits CaN from dephosphorylating a 20-residue phosphopeptide but actually increases the rate at which CaN dephosphorylates the much smaller *p*-nitrophenylphosphate by a factor of 3.

The X-ray structure of human cyclophilin · CsA–CaN (Fig. 19-52*b*), independently determined by Hengming Ke and Harrison, is strikingly similar to that of FKBP12 · FK506–CaN despite the fact that there is no structural resemblance of cyclophilin · CsA to FKBP12 · FK506. Cyclophilin binds to essentially the same region of CaNA as does FKBP12 with CsA forming an essential component of this contact, thereby limiting access to the CaN active site in the same way as does FKBP12 · FK506. However, the characteristics of rotamases that apparently uniquely suit them for their roles in CaN inhibition remain enigmatic.

4 THE PHOSPHOINOSITIDE CASCADE

Extracellular signals often cause a transient rise in the cytosolic [Ca^{2+}], which, in turn, activates a great variety of enzymes through the intermediacy of calmodulin and its homologs. An increase in cytosolic [Ca^{2+}] triggers such diverse cellular processes as glycogenolysis (Section 18-3Ce) and muscle contraction (Section 35-3C). What is the source of this Ca^{2+} and how does it enter the cytosol? In certain types of cells, neurons (nerve cells; Fig. 1-10*d*), for example, the Ca^{2+} originates in the extracellular fluid. However, the observation that the absence of extracellular Ca^{2+} does not inhibit certain Ca^{2+}-mediated processes led to the discovery that, in these cases, cytosolic Ca^{2+} is obtained from intracellular reservoirs, mostly the endoplasmic reticulum (and its equivalent in muscle, the sarcoplasmic reticulum). Extracellular stimuli leading to Ca^{2+} release must therefore be mediated by an intracellular signal.

The first clue as to the nature of this signal came from observations that the intracellular mobilization of Ca^{2+} and the turnover of **phosphatidylinositol-4,5-bisphosphate** (**PIP$_2$** or **PtdIns-4,5-P$_2$**; Fig. 19-53), which occurs mainly in the plasma membrane as a minor (<1%) component of its cytosolic leaflet, are strongly correlated. This information

Figure 19-53 **Molecular formula of the phosphatidylinositides.** The head group of these glycerophospholipids is *myo*-inositol that may be phosphorylated at its 3-, 4-, and/or 5-positions. R_1 is predominantly the hydrocarbon tail of stearic acid (an 18:0 fatty acid; Table 12-1) and R_2 is predominantly the hydrocarbon tail of arachidonic acid (a 20:4 fatty acid).

led Robert Michell to propose, in 1975, that PIP$_2$ hydrolysis is somehow associated with Ca^{2+} release.

A. Ca^{2+}, Inositol Trisphosphate, and Diacylglycerol Are Second Messengers

Investigations, notably by Mabel and Lowell Hokin, Michael Berridge, and Michell, eventually revealed that *PIP$_2$ is part of an important second messenger system, the* **phosphoinositide cascade,** *that mediates the transmission of numerous hormonal signals* including those of vasopressin, CRF, TRF (Section 19-1H), acetylcholine (a neurotransmitter; Section 20-5Cb), epinephrine (with α_1-adrenergic receptors; Section 19-1F), EGF, and PDGF. Remarkably, this system yields up to three separate types of second messengers through the following sequence of events (Fig. 19-54):

1–3. The ligand–receptor interactions described below activate a phosphoinositide-specific **phospholipase C (PLC;** Section 19-4B) to hydrolyze PIP$_2$ to **inositol-1,4,5-trisphosphate (IP$_3$ or Ins-1,4,5-P$_3$)** and *sn*-1,2-diacylglycerol **(DAG or DG)**

Figure 19-54 **Role of PIP$_2$ in intracellular signaling. (1)** The binding of a ligand to a cell-surface receptor, R, activates a phosphoinositide-specific phospholipase C through the intermediacy of what is shown here as **(2)** a G protein (G$_q$; Fig. 19-17) but in many cases is an RTK, an NRTK, or, possibly, Ca^{2+}. Phospholipase C catalyzes the hydrolysis of PIP$_2$ to IP$_3$ and DAG **(3)**. The water-soluble IP$_3$ stimulates the release of Ca^{2+} sequestered in the endoplasmic reticulum **(4),** which in turn activates numerous cellular processes through the intermediacy of calmodulin and its homologs **(5).** The nonpolar DAG remains associated with the inner leaflet of the membrane, where it activates protein kinase C to phosphorylate and thereby modulate the activities of a number of cellular proteins **(6).** This latter activation process also requires the presence of the membrane lipid phosphatidylserine (PS) and Ca^{2+}. 🐭 **See the Animated Figures**

[the stereospecific numbering (*sn*) system is described in the legend of Fig. 12-3]. PLCs catalyze the hydrolysis of the bond linking a glycerophospholipid to its phosphoryl group as indicated in Fig. 19-55 (which also shows the actions of other types of phospholipases). Note that this reaction occurs at the interface between the aqueous phase and the membrane in such a way that both PIP$_2$ and its amphipathic hydrolysis product DAG remain associated with the membrane during the catalytic reaction.

4. The water-soluble IP$_3$, acting as a second messenger, diffuses through the cytoplasm to the ER, from which it stimulates the release of Ca^{2+} into the cytoplasm by binding to and thereby opening an ER-bound transmembrane Ca^{2+}-specific ion channel known as the **IP$_3$ receptor** (ion channels are discussed in Chapter 20).

Figure 19-55 **A phospholipase is named according to the bond that it cleaves on a glycerophospholipid.** X is a phosphoinositol group for this discussion.

5. The Ca^{2+}, in turn, stimulates a variety of cellular processes, mainly through the intermediacy of calmodulin and its homologs.

6. The amphipathic DAG is constrained to remain in the inner leaflet of the plasma membrane, where it nevertheless also acts as a second messenger by activating **protein kinase C (PKC;** Section 19-4C) in the presence of Ca^{2+} and phosphatidylserine (PS; which is located exclusively on the cytosolic face of the plasma membrane). This membrane-bound enzyme (actually a family of enzymes; Section 19-4C), in turn, Ser/Thr-phosphorylates and thereby modulates the activities of several different proteins including glycogen synthase (Sections 18-3D). DAG, which predominantly has a stearoyl group at its 1-position and an arachidonoyl group at its 2-position, is further degraded in some cells by **cytosolic phospholipase A_2 (cPLA_2)** to yield arachidonate, the major substrate for the biosynthesis of prostaglandins, prostacyclins, thromboxanes, leukotrienes, and lipoxins. These paracrine hormones, as we discuss in Section 25-7, mediate or modulate a wide variety of physiological functions.

B. *The Phospholipases C*

The phosphoinositide-specific PLCs in mammals are classified in six families comprising 13 isozymes according to their sequences and modes of regulation: β(1–4), γ(1–2), δ(1, 3, 4), ε, η(1–2), and ζ (Fig. 19-56; the isozyme originally named PLC-α is actually a proteolytic fragment of PLC-δ1), several of which have splice variants. All of these PLCs require the presence of Ca^{2+} ion for enzymatic activity. The PLC-δ isozymes (~760 residues) consist of, from N- to C-terminus, an ~120-residue pleckstrin homology (PH) domain (Section 19-3Ce); an ~130-residue EF hand domain that contains four EF hand motifs (Fig. 18-18); two conserved regions known as X and Y that together form the ~250-residue PLC catalytic domain and which are separated by an ~60-residue linker; and an ~120-residue **C2 domain,** a domain that in many cases binds Ca^{2+} and which occurs in ~650 human proteins that mainly participate in signal transduction and membrane interactions. The PLC-β isozymes (~1200 residues) have an additional ~420-residue C-terminal tail that has been implicated in both membrane association and regulation by G proteins (see below). In contrast, the PLC-γ isozymes (~1270 residues) contain an ~420-residue insert between X and Y that consists of an additional PH domain that is split by two SH2 domains that are implicated in binding to activated PTKs (see below) as well as an SH3 domain. PLC-ε (~2300 residues), which is activated by several Ras-related G-proteins, has an N-terminal RasGEF domain and two C-terminal Ras-binding (RA) domains. PLC-η (~1000 residues), which occurs only in nerve tissue, contains a C-terminal Ser/Pro (S/P)-rich region of unknown function. PLC-ζ (~650 residues), the smallest of the PLCs and the only one that lacks a PH domain, occurs only in sperm. The observation that the PLCs in plants and lower eukaryotes such as yeast are

Figure 19-56 Domain organization of the six classes of phosphoinositide-specific PLCs.

δ-like suggests that the various PLC isozymes in mammals evolved from a primordial PLC-δ.

a. $G_{q\alpha}$ · GTP, $G_{\beta\gamma}$, and Rac · GTP Activate PLC-βs by Bringing Them Into Contact with the Membrane

The PLC-βs are hormonally regulated by certain G protein-coupled receptors (e.g., those for histamine, vasopressin, TSH, thromboxane A_2, and angiotensin II) via their associated heterotrimeric G proteins, as indicated in Fig. 19-54. In particular, they are activated through their interactions with the α subunits of the G_q subfamily (Section 19-2C) in complex with GTP. $G_{q\alpha}$ · GTPγS activates the PLC-β isoforms in order of potency β1 > β3 > β2, with the position of β4 in this hierarchy indeterminate because it is inhibited by GTPγS. Moreover, an important aspect of the regulation of PLC-βs by $G_{q\alpha}$ · GTP is that the PLC-βs function as GAPs to increase the GTPase activity of $G_{q\alpha}$ by >50-fold, thereby limiting the activating function of $G_{q\alpha}$.

The PLC-β isoforms are independently activated by $G_{\beta\gamma}$ complexes, which may be supplied by the dissociation of heterotrimeric G proteins other than $G_{q\alpha}G_{\beta\gamma}$. Moreover, their order of efficacy with $G_{\beta\gamma}$ differs from that with $G_{q\alpha}$ · GTP: β3 > β2 > β1, with β4 insensitive to the presence of $G_{\beta\gamma}$. Although the concentration of $G_{\beta\gamma}$ required for maximal activation of the PLC-βs is much greater than that of $G_{q\alpha}$ · GTP, their final extents of activation are similar. The sites on PLC-β2 that interact with $G_{\beta\gamma}$ are its PH domain and a 10-residue segment near the N-terminus of its Y region. The region of $G_{\beta\gamma}$ that interacts with PLC-βs overlaps the region through which it binds G_α subunits, thereby explaining why a $G_{\beta\gamma}$ cannot simultaneously bind a PLC-β and a G_α · GDP.

The Rho family small G protein named **Rac1,** in complex with GTP, activates PLC-β2, -β3, and -γ2. The X-ray structure of a complex of PLC-β2 lacking its C-terminal tail (but still catalytically active *in vitro*) with Rac1 · GTPγS was

(a)

(b)

Figure 19-57 **X-ray structure of PLC-β2 lacking its C-terminal tail in complex with Rac1 · GTPγS.** The pleckstrin homology (PH) domain of PCL-β2 is blue, its EF hand (EF) domain is orange, its TIM (α/β) barrel is red, and it's C2 domain is yellow. The Rac1 is green with its switch regions (Sw1 and Sw2) red and its bound GTPγS drawn in stick form in cyan. (*a*) View toward the cytosolic surface of the membrane. (*b*) View rotated 90° about the horizontal axis relative to Part *a*. The geranylgeranyl group at the C-terminus of Rac1 and the PIP₂ substrate (*black sticks*) are modeled as sites of membrane attachment. Note that the three loops of the TIM barrel, which form a hydrophobic ridge, are inserted into the membrane. [Courtesy of John Sondek, University of North Carolina School of Medicine. PDBid 2FJU.]

determined by John Sondek. It reveals (Fig. 19-57) that PLC-β2's remaining four domains are linked by extended polypeptide segments that form tight interactions with these domains. EF hand motifs 1 and 2 form a lobe that is closely superimposable on the N- and C-terminal lobes of calmodulin (Figs. 18-17 and 18-19), as do EF hands 3 and 4. Although EF hands are usually associated with Ca^{2+} ions, the residues necessary to do so are poorly conserved in the PLCs. Thus, although mutations to this domain are generally deleterious to a PLC's catalytic activity, its function is unclear.

The X and Y regions of PLC-β2's catalytic domain form a TIM (or α/β) barrel (Section 8-3B). The X region contributes a typical βαβαβαβ motif to form nearly half the barrel with the Y region contributing its remaining portion. The visible parts of the partially disordered ~70-residue peptide linking X and Y occlude the enzyme's active site, which as is the case for all TIM barrel enzymes, is located in

the mouth of the barrel at the C-terminal ends of its β strands. Deletion of this apparently autoinhibitory X–Y linker increases the enzyme's phospholipase activity by ~10-fold, although it can still be further activated by various G proteins. Although the X–Y linkers of the several PLCs share little sequence identity, all of them contain clusters of negatively charged residues. This suggests that when a PLC approaches the negatively charged cytosolic face of a membrane, its X–Y linker is electrostatically repelled from the PLC's active site, thereby permitting these enzymes access to their membrane-bound PIP₂ substrates.

All PLCs require a bound Ca^{2+} ion for catalytic activity. The constellation of active site residues in the PLC-β2 structure closely resemble that in the X-ray structure of PLC-δ1 in complex with IP₃ and Ca^{2+} ion. In this latter structure, a 6-coordinate Ca^{2+} is bound at the bottom of the active site, with one of its ligands contributed by the 2-hydroxyl group of IP₃ and the remainder formed by highly conserved Asp, Glu, and Asn side chains. The catalytic reaction is therefore postulated to occur via a mechanism analogous to that of the RNase A–catalyzed hydrolysis of RNA (Fig. 15-3) in which PIP₂'s 2-hydroxyl group nucleophilically attacks the neighboring 1-phosphate group to form DAG and a cyclic phosphodiester intermediate that is subsequently hydrolyzed to yield IP₃. The Ca^{2+} ion (rather than a His side chain as in RNase A) is properly positioned to promote the deprotonation of the 2-hydroxyl group so as to enhance its nucleophilicity and to subsequently help stabilize the developing negative charge on the pentavalent phosphorus in the catalytic reaction's transition state (Fig. 16-6*b*). This explains why **2-deoxy-PIP₂** is not hydrolyzed by mammalian phosphinositide-specific PLCs.

Rac1 · GTPγS binds to the PH domain of PLC-β2 exclusively via Rac1's switch regions (Fig. 19-57), which explains why Rac1 · GDP does not bind to PLC-β2. However, comparison of the X-ray structure of PLC-β2 alone with that of its Rac1 · GTPγS complex indicates that the binding of Rac1 · GTPγS does not alter the conformation of PLC-β2. How, then, does Rac1 · GTPγS activate PLC-β2? Most PH domains bind membrane-associated phosphoinositides (see below) but those of PLC-β2 and -β3 lack the residues necessary to do so. Instead, Rac1's C-terminal geranygeranyl membrane anchor (Section 12-3Ba) functions to localize PLC-β2 and probably PLC-β3 to the cytosolic face of the plasma membrane (Fig. 19-57*b*), where they can efficiently hydrolyze phosphinosites. Similarly, the membrane anchored $G_{qα}$ · GTP and $G_{βγ}$ serve to localize PLC-β2 to the membrane.

The C2 domain consists of a sandwich of two 4-stranded antiparallel β sheets with variable interstrand loops. The X-ray structure of PLC-δ1 complexed with the calcium analog lanthanum revealed that the C2 domain contains three Ca^{2+} ions sites in close proximity to one another. All of these metal ions lie in a crevice at one end of the β sandwich, where they are exposed on the surface of the enzyme. It therefore seems likely that, *in vivo,* they associate with anionic head groups such as that of phosphatidylserine on the surface of a membrane. Since the extensive interface between the C2 and catalytic domain appears to be rigid, this interaction probably helps bind the catalytic domain to the membrane such that it can productively interact with

PIP$_2$ molecules. This association appears to be supplemented through the interactions of a hydrophobic ridge comprised of three loops from one side of the active site opening that is postulated to penetrate into the membrane's nonpolar region during catalysis (Fig. 19-57*b*). This would explain how the enzyme can catalyze the hydrolysis of PIP$_2$ to DAG and IP$_3$ while the former two compounds remain associated with the membrane.

The C-terminal tail that is unique to PLC-β isoforms (Fig. 19-56) has been implicated, via studies of truncated enzymes, in binding to G$_{q\alpha}$ · GTP. The X-ray structure of this ~420-residue segment reveals that it consists almost entirely of three long helices which form a coiled coil that dimerizes along its long axis. It contains a large number of basic residues that are clustered on one face of the dimer and whose mutation results in reduced responses to G$_{q\alpha}$ · GTP. This positively charged surface is likely to interact with acidic phospholipids so as to recruit its attached PLC-β to the membrane. It therefore appears that *the activation of PLC-βs is almost entirely the result of their close association with the negatively charged membrane.*

b. The Pleckstrin Homology Domain Tethers PLC-δ1 to the Membrane

The X-ray structure of PLC-δ1's N-terminal PH domain in complex with IP$_3$, determined by Joseph Schlessinger and Sigler, reveals that the IP$_3$ binds to a positively charged surface of the protein (Fig. 19-58). This is consistent with the PH domain's proposed role as a membrane anchor, as are the observations that this PH domain binds PIP$_2$ with much greater affinity ($K_D = 1.7$ μ*M*) than does PLC-δ1's catalytic domain ($K_D > 0.1$ m*M*). Since the peptide segment linking the PH domain to the rest of the enzyme is probably flexible, it appears that the PH domain in PLC-δ1 functions to tether the enzyme to the membrane. This accounts for kinetic measurements indicating that the enzyme catalyzes multiple cycles of PIP$_2$ hydrolysis without releasing the membrane.

Despite the foregoing, the way in which the PLC-δ isozymes are regulated is poorly understood. The higher sensitivities of the PLC-δ isozymes to Ca^{2+} ion compared to those of the other PLCs suggests that the PLC-δ isozymes are regulated by changes in intracellular [Ca^{2+}]. Thus, the activation of PLC-δ isozymes may occur secondarily to the receptor-mediated activation of other PLC isozymes through their induction of the opening of Ca^{2+} channels (Fig. 19-54).

c. Protein Tyrosine Kinases Recruit PLC-γ Isozymes to the Membrane

The PLC-γ isozymes in a wide variety of cells are activated by certain protein growth factors including EGF, PDGF, FGF, and NGF. These growth factors cause their corresponding receptors, which are RTKs (Section 19-3A), to autophosphorylate at particular Tyr residues. Some of these phosphoTyr sites are specifically bound, as mutational studies indicate, by the more N-terminal SH2 (N-SH2) domain on PLC-γ1 (Figs. 19-56 and 19-27*a*) but not its C-SH2 domain. The N-SH2 domain binds peptides containing a phosphoTyr residue followed by at least 5 predominantly

Figure 19-58 **X-ray structure of the pleckstrin homology domain of PLC-δ1 in complex with PIP$_3$.** The peptide chain is shown in ribbon form colored in rainbow order from its N-terminus (*blue*) to its C-terminus (*red*). The PIP$_3$ is drawn in stick form with C green, O red, and P orange. The PH domain consists largely of the β barrel/ sandwich of 7 antiparallel strands and C-terminal α helix common to the numerous PH domains of known structure. [Based on an X-ray structure by Joseph Schlessinger, New York University Medical Center, and Paul Sigler, Yale University. PDBid 1MAI.]

hydrophobic residues, in contrast to the SH2 domain of Src, which preferentially binds pYEEI containing peptides (Section 19-3Cb).

The activated receptors for all four of the above growth factors phosphorylate PLC-γ1 at the same three Tyr residues, 771, 783 (located between the C-SH2 and SH3 domains), and 1254 (located in the C-terminal tail). In fact, mutating Tyr 783 to Phe completely blocks the activation of PLC-γ1 by PDGF, although this mutant PLC-γ1 still associates with the PDGF receptor. Conversely, mutating certain RTK autophosphorylation sites (e.g., the PDGF receptor's Tyr 1021) disrupts their binding of PLC-γ1 and hence its activation, even though these mutant receptors catalyze detectable levels of growth factor–dependent Tyr phosphorylation on PLC-γ1. Evidently, growth factor–induced activation of PLC-γ1 requires both the activating Tyr phosphorylation of PLC-γ1 and its association with the growth factor receptor, the latter presumably bringing PLC-γ1 into contact with its PIP$_2$ substrate in the cytosolic leaflet of the plasma membrane. The PLC-γ isozymes may also be activated by NRTKs, such as members of the Src and JAK families (all of which are membrane associated), that have been activated by tyrosine kinase–associated receptors (Section 19-3E). The function of the PLC-γ SH3 domain is unclear.

d. PLC-ε Is Activated by Ras · GTP

The presence of RasGEF and Ras-binding (RA) domains on PLC-ε suggests that PLC-ε is activated by Ras · GTP. This is, in fact, the case, as indicated by the observation that PLC-ε binds Ras · GTP with high affinity but does not bind Ras · GDP. Since Ras is membrane-anchored, this interaction brings PLC-ε into proximity with the membrane, much as we saw with the Rac1 · GTP activation of PLC-β2 (Section 19-4Ba). Although the growth factor–induced activation of Ras is terminated by the hydrolysis of its bound GTP to GDP, the resulting Ras · GDP may be rapidly converted to Ras · GTP by the RasGEF domain of PLC-ε, thereby prolonging the receptor-mediated activation of PLC-ε. PLC-ε may also be activated by $G_{12\alpha}$.

C. *The Protein Kinases C*

Protein kinase C (PKC), a member of the **AGC family** of protein kinases (AGC for PK*A*, PK*G*, and PK*C*), is the Ser/Thr protein kinase that transduces the numerous signals mediated by the release of DAG (Fig. 19-54). In mammals, it comprises a family of ten ~700-residue monomeric isozymes classified in three subfamilies: the "conventional" PKCs (α, βI, βII, and γ, of which βI and βII are splice variants of the same gene), the "novel" PKCs (δ, ε, η, and θ), and the "atypical" PKCs [ζ and ι(human)/λ(mouse)]. The conventional PKCs, which are activated by both DAG and Ca^{2+}, each consist of an N-terminal autoinhibitory pseudosubstrate (which resembles the enzyme's target peptide but with the Ser/Thr phosphorylation site replaced by Ala) followed by four conserved domains, C1 through C4 (C1 for *conserved region 1* of PKC, etc.). The DAG-binding **C1 domain,** which occurs in >50 other proteins including Raf (in which it does not bind DAG), consists of two tandemly repeated ~50-residue Cys-rich motifs, C1A and C1B. However, only C1B binds DAG. C2, which often binds Ca^{2+} ion, is also a component of PLC (Fig. 19-57) as well as numerous other signaling proteins. C3 and C4 form the N- and C-terminal lobes of the protein kinase, which is similar in sequence and structure to the catalytic subunit of PKA (Fig. 18-15). The protein kinase is maintained in its inactive state through its binding of the pseudosubstrate (as with MLCK; Section 18-3Ce). The novel PKCs, which are activated by DAG but not by Ca^{2+}, resemble the conventional PKCs except that their C2 domains do not bind Ca^{2+}. The atypical PKCs, which are unresponsive to both DAG and Ca^{2+}, have only one Cys-rich motif in their C1 domains and lack C2 domains.

In addition to their different regulatory properties, the various PKCs are localized to different subcellular compartments (e.g., plasma membrane, nuclear membrane, endoplasmic reticulum, Golgi apparatus, mitochondria) in ways that vary with the cell type and external stimuli. This specificity is provided both by targeting sequences on the PKC as well as by scaffolding proteins that localize individual PKCs to specific membrane microdomains in close proximity to their substrate and regulatory proteins. For example, the members of a family of membrane-associated proteins known as **RACKs** (for *r*eceptors of *a*ctivated *C*-*k*inases) each anchor a specific activated PKC to a particular subcellular location.

a. The C1 and C2 Domains Anchor PKC to the Plasma Membrane

Phorbol esters such as **12-*O*-myristoylphorbol-13-acetate**

12-*O*-Myristoylphorbol-13-acetate

(which occurs in croton seed oil and was used as a drastic purgative in folk medicine) are potent activators of protein kinase C; they structurally resemble DAG but bind to PKC with ~250-fold greater affinity. Consequently, phorbol esters are the most effective known **tumor promoters,** substances that are not in themselves carcinogenic but increase the potency of known carcinogens. They do so by inhibiting the apoptosis (programmed cell death; Section 34-4E) and stimulating the proliferation of precancerous cells (cells with only some of the mutations that make them cancerous but which nevertheless reduce the cell's sensitivity to apoptotic signals and increase it for proliferative signals) as well as stimulating metabolic processes that generate carcinogenic agents (e.g., free radicals). This increases the probability that such a cell will complete its malignant transformation. Not surprisingly, therefore, altered PKC activity is associated with various types of cancers.

The X-ray structure of the C1B motif of PKCδ in complex with 12-*O*-myristoylphorbol-13-acetate, determined by Hurley, reveals that this 50-residue motif is largely knit together by two Zn^{2+} ions, each of which is tetrahedrally liganded by one His and three Cys side chains (Fig. 19-59). The phorbol ester binds in a narrow groove between two loops that consist mainly of nonpolar residues. Since phorbol esters are also nonpolar, the entire top third of the complex, as shown in Fig. 19-59, forms a highly conserved hydrophobic surface. Very few soluble proteins have such a large fraction of their surface formed by a continuous nonpolar region. Moreover, the middle third of the protein surface, that below the nonpolar region, forms a positively charged belt about the protein. This suggests that, *in vivo*, the hydrophobic portion of the complex is inserted into the nonpolar region of its associated membrane such that the motif's positively charged belt interacts with the membrane's negatively charged head groups. This hypothesis is supported by NMR measurements indicating that residues on the ligand-binding portion of C1B interact with lipid. The fatty acyl group that is esterified to phorbol's 12-position in effective tumor promoters presumably extends into the membrane so as to help anchor the C1 domain to the membrane.

The comparison of this structure with that of C1B alone indicates that C1B does not undergo significant structural change on binding phorbol ester. Evidently, phorbol esters, and presumably DAG, activate PKC by anchoring it to the membrane rather than by an allosteric mechanism. The C2

Figure 19-59 **X-ray structure of the C1B motif of PKCδ in complex with 12-*O*-myristoylphorbol-13-acetate.** The protein tetrahedrally ligands two Zn^{2+} ions (*cyan spheres*), each via a His side chain and three Cys side chains. These side chains are shown in ball-and-stick form with C green, N blue, O red, and S yellow, as is the bound 12-*O*-myristoylphorbol-13-actetate. The C1B domains of PKCα and γ have similar structures. [Based on an X-ray structure by James Hurley, NIH. PDBid 1PTR.]

domain, as we have seen for PLC (Section 19-4Ba), also does so via Ca^{2+}-mediated binding to the membrane's phosphatidylserine head groups. These interactions are synergistic in that the greater the Ca^{2+} concentration, the lower the concentration of phorbol ester or DAG necessary to activate PKC and vice versa. Nevertheless, both the C1 and C2 domains must be membrane anchored in order to activate the protein kinase. This is because the conformation required to do so extracts the N-terminal pseudosubstrate from the protein kinase active site.

b. PKC Is Primed by Phosphorylation

The activation of all mammalian PKCs but the atypical PKCs is accompanied by their phosphorylation at three conserved Ser or Thr residues. One of these residues (Thr 500 in PKCβII) is in the protein kinase's activation loop, whereas the remaining two are in its C-terminal segment (Thr 641 and Ser 660 of the 673-residue PKCβII). In the atypical PKCs, the latter Ser/Thr residue is replaced by a phosphate-mimicking Glu residue. The sequence of events that activate PKCs, which was largely elucidated by Alexandra Newton, occurs as follows (Fig. 19-60):

1. Newly synthesized PKC binds to the membrane (or possibly to the underlying cytoskeleton), where

phosphoinositide-dependent protein kinase-1 (PDK1) phosphorylates its activation loop (at Thr 500 in PKCβII). The resulting negative charge on the activation loop is postulated to properly align the active site residues of PKC for catalysis, much as we have seen for PKA (Section 18-3Cb;

Figure 19-60 **Activation of PKC. (1)** Newly synthesized PKC is phosphorylated on its activation loop (here represented by Thr 500 of PKCβII; *yellow ball*) by phosphoinositide-dependent protein kinase-1 (PDK1), which is tethered to the membrane via its C-terminal pleckstrin homology domain (PH). **(2)** The now catalytically competent PKC autophosphorylates 2 sites on its C-terminal segment (here represented by Thr 641 and Ser 660 of PKCβII). However, the N-terminal pseudosubstrate segment now binds to PKC's active site, so that the enzyme remains inactive. **(3)** On the binding of PKC's C1 domain to membrane-bound DAG (the product of extracellular signals

inducing phosphoinositide hydrolysis) together with the Ca^{2+}-mediated binding of the C2 domain to phosphatidylserine (PS) in the membrane, the pseudosubstrate is ejected from the PKC active site, thereby yielding active enzyme. [After a drawing by Toker, A. and Newton, A.C., *Cell* **103,** 187 (2000).]

PKA's activation loop is also phosphorylated by PDK1). In fact, the mutagenic replacement of PKCα's activation loop Thr with a neutral nonphosphorylatable residue yields an inactivatable enzyme, whereas its replacement with Glu yields an enzyme that requires only DAG and Ca^{2+} for activation.

2. The now catalytically competent PKC rapidly autophosphorylates its other two phosphorylation sites. The autophosphorylation of Thr 641 appears to lock PKC into its active conformation, as suggested by the observation that in PKCβII, which has been phosphorylated at only Thr 500 and Thr 641, the selective dephosphorylation of Thr 500 yields active enzyme. The autophosphorylation of the third phosphorylation site correlates with the release of PKC into the cytosol, where PKC is maintained in its inactive state by the binding of its pseudosubstrate to its active site.

3. This autoinhibition is relieved, as described above, when PKC again binds to the membrane via DAG binding to its C1 domain and Ca^{2+}-mediated binding of its C2 domain to phosphatidylserine (PS).

The activity of PKC is regulated by the protein phosphatase **PHLPP** (for *PH* domain *l*eucine-rich *r*epeat *p*rotein *p*hosphatase), which specifically dephosphorylates its Thr 641.

D. *The Phosphoinositide 3-Kinases*

The inositol head group of phosphatidylinositol has 5 free hydroxyl groups that can be phosphorylated (Fig. 19-53). However, only its 3-, 4-, and 5-positions are known to be phosphorylated *in vivo,* and *these occur in all seven possible combinations* (Fig. 19-61), *each of which participates in signaling.* In addition to the plasma membrane, they occur in ER, Golgi, and endosome membranes, although with different distributions in each of these several subcellular compartments.

The phosphorylations of these various phosphoinositides are catalyzed by ATP-dependent enzymes known as **phosphoinositide 3-kinases (PI3Ks), phosphoinositide 4-kinases (PIP4Ks),** and **phosphoinositide 5-kinases (PIP5Ks).** Their various products function as second messengers by recruiting the proteins that bind them to the cytosolic surface of the plasma membrane (see below). The resulting colocalization of enzymes and substrates results in further signaling activity that controls such vital functions as cell survival, proliferation, cytoskeletal rearrangement, endocytosis, and vesicle trafficking.

The PI3Ks are the presently best understood phosphoinositide kinases. Consequently, in this subsection, we discuss the PI3Ks and their products as a paradigm of all phosphoinositide kinases and the signals they produce.

a. PI3Ks Have Three Classes

Mammalian PI3Ks are divided into three classes according to their structures (Fig. 19-62), substrate specificities, and modes of regulation:

1. Class I PI3Ks are heterodimeric receptor-regulated enzymes that preferentially phosphorylate PIP$_2$ (alternatively, PtdIns-4,5-P$_2$). Their ~1070-residue catalytic subunits interact with Ras · GTP via a Ras-binding domain (RBD) near their N-termini. Their regulatory subunits are adaptor proteins that link the catalytic subunits to upstream signaling events and hence form two subclasses according to the type of upstream effectors with which they interact:

(a) The class IA PI3Ks (PI3Kα, β, and δ) are activated by RTKs via the mediation of the adaptor subunit **p85** (of which there are seven isoforms), which contains SH2 and SH3 domains and may be phosphorylated on specific Tyr side chains. The gene encoding PI3Kα is one of the two most frequently mutated oncogenes in human cancers, thus indicating this enzyme's central regulatory role. Several of its most common oncogenic mutations result in enhanced enzymatic activity by altering the interactions between its kinase domain and its regulatory domains or p85.

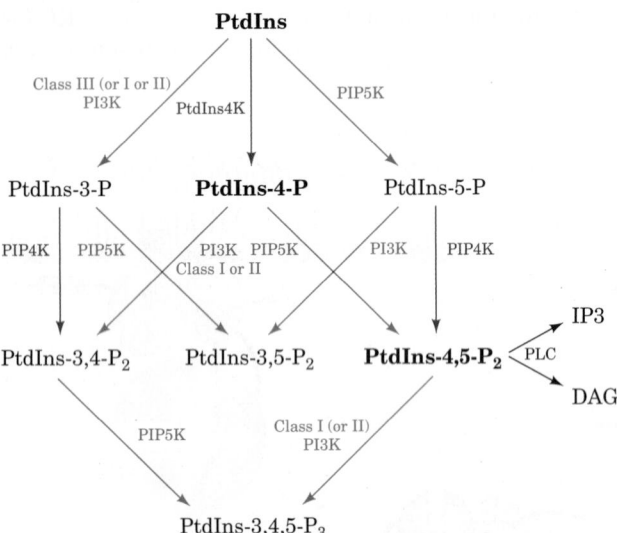

Figure 19-61   Flowchart of reactions in the synthesis of phosphoinositides in mammalian cells. PtdIns, PtdIns-4-P, and PtdIns-4,5-P$_2$ (PIP$_2$) are written in bold type to indicate their abundance: Together they comprise ~90% of the cell's total phosphoinositides. **PtdIns-3-P** and **PtdIns-5-P** each comprise 2–5% of the total, whereas the levels of **PtdIns-3,4-P$_2$** and **PtdIns-3,4,5-P$_3$** (PIP$_3$) are barely detectable in quiescent cells but rise to 1 to 3% of the total in stimulated cells. **PtdIns-3,5-P$_2$** comprises ~2% of the phosphoinositides in fibroblasts. [After Fruman, D.A., Meyers, R.E., and Cantley, L.C., *Annu. Rev. Biochem.* **67,** 501 (1998).]

Figure 19-62   Domain organization of the three classes of PI3Ks. [After Walker, E.H., Persic, O., Ried, C., Stephens, L., and Williams, R.L., *Nature* **402,** 314 (1999).]

(b) The class IB PI3K, of which PI3Kγ is its only member, is activated by the $G_{\beta\gamma}$ dimers of heterotrimeric G proteins, with its adaptor subunit **p101** rendering it far more sensitive to $G_{\beta\gamma}$.

2. Class II PI3Ks (PI3K-C2α, β, and γ) are ~1650-residue monomers that are characterized by a C-terminal C2 domain that does not bind Ca^{2+}. They preferentially phosphorylate PtdIns and **PtdIns-4-P.** Since they lack adaptors, the way in which class II PI3Ks are controlled is unknown.

3. Class III PI3K, which has one known isoform, phosphorylates only PtdIns. It is a heterodimer with an 887-residue catalytic subunit and an adaptor subunit known as **p150.** Class III PI3K is constitutively active, that is, it is unregulated and hence is thought to be the cell's main provider of **PtdIns-3-P,** whose level is essentially unaltered by cellular stimulation. It is thought to be the evolutionary predecessor of the other classes because it is the only class of PI3K present in yeast.

In addition to their lipid kinase activities, all PI3Ks have Ser/Thr protein kinase activity, although the physiological significance of this dual specificity is unclear.

b. PI3Kγ Is a Multidomain Protein

The X-ray structure of PI3Kγ · ATP, in which the PI3Kγ lacks its N-terminal 143 residues (which are important for interaction with the p101 adaptor; the analogous portion of PI3Kα interacts with its p85 adaptor), was determined by

Roger Williams. It reveals that its RBD, C2, and helical domains form a relatively compact layer that packs against the "back" of the kinase domain (Fig. 19-63). As expected, the kinase domain is grossly similar to those of protein kinases in that it is bilobal, with its N-lobe consisting largely of a 5-stranded β sheet and its C-lobe being predominantly helical. However, there are also major differences between these kinase domains, as can be seen by comparing the catalytic domain in Fig. 19-63 with that in, for example, Fig. 19-28*a*.

The RBD domain of PI3Kγ has the same fold as that of RafRBD (Fig. 19-41). Indeed, in the X-ray structure of PI3Kγ–Ras · GMPPNP (Fig. 19-64), also determined by Williams, the PI3K RBD interacts with Ras in a similar manner as we have seen that RafRBD interacts with the Ras homolog Rap1A (Fig. 19-41) in that they continue each other's central β sheets. However, Ras bound to PI3Kγ is rotated by 35° relative to Rap1A bound to RafRBD. Contacts between the Switch I region of Ras and the PI3Kγ stabilize this interaction and ensure its dependence on Ras · GTP. This complex also contains intermolecular contacts involving the Switch II region of Ras. Such an interaction had previously only been observed between Ras and its upstream effectors. Comparison of the structure of the PI3Kγ–Ras complex with that of PI3Kγ · ATP (Fig. 19-63) indicates that Ras binding induces the C-lobe of PI3Kγ's catalytic domain to pivot relative to its N-lobe in a way that substantially alters the putative binding pocket for the phosphoinositide head group. This, presumably, accounts for the ~15-fold activation of PI3Kγ on binding Ras · GTP.

Figure 19-63 **X-ray structure of PI3Kγ · ATP.** The protein is shown in ribbon form with its Ras-binding domain (RBD) green, its C2 domain magenta, its helical domain orange, the N- and C-lobes of its kinase domain pink and cyan, and interdomain segments gray. The ATP is shown in space-filling form with C green, N blue, O red, and P yellow. The protein is oriented such that its kinase domain is seen in "standard" view. The protein appears fragmented because several of its segments are disordered, including much of the kinase's activation loop. [Based on an X-ray structure by Roger Williams, MRC Laboratory of Molecular Biology, Cambridge, U.K. PDBid 1E8X.]

Figure 19-64 **X-ray structure of PI3Kγ–Ras · GMPPNP.** Here only the PI3Kγ RBD (*green*) and the Ras · GMPPNP (*gold*) are drawn, with the Switch I and Switch II regions of Ras magenta and cyan and its bound GMPPNP shown in space-filling form (C green, N blue, O red, and P yellow). The view, which is similar to that of Fig. 19-41, is related to that in Fig. 19-63 by rotating it clockwise by ~40° about its vertical axis and then turning it 180° about the axis perpendicular to the page. [Based on an X-ray structure by Roger Williams, MRC Laboratory of Molecular Biology, Cambridge, U.K. PDBid 1HE8.]

The C2 domain of PI3Kγ forms the same sandwich of two 4-stranded antiparallel β sheets seen in the C2 domain of PLC-δ1 (Section 19-4Ba). However, in contrast to the C2 domain of PLC-δ1, that of PI3Kγ does not bind Ca^{2+} ions. Nevertheless, the PI3Kγ C2 domain appears to participate in membrane association, as indicated by the observation that this isolated C2 domain binds to phospholipid vesicles with an affinity similar to that of the intact enzyme. This interaction is presumably mediated by patches of basic residues on the surface of the C2 domain.

The PI3Kγ helical domain consists of five repeating pairs of antiparallel helices that form a superhelix, which closely resembles that formed by the HEAT repeats in the A subunit of protein phosphatase 2A (PP2A; Fig. 19-51a), even though PI3Kγ does not contain a HEAT sequence motif. In analogy with the function of the A subunit of PP2A to bind other proteins (Section 19-3Fe), it is proposed that the largely solvent-exposed helical domain of PI3Kγ functions to interact with the proteins that bind PI3Kγ, such as its p101 adaptor and $G_{\beta\gamma}$.

c. Akt Activation Requires Its PH Domain–Mediated Binding to 3-Phosphoinositides

The PtdIns-3,4-P_2 and PtdIns-3,4,5-P_3 products of PI3Ks (Fig. 19-61) bind to their downstream effectors mainly via pleckstrin homology (PH) domains that preferentially bind the head groups of these 3-phosphoinositides rather than that of PIP$_2$ (as does the PH domain of PLC-δ; Fig. 19-58). Another example of a PH domain-containing protein that does so is the 556-residue phosphoinositide-dependent protein kinase-1 (PDK1), which, as we have seen, phosphorylates the activation loops of PKA and PKC (Section 19-4Cb).

PDK1 also phosphorylates the Ser/Thr protein kinase **Akt** [also called **protein kinase B (PKB)**], a proto-oncogene product that is implicated in regulating multiple biological processes including gene expression, apoptosis, glucose uptake, and cellular proliferation, and hence phosphorylates many target proteins. The ~480-residue Akt consists of an N-terminal PH domain that binds 3-phosphoinositides and a C-terminal kinase domain that is homologous to those of PKA and PKC (and is thus a member of the AGC family of protein kinases). Akt is present in multicellular organisms in three isoforms (**Akt1/PKBα, Akt2/PKBβ,** and **Akt3/PKBγ**) but is absent in yeast, which suggests that it evolved from another AGC family member coincidentally with multicellular organisms.

The full activation of Akt requires its phosphorylation at both its Ser 473 and Thr 308. Ser 473 is phosphorylated by **mTORC2** [for *m*ammalian *t*arget of *r*apomycin *c*omplex *2*; **rapamycin** is an immunosuppressant similar to FK506 (Section 9-2B)]. This stimulates PDK1 to phosphorylate Thr 308, which is located in Akt's activation loop. Mutations of the residues in Akt's PH domain responsible for lipid binding block its phosphorylation *in vitro* by PDK1. However, the deletion of Akt's PH domain overcomes this enzyme's need for binding to 3-phosphoinositides. This suggests that the binding of Akt to these membrane-bound lipids induces a conformational change that permits PDK1

to phosphorylate and hence activate Akt. It therefore appears that it is the 3-phosphoinositide-mediated colocalization of Akt and PDK1 that leads to Akt activation and hence that it is the action of PI3K that is functionally responsible for this process. In contrast, the PDK1-mediated phosphorylation of PKA and PKC, which lack PH domains, occurs in the absence of 3-phosphoinositides and is therefore constitutive. The protein phosphatase PHLPP regulates Akt activity by dephosphorylating its Ser 473, in much the same way as we have seen that PHLPP dephosphorylates PKC (Section 19-4Cb).

d. The FYVE Domain Binds the PtdIns-3-P Head Group

The singly phosphorylated PtdIns-3-P is rarely bound by PH domains. Rather, its direct effects are mediated by **FYVE domains** [named after the four proteins in which it was first identified: *F*ab1p, *Y*OTB, *V*ac1p, and **early endosome antigen 1 (EEA1)**], which have been identified in ~60 proteins. For instance, the 1410-residue eukaryotic protein EEA1, which has a 65-residue, C-terminal FYVE domain, initiates endosome fusion in eukaryotic cells (Fig. 12-91) by recruiting the membrane-anchored small G protein **Rab5** and the transmembrane SNARE protein **syntaxin** (Section 12-4Db).

The NMR structure of the EEA1 FYVE domain, determined by Michael Overduin, reveals that it assumes similar conformations in the free state, when binding dibutanoyl-PtdIns-3-P (Fig. 19-65), and when bound to dodecylphosphocholine (DPC) micelles enriched with this PtdIns-3-P. The protein is largely held together by two bound Zn^{2+} ions, each of which is tetrahedrally liganded by four conserved Cys side chains. The PtdIns-3-P head group is held in its binding pocket by a network of electrostatic, hydrogen bonding, and hydrophobic interactions involving a highly conserved (R/K)(R/K)HHCR motif (RRHHCR in EEA1).

The NMR evidence indicates that, on the addition of DPC micelles, the FYVE domain · PtdIns-3-P complex inserts a hydrophobic 5-residue loop (FSVTV; orange in Fig. 19-65), which is flanked by basic residues (blue in Fig. 19-65), into the lipid layer. This also occurs in the absence of PtdIns-3-P but to a much lesser extent. Conversely, membrane insertion increases the binding affinity of the FYVE domain for PtdIns-3-P 20-fold (from 1 µM to 50 nM). The origin of this latter effect appears to be that the 10-residue segment preceding the membrane insertion loop, the unliganded protein's most disordered region, becomes more ordered and moves toward the binding pocket on binding PtdIns-3-P. This has led to the proposal that the FYVE domain is recruited to membranes via the insertion of its hydrophobic loop into the lipid bilayer. This, in turn, primes the protein for the recognition of PtdIns-3-P, whose binding induces the protein's otherwise mobile N-terminal segment to clamp down over the PtdIns-3-P head group.

E. *Inositol Polyphosphate Phosphatases*

Signaling via the phosphoinositide cascade is terminated through the actions of a variety of inositol phosphatases that

Figure 19-65 **NMR structure of the EEA1 FYVE domain in complex with PtdIns-3-P.** The head group of PtdIns-3-P is drawn in ball-and-stick form (C green, O red, P magenta, H gray). The protein binds two Zn^{2+} ions (*cyan spheres*) that are each tetrahedrally liganded by four Cys side chains that are drawn in stick form (C green and S yellow). The 5-residue loop that inserts into DPC micelles is orange and its flanking basic residues are blue. [Based on an NMR structure by Michael Overduin, University of Colorado Health Sciences Center. PDBid 1HYI.]

are functionally classified as 1-, 3-, 4-, and 5-phosphatases. We end our consideration of the phosphoinositide cascade by discussing the characteristics of these essential enzymes.

a. The Inositol Polyphosphate 5-Phosphatases Act In Numerous Signaling Pathways

The first **inositol polyphosphate 5-phosphatases** that were studied hydrolyze IP_3 (Ins-1,4,5-P_3) to IP_2 (Ins-1,4-P_2) and thereby terminate cellular Ca^{2+} mobilization (Fig. 19-54, *bottom*). Mammals express >10 isozymes that have 5-phosphatase activity. These enzymes share a common catalytic core and have been classified according to their substrate specificities into two groups: Type I enzymes dephosphorylate inositol phosphates, whereas type II enzymes, in addition, hydrolyze the corresponding phosphoinositides.

Type I 5-phosphatases, which hydrolyze only IP_3 and **Ins-1,3,4,5-P_4,** are membrane-anchored via prenylation. That expressed in blood platelets (a type of blood cell that participates in blood clotting; Section 35-1), which is

representative of this group, forms a stoichiometric complex with **pleckstrin,** a 350-residue protein that consists largely of two PH domains. When platelets are stimulated by the proteolytic clotting enzyme **thrombin** (Section 35-1B), pleckstrin is phosphorylated on Ser and Thr residues by PKC, which in turn activates its associated 5-phosphatase. Note that PKC is activated by DAG, a product of PLC, which simultaneously generates the type I 5-phosphatase substrate IP_3 (Fig. 19-54). Hence, the PLC product IP_3 activates Ca^{2+} ion release, whereas its coproduct DAG activates type I 5-phosphatase through pleckstrin phosphorylation to terminate the Ca^{2+} signal. This termination is apparently important for normal cell growth as a decrease in the expression of type I 5-phosphatase causes increased and even uncontrolled (malignant) cell growth.

Type II 5-phosphatases share increased similarities in their catalytic cores relative to type I enzymes and, in addition, have a so-called type II domain on the N-terminal side of their catalytic cores. They occur in three main subtypes: **GIPs, SHIPs,** and **SCIPs.** GIPs are so called because they have a C-terminal GAP domain (*G*AP-containing *i*nositol *p*hosphatase), although they have no demonstrated GAP activity. GIPs hydrolyze IP_3 and Ins-1,3,4,5-P_4 and their corresponding lipids, PtdIns-4,5-P_2 and PtdIns-3,4,5-P_3, although with varying catalytic efficiencies.

There are only two known GIPs, **5-phosphatase II** and **OCRL.** OCRL is so called because its mutation causes the X-linked hereditary disease **oculocerebrorenal dystrophy** (also called **Lowe syndrome**), which is characterized by congenital cataracts, progressive retinal degeneration, mental retardation, and renal tubule defects leading to kidney failure in early adulthood. The 901-residue OCRL occurs mainly on the surface of lysosomes, where it is anchored through prenylation. Renal tubule cells from Lowe syndrome patients are deficient in PtdIns-4,5-P_2 and PtdIns-3,4,5-P_3 hydrolytic activity, whereas the corresponding inositol phosphates are hydrolyzed normally, thereby indicating that OCRL is a lipid phosphatase. PtdIns-4,5-P_2 stimulates the budding of membrane vesicles from lysosomes, so that the accumulation of this lipid probably leads to abnormally increased trafficking of enzymes from the lysosome to the extracellular space. Indeed, the lysosomal enzymes in these cells appear to be missorted (as are various lysosomal hydrolases in I-cell disease; Section 12-4Cg). It is therefore proposed that this lifelong leakage of enzymes from the lysosomes in Lowe syndrome patients causes tissue damage that eventually results in kidney failure and blindness.

SHIPs only hydrolyze substrates that also have a phosphate in their 3-positions. The two known members of this group, **SHIP** (for *S*H2-containing *i*nositol-5-*p*hosphatase) and **SHIP2,** are ~1200-residue proteins that have an N-terminal SH2 domain. Thus, these proteins can bind to PTKs and, in fact, are phosphorylated by them to yield a consensus binding sequence for PTB domains (NPXpY; Section 19-3Cc). Moreover, they also contain a C-terminal Pro-rich domain that may bind to SH3-containing proteins. Thus, it appears that SHIP activity may be under the control of several systems. Indeed, SHIP, which is expressed only in hematopoietic (blood-forming) cells, associates with the adaptor proteins Grb2 and Shc (Section 19-3Cf). It functions to hydrolyze PtdIns-3,4,5-P$_3$, which is implicated in activating Akt and PLC. SHIP2 functions similarly in nonhematopoietic cells, where it limits cellular responses to insulin, EGF, and PDGF.

SCIPs (*Sac1-*containing *inositol phosphatases*) are so named because they contain an N-terminal domain that is homologous to the yeast phosphatidylinositol phosphatase **Sac1**. The first SCIP to be characterized is named **synaptojanin1** because it was purified from synaptic vesicles and because the presence of two phosphatase domains is reminiscent of the two kinase domains in Janus kinases (JAKs; Section 19-3Eb). The 1575-residue synaptojanin1's 5-phosphatase domain hydrolyzes PIP$_3$ and PtdIns-4,5-P$_2$ and its Sac1 phosphatase domain hydrolyzes PtdIns-3-P and PtdIns-4-P. Synaptojanin1 is expressed only in neurons, where it forms complexes with the G protein dynamin (Section 12-4Cd) and thereby participates in synaptic vesicle recycling. The closely similar **synaptojanin2** is ubiquitously expressed but its functions are largely unknown.

b. Inositol Polyphosphate 1-Phosphatase Is Implicated in Bipolar Disorder

Mammals express only one type of **inositol polyphosphate 1-phosphatase,** a 399-residue enzyme that hydrolyzes **Ins-1,4-P$_2$** and Ins-1,3,4-P$_3$ (IP$_3$) but does not act on lipid substrates. This enzyme is inhibited by Li$^+$ ion. The therapeutic efficacy of Li$^+$ in controlling the incapacitating mood swings of manic-depressive individuals (those with **bipolar disorder**) therefore suggests that this mental illness is caused by an aberration of 1-phosphatase in the brain, possibly resulting in abnormal activation of Ca^{2+}-mobilizing receptors (Fig. 19-54, *bottom*). Indeed, *Drosophila* in which this 1-phosphatase has been deleted exhibit neurological deficits (the so-called "shaker" phenotype) that appear identical to those of wild-type *Drosophila* treated with Li$^+$.

c. The Inositol Polyphosphate 3-Phosphatase PTEN Is a Tumor Suppressor

The **inositol polyphosphate 3-phosphatases** undo the actions of the PI3Ks. The best characterized of these enzymes is the 403-residue **PTEN** (for *phosphatase and tensin homolog;* **tensin** is a cytoskeletal actin-binding protein), which *in vitro* dephosphorylates all 3-phosphorylated phosphoinositides and Ins-1,3,4,5-P$_4$. PTEN is a **tumor suppressor** (a protein whose loss of function is a cause of in cancer), presumably because its 3-phosphatase activity functions to downregulate the PtdIns-3,4,5-P$_3$-activated Akt. In fact, PTEN mutation or loss commonly occur in many types of cancers. PTEN can also dephosphorylate Ser-, Thr-, and Tyr-phosphorylated peptides, although this activity requires the peptides to be highly acidic.

The X-ray structure of PTEN, determined by Jack Dixon and Nikola Pavletich, reveals the protein to consist of an N-terminal phosphatase domain and a C-terminal C2 domain (Fig. 19-66). The structure of its phosphatase domain resembles that common to protein tyrosine phosphatase (PTP) domains (e.g., Fig. 19-50), but with a larger active site pocket, presumably to accommodate the large size of its PtdIns-

Figure 19-66   X-ray structure of PTEN. The protein is shown with its phosphatase domain blue, its C2 domain red, and the P loop, which interacts with the substrate, tan. The dotted line represents a 24-residue segment that was deleted from the protein to facilitate its crystallization. [Courtesy of Nikola Pavletich, Memorial Sloan-Kettering Cancer Center, New, York, New York. PDBid 1D5R.]

3,4,5-P$_3$ substrate. The C2 domain lacks bound Ca^{2+} ion as well as the ligands to bind it but, nevertheless, binds to phospholipid membranes, as does the C2 domain of PI3Kγ (Fig. 19-63). The phosphatase and C2 domains associate across an extensive interface, whose residues are frequently mutated in cancer. A similar tight interface between the C2 and kinase domain occurs in PLC-βII (Fig. 19-57). This suggests that PTEN's C2 domain functions to productively position its attached phosphatase domain at the membrane.

d. The Inositol Polyphosphate 4-Phosphatases Control the Level of PtdIns-3,4-P$_2$

There are two isoforms of inositol 4-phosphatases, **4-phosphatases I** and **II,** which catalyze the hydrolysis of **Ins-1,3,4-P$_3$, Ins-2,4-P$_2$,** and PtdIns-3,4-P$_2$. In fact, these ~940-residue proteins account for >95% of the observed PtdIns-3,4-P$_2$ phosphatase activity in many human tissues, thereby suggesting that they play an important role in the metabolism of this second messenger. This is supported by the observation that stimulating human platelets by thrombin or Ca^{2+} ion results in the inactivation of 4-phosphatase I through its proteolytic cleavage by the Ca^{2+}-dependent protease **calpain.** This inactivation of 4-phosphatase I correlates with the Ca^{2+}- and/or aggregation-dependent accumulation of PtdIns-3,4-P$_2$ characteristic of human platelets (which aggregate in the initial stages of blood clot formation; Section 35-1).

F. Epilog: Complex Systems and Emergent Properties

Complex systems are, by definition, difficult to understand and substantiate. Familiar examples include Earth's weather system, the economies of large countries, the ecologies of even small areas, and the human brain. Biological signal transduction systems, as is amply evident from a reading of this chapter, are complex systems. Thus, as we have seen, a hormonal signal is typically transduced through several intracellular signaling pathways, each of which consists of numerous components, many of which interact with components of other signaling pathways. For example, **the insulin signaling system** (Fig. 19-67), although not yet fully elucidated, is clearly highly complex. On binding insulin, the insulin receptor autophosphorylates itself at several Tyr residues (Section 19-3Ac) and then Tyr-phosphorylates its target proteins, thereby activating several signaling pathways that control a diverse array of effects:

1. Phosphorylation of Shc (Section 19-3Cc) results in stimulation of a MAP kinase cascade (Section 19-3D), ultimately affecting growth and differentiation.

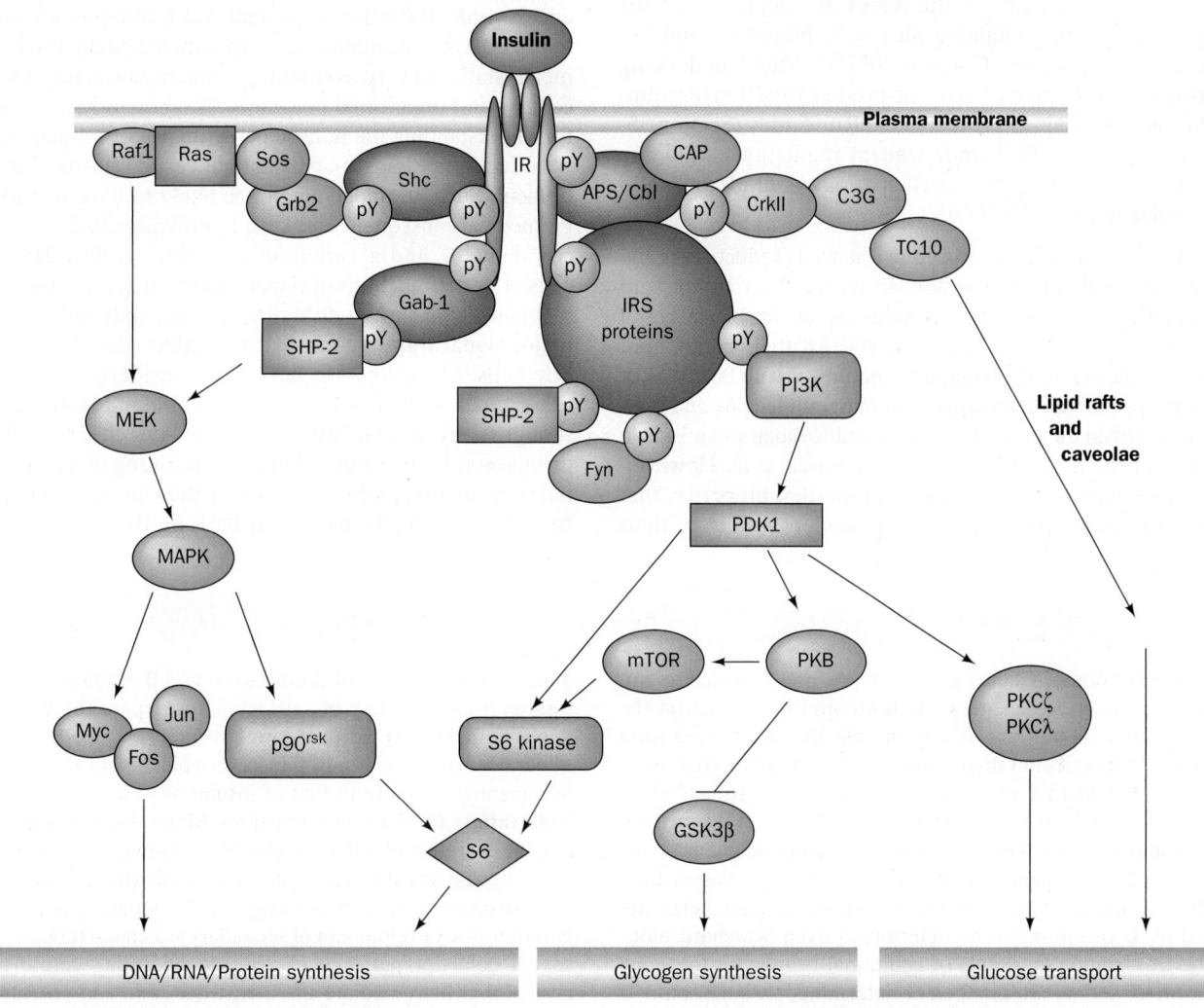

Figure 19-67   Insulin signal transduction. The binding of insulin to the insulin receptor **(IR)** induces its autophosphorylation at several Tyr residues on its β subunits. Several proteins, including Shc, Gab-1, the APS/Cbl complex, and IRS proteins, bind to these pY residues where they are Tyr-phosphorylated by the activated insulin receptor, thereby activating MAPK and PI3K phosphorylation cascades as well as a lipid raft and caveolae-associated regulation process. The MAPK cascade regulates the expression of genes involved in cellular growth and differentiation. The PI3K cascade leads to changes in the phosphorylation states of several enzymes, so as to stimulate glycogen synthesis, as well as other pathways. The PI3K cascade also participates in the control of vesicle trafficking, leading to the translocation of the **GLUT4** glucose transporter to the cell surface and thus increasing the rate of glucose transport into the cell (Section 20-2Ec). Glucose transport control is also exerted by the APS/Cbl system in a PI3K-independent manner involving lipid rafts and caveolae. Other symbols: Myc, Fos, and Jun (transcription factors; Section 19-3D), SHP-2 (an SH2-containing PTP; Section 19-3Fb), **CAP** (Cbl-associated protein), **C3G** [a guanine nucleotide exchange factor (GEF)], **CrkII** [an SH2/SH3-containing adaptor protein), PDK1 (phosphoinositide-dependent protein kinase-1; Section 19-4Cb), PKB (protein kinase B, also named Akt; Section 19-4Dc), **mTOR** [for *m*ammalian *t*arget *o*f *r*apamycin, a PI3K-related protein kinase (Section 9-2B); mTOR is also known as **FKBP12-rapamycin-associated protein (FRAP)]**, **S6** [a protein subunit of the eukaryotic ribosome's small subunit (Section 32-3Ab); its phosphorylation stimulates translation], and PKCζ and PKCλ (atypical isoforms of protein kinase C; Section 19-4C). [After Zick, Y., *Trends Cell Biol.* **11,** 437 (2001)].

2. Phosphorylation of **Gab-1 (Grb2-associated binder-1)** similarly activates this MAP kinase cascade.

3. Phosphorylation of insulin receptor substrate (IRS) proteins (Section 19-3Cg) activates a phosphoinositide cascade via a PI3K (Section 19-4Da), ultimately stimulating a variety of metabolic processes including glycogen synthesis (Section 18-3E) and glucose transport (Section 20-2E), as well as cell growth and differentiation.

4. Phosphorylation of the **APS/Cbl** complex (APS for *a*daptor *p*rotein containing *p*leckstrin homology and *S*rc homology-2 domains; Cbl is an SH2/SH3-binding docking protein that is a proto-oncogene product) leads to the stimulation of **TC10** [a G protein in the Rho family (Section 35-3E)], and to the PI3K-independent regulation of glucose transport involving the participation of lipid rafts and caveolae (Section 12-3Cb).

The predominant approach in science is reductionist: the effort to understand a system in terms of its component parts. Thus chemists and biochemists explain the properties of molecules in terms of the properties of their component atoms, cell biologists explain the nature of cells in terms of the properties of their component macromolecules, and biologists explain the characteristics of multicellular organisms in terms of the properties of their component cells. However, complex systems have **emergent properties,** properties that are not readily predicted from an understanding of their

component parts (i.e., the whole is greater than the sum of its parts). Indeed, life itself is an emergent property that arises from the numerous chemical reactions that occur in a cell.

In order to elucidate the emergent properties of a complex system, an integrative approach is required. For signal transduction systems, such an approach would entail determining how each of the components of each signaling pathway in a cell interacts with all of the other such components under the conditions that each of these components experiences within its local environment. Yet techniques for doing so are not often available. Moreover, these systems are by no means static but vary, over multiple time scales, in response to cellular and organismal programs. Consequently, the means for understanding the holistic performance of cellular signal transduction systems are only in their earliest stages of development. Such an understanding is likely to have important biomedical consequences since many diseases, including cancer, diabetes, and a variety of neurological disorders, are caused by malfunctions of signal transduction systems.

Finally, you should note that we have only outlined the major signal transduction pathways that occur in eukaryotic cells. Moreover, we have not considered numerous other such pathways that control cellular functions (although many of them are discussed in later chapters). Nevertheless, it is clear that a full understanding of a cell's signal transduction pathways and how they interact is the key to understanding the molecular basis of life.

CHAPTER SUMMARY

1 Hormones Chemical messengers are classified as autocrine, paracrine, or endocrine hormones if they act on the same cell, cells that are nearby, or cells that are distant from the cell that secreted them, respectively. The body contains a complex endocrine system that controls many aspects of its metabolism. Hormone levels may be determined through radioimmunoassays. Receptors are membrane-bound proteins that bind their ligands according to the laws of mass action. The parameters describing the binding of a radiolabeled ligand to its receptor can be determined from Scatchard plots. The dissociation constants of additional ligands for the same receptor-binding site can then be determined through competitive binding studies.

The pancreatic islet cells secrete insulin and glucagon, polypeptide hormones that induce liver and adipose tissue to store or release glucose and fat, respectively. Gastrointestinal polypeptide hormones coordinate various aspects of digestion. The thyroid hormones, T_3 and T_4, are iodinated amino acid derivatives that generally stimulate metabolism by activating cellular transcription factors. Ca^{2+} metabolism is regulated by the levels of PTH, vitamin D, and calcitonin. PTH and vitamin D induce an increase in blood $[Ca^{2+}]$ by stimulating Ca^{2+} release from bone and its absorption from kidney and intestine, whereas calcitonin has the opposite effects. Vitamin D is a steroid derivative that must be obtained in the diet or by exposure to UV radiation. Vitamin D, after being sequentially processed in the liver and kidney to $1,25(OH)_2D$, stimulates the synthesis of a Ca^{2+}-binding protein in the intestinal epithelium. The adrenal medulla secretes the catecholamines epinephrine

and norepinephrine, which bind to α- and β-adrenergic receptors on a great variety of cells so as to prepare the body for "fight or flight." The adrenal cortex secretes glucocorticoid and mineralocorticoid steroids. Glucocorticoids affect metabolism in a manner opposite to that of insulin as well as mediating a wide variety of other vital functions. Mineralocorticoids regulate the excretion of salt and water by the kidney. The gonads secrete steroid sex hormones, the androgens (male hormones) and estrogens (female hormones), which regulate sexual differentiation, the development of secondary sex characteristics, and sexual behavior patterns. Ovaries, in addition, secrete progestins that help mediate the menstrual cycle and pregnancy. Mammalian embryos develop as females unless subjected to the influence of the androgen testosterone. *SRY*, a gene that encodes a DNA-binding protein and that is normally located on the Y chromosome, induces the development of testes, which in turn secrete testosterone. The hypothalamus secretes a series of polypeptide releasing factors and release-inhibiting factors such as CRF, TRF, GnRF, and somatostatin that control the secretion of the corresponding trophic hormones from the pituitary gland's adenohypophysis. Most of these trophic hormones, such as ACTH, TSH, LH, and FSH, stimulate their target endocrine glands to secrete the corresponding hormones. However, growth hormone acts directly on tissues as well as stimulating liver to synthesize growth factors known as somatomedins. The pituitary gland's neurohypophysis secretes the polypeptides vasopressin, which stimulates the kidneys to retain water, and oxytocin, which stimulates uterine contraction. The menstrual cycle results from a complex interplay of hypothalamic, adenohypophyseal, and steroid sex hormones. A fertilized and implanted ovum secretes CG, which binds to the same receptor and has similar effects as LH, thus preventing menstruation. The binding of hGH to its receptor causes the receptor to dimerize, thereby providing the intracellular signal that the receptor has bound hCG. Many other hormonal signals are similarly mediated. The adenohypophysis also secretes opioid peptides that have opiatelike effects on the central nervous system. Nitric oxide (NO), a highly reactive radical gas, functions as a local mediator that regulates vasodilation, serves as neurotransmitter, and functions in the immune response. In mammals, it is synthesized by three isozymes of nitric oxide synthase (NOS), an enzyme that contains 5 redox-active prosthetic groups. eNOS and nNOS are activated by Ca^{2+} through their binding of Ca^{2+}–calmodulin; iNOS is transcriptionally controlled. NO activates guanylate cyclase to produce cGMP, which in turn activates cGMP-dependent protein kinase.

2 Heterotrimeric G Proteins Ligand (hormone) binding to G protein-coupled receptors (GPCRs) activates the $G_{s\alpha}$ subunit of a stimulatory G protein to replace its bound GDP with GTP, release its associated $G_{\beta\gamma}$ subunits, and activate adenylate cyclase (AC) to synthesize cAMP. Activation continues until $G_{s\alpha}$ hydrolyzes its bound GTP to GDP and recombines with $G_{\beta\gamma}$. Several types of activated hormone receptors in a cell may stimulate the same G_s protein. There are also inhibitory G proteins, which may have the same G_β and G_γ subunits as does G_s, but which have an inhibitory $G_{i\alpha}$ subunit that deactivates adenylate cyclase. Cholera toxin (CT) and heat-labile enterotoxin (LT), related bacterial AB_5 proteins, induce uncontrolled cAMP production by ADP-ribosylating $G_{s\alpha}$ so as to render it incapable of hydrolyzing GTP. Pertussis toxin, also an AB_5 protein, similarly ADP-ribosylates $G_{i\alpha}$. Biological signaling systems are subject to desensitization through the phosphorylation and endocytotic sequestering of the cell-surface receptors. The catalytic core of the numerous isoforms of AC are pseudosymmetric heterodimers that are activated, in most cases, by the binding of the Switch II region of $G_{s\alpha} \cdot$ GTP to a cleft in an AC's C_{1a} domain. cAMP and cGMP are eliminated through the actions of numerous phosphodiesterases (PDEs), whose activities are controlled by a variety of agents, thereby providing for cross talk between signaling systems.

3 Tyrosine Kinase–Based Signaling The binding of ligands such as hormones and protein growth factors activates receptor tyrosine kinases (RTKs) by inducing them to dimerize and then autophosphorylate specific Tyr residues in the activation loops of their tyrosine kinase domains. This is usually followed by the autophosphorylation of Tyr residues on other cytoplasmic domains. Cancer cells' immortality and their uncontrolled proliferation endow them with the capacity to form invasive and metastatic tumors. Rous sarcoma virus, a retrovirus causing sarcomas in chickens, carries an oncogene, v-src, that is homologous to the normal cellular gene c-src. Both genes encode a protein tyrosine kinase (PTK) that stimulates cell division. Oncogene products include analogs of growth factors, growth factor receptors, nuclear proteins that stimulate transcription and/or cell division, and G proteins.

Two-hybrid systems are used to identify interacting proteins. An autophosphorylated RTK may activate other proteins by phosphorylating them on specific Tyr side chains. It can also modulate the activities of specific proteins through the binding of an RTK's phosphoTyr-containing peptide segment to SH2 and PTB domains on these proteins or on adaptors that bind to these

proteins. Grb2, an adaptor protein, binds to certain activated RTKs in this way and simultaneously, via its SH3 domains, to Sos protein. The bound Sos, in turn, functions as a guanine nucleotide exchange factor (GEF) to induce the small G protein Ras to exchange its bound GDP for GTP. Ras is a poor GTPase but it is aided in eventually hydrolyzing its bound GTP to GDP by the GTPase activating protein (GAP) RasGAP, which insinuates a catalytically important Arg side chain into Ras's otherwise inefficient active site. Mutations that interfere with the ability of Ras–RasGAP to hydrolyze Ras's bound GTP are oncogenic.

The binding of Ras $\cdot$ GTP to Raf, a protein Ser/Thr kinase, activates Raf to phosphorylate MEK, a MAP kinase kinase (MKK), which in turn phosphorylates MAP kinase (MAPK). The activated MAPK phosphorylates various cytoplasmic and membrane-associated proteins and, in addition, is translocated to the nucleus where it phosphorylates certain transcription factors, which then induce the transcription of their target genes. The proteins of such MAP kinase cascades are organized by their binding to scaffold proteins, which also prevents the members of different MAP kinase cascades in a cell from inappropriately phosphorylating one another. However, activated members of a MAP kinase cascade may phosphorylate other regulatory proteins, thereby eliciting cross talk between different signal transduction pathways.

Tyrosine kinase–associated receptors, such as cytokine receptors, transduce the signal that they have bound effector by activating associated nonreceptor tyrosine kinases (NRTKs), many of which are members of the Src or JAK families. Activated JAK proteins phosphorylate STAT proteins, which then dimerize and are translocated to the nucleus, where they function as transcription factors. Gleevec is a highly selective Abl inhibitor that is clinically effective in the treatment of chronic myelogenous leukemia (CML). Many cancers require elevated levels of Hsp90 activity for viability because their oncogenic proteins tend to be relatively unstable. Phosphorylated proteins are deactivated by protein phosphatases. Some protein tyrosine phosphatases (PTPs) are transmembrane receptors that are deactivated by ligand-induced dimerization. Other PTPs are cytoplasmic and are activated by their binding to activated PTKs, for example, via SH2 domains, as does SHP-2.

Cells contain several types of Ser/Thr protein phosphatases: PP1 participates in the regulation of glycogen metabolism; PP2A, which participates in a wide variety of regulatory processes, is a heterotrimer with numerous variants and hence specificities and cellular locations; and calcineurin (CaN; also called PP2B) is a Ca^{2+}-activated heterodimeric phosphatase that is the target of the immunosuppressive drugs cyclosporin A and FK506 via the binding of their complexes with the rotamases cyclophilin and FKBP12 to CaN so as to prevent the binding of CaN's target phosphopeptides.

4 The Phosphoinositide Cascade PIP_2, a minor phospholipid component of the plasma membrane's inner leaflet, can yield up to three types of second messengers. Hormone–receptor interactions, through the intermediacy of a G protein or an RTK, stimulate the corresponding phospholipase C (PLC) to hydrolyze PIP_2 to the water-soluble IP_3 and the membrane-bound DAG. The IP_3 stimulates the release of Ca^{2+} from the endoplasmic reticulum through ligand-gated channels. The Ca^{2+} binds to calmodulin, which in turn activates a variety of cellular processes. The DAG activates protein kinase C (PKC) to phosphorylate and thereby modulate the activities of numerous cellular proteins. DAG may also be degraded to yield arachidonate, an obligate intermediate in the biosynthesis of prostaglandins and related compounds.

The various classes of PLCs are activated in different ways, all of which bring the PLC into contact with its PIP_2 substrate in the membrane: PLC-β's by binding $G_{q\alpha} \cdot GTP$, $G_{\beta\gamma}$, and the membrane-anchored Rac1 · GTP; PLC-γ's by binding to phosphorylated PTKs via SH2 domains followed by phosphorylation of the PLC by the PTK; PLC-δ's by Ca^{2+}; and PLC-ε by binding Ras · GTP. "Conventional" PKCs are activated by both Ca^{2+} and DAG. Phorbol esters, which are DAG mimics that activate PKC, are the most potent known tumor promoters. DAG and Ca^{2+} synergistically bind PKC to the membrane via its C1 and C2 domains, which conformationally extracts PKC's N-terminal pseudosubstrate from the kinase's active site. The kinase is catalytically activated by phosphorylation on its activation loop by PDK1 followed by autophosphorylation at two more sites.

Phosphoinositides may be phosphorylated at their inositol head group's 3-, 4-, and 5-positions in all seven combinations, yielding membrane-bound second messengers that function by recruiting the proteins that bind them to the membrane surface. Mammalian phosphoinositide 3-kinases (PI3Ks) form three classes that differ according to their structures, substrate specificities, and modes of regulation. The $PtdIns-3,4-P_2$ and $PtdIns-3,4,5-P_3$ products of PI3Ks bind to the PH domain of the proto-oncogene product Akt (PKB), thereby colocalizing Akt with PDK1, which is also tethered to the membrane via its PH domain, so that PDK1 phosphorylates and thereby activates Akt. $PtdIns-3-P$ is bound by FYVE domains, which, like PH domains, are held together by two tetrahedrally liganded Zn^{2+} ions.

The various types of inositide polyphosphate phosphatases function to terminate signaling by the phosphoinositide cascade. OCRL, a type II 5-phosphatase that participates in controlling vesicle budding from the lysosome, is mutated in oculocerebrorenal disease (Lowe syndrome). The only 1-phosphatase expressed by mammals, which hydrolyzes $Ins-1,4-P_2$ and PIP_3, is inhibited by Li^+ ion and is thereby implicated in bipolar disorder. The 3-phosphatase PTEN, a tumor suppressor whose mutant forms are common to many cancers, undoes the actions of PI3Ks. Type I 4-phosphatase in blood platelets is inactivated through proteolytic cleavage by the Ca^{2+}-activated protease calpain. Cellular signal transduction systems, such as the insulin signaling system, are complex systems with emergent properties that are, as yet, poorly understood.

REFERENCES

General

Gomperts, B.D., Tatham, P.E.R., and Kramer, I.M., *Signal Transduction,* Academic Press (2002).

Helmreich, E.J.M., *The Biochemistry of Cell Signaling,* Oxford (2001).

Krauss, G., *Biochemistry of Signal Transduction and Regulation* (4th ed.), Wiley-VCH (2008).

Marks, F., Klingmüller, U., and Müller-Decker, K., *Cellular Signal Processing. An Introduction to the Molecular Mechanisms of Signal Transduction,* Garland Science (2009).

Nelson, J., *Structure and Function in Cell Signaling,* Wiley (2008).

Science's Signal Transduction Knowledge Environment (STKE). http://stke.sciencemag.org/cm/. [A database on signaling molecules and their relationships to each other. This database is introduced in a series of authoritative articles in *Science* **296,** 1632–1657 (2002). Full access to the database requires an individual or institutional subscription.]

Hormones

Alderton, W.K., Cooper, C.E., and Knowles, R.G., Nitric oxide synthases: structure, function, and inhibition, *Biochem. J.* **357,** 593–615 (2002).

Capel, B., Sex in the 90s: *SRY* and the switch to the male pathway, *Annu. Rev. Physiol.* **60,** 497–523 (1998).

Cary, S.P.L., Winger, J.A., Derbyshire, E.R., and Marletta, M.A., Nitric oxide signaling: no longer simply on or off, *Trends Biochem. Sci.* **31,** 231–239 (2006).

DeGroot, L.J. and Jameson, J.L. (Eds.), *Endocrinology* (5th ed.), Saunders (2006). [A 3-volume compendium.]

Garcin, E.D., Bruns, C.M., Lloyd, S.J., Hosfield, D.J., Tiso, M., Gachhui, R., Stuehr, D.J., Tainer, J.A., and Getzoff, E.D., Structural basis for isozyme-specific regulation of electron transfer in nitric-oxide synthase, *J. Biol. Chem.* **36,** 37918–37927 (2004). [The X-ray structure of the nNOS reductase domain.]

Greenstein, B. and Wood, D., *The Endocrine System at a Glance* (2nd ed.), Blackwell Publishing (2006).

Hadley, M.E. and Levine, J.E., *Endocrinology* (6th ed.), Benjamin Cummings (2007).

Ignarro, L.J. (Ed.), *Nitric Oxide. Biology and Pathobiology,* Academic Press (2000).

Kossiakoff, A.A. and de Vos, A.M., Structural basis for cytokine hormone–receptor recognition and receptor activation, *Adv. Protein Chem.* **52,** 67–108 (1999).

Li, H. and Poulos, T.L., Structure–function studies on nitric oxide synthases, *J. Inorg. Biochem.* **99,** 293–305 (2005).

Ma, Y.-A., Sih, C.J., and Harms, A., Enzymatic mechanism of thyroxine biosynthesis. Identification of the "lost three-carbon fragment," *J. Am. Chem. Soc.* **121,** 8967–8968 (1999).

Murphy, K.G. and Bloom, S.R., Gut hormones and the regulation of energy homeostasis, *Nature* **444,** 854–859 (2006).

Prosser, D.E. and Jones, G., Enzymes involved in the activation and inactivation of vitamin D, *Trends Biochem. Sci.* **29,** 664–673 (2004).

Schafer, A.J. and Goodfellow, P.N., Sex determination in humans, *Bio Essays* **18,** 955–963 (1996).

Wei, C.-C., Wang, Z.-Q., Tejero, J., Yang, Y.-P., Hemann, C., Hille, R., and Steuhr, D.J., Catalytic reduction of a tetrahydrobiopterin radical with nitric-oxide synthatase, *J. Biol. Chem.* **283,** 11734–11742 (2008).

Heterotrimeric G Proteins

Cooper, D.M.F. and Crossthwaite, A.J., Higher order organization and regulation of adenylyl cyclases, *Trends Pharmacol. Sci.* **27,** 426–431 (2006).

Corbin, J.D. and Francis, S.H., Cyclic GMP phosphodiesterase-5: Target of sildenafil, *J. Biol. Chem.* **274,** 13729–13732 (1999).

Fan, E., Merritt, E.A., Verlinde, C.L.M.J., and Hol, W.G.J., AB_5 toxins: Structures and inhibitor design, *Curr. Opin. Struct. Biol.* **10,** 680–686 (2000).

Hanson, M.A. and Stevens, R.C., Discovery of new GPCR biology: one receptor structure at a time, *Structure* **17,** 8–17 (2009). [Compares the known structures of GPCRs.]

Marchese, A., Chen, C., Kim, Y.-M., and Benkovic, J.L., The in and outs of G protein-coupled receptor trafficking, *Trends Biochem. Sci.* **28**, 369–376 (2003). [Reviews mechanisms of desensitization.]

Okada, T., Sugihara, M., Bondar, A.-N., Elstner, M., Entel, P., and Buss, V., The retinal conformation and its environment in rhodopsin in light of a new 2.2 Å crystal structure, *J. Mol. Biol.* **342**, 571–583 (2004).

Oldham, W.M. and Hamm, H.E., Heterotrimeric G protein activation by G-protein-coupled receptors, *Nature Rev. Mol. Cell Biol.* **9**, 60–71 (2008); *and* Structural basis of function in heterotrimeric G proteins, *Q. Rev. Biophys.* **39**, 117–166 (2006).

Palczewski, K., G protein–coupled receptor rhodopsin, *Annu. Rev. Biochem.* **75**, 743–767 (2006).

Rasmussen, S.G.F., et al., Crystal structure of the human β₂ adrenergic G-protein-coupled receptor, *Nature*, **450**, 383–387 (2007); *and* Cherezov, V., et al., High-resolution crystal structure of an engineered human β₂ adrenergic G protein–coupled receptor, *Science* **318**, 1259–1265 (2007).

Reiter, E. and Lefkowitz, R.J., GRKs and β-arrestins: roles in receptor silencing, trafficking and signaling, *Trends Endocrinol. Metab.* **17**, 159–165 (2006).

Scheerer, P., Park, J.H., Hildebrand, P.W., Kim, Y.J., Krauss, N., Choe, H.-W., Hofmann, K.P., and Ernst, O.P., Crystal structure of opsin in its G-protein-interacting conformation, *Nature* **455**, 497–502 (2008). [Proposes a structural model for the activation by opsin of its corresponding heterotrimeric G protein.]

Soundararajan, M., et al., Structural diversity in the RGS domain and its interaction with hetertrimeric G protein α-subunits, *Proc. Natl. Acad. Sci.* **105**, 6457–6462 (2008).

Sprang, S.R. (Ed.), Mechanisms and Pathways of Heterotrimeric G Protein Signaling, *Adv. Prot. Chem.* **74** (2007).

Tesmer, J.J.G. and Sprang, S.R., The structure, catalytic mechanism and regulation of adenylyl cyclase, *Curr. Opin. Struct. Biol.* **8**, 713–719 (1998).

Tesmer, J.J.G., Sunahara, R.K., Gilman, A.G., and Sprang, S.R., Crystal structure of the catalytic domains of adenylyl cyclase in a complex with Gₛα · GTPγS, *Science* **278**, 1907–1916 (1997).

Tobin, A.B., Butcher, A.J., and Kong, K.C., Location, location, location. . . . Site-specific GPCR phosphorylation offers a mechanism for cell-type-specific signaling, *Trends Pharm. Sci.* **29**, 5–12 (2008).

Vetter, I.R., and Wittinghofer, A., The guanine nucleotide–binding switch in three dimensions, *Science* **294**, 1299–1304 (2001).

Wall, M.A., Coleman, D.E., Lee, E., Iñiguez-Lluhi, J.A., Posner, B.A., Gilman, A.G., and Sprang, S.R., The structure of the G protein heterotrimer Gᵢα₁β₁γ₂, *Cell* **83**, 1047–1058 (1995); *and* Lambright, D.G., Sondek, J., Bohm, A., Skiba, N.P., Hamm, H.E., and Sigler, P.B., The 2.0 Å crystal structure of a heterotrimeric G protein, *Nature* **379**, 311–319 (1996).

Weis, W.I., and Kobilka, B.K., Structural insights into G-protein-coupled receptor activation, *Curr. Opin. Struct. Biol.* **18**, 734–740 (2008); *and* Deupi, X. and Kobilka, B., Activation of G protein–coupled receptors, *Adv. Protein Chem.* **74**, 137–165 (2007).

Willars, G.B., Mammalian RGS proteins: Multifunctional regulators of cell signaling, *Semin. Cell Dev. Biol.* **17**, 363–376 (2006).

Zhang, R.-G., Scott, D.L., Westbrook, M.L., Nance, S., Spangler, B.D., Shipley, G.G., and Westbrook, E.M., The three-dimensional crystal structure of cholera toxin, *J. Mol. Biol.* **251**, 563–573 (1995); *and* Merrrit, E.A., Sarfaty, S., Jobling, M.G., Chang, T., Holmes, R.K., Hirst, T.R., and Hol, W.G.J., Structural studies of receptor binding by cholera toxin mutants, *Protein Sci.* **6**, 1516–1528 (1997).

Tyrosine Kinase–Based Signaling

Alonso, A., et al., Protein tyrosine phosphatases in the human genome, *Cell* **117**, 699–711 (2004). [A review that enumerates, classifies, and discusses the functions of all human PTPs.]

Beene, D.L. and Scott, J.D., A-kinase anchoring proteins take shape, *Curr. Opin. Cell Biol.* **19**, 192–198 (2007).

Bhattacharyya, R.P., Reményi, A., Yeh, B.J., and Lim, W.A., Domains, motifs and scaffolds: The role of modular interactions in the evolution and wiring of cell signaling circuits, *Annu. Rev. Biochem.* **75**, 655–680 (2006).

Boggon, T.J. and Eck, M.J., Structure and regulation of Src family kinases, *Oncogene* **23**, 7918–7927 (2004).

Bollen, M., Combinatorial control of protein phosphatase-1, *Trends Biochem. Sci.* **26**, 426–431 (2001).

Boriak-Sjodin, P.A., Margarit, S.M., Bar-Sagi, D., and Kuriyan, J., The structural basis of the activation of Ras by Sos, *Nature* **394**, 337–343 (1998).

Bos, J.L., Rehmann, H., and Wittinghofer, A., GEFs and GAPs: Critical elements in the control of small G proteins, *Cell* **129**, 865–877 (2007).

Baselga, J., Targeting tyrosine kinases in cancer: The second wave, *Science* **312**, 1175–1178 (2006).

Calderwood, S.K., Khaleque, M.A., Sawyer, D.B., and Ciocca, D.R., Heat shock proteins in cancer: chaperones of tumorigenesis, *Trends Biochem. Sci.* **31**, 164–172 (2006).

Capdeville, R., Buchdunger, E., Zimmermann, J., and Matter, A., Glivec (STI571, ImatinIB), a rationally developed targeted anticancer drug, *Nature Rev. Drug Discov.* **1**, 493–502 (2002). [Glivec was the previous name of Gleevec.]

Carlisle Michel, J.J. and Scott, J.D., AKAP mediated signal transduction, *Annu. Rev. Pharmacol. Toxicol.* **42**, 235–257 (2002).

Cho, U.S. and Xu, W., Crystal structure of a protein phosphatase 2A heterotrimeric holoenzyme, *Nature* **445**, 53–57 (2007); *and* Xu, Y., Xing, Y., Chen, Y., Chao, Y., Lin, Z., Fan, E., Yu, J., Stack, S., Jeffrey, P., and Shi, Y., Structure of the protein phosphatase 2A holoenzyme, *Cell* **127**, 1239–1251 (2006).

Chang, L. and Karin, M., Mammalian MAP kinase signaling cascades, *Nature* **410**, 37–40 (2001).

De Meyts, P., The insulin receptor: A prototype for dimeric, allosteric membrane receptors? *Trends Biochem. Sci.* **33**, 376–384 (2008).

Druker, B.J. and Lydon, N.B., Lessons learned from the development of an Abl tyrosine kinase inhibitor for chronic myelogenous leukemia, *J. Clin. Invest.* **105**, 3–7 (2000).

Garcia, K.C. (Ed.), Cell Surface Receptors, *Adv. Protein Chem.* **68** (2004).

Griffith, J.P., Kim, J.L., Kim, E.E., Sintchak, M.D., Thomson, J.A., Fitzgibbon, M.J., Fleming, M.A., Caron, P.R., Hsiao, K., and Navia, M.A., X-ray structure of calcineurin inhibited by the immunophilin-immunosuppresant FKBP12-FK506 complex, *Cell* **82**, 507–522 (1995); *and* Huai, Q., Kim, H.-Y., Liu, Y., Zhao, Y., Mondragon, A., Liu, J.O., and Ke, H., Crystal structure of calcineurin–cylophilin–cyclosporin shows common but distinct recognition of immunophilin–drug complexes, *Proc. Natl. Acad. Sci.* **99**, 12037–12042 (2002).

Groves, M.R., Hanlon, N., Turowski, P., Hemmings, B.A., and Barford, D., The structure of the protein phosphatase 2A PR65/A subunit reveals the conformation of its 15 tandemly repeated HEAT motifs, *Cell* **96**, 99–110 (1999).

Hansen, G., et al., The structure of the GM-CSF receptor complex reveals a distinct mode of cytokine receptor activation, *Cell* **134**, 496–507 (2008).

Hof, P., Pluskey, S., Dhe-Paganon, S., Eck, M.J., and Shoelson, S.E., Crystal structure of tyrosine phosphatase SHP-2, *Cell* **92,** 441–450 (1998).

Hubbard, S.R., Crystal structure of the activated insulin receptor tyrosine kinase in complex with peptide substrate and ATP analog, *EMBO J.* **16,** 5572–5581 (1997); *and* Hubbard, S.R., Wei, L., Ellis, L., and Hendrickson, W.A., Crystal structure of the tyrosine kinase domain of the human insulin receptor, *Nature* **372,** 746–753 (1994).

Hubbard, S.R. and Miller, W.T., Receptor tyrosine kinases: mechanisms of activation and signaling, *Curr. Opin. Cell Biol.* **19,** 117–123 (2007).

Karnoub, A.E. and Weinberg, R.A., Ras ongenes: split personalities, *Nature Rev. Mol. Cell Biol.* **9,** 517–531 (2008).

Kolch, W., Coordinating ERK/MAPK signaling through scaffolds and inhibitors, *Nature Rev. Mol. Cell Biol.* **6,** 827–838 (2005).

Li, L. and Dixon, J.E., Form, function, and regulation of protein tyrosine phosphatases and their involvement in human disease, *Semin. Immunol.* **12,** 75–84 (2000).

Lim, W.A., The modular logic of signaling proteins: building allosteric switches from simple binding domains, *Curr. Opin. Struct. Biol.* **12,** 61–68 (2002).

Linder, J.U and Schultz, J.E., Versatility of signal transduction encoded in dimeric adenylyl cyclases, *Curr. Opin. Struct. Biol.* **18,** 667–672 (2008).

Maignan, S., Guilloteau, J.-P., Fromage, N., Arnoux, B., Becquart, J., and Ducruix, A., Crystal structure of the mammalian Grb2 adaptor, *Science* **268,** 291–293 (1995).

McKay, M.M. and Morrison, D.K., Integrating signals from RTKs to ERK/MAPK, *Oncogene* **26,** 3113–3121 (2007).

Musacchio, A., Sareste, M., and Wilmanns, M., High-resolution crystal structures of tyrosine kinase SH3 domains complexed with proline-rich peptides, *Nature Struct. Biol.* **1,** 546–551 (1994).

Nassar, N., Horn, G., Herrmann, C., Scherer, A., McCormack, F., and Wittinghofer, A., The 2.2 Å crystal structure of the Ras-binding domain of the serine/threonine kinase c-Raf1 in complex with Rap1A and a GTP analogue, *Nature* **375,** 554–560 (1995).

Neel, B.G., Gu, H., and Pao, L., The 'Shp'ing news: SH2 domain-containing tyrosine phosphatases in cell signaling, *Trends Biochem. Sci.* **28,** 284–293 (2003).

Noble, M.E.M., Endicott, J.A., and Johnson, L.N., Protein kinase inhibitors: Insights into drug design from structure, *Science* **303,** 1800–1805 (2004).

O'Shea, J.J., Gadino, M., and Schreiber, R.D., Cytokine signaling in 2002: New surprises in the Jak/Stat pathway, *Cell* **109,** S121–S131 (2002).

Pawson, T., Dynamic control of signaling by modular adapter proteins, *Curr. Opin. Cell Biol.* **19,** 112–116 (2007).

Pawson, T. and Scott, J.D., Protein phosphorylation in signaling—50 years and counting, *Trends Biochem. Sci.* **30,** 286–290 (2005). [A historical review.]

Pellizzari, R., Guidi-Rontani, C., Vitale, G., Mock, M., and Montecucco, C., Anthrax lethal factor cleaves MKK3 in macrophages and inhibits the LPS/IFNγ-induced release of NO and TNFα, *FEBS Lett.* **462,** 199–204 (1999).

Scheffzek, K., Ahmadian, M.R., Kabsch, W., Wiesmüller, L., Lautwein, A., Schmitz, F., and Wittinghofer, A., The Ras-RasGAP complex: Structural basis for GTPase activation and its loss in oncogenic Ras mutants, *Science* **277,** 333–338 (1997).

Schindler, T., Bornmann, W., Pellicena, P., Miller, W.T., Clarkson, B., and Kuriyan, J., Structural mechanism for STI-571 inhibition of Abelson tyrosine kinase, *Science* **289,** 1938–1942 (2000). [STI-571 was the original name of Gleevec.]

Schlessinger, J., Plotnikov, A.N., Ibrahimi, O.A., Eliseenkova, A.V., Yeh, B.K., Yayon, A., Linhardt, R.J., and Mohammadi, M., Crystal structure of a ternary FGF-FGFR-heparin complex reveals a dual role for heparin in FGF binding and dimerization, *Mol. Cell* **6,** 743–750 (2000).

Sebolt-Leopold, J.S. and English, J.M., Mechanisms of drug inhibition of signaling molecules, *Nature* **441,** 457–462 (2006).

Sprang, S., GEFs: Master regulators of G-protein activation, *Trends Biochem. Sci.* **26,** 266–267 (2001).

Stoker, A.W., Protein tyrosine phosphatases and signaling, *J. Endocrinol.* **185,** 19–33 (2005).

Tiganis, T. and Bennett, A.M., Protein tyrosine kinase function: the substrate perspective, *Biochem.* **402,** 1–15 (2007).

Tonks, N.K., Protein tyrosine phosphatases: from genes, to function, to disease, *Nature Rev. Mol. Cell Biol.* **7,** 833–846 (2006).

Wang, X., Lupardus, P., La Porte, S.L., and Garcia, K.C., Structural biology of shared cytokine receptors, *Annu. Rev. Immunol.* **27,** 29–60 (2009).

Ward, C.W., Lawrence, M.C., Streltsov, V.A., Adams, T.E., and McKern, N.M., The insulin and EGF receptor structures: insights into ligand-induced receptor activation, *Trends Biochem. Sci.* **32,** 129–137 (2007).

Whitesell, L. and Lindquist, S.L., Hsp90 and the chaperoning of cancer, *Nature Rev. Cancer* **5,** 761–772 (2005).

Whitmarsh, A.J. and Davis, R.J., Structural organization of MAP-kinase signaling modules by scaffold proteins in yeast and mammals, *Trends Biochem. Sci.* **23,** 481–485 (1998).

Xu, W., Doshi, A., Lei, M., Eck, M.J., and Harrison, S.C., Crystal structures of c-Src reveal features of its autoinhibitory mechanism, *Mol. Cell* **3,** 629–638 (1999); *and* Xu, W., Harrison, S.C., and Eck, M.J., Three dimensional structure of the tyrosine kinase c-Src, *Nature* **385,** 595–602 (1995).

Yaffe, M.B., Phosphotyrosine-binding domains in tyrosine transduction, *Nature Rev. Mol. Cell Biol.* **3,** 177–186 (2002).

Young, M.A., Gonfloni, F., Superti-Furga, G., Roux, B., and Kuriyan, J., Dynamic coupling between the SH2 and SH3 domains of c-Src and Hck underlies their inactivation by C-terminal tyrosine phosphorylation, *Cell* **105,** 115–126 (2001).

Zhang, Z.-Y., Protein tyrosine phosphatases: structure and function, substrate specificity, and inhibitor development, *Annu. Rev. Pharmacol. Toxicol.* **42,** 209–234 (2002).

Zhou, M.-M., et al., Structure and ligand recognition of the phosphotyrosine binding domain of Shc, *Nature* **378,** 584–592 (1995).

The Phosphoinositide Cascade

Brazil, D.P., Yang, Z.-Z., and Hemmings, B.A., Advances in protein kinase B signaling: *AKT*ion on multiple fronts, *Trends Biochem. Sci.* **29,** 233–242 (2004).

Brognard, J. and Newton, A.C., PHLiPPing the switch on Akt and protein kinase C signaling, *Trends Endocrinol. Metab.* **19,** 223–230 (2008).

Carrasco, S. and Mérida, I., Diacylglycerol, when simplicity becomes complex, *Trends Biochem. Sci.* **32,** 27–36 (2007).

Cho, W. and Stahelin, R.V., Membrane-protein interactions in cell signaling and membrane trafficking, *Annu. Rev. Biophys. Biomol. Struct.* **34,** 119–151 (2005).

Clapham, D.E., Calcium signaling, *Cell* **131,** 1047–1058 (2007).

Cockcroft, S. (Ed.), *Biology of Phosphoinositides,* Oxford (2000).

Di Paolo, G. and De Camilli, P., Phosphoinositides in cell regulation and membrane dynamics, *Nature* **443,** 651–657 (2006).

Dekker, L.V. (Ed.), *Protein Kinase C* (2nd ed.), Kluwer Academic/Plenum Publishers (2004).

Ferguson, K.M., Lemmon, M.A., Schlessinger, M.A., and Sigler, P.B., Structure of the high affinity complex of inositol trisphosphate with a phospholipase C pleckstrin homology domain, *Cell* **83,** 1037–1046 (1995).

Gallegos, L.L. and Newton, A.C., Spaciotemporal dynamics of lipid signaling: Protein kinase C as a paradigm, *IUBMB Life* **60**, 782–786 (2008).

Harden, T.K. and Sondek, J., Regulation of phospholipase C isozymes by Ras superfamily GTPases, *Annu. Rev. Pharmacol. Toxicol.* **46**, 355–379 (2006).

Huang, C.-H., Mandelker, D., Schmidt-Kittler, O., Samuels, Y., Velculescu, V.E., Kinzler, K.W., Vogelstein, B., Gabelli, S.B., and Amzel, L.M., The structure of a human p110α/p85α complex elucidates the effects of oncogenic PI3Kα mutations, *Science* **318**, 1744–1748 (2007). [p110α is the catalytic subunit of PI3Kα.]

Hurley, J.H. and Misra, S., Signaling and subcellular targeting by membrane-binding domains, *Annu. Rev. Biophys. Biomol. Struct.* **29**, 49–79 (2000).

Jezyk, M.R., Snyder, J.T., Gershberg, S., Worthylake, D.K., Harden, T.K., and Sondek, J., Crystal structure of Rac1 bound to its effector phospholipase C-β2, *Nature Struct. Mol. Biol.* **13**, 1135–1140 (2006); *and* Hicks, S.N., Jezyk, M.R., Gershberg, S., Seifer, J.P., Harden, T.K., and Sondek, J., General and versatile autoinhibition of PLC isoenzymes, *Mol. Cell* **31**, 383–394 (2008).

Katso, R., Okkenhaug, K., Ahmadi, K., White, S., Timms, J., and Waterfield, M.D., Cellular function of phosphoinositide 3-kinases: Implications for development, immunity, homeostasis, and cancer, *Annu. Rev. Cell Dev. Biol.* **17**, 615–675 (2001).

Kok, K., Geering, B., and Vanhaesebroeck, B., Regulation of phosphoinositide 3-kinase expression in health and disease, *Trends Biochem. Sci.* **34**, 115–127 (2009).

Kutateladze, T. and Overduin, M., Structural mechanism of endosome docking by the FYVE domain, *Science* **291**, 1793–1796 (2001).

Lee, J.-O., Yang, H., Georgescu, M.-M., Di Cristofano, A., Maehama, T., Shi, Y., Dixon, J.E., Pandolfi, P., and Pavletich, N.P., Crystal structure of the PTEN tumor suppressor: Implications for its phosphoinositide phosphatase activity and membrane association, *Cell* **99**, 323–344 (1999).

Maehama, T., Taylor, G.S., and Dixon, J.E., PTEN and myotubularin: Novel phosphoinositide phosphatases, *Annu. Rev. Biochem.* **70**, 247–279 (2001).

Manning, B.D. and Cantley, L.C., AKT/PKB signaling: Navigating downstream, *Cell* **129**, 1261–1274 (2007).

Michell, R.H., Inositol derivatives: evolution and function, *Nature Rev. Mol. Cell Biol.* **9**, 151–161 (2008).

Patterson, R.L., van Rossum, D.B., Nikolaidis, N., Gill, D.L., and Snyder, S.H., Phospholipase C-γ: diverse roles in receptor-mediated calcium signaling, *Trends Biochem. Sci.* **32**, 688–697 (2005).

Rhee, S.G., Regulation of phosphoinositide-specific phospholipase C, *Annu. Rev. Biochem.* **70**, 281–312 (2001).

Salmena, L., Carracedo, A., and Pandolfi, P.P., Tenets of PTEN tumor suppression, *Cell* **133**, 403–414 (2008).

Saltiel, A.R. and Pessin, J.E., Insulin signaling pathways in time and space, *Trends Cell Biol.* **12**, 65–71 (2002).

Steinberg, S.F., Structural basis of protein kinase C isoform function, *Physiol. Rev.* **88**, 1341–1378 (2008). [A detailed review.]

Vanhaesebroek, B., Leevers, S.J., Ahmadi, K., Timms, J., Katso, R., Driscoll, P.C., Woscholski, R., Parker, P.J., and Waterfield, M.D., Synthesis and function of 3-phosphorylated inositol lipids, *Annu. Rev. Biochem.* **70**, 535–632 (2001).

Walker, E.H., Persic, O., Ried, C., Stephens, L., and Williams, R.L., Structural insights into phosphoinositide 3-kinase catalysis and signaling, *Nature* **402**, 313–320 (1999); *and* Pacold, M.E., et al., Crystal structure and functional analysis of Ras binding to its effector phosphoinositide 3-kinase γ, *Cell* **103**, 931–943 (2000).

Weng, G., Bhalla, U.S., and Iyengar, R., Complexity in biological signaling systems, *Science* **284**, 92–96 (1999).

Zick, Y., Insulin resistance: a phosphorylation-based uncoupling of insulin signaling, *Trends Cell Biol.* **11**, 437–441 (2001).

PROBLEMS

1. Explain the following observations: (a) Thyroidectomized rats, when deprived of food, survive for 20 days while normal rats starve to death within 7 days. (b) Cushing's syndrome, which results from excessive secretion of adrenocortical steroids, can be caused by a pituitary tumor. (c) **Diabetes insipidus,** which is characterized by unceasing urination and unquenchable thirst, results from an injury to the pituitary. (d) The growth of malignant tumors derived from sex organs may be slowed or even reversed by the surgical removal of the gonads and the adrenal glands.

2. How does the presence of the nonhydrolyzable GTP analog GMPPNP affect cAMP-dependent receptor systems?

3. Explain why individuals who regularly handle dynamite (which is nitroglycerin soaked into an absorbant such as wood pulp) as part of their jobs have an unusually high incidence of heart attacks on weekends.

4. A dose-dependent side effect of sildenafil (Viagra) is the transient impairment of blue/green color discrimination. What is the biochemical basis for this phenomenon?

5. Retroviruses bearing oncogenes will infect cells from their corresponding host animal but will usually not transform them. Yet these retroviruses will readily transform immortalized cells derived from the same organism. Explain.

6. Explain why mutations of the Arg residue in $G_{s\alpha}$ that is ADP-ribosylated by cholera toxin are oncogenic mutations. Why doesn't cholera toxin cause cancer?

7. Would the following alterations to Src be oncogenic? Explain. (a) The deletion or inactivation of the SH3 domain. (b) The mutation of Tyr 416 to Phe. (c) The mutation of Tyr 527 to Phe. (d) The replacement of Src residues 249 to 253 with the sequence APTMP.

8. JIP-1 was originally so named because, when it was first characterized by overexpression in mammalian cells, it appeared to act as a "*J*NK *i*nhibitor *p*rotein." What is the basis of this observation?

9. Why does pertussis toxin appear to inhibit certain isozymes of PLC? Identify these isozymes.

10. Phosphatidylethanolamine and PIP_2 containing identical fatty acyl residues can be hydrolyzed with the same efficiency by a certain phospholipase C. Will the hydrolysis products of the two lipids have the same effect on protein kinase C? Explain.

11. PKC's autoinhibitory pseudosubstrate occurs at its N-terminus, whereas that of MLCK occurs at its C-terminus (Fig. 18-20). To further investigate this phenomenon, a colleague proposes to construct a PKC with its pseudosubstrate attached to the protein's C-terminus with a sufficiently long linker so that the pseudosubstrate could bind in the enzyme's active site. Would you expect this variant PKC to be activatable? Explain.

W74'
Y41
Y6
Y118
W420
W358
F227

Transport through Membranes

CHAPTER 20

Metabolism occurs within cells that are separated from their environments by plasma membranes. Eukaryotic cells, in addition, are compartmentalized by intracellular membranes that form the boundaries and internal structures of their various organelles. The nonpolar cores of biological membranes make them highly impermeable to most ionic and polar substances, so that *these substances can traverse membranes only through the action of specific transport proteins.* Such proteins are therefore required to mediate all transmembrane movements of ions, such as Na^+, K^+, Ca^{2+}, and Cl^-, as well as metabolites such as pyruvate, amino acids, sugars, and nucleotides, and even water (despite its relatively high permeability in bilayers; Section 12-2B). Transport proteins are also responsible for all biological electrochemical phenomena such as neurotransmission. In this chapter, we discuss the thermodynamics, kinetics, and chemical mechanisms of these membrane transport

systems and end with a discussion of the mechanism of neurotransmission.

1 THERMODYNAMICS OF TRANSPORT

As we saw in Section 3-4A, the free energy of a solute, A, varies with its concentration:

$$\overline{G}_A - \overline{G}_A^{\circ\prime} = RT \ln [A] \qquad [20.1]$$

where $\overline{G}_A$ is the **chemical potential** (partial molar free energy) of A (the bar indicates quantity per mole) and $\overline{G}_A^{\circ\prime}$ is the chemical potential of its standard state. Strictly speaking, this equation applies only to ideal solutions; for nonideal (real) solutions, molar concentrations must be replaced by activities (Appendix to Chapter 3). In the dilute (millimolar) solutions that are characteristic of laboratory conditions, the activity of a substance closely approaches its molar concentration in value. However, this is not the case in the highly concentrated cellular milieu (Appendix to Chapter 3). Yet it is difficult to determine the activity of a substance in a cellular compartment. Hence, in the following derivations, we shall make the simplifying assumption that activities are equal to molar concentrations.

The diffusion of a substance between two sides of a membrane

$$A(out) \rightleftharpoons A(in)$$

thermodynamically resembles a chemical equilibration. A difference in the concentrations of the substance on two sides of a membrane generates a chemical potential difference:

$$\Delta \overline{G}_A = \overline{G}_A(in) - \overline{G}_A(out) = RT \ln\left(\frac{[A]_{in}}{[A]_{out}}\right) \quad [20.2]$$

Consequently, if the concentration of A outside the membrane is greater than that inside, $\Delta \overline{G}_A$ for the transfer of A from outside to inside will be negative and the spontaneous net flow of A will be inward. If, however, [A] is greater inside than outside, $\Delta \overline{G}_A$ is positive and an inward net flow of A can only occur if an exergonic process, such as ATP hydrolysis, is coupled to it to make the overall free energy change negative.

a. Membrane Potentials Arise from Transmembrane Concentration Differences of Ionic Substances

The permeabilities of biological membranes to ions such as H^+, Na^+, K^+, Cl^-, and Ca^{2+} are controlled by specific membrane-embedded transport systems that we shall discuss in later sections. *The resulting charge differences across a biological membrane generate an electric potential difference,* $\Delta\Psi = \Psi(in) - \Psi(out)$, *where* $\Delta\Psi$ *is termed the* **membrane potential.** Consequently, if A is ionic, Eq. [20.2] must be amended to include the electrical work required to transfer a mole of A across the membrane from outside to inside:

$$\Delta\overline{G}_A = RT \ln\left(\frac{[A]_{in}}{[A]_{out}}\right) + Z_A \mathscr{F} \Delta\psi \qquad [20.3]$$

where Z_A is the ionic charge of A; $\mathscr{F}$, the Faraday constant, is the charge of one mole of electrons ($96{,}485 \ C \cdot mol^{-1}$); and $\overline{G}_A$ is now termed the **electrochemical potential** of A.

Membrane potentials in living cells can be measured directly with microelectrodes. $\Delta\Psi$ values of $-100 \ mV$ (inside negative) are not uncommon (note that $1 \ V = 1 \ J \cdot C^{-1}$). Thus the last term of Eq. [20.3] is often significant for ionic substances.

2 KINETICS AND MECHANISMS OF TRANSPORT

Thermodynamics indicates whether a given transport process will be spontaneous but, as we saw for chemical and enzymatic reactions, provides no indication of the rates of these processes. Kinetic analyses of transport processes together with mechanistic studies have nevertheless permitted these processes to be characterized. There are two types of transport processes: **nonmediated transport** and **mediated transport.** Nonmediated transport occurs through simple diffusion. In contrast, *mediated transport occurs through the action of specific carrier proteins* that are variously called **carriers, permeases, porters, translocases, translocators,** and **transporters.** Mediated transport is further classified into two categories depending on the thermodynamics of the system:

1. Passive-mediated transport or **facilitated diffusion** in which specific molecules flow from high concentration to low concentration so as to equilibrate their concentration gradients.

2. Active transport in which specific molecules are transported from low concentration to high concentration, that is, against their concentration gradients. Such an endergonic process must be coupled to a sufficiently exergonic process to make it favorable.

In this section, we consider the nature of nonmediated transport and then compare it to passive-mediated transport as exemplified by ionophores, porins, glucose transporters, K^+ channels, Cl^- channels, and aquaporins. Active transport is examined in succeeding sections.

A. *Nonmediated Transport*

The driving force for the nonmediated flow of a substance A through a medium is A's electrochemical potential gradient. This relationship is expressed by the **Nernst–Planck equation:**

$$J_A = -[A]U_A(d\overline{G}_A/dx) \qquad [20.4]$$

where J_A is the **flux** (rate of passage per unit area) of A, x is distance, $d\overline{G}_A/dx$ is the electrochemical potential gradient of A, and U_A is its **mobility** (velocity per unit force) in the medium. If we assume, for simplicity, that A is an uncharged molecule so that $\overline{G}_A$ is given by Eq. [20.1], the Nernst–Planck equation reduces to

$$J_A = -D_A(d[A])/dx \qquad [20.5]$$

where $D_A \equiv RTU_A$ is the **diffusion coefficient** of A in the medium of interest. This is **Fick's first law of diffusion,** which states that *a substance diffuses in the direction that eliminates its concentration gradient,* $d[A]/dx$, *at a rate proportional to the magnitude of this gradient.*

For a membrane of thickness x, Eq. [20.5] is approximated by

$$J_A = \frac{D_A}{x}([A]_{out} - [A]_{in}) = P_A([A]_{out} - [A]_{in}) \quad [20.6]$$

where D_A is the diffusion coefficient of A inside the membrane and $P_A = D_A/x$ is termed the membrane's **permeability coefficient** for A. The permeability coefficient is indicative of the solute's tendency to transfer from the aqueous solvent to the membrane's nonpolar core. It should therefore vary with the ratio of the solute's solubility in a nonpolar solvent resembling the membrane's core (e.g., olive oil) to that in water, a quantity known as the solute's **partition coefficient** between the two solvents. Indeed, the fluxes of many nonelectrolytes across erythrocyte membranes vary linearly with their concentration differences across the membrane as predicted by Eq. [20.6] (Fig. 20-1). Moreover, their permeability coefficients, as obtained from the slopes of plots such as Fig. 20-1, correlate rather well

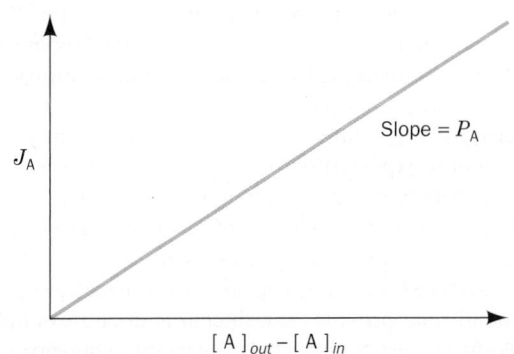

Figure 20-1 **Linear relationship between diffusional flux (J_A) and ($[A]_{out} - [A]_{in}$) across a semipermeable membrane.** See Eq. [20.6].

Figure 20-2 Permeability correlates with membrane solubility. The permeability coefficients of various organic molecules in plasma membranes from the alga *Nitella mucronata* versus their partition coefficients between olive oil and water (a measure of a molecule's polarity). This more or less linear log–log plot indicates that the rate-limiting step for the nonmediated entry of a molecule into a cell is its passage through the membrane's hydrophobic core. [Based on data from Collander, R., *Physiol. Plant.* **7**, 433–434 (1954).]

with their measured partition coefficients between nonpolar solvents and water (Fig. 20-2).

B. *Kinetics of Mediated Transport: Glucose Transport Into Erythrocytes*

Despite the success of the foregoing model in predicting the rates at which many molecules pass through membranes, there are numerous combinations of solutes and membranes that do not obey Eq. [20.6]. The flux in such a system is not linear with the solute concentration difference across the corresponding membrane (Fig. 20-3) and, furthermore, the solute's permeability coefficient is much larger than is expected on the basis of its partition coefficient. Such behavior indicates that *these solutes are conveyed across membranes in complex with carrier molecules; that is, they undergo mediated transport.*

The system that transports glucose across the erythrocyte membrane provides a well-characterized example of passive-mediated transport: It invariably transports glucose down its concentration gradient but not at the rate predicted by Eq. [20.6]. Indeed, the **erythrocyte glucose transporter** exhibits four characteristics that differentiate mediated from nonmediated transport: (1) *speed and*

Figure 20-3 Variation of glucose flux into human erythrocytes with the external glucose concentration at 5°C. The black dots are experimentally determined data points, and the solid green line is computed from Eq. [20.7] with $J_{max} = 1.0 \times 10^{-6}$ m$M \cdot$ cm $\cdot$ s^{-1} and $K_M = 0.5$ mM. The nonmediated glucose flux increases linearly with [glucose] (Fig. 20-1) but would not visibly depart from the baseline on the scale of this drawing. [Based on data from Stein, W.D., *Movement of Molecules across Membranes*, p. 134, Academic Press (1967).]

specificity, (2) saturation kinetics, (3) susceptibility to competitive inhibition, and (4) susceptibility to chemical inactivation. In the following paragraphs we shall see how the erythrocyte glucose transporter exhibits these qualities.

a. Speed and Specificity

Table 20-1 indicates that the permeability coefficients of D-glucose and D-mannitol in synthetic bilayers, and that of D-mannitol in the erythrocyte membrane, are in reasonable agreement with the values calculated from the diffusion and partition coefficients of these sugars between water and olive oil. However, the experimentally determined permeability coefficient for D-glucose in the erythrocyte membrane is four orders of magnitude greater than its predicted value. *The erythrocyte membrane must therefore contain a system that rapidly transports glucose and that can distinguish D-glucose from D-mannitol.*

b. Saturation Kinetics

The concentration dependence of glucose transport indicates that its flux obeys the relationship

$$J_A = \frac{J_{max}[A]}{K_M + [A]} \qquad [20.7]$$

This **saturation function** has a familiar hyperbolic form (Fig. 20-3). We have seen it in the equation describing the binding of O_2 to myoglobin (Eq. [10.4]) and in the Michaelis–Menten equation describing the rates of enzymatic reactions (Eq. [14.24]). Here, as before, K_M may be defined operationally as the concentration of glucose when the transport flux is half of its maximal rate, $J_{max}/2$. This observation of **saturation kinetics** for glucose transport was the first evidence that a specific, saturatable

Table 20-1 **Permeability Coefficients of Natural and Synthetic Membranes to D-Glucose and D-Mannitol at 25°C**

Membrane Preparation	Permeability Coefficient $(cm \cdot s^{-1})$	
	D-Glucose	D-Mannitol
Synthetic lipid bilayer	2.4×10^{-10}	4.4×10^{-11}
Calculated nonmediated diffusion	4×10^{-9}	3×10^{-9}
Intact human erythrocyte	2.0×10^{-4}	5×10^{-9}

Source: Jung, C.Y., *in* Surgenor, D. (Ed.), *The Red Blood Cell*, Vol. 2, p. 709, Academic Press (1975).

number of sites on the membrane were involved in the transport of any substance.

The transport process can be described by a simple four-step kinetic scheme involving binding, transport, dissociation, and recovery (Fig. 20-4). Its binding and dissociation steps are analogous to the recognition of a substrate and the release of product by an enzyme. The mechanisms of transport and recovery are discussed in Section 20-2D.

c. Susceptibility to Competitive Inhibition

Many compounds structurally similar to D-glucose inhibit glucose transport. A double-reciprocal plot (Section 14-2B) for the flux of glucose into erythrocytes in the presence or absence of 6-O-benzyl-D-galactose (Fig. 20-5) shows behavior typical of competitive inhibition of glucose transport (competitive inhibition of enzymes is discussed in Section 14-3A). *Susceptibility to competitive inhibition indicates that there is a limited number of sites available for mediated transport.*

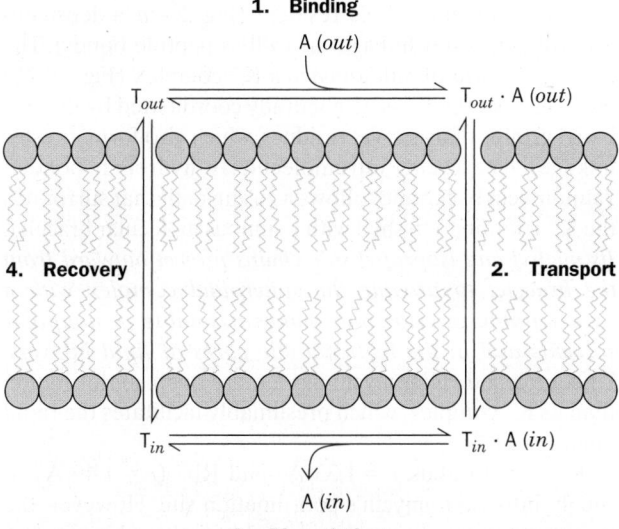

Figure 20-4 **General kinetic scheme for membrane transport.** The scheme involves four steps: binding, transport, dissociation, and recovery. T is the transport protein whose binding site for solute A is located on either the inner or the outer side of the membrane at any one time.

Figure 20-5 **Double-reciprocal plots for the net flux of glucose into erythrocytes in the presence and absence of 6-O-benzyl-D-galactose.** The pattern is that of competitive inhibition. [After Barnett, J.E.G., Holman, G.D., Chalkley, R.A., and Munday, K.A., *Biochem. J.* **145**, 422 (1975).]

d. Susceptibility to Chemical Inactivation

Treatment of erythrocytes with $HgCl_2$, which reacts with protein sulfhydryl groups

$$RSH + HgCl_2 \longrightarrow RS{-}Hg{-}Cl + HCl$$

and thus inactivates many enzymes, causes the rapid, saturatable flux of glucose to disappear so that its permeability constant approaches that of mannitol. *The erythrocyte glucose transport system's susceptibility to such protein-modifying agents indicates that it, in fact, is a protein.*

All of the above observations indicate that *glucose transport across the erythrocyte membrane is mediated by a limited number of protein carriers.* Before we discuss the mechanism of this transport system, however, we shall examine some simpler models of facilitated diffusion.

C. *Ionophores*

Our understanding of mediated transport has been enhanced by the study of **ionophores** (Greek: *phoros,* bearer), substances that vastly increase the permeability of membranes to particular ions.

a. Ionophores May Be Carriers or Channel Formers

Ionophores are organic molecules of diverse types, many of which are antibiotics of bacterial origin. Cells and organelles actively maintain concentration gradients of various ions across their membranes (Section 20-3A). The antibiotic properties of ionophores arise from their tendency to discharge these vital concentration gradients.

There are two types of ionophores:

1. *Carriers,* which increase the permeabilities of membranes to their selected ion by binding it, diffusing through the membrane, and releasing the ion on the other side (Fig. 20-6a). For net transport to occur, the uncomplexed ionophore must then return to the original side of the membrane ready to repeat the process. Carriers therefore share the common property that *their ionic complexes are soluble in nonpolar solvents.*

2. *Channel formers,* which form transmembrane channels or pores through which their selected ions can diffuse (Fig. 20-6b).

Both types of ionophores transport ions at a remarkable rate. For example, a single molecule of the carrier antibiotic **valinomycin** transports up to 10^4 K^+ ions per second across a membrane. Channel formers have an even greater ion throughput; for example, each membrane channel composed of the antibiotic **gramicidin A** permits the passage of over 10^7 K^+ ions $\cdot$ s^{-1}. Clearly, the presence of either type of ionophore, even in small amounts, greatly increases the permeability of a membrane toward the specific ions transported. However, *since ionophores passively permit ions to diffuse across a membrane in either direction, their effect can only be to equilibrate the concentrations of their selected ions across the membrane.*

Carriers and channel formers are easily distinguished experimentally through differences in the temperature

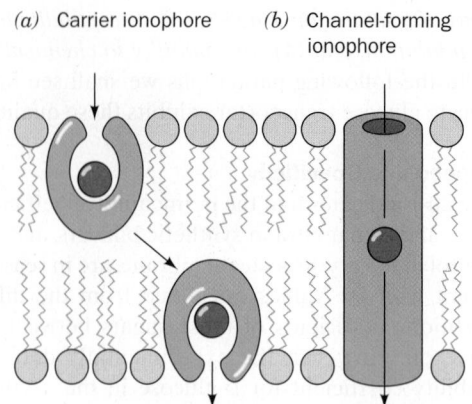

(a) Carrier ionophore   *(b)* Channel-forming ionophore

Figure 20-6   Ion transport modes of ionophores. (*a*) Carrier ionophores transport ions by diffusing through the lipid bilayer. (*b*) Channel-forming ionophores span the membrane with a channel through which ions can diffuse.

dependence of their action. Carriers depend on their ability to diffuse freely across the membrane. Consequently, cooling a membrane below its transition temperature (the temperature below which it becomes a gel-like solid; Section 12-2Cb) essentially eliminates its ionic permeability in the presence of carriers. In contrast, membrane permeability in the presence of channel formers is rather insensitive to temperature because, once in place, channel formers need not move to mediate ion transport.

b. The K^+–Valinomycin Complex Has a Polar Interior and a Hydrophobic Exterior

Valinomycin, a product of several strains of *Streptomyces* bacteria that specifically binds K^+ (and the biologically unimportant Rb^+), is perhaps the best characterized carrier ionophore. It is a **cyclic depsipeptide** that contains both D- and L-amino acid residues (Fig. 20-7a; a depsipeptide contains ester linkages as well as peptide bonds). The X-ray structure of valinomycin's K^+ complex (Fig. 20-7b) indicates that the K^+ is octahedrally coordinated by the carbonyl groups of its six Val residues, which also form its ester linkages. The cyclic, intramolecularly hydrogen bonded valinomycin backbone follows a zigzag path that surrounds the K^+ coordination shell with a sinuous molecular bracelet. *Its methyl and isopropyl side chains project outward from the bracelet to provide the spheroidal complex with a hydrophobic exterior that makes it soluble in nonpolar solvents and in the hydrophobic cores of lipid bilayers.* Uncomplexed valinomycin has a more open conformation than its K^+ complex, which presumably facilitates the rapid binding of K^+.

K^+ (ionic radius, $r = 1.33$ Å) and Rb^+ ($r = 1.49$ Å) fit snugly into valinomycin's coordination site. However, the rigidity of the valinomycin complex makes this site too large to accommodate Na^+ ($r = 0.95$ Å) or Li^+ ($r = 0.60$ Å) properly; that is, valinomycin's six carbonyl oxygen atoms cannot simultaneously coordinate these ions. Complexes of these ions with water are therefore energetically more favorable than their complexes with valinomycin. This *(a)*   **Valinomycin**

Figure 20-7   Valinomycin. (*a*) This cyclic depsipeptide (has both ester and amide bonds) contains both D- and L-amino acids. (*b*) The X-ray structure of valinomycin in complex with a K⁺ ion shown in stick form colored according to atom type (C green, H white, N blue, O red, and K⁺ purple) and embedded in its molecular surface. Note that the K⁺ ion is octahedrally coordinated by the carbonyl atoms of valinomycin's six Val residues and that the surface of the complex is largely covered with methyl groups. [Based on an X-ray structure by Max Dobler, ETH, Zürich, Switzerland.]

(b)

accounts for valinomycin's 10,000-fold greater binding affinity for K⁺ over Na⁺. No other substance discriminates more acutely between Na⁺ and K⁺. A variety of carrier ionophores with similar characteristics but with different chemical structures and metal ion specificities are known.

D. *Maltoporin: The Structural Basis of Sugar Discrimination*

The **porins** are homotrimeric transmembrane proteins that facilitate the transport of small molecules and ions across the outer membranes of gram-negative bacteria and mitochondria. Each subunit consists mainly of a 16- to 22-stranded antiparallel β barrel that forms a solvent-accessible channel along the barrel axis (Section 12-3Ad). In the *E. coli* OmpF porin (Fig. 12-27), this ∼50-Å-long channel is constricted near its center to an elliptical pore that has a minimum cross section of 7 × 11 Å. Consequently, solutes of more than ∼600 D are too large to pass through this channel.

Maltoporin is a bacterial porin that facilitates the diffusion of **maltodextrins** [the α(1 → 4)-linked glucose oligosaccharide degradation products of starch; e.g., maltose (Fig. 11-13)]. The X-ray structure of *E. coli* maltoporin (Fig. 20-8), determined by Tilman Schirmer, reveals that maltoporin is structurally similar to OmpF porin (Fig. 12-27), but with an 18-stranded rather than a 16-stranded antiparallel β barrel

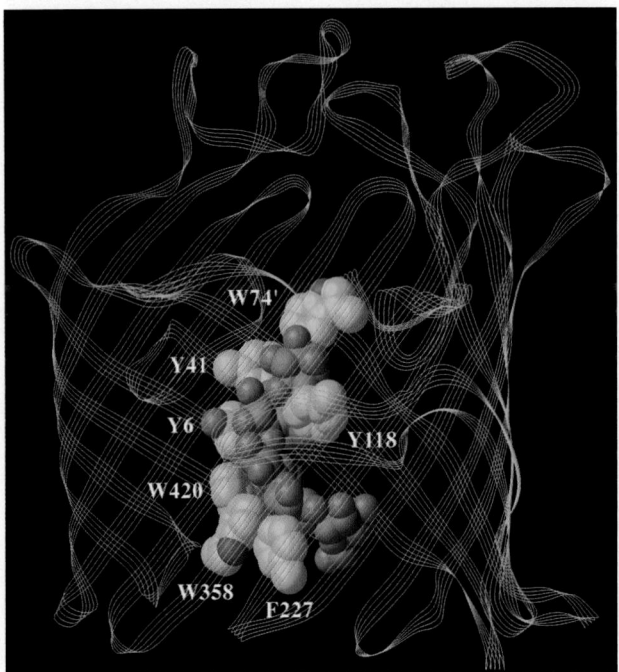

Figure 20-8   X-ray structure of a subunit of *E. coli* maltoporin in complex with a maltodextrin of six glucosyl units (Glc₆). The structure is viewed from within the bacterial outer membrane with its extracellular surface above. The polypeptide backbone is represented by a multithreaded ribbon (*cyan*). The Glc₆ (only five of whose glucosyl units are observed) and the aromatic side chains lining the constricted region of the protein's centrally located transport channel are shown in space-filling form colored according to atom type (protein side chain C gold, glucosyl C green, N blue, and O red). Note the pronounced left-handed helical twist of the Glc₆ unit. The so-called greasy slide, which consists of the aromatic side chains of six residues (W74′ is contributed by an overhanging loop from an adjacent subunit), conforms closely to this shape. The side chain of Y118 protrudes into the channel opposite the greasy slide so as to allow only the transit of near planar groups such as glucosyl residues. The maltodextrin's hydroxyl groups are arranged in two strips flanking the greasy slide (only one of which is seen here) that form an extensive hydrogen bonded network with mainly charged side chains (not shown). [Based on an X-ray structure by Tilman Schirmer, University of Basel, Switzerland. PDBid 1MPO.]

enclosing each subunit's transport channel. Three long loops from the extracellular face of each maltoporin subunit fold inward into the barrel, thereby constricting the channel near the center of the membrane to a diameter of ~5 Å (which is considerably smaller than OmpF's aperture) and giving the channel an hourglasslike cross section. The channel is lined on one side with a series of six contiguous aromatic side chains arranged in a left-handed helical path that matches the left-handed helical curvature of α-amylose (Fig. 11-18). This "greasy slide" extends from the channel's vestibule floor, through its constriction, to its periplasmic outlet.

The way in which oligosaccharides interact with malto-porin was investigated by determining the X-ray structures of maltoporin in its complexes with the maltodextrins Glc_2 (maltose), Glc_3, Glc_6, and sucrose (a glucose–fructose disaccharide; Fig. 11-13). Two Glc_2 molecules, one Glc_3 molecule, and a Glc_5 segment of Glc_6 occupied the maltoporin channel in contact and conformity with the greasy slide. Thus the hydrophobic faces of the maltodextrin's glycosyl residues stack on aromatic side chains, as is often observed in complexes of sugars with proteins. The glucose hydroxyl groups, which are arranged in two strips along opposite edges of the maltodextrins, form numerous hydrogen bonds with polar side chains that line these strips. Six of these seven polar side chains are charged, which probably strengthens their hydrogen bonds, as has also been observed in complexes of sugars with proteins. Tyr 118, which protrudes into the channel opposite the greasy slide, apparently functions as a steric barrier that permits only the passage of near-planar groups such as glucosyl residues. Thus the hook-shaped sucrose, which maltoporin transports quite slowly, binds to maltoporin with only its glucose residue inserted into the constricted part of the channel and its bulky fructose residue extending into the extracellular vestibule.

The above structures suggest a model for the selective transport of maltodextrins by maltoporin. At the start of the translocation process, the entering glucosyl residue interacts with the readily accessible end of the greasy slide in the extracellular vestibule of the channel. Further translocation along the helical channel requires the maltodextrin to follow a screwlike path that maintains the helical structure of the oligosaccharide, much like the movement of a bolt through a nut, thereby excluding molecules of comparable size that have different shapes. The translocation process is unlikely to encounter any large energy barrier due to the smooth surface of the greasy slide and the multiple polar groups at the channel constriction that would permit the essentially continuous exchange of hydrogen bonds as a maltodextrin moves through the constriction. Thus, maltoporin can be regarded as an enzyme that catalyzes the translocation of its substrate from one compartment to another.

E. *Passive-Mediated Glucose Transport*

The human erythrocyte glucose transporter is a 492-residue glycoprotein which, according to sequence hydropathy analysis (Sections 8-4C and 12-3Aa), has 12 membrane-

Figure 20-9   **Predicted secondary structure and membrane orientation of the glucose transporter.**

spanning α helices (Fig. 20-9) that are thought to form a hydrophobic cylinder. Five of these helices (3, 5, 7, 8, and 11) are amphipathic and hence most likely form a hydrophilic channel through which glucose is transported. A highly charged 66-residue domain located between helices 6 and 7, together with the 43-residue C-terminal domain, occupy the cytoplasm, whereas a 34-residue carbohydrate-bearing domain located between helices 1 and 2 is externally located. The glucose transporter accounts for 2% of erythrocyte membrane proteins and runs as band 4.5 in SDS–PAGE gels of erythrocyte membranes (Section 12-3Da; it is not visible on the gel depicted in Fig. 12-37 because the heterogeneity of its oligosaccharides makes the protein band diffuse).

a. Glucose Transport Occurs via a Gated Pore Mechanism

The erythrocyte glucose transporter has glucose binding sites on each side of the erythrocyte membrane but these have different steric requirements. Thus, John Barnett showed that 1-propylglucose will not bind to the extracellular surface of the glucose transporter but will bind to its cytoplasmic surface, whereas the converse is true of 6-propylglucose. He therefore proposed that the glucose transporter has two alternate conformations: one with the glucose binding site facing the external cell surface, requiring O1 contact and leaving O6 free, and the other with the glucose binding site facing the cytoplasm, requiring O6 contact and leaving O1 free (Fig. 20-10). *Transport apparently takes place by binding glucose to the protein on one face of the membrane, followed by a conformational change that closes the first site while exposing the other.* Glucose can then dissociate from the protein, having been translocated across the membrane. The transport cycle of this so-called **gated pore** is completed by the reversion of the glucose transporter to its initial conformation in the absence of bound glucose. Since this cycle can occur in either direction, the direction of net glucose transport is from high to low glucose concentrations. The glucose transporter thereby provides a means of equilibrating the glucose concentration across the erythrocyte

Figure 20-10   Alternating conformation model for glucose transport. Such a system is also known as a "gated pore." [After Baldwin, S.A. and Lienhard, G.E., *Trends Biochem. Sci.* **6,** 210 (1981).] 🎞 **See the Animated Figures**

membrane without any accompanying leakage of small molecules or ions.

b. Eukaryotes Express a Variety of Glucose Transporters

The erythrocyte glucose transporter, known also as **GLUT1** (for *glu*cose *t*ransporter *1*) has a highly conserved amino acid sequence (98% sequence identity between humans and rats), which suggests that all segments of this protein are functionally significant. GLUT1 is expressed in most tissues, although in liver and muscle, tissues that are highly active in glucose transport, it is present in only tiny amounts. Three other glucose transporters, **GLUT2, GLUT3,** and **GLUT4,** have been well characterized (**GLUT5** was originally thought to be a glucose transporter but was later shown to be a fructose transporter). They are 40 to 65% identical to GLUT1 but have different tissue distributions. For example, GLUT2 is prominent in pancreatic β cells (which secrete insulin in response to increased [glucose] in blood; Section 18-3F), liver (where its defects result in symptoms resembling Type I glycogen storage disease; Section 18-4), and the intestine (which absorbs dietary glucose; Section 20-4A); GLUT3 is expressed in neurons and the placenta, and GLUT4 occurs mainly in muscle and fat cells. Note that the tissue distributions of these glucose transporters correlate with the response of these tissues to insulin: Liver is unresponsive to insulin (liver functions, in part, to maintain the level of blood glucose; Section 18-3Fb), whereas muscle and fat cells take up glucose when stimulated by insulin. Analysis of the human genome has identified eight other members of the GLUT family, **GLUT6** through **GLUT12** and **HMIT** (for *H*⁺-coupled

*myo-inositol t*ransporter), although they have yet to be well characterized. All of them are members of the **major facilitator superfamily (MFS).**

c. Cellular Glucose Uptake Is Regulated through the Insulin-Sensitive Exocytosis/Endocytosis of Glucose Transporters

Insulin stimulates fat and muscle cells to take up glucose. Within 2 or 3 min after the administration of insulin to fat cells, the J_{max} for passive-mediated glucose transport into these cells increases 20- to 30-fold, whereas the K_M remains constant. On withdrawal of the insulin, the rate of glucose uptake returns to its basal level within 20 min to 2 h depending on conditions. Neither the increase nor the decrease in the rate of glucose transport is affected by the presence of protein synthesis inhibitors, so that these observations cannot be a consequence of the synthesis of new glucose transporter or of a protein that inhibits it. How, then, does insulin regulate glucose transport?

GLUT4 is the dominant glucose transporter in skeletal muscle and adipose (fat) cells. In their *basal state, these cells store most of their GLUT4 in specialized **GLUT4 storage vesicles.** On insulin stimulation, these vesicles fuse with the plasma membrane in a process known as **exocytosis** (Fig. 20-11). The consequent increased number of*

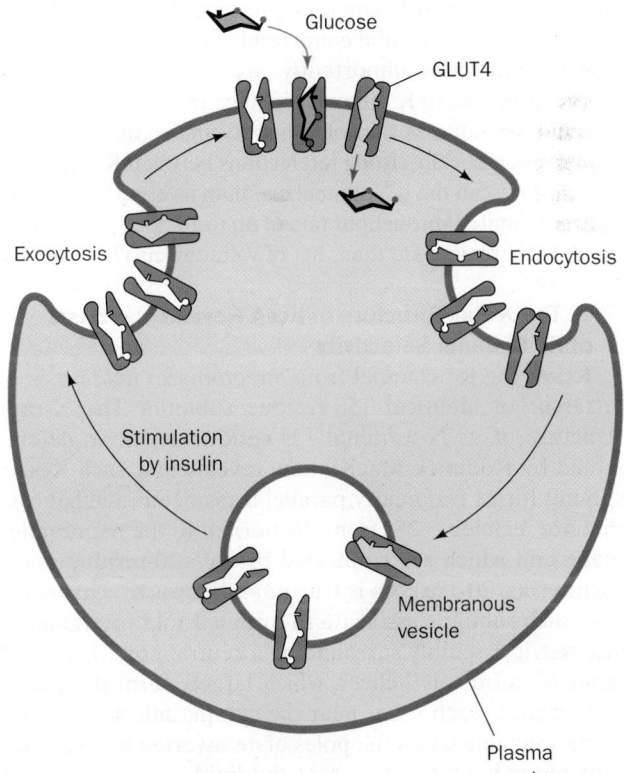

Figure 20-11   Regulation of glucose uptake in muscle and fat cells. Regulation is mediated by the insulin-stimulated exocytosis (the opposite of endocytosis; Section 12-5Bc) of membranous vesicles containing GLUT4 glucose transporters (*left*). On insulin withdrawal, the process reverses itself through endocytosis (*right*). 🎞 **See the Animated Figures**

cell-surface glucose transporters results in a proportional increase in the cell's glucose uptake rate. On insulin withdrawal, the process is reversed through the endocytosis of plasma membrane-embedded glucose transporters. The deletion or mutation of GLUT4's N-terminal eight residues, particularly Phe 5, causes this transporter to accumulate in the plasma membrane. A Leu-Leu sequence and an acidic motif near GLUT4's C-terminus are likewise essential for its sequestration by the cell's endocytotic machinery. The way in which insulin controls this system, which accounts for most of insulin's effects on muscle and fat cells, is imperfectly understood. However, it is clear that this mechanism involves a tyrosine phosphorylation cascade that is triggered by the binding of insulin to the insulin receptor (Section 19-3Ac and Fig. 19-67) and includes the activation of a class IA phosphoinositide 3-kinase (PI3K; Section 19-4Da).

F. K⁺ Channels: Ion Discrimination

Potassium ions diffuse from the cytoplasm (where $[K^+] >$ 100 mM) to the extracellular space (where $[K^+] < 5$ mM) through transmembrane proteins known as **K⁺ channels,** a process that underlies numerous important biological processes including maintenance of cellular osmotic balance, neurotransmission (Section 20-5), and signal transduction (Chapter 19). Although there is a large diversity of K⁺ channels, even within single organisms, all of them have similar sequences, exhibit comparable permeability characteristics, and most importantly, are at least 10,000-fold more permeable to K⁺ than Na⁺. Since this high selectivity (around the same as that of valinomycin; Section 20-2Cb) implies energetically strong interactions between K⁺ and the protein, how can the K⁺ channel maintain its observed nearly diffusion-limited throughput rate of up to 10^8 ions per second (a 10^4-fold greater rate than that of valinomycin)?

a. The X-Ray Structure of KcsA Reveals the Basis of K⁺ Channel Selectivity

KcsA, the K⁺ channel from *Streptomyces lividans,* is a tetramer of identical 158-residue subunits. The X-ray structure of its N-terminal 125-residue segment, determined by Roderick MacKinnon, reveals that each KcsA subunit forms two nearly parallel transmembrane helices that are inclined ~25° from the normal to the membrane plane and which are connected by an ~20-residue pore region (Fig. 20-12a). As is true of all known K⁺ channels, four such subunits associate to form a 4-fold rotationally symmetric assembly surrounding a central pore. The four inner (C-terminal) helices, which largely form the pore, pack against each other near the cytoplasmic side of the membrane much like the poles of an inverted teepee. The four outer helices, which face the lipid bilayer, buttress the inner helices but do not contact the adjacent outer helices. The pore regions, which each consist of a so-called turret, pore helix, and selectivity filter, occupy the open extracellular end of the teepee, with the pore helices fitting in between its poles. Several K⁺ ions and ordered

water molecules are seen to occupy the central pore (Figs. 20-12b and 20-13a).

The 45-Å-long central pore has variable width: It starts at its cytoplasmic side (Fig. 20-12b, *bottom*) as an ~6-Å-diameter and 18-Å-long tunnel, the so-called internal pore, whose entrance is lined with four anionic side chains that presumably help exclude anions (red area at the bottom of Fig. 20-12b). The internal pore then widens to form a cavity ~10 Å in diameter. These regions of the central pore are both wide enough so that a K⁺ ion could move through them in its hydrated state. However, the upper part of the pore, the so-called selectivity filter, narrows to 3 Å, thereby forcing a transiting K⁺ ion to shed its waters of hydration. The walls of the internal pore and the cavity are lined with hydrophobic groups that interact minimally with diffusing ions (yellow area of the pore in Fig. 20-12b). However, the selectivity filter (red area of the pore at the top of Fig. 20-12b) is lined with closely spaced main chain carbonyl oxygens of residues (Fig. 20-13a, *top*) that are highly conserved in all K⁺ channels (their so-called signature sequence, TVGYG) and whose mutations disrupt the ability of the channel to discriminate between K⁺ and Na⁺ ions.

What is the function of the cavity? Energy calculations indicate that an ion moving through a narrow transmembrane pore must surmount an energy barrier that is maximal at the center of the membrane. The existence of the cavity reduces this electrostatic destabilization by surrounding the ion with polarizable water molecules (Fig. 20-12c). In addition, the C-terminal ends of the four pore helices point directly at the center of the cavity, so that their helix dipoles impose a negative electrostatic potential on the cavity that lowers the electrostatic barrier facing a cation crossing a lipid bilayer.

Remarkably, the K⁺ ion occupying the cavity is liganded by 8 ordered water molecules located at the corners of a square antiprism (a cube with one face twisted by 45° with respect to the opposite face) in which the K⁺ ion is centered (Fig. 20-13a, *bottom*; K⁺ in aqueous solution was known to have such an inner hydration shell but it had never before been visualized). The K⁺ ion is precisely centered in the cavity but yet its liganding water molecules are not in van der Waals contact with the walls of the cavity. Indeed, there is room in the cavity for ~40 additional water molecules although they are unseen in the X-ray structure because they are disordered. This disorder arises because the cavity is lined with hydrophobic groups (mainly the side chains of Ile 100 and Phe 103; Fig. 20-13a) that interact but weakly with water molecules, thus allowing them to interact freely with the K⁺ ion so as to form an outer hydration shell. What, then, holds the hydrated K⁺ ion in place? Apparently, it is very weak indirect hydrogen bonds involving such protein groups as the hydroxyl group of Thr 107 and possibly carbonyl O atoms from the pore and inner helices. The absence of such an ordered hydration complex when Na⁺ rather than K⁺ occupies the cavity is indicative of a precise geometric match between the hydrated K⁺ and the cavity (the ionic radii of Na⁺ and K⁺ are 0.95 Å and

(a)

(b)

(c)

Figure 20-12 X-ray structure of the KcsA K⁺ channel.
(*a*) Ribbon diagram of the tetramer as viewed from within the plane of the membrane with the cytoplasm below and the extracellular region above. The protein's 4-fold axis of rotation is vertical and each of its identical subunits is differently colored. (*b*) A cutaway diagram viewed similarly to Part *a* in which the K⁺ channel is represented by its solvent-accessible surface. The surface is colored according to its physical properties, with negatively charged areas red, uncharged areas white, positively charged areas blue, and hydrophobic areas of the central pore yellow. K⁺ ions are represented by green spheres. (*c*) A schematic diagram indicating how the K⁺ channel stabilizes a cation in the center of the membrane. The central pore's 10-Å-diameter aqueous cavity (which contains ~50 water molecules) stabilizes a K⁺ ion (*green spheres*) in the otherwise hydrophobic membrane interior. In addition, the C-terminal ends of the pore helices (*red*) all point toward the K⁺ ion, thereby electrostatically stabilizing it via their dipole moments (an α helix has a strong dipole moment with its negative end pointing toward the helix's C-terminal end because the bond dipoles of its component carbonyl and N—H groups are all parallel to the helix axis with their negative ends pointing toward its C-terminal end; Fig. 8-11). This effect is magnified by the low dielectric constant at the center of the membrane interior. Electrostatic calculations indicate that the cavity is tuned to maximally stabilize monovalent cations. [Courtesy of Roderick MacKinnon, Rockefeller University. PDBid 1BL8.]

1.33 Å, respectively). The cavity thereby provides a high effective K⁺ concentration (~2*M*) at the center of the membrane and positions the K⁺ ion on the pore axis ready to enter the selectivity filter.

How does the K⁺ channel discriminate so acutely between K⁺ and Na⁺ ions? The main chain O atoms lining the selectivity filter form a stack of rings (Fig. 20-13*a, top*) that provide a series of closely spaced sites of appropriate dimensions for coordinating dehydrated K⁺ ions but not the smaller Na⁺ ions. If the observed diameter of the selectivity filter is rigidly maintained, it would make the energy of a dehydrated Na⁺ in the selectivity filter considerably

(a)

(b)

(c)

Figure 20-13 Portions of the KcsA K⁺ channel responsible for its ion selectivity viewed similarly to Fig. 20-12. (*a*) The X-ray structure of the residues forming the cavity (*bottom*) and selectivity filter (*top*) but with the front and back subunits omitted for clarity. Atoms are colored according to type, with C yellow, N blue, O red, and K⁺ ions represented by green spheres. The water and protein O atoms that ligand the K⁺ ions, including those contributed by the front and back subunits, are represented by red spheres. The coordination polyhedra formed by these O atoms are outlined by thin white lines. (*b* and *c*) Two alternative K⁺ binding states of the selectivity filter, whose superposition is presumed to be responsible for the electron density observed in the X-ray structure of KcsA. Atoms are colored as in Part *a*. Note that K⁺ ions occupying the selectivity filter are interspersed with water molecules and that the K⁺ ion immediately above the selectivity filter in Part *b* is farther above the protein than that in Part *c*. Hence these ions maintain a constant spacing while traversing the selectivity filter. [Part *a* based on an X-ray structure by, and Parts *b* and *c* courtesy of, Roderick MacKinnon, Rockefeller University. PDBid 1K4C.] 🎧 **See Interactive Exercise 14**

higher than that of hydrated Na⁺ and thus account for the K⁺ channel's high selectivity for K⁺ ions. However, proteins are not static structures. In fact, both X-ray evidence and molecular dynamics simulations (Section 9-4a) indicate that, at physiological temperatures, the atoms forming the KcsA selectivity filter undergo thermal excursions averaging ~1 Å, fluctuations sufficient to snugly cradle Na⁺ ions with little energetic cost. Instead, as free energy calculations have demonstrated, it is the electrostatic interactions of the carbonyl groups with the cation and with each other that confer specificity for binding K$^+$ ions. This is consistent with the observation that no Na$^+$-specific protein channels have evolved by refining the structure of a KcsA-like channel.

Since the selectivity filter appears designed to specifically bind K$^+$ ions, how does it support such a high throughput of these ions (up to 10^8 ions · s^{-1})? The structure in Fig. 20-13*a* shows what appear to be 4 K$^+$ ions in the selectivity filter and two more just outside it on its extracellular side. Such closely spaced positive ions would strongly repel one another and hence represent a high energy situation. However, a variety of evidence suggests that this structure is really a superposition of two sets of K$^+$ ions, one with K$^+$ ions at the topmost position in Fig. 20-13*a* and at positions 1 and 3 in the selectivity filter (Fig. 20-13*b*) and the second with K$^+$ ions at the second position from the top in Fig. 20-13*a* and at positions 2 and 4 in the selectivity filter (Fig. 20-13*c*; X-ray structures can show overlapping atoms because they are averages of many unit cells). Within the selectivity filter, the positions not occupied by K$^+$ ions are instead occupied by water molecules that coordinate the neighboring K$^+$ ions.

The electron density that is represented as the topmost 4 water molecules in Fig. 20-13*a* is highly elongated in the vertical direction in this otherwise high-resolution (2.0 Å) structure. Hence it is thought to actually arise from 8 water molecules that ligand the topmost K$^+$ ion in Fig. 20-13*b* to form an inner hydration shell similar to that of the K$^+$ in the central cavity (Fig. 20-13*a, bottom*). Moreover, the four water molecules liganding the topmost K$^+$ ion in Fig. 20-13*c* also contribute to this electron density. This latter ring of 4 waters provides half of the associated K$^+$ ion's 8 liganding O atoms. The others are contributed by the carbonyl O atoms of the 4 Gly 79 residues, which are properly oriented to do so. It therefore appears that a dehydrated K$^+$ ion transits the selectivity filter (moves to successive positions in Figs. 20-13*b,c*) by exchanging the properly spaced ligands extending from its walls and then exits into the extracellular solution by exchanging protein ligands for water molecules and hence again acquiring a hydration shell. These ligands are spaced and oriented such that there is little free energy change (estimated to be <12 kJ · mol^{-1}) along the reaction coordinate via which a K$^+$ ion transits the selectivity filter and enters the extracellular solution. The rapid dehydration of the K$^+$ ion entering the selectivity channel from the cavity is, presumably, similarly managed. The essentially level free energy landscape throughout this process is, of course, conducive to the rapid transit of K$^+$ ions through the ion channel and hence must be a product of evolutionary fine-tuning. Energy calculations indicate that mutual electrostatic repulsions between successive K$^+$ ions, whose movements are concerted, balances the attractive interactions holding these ions in the selectivity filter and hence further facilitates their rapid transit.

G. *Cl$^-$ Channels*

Cl$^-$ channels, which occur in all cell types, permit the transmembrane movement of chloride ions along their concentration gradient. In mammals, the extracellular Cl$^-$ concentration is ~120 m*M* and the intracellular concentration is ~4 m*M*.

ClC channels form a large family of Cl$^-$ channels that occur widely in all kingdoms of life. The X-ray structures of ClC channels from two species of bacteria, determined by Raimund Dutzler and MacKinnon, reveal, as biophysical measurements had previously suggested, that ClC channels are homodimers with each ~470-residue subunit forming an anion-selective pore (Fig. 20-14). Each subunit consists mainly of 18 mostly transmembrane α helices that are remarkably tilted with respect to the membrane plane and have variable lengths compared to the transmembrane helices in other integral proteins of known structures. The N- and C-terminal halves of each subunit are related by a pseudo-2-fold axis parallel to the plane of the membrane and hence these two halves have opposite orientations in

(a)

(b)

Figure 20-14 **X-ray structure of the ClC Cl$^-$ channel from** ***E. coli.*** Each subunit of the homodimer contains 18 α helices of variable lengths. The subunits are drawn in ribbon form with one colored in rainbow order from its N-terminus (*blue*) to its C-terminus (*red*) and the other pink. The two Cl$^-$ ions bound in the selectivity filter of each subunit are represented by pale green spheres. (*a*) View from within the membrane with the extracellular surface above and the 2-fold axis relating the two subunits vertical. (*b*) View from the extracellular side of the membrane along the molecular 2-fold axis. [Based on an X-ray structure by Raimund Dutzler and Roderick MacKinnon, Rockefeller University. PDBid 1OTS.]

the membrane. This suggests that the ClC channel arose through gene duplication although its two halves exhibit only weak sequence similarity. Such antiparallel architecture occurs in several types of transmembrane transport proteins.

The ClC Cl⁻ channel is located at the interface between its N- and C-terminal halves. The specificity of the ClC channel results from an electrostatic field established by basic amino acids on the protein surface, which helps funnel anions toward the pore, and by a selectivity filter formed by the dipoles of several α helices oriented with their positively charged N-terminal ends pointing toward the Cl⁻ ions (opposite to their orientation in the KcsA channel; Fig. 20-12c). This feature of the selectivity filter helps attract Cl⁻ ions, which are specifically coordinated by main chain amide nitrogens and side chain hydroxyls from Ser and Tyr residues. A positively charged residue such as Lys or Arg, if it were present in the selectivity filter, would probably bind a Cl⁻ ion too tightly to facilitate its rapid transit through the channel.

Unlike the KcsA channel, which has a central aqueous cavity (Fig. 20-12c), the Cl⁻ channel is hourglass-shaped, with its narrowest part in the center of the membrane and flanked by wider aqueous vestibules. A conserved Glu side chain projects into the pore. This group would repel other anions, suggesting that rapid Cl⁻ flux requires a protein conformational change in which the Glu side chain moves aside. Another anion could push the Glu away, which explains why some Cl⁻ channels appear to be activated by Cl⁻ ions; that is, they open in response to a certain concentration of Cl⁻ in the extracellular fluid.

H. *Aquaporins*

The observed rapid passage of water molecules across biological membranes had long been assumed to occur via simple diffusion that was made possible by the small size and high concentration of water molecules. However, certain cells, such as erythrocytes and those of the kidney, can sustain particularly rapid rates of water transport, which are reversibly inhibited by mercuric ion. This suggested the existence of previously unrecognized protein pores that conduct water through biological membranes. The first of these proteins was discovered in 1992 by Peter Agre, who named them **aquaporins.**

Aquaporins occur widely in all kingdoms of life. Plants have up to 50 different aquaporins, which is indicative of the importance of water transport to plant physiology. The 13 known mammalian aquaporins, **AQP0** through **AQP12,** are selectively expressed at high levels in tissues that rapidly transport water, such as kidneys, salivary glands, sweat glands, and lacrimal glands (which produce tears). In fact, kidneys alone employ seven different aquaporins, each with specific locations and regulatory properties. There are two subfamilies of aquaporins: those that permit only the passage of water and those that also allow the passage of small neutral molecules such as glycerol and urea and hence are named **aquaglyceroporins.** Aquaporins permit the passage of water molecules at extremely high rates (up

to ~3 × 10⁹ per second) but, quite surprisingly, not protons (really hydronium ions; H_3O^+), whose free passage would discharge the cell's membrane potential.

All known aquaporins are homotetramers, each of whose subunits contain a water-transport channel (unlike K⁺ channels, whose transport channels lie along their 4-fold axes; Section 20-2Fa). The X-ray structure of the most extensively studied aquaporin, bovine **AQP1,** reveals that each of its 271-residue subunits consists mainly of six transmembrane α helices plus two shorter helices that are components of loops that extend only to the middle of the bilayer (Fig. 20-15a). Other aquaporins of known structure have similar structures. The N- and C-terminal halves of aquaporins are ~20% identical in sequence and related by a pseudo-2-fold axis of symmetry that is parallel to the plane of the membrane (Fig. 20-15a). Evidently, these segments arose through gene duplication. ClC channels have a similar antiparallel architecture (Section 20-2G).

The helices in AQP1 surround an elongated hourglass-shaped channel through the membrane (Fig. 20-16) that at its narrowest point is ~2.8 Å wide, the diameter of a water molecule. This region is formed by the side chains of the highly conserved Phe 58, His 182, and Arg 197 (Fig. 20-15b, *lower right subunit*) and hence is known as the **ar/R constriction** (ar for *aromatic*). The side chain of Cys 191, which also forms part of the ar/R constriction, is the site of channel blockage by the binding of mercuric ion. For a water molecule to pass through the ar/R constriction, it must shed its shell of associated water molecules. This is facilitated by the side chains of His 182 and Arg 197. The water molecules then continue in single file through the ~25-Å-long and ~4-Å-wide portion of the channel, which is lined with hydrophobic groups interspersed with several hydrogen bonding groups. The water molecules' lack of interaction with the hydrophobic walls of the channel facilitates their rapid passage through the channel, whereas the hydrogen bonding groups reduce the energy barrier to water transport. It is the balancing of these opposing factors that is presumably responsible for aquaporin's selective permeability to water and its rapid transport rate.

If water were to pass through aquaporin as an uninterrupted chain of hydrogen-bonded molecules, then protons would pass even more rapidly through the channel via proton jumping (Fig. 2-10; in order for more than one such series of proton jumps to occur, each water molecule in the chain must reorient such that one of its protons forms a hydrogen bond to the next water molecule in the chain). However, aquaporin interrupts this process by forming hydrogen bonds from the side chain NH_2 groups of the highly conserved Asn 78 and Asn 194, to a water molecule that is centrally located in the channel (Fig. 20-16). Consequently, although this central water molecule can readily donate hydrogen bonds to its neighboring water molecules in the hydrogen bonded chain, it cannot accept one from them nor reorient, thereby severing the "proton-conducting wire." Both of these Asn residues occur in the sequence Asn-Pro-Ala (NPA), the signature sequence of aquaporins, in which the Ala is located at the N-terminal end of each of the half-spanning helices.

(a)

(b)

Figure 20-15 **X-ray structure of the aquaporin AQP1 from bovine erythrocytes.** (*a*) Ribbon diagram of an aquaporin subunit colored in rainbow order from its N-terminus (*blue*) to its C-terminus (*red*). The view is from within the membrane with its extracellular surface above and along the subunit's pseudo-2-fold axis of symmetry. Note that the two helices closest to the viewer (*orange* and *blue-green*) are both portions of loops that extend only to the center of the bilayer. The four water molecules that occupy the central portion of AQP1's water-transport channel are represented by red spheres. (*b*) View of the aquaporin homotetramer from the extracellular surface along its 4-fold axis. The subunit in the upper right is drawn in space-filling form with C green, N blue, and O red; that in the upper left is drawn in ribbon form colored in rainbow order from its N-terminus (*blue*) to its C-terminus (*red*), that in the lower left is represented by its solvent-accessible surface; and that in the lower right displays the side chains forming the ar/R constriction (those of Phe 58, His 182, Cys 191, and Arg 197) in stick form. Each subunit forms a water-transport channel, which is most clearly visible in the subunit drawn in space-filling form. [Based on an X-ray structure by Bing Jap, University of California at Berkeley. PDBid 1J4N.]

Figure 20-16 **Schematic drawing of the water-conducting pore of bovine aquaporin AQP1.** The pore is viewed from within the membrane with the extracellular surface above. The positions of residues critical for preventing the passage of protons, other ions, and small molecule solutes are indicated. [Courtesy of Peter Agre, Johns Hopkins School of Medicine.]

3 ATP-DRIVEN ACTIVE TRANSPORT

Mediated transport is categorized according to the stoichiometry of the transport process (Fig. 20-17):

1. A **uniport** involves the movement of a single molecule at a time. Maltoporin and GLUT1 are uniports.

2. A **symport** simultaneously transports two different molecules in the same direction.

3. An **antiport** simultaneously transports two different molecules in opposite directions.

The electrical character of ion transport is further specified as:

1. Electroneutral (electrically silent) if there is simultaneous charge neutralization, either by symport of oppositely charged ions or antiport of similarly charged ions. Aquaporin is electroneutral.

2. Electrogenic if the transport process results in a charge separation across the membrane. KcsA and ClC are electrogenic.

Since the glucose concentration in blood plasma is generally higher than that in cells, GLUT1 normally transports glucose into the erythrocyte, where it is metabolized via glycolysis. Many substances, however, are available on one side of a membrane in lower concentrations than are required on the other side of the membrane. Such substances must be actively and selectively transported across the membrane against their concentration gradients.

Active transport is an endergonic process that is often coupled to the hydrolysis of ATP. How is this coupling accomplished? In endergonic biosynthetic reactions, it often occurs through the direct phosphorylation of a substrate by ATP; for example, the formation of UTP in the synthesis of glycogen (Section 18-2B). Membrane transport, however, is usually a physical rather than a chemical process; the transported molecule is not chemically altered. Determining the mechanism by which the free energy of ATP hydrolysis is coupled to endergonic physical processes has therefore been a challenging problem.

Three types of ATP hydrolyzing, transmembrane proteins have been identified that actively transport cations:

1. P-type ATPases are located mostly in plasma membranes and are so named because they are *p*hosphorylated by ATP during the transport process. P-type ATPases are known that transport H^+, Na^+, K^+, Ca^{2+}, Cu^{2+}, Cd^{2+}, and Mg^{2+} against their concentration gradients. They are distinguished from the other types of cation-translocating ATPases by their inhibition by **vanadate** (VO_4^{3-}, a phosphate analog; see Problem 8 in this chapter).

2. F-type ATPases (F_1F_0) function to translocate protons into mitochondria and bacterial cells, which in turn powers ATP synthesis. They are discussed in Section 22-3C.

3. V-type ATPases are located in plant vacuolar membranes and acidic vesicles, such as animal lysosomes, and are homologous to the F-type ATPases.

Anions are transported by a fourth type of ATPase, the so-called **A-type ATPases.** In this section, we discuss P-type ATPases. We also examine a bacterial active transport process, in which the molecules transported are concomitantly phosphorylated, and the ABC transporters, which transport a wide variety of substances across membranes. In the next section, we study secondary active transport systems, so called because they utilize the free energy of electrochemical gradients generated by ion-pumping ATPases to transport ions and neutral molecules against their concentration gradients.

A. (Na⁺–K⁺)–ATPase of Plasma Membranes

One of the most thoroughly studied active transport systems is the **(Na⁺–K⁺)–ATPase** of plasma membranes. This transmembrane protein, which was first isolated in 1957 by Jens Skou, is often called the **(Na⁺–K⁺) pump** because *it pumps Na^+ out of and K^+ into the cell with the concomitant hydrolysis of intracellular ATP.* Unlike most P-type ATPases, which are monomeric, (Na⁺–K⁺)–ATPases consist of α and β subunits. The ~1000-residue, nonglycosylated α subunit contains the enzyme's ATP and ion binding sites. It is highly conserved (98% identical among mammals) and homologous to single-subunit P-type ATPases such as the Ca^{2+}-ATPase (Section 20-3B). The ~300-residue, glycosylated β subunit facilitates the correct insertion of the α subunit into the plasma membrane and has been implicated in K^+ transport.

The overall stoichiometry of the (Na⁺–K⁺)–ATPase reaction is

$$3Na^+ (in) + 2K^+ (out) + ATP + H_2O \rightleftharpoons$$
$$3Na^+ (out) + 2K^+ (in) + ADP + P_i$$

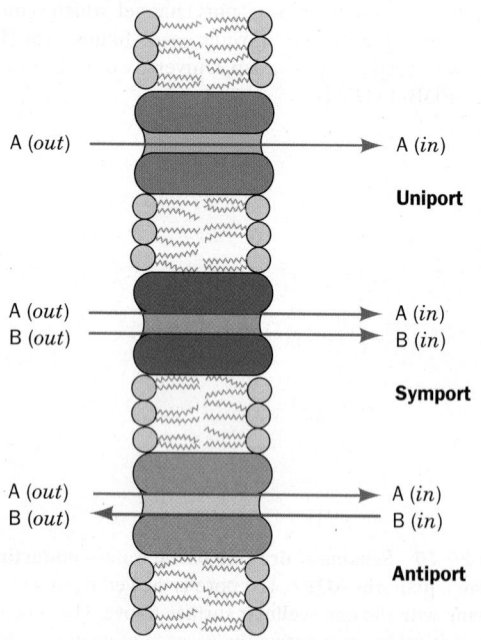

Figure 20-17 **Uniport, symport, and antiport translocation systems.**

A (*out*) → A (*in*)

Uniport

A (*out*) → A (*in*)
B (*out*) → B (*in*)

Symport

A (*out*) → A (*in*)
B (*out*) ← B (*in*)

Antiport

The $(Na^+–K^+)$–ATPase is therefore an electrogenic antiport: Three positive charges exit the cell for every two that enter. This extrusion of Na^+ enables animal cells to control their water content osmotically; without functioning $(Na^+–K^+)$ pumps, animal cells, which lack cell walls, would swell and burst (recall that lipid bilayers are permeable to H_2O; Section 12-2Ba). Moreover, the electrochemical potential gradient generated by the $(Na^+–K^+)$ pump is responsible for the electrical excitability of nerve cells (Section 20-5Ba) and provides the free energy for the active transport of glucose and amino acids into some cells (Section 20-4A). In fact, *all cells expend a large fraction of the ATP they produce (typically 30% and up to 70% in nerve cells) to maintain their required cytosolic Na^+ and K^+ concentrations.*

a. ATP Phosphorylates an Essential Asp during the Transport Process

The free energy of ATP hydrolysis powers the endergonic transport of Na^+ and K^+ against an electrochemical gradient. In coupling these two processes, a kinetic barrier must somehow be erected against the "downhill" transport of Na^+ and K^+ along their ion concentration gradients, while simultaneously facilitating their "uphill" transport. In addition, futile ATP hydrolysis must be prevented in the absence of uphill transport. How the enzyme does so is by no means well understood, although many of its mechanistic aspects have been elucidated.

Aspartyl phosphate residue

acid hydrolysis

Homoserine

Figure 20-18 **Reaction of [^{3}H]NaBH$_4$ with phosphorylated $(Na^+–K^+)$-ATPase.** The isolation of [^{3}H]homoserine following acid hydrolysis of the protein indicates that the original phosphorylated amino acid residue is Asp.

A key discovery was that the protein is phosphorylated by ATP in the presence of Na^+ during the transport process. The use of chemical trapping techniques demonstrated that this phosphorylation occurs on an Asp residue to form a highly reactive **aspartyl phosphate** intermediate. For instance, sodium borohydride reduces acyl phosphates to their corresponding alcohols. In the case of an aspartyl phosphate residue, the alcohol is **homoserine.** By use of [^{3}H]NaBH$_4$ to reduce the phosphorylated enzyme, radioactive homoserine was, in fact, isolated from the acid hydrolysate (Fig. 20-18). The phosphorylated residue, Asp 374, begins the highly conserved sequence DKTG that occurs in the central region of the polypeptide chain.

b. The $(Na^+–K^+)$-ATPase Has Two Major Conformational States

The observations that ATP phosphorylates the $(Na^+–K^+)$-ATPase only in the presence of Na^+, while the aspartyl phosphate residue is only subject to hydrolysis in the presence of K^+, led to the realization that *the enzyme has two major conformational states, E1 and E2.* These states have different tertiary structures, different catalytic activities, and different ligand specificities:

1. E1 has an inward-facing high-affinity Na^+ binding site ($K_M = 0.2$ mM, well below the intracellular [Na^+]) and reacts with ATP to form the activated product E1~P only when Na^+ is bound.

2. E2—P has an outward-facing high-affinity K^+ binding site ($K_M = 0.05M$, well below the extracellular [K^+]) and hydrolyzes to form P_i + E2 only when K^+ is bound.

c. An Ordered Sequential Kinetic Reaction Mechanism Accounts for the Coupling of Active Transport with ATP Hydrolysis

The $(Na^+–K^+)$-ATPase is thought to operate in accordance with the following ordered sequential reaction scheme (Fig. 20-19):

1. E1 · ATP, which acquired its ATP inside the cell, binds 3Na$^+$ to yield the ternary complex E1 · ATP · 3Na$^+$.

2. The ternary complex reacts to form the "high-energy" aspartyl phosphate intermediate E1~P · 3Na$^+$.

3. This "high-energy" intermediate relaxes to its "low-energy" conformation, E2—P · 3Na$^+$, and releases its bound Na$^+$ outside the cell; that is, Na$^+$ is transported through the membrane.

4. E2—P binds 2K$^+$ from outside the cell to form E2—P · 2K$^+$.

5. The phosphate group is hydrolyzed, yielding E2 · 2K$^+$.

6. E2 · 2K$^+$ changes conformation to E1, binds ATP, and releases its 2K$^+$ inside the cell, thereby completing the transport cycle.

The enzyme appears to have only one set of cation binding sites, which apparently changes both its orientation and its specificity during the course of the transport cycle.

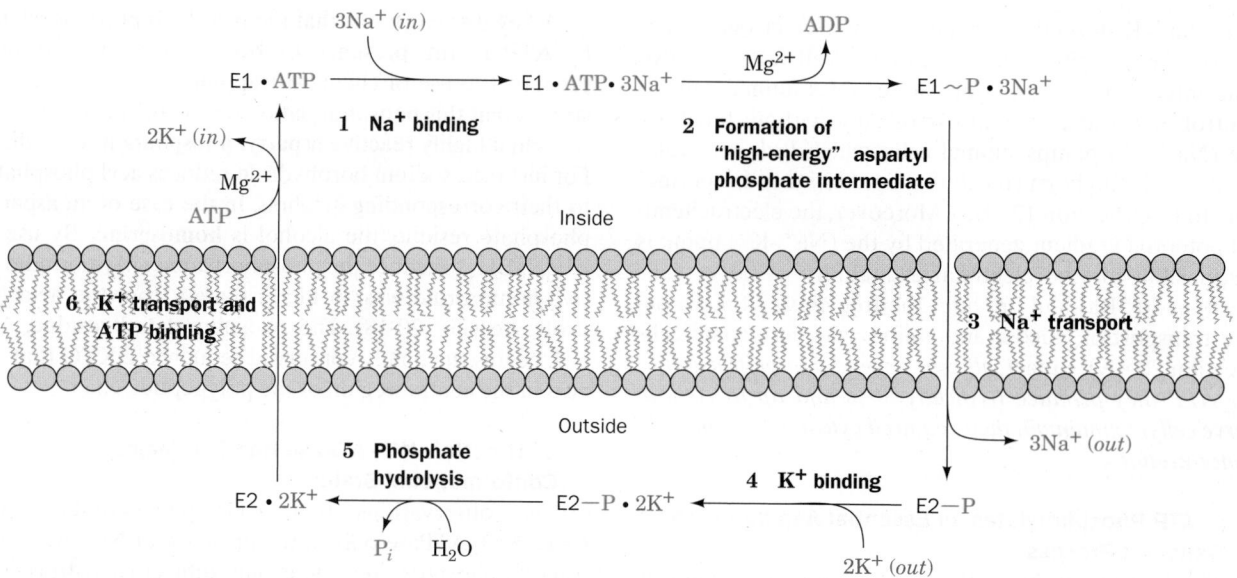

Figure 20-19 **Kinetic scheme for the active transport of Na$^+$ and K$^+$ by (Na$^+$–K$^+$)–ATPase.** Here (*in*) refers to the cytosol and (*out*) refers to the exterior of the cell.

The obligatory order of the reaction requires that ATP can be hydrolyzed only as Na$^+$ is transported "uphill." Conversely, Na$^+$ can be transported "downhill" only if ATP is concomitantly synthesized. Consequently, although each of the above reaction steps is, in fact, individually reversible, the cycle, as is diagrammed in Fig. 20-19, circulates only in the clockwise direction under normal physiological conditions; that is, ATP hydrolysis and ion transport are coupled processes. Note that the **vectorial** (unidirectional) nature of the reaction cycle results from the alternation of the steps of the exergonic ATP hydrolysis reaction (Step 2, Step 5, and ATP binding in Step 6) with the steps of the endergonic ion transport process (Step 1, Steps 3 + 4, and K$^+$ release in Step 6). Thus, neither reaction can go to completion unless the other one also does.

d. Mutual Destabilization Accounts for the Rate of Na$^+$ and K$^+$ Transport

The above ordered kinetic mechanism accounts only for the coupling of active transport with ATP hydrolysis. *In order to maintain a reasonable rate of transport, the free energies of all its intermediates must be roughly equal. If some intermediates were much more stable than others, the stable intermediates would accumulate, thereby severely reducing the overall transport rate.* For example, in order for Na$^+$ to be transported out of the cell, uphill, its binding must be strong to E1 on the inside and weak to E2 on the outside. Strong binding means greater stability and a potential bottleneck. This difficulty is counteracted by the phosphorylation of E1 · 3Na$^+$ and its subsequent conformational change to yield the low Na$^+$ affinity E2—P (Steps 2 and 3, Fig. 20-19). Likewise, the strong binding of K$^+$ to E2—P on the outside is attenuated by its dephosphorylation and conformational change to yield the low K$^+$ affinity E1 (Steps 5

and 6, Fig. 20-19). It is these mutual destabilizations that permit Na$^+$ and K$^+$ to be transported at a rapid rate.

e. The X-Ray Structure of the (Na$^+$–K$^+$)–ATPase

Chikashi Toyoshima determined the X-ray structure of shark (Na$^+$–K$^+$)–ATPase in complex with K$^+$ ions, an MgF$_4^{2-}$ ion (a P$_i$ mimic), and a 74-residue subunit named FXYD that functions as a tissue-specific regulator. This X-ray structure (Fig. 20-20) is that of the E2—P · 2K$^+$ complex (Fig. 20-19). The α subunit of this ~160-Å-long protein consists of a transmembrane domain (M) composed of 10 helices (αM1–αM10) of varied lengths and, from top to bottom in Fig. 20-20, three well-separated cytoplasmic domains: the nucleotide-binding domain (N), which binds ATP; the actuator domain (A), so named because it participates in the transmission of major conformational changes (see below); and the phosphorylation domain (P), which contains the protein's phosphorylatable Asp residue. The β subunit's single transmembrane helix is tilted ~32° from the normal to the plane of the membrane. The FXYD subunit also has a single transmembrane helix but it is nearly perpendicular to the plane of the membrane. The MgF$_4^{2-}$ ion marks the ATPase's catalytic site and is coordinated by conserved residues from both its A and P domains. Two K$^+$ ions are located ~4.1 Å apart in a common binding cavity near the center of the α subunit's transmembrane domain that is formed, in large part, by the partial unwinding of helices αM5 and αM7, and where they are each liganded by several main chain carbonyl and side chain oxygen atoms. The same cavity is implicated in binding the three Na$^+$ ions bound to the enzyme's E1 form, with two of these binding sites probably formed by the same side chains that coordinate the K$^+$ ions and the third such site formed, in part, by the side chains of the α subunit's

Figure 20-20 **X-ray structure of shark (Na⁺–K⁺)–ATPase in complex with K⁺ and MgF₄²⁻ ions.** The protein is drawn in ribbon form viewed parallel to the plane of the membrane (*gray box*) with the cytosol above. The α subunit is colored in rainbow order from its N-terminus (*blue*) to its C-terminus (*red*), the β subunit is magenta, and the FXYD subunit is brown. The α subunit's bound K⁺ and MgF₄²⁻ ions are drawn in space-filling form with K light blue, Mg pink, and F light green. [Based on an X-ray structure by Chikashi Toyoshima, University of Tokyo, Japan. PDBid 2ZXE.]

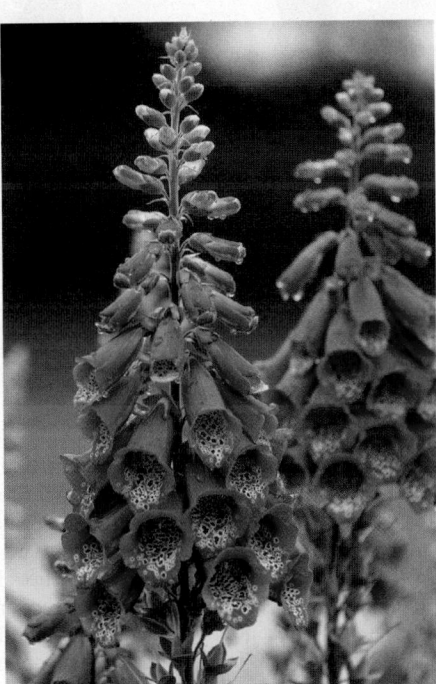

Figure 20-21 **Cardiac glycosides.** (*a*) The leaves of the purple foxglove plant are the source of the heart muscle stimulant digitalis. [iStockphoto.] (*b*) Digitalin, a major component of

small C-terminal helix. No channel leading from either side of the membrane to this binding cavity is apparent.

f. Cardiac Glycosides Specifically Inhibit the (Na⁺–K⁺)–ATPase

Study of the (Na⁺–K⁺)–ATPase has been greatly facilitated by the use of **cardiac glycosides** (also called **cardiotonic steroids**), natural products that increase the intensity of heart muscle contraction. Indeed, **digitalis,** an extract of purple foxglove leaves (Fig. 20-21*a*), which contains a mixture of cardiac glycosides including **digitalin** (Fig. 20-21*b*), has been used to treat congestive heart failure for centuries. The cardiac glycoside **ouabain** (pronounced wabane; Fig. 20-21*b*), a product of the East African ouabio tree, has been long used as an arrow poison. These two steroids, which are still among the most commonly prescribed cardiac drugs, inhibit the (Na⁺–K⁺)–ATPase by binding strongly to an externally exposed portion of the enzyme (the drugs are ineffective when injected inside cells) so as to block Step 5 in Fig. 20-19. The resultant increase in intracellular [Na⁺] stimulates the cardiac (Na⁺–Ca²⁺) antiport system, which pumps Na⁺ out of and Ca²⁺ into the cell (Section 22-1Bb). The increased cytosolic [Ca²⁺] boosts the [Ca²⁺] in other cellular organelles, principally the sarcoplasmic reticulum (**SR**). Thus, the release of Ca²⁺ to trigger

Digitalin

Ouabain

(*b*)

digitalis, and ouabain, a cardiac glycoside isolated from the East African ouabio tree, are among the most commonly prescribed cardiac drugs.

muscle contraction (Section 35-3Cb) produces a larger than normal increase in cytosolic [Ca²⁺], thereby intensifying the force of cardiac muscle contraction. Ouabain, which was once thought to be produced only by plants, is now known to also be an animal hormone that is secreted by the adrenal cortex and functions to regulate cell [Na⁺] and overall body salt and water balance.

B. *Ca²⁺–ATPase*

Calcium ion often acts as a second messenger in a manner similar to cAMP (Section 19-4). Transient increases in cytosolic [Ca²⁺] trigger numerous cellular responses, including muscle contraction (Section 35-3Ca), release of neurotransmitters (Section 20-5Cb), and, as we have seen, glycogen

Figure 20-22   Mechanism of SERCA based on four X-ray structures. (*a*) The X-ray structures of the SERCA complexes indicated in lightface are models for the conformational states indicated in boldface. These states roughly correspond to those in the corners of Fig. 20-19. The structures are represented by their ribbon diagrams embedded in their transparent surfaces (*gray*) as viewed along the plane of the membrane with the cytosol above. The A, N, and P domains are colored yellow, red, and blue, the transmembrane helices M1–2, M3–4, M5–6, and M7–10 are purple, green, tan, and gray, and bound Ca²⁺ ions are represented by gray spheres. ATP and its mimics are drawn in stick form with C green, N blue, O red, and P orange. A conserved TGES

sequence is drawn in space-filling form in magenta. In the lower right drawing, the position of the lipid bilayer is indicated by the yellow rectangle. (*b*) Schematic diagram of the conformational changes made by SERCA during its catalytic cycle. The protein components are colored as in Part *a* except that helices M5–10 are all tan and Ca²⁺ ions are represented by green spheres. In addition, protons are represented by gray spheres, and the residues that ligand Ca²⁺ ions in the binding cavity are indicated by red spheres. [Modified from drawings by Poul Nissen, University of Aarhus, Denmark. PDBids 2C88, 3BA6, 3B9B, and 3B9R.]

breakdown (Section 18-3Ce). Moreover, Ca^{2+} is an important activator of oxidative metabolism (Section 22-4).

The use of phosphate as a basic energy currency requires cells to maintain a low internal [Ca^{2+}] because, for example, Ca$_3$(PO$_4$)$_2$ has a maximum aqueous solubility of 65 μM. Thus, the [Ca^{2+}] in the cytosol ($\sim$0.1 μM) is four orders of magnitude less than it is in the extracellular spaces ($\sim$1500 μM). This large concentration gradient is maintained by the active transport of Ca^{2+} across the plasma membrane, the endoplasmic reticulum (ER; SR in muscle), and the mitochondrial inner membrane. We discuss the mitochondrial system in Section 22-1Bb. The plasma membrane and sarco(endo)plasmic reticulum each contain a P-type **Ca^{2+}–ATPase (Ca^{2+} pump)** that actively pumps Ca^{2+} out of the cytosol at the expense of ATP hydrolysis. Their kinetic mechanisms match that of the (Na$^+$–K$^+$)–ATPase (Fig. 20-19) except that two Ca^{2+} ions replace the three Na$^+$ ions, two to three H$^+$ ions replace the three K$^+$ ions, and (*out*) refers to the outside of the cell for plasma membrane Ca^{2+}–ATPase or to the lumen of the sarco(endo)plasmic reticulum for the Ca^{2+}–ATPase of that membrane.

a. X-Ray Structures of the Ca^{2+}–ATPase Suggest Its Mechanism

The X-ray structures of the 994-residue Ca^{2+}–ATPase from rabbit muscle sarcoplasmic reticulum [also known as **SERCA** for *sarco(endo)plasmic reticulum Ca^{2+}-A*TPase] in its complexes with a variety of analogs of ATP and its components have been determined, the first of which were reported by Toyoshima. SERCA, which constitutes 90% of SR membrane protein, is a 140-Å-long monomeric protein that closely resembles the α subunit of the (Na$^+$–K$^+$)–ATPase (Fig. 20-20). It contains a 10-helix transmembrane domain (M), an actuator domain (A), a nucleotide-binding domain (N) to which the ATP binds, and a phosphorylation domain (P) that contains the phosphorylatable Asp 351. The two Ca^{2+} ions are bound in a cavity, similar to that in the (Na$^+$–K$^+$)–ATPase, which is formed, in large part, by the disruption of helices M4 and M6 in this region.

Four of these structures, all determined by Poul Nissen, collectively suggest a mechanism of action of P-type ATPases resulting from a vectorial series of conformational changes driven by ATP binding, hydrolysis, and release (Figs. 20-22):

1. Let us start with the X-ray structure of SERCA in its E2 conformation in complex with the inhibitor **thapsigargin** (**TG**; Fig. 22-23*a*), which stabilizes SERCA's E2 state, and the nonhydrolyzable ATP analog **adenosine-5'-(β,γ-methylene)triphosphate** (**AMPPCP**; Fig. 22-23*b*). This structure provides a model for the E2 · ATP state (Fig. 20-22, *upper left*). The exchange of 2 to 3 protons for 2 cytosolic Ca^{2+} ions causes a rotation of the A domain that occludes the binding cavity. This activates the ATPase catalytic site to phosphorylate Asp 351 yielding ADP, which releases the N domain from the P domain due to the cleavage of the ATP's β,γ-phosphodiester bond that had previously linked these two domains. A model for the resulting Ca$_2$E1~P · ADP state is provided by the X-ray structure of SERCA in its E1 state with Asp 351

(*a*)   **Thapsigargin (TG)**

(*b*)   **Adenosine-5'-(β,γ-methylene)triphosphate (AMPPCP)**

(*c*)   **Adenosine-5'-(β-amino)diphosphate (AMPPN)**

Figure 20-23 **Molecular formulas of several SERCA inhibitors.** (*a*) Thapsigargin (TG), a product of the plant *Thapsia garganica*, (*b*) AMPPCP, and (*c*) AMPPN.

phosphorylated [via a "high energy" bond, which is represented by a squiggle ($\sim$)] and in complex with the ADP analog **adenosine-5'-(β-amino)diphosphate** (**AMPPN**; Fig. 22-23*c*).

2. The A domain rotates toward the phosphorylation site, thereby contacting the P and N domains. This movement pushes down on the M3–4 segment and pulls up on the M1–2 segment, thereby separating the lower portions of these segments from that of the M5–6 segment. This exposes the Ca^{2+}-binding cavity to the lumen of the SR and forces apart the residues that had bound the Ca^{2+} ions in the cavity, resulting in the release of these ions into the lumen. A model for this E2—P state is provided by the X-ray structure of SERCA in its E2 state in which BeF$_3^-$ is covalently linked to Asp 351 to form a mimic of a phosphate group.

3. Two or three protons bind in the lumenally exposed binding cavity. This, together with the exchange of the ADP bound to the N domain for ATP results in movements of transmembrane helices that again occludes the cavity from the lumen. A model for the resulting E2—P* · ATP state [where the star (*) is indicative of a transition state] is provided by the X-ray structure of SERCA in its E2 state in complex with AMPPCP and AlF_4^-, which is covalently linked to Asp 351 to form a trigonal bipyramidal mimic of the hydrolytic transition state (Fig. 16-6*b*).

4. The dephosphorylation of Asp 351, as stimulated by the previous binding of ATP, motivates conformational changes that opens a channel to the cytosol that permits the exchange of protons with Ca^{2+} ions, thereby completing the catalytic cycle.

This mechanism explains how P-type ATPases transport cations against their steep electrochemical gradients.

b. Calmodulin Regulates the Plasma Membrane Ca^{2+} Pump

For a cell to maintain its proper physiological state, it must regulate the activities of its ion pumps precisely. *The regulation of the Ca^{2+} pump in the plasma membrane is controlled by the level of Ca^{2+} through the mediation of calmodulin (CaM).* This ubiquitous eukaryotic Ca^{2+}-binding protein participates in numerous cellular regulatory processes including, as we have seen, the control of glycogen metabolism (Section 18-3Ce).

Ca^{2+}–Calmodulin activates the Ca^{2+}–ATPase of plasma membranes. The activation, as deduced from the study of the isolated ATPase, results in a decrease in its K_M for Ca^{2+} from 20 to 0.5 μM. Ca^{2+}–CaM activates the Ca^{2+} pump by binding to an inhibitory polypeptide segment of the pump in a manner similar to the way in which Ca^{2+}–CaM activates its target protein kinases (Section 18-3Cf). Evidence supporting this mechanism comes from proteolytically excising the Ca^{2+} pump's CaM-binding polypeptide, yielding a truncated pump that is active even in the absence of CaM. Synthetic peptides corresponding to this CaM-binding segment not only bind Ca^{2+}–CaM but inhibit the truncated pump by increasing its K_M for Ca^{2+} and decreasing its V_{max}. This suggests that, in the absence of Ca^{2+}–CaM, the CaM-binding segment of the pump interacts with the rest of the protein so as to inhibit its activity. When the Ca^{2+} concentration increases, Ca^{2+}–CaM forms and binds to the CaM-binding segment of the pump in a way that causes it to dissociate from the rest of the pump, thereby relieving the inhibition.

Now we can see how Ca^{2+} regulates its own cytoplasmic concentration: At Ca^{2+} levels below calmodulin's ~1 μM dissociation constant for Ca^{2+}, the Ca^{2+}–ATPase is relatively inactive due to autoinhibition by its CaM-binding segment. If, however, the $[Ca^{2+}]$ rises to this level, Ca^{2+} binds to calmodulin, which, in turn, binds to the CaM-binding segment so as to relieve the inhibition, thereby activating the Ca^{2+} pump:

$$Ca^{2+} + CaM \rightleftharpoons Ca^{2+}\text{–}CaM^* + \text{pump}(\textit{inactive}) \rightleftharpoons$$
$$Ca^{2+}\text{–}CaM^* \cdot \text{pump}(\textit{active})$$

(CaM* indicates activated calmodulin). This interaction decreases the pump's K_M for Ca^{2+} to below the ambient $[Ca^{2+}]$, thereby causing Ca^{2+} to be pumped out of the cytosol. When the $[Ca^{2+}]$ decreases sufficiently, Ca^{2+} dissociates from calmodulin and this series of events reverses itself, thereby inactivating the pump. The entire system is therefore analogous to a basement sump pump that is automatically activated by a float when the water reaches a preset level.

C. *(H⁺–K⁺)–ATPase of Gastric Mucosa*

Parietal cells of the mammalian gastric mucosa secrete HCl at a concentration of $0.15M$ (pH 0.8). Since the cytosolic pH of these cells is 7.4, this represents a pH difference of 6.6 units, the largest known in eukaryotic cells. The secreted protons are derived from the intracellular hydration of CO_2 by carbonic anhydrase:

$$CO_2 + H_2O \rightleftharpoons HCO_3^- + H^+$$

The secretion of H^+ involves the participation of an **(H⁺–K⁺)–ATPase,** an electroneutral antiport with structure and properties similar to that of Ca^{2+}–ATPase. Like other P-type ATPases, it is phosphorylated during the transport process. In this case, however, the K^+, which enters the cell as H^+ is pumped out, is subsequently externalized by its electroneutral cotransport with Cl^-. HCl is therefore the overall transported product.

a. Cimetidine and Omeprazole Prevent Gastric Ulcers and Heartburn

For many years, effective treatment of **peptic ulcers,** which was a frequently fatal condition caused by the attack of stomach acid on the gastric mucosa, often required surgical removal of the affected portions of the stomach or sometimes the entire stomach. The discovery, by James Black, of **cimetidine,**

Cimetidine

Histamine

which inhibits stomach acid secretion, all but eliminated the need for this dangerous and debilitating surgery. The (H⁺–K⁺)–ATPase of the gastric mucosa is activated by histamine stimulation of the **H₂-receptor** in a process mediated by cAMP. Cimetidine (trade name **Tagamet**), which became available in 1976, was the first of several widely used drugs that competitively inhibit the binding of histamine to this receptor. These drugs likewise alleviate the symptoms of **gastroesophageal reflux disease (GERD,** commonly known as **heartburn),** which is caused by the reguritation of stomach acid into the esophagus, a widespread and painful condition that, when chronic, can damage the esophagus and cause esophageal cancer. Heartburn can also be relieved by antacids but the effects of cimetidine and its analogs are longer lasting (6–10 hours vs 1–2 hours for antacids, although the former require ~30 minutes to take effect) and can be taken before meals to prevent heartburn from occurring. A drawback of most of these substances is that they inhibit several cytochrome P450s and thereby interfere with the metabolism of numerous widely used drugs (Section 15-4Bc).

Omeprazole (trade names **Prilosec** and **Nexium**),

Omeprazole

directly inhibits the $(H^+–K^+)$–ATPase by forming an adduct with the side chain of its Cys 831. Since its introduction in 1989, omeprazole has greatly supplanted the use of cimetidine and its analogs due to the irreversible nature of its inhibition (which reduces acid secretion by up to 99%) and its lack of drug–drug interactions. Hence, omeprazole is presently one of the most frequently used drugs worldwide.

D. *Group Translocation*

Group translocation is a variation of ATP-driven active transport that most bacteria use to import certain sugars. It is required for many bacterial processes, both useful and harmful (to humans), such as those that produce cheese, soy sauce, and dental cavities. *It differs from active transport in that the molecules transported are simultaneously modified chemically.* The most extensively studied example of group translocation is the **phosphoenolpyruvate-dependent phosphotransferase system (PTS)** of *E. coli* discovered by Saul Roseman in 1964. Phosphoenolpyruvate (PEP) is the phosphoryl donor for this system (recall that PEP is the "high-

Table 20-2 **Some of the Sugars Transported by the *E. coli* PEP-Dependent Phosphotransferase System (PTS)**

Glucose	Galactitol
Fructose	Mannitol
Mannose	Sorbitol
N-Acetylglucosamine	Xylitol

energy" phosphoryl donor for ATP synthesis in the pyruvate kinase reaction of glycolysis; Section 17-2J). *The PTS simultaneously transports and phosphorylates sugars. Since the cell membrane is impermeable to sugar phosphates, once they enter the cell, they remain there.* Some of the PTS-transported sugars are listed in Table 20-2.

The PTS system involves two soluble cytoplasmic proteins, **Enzyme I (EI)** and **HPr** (for *h*istidine-containing *p*hospho*c*arrier *p*rotein), which participate in the transport of all sugars (Fig. 20-24). In addition, for each sugar the system transports, there is a specific transmembrane transport protein **EII**, which consists of at least three functional components: two that are cytoplasmic, **EIIA** and **EIIB**, and a transmembrane channel, **EIIC.** These three components associate differently in different EII's. In *E. coli*, for example, EIIA, EIIB, and EIIC are separate subunits in cellobiose-specific EII; EIIB and EIIC are covalently linked and EIIA is separate in glucose-specific EII; and all three components are present on a single peptide in mannitol-specific EII.

Glucose transport, which resembles that of other sugars, involves the transfer of a phosphoryl group from PEP to glucose with net inversion of configuration about the phosphorus atom. Since each phosphoryl transfer involves inversion (Section 16-2A), an odd number of transfers must be

Figure 20-24 **Transport of glucose by the PEP-dependent phosphotransferase system (PTS).** HPr and EI are cytoplasmic proteins common to all sugars transported. EIIA^{glc} and EIIBC^{glc} are proteins specific for glucose. EIIA^{glc} inhibits non-PTS transport proteins such as the lactose permease (Section 20-4B) and enzymes such as glycerol kinase. Adenylate cyclase is activated by the presence of EIIA^{glc} ~P (or possibly inhibited by the presence of EIIA^{glc}).

involved. Four phosphorylated protein intermediates have been identified, indicative of five phosphoryl transfers:

$$PEP \rightarrow EI \rightarrow HPr \rightarrow EIIA^{glc} \rightarrow EIIBC^{glc} \rightarrow glucose$$

The transport process occurs as follows (Fig. 20-24):

1. PEP phosphorylates EI at N3 (N_ε) of His 189 to form a reactive phosphohistidine adduct.

Phosphohistidine residue

2. The phosphoryl group is transferred to N1 (N_δ) of His 15 on HPr. His is apparently a favored phosphoryl group acceptor in phosphoryl-transfer reactions. It also participates in the phosphoglycerate mutase reaction of glycolysis (Section 17-2Ha).

3. HPr~P continues the phosphoryl-transfer chain by phosphorylating EIIA^{glc} at N3 of His 90.

4. The fourth phosphoryl transfer is to Cys 421 of EIIB^{glc}.

5. The phosphoryl group is finally transferred from EIIB^{glc} to glucose, which, in the process, is transported across the membrane by EIIC^{glc}. Glucose is released into the cytoplasm only after it has been phosphorylated to glucose-6-phosphate (G6P).

Thus the transport of glucose is driven by its indirect, exergonic phosphorylation by PEP. The PTS is an energy-efficient system since only one ATP equivalent is required to both transport and phosphorylate glucose. When the active transport and phosphorylation steps occur separately, as they do in many cells, two ATPs are hydrolyzed per glucose processed.

a. Bacterial Sugar Transport Is Genetically Regulated

The PTS is more complex than the other transport systems we have encountered, probably because it is part of a complicated regulatory system governing sugar transport. When any of the sugars transported by the PTS is abundant, the active transport of sugars, which enter the cell via other transport systems, is inhibited. This inhibition, called **catabolite repression,** is mediated through the cAMP concentration (Section 31-3C). cAMP activates the transcription of genes that encode various sugar transport proteins, including **lactose permease** (Section 20-4B). The presence of glucose results in a decrease in [cAMP], which, in turn, represses the synthesis of these other sugar transport proteins. Direct inhibition of the sugar transport proteins themselves, as well as of certain enzymes, also occurs.

The mechanism for control of [cAMP] is thought to reside in EIIA^{glc}, which is transiently phosphorylated in Step 3 of the PTS transport process (Fig. 20-24). When glucose is plentiful, this enzyme is present mostly in its dephospho form since

EIIA^{glc}~P rapidly transfers its phosphoryl group through EIIBC^{glc} to glucose. Under these conditions, adenylate cyclase is inactive, although whether dephospho EIIA^{glc} inhibits this enzyme or EIIA^{glc}~P activates it is unclear. However, dephospho EIIA^{glc} binds to and inhibits many non-PTS transporters and enzymes that participate in the metabolism of sugars other than glucose (the metabolite of choice for many bacteria), including lactose permease and **glycerol kinase** (Section 17-5A). In the absence of glucose, EIIA^{glc} is converted to EIIA^{glc}~P, thereby relieving the inhibition of non-PTS transporters. In addition, adenylate cyclase is activated to produce cAMP, which, in turn, induces the increased production of some of the non-PTS transporters and enzymes that EIIA^{glc} inhibits. This is a form of energy conservation for the cell. Why synthesize the proteins required for the transport and metabolism of all sugars when the metabolism of only one sugar at a time will do?

b. The X-Ray Structure of EIIA^{glc} in Complex with Glycerol Kinase

The X-ray structures of EIIA^{glc}, both alone and in complex with one of its regulatory targets, glycerol kinase, which were determined by James Remington and Roseman, have revealed how EIIA^{glc} inhibits at least some of its targets and why EIIA^{glc}~P does not do so. EIIA^{glc} contains two His residues, His 75 and His 90, that are required for phosphoryl transfer, although only His 90 is necessary for EIIA^{glc} to accept a phosphate from HPr. The X-ray structure of *E. coli* EIIA^{glc} alone reveals that these two His residues lie in close proximity (their N3 atoms are 3.3 Å apart) in a depression on the surface of the protein that is surrounded by a remarkable ~18-Å-diameter hydrophobic ring consisting of 11 Phe, Val, and Ile side chains.

The X-ray structure of EIIA^{glc} in complex with glycerol kinase (Fig. 20-25) confirms that this hydrophobic gasket is indeed the site of interaction between the two proteins and reveals how the phosphorylation of His 90 disrupts this interaction. The two active site His residues, which are completely buried within the hydrophobic interaction surface, coordinate a previously unanticipated Zn^{2+} ion, which is additionally coordinated to Glu 478 of glycerol kinase and a water molecule. The phosphorylation of EIIA^{glc} His 90 to yield EIIA^{glc}~P no doubt disrupts this intermolecular interaction, thereby releasing glycerol kinase and reversing its inhibition.

E. ABC Transporters

The **ABC transporters,** which widely occur in all kingdoms of life, are named for their highly conserved, ~100-residue *A*TP-*b*inding *c*assette. They form a large superfamily of transmembrane proteins that collectively transport a wide variety of substances across membranes, including ions, sugars, peptides, lipids, metabolites, and numerous toxins and drugs. For example, the acquired resistance of cancer cells to chemotherapeutic agents is often due to the selection of cells that overexpress the ABC transporter named **multidrug resistance (MDR) transporter** (also called **P-glycoprotein**), which pumps a wide assortment of amphiphilic substances—including many drugs—out of the cell. Similar proteins in bacteria are, in many cases, responsible for their antibiotic resistance. The

Figure 20-25 **X-ray structure of *E. coli* EIIA^{glc} (*yellow*, a 168-residue monomer) in complex with one of its regulatory targets, glycerol kinase (*blue*, a tetramer of identical 501-residue subunits).** The two proteins associate, in part, by tetrahedrally coordinating a Zn^{2+} ion via the side chains of His 75 and His 90 of EIIA^{glc}, a carboxylate oxygen from Glu 478 of glycerol kinase, and a water molecule. These groups are shown in ball-and-stick form with C gray, N blue, O red, and Zn^{2+} white. The Zn^{2+}-mediated interaction between EIIA^{glc} and glycerol kinase inactivates glycerol kinase, presumably through an induced-fit mechanism. The phosphorylation of EIIA^{glc} at His 90 disrupts this interaction, thereby reversing the inhibition of glycerol kinase. [Courtesy of James Remington, University of Oregon. PDBid 1GLA.]

human genome encodes 48 ABC transporters and their defects are responsible for several inherited diseases including cystic fibrosis (Section 12-4Bf), Tangier disease (Section 12-5Cd), and **adrenoleukodystrophy** (Section 25-2Fa).

ABC transporters consist minimally of four modules: two highly conserved cytosolic nucleotide-binding domains, and two transmembrane domains that typically contain six transmembrane helices each. In bacteria, the four domains are contained on two or four separate polypeptides. In eukaryotes, a single polypeptide includes all four domains. Bacterial ABC transporters mediate the import as well as the export of a variety of compounds, whereas their eukaryotic counterparts function only as exporters that transport molecules out of the cell or into intracellular compartments such as the ER.

The ABC transporter of gram-negative bacteria named **MsbA,** a close homolog of the MDR transporter, functions to transport lipopolysaccharides (Section 11-3Bc) and their

glycolipid component, **lipid A,** from the cytoplasmic leaflet of the plasma membrane to its periplasmic leaflet; that is, MsbA is a flippase (Section 12-4Aa). X-ray structures of Msb, determined by Geoffrey Chang, reveal that its two identical 582-residue subunits are extensively intertwined (Fig. 20-26). In the nucleotide-free (apo) structure (Fig. 20-26*a*), each subunit contributes four of its six transmembrane helices

(*a*) **Open Apo**

(*b*) **Nucleotide Bound**

Figure 20-26 **X-ray structures of the ABC transporter MsbA.** (*a*) *E. coli* MsbA in the absence of bound nucleotide (Open Apo). (*b*) *S. typhimurium* MsbA in complex with AMPPNP (Nucleotide Bound). On the left, the structures are drawn in ribbon form viewed along the plane of the membrane with the periplasmic space above. One subunit in each homodimer is colored in rainbow order from N-terminus (*blue*) to C-terminus (*red*) and the other subunit is gray. The AMPPNP in Part *b* is drawn in stick form in blue. Transmembrane helices (TM1–6), extracellular loops (EL1–3), and intracellular helices (IH1–2) are labeled. On the upper right, the structures are drawn as semitransparent surface diagrams with one subunit cyan and the other gray, and showing the embedded transmembrane helices (1–6) of the cyan subunit. The horizontal lines delineate the lipid bilayer. On the lower right, the structures are represented by schematic diagrams with one subunit cyan and the other white and with the positions of the transmembrane helices indicated. [Modified from drawings by Geoffrey Chang, The Scripps Research Institute, La Jolla, California. PDBids 3B5W and 3B60.]

(TM1–3 and 6) to one leg of the Λ-shaped homodimer and two (TM4–5) to the other leg. The nucleotide-binding domains (NBDs; globular portions at the end of each leg) are ~50 Å apart. However, in its complex with AMPPNP (Fig. 20-26b), the two legs of the Λ have swung together by ~60° to bring the NBDs into contact, with TM3 and 6 having swung away from TM1–2 to contact TM4–5. This opens a space between the subunits on the periplasmic side of the membrane, which presumably permits a lipopolysaccharide that had bound to MsbA from the cytosolic leaflet of the plasma membrane to be released into its periplasmic leaflet. The unidirectional transport of substrate may be facilitated by the different sizes of the openings, and possibly, by a reduced affinity for substrate by MsbA's ATP-bound form.

4 ION GRADIENT–DRIVEN ACTIVE TRANSPORT

Systems such as the $(Na^+–K^+)$–ATPase discussed above utilize the free energy of ATP hydrolysis to generate electrochemical potential gradients across membranes. Conversely, *the free energy stored in an electrochemical potential gradient may be harnessed to power various endergonic physiological processes.* Indeed, ATP synthesis by mitochondria and chloroplasts is powered by the dissipation of proton gradients generated through electron transport and photosynthesis (Sections 22-3C and 24-2D). In this section we discuss active transport processes that are driven by the dissipation of ion gradients. We consider three examples: intestinal uptake of glucose by the **Na⁺–glucose symport,**

uptake of lactose by *E. coli* **lactose permease,** and the mitochondrial **ATP–ADP translocator.**

A. Na⁺–Glucose Symport

Nutritionally derived glucose is actively concentrated in **brush border cells** of the intestinal epithelium by an Na^+-dependent symport (Fig. 20-27). It is transported from these cells to the circulatory system via GLUT2 (Section 20-2Eb), a passive-mediated glucose uniport located on the capillary side of the cell. Note that *although the immediate energy source for glucose transport from the intestine is the Na^+ gradient, it is really the free energy of ATP hydrolysis that powers this process through the maintenance of the Na^+ gradient by the $(Na^+–K^+)$–ATPase.* Nevertheless, since glucose enhances Na^+ resorption, which in turn osmotically enhances water resorption, glucose (possibly as sucrose), in addition to salt and water, should be fed to individuals suffering from severe salt and water losses resulting from diarrhea (drinking only water or a salt solution is ineffective since they are rapidly excreted from the gastrointestinal tract). It is estimated that the introduction of this simple and inexpensive treatment, called **oral rehydration therapy,** has decreased the annual number of human deaths from severe diarrhea, mostly in children in less-developed countries, from 4.6 to 1.6 million.

a. Active and Passive Glucose Transporters Exhibit Differential Drug Susceptibilities

The two glucose transport systems are inhibited by different drugs:

1. Phlorizin inhibits Na^+-dependent glucose transport.

(a) Small intestine   *(b)* Villus   *(c)* Glucose transport

Figure 20-27   Glucose transport in the intestinal epithelium. The brushlike villi lining the small intestine greatly increase its surface area (*a*), thereby facilitating the absorption of nutrients. The brush border cells from which the villi are formed (*b*) actively concentrate glucose from the intestinal lumen in

symport with Na^+ (*c*), a process that is driven by the $(Na^+–K^+)$–ATPase, which is located on the capillary side of the cell and functions to maintain a low internal $[Na^+]$. The glucose is exported to the bloodstream via GLUT2, a separate passive-mediated uniport system (Section 20-2Eb).

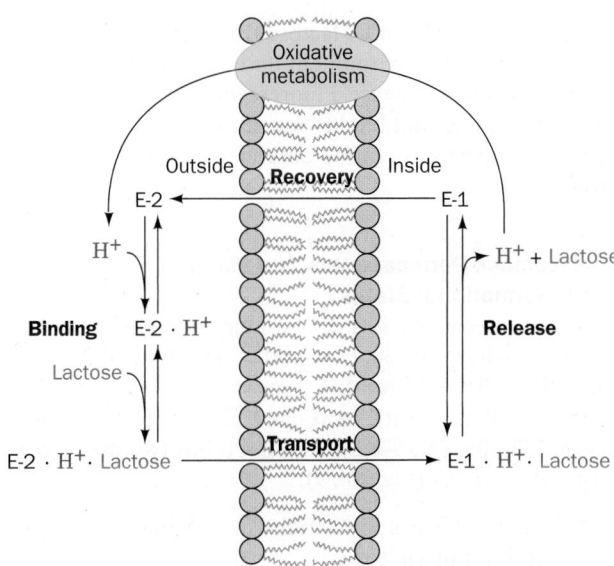

Glc $\equiv$ Glucose

Figure 20-28   The Na⁺–glucose symport system represented as a Random Bi Bi kinetic mechanism. Binding of Na⁺ increases the affinity of the transporter for glucose to such an extent that the upper pathway is heavily favored. T_o and T_i, respectively, represent the transport protein with its binding sites exposed to the outer and inner surfaces of the membrane. [After Crane, R.K. and Dorando, F.C., in Martonosi, A.N. (Ed.), *Membranes and Transport,* Vol. 2, *p.* 154, Plenum Press (1982).]

2. Cytochalasin B inhibits Na⁺-independent glucose transport.

Phlorizin

Cytochalasin B

Phlorizin binds only to the external surface of the Na⁺-dependent glucose transporter, whereas cytochalasin B binds to the cytoplasmic surface of the Na⁺-independent glucose transporter. This further indicates that these proteins are asymmetrically inserted into membranes. The use of these inhibitors permits the actions of the two glucose transporters to be studied separately in intact cells.

Kinetic studies indicate that the Na⁺–glucose symport binds its substrates, Na⁺ and glucose, in random order (Fig. 20-28), although binding of Na⁺ increases the affinity of the transporter for glucose to such an extent that the upper pathway is heavily favored. Only when both substrates are bound, however, does the protein change its conformation to expose the binding sites to the inside of the cell. This requirement for concomitant Na⁺ and glucose transport prevents the wasteful dissipation of the Na⁺ gradient.

B. *Lactose Permease*

Gram-negative bacteria such as *E. coli* contain several active transport systems for concentrating sugars. We have already discussed the PTS system. Another extensively studied system, **lactose permease** (also known as **galactoside permease**), *utilizes the proton gradient across the bacterial cell membrane to cotransport H⁺ and lactose* (Fig. 20-29). The proton gradient is metabolically generated through oxidative metabolism in a manner similar to that in mitochondria (Section 22-3B). The electrochemical potential gradient created by both these systems is used mainly to drive the synthesis of ATP.

Figure 20-29   Kinetic mechanism of lactose permease in *E. coli.* H⁺ binds first to E-2 outside the cell, followed by lactose. They are released in random order from E-1 inside the cell. E-2 must bind both lactose and H⁺ in order to change conformation to E-1, thereby cotransporting these substances into the cell. E-1 changes conformation to E-2 when neither lactose nor H⁺ is bound, thus completing the transport cycle.

How do we know that lactose transport requires the presence of a proton gradient? Ronald Kaback has established the requirement for this gradient through the following observations:

1. The rate of lactose transport into bacteria is increased enormously by the addition of D-lactate, an energy source for transmembrane proton gradient generation. Conversely, inhibitors of oxidative metabolism, such as cyanide, block both the formation of the proton gradient and lactose transport.

2. 2,4-Dinitrophenol, a proton ionophore that dissipates transmembrane proton gradients (Section 22-3D), inhibits lactose transport into both intact bacteria and membrane vesicles.

3. The fluorescence of **dansylaminoethylthiogalactoside,**

Lactose

Dansylaminoethylthiogalactoside

a competitive inhibitor of lactose transport, is sensitive to the polarity of its environment and thus changes when it binds to lactose permease. Fluorescence measurements indicate that it does not bind to membrane vesicles that contain lactose permease in the absence of a transmembrane proton gradient.

a. Lactose Permease Has Two Major Conformational States

Lactose permease is a 417-residue monomer that, like the mammalian glucose transporters (Section 20-2Eb), to which it is distantly related, consists mainly of 12 transmembrane helices with its N- and C-termini in the cytoplasm. Like the $(Na^+–K^+)$–ATPase, it has two major conformational states (Fig. 20-29):

1. E-1, which has a low-affinity lactose binding site facing the interior of the cell.

2. E-2, which has a high-affinity lactose binding site facing the exterior of the cell.

E-1 and E-2 can interconvert only when their H^+ and lactose binding sites are either both filled or both empty. This prevents not only dissipation of the H^+ gradient without cotransport of lactose into the cell, but also transport of

lactose out of the cell without cotransport of H^+ against its concentration gradient.

The X-ray structure of lactose permease in complex with a tight-binding lactose analog, determined by Kaback and So Iwata, reveals that this protein consists of two structurally similar and 2-fold pseudosymmetrically related domains containing six transmembrane helices each (Fig. 20-30a). A large internal hydrophilic cavity is open to

(a)

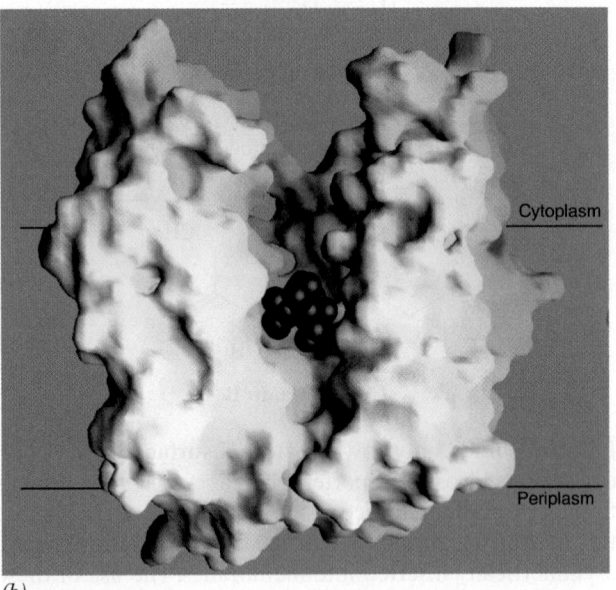

(b)

Figure 20-30 **X-ray structure of lactose permease from *E. coli.*** (a) Ribbon diagram as viewed from the membrane with the cytoplasmic side up. The protein's 12 transmembrane helices are colored in rainbow order with the N-terminus purple and the C-terminus pink. The bound lactose analog is represented by black spheres. (b) Surface model viewed as in Part a but with the two helices closest to the viewer in Part a removed to reveal the lactose-binding cavity. The surface is colored according to its electrostatic potential with positively charged areas blue, negatively charged areas red, and neutral areas white. [Courtesy of H. Ronald Kaback, UCLA. PDBid 1PV7.]

the cytoplasmic side of the membrane (Fig. 20-30*b*) so that the structure represents the E-1 state of the protein. The lactose analog is bound in the cavity at a position that is approximately equidistant from both sides of the membrane, consistent with the model that the lactose binding site is alternately accessible from each side of the membrane (e.g., Fig. 20-10). Arg, His, and Glu residues that mutational studies have implicated in proton translocation are located in the vicinity of the lactose binding site.

C. *ATP–ADP Translocator*

ATP generated in the mitochondrial matrix (its inner compartment; Section 1-2A) through oxidative phosphorylation (Section 22-3C) is largely utilized in the cytosol to drive such endergonic processes as biosynthesis, active transport, and muscle contraction. The inner mitochondrial membrane contains a transmembrane protein that exports ATP out of the matrix and then imports ADP produced in the cytosol by ATP hydrolysis. This antiport, the **ATP–ADP translocator** (also called the **ADP/ATP carrier**), is electrogenic since it exchanges ADP^{3-} for ATP^{4-}. It is the most abundant member of the **mitochondrial carrier family** of proteins that transport a variety of metabolites through the inner mitochondrial membrane; it constitutes ~10% of mitochondrial membrane proteins.

Several natural products inhibit the ATP–ADP translocator. **Atractyloside** (a poison produced by the Mediterranean thistle *Atractylis gummifera* that was known to the ancient Egyptians) and its derivative **carboxyatractyloside (CATR)** inhibit the process only from the external surface of the inner mitochondrial membrane; **bongkrekic acid** (a product of the bacterium *Pseudomonas cocovenenans*) exerts its effects only on the internal surface.

R = H **Atractyloside**
R = COOH **Carboxyatractyloside**

Bongkrekic acid

These differentially acting inhibitors have been valuable tools in the isolation of the ATP–ADP translocator and in the elucidation of its mechanism of action. For example, the translocator has been purified by affinity chromatography (Section 6-3C) using atractyloside derivatives as affinity ligands. Atractyloside binding is also a convenient means of identifying the translocator.

The ATP–ADP translocator, a dimer of identical ~300-residue subunits, has characteristics similar to those of other transport proteins. It has one binding site for which ADP and ATP compete. It has two major conformations, one with its ATP–ADP binding site facing the matrix, and the other with this site facing outward. In its X-ray structure in complex with CATR (Fig. 20-31), determined by Eva Pebay-Peyroula, each subunit's six transmembrane helices surround a deep cone-shaped cavity open to the matrix that is occupied by the CATR. The translocator is an antiport because it must bind ligand to change from one conformational state to the other at a physiologically reasonable rate.

The ATP–ADP translocator is not itself an active transport system. However, its electrogenic export of one negative charge per transport cycle in the direction of ATP export–ADP import is driven by the membrane potential difference, $\Delta\Psi$, across the inner mitochondrial membrane (positive outside). This results in the formation of gradients in ATP and ADP across the membrane.

5 NEUROTRANSMISSION

In higher animals, the most rapid and complex intercellular communications are mediated by nerve impulses. A neuron (nerve cell; e.g., Fig. 1-10*d*) electrically transmits such a signal along its highly extended length (its **axon,** which is commonly over 1 m in larger animals) as a traveling wave of ionic currents. Signal transmission between neurons, as well as between neurons and muscles or glands, is usually chemically mediated by neurotransmitters. In this section we discuss both the electrical and chemical aspects of nerve impulse transmission.

A. *Voltage-Gated Ion Channels*

Ion gradients across cell membranes, as we have seen, are generated by specific energy-driven pumps (Section 20-3). These ion gradients are, in turn, discharged through ion channels such as K^+ channels (Section 20-2F). However, the pumps cannot keep up with the massive fluxes of ions passing through the channels. Hence ion channels are normally shut and only open transiently to perform some specific task for the cell. The opening and closing of ion channels, a process known as **gating,** occurs in response to a variety of stimuli:

1. Ligand-gated channels open in response to extracellular stimuli. For example, the KcsA K^+ channel (Section 20-2F) opens when the extracellular pH is less than ~4 (the X-ray structure in Fig. 20-12 is that of its closed conformation),

(a)

(b)

Figure 20-31 **X-ray structure of bovine heart ATP–ADP translocator in complex with carboxyatractyloside.** (*a*) A subunit of the homodimeric protein, viewed from the plane of the inner membrane with the intermembrane space above, is drawn in ribbon form colored in rainbow order from its N-terminus (*blue*) to its C-terminus (*red*) and embedded in its semitransparent surface diagram. The carboxyatractyloside is drawn in space-filling form with C cyan, O red, and S yellow. Three cardiolipin molecules that cocrystallized with the protein are drawn in stick form with C green, O red, and P orange. (*b*) The molecular surface as viewed from the intermembrane space and colored according to its surface charge with blue positive, white neutral, and red negative. Note the deep positively charged cavity in which the anionic ATP binds. [Based on an X-ray structure by Eva Pebay-Peyroula, Université Joseph Fourier, Grenoble, France. PDBid 2C3E.]

whereas the ligand-gated ion channels in nerve cells open on extracellularly binding specific neurotransmitters (Section 20-5C).

2. Signal-gated channels open on intracellularly binding a second messenger such as Ca^{2+} ion or the $G_{\beta\gamma}$ subunit of a heterotrimeric G protein (Section 19-2C).

3. Mechanosensitive channels open under the influence of stretch, pressure, or displacement. In bacteria, these appear to function as safety valves to relieve high internal osmotic pressures that would otherwise burst the cell wall, whereas in animals they are important in such sensory functions as touch and hearing.

4. Voltage-gated channels open in response to a change in membrane potential. Multicellular organisms contain numerous varieties of voltage-gated channels. For example, nerve impulses arise from the sequential opening of voltage-gated channels along the length of a single nerve cell (Section 20-5Ba).

a. Voltage Gating in Kv Channels Is Triggered by the Motion of a Positively Charged Protein Helix

All voltage-gated K^+ channels are transmembrane homotetramers, each subunit of which contains an ~220-residue N-terminal cytosolic segment, an ~250-residue transmembrane segment consisting of six helices, S1 to S6, and an ~150-residue C-terminal cytosolic segment (Fig. 20-32). S5 and S6 are homologous to the KcsA channel's outer and inner helices (Fig. 20-12*a*), with their intervening so-called P loop containing the K^+ channel's TVGYG signature sequence.

Voltage-gated **Na^+ channels** and **Ca^{2+} channels** are ~2000-residue monomers that consist of four consecutive domains, each of which is homologous to the K^+ channel transmembrane domain, separated by often large cytosolic loops. These domains presumably assume a pseudotetrameric arrangement about a central pore resembling that of the subunits in voltage-gated K^+ channels. This structural homology suggests that voltage-gated ion channels

Figure 20-32 **Secondary structure and membrane orientation of voltage-gated K$^+$ channels.**

share a common architecture in which differences in ion selectivity arise from precise stereochemical variations within the central pore. However, outside of their conserved transmembrane core, voltage-gated ion channels with different ion selectivities are highly divergent. For example, voltage-gated K$^+$ channels, which are known as **Kv channels,** have a conserved ~100-residue domain, the so-called T1 domain (Fig. 20-32), that precedes the transmembrane domain and that is absent in other types of voltage-gated ion channels. The T1 domain confers specificity in subunit oligomerization: It prevents Kv subunits of different subfamilies from coassembling in the same tetramer.

What is the nature of the gating machinery in voltage-gated ion channels? The X-ray structure of the rat brain Kv

channel named **Kv1.2** in its open form, determined by MacKinnon (Fig. 20-33), reveals that its helices S1 through S4 form separate paddlelike domains that extend radially out from the channel-forming helices, S5 and S6. The 19-residue S4 helices, which each contain five positively charged residues (four Arg and one Lys) spaced every ~3 residues, act as voltage sensors. This was shown by covalently linking a dye whose fluorescence spectrum varies with the polarity of the environment to any of several residues in S4. Fluorescence measurements on each of these labeled ion channels revealed that when the membrane potential decreases (cytosol becomes more negative, which causes the channel to close), a stretch of at least 7 residues at the N-terminal end of S4 moves from a position near the extracellular environment to the center of the membrane. This suggests that the five positively charged residues on the S4 helix are drawn toward the cytosol (down in Fig. 20-34a,b) by ~15 Å, pushing down on the S4–S5 linker helix, which in turn pushes down on the S6 helix, thus collapsing together the ends of the S6 helices so as to close the cytosolic entrance to the K$^+$ channel (Fig. 20-34c,d).

b. The Kv Channel Has Two Gates

Electrophysiological measurements indicate that, a few milliseconds after opening, the Kv channel spontaneously (without a further change in membrane potential) closes, a process termed **inactivation,** and does not reopen until after the membrane has repolarized (regained its resting membrane potential). Evidently, *the Kv channel contains two voltage-sensitive gates, one to open the channel on an increase in membrane potential and one to close it a short time*

(a)

(b)

Figure 20-33 **X-ray structure of the Kv1.2 voltage-gated K$^+$ channel.** (a) View along the tetrameric protein's 4-fold axis from the extracellular side of the membrane in which it is embedded. Each of its four identical subunits is colored differently, and the T1 domain has been omitted for clarity. The fragmented appearance of the polypeptide chains is due to the high mobilities of the missing segments. The S5 and S6 helices with their intervening P loops form the pore for K$^+$ ions (delineated by green spheres). Helices S1 to S4 form a separate intramembrane voltage-sensing domain that associates with the S5 and S6

helices of the clockwise adjacent subunit. (b) View perpendicular to that in Part a with the extracellular side of the membrane above. The pore and voltage sensing domains span a distance of ~30 Å, the thickness of the membrane's hydrophobic core. The T1 domain, which occupies the cytoplasm, forms the vestibule of the transmembrane K$^+$ channel. The four large openings between the T1 domain and the K$^+$ channel are the portals through which K$^+$ ions enter the K$^+$ channel. [Based on an X-ray structure by Roderick MacKinnon, Rockefeller University. PDBid 2A79.]

Figure 20-34 **Proposed mechanism of voltage gating in the Kv1.2 voltage-gated K⁺ channel.** (*a*) The voltage-sensing domain in the X-ray structure of Kv1.2 in its open conformation as viewed from the pore parallel to the plane of the membrane with the extracellular solution above. The polypeptide chain is drawn in ribbon form with helices S1 through S4 white and the S4–S5 linker helix orange. Side chains are shown in ball-and-stick form with those of the Arg and Lys residues of S4, the so-called gating charges, cyan, those of Asp and Glu in the so-called external and internal clusters red, and that of Phe 233, which separates the external and internal clusters, green. Note that the gating charges extend upward into the extracellular solution and associate with the external cluster. (*b*) The voltage-sensing domain in the hypothetical closed conformation displayed as in Part *a*. S4, together with S3, has moved downward relative to S1 and S2 toward the now more negatively charged cytosol, thereby pivoting the S4-S5 helix in the direction of the white arrow in Part *a*. Here the gating charges extend downward toward the cytosol and associate with the internal cluster. (*c*) The S4–S6 linkers and the K⁺ pore assembly, which includes helices S5 and S6, in the X-ray structure of Kv1.2 as viewed in the opposite direction from Part *a*. The S6 helices are blue, the S4–S5 linkers, which contact S6, are orange, and the remainder of the pore assembly is green. (*d*) Hypothetical model of the S4–S5 linker in the closed conformation, displayed as in Part *c*, with the structure of the K⁺ pore assembly based on the X-ray structure of the KscA channel (Fig. 20-12*a*). [Courtesy of Roderick MacKinnon, Rockefeller University. PDBid 2R9R.]

later, a phenomenon that has important consequences for the transmission of nerve impulses (Section 20-5B). In fact, minute "gating currents" arising from the movements of these positively charged gates in opening and closing can be detected (electrical current is the movement of charge) if the much larger currents of K⁺ ions through the membrane are first blocked by plugging the Kv channel from its

cytosolic side by high concentrations of Cs⁺ or tetraethylammonium ions (which are too large to pass through the K⁺ pore but apparently become stuck within it).

c. Inactivation Occurs through the Insertion of the Kv Channel's N-Terminal Peptide into Its Central Pore

The inactivation of the Kv channel is abolished by proteolytically excising its N-terminal ~20-residue segment (its inactivation peptide), which NMR studies indicate forms a ball-like structure. However, when chemically synthesized inactivation peptide is injected into the cytosol, the truncated Kv channel inactivates at a rate proportional to the concentration of inactivation peptide. This suggests that inactivation occurs when the inactivation ball swings around at the end of the ~65-residue peptide linking it to the T1 domain so as bind to the open K⁺ pore in a way that physically blocks the passage of K⁺ ions—the so-called ball-and-chain mechanism. Indeed, the time the Kv channel stays open varies with the length of this "chain," as mutationally adjusted.

Where is the inactivation peptide's binding site? Mutational analysis by MacKinnon revealed that the hydrophobic residues lining the Kv channel's internal pore and central cavity form the receptor site for the inactivation peptide. Since the ~6-Å-diameter cytosolic entrance to the internal pore is too narrow to admit the ball, the ball peptide must unfold in order to enter the internal pore. The first 10 residues of ball peptides are predominantly hydrophobic, whereas the succeeding 10 residues are largely hydrophilic and contain several basic residues. It therefore appears that inactivation occurs through the binding of the N-terminus of the fully extended inactivation peptide inside the internal pore via hydrophobic interactions, an association that is augmented by the binding of the basic residues in the ball peptide's C-terminal segment to the acidic residues lining the entrance of the internal pore. Thus, the inactivation peptide acts more like a snake than a ball and chain.

How does the inactivation peptide gain access to the internal pore, which appears to be covered by the T1 tetramer? The passage through the center of the T1 tetramer, as seen in its X-ray structure (Fig. 20-33*b*), is too narrow to permit the passage of the inactivation peptide. Moreover, Christopher Miller eliminated the possibility that the individual T1 domains in a tetramer can ever separate far enough to admit the inactivation peptide by showing that cross-linking adjacent T1 domains by genetically engineered disulfide bonds (whose positions were selected by referring to the T1 X-ray structure) does not significantly affect the Kv channel's gating properties. This strongly suggests that the inactivation peptide gains access to the bottom of the transmembrane pore through lateral windows between the transmembrane and T1 domains, whose sides are formed by the ~35-residue peptide segment linking these domains. Presumably, K⁺ ions pass through these same windows when the Kv channel is open.

A Kv channel engineered so that only one subunit has an inactivation peptide still inactivates but at one-fourth the rate of normal Kv channels. Apparently, any of the normal Kv channel's four inactivation peptides can block the

Table 20-3   Ionic Concentrations and Membrane Permeability Coefficients in Mammals

Ion	Cell (mM)	Blood (mM)	Permeability Coefficient (cm · s⁻¹)
K⁺	139	4	5×10^{-7}
Na⁺	12	145	5×10^{-9}
Cl⁻	4	116	1×10^{-8}
X⁻ᵃ	138	9	0

ᵃX⁻ represents macromolecules that are negatively charged under physiological conditions.

Source: Darnell, J., Lodish, H., and Baltimore, D., *Molecular Cell Biology,* pp. 618 and 725, Scientific American Books (1986).

channel and it is simply a matter of chance as to which one does so. In contrast, voltage-gated Na⁺ channels have only a single inactivation peptide, which is located on the segment linking the Na⁺ channel's third and fourth homologous transmembrane domains. Consequently, a genetically engineered cut of the peptide chain in this region abolishes Na⁺ channel inactivation.

B. *Action Potentials*

Neurons, like other cells, generate ionic gradients across their plasma membranes through the actions of the corresponding ion-specific pumps. In particular, an (Na⁺–K⁺)–ATPase (Section 20-3A) pumps K⁺ into and Na⁺ out of the neuron to yield intracellular and extracellular concentrations of these ions similar to those listed in Table 20-3. The consequent membrane potential, $\Delta\Psi$, across a cell membrane is described by the **Goldman equation,** an extension of Eq. [20.3] that explicitly takes into account the various ions' different membrane permeabilities:

$$\Delta\Psi = \frac{RT}{\mathscr{F}} \ln \frac{\sum P_c[\mathrm{C}(out)] + \sum P_a[\mathrm{A}(in)]}{\sum P_c[\mathrm{C}(in)] + \sum P_a[\mathrm{A}(out)]} \quad [20.8]$$

Here, C and A represent cations and anions, respectively, and, for the sake of simplicity, we have made the physiologically reasonable assumption that only monovalent ions have significant concentrations. The quantities P_c and P_a, the respective **permeability coefficients** for the various cations and anions, are indicative of how readily the corresponding ions traverse the membrane (each is equal to the corresponding ion's diffusion coefficient through the membrane divided by the membrane's thickness; Section 20-2A). Note that Eq. [20.8] reduces to Eq. [20.3] if the permeability coefficients of all mobile ions are assumed to be equal.

Applying Eq. [20.8] to the data in Table 20-3 and assuming a temperature of 25°C yields $\Delta\Psi = -83$ mV (negative inside), which is in good agreement with experimentally measured membrane potentials for mammalian cells. This value is somewhat greater than the K⁺ equilibrium potential, the value of $\Delta\Psi = -91$ mV obtained assuming the membrane is permeable to only K⁺ ions ($P_{Na^+} = P_{Cl^-} = 0$). The membrane potential is generated by a surprisingly small imbalance in the ionic distribution across the membrane: Only ~1 ion pair per million is separated by the membrane with

the anion going to the cytosolic side and the cation going to the external side. The resulting electric field is, nevertheless, enormous by macroscopic standards: Assuming a typical membrane thickness of 50 Å, it is nearly 170,000 V · cm⁻¹.

a. Nerve Impulses Are Propagated by Action Potentials

A nerve impulse consists of a wave of transient membrane depolarization known as an **action potential** *that passes along a nerve cell.* A microelectrode implanted in an axon will record that during the first ~0.5 ms of an action potential, $\Delta\Psi$ increases from its resting potential of around –60 mV to about ~30 mV (Fig. 20-35a). This depolarization is followed by a nearly as rapid repolarization

Figure 20-35   Time course of an action potential. (*a*) The axon membrane undergoes rapid depolarization, followed by a nearly as rapid hyperpolarization and then a slow recovery to its resting potential. (*b*) The depolarization is caused by a transient increase in Na⁺ permeability (conductance), whereas the hyperpolarization results from a more prolonged increase in K⁺ permeability that begins a fraction of a millisecond later. The unit of conductance, 1 mho = 1 ohm⁻¹. [After Hodgkin, A.L. and Huxley, A.F., *J. Physiol.* **117,** 530 (1952).]

past the resting potential to the K^+ equilibrium potential (hyperpolarization) and then a slower recovery to the resting potential. What is the origin of this complicated electrical behavior? In 1952, Alan Hodgkin and Andrew Huxley demonstrated that the action potential results from a transient increase in the membrane's permeability to Na^+ (P_{Na^+}) followed, within a fraction of a millisecond, by a transient increase in its permeability to K^+ (P_{K^+}; Fig. 20-35*b*).

The ion-specific permeability changes that characterize an action potential result from the presence of Na^+- and K^+-specific voltage-gated channels. As a nerve impulse reaches a given patch of nerve cell membrane, the increased membrane potential induces the transient opening of the Na^+ channels, so that Na^+ ions diffuse into the nerve cell at the rate of ~6000 ions · ms^{-1} per channel. This increase in P_{Na^+} causes $\Delta\Psi$ to increase (Eq. [20.3]), which, in turn, induces more Na^+ channels to open, etc., leading to an explosive entry of Na^+ into the cell. Yet, before this process can equilibrate at its Na^+ equilibrium potential of around ~60 mV, the K^+ channels open (P_{K^+} increases) while the Na^+ channels close (inactivate; P_{Na^+} returns to its resting value). $\Delta\Psi$ therefore reverses sign and overshoots its resting potential to approach its K^+ equilibrium value. Eventually the K^+ channels also inactivate and the membrane patch regains its resting potential. The Na^+ channels, which remain open only 0.5 to 1.0 ms, do not reopen until the membrane has returned to its resting state, thereby limiting the axon's firing rate.

An action potential is triggered by an ~20-mV rise in $\Delta\Psi$ to about −40 mV. Action potentials therefore propagate along an axon because the initially rising value of $\Delta\Psi$ in a given patch of axonal membrane triggers the action potential in an adjacent membrane patch that does so in an adjacent membrane patch, etc. (Fig. 20-36). The nerve impulse is thereby continuously amplified so that its signal amplitude remains constant along the length of the axon (in contrast, an electrical impulse traveling down a wire dissipates as a consequence of resistive and capacitive effects). Note, however, that since the relative ion imbalance responsible for the resting membrane potential is small, only a tiny fraction of a nerve cell's Na^+–K^+ gradient is discharged by a single nerve impulse (only one K^+ ion per 3000–300,000 in the cytosol is exchanged for extracellular Na^+, as indicated by measurements with radioactive Na^+). An axon can therefore transmit a nerve impulse every few milliseconds without letup. This capacity to fire rapidly is an essential feature of neuronal communications: *Since nerve impulses all have the same amplitude, the magnitude of a stimulus is conveyed by the rate at which a nerve fires.*

b. The Voltage-Gated Na^+ Channel Is the Target of Numerous Neurotoxins

Neurotoxins have proved to be invaluable tools for dissecting the various mechanistic aspects of neurotransmission. Many neurotoxins, as we shall see, interfere with the action of neuronal voltage-gated Na^+ channels but, curiously,

Figure 20-36   Action potential propagation along an axon. Membrane depolarization at the leading edge of an action potential triggers an action potential at the immediately downstream portion of the axon membrane by inducing the opening of its voltage-gated Na^+ channels. As the depolarization wave moves farther downstream, the Na^+ channels close and the K^+ channels open to hyperpolarize the membrane. After a brief refractory period, during which the K^+ channels close and the hyperpolarized membrane recovers its resting potential, a second impulse can follow. The indicated impulse propagation speed is that measured in the giant axon of the squid, which, because of its extraordinary width (~1 mm), is a favorite experimental subject of neurophysiologists. Note that the action potential in this figure appears backward from that in Fig. 20-35 because this figure shows the distribution of the membrane potential along an axon at an instant in time, whereas Fig. 20-35 shows the membrane potential's variation with time at a fixed point on the axon.

few are known that affect voltage-gated K^+ channels. **Tetrodotoxin,**

Tetrodotoxin

a paralytic poison of enormous potency, which occurs mainly in the skin, ovaries, liver, and intestines of the puffer fish (known as fugu in Japan, where it is a delicacy that may be prepared only by chefs certified for their knowledge of puffer fish anatomy), acts by specifically blocking the Na^+ channel. The Na^+ channel is similarly blocked by **saxitoxin,**

Saxitoxin

a product of marine dinoflagellates (a type of plankton known as the "red tide") that is concentrated by filter-feeding shellfish to such an extent that a small mussel can contain sufficient saxitoxin to kill 50 people. Both of these neurotoxins have a cationic guanidino group, and both are effective only when applied to the external surface of a neuron (their injection into the cytosol elicits no response). It is therefore thought that these toxins specifically interact with an anionic carboxylate group located at the mouth of the Na^+ channel on its extracellular side.
Batrachotoxin,

Batrachotoxin

a steroidal alkaloid secreted by the skin of a Colombian

arrow-poison frog, *Phyllobates aurotaenia,* is the most potent known venom (2 $\mu g \cdot kg^{-1}$ body weight is 50% lethal in mice). This substance also specifically binds to the voltage-gated Na^+ channel but, in contrast to the actions of tetrodotoxin and saxitoxin, renders the axonal membrane highly permeable to Na^+. Indeed, batrachotoxin-induced axonal depolarization is reversed by tetrodotoxin. The observation that the repeated electrical stimulation of a neuron enhances the action of batrachotoxin indicates that this toxin binds to the Na^+ channel in its open state.

Venoms from American scorpions contain families of 60- to 70-residue protein neurotoxins that also act to depolarize neurons by binding to their Na^+ channels (the different neurotoxins in the same venom appear to be specialized for binding to the Na^+ channels in the various species the scorpion is likely to encounter). Scorpion toxins and tetrodotoxin do not, however, compete with each other for binding to the Na^+ channel and therefore must bind at separate sites.

c. Nerve Impulse Velocity Is Increased by Myelination

The axons of the larger vertebrate neurons are sheathed with **myelin,** a biological "electrical insulating tape" that is wrapped about the axon (Fig. 20-37*a*) so as to electrically isolate it from the extracellular medium. Impulses in myelinated nerves propagate with velocities of up to 100 m $\cdot$ s^{-1}, whereas those in unmyelinated nerves are no faster than 10 m $\cdot$ s^{-1} (imagine the coordination difficulties that, say, a giraffe would have if it had to rely on only unmyelinated nerves).

How does myelination increase the velocity of nerve impulses? Myelin sheaths are interrupted every millimeter or so along the axon by narrow unmyelinated gaps known as **nodes of Ranvier** (Fig. 20-37*b*), where the axon contacts the extracellular medium. Binding studies using radioactive tetrodotoxin indicate that the voltage-gated Na^+ channels of unmyelinated axons have rather sparse although uniform distributions in the axonal membrane of $\sim$20 channels $\cdot$ μm^{-2}. In contrast, the Na^+ channels of myelinated axons occur only at the nodes of Ranvier, where they are concentrated with a density of $\sim 10^4$ channels $\cdot$ μm^{-2}. The action potential of a myelinated axon evidently hops between these nodes, a process named **saltatory conduction** (Latin: *saltare,* to jump). Nerve impulse transmission between the nodes must therefore occur by the passive conduction of an ionic current, a mechanism that is inherently much faster than the continuous propagation of an action potential but that is also dissipative. The nodes act as amplification stations to maintain the intensity of the electrical impulse as it travels down the axon. Without the myelin insulation, the electrical impulse would become too attenuated through transmembrane ion leakage and capacitive effects to trigger an action potential at the next node. In fact, **multiple sclerosis,** an autoimmune disease that demyelinates nerve fibers in the brain and spinal cord, results in serious although rarely fatal neurological deficiencies.

Figure 20-37 Myelination. (*a*) An electron micrograph of myelinated nerve fibers in cross section. The myelin sheath surrounding an axon is the plasma membrane of a **Schwann cell,** which, as it spirally grows around an axon, extrudes its cytoplasm from between the layers. The resulting double bilayer, which makes between 10 and 150 turns about the axon, is a good electrical insulator because of its particularly high (79%) lipid content. [Courtesy of Cedric Raine, Albert Einstein College of Medicine of Yeshiva University.] (*b*) A schematic diagram of a myelinated axon in longitudinal section, indicating that in the nodes of Ranvier (the relatively short gaps between adjacent myelinating cells), the axonal membrane is in contact with the extracellular medium. A depolarization generated by an action potential at one node hops, via ionic conduction, down the myelinated axon (*red arrows*), to the neighboring node, where it induces a new action potential. Nerve impulses in myelinated axons are therefore transmitted by saltatory conduction.

(*a*)

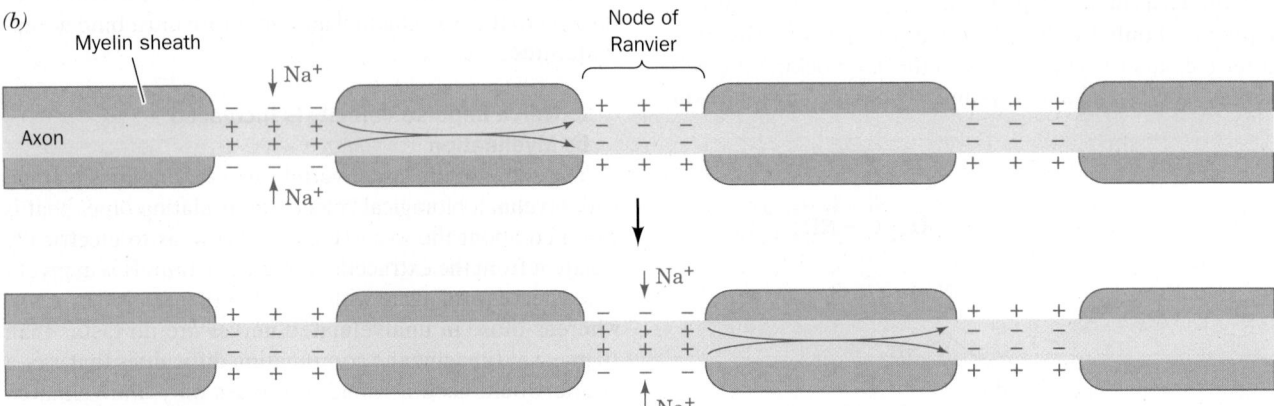

C. *Neurotransmitters and Their Receptors*

The junctions at which neurons pass signals to other neurons, muscles, or glands are called **synapses.** In **electrical synapses,** which are specialized for rapid signal transmission, the cells are separated by a gap, the **synaptic cleft,** of only 20 Å, which is spanned by gap junctions (Section 12-3F). Hence, an action potential arriving at the presynaptic side of the cleft can sufficiently depolarize the postsynaptic membrane to trigger its action potential directly. However, the >200-Å gap of most synapses is too large a distance for such direct electrical coupling. In these **chemical synapses,** the arriving action potential triggers the release from the presynaptic neuron of a specific substance known as a **neurotransmitter,** which diffuses across the cleft and binds to its corresponding receptors on the postsynaptic membrane. In **excitatory synapses,** neurotransmitter binding induces membrane depolarization, thereby triggering an action potential on the postsynaptic membrane. Conversely, neurotransmitter binding in **inhibitory synapses** alters postsynaptic membrane permeability so as to inhibit an action potential and thus attenuate excitatory signals. What is the mechanism through which an arriving action potential stimulates the release of a neurotransmitter, and by what means does its binding to a receptor alter

the postsynaptic membrane's permeability? To answer these questions let us consider the workings of **cholinergic synapses;** that is, synapses that use **acetylcholine (ACh)** as a neurotransmitter:

$$CH_3 - \overset{\overset{\displaystyle O}{\|}}{C} - O - CH_2 - CH_2 - \overset{+}{N}(CH_3)_3$$

Acetylcholine (ACh)

Nicotine

Muscarine

Two types of cholinergic synapses are known:

1. Those containing **nicotinic receptors** (receptors that respond to **nicotine**).

2. Those containing **muscarinic receptors** (receptors that respond to **muscarine,** an alkaloid produced by the poisonous mushroom *Amanita muscaria*).

In what follows, we shall focus on cholinergic synapses containing nicotinic receptors since this best characterized type of synapse occurs at all excitatory neuromuscular junctions in vertebrates and at numerous sites in the nervous system.

a. Electric Organs of Electric Rays Are Rich Sources of Cholinergic Synapses

The study of synaptic function has been greatly facilitated by the discovery that the homogenization of nerve tissue causes its presynaptic endings to pinch off and reseal to form **synaptosomes** (Section 12-4Da). The use of synaptosomes, which can be readily isolated by density gradient ultracentrifugation, has the advantage that they can be manipulated and analyzed without interference by other neuronal components.

The richest known source of cholinergic synapses is the electric organs of the freshwater electric eel *Electrophorus electricus* and saltwater electric rays of the genus *Torpedo*. Electric organs, which these organisms use to stun or kill their prey, consist of stacks of ~5000 thin flat cells called **electroplaques** that begin their development as muscle cells but ultimately lose their contractile apparatus. One side of an electroplaque is richly innervated and has high electrical resistance, whereas its opposite side lacks innervation and has low electrical resistance. Both sides maintain a resting membrane potential of around −90 mV. On neuronal stimulation, all the innervated membranes in a stack of electroplaques simultaneously depolarize to a membrane potential of around ~40 mV, yielding a potential difference across each cell of 130 mV (Fig. 20-38). Since the 5000 electroplaques in a stack are "wired" in series like the batteries in a flashlight, the total potential difference across the stack is ~5000 × 0.130 V = 650 V, enough to kill a human being.

b. Acetylcholine Is Released by the Ca²⁺-Triggered Exocytosis of Synaptic Vesicles

ACh is synthesized near the presynaptic end of a neuron by the transfer of an acetyl group from **acetyl-CoA** [the structure of coenzyme A (CoA) is given in Fig. 21-2] to **choline** in a reaction catalyzed by **choline acetyltransferase.**

$$\underset{\textbf{Acetyl-CoA}}{H_3C-\overset{\overset{\textstyle O}{\|}}{C}-S-CoA} \ + \ \underset{\textbf{Choline}}{HO-CH_2-CH_2-\overset{+}{N}(CH_3)_3}$$

$$\downarrow \text{choline acetyltransferase}$$

$$\underset{\textbf{Acetylcholine}}{H_3C-\overset{\overset{\textstyle O}{\|}}{C}-O-CH_2-CH_2-\overset{+}{N}(CH_3)_3} \ + \ HS-CoA$$

Much of this ACh is sequestered in ~400-Å-diameter membrane-enveloped **synaptic vesicles,** which typically contain ~10^4 ACh molecules each.

*The arrival of an action potential at the presynaptic membrane triggers the opening of **voltage-gated Ca²⁺** channels, which transiently raises the local [Ca²⁺] from its resting level of 0.1 μM to 10 to 100 μM. The resulting influx of extracellular Ca²⁺, in turn, stimulates the exocytosis of the synaptic vesicles in the vicinity of the Ca²⁺ channel so that they release their packets of ACh into the synaptic cleft (Fig. 12-73).* The mechanism by which synaptic vesicles fuse with the presynaptic membrane is discussed in Section 12-4D.

The mechanism through which Ca²⁺ induces synaptic vesicle exocytosis is beginning to come into focus. The major Ca²⁺-sensing protein appears to be **synaptotagmin I,** a protein with a single helix passing through the synaptic vesicle membrane, whose cytosolic domain contains four Ca²⁺ binding sites. At resting levels of Ca²⁺, synaptotagmin I binds to the Q-SNARE syntaxin (Section 12-4Db) so as to block its binding to the R-SNARE synaptobrevin and the Q-SNARE SNAP25, thereby preventing vesicle fusion. However, on binding Ca²⁺, synaptotagmin I releases syntaxin, permitting vesicle fusion to commence.

Once triggered, the fusion of the synaptic vesicles with the presynaptic membrane occurs very rapidly (in <0.3 ms)

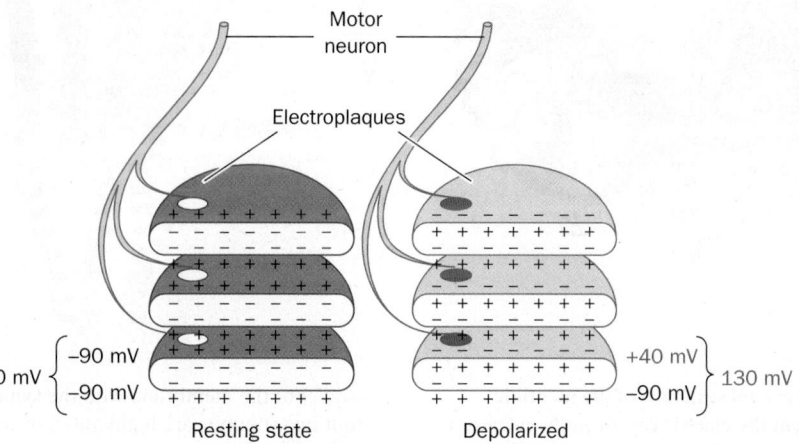

Figure 20-38 **The simultaneous depolarization (*red, right*) of the innervated membranes in a stack of electroplaques "wired" in series results in a large voltage difference between the two ends of the stack.** This is because the total voltage across the stack is the sum of the voltages generated by each of its numerous electroplaques.

because many synaptic vesicles are already docked with the presynaptic membrane. Each Ca^{2+} pulse triggers the exocytosis of ~10% of these docked vesicles. However, they are rapidly replaced because most of the remaining synaptic vesicles are held in reserve in a so-called active zone within 20 nm of the presynaptic membrane. Vesicles are held in the active zone by a fibrous phosphoprotein named **synapsin I** that also binds the cytoskeletal proteins actin and spectrin (Section 12-3Db). Synapsin I is a substrate for calmodulin-dependent protein kinase (Section 18-3Ce), so that a rise in $[Ca^{2+}]$ causes its phosphorylation. This apparently releases the synaptic vesicles from the active zone, thereby permitting them to dock to the presynaptic membrane in preparation for exocytosis. The 20-nm distance between the active zone and the presynaptic membrane may be close enough for Q- and R-SNAREs to initiate the formation of their coiled coil complex (Fig. 12-74), which may thereby facilitate the docking process.

The black widow spider takes advantage of this system: Its highly neurotoxic venom protein, **α-latrotoxin** (130 kD), causes massive release of ACh at the neuromuscular junction, in part by forming homotetrameric transmembrane channels through the presynaptic membrane that act as Ca^{2+} ionophores. In contrast, **botulinus toxin,** as we have seen (Section 12-4Dd), interferes with the exocytosis of synaptic vesicles by proteolytically cleaving specific SNARE proteins, thereby preventing ACh release.

Exocytosed synaptic vesicle proteins are rapidly recovered from the presynaptic membrane via endocytosis mainly in clathrin-coated vesicles (Section 12-4C). However, once the resulting endocytotic vesicles lose their clathrin coats, they do not fuse with endosomes, as is usually the case (Fig. 12-91). Rather, they are immediately refilled with ACh by an H^+–ACh antiport, which is driven by the protons pumped into the vesicle by a V-type ATPase (Section 20-3), and then translocated to the active zone. This rapid recycling of the synaptic vesicles (which takes <1 min) permits neurons to fire continuously at a rate of ~50 times per second.

c. The Acetylcholine Receptor Is a Ligand-Gated Cation Channel

After its release at the presynaptic membrane, acetylcholine quickly diffuses across the synaptic cleft to the postsynaptice membrane, where it binds to the 290-kD **acetylcholine receptor (AChR),** an $\alpha_2\beta\gamma\delta$ transmembrane glycoprotein, whose four different ~490-residue subunits are homologous. *The binding of two ACh molecules to the AChR allosterically induces the opening of a channel through the AChR that permits Na^+ ions to diffuse into the cell at a peak rate of ~30,000 ions per millisecond. The*

(a)　　　　　　　　　(b)

Figure 20-39 **Electron crystal structure of the nicotinic acetylcholine receptor from the electric ray *Torpedo marmorata.*** (*a*) View from the synaptic cleft. The extracellular domains of each type of subunit are differently colored and their remaining portions are gray. The side chains of αTrp 149, which mark the ACh-binding sites, are drawn in stick form (*gold*). (*b*) View parallel to the membrane with the synaptic cleft above. Only the front two subunits are highlighted in color. The horizontal gray bars delineate the position of the postsynaptic membrane. [Courtesy of Nigel Unwin, MRC Laboratory of Molecular Biology, Cambridge, U.K. PDBid 2BG9.]

resulting depolarization of the postsynaptic membrane initiates a new action potential. The AChR is also permeable to K$^+$ and Ca^{2+} ions. However, since the thermodynamic driving forces exerted by the resting membrane potential and the K$^+$ concentration gradient oppose each other and largely balance, relatively few K$^+$ ions are transferred (for Na$^+$ ions, these forces reinforce each other). The extracellular concentration of Ca^{2+} is so much less than that of Na$^+$ that it makes only a negligible contribution to the inward ionic current. After 1 to 2 ms, the ACh spontaneously dissociates from the receptor and the channel closes.

The electron crystallography–based structure (Section 12-3Ab) of the AChR in its closed (unliganded) form, determined by Nigel Unwin, when viewed from the synaptic cleft, is an 80-Å-diameter 5-fold pseudosymmetric assembly whose subunits are arranged in the clockwise order -α-β-δ-α-γ- (Fig. 20-39a). When viewed parallel to the plane of the postsynaptic membrane, it is a 160-Å-long wedge-shaped particle (Fig. 20-39b). Each of its subunits consists of three domains: an N-terminal extracellular assembly that consists mainly of a 10-stranded β barrel; a transmembrane 4-helix down–up–down–up bundle (M1–M4); and an intracellular component composed mainly of a disordered segment extending from M3 followed by a curved helix (MA) that precedes M4. The ACh-binding sites are located between the α–γ and α–δ interfaces of their extracellular domains.

The ACh receptor's most striking structural feature is an ~20-Å-diameter and ~65-Å-long water-filled central channel that extends from the receptor's synaptic entrance to the level of the lipid bilayer. There it forms a more constricted ~30-Å-long pore, constructed from the ring of the five M2 helices, that is blocked near the middle of the bilayer. This blockage, which is presumably the channel's gate, consists of rings of conserved Leu 264 and Val/Ile 268 side chains that project inward from the M2 helices. Unwin has proposed that the binding of ACh to the AChR allosterically induces clockwise rotations (viewed from the synaptic cleft) of its α subunit's M2 helices that results in the opening of the gate.

The intracellular end of the AChR contains an ~20-Å-diameter central cavity that is ~20 Å long and is connected to the cytosol via lateral openings between adjacent MA helices at a level that is ~30 Å below the membrane surface. These openings are largely formed by anionic residues and have widths of ~8 Å, which is comparable to the diameters of K$^+$ and Na$^+$ ions surrounded by their first hydration shells. Thus, they likely serve as filters that prevent the passage of cytoplasmic anions and large cations. Recall that Kv channels have similar lateral openings to the cytoplasm (Section 20-5Ac).

The ACh receptor is the target of some of the most deadly known neurotoxins (death occurs through respiratory arrest), whose use has greatly aided in the elucidation of receptor function. **Histrionicatoxin,** an alkaloid secreted by the skin of the Colombian arrow-poison frog *Dendrobates histrionicus,* and *d*-**tubocurarine,** the active ingredient of the Amazonian arrow-poison **curare** as well as a med-

ically useful paralytic agent, are both ACh antagonists that prevent ACh receptor channel opening:

Histrionicatoxin

d-**Tubocurarine**

Similarly, a family of homologous 7- to 8-kD venom proteins from some of the world's most poisonous snakes, including **α-bungarotoxin** from snakes of the genus *Bungarus,* **erabutoxin** from sea snakes, and **cobratoxin** from cobras, prevent ACh receptor channel opening by binding specifically and all but irreversibly to its α subunits. Indeed, detergent-solubilized ACh receptor has been purified by affinity chromatography on a column containing covalently attached cobra toxin. **Myasthenia gravis** (Greek: *mys,* muscle + *astheneia,* weakness and Latin: *gravis,* serious), an autoimmune disease characterized by muscle weakness, is caused by circulating antibodies that bind to the AChR so as to lead to its destruction or block its binding of ACh.

d. Acetylcholine Is Rapidly Degraded by Acetylcholinesterase

An ACh molecule that participates in the transmission of a given nerve impulse must be degraded in the few milliseconds before the potential arrival of the next nerve impulse. This essential task is accomplished by **acetylcholinesterase (AChE),** a 75-kD fast-acting enzyme that is GPI anchored (Section 12-3B) to the surface of the postsynaptic membrane:

(the turnover number of AChE, $k_{cat} = 14{,}000$ s^{-1}; the enzyme's catalytic efficiency, $k_{cat}/K_M = 1.5 \times 10^8 M^{-1} \cdot$ s^{-1}, is close to the diffusion-controlled limit, so that it is a nearly perfect catalyst; Section 14-2Bb). The resulting choline is taken up by the presynaptic cell via an Na$^+$–choline symport for use in the resynthesis of ACh. The operation of this transporter is similar to that of the Na$^+$–glucose symport of intestinal brush border cells (Section 20-4A).

AChE is a serine esterase; that is, its catalytic mechanism resembles that of serine proteases such as trypsin. These enzymes, as we have seen in Section 15-3Ab, are irreversibly inhibited by alkylphosphofluoridates such as diisopropylphosphofluoridate (DIPF). Indeed, related compounds such as **tabun** and **sarin**

(CH$_3$)$_2$N   O

(CH$_3$)$_2$CH

P

O

CH$_3$—CH$_2$—O   CN

O

P

O

H$_3$C   F

Tabun   **Sarin**

are military nerve gases because their efficient inactivation of human AChE by reaction with the active site Ser causes paralysis stemming from cholinergic nerve impulse blockade and thus death by suffocation. **Succinylcholine,**

O
‖
H$_2$C—C—O—CH$_2$—CH$_2$—$\overset{+}{\text{N}}$(CH$_3$)$_3$
|
H$_2$C—C—O—CH$_2$—CH$_2$—$\overset{+}{\text{N}}$(CH$_3$)$_3$
‖
O

Succinylcholine

which is used as a muscle relaxant during surgery, is an ACh agonist that, although rapidly released by the ACh receptor, is but slowly hydrolyzed by AChE. Succinylcholine therefore produces persistent depolarization of the postsynaptic membrane. Its effects are short-lived, however, because it is rapidly hydrolyzed by the relatively nonspecific liver and plasma enzyme **butyrylcholinesterase.** Certain snake venoms, such as that of the green mamba snake, inactivate AChE, although they do so by binding to a site on AChE distinct from its active site.

e. X-Ray Structure of Acetylcholinesterase

The X-ray structure of the 537-residue AChE from the electric ray *Torpedo californica*, determined by Joel Sussman, Israel Silman, and Michal Harel, confirms that the previously identified Ser 200 and His 440 are members of AChE's catalytic triad. The structure further reveals that the third member of AChE's catalytic triad is Glu 327 rather than an Asp residue, only the second instance of a Glu in this position among the many serine proteases, lipases, and esterases of known structure. AChE's catalytic triad is arranged in what appears to be the mirror image of the catalytic triads in trypsin and subtilisin, for example (Figure 15-20), although, of course, this is not actually the case since all proteins consist of L-amino acid residues.

Figure 20-40   **X-ray structure of *Torpedo californica* acetylcholinesterase in complex with its inhibitor decamethonium ion [(CH$_3$)$_3$N$^+$(CH$_2$)$_{10}$N$^+$(CH$_3$)$_3$].** In this cutaway drawing through the enzyme's active site gorge, the protein's ribbon diagram is embedded in its semitransparent surface diagram, which is tinted blue with the exception of the aromatic residues (Phe, Trp, and Tyr) lining the active site gorge, which are yellow. The decamethonium ion that is bound in the active site gorge is drawn in stick form with C green and N blue. [Based on an X-ray structure by Joel Sussman, Israel Silman, and Michal Harel, The Weizmann Institute of Science, Rehovot, Israel. PDBid 1ACL.]

AChE's catalytic site is near the bottom of a narrow and 20-Å-deep gorge that extends halfway through the protein and widens out near its base (Fig. 20-40). The sides of this so-called active site gorge are lined with the side chains of 14 aromatic residues that comprise 40% of its surface area. Since the side chain O atom of the active site Ser is only 4 Å from the bottom of the gorge, ACh must bind in the gorge with its positively charged trimethylammonium group surrounded by aromatic side chains. This conclusion came as a surprise since it had been understandably expected that the trimethylammonium group would be bound at an anionic site. Perhaps the weak binding provided by the interactions of the trimethylammonium group with the π electrons of the aromatic rings facilitates the rapid diffusion of ACh to the bottom of the gorge, thereby accounting for the enzyme's high turnover number. In fact, model aromatic compounds have been synthesized that also bind quaternary ammonium compounds.

f. Amino Acids and Their Derivatives Function as Neurotransmitters

The mammalian nervous system employs well over 30 substances as neurotransmitters. Some of these substances, such as glycine and glutamate, are amino acids; many others are amino acid decarboxylation products or their derivatives (often referred to as **biogenic amines**). For example, as we shall see in Section 26-4B, **dopamine,** norepinephrine, and epinephrine [which are collectively termed catecholamines because they are derivatives of catechol (1,2-dihydroxybenzene)] are sequentially synthesized from tyrosine, whereas **γ-aminobutyric acid (GABA), histamine,** and **serotonin** are derived from glutamate, histidine, and tryptophan, respectively (Fig. 20-41). Many of these compounds are hormonally active substances that are present in the bloodstream. However, since the brain is largely isolated from the general circulation by a selective filtration system known as the **blood–brain barrier** (Section 15-4Ba), the presence of these substances in the blood has no direct effect on the brain. The use of the same compounds as hormones and neurotransmitters apparently has no physiological significance but, rather, is thought to reflect evolutionary opportunism in adapting already available systems to new roles.

The use of selective staining techniques has established that each of the different neurotransmitters is used in discrete and often highly localized regions of the nervous system. The various neurotransmitters are, nevertheless, not simply functional equivalents of acetylcholine. Rather, many of them have distinctive physiological roles. For example, both GABA and glycine are inhibitory rather than excitatory neurotransmitters. The receptors for these substances are ligand-gated channels that are selectively permeable to Cl^-. Hence, their opening tends to hyperpolarize the membrane (make its membrane potential more negative) rather than depolarize it. A neuron inhibited in this manner must therefore be more intensely depolarized than otherwise to trigger an action potential (note that these neurons respond to more than one type of neurotransmitter). Thus, anion channels are inhibitory, whereas cation channels are excitatory. Ethanol, the oldest and most widely used psychoactive drug, is thought to act by inducing GABA receptors in the brain to open their Cl^- channels.

The subunits of the various neurotransmitter-gated cation channels have 20 to 40% sequence identity, as do those of the anion channels. However, the two families of channel proteins appear to be unrelated. Despite this lack of homology, the sequences of the two types of channels suggest that they have considerable structural similarity.

The actual nature of a neuron's response to a neurotransmitter depends more on the characteristics of the corresponding receptor than on the neurotransmitter's identity. Thus, as we have seen, nicotinic ACh receptors, which trigger the rapid contraction of skeletal muscles, respond to ACh within a few milliseconds by depolarizing the postsynaptic membrane. In contrast, the binding of ACh to muscarinic ACh receptors in heart muscle inhibits muscle contraction over a period of several seconds (several heartbeats). This is accomplished by hyperpolarizing the postsynaptic membrane through the closure of otherwise open K^+ channels. Slow-acting neurotransmitters may act by inducing the formation of a second messenger such as cAMP. In fact, the brain has the highest concentration of cAMP-dependent kinases in the body. The binding of catecholamines to their respective neuronal receptors, through the intermediacy of adenylate cyclase and cAMP, activates protein kinases to phosphorylate ion channels so as to alter the neuron's electrical properties. The ultimate effect of this process can be either excitatory or inhibitory. Thus catecholamines, whether acting as hormones (Section 19-1F) or as neurotransmitters, have similar mechanisms of receptor activation.

g. Neuropeptides Are Neurotransmitters

A large and growing list of hormonally active polypeptides known as **neuropeptides** also act as neurotransmitters. Not surprisingly, perhaps, the opioid peptides β-endorphin, met-enkephalin, and leu-enkephalin (Section 19-1K), as well as the hypothalamic releasing factors TRF, GnRF, and somatostatin (Section 19-1H), are in this category. What is less expected is that several gastrointestinal polypeptides, including the hormones gastrin, secretin, and cholecystokinin

Figure 20-41 A selection of neurotransmitters.

(CCK; Section 19-1C), may also act as neurotransmitters in discrete regions of the brain, as do the pituitary hormones oxytocin and vasopressin (Section 19-1Ha). Such neuropeptides differ from the simpler neurotransmitters in that they seem to elicit complex behavior patterns. For example, intracranially injecting rats with a nanogram of vasopressin greatly enhances their ability to learn and remember new tasks. Similarly, injecting a male or a female rat with GnRF evokes the respective postures they require for copulation. Just how these neuropeptides operate is but one of the many enigmas of brain function and organization.

CHAPTER SUMMARY

1 Thermodynamics of Transport Polar molecules and ions are transported across biological membranes by specific transmembrane transport proteins. The free energy change of the species transported depends on the ratio of its concentrations on the two sides of the membrane and, if the species is charged, on the membrane potential, $\Delta\Psi$.

2 Kinetics and Mechanisms of Transport The rate of nonmediated diffusion across a membrane is a linear function of the difference in concentration of the species on the two sides of the membrane as governed by Fick's first law of diffusion. Mediated transport is characterized by rapid saturation kinetics and specificity for the substance transported. It is also subject to competitive inhibition and chemical inactivation. Ionophores transport ions through membranes. Carrier ionophores, such as valinomycin, do so by wrapping a specific ion in a hydrophobic, membrane-soluble coat that can freely diffuse through the membrane. Maltoporin is specific for the passage of maltodextrins because its transport channel matches their left-handed helical shape and is lined with aromatic side chains that form a so-called greasy slide. Glucose transport across erythrocyte membranes is mediated by dimeric transmembrane glycoproteins that can assume two conformations: one with a glucose binding site facing the external cell surface and the other with the glucose site facing the cytosol. Transport occurs by the binding of glucose to the protein on one face of the membrane, followed by a conformational change that closes this site and exposes the other (a gated pore). GLUT4 is stored in specialized vesicles, which on insulin stimulation fuse with the plasma membrane.

The KcsA K^+ channel, a transmembrane homotetramer, permits the rapid passage of K^+ ions, for which it is highly selective. It does so, in part, because it forms an aqueous cavity surrounded by the negative ends of helix dipoles that stabilizes K^+ ions in the middle of the bilayer. K^+ ions, but not the smaller Na^+ ions, are transported because the K^+ channel's selectivity filter selectively coordinates K^+ ions by rings of O atoms in a way that allows their dehydration, passage, and subsequent hydration without significant activation barriers. ClC Cl^- channels are TM homodimers, each subunit of which contains an hourglass-shaped channel. Aquaporins are TM homotetramers with an hourglass-shaped channel through each subunit. They facilitate the rapid passage of water molecules through membranes but inhibit the passage of protons via proton jumping by preventing the reorientation of a centrally located water molecule.

3 ATP-Driven Active Transport Active transport of molecules or ions against a concentration gradient requires an input of free energy. The free energy of ATP hydrolysis is coupled to the transport of three Na^+ ions out of and two K^+ ions into the cell by the $(Na^+–K^+)$–ATPase. This electrogenic process involves phosphorylation of an Asp residue (by ATP) in the presence of Na^+ and its dephosphorylation (hydrolysis) in the presence of K^+. Phosphorylation and dephosphorylation are accompanied by conformational changes that ensure rapid interconversion of all intermediates along the transport pathway. The X-ray structure of the $(Na^+–K^+)$–ATPase reveals that it has a 10-helix transmembrane domain that binds 3 Na^+ ions or 2 K^+ ions near its center and three well-separated cytoplasmic domains. The X-ray structures of four different conformational states of the sarcoplasmic reticulum Ca^{2+}–ATPase (SERCA) indicate its transport mechanism and that of P-type ATPases in general. Plasma membrane Ca^{2+}–ATPase is activated by Ca^{2+}–calmodulin. The $(H^+–K^+)$–ATPase in the stomach is indirectly inhibited by cimetidine and its analogs, and directly inhibited by omeprazole. Bacteria transport sugars by group translocation, a process in which the transported substance is chemically modified. The PTS system, which has important regulatory functions, phosphorylates sugars as they are transported by utilizing phosphoenolpyruvate as a phosphoryl donor. ABC transporters use ATP to power conformational changes that transport a wide variety of substances across membranes.

4 Ion Gradient–Driven Active Transport Active transport may be driven by the free energy stored in ion gradients (secondary active transport). Glucose is transported into intestinal epithelial cells against its concentration gradient by an Na^+–glucose symport. This process is ultimately powered by the free energy of ATP hydrolysis since the Na^+ gradient is constantly being replenished via the $(Na^+–K^+)$–ATPase. The system conforms to a Random Bi Bi kinetic mechanism, implying that both Na^+ and glucose must be bound for the transport-producing conformational change to occur. Lactose is transported into *E. coli* by lactose permease, an H^+–lactose symport. This process is driven by the cell's electrochemical H^+ gradient, which is, in turn, maintained by a proton pump coupled with oxidative metabolism. The mitochondrial ATP–ADP translocator also interacts with the membrane potential in the asymmetric transport of ATP out of and ADP into the mitochondrion.

5 Neurotransmission Voltage-gated cation channels such as Kv channels open in response to the membrane potential and close a short time later through the action of a second gate that functions via a modified "ball-and-chain" mechanism. Nerve impulses are traveling waves of electrical excitation along axon plasma membranes known as action potentials that are generated by the transient opening of voltage-gated Na^+ channels to let Na^+ ions into the cell followed a short time later by the transient opening of voltage-gated K^+ channels to let K^+ ions out of the cell. Nerve impulses are chemically transmitted across most synapses by the release of neurotransmitters. Acetylcholine (ACh), the best characterized neurotransmitter, is packaged in synaptic vesicles that are exocytotically released into the synaptic cleft. This process is triggered by an increase in cytosolic $[Ca^{2+}]$ resulting from the arriving action potential's opening of voltage-gated Ca^{2+} channels. The ACh diffuses across the synaptic cleft, where it binds to the ACh receptor, a transmembrane cation channel that opens in response to ACh binding. The resultant flow of Na^+ into and K^+ out of the post-synaptic cell depolarizes the postsynaptic membrane, which, if sufficient neurotransmitter has been released, triggers a postsynaptic action potential. The ACh receptor is the target of numerous deadly neurotoxins, including histrionicatoxin, *d*-tubocurarine, and cobra toxin, which all bind to the ACh receptor so as to prevent its opening. The ACh is rapidly degraded, before the possible arrival of the next nerve impulse, through the action of acetylcholinesterase, a fast-acting serine esterase that has an unusual aromatic side chain–lined active site gorge. Nerve gases and succinylcholine inhibit acetyl-cholinesterase and therefore block nerve impulse transmission at cholinergic synapses.

Many specific regions of the nervous system employ neurotransmitters other than ACh. Most of these neurotransmitters are amino acids, such as glycine and glutamate, or their decarboxylation products and their derivatives, including catecholamines, GABA, histamine, and serotonin. Many of these compounds are also hormonally active, but they are excluded from the brain by the blood–brain barrier. Although many neurotransmitters, such as ACh, are excitatory, others are inhibitory. The latter stimulate the opening of anion (Cl^-) channels, thereby causing the postsynaptic membrane to become hyperpolarized, so that it must be more highly depolarized than otherwise to trigger an outgoing action potential. There is also a growing list of polypeptide neurotransmitters, many of which are also polypeptide hormones, that elicit complex behavior patterns.

REFERENCES

General

Alberts, B., Johnson, A., Lewis, J., Raff, M., Roberts, K., and Walter, P., *The Molecular Biology of the Cell* (5th ed.), Chapter 11, Garland Science (2008).

Ashcroft, F.M., From molecule to malady, *Nature* **440,** 440–447 (2006). [Reviews ion channels and their defects.]

Busch, W. and Saier, M.H., Jr., The transporter classification (TC) system, 2002, *Crit. Rev. Biochem. Mol. Biol.* **37,** 287–337 (2002). [Summarizes the classification of the nearly 400 families of transport systems and their distribution among the three domains of life.]

Gadsby, D.C., Ion channels versus ion pumps: the principal difference, in principle, *Nature Struct. Mol. Biol.* **10,** 344–352 (2009).

Gouaux, E. and MacKinnon, R., Principles of selective ion pump transport in channels and pumps, *Science* **310,** 1461–1465 (2005). [Compares several transport proteins of known structure and discusses the selectivity of Na^+, K^+, Ca^{2+}, and Cl^- transport.]

Lodish, H., Berk, A., Kaiser, C.A., Krieger, M., Scott, M.P., Bretscher, A., Ploegh, H., and Matsudaira, P., *Molecular Cell Biology* (6th ed.), Chapters 11 and 23, Freeman (2008).

Kinetics and Mechanism of Transport

Clapham, D.E., Unlocking family secrets: K^+ channel transmembrane domains, *Cell* **97,** 547–550 (1999).

Dutzler, R., Wang, Y.-F., Rizkallah, P.J., Rosenbusch, J.P., and Schirmer, T., Crystal structures of various maltooligosaccharides reveal a specific sugar translocation pathway, *Structure* **4,** 127–134 (1996); *and* Dutzler, R., Schirmer, T., Karplus, M., and Fischer, S., Translation mechanism of long sugar chains across the maltoporin membrane channel, *Structure* **10,** 1273–1284 (2002).

Glucose Transport

Barnett, J.E.G., Holman, G.D., Chalkley, R.A., and Munday, K.A., Evidence for two asymmetric conformational states in the human erythrocyte sugar transport system, *Biochem. J.* **145,** 417–429 (1975).

Huang, S. and Czech, M.P., The GLUT4 glucose transporter, *Cell. Metab.* **5,** 237–252 (2007).

Watson, R.T. and Pessin, J.E., GLUT4 translocation: The last 200 nanometers, *Cell. Signal.* **19,** 2209–2217 (2007); *and* Bridging the GAP between insulin signaling and GLUT4 translocation, *Trends Biochem. Sci.* **31,** 215–222 (2006).

K^+ Channels

Roux, B., Ion conduction and selectivity in K^+ channels, *Annu. Rev. Biophys. Biomol. Struct.* **34,** 153–171 (2005).

Zhou, Y., Morais-Cabral, J.H., Kaufman, A., and MacKinnon, R., Chemistry of ion coordination and hydration revealed by a K^+ channel–Fab complex at 2.0 Å resolution, *Nature* **414,** 43–48 (2001); *and* Doyle, D.A., Morais-Cabral, J.M., Pfuetzner, R.A., Kuo, A., Gulbis, J.M., Cohen, S.L., Chait, B.T., and MacKinnon, R., The structure of the potassium channel: Molecular basis of K^+ conduction and selectivity, *Science* **280,** 69–77 (1998). [High and medium resolution X-ray structures of the KcsA channel.]

Cl^- Channels

Dutzler, R., The ClC family of chloride channels and transporters, *Curr. Opin. Struct. Biol.* **16,** 439–446 (2006).

Dutzler, R., Campbell, E.B., Cadene, M., Chait, B.T., and MacKinnon, R., X-ray structure of a ClC chloride channel at 3.0 Å reveals the molecular basis of anion selectivity, *Nature* **415,** 287–294 (2002).

Lobet, S. and Dutzler, R., Ion-binding properties of the ClC chloride selectivity filter, *EMBO J.* **25,** 24–33 (2006).

Aquaporins

Fu, D. and Lu, M., The structural basis of water permeation and proton exclusion in aquaporins, *Mol. Memb. Biol.* **24,** 366–374 (2007). [A review.]

Gonen, T. and Walz, T., The structure of aquaporins, *Q. Rev. Biophys.* **39,** 361–396 (2006).

King, L.S., Kozono, D., and Agre, P., From structure to disease: The evolving tale of aquaporin biology, *Nature Rev. Mol. Cell Biol.* **5,** 687–698 (2004).

Sui, H., Han, B.-G., Lee, J.K., and Jap, B.K., Structural basis of water-specific transport through the AQP1 water channel, *Nature* **414,** 872–878 (2001).

$(Na^+–K^+)$–ATPase

Blaustein, M.P., Physiological effects of endogenous ouabain: Control of intracellular Ca^{2+} stores and cell responsiveness, *Am. J. Physiol.* **264,** C1367–C1378 (1993).

Kaplan, J.H., Biochemistry of the Na,K-ATPase, *Annu. Rev. Biochem.* **71,** 511–535 (2002).

Morph, J.P., Pedersen, B.P., Toustrop-Jensen, M.S., Sørensen, T.L.-M., Petersen, J., Andersen, J.P., Vilsen, B., and Nissen, P., Crystal structure of the sodium–potassium pump, *Nature* **450,** 1043–1049 (2007). [The 3.4-Å resolution structure of the porcine enzyme.]

Shinoda, T., Ogawa, H., Cornelius, F., and Toyoshima, C., Crystal structure of the sodium–potassium pump at 2.4 Å resolution, *Nature* **459,** 446–450 (2009). [The shark enzyme.]

Ca^{2+} –ATPase

Enyedi, A., Vorherr, T., James, P., McCormick, D.J., Filoteo, A.G., Carafoli, E., and Penniston, J.T., The calmodulin binding domain of the plasma membrane Ca^{2+} pump interacts both with calmodulin and with another part of the pump, *J. Biol. Chem.* **264,** 12313–12321 (1989).

Jencks, W.P., Coupling of hydrolysis of ATP and the transport of Ca^{2+} by the calcium ATPase of sarcoplasmic reticulum, *Biochem. Soc. Trans.* **20,** 555–559 (1992). [An insightful discussion of the mechanism of coupling chemical energy to the vectorial transport of ions against a concentration gradient.]

Olesen, C., Picard, M., Winther, A.-M.L., Gyrup, C., Morth, J.P., Oxvig, C., Møller, J.V., and Nissen, P., The structural basis of calcium transport by the calcium pump, *Nature* **450,** 1036–1042 (2007).

Toyoshima, C. and Inesi, G., Structural basis of ion pumping by Ca^{2+}-ATPase of the sarcoplasmic reticulum, *Annu. Rev. Biochem.* **73,** 269–292 (2004).

Toyoshima, C., Nakasako, M., and Ogawa, H., Crystal structure of the calcium pump of sacroplasmic reticulum at 2.6 Å resolution, *Nature* **405,** 647–655 (2000).

(H$^+$–K$^+$)–ATPase

Besanèon, M., Shin, J.M., Mercier, F., Munson, K., Rabon, E., Hersey, S., and Sachs, G., Chemomechanical coupling in the gastric H,K ATPase, *Acta Physiol. Scand.* **146,** 77–88 (1992).

Pedersen, B.P., Buch-Pedersen, M.J., Morth, J.P., Palmgren, M.G., and Nissen, P., Crystal structure of the plasma membrane proton pump, *Nature* **450,** 1111–1114 (2007).

PEP-Dependent Phosphotransferase System

Herzberg, O. and Klevit, R., Unraveling a bacterial hexose transport pathway, *Curr. Opin. Struct. Biol.* **4,** 814–822 (1994).

Hurley, J.H., Faber, H.R., Worthylake, D., Meadow, N.D., Roseman, S., Pettigrew, D.W., and Remington, S.J., Structure of the regulatory complex of *Escherichia coli* E III^{glc} with glycerol kinase, *Science* **259,** 673–677 (1993). [EIIA was previously named EIII.]

Meadow, N.D., Fox, D.K., and Roseman, S., The bacterial phosphoenolpyruvate:glucose phosphotransferase system, *Annu. Rev. Biochem.* **59,** 497–542 (1990); *and* Feese, M., Pettigrew, D.W., Meadow, N.D., Roseman, S., and Remington, S.J., Cation promoted association (CPA) of a regulatory and target protein is controlled by protein phosphorylation, *Proc. Natl. Acad Sci.* **91,** 3544–3548 (1994).

Saier, M.H., Jr., Chauvaux, S., Deutscher, J., Reizer, J., and Ye, J.-J., Protein phosphorylation and regulation of carbon metabolism in gram-negative versus gram-positive bacteria, *Trends Biochem. Sci.* **20,** 267–271 (1995).

ABC Transporters

Higgins, C.F., Multiple molecular mechanisms for multidrug resistance transporters, *Nature* **446,** 749–757 (2007).

Hollerstein, K., Dawson, R.J.P., and Locher, K.P., Structure and mechanism of ABC transporter proteins, *Curr. Opin. Struct. Biol.* **17,** 412–418 (2007).

Jones, P.M., O'Mara, M.L., and George, A.M., ABC transporters: a riddle wrapped in a mystery inside an enigma, *Trends Biochem. Sci.* **34,** 520–531 (2009).

Oldham, M.L., Davidson, A.L., and Chen, J., Structural insights into ABC transporter mechanism, *Curr. Opin. Struct. Biol.* **18,** 726–733 (2008).

Rees, D.C., Johnson, E., and Lewinson, O., ABC transporters: the power to change, *Nature Rev. Mol. Cell Biol.* **10,** 218–227 (2009).

Ward, A., Reyes, C.L., Yu, J., Roth, C.B., and Chang, G., Flexibility in the ABC transporter MsbA: Alternating access with a twist, *Proc. Natl. Acad Sci.* **104,** 19005–19010 (2007).

Na$^+$–Glucose Symport

Faham, S., Watanabe, A., Besserer, G.M., Cascio, D., Specht, A., Hirayama, B.A., Wright, E.M., and Abramson, J., The crystal structure of a sodium galactose transporter reveals insights into Na$^+$/sugar symport, *Science* **321,** 810–814 (2008).

Wright, E.M., Hirayama, B.A., and Loo, D.F., Active sugar transport in health and disease, *J. Intern. Med.* **261,** 32–43 (2007).

Lactose Permease

Abramson, J., Smirnova, I., Kasho, V., Verner, G., Kaback, H.R., and Iwata, S., Structure and mechanism of the lactose permease of *Escherichia coli*, *Science* **301,** 610–615 (2003).

Guan, L. and Kaback, H.R., Lessons from lactose permease, *Annu. Rev. Biophys. Biomol. Struct.* **35,** 67–91 (2006).

ATP–ADP Translocator

Klingenberg, M., Molecular aspects of the adenine nucleotide carrier from mitochondria, *Arch. Biochem. Biophys.* **270,** 1–14 (1989).

Nury, H., Dahout-Gonzalez, C., Trézéguet, V., Lauquin, G.J.M., Brandolin, G., and Pebay-Peyroula, E., Relations between structure and function of the mitochondrial ADP/ATP carrier, *Annu. Rev. Biochem.* **75,** 713–741 (2006).

Neurotransmission

Bezanilla, F., How membrane proteins sense voltage, *Nature Rev. Mol. Cell Biol.* **9,** 323–332 (2008).

Catterall, W.A., Structure and regulation of voltage-gated Ca^{2+} channels, *Annu. Rev. Cell Dev. Biol.* **16,** 521–555 (2000).

Geppert, M. and Südhof, T.C., Rab3 and synaptotagmin. The yin and yang of synaptic transmission, *Annu. Rev. Neurosci.* **21,** 75–95 (1998).

Gulbis, J.M. and Doyle, D.A., Potassium channel structures: do they conform? *Curr. Opin. Struct. Biol.* **14,** 440–446 (2004).

Hille, B., *Ionic Channels of Excitable Membranes* (3rd ed.), Sinauer Associates (2001).

Lin, R.C. and Scheller, R.H., Mechanisms of synaptic vesicle exocytosis, *Annu. Rev. Cell Dev. Biol.* **16,** 19–49 (2000).

Long, S.B., Campbell, E.B., and MacKinnon, R., Crystal structure of a mammalian voltage-dependent *Shaker* family K$^+$ channel; *and* Voltage sensor of Kv1.2: Structural basis of electromechanical coupling, *Science* **309,** 897–903; *and* 903–908 (2005).

Long, S.B., Tao, X., Campbell, E.B., and MacKinnon, R., Atomic structure of a voltage-dependent K$^+$ channel in a lipid membrane-like environment, *Nature* **450,** 376–382 (2007).

Orlova, E.V., Rahman, M.A., Gowen, B., Volynski, K.E., Ashton, A.C., Manser, C., van Heel, M., and Ushkaryov, Y.A., Structure of α-latrotoxin oligomers reveals that divalent cation-dependent tetramers form membrane pores, *Nature Struct. Biol.* **7,** 48–53 (2000).

Roosild, T.P., Lê, K.-T., and Choe, S., Cytoplasmic gatekeepers of K$^+$-channel flux: a structural perspective, *Trends Biochem. Sci.* **29,** 39–45 (2004).

Sussman, J.L., Harel, M., Frolow, F., Oefner, C., Goldman, A., Toker, L., and Silman, I., Atomic structure of acetylcholinesterase from *Torpedo californica*: A prototypic acetylcholine-binding protein, *Science* **253,** 872–879 (1991).

Swartz, K.J., Sensing voltage across lipid membranes, *Nature* **456**, 891–897 (2008).

Unwin, N., Refined structure of the nicotinic acetylcholine receptor at 4 Å resolution, *J. Mol. Biol.* **346**, 967–989 (2004); *and* Miyazawa, A., Fujiyoshi, Y., and Unwin, N., Structure and gating mechanism of the acetylcholine receptor pore, *Nature* **423**, 949–955 (2003).

Yellin, G., The voltage-gated potassium channels and their relatives, *Nature* **419**, 35–42 (2002).

Zhou, M., Morais-Cabral, J.H., Mann, S., and MacKinnon, R., Potassium channel receptor site for the inactivation gate and quaternary amine inhibitors, *Nature* **411**, 657–661 (2001).

PROBLEMS

1. If the glucose concentration outside a cell is 10 mM but that inside a cell is 0.1 mM, what is glucose's chemical potential difference across the membrane at 37°C?

*2. If a solution of an ionic macromolecule is equilibrated with a salt solution from which it is separated by a membrane through which the salt ions but not the macromolecule can pass, a membrane potential is generated across the membrane. This so-called **Donnan equilibrium** arises because the impermeability of the membrane to some ions but not others prevents the equalization of the ionic concentrations on the two sides of the membrane. To demonstrate this effect, assume that the Cl^- salt of a monocationic protein, P^+, is dissolved in water to the extent that $[Cl^-] = 0.1M$ and is separated by a membrane impermeable to the protein but not NaCl from an equal volume of $0.1M$ NaCl solution. Assuming no volume change in either compartment, what are the concentrations of the various ionic species on either side of the membrane after the system has equilibrated? What is the membrane potential across the membrane? (*Hint:* Mass is conserved and the solution on each side of the membrane must be electrically neutral. At equilibrium, $\Delta G_{Na^+} + \Delta G_{Cl^-} = 0$.)

3. Gramicidin A, a dimer of 15-residue polypeptides that forms a hollow helix, is a channel-forming ionophore that permits the passage of ~10^7 alkalai metal ions per second through a membrane. How long would it take one molecule of gramicidin A to transport enough Na^+ to change the concentration inside an erythrocyte of volume 80 μm^3 by 10 mM? Assume the erythrocyte's Na^+ pumps are inoperative and that gramicidin A does not also transport ions out of the cell, which it really does.

4. Predict whether the following compounds can cross a membrane without mediation or will require facilitation. Indicate the criteria you used to make these predictions. (a) Ethanol, (b) glycine, (c) cholesterol, and (d) ATP.

5. The rate of movement (flux) of a substance X into cells was measured at different concentrations of X to construct the graph below.

(a) Does this information suggest that the movement of X into the cells is mediated by a protein transporter? Explain.

(b) What additional experiment could you perform to verify that a transport protein is or is not involved?

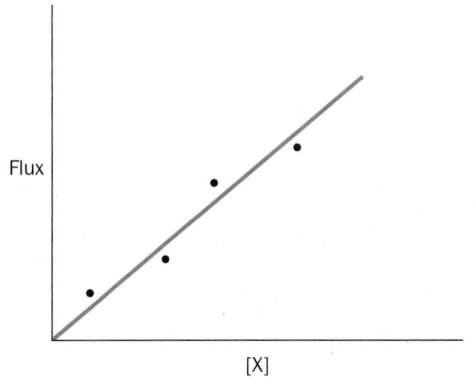

6. You have isolated a new strain of bacteria and would like to know whether leucine and ethylene glycol enter the cells by mediated diffusion or only by a nonmediated route. To do this you measure the initial rates of uptake of these molecules as a function of external concentration and obtain the data in the following table.

Compound	Concentration (M)	Initial Uptake Rate (arbitrary units)
Leucine	1×10^{-6}	110
	2×10^{-6}	220
	5×10^{-6}	480
	1×10^{-5}	830
	3×10^{-5}	1700
	1×10^{-4}	2600
	5×10^{-4}	3100
	1×10^{-3}	3200
Ethylene glycol	1×10^{-3}	1
	5×10^{-3}	5
	0.01	10
	0.05	50
	0.1	100
	0.5	500
	1.0	1000

Which compound(s) enters by a mediated route? What criteria did you use for this decision?

7. Aquaporin AQP1 (Fig. 20-15) forms an hourglass-shaped pore that, in its narrowest region, is 3 Å in diameter. AQP1 is impermeable to glycerol [CHOH(CH$_2$OH)$_2$]. However, a homologous and structurally similar aquaglyceroporin, which is minimally 3.4 Å wide, permits the passage of glycerol but is only poorly permeable to water. Discuss the possible differences between these channels that would account for their different permeabilities.

8. The (Na^+–K^+)–ATPase is inhibited by nanomolar concentrations of vanadate, which forms a pentavalent ion, VO_5^{5-}, with trigonal bipyramidal symmetry. Explain the mechanism of this inhibition. (*Hint:* See Section 16-2B.)

9. What function might the synthesis of digitalis serve in the purple foxglove plant?

10. The (H^+–K^+)–ATPase secretes H^+ at a concentration of $0.18M$ from cells that have an internal pH of 7. What is the ΔG required for the transport of 1 mol of H^+ under these conditions? Assuming that the ΔG for ATP hydrolysis is -31.5 kJ · mol^{-1} under these conditions, and that the membrane potential is 0.06 V, inside negative, how much ATP must be hydrolyzed per mole of H^+ transported in order to make this transport exergonic?

11. A 100-Å-thick membrane has a membrane potential of 100 mV. What is the magnitude of this potential difference in $V \cdot cm^{-1}$? Comment on the magnitude of this potential field in macroscopic terms.

12. The resting membrane potential ($\Delta\Psi$) of a neuron at 37°C is −60 mV (inside negative). If the free energy change associated with the transport of one Na^+ ion from outside to inside is −11.9 $kJ \cdot mol^{-1}$, and [Na^+] outside the cell is 260 mM, what is [Na^+] inside the cell?

13. Write a kinetic scheme for the (H^+–K^+)–ATPase that provides for coupled ATP hydrolysis with H^+ transport. Discuss the order of substrate addition required for coupling. Identify the steps in which mutual destabilization results in reasonable rates of transport.

14. If the ATP supply in the cell shown in Fig. 20-27c suddenly vanished, would the intracellular glucose concentration increase, decrease, or remain the same?

15. Endothelial cells and pericytes in the retina of the eye have different mechanisms for glucose uptake. The figure below shows the rate of glucose uptake for each type of cell in the presence of increasing amounts of sodium. What do these results reveal about the glucose transporter in each cell type?

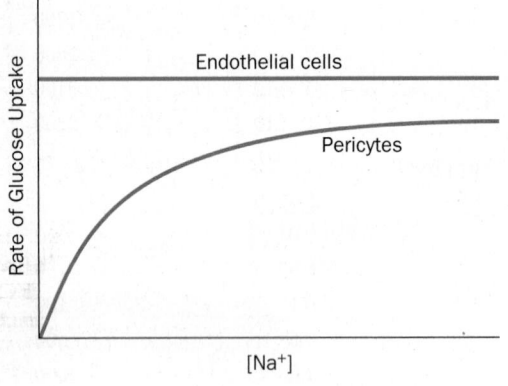

16. Why don't nerve impulses propagate in the reverse direction?

17. What is the resting membrane potential across an axonic membrane at 25°C (a) in the presence of tetrodotoxin or (b) with a high concentration of Cs^+ inside the axon (use the data in Table 20-3)? How do these substances affect the axon's action potential?

18. Decamethonium ion [$(CH_3)_3N^+(CH_2)_{10}N^+(CH_3)_3$] is a synthetic muscle relaxant. What is its mechanism of action?

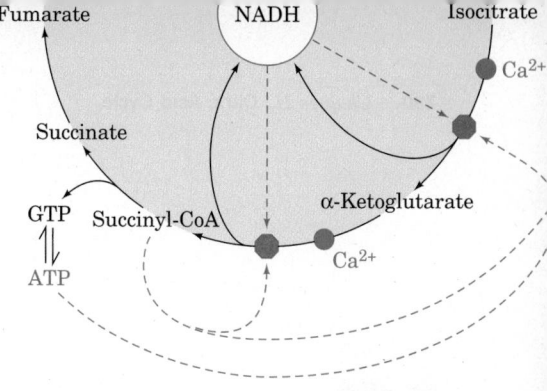

Citric Acid Cycle

CHAPTER 21

In this chapter we continue our metabolic explorations by examining the **citric acid cycle,** the common mode of oxidative degradation in eukaryotes and prokaryotes. This cycle, which is alternatively known as the **tricarboxylic acid (TCA) cycle** and the **Krebs cycle,** marks the "hub" of the metabolic system: *It accounts for the major portion of carbohydrate, fatty acid, and amino acid oxidation and generates numerous biosynthetic precursors.* The citric acid cycle is therefore **amphibolic,** that is, it operates both catabolically and anabolically.

We begin our study of the citric acid cycle with an overview of its component reactions and a historical synopsis of its elucidation. Next, we explore the origin of the cycle's starting compound, **acetyl-coenzyme A (acetyl-CoA),** the common intermediate formed by the breakdown of most metabolic fuels. Then, after discussing the reaction mechanisms of the enzymes that catalyze the cycle, we consider the various means by which it is regulated. Finally, we deal with the citric acid cycle's amphibolic nature by examining its interrelationships with other metabolic pathways.

1 CYCLE OVERVIEW

See Guided Exploration 18: Citric acid cycle overview *The citric acid cycle (Fig. 21-1) is an ingenious series of reactions that oxidizes the acetyl group of acetyl-CoA to two molecules of CO_2 in a manner that conserves the liberated free energy for utilization in ATP generation.* Before we study these reactions in detail, let us consider the cycle's chemical strategy by "walking" through the cycle and noting the fate of the acetyl group at each step. Following this preview, we shall consider some of the major discoveries that led to our present understanding of the citric acid cycle.

A. Reactions of the Cycle

The eight enzymes of the citric acid cycle (Fig. 21-1) catalyze a series of well-known organic reactions that cumulatively oxidize an acetyl group to two CO_2 molecules with the concomitant generation of three NADHs, one FADH$_2$, and one GTP:

1. Citrate synthase catalyzes the condensation of acetyl-CoA and **oxaloacetate** to yield **citrate,** giving the cycle its name.

2. The strategy of the cycle's next two steps is to rearrange citrate to a more easily oxidized isomer and then oxidize it. **Aconitase** isomerizes citrate, a not readily oxidized tertiary alcohol, to the easily oxidized secondary alcohol **isocitrate.** The reaction sequence involves a dehydration, producing enzyme-bound *cis*-**aconitate,** followed by a hydration, so that citrate's hydroxyl group is, in effect, transferred to an adjacent carbon atom.

3. Isocitrate dehydrogenase oxidizes isocitrate to the β-keto acid intermediate **oxalosuccinate** with the coupled reduction of NAD$^+$ to NADH; oxalosuccinate is then decarboxylated, yielding **α-ketoglutarate.** This is the first step in which oxidation is coupled to NADH production and also the first CO_2-generating step.

4. The multienzyme complex **α-ketoglutarate dehydrogenase** oxidatively decarboxylates α-ketoglutarate to **succinyl-coenzyme A.** The reaction involves the reduction of a second NAD$^+$ to NADH and the generation of a second molecule of CO_2. At this point in the cycle, two molecules of CO_2 have been produced, so that the net oxidation of the acetyl group is complete. Note, however, that it is not the carbon atoms of the entering acetyl-CoA that have been oxidized.

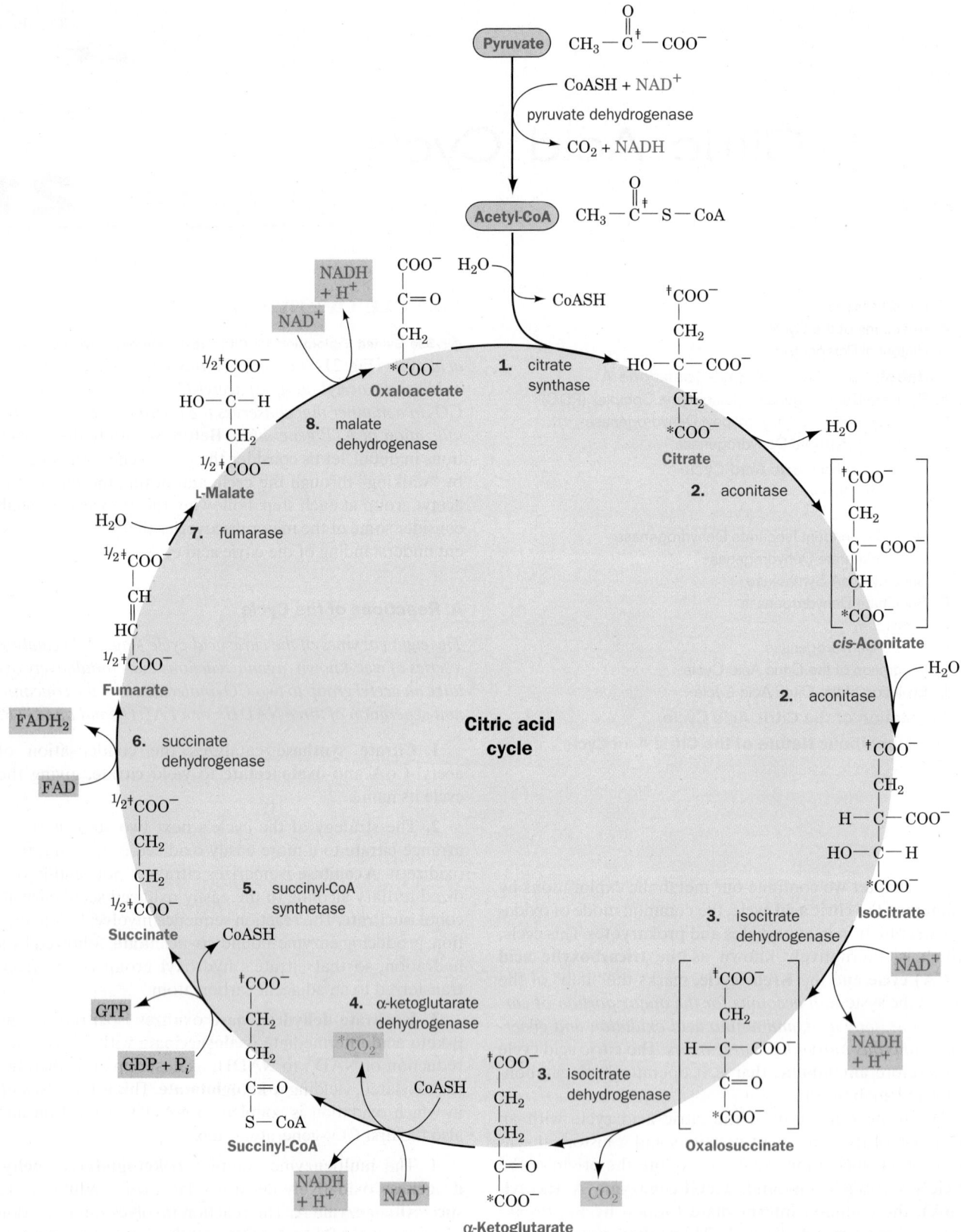

Figure 21-1 **Reactions of the citric acid cycle.** The reactants and products of this catalytic cycle are boxed. The pyruvate → acetyl-CoA reaction (*top*) supplies the cycle's substrate via carbohydrate metabolism but is not considered to be part of the cycle. The bracketed compounds are enzyme-bound intermediates. An isotopic label at C4 of oxaloacetate (*) becomes C1 of α-ketoglutarate and is released as CO_2 in Reaction 4. An isotopic label at C1 of acetyl-CoA (‡) becomes C5 of α-ketoglutarate and is scrambled in Reaction 5 between C1 and C4 of succinate (1/2‡). See the Animated Figures

5. Succinyl-CoA synthetase converts succinyl-coenzyme A to **succinate.** The free energy of the thioester bond is conserved in this reaction by the formation of "high-energy" GTP from GDP + P_i.

6. The remaining reactions of the cycle serve to oxidize succinate back to oxaloacetate in preparation for another round of the cycle. **Succinate dehydrogenase** catalyzes the oxidation of succinate's central single bond to a trans double bond, yielding **fumarate** with the concomitant reduction of the redox coenzyme FAD to $FADH_2$ (the molecular formulas of FAD and $FADH_2$ and the reactions through which they are interconverted are given in Fig. 16-8).

7. Fumarase then catalyzes the hydration of fumarate's double bond to yield **malate.**

8. Finally, **malate dehydrogenase** reforms oxaloacetate by oxidizing malate's secondary alcohol group to the corresponding ketone with concomitant reduction of a third NAD^+ to NADH. Acetyl groups are thereby completely oxidized to CO_2 with the following stoichiometry:

$$3NAD^+ + FAD + GDP + P_i + \text{acetyl-CoA} \longrightarrow$$
$$3NADH + FADH_2 + GTP + CoA + 2CO_2$$

The citric acid cycle functions catalytically as a consequence of its regeneration of oxaloacetate: An endless number of acetyl groups can be oxidized through the agency of a single oxaloacetate molecule.

NADH and $FADH_2$ are vital products of the citric acid cycle. Their reoxidation by O_2 through the mediation of the electron-transport chain and oxidative phosphorylation (Chapter 22) completes the breakdown of metabolic fuel in a manner that drives the synthesis of ATP. Other functions of the cycle are discussed in Section 21-5.

B. *Historical Perspective*

The citric acid cycle was proposed in 1937 by Hans Krebs, a contribution that ranks as one of the most important achievements of metabolic chemistry. We therefore outline the intellectual history of this cycle's discovery.

By the early 1930s, significant progress had been made in elucidating the glycolytic pathway (Section 17-1A). Yet the mechanism of glucose oxidation and its relationship to cellular respiration (oxygen uptake) was still a mystery. Nevertheless, the involvement of several metabolites in cellular oxidative processes was recognized. It was well known, for example, that in addition to lactate and acetate, the dicarboxylates succinate, malate, and α-ketoglutarate, as well as the tricarboxylate citrate, are rapidly oxidized by muscle tissue during respiration. It had also been shown that **malonate** (Section 21-3F), a potent inhibitor of succinate oxidation to fumarate, also inhibits cellular respiration, thereby suggesting that succinate plays a central role in oxidative metabolism rather than being just another metabolic fuel.

In 1935, Albert Szent-Györgyi demonstrated that cellular respiration is dramatically accelerated by catalytic amounts of succinate, fumarate, malate, and oxaloacetate;

that is, *the addition of any of these substances to minced muscle tissue stimulates O_2 uptake and CO_2 production far in excess of that required to oxidize the added dicarboxylic acid.* Szent-Györgyi further showed that these compounds were interconverted according to the reaction sequence:

$$\text{Succinate} \longrightarrow \text{fumarate} \longrightarrow \text{malate} \longrightarrow \text{oxaloacetate}$$

Shortly afterward, Carl Martius and Franz Knoop demonstrated that citrate is rearranged, via *cis*-aconitate, to isocitrate and then dehydrogenated to α-ketoglutarate. α-Ketoglutarate was already known to undergo oxidative decarboxylation to succinate and CO_2. This extended the proposed reaction sequence to

Citrate $\longrightarrow$ *cis*-aconitate $\longrightarrow$ isocitrate
$\longrightarrow$ α-ketoglutarate $\longrightarrow$ succinate $\longrightarrow$ fumarate
$\longrightarrow$ malate $\longrightarrow$ oxaloacetate

What was necessary to close the circle so as to make the system catalytic was to establish that oxaloacetate is converted to citrate. In 1936, Martius and Knoop demonstrated that citrate could be formed nonenzymatically from oxaloacetate and pyruvate by treatment with hydrogen peroxide under basic conditions. Krebs used this chemical model as the point of departure for the biochemical experiments that led to his proposal of the citric acid cycle.

Krebs' hypothesis was based on his investigations, starting in 1936, on respiration in minced pigeon breast muscle (which has a particularly high rate of respiration). The idea of a catalytic cycle was not new to him: In 1932, he and Kurt Henseleit had elucidated the outlines of the **urea cycle,** a process in which ammonia and CO_2 are converted to urea (Section 26-2). The most important observations Krebs made in support of the existence of the citric acid cycle were as follows:

1. Succinate is formed from fumarate, malate, or oxaloacetate in the presence of the metabolic inhibitor malonate. Since malonate inhibits the direct reduction of fumarate to succinate, the succinate must be formed by an oxidative cycle.

2. Pyruvate and oxaloacetate can form citrate enzymatically. Krebs therefore suggested that the metabolic cycle is closed with the reaction:

$$\text{Pyruvate} + \text{oxaloacetate} \longrightarrow \text{citrate} + CO_2$$

3. The interconversion rates of the cycle's individual steps are sufficiently rapid to account for observed respiration rates, so it must be (at least) the major pathway for pyruvate oxidation in muscle.

Although Krebs had established the existence of the citric acid cycle, some major gaps still remained in its complete elucidation. The mechanism of citrate formation did not become clear until Nathan Kaplan and Fritz Lipmann discovered **coenzyme A** in 1945 (Section 21-2), and Severo Ochoa and Feodor Lynen established, in 1951, that acetyl-CoA is the intermediate that condenses with oxaloacetate to form citrate. Oxidative decarboxylation of α-ketoglutarate to succinate was also shown to involve coenzyme A with succinyl-CoA as an intermediate.

The elucidation of the citric acid cycle was a major achievement and, like all achievements of this magnitude, required the efforts of numerous investigators. Indeed, biochemists are still working to understand the cycle on a molecular and enzymatic level. We shall study the eight enzymes that catalyze the cycle after first discussing the cycle's major fuel, acetyl-CoA, and its formation from pyruvate.

2 METABOLIC SOURCES OF ACETYL-COENZYME A

Acetyl groups enter the citric acid cycle as **acetyl-coenzyme A (acetyl-SCoA or acetyl-CoA;** *Fig. 21-2), the common product of carbohydrate, fatty acid, and amino acid breakdown.* **Coenzyme A (CoASH or CoA)** consists of a β-mercaptoethylamine group bonded through an amide linkage to the vitamin **pantothenic acid,** which, in turn, is attached to a 3′-phosphoadenosine moiety via a pyrophosphate bridge. The acetyl group of acetyl-CoA is bonded as a thioester to the sulfhydryl portion of the β-mercaptoethylamine group. *CoA thereby functions as a carrier of acetyl and other acyl groups (the A in CoA stands for "Acetylation").*

Acetyl-CoA is a "high-energy" compound: The $\Delta G^{\circ\prime}$ for the hydrolysis of its thioester bond is -31.5 kJ · mol^{-1}, which makes this reaction slightly (1 kJ · mol^{-1}) more exergonic than that of ATP hydrolysis (Section 16-4A). The formation of this thioester bond in a metabolic intermediate therefore conserves a portion of the free energy of oxidation of a metabolic fuel.

A. *Pyruvate Dehydrogenase Multienzyme Complex (PDC)*

The immediate precursor to acetyl-CoA from carbohydrate sources is the glycolytic product pyruvate. As we saw in Section 17-3, under anaerobic conditions the NADH produced by glycolysis is reoxidized with concomitant reduction of pyruvate to lactate (in muscle) or ethanol (in yeast). Under aerobic conditions, however, NADH is reoxidized by the mitochondrial electron-transport chain (Section 22-2), so that pyruvate, which enters the mitochondrion via a specific pyruvate–H$^+$ symport (membrane transport nomenclature is discussed in Section 20-3), can undergo further oxidation. (The formation of acetyl-CoA from fatty acids and amino acids is discussed in Sections 25-2 and 26-3.)

Acetyl-CoA is synthesized from pyruvate and CoA through oxidative decarboxylation by a **multienzyme complex** named **pyruvate dehydrogenase.** In general, multienzyme complexes are groups of noncovalently associated enzymes that catalyze two or more sequential steps in a metabolic pathway. The **pyruvate dehydrogenase multienzyme complex (PDC)** consists of three enzymes: **pyruvate dehydrogenase (E1), dihydrolipoyl transacetylase (E2),** and **dihydrolipoyl dehydrogenase (E3).** The *E. coli* pyruvate dehydrogenase complex, whose characterization was pioneered by Lester Reed, is an ~4600-kD polyhedral particle that is ~300 Å in diameter (Fig. 21-3a). Isolated *E. coli* E2 forms a particle with 24 identical subunits, which electron micrographs (Figs. 21-3b and 21-4a), together with an X-ray structure by Wim Hol (Fig. 21-5a), indicate are arranged with cubic symmetry. The E1 subunits form dimers that associate with the E2 cube at the centers of the cube's 12 edges (Fig. 21-4b,c), whereas the E3 subunits form dimers that are located at the centers of this cube's 6 faces. We discuss the X-ray structures of the E1, E2, and E3 subunits below.

a. Some PDCs Have a Dodecahedral Form

Although all PDCs catalyze the same reactions by similar mechanisms, they may have different quaternary structures. Whereas the PDCs of *E. coli* and most other gram-negative bacteria have the foregoing cubic symmetry, those of

Acetyl-coenzyme A (acetyl-CoA)

Figure 21-2   Chemical structure of acetyl-CoA. The thioester bond is drawn with a squiggle ($\sim$) to indicate that it is a "high-energy" bond. In CoA, the acetyl group is replaced by a hydrogen atom.

Figure 21-3 **Electron micrographs of the *E. coli* pyruvate dehydrogenase multienzyme complex.** (*a*) The intact complex. (*b*) The dihydrolipoyl transacetylase (E2) "core" complex. [Courtesy of Lester Reed, University of Texas.]

Figure 21-4 **Structural organization of the *E. coli* PDC.** (*a*) The dihydrolipoyl transacetylase (E2) "core." Its 24 subunits (*green spheres*) associate as trimers located at the corners of a cube to form a particle that has cubic symmetry (*O* symmetry; Section 8-5B). (*b*) The 24 pyruvate dehydrogenase (E1) subunits (*orange spheres*) form dimers that associate with the E2 core (*shaded cube*) at the centers of each of its 12 edges, whereas the 12 dihydrolipoyl dehydrogenase (E3) subunits (*purple spheres*) form dimers that attach to the E2 cube at the centers of each of its 6 faces. (*c*) Parts *a* and *b* combined to form the entire 60-subunit complex.

(*a*)

(*b*)

Figure 21-5 **Comparison of the X-ray structures of the dihydrolipoyl transacetylase (E2) cores of PDCs.** Both complexes are drawn mainly in space-filling representation and viewed along their respective 2-fold axes of symmetry. (*a*) The 125-Å high cubic (*O* symmetry; Fig. 8-65*c*) E2 core from the gram-negative bacterium *Azotobacter vinelandii*. It consists of 24 subunits that form 8 trimers, shown here in different colors. The positions of a 2-fold, a 3-fold, and a 4-fold axis of symmetry are indicated. The inset (*upper right*) is a perspective drawing of a cube. (*b*) The 237-Å-diameter dodecahedral (*I* symmetry; Fig. 8-65*c*) E2 core from the gram-positive bacterium *Bacillus stearothermophilus*. It consists of 60 subunits that form 20 trimers, shown here in different colors. The positions of a 2-fold, a 3-fold, and a 5-fold axis of symmetry are indicated. The inset (*upper right*) is a perspective drawing of a dodecahedron. The rear portion of each complex, which is largely eclipsed by the front portion, is gray. The subunits forming the trimers closest to the viewer in each drawing are individually colored with the lower trimer drawn in ribbon form. Note that the subunits in each trimer are extensively associated and that the trimers in the two complexes have closely similar structures. In contrast, the interactions between contacting trimers in both types of complexes are relatively tenuous. Also note that these contacting trimers form the 4- and 5-membered rings that comprise the square and pentagonal faces of the cubic and dodecahedral complexes, respectively. [Based on X-ray structures by Wim Hol, University of Washington. PDBids 1EAB and 1B5S.]

eukaryotes and some gram-positive bacteria have an analogous dodecahedral form [Fig. 21-5*b;* a dodecahedron is a regular polyhedron with *I* symmetry (Section 8-5B) that has 20 vertices, each lying on a 3-fold axis, and 12 pentagonal faces having an aggregate of 30 edges]. Thus, the mitochondrially located ~10,000-kD eukaryotic complex, the largest known multienzyme complex, consists of a dodecahedral core of 20 E2 trimers (one centered on every vertex) surrounded by 30 E1 $\alpha_2\beta_2$ heterotetramers (one centered on every edge) and 12 E3 dimers (one centered in every face).

b. Multienzyme Complexes Are Catalytically Efficient

Multienzyme complexes are a step forward in the evolution of catalytic efficiency. They offer the following mechanistic advantages:

1. Enzymatic reaction rates are limited by the frequency at which enzymes collide with their substrates (Section 14-2Ba). If a series of reactions occurs within a multienzyme complex, the distance that substrates must diffuse between active sites is minimized, thereby achieving a rate enhancement.

2. Complex formation provides the means for **channeling** (passing) metabolic intermediates between successive enzymes in a metabolic pathway, thereby minimizing side reactions.

3. The reactions catalyzed by a multienzyme complex may be coordinately controlled.

c. Acetyl-CoA Formation Occurs in Five Reactions

The PDC catalyzes five sequential reactions (Fig. 21-6) with the overall stoichiometry:

Pyruvate + CoA + NAD$^+$ ⟶

acetyl-CoA + CO$_2$ + NADH

The coenzymes and prosthetic groups required in this reaction sequence are thiamine pyrophosphate (TPP; Fig. 17-26), flavin adenine dinucleotide (FAD; Fig. 16-8), nicotinamide adenine dinucleotide (NAD$^+$; Fig. 13-2), and **lipoamide**

(Fig. 21-7); their functions are listed in Table 21-1. Lipoamide consists of **lipoic acid** joined in amide linkage to the ε-amino group of a Lys residue. Reduction of its cyclic disulfide to a dithiol, **dihydrolipoamide,** and its reoxidation (Fig. 21-7) are the "business" of this prosthetic group.

The five reactions catalyzed by the PDC are as follows (Fig. 21-6):

1. Pyruvate dehydrogenase (E1), a TPP-requiring enzyme, decarboxylates pyruvate, with the intermediate formation of hydroxyethyl-TPP. This reaction is identical with that catalyzed by yeast pyruvate decarboxylase (Section 17-3B):

Figure 21-6 **The five reactions of the PDC.** E1 (pyruvate dehydrogenase) contains TPP and catalyzes Reactions 1 and 2. E2 (dihydrolipoyl transacetylase) contains lipoamide and catalyzes Reaction 3. E3 (dihydrolipoyl dehydrogenase) contains FAD and a redox-active disulfide and catalyzes Reactions 4 and 5.

2. Unlike pyruvate decarboxylase, however, pyruvate dehydrogenase does not convert the hydroxyethyl-TPP intermediate into acetaldehyde and TPP. Instead, the hydroxyethyl group is transferred to the next enzyme in the multienzyme sequence, dihydrolipoyl transacetylase (E2). The reaction occurs by attack of the hydroxyethyl group carbanion on the lipoamide disulfide, followed by the elimination of TPP from the intermediate adduct to form acetyl-dihydrolipoamide and regenerate active E1. The hydroxyethyl carbanion is thereby oxidized to an acetyl group by the concomitant reduction of the lipoamide disulfide bond:

3. E2 then catalyzes the transfer of the acetyl group to CoA, yielding acetyl-CoA and dihydrolipoamide–E2:

This is a transesterification in which the sulfhydryl group of CoA attacks the acetyl group of acetyl-dihydrolipoamide–E2 to form a tetrahedral intermediate (not shown), which decomposes to acetyl-CoA and dihydrolipoamide–E2.

Figure 21-7 **Interconversion of lipoamide and dihydrolipoamide.** Lipoamide is lipoic acid covalently joined to the ε-amino group of a Lys residue via an amide linkage.

Table 21-1 **The Coenzymes and Prosthetic Groups of Pyruvate Dehydrogenase**

Cofactor	Location	Function
Thiamine pyrophosphate (TPP)	Bound to E1	Decarboxylates pyruvate, yielding a hydroxyethyl-TPP carbanion
Lipoic acid	Covalently linked to a Lys on E2 (lipoamide)	Accepts the hydroxyethyl carbanion from TPP as an acetyl group
Coenzyme A (CoA)	Substrate for E2	Accepts the acetyl group from acetyl-dihydrolipoamide
Flavin adenine dinucleotide (FAD)	Bound to E3	Reduced by dihydrolipoamide
Nicotinamide adenine dinucleotide (NAD$^+$)	Substrate for E3	Reduced by FADH$_2$

4. Dihydrolipoyl dehydrogenase (E3; also called **dihydrolipoamide dehydrogenase**) reoxidizes dihydrolipoamide, thereby completing the catalytic cycle of E2:

Oxidized E3 contains a reactive disulfide group and a tightly bound FAD. The oxidation of dihydrolipoamide is a disulfide interchange reaction (Section 9-1A): The lipoamide disulfide bond forms with concomitant reduction of E3's reactive disulfide to two sulfhydryl groups.

5. Reduced E3 is reoxidized by NAD^+:

The enzyme's active sulfhydryl groups are reoxidized by the enzyme-bound FAD, which is thereby reduced to $FADH_2$. The $FADH_2$ is then reoxidized to FAD by NAD^+, producing NADH.

d. The Structure of E2

Dihydrolipoyl transacetylase (E2) consists of several domains (Fig. 21-8): one to three N-terminal lipoyl domains (~80 residues) that each covalently bind a lipoyl group; a **peripheral subunit-binding domain (PSBD; ~35 residues)** that binds to both E1 and E3 and hence holds the complex together; and a C-terminal catalytic domain (~250 residues) that contains the enzyme's catalytic center and its intersubunit binding sites. These domains are linked by 20- to 40-residue Pro- and Ala-rich segments that are largely extended and highly flexible and thereby provide the lipoyl domains with the mobility they require to interact with E1 and E3 and with neighboring E2 subunits (see below).

The hollow cagelike structures formed by the E2 catalytic domains (Fig. 21-5) contain channels large enough to allow substrates to diffuse in and out. In fact, coenzyme A and lipoamide bind in extended conformations at opposite ends of a 30-Å-long channel that is located at the interface between each pair of the subunits in a trimer (Fig. 21-9). This arrangement requires CoA to approach its binding site from inside the cage.

The NMR structures of the lipoyl domains of E2 from several sources verify the exposed nature of the site of the lipoyl group. They consist of a β barrel/sandwich containing a 4-stranded and a 3-stranded antiparallel β sheet, with the Lys to which the lipoyl group would be attached extending from an exposed position on a type I β bend linking two of the β strands in the 4-stranded sheet (Fig. 21-10*a*).

The NMR structure of the PSBD from *B. stearothermophilus*, determined by Richard Perham, reveals that its ~35-residue ordered region consists of two parallel helices separated by a loop that form a close-packed hydrophobic

Figure 21-8 **Domain structure of the dihydrolipoyl transacetylase (E2) subunit of the PDC.** The number of lipoyl domains, *n*, is species dependent: *n* = 3 for *E. coli* and *A. vinelandii*, *n* = 2 for mammals and *Streptococcus faecalis*, and *n* = 1 for *B. stearothermophilus* and yeast.

Figure 21-9 **X-ray structure of a trimer of *A. vinelandii* dihydrolipoyl transacetylase (E2) catalytic domains in complex with CoA and lipoamide.** The trimer, a portion of the cubic complex (Fig. 21-5*a*), is drawn in ribbon form and viewed along its 3-fold axis (the cubic complex's body diagonal) from inside the cube. The three identical subunits have different colors with the upper subunit colored in rainbow order from its N-terminus (*blue*) to its C-terminus (*red*). The CoA and lipoamide bound to the upper and lower left subunits are drawn in space-filling form and those bound to the lower right subunit are drawn in stick form, with CoA C green, lipoamide C cyan, N blue, O red, P orange, and S yellow. Note how the N-terminal "elbow" of each subunit extends behind the counterclockwise adjacent subunit; its deletion greatly destabilizes the complex. [Based on an X-ray structure by Wim Hol, University of Washington. PDBid 1EAB.]

(a)   (b)

Figure 21-10 **NMR structures of the lipoyl and peripheral subunit-binding domains of dihydrolipoyl transacetylase (E2).** (*a*) The NMR structure of the *A. vinelandii* dihydrolipoyl transacetylase (E2) lipoyl domain. The polypeptide chain is colored in rainbow order from its N-terminus (*blue*) to its C-terminus (*red*). The Lys side chain to which the lipoyl group would be bound is shown in stick form colored according to atom type (C green, H white, and N blue) and is located in a type I β turn. [Based on an NMR structure by Aart de Kok, Wageningen Agricultural University, Wageningen, Netherlands. PDBid 1IYU.] (*b*) The NMR structure of the peripheral subunit-binding domain (PSBD) from *B. stearothermophilus* E2. The polypeptide chain is colored in rainbow order from its N-terminus (*blue*) to its C-terminus (*red*). [Based on an NMR structure by Richard Perham, Cambridge University, U.K. PDBid 2PDD.]

core (Fig. 21-10*b*). It is one of the smallest known polypeptides that have a globular structure but lack disulfide bonds or prosthetic groups.

e. Intermediates Are Transferred between Enzyme Subunits by Flexible Tethers

How are reaction intermediates transferred between the component enzymes of the PDC? The group between the lipoamide disulfide bond and the E2 polypeptide backbone, the so-called **lipoyllysyl arm,** has a fully extended length of 14 Å:

**Lipoyllysyl arm
(fully extended)**

Evidently, *the lipoyllysyl arm, in combination with the >140-Å-long outer linker (Fig. 21-8), acts as a long flexible tether that swings its attached lipoyl group among the active sites of E1, E2, and E3.* Moreover, there is rapid interchange of acetyl groups among the lipoyl groups of the E2 core ($3 \times 24 = 72$ lipoyl groups in an *E. coli* PDC; $2 \times 60 = 120$ in mammals); the tethered arms also swing among themselves, exchanging both acetyl groups and disulfides:

One E1 subunit can therefore acetylate numerous E2 subunits and one E3 subunit can reoxidize several dihydrolipoamide groups.

f. Mammalian and Yeast PDCs Contain Additional Subunits

In mammals and yeast, the dodecahedral PDC's already complicated structure has a further level of complexity in that it contains additional subunits: About 12 copies of **E3 binding protein (E3BP)** facilitate the binding of E3 to the E2 core of the eukaryotic dodecahedral complex. E3BP has a lipoyllysine-containing domain similar to E2 and can accept an acetyl group, but its C-terminal domain has no catalytic activity, and the removal of its lipoyllysine domain does not diminish the catalytic activity of the complex. E3BP's main role seems to be to aid in the binding of E3, since limited proteolysis of E3BP decreases E3's binding ability.

g. Cryoelectron Microscopy Reveals the Structure of the Dodecahedral PDC

James Stoops and Reed have determined the organization of the bovine kidney dodecahedral complex through cryoelectron microscopy (Fig. 21-11). As expected, the E2 subunits form a dodecahedral core that is surrounded by a concentric dodecahedron of E1 subunits (Fig. 21-11a,b). The E2 subunits bind the E1 subunits via the radially extending, ~50-Å-long inner linkers that precede the PSBD of E2 (Fig. 21-8; these linkers are not present in Fig. 21-5b because that X-ray structure only contains E2's catalytic domain). Although the bovine kidney PDC used in the structure determination lacked sufficient E3BP · E3 to be visible, its position in the electron microscopy–based structure of yeast PDC indicates that an E3 dimer occupies and largely fills each pentagonal opening of the E2 core (Fig. 21-11c). Hence the E3 subunits are not arranged with dodecahedral symmetry (similarly, the E3 subunits in the cubic *E. coli* PDC are not arranged with cubic symmetry;

(a)

100 Å

(b)

Figure 21-11   Electron microscopy–based images of the bovine kidney pyruvate dehydrogenase complex at ~35 Å resolution.
(*a*) The entire particle as viewed along its 3-fold axis of symmetry. E1 is yellow, the E2 catalytic core is green, and the inner linkers that connect E2's catalytic domains to its E1-binding domains, the PSBDs (Fig. 21-8), are cyan. The particle is viewed along a 3-fold axis of symmetry with the positions of a 5-fold axis and a 2-fold axis also marked. (*b*) A cutaway diagram viewed and colored as in Part *a* but with the particle's closest half removed to reveal the E2 catalytic core and the inner linkers. Compare the green portion with the X-ray structure of a dodecahedral E2 catalytic core as viewed along a 2-fold axis (Fig. 21-5b). (*c*) A cutaway diagram as in Part *b* but with E3 dimers (Fig. 21-13a) shown at 20 Å resolution (*red*) modeled into the pentagonal openings of the E2 core. The position of a peripheral subunit-binding site at the end of an E2 inner linker is marked by an asterisk (*). [From Zhou, Z.H., McCarthy, D.B., O'Connor, C.M., Reed, L.J., and Stoops, J.K., *Proc. Natl. Acad. Sci.* **98,** 14802 (2001).]

(c)

Fig. 21-4*b*). The PSBD, which is located at the end of the E2 inner linker (* in Fig. 21-11*c*), is the point from which the E2 lipoyl domains (Fig. 21-10*a*) swing. It is ~50 Å from the nearest active sites of E1, E2, and E3.

In addition to E3BP, the mammalian PDC contains one to three copies each of **pyruvate dehydrogenase kinase** and **pyruvate dehydrogenase phosphatase.** The kinase and phosphatase function to regulate the catalytic activity of the complex (Section 21-2Cb).

h. Two Other Multienzyme Complexes Closely Resemble the PDC

In addition to the PDC, most cells contain two other closely related multienzyme complexes: the **α-ketoglutarate dehydrogenase complex** [also called **2-oxoglutarate dehydrogenase (OGDH),** which catalyzes Reaction 4 of the citric acid cycle; Fig. 21-1] and the **branched-chain α-keto acid dehydrogenase complex (BCKDH;** which participates in the degradation of isoleucine, leucine, and valine; Section 26-3Ee). These multienzyme complexes all catalyze similar reactions: the NAD⁺-linked oxidative decarboxylation of an α-keto acid with the transfer of the resulting acyl group to CoA. In fact, all three members of this **2-ketoacid dehydrogenase** family (alternatively, the **2-oxo acid dehydrogenase** family) of multienzyme complexes share the same E3 subunit, and their E1 and E2 subunits, which are specific for their corresponding substrates, are homologous and use identical cofactors. Thus, to differentiate them, the E1s for PDC, OGDH, and BCKDH are often referred to as **E1p, E1o,** and **E1b,** respectively, and likewise, their E3s are called **E3p, E3o,** and **E3b.**

i. Arsenic Compounds Are Poisonous because They Sequester Lipoamide

Arsenic has been known to be a poison since ancient times. As(III) compounds, such as **arsenite** (AsO_3^{3-}) and organic arsenicals, are toxic because of their ability to covalently bind sulfhydryl compounds. This is particularly true of vicinal (adjacent) sulfhydryls such as those of lipoamide because they can form bidentate adducts:

Arsenite   Dihydrolipoamide

Organic arsenical

The resultant inactivation of lipoamide-containing enzymes, especially pyruvate dehydrogenase and α-ketoglutarate dehydrogenase (Section 21-3D), brings respiration to a halt.

Organic arsenicals are more toxic to microorganisms than to humans, apparently because of differences in the sensitivities of their various enzymes to these compounds. This differential toxicity is the basis for the early twentieth century use of organic arsenicals in the treatment of **syphilis** (now superseded by penicillin) and **trypanosomiasis** (typanosomes are parasitic protozoa that cause several diseases including **African sleeping sickness** and **Chagas-Cruz disease**). These compounds were really the first antibiotics, although, not surprisingly, they had severe side effects.

j. Arsenic Poisoning, Napoleon, and Darwin

Arsenic is often suspected as a poison in untimely deaths. In fact, it has long been thought that Napoleon Bonaparte died from arsenic poisoning while in exile on St. Helena, an island in the Atlantic Ocean. This suspicion, and the chemical analyses it sparked, makes a fascinating chemical anecdote. The finding that a lock of Napoleon's hair indeed contains a high level of arsenic strongly supports the notion that arsenic poisoning at least contributed to his death. But was it murder or environmental pollution? A sample of the wallpaper from Napoleon's drawing room was found to contain the commonly used (at the time) green pigment copper arsenate ($CuHAsO_4$). It was eventually determined that in a damp climate, as occurs on St. Helena, fungi growing on the wallpaper eliminate the arsenic by converting it to the volatile and highly toxic trimethyl arsine [$(CH_3)_3As$]. Indeed, Napoleon's regular visitors also suffered from symptoms of arsenic poisoning (e.g., gastrointestinal disturbances), which appeared to moderate when they spent much of their time outdoors. Thus, Napoleon's arsenic poisoning may have been unintentional.

Retrospective detective work also suggests that Charles Darwin was a victim of chronic arsenic poisoning. For most of his life after he returned from his epic voyage, Darwin complained of numerous ailments, including eczema, vertigo, headaches, arthritis, gout, palpitations, and nausea, all symptoms of arsenic poisoning. Fowler's solution, a common nineteenth century tonic, contained 10 mg of arsenite · mL^{-1}. Many individuals, quite possibly Darwin himself, took this "medication" for years.

k. The Structure of E1

In dodecahedral PDCs, E1 (pyruvate dehydrogenase) is a 2-fold symmetric $\alpha_2\beta_2$ heterotetramer. However, in cubic PDCs and in α-ketoglutarate dehydrogenase, which is also cubic, the α and β subunits are genetically fused to form a single polypeptide and hence a homodimeric enzyme.

The X-ray structure of E1 from the (dodecahedral) *B. stearothermophilus* PDC in complex with TPP and a

PSBD (Fig. 21-12*a*), determined by Perham and Ben Luisi, reveals a tightly associated $\alpha_2\beta_2$ heterotetramer, whose 368-residue α subunits and 324-residue β sub-

(a)

(b)

Figure 21-12   **X-ray structure of *B. stearothermophilus* E1 in complex with TPP and the peripheral subunit-binding domain (PSBD).** (*a*) The E1 $\alpha_2\beta_2$ heterotetramer, viewed with its 2-fold axis of symmetry vertical and the E2 dodecahedron below, is drawn with its α subunits in worm form and its β subunits in ribbon form. The α and β subunits that are closest to the viewer and predominantly on the left are each colored in rainbow order from their N-termini (*blue*) to their C-termini (*red*); the remaining α and β subunits are pale cyan and pale green. The PSBD (*below*) is light blue. The TPP is shown in space-filling form with C green, N blue, O red, P orange, and S yellow. The entrances to the enzyme's active site channels are indicated by black arrows. (*b*) A cutaway surface diagram of the solvent-filled tunnel connecting the enzyme's two active sites. The hydroxyethyl-TPPs bound at the enzyme's active sites and the Asp and Glu side chains lining the tunnel are drawn in stick form with C gray, N blue, O red, P green, and S yellow. Water molecules and Mg^{2+} ions are represented by red and magenta spheres. [Part *a* based on an X-ray structure by and Part *b* courtesy of Richard Perham and Ben Luisi, University of Cambridge, U.K. PDBid 1W85.]

units each consist of two domains. A single PSBD binds to the C-terminal domains of the β subunits along the 2-fold axis relating them in a way that sterically prevents the binding of a second PSBD. Hence the PSBD interacts asymmetrically with the two chemically identical domains in a manner reminiscent of the way that human growth hormone interacts with its homodimeric receptor (Section 19-1J). E1's conserved ~310-residue core occurs in other TPP-utilizing enzymes of known structure, including pyruvate decarboxylase (Section 17-3Ba). Each TPP is bound between the N-terminal domains of an α and a β subunit at the end of an ~21-Å-deep funnel-shaped channel with its thiazolium ring (its reactive group) closest to the channel entrance. Apparently, the lipoyllysyl arm at the end of an E2 lipoyl domain is inserted into this channel in an extended conformation for the transfer of the TPP's hydroxyacyl substituent to lipoamide.

I. E1 Active Sites Are Coordinated via a Proton Wire

When apo-E1 is mixed with TPP, the first TPP to bind to this apparently 2-fold symmetric protein does so orders of magnitude faster than the second TPP. This indicates that these two binding sites, which are ~20 Å apart, somehow communicate. In fact, the enzyme's two active sites in the foregoing X-ray structure have different structures in that two conserved loops (residues 203–212 and 276–287 of the α subunits) at the entrance to the channel leading to TPP are ordered in one subunit in a way that blocks the active site but are disordered in the other subunit. That this is not an artifact of crystallization was demonstrated by the observation that limited proteolysis cleaves the more C-terminal of these loops on one subunit, but not the other. Similar asymmetries have been observed in all TPP-dependent enzymes of known structure.

The X-ray structure of E1 reveals a solvent-filled tunnel linking its two active sites that is lined with 10 conserved acidic residues (Fig. 21-12*b*). Similar acidic tunnels are present in all TPP-dependent enzymes of known structure, all of which are either dimeric or tetrameric. This led Luisi and Perham to propose that as pyruvate proceeds through the E1-catalyzed Reactions 1 and 2 discussed in Section 21-2Ac, the proton required by Reaction 1 in one active site is supplied by the proton released in Reaction 2 in the other active site. The protons are presumably transferred between the two active sites via series of proton jumps (Section 2-1C) through the acidic tunnel, which thereby functions as a **proton wire.** Consequently, as the two active sites progress through their catalytic cycles, they must remain out of phase with one another; that is, one site requires a general base when the other site requires a general acid and vice versa so that they reciprocate their catalytic needs via the proton wire. This explains the observed negative cooperativity of substrate binding exhibited by E1. Structurally, this is rationalized by the alternate closing and opening of the active sites through the out of phase ordering and disordering of the loops at the entrance of each active site channel.

B. *The Mechanism of Dihydrolipoyl Dehydrogenase*

The reaction catalyzed by dihydrolipoyl dehydrogenase (E3) is more complex than Reactions 4 and 5 in Fig. 21-6 suggest. Vincent Massey demonstrated that *oxidized dihydrolipoyl dehydrogenase contains a "redox-active" disulfide bond, which in the enzyme's reduced form has accepted an electron pair through bond cleavage to form a dithiol:*

He established this through the following observations involving arsenite (which, as we saw in Section 21-2Ai, reacts with vicinal sulfhydryl groups but not with disulfides).

1. The spectrum of oxidized dihydrolipoyl dehydrogenase (E) is unaffected by arsenite.

2. When NADH reacts with the oxidized enzyme in the presence of arsenite, the resulting reduced enzyme (EH$_2$) binds arsenite to form an enzymatically inactive species.

3. The oxidation state of the flavin in a **flavoprotein** (flavin-containing protein) is readily established from its characteristic UV–visible spectrum: FAD is an intense yellow, whereas FADH$_2$ is pale yellow. The spectrum of the arsenite-inactivated EH$_2$ indicates that its FAD prosthetic group is fully oxidized.

Oxidized dihydrolipoyl dehydrogenase must therefore have a second electron acceptor in addition to FAD; arsenite's known specificity suggests that the acceptor is a disulfide. Dihydrolipoyl dehydrogenase's amino acid sequence indicates that its redox-active disulfide bond forms between its Cys 43 and Cys 48, which occur on a highly conserved segment of the enzyme's polypeptide chain.

a. The X-Ray Structures of Dihydrolipoyl Dehydrogenase and Glutathione Reductase

The X-ray structures of dihydrolipoyl dehydrogenase from several microorganisms, mostly determined by Hol, reveal that each ~470-residue subunit of these homodimeric enzymes is folded into four domains, all of which participate in forming the subunit's catalytic center (Fig. 21-13*a*). An E2 PSBD (not present in Fig. 21-13) binds to the bottom of the enzyme along its 2-fold axis in much the same way that it binds to E1 (Fig. 21-12). The flavin is almost completely buried in the protein, which prevents the surrounding solution from interfering with the electron-transfer reaction catalyzed by the enzyme (FADH$_2$, but not NADH or thiol, is rapidly oxidized by O$_2$). The redox-active disulfide, which is located on the opposite side of the flavin ring from the nicotinamide ring (Fig. 21-13*b*), links successive turns in a distorted segment of an α helix (in an undistorted helix, the C$_\alpha$ atoms of Cys 43 and Cys 48 would be too far apart to permit the disulfide bond to form).

(a)

(b)

Figure 21-13 **X-ray structure of dihydrolipoyl dehydrogenase (E3) from the gram-negative bacterium *Pseudomonas putida* in complex with FAD and NAD$^+$.** (*a*) The homodimeric enzyme is viewed with its 2-fold axis of symmetry vertical and the E2 dodecahedron below. One subunit is gray and the other is colored according to domain, with its FAD-binding domain (residues 1–142) light orange, its NAD$^+$-binding domain (residues 143–268) purple, its central domain (residues 269–337) yellow-green, and its interface domain (residues 338–458) pink. The NAD$^+$ and FAD in both subunits are shown in stick form with FAD C green, NAD$^+$ C cyan, N blue, O red, and P orange. (*b*) The active site region of the left subunit in Part *a* in which the central and interface domains have been deleted for clarity. The view is very roughly in the same direction as in Part *a*. The redox-active portions of the bound NAD$^+$ and FAD cofactors, the side chains of Cys 43 and Cys 48 forming the redox-active disulfide bond, and the side chain of Tyr 181 are shown in ball-and-stick form with FAD C green, NAD$^+$ C cyan, side chain C magenta, N blue, O red, P magenta, and S yellow. Note that the side chain of Tyr 181 is interposed between the flavin and the nicotinamide rings. [Based on an X-ray structure by Wim Hol, University of Washington. PDBid 1LVL.]

E3's catalytic mechanism has been largely determined in analogy with that of the homologous (~33% identical) but structurally more extensively characterized **glutathione reductase (GR).** This nearly ubiquitous enzyme catalyzes

Stage II:
NAD⁺
reduction

Stage I:
Dihydrolipoamide
oxidation

Figure 21-14 **Catalytic reaction cycle of dihydrolipoyl dehydrogenase.** The catalytic center is surrounded by protein so that the NAD⁺ and dihydrolipoamide binding sites are in deep pockets. The catalytic cycle consists of six steps: **(1)** The oxidized enzyme, E, which contains a redox-active disulfide bond between Cys 43 and Cys 48, binds dihydrolipoamide, LH₂, the enzyme's first substrate, to form an enzyme–substrate complex, E · LH₂. **(2)** A substrate S atom nucleophilically attacks S₄₃ to yield a disulfide bond and release S₄₈ as a thiolate ion. The proton on the substrate's second thiol group is abstracted by His 451′ to yield a second thiolate ion, E–S–S–L · S⁻. **(3)** The substrate thiolate ion nucleophilically displaces S₄₃ aided by general acid catalysis by

His 451′ to yield the enzyme's lipoamide product in its complex with the stable reduced enzyme, EH₂ · L, in which S₄₈ forms a charge-transfer complex with the flavin ring (*dotted red line to the red flavin ring*). **(4)** The lipoamide product is released, yielding EH₂. The phenol side chain of Tyr 181 continues to block access to the flavin ring of FAD, so as to prevent the enzyme's oxidation by O₂. **(5)** The enzyme's second substrate, NAD⁺, binds to EH₂ to form EH₂ · NAD⁺. The phenol side chain of Tyr 181 is pushed aside by the nicotinamide ring of NAD⁺. **(6)** The catalytic cycle is then closed by the reduction of the NAD⁺ by EH₂ to reform the oxidized enzyme E and yield the enzyme's second product, NADH.

the NADPH-dependent reduction of **glutathione disulfide (GSSG)** to the intracellular reducing agent **glutathione (GSH;** its physiological function is discussed in Sections 23-4E and 26-4C):

$$H_3\overset{+}{N}-CH-CH_2-CH_2-\overset{O}{\overset{\|}{C}}-NH-CH-\overset{O}{\overset{\|}{C}}-NH-CH_2-COO^-$$

with side chains COO⁻, and CH₂–S–S–CH₂

Second molecule:
$$H_3\overset{+}{N}-CH-CH_2-CH_2-\overset{O}{\overset{\|}{C}}-NH-CH-\overset{O}{\overset{\|}{C}}-NH-CH_2-COO^-$$
with side chain COO⁻

Glutathione disulfide (GSSG)

NADPH + H⁺ ⟶ NADP⁺ | glutathione reductase

$$2\ H_3\overset{+}{N}-CH-CH_2-CH_2-\overset{O}{\overset{\|}{C}}-NH-CH-\overset{O}{\overset{\|}{C}}-NH-CH_2-COO^-$$
with side chains COO⁻ and CH₂–SH

Glutathione (GSH)
(γ-L-Glutamyl-L-cysteinylglycine)

Figure 21-15   The complex of the flavin ring, the nicotinamide ring, and the side chain of Cys 63 observed in the X-ray structure of human erythrocyte glutathione reductase. The view is perpendicular to the flavin ring. Atoms are colored according to type with nicotinamide C cyan, flavin C green, Cys 63 C purple, N blue, O red, and S yellow. The two planar heterocycles are parallel and in van der Waals contact. The Cys 63 S atom (equivalent to S_{48} of dihydrolipoyl dehydrogenase), a member of the redox-active disulfide, is also in van der Waals contact with the flavin ring on the opposite side of it from the nicotinamide ring. [Based on an X-ray structure by Andrew Karplus and Georg Schulz, Institut für Organische Chemie und Biochemie, Freiburg, Germany. PDBid 1GRB.]

Note that this reaction is analogous to that catalyzed by dihydrolipoyl dehydrogenase, but that these two reactions normally occur in opposite directions; that is, dihydrolipoyl dehydrogenase normally uses NAD^+ to oxidize two thiol groups to a disulfide (Fig. 21-6), whereas GR uses NADPH to reduce a disulfide to two thiol groups. Nevertheless, the active sites of these two enzymes are closely superimposable.

The arrangement of the groups in dihydrolipoyl dehydrogenase's catalytic center and its reaction sequence are diagrammed in Fig. 21-14. The substrate-binding site is located in the interface between the enzyme's two subunits in the vicinity of the redox-active disulfide. In the absence of NAD^+, the phenol side chain of its Tyr 181 covers the nicotinamide-binding pocket so as to shield the flavin from contact with the solution. Indeed, Fig. 21-13*b* shows an enzyme–product complex of oxidized dihydrolipoyl dehydrogenase with NAD^+ in which the side chain of Tyr 181 is interposed between the nicotinamide ring and its binding site at the enzyme's catalytic center. However, in the X-ray structure of GR in complex with NADPH, determined by Georg Schulz and Heiner Schirmer, this side chain has moved aside such that the reduced nicotinamide ring binds parallel to and in van der Waals contact with the flavin ring of the fully oxidized enzyme, E (Fig. 21-15). The H_S substituent to this reduced nicotinamide's prochiral C4 atom (that facing the flavin), the H atom that is known to be lost in the GR reaction, lies near flavin atom N5, the position through which electrons often enter a flavin ring on its reduction. This positioning is particularly significant in view of the catalytic mechanism described below.

b. Catalytic Mechanism

Protein X-ray structures neither reveal H atom positions (except in the most highly resolved structures) nor indicate pathways of electron transfer. The electron-transfer pathway in the glutathione reductase reaction has nevertheless been inferred from the X-ray structures of a series of stable enzymatic reaction intermediates as augmented with a variety of enzymological data. Here we present this mechanism in terms of the reaction catalyzed by the closely similar dihydrolipoyl dehydrogenase (Fig. 21-14).

Dihydrolipoyl dehydrogenase has a Ping Pong mechanism (Section 14-5Aa); each of its two substrates, dihydrolipoamide and NAD^+, react in the absence of the other (Fig. 21-6, Reactions 4 and 5). Stage I of the catalytic reaction (Fig. 21-14, Steps 1–4) involves a disulfide interchange reaction between the first substrate, dihydrolipoamide (LH_2), and the redox-active disulfide on the enzyme, a process in which His 451′ functions as a general acid–base (primed residues refer to the opposite subunit). In fact, Glu 456′, His 451′, and Cys 43 are arranged much like the catalytic triad of serine proteases (Section 15-3B), with the Cys SH replacing the Ser OH. The importance of His 451′ as an acid–base catalyst was established by Charles Williams through his observation that mutationally changing it to Gln yields an enzyme that retains only ~0.4% of the wild-type catalytic activity.

The thiolate anion formed from Cys 48 in Step 2 forms a **charge-transfer complex** with the flavin in which S_{48}^- (the S atom of Cys 48) contacts the flavin ring near its 4a position (a charge-transfer complex is a noncovalent interaction in which an electron pair is partially transferred from a donor, in this case S_{48}^-, to an acceptor, in this case the oxidized flavin ring; the red color of this complex is indicative of the formation of the charge-transfer complex).

Stage II of this Ping Pong reaction (Fig. 21-14, Steps 5 and 6) involves the binding and reduction of NAD^+ followed by the release of NADH to regenerate the oxidized enzyme. The path of the electrons from the reactive disulfide in its reduced form through the FAD to NAD^+ has been elucidated by spectroscopic studies and the chemistry of model compounds. These indicate that an electron pair is rapidly transferred from S_{48}^- to the flavin ring through the transient formation of a covalent bond from S_{48} to flavin atom 4a (Fig. 21-16, Step 1). His 451' then abstracts the proton from the S_{43} thiol to form a thiolate ion, which nucleophilically attacks S_{48} to reform the redox-active disulfide group (Fig. 21-16, Step 2). The resulting reduced flavin anion ($FADH^-$) has but transient existence. The H atom substituent to its N5 is immediately transferred (formally as a hydride ion) to the juxtaposed C4 atom of the nicotinamide ring (Fig. 21-15), yielding FAD and the reaction's second product, NADH, thereby completing the catalytic cycle. Thus, the FAD appears to function more like an electron conduit between the reduced form of the redox-active disulfide and NAD^+ than as a source or sink of electrons.

C. *Control of Pyruvate Dehydrogenase*

The PDC regulates the entrance of acetyl units derived from carbohydrate sources into the citric acid cycle. The decarboxylation of pyruvate by E1 is irreversible and, since there are no other pathways in mammals for the synthesis of acetyl-CoA from pyruvate, it is crucial that the reaction be carefully controlled. Two regulatory systems are employed:

1. Product inhibition by NADH and acetyl-CoA (Fig. 21-17*a*).

2. Covalent modification by phosphorylation/dephosphorylation of the pyruvate dehydrogenase (E1) subunit (Fig. 21-17*b*; enzymatic regulation by covalent modification is discussed in Section 18-3Ba).

a. Control by Product Inhibition
NADH and acetyl-CoA compete with NAD^+ and CoA for binding sites on their respective enzymes. They also drive the reversible transacetylase (E2) and dihydrolipoyl dehydrogenase (E3) reactions backward (Fig. 21-17a). High ratios of [NADH]/[NAD^+] and [acetyl-CoA]/[CoA] therefore maintain E2 in the acetylated form, incapable of accepting the hydroxyethyl group from the TPP on E1. This, in turn, ties up the TPP on the E1 subunit in its hydroxyethyl form, decreasing the rate of pyruvate decarboxylation.

b. Control by Phosphorylation/Dephosphorylation
Control by phosphorylation/dephosphorylation occurs only in eukaryotic enzyme complexes. These complexes contain pyruvate dehydrogenase kinase and pyruvate

Figure 21-16 **The reaction transferring an electron pair from dihydrolipoyl dehydrogenase's redox-active disulfide in its reduced form to the enzyme's bound flavin ring. (1)** The collapse of the charge-transfer complex between the Cys 48 thiolate ion and the flavin ring (*dashed red line*) to form a covalent bond between S_{48} and flavin atom C4a. S_{48} is located out of the plane of the flavin, as Fig. 21-15 indicates. Flavin atom N5 acquires a proton, possibly from S_{43}, which becomes a thiolate ion. **(2)** The S_{43} thiolate nucleophilically attacks S_{48} to form the redox-active disulfide bond, thereby releasing the reduced flavin anion $FADH^-$.

(a) **Product inhibition**

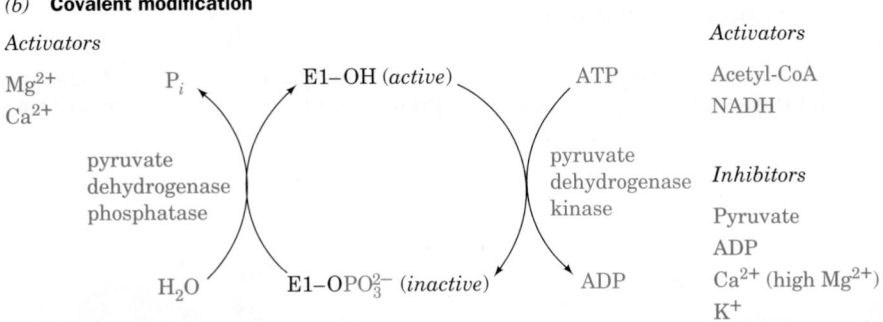

(b) **Covalent modification**

Activators   *Activators*

Mg^{2+} Acetyl-CoA
Ca^{2+} NADH

pyruvate   pyruvate   *Inhibitors*
dehydrogenase   dehydrogenase
phosphatase   kinase   Pyruvate
                                                                ADP
                                                                Ca^{2+} (high Mg^{2+})
                                                                K^+

Figure 21-17   Factors controlling the activity of the PDC.
(*a*) Product inhibition. NADH and acetyl-CoA, respectively,
compete with NAD^+ and CoA in Reactions 3 and 5 of the
pyruvate dehydrogenase reaction sequence. When the relative
concentrations of NADH and acetyl-CoA are high, the
reversible reactions catalyzed by E2 and E3 are driven
backward (*red arrows*), thereby inhibiting further formation of
acetyl-CoA. (*b*) Covalent modification in the eukaryotic

complex. Pyruvate dehydrogenase (E1) is inactivated by the
specific phosphorylation of one of its Ser residues in a reaction
catalyzed by pyruvate dehydrogenase kinase (*right*). This
phosphoryl group is hydrolyzed through the action of pyruvate
dehydrogenase phosphatase (*left*), thereby reactivating E1. The
activators and inhibitors of the kinase are listed on the right and
the activators of the phosphatase are listed on the left.

dehydrogenase phosphatase bound to the dihydrolipoyl
transacetylase (E2) core. *The kinase inactivates the pyruvate dehydrogenase (E1) subunit by catalyzing the ATP-
dependent phosphorylation of a Ser residue on the more
C-terminal loop at the entrance of the active site channel
(Fig. 21-17b).* Moreover, phosphorylation at only one of
E1's two active sites inactivates the entire enzyme, thereby
further demonstrating the obligatory nature of the out of
phase coupling between these two active sites (Section 21-
2AI). *Hydrolysis of this phosphoSer residue by the phosphatase reactivates the complex.*

Pyruvate dehydrogenase kinase is activated through its
interaction with the acetylated form of E2. Consequently,
the products of the reaction, NADH and acetyl-CoA, in
addition to their direct effects on the PDC, indirectly activate pyruvate dehydrogenase kinase. The resultant phosphorylation inactivates the complex just as the products
themselves inhibit it. Acetyl-CoA and NADH are products
of fatty acid oxidation (Section 25-2) so that this inhibition

of PDC serves to preserve carbohydrate stores when lipid
fuels are available.

Calcium ion is an important second messenger signaling
the need for increased energy (e.g., for muscle contraction). Increasing $[Ca^{2+}]$ enhances pyruvate dehydrogenase
phosphatase activity, thus activating PDC.

Insulin is involved in the control of this system through
its indirect activation of pyruvate dehydrogenase phosphatase. Recall that insulin activates glycogen synthesis as
well by activating phosphoprotein phosphatase-1 (Section
18-3Cg). Insulin, in response to increases in blood glucose,
is now seen as promoting the synthesis of acetyl-CoA as
well as glycogen. As we shall see in Section 25-4, acetyl-
CoA is the precursor to fatty acids in addition to being the
fuel for the citric acid cycle. Various other activators and
inhibitors regulate the pyruvate dehydrogenase system
(Fig. 21-17b); in contrast to the glycogen metabolism control system (Section 18-3), however, it is unaffected by
cAMP.

3 ENZYMES OF THE CITRIC ACID CYCLE

In this section we discuss the reaction mechanisms of the eight citric acid cycle enzymes. Our knowledge of these mechanisms rests on an enormous amount of experimental work; as we progress, we shall pause to examine some of these experimental details. Consideration of how this cycle is regulated and its relationship to cellular metabolism are the subjects of the following sections.

A. *Citrate Synthase*

Citrate synthase (originally named **citrate condensing enzyme***) catalyzes the condensation of acetyl-CoA and oxaloacetate (Reaction 1 of Fig. 21-1).* This initial reaction of the citric acid cycle is the point at which carbon atoms are "fed into the furnace" as acetyl-CoA. The citrate synthase reaction proceeds via an Ordered sequential kinetic mechanism (Section 14-5Aa), with oxaloacetate adding to the enzyme before acetyl-CoA.

The X-ray structure of the free dimeric enzyme, determined by James Remington and Robert Huber, shows it to adopt an "open form" in which the two domains of each subunit form a deep cleft that contains the oxaloacetate binding site (Fig. 21-18a). On binding oxaloacetate, however, the smaller domain undergoes a remarkable 18° rotation relative to the larger domain, which closes the cleft (Fig. 21-18b).

The X-ray structures of two inhibitors of citrate synthase in ternary complex with the enzyme and oxaloacetate have also been determined. **Acetonyl-CoA,** an inhibitory analog of acetyl-CoA in the ground state, and **carboxymethyl-CoA,** a proposed transition state analog (see below),

Acetyl-CoA

Proposed enol intermediate

Acetonyl-CoA (ground-state analog)

Carboxymethyl-CoA (transition state analog)

Figure 21-18 **Conformational changes in citrate synthase.**
(*a*) The open conformation. (*b*) The closed, substrate-binding conformation. In both forms, the homodimeric protein is viewed along its 2-fold axis and is represented by its transparent molecular surface with its polypeptide chains drawn in worm form colored in rainbow order from blue at their N-termini to red at their C-termini. The reaction product citrate, which is bound to both enzymatic forms, and coenzyme A, which is also bound to the closed form, are shown in space-filling form with citrate C cyan, CoA C green, N blue, O red, and P orange. The large conformational shift between the open and closed forms entails 18° rotations of the small domains (*upper left and lower right*) relative to the large domains resulting in relative interatomic movements of up to 15 Å. [Based on X-ray structures by James Remington and Robert Huber, Max-Planck-Institut für Biochemie, Martinsried, Germany. PDBids 1CTS and 2CTS.] *See* **Interactive Exercise 15**

bind to the enzyme in its "closed" form, thereby identifying the acetyl-CoA binding site. The existence of the "open" and "closed" forms explains the enzyme's ordered sequential kinetic behavior: *The conformational change induced by oxaloacetate binding generates the acetyl-CoA binding site while sealing off the solvent's access to the bound oxaloacetate.* This is a classic example of the induced-fit model of substrate binding (Section 10-4C). Hexokinase exhibits similar behavior (Section 17-2Aa).

The citrate synthase reaction is a mixed aldol–Claisen ester condensation, subject to general acid–base catalysis and the intermediate participation of the enol(ate) form of acetyl-CoA. The X-ray structure of the enzyme in ternary complex with oxaloacetate and carboxymethyl-CoA reveals that three of its ionizable side chains are properly oriented to play catalytic roles: His 274, Asp 375, and His 320. The N1 atoms in both of these His side chains are hydrogen bonded to two backbone NH groups indicating that these N1 atoms are not protonated. The participation of His 274, Asp 375, and His 320 has been confirmed by kinetic studies of mutant enzymes generated by site-directed mutagenesis. The following three-step mechanism, formulated mainly by Remington, takes into account these observations (Fig. 21-19):

1. The enolate of acetyl-CoA is generated in the rate-limiting step of the reaction, with the catalytic participation of Asp 375 acting as a base to remove a proton from the methyl group and His 274, in its neutral form, H-bonding the enolate oxygen (whose protonated carbonyl form is normally far more acidic than a neutral His side chain). Much as we discussed for a similar step in the triose phosphate isomerase reaction (Section 17-2Eb), the pK of the protonated form of the substrate thioester carbonyl oxygen increases to ~14 on enolization, which approximates that of neutral His 274. Hence, it has been proposed that this step is facilitated through the formation of a low-barrier hydrogen bond (which, it will be recalled, is a particularly strong form of hydrogen bonding interaction in which the hydrogen atom is more or less equally shared between the "donor" and "acceptor" atoms; Section 15-3Dd). However, whether or not this low-barrier hydrogen bond actually forms is a subject of controversy. The presence of the thioester bond to CoA facilitates the enolization; enolization of acetate alone would require the generation of a much more highly charged and therefore less stable intermediate.

2. Citryl-CoA is formed in a second acid–base-catalyzed reaction step in which the acetyl-CoA enolate nucleophilically attacks oxaloacetate while His 320, also in its neutral form, donates a proton to oxaloacetate's carbonyl group. The citryl-CoA remains bound to the enzyme.

3. Citryl-CoA is hydrolyzed to citrate and CoA, with His 320, now in the anionic form, abstracting a proton from water as it attacks the carbonyl group of the citryl-CoA and the acidic form of Asp 375 donating a proton to the CoA as it leaves. This hydrolysis provides the reaction's thermodynamic driving force ($\Delta G^{\circ\prime} = -31.5$ kJ · mol^{-1}). We shall see presently why the reaction requires such a large, seemingly wasteful, release of free energy.

Figure 21-19 **Mechanism and stereochemistry of the citrate synthase reaction.** His 274 and His 320 in their neutral forms and Asp 375 have been implicated as general acid–base catalysts. The overall reaction's rate-limiting step is the formation of the acetyl-CoA enolate, which may be stabilized by a low-barrier hydrogen bond to His 274. The acetyl-CoA enolate then nucleophilically attacks the *si* face of oxaloacetate's carbonyl carbon. The resulting intermediate, (*S*)-citryl-CoA, is hydrolyzed to yield citrate and CoA. [Mostly after Remington, S.J., *Curr. Opin. Struct. Biol.* **2,** 732 (1992)].

Enzyme-catalyzed reactions, as we have previously noted, are stereospecific. The aldol–Claisen condensation that occurs here involves attack of the acetyl-CoA enolate exclusively at the *si* face of oxaloacetate's carbonyl carbon atom, thereby yielding (*S*)-citryl-CoA (chirality nomenclature is presented in Section 4-2C). The acetyl group of acetyl-CoA thereby only forms citrate's *pro-S* carboxymethyl group.

B. Aconitase

Aconitase catalyzes the reversible isomerization of citrate and isocitrate with cis-aconitate as an intermediate (Reaction 2 of Fig. 21-1). Although citrate has a plane of symmetry and is therefore not optically active, it is nevertheless prochiral; aconitase can distinguish between citrate's *pro-R* and *pro-S* carboxymethyl groups.

A combination of X-ray crystallography and site-directed mutagenesis by Helmut Beinert and David Stout identified several amino acid residues that participate in catalysis. Formation of the *cis*-aconitate intermediate involves a dehydration in which Ser 642 alkoxide, acting as a general base, abstracts the *pro-R* proton at C2 of citrate's *pro-R* carboxymethyl group (Fig. 21-20, *top*). [The O_γ of Ser 642 occupies a sort of oxyanion hole (Section 15-3Db) that apparently stabilizes its otherwise highly basic alkoxide form. The carbanion formed in the transition state of this reaction is not shown.] This is followed by the loss of the OH group at C3 in a trans elimination of H_2O to form the *cis*-aconitate intermediate. The latter reaction step is facilitated through general acid catalysis by His 101, whose imidazolium group is polarized through ion pairing to the side chain of Asp 100. This model is corroborated by the observations that the mutagenesis of Asp 100, His 101, or Ser 642 results in a 10^3- to 10^5-fold reduction in aconitase's catalytic activity without greatly affecting its substrate binding affinity.

Figure 21-20   **Mechanism and stereochemistry of the aconitase reaction.** Fe_a of the enzyme's [4Fe–4S] cluster coordinates the citrate hydroxyl and central carboxyl groups; Arg 580 forms a salt bridge with the *pro-S* carboxyl group; Ser 642, in its alkoxide form, acts as a general base; and the Asp 100-polarized His 101 acts as a general acid in the elimination of water to form *cis*-aconitate. Note the unusual 180° flip that *cis*-aconitate apparently undergoes, possibly while remaining bound to the active site. Thus rehydration takes place on the opposite face of the substrate from which dehydration occurred, yielding (2*R*,3*S*)-isocitrate.

Aconitase contains a covalently bound **[4Fe–4S] iron–sulfur cluster,** which is required for catalytic activity (the properties of iron–sulfur clusters are discussed in Section 22-2Ca). A specific Fe(II) atom in this cluster, the so-called Fe_a atom, is thought to coordinate the OH group of the substrate so as to facilitate its elimination. Iron–sulfur clusters are almost always associated with redox processes although, intriguingly, not in the case of aconitase. It has been postulated that the electronic properties of the [4Fe–4S] cluster permit the Fe_a atom to expand its coordination shell from the four ligands observed in the X-ray structure of the free enzyme (three S^{2-} ions and a OH^- ion) to the six octahedrally arranged ligands observed in the enzyme–substrate complex. A single metal ion, such as Zn^{2+} or Cu^{2+}, is unable to do so, and hence would require that some of its ligands be displaced on binding substrate.

The second stage of the aconitase reaction is rehydration of *cis*-aconitate's double bond to form isocitrate (Fig. 21-20, *bottom*). The nonenzymatic addition of H_2O across the double bond of *cis*-aconitate would yield four stereoisomers. Aconitase, however, catalyzes the stereospecific trans addition of OH^- and H^+ across the double bond to form only (2R,3S)-isocitrate in the forward reaction and citrate in the reverse reaction.

Although citrate's OH group is lost to the solvent in the aconitase reaction, the abstracted H^+ is retained by Ser 642. Remarkably, it adds to the opposite faces of *cis*-aconitate's double bond in forming citrate and isocitrate. The *cis*-aconitate must expose a different face to the sequestered H^+ for the formation of isocitrate. This is thought to occur by a 180° flip of the intermediate on the enzyme's surface while maintaining its association with Arg 580 or else by dissociation from the enzyme and replacement by another *cis*-aconitate in the "flipped" orientation.

a. Fluorocitrate Inhibits Aconitase

Fluoroacetate, one of the most toxic small molecules known ($LD_{50} = 0.2$ mg · kg^{-1} of body weight in rats), occurs in the leaves of certain African, Australian, and South American poisonous plants. Interestingly, fluoroacetate itself has little toxic effect on cells; instead, cells enzymatically convert it first to fluoroacetyl-CoA and then to **(2R,3R)-2-fluorocitrate,** which specifically inhibits aconitase (see Problem 9):

It is not clear, however, that the inhibition of aconitase fully accounts for fluorocitrate's high toxicity. Indeed, fluorocitrate also inhibits the transport of citrate across the mitochondrial membrane.

C. *NAD⁺-Dependent Isocitrate Dehydrogenase*

Isocitrate dehydrogenase catalyzes the oxidative decarboxylation of isocitrate to α-ketoglutarate to produce the citric acid cycle's first CO_2 and NADH (Reaction 3 of Fig. 21-1). Mammalian tissues contain two isoforms of this enzyme. Although they both catalyze the same reaction, one isoform participates in the citric acid cycle, is located entirely in the mitochondrion, and utilizes NAD^+ as a cofactor. The other isoform occurs in both the mitochondrion and the cytosol, utilizes $NADP^+$ as a cofactor, and generates NADPH for use in reductive biosynthesis.

NAD^+-dependent isocitrate dehydrogenase, which requires an Mn^{2+} or Mg^{2+} cofactor, catalyzes the oxidation of a secondary alcohol (isocitrate) to a ketone (oxalosuccinate) followed by the decarboxylation of the carboxyl group β to the ketone (Fig. 21-21). In this sequence, the keto group β to the carboxyl group facilitates the decarboxylation by acting as an electron sink. The oxidation occurs with the stereospecific reduction of NAD^+ at its *re* face (A-side addition; Section 13-2Aa). Mn^{2+} coordinates the newly formed carbonyl group so as to polarize its electronic charge.

Figure 21-21 **Probable reaction mechanism of isocitrate dehydrogenase.** Oxalosuccinate is shown in brackets because it does not dissociate from the enzyme.

Although the intermediate formation of oxalosuccinate is a logical chemical prediction, evidence for it has been difficult to obtain because it has only a transient existence in reactions catalyzed by the wild-type enzyme. However, an enzymatic reaction rate can be slowed by the mutation of particular catalytically important residues, resulting in the accumulation of specific intermediates. Thus, when Lys 230, which facilitates the decarboxylation of the oxalosuccinate intermediate (Fig. 21-21) is mutated to Met in $NADP^+$-dependent isocitrate dehydrogenase, the oxalosuccinate intermediate accumulates. This accumulated intermediate was directly visualized in the X-ray structure of the mutant enzyme, in the presence of a steady-state flow of substrate, through the use of fast X-ray intensity measurements.

D. α-*Ketoglutarate Dehydrogenase*

α-Ketoglutarate dehydrogenase catalyzes the oxidative decarboxylation of an α-keto acid (α-ketoglutarate), releasing the citric acid cycle's second CO_2 and NADH (Reaction 4 of Fig. 21-1). The overall reaction, which chemically resembles that catalyzed by the PDC (Fig. 21-6), is mediated by a homologous multienzyme complex consisting of **α-ketoglutarate dehydrogenase (E1o), dihydrolipoyl transsuccinylase (E2o),** and dihydrolipoyl dehydrogenase (E3) in which the E3 subunits are identical to those in the PDC (Section 21-2A).

Individual reactions catalyzed by the complex occur by mechanisms identical to those of the pyruvate dehydrogenase reaction (Section 21-2A), the product likewise being a "high-energy" thioester, in this case succinyl-CoA. There are no covalent modification enzymes in the α-ketoglutarate dehydrogenase complex, however.

E. *Succinyl-CoA Synthetase*

*Succinyl-CoA synthetase (also called **succinate thiokinase**) hydrolyzes the "high-energy" compound succinyl-CoA with the coupled synthesis of a "high-energy" nucleoside triphosphate (Reaction 5 of Fig. 21-1).* (Note: Enzyme names can refer to either the forward or the reverse reaction; in this case, succinyl-CoA synthetase and succinate thiokinase refer to the reverse reaction.) GTP is synthesized from GDP + P_i by the mammalian enzyme; plant and bacterial enzymes utilize ADP + P_i to form ATP. These reactions are nevertheless equivalent since ATP and GTP are rapidly interconverted through the action of nucleoside diphosphate kinase (Section 16-4C):

$$GTP + ADP \rightleftharpoons GDP + ATP \qquad \Delta G^{\circ\prime} = 0$$

a. The Succinyl-CoA Thioester Bond Energy Is Preserved through the Formation of a Series of "High-Energy" Phosphates

How does succinyl-CoA synthetase couple the exergonic hydrolysis of succinyl-CoA ($\Delta G^{\circ\prime} = -32.6$ kJ · mol^{-1}) to the endergonic formation of a nucleoside triphosphate ($\Delta G^{\circ\prime} = 30.5$ kJ · mol^{-1})? This question was answered

through the creative use of isotope tracers. In the absence of succinyl-CoA, the spinach enzyme (which utilizes adenine nucleotides) catalyzes the transfer of ATP's γ phosphoryl group to ADP as detected by ^{14}C-labeling ADP and observing the label to appear in ATP. Such an isotope exchange reaction (Section 14-5D) suggests the participation of a phosphoryl–enzyme intermediate that mediates the reaction sequence:

Indeed, this information led to the isolation of a kinetically active phosphoryl–enzyme in which the phosphoryl group is covalently linked to the N3 position of a His residue.

When the succinyl-CoA synthetase reaction, which is freely reversible, is run in the direction of succinyl-CoA synthesis (opposite to its direction in the citric acid cycle) using [^{18}O]succinate as a substrate, ^{18}O is transferred from succinate to phosphate. Evidently, succinyl phosphate, a "high-energy" mixed anhydride, is transiently formed during the reaction.

These observations suggest the following three-step sequence for the mammalian succinyl-CoA synthetase reaction (Fig. 21-22):

1. Succinyl-CoA reacts with P_i to form succinyl phosphate and CoA (accounting for the ^{18}O-exchange reaction).

2. Succinyl phosphate's phosphoryl group is transferred to an enzyme His residue, releasing succinate (accounting for the 3-phosphoHis residue).

3. The phosphoryl group on the enzyme is transferred to GDP, forming GTP (accounting for the nucleoside diphosphate exchange reaction).

Note how in each of these steps *the "high-energy" succinyl-CoA's free energy of hydrolysis is conserved through the successive formation of "high-energy" compounds: First succinyl phosphate, then a 3-phosphoHis residue, and finally GTP.* The process is reminiscent of passing a hot potato.

b. A Pause for Perspective

Up to this point in the cycle, one acetyl equivalent has been completely oxidized to two CO_2. Two NADHs and one GTP (in equilibrium with ATP) have also been generated. In order to complete the cycle, succinate must be converted back to oxaloacetate. This is accomplished by the cycle's remaining three reactions.

Figure 21-22 Reactions catalyzed by succinyl-CoA synthetase.
(1) Formation of succinyl phosphate, a "high-energy" mixed anhydride. **(2)** Formation of phosphoryl–His, a "high-energy" inter-

mediate. **(3)** Transfer of the phosphoryl group to GDP, forming
GTP. The symbol ‡ represents ^{18}O in isotopic labeling reactions.

F. *Succinate Dehydrogenase*

*Succinate dehydrogenase catalyzes stereospecific dehydro-
genation of succinate to fumarate (Reaction 6 of Fig. 21-1):*

The enzyme is strongly inhibited by **malonate,** a structural
analog of succinate and a classic example of a competitive
inhibitor:

Recall that malonate inhibition of cellular respiration was
one of the observations that led Krebs to hypothesize the
citric acid cycle (Section 21-1B).

Succinate dehydrogenase contains an FAD, the reaction's electron acceptor. In general, FAD functions biochemically to oxidize alkanes to alkenes, whereas NAD^+
oxidizes alcohols to aldehydes or ketones. This is because
the oxidation of an alkane (such as succinate) to an alkene
(such as fumarate) is sufficiently exergonic to reduce FAD
to $FADH_2$ but not to reduce NAD^+ to NADH. Alcohol oxidation, in contrast, can reduce NAD^+ (Table 16-4).

Succinate dehydrogenase's FAD is covalently bound via
its C8a atom to an enzyme His residue (Fig. 21-23). A covalent link between FAD and a protein is unusual; in most
cases FAD is noncovalently although tightly bound to its
associated enzyme (e.g., to dihydrolipoyl dehydrogenase;
Section 21-2Ba). How does succinate dehydrogenase's
$FADH_2$ become reoxidized? Being permanently linked to
the enzyme, this prosthetic group cannot function as a
metabolite as does NADH. Rather, *succinate dehydrogenase
(also known as Complex II) is reoxidized by the membrane-
soluble coenzyme Q in the electron-transport chain,* an aspect of its function that we discuss in Section 22-2C. This
rationalizes why succinate dehydrogenase is the only citric
acid cycle enzyme that is anchored in the inner mitochondrial membrane; all the others are dissolved in the mitochondrial matrix (mitochondrial anatomy is described in
Section 22-1A).

Figure 21-23   **Covalent attachment of FAD to a His residue of succinate dehydrogenase.**

FAD

G. Fumarase

Fumarase (fumarate hydratase) catalyzes the hydration of fumarate's double bond to form (S)-malate (L-malate) (Reaction 7 of Fig. 21-1). Consideration of experiments that have contributed to our understanding of the fumarase mechanism illustrates the role played by independent investigations.

a. Conflicting Mechanistic Evidence: What Is the Sequence of H$^+$ and OH$^-$ Addition?

Experiments designed to establish whether the fumarase reaction occurs by a carbanion (OH$^-$addition first) or carbocation (H$^+$ addition first) mechanism have provided contradictory information (Fig. 21-24). Evidence favoring the carbocation mechanism was obtained by studying the dehydration of (S)-malate (the fumarase reaction run in reverse) in H$_2{}^{18}$O. [^{18}O]Malate appears in the reaction mixture more rapidly than it would if the ^{18}O were incorporated via a back reaction of the newly formed fumarate. This suggests the rapid formation of a carbocation intermediate at C2, from which OH$^-$ could exchange with ^{18}OH$^-$, followed by slow hydrogen removal at C3 (Fig. 21-24, *lower route*).

Other observations, however, indicate that the reaction occurs via the formation of a carbanion intermediate at C3 (Fig. 21-24, *upper route*). David Porter synthesized **3-nitro-2-(S)-hydroxypropionate,** which sterically resembles (S)-malate:

(S)-Malate

3-Nitro-2-(S)-hydroxypropionate

Carbanion transition state

Fumarate

(S)-Malate

Carbocation transition state

Figure 21-24   **Possible mechanisms for the hydration of fumarate as catalyzed by fumarase.**

The nitro group's electron-withdrawing character renders the C3 protons relatively acidic (pK $\approx$ 10). The resulting anion is an analog of the postulated C3 carbanion transition state of the fumarase reaction but not of the C2 carbocation transition state (Fig. 21-24; transition state analogs are discussed in Section 15-1Fa). This anion is, in fact, an excellent inhibitor of fumarase: It has an 11,000-fold higher binding affinity for the enzyme than does (S)-malate.

If the fumarase reaction proceeds by a carbanion mechanism, how can the rapid OH^- exchange be explained? Conversely, if it has a carbocation mechanism, why is the nitro anion such an effective inhibitor? This contradictory set of observations makes only one thing clear: *When studying enzyme reaction mechanisms, it is always necessary to approach the problem from several different directions. One set of experiments should never be taken as proof, and interpretation should never be taken as fact.* Indeed, reinterpretation of the ^{18}O exchange experiment makes it consistent with a carbanion mechanism. It turns out that product release is the enzyme's rate-determining step:

$$E + Mal \underset{1}{\rightleftharpoons} E \cdot Mal \underset{2}{\rightleftharpoons} EH^+ \cdot Mal^- \underset{3}{\rightleftharpoons} EH^+ \cdot Fum \cdot OH^-$$

Carbanion intermediate

$$4 \;\; rapid$$

$$^{18}OH^- \longrightarrow OH^-$$

$$E + H^+ + Fum \underset{5}{\rightleftharpoons} E \cdot H^+ \cdot Fum$$
$$slow$$

Malate binds to fumarase **(1)**, forms a carbanion **(2)**, eliminates OH^- to form fumarate **(3)**, and rapidly releases OH^- from the enzyme surface **(4)**. Release of the other products **(5)** is slow. $^{18}OH^-$ can therefore exchange with OH^- to produce $[^{18}O]$malate more rapidly than the overall rate of the fumarase reaction.

H. *Malate Dehydrogenase*

Malate dehydrogenase catalyzes the final reaction of the citric acid cycle, the regeneration of oxaloacetate (Reaction 8 of Fig. 21-1). This occurs through the oxidation of (S)-malate's hydroxyl group to a ketone in an NAD^+-dependent reaction:

(S)-Malate

Oxaloacetate

The hydride ion released by the alcohol in this reaction is transferred to the *re* face of NAD^+, the same face acted on by lactate dehydrogenase (Section 17-3A) and alcohol dehydrogenase (Section 17-3Bc). In fact, *X-ray crystallographic comparisons of the NAD^+-binding domains of these three dehydrogenases indicate that they are remarkably similar and have led to the proposal that all NAD^+-binding domains have evolved from a common ancestor (Section 9-3B).*

The $\Delta G^{\circ\prime}$ for the malate dehydrogenase reaction is $+29.7 \; kJ \cdot mol^{-1}$; the concentration of oxaloacetate formed at equilibrium is consequently very low. Recall, however, that the reaction catalyzed by citrate synthase, the first enzyme in the cycle, is highly exergonic ($\Delta G^{\circ\prime} = -31.5 \; kJ \cdot mol^{-1}$; Section 21-3A) because of the hydrolysis of citryl-CoA's thioester bond. We can now understand the necessity for such a seemingly wasteful process. It allows citrate formation to be exergonic at even low physiological concentrations of oxaloacetate, with the resultant initiation of another turn of the cycle.

I. *Integration of the Citric Acid Cycle*

a. The Impact of the Citric Acid Cycle on ATP Production

The foregoing discussion indicates that one turn of the citric acid cycle results in the following chemical transformations (Fig. 21-1):

1. One acetyl group is oxidized to two molecules of CO_2, a four-electron pair process (although, as we discuss below, it is not the carbon atoms of the entering acetyl group that are oxidized).

2. Three molecules of NAD^+ are reduced to NADH, which accounts for three of the electron pairs.

3. One molecule of FAD is reduced to $FADH_2$, which accounts for the fourth electron pair.

4. One "high-energy" phosphate group is produced as GTP (or ATP).

The eight electrons abstracted from the acetyl group in the citric acid cycle subsequently pass into the electron-transport chain, where they ultimately reduce two molecules of O_2 to H_2O. The three electron pairs from NADH each produce ~2.5 ATPs via oxidative phosphorylation, whereas the one electron pair of $FADH_2$ produces ~1.5 ATPs. *One turn of the citric acid cycle therefore gives rise to the generation of ~10 ATPs.* Electron transport and oxidative phosphorylation are the subject of Chapter 22.

b. Isotopic Tests of Citric Acid Cycle Stereochemistry

The reactions of the citric acid cycle have been confirmed through the use of radioactive tracer experiments that became possible in the late 1930s and early 1940s. At that time, compounds could be synthesized enriched with the stable isotope ^{13}C (detectable at the time by mass spectrometry and now also by NMR) or with the radioactive isotope ^{11}C, which has a half-life of only 20 min. ^{14}C, whose use was pioneered in the late 1940s by Samuel Ruben and Martin Kamen, has the advantage over ^{11}C that its half-life is 5715 years.

In one landmark experiment, [4-^{11}C]oxaloacetate was generated from $^{11}CO_2$ and pyruvate,

$$^*CO_2 + CH_3-\overset{\overset{\text{O}}{\|}}{C}-CO_2^-$$

Pyruvate

ATP $\longrightarrow$ | pyruvate
ADP + P$_i$ $\longleftarrow$ | carboxylase

$$^-OOC^*-CH_2-\overset{\overset{\text{O}}{\|}}{C}-CO_2^-$$

Oxaloacetate

reacted in the citric acid cycle of metabolizing muscle cells, and the resulting cycle intermediates were isolated. Identification of the labeled position in the isolated α-ketoglutarate caused a furor. As we have seen in Sections 21-3A and 21-3B, citrate synthase and aconitase catalyze stereospecific reactions in which citrate synthase can distinguish between the two faces of oxaloacetate's carbonyl group and aconitase can distinguish between citrate's *pro-R* and *pro-S* carboxymethyl groups. In the early 1940s, however, the concept of prochirality had not been established; it was assumed that the two halves of citrate were indistinguishable (in nonenzymatic systems, this is, in effect, the case). It was therefore assumed that the radioactivity originally located at C4 in oxaloacetate (* in Fig. 21-1) would be scrambled in citrate so that its C1 and C6 atoms would be equally labeled, resulting in α-ketoglutarate labeled at both C1 and C5. In fact, only C1, the carboxyl group α to the keto group of α-ketoglutarate, was found to be radioactive (Fig. 21-1). This result threw the identity of the condensation product of oxaloacetate and acetyl-CoA into doubt. How could it be the symmetrical citrate molecule in light of such "conclusive" labeling experiments? This problem of which tricarboxylic acid was the cycle's original condensation product resulted in a name change from the citric acid cycle (proposed by Krebs) to the tricarboxylic acid (TCA) cycle.

In 1948, Alexander Ogston pointed out that citric acid, while symmetrical, is prochiral and thus can interact asymmetrically with the surface of aconitase (Section 13-2A). Even though citrate is now accepted as a cycle intermediate, the duality of the cycle's name persists.

While the net reaction of the cycle is oxidation of the carbon atoms of acetyl units to CO_2, the CO_2 lost in a given turn of the cycle is derived from the carbon skeleton of oxaloacetate. This can be shown by following the fate of isotopically labeled C4 of oxaloacetate (* in Fig. 21-1) through the citric acid cycle. We can see (Fig. 21-1) that it is lost as CO_2 at the α-ketoglutarate dehydrogenase reaction.

Experiments have been performed using [1-^{14}C]acetate, which cells convert to [1-^{14}C]acetyl-CoA:

$$CH_3-{}^*COO^- + CoASH \xrightarrow[\text{acetate} \atop \text{thiokinase}]{ATP \quad AMP + PP_i} CH_3-\overset{\overset{\text{O}}{\|}}{C}{}^*-SCoA$$

Acetate   **Acetyl-CoA**

Tracing the path of this label (‡ in Fig. 21-1) allows the reader to ascertain that the label is not lost as CO_2. It becomes scrambled (1/2‡ in Fig. 21-1) during the cycle, but not until the formation of succinate, the cycle's first rotationally symmetric (nonprochiral) intermediate.

J. Evolution of the Citric Acid Cycle

The citric acid cycle is ubiquitous in aerobic organisms, where it has a central role in energy metabolism. However, an eight-step catalytic cycle such as the citric acid cycle could not have arisen as a whole and must therefore have evolved from simpler sets of enzyme-catalyzed reactions. Clues to the citric acid cycle's origins can be found by examining the metabolism of cells that resemble early lifeforms. Such organisms emerged over 3 billion years ago (Section 1-5), before significant quantities of atmospheric O_2 were available. These cells probably used sulfur as their ultimate oxidizing agent, reducing it to H_2S. Their modern counterparts are anaerobic autotrophs (cells that synthesize all their components from simple molecules such as H_2O, CO_2, NH_3, and H_2S) that harvest free energy by pathways, which are independent of those that oxidize carbon-containing compounds. These organisms therefore do not use the citric acid cycle to generate reduced cofactors such as NADH that are subsequently oxidized by O_2.

The task of divining an organism's metabolic capabilities has been facilitated by bioinformatics. By comparing the sequences of prokaryotic genomes and assigning functions to various homologous genes, it is possible to reconstruct their central metabolic pathways. This approach has been fruitful because the genes that encode the enzymes that generate free energy and a cell's most common metabolites are highly conserved among different species and hence are relatively easy to recognize.

Many prokaryotes lack the citric acid cycle. However, these organisms harbor genes for some citric acid cycle enzymes. The last three reactions of the cycle, leading from succinate to oxaloacetate, appear to be the most highly conserved. This pathway fragment constitutes a mechanism for accepting electrons that are released during sugar fermentation. For example, the reverse of this pathway could regenerate NAD^+ from the NADH produced by the glyceraldehyde-3-phosphate dehydrogenase step of glycolysis.

The resulting succinate could then be used as a starting material for the biosynthesis of other compounds.

Many archaeal cells have a **pyruvate:ferredoxin oxidoreductase** that converts pyruvate to acetyl-CoA (but without producing NADH). In a primitive cell, the resulting acetyl groups could have condensed with oxaloacetate (by the action of a citrate synthase), eventually giving rise to an oxidative sequence of reactions resembling the first few steps of the modern citric acid cycle.

Acetyl-CoA

Oxaloacetate

Citrate

Isocitrate

NAD$^+$

CO$_2$ NADH

α-Ketoglutarate

The α-ketoglutarate produced in this way can be converted to glutamate and other amino acids (Section 26-5Ab).

The reductive and oxidative branches of the citric acid cycle outlined so far function in modern bacterial cells such as *E. coli* cells when they are growing anaerobically, suggesting that similar pathways could have filled the metabolic needs of primordial cells. The evolution of a complete citric acid cycle in which the two branches are linked and both proceed in an oxidative direction (clockwise) would have required an enzyme such as **α-ketoglutarate: ferredoxin reductase** (a homolog of pyruvate:ferredoxin oxidoreductase) to link α-ketoglutarate and succinate.

Interestingly, a primitive citric acid cycle that operated in the reverse (counterclockwise) direction could have provided a pathway for incorporating CO$_2$ into biological molecules.

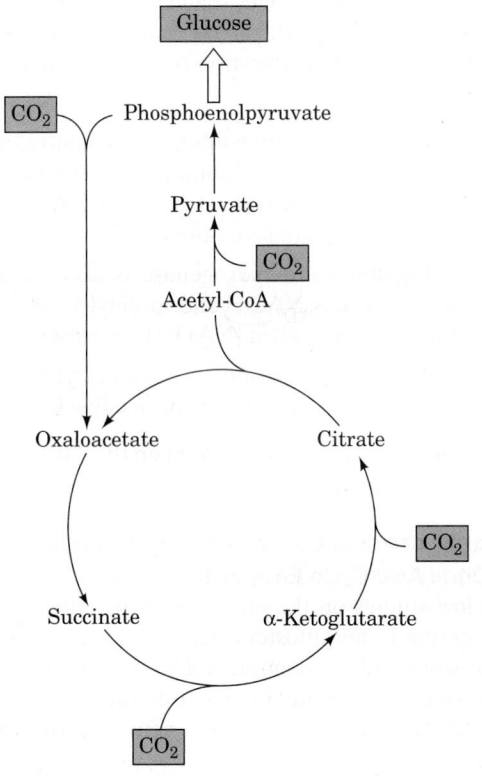

The genes encoding enzymes that catalyze the steps of such a pathway have been identified in several modern autotrophic bacteria. This reductive pathway, which occurs in some deeply rooted archaeal species, possibly predates the CO$_2$-fixing pathway used in some photosynthetic bacteria and in the chloroplasts of green plants (Section 24-3A).

4 REGULATION OF THE CITRIC ACID CYCLE

In this section we consider how metabolic flux through the citric acid cycle is regulated. In our discussions of metabolic flux control (Section 17-4), we established that to understand how a metabolic pathway is controlled, we must identify the enzyme(s) that catalyzes its rate-determining step(s), the *in vitro* effectors of these enzymes, and the *in vivo* concentrations of these substances. *A proposed mechanism of flux control must demonstrate that an increase or decrease in flux is correlated with an increase or decrease in the concentration of the proposed effector.*

a. Citrate Synthase, Isocitrate Dehydrogenase, and α-Ketoglutarate Dehydrogenase Are the Citric Acid Cycle's Rate-Controlling Enzymes

Establishing the rate-determining steps of the citric acid cycle is more difficult than it is for glycolysis because most of the cycle's metabolites are present in both mitochondria and cytosol and we do not know their distribution between these two compartments [recall that identifying a pathway's rate-determining step(s) requires determining the ΔG of each of its reactions from the concentrations of its substrates and products]. If, however, we assume equilibrium between the two compartments, we can use the total cell contents of these substances to estimate their mitochondrial concentrations. Table 21-2 gives the standard free energy changes for the eight citric acid cycle enzymes

Table 21-2 **Standard Free Energy Changes ($\Delta G°'$) and Physiological Free Energy Changes (ΔG) of Citric Acid Cycle Reactions**

Reaction	Enzyme	$\Delta G°'$ (kJ · mol^{-1})	ΔG (kJ · mol^{-1})
1	Citrate synthase	−31.5	Negative
2	Aconitase	~5	~0
3	Isocitrate dehydrogenase	−21	Negative
4	α-Ketoglutarate dehydrogenase multienzyme complex	−33	Negative
5	Succinyl-CoA synthetase	−2.1	~0
6	Succinate dehydrogenase	+6	~0
7	Fumarase	−3.4	~0
8	Malate dehydrogenase	+29.7	~0

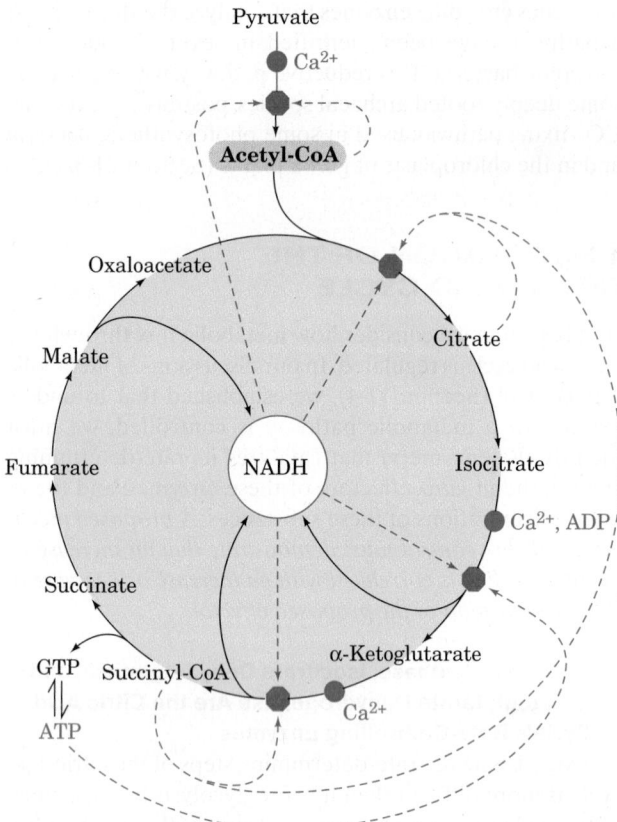

Figure 21-25   Regulation of the citric acid cycle. This diagram of the citric acid cycle and the pyruvate dehydrogenase reaction indicates their points of inhibition (*red octagons*) and the pathway intermediates that function as inhibitors (*dashed red arrows*). ADP and Ca²⁺ (*green dots*) are activators. *See the* **Animated Figures**

and estimates of the physiological free energy changes for the reactions in heart muscle or liver tissue. We can see that three of the enzymes are likely to function far from equilibrium under physiological conditions (negative ΔG): citrate synthase, NAD^+-dependent isocitrate dehydrogenase, and α-ketoglutarate dehydrogenase. We shall therefore focus our discussion on how these enzymes are regulated (Fig. 21-25).

b. The Citric Acid Cycle Is Largely Regulated by Substrate Availability, Product Inhibition, and Inhibition by Other Cycle Intermediates

In heart muscle, where the citric acid cycle functions mainly to generate ATP for use in muscle contraction, the enzymes of the cycle almost always act as a functional unit, with their metabolic flux proportional to the rate of cellular oxygen consumption. *Since oxygen consumption, NADH reoxidation, and ATP production are tightly coupled (Section 22-4), the citric acid cycle must be regulated by feedback mechanisms that coordinate its NADH production with energy expenditure.* Unlike the rate-limiting enzymes of glycolysis and glycogen metabolism, which utilize elaborate systems of allosteric control, substrate cycles, and

covalent modification as flux control mechanisms, the regulatory enzymes of the citric acid cycle seem to be controlled almost entirely in three simple ways: (1) substrate availability, (2) product inhibition, and (3) competitive feedback inhibition by intermediates farther along the cycle. We shall encounter several examples of these straightforward mechanisms in the following discussion.

Perhaps the most crucial regulators of the citric acid cycle are its substrates, acetyl-CoA and oxaloacetate, and its product NADH. Both acetyl-CoA and oxaloacetate are present in mitochondria at concentrations that do not saturate citrate synthase. The metabolic flux through the enzyme therefore varies with substrate concentration and is subject to control by substrate availability. The production of acetyl-CoA from pyruvate is regulated by the activity of pyruvate dehydrogenase (Section 21-2C). Oxaloacetate is in equilibrium with malate, its concentration fluctuating with the [NADH]/[NAD⁺] ratio according to the equilibrium expression

$$K = \frac{[\text{oxaloacetate}][\text{NADH}]}{[\text{malate}][\text{NAD}^+]}$$

In the transition from low to high work and respiration rates, mitochondrial [NADH] decreases. The consequent increase in [oxaloacetate] stimulates the citrate synthase reaction, which controls the rate of citrate formation.

The observation that [citrate] invariably falls as the workload increases indicates that the rate of citrate removal increases more than its rate of formation. The rate of citrate removal is governed by NAD^+-dependent isocitrate dehydrogenase (aconitase functions close to equilibrium), which is strongly inhibited *in vitro* by NADH (product inhibition). Citrate synthase is also inhibited by NADH. Evidently, NAD^+-dependent isocitrate dehydrogenase is more sensitive to [NADH] changes than citrate synthase.

The decrease in [citrate] that occurs on transition from low to high work and respiration rates results in a domino effect:

1. Citrate is a competitive inhibitor of oxaloacetate for citrate synthase (product inhibition); the fall in [citrate] caused by increased isocitrate dehydrogenase activity increases the rate of citrate formation.

2. α-Ketoglutarate dehydrogenase is also strongly inhibited by its products, NADH and succinyl-CoA. Its activity therefore increases when [NADH] decreases.

3. Succinyl-CoA also competes with acetyl-CoA in the citrate synthase reaction (competitive feedback inhibition).

This interlocking system serves to keep the citric acid cycle coordinately regulated.

c. ADP, ATP, and Ca²⁺ Are Allosteric Regulators of Citric Acid Cycle Enzymes

In vitro studies on the enzymes of the citric acid cycle have identified a few allosteric activators and inhibitors. Increased workload is accompanied by increased [ADP] resulting from the consequent increased rate of ATP hydrolysis. ADP acts as an allosteric activator of isocitrate dehydrogenase by decreasing its apparent K_M for isocitrate. ATP, which builds up when muscle is at rest, inhibits this enzyme.

Ca^{2+}, among its many biological functions, is an essential metabolic regulator. It stimulates glycogen breakdown (Section 18-3Ce), triggers muscle contraction (Section 35-3Ca), and mediates many hormonal signals as a second messenger (Section 19-4A). Ca^{2+} also plays an important role in the regulation of the citric acid cycle (Fig. 21-25). It activates pyruvate dehydrogenase phosphatase and inhibits pyruvate dehydrogenase kinase, thereby activating the PDC to produce acetyl-CoA (Fig. 21-17b). In addition, Ca^{2+} activates both isocitrate dehydrogenase and α-ketoglutarate dehydrogenase. Thus, the same signal stimulates muscle contraction and the production of the ATP to fuel it.

In the liver, the role of the citric acid cycle is more complex than in heart muscle. The liver synthesizes many substances required by the body including glucose, fatty acids, cholesterol, amino acids, and porphyrins. Reactions of the citric acid cycle play a part in many of these biosynthetic pathways in addition to their role in energy metabolism. In the next section, we discuss the contribution of the citric acid cycle to these processes.

d. Are the Enzymes of the Citric Acid Cycle Organized Into a Metabolon?

Considerable efficiency can be gained by organizing the enzymes of a metabolic pathway such that the enzymes catalyzing its sequential steps interact to channel intermediates between them. Indeed, we saw this to be the case with the PDC (Section 21-2A). The advantages of such an assembly, termed a **metabolon,** include the protection of labile intermediates and the increase of their local concentration for more efficient catalysis. Considerable effort has been expended to obtain evidence for such interactions in the major metabolic pathways including glycolysis and the citric acid cycle. However, since citric acid cycle intermediates must be available for use in other metabolic pathways, any complexes between citric acid cycle enzymes are likely to be weak and therefore unable to withstand the laboratory manipulations necessary to isolate them.

Despite the foregoing, specific interactions have been demonstrated *in vivo* between the members of several pairs of citric acid cycle enzymes, including those between citrate synthase and malate dehydrogenase. For example, Paul Srere isolated the gene for a mutant citrate synthase in yeast that he termed an "assembly mutant" because it has normal enzymatic activity *in vitro* but nevertheless causes a citric acid cycle deficiency *in vivo*. This mutation occurs in a highly conserved 13-residue segment of the enzyme (Pro 354–Pro 366 in yeast), which forms a solvent-exposed loop that could interact with other proteins. To further investigate this phenomenon, Srere constructed a plasmid expressing a fusion protein (Section 5-5Ga) consisting of the wild-type citrate synthase peptide grafted to the C-terminal end of green fluorescent protein (GFP; Section 5-5Gd). If the peptide actually binds to malate dehydrogenase, the expression of this enzymatically inactive fusion protein in yeast would be expected to inhibit the citric

acid cycle. This is, in fact, the case as judged by the severely decreased ability of such yeast to grow on acetate, a metabolite that can only be metabolized via the citric acid cycle. Moreover, replacing the citrate synthase peptide in the fusion protein with an unrelated peptide only slightly decreases the normal growth rate on acetate, thereby demonstrating that the specific sequence of the citrate synthase peptide is the major cause of this growth inhibition. If the decrease in growth rate is actually caused by competition of the peptide with citrate synthase for an interacting partner in a metabolon, then this should be overcome by the overexpression of either citrate synthase or its interaction partner. Indeed, when either citrate synthase or malate dehydrogenase was overexpressed in yeast expressing the GFP–citrate synthase peptide fusion protein, the growth rate on acetate was restored. However, the overexpression of aconitase did not overcome this growth inhibition. These observations support the hypothesis that malate dehydrogenase and citrate synthase must interact for optimal citric acid cycle function and identify the interaction site for malate dehydrogenase on citrate synthase as the 13-residue Pro 354–Pro 366 peptide.

e. A Bacterial Isocitrate Dehydrogenase Is Regulated by Phosphorylation

Escherichia coli isocitrate dehydrogenase is a dimer of identical 416-residue subunits that is inactivated by phosphorylation of its Ser 113, an active site residue. In contrast, most other enzymes that are known to be subject to covalent modification/demodification, for example, glycogen phosphorylase (Section 18-3), are phosphorylated at allosteric sites. In the case of isocitrate dehydrogenase, phosphorylation renders the enzyme unable to bind its substrate, isocitrate.

Comparison of the X-ray structures, determined by Daniel Koshland and Robert Stroud, of isocitrate dehydrogenase alone, in its phosphorylated form, and with bound isocitrate reveals only small conformational differences, suggesting that electrostatic repulsions between the anionic isocitrate and Ser phosphate groups prevent the enzyme from binding substrate. Evidently, phosphorylation can regulate enzyme activity by directly interfering with active site ligand binding as well as by inducing a conformational change from an allosteric site.

5 THE AMPHIBOLIC NATURE OF THE CITRIC ACID CYCLE

Ordinarily one thinks of a metabolic pathway as being either catabolic with the release (and conservation) of free energy, or anabolic with a requirement for free energy. The citric acid cycle is, of course, catabolic because it involves degradation and is a major free energy conservation system in most organisms. Cycle intermediates are only required in catalytic amounts to maintain the degradative function of the cycle. However, several biosynthetic pathways utilize citric acid cycle intermediates as starting materials (anabolism). The citric acid cycle is therefore **amphibolic** (both anabolic and catabolic).

All of the biosynthetic pathways that utilize citric acid cycle intermediates also require free energy. Consequently, the catabolic function of the cycle cannot be interrupted; *cycle intermediates that have been siphoned off must be replaced.* Although the mechanistic aspects of the enzymes involved in the pathways that utilize and replenish citric acid cycle intermediates are discussed in subsequent chapters, it is useful to briefly mention these metabolic interconnections here (Fig. 21-26).

a. Pathways That Utilize Citric Acid Cycle Intermediates

Reactions that utilize and therefore drain citric acid cycle intermediates are called **cataplerotic reactions** (emptying; Greek: *cata,* down + *plerotikos,* to fill). These reactions serve not only to synthesize important products but also to avoid the inappropriate buildup of citric acid cycle intermediates in the mitochondrion, for example, when there is a high rate of breakdown of amino acids to citric acid cycle intermediates (Section 26-3). Cataplerotic reactions occur in the following pathways:

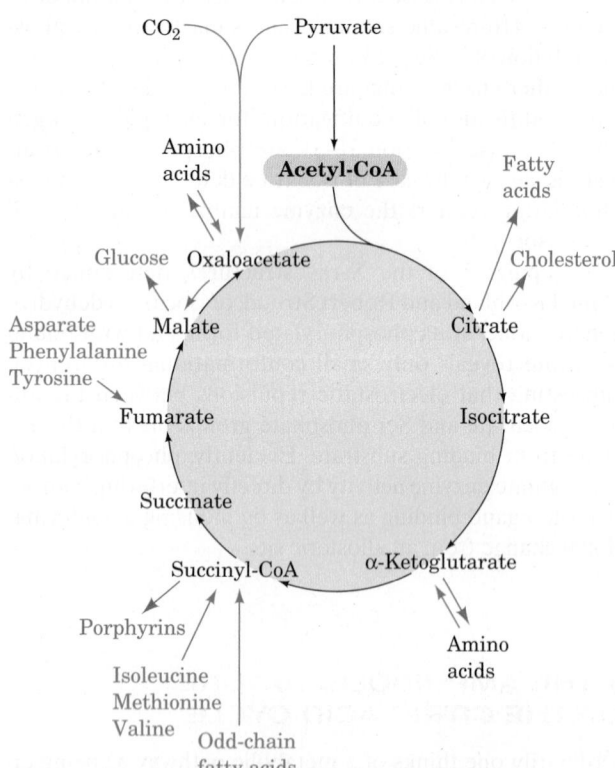

Figure 21-26 Amphibolic functions of the citric acid cycle. This diagram indicates the positions at which intermediates are cataplerotically drawn off for use in anabolic pathways (*red arrows*) and the points where anaplerotic reactions replenish depleted cycle intermediates (*green arrows*). Reactions involving amino acid transamination and deamination are reversible, so their direction varies with metabolic demand. ✐ **See the Animated Figures**

1. Glucose biosynthesis (gluconeogenesis; Section 23-1), which occurs in the cytosol, utilizes oxaloacetate as its starting material. Oxaloacetate is not transported across the mitochondrial membrane, but malate is. Malate that has been transported across the mitochondrial membrane is converted to oxaloacetate in the cytosol for gluconeogenesis.

2. Lipid biosynthesis, which includes **fatty acid biosynthesis** (Section 25-4) and **cholesterol biosynthesis** (Section 25-6A), is a cytosolic process that requires acetyl-CoA. Acetyl-CoA is generated in the mitochondrion and is not transported across the inner mitochondrial membrane. Cytosolic acetyl-CoA is therefore generated by the breakdown of citrate, which can cross the inner mitochondrial membrane, in a reaction catalyzed by **ATP-citrate lyase** (Section 25-4D):

$$\text{ATP} + \text{citrate} + \text{CoA} \rightleftharpoons$$
$$\text{ADP} + \text{P}_i + \text{oxaloacetate} + \text{acetyl-CoA}$$

3. Amino acid biosynthesis utilizes citric acid cycle intermediates in two ways. α-Ketoglutarate is converted to glutamate in a reductive amination reaction involving either NAD^+ or $NADP^+$ catalyzed by **glutamate dehydrogenase** (Section 26-1):

$$\alpha\text{-Ketoglutarate} + \text{NAD(P)H} + \text{NH}_4^+ \rightleftharpoons$$
$$\text{glutamate} + \text{NAD(P)}^+ + \text{H}_2\text{O}$$

α-Ketoglutarate and oxaloacetate are also used to synthesize glutamate and aspartate in transamination reactions (Section 26-1):

$$\alpha\text{-Ketoglutarate} + \text{alanine} \rightleftharpoons \text{glutamate} + \text{pyruvate}$$

and

$$\text{Oxaloacetate} + \text{alanine} \rightleftharpoons \text{aspartate} + \text{pyruvate}$$

4. Porphyrin biosynthesis (Section 26-4A) utilizes succinyl-CoA as a starting material.

5. Complete oxidation of amino acids requires that the citric acid cycle intermediates to which the amino acids are degraded be converted first to PEP [in a reaction catalyzed by **phosphoenolpyruvate carboxykinase (PEPCK;** Section 23-1): Oxaloacetate + GTP ⇋ PEP + GDP], then to pyruvate by pyruvate kinase (Section 17-2J), and finally to acetyl-CoA by pyruvate dehydrogenase (Section 21-2A).

b. Reactions That Replenish Citric Acid Cycle Intermediates

Reactions that replenish citric acid cycle intermediates are called **anaplerotic reactions** (filling up, Greek: *ana,* up). The main reaction of this type is catalyzed by **pyruvate carboxylase,** which produces oxaloacetate (Section 23-1Aa):

$$\text{Pyruvate} + \text{CO}_2 + \text{ATP} + \text{H}_2\text{O} \rightleftharpoons$$
$$\text{oxaloacetate} + \text{ADP} + \text{P}_i$$

This enzyme "senses" the need for more citric acid cycle intermediates through its activator, acetyl-CoA. Any decrease in the rate of the cycle caused by insufficient oxaloacetate or other cycle intermediates results in an increased level of acetyl-CoA because of its underutilization. This activates pyruvate carboxylase, which replenishes oxaloacetate, increasing the rate of the cycle. Of course, if the citric acid cycle is inhibited at some other step, by high NADH concentration, for example, increased oxaloacetate concentration will not activate the cycle. The excess oxaloacetate instead equilibrates with malate, which is transported out of the mitochondria for use in gluconeogenesis.

Degradative pathways generate citric acid cycle intermediates:

1. Oxidation of odd-chain fatty acids (Section 25-2E) leads to the production of succinyl-CoA.

2. Breakdown of the amino acids isoleucine, methionine, and valine (Section 26-3E) also leads to the production of succinyl-CoA.

3. Transamination and deamination of amino acids lead to the production of α-ketoglutarate and oxaloacetate. These reactions are reversible and, depending on metabolic demand, serve to remove or replenish these citric acid cycle intermediates.

The citric acid cycle is truly at the center of metabolism (see Fig. 16-1). Its reduced products, NADH and $FADH_2$, are reoxidized by the electron-transport chain during oxidative phosphorylation and the free energy released is coupled to the biosynthesis of ATP. Citric acid cycle intermediates are utilized in the biosynthesis of many vital cellular constituents. In the next few chapters we shall explore the interrelationships of these pathways in more detail.

CHAPTER SUMMARY

1 Cycle Overview The citric acid cycle, the common mode of oxidative metabolism in most organisms, is mediated by eight enzymes that collectively convert 1 acetyl-CoA to 2 CO_2 molecules so as to yield 3 NADHs, 1 $FADH_2$, and 1 GTP (or ATP). The NADH and $FADH_2$ are oxidized by O_2 in the electron-transport chain with the concomitant synthesis of around 11 more ATPs, yielding a total of about 12 ATPs for one turn of the citric acid cycle.

2 Metabolic Sources of Acetyl-Coenzyme A Pyruvate, the end product of glycolysis under aerobic conditions, is converted to acetyl-CoA by the pyruvate dehydrogenase multienzyme complex (PDC), a cubic or dodecahedral cluster of three enzymes: pyruvate dehydrogenase (E1), dihydrolipoyl transacetylase (E2), and dihydrolipoyl dehydrogenase (E3). The pyruvate dehydrogenase subunit catalyzes the conversion of pyruvate to CO_2 and a hydroxyethyl-TPP intermediate. The latter is channeled to dihydrolipoyl transacetylase, which oxidizes the hydroxyethyl group to acetate and transfers it to CoA to form acetyl-CoA. The lipoamide prosthetic group, which is reduced to the dihydro form in the process, is reoxidized by dihydrolipoyl dehydrogenase in a reaction involving bound FAD that reduces NAD^+ to NADH. Dihydrolipoyl transacetylase is inactivated by the formation of a covalent adduct between lipoamide and As(III) compounds. Pyruvate dehydrogenase's two active sites are connected by an acidic tunnel through which they alternately supply and abstract protons from one another and thereby keep their reactions out of phase.

Dihydrolipoyl dehydrogenase, which closely resembles glutathione reductase, catalyzes a two-stage reaction. In the first stage, dihydrolipoamide reduces the enzyme's redox-active disulfide group, yielding the reaction's first product, lipoamide. In the second stage, NAD^+ reoxidizes the reduced enzyme through the intermediacy of the enzyme's FAD prosthetic group, thereby closing the catalytic cycle and yielding the enzyme's second product, NADH. The activity of the pyruvate dehydrogenase complex varies with the [NADH]/[NAD^+] and [acetyl-CoA]/[CoA] ratios. In eukaryotes, the pyruvate dehydrogenase subunit is also inactivated by phosphorylation of a specific Ser residue and is reactivated by its removal. These modifications are mediated, respectively, by pyruvate dehydrogenase kinase and pyruvate dehydrogenase phosphatase, which are components of the multienzyme complex and respond to the levels of metabolic intermediates such as NADH and acetyl-CoA.

3 Enzymes of the Citric Acid Cycle Citrate is formed by the condensation of acetyl-CoA and oxaloacetate by citrate synthase. The citrate is dehydrated to cis-aconitate and then rehydrated to isocitrate in a stereospecific reaction catalyzed by aconitase. This enzyme is specifically inhibited by (2R,3R)-2-fluorocitrate, which is enzymatically synthesized from fluoroacetate and oxaloacetate. Isocitrate is oxidatively decarboxylated to α-ketoglutarate by isocitrate dehydrogenase, which produces NADH and CO_2. The α-ketoglutarate, in turn, is oxidatively decarboxylated by α-ketoglutarate dehydrogenase, a multienzyme complex homologous to the PDC. This reaction generates the second NADH and CO_2. The resulting succinyl-CoA is converted to succinate with the generation of GTP (ATP in plants and bacteria) by succinyl-CoA synthetase. The succinate is stereospecifically dehydrogenated to fumarate by succinate dehydrogenase in a reaction that generates $FADH_2$. The final two reactions of the citric acid cycle, which are catalyzed by fumarase and malate dehydrogenase, in turn hydrate fumarate to (S)-malate and oxidize this alcohol to its corresponding ketone, oxaloacetate, with concomitant production of the pathway's third and final NADH. The citric acid cycle appears to have evolved through the piecing together of enzymatic reactions that originally carried out other metabolic functions.

4 Regulation of the Citric Acid Cycle The enzymes of the citric acid cycle act as a functional unit that keeps pace with the metabolic demands of the cell. The flux-controlling enzymes appear to be citrate synthase, isocitrate dehydrogenase, and α-ketoglutarate dehydrogenase. Their activities are controlled by substrate availability, product inhibition, inhibition by cycle intermediates, and activation by Ca^{2+}. The enzymes of the citric acid cycle may be organized, in vivo, into a metabolon for channeling the products of one enzyme to the next enzyme in the cycle.

5 The Amphibolic Nature of the Citric Acid Cycle   Several anabolic pathways cataplerotically utilize citric acid cycle intermediates as starting materials. These essential substances are replaced by anaplerotic reactions of which the major one is synthesis of oxaloacetate from pyruvate and CO_2 by pyruvate carboxylase.

REFERENCES

History

Holmes, F.L., *Hans Krebs:* Vol. 1: *The Formation of a Scientific Life, 1900–1933; and* Vol. 2: *Architect of Intermediary Metabolism, 1933–1937,* Oxford University Press (1991 and 1993). [The biography of the discoverer of the citric acid cycle through the time of its discovery.]

Kornberg, H.L., Tricarboxylic acid cycles, *BioEssays* **7**, 236–238 (1987). [A historical synopsis of the intellectual background leading to the discovery of the citric acid cycle.]

Krebs, H.A., The history of the tricarboxylic acid cycle, *Perspect. Biol. Med.* **14**, 154–170 (1970).

Pyruvate Dehydrogenase Multienzyme Complex

Frank, R.A.W., Pratap, J.V., Pei, X.Y., Perham, R.N., and Luisi, B.F., The molecular origins of specificity in the assembly of a multienzyme complex, *Structure* **13**, 1119–1130 (2005). [The X-ray structure of *B. stearothermophilus* E1 · TPP · PSBD.]

Frank, R.A.W., Titman, C.M., Pratap, J.V., Luisi, B.F., and Perham, R.N., A molecular switch and proton wire synchronize the active sites in thiamine enzymes, *Science* **306**, 872–876 (2004).

Izard, T., Ævarsson, A., Allen, M.D., Westphal, A.H., Perham, R.N., de Kok, A., and Hol, W.G.J., Principles of quasi-equivalence and euclidean geometry govern the assembly of cubic and dodecahedral cores of pyruvate dehydrogenase, *Proc. Natl. Acad. Sci.* **96**, 1240–1245 (1999).

Karplus, P.A. and Schulz, G.E., Refined structure of glutathione reductase at 1.54 Å resolution, *J. Mol. Biol.* **195**, 701–729 (1987).

Karplus, P.A. and Schulz, G.E., Substrate binding and catalysis by glutathione reductase derived from refined enzyme:substrate crystal structures at 2Å resolution, *J. Mol. Biol.* **210**, 163–180 (1989).

Lengyel, J.S., Stott, K.M., Wu, X., Brooks, B.R., Balbo, A., Schuck, P., Perham, R.N., Subramaniam, S., and Milne, J.L.S., Extended polypeptide linkers establish the spatial architecture of a pyruvate dehydrogenase multienzyme complex, *Structure* **16**, 93–103 (2008).

Mande, S.S., Sarfaty, S., Allen, M.D., Perham, R.N., and Hol, W.G.J., Protein–protein interactions in the pyruvate dehydrogenase complex: dihydrolipoamide dehydrogenase complexed with the binding domain of dihydrolipoamide acetyltransferase, *Structure* **4**, 277–286 (1996).

Mattevi, A., Obmolova, G., Sokatch, J.R., Betzel, C., and Hol, W.G.J., The refined crystal structure of *Pseudomonas putida* lipoamide dehydrogenase complexed with NAD⁺ at 2.45 Å resolution, *Proteins* **13**, 336–351 (1992); *and* Mattevi, A., Schierbeek, A.J., and Hol, W.G.J., Refined crystal structure of lipoamide dehydrogenase from *Azotobacter vinelandii* at 2.2 Å resolution. A comparison with the structure of glutathione reductase, *J. Mol. Biol.* **220**, 975–994 (1991).

Milne, J.L.S., Wu, X., Borgnia, M.J., Lengyel, J.S., Brooks, B.R., Shi, D., Perham, R.N., and Subramaniam, S., Molecular structure of a 9-MDa icosahedral pyruvate dehydrogenase subcomplex containing the E2 and E3 enzymes using cryoelectron microscopy, *J. Biol. Chem.* **281**, 4364–4370 (2006).

Patel, M.S. and Korotchkina, L.G., The biochemistry of the pyruvate dehydrogenase complex, *Biochem. Mol. Biol. Educ.* **31**, 5–15 (2003).

Perham, R.N., Swinging arms and swinging domains in multifunctional enzymes: Catalytic machines for multistep reactions, *Annu. Rev. Biochem.* **69**, 961–1004 (2000). [An authoritative review on multienzyme complexes.]

Reed, L.J., A trail of research from lipoic acid to α-keto acid dehydrogenase complexes, *J. Biol. Chem.* **276**, 38329–38336 (2001). [A scientific memoir.]

Roche, T.E., Baker, J.C., Yan, X., Hiromasa, Y., Gong, X., Peng, T., Dong, J., Turkan, A., and Kasten, S.E., Distinct regulatory properties of pyruvate dehydrogenase kinase and phosphatase isoforms, *Prog. Nucleic Acid Res. Mol. Biol.* **70**, 33–75 (2001).

Williams, C.H., Jr., Lipoamide dehydrogenase, glutathione reductase, thioredoxin reductase, and mercuric ion reductase—A family of flavoenzyme transhydrogenases, *in* Müller, F. (Ed.), *Chemistry and Biochemistry of Flavoenzymes,* Vol. III, pp. 121–211, CRC Press (2000).

Zhou, Z.H., McCarthy, D.B., O'Connor, C.M., Reed, L.J., and Stoops, J.K., The remarkable structural and functional organization of the eukaryotic pyruvate dehydrogenase complexes, *Proc. Natl. Acad. Sci.* **98**, 14802–14807 (2001).

Enzymes of the Citric Acid Cycle

Beinert, H., Kennedy, M.C., and Stout, D.C., Aconitase as iron-sulfur protein, enzyme, and iron-regulatory protein, *Chem. Rev.* **96**, 2335–2374 (1996).

Bolduc, J.M., Dyer, D.H., Scott, W.G., Singer, P., Sweet, R.M., Koshland, D.E., Jr., and Stoddard, B.L., Mutagenesis and Laue structures of enzyme intermediates: Isocitrate dehydrogenase, *Science* **268**, 1312–1318 (1995).

Cleland, W.W. and Kreevoy, M.M., Low-barrier hydrogen bonds and enzymic catalysis, *Science* **264**, 1887–1890 (1994).

Frey, P.A. and Hegeman, A.D., *Enzymatic Reaction Mechanisms,* Oxford (2007). [Contains discussions of the mechanisms of various citric acid cycle enzymes.]

Huynen, M.A., Dandekar, T., and Bork, P., Variation and evolution of the citric-acid cycle: a genomic perspective, *Trends Microbiol.* **7**, 281–291 (1999). [Discusses how genomic studies can allow reconstruction of metabolic pathways, even when some enzymes appear to be missing.]

Karpusas, M., Branchaud, B., and Remington, S.J., Proposed mechanism for the condensation reaction of citrate synthase: 1.9-Å structure of the ternary complex with oxaloacetate and carboxymethyl coenzyme A, *Biochemistry* **29**, 2213–2219 (1990).

Kurz, L.C., Nakra, T., Stein, R., Plungkhen, W., Riley, M., Hsu, F., and Drysdale, G.R., Effects of changes in three catalytic residues on the relative stabilities of some of the intermediates and transition states in the citrate synthase reaction, *Biochemistry* **37**, 9724–9737 (1998).

Lauble, H., Kennedy, M.C., Beinert, H., and Stout, D.C., Crystal structures of aconitase with isocitrate and nitroisocitrate bound, *Biochemistry* **31**, 2735–2748 (1992).

Mulholland, A.J., Lyne, P.D., and Karplus, M., Ab initio QM/MM study of the citrate synthase mechanism. A low-barrier hydrogen bond is not involved, *J. Am. Chem. Soc.* **122**, 534–535 (2000).

Porter, D.J.T. and Bright, H.J., 3-Carbanionic substrate analogues bind very tightly to fumarase and aspartase, *J. Biol. Chem.* **255**, 4772–4780 (1980).

Remington, S.J., Structure and mechanism of citrate synthase, *Curr. Top. Cell Regul.* **33**, 202–229 (1992); *and* Mechanisms of citrate synthase and related enzymes (triose phosphate isomerase and mandelate racemase), *Curr. Opin. Struct. Biol.* **2**, 730–735 (1992).

Wolodk, W.T., Fraser, M.E., James, M.N.G., and Bridger, W.A., The crystal structure of succinyl-CoA synthetase from *Escherichia coli* at 2.5 Å resolution, *J. Biol. Chem.* **269**, 10883–10890 (1994).

Zheng, L., Kennedy, M.C., Beinert, H., and Zalkin, H. Mutational analysis of active site residues in pig heart aconitase, *J. Biol. Chem.* **267**, 7895–7903 (1992).

Metabolic Poisons

Gibble, G.W., Fluoroacetate toxicity, *J. Chem. Educ.* **50**, 460–462 (1973).

Jones, D.E.H. and Ledingham, K.W.D., Arsenic in Napoleon's wallpaper, *Nature* **299**, 626–627 (1982).

Lauble, H., Kennedy, M.C., Emptage, M.H., Beinert, H., and Stout, C.D., The reaction of fluorocitrate with aconitase and the crystal structure of the enzyme-inhibitor complex, *Proc. Natl. Acad. Sci.* **93**, 13699–13703 (1996).

Winslow, J.H., *Darwin's Victorian Malady,* American Philosophical Society (1971).

Control Mechanisms

Hurley, J.H., Dean, A.M., Sohl, J.L., Koshland, D.E., Jr., and Stroud, R.M., Regulation of an enzyme by phosphorylation at the active site, *Science* **249**, 1012–1016 (1990).

Owen, O.E., Kalhan, S.C., and Hanson, R.W., The key role of anaplerosis and cataplerosis for citric acid cycle function, *J. Biol. Chem.* **277**, 30409–30412 (2002).

Reed, L.J., Damuni, Z., and Merryfield, M.L., Regulation of mammalian pyruvate and branched-chain α-keto-acid dehydrogenase complexes by phosphorylation and dephosphorylation, *Curr. Top. Cell. Regul.* **27**, 41–49 (1985).

Srere, P.A., Sherry, A.D., Malloy, C.R., and Sumegi, B., Channelling in the Krebs tricarboxylic acid cycle, *in* Agius, L. and Sherratt, H.S.A. (Eds.), *Channelling in Intermediary Metabolism, pp.* 201–217, Portland Press (1997).

Stroud, R.M., Mechanisms of biological control by phosphorylation, *Curr. Opin. Struct. Biol.* **1**, 826–835 (1991). [Reviews, among other things, the inactivation of isocitrate dehydrogenase by phosphorylation.]

Vélot, C., Mixon, M.B., Teige, M., and Srere, P.A., Model of a quinary structure between Krebs TCA cycle enzymes: A model for the metabolon, *Biochemistry* **36**, 14271–14276 (1997).

Vélot, C. and Srere, P.A., Reversible transdominant inhibition of a metabolic pathway. *In vivo* evidence of interaction between two sequential tricarboxylic acid cycle enzymes in yeast. *J. Biol. Chem.* **275**, 12926–12933 (2000).

PROBLEMS

1. Trace the course of the radioactive label in [2-^{14}C]glucose through glycolysis and the citric acid cycle. At what point(s) in the cycle will the radioactivity be released as $^{14}CO_2$? How many turns of the cycle will be required for complete conversion of the radioactivity to CO_2? Repeat this problem for pyruvate that is ^{14}C-labeled at its methyl group.

2. The reaction of glutathione reductase with an excess of NADPH in the presence of arsenite yields a nonphysiological four-electron reduced form of the enzyme. What is the chemical nature of this catalytically inactive species?

3. Two-electron reduced dihydrolipoyl dehydrogenase (EH$_2$), but not the oxidized enzyme (E), reacts with iodoacetate (ICH$_2$COO$^-$) to yield an inactive enzyme. Explain.

4. Given the following information, calculate the physiological ΔG of the isocitrate dehydrogenase reaction at 25°C and pH 7.0: [NAD$^+$]/[NADH] = 8; [α-ketoglutarate] = 0.1 mM; [isocitrate] = 0.02 mM; assume standard conditions for CO$_2$ (ΔG^σ is given in Table 21-2). Is this reaction a likely site for metabolic control? Explain.

5. The oxidation of acetyl-CoA to two molecules of CO$_2$ involves the transfer of four electron pairs to redox coenzymes. In which of the cycle's reactions do these electron transfers occur? Identify the redox coenzyme in each case. For each reaction, draw the structural formulas of the reactants, intermediates, and products and show, using curved arrows, how the electrons are transferred.

6. The citrate synthase reaction has been proposed to proceed via the formation of the enol(ate) form of acetyl-CoA. How, then, would you account for the observation that ^{3}H is not incorporated into acetyl-CoA when acetyl-CoA is incubated with citrate synthase in ^3H_2O?

7. Malonate is a competitive inhibitor of succinate in the succinate dehydrogenase reaction. (a) Sketch the graphs that would be obtained on plotting 1/v versus 1/[succinate] at three different malonate concentrations. Label the lines for low, medium, and high [malonate]. (b) Explain why increasing the oxaloacetate concentration in a cell can overcome malonate inhibition.

8. Krebs found that malonate inhibition of the citric acid cycle could be overcome by raising the oxaloacetate concentration. Explain the mechanism of this process in light of your findings in Problem 7.

*9. (2R,3R)-2-Fluorocitrate contains F in the *pro-S* carboxymethyl arm of citrate [note that the rules of organic nomenclature require that atom C2 in citrate (Fig. 21-20) be renumbered as C4 in (2R,3R)-2-fluorocitrate]. This compound, but not its diastereomer, is a potent inhibitor of aconitase. (a) Draw the aconitase-catalyzed reaction pathway of (2R,3R)-2-fluorocitrate assuming it follows the same reaction pathway as citrate (Fig. 21-20). (b) Aconitase, in fact, does not catalyze the foregoing reaction with (2R,3R)-2-fluorocitrate but, rather, yields the following tight-binding inhibitor:

Draw an alternative aconitase-catalyzed reaction that would generate this inhibitor. (c) Draw the aconitase-catalyzed reaction of

(2*S*,3*R*)-3-fluorocitrate, the diastereomer of (2*R*,3*R*)-2-fluorocitrate (fluorocitrate containing F in the *pro-R* carboxymethyl arm of citrate; here the atom numbering scheme is the same as that in Fig. 21-20). Would a tight-binding inhibitor be formed?

10. Which of the following metabolites undergo net oxidation by the citric acid cycle: (a) α-ketoglutarate, (b) succinate, (c) citrate, and (d) acetyl-CoA?

11. Although there is no net synthesis of intermediates by the citric acid cycle, citric acid cycle intermediates are used in biosynthetic reactions such as the synthesis of porphyrins from succinyl-CoA. Write a reaction for the net synthesis of succinyl-CoA from pyruvate.

12. Oxaloacetate and α-ketoglutarate are precursors of the amino acids aspartate and glutamate as well as being catalytic intermediates in the citric acid cycle. Describe the net synthesis of α-ketoglutarate from pyruvate in which no citric acid cycle intermediates are depleted.

13. Lipoic acid is bound to enzymes that catalyze oxidative decarboxylation of α-keto acids. (a) What is the chemical mode of attachment of lipoic acid to enzymes? (b) Using chemical structures, show how lipoic acid participates in the oxidative decarboxylation of α-keto acids.

14. British anti-lewisite (BAL), which was designed to counter the effects of the arsenical war gas **lewisite,** is useful in treating arsenic poisoning. Explain.

$$CH_2 - SH$$
$$|$$
$$CH - SH \qquad Cl - CH = CH - AsCl_2$$
$$|$$
$$CH_2 - OH$$

British anti-lewisite   **Lewisite**
(BAL)

15. The first organisms on Earth may have been chemoautotrophs in which the citric acid cycle operated in reverse to "fix" atmospheric CO_2 in organic compounds. Complete a catalytic cycle that begins with the hypothetical overall reaction succinate + $2 CO_2 \rightarrow$ citrate.

16. Why is it advantageous for citrate, the product of Reaction 1 of the citric acid cycle, to inhibit phosphofructokinase, which catalyzes the third reaction of glycolysis?

17. Anaplerotic reactions permit the citric acid cycle to supply intermediates to biosynthetic pathways while maintaining the proper levels of cycle intermediates. Write the equation for the net synthesis of citrate from pyruvate.

18. Many amino acids are broken down to intermediates of the citric acid cycle. (a) Why can't these amino acid "remnants" be directly oxidized to CO_2 by the citric acid cycle? (b) Explain why amino acids that are broken down to pyruvate can be completely oxidized by the citric acid cycle.

Electron Transport and Oxidative Phosphorylation

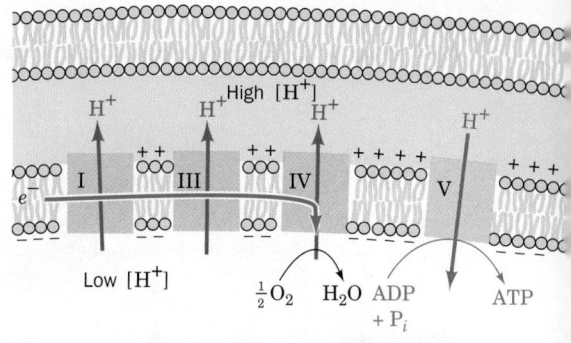

In 1789, Armand Séguin and Antoine Lavoisier (the father of modern chemistry) wrote:

> . . . in general, respiration is nothing but a slow combustion of carbon and hydrogen, which is entirely similar to that which occurs in a lamp or lighted candle, and that, from this point of view, animals that respire are true combustible bodies that burn and consume themselves.

Lavoisier had by this time demonstrated that living animals consume oxygen and generate carbon dioxide. It was not until the early twentieth century, however, after the rise of enzymology, that it was established, largely through the work of Otto Warburg, that biological oxidations are catalyzed by intracellular enzymes. As we have seen, glucose is completely oxidized to CO_2 through the enzymatic reactions of glycolysis and the citric acid cycle. In this chapter we shall examine the fate of the electrons that are removed from glucose by this oxidation process.

The complete oxidation of glucose by molecular oxygen is described by the following redox equation:

$$C_6H_{12}O_6 + 6O_2 \longrightarrow 6CO_2 + 6H_2O$$
$$\Delta G^{\circ\prime} = -2823 \text{ kJ} \cdot \text{mol}^{-1}$$

To see more clearly the transfer of electrons, let us break this equation down into two half-reactions. In the first half-reaction the glucose carbon atoms are oxidized:

$$C_6H_{12}O_6 + 6H_2O \longrightarrow 6CO_2 + 24H^+ + 24e^-$$

and in the second, molecular oxygen is reduced:

$$6O_2 + 24H^+ + 24e^- \longrightarrow 12H_2O$$

In living systems, the electron-transfer process connecting these half-reactions occurs through a multistep pathway that harnesses the liberated free energy to form ATP.

The 12 electron pairs involved in glucose oxidation are not transferred directly to O_2. Rather, as we have seen, *they are transferred to the coenzymes NAD^+ and FAD to form 10 NADH and 2 $FADH_2$* (Fig. 22-1) in the reactions catalyzed by the glycolytic enzyme glyceraldehyde-3-phosphate dehydrogenase (Section 17-2F), pyruvate dehydrogenase (Section 21-2A), and the citric acid cycle enzymes isocitrate dehydrogenase, α-ketoglutarate dehydrogenase, succinate dehydrogenase (the only FAD reduction), and malate dehydrogenase (Section 21-3). *The electrons then pass into the* **electron-transport chain** *(alternatively, the* **respiratory chain***), where, through reoxidation of NADH and $FADH_2$, they participate in the sequential oxidation–reduction of over 10 redox centers before reducing O_2 to H_2O. In this process, protons are expelled from the mitochondrion. The free energy stored in the resulting pH gradient drives the synthesis of ATP from ADP and P_i through* **oxidative phosphorylation.** Reoxidation of each NADH results in the synthesis of ~2.5 ATP, and reoxidation of $FADH_2$ yields ~1.5 ATP for a total of ~32 ATP for each glucose completely oxidized to CO_2 and H_2O (including the 2 ATP made in glycolysis and the 2 ATP made in the citric acid cycle).

In this chapter we explore the mechanisms of electron transport and oxidative phosphorylation and their regulation. We begin with a discussion of mitochondrial structure and transport systems.

1 THE MITOCHONDRION

The mitochondrion (Section 1-2Ac) is the site of eukaryotic oxidative metabolism. It contains, as Albert Lehninger and Eugene Kennedy demonstrated in 1948, the enzymes that mediate this process, including pyruvate dehydrogenase, the citric acid cycle enzymes, the enzymes catalyzing fatty acid oxidation (Section 25-2C), and the enzymes and redox proteins involved in electron transport and oxidative phosphorylation. It is therefore with good reason that the mitochondrion is described as the cell's "power plant."

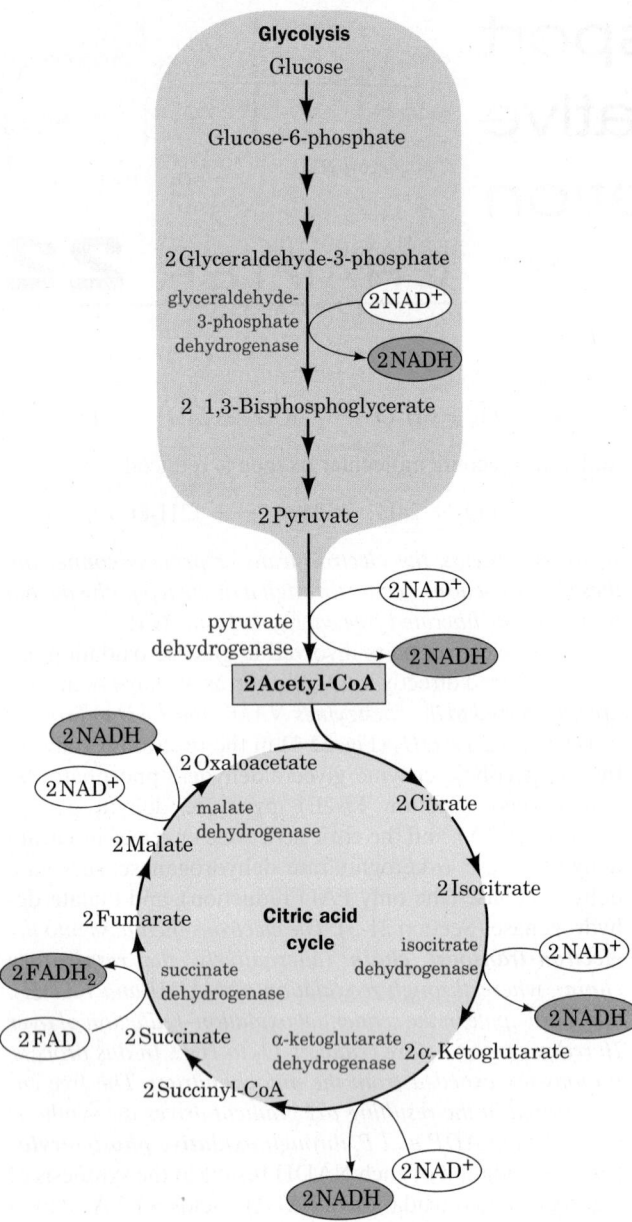

Figure 22-1 **The sites of electron transfer that form NADH and FADH$_2$ in glycolysis and the citric acid cycle.**

A. Mitochondrial Anatomy

Mitochondria vary considerably in size and shape depending on their source and metabolic state. They are typically ellipsoids ~0.5 μm in diameter and 1 μm in length (about the size of a bacterium; Fig. 22-2). The mitochondrion is bounded by a smooth outer membrane and contains an extensively invaginated inner membrane. The number of invaginations, called **cristae,** varies with the respiratory activity of the particular type of cell. This is because the proteins mediating electron transport and oxidative phosphorylation are bound to the inner mitochondrial membrane, so that the respiration rate varies with membrane surface

area. Liver, for example, which has a relatively low respiration rate, contains mitochondria with relatively few cristae, whereas those of heart muscle contain many. Nevertheless, the aggregate area of the inner mitochondrial membranes in a liver cell is ~15-fold greater than that of its plasma membrane.

The inner mitochondrial compartment consists of a gel-like substance of <50% water, named the **matrix,** which contains remarkably high concentrations of the soluble enzymes of oxidative metabolism (e.g., citric acid cycle enzymes), as well as substrates, nucleotide cofactors, and inorganic ions. The matrix also contains the mitochondrial genetic machinery—DNA, RNA, and ribosomes—that in mammals expresses only 13 mitochondrial inner membrane proteins together with 22 tRNAs and two ribosomal RNAs.

a. The Inner Mitochondrial Membrane and Cristae Compartmentalize Metabolic Functions

The outer mitochondrial membrane contains **porin,** a protein that forms nonspecific pores that permit free diffusion of up to 10-kD molecules (the X-ray structures of bacterial porins are discussed in Sections 12-3Ad and 20-2D). The inner membrane, which is ~75% protein by mass, is considerably richer in proteins than the outer membrane (Fig. 22-3). It is freely permeable only to O_2, CO_2, and H_2O and contains, in addition to respiratory chain proteins, numerous transport proteins that control the passage of metabolites such as ATP, ADP, pyruvate, Ca^{2+}, and phosphate (see below). *This controlled impermeability of the inner mitochondrial membrane to most ions, metabolites, and low molecular mass compounds permits the generation of ionic gradients across this barrier and results in the compartmentalization of metabolic functions between cytosol and mitochondria.*

Two-dimensional electron micrographs of mitochondria such as Fig. 22-2a suggest that cristae resemble baffles and that the intercristal spaces communicate freely with the mitochondrion's intermembrane space, as Fig. 22-2b implies. However, electron microscopy–based three-dimensional image reconstruction methods have revealed that cristae can vary in shape from simple tubular entities to more complicated lamellar assemblies that merge with the inner membrane via narrow tubular structures (Fig. 22-4). Evidently, cristae form microcompartments that restrict the diffusion of substrates and ions between the intercristal and intermembrane spaces. This has important functional implications because it would result in a locally greater pH gradient across cristal membranes than across inner membranes that are not part of cristae, thereby significantly influencing the rate of oxidative phosphorylation (Section 22-3).

B. Mitochondrial Transport Systems

The inner mitochondrial membrane is impermeable to most hydrophilic substances. It must therefore contain specific transport systems to permit the following processes:

(a)   *(b)*

Figure 22-2   **Mitochondria.** (*a*) An electron micrograph of an animal mitochondrion. [K.R. Porter/Photo Researchers, Inc.] (*b*) Cutaway diagram of a mitochondrion.

1. Glycolytically produced cytosolic NADH must gain access to the electron-transport chain for aerobic oxidation.

2. Mitochondrially produced metabolites such as oxaloacetate and acetyl-CoA, the respective precursors for cytosolic glucose and fatty acid biosynthesis, must reach their metabolic destinations.

3. Mitochondrially produced ATP must reach the cytosol, where most ATP-utilizing reactions take place, whereas ADP and P_i, the substrates for oxidative phosphorylation, must enter the mitochondrion.

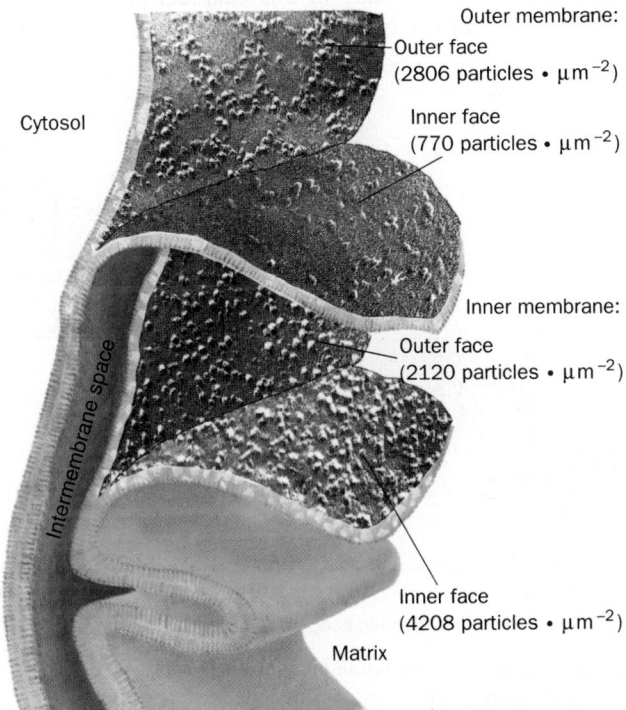

Figure 22-3   **Freeze-fracture and freeze-etch electron micrographs of the inner and outer mitochondrial membranes.** The inner membrane contains about twice the density of embedded particles as does the outer membrane. [Courtesy of Lester Packer, University of California at Berkeley.]

Figure 22-4   **Electron microscopy–based three-dimensional image reconstruction of a rat liver mitochondrion.** The outer membrane (OM) is red, the inner membrane (IM) is yellow, and the cristae (C) are green. The arrowheads point to tubular regions of the cristae that connect them to the inner membrane and to each other. [Courtesy of Carmen Mannella, Wadsworth Center, Albany, New York.]

We have already studied the ADP–ATP translocator and its dependence on $\Delta\Psi$, the electric potential difference across the mitochondrial membrane (Section 20-4C). The export mechanisms of oxaloacetate and acetyl-CoA from the mitochondrion are, respectively, discussed in Sections 23-1Ag and 25-4D. In the remainder of this section we examine the mitochondrial transport systems for P_i and Ca^{2+} and the shuttle systems for NADH.

a. P_i Transport

ATP is generated from ADP + P_i in the mitochondrion but is utilized in the cytosol. The P_i produced is returned to the mitochondrion by the **phosphate carrier,** an electroneutral P_i–H^+ symport that is driven by ΔpH. The proton that accompanies the P_i into the mitochondrion had, in effect, been previously expelled from the mitochondrion by the redox-driven pumps of the electron-transport chain (Section 22-3B). The electrochemical potential gradient generated by these proton pumps is therefore responsible for maintaining high mitochondrial ADP and P_i concentrations in addition to providing the free energy for ATP synthesis.

b. Ca^{2+} Transport

Since Ca^{2+}, like cAMP, functions as a second messenger (Section 18-3Ce), its concentrations in the various cellular compartments must be precisely controlled. The mitochondrion, endoplasmic reticulum, and extracellular spaces act as Ca^{2+} storage tanks. We studied the Ca^{2+}–ATPases of the plasma membrane, endoplasmic reticulum, and sarcoplasmic reticulum in Section 20-3B. Here we consider the mitochondrial Ca^{2+} transport systems.

Mitochondrial inner membrane systems separately mediate the influx and the efflux of Ca^{2+} (Fig. 22-5). The Ca^{2+} influx is driven by the inner mitochondrial membrane's membrane potential ($\Delta\Psi$, inside negative), which attracts positively charged ions. The rate of influx varies with the external $[Ca^{2+}]$ because the K_M for Ca^{2+} transport by this system is greater than the cytosolic Ca^{2+} concentration.

In heart, brain, and skeletal muscle mitochondria especially, Ca^{2+} efflux is independently driven by the Na^+ gradient across the inner mitochondrial membrane. Ca^{2+} exits the matrix only in exchange for Na^+, so that this system is an antiport. This exchange process normally operates at its maximal velocity. *Mitochondria (as well as the endoplasmic and sarcoplasmic reticulum) therefore can act as a "buffer" for cytosolic Ca^{2+}* (Fig. 22-6): If cytosolic $[Ca^{2+}]$ rises, the rate of mitochondrial Ca^{2+} influx increases while that of Ca^{2+} efflux remains constant, causing the mitochondrial $[Ca^{2+}]$ to increase while the cytosolic $[Ca^{2+}]$ decreases to its original level (its set point). Conversely, a decrease in cytosolic $[Ca^{2+}]$ reduces the influx rate, causing net efflux of $[Ca^{2+}]$ and an increase of cytosolic $[Ca^{2+}]$ back to the set point.

Oxidation carried out by the citric acid cycle in the mitochondrial matrix is controlled by the matrix $[Ca^{2+}]$ (Section 21-4c). It is interesting to note, therefore, that in response to increases in cytosolic $[Ca^{2+}]$ caused by increased muscle activity, the matrix $[Ca^{2+}]$ increases, thereby activating

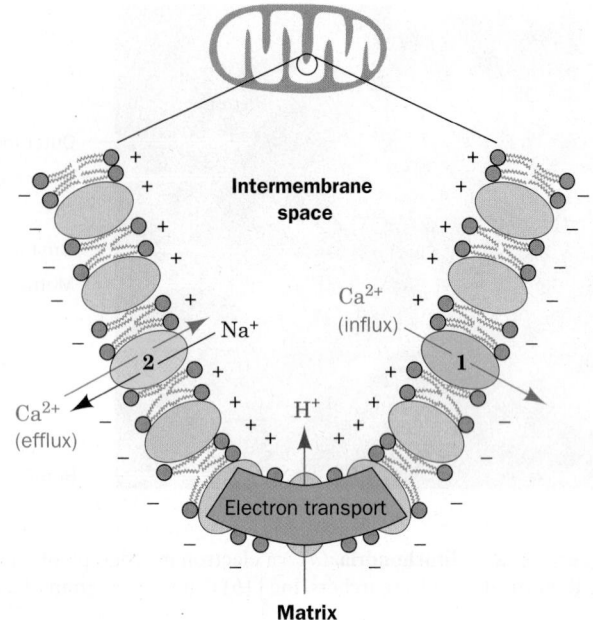

Figure 22-5 **The two mitochondrial Ca^{2+} transport systems.** System 1 mediates Ca^{2+} influx to the matrix in response to the membrane potential (negative inside). System 2 mediates Ca^{2+} efflux in exchange for Na^+.

the enzymes of the citric acid cycle. This leads to an increase in [NADH], whose reoxidation by oxidative phosphorylation (as we study in this chapter) generates the ATP needed for this increased muscle activity.

Figure 22-6 **The regulation of cytosolic $[Ca^{2+}]$.** The efflux pathway operates at a constant rate independent of $[Ca^{2+}]$, whereas the activity of the influx pathway varies with $[Ca^{2+}]$. At the set point, the activities of the two pathways are equal and there is no net Ca^{2+} flux. An increase in cytosolic $[Ca^{2+}]$ results in net mitochondrial influx, and a decrease in cytosolic $[Ca^{2+}]$ results in net mitochondrial efflux. Both effects lead to the restoration of the cytosolic $[Ca^{2+}]$. [After Nicholls, D., *Trends Biochem. Sci.* **6,** 37 (1981).]

c. Cytoplasmic Shuttle Systems "Transport" NADH Across the Inner Mitochondrial Membrane

Although most of the NADH generated by glucose oxidation is formed in the mitochondrial matrix via the citric acid cycle, that generated by glycolysis occurs in the cytosol. Yet the inner mitochondrial membrane lacks an NADH transport protein. *Only the electrons from cytosolic NADH are transported into the mitochondrion by one of several ingenious "shuttle" systems.* In the **malate–aspartate shuttle** (Fig. 22-7), which functions in heart, liver, and kidney, mitochondrial NAD$^+$ is reduced by cytosolic NADH through the intermediate reduction and subsequent regeneration of oxaloacetate. This process occurs in two phases of three reactions each:

Phase A (transport of electrons into the matrix):

1. In the cytosol, NADH reduces oxaloacetate to yield NAD$^+$ and malate in a reaction catalyzed by cytosolic malate dehydrogenase.

2. The **malate–α-ketoglutarate carrier** transports malate from the cytosol to the mitochondrial matrix in exchange for α-ketoglutarate from the matrix.

3. In the matrix, NAD$^+$ reoxidizes malate to yield NADH and oxaloacetate in a reaction catalyzed by mitochondrial malate dehydrogenase (Section 21-3H).

Phase B (regeneration of cytosolic oxaloacetate):

4. In the matrix, a transaminase (Section 26-1A) converts oxaloacetate to aspartate with the concomitant conversion of glutamate to α-ketoglutarate.

5. The **glutamate–aspartate carrier** transports aspartate from the matrix to the cytosol in exchange for cytosolic glutamate.

6. In the cytosol, a transaminase converts aspartate to oxaloacetate with the concomitant conversion of α-ketoglutarate to glutamate.

The electrons of cytosolic NADH are thereby transferred to mitochondrial NAD$^+$ to form NADH, which is subject to reoxidation via the electron-transport chain. *The malate–aspartate shuttle yields ~2.5 ATPs for every cytosolic NADH.* Note, however, that every NADH that enters the matrix is accompanied by a proton which, as we shall

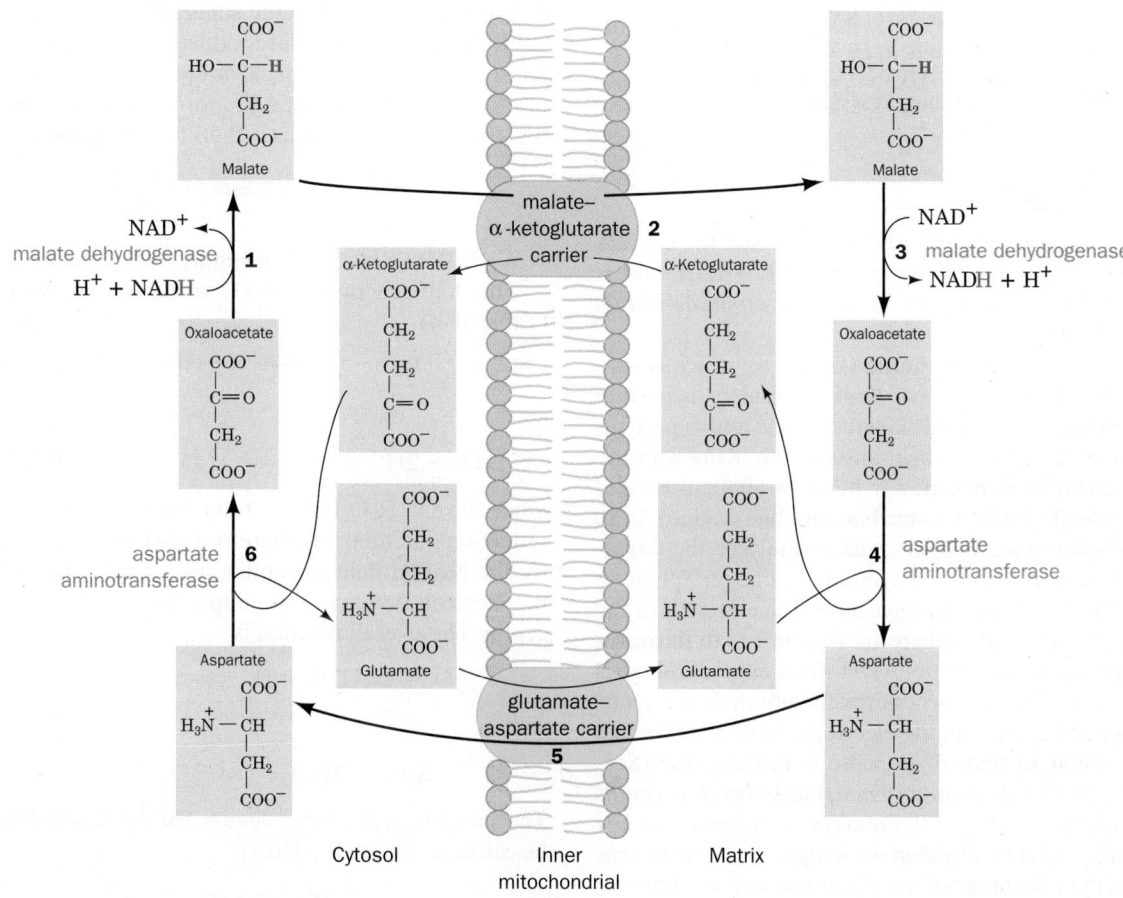

Figure 22-7 **The malate–aspartate shuttle.** The electrons of cytosolic NADH are transported to mitochondrial NADH (shown in red as hydride transfers) in Steps **1** to **3**. Steps **4** to **6** then serve to regenerate cytosolic oxaloacetate.

Inner
mitochondrial
Cytosol membrane Matrix

**Dihydroxyacetone
phosphate**

$$H_2C—OH$$
$$C=O$$
$$CH_2OPO_3^{2-}$$

$H^+ + NADH$

glycerol-3-phosphate
dehydrogenase

NAD^+

$$H_2C—OH$$
$$HO—C—H$$
$$CH_2—OPO_3^{2-}$$

Glycerol-3-phosphate

Electron-
transport
chain

$2e^-$
| 3
$FADH_2$

flavoprotein
dehydrogenase

FAD

Figure 22-8 **The glycerophosphate shuttle.** The electrons of cytosolic NADH are transported to the mitochondrial electron-transport chain in three steps (shown in red as hydride transfers): **(1)** Cytosolic oxidation of NADH by dihydroxyacetone phosphate catalyzed by glycerol-3-phosphate dehydrogenase. **(2)** Oxidation of glycerol-3-phosphate by flavoprotein dehydrogenase with the reduction of FAD to $FADH_2$. **(3)** Reoxidation of $FADH_2$ with the passage of electrons into the electron-transport chain. Note that the glycerophosphate shuttle is not a membrane transport system.

see (Section 22-3C), would otherwise be used to generate ~0.3 ATP. Consequently, *every cytosolic NADH that is translocated to the matrix by the malate–aspartate shuttle yields ~2.2 ATPs.*

The **glycerophosphate shuttle** (Fig. 22-8), which is simpler but less energy efficient than the malate–aspartate shuttle, occurs in brain and skeletal muscle and is particularly prominent in insect flight muscle (the tissue with the largest known sustained power output—about the same power-to-weight ratio as a small automobile engine). In it, **glycerol-3-phosphate dehydrogenase** catalyzes the oxidation of cytosolic NADH by dihydroxyacetone phosphate to yield NAD^+, which reenters glycolysis. The electrons of the resulting **glycerol-3-phosphate** are transferred to **flavoprotein dehydrogenase** to form $FADH_2$. This enzyme, which is situated on the inner mitochondrial membrane's outer surface, supplies electrons to the electron-transport chain in a manner similar to that of succinate dehydrogenase (Section 22–2C2). *The glycerophosphate shuttle therefore results in the synthesis of ~1.5 ATPs for every cytoplasmic NADH reoxidized, ~0.7 ATP less than the malate–aspartate shuttle.* However, the advantage of the glycerophosphate shuttle is that, being essentially irreversible, it operates efficiently even when the cytoplasmic NADH concentration is low relative to that of NAD^+, as occurs in rapidly metabolizing

tissues. In contrast, the malate–aspartate shuttle is reversible and hence is driven by concentration gradients.

2 ELECTRON TRANSPORT

In the electron-transport process, the free energy of electron transfer from NADH and $FADH_2$ to O_2 via protein-bound redox centers is coupled to ATP synthesis. We begin our study of this process by considering its thermodynamics. We then examine the path of electrons through the redox centers of the system and discuss the experiments used to unravel this pathway. Finally, we study the four complexes that make up the electron-transport chain. In the next section we discuss how the free energy harvested by the electron-transport process is coupled to ATP synthesis.

A. *Thermodynamics of Electron Transport*

We can estimate the thermodynamic efficiency of electron transport through knowledge of standard reduction potentials. As we have seen in our thermodynamic considerations of oxidation–reduction reactions (Section 16-5), an oxidized substrate's affinity for electrons increases with its standard reduction potential, $\mathscr{E}°'$ [the voltage generated by the reaction of the half-cell under standard biochemical conditions ($1M$ reactants and products with $[H^+]$ defined as 1 at pH 7) relative to the standard hydrogen electrode; Table 16-4 lists the standard reduction potentials of several half-reactions of biochemical interest]. The standard reduction potential difference, $\Delta\mathscr{E}°'$, for a redox reaction involving any two half-reactions is therefore expressed:

$$\Delta\mathscr{E}°' = \mathscr{E}°'_{(e^- \text{ acceptor})} - \mathscr{E}°'_{(e^- \text{ donor})}$$

a. NADH Oxidation Is a Highly Exergonic Reaction

The half-reactions for O_2 oxidation of NADH are (Table 16-4)

$$NAD^+ + H^+ + 2e^- \rightleftharpoons NADH \qquad \mathscr{E}°' = -0.315 \text{ V}$$

and

$$\tfrac{1}{2}O_2 + 2H^+ + 2e^- \rightleftharpoons H_2O \qquad \mathscr{E}°' = 0.815 \text{ V}$$

Since the O_2/H_2O half-reaction has the greater standard reduction potential and therefore the higher electron affinity, the NADH half-reaction is reversed, so that NADH is the electron donor in this couple and O_2 the electron acceptor. The overall reaction is

$$\tfrac{1}{2}O_2 + NADH + H^+ \rightleftharpoons H_2O + NAD^+$$

so that

$$\Delta\mathscr{E}°' = 0.815 - (-0.315) = 1.130 \text{ V}$$

The standard free energy change for the reaction can then be calculated from Eq. [16.7]:

$$\Delta G°' = -n\mathscr{F}\,\Delta\mathscr{E}°'$$

where $\mathscr{F}$, the Faraday constant, is 96,485 C · mol^{-1} of electrons and n is the number of electrons transferred per mole of reactants. Thus, since $1\ V = 1\ J \cdot C^{-1}$, for NADH oxidation:

$$\Delta G^{\circ\prime} = -2\ \frac{\text{mol } e^-}{\text{mol reactant}} \times 96{,}485\ \frac{C}{\text{mol } e^-} \times 1.13\ J \cdot C^{-1}$$
$$= -218\ kJ \cdot mol^{-1}$$

In other words, the oxidation of 1 mol of NADH by O_2 (the transfer of $2e^-$) under standard biochemical conditions is associated with the release of 218 kJ of free energy.

b. Electron Transport Is Thermodynamically Efficient

The standard free energy required to synthesize 1 mol of ATP from ADP + P_i is 30.5 kJ. The standard free energy of oxidation of NADH by O_2, if coupled to ATP synthesis, is therefore sufficient to drive the formation of several moles of ATP. This coupling, as we shall see, is achieved by an electron-transport chain in which electrons are passed through three protein complexes containing redox centers with progressively greater affinity for electrons (increasing standard reduction potentials) instead of directly to O_2. *This allows the large overall free energy change to be broken up into three smaller packets, each of which is coupled with ATP synthesis in a process called* **oxidative phosphorylation.** *Oxidation of 1 NADH therefore results in the synthesis of ~2.5 ATP.* (Oxidation of $FADH_2$, whose entrance into the electron-transport chain is regulated by a fourth protein complex, is similarly coupled to the synthesis of ~1.5 ATP.) The thermodynamic efficiency of oxidative phosphorylation is therefore $2.5 \times 30.5\ kJ \cdot mol^{-1} \times 100/218\ kJ \cdot mol^{-1} = 35\%$ under standard biochemical conditions. However, under physiological conditions in active mitochondria (where the reactant and product concentrations as well as the pH deviate from standard conditions), this thermodynamic efficiency is thought to be ~70%. In comparison, the energy efficiency of a typical automobile engine is <30%.

B. *The Sequence of Electron Transport*

See Guided Exploration 19: Electron transport and oxidative phosphorylation overview *The free energy necessary to generate ATP is extracted from the oxidation of NADH and $FADH_2$ by the electron-transport chain, a series of four protein complexes through which electrons pass from lower to higher standard reduction potentials* (Fig. 22-9). Electrons are carried from **Complexes I** and **II** to **Complex III** by **coenzyme Q** (**CoQ** or **ubiquinone;** so named because of its ubiquity in respiring organisms), and from Complex III to **Complex IV** by the peripheral membrane protein **cytochrome c** (Sections 7-3B and 9-6A).

Figure 22-9 The mitochondrial electron-transport chain. The standard reduction potentials of its most mobile components (*green*) are indicated, as are the points where sufficient free energy is harvested to synthesize ATP (*blue*) and the sites of action of several respiratory inhibitors (*red*). (Note that Complexes I, III, and IV do not directly synthesize ATP but, rather, sequester the free energy necessary to do so by pumping protons outside the mitochondrion to form a proton gradient; Section 22-3.)

Complex I catalyzes oxidation of NADH by CoQ:

$$NADH + CoQ\ (oxidized) \longrightarrow$$
$$NAD^+ + CoQ\ (reduced)$$
$$\Delta\mathscr{E}^{\circ\prime} = 0.360\ V \qquad \Delta G^{\circ\prime} = -69.5\ kJ \cdot mol^{-1}$$

Complex III catalyzes oxidation of CoQ (reduced) by cytochrome c:

$$CoQ\ (reduced) + 2\ cytochrome\ c\ (oxidized) \longrightarrow$$
$$CoQ\ (oxidized) + 2\ cytochrome\ c\ (reduced)$$
$$\Delta\mathscr{E}^{\circ\prime} = 0.190\ V \qquad \Delta G^{\circ\prime} = -36.7\ kJ \cdot mol^{-1}$$

Complex IV catalyzes oxidation of cytochrome c (reduced) by O_2, the terminal electron acceptor of the electron-transport process:

$$2\ cytochrome\ c\ (reduced) + \tfrac{1}{2}O_2 \longrightarrow$$
$$2\ cytochrome\ c\ (oxidized) + H_2O$$
$$\Delta\mathscr{E}^{\circ\prime} = 0.580\ V \qquad \Delta G^{\circ\prime} = -112\ kJ \cdot mol^{-1}$$

The changes in standard reduction potential of an electron pair as it successively traverses Complexes I, III, and IV correspond, at each stage, to sufficient free energy to power the synthesis of nearly one ATP molecule.

Complex II catalyzes the oxidation of $FADH_2$ by CoQ.

$$FADH_2 + CoQ\ (oxidized) \longrightarrow FAD + CoQ\ (reduced)$$
$$\Delta\mathscr{E}^{\circ\prime} = 0.085\ V \qquad \Delta G^{\circ\prime} = -16.4\ kJ \cdot mol^{-1}$$

This redox reaction does not release sufficient free energy to synthesize ATP; it functions only to inject the electrons from $FADH_2$ into the electron-transport chain.

a. The Workings of the Electron-Transport Chain Have Been Elucidated through the Use of Inhibitors

Our understanding of the sequence of events in electron transport is largely based on the use of specific inhibitors. This sequence has been corroborated by measurements of the standard reduction potentials of the redox components of each of the complexes as well as by determining the stoichiometry of electron transport and the coupled ATP synthesis.

The rate at which O_2 is consumed by a suspension of mitochondria is a sensitive measure of the functioning of the electron-transport chain. It is conveniently measured with an **oxygen electrode** (Fig. 22-10). Compounds that inhibit electron transport, as judged by their effect on O_2 disappearance in such an experimental system, have been invaluable experimental probes in tracing the path of electrons through the electron-transport chain and in determining the points of entry of electrons from various substrates. Among the most useful such substances are **rotenone** (a plant toxin used by Amazonian Indians to poison fish and which is also used as an insecticide), **amytal** (a barbiturate), **antimycin** (an antibiotic), and **cyanide:**

Rotenone

Amytal

Cyanide

Antimycin

Figure 22-10 The oxygen electrode. This electrode consists of an Ag/AgCl reference electrode and a Pt electrode, both immersed in a KCl solution and in contact with the sample chamber through an O_2-permeable Teflon membrane. O_2 is reduced to H_2O at the Pt electrode, thereby generating a voltage with respect to the Ag/AgCl electrode that is proportional to the O_2 concentration in the sealed sample chamber. [After Cooper, T.G., *The Tools of Biochemistry*, p. 69, Wiley (1977).]

The following experiment illustrates the use of these inhibitors:

A buffered solution containing excess ADP and P_i is equilibrated in the reaction vessel of an oxygen electrode.

Mitochondria + β-hydroxybutyrate added,
NAD⁺-linked oxidation commences

Rotenone or amytal added,
inhibiting NAD⁺-linked oxidation

Succinate added, FAD-linked
oxidation begins

Antimycin added, inhibiting
FAD-linked oxidation

TMPD/ascorbate added,
oxidation resumes

CN⁻ added,
oxidation terminates

$[O_2]$

Time

Figure 22-11   Effect of inhibitors on electron transport. This diagram shows an idealized oxygen electrode trace of a mitochondrial suspension containing excess ADP and P_i. At the numbered points, the indicated reagents are injected into the sample chamber and the resulting changes in $[O_2]$ are recorded. The numbers refer to the discussion in the text. [After Nicholls, D.G., *Bioenergetics, p.* 110, Academic Press (1982).]

Reagents are then injected into the chamber and the O_2 consumption recorded (Fig. 22-11):

1. Mitochondria and **β-hydroxybutyrate** are injected into the chamber. Mitochondria mediate the NAD⁺-linked oxidation of β-hydroxybutyrate (Section 25-3):

$$CH_3-\overset{\overset{\displaystyle OH}{|}}{CH}-CH_2-CO_2^-$$

β-Hydroxybutyrate

NAD⁺ ⟍
          ⟩ β-hydroxybutyrate
NADH + H⁺ ⟋   dehydrogenase

$$CH_3-\overset{\overset{\displaystyle O}{\|}}{C}-CH_2-CO_2^-$$

Acetoacetate

As the resulting NADH is oxidized by the electron-transport chain with O_2 as the terminal electron acceptor, the O_2 concentration in the reaction mixture decreases.

2. Addition of rotenone or amytal completely stops the β-hydroxybutyrate oxidation.

3. Addition of succinate, which undergoes FAD-linked oxidation, causes the $[O_2]$ to resume its decrease. Electrons from FADH₂ are therefore still able to reduce O_2 in the presence of rotenone; that is, *electrons from FADH₂ enter the electron-transport chain after the rotenone-blocked step.*

4. Addition of antimycin inhibits electron transport from FADH₂.

5. Although NADH and FADH₂ are the electron-transport chain's two physiological electron donors,

nonphysiological reducing agents can also be used to probe the flow of electrons. **Tetramethyl-*p*-phenylenediamine (TMPD)** is an ascorbate-reducible redox carrier that transfers electrons directly to cytochrome *c:*

Tetramethyl-*p*-phenylenediamine
(TMPD), oxidized form

Ascorbic acid

TMPD, reduced form

Dehydroascorbic acid

Addition of TMPD and ascorbate to the antimycin-inhibited reaction mixture results in resumption of oxygen consumption; evidently *there is a third point at which electrons can enter the electron-transport chain.*

6. The addition of CN⁻ completely inhibits oxidation of all three electron donors, indicating that it blocks the electron-transport chain after the third point of entry of electrons.

Experiments such as these established the order of electron flow through the electron-transport chain complexes and the positions blocked by various electron-transport inhibitors (Fig. 22-9). This order was confirmed and extended by observations that the standard reduction potentials of the redox carriers forming the electron-transport chain complexes are very close to the standard reduction potentials of their electron donor substrates (Table 22-1). *The three jumps in reduction potential between NADH, CoQ, cytochrome c, and O_2 are each of sufficient magnitude to drive ATP synthesis.* Indeed, these redox potential jumps correspond to the points of inhibition of rotenone (or amytal), antimycin, and CN⁻.

b. Phosphorylation and Oxidation Are Rigidly Coupled

The foregoing thermodynamic studies suggest that oxidation of NADH, FADH₂, and ascorbate by O_2 is associated with the synthesis of about 2.5, 1.5, and 1 ATP, respectively. This stoichiometry, called the **P/O ratio** [the ratio of ATP synthesized to O atoms reduced (electron pairs taken up)], has been confirmed experimentally through measurements of O_2 uptake by resting and active mitochondria. An example of a typical experiment used to determine the P/O ratio is as follows: A suspension of mitochondria (isolated by differential centrifugation after cell disruption; Section 6-1B) containing an excess of P_i but no ADP is incubated in an oxygen electrode reaction chamber. *Oxidation and*

Reduction Potentials of Electron-Transport Chain Components in Resting Mitochondria

Component	$\mathscr{E}°'$(V)
NADH	−0.315
Complex I (NADH:CoQ oxidoreductase; ~900 kD, 45 subunits):	
FMN	−0.380
[2Fe–2S]N1a	−0.370
[2Fe–2S]N1b	−0.250
[4Fe–4S]N3, 4, 5, 6a, 6b, 7	−0.250
[4Fe–4S]N2	−0.150
Succinate	0.031
Complex II (succinate–CoQ oxidoreductase; ~420 kD, 4 subunits):	
FAD	−0.040
[2Fe–2S]	−0.030
[4Fe–4S]	−0.245
[3Fe–4S]	−0.060
Heme b_{560}	−0.080
Coenzyme Q	0.045
Complex III (CoQ–cytochrome *c* oxidoreductase; ~450 kD, 9–11 subunits):	
Heme b_H (b_{562})	0.030
Heme b_L (b_{566})	−0.030
[2Fe–2S]	0.280
Heme c_1	0.215
Cytochrome *c*	0.235
Complex IV (cytochrome *c* oxidase; ~410 kD, 8–13 subunits):	
Heme *a*	0.210
Cu_A	0.245
Cu_B	0.340
Heme a_3	0.385
O_2	0.815

Source: Mainly Wilson, D.F., Erecinska, M., and Dutton, P.L., *Annu. Rev. Biophys. Bioeng.* **3**, 205 and 208 (1974); *and* Wilson, D.F., *In* Bittar, E.E. (Ed.), *Membrane Structure and Function*, Vol. 1, p. 160, Wiley (1980).

phosphorylation are closely coupled in well-functioning mitochondria, so electron transport can occur only if ADP is being phosphorylated (Section 22-3). Indeed, mitochondrial metabolism is so tightly regulated that even the appearances of actively respiring and resting mitochondria are greatly different (Fig. 22-12). Since no ADP is present in the reaction mixture, the mitochondria are resting and the O_2 consumption rate is minimal (Fig. 22-13; Region 1). The system is then manipulated as follows:

(a) ADP (75 μmol) and an excess of β-hydroxybutyrate (an NAD⁺-linked substrate) are added. The mitochondria immediately enter the active state and the rate of oxygen consumption increases (Fig. 22-13, Region 2) and is maintained at this elevated

level until all the ADP is phosphorylated. The mitochondria then return to the resting state (Fig. 22-13, Region 3). Phosphorylation of 75 μmol of ADP under these conditions consumes 15 μmol of O_2. Since the oxidation of NADH by O_2 consumes twice as many moles of NADH (i.e., electron pairs) as of O_2, the P/O ratio for NADH reoxidation at Region 2 is 75 μmol of ADP/(2 × 15 μmol of O_2) = 2.5; that is, *2.5 mol of ADP are phosphorylated per mole of NADH oxidized.*

(b) The experiment is continued by inhibiting electron transfer from NADH by rotenone and adding an additional 75 μmol of ADP (Fig. 22-13, Region 4), this time together with an excess of the FAD-linked substrate succinate. Oxygen consumption again continues until all the ADP is phosphorylated, and the system again returns to the resting state (Fig. 22-13, Region 5). Calculation of the P/O ratio for $FADH_2$ oxidation yields the value 1.5; that is, *1.5 mol of ADP are phosphorylated per mole of $FADH_2$ oxidized.*

(c) In the same manner, *the oxidation of ascorbate/TMPD yields a P/O ratio of 1 (Fig. 22-13, Regions 6 and 7).*

These conclusions agree with the inhibitor studies indicating that there are three entry points for electrons into the electron-transport chain and with the standard reduction potential measurements exhibiting three potential jumps, each sufficient to provide the free energy for ATP synthesis (Fig. 22-9).

c. The P/O Ratios Are Difficult to Measure Accurately

Measurements of P/O ratios are subject to systematic experimental errors for which it is difficult to correct, such as inaccuracies in the measurement of the oxygen concentration, the presence of AMP, and proton leakage through the inner mitochondrial membrane. In the older literature, these values were taken to be 3, 2, and 1. However, careful measurements by Peter Hinkle have yielded values close to 2.5, 1.5, and 1 for these quantities (we shall see in Section 22-3 that the mechanism of oxidative phosphorylation does not require that P/O ratios have integer values). If these latter values are correct, then the number of ATP molecules that are synthesized per molecule of glucose oxidized is 2.5 ATP/NADH × 10 NADH/glucose + 1.5 ATP/$FADH_2$ × 2 $FADH_2$/glucose + 2 ATP/glucose from the citric acid cycle + 2 ATP/glucose from glycolysis = 32 ATP/glucose (rather than the value of 38 ATP/glucose implied by P/O ratios of 3, 2, and 1).

The determination of the *in vivo* yield of ATP/glucose is even more problematic. The reducing equivalents from the GAPDH-generated NADH may be imported into the mitochondrion via the glycerophosphate shuttle, which yields $FADH_2$, or via the malate–aspartate shuttle, which yields NADH but passes a proton into the matrix (Section 22-1Bc). The mix of these two shuttle systems varies from tissue to tissue and hence so does the ATP yield. In addition, the rate of proton leakage back across the inner mitochondrial membrane, which is significant, may vary with

(a)

(b)

Figure 22-12   Electron micrographs of mouse liver mitochondria. (*a*) In the actively respiring state and (*b*) in the resting state. The cristae in actively respiring mitochondria are far more condensed than they are in resting mitochondria. [Courtesy of Charles Hackenbrock, University of North Carolina Medical School.]

conditions and cellular identity. Hence, the *in vivo* yield of ATP per glucose is likely to be significantly less than the above figures of 32 ATP/glucose.

How the free energy of electron transport is actually coupled to ATP synthesis, a subject of active research, is discussed in Section 22-3. We first examine the structures of the four respiratory complexes in order to understand how they are related to the function of the electron-transport chain. Keep in mind, however, that as in most areas of biochemistry, this field is under intense scrutiny and much of the information we need for a complete understanding of these relationships has yet to be elucidated.

C. Components of the Electron-Transport Chain

Many of the proteins embedded in the inner mitochondrial membrane are organized into the four respiratory complexes of the electron-transport chain. Each complex consists of several protein components that are associated with a variety of redox-active prosthetic groups with successively

Figure 22-13   Determination of the stoichiometry of coupled oxidation and phosphorylation (the P/O ratio) with different electron donors. Mitochondria are incubated in excess phosphate buffer in the sample chamber of an oxygen electrode. (*a*) Next, 75 μmol of ADP and excess β-hydroxybutyrate are added. Respiration continues until all the ADP is phosphorylated. ΔO_2 in Region 2 is 15 μmol, corresponding to 30 μmol of NADH oxidized; thus, P/O = 75/30 = 2.5. (*b*) Then 75 μmol of ADP and excess succinate are added together with rotenone to inhibit electron transfer from NADH. ΔO_2 in Region 4 is 25 μmol, corresponding to 50 μmol of $FADH_2$ oxidized; P/O = 75/50 = 1.5. (*c*) Finally, 75 μmol of ADP and excess TMPD/ascorbate are added with antimycin to inhibit electron transfer from $FADH_2$. ΔO_2 in Region 6 is 37.5 μmol, corresponding to 75 μmol of ascorbate oxidized; P/O = 75/75 = 1.

Figure 22-14 **The mitochondrial electron-transport chain.** The pathways of electron transfer (*black*) and proton pumping (*red*) are indicated. Electrons are transferred between Complexes I and III by membrane-soluble CoQ (Q) and between Complexes III and IV by the peripheral membrane protein cytochrome *c* (Cyt *c*). Complex II (not shown) transfers electrons from succinate to CoQ. 🔊 **See the Animated Figures**

increasing reduction potentials (Table 22-1). In the following paragraphs, we examine their structures and the agents that transfer electrons between them. Their relationships are summarized in Fig. 22-14.

1. Complex I (NADH Dehydrogenase or NADH–Coenzyme Q Oxidoreductase)

*Complex I (also called **NADH dehydrogenase**) passes electrons from NADH to CoQ.* This probably largest protein component of the inner mitochondrial membrane (~980 kD in mammals and ~700 kD in *Neurospora crassa*) contains one molecule of **flavin mononucleotide** (**FMN;** a redox-active prosthetic group that differs from FAD only by the absence of the AMP group) and eight or nine **iron–sulfur clusters** that participate in the electron-transport process (Table 22-1). In mammals, 7 of its 45 subunits, its most hydrophobic subunits, which form the core of its transmembrane region, are encoded by mitochondrial genes, with the remainder encoded by nuclear genes. Most of its subunits are homologous to soluble redox center-containing proteins, which strongly suggests that Complex I arose through the evolutionary aggregation of these previously existing proteins.

a. Iron–Sulfur Clusters Are Redox Active

Iron–sulfur clusters, first discovered by Helmut Beinert, commonly occur as prosthetic groups in **iron–sulfur proteins.** There are four common types of iron–sulfur clusters (Fig. 22-15). Those designated **[2Fe–2S]** and **[4Fe–4S]**

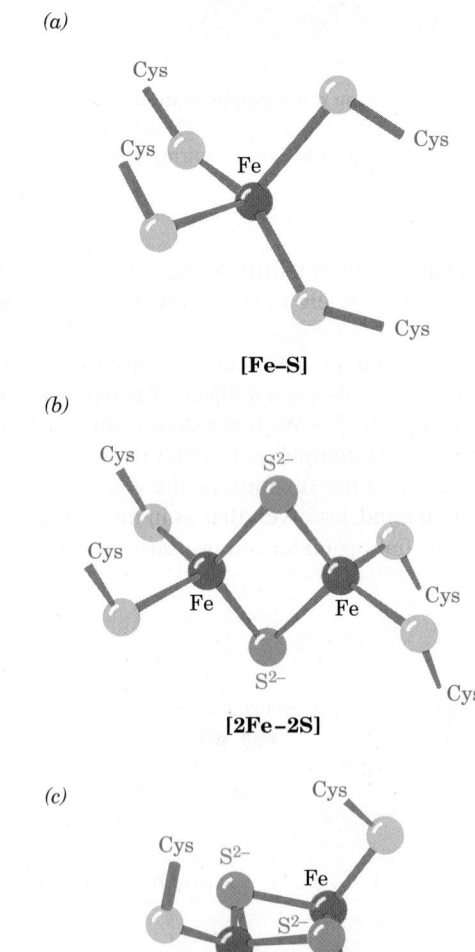

(a)

(b)

(c)

Figure 22-15 **Structures of the common iron–sulfur clusters.** (*a*) An [Fe–S] cluster, (*b*) a [2Fe–2S] cluster, and (*c*) a [4Fe–4S] cluster. The [3Fe–4S] cluster resembles the [4Fe–4S] cluster with one of its Fe ions removed. Note that, whereas the Fe and S^{2-} ions of [4Fe–4S] clusters form what appears to be a distorted cube, the structure is really two interpenetrating tetrahedra of Fe ions and S^{2-} ions.

clusters consist of equal numbers of iron and sulfide ions and are both coordinated to four protein Cys sulfhydryl groups. The **[3Fe–4S] cluster** is essentially a [4Fe–4S] cluster that lacks one Fe. One means of identifying these clusters utilizes the fact that their sulfide ions are acid labile: They are released as H_2S near pH 1. The **[Fe–S] cluster,** which occurs only in bacteria, consists of a single Fe atom liganded to four Cys residues. Note that the Fe ions in all four types of clusters are each coordinated by four S atoms, which are more or less tetrahedrally disposed around the Fe. However, in **Rieske iron–sulfur proteins** (named after their discoverer, John Rieske), one of the Fe atoms in a [2Fe–2S] cluster is coordinated by two His residues rather than two Cys residues.

The oxidized and reduced states of all iron–sulfur clusters differ by one formal charge regardless of their number of Fe ions. This is because the Fe ions in each cluster form a conjugated system and thus can have oxidation states between the +2 and +3 values possible for individual Fe ions. For example, each of the two [4Fe–4S] clusters in the protein **ferredoxin** (Fig. 22-16) contains one Fe(II) and three Fe(III) in its oxidized form and two Fe(II) and two Fe(III) in its reduced form. The standard reduction potential of a given type of iron–sulfur cluster depends on its interaction with its associated protein as well as on its oxidation state.

Figure 22-16 **X-ray structure of ferredoxin from *Peptostreptococcus asaccharolyticus.*** This monomeric 55-residue protein is drawn in ribbon form colored in rainbow order from its N-terminus (*blue*) to its C-terminus (*red*). Its two [4Fe–4S] clusters are drawn in ball-and-stick form with S yellow and Fe red-brown. The Cys side chains that ligand each Fe atom are drawn in stick form with C green and S yellow. [Based on an X-ray structure by Elinor Adman, Larry Sieker, and Lyle Jensen, University of Washington. PDBid 1DUR.] 🔖 **See Interactive Exercise 16**

Iron–sulfur proteins also occur in the photosynthetic electron-transport chains of plants and bacteria (Section 24-2); indeed, photosynthetic electron-transport chains are thought to be the evolutionary precursors of oxidative electron-transport chains (Section 1-5Cb).

b. The Coenzymes of Complex I

FMN and ubiquinone (CoQ), the coenzymes of Complex I, can each adopt three oxidation states (Fig. 22-17). Although NADH can only participate in a two-electron transfer, both FMN and CoQ are capable of accepting and donating either one or two electrons because their semiquinone forms are stable. In contrast, the cytochromes of Complex III (see below), to which reduced CoQ passes its electrons, are only capable of one-electron reductions. *FMN and CoQ therefore provide an electron conduit between the two-electron donor NADH and the one-electron acceptors, the cytochromes.*

CoQ's hydrophobic tail makes it soluble in the inner mitochondrial membrane's lipid bilayer. In mammals, this tail consists of ten C_5 isoprenoid units and hence the coenzyme is designated $\mathbf{Q_{10}}$. In other organisms, CoQ may have only six ($\mathbf{Q_6}$) or eight ($\mathbf{Q_8}$) isoprenoid units.

c. Electrons Follow a Multistep Path through Complex I

Complex I from prokaryotes consists of 13 to 15 subunits with an aggregate molecular mass of ~550 kD. Homologs of all of its conserved subunits occur in the mitochondrial enzyme and they contain equivalent redox components. Evidently, the prokaryotic enzyme is a stripped down version of mitochondrial Complex I. The functions of the 30 to 32 additional "accessory" subunits that are components of mitochondrial Complex I are largely unknown.

The low resolution X-ray structure of the 15-subunit Complex I from the thermophilic bacterium *Thermus thermophilus* (Fig. 22-18), determined by Leonid Sazanov, reveals an L-shaped protein with one arm of the L, the transmembrane arm, immersed in the plasma membrane (the inner mitochondrial membrane for eukaryotes) and the other, the peripheral arm, extending into the cytosol (the matrix for eukaryotes). The transmembrane arm, consists of seven subunits that collectively have 63 transmembrane helices. The three largest transmembrane subunits, Nqo12, 13, and 14, are structurally similar to each other and to previously determined structures of Na^+/H^+ antiporters. However, the most unusual feature of the transmembrane portion is a 110-Å long amphipathic α helix that extends from Nqo12 parallel to the plane of the membrane so as to span Nqo13 and 14.

The high resolution X-ray structure of the 8-subunit, hydrophilic, peripheral arm of Complex I from the thermophilic bacterium. *Thermophilus*, also determined by Sazanov, reveals a Y-shaped assembly that is 140 Å high (Fig. 22-19a). This subcomplex contains all of the enzyme's redox centers: an FMN, seven [4Fe–4S] clusters, and two [2Fe–2S] clusters. The FMN is located at the end of a solvent-exposed cavity that presumably forms the NADH

(a)

Flavin mononucleotide (FMN)
(oxidized or quinone form)

[H•]

FMNH• (radical or semiquinone form)

[H•]

FMNH$_2$ (reduced or hydroquinone form)

(b)

Isoprenoid units

Coenzyme Q (CoQ) or ubiquinone
(oxidized or quinone form)

[H•]

Coenzyme QH• or ubisemiquinone
(radical or semiquinone form)

[H•]

Coenzyme QH$_2$ or ubiquinol
(reduced or hydroquinone form)

Figure 22-17   Oxidation states of the organic coenzymes of complex I. (*a*) FMN and (*b*) CoQ. Both coenzymes form stable semiquinone free radical states.

Figure 22-18   Low (4.5-Å) resolution X-ray structure of Complex I from *T. thermophilus*. The structure is viewed parallel to the plane of the plasma membrane with the cytoplasm above and the periplasm below. Its 15 subunits are drawn in ribbon form in different colors with the 110-Å-long horizontal helix of Nqo12 purple. The nine Fe–S clusters in the peripheral arm are drawn in space-filling form with S yellow and Fe red-brown. At the structure's low resolution, only the transmembrane helices of the transmembrane arm can be seen; their connecting loops are not visible, which accounts for the apparent gap between the transmembrane and peripheral arms. However, the loops of the peripheral arm are visible because that part of the structure was based on the higher resolution X-ray structure of the peripheral arm alone (see Fig. 22-19). [Courtesy of Leonid Sazanov, Medical Research Council, Cambridge, U.K. PDBid 3M9S.]

(a)

(b)

Figure 22-19 **X-ray structure of the peripheral arm of Complex I from *T. thermophilus*.** (*a*) The protein is viewed parallel to the plane of the plasma membrane but oriented differently to that in Fig. 22-18. Its eight subunits are drawn in semitransparent ribbon form in different colors (which differ from those in Fig. 22-18). The FMN (*upper left*) and the nine Fe–S clusters are shown in space-filling form with C green, N blue, O red, P orange, S yellow, and Fe red-brown. The likely CoQ binding site (Q) is indicated by a gray arrow. (*b*) The arrangement of redox groups viewed similarly to Part *a*. The FMN (stick model with C yellow and the two [Fe–S] and seven

[4Fe–4S] clusters (space-filling models) are shown together with their center-to-center distances in angstroms (shortest edge-to-edge distances are indicated in parentheses). Blue arrows represent the ~94-Å-long main path of electrons after their transfer from NADH to FMN. Fe–S clusters N1a and N7 lie too far off this path for electrons to pass through them but perhaps they function to fine-tune its electronic properties. Spectroscopic measurements suggest that the CoQ binding site is ~12 Å distant from N2. [Part *a* based on an X-ray structure by and Part *b* courtesy of Leonid Sazanov, Medical Research Council, Cambridge, U.K. PDBid 2FUG.]

binding site. Model building suggests that bound NADH protects the FMN from solvent. The CoQ binding site appears to be in a cavity below the [4Fe–4S] cluster N2 that is located at the interface with Complex I's transmembrane arm.

The transit of electrons from NADH to CoQ in Complex I begins with a two-electron reduction in the form of a hydride ion (H⁻) transfer from NADH to FMN to yield NAD⁺ and FMNH⁻ (similar to the reverse of the final step of the dihydrolipoyl dehydrogenase reaction; Section 21-2Bb). The electrons are then transferred, one at a time, to Fe–S cluster N3 and then between Fe–S clusters along a gradient of increasing redox potential (Table 22-1) until they reach Fe–S cluster N2. This "downhill" process involves the transient reduction of each Fe–S cluster as it binds electrons and its reoxidation when it passes the electrons to the next cluster. Their spatial arrangement indicates the probable path of the electrons (Fig. 22-19*b*). Note that redox centers need not be in contact in order to transfer electrons. Finally, the electrons are passed from N2 to CoQ, which is sequentially reduced to CoQH· and then to CoQH₂, which is released into the membrane and replaced by CoQ, thereby completing the catalytic cycle.

In translocating an electron pair from NADH to CoQ, Complex I pumps four protons out of the mitochondrial

matrix (the bacterial cell). How complex I accomplishes this task is discussed in Section 22-3Bj.

2. Complex II [Succinate Dehydrogenase or Succinate–Coenzyme Q Oxidoreductase (SQR)]

Complex II, which is also the citric acid cycle enzyme succinate dehydrogenase (Section 21-3F), passes electrons from succinate to CoQ to yield fumarate and CoQH₂. It does so with the participation of a covalently bound FAD, a [2Fe–2S] cluster, a [4Fe–4S] cluster, a [3Fe–4S] cluster, and one heme b_{560} (Table 22-1). All of its four subunits are encoded by nuclear genes.

The standard redox potential difference for electron transfer from succinate to CoQ (Fig. 22-9) is insufficient to provide the free energy necessary to drive ATP synthesis. Complex II is, nevertheless, important because it injects these relatively high-potential electrons into the electron-transport chain. Two other enzymes, in addition to Complexes I and II, synthesize and release CoQH₂ in the inner mitochondrial membrane and thereby power oxidative phosphorylation via the actions of Complexes III and IV. They are glycerol-3-phosphate dehydrogenase of the glycerophosphate shuttle (Fig. 22-8) and **ETF–ubiquinone oxidoreductase,** which participates in fatty acid oxidation (Section 25-2Ca; ETF stands for *electron-transfer flavoprotein*).

a. Cytochromes Are Heme Proteins That Transport Electrons

Cytochromes, whose function was elucidated in 1925 by David Keilin, are redox-active proteins that occur in all organisms except a few types of obligate anaerobes. These proteins contain heme groups that reversibly alternate between their Fe(II) and Fe(III) oxidation states during electron transport.

The heme groups of the reduced [Fe(II)] cytochromes have prominent visible absorption spectra consisting of three peaks: the α, β, and γ **(Soret)** bands (Fig. 22-20a). The wavelength of the α peak, which varies characteristically with the particular reduced cytochrome species (it is absent in oxidized cytochromes), is useful for differentiating the various cytochromes. Accordingly, the spectra of mitochondrial membranes (Fig. 22-20b) indicate that they contain three cytochrome types, **cytochromes a, b,** and **c.**

Within each type of cytochrome, different heme group environments may be characterized by slightly different α peak wavelengths. For example, Complex III (see below) has two b-type hemes: That absorbing maximally at 562 nm is referred to as **heme b_{562}** or b_H (for *h*igh potential), whereas that absorbing maximally at 566 nm is referred to as **heme b_{566}** or b_L (for *l*ow potential).

Each type of cytochrome contains a differently substituted porphyrin ring (Fig. 22-21a) coordinated with the redox-active iron atom. A b-type cytochrome contains **protoporphyrin IX,** which also occurs in hemoglobin and myoglobin (Section 10-1A). The heme group of a c-type cytochrome differs from protoporphyrin IX in that its vinyl groups have added the Cys sulfhydryls in the sequence Cys-X-Y-Cys-His across their double bonds to form thioether linkages to the protein (Fig. 9-39). Heme a con-

tains a long hydrophobic tail of three isoprene units (a **farnesyl** group) linked to the porphyrin via a hydroxyethyl group, as well as a formyl group in place of a methyl substituent.

The axial ligands of the heme iron also vary with the cytochrome type. In cytochromes a and b, both ligands are His residues, whereas in cytochrome c, one is the His in the Cys-X-Y-Cys-His sequence and the other is Met (Fig. 22-21b). Note that the designation of a given cytochrome type refers only to the identity of the cytochrome's heme prosthetic group(s); a given cytochrome type may have any of several unrelated protein folds.

b. Complex II Contains a Linear Chain of Redox Cofactors

Both bacterial and mitochondrial Complex II's consist of four subunits: two water-soluble subunits, a flavoprotein (Fp, ~665 residues) that binds the enzyme's substrate succinate and FAD (which is covalently linked to a specific His side chain via a bond to FAD atom C8a; Section 21-3F), and an iron–sulfur protein (Ip, ~290 residues) that binds the three Fe–S clusters; and two transmembrane subunits, CybL (~170 residues) and CybS (~160 residues) that collectively bind the heme b_{560} and the CoQ. Fp and Ip are highly conserved, whereas the sequences of CybS and CybL vary among different organisms.

The X-ray structures of porcine, chicken, and *E. coli* Complex II were independently determined by Zihe Rao, Edward Berry, and So Iwata. In all cases, it has the shape of the letter "q," with its top lobe containing Fp and Ip and its tail composed of the transmembrane subunits (Fig. 22-22a). The complex is oriented in the inner mitochondrial membrane such that Fp and Ip extend into the matrix. The

Figure 22-20 Visible absorption spectra of cytochromes.
(*a*) Absorption spectrum of reduced cytochrome c showing its characteristic α, β, and γ (Soret) absorption bands. The absorption maxima for cytochromes a, b, c, and c_1 are listed.
(*b*) The three separate α bands in the visible absorption

spectrum of beef heart mitochondrial membranes (*below*) indicate the presence of cytochromes a, b, and c. The spectrum of purified cytochrome c (*above*) is provided for reference. [After Nicholls, D.G. and Ferguson, S.J., *Bioenergetics 3*, p. 96, Academic Press (2002).]

Heme *a*

Heme *b*
(iron–protoporphyrin IX)

Heme *c*

Figure 22-21   Porphyrin rings in cytochromes. The (*a*) chemical structures and (*b*) axial liganding of the heme groups contained in cytochromes *a*, *b*, and *c* are shown.

Hemes *a* and *b*

Met   His
Heme *c*

(a)

Fp

82 Å

Ip

CybL

Matrix

Membrane

45 Å

CybS

Figure 22-22   X-ray structure of chicken Complex II in complex with its inhibitor oxaloacetate and CoQ. (*a*) A semitransparent ribbon diagram viewed parallel to the inner mitochondrial membrane with the matrix above and in which the enzyme's four subunits are drawn in different colors. The oxaloacetate (OAA),

(b)
FAD
OAA
12.2 Å
[2Fe–2S]
9.8 Å
[4Fe–4S]
8.9 Å
[3Fe–4S]   7.1 Å
13.3 Å   CoQ
11.4 Å
Heme *b*

FAD, three Fe–S clusters, CoQ, and heme *b* are shown in space-filling form with OAA C magenta, FAD and heme *b* C green, CoQ C cyan, N blue, O red, P orange, S yellow, and Fe red-brown. The inferred position of the inner mitochondrial membrane is indicated. (*b*) The ligands and redox cofactors viewed and colored as in Part *a* but drawn in stick form. Their closest edge-to-edge distances are indicated. [Based on an X-ray structure by Edward Berry, Lawrence Berkeley National Laboratory, Berkeley, California. PDBid 1YQ3.]

substrate and CoQ binding sites are connected by a nearly linear ~40-Å-long chain of redox centers with the sequence substrate–FAD—[2Fe–2S]—[4Fe–4S]—[3Fe–4S]—CoQ (top to bottom in Fig. 22-22*b*). Thus, despite the measured –0.245 V reduction potential of the [4Fe–4S] cluster (Table 22-1), which would appear to be too low for it to accept electrons from succinate in the succinate → fumarate reaction, the [4Fe–4S] cluster probably participates in the electron-transfer process.

c. Complex II Suppresses the Formation of Reactive Oxygen Species

The bacterial respiratory complex **quinol–fumarate reductase (QFR)** is a homolog of Complex II that functions in anaerobic organisms that use fumarate as a terminal electron acceptor. There it catalyzes the same reaction as does Complex II but in the opposite direction, that is, it uses a quinol to reduce fumarate to succinate.

Even though QFR and Complex II have similar structures, organisms such as *E. coli* that are capable of both aerobic and anaerobic metabolism employ the two different complexes for these different purposes. This appears to be because under aerobic conditions, QFR produces 25 times as much superoxide radical ($O_2^{-\cdot}$) as does *E. coli* Complex II as well as H_2O_2, which Complex II does not produce, presumably by leaking electrons to O_2. As we discuss in Section 22-4Cg, these **reactive oxygen species (ROS)** are highly destructive. Comparison of the X-ray structure of QFR with that of the closely similar *E. coli* Complex II indicates that the electron distributions about their various redox centers greatly favors the ROS-generating side reactions of O_2 with the flavin ring of QFR relative to that of Complex II. Indeed, the heme b_{560} of Complex II, which QFR lacks and which is not in Complex II's direct electron transfer pathway (Fig. 22-22*b*), appears to fine-tune its electronic properties so as to suppress the side reactions that generate the ROS. This suggests that mutations of the genes encoding human Complex II, which cause a wide variety of disorders including tumor formation, neurological defects, and premature aging, result from ROS generation.

3. Complex III (Coenzyme Q–Cytochrome c Oxidoreductase or Cytochrome bc_1 Complex)

Complex III passes electrons from reduced CoQ to cytochrome c. It contains four redox cofactors: two *b*-type hemes, a *c*-type heme, and one [2Fe–2S] cluster (Table 22-1).

a. X-Ray Structure of the Cytochrome bc_1 Complex

All known cytochrome bc_1 complexes contain three common subunits: **cytochrome b,** which binds both the b_H and b_L hemes, **cytochrome c_1,** which contains a single *c*-type heme, and a Rieske iron–sulfur protein **(ISP),** which contains a [2Fe–2S] cluster. The bovine bc_1 complex contains 8 additional subunits for a total of 11 different subunits that combine to form a 2166-residue (243-kD) protomer that dimerizes. Of these, only cytochrome *b* is encoded by a mitochondrial gene.

The X-ray structures of bovine, chicken, and yeast cytochrome bc_1 complexes were independently determined by Johann Deisenhofer, by Iwata and Bing Jap, by Berry, Antony Crofts, and Sung-Hou Kim, and by Hartmut Michel. All of these structures reveal a 2-fold symmetric pear-shaped molecule of maximum diameter ~130 Å and height ~150 Å, whose wide end extends ~75 Å into the matrix space and whose narrow end extends ~35 Å into the intermembrane space (Fig. 22-23). Its membrane-spanning region is ~40 Å thick and consists of 13 transmembrane helices in each protomer (12 in yeast, which consists of 9 different subunits). Eight of these transmembrane helices are contributed by the cytochrome *b* subunit, which binds both heme b_H and heme b_L within its transmembrane region, with heme b_L being closest to the intermembrane space. One of the remaining transmembrane helices is the membrane anchor of cytochrome c_1, the rest of which is a globular domain that extends into the intermembrane space. This is the portion of the complex

Figure 22-23   **X-ray structure of the yeast Complex III homodimer in complex with cytochrome c and the inhibitor stigmatellin.** The view is perpendicular to its 2-fold axis and parallel to the membrane with the matrix below. The nine different subunits in each protomer of Complex III, which collectively have 12 transmembrane helices, are drawn in semitransparent ribbon form and are differently colored with cytochrome *b* pale yellow, cytochrome c_1 light purple, the ISP tan, Core 1 cyan, and Core 2 orange. The bound cytochrome *c* is pink. The four different heme groups, the [2Fe–2S] cluster, and stigmatellin are drawn in space-filling form with heme C green, stigmatellin C magenta, N blue, O red, S yellow, and Fe red-brown. The horizontal lines delineate the inferred position of the membrane. Note that only one cytochrome *c* is bound to the homodimeric Complex III. [Based on an X-ray structure by Carola Hunte, Max Planck Institute for Biophysics, Frankfurt am Main, Germany. PDBid 1KYO.] **See Interactive Exercise 17.**

that contains heme c_1 and to which cytochrome c docks (Figure 22-23). The ISP is likewise anchored by a single transmembrane helix and extends into the intermembrane space. Interestingly, the two ISPs of the dimeric complex are domain swapped (intertwined) such that the [2Fe–2S] cluster-containing domain of one protomer interacts with the cytochrome b and cytochrome c_1 subunits of the other protomer. The distances between the various metal centers are all quite large, ranging from 21 to 34 Å. The portion of the complex that occupies the matrix, which accounts for more than half the mass of cytochrome bc_1, consists largely of the structurally homologous **Core 1** and **Core 2** proteins.

The route of electrons through the cytochrome bc_1 complex is discussed in Section 22-3B, together with the mechanism by which the complex preserves the free energy of electron transfer from $CoQH_2$ to cytochrome c for ATP synthesis.

4. Cytochrome c

Cytochrome c, whose evolution we discussed in Section 7-3B, is a peripheral membrane protein of known crystal structure (Figs. 8-42 and 9-41c) that is loosely bound to the outer surface of the inner mitochondrial membrane. *It alternately binds to cytochrome c_1 (of Complex III) and to cytochrome c oxidase (Complex IV) and thereby functions to shuttle electrons between them.*

The X-ray structure of yeast cytochrome bc_1 in complex with cytochrome c, determined by Carola Hunte, reveals, as expected, that cytochrome c binds to the cytochrome c_1 subunit of cytochrome bc_1 (Fig. 22-23). This association appears to be particularly tenuous because its interfacial area (880 Å^2) is significantly less than that exhibited by protein–protein complexes known to have low stability (typically <1600 Å^2). Such a small interface is well suited for fast binding and release. The closest approach between the heme groups of the contacting proteins is 4.5 Å between atoms of their respective thioether-bonded substituents, and their Fe–Fe distance is 17.4 Å. This accounts for the $8.3 \times 10^6 \, s^{-1}$ rate of electron transfer between these two redox centers (see below).

a. Protein Structure Influences the Rate of Electron Transfer

Reduced hemes are highly reactive entities; they can transfer electrons over distances of 10 to 20 Å at physiologically significant rates. Hence cytochromes, in a sense, have the opposite function of enzymes: Instead of persuading unreactive substrates to react, they must prevent their hemes from transferring electrons nonspecifically to other cellular components. This, no doubt, is why these hemes are almost entirely enveloped by protein. However, cytochromes must also provide a path for electron transfer to an appropriate partner.

In proteins, electron transfers occur between redox-active cofactors such as hemes and/or Fe–S clusters. However, a survey of proteins of known structure that function in electron transfer reveals that electrons travel no more than 14 Å between protein-embedded redox centers and that transfers over longer distances always involve chains of redox-active cofactors (e.g., Figs. 22-19b and 22-22b). Electron transfers occur far more efficiently through bonds than through space via a quantum mechanical process known as **electron tunneling.** Thus, as Harry Gray has experimentally demonstrated, electron tunneling within proteins occurs largely through the polypeptide chains between bound redox-active groups and the rate of electron transfer varies with the structure of the intervening polypeptide. Moreover, tunneling across protein–protein interfaces is largely mediated by van der Waals interactions and water-bridged hydrogen bonds. Nevertheless, Leslie Dutton has shown that the experimentally measured electron transfer rates within proteins vary only with the distance between the electron donor and the electron acceptor and fall off with an ~10-fold decrease in rate for each 1.7 Å increase in this distance.

5. Complex IV (Cytochrome c Oxidase)

Cytochrome c oxidase **(COX** or **CcO),** *the terminal enzyme of the electron-transport chain, catalyzes the one-electron oxidations of four consecutive reduced cytochrome c molecules and the concomitant four-electron reduction of one O_2 molecule to yield $2H_2O$:*

$$4 \text{ cytochrome } c^{2+} + 4H^+ + O_2 \longrightarrow$$
$$4 \text{ cytochrome } c^{3+} + 2H_2O$$

Eukaryotic COX is an ~200-kD transmembrane protein composed of 8 to 13 subunits, whose largest and most hydrophobic chains, Subunits I, II, and III, are encoded by mitochondrial genes, with the remaining subunits encoded by nuclear genes. Eukaryotic COX exists in membranes as a dimer. Subunits I and II of this complex contain all four of its redox-active centers: two a-type hemes, a and a_3, and two Cu-containing centers, Cu_A and Cu_B. The Cu_A center and heme a are of low potential (0.245 and 0.210 V; Table 22-1), whereas Cu_B and heme a_3 are of higher potential (0.340 and 0.385 V). Spectroscopic studies indicate that electrons are passed from cytochrome c to the Cu_A center, then to the heme a, and finally to a binuclear complex of heme a_3 and Cu_B. O_2 binds to this binuclear complex and is reduced to H_2O in a complex four-electron reaction (see below).

a. X-Ray Structures of Cytochrome c Oxidase

The X-ray structures of three species of cytochrome c oxidase have been determined: two relatively simple forms from the soil bacterium *Paracoccus denitrificans* (1106 residues) by Michel and the purple photosynthetic bacterium *Rhodobacter sphaeroides* by Iwata, and a more complex form from bovine heart (1806 residues) by Shinya Yoshikawa. Each protomer of the 2-fold symmetric dimeric bovine COX has an ellipsoidal (potatolike) shape comprised of a 48-Å-thick transmembrane portion and hydrophilic portions that protrude 32 and 37 Å into the mitochondrial matrix and the intermembrane space,

Figure 22-24   X-ray structure of the bovine heart cytochrome *c* oxidase homodimer. The view is perpendicular to its 2-fold axis and parallel to the membrane with the matrix below. The 13 subunits in each protomer, which collectively have 28 transmembrane helices, are drawn in semitransparent ribbon form and differently colored according to type with Subunit I pale green, Subunit II pink, and Subunit III pale yellow. The protein's bound heme groups and Cu ions are drawn in space-filling form with C green, N blue, O red, Fe red-brown, and Cu cyan. The horizontal black lines delineate the inferred position of the inner mitochondrial membrane. [Based on an X-ray structure by Shinya Yoshikawa, Himeji Institute of Technology, Hyogo, Japan. PDBid 1V54.] *See Interactive Exercise 18.*

respectively (Fig. 22-24). These protomers each consist of 13 different subunits that mainly form 28 transmembrane helices. The protomer surfaces that comprise the dimer interface are concave (Fig. 22-24) and hence their relatively tenuous contacts enclose a lipid-filled cavity. Moreover, mitochondrial COX is active as a monomer. Thus a mechanistic role for dimer formation seems unlikely. The *P. denitrificans* and *Rb. sphaeroides* COXs are monomeric complexes, each of which consists of only 4 subunits that collectively contain 22 transmembrane helices.

The structures of bovine COX Subunits I (12 transmembrane helices) and II (2 transmembrane helices), which bind all of the complex's four redox centers, are closely similar to those of *P. denitrificans* and *Rb. sphaeroides* COXs. These are the subunits that carry out the main functions of the complex: transporting electrons from cytochrome *c* to O_2 to yield water, while pumping protons from inside (as we shall refer to the mitochondrial matrix and the bacterial cytoplasm; also called the **N-side** due to its *n*egative charge) to outside (as we shall refer to the mitochondrial intermembrane space and the bacterial periplasmic space; also called the **P-side** due to its *p*ositive charge). Bovine Subunit III (7 transmembrane helices), whose structure also resembles that of its *P. denitrificans* and *Rb. sphaeroides* counterparts, does not appear to directly participate in electron transfer or proton translocation. Indeed, a complex consisting of only *P. denitrificans* Subunits I and II can actively transport electrons and pump protons. Thus the function of Subunit III is unknown,

although there is some evidence that it facilitates the assembly of Subunits I and II to form the active complex. Note, however, that Subunit III does not contact Subunit II. None of the 10 nuclear-encoded subunits of bovine COX resemble Subunit IV of *P. denitrificans* and *Rb. sphaeroides* COXs (1 transmembrane helix). Seven of these bovine subunits each have one transmembrane helix, all oriented with their N-terminal ends on the matrix side of the membrane. These helices are distributed about the periphery of the dimeric core formed by Subunits I, II, and III. The remaining three bovine subunits are globular and associate entirely with the extramembrane portions of the complex. The X-ray structure of bovine COX provides little indication of the function of any of its nuclear-encoded subunits. Perhaps they have regulatory roles.

Subunit I binds heme *a* and the heme a_3–Cu_B binuclear center (Fig. 22-25), whose metal ions are all located ~13 Å below the membrane surface on its intermembrane/periplasmic side (Fig. 22-24). The heme a_3 Fe has one axial His ligand, the heme *a* Fe has two axial His ligands

Figure 22-25   The redox centers in the X-ray structure of bovine heart cytochrome *c* oxidase. The view is similar to that of the left protomer in Fig. 22-24. The Fe and Cu ions are represented by orange and cyan spheres. Their liganding heme and protein groups (from subunit II for the Cu_B center and subunit I for the others) are drawn in stick form colored according to atom type with heme C magenta, protein C green, N blue, O red, and S yellow. The peroxy group that bridges the Cu_B and heme a_3 Fe ions is shown in ball-and-stick form in red. Coordination bonds are drawn as thin gray lines. Note that the side chains of His 240 and Tyr 244 are joined by a covalent bond (*lower right*). [Based on an X-ray structure by Shinya Yoshikawa, Himeji Institute of Technology, Hyogo, Japan. PDBid 2OCC.]

(as in Fig. 22-21*b, top*), and the Cu_B atom has three His ligands, whose coordinating N atoms are arranged in an equilateral triangle that is centered on Cu_B and is parallel to heme a_3. The X-ray structure of fully oxidized bovine COX reveals a peroxide group (O_2^{2-}; Fig. 22-25) that bridges the heme a_3 Fe atom and Cu_B (which are separated by 4.9 Å), thereby liganding Cu_B via a distorted square-planar arrangement, a stable coordination geometry for Cu(II). However, in the X-ray structure of the fully reduced form of bovine COX (in which the heme a_3 Fe—Cu_B distance is 5.2 Å), this ligand is absent, so that Cu_B is trigonally liganded, a stable coordination geometry for Cu(I). The closest approach of the two heme groups is 4 Å, and the distance between their Fe atoms is 13.2 Å.

Subunit II, in addition to its two transmembrane helices, has a globular domain on the outside surface that binds the Cu_A center and largely consists of a 10-stranded β barrel (Fig. 22-24). The Cu_A center is located ~8 Å above the outside membrane surface. Although the Cu_A center was, for many years, widely believed to contain only one Cu atom, the X-ray structures of COX clearly indicate that Cu_A contains two Cu atoms (Fig. 22-25). These are bridged by two Cys S atom ligands and have two additional protein ligands each to form an arrangement similar to that of a [2Fe–2S] cluster (Fig. 22-15*b*) in which the two Cu atoms are 2.4 Å apart. Spectroscopic measurements indicate that in the Cu_A center's reduced form, both of its Cu atoms are in their Cu(I) states, whereas in its fully oxidized form, an electron appears to be delocalized between the two Cu atoms such that they assume the $[Cu^{1.5+} \cdots Cu^{1.5+}]$ state.

b. Electron and Proton Acquisition

COX's cytochrome *c* binding site is postulated to be in a corner formed by the globular domain of Subunit II and the outside surface of Subunit I, since this region is close to the Cu_A site and contains 10 acidic side chains that could interact with a ring of Lys side chains that surrounds cytochrome *c*'s heme crevice. Indeed, differential labeling of cytochrome *c* oxidase's carboxyl groups in the presence and absence of cytochrome *c* demonstrated that cytochrome *c* shields the invariant residues Asp 112, Glu 114, and Glu 198 of Subunit II (bovine numbering). Glu 198 is located between the two Cys residues of Subunit II that ligand Cu_A (Fig. 22-25). This observation supports the spectroscopic evidence that places the cytochrome *c* binding site on Subunit II in close proximity to Cu_A. Cross-linking studies have additionally shown that the cytochrome *c* surface opposite the electron-transferring site interacts with Subunit III, suggesting that Subunit III also participates in binding cytochrome *c*.

Time-resolved spectroscopic studies indicate that an electron obtained from cytochrome *c* is first acquired by the Cu_A center and is then transferred to heme *a* rather than heme a_3, probably because the shortest $Cu_A \cdots$ heme *a* distance of 11.7 Å is less than the shortest $Cu_A \cdots$ heme a_3 distance of 14.7 Å. The electron is then rapidly transferred to the heme a_3–Cu_B binuclear center, where it participates in reducing the bound O_2 to H_2O. Note that the fifth ligand

Figure 22-26 **Synthetic model of the cytochrome a_3–Cu_B binuclear complex.** This structure efficiently reduces O_2 to H_2O when attached to an electrode. The pyridine group that axially ligands the Fe ion (*bottom*) can be replaced by an imidazole group.

of heme *a*, His 378, is separated from the fifth ligand of heme a_3, His 376, by only one residue, and hence there is a relatively short through-the-bonds electron transfer pathway between heme *a* and a_3 (Fig. 22-25, *lower left*). In addition, the closest approach of the two hemes is 4 Å.

COX must acquire four so-called **chemical** or **scalar protons** from the inside for every molecule of O_2 it reduces to H_2O. This four-electron process is coupled to the translocation of up to four so-called **pumped** or **vectorial protons** from the inside to the outside, thereby contributing to the proton gradient that powers ATP synthesis (Section 22-3C). Note that for each turnover of the enzyme,

$$8H_{inside}^+ + 4 \text{ cyt } c^{2+} + O_2 \longrightarrow$$
$$4 \text{ cyt } c^{3+} + 2H_2O + 4H_{outside}^+$$

a total of eight positive charges are transported across the membrane, thereby contributing to its membrane potential.

c. The Reduction of O_2 by Cytochrome *c* Oxidase Occurs in Multiple Steps

The reduction of O_2 to $2H_2O$ by cytochrome *c* oxidase takes place on the cytochrome a_3–Cu_B binuclear complex (Fig. 22-25). Indeed, a synthetic model of this binuclear complex (Fig. 22-26), synthesized by James Collman, efficiently catalyzes the reduction of O_2 to H_2O when attached to an electrode.

The COX-mediated reduction of O_2 requires, as we shall see, the nearly simultaneous input of four electrons. However, the fully reduced a_3^{2+}–Cu_B^{1+} binuclear complex can readily contribute only three electrons to its bound O_2 in reaching its fully oxidized a_3^{4+}–Cu_B^{2+} state [cytochrome a_3 transiently assumes its Fe(IV) or **ferryl** oxidation state during the reduction of O_2; see below]. What is the source of the fourth electron?

The X-ray structures of both bovine and *P. denitrificans* COX clearly indicate that the His 240 ligand of Cu_B (bovine numbering) is covalently cross-linked to the side chain of the conserved Tyr 244 (Fig. 22-25, *lower right*). This places Tyr 244's phenolic —OH group in close proximity to the heme a_3-ligated O_2 such that Tyr 244 can supply the fourth electron by transiently forming a tyrosyl radical (TyrO·). In fact, adding peroxide to the resting enzyme generates a tyrosyl radical, whereas mutating Tyr 244 to Phe inactivates the enzyme. Moreover, tyrosyl radicals have been implicated in several enzymatically mediated redox processes, including the generation of O_2 from H_2O in photosynthesis (in a sense, the reverse of the COX reaction; Section 24-2Cd), and in the **ribonucleotide reductase** reaction (which converts NDP to dNDP; Section 28-3Aa). Tyr 244's phenolic—OH group is within hydrogen bonding distance of the COX-bound O_2 and hence is a likely H^+ donor during O—O bond cleavage. The formation of the covalent cross-link is expected to lower both the reduction potential and the p*K* of Tyr 244, thereby facilitating both radical formation and proton donation (the synthetic binuclear complex in Fig. 22-26 can function without an associated tyrosyl radical, presumably because its associated electrode can supply it with electrons much faster than cytochrome *c* can supply them to COX).

The COX reaction, elucidated in large part by Mårten Wikström and Gerald Babcock using a variety of spectroscopic techniques, involves four consecutive one-electron transfers from the Cu_A and cytochrome *a* sites and occurs as follows (Fig. 22-27):

1 and 2. The oxidized binuclear complex [Fe(III)$_{a3}$—OH$^-$ Cu(II)$_B$] is reduced to its [Fe(II)$_{a3}$ Cu(I)$_B$] state by two con-

secutive one-electron transfers from cytochrome *c* via cytochrome *a* and Cu_A. A proton from the matrix is concomitantly acquired and an H_2O is released in this process. Tyr 244 (Y—OH) is in its phenolic state.

3. O_2 binds to the reduced binuclear complex so as to ligand its Fe(II)$_{a3}$ atom. It binds to the heme with much the same configuration it has in oxymyoglobin (Fig. 10-12).

4. Internal electron redistribution rapidly yields the oxyferryl complex [Fe(IV)═O^{2-} HO$^-$—Cu(II)] in which Tyr 244 has donated an electron and a proton to the complex and thereby assumed its neutral radical state (Y—O·). This is known as compound P because it was originally thought that this spectroscopically identified state was a peroxy complex. However, it has since been shown that a peroxy compound is not on the reaction pathway. The peroxy complex displayed in Fig. 22-25 is a two-electron reduced "mixed valence" state of the enzyme that cannot reduce O_2 past its peroxy form.

5. A third one-electron transfer from cytochrome *c* together with the acquisition of two protons reconverts Tyr 244 to its phenolic state, yielding compound F (for ferryl) and releasing an H_2O.

6. A fourth and final electron transfer and proton acquisition yields the oxidized [Fe(III)$_{a3}$—OH$^-$ Cu(II)$_B$] complex, thereby completing the catalytic cycle.

COX typically undergoes 100 to 200 turnovers per second so that one catalytic cycle takes only a few milliseconds. Note that the COX reaction proceeds without the release of the destructive partially reduced reactive oxygen species (ROS) from its active site. The positions in this proposed catalytic cycle at which protons appear to be

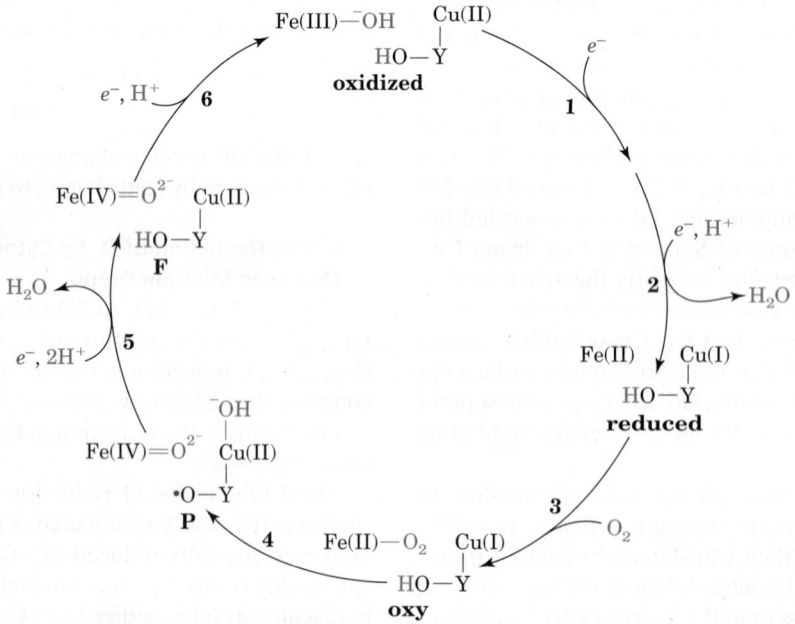

Figure 22-27 **Proposed reaction sequence for the reduction of O_2 by the cytochrome a_3–Cu_B binuclear complex of cytochrome *c* oxidase.** The numbered steps are discussed in the text. The entire reaction is extremely fast; it goes to completion in ~1 ms at room temperature. [Modified from Babcock, G.T., *Proc. Natl. Acad. Sci.* **96**, 12971 (1999).]

pumped from the matrix (bacterial cytoplasm) to the intermembrane (periplasmic) space are discussed in Section 22-3B. Keep in mind, however, that aspects of this cycle are uncertain and/or disputed and hence it remains under investigation.

d. Complexes I, III, and IV Form Supercomplexes

For many years, it had been widely assumed that the complexes of the respiratory chain were laterally mobile within the inner mitochondrial membrane and hence did not associate. However, the development of gentle methods of separating the components of the inner mitochondrial membrane has made it increasingly clear that these already large protein complexes form supercomplexes. In fact, a variety of supercomplexes from several organisms have been characterized, including yeast III_2IV_2, bovine and *Arabadopsis thaliana* (a plant) I_1III_2, and bovine $I_1III_2IV_1$ (in which, for example, IV_1 represents a protomer of Complex IV). Electron microscopy–based images such as Fig. 22-28 indicate that Complex IV monomers associate with Complex III dimers, which limits the distance that cytochrome *c* must travel to transport an electron from Complex III to Complex IV to <40 Å. Similarly, the association of Complex III with Complex I reduces the distance over which $CoQH_2$ must diffuse between these two complexes. Apparently, these supermolecular complexes function to increase the efficiency of electron transport via the channeling of their intermediates, a conclusion supported by kinetic measurements. Note that none of these supercomplexes contain Complex II, also in agreement with kinetic measurements.

Figure 22-28   Electron microscopy–based image of the bovine supercomplex $I_1III_2IV_1$ at 32 Å resolution. The supercomplex is represented by its semitransparent surface (*blue*) fitted with the EM-based structure of Complex I (*yellow*) and the X-ray structures of the Complex III dimer (*red*) and the Complex IV monomer (*green*). The view is parallel to the plane of the membrane with the matrix above. [Courtesy of Eva Schäfer, Birbeck College, London, U.K.]

3 OXIDATIVE PHOSPHORYLATION

The endergonic synthesis of ATP from ADP and P_i in mitochondria, which, as we shall see, is catalyzed by **proton-translocating ATP synthase (Complex V)**, is driven by the electron-transport process. Yet, since Complex V is physically distinct from the proteins mediating electron transport (Complexes I–IV), *the free energy released by electron transport must be conserved in a form that ATP synthase can utilize.* Such energy conservation is referred to as **energy coupling** or **energy transduction**.

The physical characterization of energy coupling proved to be surprisingly elusive; many sensible and often ingenious ideas have failed to withstand the test of experimental scrutiny. In this section we first examine some of the hypotheses that have been formulated to explain the coupling of electron transport and ATP synthesis. We shall then explore the coupling mechanism that has garnered the most experimental support, analyze the mechanism by which ATP is synthesized by ATP synthase, and, finally, discuss how electron transport and ATP synthesis can be uncoupled.

A. Energy Coupling Hypotheses

In the more than 70 years that electron transport and oxidative phosphorylation have been studied, numerous mechanisms have been proposed to explain how these processes are coupled. In the following paragraphs, we examine the mechanisms that have received the greatest experimental attention.

1. The chemical coupling hypothesis. In 1953, Edward Slater formulated the **chemical coupling hypothesis,** in which he proposed that electron transport yielded reactive intermediates whose subsequent breakdown drove oxidative phosphorylation. We have seen, for example, that such a mechanism is responsible for ATP synthesis in glycolysis (Sections 17-2G and 17-2J). Thus, the exergonic oxidation of glyceraldehyde-3-phosphate by NAD^+ yields 1,3-bisphosphoglycerate, a reactive ("high-energy") acyl phosphate whose phosphoryl group is then transferred to ADP to form ATP in the phosphoglycerate kinase reaction. The difficulty with such a mechanism for oxidative phosphorylation, which has caused it to be abandoned, is that despite intensive efforts in numerous laboratories over many years, no appropriate reactive intermediates have been identified.

2. The conformational coupling hypothesis. The **conformational coupling hypothesis,** which Paul Boyer formulated in 1964, proposes that electron transport causes proteins of the inner mitochondrial membrane to assume "activated" or "energized" conformational states. These proteins are somehow associated with ATP synthase such that their relaxation back to the deactivated conformation drives ATP synthesis. As with the chemical coupling hypothesis, the conformational coupling hypothesis has found little experimental support. However, conformational coupling of a different sort appears to be involved in ATP synthesis (Section 22-3Cd).

3. The chemiosmotic hypothesis. The **chemiosmotic hypothesis,** proposed in 1961 by Peter Mitchell, has spurred considerable controversy, as well as much research, and is now the model most consistent with the experimental evidence. It postulates that *the free energy of electron transport is conserved by pumping H⁺ from the mitochondrial matrix to the intermembrane space so as to create an electrochemical H⁺ gradient across the inner mitochondrial membrane. The electrochemical potential of this gradient is harnessed to synthesize ATP* (Fig. 22-29).

Several key observations are explained by the chemiosmotic hypothesis:

(a) Oxidative phosphorylation requires an intact inner mitochondrial membrane.

(b) The inner mitochondrial membrane is impermeable to ions such as H⁺, OH⁻, K⁺, and Cl⁻, whose free diffusion would discharge an electrochemical gradient.

(c) Electron transport results in the transport of H⁺ out of intact mitochondria, thereby creating a measurable electrochemical gradient across the inner mitochondrial membrane.

(d) Compounds that increase the permeability of the inner mitochondrial membrane to protons, and thereby dissipate the electrochemical gradient, allow electron transport (from NADH and succinate oxidation) to continue but inhibit ATP synthesis; that is, they "uncouple" electron transport from oxidative phosphorylation. Conversely, increasing the acidity outside the inner mitochondrial membrane stimulates ATP synthesis.

In the remainder of this section we examine the mechanisms through which electron transport can result in proton translocation and how an electrochemical gradient can interact with ATP synthase to drive ATP synthesis.

B. *Proton Gradient Generation*

Electron transport, as we shall see, causes Complexes I, III, and IV to transport protons across the inner mitochondrial membrane from the matrix, a region of low [H⁺] and negative electrical potential, to the intermembrane space (which is in contact with the cytosol), a region of high [H⁺] and positive electrical potential (Fig. 22-14). The free energy sequestered by the resulting electrochemical gradient [which, in analogy to the term electromotive force (emf), is called **proton-motive force (pmf)**] powers ATP synthesis.

a. Proton Pumping Is an Endergonic Process

The free energy change of transporting a proton out of the mitochondrion against an electrochemical gradient is expressed by Eq. [20.3], which, in terms of pH, is

$$\Delta G = 2.3\, RT[\mathrm{pH}(in) - \mathrm{pH}(out)] + Z\mathscr{F}\,\Delta\Psi \quad [22.1]$$

where Z is the charge on the proton (including sign), $\mathscr{F}$ is the Faraday constant, and ΔG is the membrane potential. The sign convention for ΔG is that when a positive ion is transported from negative to positive, ΔG is positive. Since $\mathrm{pH}(out)$ is less than $\mathrm{pH}(in)$, the export of protons from the mitochondrial matrix (against the proton gradient) is an endergonic process. In addition, *proton transport out of the matrix makes the inner membrane's internal surface more negative than its external surface.* Outward transport of a positive ion is consequently associated with a positive ΔG and an increase in free energy (endergonic process), whereas the outward transport of a negative ion yields the opposite result. Clearly, it is always necessary to describe membrane polarity when specifying a membrane potential.

The measured membrane potential across the inner membrane of a liver mitochondrion, for example, is 0.168 V

Figure 22-29 **Coupling of electron transport (*green arrow*) and ATP synthesis.** H⁺ is pumped out of the mitochondrion by Complexes I, III, and IV of the electron-transport chain (*blue arrows*), thereby generating an electrochemical gradient across the inner mitochondrial membrane. The exergonic return of these protons to the matrix powers the synthesis of ATP (*red arrow*). Note that the outer mitochondrial membrane is permeable to small molecules and ions, including H⁺. 🎧 **See the Animated Figures**

(inside negative; this corresponds to an $\sim$210,000 V $\cdot$ cm^{-1} electric field across its $\sim$80 Å thickness). The pH of the matrix is 0.75 units higher than that of the intermembrane space. ΔG for proton transport out of this mitochondrial matrix is therefore 21.5 kJ $\cdot$ mol^{-1}.

b. The Passage of About Three Protons Is Required to Synthesize One ATP

An ATP molecule's estimated physiological free energy of synthesis, around +40 to +50 kJ $\cdot$ mol^{-1}, is too large to be driven by the passage of a single proton back into the mitochondrial matrix; at least two protons are required. This number is difficult to measure precisely, in part because transported protons tend to leak back across the mitochondrial membrane. However, most estimates indicate that around three protons are passed per ATP synthesized.

c. Two Mechanisms of Proton Transport Have Been Proposed

Three of the four electron-transport complexes, Complexes I, III, and IV, are involved in proton translocation. Two mechanisms have been entertained that would couple the free energy of electron transport with the active transport of protons: the **redox loop mechanism** and the **proton pump mechanism.**

d. The Redox Loop Mechanism

This mechanism, proposed by Mitchell, requires that the redox centers of the respiratory chain (FMN, CoQ, cytochromes, and iron–sulfur clusters) be so arranged in the membrane that reduction would involve a redox center simultaneously accepting e^- and H$^+$ from the matrix side of the membrane. Reoxidation of this redox center by the next center in the chain would involve release of H$^+$ on the cytosolic side of the membrane together with the transfer of electrons back to the matrix side (Fig. 22-30). Electron flow from one center to the next would therefore yield net translocation of H$^+$ and the creation of an electrochemical gradient (ΔG and ΔpH).

The redox loop mechanism requires that the first redox carrier contain more hydrogen atoms in its reduced state than in its oxidized state and that the second redox carrier have no difference in its hydrogen atom content between its reduced and oxidized states. Are these requirements met in the electron-transport chain? Some of the redox carriers, FMN and CoQ, in fact, contain more hydrogen atoms in their reduced state than in their oxidized state and thus can qualify as proton carriers as well as electron carriers. If these centers were spatially alternated with pure electron carriers (cytochromes and iron–sulfur clusters), such a mechanism could well be accommodated.

The main difficulty with the redox loop mechanism involves the deficiency of (H$^+$ + e^-) carriers that can alternate with pure e^- carriers. Whereas the electron-transport chain has as many as 15 pure e^- carriers (up to 8 iron–sulfur proteins, 5 cytochromes, and 2 Cu centers), it has only 2

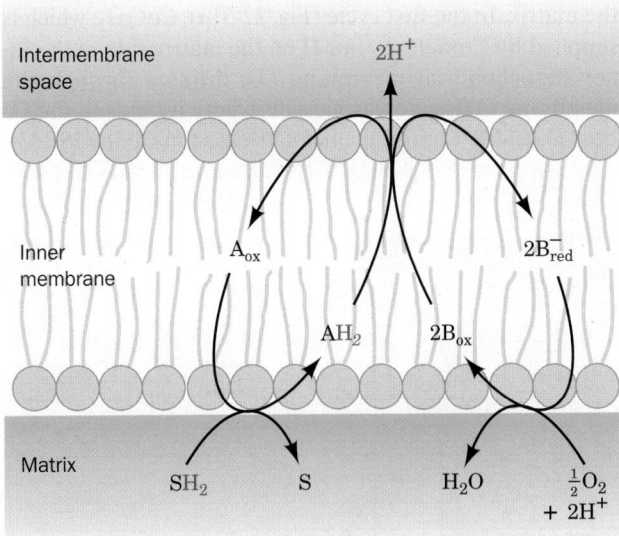

Figure 22-30   The redox loop mechanism for electron transport–linked H$^+$ translocation. AH$_2$ represents (H$^+$ + e^-) carriers such as FMNH$_2$ and CoQH$_2$, whereas B represents pure e^- carriers such as iron–sulfur clusters and the cytochromes. These components are so arranged as to require that electron transport be accompanied by H$^+$ translocation.

(H$^+$ + e^-) carriers. The fact that there are three complexes with standard reduction potential changes large enough to provide free energy for ATP synthesis suggests the need for at least three proton-transport redox carriers. As we shall see, however, there are, in fact, three proton-transport sites but only two proton-transport redox carriers: Both the redox loop mechanism and the proton pump mechanism (discussed below) are employed.

e. Complex III Pumps Protons via the Q Cycle, a Type of Redox Loop

See Guided Exploration 20: The Q cycle Mitchell postulated that Complex III functions in a way that permits one molecule of CoQH$_2$, the two-electron carrier, to sequentially reduce two molecules of cytochrome c, a one-electron carrier, while transporting four protons. This occurs via a modified redox loop mechanism involving a remarkable bifurcation of the flow of electrons from CoQH$_2$ to cytochrome c_1 and to cytochrome b. It is through this so-called **Q cycle** that Complex III pumps protons from the matrix to the intermembrane space.

The essence of the Q cycle is that *CoQH$_2$ undergoes a two-cycle reoxidation in which the semiquinone Q$^{\cdot-}$ is a stable intermediate.* This involves two independent binding sites for coenzyme Q: Q$_o$, which binds CoQH$_2$ and is located between the ISP and heme b_L in proximity to the intermembrane space (Fig. 22-23); and Q$_i$, which binds both Q$^{\cdot-}$ and CoQ and is located near heme b_H in proximity to the matrix. In the first cycle (Fig. 22-31*a*), CoQH$_2$, which is supplied by Complexes I or II on the matrix side of the inner mitochondrial membrane (**1**), diffuses through the membrane to its cytoplasmic side, where it binds to the Q$_o$ site (**2**). There it transfers one of its electrons to the ISP (**3**),

releasing its two protons into the intermembrane space and yielding Q$^{\cdot-}$. The ISP then reduces cytochrome c_1, whereas the Q$^{\cdot-}$ transfers its remaining electron to heme b_L (**4**), yielding fully oxidized CoQ. Heme b_L then reduces heme b_H (**6**). The CoQ from Step 4 is released from the Q$_o$ site and diffuses back through the membrane to rebind at the Q$_i$ site (**5**), where it picks up the electron from heme b_H (**7**), reverting to the semiquinone form, Q$^{\cdot-}$. Thus, the reaction for this first cycle is

$$CoQH_2 + \text{cytochrome } c_1(Fe^{3+}) \longrightarrow$$
$$Q^{\cdot-} + \text{cytochrome } c_1(Fe^{2+}) + 2H^+(outside)$$

In the second cycle (Fig. 22-31*b*), another CoQH$_2$ repeats Steps 1 through 6: One electron reduces the ISP and then cytochrome c_1, and the other electron sequentially reduces heme b_L and then heme b_H. This second electron then reduces the Q$^{\cdot-}$ at the Q$_i$ site produced in the first cycle (**8**), yielding CoQH$_2$. The protons taken up in this last step originate in the mitochondrial matrix. The reaction for the second cycle is therefore

$$CoQH_2 + Q^{\cdot-} + \text{cytochrome } c_1(Fe^{3+}) + 2H^+(matrix) \longrightarrow$$
$$CoQ + CoQH_2 + \text{cytochrome } c_1(Fe^{2+}) + 2H^+(outside)$$

For every two CoQH$_2$ that enter the Q cycle, one CoQH$_2$ is regenerated. The combination of both cycles, in which two electrons are transferred from CoQH$_2$ to cytochrome c_1, results in the overall reaction

$$CoQH_2 + 2 \text{ cytochrome } c_1(Fe^{3+}) + 2H^+(matrix) \longrightarrow$$
$$CoQ + 2 \text{ cytochrome } c_1(Fe^{2+}) + 4H^+(outside)$$

X-ray studies of Complex III provide direct evidence for the independent existence of the Q$_o$ and Q$_i$ sites. The antifungal agents **myxothiazol** and **stigmatellin**,

Figure 22-31   The Q cycle. The Q cycle is an electron-transport cycle in Complex III that accounts for H$^+$ translocation during the transport of electrons from cytochrome b to cytochrome c. The overall cycle is actually two cycles, the first (*a*) requiring reactions 1 through 7 and the second (*b*) requiring reactions 1 through 6 and 8. (**1**) Coenzyme QH$_2$ is supplied by Complex I on the matrix side of the membrane. (**2**) QH$_2$ diffuses to the outside of the membrane. (**3**) QH$_2$ reduces the Rieske iron–sulfur protein (ISP) forming Q$^{\cdot-}$ semiquinone and releasing 2H$^+$. The ISP goes on to reduce cytochrome c_1. (**4**) Q$^{\cdot-}$ reduces heme b_L to form coenzyme Q. (**5**) Q diffuses to the matrix side. (**6**) Heme b_L reduces heme b_H. (**7**, cycle 1 only) Q is reduced to Q$^{\cdot-}$ by heme b_H. (**8**, cycle 2 only) Q$^{\cdot-}$ is reduced to QH$_2$ by heme b_H. [After Trumpower, B.L., *J. Biol. Chem.* **265**, 11410 (1990).]

Myxothiazol

Stigmatellin

which both block electron flow from $CoQH_2$ to the ISP and to heme b_L (Steps 3 and 4 of both cycles), bind in a pocket within cytochrome b between the ISP and heme b_L (Fig. 22-23). Evidently, this binding pocket overlaps the Q_o site. Similarly, antimycin (Section 22-2Ba), which blocks electron flow from heme b_H to CoQ and $Q^{\overline{}}$ (Step 7 of Cycle 1 and Step 8 of Cycle 2), binds in a pocket near heme b_H, thereby identifying this pocket as site Q_i.

The circuitous route of electron transfer in Complex III is tied to the ability of coenzyme Q to diffuse within the hydrophobic core of the membrane in order to bind to both the Q_o and Q_i sites. This process is facilitated by an indentation in the surface of cytochrome b's transmembrane region that contains Q_o from one protomer and Q_i from the other. *When CoQH$_2$ is oxidized, two reduced cytochrome c molecules and four protons appear on the outer side of the membrane.* Proton transport by the Q cycle thus follows the redox loop mechanism of proton transport in which a redox center itself (CoQ) is the proton carrier. As we shall see below, however, Complexes I and IV follow a different mechanism of proton transport, the proton pump mechanism.

f. The Bifurcation of the Q Cycle's Electron Flow Occurs via Domain Movement

Why does Q_o-bound $Q^{\overline{}}$ exclusively reduce heme b_L (Step 4 of the Q cycle) rather than the Rieske [2Fe–2S] cluster of the ISP, despite the greater reduction potential difference ($\Delta\mathscr{E}$) favoring the latter reaction (Table 22-1)?

The remarkable answer to this question provides a fascinating insight into the inner workings of Complex III. Although the binding of stigmatellin and myxothiazol to Q_o are mutually exclusive, these inhibitors affect this site differently: Stigmatellin perturbs the spectrum and redox properties of the ISP's Rieske [2Fe–2S] cluster as well as prevents it from oxidizing cytochrome c_1, whereas myxothiazol does not interact with the ISP but, instead, shifts the spectrum of heme b_L. Evidently, stigmatellin is a $CoQH_2$ analog and myxothiazol is a $Q^{\overline{}}$ mimic.

X-ray structures of Complex III reveal that its Q_o site is a bifurcated pocket in which stigmatellin binds close to the ISP docking interface (see below), whereas myxothiazol binds in the vicinity of heme b_L (Fig. 22-32; their binding to Q_o is mutually exclusive because their hydrophobic tails would overlap). Furthermore, the globular Rieske [2Fe–2S] cluster–containing domain of the ISP (Fig. 22-23, *top*) is conformationally mobile and assumes a conformational state that is controlled by the ligand-binding state of Q_o: It binds to cytochrome b near the heme b_L site when stigmatellin is bound to the Q_o site, but has swung around by ~20 Å (via an ~57° hinge motion that leaves intact its tertiary structure) to bind to the cytochrome c_1 near its heme c when myxothiazol is bound to Q_o. Apparently the globular domain of the ISP functions to shuttle an electron from $CoQH_2$, which is bound to Q_o near the ISP docking interface, to heme c_1 by mechanically swinging the reduced Rieske [2Fe–2S] cluster between these sites. The resulting $Q^{\overline{}}$ shifts to the position near heme b_L (probably via a

(a)

(b)

Figure 22-32   X-ray structures of the Q_o binding site of the chicken cytochrome bc_1 complex occupied by inhibitors. The structures show (*a*) its complex with stigmatellin and (*b*) its complex with myxothiazol. The protein surface (*white*) has been cut away to show the Q_o pocket. Heme b_L (*upper right*) is shown in ball-and-stick form with C gray, N blue, O red, and Fe tan. In Part *a*, stigmatellin is drawn as a stick model with C yellow and its volume represented by the dotted surface. The Rieske [2Fe–2S] cluster (*lower left*) is represented by gold spheres, and the ISP

domain to which it is bound is drawn as a cyan ribbon. Note that His 161, which is a ligand of the Rieske [2Fe–2S] cluster, forms a hydrogen bond to stigmatellin. In Part *b*, the myxothiazol is drawn as a stick model with C orange. Note that the semiquinone-mimicking portion of myxothiazol binds to Q_o in proximity to heme b_L, whereas its hydrophobic tail occupies the same position as that of stigmatellin. Also note that the [2Fe–2S] cluster–containing ISP domain is not visible in this diagram; it has rotated into proximity with cytochrome c_1. [Courtesy of Antony Crofts, University of Illinois at Urbana–Champaigne, and Edward Berry, University of California at Berkeley. PDBid 3BCC.]

rotation about a bond connecting the semiquinone ring to its nonpolar tail), which it then reduces. Thus, Q⁻ is unable to reduce the ISP (after it has reduced cytochrome c_1) because it is too far away to do so. This novel mechanism is supported by the observation that mutagenically inserting a disulfide bond either within the hinge of the ISP or between it and cytochrome *b* greatly diminishes the activity of Complex III but that its activity is restored on exposure to reducing agents.

g. The Proton Pump Mechanism

Complex IV (COX) transports four protons from the matrix to the intermembrane space for each O_2 it reduces ($2H^+$ per electron pair; Fig. 22-14). It contains no ($H^+ + e^-$) carriers and hence cannot do so via a redox loop (Q cycle–like) system. Rather, as we shall see, it does so via the proton pump mechanism (Fig. 22-33), which does not require that the redox centers themselves be H^+ carriers. In this model, *the transfer of electrons results in conformational changes to the complex. The unidirectional translocation of protons occurs as a result of the influence of these conformational changes on the pK's of amino acid side chains and their alternate exposure to the internal and external side of the membrane.* We have previously seen that conformation can influence pK. The Bohr effect in hemoglobin, for example, results from conformational changes induced by O_2 binding, which causes pK changes in protein acid–base groups (Section 10-2E). If such a protein were located in a membrane and if, in addition to pK changes, the conformational changes altered the side of the membrane to which the affected amino acid side chains were exposed, the result would be H^+ transport and the system would be a proton pump.

Keep in mind that protons, being atomic nuclei, must always be associated with molecules or ions. Consequently, a proton cannot be transported across a membrane in the same way that, say, a K^+ ion is. Rather, protons are translocated by hopping along chains of hydrogen bonded groups

in the transport protein in much the same way that hydronium ions migrate through an aqueous solution (Fig. 2-10), that is, they move along a proton wire (Section 21-2Al). However, unlike a wire in an electrical circuit, all elements of a proton wire need not be connected at the same time and, moreover, internal water molecules, which are not always apparent in protein X-ray structures, are likely to be integral parts of the proton wire. Hence, elucidating the precise pathway of proton transport through a protein is a difficult and uncertain task. Note that proton pumps, like other active transport systems, must be gated to prevent protons from leaking back through the pump and thus short-circuiting the system.

h. Bacteriorhodopsin Is a Light-Driven Proton Pump

The simplest known and best characterized proton pump is the intrinsic membrane protein **bacteriorhodopsin** of *Halobacterium halobium*. It consists mainly of seven transmembrane helices, A through G, that form a central polar channel (Section 12-3Ab). This channel contains a retinal prosthetic group that is covalently linked, via a protonated Schiff base, to Lys 216, which extends from helix G (Fig. 12-24). The protein obtains the free energy required for unidirectionally pumping protons from the absorption of a photon by the retinal. This initiates a sequence of events in which the protein conformationally adjusts through successive spectroscopically characterized intermediates, designated J, K, L, M, N, and O, as the system decays back to its ground state over a period of ~10 ms. The result of this cycle is the net translocation of one proton from the cytoplasm to the extracellular medium, thereby converting light energy to proton-motive force. The as yet incompletely understood mechanism of this process, which was elucidated through detailed structural, mutational, and time-resolved spectroscopic studies carried out in several laboratories, is outlined in Fig. 22-34:

1. On absorbing a photon, the ground state all-*trans*-retinal photoisomerizes to its 13-cis form. This is a multi-step process that rapidly passes (in ~3 ps) through the J and K states. The free end of the retinal, which is now twisted around the newly cis double bond, moves relative to the protein scaffold such that retinal's C13 methyl group and its C14 atom shift toward the inside by 1.3 and 1.7 Å, respectively. This yields the L state.

2. Further conformational adjustments yield the M state. Here the N atom of the Schiff base has rotated and shifted from its ground state position, in which it is hydrogen bonded to an internal water molecule, to one in which it points toward the inside face of the protein in the vicinity of the hydrophobic side chains of Val 49 and Leu 93. This reduces the pK of the protonated Schiff base. In contrast, the pK of Asp 85 increases. This is because, in the ground state, Asp 85 effectively serves as the counterion of the protonated Schiff base and participates in a hydrogen bonded network with three internal water molecules, but in the M state, it is only associated with a single water molecule. Consequently, the Schiff base protonates Asp 85. This process is facilitated by a slight movement of helix C that

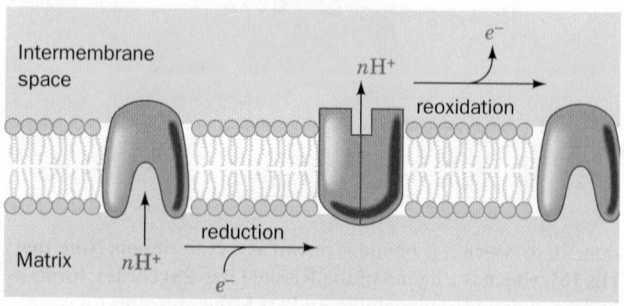

Figure 22-33 **Proton pump mechanism of electron transport–linked proton translocation.** At each H^+ translocation site, *n* protons bind to amino acid side chains on the matrix side (inside) of the membrane. Reduction causes a conformational change that decreases the pK's of these side chains and exposes them to the cytosolic side (outside) of the membrane, where the protons dissociate. Reoxidation results in a conformational change that restores the pump to its original conformation.

Ground state
all trans retinal protonated

L intermediate
13-cis retinal protonated

Inside

light

Outside

Membrane

N intermediate
13-cis retinal protonated

Late M intermediate
13-cis retinal neutral

Figure 22-34 **Proton pump of bacteriorhodopsin.** See the text for a description of this mechanism. The protein is represented by its seven transmembrane helices, A through G (with helices D and E omitted for clarity in all but the upper left panel), and several mechanistically important side chains. The retinal is drawn with the approximate color of the complex in its various spectroscopically characterized states. Red arrows indicate proton movements, blue arrows indicate the movements of groups of atoms, and the "paddle" attached to helix F represents the bulky side chains that must move aside to open the cytoplasmic channel. [After Kühlbrandt, W., *Nature* **406**, 569 (2000).]

brings Asp 85 closer to the Schiff base N atom. The deprotonated retinal straightens and, in doing so, moves upward (toward the inside) by 0.7 to 1.0 Å. It thereby pushes against the F helix, causing its inside (cytoplasmic) end to tilt outward from the channel by ~3.5 Å and the G helix to partially replace it.

3. The movement of the F helix opens up the central channel on the inside of the membrane, admitting several water molecules that form a hydrogen bonded chain between Asp 96 and the Schiff base. One of these water molecules hydrogen bonds with Asp 96 so as to lower its pK. This permits Asp 96 to protonate the Schiff base via the intermediacy of the chain of hydrogen bonded water molecules, thereby yielding the N state.

4. Asp 96 is reprotonated by the cytoplasmic solution. The loops forming bacteriorhodopsin's inside surface bear numerous charged residues and hence it appears that they function as "antennas" to capture protons from the alkaline cytoplasmic medium. Asp 85 transfers its proton to the extracellular medium via a hydrogen bonded network that includes several bound water molecules. This process is facilitated by a preceding 1.6-Å displacement of the Arg 82 side chain toward a complex of residues that includes Glu 194 and Glu 204, which reduces the pK of this complex. The retinal then relaxes, via the O state, to its original all-trans form and helices F and G return to their original positions, thereby re-forming the ground state of the protein and completing the catalytic cycle.

The retinal, which occupies the center of the protein channel, thereby acts as a one-way proton valve. The vectorial nature of this process arises from the unidirectional series of conformational changes made by the photoexcited retinal as it relaxes to its ground state. The protein's principal proton pumping motions are remarkably small, involving group movements of ~1 Å or less in response to the light-induced flexing of the retinal. These, nevertheless, cause pK changes in various residues that facilitate proton transfer as well as making and breaking hydrogen bonded networks of protein groups and water molecules in the proper sequence to transport a proton. A similar mechanism is likely to operate in COX, but with its conformational changes motivated by redox reactions rather than by photoexcitation.

i. COX Has Two Proton Translocation Channels

Two channels that are candidates for translocating protons from the inside to the vicinity of the O_2-reducing center have been described in bovine, *P. denitrificans,* and *Rb. sphaeroides* COX (Fig. 22-35). These channels, which are both contained in Subunit I, are named the K- and D-channels after their respective key residues (K319 and D91 in the bovine numbering scheme, which we shall use in the subsequent discussion). Both putative channels are similar in character to that present in bacteriorhodopsin in that they consist of chains of hydrogen bonded and potentially hydrogen bonded protein groups, bound water molecules, and water-filled cavities.

The K-channel leads from K319, which is exposed to the inside, to Y244, the residue that is the proposed substrate electron and proton donor in the reaction forming the P state (Step 4 of Fig. 22-27). The K319M mutant has an extremely low activity (<0.05% of wild type), which is not increased by supplying additional protons from the outside. Hence it appears that the K-channel is not connected to the putative exit channel (Fig. 22-35) that leads to the outside. It therefore seems likely that the K-channel only functions to supply chemical protons to the O_2-reduction center.

The entrance to the D-channel is within a region on the protein surface that appears likely to act as a proton-gathering antenna. The mutation of D91 to any noncarboxylate residue eliminates proton pumping but reduces the rate of O_2 reduction to only 45% of wild type (in

Outside

exit channel

Asp 369   His 368

Arg 439   Asp 364

Arg 438

His 291

heme *a*

D

heme *a₃*

hydrophobic
cavity

copper B

Tyr 244

Glu 242

Thr 316

Ser 157

Ser 156

Lys 319

Ser 255

Asp 91

D-channel   K-channel

Inside

Figure 22-35 **The proton-translocating channels in bovine COX.** The enzyme is viewed parallel to the membrane with the matrix below. The four rectangles delineate the proposed proton entry and exit conduits. Single and double circles, respectively, represent water molecules that are observed in X-ray structures or that theoretical methods suggest are likely to be present. [After a drawing by Mårten Wikström, University of Helsinki, Helsinki, Finland.]

E. coli). Evidently, the D-channel, in series with the exit channel, is the proton pumping channel. Moreover, the D-channel, which extends to the vicinity of the heme a_3–Cu_B binuclear center, is also the conduit for the chemical protons required for the second part of the reaction cycle (Steps 5 and 6 of Fig. 22-27).

What is the mechanism that couples O_2 reduction to proton pumping in COX? Unfortunately, X-ray crystallography has provided little guidance in answering this question because the X-ray structures of only a few different states of COX have as yet been determined and because the resolution of several of these structures is too low to reliably reveal small structural differences between them. Consequently, the mechanisms that have been proposed to explain how COX pumps protons are largely inferences based on limited structural information, interpretation of site-directed mutagenesis experiments, spectroscopic data, theoretical considerations, and chemical intuition. Several ingenious although largely phenomenological models of how COX pumps protons have been proposed. These mod-

els agree that at least one proton is pumped during each of Steps 5 and 6 in the COX reaction cycle (Fig. 22-27) but disagree as to where in the reaction cycle the remaining two protons are pumped and how the protein actuates this process. Clearly there is still much to learn about how COX carries out its function.

j. The Horizontal Helix of Complex I's Peripheral Arm Functions Like a Piston.

Complex I pumps four protons out of the matrix for every electron pair it translocates from NADH to CoQ. One of these protons appears to be transported at the interface between the peripheral and transmembrane arms of Complex I (Fig. 2-18) in a way that presumably is driven by the conformational changes in the peripheral arm as it translocates electrons from NADH to CoQ. But how are these conformational changes coupled to the far more distant antiporter-like subunits, Nqo12, 13, and 14, which presumably pump one proton each out of the matrix? The low resolution of the Complex I X-ray structure (Fig. 22-18) precludes the visualization of the membrane arm's side chains and hence a detailed mechanism for proton transport. However, the unusual 110-Å-long horizontal helix that appears to link the antiporter-like subunits to the conformational changes in the peripheral arm has led Sazanov to postulate that this helix functions like a piston to couple conformational changes in the peripheral arm to those in the antiporter-like channels.

C. *Mechanism of ATP Synthesis*

See Guided Exploration 21: F₁F₀–ATP synthase and the binding change mechanism *The proton-motive force across the mitochondrial membrane is harnessed in the synthesis of ATP by **proton-translocating ATP synthase** (also known as F₁F₀–ATPase, Complex V, and F-type H⁺–ATPase).* In the following subsections we discuss the location and structure of this ATP synthase and the mechanism by which it harnesses proton flux to drive ATP synthesis.

a. Proton-Translocating ATP Synthase Is a Multisubunit Transmembrane Protein

Proton-translocating ATP synthase consists of two major substructures comprising 8 to 13 different subunits. Electron micrographs of mitochondria (Fig. 22-36) show lollipop-shaped structures studding the matrix surface of the inner mitochondrial membrane (Fig. 22-36*a*). Similar entities have been observed lining the inner surface of the bacterial plasma membrane and in chloroplasts (Section 24-2Da). Sonication of the inner mitochondrial membrane yields sealed vesicles, **submitochondrial particles,** from which the "lollipops" project (Fig. 22-36*b*) and which can carry out ATP synthesis.

Efraim Racker discovered that the proton-translocating ATP synthase from submitochondrial particles comprises two functional units, **F₀** and **F₁**. F₀ is a water-insoluble transmembrane protein composed of as many as eight different types of subunits (although only three in *E. coli*) that contains a proton translocation channel. F₁ is a water-soluble

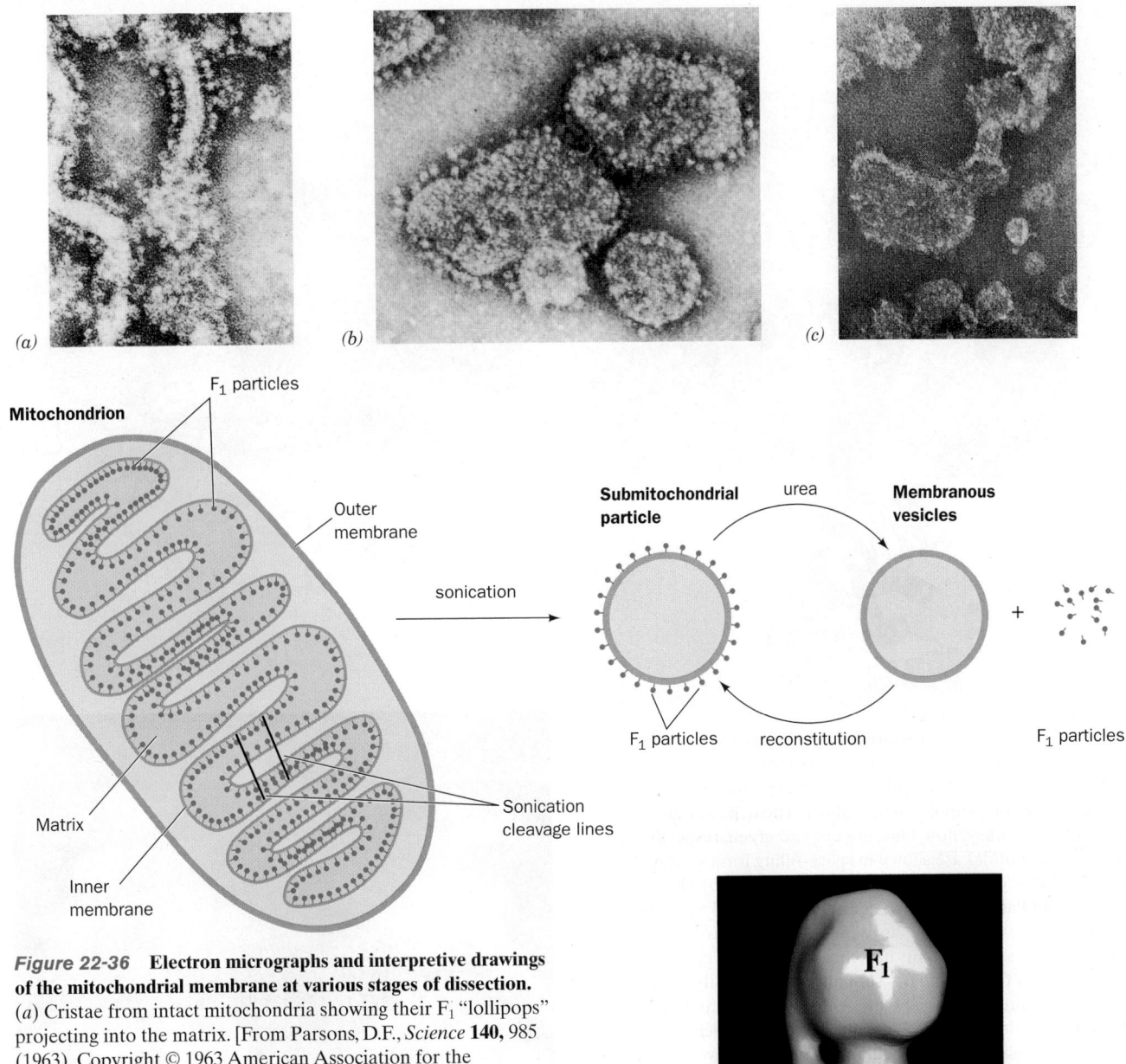

Figure 22-36 **Electron micrographs and interpretive drawings of the mitochondrial membrane at various stages of dissection.** (*a*) Cristae from intact mitochondria showing their F_1 "lollipops" projecting into the matrix. [From Parsons, D.F., *Science* **140**, 985 (1963). Copyright © 1963 American Association for the Advancement of Science. Used by permission.] (*b*) Submitochondrial particles, showing their outwardly projecting F_1 lollipops. Submitochondrial particles are prepared by the sonication (ultrasonic disruption) of inner mitochondrial membranes. [Courtesy of Peter Hinkle, Cornell University.] (*c*) Submitochondrial particles after treatment with urea. [Courtesy of Efraim Racker, Cornell University.]

Figure 22-37 **Cryoelectron microscopy–based image of F_1F_0–ATPase from bovine heart mitochondria.** [Courtesy of John Rubinstein, University of Toronto, Canada, and John Walker and Richard Henderson, MRC Laboratory of Molecular Biology, Cambridge, U.K.]

peripheral membrane protein, composed of five types of subunits, that is easily dissociated from F_0 by treatment with urea. Solubilized F_1 can hydrolyze ATP but cannot synthesize it (hence the name ATPase). Submitochondrial particles from which F_1 has been removed by urea treatment no longer exhibit the lollipops in their electron micrographs (Fig. 22-36*c*) and lack the ability to synthesize ATP. If, however, F_1 is added back to these F_0-containing submitochondrial particles, their ability to synthesize ATP is restored and their electron micrographs again exhibit the

lollipops. Thus *the lollipops are the F_1 particles.* Higher resolution cryoelectron micrograph–based images of the ATP synthase from bovine heart mitochondria reveal that its F_1 and F_0 components are joined by both an ~50-Å-long central stalk and a less substantial peripheral stalk (Fig. 22-37). Certain bacterial F_1F_0–ATPases translocate Na^+ ions rather than protons.

(a)

(b)

(c)

Figure 22-38   X-ray structures of F₁–ATPase from bovine heart mitochondria. (*a*) F₁–ATPase in complex with ADP and ATP drawn as a semitransparent ribbon diagram viewed parallel to the membrane with the matrix above. The α, β, γ, δ, and ε subunits are pink, yellow, blue, orange, and green, respectively, and the nucleotides are shown in space-filling form with ATP C green, ADP C cyan, N blue, O red, and P orange. (*b*) The structure in Part *a* rotated 90° about the horizontal axis such that the protein is viewed along its pseudo-3-fold axis from the matrix. The δ and ε subunits have been deleted for clarity. Note that although the α_{TP} and β_{TP} subunits both normally bind ATP, in this structure they have bound ADP. (*c*) A surface diagram of the inner portion of the $\alpha_3\beta_3$ assembly through which the C-terminal helix of the γ subunit penetrates as viewed from the matrix. The surface is colored according to its electrical potential, with positive potentials blue, negative potentials red, and neutral white. Note the absence of charge on the inner surface of this sleeve. That part of the γ subunit's C-terminal helix which contacts this sleeve is similarly devoid of charge. [Parts *a* and *b* based on an X-ray structure by Andrew Leslie and John Walker, MRC Laboratory of Molecular Biology, Cambridge, U.K. PDBid 1E79. Part *c* from Abrahams, J.P., Leslie, A.G.W., Lutter, R., and Walker, J.E., *Nature* **370,** 621 (1994). PDBid 1BMF.]

📙 **See Interactive Exercise 19**

b. The X-Ray Structure of F₁ Reveals the Basis of Its Lollipoplike Structure

Mitochondrial F₁ is an $\alpha_3\beta_3\gamma\delta\varepsilon$ nonamer in which the β subunits contain the catalytic sites for ATP synthesis and the δ subunit is required for binding of F₁ to F₀. The X-ray structure of F₁ from bovine heart mitochondria, determined by John Walker and Andrew Leslie, reveals that this ~400-kD protein is a 100-Å-high and 100-Å-wide spheroid

that is mounted on a 50-Å-long stem (Fig. 22-38*a,b*). F₁'s α and β subunits (553 and 528 residues), which are 20% identical in sequence and have nearly identical folds, are arranged alternately, like the segments of an orange, about the upper portion of a 114-Å-long, 74-residue α helix formed by the γ subunit (298 residues). The C-terminus of this helix protrudes into a 15-Å-deep dimple that is centrally located at the top of the spheroid. The lower half of

Cytosol

(a)

Periplasm

(b)

Figure 22-39   X-ray structure of the *c*₁₁ assembly of the *I.* *tartaricus* Na⁺-translocating F₁F₀–ATPase. (*a*) Ribbon diagram as viewed along the plane of the plasma membrane with the cytosol above. Each subunit is colored in rainbow order from its N-terminus (*blue*) to its C-terminus (*red*). Bound Na⁺ ions are represented by purple spheres. The parallel lines delineate the inferred position of the plasma membrane. (*b*) View from the cytosol parallel to the assembly's 11-fold axis in which each subunit is drawn in worm form and colored as in Part *a*. The minimally ~20-Å-diameter central channel presumably contains a lipid bilayer. [Based on an X-ray structure by Peter Dimroth, ETH, Zürich, Switzerland. PDBid 1YCE.]

the helix forms a bent left-handed antiparallel coiled coil with the N-terminal segment of the γ subunit. This coiled coil forms much of the ~50-Å-long central stalk seen in cryoEM-based images of the F₁F₀–ATPase such as Fig. 22-37.

The cyclical arrangement and structural similarities of F₁'s α and β subunits gives it both pseudo-3-fold and pseudo-6-fold rotational symmetry. Nevertheless, the protein is asymmetric. This is in part due to the presence of the γ, δ, and ε subunits but, more importantly, because each of the α and β subunits have a somewhat different conformation. Thus, one β subunit (designated β_TP) normally binds a molecule of ATP, the second (β_DP) binds ADP, and the third (β_E) has an empty and distorted binding site. The α subunits all normally bind ATP, although they also differ conformationally from one another (however, α_TP and β_TP both bind ADP in Fig. 22-38*a,b*). The ATP and ADP binding sites each lie at a radius of ~20 Å near an interface between adjacent α and β subunits and, in fact, all incorporate a few residues from the adjacent subunit. The α and β subunits both contain two sequence motifs, the **Walker A motif** (GXXXXGKT/S, where X is any residue) and the **Walker B motif** (R/KXXXGXXL/VhhD, where h is a hydrophobic residue), that participate in ATP binding and which occur widely in nucleotide-binding proteins of all descriptions.

The δ and ε subunits (168 and 51 residues) are wrapped about the base of the γ subunit's coiled coil. The X-ray structure of *E. coli* F₁ resembles its bovine counterpart. Note, however, that in an unfortunate confusion of nomenclature, the *E. coli* ε subunit is the homolog of the mitochondrial δ subunit; the *E. coli* δ subunit is the counterpart of the mitochondrial **oligomycin-sensitivity conferral protein (OSCP;** see Problem 11), and the mitochondrial ε subunit has no counterpart in either bacterial or chloroplast ATP synthases.

c. The *c* Subunits of F₀ Form a Transmembrane Ring

The F₀ component of the F₁F₀–ATPase from *E. coli* consists of three transmembrane subunits, *a, b,* and *c* (271, 161, and 89 residues) that form an $a_1b_2c_{12}$ complex. However, the number *c* subunits per F₀ assembly varies from 10 to 15, depending on the species. Mitochondrial F₀ additionally contains one copy each of three different subunits, *d,* F₆, and OSCP (256, 108, and 213 residues), as well as several "minor" subunits, *e, f, g,* and A6L, of unknown function. Subunits *a* and A6L are encoded by mitochondrial genes. A variety of evidence indicates that the hydrophobic *c* subunits associate to form a ring with the *ab*₂ unit located at its periphery (see below). The sequence of the *a* subunit suggests that this highly hydrophobic peptide forms five transmembrane helices.

The X-ray structure of the *c*₁₁ unit from the Na⁺-translocating F₁F₀–ATPase of the gram-negative bacterium *Ilyobacter tartaricus,* determined by Peter Dimroth, reveals an 11-fold symmetric cylindrical assembly that is 70 Å high and ~50 Å in diameter (Fig. 22-39). Each of its identical, largely hydrophobic subunits consists almost entirely of an inner helix that defines the length of the cylinder and an outer helix that is somewhat shorter. Both helices are bent in the vicinity of the Na⁺ ion-binding

Figure 22-40   X-ray structure of a portion of the peripheral stalk of bovine F_1F_0–ATPase. This protein fragment is drawn in ribbon form embedded in its semitransparent molecular surface, with *b* (residues 79–183) magenta, *d* (residues 3–123) cyan, and F_6 (residues 5–70) orange. The N- and C-termini of each subunit are indicated. [Based on an X-ray structure by Andrew Leslie and John Walker, MRC Laboratory of Molecular Biology, Cambridge, U.K. PDBid 2CLY.]

Figure 22-41   A composite model of the *E. coli* F_1F_0–ATPase. This model is based on the X-ray structure of the *E. coli* F_1 subunit (PDBid 1JNV), which resembles that of bovine F_1 (Fig. 22-38), the structures displayed in Figs. 22-39 and 22-40, and the NMR structures of δ and the transmembrane segment of the *b* from *E. coli* (PDBids 2A7U and 1B9U). The structures of *a* and the so-called hinge region of *b* are unknown. [Courtesy of Peter Dimroth, ETH, Zürich, Switzerland.]

site, thereby conferring an hourglasslike shape on the c_{11} assembly (Fig. 22-39*a*). Each Na^+ ion is liganded by residues from both helices of a given subunit as well as those of the clockwise neighboring outer helix as seen in Fig. 22-29*b*.

In vertebrates, the peripheral stalk (Fig. 22-37) consists of the OSCP, *b, d,* and F_6 subunits, whereas in most prokaryotes, it consists of the OSCP homolog δ and two copies of *b*. The X-ray structure of most of the cytosolic portion of a complex of bovine subunits *b, d,* and F_6, constituting 54% of the extramembrane segment of the peripheral stalk, determined by Leslie and Walker, reveals that this *b* fragment forms a 160-Å-long and curved α helix, and that the other subunit fragments are also mainly helical (Fig. 22-40). The curvature of this assembly closely matches that of the peripheral stalk in cryoEM-based images of the F_1F_0–ATPase (e.g., Fig. 22-37). Such cryoEM-based images, together with the foregoing X-ray structures and the NMR structures of δ and the transmembrane segment of *b*, both from *E. coli*, has enabled the construction of a composite model of the *E. coli* F_1F_0–ATPase (Fig. 22-41), which, of course, resembles the mitochondrial assembly. Note the extensive contacts between γ and ε and the top of the *c* cylinder.

d. The Binding Change Mechanism: Proton-Translocating ATP Synthase Is Driven by Conformational Changes

The mechanism of ATP synthesis by proton-translocating ATP synthase can be conceptually broken down into three phases:

1. Translocation of protons carried out by F_0.

2. Catalysis of formation of the phosphoanhydride bond of ATP carried out by F_1.

3. Coupling of the dissipation of the proton gradient with ATP synthesis, which requires interaction of F_1 and F_0.

The available evidence supports a mechanism for ATP formation, proposed by Boyer, that resembles the conformational coupling hypothesis of oxidative phosphorylation (Section 22-3A). However, the conformational changes in the ATP synthase that power ATP formation are generated by proton translocation rather than by direct electron transfer, as proposed in the original formulation of the conformational coupling hypothesis.

F_1 is proposed to have three interacting catalytic protomers, each in a different conformational state: one that binds substrates and products loosely (L state), one that binds them tightly (T state), and one that does not bind them at all (open or O state). The free energy released on proton translocation is harnessed to interconvert these three states. The phosphoanhydride bond of ATP is synthesized only in the T state and ATP is released only in the O state. The reaction involves three steps (Fig. 22-42):

1. Binding of ADP and P_i to the "loose" (L) binding site.

2. A free energy–driven conformational change that converts the L site to a "tight" (T) binding site that catalyzes the formation of ATP. This step also involves conformational changes of the other two subunits that convert

Figure 22-42 **Energy-dependent binding change mechanism for ATP synthesis by proton-translocating ATP synthase.** F_1 has three chemically identical but conformationally distinct interacting αβ protomers: O, the open conformation, has very low affinity for ligands and is catalytically inactive; L has loose binding for ligands and is catalytically inactive; T has tight binding for ligands and is catalytically active. ATP synthesis occurs in three steps. **(1)** Binding of ADP and P_i to site L. **(2)** Energy-dependent conformational change converting

binding site L to T, T to O, and O to L. **(3)** Synthesis of ATP at site T and release of ATP from site O. The enzyme returns to its initial state after two more passes of this reaction sequence. The energy that drives the conformational change is apparently transmitted to the catalytic $\alpha_3\beta_3$ assembly via the rotation of the γε assembly (in *E. coli;* γδε in mitochondria), here represented by the centrally located asymmetric pointer (*green*). [After Cross, R.L., *Annu. Rev. Biochem.* **50,** 687 (1980).] **See the Animated Figures**

the ATP-containing T site to an "open" (O) site and convert the O site to an L site.

3. ATP is synthesized at the T site on one subunit while ATP dissociates from the O site on another subunit. On the surface of the active site, the formation of ATP from ADP and P_i entails little free energy change, that is, the reaction is essentially at equilibrium. Consequently, the free energy supplied by the proton flow primarily facilitates the release of the newly synthesized ATP from the enzyme; that is, it drives the T → O transition, thereby disrupting the enzyme–ATP interactions that had previously promoted the spontaneous formation of ATP from ADP + P_i in the T site.

How is the free energy of proton transfer coupled to the synthesis of ATP? Boyer proposed that *the binding changes are driven by the rotation of the catalytic assembly, $\alpha_3\beta_3$ with respect to other portions of the F_1F_0–ATPase.* This hypothesis is supported by the X-ray structure of F_1. Thus, the closely fitting nearly circular arrangement of the α and β subunits' inner surface about the γ subunit's helical C-terminus is reminiscent of a cylindrical bearing rotating in a sleeve (Fig. 22-38c). Indeed, the contacting hydrophobic surfaces in this assembly are devoid of the hydrogen bonding and ionic interactions that would interfere with their free rotation; that is, the bearing and sleeve appear to be "lubricated." Moreover, the central cavity in the $\alpha_3\beta_3$ assembly (Fig. 22-38a) would permit the passage of the γ subunit's N-terminal helix within the core of this particle during rotation. Finally, the conformational differences between F_1's three catalytic sites appear to be correlated with the rotational position of the γ subunit. *Apparently the γ subunit, which is thought to rotate within the fixed $\alpha_3\beta_3$ assembly, acts as a molecular cam shaft in linking the proton-motive force–driven rotational motor to the conformational changes in the catalytic sites of F_1.* This concept is also supported by molecular dynamics simulations (Section 9-4) by Leslie, Walker, and Martin Karplus, which indicate that the conformational changes in the β subunits arise from both steric and electrostatic interactions with the rotating γ subunit.

Rotating assemblies are not unprecedented in biological systems. Bacterial flagella, which function as propellers, had previously been shown to be membrane-mounted rotary engines that are driven by the discharge of a proton gradient (Section 35-3Ib).

e. The F_1F_0–ATPase Is a Rotary Engine

In the F_1F_0–ATPase, the rotor is proposed to be an assembly of the *c* ring and its associated γ and (*E. coli*) ε subunits, whereas the ab_2 unit and the (*E. coli*) δ subunit together with the $\alpha_3\beta_3$ spheroid form the stator (Fig. 22-41). The rotation of the *c* ring in the membrane relative to the stationary *a* subunit is driven by the migration of protons from the outside to the inside, as we discuss below. The peripheral arm ($b_2\delta$) presumably functions to hold the $\alpha_3\beta_3$ spheroid in place while the γ subunit rotates inside it.

The rotation of the *E. coli* γε–*c*-ring rotor with respect to the $ab_2\delta$–$\alpha_3\beta_3$ stator has been ingeniously demonstrated by Masamitsu Futai using techniques developed by Kazuhiko Kinosita Jr. and Masasuke Yoshida (Fig. 22-43a). The $\alpha_3\beta_3$ spheroid of *E. coli* F_1F_0–ATPase was fixed, head down, to a glass surface as follows: Six consecutive His residues (a so-called **His Tag;** Section 6-3Dg) were mutagenically appended to the N-terminus of the α subunit, which is located at the top of the $\alpha_3\beta_3$ spheroid as it is drawn in Fig. 22-38a. The His-tagged assembly was applied to a glass surface coated with horseradish peroxidase (which, like most proteins, sticks to glass) conjugated with Ni^{2+}-**nitriloacetic acid** $[N(CH_2COOH)_3$, which tightly binds His tags], thereby binding the F_1F_0–ATPase with its F_0 side facing away from the surface. The Glu 2 residues of this assembly's *c* subunits, which are located on the side of the *c* ring facing away from F_1, had been mutagenically replaced by Cys residues, which were then covalently linked to **biotin** (a coenzyme that normally participates in carboxylation reactions; Section 23-1Ab). A fluorescently labeled and biotinylated (at one end) filament of the muscle protein **actin** (Section 35-3Ac) was then attached to the *c* subunit through the addition of a bridging molecule of **streptavidin**, a protein that avidly binds biotin to each of four binding sites (Cys 193 of

(a)

(b)

Figure 22-43 **Rotation of the *c* ring in *E. coli* F₁F₀–ATPase.**
(a) The experimental system used to observe the rotation. See
the text for details. The blue arrow indicates the observed
direction of rotation of the fluorescently labeled actin filament
that was linked to the *c* ring. (b) The rotation of a 3.6-μm-long
actin filament in the presence of 5 m*M* MgATP as seen in
successive video images taken through a fluorescence microscope. [Courtesy of Masamitsu Futai, Osaka University, Osaka,
Japan.]

the γ subunit, the only other Cys residue in the rotor, was
mutagenically replaced by Ala to prevent it from being
linked to an actin filament).

E. coli F₁F₀–ATPase can work in reverse, that is, it can
pump protons from the inside (cytoplasm) to the outside
(periplasm) at the expense of ATP hydrolysis (this enables
the bacterium to maintain its proton gradient under anaerobic conditions, which it uses to drive various processes
such as flagellar rotation). Thus, the foregoing preparation
was observed under a fluorescence microscope as a 5 m*M*
MgATP solution was infused over it. *Many of the actin filaments were seen to rotate (Fig. 22-43b), and always in a
counterclockwise direction when viewed looking down on
the glass surface (from the outside).* This would permit the γ
subunit to sequentially interact with the β subunits in the
direction

$$\beta_E(\text{O state}) \longrightarrow \beta_{TP}(\text{T state}) \longrightarrow \beta_{DP}(\text{L state})$$

(Figs. 22-38*b* and 22-42), the direction expected for ATP
hydrolysis.

a [ATP] = 0.02 μM, Actin length = 1.1 μm

Figure 22-44 **Stepwise rotation of the γ subunit of F₁ relative
to an immobilized α₃β₃ unit at low ATP concentration as
observed by fluorescence microscopy.** The graph plots the
cumulative number of rotations made by a fluorescently labeled
actin filament that was linked at one end to the γ subunit in a
preparation similar to that diagrammed in Fig. 22-43*a* (but
lacking F₀, δ, and ε). Note that the actin filament rotates in
increments of 120°. This is also evident in the inset, which shows
the superposition of the centers of the actin images (the δ₃β₃γ
assembly is fixed in the center). [Courtesy of Kazuhiko Kinosita,
Jr., Keio University, Yokohama, Japan.]

In a variation of the above experiment, the γ subunit of
the α₃β₃γ complex was directly cross-linked, via its Cys 193,
to a fluorescently labeled actin filament and, in this case,
the β subunits were immobilized by appended His tags. At
very low ATP concentrations (e.g., 0.02 μ*M*), video images
(Fig. 22-44) revealed that the fluorescent actin filament rotated counterclockwise in discrete steps of 120°, as the
binding change mechanism predicts. Moreover, the calculated frictional work done in each rotational step is very
nearly equal to the energy available from the hydrolysis of
one ATP molecule, that is, *the F₁F₀–ATPase converts chemical to mechanical energy with nearly 100% efficiency.*
    The foregoing system also works in reverse. An ~0.7-
μm-diameter magnetic bead that was coated with streptavidin was attached to the biotinylated γ subunit of an immobilized α₃β₃γ complex. When the resulting assembly was
placed in a rotating magnetic field in the presence of ADP
and P₁, ATP was produced when the magnetic field rotated
in the clockwise direction but hydrolyzed when it rotated
in the counterclockwise direction. This further demonstrates that F₁ is a device that interconverts mechanical and
chemical energy.

f. *c*-Ring Rotation Is Impelled by H⁺-Induced Conformational Changes

The foregoing structural and biochemical information
has led to the model for proton-driven rotation of the F₀
subunit that is diagrammed in Fig. 22-45. Protons from the
outside enter a hydrophilic channel between the *a* subunit
and the *c* ring, where they bind to a *c* subunit. The *c* ring

Figure 22-45   Schematic diagram of the action of the *E. coli* F_1F_0-ATPase. The $\gamma\varepsilon$–c_{12} ring complex is the rotor and the ab_2–$\alpha_3\beta_3\delta$ complex is the stator. Rotational motion is imparted to the rotor by the passage of protons from the outside (periplasm) to the inside (cytoplasm). Protons entering from the outside bind to a *c* subunit where it interacts with the *a* subunit, and exit to the inside after the *c* ring has made a nearly full rotation as indicated (*black arrows*), so that the *c* subunit again contacts the *a* subunit. The $b_2\delta$ complex presumably functions to prevent the $\alpha_3\beta_3$ assembly from rotating with the γ subunit. [Courtesy of Richard Cross, State University of New York, Syracuse, New York.]

Figure 22-46   NMR structures of the *c* subunit of *E. coli* F_1F_0-ATPase. The structures, which closely resemble that in Fig. 22-39, were determined in chloroform–methanol–water (4:4:1) solution at (*a*) pH 8 (at which D61 is deprotonated) and (*b*) pH 5 (at which D61 is protonated). Selected side chains are shown to aid in the comparison of the two structures. Note that the C-terminal helix in the pH 8 structure has rotated by 140° clockwise, as viewed from the top of the drawing, relative to that in the pH 5 structure. [Courtesy of Mark Girvin, Albert Einstein College of Medicine. PDBids (*a*) 1C99 and (*b*) 1C0V.]

then rotates nearly a full turn (while protons bind to successive *c* subunits as they pass this input channel) until the subunit reaches a second hydrophilic channel between the *a* subunit and the *c*-ring that opens into the inside. There the proton is released. Thus, *E. coli* F_1F_0-ATPase, which has 12 *c* subunits in its F_0 assembly and generates 3ATP per turn, ideally forms 3/12 = 0.25 ATP for every proton it passes from the periplasmic space (outside) to the cytosol (inside). Organisms with more/less *c* subunits in their *c*-rotor tend to have lesser/greater values of proton-motive force across their membranes and hence have less/more impetus imparted to their *c*-rotor per proton passed. Hence it takes the passage of proportionately more/less protons to generate each ATP, as the first law of themodynamics requires.

How does the passage of protons through this system induce the rotation of the *c* ring and hence the synthesis of ATP? The mutation of the *c* subunit's conserved Asp 61 to Asn inactivates *E. coli* F_1F_0-ATPase. The *a* subunit's invariant Arg 210 (*E. coli* numbering) has been similarly impli-

cated in proton translocation. Through the mutagenic conversion of selected residues on the *a* and *c* subunits to Cys, Robert Fillingame has shown that the outer (C-terminal) helix of *E. coli* subunit *c* (Fig. 22-39), which contains Asp 61, can be disulfide-cross-linked to the putative fourth helix of subunit *a*, which contains Arg 210. Evidently, these helices are juxtaposed at some point in the *c* ring's rotation cycle. Thus, it is postulated that the protonation of Asp 61 releases its attraction to Arg 210, thereby permitting the *c* ring to rotate.

Comparison of the NMR structures of subunit *c* at pH 8 and pH 5 (Fig. 22-46), at which Asp 61 is, respectively, deprotonated and protonated, reveals that its main conformational change on protonation is an ~140° clockwise rotation (as viewed from F_1) of its Asp 61-containing C-terminal helix with respect to its N-terminal helix. Since the C-terminal helix is the *c* ring's outer helix (Fig. 22-39*b*), this suggests that, on protonation, the rotation of the C-terminal helix mechanically pushes against the juxtaposed *a* subunit so as to rotate the *c* ring in the direction indicated in Fig. 22-45.

D.   *Uncoupling of Oxidative Phosphorylation*

Electron transport (the oxidation of NADH and FADH$_2$ by O$_2$) and oxidative phosphorylation (the synthesis of ATP) are normally tightly coupled due to the impermeability of the inner mitochondrial membrane to the passage of protons. Thus the only way for H$^+$ to reenter the matrix is through the F_0 portion of the proton-translocating ATP synthase. In the resting state, when oxidative phosphorylation is minimal, the proton-motive force across the inner mitochondrial membrane builds up to the extent that the free energy to pump additional protons is greater than the electron-transport chain can muster, thereby inhibiting further electron transport. However, many compounds, including **2,4-dinitrophenol (DNP)** and **carbonylcyanide-*p*-trifluoromethoxyphenylhydrazone (FCCP),** have been found to "uncouple" these processes. The chemiosmotic hypothesis has provided a rationale for understanding the mechanism by which these uncouplers act.

The presence in the inner mitochondrial membrane of an agent that renders it permeable to H⁺ uncouples oxidative phosphorylation from electron transport by providing a route for the dissipation of the proton-motove force that does not require ATP synthesis. Uncoupling therefore allows electron transport to proceed unchecked even when ATP synthesis is inhibited. DNP and FCCP are lipophilic weak acids that therefore readily pass through membranes. In a pH gradient, they bind protons on the acidic side of the membrane, diffuse through, and release them on the alkaline side, thereby dissipating the gradient (Fig. 22-47). Thus, *such uncouplers are proton-transporting ionophores* (Section 20-2C).

Even before the mechanism of uncoupling was known, it was recognized that metabolic rates were increased by such compounds. Studies at Stanford University in the early part of the twentieth century documented an increase in respiration and weight loss caused by DNP. The compound was even used as a "diet pill" for several years. In the words of Efraim Racker (*A New Look at Mechanisms in Bioenergetics,* p. 155):

> *In spite of warnings from the Stanford scientists, some enterprising physicians started to administer dinitrophenol to obese patients without proper precautions. The results were striking. Unfortunately in some cases the treatment eliminated not only the fat but also the patients, and several fatalities were reported in the Journal of the American Medical Association in 1929. This discouraged physicians for a while....*

a. Hormonally Controlled Uncoupling in Brown Adipose Tissue Functions to Generate Heat

The dissipation of an electrochemical H⁺ gradient, which is generated by electron transport and uncoupled from ATP synthesis, produces heat. Heat generation is the physiological function of **brown adipose tissue (brown fat).** This tissue is unlike typical (white) adipose tissue in that, besides containing large amounts of triacylglycerols, it contains numerous mitochondria whose cytochromes color it brown. Newborn mammals that lack fur, such as humans, as well as hibernating mammals, contain brown fat in their neck and upper back that functions in **nonshivering thermogenesis,** that is, as a "biological heating pad." (The ATP hydrolysis that occurs during the muscle contractions of shivering—or

Figure 22-47   Uncoupling of oxidative phosphorylation. The proton-transporting ionophores DNP and FCCP uncouple oxidative phosphorylation from electron transport by discharging the electrochemical proton gradient generated by electron transport.

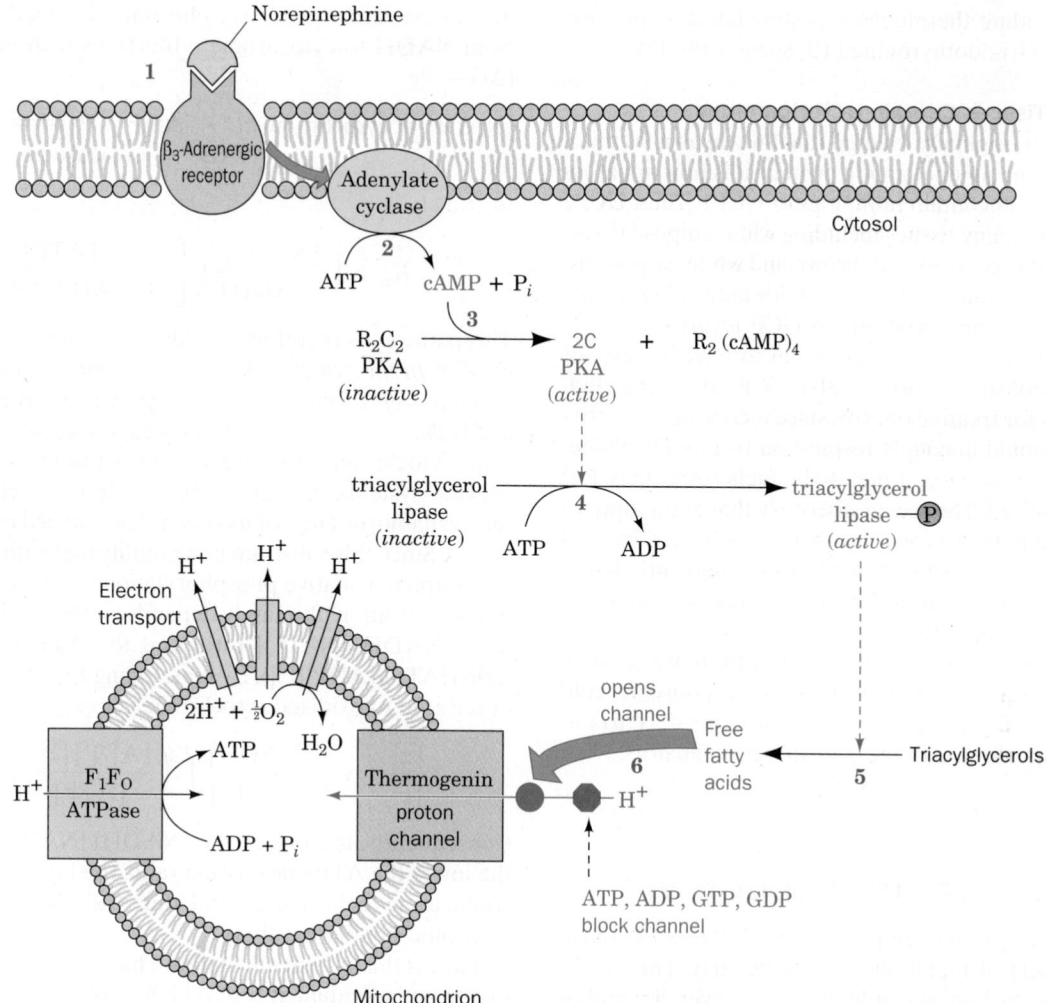

Figure 22-48 **Mechanism of hormonally induced uncoupling of oxidative phosphorylation in brown fat mitochondria.** **(1)** Norepinephrine binds to a β_3-adrenergic receptor. **(2)** This stimulates the associated heterotrimeric G protein to activate adenylate cyclase (*upper green arrow*) to synthesize cAMP. **(3)** cAMP binding activates protein kinase A (PKA). **(4)** PKA phosphorylates hormone-sensitive triacylglycerol lipase, thereby activating it. **(5)** Triacylglycerols are hydrolyzed, yielding free fatty acids. **(6)** Free fatty acids overcome the purine nucleotide block of thermogenin's proton channel (*lower green arrow*), allowing H^+ to enter the mitochondrion uncoupled from ATP synthesis.

any other movement—also produces heat. Nonshivering thermogenesis through substrate cycling is discussed in Section 17-4Fi.)

The mechanism of heat generation in brown fat involves the regulated uncoupling of oxidative phosphorylation in their mitochondria. These mitochondria contain the protein **thermogenin** [also called **uncoupling protein (UCP)**], a transmembrane homodimer of 307-residue subunits that acts as a channel to control the permeability of the inner mitochondrial membrane to protons. In cold-adapted animals, thermogenin constitutes up to 15% of brown fat inner mitochondrial membrane proteins. The flow of protons through this channel protein is inhibited by physiological concentrations of purine nucleotides (ADP, ATP, GDP, GTP), but this inhibition can be overcome by free fatty acids. The components of this system interact under hormonal control.

Thermogenesis in brown fat mitochondria is activated by free fatty acids. These counteract the inhibitory effects of purine nucleotides, thereby stimulating the flux through the proton channel and uncoupling electron transport from oxidative phosphorylation. *The concentration of fatty acids in brown adipose tissue is controlled by the adrenal hormone norepinephrine (noradrenaline; Section 18-3E) with cAMP acting as a second messenger* (Section 18-3). Norepinephrine binds to the **β_3-adrenergic receptor,** a G-protein coupled receptor (GPCR) that, via an associated heterotrimeric G protein, stimulates adenylate cyclase to synthesize cAMP (Fig. 22-48), as described in Section 19-2. The cAMP, in turn, activates protein kinase A (PKA), which activates **hormone-sensitive triacylglycerol lipase** by phosphorylating it (Section 25-5). Finally, the activated lipase hydrolyzes triacylglycerols to yield the free fatty acids that open thermogenin's proton channel. The transcription of the gene encoding thermogenin is stimulated by the thyroid hormone triiodothyronine (T3; Section 19-1D).

b. Other Tissues Contain UCP Homologs

Although it originally seemed that only brown fat mitochondria contain an uncoupling protein, it is now apparent that other tissues contain homologs of UCP1. Thus, **UCP2** is expressed in many tissues including white adipose tissue, whereas **UCP3** occurs in both brown and white adipose tissues as well as in muscle. These proteins may help regulate metabolic rates, and variations in UCP levels or activity might explain why some people seem to have a "fast" or "slow" metabolism (Section 27-3E). UCPs are being studied as targets for treating obesity, since increasing the activity of UCPs could uncouple respiration from ATP synthesis, thus permitting stored metabolic fuels (especially fat) to be metabolized. The recent discovery that adult humans have small depots of brown fat that are activated by cold has made this an attractive weight loss strategy, although it is possible that stimulating UCPs will cause a compensatory increase in appetite.

Uncoupling proteins are not limited to animals. Some plants express uncoupling proteins in response to cold stress or to increase flower temperature, possibly to enhance the vaporization of scent to attract pollinators.

4 CONTROL OF ATP PRODUCTION

A typical adult woman requires some 1500 to 1800 kcal (6300–7500 kJ) of metabolic energy per day. This corresponds to the free energy of hydrolysis of over 200 mol of ATP to ADP and P_i. Yet the total amount of ATP present in the body at any one time is <0.1 mol; obviously, this sparse supply of ATP must be continually recycled. As we have seen, when carbohydrates serve as the energy supply and aerobic conditions prevail, this recycling involves glycogenolysis, glycolysis, the citric acid cycle, and oxidative phosphorylation.

Of course the need for ATP is not constant. There is a 100-fold change in ATP utilization between sleep and vigorous activity. *The activities of the pathways that produce ATP are under strict coordinated control so that ATP is never produced more rapidly than necessary.* We have already discussed the control mechanisms of glycolysis, glycogenolysis, and the citric acid cycle (Sections 17-4, 18-3, and 21-4). In this section we discuss the mechanisms through which oxidative phosphorylation is controlled and observe how all four systems are synchronized to produce ATP at precisely the rate required at any particular moment.

A. *Control of Oxidative Phosphorylation*

In our discussion of the control of glycolysis, we saw that most of the reactions in a metabolic pathway function close to equilibrium. *The few irreversible reactions constitute the potential control points of the pathway and usually are catalyzed by regulatory enzymes that are under allosteric con-*

trol. In the case of oxidative phosphorylation, the pathway from NADH to cytochrome *c* functions near equilibrium ($\Delta G \approx 0$):

$$\tfrac{1}{2}NADH + \text{cytochrome } c^{3+} + ADP + P_i \rightleftharpoons$$
$$\tfrac{1}{2}NAD^+ + \text{cytochrome } c^{2+} + ATP$$

for which

$$K_{eq} = \left(\frac{[NAD^+]}{[NADH]}\right)^{\tfrac{1}{2}}\frac{[c^{2+}]}{[c^{3+}]}\frac{[ATP]}{[ADP][P_i]} \quad [22.2]$$

This pathway is therefore readily reversed by the addition of ATP. *In the cytochrome c oxidase reaction, however, the terminal step of the electron-transport chain is irreversible and is thus one of the important regulatory sites of the pathway.* Cytochrome *c* oxidase, in contrast to most regulatory enzyme systems, appears to be controlled exclusively by the availability of one of its substrates, reduced cytochrome *c* (c^{2+}). Since this substrate is in equilibrium with the rest of the coupled oxidative phosphorylation system (Eq. [22.2]), its concentration ultimately depends on the intramitochondrial [NADH]/[NAD$^+$] ratio and the **ATP mass action ratio** ([ATP]/[ADP][P$_i$]). By rearranging Eq. [22.2], the ratio of reduced to oxidized cytochrome *c* is expressed

$$\frac{[c^{2+}]}{[c^{3+}]} = \left(\frac{[NADH]}{[NAD^+]}\right)^{1/2}\left(\frac{[ADP][P_i]}{[ATP]}\right)K_{eq} \quad [22.3]$$

Consequently, the higher the [NADH]/[NAD$^+$] ratio and the lower the ATP mass action ratio, the higher is the [c^{2+}] (reduced cytochrome *c*) and thus the higher is the cytochrome *c* oxidase activity.

How is this system affected by changes in physical activity? In an individual at rest, ATP hydrolysis to ADP and P$_i$ is minimal and the ATP mass action ratio is high; the concentration of reduced cytochrome *c* is therefore low and oxidative phosphorylation is minimal. Increased activity results in hydrolysis of ATP to ADP and P$_i$, thereby decreasing the ATP mass action ratio and increasing the concentration of reduced cytochrome *c*. This results in an increase in the electron-transport rate and its coupled phosphorylation. Such control of oxidative phosphorylation by the ATP mass action ratio is called **acceptor control** because the rate of oxidative phosphorylation increases with the concentration of ADP, the phosphoryl group acceptor. In terms of a supply–demand system (Section 17-4D), acceptor control is understood as control by the demand block.

The compartmentalization of the cell into mitochondria, where ATP is synthesized, and cytoplasm, where ATP is utilized, presents an interesting control problem: Is it the ATP mass action ratio in the cytosol or in the mitochondrial matrix that ultimately controls oxidative phosphorylation? Clearly the ATP mass action ratio that exerts direct control must be that of the mitochondrial matrix where ATP is synthesized. However, the inner mitochondrial membrane, which is impermeable to adenine nucleotides and P$_i$, depends on specific transport systems to maintain communication between the two compartments (Section 20-4C). This organization makes it possible for the transport of adenine nucleotides or P_i to participate in the control of oxidative phosphorylation.

Considerable research effort has been aimed at determining how oxidative phosphorylation is controlled in terms of metabolic control analysis. For example, Hans Westerhoff and Martin Kushmerick employed ^{31}P NMR to measure the ATP/ADP ratios in human forearm muscle at rest and during twitch contractions caused by external electrical stimulation (the ^{31}P NMR spectrum of ATP is shown in Fig. 16-15). Under conditions of low to moderate ATP demand, the cytosolic mass action ratio as controlled by the demand block of the system appears to be the major control factor for mitochondrial oxidation. However, as other laboratories have shown, as the demand for ATP increases, the ADP–ATP translocator exerts greater control until finally, when the demand for ATP is high, control shifts to the supply block of the system, oxidative phosphorylation itself.

B. *Coordinated Control of ATP Production*

Glycolysis, the citric acid cycle, and oxidative phosphorylation constitute the major pathways for cellular ATP production. Control of oxidative phosphorylation by the ATP mass action ratio depends, of course, on an adequate supply of electrons to fuel the electron-transport chain. This aspect of the system's control is, in turn, dependent on the $[NADH]/[NAD^+]$ ratio (Eq. [22.3]), which is maintained high by the combined action of glycolysis and the citric acid cycle in converting 10 molecules of NAD^+ to NADH per molecule of glucose oxidized (Fig. 22-1). It is clear, therefore, that coordinated control is necessary for the three processes. This is provided by the regulation of each of the control points of glycolysis [hexokinase, phosphofructokinase (PFK), and pyruvate kinase] and the citric acid cycle (pyruvate dehydrogenase, citrate synthase, isocitrate dehydrogenase, and α-ketoglutarate dehydrogenase) by adenine nucleotides or NADH or both as well as by certain metabolites (Fig. 22-49).

a. Citrate Inhibits Glycolysis

The main control points of glycolysis and the citric acid cycle are regulated by several effectors besides adenine nucleotides or NADH (Fig. 22-49). This is an extremely complex system with complex demands. Its many effectors, which are involved in various aspects of metabolism, increase its regulatory sensitivity. One particularly interesting regulatory effect is the inhibition of PFK by citrate. When demand for ATP decreases, [ATP] increases and [ADP] decreases. The

Figure 22-49 **Schematic diagram depicting the coordinated control of glycolysis and the citric acid cycle by ATP, ADP, AMP, P_i, Ca^{2+}, and the $[NADH]/[NAD^+]$ ratio (the vertical arrows indicate increases in this ratio).** Here a green dot signifies activation and a red octagon represents inhibition. [After Newsholme, E.A. and Leech, A.R., *Biochemistry for the Medical Sciences,* pp. 316 and 320, Wiley (1983).] 🔗 **See the Animated Figures**

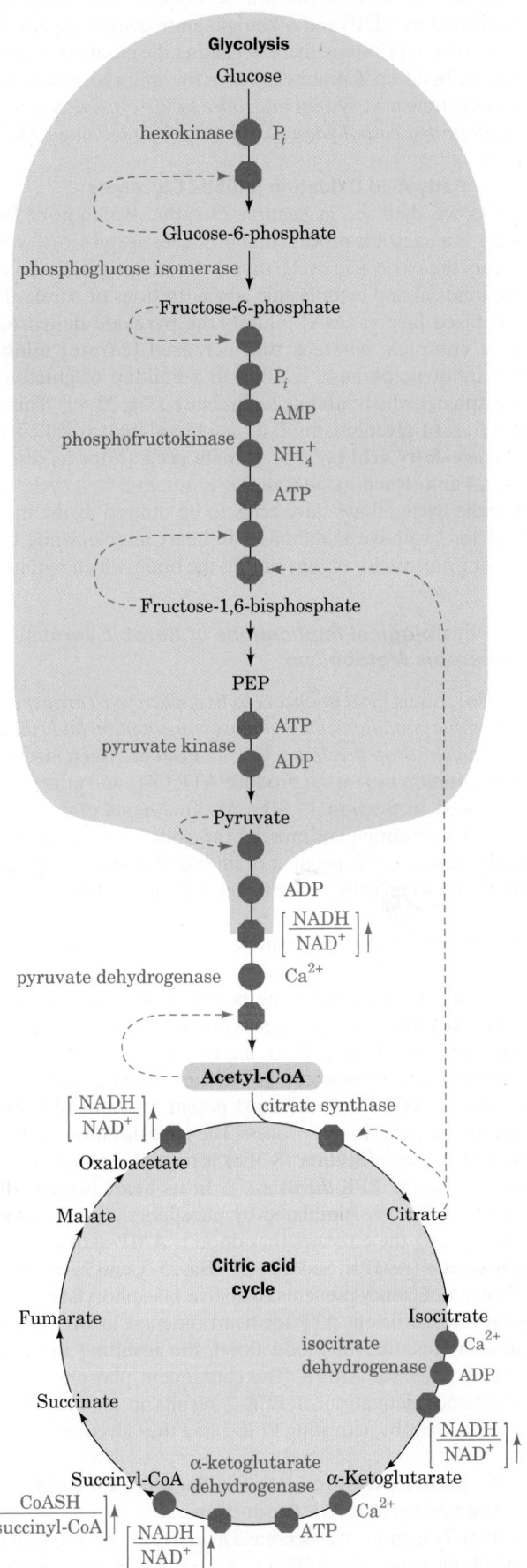

citric acid cycle slows down at its isocitrate dehydrogenase (activated by ADP) and α-ketoglutarate dehydrogenase (inhibited by ATP) steps, thereby causing the citrate concentration to build up. Citrate can leave the mitochondrion via a specific transport system and, *once in the cytosol, acts to restrain further carbohydrate breakdown by inhibiting PFK.*

b. Fatty Acid Oxidation Inhibits Glycolysis

As we shall see in Section 25-2, the oxidation of fatty acids is an aerobic process that produces acetyl-CoA, which enters the citric acid cycle, thereby increasing both the mitochondrial and cytoplasmic concentrations of citrate. The increased [acetyl-CoA] inhibits the pyruvate dehydrogenase complex, whereas the increased [citrate] inhibits phosphofructokinase, leading to a buildup of glucose-6-phosphate, which inhibits hexokinase (Fig. 22-49). This inhibition of glycolysis by fatty acid oxidation is called the **glucose–fatty acid cycle** or **Randle cycle** (after its discoverer, Philip Randle), although it is not, in fact, a cycle. The Randle cycle allows fatty acids to be utilized as the major fuel for oxidative metabolism in heart muscle, while conserving glucose for organs such as the brain, which require it.

C. *Physiological Implications of Aerobic versus Anaerobic Metabolism*

In 1861, Louis Pasteur observed that *when yeast are exposed to aerobic conditions, their glucose consumption and ethanol production drop precipitously* (the **Pasteur effect;** alcoholic fermentation in yeast to produce ATP, CO_2, and ethanol are discussed in Section 17-3B). An analogous effect is observed in mammalian muscle; the concentration of lactic acid, the anaerobic product of muscle glycolysis, drops dramatically when cells switch to aerobic metabolism.

a. Hypoxia Causes an Increase in Glycolysis

In the presence of sufficient oxygen, oxidative phosphorylation supplies most of the body's ATP needs. However, during **hypoxia** (when oxygen is limiting), glycolysis must be stimulated (with its inherent increased rate of glucose consumption; the reverse of the Pasteur effect) to supply the necessary ATP. F2,6P, the most potent activator of PFK-1, also participates in this process. The concentration of F2,6P, as we have seen (Section 18-3Fc), is regulated by the bifunctional enzyme PFK-2/FBPase-2. In its heart isozyme, the PFK-2 activity is stimulated by phosphorylation at its Ser 466. Among the enzymes that do so is **AMP-activated protein kinase** (AMPK; Sections 25-4Ba, 25-5, and 27-1). When oxygen deficiency prevents oxidative phosphorylation from providing sufficient ATP for heart function, as occurs in **ischemia** (insufficient blood flow), the resulting increased [AMP] activates AMPK. The consequent phosphorylation and hence activation of PFK-2 results in an increase of [F2,6P], thereby activating PFK-1 and thus glycolysis.

b. Aerobic ATP Production Is Far More Efficient than Anaerobic ATP Production

One reason for the decrease in glucose consumption on switching from anaerobic to aerobic metabolism is clear from an examination of the stoichiometries of anaerobic and aerobic breakdown of glucose ($C_6H_{12}O_6$).

Anaerobic glycolysis:

$$C_6H_{12}O_6 + 2ADP + 2P_i \longrightarrow$$
$$2 \text{ lactate} + 2H^+ + 2H_2O + 2ATP$$

Aerobic metabolism of glucose:

$$C_6H_{12}O_6 + 32ADP + 32P_i + 6O_2 \longrightarrow$$
$$6CO_2 + 38H_2O + 32ATP$$

(2.5 ATP for each of the 10 NADH generated per glucose oxidized, 1.5 ATP for each of the 2 $FADH_2$ generated, 2 ATP produced in glycolysis, and 2GTP $\rightleftharpoons$ 2ATP produced in the citric acid cycle.) Thus *aerobic metabolism is 16 times more efficient than anaerobic glycolysis in producing ATP.* The switch to aerobic metabolism therefore rapidly increases the ATP mass action ratio. As the ATP mass action ratio increases, the rate of electron transport decreases, which has the effect of increasing the [NADH]/[NAD$^+$] ratio. The increases in [ATP] and [NADH] inhibit their target enzymes in the citric acid cycle and in the glycolytic pathway. *The activity of PFK, which is citrate- and adenine nucleotide-regulated and one of the rate-controlling enzymes of glycolysis, decreases manyfold on switching from anaerobic to aerobic metabolism. This accounts for the dramatic decrease in glycolysis.*

c. Anaerobic Glycolysis Has Advantages as Well as Limitations

Animals can sustain anaerobic glycolysis for only short periods of time. This is because PFK, which cannot function effectively much below pH 7, is inhibited by the acidification arising from lactic acid production. Despite this limitation and the low efficiency of glycolytic ATP production, *the enzymes of glycolysis are present in such great concentrations that when they are not inhibited, ATP can be produced much more rapidly than through oxidative phosphorylation.*

The different characteristics of aerobic and anaerobic metabolism permit us to understand certain aspects of cancer cell metabolism and cardiovascular disease.

d. Cancer Cell Metabolism

As Warburg first noted in 1926, certain cancer cells produce more lactic acid under aerobic conditions than do normal cells. This is because the glycolytic pathway in these cells produces pyruvate more rapidly than the citric acid cycle can accommodate. How can this happen given the interlocking controls on the system? One explanation is that these controls have broken down in cancer cells. Another is that their ATP utilization occurs at rates too rapid to be replenished by oxidative phosphorylation. This would alter the ratios of adenine nucleotides so as to relieve the inhibition of PFK-1. In addition, many cancer cell lines have a much larger [F2,6P] than do normal cells. These cells contain an inducible isozyme of PFK-2/FBPase-2 that has an AMPK-phosphorylatable site for activating PFK-2. Consequently, an [AMP] increase in these cells results in an increase in their [F2,6P], which further activates PFK-1 and glycolysis. Efforts to understand the metabolic differences between cancer cells and normal cells may eventually lead to a treatment of certain forms of this devastating disease.

e. Cardiovascular Disease

Oxygen deprivation of certain tissues resulting from cardiovascular disease is of major medical concern. For example, two of the most common causes of human death, **myocardial infarction** (heart attack) and **stroke,** are caused by interruption of the blood (O_2) supply to a portion of the heart or the brain, respectively. It seems obvious why this should result in a cessation of cellular activity, but why does it cause cell death?

In the absence of O_2, a cell, which must then rely only on glycolysis for ATP production, rapidly depletes its stores of phosphocreatine (a source of rapid ATP production; Section 16-4Cd) and glycogen. As the rate of ATP production falls below the level required by membrane ion pumps for the maintenance of proper intracellular ionic concentrations, the osmotic balance of the system is disrupted, so that the cell and its membrane-enveloped organelles begin to swell. The resulting overstretched membranes become permeable, thereby leaking their enclosed contents. [In fact, a useful diagnostic criterion for myocardial infarction is the presence in the blood of heart-specific enzymes, such as the H-type isozyme of lactate dehydrogenase (vs the M-type isozyme, which predominates in skeletal muscle; Section 17-3A), which leak out of necrotic (dead) heart tissue.] Moreover, the decreased intracellular pH that accompanies anaerobic glycolysis (because of lactic acid production; Section 17-3A) permits the released lysosomal enzymes (which are active only at acidic pH's) to degrade the cell contents. Thus, the cessation of metabolic activity results in irreversible cell damage. Rapidly respiring tissues, such as those of heart and brain, are particularly susceptible to such damage.

f. IF$_1$ Inhibits F$_1$F$_0$–ATPase during Hypoxia

Under hypoxic conditions, the proton-motive force across the inner mitochondrial membrane is reduced to the point that F_1F_0–ATPase would switch from the synthesis of ATP to its hydrolysis, resulting in a catastrophic loss of ATP. This is prevented through the interaction of the F_1F_0–ATPase with an 84-residue regulatory protein named **IF$_1$.** Under normal physiological conditions, IF_1 forms inactive tetramers and higher order oligomers. However, when the pH drops below 6.5, which occurs under anaerobic conditions due to lactic acid production, IF_1 forms dimers in which its almost entirely α helical subunits associate via an antiparallel coiled coil involving its residues 48 to 84. The X-ray structure of F_1 in complex with AMPPNP and IF_1, determined by Leslie and Walker, reveals that each N-terminal segment of the IF_1 dimer has bound to the α_{DP}–β_{DP} interface (Fig. 22-38b) of a separate F_1. This traps AMPPNP and presumably ATP in the β_{DP} binding site, which would prevent it from hydrolyzing ATP (which, since AMPPNP rather than ADP is bound to β_{DP}, suggests that this structure is that of a prehydrolysis step in the catalytic reaction). When oxygen becomes available, the cell re-energizes and its pH increases, thereby causing IF_1 to dissociate from F_1F_0, which then commences synthesizing ATP.

g. Partial Oxygen Reduction Produces Reactive Oxygen Species (ROS)

Although the four-electron reduction of O_2 by cytochrome c oxidase normally goes to completion, the enzyme infrequently releases partially reduced reactive oxygen species (ROS) that readily react with a variety of cellular components. The best known ROS is the superoxide radical, $O_2^{\cdot-}$. It is also produced by the occasional leakage of electrons from Complexes I and III:

$$O_2 + e^- \longrightarrow O_2^{\cdot-}$$

Its production is enhanced under hypoxic conditions.

Superoxide radical is a precursor of other reactive species. Protonation of $O_2^{\cdot-}$ yields **HO$_2$·**, a much stronger oxidant than $O_2^{\cdot-}$. The most potent oxygen species in biological systems is probably the **hydroxyl radical,** which forms from the relatively harmless hydrogen peroxide (H_2O_2):

$$H_2O_2 + Fe^{2+} \longrightarrow \cdot OH + OH^- + Fe^{3+}$$

The hydroxyl radical also forms through the reaction of superoxide with H_2O_2:

$$O_2^{\cdot-} + H_2O_2 \longrightarrow O_2 + OH^- + \cdot OH$$

ROS readily extract electrons from other molecules, converting them to free radicals and thereby initiating a chain reaction.

The random nature of ROS attacks makes it difficult to characterize their reaction products, but all classes of biological molecules are susceptible to oxidative damage caused by free radicals. The oxidation of polyunsaturated lipids in cells may disrupt the structures of membranes, and oxidative damage to DNA may result in point mutations. Enzyme function may also be compromised through radical reactions with amino acid side chains. Because the mitochondrion is the site of the bulk of the cell's oxidative metabolism, its lipids, DNA, and proteins bear the brunt of free radical–related damage.

Several degenerative diseases, including Parkinson's, Alzheimer's, and Huntington's diseases, are associated with oxidative damage to mitochondria. Such observations have led to the free-radical theory of aging, which holds that *free-radical reactions arising during the course of normal oxidative metabolism are at least partially responsible for the aging process.* In fact, individuals with congenital defects in their mitochondrial DNA suffer from a variety of symptoms typical of old age, including neuromotor difficulties, deafness, and dementia. These genetic defects may increase the susceptibility of mitochondria to ROS-generated damage.

h. Cells Are Equipped with Antioxidant Mechanisms

Antioxidants eliminate oxidative free radicals such as $O_2^{\cdot-}$ and ·OH. In 1969, Irwin Fridovich discovered that the enzyme **superoxide dismutase (SOD),** which is present in nearly all cells, catalyzes the conversion of $O_2^{\bar{\;}}$ to H_2O_2:

$$2O_2^{\bar{\;}} + 2H^+ \longrightarrow H_2O_2 + O_2$$

Mitochondrial and bacterial SOD are both Mn^{2+}-containing tetramers; eukaryotic cytosolic SOD is a dimer that contains both Cu^{2+} and Zn^{2+} ions. Although the rate of nonenzymatic superoxide breakdown is $\sim 2 \times 10^5 \, M^{-1} \cdot s^{-1}$, that of the Cu,Zn-SOD–catalyzed reaction is $\sim 2 \times 10^9 \, M^{-1} \cdot s^{-1}$, close to the diffusion-controlled limit (Section 14-2Bb). This is apparently accomplished by electrostatic guidance of the negatively charged superoxide substrate into the enzyme's active site (Fig. 14-10).

H_2O_2 is degraded to water and oxygen by enzymes such as **catalase,** which catalyzes the reaction

$$2H_2O_2 \longrightarrow 2H_2O + O_2$$

and **glutathione peroxidase,** which uses glutathione (GSH; Section 21-2Ba) as a reducing agent:

$$2GSH + H_2O_2 \longrightarrow GSSG + 2H_2O$$

The latter enzyme also catalyzes the breakdown of organic hydroperoxides. Some types of glutathione peroxidase require Se for activity, which is one reason why Se appears to have antioxidant activity.

Other potential antioxidants are plant-derived compounds such as ascorbic acid (vitamin C; Section 11-1Cb) and **vitamin E,** a group of compounds whose most prominent member is **α-tocopherol.**

α-Tocopherol (vitamin E)

These compounds may help protect plants from oxidative damage during photosynthesis, a process in which H_2O is oxidized to O_2 (Section 24-2). However, clinical trials indicate that their use does not contribute to longevity in humans.

CHAPTER SUMMARY

1 The Mitochondrion Oxidative phosphorylation is the process through which the NADH and $FADH_2$ produced by nutrient oxidation are oxidized with the concomitant formation of ATP. The process takes place in the mitochondrion, an ellipsoidal organelle that is bounded by a permeable outer membrane and contains an impermeable and highly invaginated inner membrane that encloses the matrix. Enzymes of oxidative phosphorylation are embedded in the inner mitochondrial membrane. P_i is imported into the mitochondrion by a specific transport protein. Ca^{2+} import and Ca^{2+} export proteins operate to maintain a constant cytosolic $[Ca^{2+}]$. NADH's electrons are imported into the mitochondrion by shuttle systems such as the glycerophosphate shuttle and the malate–aspartate shuttle.

2 Electron Transport The standard free energy change for the oxidation of NADH by O_2 is $\Delta G^{\circ\prime} = -218 \, kJ \cdot mol^{-1}$, whereas that for the synthesis of ATP from ADP and P_i is $\Delta G^{\circ\prime} = 30.5 \, kJ \cdot mol^{-1}$. Consequently, the molar free energy of oxidation of NADH by O_2 is sufficient to power the synthesis of several moles of ATP under standard conditions. The electrons generated by oxidation of NADH and $FADH_2$ pass through four protein complexes, the electron-transport chain, with the coupled synthesis of ATP. Complexes I, III, and IV participate in the oxidation of NADH, producing ~ 2.5 ATPs per NADH, whereas $FADH_2$ oxidation, which involves Complexes II, III, and IV, produces only ~ 1.5 ATPs per $FADH_2$. Thus, the ratio of moles of ATP produced per mole of coenzyme oxidized by O_2, the P/O ratio, is ~ 2.5 for NADH oxidation and ~ 1.5 for $FADH_2$ oxidation. The route taken by electrons through the electron-transport chain was elucidated, in part, through the use of electron-transport inhibitors. Rotenone and amytal inhibit Complex I, antimycin inhibits Complex III, and CN^- inhibits Complex IV. Also involved were measurements of the reduction potentials of the electron-carrying prosthetic groups contained in the electron-transport complexes.

Complex I contains FMN and nine iron–sulfur clusters in a 45-subunit (in mammals) transmembrane protein complex. This L-shaped complex passes electrons from NADH to CoQ, a nonpolar small molecule that diffuses freely within the membrane. Complex II, which is also the citric acid cycle enzyme succinate dehydrogenase, also passes electrons to CoQ, in this case from succinate through FAD and three iron–sulfur clusters. The X-ray structure of Complex II indicates that its redox cofactors are arranged in a linear chain. $CoQH_2$ passes electrons to Complex III (cytochrome bc_1), a homodimeric complex whose protomers each contain two b-type hemes bound to a cytochrome b subunit, a Rieske iron–sulfur protein (ISP), and a cytochrome c_1. An electron from cytochrome c_1 of Complex III is passed to the Cu_A center of Complex IV (cytochrome c oxidase) via the peripheral membrane protein cytochrome c. This electron is then passed to cytochrome a, which, in turn, passes it to a binuclear center composed of heme a_3 and Cu_B, which reduces O_2 to H_2O. This process occurs in four 1-electron steps that pump four protons from the mitochondrial matrix/bacterial cytoplasm to the intermembrane space/periplasm. Complexes I, III, and IV form supercomplexes that increase the efficiency of electron transport.

3 Oxidative Phosphorylation The mechanism by which the free energy released by the electron-transport chain is stored and utilized in ATP synthesis is described by the chemiosmotic hypothesis. This hypothesis states that the free energy released by electron transport is conserved by the generation of an electrochemical proton gradient across the inner mitochondrial membrane (bacterial cell membrane; outside positive and acidic), which is harnessed to synthesize ATP. The proton gradient is created and maintained by the obligatory outward translocation of H^+ across the inner mitochondrial membrane as electrons travel through Complexes I, III, and IV.

Complex III pumps protons via a redox loop mechanism called the Q cycle, a bifurcated double cycle in which one molecule of $CoQH_2$ is oxidized to CoQ and then is rereduced to $CoQH_2$ by a second molecule of $CoQH_2$ in a process that collectively transfers four protons from the inside to the outside while oxidizing one molecule of $CoQH_2$ to CoQ. Electrons are transferred between the two CoQ's, which are bound at different sites, Q_o and Q_i, as well as between the $CoQH_2$ bound at Q_o and cytochrome c_1 via the ISP, which undergoes a conformational change in doing so. Complex IV contains no $(H^+ + e^-)$ carriers such as $CoQH_2$ and hence translocates proteins via a proton pump mechanism. Bacteriorhodopsin, the best characterized proton pump, translocates protons in a light driven process. This involves a trans to cis isomerization of bacteriorhodopsin's retinal prosthetic group on absorbing a photon, followed by the translocation of a proton through the hydrophilic central channel of this transmembrane protein via a process that involves conformational and pK changes of the polar groups lining the channel as the retinal relaxes to its ground state. Complex IV is thought to pump protons via a similar mechanism that is driven by the changes in the redox state of its heme a_3–Cu_B binuclear center as it reduces O_2 to H_2O.

The energy stored in the electrochemical proton gradient is utilized by proton-translocating ATP synthase (Complex V, F_1F_0–ATPase) in the synthesis of ATP via the binding change mechanism, by coupling this process to the exergonic transport of H^+ back to the inside. Mitochondrial proton-translocating ATP synthase consists of two oligomeric components: F_1 ($\alpha_3\beta_3\gamma\delta\varepsilon$), a peripheral membrane protein that appears as "lollipops" in electron micrographs of the inner mitochondrial membrane, and F_0 (ab_2c_{12} in *E. coli*), an integral membrane protein that contains the proton channel. The conformational changes that promote the synthesis of ATP from ADP + P_i arise through the demonstrated rotation of the γ subunit relative to the catalytic $\alpha_3\beta_3$ assembly that contains the enzyme's three active sites. The γ subunit is attached to a ring of c subunits in F_0, whose rotation is driven by the passage of protons between it and the a subunit.

Compounds such as 2,4-dinitrophenol are uncouplers of oxidative phosphorylation because they carry H^+ across the mitochondrial membrane, thereby dissipating the proton gradient and allowing electron transport to continue without concomitant ATP synthesis. Brown fat mitochondria contain a regulated uncoupling system that, under hormonal control, generates heat instead of ATP.

4 Control of ATP Production Under aerobic conditions, the rate of ATP synthesis by oxidative phosphorylation is regulated, in a phenomenon known as acceptor control, by the ATP mass action ratio. ATP synthesis is tightly coupled to the oxidation of NADH and $FADH_2$ by the electron-transport chain. Glycolysis and the citric acid cycle are coordinately controlled so as to produce NADH and $FADH_2$ only at a rate required to meet the system's demand for ATP. IF_1 inhibits the ATP hydrolysis by mitochondrial F_1F_0–ATPase that would otherwise occur under hypoxic conditions. Incomplete and side reactions of Complexes I, III, and IV produce damaging reactive oxygen species (ROS) that are largely eliminated through the actions of several cellular enzymes, most notably superoxide dismutase.

REFERENCES

Historical Overview

Ernster, L. and Schatz, G., Mitochondria: A historical review, *J. Cell Biol.* **91,** 227s–255s (1981).

Fruton, J.S., *Molecules and Life, pp.* 262–396, Wiley–Interscience (1972).

Krebs, H., *Otto Warburg. Cell Physiologist, Biochemist, and Eccentric,* Clarendon Press (1981). [A biography of one of the pioneers in the biochemical study of respiration, by a distinguished student.]

Prebble, J., Peter Mitchell and the ox phos wars, *Trends Biochem. Sci.* **27,** 209–212 (2002).

Racker, E., *A New Look at Mechanisms in Bioenergetics,* Academic Press (1976). [A fascinating personal account by one of the outstanding contributors to the field.]

General

Nicholls, D.G. and Ferguson, S.J., *Bioenergetics* (3rd ed.), Academic Press (2002). [An authoritative monograph devoted almost entirely to the mechanism of oxidative phosphorylation and the techniques used to elucidate it.]

Schäfer, G. and Penefsky, H. (Eds.), *Bioenergetics,* Springer (2008).

Schultz, B.E. and Chan, S.I., Structures and proton-pumping strategies of mitochondrial respiratory enzymes, *Annu. Rev. Biophys. Biomol. Struct.* **30,** 23–65 (2001).

Mitochondria

Goodsell, D.S., Mitochondrion, *Biochem. Mol. Biol. Educ.* **38,** 134–140 (2010). [An illustrated guide to the mitochondrion.]

Frey, T.G. and Mannella, C.A., The internal structure of mitochondria, *Trends Biochem. Sci.* **23,** 319–324 (2000).

Frey, T.G., Perkins, G.A., and Ellisman, M.H., Electron tomography of membrane-bound cellular organelles, *Annu. Rev. Biophys. Biomol. Struct.* **35,** 199–224 (2006).

Logan, D.C., The mitochondrial compartment, *J. Exp. Botany* **57,** 1225–1243 (2006).

Scheffler, I.E., *Mitochondria* (2nd ed.), Wiley-Liss (2008).

Electron Transport

Beinert, H., Holm, R.H., and Münck, E., Iron-sulfur clusters: Nature's modular, multipurpose structures, *Science* **277,** 653–659 (1997).

Belevich, I. and Verkhovsky, M.I., Molecular mechanism of proton translocation by cytochrome c oxidase, *Antioxidants Redox Signaling* **10,** 1–29 (2008). [A comprehensive review.]

Brandt, U., Energy converting NADH:quinone oxidoreductase (Complex I), *Annu. Rev. Biochem.* **75,** 69–92 (2006).

Collman, J.P., Rapta, M., Bröring, M., Raptova, L., Schwenninger, R., Boitrel, B., Fu, L., and L'Her, M., Close structural analogues of the cytochrome c oxidase Fe_{a3}/Cu_B center show clean $4e^-$ electroreduction of O_2 to H_2O at physiological pH, *J. Am. Chem. Soc.* **121,** 1387–1388 (1999).

Crofts, A.R., The cytochrome bc_1 complex: Function in the context of structure, *Annu. Rev. Physiol.* **66,** 689–733 (2004).

Efremov, R.G., Baradaran, R., and Sazanov, L.A., The architecture of respiratory complex I, *Nature* **465,** 441–445 (2010).

Hinkle, P.C., P/O ratios of mitochondrial oxidative phosporylation, *Biochim. Biophys. Acta* **1706,** 1–11 (2005).

Hosler, J.P., Ferguson-Miller, S., and Mills, D.A., Energy transduction: Proton transfer through the respiratory complexes, *Annu. Rev.*

Biochem. **75,** 165–187 (2006). [A review that focuses on cytochrome *c* oxidase.]

Huang, L., Sun, G., Cobessi, D., Wang, A.C., Shen, J.T., Tung, E.Y., Anderson, V.E., and Berry, E.A., 3-Nitropropionic acid is a suicide inhibitor of mitochondrial respiration that, upon oxidation by Complex II, forms a covalent adduct with a catalytic base arginine in the active site of the enzyme, *J. Biol. Chem.* **281,** 5965–5972 (2006). [X-ray structure of chicken Complex II.]

Hunte, C., Koepke, J., Lange, C., Rossmanith, T., and Michel, H., Structure at 2.3 Å resolution of the cytochrome bc_1 complex from the yeast *Saccharomyces cerevisiae* co-crystallized with an antibody Fv fragment, *Structure* **8,** 669–684 (2000).

Iwata, S., Ostermeier, C., Ludwig, B., and Michel, H., Structure at 2.8 Å resolution of cytochrome *c* oxidase from *Paracoccus denitrificans, Nature* **376,** 660–669 (1995).

Iwata, S., Lee, J.W., Okada, K., Lee, J.K., Iwata, M., Rasmussen, B., Link, T.A., Ramaswamy, S., and Jap, B.K., Complete structure of the 11-subunit bovine mitochondrial cytochrome bc_1 complex, *Science* **281,** 64–71 (1998).

Johnson, D.C., Dean, D.R., Smith, A.D., and Johnson, M.K., Structure, function, and formation of biological iron–sulfur clusters, *Annu. Rev. Biochem.* **74,** 247–281 (2005).

Lenaz, G., Fato, R., Genova, M.L., Bergamini, C., Bianchi, C., and Biondi, A., Mitochondrial complex I: Structural and functional aspects, *Biochim. Biophys. Acta* **1757,** 1406–1420 (2006).

Michel, H., Behr, J., Harrenga, A., and Kannt, A., Cytochrome *c* oxidase: Structure and spectroscopy, *Annu. Rev. Biophys. Biomol. Struct.* **27,** 329–356 (1998).

Moser, C.C., Keske, J.M., Warncke, K., Farid, R.S., and Dutton, L.S., Nature of biological electron transfer, *Nature* **355,** 796–802 (1992).

Osyczka, A., Moser, C.C., and Dutton, P.L., Fixing the Q cycle, *Trends Biochem. Sci.* **30,** 176–182 (2005).

Radermacher, M., Ruiz, T., Clason, T., Benjamin, S., Brandt, U., and Zickerman, V., The three dimensional structure of complex I from *Yarrowia lipolytica:* A highly dynamic enzyme, *J. Struct. Biol.* **154,** 269–279 (2006).

Sazanov, L.A. and Hinchliffe, P., Structure of the hydrophilic domain of respiratory Complex I from *Thermus thermophilus, Science* **311,** 1430–1436 (2006); *and* Sazanov, L.A., Respiratory Complex I: Mechanistic and structural insights provided by the crystal structure of the hydrophilic domain, *Biochemistry* **46,** 2275–2288 (2007).

Schäfer, E., Dencher, N.A., Vonck, J., and Parcej, D.N., Three dimensional structure of the respiratory supercomplex $I_1III_2IV_1$ from bovine heart mitochondria, *Biochemistry* **46,** 12579–12585 (2007); *and* Vonck, J., and Schäfer, E., Supramolecular organization of protein complexes in the mitochondrial inner membrane, *Biochim. Biophys. Acta* **1793,** 117–124 (2009).

Solmaz, S.R.N. and Hunte, C., Structure of Complex III with bound cytochrome *c* in reduced state and definition of a minimal core interface for electron transfer, *J. Biol. Chem.* **283,** 17542–17549 (2008); *and* Lange, C. and Hunte, C., Crystal structure of the yeast cytochrome bc_1 complex with its bound substrate cytochrome *c, Proc. Natl. Acad. Sci.* **99,** 2800–2805 (2002).

Sun, F., Huo, X., Zhai, Y., Wang, A., Xu, J., Su, D., Bartlam, M., and Rao, Z., Crystal structure of mitochondrial respiratory protein Complex II, *Cell* **121,** 1043–1057 (2005).

Tsukihara, T., Aoyama, H., Yamashita, E., Tomizaki, T., Yamaguchi, H., Shinzawa-Itoh, K., Nakashima, R., Yaono, R., and Yoshikawa, S., The whole structure of the 13-subunit oxidized cytochrome *c* oxidase at 2.8 Å, *Science* **272,** 1136–1144 (1996).

Xia, D., Yu, C.-A., Kim, H., Xia, J.-Z., Kachurin, A.M., Zhang, L., Yu,

L., and Deisenhofer, J., Crystal structure of the cytochrome bc_1 complex from heart mitochondria, *Science* **277,** 60–66 (1997).

Yankovskaya, V., Horsefield, R., Törnroth, S., Luna-Chavez, C., Miyoshi, H., Légar, C., Byrne, B., Cecchini, G., and Iwata, S., Architecture of succinate dehydrogenase and reactive oxygen species generation, *Science* **299,** 700–704 (2003).

Yoshikawa, S., et al., Redox-coupled crystal structural changes in bovine heart cytochrome *c* oxidase, *Science* **280,** 1723–1729 (1998); *and* Yoshikawa, S., Beef heart cytochrome *c* oxidase, *Curr. Opin. Struct. Biol.* **7,** 574–579 (1997).

Zhang, Z., Huang, L., Shulmeister, V.M., Chi, Y.-I., Kim, K.K., Huang, L.-W., Crofts, A.R., Berry, E.A., and Kim, S.H., Electron transfer by domain movement in cytochrome bc_1, *Nature* **392,** 677–684 (1998).

Bacteriorhodopsin

Heberle, J., Proton transfer reactions across bacteriorhodopsin and along the membrane, *Biochim. Biophys. Acta* **1458,** 135–147 (2000).

Kühlbrandt, W., Bacteriorhodopsin—the movie, *Nature* **406,** 569–570 (2000).

Lanyi, J.K., Bacteriorhodopsin, *Annu. Rev. Physiol.* **66,** 665–688 (2004).

Oxidative Phosphorylation

Abrahams, J.P., Leslie, A.G.W., Lutter, R., and Walker, J.E., Structure at 2.8 Å resolution of F_1-ATPase from bovine heart mitochondria, *Nature* **370,** 621–628 (1994).

Boyer, P.D., The binding change mechanism for ATP synthase—some probabilities and possibilities, *Biochim. Biophys. Acta* **1140,** 215–250 (1993).

Boyer, P.D., The ATP synthase—a splendid molecular machine, *Annu. Rev. Biochem.* **66,** 717–749 (1997).

Capaldi, R. and Aggeler, R., Mechanism of F_1F_0-type ATP synthase, a biological rotary motor, *Trends Biochem. Sci.* **27,** 154–160 (2002).

Dickson, V.K., Silvester, J.A., Fearnley, I.M., Leslie, A.G.W., and Walker, J.E., On the structure of the stator of the mitochondrial ATP synthase, *EMBO J.* **25,** 2911–2918 (2006).

Gibbons, C., Montgomery, M.G., Leslie, A.G.W., and Walker, J.E., The structure of the central stalk in bovine F_1-ATPase at 2.4 Å resolution, *Nature Struct. Biol.* **7,** 1055–1061 (2000).

Hausrath, A.C., Capaldi, R.A., and Matthews, B.M., The conformation of the ε- and γ-subunits of *Escherichia coli* F_1 ATPase, *J. Biol. Chem.* **276,** 47227–47232 (2001). [The X-ray structure of the *E. coli* F_1.]

Itoh, H., Takahashi, A., Adachi, K., Noji, H., Yasuda, R., Yoshida, M., and Kinosita, K., Jr., Mechanically driven ATP synthesis by F_1-ATPase, *Nature* **427,** 465–468 (2004).

Klingenberg, M., Mechanism and evolution of the uncoupling protein of brown adipose tissue, *Trends Biochem. Sci.* **15,** 108–112 (1990).

Ma, J., Flynn, T.C., Cui, Q., Leslie, A.G.W., Walker, J.E., and Karplus, M., A dynamic analysis of the rotation mechanism for the conformational change in F_1-ATPase, *Structure* **10,** 921–931 (2002).

Meier, T., Polzer, P., Diederichs, K., Welte, W., and Dimroth, P., Structure of the rotor ring of F-type Na^+-ATPase from *Ilyobacter tartaricus, Science* **308,** 659–662 (2005).

Mitchell, P., Vectorial chemistry and the molecular mechanics of chemiosmotic coupling: Power transmission by proticity, *Biochem. Soc. Trans.* **4,** 398–430 (1976).

Nicholls, D.G. and Rial, E., Brown fat mitochondria, *Trends Biochem. Sci.* **9,** 489–491 (1984).

Noji, H. and Yoshida, M., The rotary engine in cell ATP synthase, *J. Biol. Chem.* **276**, 1665–1668 (2001).

Rastogi, V.K. and Girvin, M.E., Structural changes linked to proton translocation by subunit c of the ATP synthase, *Nature* **402**, 262–268 (1999); *and* Girvin, M.E., Rastogi, V.K., Abildgaard, F., Markley, J.L., and Fillingame, R.H., Solution structure of the transmembrane H^+-transporting subunit c of the ATP synthase, *Biochemistry* **37**, 8817–8824 (1998).

Rubinstein, J.L., Walker, J.E., and Henderson, R., Structure of the mitochondrial ATP synthase by electron cryomicroscopy, *EMBO J.* **22**, 6182–6192 (2003).

Sambongi, Y., Iko, Y., Tanabe, M., Omote, H., Iwamoto-Kihara, A., Ueda, I., Yanagida, T., Wada, Y., and Futai, M., Mechanical rotation of the c subunit oligomer in ATP synthase (F_0F_1): Direct observation, *Science* **286**, 1722–1724 (1999).

Stock, D., Gibbons, C., Arechaga, I., Leslie, A.G.W., and Walker, J.E., The rotary mechanism of ATP synthase, *Curr. Opin. Struct. Biol.* **10**, 672–679 (2000).

Verkhovsky, M.I., Jasaitis, A., Verkhovskaya, M.L., Morgan, J.E., and Wikström, M., Proton translocation by cytochrome c oxidase, *Nature* **400**, 480–483 (1999).

von Ballamoos, C., Wiedenmann, A., and Dimroth, P., Essentials for ATP synthesis by F_1F_0 ATP synthases, *Annu. Rev. Biochem.* **78**, 649–672 (2009); *and* von Ballamoos, C., Cook, G.M., and Dimroth, P., Unique rotary ATP synthase and its biological diversity, *Annu. Rev. Biophys.* **37**, 43–64 (2008).

Wilkens, S., Rotary molecular motors, *Adv. Prot. Chem.* **71**, 345–382 (2005).

Yasuda, R., Noji, H., Kinosita, K., Jr., and Yoshida, M., F_1-ATPase is a highly efficient molecular motor that rotates with discrete 120° steps, *Cell* **93**, 1117–1124 (1998).

Yoshida, M., Muneyuki, E., and Hisabori, T., ATP synthase—a marvelous rotary engine of the cell, *Nature Rev. Mol. Cell Biol.* **2**, 669–677 (2001); *and* Noji, H. and Yoshida, M., The rotary machine in the cell ATP synthase, *J. Biol. Chem.* **276**, 1665–1668 (2001).

Control of ATP Production

Brown, G.C., Control of respiration and ATP synthesis in mammalian mitochondria and cells, *Biochem. J.* **284**, 1–13 (1992).

Cabezón, E., Montgomery, M.G., Leslie, A.G.W., and Walker, J.E., The structure of bovine F_1-ATPase in complex with its regulatory protein IF_1, *Nature Struct. Biol.* **10**, 744–750 (2003).

Celi, F.S., Brown adipose tissue—when it pays to be inefficient, *New Engl. J. Med.* **360** 1553–1556 (2009).

Chesney J., Mitchell, R., Benigni, F., Bacher, M., Spiegel, L., Al-Abed, Y., Han, J.H., Metz, C., and Bucala, R., An inducible gene product for 6-phosphofructo-2-kinase with an AU-rich instability element: role in tumor cell glycolysis and the Warburg effect, *Proc. Natl. Acad. Sci.* **96**, 3047–3052 (2000).

Harris, D.A. and Das, A.M., Control of mitochondrial ATP synthesis in the heart, *Biochem. J.* **280**, 561–573 (1991).

Jeneson, J.A.L., Westerhoff, H.V., and Kushmerick, M.J., A metabolic control analysis of kinetic controls in ATP free energy metabolism in contracting skeletal muscle, *Am. J. Physiol. Cell Physiol.* **279**, C813–C832 (2000).

Marsin, A-S., Bertrand, L., Rider, M.H., Deprez, J., Beauloye, C., Vincent, M.F., Van den Berghe, G., Carling, D., and Hue, L., Phosphorylation and activation of heart PFK-2 by AMPK has a role in the stimulation of glycolysis during ischaemia, *Curr. Biol.* **10**, 1247–1255 (2000).

Marsin, A-S., Bouzin, C., Bertrand, L., and Hue, L., The stimulation of glycolysis by hypoxia in activated monocytes is mediated by AMP-activated protein kinase and inducible 6-phosphofructo-3-kinase, *J. Biol. Chem.* **277**, 30778–30783 (2002).

Randle, P. J., Regulatory interactions between lipids and carbohydrates: The glucose fatty acid cycle after 35 years. *Diabetes/Metab. Rev.* **14**, 263–283 (1998).

Ricquier, D. and Bouillaud, F., The mitochondrial uncoupling protein: Structural and genetic studies, *Prog. Nucleic Acid Res. Mol. Biol.* **56**, 83–108 (1997).

PROBLEMS

1. Rank the following redox-active coenzymes and prosthetic groups of the electron-transport chain in order of increasing affinity for electrons: cytochrome *a*, CoQ, FAD, cytochrome *c*, NAD^+.

2. Why is the oxidation of succinate to fumarate only associated with the production of two ATPs during oxidative phosphorylation, whereas the oxidation of malate to oxaloacetate is associated with the production of three ATPs?

3. What is the thermodynamic efficiency of oxidizing $FADH_2$ so as to synthesize two ATPs under standard biochemical conditions?

4. Sublethal cyanide poisoning may be reversed by the administration of nitrites. These substances oxidize hemoglobin, which has a relatively low affinity for CN^-, to methemoglobin, which has a relatively high affinity for CN^-. Why is this treatment effective?

5. Match the compound with its behavior: (1) rotenone, (2) dinitrophenol, and (3) antimycin. (a) Inhibits oxidative phosphorylation when the substrate is pyruvate but not when the substrate is succinate. (b) Inhibits oxidative phosphorylation when the substrate is either pyruvate or succinate. (c) Allows pyruvate to be oxidized by mitochondria even in the absence of ADP.

6. Nigericin is an ionophore (Section 20-2C) that exchanges K^+ for H^+ across membranes. Explain how the treatment of functioning mitochondria with nigericin uncouples electron transport from oxidative phosphorylation. Does valinomycin, an ionophore that transports K^+ but not H^+, do the same? Explain.

7. Why is it possible for electrons in an electron-transfer complex to flow from a redox center to one with a lesser value of $\mathscr{E}°'$?

8. How do the P/O ratios for NADH differ in ATP synthases that contain 10 and 15 *c* subunits?

9. The difference in pH between the internal and external surfaces of the inner mitochondrial membrane is 1.4 pH units (external side acidic). If the membrane potential is 0.06 V (inside negative), what is the free energy released on transporting 1 mol of protons back across the membrane? How many protons must be transported to provide enough free energy for the synthesis of 1 mol of ATP (assume standard biochemical conditions)?

*10. (a) A simplistic interpretation of the Q cycle would predict that the proton pumping efficiency of cytochrome bc_1 would be reduced by no more than 50% in the presence of saturating amounts of antimycin. Explain. (b) Indicate why cytochrome bc_1 is nearly 100% inhibited by antimycin.

11. The antibiotic oligomycin B

Oligomycin B

binds to the F_0 subunit of the mitochondrial F_1F_0–ATPase and thereby prevents it from synthesizing ATP [note that oligomycin-sensitivity conferral protein (OSCP), the mitochondrial counterpart of the *E. coli* δ subunit (Fig. 22-41), does not bind oligomycin B.] Explain why: (a) Submitochondrial particles from which F_1 has been removed are permeable to protons. (b) Addition of oligomycin B to F_1-depleted submitochondrial particles decreases this permeability severalfold.

12. Oligomycin B (see Problem 11) and cyanide both inhibit oxidative phosphorylation when the substrate is either pyruvate or succinate. Dinitrophenol can be used to distinguish between these inhibitors. Explain.

13. The *E. coli* F_1F_0–ATPase cannot synthesize ATP when Met 23 of its γ subunit is mutated to Lys. Yet the F_1 component of this complex still exhibits rotation of its γ subunit relative to its $α_3β_3$ spheroid when it is supplied with ATP. Suggest a reason for these effects.

14. For the oxidation of a given amount of glucose, does non-shivering thermogenesis by brown fat or shivering thermogenesis by muscle produce more heat?

15. What is the advantage of hormones activating a lipase to stimulate nonshivering thermogenesis in brown fat rather than activating UCP1 directly?

16. How does atractyloside affect mitochondrial respiration? (*Hint:* See Section 20-4C.)

17. Certain unscrupulous operators offer, for a fee, to freeze recently deceased individuals in liquid nitrogen until medical science can cure the disease from which they died. What is the biochemical fallacy of this procedure?

Other Pathways of Carbohydrate Metabolism

trans Golgi

Polypeptide

CHAPTER 23

Heretofore, we have dealt with many aspects of carbohydrate metabolism. We have seen how the free energy of glucose oxidation is sequestered in ATP through glycolysis, the citric acid cycle, and oxidative phosphorylation. We have also studied the mechanism by which glucose is stored as glycogen for future use and how glycogen metabolism is controlled in response to the needs of the organism. In this chapter, we examine several other carbohydrate metabolism pathways of importance:

1. Gluconeogenesis, through which noncarbohydrate precursors such as lactate, pyruvate, glycerol, and amino acids are converted to glucose.

2. The **glyoxylate cycle,** through which plants convert acetyl-CoA to glucose.

3. Oligosaccharide and glycoprotein biosynthesis, through which oligosaccharides are synthesized and added to specific amino acid residues of proteins.

4. The **pentose phosphate pathway,** an alternate pathway of glucose degradation, which generates **NADPH,** the source of reducing equivalents in reductive biosynthesis, and **ribose-5-phosphate,** the sugar precursor of the nucleic acids.

This chapter completes our study of carbohydrate metabolism in animals; photosynthesis, which occurs only in plants and certain bacteria, is the subject of Chapter 24.

1 GLUCONEOGENESIS

Glucose occupies a central role in metabolism, both as a fuel and as a precursor of essential structural carbohydrates and other biomolecules. The brain and red blood cells are almost completely dependent on glucose as an energy source. Yet the liver's capacity to store glycogen is only sufficient to supply the brain with glucose for about half a day under fasting or starvation conditions. Thus, *when fasting, most of the body's glucose needs must be met by gluconeogenesis (literally, new glucose synthesis), the biosynthesis of glucose from noncarbohydrate precursors.* Indeed, isotopic labeling studies determining the source of glucose in the blood during a fast showed that gluconeogenesis is responsible for 64% of total glucose production over the first 22 hours of the fast and accounts for almost all the glucose production by 46 hours. Thus, gluconeogenesis provides a substantial fraction of the glucose produced in fasting humans, even after a few hours' fast. Gluconeogenesis occurs in liver and, to a smaller extent, in kidney.

The noncarbohydrate precursors that can be converted to glucose include the glycolysis products lactate and pyruvate, citric acid cycle intermediates, and the carbon skeletons of most amino acids. First, however, all these substances must be converted to oxaloacetate, the starting material for gluconeogenesis (Fig. 23-1). The only amino acids that cannot be converted to oxaloacetate in animals are leucine and lysine because their breakdown yields only acetyl-CoA (Section 26-3F). *There is no pathway in animals for the net conversion of acetyl-CoA to oxaloacetate.* Likewise, fatty acids cannot serve as glucose precursors in animals because most fatty acids are degraded completely to acetyl-CoA (Section 25-2C). Unlike animals, however, plants do contain a pathway for the conversion of acetyl-CoA to oxaloacetate, the **glyoxylate cycle** (Section 23-2), so that fatty acids can serve as a plant cell's only carbon source. Glycerol, a triacylglycerol breakdown product, is converted to glucose via synthesis of the glycolytic intermediate dihydroxyacetone phosphate, as described in Section 25-1.

Figure 23-1 **Pathways converting lactate, pyruvate, and citric acid cycle intermediates to oxaloacetate.** The carbon skeletons of all amino acids but leucine and lysine may be, at least in part, converted to oxaloacetate and thus to glucose by these reactions.

A. *The Gluconeogenesis Pathway*

Gluconeogenesis utilizes glycolytic enzymes. Yet three of these enzymes, hexokinase, phosphofructokinase (PFK), and pyruvate kinase, catalyze reactions with large negative free energy changes in the direction of glycolysis. These reactions must therefore be replaced in gluconeogenesis by reactions that make glucose synthesis thermodynamically favorable. Here, as in glycogen metabolism (Section 18-1D), we see the recurrent theme that *biosynthetic and degradative pathways differ in at least one reaction. This not only permits both directions to be thermodynamically favorable under the same physiological conditions but allows the pathways to be independently controlled so that one direction can be activated while the other is inhibited.*

a. Pyruvate Is Converted to Oxaloacetate before Conversion to Phosphoenolpyruvate

The formation of phosphoenolpyruvate (PEP) from pyruvate, the reverse of the pyruvate kinase reaction, is endergonic and therefore requires free energy input. This is accomplished by first converting the pyruvate to oxaloacetate. Oxaloacetate is a "high-energy" intermediate whose exergonic decarboxylation provides the free energy necessary for PEP synthesis. The process requires the participation of two enzymes (Fig. 23-2):

1. Pyruvate carboxylase catalyzes the ATP-driven formation of oxaloacetate from pyruvate and HCO_3^-.

2. PEP carboxykinase (PEPCK) converts oxaloacetate

Figure 23-2 **Conversion of pyruvate to oxaloacetate and then to phosphoenolpyruvate.** The enzymes involved are **(1)** pyruvate carboxylase and **(2)** PEP carboxykinase (PEPCK).

to PEP in a reaction that uses GTP as a phosphorylating agent.

b. Pyruvate Carboxylase Has a Biotin Prosthetic Group

Pyruvate carboxylase, discovered in 1959 by Merton Utter, is a tetrameric protein of identical ~130-kD subunits, each of which has a **biotin** prosthetic group. *Biotin* (Fig. 23-3a) *functions as a CO$_2$ carrier by acquiring a carboxyl substituent at its **ureido group*** (Fig. 23-3b). Biotin is covalently bound to the enzyme by an amide linkage between the carboxyl group of its valerate side chain and the ε-amino group of an enzyme Lys residue to form a **biocytin**

Figure 23-3 **Biotin and carboxybiotinyl–enzyme.** (a) Biotin consists of an imidazoline ring that is cis-fused to a tetrahydrothiophene ring bearing a valerate side chain. The chirality at each of its three asymmetric centers is indicated. Positions 1, 2, and 3 constitute a ureido group. (b) In carboxybiotinyl–enzyme, N1 of the biotin ureido group is the carboxylation site. Biotin is covalently attached to carboxylases by an amide linkage between its valeryl carboxyl group and the ε-amino group of an enzyme Lys side chain to form biocytin.

(alternatively, **biotinyllysine**) residue (Fig. 23-3b). The biotin ring system is therefore at the end of a 16-Å-long flexible arm, much like that of the lipoic acid prosthetic group in the pyruvate dehydrogenase multienzyme complex (Section 21-2Ac).

Biotin, which was first identified in 1935 as a growth factor in yeast, is an essential human nutrient. Its nutritional deficiency is rare, however, because it occurs in many foods and is synthesized by intestinal bacteria. Human biotin deficiency almost always results from the consumption of large quantities of raw eggs. This is because egg whites contain a protein, **avidin,** that binds biotin so tightly (dissociation constant, $K = 10^{-15}$ M) as to prevent its intestinal absorption (cooked eggs do not cause this problem because cooking denatures avidin). The presence of avidin in eggs is thought to inhibit the growth of microorganisms in this highly nutritious environment. The avidin homolog **streptavidin,** which is secreted by *Streptomyces avidinii,* is used as a linking agent in numerous biotechnological applications (e.g., Section 22-3Ce) because of its particularly high affinity for biotin.

c. The Pyruvate Carboxylase Reaction

The pyruvate carboxylase reaction occurs in two phases (Fig. 23-4):

Phase I Biotin is carboxylated at its N1 atom by bicarbonate ion in a three-step reaction in which the hydrolysis of ATP to ADP + P$_i$ functions, via the intermediate formation of **carboxyphosphate,** to dehydrate bicarbonate. This yields free CO$_2$, which has sufficient free energy to carboxylate biotin. The resulting carboxyl group is activated relative to bicarbonate ($\Delta G^{o\prime}$ for its cleavage is -19.7 kJ · mol^{-1}) and can therefore be transferred without further free energy input.

Phase II The activated carboxyl group is transferred from carboxybiotin to pyruvate in a three-step reaction to form oxaloacetate.

These two reaction phases occur on different active sites of the same enzyme.

d. Acetyl-CoA Regulates Pyruvate Carboxylase

Oxaloacetate synthesis is an anaplerotic (filling up) reaction that increases citric acid cycle activity (Section 21-5b). Accumulation of the citric acid cycle substrate acetyl-CoA is therefore indicative of the need for more oxaloacetate.

Phase I

Phase II

Figure 23-4 **Two-phase reaction mechanism of pyruvate carboxylase. Phase I** is a three-step reaction in which carboxyphosphate is formed from bicarbonate and ATP, followed by the generation of CO_2 on the enzyme, which then carboxylates biotin. **Phase II** is a three-step reaction in which CO_2 is produced at the active site via the elimination of the biotinyl enzyme, which accepts a proton from pyruvate to generate pyruvate enolate. This, in turn, nucleophilically attacks the CO_2, yielding oxaloacetate. [After Knowles, J.R., *Annu. Rev. Biochem.* **58,** 217 (1989).]

Indeed, acetyl-CoA is a powerful allosteric activator of pyruvate carboxylase; the enzyme is all but inactive without bound acetyl-CoA. *If, however, the citric acid cycle is inhibited (by ATP and NADH, whose presence in high concentrations indicates a satisfied demand for oxidative phosphorylation; Section 21-4), oxaloacetate instead undergoes gluconeogenesis.*

e. X-Ray Structure of Pyruvate Carboxylase Suggests a Mechanism for Carboxyl Group Transfer

The X-ray structure of pyruvate carboxylase from the soil bacterium *Rhizobium etli* in complex with ATPγS and **ethyl-CoA** (acetyl-CoA with an ethyl group in place of its acetyl group), determined by Ivan Rayment, reveals that each subunit of this homotetrameric, 1154-residue protein consists of four domains (Fig. 23-5a): a biotin carboxylation (BC) domain (residues 1–465) that carries out Phase I of

the pyruvate carboxylase reaction (Fig. 23-4); an allosteric domain (residues 466–510 and 1001–1073) that binds acetyl-CoA; a carboxyltransferase (CT) domain (residues 511–1000) that catalyzes Phase II of the pyruvate carboxylase reaction (Fig. 23-4); and a domain named the biotin carboxyl carrier protein (BCCP; residues 1074–1154) to which the enzyme's biotin prosthetic group is covalently linked via Lys 1119 (Fig. 23-3). The active site of the BC domain is marked by its bound ATPγS, the CT domain is mainly an α/β barrel (viewed from the side in Fig. 23-5a) whose active site is located in the mouth of the barrel and is marked by its bound Zn^{2+} ion, and the allosteric domain binds the ethyl-CoA, which like acetyl-CoA, activates the enzyme.

The biotin prosthetic group on the BCCP domain is disordered and hence not visible. Nevertheless it is clear that the ~80-Å distance between a subunit's active sites is too

(a)

Biotin Carboxylation (BC) Domain

Carboxyltransferase (CT) Domain

(b)

(c)

Figure 23-5 **X-ray structure of *R. etli* pyruvate carboxylase.**
(*a*) Ribbon diagram in which the BC and CT domains are each
colored in rainbow order from their N-termini (*blue*) to their
C-termini (*red*), the allosteric domain is light green, and the
BCCP domain is pink. The ATPγS bound at the active site of the
BC domain and the ethyl-CoA bound to the allosteric domain
(with only its nucleotide portion visible) are drawn in space-
filling form with ATPγS C green, ethyl-CoA C cyan, N blue, O
red, and P orange. The Zn^{2+} ion bound at the CT domain active
site is represented by a purple sphere. (*b*) Surface representation
of the tetramer viewed along its 2-fold axis with the two active
subunits closest to the viewer. The BC domain is purple, the
allosteric domain is light green, the CT domain is yellow, and the
BCCP domain is red. For clarity, one of the subunits is outlined
in black. The distance between the ATPγS in the BC active site
and the Zn^{2+} ion in the CT active site is 65 Å. (*c*) View relative to
Part *b* by a 180° rotation about the vertical axis. The top pair of
subunits have undergone a conformational change relative to
the top pair in Part *b* such that the ATPγS–Zn^{2+} distance
between neighboring subunits is 80 Å. In addition, the BCCP
domains in the top pair here are disordered. (*d*) Model of the
tetramer indicating how the BCCP domain transfers a carboxyl
group between the BC domain on the same subunit and the CT

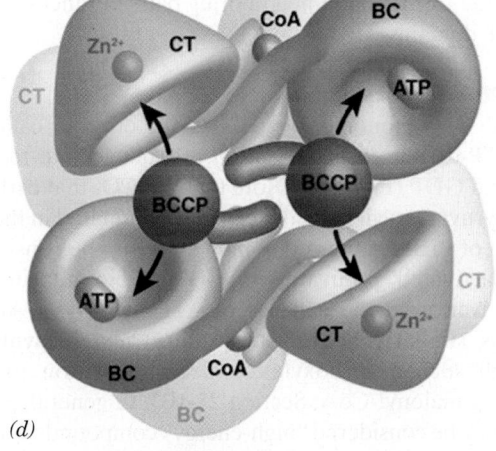

(d)

domain of its neighboring subunit. The view and domain colors
are the same as in Part *b*. [Part *a* based on an X-ray structure by
and Parts *b*, *c*, and *d* courtesy of Ivan Rayment, University of
Wisconsin, Madison, Wisconsin. PDBid 2QF7.]

large to be bridged by the 16-Å-long carboxybiotinyl arm (Fig. 23-3*b*). However, the BCCP domain is attached to the enzyme by a flexible polypeptide linker that is 34 Å long, much like that linking the lipoyl domain(s) to each dihydrolipoyl transacetylase (E2) subunit of the pyruvate dehydrogenase complex (Section 21-2Ae). Even so, it would require a dramatic movement of the entire BCCP domain to transfer substrate between the two active sites of a single subunit. How, then, does a BCCP domain translocate its carboxybiotin group between the active sites of a BC domain and a CT domain?

Pyruvate carboxylase's homotetrameric structure is required for its enzymatic activity; isolated subunits are catalytically inactive. However, the tetramer has only 2-fold symmetry because the top pair of subunits in Fig. 23-5*b* differ in conformation from the top pair in Fig. 23-5*c* by a 40° rotation and a 40-Å translocation of the BC domain relative to the CT domain of the same subunit. Indeed, the BCCP domains in the top pair of subunits in Fig. 23-5*c* are disordered, probably because the allosteric domains on these subunits do not bind ethyl-CoA. Consequently, the distance between active sites from adjacent subunits is 65 Å for the top pair in Fig. 23-5*b*, whereas it is 80 Å for the top pair in Fig. 23-5*c*. This suggests the model drawn in Fig. 23-5*d* in which each BCCP domain at the top of Fig. 23-5*b* shuttles CO_2 in the form of carboxybiotin from the active site of the BC domain on the same subunit to the CT domain on the adjacent subunit, whereas the other two subunits are inactive. This is an unusual example of allosteric activation coupled with negative cooperativity. It may permit pyruvate carboxylase to carry out efficient catalysis in association with other metabolic enzymes.

The foregoing model is supported by experiments involving two mutant forms of pyruvate carboxylase: K1119Q, which eliminates the biotinylation of the BC domain; and K718Q, which impairs the Phase II reaction. Tetramers of each of these mutant subunits exhibited 0.1% and 4% of the wild-type enzymatic activity, respectively. However, mixed tetramers exhibit 20% activity, thus indicating the formation of neighboring pairs of functional BC and CT domains.

f. PEP Carboxykinase

PEPCK, a monomeric ~630-residue enzyme, catalyzes the GTP-driven decarboxylation of oxaloacetate to form PEP and GDP (Fig. 23-6). Note that the CO_2 that carboxylates pyruvate to yield oxaloacetate is eliminated in the formation of PEP. Oxaloacetate may therefore be considered to be "activated" pyruvate, with CO_2 and biotin facilitating the activation at the expense of ATP hydrolysis. Acetyl-CoA is similarly activated for fatty acid biosynthesis through such a carboxylation decarboxylation process (forming malonyl-CoA; Section 25-4B). In general, β-keto acids may be considered "high-energy" compounds because of the high free energy of decarboxylation of the β-carboxyl group. The enolates they generate are used to form carbon–carbon bonds in fatty acid biosynthesis or phosphoenolpyruvate here in gluconeogenesis.

Figure 23-6 **The PEPCK mechanism.** Decarboxylation of oxaloacetate (a β-keto acid) forms a resonance-stabilized enolate anion whose oxygen atom attacks the γ phosphoryl group of GTP forming PEP and GDP.

g. Gluconeogenesis Requires Metabolite Transport between Mitochondria and Cytosol

The generation of oxaloacetate from pyruvate or citric acid cycle intermediates occurs only in the mitochondrion, whereas the enzymes that convert PEP to glucose are cytosolic. The cellular location of PEPCK varies with the species. In mouse and rat liver it is located almost exclusively in the cytosol, in pigeon and rabbit liver it is mitochondrial, and in guinea pig and humans it is more or less equally distributed between both compartments. In order for gluconeogenesis to occur, either oxaloacetate must leave the mitochondrion for conversion to PEP or the PEP formed there must enter the cytosol.

PEP is transported across the mitochondrial membrane by specific membrane transport proteins. There is, however, no such transport system for oxaloacetate. It must first be converted either to aspartate (Fig. 23-7, Route 1) or to malate (Fig. 23-7, Route 2), for which mitochondrial transport systems exist (Section 22-1B). The difference between these two routes involves the transport of NADH reducing equivalents. The **malate dehydrogenase** route (Route 2) results in the transport of reducing equivalents from the mitochondrion to the cytosol, since it utilizes mitochondrial NADH and produces cytosolic NADH. The **aspartate aminotransferase** route (Route 1) does not involve NADH. Cytosolic NADH is required for gluconeogenesis so, under most conditions, the route through malate is a necessity. If the gluconeogenic precursor is lactate, however (Section 23-1C), its oxidation to pyruvate generates cytosolic NADH, so that either transport route may then be used. Of course, as we have seen, during oxidative metabolism

Figure 23-7 **Transport of PEP and oxaloacetate from the mitochondrion to the cytosol.** PEP is directly transported between these compartments. Oxaloacetate, however, must first be converted to either aspartate through the action of **aspartate aminotransferase** (Route 1) or to malate by malate

dehydrogenase (Route 2). Route 2 involves the mitochondrial oxidation of NADH followed by the cytosolic reduction of NAD^+ and therefore also transfers NADH reducing equivalents from the mitochondrion to the cytosol. 🎧 **See the Animated Figures**

the two routes may also alternate (with Route 2 reversed) to form the malate–aspartate shuttle, which transports NADH reducing equivalents into the mitochondrion (Section 22-1Bc).

In the liver, where the urea cycle occurs (Section 26-2), a third route, a modification of Route 1, may be followed for transporting oxaloacetate into the cytosol. The aspartate that enters the cytosol by Route 1 may be converted to fumarate as part of the urea cycle (Fig. 26-8), instead of being transaminated. Fumarate is then hydrated to malate and dehydrogenated to oxaloacetate by cytosolic equivalents of citric acid cycle enzymes. This third route generates cytosolic NADH in the same way as does Route 2.

h. Hydrolytic Reactions Bypass PFK and Hexokinase
The opposing pathways of gluconeogenesis and glycolysis utilize many of the same enzymes (Fig. 23-8). However, the free energy change is highly unfavorable in the gluconeogenic

direction at two other points in the pathway in addition to the pyruvate kinase reaction: the PFK reaction and the hexokinase reaction. At these points, instead of generating ATP by reversing the glycolytic reactions, FBP and G6P are hydrolyzed, releasing P_i in exergonic processes catalyzed by **fructose-1,6-bisphosphatase (FBPase)** and **glucose-6-phosphatase,** respectively. *Glucose-6-phosphatase is unique to liver and kidney, permitting them to supply glucose to other tissues.*

Because of the presence of separate gluconeogenic enzymes at the three irreversible steps in the glycolytic conversion of glucose to pyruvate, both glycolysis and gluconeogenesis are rendered thermodynamically favorable. This is accomplished at the expense of the free energy of hydrolysis of two molecules each of ATP and GTP per molecule of glucose synthesized by gluconeogenesis in addition to that which would be consumed by the direct reversal of glycolysis.

Glycolysis:

Glucose + 2NAD$^+$ + 2ADP + 2P$_i$ $\longrightarrow$

    2 pyruvate + 2NADH + 4H$^+$ + 2ATP + 2H$_2$O

Gluconeogenesis:

2 Pyruvate + 2NADH + 4H$^+$ + **4ATP** + **2GTP** + 6H$_2$O

   $\longrightarrow$ glucose + 2NAD$^+$ + 4ADP + 2GDP + 6P$_i$

Overall:

  2ATP + 2GTP + 4H$_2$O $\longrightarrow$ 2ADP + 2GDP + 4P$_i$

Such free energy losses in a cyclic process are thermodynamically inescapable. They are the price that must be paid to maintain independent regulation of the two pathways.

B. *Regulation of Gluconeogenesis*

If both glycolysis and gluconeogenesis were to proceed in an uncontrolled manner, the net effect would be a futile cycle wastefully hydrolyzing ATP and GTP. This does not occur. Rather, *these pathways are reciprocally regulated so as to meet the needs of the organism.* In the fed state, when the blood glucose level is high, the liver is geared toward fuel conservation: Glycogen is synthesized and the glycolytic pathway and pyruvate dehydrogenase are activated, breaking down glucose to acetyl-CoA for fatty acid biosynthesis and fat storage. In the fasted state, however, the liver maintains the blood glucose level both by glycogen breakdown and by reversing the flux through glycolysis toward gluconeogenesis [using mainly protein degradation products via the **glucose–alanine cycle** (Section 26-1Ad) and glycerol from triacylglycerol hydrolysis (Section 25-1e)].

a. Glycolysis and Gluconeogenesis Are Controlled by Allosteric Interactions and Covalent Modifications

The rate and direction of glycolysis and gluconeogenesis are controlled at the points in these pathways where the forward and reverse directions can be independently regulated: the reactions catalyzed by (1) hexokinase/glucose-6-phosphatase, (2) PFK/FBPase, and (3) pyruvate kinase/pyruvate carboxylase–PEPCK (Fig. 23-8). Table 23-1 lists these regulatory enzymes and their regulators. The dominant mechanisms are allosteric interactions and cAMP-dependent covalent modifications (phosphorylation/dephosphorylation; Section 18-3). cAMP-dependent covalent modification renders this system sensitive to control by glucagon and other hormones that alter cAMP levels.

 One of the most important allosteric effectors involved in the regulation of glycolysis and gluconeogenesis is fructose-2,6-bisphosphate (F2,6P), which activates PFK and inhibits FBPase (Section 18-3F). The concentration of F2,6P is controlled by its rates of synthesis and breakdown by phosphofructokinase-2 (PFK-2) and fructose bisphosphatase-2 (FBPase-2), respectively. Control of the activities of PFK-2 and FBPase-2 is therefore an important aspect of gluconeogenic regulation even though these

enzymes do not catalyze reactions of the pathway. PFK-2 and FBPase-2 activities, which occur on separate domains of the same bifunctional enzyme, are subject to allosteric

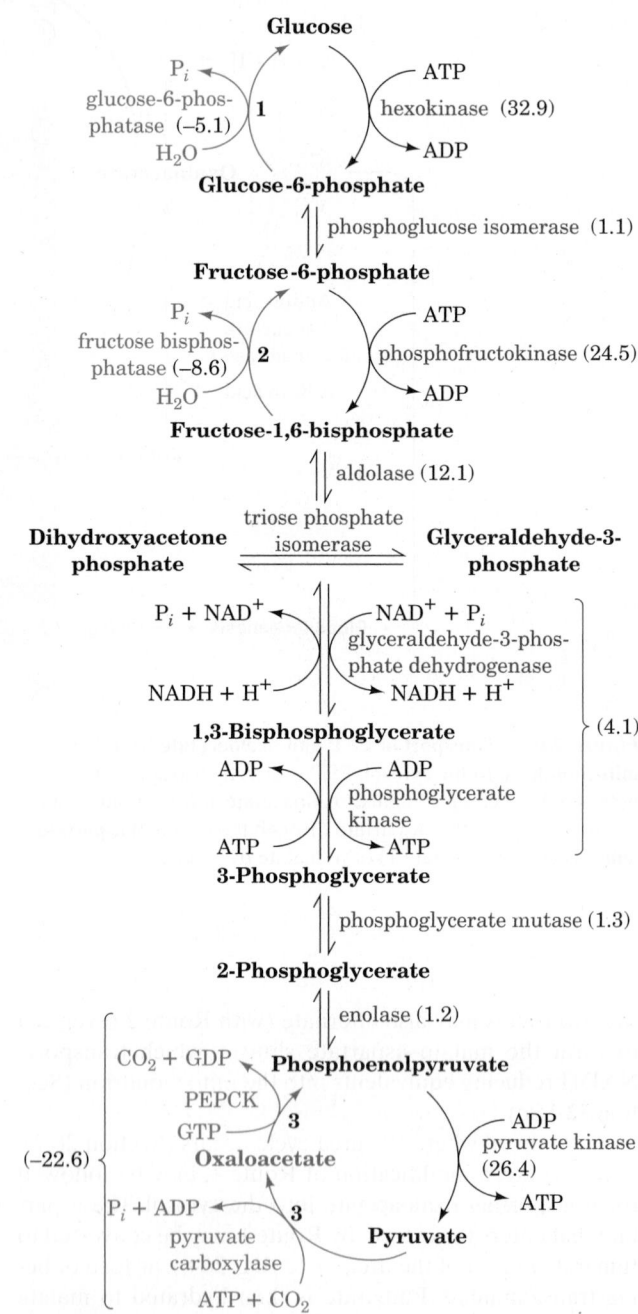

Figure 23-8 **Pathways of gluconeogenesis and glycolysis.** The three numbered steps, which are catalyzed by different enzymes in gluconeogenesis, have red arrows. The ΔG's for the reactions in the direction of gluconeogenesis under physiological conditions in liver are given in parentheses in kJ · mol^{-1}. [ΔG's obtained from Newsholme, E.A. and Leech, A.R., *Biochemistry for the Medical Sciences,* p. 448, Wiley (1983).] **See the Animated Figures**

Table 23-1   **Regulators of Gluconeogenic Enzyme Activity**

Enzyme	Allosteric Inhibitors	Allosteric Activators	Enzyme Phosphorylation	Protein Synthesis
PFK	ATP, citrate	AMP, F2,6P		
FBPase	AMP, F2,6P			
Pyruvate kinase	Alanine	F1,6P	Inactivates	
Pyruvate carboxylase		Acetyl-CoA		
PEPCK				Stimulated by glucagon, thyroid hormone, and glucocorticoids, and inhibited by insulin
PFK-2	Citrate	AMP, F6P, P_i	Inactivates	
FBPase-2	F6P	Glycerol-3-P	Activates	

regulation as well as control by covalent modifications (Table 23-1). Low levels of blood glucose result in hormonal activation of gluconeogenesis through regulation of [F2,6P] (Fig. 23-9).

Activation of gluconeogenesis in liver also involves inhibition of glycolysis at the level of pyruvate kinase. *Liver pyruvate kinase is inhibited both allosterically by alanine (a pyruvate precursor; Section 26-1Ad) and by phosphorylation.* Glycogen breakdown, in contrast, is stimulated by phosphorylation (Section 18-3C). Both pathways then flow toward G6P, which is converted to glucose for export to muscle and brain. Muscle pyruvate kinase, an isozyme of the liver enzyme, is not subject to these controls. Indeed, such controls would be counterproductive in muscle since this tissue lacks glucose-6-phosphatase and thus the ability to synthesize glucose via gluconeogenesis.

Low blood [glucose]

↓

Increased glucagon secretion

↓

Increased [cAMP]

↓

Increased enzyme phosphorylation

↓

Activation of FBPase-2 and inactivation of PFK-2

↓

Decreased [F2,6P]

↓

Inhibition of PFK and activation of FBPase

↓

Increased gluconeogenesis

Figure 23-9   Hormonal regulation of [F2,6P]. This process activates gluconeogenesis in liver in response to low blood [glucose].

b. PEPCK Concentration Is Transcriptionally Controlled

PEPCK is the enzyme that catalyzes the first committed reaction of gluconeogenesis. It is therefore of interest (Table 23-1) that PEPCK's activity is controlled solely through the transcriptional regulation of the gene encoding it (transcriptional regulation is outlined in Section 5-4Aa and discussed in detail in Sections 31-3 and 34-3). In particular, the transcription of the PEPCK gene is stimulated by glucagon, glucocorticoids, and thyroid hormones, and is inhibited by insulin. For instance, the cAMP that is produced in response to stimulation of the liver by glucagon, in addition to its initiation of phosphorylation cascades (Section 18-3), induces the transcription of the PEPCK gene. Richard Hanson has shown that this occurs because the PEPCK gene **promoter** (a control region that precedes the transcriptional initiation site of genes encoding proteins; Section 5-4Aa) contains a specific DNA sequence called the **cAMP response element (CRE)** that is bound by a **transcription factor** named **CRE binding protein (CREB),** but only when CREB is also binding cAMP (recall that a transcription factor is a protein that binds to a specific segment of its target promoter and, in doing so, activates RNA polymerase to initiate the transcription of the associated gene; Section 5-4Aa). However, the PEPCK gene promoter contains numerous other binding sites for specific transcription factors. Among them are the **thyroid hormone response element (TRE),** which is bound by thyroid hormone receptor in complex with thyroid hormone (Section 19-1D), and the **glucocorticoid hormone response element (GRE),** which is bound by the **glucocorticoid receptor** in complex with a glucocorticoid hormone (Sections 19-1G and 34-3Bn). In contrast, PEPCK gene transcription is strongly repressed by protein factors phosphorylated by the PI3K signaling cascade (Section 19-4D) initiated by the binding of insulin to the insulin receptor (these protein factors may repress transcription by interfering with the binding of the above transcription factors; the mechanism of insulin signaling is discussed in Sections 19-3Ac, 19-3Cg, and 19-4F). The rate of PEPCK mRNA production is determined by the integration of these various interactions and hence of the signals that caused them.

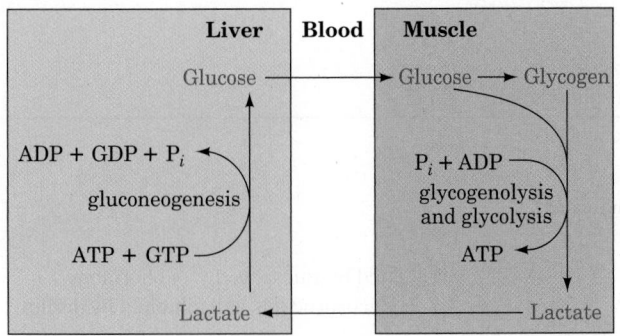

Figure 23-10   **The Cori cycle.** Lactate produced by muscle glycolysis is transported by the bloodstream to the liver, where it is converted to glucose by gluconeogenesis. The bloodstream carries the glucose back to the muscles, where it may be stored as glycogen. **See the Animated Figures**

C. *The Cori Cycle*

Muscle contraction is powered by hydrolysis of ATP, which is then regenerated through oxidative phosphorylation in the mitochondria of slow-twitch (red) muscle fibers and by glycolysis yielding lactate in fast-twitch (white) muscle fibers. Slow-twitch fibers also produce lactate when ATP demand exceeds oxidative flux. The lactate is transferred, via the bloodstream, to the liver, where it is reconverted to pyruvate by lactate dehydrogenase and then to glucose by gluconeogenesis. Thus, through the intermediacy of the bloodstream, liver and muscle participate in a metabolic cycle known as the **Cori cycle** (Fig. 23-10) in honor of Carl and Gerty Cori, who first described it. This is the same ATP-consuming glycolysis/gluconeogenesis "futile cycle" we discussed above. Here, however, instead of occurring in the same cell, the two pathways occur in different organs. Liver ATP is used to resynthesize glucose from lactate produced in muscle. The resynthesized glucose is returned to the muscle, where it is stored as glycogen and used, on demand, to generate ATP for muscle contraction. The ATP utilized by the liver for this process is regenerated by oxidative phosphorylation. After vigorous exertion, it often takes at least 30 min for all of the lactate so produced to be converted to glycogen and the oxygen consumption rate to return to its resting level, a phenomenon known as **oxygen debt.**

2 THE GLYOXYLATE CYCLE

Plants, but not animals, possess enzymes that mediate the net conversion of acetyl-CoA to succinate, which is then converted, via malate, to oxaloacetate. This is accomplished via the **glyoxylate cycle** (Fig. 23-11), a pathway involving enzymes of the **glyoxysome** (a membranous plant organelle; Section 1-2Ad). The glyoxylate cycle involves five enzymes, three of which also participate in the citric acid cycle: citrate synthase, aconitase, and malate dehydrogenase.

The two other enzymes, isocitrate lyase and malate synthase, are unique to the cycle.

The glyoxalate cycle consists of five reactions (Fig. 23-11):

Reactions 1 and 2. Glyoxysomal oxaloacetate is condensed with acetyl-CoA to form citrate, which is isomerized to isocitrate as in the citric acid cycle. Since the glyoxysome contains no aconitase, Reaction 2 presumably takes place in the cytosol.

Reaction 3. Glyoxysomal **isocitrate lyase** cleaves the isocitrate to succinate and **glyoxylate** (hence the cycle's name).

Reaction 4. Malate synthase, a glyoxysomal enzyme, condenses glyoxylate with a second molecule of acetyl-CoA to form malate.

Reaction 5. Glyoxysomal malate dehydrogenase catalyzes the oxidation of malate to oxaloacetate by NAD^+, thereby completing the cycle.

The glyoxylate cycle therefore results in the net conversion of two acetyl-CoA to succinate instead of to four molecules of CO_2, as would occur in the citric acid cycle. The succinate produced in Reaction 3 is transported to the mitochondrion, where it enters the citric acid cycle and is converted to malate, which has two alternative fates: (1) It can be converted to oxaloacetate in the mitochondrion, continuing the citric acid cycle and thereby making the glyoxylate pathway an anaplerotic process (Section 21-5b); or (2) it can be transported to the cytosol, where it is converted to oxaloacetate for entry into gluconeogenesis.

The overall reaction of the glyoxylate cycle can be considered to be the formation of oxaloacetate from two molecules of acetyl-CoA.

$$2 \text{ Acetyl-CoA} + 2NAD^+ + FAD \longrightarrow$$
$$\text{oxaloacetate} + 2CoA + 2NADH + FADH_2 + 2H^+$$

Isocitrate lyase and malate synthase, the only enzymes of the glyoxylate pathway unique to plants, enable germinating seeds to convert their stored triacylglycerols, through acetyl-CoA, to glucose. It had long been assumed that this was a requirement of germination. However, a mutant of *Arabidopsis thaliana* (an oilseed plant) lacking isocitrate lyase, and hence unable to convert lipids to carbohydrate, nevertheless germinated. This process was only inhibited when the mutant plants were subjected to low light conditions. It therefore appears that the glyoxylate cycle's importance in seedling growth is its anaplerotic function in providing 4-carbon units to the citric acid cycle, which can then oxidize the triacylglycerol-derived acetyl-CoA.

3 BIOSYNTHESIS OF OLIGOSACCHARIDES AND GLYCOPROTEINS

Oligosaccharides consist of monosaccharide units joined together by glycosidic bonds (linkages between C1, the anomeric carbon, of one unit and an OH group of a second

Figure 23-11   The glyoxylate cycle. The cycle results in the net conversion of two acetyl-CoA to succinate in the glyoxysome, which can be converted to malate in the mitochondrion for use in gluconeogenesis. Isocitrate lyase and malate synthase, enzymes unique to glyoxysomes (which occur only in plants), are boxed in blue. **(1)** Glyoxysomal citrate synthase catalyzes the condensation of oxaloacetate with acetyl-CoA to form citrate. **(2)** Cytosolic aconitase catalyzes the conversion of citrate to isocitrate. **(3)** Isocitrate lyase catalyzes the cleavage of isocitrate to succinate and glyoxylate. **(4)** Malate synthase catalyzes the condensation of glyoxylate with acetyl-CoA to form malate. **(5)** Glyoxysomal malate dehydrogenase catalyzes the oxidation of malate to oxaloacetate, completing the cycle. **(6)** Succinate is transported to the mitochondrion, where it is converted to malate via the citric acid cycle. **(7)** Malate is transported to the cytosol, where malate dehydrogenase catalyzes its oxidation to oxaloacetate, which can then be used in gluconeogenesis. **(8)** Alternatively, malate can continue in the citric acid cycle, making the glyoxylate cycle anaplerotic.

unit; Section 11-1C). About 80 different kinds of naturally occurring glycosidic linkages are known, most of which involve mannose, *N*-acetylglucosamine, *N*-acetylmuramic acid, glucose, galactose, fucose (6-deoxygalactose), *N*-acetylneuraminic acid (sialic acid), and *N*-acetylgalactosamine (Section 11-1C). Glycosidic linkages also occur to lipids (e.g., glycosphingolipids; Section 12-1D) and proteins (glycoproteins; Section 11-3C).

Glycosidic bond formation requires free energy input under physiological conditions ($\Delta G^{\circ\prime} = 16$ kJ · mol^{-1}). This free energy, as we have seen in the case of glycogen synthesis (Section 18-2B), is acquired through the conversion of monosaccharide units to nucleotide sugars. A nucleotide at a sugar's anomeric carbon atom is a good leaving group and thereby facilitates formation of a glycosidic bond to a second sugar unit via reactions catalyzed by **glycosyltransferases**

(Fig. 23-12). The nucleotides that participate in monosaccharide transfers are UDP, GDP, and CMP; a given sugar is associated with only one of these nucleotides (Table 23-2).

A. *Lactose Synthesis*

Several disaccharides are synthesized for future use as metabolic fuels. In plants, the major fuel disaccharide is sucrose (Section 11-2B), whose synthesis is discussed in Section 24-3Ad. Typical of mammalian disaccharides is lactose [β-galactosyl-(1 → 4)-glucose; milk sugar], which is synthesized in the mammary gland by **lactose synthase** (Fig. 23-13). The donor sugar is UDP–galactose, which is formed by epimerization of UDP–glucose (Section 17-5B). The acceptor sugar is glucose.

Lactose synthase consists of two subunits:

1. Galactosyltransferase, the catalytic subunit, which occurs in many tissues, where it catalyzes the reaction of UDP–galactose and N-acetylglucosamine to yield N-acetyllactosamine, a constituent of many complex oligosaccharides (see, e.g., Fig. 23-20, Reaction 6).

2. α-Lactalbumin, a mammary gland protein with no catalytic activity, which alters the specificity of galactosyltransferase such that it utilizes glucose as an acceptor, rather than N-acetylglucosamine, to form lactose instead of N-acetyllactosamine.

B. *Glycoprotein Synthesis*

Eukaryotic proteins destined for secretion, incorporation into membranes, or localization inside membranous organelles contain carbohydrates and are therefore classified as glycoproteins. *Glycosylation and oligosaccharide processing play an indispensable role in the sorting and the distribution of these proteins to their proper cellular destinations.* Their polypeptide components are ribosomally synthesized and processed by addition and modification of oligosaccharides.

The oligosaccharide portions of glycoproteins, as we have seen in Sections 11-3C and 12-3Bc, are classified into three groups:

1. *N*-Linked oligosaccharides, which are attached to their polypeptide chain by a β-*N*-glycosidic bond to the side chain N of an Asn residue in the sequence Asn-X-Ser

Table 23-2 Sugar Nucleotides and Their Corresponding Monosaccharides in Glycosyltransferase Reactions

UDP	GDP	CMP
N-Acetylgalactosamine	Fucose	Sialic acid
N-Acetylglucosamine	Mannose	
N-Acetylmuramic acid		
Galactose		
Glucose		
Glucuronic acid		
Xylose		

or Asn-X-Thr, where X is any amino acid residue except Pro (Fig. 23-14*a*).

2. *O*-Linked oligosaccharides, which are attached to their polypeptide chain through an α-*O*-glycosidic bond to the side chain O of a Ser or Thr residue (Fig. 23-14*b*) or, only in collagens (Section 8-2Bb), to that of a 5-hydroxylysine (Hyl) residue (Fig. 23-14*c*).

3. Glycosylphosphatidylinositol (GPI) membrane anchors, which are attached to their polypeptide chain through an amide bond between mannose-6-phosphoethanolamine and the C-terminal carboxyl group (Fig. 23-14*d*).

We shall consider the synthesis of these three types of oligosaccharides in turn.

a. *N*-Linked Glycoproteins Are Synthesized in Four Stages

N-Linked glycoproteins are formed in the endoplasmic reticulum and further processed in the Golgi apparatus. Synthesis of their carbohydrate moieties occurs in four stages:

1. Synthesis of a lipid-linked oligosaccharide precursor.

2. Transfer of this precursor to the side chain N of an Asn residue on a growing polypeptide.

3. Removal of some of the precursor's sugar units.

4. Addition of sugar residues to the remaining core oligosaccharide.

Figure 23-12 Role of nucleotide sugars. These compounds are the glycosyl donors in oligosaccharide biosynthesis catalyzed by glycosyltransferases.

Figure 23-13   Lactose synthase. This enzyme catalyzes the formation of lactose from UDP–galactose and glucose.

We shall discuss these stages in order.

b. N-Linked Oligosaccharides Are Constructed on Dolichol Carriers

N-Linked oligosaccharides are initially synthesized as lipid-linked precursors. The lipid component in this process is **dolichol,** a long-chain polyisoprenol of 14 to 24 isoprene units (17–21 units in animals and 14–24 units in fungi and plants; isoprene units are C_5 units with the carbon skeleton of isoprene; Section 25-6A), which is linked to the oligosaccharide precursor via a pyrophosphate bridge (Fig. 23-15). Dolichol apparently anchors the growing oligosaccharide to the endoplasmic reticulum membrane. Involvement of lipid-linked oligosaccharides in *N*-linked glycoprotein synthesis was first demonstrated in 1972 by Armando Parodi and Luis Leloir, who showed that, when a lipid-linked oligosaccharide containing [^{14}C]glucose is incubated with rat liver **microsomes** (vesicular fragments of isolated endoplasmic reticulum), the radioactivity becomes associated with protein.

c. N-Linked Glycoproteins Have a Common Oligosaccharide Core

The pathway of dolichol-PP-oligosaccharide synthesis involves stepwise addition of monosaccharide units to the

Figure 23-14   Types of saccharide–polypeptide linkages in glycoproteins. (*a*) An *N*-linked glycosidic bond to an Asn residue in the sequence Asn-X-Ser/Thr. (*b*) An *O*-linked glycosidic bond to a Ser (or Thr) residue. (*c*) An *O*-linked glycosidic bond to a 5-hydroxylysine residue in collagen. (*d*) An amide bond between the C-terminal amino acid of a protein and the phosphoethanolamine bridge to the 6 position of mannose in the glycophosphatidylinositol (GPI) anchor. The X group (*green*) denotes the rest of the GPI anchor (Fig. 12-30).

Figure 23-15   Dolichol pyrophosphate glycoside. The carbohydrate precursors of *N*-linked glycosides are synthesized as dolichol pyrophosphate glycosides. Dolichols are long-chain polyisoprenols ($n = 14$–24) in which the α-isoprene unit is saturated.

growing glycolipid by specific glycosyltransferases to form a common "core" structure. Each monosaccharide unit is added by a unique glycosyltransferase (Fig. 23-16). For example, in Reaction 2 of Fig. 23-16, five mannosyl units are added through the action of five different mannosyltransferases, each with a different oligosaccharide-acceptor specificity. The oligosaccharide core, the product of Reaction 9 in Fig. 23-16, has the composition (*N*-acetylglucosamine)$_2$(mannose)$_9$(glucose)$_3$.

Although nucleotide sugars are the most common monosaccharide donors in glycosyltransferase reactions, *several mannosyl and glucosyl residues are transferred to*

the growing dolichol-PP-oligosaccharide from their corresponding dolichol-P derivatives. This requirement for **dolichol-P-mannose** was discovered by Stuart Kornfeld, who found that mutant mouse lymphoma cells (lymphoma is a type of cancer) that are unable to synthesize normal lipid-linked oligosaccharides formed a defective, smaller glycolipid. These cells contain all the requisite glycosyltransferases but are unable to synthesize dolichol-P-mannose (Reaction 4 in Fig. 23-16 is blocked). When this substance is supplied to the mutant cells, mannosyl units are added to the defective dolichol-PP-oligosaccharide.

♦ = Glucose   ▼ = Mannose

● = *N*-Acetylglucosamine   ∿-P = Dolichol phosphate

Figure 23-16 **Pathway of dolichol-PP-oligosaccharide synthesis.**
(1) Addition of *N*-acetylglucosamine-1-P and a second *N*-acetylglucosamine to dolichol-P. **(2)** Addition of five mannosyl residues from GDP–mannose in reactions catalyzed by five different mannosyltransferases. **(3)** Membrane translocation of dolichol-PP-(*N*-acetylglucosamine)$_2$(mannose)$_5$ to the lumen of the endoplasmic reticulum (ER). **(4)** Cytosolic synthesis of dolichol-P-mannose from GDP–mannose and dolichol-P. **(5)** Membrane translocation of dolichol-P-mannose to the lumen of the ER. **(6)** Addition of four mannosyl residues from dolichol-P-mannose in reactions catalyzed by four different mannosyltransferases. **(7)** Cytosolic synthesis of dolichol-P-

glucose from UDP–glucose and dolichol-P. **(8)** Membrane translocation of dolichol-P-glucose to the lumen of the ER. **(9)** Addition of three glucosyl residues from dolichol-P-glucose. **(10)** Transfer of the oligosaccharide from dolichol-PP to the polypeptide chain at an Asn residue in the sequence Asn-X-Ser/Thr, releasing dolichol-PP. **(11)** Translocation of dolichol-PP to the cytoplasmic surface of the ER membrane. **(12)** Hydrolysis of dolichol-PP to dolichol-P. **(13)** Dolichol-P can also be formed by phosphorylation of dolichol by CTP. [Modified from Abeijon, C. and Hirschberg, C.B., *Trends Biochem. Sci.* **17,** 34 (1992).]
🎬 **See the Animated Figures**

d. Dolichol-PP-Oligosaccharide Synthesis Involves Topological Changes of the Intermediates

Reactions 1, 2, 4, and 7 of Fig. 23-16 all occur on the cytoplasmic side of the endoplasmic reticulum (ER) membrane. This was determined by using "right-side-out" rough ER vesicles and showing that various membrane-impermeant reagents can disrupt one or another of these reactions. Reactions 6, 9, and 10 occur in the lumen of the ER as judged by the inability of concanavalin A, a **lectin** (carbohydrate-binding protein), to bind to the products of these reactions until the membrane is permeabilized. The (mannose)$_5$ (*N*-acetylglucosamine)$_2$-PP-dolichol product of Reaction 2, the dolichol-P-mannose product of Reaction 4, and the dolichol-P-glucose product of Reaction 7 must therefore be translocated across the ER membrane (Reactions 3, 5, and 8) such that they extend from its luminal surface in order for the synthesis of *N*-linked oligosaccharides to continue. The translocations are mediated by specific ATP-independent flippases.

e. *N*-Linked Oligosaccharides Are Cotranslationally Added to Proteins

Vesicular stomatitis virus (VSV), which infects cattle, producing influenza-like symptoms, provides an excellent model system for studying *N*-linked glycoprotein processing. The VSV coat consists of host-cell membrane in which a single viral glycoprotein, the **VSV G-protein** (not to be confused with the GTPases involved in signal transduction; Chapter 19), is embedded. Since a viral infection almost totally usurps an infected cell's protein synthesizing machinery, a VSV-infected cell's Golgi apparatus, which normally contains hundreds of different types of glycoproteins, contains virtually no other glycoprotein but G-protein. Consequently, the maturation of the G-protein is relatively easy to follow.

Of the Asn-X-Ser/Thr sites in mature eukaryotic proteins, 70 to 90% are *N*-glycosylated. Studies of VSV-infected cells indicate that the *transfer of the lipid-linked oligosaccharide to a polypeptide chain occurs while the polypeptide chain is still being synthesized.* Structural predictions (Section 9-3A), together with glycosylation studies of model polypeptides, suggest that the amino acid sequences flanking known *N*-glycosylation sites occur at β turns or loops in which Asn's backbone N—H group is hydrogen bonded to the Ser/Thr hydroxyl O atom (Fig. 23-17a). This explains why Pro cannot occupy the X position; it would prevent Asn-X-Ser/Thr from assuming the putative required hydrogen bonded conformation.

VSV G-protein is *N*-glycosylated by **oligosaccharyltransferase (OST),** a membrane-bound, ~300-kD, 8-subunit enzyme that recognizes the amino acid sequence Asn-X-Ser/Thr (Fig. 23-16, Reaction 10). Ernst Bause has proposed a catalytic mechanism for OST in which an enzyme base

Figure 23-17 The oligosaccharyltransferase (OST) reaction. (*a*) The Asn-X-Thr component of a hexapeptide model substrate forms a ring that is closed by a hydrogen bond from the Asn amide group to the Thr hydroxyl group. A base on the enzyme facilitates the nucleophilic displacement of dolichol pyrophosphate from the oligosaccharide (Sac) by the amide nitrogen. (*b*) The inactivation of the OST by reacting it with a hexapeptide containing Asn-Gly-epoxyethylGly in the presence of dolichol-PP-oligosaccharide. This chemically labels the base with the oligopeptide to which the oligosaccharide has become covalently linked.

abstracts a proton from the Ser/Thr hydroxyl group, which in turn abstracts a proton from the Asn NH$_2$ group, thereby promoting its nucleophilic attack on the oligosaccharide (Sac), which then displaces the dolichol pyrophosphate (Fig. 23-17*a*). This mechanism is supported by the observation that reacting the OST with dolichol-PP-oligosaccharide and a hexapeptide model substrate containing the sequence Asn-X-epoxyethylGly (rather than Asn-X-Ser/Thr) irreversibly inactivates the enzyme by covalently linking it to the now glycosylated hexapeptide (Fig. 23-17*b*).

f. The Calnexin/Calreticulin Cycle Facilitates Glycoprotein Folding

The processing of an N-linked core oligosaccharide begins in the endoplasmic reticulum by the enzymatic trimming (removal) of its three glucose residues (Fig. 23-18, Reactions 2 and 3) and one of its mannose residues (Fig. 23-18, Reaction 4) before the protein has folded to its native conformation. This is not a straightforward process, however, because **UDP–glucose:glycoprotein glucosyltransferase (GT),** a 1513-residue soluble protein, reglucosylates the oligosaccharides of partially folded glycoproteins, a reaction that reverses the removal of the last of the three glucose residues by **glucosidase II** (Fig. 23-18, Reaction 3). This futile cycle (most glycoproteins undergo reglucosylation at least

once) is part of a chaperone-mediated glycoprotein folding process called the **calnexin/calreticulin cycle. Calnexin (CNX;** ~570 residues), which is membrane bound, and **calreticulin (CRT;** ~400 residues), its soluble homolog, are ER-resident lectins that bind partially folded glycoproteins bearing a monoglucosylated oligosaccharide in a way that protects the glycoprotein from degradation and premature transfer to the Golgi apparatus. If the glycoprotein is released and deglucosylated before it has correctly folded, GT, which recognizes only non-native glycoproteins, reglucosylates it so that the CNX/CRT cycle can repeat. CNX and CRT both also bind **ERp57,** a 481-residue thiol oxidoreductase homologous to protein disulfide isomerase (PDI; Section 9-2A). While the partially folded glycoprotein is bound to the complex, ERp57 catalyzes disulfide interchange reactions to facilitate the formation of the correctly paired disulfide bonds. The CNX/ ERp57 and CRT/ERp57 complexes are therefore responsible for the correct folding and disulfide bond formation of the glycoproteins in the ER. The importance of this process is demonstrated by the observation that knockout mice lacking the gene for CRT die *in utero.*

The X-ray structure of the luminal domain of calnexin (residues 61–458), determined by Miroslaw Cygler, reveals a most unusual structure (Fig. 23-19): a compact globular domain (residues 61–262 and 415–458) from which extends

= Glucose

= Mannose

= *N*-Acetylglucosamine

Transfer via vesicles to the cis Golgi network

Figure 23-18 **The calnexin/calreticulin cycle for glycoprotein folding in the endoplasmic reticulum.** The reactions are catalyzed by: **(1)** oligosaccharyltransferase (OST); **(2)** α-glucosidase I; **(3)** α-glucosidase II, UDP–glucose:glycoprotein glucosyltransferase (GT), calreticulin (CRT), calnexin (CNX), and the thiol oxidoreductase ERp57; and **(4)** ER α-1,2-mannosidase. [After Helenius, A. and Aebi, M., *Science* **291,** 2367 (2001).]

a 145-Å-long arm (residues 270–414). The globular domain forms a sandwich of a 6-stranded and a 7-stranded antiparallel β sheet that binds a Ca^{2+} ion and which resembles legume lectins such as concanavalin A (Fig. 8-40). This domain binds glucose on its concave (*blue*) surface, which is lined by hydrogen bonding groups that model building suggests binds the $(glucose)_1(mannose)_3$ portion of calnexin's natural $(glucose)_1(mannose)_9$ substrate. The long arm, which consists of an extended hairpin, is known as the P domain because it has four copies each of two different Pro-rich motifs arranged in the sequence 11112222, with each ~18-residue motif 1 in antiparallel association with an ~14-residue motif 2 on the opposite strand of the hairpin. Each of these motif pairs has a similar structure, with its conserved residues maintaining identical interactions in each pair. The P domain has been shown to form the binding site for ERp57 in both calnexin and calreticulin.

g. Glycoprotein Processing is Completed in the Golgi Apparatus

Once a glycoprotein has folded to its native conformation and ER α-1,2-mannosidase has removed one of its

mannosyl residues (Fig. 23-18, Step 4), the glycoprotein is transported, in membranous vesicles, to the Golgi apparatus, where it is further processed (Fig. 23-20). The Golgi apparatus (Fig. 12-58), as we discussed in Section 12-4C, consists of, from opposite the ER outward, the cis Golgi

Figure 23-19 **X-ray structure of the luminal portion of canine calnexin.** The protein is drawn in ribbon form embedded in its semitransparent molecular surface. The 6- and 7-stranded antiparallel β sheets of its globular domain are colored orange and blue, with its remaining portions gray and its bound Ca^{2+} ion represented by a blue-green sphere. In the P domain, motifs 1 are alternately colored green and yellow and motifs 2 are alternately colored magenta and cyan. [Based on an X-ray structure by Miroslaw Cygler, Biotechnology Research Institute, NRC, Montreal, Quebec, Canada. PDBid 1JHN.]

Figure 23-20 **Oligosaccharide processing of VSV G-protein in the Golgi network.** The reactions are catalyzed by: **(1)** Golgi α-mannosidase I, **(2)** *N*-acetylglucosaminyltransferase I, **(3)** Golgi α-mannosidase II, **(4)** *N*-acetylglucosaminyltransferase II, **(5)** fucosyltransferase, **(6)** galactosyltransferase, and **(7)** sialyltransferase. Lysosomal proteins are modified by: **(I)** *N*-acetylglucosaminyl phosphotransferase and **(II)** *N*-acetylglucosamine-1-phosphodiester α-*N*-acetylglucosaminidase. [Modified from Kornfeld, R. and Kornfeld, S., *Annu. Rev. Biochem.* **54,** 640 (1985).]

network, through which glycoproteins enter the Golgi apparatus; a stack of at least three different types of sacs, the cis, medial, and trans cisternae; and the trans Golgi network, through which proteins exit the Golgi apparatus. Glycoproteins traverse the Golgi stack, from the cis to the medial to the trans cisternae, each of which, as shown by James Rothman and Kornfeld, contains different sets of glycoprotein processing enzymes. As this occurs, mannose residues are trimmed from each oligosaccharide group and *N*-acetylglucosamine, galactose, fucose, and/or sialic acid residues are added to complete the processing of the glycoprotein (Fig. 23-20; Reactions 1–7). The glycoproteins are then sorted in the trans Golgi network for transport to their respective cellular destinations via membranous vesicles (Sections 12-4C and 12-4D).

There is enormous diversity among the different oligosaccharides of *N*-linked glycoproteins, as is indicated, for example, in Fig. 11-32*c*. Indeed, *even glycoproteins with a given polypeptide chain exhibit considerable microheterogeneity* (Section 11-3C), presumably as a consequence of incomplete glycosylation and lack of absolute specificity on the part of glycosyltransferases and glycosylases.

The processing of all *N*-linked oligosaccharides is identical through Reaction 4 of Fig. 23-18, so that all of them have a common (*N*-acetylglucosamine)$_2$(mannose)$_3$ core (five "noncore" mannose residues are subsequently trimmed from VSV G-protein; Fig. 23-20, Reactions 1 and 3). The diversity of the *N*-linked oligosaccharides therefore arises through divergence from this sequence after Fig. 23-20, Reaction 3. The resulting oligosaccharides are classified into three groups:

1. High-mannose oligosaccharides (Fig. 23-21*a*), which contain 2 to 9 mannose residues appended to the common pentasaccharide core (red residues in Fig. 23-21).

2. Complex oligosaccharides (Fig. 23-21*b*), which contain variable numbers of *N*-acetyllactosamine units as well as sialic acid and/or fucose residues linked to the core.

3. Hybrid oligosaccharides (Fig. 23-21*c*), which contain elements of both high-mannose and complex chains.

It is unclear how different types of oligosaccharides are related to the functions and/or final cellular locations of their glycoproteins. Lysosomal glycoproteins, however, appear to be of the high-mannose variety.

h. Inhibitors Have Aided the Study of *N*-Linked Glycosylation

Elucidation of the events in the glycosylation process has been greatly facilitated through the use of inhibitors that block specific glycosylation enzymes. Two of the most useful are the antibiotics **tunicamycin** (Fig. 23-22*a*), a hydrophobic analog of UDP–*N*-acetylglucosamine, and **bacitracin** (Fig. 23-23), a cyclic polypeptide. Both were discovered because of their ability to inhibit bacterial cell wall biosynthesis, a process that also involves the participation of lipid-linked oligosaccharides. Tunicamycin blocks the formation of dolichol-PP-oligosaccharides by inhibiting the synthesis of dolichol-PP-*N*-acetylglucosamine from dolichol-P and UDP–*N*-acetylglucosamine (Fig. 23-16, Reaction 1). Tunicamycin resembles an adduct of these reactants (Fig. 23-22*b*) and, in fact, binds to the enzyme with a dissociation constant of $7 \times 10^{-9} M$.

Figure 23-21   Types of *N*-linked oligosaccharides. Typical primary structures of (*a*) high-mannose, (*b*) complex, and (*c*) hybrid *N*-linked oligosaccharides. The pentasaccharide core common to all *N*-linked oligosaccharides is indicated in red. [After Kornfeld, R. and Kornfeld, S., *Annu. Rev. Biochem.* **54,** 633 (1985).]

Figure 23-22 Chemical structure of tunicamycin. The structure of (*a*) the glycosylation inhibitor tunicamycin is compared to that of (*b*) dolichol-P + UDP–*N*-acetylglucosamine.

(*a*)

Tunicamycin

n = 8,9,10, or 11

(*b*)

Dolichol phosphate

UDP-*N*-Acetylglucosamine

Bacitracin forms a complex with dolichol-PP that inhibits its dephosphorylation (Fig. 23-16, Reaction 12), thereby preventing glycoprotein synthesis from lipid-linked oligosaccharide precursors. Bacitracin is clinically useful because it destroys bacterial cell walls but does not affect animal cells because it cannot cross cell membranes (bacterial cell wall biosynthesis is an extracellular process).

Bacitracin

Figure 23-23 Chemical structure of bacitracin. Note that this dodecapeptide has four D-amino acid residues and two unusual intrachain linkages. "Orn" represents the nonstandard amino acid residue ornithine (Fig. 26-7).

i. *O*-Linked Oligosaccharides Are Post-Translationally Formed

The study of the biosynthesis of **mucin,** an *O*-linked glycoprotein secreted by the submaxillary salivary gland, indicates that *O-linked oligosaccharides are synthesized in the Golgi apparatus by serial addition of monosaccharide units to a completed polypeptide chain* (Fig. 23-24). Synthesis starts with the transfer of *N*-acetylgalactosamine (GalNAc) from UDP–GalNAc to a Ser or Thr residue on the polypeptide by **GalNAc transferase.** In contrast to *N*-linked oligosaccharides, which are transferred to an Asn in a specific amino acid sequence, the *O*-glycosylated Ser and Thr residues are not members of any common sequence. Rather, it appears that the location of glycosylation sites is specified only by the secondary or tertiary structure of the polypeptide. Glycosylation continues with stepwise addition of galactose, sialic acid, *N*-acetylglucosamine, and/or fucose by the corresponding glycosyltransferases.

j. Oligosaccharides on Glycoproteins Act as Recognition Sites

Glycoproteins that are synthesized in the endoplasmic reticulum and processed in the Golgi apparatus are targeted for secretion, insertion into cell membranes, or incorporation into cellular organelles such as lysosomes. This suggests that *oligosaccharides serve as recognition markers for this sorting process.* For example, the study of I-cell disease (Section 12-4Cg) demonstrated that in glycoprotein enzymes destined for the lysosome, a mannose residue is converted to mannose-6-phosphate (M6P) in the cis cisternae of the Golgi. The process involves two enzymes (Fig. 23-20, Reactions I and II), which are thought to recognize lysosomal protein precursors by certain structural features on these proteins rather than a specific amino acid sequence. In the trans Golgi network, M6P-bearing glycoproteins are sorted into lysosome-bound coated vesicles through their specific binding to one of two M6P receptors, one of which is a 275-kD membrane glycoprotein called the **M6P/IGF-II receptor** (because it has been found that this M6P receptor and the **insulinlike growth factor II receptor** are the same protein). Individuals with I-cell disease lack the enzyme catalyzing mannose phosphorylation (Fig. 23-20, Reaction I), resulting in the secretion of the normally lysosome-resident enzymes.

ABO blood group antigens (Section 12-3E) are *O*-linked glycoproteins. Their characteristic oligosaccharides are components of both cell-surface lipids and of proteins that occur in various secretions such as saliva. These oligosaccharides form antibody recognition sites.

Glycoproteins are believed to mediate cell–cell recognition. For example, an *O*-linked oligosaccharide on a glycoprotein that coats the mouse ovum surface (zona pellucida) acts as the sperm receptor. Even when this oligosaccharide is separated from its protein, it retains the ability to bind mouse sperm.

k. GPI-Linked Proteins

Glycosylphosphatidylinositol (GPI) groups function to anchor a wide variety of proteins to the exterior surface of the eukaryotic plasma membrane, thus providing an alternative to transmembrane polypeptide domains (Section

Figure 23-24 **Proposed synthesis pathway for the carbohydrate moiety of an *O*-linked oligosaccharide chain of canine submaxillary mucin.** SA and Fuc represent sialic acid and fucose.

12-3Bc; Fig. 12-30). This anchoring results from transamidation of a preformed GPI glycolipid within 1 min of the synthesis and transfer of a target protein to the ER. Biosynthesis of the GPI core structure (Fig. 23-25a) begins on the cytoplasmic side of the ER with the transfer of

Figure 23-25 **GPI anchors.** (*Opposite*) (*a*) The pathway of synthesis of the tetrasaccharide core of glycophosphatidylinositol (GPI). The following enzymes and steps are involved: **(1)** UDP–GlcNAc:PI α1 → 6 *N*-acetylglucosaminyltransferase complex, **(2)** GlcNAc–PI de-*N*-acetylase, **(3)** inositol acyltransferase, **(4)** Dol-P-Man:GlcN–PI/GlcN–(acyl)PI α1 → 4 mannosyltransferase (MT-I), **(5)** an ethanolamine phosphotransferase, **(6)** Dol-P-Man:Man$_1$GlcN–(acyl)PI α1 → 6 mannosyltransferase (MT-II), **(7)** Dol-P–Man:Man$_2$GlcN–(acyl)PI α1 → 2 mannosyltransferase (MT-III), **(8)** lipid remodeling (replacement of the fatty acyl groups on PI), and **(9)** transfer of phosphoethanolamine from phosphatidylethanolamine to the 6-hydroxyl group of the terminal mannose residue of the core tetrasaccharide by an ethanolamine phosphotransferase. (*b*) Transamidation of the target protein, resulting in a C-terminal amide link to the GPI anchor.

(a)

UDP–GlcNac   **Phosphatidylinositol (PI)**   **GlcNac–PI**

Phosphatidyl
ethanolamine   **5**

GlcNH₂ — PI

CoA   Acyl-CoA

HOAc

lipid   Phosphatidyl
remodeling   ethanolamine

8 9

H₂NCH₂CH₂O—P=O

(b)

TARGET PROTEIN—C—HN—C-TERMINAL PEPTIDE + H₂NCH₂CH₂O—P=O

C-TERMINAL PEPTIDE

TARGET PROTEIN—C—NHCH₂CH₂—O—P=O

P Dolichol phosphate
▼ mannose
○ Glucosamine

Phosphalidylinositol (PI)

Acyl group

N-acetylglucosamine from UDP–*N*-acetylglucosamine (UDP–GlcNAc) to the 6 hydroxyl of the inositol of phosphatidylinositol, followed by the removal of the acetyl group. The mammalian pathway then continues with the 2-acylation of inositol, translocation to the luminal side of the ER membrane, and the addition of mannose from dolichol-P-mannose (Dol-P-Man; Fig. 23-16) and phosphoethanolamine from phosphatidylethanolamine (Table 12-2), as indicated in Fig. 23-25*a*. This core is modified with a variety of additional sugar residues, depending on the species and the protein to which it is attached. There is considerable diversity in the fatty acid residues of GPI anchors due to the extensive lipid remodeling that occurs during anchor synthesis. Target proteins become anchored to the membrane surface when the amino group of the GPI phosphoethanolamine nucleophilically attacks a specific amino acyl group of the protein near its C-terminus, resulting in a transamidation that releases a 20- to 30-residue hydrophobic C-terminal signal peptide (Fig. 23-25*b*). Since GPI groups are appended to proteins on the luminal surface of the RER, GPI-anchored proteins occur on the exterior surface of the plasma membrane (Fig. 12-60). However, they are distributed unevenly in the outer leaflet of the plasma membrane because they prefer to associate with sphingolipid–cholesterol rafts (Section 12-3Cb).

The core GPI structure is evolutionarily conserved among all eukaryotes, although there are differences between species in its synthesis. For example, the cell surface of the trypanosomes that cause African sleeping sickness (a debilitating and often fatal disease that afflicts millions of people in sub-Saharan Africa) has a dense coating of **variant surface glycoprotein (VSG)** that is GPI-anchored to its plasma membrane. The VSG coating conceals the trypanosome's plasma membrane from the host's immune system although it recognizes and attacks the VSG itself. The parasite is nevertheless able to evade the host's immunological defenses because it has a genetic repertoire of about a thousand immunologically distinct VSGs. An individual trypanosome expresses only one of its VSG genes and hence the host can mount an effective immunological attack against the prevailing population of VSGs, a process that takes around 1 week (Section 35-2A). However, by switching VSG genes, a new population of trypanosomes arises that replicates unchecked until the host can mount a new immune response, a cycle that repeats until the death of the host. The comparison of the GPI biosynthetic pathway in trypanosomes with that in mammalian systems has revealed several differences in the pathway order. For example, Steps 3 and 4 of Fig. 23-25*a* are reversed in trypanosomes. This and other differences in the substrate specificities of the enzymes catalyzing this pathway have brought to light several promising drug targets for the treatment of African sleeping sickness.

4 THE PENTOSE PHOSPHATE PATHWAY

ATP is the cell's "energy currency"; its exergonic hydrolysis is coupled to many otherwise endergonic cell functions. *Cells have a second currency, reducing power.* Many endergonic reactions, notably the reductive biosynthesis of fatty acids

(Section 25-4) and cholesterol (Section 25-6A), as well as photosynthesis (Section 24-3A), require NADPH in addition to ATP. Despite their close chemical resemblance, *NADPH and NADH are not metabolically interchangeable* (recall that these coenzymes differ only by a phosphate group at the 2′-OH group of NADPH's adenosine moiety; Fig. 13-2). Whereas NADH participates in utilizing the free energy of metabolite oxidation to synthesize ATP (oxidative phosphorylation), *NADPH is involved in utilizing the free energy of metabolite oxidation for otherwise endergonic reductive biosynthesis.* This differentiation is possible because the dehydrogenase enzymes involved in oxidative and reductive metabolism exhibit a high degree of specificity toward their respective coenzymes. Indeed, cells normally maintain their [NAD$^+$]/[NADH] ratio near 1000, which favors metabolite oxidation, while keeping their [NADP$^+$]/[NADPH] ratio near 0.01, which favors metabolite reduction.

NADPH is generated by the oxidation of G6P via an alternative pathway to glycolysis, the **pentose phosphate pathway** *[also called the* **hexose monophosphate (HMP) shunt** *and the* **phosphogluconate pathway;** *Fig. 23-26]. The pathway also produces ribose-5-phosphate (R5P), an essential precursor in nucleotide biosynthesis (Sections 28-1, 28-2, and 28-5). The first evidence of this pathway's existence was obtained in the 1930s by Otto Warburg, who discovered NADP$^+$ through his studies on the oxidation of G6P to 6-phosphogluconate. Further indications came from the observation that tissues continue to respire in the presence of high concentrations of fluoride ion, which, it will be recalled, blocks glycolysis by inhibiting enolase (Section 17-2I). It was not until the 1950s, however, that the pentose phosphate pathway was elucidated by Frank Dickens, Bernard Horecker, Fritz Lipmann, and Efraim Racker. Tissues most heavily involved in fatty acid and cholesterol biosynthesis (liver, mammary gland, adipose tissue, and adrenal cortex) are rich in pentose phosphate pathway enzymes. Indeed, some 30% of the glucose oxidation in liver occurs via the pentose phosphate pathway.

The overall reaction of the pentose phosphate pathway is

$$3G6P + 6NADP^+ + 3H_2O \rightleftharpoons$$
$$6NADPH + 6H^+ + 3CO_2 + 2F6P + GAP$$

However, the pathway may be considered to have three stages:

1. Oxidative reactions (Fig. 23-26, Reactions 1–3), which yield NADPH and **ribulose-5-phosphate (Ru5P).**

$$3G6P + 6NADP^+ + 3H_2O \longrightarrow$$
$$6NADPH + 6H^+ + 3CO_2 + 3Ru5P$$

2. Isomerization and epimerization reactions (Fig. 23-26, Reactions 4 and 5), which transform Ru5P either to **ribose-5-phosphate (R5P)** or to **xylulose-5-phosphate (Xu5P).**

$$3Ru5P \rightleftharpoons R5P + 2Xu5P$$

3. A series of C—C bond cleavage and formation reactions (Fig. 23-26, Reactions 6–8) that convert two molecules of Xu5P and one molecule of R5P to two molecules

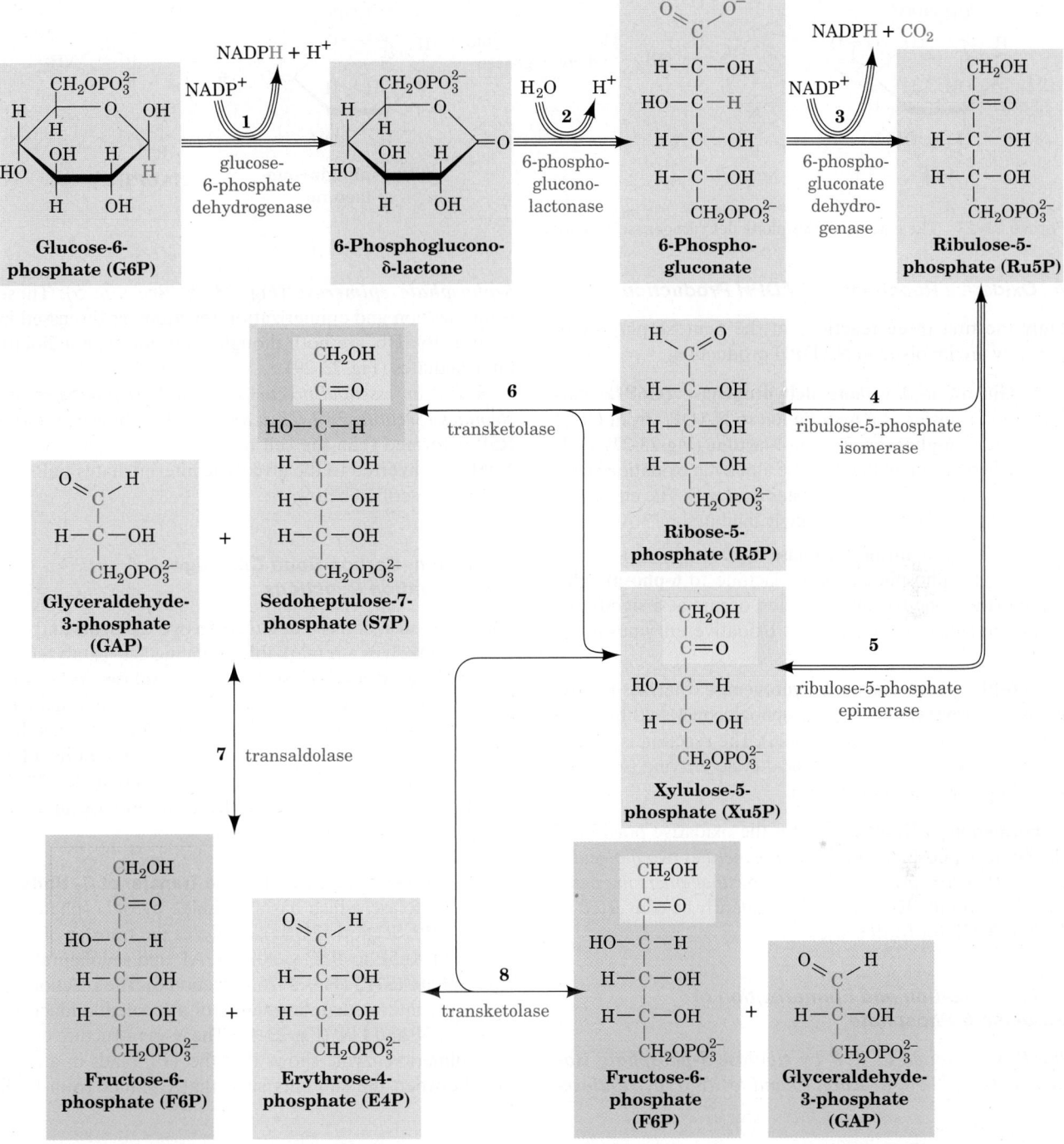

Figure 23-26 **The pentose phosphate pathway.** The number of lines in an arrow represents the number of molecules reacting in one turn of the pathway so as to convert three G6P to three CO_2, two F6P, and one GAP. For the sake of clarity, sugars from Reaction 3 onward are shown in their linear forms. The carbon skeleton of R5P and the atoms derived from it are drawn in red and those from Xu5P are drawn in green. The C_2 units transferred by transketolase are shaded in green and the C_3 units transferred by transaldolase are shaded in blue. Double-headed arrows indicate reversible reactions.

of fructose-6-phosphate (F6P) and one of glyceraldehyde-3-phosphate (GAP).

$$R5P + 2Xu5P \rightleftharpoons 2F6P + GAP$$

The reactions of Stages 2 and 3 are freely reversible so that the products of the pathway vary with the needs of the cell.

For example, when R5P is required for nucleotide biosynthesis, Stage 3 works in reverse, producing R5P from F6P and GAP nonoxidatively. In this section, we discuss the three stages of the pentose phosphate pathway and how this pathway is controlled. We close by considering the consequences of one of its abnormalities.

Figure 23-27   **The glucose-6-phosphate dehydrogenase reaction.**

A. *Oxidative Reactions of NADPH Production*

Only the first three reactions of the pentose phosphate pathway are involved in NADPH production.

1. Glucose-6-phosphate dehydrogenase (G6PD) catalyzes net transfer of a hydride ion to $NADP^+$ from C1 of G6P to form **6-phosphoglucono-δ-lactone** (Fig. 23-27). G6P, a cyclic hemiacetal with C1 in the aldehyde oxidation state, is thereby oxidized to a cyclic ester (lactone). The enzyme is specific for $NADP^+$ and is strongly inhibited by NADPH.

2. 6-Phosphogluconolactonase increases the rate of hydrolysis of 6-phosphoglucono-δ-lactone to **6-phosphogluconate** (the nonenzymatic reaction occurs at a significant rate), the substrate of the next oxidative enzyme in the pathway.

3. 6-Phosphogluconate dehydrogenase catalyzes the oxidative decarboxylation of 6-phosphogluconate, a β-hydroxy acid, to Ru5P and CO_2 (Fig. 23-28). The reaction is similar to that catalyzed by the citric acid cycle enzyme isocitrate dehydrogenase (Section 21-3C).

Formation of Ru5P completes the oxidative portion of the pentose phosphate pathway. *It generates two molecules of NADPH for each molecule of G6P that enters the pathway.* The product Ru5P must subsequently be converted to R5P or Xu5P for further use.

B. *Isomerization and Epimerization of Ribulose-5-Phosphate*

Ru5P is converted to R5P by **ribulose-5-phosphate isomerase** *(Fig. 23-26, Reaction 4) and to Xu5P by* **ribulose-**

5-phosphate epimerase *(Fig. 23-26, Reaction 5).* These isomerization and epimerization reactions, as discussed in Section 16-2Db, are both thought to occur via enediolate intermediates (Fig. 23-29).

R5P is an essential precursor in the biosynthesis of nucleotides (Sections 28-1, 28-2, and 28-5). If, however, more R5P is formed than the cell needs, the excess, along with Xu5P, is converted to the glycolytic intermediates F6P and GAP as described below.

C. *Carbon–Carbon Bond Cleavage and Formation Reactions*

The conversion of three C_5 sugars to two C_6 sugars and one C_3 sugar involves a remarkable "juggling act" catalyzed by two enzymes, **transaldolase** and **transketolase**. As we discussed in Section 16-2E, enzymatic reactions that make or break carbon–carbon bonds usually have mechanisms that involve generation of a stabilized carbanion and its addition to an electrophilic center such as an aldehyde. This is the dominant theme of both the transaldolase and the transketolase reactions.

a. Transketolase Catalyzes the Transfer of C_2 Units

Transketolase, which has a thiamine pyrophosphate cofactor (TPP; Section 17-3Ba), catalyzes the transfer of a C_2 unit from Xu5P to R5P, yielding GAP and **sedoheptulose-7-phosphate (S7P)** (Fig. 23-26, Reaction 6). The reaction involves the intermediate formation of a covalent adduct between Xu5P and TPP (Fig. 23-30). The X-ray structure of this homodimeric enzyme shows that the TPP binds in a deep cleft between the subunits such that residues from both

Figure 23-28   **The phosphogluconate dehydrogenase reaction.** Oxidation of the OH group forms an easily decarboxylated

β-keto acid (although the proposed intermediate has not been isolated).

Figure 23-29   Ribulose-5-phosphate isomerase and ribulose-5-phosphate epimerase. The reactions catalyzed by both these enzymes involve enediolate intermediates. In the isomerase reaction (*right*), a base on the enzyme removes a proton from C1 of Ru5P to form a 1,2-enediolate and then adds a proton at C2 to form R5P. In the epimerase reaction (*left*), a base on the enzyme removes a C3 proton to form a 2,3-enediolate. A proton is then added to the same carbon atom but with inversion of configuration to yield Xu5P.

Figure 23-30   Mechanism of transketolase. Transketolase utilizes the coenzyme thiamine pyrophosphate to stabilize the carbanion formed on cleavage of the C2—C3 bond of Xu5P. The reaction occurs as follows: **(1)** The TPP ylid attacks the carbonyl group of Xu5P. **(2)** C2—C3 bond cleavage yields GAP and enzyme-bound 2-(1,2-dihydroxyethyl)-TPP, a resonance-stabilized carbanion. **(3)** The C2 carbanion attacks the aldehyde carbon of R5P forming an S7P–TPP adduct. **(4)** TPP is eliminated yielding S7P and the regenerated TPP–enzyme.

subunits participate in its binding, just as in pyruvate decarboxylase (another TPP-requiring enzyme; Figure 17-28). In fact the structures are so similar that it is likely that they diverged from a common ancestor.

b. Transaldolase Catalyzes the Transfer of C₃ Units

Transaldolase catalyzes the transfer of a C_3 unit from S7P to GAP, yielding **erythrose-4-phosphate (E4P)** and F6P (Fig. 23-26, Reaction 7). The reaction occurs by aldol cleavage, which begins with the formation of a Schiff base between an ε-amino group of an essential enzyme Lys residue and the carbonyl group of S7P (Fig. 23-31). Transaldolase and Class I aldolase (Section 17-2Da) share a common reaction mechanism and may also share a common ancestor, despite their lack of significant sequence identity. Both are α/β barrel proteins (Section 8-3Bh), but while the Schiff base–forming Lys is on β4 (the fourth β strand from the N-terminus) of transaldolase, it is on β6 of Class I aldolase. Superimposing the barrel structures of these two enzymes while maintaining the alignment of the β strands bearing the Schiff base–forming Lys residues results in a significantly better fit than doing so while maintaining the alignment of their entire α/β barrels. Moreover, five of the pairs of matched active site residues in the former superposition are identical. This suggests that, during evolution, the DNA sequence for two α/β units was transferred from the N-terminus to the C-terminus of the evolving Class I aldolase, moving the active site Lys from β6 to β4. Such a circular permutation of an α/β barrel's structural elements does not greatly change its structure.

c. A Second Transketolase Reaction Yields GAP and a Second F6P Molecule

In a second transketolase reaction, a C_2 unit is transferred from a second molecule of Xu5P to E4P to form GAP and another molecule of F6P (Fig. 23-26, Reaction 8). The third phase of the pentose phosphate pathway thus transforms two molecules of Xu5P and one of R5P to two molecules of F6P and one molecule of GAP. These carbon skeleton transformations (Fig. 23-26, Reactions 6–8) are summarized in Fig. 23-32.

D. *Control of the Pentose Phosphate Pathway*

The principal products of the pentose phosphate pathway are R5P and NADPH. The transaldolase and transketolase reactions serve to convert excess R5P to glycolytic intermediates

Figure 23-31   Mechanism of transaldolase. Transaldolase contains an essential Lys residue that forms a Schiff base with S7P to facilitate an aldol cleavage reaction. The reaction occurs as follows: **(1)** The ε-amino group of an essential Lys residue forms a Schiff base with the carbonyl group of S7P. **(2)** A Schiff base–stabilized C3 carbanion is formed in an aldol cleavage reaction between C3 and C4 that eliminates E4P. **(3)** The enzyme-bound resonance-stabilized carbanion adds to the carbonyl C atom of GAP, forming F6P linked to the enzyme via a Schiff base. **(4)** The Schiff base hydrolyzes, regenerating active enzyme and releasing F6P.

(6) $\quad C_5 + C_5 \rightleftharpoons C_7 + C_3$

(7) $\quad C_7 + C_3 \rightleftharpoons C_6 + C_4$

(8) $\quad \underline{C_5 + C_4 \rightleftharpoons C_6 + C_3}$

(Sum) $\quad 3\,C_5 \rightleftharpoons 2\,C_6 + C_3$

Figure 23-32 Summary of carbon skeleton rearrangements in the pentose phosphate pathway. A series of carbon–carbon bond formations and cleavages convert three C_5 sugars to two C_6 and one C_3 sugar. The number to the left of each reaction is keyed to the corresponding reaction in Fig. 23-26.

when the metabolic need for NADPH exceeds that of R5P in nucleotide biosynthesis. The resulting GAP and F6P can be consumed through glycolysis and oxidative phosphorylation or recycled by gluconeogenesis to form G6P. *In the latter case, 1 molecule of G6P can be converted, via six cycles of the pentose phosphate pathway and gluconeogenesis, to 6 CO_2 molecules with the concomitant generation of 12 NADPH molecules.* When the need for R5P outstrips that for NADPH, F6P and GAP can be diverted from the glycolytic pathway for use in the synthesis of R5P by reversal of the transaldolase and transketolase reactions. In fact, mass spectral analysis of the ^{13}C-labeled carbons from [1,2-^{13}C]glucose incorporated into RNA in rapidly proliferating cancer cells has shown that more than ~70% of the *de novo* ribose synthesis arises through this nonoxidative reversal of the pentose phosphate pathway (rather than its forward direction).

Flux through the oxidative pentose phosphate pathway and thus the rate of NADPH production is controlled by the rate of the glucose-6-phosphate dehydrogenase reaction (Fig. 23-26, Reaction 1). The activity of this enzyme, which catalyzes the pentose phosphate pathway's first committed step ($\Delta G = -17.6$ kJ · mol^{-1} in liver), is regulated by the NADP$^+$ concentration (substrate availability). When the cell consumes NADPH, the NADP$^+$ concentration rises, increasing the rate of the glucose-6-phosphate dehydrogenase reaction, thereby stimulating NADPH regeneration.

E. *Glucose-6-Phosphate Dehydrogenase Deficiency*

NADPH is required for several reductive processes in addition to biosynthesis. For example, erythrocyte membrane integrity requires a plentiful supply of reduced glutathione (GSH), a Cys-containing tripeptide (Sections 21-2Ba and 26-4C). A major function of GSH in the erythrocyte is to eliminate H_2O_2 and organic hydroperoxides. H_2O_2, a toxic product of various oxidative processes (Section 22-4Cg), reacts with double bonds in the fatty acid residues of the erythrocyte cell membrane to form organic hydroperoxides. These, in turn, react to cleave fatty acid C—C bonds, thereby damaging the membrane. In erythrocytes, the unchecked buildup of peroxides results in premature cell lysis. Peroxides are eliminated through the action of **glutathione peroxidase,** one of the handful of

enzymes with a selenium cofactor, yielding glutathione disulfide (GSSG).

$$2GSH + R{-}O{-}O{-}H \xrightarrow{\text{glutathione peroxidase}} GSSG + ROH + H_2O$$
Organic hydroperoxide

GSH is subsequently regenerated by the NADPH reduction of GSSG catalyzed by glutathione reductase (Section 21-2Ba).

$$GSSG + NADPH + H^+ \xrightarrow{\text{glutathione reductase}} 2GSH + NADP^+$$

A steady supply of NADPH is therefore vital for erythrocyte integrity.

a. Primaquine Causes Hemolytic Anemia in Glucose-6-Phosphate Dehydrogenase Mutants

A genetic defect, common in African, Asian, and Mediterranean populations, results in severe hemolytic anemia on infection or on the administration of certain drugs including the antimalarial agent **primaquine.**

Primaquine

Similar effects, which go by the name of **favism,** occur when individuals bearing this trait eat **fava beans (broad beans,** *Vicia faba*), a staple Middle Eastern vegetable that contains small quantities of toxic glycosides (the Greek philosopher and mathematician Pythagoras, who lived in the sixth century BCE, forbade his followers from eating fava beans, possibly because of their deleterious effects). This trait has been traced to an altered gene for glucose-6-phosphate dehydrogenase (G6PD). Under most conditions, mutant erythrocytes have sufficient enzyme activity for normal function. Agents such as primaquine and fava beans, however, stimulate peroxide formation, thereby increasing the demand for NADPH to a level that mutant cells cannot meet.

The major reason for low enzymatic activity in affected cells appears to be an accelerated rate of breakdown of the mutant enzyme (protein degradation is discussed in Section 32-6). This explains why patients with G6PD deficiency react to primaquine with hemolytic anemia but recover within a week despite continued primaquine treatment. Mature erythrocytes lack a nucleus and protein synthesizing machinery and therefore cannot synthesize new enzyme molecules to replace degraded ones (they likewise cannot synthesize new membrane components, which is why they are so sensitive to membrane damage in the first place). The initial primaquine treatments result in the lysis of old red blood cells whose defective G6PD has been largely degraded. Lysis products stimulate the release of young cells that contain more enzyme and are therefore better able to cope with primaquine stress.

It is estimated that over 400 million people are deficient in G6PD, which makes this condition the most common human enzymopathy. Indeed, ~400 G6PD variants have been reported and at least 140 of them have been characterized at the molecular level. G6PD is active in a dimer–tetramer equilibrium. Many of the mutation sites in individuals with the most severe G6PD deficiency are at the dimer interface, shifting the equilibrium toward the inactive and unstable monomer.

Several G6PD variants occur with high incidence. For example, the so-called type A⁻ deficiency, which exhibits ~10% of the normal G6PD activity, has an incidence of 11% among African Americans. This variant is also the most common form of G6PD deficiency in sub-Saharan Africa. The variant "Mediterranean" is found throughout the Mediterranean and Middle East regions, and occurs in

65% of Kurdish Jews, the population with the highest known incidence of this trait. The high prevalence of defective G6PD in malarial areas of the world suggests that such mutations confer resistance to the malarial parasite, *Plasmodium falciparum* (as we likewise saw to be the case for the sickle-cell trait; Section 7-3Ab). Indeed, two epidemiological studies involving over 2000 African children with A⁻ G6PD deficiency indicate that this form is associated with an ~50% reduction in the risk of severe malaria for both female heterozygotes and male hemizygotes (G6PD deficiency is an X-linked trait).

In vitro studies indicate that erythrocytes with G6PD deficiency are less suitable hosts for plasmodia than are normal cells. This is presumably because the parasite requires the products of the pentose phosphate pathway and/or because the erythrocyte is lysed before the parasite has had a chance to mature. Thus, like the sickle-cell trait, *a defective G6PD confers a selective advantage on individuals living where malaria is endemic.*

CHAPTER SUMMARY

1 Gluconeogenesis Lactate, pyruvate, citric acid cycle intermediates, and many amino acids may be converted, by gluconeogenesis, to glucose via the formation of oxaloacetate. For this to occur, the three irreversible steps of glycolysis must be bypassed. The pyruvate kinase reaction is bypassed by converting pyruvate to oxaloacetate in an ATP-driven reaction catalyzed by the biotinyl-containing enzyme pyruvate carboxylase. The two phases of the pyruvate decarboxylase reaction are catalyzed on different active sites of the homotetrameric enzyme, which translocates its covalently linked carboxybiotinyl group from its BC domain to the CT domain of a neighboring subunit. The oxaloacetate is subsequently decarboxylated and phosphorylated by GTP to form PEP in a reaction catalyzed by PEPCK. For this to happen in species in which PEPCK is a cytosolic enzyme, the oxaloacetate must be transported from the mitochondrion to the cytosol via its interim conversion to either malate or aspartate. Conversion to malate concomitantly transports reducing equivalents to the cytosol in the form of NADH. The two other irreversible steps of glycolysis, the PFK reaction and the hexokinase reaction, are bypassed by simply hydrolyzing their products, FBP and G6P, by FBPase and glucose-6-phosphatase, respectively. A glucose molecule may therefore be synthesized from pyruvate at the expense of four ATPs more than are generated by the reverse process.

Glycolysis and gluconeogenesis are reciprocally regulated so as to consume glucose when the demand for ATP is high and synthesize it when the demand is low. The control points in these processes are at pyruvate kinase/pyruvate carboxylase–PEPCK, PFK/FBPase, and hexokinase/glucose-6-phosphatase. Regulation of these enzymes is exerted largely through allosteric interactions, cAMP-dependent enzyme modifications, and, for PEPCK, gene expression. Muscle, which is incapable of gluconeogenesis, transfers much of the lactate it produces to the liver via the blood for conversion to glucose and return to the muscle. This Cori cycle shifts the metabolic burden of oxidative ATP generation for gluconeogenesis from muscle to liver.

2 The Glyoxylate Cycle Animals cannot convert fatty acids to glucose because they lack the enzymes necessary to synthesize oxaloacetate from acetyl-CoA. Plants, however, can do so via the glyoxylate cycle, a glyoxysomal process that converts two molecules of acetyl-CoA to one molecule of succinate via the intermediate formation of glyoxylate. Succinate is converted to oxaloacetate for use in gluconeogensis or the citric acid cycle.

3 Biosynthesis of Oligosaccharides and Glycoproteins Glycosidic bonds are formed by transfer of the monosaccharide unit of a sugar nucleotide to a second sugar unit. Such reactions occur in the synthesis of disaccharides such as lactose and in the synthesis of the carbohydrate components of glycoproteins. In *N*-linked glycoproteins, the carbohydrate component is attached to the protein via an *N*-glycosidic bond to an Asn residue in the sequence Asn-X-Ser/Thr. In *O*-linked glycoproteins, the carbohydrate attachment is an *O*-glycosidic bond to Ser or Thr or, in collagens, to 5-hydroxylysine. In GPI-anchored proteins a glycosylphosphatidylinositol group is linked to the protein through an intermediary phosphoethanolamine bridge, which forms an amide bond to the protein's C-terminal amino acid residue.

Synthesis of *N*-linked oligosaccharides begins in the endoplasmic reticulum with the multistep formation of a lipid-linked precursor consisting of dolichol pyrophosphate bonded to a common 14-residue core oligosaccharide. The carbohydrate is then transferred to an Asn residue of a growing polypeptide chain. The correct folding of the immature *N*-linked glycoprotein is assisted via the calnexin/calreticulin cycle and it is subsequently transferred, via a membranous vesicle, to the cis Golgi network of the Golgi apparatus. Processing is completed by the trimming of mannose residues followed by attachment of a variety of other monosaccharides as catalyzed by specific enzymes in the cis, medial, and trans Golgi cisternae. Completed *N*-linked glycoproteins are sorted in the trans Golgi network according to the identities of their carbohydrate components for transport, via membranous vesicles, to their final cellular destinations. Three major types of N-linked oligosaccharides have been identified, high mannose, complex, and hybrid oligosaccharides, all of which contain a common pentasaccharide core. Studies of glycoprotein formation have been facilitated by the use of antibiotics, such as tunicamycin and bacitracin, which inhibit specific enzymes involved in the synthesis of these oligosaccharides.

O-Linked oligosaccharides are synthesized in the Golgi apparatus by sequential attachments of specific monosaccharide units to certain Ser or Thr residues. Carbohydrate components of glycoproteins are thought to act as recognition markers for the transport of glycoproteins to their proper cellular destinations and for cell–cell and antibody recognition. The GPI membrane anchor is appended to proteins on the luminal surface of the endoplasmic reticulum, thereby targeting GPI-anchored proteins to the external surface of the plasma membrane.

4 The Pentose Phosphate Pathway The cell uses NAD^+ in oxidative reactions and employs NADPH in reductive biosynthesis. NADPH is synthesized by the pentose phosphate pathway, an alternate mode of glucose oxidation. This pathway also synthesizes R5P for use in nucleotide biosynthesis. The first three reactions of the pentose phosphate pathway involve oxidation of G6P to Ru5P with release of CO_2 and formation of two NADPH molecules. This is followed by reactions that either isomerize Ru5P to R5P or epimerize it to Xu5P. Each molecule of R5P not required for nucleotide biosynthesis, together with two Xu5P, is converted to two molecules of F6P and one molecule of GAP via the sequential actions of transketolase, transaldolase, and, again, transketolase. The products of the pentose phosphate pathway depend on the needs of the cell. The F6P and GAP may be metabolized through glycolysis and the citric acid cycle or recycled via gluconeogenesis. If NADPH is in excess, the latter portion of the pentose phosphate pathway may be reversed to synthesize R5P from glycolytic intermediates. The pentose phosphate pathway is controlled at its first committed step, the glucose-6-phosphate dehydrogenase reaction, by the $NADP^+$ concentration. A genetic deficiency in glucose-6-phosphate dehydrogenase leads to hemolytic anemia on administration of the antimalarial drug primaquine. This X-linked deficiency, which results from the accelerated degradation of the mutant enzyme, provides resistance against severe malaria to female heterozygotes and male hemizygotes for this sex-linked trait.

REFERENCES

Gluconeogenesis

Croniger, C.M., Olswang, Y., Reshef, L., Kalhan, S.C., Tilghman, S.M., and Hanson, R.W., Phosphoenolpyruvate carboxykinase revisited. Insights into its metabolic role, *Biochem. Mol. Biol. Educ.* **30,** 14–20 (2002); *and* Croniger, C.M., Chakravarty, K., Olswang, Y., Cassuto, H., Reshef, L., and Hanson, R.W., Phosphoenolpyruvate carboxykinase revisited. II. Control of PEPCK-C gene expression, *Biochem. Mol. Biol. Educ.* **30,** 353–362 (2002).

Knowles, J.R., The mechanism of biotin-dependent enzymes, *Annu. Rev. Biochem.* **58,** 195–221 (1989).

Matte, A., Tari, L.W., Goldie, H., and Delbaere, T.J., Structure and mechanism of phosphoenolpyruvate carboxykinase, *J. Biol. Chem.* **272,** 8105–8108 (1997).

Pilkis, S.J., Mahgrabi, M.R., and Claus, T.H., Hormonal regulation of hepatic gluconeogenesis and glycolysis, *Annu. Rev. Biochem.* **57,** 755–783 (1988).

Rothman, D.L., Magnusson, I., Katz, L.D., Shulman, R.G., and Shulman, G.I., Quantitation of hepatic gluconeogenesis in fasting humans with ^{13}C NMR, *Science* **254,** 573–576 (1991).

St. Maurice, M., Reinhardt, L., Surinya, K.H., Attwood, P.V., Wallace, J.C., Cleland, W.W., and Rayment, I., Domain architecture of pyruvate carboxylase, a biotin-dependent multifunctional enzyme, *Science* **317,** 1076–1079 (2007).

Van Schaftingen, E., and Gerin, I., The glucose-6-phosphatase system, *Biochem. J.* **362,** 513–532 (2002).

Yang, J., Kalhan, S. C., and Hanson, R.W., What is the metabolic role of phosphoenolpyruvate carboxykinase? *and* Yang, J., Rashef, L., Cassuto, H., Aleman, G., and Hanson, R. W., Aspects of control of phosphoenolpyruvate carboxykinase gene transcription, *J. Biol. Chem.* **284,** 27025–27029 *and* 27031–27035 (2009).

The Glyoxylate Cycle

Eastmond, P.J. and Graham, I.A., Re-examining the role of the glyoxylate cycle in oilseeds, *Trends Plant Sci.* **6,** 72–77 (2001).

Oligosaccharide Biosynthesis

Abeijon, C. and Hirschberg, C.B., Topography of glycosylation reactions in the endoplasmic reticulum, *Trends Biochem. Sci.* **17,** 32–36 (1992).

Aebi, C., Bernasconi, R., Clerc, S., and Molinari, M., N-glycan structures: recognition and processing in the ER, *Trends Biochem. Sci.* **35,** 74–82 (2010).

Bause, E., Wesemann, M., Bartoschek, A., and Breuer, W., Epoxyethylglycyl peptides as inhibitors of oligosaccharyltransferase: double-labeling of the active site, *Biochem. J.* **322,** 95–102 (1997).

Burda, P. and Aebi, M., The dolichol pathway of N-linked glycosylation, *Biochim. Biophys. Acta* **1426,** 239–257 (1999).

Elbein, A.D., Inhibitors of the biosynthesis and processing of N linked oligosaccharide chains, *Annu. Rev. Biochem.* **56,** 497–534 (1987).

Englund, P.T., The structure and biosynthesis of glycosyl phosphatidylinositol protein anchors, *Annu. Rev. Biochem.* **62,** 65–100 (1993).

Ferguson, M.A.J., Brimacombe, J.S., Brown, J.R., Crossman, A., Dix, A., Field, R.A., Güther, M.L.S., Milne, K.G., Sharma, D.K., and Smith, T.K., The GPI biosynthetic pathway as a therapeutic target for African sleeping sickness, *Biochim. Biophys. Acta* **1455,** 327–340 (1999).

Florman, H.M. and Wasserman, P.M., *O*-Linked oligosaccharides of mouse egg ZP3 account for its sperm receptor activity, *Cell* **41,** 313–324 (1985).

Helenius, A. and Aebi, M., Intracellular functions of N-linked glycans, *Science* **291,** 2364–2369 (2001).

Helenius, A., Trombetta, E.S., Hebert, J.N., and Simons, J.F., Calnexin, calreticulin and the folding of glycoproteins, *Trends Cell Biol.* **7,** 193–200 (1997).

Hirschberg, C.B. and Snider, M.D., Topography of glycosylation in the rough endoplasmic reticulum and the Golgi apparatus, *Annu. Rev. Biochem.* **56,** 63–87 (1987).

Kornfeld, R. and Kornfeld, S., Assembly of asparagine-linked oligosaccharides, *Annu. Rev. Biochem.* **54,** 631–664 (1985).

Lairson, L.L., Henrissat, B., Davies, G.J., and Withers, S.G., Glycosyltransferases: Structures, functions, and mechanisms, *Annu. Rev. Biochem.* **77,** 521–555 (2008).

Maeda, Y., Watanabe, R., Harris, C.L., Hong, Y., Ohishi, K., Kinoshita, K., and Kinoshita, T., PIG-M transfers the first mannose to glycosylphosphatidylinositol on the lumenal side of the ER, *EMBO J.* **20,** 250–261 (2001).

Parodi, A.J., Role of *N*-oligosaccharide endoplasmic reticulum processing reactions in glycoprotein folding and degradation, *Biochem. J.* **348,** 1–13 (2000); *and* Protein glucosylation and its role in protein folding, *Annu. Rev. Biochem.* **69,** 69–93 (2000).

Sanyal, S., Frank, C.G., and Menon, A.K., Distinct flippases translocate glycerophospholipids and oligosaccharide diphosphate dolichols across the endoplasmic reticulum, *Biochemistry* **47,** 7937–7946 (2008).

Schachter, H., Enzymes associated with glycosylation, *Curr. Opin. Struct. Biol.* **1,** 755–765 (1991).

Schrag, J.D., Bereron, J.J.M., Li, Y., Borisova, S., Hahn, M., Thomas, D.Y., and Cygler, M., The structure of calnexin, an ER chaperone involved in quality control of protein folding, *Mol. Cell* **8,** 633–644 (2001).

Shaper, J.H. and Shaper, N.L., Enzymes associated with glycosylation, *Curr. Opin. Struct. Biol.* **2,** 701–709 (1992).

Tartakoff, A.M. and Singh, N., How to make a glycoinositol phospholipid anchor, *Trends Biochem. Sci.* **17,** 470–473 (1992).

Taylor, M.E. and Drickamer, K., *Introduction to Glycobiology* (2nd ed.), Oxford University Press (2006).

von Figura, K. and Hasilik, A., Lysosomal enzymes and their receptors, *Annu. Rev. Biochem.* **55,** 167–193 (1986).

The Pentose Phosphate Pathway

Adams, M.J., Ellis, G.H., Gover, S., Naylor, C.E., and Phillips, C., Crystallographic study of coenzyme, coenzyme analogue and substrate binding in 6-phosphogluconate dehydrogenase: Implications for NADP specificity and enzyme mechanism, *Structure* **2,** 651–668 (1994).

Au, S.W.N., Gover, S., Lam, V.M.S., and Adams, M.J., Human glucose-6-phosphate dehydrogenase: The crystal structure reveals a structural NADP$^+$ molecule and provides insights into enzyme deficiency, *Structure* **8,** 293–303 (2000).

Beutler, E., The molecular biology of G6PD variants and other red cell enzyme defects, *Annu. Rev. Med.* **43,** 47–59 (1992).

Cappellini, M.D. and Fiorelli, G., Glucose-6-phosphate dehydrogenase deficiency, *Lancet* **375,** 64–74 (2008).

Jia, J., Huang, W., Schörken, U., Sahm, H., Sprenger, G.A., Lindqvist, Y., and Schneider, G., Crystal structure of transaldolase B from Escherichia coli suggests a circular permutation of the α/β barrel within the class I aldolase family, *Structure* **4,** 715–724 (1996).

Lindqvist, Y. and Schneider, G., Thiamin diphosphate dependent enzymes: transketolase, pyruvate oxidase and pyruvate decarboxylase, *Curr. Opin. Struct. Biol.* **3,** 896–901 (1993); *and* Muller, Y.A., Lindqvist, Y., Furey, W., Schulz, G.E., Jordan, F., and Schneider, G., A thiamin diphosphate binding fold revealed by comparison of the crystal structures of transketolase, pyruvate oxidase and pyruvate decarboxylase, *Structure* **1,** 95–103 (1993).

Ruwende, C., et al., Natural selection of hemi- and heterozygotes for G6PD deficiency in Africa by resistance to severe malaria, *Nature* **376,** 246–249 (1995).

Vulliamy, T., Mehta, A., and Luzzato, L., Glucose-6-phosphate dehydrogenase deficiency, Chap. 179 *in* Valle, D. (Ed.), *The Online Metabolic & Molecular Bases of Inherited Disease* http://www.ommbid.com/

Wood, T., *The Pentose Phosphate Pathway*, Academic Press (1985).

PROBLEMS

1. Compare the relative energetic efficiencies, in ATPs per mole of glucose oxidized, of glucose oxidation via glycolysis + the citric acid cycle versus glucose oxidation via the pentose phosphate pathway + gluconeogenesis. Assume that NADH and NADPH are each energetically equivalent to 2.5 ATP.

2. Although animals cannot synthesize glucose from acetyl-CoA, if a rat is fed ^{14}C-labeled acetate, some of the label will appear in the glycogen extracted from its muscles. Explain.

3. Substances that inhibit specific trimming steps in the processing of *N*-linked glycoproteins have been useful tools in elucidating the pathway of this process. Explain.

4. Through clever genetic engineering you have developed an unregulatable enzyme that can interchangeably use NAD$^+$ or NADP$^+$ in a redox reaction. What would be the physiological consequence(s) on an organism of having such an enzyme?

5. What is the free energy change of the reaction

$$\text{NADH} + \text{NADP}^+ \rightleftharpoons \text{NAD}^+ + \text{NADPH}$$

under physiological conditions? Assume that $\Delta G^{\circ\prime} = 0$ for this reaction and that $T = 37°C$.

6. If G6P is ^{14}C-labeled at its C2 position, what is the distribution of the radioactive label in the products of the pentose phosphate pathway after one turnover of the pathway? What is the distribution of the label after passage of these products through gluconeogenesis followed by a second round of the pentose phosphate pathway?

7. After feeding rapidly growing and proliferating cells [1,2-^{13}C]glucose and isolating the RNA, you find that both the C1 and C2 atoms of the ribosyl units are labeled. Show, using chemical structures and the appropriate enzymes, how the pentose phosphate pathway can yield this distribution of the label.

8. The relative metabolic activities in an organism of glycolysis + the citric acid cycle versus the pentose phosphate pathway + gluconeogenesis can be measured by comparing the rates of $^{14}CO_2$ generation on administration of glucose labeled with ^{14}C at C1 with that of glucose labeled at C6. Explain.

9. (a) Describe the lengths of the products of the transketolase reaction when the two substrates are both five-carbon sugars. (b) Describe the products of the reaction when the substrates are a five-carbon aldose and a six-carbon ketose. Does it matter which of the substrates binds to the enzyme first?

10. In light of the finding that an otherwise benign or even advantageous mutation leads to abnormal primaquine sensitivity combined with the fact that human beings have enormous genetic complexity, comment on the possibility of developing drugs that exhibit no atypical side effects in any individual.

11. Glucose-6-phosphatase is located inside the endoplasmic reticulum. Describe the probable symptoms of a defect in G6P transport across the endoplasmic reticulum membrane.

Photosynthesis

Life on Earth depends on the sun. *Plants and cyanobacteria chemically sequester light energy through photosynthesis, a light-driven process in which CO_2 is "fixed" to yield carbohydrates (CH_2O).*

$$CO_2 + H_2O \xrightarrow{\text{light}} (CH_2O) + O_2$$

This process, in which CO_2 is reduced and H_2O is oxidized to yield carbohydrates and O_2, is essentially the reverse of oxidative carbohydrate metabolism. Photosynthetically produced carbohydrates therefore serve as an energy source for the organism that produced them as well as for nonphotosynthetic organisms that directly or indirectly consume photosynthetic organisms. In fact, even modern industry is highly dependent on the products of photosynthesis because coal, oil, and gas (the so-called fossil fuels) are thought to be the remains of ancient organisms. It is estimated that photosynthesis annually fixes $\sim 10^{11}$ tons of carbon, which represents the storage of over 10^{18} kJ of energy. Moreover, photosynthesis, over the eons, has produced the O_2 in Earth's atmosphere (Section 1-5Cb).

The notion that plants obtain nourishment from such insubstantial things as light and air took nearly two centuries to develop. In 1648, the Flemish physician Jean Baptiste van Helmont reported that growing a potted willow tree from a shoot caused an insignificant change in the weight of the soil in which the tree had been rooted. Although another century was to pass before the law of conservation of matter was formulated, van Helmont attributed the tree's weight gain to the water it had taken up. This idea was extended in 1727 by Stephen Hales, who proposed that plants extract some of their matter from the air.

The first indication that plants produce oxygen was found by the English clergyman and pioneering chemist Joseph Priestley, who reported:

Finding that candles burn very well in air in which plants had grown a long time, and having some reason to think, that there was something attending vegetation, which restored air that had been injured by respiration, I thought it was possible that the same process might also restore the air that had been injured by the burning of candles. Accordingly, on the 17th of August, 1771, I put a sprig of mint into a quantity of air, in which a wax candle had burned out, and found that, on the 27th of the same month, another candle burned perfectly well in it.

Although Priestley later discovered oxygen, which he named "dephlogisticated air," it was Antoine Lavoisier who elucidated its role in combustion and respiration. Nevertheless, Priestley's work inspired the Dutch physician Jan Ingenhousz, who in 1779 demonstrated that the "purifying" power of plants resides in the influence of sunlight on their green parts. In 1782, the Swiss pastor Jean Senebier showed that CO_2, which he called "fixed air," is taken up during photosynthesis. His compatriot Nicolas-Théodore de Saussure found, in 1804, that the combined weights of the organic matter produced by plants and the oxygen they evolve is greater than the weight of the CO_2 they consume. He therefore concluded that water, the only other substance he added to his system, was also necessary for photosynthesis. The final ingredient in the overall photosynthetic recipe was established in 1842 by the German physiologist Robert Mayer, one of the formulators of the first law of thermodynamics, who concluded that plants convert light energy to chemical energy.

1 CHLOROPLASTS

*The site of photosynthesis in eukaryotes (algae and higher plants) is the **chloroplast** (Section 1-2Ag), a member of the membranous subcellular organelles peculiar to plants known as **plastids.*** The first indication that chloroplasts have a photosynthetic function was Theodor Englemann's observation, in 1882, that small, motile, O_2-seeking bacteria congregate at the surface of the alga *Spirogyra*, overlying its single chloroplast, but only while the chloroplast is illuminated. Chloroplasts must therefore be the site of light-induced O_2 evolution, that is, photosynthesis. Chloroplasts, of which there are 1 to 1000 per cell, vary considerably in size and

Figure 24-1   Chloroplast from corn. (*a*) An electron micrograph. (*b*) Schematic diagram. [©George Chapman/Visuals Unlimited.]

shape but are typically ~5-μm-long ellipsoids. Like mitochondria, which they resemble in many ways, chloroplasts have a highly permeable outer membrane and a nearly impermeable inner membrane separated by a narrow intermembrane space (Fig. 24-1). The inner membrane encloses the **stroma,** a concentrated solution of enzymes much like the mitochondrial matrix, that also contains the DNA, RNA, and ribosomes involved in the synthesis of several chloroplast proteins. The stroma, in turn, surrounds a third membranous compartment, the **thylakoid** (Greek: *thylakos,* a sac or pouch). The thylakoid is probably a single highly folded vesicle, although in most organisms it appears to consist of stacks of disklike sacs named **grana,** which are interconnected by unstacked **stromal lamellae.** A chloroplast usually contains 10 to 100 grana. Thylakoid membranes arise from invaginations in the inner membrane of developing chloroplasts and therefore resemble mitochondrial cristae.

The lipids of the thylakoid membrane have a distinctive composition. They consist of only ~10% phospholipids; the majority, ~80%, are uncharged **mono-** and **digalactosyl diacylglycerols,** and the remaining ~10% are the sulfolipids **sulfoquinovosyl diacylglycerols (quinovose** is 6-deoxyglucose):

The acyl chains of these lipids have a high degree of unsaturation, which gives the thylakoid membrane a highly fluid character.

Photosynthesis occurs in two distinct phases:

1. The **light reactions,** which use light energy to generate NADPH and ATP.

2. The **dark reactions,** actually light-independent reactions, which use NADPH and ATP to drive the synthesis of carbohydrate from CO_2 and H_2O.

The light reactions occur in the thylakoid membrane and involve processes that resemble mitochondrial electron transport and oxidative phosphorylation (Sections 22-2 and 22-3). In photosynthetic prokaryotes, which lack chloroplasts, the light reactions take place in the cell's plasma (inner) membrane or in highly invaginated structures derived from it called **chromatophores** (e.g., Fig. 24-2; recall that chloroplasts evolved from cyanobacteria that assumed a symbiotic relationship with a nonphotosynthetic

X = OH   **Galactosyl diacylglycerol**

Digalactosyl diacylglycerol

X = SO_3^-
* = C atom inverted   **Sulfoquinovosyl diacylglycerol**

Figure 24-2   Electron micrograph of a section through the purple photosynthetic bacterium *Rhodobacter sphaeroides.* Its plasma membrane invaginates to form externally connected tubules known as chromatophores (*arrows;* seen here in circular cross section) that are the sites of photosynthesis. [Courtesy of Gerald A. Peters, Virginia Commonwealth University.]

eukaryote; Section 1-2Ag). In eukaryotes, the dark reactions occur in the stroma through a cyclic series of enzyme-catalyzed reactions. In the following sections, we consider the light and dark reactions in detail.

2 LIGHT REACTIONS

In the first decades of the twentieth century, it was generally assumed that light, as absorbed by photosynthetic pigments, directly reduced CO_2, which, in turn, combined with water to form carbohydrate. In this view, CO_2 is the source of the O_2 generated by photosynthesis. In 1931, however, Cornelis van Niel showed that green photosynthetic bacteria, anaerobes that use H_2S in photosynthesis, generate sulfur:

$$CO_2 + 2H_2S \xrightarrow{\text{light}} (CH_2O) + 2S + H_2O$$

The chemical similarity between H_2S and H_2O led van Niel to propose that the general photosynthetic reaction is

$$CO_2 + 2H_2A \xrightarrow{\text{light}} (CH_2O) + 2A + H_2O$$

where H_2A is H_2O in green plants and cyanobacteria and H_2S in photosynthetic sulfur bacteria. This suggests that photosynthesis is a two-stage process in which light energy is harnessed to oxidize H_2A (the light reactions):

$$2H_2A \xrightarrow{\text{light}} 2A + 4[H]$$

and the resulting reducing agent [H] subsequently reduces CO_2 (the dark reactions):

$$4[H] + CO_2 \longrightarrow (CH_2O) + H_2O$$

Thus, in aerobic photosynthesis, H_2O, not CO_2, is photolyzed (split by light).

The validity of van Niel's hypothesis was established unequivocally by two experiments. In 1937, Robert Hill discovered that when isolated chloroplasts that lack CO_2 are illuminated in the presence of an artificial electron acceptor such as ferricyanide $[Fe(CN)_6^{3-}]$, O_2 is evolved with concomitant reduction of the acceptor [to ferrocyanide, $Fe(CN)_6^{4-}$, in our example]. This so-called **Hill reaction** demonstrates that CO_2 does not participate directly in the O_2-producing reaction. It was discovered eventually that the natural photosynthetic electron acceptor is $NADP^+$ (Fig. 13-2), whose reduction product, NADPH, is utilized in the dark reactions to reduce CO_2 to carbohydrate (Section 24-3A). In 1941, when the oxygen isotope ^{18}O became available, Samuel Ruben and Martin Kamen directly demonstrated that the source of the O_2 formed in photosynthesis is H_2O:

$$H_2^{18}O + CO_2 \xrightarrow{\text{light}} (CH_2O) + {}^{18}O_2$$

This section discusses the major aspects of the light reactions.

A. *Absorption of Light*

The principal photoreceptor in photosynthesis is **chlorophyll.** This cyclic tetrapyrrole, like the heme group of globins and cytochromes (Sections 10-1A and 22-2C), is derived biosynthetically from protoporphyrin IX. Chlorophyll, however, differs from heme in four major respects (Fig. 24-3):

1. Its central metal ion is Mg^{2+} rather than Fe(II) or Fe(III).

2. It has a cyclopentenone ring, Ring V, fused to pyrrole Ring III.

3. Pyrrole Ring IV is partially reduced in **chlorophyll *a*** **(Chl *a*)** and **chlorophyll *b* (Chl *b*),** the two major chlorophyll varieties in eukaryotes and cyanobacteria, whereas in **bacteriochlorophyll *a* (BChl *a*)** and **bacteriochlorophyll *b*** **(BChl *b*),** the principal chlorophylls of photosynthetic bacteria, Rings II and IV are partially reduced.

4. The propionyl side chain of Ring IV is esterified to a tetraisoprenoid alcohol. In Chl *a* and *b* as well as in BChl *b* it is **phytol** but in BChl *a* it is either phytol or **geranylgeraniol,** depending on the bacterial species.

In addition, Chl *b* has a formyl group in place of the methyl substituent to atom C3 of Ring II of Chl *a*. Similarly, BChl *a* and BChl *b* have different substituents to atom C4.

a. Light and Matter Interact in Complex Ways

As photosynthesis is a light-driven process, it is worthwhile reviewing how light and matter interact. Electromagnetic radiation is propagated as discrete **quanta (photons)** whose energy E is given by **Planck's law:**

$$E = h\nu = \frac{hc}{\lambda} \qquad [24.1]$$

where h is **Planck's constant** $(6.626 \times 10^{-34}\ \text{J} \cdot \text{s})$, c is the speed of light $(2.998 \times 10^8\ \text{m} \cdot \text{s}^{-1}$ in a vacuum), ν is the frequency of the radiation, and λ is its wavelength (visible light ranges in wavelength from 400 to 700 nm). Thus red light with $\lambda = 680$ nm has an energy of 176 kJ · einstein^{-1} (an **einstein** is a mole of photons).

Molecules, like atoms, have numerous electronic quantum states of differing energies. Moreover, because molecules contain more than one nucleus, each of their electronic states has an associated series of vibrational and rotational substates that are closely spaced in energy (Fig. 24-4). Absorption of light by a molecule usually occurs through the promotion of an electron from its ground (lowest energy) state molecular orbital to one of higher energy. However, *a given molecule can only absorb photons of certain wavelengths because, as is required by the law of conservation of energy, the energy difference between the two states must exactly match the energy of the absorbed photon.*

Chlorophyll   **Iron–protoporphyrin IX**

	R_1	R_2	R_3	R_4
Chlorophyll *a*	$-CH=CH_2$	$-CH_3$	$-CH_2-CH_3$	P
Chlorophyll *b*	$-CH=CH_2$	$\overset{O}{\underset{\|\|}{-C}}-H$	$-CH_2-CH_3$	P
Bacteriochlorophyll *a*	$\overset{O}{\underset{\|\|}{-C}}-CH_3$	$-CH_3^{\,a}$	$-CH_2-CH_3^{\,a}$	P or G
Bacteriochlorophyll *b*	$\overset{O}{\underset{\|\|}{-C}}-CH_3$	$-CH_3^{\,a}$	$=CH-CH_3^{\,a}$	P

[a] No double bond between positions C3 and C4.

Figure 24-3   Chlorophyll structures. The molecular formulas of chlorophylls *a* and *b* and bacteriochlorophylls *a* and *b* are compared to that of iron protoporphyrin IX (heme). The starred atom has the opposite stereochemistry in **chlorophyll *a′* (Chl *a′*).** The isoprenoid phytyl and geranylgeranyl tails presumably increase the chlorophylls' solubility in nonpolar media.

$P = -CH_2$

Phytyl side chain

$G = -CH_2$

Geranylgeranyl side chain

The various chlorophylls are highly conjugated molecules (Fig. 24-3). It is just such molecules that strongly absorb visible light (the spectral band in which the solar radiation reaching Earth's surface is of peak intensity). In fact, the peak molar extinction coefficients of the various chlorophylls, $>10^5$ $M^{-1} \cdot cm^{-1}$, are among the highest known for organic molecules (Fig. 24-5; absorbance spectra are discussed in Section 5-3Ca). Yet the relatively small chemical differences among the various chlorophylls greatly affect their absorption spectra. These spectral differences, as we shall see, are functionally significant.

An electronically excited molecule can dissipate its excitation energy in many ways. Those modes with the greatest photosynthetic significance are as follows (Fig. 24-4):

1. Internal conversion, a common mode of decay in which electronic energy is converted to the kinetic energy of molecular motion, that is, to heat. This process occurs very rapidly, being complete in $<10^{-11}$ s. Many molecules relax in this manner to their ground states. Chlorophyll molecules, however, usually relax only to their lowest excited states. Therefore, *the photosynthetically applicable excitation energy of a chlorophyll molecule that has absorbed a photon in its short-wavelength band, which corresponds to its second excited state, is no different than if it had absorbed a photon in its less energetic long-wavelength band.*

2. Fluorescence, in which an electronically excited molecule decays to its ground state by emitting a photon. Such a process requires $\sim 10^{-8}$ s, so it occurs much more slowly than internal conversion. Consequently, a fluorescently emitted photon generally has a longer wavelength (lower energy) than that initially absorbed. Fluorescence accounts for the dissipation of only 3 to 6% of the light energy absorbed by living plants. However, chlorophyll in solution,

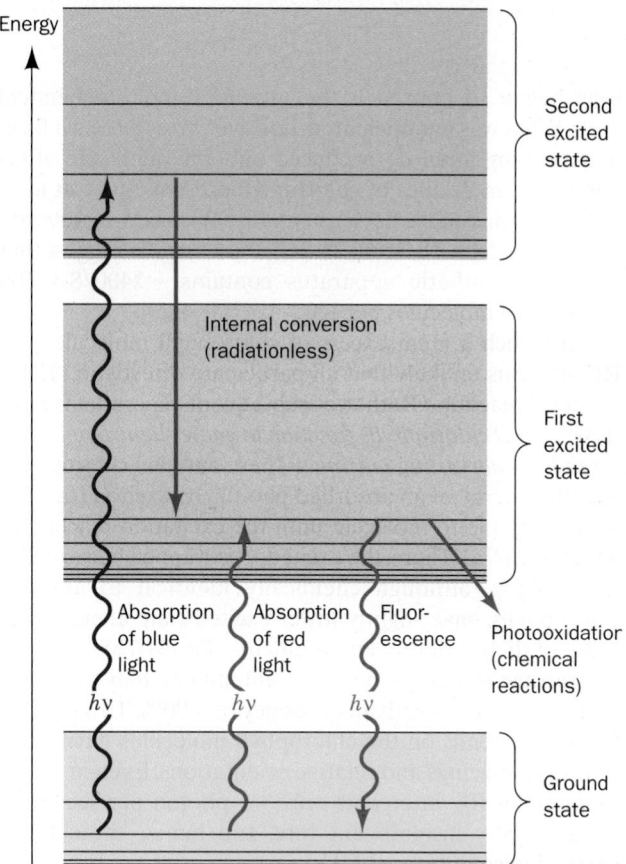

Figure 24-4   Energy diagram indicating the electronic states of chlorophyll and their most important modes of interconversion. The thin black lines denote different vibrational and rotational substates of each electronic state. The wiggly arrows represent the absorption of photons or their fluorescent emission. Excitation energy may also be dissipated in radiationless processes such as internal conversion (heat production) or chemical reactions.

See the Animated Figures

where of course the photosynthetic uptake of this energy cannot occur, has an intense red fluorescence.

3. Exciton transfer (also known as **resonance energy transfer**), in which an excited molecule directly transfers its excitation energy to nearby unexcited molecules with similar electronic properties (the basis of FRET; Section 9-1Cd). This process occurs through interactions between the molecular orbitals of the participating molecules in a manner analogous to the interactions between mechanically coupled pendulums of similar frequencies. An exciton (excitation) may be serially transferred between members of a group of molecules or, if their electronic coupling is strong enough, the entire group may act as a single excited "supermolecule." We shall see that *exciton transfer is of particular importance in funneling light energy to photosynthetic reaction centers.*

4. Photooxidation, in which a light-excited donor molecule is oxidized by transferring an electron to an acceptor molecule, which is thereby reduced. This process occurs because the transferred electron is less tightly bound to the donor in its excited state than it is in the ground state. In photosynthesis, excited chlorophyll (Chl*) is such a donor. *The energy of the absorbed photon is thereby chemically transferred to the photosynthetic reaction system.* Photooxidized chlorophyll, Chl^+, a cationic free radical, eventually returns to its ground state by oxidizing some other molecule.

b. Light Absorbed by Antenna Chlorophylls Is Transferred to Photosynthetic Reaction Centers

The primary reactions of photosynthesis, as is explained in Sections 24-2B and 24-2C, take place at **photosynthetic reaction centers (RCs).** Yet *photosynthetic organelles contain far more chlorophyll molecules than RCs.* This was

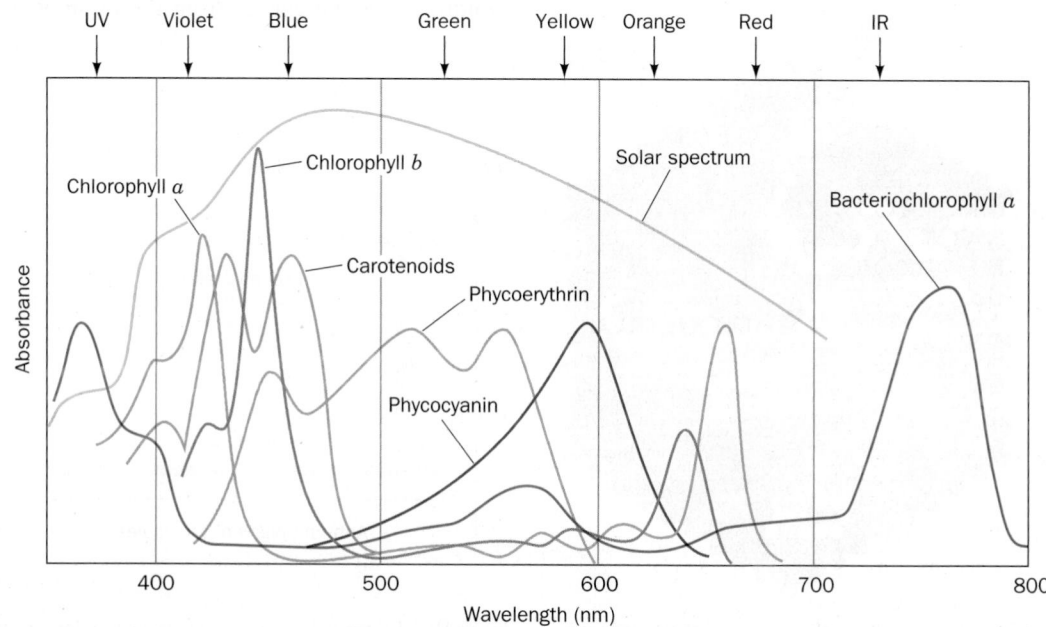

Figure 24-5   Absorption spectra of various photosynthetic pigments. The chlorophylls each have two absorption bands, one in the red and one in the blue. Phycoerythrin absorbs blue and

green light, whereas phycocyanin absorbs yellow light. Together, these pigments absorb most of the visible light in the solar spectrum. [After a drawing by Govindjee, University of Illinois.]

Figure 24-6 **The amount of O_2 evolved by *Chlorella* algae versus the intensity of light flashes.** Flashes are separated by dark intervals of >20 ms.

demonstrated in 1932 by Robert Emerson and William Arnold in their studies of O_2 production by the green alga *Chlorella* (a favorite experimental subject), which had been exposed to repeated brief (10-μs) flashes of light. The amount of O_2 generated per flash was maximal when the interval between flashes was at least 20 ms. Evidently, this is the time required for a single turnover of the photosynthetic reaction cycle. Emerson and Arnold then measured the variation of O_2 yield with flash intensity when the flash interval was the optimal 20 ms. With weak flashes, the O_2 increased linearly with flash intensity such that about one molecule of O_2 was generated per eight photons absorbed (Fig. 24-6). With increasing flash intensity the efficiency of this process fell off, no doubt because the number of pho-

tons began to approach the number of photochemical units. What was unanticipated, however, was that each flash of saturating intensity produced only one molecule of O_2 per ~2400 molecules of chlorophyll present. Since at least eight photons must be sequentially absorbed to liberate one O_2 molecule (Section 24-2C), these results suggest that the photosynthetic apparatus contains ~2400/8 = 300 chlorophyll molecules per RC.

With such a great excess of chlorophyll molecules per RC, it seems unlikely that all participate directly in photochemical reactions. Rather, as subsequent experiments have shown, *most chlorophylls function to gather light; that is, they act as light-harvesting antennas.* These **antenna chlorophylls** pass the energy of an absorbed photon, by exciton transfer, from molecule to molecule until the excitation reaches an RC (Fig. 24-7a). There, the excitation is trapped because RC chlorophylls, although chemically identical to antenna chlorophylls, have slightly lower excited state energies because of their different environments (Fig. 24-7b).

Transfer of energy from the antenna system to an RC occurs in $<10^{-10}$ s with an efficiency of >90%. This high efficiency depends on the chlorophyll molecules having appropriate spacings and relative orientations. Even in bright sunlight, an RC intercepts only ~1 photon per second, a metabolically insignificant rate, and hence, these **light-harvesting complexes (LHCs)** serve an essential function.

c. The LHCs of Purple Photosynthetic Bacteria Contain Multiple Symmetrically Arranged Light-Absorbing Molecules

Most **purple photosynthetic bacteria,** which are among the simplest photosynthetic organisms, have two types of LHCs, **LH1** and **LH2,** that are transmembrane proteins but have different spectral and biochemical properties. LH2, which absorbs light at shorter wavelengths than LH1, rapidly passes the energy from the photons it absorbs to

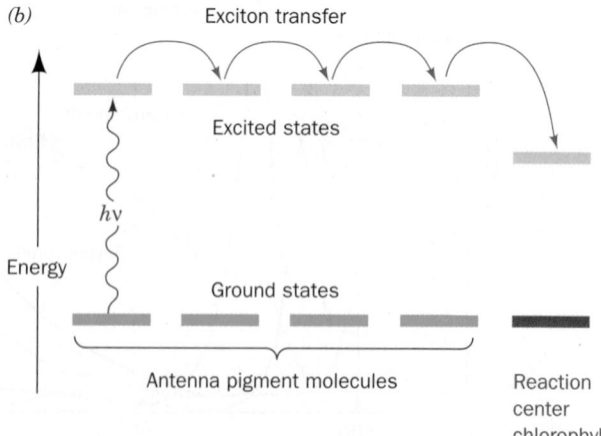

Figure 24-7 **Flow of energy through a photosynthetic antenna complex.** (*a*) The excitation resulting from photon absorption randomly migrates by exciton transfer among the molecules of the antenna complex (*light green circles*) until it is either trapped

by an RC chlorophyll (*dark green circles*) or, less frequently, fluorescently reemitted. (*b*) The excitation is trapped by the RC chlorophyll because its lowest excited state has a lower energy than those of the antenna pigment molecules.

LH1, which, in turn, passes it to the RC. The X-ray structure of LH2 from the purple photosynthetic bacterium *Rhodospirillum (Rs.) molischianum* (Fig. 24-8), determined by Hartmut Michel, reveals that this protein is an 8-fold rotationally symmetric $\alpha_8\beta_8$ 16-mer that binds 24 bacteriochlorophyll *a* (BChl *a*) molecules and 8 **lycopene** molecules (a **carotenoid;** see below):

Lycopene

The α and β subunits (56 and 45 residues, respectively) both consist largely of single helices that are aligned nearly perpendicularly to the plane of the membrane in which they are embedded. The eight α subunits pack side by side to form a hollow cylinder of diameter ~31 Å (as measured between helix axes). Each of the eight β subunits occupies a position radially outward from an α subunit to form a concentric cylinder of diameter ~62 Å. Sixteen of the BChl *a* molecules are packed between these rings of helices in an arrangement resembling a 16-bladed turbine: Successive nearly parallel BChl *a* ring systems are in partial van der Waals contact (their Mg^{2+} ions are ~9 Å apart) with their planes perpendicular to the plane of the membrane. Their Mg^{2+} atoms are each singly axially liganded by His side

chains [much like the Fe(II) in deoxyhemoglobin] that alternately extend from an α and a β subunit around the lower end of the cylinder. The remaining eight BChl *a* molecules, which are each singly axially liganded by a side chain of Asp 6α near the upper end of the cylinder, are arranged in an 8-fold symmetric ring between successive β subunit helices and are oriented with the planes of their ring systems tilted by ~35° relative to the plane of the membrane. The eight lycopene molecules are sandwiched between the α and β subunits and extend along much of their lengths, thereby contacting both sets of BChl *a* molecules. The LH2 from *Rhodopseudomonas (Rps.) acidophila,* another purple photosynthetic bacterium, is an $\alpha_9\beta_9$ 18-mer but otherwise has a similar structure in its transmembrane region to that of *Rs. molischianum,* even though their α and β subunits are only 26 and 31% identical.

Spectroscopic measurements indicate that an LH2's His-liganded and closely associated BChl *a* molecules maximally absorb radiation at a wavelength of 850 nm (and hence are called B850) and are strongly coupled, that is, they absorb radiation almost as a unit. The other, more loosely associated BChl *a* molecules (B800) maximally absorb radiation at 800 nm, largely as individual molecules (BChl *a*'s local environment in the protein alters its spectrum from that in solution; Fig. 24-5). When a B800 BChl *a* absorbs a photon, the excitation is rapidly [in ~700 femtoseconds (fs); 1 fs = 10^{-15} s] transferred to a lower energy B850 BCl *a* (which may independently absorb a photon),

(a)

(b)

Figure 24-8   X-ray structure of LH2 from *Rs. molischianum.* The α subunits are purple and the β subunits are pink. The bound chromophores are drawn in stick form with the BChl *a*'s green and the lycopenes yellow. The phytyl tails of the BChl *a*'s have been truncated for clarity. (*a*) View perpendicular to the bacterial membrane from the cytoplasm. The polypeptide chains are drawn in worm form. (*b*) View parallel to the membrane with

the cytoplasm above. The protein subunits are represented by only their helices, which are shown as cylinders. The Mg^{2+} ions are represented by white spheres. [Courtesy of Juergen Koepke and Hartmut Michel, Max-Planck-Institut für Biochemie, Frankfurt, Germany. PDBid 1LGH.] 🐾 **See Interactive Exercise 20**

Figure 24-9 **Model of the light-absorbing antenna system of purple photosynthetic bacteria.** Several LH2s associate with each other and with LH1, which surrounds the photosynthetic reaction center (RC). The BChl *a*'s of LH2 B850 and LH1 are green, those of LH2 B800 are purple, and the light-absorbing pigments of the RC (see below) are red and black. Light absorbed by the Bchl *a* and lycopene molecules of an LH2 is rapidly transferred (*curved arrows*), often via other contacting LH2s, to LH1, which, in turn, transfers the excitation to its enclosed RC. [From Bhattarchardee, Y., *Nature* **412**, 474 (2001).]

which even more rapidly (in ~100 fs) exchanges the excitation among the other B850 BChl *a* molecules. Hence, the B850 system acts as a kind of energy storage ring that delocalizes the excitation over a large region. The carotenoid molecules in this system absorb visible light (<800 nm) and may also facilitate the transmission of excitation between the rather distantly separated (19 Å between Mg atoms) nearest-neighbor B850 and B800 BChl *a* molecules.

LH1, like LH2, has α and β subunits of ~50 residues each. The low (8.5-Å) resolution structure of LH1 from *Rs. rubrum,* as determined by electron crystallography, reveals that it resembles LH2 but with 16-fold rotational symmetry, and forms a 116-Å-diameter cylinder with a 68-Å-diameter hole down its center. This hole is of sufficient size to contain an RC (see below), as electron microscopy studies indicate is, in fact, the case (Fig. 24-9). LH1's BChl *a* molecules absorb radiation at a longer wavelength than those of LH2 and consequently, when these two assemblies are in contact, excitation is rapidly [in 1–5 picoseconds (ps); 1 ps = 10^{-12} s] transferred from LH2 to LH1 and then (in 20–40 ps) to LH1's enclosed RC. Excitations may also be rapidly exchanged between contacting LH2s. Thus, this antenna system transfers virtually all of the radiation energy it absorbs to the RC in far less than the few nanoseconds (ns; 1 ns = 10^{-9} s) over which these excitations would otherwise decay. It should be noted that this complicated arrangement of **chromophores** (light-absorbing molecules) is among the simplest known; those of the light-harvesting systems of plants are even more elaborate (see below).

d. LHCs Contain Accessory Pigments

Most LHCs contain organized arrays of other light-absorbing substances in addition to chlorophyll. These **accessory pigments** function to fill in the absorption spectra of the antenna complexes in spectral regions where chlorophylls do not absorb strongly (Fig. 24-5). **Carotenoids,** which are C_{40}, largely linear polyenes such as lycopene and **β-carotene,**

β-Carotene

are components of all green plants and many photosynthetic bacteria and are therefore the most common accessory pigments. They are largely responsible for the brilliant fall colors of deciduous trees as well as for the orange color of carrots (after which carotenoids are named).

[Ross M. Horowitz/The Image Bank/Getty Images.]

The light-harvesting protein **LHC-II** of green plants comprises ~30% of the protein in chloroplast membranes, which makes it the most abundant membrane protein in nature. Each subunit of this highly conserved, 232-residue, trimeric protein binds eight Chl *a*'s, six Chl *b*'s, and four carotenoids (Fig. 24-10), thereby accounting for around half the chlorophyll in the biosphere. The orientations of the 42 chlorophylls in each LHC-II trimer evenly sample nearly all directions in space, thus maximizing the efficiency of light harvesting.

Carotenoids serve an additional function besides that of light-gathering antennas: Through electronic interactions, they prevent their associated light-excited chlorophyll molecules from transferring this excitation to O_2, which would otherwise yield highly destructive reactive oxygen species (ROS; Section 22-4Cg). This is particularly important under full sunlight, when the rate that light energy is absorbed exceeds the rate that it can be used in photosynthesis. Then, the excess energy is dissipated as heat through internal

(a)

(b)

stromal surface

lumenal surface

Figure 24-10   X-ray structure of the homotrimeric protein LHC-II from pea chloroplasts. The protein is drawn in ribbon form (*gray*) viewed (*a*) perpendicular to the thylakoid membrane from the stroma and along its 3-fold axis; and (*b*) parallel to the membrane (*gray band*) with the stroma above. Its bound carotenoids and chlorophylls are drawn in stick form with Chl *a* cyan, Chl *b* green, carotenoids orange, and lipids magenta. Mg^{2+} ions are represented by light pink spheres. Each subunit has three transmembrane helices oriented with its N-terminus on the stromal surface. Note the unusually high density of cofactors: Nearly 40% of this protein's nonhydrogen atoms comprise its chlorophylls and carotenoids. [Courtesy of Werner Kühlbrandt, Max Planck Institute of Biophysics, Frankfurt, Germany. PDBid 2BHW.]

conversion by carotenoids, thereby minimizing the irreversible damage to the photosynthetic system that would otherwise occur. In fact, the acidification of the thylakoid lumen resulting from high photosynthetic activity (see below) induces a conformational change in LHC-II. This converts it to a dissipative state by twisting one of its carotenoids and hence changing its electronic properties.

Aquatic photosynthetic organisms, which are responsible for nearly half of the photosynthesis on Earth, additionally contain other types of accessory pigments. This is because light outside the wavelengths 450 to 550 nm (blue and green light) is absorbed almost completely by passage through more than 10 m of water. In red algae and cyanobacteria, Chl *a* is therefore replaced as an antenna pigment by a

series of linear tetrapyrroles, notably the red **phycoerythrobilin** and the blue **phycocyanobilin:**

Peptide-linked phycoerythrobilin and phycocyanobilin

The lowest excited states of these so-called **bilins** have higher energies than those of the chlorophylls, thereby facilitating energy transfer to the RC. The bilins are covalently linked via Cys S atoms to **phycobiliproteins** to form **phycoerythrin** and **phycocyanin** (spectra in Fig. 24-5). These, in turn, are organized in high molecular mass particles called **phycobilisomes** that are bound to the outer faces of photosynthetic membranes so as to funnel excitation energy to RCs over long distances with >90% efficiency.

B. *Electron Transport in Purple Photosynthetic Bacteria*

Photosynthesis is a process in which electrons from excited chlorophyll molecules are passed through a series of acceptors that convert electronic energy to chemical energy. Thus two questions arise: (1) What is the mechanism of energy transduction; and (2) how do photooxidized chlorophyll molecules regain their lost electrons? We shall see that photosynthetic bacteria solve these problems somewhat differently from cyanobacteria and plants. We first discuss these mechanisms in photosynthetic bacteria, where they are simpler and better understood. Electron transport in cyanobacteria and plants is the subject of Section 24-2C.

a. The Photosynthetic Reaction Center Is a Transmembrane Protein Containing a Variety of Chromophores

The first indication that chlorophyll undergoes direct photooxidation during photosynthesis was obtained by Louis Duysens in 1952. He observed that illumination of membrane preparations from the purple photosynthetic bacterium *Rs. rubrum* caused a slight (~2%) bleaching of their absorbance at 870 nm, which returned to their original levels in the dark. Duysens suggested that this bleaching is caused by photooxidation of a bacteriochlorophyll complex that he named **P870** (P for pigment and 870 nm for the position of the major long-wavelength absorption band of BChl *a*; photosynthetic bacteria tend to inhabit murky stagnant ponds, so that they require an infrared-absorbing species of chlorophyll). The ability to detect the presence of P870 eventually led to the purification and characterization of the RC to which it is bound.

RC particles from several species of purple photosynthetic bacteria (**PbRCs**) have similar compositions. That from *Rps. viridis* consists of three hydrophobic subunits: H (258 residues), L (273 residues), and M (323 residues). The L and M subunits of this membrane-spanning protein collectively bind four molecules of BChl *b* (which maximally absorbs light at 960 nm), two molecules of **bacteriopheophytin *b*** (**BPheo *b*;** BChl *b* in which the Mg^{2+} is replaced by two protons), one nonheme/non-Fe–S Fe(II) ion, one molecule of the redox coenzyme ubiquinone (Fig. 22-17*b*), and one molecule of the related **menaquinone**

Menaquinone

(**vitamin K$_2$,** a substance required for proper blood clotting; Section 35-1Ba). In many PbRCs, however, the BChl *b*, BPheo *b*, and menaquinone are replaced by BChl *a*, **BPheo *a*,** and a second ubiquinone, respectively.

The RC of *Rps. viridis,* whose X-ray structure was determined by Johann Deisenhofer, Robert Huber, and Hartmut Michel in 1984, was the first transmembrane protein to be described in atomic detail (Fig. 12-26). *The protein's transmembrane portion consists of 11 α helices that form a 45-Å-long flattened cylinder with the expected hydrophobic surface.* A *c*-type cytochrome containing four hemes, which is an integral constituent of the PbRC complex in only some photosynthetic bacteria, binds to the PbRC on the external side of the plasma membrane. In fact, the PbRC from another bacterial species, *Rhodobacter (Rb.) sphaeroides,* whose X-ray structure (Fig. 24-11) was independently determined by Marianne Schiffer and by Douglas Rees and George Feher, is nearly identical to that of *Rps. viridis* but lacks such a bound cytochrome.

b. Two BChl Molecules Form a "Special Pair"

The most striking aspect of the PbRC is that its chromophoric prosthetic groups are arranged with nearly perfect 2-fold symmetry (Fig. 24-12*a*). This symmetry arises because the L and M subunits, with which these prosthetic groups are exclusively associated, have homologous sequences and similar folds. Two of the BChl *b* molecules in the *Rps. viridis* PbRC, the so-called **special pair,** are closely associated; they are nearly parallel and have an Mg—Mg distance of ~7 Å. The special pair occupies a predominantly hydrophobic region of the protein and each of its Mg^{2+} ions has a His side chain as a fifth ligand. Each member of the special pair is in contact with another His-liganded BChl *b* molecule, which, in turn, is associated with a BPheo *b* molecule. The menaquinone is in close association with the L subunit BPheo *b* (Fig. 24-12*a, right*), whereas the ubiquinone, which is but loosely bound to the protein,

(a)

(b)

Figure 24-11 **A ribbon diagram of the photosynthetic reaction center (RC) from *Rb. sphaeroides*.** (*a*) The H, M, and L subunits, as viewed from within the plane of the plasma membrane with the cytoplasm below, are magenta, cyan, and orange, respectively. The prosthetic groups are drawn in stick form with C green, N blue, and O red. The Fe(II) atom is represented by a yellow sphere. The 11 largely vertical helices that form the central portion of the protein constitute its transmembrane region. Compare this structure with that of the RC from *Rps. viridis* (Fig. 12-26), whose H, M, and L subunits are 39, 50, and 59% identical to those of *Rb. sphaeroides*. Note that the *Rb. sphaeroides* protein lacks the four-heme *c*-type cytochrome (*green* in Fig. 12-26) on its periplasmic surface and that the Q_A prosthetic group, whose quinone ring lies to the right of the Fe(II), is ubiquinone in *Rb. sphaeroides* but menaquinone in *Rps. viridis*. (*b*) View from the extracellular side of the membrane. Note how the transmembrane portions of the M and L subunits are related by a pseudo-2-fold axis passing through the Fe(II) ion and that the prosthetic groups are sandwiched between these two subunits. [Based on an X-ray structure by Marianne Schiffer, Argonne National Laboratory. PDBid 2RCR.] ✏️ **See Interactive Exercise 21 and Kinemage Exercise 8-2**

(a) 0 s

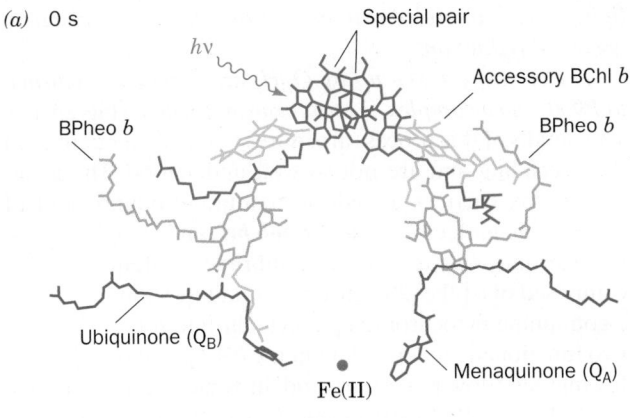

(b) 3×10^{-12} s

(c) 200×10^{-12} s

(d) 100×10^{-6} s

Figure 24-12 **Sequence of excitations in the bacterial RC of** ***Rps. viridis.*** The RC chromophores are shown in the same view as in Fig. 12-26*a*, which resembles that in Fig. 24-11*a*. Note that their rings, but not their aliphatic side chains, are arranged with close to 2-fold symmetry. (*a*) At zero time, a photon is absorbed by the "special pair" of BChl *b* molecules, thereby collectively raising them to an excited state [in each step, the excited molecule(s) is shown in red]. (*b*) Within 3 ps, an excited electron has passed to the BPheo *b* of the L subunit (right arm of the system) without becoming closely associated with the accessory BChl *b*. The special pair is thereby left with a positive charge. (*c*) Some 200 ps later, the excited electron has transferred to the menaquinone (Q_A, which is ubiquinone in *Rb. sphaeroides*). (*d*) Within the next 100 μs, the special pair has been reduced (via an electron-transport chain discussed in the text), thereby eliminating its positive charge, while the excited electron migrates to the ubiquinone (Q_B). After a second such electron has been transferred to Q_B, it picks up two protons from solution and exchanges with the membrane-bound ubiquinone pool.

See Kinemage Exercise 8-2

described below. The Fe(II) is positioned between the menaquinone and ubiquinone rings and is octahedrally liganded by four His side chains and the two carboxyl oxygen atoms of a Glu side chain. Curiously, the two symmetry related groups of chromophores are not functionally equivalent; electrons, as we shall see, are almost exclusively transferred through the L subunit (the right sides of Figs. 24-11 and 24-12). This effect is generally attributed to subtle structural and electronic differences between the L and M subunits.

c. The Electronic States of Molecules Undergoing Fast Reactions Can Be Monitored by EPR and Laser Spectroscopy Techniques

The turnover time of a photosynthetic reaction cycle, as we have seen, is only a few milliseconds. Its sequence of reactions can therefore only be traced by measurements that can follow extremely rapid electronic changes in molecules. Two techniques are well suited to this task:

1. Electron paramagnetic resonance (EPR) spectroscopy [also called **electron spin resonance (ESR) spectroscopy**], which detects the spins of unpaired electrons in a manner analogous to the detection of nuclear spins in NMR spectroscopy. A molecular species with unpaired electrons, such as an organic radical or a transition metal ion, has a characteristic EPR spectrum because its unpaired electrons interact with the magnetic fields generated by the nuclei and the other electrons of the molecule. Paramagnetic species as short lived as 10 ps can exhibit definitive EPR spectra.

2. Optical spectroscopy using pulsed lasers. Laser flashes as brief as 20 attoseconds (as; 1 as = 10^{-18} s) have been generated. By monitoring the bleaching (disappearance) of certain absorption bands and the emergence of others, laser spectroscopy can track the time course of a fast reaction process.

associates with the M subunit BPheo *b* (Fig. 24-12*a, left*). These various chromophores are closely associated with a number of protein aromatic rings, which are therefore also thought to participate in the electron-transfer process

d. Photon Absorption Rapidly Photooxidizes the Special Pair

The sequence of photochemical events mediated by the photosynthetic reaction center is diagrammed in Fig. 24-12:

(a) The primary photochemical event of bacterial photosynthesis is absorption of a photon by the special pair (P870 or **P960** depending on whether it consists of BChl *a* or *b*; here, for argument's sake, we assume it to be P960). This event is nearly instantaneous; it occupies the ~3-fs oscillation time of a light wave. EPR measurements established that P960 is, in fact, a pair of BChl *b* molecules and indicated that the excited electron is delocalized over both of them.

(b) P960*, the excited state of P960, has but a fleeting existence. Laser spectroscopy has demonstrated that within ~3 ps after its formation, P960* has transferred an electron to the BPheo *b* on the right in Fig. 24-12*b* to yield P960$^+$ BPheo b^-. In forming this radical pair, the transferred electron must pass near but seems not to reduce the intervening BChl *b* (which is therefore termed an accessory chlorophyll), although its position strongly suggests that it has an important role in conveying electrons.

(c) By some 200 ps later, the electron has further migrated to the menaquinone (or, in many species, the second ubiquinone), designated Q_A, to form the anionic semiquinone radical Q_A^- All these electron transfers, as diagrammed in Fig. 24-13, are to progressively lower energy states, which makes this process all but irreversible.

Rapid removal of the excited electron from the vicinity of P960$^+$ is an essential feature of the PbRC; this prevents back reactions that would return the electron to P960$^+$ so as to provide the time required for the wasteful internal conversion of its excitation energy to heat. In fact, *this sequence of electron transfers is so efficient that its overall* **quantum yield** *(ratio of molecules reacted to photons absorbed) is virtually 100%.* No man-made device has yet approached this level of efficiency.

e. Electrons Are Returned to the Photooxidized Special Pair via an Electron-Transport Chain

The remainder of the photosynthetic electron-transport process occurs on a much slower timescale. Within ~100 μs after its formation, Q_A^-, which occupies a hydrophobic pocket in the protein, transfers its excited electron to the more solvent-exposed ubiquinone, Q_B, to form Q_B^- (Fig. 24-12*d*). The nonheme Fe(II) is not reduced in this process and, in fact, its removal only slightly affects the electron transfer rate, so that the Fe(II) probably functions to fine-tune the PbRC's electronic character. Q_A never becomes fully reduced; it shuttles between its oxidized and semiquinone forms. Moreover, the lifetime of Q_A^- is so short that it never becomes protonated. In contrast, once the PbRC again becomes excited, it transfers a second electron to Q_B^- to form the fully reduced Q_B^{2-}. This anionic quinol takes up two protons from the solution on the cytoplasmic side of the plasma membrane to form Q_BH_2. Thus Q_B *is a molecular transducer that converts two* *light-driven one-electron excitations to a two-electron chemical reduction.*

The electrons taken up by Q_BH_2 are eventually returned to P960$^+$ via a complex electron-transport chain (Fig. 24-13). The details of this process are more species dependent than the preceding and are not so well understood. The available redox carriers include a membrane-bound pool of ubiquinone molecules, **cytochrome bc_1,** and **cytochrome c_2.** Cytochrome bc_1 is a transmembrane protein complex composed of a [2Fe–2S] cluster–containing subunit; a heme *c*-containing cytochrome c_1; a cytochrome *b* that contains two functionally inequivalent heme *b*'s, b_H and b_L (H and L for high and low potential); and, in some species, a fourth subunit. Note that cytochrome bc_1 is strikingly similar to the proton-translocating Complex III of mitochondria (Section 22-2C3a), which is also called cytochrome bc_1. The electron-transport pathway leads from Q_BH_2 on the cytoplasmic side of the plasma membrane, through the ubiquinone pool, with which Q_BH_2 exchanges, to cytochrome bc_1, and then to cytochrome c_2 on the external (periplasmic) side of the plasma membrane. The reduced cytochrome c_2, which, as its name implies, closely resembles mitochondrial cytochrome *c*, diffuses along the external membrane surface until it reacts with the membrane-spanning PbRC to transfer an electron to P960$^+$ (the structures of several *c*-type cytochromes, including that of cytochrome c_2 from *Rs. rubrum*, are diagrammed in Fig. 9-41). In *Rps. viridis,* the four-heme *c*-type cytochrome bound to the PbRC complex on the external side of the plasma membrane (Fig. 12-26) is interposed between cytochrome c_2 and P960$^+$. Note that one of this *c*-type cytochrome's hemes is positioned to reduce the photooxidized special pair. The PbRC is thereby prepared to absorb another photon.

f. Photosynthetic Electron Transport Drives the Formation of a Proton Gradient

Since electron transport in PbRCs is a cyclic process (Fig. 24-13), *it results in no net oxidation–reduction. Rather, it functions to translocate the cytoplasmic protons acquired by Q_BH_2 across the plasma membrane, thereby making the cell alkaline relative to its environment.* The mechanism of this process is essentially identical to that of proton transport in mitochondrial Complex III (Section 22-3Be); that is, in addition to the translocation of the two H$^+$ resulting from the two-electron reduction of Q_B to QH$_2$, a Q cycle mediated by cytochrome bc_1 translocates two H$^+$ for a total of four H$^+$ translocated per two photons absorbed (Fig. 24-13*a*; also see Fig. 22-31). *Synthesis of ATP, a process known as* **photophosphorylation,** *is driven by the dissipation of the resulting pH gradient in a manner that closely resembles ATP synthesis in oxidative phosphorylation (Section 22-3C).* We further discuss the mechanism of photophosphorylation in Section 24-2D.

Photosynthetic bacteria use photophosphorylation-generated ATP to drive their various endergonic processes. However, unlike cyanobacteria and plants, which generate their required reducing equivalents by the light-driven oxidation of H$_2$O (see below), photosynthetic bacteria must obtain their reducing equivalents from the environment.

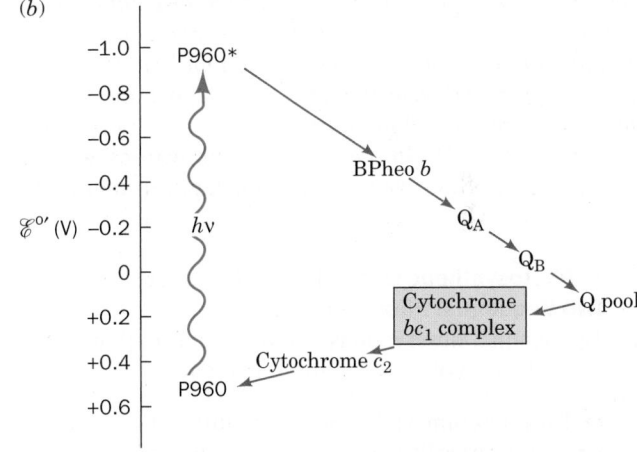

Figure 24-13   Photosynthetic electron-transport system of purple photosynthetic bacteria. (*a*) Schematic diagram indicating the arrangement of the system components in the bacterial plasma membrane and the flows of electrons (*black arrows*) and protons (*blue arrows*) that photon (*hv*) absorption promotes through them. The system contains two protein complexes, the RC and cytochrome bc_1. Two electrons liberated from the special pair, here P870 (as in *Rb. sphaeroides*), by the consecutive absorption of two photons are taken up by ubiquinone (Q_B) together with two protons from the cytoplasm to yield ubiquinol (QH_2). The QH_2 is released from the RC and diffuses (*dotted arrows*) through the membrane to cytochrome bc_1, which, in a two-electron reaction, oxidizes it to ubiquinone with the concomitant liberation of its two protons to the external medium. One of the two electrons is passed, via the [2Fe–2S] cluster and cytochrome c_1, to cytochrome c_2, a peripheral membrane protein that then diffuses across the external surface of the membrane so as to return the electron to P870 of the RC. The second electron from QH_2 passes, via a Q cycle, through hemes b_L and b_H of cytochrome bc_1 and then contributes to the reduction of a molecule of ubiquinone (Q) with the concomitant uptake of two more cytoplasmic protons (two rounds of a Q cycle are required for the reduction of one molecule of Q to QH_2; Fig. 22-31). The resulting QH_2 diffuses back to cytochrome bc_1. There it is again oxidized, with the liberation of its two protons to the exterior and the return of one of its two electrons, via cytochrome c_2, to P870, thereby completing the electrical circuit. Note that in every turn of a Q cycle, half the electrons liberated by the oxidation of

QH_2 to Q are used to reduce Q to QH_2, so that, after a large number of turns, an electron that enters the Q cycle, on average, passes through it twice before being returned to P870. Thus, the net result of the absorption of two photons by the RC is the translocation of four H^+ from the cytoplasm to the external medium. (*b*) The approximate standard reduction potentials of the photosynthetic electron-transport system's various components. The overall process is essentially irreversible because electrons are transferred to progressively lower energy states (more positive standard reduction potentials).

Various substances, such as H_2S, S, $S_2O_3^{2-}$, H_2, and many organic compounds, function in this capacity depending on the bacterial species.

Modern photosynthetic bacteria are thought to resemble the original photosynthetic organisms. These presumably arose very early in the history of cellular life when environmentally supplied sources of "high-energy" compounds were dwindling but reducing agents were still plentiful (Section 1-5Cb). During this era, photosynthetic bacteria were no doubt the dominant form of life. However, their very success eventually caused them to exhaust the available reductive resources. The ancestors of modern cyanobacteria adapted to this situation by evolving a photosynthetic sys-

tem with sufficient electromotive force to abstract electrons from H_2O. The gradual accumulation of the resulting toxic waste product, O_2, forced photosynthetic bacteria, which cannot photosynthesize in the presence of O_2 (although some species have evolved the ability to respire), into the narrow ecological niches to which they are presently confined (Section 1-1Ab).

C. Two-Center Electron Transport

See Guided Exploration 22: Two-center photosynthesis (Z-scheme) **overview** *Plants and cyanobacteria use the reducing power generated by the light-driven oxidation of H_2O to produce*

NADPH. The component half-reactions of this process, together with their standard reduction potentials, are

$$O_2 + 4e^- + 4H^+ \rightleftharpoons 2H_2O \qquad \mathscr{E}^{\circ\prime} = +0.815V$$

and

$$NADP^+ + H^+ + 2e^- \rightleftharpoons NADPH \qquad \mathscr{E}^{\circ\prime} = -0.320V$$

Hence, the overall four-electron reaction and its standard redox potential is

$$2NADP^+ + 2H_2O \rightleftharpoons 2NADPH + O_2 + 2H^+$$
$$\mathscr{E}^{\circ\prime} = -1.135V$$

This latter quantity corresponds (Eq. [16.5]) to a standard free energy change of $\Delta G^{\circ\prime} = 438$ kJ · mol^{-1}, which Eq. [24.1] indicates is the energy of one einstein of 223-nm photons (UV light). Clearly, *even if photosynthesis were 100% efficient, which it is not, it would require more than one photon of visible light to generate a molecule of O$_2$. In fact, experimental measurements indicate that algae minimally require 8 to 10 photons of visible light to produce one molecule of O$_2$.* In the following subsections, we discuss how plants and cyanobacteria manage this multiphoton process.

a. Photosynthetic O$_2$ Production Requires Two Sequential Photosystems

Two seminal observations led to the elucidation of the basic mechanism of photosynthesis in plants:

1. The quantum yield for O$_2$ evolution by *Chlorella pyrenoidosa* varies little with the wavelength of the illuminating light between 400 and 675 nm, but decreases precipitously above 680 nm (Fig. 24-14, *lower curve*). This phenomenon, the "red drop," was unexpected because Chl *a* absorbs such far-red light (Fig. 24-5).

2. Shorter wavelength light, such as yellow-green light, enhances the photosynthetic efficiency of 700-nm light well in excess of the energy content of the shorter wavelength light; that is, *the rate of O$_2$ evolution by both lights is greater than the sum of the rates for each light acting alone (Fig. 24-14, upper curve).* Moreover, this enhancement still occurs if the yellow-green light is switched off several seconds before the red light is turned on and vice versa.

These observations clearly indicate that two processes are involved. They are explained by a mechanistic model, the **Z-scheme,** which postulates that *O$_2$-producing photosynthesis occurs through the actions of two photosynthetic RCs that are connected essentially in series* (Fig. 24-15).

1. Photosystem I (PSI) generates a strong reductant capable of reducing NADP$^+$ and, concomitantly, a weak oxidant.

2. Photosystem II (PSII) generates a strong oxidant capable of oxidizing H$_2$O and, concomitantly, a weak reductant.

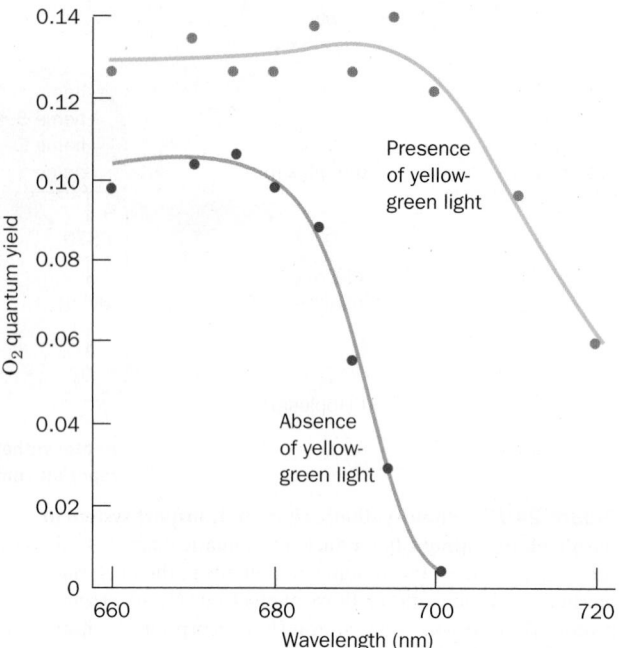

Figure 24-14 **Quantum yield for O$_2$ production by *Chlorella* algae as a function of the wavelength of the incident light.** The experiment was conducted in the absence (*lower curve*) and the presence (*upper curve*) of supplementary yellow-green light. The upper curve has been corrected for the amount of O$_2$ production stimulated by the supplementary light alone. Note that the lower curve falls off precipitously above 680 nm (the red drop). However, the supplementary light greatly increases the quantum yield in the wavelength range above 680 nm (far-red) in which the algae absorb light. [After Emerson, R., Chalmers, R., and Cederstrand, C., *Proc. Natl. Acad. Sci.* **49,** 137 (1957).]

The weak reductant reduces the weak oxidant, so that *PSI and PSII form a two-stage electron "energizer." Both photosystems must therefore function for photosynthesis (electron transfer from H$_2$O to NADP$^+$, forming O$_2$ and NADPH) to occur.*

The red drop is explained in terms of the Z-scheme by the observation that PSII is only poorly activated by 680-nm light. In the presence of only this far-red light, PSI is activated but is unable to obtain more than a few of the electrons it is capable of energizing. Yellow-green light, however, efficiently stimulates PSII to supply these electrons. The observation that the far-red and yellow-green lights can be alternated indicates that both photosystems remain activated for a time after the light is switched off.

The validity of the Z-scheme was established as follows. The oxidation state of **cytochrome f,** a *c*-type cytochrome of the electron-transport chain connecting PSI and PSII (see below), can be spectroscopically monitored. Illumination of algae with 680-nm (far-red) light results in the oxidation of cytochrome *f* (Fig. 24-16). However, the additional imposition of a 562-nm (yellow-green) light results in this

Figure 24-15 **The Z-scheme for photosynthesis in plants and cyanobacteria.** Two photosystems, PSI and PSII, function to drive electrons from H₂O to NADPH. The reduction potential increases downward so that electron flow occurs spontaneously in this direction. The herbicide DCMU (see text) blocks photosynthetic electron transport from PSII to cytochrome *f*.

Figure 24-16 **The oxidation state of cytochrome *f* in *Porphyridium cruentum* algae as monitored by a weak beam of 420-nm (blue-violet) light.** An increase in the transmitted light signals the oxidation of cytochrome *f*. In the upper curve, strong light at 680 nm (far-red) causes the oxidation of the cytochrome *f* but the superposition of 562-nm (yellow-green) light causes its partial re-reduction. In the lower curve, the presence of the herbicide DCMU, which inhibits photosynthetic electron transport, causes 562-nm light to further oxidize, rather than reduce, the cytochrome *f*.

protein's partial re-reduction. In the presence of the herbicide **3-(3,4-dichlorophenyl)-1,1-dimethylurea (DCMU),**

3-(3,4-Dichlorophenyl)-1,1-dimethylurea (DCMU)

which abolishes photosynthetic oxygen production, 680-nm light still oxidizes cytochrome *f* but simultaneous 562-nm light only oxidizes it further. The explanation for these effects is that 680-nm light, which efficiently activates only PSI, causes it to withdraw electrons from (oxidize) cytochrome *f*. The 562-nm light also activates PSII, which thereby transfers electrons to (reduces) cytochrome *f*. DCMU blocks electron flow from PSII to cytochrome *f* (Fig. 24-15), so an increased intensity of light, whatever its wavelength, only serves to activate PSI further.

b. O₂-Producing Photosynthesis Is Mediated by Three Transmembrane Protein Complexes Linked by Mobile Electron Carriers

The components of the Z-scheme, which mediate electron transport from H₂O to NADPH, are largely organized into three thylakoid membrane-bound particles (Fig. 24-17): *(1)*

PSII, (2) the **cytochrome b₆f complex,** *and (3) PSI.* As in oxidative phosphorylation, electrons are transferred between these complexes via mobile electron carriers. The ubiquinone analog **plastoquinone (Q),** via its reduction to **plastoquinol (QH₂),**

Plastoquinone

$$\Updownarrow 2 \, [H\cdot]$$

Plastoquinol

Figure 24-17 **Schematic representation of the thylakoid membrane showing the components of its electron-transport chain.** The system consists of three protein complexes: PSII, the cytochrome $b_6 f$ complex, and PSI, which are electrically "connected" by the diffusion of the electron carriers plastoquinol (Q) and plastocyanin (PC). Light-driven transport of electrons (*black arrows*) from H_2O to $NADP^+$ forming NADPH motivates the transport of protons (*red arrows*) into the thylakoid space (Fd is ferredoxin). Additional protons are split off from water by the oxygen-evolving complex (OEC), yielding O_2. The resulting proton gradient powers the synthesis of ATP by the CF_1CF_0 proton-translocating ATP synthase [CF_1 and CF_0 are chloroplast (C) analogs of mitochondrial F_1 and F_0]. The membrane also contains light-harvesting complexes whose component chlorophylls and other chromophores transfer their excitations to PSI and PSII. [After Ort, D.R. and Good, N.E., *Trends Biochem. Sci.* **13,** 469 (1988).]

links PSII to the cytochrome $b_6 f$ complex, which, in turn, interacts with PSI through the mobile Cu-containing redox protein **plastocyanin (PC).** In what follows, we trace the electron pathway through this chloroplast system from H_2O to $NADP^+$ (Fig. 24-18).

c. PSII Resembles the PbRC

PSII from the thermophilic cyanobacterium *Thermosynechococcus elongatus* consists of 20 subunits, 14 of which occupy the photosynthetic membrane. These transmembrane subunits include the reaction center proteins **D1 (PsbA)** and **D2 (PsbD),** the chlorophyll-containing inner-antenna subunits **CP43 (PsbC)** and **CP47 (PsbB),** and **cytochrome b_{559}.** The X-ray structure of this PSII (Fig. 24-19), independently determined by James Barber and So Iwata and by Wolfram Saenger, reveals that this ~340-kD protein is a symmetric dimer, whose protomeric units each contain 35 transmembrane helices. Each protomer, which has pseudo-2-fold symmetry, binds 36 Chl *a*'s, 2 **pheophytin *a*'s (Pheo *a*'s;** Chl *a* with its Mg^{2+} replaced by two protons), one heme *b*, one heme *c*, 2 plastoquinones, one nonheme

Fe, 12 all-trans carotenoids presumed to be β-carotene, one HCO_3^- ion, and one Mn_4CaO_4 complex known as the **oxygen-evolving center [OEC;** alternatively, the **water-oxidizing complex (WOC)].** In higher plants, the PSII protomer contains ~25 subunits and forms an ~1000-kD transmembrane supercomplex with several antenna proteins. The arrangement of the 5 transmembrane helices in both D1 and D2 resembles that in the L and M subunits of the PbRC (Fig. 24-11). Indeed, these two sets of subunits have similar sequences, thereby indicating that they arose from a common ancestor.

The cofactors of PSII's RC (Fig. 24-20) are organized similarly to those of the bacterial system (Fig. 24-12): They have essentially the same components (with Chl *a*, Pheo *a*, and plastoquinone replacing BChl *b*, BPheo *b*, and menaquinone, respectively) and are symmetrically organized along the complex's pseudo-2-fold axis. The two Chl *a* rings labeled P_{D1} and P_{D2} in Fig. 24-20 are positioned analogously to the BChl *b*'s of P960's "special pair" and are therefore presumed to form PSII's primary electron donor, **P680** (named after the wavelength at which its absorbance

Figure 24-18 Detailed diagram of the Z-scheme of photosynthesis. Electrons ejected from P680 by the absorption of photons are replaced with electrons abstracted from H_2O by an Mn complex (OEC), thereby forming O_2 and four H^+. Each ejected electron is passed through a chain of electron carriers to a pool of plastoquinone molecules (Q). The resulting plastoquinol, in turn, reduces the cytochrome $b_6 f$ particle (*yellow box*) that transfers electrons with the concomitant translocation of protons, via a Q cycle, into the thylakoid lumen. Cytochrome

$b_6 f$ then transfers the electrons to plastocyanin (PC). The plastocyanin regenerates photooxidized P700. The electron ejected from P700, through the intermediacy of a chain of electron carriers (A_0, A_1, F_X, F_A, F_B, and Fd), reduces $NADP^+$ to NADPH in noncyclic electron transport. Alternatively, the electron may be returned to the cytochrome $b_6 f$ complex in a cyclic process that only translocates protons into the thylakoid lumen.

maximally decreases on photooxidation). The electron ejected from P680 follows a similar asymmetric course as that in the PbRC even though the two systems operate over different ranges of reduction potential (compare Figs. 24-13b and 24-18). As indicated in the central part of Fig. 24-18, the electron is transferred to a molecule of Pheo *a* (Pheo$_{D1}$ in Fig. 24-20), probably via a Chl *a* molecule (Chl$_{D1}$), and then to a bound plastoquinone (Q_A). The electron is subsequently transferred to a second plastoquinone molecule, Q_B, which after it receives a second electron in a like manner, takes up two protons at the stromal (cytosolic in cyanobacteria) surface of the thylakoid membrane. The resulting plastoquinol, $Q_B H_2$, then exchanges with a membrane-bound pool of plastoquinone molecules. DCMU as well as many other commonly used herbicides compete with plastoquinone for the Q_B-binding site on PSII, which explains how they inhibit photosynthesis.

Two "extra" Chl *a* molecules, Chlz$_{D1}$ and Chlz$_{D2}$, lie on the periphery of the RC, where they are postulated to func-

tion in the transfer of excitation from the antenna systems to P680. Cytochrome b_{559}, whose function is unclear, breaks the pseudosymmetry of the PSII protomer as does the Mn cluster, whose function we now discuss.

d. O_2 Is Generated in a Five-Stage Water-Splitting Reaction Mediated by an Mn-Containing Protein Complex

The oxidation by the OEC of two molecules of H_2O to form one molecule of O_2 requires four electrons. Since transfer of a single electron from H_2O to $NADP^+$ requires two photochemical events, this accounts for the observed minimum of 8 to 10 photons absorbed per molecule of O_2 produced.

Must the four electrons necessary to produce a given O_2 molecule be removed by a single photosystem or can they be extracted by several different photosystems? Pierre Joliet and Bessel Kok answered this question by analyzing the rate at which dark-adapted chloroplasts produce O_2

(a)

(b)

Figure 24-19 **X-ray structure of PSII from *T. elongatus*.**
(*a*) The PSII dimer is viewed from within the plane of the
membrane with the stroma above. Its transmembrane subunits
include D1 (*yellow*), D2 (*orange*), CP47 (*red*), CP43 (*green*), and
cytochrome b_{559} (*magenta*). Other transmembrane subunits are
colored light blue and blue-gray. Its membrane-extrinsic proteins
are PsbO (*dark blue*), PsbU (*purple*), and PsbV (*light green*). The
various cofactors are drawn in stick form with the chlorophylls of
the D1/D2 reaction center light green, those of the antenna
complexes dark green, pheophytins dark blue, hemes red,
β-carotenes orange, Q_A and Q_B purple, and the nonheme Fe
represented by a red sphere. The inferred position of the
membrane is indicated by the light blue band. (*b*) View of a PSII
protomer perpendicular to the membrane from the thylakoid
lumen showing only the transmembrane portions of the complex
and colored as in Part *a*. A portion of the other protomer in the
PSII dimer is shown in muted colors with the dashed line
indicating the region of monomer–monomer interactions and the
black ellipse indicating the position of the 2-fold axis. The
pseudo-2-fold axis, which is perpendicular to the membrane and
passes through the nonheme Fe, relates the transmembrane
helices of the D1/D2 heterodimer, CP43 and CP47, and PsbI and
PsbX as emphasized by the black lines encircling these subunits.
[Courtesy of James Barber and So Iwata, Imperial College
London, U.K. PDBid 1S5L.]

when exposed to a series of short flashes. O_2 was evolved
with a peculiar oscillatory pattern (Fig. 24-21). There is virtually no O_2 evolved by the first two flashes. The third flash
results in the maximal O_2 yield. Thereafter, the amount of
O_2 produced peaks with every fourth flash until the oscillations damp out to a steady state. This periodicity indicates
that each OEC cycles through five different states, S_0
through S_4 (Fig. 24-22). Each of the transitions between S_0
and S_4 is a photon-driven redox reaction; that from S_4 to S_0

results in the release of O_2. Thus, *each O_2 molecule must be
produced by a single photosystem.* The observation that O_2
evolution peaks at the third rather than the fourth flash indicates that the OEC's resting state is predominantly S_1
rather than S_0. The oscillations gradually damp out because
a small fraction of the RCs fail to be excited or become
doubly excited by a given flash of light, so that they eventually lose synchrony. The five reaction steps release a total of
four water-derived protons into the inner thylakoid space

Figure 24-20 The arrangement of electron-transfer cofactors in PSII from *T. elongatus.* The complex is viewed along the membrane plane with the thylakoid lumen below. The cofactors are colored as in Fig. 24-19 but with Mg^{2+} yellow, N blue, and O red. The phytyl tails of the chlorophylls and pheophytins have been removed for clarity. The side chain C atoms of Tyr_Z (D1 Tyr 161) and D1 His 190 are yellow, and those of Tyr_D (D2 Tyr 160) and D2 His 189 are orange. The OEC is drawn in space-filling form with Mn purple, Ca^{2+} cyan, and O red. The numbers indicate the center-to-center distances, in angstroms, between the cofactors spanned by the accompanying thin black lines. Compare this figure to Fig. 24-12 (which is drawn upside down relative to this figure). [Courtesy of James Barber and So Iwata, Imperial College London, U.K. PDBid 1S5L.]

(lumen) in a stepwise manner (Fig. 24-22). These protons contribute to the transmembrane proton gradient.

Since the OEC abstracts electrons from H_2O, its five states must have extraordinarily high reduction potentials (recall from Table 22-1 that the O_2/H_2O half-reaction has a standard reduction potential of 0.815 V). PSII must also stabilize the highly reactive intermediates for extended periods (as much as minutes) in close proximity to water.

The OEC, which is located at the lumenal surface of the D1 subunit (Fig. 24-20), is a Mn_4CaO_4 or Mn_4CaO_5 complex in which the O atoms bridge neighboring Mn atoms. The structure of the OEC remains elusive due to PSII's relatively poorly resolved X-ray structures and the observation

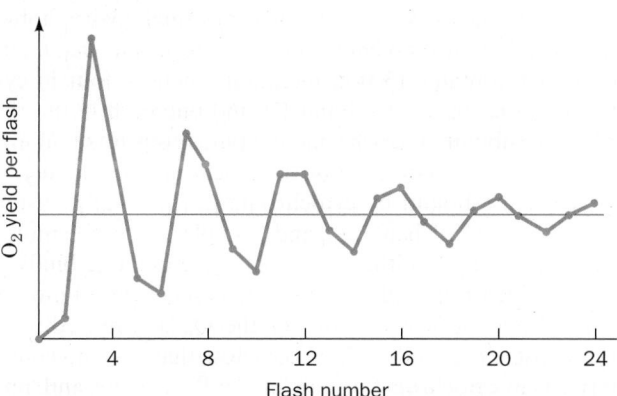

Figure 24-21 The O_2 yield per flash in dark-adapted spinach chloroplasts. Note that the yield peaks on the third flash and then on every fourth flash thereafter until the curve eventually damps out to its average value. [After Forbush, B., Kok, B., and McGloin, M.P., *Photochem. Photobiol.* **14**, 309 (1971).]

Figure 24-22 Schematic mechanism of O_2 generation in chloroplasts. Four electrons are stripped, one at a time in light-driven reactions ($S_0 \rightarrow S_4$), from two bound H_2O molecules. In the recovery step ($S_4 \rightarrow S_0$), which is light independent, O_2 is released and two more H_2O molecules are bound. Three of these five steps release protons into the thylakoid lumen.

Figure 24-23   A model of the OEC. This Mn_4CaO_5 complex is shown in ball-and-stick form with Mn ions purple, the Ca^{2+} ion cyan, and O red. The bonds between the Ca and Mn ions are drawn in gray to indicate that the position of the Ca ion is relatively poorly defined. Presumably, numerous protein side chains and water molecules ligand the Ca and Mn ions. Several related models are also compatible with the structural data. [Based on a model by Vittal Yachandra, Lawrence Berkeley National Laboratory, Berkeley, California.]

that the OEC decomposes when illuminated with X-rays at the intensities used in X-ray structure determinations. However, the use of X-ray spectroscopy techniques of lower intensity that are sensitive to bond lengths have led to the formulation of several related models for the OEC that are compatible with the X-ray structure of PSII. One of these models is shown in Fig. 24-23.

The water-splitting reaction is driven by the excitation of the PSII RC. A variety of evidence indicates that the Mn ions in the OEC's various S states (Fig. 24-22) cycle through specific combinations of Mn(II), Mn(III), Mn(IV), and Mn(V) while abstracting protons and electrons from two H_2O molecules to yield O_2, which is released into the thylakoid lumen. However, the mechanism whereby this occurs, that is, the nature of the five S states, remains unknown due to the lack of structural information concerning these states.

The next link in the PSII electron transport chain is an entity, originally named Z (Fig. 24-18), which relays electrons from the OEC to P680. The existence of Z is signaled by a transient EPR spectrum of illuminated chloroplasts that parallels the S-state transitions. The change in this spectrum on supplying deuterated tyrosine to cyanobacteria indicates that Z^+ is a tyrosyl radical (TyrO·; EPR spectra reflect the nuclear spins of the atoms with which the unpaired electrons interact). It has been identified as Tyr_Z in PSII (Fig. 24-20) due to its position between the Mn cluster and P680's chlorophyll P_{D1}. Recall that a tyrosyl radical has also been implicated in the reduction of O_2 to 2 H_2O by cytochrome *c* oxidase (Complex IV) in the respiratory electron-transport chain (Section 22-2C5c).

e. Electron Transport through the Cytochrome b_6f Complex Generates a Proton Gradient

From the plastoquinone pool, electrons pass through the cytochrome b_6f complex. This integral membrane assembly resembles cytochrome bc_1, its purple bacterial counterpart (Section 24-2Be), as well as Complex III of the mitochondrial electron-transport chain (also called cytochrome bc_1; Section 22-2C3a). Electron flow through the cytochrome b_6f complex occurs through a Q cycle (Fig. 22-31) in which plastoquinone is the $(H^+ + e^-)$ carrier. Accordingly, two protons are translocated across the thylakoid membrane for every electron transported. The four electrons abstracted from 2 H_2O by the OEC therefore lead to the translocation of eight H^+ from the stroma to the thylakoid lumen. *Electron transport via the cytochrome b_6f complex generates much of the electrochemical proton gradient that drives the synthesis of ATP in chloroplasts.*

The X-ray structure of cytochrome b_6f (Fig. 24-24) was independently determined by Janet Smith and William Cramer and by Jean-Luc Popot and Daniel Picot. Cytochrome b_6f is a dimer of ~109-kD protomers, each containing four large subunits (18–32 kD) that have counterparts in cytochrome bc_1: **cytochrome b_6,** a homolog of the N-terminal half of cytochrome *b*; **subunit IV,** a homolog of the C-terminal half of cytochrome *b*; a Rieske iron–sulfur protein (ISP), which is also present in cytochrome bc_1; and **cytochrome *f*** (*f* for *feuille,* French for leaf), a *c*-type cytochrome that is a functional analog of cytochrome c_1, although the two are unrelated in structure or sequence. In fact, cytochrome *f* is an elongated, two-domain protein that is dominated by β sheets and hence has an entirely different fold from those of other *c*-type cytochromes of known structure. Cytochrome *f*'s single heme *c* is, nevertheless, covalently linked to the protein's larger domain via the two Cys residues in a Cys-X-Y-Cys-His sequence that is characteristic of *c*-type cytochromes (Fig. 9-41) and whose His residue forms one of the Fe ion's two axial ligands (Fig. 22-21). Intriguingly, however, the second axial ligand is not a Met S atom, as occurs in most *c*-type cytochromes, but instead is the protein's N-terminal amino group, a group that had previously not been observed to be a heme ligand.

In addition, cytochrome b_6f has four small hydrophobic subunits that have no equivalents in cytochrome bc_1. Each protomer contains 13 transmembrane helices, four in cytochrome b_6, three in subunit IV, and one each in the remaining subunits. Cytochrome b_6f binds cofactors that are the equivalents of all of those in cytochrome bc_1: **heme *f*,** a *c*-type heme bound by cytochrome *f*; a [2Fe–2S] cluster bound by the ISP; hemes b_H and b_L; a plastoquinone molecule that occupies either the Q_i site (the quinone-binding site at which fully reduced quinone is regenerated during the Q cycle; Section 22-3Be) or the Q_o site. In addition, cytochrome b_6f binds several cofactors that have no counterparts in cytochrome bc_1: a Chl *a,* a β-carotene, and, unexpectedly, a novel heme named **heme *x*** (alternatively, **heme c_i**), which is covalently linked to the protein via a single thioether bond to Cys 35 of cytochrome b_6, and whose only axial ligand is a water molecule (compare with hemes *a, b,* and *c*; Fig. 22-21).

Figure 24-24 **X-ray structure of the cytochrome b_6f complex from the thermophilic cyanobacterium *Mastigocladus laminosus*.** A ribbon diagram of the dimeric complex is drawn on the left with cytochrome b_6 blue, subunit IV purple, cytochrome *f* red, the iron–sulfur protein (ISP) yellow, and the other subunits green. The inferred position of the lipid bilayer is indicated by a yellow band. Compare this figure to Fig. 22-23 (which is upside down relative to this figure). The paths of electron and proton transfer through the complex and the distances, in angstroms, between redox centers are shown on the right. [Modified from a drawing by William A. Cramer and Janet Smith, Purdue University. PDBid 1UM3.]

f. Plastocyanin Transports Electrons from Cytochrome b_6f to PSI

Electron transfer between cytochrome *f*, the terminal electron carrier of the cytochrome b_6f complex, and PSI is mediated by **plastocyanin (PC),** a 99-residue, monomeric, Cu-containing, peripheral membrane protein located on the thylakoid luminal surface (Fig. 24-17). Thus PC is the functional analog of cytochrome *c*, which transfers electrons from Complex III to Complex IV in the mitochondrial electron-transport chain (Section 22-2C4).

PC's redox center cycles between its Cu(I) and Cu(II) oxidation states. The X-ray structure of PC from poplar leaves, determined by Hans Freeman, shows that its single Cu atom is coordinated with distorted tetrahedral geometry by a Cys, a Met, and two His residues (Fig. 24-25). Cu(II) complexes with four ligands normally adopt a square planar coordination geometry, whereas those of Cu(I) are generally tetrahedral. Evidently, the strain of Cu(II)'s protein-imposed tetrahedral coordination in PC promotes its

Figure 24-25 **X-ray structure of plastocyanin (PC) from poplar leaves.** This 99-residue monomeric protein, a member of the family of **blue copper proteins** (as is the globular domain of Complex IV's Subunit II, which binds the Cu_A center; Section 22-2C5a), folds into a β sandwich. Its Cu atom (*orange sphere*), which alternates between its Cu(I) and Cu(II) oxidation states, is tetrahedrally coordinated by the side chains of His 37, Cys 84, His 87, and Met 92, which are shown in stick form with their C, N, and S atoms green, blue, and yellow. Seven conserved Asp and Glu residues (*red*) form a negatively charged patch on the surface of PC that has been implicated in electrostatically binding to a positively charged patch on the surface of cytochrome *f* formed by five Lys and Arg residues. [Based on an X-ray structure by Mitchell Guss and Hans Freeman, University of Sydney, Australia. PDBid 1PLC.]

reduction to Cu(I). This hypothesis accounts for PC's high standard reduction potential (0.370 V) compared to that of the normal Cu(II)/Cu(I) half-reaction (0.158 V). This is an example of how proteins modulate the reduction potentials of their redox centers so as to match them to their function— in the case of plastocyanin, the efficient transfer of electrons from the cytochrome b_6f complex to PSI.

g. PSI Resembles Both PSII and the PbRC

Cyanobacterial PSIs are trimers of protomers that each consist of at least 11 different protein subunits coordinating >100 cofactors. The X-ray structure of PSI from *T. elongatus* (Fig. 24-26), determined at 2.5-Å resolution by Norbert Krauss, Saenger, and Petra Fromme, reveals that each of its 356-kD protomers contains nine transmembrane subunits **(PsaA, PsbB, PsaF, PsaI–M,** and **PsaX)** and three stromal (cytoplasmic) subunits **(PsaC–E),** which collectively bind 127 cofactors that comprise 30% of PSI's mass. The cofactors forming the PSI RC are all bound by the homologous subunits PsaA (755 residues) and PsaB (740 residues), whose 11 transmembrane helices each are arranged in a manner resembling those in the L and M subunits of the PbRC (Fig. 24-11) and the D1 and D2 subunits of PSII (Fig. 24-19), thus supporting the hypothesis that all

RCs arose from a common ancestor. PsaA and PsaB, together with other transmembrane subunits, also bind the cofactors of the core antenna system (see below).

Figure 24-27 indicates that PSI's RC consists of six molecules of chlorophyll and two molecules of **phylloquinone** (**vitamin K₁;** note that it has the same phytyl side chain as chlorophylls; Fig. 24-3),

Phylloquinone

all arranged in two pseudosymmetrically related branches, followed by three [4Fe–4S] clusters. The primary electron donor of this system, **P700,** consists of a Chl *a'* (Fig. 24-3) and a Chl *a* (A1 and B1, respectively), whose rings are

(a)

Figure 24-26 **X-ray structure of PSI from *T. elongatus.***
(*a*) View of the trimer perpendicular to the membrane from its stromal side. The stromal subunits have been removed for clarity. PSI's 3-fold axis of symmetry is represented by the small black triangle. Different structural elements are shown for each of the three protomers (I, II, and III). I shows the arrangement of transmembrane helices (*cylinders*), which are differently colored for each subunit. The transmembrane helices of both PsaA (*blue*) and PsaB (*red*) are named a through k from their N- to C-termini. The six helices in extramembranous loop regions are drawn as spirals. II shows the transmembrane helices as cylinders with the stromal and lumenal loop regions drawn in ribbon form. III shows the transmembrane helices as cylinders together with

all cofactors. The RC Chl *a*'s and quinones, drawn in stick form, are purple, the Fe and S atoms of the [4Fe–4S] clusters are drawn as orange and yellow spheres, the antenna system Chl *a*'s (whose side chains have been removed for clarity) are yellow, the carotenoids are black, and the bound lipids are light green.
(*b*) One protomer as viewed parallel to the membrane along the arrow in Part *a* with the stroma above. The transmembrane subunits are colored as in Part *a* with the stromal subunits PsaC, PsaD, and PsaE pink, cyan, and light green. The vertical line and triangle mark the trimer's 3-fold axis of symmetry. [Courtesy of Wolfram Saenger, Freie Universität Berlin, Germany. PDBid 1JB0.]

Figure 24-27 Cofactors of the PSI RC and PsaC. The structure is viewed parallel to the membrane plane with the stroma above. The Chl *a* and phylloquinone molecules are arranged in two branches that are related by PSI's 2-fold axis of pseudosymmetry, which is vertical in this drawing. The Chl *a*'s are labeled A or B to indicate that their Mg^{2+} ions are liganded by the side chains of PsaA or PsaB, respectively, and, from the luminal side upward, by different colors and numbers, 1 to 3. The phylloquinones are named Q_K-A and Q_K-B. PsaC is shown in ribbon form with those portions resembling segments in bacterial 2[4Fe–4S] ferredoxins pink and with insertions and extensions green. The three [4Fe–4S] clusters are shown in ball-and-stick form and labeled according to their spectroscopic identities F_X, F_A, and F_B. The center-to-center distances between cofactors (*vertical black lines*) are given in angstroms. Compare this figure with Figs. 24-20 and 24-12. [Courtesy of Wolfram Saenger, Freie Universität Berlin, Germany. PDBid 1JB0.]

parallel and whose Mg^{2+} ions are separated by 6.3 Å. Thus P700 resembles the special pair in the PbRC. However, EPR studies indicate that ~80% of the unpaired electron associated with photooxidized $P700^+$ resides on Chl *a* B1. A1 is followed in the left branch of Fig. 24-27 by two more Chl *a* rings, B2 and A3, and B1 is followed by A2 and B3 in the right branch. One or both of the third pair of Chl *a* molecules, A3 and B3, probably form the spectroscopically identified primary electron acceptor A_0 (right side of Fig. 24-18). The Mg^{2+} ions of A3 and B3 are each axially liganded by the S atom of a Met residue rather than by a His side chain (thereby forming the only known biological examples of Mg^{2+}—S coordination). All of the residues involved in Mg^{2+} coordination and hydrogen bonding to these second and third Chl *a*'s are strictly conserved in PSI's, from cyanobacteria to higher plants, thereby suggesting that all of these interactions are important for fine-tuning their redox potentials. Electrons are passed from A3 and B3 to the phylloquinones, Q_K-A and Q_K-B, which almost certainly correspond to the spectroscopically identified electron acceptor A_1. Spectroscopically based kinetic investigations indicate that, in contrast to the case for the PbRC, electrons pass through both branches of the PSI

RC, although at different rates: 35×10^6 s^{-1} for the branch ending in Q_K-B and 4.4×10^6 s^{-1} for that ending in Q_K-A. Indeed, the PSI RC is most closely related to the RC of **green sulfur bacteria** (a second class of photosynthetic bacteria), which is a true homodimer.

Up until this point, PSI's RC resembles those of PSII and purple photosynthetic bacteria. However, rather than the reduced forms of either Q_K-A or Q_K-B dissociating from PSI, both of these quinones directly pass their photoexcited electron to a chain of three spectroscopically identified [4Fe–4S] clusters designated F_X, F_A, and F_B (right side of Fig. 24-18). F_X, which lies on the pseudo-2-fold axis relating PsaA and PsaB, is coordinated by two Cys residues from each of these subunits. F_A and F_B are bound to the stromal subunit PsaC, which structurally resembles bacterial 2[4Fe–4S] ferredoxins (e.g., Fig. 22-16). Mutational studies on the Cys residues of PsaC that coordinate its two [4Fe–4S] clusters indicate that the cluster that lies closer to F_X is F_A and the more distant cluster is F_B (Fig. 24-27). The observation that both branches of PSI's electron-transfer pathways are active, in contrast to only one active branch in PSII and the PbRC, is rationalized by the fact that the two quinones at the ends of each branch are functionally equivalent in PSI but functionally different in PSII and the PbRC.

PSI's core antenna system consists of 90 Chl *a* molecules and 22 carotenoids (Fig. 24-26*a*). The Mg^{2+} ions of 79 of these Chl *a* molecules are axially liganded by residues of PsaA and PsaB (mostly His side chains or protein-bound water molecules), whereas the remaining 11 are so liganded by the smaller subunits PsaJ through M and PsaX. The spatial distribution of these antenna Chl *a*'s resembles that in the core antenna subunits CP43 and CP47 of PSII. Indeed, the N-terminal domains of PsaA and PsaB are similar in sequence to those of CP43 and CP47 and fold into similar structures containing six transmembrane helices each. The carotenoids, which are mostly β-carotenes, are deeply buried in the membrane, where they are in van der Waals contact with Chl *a* rings. This permits efficient energy transfer from photoexcited carotenoids to Chl *a* as well as protects PSI from photooxidative damage. PSI also tightly binds four lipid molecules such that their fatty acyl groups are embedded among the complex's transmembrane helices. This strongly suggests that these lipids have specific structural and/or functional roles rather than being artifacts of preparation. Indeed, the head group of one of them, a phospholipid, coordinates the Mg^{2+} of an antenna Chl *a*, an unprecedented interaction.

PSIs from higher plants are monomers rather than trimers as are cyanobacterial PSIs. Nevertheless, the X-ray structure of PSI from peas, determined by Nathan Nelson, reveals that the positions and orientations of the chlorophylls in both species of PSIs are nearly identical, a remarkable finding considering the >1 billion years since chloroplasts diverged from their cyanobacterial ancestors. However, pea PSI has four antenna proteins not present in cyanobacterial PSI that are arranged in a crescent-shaped transmembrane belt around one side of its RC and which collectively bind 56 chlorophyll molecules.

h. PSI-Activated Electrons May Reduce NADP⁺ or Motivate Proton Gradient Formation

Electrons ejected from F_B in PSI may follow either of two alternative pathways (Fig. 24-18):

1. Most electrons follow a noncyclic pathway by passing to an ~100-residue, single [2Fe–2S]-containing, soluble ferredoxin **(Fd)** that is located in the stroma. Reduced Fd, in turn, reduces NADP⁺ in a reaction mediated by the ~310-residue, monomeric, FAD-containing **ferredoxin–NADP⁺ reductase** (**FNR,** Fig. 24-28*a*), to yield the final product of the chloroplast light reaction, NADPH. Two reduced Fd molecules successively deliver one electron each to the FAD of FNR, which thereby sequentially assumes the neutral semiquinone and fully reduced states before transferring the two electrons and a proton to the NADP⁺ via what is formally a hydride ion transfer. The X-ray structure of the complex between Fd and FNR from maize leaf (Fig. 24-28*b*), determined by Genji Kurisu, reveals that the shortest interatomic approach between Fd's [2Fe–2S] cluster and FNR's FAD is the 6.0 Å between an Fe atom and FAD atom C8a (the methyl C closest to its ribitol residue; Fig. 16-8). This is

sufficiently close for direct electron transfer through space between these prosthetic groups. The complex is stabilized by five salt bridges, as similarly appears to be the case for the interaction between cytochrome *f* and PC.

2. Some electrons are returned from PSI, via cytochrome b_6, to the plastoquinone pool, thereby traversing a cyclic pathway that translocates protons across the thylakoid membrane. A mechanism that has been proposed for this process is that Fd transfers an electron to heme *x* of cytochrome b_6 (Fig. 24-24) rather than to FNR. Since heme *x* contacts heme b_L at the periphery of cytochrome b_6f's Q_i site, an electron injected into heme *x* would be expected to reduce plastoquinone via a Q cycle-like mechanism (Fig. 22-31). Note that the cyclic pathway is independent of the action of PSII and hence does not result in the evolution of O₂. This accounts for the observation that chloroplasts absorb more than eight photons per O₂ molecule evolved.

The cyclic electron flow presumably functions to increase the amount of ATP produced relative to that of

(a)

(b)

Figure 24-28 **Ferredoxin–NADP⁺ reductase.** (*a*) The X-ray structure of the Y308S mutant form of pea ferredoxin–NADP⁺ reductase (FNR) in complex with FAD and NADP⁺. This 308-residue protein has two domains: The N-terminal domain (*gold*), which forms the FAD binding site, folds into an antiparallel β barrel, whereas the C-terminal domain (*magenta*), which provides the NADP⁺ binding site, forms a dinucleotide-binding fold (Section 8-3Bi). The FAD and NADP⁺ are shown in stick form with NADP⁺ C green, FAD C cyan, N blue, O red, and P yellow. The flavin and nicotinamide rings are in opposition with C4 of the nicotinamide ring and C5 of the flavin ring 3.0 Å apart, an arrangement that is consistent with direct hydride transfer as also occurs in glutathione reductase and dihydrolipoyl dehydrogenase (Section 21-2Ba). However, in contrast to these

latter enzymes, whose bound flavin and nicotinamide rings are parallel, those in FNR are inclined by ~30°, a heretofore unobserved binding mode. [Based on an X-ray structure by Andrew Karplus, Cornell University. PDBid 1QFY.] (*b*) The X-ray structure of the complex between Fd (*red*) and FNR (*blue*) from maize leaf with both proteins drawn in ribbon form embedded in their transparent solvent-accessible surfaces. The [2Fe–2S] cluster of Fd (*green*) and the FAD of FNR (*yellow*) are drawn in ball-and-stick form. The Fd binds in a hollow between FNR's two domains (Part *a*) such that the line joining the two Fe's of the [2Fe–2S] cluster lies roughly in the plane of the flavin ring. [Courtesy of Genji Kurisu, Osaka University, Osaka, Japan. PDBid 1GAQ.] **See Interactive Exercise 22**

NADPH and thus permits the cell to adjust the relative amounts of these two substances produced according to its needs. However, the mechanism that apportions electrons between the cyclic and noncyclic pathways is unknown.

i. PSI and PSII Occupy Different Parts of the Thylakoid Membrane

Freeze-fracture electron microscopy (Section 12-3Ca) revealed that the protein complexes of the thylakoid membrane have characteristic distributions (Fig. 24-29):

1. PSI occurs mainly in the unstacked stroma lamellae, in contact with the stroma, where it has access to $NADP^+$.

2. PSII is located almost exclusively between the closely stacked grana, out of direct contact with the stroma.

3. Cytochrome b_6f is uniformly distributed throughout the membrane.

The high mobilities of plastoquinone and plastocyanin, the electron carriers that shuttle electrons between these complexes, permits photosynthesis to proceed at a reasonable rate.

What function is served by the segregation of PSI and PSII, which are typically present in chloroplasts in equimolar amounts? If these two photosystems were in close proximity, the higher excitation energy of PSII (P680 vs P700) would cause it to pass a large fraction of its absorbed photons to PSI via exciton transfer; that is, PSII would act as a light-harvesting antenna for PSI (Fig. 24-7b). The separation of these particles by around 100 Å eliminates this difficulty.

The physical separation of PSI and PSII also permits the chloroplast to respond to changes in illumination. The relative amounts of light absorbed by the two photosystems vary with how the light-harvesting complexes (LHCs) are distributed between the stacked and unstacked portions of the thylakoid membrane. Under high illumination (normally direct sunlight, which contains a high proportion of short-wavelength blue light), all else being equal, PSII absorbs more light than PSI. PSI is then unable to take up electrons as fast as PSII can supply them, so the plastoquinone is predominantly in its reduced state. The reduced plastoquinone activates a protein kinase to phosphorylate specific Thr residues of the LHCs, which, in response, migrate to the unstacked regions of the thylakoid membrane, where they bind to PSI. A greater fraction of the incident light is thereby funneled to PSI. Under low illumination (normally shady light, which contains a high proportion of long-wavelength red light), PSI takes up electrons faster than PSII can provide them so that plastoquinone predominantly assumes its oxidized form. The LHCs are consequently dephosphorylated and migrate to the stacked portions of the thylakoid membrane, where they drive PSII. The chloroplast therefore maintains the balance between its two photosystems by a light-activated feedback mechanism.

D. *Photophosphorylation*

Chloroplasts generate ATP in much the same way as mitochondria, that is, by coupling the dissipation of a proton gradient to the enzymatic synthesis of ATP (Section 22-3C). This was clearly demonstrated by the imposition of an artificially produced pH gradient across the thylakoid membrane. Chloroplasts were soaked, in the dark, for several hours in a succinic acid solution at pH 4 so as to bring the thylakoid lumen to this pH (the thylakoid membrane is permeable to unionized succinic acid). The abrupt transfer of these chloroplasts to an $ADP + P_i$–containing buffer at pH 8 resulted in an impressive burst of ATP synthesis: About 100 ATPs were synthesized per molecule of cytochrome f present. Moreover, the amount of ATP

Figure 24-29 **Segregation of PSI and PSII.** The distribution of photosynthetic protein complexes between the stacked (grana) and the unstacked (stroma exposed) regions of the thylakoid membrane is shown. [After Anderson, J.M. and Anderson, B., *Trends Biochem. Sci.* **7,** 291 (1982).]

synthesized was unaffected by the presence of electron-transport inhibitors such as DCMU. This, together with the observations that photophosphorylation requires an intact thylakoid membrane and that proton translocators such as 2,4-dinitrophenol (Section 22-3D) uncouple photophosphorylation from light-driven electron transport, provides convincing evidence favoring the chemiosmotic hypothesis (Section 22-3A).

a. Chloroplast Proton-Translocating ATP Synthase Resembles That of Mitochondria

Electron micrographs of thylakoid membrane stromal surfaces and bacterial plasma membrane inner surfaces reveal lollipop-shaped structures (Fig. 24-30). These closely resemble the F_1 units of the proton-translocating ATP synthase studding the matrix surfaces of inner mitochondrial membranes (Fig. 22-36a). In fact, the chloroplast ATP synthase, which is termed the **CF_1CF_0 complex** (C for chloroplast), has remarkably similar properties to the mitochondrial F_1F_0 complex (Section 22-3C). For example,

1. Both F_0 and CF_0 units are hydrophobic transmembrane proteins that contain a proton-translocating channel.

2. Both F_1 and CF_1 are hydrophilic peripheral membrane proteins of subunit composition $\alpha_3\beta_3\gamma\delta\epsilon$, of which β is a reversible ATPase.

3. Both ATP synthases are inhibited by oligomycin.

4. Chloroplast ATP synthase translocates protons out of the thylakoid lumen into the stroma (Fig. 24-17), and mitochondrial ATP synthase conducts them into the matrix space (the mitochondrial equivalent of the stroma) from the intermembrane space (Section 22-3A).

Figure 24-30 **Electron micrograph of thylakoids.** The CF_1 "lollipops" of their ATP synthases project from their stromal surfaces. Compare this with Fig. 22-36a. [Courtesy of Peter Hinkle, Cornell University.]

Clearly, proton-translocating ATP synthases must have evolved very early in the history of cellular life. Chloroplast ATP synthase is located in the unstacked portions of the thylakoid membrane, in contact with the stroma, where there is room for the bulky CF_1 globule and access to ADP (Fig. 24-29).

b. Photosynthesis with Noncyclic Electron Transport Produces Around 1.25 ATP Equivalents per Absorbed Photon

At saturating light intensities, chloroplasts generate proton gradients of ~3.5 pH units across their thylakoid membranes. This, as we have seen (Figs. 24-17 and 24-18), arises from two sources:

1. The evolution of a molecule of O_2 from two H_2O molecules releases four protons into the thylakoid lumen. These protons should be considered as being supplied from the stroma by the protons and H atoms taken up in the synthesis of NADPH.

2. The transport of the liberated four electrons through the cytochrome b_6f complex occurs with the translocation of what is estimated to be eight protons from the stroma to the thylakoid lumen.

Altogether ~12 protons are translocated per molecule of O_2 produced by noncyclic electron transport.

The thylakoid membrane, in contrast to the inner mitochondrial membrane, is permeable to ions such as Mg^{2+} and Cl^-. Translocation of protons and electrons across the thylakoid membrane is consequently accompanied by the passage of these ions so as to maintain electrical neutrality (Mg^{2+} out and Cl^- in). This all but eliminates the membrane potential, $\Delta\Psi$ (Eq. [22.1]). *The electrochemical gradient in chloroplasts is therefore almost entirely a result of the pH gradient.*

Chloroplast ATP synthase, according to most estimates, produces one ATP for every three protons it transports out of the thylakoid lumen. Noncyclic electron transport in chloroplasts therefore results in the production of ~12/3 = 4 molecules of ATP per molecule of O_2 evolved (although this quantity is subject to revision) or around half an ATP per photon absorbed. Cyclic electron transport is a more productive ATP generator since it yields two-thirds of an ATP (two protons transported) per absorbed photon. The noncyclic process, of course, also yields NADPH, each molecule of which has the free energy to produce three ATPs (Section 22-2A; although this does not normally occur), for a total of six more ATP equivalents per O_2 produced. Consequently, the energetic efficiency of the noncyclic process is 4/8 + 6/8 = 1.25 ATP equivalents per absorbed photon.

3 DARK REACTIONS

In the previous section we saw how light energy is harnessed to generate ATP and NADPH. In this section we discuss how these products are used to synthesize carbohydrates and other substances from CO_2.

A. *The Calvin Cycle*

The metabolic pathway by which plants incorporate CO_2 into carbohydrates was elucidated between 1946 and 1953 by Melvin Calvin, James Bassham, and Andrew Benson. They did so by tracing the metabolic fate of the radioactive label from $^{14}CO_2$ as it passed through a series of photosynthetic intermediates. The basic experimental strategy they used was to expose growing cultures of algae, such as *Chlorella,* to $^{14}CO_2$ for varying times and under differing illumination conditions and then to drop the cells into boiling alcohol so as to disrupt them while preserving their labeling pattern. The radioactive products were subsequently separated and identified (an often difficult task) through the use of the then recently developed technique of two-dimensional paper chromatography (Section 6-3Dc) coupled with autoradiography. The overall pathway, diagrammed in Fig. 24-31, is known as the **Calvin cycle** or the **reductive pentose phosphate cycle.**

Some of Calvin's earliest experiments indicated that algae exposed to $^{14}CO_2$ for a minute or more had synthesized a complex mixture of labeled metabolic products, including sugars and amino acids. By inactivating the algae within 5 s of their exposure to $^{14}CO_2$, however, it was shown that *the first stable radioactively labeled compound formed is 3-phosphoglycerate (3PG), which is initially labeled only in its carboxyl group.* This result immediately suggested, in analogy with most biochemical experience, that the 3PG was formed by the carboxylation of a C_2 compound. Yet the failure to find any such precursor eventually forced this hypothesis to be abandoned. The actual carboxylation reaction was discovered through an experiment in which illuminated algae had been exposed to $^{14}CO_2$ for ~10 min so that the levels of their labeled photosynthetic intermediates had reached a steady state. The CO_2 was then withdrawn. As expected, the carboxylation product, 3PG, decreased in concentration (Fig. 24-32) because it was depleted by reactions farther along the pathway. The concentration of **ribulose-5-phosphate (Ru5P),**

$$
\begin{array}{c}
CH_2OH \\
| \\
C{=}O \\
| \\
H{-}C{-}OH \\
| \\
H{-}C{-}OH \\
| \\
CH_2OPO_3^{2-}
\end{array}
$$

Ribulose-5-phosphate (Ru5P)

however, simultaneously increased. Evidently, Ru5P is the Calvin cycle's carboxylation substrate. If so, the resulting C_6 carboxylation product must split into two C_3 compounds, one of which is 3PG (Fig. 24-31, Reaction 2). A consideration of the oxidation states of Ru5P and CO_2 indicates that, in fact, both C_3 compounds must be 3PG and that the carboxylation reaction requires no external redox source.

While the search for the carboxylation substrate was going on, several other photosynthetic intermediates had

been identified and, through chemical degradation studies, their labeling patterns had been elucidated. For example, the hexose fructose-1,6-bisphosphate (FBP) is initially labeled only at its C3 and C4 positions (Fig. 24-31) but later becomes labeled to a lesser degree at its other atoms. Similarly, a series of tetrose, pentose, hexose, and heptose phosphates were isolated that had the identities and initial labeling patterns indicated in Fig. 24-31. A consideration of the flow of the labeled atoms through these various intermediates led, in what was a milestone of metabolic biochemistry, to the deduction of the Calvin cycle as is diagrammed in Fig. 24-31. The existence of many of its postulated reactions was eventually confirmed by *in vitro* studies using purified enzymes.

a. The Calvin Cycle Generates GAP from CO_2 via a Two-Stage Process

The Calvin cycle may be considered to have two stages:

Stage 1 The production phase (top line of Fig. 24-31), in which three molecules of Ru5P react with three molecules of CO_2 to yield six molecules of glyceraldehyde-3-phosphate (GAP) at the expense of nine ATP and six NADPH molecules. *The cyclic nature of the pathway makes this process equivalent to the synthesis of one GAP from three CO_2 molecules.* Indeed, at this point, one GAP can be bled off from the cycle for use in biosynthesis (see Stage 2).

Stage 2 The recovery phase (bottom lines of Fig. 24-31), in which the carbon atoms of the remaining five GAPs are shuffled in a remarkable series of reactions, similar to those of the pentose phosphate pathway (Section 23-4), to reform the three Ru5Ps with which the cycle began. Indeed, the elucidation of the pentose phosphate pathway at about the same time that the Calvin cycle was being worked out provided much of the biochemical evidence in support of the Calvin cycle. This stage can be conceptually decomposed into four sets of reactions (with the numbers keyed to the corresponding reactions in Fig. 24-31):

6. $C_3 + C_3 \longrightarrow C_6$

8. $C_3 + C_6 \longrightarrow C_4 + C_5$

9. $C_3 + C_4 \longrightarrow C_7$

11. $C_3 + C_7 \longrightarrow C_5 + C_5$

The overall stoichiometry for this process is therefore

$$5C_3 \longrightarrow 3C_5$$

Note that this stage of the Calvin cycle occurs without further input of free energy (ATP) or reducing power (NADPH).

b. Most Calvin Cycle Reactions Also Occur in Other Metabolic Pathways

The types of reactions in the Calvin cycle are all familiar (Section 23-4), with the exception of the carboxylation reaction. This first stage of the Calvin cycle begins with the phosphorylation of Ru5P by **phosphoribulokinase** to form **ribulose-1, 5-bisphosphate (RuBP).** Following the

Figure 24-31 The Calvin cycle. (*Opposite*) The number of lines in an arrow indicates the number of molecules reacting in that step for a single turn of the cycle that converts three CO_2 molecules to one GAP molecule. For the sake of clarity, the sugars are all shown in their linear forms, although the hexoses and heptoses predominantly exist in their cyclic forms (Section 11-1B). The ^{14}C-labeling patterns generated in one turn of the cycle through the use of $^{14}CO_2$ are indicated in red. Note that two of the Ru5Ps are labeled only at C3, whereas the third Ru5P is equally labeled at C1, C2, and C3. **See the Animated Figures**

carboxylation step, which is discussed below, the resulting 3PG is converted first to 1,3-bisphosphoglycerate (BPG) and then to GAP. This latter sequence is the reverse of two consecutive glycolytic reactions (Sections 17-2G and 17-2F) except that the Calvin cycle reaction involves NADPH rather than NADH.

The second stage of the Calvin cycle begins with the reverse of a familiar glycolytic reaction, the isomerization of GAP to dihydroxyacetone phosphate (DHAP) by triose phosphate isomerase (Section 17-2E). Following this, DHAP is directed along two analogous paths (Fig. 24-31): Reactions 6–8 or Reactions 9–11. Reactions 6 and 9 are

aldolase-catalyzed aldol condensations in which DHAP is linked to an aldehyde (aldolase is specific for DHAP but accepts a variety of aldehydes). Reaction 6 is also the reverse of a glycolytic reaction (Section 17-2D). Reactions 7 and 10 are phosphate hydrolysis reactions that are catalyzed, respectively, by fructose bisphosphatase (FBPase, which we previously encountered in our discussion of glycolytic substrate cycles and gluconeogenesis; Sections 17-4Ff and 23-1Ah) and **sedoheptulose bisphosphatase (SBPase).** The remaining Calvin cycle reactions are catalyzed by enzymes that also participate in the pentose phosphate pathway. In Reactions 8 and 11, both catalyzed by **transketolase,** a C_2 keto unit (shaded in green in Fig. 24-31) is transferred from a ketose to GAP to form **xylulose-5-phosphate (Xu5P)** and leave the aldoses **erythrose-4-phosphate (E4P)** in Reaction 8 and **ribose-5-phosphate (R5P)** in Reaction 11. The E4P produced by Reaction 8 feeds into Reaction 9. The Xu5Ps produced by Reactions 8 and 11 are converted to Ru5P by **phosphopentose epimerase** in Reaction 12. The R5P from Reaction 11 is also converted to Ru5P by **ribose phosphate isomerase** in Reaction 13, thereby completing a turn of the Calvin cycle. Thus only 3 of the 11 Calvin cycle enzymes, phosphoribulokinase, the carboxylation enzyme **ribulose bisphosphate carboxylase,** and SBPase, have no equivalents in animal tissues.

c. RuBP Carboxylase Catalyzes CO_2 Fixation in an Exergonic Process

The enzyme that catalyzes CO_2 fixation, ribulose bisphosphate carboxylase **(RuBP carboxylase),** is arguably the world's most important enzyme, since nearly all life on Earth ultimately depends on its action. This protein, presumably as a consequence of its particularly low catalytic efficiency ($k_{cat} = \sim3\ s^{-1}$), comprises up to 50% of leaf proteins and is therefore the most abundant protein in the biosphere (it is estimated to be synthesized at the rate of $\sim4 \times 10^9$ tons/year, which fixes $\sim10^{11}$ tons of CO_2/year; in comparison crude oil is consumed at the rate of $\sim3 \times 10^9$ tons/year). RuBP carboxylase from higher plants and most photosynthetic microorganisms consists of eight large (L) subunits (477 residues in tobacco leaves) encoded by chloroplast DNA and eight small (S) subunits (123 residues) specified by a nuclear gene (the RuBP carboxylase from certain photosynthetic bacteria is an L_2 dimer whose L subunit has 28% sequence identity with and is structurally similar to that of the L_8S_8 enzyme). X-ray studies by Carl-Ivar Brändén and by David Eisenberg demonstrated that the L_8S_8 enzyme has the D_4 symmetry of a square prism (Fig. 24-33*a,b*). The L subunit contains the enzyme's catalytic site, as is demonstrated by its enzymatic activity in the absence of the S subunit. It consists of two domains (Fig. 24-33*c*): Residues 1 to 150 form a mixed five-stranded β sheet and residues 151 to 477 fold into an α/β barrel (Fig. 8-19*b,c*) which, as do nearly all known α/β barrel enzymes (Section 8-3Bh), contains the enzyme's active site at the mouth of the barrel near the C-terminus of its β strands. The function of the S subunit is unknown; attempts to show that it has a regulatory role, in analogy with other enzymes, have been unsuccessful.

Figure 24-32 Algal 3PG and RuBP levels on removal of CO_2. The time course of the levels of 3PG (*purple curve*) and RuBP (*green curve*) in steady-state $^{14}CO_2$-labeled, illuminated algae is shown during a period in which the CO_2 (*orange curve*) is abruptly withdrawn. In the absence of CO_2, the 3PG concentration rapidly decreases because it is taken up by the reactions of the Calvin cycle but cannot be replenished by them. Conversely, the RuBP concentration transiently increases as it is synthesized from the residual pool of Calvin cycle intermediates but, in the absence of CO_2, cannot be used for their regeneration.

(a)

(b)

Figure 24-33 **X-ray structure of tobacco RuBP carboxylase in complex with the transition state inhibitor 2-carboxyarabinitol-1,5-bisphosphate.** The ~535-kD L_8S_8 protein has D_4 symmetry (the rotational symmetry of a square prism; Fig. 8-65*b*). (*a*) Surface diagram viewed along the protein's 4-fold axis, and (*b*) along a 2-fold axis. Parts *a* and *b* are related by a 90° rotation about the horizontal axis. The elongated L subunits can be considered to associate as two interdigitated tetramers, with that extending from the top in Part *b* cyan and that extending from the bottom green. The S subunits, which form C_4 tetramers that cap the top and bottom of the complex, are alternately yellow and orange. The transition state inhibitor **2-carboxyarabinitol-1,5-bisphosphate** (**CABP;** see text), which is partially visible in Part *b*, is drawn in stick form with C green, O red, and P orange. (*c*) Ribbon diagram of an L subunit oriented as is the central green subunit in Part *b* and colored in rainbow order from its N-terminus (*blue*) to its C-terminus (*red*). The 2-carboxyarabinitol-1,5-bisphosphate is bound in the mouth of the enzyme's α/β barrel. [Based on an X-ray structure by David Eisenberg, UCLA. PDBid 1RLC.]

(c)

The accepted mechanism of RuBP carboxylase, which was largely formulated by Calvin, is indicated in Fig. 24-34. Abstraction of the C3 proton of RuBP, the reaction's rate-determining step, generates an enediolate that nucleophilically attacks CO_2 (not HCO_3^-). The resulting β-keto acid is rapidly attacked at its C3 position by H_2O to yield an adduct that splits, by a reaction similar to aldol cleavage, to yield the two product 3PG molecules. The following evidence favors this mechanism:

1. The C3 proton of enzyme-bound RuBP exchanges with solvent, an observation compatible with the existence of the enediolate intermediate.

2. The C2 and C3 oxygen atoms remain attached to their respective C atoms, which eliminates mechanisms involving a covalent adduct such as a Schiff base between RuBP and the enzyme.

3. The trapping of the proposed β-keto acid intermediate by borohydride reduction and the tight enzymatic binding of its analogs, such as **2-carboxyarabinitol-1-phosphate**

(**CA1P**) and **2-carboxyarabinitol-1,5-bisphosphate** (**CABP**),

$$
\begin{array}{cc}
CH_2OPO_3^{2-} & CH_2OPO_3^{2-} \\
| & | \\
HO-C-CO_2^- & HO-C-CO_2^- \\
| & | \\
H-C-OH & H-C-OH \\
| & | \\
H-C-OH & H-C-OH \\
| & | \\
CH_2OH & CH_2OPO_3^{2-}
\end{array}
$$

2-Carboxyarabinitol-1-phosphate (CA1P)   **2-Carboxyarabinitol-1,5-bisphosphate (CABP)**

provide strong evidence for the existence of this intermediate.

The driving force for the overall reaction, which is highly exergonic ($\Delta G^{\circ\prime} = -35.1 \ kJ \cdot mol^{-1}$), *is provided by the*

Figure 24-34 **Probable mechanism of the carboxylation reaction catalyzed by RuBP carboxylase.** The reaction proceeds via an enediolate intermediate that nucleophilically attacks CO_2 to form a β-keto acid. This intermediate reacts with water to yield two molecules of 3PG. See the Animated Figures

cleavage of the β-keto acid intermediate to yield an additional resonance-stabilized carboxylate group.

RuBP carboxylase activity requires a bound divalent metal ion, physiologically Mg^{2+}, which acts to stabilize the developing negative charges during catalysis. The Mg^{2+} is, in part, bound to the enzyme by a catalytically essential carbamate group that is generated by the reaction of a non-substrate CO_2 with the ε-amino group of Lys 201 ($R-NH_2 + CO_2 \rightarrow R-NH-COO^- + H^+$). Although the *in vitro* activation reaction occurs spontaneously in the presence of Mg^{2+} and HCO_3^-, it is blocked *in vivo* by the particularly tight binding of RuBP to active sites lacking carbamate. This inhibition is relieved, however, by the release of RuBP in an ATP-driven process catalyzed by **RuBP carboxylase activase.**

d. Calvin Cycle Products Are Converted to Starch, Sucrose, and Cellulose

The overall stoichiometry of the Calvin cycle is

$$3CO_2 + 9ATP + 6NADPH \longrightarrow$$
$$GAP + 9ADP + 8P_i + 6NADP^+$$

GAP, the primary product of photosynthesis, is used in a variety of biosynthetic pathways, both inside and outside the chloroplast. For example, fatty acids and amino acids are synthesized from GAP as is described, respectively, in Sections 25-4 and 26-5. GAP can also be converted to fructose-

6-phosphate by the further action of Calvin cycle enzymes and then to glucose-1-phosphate (G1P) by phosphoglucose isomerase (Section 17-2B) and phosphoglucomutase (Section 18-1B). *G1P is the precursor of the higher carbohydrates characteristic of plants.*

The polysaccharide α-amylose, a major component of starch (Section 11-2D), is synthesized in the chloroplast stroma as a temporary storage depot for glucose units. It is also synthesized as a long-term storage molecule elsewhere in the plant, including leaves, seeds, and roots. G1P is first activated by its reaction with ATP to form ADP–glucose as catalyzed by **ADP–glucose pyrophosphorylase. Starch synthase** then transfers the glucose residue to a nonreducing end of an α-amylose or amylopectin molecule, forming a new glycosidic linkage (Fig. 24-35). The overall reaction is driven by the exergonic hydrolysis of the PP_i released in the formation of ADP–glucose. A similar reaction sequence occurs in glycogen synthesis, which uses UDP–glucose (Section 18-2A). The $\alpha(1 \rightarrow 6)$ branches of amylopectin (Section 11-2D) are made by **starch-branching enzyme,** which functions similarly to glycogen branching enzyme (Section 18-2C).

Sucrose, a disaccharide of glucose and fructose (Section 11-2B), is the major transport sugar for delivering carbohydrates to nonphotosynthesizing cells and hence is the major photosynthetic product of green leaves. Since sucrose is synthesized in the cytosol, either glyceraldehyde-3-phosphate

Figure 24-35 **Starch synthesis.** ADP–glucose is formed from G1P and ATP in a phosphoanhydride exchange reaction (**1**). The PP$_i$ product is rapidly hydrolyzed. ADP–glucose is the substrate for starch synthase (**2**), which adds the glucose residue to a nonreducing end of an existing polysaccharide, releasing ADP.

α-Amylose (*n* residues)

α-Amylose (*n* + 1 residues)

or dihydroxyacetone phosphate is transported out of the chloroplast by an antiporter that exchanges phosphate for a triose phosphate. Two trioses combine to form fructose-6-phosphate (F6P) and subsequently glucose-1-phosphate (G1P), which is then activated by UTP to form UDP–glucose. Next, sucrose-6-phosphate is produced in a reaction catalyzed by **sucrose-phosphate synthase.** Finally, sucrose-6-phosphate is hydrolyzed by **sucrose-phosphate phosphatase** to yield sucrose (Fig. 24-36), which is then exported to other plant tissues.

Cellulose, which consists of long chains of β(1 → 4)-linked glucose units and is the major polysaccharide of plants, is also synthesized from UDP–glucose. Plant cell walls consist of almost crystalline cables containing 36 par-

UDP–glucose + F6P

| sucrose-phosphate
UDP ← | synthase

Sucrose-6-phosphate

H$_2$O ⟍ | sucrose-phosphate
P$_i$ ← | phosphatase

Sucrose

Figure 24-36 **The synthesis of sucrose.**

allel cellulose chains, each of 500 to 15,000 glucose units, all embedded in an amorphous matrix of other polysaccharides and lignin (Section 11-2C). Unlike starch in plants or glycogen in mammals, cellulose is synthesized by multisubunit enzyme complexes in the plant plasma membrane and extruded into the extracellular space.

B. *Control of the Calvin Cycle*

During the day, plants satisfy their energy needs via the light and dark reactions of photosynthesis. At night, however, like other organisms, they must use their nutritional reserves to generate their required ATP and NADPH through glycolysis, oxidative phosphorylation, and the pentose phosphate pathway. Since the stroma contains the enzymes of glycolysis and the pentose phosphate pathway as well as those of the Calvin cycle, *plants must have a light-sensitive control mechanism to prevent the Calvin cycle from consuming this catabolically produced ATP and NADPH in a wasteful futile cycle.*

As we saw in Section 17-4F, the control of flux in a metabolic pathway occurs at enzymatic steps that are far from equilibrium; that is, those that have a large negative value of ΔG. Inspection of Table 24-1 indicates that the four best candidates for flux control in the Calvin cycle are the reactions catalyzed by phosphoribulokinase, RuBP carboxylase, FBPase, and SBPase (Reactions 1, 2, 7, and 10, Fig. 24-31). In fact, the catalytic efficiencies of these four enzymes all vary, *in vivo,* with the level of illumination.

Table 24-1 **Standard and Physiological Free Energy Changes for the Reactions of the Calvin Cycle**

Step[a]	Enzyme	$\Delta G^{\circ\prime}$ (kJ $\cdot$ mol^{-1})	ΔG (kJ $\cdot$ mol^{-1})
1	Phosphoribulokinase	−21.8	−15.9
2	Ribulose bisphosphate carboxylase	−35.1	−41.0
3 + 4	Phosphoglycerate kinase + glyceraldehyde-3-phosphate dehydrogenase	+18.0	−6.7
5	Triose phosphate isomerase	−7.5	−0.8
6	Aldolase	−21.8	−1.7
7	Fructose bisphosphatase	−14.2	−27.2
8	Transketolase	+6.3	−3.8
9	Aldolase	−23.4	−0.8
10	Sedoheptulose bisphosphatase	−14.2	−29.7
11	Transketolase	+0.4	−5.9
12	Phosphopentose epimerase	+0.8	−0.4
13	Ribose phosphate isomerase	+2.1	−0.4

[a]Refer to Fig. 24-31.

Source: Bassham, J.A. and Buchanan, B.B., *in* Govindjee (Ed.), *Photosynthesis,* Vol. II, *p.* 155, Academic Press (1982).

The activity of RuBP carboxylase responds to three light-dependent factors:

1. It varies with pH. On illumination, the pH of the stroma increases from around 7.0 to about 8.0 as protons are pumped from the stroma into the thylakoid lumen. RuBP carboxylase has a sharp pH optimum near pH 8.0.

2. It is stimulated by Mg^{2+}. Recall that the light-induced influx of protons to the thylakoid lumen is accompanied by the efflux of Mg^{2+} to the stroma (Section 24-2Db).

3. It is strongly inhibited by its transition state analog 2-carboxyarabinitol-1-phosphate (CA1P; Section 24-3Ac), which many plants synthesize only in the dark. RuBP carboxylase activase (Section 24-3Ac) also facilitates the release of the tight-binding CA1P from RuBP carboxylase.

FBPase and SBPase are also activated by increased pH and Mg^{2+}, and by NADPH as well. The action of these factors is complemented by a second regulatory system that responds to the redox potential of the stroma. **Thioredoxin (Trx),** an ~105-residue protein that occurs in many types of cells, contains a redox-active disulfide group. Reduced Trx activates five Calvin cycle enzymes by disulfide interchange reactions (Fig. 24-37): phosphoribulokinase, glyceraldehyde-3-phosphate dehydrogenase, FBPase, SBPase, and RuBP carboxylase activase. This explains why these enzymes are activated by reduced disulfide reagents such as dithiothreitol. The redox level of Trx is maintained by **ferredoxin–thioredoxin reductase (FTR),** which contains a redox-active disul-

fide that is closely associated with a [4Fe–4S] cluster through which the protein directly responds to the redox state of soluble ferredoxin (Fd) in the stroma. This, as we have seen (Section 24-2Ch), varies with the illumination level.

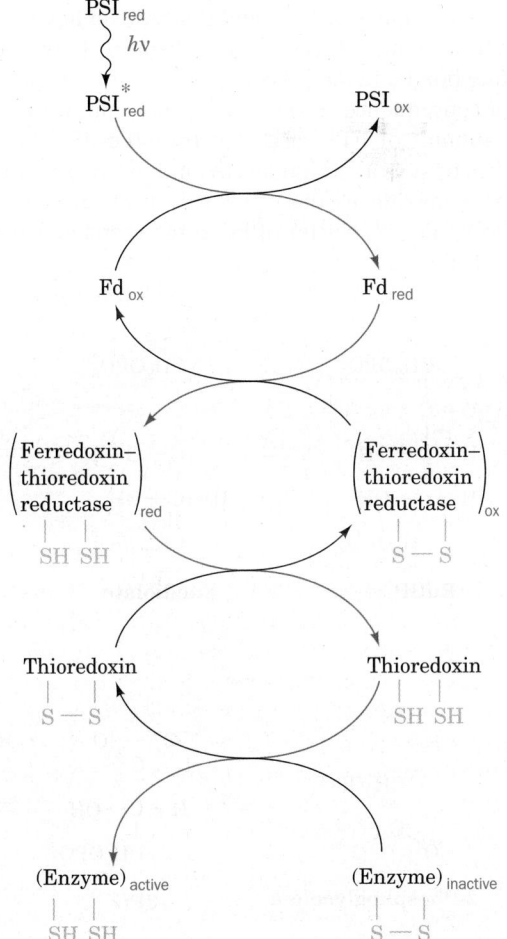

Figure 24-37 **Light-activation mechanism of chloroplast enzymes.** Photoactivated PSI reduces soluble ferredoxin (Fd), which reduces ferredoxin–thioredoxin reductase, which, in turn, reduces the disulfide linkage of thioredoxin. Reduced thioredoxin reacts with its target enzymes by disulfide interchange, thereby activating or deactivating these enzymes. In the dark, these processses are quickly reversed by reaction with oxygen.

Figure 24-38 **X-ray structure of a Fd–FTR–Trx complex.** The subunits are drawn in ribbon form with Fd green, the catalytic subunit of FTR blue, its variable subunit cyan, and Trx magenta. The [2Fe–2S] cluster of Fd and the [4Fe–4S] cluster of FTR are drawn in space-filling form with S yellow and Fe red-brown. The two Cys side chains forming the disulfide bond between FTR and Trx are shown in stick form with C green and S yellow. [Based on an X-ray structure by Hans Eklund, Swedish University of Agricultural Sciences, Uppsala, Sweden. PDBid 2PVO.]

The X-ray structure of a Fd–FTR–Trx complex (Fig. 24-38), determined by Hans Eklund, reveals that redox-active Cys residues of FTR and thioredoxin have formed a disulfide bond. Moreover, the distance between this disulfide bond and the [2Fe–2S] cluster of Fd is only ~20 Å as a consequence of the remarkably thin disklike catalytic subunit of FTR, and that the [4Fe–4S] cluster of FTR lies on a straight line between these two redox centers. This maximizes the efficiency of electron transfer from the [2Fe–2S] cluster of Fd to the redox active disulfide of Trx.

Reduced Trx also deactivates the chloroplastic enzymes phosphofructokinase (PFK), the main flux-generating enzyme of glycolysis (Section 17-4Fb), and glucose-6-phosphate dehydrogenase, the first enzyme in the pentose phosphate pathway (Section 23-4A), whose products, ATP and NADPH, would otherwise be used by the Calvin cycle in a futile cycle. Thus in plants, *light stimulates the Calvin cycle while deactivating glycolysis and the pentose phosphate pathway, whereas darkness has the opposite effect (that is, the so-called dark reactions do not occur in the dark).* Moreover, chloroplast ATP synthase is activated by reduced Trx, thus preventing it from uselessly hydrolyzing glycolytically produced ATP in the dark. Indeed, the redox state of Trx regulates a great variety of plant metabolic processes.

We have seen that ferredoxin reduces ferredoxin–NADP$^+$ reductase (Section 24-2Ch) and FTR, as well as supplying electrons to the cyclic pathway of chloroplast photosynthesis (Section 24-2Ch). In addition, ferredoxin is the reducing agent for three metabolically essential chloroplast enzymes: **sulfite reductase** (which reduces SO_3^{2-} to S_2^{2-}), **nitrite reductase** (which reduces NO_2^- to NH_4^+), and **glutamate synthase** (which catalyzes the reaction of α-ketoglutarate and NH_4^+ to form glutamate; Section 26-5Aa). Thus Fd stands at the center of a complex web of enzymatic and regulatory processes.

C. *Photorespiration and the C₄ Cycle*

It has been known since the 1960s that *illuminated plants consume O_2 and evolve CO_2 in a pathway distinct from oxidative phosphorylation. In fact, at low CO_2 and high O_2 levels, this photorespiration process can outstrip photosynthetic CO_2 fixation.* The basis of photorespiration was unexpected: O_2 competes with CO_2 as a substrate for RuBP carboxylase (RuBP carboxylase is therefore also called **RuBP carboxylase–oxygenase** or **RuBisCO**). In the oxygenase reaction, O_2 reacts with RuBisCO's second substrate, RuBP, to form 3PG and **2-phosphoglycolate** (Fig. 24-39). The 2-phosphoglycolate

Figure 24-39 **Probable mechanism of the oxygenase reaction catalyzed by RuBP carboxylase–oxygenase.** Note the similarity of this mechanism to that of the carboxylase reaction catalyzed by the same enzyme (Fig. 24-34).

is hydrolyzed to **glycolate** by **glycolate phosphatase** and, as described below, is partially oxidized to yield CO_2 by a series of enzymatic reactions that occur in the peroxisome and the mitochondrion. Thus photorespiration is a seemingly wasteful process that undoes some of the work of photosynthesis. In the following subsections we discuss the biochemical basis of photorespiration, its significance, and how certain plants manage to evade its deleterious effects.

a. Photorespiration Dissipates ATP and NADPH

The photorespiration pathway is outlined in Fig. 24-40. Glycolate is exported from the chloroplast to the peroxisome (also called the glyoxisome, Sections 1-2Ad and 23-2), where it is oxidized by **glycolate oxidase** to **glyoxylate** and H_2O_2. The H_2O_2, a powerful and potentially harmful oxidizing agent, is disproportionated to H_2O and O_2 in the peroxisome by the heme-containing enzyme **catalase.** Some of the glyoxylate is further oxidized by glycolate oxidase to oxalate. The remainder is converted to glycine in a **transamination reaction,** as discussed in Section 26-1A, and exported to the mitochondrion. There, two molecules of glycine are converted to one molecule of serine and one of CO_2 by a reaction described in Section 26-3B. *This is the origin of the CO_2 generated by photorespiration.* The serine is transported back to the peroxisome, where a transamination reaction converts it to **hydroxypyruvate.** This substance is reduced to **glycerate** and phosphorylated in the cytosol to 3PG, which reenters the chloroplast, where it is reconverted to RuBP in the Calvin cycle. *The net result of this complex photorespiration cycle is that some of the ATP and NADPH generated by the light reactions is uselessly dissipated.*

Although photorespiration has no known metabolic function, the RuBisCOs from the great variety of photosynthetic organisms so far tested all exhibit oxygenase activity. Yet, over the eons, the forces of evolution must have optimized the function of this important enzyme. It is thought that photosynthesis evolved at a time when Earth's atmosphere contained large quantities of CO_2 and very little O_2, so that photorespiration was of no consequence. It has therefore been suggested that the RuBisCO reaction has an obligate intermediate that is inherently autooxidizable. Another possibility is that photorespiration protects the photosynthetic apparatus from photooxidative damage when insufficient CO_2 is available to otherwise dissipate its absorbed light energy. This hypothesis is supported by the observation that when chloroplasts or leaf cells are brightly illuminated in the absence of both CO_2 and O_2, their photosynthetic capacity is rapidly and irreversibly lost.

b. Photorespiration Limits the Growth Rates of Plants

The steady-state CO_2 concentration attained when a photosynthetic organism is illuminated in a sealed system is named its **CO_2 compensation point.** For healthy plants, this is the CO_2 concentration at which the rates of photosynthesis and photorespiration are equal. For many species it is ~40 to 70 ppm (parts per million) CO_2 (the normal

Figure 24-40 **Photorespiration.** This pathway metabolizes the phosphoglycolate produced by the RuBP carboxylase-catalyzed oxidation of RuBP. The reactions occur, as indicated, in the chloroplast, the peroxisome, the mitochondrion, and the cytosol. Note that two glycines are required to form serine + CO_2 (Section 26-3B).

atmospheric concentration of CO_2 is 330 ppm), so their photosynthetic CO_2 fixation usually dominates their photorespiratory CO_2 release. However, the CO_2 compensation point increases with temperature because the oxygenase activity of RuBisCO increases more rapidly with temperature than does its carboxylase activity. Thus, *on a hot bright day, when photosynthesis has depleted the level of CO_2 at the chloroplast and raised that of O_2, the rate of photorespiration may approach that of photosynthesis. This phenomenon is, in fact, a major limiting factor in the growth of many plants.* Indeed, plants possessing a RuBisCO with significantly less oxygenase activity would not only have increased photosynthetic efficiency but would need less water because they could spend less time with their **stomata** (the pores leading to their internal leaf spaces) open acquiring CO_2 and would have a reduced need for fertilizer because they would require less RuBisCO. The control of photorespiration is therefore an important unsolved agricultural problem that is presently being attacked through genetic engineering studies (Section 5-5).

c. C_4 Plants Concentrate CO_2

Certain species of plants, such as sugarcane, corn, and most important weeds, have a metabolic cycle that concentrates CO_2 in their photosynthetic cells, thereby almost totally preventing photorespiration (their CO_2 compensation points are in the range 2 to 5 ppm). The leaves of plants that have this so-called **C_4 cycle** have a characteristic anatomy. Their fine veins are concentrically surrounded by a single layer of so-called **bundle-sheath cells,** which in turn are surrounded by **mesophyll cells.**

The C_4 cycle (Fig. 24-41) was elucidated in the 1960s by Marshall Hatch and Roger Slack. It begins with the uptake of atmospheric CO_2 by the mesophyll cells, which, lacking RuBisCO in their chloroplasts, do so by condensing it as HCO_3^- with phosphoenolpyruvate (PEP) to yield oxaloacetate. The oxaloacetate is reduced by NADPH to **malate,** which is exported to the bundle-sheath cells (the name C_4 refers to these four-carbon acids). There the malate is oxidatively decarboxylated by $NADP^+$ to form CO_2, pyruvate, and NADPH. The CO_2, which has been concentrated by this process, enters the Calvin cycle. The pyruvate is returned to the mesophyll cells, where it is phosphorylated to again form PEP. The enzyme that mediates this reaction, **pyruvate-phosphate dikinase,** has the unusual action of activating a phosphate group through the hydrolysis of ATP to AMP $+$ PP_i. This PP_i is further hydrolyzed to two P_i, which is tantamount to the consumption of a second ATP. *CO_2 is thereby concentrated in the bundle-sheath cells at the expense of two ATPs per CO_2. The dark reactions of photosynthesis in C_4 plants therefore consume a total of five ATPs per CO_2 fixed versus the three ATPs required by the Calvin cycle alone.* The additional ATP is presumably generated through the cyclic flow of electrons in the light reactions (Section 24-2Ch).

C_4 plants, which comprise ~5% of terrestrial plants, occur largely in unshaded areas of tropical regions because they grow faster under hot and sunny conditions than other, so-called **C_3 plants** (so named because they initially fix CO_2 in the form of three-carbon acids). In cooler climates, where photorespiration is less of a burden, C_3 plants have the advantage because they require less energy to fix CO_2.

Figure 24-41   **The C_4 pathway.** CO_2 is concentrated in the mesophyll cells and transported to the bundle-sheath cells for entry into the Calvin cycle.

d. CAM Plants Store CO_2 through a Variant of the C_4 Cycle

Desert-adapted plants known as succulents (e.g., cacti) employ a variant of the C_4 cycle that separates CO_2 acquisition and the Calvin cycle in time rather than in space. If, as most plants, they opened their stomata by day to acquire CO_2, they would simultaneously transpire (lose by evaporation) what for them would be unacceptable amounts of water. To minimize this loss, these succulents only absorb CO_2 at night when the temperature is relatively cool. They store this CO_2, in a process known as **crassulacean acid metabolism (CAM;** so named because it was first discovered in plants of the family Crassulaceae), by the synthesis of malate through the reactions of the C_4 pathway (Fig. 24-41). The large amount of PEP necessary to store a day's supply of CO_2 is obtained by the breakdown of starch via glycolysis. During the course of the day, this malate is broken down to CO_2, which enters the Calvin cycle, and pyruvate, which is used to resynthesize starch. CAM plants are able, in this way, to carry out photosynthesis with minimal water loss.

CHAPTER SUMMARY

1 Chloroplasts Photosynthesis is the light-driven fixation of CO_2 to form carbohydrates and other biological molecules. In plants, photosynthesis takes place in the chloroplast, which consists of an inner and an outer membrane surrounding the stroma, a concentrated enzyme solution in which the thylakoid membrane system is immersed. Photosynthesis occurs in two stages, the so-called light reactions in which light energy is harnessed to synthesize ATP and NADPH, and the dark reactions in which these products are used to drive the synthesis of carbohydrates from CO_2 and H_2O. The thylakoid membrane is the site of the photosynthetic light reactions, whereas the dark reactions take place in the stroma. The counterpart of the thylakoid in photosynthetic bacteria is a specialized portion of the plasma membrane named the chromatophore.

2 Light Reactions Chlorophyll is the principal photoreceptor of photosynthesis. Light is absorbed initially by light-harvesting complexes (LHCs) that contain chlorophyll and accessory pigments such as carotenoids. The resulting excitation then migrates via exciton transfer until it reaches the reaction center chlorophyll, where it is trapped. LH2 from purple photosynthetic bacteria is a transmembrane protein that consists of eight or nine rotationally related subunits that each bind three BChl a molecules and one carotenoid. LH1, which is similarly arranged but 16-fold symmetric, contains a central hole that binds a photosynthetic reaction center (RC). Light energy absorbed by LH2 is transmitted to LH1, which, in turn, transmits it to the RC.

The purple photosynthetic bacterial RC (PbRC) is a protein that consists of three subunits and several redox-active small molecules that are arranged as two pseudosymmetrically related chains of electron carriers. The primary photon absorbing species of the *Rps. viridis* bacterial reaction center is a special pair of BChl b molecules known as P960. By rapid measurement techniques it has been determined that the electron ejected by P960* passes by a third BChl b to a BPheo b molecule located in only one of the two chains (the other is apparently nonfunctional) and then sequentially to a menaquinone (Q_A) and a ubiquinone (Q_B). The resulting Q_B^- is subsequently further reduced in a second one-electron transfer process and then takes up two protons from the cytosol to form Q_BH_2. The electrons taken up by this species are returned to P960 via a cytochrome bc_1 complex, cytochrome c_2, and, in some purple photosynthetic bacteria, a four-heme c-type cytochrome associated with the photosynthetic reaction center. This cyclic electron-transport process functions to translocate protons, via a Q cycle mediated by the cytochrome bc_1, from the cytoplasm to the outside of the cell. The resulting proton gradient, in a process known as photophosphorylation, drives the synthesis of ATP. Since bacterial photosynthesis does not generate the reducing equivalents needed in many biosynthetic processes, photosynthetic bacteria require an outside source of reducing agents such as H_2S.

In plants and cyanobacteria, the light reactions occur in two reaction centers, those of PSI and PSII, which are electrically "connected" in series. This enables the system to generate sufficient electromotive force to form NADPH by oxidizing H_2O in a noncyclic pathway known as the Z-scheme. PSI and PSII both contain core antenna systems and their RCs are evolutionarily related to each other and to the PbRC. PSII contains an Mn_4CaO_4 complex that oxidizes two H_2O molecules to four H^+ and O_2 in four one-electron steps. The electrons are passed singly, through a Tyr side chain named Z, to photooxidized P680, the reaction center's photon-absorbing species, a special pair that consists of two Chl a molecules. The electron previously ejected from P680* passes through a series of carriers resembling those of the PbRC to a pool of plastoquinone molecules. The electrons then enter the cytochrome b_6f complex, which transports protons, via a Q cycle, from the stroma to the thylakoid space. These electrons are transferred individually, by a plastocyanin carrier, directly to PSI's photooxidized photon-absorbing pigment, P700, a pair of Chl a's that resembles the PbRC's special pair. The electron that had been previously ejected from P700* migrates through both sides of a bifurcated chain of Chl a molecules and then through a chain of three [4Fe–4S] clusters to a soluble ferredoxin (Fd) that contains a [2Fe–2S] cluster. The electron then reduces NADP$^+$ in a noncyclic process mediated by ferredoxin–NADP$^+$ reductase. Alternatively, it may be returned, presumably via ferredoxin–plastoquinone reductase, to the plastoquinone pool in a cyclic process that does not require electron input from PSII and only translocates protons across the thylakoid membrane. ATP is synthesized by the CF_1CF_0-ATP synthase, which closely resembles the analogous mitochondrial complex, in a reaction driven by the dissipation of the proton gradient across the thylakoid membrane.

3 Dark Reactions CO_2 is fixed in the photosynthetic dark reactions of plants and cyanobacteria by the reactions of the Calvin cycle. The first stage of the Calvin cycle, in sum, mediates the reaction $3RuBP + 3CO_2 \rightarrow 6GAP$ with the consumption of 9 ATP and 6 NADPH generated by the light reactions. The second stage reshuffles the atoms of five GAPs to reform the three RuBPs with which the cycle began, a process that requires no further input of free energy or reduction equivalents. The sixth GAP, the product of the Calvin cycle, is used to synthesize carbohydrates, amino acids, and fatty acids. The flux-controlling enzymes of the Calvin cycle are activated in the light through variations in the pH and the Mg^{2+} and NADPH concentrations, and by the redox level of thioredoxin. The central enzyme of the Calvin cycle, RuBP carboxylase, catalyzes both a carboxylase and an oxygenase reaction with RuBP. The latter reaction is the first step in the photorespiration cycle that liberates CO_2. The rate of photorespiration increases with temperature and decreases with CO_2 concentration, so photorespiration constitutes a significant energetic drain on most plants on hot bright days. Calvin cycle products are converted to sucrose, starch, and cellulose, as well as fatty acids and amino acids. C_4 plants, which are most common in the tropics, have a system for concentrating CO_2 in their photosynthetic cells so as to minimize the effects of photorespiration but at the cost of 2 ATP per CO_2 fixed. Certain desert plants conserve water by absorbing CO_2 at night and releasing it to the Calvin cycle by day. This crassulacean acid metabolism (CAM) occurs through a process similar to the C_4 cycle.

REFERENCES

General

Blankenship, R.E., *Molecular Mechanisms of Photosynthesis,* Blackwell Science (2002).

Buchanan, B.B., Gruissem, W., and Jones, R.L. (Eds.), *Biochemistry and Molecular Biology of Plants,* American Society of Plant Physiologists (2000).

Hall, D.O. and Rao, K.K., *Photosynthesis* (6th ed.), Cambridge (1999).

Heldt, H.-W., *Plant Biochemistry,* Elsevier (2005).

Lawlor, D.W., *Photosynthesis* (3rd ed.), BIOS Scientific Publishers Ltd. (2001).

Nicholls, D.G. and Ferguson, S.J., *Bioenergetics 3,* Chapter 6, Academic Press (2002).

Chloroplasts

Bogorad, L. and Vasil, I.K. (Eds.), *The Molecular Biology of Plastids,* Academic Press (1991).

Hoober, J.K., *Chloroplasts,* Plenum Press (1984).

Light Reactions

Amunts, A., Drory, O., and Nelson, N., The structure of plant photosystem I at 3.4 Å resolution, *Nature* **447,** 58–63 (2007).

Barber, J., Photosystem II: a multisubunit membrane protein that oxidises water, *Curr. Opin. Struct. Biol.* **12,** 523–530 (2002).

Chitnis, P.R., Photosystem I: Function and physiology, *Annu. Rev. Plant Physiol. Plant Biol.* **52,** 593–626 (2001).

Deisenhofer, J., Epp, O., Sinning, I., and Michel, H., Crystallographic refinement at 2.3 Å resolution and refined model of the photosynthetic reaction centre from *Rhodopseudomonas viridis, J. Mol. Biol.* **246,** 429–457 (1995).

Deisenhofer, J. and Michel, H., High-resolution structures of photosynthetic reaction centers, *Annu. Rev. Biophys. Biophys. Chem.* **20,** 247–266 (1991); *and* Structures of bacterial photosynthetic reaction centers, *Annu. Rev. Cell Biol.* **7,** 1–23 (1991).

Deng, Z., Aliverti, A., Zanetti, G., Arakaki, A.K., Ottado, J., Orellano, E.G., Calcaterra, N.B., Ceccarelli, E.A., Carrillo, N., and Karplus, P.A., A productive NADP⁺ binding mode of ferredoxin–NADP⁺ reductase revealed by protein engineering and crystallographic studies, *Nature Struct. Biol.* **6,** 847–853 (1999); *and* Bruns, C.M. and Karplus, P.A., Refined crystal structure of spinach ferredoxin reductase at 1.7 Å resolution: Oxidized, reduced, and 2′-phospho-5′-AMP bound states, *J. Mol. Biol.* **247,** 125–145 (1995).

Diner, B.A. and Rappaport, F., Structure, dynamics, and energetics of the primary photochemistry of photosystem II of oxygenic photosynthesis, *Annu. Rev. Plant Biol.* **53,** 551–580 (2002).

Eberhard, S., Finazzi, G., and Wollman, F.-A., The dynamics of photosynthesis, *Annu. Rev. Genet.* **42,** 463–515 (2008).

El-Kabbani, O., Chang, C.-H., Tiede, D., Norris, J., and Schiffer, M., Comparison of reaction centers from *Rhodobacter sphaeroides* and *Rhodopseudomonas viridis:* Overall architecture and protein-pigment interactions, *Biochemistry* **30,** 5361–5369 (1991).

Fromme, P. (Ed.), *Photosynthetic Protein Complexes. A Structural Approach,* Wiley–Blackwell (2008).

Guskov, A., Kern, J., Gabdulkhakov, A., Broser, M., Zouni, A., and Saenger, W., Cyanobacterial photosystem II at 2.9-Å resolution and the roles of quinones, lipids, channels, and chloride, *Nature Struct. Mol. Biol.* **16,** 334–342 (2009); Loll, B., Kern, J., Saenger, W., Zouni, A., and Biesiadka, J., Towards complete cofactor arrangement in the 3.0 Å resolution structure of photosystem II, *Nature* **438,** 1040–1044 (2005); *and* Ferreira, K.N., Iverson, T.M., Maghlaoui, K., Barber, J., and Iwata, S., Architecture of the photosynthetic oxygen-evolving center, *Science* **303,** 1831–1838 (2004). [X-ray structures of PSII.]

Heathcote, P., Fyfe, P.K., and Jones, M.R., Reaction centers: The structure and evolution of biological solar power, *Trends Biochem. Sci.* **27,** 79–87 (2002).

Jordan, P., Fromme, P., Witt, H.T., Klukas, O., Saenger, W., and Krauss, N., Three-dimensional structure of cyanobacterial photosystem I at 2.5 Å resolution, *Nature* **411,** 909–917 (2001).

Koepke, J., Hu, X., Muenke, C., Schulen, K., and Michel, H., The crystal structure of the light-harvesting complex II (B800–850) from *Rhodospirillum molischianum, Structure* **4,** 581–597 (1996); *and* McDermott, G., Prince, S.M., Freer, A.A., Horthornthwaite-Lawless, A.M., Papiz, M.Z., Cogdell, R.J., and Isaacs, N.W., Crystal structure of an integral membrane light-harvesting complex from photosynthetic bacteria, *Nature* **374,** 517–521 (1995). [The X-ray structures of LH2s.]

Kurisu, G., Kusunoki, M., Katoh, E., Yamazaki, T., Teshima, K., Onda, Y., Kimata-Ariga, Y., and Hase, T., Structure of the electron transfer complex between ferredoxin and ferredoxin-NADP⁺ reductase, *Nature Struct. Biol.* **8,** 117–121 (2001).

Kurisu, G., Zhang, H., Smith, J.L., and Cramer, W.A., Structure of the cytochrome b_6f complex of oxygenic photosynthesis: Tuning the cavity, *Science* **302,** 1009–1014 (2003); Stroebel, D., Choquet, Y., Popot, J.-L., and Picot, D., An atypical haem in the cytochrome b_6f complex, *Nature* **426,** 413–418 (2003); *and* Cramer, W.A., Zhang, H., Yan, J., Kurisu, G., and Smith, J.L., Transmembrane traffic in the cytochrome b_6f complex, *Annu. Rev. Biochem.* **75,** 769–790 (2006).

Nelson, N. and Yocum, C.F., Structure and function of photosystems I and II, *Annu. Rev. Plant Biol.* **57,** 521–565 (2006).

Renger, G. and Tenger, T., Photosystem II: The machinery of photosynthetic water splitting, *Photosyn. Res.* **98,** 53–80 (2008).

Ruban, A.V., et al., Identification of a mechanism of photoprotective energy dissipation in higher plants, *Nature* **450,** 575–578 (2007).

Standfuss, J., van Scheltinga, A.C.T., Lamborghini, M., and Kühlbrandt, W., Mechanisms of photoprotection and nonphotochemical quenching in pea light-harvesting complex at 2.5 Å resolution, *EMBO J.* **24,** 919–928 (2005). [The X-ray structure of LHC-II.]

Yano, J., et al., Where water is oxidized to dioxygen: Structure of the photosynthetic Mn_4Ca cluster, *Science* **314,** 821–825 (2006).

Dark Reactions

Black, C.C. and Osmond, C.B., Crassulacean acid metabolism pnotosynthesis: 'working the night shift,' *Photosynthesis Res.* **76,** 329–31 (2003).

Cushman, J.C. and Bohnert, H.J., Crassalacean acid metabolism: Molecular genetics, *Annu. Rev. Plant Physiol. Plant Mol. Biol.* **50,** 305–332 (1999).

Dai, S., Friemann, R., Glauser, D.A., Bourquin, F., Manieri, W., Schürmann, P., and Eklund, H., Structural snapshots along the reaction pathway of ferredoxin–thioredoxin reductase, *Nature* **448,** 92–96 (2007).

Hartman, F.C. and Harpel, M.R., Chemical and genetic probes of the active site of D-ribulose-1,5-bisphosphate carboxylase/oxygenase: A retrospective based on the three-dimensional structure, *Adv. Enzymol. Relat. Areas Mol. Biol.* **67,** 1–75 (1993).

Hatch, M.D., C_4 photosynthesis: A unique blend of modified biochemistry, anatomy, and ultrastructure, *Biochim. Biophys. Acta* **895,** 81–106 (1987).

Portis, A.R., Jr., Regulation of ribulose 1,5-bisphosphate carboxylase/oxygenase activity, *Annu. Rev. Plant Physiol. Plant Mol. Biol.* **43,** 415–437 (1992); *and* Rubisco activase, *Biochim. Biophys. Acta* **1015,** 15–28 (1990).

Saxena, I.M. and Brown, R.M., Jr., Cellulose biosynthesis: current views and evolving concepts, *Ann. Bot.* **96,** 9–21 (2005).

Schneider, G., Lindqvist, Y., and Branden, C.-I., RUBISCO: Structure and mechanism, *Annu. Rev. Biophys. Biomol. Struct.* **21,** 119–143 (1992).

Schreuder, H.A., Knight, S., Curmi, P.M.G., Andersson, I., Cascio, D., Sweet, R.M., Brändén, C.-I., and Eisenberg, D., Crystal structure of activated tobacco rubisco complexed with the reaction-intermediate analogue 2-carboxy-arabinitol 1,5-bisphosphate, *Protein Sci.* **2,** 1136–1146 (1993).

Schürmann, P., Redox signaling in the chloroplast: The ferredoxin/thioredoxin system, *Antioxidants Redox Signaling* **5,** 69–79 (2003).

Spreitzer, R.J. and Salvucci, M.E., Rubisco: structure, regulatory interactions, and possibilities for a better enzyme, *Annu. Rev. Plant Biol.* **53,** 449–475 (2002).

Taylor, T.C. and Andersson, I., The structure of the complex between rubisco and its natural substrate ribulose 1,5-bisphosphate, *J. Mol. Biol.* **265,** 432–444 (1997).

PROBLEMS

1. Why is chlorophyll green in color when it absorbs in the red and the blue regions of the spectrum (Fig. 24-5)?

2. The "red tide" is a massive proliferation of certain algal species that cause seawater to become visibly red. Describe the spectral characteristics of the dominant photosynthetic pigments in these algae.

3. $H_2^{18}O$ is added to a suspension of chloroplasts capable of photosynthesis. Where does the label appear when the suspension is exposed to light?

4. Indicate, where appropriate, the analogous components in the photosynthetic electron-transport chains of purple photosynthetic bacteria and chloroplasts.

5. Antimycin inhibits photosynthesis in chloroplasts. Indicate its most likely site of action and explain your reasoning.

6. Calculate the energy efficiency of cyclic and noncyclic photosynthesis in chloroplasts using 680-nm light. What would this efficiency be with 500-nm light? Assume that ATP formation requires $59 kJ \cdot mol^{-1}$ under physiological conditions.

*7. What is the minimum pH gradient required to synthesize ATP from $ADP + P_i$? Assume $[ATP]/([ADP][P_i]) = 10^3$, $T = 25°C$, and that three protons must be translocated per ATP generated. (See Table 16-3 for useful thermodynamic information.)

8. Indicate the average Calvin cycle labeling pattern in ribulose-5-phosphate after two rounds of exposure to $^{14}CO_2$.

9. Chloroplasts are illuminated until the levels of their Calvin cycle intermediates reach a steady state. The light is then turned off. How do the levels of RuBP and 3PG vary after this time?

10. What is the energy efficiency of the Calvin cycle combined with glycolysis and oxidative phosphorylation; that is, what percentage of the input energy can be metabolically recovered in synthesizing starch from CO_2 using photosynthetically produced NADPH and ATP rather than somehow directly storing these "high-energy" intermediates? Assume that each NADPH is energetically equivalent to three ATPs and that starch synthesis and breakdown are energetically equivalent to glycogen synthesis and breakdown.

11. Predict the effect of an uncoupler such as dinitrophenol (Fig. 22-47) on production of (a) ATP and (b) NADPH in a chloroplast.

12. Describe the effects of an increase in oxygen pressure on the dark reactions of photosynthesis.

13. If a C_3 plant and a C_4 plant are placed together in a sealed illuminated box with sufficient moisture, the C_4 plant thrives while the C_3 plant sickens and eventually dies. Explain.

14. The leaves of some species of desert plants taste sour in the early morning but, as the day wears on, they become tasteless and then bitter. Explain.

Lipid Metabolism

CHAPTER 25

Lipids play indispensable roles in cell structure and metabolism. For example, triacylglycerols are the major storage form of metabolic energy in animals; cholesterol is a vital component of cell membranes and a precursor of the steroid hormones and bile salts; arachidonate, a C_{20} unsaturated fatty acid, is the precursor of the prostaglandins, prostacyclins, thromboxanes, leukotrienes, and lipoxins, potent intercellular mediators that control a variety of complex processes; and complex glycolipids and phospholipids are major components of biological membranes. We dis-

cussed the structures of simple and complex lipids in Section 12-1. In the first half of this chapter, we consider the metabolism of fatty acids and triacylglycerols, including their digestion, oxidation, and biosynthesis. We then consider how cholesterol is synthesized and utilized, and how prostaglandins, prostacyclins, thromboxanes, leukotrienes, and lipoxins are synthesized. We end by studying how complex glycolipids and phospholipids are synthesized from their simpler lipid and carbohydrate components.

1 LIPID DIGESTION, ABSORPTION, AND TRANSPORT

Triacylglycerols *(also called **fats** or **triglycerides**) constitute ~90% of the dietary lipid and are the major form of metabolic energy in humans*. Triacylglycerols consist of glycerol triesters of fatty acids such as palmitic and oleic acids

$$
\begin{aligned}
&H_2C^1{-}O{-}C^1 \diagup\diagdown\diagup\diagdown\diagup\diagdown \,16 \\
&\qquad\qquad\;\; \| \\
&\qquad\qquad\;\; O \\
&HC^2{-}O{-}C^1 \diagup\diagdown\diagup\diagdown\diagup\diagdown \,18 \\
&\qquad\qquad\;\; \| \qquad\quad 9 \\
&\qquad\qquad\;\; O \\
&H_2C^3{-}O{-}C^1 \diagup\diagdown\diagup\diagdown\diagup\diagdown \,18 \\
&\qquad\qquad\;\; \| \qquad\quad 9 \\
&\qquad\qquad\;\; O
\end{aligned}
$$

1-Palmitoyl-2,3-dioleoyl-glycerol

(the names and structural formulas of some biologically common fatty acids are listed in Table 12-1). Like glucose, they are metabolically oxidized to CO_2 and H_2O. Yet, since most carbon atoms of triacylglycerols have lower oxidation states than those of glucose, *the oxidative metabolism of fats yields over twice the energy of an equal weight of dry carbohydrate or protein* (Table 25-1). Moreover, fats, being nonpolar, are stored in an anhydrous state, whereas glycogen, the storage form of glucose, is polar and is consequently stored in a hydrated form that contains about twice its dry weight of water. Fats therefore provide up to six times the metabolic energy of an equal weight of hydrated glycogen.

a. Lipid Digestion Occurs at Lipid–Water Interfaces
Since triacylglycerols are water insoluble, whereas digestive enzymes are water soluble, *triacylglycerol digestion takes place at lipid–water interfaces*. The rate of triacylglyc-

Table 25-1 Energy Content of Food Constituents

Constituent	ΔH (kJ · g^{-1} dry weight)
Carbohydrate	16
Fat	37
Protein	17

Source: Newsholme, E.A. and Leech, A.R., *Biochemistry for the Medical Sciences*, p. 16, Wiley (1983).

erol digestion therefore depends on the surface area of the interface, a quantity that is greatly increased by the churning peristaltic movements of the intestine combined with the emulsifying action of **bile salts.** These are powerful digestive detergents that, as we shall see in Section 25-6C, are synthesized by the liver and secreted via the gallbladder into the small intestine where lipid digestion and absorption mainly take place.

b. Pancreatic Lipase Requires Activation and Has a Catalytic Triad

Pancreatic **lipase (triacylglycerol lipase)** catalyzes the hydrolysis of triacylglycerols at their 1 and 3 positions to form sequentially **1,2-diacylglycerols** and **2-acylglycerols,** together with the Na$^+$ and K$^+$ salts of fatty acids (soaps). These soaps, being amphipathic, aid in the lipid emulsification process.

The enzymatic activity of pancreatic lipase greatly increases when it contacts the lipid–water interface, a phenomenon known as **interfacial activation.** Binding to the lipid–water interface requires the presence of mixed micelles of phosphatidylcholine (Fig. 12-4) and bile salts, as well as the pancreatically produced protein named **colipase,** which forms a 1:1 complex with lipase. This complex aids in the adsorption of the enzyme to emulsified oil droplets as well as stabilizes the enzyme in an active conformation. The X-ray structures, determined by Christian Cambillau, of pancreatic lipase–colipase complexes, alone and cocrystallized with mixed micelles of phosphatidylcholine and bile salts, have revealed the structural basis of lipase activation as well as how colipase and micelles aid lipase in binding to the lipid–water interface (Fig. 25-1).

The active site of the 449-residue pancreatic lipase, which is contained in the enzyme's N-terminal domain, has a catalytic triad that closely resembles that in the serine proteases (Section 15-3B; recall that ester hydrolysis is mechanistically similar to peptide hydrolysis). In aqueous solution (Fig. 25-1a), the lipase's active site is covered by a 26-residue helical lid. However, in the presence of the mixed micelles (Fig. 25-1b), the lid undergoes a complex structural reorganization that exposes the active site; causes a contacting 10-residue loop, the β5 loop, to change conformation in a way that forms the active enzyme's oxyanion hole; and generates a hydrophobic surface about the entrance to the active site. Indeed, the active site of the mixed micelle–containing complex contains a long rod of electron density that contacts the catalytic triad's Ser residue and appears to be a phosphatidylcholine molecule.

(a)

(b)

Figure 25-1 **X-ray structures of pancreatic lipase in complex with colipase.** (*a*) In aqueous solution, and (*b*) cocrystallized with mixed micelles of phosphatidylcholine and bile salts. The lipase is drawn in ribbon form with its N-terminal domain (residues 1–336) cyan, its C-terminal domain (residues 337–449) green, the lid (residues 237–262) magenta, and the β5 loop (residues 76–85) orange. The colipase is yellow. A phosphatidylcholine molecule that is bound in the lipase active site in Part *b* is shown in stick form with C green, O red, and P orange. The micelles, which have irregular structures (e.g., Fig. 2-9), are not visible. [Based on X-ray structures by Christian Cambillau, LCCMB-CNRS, Marseille, France. PDBids 1N8S and 1LPB.]

Colipase binds to the C-terminal domain of lipase such that the hydrophobic tips of the three loops that comprise much of this 90-residue protein extend from the complex on the same face as lipase's active site. A continuous hydrophobic plateau is thereby created that extends over a distance of >50 Å past the active site (bottom of Fig. 25-1b) and that, presumably, helps bind the complex to the lipid surface. In the presence of the mixed micelles, colipase changes conformation so as to form three hydrogen bonds to the opened lid, thereby stabilizing it in this conformation.

The mixed micelles are not visible in the X-ray structure. However, neutron diffraction studies, by Juan Fontecilla-Camps, of crystals of a lipase–colipase–micelle complex in which the lipase is in its active conformation reveal that the activating micelle interacts, not with the substrate site, but with the concave face of colipase and the adjacent tip of the lipase's C-terminal domain (left side of

phospholipase A$_1$

phospholipase C   phospholipase D

Phospholipid

Lysophospholipid

phospholipase A$_2$

H$_2$O

Figure 25-2   Catalytic action of phospholipase A$_2$. Phospholipase A$_2$ hydrolytically excises the C2 fatty acid residue from a phospholipid to yield the corresponding lysophospholipid. The bonds hydrolyzed by other types of phospholipases, which are named according to their specificities, are also indicated.

Fig. 25-1*b*). Apparently, micelle binding and substrate binding involve different regions of the lipase–colipase complex. Hence, strictly speaking, lipase activation appears not to be interfacial but, instead, occurs in the aqueous phase and requires the binding of colipase and a micelle.

c. Pancreatic Phospholipase A$_2$ Has a Modified Catalytic Triad

Phospholipids are degraded by pancreatic **phospholipase A$_2$**, which hydrolytically excises the fatty acid residue at C2 to yield the corresponding **lysophospholipids** (Fig. 25-2), which are also powerful detergents. Indeed, the phospholipid lecithin (phosphatidylcholine) is secreted in the bile, presumably to aid in lipid digestion.

Phospholipase A$_2$, as does triacylglycerol lipase, preferentially catalyzes reactions at interfaces. However, as Paul Sigler's determinations of the X-ray structures of the phospholipases A$_2$ from cobra venom and bee venom revealed, its mechanism of interfacial activation differs from that of triacylglycerol lipase in that it does not change its conformation. Instead, phospholipase A$_2$ contains a hy-

drophobic channel that provides the substrate with direct access from the phospholipid aggregate (micelle or membrane) surface to the bound enzyme's active site. Hence, on leaving its micelle to bind to the enzyme, the substrate need not become solvated and then desolvated (Fig. 25-3). In contrast, soluble and dispersed phospholipids must first surmount these significant kinetic barriers in order to bind to the enzyme.

The catalytic mechanism of phospholipase A$_2$ also differs substantially from that of triacylglycerol lipase. Although the phospholipase A$_2$ active site contains the His and Asp components of a catalytic triad, an enzyme-bound water molecule occupies the position expected for an active site Ser. Moreover, the active site contains a bound Ca^{2+} ion and does not form an acyl–enzyme intermediate. Sigler therefore proposed that phospholipase A$_2$ catalyzes the direct hydrolysis of phospholipid with a His–Asp "catalytic dyad" activating an active site water molecule for nucleophilic attack on the ester, and with the Ca^{2+} ion stabilizing the oxyanion transition state. However, the subsequently determined X-ray structure, by Mahendra Jain and

(a)

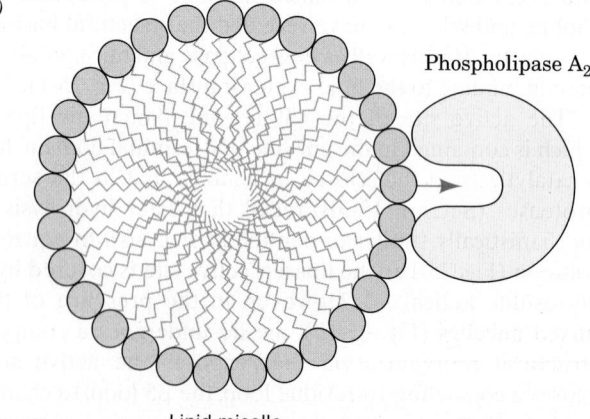

(b)

Phospholipase A$_2$

Lipid micelle

Figure 25-3   Substrate binding to phospholipase A$_2$. (*a*) A hypothetical model of phospholipase A$_2$ in complex with a micelle of lysophosphatidylethanolamine as shown in cross section. The protein is drawn in cyan, the phospholipid head groups are yellow, and their hydrocarbon tails are blue. The calculated atomic motions of the assembly are indicated through a series of superimposed images taken at 5-ps intervals. [Courtesy of Raymond Salemme, E.I. du Pont de Nemours & Company.] (*b*) Schematic diagram of a productive interaction between phospholipase A$_2$ and a phospholipid contained in a micelle.

Brian Bahnson, of phospholipase A_2 in complex with the tetrahedral intermediate mimic **MJ33**

MJ33 [1-Hexadecyl-3-(trifluoroethyl)-*sn*-glycero-2-phosphomethanol]

suggests that a second, previously unobserved water molecule, which is liganded by the Ca^{2+} ion, is the attacking nucleophile (Fig. 25-4*a*). This has led to the formulation of a reaction mechanism (Fig. 25-4*b*) in which the Asp–His–water catalytic triad and the Ca^{2+} ion both activate the second water molecule, with the Ca^{2+} ion also stabilizing the resulting tetrahedral intermediate.

d. Bile Salts and Fatty Acid–Binding Protein Facilitate the Intestinal Absorption of Lipids

The mixture of fatty acids and mono- and diacylglycerols produced by lipid digestion is absorbed by the cells lining the small intestine (the intestinal mucosa) in a process facilitated by bile salts. The micelles formed by the bile salts take up the nonpolar lipid degradation products so as to permit their transport across the unstirred aqueous boundary layer at the intestinal wall. The importance of this process is demonstrated in individuals with obstructed bile ducts: They absorb little of their dietary lipids but, rather, eliminate them in hydrolyzed form in their feces **(steatorrhea).** Evidently, *bile salts are not only an aid to lipid digestion but are essential for the absorption of lipid digestion products.* Bile salts are likewise required for the efficient intestinal absorption of the lipid-soluble vitamins A, D, E, and K.

Inside the intestinal cells, fatty acids form complexes with **intestinal fatty acid–binding protein (I-FABP),** a cyto-

Figure 25-4   **Structure and mechanism of phospholipase A_2.** (*a*) The X-ray structure of the 124-residue monomeric porcine phospholipase A_2 (*lavender*) in complex with the tetrahedral intermediate mimic MJ33. The enzyme's active site contains a catalytic triad similar to those of the serine proteases (Fig. 15-20) with a water molecule replacing the catalytic Ser. The His 48 and Asp 99 side chains of the catalytic triad together with MJ33 are drawn in stick form colored according to atom type (C gray, N blue, O red, F green, and P yellow). The water molecule of the catalytic triad and the catalytically important Ca^{2+} ion are represented by red and cyan spheres. Catalytically important hydrogen bonds and Ca^{2+} liganding interactions are represented by thin black lines. The tetrahedral phosphoryl group of MJ33 presumably occupies the site of the unobserved H_2O. Residues 65 to 74 of the protein have been deleted for clarity. [Based on an X-ray structure by Mahendra Jain and Brian Bahnson, University of Delaware. PDBid 1FXF.] (*b*) The catalytic mechanism of phospholipase A_2. **(1)** The catalytic triad activates a second water molecule to attack the scissile carbonyl carbon with Ca^{2+} coordinating the activated water molecule as well as electrostatically stabilizing the resulting tetrahedral intermediate (rather than doing so via nucleophilic catalysis as occurs in the serine proteases; Fig. 15-23). **(2)** The tetrahedral intermediate decomposes to yield products. [After Berg, O.G., Gelb, M.H., Tsai, M.-D., and Jain, M.K., *Chem. Rev.* **101,** 2638 (2001).]

Figure 25-5 **X-ray structure of rat intestinal fatty acid–binding protein in complex with palmitate.** The protein is drawn in ribbon form colored in rainbow order from its N-terminus (*blue*) to its C-terminus (*red*). The palmitate is shown in space-filling form with C green and O red. [Based on an X-ray structure by James Sacchettini, Albert Einstein College of Medicine. PDBid 2IFB.]

plasmic protein, which serves to increase the effective solubility of these water-insoluble substances and also to protect the cell from their detergent-like effects (recall that soaps are fatty acid salts). The X-ray structures of rat I-FABP, both alone and in complex with a single molecule of palmitate, were determined by James Sacchettini. This monomeric, 131-residue protein consists largely of 10 antiparallel β strands organized into a stack of two approximately orthogonal β sheets (Fig. 25-5). The palmitate occupies a gap between two of the β strands such that it lies between the β sheets with an orientation that, over much of its length, is more or less parallel to the gapped β strands (this structure has therefore been described as forming a "β-clam"). The palmitate's carboxyl group interacts with Arg 106, Gln 115, and two bound water molecules, whereas the methylene chain is encased by the side chains of several hydrophobic, mostly aromatic, residues.

e. Lipids Are Transported in Lipoprotein Complexes

The lipid digestion products absorbed by the intestinal mucosa are converted by these tissues to triacylglycerols (Section 25-4F) and then packaged into lipoprotein particles called **chylomicrons.** These, in turn, are released into the bloodstream via the lymph system for delivery to the tissues. Similarly, triacylglycerols synthesized by the liver are packaged into **very low density lipoproteins (VLDL)** and released directly into the blood. These lipoproteins, whose origins, structures, and functions are discussed in Section 12-5, maintain their otherwise insoluble lipid components in aqueous solution.

The triacylglycerol components of chylomicrons and VLDL are hydrolyzed to free fatty acids and glycerol in the capillaries of adipose tissue and skeletal muscle by **lipoprotein lipase** (Section 12-5Ba). The resulting free fatty acids are taken up by these tissues while the glycerol is transported to the liver or kidneys. There it is converted to the glycolytic intermediate dihydroxyacetone phosphate by the sequential actions of **glycerol kinase** and **glycerol-3-phosphate dehydrogenase** (Fig. 25-6).

Mobilization of triacylglycerols stored in adipose tissue involves their hydrolysis to glycerol and free fatty acids by **hormone-sensitive triacylglycerol lipase** (or just **hormone-sensitive lipase).** The free fatty acids are released into the bloodstream, where they bind to **serum albumin** (or just **albumin**), a soluble 585-residue monomeric protein that comprises about half of the blood serum protein. In the absence of albumin, the maximum solubility of free fatty acids is $\sim 10^{-6}$ M. Above this concentration, free fatty acids form micelles that act as detergents to disrupt protein and membrane structure and would therefore be toxic. However, the effective solubility of fatty acids in fatty acid–albumin complexes is as much as 2 mM. Nevertheless, those rare individuals with **analbuminemia** (severely depressed levels of albumin) suffer no apparent adverse symptoms; evidently, their fatty acids are transported in complex with other serum proteins.

The X-ray structure of human serum albumin in its complexes with a variety of common fatty acids, determined by Stephen Curry, reveals that each albumin molecule can bind up to seven fatty acid molecules (Fig. 25-7). However, these binding sites have different fatty acid–binding affinities so that, under normal physiological conditions, albumin carries between 0.1 and 2 fatty acid molecules per protein molecule. Albumin also binds an extraordinarily broad range of drugs and is thereby a major and usually unpredictable influence on their pharmacokinetics (Section 15-4Ba). Indeed, the large amounts of fatty acids in the blood after meals can significantly affect the pharmacokinetics of a drug through competitive and/or cooperative interactions.

Figure 25-6 Conversion of glycerol to the glycolytic intermediate dihydroxyacetone phosphate.

Figure 25-7 **X-ray structure of human serum albumin in complex with 7 molecules of palmitic acid.** The protein is drawn in semitransparent ribbon form colored in rainbow order from its N-terminus (*blue*) to its C-terminus (*red*). The fatty acids are shown in space-filling form with C green and O red. [Based on an X-ray structure by Stephen Curry, Imperial College of Science, Technology, and Medicine, London, U.K. PDBid 1E7H.]

2 FATTY ACID OXIDATION

The biochemical strategy of fatty acid oxidation was understood long before the advent of biochemical techniques involving enzyme purification or the use of radioactive tracers. In 1904, Franz Knoop, in the first use of chemical labels to trace metabolic pathways, fed dogs fatty acids labeled at their ω (last) carbon atom by a benzene ring and isolated the phenyl-containing metabolic products from their urine. Dogs fed labeled odd-chain fatty acids excreted **hippuric acid,** the glycine amide of **benzoic acid,** whereas those fed labeled even-chain fatty acids excreted **phenylaceturic acid,** the glycine amide of **phenylacetic acid** (Fig. 25-8). Knoop therefore deduced that the oxidation of the carbon atom β to the carboxyl group is involved in fatty acid breakdown. Otherwise, the phenylacetic acid would be further oxidized to benzoic acid. Knoop proposed that this breakdown occurs by a mechanism known as **β oxidation** in which the fatty acid's C_β atom is oxidized. It was not until after 1950, following the discovery of coenzyme A, that the enzymes of fatty acid oxidation were isolated and their reaction mechanisms elucidated. This work confirmed Knoop's hypothesis.

A. *Fatty Acid Activation*

Before fatty acids can be oxidized, they must be "primed" for reaction in an ATP-dependent acylation reaction to form fatty acyl-CoA. This activation process is catalyzed by a family of at least three **acyl-CoA synthetases** (also called **thiokinases**) that differ according to their chain-length specificities. These enzymes, which are associated with either the endoplasmic reticulum (ER) or the outer mitochondrial membrane, all catalyze the reaction

$$\text{Fatty acid} + \text{CoA} + \text{ATP} \rightleftharpoons$$
$$\text{fatty acyl-CoA} + \text{AMP} + PP_i$$

In the activation of ^{18}O-labeled palmitate by a long-chain acyl-CoA synthetase, both the AMP and the acyl-CoA products become ^{18}O labeled. This observation

Fatty acid fed	Breakdown product	Excretion product

Odd-chain fatty acid $\quad\quad\quad$ **Benzoic acid** $\quad\quad\quad$ **Hippuric acid**

Even-chain fatty acid $\quad\quad\quad$ **Phenylacetic acid** $\quad\quad\quad$ **Phenylaceturic acid**

Figure 25-8 **Franz Knoop's classic experiment indicating that fatty acids are metabolically oxidized at their β-carbon atom.** ω-Phenyl-labeled fatty acids containing an odd number of carbon atoms are oxidized to the phenyl-labeled C_1 product, benzoic acid, whereas those with an even number of carbon atoms are oxidized to the phenyl-labeled C_2 product, phenylacetic acid.

These products are excreted as their respective glycine amides, hippuric and phenylaceturic acids. The vertical arrows indicate the deduced sites of carbon oxidation. The intermediate C_2 products are oxidized to CO_2 and H_2O and were therefore not isolated.

$$R-C \overset{O^*}{\underset{O^*}{\big<}} \quad + \quad {}^-O-\overset{O}{\underset{O^-}{P}}-O-\overset{O}{\underset{O^-}{P}}-O-\overset{O}{\underset{O^-}{P}}-O-Adenosine$$

Fatty acid   **ATP**

$$\downarrow$$

$$\xrightarrow{\quad} PP_i \xrightarrow[\text{H}_2\text{O}]{\substack{\text{inorganic} \\ \text{pyrophosphatase}}} 2P_i$$

$$\downarrow$$

$$R-\overset{O^*}{\overset{\|}{C}}-O^*-\overset{O}{\overset{\|}{P}}-O-Adenosine \\ \underset{O^-}{|}$$

**Acyladenylate
mixed anhydride**

$$H-SCoA \overset{\curvearrowleft}{\longrightarrow}$$

$$\downarrow$$

$$R-\overset{O^*}{\overset{\|}{C}}-SCoA \quad + \quad {}^-O^*-\overset{O}{\overset{\|}{P}}-O-Adenosine \\ \underset{O^-}{|}$$

Acyl-CoA   **AMP**

**Figure 25-9   Mechanism of fatty acid activation catalyzed by
acyl-CoA synthetase.** Experiments utilizing ^{18}O-labeled fatty
acids (*) demonstrate that the formation of acyl-CoA involves an
intermediate acyladenylate mixed anhydride.

$$(CH_3)_3\overset{+}{N}-CH_2-\overset{OH}{\overset{|}{CH}}-CH_2-COO^- \quad + \quad R-\overset{O}{\overset{\|}{C}}-SCoA$$

**Carnitine (4-trimethylamino-
3-hydroxybutyrate)**

$$\updownarrow \text{ carnitine palmitoyltransferase}$$

$$(CH_3)_3\overset{+}{N}-CH_2-\overset{\overset{\displaystyle O}{\overset{\displaystyle \|}{\overset{\displaystyle R-C-O}{|}}}}{CH}-CH_2-COO^- \quad + \quad H-SCoA$$

Acyl-carnitine

**Figure 25-10   Acylation of carnitine catalyzed by carnitine
palmitoyltransferase.**

indicates that the reaction has an acyladenylate mixed anhydride intermediate that is attacked by the sulfhydryl group
of CoA to form the thioester product (Fig. 25-9). The reaction involves both the cleavage and the synthesis of bonds
with large negative free energies of hydrolysis so that the
free energy change associated with the overall reaction is
close to zero. The reaction is driven to completion in the
cell by the highly exergonic hydrolysis of the product pyrophosphate (PP$_i$) catalyzed by the ubiquitous **inorganic
pyrophosphatase.** Thus, as commonly occurs in metabolic
pathways, *a reaction forming a "high-energy" bond through
the hydrolysis of one of ATP's phosphoanhydride bonds is
driven to completion by the hydrolysis of its second such
bond.*

B. *Transport Across the Mitochondrial Membrane*

Although fatty acids are activated for oxidation in the cytosol, they are oxidized in the mitochondrion as Eugene
Kennedy and Albert Lehninger established in 1950. We
must therefore consider how fatty acyl-CoA is transported
across the inner mitochondrial membrane. A long-chain
fatty acyl-CoA cannot directly cross the inner mitochondrial membrane. Rather, its acyl portion is first transferred to
carnitine (Fig. 25-10), a compound that occurs in both plant
and animal tissues. This transesterification reaction has an
equilibrium constant close to 1, which indicates that the
O-acyl bond of **acyl-carnitine** has a free energy of hydrolysis
similar to that of the thioester. **Carnitine palmitoyltrans-
ferases I** and **II,** which can transfer a variety of acyl groups,
are located, respectively, on the external and internal surfaces
of the inner mitochondrial membrane. The translocation
process itself is mediated by a specific carrier protein that
transports acyl-carnitine into the mitochondrion while transporting free carnitine in the opposite direction. Acyl-CoA
transport therefore occurs via four reactions (Fig. 25-11):

1. The acyl group of a cytosolic acyl-CoA is transferred
to carnitine, thereby releasing the CoA to its cytosolic pool.

2. The resulting acyl-carnitine is transported into the
mitochondrial matrix by the transport system.

3. The acyl group is transferred to a CoA molecule
from the mitochondrial pool.

4. The product carnitine is returned to the cytosol.

Figure 25-11   Transport of fatty acids into the mitochondrion.

Figure 25-12 The β-oxidation pathway of fatty acyl-CoA.
🖱 See the Animated Figures

The cell thereby maintains separate cytosolic and mitochondrial pools of CoA. The mitochondrial pool functions in the oxidative degradation of pyruvate (Section 21-2A) and certain amino acids (Sections 26-3E–G) as well as fatty acids, whereas the cytosolic pool supplies fatty acid biosynthesis (Section 25-4). The cell similarly maintains separate cytosolic and mitochondrial pools of ATP and NAD^+.

C. β Oxidation

Fatty acids are dismembered through the β oxidation of fatty acyl-CoA, a process that occurs in four reactions (Fig. 25-12):

1. Formation of a trans-α,β double bond through dehydrogenation by the flavoenzyme **acyl-CoA dehydrogenase (AD).**

2. Hydration of the double bond by **enoyl-CoA hydratase (EH)** to form a **3-L-hydroxyacyl-CoA.**

3. NAD^+-dependent dehydrogenation of this β-hydroxyacyl-CoA by **3-L-hydroxyacyl-CoA dehydrogenase (HAD)** to form the corresponding β-ketoacyl-CoA.

4. C_α—C_β cleavage in a thiolysis reaction with CoA as catalyzed by **β-ketoacyl-CoA thiolase (KT;** also called just **thiolase)** to form acetyl-CoA and a new acyl-CoA containing two less C atoms than the original one.

The first three steps of this process chemically resemble the citric acid cycle reactions that convert succinate to oxaloacetate (Sections 21-3F–H):

Figure 25-13 **Ribbon diagram of the active site region in a subunit of medium-chain acyl-CoA dehydrogenase from pig liver mitochondria in complex with octanoyl-CoA.** The enzyme is a tetramer of identical 385-residue subunits, each of which binds an FAD prosthetic group (*green*) and its octanoyl-CoA substrate (whose octanoyl and CoA moieties are blue and white) in largely extended conformations. The octanoyl-CoA binds such that its C_α—C_β bond is sandwiched between the carboxylate group of Glu 376 (*red*) and the flavin ring (*green*), consistent with the proposal that Glu 376 is the general base that abstracts the α proton in the α,β dehydrogenation reaction catalyzed by the enzyme. [Based on an X-ray structure by Jung-Ja Kim, Medical College of Wisconsin. PDBid 3MDE.] *See Interactive Exercise 23*

Mitochondria contain four acyl-CoA dehydrogenases, with specificities for short- (C_4 to C_6), medium- (C_6 to C_{10}), long- (between medium and very long), and very long-chain (C_{12} to C_{18}) fatty acyl-CoAs. The reaction catalyzed by these enzymes is thought to involve removal of a proton at C_α and transfer of a hydride ion equivalent from C_β to FAD (Fig. 25-12, Reaction 1). The X-ray structure of the **medium-chain acyl-CoA dehydrogenase (MCAD)** in complex with **octanoyl-CoA,** determined by Jung-Ja Kim, clearly shows how the enzyme orients the enzyme's base (Glu 376), the substrate C_α—C_β bond, and the FAD prosthetic group for reaction (Fig. 25-13).

a. Acyl-CoA Dehydrogenase Is Reoxidized via the Electron-Transport Chain

The $FADH_2$ resulting from the oxidation of the fatty acyl-CoA substrate is reoxidized by the mitochondrial electron-transport chain through the intermediacy of a series of electron-transfer reactions. **Electron-transfer flavoprotein (ETF)** transfers two electrons from $FADH_2$ to the flavoiron–sulfur protein **ETF:ubiquinone oxidoreductase,** which in turn transfers two electrons to the mitochondrial electron-transport chain by reducing coenzyme Q (CoQ; Fig. 25-12, Reactions 5–8). Reduction of O_2 to H_2O by the electron-transport chain beginning at the CoQ stage re-

sults in the synthesis of 1.5 ATPs per two electrons transferred (Section 22-2Bc).

b. Acyl-CoA Dehydrogenase Deficiency Has Fatal Consequences

The unexpected death of an apparently healthy infant, often overnight, has been, for lack of any real explanation, termed **sudden infant death syndrome (SIDS).** MCAD has been shown to be deficient in up to 10% of these infants, making this genetic disease more prevalent than **phenylketonuria (PKU)** (Section 26-3Hd), a genetic defect in phenylalanine degradation for which babies born in the United States are routinely tested. Glucose is the principal energy metabolism substrate just after eating, but when the glucose level later decreases, the rate of fatty acid oxidation must correspondingly increase. The sudden death of infants lacking MCAD may be caused by the imbalance between glucose and fatty acid oxidation.

Lys 304, which becomes Glu in the most prevalent mutation among individuals with MCAD deficiency, is ~20 Å distant from the enzyme's active site and hence cannot participate in binding substrate or FAD. However, since the side chains of Asp 300 and Asp 346 lie within 6 Å of Glu 304, near a subunit–subunit interface, it seems likely that the high concentration of negative charges resulting from the Lys 304 → Glu mutation structurally destabilizes the enzyme.

Deficiency of acyl-CoA dehydrogenase has also been implicated in **Jamaican vomiting sickness,** whose victims suffer violent vomiting followed by convulsions, coma, and death. Severe hypoglycemia is observed in most cases. This condition results from eating unripe **ackee fruit,** which contains **hypoglycin A,** an unusual amino acid, which is metabolized to **methylenecyclopropylacetyl-CoA (MCPA-CoA; Fig. 25-14).** MCPA-CoA, a substrate for acyl-CoA dehydrogenase, is thought to undergo the first step of the reaction that this enzyme catalyzes, removal of a proton from C_α, to form a reactive intermediate that covalently modifies the enzyme's FAD prosthetic group (Fig. 25-14). Since a normal step in the enzyme's reaction mechanism generates the reactive intermediate, MCPA-CoA is said to be a **mechanism-based inhibitor.**

c. Long-Chain Enoyl-CoAs Are Converted to Acetyl-CoA and a Shorter Acyl-CoA by Mitochondrial Trifunctional Protein

The products of acyl-CoA dehydrogenases are 2-enoyl-CoAs. Depending on their chain lengths their processing is continued by one of three systems (Fig. 25-12): the short-chain, medium-chain, or long-chain 2-enoyl-CoA hydratases (EHs), hydroxyacyl-CoA dehydrogenases (HADs), and β-ketoacyl-CoA thiolases (KTs). The long-chain (LC) versions of these enzymes are contained on one $\alpha_4\beta_4$ octameric protein, **mitochondrial trifunctional protein,** located in the inner mitochondrial membrane. LCEH and LCHAD are contained on the α subunits while LCKT is located on the β subunits. The protein is therefore a combination multifunctional protein (more than one enzyme activity on a sin-

Figure 25-14 Metabolic conversions of hypoglycin A to yield a product that inactivates acyl-CoA dehydrogenase. Spectral changes suggest that the enzyme's FAD prosthetic group has been modified.

gle polypeptide chain)–multienzyme complex (a complex of polypeptides catalyzing more than one reaction). The advantage of such a trifunctional enzyme is the ability to channel the intermediates toward the final product. Indeed, no long-chain hydroxyacyl-CoA or ketoacyl-CoA intermediates are released into solution by this system.

d. The Thiolase Reaction Occurs via Claisen Ester Cleavage

The final stage of the fatty acid β-oxidation process, the thiolase reaction, forms acetyl-CoA and a new acyl-CoA which is two carbon atoms shorter than the one that began the cycle. This occurs in five reaction steps (Fig. 25-15):

1. An active site thiol is added to the substrate β-keto group.

2. Carbon–carbon bond cleavage forms an acetyl-CoA carbanion intermediate that is stabilized by electron withdrawal into this thioester's carbonyl group. Such a reaction is known as a Claisen ester cleavage (the reverse of a Claisen condensation). The citric acid cycle enzyme citrate synthase also catalyzes a reaction that involves a stabilized acetyl-CoA carbanion intermediate (Section 21-3A).

3. The acetyl-CoA carbanion intermediate is protonated by an enzyme acid group, yielding acetyl-CoA.

4 and 5. Finally, CoA displaces the enzyme thiol group from the enzyme–thioester intermediate, yielding acyl-CoA.

The formation of an enzyme–thioester intermediate involving an active site thiol group is based on the observation that incubation of the enzyme with [^{14}C]acetyl-CoA

Figure 25-15 Mechanism of action of β-ketoacyl-CoA thiolase. An active site Cys residue participates in the formation of an enzyme–thioester intermediate.

yields a specifically labeled enzyme Cys residue (the reverse of steps 4 and 5):

$$E-SH + CH_3-\overset{\overset{\text{O}}{\|}}{\underset{*}{C}}-SCoA$$

Thiolase   Acetyl-CoA

↓ ↘ CoASH

$$E-S-\overset{\overset{\text{O}}{\|}}{\underset{*}{C}}-CH_3$$

↓ trypsin degradation

$$Val-\underset{\underset{\underset{\underset{CH_3}{|}}{\underset{*}{C=O}}}{\underset{|}{S}}}{Cys}-Ala-Ser-Gly-Met-Lys$$

e. Fatty Acid Oxidation Is Highly Exergonic

The function of fatty acid oxidation is, of course, to generate metabolic energy. Each round of β oxidation produces one NADH, one FADH$_2$, and one acetyl-CoA. Oxidation of acetyl-CoA via the citric acid cycle generates additional FADH$_2$ and NADH, which are reoxidized through oxidative phosphorylation to form ATP. Complete oxidation of a fatty acid molecule is therefore a highly exergonic process, which yields numerous ATPs. For example, oxidation of palmitoyl-CoA (which has a C$_{16}$ fatty acyl group) involves seven rounds of β oxidation, yielding 7 FADH$_2$, 7 NADH, and 8 acetyl-CoA. Oxidation of the 8 acetyl-CoA, in turn, yields 8 GTP, 24 NADH, and 8 FADH$_2$. Since oxidative phosphorylation of the 31 NADH molecules yields 77.5 ATP and that of the 15 FADH$_2$ yields 22.5 ATPs, subtracting the 2 ATP equivalents required for fatty acyl-CoA formation (Section 25-2A), *the oxidation of one palmitate molecule has a net yield of 106 ATP.*

D. *Oxidation of Unsaturated Fatty Acids*

Almost all unsaturated fatty acids of biological origin (Section 12-1A) contain only cis double bonds, which most often begin between C9 and C10 (referred to as a Δ^9 or 9-double bond; Table 12-1). Additional double bonds, if any, occur at three-carbon intervals and are therefore never conjugated. Two examples of unsaturated fatty acids are oleic acid and linoleic acid (Fig. 25-16). Note that one of the double bonds in linoleic acid is at an odd-numbered carbon atom and the other is at an even-numbered carbon atom. Double bonds at these positions in fatty acids pose three problems for the β-oxidation pathway that are solved through the actions of four additional enzymes (Fig. 25-17):

Problem 1: A β,γ Double Bond

The first enzymatic difficulty occurs on the left-hand pathway in Fig. 25-17 after the third round of β oxidation:

Oleic acid
(9-*cis*-Octadecenoic acid)

Linoleic acid
(9,12-*cis*-Octadecadienoic acid)

Figure 25-16 **Structures of two common unsaturated fatty acids.** Most unsaturated fatty acids contain unconjugated cis double bonds.

The resulting cis-β,γ double bond–containing enoyl-CoA is not a substrate for enoyl-CoA hydratase. **Enoyl-CoA isomerase,** however, mediates conversion of the cis-Δ^3 double bond to the more stable, ester-conjugated trans-Δ^2 form:

Such compounds are normal substrates of enoyl-CoA hydratase so that β oxidation can then continue.

Problem 2: A Δ^4 Double Bond Inhibits Hydratase Action

The next difficulty arises on the left-hand pathway in Fig. 25-17 in the fifth round of β oxidation. The presence of a double bond at an even-numbered carbon atom results in the formation of 2,4-dienoyl-CoA, which is a poor substrate for enoyl-CoA hydratase. However, NADPH-dependent

Figure 25-17 (*Opposite*) **Problems in the oxidation of unsaturated fatty acids and their solutions.** Linoleic acid is used as an example. The first problem, the presence of a β,γ double bond seen in the left-hand pathway, is solved by the bond's enoyl-CoA isomerase–catalyzed conversion to a trans-α,β double bond. The second problem in the left-hand pathway, that a 2,4-dienoyl-CoA is not a substrate for enoyl-CoA hydratase, is eliminated by the NADPH-dependent reduction of the Δ^4 bond by 2,4-dienoyl-CoA reductase to yield the β-oxidation substrate *trans*-2-enoyl-CoA in *E. coli* but *trans*-3-enoyl-CoA in mammals. Mammals therefore also have 3,2-enoyl-CoA isomerase, which converts the *trans*-3-enoyl-CoA to *trans*-2-enoyl-CoA. The third problem, the isomerization of 2,5-dienoyl-CoA (originating from the oxidation of unsaturated fatty acids with double bonds at odd-numbered C atoms) to 3,5-dienoyl-CoA by 3,2-enoyl-CoA isomerase, is solved by 3,5–2,4-dienoyl-CoA isomerase, which converts the 3,5-dienoyl-CoA to 2,4-dienoyl-CoA, a substrate for 2,4-dienoyl-CoA reductase.

Linoleic acid

$2NAD^+ + 2FAD + 2CoASH$

$2NADH + 2FADH_2 + 2acetyl\text{-}CoA$

2 rounds of β oxidation

2,5,8-Trienoyl-CoA

FAD

$FADH_2$

acyl-CoA dehydrogenase

$NAD^+ + CoASH$

$NADH + acetyl\text{-}CoA$

completion of β-oxidation round

3,2-enoyl-CoA isomerase

Problem 3: Isomerization

3,5,8-Trienoyl-CoA

3,5–2,4-dienoyl-CoA isomerase

$\Delta^2, \Delta^4, \Delta^8$**-Trienoyl-CoA**

$NADPH + H^+$

$NADP^+$

2,4-dienoyl-CoA reductase

3,2-enoyl-CoA isomerase

Problem 1:
β,γ double bond

enoyl-CoA isomerase

$NAD^+ + FAD + CoASH$

$NADH + FADH_2 + Acetyl\text{-}CoA$

one round of β oxidation + the first oxidation of the next round

Problem 2:
Δ^4 double bond

$NADPH + H^+$

$NADP^+$

2,4-dienoyl-CoA reductase (mammalian)

2,4-dienoyl-CoA reductase (*E. coli*)

3,2-enoyl-CoA isomerase (mammalian)

$2NADH + 2FADH_2 + 2Acetyl\text{-}CoA$

$2NAD^+ + 2FAD + 2CoA$

2 rounds of β-oxidation

Continuation of β oxidation

2,4-dienoyl-CoA reductase reduces the Δ^4 double bond. The *E. coli* reductase produces *trans*-2-enoyl-CoA, a normal substrate of β oxidation. The mammalian reductase, however, yields *trans*-3-enoyl-CoA, which, to proceed along the β-oxidation pathway, must first be isomerized to *trans*-2-enoyl-CoA by **3,2-enoyl-CoA isomerase.**

Problem 3: The Unanticipated Isomerization of 2,5-Enoyl-CoA by 3,2-Enoyl-CoA Isomerase

Mammalian 3,2-enoyl-CoA isomerase catalyzes a reversible reaction that interconverts Δ^2 and Δ^3 double bonds. A carbonyl group is stabilized by being conjugated to a Δ^2 double bond. However, the presence of a Δ^5 double bond (originating from an unsaturated fatty acid with a double bond at an odd-numbered C atom such as the Δ^9 double bond of linoleic acid) is likewise stabilized by being conjugated with a Δ^3 double bond (right-hand pathway of Fig. 25-17). If a 2,5-enoyl-CoA is converted by 3,2-enoyl-CoA isomerase to 3,5-enoyl-CoA, which occurs up to 20% of the time, another enzyme is necessary to continue the oxidation: **3,5–2,4-Dienoyl-CoA isomerase** isomerizes the 3,5-diene to a 2,4-diene, which is then reduced by 2,4-dienoyl-CoA reductase and isomerized by 3,2-enoyl-CoA isomerase as in Problem 2 above. After two more rounds of β oxidation, the cis-Δ^4 double bond originating from the cis-Δ^{12} double bond of linoleic acid is also dealt with as in Problem 2.

E. Oxidation of Odd-Chain Fatty Acids

Most fatty acids have even numbers of carbon atoms and are therefore completely converted to acetyl-CoA. Some plants and marine organisms, however, synthesize fatty acids with an odd number of carbon atoms. *The final round of β oxidation of these fatty acids forms propionyl-CoA, which, as we shall see, is converted to succinyl-CoA for entry into the citric acid cycle.* Propionate or propionyl-CoA is also produced by oxidation of the amino acids isoleucine, valine, and methionine (Section 26-3E). Furthermore, ruminant animals such as cattle derive most of their caloric intake from the acetate and propionate produced in their rumen (stomach) by bacterial fermentation of carbohydrates. These products are absorbed by the animal and metabolized after conversion to the corresponding acyl-CoA.

a. Propionyl-CoA Carboxylase Has a Biotin Prosthetic Group

The conversion of propionyl-CoA to succinyl-CoA involves three enzymes (Fig. 25-18). The first reaction is that of **propionyl-CoA carboxylase,** a biotin-dependent enzyme (Section 23-1Ab) with subunit composition $\alpha_6\beta_6$. The reaction, which resembles that catalyzed by the homologous biotin-containing enzyme pyruvate decarboxylase (Section 23-1Ac), occurs in two steps (Fig. 25-19):

1. Carboxylation of biotin at N1′ by bicarbonate ion as in the pyruvate carboxylase reaction (Fig. 23-4). This step,

Figure 25-18   Conversion of propionyl-CoA to succinyl-CoA.

which is driven by the concomitant hydrolysis of ATP to ADP and P_i, activates the resulting carboxyl group for transfer without further free energy input.

2. Stereospecific transfer of the activated carboxyl group from carboxybiotin to propionyl-CoA to form **(S)-methylmalonyl-CoA.** This step occurs via nucleophilic attack on carboxybiotin by a carbanion at C2 of propionyl-CoA (see below).

These two reaction steps occur at different catalytic sites on propionyl-CoA carboxylase. It therefore appears that the biotinyllysine linkage attaching the biotin ring to the enzyme forms a flexible tether that permits the efficient transfer of the biotin ring between these two active sites as occurs in pyruvate carboxylase (Section 23-1Ae).

Formation of the C2 carbanion in the second stage of the propionyl-CoA carboxylase reaction involves removal of a proton α to a thioester. This proton is relatively acidic since, as we have seen in Section 25-2Cd, the negative charge on a carbanion α to a thioester can be delocalized into the thioester's carbonyl group. This explains the relatively convoluted path taken in the conversion of propionyl-CoA to succinyl-CoA (Fig. 25-18). It would seem simpler, at least on paper, for this process to occur in one step, with carboxylation occurring on C3 of propionyl-CoA so as to form succinyl-CoA directly. Yet, the C3 carbanion required for

Figure 25-19 The propionyl-CoA carboxylase reaction. (1)
The carboxylation of biotin with the concomitant hydrolysis of
ATP is followed by **(2)** the carboxylation of a propionyl-CoA
carbanion by its attack on carboxybiotin. Each reaction step
probably involves the intermediate formation of CO_2 as occurs
in the pyruvate carboxylase reaction (Fig. 23-4).

such a carboxylation has a high free energy of formation.
Nature has instead chosen a more facile, albeit less direct
route, which carboxylates propionyl-CoA at a more reactive
position and then rearranges the C_4 skeleton to form the desired product.

b. Methylmalonyl-CoA Mutase Contains a Coenzyme B_{12} Prosthetic Group

Methylmalonyl-CoA mutase, which catalyzes the third
reaction of the propionyl-CoA to succinyl-CoA conversion
(Fig. 25-18), is specific for (*R*)-methylmalonyl-CoA even
though propionyl-CoA carboxylase stereospecifically synthesizes (*S*)-methylmalonyl-CoA. This diversion is rectified by **methylmalonyl-CoA racemase,** which interconverts
the (*R*) and (*S*) configurations of methylmalonyl-CoA,
presumably by promoting the reversible dissociation of its
acidic α-H via formation of a resonance-stabilized carbanion intermediate:

**Resonance-stabilized carbanion
intermediate**

AdoCbl's reactive C—Co bond participates in two types of enzyme-catalyzed reactions:

1. Rearrangements in which a hydrogen atom is directly transferred between two adjacent carbon atoms with concomitant exchange of the second substituent, X:

where X may be a carbon atom with substituents, an oxygen atom of an alcohol, or an amine.

2. Methyl group transfers between two molecules.

There are about a dozen known cobalamin-dependent enzymes. However, only two occur in mammalian systems: (1) methylmalonyl-CoA mutase, which catalyzes a carbon skeleton rearrangement (the X group in the rearrangement is —COSCoA; Fig. 25-20) and is the only B$_{12}$-containing enzyme that occurs in both eukaryotes and prokaryotes; and (2) **methionine synthase,** a methyl transfer enzyme that participates in methionine biosynthesis (Sections 26-3Ec and 26-5B). Defects in methylmalonyl-CoA mutase result in **methylmalonic aciduria,** a condition that is often fatal in infancy due to **acidosis** (low blood pH) without a diet de-

void of odd-chain fatty acids and low in the amino acid residues that are degraded to propionyl-CoA (Ile, Val, and Met; Section 26-3E).

c. The Methylmalonyl-CoA Mutase Reaction Occurs via a Free Radical Mechanism

Methylmalonyl-CoA mutase from *Propionibacterium shermanii* is an αβ heterodimer whose catalytically active 728-residue α subunit is 24% identical to its catalytically inactive 638-residue β subunit. In contrast, the human enzyme is a homodimer whose subunits are 60% identical in sequence to *P. shermanii*'s α subunit. Hence *P. shermanii*'s β subunit is thought to be an evolutionary fossil.

The X-ray structure of methylmalonyl-CoA mutase from *P. shermanii* in complex with the substrate analog **2-carboxypropyl-CoA** (which lacks methylmalonyl-CoA's thioester oxygen atom) was determined by Philip Evans. Its AdoCbl cofactor is sandwiched between the α subunit's two domains: a 559-residue N-terminal α/β barrel (TIM barrel, the most common enzymatic motif; Section 8-3B) and a 169-residue C-terminal α/β domain that resembles a Rossmann fold (Section 8-3Bh). The structure of the α/β barrel contains several surprising features (Fig. 25-22):

1. The active sites of nearly all α/β barrel enzymes are located at the C-terminal ends of the barrel's β strands.

Figure 25-22 **X-ray structure of *P. shermanii* methylmalonyl-CoA mutase in complex with 2-carboxypropyl-CoA and AdoCbl.** (*a*) The catalytically active α subunit in which the N-terminal domain is cyan with the β strands of its α/β barrel orange, and the C-terminal domain is pink. The 2-carboxypropyl-CoA (*magenta*) and AdoCbl (*green*) are drawn in space-filling form. The 2-carboxypropyl-CoA passes through the center of the α/β barrel and is oriented such that the methylmalonyl group of methylmalonyl-CoA would contact the corrin ring of the AdoCbl, which is sandwiched between the enzyme's N- and C-terminal domains. (*b*) The arrangement of the AdoCbl and 2-carboxypropyl-CoA molecules which, together with the side chain of His 610, are represented in stick form colored according to atom type (AdoCbl and His C green, 2-carboxypropyl-CoA C cyan, N blue, O red, P magenta, and S yellow). The corrin ring's Co atom is represented by a lavender sphere and the α/β barrel's β strands are represented by orange ribbons. The view is similar to that in Part *a*. Note that the DMB group (*bottom*) has swung away from the corrin ring (seen edgewise) to be replaced by the side chain of His 610 from the C-terminal domain and that the 5'-deoxyadenosyl group is unseen (due to disorder). [Based on an X-ray structure by Philip Evans, MRC Laboratory of Molecular Biology, Cambridge, U.K. PDBid 7REQ.]

See Interactive Exercise 24

However, in methylmalonyl-CoA mutase, the AdoCbl is packed against the N-terminal ends of the barrel's β strands.

2. In free AdoCbl, the Co atom is axially liganded by an N atom of its DMB group and by the adenosyl residue's 5′-CH_2 group (Fig. 25-21). However, in the enzyme, the DMB has swung aside to bind in a separate pocket and has been replaced by the side chain of His 610 from the C-terminal domain. The adenosyl group is not visible in the structure due to disorder and hence has probably also swung aside.

3. In nearly all other α/β barrel–containing enzymes, the center of the barrel is occluded by large, often branched, hydrophobic side chains. However, in methylmalonyl-CoA mutase, the 2-carboxypropyl-CoA's pantetheine group binds in a narrow tunnel through the center of the α/β barrel so as to put the methylmalonyl group of an intact substrate in close proximity to the unliganded face of the cobalamin ring. This tunnel provides the only direct access to the active site cavity, thereby protecting the reactive free radical intermediates that are produced in the catalytic reaction from side reactions (see below). The tunnel is lined by small hydrophilic residues (Ser and Thr).

Methylmalonyl-CoA mutase's substrate binding mode resembles those of several other AdoCbl-containing enzymes of known structure, which are collectively unique among α/β barrel–containing proteins.

The proposed methylmalonyl-CoA mutase reaction mechanism (Fig. 25-23) begins with **homolytic cleavage** of the cobalamin C—Co(III) bond (the C and Co atoms each acquire one of the electrons that formed the cleaved electron pair bond). The Co ion therefore fluctuates between its Co(III) and Co(II) oxidation states [the two states are spectroscopically distinguishable: Co(III) is red and diamagnetic (no unpaired electrons), whereas Co(II) is yellow and paramagnetic (unpaired electrons)]. Note that a homolytic cleavage reaction is unusual in biology; most other biological bond cleavage reactions occur via **heterolytic cleavage** (in which the electron pair forming the cleaved bond is fully acquired by one of the separating atoms).

The role of AdoCbl in the catalytic process is that of a reversible free radical generator. The C—Co(III) bond is well suited to this function because it is inherently weak (dissociation energy = 109 kJ · mol⁻¹) and appears to be further weakened through steric interactions with the enzyme. Indeed, as Fig. 25-22 indicates, the Co atom in methylmalonyl-CoA mutase has no sixth ligand and hence, as confirmed by spectroscopic measurements, is in its Co(II) state. The His N—Co bond is extremely long (2.5 Å vs 1.9–2.0 Å in various other B_{12}-containing structures). It is proposed that this strained and hence weakened bond stabilizes the Co(II) state with respect to the Co(III) state, thus favoring the formation of the adenosyl radical and facilitating the homolytic cleavage through which the catalyzed reaction occurs (Fig. 23-23). The adenosyl radical presumably abstracts a hydrogen atom from the substrate, thereby facilitating the rearrangement reaction through the intermediate formation of a cyclopropyloxy radical.

d. Succinyl-CoA Cannot Be Directly Consumed by the Citric Acid Cycle

Methylmalonyl-CoA mutase catalyzes the conversion of a metabolite to a C_4 citric acid cycle intermediate, not acetyl-CoA. The route of succinyl-CoA oxidation is therefore not as simple as it may first appear. The citric acid cycle regenerates all of its C_4 intermediates so that these compounds are really catalysts, not substrates. Consequently, succinyl-CoA cannot undergo net degradation by citric acid cycle enzymes alone. Rather, *in order for a metabolite to undergo net oxidation by the citric acid cycle, it must first be converted either to pyruvate or directly to acetyl-CoA.* Net degradation of succinyl-CoA begins with its conversion, via the citric acid cycle, to malate. At high concentrations, malate is transported, by a specific transport protein, to the cytosol, where it may be oxidatively decarboxylated to pyruvate and CO_2 by **malic enzyme (malate dehydrogenase, decarboxylating):**

Malate   **Pyruvate**

(We previously encountered this enzyme in the C_4 cycle of photosynthesis; Fig. 24-41.) Pyruvate is then completely oxidized via pyruvate dehydrogenase and the citric acid cycle.

e. Pernicious Anemia Results from Vitamin B_{12} Deficiency

The existence of **vitamin B_{12}** came to light in 1926 when George Minot and William Murphy discovered that **pernicious anemia,** a rare but often fatal disease of the elderly characterized by decreased numbers of red blood cells, low hemoglobin levels (for reasons explained in Section 26-4D), and progressive neurological deterioration (caused by the accumulation of odd-chain fatty acid residues in neuronal membranes), can be treated by the daily consumption of large amounts of raw liver (a treatment that some patients considered worse than the disease). It was not until 1948, however, after a bacterial assay for antipernicious anemia factor had been developed, that vitamin B_{12} was isolated.

Vitamin B_{12} is synthesized by neither plants nor animals but only by a few species of bacteria. Herbivores obtain their vitamin B_{12} from the bacteria that inhabit their gut (in fact, some animals, such as rabbits, must periodically eat some of their feces to obtain sufficient amounts of this essential substance). Humans, however, obtain almost all their vitamin B_{12} directly from their diet, particularly from meat. The vitamin is specifically bound in the intestine by the glycoprotein **intrinsic factor** that is secreted by the stomach. This complex is absorbed by a specific receptor in the intestinal mucosa, where the complex is dissociated and the liberated vitamin B_{12} transported to the bloodstream.

There it is bound by at least three different plasma globulins, called **transcobalamins,** which facilitate its uptake by the tissues.

Pernicious anemia is not usually a dietary deficiency disease but, rather, results from insufficient secretion of intrinsic factor, often due to an autoimmune attack against the cells that produce it. The normal human requirement for cobalamin is very small, ~ 3 μg $\cdot$ day^{-1}, and the liver stores a 3- to 5-year supply of this vitamin. This accounts for the insidious onset of pernicious anemia and the fact that true dietary deficiency of vitamin B$_{12}$, even among strict vegetarians, is extremely rare.

Figure 25-23 **Proposed mechanism of methylmalonyl-CoA mutase. (1)** The homolytic cleavage of the C—Co(III) bond yielding a 5'-deoxyadenosyl radical and cobalamin in its Co(II) oxidation state. **(2)** Abstraction of a hydrogen atom from the methylmalonyl-CoA by the 5'-deoxyadenosyl radical, thereby generating a methylmalonyl-CoA radical. **(3)** Carbon skeleton rearrangement to form a succinyl-CoA radical via a proposed cyclopropyloxy radical intermediate. **(4)** Abstraction of a hydrogen atom from 5'-deoxyadenosine by the succinyl-CoA radical to form succinyl-CoA and regenerate the 5'-deoxyadenosyl radical. **(5)** Release of succinyl-CoA and reformation of the coenzyme.

F. *Peroxisomal β Oxidation*

In mammalian cells, the bulk of β oxidation occurs in the mitochondria, but peroxisomes (Fig. 25-24) also oxidize fatty acids, particularly those with very long chains or branched chains. *Peroxisomal β oxidation in animals functions to shorten very long chain fatty acids (>22 C atoms) so as to facilitate their degradation by the mitochondrial β-oxidation system. In yeast and plants, fatty acid oxidation occurs exclusively in the peroxisomes and glyoxysomes (specialized peroxisomes, Sections 23-2 and 1-2Ad).*

The peroxisomal pathway results in the same chemical changes to fatty acids as does the mitochondrial pathway, although the enzymes in these two organelles are different. The protein that transports very long-chain fatty acids into the peroxisome, **ALD protein** (see below), does not have a carnitine requirement. The very long-chain fatty acids that enter this compartment are activated by a peroxisomal very long-chain acyl-CoA synthetase to form their CoA esters, and are oxidized directly. The shorter chain acyl products of this β-oxidation process are then linked to carnitine for transport into mitochondria for further oxidation.

a. Adrenoleukodystrophy Is Caused by a Defect in ALD Protein

Adrenoleukodystrophy (ALD) is a rare X-linked inherited disease that results in progressive brain damage and adrenal gland failure. It causes very long-chain saturated fatty acids to accumulate in the blood and destroy the insulating myelin sheath surrounding the axons of many neurons (Section 20-5Bc). Its varied neurological symptoms present (become evident) between the ages of 4 and 10 years and are usually fatal within 1 to 10 years (except after a successful bone marrow transplant). ALD is caused by a defective ALD protein, an ABC transporter (Section 20-3E). Thus in ALD patients, **lignoceric acid** (24:0; recall that the symbol *n:m* indicates a C_n fatty acid with *m* double bonds) is converted to lignoceroyl-CoA at only 13% of the normal rate, although once formed, it undergoes β oxidation at the normal rate.

b. Peroxisomal β Oxidation Differs in Detail from Mitochondrial β Oxidation

The β-oxidation pathway in peroxisomes differs from that in mitochondria as follows:

1. The first enzyme in the peroxisomal pathway, **acyl-CoA oxidase,** catalyzes the reaction

$$\text{Fatty acyl-CoA} + O_2 \longrightarrow \textit{trans-}\Delta^2\text{-enoyl-CoA} + H_2O_2$$

This reaction involves participation of an FAD cofactor but differs from its mitochondrial counterpart in that the abstracted electrons are transferred directly to O_2 rather than passing through the electron-transport chain with its concomitant oxidative phosphorylation (Fig. 25-12). Peroxisomal fatty acid oxidation is therefore less efficient than the mitochondrial process by two ATPs for each C_2 cycle. The H_2O_2 produced is disproportionated to H_2O and O_2

Figure 25-24   Peroxisomes. These membrane-bounded organelles perform a variety of metabolic functions, including the oxidation of very long chain fatty acids. [© Donald Fawcett/ Visuals Unlimited.]

through the action of peroxisomal catalase (Section 1-2Ad).

2. Peroxisomal enoyl-CoA hydratase and 3-L-hydroxy-acyl-CoA dehydrogenase are activities that occur on a single polypeptide and therefore join the growing list of multifunctional enzymes. The reactions catalyzed are identical to those of the mitochondrial system (Fig. 25-12).

3. Peroxisomal thiolase has a different chain-length specificity than its mitochondrial counterpart. It is almost inactive with acyl-CoAs of length C_8 or less so that fatty acids are incompletely oxidized by peroxisomes.

Although peroxisomal β oxidation is not dependent on the transport of acyl groups into the peroxisome as their carnitine esters, the peroxisome contains carnitine acyltransferases. Acyl-CoAs that have been chain-shortened by peroxisomal β oxidation are thereby converted to their carnitine esters. These substances, for the most part, passively diffuse out of the peroxisome to the mitochondrion, where they are oxidized further.

G. *Minor Pathways of Fatty Acid Oxidation*

β Oxidation is blocked by an alkyl group at the C_β of a fatty acid, and thus at any odd-numbered carbon atom. One such branched-chain fatty acid, a common dietary component, is **phytanic acid.** This metabolic breakdown product of chlorophyll's phytyl side chain (Section 24-2A) is present in dairy products, ruminant fats, and fish although, surprisingly, chlorophyll itself is but a poor dietary source of phytanic acid for humans. The oxidation of branched-chain fatty acids such as phytanic acid is facilitated by α **oxidation** (Fig. 25-25). In this process, the fatty acid is converted to its CoA thioester and its C_α is hydroxylated by the Fe^{2+}-containing **phytanoyl-CoA hydroxylase.** The resulting

Figure 25-25 Pathway of α oxidation of fatty acids. Phytanic acid, a degradation product of the phytyl side chain of chlorophyll, is metabolized through α oxidation to **pristanic acid** followed by β oxidation.

CoA thioester is, in effect, oxidatively decarboxylated to yield a new fatty acid with an unsubstituted C_β. Further degradation of the molecule can then continue via six cycles of normal β oxidation to yield three propionyl-CoAs, three acetyl-CoAs, and one 2-methylpropionyl-CoA (which is converted to succinyl-CoA).

A rare genetic defect, **Refsum's disease** or **phytanic acid storage syndrome,** results from the accumulation of this metabolite throughout the body. The disease, which is characterized by progressive neurological difficulties such as tremors, unsteady gait, and poor night vision, results from a defective phytanoyl-CoA hydroxylase. Its symptoms can therefore be attenuated by a diet that restricts the intake of phytanic acid–containing foods.

Medium- and long-chain fatty acids are converted to dicarboxylic acids through **ω oxidation** (oxidation of the last carbon atom). This process, which is catalyzed by enzymes of the ER, involves hydroxylation of a fatty acid's C_ω atom by a **cytochrome P450,** a monooxygenase that utilizes NADPH and O_2 (Section 15-4Bc). This CH_2—OH group is then oxidized to a carboxyl group, converted to a CoA derivative at either end, and oxidized via the β-oxidation pathway. ω Oxidation is probably of only minor significance in fatty acid oxidation.

3 KETONE BODIES

Acetyl-CoA produced by oxidation of fatty acids in liver mitochondria can be further oxidized via the citric acid cycle as is discussed in Chapter 21. A significant fraction of this acetyl-CoA has another fate, however. *By a process known as* **ketogenesis,** *which occurs primarily in liver mitochondria, acetyl-CoA is converted to* **acetoacetate** *or* **D-β-hydroxybutyrate.** *These compounds, which together with* **acetone** *are somewhat inaccurately referred to as* **ketone bodies,**

serve as important metabolic fuels for many peripheral tissues, particularly heart and skeletal muscle. The brain, under normal circumstances, uses only glucose as its energy source (fatty acids are unable to pass the blood–brain barrier), but during starvation, ketone bodies become the brain's major fuel source (Section 27-4A). Ketone bodies are water-soluble equivalents of fatty acids.

Acetoacetate formation occurs in three reactions (Fig. 25-26):

1. Two molecules of acetyl-CoA are condensed to **acetoacetyl-CoA** by thiolase (also called **acetyl-CoA acetyltransferase**) working in the reverse direction from the way it does in the final step of β oxidation (Section 25-2Cd).

2. Condensation of the acetoacetyl-CoA with a third acetyl-CoA by **HMG-CoA synthase** forms **β-hydroxy-β-methylglutaryl-CoA (HMG-CoA)**. The mechanism of this reaction resembles the reverse of the thiolase reaction (Fig. 25-15) in that an active site thiol group forms an acyl–thioester intermediate.

3. Degradation of HMG-CoA to acetoacetate and acetyl-CoA in a mixed aldol–Claisen ester cleavage is catalyzed by **HMG-CoA lyase.** The mechanism of this reaction is analogous to the reverse of the citrate synthase reaction (Section 21-3A). (HMG-CoA is also a precursor in cholesterol biosynthesis and hence may be diverted to this purpose as is discussed in Section 25-6A.)

The overall reaction catalyzed by HMG-CoA synthase and HMG-CoA lyase is

$$\text{Acetoacetyl-CoA} + H_2O \longrightarrow \text{acetoacetate} + \text{CoA}$$

One may well ask why this apparently simple hydrolysis reaction occurs in such an indirect manner. The answer is unclear but may lie in the regulation of the process.

Acetoacetate may be reduced to D-β-hydroxybutyrate by **β-hydroxybutyrate dehydrogenase:**

Acetoacetate   **D-β-Hydroxybutyrate**

Note that this product is the stereoisomer of the L-β-hydroxyacyl-CoA that occurs in the β-oxidation pathway. Acetoacetate, being a β-keto acid, also undergoes relatively facile nonenzymatic decarboxylation to acetone and

Figure 25-26 Ketogenesis: the enzymatic reactions forming acetoacetate from acetyl-CoA. (1) Two molecules of acetyl-CoA condense to form acetoacetyl-CoA in a thiolase-catalyzed reaction. **(2)** A Claisen ester condensation of the acetoacetyl-CoA with a third acetyl-CoA to form β-hydroxy-β-methylglutaryl-CoA (HMG-CoA) as catalyzed by HMG-CoA synthase. **(3)** The degradation of HMG-CoA to acetoacetate and acetyl-CoA in a mixed aldol–Claisen ester cleavage catalyzed by HMG-CoA lyase.

Figure 25-27 The metabolic conversion of ketone bodies to acetyl-CoA.

Figure 25-28 **Proposed mechanism of 3-ketoacyl-CoA transferase involving an enzyme–CoA thioester intermediate.**

step and therefore "costs" the free energy of GTP hydrolysis. The liver lacks 3-ketoacyl-CoA transferase, which permits it to supply ketone bodies to other tissues.

4 FATTY ACID BIOSYNTHESIS

Fatty acid biosynthesis occurs through condensation of C_2 units, the reverse of the β-oxidation process. Through isotopic labeling techniques, David Rittenberg and Konrad Bloch demonstrated, in 1945, that these condensation units are derived from acetic acid. Acetyl-CoA was soon proven to be a precursor of the condensation reaction, but its mechanism remained obscure until the late 1950s when Salih Wakil discovered a requirement for bicarbonate in fatty acid biosynthesis and malonyl-CoA was shown to be an intermediate. In this section, we discuss the reactions of fatty acid biosynthesis.

A. Pathway Overview

The pathway of fatty acid synthesis differs from that of fatty acid oxidation. This situation, as we saw in Section 18-1D, is typically the case of opposing biosynthetic and degradative pathways because it permits them both to be thermodynamically favorable and independently regulated under similar physiological conditions. Figure 25-29 outlines fatty acid oxidation and synthesis with emphasis on the differences between these pathways. Whereas fatty acid oxidation occurs in the mitochondrion and utilizes fatty acyl-CoA esters, fatty acid biosynthesis occurs in the cytosol with, as Roy Vagelos discovered, the growing fatty acids esterified to **acyl-carrier protein** (**ACP;** Fig. 25-30). ACP, like CoA, contains a phosphopantetheine group that forms thioesters with acyl groups. The phosphopantetheine phosphoryl group is esterified to a Ser OH group of ACP, whereas in CoA it is esterified to AMP. In animals, ACP is part of a large multifunctional protein (Type I ACP; see below), whereas in *E. coli* it is a 125-residue polypeptide (Type II ACP). The phosphopantetheine group is transferred from CoA to apo-ACP to form the active holo-ACP by **phosphopantetheine transferase** (alternatively, **ACP synthase**).

The redox coenzymes of the animal fatty acid oxidative and biosynthetic pathways differ (NAD^+ and FAD for oxidation; NADPH for biosynthesis) as does the stereochemistry of their intermediate steps, but their main difference is the manner in which C_2 units are removed from or added to the fatty acyl thioester chain. In the oxidative pathway, β-ketothiolase catalyzes the cleavage of the C_α—C_β bond of β-ketoacyl-CoA so as to produce acetyl-CoA and a new fatty acyl-CoA, which is shorter by a C_2 unit. The $\Delta G^{\circ\prime}$ of this reaction is very close to zero so it can also function in the reverse direction (ketone body formation). In the biosynthetic pathway, the condensation reaction is coupled to the hydrolysis of ATP, thereby driving the reaction to completion. This process involves two steps: (1) the ATP-dependent carboxylation of acetyl-CoA by **acetyl-CoA carboxylase (ACC)** to form **malonyl-CoA,** and (2) the

CO_2. Indeed, the breath of individuals with **ketosis** (also called **ketoacidosis**), a potentially pathological condition in which acetoacetate is produced faster than it can be metabolized (a symptom of diabetes; Section 27-4B), has the characteristic sweet smell of acetone.

The liver releases acetoacetate and β-hydroxybutyrate, which are carried by the bloodstream to the peripheral tissues for use as alternative fuels. There, these products are converted to acetyl-CoA as is diagrammed in Fig. 25-27. The proposed reaction mechanism of **3-ketoacyl-CoA transferase** (Fig. 25-28), which catalyzes this pathway's second step, involves the participation of an active site carboxyl group both in an enzyme–CoA thioester intermediate and in an unstable anhydride. Succinyl-CoA, which acts as the CoA donor in this reaction, can also be converted to succinate with the coupled synthesis of GTP in the succinyl-CoA synthetase reaction of the citric acid cycle (Section 21-3Ea). The "activation" of acetoacetate bypasses this

Figure 25-29 **Comparison of fatty acid β oxidation and fatty acid biosynthesis.** Differences occur in **(1)** cellular location, **(2)** acyl group carrier, **(3)** electron acceptor/donor, **(4)** stereo- chemistry of the hydration/dehydration reaction, and **(5)** the form in which C$_2$ units are produced/donated. 🔊 **See the Animated Figures**

exergonic decarboxylation of the malonyl group in the condensation reaction catalyzed by **fatty acid synthase.** These enzymes are discussed below.

B. *Acetyl-CoA Carboxylase*

ACC is a biotin-dependent enzyme that catalyzes the first committed step of fatty acid biosynthesis and one of its rate-controlling steps. It is a member of a family of biotin-dependent carboxylases that, in humans, has only three members besides ACC: propionyl-CoA carboxylase (Section 25-2Ea), pyruvate carboxylase (Fig. 23-4), and **β-methylcrotonyl-CoA carboxylase** (which participates in

the degradation of leucine; Section 26-3F). The ACC reaction, like those of the other biotin-dependent carboxylases, occurs in two steps, a CO$_2$ activation and a carboxylation:

E—biotin
Biotinyl–enzyme

^-O_2C—CH$_2$—C(=O)—SCoA + E—biotin
Malonyl-CoA

HCO$_3^-$ + ATP → ADP + P$_i$

CH$_3$—C(=O)—SCoA
Acetyl-CoA

E—biotin—CO$_2^-$
Carboxybiotinyl–enzyme

HS—CH$_2$—CH$_2$—N(H)—C(=O)—CH$_2$—CH$_2$—N(H)—C(=O)—C(OH)(H)—C(CH$_3$)(CH$_3$)—CH$_2$—O—P(=O)(O$^-$)—O—CH$_2$—Ser—ACP

Cysteamine

Phosphopantetheine prosthetic group of ACP

HS—CH$_2$—CH$_2$—N(H)—C(=O)—CH$_2$—CH$_2$—N(H)—C(=O)—C(OH)(H)—C(CH$_3$)(CH$_3$)—CH$_2$—O—P(=O)(O$^-$)—O—P(=O)(O$^-$)—O—CH$_2$—O Adenine

Cysteamine

$^{2-}O_3PO$ OH

Phosphopantetheine group of CoA

Figure 25-30 **The phosphopantetheine group in acyl-carrier protein (ACP) and in CoA.**

E. coli ACC is a multienzyme complex in which these steps are catalyzed by separate subunits: the homodimer biotin carboxylase (BC; 456 residues) and the $\alpha_2\beta_2$ heterotetramer carboxyltransferase (CT; 319 and 304 residues). In addition, the biotin is linked as a biocytin residue (Fig. 23-3*b*) to the biotin carboxyl-carrier protein (BCCP; 156 residues), which forms homodimers. In contrast, mammalian and avian ACCs contain both enzymatic activities as well as the biotin carboxyl carrier on a single 2346-residue polypeptide chain in the order BC–BCCP–CT (which differs from the order in pyruvate carboxylase, which is BC–CT–BCCP; Section 23-1Ae). The structure of an intact ACC has not been determined, although the X-ray structure the *E. coli* BC subunit closely resembles the BC domain of pyruvate carboxylase (Fig. 23-5*a*). Interestingly, however, the CT domains of the various biotin-dependent carboxylases differ greatly in sequence and structure.

a. Acetyl-CoA Carboxylase Is Regulated by Hormonally Controlled Reversible Phosphorylation

ACC is subject to hormonal regulation. Glucagon as well as epinephrine and norepinephrine (adrenaline and noradrenaline; Section 18-3Ea) trigger the enzyme's cAMP-dependent increase in phosphorylation, which inactivates the enzyme. Insulin, on the other hand, stimulates enzyme dephosphorylation and thus its activation.

The mechanism by which cAMP causes an increase in the phosphorylation state of ACC is interesting. ACC is phosphorylated, *in vitro*, by two different kinases, the cAMP-dependent protein kinase A (PKA; Section 18-3Cb) at Ser 77 and **AMP-dependent protein kinase** (**AMPK;** Sections 25-5a and 27-1**)** (which is cAMP independent) at Ser 79, 1200, and 1215. Yet, when liver cells are incubated with cAMP-elevating hormones in the presence of ^{32}P-ATP, Ser 77 is found not to be labeled. Evidently, an increase in [cAMP] results in a phosphorylation increase at sites modified by AMPK rather than by PKA. How can this be? It appears that, *in vivo*, the cAMP-dependent increase in phosphorylation occurs not through the phosphorylation of new sites but, rather, through the inhibition of dephosphorylation of previously phosphorylated positions. We have already seen such a mechanism in operation in the control of glycogen metabolism, where the cAMP-dependent phosphorylation of phosphoprotein phosphatase inhibitor-1 causes the inhibition of dephosphorylation (Section 18-3C). In the case of ACC, however, dephosphorylation is catalyzed by **phosphoprotein phosphatase-2A,** which is not affected by phosphoprotein phosphatase inhibitor-1. The mechanism by which PKA causes the increase in phosphorylation associated with AMPK activity is, as yet, unknown.

b. Avian and Mammalian Acetyl-CoA Carboxylases Undergo Enzyme Polymerization on Activation

Electron microscopy reveals that both avian and mammalian ACCs form long filaments of 20 to 40 protomers (Fig. 25-31). *This polymeric form of the enzyme is catalytically active but the protomer is not.* The rate of fatty acid

Figure 25-31   **Association of acetyl-CoA carboxylase protomers.** An electron micrograph with an accompanying interpretive drawing indicates that filaments of avian liver acetyl-CoA carboxylase consist of linear chains of flat rectangular protomers. [Courtesy of Malcolm Lane, The Johns Hopkins University School of Medicine.]

biosynthesis is therefore controlled by the position of the equilibrium between these forms:

$$\text{Protomer}(\textit{inactive}) \rightleftharpoons \text{polymer}(\textit{active})$$

Phosphorylation favors the inactive protomer while dephosphorylation favors the active polymer. Several metabolites also affect the activity of acetyl-CoA carboxylase. Citrate promotes the polymerization of ACC, whereas palmitoyl-CoA and other fatty acyl-CoA's promote its depolymerization. Thus, cytosolic citrate, whose concentration increases when the mitochondrial acetyl-CoA concentration builds up (Section 25-4D), activates fatty acid biosynthesis and hence is a feedforward activator, whereas palmitoyl-CoA, the pathway product, is a feedback inhibitor.

c. Mammalian Acetyl-CoA Carboxylase Has Two Major Isoforms

There are two major isoforms of mammalian ACC. **ACC1** occurs in adipose tissue and **ACC2** occurs in tissues that oxidize but do not synthesize fatty acids, such as heart muscle. Tissues that both synthesize and oxidize fatty acids, such as liver, contain both isoforms, which are homologous although the genes encoding them are located on different chromosomes. What is the function of ACC2? The product of the ACC-catalyzed reaction, malonyl-CoA, strongly inhibits

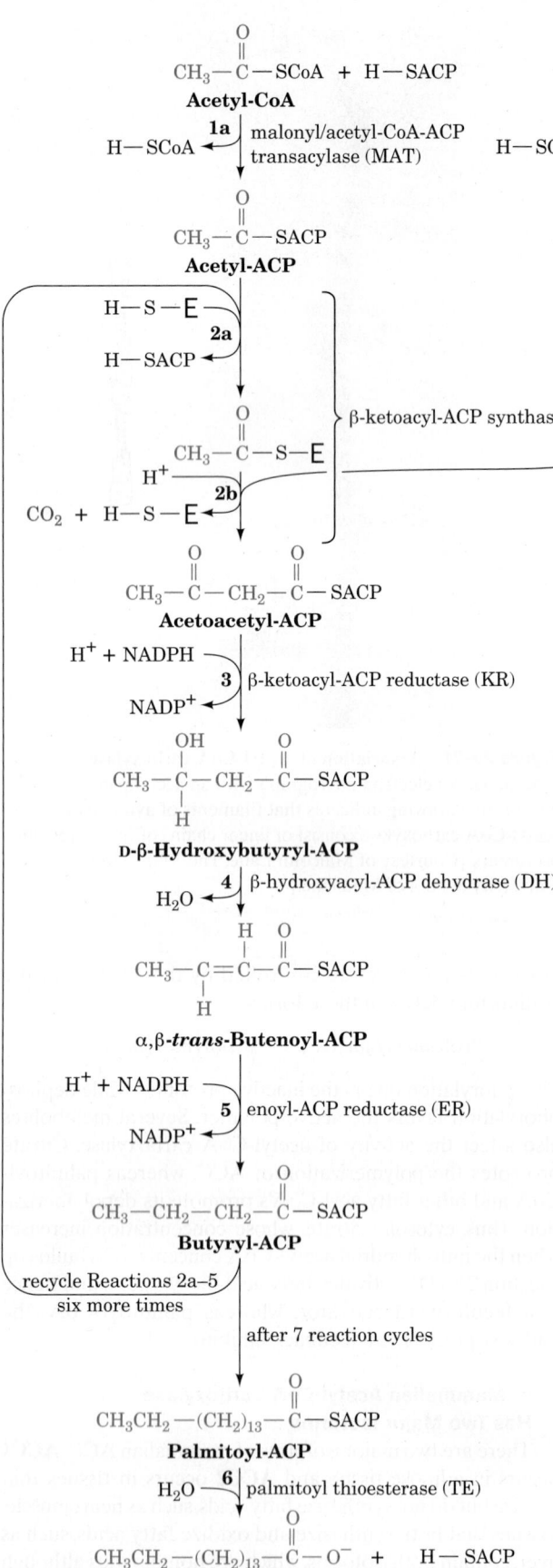

Figure 25-32   Reaction cycle for the biosynthesis of fatty acids. The biosynthesis of palmitate requires seven cycles of C_2 elongation followed by a final hydrolysis step. See the Animated Figures

the mitochondrial import of fatty acyl-CoA for fatty acid oxidation, the major control point for this process. Thus it appears that ACC2 has a regulatory function (Section 25-5).

Prokaryotic acetyl-CoA carboxylases are not subject to any of these controls. This is because fatty acids in these organisms are not stored as fats but function largely as phospholipid precursors. The *E. coli* enzyme is instead regulated by guanine nucleotides so that fatty acids are synthesized in response to the cell's growth requirements.

C. Fatty Acid Synthase

The synthesis of fatty acids from acetyl-CoA and malonyl-CoA involves seven enzymatic reactions that yield mainly palmitic acid (see below). These reactions were first studied in cell-free extracts of *E. coli*, in which they are catalyzed by independent enzymes together with ACP. These proteins are collectively known as **type II fatty acid synthase (FAS-II).** FAS-II also occurs in chloroplasts, the only site of fatty acid synthesis in plants, and in vertebrate mitochondria, whose components are all nuclear encoded and closely resemble their bacterial counterparts. In fungi, however, fatty acids are synthesized by a 2600-kD $\alpha_6\beta_6$ multifunctional enzyme, whereas in animals they are synthesized by a multifunctional enzyme consisting of two identical ~275-kD polypeptide chains, each containing all seven activities plus ACP and known as **type I fatty acid synthase (FAS-I).** Most of the domains catalyzing these reactions in these so-called **megasynthases** are homologous to the corresponding proteins that constitute FAS-II, which indicates that they arose through the joining of what were previously independent proteins.

The reactions catalyzed by FAS-I to synthesize palmitate are diagrammed in Fig. 25-32. The long flexible phosphopantetheine chain of ACP (Fig. 25-30) functions to transport the substrate between the protein's various enzymatic domains:

1a. The transfer of the acetyl group from acetyl-CoA to ACP to yield acetyl-ACP as catalyzed by **malonyl/CoA-ACP transacylase (MAT).**

Malonyl-ACP

CO_2

$CH_2=\overset{O^-}{\underset{|}{C}}-S-ACP$

$CH_2-\overset{O}{\underset{||}{C}}-S-ACP$

$H_3C-\overset{O}{\underset{||}{C}}-S-Cys-KS$

H^+

$HS-Cys-KS$

$H_3C-\overset{O}{\underset{||}{C}}-CH_2-\overset{O}{\underset{||}{C}}-S-ACP$

Acetoacetyl-CoA

Figure 25-33 **The mechanism of carbon–carbon bond formation in fatty acid biosynthesis.** The condensation of an acetyl group on the active site Cys of β-ketoacyl-ACP synthase (KS) with a malonyl group on the phosphopantetheine arm of ACP forms a β-ketoacyl-ACP. The reaction is driven by the exergonic elimination of CO_2 from the malonyl group to generate a resonance-stabilized acetyl-ACP carbanion intermediate that functions as a good nucleophile.

2a. The loading of **β-ketoacyl-ACP synthase (KS;** also known as **condensing enzyme)** by the transfer of the acetyl group from ACP to a KS Cys residue, thus maintaining the acetyl group's thioester linkage.

1b. The formation of malonyl-ACP in a reaction analogous to that of Reaction 1a, which in animals is catalyzed by the same enzyme, MAT.

2b. The coupling of the acetyl group to the C_β of the malonyl group on the ACP with the malonyl group's accompanying decarboxylation so as to form acetoacetyl-ACP and free the KS active site Cys-SH group (Fig. 25-33). *Consequently, the CO_2 taken up in the acetyl-CoA carboxylase reaction (Section 25-4B) does not appear in the product fatty acid. Rather, the decarboxylation functions to drive carbon–carbon bond formation in the condensation reaction, which through the acetyl-CoA carboxylase reaction, is coupled to ATP hydrolysis.*

3–5. The reduction, dehydration, and further reduction of acetoacetyl-ACP so as to form **butyryl-ACP** as sequentially catalyzed by **β-ketoacyl-ACP reductase (KR), β-hydroxyacyl-ACP dehydrase (DH),** and **enoyl-ACP reductase (ER).** The coenzyme in both reductive steps is NADPH, whereas in β oxidation, the analogs of Reactions 3 and 5, respectively, use NAD^+ and FAD (Fig. 25-29). Moreover, Reaction 3 produces and Reaction 4 requires a D-β-hydroxyacyl group, whereas the analogous reactions in β oxidation involve the corresponding L isomer.

2a to 5 Repeat. The butyryl group from the butyryl-ACP is transferred to the Cys-SH of KS. Thus the acetyl group with which the system was initially loaded has been elongated by a C_2 unit. The ACP is "reloaded" with a malonyl group (Step 1b), and another cycle of C_2 elongation occurs. This process occurs altogether seven times to yield **palmitoyl-ACP.**

6. The palmitoyl-ACP thioester bond is hydrolyzed by **palmitoyl thioesterase (TE),** yielding palmitate, the normal product of the fatty acid synthase pathway, and regenerating the enzyme for a new round of synthesis.

The stoichiometry of palmitate synthesis therefore is

Acetyl-CoA + 7 malonyl-CoA + 14NADPH + $7H^+ \longrightarrow$
Palmitate + $7CO_2$ + $14NADP^+$ + 8CoA + $6H_2O$

Since the 7 malonyl-CoA are derived from acetyl-CoA as follows:

7 Acetyl-CoA + $7CO_2$ + 7ATP $\longrightarrow$
7 malonyl-CoA + 7ADP + $7P_i$ + $7H^+$

the overall stoichiometry for palmitate biosynthesis is

8 Acetyl-CoA + 14NADPH + 7ATP $\longrightarrow$
Palmitate + $14NADP^+$ + 8CoA + $6H_2O$ + 7ADP + $7P_i$

a. Animal FAS-I Is a Flexible X-Shaped Dimer

Most but not all of the enzymatic activities of animal FAS-I remain functional when this dimeric enzyme is dissociated into monomers. Moreover, fragments resulting from the limited proteolysis of animal FAS-I exhibit many of the enzymatic activities of the intact protein. Apparently, *contiguous stretches of its polypeptide chain fold to form a series of autonomous domains, each with a specific but different catalytic activity.* Several other enzymes, for example, mammalian acetyl-CoA carboxylase (Section 25-4B), exhibit similar multifunctionality but none has as many separate catalytic activities as does animal FAS-I. The order of the domains along the animal FAS-I polypeptide chain is indicated in Fig. 25-34. Three of these domains besides ACP lack enzymatic activity: a linker domain

Figure 25-34 **Domain organization of porcine FAS-I at approximate sequence scale.** [Modified from a drawing by Timm Maier and Nenad Ban, ETH Zurich, Switzerland.]

Figure 25-35 **X-ray structure of porcine FAS-I in complex with NADP⁺.** (*a*) The ~190-Å wide homodimer, as viewed perpendicular to its pseudo-2-fold axis (*vertical black arrow*), with its various domains colored as in Fig. 25-34 and its linkers gray. The bound NADP⁺ cofactors are drawn in space-filling form in blue. The attachment sites of the disordered ACP/TE domains are indicated by black dots. The domain names of the second subunit have appended primes. (*b*) A corresponding schematic diagram indicating how the various domains are linked. [Courtesy of Timm Maier and Nenad Ban, ETH Zurich, Switzerland. PDBid 2VZ9.]

(LD) that bridges the KS and MAT domains; a pseudo-methyltransferase (ΨME) domain, so named because it is homologous to the methyltransferase family; and a pseudo-ketoreductase (ΨKR) domain, so named because it is a truncated form of the KR domain.

The X-ray structure of porcine FAS-I in complex with NADP⁺, determined by Nenad Ban, reveals a pseudo-2-fold symmetric, X-shaped dimer (Fig. 25-35). Its two subunits associate via an extensive interface involving over 150 residues per chain from the KS, ER, and DH domains. However the upper and lower portions of the X are only loosely connected. The lower portion of the X contains the enzyme's two condensing activities, MAT and KS (which

catalyze reactions 1 and 2 in Fig. 25-32), whereas the upper portion of the X contains the enzyme's β-carbon modifying activities, DH, KR, and ER (which catalyze reactions 3–5 in Fig. 25-32). The C-terminal ACP and TE domains are flexibly tethered to the enzyme and are therefore unobserved. The KR–ACP linker consists of ~13 residues and can therefore span a distance of up to ~40 Å, whereas the ACP–TE linker has a length of ~25 residues and can therefore extend ~80 Å.

Each of the two reaction chambers of porcine FAS-I is lined by the full set of catalytic domains from one subunit, all of which must be visited by the phosphopan-tethiene arm of an ACP to carry out all the FAS reac-

Figure 25-36 **Domain organization of fungal FAS-I at approximate sequence scale.** [Modified from a drawing by Simon Jenni and Nanad Ban, ETH Zurich, Switzerland.]

tions. The catalytic sites of these domains occur on both the front and back faces of the protein and their distribution has no discernable relationship to the order of the reactions in the catalytic cycle. In the structure shown in Fig. 25-35a, the length of the KR–ACP linker would allow the ACP to reach all the catalytic sites in the same reaction chamber but not those in the opposite reaction chamber (e.g., the distance between the ACP attachment site on one subunit and the MAT domain in the other is ~135 Å). Nevertheless, mutational studies by Stuart Smith in which ACP on one subunit and MAT or KS on the other subunit were inactivated yielded functional enzymes albeit with reduced activities, thus indicating that an ACP domain can service both MAT domains and both KS domains. The most plausible explanation for these observations is that the upper portion of the X can rotate 180° with respect to the lower portion. This hypothesis is supported by the observation that the quite tenuous DH/KS contact that joins the top and bottom portions of the X lacks perfect 2-fold symmetry in Fig. 25-35a, thus indicating that this joint is flexible.

b. Fungal FAS-I Has a Barrel-Like Shape

The reactions catalyzed by $\alpha_6\beta_6$ fungal FAS-I differ from those mediated by animal FAS-I in several respects:

1. The bifunctional MAT activity of animal FAS-I transfers both acetyl and malonyl groups from CoA to ACP (Reactions 1a and 1b of Fig. 25-32). However, fungal FAS-I employs a monofunctional **acetyl transferase (AT)** activity to transfer the incoming acetyl group from CoA to ACP and a **malonyl/palmitoyl transferase (MPT)** activity to do so for malonyl groups.

2. Animal FAS-I synthesizes an ACP-linked palmitoyl group, which its TE activity hydrolytically releases as palmitate (Reaction 6 of Fig. 25-32). In contrast, fungal FAS-I synthesizes ACP-linked palmitoyl (C_{16}) and stearoyl (C_{18}) groups in a ratio of roughly 2:3, which its bifunctional MPT activity then transfers to CoA to yield the pathway's final products, palmitoyl-CoA and stearoyl-CoA.

3. The ER activity in animal FAS-I uses NADPH to directly reduce the C≡C double bond in Reaction 5 of Fig. 25-32. However, in fungal ER, NAPDH reduces FMN to $FMNH_2$, which in turn, reduces the double bond.

4. Fungal FAS-I has a **phosphopantetheine transferase (PPT)** activity that attaches the phosphopantetheine group to ACP. This activity is not part of animal FAS-I.

The distribution of the various enzymatic activities along the α and β chains of fungal FAS-I (Fig. 25-36) bears no resemblance to that of animal FAS-I (Fig. 25-34). Note that the MPT activity is shared by both the α and β chains.

The X-ray structure of FAS-I from the thermophilic fungus *Thermomyces lanuginosus* (Fig. 25-37), determined by Ban, reveals that it forms a hollow barrel-shaped protein with D_3 symmetry (the rotational symmetry of a trigonal prism; Fig. 8-65b). The six α chains form a D_3-symmetric wheel that is capped on each side by a 3-fold symmetric dome that is predominantly formed by the β chains. The central wheel splits the hollow interior of the barrel into two identical reaction chambers, each of which has several openings in its side walls through which small molecules can enter. The active sites of all the enzymatic activities face the interiors of the reaction chambers. Clearly, animal and fungal FAS-I's have divergently evolved from FAS-II.

Figure 25-37 **X-ray structure of *T. lanuginosus* FAS.** The ~270-Å-high by ~250-Å-diameter $\alpha_6\beta_6$ heterododecamer is viewed along a 2-fold axis (*ellipsoid*) and perpendicular to its 3-fold axis (*triangle*) with its various domains colored as in Fig. 25-36 and its linker domains gray. The attachment sites of the disordered ACP domains are indicated by black dots. [Courtesy of Simon Jenni and Nenad Ban, ETH Zurich, Switzerland. PDBids 2UV9 and 2UVA.]

The six ACP domains, which are N-terminally anchored to the chamber walls and C-terminally anchored to the middle of the central wheel, are disordered in the X-ray structure of *T. lanuginosus* FAS-I. However, they are visible in the otherwise closely similar X-ray structure of yeast FAS-I, which was independently determined by Ban and by Thomas Steitz. Structural considerations suggest that each doubly tethered ACP domain can swing to visit the required six catalytic centers, which in the case of the ACP tethered to subchamber 1 are KR from α1, KS from α2, MPT and DH from β1, and AT and ER from β2, where subchamber 2 is on the clockwise side of subchamber 1 as viewed from the top of the dome.

The PPT domains are located on the outside of the barrel where they cannot interact with the ACP domains. This suggests that they attach the phosphopantetheine groups to the ACPs before the barrel has fully assembled.

c. Fatty Acid Synthase Inhibitors Are Drug Candidates

In well-nourished individuals, fatty acid synthesis proceeds at a low rate. However, certain tissues, particularly cancers, express high levels of FAS-I and produce fatty acids at a high rate. Consequently, inhibitors of animal FAS-I are being investigated as possible anticancer agents. Moreover, the differences between the enzymatic activities of the various types of FAS's, particularly their ER activities, makes FAS-II and fungal FAS-I targets for the development of novel antibiotics.

d. Variations on a Theme: Polyketide Biosynthesis

Polyketides are a family of >10,000 diverse and structurally complex natural products, many of which have antibacterial, antifungal, antitumor, and immunosuppressive properties, that are synthesized by bacteria, fungi, plants, and certain marine animals. They are made by the modular condensation of acyl-CoA monomers such as acetyl-CoA and propionyl-CoA with malonyl-CoA and methylmalonyl-CoA extender units whose decarboxylation drives the condensation reaction. The name polyketide comes from the fact that the primary condensation products have β-keto functional groups. Palmitate is an example of a polyketide since it is formed by the condensation of one acetyl-CoA primer and seven malonyl-CoA extender units. Following each condensation reaction, the new β-keto group may be reduced, dehydrated, and reduced again as with fatty acids, or may undergo only partial modification.

Polyketides are synthesized by megasynthases. We have already seen that animal FAS-I contains seven enzymatic activities as well as ACP. Another example of a polyketide is **6-deoxyerythronolide B (6dEB),** the parent **macrolactone** of the antibiotic **erythromycin A** (Section 32-3G), which is synthesized in the soil bacterium *Saccharopolyspora erythraea* from one propionyl-CoA primer and six (*S*)-methylmalonyl-CoA extenders by **deoxyerythronolide B synthase (DEBS;** Fig. 25-38). DEBS is a 2000-kD, $\alpha_2\beta_2\gamma_2$ complex of >3000-residue subunits

whose three homodimeric units each catalyze two elongation/modification cycles. Unlike FAS-I, which catalyzes several cycles of elongation/modification with the same active sites, DEBS catalyzes each elongation/modification cycle on a different module, which permits the differences in the modifications that occur at each cycle. Thus, DEBS, which has 28 different active sites, functions much like an assembly line. Module 4, as Fig. 25-38 indicates, is almost identical in function to FAS-I, containing KS, AT, ACP, KR, DH, and ER, and reducing its primary β-ketone condensation product to a methylene group. However, it does not contain TE because the elongation process is not complete after this phase. Module 3 contains only ACP, KS, and AT, the minimal set of activities for a module, and passes its β-ketone condensation product to module 4 without further modification. Modules 1, 2, 5, and 6 contain only ACP, AT, KS, and KR, the sites necessary for the condensation and ketone reduction steps, thereby generating hydroxy products. The overall organization of the modules therefore creates a polyhydroxy product containing one keto group and one methylene group in the chain. The DEBS final product, 6dEB, is a lactone produced by the reaction of the terminal hydroxyl group with the thioester anchoring the growing chain to the synthase. The various polyketide synthases have different organizations of modules, and consequently synthesize a multitude of different compounds.

D. Transport of Mitochondrial Acetyl-CoA Into the Cytosol

Acetyl-CoA is generated in the mitochondrion by the oxidative decarboxylation of pyruvate as catalyzed by pyruvate dehydrogenase (Section 21-2A) as well as by the oxidation of fatty acids. When the need for ATP synthesis is low, so that the oxidation of acetyl-CoA via the citric acid cycle and oxidative phosphorylation is minimal, this mitochondrial acetyl-CoA may be stored for future use as fat. Fatty acid biosynthesis occurs in the cytosol but the mitochondrial membrane is essentially impermeable to acetyl-CoA. *Acetyl-CoA enters the cytosol in the form of citrate via the* **tricarboxylate transport system** (Fig. 25-39). Cytosolic **ATP-citrate lyase** then catalyzes the reaction

Citrate + CoA + ATP $\rightleftharpoons$

acetyl-CoA + oxaloacetate + ADP + P$_i$

which resembles the reverse of the citrate synthase reaction (Section 21-3A) except that ATP hydrolysis is required to drive the intermediate synthesis of the "high-energy" citryl-CoA, whose hydrolysis drives the citrate synthase reaction to completion. ATP hydrolysis is therefore required in the ATP-citrate lyase reaction to power the resynthesis of this thioester bond. Oxaloacetate is reduced to malate by malate dehydrogenase. Malate may be oxidatively decarboxylated to pyruvate by malic enzyme (Section 25-2Ed) and be returned in this form to the mitochondrion. The malic enzyme reaction resembles that

Figure 25-38   An example of polyketide biosynthesis: the synthesis of erythromycin A. [After Pfeifer, B.A., Admiraal, S.J., Gramajo, H., Cane, D.E., and Khosla, C., *Science* **291,** 1790 (2001).]

of isocitrate dehydrogenase in which a β-hydroxy acid is oxidized to a β-keto acid, whose decarboxylation is strongly favored (Section 21-3C). Malic enzyme's coenzyme is $NADP^+$, so when this route is used NADPH is produced for use in the reductive reactions of fatty acid biosynthesis.

Citrate transport out of the mitochondrion must be balanced by anion transport into the mitochondrion. Malate, pyruvate, and P_i can act in this capacity. Malate may therefore also be transported directly back to the mitochondrion without generating NADPH. As we have seen in Section 25-4C, synthesis of each palmitate ion requires 8 molecules of acetyl-CoA and 14 molecules of NADPH. As many as 8 of these NADPH molecules may be supplied with the 8 molecules of acetyl-CoA if all the malate produced in the cytosol is oxidatively decarboxylated. The re-

maining NADPH is provided through the pentose phosphate pathway (Section 23-4).

E. Elongases and Desaturases

*Palmitate (16:0), the normal product of the animal fatty acid synthase pathway, is the precursor of longer chain saturated and unsaturated fatty acids through the actions of **elongases** and **desaturases.*** Elongases are present in both the mitochondrion and the ER but the mechanisms of elongation at the two sites differ. Mitochondrial elongation (a process independent of the fatty acid synthase pathway) occurs by successive addition and reduction of acetyl units in a reversal of fatty acid oxidation; the only chemical difference between these two pathways occurs in the final reduction step in which NADPH takes the place of $FADH_2$ as the terminal

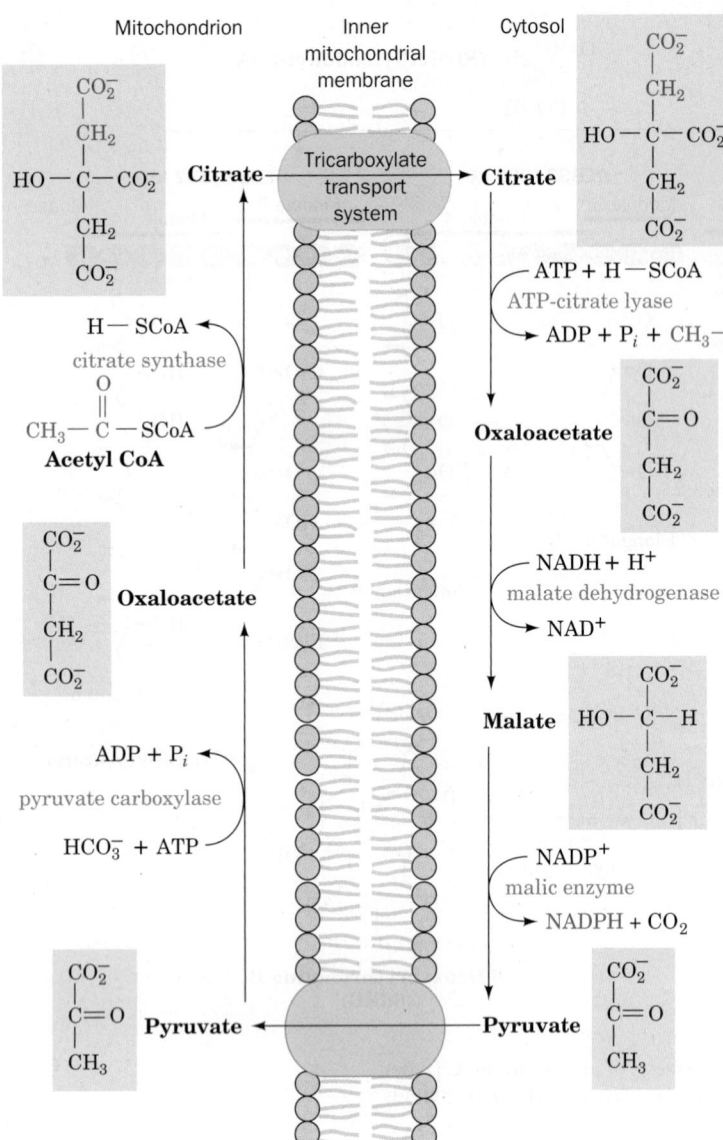

Figure 25-39 Transfer of acetyl-CoA from mitochondrion to cytosol via the tricarboxylate transport system.

redox coenzyme (Fig. 25-40). Elongation in the ER involves the successive condensations of malonyl-CoA with acyl-CoA. These reactions are each followed by NADPH-associated reductions similar to those catalyzed by FAS-I, the only difference being that the fatty acid is elongated as its CoA derivative rather than as its ACP derivative.

Unsaturated fatty acids are produced by **terminal desaturases.** Mammalian systems contain four terminal desaturases of broad chain-length specificities designated Δ^9-, Δ^6-, Δ^5-, and Δ^4-**fatty acyl-CoA desaturases.** These

Figure 25-40 Mitochondrial fatty acid elongation. Elongation occurs by the reversal of fatty acid oxidation with the exception that the final reaction employs NADPH rather than $FADH_2$ as its redox coenzyme.

membrane-bound, nonheme iron–containing enzymes catalyze the general reaction

$$CH_3-(CH_2)_x-\overset{\overset{\displaystyle H}{|}}{\underset{\underset{\displaystyle H}{|}}{C}}-\overset{\overset{\displaystyle H}{|}}{\underset{\underset{\displaystyle H}{|}}{C}}-(CH_2)_y-\overset{\overset{\displaystyle O}{||}}{C}-SCoA + NADH + H^+ + O_2$$

$$\downarrow$$

$$CH_3-(CH_2)_{\overline{x}}-\underset{\underset{\displaystyle H}{|}}{C}=\underset{\underset{\displaystyle H}{|}}{C}-(CH_2)_y-\overset{\overset{\displaystyle O}{||}}{C}-SCoA + 2H_2O + NAD^+$$

where x is at least 5 and where $(CH_2)_x$ can contain one or more double bonds. The $(CH_2)_y$ portion of the substrate is always saturated. Double bonds are inserted between existing double bonds in the $(CH_2)_x$ portion of the substrate and the CoA group such that the new double bond is three carbon atoms closer to the CoA group than the next double bond (not conjugated to an existing double bond) and, in animals, never at positions beyond C9. Mammalian terminal desaturases are components of mini-electron-transport systems that contain two other proteins: **cytochrome b_5** and **NADH–cytochrome b_5 reductase.** The electron-transfer reactions mediated by these complexes occur at the inner surface of the ER membrane (Fig. 25-41) and are therefore not associated with oxidative phosphorylation.

a. Some Unsaturated Fatty Acids Must Be Obtained in the Diet

A variety of unsaturated fatty acids may be synthesized by combinations of elongation and desaturation reactions. However, since palmitic acid is the shortest available fatty acid in animals, the above rules preclude the formation of the Δ^{12} double bond of linoleic acid [$\Delta^{9,12}$-octadecadienoic acid; 18:2n–6 (this nomenclature is explained in Table 12-1)], a required precursor of **prostaglandins.** *Linoleic acid must consequently be obtained in the diet (ultimately from plants that have Δ^{12}- and Δ^{15}-desaturases; it is abundant in most vegetable oils) and is therefore termed an **essential fatty acid.*** Indeed, animals maintained on a fat-free diet develop an ultimately fatal condition that is initially characterized by poor growth, poor wound healing, and dermatitis. Linoleic acid is also an important constituent of epidermal sphingolipids that function as the skin's water-permeability barrier.

Because of the inability of animal desaturases to add double bonds to positions beyond C9, another essential fatty acid is **α-linolenic acid [ALA; $\Delta^{9,12,15}$**-octadecatrienoic acid (18:3n–3, an ω–3 fatty acid)]. This fatty acid is a precursor to **EPA** ($\Delta^{5,8,11,14,17}$**-eicosapentaenoic acid; 20:5n–3**) and **DHA** ($\Delta^{4,7,10,13,16,19}$**-docosahexaenoic acid; 22:6n–3**), polyunsaturated ω–3 fatty acids recently found to be important dietary constituents (present in fish oils) that improve cognitive function and vision, and contribute to protection against inflammation and cardiovascular disease. DHA is, among other things, the predominant fatty acid in the phospholipids of retinal rod outer segments. Substitution of DHA with otherwise identical ω–6 fatty acids in phospholipids results in impaired visual acuity. Deficiency of ω–3 polyunsaturated fatty acids in brain phospholipids is associated with memory loss and diminished cognitive function.

F. Synthesis of Triacylglycerols

Triacylglycerols are synthesized from fatty acyl-CoA esters and glycerol-3-phosphate or dihydroxyacetone phosphate (Fig. 25-42). The initial step in this process is catalyzed either by **glycerol-3-phosphate acyltransferase** in mitochondria and the ER, or by **dihydroxyacetone phosphate acyltransferase** in the ER or peroxisomes. In the latter case, the product **acyl-dihydroxyacetone phosphate** is reduced to the corresponding **lysophosphatidic acid** by an NADPH-dependent reductase. The lysophosphatidic acid is converted to a triacylglycerol by the successive actions of **1-acylglycerol-3-phosphate acyltransferase, phosphatidic acid phosphatase,** and **diacylglycerol acyltransferase.** The intermediate phosphatidic acid and 1,2-diacylglycerol (DAG) can also be converted to phospholipids by the pathways described in Section 25-8. The acyltransferases are not completely specific for particular fatty acyl-CoAs, either in chain length or in degree of unsaturation, but in triacylglycerols of human adipose tissue, palmitate tends to be concentrated at position 1 and oleate at position 2.

a. Glyceroneogenesis Is Important for Triacylglycerol Biosynthesis

The dihydroxyacetone phosphate used to make glycerol-3-phosphate for triacylglycerol synthesis comes either from glucose via the glycolytic pathway (Fig. 17-3) or

Figure 25-41 **The electron-transfer reactions mediated by the Δ^9-fatty acyl-CoA desaturase complex.** Its three proteins, desaturase, cytochrome b_5, and NADH–cytochrome b_5 reductase, are situated in the endoplasmic reticulum membrane. [After Jeffcoat, R., *Essays Biochem.* **15,** 19 (1979).]

Figure 25-42 **The reactions of triacylglycerol biosynthesis.**

from oxaloacetate via an abbreviated version of gluconeogenesis (Fig. 23-8) termed **glyceroneogenesis.** Glyceroneogenesis is necessary in times of starvation, since approximately 30% of the fatty acids that enter the liver during a fast are reesterified to triacylglycerol and exported as VLDL (Section 25-1 and 25-6A). Adipocytes also carry out glyceroneogenesis in times of starvation. They do not carry out gluconeogenesis but contain the gluconeogenic enzyme phosphoenolpyruvate carboxykinase (PEPCK), which is upregulated when glucose concentration is low, and participates in the glyceroneogenesis required for triacylglycerol biosynthesis.

5 REGULATION OF FATTY ACID METABOLISM

Discussions of metabolic control are usually concerned with the regulation of metabolite flow through a pathway in response to the differing energy needs and dietary states of an organism. For example, the difference in the energy requirement of muscle between rest and vigorous exertion may be as much as 100-fold. Such varying demands may be placed on the body when it is in either a fed or a fasted state. For instance, Eric Newsholme, an authority on the biochemistry of exercise, enjoys a 2-hour run before breakfast. Others might wish for no greater exertion than the motion of hand to mouth. In both individuals, glycogen and triacylglycerols serve as primary fuels for energy-requiring processes and are synthesized in times of quiet plenty for future use.

a. Hormones Regulate Fatty Acid Metabolism

Synthesis and breakdown of glycogen and triacylglycerols, as detailed in Chapter 18 and above, are processes that concern the whole organism, with its organs and tissues forming an interdependent network connected by the bloodstream. The blood carries the metabolites responsible for energy production: triacylglycerols in the form of chylomicrons and VLDL (Section 12-5A), fatty acids as their albumin complexes (Section 25-1e), ketone bodies, amino acids, lactate, and glucose. The pancreatic α and β cells sense the organism's dietary and energetic state mainly through the glucose concentration in the blood. The α cells respond to the low blood glucose concentration of the fasting and energy-demanding states by secreting glucagon. The β cells respond to the high blood glucose concentration of the fed and resting states by secreting insulin. We have previously discussed (Sections 18-3E and 18-3F) how these hormones are involved in glycogen metabolism. *They also regulate the rates of the opposing pathways of lipid metabolism and therefore control whether fatty acids will be oxidized or synthesized.* Their targets are the regulatory (flux-generating) enzymes of fatty acid synthesis and breakdown in specific tissues (Fig. 25-43).

We are already familiar with most of the mechanisms by which the catalytic activities of regulatory enzymes may be controlled: substrate availability, allosteric interactions, and covalent modification (phosphorylation). These are examples of **short-term regulation,** regulation that occurs with a response time of minutes or less. *Fatty acid synthesis is controlled, in part, by short-term regulation.* Acetyl-CoA carboxylase, which catalyzes the first committed step of this pathway, is inhibited by palmitoyl-CoA and by the glucagon-stimulated cAMP-dependent increase in phosphorylation, and is activated by citrate and by insulin-stimulated dephosphorylation (Section 25-4B).

Another mechanism exists for controlling a pathway's regulatory enzymes: alteration of the amount of enzyme present by changes in the rates of protein synthesis and/or breakdown. This process requires hours or days and is therefore called **long-term regulation** (the control of protein synthesis and breakdown is discussed in Chapters 31 and 32). *Lipid biosynthesis is also controlled by long-term regulation,* with insulin stimulating and starvation inhibiting the synthesis of acetyl-CoA carboxylase and fatty acid synthase. The presence in the diet of polyunsaturated fatty acids also decreases the concentrations of these enzymes. The amount of adipose tissue lipoprotein lipase, the enzyme that initiates the entry of lipoprotein-packaged fatty acids into adipose tissue for storage (Section 12-5Ba), is also increased by insulin and decreased by starvation. In contrast, the concentration of heart lipoprotein lipase, which controls the entry of fatty acids from lipoproteins into heart tissue for oxidation rather than storage, is decreased by insulin and increased by starvation. *Starvation and/or regular exercise, by decreasing the glucose concentration in the blood, change the body's hormone balance. This situation results in long-term changes in gene expression that increase the levels of fatty acid oxidation enzymes and decrease those of lipid biosynthesis.*

Fatty acid oxidation is regulated largely by the concentration of fatty acids in the blood, which is, in turn, controlled by the hydrolysis rate of triacylglycerols in adipose tissue by **hormone-sensitive triacylglycerol lipase.** This enzyme is so named because it is susceptible to regulation by phosphorylation and dephosphorylation in response to hormonally controlled cAMP levels. Epinephrine and norepinephrine, as does glucagon, act to increase adipose tissue cAMP concentrations. cAMP allosterically activates protein kinase A (PKA) which, in turn, increases the phosphorylation levels of susceptible enzymes. Phosphorylation activates hormone-sensitive triacylglycerol lipase, thereby stimulating lipolysis in adipose tissue, raising blood fatty acid levels, and ultimately activating the β-oxidation pathway in other tissues such as liver and muscle. In liver, this process leads to the production of ketone bodies that are secreted into the bloodstream for use by peripheral tissues as an alternative fuel to glucose. PKA, acting in concert with AMP-dependent protein kinase (AMPK), also causes the inactivation of acetyl-CoA carboxylase (Section 25-4B), one of the rate-determining enzymes of fatty acid synthesis, so that *cAMP-dependent phosphorylation simultaneously stimulates fatty acid oxidation and inhibits fatty acid synthesis.*

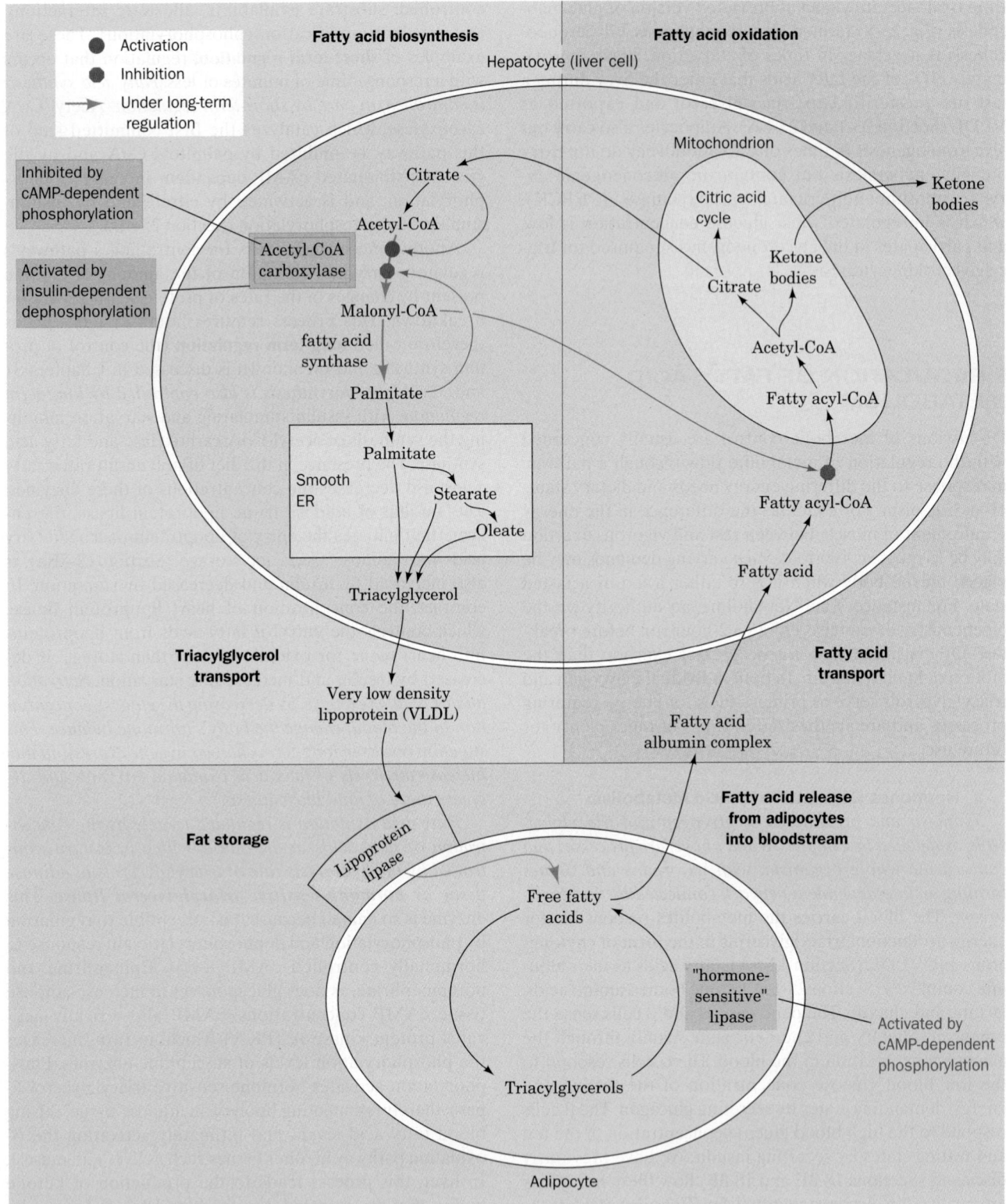

Figure 25-43 **Sites of regulation of fatty acid metabolism.**

Insulin has the opposite effect of glucagon and epineph-rine: It stimulates the formation of glycogen and triacyl-glycerols. This protein hormone, which is secreted in re-sponse to high blood glucose concentrations, triggers a highly complex signal transduction network (Section 19-4F)

that induces the long-term regulation of numerous enzymes as well as decreasing cAMP levels. This latter situation leads to the dephosphorylation and thus the inactivation of hormone-sensitive triacylglycerol lipase, thereby reducing the amount of fatty acid available for oxidation. Insulin also stimulates the dephosphorylation of acetyl-CoA carboxylase, thereby activating this enzyme (Section 25-4Ba). *The glucagon–insulin ratio is therefore of prime importance in determining the rate and direction of fatty acid metabolism.*

Another control point that inhibits fatty acid oxidation when fatty acid synthesis is stimulated is the inhibition of carnitine palmitoyltransferase I by malonyl-CoA. This inhibition keeps the newly synthesized fatty acids out of the mitochondrion (Section 25-2B) and thus away from the β-oxidation system. As we have seen (Section 25-4Bc), heart muscle, an oxidative tissue that does not carry out fatty acid biosynthesis, contains an isoform of acetyl-CoA carboxylase, ACC2, whose sole function appears to be the synthesis of malonyl-CoA to regulate fatty acid oxidation.

AMPK may itself be an important regulator of fatty acid metabolism. This phosphorylating enzyme is activated by AMP and inhibited by ATP and thus has been proposed to serve as a fuel gauge for the cell. When ATP levels are high, signaling the fed and rested state, this kinase is inhibited, allowing ACC to become dephosphorylated (activated) so as to stimulate malonyl-CoA production for fatty acid synthesis in adipose tissue and for inhibition of fatty acid oxidation in muscle cells. When activity levels increase causing ATP levels to decrease with a concomitant increase in AMP levels, AMPK is activated to phosphorylate (inac-

tivate) ACC. The resulting decrease in malonyl-CoA levels causes fatty acid biosynthesis to decrease in adipose tissue while fatty acid oxidation increases in muscle to provide the ATP for continued activity.

6 CHOLESTEROL METABOLISM

Cholesterol is a vital constituent of cell membranes and the precursor of steroid hormones and bile salts. It is clearly essential to life, yet its deposition in arteries has been associated with cardiovascular disease and stroke, two leading causes of death in humans. In a healthy organism, an intricate balance is maintained between the biosynthesis, utilization, and transport of cholesterol, keeping its harmful deposition to a minimum. In this section, we study the pathways of cholesterol biosynthesis and transport and how they are controlled. We also examine how cholesterol is utilized in the biosynthesis of steroid hormones and bile salts.

A. *Cholesterol Biosynthesis*

All of the carbon atoms of cholesterol are derived from acetate (Fig. 25-44). Observation of their pattern of incorporation into cholesterol led Konrad Bloch to propose that acetate was first converted to **isoprene units,** C_5 units that have the carbon skeleton of **isoprene:**

Isoprene
(2-Methyl-1,3-butadiene)

An isoprene unit

Isoprene units are condensed to form a linear precursor to cholesterol, and then cyclized.

Squalene, a polyisoprenoid hydrocarbon (Fig. 25-45*a*), was demonstrated to be the linear intermediate in cholesterol biosynthesis by the observation that feeding isotopically labeled squalene to animals yields labeled cholesterol. Squalene may be folded in several ways that would enable it to cyclize to the four-ring sterol nucleus (Section 12-1E). The folding pattern proposed by Bloch and Robert B. Woodward (Fig. 25-45*b*) proved to be correct.

Figure 25-44 **All of cholesterol's carbon atoms are derived from acetate.**

Figure 25-45 **Squalene.** (*a*) Extended conformation. Each box contains one isoprene unit. (*b*) Folded in preparation for cyclization as predicted by Bloch and Woodward.

Figure 25-46   The branched pathway of isoprenoid metabolism in mammalian cells. The pathway produces ubiquinone, dolichol, farnesylated and geranylgeranylated proteins, and isopentenyl adenosine, a modified tRNA base, in addition to cholesterol.

Bloch's outline for the major stages of cholesterol biosynthesis was

Acetate ⟶ isoprenoid intermediate ⟶
   squalene ⟶ cyclization product ⟶ cholesterol

This pathway has been experimentally verified and its details elaborated. It is now known to be part of a branched pathway (Fig. 25-46) that produces several other essential isoprenoids in addition to cholesterol, namely, ubiquinone (CoQ; Fig. 22-17*b*), dolichol (Fig. 23-15), farnesylated and geranylgeranylated proteins (Fig. 12-29), and **isopentenyl-adenosine** (a modified base of tRNA; Fig. 32-10). We shall examine in detail the portion of this pathway that synthesizes cholesterol. Note, however, that >25,000 isoprenoids (also known as **terpenoids**), mostly of plant, fungal, and bacterial origin, have been characterized. These serve as membrane constituents (e.g., cholesterol), hormones

(steroids), pheromones, defensive agents, photoprotective agents (e.g., β-carotene; Section 24-2Ad), and visual pigments (e.g., retinal; Section 12-3Ab), to name only a few of their many biological functions.

a. HMG-CoA Is a Key Cholesterol Precursor

Acetyl-CoA is converted to isoprene units by a series of reactions that begins with formation of hydroxymethylglutaryl-CoA (HMG-CoA; Fig. 25-26), a compound we previously encountered as an intermediate in ketone body biosynthesis (Section 25-3). HMG-CoA synthesis requires the participation of two enzymes: thiolase and HMG-CoA synthase. The enzymes forming the HMG-CoA leading to ketone bodies occur in the mitochondria, whereas those responsible for the synthesis of the HMG-CoA that is destined for cholesterol biosynthesis are located in the cytosol. Their catalytic mechanisms, however, are identical.

HMG-CoA is the precursor of two isoprenoid intermediates, **isopentenyl pyrophosphate** and **dimethylallyl pyrophosphate:**

Isopentenyl pyrophosphate

Dimethylallyl pyrophosphate

The formation of isopentenyl pyrophosphate involves four reactions (Fig. 25-47):

1. The CoA thioester group of HMG-CoA is reduced to an alcohol in an NADPH-dependent four-electron reduction catalyzed by **HMG-CoA reductase,** yielding **mevalonate.**

2. The new OH group is phosphorylated by **mevalonate-5-phosphotransferase.**

3. The phosphate group is converted to a pyrophosphate by **phosphomevalonate kinase.**

4. The molecule is decarboxylated and the resulting alcohol dehydrated by **pyrophosphomevalonate decarboxylase.**

HMG-CoA reductase mediates the rate-determining step of cholesterol biosynthesis and is the most elaborately regulated enzyme of this pathway. This 888-residue ER membrane-bound enzyme is regulated, as we shall see in Section 25-6Bb, by competitive and allosteric mechanisms, phosphorylation/dephosphorylation, and long-term regulation. Cholesterol itself is an important feedback regulator of the enzyme.

b. Pyrophosphomevalonate Decarboxylase Catalyzes an Apparently Concerted Reaction

5-Pyrophosphomevalonate is converted to isopentenyl pyrophosphate by an ATP-dependent dehydration–decarboxylation reaction catalyzed by **pyrophosphomevalonate decarboxylase** (Fig. 25-48). When [3-^{18}O]5-pyrophosphomevalonate (*O in Fig. 25-48) is used as a substrate, the labeled oxygen appears in P_i. This observation suggests that 3-phospho-5-pyrophosphomevalonate is a reaction intermediate. Since all attempts to isolate this intermediate have failed, however, it has been proposed that phosphorylation, the α,β elimination of CO_2, and the elimination of P_i occur in a concerted reaction.

HMG-CoA

Figure 25-47 Formation of isopentenyl pyrophosphate from HMG-CoA.

Isopentenyl pyrophosphate

5-Pyrophosphomevalonate

Isopentenyl pyrophosphate

Figure 25-48   Action of pyrophosphomevalonate decarboxylase. The enzyme catalyzes an ATP-dependent concerted dehydration–decarboxylation of pyrophosphomevalonate, yielding isopentenyl pyrophosphate.

The equilibration between isopentenyl pyrophosphate and dimethylallyl pyrophosphate is catalyzed by **isopentenyl pyrophosphate isomerase.** The reaction appears to occur via a protonation/deprotonation reaction involving the intermediacy of a tertiary carbocation intermediate. Cys and Glu residues have been implicated as the general acid and base catalysts, respectively (Fig. 25-49), as supported by site-directed mutagenesis and the X-ray structure of the enzyme. The carbocation is thought to be stabilized through interactions with the aromatic π cloud of an adjacent Trp residue. Aromatic residues provide electron-rich interactions with positively charged groups without forming covalent bonds that would destroy the intermediate.

c. Squalene Is Formed by the Condensation of Six Isoprene Units

Four isopentenyl pyrophosphates and two dimethylallyl pyrophosphates condense to form the C_{30} cholesterol precursor squalene in three reactions catalyzed by two enzymes (Fig. 25-50):

1. Prenyltransferase (farnesyl pyrophosphate synthase) catalyzes the head-to-tail (1'–4) condensation of dimethylallyl pyrophosphate and isopentenyl pyrophosphate to yield **geranyl pyrophosphate.**

2. Prenyltransferase catalyzes a second head-to-tail condensation of geranyl pyrophosphate and isopentenyl pyrophosphate to yield **farnesyl pyrophosphate (FPP).**

3. Squalene synthase (SQS) then catalyzes the head-to-head (1–1') condensation of two farnesyl pyrophosphate molecules to form squalene. Farnesyl pyrophosphate is also a precursor to dolichol, farnesylated and geranylgeranylated proteins, and ubiquinone (Fig. 25-46).

Prenyltransferase catalyzes the condensation of isopentenyl pyrophosphate with an allylic (conjugated to a C=C double bond) pyrophosphate. It is specific for isopentenyl pyrophosphate but can use either the 5-carbon dimethylallyl pyrophosphate or the 10-carbon **geranyl pyrophosphate** as its allylic substrate. The prenyltransferase-catalyzed condensation mechanism is particularly interesting since it is one of the few known enzyme-catalyzed reactions that pro-

Isopentenyl pyrophosphate

Tertiary carbocation

Dimethylallyl pyrophosphate

Figure 25-49   Mechanism of isopentenyl pyrophosphate isomerase. The enzyme interconverts isopentenyl pyrophosphate and dimethylallyl pyrophosphate by a protonation/deprotonation reaction involving a carbocation intermediate in which a Cys and a Glu residue act as a proton donor and acceptor. The carbocation intermediate appears to be stabilized by π interactions with a nearby Trp side chain.

Figure 25-50 **Formation of squalene from isopentenyl pyrophosphate and dimethylallyl pyrophosphate.** The pathway involves two head-to-tail condensations catalyzed by prenyltransferase and a head-to-head condensation catalyzed by squalene synthase.

ceed via a carbocation intermediate. Two possible condensation mechanisms can be envisioned (Fig. 25-51):

Scheme I An S_N1 mechanism in which an allylic carbocation forms by the elimination of PP_i. Isopentenyl pyrophosphate then condenses with this carbocation, forming a new carbocation that eliminates a proton to form product.

Scheme II An S_N2 reaction in which the allylic PP_i is displaced in a concerted manner. In this case, an enzyme nucleophile, X, assists in the reaction. This group is eliminated in the second step with the loss of a proton to form product.

Dale Poulter and Hans Rilling used chemical logic to differentiate between these two mechanisms. Capitalizing

on the observation that S_N1 reactions are much more sensitive to electron-withdrawing groups than S_N2 reactions, they synthesized a geranyl pyrophosphate derivative in which the H at C2 is replaced by the electron-withdrawing group F. This allylic substrate for the second (1′–4) condensation catalyzed by prenyltransferase, not surprisingly, has the same K_M as the natural substrate (F and H have similar atomic radii):

It is, however, the V_{max} of this reaction that tells the story. If the reaction is an S_N2 displacement, the fluoro derivative

Scheme I
Ionization–condensation–elimination

S$_N$1

PP$_i$

H$^+$

Scheme II
Condensation–elimination

S$_N$2

PP$_i$

H$^+$

***Figure 25-51*   Two possible mechanisms for the prenyltransferase reaction.** Scheme I involves the formation of a carbocation intermediate, whereas Scheme II involves the participation of an enzyme nucleophile, X.

should react at a rate similar to that of the natural substrate. If, instead, the reaction has an S$_N$1 mechanism, the fluoro derivative should react orders of magnitude more slowly than the natural substrate. In fact, 3-fluorogeranyl pyrophosphate forms product at <1% of the rate of the natural substrate, strongly supporting an S$_N$1 mechanism with a carbocation intermediate.

Carbocations are now known to participate in several reactions of isoprenoid biosynthesis. The enzymes are classified according to how they generate these carbocations. Class I enzymes do so via the release of pyrophosphate, as we have seen for prenyltransferase. Class II enzymes do so by protonating a double bond, as does isopentenyl pyrophosphate isomerase (Fig. 25-49), or an epoxide, as we shall see below for oxidosqualene cyclase.

Squalene, the immediate sterol precursor, is formed by the head-to-head condensation of two FPP molecules by SQS. Although the enzyme is a Class I enzyme that is structurally related to prenyltransferase and generates carbocations by the release of pyrophosphate, the reaction is not a simple head-to-tail condensation, as might be expected, but, rather, proceeds via a complex two-step mechanism with each step catalyzed by a different active site on the enzyme (Fig. 25-52):

Step I The reaction of two FPP molecules to yield the stable intermediate **presqualene pyrophosphate.** This reac-

Farnesyl pyrophosphate

Farnesyl pyrophosphate

Presqualene pyrophosphate

Squalene

***Figure 25-52   Action of squalene synthase.** The enzyme catalyzes the head-to-head condensation of two farnesyl pyrophosphate molecules to form squalene.

tion is initiated by the elimination of PP$_i$ from one farnesyl pyrophosphate molecule to form an allylic carbocation at C1 that is stabilized by a π interaction with an essential Tyr

Farnesyl pyrophosphate

Farnesyl pyrophosphate

Allylic carbocation

Tertiary carbocation

Presqualene pyrophosphate

Figure 25-53 **Proposed mechanism for the formation of presqualene pyrophosphate from two farnesyl pyrophosphate molecules by squalene synthase (Fig. 25-52, Step I). (1)** The pyrophosphate group on one farnesyl pyrophosphate leaves, yielding an allylic carbocation. This reaction step is facilitated by proton donation from the side chain of an essential Tyr residue, which then stabilizes the allylic cation via π–cation interactions. **(2)** The C2═C3 double bond of the second farnesyl pyrophosphate nucleophilically attacks the allylic carbocation to form a tertiary carbocation at C3. **(3)** The abstraction of the pro-*S* proton at C1′ by the phenolate group of the essential Tyr residue results in the formation of a C1′—C3 bond yielding presqualene pyrophosphate.

residue (Fig. 25-53). The highly reactive electron-deficient carbocation inserts into the electron-rich C2═C3 double bond of the second molecule, yielding presqualene pyrophosphate, a cyclopropylcarbinyl pyrophosphate.

Step II The rearrangement and reduction of presqualene pyrophosphate by NADPH to form squalene. This reaction involves the formation and rearrangement of a cyclopropylcarbinyl cation in a complex reaction sequence called a **1′–2–3 process** (Fig. 25-54).

SQS, a monomeric protein, is anchored to the ER membrane via a short C-terminal transmembrane domain, with its active site facing the cytosol. This allows it to accept its water-soluble substrates, farnesyl pyrophosphate and NADPH, from the cytosol and release its hydrophobic product, squalene, in the ER membrane.

Presqualene pyrophosphate

Primary cyclopropylcarbinyl cation

Tertiary cyclopropylcarbinyl cation

Squalene

Figure 25-54 **Mechanism of rearrangement and reduction of presqualene pyrophosphate to squalene as catalyzed by squalene synthase (Fig. 25-52, Step II). (1)** Presqualene's pyrophosphate group leaves, yielding a primary carbocation at C1. **(2)** The electrons forming the C1′—C3 bond migrate to C1, forming squalene's C1—C1′ bond and a tertiary carbocation at C3. **(3)** The process is completed by the addition of an NADPH-supplied hydride ion to C1′ and the formation of the C2═C3 double bond.

The X-ray structure of the 417-residue human SQS, in complex with the inhibitor **CP-320473,**

CP-320473

determined by Jayvardhan Pandit, reveals that the protein folds as a single domain with a large channel across one face into which CP-320473 binds (Fig. 25-55). The channel is lined with Asp and Arg residues that mutagenesis studies indicate are involved in FPP binding. Of these, the conserved Asp 80 and Asp 84 are implicated in binding Mg^{2+} ions that ligand an FPP pyrophosphate group. These Asp residues are adjacent to Tyr 171, which forms the base of the channel and which mutagenesis studies have identified as the essential Tyr that is implicated in stabilizing the allylic carbocation intermediate in Step I of the SQS reaction. Step II of the SQS reaction requires that its highly reactive carbocation intermediates be shielded from contact with the aqueous solvent to prevent it from quenching the reaction. This suggests that for Step II, the presqualene pyrophosphate product of Step I moves deeper into the channel into a pocket that is lined with hydrophobic groups, including Phe 288, whose mutation inactivates the enzyme. This further suggests that Phe 288 functions to stabilize one of the cationic intermediates in Step II (Fig. 25-54) through π–cation interactions.

d. Lanosterol Is Produced by Squalene Cyclization

Squalene, an open-chain C_{30} hydrocarbon, is cyclized to form the tetracyclic steroid skeleton in two steps. **Squalene epoxidase** catalyzes the oxidation of squalene to form **2,3-oxidosqualene** (Fig. 25-56). **Oxidosqualene cyclase** (alternatively, **lanosterol synthase**) converts this epoxide to **lanosterol,** the sterol precursor of cholesterol. The highly exergonic reaction is a complex process involving cyclization of 2,3-oxidosqualene to a **protosterol** cation, via a Class II mechanism involving protonation of the epoxide, and rearrangement of this cation to lanosterol by a series of 1,2 hydride and methyl shifts (Fig. 25-57). Note that this reaction generates six of lanosterol's seven chiral centers.

The X-ray structure of human oxidosqualene cyclase in complex with lanosterol, determined by Armin Ruf, reveals that this 732-residue monomeric and monotopic [integral but not transmembrane] protein contains two structurally similar domains named **α/α barrels** (Fig. 25-58a). An α/α barrel consists of two concentric barrels of 6 helices each, with the helices of the inner barrel largely parallel to each other and antiparallel to those of the outer barrel (much like an α/β barrel with the β strands of the inner barrel replaced by helices but with only 6 α/α units rather than 8 α/β units). The enzyme's active site is located inside an elongated central cavity (Fig. 25-58b) to which the lanosterol is bound and which is accessible from the membrane via a nonpolar channel through the enzyme's membrane-immersed part.

The interactions of 2,3-oxidosqualene with oxidosqualene cyclase cause it to fold and react such that it only forms lanosterol. The active site cavity is lined with several conserved aromatic side chains that are suitably positioned to stabilize the catalyzed reaction's intermediate carbocations, while shielding the cationic intermediates from premature quenching by either enzyme nucleophiles or water. The cationic cascade (Fig. 25-57) is initiated by proton donation by Asp 455 to the epoxide ring of the prefolded 2,3-oxidosqualene, and is quenched by His 232 acting as a general base.

The importance of the proper placement of residues for the formation of the correct product was demonstrated by site-directed mutagenesis: The conversion of Thr 384 in

Figure 25-55 **X-ray structure of human squalene synthase (SQS) in complex with the inhibitor CP-320473.** This inhibitor together with the side chains of D80, D84, Y171, and F288 are drawn in ball-and-stick form colored according to atom type (inhibitor C green, protein C cyan, N blue, O red, and Cl magenta). The protein is viewed looking into its central channel with the putative active sites for Steps I and II of the catalyzed reaction at the bottom and top of the cleft, respectively. The protein's C-terminal transmembrane segment (residues 371–417) together with its N-terminal 30 residues were excised to facilitate its crystallization, which did not affect its *in vitro* catalytic activity. [Based on an X-ray structure by Jayvardhan Pandit, Pfizer Central Research, Groton, Connecticut. PDBid 1EZF.]

Figure 25-56   The squalene epoxidase reaction.

oxidosqualene cyclase to Tyr causes some misplacement of the C8=C9 double bond, with 11% of the product having a C9=C11 double bond and 10% having a 9-hydroxy group; the double mutation, T384Y/V454I, increases the yield of the C9=C11 double bond to 64%. This, together

with different ways of substrate folding, explains how enzymes homologous to oxidosqualene cyclase cause oxidosqualene and squalene to form different products. Indeed, the bacterial enzyme **squalene–hopene synthase,** a homolog of oxidosqualene cyclase, synthesizes the C_{30} compound **hopene**

Hopene

from squalene. Its X-ray structure in complex with 2-azasqualene (squalene with its C2 atom replaced N), determined by Georg Schulz, reveals that the enzyme, which structurally resembles oxidosqualene cyclase, folds squalene in the configuration that is predicted to yield hopene. In this case, the cationic cyclase cascade is initiated by the protonation of a double bond instead of an epoxide.

Figure 25-57   The oxidosqualene cyclase reaction. (1) 2,3-Oxidosqualene is cyclized to the protosterol cation in a process that is initiated by the enzyme-mediated protonation of the squalene epoxide oxygen while this extended molecule is folded in the manner predicted by Bloch and Woodward. The opening of the epoxide leaves an electron-deficient center whose migration drives the series of cyclizations that ultimately form the protosterol cation. **(2)** A series of methyl and hydride migrations yields a hypothesized intermediate carbocation at C8, which then eliminates a proton from C9 to form the C8=C9 double bond of lanosterol.

(a)

(b)

Figure 25-58   X-ray structure of human oxidosqualene cyclase in complex with its product lanosterol. The monomeric protein is shown in ribbon form (*a*) viewed along the axis of its α/α barrels, and (*b*) rotated 90° about the horizontal axis with respect to Part *a*. The inner helices of its two α/α barrels are cyan, its outer helices are purple, its nonpolar membrane-immersed portion is orange, and the remainder of the protein is pink. The lanosterol, which occupies the enzyme's centrally located active site cavity, is shown in Part *a* in space-filling form and in Part *b* in stick form embedded in its semitransparent space-filling form, both with C

green and O red. The catalytically important side chains of His 232 and Asp 455, together with molecules of the detergent β-octylglucoside and fragments of lipids that coat the enzyme's membrane-immersed portion, are drawn in stick form with side chain C magenta, lipid and detergent C yellow, and O red. The arrow in Part *b* points to the membrane-immersed opening of the enzyme's active site cavity. [Based on an X-ray structure by Armin Ruf, F. Hoffmann-La Roche AG, Basel, Switzerland. PDBid 1W6K.]

e. Cholesterol Is Synthesized from Lanosterol

Conversion of lanosterol to cholesterol (Fig. 25-59) is a 19-step process that we shall not explore in detail. It involves an oxidation and loss of three methyl groups. The first methyl group is removed as formate and the other two are eliminated as CO_2 in reactions that all require NADPH and O_2. The enzymes involved in this process are embedded in the ER membrane.

f. Cholesterol Is Transported in the Blood and Taken Up by Cells in Lipoprotein Complexes

Transport and cellular uptake of cholesterol are described in Section 12-5. To recapitulate, cholesterol synthesized by the liver is either converted to bile salts for use in the digestive process (Section 25-1) or esterified by **acyl-**

CoA:cholesterol acyltransferase (ACAT) to form **cholesteryl esters**

Cholesteryl ester

which are secreted into the bloodstream as part of the lipoprotein complexes called **very low density lipoproteins (VLDL).** As the VLDL circulate, their component triacyl-

Figure 25-59 **The 19-reaction conversion of lanosterol to cholesterol.** [After Rilling, H.C. and Chayet, L.T., *in* Danielsson, H. and Sjövall, J. (Eds.), *Sterols and Bile Acids, p.* 33, Elsevier (1985), as modified by Bae, S.-H. and Paik, Y.-K., *Biochem. J.* **326,** 609–616 (1997).]

glycerols and most types of their **apolipoproteins** (Table 12-6) are removed in the capillaries of muscle and adipose tissues, sequentially converting the VLDL to **intermediate-density lipoproteins (IDL)** and then to **low-density lipoproteins (LDL).** Peripheral tissues normally obtain most of their exogenous cholesterol from LDL by receptor-mediated endocytosis (Fig. 25-60; Section 12-5Bc). Inside the cell, cholesteryl esters are hydrolyzed by a lysosomal lipase to free cholesterol, which is either incorporated into cell membranes or reesterified by ACAT for storage as cholesteryl ester droplets.

Dietary cholesterol, cholesteryl esters, and triacylglycerols are transported in the blood by intestinally synthesized lipoprotein complexes called **chylomicrons.** After removal of their triacylglycerols at the peripheral tissues, the resulting **chylomicron remnants** bind to specific liver cell remnant receptors and are taken up by receptor-mediated endocytosis in a manner similar to that of LDL. In the liver, dietary cholesterol is either used in bile salt biosynthesis (Section 25-6C) or packaged into VLDL for export. *Liver and peripheral tissues therefore have two ways of obtaining cholesterol: They may either synthesize it from acetyl-CoA*

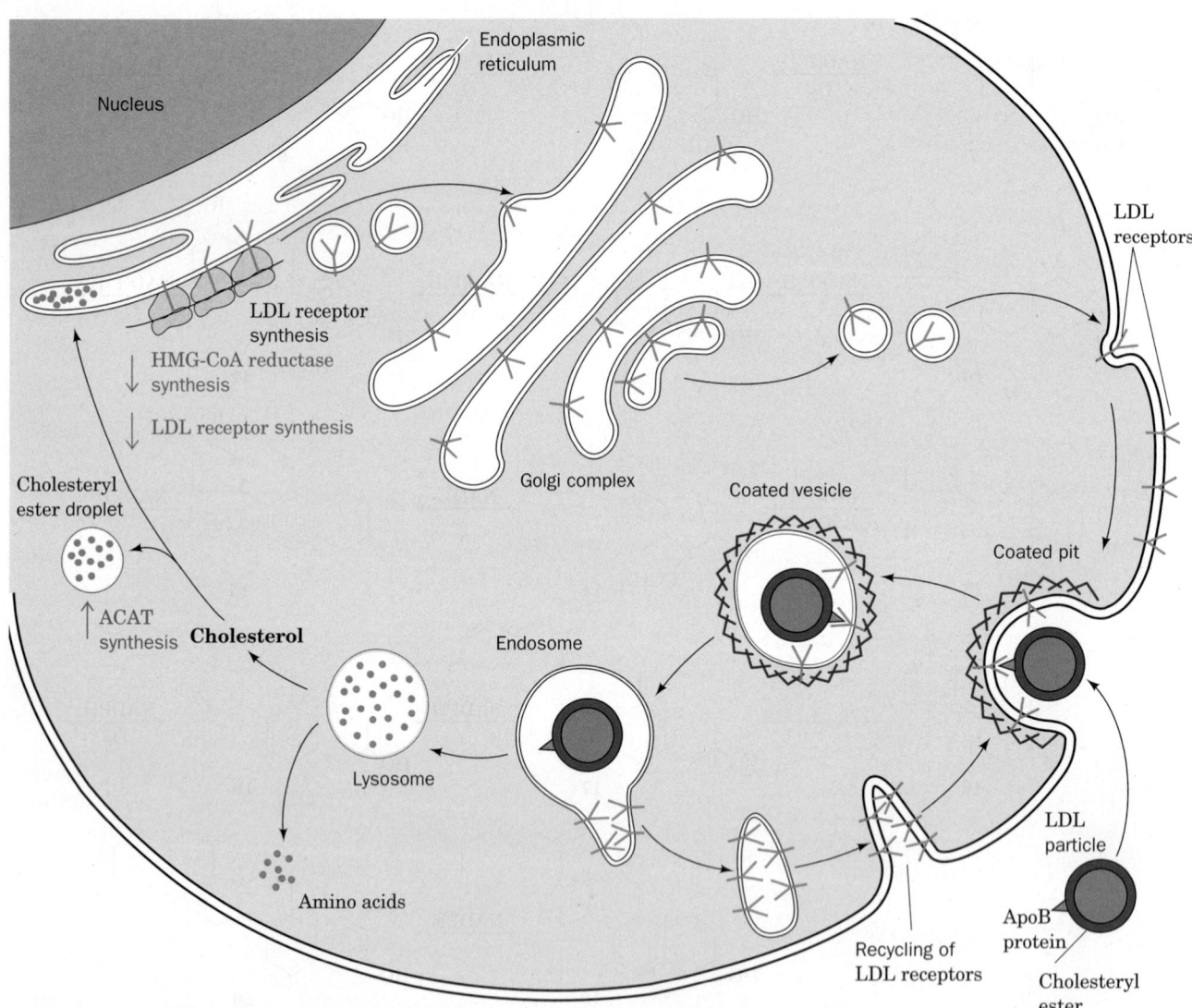

Figure 25-60   LDL receptor-mediated endocytosis in mammalian cells. LDL receptor is synthesized on the endoplasmic reticulum, processed in the Golgi complex, and inserted into the plasma membrane as a component of coated pits. LDL is specifically bound by the LDL receptor on the coated pit and brought into the cell in endosomes that deliver LDL to lysosomes while recycling LDL receptor to the plasma membrane (Section 12-5Bc). Lysosomal degradation of LDL releases cholesterol, whose presence decreases the rate of synthesis of HMG-CoA reductase and LDL receptors (*down arrows*) while increasing that of acyl-CoA:cholesterol acyltransferase (ACAT; *up arrow*). [After Brown, M.S. and Goldstein, J.L., *Curr. Top. Cell. Reg.* **26,** 7 (1985).] **See the Animated Figures**

by the de novo pathway we have just discussed, or they may obtain it from the bloodstream by receptor-mediated endocytosis. A small amount of cholesterol also enters cells by a non-receptor-mediated pathway. Note however, that the brain, which comprises ~2% of the human body mass but contains ~30% of its cholesterol, must synthesize all of its cholesterol because cholesterol cannot pass the blood-brain barrier.

Cholesterol actually circulates back and forth between the liver and peripheral tissues. While LDL transports cholesterol from the liver, cholesterol is transported back to the liver by **high-density lipoproteins (HDL).** Surplus cholesterol is disposed of by the liver as bile salts, thereby protecting the body from an overaccumulation of this water-insoluble substance.

B. *Control of Cholesterol Biosynthesis and Transport*

Cholesterol biosynthesis and transport must be tightly regulated. There are three ways in which the cellular cholesterol supply is maintained:

1. By regulating the activity of HMG-CoA reductase, the enzyme catalyzing the rate-limiting step in the *de novo* cholesterol biosynthesis pathway. This is accomplished in two ways:

(i) Short-term regulation of the enzyme's catalytic activity by (a) competitive inhibition, (b) allosteric effects, and (c) covalent modification involving reversible phosphorylation.

(ii) Long-term regulation of the enzyme's concentration by modulating its rates of synthesis and degradation.

2. By regulating the rate of LDL receptor synthesis, and therefore the rate of cholesterol uptake. High intracellular concentrations of cholesterol suppress LDL receptor synthesis, whereas low cholesterol concentrations stimulate it.

3. By regulating the rate of esterification and hence the removal of free cholesterol. ACAT, the enzyme that catalyzes intracellular cholesterol esterification, is regulated by reversible phosphorylation and by long-term control.

a. HMG-CoA Reductase Is the Primary Control Site for Cholesterol Biosynthesis

HMG-CoA reductase is the rate-limiting enzyme in cholesterol biosynthesis and, as therefore might be expected, constitutes the pathway's main regulatory site. The pathway branches after this reaction, however (Fig. 25-46); ubiquinone, dolichol, farnesylated and geranylgeranylated proteins, and isopentenyl adenosine are also essential, albeit minor, products. HMG-CoA is therefore subject to "multivalent" control, both long-term and short-term, in order to coordinate the synthesis of all of these products.

b. Long-Term Feedback Regulation of HMG-CoA Reductase Is Its Primary Means of Control

The main way in which HMG-CoA reductase is controlled is by long-term feedback control of the amount of enzyme present in the cell. When either LDL–cholesterol or

mevalonate levels fall, the amount of HMG-CoA reductase present in the cell can rise as much as 200-fold, due to an increase in enzyme synthesis combined with a decrease in its degradation. When LDL–cholesterol or **mevalonolactone** (an internal ester of mevalonate that is hydrolyzed to mevalonate and metabolized in the cell)

Mevalonolactone

are added back to a cell, these effects are reversed.

The mechanism by which cholesterol serves to control the expression of the >20 genes involved in its biosynthesis and uptake, such as those encoding HMG-CoA reductase and the LDL receptor, has been elucidated by Michael Brown and Joseph Goldstein. These genes all contain a DNA sequence upstream from the transcription initiation site called the **sterol regulatory element (SRE).** In order for these genes to be transcribed, a specific transcription factor, the **sterol regulatory element binding protein (SREBP),** must bind to the SRE (eukaryotic gene expression is discussed in Section 34-3). SREBP is synthesized as an integral membrane protein that, when the cholesterol concentration is sufficiently high (Fig. 25-61a), resides in the ER membrane in complex with **SREBP cleavage-activating protein (SCAP)** and a protein named **Insig.** SREBP (~1160 residues) consists of three domains (Fig. 25-61a): (1) an ~480-residue cytosolic N-terminal domain that is a member of the **basic helix–loop–helix/leucine zipper (bHLH/Z)** family of transcription factors (Section 34-3Br), which specifically binds to SREs; (2) an ~90-residue transmembrane (TM) domain consisting of two TM helices connected by an ~30-residue hydrophilic luminal loop; and (3) an ~590-residue cytosolic C-terminal regulatory domain. SCAP (1276 residues) consists of two domains: (1) a 730-residue N-terminal domain that contains eight TM helices; and (2) a 546-residue cytosolic C-terminal domain that contains five copies of the protein–protein interaction motif known as a **WD repeat** (also called a WD40 sequence motif because it is ~40 residues long; Section 19-2Cc) and which presumably forms a 5-bladed β propeller similar to the 7-bladed β propeller of the G_β subunit (Fig. 19-19b). Insig (named to denote *insulin-induced gene*, although the insulin effect was later found to be indirect) consists mainly of six TM helices. SCAP and SREBP associate through the interaction of the SCAP's regulatory domain with SREBP's WD domain (Fig. 25-61a).

SCAP functions as a sterol sensor. An ~170-residue segment of its TM domain, the **sterol-sensing domain,** interacts with sterols although how it does so is unknown. When the cholesterol in the ER membrane is depleted (Fig. 25-61b), SCAP changes conformation and then

Figure 25-61   **The cholesterol-mediated proteolytic activation of SREBP.** (*a*) When cholesterol levels in the cell are high, the Insig–SCAP–SREBP complex resides in the ER membrane. (*b*) When cholesterol levels are low, SCAP releases Insig and binds COPII protein. The SCAP–SREBP complex is then transported, via COPII-coated vesicles, to the Golgi apparatus, where SREBP undergoes sequential proteolytic cleavage by the membrane-bound proteases S1P and S2P. This releases SREBP's bHLH/Z-containing N-terminal domain, which enters the nucleus where it binds to the SREs of its target genes, thereby inducing their transcription. [After Goldstein, J., Rawson, R.B., and Brown, M., *Arch. Biochem. Biophys.* **397,** 139 (2002).]

releases Insig, which allows SCAP to bind to the Sec24 cargo-binding subunit of the COPII coated vesicle protein (Section 12-4Ca). The COPII vesicle then escorts its bound SREBP–SCAP complex to the Golgi apparatus.

In the Golgi apparatus, SREBP is sequentially cleaved by two membrane-bound proteases (Fig. 25-61*b*). **Site-1 protease (S1P),** a serine protease of the subtilisin family, cleaves SREBP in the luminal loop that connects its two transmembrane helices but only when it is associated with SCAP. This cleavage exposes a peptide bond located 3 residues from the beginning of SREBP's N-terminal TM helix to cleavage by **site-2 protease (S2P),** a zinc metalloprotease. This releases the bHLH/Z domain to migrate to the nucleus where it activates the transcription of its target genes. The cholesterol level in the cell thereby rises until SCAP no longer induces the translocation of SREBP to the Golgi, a classic case of feedback inhibition.

This complex regulatory pathway was elucidated, in part, through the generation of several lines of transgenic mice that overexpress one or another of the foregoing proteins, and knockout mice that lack one or another of these proteins. For example, knockout mice lacking either SCAP or S1P in their livers have decreased expression of both HMG-CoA reductase and LDL receptor, even when fed a cholesterol-deficient diet. In contrast, transgenic mice that overexpress SREBP or SCAP have greatly increased expression of the foregoing proteins. In fact, animals overproducing only the bHLH/Z domain of SREBP have massively enlarged livers (up to 4-fold larger than normal) due to engorgement with triacylglycerols and cholesteryl esters and yet they continue to transcribe SREBP's target genes such that their mRNA levels are up to 75-fold greater than normal. Many individuals suffering from obesity or diabetes caused by insulin resistance (type 2 diabetes; Section 27-4B) have fatty livers, which in some cases leads to liver failure. Fatty livers due to insulin resistance appear to be caused by elevated levels of SREBP in response to elevated insulin levels.

The level of HMG-CoA reductase also responds to the level of the cholesterol precursor lanosterol (Section 25-6Ad). HMG-CoA reductase's ER membrane-bound N-terminal domain contains eight TM helices, whereas its C-terminal domain, which contains its active site and is linked to the N-terminal domain via a flexible Pro-rich sequence, projects into the cytosol. Insig binds to an enzymatic complex that marks proteins for degradation by covalently linking them to the protein **ubiquitin** (Section 32-6B). When lanosterol accumulates in the ER membrane, the N-terminal domain of HMG-CoA reductase also binds to Insig, and is thus marked for destruction. Consequently, the >12-hour half-life of HMG-CoA reductase in sterol-deprived cells decreases to <1 hour when sterols are plentiful.

c. Regulation of HMG-CoA Reductase by Covalent Modification Is a Means of Cellular Energy Conservation

HMG-CoA reductase exists in interconvertible more active and less active forms, as do glycogen phosphorylase (Section 18-3Ca), glycogen synthase (Section 18-3D), pyruvate dehydrogenase (Section 21-2Cb), and acetyl-CoA carboxylase (Section 25-4Ba), among others. The unmodified form of HMG-CoA reductase is more active and the phosphorylated form is less active. HMG-CoA reductase is phosphorylated (inactivated) at its Ser 871 in a bicyclic cascade system by the covalently modifiable enzyme AMP-dependent protein kinase (AMPK), which, as we saw in Section 25-4Ba, also acts on acetyl-CoA carboxylase [in this context, this enzyme was originally named **HMG-CoA reductase kinase (RK),** until it was found to be identical to AMPK]. It appears that this control is exerted to conserve energy when ATP levels fall and AMP levels rise, by inhibiting biosynthetic pathways. This hypothesis was tested by Brown and Goldstein, who used genetic engineering techniques to produce hamster cells containing a mutant HMG-CoA reductase with Ala replacing Ser 871 and therefore incapable of phosphorylation control. These cells respond normally to feedback regulation of cholesterol biosynthesis by LDL–cholesterol and mevalonate but, un-

like normal cells, do not decrease their synthesis of cholesterol on ATP depletion, supporting the idea that control of HMG-CoA reductase by phosphorylation is involved in energy conservation.

d. LDL Receptor Activity Controls Cholesterol Homeostasis

LDL receptors clearly play an important role in the maintenance of plasma LDL–cholesterol levels. In normal individuals, about half of the IDL formed from the VLDL reenters the liver through LDL receptor-mediated endocytosis (IDL and LDL both contain apolipoproteins that specifically bind to the LDL receptor; Section 12-5Bc). The remaining IDL are converted to LDL (Fig. 25-62a). *The serum concentration of LDL therefore depends on the rate at which liver removes IDL from the circulation, which, in turn, depends on the number of functioning LDL receptors on the liver cell surface.*

High blood cholesterol **(hypercholesterolemia),** which results from the overproduction and/or underutilization of LDL, is known to be caused by either of two metabolic irregularities: (1) the genetic disease **familial hypercholesterolemia (FH)** or (2) the consumption of a high-cholesterol diet. FH is a dominant genetic defect that results in a deficiency of functional LDL receptors (Section 12-5Ca). Homozygotes for this disorder lack functional LDL receptors, so their cells can absorb neither IDL nor LDL by receptor-mediated endocytosis. The increased concentration of IDL in the bloodstream leads to a corresponding increase in LDL, which is, of course, underutilized since it cannot be taken up by the cells (Fig. 25-62b). FH homozygotes therefore have plasma LDL–cholesterol levels three to five times higher than average. FH heterozygotes, which are far more common, have about half of the normal number of functional LDL receptors and plasma LDL–cholesterol levels about twice the average.

The long-term ingestion of a high-cholesterol diet has an effect similar, although not as extreme, as FH (Fig. 25-62c). Excessive dietary cholesterol enters the liver cells in chylomicron remnants and represses the synthesis of LDL receptor protein. The resulting insufficiency of LDL receptors on the liver cell surface has consequences similar to those of FH.

LDL receptor deficiency, whether of genetic or dietary origin, raises the LDL level by two mechanisms: (1) increased LDL production resulting from decreased IDL uptake and (2) decreased LDL uptake. Two strategies for reversing these conditions (besides maintaining a low-cholesterol diet) are being used in humans:

1. *Ingestion of anion exchange resins (Section 6-3A) that bind bile salts, thereby preventing their intestinal absorption* (resins are insoluble in water). Bile salts, which are derived from cholesterol, are normally efficiently recycled by the liver (Section 25-6C). Elimination of resin-bound bile salts in the feces forces the liver to convert more cholesterol to bile salts than otherwise. The consequent decrease in the serum cholesterol concentration induces synthesis of LDL receptors (of course, not in FH homozygotes). Unfortunately, the

(a) **Normal**

(b) **Familial hypercholesterolemia**

(c) **High cholesterol diet**

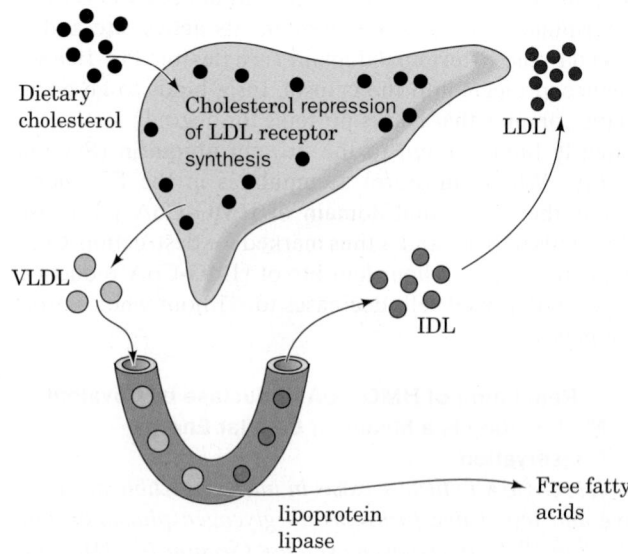

Figure 25-62 **Control of plasma LDL production and uptake by liver LDL receptors.** (*a*) In normal human subjects, VLDL is secreted by the liver and converted to IDL in the capillaries of the peripheral tissues. About half of the plasma IDL particles bind to the LDL receptor and are taken up by the liver. The remainder are converted to LDL at the peripheral tissues. (*b*) In individuals with familial hypercholesterolemia (FH), liver LDL receptors are diminished or eliminated because of a genetic defect. (*c*) In normal individuals who have a long-term high-cholesterol diet, the liver is filled with cholesterol, which represses the rate of LDL receptor production. Receptor deficiency, whether of genetic or dietary cause, raises the plasma LDL level by increasing the rate of LDL production and decreasing the rate of LDL uptake. [After Goldstein, J.L. and Brown, M.S., *J. Lipid Res.* **25,** 1457 (1984).]

decreased serum cholesterol level also induces the synthesis of HMG-CoA reductase, which increases the rate of cholesterol biosynthesis. Ingestion of bile salt–binding resins such as **cholestyramine** (sold as **Questran**) therefore provides only a 15 to 20% drop in serum cholesterol levels.

2. *Treatment with competitive inhibitors of HMG-CoA reductase.* These include (Fig. 25-63) the fungal derivatives **lovastatin** (also called **mevinolin** and sold as **Mevacor**), **pravastatin (Pravachol),** and **simvastatin (Zocor)** as well as the synthetic inhibitor **atorvastatin (Lipitor),** compounds

X = H   R = CH₃   **Lovastatin (Mevacor)**
X = H   R = OH   **Pravastatin (Pravachol)**
X = CH₃   R = CH₃   **Simvastatin (Zocor)**

Atorvastatin (Lipitor)

HMG-CoA

Mevalonate

Figure 25-63 **Some competitive inhibitors of HMG-CoA reductase used for the treatment of hypercholesterolemia.** The molecular formulas of lovastatin (Mevacor), pravastatin (Pravachol), simvastatin (Zocor), and atorvastatin (Lipitor), all of which are potent competitive inhibitors of HMG-CoA reductase, are given. The structures of HMG-CoA and mevalonate are shown for comparison. Note that lovastatin, pravastatin, and simvastatin are lactones, whereas atorvastatin and mevalonate are hydroxy acids. The lactones are hydrolyzed enzymatically *in vivo* to their active hydroxy-acid forms.

that are collectively known as **statins.** Indeed, Lipitor is presently one of the most widely prescribed drugs in the United States. The initial decreased cholesterol supply in the cell caused by the presence of statins is again met by induction of LDL receptors and HMG-CoA reductase so that, at the new steady state, the HMG-CoA reductase level is almost that of the predrug state. However, the increased number of LDL receptors causes increased removal of both LDL and IDL (the apoB-containing precursor to LDL), decreasing serum LDL levels appreciably. Lipitor-treated FH heterozygotes routinely show a serum cholesterol decrease of 40–50%.

The combined use of these agents, moreover, results in a clinically dramatic 50 to 60% decrease in serum cholesterol levels.

e. Overexpression of LDL Receptor Prevents Diet-Induced Hypercholesterolemia

Experiments are well underway toward the treatment of hypercholesterolemic individuals by **gene therapy**

(Section 5-5Hb). A line of transgenic mice has been developed that overproduce the human LDL receptor. When fed a diet high in cholesterol, fat, and bile salts, these transgenic animals did not develop a detectable increase in plasma LDL. In contrast, normal mice fed the same diet exhibited large increases in plasma LDL levels. Evidently, the unregulated overexpression of LDL receptors can prevent diet-induced hypercholesterolemia, at least in mice.

C. *Cholesterol Utilization*

Cholesterol is the precursor of steroid hormones and bile salts. Steroid hormones, which are grouped into five categories, **progestins, glucocorticoids, mineralocorticoids, androgens,** and **estrogens,** mediate a wide variety of vital physiological functions (Section 19-1G). All contain the four-ring structure of the sterol nucleus and are remarkably similar in structure, considering the enormous differences in their physiological effects. A simplified

Figure 25-64 **Simplified scheme of steroid biosynthesis.** The enzymes involved are (**1**) the cholesterol side chain cleavage enzyme, (**2**) steroid C17 hydroxylase, (**3**) steroid C17, C20 lyase,
(**4**) steroid C21 hydroxylase, (**5**) steroid 11β-hydroxylase, (**6**) steroid C18 hydroxylase, (**7**) 18-hydroxysteroid oxidase, and (**8**) aromatase.

biosynthetic scheme (Fig. 25-64) indicates their structural similarities and differences. We shall not discuss the details of these pathways.

*The quantitatively most important pathway for the excretion of cholesterol in mammals is the formation of **bile acids.*** The major bile acids, **cholate** and **chenodeoxycholate,**

	$R_1 = OH$	$R_1 = H$
$R_2 = OH$	**Cholic acid**	**Chenodeoxycholic acid**
$R_2 = NH-CH_2-COOH$	**Glycocholic acid**	**Glycochenodeoxycholic acid**
$R_2 = NH-CH_2-CH_2-SO_3H$	**Taurocholic acid**	**Taurochenodeoxycholic acid**

Figure 25-65   **Structures of the major bile acids and their glycine and taurine conjugates.**

are synthesized in the liver and secreted as their glycine or **taurine** conjugates (Fig. 25-65), which are known as bile salts, into the gallbladder. From there, they are secreted into the small intestine, where they act as emulsifying agents in the digestion and absorption of fats and fat-soluble vitamins (Section 25-1). An efficient recycling system allows the bile salts to reenter the bloodstream and return to the liver for reuse several times each day. The $<1 \ g \cdot day^{-1}$ of bile salts that normally escape this recycling system are further metabolized by microorganisms in the large intestine and excreted. *This is the body's only route for cholesterol excretion.*

Comparison of the structures of cholesterol and the bile acids (Figs. 25-44 and 25-65) indicates that biosynthesis of bile acids from cholesterol involves (1) saturation of the 5,6 double bond, (2) epimerization of the 3β-OH group, (3) introduction of OH groups into the 7α and 12α positions, (4) oxidation of C24 to a carboxylic acid, and (5) conjugation of this side chain carboxylic acid with glycine or taurine. **Cholesterol 7α-hydroxylase** catalyzes the first and rate-limiting step in bile acid synthesis and is closely regulated.

7 EICOSANOID METABOLISM: PROSTAGLANDINS, PROSTACYCLINS, THROMBOXANES, LEUKOTRIENES, AND LIPOXINS

Prostaglandins (PGs) were first identified in human semen by Ulf von Euler in the early 1930s through their ability to stimulate uterine contractions and lower blood pressure. von Euler thought that these compounds originated in the prostate gland (hence their name) but they were later shown to be synthesized in the seminal vesicles. By the time the mistake was realized, the name was firmly entrenched. In the mid-1950s, crystalline materials were isolated from biological fluids and called PGE (ether-soluble) and PGF

(phosphate buffer–soluble; *fosfat* in Swedish). This began an explosion of research on these potent substances.

Almost all mammalian cells except red blood cells produce prostaglandins and their related compounds, the **prostacyclins, thromboxanes, leukotrienes** *and* **lipoxins,** *known collectively as* **eicosanoids** *since they are all C_{20} compounds; Greek: eikosi, twenty). The eicosanoids, like endocrine hormones, have profound physiological effects at extremely low concentrations.* For example, they mediate the following: (1) the inflammatory response, notably as it involves the joints (rheumatoid arthritis), skin (psoriasis), and eyes; (2) the production of pain and fever; (3) the regulation of blood pressure; (4) the induction of blood clotting; (5) the control of several reproductive functions such as the induction of labor; and (6) the regulation of the sleep/wake cycle. The enzymes that synthesize these compounds and the receptors to which they bind are therefore the targets of intensive pharmacological research.

The eicosanoids are also hormonelike in that they bind to G-protein-coupled receptors (Section 19-2B), and many of their effects are intracellularly mediated by cAMP. Unlike endocrine hormones, however, they are not transported in the bloodstream to their sites of action. Rather, these chemically and biologically unstable substances (some decompose within minutes or less *in vitro*) are local mediators (paracrine hormones; Section 19-1); that is, *they act in the same environment in which they are synthesized.*

In this section, we discuss the structures of the eicosanoids and outline their biosynthetic pathways and modes of action. As we do so, note the great diversity of their structures and functions, a phenomenon that makes the elucidation of the physiological roles of these potent substances a challenging research area.

A. Background

Prostaglandins are all derivatives of the hypothetical C_{20} fatty acid **prostanoic acid** *in which carbon atoms 8 to 12*

(a)

Prostanoic acid

(b)

(c)

PGE₁

PGE₂

PGF₂ₐ

Figure 25-66 **Prostaglandin structures.** (*a*) The carbon skeleton of prostanoic acid, the prostaglandin parent compound. (*b*) Structures of prostaglandins A through I. (*c*) Structures of prostaglandins E₁, E₂, and F₂ₐ (the first prostaglandins to be identified).

form a cyclopentane ring (Fig. 25-66a). Prostaglandins A through I differ in the substituents on the cyclopentane ring (Fig. 25-66b): **PGAs** are α,β-unsaturated ketones, **PGEs** are β-hydroxy ketones, **PGFs** are 1,3-diols, etc. In **PGF**$_\alpha$, the C9 OH group is on the same side of the ring as R₁; it is on the opposite side in **PGF**$_\beta$. The numerical subscript in the name refers to the number of double bonds contained on the side chains of the cyclopentane ring (Fig. 25-66c).

In humans, the most prevalent prostaglandin precursor is **arachidonic acid (5,8,11,14-eicosatetraenoic acid), a C₂₀** *polyunsaturated fatty acid that has four nonconjugated double bonds.* The double bond at C14 is six carbon atoms from the terminal carbon atom (the ω carbon atom), making arachidonic acid an ω–6 fatty acid. As Sune Bergström and Bengt Samuelsson demonstrated, arachidonic acid is syn-

thesized from the essential fatty acid linoleic acid (also an ω–6 fatty acid). This occurs via its desaturation with a Δ^6-desaturase to yield **γ-linolenic acid (GLA),** followed by elongation and a second desaturation, this time with a Δ^5-desaturase (Fig. 25-67; Section 25-4E). Prostaglandins with the subscript 1 (the "series-1" prostaglandins) are synthesized from **dihomo-γ-linolenic acid (DGLA; 8,11,14-eicosatrienoic acid),** whereas "series-2" prostaglandins are synthesized from arachidonic acid. α-Linolenic acid (ALA), another essential fatty acid since the Δ^{15}-desaturase required for its synthesis occurs only in plants, is a precursor of 5,8,11,14,17-eicosapentaenoic acid (EPA) and the "series-3" prostaglandins. Since arachidonate is the primary prostaglandin precursor in humans, we shall mostly refer to the series-2 prostaglandins in our examples. Note, however, that when dietary linoleic acid and α-linolenic acid are equally available, the relative activities of the Δ^5- and Δ^6-desaturases are important in determining the relative amounts of these prostaglandin precursors.

a. Arachidonate Is Generated by Phospholipid Hydrolysis

Arachidonate is stored in cell membranes esterified at glycerol C2 of phosphatidylinositol and other phospholipids. The production of arachidonate metabolites is controlled by the rate of arachidonate release from these phospholipids through three alternative pathways (Fig. 25-68):

1. Phospholipase A₂ hydrolyzes acyl groups at C2 of phospholipids (Fig. 25-68b, left).

2. Phospholipase C (Section 19-4B) specifically hydrolyzes the phosphatidylinositol head group to yield a **1,2-diacylglycerol (DAG)** and **phosphoinositol.** DAG is phosphorylated by **diacylglycerol kinase** to phosphatidic acid, a phospholipase A₂ substrate (Fig. 25-68b, center). (Recall that DAG and the various phosphorylated forms of phosphoinositol are also important signaling molecules in that they mediate the phosphoinositide cascade; Section 19-4.)

3. The DAG also may be hydrolyzed directly by **diacylglycerol lipase** (Fig. 25-68b, right). Corticosteroids are used as anti-inflammatory agents because they inhibit phospholipase A₂, reducing the rate of arachidonate production.

b. Aspirin Inhibits Prostaglandin Synthesis

The use of **aspirin** as an analgesic (pain-relieving), antipyretic (fever-reducing), and anti-inflammatory agent has been widespread since the nineteenth century. Yet, it was not until 1971 that John Vane discovered its mechanism of action. *Aspirin, as do other* **nonsteroidal anti-inflammatory drugs (NSAIDs),** *inhibits the synthesis of prostaglandins from eicosanoid precursors (Section 25-7Ba).* These inhibitors have therefore proved to be valuable tools in the elucidation of prostaglandin biosynthesis pathways and have provided a starting point for the rational synthesis of new anti-inflammatory drugs.

Figure 25-67   Synthesis of prostaglandin precursors. The linoleic acid derivatives dihomoGLA (DGLA), arachidonic acid, and 5,8,11,14,17-eicosapentaenoic acid (EPA) are the respective precursors of the series-1, series-2, and series-3 prostaglandins.

c. Arachidonic Acid Is a Precursor of Leukotrienes, Thromboxanes, and Prostacyclins

Arachidonic acid also serves as a precursor to compounds whose synthesis is not inhibited by aspirin. In fact, there are two main pathways of eicosanoid metabolism. The so-called cyclic pathway, which is inhibited by NSAIDs, forms prostaglandin's characteristic cyclopentane ring, whereas the so-called linear pathway, which is

Figure 25-68   Release of arachidonic acid by phospholipid hydrolysis. (*a*) The sites of hydrolytic cleavage mediated by phospholipases A$_2$ and C. The polar head group, X, is often inositol and its various phosphorylated forms (Section 19-4D). (*b*) Pathways of arachidonic acid liberation from phospholipids.

Figure 25-69 The cyclic and linear pathways of arachidonic acid metabolism.

COOH

Arachidonic acid

PGH synthase ● ← (Aspirin inhibits)

COOH

PGI₂ (unstable)

prostacyclin synthase

thromboxane synthase

TxA₂ (unstable)

HO   OH

COOH

Prostacyclins

PGH₂

OH

Thromboxanes

HO   O   OH

TxB₂

HO   OH

6-Oxo-PGF₁ₐ

COOH

COOH

COOH

HO   OH
PGE₂

HO   OH
PGF₂ₐ

O   OH
PGD₂

Prostaglandins

Figure 25-70 The cyclic pathway of arachidonic acid metabolism. This pathway's branches lead
to prostaglandins, prostacyclins, and thromboxanes.

not inhibited by these agents, leads to the formation of the **leukotrienes** and **HPETEs** (Fig. 25-69; Section 25-7C).

Studies using NSAIDs helped demonstrate that two structurally related and highly short-lived classes of compounds, the prostacyclins and the thromboxanes (Fig. 25-70), are also products of the cyclic pathway of eicosanoid metabolism. The specific products produced by this branched pathway depend on the tissue involved. For example, blood platelets (thrombocytes) produce thromboxanes almost exclusively; vascular endothelial cells, which make up the walls of veins and arteries, predominantly synthesize the prostacyclins; and heart muscle makes PGI_2, PGE_2, and $PGF_{2\alpha}$ in more or less equal quantities. In the remainder of this section, we study the cyclic and the linear pathways of eicosanoid metabolism.

B. The Cyclic Pathway of Eicosanoid Metabolism: Prostaglandins, Prostacyclins, and Thromboxanes

The first step in the cyclic pathway of eicosanoid metabolism is catalyzed by **PGH synthase** (**PGHS;** also called **prostaglandin H synthase** and **prostaglandin endoperoxide synthase;** Fig. 25-71). This heme-containing enzyme contains two catalytic activities: a cyclooxygenase activity and a peroxidase activity. The former catalyzes the tyrosyl radical-mediated addition of two molecules of O_2 to arachidonic acid, forming **PGG_2.** The latter converts the hydroperoxy function of PGG_2 to an OH group, yielding **PGH_2.** *PGH$_2$ is the immediate precursor of all series-2 prostaglandins, prostacyclins, and thromboxanes (Fig. 25-70).* The cyclooxygenase activity of the enzyme gives it its common name, **COX** [not to be confused with cytochrome *c* oxidase, which is also called COX (Section 22-2C5)].

PGHS, a homodimeric glycoprotein of 576-residue subunits, is a monotopic membrane protein that extends into the lumen of the endoplasmic reticulum. Its X-ray structure, determined by Michael Garavito, reveals that each of its subunits folds into three domains (Fig. 25-72*a*): an N-terminal module that structurally resembles **epidermal growth factor** (**EGF;** a hormonally active polypeptide that stimulates cell proliferation; Section 12-3Ae); a central membrane-binding motif; and a C-terminal enzymatic domain. The 44-residue membrane-binding motif has a hydrophobic surface that faces away from the body of the protein [as is also true of oxidosqualene cyclase (Fig. 25-58*b*) and fatty acid amide hydrolase (Fig. 12-28)].

The peroxidase active site of PGHS occurs at the interface between the large and small lobes of the catalytic domain, in a shallow cleft that contains the enzyme's Fe(III)–heme prosthetic group. The cleft exposes a large portion of the heme to solvent and is therefore thought to comprise the substrate binding site.

The cyclooxygenase active site lies on the opposite side of the heme at the end of a long narrow hydrophobic channel (~8 × 25 Å) extending from the outer surface of the membrane-binding motif to the center of each subunit (Fig. 25-72*b*). This channel allows access of the membrane-associated substrate to the active site. Tyr 385, which lies near the top of the channel, just beneath the heme, has

Figure 25-71 Reactions catalyzed by PGH synthase (PGHS). The enzyme contains two activities: a cyclooxygenase, which catalyzes Steps 1 to 3 and is inhibited by aspirin; and a peroxidase, which catalyzes Step 4. **(1)** A radical at Tyr 385 that is generated by the enzyme's heme cofactor stereospecifically abstracts a hydrogen atom from C13 of arachidonic acid, which then rearranges so that the radical is on C11. **(2)** The radical reacts with O_2 to yield a hydroperoxide radical. **(3)** The radical cyclizes and reacts with a second O_2 molecule at C15 to yield a peroxide in a process that regenerates the Tyr radical. **(4)** The enzyme's peroxidase activity converts the peroxide at C15 to a hydroxyl group.

been shown to form a transient radical during the cyclooxygenase reaction as does, for example, Tyr 244 of cytochrome *c* oxidase (Section 22-2C5c). Indeed, the mutagenic replacement of PGHS's Tyr 385 by Phe abolishes its cyclooxygenase activity. The Tyr 385 radical is generated via an intramolecular oxidation by the heme cofactor.

ER lumen

(a)

(b)

Figure 25-72 **X-ray structure of PGH synthase (PGHS) from sheep seminal vesicles in complex with the NSAID flurbiprofen.** (*a*) This homodimeric monotopic membrane protein is viewed parallel to the plane of the ER membrane with its 2-fold axis of symmetry vertical. The EGF-like module is green, the membrane-binding motif is orange, and the catalytic domain is blue. The heme (*red*); flurbiprofen (*yellow*); Tyr 385 (*magenta*), which forms a transient radical during the cyclooxygenase reaction; and Arg 120 (*green*), which forms an ion pair with flurbiprofen, are drawn in space-filling form. (*b*) A C_α diagram of a PGHS subunit (*green*), the left subunit in Part *a* as viewed from 30° to the left. The peroxidase active site is located above the heme (*pink*). The hydrophobic channel, which penetrates the subunit from the membrane-binding motif at the bottom of the figure to the cyclooxygenase active site below the heme, is represented by its van der Waals surface (*blue dots*). The three residues in the channel that are shown in stick form in orange are, from top to bottom: Tyr 385, Ser 530, which is acetylated by aspirin, and Arg 120. [Courtesy of Michael Garavito, Michigan State University. PDBid 1CQE.]

The fate of PGH_2 depends on the relative activities of the enzymes catalyzing the specific interconversions (Fig. 25-70). Platelets contain **thromboxane synthase,** which mediates the formation of **thromboxane A_2 (TxA$_2$),** a vasoconstrictor and stimulator of platelet aggregation (an initial step in blood clotting; Section 35-1). Vascular endothelial cells contain **prostacyclin synthase,** which catalyzes the synthesis of **prostacyclin I_2 (PGI$_2$),** a vasodilator and inhibitor of platelet aggregation. These two substances act in opposition, maintaining a balance in the cardiovascular system.

a. NSAIDs Inhibit PGH Synthase

Nonsteroidal anti-inflammatory drugs (NSAIDs; Fig. 25-73) inhibit the synthesis of the prostaglandins, prostacyclins, and thromboxanes by inhibiting or inactivating the cyclooxygenase activity of PGHS. Aspirin **(acetylsalicylic acid),** for example, acetylates this enzyme: If [^{14}C-*acetyl*] acetylsalicylic acid is incubated with the enzyme, radioactivity becomes irreversibly associated with the inactive enzyme as Ser 530 becomes acetylated (Fig. 25-74). The X-ray structure of PGHS reveals that Ser 530, which is not implicated in catalysis, extends into the cyclooxygenase channel just below Tyr 385 such that its acetylation would block arachidonic acid's access to the active site (Fig. 25-72*b*). The structure of PGHS, which was crystallized with the NSAID **flurbiprofen** (Fig. 25-73), indicates that this drug binds in the

cyclooxygenase channel, with its carboxyl group forming an ion pair with Arg 120 (Fig. 25-72*a*). Evidently, flurbiprofen, and by implication other NSAIDs, inhibits the cyclooxygenase activity of PGHS by blocking its active site channel.

Low doses of aspirin, ~80 mg (baby aspirin) every day, significantly reduce the long-term incidence of heart attacks and strokes. Such low doses selectively inhibit platelet aggregation and thus blood clot formation because these enucleated cells, which have a lifetime in the circulation of ~10 days, cannot resynthesize their inactivated enzymes. Vascular endothelial cells are not so drastically affected since, for the most part, they are far from the site where aspirin is absorbed, are exposed to lesser concentrations of aspirin and, in any case, can synthesize additional PGHS.

b. COX-2 Inhibitors Lack the Side Effects of Other NSAIDs

PGHS has two isoforms, **COX-1** and **COX-2,** that share a high degree (60%) of sequence identity and structural homology. COX-1 is constitutively (without regulation) expressed in most, if not all, mammalian tissues, thereby supporting levels of prostaglandin synthesis necessary to maintain organ and tissue homeostasis such as that of the gastrointestinal mucosa. In contrast, COX-2 is only expressed in certain tissues in response to inflammatory stim-

**Aspirin
(acetylsalicylic acid)**

Indomethacin

Ibuprofen

Flurbiprofen

Acetaminophen

Naproxen

Phenylbutazone

Figure 25-73   Some nonsteroidal anti-inflammatory drugs (NSAIDs).

uli such as cytokines, protein growth factors, and endotoxins, and hence is responsible for the elevated prostaglandin levels that cause inflammation. The NSAIDs in Fig. 25-73 are relatively nonspecific and therefore can have adverse side effects, most notably gastrointestinal ulceration, when used to treat inflammation or fever. A structure-based drug design program (Section 15-4Ad) was therefore instituted to create inhibitors that would target COX-2 but not COX-1. The three-dimensional structures of COX-1 and COX-2 are almost identical. However, their amino acid differences, specifically I523V, I434V, and H513R (COX-1 amino acid on the left and COX-2 amino acid on the right), make COX-2's active site channel ~20% larger in volume

than that of COX-1. In addition, the fourth helix of the membrane-binding domain is oriented slightly differently so as to provide a larger opening to the channel. Medicinal chemists therefore synthesized inhibitors, collectively known as **coxibs,** that could enter the COX-2 channel but are excluded from that of COX-1. Two of these inhibitors, **rofecoxib (Vioxx)** and **celecoxib (Celebrex;** Fig. 25-75), became major drugs for the treatment of inflammatory diseases such as arthritis because they lack the major side

Rofecoxib (Vioxx)

Celecoxib (Celebrex)

Figure 25-75   COX-2 inhibitors. Rofecoxib and celecoxib are specific inhibitors of COX-2 **(PGH synthase-2).**

Aspirin

PGH synthase
(*active*)

**Salicylic
acid**

PGH synthase
(*inactive*)

Figure 25-74   Inactivation of PGH synthase by aspirin. Aspirin acetylates Ser 530 of PGH synthase, thereby blocking the enzyme's cyclooxygenase activity.

effects of the nonspecific NSAIDs. However, in 2004, Vioxx was withdrawn from the pharmaceutical market because of unanticipated cardiac side effects arising from its attenuation of PGI$_2$ formation.

c. COX-3 May Be the Target of Acetaminophen

Acetaminophen, which is among the most widely used analgesic/antipyretic drugs (but possesses little anti-inflammatory activity, so that it is not really an NSAID), does not significantly bind to either COX-1 or COX-2. Thus its mechanism of action remained a mystery until the discovery by Daniel Simmons of a third COX isozyme, **COX-3,** a splice variant of COX-1, that is selectively inhibited by acetaminophen as well as by certain NSAIDs. This suggests that COX-3 is the primary target of drugs that decrease pain and fever.

C. *The Linear Pathway of Eicosanoid Metabolism: Leukotrienes and Lipoxins*

Arachidonic acid can be converted by a linear pathway to several different **hydroperoxyeicosatetraenoic acids (HPETEs)** by the **5-, 12-,** and **15-lipoxygenases (5-, 12-,** and **15-LOs;** Fig. 25-69). **Hepoxilins** are hydroxy epoxy derivatives of **12-HPETE** whose functions are not as yet well understood. **Lipoxins,** the products of a second lipoxygenase acting on **15-HPETE,** are anti-inflammatory substances. Leukotrienes, derived from the 5-LO reaction, are synthesized by a variety of white blood cells, mast cells (connective tissue cells derived from the blood-forming tissues that secrete substances which mediate inflammatory and allergic reactions), as well as lung, spleen, brain, and heart. **Peptidoleukotrienes (LTC$_4$, LTD$_4$,** and **LTE$_4$)** are now recognized to be the components of the **slow reacting substances of anaphylaxis (SRS-A;** anaphalaxis is a violent and potentially fatal allergic reaction) released from sensitized lung after immunological challenge. These substances act at very low concentrations (as little as 10^{-10} *M*) to contract vascular, respiratory, and intestinal smooth muscle. Peptidoleukotrienes, for example, are ~10,000-fold more potent than histamine, a well-known stimulant of allergic reactions. In the respiratory system, they constrict bronchi, especially the smaller airways; increase mucus secretion; and are thought to be the mediators in asthma. They are also implicated in immediate hypersensitivity (allergic) reactions, inflammatory reactions, and heart attacks.

a. Leukotriene Synthesis

The first two reactions in the conversion of arachidonic acid to leukotrienes are both catalyzed by 5-LO, which contains a nonheme, non-[Fe–S] cluster iron atom that must be in its Fe(III) state to be active. These reactions occur as follows (Fig. 25-76):

1. The oxidation of arachidonic acid to form 5-HPETE, a substance that, in itself, is not a physiological mediator. This reaction occurs in three steps:

(a) The active site iron atom, in its active Fe(III) state,

Figure 25-76 The 5-LO-catalyzed oxidation of arachidonic acid to LTA$_4$ via the intermediate 5-HPETE.

abstracts an electron from the central methylene group of the 5,8-pentadiene moiety of arachidonate and the resulting free radical loses a proton to an enzymatic base.

(b) The free radical rearranges and adds O$_2$ to form a hydroperoxide radical.

(c) The hydroperoxide radical reacts with the active site iron, now in its Fe(II) form, to yield the hydroperoxide in its anionic form, which the enzyme then protonates to yield the hydroperoxide product, regenerating the active Fe(III) enzyme.

2. The base-catalyzed elimination of water to form the unstable epoxide **leukotriene A$_4$ (LTA$_4$;** the subscript indicates the number of carbon–carbon double bonds in the molecule, which is also its series number).

The X-ray structure of the rabbit reticulocyte 15-LO, a homolog of 5-LO, in complex with the competitive inhibitor **RS75091**

COOH

CH₃

RS75091

was determined by Michelle Browner. This 663-residue monomeric protein consists of an N-terminal 8-stranded β barrel domain and a C-terminal catalytic domain (Fig. 25-77). Its active site Fe atom is coordinated by four invariant His residues and by a C-terminal carboxylate oxygen in a liganding arrangement that is best described as a distorted octahedron with one of its six vertices unoccupied. The Fe, which is well below the protein surface, faces an internal cavity occupied by RS75091. This identifies the substrate binding cavity, which is lined with mostly hydrophobic residues and follows an irregular pathway past the Fe atom to the protein surface. Intriguingly, 15-LO (as does soybean **lipoxygenase-1**) contains two rarely observed π helices (Fig. 8-14*c*), each of which contains two of the Fe-liganding His residues. Each of these π helices is embedded in a longer helix rather than being at the end of an α helix as is the case for all previously observed π helices.

The sizes of the substrate-binding cavities of the 5- and 12-LOs have been predicted through their homology modeling (Section 9-3B) with 15-LO. 5-LO and 12-LO have smaller amino acids substituted for those in 15-LO, such that, for example, 5-LO is predicted to have a cavity with ~20% greater volume than that of 15-LO. The mutagenesis of 5-LO by Harmut Kuhn so as to decrease the size of its cavity yielded an enzyme with the specificity of 15-LO, thus supporting the proposal that it is the size of the cavity that determines lipoxygenase specificity.

b. Peptidoleukotrienes

LTA₄ is converted to peptidoleukotrienes by reaction with **LTC₄ synthase**, a **glutathione-*S*-transferase** that catalyzes the addition of the glutathione sulfhydryl group to the LTA₄ epoxide, forming the first of the peptidoleukotrienes, **leukotriene C₄ (LTC₄,** Fig. 25-78). **γ-Glutamyltransferase** removes glutamic acid, converting LTC₄ to **leukotriene D₄ (LTD₄).** LTD₄ is converted to **leukotriene E₄ (LTE₄)** by a dipeptidase that removes glycine. LTA₄ can also be hydrolyzed to **leukotriene B₄ (LTB₄),** a potent chemotactic agent (a substance that attracts motile cells) involved in attracting certain types of white blood cells to fight infection.

Various inflammatory and hypersensitivity disorders (such as asthma) are associated with elevated levels of leukotrienes. The development of drugs that inhibit leukotriene synthesis has therefore been an active field of

Figure 25-77 **X-ray structure of rabbit reticulocyte 15-lipoxygenase (15-LO) in complex with its competitive inhibitor RS75091.** The N-terminal β barrel domain is gold and the C-terminal catalytic domain is light blue with its two Fe-liganding π helical segments magenta. The Fe is represented by an orange sphere and the RS75091 is drawn in space-filling form with C green and O red. [Based on an X-ray structure by Michelle Browner, Roche Bioscience, Palo Alto, California. PDBid 1LOX.]

research. 5-LO activity requires the presence of **5-lipoxygenase-activating protein (FLAP),** a homotrimeric integral membrane protein of 161-residue subunits that is located in both the nuclear and ER membranes. FLAP, which has no enzymatic activity, binds the arachidonic acid substrate of 5-LO and facilitates enzyme–substrate binding as well as 5-LO's interaction with the membrane. Several inhibitors of leukotriene synthesis, such as **MK-591,**

S

N

HO

O

Cl

MK-591

bind to FLAP so as to inhibit both of its functions.

Leukotriene A₄ (LTA₄)

H_2O → leukotriene A₄ hydrolase → **Leukotriene B₄ (LTB₄)**

Glutathione → LTC₄ synthase (a glutathione-S-transferase)

Leukotriene C₄ (LTC₄)

γ-glutamyltransferase → $HOOC—CH_2—CH_2—CH—COOH$ with NH_2

Glutamic acid

Leukotriene D₄ (LTD₄)

$H_2N—CH_2—COOH$ ← dipeptidase

Glycine

Leukotriene E₄ (LTE₄)

Figure 25-78 Formation of the leukotrienes from LTA₄.

Figure 25-79 X-ray structure of human 5-lipoxygenase-activating protein (FLAP) in complex with MK-591. The protein is drawn in ribbon form and viewed from within the membrane with its approximate molecular 3-fold axis vertical. The subunit closest to the viewer of this homotrimeric protein is colored in rainbow order from its N-terminus (*blue*) to its C-terminus (*red*). The other subunits are pale blue and pale orange. The bound MK-591 is drawn in stick form with C magenta, N blue, O red, S yellow, and C green. The inferred position of the membrane is indicated by the horizontal black lines. [Based on an X-ray structure by Joseph Becker, Merck Research Laboratories, Rahway, New Jersey. PDBid 2Q7M.]

The X-ray structure of human FLAP in complex with MK-591 (Fig. 25-79), determined by Joseph Becker, reveals that each subunit forms an up–down–up–down four-helix bundle whose N- and C-termini are on the luminal side of the membrane, and that its three identical subunits have extensive intersubunit contacts, thereby forming a 12-helix bundle. MK-591 binds in lipid-exposed grooves between pairs of subunits. Moreover, the trimer forms an elongated pocket along its 3-fold axis that is open to the lumen. This, together with biochemical data, suggests that one 5-LO molecule binds to the cytosolic surface of a FLAP trimer where, through conformational changes in FLAP, it accepts arachidonic acid molecules that FLAP has extracted from the membrane. Inhibitors such as MK-591 presumably block this extraction process.

c. Diets Rich in Marine Lipids May Decrease Cholesterol, Prostaglandin, and Leukotriene Levels

Greenland Eskimos have a very low incidence of coronary heart disease and thrombosis despite their high dietary intake of cholesterol and fat. Their consumption of marine animals provides them with a higher proportion of unsaturated fats than the typical American diet. A major unsaturated component of marine lipids is 5,8,11,14,17-

eicosapentaenoic acid (EPA; Fig. 25-67), an ω–3 fatty acid, rather than the arachidonic acid precursor linoleic acid, an ω–6 fatty acid. EPA inhibits formation of TxA$_2$ (Fig. 25-70) and is a precursor of the **series-5 leukotrienes,** compounds with substantially lower physiological activities than their arachidonate-derived (series-4) counterparts. This suggests that a diet containing marine lipids should decrease the extent of prostaglandin- and leukotriene-mediated inflammatory responses. Indeed, dietary enrichment with EPA inhibits the *in vitro* chemotactic and aggregating activities of neutrophils (a type of white blood cell). Moreover, an EPA-rich diet decreases the cholesterol and triacylglycerol levels in the plasma of hypertriacylglycerolemic patients.

d. Lipoxins and Aspirin-Induced *epi*-Lipoxins Have Anti-Inflammatory Properties

Eicosanoids are usually associated with the inflammatory response. However, some eicosanoids have anti-inflammatory properties. The **lipoxins (LXs),** products of the 12- and 15-LO pathways (sometimes also involving 5-LO), are so named because they are synthesized through *lipox*ygenase *in*teractions. Their activities appear to inhibit those of leukotrienes and hence they are anti-inflammatory. There are many ways for LXs to be synthesized by combinations of the actions of 5-, 12-, and 15-LOs. Here we discuss only one such route (Fig. 25-80, *left*). **Lipoxin A$_4$ (LXA$_4$)** synthesis from arachidonic acid begins in endothelial and epithelial cells by the 15-LO-catalyzed synthesis of **(15S)-hydroperoxyeicosatetraenoic acid [(15S)-HPETE],** which is reduced by **glutathione peroxidase** to **(15S)-hydroxyeicosatetraenoic acid [(15S)-HETE].** The (15S)-HETE then makes its way to leukocytes where it is converted to LXA$_4$ by 5-LO and a hydrolase.

Charles Serhan discovered an additional pathway for the anti-inflammatory action of aspirin that also involves lipoxin production. As we have previously seen (Fig. 25-74), aspirin covalently inhibits the cyclooxygenase activity of

Figure 25-80   Lipoxin biosynthesis. The biosynthesis of the lipoxin LXA$_4$ (*left*) and the aspirin-triggered *epi*-lipoxin (ATL) 15-*epi*-LXA$_4$ (*right*). In endothelial and epithelial cells, arachidonic acid is converted by 15-LO and glutathione peroxidase to (15S)-HETE or by aspirin-acetylated COX-2 to (15R)-HETE. After transfer to leukocytes, 5-LO and a hydrolase convert these intermediate products to LXA$_4$ and 15-*epi*-LXA$_4$.

PGHS (COX). However, aspirin-acetylated COX-2 retains a residual 15-LO activity (Steps 3 and 4 in Fig. 25-71) through which it initiates a pathway that converts arachidonic acid to the anti-inflammatory agents called **aspirin-triggered *epi*-lipoxins (ATLs;** Fig. 25-80, *right*). This pathway begins in endothelial and epithelial cells with the aspirin-acetylated COX-2-catalyzed conversion of arachidonic acid to **(15*R*)-hydroxyeicosatetraenoic acid [(15*R*)-HETE],** the epimer of (15*S*)-HETE. In leukocytes, 5-LO and a hydrolase then convert (15*R*)-HETE to the anti-inflammatory agent **15-*epi*-lipoxin A₄ (15-*epi*-LXA₄).**

These are indeed exciting times in the study of eicosanoid metabolism and its physiological manifestations. As the mechanisms of action of the prostaglandins, prostacyclins, thromboxanes, leukotrienes, and lipoxins are becoming better understood, they are providing the insights required for the development of new and improved therapeutic agents.

8 PHOSPHOLIPID AND GLYCOLIPID METABOLISM

The "complex lipids" are dual-tailed amphipathic molecules composed of either 1,2-diacyl-sn-glycerol or N-acylsphingosine (ceramide) linked to a polar head group that is either a carbohydrate or a phosphate ester (Fig. 25-81; Sections 12-1C and 12-1D; sn stands for stereospecific numbering, which assigns the 1 position to the group occupying the pro-S position of a prochiral center). Hence, there are two categories of phospholipids, **glycerophospholipids** and **sphingophospholipids,** and two categories of glycolipids, **glyceroglycolipids** and **sphingoglycolipids** (also called **glycosphingolipids; GSLs**). In this section we describe the biosynthesis of the complex lipids from their simpler components. We shall see that the great variety of these substances is matched by the numerous enzymes required for their specific syntheses. Note also that these substances are synthesized in membranes, mostly on the cytosolic face of the endoplasmic reticulum, and from there are transported to their final cellular destinations as indicated in Sections 12-4B–D.

A. Glycerophospholipids

Glycerophospholipids have significant asymmetry in their C1- and C2-linked fatty acyl groups: C1 substituents are mostly saturated fatty acids, whereas those at C2 are by and large unsaturated fatty acids. We shall examine the major pathways of biosynthesis and metabolism of the glycerophospholipids with an eye toward understanding the origin of this asymmetry.

a. Biosynthesis of Diacylglycerophospholipids

The triacyglycerol precursors 1,2-diacyl-sn-glycerol and phosphatidic acid are also the precursors of certain glycerophospholipids (Figs. 25-42 and 25-81). Activated phosphate esters of the polar head groups (Table 12-2) react with the C3 ÔH group of 1,2-diacyl-*sn*-glycerol to form the phospholipid's phosphodiester bond. In some cases the phosphoryl group of phosphatidic acid is activated and reacts with the unactivated polar head group.

The mechanism of activated phosphate ester formation is the same for both the polar head groups **ethanolamine** and **choline** (Fig. 25-82):

1. ATP first phosphorylates the OH group of choline or ethanolamine.

2. The phosphoryl group of the resulting **phosphoethanolamine** or **phosphocholine** then attacks CTP, displacing PP$_i$, to form the corresponding CDP derivatives, which are activated phosphate esters of the polar head group.

3. The C3 OH group of 1,2-diacyl-*sn*-glycerol attacks the phosphoryl group of the activated CDP–ethanolamine or CDP–choline, displacing CMP to yield the corresponding glycerophospholipid.

The liver also converts phosphatidylethanolamine to phosphatidylcholine by trimethylating its amino group, using **S-adenosylmethionine** (Section 26-3Ea) as the methyl donor.

Phosphatidylserine is synthesized from phosphatidylethanolamine by a head group exchange reaction catalyzed by **phosphatidylethanolamine:serine transferase** in which serine's OH group attacks the donor's phosphoryl group (Fig. 25-83). The original head group is then eliminated, forming phosphatidylserine.

In the synthesis of **phosphatidylinositol** and **phosphatidylglycerol,** the hydrophobic tail is activated rather than the polar head group. Phosphatidic acid, the precursor of 1,2-diacyl-*sn*-glycerol (Fig. 25-42), attacks the α-phosphoryl group of CTP to form the activated **CDP–**

Glycerolipid	**Sphingolipid**

X = H	**1,2-Diacylglycerol**	**N-Acylsphingosine (ceramide)**
X = Carbohydrate	**Glyceroglycolipid**	**Sphingoglycolipid (glycosphingolipid)**
X = Phosphate ester	**Glycerophospholipid**	**Sphingophospholipid**

Figure 25-81 **The glycerolipids and sphingolipids.** The structures of the common head groups, X, are presented in Table 12-2.

HO—CH$_2$—CH$_2$—NR$_3'^+$

R' = H   **Ethanolamine**
R' = CH$_3$ **Choline**

ethanolamine kinase
or choline kinase | **1** ⌐ ATP
                              └→ ADP

$$\begin{array}{c} O \\ \parallel \\ ^-O-P-O-CH_2-CH_2-NR_3'^+ \\ \mid \\ O^- \end{array}$$

R' = H   **Phosphoethanolamine**
R' = CH$_3$ **Phosphocholine**

CTP:phosphoethanolamine
cytidyltransferase | **2** ⌐ CTP
or CTP:phosphocholine   └→ PP$_i$
cytidyltransferase

$$\text{Cytidine}\begin{array}{cc} O & O \\ \parallel & \parallel \\ -P-O-P- \\ \mid & \mid \\ O^- & O^- \end{array}O-CH_2-CH_2-NR_3'^+$$

R' = H   **CDP–ethanolamine**
R' = CH$_3$ **CDP–choline**

CDP–ethanolamine:1,2-diacylglycerol
phosphoethanolamine transferase | **3** ⌐ 1,2-Diacylglycerol
or CDP–choline:1,2-diacylglycerol   └→ CMP
phosphocholine transferase

$$R_2-\overset{O}{\overset{\parallel}{C}}-O-\overset{\displaystyle CH_2-O-\overset{O}{\overset{\parallel}{C}}-R_1}{\underset{\displaystyle CH_2-O-\overset{O}{\overset{\parallel}{P}}-O-CH_2-CH_2-NR_3'^+}{C-H}}$$

R' = H   **Phosphatidylethanolamine**
R' = CH$_3$ **Phosphatidylcholine (lecithin)**

***Figure 25-82* The biosynthesis of phosphatidylethanolamine and phosphatidylcholine.** In mammals, CDP–ethanolamine and CDP–choline are the precursors of the head groups.

diacylglycerol and PP$_i$ (Fig. 25-84). Phosphatidylinositol results from the attack of inositol on CDP–diacylglycerol. Phosphatidylglycerol is formed in two reactions: (1) attack of the C1 —OH group of *sn*-glycerol-3-phosphate on CDP–diacylglycerol, yielding **phosphatidylglycerol phosphate**; and (2) hydrolysis of the phosphoryl group to form phosphatidylglycerol.

Cardiolipin, an important phospholipid first isolated from heart tissue, is synthesized from two molecules of phosphatidylglycerol (Fig. 25-85). The reaction occurs by the attack of the C1 OH group of one of the phosphatidylglycerol molecules on the phosphoryl group of the other, displacing a molecule of glycerol.

Enzymes that synthesize phosphatidic acid have a general preference for saturated fatty acids at C1 and for unsaturated fatty acids at C2. Yet, this general preference cannot account, for example, for the observations that ~80%

$$R_2-\overset{O}{\overset{\parallel}{C}}-O-\overset{\displaystyle CH_2-O-\overset{O}{\overset{\parallel}{C}}-R_1}{\underset{\displaystyle CH_2-O-\overset{O}{\overset{\parallel}{P}}-O-CH_2CH_2NH_3^+}{CH}}\overset{\mid}{\underset{O^-}{}}$$

Phosphatidylethanolamine

+

HO—CH$_2$—CH—COO$^-$
         |
        NH$_3^+$

Serine

phosphatidylethanolamine:
serine transferase | └→ HO—CH$_2$—CH$_2$—NH$_3^+$

$$R_2-\overset{O}{\overset{\parallel}{C}}-O-\overset{\displaystyle CH_2-O-\overset{O}{\overset{\parallel}{C}}-R_1}{\underset{\displaystyle CH_2-O-\overset{O}{\overset{\parallel}{P}}-O-CH_2-CH-COO^-}{CH}}\overset{}{\underset{O^- \qquad\quad NH_3^+}{}}$$

Phosphatidylserine

***Figure 25-83* Phosphatidylserine synthesis.** Serine replaces ethanolamine in phosphatidylethanolamine by a head group exchange reaction.

of brain phosphatidylinositol has a stearoyl group (18:0) at C1 and an arachidonoyl group (20:4) at C2, and that ~40% of lung phosphatidylcholine has palmitoyl groups (16:0) at both positions (this latter substance is the major component of the surfactant that prevents the lung from collapsing when air is expelled; its deficiency is responsible for **respiratory distress syndrome** in premature infants). William Lands showed *that such side chain specificity results from "remodeling" reactions in which specific acyl groups of individual glycerophospholipids are exchanged by specific phospholipases and acyltransferases.*

b. Biosynthesis of Plasmalogens and Alkylacylglycerophospholipids

Eukaryotic membranes contain significant amounts of two other types of glycerophospholipids:

1. Plasmalogens, which contain a hydrocarbon chain linked to glycerol C1 via a vinyl ether linkage:

$$R_2-\overset{O}{\overset{\parallel}{C}}-O-\overset{\displaystyle CH_2-O-CH=CH-R_1}{\underset{\displaystyle CH_2-O-\overset{O}{\overset{\parallel}{P}}-O-X}{CH}}\overset{}{\underset{O^-}{}}$$

A plasmalogen

Figure 25-84 **The biosynthesis of phosphatidylinositol and phosphatidylglycerol.** In mammals, this process involves a CDP–diacylglycerol intermediate.

Figure 25-85 **The formation of cardiolipin.**

Figure 25-86 **The biosynthesis of ethanolamine plasmalogen via a pathway in which 1-alkyl-2-acyl-*sn*-glycerophosphoethanolamine is an intermediate.** The participating enzymes are **(1)** alkyl-DHAP synthase, **(2)** 1-alkyl-*sn*-glycerol-3-phosphate dehydrogenase, **(3)** acyl-CoA:1-alkyl-*sn*-glycerol-3-phosphate acyltransferase, **(4)** 1-alkyl-2-acyl-*sn*-glycerol-3-phosphate phosphatase, **(5)** CDP–ethanolamine: 1-alkyl-2-acyl-*sn*-glycerophosphoethanolamine transferase, and **(6)** 1-alkyl-2-acyl-*sn*-glycerophosphoethanolamine desaturase.

2. Alkylacylglycerophospholipids, in which the alkyl substituent at glycerol C1 is attached via an ether linkage:

**An alkylacyl-
glycerophospholipid**

About 20% of mammalian glycerophospholipids are plasmalogens. The exact percentage varies both from species to species and from tissue to tissue within a given organism. While plasmalogens comprise only 0.8% of the phospholipids in human liver, they account for 23% of those in human nervous tissue. The alkylacylglycerophospholipids are less abundant than the plasmalogens; for instance, 59% of the ethanolamine glycerophospholipids of human heart are plasmalogens, whereas only 3.6% are alkylacylglycerophospholipids. However, in bovine erythrocytes, 75% of the ethanolamine glycerophospholipids are of the alkylacyl type.

The pathway forming ethanolamine plasmalogens and alkylacylglycerophospholipids involves several reactions (Fig. 25-86):

1. Exchange of the acyl group of **1-acyldihydroxyacetone phosphate** for an alcohol.

2. Reduction of the ketone to **1-alkyl-*sn*-glycerol-3-phosphate.**

3. Acylation of the resulting C2 OH group by acyl-CoA.

4. Hydrolysis of the phosphoryl group to yield an alkylacylglycerol.

5. Attack by the new OH group of alkylacylglycerol on CDP–ethanolamine to yield **1-alkyl-2-acyl-*sn*-glycerophosphoethanolamine.**

6. Introduction of a double bond into the alkyl group to form the plasmalogen by a desaturase having the same cofactor requirements as the fatty acid desaturases (Section 25-4E).

Recall that the precursor–product relationship between the alkylacylglycerophospholipid and the plasmalogen was established through studies using [^{14}C]ethanolamine (Section 16-3Be).

The alkylacylglycerophospholipid with an acetyl group at R$_2$ and a choline polar head group (X), **1-*O*-hexadecyl-2-acetyl-*sn*-glycero-3-phosphocholine,** is known as **platelet-activating factor (PAF).** This molecule has diverse functions and acts at very low concentrations (10^{-10} *M*) to lower blood pressure and to cause blood platelets to aggregate.

B. *Sphingophospholipids*

Only one major phospholipid contains ceramide (*N*-acylsphingosine) as its hydrophobic tail: **sphingomyelin (*N*-acylsphingosine phosphocholine;** Section 12-1D), an important structural lipid of nerve cell membranes. The molecule was once thought to be synthesized from *N*-acylsphingosine and CDP–choline. However, it is now known that the main route of sphingomyelin synthesis occurs through donation of the phosphocholine group of phosphatidylcholine to *N*-acylsphingosine (Fig. 25-87). These pathways were differentiated by establishing the precursor–product relationships between CDP–choline, phosphatidylcholine, and sphingomyelin (Section 16-3Be). Mouse liver microsomes were isolated and incubated for a short time with [^{3}H]choline. Radioactivity appeared in sphingomyelin only after first appearing in both CDP–choline and phosphatidylcholine, ruling out the direct transfer of phosphocholine from CDP–choline to *N*-acylsphingosine.

The most prevalent acyl groups of sphingomyelin are palmitoyl (16:0) and stearoyl (18:0) groups. Longer chain fatty acids such as nervonic acid (24:1) and behenic acid (22:0) occur with lesser frequency in sphingomyelins.

C. *Sphingoglycolipids*

Most sphingolipids are sphingoglycolipids, that is, their polar head groups consist of carbohydrate units (Section 12-1D). The principal classes of sphingoglycolipids, as indicated in Fig. 25-88, are **cerebrosides** (ceramide monosaccharides), **sulfatides** (ceramide monosaccharide sulfates), **globosides** (neutral ceramide oligosaccharides), and **gangliosides** (acidic, sialic acid–containing ceramide oligosaccharides). The carbohydrate unit is glycosidically attached to the *N*-acylsphingosine at its C1 OH group (Fig. 25-81).

The lipids providing the carbohydrate that covers the external surfaces of eukaryotic cells are sphingoglycolipids. Along with glycoproteins (Section 23-3), they are biosynthesized on the luminal surfaces of the endoplasmic reticulum and the Golgi apparatus and reach the plasma membrane through vesicle flow (Sections 12-4C and 12-4D), where membrane fusion results in their facing the external surface of the lipid bilayer (Fig. 12-60). Degradation of sphingoglycolipids occurs in the lysosomes after endocytosis from the plasma membrane.

In the following subsections, we discuss the biosynthesis and breakdown of *N*-acylsphingosine and sphingoglycolipids and consider the diseases caused by deficiencies in their degradative enzymes.

a. Biosynthesis of Ceramide (*N*-Acylsphingosine)

Biosynthesis of *N*-acylsphingosine occurs in four reactions from the precursors palmitoyl-CoA and serine (Fig. 25-89):

1. 3-Ketosphinganine synthase (serine palmitoyltransferase), a pyridoxal phosphate–dependent enzyme, catalyzes the condensation of palmitoyl-CoA with serine yielding **3-ketosphinganine** (pyridoxal phosphate–dependent reactions are discussed in Section 26-1A).

Figure 25-87 **The synthesis of sphingomyelin from *N*-acylsphingosine and phosphatidylcholine.**

Cerebrosides

Glucocerebroside

Galactocerebroside

Sulfatide

OSO_3^-

Globosides

Lactosyl ceramide

Trihexosyl ceramide

Globoside

Gangliosides

G_{M3}

NANA

G_{M2}

NANA

G_{M1}

NANA

◯ = glucose ⬢ = *N*-acetylgalactosamine

◯ = galactose ▭ = ceramide

NANA = *N*-acetylneuraminic acid (sialic acid)

Figure 25-88 **Diagrammatic representation of the principal classes of sphingoglycolipids.** The G_M ganglioside structures are presented in greater detail in Fig. 12-7.

2. 3-Ketosphinganine reductase catalyzes the NADPH-dependent reduction of 3-ketosphinganine's keto group to form **sphinganine (dihydrosphingosine).**

3. Dihydroceramide is formed by transfer of an acyl group from an acyl-CoA to the sphinganine's 2-amino group, forming an amide bond.

4. Dihydroceramide dehydrogenase converts dihydroceramide to ceramide by an FAD-dependent oxidation reaction.

b. Biosynthesis of Cerebrosides (Glycosylceramides)

Galactocerebroside (1-β-galactosylceramide) and **glucocerebroside (1-β-glucosylceramide)** are the two most common cerebrosides. In fact, the term cerebroside is often

$$CoA-S-\overset{O}{\overset{\|}{C}}-CH_2-CH_2-(CH_2)_{12}-CH_3 \quad + \quad H_2N-\overset{CO_2^-}{\underset{CH_2OH}{\overset{|}{C}}}-H$$

Palmitoyl-CoA   **Serine**

1   3-ketosphinganine synthase

$\longrightarrow CO_2^- + CoASH$

$$\overset{O}{\overset{\|}{C}}-CH_2-CH_2-(CH_2)_{12}-CH_3$$
$$H_2N-\overset{|}{C}-H$$
$$\overset{|}{C}H_2OH$$

**3-Ketosphinganine
(3-ketodihydrosphingosine)**

2   NADPH + H⁺   3-ketosphinganine reductase

$\longrightarrow NADP^+$

$$\overset{OH}{\overset{|}{C}H}-CH_2-CH_2-(CH_2)_{12}-CH_3$$
$$H_2N-\overset{|}{C}-H$$
$$\overset{|}{C}H_2OH$$

**Sphinganine
(dihydrosphingosine)**

3   $R-\overset{O}{\overset{\|}{C}}-SCoA$ acyl-CoA transferase

$\longrightarrow CoASH$

$$\overset{OH}{\overset{|}{C}H}-CH_2-CH_2-(CH_2)_{12}-CH_3$$
$$R-\overset{O}{\overset{\|}{C}}-NH-\overset{|}{C}-H$$
$$\overset{|}{C}H_2OH$$

**Dihydroceramide
(*N*-acylsphinganine)**

4   FAD   dihydroceramide dehydrogenase

$\longrightarrow FADH_2$

$$\overset{OH}{\overset{|}{C}H}-\overset{H}{\overset{|}{C}}=C-(CH_2)_{12}-CH_3$$
$$R-\overset{O}{\overset{\|}{C}}-NH-\overset{|}{C}-H \quad H$$
$$\overset{|}{C}H_2OH$$

**Ceramide
(*N*-acylsphingosine)**

Figure 25-89 **The biosynthesis of ceramide (*N*-acylsphingosine).**

used synonymously with galactocerebroside. Both are synthesized from ceramide by addition of a glycosyl unit from the corresponding UDP–hexose (Fig. 25-90). Galactocerebroside is a common component of brain lipids. Glucocerebroside, although relatively uncommon, is the precursor of globosides and gangliosides.

Glucocerebroside (1-β-D-glucosylceramide)

UDP glucose:ceramide glycosyltransferase (glucosylceramide synthase)

UDP

UDP–glucose

Ceramide

UDP galactose:ceramide glycosyltransferase

UDP–galactose

UDP

Galactocerebroside (1-β-D-galactosylceramide)

Figure 25-90 **The biosynthesis of cerebrosides.**

c. Biosynthesis of Sulfatides

Sulfatides (galactocerebroside-3-sulfate) account for 15% of the lipids of white matter in the brain. They are formed by transfer of an "activated" sulfate group from **3'-phosphoadenosine-5'-phosphosulfate (PAPS)** to the C3 OH group of galactose in galactocerebroside (Fig. 25-91).

d. Biosynthesis of Globosides and Gangliosides

Biosynthesis of both globosides (neutral ceramide oligosaccharides) and gangliosides (acidic, sialic acid–containing ceramide oligosaccharides) is catalyzed by a series of **glycosyltransferases.** While the reactions are chemically similar, each is catalyzed by a specific enzyme. The pathways begin with transfer of a galactosyl unit from UDP–Gal to glucocerebroside to form a β(1 → 4) linkage (Fig. 25-92). Since this bond is the same as that linking glucose and galactose in lactose, this glycolipid is often referred to as **lactosyl ceramide.** Lactosyl ceramide is the precursor of both globosides and gangliosides. To form a globoside, one galactosyl and one *N*-acetylgalactosaminyl unit are sequentially added to lactosyl ceramide from UDP–Gal and UDP–GalNAc, respectively. The G_M gangliosides are formed by addition of **N-acetylneuraminic acid (NANA, sialic acid)**

N-Acetylneuraminic acid (NANA, sialic acid)

from CMP–NANA to lactosyl ceramide in α(2 → 3) linkage yielding G_{M3}. The sequential additions to G_{M3} of the *N*-acetylgalactosamine and galactose units from UDP–GalNAc and UDP–Gal yield gangliosides G_{M2} and G_{M1}. Other gangliosides are formed by adding a second

Figure 25-91 **The biosynthesis of sulfatides.**

Sulfatide (galactocerebroside-3-sulfate)

Figure 25-92 The biosynthesis of globosides and G_M gangliosides.

NANA group to G_{M3}, forming G_{D3}, or by adding an *N*-acetyl-glucosamine unit to lactosyl ceramide before NANA addition, forming G_{A2}. Hundreds of different gangliosides are known.

e. Sphingoglycolipid Degradation and Lipid Storage Diseases

Sphingoglycolipids are lysosomally degraded by a series of enzymatically mediated hydrolytic reactions (Fig. 25-93). These reactions are catalyzed at the lipid–water interface by soluble enzymes, often with the aid of **sphingolipid activator proteins (SAPs;** including **saposins, G_{M2}-activator protein,** and **SAP-A** through **SAP-D).** These nonenzymatic ancillary proteins are thought to increase the accessibility of the carbohydrate moiety of the sphingolipid to the degradation enzyme. For example, G_{M2}-activator binds G_{M2} and helps expose it to the surface of the membrane. The G_{M2}-activator–G_{M2} complex can then bind **hexosaminidase A,** an $\alpha\beta$ dimer that hydrolyzes *N*-acetylgalactosamine from G_{M2} at the lipid–water interface (Fig. 25-94).

The hereditary absence of one of the sphingolipid hydrolases or a SAP results in a **sphingolipid storage disease** (Table 25-2). One of the most common such conditions is **Tay–Sachs disease,** an autosomal recessive deficiency in hexosaminidase A. The absence of hexosaminidase A activity results in the neuronal accumulation of G_{M2} as shell-like inclusions (Fig. 25-95).

Although infants born with Tay–Sachs disease at first appear normal, by ~1 year of age, when sufficient G_{M2} has accumulated to interfere with neuronal function, they be-

come progressively weaker, retarded, and blinded until they die, usually by the age of 3 years. It is possible, however, to screen potential carriers of this disease by a simple serum assay. It is also possible to detect the disease *in utero* by assay of amniotic fluid or amniotic cells obtained by amniocentesis. The assay involves use of an artificial hexosaminidase substrate, **4-methylumbelliferyl-β-D-*N*-acetylglucosamine,** which yields a fluorescent product on hydrolysis:

4-Methylumbelliferyl-β-D-*N*-acetylglucosamine

hexosaminidase H_2O → GlcNAc

4-Methylumbelliferone (fluorescent in alkaline medium)

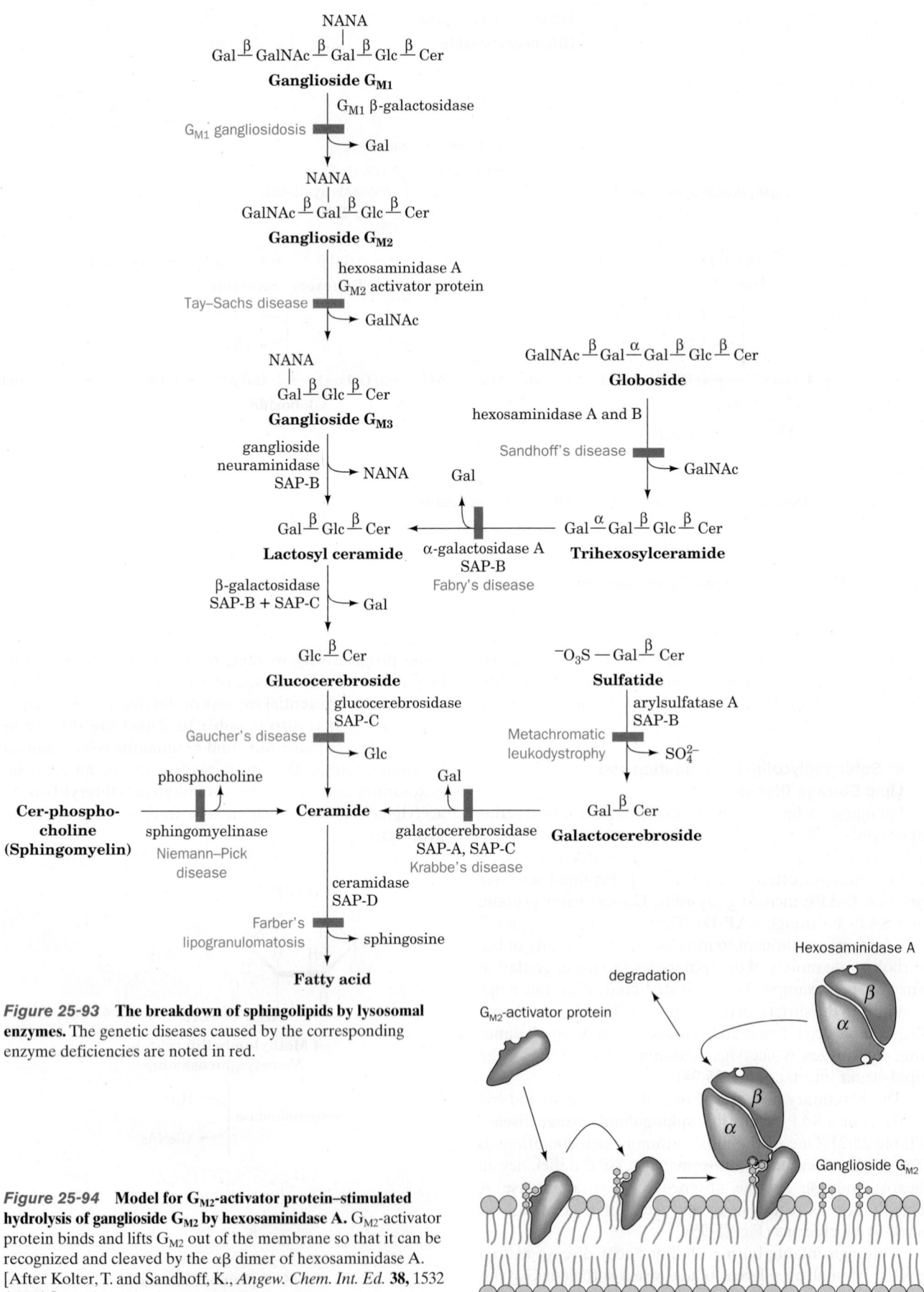

NANA
|
Gal $\xrightarrow{\beta}$ GalNAc $\xrightarrow{\beta}$ Gal $\xrightarrow{\beta}$ Glc $\xrightarrow{\beta}$ Cer

Ganglioside G$_{M1}$

G$_{M1}$ β-galactosidase

G$_{M1}$ gangliosidosis → Gal

NANA
|
GalNAc $\xrightarrow{\beta}$ Gal $\xrightarrow{\beta}$ Glc $\xrightarrow{\beta}$ Cer

Ganglioside G$_{M2}$

hexosaminidase A
G$_{M2}$ activator protein

Tay–Sachs disease → GalNAc

NANA
|
Gal $\xrightarrow{\beta}$ Glc $\xrightarrow{\beta}$ Cer

Ganglioside G$_{M3}$

GalNAc $\xrightarrow{\beta}$ Gal $\xrightarrow{\alpha}$ Gal $\xrightarrow{\beta}$ Glc $\xrightarrow{\beta}$ Cer

Globoside

hexosaminidase A and B

Sandhoff's disease → GalNAc

ganglioside
neuraminidase
SAP-B → NANA

Gal

Gal $\xrightarrow{\beta}$ Glc $\xrightarrow{\beta}$ Cer ← α-galactosidase A → Gal $\xrightarrow{\alpha}$ Gal $\xrightarrow{\beta}$ Glc $\xrightarrow{\beta}$ Cer

Lactosyl ceramide   α-galactosidase A
SAP-B
Fabry's disease   **Trihexosylceramide**

β-galactosidase
SAP-B + SAP-C → Gal

Glc $\xrightarrow{\beta}$ Cer

Glucocerebroside

^-O_3S — Gal $\xrightarrow{\beta}$ Cer

Sulfatide

glucocerebrosidase
SAP-C

arylsulfatase A
SAP-B

Gaucher's disease → Glc

Metachromatic
leukodystrophy → SO_4^{2-}

phosphocholine

Gal

**Cer-phospho-
choline
(Sphingomyelin)** ⟶ **Ceramide** ← Gal $\xrightarrow{\beta}$ Cer

sphingomyelinase

Niemann–Pick
disease

galactocerebrosidase
SAP-A, SAP-C

Krabbe's disease

Galactocerebroside

ceramidase
SAP-D

Farber's
lipogranulomatosis → sphingosine

Fatty acid

Figure 25-93 **The breakdown of sphingolipids by lysosomal enzymes.** The genetic diseases caused by the corresponding enzyme deficiencies are noted in red.

Hexosaminidase A

degradation

G$_{M2}$-activator protein

Ganglioside G$_{M2}$

Figure 25-94 **Model for G$_{M2}$-activator protein–stimulated hydrolysis of ganglioside G$_{M2}$ by hexosaminidase A.** G$_{M2}$-activator protein binds and lifts G$_{M2}$ out of the membrane so that it can be recognized and cleaved by the αβ dimer of hexosaminidase A. [After Kolter, T. and Sandhoff, K., *Angew. Chem. Int. Ed.* **38**, 1532 (1999).]

Table 25-2   **Sphingolipid Storage Diseases**

Disease	Enzyme Deficiency	Principal Storage Substance	Major Symptoms
G_{M1} Gangliosidosis	G_{M1} β-galactosidase	Ganglioside G_{M1}	Mental retardation, liver enlargement, skeletal involvement, death by age 2
Tay–Sachs disease	Hexosaminidase A	Ganglioside G_{M2}	Mental retardation, blindness, death by age 3
Fabry's disease	α-Galactosidase A	Trihexosylceramide	Skin rash, kidney failure, pain in lower extremities
Sandhoff's disease	Hexosaminidases A and B	Ganglioside G_{M2} and globoside	Similar to Tay–Sachs disease but progressing more rapidly
Gaucher's disease	Glucocerebrosidase	Glucocerebroside	Liver and spleen enlargement, erosion of long bones, mental retardation in infantile form only
Niemann–Pick disease	Sphingomyelinase	Sphingomyelin	Liver and spleen enlargement, mental retardation
Farber's lipogranulomatosis	Ceramidase	Ceramide	Painful and progressively deformed joints, skin nodules, death within a few years
Krabbe's disease	Galactocerebrosidase	Deacylated galactocerebroside	Loss of myelin, mental retardation, death by age 2
Metachromatic leukodistrophy (Sulfatide lipidosis)	Arylsulfatase A	Sulfatide	Mental retardation, death in first decade

Figure 25-95   **Cytoplasmic membranous body in a neuron affected by Tay–Sachs disease.** [Courtesy of John S. O'Brien, University of California at San Diego Medical School.]

Since this substrate is also recognized by **hexosaminidase B,** which is unaffected in Tay–Sachs disease, the hexosaminidase B is first heat inactivated since it is more heat labile than hexosaminidase A. As a result of mass screening efforts, the tragic consequences of this genetic enzyme deficiency are being averted. The other sphingolipid storage diseases, although less common, have similar consequences (Table 25-2). In recent years, DNA sequencing techniques that screen potential carriers for the most common mutations have become economically feasible.

CHAPTER SUMMARY

1 Lipid Digestion, Absorption, and Transport Triacylglycerols, the storage form of metabolic energy in animals, provide up to six times the metabolic energy of an equal weight of hydrated glycogen. Dietary lipids are digested by pancreatic digestive enzymes such as lipase and phospholipase A_2 that are active at the lipid–water interface of bile salt–stabilized emulsions. Bile salts are also essential for the intestinal absorption of dietary lipids, as is fatty acid–binding protein. Dietary triacylglycerols and those synthesized by the liver are transported in the blood as chylomicrons and VLDL, respectively. Triacylglycerols present in these lipoproteins are hydrolyzed by lipoprotein lipase outside the cells and enter them as free fatty acids. Fatty acids resulting from hydrolysis of adipose tissue triacylglycerols by hormone-sensitive triacylglycerol lipase are transported in the bloodstream as fatty acid–albumin complexes.

2 Fatty Acid Oxidation Before fatty acids are oxidized, they are converted to their acyl-CoA derivatives by acyl-CoA synthase in an ATP-requiring process, transported into mitochondria as carnitine esters, and reconverted inside the mitochondrial matrix to acyl-CoA. β Oxidation of fatty acyl-CoA occurs in 2-carbon increments so as to convert even-chain fatty acyl-CoAs completely to acetyl-CoA. The pathway involves FAD-dependent dehydrogenation of an alkyl group, hydration of the resulting double bond, NAD^+-dependent oxidation of this alcohol to a ketone, and C—C bond cleavage to form acetyl-CoA and a new fatty acyl-CoA with two fewer carbon atoms. The process then repeats itself. Complete oxidation of the acetyl-CoA, NADH, and $FADH_2$ is achieved by the citric acid cycle and oxidative phosphorylation. Oxidation of unsaturated fatty acids and odd-chain fatty acids also occurs by β oxidation but requires the participation of additional enzymes. Odd-chain fatty acid oxidation generates propionyl-CoA, whose further metabolism requires the participation of (1) propionyl-CoA carboxylase, which has a biotin prosthetic group, (2) methylmalonyl-CoA racemase, and (3) methylmalonyl-CoA mutase, which contains AdoCbl (coenzyme B_{12}). Methylmalonyl-CoA mutase catalyzes a carbon skeleton rearrangement reaction via a free radical mechanism in which the free radical is generated by the homolytic cleavage of AdoCbl's C—Co(III) bond.

β Oxidation of fatty acids takes place in the peroxisomes in addition to the mitochondrion. The peroxisomal pathway differs from the mitochondrial pathway in that the $FADH_2$ produced in the first step, rather than generating ATP by oxidative phosphorylation, is directly oxidized by O_2 to produce H_2O_2. Peroxisomal enzymes are specific for long-chain fatty acids and function in a chain-shortening process via β oxidation. The resultant intermediate chain-length products are transferred to the mitochondrion for complete oxidation.

3 Ketone Bodies A significant fraction of the acetyl-CoA produced by fatty acid oxidation in the liver is converted to acetoacetate and D-β-hydroxybutyrate, which, together with acetone, are referred to as ketone bodies. The first two compounds serve as important fuels for the peripheral tissues.

4 Fatty Acid Biosynthesis Fatty acid biosynthesis differs from fatty acid oxidation in several respects. Whereas fatty acid oxidation occurs in the mitochondrion utilizing fatty acyl-CoA esters, fatty acid biosynthesis occurs in the cytosol with the growing fatty acids esterified to acyl-carrier protein (ACP). The redox coenzymes differ (FAD and NAD^+ for oxidation; NADPH for biosynthesis), as does the stereochemistry of the pathway's intermediate steps. Oxidation produces acetyl-CoA, whereas malonyl-CoA is the immediate precursor in biosynthesis. Malonyl-CoA is produced by the ATP-driven reaction of HCO_3^- with acetyl-CoA as catalyzed by the biotin-containing enzyme acetyl-CoA carboxylase, although this HCO_3^- is not incorporated into the final fatty acid product. In bacteria, fatty acids synthesis is carried out by a series of independent enzymes that are collectively known as FAS-I. In animals, fatty acid synthesis yielding palmitate occurs on a single X-shaped homodimeric protein known as FAS-II, each of whose subunits contain all six activities required to do so on

separate domains that are serviced by ACP. In fungi, FAS-II is an $\alpha_6\beta_6$ heterododecamer that forms a barrel-shaped protein that contains two equivalent reaction chambers. Similar but more extensive systems are used in the synthesis of the various polyketides. Acetyl-CoA is transferred from the mitochondrion to the cytosol as citrate via the tricarboxylate transport system and released by citrate cleavage to yield acetyl-CoA and oxaloacetate. Oxaloacetate is converted to malate and then to pyruvate for transport back to the mitochondrion, a process that also generates some of the NADPH required for biosynthesis. Palmitate is the primary product of fatty acid biosynthesis in animals. Longer chain fatty acids and unsaturated fatty acids are synthesized from palmitate by elongation and desaturation reactions. Certain essential unsaturated fatty acids cannot be synthesized by animals and hence must be obtained from the diet. Triacylglycerols are synthesized from fatty acyl-CoA esters and glycerol-3-phosphate.

5 Regulation of Fatty Acid Metabolism Fatty acid metabolism is regulated through the allosteric control of hormone-sensitive triacylglycerol lipase and acetyl-CoA carboxylase, phosphorylation/dephosphorylation, and/or changes in the rates of protein synthesis and breakdown. This regulation is mediated by the hormones glucagon, epinephrine, and norepinephrine, which activate degradation, and by insulin, which activates biosynthesis. These hormones interact to control the cAMP concentration, which in turn controls phosphorylation/dephosphorylation ratios via PKA. AMPK, which senses the level of ATP, is also an important regulator of fatty acid metabolism.

6 Cholesterol Metabolism Cholesterol is a vital constituent of cell membranes and is the precursor of the steroid hormones and bile salts. Its biosynthesis, transport, and utilization are rigidly controlled. Cholesterol is synthesized in the liver from acetate in a pathway that involves formation of HMG-CoA from three molecules of acetate followed by reduction, phosphorylation, decarboxylation, and dehydration to the isoprene units isopentenyl pyrophosphate and dimethylallyl pyrophosphate. Four of these isoprene units are then condensed via cationic mechanisms to form squalene, which, in turn, undergoes a cyclization reaction, via a cationic cascade, to form lanosterol, the sterol precursor to cholesterol.

The pathway's major control point is at HMG-CoA reductase. This enzyme is regulated by competitive and allosteric mechanisms, by phosphorylation/dephosphorylation, and, most importantly, by long-term control of the rates of enzyme synthesis and degradation. Long-term control is mediated by the integral membrane protein SREBP, which when the cholesterol level is low, is escorted by SCAP to the Golgi apparatus via COPII-coated vesicles. There it is sequentially cleaved by the proteases S1P and S2P, thereby releasing its soluble bHLH/Z domain to travel to the nucleus where it induces the transcription of SRE-containing genes such as those encoding HMG-CoA reductase and the LDL receptor. In addition, when sterol levels are high, the rate at which HMG-CoA reductase is proteolytically degraded is greatly increased.

The liver secretes cholesterol into the bloodstream in esterified form as part of the VLDL. This complex is sequentially converted to IDL and then to LDL. LDL, which is brought into the cells by receptor-mediated endocytosis, carries the major portion of cholesterol to peripheral tissues for utilization. Excess cholesterol is returned to the liver from peripheral tissues by HDL. The cellular supply of cholesterol is controlled by three mechanisms: (1) long- and short-term regulation of HMG-CoA reductase; (2) control of LDL receptor synthesis by cholesterol concentration; and (3) long- and short-term regulation of acyl-CoA:cholesterol acyltransferase (ACAT), which mediates cholesterol esterification. Cholesterol is the precursor to the steroid hormones, which are classified as progestins, glucocorticoids, mineralocorticoids, androgens, and estrogens. The only pathway for the excretion of cholesterol in mammals is through the formation and elimination of bile salts.

7 Eicosanoid Metabolism: Prostaglandins, Prostacyclins, Thromboxanes, Leukotrienes, and Lipoxins Prostaglandins, prostacyclins, thromboxanes, leukotrienes, and lipoxins are eicosanoid products produced largely by the metabolism of arachidonate. These highly unstable compounds have profound physiological effects at extremely low concentrations. They are involved in the inflammatory response, the production of pain and fever, the regulation of blood pressure, and many other important physiological processes. Arachidonate is synthesized from linoleic acid, an essential fatty acid, and stored as phosphatidylinositol and other phospholipids. Prostaglandins, prostacyclins, and thromboxanes are synthesized via the cyclic pathway, whereas leukotrienes and lipoxins are synthesized via the linear pathway. Aspirin and other nonsteroidal anti-inflammatory drugs (NSAIDs) inhibit the cyclic pathway but not the linear pathway. COX-2 inhibitors are NSAIDs that bind to COX-2 but not COX-1, thereby eliminating the side effects of other NSAIDs. Peptidoleukotrienes have been identified as the slow reacting substances of anaphylaxis (SRS-A) released from sensitized lung after immuno-logical challenge. Lipoxins and aspirin-induced *epi*-lipoxins have anti-inflammatory properties.

8 Phospholipid and Glycolipid Metabolism Complex lipids have either a phosphate ester or a carbohydrate as their polar head group and either 1,2-diacyl-*sn*-glycerol or ceramide (*N*-acylsphingosine) as their hydrophobic tail. Phospholipids are either glycerophospholipids or sphingophospholipids, whereas glycolipids are either glyceroglycolipids or sphingoglycolipids. The polar head groups of glycerophospholipids, which are phosphate esters of either ethanolamine, serine, choline, inositol, or glycerol, are attached to 1,2-diacyl-*sn*-glycerol's C3 OH group by means of CTP-linked transferase reactions. The specific long-chain fatty acids found at the C1 and C2 positions are incorporated by "remodeling reactions" after the addition of the polar head group. Plasmalogens and alkylacylglycerophospholipids, respectively, contain a long-chain alkyl group in a vinyl ether linkage or in an ether linkage to glycerol's C1 OH group. Platelet-activating factor (PAF) is an important alkylacylglycerophospholipid. The only major sphingophospholipid is sphingomyelin (*N*-acylsphingosine phosphocholine), an important structural lipid of nerve cell membranes. Most sphingolipids contain polar head groups composed of carbohydrate units and are therefore referred to as sphingoglycolipids. The principal classes of sphingoglycolipids are cerebrosides, sulfatides, globosides, and gangliosides. Their carbohydrate units, which are attached to *N*-acylsphingosine's C1 OH group by glycosidic linkages, are formed by stepwise addition of activated monosaccharide units. Several lysosomal sphingolipid storage diseases, including Tay–Sachs disease, result from deficiencies in the enzymes that degrade sphingoglycolipids.

REFERENCES

General

Newsholme, E.A. and Leech, A.R., *Biochemistry for the Medical Sciences,* Wiley (1983). [Chapters 6–8 contain a wealth of information on the control of fatty acid metabolism and its integration into the overall scheme of metabolism.]

Valle, D. (Ed.), *The Online Metabolic & Molecular Bases of Inherited Disease,* http://www.ommbid.com/. [Contain numerous chapters on defects in lipid metabolism.]

Vance, D.E. and Vance J.E. (Eds.), *Biochemistry of Lipids, Lipoproteins, and Membranes* (5th ed.), Elsevier (2008).

Lipid Digestion

Berg, O.G., Gelb, M.H., Tsai, M.-D., and Jain, M.K., Interfacial enzymology: the secreted phospholipase A$_2$-paradigm, *Chem. Rev.* **101,** 2613–2653 (2001).

Bhattacharya, A.A., Grüne, T., and Curry, S., Crystallographic analysis reveals common modes of binding of medium and long-chain fatty acids to human serum albumin, *J. Mol. Biol.* **303,** 721–732 (2002).

Borgström, B., Barrowman, J.A., and Lindström, M., Roles of bile acids in intestinal lipid digestion and absorption, *in* Danielsson, H. and Sjövall, J. (Eds.), *Sterols and Bile Acids,* pp. 405–425, Elsevier (1985).

Brady, L., Brzozowski, A.M., Derewenda, Z.S., Dodson, E., Dodson, G., Tolley, S., Turkenburg, J.P., Christiansen, L., Huge-Jensen, B., Norskov, L., Thim, L., and Menge, U., A serine protease triad forms the catalytic centre of a triacylglycerol lipase, *Nature* **343,** 767–770 (1990).

Hermoso, J., Pignol, D., Penel, S., Roth, M., Chapus, C., and Fontecilla-Camps, J.C., Neutron crystallographic evidence of lipase–colipase complex activation by a micelle, *EMBO J.* **16,** 5531–5536 (1997).

Scapin, G., Gordon, J.I., and Sacchettini, J.C., Refinement of the structure of recombinant rat intestinal fatty acid-binding apoprotein at 1.2-Å resolution, *J. Biol. Chem.* **267,** 4253–4269 (1992).

van Tilbeurgh, H., Bezzine, S., Cambillau, C., Verger, R., and Carriére, F., Colipase: structure and interaction with pancreatic lipase, *Biochim. Biophys. Acta* **1441,** 173–184 (1999).

Fatty Acid Oxidation

Bannerjee, R., Radical peregrinations catalyzed by coenzyme B$_{12}$-dependent enzymes, *Biochem.* **40,** 6191–6198 (2001).

Bartlett, K. and Eaton, S., Mitochondrial β-oxidation, *Eur. J. Biochem.* **271,** 462–469 (2004). [Discusses the reactions and enzymes of mitochondrial β oxidation as well as their regulation.]

Bieber, L.L., Carnitine, *Annu. Rev. Biochem.* **88,** 261–283 (1988).

Brownsey, R.W., Boone, A.N., Elliott, J.E., Kulpa, J.E., and Lee, W.M., Regulation of acetyl-CoA carboxylase, *Biochem. Soc. Trans.* **34,** 223–227 (2006).

Chou, C.-Y., Yu, L.P.C., and Tong, L., Crystal structure of biotin carboxylase in complex with substrates and implications for its catalytic mechanism, *J. Biol. Chem.* **284,** 11690–11697 (2009).

Kim, J.-J.P. and Battaile, K.P., Burning fat: The structural basis of fatty acid β-oxidation, *Curr. Opin. Struct. Biol.* **12,** 721–728 (2002).

Kim, J.-J.P., Wang, M., and Pashke, R., Crystal structures of medium-chain acyl-CoA dehydrogenase from pig liver mitochondria with and without substrate, *Proc. Natl. Acad. Sci.* **90,** 7523–7527 (1993).

Kindl, H., Fatty acid degradation in plant peroxisomes: Function and biosynthesis of the enzymes involved, *Biochimie* **75,** 225–230 (1993).

Mancia, R., Smith, G.A., and Evans, P.R., Crystal structure of substrate complexes of methylmalonyl-CoA mutase, *Biochem.* **38,** 7999–8005 (1999).

Marsh, E.N.G. and Drennan, C.L., Adenosylcobalamin-dependent isomerases: new insights into structure and mechanism, *Curr. Opin. Chem. Biol.* **5,** 499–505 (2001).

Rinaldo, P., Matern, D., and Bennett, M.J., Fatty acid oxidation disorders, *Annu. Rev. Physiol.* **64,** 477–502 (2002).

Shoukry, K. and Schulz, H., Significance of the reductase-dependent pathway for the β-oxidation of unsaturated fatty acids with odd-numbered double bonds: mitochondrial metabolism of 2-*trans*-5-*cis*-octadienoyl-CoA, *J. Biol. Chem.* **273,** 6892–6899 (1998).

Sudden infant death and inherited disorders of fat oxidation, *Lancet,* 1073–1075, Nov. 8, 1986.

Tong, L., Acetyl-coenzyme A carboxylase: crucial metabolic enzyme and attractive target for drug discovery, *Cell. Mol. Life Sci.* **62,** 1784–1893 (2005).

van den Bosch, H., Schutgens, R.B.H., Wanders, R.J.A., and Tager, J.M., Biochemistry of peroxisomes, *Annu. Rev. Biochem.* **61,** 157–197 (1992).

Watkins, P.A., Very-long-chain acyl-CoA synthetases, *J. Biol. Chem.* **283,** 1773–1777 (2008).

Fatty Acid Biosynthesis

Bartlett, K. and Eaton, S., Mitochondrial β-oxidation, *Eur. J. Biochem.* **271,** 462–469 (2004). [Discusses the reactions and enzymes of mitochondrial β oxidation as well as their regulation.]

Brownsey, R.W. and Denton, R.M., Acetyl-coenzyme A carboxylase, *in* Boyer, P.D. and Krebs, E.G. (Eds.), *The Enzymes* (3rd ed.), Vol. 18, *pp.* 123–146, Academic Press (1987).

Brownsey, R.W., Zhande, R., and Boone, A.N., Isoforms of acetyl-CoA carboxylase: structures, regulatory properties and metabolic functions, *Biochem. Soc. Trans.* **25,** 1232–1238 (1997).

Haapalainen, A.M., Meriläinen, G., and Wierenga, R.K., The thiolase superfamily: condensing enzymes with diverse reaction specificities, *Trends Biochem. Sci.* **31,** 64–71 (2006).

Jenni, S., Leibundgut, M., Boehringer, D., Frick, C., Mikolásek, B., and Ban, N., Structure of fungal fatty acid synthase and implications for iterative substrate shuttling, *Science* **316,** 254–261 (2007).

Jump, D.B., The biochemistry of *n*–3 polyunsaturated fatty acids, *J. Biol. Chem.* **277,** 8755–8758 (2002).

Khosla, C., Tang, Y., Chen, A.Y., Schnarr, N.A., and Cane, D.E., Structure and mechanism of the 6-deoxyerythronolide B synthase, *Annu. Rev. Biochem.* **76,** 195–221 (2007).

Leibundgut, M., Maier, T., Jenni, S., and Ban, N., The multienzyme architecture of eukaryotic fatty acid synthases, *Curr. Opin. Struct. Biol.* **18,** 714–725 (2008).

Lomakin, I.B., Xiong, Y., and Steitz, T.A., The crystal structure of yeast fatty acid synthase, a cellular machine with eight active sites working together, *Cell* **129,** 319–332 (2007); *and* Leibundgut, M., Jenni, S., Frick, C., and Ban, N., Structural basis for substrate delivery by acyl carrier protein in the yeast fatty acid synthase, *Science* **316,** 288–290 (2007).

Los, D.A. and Murata, N., Structure and expression of fatty acid desaturases, *Biochim. Biophys. Acta* **1394,** 3–15 (1998).

Maier, T., Leibundgut, M., and Ban, N., The crystal structure of a mammalian fatty acid synthase, *Science* **321,** 1315–1322 (2008).

Smith, S. and Tsai, S.-C., The type I fatty acid and polyketide synthetases: a tale of two megasynthases, *Nat. Prod. Rep.* **24,** 1041–1072 (2007).

Wakil, S.J., Fatty acid synthase, a proficient multifunctional enzyme, *Biochemistry* **28,** 4523–4530 (1989).

Walsh, C.T., Polyketide and nonribosomal peptide antibiotics: Modularity and versatility, *Science* **303,** 1805–1810 (2004).

White, S.W., Zheng, J., Zhang, Y.-M., and Rock, C.O., The structural biology of type II fatty acid biosynthesis, *Annu. Rev. Biochem.* **74,** 791–831 (2005).

Regulation of Fatty Acid Metabolism

Eaton, S., Control of mitochondrial β-oxidation flux, *Prog. Lipid Res.* **41,** 197–239 (2002).

Hardie, D.G. and Carling, D., The AMP-activated protein kinase: fuel gauge of the mammalian cell? *Eur. J. Biochem.* **246,** 259–273 (1997).

Hardie, D.G., Carling, D., and Carlson, M., The AMP-activated/SNF1 protein kinase subfamily: metabolic sensors of the eukaryotic cell? *Annu. Rev. Biochem.* **67,** 821–855 (1998).

Munday M.R. and Hemingway C.J., The regulation of acetyl-CoA carboxylase—A potential target for the action of hypolipidemic agents, *Adv. Enzym. Regul.* **39,** 205–234 (1999).

Witters, L.A., Watts, T.D., Daniels, D.L., and Evans, J.L., Insulin stimulates the dephosphorylation and activation of acetyl-CoA carboxylase, *Proc. Natl. Acad. Sci.* **85,** 5473–5477 (1988).

Cholesterol Metabolism

Bloch, K., The biological synthesis of cholesterol, *Science* **150,** 19–28 (1965).

Chang, T.Y., Chang, C.C.Y., and Cheng, D., Acyl-coenzyme A:cholesterol acyltransferase, *Annu. Rev. Biochem.* **66,** 613–638 (1997).

Durbecq, V., et al., Crystal structure of isopentenyl diphosphate:dimethylallyl diphosphate isomerase, *EMBO J.* **20,** 1530–1537 (2001).

Edwards, P.A., Sterols and isoprenoids: signaling molecules derived from the cholesterol biosynthetic pathway, *Annu. Rev. Biochem.* **68,** 157–185 (1999).

Goldstein, J.L. and Brown, M.S., Regulation of the mevalonate pathway, *Nature* **343,** 425–430 (1990).

Goldstein, J.L., DeBose-Boyd, R.A., and Brown, M.S., Protein sensors for membrane sterols, *Cell* **124,** 35–46 (2006).

Goldstein, J.L., Rawson, R.B., and Brown, M.S., Mutant mammalian cells as tools to delineate the sterol regulatory element binding protein pathway for feedback regulation of lipid synthesis, *Arch. Biochem. Biophys.* **397,** 139–148 (2002).

Ikonen, E., Cellular cholesterol trafficking and compartmentalization, *Nature Rev. Mol. Cell Biol.* **9,** 125–138 (2008).

Istvan, E.S., Bacterial and mammalian HMG-CoA reductases: related enzymes with distinct architectures, *Curr. Opin. Struct. Biol.* **11,** 746–751 (2001).

Istvan, E.S. and Deisenhofer, J., Structural mechanism for statin inhibition of HMG-CoA reductase, *Science* **292,** 1160–1164 (2001).

Knopp, R.H., Drug therapy: drug treatment of lipid disorders, *New Engl. J. Med.* **341,** 498–511 (1999).

Meyer, M.M., Segura, M.J.R., Wilson, W.K., and Matsuda, S.P.T., Oxidosqualene cyclase residues that promote formation of cycloartenol, lanosterol and parkeol, *Angew. Chem. Int. Ed.* **39,** 4090–4092 (2000).

Reinert, D.J., Balliano, G., and Schulz, G.E., Conversion of squalene to the pentacarbocyclic hopene, *Chem. Biol.* **11**, 121–126 (2004). [The X-ray structure of squalene-hopene cyclase in complex with 2-azasqualene.]

Russell, D.W. and Setchell, K.D.R., Bile acid biosynthesis, *Biochemistry* **31**, 4737–4749 (1992).

Thoma, R., Schulz-Gasch, T., D'Arcy, B., Benz, J., Aebl, J., Dehmiow, H., Hennig, M., Stihle, M., and Ruf, A., Insight into steroid scaffold formation from the structure of human oxidosqualene cyclase, *Nature* **432**, 118–122 (2004).

Wang, K.C. and Ohnuma, S.-I., Isoprenyl diphosphate synthases, *Biochim. Biophys. Acta* **1529**, 33–48 (2000).

Wendt, K.U., Schulz, G.E., Corey, E.J. and Liu, D.R., Enzyme mechanisms for polycyclic triterpene formation, *Angew. Chem. Int. Ed.* **39**, 2812–2833 (2000).

Yokode, M., Hammer, R.E., Ishibashi, S., Brown, M.S., and Goldstein, J.L., Diet-induced hypercholesterolemia in mice: Prevention by overexpression of LDL receptors, *Science* **250**, 1273–1275 (1990).

Eicosanoid Metabolism

Abramovitz, M., Wong, E., Cox, M.E., Richardson, C.D., Li, C., and Vickers, P.J., 5-Lipoxygenase-activating protein stimulates the utilization of arachidonic acid by 5-lipoxygenase, *Eur. J Biochem.* **215**, 105–111 (1993).

Chandrasekharan, N.V., Dai, H., Roos, K.L.T., Evanson, N.K., Tomsik, J., Elton, T.S., and Simmons, D.L., COX-3, a cyclooxygenase-1 variant inhibited by acetaminophen and other analgesic/antipyretic drugs: Cloning, structure, and expression, *Proc. Natl. Acad. Sci.* **99**, 13926–13931 (2002).

Ferguson, A.D., et al., Crystal structure of inhibitor-bound human 5-lipoxygenase–activating protein, *Science* **317**, 510–512 (2007).

Ford-Huchinson, A.W., Gresser, M., and Young, R.N., 5-Lipoxygenase, *Annu. Rev. Biochem.* **63**, 383–417 (1994).

Gillmor, S.A., Villaseñor, A., Fletterick, R., Sigal, E., and Browner, M.F., The structure of mammalian 15-lipoxygenase reveals similarity to the lipases and the determinants of substrate specificity, *Nature Struct. Biol.* **4**, 1003–1009 (1997).

Kurumbail, R.G., Kiefer, J.R., and Marnett, L.J., Cyclooxygenase enzymes: catalysis and inhibition, *Curr. Opin. Struct. Biol.* **11**, 752–760 (2001).

Phillipson, B.E., Rothrock, D.W., Conner, W.E., Harris, W.S., and Illingworth, D.R., Reduction of plasma lipids, lipoproteins and apoproteins by dietary fish oils in patients with hypertriglyceridemia, *New Engl. J. Med.* **312**, 1210–1216 (1985).

Picot, D., Loll, P.J., and Garavito, R.M., The X-ray crystal structure of the membrane protein prostaglandin H₂ synthase-1, *Nature* **367**, 243–249 (1994).

Rådmark, O., Werz, O., Steinhilber, D., and Samuelsson, B., 5-Lipoxygenase: regulation of expression and enzyme activity, *Trends Biochem. Sci.* **32**, 332–341 (2007).

Samuelsson, B. and Funk, C.D., Enzymes involved in the biosynthesis of leukotriene B4, *J. Biol. Chem.* **264**, 19469–19472 (1989).

Schwarz, K., Walther, M., Anton, M., Gerth, C., Feussner, I., and Kuhn, H., Structural basis for lipoxygenase specificity: conversion of the human leukocyte 5-lipoxygenase to a 15-lipoxygenating enzyme species by site-directed mutagenesis, *J. Biol. Chem.* **276**, 773–339 (2001).

Serhan, C.N., Lipoxins and novel aspirin-triggered 15-*epi*-lipoxins (ATL): a jungle of cell-cell interactions or a therapeutic opportunity? *Prostaglandins* **53**, 107–137 (1997).

Smith, W.L., Nutritionally essential fatty acids and biologically indispensible cyclooxygenases, *Trends Biochem. Sci.* **33**, 27–37 (2003).

Smith, W.L., DeWitt, D.L., and Garavito, R.M., Cyclooxygenases: structural, cellular and molecular biology, *Annu. Rev. Biochem.* **69**, 145–182 (2000).

Turini, M.E. and DuBois, R.N., Cyclooxygenase-2: a therapeutic target, *Annu. Rev. Med.* **53**, 35–57 (2002).

Warner, T.D. and Mitchell, J.A., Cyclooxygenases: new forms, new inhibitors, and lessons from the clinic, *FASEB J.* **18**, 790–804 (2004).

Phospholipid and Glycolipid Metabolism

Conzelmann, E. and Sandhoff, K., Glycolipid and glycoprotein degradation, *Adv. Enzymol.* **60**, 89–216 (1987).

Dowhan, W., Molecular basis for membrane diversity: Why are there so many lipids? *Annu. Rev. Biochem.* **66**, 199–232 (1997).

Kent, C., Eukaryotic phospholipid synthesis, *Annu. Rev. Biochem.* **64**, 315–342 (1995).

Kolter, T. and Sandhoff, K., Sphingolipids—their metabolic pathways and the pathobiochemistry of neurodegenerative diseases, *Angew. Chem. Int. Ed.* **38**, 1532–1568 (1999).

Neufield, E.F., Natural history and inherited disorders of a lysosomal enzyme, β-hexosaminidase, *J. Biol. Chem.* **264**, 10927–10930 (1989).

Prescott, S.M., Zimmerman, G.A., and McIntire, T.M., Platelet-activating factor, *Biol. Chem.* **265**, 17381–17384 (1990).

Tifft, C.J. and Proila, R.L., Stemming the tide: glycosphingloipid synthesis inhibitors as therapy for storage diseases, *Glycobiology* **10**, 1249–1258 (2000).

van Echten, G. and Sandhoff, K., Ganglioside metabolism, *J. Biol. Chem.* **268**, 5341–5344 (1993).

PROBLEMS

1. The venoms of many poisonous snakes, including rattlesnakes, contain a phospholipase A_2 that causes tissue damage that is seemingly far out of proportion to the small amount of enzyme injected. Explain.

2. Explain why individuals with a hereditary deficiency of carnitine palmitoyltransferase II have muscle weakness. Why are these symptoms more severe during fasting?

3. Why are the livers of Jamaican vomiting sickness victims usually depleted of glycogen?

4. Compare the metabolic efficiencies, in moles of ATP produced per gram, of completely oxidized fat (tripalmitoyl glycerol) versus glucose derived from glycogen. Assume that the fat is anhydrous and the glycogen is stored with twice its weight of water.

5. Methylmalonyl-CoA mutase is incubated with deuterated methylmalonyl-CoA. The coenzyme B_{12} extracted from this mutase is found to contain deuterium at its 5′-methylene group. Account for the transfer of label from substrate to coenzyme.

6. What is the energetic price, in ATP units, of converting acetoacetyl-CoA to acetoacetate and then resynthesizing acetoacetyl-CoA?

7. A fasting animal is fed palmitic acid that has a ^{14}C-labeled carboxyl group. (a) After allowing sufficient time for fatty acid breakdown and resynthesis, what would be the ^{14}C-labeling pattern in the animal's palmitic acid residues? (b) The animal's liver glycogen becomes ^{14}C labeled although there is no net increase in the amount of this substance present. Indicate the sequence of reactions whereby the glycogen becomes labeled. Why is there no net glycogen synthesis?

8. What is the ATP yield from the complete oxidation of a molecule of (a) α-linolenic acid (9,12,15-octadecatrienoic acid, 18:3n–3) and (b) **margaric acid** (heptadecanoic acid, 17:0)? Which has the greater amount of available biological energy on a per carbon basis?

*9. The role of coenzyme B$_{12}$ in mediating hydrogen transfer was established using the coenzyme B$_{12}$-dependent bacterial enzyme **dioldehydrase,** which catalyzes the reaction:

$$CH_3-\underset{\underset{\displaystyle OH}{|}}{CH}-\underset{\underset{\displaystyle H}{|}}{CH}-OH \longrightarrow CH_3-\underset{\underset{\displaystyle H}{|}}{CH}-\underset{\underset{\displaystyle OH}{|}}{CH}-OH$$

1,2-Propanediol

$$\downarrow \searrow H_2O$$

$$CH_3-CH_2-\overset{\overset{\displaystyle O}{\|}}{CH}$$

Propionaldehyde

The enzyme converts [1-^{3}H$_2$]1,2-propanediol to [1,2-^{3}H]propionaldehyde with the incorporation of tritium into both C5′ positions of 5′-deoxyadenosylcobalamin's 5′-deoxyadenosyl residue. Suggest the mechanism of this reaction. What would be the products of the dioldehydrase reaction if the enzyme was supplied with [5′-^{3}H]deoxyadenosylcobalamin and unlabeled 1,2-propanediol?

10. Why is it important that liver cells lack 3-ketoacyl-CoA transferase (Fig. 25-27)?

11. What is the energetic price, in ATP equivalents, of breaking down palmitic acid to acetyl-CoA and then resynthesizing it?

12. Is the fatty acid shown below likely to be synthesized in animals? Explain.

$$CH_3-(CH_2-CH=CH)_3-(CH_2)_7-\overset{\overset{\displaystyle O}{\|}}{C}\diagdown_{O^-}$$

13. What is the energetic price, in ATP equivalents, of synthesizing cholesterol from acetyl-CoA?

14. What would be the ^{14}C-labeling pattern in cholesterol if it were synthesized from HMG-CoA that was ^{14}C labeled (a) at C5, its carboxyl carbon atom, or (b) C1, its thioester carbon atom?

15. Hypercholesterolemic individuals taking statins are sometimes advised to take supplements of coenzyme Q. Explain.

*16. A child suffering from severe abdominal pain is admitted to the hospital several hours after eating a meal consisting of hamburgers, fried potatoes, and ice cream. Her blood has the appearance of "creamed tomato soup" and on analysis is found to contain massive quantities of chylomicrons. As the attending physician, what is your diagnosis of the patient's difficulty (the cause of the abdominal pain is unclear)? What treatment would you prescribe to alleviate the symptoms of this inherited disease?

17. Although linoleic acid is an essential fatty acid in animals, it is not required by animal cells in tissue culture. Explain.

18. The inactivation of the peroxidase function of prostaglandin H synthase (PGHS) also inactivates its cyclooxygenase function but not vice versa. Explain.

Amino Acid Metabolism

CHAPTER 26

α-Amino acids, in addition to their role as protein monomeric units, are energy metabolites and precursors of many biologically important nitrogen-containing compounds, notably heme, physiologically active amines, glutathione, nucleotides, and nucleotide coenzymes. Amino acids are classified into two groups: **essential** and **nonessential.** Mammals synthesize the nonessential amino acids from metabolic precursors but must obtain the essential amino acids from their diet. Excess dietary amino acids are neither stored for future use nor excreted. Rather, they are converted to common metabolic intermediates such as pyruvate, oxaloacetate, acetyl-CoA, and α-keto-glutarate. Consequently, *amino acids are also precursors of glucose, fatty acids, and ketone bodies and are therefore metabolic fuels.*

In this chapter, we consider the pathways of amino acid breakdown, synthesis, and utilization. We begin by examining the three common stages of amino acid breakdown:

1. Deamination (amino group removal), whereby amino groups are converted either to ammonia or to the amino group of aspartate.

2. Incorporation of ammonia and aspartate nitrogen atoms into urea for excretion.

3. Conversion of amino acid carbon skeletons (the α-keto acids produced by deamination) to common metabolic intermediates.

Many of these reactions are similar to those we have considered in other pathways. Others employ enzyme cofactors we have not previously encountered. One of our goals in studying amino acid metabolism is to understand the mechanisms of action of these cofactors.

After our discussion of amino acid breakdown, we examine the pathways by which amino acids are utilized in the biosynthesis of heme, physiologically active amines, and glutathione (the synthesis of nucleotides and nucleotide coenzymes is the subject of Chapter 28). Next, we study amino acid biosynthesis pathways. The chapter ends with a discussion of nitrogen fixation, a process that converts atmospheric N_2 to ammonia and is therefore the ultimate biological source of metabolically useful nitrogen.

1 AMINO ACID DEAMINATION

The first reaction in the breakdown of an amino acid is almost always removal of its α-amino group with the object of excreting excess nitrogen and degrading the remaining carbon skeleton or converting it to glucose. Urea, the predominant nitrogen excretion product in terrestrial mammals, is synthesized from ammonia and aspartate. Both of these latter substances are derived mainly from glutamate, a product of most deamination reactions. In this section we examine the routes by which α-amino groups are incorporated into glutamate and then into aspartate and ammonia. In Section 26-2, we discuss urea biosynthesis from these precursors.

Most amino acids are deaminated by **transamination,** the transfer of their amino group to an α-keto acid to yield the α-keto acid of the original amino acid and a new amino acid, in reactions catalyzed by **aminotransferases** (alternatively, **transaminases**). The predominant amino group acceptor is α-ketoglutarate, producing glutamate as the new amino acid:

Amino acid + α-ketoglutarate $\rightleftharpoons$
$$\text{α-keto acid + glutamate}$$

Glutamate's amino group, in turn, is transferred to oxaloacetate in a second transamination reaction, yielding aspartate:

Glutamate + oxaloacetate $\rightleftharpoons$
$$\text{α-ketoglutarate + aspartate}$$

Transamination, of course, does not result in any net deamination. Deamination occurs largely through the oxidative deamination of glutamate by **glutamate dehydrogenase (GDH),** yielding ammonia. The reaction requires NAD^+ or $NADP^+$ as an oxidizing agent and regenerates α-ketoglutarate for use in additional transamination reactions:

Glutamate + $NAD(P)^+$ + H_2O $\rightleftharpoons$
$$\text{α-ketoglutarate + } NH_4^+ + NAD(P)H$$

The mechanisms of transamination and oxidative deamination are the subjects of this section. We also consider other means of amino group removal from specific amino acids.

A. Transamination

a. Aminotransferase Reactions Occur in Two Stages

1. The amino group of an amino acid is transferred to the enzyme, producing the corresponding keto acid and the aminated enzyme.

Amino acid + enzyme $\rightleftharpoons$
$$\text{α-keto acid + enzyme—} NH_2$$

2. The amino group is transferred to the keto acid acceptor (e.g., α-ketoglutarate), forming the amino acid product (e.g., glutamate) and regenerating the enzyme.

α-Ketoglutarate + enzyme—NH_2 $\rightleftharpoons$
$$\text{enzyme + glutamate}$$

To carry the amino group, aminotransferases require participation of an aldehyde-containing coenzyme, pyridoxal-5'-phosphate (PLP), a derivative of pyridoxine (vitamin B_6; Fig. 26-1a,b). The amino group is accommodated by conversion of this coenzyme to **pyridoxamine-5'-phosphate (PMP;** Fig. 26-1c). PLP is covalently attached to the enzyme via a Schiff base (imine) linkage formed by the

Figure 26-1 Forms of pyridoxal-5'-phosphate.
(*a*) Pyridoxine (vitamin B_6). (*b*) Pyridoxal-5'-phosphate (PLP). (*c*) Pyridoxamine-5'-phosphate (PMP). (*d*) The Schiff base that forms between PLP and an enzyme ε-amino group.

condensation of its aldehyde group with the ε-amino group of an enzymatic Lys residue (Fig. 26-1*d*). This Schiff base, which is conjugated to the coenzyme's pyridinium ring, is the focus of the coenzyme's activity.

Esmond Snell, Alexander Braunstein, and David Metzler demonstrated that the aminotransferase reaction occurs via a Ping Pong Bi Bi mechanism whose two stages consist of three steps each (Fig. 26-2):

Figure 26-2 The mechanism of PLP-dependent enzyme-catalyzed transamination. The first stage of the reaction, in which the α-amino group of an amino acid is transferred to PLP yielding an α-keto acid and PMP, consists of three steps: **(1)** transimination; **(2)** tautomerization, in which the Lys released during the transimination reaction acts as a general acid–base catalyst; and **(3)** hydrolysis. The second stage of the reaction, in which the amino group of PMP is transferred to a different α-keto acid to yield a new α-amino acid and PLP, is essentially the reverse of the first stage: Steps 3′, 2′, and 1′ are, respectively, the reverse of Steps 3, 2, and 1. **See the Animated Figures**

b. Stage I: Conversion of an Amino Acid to an α-Keto Acid

Step 1. The amino acid's nucleophilic amino group attacks the enzyme–PLP Schiff base carbon atom in a **transimination (trans-Schiffization)** reaction to form an amino acid–PLP Schiff base (aldimine), with concomitant release of the enzyme's Lys amino group. This Lys is then free to act as a general base at the active site.

Step 2. The amino acid–PLP Schiff base tautomerizes to an α-keto acid–PMP Schiff base by the active site Lys–catalyzed removal of the amino acid α hydrogen and protonation of PLP atom C4' via a resonance-stabilized carbanion intermediate. This resonance stabilization facilitates the cleavage of the C_α—H bond.

Step 3. The α-keto acid–PMP Schiff base is hydrolyzed to PMP and an α-keto acid.

c. Stage II: Conversion of an α-Keto Acid to an Amino Acid

To complete the aminotransferase's catalytic cycle, the coenzyme must be converted from PMP back to the enzyme–PLP Schiff base. This involves the same three steps as above, but in reverse order:

Step 3'. PMP reacts with an α-keto acid to form a Schiff base.

Step 2'. The α-keto acid–PMP Schiff base tautomerizes to form an amino acid–PLP Schiff base.

Step 1'. The ε-amino group of the active site Lys residue attacks the amino acid–PLP Schiff base in a transimination reaction to regenerate the active enzyme–PLP Schiff base, with release of the newly formed amino acid.

The reaction's overall stoichiometry therefore is

Amino acid 1 + α-keto acid 2 $\rightleftharpoons$
α-keto acid 1 + amino acid 2

Examination of the amino acid–PLP Schiff base's structure (Fig. 26-2, Step 1) reveals why this system is called "an electron-pusher's delight." *Cleavage of any of the amino acid C_α atom's three bonds (labeled a, b, and c) produces a resonance-stabilized C_α carbanion whose electrons are delocalized all the way to the coenzyme's protonated pyridinium nitrogen atom; that is, PLP functions as an electron sink.* For transamination reactions, this electron-withdrawing capacity facilitates removal of the α proton (a bond cleavage) in the tautomerization of the Schiff base. PLP-dependent reactions involving b bond cleavage (amino acid decarboxylation) and c bond labilization are discussed in Section 26-4B and in Sections 26-3Bb and 26-3G, respectively.

Aminotransferases differ in their specificity for amino acid substrates in the first stage of the transamination reaction, thereby producing the correspondingly different α-keto acid products. Most aminotransferases, however, accept only α-ketoglutarate or (to a lesser extent) oxaloacetate as the α-keto acid substrate in the second stage of the reaction, thereby yielding glutamate or aspartate as their only amino acid products. *The amino groups of most amino acids are consequently funneled into the formation of glutamate or aspartate, which are themselves interconverted by* **glutamate–aspartate aminotransferase:**

Glutamate + oxaloacetate $\rightleftharpoons$
α-ketoglutarate + aspartate

Oxidative deamination of glutamate (Section 26-1B) yields ammonia and regenerates α-ketoglutarate for another round of transamination reactions. Ammonia and aspartate are the two amino group donors in the synthesis of urea.

d. The Glucose–Alanine Cycle Transports Nitrogen to the Liver

An important exception to the foregoing is a group of muscle aminotransferases that accept pyruvate as their α-keto acid substrate. The product amino acid, alanine, is released into the bloodstream and transported to the liver, where it undergoes transamination to yield pyruvate for use in gluconeogenesis (Section 23-1A). The resulting glucose is returned to the muscles, where it is glycolytically degraded to pyruvate. This is the **glucose–alanine cycle** (Fig. 26-3). The amino group ends up in either ammonium ion or aspartate for urea biosynthesis. Evidently, the glucose–alanine cycle functions to transport nitrogen from muscle to liver.

During starvation the glucose formed in the liver by this route is also used by the other peripheral tissues, breaking the

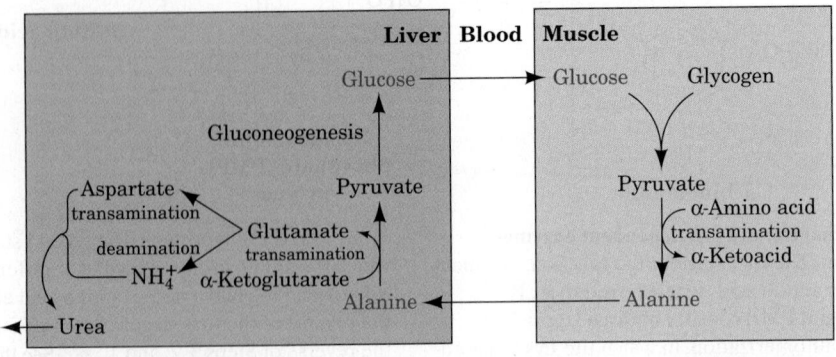

Figure 26-3 **The glucose–alanine cycle.** 🔊 **See the Animated Figures**

cycle. Under these conditions both the amino group and the pyruvate originate from muscle protein degradation, providing a pathway yielding glucose for other tissue use (recall that muscle is not a gluconeogenic tissue; Section 23-1).

Nitrogen is also transported to the liver in the form of glutamine, synthesized from glutamate and ammonia in a reaction catalyzed by **glutamine synthetase** (Section 26-5Ab). The ammonia is released for urea synthesis in liver mitochondria or for excretion in the kidney through the action of **glutaminase** (Section 26-3D).

B. *Oxidative Deamination: Glutamate Dehydrogenase*

Glutamate is oxidatively deaminated in the mitochondrial matrix by glutamate dehydrogenase (GDH), the only known enzyme that, in at least some organisms, can accept either NAD^+ or $NADP^+$ as its redox coenzyme. Oxidation is thought to occur with transfer of a hydride ion from glutamate's C_α to $NAD(P)^+$, thereby forming α-iminoglutarate, which is hydrolyzed to α-ketoglutarate and ammonia (Fig. 26-4). GDH is allosterically inhibited by GTP, NADH, and nonpolar compounds such as palmitoyl-CoA and steroid hormones. It is activated by ADP, NAD^+, and leucine (the most abundant amino acid in proteins; Table 4-1).

a. The X-Ray Structures of GDH Reveal its Allosteric Mechanism

The X-ray structures of homohexameric GDH from bovine and human liver mitochondria, determined by Thomas Smith, reveal that each monomer has three domains, a substrate domain, a coenzyme domain, and an antenna domain. The protein, which has D_3 symmetry, can be considered to be a dimer of trimers, with the antenna domains of each trimer wrapping around each other about the 3-fold axis (Fig. 26-5a). Structural comparison of a 501-residue monomer of the bovine GDH–glutamate–NADH–GTP complex (Fig. 26-5b) with that of the 96% identical human **apoenzyme** (no active site or regulatory ligands bound; Fig. 26-5c) reveals that, on binding ligands, the coenzyme binding domain rotates about the so-called pivot helix so as to close the cleft between the coenzyme and substrate domains. Simultaneously, the antenna domain twists in a way that unwinds one turn of the antenna helix that is connected to the pivot helix. Although the closed form is required for catalysis, the open form favors the association and dissociation of substrates and products. In the open state, Arg 463 (human numbering) in the center of the pivot helix interacts with the activator ADP (whose binding site in the bovine complex is occupied by the ADP moiety of an NADH; Fig. 26-5b), whereas in the closed state, the side chain of His 454 hydrogen-bonds to the γ-phosphate of the inhibitor GTP. The GTP binding site is distorted and blocked in the open state so that GTP binding favors the closed form of the enzyme. This results in tight binding of substrates and products and hence inhibition of the enzyme. ADP binding favors the open form, allowing product dissociation, and therefore activates the enzyme. Allosteric interactions appear to be communicated between subunits through the interactions of the antenna domains. In fact, bacterial GDHs, which lack allosteric regulation, differ from mammalian GDHs mainly by the absence of antenna domains.

b. Hyperinsulinism/Hyperammonemia (HI/HA) Is Caused by Uncontrolled GDH Activity

Charles Stanley has reported a new form of congenital hyperinsulinism that is characterized by hypoglycemia and **hyperammonemia** (**HI/HA**; hyperammonemia is elevated levels of ammonia in the blood) and has shown that it is caused by mutations in GDH at the N-terminal end of its pivot helix in the GTP binding site or in the antenna domain near its joint with the pivot helix. The mutant enzymes have reduced sensitivity to GTP inhibition but retain their ability to be activated by ADP. The GDH mutants S448P, H454Y, and R463A, which were respectively designed to affect the antenna region, the GTP binding site, and the ADP binding site (Fig. 26-5b), all have decreased sensitivity to GTP inhibition (Fig. 26-6), with H454Y and S448P, which were previously known to be associated with HI/HA, conferring the most resistance to GTP inhibition. The hypoglycemia and hyperammonemia in HI/HA patients arises from the increased activity of the GDH mutants in the breakdown direction, producing increased amounts of α-ketoglutarate and NH_3. The increased levels of α-ketoglutarate stimulate the citric acid cycle and oxidative phosphorylation, which has been shown to lead to increased insulin secretion and hypoglycemia, thereby producing the symptoms of the disease. The NH_4^+ produced is usually converted to urea (Section 26-2) but can also be exported to the bloodstream.

If this scenario for the cause of HI/HA is correct, it requires a reassessment of the role of GDH in ammonia homeostasis. The equilibrium position of the GDH reaction greatly favors the synthesis of Glu ($\Delta G^{\circ\prime} \approx 30\ kJ \cdot mol^{-1}$ for the reaction as written in Fig. 26-4), but studies of cellular

$$^-OOC-CH_2-CH_2-\underset{\underset{H}{|}}{\overset{\overset{NH_3^+}{|}}{C}}-COO^- + NAD(P)^+$$

Glutamate

$$\left[^-OOC-CH_2-CH_2-\overset{\overset{NH_2^+}{\|}}{C}-COO^- \right] + NAD(P)H + H^+$$

α-Iminoglutarate

$$\Big\updownarrow \; -\; H_2O$$

$$^-OOC-CH_2-CH_2-\overset{\overset{O}{\|}}{C}-COO^- + NH_4^+$$

α-Ketoglutarate

Figure 26-4 **The oxidative deamination of glutamate by glutamate dehydrogenase.** This reaction involves the intermediate formation of α-iminoglutarate.

Figure 26-5 X-ray structures of glutamate dehydrogenase (GDH).
(*a*) Bovine GDH in complex with glutamate, NADH, and GTP. The homohexameric enzyme, which has D_3 symmetry, is viewed along one of its 2-fold axes with its 3-fold axis vertical. Each of its subunits is differently colored. The bound substrates and ligands are shown in space-filling form with glutamate orange, the substrate NADH pink, the NADH bound at the ADP effector site brown, and the GTP effector gray. (*b*) One subunit of the bovine GDH–glutamate–NADH–GTP complex drawn with the coenzyme binding domain magenta, the substrate binding domain orange, the antenna domain green, and the pivot helix cyan. The substrates and ligands are shown in space-filling form colored according to atom type with glutamate C green, substrate NADH C gold, ADP site-bound NADH C pink, GTP C cyan, N blue, O red, and P magenta. The C_α atoms of Ser 448, His 454, and Arg 463 (human numbering) are represented by yellow spheres. (*c*) One subunit of human apoGDH with the protein colored as and viewed similarly to Part *b*. [Based on X-ray structures by Thomas Smith, Donald Danforth Plant Science Center, St. Louis, Missouri. PDBids (*a* and *b*) 1HWX and (*c*) 1L1F.]

Ser 448
His 454
Arg 463

Closed active site cleft

(*b*)

Ser 448
His 454
Arg 463

Open active site cleft

(*c*)

W/T
R463A
S448P
H454Y

GDH activity (% basal)

100

50

0

0.01   0.1   1   10   100   1000   10,000   100,000

GTP (μmol/L)

Figure 26-6 Inhibition of human glutamate dehydrogenase (GDH) by GTP. Human wild-type and mutant GDHs were expressed in *E. coli* and assayed for sensitivity to GTP inhibition. The midpoint of each curve corresponds to the concentration of GTP causing 50% inhibition. [After Fang, J., Hsu, B.Y.L., MacMullen, C.M., Poncz, M., Smith, T.J., and Stanley, C.A., *Biochem. J.* **363,** 81 (2002).]

substrate and product concentrations suggested that the enzyme functions close to equilibrium ($\Delta G \approx 0$) *in vivo*. It was therefore widely accepted that increases in [NH_3], high levels of which are toxic, would cause GDH to act in reverse, removing NH_3 and hence preventing its buildup to toxic levels. However, since HI/HA patients have increased GDH activity yet have higher levels of NH_3 than normal, this accepted role of GDH cannot be correct. Indeed, if GDH functioned close to equilibrium, changes in its activity resulting from allosteric interactions would not result in significant flux changes.

C. *Other Deamination Mechanisms*

Two nonspecific amino acid oxidases, **L-amino acid oxidase** and **D-amino acid oxidase,** catalyze the oxidation of L- and D-amino acids, utilizing FAD as their redox coenzyme [rather than $NAD(P)^+$]. The resulting $FADH_2$ is reoxidized by O_2.

$$\text{Amino acid} + FAD + H_2O \longrightarrow$$
$$\alpha\text{-keto acid} + NH_3 + FADH_2$$
$$FADH_2 + O_2 \longrightarrow FAD + H_2O_2$$

D-Amino acid oxidase occurs mainly in kidney. Its function is an enigma since D-amino acids are associated mostly with bacterial cell walls (Section 11-3Ba). A few amino acids, such as serine and histidine, are deaminated nonoxidatively (Sections 26-3B and 26-3D).

2 THE UREA CYCLE

Living organisms excrete the excess nitrogen resulting from the metabolic breakdown of amino acids in one of three ways. Many aquatic animals simply excrete ammonia. Where water is less plentiful, however, processes have evolved that convert ammonia to less toxic waste products that therefore require less water for excretion. One such product is urea, which is excreted by most terrestrial vertebrates; another is **uric acid,** which is excreted by birds and terrestrial reptiles:

| Ammonia | Urea | Uric acid |

Accordingly, living organisms are classified as being either **ammonotelic** (ammonia excreting), **ureotelic** (urea excreting), or **uricotelic** (uric acid excreting). Some animals can shift from ammonotelism to ureotelism or uricotelism if their water supply becomes restricted. Here we focus our attention on urea formation. Uric acid biosynthesis is discussed in Section 28-4A.

Urea is synthesized in the liver by the enzymes of the **urea cycle.** It is then secreted into the bloodstream and sequestered by the kidneys for excretion in the urine. The urea cycle was elucidated in outline in 1932 by Hans Krebs and Kurt Henseleit (the first known metabolic cycle; Krebs did not elucidate the citric acid cycle until 1937). Its individual reactions were later described in detail by Sarah Ratner and Philip Cohen. The overall urea cycle reaction is

Thus, the two urea nitrogen atoms are contributed by NH_3 and aspartate, whereas the carbon atom comes from HCO_3^-. Five enzymatic reactions are involved in the urea cycle, two of which are mitochondrial and three cytosolic (Fig. 26-7). In this section, we examine the mechanisms of these reactions and their regulation.

A. *Carbamoyl Phosphate Synthetase: Acquisition of the First Urea Nitrogen Atom*

Carbamoyl phosphate synthetase (CPS) is technically not a urea cycle enzyme. It catalyzes the condensation and activation of NH_3 and HCO_3^- to form **carbamoyl phosphate,** the first of the cycle's two nitrogen-containing substrates, with the concomitant hydrolysis of two ATPs. Eukaryotes have two forms of CPS:

1. Mitochondrial **CPS I** uses NH_3 as its nitrogen donor and participates in urea biosynthesis.

2. Cytosolic **CPS II** uses glutamine as its nitrogen donor and is involved in pyrimidine biosynthesis (Section 28-2A). The reaction catalyzed by CPS I involves three steps (Fig. 26-8):

1. Activation of HCO_3^- by ATP to form **carboxyphosphate** and ADP.

2. Nucleophilic attack of NH_3 on carboxyphosphate, displacing the phosphate to form **carbamate** and P_i.

3. Phosphorylation of carbamate by the second ATP to form carbamoyl phosphate and ADP.

The reaction is essentially irreversible and is the rate-limiting step of the urea cycle. CPS I is subject to allosteric activation by **N-acetylglutamate** as is discussed in Section 26-2F.

E. coli contains only one type of CPS, which is homologous to both CPS I and CPS II. The enzyme is a heterodimer but when allosterically activated by ornithine (a urea cycle intemediate), it forms a tetramer of heterodimers, $(\alpha\beta)_4$. Its small subunit (382 residues) functions to hydrolyze glutamine and deliver the resulting NH_3 to its large subunit (1073 residues). However, if the enzyme's **glutaminase (glutamine**

$$H_3C - \underset{\underset{O}{\|}}{C} - COO^-$$

Pyruvate

Glutamate

glutamate
dehydrogenase

transaminase

$NAD(P)^+$

$$H_3C - \underset{\underset{NH_3^+}{|}}{CH} - COO^-$$

Alanine

$H^+ + \boxed{NAD(P)H}$ ⟶ α-Ketoglutarate

Mitochondrion

$$2ATP + HCO_3^- + NH_3 \xrightarrow{\textbf{1}} H_2N - \underset{\underset{O}{\|}}{C} - OPO_3^{2-} + 2ADP + P_i$$

Carbamoyl phosphate

P_i

$$O = C - NH_2$$
$$|$$
$$NH$$
$$|$$
$$(CH_2)_3$$
$$|$$
$$HC - NH_3^+$$
$$|$$
$$COO^-$$

Ornithine

2

Citrulline

Pyruvate

$CO_2 + ATP$

pyruvate
carboxylase

$ADP + P_i$

Oxaloacetate

trans-
aminase

α-Ketoglutarate

Glutamate

Aspartate

Glutamate

$$\begin{array}{l} NH_3^+ \\ | \\ (CH_2)_3 \\ | \\ H - C - NH_3^+ \\ | \\ COO^- \end{array}$$

Ornithine

Citrulline

ATP

3

α-Keto acid
transaminase
Amino acid

α-Ketoglutarate

$$\begin{array}{l} COO^- \\ | \\ CH_2 \\ | \\ HC - NH_3^+ \\ | \\ COO^- \end{array}$$

Aspartate

$$\begin{array}{l} O \\ \| \\ H_2N - C - NH_2 \end{array}$$

Urea

**Urea
cycle**

AMP + PP$_i$

5

H_2O

Arginine

$$\begin{array}{l} H_2N \quad \overset{+}{NH_2} \\ \diagdown \; / \\ C \\ | \\ NH \\ | \\ (CH_2)_3 \\ | \\ H - C - NH_3^+ \\ | \\ COO^- \end{array}$$

**Arginino-
succinate**

4

$$\begin{array}{l} COO^- \\ | \\ CH_2 \\ | \\ HC - \underset{\underset{H}{|}}{N} - \overset{\overset{+}{NH_2}}{\underset{\underset{(CH_2)_3}{|}}{\underset{\underset{HC - NH_3^+}{|}}{\underset{|}{NH}}}} \\ COO^- \qquad\qquad COO^- \end{array}$$

Cytosol

$$\begin{array}{l} COO^- \\ | \\ HC \\ \| \\ CH \\ | \\ COO^- \end{array}$$

Fumarate

H_2O | fumarase

$$\begin{array}{l} COO^- \\ | \\ CH_2 \\ | \\ CH - OH \\ | \\ COO^- \end{array}$$

Malate

NAD^+

malate dehydrogenase

$\boxed{NADH} + H^+$

$$\begin{array}{l} COO^- \\ | \\ CH_2 \\ | \\ C = O \\ | \\ COO^- \end{array}$$

Oxaloacetate

⟶ $\boxed{\text{Gluconeogenesis}}$

Figure 26-7 *(Opposite)* **The urea cycle.** Its five enzymes are (1) carbamoyl phosphate synthetase, (2) ornithine transcarbamoylase, (3) argininosuccinate synthetase, (4) argininosuccinase, and (5) arginase. The reactions occur in part in the mitochondrion and in part in the cytosol with ornithine and citrulline being transported across the mitochondrial membrane by specific transport systems (*yellow circles*). One of the urea amino groups (*green*) originates as the NH_3 product of the glutamate dehydrogenase reaction (*top*). The other amino group (*red*) is obtained from aspartate through the transfer of an amino acid to oxaloacetate via transamination (*right*). The fumarate product of the argininosuccinase reaction is converted to oxaloacetate for entry into gluconeogenesis via the same reactions that occur in the citric acid cycle but take place in the cytosol (*bottom*). The ATP utilized in Reactions 1 and 3 of the cycle can be regenerated by oxidative phosphorylation from the NAD(P)H produced in the glutamate dehydrogenase (*top*) and malate dehydrogenase (*bottom*) reactions. 🐭**See the Animated Figures**

amidotransferase) activity is eliminated (e.g., by site-directed mutagenesis), the large subunit can still produce carbamoyl phosphate if NH_3 is supplied in high enough concentration. The large subunit is composed of two nearly superimposable halves that have 40% sequence identity. The N-terminal half contains the carboxyphosphate synthetic component and an oligomerization domain while the C-terminal half contains the carbamoyl phosphate synthetic component and an allosteric binding domain.

a. *E. coli* CPS Contains an Extraordinarily Long Tunnel

The X-ray structure of *E. coli* CPS in complex with Mn^{2+}, ADP, P_i, and ornithine, determined by Hazel Holden and Ivan Rayment, reveals that the active site for synthesis of the carboxyphosphate intermediate is ~45 Å away from the ammonia synthesis site and also ~35 Å away from the carbamoyl phosphate synthesis active site. Astonishingly, the three sites are connected by a narrow 96-Å-long molecular tunnel that runs nearly the length of the elongated protein molecule (Fig. 26-9). It therefore appears that CPS guides its intermediate products from the active site in which they are formed to that in which they are utilized. This phenomenon, in which the intermediate of two reactions is directly transferred from one enzyme active site to

Figure 26-9 **X-ray structure of *E. coli* carbamoyl phosphate synthetase (CPS).** The protein is represented by its C_α backbone. The small subunit (*magenta*) contains the glutamine binding site where NH_3 is produced or bound. The large subunit consists of the carboxyphosphate domain (*green*), the oligomerization domain (*yellow*), the carbamoyl phosphate domain (*blue*), and the allosteric binding domain (*orange*). The 96-Å-long tunnel connecting the three active sites is outlined in red. [Courtesy of Hazel Holden and Ivan Rayment, University of Wisconsin. PDBid 1JDB.]

Figure 26-8 **The mechanism of action of CPS I.** (1) Activation of HCO_3^- by phosphorylation forms the intermediate, carboxyphosphate; (2) nucleophilic attack on carboxyphosphate by NH_3 forms the reaction's second intermediate, carbamate; and (3) phosphorylation of carbamate by ATP yields the reaction product carbamoyl phosphate.

another, is called **channeling** (the term "tunneling" is reserved for certain quantum mechanical phenomena).

Channeling increases the rate of a metabolic pathway by preventing the loss of its intermediate products as well as protecting the intermediate from degradation. NH_3 must travel ~45 Å down the CPS tunnel to react with carboxyphosphate to form the next intermediate, carbamate. The carbamate, in turn, must travel an additional ~35 Å to the site where it is phosphorylated by ATP to form the final product carbamoyl phosphate. The NH_3 transfer tunnel is lined with polar groups capable of forming hydrogen bonds with NH_3, whereas the tunnel through which carbamate travels is lined with backbone atoms and lacks charged groups that might induce its hydrolysis as it diffuses between active sites. Shielding and channeling are necessary because the intermediates carboxyphosphate and carbamate are extremely reactive, having half-lives of 28 and 70 ms, respectively, at neutral pH. Also, channeling allows the local concentration of NH_3 to reach a higher value than is present in the cellular medium. We shall encounter several other examples of channeling in our studies of metabolic enzymes, but the CPS tunnel is far longer than that in any other known enzyme.

B. Ornithine Transcarbamoylase

Ornithine transcarbamoylase transfers the carbamoyl group of carbamoyl phosphate to **ornithine**, yielding **citrulline** (Fig. 26-7, Reaction 2; note that both of these compounds are "nonstandard" α-amino acids in that they do not occur in proteins). The reaction occurs in the mitochondrion so that ornithine, which is produced in the cytosol, must enter the mitochondrion via a specific transport system. Likewise, since the remaining urea cycle reactions occur in the cytosol, citrulline must be exported from the mitochondrion.

C. Argininosuccinate Synthetase: Acquisition of the Second Urea Nitrogen Atom

Urea's second nitrogen atom is introduced in the urea cycle's third reaction by the condensation of citrulline's ureido group with an aspartate amino group by **argininosuccinate synthetase** (Fig. 26-10). The ureido oxygen atom is activated as a leaving group through formation of a citrullyl–AMP intermediate, which is subsequently displaced by the aspartate amino group. Support for the existence of the citrullyl–AMP intermediate comes from experiments using ^{18}O-labeled citrulline (* in Fig. 26-10). The label was isolated in the AMP produced by the reaction, demonstrating that at some stage of the reaction, AMP and citrulline are linked covalently through the ureido oxygen atom.

D. Argininosuccinase

With formation of argininosuccinate, all of the urea molecule components have been assembled. However, the amino group donated by aspartate is still attached to the aspartate carbon skeleton. This situation is remedied by the **argininosuccinase**-catalyzed elimination of arginine from the aspartate carbon skeleton forming fumarate (Fig. 26-7, Reaction 4). Arginine is urea's immediate precursor. The fumarate produced in the argininosuccinase reaction reacts via the fumarase and malate dehydrogenase reactions to form oxaloacetate (Fig. 26-7, *bottom*), which is then used in gluconeogenesis (Section 23-1).

E. Arginase

The urea cycle's fifth and final reaction is the **arginase**-catalyzed hydrolysis of arginine to yield urea and regenerate ornithine (Fig. 26-7, Reaction 5). Ornithine is then returned to the mitochondrion for another round of the cycle. The urea cycle thereby converts two amino groups, one from NH_3 and one from aspartate, and a carbon atom from HCO_3^- to the relatively nontoxic excretion product urea at the cost of four "high-energy" phosphate bonds (three ATP hydrolyzed to two ADP, two P_i, AMP, and PP_i, followed by rapid PP_i hydrolysis). This energetic cost, together with that of gluconeogenesis, is supplied by the oxidation of the acetyl-CoA formed by the breakdown of amino acid carbon skeletons (e.g., threonine, Fig. 26-12). Indeed, half the oxygen that the liver consumes is used to provide this energy.

F. Regulation of the Urea Cycle

Carbamoyl phosphate synthetase I, the mitochondrial enzyme that catalyzes the first committed reaction of the urea

Figure 26-10 **The mechanism of action of argininosuccinate synthetase.** The steps involved are **(1)** activation of the ureido oxygen of citrulline through the formation of citrullyl–AMP and **(2)** displacement of AMP by the α-amino group of aspartate. The asterisk (*) traces the fate of ^{18}O originating in citrulline's ureido group.

cycle, is allosterically activated by *N*-acetylglutamate:

$$
\begin{array}{c}
\text{COO}^- \\
| \\
(\text{CH}_2)_2 \quad \text{O} \\
| \qquad \quad \| \\
\text{H}\!-\!\text{C}\!-\!\text{N}\!-\!\text{C}\!-\!\text{CH}_3 \\
| \quad \ \ \text{H} \\
{}^-\text{OOC}
\end{array}
$$

N-Acetylglutamate

This metabolite is synthesized from glutamate and acetyl-CoA by **N-acetylglutamate synthase** and hydrolyzed by a specific hydrolase. The rate of urea production by the liver is, in fact, correlated with the *N*-acetylglutamate concentration. Increased urea synthesis is required when amino acid breakdown rates increase, generating excess nitrogen that must be excreted. Increases in these breakdown rates are signaled by an increase in glutamate concentration through transamination reactions (Section 26-1). This situation, in turn, causes an increase in *N*-acetylglutamate synthesis, stimulating carbamoyl phosphate synthetase and thus the entire urea cycle.

The remaining enzymes of the urea cycle are controlled by the concentrations of their substrates. Thus, inherited deficiencies in urea cycle enzymes other than arginase do not result in significant decreases in urea production (the total lack of any urea cycle enzyme results in death shortly after birth). Rather, the deficient enzyme's substrate builds up, increasing the rate of the deficient reaction to normal. The anomalous substrate buildup is not without cost, however. The substrate concentrations become elevated all the way back up the cycle to NH_3, resulting in hyperammonemia. Although the root cause of NH_3 toxicity is not completely understood, high $[NH_3]$ puts an enormous strain on the NH_3-clearing system, especially in the brain (symptoms of urea cycle enzyme deficiencies include mental retardation and lethargy). This clearing system has been proposed to involve glutamate dehydrogenase (working in reverse) and **glutamine synthetase,** which decrease the α-ketoglutarate and glutamate pools (Sections 26-1 and 26-5Ab). The brain is most sensitive to the depletion of these pools. Depletion of α-ketoglutarate decreases the rate of the energy-generating citric acid cycle, whereas decreasing the glutamate concentration disturbs neuronal function, since it is both a neurotransmitter and a precursor to γ-aminobutyrate (GABA), another neurotransmitter (Section 20-5Cf). Glutamate depletion would also decrease the functioning of the urea cycle, since it is also the precursor to *N*-acetylglutamate, the major regulator of the cycle. The involvement of GDH in NH_3 clearance is a subject of debate in light of the observation that HI/HA involves deinhibition of GDH (Section 26-1Bb), suggesting that increased GDH activity increases the NH_3 concentration rather than decreasing it.

3 METABOLIC BREAKDOWN OF INDIVIDUAL AMINO ACIDS

The degradation of amino acids converts them to citric acid cycle intermediates or their precursors so that they can be metabolized to CO_2 and H_2O or used in gluconeogenesis.

Indeed, oxidative breakdown of amino acids typically accounts for 10 to 15% of the metabolic energy generated by animals. In this section we consider how amino acid carbon skeletons are catabolized. The 20 "standard" amino acids (the amino acids of proteins) have widely differing carbon skeletons, so their conversions to citric acid cycle intermediates follow correspondingly diverse pathways. We shall not describe all of the many reactions involved in detail. Rather, we shall consider how these pathways are organized and focus on a few reactions of chemical and/or medical interest.

A. Amino Acids Can Be Glucogenic, Ketogenic, or Both

"Standard" amino acids are degraded to one of seven metabolic intermediates: pyruvate, α-ketoglutarate, succinyl-CoA, fumarate, oxaloacetate, acetyl-CoA, or acetoacetate (Fig. 26-11). The amino acids may therefore be divided into two groups based on their catabolic pathways (Fig. 26-11):

1. Glucogenic amino acids, whose carbon skeletons are degraded to pyruvate, α-ketoglutarate, succinyl-CoA, fumarate, or oxaloacetate and are therefore glucose precursors (Section 23-1A).

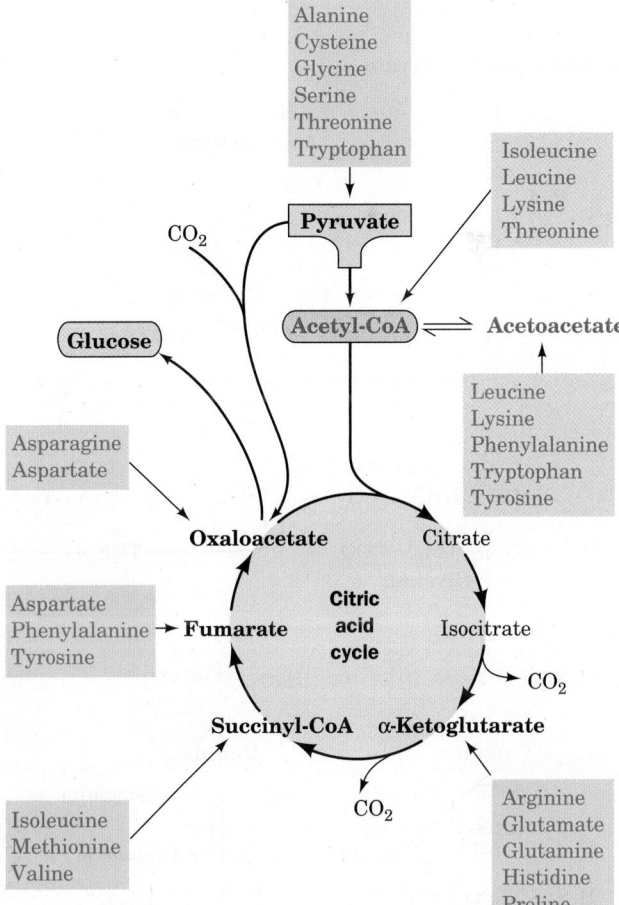

Figure 26-11 **Degradation of amino acids to one of seven common metabolic intermediates.** Glucogenic and ketogenic degradations are indicated in green and red, respectively.

2. Ketogenic amino acids, whose carbon skeletons are broken down to acetyl-CoA or acetoacetate and can thus be converted to ketone bodies or fatty acids (Sections 25-3 and 25-4).

For example, alanine is glucogenic because its transamination product, pyruvate (Section 26-1A), can be converted to glucose via gluconeogenesis (Section 23-1A). Leucine, on the other hand, is ketogenic; its carbon skeleton is converted to acetyl-CoA and acetoacetate (Section 26-3F). Since animals lack any metabolic pathway for the net conversion of acetyl-CoA or acetoacetate to gluconeogenic precursors, no net synthesis of carbohydrates is possible from leucine, or from lysine, the only other purely ketogenic amino acid. Isoleucine, phenylalanine, threonine, tryptophan, and tyrosine, however, are both glucogenic and ketogenic; isoleucine, for example, is broken down to succinyl-CoA and acetyl-CoA and hence is a precursor of both carbohydrates and ketone bodies (Section 26-3Ed). The remaining 13 amino acids are purely glucogenic.

In studying the specific pathways of amino acid breakdown, we shall organize the amino acids into groups that are degraded into each of the seven metabolic intermediates mentioned above: pyruvate, oxaloacetate, α-ketoglutarate, succinyl-CoA, fumarate, acetyl-CoA, and acetoacetate. When acetoacetyl-CoA is a product in amino acid degradation, it can, of course, be directly converted to acetyl-CoA (Section 25-3). We also discuss the pathway by which, in liver, it is converted instead to acetoacetate for use as an alternative fuel source in peripheral tissues (Section 25-3).

B. Alanine, Cysteine, Glycine, Serine, and Threonine Are Degraded to Pyruvate

Five amino acids, alanine, cysteine, glycine, serine, and threonine, are broken down to yield pyruvate (Fig. 26-12). Tryptophan should also be included in this group since one of its breakdown products is alanine (Section 26-3G), which, as we have seen (Section 26-1Ad), is transaminated to pyruvate.

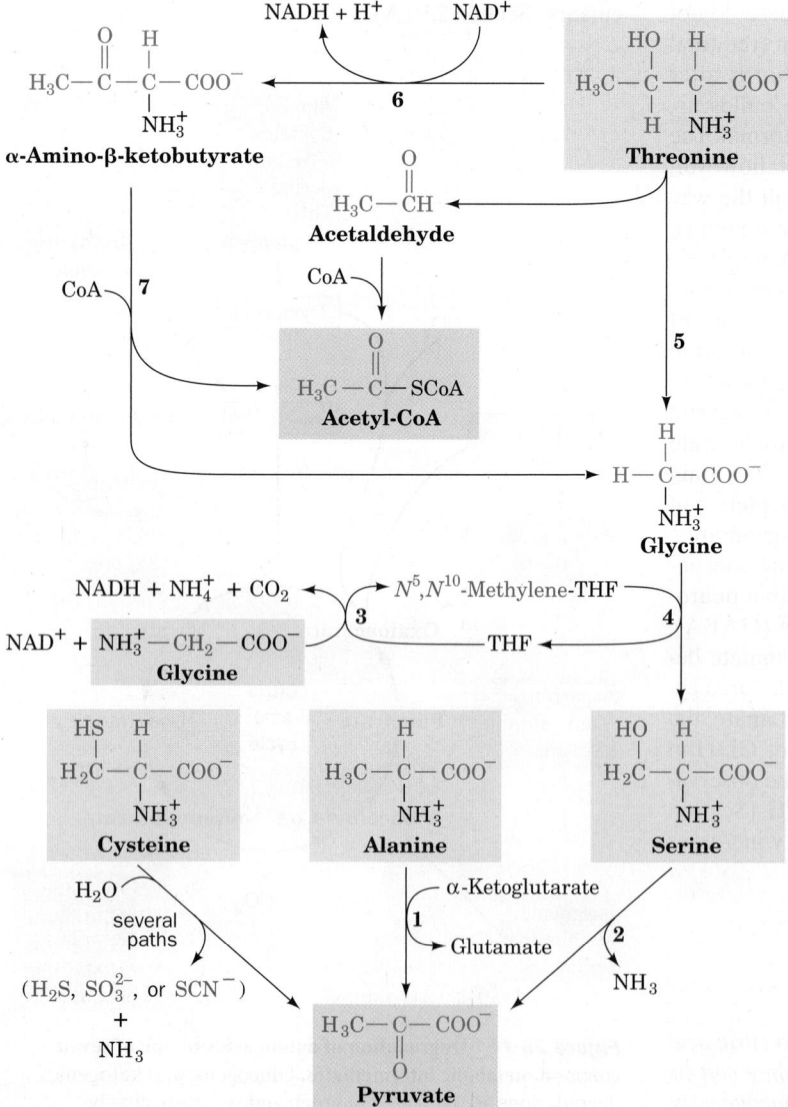

Figure 26-12 **The pathways converting alanine, cysteine, glycine, serine, and threonine to pyruvate.** The enzymes involved are **(1)** alanine aminotransferase, **(2)** serine dehydratase, **(3)** glycine cleavage system, **(4)** and **(5)** serine hydroxymethyltransferase, **(6)** threonine dehydrogenase, and **(7)** α-amino-β-ketobutyrate lyase.

Figure 26-13 **The serine dehydratase reaction.** This PLP-dependent enzyme catalyzes the elimination of water from serine. The steps in the reaction are **(1)** formation of a serine–PLP Schiff base, **(2)** removal of the α-H atom of serine to form a resonance-stabilized carbanion, **(3)** β elimination of OH^-, **(4)** hydrolysis of the Schiff base to yield the PLP–enzyme and aminoacrylate, **(5)** nonenzymatic tautomerization to the imine, and **(6)** nonenzymatic hydrolysis to form pyruvate and ammonia.

Pyruvate

Aminoacrylate

Serine is converted to pyruvate through dehydration by **serine dehydratase.** This PLP–enzyme, like the aminotransferases (Section 26-1), functions by forming a PLP–amino acid Schiff base, which facilitates the removal of the amino acid's α-hydrogen atom. In the serine dehydratase reaction, however, the C_α carbanion breaks down with the elimination of the amino acid's C_β OH, rather than with tautomerization (Fig. 26-2, Step 2), so that the substrate undergoes α,β elimination of H_2O rather than deamination (Fig. 26-13). The product of the dehydration, the enamine **aminoacrylate,** tautomerizes nonenzymatically to the corresponding imine, which spontaneously hydrolyzes to pyruvate and ammonia.

Cysteine may be converted to pyruvate via several routes in which the sulfhydryl group is released as H_2S, SO_3^{2-} or SCN^-.

Glycine is converted to serine by the enzyme **serine hydroxymethyltransferase,** another PLP-containing enzyme (Fig. 26-12, Reaction 4). This enzyme utilizes N^5,N^{10}-**methylene-tetrahydrofolate (N^5,N^{10}-methylene-THF)** as a cofactor to provide the C_1 unit necessary for this conversion. We shall defer a detailed discussion of THF cofactors until Section 26-4D.

a. The Glycine Cleavage System Is a Multienzyme Complex

The methylene group of the N^5,N^{10}-methylene-THF utilized in the conversion of glycine to serine is obtained from the methylene group of a second glycine via a reaction in which this glycine's remaining atoms are released as CO_2 and NH_4^+ (Fig. 26-12, Reaction 3). This reaction is catalyzed by the **glycine cleavage system** (also called the **glycine decarboxylase multienzyme system** in plants and **glycine synthase** when acting in the reverse direction; Section 26-5Ae), a complex resembling the pyruvate dehydrogenase complex (Section 21-2A) that consists of four proteins (Fig. 26-14):

1. A PLP-dependent glycine decarboxylase **(P-protein).**

2. A lipoamide-containing aminomethyl carrier **(H-protein),** which carries the aminomethyl group remaining after glycine decarboxylation.

Figure 26-14 **The reactions catalyzed by the glycine cleavage system, a multienzyme complex.** The enzymes involved are **(1)** a PLP-dependent glycine decarboxylase (P-protein), **(2)** a lipoamide-containing protein (H-protein), **(3)** a THF-requiring enzyme (T-protein), and **(4)** an NAD^+-dependent, FAD-requiring dihydrolipoyl dehydrogenase (L-protein).

3. An N^5,N^{10}-methylene-THF synthesizing enzyme **[T-protein;** alternatively **aminomethyltransferase (AMT)],** which accepts a methylene group from the aminomethyl carrier (H-protein; the amino group is released as ammonia).

4. An NAD$^+$-dependent, FAD-requiring dihydrolipoyl dehydrogenase **(L-protein),** a protein shared by and known as E3 in the pyruvate dehydrogenase complex (Section 21-2A).

Unlike the pyruvate dehydrogenase complex, the glycine cleavage system components are only loosely associated and hence are isolated as individual proteins. Nevertheless, H-protein has the central role in this multienzyme system: Its oxidized lipoyllysyl arm (Section 21-2Ad) is reduced as it accepts an aminomethyl group from P-protein (Fig. 26-14, Step 2), it donates the methylene group to THF in complex with T-protein as ammonia is released (Fig. 26-14, Step 3), and is then reoxidized by L-protein (Fig. 26-14, Step 4). The X-ray structure of pea leaf H-protein (Fig. 26-15), determined by Roland Douce, reveals that it is largely composed of a sandwich of 3-stranded and 6-stranded antiparallel β sheets that structurally resembles the lipoyl domain of E2 in the pyruvate dehydrogenase complex (Fig. 21-10a).

The aminomethylthio group is unstable and ordinarily is rapidly hydrolyzed to formaldehyde and. NH$_4^+$ However, on its aminomethylation, the previously exposed lipoyl group (Fig. 26-15a) inserts into a hydrophobic cleft

in the H-protein where its amino group hydrogen-bonds to Glu 14 (Fig. 26-15b), thereby shielding the aminomethyl group from hydrolysis. Indeed, the replacement of Glu 14 by Ala results in the rapid hydrolysis of the aminomethyl group. It therefore appears that the T-protein in the THF · T · H complex functions to release the lipoyl group from the H-protein cleft and to orient the THF for approach to the methylene C of the aminomethyl group for reaction.

Two observations indicate that the above pathway is the major route of glycine degradation in mammalian tissues:

1. The serine isolated from an animal that has been fed [2-^{14}C]glycine is ^{14}C labeled at both C2 and C3. This observation indicates that the methylene group of the N^5,N^{10}-methylene-THF utilized by serine hydroxymethyltransferase is obtained from glycine C2.

2. The inherited human disease **nonketotic hyperglycinemia,** which is characterized by mental retardation and accumulation of large amounts of glycine in body fluids, results from the absence of one of the components of the glycine cleavage system.

The glycine cleavage system and serine hydroxymethyltransferase occupy a vital role in green leaves, catalyzing the rapid destruction of the huge amounts of glycine produced by photorespiration (Section 24-3Ca). In fact, these

(a)

Lys 63

(b)

Lys 63

Figure 26-15 **X-ray structure of H-protein from the pea leaf glycine cleavage system.** (*a*) The oxidized lipoamide-containing form in which the side chain of Glu 14 together with that of Lys 63 with its covalently linked lipoyl group are represented in ball-and-stick form colored according to atom type (Glu 14 C gold, Lys 63 C cyan, lipoyl C green, N blue, O red, and S yellow). (*b*) The reduced aminomethyl-dihydrolipoamide form of H-protein viewed and colored as in Part *a*. The dot surface represents the protein's solvent-accessible surface. Note how the

aminomethyl-dihydrolipoamide has changed conformation relative to the lipoamide in Part *a* so as to bind in a hydrophobic cleft in the protein where its amino group is hydrogen bonded to Glu 14 (*dashed black bond*). This protects the aminomethyl group from hydrolysis. [Based on X-ray structures by Roland Douce, Centre National de la Recherche Scientifique et Commisariat à l'Energy Atomique, Grenoble, France. PDBids (*a*) 1HPC and (*b*) 1HTP.]

enzymes comprise about half the proteins present in the mitochondria from pea and spinach leaves.

Threonine is both glucogenic and ketogenic, since one of its degradation routes produces both pyruvate and acetyl-CoA (Fig. 26-12, Reactions 6 and 7). Its major route of breakdown is through **threonine dehydrogenase,** producing **α-amino-β-ketobutyrate,** which is converted to acetyl-CoA and glycine by **α-amino-β-ketobutyrate lyase.** The glycine may be converted, through serine, to pyruvate.

b. Serine Hydroxymethyltransferase Catalyzes PLP-Dependent C_α—C_β Bond Cleavage

Threonine may also be converted directly to glycine and acetaldehyde (the latter being subsequently oxidized to acetyl-CoA), at least *in vitro,* via Reaction 5 of Fig. 26-12. Surprisingly, this reaction is catalyzed by serine hydroxymethyltransferase. We have heretofore considered PLP-catalyzed reactions that begin with the cleavage of an amino acid's C_α—H bond (Fig. 26-2). Degradation of threonine to glycine and acetaldehyde by serine hydroxymethyltransferase demonstrates that PLP also facilitates cleavage of an amino acid's C_α—C_β bond by delocalizing the electrons of the resulting carbanion into the conjugated PLP ring:

c. PLP Facilitates the Cleavage of Different Bonds in Different Enzymes

How can the same amino acid–PLP Schiff base be involved in the cleavage of the different bonds to an amino acid C_α in different enzymes? The answer to this conundrum was suggested by Harmon Dunathan. For electrons to be withdrawn into the conjugated ring system of PLP, the π-orbital system of PLP must overlap with the bonding orbital containing the electron pair being delocalized. This is possible only if the bond being broken lies in the plane perpendicular to the plane of the PLP π-orbital system (Fig. 26-16*a*). Different bonds to C_α can be placed in this plane by rotation about the C_α—N bond. Indeed, the X-ray structure of aspartate aminotransferase reveals that the C_α—H of its aspartate substrate assumes just this

Amino acid–PLP Schiff base

Delocalized α carbanion

(a)

(b)

Figure 26-16 **Bond orientation in a PLP–amino acid Schiff base.** (*a*) The π-orbital framework of a PLP–amino acid Schiff base. The bond to C_α in the plane perpendicular to the PLP π-orbital system (from X in the illustration) is labile as a consequence of its overlap with the π system, which permits the broken bond's electron pair to be delocalized over the conjugated molecule. (*b*) The Schiff base complex of the inhibitor **α-methylaspartate** with PLP in the X-ray structure of porcine aspartate aminotransferase as viewed normal to the pyridoxal ring. This inhibitor is drawn in ball-and-stick form with C green, N blue, O red, and P gold, with the exception that the methyl C atom and the bond linking it to the aspartate residue are magenta. Here the methyl C occupies the position of the H atom that the enzyme normally excises from aspartate. Note that the bond linking the methyl C to aspartate is in the plane perpendicular to the pyridoxal ring and is thus ideally oriented for bond cleavage. [Part *b* based on an X-ray structure by David Metzler and Arthur Arnone, University of Iowa. PDBid 1AJS.]

conformation (Fig. 26-16*b*). Evidently, *each enzyme specifically cleaves its corresponding bond because the enzyme binds the amino acid–PLP Schiff base adduct with this bond in the plane perpendicular to that of the PLP ring.* This is an example of stereoelectronic assistance (Section 15-1Eb): *The enzyme binds substrate in a conformation that minimizes the electronic energy of the transition state.*

C. Asparagine and Aspartate Are Degraded to Oxaloacetate

Transamination of aspartate leads directly to oxaloacetate:

Aspartate

Oxaloacetate

Asparagine is also converted to oxaloacetate in this manner after its hydrolysis to aspartate by L-**asparaginase**:

Asparagine

Aspartate

Interestingly, L-asparaginase is an effective chemotherapeutic agent in the treatment of cancers that must obtain asparagine from the blood, particularly **acute lymphoblastic leukemia.** The cancerous cells express particularly low levels of the enzyme asparagine synthetase (Section 26-5Ab) and hence die without an external source of asparagine. However, L-asparaginase treatment may select for cells with increased levels of asparagine synthetase expression, and hence, in these cases, the surviving cancer cells are resistant to this treatment.

D. Arginine, Glutamate, Glutamine, Histidine, and Proline Are Degraded to α-Ketoglutarate

Arginine, glutamine, histidine, and proline are all degraded by conversion to glutamate (Fig. 26-17), which in turn is oxidized to α-ketoglutarate by glutamate dehydrogenase (Section 26-1). Conversion of glutamine to glutamate involves only one reaction: hydrolysis by **glutaminase.** Histidine's conversion to glutamate is more complicated: It is nonoxidatively deaminated, then it is hydrated, and its imidazole ring is cleaved to form *N*-**formiminoglutamate.** The formimino group is then transferred to tetrahydrofolate forming glutamate and N^5-**formimino-tetrahydrofolate** (Section 26-4D). Both arginine and proline are converted to glutamate through the intermediate formation of **glutamate-5-semialdehyde.**

E. Isoleucine, Methionine, and Valine Are Degraded to Succinyl-CoA

Isoleucine, methionine, and valine have complex degradative pathways that all yield propionyl-CoA. Propionyl-CoA, which is also a product of odd-chain fatty acid degradation, is converted, as we have seen, to succinyl-CoA by a series of reactions involving the participation of biotin and coenzyme B_{12} (Section 25-2E).

a. Methionine Breakdown Involves Synthesis of *S*-Adenosylmethionine and Cysteine

Methionine degradation (Fig. 26-18) begins with its reaction with ATP to form *S*-**adenosylmethionine (SAM;** alternatively **AdoMet).** *This sulfonium ion's highly reactive methyl group makes it an important biological methylating agent.* For instance, we have already seen that SAM is the methyl donor in the synthesis of phosphatidylcholine from phosphatidylethanolamine (Section 25-8Aa). It is also the methyl donor in the conversion of norepinephrine to epinephrine (Section 26-4B).

Methylation reactions involving SAM yield *S*-**adenosylhomocysteine** in addition to the methylated acceptor. The former product is hydrolyzed to adenosine and **homocysteine** in the next reaction of the methionine degradation pathway. The homocysteine may be methylated to form methionine via a B_{12}-requiring reaction in which N^5-**methyl-THF** is the methyl donor. Alternatively, the homocysteine may combine with serine to yield **cystathionine** in a PLP-requiring reaction, which subsequently forms cysteine (cysteine biosynthesis) and **α-ketobutyrate.** The α-ketobutyrate continues along the degradative pathway to propionyl-CoA and then succinyl-CoA.

b. Hyperhomocysteinemia Is Associated with Disease

Imbalance between the rate of production of homocysteine through methylation reactions utilizing SAM (Fig. 26-18, Reactions 2 and 3) and its rate of breakdown by either remethylation to form methionine (Fig. 26-18, Reaction 4) or reaction with serine to form cystathionine in the

Figure 26-17 **Degradation pathways of arginine, glutamate, glutamine, histidine, and proline to α-ketoglutarate.** The enzymes catalyzing the reactions are **(1)** glutamate dehydrogenase, **(2)** glutaminase, **(3)** arginase, **(4)** ornithine-δ-aminotransferase, **(5)** glutamate-5-semialdehyde dehydrogenase, **(6)** proline oxidase, **(7)** spontaneous, **(8)** histidine ammonia lyase, **(9)** urocanate hydratase, **(10)** imidazolone propionase, and **(11)** glutamate formiminotransferase.

cysteine biosynthesis pathway (Fig. 26-18, Reaction 5) can result in an increase in the release of homocysteine to the extracellular medium and ultimately the plasma and urine. Moderately elevated concentrations of homocysteine in the plasma, **hyperhomocysteinemia,** for reasons that are poorly understood, are closely associated with cardiovascular disease, cognitive impairment, and **neural tube defects** [the cause of a variety of severe birth defects including **spina bifida** (defects in the spinal column that often result in paralysis) and **anencephaly** (the invariably fatal failure of the brain to develop, which is the leading cause of infant death due to congenital anomalies)]. Hyperhomo-

cysteinemia is readily controlled by ingesting the vitamin precursors of the coenzymes that participate in homocysteine breakdown, namely, B$_6$ (pyridoxine, the PLP precursor; Fig. 26-1), B$_{12}$ (Fig. 25-21), and folate (Section 26-4D). Folate, especially, appears to alleviate hyperhomocysteinemia; its administration to pregnant women dramatically reduces the incidence of neural tube defects in newborns. This has led to the discovery that 10% of the population is homozygous for the A222V mutation in N^5,N^{10}-**methylenetetrahydrofolate reductase** (**MTHFR;** Fig. 26-18, Reaction 12; Section 26-4D), the enzyme that generates N^5-methyl-THF for the methionine synthase reaction (Fig. 26-18,

Figure 26-18 **The pathway of methionine degradation, yielding cysteine and succinyl-CoA as products.** The enzymes involved are **(1)** methionine adenosyltransferase in a reaction that yields the biological methylating agent *S*-adenosylmethionine (SAM), **(2)** methyltransferase, **(3)** adenosylhomocysteinase, **(4)** methionine synthase (a coenzyme B$_{12}$–dependent enzyme), **(5)** cystathionine β-synthase (a PLP-dependent enzyme), **(6)** cystathionine γ-lyase (a PLP-dependent enzyme), **(7)** α-keto acid dehydrogenase, **(8)** propionyl-CoA carboxylase, **(9)** methylmalonyl-CoA racemase, **(10)** methylmalonyl-CoA mutase (a coenzyme B$_{12}$–dependent enzyme; Reactions 8–10 are discussed in Section 25-2E), **(11)** glycine cleavage system (Figs. 26-12 and 26-14) or serine hydroxymethyltransferase (Fig. 26-12), **(12)** N^5,N^{10}-methylene-tetrahydrofolate reductase (a coenzyme B$_{12}$– and FAD-dependent enzyme; Figs. 26-19 and 26-49).

Reaction 4). This mutation does not affect this homo-tetrameric enzyme's reaction kinetics but instead increases the rate at which it dissociates into dimers that readily lose their essential flavin cofactor. Folate derivatives that bind to the enzyme decrease its rate of dissociation and

flavin loss, thus increasing the mutant enzyme's overall activity and decreasing the homocysteine concentration.

The X-ray structure of *E. coli* MTHFR (which is 30% identical with the catalytic domain of human MTHFR), determined by Rowena Matthews and Martha Ludwig,

Figure 26-19 **X-ray structure of *E. coli* N^5,N^{10}-methylene-tetrahydrofolate reductase (MTHFR).** The structure is viewed along the axis of its α/β barrel looking toward the C-terminal ends of its β strands. The protein is colored according to its secondary structure with β strands yellow and α helices cyan except for helix $\alpha5$, which is red. The enzyme's bound FAD is drawn in ball-and-stick form with C yellow, N blue, O red, and P green. Note that the AMP moiety of the FAD is in contact with helix $\alpha5$. [Courtesy of Rowena Matthews and Martha Ludwig, University of Michigan. PDBid 1B5T.]

reveals that this 296-residue enzyme forms an α/β barrel. The FAD cofactor binds at the C-terminal ends of barrel strands $\beta3$, $\beta4$, and $\beta5$ and along helix $\alpha5$ (Fig. 26-19). Ala 177, which corresponds to Ala 222 in the mammalian enzyme, does not interact directly with the active site FAD. Instead it occupies a position flush against helix $\alpha5$ (which ends with residue 176). It is postulated that the replacement of Ala 177 by a bulkier Val residue would force helix $\alpha5$ to reorient. Since this helix appears to be involved in the subunit interface as well as in FAD binding, its reorientation is likely to decrease the strength of subunit and FAD interactions.

Why should this mutation be so prevalent in the human population? What selective advantage, if any, might it confer? We have seen that the gene for sickle-cell anemia provides a selective advantage against malaria (Section 7-3Ab). However, the selective advantage of the A222V mutation in human MTHFR is as yet a matter of speculation.

c. Methionine Synthase Is a Coenzyme B₁₂–Dependent Enzyme

Methionine synthase (alternatively **homocysteine methyltransferase**), the enzyme that catalyzes Reaction 4 in Fig. 26-18, is the only coenzyme B$_{12}$–associated enzyme in mammals besides methylmalonyl-CoA mutase (Section 25-2Eb). However, in methionine synthase, the cobalamin Co ion is axially liganded by a methyl group forming **methylcobalamin** rather than by a 5′-deoxyadenosyl group as occurs in methylmalonyl-CoA mutase (Fig. 25-21). This

is because the cobalamin functions to accept the methyl group from N^5-methyl-THF to yield methylcobalamin (and THF), which, in turn, donates the methyl group to homocysteine to yield methionine.

The X-ray structure of the 246-residue methylcobalamin-binding portion of the 1227-residue monomeric *E. coli* methionine synthase, also determined by Matthews and Ludwig, reveals that it consists of two domains, an N-terminal helical domain and a C-terminal Rossmann fold-like α/β domain, with the corrin ring sandwiched between them (Fig. 26-20). The α/β domain resembles the corrin-binding α/β domain in methylmalonyl-CoA mutase (Fig. 25-22) and, in fact, sequence homologies suggest that this domain is a common binding motif in B$_{12}$-associated enzymes. The Co ion's second axial ligand is a His side chain as is also the case in methylmalonyl-CoA mutase; the coenzyme's 5,6-dimethylbenzamidazole (DMB) moiety, which ligands the Co ion in free methylcobalamin, has swung aside to become anchored to the protein at some distance from the corrin ring.

Figure 26-20 **X-ray structure of the B₁₂-binding domains of *E. coli* methionine synthase.** Its N-terminal helical domain (residues 651–743) is cyan and its C-terminal α/β domain (residues 744–896) is pink. The methylcobalamin cofactor and its axially liganded His 759 side chain are drawn in stick form with cobalamin C green, His C gold, N blue, O red, and the Co ion and its axially liganded methyl group represented by light blue and orange spheres, respectively. [Based on an X-ray structure by Rowena Matthews and Martha Ludwig, University of Michigan. PDBid 1BMT.]

Figure 26-21   The degradation of the branched-chain amino acids (A) isoleucine, (B) valine, and (C) leucine. The first three reactions of each pathway utilize the common enzymes **(1)** branched-chain amino acid aminotransferase, **(2)** branched-chain α-keto acid dehydrogenase (BCKDH), and **(3)** acyl-CoA dehydrogenase. Isoleucine degradation then continues (*left*) with **(4)** enoyl-CoA hydratase, **(5)** β-hydroxyacyl-CoA dehydrogenase, and **(6)** acetyl-CoA acetyltransferase to yield acetyl-CoA and the succinyl-CoA precursor propionyl-CoA. Valine degradation (*center*) continues with **(7)** enoyl-CoA hydratase, **(8)** β-hydroxyisobutyryl-CoA hydrolase, **(9)** β-hydroxyisobutyrate dehydrogenase, and **(10)** methylmalonate semialdehyde dehydrogenase to also yield propionyl-CoA. Leucine degradation (*right*) continues with **(11)** β-methylcrotonyl-CoA carboxylase (a biotin-dependent enzyme), **(12)** β-methylglutaconyl-CoA hydratase, and **(13)** HMG-CoA lyase to yield acetyl-CoA and acetoacetate.

d. Branched-Chain Amino Acid Degradation Pathways Contain Themes Common to All Acyl-CoA Oxidations

Degradation of the branched-chain amino acids isoleucine, leucine, and valine begins with three reactions that employ common enzymes (Fig. 26-21, *top*): (1) transamination to the corresponding α-keto acid, (2) oxidative decarboxylation to the corresponding acyl-CoA, and (3) dehydrogenation by FAD to form a double bond.

The remainder of the isoleucine degradation pathway (Fig. 26-21, *left*) is identical to that of fatty acid oxidation (Section 25-2C): (4) double-bond hydration, (5) dehydrogenation by NAD^+, and (6) thiolytic cleavage yielding acetyl-CoA and propionyl-CoA, which is subsequently converted to succinyl-CoA. Valine degradation is a variation on this theme (Fig. 26-21, *center*): Following (7) double-bond hydration, (8) the CoA thioester bond is hydrolyzed before (9) the second dehydrogenation reaction. The thioester bond is then regenerated as propionyl-CoA in the sequence's last reaction (10), an oxidative decarboxylation rather than a thiolytic cleavage.

e. Maple Syrup Urine Disease Results from a Defect in Branched-Chain Amino Acid Degradation

Branched-chain α-keto acid dehydrogenase (BCKDH; also known as α-ketoisovalerate dehydrogenase), which catalyzes Reaction 2 of branched-chain amino acid degradation (Fig. 26-21), is a multienzyme complex containing three enzymatic components, E1, E2, and E3, together with **BCKDH kinase** (phosphorylation inactivates) and **BCKDH phosphatase** (dephosphorylation activates), which impart control by covalent modification. This complex closely resembles the pyruvate dehydrogenase and α-ketoglutarate dehydrogenase multienzyme complexes (Sections 21-2A and 21-3D). Indeed, all three of these multienzyme complexes share a common protein component, E3 (dihydrolipoyl dehydrogenase), and employ the coenzymes thiamine pyrophosphate (TPP), lipoamide, and FAD in addition to their terminal oxidizing agent, NAD^+.

A genetic deficiency in BCKDH causes **maple syrup urine disease (MSUD),** so named because the consequent buildup of branched-chain α-keto acids imparts the urine with the characteristic odor of maple syrup. Unless promptly treated by a diet low in branched-chain amino acids (but not too low because they are essential amino acids; Section 26-5), MSUD is rapidly fatal.

MSUD is an autosomal recessive disorder that is caused by defects in any of four of the complex's six subunits, E1α, E1β, E2, or E3 (E1 is an $α_2β_2$ heterotetramer). The determination of the X-ray structure of human BCKDH E1 by Wim Hol (Fig. 26-22) has enabled the interpretation of several of the mutations causing MSUD. The most common mutation is Y393N-α, the so-called Mennonite mutation, which occurs once in every 176 live births in the Old Order Mennonite population (versus 1 in 185,000 worldwide). This mutation is so common among Old Order Mennonites that it has been attributed to a founder effect, that is, a mutation that originated in one of the handful of founders of this isolated community. The E1 tetramer can be considered to be a dimer of αβ heterodimers with a TPP cofactor at the interface between an α subunit and a β subunit and with each α subunit

contacting both the β and β′ subunits (Fig. 26-22*a*). The amino acid change in the Mennonite mutation occurs at the α–β′ interface: Tyr 393α is hydrogen bonded to both His 385α and Asp 328β′ (Fig. 26-22*b*). Its mutation to Asn disrupts these interactions and thereby impedes tetramerization.

(a)

(b)

Figure 26-22 **X-ray structure of the E1 component of the human branched-chain α-keto acid dehydrogenase multienzyme complex.** (*a*) The $α_2β_2$ heterotetramer. The α subunits are colored cyan and orange, and the β subunits are blue and pink. The thiamine pyrophosphate (TPP) cofactor and Tyr 393α (which is mutated to Asn in the Mennonite mutation, causing maple syrup urine disease) are shown in space-filling form with TPP C green, Tyr 393α C gold, N blue, O red, S yellow, and P magenta. Note the similarity of this structure to that of the E1 component of the pyruvate dehydrogenase multienzyme complex (Fig. 21-12*a*). (*b*) The α–β′ interface colored as in Part *a* and showing the interactions of Tyr 393α with His 385α and Asp 328β′. The side chains of these residues are drawn in ball-and-stick form with C green, N blue, and O red and with the hydrogen bonds between them represented by dashed lines. [Based on an X-ray structure by Wim Hol, University of Washington. PDBid 1DTW.]

F. *Leucine and Lysine Are Degraded to Acetoacetate and/or Acetyl-CoA*

Leucine is oxidized by a combination of reactions used in β oxidation and ketone body synthesis (Fig. 26-21, *right*). The first dehydrogenation and the hydration reactions are interspersed by (11) a carboxylation reaction catalyzed by the biotin-containing enzyme **β-methylcrotonyl-CoA carboxylase.** The hydration reaction (12) then produces **β-hydroxy-β-methylglutaryl-CoA (HMG-CoA),** which is cleaved by HMG-CoA lyase to form acetyl-CoA and the

ketone body acetoacetate (13) (which, in turn, may be converted to 2 acetyl-CoA; Section 25-3).

Although there are several pathways for lysine degradation, the one that proceeds via formation of the α-ketoglutarate–lysine adduct **saccharopine** predominates in mammalian liver (Fig. 26-23). This pathway is of interest because we have encountered 7 of its 11 reactions in other pathways. Reaction 4 is a PLP-dependent transamination. Reaction 5 is the oxidative decarboxylation of an α-keto acid by a multienzyme complex similar to pyruvate dehydrogenase

Figure 26-23 **The pathway of lysine degradation in mammalian liver.** The enzymes involved are **(1)** saccharopine dehydrogenase (NADP⁺, lysine forming), **(2)** saccharopine dehydrogenase (NAD⁺, glutamate forming), **(3)** aminoadipate semialdehyde dehydrogenase, **(4)** aminoadipate aminotransferase (a PLP enzyme), **(5)** α-keto acid dehydrogenase, **(6)** glutaryl-CoA dehydrogenase, **(7)** decarboxylase, **(8)** enoyl-CoA hydratase, **(9)** β-hydroxyacyl-CoA dehydrogenase, **(10)** HMG-CoA synthase, and **(11)** HMG-CoA lyase. Reactions 10 and 11 are discussed in Section 25-3.

and α-ketoglutarate dehydrogenase (Sections 21-2A and 21-3D). Reactions 6, 8, and 9 are standard reactions of fatty acyl-CoA oxidation: dehydrogenation by FAD, hydration, and dehydrogenation by NAD^+. Reactions 10 and 11 are standard reactions in ketone body formation. Two molecules of CO_2 are produced at Reactions 5 and 7 of the pathway.

The saccharopine pathway is thought to predominate in mammals because a genetic defect in the enzyme that catalyzes Reaction 1 in the sequence results in **hyperlysinemia** and **hyperlysinuria** (elevated levels of lysine in the blood and urine, respectively) along with mental and physical retardation. This is yet another example of how the study of rare inherited disorders has helped to trace metabolic pathways.

Leucine's carbon skeleton, as we have seen, is converted to one molecule each of acetoacetate and acetyl-CoA, whereas that of lysine is converted to one molecule of acetoacetate and two of CO_2. Since neither acetoacetate nor acetyl-CoA can be converted to glucose in animals, leucine and lysine are purely ketogenic amino acids.

G. *Tryptophan Is Degraded to Alanine and Acetoacetate*

The complexity of the major tryptophan degradation pathway (Fig. 26-24) precludes detailed discussion of all of its reactions. However, one reaction in the pathway is of particular interest. Reaction 4, cleavage of **3-hydroxykynurenine**

Figure 26-24 **The pathway of tryptophan degradation.** The enzymes involved are **(1)** tryptophan-2,3-dioxygenase, **(2)** formamidase, **(3)** kynurenine-3-monooxygenase, **(4)** kynureninase (PLP dependent), **(5)** 3-hydroxyanthranilate-3,4-dioxygenase, **(6)** amino carboxymuconate semialdehyde decarboxylase, **(7)** aminomuconate semialdehyde dehydrogenase,

(8) hydratase, **(9)** dehydrogenase, and **(10–16)** enzymes of Reactions 5 through 11 in lysine degradation (Fig. 26-23). 2-Amino-3-carboxymuconate-6-semialdehyde, in addition to undergoing Reaction 6, spontaneously forms **quinolinate,** an NAD^+ and $NADP^+$ precursor (Section 28-5A).

Figure 26-25 **Proposed mechanism for the PLP-dependent kynureninase-catalyzed C$_\beta$—C$_\gamma$ bond cleavage of 3-hydroxykynurenine.** The reaction occurs in eight steps: **(1)** transimination; **(2)** tautomerization; **(3)** attack of an enzyme nucleophile, X; **(4)** C$_\beta$—C$_\gamma$ bond cleavage with formation of an acyl–enzyme intermediate; **(5)** acyl–enzyme hydrolysis; **(6)** and **(7)** tautomerization; and **(8)** transimination.

to alanine and **3-hydroxyanthranilate,** is catalyzed by **kynureninase,** a PLP-dependent enzyme. The reaction further demonstrates the enormous versatility of PLP. We have seen how PLP can labilize an α-amino acid's C$_\alpha$—H and C$_\alpha$—C$_\beta$ bonds (Fig. 26-16). Here we see the facilitation of C$_\beta$—C$_\gamma$ bond cleavage. The reaction follows the same steps as transamination reactions but does not hydrolyze the tautomerized Schiff base (Fig. 26-25). The proposed reaction mechanism involves an attack of an enzyme nucleophile on the carbonyl carbon (C$_\gamma$) of the tautomerized

3-hydroxykynurenine–PLP Schiff base (Fig. 26-25, Step 3). This is followed by C$_\beta$—C$_\gamma$ bond cleavage to generate an acyl–enzyme intermediate together with a tautomerized alanine–PLP adduct (Fig. 26-25, Step 4). Hydrolysis of the acyl–enzyme then yields 3-hydroxyanthranilate, whose further degradation yields **α-ketoadipate** (Fig. 26-24, Reactions 5–9). α-Ketoadipate is also an intermediate in lysine breakdown (Fig. 26-23, Reaction 4) so that the last seven reactions in the degradation of both these amino acids are identical, forming acetoacetate and two molecules of CO_2.

Figure 26-26 The pathway of phenylalanine degradation. The enzymes involved are **(1)** phenylalanine hydroxylase, **(2)** aminotransferase, **(3)** *p*-hydroxyphenylpyruvate dioxygenase, **(4)** homogentisate dioxygenase, **(5)** maleylacetoacetate isomerase, and **(6)** fumarylacetoacetase. The symbols labeling the various carbon atoms serve to indicate the group migration that occurs in Reaction 3 of the pathway (see Fig. 26-31).

H. *Phenylalanine and Tyrosine Are Degraded to Fumarate and Acetoacetate*

Since the first reaction in phenylalanine degradation is its hydroxylation to tyrosine, a single pathway (Fig. 26-26) is responsible for the breakdown of both of these amino acids. The final products of the six-reaction degradation are fumarate, a citric acid cycle intermediate, and acetoacetate, a ketone body.

a. Pterins Are Redox Cofactors

The hydroxylation of phenylalanine by the nonheme iron–containing homotetrameric enzyme **phenylalanine hydroxylase (PAH)** requires O_2 and that the iron be in the Fe(II) state. The enzyme also requires the participation of **biopterin,** a **pterin** derivative. Pterins are compounds that contain the **pteridine** ring (Fig. 26-27). Note the resemblance between the pteridine ring and the isoalloxazine ring of the flavin coenzymes; the positions of the nitrogen atoms in pteridine are identical with those of the B and C rings of isoalloxazine. Folate derivatives also contain the pterin ring (Section 26-4D). Pterins, like flavins, participate in biological oxidations. The active form of biopterin is the fully reduced form, **5,6,7,8-tetrahydrobiopterin (BH₄).** It is produced from **7,8-dihydrobiopterin** and NADPH, in what

Figure 26-27 The pteridine ring, the nucleus of biopterin and folate. Note the similar structures of pteridine and the isoalloxazine ring of flavin coenzymes.

may be considered a priming reaction, by **dihydrofolate reductase** (Fig. 26-28).

Each 452-residue subunit of the PAH homotetramer contains three domains, an N-terminal regulatory domain, a catalytic domain, and a C-terminal tetramerization domain. However, the 325-residue catalytic domain alone forms catalytically competent dimers. The X-ray structure of the catalytic domain of PAH in its Fe(II) state in complex with BH$_4$, determined by Edward Hough, reveals that the Fe(II) is octahedrally coordinated by His 285, His 290, Glu 330, and three water molecules, and that atom O4 of BH$_4$ is hydrogen bonded to two of these waters (Fig. 26-29).

In the phenylalanine hydroxylase reaction, 5,6,7,8-tetrahydrobiopterin is hydroxylated to **pterin-4a-carbinolamine** (Fig. 26-28), which is converted to **7,8-dihydrobiopterin (quinoid form)** by **pterin-4a-carbinolamine dehydratase.** The quinoid is subsequently reduced by the NAD(P)H-requiring enzyme **dihydropteridine reductase** to regenerate the active cofactor. Note that although dihydrofolate reductase and dihydropteridine reductase produce the same product, they utilize different tautomers of the substrate. Although this suggests that these enzymes may be evolutionarily related, the comparison of their X-ray structures indicates that this is not the case: Dihydropteridine reductase resembles nicotinamide

Figure 26-28 The formation, utilization, and regeneration of 5,6,7,8-tetrahydrobiopterin (BH$_4$) in the phenylalanine hydroxylase reaction.

Figure 26-29 **The active site of the Fe(II) form of phenylalanine hydroxylase (PAH) in complex with 5,6,7,8-tetrahydrobiopterin (BH₄).** The Fe(II) (*orange sphere*) is octahedrally coordinated (*gray lines*) by His 285, His 290, and Glu 330 (C green, N blue, and O red) and three water molecules (*red spheres*). BH₄ atom O4 is hydrogen bonded (*black dashed lines*) to two of these water molecules. [Based on an X-ray structure by Edward Hough, University of Tromsø, Norway. PDBid 1J8U.]

coenzyme–requiring flavin-dependent enzymes such as glutathione reductase and dihydrolipoyl dehydrogenase (Section 21-2B).

b. Phenylalanine Hydroxylase Is Controlled by Phosphorylation and by Allosteric Interactions

PAH initiates the detoxification of high concentrations of phenylalanine as well as the synthesis of the catecholamine hormones and neurotransmitters (Section 26-4B). It is allosterically activated by its substrate, phenylalanine, and by phosphorylation at its Ser 16 by the cAMP-dependent protein kinase A (PKA; Section 18-3Cb). Its second substrate, BH_4, allosterically inhibits the enzyme.

c. The NIH Shift

An unexpected aspect of the PAH reaction is that a ^3H atom, which begins on C4 of phenylalanine's phenyl ring, ends up on C3 of this ring in tyrosine (Fig. 26-28, *right*) rather than being lost to the solvent by replacement with the OH group. The mechanism postulated to account for this **NIH shift** (so called because it was first characterized by chemists at the *N*ational *I*nstitutes of *H*ealth) involves the activation of oxygen by the pterin and Fe cofactors to form the pterin-4a-carbinolamine and a reactive oxyferryl group [Fe(IV)=O^{2-}; Fig. 26-30, Steps 1 and 2] that reacts with the substrate to form an epoxide across the phenyl ring's 3,4 bond (Fig. 26-30, Step 3). This is followed by epoxide opening to form a carbocation at C3 (Fig. 26-30, Step 4). Migration of a hydride from C4 to C3 forms a more stable carbocation (an oxonium ion; Fig. 26-30, Step 5). This migration is followed by ring aromatization to form tyrosine

(Fig. 26-30, Step 6). **Tyrosine hydroxylase** and **tryptophan hydroxylase** (Section 26-4B) are both homologous to phenylalanine hydroxylase and utilize this same NIH shift reaction mechanism, although there may not be an epoxide intermediate in these cases.

Reaction 3 in the phenylalanine degradation pathway (Fig. 26-26) provides another example of an NIH shift. This reaction, which is catalyzed by the Fe(II)-containing *p*-**hydroxyphenylpyruvate dioxygenase,** involves the oxidative decarboxylation of an α-keto acid as well as ring hydroxylation. In this case, the NIH shift involves migration of an alkyl group rather than of a hydride ion to form a more stable carbocation (Fig. 26-31). This shift, which has been demonstrated through isotope-labeling studies (represented by the different symbols in Figs. 26-26 and 26-31), accounts for the observation that C3 is bonded to C4 in *p*-**hydroxyphenylpyruvate** but to C5 in **homogentisate.**

d. Alkaptonuria and Phenylketonuria Result from Defects in Phenylalanine Degradation

Archibald Garrod realized in the early 1900s that human genetic diseases result from specific enzyme deficiencies. We have repeatedly seen how this realization has contributed to the elucidation of metabolic pathways. The first such disease to be recognized was **alkaptonuria,** which, Garrod observed, resulted in the excretion of large quantities of homogentisic acid. This condition results from deficiency of **homogentisate dioxygenase** (Fig. 26-26, Reaction 4). Alkaptonurics suffer no ill effects other than arthritis later in life (although their urine darkens alarmingly because of the rapid air oxidation of the homogentisate they excrete).

Individuals suffering from **phenylketonuria (PKU)** are not so fortunate. Severe mental retardation occurs within a few months of birth if the disease is not detected and treated immediately (see below). Indeed, ~1% of the patients in mental institutions were, at one time (before routine screening), phenylketonurics. PKU is caused by the inability to hydroxylate phenylalanine (Fig. 26-26, Reaction 1) and therefore results in increased blood levels of phenylalanine **(hyperphenylalaninemia).** The excess phenylalanine is transaminated to **phenylpyruvate**

$$\text{CH}_2-\overset{\displaystyle O}{\overset{\|}{\text{C}}}-\text{COO}^-$$

Phenylpyruvate

by an otherwise minor pathway. The "spillover" of phenylpyruvate (a phenylketone) into the urine was the first observation connected with the disease and gave the disease its name, although it has since been demonstrated that it is the high concentration of phenylalanine itself that gives rise to brain dysfunction. All babies born in the United States are now screened for PKU immediately after birth by testing for elevated levels of phenylalanine in the blood.

Classic PKU results from a deficiency in phenylalanine hydroxylase (PAH). When this was established in 1947, it was the first human inborn error of metabolism whose basic biochemical defect had been identified. Since then, over

Figure 26-30   Proposed mechanism of the NIH shift in the phenylalanine hydroxylase reaction. The mechanism involves (**1** and **2**) activation of oxygen by the enzyme's BH$_4$ and Fe(II) cofactors to yield pterin-4a-carbinolamine and a reactive oxyferryl species [Fe(IV)=O^{2-}]; (**3**) reaction of the Fe(IV)=O^{2-} with the phenylalanine substrate to form an epoxide across its phenyl ring's 3,4 bond; (**4**) epoxide opening to form a carbocation at C3; (**5**) migration of a hydride from C4 to C3 to form a more stable carbocation (an oxonium ion); and (**6**) ring aromatization to form tyrosine.

400 mutations have been identified in PAH. Because all of the tyrosine breakdown enzymes are normal, treatment consists in providing the patient with a low-phenylalanine diet and monitoring the blood level of phenylalanine to ensure that it remains within normal limits for the first 5 to 10 years of life (the adverse effects of hyperphenylalaninemia seem to disappear after that age). PAH deficiency also accounts for another common symptom of PKU: Its victims have lighter hair and skin color than their siblings. This is because tyrosine hydroxylation, the first reaction in the formation of the black skin pigment **melanin** (Section 26-4B), is inhibited by elevated phenylalanine levels.

Other causes of hyperphenylalaninemia have been discovered since the introduction of infant screening techniques. These result from deficiencies in the enzymes catalyzing the formation or regeneration of 5,6,7,8-tetrahydrobiopterin (BH$_4$), the PAH cofactor (Fig. 26-28). In such cases, patients must also be supplied with **L-3,4-dihydroxyphenylalanine (L-DOPA)** and **5-hydroxytryptophan,** metabolic precursors of the neurotransmitters **norepinephrine** and **serotonin,** respectively, since tyrosine hydroxylase and tryptophan hydroxylase, the PAH homologs that produce these physiologically active amines, also require 5,6,7,8-tetrahydrobiopterin (Section 26-4B). Unfortunately, simply adding BH$_4$ to the diet of

p-Hydroxyphenypyruvate

Figure 26-31 **The NIH shift in the *p*-hydroxyphenylpyruvate dioxygenase reaction.** Carbon atoms are labeled as an aid to following the group migration constituting the shift.

an affected individual is not an effective treatment because BH_4 is unstable and cannot cross the blood–brain barrier.

4 AMINO ACIDS AS BIOSYNTHETIC PRECURSORS

Certain amino acids, in addition to their major function as protein building blocks, are essential precursors of a variety of important biomolecules, including nucleotides and nucleotide coenzymes, heme, various hormones and neurotransmitters, and glutathione. In this section, we consider the pathways producing some of these substances. We begin by discussing the biosynthesis of heme from glycine and succinyl-CoA. We then examine the pathways by which tyrosine, tryptophan, glutamate, and histidine are converted to various neurotransmitters and study certain aspects of glutathione biosynthesis and the involvement of this tripeptide in amino acid transport and other processes. Finally, we consider the role of folate derivatives in the biosynthetic transfer of C_1 units. The biosynthesis of nucleotides and nucleotide coenzymes is the subject of Chapter 28.

A. *Heme Biosynthesis and Degradation*

Heme (Fig. 26-32), as we have seen, is an Fe-containing prosthetic group that is an essential component of many proteins, notably hemoglobin, myoglobin, and the cytochromes. The initial reactions of heme biosynthesis are common to the formation of other tetrapyrroles including

Heme

Figure 26-32 **Structure of heme.** Heme's C and N atoms are derived from those of glycine and acetate.

chlorophyll in plants and bacteria (Section 24-2A) and coenzyme B_{12} in bacteria (Section 25-2Eb).

a. Porphyrins Are Derived from Succinyl-CoA and Glycine

Elucidation of the heme biosynthesis pathway involved some interesting detective work. David Shemin and David Rittenberg, who were among the first to use isotopic tracers in the elucidation of metabolic pathways, demonstrated, in 1945, that *all of heme's C and N atoms can be derived from acetate and glycine.* Only glycine, out of a variety of ^{15}N-labeled metabolites they tested (including ammonia, glutamate, leucine, and proline), yielded ^{15}N-labeled heme in the hemoglobin of experimental subjects to whom these metabolites were administered. Similar experiments, using acetate labeled with ^{14}C in its methyl or carboxyl groups, or $[^{14}C_\alpha]$glycine, demonstrated that 24 of heme's 34 carbon atoms are derived from acetate's methyl carbon, 2 from acetate's carboxyl carbon, and 8 from glycine's C_α atom (Fig. 26-32). None of the heme atoms is derived from glycine's carboxyl carbon atom.

Figure 26-32 indicates that heme C atoms derived from acetate methyl groups occur in groups of three linked atoms. Evidently, acetate is first converted to some other metabolite that has this labeling pattern. Shemin and Rittenberg postulated that this metabolite is succinyl-CoA based on the following reasoning (Fig. 26-33):

1. Acetate is metabolized via the citric acid cycle (Section 21-1B).

2. Labeling studies indicate that atom C3 of the citric acid cycle intermediate succinyl-CoA is derived from acetate's methyl C atom, whereas atom C4 comes from acetate's carboxyl C atom.

3. After many turns of the citric acid cycle, C1 and C2 of succinyl-CoA likewise become fully derived from acetate's methyl C atom.

We shall see that this labeling pattern indeed leads to that of heme.

In the mitochondria of yeast and animals as well as in some bacteria, the first phase of heme biosynthesis is a condensation of succinyl-CoA with glycine followed by decarboxylation to form **δ-aminolevulinic acid (ALA)** as catalyzed by the PLP-dependent enzyme **δ-aminolevulinate synthase (ALA synthase or ALAS;** Fig. 26-34). The carboxyl group lost in the decarboxylation (Fig. 26-34, Reaction 5) originates in glycine, which is why heme contains no label from this group.

b. The Pyrrole Ring Is the Product of Two ALA Molecules

The pyrrole ring is formed in the next phase of the pathway through linkage of two molecules of ALA to yield **porphobilinogen (PBG).** The reaction is catalyzed by **porphobilinogen synthase [PBGS; alternatively, δ-aminolevulinic acid dehydratase (ALAD)],** which in yeast and mammals, is Zn^{2+}-dependent and involves Schiff base formation of one of the substrate molecules with an enzyme amine

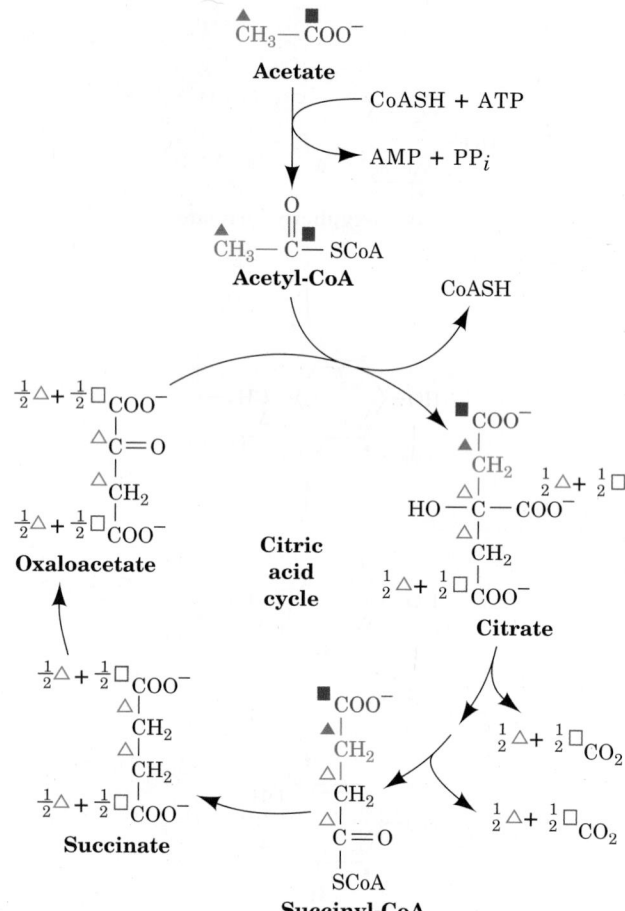

Figure 26-33 The origin of the C atoms of succinyl-CoA as derived from acetate via the citric acid cycle. C atoms labeled with triangles and squares are derived, respectively, from acetate's methyl and carboxyl C atoms. Filled symbols label atoms derived from acetate in the present round of the citric acid cycle, whereas open symbols label atoms derived from acetate in previous rounds of the citric acid cycle. Note that the C1 and C4 atoms of succinyl-CoA are scrambled on forming the 2-fold symmetric succinate.

group (in some bacteria and all plants, Mg^{2+} substitutes for Zn^{2+}). One possible mechanism of this condensation–elimination reaction involves formation of a second Schiff base between the ALA–enzyme Schiff base and the second ALA molecule (Fig. 26-35). At this point, if we continue tracing the acetate and glycine labels through the PBG synthase reaction (Fig. 26-35), we can begin to see how heme's labeling pattern arises.

The X-ray structure of human PBGS in covalent complex with its product PBG, determined by Jonathan Cooper, indicates that this enzyme is a homooctamer with D_4 symmetry. Each of its 330-residue subunits consists of an α/β barrel and a 39-residue N-terminal tail that wraps around a neighboring monomer (related to it by 2-fold symmetry) so that the protein is better described as a relatively loosely organized tetramer of compact dimers. As is the case with nearly all α/β barrel enzymes, PBGS's active site (Fig. 26-36a) lies at the mouth of the barrel at the

Figure 26-34 **The mechanism of action of the PLP-dependent enzyme δ-aminolevulinate synthase (ALAS).** The reaction steps are **(1)** transimination, **(2)** PLP-stabilized carbanion formation, **(3)** C—C bond formation, **(4)** CoA elimination, **(5)** decarboxylation facilitated by the PLP–Schiff base, and **(6)** transimination yielding ALA and regenerating the PLP–enzyme.

C-terminal ends of its β strands. The active site is covered by a loop that comparison with other PBGS structures indicates forms a flexible lid over the substrate, an arrangement that is reminiscent of the glycolytic enzyme triose phosphate isomerase (TIM; Fig. 17-11). PBG is covalently bound to Lys 252 and its free amino group is coordinated to the active site Zn^{2+} ion. Lys 199 appears to be properly positioned to act as a general acid–base catalyst.

Inhibition of PBG synthase by Pb^{2+} (a competitor of its active site Zn^{2+} ion) is one of the major manifestations of lead poisoning, which is among the most common acquired environmental diseases. Indeed, it has been suggested that the accumulation, in the blood, of ALA, which resembles the neurotransmitter **γ-aminobutyric acid** (Section 26-4B), is responsible for the psychosis that often accompanies lead poisoning.

Figure 26-35   A possible mechanism for porphobilinogen synthase. The reaction involves **(1)** Schiff base formation, **(2)** second Schiff base formation, **(3)** formation of a carbanion α to a Schiff base, **(4)** cyclization by an aldol-type condensation, **(5)** elimination of the enzyme—NH_2 group, and **(6)** tautomerization.

ALA

Porphobilinogen
(PBG)

c. PBGS Has Two Quaternary States with Different Enzymatic Activities

Although the X-ray structure of wild-type human PBGS indicates that it is a D_4-symmetric homooctamer, the X-ray structure of its rare F12L mutant form, determined by Eileen Jaffe, is a D_3-symmetric homohexamer (Fig. 26-36b). Moreover, the F12L mutant has far less activity than does wild-type PBGS, even though residue 12 is far from the enzyme's active site in both proteins. This is apparently due to a conformational change in PBGS's N-terminal arm: In the octamer, the N-terminal arms of two adjacent monomers wrap around each other's barrel to form a so-called hugging dimer, whereas in the hexamer, the N-terminal arms extend away from the core of the protein to form a so-called detached dimer (Fig. 26-36b). Nevertheless, the α/β barrels of these two oligomeric forms are closely superimposable. The difference in the activities of these two quaternary forms is caused by the binding of an allosteric Mg^{2+} ion in the octamer's "hugging" interface that is absent in the detached dimer.

The alternate oligomeric forms have been observed in solutions of wild-type PBGS and the equilibrium between these forms can be varied by changing the pH, the enzyme concentration, and the substrate concentration, as well as through mutation. Jaffe has therefore proposed that this quaternary structural change is an allosteric mechanism for controlling the activity of the enzyme. In contrast, in the symmetry and sequential models of allosterism (Sections 10-4B and 10-4C), the quaternary state of the enzyme does not change during an allosteric transition.

Jaffe has termed enzymes that control their activities by changing their oligomeric states **morpheeins** [pronounced morph-ee′-in; derived from the verb "to morph" and the classic pronunciation of the word protein (pro-tee′-in)]. Although PBGS is the first known morpheein, several others have since been identified (e.g., ribonucleotide reductase; Section 28-3Ad). In fact, it may be that morpheeins are relatively common because when an enzyme is purified, inactive fractions are usually assumed to be denatured and are therefore discarded rather than being characterized.

(a)

(b)

Figure 26-36 X-ray structure of human porphobilinogen synthase (PBGS). (a) PBGS in covalent complex with its porphobiligen (PBG) product. A monomer of this homooctameric protein is viewed perpendicular to the axis of its α/β barrel and is drawn in gray with its β strands cyan and the loop forming its flexible lid (residues 201–222) magenta. The PBG, product, Lys 252 to which it is covalently linked, and the three Cys side chains that ligand the active site Zn^{2+} ion (*blue sphere*) are shown in stick form with PBG C pink, side chain C green, N blue, O red, S yellow, and the N—C bond linking Lys 252 to PBG gold. The active site Zn^{2+} ion is liganded (*black lines*) by the S atoms of Cys 122, Cys 124, Cys 132, and the PBG amino group. Lys 199, which lies directly behind Lys 252 in this view, appears to be properly positioned to act as an acid–base catalyst. [Based on an X-ray structure by Jonathan Cooper,

University of Southampton, U.K. PDBid 1E51.] (*b*) Quaternary structural changes between the low activity hexameric state of the F12L mutant of human PBGS (*blue*) and the high activity octameric state of the wild-type enzyme (*red*). In the upper panel, the proteins are viewed along their 3-fold and 4-fold axes with the subunits closest to the viewer more darkly colored. In the lower panel, the outer drawings are viewed along their 2-fold axes with one dimer drawn in ribbon form and the others in space-filling form. The inner drawings show only the dimers into which the oligomers are assumed to dissociate before reassembling to an alternate quaternary state. [Courtesy of Sarah Lawrence and Eileen Jaffe, The Fox Chase Cancer Center, Philadelphia, Pennsylvania. The X-ray structure of the F12L mutant was determined by Eileen Jaffe. PDBid 1PV8.]

Figure 26-37 **The synthesis of uroporphyrinogen III from PBG as catalyzed by porphobilinogen deaminase and uroporphyrinogen III synthase. (1a)** General base-catalyzed elimination of NH_3 to form a **methylene pyrrolinene** intermediate. **(1b)** Addition to the methylene pyrrolinene intermediate of the enzyme's covalently linked dipyrromethane cofactor to form a covalent adduct. **(2–4)** Sequential addition of a second, third, and fourth PBG through successive NH_3 eliminations from PBG to form methylene pyrrolinene, as in Reaction 1a, followed by addition of a pyrrole ring carbon atom from the growing chain, as in Reaction 1b. **(5)** Hydrolysis of the methylbilane–enzyme to yield hydroxymethylbilane and regenerate the free enzyme–dipyrromethane complex. **(6)** Synthesis of uroporphyrinogen III via a spiro intermediate by porphobilinogen deaminase and uroporphyrinogen III synthase. **(7)** Spontaneous cyclization of hydroxymethylbilane in the absence of uroporphyrinogen III synthase. A and P represent acetyl and propionyl groups.

$A = -CH_2-COO^-$

$P = -CH_2-CH_2-COO^-$

d. The Porphyrin Ring Is Formed from Four PBG Molecules

The next phase of heme biosynthesis is the condensation of four PBG molecules to form **uroporphyrinogen III,** the porphyrin nucleus, in a series of reactions catalyzed by **porphobilinogen deaminase** (alternatively, **hydroxymethylbilane synthase** or **uroporphyrinogen synthase**) and **uroporphyrinogen III synthase.** The reaction (Fig. 26-37) begins with the enzyme's displacement of the amino group in PBG to form a covalent adduct. A second, third, and fourth PBG then sequentially add through the displacement of the primary amino group on one PBG by a carbon atom on the pyrrole ring of the succeeding PBG to yield a linear tetrapyrrole that is hydrolyzed and released from the enzyme as **hydroxymethylbilane** (also called **preuroporphyrinogen**).

e. Porphobilinogen Deaminase Has a Dipyrromethane Cofactor

Peter Shoolingin-Jordan and Alan Battersby independently showed that porphobilinogen deaminase contains a unique **dipyrromethane** cofactor (two pyrroles linked by a methylene bridge; rings C_1 and C_2 in Fig. 26-37), which is covalently linked to the enzyme via a C—S bond to an enzyme Cys residue. Thus, the methylbilane–enzyme complex really contains a linear hexapyrrole. The subsequent reaction step, also catalyzed by porphobilinogen deaminase (Step 5 in Fig. 26-37), is the hydrolysis of the bond linking the second and third pyrrole units of the hexapyrrole to yield hydroxymethylbilane and the dipyrromethane cofactor. This cofactor is still linked to the enzyme, which is therefore ready to catalyze a new round of hydroxymethylbilane synthesis.

How is the dipyrromethane cofactor assembled? Shoolingin-Jordan has shown that porphobilinogen deaminase synthesizes its own cofactor from two PBG units using, it appears, the same catalytic machinery with which it synthesizes methylbilane. However, the enzyme Cys reacts much more rapidly with presynthesized hydroxymethylbilane to form a reaction intermediate (the product of Step 2 in Fig. 26-37) that continues to add two more PBG units. When hydroxymethylbilane is released, the enzyme retains its dipyrromethane cofactor.

The X-ray structure of human porphobilinogen deaminase (whose sequence is >45% identical to those of the *E. coli* enzyme), in covalent complex with its dipyrromethane cofactor, indicates that this monomeric, 364-residue protein folds into three nearly equal sized domains (Fig 26-38). The dipyrromethane cofactor lies deep in a cleft between domains 1 and 2 such that there is still considerable unoccupied space in the cleft. Although the enzyme sequentially appends four PBG residues to the cofactor, it has only one catalytic site.

If the enzyme has only one catalytic site, how does it reposition the polypyrrole chain after each catalytic cycle so that it can further extend this chain? One possibility is that the polypyrrole chain fills the cavity next to the cofactor. This model provides a simple steric rationale for why the length of the polypyrrole chain is limited to six residues (the final four of which are hydrolytically cleaved away by the enzyme to yield the hydroxymethylbilane product and regenerate the dipyrromethane cofactor).

Figure 26-38   X-ray structure of human porphobilinogen deaminase in covalent complex with its dipyrromethane cofactor. The protein is shown in ribbon form with domain 1 (residues 1–116 and 216–239) magenta, domain 2 (residues 117–215) cyan, and domain 3 (residues 240–364) orange. The dipyrromethane cofactor and the Cys side chain to which it is covalently linked are drawn in stick form with C green, N blue, O red, and C yellow. [Based on an X-ray structure by Zhi-Jie Liu, Institute of Biophysics, Beijing, China. PDBid 3ECR.]

f. Protoporphyrin IX Biosynthesis Requires Four More Reactions

Cyclization of the hydroxymethylbilane product requires **uroporphyrinogen III synthase** (Fig. 26-37). In the absence of this enzyme, hydroxymethylbilane is released from the synthase and rapidly cyclizes nonenzymatically to the symmetric **uroporphyrinogen I.** Heme, however, is an asymmetric molecule; the methyl substituent of pyrrole ring D has an inverted placement compared to those of rings A, B, and C (Fig. 26-32). This ring reversal to yield uroporphyrinogen III has been shown by Battersby to proceed through attachment of the methylenes from rings A and C to the same carbon of ring D so as to form a spiro compound (a bicyclic compound with a carbon atom common to both rings; Fig. 26-37).

Heme biosynthesis takes place partly in the mitochondrion and partly in the cytosol (Fig. 26-39). ALA is mitochondrially synthesized and is transported to the cytosol for conversion to PBG and then to uroporphyrinogen III. **Protoporphyrin IX,** to which Fe is added to form heme, is produced from uroporphyrinogen III in a series of reactions catalyzed by (1) **uroporphyrinogen decarboxylase,** which decarboxylates all four acetate side chains (A) to form methyl groups (M); (2) **coproporphyrinogen oxidase,** which oxidatively decarboxylates two of the propionate side chains (P) to vinyl groups (V); and (3) **protoporphyrinogen**

Figure 26-39   The overall pathway of heme biosynthesis.
δ-Aminolevulinic acid (ALA) is synthesized in the mitochondrion
by ALA synthase. ALA (*left*) leaves the mitochondrion and is
converted to PBG, four molecules of which condense to form a
porphyrin ring. The next two reactions involve oxidation of the
pyrrole ring substituents yielding protoporphyrinogen IX whose

formation is accompanied by its transport back into the
mitochondrion. After oxidation of the methylene groups linking
the pyrroles to yield protoporphyrin IX, ferrochelatase catalyzes
the insertion of Fe^{2+} to yield heme. A, P, M, and V, respectively,
represent acetyl, propionyl, methyl, and vinyl ($-CH_2\!=\!CH_2$) groups.
C atoms originating as the carboxyl group of acetate are red.

oxidase, which oxidizes the methylene groups linking the pyrrole rings to methenyl groups. Altogether, six carboxyl groups originally from carboxyl-labeled acetate are lost as CO_2. The only remaining C atoms from carboxyl-labeled acetate are the carboxyl groups of heme's two propionate side chains (P). During the coproporphyrinogen oxidase reaction, the macrocycle is transported back into the mitochondrion for the pathway's final reactions.

g. Ferrochelatase Catalyzes the Insertion of Fe(II) Into Protoporphyrin IX to Form Heme

Protoporphyrin IX is converted to heme by the insertion of Fe(II) into the tetrapyrrole nucleus by **ferrochelatase,** a protein that is associated with the inner mitochondrial membrane on the matrix side. The X-ray structure of human ferrochelatase in complex with its heme product (Fig. 26-40), determined by Harry Dailey and William Lanzilotta, reveals that the 369-residue subunits of this homodimeric protein consist of two structurally similar domains and a C-terminal extension that occurs only in animal ferrochelatases (Fig. 26-40). This C-terminal extension participates in hydrogen bonding between the monomers; bacterial ferrochelatases, which lack this extension, are monomeric although they are otherwise structurally similar to the human enzyme despite their only ~10% sequence identity. In addition, the C-terminal extension is bound to the N-terminal domain by an unusual [2Fe–2S] cluster that is coordinated by C196 of the N-terminal domain and C403, C406, and C411 of the C-terminal extension. The function of this [2Fe–2S] cluster, which is distant from the active site, is unclear although it appears likely that it only has a structural role. Nevertheless, three mutations, C406Y, C406S, and C411G, that inactivate the enzyme and thereby cause the rare inherited disease **erythropoietic protoporphyria** (see below) demonstrate the importance of the [2Fe–2S] cluster for activity.

The ferrochelatase active site (Fig. 26-40) consists of two hydrophobic lips that, it has been proposed, participate in the enzyme's association with the inner mitochondrial membrane. The ferrochelatase reaction follows an ordered mechanism in which the Fe(II) binds to the enzyme before the porphyrin. The reaction requires that the two pyrrole NH protons be removed from the porphyrin prior to the binding of the Fe(II) (Fig. 26-39). The invariant H263 appears properly positioned to abstract these protons from the porphyrin (Fig. 26-40), a hypothesis that is supported by mutagenesis studies. Structural and spectroscopic studies indicate that the metalation reaction is accompanied by the folding of the porphyrin by ~12° to a nonplanar conformation. Product release is then facilitated by the partial unwinding of a structurally conserved π helix (Fig. 26-40; π helices are discussed in Section 8-1Bb).

h. Heme Biosynthesis Is Regulated Differently in Erythroid and Liver Cells

The two major sites of heme biosynthesis are erythroid cells, which synthesize ~85% of the body's heme groups, and the liver, which synthesizes ~80% of the remainder. An important function of heme in liver is as the prosthetic groups of the **cytochromes P450,** a family of oxidative enzymes involved in detoxification (Section 15-4Bc), whose members are required throughout a liver cell's lifetime in amounts that vary with conditions. In contrast, erythroid cells, in which heme is, of course, a hemoglobin component, engage in heme synthesis only on differentiation, when they synthesize hemoglobin in vast quantities. This is a one-time synthesis; the heme must last the erythrocyte's lifetime (normally 120 days) since heme and hemoglobin synthesis stop on red cell maturation (protein synthesis stops on the loss of nuclei and ribosomes). The different ways in which heme biosynthesis is regulated in liver and in erythroid cells reflect these different demands: In liver, heme biosynthesis must really be "controlled," whereas in erythroid cells, the process is more like breaking a dam.

In liver, the main control target in heme biosynthesis is ALA synthase, the enzyme catalyzing the pathway's first committed step. Heme, or its Fe(III) oxidation product **hemin,** controls this enzyme's activity through three mechanisms: (1) feedback inhibition, (2) inhibition of the transport of ALA synthase (ALAS) from its site of synthesis in the cytosol to its reaction site in the mitochondrion (Fig. 26-39), and (3) repression of ALAS synthesis.

π Helix

Heme

His 263

[2Fe–2S]

Figure 26-40 **X-ray structure of human ferrochelatase in complex with its heme product.** A subunit of this homodimeric protein is drawn in ribbon form colored in rainbow order from its N-terminus (*blue*) to its C-terminus (*red*), but with its conserved π helix (residues 340–360) magenta. The bound heme and [2Fe–2S] cluster are shown in space-filling form with C green, N blue, O red, S yellow, and Fe orange. The side chain of the catalytically essential His 263 is drawn in stick form with C cyan and N blue. [Based on an X-ray structure by Harry Dailey and William Lanzilotta, University of Georgia. PDBid 2QD3.]

In erythroid cells, heme exerts quite a different effect on its biosynthesis. Heme induces, rather than represses, protein synthesis in reticulocytes (immature erythrocytes). Although the vast majority of the protein synthesized by reticulocytes is globin, heme may also induce these cells to synthesize the enzymes of the heme biosynthesis pathway. Moreover, the rate-determining step of heme biosynthesis in erythroid cells may not be the ALA synthase reaction, which in mammalian reticulocytes is catalyzed by a different isozyme (**ALAS-2**) than the ALA synthase that is expressed in other cells (**ALAS-1**). Experiments on various systems of differentiating erythroid cells implicate ferrochelatase and porphobilinogen deaminase in the control of heme biosynthesis in these cells. There are also indications that cellular uptake of iron may be rate limiting. Iron is transported in the plasma complexed with the iron transport protein **transferrin.** The rate at which the iron–transferrin complex enters most cells, including those of liver, is controlled by receptor-mediated endocytosis (Section 12-5Bc). However, lipid-soluble iron complexes that diffuse directly into reticulocytes stimulate *in vitro* heme biosynthesis. The existence of several control points supports the supposition that when erythroid heme biosynthesis is "switched on," all of its steps function at their maximal rates rather than any one step limiting the flow through the pathway. Heme-stimulated synthesis of globin also ensures that heme and globin are synthesized in the correct ratio for assembly into hemoglobin (Section 32-4Aa).

i. Porphyrias Have Bizarre Symptoms

Seven sets of genetic defects in heme biosynthesis, in liver or erythroid cells, are recognized. All involve the accumulation of porphyrin and/or its precursors and are therefore known as **porphyrias** (Greek: *porphyra,* purple). Two such defects are known to affect erythroid cells: uroporphyrinogen III synthase deficiency (**congenital erythropoietic porphyria**) and ferrochelatase deficiency (**erythropoietic protoporphyria**). The former results in accumulation of uroporphyrinogen I and its decarboxylation product **coproporphyrinogen I.** Excretion of these compounds colors the urine red, their deposition in the teeth turns them a fluorescent reddish brown, and their accumulation in the skin renders it extremely photosensitive such that it ulcerates and forms disfiguring scars. Increased hair growth is also observed in afflicted individuals such that fine hair may cover much of the face and extremities. These symptoms have prompted speculation that the werewolf legend has a biochemical basis.

The most common porphyria that primarily affects liver is porphobilinogen deaminase deficiency (**acute intermittent porphyria**). This disease is marked by intermittent attacks of abdominal pain and neurological dysfunction, often brought about by infection, fasting, certain drugs, alcohol, steroids, and other chemicals, all of which induce the expression of ALAS-1. Excessive amounts of ALA and PBG are excreted in the urine during and after such attacks. The urine may become red resulting from the excretion of excess porphyrins synthesized from PBG in nonhepatic cells although the skin does not become unusually photosensitive. King George III, who ruled England during the American Revolution, and who has been widely portrayed as being mad, in fact had attacks characteristic of acute intermittent porphyria, was reported to have urine the color of port wine, and had several descendants who were diagnosed as having this disease. American history might have been quite different had George III not inherited this metabolic defect.

j. Heme Is Degraded to Bile Pigments

At the end of their lifetime, red cells are removed from the circulation and their components degraded. Heme catabolism (Fig. 26-41) begins with oxidative cleavage, by **heme oxygenase,** of the porphyrin between rings A and B to form **biliverdin,** a green linear tetrapyrrole. Biliverdin's central methenyl bridge (between rings C and D) is then reduced to form the red-orange **bilirubin.** The changing colors of a healing bruise are a visible manifestation of heme degradation.

The highly lipophilic bilirubin is insoluble in aqueous solutions. Like other lipophilic metabolites, such as free fatty acids, it is transported in the blood in complex with serum albumin. In the liver, its aqueous solubility is increased by esterification of its two propionate side groups with glucuronic acid, yielding **bilirubin diglucuronide,** which is secreted into the bile. Bacterial enzymes in the large intestine hydrolyze the glucuronic acid groups and, in a multistep process, convert bilirubin to several products, most notably **urobilinogen.** Some urobilinogen is reabsorbed and transported via the bloodstream to the kidney, where it is converted to the yellow **urobilin** and excreted, thus giving urine its characteristic color. Most of the urobilinogen, however, is microbially converted to the deeply red-brown **stercobilin,** the major pigment of feces.

When the blood contains excessive amounts of bilirubin, the deposition of this highly insoluble substance colors the skin and the whites of the eyes yellow. This condition, called **jaundice** (French: *jaune,* yellow), signals either an abnormally high rate of red cell destruction, liver dysfunction, or bile duct obstruction. Newborn infants, particularly when premature, often become jaundiced because their livers do not yet make sufficient **bilirubin UDP-glucuronosyltransferase** to glucuronidate the incoming bilirubin. Jaundiced infants are treated by bathing them with light from a fluorescent lamp; this photochemically converts bilirubin to more soluble isomers that the infant can degrade and excrete.

k. Hemoglobin's Reduced Affinity for CO Prevents Asphyxiation

In the reaction forming biliverdin, the methenyl bridge carbon between porphyrin rings A and B is released as CO (Fig. 26-41, *top*), which, we have seen, is a tenacious heme ligand (with 200-fold greater affinity for hemoglobin and myoglobin than O_2; Section 10-1A). Consequently, ~1% of hemoglobin's O_2-binding sites are blocked by CO, even in the absence of air pollution. However, free heme in solution binds CO with 20,000-fold greater affinity than it binds O_2. Thus, the globin (protein) portion of hemoglobin (and likewise myoglobin) somehow lowers the affinity of its bound heme for CO, thereby making O_2 transport possible. How does the globin do so?

Figure 26-41 **The heme degradation pathway.** M, V, P, and E, respectively, represent methyl, vinyl, propionyl, and ethyl groups.

Early X-ray structures of **carboxymyoglobin** (myoglobin with a CO ligand) indicated that the bound CO was inclined from the normal to the heme plane by 40° to 60° (the Fe—C—O bond angle appeared to be 120° to 140°), approximately the same angle with which O_2 binds to heme (Fig. 10-12). Yet, in complexes of CO with porphyrins in the absence of protein, the CO is normal to the heme plane. This suggested that the globin (in both myoglobin and hemoglobin) sterically bends the bound CO away from its preferred linear geometry, thereby reducing its affinity for CO and hence permitting the CO to be slowly exhaled. However, a variety of spectroscopic investigations together with highly accurate X-ray structures of carboxymyoglobin revealed that the bound CO is, in fact, inclined from the normal to the heme plane by ~7°, a distortion that is too small to explain the reduced affinity of myoglobin for CO.

Of course, this reduced affinity might instead be explained by the distortions that the upright CO ligand imposes on the globin, presumably via the distal His (E7, the His residue that hydrogen bonds to the bound O_2; Section 10-2A). However, studies of the energetics of binding of CO and O_2 to myoglobins in which His E7 has been mutated to nonpolar residues of comparable bulk (e.g., Leu) indicate that this is not the main determinant of the ligand affinity changes. Rather, the reduction in affinity of myoglobin for CO relative to that for O_2 has been shown to arise from the greater hydrogen bonding affinity that His E7 has for O_2 relative to CO together with electrostatic effects due to the differing charge distributions in the O_2 and CO ligands.

I. Chloroquine Prevents Malaria by Inhibiting Plasmodial Heme Sequestration

Malaria is caused by the mosquito-borne parasite *Plasmodium falciparum* (Section 7-3Ab), which multiplies within and destroys red blood cells in a 2-day cycle. During the intraerythrocytic stages of its life cycle, the parasite partially meets its nutritional needs by proteolyzing up to ~80% of the host cell's hemoglobin in its so-called acid food vacuole, whose pH is 4.7. This process releases heme, which in its soluble form, is toxic to the parasite because it damages cell membranes and inhibits a variety of enzymes. Since, unlike their human hosts, plasmodia cannot degrade heme, they sequester it within their food vacuoles in the form of harmless dark brown granules known as **hemozoin,** which consist of crystals of dimerized hemes linked together by reciprocal iron–carboxylate bonds between the ferric ions and the propionate side chains of adjacent molecules. Hemozoin has been found to be identical to **β-hematin,**

β-Hematin (hemozoin)

whose X-ray structure has been determined. Dimers interact in the crystals through hydrogen bonds between the remaining carboxyl groups.

Chloroquine,

Chloroquine

Quinine

a member of the quinoline ring–containing family of antimalarials, which includes **quinine,** is one of the most successful antimicrobial agents that has been produced. It is effective against plasmodia only during their intraerythrocytic stages. This drug, being a weak base that can readily pass through membranes in its uncharged form, accumulates in the plasmodial acid food vacuole in its acidic (charged) form in millimolar concentrations. Chloroquine and several other quinoline-containing antimalarials inhibit the crystallization of hemes to form hemozoin. This inhibition *in vivo* is almost certainly responsible for the antimalarial properties of these drugs. The mechanism of inhibition is as yet unclear although a plausible hypothesis is that the drug adsorbs onto crystallized hemozoin, inhibiting further crystallization.

The massive use of chloroquine has, unfortunately, led to the appearance of chloroquine-resistant plasmodia in nearly every malarial region of the world. Resistant plasmodia do not concentrate chloroquine in their food vacuoles to the high levels found in sensitive parasites. Rather, they export this drug out of their food vacuoles at an ~50-fold higher rate than do sensitive organisms. Since chloroquine activity and chloroquine resistance have different mechanisms, it has been possible to modify existing quinoline-containing structures and to develop new hemozoin crystallization inhibitors that are effective antimalarial agents but to which plasmodia are not (yet) resistant.

B. *Biosynthesis of Physiologically Active Amines*

Epinephrine, norepinephrine, dopamine, serotonin (5-hydroxytryptamine), γ-aminobutyric acid (GABA), and histamine

X = OH, R = CH$_3$ **Epinephrine (Adrenalin)**
X = OH, R = H   **Norepinephrine**
X = H,   R = H   **Dopamine**

Serotonin
(5-hydroxytryptamine)

$$^-OOC-CH_2-CH_2-CH_2-NH_3^+$$

γ-Aminobutyric acid (GABA)

Histamine

are hormones and/or neurotransmitters derived from amino acids. For instance, epinephrine, as we have seen, activates muscle adenylate cyclase, thereby stimulating glycogen breakdown (Section 18-3E); deficiency in dopamine production is associated with **Parkinson's disease,** a degenerative condition causing "shaking palsy"; serotonin causes smooth muscle contraction; GABA is one of the brain's major inhibitory neurotransmitters (Section 20-5Cf), being released at 30% of its synapses; and histamine is involved in allergic responses (as allergy sufferers who take antihistamines will realize), as well as in the control of acid secretion by the stomach (Section 20-3C).

The biosynthesis of each of these physiologically active amines involves decarboxylation of the corresponding precursor amino acid. Amino acid decarboxylases are PLP-dependent enzymes that form a PLP–Schiff base with the

substrate so as to stabilize the C_α carbanion formed on C_α—COO$^-$ bond cleavage (Section 26-1Aa):

Formation of histamine and GABA are one-step processes (Fig. 26-42). In the synthesis of serotonin from tryptophan, the decarboxylation is preceded by a hydroxylation (Fig. 26-43) by **tryptophan hydroxylase,** one of three mammalian enzymes that has a 5,6,7,8-tetrahydrobiopterin cofactor (Section 26-3Ha). This hydroxylation involves an NIH shift similar to that occurring in phenylalanine hydroxylase (Fig. 26-30), although no epoxide intermediate has been observed in this case. Dopamine, norepinephrine, and epinephrine are all termed **catecholamines** because they are amine derivatives of **catechol:**

Catechol

The conversion of tyrosine to these various catecholamines occurs as follows (Fig. 26-44):

1. Tyrosine is hydroxylated to **3,4-dihydroxyphenylalanine (L-DOPA)** by **tyrosine hydroxylase,** another 5,6,7,8-tetrahydrobiopterin-requiring enzyme.

2. L-DOPA is decarboxylated to dopamine.

3. A second hydroxylation yields norepinephrine.

Figure 26-42   The formation of γ-aminobutyric acid (GABA) and histamine. The reactions involve the decarboxylations of glutamate to form GABA and of histidine to form histamine.

Figure 26-43 The formation of serotonin. The biosynthesis involves the hydroxylation and subsequent decarboxylation of tryptophan.

4. Methylation of norepinephrine's amino group by *S*-adenosylmethionine (SAM; Section 26-3Ea) produces epinephrine.

The specific catecholamine that a cell produces depends on which enzymes of the pathway are present. In adrenal medulla, which functions to produce hormones (Section 19-1F), epinephrine is the predominant product. In some areas of the brain, norepinephrine is more common. In other areas, most prominently the **substantia nigra,** the pathway stops at dopamine synthesis. Indeed, Parkinson's disease, which is caused by degeneration of the substantia nigra, has been treated with some success by the administration of L-DOPA, dopamine's immediate precursor. Dopamine itself is ineffective because it cannot cross the blood–brain barrier. L-DOPA, however, is able to get to its sites of action where it is decarboxylated to dopamine. The enzyme catalyzing this reaction, **aromatic amino acid decarboxylase,** decarboxylates all aromatic amino acids and is therefore also responsible for serotonin formation. L-DOPA is also a precursor of the black skin pigment melanin.

C. Glutathione

Glutathione (GSH; γ-glutamylcysteinylglycine),

Glutathione (GSH; γ-glutamylcysteinylglycine)

a tripeptide that contains an unusual γ-amide bond, participates in a variety of detoxification, transport, and metabolic processes (Fig. 26-45). For instance, it is a substrate for

Figure 26-44 The sequential synthesis of L-DOPA, dopamine, norepinephrine, and epinephrine from tyrosine. L-DOPA is also the precursor of the black skin pigment melanin, an oxidized polymeric material.

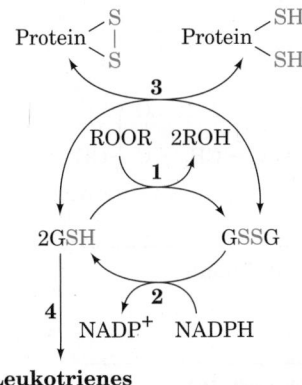

Figure 26-45 **Some reactions involving glutathione.** The reactions and enzymes are **(1)** peroxide detoxification by **glutathione peroxidase, (2)** regeneration of GSH from GSSG by glutathione reductase (Section 21-2Ba), **(3)** thiol transferase modulation of protein thiol–disulfide balance, and **(4)** leukotriene biosynthesis by a glutathione-*S*-transferase.

peroxidase reactions, helping to destroy peroxides generated by oxidases; it is involved in leukotriene biosynthesis (Section 25-7Cb); and the balance between its reduced (GSH) and oxidized (GSSG) forms maintains the sulfhydryl groups of intracellular proteins in their correct oxidation states.

The **γ-glutamyl cycle,** which was elucidated by Alton Meister, *provides a vehicle for the energy-driven transport*

of amino acids into cells through the synthesis and break-down of GSH (Fig. 26-46). GSH is synthesized from glutamate, cysteine, and glycine by the consecutive action of **γ-glutamylcysteine synthetase** and **GSH synthetase** (Fig. 26-46, Reactions 1 and 2). ATP hydrolysis provides the free energy for each reaction. The carboxyl group is activated for peptide bond synthesis by formation of an acyl phosphate intermediate:

The breakdown of GSH is catalyzed by **γ-glutamyl transpeptidase, γ-glutamyl cyclotransferase, 5-oxoprolinase,** and an intracellullar protease (Fig. 26-46, Reactions 3–6).

Amino acid transport occurs because, whereas GSH is synthesized intracellularly and is located largely within the cell, γ-glutamyl transpeptidase, which catalyzes GSH breakdown (Fig. 26-46, Reaction 3), is situated on the cell membrane's external surface and accepts amino acids, notably cysteine and methionine. GSH is first transported to the

Figure 26-46 **Glutathione synthesis as part of the γ-glutamyl cycle of glutathione metabolism.** The cycle's reactions are catalyzed by **(1)** γ-glutamylcysteine synthetase, **(2)** glutathione synthetase, **(3)** γ-glutamyl transpeptidase, **(4)** γ-glutamyl cyclotransferase, **(5)** 5-oxoprolinase, and **(6)** an intracellular protease.

Figure 26-47 **Tetrahydrofolate (THF).**

external surface of the cell membrane, where the transfer of the γ-glutamyl group from GSH to an external amino acid occurs. The γ-glutamyl amino acid is then transported back into the cell and converted to glutamate by a two-step process in which the transported amino acid is released and **5-oxoproline** is formed as an intermediate. The last step in the cycle, the hydrolysis of 5-oxoproline, requires ATP hydrolysis. This surprising observation (amide bond hydrolysis is almost always an exergonic process) is a consequence of 5-oxoproline's unusually stable internal amide bond.

D. *Tetrahydrofolate Cofactors: The Metabolism of C₁ Units*

Many biosynthetic processes involve the addition of a C_1 unit to a metabolic precursor. A familiar example is carboxylation. For instance, gluconeogenesis from pyruvate begins with the addition of a carboxyl group to form oxaloacetate (Section 23-1Aa). The coenzyme involved in this and most other carboxylation reactions is biotin (Section 23-1Ab). In contrast, *S*-adenosylmethionine functions as a methylating agent (Section 26-3Ea).

Tetrahydrofolate (THF) is more versatile than the above cofactors in that it functions to transfer C_1 units in several oxidation states. THF is a 6-methylpterin derivative linked in sequence to *p*-aminobenzoic acid and Glu residues (Fig. 26-47). Up to five additional Glu residues may be linked to the first glutamate via isopeptide bonds to form a polyglutamyl tail.

THF is derived from **folic acid** (Latin: *folium,* leaf), a doubly oxidized form of THF that must be enzymatically reduced before it becomes an active coenzyme (Fig. 26-48). Both reductions are catalyzed by **dihydrofolate reductase (DHFR).** Mammals cannot synthesize folic acid and so must obtain it from their diets or from intestinal microorganisms.

C_1 units are covalently attached to THF at its positions N5, N10, or both N5 and N10. These C_1 units, which may be at the oxidation levels of formate, formaldehyde, or methanol (Table 26-1), are all interconvertible by enzymatic redox reactions (Fig. 26-49).

The main entry of C_1 units into the THF pool is as N^5,N^{10}-**methylene-THF** through the conversion of serine to glycine by serine hydroxymethyltransferase (Sections 26-3Bb and 26-5Ae) and the cleavage of glycine by glycine synthase (the glycine cleavage system; Section 26-3Ba, Fig. 26-14). Histidine also contributes C_1 units through its degradation with the formation of N^5-**formimino-THF** (Fig. 26-17, Reaction 11).

A C_1 unit in the THF pool can have several fates (Fig. 26-50):

1. It may be used directly as N^5,N^{10}-methylene-THF in the conversion of the deoxynucleotide dUMP to dTMP by **thymidylate synthase** (Section 28-3Bb).

2. It may be reduced to N^5-**methyl-THF** for the synthesis of methionine from homocysteine (Section 26-3Ea).

3. It may be oxidized through N^5,N^{10}-methenyl-THF to N^{10}-**formyl-THF** for use in the synthesis of purines (Section

Figure 26-48 **The two-stage reduction of folate to THF.** Both reactions are catalyzed by dihydrofolate reductase (DHFR).

Table 26-1 **Oxidation Levels of C_1 Groups Carried by THF**

Oxidation Level	Group Carried	THF Derivative(s)
Methanol	Methyl ($-CH_3$)	N^5-Methyl-THF
Formaldehyde	Methylene ($-CH_2-$)	N^5,N^{10}-Methylene-THF
Formate	Formyl ($-CH=O$)	N^5-Formyl-THF, N^{10}-formyl-THF
	Formimino ($-CH=NH$)	N^5-Formimino-THF
	Methenyl ($-CH=$)	N^5,N^{10}-Methenyl-THF

28-1A). Since the purine ring of ATP is involved in histidine biosynthesis in microorganisms and plants (Section 26-5Be), N^{10}-formyl-THF is indirectly involved in this pathway as well. Prokaryotes use N^{10}-formyl-THF in a formylation reaction yielding **formylmethionyl-tRNA,** which they require for the initiation of protein synthesis (Section 32-3Ca).

Figure 26-49 **Interconversion of the C_1 units carried by THF.**

N^5-Methyl-THF $\overset{2}{\rightleftharpoons}$ methionine $\Longrightarrow$ SAM $\Longrightarrow$ epinephrine

SAM $\Longrightarrow$ phosphatidylcholine

N^5,N^{10}-Methylene-THF $\overset{1}{\rightleftharpoons}$ thymidylate (dTMP)

N^5,N^{10}-Methenyl-THF

N^{10}-Formyl-THF $\overset{3}{\Longleftarrow}$ formylmethionyl-tRNA

purines $\Longrightarrow$ histidine

Figure 26-50 **The biosynthetic fates of the C_1 units in the THF pool.**

A deficiency in folic acid results in **megloblastic anemia,** a condition in which erythrocytes in the blood are replaced by fewer abnormally large erythrocytes known as **macrocytes.** This occurs as a consequence of the lack of N^5,N^{10}-methylene-THF, which is required for the synthesis of the DNA precursor deoxythymidine monophosphate from deoxyuridine monophosphate (Section 28-3Bb). Pernicious anemia (vitamin B_{12} deficiency; Section 25-2Ee) has similar anemic symptoms. This is because the irreversibility of the reaction forming N^5-methyl-THF and the now inactive coenzyme B_{12}–dependent methionine synthase (top of Fig. 26-49), the only mammalian enzyme that utilizes N^5-methyl-THF, result in folates becoming trapped in their N^5-methyl-THF form. Consequently, the anemic symptoms of pernicious anemia may be alleviated by the administration of folic acid but this does not relieve its neurological symptoms.

Sulfonamides (sulfa drugs) such as **sulfanilamide** are antibiotics that are structural analogs of the *p*-aminobenzoic acid constituent of THF:

Sulfonamides
(R = H, sulfanilamide)

***p*-Aminobenzoic acid**

They competitively inhibit bacterial synthesis of THF at the *p*-aminobenzoic acid incorporation step, thereby blocking the above THF-requiring reactions. The inability of mammals to synthesize folic acid leaves them unaffected by sulfonamides, which accounts for the medical utility of these widely used antibacterial agents.

5 AMINO ACID BIOSYNTHESIS

Many amino acids are synthesized by pathways that are present only in plants and microorganisms. Since mammals must obtain these amino acids in their diets, these substances are known as **essential amino acids.** The other amino acids, which can be synthesized by mammals from common intermediates, are termed **nonessential amino**

acids. Their α-keto acid carbon skeletons are converted to amino acids by transamination reactions (Section 26-1A) utilizing the preformed α-amino nitrogen of another amino acid, usually glutamate. Yet, although it was originally presumed that glutamate can be synthesized from ammonia and α-ketoglutarate by glutamate dehydrogenase acting in reverse, it now appears that the predominant physiological direction of this enzyme is glutamate breakdown (Section 26-1B). Consequently, *preformed α-amino nitrogen should also be considered to be an essential nutrient.* In this context, it is interesting to note that, in addition to the four well-known taste receptors, those for sweet, sour, salty, and bitter tastes, a fifth taste receptor has been characterized, that for the meaty taste of **monosodium glutamate (MSG),** which is known as **umami** (Japanese: flavor).

The essential and nonessential amino acids for humans are listed in Table 26-2. Arginine is classified as essential, even though it is synthesized by the urea cycle (Section 26-2D), because it is required in greater amounts than can be produced by this route during the normal growth and development of children (but not adults). The essential amino acids occur in animal and vegetable proteins. Different proteins, however, contain different proportions of the essential amino acids. Milk proteins, for example, contain them all in the proportions required for proper human nutrition. Bean protein, on the other hand, contains an abundance of lysine but is deficient in methionine, whereas wheat is deficient in lysine but contains ample methionine. A balanced protein diet therefore must contain a variety of different protein sources that complement each other to supply the proper proportions of all the essential amino acids.

In this section we study the pathways involved in the formation of the nonessential amino acids. We also briefly consider such pathways for the essential amino acids as they occur in plants and microorganisms. You should note, however, that although we discuss some of the most common pathways for amino acid biosynthesis, there is considerable variation in these pathways among different species. In contrast, as we have seen, the basic pathways of carbohydrate and lipid metabolism are all but universal.

A. Biosynthesis of the Nonessential Amino Acids

All the nonessential amino acids except tyrosine are synthesized by simple pathways leading from one of four common metabolic intermediates: pyruvate, oxaloacetate, α-ketoglutarate, and 3-phosphoglycerate. Tyrosine, which is really misclassified as nonessential, is synthesized by the one-step hydroxylation of the essential amino acid phenylalanine (Section 26-3H). Indeed, the dietary requirement for phenylalanine reflects the need for tyrosine as well. The presence of dietary tyrosine therefore decreases the need for phenylalanine. Since preformed α-amino nitrogen in the form of glutamate is an essential nutrient for nonessential

Table 26-2   Essential and Nonessential Amino Acids in Humans

Essential	Nonessential
Arginine[a]	Alanine
Histidine	Asparagine
Isoleucine	Aspartate
Leucine	Cysteine
Lysine	Glutamate
Methionine	Glutamine
Phenylalanine	Glycine
Threonine	Proline
Tryptophan	Serine
Valine	Tyrosine

[a]Although mammals synthesize arginine, they cleave most of it to form urea (Sections 26-2D and 26-2E).

amino acid biosynthesis, we first discuss its production by plants and microorganisms.

a. Glutamate Is Synthesized by Glutamate Synthase

Glutamate synthase, an enzyme that occurs only in microorganisms, plants, and lower animals, converts

α-ketoglutarate and ammonia originating in glutamine to glutamate. The electrons required for this reductive amination come from NADPH or ferredoxin, depending on the organism. The NADPH-dependent glutamate synthase from the nitrogen-fixing bacterium *Azospirillum brasilense,* the best characterized such enzyme, is an $\alpha_2\beta_2$ heterotetramer that binds an FAD and two [4Fe–4S] clusters on each β subunit, and an FMN and a [3Fe–4S] cluster on each α subunit. The overall reaction is

$$\text{NADPH} + \text{H}^+ + \text{glutamine} + \alpha\text{-ketoglutarate} \longrightarrow$$
$$2 \text{ glutamate} + \text{NADP}^+$$

and involves five steps that occur at three distinct active sites (Fig. 26-51):

1. Electrons are transferred from NADPH to FAD at active site 1 on the β subunit to yield $FADH_2$.

2. The electrons are transferred from the $FADH_2$ to FMN at site 2 on a specific α subunit through the iron–sulfur clusters to yield $FMNH_2$.

3. Glutamine is hydrolyzed to α-glutamate and ammonia at site 3 on the α subunit.

4. The ammonia produced is transferred to site 2 where it reacts with α-ketoglutarate to form α-iminoglutarate.

Overall: $\text{NADPH} + \text{H}^+ + \text{glutamine} + \alpha\text{-ketoglutarate} \longrightarrow 2 \text{ glutamate} + \text{NADP}^+$

Figure 26-51   The sequence of reactions catalyzed by glutamate synthase.

5. The α-iminoglutarate is reduced by FMNH$_2$ to form glutamate.

In the absence of the β subunit, the α subunit can synthesize glutamate from glutamine and α-ketoglutarate using an artificial electron donor; moreover, it is homologous and functionally similar to ferredoxin-dependent glutamine synthases. The *A. brasilense* α subunit is therefore considered to be the enzyme's catalytic core.

The X-ray structure of the 1479-residue α subunit of *A. brasilense* glutamate synthase in complex with a [3Fe–4S] cluster, an FMN, an α-ketoglutarate substrate, and a **methionine sulfone** inhibitor (a tetrahedral transition state analog)

Glutamine

Methionine sulfone

was determined by Andrea Mattevi (Fig. 26-52). The subunit has four domains, an N-terminal glutamine amidotransferase domain, a central domain, an FMN-binding domain, and a C-terminal **β-helix** domain (see below). The methionine sulfone binds in the N-terminal amidotransferase domain (site 3), where glutamine is normally hydrolyzed via the nucleophilic attack of the Cys 1 sulfhydryl group on the glutamine C$_\gamma$ atom to transiently form a tetrahedral intermediate that the tetrahedral sulfone group mimics. The FMN-binding domain consists in large part of an α/β barrel, at the mouth of which the α-ketoglutarate and the [3Fe–4S] cluster bind. The distance between the methyl group on methionine sulfone (the analog of the glutamine amido group) and the α-keto group of the α-ketoglutarate is 31 Å. These two sites are connected by a tunnel through which ammonia must diffuse in order to react with α-ketoglutarate (Reaction 4, Fig. 26-51). However, the tunnel is blocked by the main chain atoms of four residues that protrude into the tunnel. Consequently, there must be at least a 2- to 3-Å shift in the structure during the reaction to permit ammonia to diffuse between these sites. Such gating of the tunnel may well be crucial for the control of enzyme function so as to avoid the wasteful hydrolysis of glutamine. In fact, glutamine is hydrolyzed only when the enzyme has bound α-ketoglutarate and reducing equivalents are available for iminoglutarate reduction.

The C-terminal domain of glutamate synthase consists largely of a 7-turn, right-handed β helix (Fig. 26-53). In this unusual motif, the polypeptide chain is wrapped in a wide helix such that neighboring turns of chain interact as do the strands of a parallel β sheet. The 43-Å-long glutamate synthase β helix has an elliptical cross section of 16 by 23 Å. β Helices have been observed in only a handful of mainly bacterial enzymes. The β helix of glutamate synthase does

Figure 26-52 **X-ray structure of the α subunit of *A. brasilense* glutamate synthase as represented by its C$_\alpha$ backbone.** The subunit has four domains, an N-terminal glutamine amidotransferase domain (*blue*) to which methionine sulfone (a glutamine tetrahedral transition state analog) is bound; a central domain (*red*); an FMN-binding domain (*green*) to which an FMN, a [3Fe–4S] cluster, and an α-ketoglutarate are bound; and a β helix domain (*purple*). The foregoing ligands are drawn in black in ball-and-stick form. The 31-Å-long tunnel from the methyl group on the methionine sulfone (analogous to the amido group of glutamine) to the α-keto group of the α-ketoglutarate is outlined by a gray surface. The tunnel is blocked in this structure (it is divided into two cavities) by the main chain atoms of four residues that protrude into the tunnel. [Courtesy of Andrea Mattevi, Universitá degli Studi di Pavia, Italy. PDBid 1EA0.]

not contain a residue that is involved in catalysis or electron transfer. However, it does appear to have an important structural role because some of its residues line the tunnel through which ammonia passes.

Glutamine amidotransferase domains or subunits are part of several protein structures in which glutamine is the ammonia donor for a further reaction. In glutamate synthase, it is a domain within the α subunit (Fig. 26-52). In *E. coli* carbamoyl phosphate synthetase (CPS; Section 26-2Aa), it is a complete subunit within a heterodimer. These domains or subunits occur in one of two families that are differentiated by their active site structures. In CPS, it belongs to the **triad family,** so called because it contains an active site Cys in a catalytic triad reminiscent of the catalytic triad in serine proteases (Section 15-3). The glutamine amidotransferase domain of glutamate synthase belongs to the **N-terminal nucleophile (Ntn) family,** which, as we have seen, has an N-terminal Cys that acts as the active site nucleophile. Other enzymes involved in

Figure 26-53 **The β helix of *A. brasilense* glutamate synthase.** The polypeptide backbone (residues 1225–1416) is colored in rainbow order from its N-terminus (*blue*) to its C-terminus (*red*). Neighboring turns of polypeptide chain within the β helix interact as do the strands of parallel β sheets. Nevertheless, the conformations of many of these segments lie outside the normal range for β strands, and hence they are drawn in coil form. [Based on an X-ray structure by Andrea Mattevi, Universitá degli Studi di Pavia, Italy. PDBid 1EA0.]

amino acid biosynthesis and having a glutamine amidotransferase domain include asparagine synthetase (Fig. 26-54, Reaction 4; see below), a member of the Ntn family, and imidazole glycerol phosphate synthase (Fig. 26-65, Reaction 5), which belongs to the triad family. All of these enzymes have an ammonia-channeling tunnel that connects the amidotransferase site with the ammonia-utilizing site.

b. Alanine, Asparagine, Aspartate, Glutamate, and Glutamine Are Synthesized from Pyruvate, Oxaloacetate, and α-Ketoglutarate

Pyruvate, oxaloacetate, and α-ketoglutarate are the keto acids that correspond to alanine, aspartate, and glutamate, respectively. Indeed, as we have seen (Section 26-1), the synthesis of each of these amino acids is a one-step

Figure 26-54 **The syntheses of alanine, aspartate, glutamate, asparagine, and glutamine.** These reactions involve the transaminations of (**1**) pyruvate, (**2**) oxaloacetate, and (**3**) α-ketoglutarate, and the amidations of (**4**) aspartate and (**5**) glutamate.

transamination reaction (Fig. 26-54, Reactions 1–3). As-paragine and glutamine are, respectively, synthesized from aspartate and glutamate by amidation (Fig. 26-54, Reactions 4 and 5). **Glutamine synthetase** catalyzes the formation of glutamine in a reaction in which ATP is hydrolyzed to ADP

and P_i via the intermediacy of **γ-glutamylphosphate** and NH_4^+ is the amino group donor (Fig. 26-54, Reaction 5). Curiously, aspartate amidation by **asparagine synthetase** to form asparagine follows a different route; it utilizes glutamine as its amino group donor and hydrolyzes ATP to $AMP+PP_i$ (Fig. 26-54, Reaction 4). This enzyme is composed of a glutamine amidotransferase domain of the Ntn family (see above) and a second domain in which β-aspartyl-AMP

β-Aspartyl-AMP

is synthesized from Asp and ATP and then reacts with ammonia to form Asn. As with other glutamine amidotransferase–containing enzymes, the two domains are connected by a tunnel that channels ammonia between their two active sites.

c. Glutamine Synthetase Is a Central Control Point in Nitrogen Metabolism

Glutamine, as we have seen, is the amino group donor in the formation of many biosynthetic products, as well as being a storage form of ammonia. Glutamine synthetase's consequent pivotal position in nitrogen metabolism makes its control a vital aspect of this process. In fact, mammalian glutamine synthetases are activated by α-ketoglutarate, the product of glutamate's oxidative deamination. This control presumably prevents the accumulation of the ammonia produced by that reaction.

Bacterial glutamine synthetase, as Earl Stadtman showed, has a much more elaborate control system. This enzyme, which consists of 12 identical 469-residue subunits arranged with D_6 symmetry (Fig. 26-55), is regulated by several effectors as well as by covalent modification. Although a complete description of this complex enzyme is not given here, several aspects of its catalytic and control systems bear note.

The X-ray structure of *Salmonella typhimurium* glutamine synthetase in complex with the glutamate structural analog **phosphinothricin,**

Phosphinothricin   **Glutamate**

determined by David Eisenberg, reveals that its catalytic sites occur at the interface between the C-terminal domain

Figure 26-55 **X-ray structure of *S. typhimurium* glutamine synthetase.** The enzyme consists of 12 identical subunits, here represented by their C_α backbones, arranged with D_6 symmetry (the symmetry of a hexagonal prism). (*a*) View down the 6-fold axis of symmetry showing only the six subunits of the upper ring in alternating blue and green. The subunits of the lower ring are roughly directly below those of the upper ring. The protein, including its side chains (not shown), has a diameter of 143 Å. The six active sites shown are marked by the pairs of Mn^{2+} ions (*magenta spheres*; divalent metal ions, physiologically Mg^{2+}, are required for enzymatic activity). Each adenylylation site, Tyr 397 (*yellow*), lies between two subunits at a higher radius than the corresponding active site. Also drawn in one active site are ADP (*cyan*) and phosphinothricin (*orange*), a competitive inhibitor of glutamate. (*b*) Side view along one of the enzyme's 2-fold axes showing only the eight nearest subunits. The molecule extends 103 Å along the 6-fold axis, which is vertical in this view. [Based on an X-ray structure by David Eisenberg, UCLA. PDBid 1FPY.]

of one subunit and the N-terminal domain of an adjacent subunit. These catalytic sites have a shape described as a "bifunnel" that opens at both the exposed top (ATP binding) and bottom (Glu and NH_4^+ binding) of the molecule (between the two hexameric rings) and is narrow in the plane of its essential metal ions (two per subunit). Nucleotide binding induces conformational changes that increase the enzyme's affinity for glutamate and ammonium ion, leading to an ordered sequential mechanism.

Nine feedback inhibitors cumulatively control the activity of bacterial glutamine synthetase: histidine, tryptophan, carbamoyl phosphate (as synthesized by carbamoyl phosphate synthetase II), glucosamine-6-phosphate, AMP, and CTP are all end products of pathways leading from glutamine, whereas alanine, serine, and glycine reflect the cell's nitrogen

level. Several of these inhibitors act in a competitive manner, binding either to the glutamate binding site (serine, glycine, and alanine) or to the ATP binding site (AMP and CTP).

E. coli glutamine synthetase is covalently modified by adenylylation of a specific Tyr residue (Fig. 26-56). The enzyme's susceptibility to cumulative feedback inhibition increases, and its activity therefore decreases, with its degree of adenylylation. The level of adenylylation is controlled by a complex metabolic cascade that is conceptually similar to the one controlling glycogen phosphorylase (although the type of covalent modification differs in that glycogen phosphorylase is phosphorylated at a specific Ser residue; Section 18-3Ca). Both adenylylation and deadenylylation of glutamine synthetase are catalyzed by **adenylyltransferase** in complex with a tetrameric regulatory protein, **P_{II}.** This

Figure 26-56 The regulation of bacterial glutamine synthetase. The adenylylation/deadenylylation of a specific Tyr residue is controlled by the level of uridylylation of a specific adenylyltransferase · P_{II} Tyr residue. This uridylylation level, in turn, is controlled by the relative activities of uridylyltransferase, which is sensitive to the levels of a variety of nitrogen metabolites, and uridylyl-removing enzyme, whose activity is independent of these metabolite levels.

Figure 26-57   The biosynthesis of the "glutamate family" of amino acids: arginine, ornithine, and proline. The enzymes catalyzing proline biosynthesis are **(1)** γ-glutamyl kinase, **(2)** dehydrogenase, **(3)** nonenzymatic, and **(4)** pyrroline-5-carboxylate reductase. The enzymes catalyzing ornithine biosynthesis are **(5)** *N*-acetylglutamate synthase, **(6)** acetylglutamate kinase, **(7)** *N*-acetyl-γ-glutamyl phosphate reductase, **(8)** *N*-acetylornithine-δ-aminotransferase, and **(9)** acetylornithine deacetylase. An alternate pathway to ornithine is through Reaction **10**, catalyzed by ornithine-δ-aminotransferase. Ornithine is converted to arginine **(11)** via the urea cycle (Fig. 26-7, Reactions **2–4**).

complex deadenylylates glutamine synthetase when P_{II} is uridylylated (also at a Tyr residue) and adenylylates glutamine synthetase when P_{II} lacks UMP residues. The level of P_{II} uridylylation, in turn, depends on the relative activities of two enzymatic activities located on the same protein: a **uridylyltransferase** that uridylylates P_{II} and a **uridylyl-removing enzyme** that hydrolytically excises the attached UMP groups of P_{II} (Fig. 26-56). The uridylyltransferase is activated by α-ketoglutarate and ATP and inhibited by glutamine and P_i, whereas uridylyl-removing enzyme is insensitive to these metabolites. This complex metabolic cascade therefore renders the activity of *E. coli* glutamine synthetase extremely responsive to the cell's nitrogen requirements.

d. Glutamate Is the Precursor of Proline, Ornithine, and Arginine

Conversion of glutamate to proline (Fig. 26-57, Reactions 1–4) involves the reduction of the γ-carboxyl group to an aldehyde followed by the formation of an internal Schiff base whose further reduction yields proline. Reduction of the glutamate γ-carboxyl group to an aldehyde is an endergonic process that is facilitated by the carboxyl group's prior phosphorylation by **γ-glutamyl kinase.** The unstable product, **glutamate-5-phosphate,** has not been isolated from reaction mixtures but is presumed to be the substrate for the reduction that follows. The resulting **glutamate-5-semialdehyde** cyclizes spontaneously to form the internal Schiff base **Δ^1-pyrroline-5-carboxylate.** The final reduction to proline is catalyzed by **pyrroline-5-carboxylate reductase.** Whether the enzyme requires NADH or NADPH is unclear.

The *E. coli* pathway from glutamate to ornithine and hence to arginine likewise involves the ATP-driven reduction of the glutamate γ-carboxyl group to an aldehyde (Fig. 26-57, Reactions 6 and 7). Spontaneous cyclization of this intermediate, **N-acetylglutamate-5-semialdehyde,** is prevented by prior acetylation of its amino group by **N-acetylglutamate synthase** to form **N-acetylglutamate** (Fig. 26-57, Reaction 5). *N*-Acetylglutamate-5-semialdehyde, in turn, is converted to the corresponding amine by transamination (Fig. 26-57, Reaction 8). Hydrolysis of the acetyl protecting group finally yields ornithine, which, as we have seen (Section 26-2), is converted to arginine via the urea cycle. In humans, however, the pathway to ornithine is more direct. The *N*-acetylation of glutamate that protects it from cyclization does not occur. Rather, glutamate-5-semialdehyde, which is in equilibrium with Δ^1-pyrroline-5-carboxylate, is directly transaminated to yield ornithine in a reaction catalyzed by **ornithine-δ-aminotransferase** (Fig. 26-57, Reaction 10).

e. Serine, Cysteine, and Glycine Are Derived from 3-Phosphoglycerate

Serine is formed from the glycolytic intermediate 3-phosphoglycerate in a three-reaction pathway (Fig. 26-58):

1. Conversion of 3-phosphoglycerate's 2-OH group to a ketone yielding **3-phosphohydroxypyruvate,** serine's phosphorylated keto acid analog.

2. Transamination of 3-phosphohydroxypyruvate to phosphoserine.

3. Hydrolysis of phosphoserine to yield serine.

Figure 26-58 **The conversion of 3-phosphoglycerate to serine.** The pathway enzymes are **(1)** 3-phosphoglycerate dehydrogenase, **(2)** a PLP-dependent aminotransferase, and **(3)** phosphoserine phosphatase.

Serine participates in glycine synthesis in two ways (Section 26-3B):

1. Direct conversion of serine to glycine by serine hydroxymethyl transferase in a reaction that also yields N^5,N^{10}-methylene-THF (Fig. 26-12, Reaction 4 in reverse).

2. Condensation of the N^5,N^{10}-methylene-THF with CO_2 and NH_4^+ by the glycine cleavage system (Fig. 26-12, Reaction 3 in reverse).

We have already discussed the synthesis, in animals, of cysteine from serine and homocysteine, a breakdown product of methionine (Section 26-3Ea). Homocysteine combines with serine to yield cystathionine, which subsequently forms cysteine and α-ketobutyrate (Fig. 26-18, Reactions 5 and 6). Since cysteine's sulfhydryl group is derived from the essential amino acid methionine, cysteine is really an essential amino acid. In plants and microorganisms, however, cysteine is synthesized from serine in a two-step reaction involving the activation of the serine —OH by converting it to **O-acetylserine** followed by the displacement of acetate by sulfide (Fig. 26-59a). The sulfide required is produced from sulfate in an 8-electron reduction that occurs in *E. coli* as shown in Fig. 26-59b. Sulfate is first activated by the

(a)

(b)

Figure 26-59   Cysteine biosynthesis. (*a*) The synthesis of cysteine from serine in plants and microorganisms. (*b*) The 8-electron reduction of sulfate to sulfide in *E. coli.*

enzymes **ATP sulfurylase** (which is used in the pyrosequencing of DNA; Section 7-2Ca) and **adenosine-5′-phosphosulfate (APS) kinase.** The activated sulfate is then reduced to sulfite by **3′-phosphoadenosine-5′-phosphosulfate (PAPS) reductase** and to sulfide by **sulfite reductase.**

B. *Biosynthesis of the Essential Amino Acids*

Essential amino acids, like nonessential amino acids, are synthesized from familiar metabolic precursors. Their synthetic pathways are present only in microorganisms and plants, however, and usually involve more steps than those of the nonessential amino acids. For example, lysine, methionine, and threonine are all synthesized from aspartate in pathways whose common first reaction is catalyzed by **aspartokinase,** an enzyme that is present only in plants and microorganisms. Similarly, valine and leucine are formed from pyruvate; isoleucine is formed from pyruvate and α-ketobutyrate; and tryptophan, phenylalanine, and tyrosine are formed from phosphoenolpyruvate and erythrose-4-phosphate. The enzymes that synthesize essential amino acids were apparently lost early in animal evolution, possibly because of the ready availability of these amino acids in the diet.

Time and space prevent a detailed discussion of the many interesting reactions that occur in these pathways. The biosynthetic pathways of the aspartate family of amino acids, the pyruvate family, the aromatic family, and histidine are presented in Figs. 26-60 through 26-63 and 26-65 together with lists of the enzymes involved. Several agriculturally useful

Figure 26-60 (*Opposite*) **The biosynthesis of the "aspartate family" of amino acids: lysine, methionine, and threonine.** The pathway enzymes are **(1)** aspartokinase, **(2)** β-aspartate semialdehyde dehydrogenase, **(3)** homoserine dehydrogenase, **(4)** homoserine kinase, **(5)** threonine synthase (a PLP enzyme), **(6)** homoserine acyltransferase, **(7)** cystathionine γ-synthase, **(8)** cystathionine β-lyase, **(9)** methionine synthase (alternatively homocysteine methyltransferase, which also occurs in mammals; Section 26-3Ec), **(10)** dihydrodipicolinate synthase, **(11)** dihydrodipicolinate reductase, **(12)** *N*-succinyl-2-amino-6-ketopimelate synthase, **(13)** succinyl-diaminopimelate aminotransferase (a PLP enzyme), **(14)** succinyl-diaminopimelate desuccinylase, **(15)** diaminopimelate epimerase, and **(16)** diaminopimelate decarboxylase.

Aspartate

$$O=C-CH_2-\overset{\overset{\displaystyle H}{|}}{\underset{\underset{\displaystyle NH_3^+}{|}}{C}}-COO^-$$

1 ATP → ADP

Aspartyl-β-phosphate

$$^{-2}O_3P-O-C-CH_2-\overset{\overset{\displaystyle H}{|}}{\underset{\underset{\displaystyle NH_3^+}{|}}{C}}-COO^-$$

2 NADPH → NADP$^+$ + P$_i$

β-Aspartate-semialdehyde

$$\underset{H}{O}=C-CH_2-\overset{\overset{\displaystyle H}{|}}{\underset{\underset{\displaystyle NH_3^+}{|}}{C}}-COO^-$$

Pyruvate 10

NADPH → NADP 3

4-Hydroxy-tetrahydrodipicolinate

OH, $^-$OOC, N, COO$^-$

spontaneous → H$_2$O

Dihydrodipicolinate

$^-$OOC, N, COO$^-$

NADPH → NADP$^+$ 11

Tetrahydrodipicolinate

$^-$OOC, N, COO$^-$

Succinyl-CoA + H$_2$O 12 → CoASH

N-Succinyl-2-amino-6-keto-L-pimelate

$^-$OOC, O, NH, Succ, COO$^-$

α-Ketoglutarate ← Glutamate 13

N-Succinyl-L,L-α,ε-diamino-pimelate

$^-$OOC, NH$_2$, NH$_3^+$, COO$^-$, Succ

14 H$_2$O → Succinate

L,L-α,ε-Diamino-pimelate

$$NH_3^+-\overset{COO^-}{\underset{(CH_2)_3}{C}}-H$$
$$H-C-NH_3^+$$
$$COO^-$$

15 →

meso-α,ε-Diamino-pimelate

$$H-\overset{COO^-}{\underset{(CH_2)_3}{C}}-NH_3^+$$
$$H-C-NH_3^+$$
$$COO^-$$

H$^+$ → CO$_2$ 16

Lysine

$$CH_2-NH_3^+$$
$$(CH_2)_3$$
$$H-C-NH_3^+$$
$$COO^-$$

Threonine

$$H_3C-\overset{}{\underset{OH}{CH}}-\overset{\overset{\displaystyle H}{|}}{\underset{\underset{\displaystyle NH_3^+}{|}}{C}}-COO^-$$

5 → P$_i$, H$_2$O

Phosphohomoserine

$$CH_2-CH_2-\overset{\overset{\displaystyle H}{|}}{\underset{\underset{\displaystyle NH_3^+}{|}}{C}}-COO^-$$
$$O-PO_3^{2-}$$

4 → ADP, ← ATP

Homoserine

$$CH_2-CH_2-\overset{\overset{\displaystyle H}{|}}{\underset{\underset{\displaystyle NH_3^+}{|}}{C}}-COO^-$$
$$OH$$

Succinyl-CoA → CoASH 6

O-Succinylhomoserine

$$CH_2-CH_2-\overset{\overset{\displaystyle H}{|}}{\underset{\underset{\displaystyle NH_3^+}{|}}{C}}-COO^-$$
$$O-Succ$$

Cysteine 7 → Succinate

Cystathionine

$$H_2C-S-CH_2-CH_2-\overset{\overset{\displaystyle H}{|}}{\underset{\underset{\displaystyle NH_3^+}{|}}{C}}-COO^-$$
$$H-C-NH_3^+$$
$$COO^-$$

8 → Pyruvate + NH$_3$

Homocysteine

$$HS-CH_2-CH_2-\overset{\overset{\displaystyle H}{|}}{\underset{\underset{\displaystyle NH_3^+}{|}}{C}}-COO^-$$

N^5-Methyl-THF 9 → THF

Methionine

$$H_3C-S-CH_2-CH_2-\overset{\overset{\displaystyle H}{|}}{\underset{\underset{\displaystyle NH_3^+}{|}}{C}}-COO^-$$

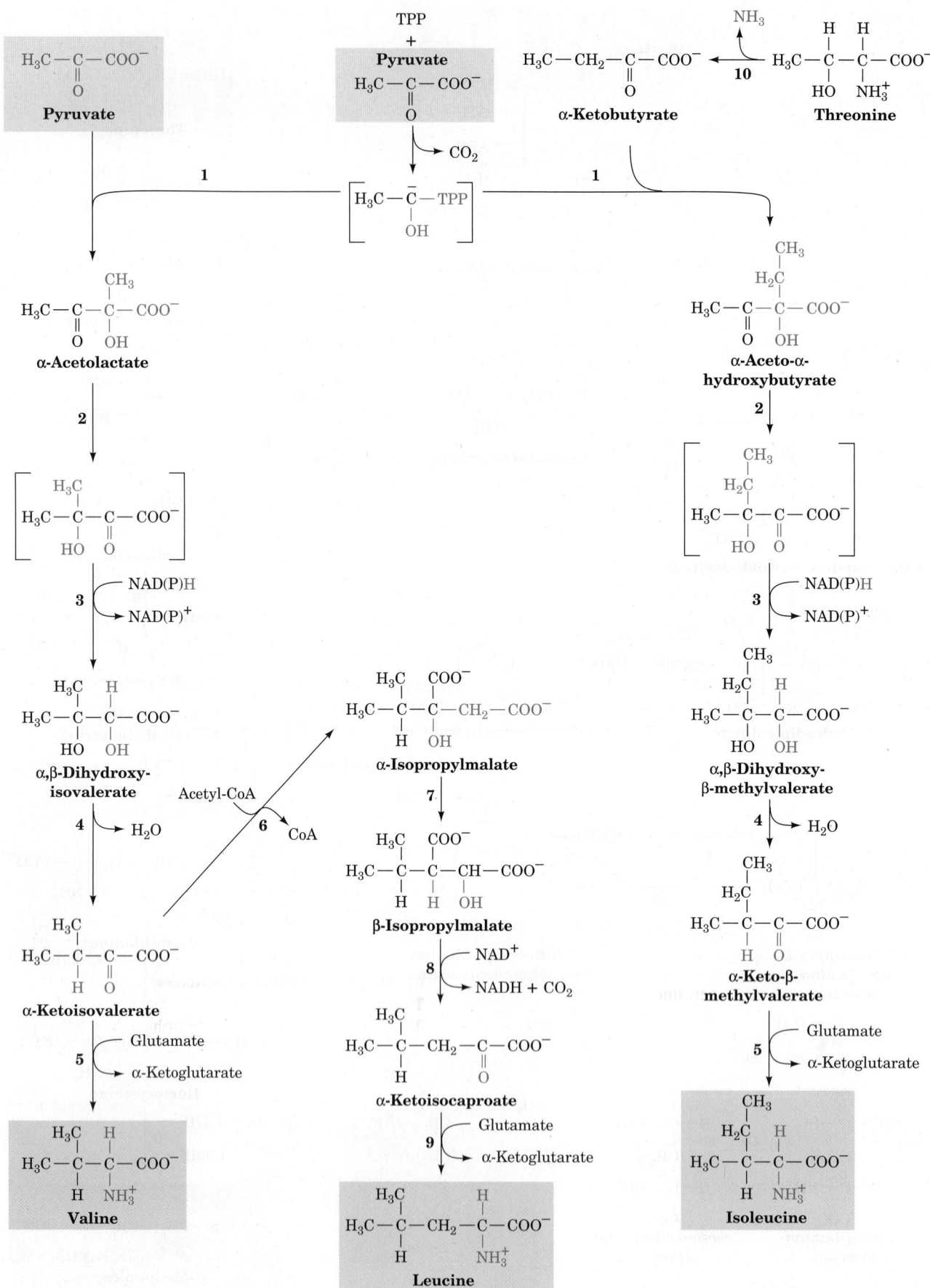

herbicides are specific inhibitors of some of these enzymes. Such herbicides have little toxicity toward animals and hence pose minimal risk to human health and the environment.

a. The Aspartate Family: Lysine, Methionine, and Threonine

In bacteria, aspartate is the common precursor of lysine, methionine, and threonine (Fig. 26-60). The biosyntheses of these essential amino acids all begin with the aspartokinase-catalyzed phosphorylation of aspartate to yield **aspartyl-β-phosphate.** We have seen that the control of metabolic pathways commonly occurs at the first committed step of the pathway. One might therefore expect lysine, methionine, and threonine biosynthesis to be controlled as a group. Each of these pathways is, in fact, independently controlled. *E. coli* does so via three isozymes of aspartokinase that respond differently to the three amino acids in terms both of feedback inhibition of enzyme activity and repression of enzyme synthesis. Table 26-3 summarizes this differential control. In addition, the pathway direction is controlled by feedback inhibition at the branch points by the individual amino acids. Thus methionine inhibits the *O*-acylation of homoserine (Fig. 26-60, Reaction 6), and lysine inhibits dihydrodipicolinate synthase (Fig. 26-60, Reaction 10).

b. The Pyruvate Family: Leucine, Isoleucine, and Valine

Valine and isoleucine are both synthesized via the same five-step pathway (Fig. 26-61), the only difference being in the first step of the series. In this TPP-dependent reaction, which resembles those catalyzed by pyruvate decarboxylase (Section 17-3Ba) and transketolase (Section 23-4Ca), pyruvate forms an adduct with TPP, which is decarboxylated to hydroxyethyl-TPP. This resonance-stabilized carbanion adds either to the keto group of a second pyruvate to form **acetolactate** on the way to valine, or to the keto group of threonine-derived **α-ketobutyrate** to form **α-aceto-α-hydroxybutyrate** on the way to isoleucine. The leucine biosynthetic pathway branches off from the valine pathway at α-ketoisovalerate (Fig. 26-61, Reaction 6). Reactions 6 to 8 in Fig. 26-61 are reminiscent of the first three reactions of the citric acid cycle (Sections 21-3A–C). Here, acetyl-CoA condenses with **α-ketoisovalerate** to form **α-isopropylmalate,** which then undergoes a dehydration/

Table 26-3 Differential Control of Aspartokinase Isoenzymes in *E. Coli*

Enzyme	Feedback Inhibitor	Corepressor(s)[a]
Aspartokinase I	Threonine	Threonine and isoleucine
Aspartokinase II	None	Methionine
Aspartokinase III	Lysine	Lysine

[a]Compounds whose presence results in the transcriptional repression of enzyme synthesis (Section 31-3G).

hydration reaction, followed by oxidative decarboxylation and transamination, to yield leucine.

c. The Aromatic Amino Acids: Phenylalanine, Tyrosine, and Tryptophan

The precursors to the aromatic amino acids are the glycolytic intermediate phosphoenolpyruvate (PEP) and erythrose-4-phosphate (an intermediate in the pentose phosphate pathway; Section 23-4Cb). Their condensation forms **2-keto-3-deoxy-D-arabinoheptulosonate-7-phosphate,** a C_7 compound that cyclizes and is ultimately converted to **chorismate** (Fig. 26-62), the branch point for tryptophan biosynthesis. Chorismate is converted either to **anthranilate** and then on to tryptophan, or to **prephenate** and on to either tyrosine or phenylalanine (Fig. 26-63). Although mammals synthesize tyrosine by the hydroxylation of phenylalanine (Section 26-3Ha), many microorganisms synthesize it directly from prephenate.

Since the synthesis of aromatic amino acids only occurs in plants and microorganisms, this pathway is a natural target for herbicides that will not be toxic to animals. For example, **glyphosate,**

$$^{-2}O_3P - CH_2 - NH - CH_2 - COO^-$$
Glyphosate

the active ingredient in one of the most widely used weed killers, Roundup, is a competitive inhibitor with respect to PEP in the **5-enolpyruvylshikimate-3-phosphate (EPSP) synthase** reaction (Reaction 6 of Fig. 26-62).

d. A Protein Tunnel Channels the Intermediate Product of Tryptophan Synthase between Two Active Sites

The final two reactions in tryptophan biosynthesis, Reactions 5 and 6 in Fig. 26-63, are both catalyzed by **tryptophan synthase:**

1. The α subunit (268 residues) of this $\alpha_2\beta_2$ bifunctional enzyme cleaves **indole-3-glycerol phosphate,** yielding **indole** and glyceraldehyde-3-phosphate (Reaction 5).

2. The β subunit (396 residues) joins indole with L-serine in a PLP-dependent reaction to form L-tryptophan (Reaction 6).

Either subunit alone is enzymatically active, but when they are joined in the $\alpha_2\beta_2$ tetramer, the rates of both reactions

Figure 26-61 *(Opposite)* **The biosynthesis of the "pyruvate family" of amino acids: isoleucine, leucine, and valine.** The pathway enzymes are **(1)** acetolactate synthase (a TPP enzyme), **(2)** acetolactate mutase, **(3)** reductase, **(4)** dihydroxy acid dehydratase, **(5)** valine aminotransferase (a PLP enzyme), **(6)** α-isopropylmalate synthase, **(7)** α-isopropylmalate dehydratase, **(8)** isopropylmalate dehydrogenase, **(9)** leucine aminotransferase (a PLP enzyme), and **(10)** threonine deaminase (serine dehydratase, a PLP enzyme).

Figure 26-62   The biosynthesis of chorismate, the aromatic amino acid precursor. The pathway enzymes are **(1)** 2-keto-3-deoxy-D-arabinoheptulosonate-7-phosphate synthase, **(2)** dehydroquinate synthase (an NAD$^+$-requiring reaction that yields an unchanged NAD$^+$ product and is thereby indicative of an oxidized intermediate as similarly occurs in the UDP–galactose-4-epimerase reaction; Section 17-5B), **(3)** 5-dehydroquinate dehydratase, **(4)** shikimate dehydrogenase, **(5)** shikimate kinase, **(6)** 5-enolpyruvylshikimate-3-phosphate synthase, and **(7)** chorismate synthase.

and their substrate affinities are increased by 1 to 2 orders of magnitude. Indole, the intermediate product, does not appear free in solution; the enzyme apparently sequesters it.

The X-ray structure of tryptophan synthase from *Salmonella typhimurium*, determined by Craig Hyde, Edith Miles, and David Davies, explains the latter observation. The protein forms a 150-Å-long, 2-fold symmetric α–β–β–α complex (Fig. 26-64) in which the active sites of neighboring α and β subunits are separated by ~25 Å. *These active sites are joined by a solvent-filled tunnel that is* *wide enough to permit the passage of the intermediate substrate, indole.* This structure, the first in which the presence of a tunnel between active sites was observed, suggests the following series of events. The indole-3-glycerol phosphate substrate binds to the α subunit through an opening into its active site, its "front door," and the glyceraldehyde-3-phosphate product leaves via the same route. Similarly, the β subunit active site has a "front door" opening to the solvent through which serine enters and tryptophan leaves. Both active sites also have "back doors" that are connected

Figure 26-63 The biosynthesis of phenylalanine, tryptophan, and tyrosine from chorismate. The pathway enzymes are (1) anthranilate synthase, (2) anthranilate phosphoribosyltransferase, (3) N-(5'-phosphoribosyl) anthranilate isomerase, (4) indole-3-glycerol phosphate synthase, (5) tryptophan synthase, α subunit, (6) tryptophan synthase, β subunit, (7) chorismate mutase, (8) prephenate dehydrogenase, (9) aminotransferase, (10) prephenate dehydratase, and (11) aminotransferase.

by the tunnel. The indole intermediate presumably diffuses between the two active sites via the tunnel and hence does not escape to the solvent.

Allosteric interactions between the subunits to control the activity of the α subunit also serve to ensure that indole is only released when the β subunit is ready to accept it.

Figure 26-64 **A ribbon diagram of the bifunctional enzyme tryptophan synthase from *S. typhimurium.*** Only one αβ protomer of this 2-fold symmetric αββα heterotetramer is shown. The α subunit is blue, the β subunit's N-terminal domain is orange, its C-terminal domain is red-orange, and all β sheets are tan. The active site of the α subunit is located by its bound competitive inhibitor, **indolepropanol phosphate** (**IPP;** *red ball-and-stick model*), whereas that of the β subunit is marked by its PLP coenzyme (*yellow ball-and-stick model*). The solvent-accessible surface of the ~25-Å-long "tunnel" connecting the active sites of the α and β subunits is outlined by a yellow dot surface. Several indole molecules (*green ball-and-stick models*) have been modeled into the tunnel in head to tail fashion, thereby demonstrating that the tunnel has sufficient width to permit the indole product of the α subunit to pass through the tunnel to the β subunit's active site. [Courtesy of Craig Hyde, Edith Miles, and David Davies, National Institutes of Health.]
See Interactive Exercise 25

Michael Dunn has shown that the elimination of water from the serine–PLP Schiff base on the β subunit to form an aminoacrylate–PLP Schiff base intermediate

Aminoacrylate–PLP Schiff base

triggers a conformational change that activates the α subunit to produce indole. The diffusion of the indole to the β

Figure 26-65 (*Opposite*) **The biosynthesis of histidine.** The pathway enzymes are (**1**) ATP phosphoribosyltransferase, (**2**) pyrophosphohydrolase, (**3**) phosphoribosyl–AMP cyclohydrolase, (**4**) phosphoribosylformimino-5-aminoimidazole carboxamide ribonucleotide isomerase, (**5**) imidazole glycerol phosphate synthase (a glutamine amidotransferase), (**6**) imidazole glycerol phosphate dehydratase, (**7**) L-histidinol phosphate aminotransferase, (**8**) histidinol phosphate phosphatase, and (**9**) histidinol dehydrogenase.

subunit to react with this intermediate then results in the formation of tryptophan.

Channeling may be particularly important for indole since this nonpolar molecule otherwise can escape the bacterial cell by diffusing through its plasma and outer membranes. We have seen similar phenomena in reactions involving glutamine amidotransferases (Sections 26-2Aa and 26-5Aa), as well as in the series of reactions catalyzed by fatty acid synthase, in which the growing product is kept in the vicinity of the multifunctional enzyme's active site by covalent attachment to the enzyme's flexible phosphopantetheine arm (Section 25-4Ca). Channeling is also implicated in the multistep biosyntheses of purines and pyrimidines (Sections 28-1A and 28-2A).

e. Histidine Biosynthesis

Five of histidine's six C atoms are derived from **5-phosphoribosyl-α-pyrophosphate** (**PRPP;** Fig. 26-65), an intermediate also involved in the biosynthesis of tryptophan (Fig. 26-63, Reaction 2), purine nucleotides (Section 28-1A), and pyrimidine nucleotides (Section 28-2A). The histidine's sixth carbon originates from ATP. The ATP atoms that are not incorporated into histidine are eliminated as **5-aminoimidazole-4-carboxamide ribonucleotide** (**AICAR;** Fig. 26-65, Reaction 5), which is also an intermediate in purine biosynthesis (Section 28-1A).

The unusual biosynthesis of histidine from a purine has been cited as evidence supporting the hypothesis that life was originally RNA based (Section 1-5Ca). His residues, as we have seen, are often components of enzyme active sites, where they act as nucleophiles and/or general acid–base catalysts. The discovery that RNA can have catalytic properties (Section 31-4Ae) therefore suggests that the imidazole moiety of purines plays a similar role in these RNA enzymes (**ribozymes**). This further suggests that the histidine biosynthesis pathway is a "fossil" of the transition to more efficient protein-based life-forms.

6 NITROGEN FIXATION

The most prominent chemical elements in living systems are O, H, C, N, and P. The elements O, H, and P occur widely in metabolically available forms (e.g., H_2O, O_2, and P_i). However, the major available forms of C and N, CO_2 and N_2, are extremely stable (unreactive); for example, the $N\equiv N$ triple bond has a bond energy of 945 kJ · mol^{-1}

5-Phosphoribosyl-
α-pyrophosphate (PRPP)

ATP

1 → PP$_i$

N^1-5′-Phosphoribosyl-ATP

PP$_i$

2

N^1-5′-Phosphoribosyl-AMP

3 → H$_2$O

N^1-5′-Phosphoribosylformimino-
5-aminoimidazole-4-
carboxamide ribonucleotide

To purine biosynthesis

5-Aminoimidazole-
4-carboxamide
ribonucleotide (AICAR)

4

N^1-5′-Phosphoribulosylformimino-
5-aminoimidazole-4-
carboxamide ribonucleotide

Glutamine

Glutamate

5

Imidazole glycerol
phosphate

6 → H$_2$O

Imidazole acetol
phosphate

Glutamate   α-Ketoglutarate

7

L-Histidinol
phosphate

H$_2$O   P$_i$

8

L-Histidinol

2NAD$^+$ 2NADH

9

Histidine

(versus 351 kJ · mol^{-1} for a C—O single bond). CO_2, with only minor exceptions, is metabolized (fixed) only by photosynthetic organisms (Chapter 24). *N_2 fixation is even less common; this element is converted to metabolically useful forms by only a few strains of bacteria, named **diazatrophs.***

Diazatrophs of the genus *Rhizobium* live in symbiotic relationship with root nodule cells of legumes (plants belonging to the pea family, including beans, clover, and alfalfa; Fig. 26-66) where they convert N_2 to NH_3:

$$N_2 + 8H^+ + 16ATP + 16H_2O + 8e^- \longrightarrow$$
$$2NH_3 + H_2 + 16ADP + 16P_i$$

The NH_3 thus formed can be incorporated either into glutamate by glutamate dehydrogenase (Section 26-1B) or into glutamine by glutamine synthetase (Section 26-5Ab). This nitrogen-fixing system produces more metabolically useful nitrogen than the legume needs; the excess is excreted into the soil, enriching it. It is therefore common agricultural practice to plant a field with alfalfa every few years to build up the supply of usable nitrogen in the soil for later use in growing other crops.

a. Nitrogenase Contains Novel Redox Centers

Nitrogenase, which catalyzes the reduction of N_2 to NH_3, is a complex of two proteins:

1. The **Fe-protein,** a homodimer that contains one [4Fe–4S] cluster and two ATP binding sites.

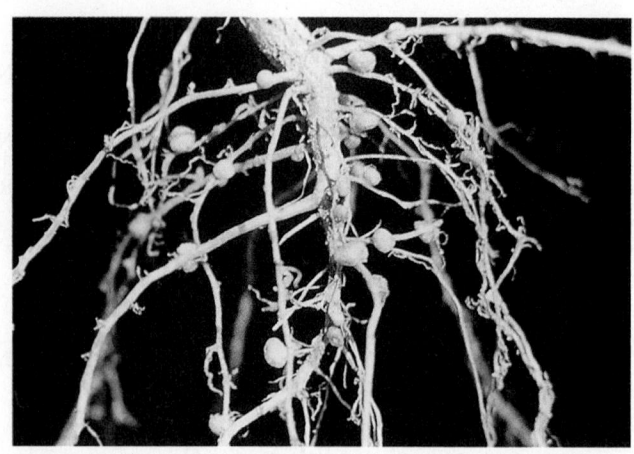

Figure 26-66 **Photograph showing the root nodules of the legume bird's-foot trefoil.** [Vu/Cabisco/Visuals Unlimited.]

2. The **MoFe-protein,** an $\alpha_2\beta_2$ heterotetramer that contains Fe and Mo.

The X-ray structure of *Azotobacter vinelandii* nitrogenase in complex with the inhibitor ADP · AlF$_4^-$ (which mimics the transition state in ATP hydrolysis), determined by Douglas Rees, reveals that each MoFe-protein associates with two molecules of Fe-protein (Fig. 26-67).

Each Fe-protein dimer's single [4Fe–4S] cluster is located in a solvent-exposed cleft between the two subunits

Figure 26-67 **X-ray structure of the A. vinelandii nitrogenase in complex with ADP · AlF$_4^-$.** The enzyme, which is viewed along its molecular 2-fold axis, is an $(\alpha\beta\gamma_2)_2$ heterooctamer in which the β–α–α–β assembly, the MoFe-protein, is flanked by two γ_2 Fe-proteins whose 289-residue subunits are related by local 2-fold symmetry. The homologous α subunits (*cyan and magenta;* 491 residues) and β subunits (*light red and light blue;* 522 residues) are related by pseudo-2-fold symmetry. The two γ subunits forming each Fe-protein (*pink and green with their* *Switch I and Switch II segments red and blue*) bind to the MoFe-protein with the 2-fold axis relating them coincident with the pseudo-2-fold axis relating the MoFe-protein's α and β subunits. The ADP · AlF$_4^-$, [4Fe–4S] cluster, FeMo-cofactor, and P-cluster are drawn in space-filling form with C green, N blue, O red, S yellow, Fe orange, Mo pink, and the AlF$_4^-$ ion purple. [Based on an X-ray structure by Douglas Rees, California Institute of Technology. PDBid 1N2C.] **See Interactive Exercise 26**

and is symmetrically linked to Cys 97 and Cys 132 from both subunits such that an Fe-protein resembles an "iron butterfly" with the [4Fe–4S] cluster at its head. Its two identical nucleotide binding sites are located at the interface between its two subunits. The [4Fe–4S] cluster cycles between its +1 and +2 oxidation states.

The MoFe-protein's α and β subunits assume similar folds and extensively associate to form a pseudo-2-fold symmetric αβ dimer, two of which more loosely associate to form the 2-fold symmetric $\alpha_2\beta_2$ tetramer (Fig. 26-67). Each αβ dimer has two bound redox centers:

1. The **P-cluster** (Fig. 26-68a,b), which consists of two [4Fe–3S] clusters linked through an additional sulfide ion forming the eighth corner of each of the clusters to make cubane-like structures, and bridged by two Cys thiol ligands, each coordinating one Fe from each cluster. Four additional Cys thiols coordinate the remaining four Fe atoms. The positions of two of the Fe atoms in one of the [4Fe–3S] clusters change on oxidation, rupturing the bonds from these Fe atoms to the linking sulfide ion. These bonds are replaced in the oxidized state by a Ser oxygen ligand to one of the Fe atoms, and by a bond to the amide N of a Cys from the other Fe atom.

2. The **FeMo-cofactor** (Fig. 26-68c), which consists of a [4Fe–3S] cluster and a [1Mo–3Fe–3S] cluster bridged by

three sulfide ions. The FeMo-cofactor's Mo atom is approximately octahedrally coordinated by three cofactor sulfide ions, a His imidazole nitrogen, and two oxygens from a bound **homocitrate** ion:

$$^-COO-CH_2-CH_2-\overset{\overset{\displaystyle OH}{|}}{\underset{\underset{\displaystyle COO^-}{|}}{C}}-CH_2-COO^-$$

Homocitrate

(an essential component of the FeMo-cofactor). The FeMo-cofactor contains a central cavity that a high resolution (1.16 Å) X-ray structure of *A. vinelandii* MoFe-protein, also determined by Rees, reveals contains what most probably is a nitrogen atom (although a C or an O atom cannot be ruled out). This putative N atom is liganded to the FeMo-cofactor's central six Fe atoms such that it completes the approximate tetrahedral coordination environment of each of these Fe atoms.

The FeMo-cofactor is located ~10 Å below the α subunit surface, and hence the N_2 is thought to gain access to its binding site through conformational fluctuations of the protein (recall that myoglobin and hemoglobin likewise have no clear path for O_2 to approach its heme binding

Figure 26-68 **The prosthetic groups of the nitrogenase MoFe-protein.** The molecules are drawn in ball-and-stick form with C green, N blue, O red, S yellow, Fe orange, and Mo pink. (*a*) The reduced *Klebsiella pneumoniae* P-cluster. It consists of two [4Fe–3S] complexes linked by an additional sulfide ion forming the eighth corner of each cubane-like structure, and bridged by two Cys thiol ligands, each coordinating one Fe from each cluster. Four additional Cys thiols coordinate the remaining 4 Fe atoms. (*b*) The 2-electron-oxidized *K. pneumoniae* P-cluster. In comparison with the reduced complex in Part *a*, two of the Fe—S bonds from the centrally located sulfide ion that bridges the two [4Fe–3S] clusters have been replaced by ligands from the Cys 87α amide N and the Ser 186β side chain O yielding a [4Fe–3S] cluster (*left*) and a [4Fe–4S] cluster (*right*) that remain linked by a direct Fe—S bond and two bridging Cys thiols. (*c*) The *A. vinelandi* FeMo-cofactor. It consists of a [4Fe–3S] cluster and a [1Mo–3Fe–3S] cluster that are bridged by three sulfide ions. The FeMo-cofactor is linked to the protein by only two ligands at its opposite ends, one from His 442α to the Mo atom and the other from Cys 275α to an Fe atom. The Mo atom is additionally doubly liganded by homocitrate. What is most likely an N atom (*blue sphere*) is liganded to the FeMo-cluster's six central Fe atoms (*dashed black lines*). [Parts *a* and *b* based on X-ray structures by David Lawson, John Innes Centre, Norwich, U.K. Part *c* based on an X-ray structure by Douglas Rees, California Institute of Technology. PDBids (*a*) 1QGU, (*b*) 1QH1, and (*c*) 1M1N.]

sites in these proteins; Section 10-2). The P-cluster, which is also ~10 Å below the protein surface, is at the interface between the α and β subunits on the pseudo-2-fold axis that roughly relates these two subunits. The 2-fold axis of the Fe-protein and the pseudo-2-fold axis of the MoFe-proteins coincide in their complex.

The Fe-protein hydrolyzes two ATP molecules for each electron it transfers from its $[4Fe–4S]^{+1}$ cluster to the P-cluster. Since the nucleotide binding sites and the [4Fe–4S] cluster on the Fe-protein are separated by ~20 Å, a distance too large for direct coupling between electron transfer and ATP hydrolysis, it appears that these processes are allosterically coupled through conformational changes at the subunit interface. Indeed, portions of the Fe-protein resemble those of G-proteins, in which nucleotide hydrolysis is coupled to conformational changes controlling the protein's actions (Sections 19-2Cb and 19-3Cf). Specifically, two regions of the Fe-protein, designated Switch I and Switch II (Fig. 26-67), are homologous with those of Ras (Section 19-3Cf). The binding of $ADP \cdot AlF_4^-$ to Fe-protein induces conformational changes in Switch I that affect the interactions between the Fe-protein and the MoFe-protein, and in Switch II that affect the environment of the [4Fe–4S] cluster.

In nitrogenase, the [4Fe–4S] cluster of the Fe-protein approaches within ~14 Å of the P-cluster in the MoFe-protein, whereas the P-cluster and the FeMo-cofactor are ~13 Å apart. Hence, the sequence of the electron-transfer steps in the nitrogenase reaction appears to be

$$[4Fe–4s]\text{ cluster} \longrightarrow P\text{-cluster} \longrightarrow$$
$$FeMo\text{-cofactor} \longrightarrow N_2$$

It therefore seems that the role of ATP hydrolysis is to stabilize a conformation in the Fe-protein that it cannot achieve on its own and which facilitates electron transfer from the [4Fe–4S] cluster on the Fe-protein to the P-cluster on the MoFe-protein.

b. N₂ Reduction Is Energetically Costly

Nitrogen fixation requires two participants in addition to N₂ and nitrogenase: (1) a source of electrons and (2) ATP. Electrons are generated either oxidatively or photosynthetically, depending on the organism. These electrons are transferred to ferredoxin (Section 22-2C1a), a [4Fe–4S]-containing electron carrier that transfers an electron to the Fe-protein of nitrogenase, beginning the nitrogen fixation process (Fig. 26-69). Two molecules of ATP bind to the reduced Fe-protein and are hydrolyzed as each electron is passed from the Fe-protein to the MoFe-protein. The ATP hydrolysis-induced conformational change in the Fe-protein alters its redox potential from −0.29 to −0.40 V, making the electron capable of N₂ reduction ($\mathscr{E}°' = -0.34$ V for the half-cell $N_2 + 6H^+ + 6e^- \rightleftharpoons 2NH_3$).

The actual reduction of N₂ occurs on the MoFe-protein in three discrete steps, each involving an electron pair:

$$N{\equiv}N \xrightarrow{2H^+ + 2e^-} H–N{=}N–H \xrightarrow{2H^+ + 2e^-} \begin{matrix} H & & H \\ \backslash & & / \\ N{-}N \\ / & & \backslash \\ H & & H \end{matrix} \xrightarrow{2H^+ + 2e^-} 2NH_3$$

$$\textbf{Diimine} \qquad \textbf{Hydrazine}$$

An electron transfer must occur six times per N₂ molecule fixed so that a total of 12 ATPs are required to fix one N₂ molecule. Although the N₂ binding site is almost certainly the FeMo-cofactor, exactly how the N₂ is bound and reduced are largely a matter of speculation. Theoretical studies suggest that the FeMo-cofactor's prismatically arranged Fe atoms provide favorable interaction sites for N₂ and its reduction products. Indeed, it seems highly likely that the putative N atom that is liganded to the FeMo-cofactor (Fig. 26-68c) participates in N₂ reduction.

Nitrogenase also reduces H₂O to H₂, which in turn reacts with **diimine** to reform N₂:

$$HN{=}NH + H_2 \longrightarrow N_2 + 2H_2$$

The resulting futile cycle is favored when the ATP level is low and/or the reduction of the Fe-protein is sluggish. Even when ATP is plentiful, however, the cycle cannot be suppressed beyond about one H₂ molecule produced per N₂ reduced and hence appears to be a requirement of the nitrogenase reaction. The total cost of N₂ reduction is therefore 8 electrons transferred and 16 ATPs hydrolyzed (physiologically, 20–30 ATPs). Hence nitrogen fixation is an energetically expensive process; indeed, the nitrogen-fixing bacteria in the root nodules of pea plants consume nearly 20% of the ATP that the plant produces.

c. Leghemoglobin Protects Nitrogenase from Oxygen Inactivation

Nitrogenase is rapidly inactivated by O₂, so the enzyme must be protected from this reactive substance. Cyanobacteria

Figure 26-69 The flow of electrons in the nitrogenase-catalyzed reduction of N₂.

(photosynthetic oxygen-evolving bacteria; Section 1-1Ab) provide protection by carrying out nitrogen fixation in specialized nonphotosynthetic cells called **heterocysts,** which have Photosystem I but lack Photosystem II (Section 24-2Ca). In the root nodules of legumes (Fig. 26-66), however, protection is afforded by the symbiotic synthesis of **leghemoglobin.** The globin portion of this ~145-residue monomeric oxygen-binding protein is synthesized by the plant (an evolutionary curiosity since globins are otherwise known to occur only in animals), whereas the heme is synthesized by the *Rhizobium.* Leghemoglobin has a very high O_2 affinity, thus keeping the pO_2 low enough to protect the nitrogenase while providing passive O_2 transport for the aerobic bacterium.

d. Installing the Nitrogen Fixation Machinery in Nonleguminous Plants Would Revolutionize Agriculture

Although atmospheric N_2 is the ultimate nitrogen source for all living things, most plants do not support the symbiotic growth of nitrogen-fixing bacteria. They must therefore depend on a source of "prefixed" nitrogen such as nitrate or ammonia. These nutrients come from lightning discharges (the source of ~10% of naturally fixed N_2), decaying organic matter in the soil, or from fertilizer applied to it. The Haber process, which was invented by Fritz Haber in 1910, is a chemical process for N_2 fixation that is still widely used in fertilizer manufacture. This direct reduction of N_2 by H_2 to form NH_3 requires temperatures of 300 to 500°C, pressures of >300 atm, and an Fe catalyst. Intriguingly, the spacing of the Fe atoms on the surface of this catalyst resembles that of the FeMo-cofactor's central Fe atoms (Fig. 26-68c).

One of the major long-term goals of genetic engineering is to induce agriculturally useful nonleguminous plants to fix their own nitrogen, a complex undertaking in which the plant must be made to provide a hospitable environment for nitrogen fixation as well as to acquire the enzymatic machinery to do so. This would free farmers, particularly those in developing countries, from the need for either purchasing fertilizers, periodically letting their fields lie fallow (giving legumes the opportunity to grow), or following the slash-and-burn techniques that are rapidly destroying the world's tropical forests and contributing significantly to the greenhouse effect (atmospheric CO_2 pollution causing long-term global warming).

e. The Nitrogen Cycle Describes the Interconversion of Nitrogen in the Biosphere

The ammonia produced by the nitrogenase reaction and incorporated into amino acids is eventually recycled in the biosphere as described by the **nitrogen cycle** (Fig. 26-70). Nitrate is produced by certain bacteria that oxidize NH_3 to NO_2^- and then NO_3^-, a process called **nitrification.** Still other organisms convert nitrate back to N_2, which is known as **denitrification.** In addition, nitrate is reduced to NH_3 by plants, fungi, and many bacteria, a process called **ammonification** in which **nitrate reductase** catalyzes the two-electron reduction of nitrate to nitrite (NO_2^-):

$$NO_3^- + 2H^+ + 2e^- \longrightarrow NO_2^- + H_2O$$

and then **nitrite reductase** converts nitrite to ammonia,

$$NO_2^- + 7H^+ + 6e^- \longrightarrow NH_3 + 2H_2O$$

The direct anaerobic oxidation of NH_3 back to N_2 without the intermediacy of nitrate, the reverse of nitrogen fixation, has recently been discovered in certain bacteria.

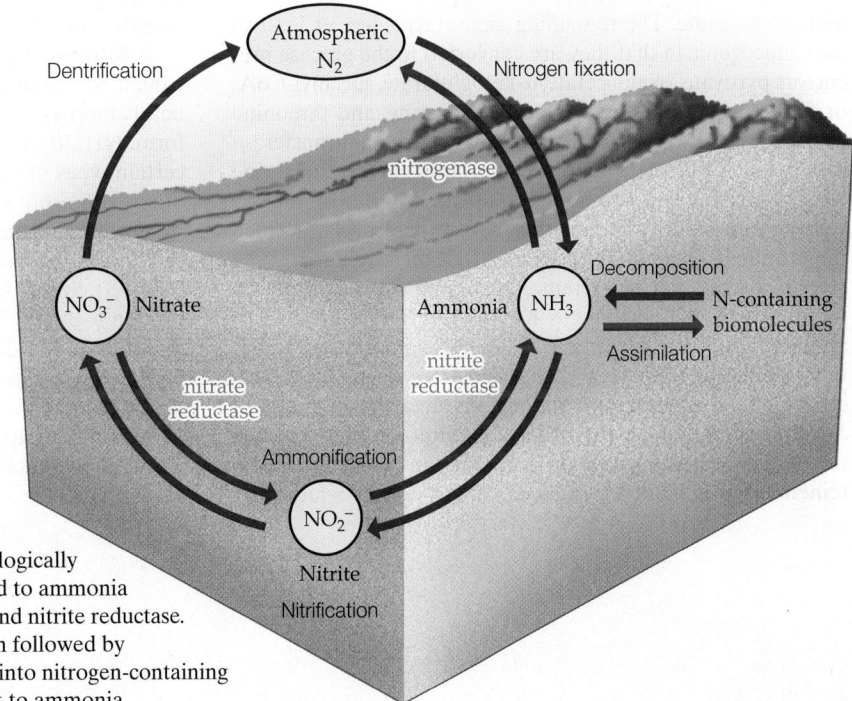

Figure 26-70   The nitrogen cycle. Nitrogen fixation by nitrogenase converts N_2 to the biologically useful ammonia. Nitrate can also be converted to ammonia by the sequential actions of nitrate reductase and nitrite reductase. Ammonia is transformed to N_2 by nitrification followed by denitrification. Ammonia may be assimilated into nitrogen-containing biomolecules, which may be decomposed back to ammonia.

CHAPTER SUMMARY

1 Amino Acid Deamination Amino acids are the precursors for numerous nitrogen-containing compounds such as heme, physiologically active amines, and glutathione. Excess amino acids are converted to common metabolic intermediates for use as fuels. The first step in amino acid breakdown is removal of the α-amino group by transamination. Transaminases require pyridoxal phosphate (PLP) and convert amino acids to their corresponding α-keto acids. The amino group is transferred to α-ketoglutarate to form glutamate, oxaloacetate to form aspartate, or pyruvate to form alanine. Glutamate is subsequently oxidatively deaminated by glutamate dehydrogenase (GDH) to form ammonia and regenerate α-ketoglutarate. Hyperinsulinism/hyperammonemia (HI/HA), a genetic disease, is caused by a mutation of the GDH gene that decreases GTP's ability to inhibit GDH.

2 The Urea Cycle In the urea cycle, amino groups from NH_3 and aspartate combine with HCO_3^- to form urea. This pathway takes place in the liver, partially in the mitochondrion and partially in the cytosol. It begins with the ATP-dependent condensation of NH_3 and HCO_3^- by carbamoyl phosphate synthetase, an enzyme with a 96-Å-long tunnel connecting its three active sites through which its highly reactive intermediate products are channeled. The resulting carbamoyl phosphate then combines with ornithine to yield citrulline, which combines with aspartate to form argininosuccinate, which in turn is cleaved to fumarate and arginine. The arginine is then hydrolyzed to urea, which is excreted, and ornithine, which reenters the urea cycle. *N*-Acetylglutamate regulates the urea cycle by activating carbamoyl phosphate synthetase allosterically.

3 Metabolic Breakdown of Individual Amino Acids The α-keto acid products of transamination reactions are degraded to citric acid cycle intermediates or their precursors. The amino acids leucine and lysine are ketogenic in that they are converted only to the ketone body precursors acetyl-CoA and acetoacetate. The remaining amino acids are, at least in part, glucogenic in that they are converted to the glucose precursors pyruvate, oxaloacetate, α-ketoglutarate, succinyl-CoA, or fumarate. Alanine, cysteine, glycine, serine, and threonine are converted to pyruvate. Serine hydroxymethyltransferase catalyzes the PLP-dependent C_α—C_β bond cleavage of serine to form glycine. This reaction requires the transfer of a methylene group from N^5,N^{10}-methylene-tetrahydrofolate, which the tetrahydrofolate (THF) obtains from the glycine cleavage system, a multienzyme system. Asparagine and aspartate are converted to oxaloacetate. α-Ketoglutarate is a product of arginine, glutamate, glutamine, histidine, and proline degradation. Methionine, isoleucine, and valine are degraded to succinyl-CoA. Methionine breakdown involves the synthesis of *S*-adenosylmethionine (SAM), a sulfonium ion that acts as a methyl donor in many biosynthetic reactions. Hyperhomocysteinemia, a risk factor for cardiovascular disease, cognitive impairment, and neural tube defects, is caused by a deficiency in its folate-dependent degradation. Maple syrup urine disease (MSUD) is caused by an inherited defect in branched-chain amino acid degradation. Branched-chain amino acid degradation pathways contain reactions common to all acyl-CoA oxidations. Tryptophan is degraded to alanine and acetoacetate. Phenylalanine and tyrosine are degraded to fumarate and acetoacetate. Most individuals with the hereditary disease phenylketonuria lack phenylalanine hydroxylase (PAH), which converts phenylalanine to tyrosine.

4 Amino Acids as Biosynthetic Precursors Heme is synthesized from glycine and succinyl-CoA. These precursors condense to form δ-aminolevulinic acid (ALA), which cyclizes to form the pyrrole porphobilinogen (PBG). Four molecules of PBG condense to form uroporphyrinogen III, which then goes on to form heme, with the final reaction, the insertion of Fe(II) into protoporphyrin IX, catalyzed by ferrochelatase. Defects in heme biosynthesis, which are known as porphyrias, have a variety of bizarre symptoms. Heme is degraded to form linear tetrapyrroles, which are subsequently excreted as bile pigments. The hormones and neurotransmitters L-DOPA, epinephrine, norepinephrine, serotonin, γ-aminobutyric acid (GABA), and histamine are all synthesized from amino acid precursors. Glutathione, a tripeptide that is synthesized from glutamate, cysteine, and glycine, is involved in a variety of protective, transport, and metabolic processes. Tetrahydrofolate is a coenzyme that participates in the transfer of C_1 units.

5 Amino Acid Biosynthesis Amino acids are required for many vital functions of an organism. Those amino acids that mammals can synthesize from common α-keto acid carbon skeletons and preformed α-amino nitrogen such as that of glutamate are known as nonessential amino acids; those that mammals must obtain from their diets are called essential amino acids. The biosynthesis of nonessential amino acids involves relatively simple pathways, whereas those forming the essential amino acids are generally more complex.

6 Nitrogen Fixation Although the ultimate source of nitrogen for amino acid biosynthesis is atmospheric N_2, this nearly inert gas must first be reduced to a metabolically useful form, NH_3, by nitrogen fixation. This process occurs only in certain types of bacteria, one genus of which occurs in symbiotic relationship with legumes. N_2 is fixed in these organisms by an oxygen-sensitive enzyme, nitrogenase, that consists of two proteins: the Fe-protein dimer, which contains one [4Fe–4S] cluster and two ATP binding sites, and the MoFe-protein $\alpha_2\beta_2$ tetramer, which contains one P-cluster (consisting of two [4Fe–3S] clusters linked by a sulfide ion) and one FeMo-cofactor (a [4Fe–3S] cluster and a [1Mo–3Fe–3S] cluster bridged by three sulfide ions and coordinated with homocitrate) in each αβ dimer. These cofactors each function as two-electron carriers for the ATP-driven reduction of N_2 to NH_3.

REFERENCES

General

Bender, D.A., *Amino Acid Metabolism*, Wiley (1985).

Frey, P.A. and Hegeman, A.D., *Enzymatic Reaction Mechanisms*, Oxford University Press (2007). [Describes the mechanisms of many of the enzymes discussed in this chapter.]

Valle, D. (Ed.), *The Online Metabolic & Molecular Bases of Inherited Disease*, http://www.ommbid.com/. [Part 8 contains numerous chapters on defects in amino acid metabolism.]

Amino Acid Deamination and the Urea Cycle

Eliot, A.C. and Kirsch, J.F., Pyridoxal phosphate enzymes, *Annu. Rev. Biochem.* **73**, 383–415 (2004).

Holden, H.M., Thoden, J.B., and Raushel, F.M., Carbamoyl phosphate synthetase: a tunnel runs through it, *Curr. Opin. Struct. Biol.* **8**, 679–685 (1998); *and* Thoden, J.B., Holden, H.M., Wesenberg, G., Raushel, F.M., and Rayment, I., Structure of carbamoyl phosphate synthetase: A journey of 96 Å from substrate to product, *Biochemistry* **36**, 6305–6316 (1997).

Jansonius, J.N., Structure, evolution and action of vitamin B_6-dependent enzymes, *Curr. Opin. Struct. Biol.* **8**, 759–769 (1998).

Jungas, R.L., Halperin, M.L., and Brosnan, J.T., Quantitative analysis of amino acid oxidation and related gluconeogenesis in humans, *Physiol. Rev.* **72**, 419–448 (1992).

Miles, E.W., Rhee, S., and Davies, D.R., The molecular basis of substrate channeling, *Biochemistry* **274**, 12193–12196 (1999).

Saeed-Kothe, A. and Powers-Lee, S.G., Specificity determining residues in ammonia- and glutamine-dependent carbamoyl phosphate synthetases, *J. Biol. Chem.* **277**, 7231–7238 (2002).

Smith, T.J., Schmidt, T., Fang, J., Wu, J., Siuzdak, G., and Stanley, C.A., The structure of apo human glutamate dehydrogenase details subunit communication and allostery, *J. Mol. Biol.* **318**, 765–777 (2002).

Smith, T.J. and Stanley, C.A., Untangling the glutamate dehydrogenase allosteric nightmare, *Trends Biochem. Sci.* **33**, 557–564 (2008).

Stipanuk, M.H. and Watford, M., Amino acid metabolism, Chap. 11, *in* Stipanuk, M.H. (Ed.), *Biochemical and Physiological Basis of Nutrition*, Saunders (2000).

Torchinsky, Yu.M., Transamination: its discovery, biological and chemical aspects (1937–1987), *Trends Biochem. Sci.* **12**, 115–117 (1987).

Metabolic Breakdown of Individual Amino Acids

Anderson, O.A., Flatmark, T., and Hough, E., High resolution crystal structures of the catalytic domain of human phenylalanine hydroxylase in its catalytically active Fe(II) form and binary complex with tetrahydrobiopterin, *J. Mol. Biol.* **314**, 279–291 (2001).

Ævarsson, A., Chuang, J.L., Wynn, R.M., Turley, S., Chuang, D.T., and Hol, W.G.J., Crystal structure of human branched-chain α-ketoacid dehydrogenase and the molecular basis of multienzyme complex deficiency in maple syrup urine disease, *Structure* **8**, 277–291 (2000); *and* Wynn, R.M., Davie, J.R., Chuang, J.L., Cote, C.D., and Chuang, D.T., Impaired assembly of E1 decarboxylase of the branched-chain α-ketoacid dehydrogenase complex in type IA maple syrup urine disease, *J. Biol. Chem.* **273**, 13110–13118 (1998).

Binda, C., Bossi, R.T., Wakatsuki, S., Arzt, S., Coda, A., Curti, B., Vanoni, M.A., and Mattevi, A., Cross-talk and ammonia channeling between active centers in the unexpected domain arrangement of glutamate synthase, *Structure* **8**, 1299–1308 (2000).

Douce, R., Bourguignon, J., Neuburger, M., and Rébeillé, F., The glycine decarboxylase system: a fascinating complex, *Trends Plant Sci.* **6**, 167–176 (2001).

Drennen, C.L., Huang, S., Drumond, J.T., Matthews, R., and Ludwig, M.L., How a protein binds B_{12}: A 3.0 Å X-ray structure of B_{12}-binding domains of methionine synthase, *Science* **266**, 1669–1674 (1994).

Faure, M., Rourguignon, J., Neuburger, M., Macherel, D., Sieker, L., Ober, R., Kahn, R., Cohen-Addad, C., and Douce, R., Interaction between the lipoamide-containing H-protein and the lipoamide dehydrogenase (L-protein) of the glycine decarboxylase multienzyme system, 2. Crystal structures of H- and L-proteins, *Eur. J. Biochem.* **267**, 2890–2898 (2000).

Guenther, B.D., Sheppard, C.A., Tran, P., Rozen, R., Matthews, R.G., and Ludwig, M.L., The structure and properties of methylenetetrahydrofolate reductase from *Escherichia coli* suggest how folate ameliorates human hyperhomocysteinemia, *Nature Struct. Biol.* **6**, 359–365 (1999).

Guilhaudis, L., Simorre, J.-P., Blackledge, M., Marion, D., Gans, P., Neuburger, M., and Douce, R., Combined structural and biochemical analysis of the H–T complex in the glycine decarboxylase cycle: evidence for a destabilization mechanism of the H-protein, *Biochemistry* **39**, 4259–4266 (2000).

Huang, X., Holden, H.M., and Raushel, F.M., Channeling of substrates and intermediates in enzyme-catalyzed reactions, *Annu. Rev. Biochem.* **70**, 149–180 (2001).

Kelly, A. and Stanley, C.A., Disorders of glutamate metabolism, *Mental Retard. Devel. Dis. Res. Rev.* **7**, 287–295 (2001).

Ludwig, M.L. and Matthews, R.G., Structure-based perspectives on B_{12}-dependent enzymes, *Annu. Rev. Biochem.* **66**, 269–313 (1997).

Matthews, R.G., Koutmos, M., and Datta, S., Cobalamin-dependent and cobamide-dependent methyltransferases, *Curr. Opin. Struct. Biol.* **18**, 658–666 (2008).

Medina, M.Á., Urdiales, J.L., and Amores-Sánchez, M.I., Roles of homocysteine in cell metabolism: Old and new functions, *Eur. J. Biochem.* **268**, 3871–3882 (2001).

Spiro, T.G. and Kozlowski, P.M., Is the CO adduct of myoglobin bent, and does it matter? *Acc. Chem. Res.* **34**, 137–144 (2001).

Swain, A.L., Jaskólski, M., Housset, D., Rao, J.K.M., and Wladower, A., Crystal structure of *Escherichia coli* L-asparaginase, an enzyme used in cancer therapy, *Proc. Natl. Acad. Sci.* **90**, 1474–1478 (1993).

Varughese, K.I., Skinner, M.M., Whiteley, J.M., Matthews, D.A., and Xuong, N.H., Crystal structure of rat liver dihydropteridine reductase, *Proc. Natl. Acad. Sci.* **89**, 6080–6084 (1992).

Zalkin, H. and Smith, J.L. Enzymes utilizing glutamine as an amide donor, *Adv. Enzymol.* **72**, 87–144 (1998).

Amino Acids as Biosynthetic Precursors

Ajioka, R.S., Phillips, J.D., and Kushner, J.P., Biosynthesis of heme in mammals, *Biochim. Biophys. Acta* **1763**, 723–736 (2006).

Al-Karadaghi, S., Franco, R., Hansson, M., Shelnutt, J.A., Isaya, G., and Ferreira, G.C., Chelatases: distort to select, *Trends Biochem. Sci.* **31**, 135–142 (2006).

Battersby, A.R., Tetrapyrroles: the pigments of life, *Nat. Prod. Rep.* **17**, 507–526 (2000).

Erskine, P.T., Newbold, R., Brindley, A.A., Wood, S.P., Shoolingin-Jordan, P.M., Warren, M.J., and Cooper J.B., The X-ray structure of yeast 5-aminolaevulinic acid dehydratase complexed with substrate and three inhibitors, *J. Mol. Biol.* **312**, 133–141 (2001).

Fitzpatrick, P.F., Tetrahydropterin-dependent amino acid hydroxylases, *Annu. Rev. Biochem.* **68,** 355–381 (1999).

Jaffe, E.K., Martins, J., Li, J., Kervinen, J., and Dunbrack, R.L., Jr., The molecular mechanism of lead inhibition of human porphobilinogen synthase, *J. Biol. Chem.* **276,** 1531–1537 (2001).

Jaffe, E.K., Morpheeins—a new structural paradigm for allosteric regulation, *Trends Biochem. Sci.* **39,** 490–497 (2005); *and* Lawrence, S.H. and Jaffe, E.K., Expanding the concepts in protein structure-function relationships and enzyme kinetics: Teaching using morpheeins, *Biochem. Mol. Biol. Educ.* **36,** 274–283 (2008).

Kauppinen, R., Porphyrias, *Lancet* **365,** 241–252 (2005).

Louie, G.V., Brownlie, P.D., Lambert, R., Cooper, J.B., Blundell, T.L., Wood, S.P., Warren, M.J., Woodcock, S.C., and Jordan, P.M., Structure of porphobilinogen deaminase reveals a flexible multidomain polymerase with a single catalytic site, *Nature* **359,** 33–39 (1992).

Medlock, A.E., Dailey, T.A., Ross, T.A., Dailey, H.A., and Lanzilotta, W.N., A π-helix switch selective for porphyrin deprotonation and product release in human ferrochelatase, *J. Mol. Biol.* **373,** 1006–1016 (2007).

Pagola, S., Stephens, P.W., Bohle, D.S., Kosar, A.D., and Madsen, S.K., The structure of malaria pigment β-haematin, *Nature* **404,** 307–310 (2000).

Schneider-Yin, X., Gouya, L., Dorsey, M., Rüfenacht, U., Deybach, J.-C., and Ferreira, G.C., Mutations in the iron-sulfur cluster ligands of the human ferrochelatase lead to erythropoietic protoporphyria, *Blood* **96,** 1545–1549 (2000).

Sellers, V.M., Wu, C.-K., Dailey, T.A., and Dailey, H.A., Human ferrochelatase: Characterization of substrate-iron binding and proton-abstracting residues, *Biochemistry* **40,** 9821–9827 (2001).

Shoolingin-Jordan, P.M., Warren, M.J., and Awan, S.J., Discovery that the assembly of the dipyrromethane cofactor of porphobilinogen deaminase holoenzyme proceeds initially by the reaction of preuroporphyrinogen with the apoenzyme, *Biochem. J.* **316,** 373–376 (1996).

Song, G., Li, Y., Cheng, C., Zhao, Y., Gao, A., Zhang, R., Joachimiak, A., Shaw, N., and Li, Z.-J., Structural insight into acute intermittent porphyria, *FASEB J.* **23,** 396–404 (2009). [Reports the X-ray structure of human porphobilinogen deaminase.]

Thunell, S., Porphyrins, porphyrin metabolism and porphyrias. I. Update, *Scand. J. Clin. Lab. Invest.* **60,** 509–540 (2000).

Wellems, T.E., How chloroquine works, *Nature* **355,** 108–109 (1992).

Wu, C.-K., Dailey, H.A., Rose, J.P., Burden, A., Sellers, V.M., and Wang, B.-C., The 2.0 Å structure of human ferrochelatase, the terminal enzyme of heme biosynthesis, *Nature Struct. Biol.* **8,** 156–160 (2001).

Amino Acids Biosynthesis

Chaudhuri, B.N., Lange, S.C., Myers, R.S., Chittur, S.V., Davisson, V.J., and Smith, J.L., Crystal structure of imidazole glycerol phosphate synthase: a tunnel through a $(\beta/\alpha)_8$ barrel joins two active sites, *Structure* **9,** 987–997 (2001).

Dunn, M.F., Niks, D., Ngo, H., Barends, T.R.M., and Schlichting, I., Tryptophan synthase: the workings of a channeling nanomachine, *Trends Biochem. Sci.* **33,** 254–264 (2008).

Eisenberg, D., Gill, H.S., Pfluegl, M.U., and Rotstein, S.H., Structure–function relationships of glutamine synthetases, *Biochim. Biophys. Acta* **1477,** 122–145 (2000); *and* Gill, H.S. and Eisenberg, D., The crystal structure of phosphinothricin in the active site of glutamine synthetase illuminates the mechanism of enzymatic inhibition, *Biochemistry* **40,** 1903–1912 (2001).

Hyde, C.C., Ahmed, S.A., Padlan, E.A., Miles, E.W., and Davies, D.R., Three-dimensional structure of the tryptophan synthase $\alpha_2\beta_2$ multienzyme complex from *Salmonella typhimurium*, *J. Biol. Chem.* **263,** 17857–17871 (1988).

Katagiri, M. and Nakamura, M., Animals are dependent on preformed α-amino nitrogen as an essential nutrient, *Life* **53,** 125–129 (2002).

Kishore, G.M. and Shah, D.M., Amino acid biosynthesis inhibitors as herbicides, *Annu. Rev. Biochem.* **57,** 627–663 (1988). [Discusses the biosynthesis of the essential amino acids.]

Larsen, T.M., Boehlein, S.K., Schuster, S.M., Richards, N.G.J., Thoden, J.B., Holden, H.M., and Rayment, I., Three-dimensional structure of *Escherichia coli* asparagine synthetase B: a short journey from substrate to product, *Biochemistry* **38,** 16146–16167 (1999).

Stadtman, E.R., The story of glutamine synthetase regulation, *J. Biol. Chem.* **276,** 44357–44364 (2001).

Stallings, W.C., Abdel-Meguid, S.S., Lim, L.W., Shieh, H.-S., Dayringer, H.E., Leimgruber, N.K., Stegeman, R.A., Anderson, K.S., Sikorski, J.A., Padgette, S.R., and Kishore, G.M., Structure and topological symmetry of the glyphosate target 5-*enol*-pyruvylshikimate-3-phosphate synthase: A distinctive protein fold, *Proc. Natl. Acad. Sci.* **88,** 5046–5050 (1991). [The enzyme that catalyzes Reaction 6 of Fig. 26-62 in complex with glyphosate, an inhibitor that is a broad-spectrum herbicide.]

Weeks, A., Lund, L., and Raushel, F.M., Tunneling of intermediates in enzyme-catalyzed reactions, *Curr. Opin. Chem. Biol.* **10,** 465–472 (2006).

Nitrogen Fixation

Einsle, O., Tezcan, F.A., Andrade, A.L.A., Schmidt, B., Yoshida, M., Howard, J.B., and Rees, D.C., Nitrogense MoFe-protein at 1.16 Å resolution: A central ligand in the FeMo-cofactor, *Science* **297,** 1696–1700 (2002).

Fisher, R.F. and Long, S.R., *Rhizobium*–plant signal exchange, *Nature* **357,** 655–660 (1992). [Discusses the signals through which Rhizobiaceae and legumes communicate to symbiotically generate the root nodules in which nitrogen fixation occurs.]

Jang, S.B., Seefeldt, L.C., and Peters, J.W., Insights into nucleotide signal transduction in nitrogenase: Structure of an iron protein with MgADP bound, *Biochemistry* **39,** 14745–14752 (2000).

Lawson, D.M. and Smith, B.E., Molybdenum nitrogenases: a crystallographic and mechanistic view, *Metal Ions Biol. Sys.* **39,** 75–120 (2002).

Mayer, S.M., Lawson, D.M., Gormal, C.A., Roe, S.M., and Smith, B.E., New insights into structure-function relationships in nitrogenase: A 1.6 Å resolution X-ray crystallographic study of *Klebsiella pneumoniae* MoFe-protein, *J. Mol. Biol.* **292,** 871–891 (1999).

Peters, J.W., Stowell, M.H.B., Soltis, S.M., Finnegan, M.G., Johnson, M.K., and Rees, D.C., Redox-dependent structural changes in the nitrogenase P-cluster, *Biochemistry* **36,** 1181–1187 (1997).

Peters, J.W. and Szilagyi, R.K., Exploring new frontiers of nitrogenase structure and mechanism, *Curr. Opin. Chem. Biol.* **10,** 101–108 (2006).

Rees, D.C., Tezcan, F.A., Haynes, C.A., Walton, M.Y., Andrade, S., Einsle, O., and Howard, J.B., Structural basis of biological nitrogen fixation, *Philos. Trans. Roy. Soc. A* **363,** 971–984 (2005); *and* Howard, J.B. and Rees, D.C., How many metals does it take to fix N_2? A mechanistic overview of biological nitrogen fixation, *Proc. Natl. Acad. Sci.* **103,** 17088–17093 (2006).

Schindelin, H., Kisker, C., Schlessman, J.L., Howard, J.B., and Rees., D.C., Structure of ADP · AIF$_4^-$ stabilized nitrogenase complex and its implications for signal transduction, *Nature* **387,** 370–376 (1997).

Seefeldt, L.C., Hoffman, B.M., and Dean, D.R., Mechanism of Mo-dependent nitrogenase, *Annu. Rev. Biochem.* **78,** 701–722 (2009).

PROBLEMS

1. Write the reaction for the transamination of an amino acid in terms of Cleland notation (Section 14-5A).

2. Explain why the symptoms of the partial deficiency of a urea cycle enzyme may be attenuated by a low-protein diet.

3. Why are people on a high-protein diet instructed to drink lots of water?

4. A student on a particular diet expends $10,000 \text{ kJ} \cdot \text{day}^{-1}$ while excreting 40 g of urea. Assuming that protein is 16% N by weight and that its metabolism yields $18 \text{ kJ} \cdot \text{g}^{-1}$, what percentage of the student's energy requirement is met by protein?

5. Production of the enzymes that catalyze the reactions of the urea cycle can increase or decrease according to the metabolic needs of the organism. High levels of these enzymes are associated with high-protein diets as well as starvation. Explain this apparent paradox.

6. *Helicobacter pylori,* the bacterium responsible for gastric ulcers, can survive in the stomach (where the pH is as low as 1.5) in part because it synthesizes large amounts of the enzyme urease. (a) Write the reaction for urea hydrolysis by urease. (b) Explain why this reaction could help establish a more hospitable environment for *H. pylori,* which tolerates acid but prefers to grow at near-neutral pH.

7. Why are phenylketonurics warned against eating products containing the artificial sweetener **aspartame** (**NutraSweet;** chemical name L-aspartyl-L-phenylalanine methyl ester)?

8. Demonstrate that the synthesis of heme from PBG as labeled in Fig. 26-35 results in the heme-labeling pattern given in Fig. 26-32.

9. Explain why certain drugs and other chemicals can precipitate an attack of acute intermittent porphyria.

10. Heterozygotes for erythropoietic protoporphyria show only 20 to 30% residual ferrochelatase activity rather than the 50% that is normally expected for an autosomal dominant inherited disease. Provide a plausible explanation for this observation.

11. One of the symptoms of **kwashiorkor,** the dietary protein deficiency disease in children, is the depigmentation of the skin and hair. Explain the biochemical basis of this symptom.

12. What are the metabolic consequences of a defective uridylyl-removing enzyme in *E. coli*?

13. Figure 26-60, Reaction 9, indicates that methionine is synthesized in microorganisms by the methylation of homocysteine in a reaction in which N^5-methyl-THF is the methyl donor. Yet, in the breakdown of methionine (Fig. 26-18), its demethylation occurs in three steps in which SAM is an intermediate. Discuss why this reaction does not occur via the simpler one-step reversal of the methylation reaction.

*14. In the glucose–alanine cycle (Fig. 26-3), glycolytically derived pyruvate is transaminated to alanine and exported to the liver for conversion to glucose and return to the cell. Explain how a muscle cell is able to participate in this cycle under anaerobic (vigorously contracting) conditions. (*Hint:* The breakdown of many amino acids yields NH_3.)

15. Draw the activated intermediates involved in (a) glutamine and (b) asparagine biosynthesis from glutamate and aspartate, respectively. (c) Provide an example of another metabolic activation of a carboxylic acid group analogous to each of these reactions.

16. The $\alpha_2\beta_2$ tetramer of tryptophan synthase catalyzes the PLP-dependent reaction of indole-3-glycerol phosphate and serine to form tryptophan (Fig. 26-63, Reactions 5 and 6). Draw the chemical reactions involved in this synthesis, including the participation of PLP, and use curved arrows to show the flow of electrons. What role does PLP play in the reaction?

17. Suggest a reason why the nitrogen-fixing heterocysts of cyanobacteria have lost Photosystem II but retain Photosystem I.

Energy Metabolism: Integration and Organ Specialization

At this point in our narrative we have studied all of the major pathways of energy metabolism. Consequently, we are now in a position to consider how organisms, mammals in particular, orchestrate the metabolic symphony to meet their energy needs. This chapter therefore begins with a recapitulation of the major metabolic pathways and their control systems, then considers how these processes are apportioned among the various organs of the body, and ends with a discussion of metabolic adaptation, including how the body maintains energy balance (homeostasis), how it deals with the metabolic challenges of starvation and obesity, and how it responds to the loss of control resulting from diabetes mellitus.

1 MAJOR PATHWAYS AND STRATEGIES OF ENERGY METABOLISM: A SUMMARY

Figure 27-1 indicates the interrelationships among the major pathways involved in energy metabolism. Let us review these pathways and their control mechanisms.

1. Glycolysis (Chapter 17) The metabolic degradation of glucose begins with its conversion to two molecules of pyruvate with the net generation of two molecules each of ATP and NADH. Under anaerobic conditions, pyruvate is converted to lactate (or, in yeast, to ethanol) so as to recycle the NADH. Under aerobic conditions, however, when glycolysis serves to prepare glucose for further oxidation, the NAD^+ is regenerated through oxidative phosphorylation (see below). The flow of metabolites through the glycolytic pathway is largely controlled by the activity of phosphofructokinase (PFK). This enzyme is activated by AMP and ADP, whose concentrations rise as the need for energy metabolism increases, and is inhibited by ATP and citrate, whose concentrations increase when the demand for energy metabolism has slackened. Citrate, a citric acid cycle intermediate, also inhibits PFK and glycolysis when aerobic metabolism takes over from anaerobic metabolism, making glucose oxidation more efficient (the Pasteur effect; Section 22-4C), and when fatty acid and/or ketone body oxidation (which are also aerobic pathways) are providing for energy needs (the glucose–fatty acid or Randle cycle; Section 22-4Bb). PFK is also activated by fructose-2,6-bisphosphate, whose concentration is regulated by the levels of glucagon, epinephrine, and norepinephrine through the intermediacy of cAMP (Section 18-3Fc). Liver and heart muscle F2,6P levels are regulated oppositely: A [cAMP] increase causes an [F2,6P] decrease in liver and an [F2,6P] increase in heart muscle. However, skeletal muscle [F2,6P] does not respond to changes in [cAMP].

2. Gluconeogenesis (Section 23-1) Mammals can synthesize glucose from a variety of precursors, including pyruvate, lactate, glycerol, and glucogenic amino acids (but not fatty acids), through pathways that occur mainly in liver and kidney. Many of these precursors are converted to oxaloacetate which, in turn, is converted to phosphoenolpyruvate and then, through a series of reactions that largely reverse the path of glycolysis, to glucose. The irreversible steps of glycolysis, those catalyzed by PFK and hexokinase, are bypassed in gluconeogenesis by hydrolytic reactions catalyzed, respectively, by fructose-1,6-bisphosphatase (FBPase) and glucose-6-phosphatase. FBPase and PFK may both be at least partially active simultaneously, creating a substrate cycle. This cycle, and the reciprocal regulation of PFK and FBPase, are important in regulating both the rate and direction of flux through glycolysis and

Figure 27-1 **The major energy metabolism pathways.**

3. Glycogen degradation and synthesis (Chapter 18)
Glycogen, the storage form of glucose in animals, occurs mostly in liver and muscle. Its conversion to glucose-6-phosphate (G6P) for entry into glycolysis in muscle and its conversion to glucose in liver is catalyzed, in part, by glycogen phosphorylase, whereas the opposing synthetic pathway is mediated by glycogen synthase. These enzymes are reciprocally regulated through phosphorylation/dephosphorylation reactions as catalyzed by amplifying cascades that respond to the levels of the hormones glucagon and epinephrine through the intermediacy of cAMP, and by insulin (Sections 18-3E and 19-4F). *The glucagon–insulin ratio is therefore a crucial factor in determining the rate and direction of glycogen metabolism.*

4. Fatty acid degradation and synthesis (Sections 25-1 through 25-5) Fatty acids are broken down in increments of C_2 units through β oxidation to form acetyl-CoA. They are synthesized from this compound via a separate pathway. The activity of the β-oxidation pathway varies with the fatty acid concentration. This, in turn, depends on the activity of "hormone-sensitive" triacylglycerol lipase in adipose tissue that is stimulated, through cAMP-regulated phosphorylation/dephosphorylation reactions, by glucagon and epinephrine but inhibited by insulin. The fatty acid synthesis rate varies with the activity of acetyl-CoA carboxylase,

gluconeogenesis (Sections 17-4F and 23-1B). Fatty acid and ketone body oxidation can increase the rate of gluconeogenesis in liver by decreasing the concentration of F2,6P (Section 18-3Fc). This occurs because the increased citrate concentration accompanying activation of the citric acid cycle during fatty acid oxidation inhibits PFK-2 as well as PFK (Table 23-1). Phosphoenolpyruvate carboxykinase (PEPCK) bypasses the third irreversible reaction of glycolysis, that catalyzed by pyruvate kinase (PK), and is controlled exclusively by long-term transcriptional regulation.

which is activated by citrate and insulin-dependent dephosphorylation, and inhibited by the pathway product palmitoyl-CoA and by cAMP- and AMP-dependent phosphorylation. Fatty acid synthesis is also subject to long-term regulation through alterations in the rates of synthesis of the enzymes mediating this process as stimulated by insulin and inhibited by fasting. *The glucagon–insulin ratio is therefore of prime importance in determining the rate and direction of fatty acid metabolism.*

5. Citric acid cycle (Chapter 21)   The citric acid cycle oxidizes acetyl-CoA, the common degradation product of glucose, fatty acids, ketone bodies, and ketogenic amino acids, to CO_2 and H_2O with the concomitant production of NADH and $FADH_2$. Many glucogenic amino acids can also be oxidized via the citric acid cycle through their breakdown, ultimately to pyruvate and then to acetyl-CoA, sometimes via the **cataplerosis** (using up) of a citric acid cycle intermediate (Section 21-5). The activities of the citric acid cycle regulatory enzymes citrate synthase, isocitrate dehydrogenase, and α-ketoglutarate dehydrogenase are controlled by substrate availability and feedback inhibition by cycle intermediates, NADH, and ATP.

6. Oxidative phosphorylation (Chapter 22)   This mitochondrial pathway oxidizes NADH and $FADH_2$ to NAD^+ and FAD with the coupled synthesis of ATP. The rate of oxidative phosphorylation, which is tightly coordinated with the metabolic fluxes through glycolysis and the citric acid cycle, is largely dependent on the concentrations of ATP, ADP, and P_i, as well as O_2.

7. Pentose phosphate pathway (Section 23-4)   This pathway functions to generate NADPH for use in reductive biosynthesis, as well as the nucleotide precursor ribose-5-phosphate, through the oxidation of G6P. Its flux-generating step is catalyzed by glucose-6-phosphate dehydrogenase, which is controlled by the level of $NADP^+$. *The ability of enzymes to distinguish between NADH, which is mainly utilized in energy metabolism, and NADPH permits energy metabolism and biosynthesis to be regulated independently.*

8. Amino acid degradation and synthesis (Sections 26-1 through 26-5)   Excess amino acids may be degraded to common metabolic intermediates. Most of these pathways begin with an amino acid's transamination to its corresponding α-keto acid with the eventual transfer of the amino group to urea via the urea cycle. Leucine and lysine are ketogenic amino acids in that they can be converted only to acetyl-CoA or acetoacetate and hence cannot be glucose precursors. The other amino acids are glucogenic in that they may be, at least in part, converted to one of the glucose precursors pyruvate, oxaloacetate, α-ketoglutarate, succinyl-CoA, or fumarate. Five amino acids are both ketogenic and glucogenic. Essential amino acids are those that an animal cannot synthesize itself; they must be obtained from plant and microbial sources. Nonessential amino acids can be synthesized by animals utilizing preformed amino groups via pathways that are generally simpler than those synthesizing essential amino acids.

Two compounds lie at the crossroads of the foregoing metabolic pathways: acetyl-CoA and pyruvate (Fig. 27-1).

Acetyl-CoA is the common degradation product of most metabolic fuels, including polysaccharides, lipids, and proteins. Its acetyl group may be oxidized to CO_2 and H_2O via the citric acid cycle and oxidative phosphorylation or used to synthesize fatty acids. Pyruvate is the product of glycolysis, the dehydrogenation of lactate, and the breakdown of certain glucogenic amino acids. It may be oxidatively decarboxylated to yield acetyl-CoA, thereby committing its atoms either to oxidation or to the biosynthesis of fatty acids. Alternatively, it may be carboxylated via the pyruvate carboxylase reaction to form oxaloacetate, which, in turn, either replenishes citric acid cycle intermediates or enters gluconeogenesis via phosphoenolpyruvate, thereby bypassing an irreversible step in glycolysis. Pyruvate is therefore a precursor of several amino acids as well as of glucose.

The foregoing pathways occur in specific cellular compartments. Glycolysis, glycogen synthesis and degradation, fatty acid synthesis, and the pentose phosphate pathway are largely or entirely cytosolically based, whereas fatty acid degradation, the citric acid cycle, and oxidative phosphorylation occur in the mitochondrion. Different phases of gluconeogenesis and amino acid degradation occur in each of these compartments. *The flow of metabolites across compartment membranes is mediated, in most cases, by specific carriers that are also subject to regulation.*

The enormous number of enzymatic reactions that simultaneously occur in every cell (Fig. 16-1) must be coordinated and strictly controlled to meet the cell's needs. Such regulation occurs on many levels. Intercellular communications regulating metabolism occur via certain hormones, including epinephrine, norepinephrine, glucagon, and insulin, as well as through a series of steroid hormones known as **glucocorticoids** (whose effects are discussed in Section 19-1Ga). These hormonal signals trigger a variety of cellular responses, including the synthesis of second messengers such as cAMP in the short term and the modulation of protein synthesis rates in the long term. On the molecular level, enzymatic reaction rates are controlled by phosphorylation/dephosphorylation via amplifying reaction cascades, by allosteric responses to the presence of effectors, which are usually precursors or products of the reaction pathway being controlled, and by substrate availability. The regulatory machinery of opposing catabolic and anabolic pathways is generally arranged such that these pathways are reciprocally regulated.

2 ORGAN SPECIALIZATION

Different organs have different metabolic functions and capabilities. In this section we consider how the special needs of the mammalian body organs are met and how their metabolic capabilities are coordinated to meet these needs. In particular, we discuss brain, muscle, adipose tissue, liver, and kidney (Fig. 27-2).

A. Brain

Brain tissue has a remarkably high respiration rate. For instance, the human brain constitutes only ~2% of the adult

Figure 27-2   The metabolic interrelationships among brain, adipose tissue, muscle, liver, and kidney. The red-outlined arrows indicate pathways that predominate in the well-fed state when glucose, amino acids, and fatty acids are directly available from the intestines.

body mass but is responsible for ~20% of its resting O_2 consumption. This consumption, moreover, is independent of the state of mental activity; it varies little between sleep and the intense concentration required of, say, the study of biochemistry. Most of the brain's energy production serves to power the plasma membrane $(Na^+–K^+)$–ATPase (Section 20-3A), which maintains the membrane potential required for nerve impulse transmission (Section 20-5). In fact, the respiration of brain slices is over 50% reduced by the $(Na^+–K^+)$–ATPase inhibitor ouabain (Section 20-3Af).

Under usual conditions, glucose serves as the brain's only fuel (although, with extended fasting, the brain gradually switches to ketone bodies; Section 27-4A). Indeed, since brain cells store very little glycogen, they require a steady supply of glucose from the blood. A blood glucose concentration of less than half of the normal value of ~5 m*M* results in brain dysfunction. Levels much below this, for example, caused by severe insulin overdose, result in coma, irreversible damage, and ultimately death. One of the liver's major functions, therefore, is to maintain the blood glucose level (Sections 18-3F and 27-2D).

B. Muscle

Muscle's major fuels are glucose from glycogen, fatty acids, and ketone bodies. Rested, well-fed muscle, in contrast to brain, synthesizes a glycogen store comprising 1 to 2% of its mass. The glycogen serves muscle as a readily available fuel depot since it can be rapidly converted to G6P for entry into glycolysis (Section 18-1).

Muscle cannot export glucose because it lacks glucose-6-phosphatase. Nevertheless, muscle serves the body as an energy reservoir because, during the fasting state, its proteins are degraded to amino acids, many of which are converted to pyruvate, which in turn, is transaminated to alanine. The alanine is then exported via the bloodstream to the liver, which transaminates it back to pyruvate, a glucose precursor. This process is known as the glucose–alanine cycle (Section 26-1Ad).

Since muscle does not participate in gluconeogenesis, it lacks the machinery that regulates this process in such gluconeogenic organs as liver and kidney. Muscle does not have receptors for glucagon, which, it will be recalled, stimulates an increase in blood glucose levels (Section 18-3F). However, muscle possesses epinephrine receptors (β-adrenergic receptors; Section 19-1F), which through the intermediacy of cAMP control the phosphorylation/dephosphorylation cascade system that regulates glycogen breakdown and synthesis (Section 18-3). This is the same cascade system that controls the competition between glycolysis and gluconeogenesis in liver in response to glucagon.

Heart muscle and skeletal muscle contain different isozymes of PFK-2/FBPase-2. The heart muscle isozyme is controlled by phosphorylation oppositely to that in liver, whereas skeletal muscle PFK-2/FBPase-2 is not controlled by phosphorylation at all (Section 18-3Fc). Thus the concentration of F2,6P rises in heart muscle but falls in liver in response to an increase in [cAMP]. Moreover the muscle isozyme of pyruvate kinase, which, it will be recalled, catalyzes the final step of glycolysis, is not subject to phosphorylation/dephosphorylation as is the liver isozyme (Section 23-1Ba). Thus, *whereas an increase in liver cAMP stimulates glycogen breakdown and gluconeogenesis, resulting in glucose export, an increase in heart muscle cAMP activates glycogen breakdown and glycolysis, resulting in glucose consumption. Consequently, epinephrine, which prepares the organism for action (fight or flight), acts independently of glucagon which, acting reciprocally with insulin, regulates the general level of blood glucose.*

a. Muscle Contraction Is Anaerobic Under Conditions of High Exertion

Muscle contraction is driven by ATP hydrolysis (Section 35-3Bb) and is therefore ultimately dependent on respiration. Skeletal muscle at rest utilizes ~30% of the O_2 consumed by the human body. A muscle's respiration rate may increase in response to a heavy workload by as much as 25-fold. Yet, its rate of ATP hydrolysis can increase by a much greater amount. The ATP is initially regenerated by the reaction of ADP with phosphocreatine as catalyzed by creatine kinase (Section 16-4Cd):

$$\text{Phosphocreatine} + \text{ADP} \rightleftharpoons \text{creatine} + \text{ATP}$$

(phosphocreatine is resynthesized in resting muscle by the reversal of this reaction). Under conditions of maximum exertion, however, such as occurs in a sprint, a muscle has

only an ~5-s supply of phosphocreatine. It must then shift to ATP production via glycolysis of G6P resulting from glycogen breakdown, a process whose maximum flux greatly exceeds those of the citric acid cycle and oxidative phosphorylation. Much of this G6P is therefore degraded anaerobically to lactate (Section 17-3A) which, in the Cori cycle (Section 23-1C), is exported via the bloodstream to the liver, where it is reconverted to glucose through gluconeogenesis. Gluconeogenesis requires ATP generated by oxidative phosphorylation. Muscles thereby shift much of their respiratory burden to the liver and consequently also delay the O_2-consumption process, a phenomenon known as oxygen debt. The source of ATP during exercise of varying duration is summarized in Fig. 27-3.

b. Muscle Fatigue Has a Protective Function

Muscle fatigue, defined as the inability of a muscle to maintain a given power output, occurs in ~20 s under conditions of maximum exertion. Such fatigue is not caused by the exhaustion of the muscle's glycogen supply. Rather, it may result from glycolytic proton generation that can drop the intramuscular pH from its resting value of 7.0 to as low as 6.4 (fatigue does not, as is widely believed, result from the buildup of lactate itself, as is demonstrated by the observation that muscles can sustain a large power output under high lactate concentrations if the pH is maintained near 7.0). Nevertheless, how acidification might cause muscle fatigue is unclear. Two other proposed causes for muscle fatigue are (1) the increased [P_i] arising largely from the

Figure 27-3 Source of ATP during exercise in humans. The supply of endogenous ATP is extended for a few seconds by phosphocreatine, after which anaerobic glycolysis generates ATP. The shift from anaerobic to aerobic metabolism (oxidative phosphorylation) occurs after about 90 s, or slightly later in trained athletes. [Adapted from McArdle, W.D., Katch, F.I., and Katch, V.L., *Exercise Physiology*, 2nd ed., Lea & Febiger (1986), p. 348.]

utilization of ATP may precipitate Ca^{2+} as calcium phosphate (which is highly insoluble), thereby decreasing contractile force (muscle contraction is triggered by the release of Ca^{2+} ion; Section 35-3Cb); and (2) the K^+ ion known to be released from contracting muscle cells may result in their depolarization (Section 20-5Ba) and hence a reduction in their contraction. Whatever its cause(s), it seems likely that muscle fatigue is an adaptation that prevents muscle cells from committing suicide by exhausting their ATP supply (recall that glycolysis and other ATP-generating pathways must be primed by ATP).

c. The Heart Is a Largely Aerobic Organ

The heart is a muscular organ but one that must maintain continuous rather than intermittent activity. Thus heart muscle, except for short periods of extreme exertion, relies entirely on aerobic metabolism. It is therefore richly endowed with mitochondria; they comprise up to 40% of its cytoplasmic space, whereas some types of skeletal muscle are nearly devoid of mitochondria. The heart can metabolize fatty acids, ketone bodies, glucose, pyruvate, and lactate. Fatty acids are the resting heart's fuel of choice but, on the imposition of a heavy workload, the heart greatly increases its rate of consumption of glucose, which is derived mostly from its relatively limited glycogen store.

C. *Adipose Tissue*

Adipose tissue, which consists of cells known as adipocytes (Fig. 12-2), is widely distributed about the body but occurs most prominently under the skin, in the abdominal cavity, in skeletal muscle, around blood vessels, and in mammary gland. The adipose tissue of a normal 70-kg man contains ~15 kg of fat. This amount represents some 590,000 kJ of energy (141,000 dieter's Calories), which is sufficient to maintain life for ~3 months. Yet, adipose tissue is by no means just a passive storage depot. In fact, it is second in importance only to liver in the maintenance of metabolic homeostasis (Section 27-3).

Adipose tissue obtains most of its fatty acids from the liver or from the diet as described in Section 25-1. Fatty acids are activated by the formation of the corresponding fatty acyl-CoA and then esterified with glycerol-3-phosphate to form the stored triacylglycerols (Section 25-4F). The glycerol-3-phosphate arises from the reduction of dihydroxyacetone phosphate, which must be glycolytically generated from glucose or gluconeogenically generated from pyruvate or oxaloacetate (a process called **glyceroneogenesis;** Section 25-4Fa) because adipocytes lack a kinase that phosphorylates endogenous glycerol.

Adipocytes hydrolyze triacylglycerols to fatty acids and glycerol in response to the levels of glucagon, epinephrine, and insulin through a reaction catalyzed by hormone-sensitive triacylglycerol lipase (Section 25-5). If glycerol-3-phosphate is abundant, many of the fatty acids so formed are reesterified to triacylglycerols. Indeed, the average turnover time for triacylglycerols in adipocytes is only a few days. If, however, glycerol-3-phosphate is in short supply, the fatty acids are released into the bloodstream. *The rate of glucose uptake by adipocytes, which is regulated by insulin as well as by glucose availability, is therefore also an important factor in triacylglycerol formation and mobilization.* However, glycerol-3-phosphate is also produced via glyceroneogenesis under the control of PEPCK, allowing triacylglycerol turnover even when glucose concentration is low.

a. Obesity Results from Aberrant Metabolic Control

The human body regulates glycogen and protein levels within relatively narrow limits, but fat reserves, which are much larger, can become enormous. The accumulation of fatty acids as triacylglycerols in adipose tissue is largely a result of excess fat or carbohydrate intake compared to energy expenditure. Fat synthesis from carbohydrates occurs when the carbohydrate intake is high enough that glycogen stores, to which excess carbohydrate is normally directed, approach their maximum capacity.

Obesity is one of the major health-related problems in industrial countries. An estimated 30% of adults in the United States are obese (are at least 20% above their desirable weights) and another 35% are overweight. Most obese people find it inordinately difficult to lose weight or, having done so, to keep it off. Yet most animals, including humans, tend to have stable weights; that is, if they are given free access to food, they eat just enough to maintain this so-called set point weight. The nature of the regulatory machinery that controls the set point, which in obese individuals seems to be aberrantly high, is just beginning to come to light (see Section 27-3).

Formerly grossly obese individuals who have lost at least 100 kg to reach their normal weights exhibit some of the metabolic symptoms of starvation: they are obsessed with food, have low heart rates, are cold intolerant, and require 25% less caloric intake than normal individuals of similar heights and weights. In both normal and obese individuals, some 50% of the fatty acids liberated by the hydrolysis of triacylglycerols are reesterified before they can leave the adipocytes. In formerly obese subjects, this reesterification rate is only 35 to 40%, a level similar to that observed in normal individuals after a several day fast. The fat cells in normal and obese individuals, moreover, are of roughly the same size; obese people just have more of them. In fact, adipocyte precursor cells from massively obese individuals proliferate excessively in tissue culture compared to those from normal or even moderately obese subjects (adipocytes themselves do not replicate). Since fat cells, once gained, are never lost, this suggests that adipocytes, although highly elastic in size, tend to maintain a certain fixed volume and in doing so influence the metabolism and thus the appetite. This insight, unfortunately, has not yet led to a method for lowering the set points of individuals with a tendency toward obesity.

D. *Liver*

The liver is the body's central metabolic clearinghouse. It functions to maintain the proper levels of nutrients in the blood for use by the brain, muscles, and other tissues. The liver is uniquely situated to carry out this task because all the nutrients absorbed by the intestines except fatty acids are released into the portal vein, which drains directly into the liver.

One of the liver's major functions is to act as a blood glucose "buffer." It does so by taking up or releasing glucose in response to the levels of glucagon, epinephrine, and insulin as well as to the concentration of glucose itself. After a carbohydrate-containing meal, when the blood glucose concentration reaches ~6 mM, the liver takes up glucose by converting it to G6P. The process is catalyzed by glucokinase (Section 18-3Fa), which differs from hexokinase, the analogous glycolytic enzyme in other cells, in that glucokinase has a much lower affinity for glucose (glucokinase reaches half-maximal velocity at ~5 mM glucose vs <0.1 mM glucose for hexokinase) and is not inhibited by G6P. Liver cells, in contrast to muscle and adipose cells, are permeable to glucose, and thus insulin has no direct effect on their glucose uptake. Since the blood glucose concentration is normally less than glucokinase's K_M, the rate of glucose phosphorylation in the liver is more or less proportional to the blood glucose concentration. The other intestinally absorbed sugars, mostly fructose, galactose, and mannose, are also converted to G6P in the liver (Section 17-5). After an overnight fast, the blood glucose level drops to ~4 mM. The liver keeps it from dropping below this level by releasing glucose into the blood as is described below. In addition, lactate, the product of anaerobic glucose metabolism in the muscle, is taken up by the liver for use in gluconeogenesis and lipogenesis as well as in oxidative phosphorylation (the Cori cycle; Section 23-1C). Alanine produced in the muscle is taken up by the liver and converted to pyruvate for gluconeogenesis as well (the glucose–alanine cycle; Section 26-1Ad).

a. The Fate of Glucose-6-Phosphate Varies with Metabolic Requirements

G6P is at the crossroads of carbohydrate metabolism; it can have several alternative fates depending on the glucose demand (Fig. 27-1):

1. G6P can be converted to glucose by the action of glucose-6-phosphatase for transport via the bloodstream to the peripheral organs.

2. G6P can be converted to glycogen (Section 18-2) when the body's demand for glucose is low. Yet, increased glucose demand, as signaled by higher levels of glucagon and/or epinephrine, reverses this process (Section 18-1).

3. G6P can be converted to acetyl-CoA via glycolysis and the action of pyruvate dehydrogenase (Chapter 17 and Section 21-2). Most of this glucose-derived acetyl-CoA is used in the synthesis of fatty acids (Section 25-4), whose fate is described below, and in the synthesis of phospholipids (Section 25-8) and cholesterol (Section 25-6A). Cholesterol, in turn, is a precursor of bile salts, which are produced by the liver (Section 25-6C) for use as emulsifying agents in the intestinal digestion and absorption of fats (Section 25-1).

4. G6P can be degraded via the pentose phosphate pathway (Section 23-4) to generate the NADPH required

for fatty acid biosynthesis and the liver's many other biosynthetic functions, as well as ribose-5-phosphate (R5P) for nucleotide biosynthesis (Sections 28-1A and 28-2A).

b. The Liver Can Synthesize or Degrade Triacylglycerols

Fatty acids are also subject to alternative metabolic fates in the liver (Fig. 27-1):

1. When the demand for metabolic fuels is high, fatty acids are degraded to acetyl-CoA and then to ketone bodies (Section 25-3) for export via the bloodstream to the peripheral tissues.

2. When the demand for metabolic fuels is low, fatty acids are used to synthesize triacylglycerols that are secreted into the bloodstream as VLDL for uptake by adipose tissue. Fatty acids may also be incorporated into phospholipids (Section 25-8).

Since the rate of fatty acid oxidation varies only with fatty acid concentration (Section 25-5), fatty acids produced by the liver might be expected to be subject to reoxidation before they can be exported. Such a futile cycle is prevented by the compartmentalization of fatty acid oxidation in the mitochondrion and fatty acid synthesis in the cytosol. Carnitine palmitoyltransferase I, a component of the system that transports fatty acids into the mitochondrion (Section 25-2B), is inhibited by malonyl-CoA, the key intermediate in fatty acid biosynthesis (Section 25-4A). Hence, when the demand for metabolic fuels is low so that fatty acids are being synthesized, they cannot enter the mitochondrion for conversion to acetyl-CoA. Rather, the liver's biosynthetic demand for acetyl-CoA is met through the degradation of glucose.

When the demand for metabolic fuel rises so as to inhibit fatty acid biosynthesis, however, fatty acids are transported into the liver mitochondria for conversion to ketone bodies. Under such conditions of low blood glucose concentrations, glucokinase has reduced activity so that there is net glucose export (there is, however, always a futile cycle between the reactions catalyzed by glucokinase and glucose-6-phosphatase; Section 18-3Fb). The liver cannot use ketone bodies for its own metabolic purposes because liver cells lack 3-ketoacyl-CoA transferase (Section 25-3). Fatty acids rather than glucose or ketone bodies are therefore the liver's major acetyl-CoA source under conditions of high metabolic demand. The liver generates its ATP from this acetyl-CoA through the citric acid cycle and oxidative phosphorylation. The aerobic oxidation of fatty acids inhibits glucose utilization since activation of the citric acid cycle and oxidative phosphorylation increases the concentration of citrate, which inhibits glycolysis (the glucose–fatty acid or Randle cycle; Section 22-4Bb).

c. Amino Acids Are Important Metabolic Fuels

The liver degrades amino acids to a variety of metabolic intermediates (Section 26-3). These pathways mostly begin with amino acid transamination to yield the corresponding α-keto acid (Section 26-1A) with the amino group being ultimately converted, via the urea cycle (Section 26-2), to the subsequently excreted urea. Glucogenic amino acids can be converted in this manner to pyruvate or citric acid cycle intermediates such as oxaloacetate and are thereby gluconeogenic precursors (Section 23-1). Ketogenic amino acids, many of which are also glucogenic, may be converted to ketone bodies.

The liver's glycogen store is insufficient to supply the body's glucose needs for more than ~6 h after a meal. After that, glucose is supplied through gluconeogenesis from amino acids arising mostly from muscle protein degradation to alanine (the glucose–alanine cycle; Section 26-1Ad) and glutamine (the transport form of ammonia; Section 26-1B). Thus proteins, in addition to their structural and functional roles, are important fuel resources. (Animals cannot convert fat to glucose because they lack a pathway for the net conversion of acetyl-CoA to oxaloacetate; Section 23-2).

d. The Liver Is the Body's Major Metabolic Processing Unit

The liver has numerous specialized biochemical functions in addition to those already mentioned. Prominent among them are the synthesis of blood plasma proteins, the degradation of porphyrins (Section 26-4A) and nucleic acid bases (Section 28-4), the storage of iron, and the detoxification of biologically active substances such as drugs, poisons, and hormones by a variety of oxidation (e.g., by cytochromes P450; Section 15-4Bc), reduction, hydrolysis, conjugation, and methylation reactions.

E. *Kidney*

The kidney functions to filter out the waste product urea from the blood and concentrate it for excretion, to recover important metabolites such as glucose, and to maintain the blood's pH. Blood pH is maintained by regenerating depleted blood buffers such as bicarbonate (lost by the exhalation of CO_2) and by removing for excretion excess H^+ together with the conjugate bases of excess metabolic acids such the ketone bodies acetoacetate and β-hydroxybutyrate. Phosphate, the major buffer in urine for moderate acid excretion, is accompanied by equivalent quantities of cations such as Na^+ and K^+. However, large losses of Na^+ and K^+ would upset the body's electrolyte balance, so on the production of large amounts of acids such as lactic acid or ketone bodies, the kidney produces NH_4^+ to aid in the excretion of the excess H^+ (utilizing Cl^- or the conjugate base of a metabolic acid as the counterion). This NH_4^+ is generated from glutamine, which is converted first to glutamate and then to α-ketoglutarate by glutaminase and glutamate dehydrogenase. The overall reaction is

$$\text{Glutamine} \longrightarrow \text{α-ketoglutarate} + 2NH_4^+$$

The α-ketoglutarate is converted to malate by the citric acid cycle and then is exported from the mitochondrion and converted either to pyruvate, which is oxidized completely to CO_2, or via oxaloacetate to PEP and then to glucose via gluconeogenesis. High fat diets, which produce high blood concentrations of free fatty acids and ketone bodies and hence high acidic loads, cause α-ketoglutarate to be converted completely to CO_2, and then to bicarbonate, thereby increasing the blood's buffering capacity. During starvation, the α-ketoglutarate enters gluconeogenesis, to the extent that the kidneys generate as much as 50% of the body's glucose supply.

3 METABOLIC HOMEOSTASIS: REGULATION OF APPETITE, ENERGY EXPENDITURE, AND BODY WEIGHT

When a normal animal overeats, the resulting additional fat somehow signals the brain to induce the animal to eat less and to expend more energy. Conversely, the loss of fat stimulates increased eating until the lost fat is replaced. Evidently, animals have a "lipostat" that can keep the amount of body fat constant to within 1% over many years. At least a portion of the lipostat resides in the hypothalamus (a part of the brain that hormonally controls numerous physiological functions; Section 19-1H), since damaging it can yield a grossly obese animal.

Despite this obvious set of controls in animals, there has been an explosion of obesity in many industrial nations. It has, in fact, become a world health problem, leading to diabetes and heart disease. As a result of numerous studies in recent years, researchers have been able to outline the mechanisms involved in **metabolic homeostasis,** the balance between energy influx and energy expenditure, and to identify some of the irregularities that lead to obesity. A variety of mutant strains of rodents have been generated that cause obesity. The study of these mutants has resulted in the identification of several hormones that act in a coordinated manner to regulate appetite.

A. *AMP-Dependent Protein Kinase Is the Cell's Fuel Gauge*

All of the metabolic pathways discussed in Section 27-1 are affected in one way or another by the need for ATP, as is indicated by the cell's AMP-to-ATP ratio (Section 17-4Fd). Several enzymes are either activated or inhibited allosterically by AMP, and several others are phosphorylated by **AMP-dependent protein kinase (AMPK),** a major regulator of metabolic homeostasis. *AMPK activates metabolic breakdown pathways that generate ATP while inhibiting biosynthetic pathways so as to conserve ATP for more vital processes.* AMPK is an αβγ heterotrimer found in all eukaryotic organisms from yeast to humans. The α subunit contains a Ser/Thr protein kinase domain, and the γ subunit contains sites for allosteric activation by AMP and inhibition by ATP. Like other protein kinases, AMPK's kinase domain must be phosphorylated for activity. Binding of AMP to the γ subunit causes a conformational change that exposes Thr 172 in the activation loop of the α subunit, promoting its phosphorylation and increasing its activity at least 100-fold. AMP can activate the phosphorylated enzyme up to 5-fold more. There are two isoforms of the α subunit, two of the β subunit, and three of the γ subunit, giving rise to 12 possible heterotrimeric combinations, with splice variants yielding further diversity. The major kinase that phosphorylates AMPK is named **LKB1.** The knockout of LKB1 in mouse liver results in the loss of the phosphorylated form of AMPK.

a. AMPK Activates Glycolysis in Ischemic Cardiac Muscle

AMPK's targets include the heart isozyme of the bifunctional enzyme PFK-2/FBPase-2, which controls the fructose-2,6-bisphosphate (F2,6P) concentration (Section 18-3Fc). The phosphorylation of this isozyme activates the PFK-2 activity, increasing [F2,6P], which in turn activates PFK-1, the rate-determining enzyme of glycolysis (Section 17-4Fb). Consequently, in ischemic (blood-starved) heart muscle cells, which receive insufficient oxygen for oxidative phosphorylation to maintain adequate concentrations of ATP, the resulting AMP buildup causes the cells to switch to anaerobic glycolysis for ATP production.

b. AMPK Inhibits Lipogenesis, Cholesterol Synthesis, and Gluconeogenesis in Liver

AMPK-mediated phosphorylation also inhibits acetyl-CoA carboxylase (ACC), which catalyzes the first committed step of fatty acid synthesis (Section 25-4B), and hydroxymethylglutaryl-CoA reductase (HMG-CoA reductase), which catalyzes the rate-determining step in cholesterol biosynthesis (Section 25-6Aa). Activated AMPK inhibits gluconeogenesis in a more complicated way: It phosphorylates and thereby inactivates the transcriptional coactivator **TORC2** (for *transducer* of *regulated* CREB activity-2), which in concert with the transcriptional activator CREB, would otherwise induce the transcription of the gene encoding PEP carboxykinase (PEPCK), the enzyme that catalyzes the rate-determining step of gluconeogenesis (Sections 23-1Af and 23-1Bb). Consequently, when the rate of ATP production is inadequate, these biosynthetic pathways are turned off, thereby conserving ATP for more vital cellular functions.

c. AMPK Promotes Fatty Acid Oxidation and Glucose Uptake but Inhibits Glycogen Synthesis in Skeletal Muscle

The inhibition of ACC results in a decrease in the concentration of malonyl-CoA, the starting material for fatty acid biosynthesis. Malonyl-CoA has an additional role, however. It is an inhibitor of carnitine palmitoyltransferase I (Section 25-2B), which is required to transfer cytosolic palmitoyl-CoA into mitochondria for oxidation. The decrease in malonyl-CoA concentration therefore allows more palmitoyl-CoA to be oxidized. AMPK also increases the recruitment of GLUT4 to muscle cell plasma membranes (Section 20-2Ec), as well as stimulating its expression, thus facilitating the insulin-independent entry of glucose into these cells. In addition, AMPK inhibits glycogen synthase (which catalyzes the rate-limiting reaction in glycogen synthesis; Section 18-3B). In fact, the β subunit of AMPK has a glycogen-binding domain that presumably recruits AMPK to the vicinity of glycogen synthase.

d. AMPK Inhibits Fatty Acid Synthesis and Lipolysis in Adipocytes

AMPK inhibits fatty acid synthesis in adipocytes by phosphorylating ACC as described above. Moreover, AMPK phosphorylates hormone-sensitive triacylglycerol lipase in adipose tissue (Section 25-5). This phosphorylation inhibits rather than activates the enzyme, in part by preventing the relocation of the enzyme to the lipid droplet, the cellular location of lipolysis. As a result, fewer triacylglycerol molecules are broken down so that fewer fatty acids are exported to the bloodstream. This latter process seems paradoxical (fatty acid oxidation would help relieve an ATP deficit), although it has been speculated that it prevents the cellular buildup of fatty acids to toxic levels. The major effects of AMPK activation on glucose and lipid metabolism in liver, skeletal muscle, heart muscle, and adipose tissue are diagrammed in Fig. 27-4.

B. Adiponectin Regulates AMPK Activity

Adiponectin is a 247-residue protein hormone, secreted exclusively by adipocytes, that helps regulate energy homeostasis and glucose and lipid metabolism by controlling AMPK activity. Its monomers consist of an N-terminal collagenlike domain and a C-terminal globular domain. Adiponectin occurs in the bloodstream in several forms: a low molecular weight (LMW) trimer formed by the coiling of its collagenlike domains into a triple helix (Section 8-2Ba) as well as hexamers (MMW) and multimers (HMW) that form disulfide cross-linked bouquets (Fig. 27-5). In addition, globular adiponectin, formed by the cleavage of the collagenlike domain to release globular monomers, occurs in lower concentrations.

The binding of adiponectin to **adiponectin receptors,** which occur on the surfaces of both liver and muscle cells, acts to increase the phosphorylation and activity of AMPK. This, as we have seen (Section 27-3A), inhibits gluconeogenesis and stimulates fatty acid oxidation in liver and stimulates glucose uptake and glucose and fatty acid oxidation in muscle. All of these effects act to increase insulin sensitivity, in part because adiponectin and insulin elicit similar responses in tissues such as liver. Decreased adiponectin is associated with insulin resistance (Section 27-4B). Paradoxically, the blood concentration of adiponectin, which is secreted by adipocytes, decreases with increased amounts of adipose tissue. This may be because increased adipose tissue is also associated with increased production of **tumor necrosis factor-α (TNF-α),** a cytokine that decreases both the expression and secretion of adiponectin from adipose tissue (Section 19-3Db).

C. Leptin

Two of the genes whose mutations cause obesity in mice are known as *obese* (*ob*) and *diabetes* (*db;* the wild-type genes are designated *OB* and *DB*). Homozygotes for defects in either of these recessive genes, *ob/ob* or *db/db,* are grossly obese and have nearly identical phenotypes (Fig. 27-6). Indeed, the way in which these phenotypes

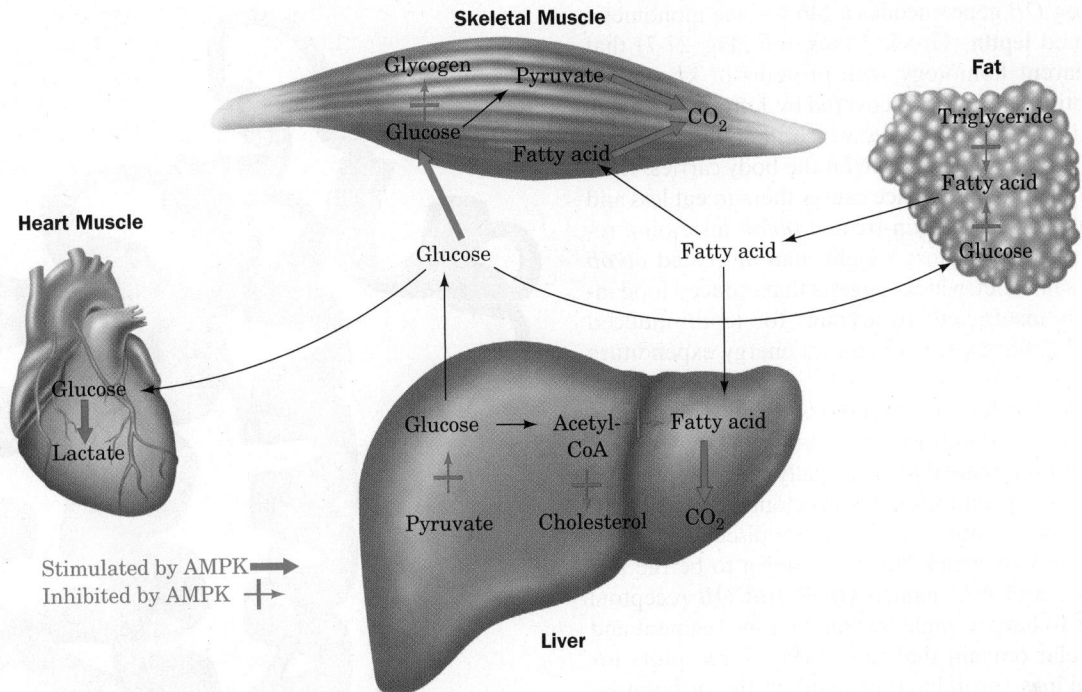

Figure 27-4   Major effects of AMP-activated protein kinase (AMPK) on glucose and lipid metabolism in liver, muscle, and adipose tissue. In skeletal muscle, AMPK stimulates glucose and fatty acid oxidation while inhibiting glycogen synthesis. In heart muscle, AMPK stimulates glycolysis. In liver, AMPK inhibits lipid biosynthesis and gluconeogenesis while activating fatty acid oxidation. In adipose tissue, AMPK inhibits fatty acid biosynthesis, lipolysis, and fatty acid export. [After Towler, M.C. and Hardie, D.G., *Circ. Res.* **100,** 328 (2007).]

were distinguished was by surgically linking the circulation of a mutant mouse to that of a normal (*OB/OB*) mouse, a phenomenon named **parabiosis.** *ob/ob* mice so linked exhibit normalization of body weight and reduced food intake, whereas *db/db* mice do not do so. This suggests that *ob/ob* mice are deficient in a circulating factor that regulates appetite and metabolism, whereas *db/db* mice are defective in the receptor for this circulating factor.

Figure 27-5   Adiponectin trimers, hexamers, and multimers. These complexes are referred to as low molecular weight (LMW), medium molecular weight (MMW), and high molecular weight (HMW) forms. [After Kadowaki, T., and Yamauchi, T., *Endocr. Rev.* **26,** 439 (2005).]

Figure 27-6   Normal (*OB/OB, left*) and obese (*ob/ob, right*) mice. [Courtesy of Richard D. Palmiter, University of Washington.]

The mouse *OB* gene encodes a 146-residue monomeric protein named **leptin** (Greek: *leptos,* thin; Fig. 27-7) that has no apparent homology with proteins of known sequence. Leptin, which was discovered by Jeffrey Friedman, is expressed only by adipocytes, which in doing so appear to inform the brain of how much fat the body carries. Thus, injecting leptin into *ob/ob* mice causes them to eat less and to lose weight. In fact, leptin-treated *ob/ob* mice on a restricted diet lost 50% more weight than untreated *ob/ob* mice on the same diet, which suggests that reduced food intake alone is insufficient to account for leptin-induced weight loss. Leptin appears to control energy expenditure as well.

Leptin injection has no effect on *db/db* mice. The leptin receptor gene was identified by making a cDNA library from mouse brain tissue that specifically bound leptin and then identifying a receptor-expressing clone by its ability to bind leptin (gene cloning techniques are discussed in Section 5-5). This gene, which has been shown to be the *DB* gene, encodes a protein named **OB-R** (for *OB* receptor) that appears to have a single transmembrane segment and an extracellular domain that resembles the receptors for certain cytokines (proteins that regulate the differentiation, proliferation, and activities of various blood cells; Section 19-3Eb).

OB-R protein, which was discovered by Louis Tartaglia, has at least six alternatively spliced forms that appear to be expressed in a tissue-specific manner (alternative gene splicing is discussed in Section 31-4Am). In normal mice, the hypothalamus expresses high levels of a splice variant of OB-R that has a 302-residue cytoplasmic segment. However, in *db/db* mice, this segment has an abnormal splice site that truncates it to only 34 residues, which almost certainly renders this OB-R variant unable to transmit leptin signals. Thus, it appears that leptin's weight-controlling effects are mediated by signal transduction resulting from its binding to the OB-R protein in the hypothalamus.

Human leptin is 84% identical in sequence to that of mice. The use of a radioimmunoassay (Section 19-1A) to measure the serum levels of leptin in normal-weight and obese humans established that in both groups serum leptin concentrations increase with their percentage of body fat as does the *ob* mRNA content of their adipocytes. Moreover, after obese individuals had lost weight, their serum leptin concentrations and adipocyte *ob* mRNA content declined. This suggests that most obese persons produce sufficient amounts of leptin but have developed "leptin resistance." Since leptin must cross the blood–brain barrier in order to exert its effects on the hypothalamus, it has been suggested that this crossing is somehow saturatable, thus limiting the concentration of leptin in the brain. The high concentration of leptin in obese individuals is not without affect, however. OB-R is also expressed in peripheral tissues where leptin has been shown to function as well. While not preventing obesity, the hormone has been shown to directly stimulate the oxidation of fatty acids as well as to inhibit the accumulation of lipids in non-adipose tissue. It does so by activating AMP-dependent protein kinase (AMPK), which in turn phosphorylates and thereby inacti-

Figure 27-7 **X-ray structure of human leptin-E100.** This mutant form of leptin (W100E) has comparable biological activity to the wild-type protein but crystallizes more readily. The protein, which is colored in rainbow order from its N-terminus (*blue*) to its C-terminus (*red*), forms a four-helix bundle, as do many protein growth factors (e.g., human growth hormone; Fig. 19-10). Residues 25 to 38 are not visible in the X-ray structure. [Based on an X-ray structure by Faming Zhang, Eli Lilly & Co., Indianapolis, Indiana. PDBid 1AX8.]
See Interactive Exercise 27.

vates acetyl-CoA carboxylase (ACC). This reduces the malonyl-CoA concentration, thereby decreasing its inhibition of carnitine palmitoyltransferase I, which then transports fatty acyl-CoA into the mitochondrion for oxidation (Section 25-5). We discuss the function of leptin in peripheral tissues in Section 27-3H.

A small minority of obese individuals have been found to be leptin deficient in a manner similar to *ob/ob* mice. Two grossly obese children who are members of the same highly consanguineous (descended from the same ancestors) family (they are cousins and both sets of parents are cousins) have been shown to be homozygous for a defective OB gene. The children, at the ages of 8 and 2 years old, respectively, weighed 86 and 29 kg and were noted to have remarkably large appetites. Their OB genes have a deletion of a single guanine nucleotide in codon 133, thereby causing a frameshift mutation that, it is likely, renders the mutant leptin biologically inactive. Moreover, their leptin serum levels were only ~10% of normal. Leptin injections have relieved their symptoms.

D. *Insulin*

We have discussed the insulin signaling cascade (Section 19-4F) and the role of insulin in peripheral tissues such as muscle and adipose tissue in stimulating the uptake of glucose (Fig. 20-11) and its storage as glycogen (Section 18-3) or fat (Section 25-5). Insulin receptors also occur in the hypothalamus. Consequently, the infusion of insulin into rats with insulin-deficient diabetes inhibits food intake, reversing the overeating behavior characteristic of the disease. Knock-out mice have been developed with a central nervous system–specific disruption of the insulin receptor gene. These mice have no alteration in brain development or survival but become obese, with increased body fat, increased leptin levels, increased serum triacylglycerol, and the elevated plasma insulin levels characteristic of insulin resistance (Section 27-4B). Evidently, insulin also plays a role in the neuronal regulation of food intake and body weight. As we discuss in Section 27-3F, insulin and leptin both act through receptors in the hypothalamus to decrease food intake.

E. *Ghrelin and PYY$_{3-36}$*

a. Ghrelin and PYY$_{3-36}$ Act as Short-Term Regulators of Appetite

Ghrelin, which was discovered by Masayasu Kojima and Kenji Kanagawa, is an appetite-stimulating gastric peptide that is secreted by the empty stomach. This 28-residue peptide was first discovered and named for its function as a growth hormone–releasing peptide (ghrelin is an abbreviation for *growth-hormone-rel*ease). Octanoylation of its Ser 3 is required for activity.

$$\overset{10}{\text{GSXFLSPEHQ}}\ \overset{20}{\text{RVQQRKESKK}}\ \overset{28}{\text{PPAKLQPR}}$$

Human ghrelin

X = Ser modified with *n*-octanoic acid

Injection of ghrelin has been shown to induce adiposity (increased adipose tissue) in rodents by stimulating an increase in food intake while reducing fat utilization. In humans in states of positive energy balance such as obesity or high caloric intake, circulating ghrelin levels are decreased, whereas during fasting, circulating ghrelin levels increase.

PYY$_{3-36}$

$$\overset{3}{\text{IKPEAPGE}}\ \overset{10}{\text{DASPEELNRY}}\ \overset{20}{\text{YASLRHYLNL}}\ \overset{30}{\text{VTRQRY}}\overset{36}{}$$

Human PYY$_{3-36}$

is a peptide secreted by the gastrointestinal tract in proportion to the caloric intake of a meal, which acts to inhibit further food intake. Both rodents and humans have been shown to respond to the presence of this peptide by decreasing their food intake for up to 12 hours. Human subjects receiving a 90-minute infusion of PYY$_{3-36}$ ate only 1500 kcal of food during the next 24-hour period, whereas those receiving saline controls ate 2200 kcal during the same period.

F. *Hypothalamic Integration of Hormonal Signals*

a. Neurons of the Arcuate Nucleus Region of the Hypothalamus Integrate and Transmit Hunger Signals

About half of the length of the hypothalamus is taken up by the **arcuate nucleus,** a collection of neuronal cell bodies consisting of two cell types: the **NPY/AgRP** cell type and the **POMC/CART** cell type. These cell types are named after the neuropeptides they secrete. **Neuropeptide Y (NPY)**

$$\overset{1}{\text{YPSKPDNPGE}}\ \overset{20}{\text{DAPAEDMARY}}\ \overset{30}{\text{YSALRHYINL}}\ \overset{36}{\text{ITRQRY}}\text{–NH}_2$$

Neuropeptide Y

The C-terminal carboxyl is amidated

is a potent stimulator of food intake and an inhibitor of energy expenditure, as is **Agouti related peptide (AgRP).** **Pro-opiomelanocortin (POMC)** is post-translationally processed in the hypothalamus to release **α-melanocyte stimulating hormone (α-MSH;** Section 34-3C). **Cocaine and amphetamine-regulated transcript (CART)** and α-MSH are both inhibitors of food intake and stimulators of energy expenditure.

The balance of the secretions from these two cell types is controlled by leptin, insulin, ghrelin, and PYY$_{3-36}$ (Fig. 27-8). Leptin and insulin signal satiety and therefore decrease appetite by diffusing across the blood–brain barrier to the arcuate nucleus, where they stimulate POMC/CART neurons to produce CART and α-MSH, while inhibiting the production of NPY from NPY/AgRP neurons. Leptin receptors act through the JAK–STAT signal transduction pathway (Section 19-3Eb). Ghrelin has receptors on NPY/AgRP neurons that stimulate the secretion of NPY and AgRP to increase appetite. Interestingly, PYY$_{3-36}$, a peptide that is homologous to NPY, binds specifically to NPY receptor subtype Y2R on NPY/AgRP neurons. This subtype is an inhibitory receptor, however, so binding of PYY$_{3-36}$ causes a decrease in secretion from NPY/AgRP neurons. The integrated stimuli of all these secretions from the arcuate nucleus control appetite.

G. *Control of Energy Expenditure by Adaptive Thermogenesis*

The energy content of food is utilized by an organism either in the performance of work or the generation of heat. Excess energy is stored as glycogen or fat for future use. In well-balanced individuals, the storage of excess fuel remains constant over many years. However, when energy consumed is consistently greater than energy expended, obesity results. The body has several mechanisms for preventing obesity. One of them, as discussed above, is appetite control. The other is **diet-induced thermogenesis,** a form of **adaptive thermogenesis** (heat production in

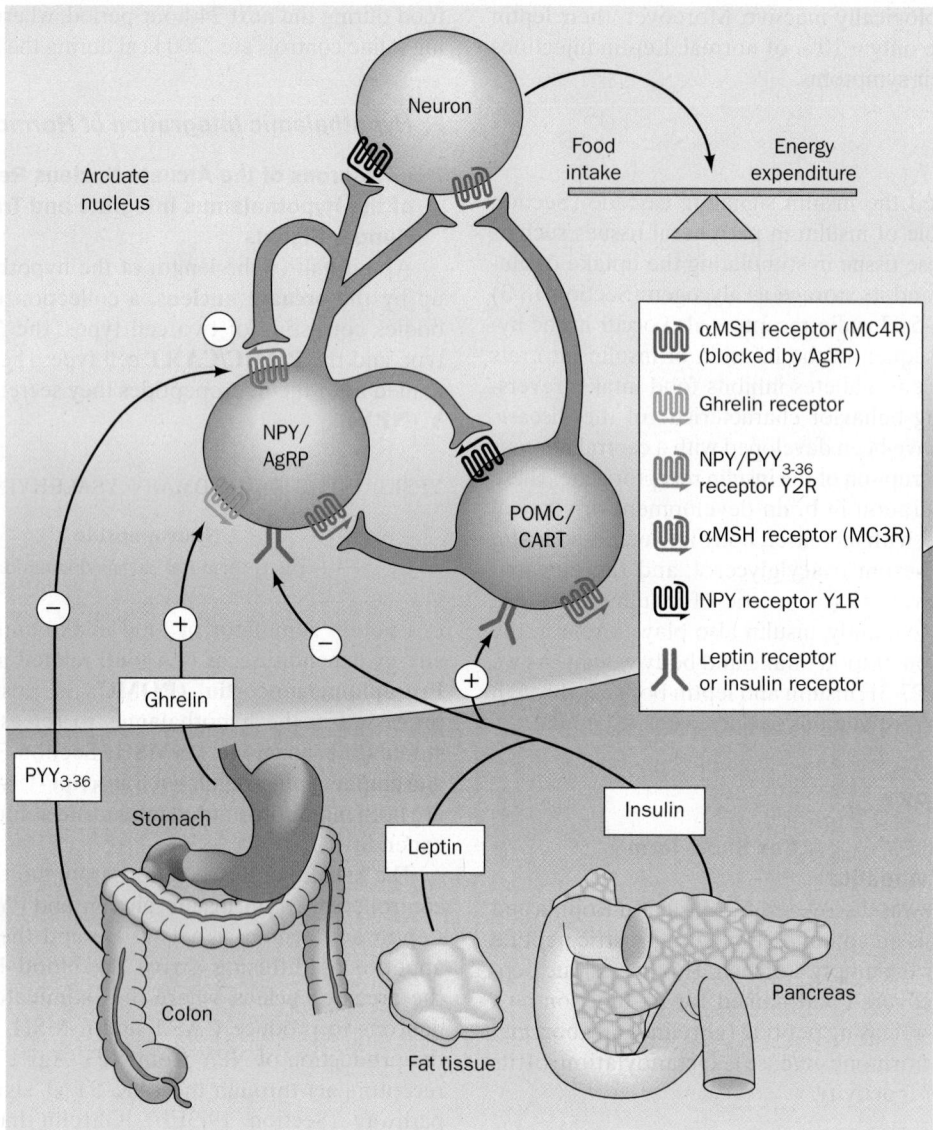

Figure 27-8 **Hormones that control the appetite.** Leptin and insulin (*bottom*) circulate in the blood at concentrations proportional to body-fat mass. They decrease appetite by inhibiting NPY/AgRP neurons (*center*) while stimulating melanocortin-producing neurons in the arcuate nucleus region of the hypothalamus. NPY and AgRP increase the appetite, and melanocortins decrease the appetite, via other neurons (*top*). Activation of NPY/AgRP-expressing neurons inhibits melanocortin-producing neurons. The gastric hormone ghrelin stimulates appetite by activating the NPY/AgRP-expressing neurons. PYY$_{3-36}$, released from the gastrointestinal tract, inhibits NPY/AgRP-expressing neurons and thereby decreases the appetite. PYY$_{3-36}$ works in part through the autoinhibitory NPY receptor subtype Y2R. [After Schwartz, M.W. and Morton, G.J., *Nature* **418,** 596 (2002).]

response to environmental stress). We have previously discussed adaptive thermogenesis in response to cold, which occurs in rodents and newborn humans through the uncoupling of oxidative phosphorylation in brown adipose tissue (Section 22-3Da). The mechanism of this thermogenesis involves the release of norepinephrine from the brain in response to cold, its binding to β-adrenergic receptors on brown adipose tissue inducing an increase in [cAMP], which in turn initiates an enzymatic phosphorylation cascade that activates hormone-sensitive triacylglycerol lipase. The resulting increase in the concentration of free fatty acids provides fuel for oxidation as well as inducing

the opening of a proton channel, called uncoupling protein-1 (UCP1) or thermogenin, in the inner mitochondrial membrane. The opening of UCP1 discharges the proton gradient across the inner mitochondrial membrane, thus uncoupling electron transport from ATP production. The energy that would otherwise have been used to drive ATP synthesis is thereby released as heat.

Although metabolic measurements in adult humans clearly demonstrate that an increase in energy intake causes an increase in metabolic rate and thermogenesis, the cause of this increase is unclear. Adult humans have little brown adipose tissue. However, skeletal muscle represents up to 40% of their total body weight and has high mitochondrial capacity. Homologs of UCP1 have been identified: **UCP2** occurs in many tissues including white adipose tissue, whereas **UCP3** occurs in brown adipose tissue, white adipose tissue, and muscle. Leptin has been shown to up-regulate UCP2. However, it has yet to be demonstrated that UCP3 in muscle participates in diet-induced thermogenesis. ATP-hydrolyzing substrate cycles such as that between fatty acids and triacylglycerol in adipose tissue (Section 27-2C) may also be involved.

H. *Did Leptin Evolve as a Thrifty Gene?*

The unusual behavior of leptin, which serves to control weight in normal-weight individuals while its concentration continues to climb without apparent effect in obese individuals, has led to the proposal that leptin evolved as a "thrifty gene." In hunter-gatherer societies, it was a distinct advantage to be able to survive intermittent famines. In order to do this, fat must be stored in adipose tissue in times of plenty, making short-term obesity advantageous. However, the accumulation of fatty acids and lipids in non-adipose tissue results in coronary artery disease, insulin resistance, and diabetes (Section 27-4B). Leptin, by directly stimulating the oxidation of fatty acids as well as inhibiting the accumulation of lipids in non-adipose tissue, is thought to protect against these diseases during short-term obesity, thereby providing an evolutionary advantage. However, in recent times in industrialized nations, the unprecedented availability of food and lack of famine has made obesity a long-term rather than a short-term condition, which is a liability rather than a benefit.

4 METABOLIC ADAPTATION

In this section we consider the body's responses to two metabolically abnormal situations: (1) starvation and (2) the disease diabetes mellitus.

A. *Starvation*

Glucose is the metabolite of choice of both brain and working muscle. Yet, the body stores less than a day's supply of carbohydrate (Table 27-1). Thus, the low blood sugar caused by even an overnight fast results, through an increase in glucagon secretion and a decrease in insulin secretion, in the mobilization of fatty acids from adipose tissue (Section 25-5). The diminished insulin level also inhibits glucose uptake by muscle tissue. Muscles therefore switch from glucose to fatty acid metabolism for energy production. The brain, however, still remains heavily dependent on glucose.

In animals, glucose cannot be synthesized from fatty acids. This is because neither pyruvate nor oxaloacetate, the precursors of glucose in gluconeogenesis (Section 23-1), can be synthesized from acetyl-CoA (the oxaloacetate in the citric acid cycle is derived from acetyl-CoA but the cyclic nature of this process requires that the oxaloacetate be consumed as fast as it is synthesized; Section 21-1A).

Table 27-1 Fuel Reserves for a Normal 70-kg Man

Fuel	Mass (kg)	Calories[a]
Tissues		
Fat (adipose triacyglycerols)	15	141,000
Protein (mainly muscle)	6	24,000
Glycogen (muscle)	0.150	600
Glycogen (liver)	0.075	300
Circulating fuels		
Glucose (extracellular fluid)	0.020	80
Free fatty acids (plasma)	0.0003	3
Triacylglycerols (plasma)	0.003	30
Total		166,000

[a]One (dieter's) Calorie = 1 kcal = 4.184 kJ.

Source: Cahill, G.F., Jr., *New Engl. J. Med.* **282**, 669 (1970).

During starvation, glucose must therefore be synthesized from the glycerol product of triacylglycerol breakdown and, more importantly, from the amino acids derived from the proteolytic degradation of proteins, the major source of which is muscle. Thus, after a 40-hour fast, gluconeogenesis supplies ~96% of the glucose produced by the liver. However, the continued breakdown of muscle during prolonged starvation would ensure that this process became irreversible since a large muscle mass is essential for an animal to move about in search of food. The organism must therefore make alternate metabolic arrangements.

After several days of starvation, gluconeogenesis has so depleted the liver's oxaloacetate supply that this organ's ability to metabolize acetyl-CoA via the citric acid cycle is greatly diminished. Rather, the liver converts the acetyl-CoA to ketone bodies (Section 25-3), which it releases into the blood. The brain gradually adapts to using ketone bodies as fuel through the synthesis of the appropriate enzymes: After a 3-day fast, only about one-third of the brain's energy requirements are satisfied by ketone bodies but after 40 days of starvation, ~70% of its energy needs are so met. The rate of muscle breakdown during prolonged starvation consequently decreases to ~25% of its rate after a several-day fast. The survival time of a starving individual is therefore much more dependent on the size of his or her fat reserves than it is on his or her muscle mass. Indeed, highly obese individuals can survive for over a year without eating (and have occasionally done so in clinically supervised weight reduction programs).

a. Caloric Restriction May Increase Longevity

Caloric restriction is a modified form of starvation whereby energy intake is reduced 30–40%, while micronutrient (vitamin and mineral) levels are maintained. Rodents kept on such a diet live up to 50% longer than rodents on normal diets and exhibit fewer of the debilitating symptoms of old age. The life spans of a large range of organisms from yeast to primates are similarly extended. Considerable research effort is being expended to determine the biochemical basis of these observations.

B. *Diabetes Mellitus*

The polypeptide hormone insulin acts mainly on muscle, liver, and adipose tissue cells to stimulate the synthesis of glycogen, fats, and proteins while inhibiting the breakdown of these metabolic fuels. In addition, insulin stimulates the uptake of glucose by most cells, with the notable exception of brain and liver cells. Together with glucagon, which has largely opposite effects, insulin acts to maintain the proper level of blood glucose.

In the disease **diabetes mellitus,** which is the third leading cause of death in the United States after heart disease and cancer, insulin either is not secreted in sufficient amounts or does not efficiently stimulate its target cells. As a consequence, blood glucose levels become so elevated that the glucose "spills over" into the urine, providing a convenient diagnostic test for the disease. Yet, despite these high blood glucose levels, cells "starve" since insulin-stimulated glucose entry into cells is impaired. Triacylglycerol hydrolysis, fatty acid oxidation, gluconeogenesis, and ketone body formation are accelerated and, in a condition termed **ketoacidosis,** ketone body levels in the blood become abnormally high. Since ketone bodies are acids, their high concentration puts a strain on the buffering capacity of the blood and on the kidney, which controls blood pH by excreting the excess H^+ into the urine (Section 27-2E). This unusually high excess H^+ excretion is accompanied by NH_4^+, Na^+, K^+, P_i, and H_2O excretion, causing severe dehydration (which compounds the dehydration resulting from the osmotic effect of the high glucose concentration in the blood; excessive thirst is a classic symptom of diabetes) and a decrease in blood volume—ultimately life-threatening situations.

There are two major forms of diabetes mellitus:

1. Insulin-dependent, type 1, or **juvenile-onset diabetes mellitus,** which most often strikes suddenly in childhood.

2. Noninsulin-dependent, type 2, or **maturity-onset diabetes mellitus,** which usually develops rather gradually after the age of 40.

a. Insulin-Dependent Diabetes Is Caused by a Deficiency of Pancreatic β Cells

In insulin-dependent (type 1) diabetes mellitus, insulin is absent or nearly so because the pancreas lacks or has defective β cells. This condition results, in genetically susceptible individuals (see below), from an autoimmune response that selectively destroys their β cells. Individuals with insulin-dependent diabetes, as Frederick Banting and George Best first demonstrated in 1921, require daily insulin injections to survive and must follow carefully balanced diet and exercise regimens. Their life spans are, nevertheless, reduced by up to one-third as a result of degenerative complications such as kidney malfunction, nerve impairment, and cardiovascular disease, which apparently arise from the imprecise metabolic control provided by periodic insulin injections. The **hyperglycemia** (high blood [glucose]) of diabetes mellitus also leads to blindness through retinal degeneration and the glucosylation of lens proteins, which causes cataracts (Fig. 27-9).

Figure 27-9 **Photo of a diabetic cataract.** The accumulation of glucose in the lens leads to swelling and precipitation of lens proteins. The resulting opacification causes blurred vision and ultimately complete loss of sight. [© Sue Ford/Photo Researchers.]

Perhaps newly developed systems that monitor blood glucose levels and continuously deliver insulin in the required amounts will rectify this situation.

The usually rapid onset of the symptoms of insulin-dependent diabetes had suggested that the autoimmune attack on the pancreatic β cells responsible for this disease is one of short duration. Typically, however, the disease "brews" for several years as the aberrantly aroused immune system slowly destroys the β cells. Only when >80% of these cells have been eliminated do the classic symptoms of diabetes suddenly emerge. Consequently, one of the most successful treatments for insulin-dependent diabetes is a β-cell transplant, a procedure that became possible with the development of relatively benign immunosuppressive drugs.

Why does the immune system attack the pancreatic β cells? It has long been known that certain alleles (genetic variants) of the **Class II major histocompatibility complex (MHC) proteins** are particularly common in insulin-dependent diabetics [MHC proteins are highly polymorphic (variable within a species) immune system components to which cell-generated antigens such as viral proteins must bind in order to be recognized as foreign; Sections 35-2Aa and 35-2E]. It is thought that autoimmunity against β cells is induced in a susceptible individual by a foreign antigen, perhaps a virus, which immunologically resembles some β cell component. The Class II MHC protein that binds this antigen does so with such tenacity that it stimulates the immune system to launch an unusually vigorous and prolonged attack on the antigen. Some of the activated immune system cells eventually make their way to the pancreas, where they initiate an attack on the β cells due to the close resemblance of the β cell component to the foreign antigen.

b. Noninsulin-Dependent Diabetes Is Characterized by Insulin Resistance as Well as Impaired Insulin Secretion

Noninsulin-dependent (type 2) diabetes mellitus **(NIDDM),** which accounts for over 90% of the diagnosed cases of diabetes and affects 18% of the population over 65 years of age, usually occurs in obese individuals with a genetic predisposition for this condition (although one that differs from that associated with insulin-dependent diabetes). These individuals may have normal or even greatly elevated insulin levels. Their symptoms arise from **insulin resistance,** an apparent lack of sensitivity to insulin in normally insulin-responsive cells.

The hyperglycemia that accompanies insulin resistance induces the pancreatic β cells to increase their production of insulin. Yet the high basal level of insulin secretion diminishes the ability of the β cells to respond to further increases in blood glucose. Consequently, the hyperglycemia and its attendant complications tend to worsen over time.

A small percentage of cases of type 2 diabetes result from mutations in the insulin receptor that affect its insulin-binding ability or tyrosine kinase activity. However, a clear genetic cause has not been identified in the vast majority of cases. It is therefore likely that many factors play a role in the development of this disease. For example, the increased insulin production resulting from overeating may eventually suppress the synthesis of insulin receptors. This hypothesis accounts for the observation that diet alone is often sufficient to control this type of diabetes.

Insulin resistance, which may precede NIDDM by as much as 10 to 20 years, appears to be caused by an interruption in the insulin signaling pathway (Section 19-4F). Gerald Shulman has proposed that this interruption is caused by a Ser/Thr kinase cascade that phosphorylates

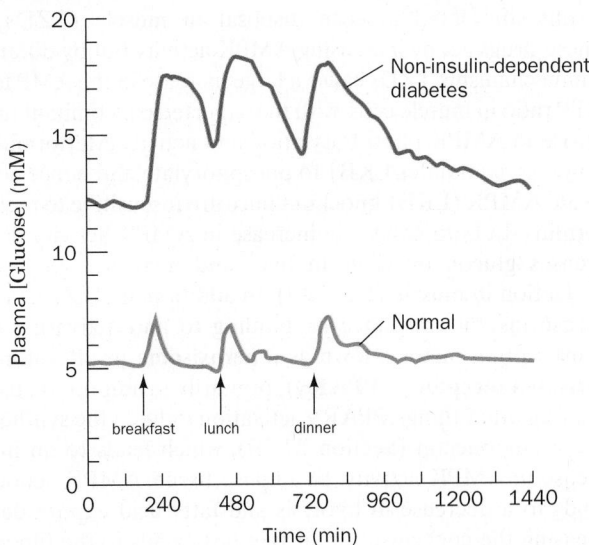

Figure 27-11 **Twenty-four-hour plasma glucose profiles in normal and noninsulin-dependent diabetic subjects.** The basal level of glucose and the peaks following meals are higher in the diabetic individuals. [After Bell, G.I., Pilkis, S.J., Weber, I.T., and Polonsky, K.S., *Annu. Rev. Physiol.* **58**, 178 (1996).]

proteins known as **insulin receptor substrates (IRSs;** Section 19-3Cg) so as to decrease their ability to be phosphorylated on their Tyr residues by activated insulin receptor. Tyrosine phosphorylation is required for IRS activation and communication with phosphoinositide 3-kinase (PI3K; Section 19-4D), which subsequently activates the translocation of GLUT4-containing vesicles to the cell surface for increased glucose transport into cells (Section 20-2Ec). The original Ser/Thr kinase cascade is triggered by the activation of an isoform of protein kinase C (PKC; Section 19-4C) caused by an increase in fatty acyl-CoA, diacylglycerol, and ceramides (Section 12-1D) resulting from elevated free fatty acids (Fig. 27-10). The failure to activate IRSs decreases the cell's response to insulin (Fig. 27-11).

c. Substances That Activate AMPK Attenuate the Symptoms of Noninsulin-Dependent Diabetes

Other treatments for noninsulin-dependent diabetes are drugs such as **metformin** and the **thiazolidinediones (TZDs),**

Metformin

A thiazolidinedione (TZD)

which decrease insulin resistance by either suppressing glucose release by the liver (metformin) or promoting

Figure 27-10 **The mechanism through which high concentrations of free fatty acids cause insulin resistance.** Elevated concentrations of free fatty acids in the blood diffuse into muscle cells where they are converted to fatty acyl-CoA, diacylglycerols, and ceramides. These lipotoxic substances activate an isoform of protein kinase C (PKC), triggering a Ser/Thr kinase cascade that results in the phosphorylation of IRS-1 and IRS-2. This phosphorylation inhibits the Tyr phosphorylation required for transmission of the insulin signal, thereby decreasing the activation of PI3K, which decreases the rate of fusion of GLUT4-containing vesicles with the plasma membrane and hence the amount of glucose entering the cell. [Modified from Shulman, G.I., *J. Clin. Invest.* **106**, 173 (2000).]

insulin-stimulated glucose disposal in muscle (TZDs). These drugs act by increasing AMPK activity but by different mechanisms. TZDs cause a large increase in the AMP to ATP ratio in muscle cells, with the expected concomitant increase in AMPK phosphorylation and activity. Metformin, however, stimulates LKB1 to phosphorylate and hence activate AMPK (LKB1 knockout mice are insensitive to metformin). In both cases, the increase in AMPK activity decreases gluconeogenesis in liver and increases glucose utilization in muscle (Fig. 27-4). In addition, the TZDs decrease insulin resistance by binding to and activating a transcription factor known as a **peroxisome proliferator–activated receptor γ (PPARγ),** primarily in adipose tissue. Among other things, PPARγ activation induces the synthesis of adiponectin (Section 27-3B), which leads to an increase in AMPK activity. In adipose tissue, AMPK action leads to a decrease in lipolysis and fatty acid export, decreasing the concentration of free fatty acids in the blood and therefore decreasing insulin resistance (see above).

Intriguingly, Ronald Evans has shown that transgenic mice expressing an activated form of PPARγ in their skeletal muscles can run around twice the distance of wild-type mice and are resistant to weight gain, even on a high fat diet. This activated PPARγ induces an increase in the number of the aerobic and hence fatty acid–oxidizing slow-twitch (Type I) muscle fibers (Section 17-3Ca) relative to the largely anaerobic and hence less energy-efficient fast-twitch (Type II) muscle fibers.

Rodent adipocytes secrete a 108-residue polypeptide hormone called **resistin.** The hormone is named for its ability to block the action of insulin on adipocytes. In fact, resistin production is decreased by TZDs, a phenomenon that led to the discovery of resistin. Overproduction of resistin was proposed to contribute to the development of noninsulin-dependent diabetes. An interesting difference between rodents and humans is that in humans, resistin is produced by macrophages, a divergence whose evolutionary and functional implications are unclear.

d. Obesity Is a Contributing Factor in Metabolic Syndrome

Metabolic syndrome is a disturbance in metabolism characterized by insulin resistance, inflammation, and a predisposition to several disorders including type 2 diabetes, hypertension, and atherosclerosis. These disorders are accompanied by an increase in coronary heart disease. Obesity, physical inactivity, and possibly genetic determinants have been implicated in its occurrence, which affects as many as 65 million people in the United States alone. Exercise, calorie/weight reduction, adiponectin, leptin, metformin, and TZDs have all been successfully used to treat metabolic syndrome. Similarly, the PPARγ agonist known as **GW1516**

GW1516

alleviates the symptoms of metabolic syndrome in obese men, probably by stimulating fatty acid oxidation.

Evans has shown that GW1516 greatly increases exercise endurance in mice, particularly when it is administered together with the AMPK agonist **5-aminoimidazole-4-carboxamide ribotide [AICAR];**

5-Aminoimidazole-4-carboxamide ribotide (AICAR)

which is also a product of histidine biosynthesis (Section 26-5Be) and an intermediate in purine ribonucleotide biosynthesis (Section 28-1A)]. This treatment mimics the effects of the expression of activated PPARγ, which suggests that the administration of GW1516 and AICAR can confer some of the benefits of exercise without actually exercising. Indeed, the World Anti-Doping Agency has placed these compounds on the list of performance-enhancing drugs that athletes are forbidden from taking.

e. DNA Chip Technology Permits the Integrated Study of Metabolic Regulation

Our ability to understand the integrated nature of metabolism and its genetic regulation in health and disease has taken a giant step forward with the advent of DNA chips (microarrays; Section 7-6B). For example, Ronald Kahn has used this technology to study the genetic basis of the metabolic abnormalities underlying both obesity and diabetes. To do so, he isolated the mRNA from the skeletal muscle of normal, diabetic, and insulin-treated diabetic mice and reverse-transcribed it to cDNA (Section 5-5Fa), which was then hybridized to oligonucleotide microarrays that represented 14,288 mouse genes. Thereby, 129 up-regulated and 106 down-regulated genes were identified in diabetic mice. Not surprisingly, the expression of the mRNAs encoding enzymes of the fatty acid β-oxidation pathway were increased, whereas those for GLUT4, glucokinase, the E1 component of the pyruvate dehydrogenase multienzyme complex, and the subunits of all four mitochondrial electron-transport chain complexes were coordinately decreased. Intriguingly, only about half of these changes in gene expression could be reversed by insulin treatment. Thus, the post-genomic era will almost certainly witness an explosion in our knowledge of metabolic regulation that should yield major health benefits. Nevertheless, our ability to sensibly interpret this huge influx of information may prove to be the greatest challenge.

CHAPTER SUMMARY

1 Major Pathways and Strategies of Energy Metabolism: A Summary The complex network of processes involved in energy metabolism are distributed among different compartments within cells and in different organs of the body. These processes function to generate ATP "on demand," to generate and store glucose, triacylglycerols, and proteins in times of plenty for use when needed, and to keep the concentration of glucose in the blood at the proper level for use by organs such as the brain, whose sole fuel source, under normal conditions, is glucose. The major energy metabolism pathways include glycolysis, glycogen degradation and synthesis, gluconeogenesis, the pentose phosphate pathway, and triacylglycerol and fatty acid synthesis, which are cytosolically based, and fatty acid oxidation, the citric acid cycle, and oxidative phosphorylation, which are confined to the mitochondrion. Amino acid degradation occurs, in part, in both compartments. The mediated membrane transport of metabolites therefore also plays an essential metabolic role.

2 Organ Specialization The brain normally consumes large amounts of glucose. Muscle, under intense ATP demand such as in sprinting, degrades glucose and glycogen anaerobically, thereby producing lactate, which is exported via the blood to the liver for reconversion to glucose through gluconeogenesis. During moderate activity, muscle generates ATP by oxidizing glucose from glycogen, fatty acids, and ketone bodies completely to CO_2 and H_2O via the citric acid cycle and oxidative phosphorylation. Adipose tissue stores triacylglycerols and releases fatty acids into the bloodstream in response to the organism's metabolic needs. These metabolic needs are communicated to adipose tissue by means of the hormones insulin, which indicates a fed state in which storage is appropriate, and glucagon, epinephrine, and norepinephrine, which signal a need for fatty acid release to provide fuel for other tissues. The liver, the body's central metabolic clearinghouse, maintains blood glucose concentrations by storing glucose as glycogen in times of plenty and releasing glucose in times of need both by glycogen breakdown and by gluconeogenesis. It also converts fatty acids to ketone bodies for use by peripheral tissues. During a fast, it breaks down amino acids resulting from protein degradation to metabolic intermediates that can be used to generate glucose. The kidney filters out urea from the blood, recovers important metabolites, and maintains pH balance. To do so, glutamine is broken down to produce NH_4^+ for H^+ excretion. The resulting α-ketoglutarate product is converted to CO_2 to resupply HCO_3^- to the blood to maintain its buffering capacity. During starvation, the kidney uses the α-ketoglutarate from glutamine breakdown for gluconeogenesis.

3 Metabolic Homeostasis: Regulation of Appetite, Energy Expenditure, and Body Weight AMP-dependent protein kinase (AMPK), the cell's fuel gauge, senses the cell's need for ATP and activates metabolic breakdown pathways while inhibiting biosynthetic pathways. Adiponectin, an adipocyte hormone that increases insulin sensitivity, acts by activating AMPK. Appetite is suppressed by the actions of leptin, a hormone produced by adipose tissue, insulin, produced by the β cells of the pancreas, and PYY_{3-36}, produced by the gastrointestinal tract, which act in the hypothalamus to inhibit the secretion of neuropeptide Y (NPY) and stimulate the secretion of α-MSH and CART. This decreases the appetite and hence food intake. Ghrelin, a hormone secreted by the empty stomach, opposes the actions of leptin, insulin, and PYY_{3-36}, stimulating appetite and food intake. Leptin also acts in peripheral tissues to stimulate energy expenditure by fatty acid oxidation and thermogenesis.

4 Metabolic Adaptation During prolonged starvation, the brain slowly adapts from the use of glucose as its sole fuel source to the use of ketone bodies, thereby shifting the metabolic burden from protein breakdown to fat breakdown. Diabetes mellitus is a disease in which insulin either is not secreted or does not efficiently stimulate its target tissues, leading to high concentrations of glucose in the blood and urine. Cells "starve" in the midst of plenty since they cannot absorb blood glucose and their hormonal signals remain those of starvation. Abnormally high production of ketone bodies is one of the most dangerous effects of uncontrolled diabetes. Metabolic syndrome is caused by obesity, physical inactivity, and possibly genetic determinants. It symptoms can be relieved by substances that activate AMPK.

REFERENCES

Chapters 17 to 26 of this text.

Batterham, R.L., et al., Gut hormone PYY_{3-36} physiologically inhibits food intake, *Nature* **418,** 650–654 (2002).

Brüning, J.C., Gautam, D., Burks, D.J., Gillette, J., Schubert, M., Orban, P.C., Klein, R., Krone, W., Müller-Weiland, D., and Kahn, C.R., Role of brain insulin receptor in control of body weight and reproduction, *Science* **289,** 2122–2125 (2000).

Carling, D., The AMP-activated protein kinase cascade—a unifying system for energy control, *Trends Biochem. Sci.* **29,** 18–24 (2004).

Coll, A.P., Farooqi, I.S., and O'Rahilly, S., The hormonal control of food intake, *Cell* **129,** 251–262 (2007).

Evans, J.L., Goldfine, I.D., Maddux, B.A., and Grodsky, G.M., Oxidative stress and stress-activated signaling pathways: a unifying hypothesis of type 2 diabetes, *Endocrine Rev.* **23,** 599–622 (2002).

Flier, J.S., Obesity wars: Molecular progress confronts an expanding epidemic, *Cell* **116,** 337–350 (2004).

Kadowaki, T. and Yamauchi, T., Adiponectin and adiponectin receptors, *Endocrine Rev.* **26,** 439–451 (2005).

Lowell, B.B. and Spiegelman, B.M., Towards a molecular understanding of adaptive thermogenesis, *Nature* **404,** 652–660 (2000).

Montague, C.T., et al., Congenital leptin deficiency is associated with severe early-onset obesity in humans, *Nature* **387,** 903–908 (1997).

Moreno-Aliaga, M.J., Marti, A., García-Foncillas, J. and Martínes, J.A., DNA hybridization arrays: a powerful technology for nutritional and obesity research, *Br. J. Nutr.* **86,** 119–122 (2001).

Nakar, V.A., et al., AMPK and PPARγ agonists are exercise mimetics, *Cell* **134**, 405–415 (2008).

Nakazato, M., Murakami, N., Date, Y., Kojima, M., Matsuo, H., Kangawa, K., and Matsukara, S., A role for ghrelin in the central regulation of feeding, *Nature* **409**, 194–198 (2001).

Schwartz, M.W. and Morton, G.J., Keeping hunger at bay, *Nature* **418**, 595–597 (2002).

Shaw, R.J., Lamia, K.A., Vasquez, D., Koo, S.-H., Bardeesy, N., DePinho, R.A., Montminy, M., and Cantley, L.C., The kinase LKB1 mediates glucose homeostasis in liver and therapeutic effects of metformin, *Science* **310**, 1642–1646 (2005).

Shulman, G.I., Cellular mechanisms of insulin resistance, *J. Clin. Invest.* **106**, 171–176 (2000).

Towler, M.C. and Hardie, D.G., AMP-Activated protein kinase in metabolic control and insulin signaling, *Circ. Res.* **100**, 328–341 (2007).

Tshöp, M., Smiley, D.L., and Heiman, M.L., Ghrelin induces adiposity in rodents, *Nature* **407**, 908–913 (2000).

Type 2 Diabetes, *Science* **307**, 369–387 (2005). [A series of informative articles on the origins of type 2 diabetes and its relationship to obesity.]

Unger, R.H., Leptin physiology: a second look, *Regul. Pept.* **92**, 87–95 (2000).

Wang, Y.-X., Zhang, C.L., Yu, R.T., Cho, H.K., Nelson, M.C., Bayuga-Ocampo, C.R., Ham, J., Kang, H., and Evans, R.M., Regulation of muscle fiber type and running endurance by PPARγ, *PLoS Biol.* **2**, e294 (2004).

Yechoor, V.K., Patti, M.-E., Saccone, R., and Kahn, C.R., Coordinated patterns of gene expression for substrate and energy metabolism in skeletal muscle of diabetic mice, *Proc. Natl. Acad. Sci.* **99**, 10587–10592 (2002).

Zhang, F., et al., Crystal structure of the *obese* protein leptin-E100, *Nature* **387**, 206–209 (1997).

Zick, Y., Insulin resistance: a phosphorylation-based uncoupling of insulin signaling, *Trends Cell Biol.* **11**, 437–441 (2001).

PROBLEMS

1. Describe the metabolic effects of liver failure.

2. What is the basis of the hypothesis that athletes' muscles are more heavily buffered than those of normal individuals?

3. Experienced runners know that it is poor practice to ingest large amounts of glucose prior to running a long-distance race such as a marathon. What is the metabolic basis of this apparent paradox?

4. Explain why urea output is vastly decreased during starvation.

5. Explain why people survive longer by total fasting than on a diet consisting only of carbohydrates.

6. Explain why the breath of an untreated diabetic smells of acetone.

7. Among the many eat-all-you-want-and-lose-weight diets that have been popular for a time is one that eliminates all carbohydrates but permits the consumption of all the protein and fat desired. Would such a diet be effective? (*Hint:* Individuals on such a diet often complain that they have bad breath.)

8. Pancreatic β cells express a receptor for fatty acids. Fatty acid binding to this protein appears to stimulate insulin secretion. (a) Does this phenomenon make metabolic sense? (b) Fatty acids appear to stimulate insulin secretion to a much greater extent when glucose is also present. Why is this significant?

9. High concentrations of free fatty acids in the blood are known to cause insulin resistance in muscle, but only after 5 hours. This suggests that a metabolite of these fatty acids may be responsible for this phenomenon. It is also known that an isoform of protein kinase C is activated during the process and that high concentrations of free fatty acids result in intramuscular accumulation of triacylglycerols. With this information, review the mechanism of activation of PKC and the pathway of triacylglycerol biosynthesis and suggest a metabolite that may be responsible for PKC activation.

10. Discuss, in molecular terms, how physical inactivity might lead to insulin resistance.

Nucleotide Metabolism

CHAPTER 28

Nucleotides, as we have seen, are biologically ubiquitous substances that participate in nearly all biochemical processes: They are the monomeric units of DNA and RNA; the hydrolysis of ATP and GTP drives many free energy–requiring processes; the levels of ATP, ADP, and AMP regulate numerous metabolic pathways; cAMP and cGMP mediate hormonal signals; and NAD^+, $NADP^+$, FMN, FAD, and coenzyme A are essential coenzymes in a great variety of enzymatic reactions. The importance of nucleotides in cellular metabolism is indicated by the observation that nearly all cells can synthesize them both *de novo* (anew) and from the degradation products of nucleic acids. In this chapter, we consider the nature of these biosynthetic pathways. In doing so, we shall examine how they are regulated and the consequences of their blockade, both by genetic defects and through the administration of chemotherapeutic agents. We then discuss how nucleotides are degraded. Finally, we outline the biosynthesis of the nucleotide coenzymes.

1 SYNTHESIS OF PURINE RIBONUCLEOTIDES

In this section we commence our considerations of how nucleic acids and their components are synthesized by describing the synthesis of purine ribonucleotides. In 1948, John Buchanan obtained the first clues as to how this process occurs *de novo* by feeding a variety of isotopically labeled compounds to pigeons and chemically determining the positions of the labeled atoms in their excreted **uric acid** (a purine).

Uric acid

He used birds in these experiments because they excrete waste nitrogen almost entirely as uric acid, a water-insoluble and therefore easily isolated substance. The results of his studies, which are summarized in Fig. 28-1, demonstrated that N1 of purines arises from the amine group of aspartate; C2 and C8 originate from formate; N3 and N9 are contributed by the amide group of glutamine; C4, C5, and N7 are derived from glycine (strongly suggesting that this molecule is wholly incorporated into the purine ring); and C6 comes from HCO_3^- (CO_2).

Figure 28-1 The biosynthetic origins of purine ring atoms. Note that C4, C5, and N7 come from a single glycine molecule but each of the other atoms is derived from an independent precursor.

The actual pathway by which these precursors are incorporated into the purine ring, the subject of Section 28-1A, was elucidated in subsequent investigations performed largely by Buchanan and by G. Robert Greenberg. These investigations showed that the initially synthesized purine derivative is **inosine monophosphate (IMP),**

Inosine monophosphate (IMP)

the nucleotide of the base **hypoxanthine.** AMP and GMP are subsequently synthesized from this intermediate via separate pathways (Section 28-1B). Thus, contrary to naive expectation, purines are initially formed as ribonucleotides rather than as free bases. Additional studies have demonstrated that such widely divergent organisms as *E. coli,* yeast, pigeons, and humans have virtually identical pathways for the biosynthesis of purine nucleotides, thereby further demonstrating the biochemical unity of life.

A. *Synthesis of Inosine Monophosphate*

IMP is synthesized in a pathway comprising 11 reactions (Fig. 28-2):

1. Activation of ribose-5-phosphate. The starting material for purine biosynthesis is α-D-ribose-5-phosphate (R5P), a product of the pentose phosphate pathway (Section 23-4). In the first step of *de novo* purine biosynthesis, **ribose phosphate pyrophosphokinase** (also known as **phosphoribosylpyrophosphate synthetase**) activates R5P by reacting it with ATP to form **5-phosphoribosyl-α-pyrophosphate (PRPP).** This reaction, which occurs via the nucleophilic attack of the R5P's C1—H group on the P_β of ATP, is unusual in that a pyrophosphoryl group is directly transferred from ATP to C1 of R5P and that the product has the α anomeric configuration. PRPP is also a precursor in the biosynthesis of pyrimidines (Section 28-2A) and the amino acids tryptophan and histidine (Section 26-5Bd,e). Thus, as is expected for an enzyme at such an important biosynthetic crossroads, the activity of ribose phosphate pyrophosphokinase varies with the concentrations of numerous metabolites, including PP_i and 2,3-bisphosphoglycerate, which are activators, and ADP and GDP, which are mixed inhibitors (Section 14-3C). The regulation of purine nucleotide biosynthesis is further discussed in Section 28-1C.

2. Acquisition of purine atom N9. Amidophosphoribosyltransferase (alternatively, **glutamine PRPP aminotransferase** or **PurF;** the latter being named for the *E. coli* gene encoding it, *purF*) catalyzes the displacement of PRPP's

pyrophosphate group by glutamine's amide nitrogen to yield **β-5-phosphoribosylamine (PRA).** This is the first reaction in the pathway that is unique to *de novo* purine biosynthesis (and hence some sources refer to it as the first reaction of the pathway, which is then said to consist of 10 reactions). This process occurs in two consecutive reactions that take place on separate active sites on the enzyme:

1. Glutamine + $H_2O \rightarrow$ glutamic acid + NH_3

2. NH_3 + PRPP $\rightarrow$ PRA + PP_i

Step 1 is catalyzed by a member of the N-terminal nucleophile (Ntn) amidotransferase family (Section 26-5Aa). Step 2 occurs with inversion of configuration about ribose C1 and hence establishes the anomeric form of the future nucleotide. The NH_3 passes between the two active sites through a 20-Å-long tunnel that is lined with conserved

Figure 28-2 (*Opposite*) **The metabolic pathway for the *de novo* biosynthesis of IMP.** The purine residue is built up on a ribose ring in 11 enzymatically catalyzed reactions. The X-ray structures for all enzymes are shown to the outside of the corresponding reaction arrow. The peptide chains of monomeric enzymes are colored in rainbow order from their N-termini (*blue*) to their C-termini (*red*). The oligomeric enzymes, all of which consist of identical polypeptide chains, are viewed along a rotation axis with their various chains differently colored. Bound ligands are shown in space-filling form. Enzyme 1, determined by Sine Larsen, University of Copenhagen, Denmark, is a D_3 hexamer from *B. subtilis* that binds α,β-methylene-ADP at its catalytic (*red*) and allosteric (*blue*) sites; PDBid 1DKU. Enzyme 2, determined by Janet Smith, Purdue University, is a D_2 tetramer from *B. subtilis* that binds GMP (*blue*), ADP (*red*), and a [4Fe–4S] cluster (*orange,* which appears to have a regulatory rather than a redox function); PDBid 1AO0. Enzymes 3 and 6, both from *E. coli,* were determined by JoAnne Stubbe, MIT, and Steven Ealick, Cornell University; PDBids 1GSO and 1CLI. Enzyme 4, from *E. coli,* determined by Robert Almassy, Agouron Pharmaceuticals, San Diego, California, binds GAR (*cyan*) and 5-deazatetrahydrofolate (*red*); PDBid 1CDE. Enzyme 5, from *Thermatoga maritima, was* determined by Ian Wilson, Scripps Research Institute, La Jolla, California; PDBid 1VK3. Reaction 7, in *E. coli,* is catalyzed by two sequentially acting enzymes, Class I PurE (*above*) and PurK (*below*). Class I PurE, determined by JoAnne Stubbe, MIT, and Steven Ealick, Cornell University, is a D_4 octamer that binds AIR (*red*); PDBid 1D7A. PurK, determined by JoAnne Stubbe, MIT, and Hazel Holden, University of Wisconsin, is a C_2 dimer that binds ADP (*red*); PDBid 1B6S. Enzyme 8, from yeast, was determined by Victor Lamzin, Academy of Sciences, Moscow, Russia, and Keith Wilson, EMBL, Hamburg, Germany; PDBid 1A48. Enzyme 9, from *Thermatoga maritima,* determined by Todd Yeates, UCLA, is a D_2 tetramer; PDBid 1C3U. Reactions 10 and 11 in chicken are catalyzed by a bifunctional enzyme that was determined by Stephen Benkovic, Pennsylvania State University, and Ian Wilson, The Scripps Research Institute, La Jolla, California. It forms a C_2 dimer shown with its AICAR transformylase function above and its IMP cyclohydrolase function, which binds GMP (*purple*), below; PDBid 1G8M. ♻ **See the Animated Figures**

α-D-Ribose-5-phosphate (R5P)

ATP → ribose phosphate pyrophosphokinase, **1**, AMP

5-Phosphoribosyl-α-pyrophosphate (PRPP)

Glutamine + H_2O → amidophosphoribosyl transferase, **2**, Glutamate + PP_i

β-5-Phosphoribosylamine (PRA)

Glycine + ATP → GAR synthetase, **3**, ADP + P_i

Glycinamide ribotide (GAR)

N^{10}-Formyl-THF → GAR transformylase, **4**, THF

Formylglycinamide ribotide (FGAR)

ATP + Glutamine + H_2O → FGAM synthetase, **5**, ADP + Glutamate + P_i

Formylglycinamidine ribotide (FGAM)

ATP → AIR synthetase, **6**, ADP + P_i

6 →

5-Aminoimidazole ribotide (AIR)

ATP + HCO_3^- → AIR carboxylase, **7**, ADP + P_i

Carboxyaminoimidazole ribotide (CAIR)

Aspartate + ATP → SAICAR synthetase, **8**, ADP + P_i

5-Aminoimidazole-4-(N-succinylocarboxamide) ribotide (SAICAR)

Fumarate ← adenylosuccinate lyase, **9**

5-Aminoimidazole-4-carboxamide ribotide (AICAR)

N^{10}-Formyl-THF → AICAR transformylase, **10**, THF

5-Formaminoimidazole-4-carboxamide ribotide (FAICAR)

H_2O ← IMP cyclohydrolase, **11**

Inosine monophosphate (IMP)

nonpolar residues that lack hydrogen bonding groups and hence do not impede the diffusion of the NH_3 [we have seen that NH_3 generated by glutamine hydrolysis is similarly channeled to the active site that uses it in carbamoyl phosphate synthetase (Section 26-2Aa) and glutamate synthetase (Section 26-5Aa)]. These reactions, which are driven to completion by the subsequent hydrolysis of the released PP_i, constitute the pathway's flux-generating step. Not surprisingly, therefore, amidophosphoribosyltransferase is subject to feedback inhibition by purine nucleotides (Section 28-1C).

3. Acquisition of purine atoms C4, C5, and N7. Glycine's carboxyl group forms an amide with the amino group of PRA, yielding **glycinamide ribotide (GAR)** in a reaction, catalyzed by **GAR synthetase (PurD),** that occurs via the intermediate phosphorylation of glycine's carboxyl group. The reaction, which is reversible despite its concomitant hydrolysis of ATP to ADP + P_i, is the only step of the purine biosynthesis pathway in which more than one purine ring atom is acquired. The observation that PRA is chemically unstable (it is hydrolyzed to R5P and NH_3 with a half-life of 5 s at 37°C) suggests that GAR synthetase and amidophosphoribosyltransferase associate in a way that channels PRA between them. Indeed, a sterically and electrostatically plausible model of such a complex has been built based on the X-ray structures of these two enzymes.

4. Acquisition of purine atom C8. GAR's primary α-amino group is formylated by **GAR transformylase (PurN)** to yield **formylglycinamide ribotide (FGAR).** The formyl donor in this reaction is N^{10}-formyltetrahydrofolate (N^{10}-formyl-THF), a cofactor that transfers C_1 units from such donors as serine, glycine, and formate to various acceptors in biosynthetic reactions (Section 26-4D). Structural and enzymological studies indicate that the reaction proceeds via the nucleophilic attack of the GAR primary amine group on the formyl carbon of N^{10}-formyl-THF to yield a tetrahedral intermediate.

5. Acquisition of purine atom N3. The amide amino group of a second glutamine is transferred to the growing purine ring to form **formylglycinamidine ribotide (FGAM).** This reaction, which is catalyzed by **FGAM synthetase (PurL),** is driven by the coupled hydrolysis of ATP to ADP + P_i. It is thought to proceed by the mechanism diagrammed in Fig. 28-3. Here the oxygen of the FGAR isoamide form reacts with ATP to yield a phosphoryl ester intermediate. This intermediate then reacts with NH_3 (the glutamine amide nitrogen as labilized through the transient formation of an enzyme thioester) to form a tetrahedral adduct. The adduct then eliminates P_i to yield the imine product, FGAM. Such reactions, in which a carboxamide oxygen is replaced by an imino group, are common in the biosynthesis of nucleotides. For example, Reaction 6 of this pathway and the reactions converting IMP to AMP (Section 28-1B) and UTP to CTP (Section 28-2B) follow similar mechanisms, that is, conversion of a carboxamide oxygen to a phosphoryl ester that is nucleophilically

Figure 28-3   The proposed mechanism of formylglycinamide ribotide (FGAM) synthetase. The glutaminase domain of the enzyme contains an active site Cys residue that catalyzes the release of NH_3 with the transient formation of an enzyme thioester (not shown) whose hydrolysis produces glutamate. The isoamide form of FGAR is phosphorylated by ATP and then reacts with "NH_3" to form a tetrahedral intermediate whose collapse yields FGAM + P_i.

attacked by an amine nitrogen atom to yield a tetrahedral adduct that, in turn, expels P_i to form product.

6. Formation of the purine imidazole ring. The purine imidazole ring is closed in an ATP-requiring intramolecular condensation that yields **5-aminoimidazole ribotide (AIR)** in a reaction catalyzed by **AIR synthetase (PurM).**

The aromatization of the imidazole ring is facilitated by the tautomeric shift of the reactant from its imine to its enamine form.

7. Acquisition of C6. In higher eukaryotes, purine C6 is introduced as HCO_3^- (CO_2) in an ATP-dependent reaction catalyzed by **AIR carboxylase** that yields **carboxyaminoimidazole ribotide (CAIR)** and ADP + P_i. However, in yeast, plants, and most prokaryotes (including *E. coli*), this overall reaction occurs in two steps that are mediated by separate enzymatic activities: **PurK** and **Class I PurE.**

PurK catalyzes the ATP-dependent carboxylation of AIR to yield N^5-**CAIR,** which Class I PurE rearranges to yield CAIR. Class I PurE is homologous to AIR carboxylase, which is therefore also called **Class II PurE.** Class I PurE alone can catalyze the AIR carboxylase reaction but since its K_M for HCO_3^- is 110 mM, it requires an unphysiologically high ($\sim$100 mM) HCO_3^- concentration to do so at a significant rate. However, the action of PurK decreases the HCO_3^- concentration required for the PurE-catalyzed reaction by >1000-fold, presumably through the ATP-driven formation of carbonyl phosphate, as is also postulated to occur in the carbamoyl phosphate synthetase reaction (Section 26-2A). The observation that N^5-CAIR is chemically unstable (it decomposes to AIR with a half-life of 15 s at pH 7.5 and 25°C) suggests that N^5-CAIR is channeled between PurK and Class I PurE. In fact, in yeast and plants, the N-terminus of Class I PurE is fused to the C-terminus of PurK. However, in *E. coli*, these two enzymatic activities occur on separate proteins for which there is no evidence of association.

8. Acquisition of N1. Purine atom N1 is contributed by aspartate in an amide-forming condensation reaction yielding **5-aminoimidazole-4-(*N*-succinylocarboxamide) ribotide (SAICAR)** that is catalyzed by **SAICAR synthetase (PurC).** The reaction, which is driven by the hydrolysis of ATP to ADP+P_i, chemically resembles Reaction 3.

9. Elimination of fumarate. SAICAR is cleaved with the release of fumarate, yielding **5-aminoimidazole-4-carboxamide ribotide (AICAR)** in a reaction catalyzed by **adenylosuccinate lyase (PurB).** Reactions 8 and 9 chemically

resemble the reactions in the urea cycle in which citrulline is aminated to form arginine (Sections 26-2C and 26-2D). In both pathways, aspartate's amino group is transferred to an acceptor through an ATP-driven coupling reaction followed by the elimination of the aspartate carbon skeleton as fumarate. In plants and microorganisms, AICAR is also formed in the biosynthesis of histidine (Section 26-5Be) but since in that process the AICAR is derived from ATP, it provides for no net purine biosynthesis.

10. Acquisition of C2. The final purine ring atom is acquired through formylation by N^{10}-formyltetrahydrofolate, yielding **5-formaminoimidazole-4-carboxamide ribotide (FAICAR)** in a reaction catalyzed by **AICAR transformylase (PurH).** In bacteria, this reaction and that of Reaction 4 are indirectly inhibited by sulfonamides, which, it will be recalled, prevent the synthesis of folate by competing with its *p*-aminobenzoate component (Section 26-4D). Animals, including humans, must acquire folate through the diet, since they are incapable of synthesizing it. They are therefore unaffected by sulfonamides. The antibiotic properties of sulfonamides are therefore largely a result of their inhibition of nucleic acid biosynthesis in susceptible bacteria.

11. Cyclization to form IMP. The final reaction in the pathway, ring closure to form IMP, occurs through the elimination of water as catalyzed by **IMP cyclohydrolase (PurJ).** In contrast to Reaction 6, the cyclization that forms the imidazole ring, this reaction does not entail ATP hydrolysis.

In animals, the activities catalyzing Reactions 3, 4, and 6, Reactions 7 and 8, and Reactions 10 and 11 occur on single polypeptides. The intermediate products of these multifunctional enzymes are not readily released to the medium but are channeled to the succeeding enzymatic activities of the pathway, thereby increasing the overall rates of these multistep processes and protecting the intermediates from degradation by other cellular enzymes. We have previously seen, for example, that the formation of acetyl-CoA from pyruvate takes place on the pyruvate dehydrogenase multienzyme complex, which contains three enzymes catalyzing five consecutive reactions (Section 21-2A); that all seven enzymatic activities catalyzing fatty acid synthesis in animals occur on a single protein molecule (Section 25-4Ca,b); and that the multifunctional enzymes carbamoyl phosphate synthase I (Section 26-2Aa), glutamate synthase (Section 26-5Aa), tryptophan synthase (Section 26-5Bd), and amidophosphoribosyltransferase (see above) pass reactive intermediate products between their active sites via protein tunnels. It is becoming increasingly apparent that the association of functionally related enzymes is a widespread phenomenon.

B. Synthesis of Adenine and Guanine Ribonucleotides

IMP does not accumulate in the cell but is rapidly converted to AMP and GMP. AMP, which differs from IMP only in the replacement of its 6-keto group by an amino

Figure 28-4 **IMP is converted to AMP or GMP in separate two-reaction pathways.** The X-ray structures for all of the enzymes catalyzing these reactions are shown to the outside of the corresponding reaction arrow. The X-ray structures for these homooligomers are shown as described in the legend to Fig. 28-2. Adenylosuccinate synthetase from *E. coli,* determined by Herbert Fromm and Richard Honzatko, Iowa State University, is a C_2 dimer in complex with IMP (*green*), GDP (*red*), and **hadacidin** (*magenta;* a competitive inhibitor of aspartate); PDBid

1GIM. Adenylosuccinate lyase, from *Thermatoga maritima,* determined by Todd Yeates, UCLA, is a D_2 tetramer; PDBid 1C3U. IMP dehydrogenase from Chinese hamsters, determined by Keith Wilson, Vertex Pharmaceuticals, Cambridge, Massachusetts, is a C_4 tetramer in complex with oxidized IMP (*red*) and MPA (*purple*); PDBid 1JR1. GMP synthetase from *E. coli,* determined by Janet Smith, Purdue University, is a D_2 tetramer in complex with AMP (*red*), pyrophosphate (*blue*), and citrate (*purple*); PDBid 1GPM.

group, is synthesized in a two-reaction pathway (Fig. 28-4, *left*). In the first reaction, aspartate's amino group is linked to IMP in a reaction driven by the hydrolysis of GTP to GDP + P_i to yield **adenylosuccinate.** In the second reaction, **adenylosuccinate lyase** eliminates fumarate from adenylosuccinate to form AMP. This enzyme also catalyzes Reaction 9 of the IMP pathway (Fig. 28-2).

GMP is also synthesized from IMP in a two-reaction pathway (Fig. 28-4, *right*). In the first reaction, **IMP dehydrogenase** catalyzes the NAD^+-dependent oxidation of IMP to form **xanthosine monophosphate (XMP;** the ribonucleotide of the base **xanthine**). XMP is then converted to GMP by the replacement of its 2-keto group with glutamine's amide nitrogen in a reaction driven by the hydrolysis of ATP to AMP + PP_i (and subsequently to 2 P_i).

IMP dehydrogenase, a homotetramer of 514-residue subunits, was incubated with IMP, NAD^+, and the fungally produced inhibitor **mycophenolic acid (MPA).**

Mycophenolic acid (MPA)

The X-ray structure of the resulting complex, determined by Keith Wilson, reveals that the enzyme had bound MPA together with a reaction intermediate in which IMP atom

C2 had become covalently linked to the Cys 331 S atom and then dehydrogenated by NAD^+ to yield a thioimidate ester:

Enzyme–product thioimidate ester

The mutagenic replacement of Cys 331 by Ala inactivates the enzyme. These observations strongly support a catalytic mechanism in which the Cys 331 thiol group nucleophilically attacks IMP's C2 atom, followed by hydride transfer to NAD^+ to yield the above covalently bound intermediate, which is subsequently hydrolyzed to yield XMP. The MPA binds to the enzyme with its bicyclic ring stacked on the purine ring (as would be expected for NAD^+'s nicotinamide ring) and with its phenolic hydroxyl group in the proposed hydrolytic water site. This blocks the hydrolysis of the thioimidate ester, thereby inactivating the enzyme.

IMP dehydrogenase activity is essential to the immune response (Section 35-2) because it is required by the immune system cells known as B and T lymphocytes to generate the guanosine nucleotides they need to proliferate. Moreover, certain cancer cells have increased IMP dehydrogenase activity. Hence, IMP dehydrogenase is a target for both immunosuppressive therapy and cancer chemotherapy. Indeed, MPA is in clinical use to prevent the rejection of transplanted kidneys.

a. Nucleoside Diphosphates and Triphosphates Are Synthesized by the Phosphorylation of Nucleoside Monophosphates

In order to participate in nucleic acid synthesis, nucleoside monophosphates must first be converted to the corresponding nucleoside triphosphates. In the first of the two sequential phosphorylation reactions that do so, nucleoside diphosphates are synthesized from the corresponding nucleoside monophosphates by base-specific **nucleoside monophosphate kinases.** For example, adenylate kinase (Section 17-4Fd) catalyzes the phosphorylation of AMP to ADP:

$$AMP + ATP \rightleftharpoons 2ADP$$

Similarly, GDP is produced by a guanine-specific enzyme:

$$GMP + ATP \rightleftharpoons GDP + ADP$$

These nucleoside monophosphate kinases do not discriminate between ribose and deoxyribose in the substrate.

Nucleoside diphosphates are converted to the corresponding triphosphates by **nucleoside diphosphate kinase;** for instance,

$$ATP + GDP \rightleftharpoons ADP + GTP$$

Although this reaction is written with ATP as the phosphoryl donor and GDP as the acceptor, nucleoside diphosphate kinase is nonspecific as to the bases on either of its

substrates and as to whether their sugar residues are ribose or deoxyribose. The reaction occurs via a Ping Pong mechanism in which the substrate NTP phosphorylates an enzyme His residue, which in turn, phosphorylates the substrate NDP. The phosphoglycerate mutase reaction of glycolysis also has a phospho-His intermediate (Section 17-2H). The nucleoside diphosphate kinase reaction, as might be expected from the nearly identical structures of its substrates and products, normally operates close to equilibrium ($\Delta G \approx 0$). ADP is, of course, also converted to ATP by a variety of energy-releasing reactions such as those of glycolysis and oxidative phosphorylation. Indeed, it is these reactions that ultimately drive the foregoing kinase reactions.

C. *Regulation of Purine Nucleotide Biosynthesis*

The pathways involved in nucleic acid metabolism are tightly regulated, as is evidenced, for example, by the increased rates of nucleotide synthesis during cell proliferation. In fact, the pathways synthesizing IMP, ATP, and GTP are individually regulated in most cells so as not only to control the total amounts of purine nucleotides produced but also to coordinate the relative amounts of ATP and GTP. This control network is diagrammed in Fig. 28-5.

The IMP pathway is regulated at its first two reactions: those catalyzing the synthesis of PRPP and 5-phosphoribo-

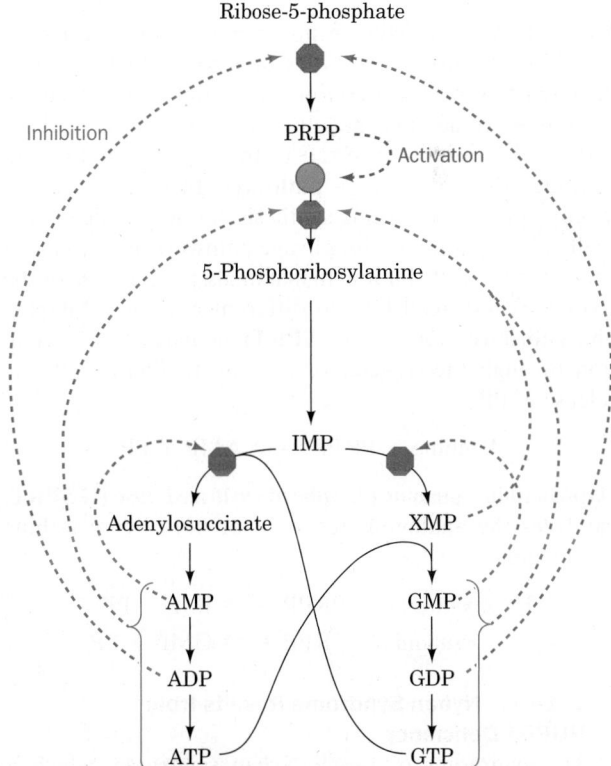

Figure 28-5 **Control network for the purine biosynthesis pathway.** Red octagons and green dots indicate control points. Feedback inhibition is indicated by dashed red arrows and feedforward activation is represented by dashed green arrows. **See the Animated Figures**

sylamine. We have already seen that ribose phosphate pyrophosphokinase, the enzyme catalyzing Reaction 1 of the IMP pathway, is inhibited by both ADP and GDP (Section 28-1A). Amidophosphoribosyltransferase, the enzyme catalyzing the first committed step of the IMP pathway (Reaction 2), is likewise subject to feedback inhibition. In this case, however, the enzyme binds ATP, ADP, and AMP at one inhibitory site and GTP, GDP, and GMP at another. *The rate of IMP production is consequently independently but synergistically controlled by the levels of adenine nucleotides and guanine nucleotides.* Moreover, amidophosphoribosyltransferase is allosterically stimulated by PRPP (feedforward activation).

A second level of regulation occurs immediately below the branch point leading from IMP to AMP and GMP (Fig. 28-4). AMP and GMP are each competitive inhibitors of IMP in their own synthesis, so that excessive buildup of these products is impeded. In addition, the synthesis rates of adenine and guanine nucleotides are coordinated because GTP powers the synthesis of AMP from IMP, whereas ATP powers the synthesis of GMP from IMP. This reciprocity serves to balance the production of AMP and GMP (which are required in roughly equal amounts in nucleic acid biosynthesis): *The rate of synthesis of GMP increases with [ATP], whereas that of AMP increases with [GTP].*

D. *Salvage of Purines*

Most cells have an active turnover of many of their nucleic acids (particularly some types of RNA) which, through degradative processes described in Section 28-4A, result in the release of adenine, guanine, and hypoxanthine. These free purines are reconverted to their corresponding nucleotides through **salvage pathways.** In contrast to the *de novo* purine nucleotide synthesis pathway, which is virtually identical in all cells, salvage pathways are diverse in character and distribution. In mammals, purines are, for the most part, salvaged by two different enzymes. **Adenine phosphoribosyltransferase (APRT)** mediates AMP formation through the transfer of adenine to PRPP with the release of PP_i:

$$\text{Adenine} + \text{PRPP} \rightleftharpoons \text{AMP} + PP_i$$

Hypoxanthine–guanine phosphoribosyltransferase (HGPRT) catalyzes the analogous reaction for both hypoxanthine and guanine:

$$\text{Hypoxanthine} + \text{PRPP} \rightleftharpoons \text{IMP} + PP_i$$
$$\text{Guanine} + \text{PRPP} \rightleftharpoons \text{GMP} + PP_i$$

a. Lesch–Nyhan Syndrome Results from HGPRT Deficiency

The symptoms of **Lesch–Nyhan syndrome,** which is caused by a severe HGPRT deficiency, indicate that purine salvage reactions have functions other than conservation of the energy required for *de novo* purine biosynthesis. This sex-linked congenital defect (affects almost only males) results in excessive uric acid production (uric acid is a purine

degradation product; Section 28-4A) and neurological abnormalities such as spasticity, mental retardation, and highly aggressive and destructive behavior, including a bizarre compulsion toward self-mutilation. For example, many children with Lesch–Nyhan syndrome have such an irresistible urge to bite their lips and fingers that they must be restrained. If the restraints are removed, communicative patients will plead that the restraints be replaced even as they attempt to injure themselves.

The excessive uric acid production in patients with Lesch–Nyhan syndrome is readily explained. The lack of HGPRT activity leads to an accumulation of the PRPP that would normally be used in the salvage of hypoxanthine and guanine. The excess PRPP activates amidophosphoribosyltransferase (which catalyzes Reaction 2 of the IMP biosynthesis pathway; Fig. 28-2), thereby greatly increasing the rate of synthesis of purine nucleotides and consequently that of their degradation product, uric acid. Yet the physiological basis of the associated neurological abnormalities remains obscure. That a defect in a single enzyme can cause such profound but well-defined behavioral changes nevertheless has important psychiatric implications.

2 SYNTHESIS OF PYRIMIDINE RIBONUCLEOTIDES

The biosynthesis of pyrimidines is a simpler process than that of purines. Isotopic labeling experiments have shown that atoms N1, C4, C5, and C6 of the pyrimidine ring are all derived from aspartic acid, C2 arises from HCO_3^-, and N3 is contributed by glutamine (Fig. 28-6). In this section we discuss the pathways for pyrimidine ribonucleotide biosynthesis and how these processes are regulated.

A. *Synthesis of UMP*

The major breakthrough in the determination of the pathway for the *de novo* biosynthesis of pyrimidine ribonucleotides was the observation that mutants of the bread mold *Neurospora crassa,* which are unable to synthesize pyrimidines and therefore require both cytosine and uracil in their growth medium, grow normally when supplied instead with the pyrimidine **orotic acid** (uracil-6-carboxylic acid).

Orotic acid (uracil-6-carboxylic acid)

This observation led to the elucidation of the following six-reaction pathway for the biosynthesis of UMP (Fig. 28-7). Note that, in contrast to the case for purine nucleotides, the pyrimidine ring is coupled to the ribose-5-phosphate moiety *after* the ring has been synthesized.

Glutamine amide → N_3 $\overset{\displaystyle 4}{\underset{\displaystyle}{C}}$ C ← Aspartate

HCO_3^- → $\underset{\displaystyle N}{C}$ $\overset{\displaystyle 2}{\underset{\displaystyle 1}{}}$ C

Figure 28-6   **The biosynthetic origins of pyrimidine ring atoms.**

1. Synthesis of carbamoyl phosphate. The first reaction of pyrimidine biosynthesis is the synthesis of **carbamoyl phosphate** from HCO_3^- and the amide nitrogen of glutamine by the cytosolic enzyme **carbamoyl phosphate synthetase II (CPS II).** This reaction is unusual in that it does not use biotin and consumes two molecules of ATP: One provides a phosphate group and the other energizes the reaction. We have previously discussed the synthesis of carbamoyl phosphate in connection with the formation of arginine (Section 26-2A). The carbamoyl phosphate that is used to

Figure 28-7   Metabolic pathway for the *de novo* synthesis of UMP. The pathway consists of six enzymatically catalyzed reactions. Note that, in contrast to the case for purine biosynthesis (Fig. 28-2), the pyrimidine ring is formed before its attachment to a ribose ring. The X-ray structures for the enzymes are drawn as described in the legend to Fig. 28-2. Enzyme 1, from *E. coli,* determined by Hazel Holden, University of Wisconsin, is an $\alpha_4\beta_4$ heterooctamer with D_2 symmetry, whose large subunits each bind two ADPNPs (*red*) and one ornithine (*purple*); PDBid 1BXR. Enzyme 2, from *E. coli,* determined by William Lipscomb, Harvard University, is a c_6r_6 heterododecamer with D_3 symmetry, whose regulatory (*r*) subunits each bind a CTP (*green*); PDBid

5AT1. Enzyme 3 from *E. coli,* determined by Hazel Holden, University of Wisconsin, is a C_2 dimer that binds carbamoyl aspartate (*purple*) in one subunit and orotate (*red*) in the other; PDBid 1J79. Enzyme 4, from humans, determined by Jon Clardy, Cornell University, is a monomer that binds orotate (*yellow*), FMN (*magenta*), and A77 1726 (*green*); PDBid 1D3H. Enzyme 5, from *Salmonella typhimurium,* determined by James Sacchettini, Albert Einstein College of Medicine, is a C_2 dimer that binds orotate (*orange*) and PRPP (*purple*); PDBid 1OPR. Enzyme 6, from *B. subtilis,* determined by Steven Ealick, Cornell University, is a C_2 dimer that binds UMP (*green*); PDBid 1DBT. See the Animated Figures synthesize arginine via the urea cycle is synthesized by a separate mitochondrial enzyme, **carbamoyl phosphate synthetase I (CPS I),** which uses ammonia as its nitrogen source. Prokaryotes only have one carbamoyl phosphate synthetase, which supplies both pyrimidine and arginine biosynthesis and utilizes glutamine. This latter enzyme, as we have seen, contains three different active sites that are connected by a remarkable 96-Å-long tunnel through which intermediate products diffuse (Fig. 26-9).

The pyrimidine biosynthetic pathway is a target for antiparasitic drugs. For example, the parasitic protozoan *Toxoplasma gondii*, which infects most mammals, causes **toxoplasmosis,** a disease whose complications include blindness, neurological dysfunction, and death in immunocompromised individuals (e.g., those with AIDS). Most parasites have evolved to take advantage of nutrients supplied by their hosts. However, *T. gondii* is unable to meet its needs exclusively through nucleotide salvage pathways and retains the ability to synthesize uracil *de novo*. Drugs that target the parasite's carbamoyl phosphate synthetase II (an enzyme whose structure and kinetics distinguish it from its mammalian counterpart) could therefore prevent *T. gondii* growth. Moreover, there is evidence that *T. gondii* strains that have been engineered to lack carbamoyl phosphate synthetase II are avirulent and could be useful as vaccines in humans and livestock.

2. Synthesis of carbamoyl aspartate. Condensation of carbamoyl phosphate with aspartate to form **carbamoyl aspartate** is catalyzed by **aspartate transcarbamylase (ATCase).** This reaction, the pathway's flux-generating step, occurs without need of ATP because carbamoyl phosphate is intrinsically activated. The structure and regulation of *E. coli* ATCase is discussed in Section 13-4.

3. Ring closure to form dihydroorotate. The third reaction of the pathway was elucidated by Arthur Kornberg following his observation that microorganisms made to utilize orotic acid as a carbon source first reduce it to **dihydroorotate.** The reaction forming the pyrimidine ring yields dihydroorotate in an intramolecular condensation catalyzed by the zinc metalloenzyme **dihydroorotase.**

4. Oxidation of dihydroorotate. Dihydroorotate is irreversibly oxidized to orotate by **dihydroorotate dehydrogenase (DHODH).** The eukaryotic enzyme, which contains FMN, is an integral membrane protein that is located on the outer surface of the inner mitochondrial membrane, where ubiquinone supplies its oxidizing power. The other five enzymes of pyrimidine nucleotide biosynthesis are cytosolic in animal cells. Many bacterial dihydroorotate dehydrogenases are NAD^+-linked flavoproteins that contain FMN, FAD, and a [2Fe–2S] cluster. These enzymes normally function degradatively, that is, in the direction orotate → dihydroorotate, thereby permitting these bacteria to metabolize orotate and accounting for Kornberg's observation. The reaction mediated by eukaryotic DHODH involves two redox steps, as is indicated in Fig. 28-8. The X-ray structure of human DHODH in complex with orotate, determined by Jon Clardy, reveals that the pyrimidine ring of orotate is stacked over the FMN's flavin ring with

the orotate C6 and FMN N5 separated by 3.6 Å, a distance that is compatible with direct hydride transfer between these two centers. A tunnel leads from the opposite side of the flavin ring to a hydrophobic region on the enzyme surface. The enzyme presumably binds to the mitochondrial membrane surface via this hydrophobic patch, thereby permitting ubiquinone, which readily diffuses within the mitochondrial membrane, to approach and reoxidize the enzyme's bound $FMNH_2$. In the X-ray structure, this tunnel contains a tightly bound molecule named **A77 1726,** which is the primary metabolite of **leflunomide** (trade name **Arava**),

A77 1726

Leflunomide

a compound that is in clinical use for the treatment of rheumatoid arthritis. A77 1726 attenuates this autoimmune disease by blocking pyrimidine biosynthesis in T lymphocytes, thereby reducing their inappropriate proliferation. However, A77 1726 does not inhibit bacterial DHODHs.

5. Acquisition of the ribose phosphate moiety. Orotate reacts with PRPP to yield **orotidine-5′-monophosphate (OMP)** in a reaction catalyzed by **orotate phosphoribosyltransferase** and driven by hydrolysis of the eliminated PP_i. This reaction fixes the anomeric form of pyrimidine nucleotides in the β configuration. Orotate phosphoribosyltransferase also acts to salvage other pyrimidine bases, such as uracil and cytosine, by converting them to their corresponding nucleotides. Although the various phosphoribosyltransferases, including HGPRT, exhibit little sequence similarity, their X-ray structures indicate that they contain a common structural core that resembles the dinucleotide binding fold (Section 8-3Bi) but lacks one of its β strands.

6. Decarboxylation to form UMP. The final reaction of the pathway is the decarboxylation of OMP by **OMP decarboxylase (ODCase)** to form UMP. ODCase enhances the rate (k_{cat}/K_M) of OMP decarboxylation by a factor of 2×10^{23} over that of the uncatalyzed reaction, making it the most catalytically proficient enzyme known. Yet ODCase has no cofactors to help stabilize the reaction's putative carbanion intermediate. How is it able to do so? The X-ray structure, by Steven Ealick, of ODCase from *B. subtilis* in complex with UMP indicates that a bound OMP's C6

Figure 28-8 **Reactions catalyzed by eukaryotic dihydroorotate dehydrogenase.** The reaction is initiated by the enzyme-mediated abstraction of a proton from C5 of dihydroorotate followed by the direct hydride transfer from C6 of dihydroorotate to N5 of FMN to yield orotate and FMNH⁻, which may then be

protonated to yield FMNH₂. The FMNH₂ (or FMNH⁻) then reacts with coenzyme Q acquired from the inner mitochondrial membrane to regenerate the enzyme in its FMN form and yield coenzyme QH₂, which then re-enters the inner mitochondrial membrane.

carboxyl group that is coplanar with its pyrimidine ring would be in close proximity to the side chains of both Asp 60 and Lys 62. Ealick has therefore proposed a mechanism (Fig. 28-9) in which the electrostatic interactions between the closely spaced carboxyl groups of OMP and Asp 60 destabilize OMP's ground state. This destabilization would be reduced in the transition state by the shift of OMP's negative charge from its carboxyl group toward C6, where it would be stabilized by the adjacent positively charged side chain of Lys 62. This side chain is also proposed to protonate the fragmenting C—C bond when it becomes sufficiently basic to accept the proton, thus avoiding the formation of a high-energy carbanion intermediate. The

unfavorable electrostatic interaction between OMP and Asp 60 occurs because the enzyme tightly binds OMP through extensive interactions with its other functional groups. Indeed, the removal of OMP's phosphate group, which is quite distant from the C6 carboxyl group, decreases the catalytic reaction's k_{cat}/K_M by a factor of 7×10^7, thus providing a striking example of how binding energy can be applied to catalysis (preferential transition state binding).

In bacteria, the six enzyme activities mediating UMP biosynthesis occur on independent proteins (Fig. 28-7). In animals, however, as Mary Ellen Jones demonstrated, the first three enzymatic activities of the pathway, *carbamoyl*

Figure 28-9 **Proposed catalytic mechanism for OMP decarboxylase.** [After Appleby, T.C., Kinsland, C., Begley, T.P., and Ealick, S.E., *Proc. Natl. Acad. Sci.* **97**, 2005 (2000).]

Figure 28-10 **Synthesis of CTP from UTP.**

phosphate synthetase II, ATCase, and *d*ihydroorotase, occur on a single 2225-residue polypeptide chain known as **CAD,** which forms 1400-kD homohexamers. Similarly, Reactions 5 and 6 of the animal pyrimidine pathway are catalyzed by a single 480-residue polypeptide named **UMP synthase** that forms 102-kD homodimers.

B. *Synthesis of UTP and CTP*

The synthesis of UTP from UMP is analogous to the synthesis of purine nucleotide triphosphates (Section 28-1B). The process occurs by the sequential actions of a nucleoside monophosphate kinase and nucleoside diphosphate kinase:

$$\text{UMP} + \text{ATP} \rightleftharpoons \text{UDP} + \text{ADP}$$
$$\text{UDP} + \text{ATP} \rightleftharpoons \text{UTP} + \text{ADP}$$

CTP is formed by amination of UTP by **CTP synthetase** (Fig. 28-10). In animals, the amino group is donated by glutamine, whereas in bacteria it is supplied directly by ammonia.

C. *Regulation of Pyrimidine Nucleotide Biosynthesis*

In bacteria, the pyrimidine biosynthesis pathway is primarily regulated at Reaction 2, the ATCase reaction (Fig. 28-11*a*). In *E. coli,* control is exerted there through the allosteric stimulation of ATCase by ATP and its inhibition by CTP (Section 13-4). In many bacteria, however, UTP is the major ATCase inhibitor.

In animals, ATCase is not a regulatory enzyme. Rather, pyrimidine biosynthesis is controlled by the activity of carbamoyl phosphate synthetase II, which is inhibited by UDP and UTP and activated by ATP and PRPP (Fig. 28-11*b*). A second level of control in the mammalian pathway occurs at OMP decarboxylase, for which UMP and to a lesser extent CMP are competitive inhibitors. In all organisms, the rate of OMP production varies with the availability of its precursor, PRPP. The PRPP level, it will be recalled, depends on the activity of ribose phosphate pyrophosphokinase, which is inhibited by ADP and GDP (Section 28-1A).

a. **Orotic Aciduria Results from an Inherited Enzyme Deficiency**

Orotic aciduria, an inherited human disease, is characterized by the excretion of large amounts of orotic acid in the urine, retarded growth, and severe anemia. It results from a deficiency in the bifunctional enzyme catalyzing

Figure 28-11 **Regulation of pyrimidine biosynthesis.** The control networks are shown for (*a*) *E. coli* and (*b*) animals. Red octagons and green dots indicate control points. Feedback inhibition is represented by dashed red arrows and activation is indicated by dashed green arrows. 🔊 **See the Animated Figures**

Reactions 5 and 6 of pyrimidine nucleotide biosynthesis. Consideration of the biochemistry of this situation led to its effective treatment: the administration of uridine and/or cytidine. The UMP formed through the phosphorylation of these nucleosides, besides replacing that normally synthesized, inhibits carbamoyl phosphate synthetase II so as to attenuate the rate of orotic acid synthesis. Few other genetic deficiencies in pyrimidine nucleotide biosynthesis are known in humans, presumably because most such defects are lethal *in utero.*

3 FORMATION OF DEOXYRIBONUCLEOTIDES

DNA differs chemically from RNA in two major respects: (1) its nucleotides contain 2'-deoxyribose residues rather than ribose residues, and (2) it contains the base thymine (5-methyluracil) rather than uracil. In this section we consider the biosynthesis of these DNA components.

A. *Production of Deoxyribose Residues*

Deoxyribonucleotides are synthesized from their corresponding ribonucleotides by the reduction of their C2' position rather than by their de novo synthesis from deoxyribose-containing precursors.

NDP

dNDP

This pathway was established through Irwin Rose's study of how rats metabolize cytidine that is ^{14}C-labeled in both its base and ribose components. The dCMP recovered from the rats' DNA had the same labeling ratio in its cytosine and deoxyribose residues as had the original cytidine, indicating that the DNA's components remained linked during DNA synthesis. If the cytosine and the ribose residues had become separated, dilution of the labeled cytosine and ribose residues with unlabeled residues, which are present in rat tissues in different amounts, would have altered this ratio.

The enzymes that catalyze the formation of deoxyribonucleotides by the reduction of the corresponding ribonucleotides are named **ribonucleotide reductases (RNRs).** Three classes of RNRs are known that differ in their

substrates (NDP or NTP), the cofactors they employ, and in the way they obtain reducing equivalents (see below). Class I and II RNRs are widely distributed among prokaryotes; some species have a Class I RNR, whereas other, sometimes related species have a Class II RNR. However, all eukaryotes except a few unicellular species have Class I RNRs. Class III RNRs occur in prokaryotes that can grow anaerobically. (Class III RNRs are O_2-sensitive whereas Class I RNRs require O_2 for activation; see below.) In fact, *E. coli*, which can grow both aerobically and anaerobically, expresses a Class I and a Class III RNR. In what follows, we shall mainly discuss the mechanism of Class I RNRs but end with a consideration of the evolutionary relationships among the different classes of RNRs.

a. Class I Ribonucleotide Reductase: Structure and Mechanism

The *E. coli* Class I RNR, as Peter Reichard demonstrated, is mainly present *in vitro* as a heterotetramer that can be decomposed to two catalytically inactive homodimers, R1$_2$ (761-residue subunits) and R2$_2$ (375-residue subunits), which together form the enzyme's two active sites (Fig. 28-12*a*). Each R1 subunit contains a substrate binding site as well as three independent effector binding sites that control both the enzyme's catalytic activity and its substrate specificity (see below). R1's catalytic residues include several redox-active thiol groups.

The X-ray structure of R2$_2$ (Fig. 28-12*b*), determined by Hans Eklund, reveals that each of its subunits contains a novel binuclear Fe(III) prosthetic group whose two Fe(III) ions are bridged by both an O^{2-} ion (a μ-oxo bridge) and the carboxyl group of Glu 115 (Fig. 28-12*c*). Each Fe(III) is further liganded by two carboxyl O atoms from Asp or Glu residues, a His N_δ atom, and a water molecule. The Fe(III) complex interacts with Tyr 122 to form, as EPR measurements indicate, an unusual tyrosyl free radical (TyrO·) that is 5 Å from the closest Fe atom and is buried 10 Å beneath the surface of the protein, where it is out of contact with solvent and any oxidizable side chain [tyrosyl radicals have also been observed in cytochrome *c* oxidase (Section 22-2C5c) and in Photosystem II (Section 24-2Cd)].

The *E. coli* RNR is inhibited by **hydroxyurea,** which specifically quenches (destroys) the tyrosyl radical, and by **8-hydroxyquinoline,** which chelates Fe^{3+} ions.

Hydroxyurea   **8-Hydroxyquinoline**

Mammalian RNRs have similar characteristics to the *E. coli* enzyme. Indeed, hydroxyurea is in clinical use as an antitumor agent.

If *E. coli* RNR is incubated with [3'-^{3}H]UDP, a small but reproducible fraction of the ^{3}H is released as ^{3}H$_2$O. This observation, together with kinetic, spectroscopic, and site-directed mutagenesis studies, led JoAnne Stubbe to

Figure 28-12 (*Opposite*) **Class I ribonucleotide reductase from *E. coli.*** (*a*) A schematic diagram of its quaternary structure. The enzyme consists of two pairs of identical subunits, R1$_2$ and R2$_2$. Each R2 subunit contains a binuclear Fe(III) complex that generates a phenoxy radical at its Tyr 122. The R1 subunits each contain three different allosteric effector sites and five catalytically important Cys residues. The enzyme's two active sites are located near the interface between neighboring R1 and R2 subunits. (*b*) The X-ray structure of R2$_2$ as viewed perpendicular to its 2-fold axis with the dimer's longest dimension in the horizontal plane. One subunit of the homodimeric protein is shown in blue and the other in yellow. The Fe(III) ions of its binuclear Fe complexes are represented by orange spheres and the radical-harboring Tyr 122 side chains are drawn in space-filling form with their C and O atoms green and red. Note that each subunit consists mainly of a bundle of eight unusually long helices. (*c*) The binuclear Fe(III) complex of R2.

Each Fe(III) ion is octahedrally coordinated by a His N$_\delta$ atom and five O atoms, including those of the O^{2-} ion and the Glu carboxyl group that bridge the two Fe(III) ions. (*d*) The X-ray structure of the R1 dimer, each subunit of which is in complex with the 20-residue C-terminal peptide of R2 together with GDP in the active site and dTTP in the specificity site. The ATP analog AMPPNP bound in the activity site of the closely similar complex of R1 with the 20-residue peptide and AMPPNP has been superimposed on this structure. The structure is viewed along its 2-fold axis with its two subunits lavender and light green, the two R2 peptides cyan and magenta, and the GDP, dTTP, and ATP shown in space-filling form colored according to atom type (C green, N blue, O red, and P gold). [Parts *b* and *d* based on X-ray structures by Hans Eklund, Swedish University of Agricultural Sciences, Uppsala, Sweden. PDBids (*b*) 1RIB and (*d*) 3R1R and 4R1R.] 🖱 **See the Animated Figures and Interactive Exercise 28** formulate the following catalytic mechanism for *E. coli* RNR (Fig. 28-13):

1. RNR's free radical (X·) abstracts an H atom from C3′ of the substrate in the reaction's rate-determining step.

2 and 3. Acid-catalyzed cleavage of the C2′—OH bond releases H_2O to yield a radical cation intermediate. The radical mediates the stabilization of the C2′ cation by the 3′—OH group's unshared electron pair, thereby accounting for the radical's catalytic role.

4. The radical cation intermediate is reduced by the enzyme's redox-active sulfhydryl pair to yield a 3′-deoxynucleotide radical and a protein disulfide group.

5. The 3′ radical reabstracts an H atom from the protein to yield the product deoxynucleoside diphosphate and restore the enzyme to its radical state. A small fraction of the originally abstracted H atom exchanges with solvent before it can be replaced, thus accounting for the release of ^3H on reduction of [3′-^3H]UDP.

The Tyr 122 radical in R2 is too far away (>10 Å) from the enzyme's catalytic site to abstract an electron directly from the substrate. Evidently, the protein mediates electron transfer from this tyrosyl radical to some other group (X· in Fig. 28-13) that is in close proximity to the substrate C3′—H group. Site-directed mutagenesis studies suggest that Cys 439 of R1, in its thiyl radical form (—S·), is the most plausible candidate for X· (which makes RNR the only enzyme in which a Cys residue is known to reduce a carbohydrate substrate). Similar studies suggest that Cys 225 and Cys 462 of R1 form the redox-active sulfhydryl pair that directly reduces substrate. Moreover, the resulting disulfide bond is subsequently reduced to regenerate active enzyme via disulfide interchange with Cys 754 and Cys 759 on R1, which are apparently positioned to accept electrons from external reducing agents (see below). Thus, each R1 subunit contains at least five Cys residues that chemically participate in nucleotide reduction.

These observations are confirmed by the X-ray structure of R1 in complex with R2's 20-residue C-terminal

Figure 28-13 **Enzymatic mechanism of ribonucleotide reductase.** The reaction occurs via a free radical–mediated process in which reducing equivalents are supplied by the formation of an enzyme disulfide bond. [After Stubbe, J.A., *Biol. Chem.* **265,** 5330 (1990).]

polypeptide (R1 does not crystallize satisfactorily in the absence of this polypeptide), also determined by Eklund (Fig. 28-12*d*). The central domain of the three-domain R1 monomer consists of a novel 10-stranded α/β barrel that is formed by the antiparallel joining of two topologically similar half-barrels, each comprising five parallel β strands connected by four α helices. As with the similar 8-stranded α/β barrels that form the active sites of numerous enzymes (Section 8-3Bh), R1's active site Cys residues (439, 225, 462) are located in the mouth of the 10-stranded α/β barrel.

The two R1 Cys residues, 754 and 759, that are implicated in the regeneration of the active enzyme are components of R1's C-terminal segment, which is not visible in the X-ray structure of R1 and is presumably disordered. This observation supports the hypothesis that this C-terminal segment acts to flexibly shuttle reducing equivalents from the enzyme surface to its active site.

b. Radical Generation in Class I RNR Requires the Presence of O_2

One of the most remarkable aspects of Class I RNR is its ability to stabilize its normally highly reactive TyrO· radical (its half-life is 4 days in the protein vs milliseconds in solution). Yet quenching the radical, say, by hydroxyurea, inactivates the enzyme. How, then, is the radical generated in the first place? The radical may be restored *in vitro* by simply treating the inactive enzyme with Fe(II) and a reducing agent in the presence of O_2.

This is a four-electron reduction of O_2 in which the reducing agent that supplies the electron represented by e^- may be ascorbate or even excess Fe^{2+}.

c. The Inability of Oxidized RNR to Bind Substrate Serves an Important Protective Function

Comparison of the X-ray structures of reduced R1 (in which the redox-active Cys 225 and Cys 462 residues are in their SH forms) with that of oxidized R1 (in which Cys 225 and Cys 462 are disulfide linked) reveals that Cys 462 in reduced R1 has rotated away from its position in oxidized R1 to become buried in a hydrophobic pocket, whereas Cys 225 moves into the region formerly occupied by Cys 462. The distance between the formerly disulfide-linked S atoms thereby increases from 2.0 to 5.7 Å. These movements are accompanied by small shifts of the surrounding polypeptide chain. Oxidized RNR does not bind substrate because its R1 Cys

225 would prevent the binding of substrate through steric interference of its S atom with the substrate NDP's O2′ atom.

The inability of oxidized RNR to bind substrate has functional significance. In the absence of substrate, the enzyme's free radical is stored in the interior of the R2 protein, close to its dinuclear iron center. When substrate is bound, the radical is presumably transferred to it via a series of protein side chains in both R2 and R1. If the substrate is unable to properly react after accepting this free radical, as would be the case if RNR were in its oxidized state, this could result in the destruction of the substrate and/or the enzyme. Indeed, the mutation of the redox-active Cys 225 to Ser results in an enzyme that permits the formation of the substrate radical (Fig. 28-13); however, since the mutant enzyme is incapable of reducing it, the substrate radical instead decomposes followed by the release of its base and phosphate moieties. More importantly, a transient peptide radical forms, which cleaves and inactivates the R1 polypeptide chain while consuming the radical and thereby inactivating R2. Thus, an important role of the enzyme is to control the release of the radical's powerful oxidizing capability. It does so in part by preventing the binding of substrate while the enzyme is in its oxidized form.

d. Ribonucleotide Reductase Is Regulated by Effector-Induced Oligomerization

The synthesis of the four dNTPs in the amounts required for DNA synthesis is accomplished through feedback control. The maintenance of the proper intracellular ratios of dNTPs is essential for normal growth. Indeed, *a deficiency of any dNTP is lethal, whereas an excess is mutagenic because the probability that a given dNTP will be erroneously incorporated into a growing DNA strand increases with its concentration relative to those of the other dNTPs.*

The activities of both *E. coli* and mammalian Class I RNRs are allosterically responsive to the concentrations of various (d)NTPs. Thus, as Reichard has shown, ATP induces the reduction of CDP and UDP; dTTP induces the reduction of GDP and inhibits the reduction of CDP and UDP; dGTP induces the reduction of ADP and, in mammals but not *E. coli*, inhibits the reduction of CDP and UDP; and dATP inhibits the reduction of all NDPs.

Barry Cooperman has shown that the catalytic activity of mouse RNR varies with its state of oligomerization (that is, RNR is a morpheein; Section 26-4Ac), which in turn is governed by the binding of nucleotide effectors to three independent allosteric sites on R1: (1) the specificity site, which binds ATP, dATP, dGTP, and dTTP; (2) the activity site, which binds ATP and dATP; and (3) the hexamerization site, which binds only ATP. On the basis of molecular mass, ligand binding, and activity studies on mouse RNR, Cooperman formulated a model that quantitatively accounts for the allosteric regulation of Class I RNR. It has the following features (Fig. 28-14*a*):

1. The binding of ATP, dATP, dGTP, or dTTP to the specificity site induces the catalytically inactive R1 monomers to form a catalytically active dimer, $R1_2$.

Figure 28-14 **Ribonucleotide reductase regulation.** (*a*) A model for the allosteric regulation of Class I RNR via its oligomerization. States shown in green have high activity and those shown in red have little or no activity. R2 has been omitted for simplicity. [After Kashlan, O.B., Scott, C.P., Lear, J.D., and Cooperman, B.S., *Biochemistry* **41,** 461 (2002).] (*b*) The X-ray structure of the R1 hexamer, which has D_3 symmetry, in complex with AMPPNP as viewed along its 3-fold axis. Each of its three dimers are differently colored (the X-ray structure of a dimer is shown in Fig. 28-12*d*). The AMPPNP, which binds to the enzyme's activity sites, is drawn in space-filling form with C green, N blue, O red, and P gold. The black arrows point along the R1 dimers' 2-fold axes and indicate the probable docking sites for the binding of R2 dimers. (*c*) The R1 · AMPPNP hexamer as viewed along the vertical 2-fold axis in Part *b*. [Parts *b* and *c* based on an X-ray structure by Hans Eklund, Swedish University of Agricultural Sciences, Uppsala, Sweden. PDBid 3R1R.]

2. The binding of dATP or ATP to the activity site causes the dimers to form catalytically active tetramers, R1$_{4a}$, that slowly but reversibly change conformation to a catalytically inactive state, R1$_{4b}$.

3. The binding of ATP to the hexamerization site induces the tetramers to further aggregate to form catalytically active hexamers, R1$_6$, RNR's major active form.

The concentration of ATP in a cell is such that, *in vivo,* R1 is almost entirely in its tetrameric or hexameric forms.

As a consequence, ATP couples the overall rate of DNA synthesis to the cell's energy state.

The specificity and activity sites have been located in X-ray structures of *E. coli* R1 (Fig. 28-12*d*); the hexamerization site has not yet been identified. The R1 hexamer had, in fact, been previously observed in the X-ray structures of R1 (Fig. 28-14*b,c*), but the interactions between its contacting dimers are so tenuous that it was assumed that they are merely artifacts of crystallization with no physiological significance. Yet, since the activity site is located at this contact site, it now seems likely that its binding of (d)ATP induces R1 oligomerization through local conformational changes.

The foregoing model has, for simplicity, neglected the presence of R2 subunits although, of course, R1 and R2 must be present in equimolar amounts in the active enzyme. Presumably, the R1 and R2 dimers bind to one another such that their 2-fold axes coincide. The lack of space on the inside of the R1 hexamer dictates that the R2 dimers must contact the R1 dimers from outside the hexamer (Fig. 28-14*b*).

dCTP is not an effector of RNR. This is presumably because the intracellular balance between dCTP and dTTP is not controlled by RNR but, rather, is maintained by **deoxycytidine deaminase,** which converts dCTP to dUMP, the precursor of dTTP. This enzyme is activated by dCTP and inhibited by dTTP.

e. Thioredoxin and Glutaredoxin Are Class I Ribonucleotide Reductase's Physiological Reducing Agents

The final step in the RNR catalytic cycle is the reduction of the enzyme's newly formed disulfide bond to reform its redox-active sulfhydryl pair. Dithiols such as that of 2-mercaptoethanol (Section 7-1B) can serve as the reducing agent for this process *in vitro* through a disulfide interchange reaction. One of the enzyme's physiological reducing agents, however, is **thioredoxin (Trx),** a ubiquitous monomeric 105-residue protein that has a pair of closely proximal redox-active Cys residues, Cys 32 and Cys 35 (we have previously encountered thioredoxin in our study of the light-induced activation of the Calvin cycle; Section 24-3B). Thioredoxin reduces oxidized RNR via disulfide interchange.

The X-ray structure of reduced *E. coli* Trx (Fig. 28-15) reveals that the side chain of the redox-active Cys 32 is exposed on the protein's surface, where it is available for oxidation. Oxidized thioredoxin is, in turn, reduced by

Figure 28-15   X-ray structure of human thioredoxin in its reduced (sulfhydryl) state. The 105-residue polypeptide chain is drawn in ribbon form colored in rainbow order from its N-terminus (*blue*) to its C-terminus (*red*). The side chains of the redox-active residues, Cys 32 and Cys 35, are shown in space-filling form with C green and S yellow. This structure closely resembles those of the homologous a and a′ domains of protein disulfide isomerase (PDI, Fig. 9-16). [Based on an X-ray structure by William Montfort, University of Arizona. PDBid 1ERT.]

NADPH in a reaction mediated by the flavoprotein **thioredoxin reductase.** NADPH therefore serves as the terminal reducing agent in the RNR-mediated reduction of NDPs to dNDPs (Fig. 28-16).

The existence of a viable *E. coli* mutant devoid of thioredoxin indicates that this protein is not the only substance capable of reducing oxidized RNR *in vivo.* This observation led to the discovery of **glutaredoxin,** a disulfide-containing, monomeric, 85-residue protein that can also reduce RNR (mutants devoid of both thioredoxin and glutaredoxin are nonviable). Oxidized glutaredoxin is reduced, via disulfide interchange, by the Cys-containing tripeptide glutathione which, in turn, is reduced by NADPH as catalyzed by glutathione reductase (GR; Section 21-2Ba). The relative

Figure 28-16   Electron-transfer pathway for nucleoside diphosphate (NDP) reduction. NADPH provides the reducing equivalents for this process through the intermediacy of thioredoxin reductase, thioredoxin, and ribonucleotide reductase.

importance of thioredoxin and glutaredoxin in the reduction of RNRs remains to be established.

f. Thioredoxin Reductase Alternates Its Conformation with Its Redox State

Thioredoxin reductase **(TrxR)**, a homodimer of 316-residue subunits, is a homolog of GR that catalyzes a similar reaction: the reduction of a substrate disulfide bond by NADPH as mediated by an FAD prosthetic group and a redox active sulfhydryl pair (Cys 135 and Cys 138). However, the X-ray structure of the C138S mutant of *E. coli* TrxR in complex with $NADP^+$ (Fig. 28-17*a*), determined by Charles Williams and John Kuriyan, reveals that TrxR and GR differ in their active site arrangements such that their redox-active sulfhydryl pairs are on opposite sides of the flavin rings in the two enzymes. Nevertheless, TrxR's redox-active sulfhydryl pair appears properly positioned to reduce the flavin ring. However, the $NADP^+$'s nicotinamide ring is >17 Å from the flavin ring and the redox-active sulfhydryl pair is buried such that it could not react with the enzyme's Trx substrate. How then does TrxR manage to transfer an electron pair from its bound NADPH via its flavin ring and redox-active sulfhydryl pair to Trx?

This question was answered by Williams and Martha Ludwig through their X-ray structure determination of the C135S mutant of TrxR, whose Cys 138 is disulfide-linked to Cys 32 of the C35S mutant of *E. coli* Trx (probably the physiologically relevant disulfide bond) and which is in complex with the $NADP^+$ analog **3-aminopyridine adenine dinucleotide phosphate (AADP⁺)**. In this complex (Fig. 28-17*b*), TrxR's $NADP^+$-binding domain has rotated by 67°

relative to the rest of the protein compared to its position in TrxR alone (Fig. 28-17*a*). This positions the $AADP^+$'s pyridine ring to react with the flavin ring and positions TrxR's redox-active sulfhydryl pair to undergo a disulfide interchange reaction with that of Trx. Moreover, in this latter conformation, the $NADP^+$-binding domain appears to provide the recognition site for the substrate Trx. Evidently, TrxR alternates its conformation with each successive step in the process of transferring an electron pair from NADPH to the flavin to its redox-active sulfhydryl pair to its bound Trx substrate. This added mechanistic complication relative to that of GR, which does not undergo a significant conformational change in reducing glutathione disulfide (Section 21-2Ba), has apparently evolved to permit TrxR to reduce its protein substrate: Trx would be too large for its redox-active sulfhydryl pair to properly approach the active site sulfhydryl pair in a GR-like enzyme.

g. The Three Classes of Ribonucleotide Reductases Are Evolutionarily Related

We have seen that the active forms of Class I RNRs are $R1_2R2_2$, $R1_4R2_4$, and $R1_6R2_6$ oligomers that have mechanistically essential tyrosyl radicals that are stabilized by oxo-bridged binuclear Fe(III) complexes, have NDPs for substrates, and obtain their reducing equivalents from thioredoxin and glutaredoxin. In contrast, Class II RNRs, which are α monomers or $α_2$ dimers, utilize a 5′-deoxyadenosylcobalamin cofactor (coenzyme B_{12}; Section 25-2Eb) for radical generation, have NDPs for substrates, and are reduced by thioredoxin and glutaredoxin; whereas Class III RNRs,

Figure 28-17 **X-ray structures of *E. coli* thioredoxin reductase (TrxR).** (*a*) The C138S mutant TrxR in complex with $NADP^+$. The protein is shown in ribbon form colored according to its secondary structure. The $NADP^+$, the FAD, and the side chains of Cys 135 and Ser 138 are drawn in ball-and-stick form with C yellow, N blue, O red, S green, and P magenta. (*b*) The C135S mutant TrxR in complex with $AADP^+$ and covalently linked to the C35S mutant of Trx via a disulfide bond between TrxR Cys 138 and Trx Cys 32. The TrxR is represented as in Part *a*, the Trx ribbon is blue-gray, and its Cys 32 and Ser 35 side chains are drawn in ball-and-stick form. Comparison of these two structures reveals that TrxR's $NADP^+$-binding domain (residues 120–243) undergoes a 67° rotation about the axis drawn in blue relative to the rest of the protein, which is shown in the same orientation in both structures. [Courtesy of Martha Ludwig, University of Michigan. PDBids (*a*) 1TDF and (*b*) 1F6M.]

(a)   *(b)*

which are α_2 dimers that interact with a radical-generating protein β_2 that contains a [4Fe–4S] cluster and requires *S*-adenosylmethionine (SAM; Section 26-3Ea) and NADPH for activity, have NTPs for substrates, and their reducing equivalents are provided by the oxidation of formate to CO_2.

Since all known cellular life synthesizes its deoxyribonucleotides from ribonucleotides, the rise of an RNR must have preceded the evolutionary transition from the RNA world (Section 1-5Ca) to DNA-based life-forms. Did the three classes of RNRs arise independently or are they evolutionarily related? Despite the seemingly large differences between these different classes of RNRs, the reactions they catalyze are surprisingly similar. All replace the 2′ OH group of ribose with H via a free radical mechanism involving a thiyl radical with the reducing equivalents provided by a Cys sulfhydryl group (Fig. 28-13; the second Cys residue of the redox-active sulfhydryl pair in Class I and II RNRs is replaced by formate in Class III RNRs). They differ mainly in the way they generate the free radical. [In Class II RNRs, the radical is generated by the homolytic cleavage of its 5′-deoxyadenosylcobalamin cofactor's C—Co(III) bond (Section 25-2Ec). In Class III RNRs, it is generated by the NADPH-supplied and [4Fe–4S] cluster–mediated one-electron reductive cleavage of SAM by the β_2 protein to yield methionine and the 5′-deoxyadenosyl radical (the same radical generated by the homolytic cleavage of 5′-deoxyadenosylcobalamin), which then abstracts the H atom from a C_α—H group of a specific Gly on the α subunit to yield 5′-deoxyadenosine and a stable but O_2-sensitive glycyl radical.] Moreover, the X-ray structures of both Class II and Class III RNRs reveal that their active sites are formed by 10-stranded α/β barrels that have the same connectivity as and are closely superimposable on that of Class I RNRs. It therefore appears that all three classes of RNRs are evolutionarily related. Reichard has proposed that, since life arose under anaerobic conditions and that formate, one of simplest organic reductants, was probably widely available on primitive Earth (Section 1-5B), the primordial RNR was a Class III–like enzyme. The rise of photosynthetic organisms that generated O_2 then promoted the evolution of Class II RNRs, which can function under both anaerobic and aerobic conditions. Class I RNRs, which require the presence of O_2 for activation, evolved last, presumably from a Class II RNR.

h. dNTPs Are Produced by Phosphorylation of dNDPs

In pathways involving Class I and Class II RNRs, the final step in the production of dNTPs is the phosphorylation of the corresponding dNDPs:

$$dNDP + ATP \rightleftharpoons dNTP + ADP$$

This reaction is catalyzed by nucleoside diphosphate kinase, the same enzyme that phosphorylates NDPs (Section 28-1Ba). As before, the reaction is written with ATP as the phosphoryl donor, although any NTP or dNTP can function in this capacity. In pathways involving Class III RNRs, the production of NTPs from NDPs precedes the reduction of NTPs to dNTPs.

B. *Origin of Thymine*

a. dUTP Diphosphohydrolase

The dTMP component of DNA is synthesized, as we discuss below, by methylation of dUMP. The dUMP is generated through the hydrolysis of dUTP by **dUTP diphosphohydrolase** (**dUTPase**; also called **dUTP pyrophosphatase**):

$$dUTP + H_2O \rightleftharpoons dUMP + PP_i$$

The reason for this apparently energetically wasteful process (dTMP, once formed, is rephosphorylated to dTTP) is that cells must minimize their concentration of dUTP in order to prevent incorporation of uracil into their DNA. This is because, as we discuss in Section 30-5Bd, DNA polymerase does not discriminate between dUTP and dTTP.

The X-ray structure of human dUTPase, determined by John Tainer, reveals the basis for this enzyme's exquisite specificity for dUTP. This homotrimer of 141-residue subunits binds dUTP in a snug-fitting cavity that sterically excludes thymine's C5 methyl group via the side chains of conserved residues (Fig. 28-18*a*). It differentiates uracil from cytosine via a set of hydrogen bonds from the protein backbone that mimic adenine's base-pairing interaction (Fig. 28-18*b*), and it differentiates dUTP from UTP by the steric exclusion of ribose's 2′ OH group by the side chain of a conserved Tyr.

b. Thymidylate Synthase

dTMP is synthesized from dUMP by **thymidylate synthase (TS)** *with* N^5,N^{10}-*methylenetetrahydrofolate* (N^5,N^{10}-*methylene-THF*) *as the methyl donor:*

(THF cofactors are discussed in Section 26-4D). Note that the transferred methylene group (in which the carbon has the oxidation state of formaldehyde) is reduced to a methyl group (which has the oxidation state of methanol) at the expense of the oxidation of the THF cofactor to dihydrofolate (DHF).

The catalytic mechanism of TS, a highly conserved homodimeric protein (with 264-residue subunits in *E. coli*), has been extensively investigated. On incubation of the enzyme with N^5,N^{10}-methylene-[6-^{3}H]THF and dUMP, the ^{3}H is quantitatively transferred to the methyl group of the product dTMP. When [5-^{3}H]dUMP is the substrate, however, the ^{3}H is released into the aqueous solvent. Such information, together with the knowledge that uracil C6, which occupies the β position of an α,β-unsaturated ketone, is susceptible to nucleophilic attack, led Daniel Santi to propose the following mechanistic scheme for the TS reaction (Fig. 28-19):

1. An enzyme nucleophile, identified as the thiolate group of Cys 146, attacks C6 of dUMP to form a covalent adduct.

2. C5 of the resulting enolate ion attacks the CH$_2$ group of the iminium cation in equilibrium with N^5,N^{10}-methylene-THF to form an enzyme–dUMP–THF ternary covalent complex.

3. An enzyme base abstracts the acidic proton at the C5 position of the enzyme-bound dUMP, forming an exocyclic methylene group and eliminating the THF cofactor. The abstracted proton subsequently exchanges with solvent.

4. The redox change occurs via the migration of the N6—H atom of THF as a hydride ion to the exocyclic methylene group, converting it to a methyl group (thus

accounting for the above described transfer of ^{3}H) and yielding DHF. This reduction promotes displacement of the Cys thiolate group from the intermediate so as to release product, dTMP, and reform active enzyme.

c. 5-Fluorodeoxyuridylate Is a Potent Antitumor Agent

The above mechanism is supported by the observation that **5-fluorodeoxyuridylate (FdUMP)**

5-Fluorodeoxyuridylate (FdUMP)

is an irreversible inhibitor of TS. This substance, like dUMP, binds to the enzyme (an F atom has approximately the same radius as an H atom) and undergoes the first two steps of the normal enzymatic reaction. In Step 3, however, the enzyme cannot abstract the F atom as F$^+$ (recall that F is the most electronegative element), so that the enzyme is all but permanently immobilized as the enzyme–FdUMP–THF

(a)

(b)

Figure 28-18 **X-ray structure of human dUTPase.** (*a*) The molecular surface at the substrate binding site showing how the enzyme differentiates uracil from thymine. Bound dUTP is drawn in stick form with its N, O, and P atoms represented by blue, red, and yellow spheres. Mg^{2+} ions that were modeled into the structure are represented by green spheres. The protein's molecular surface is colored according to its electrostatic potential with positive, negative, and near neutral regions blue, red, and white, respectively. Note how the snug fit of the uracil ring into its binding site would sterically exclude thymine's C5 methyl group. (*b*) The substrate binding site indicating how the enzyme differentiates uracil from cytosine and 2′-deoxyribose from ribose. dUMP bound at the active site is drawn as in Part *a*. The protein, mainly the backbone of a β hairpin motif, is similarly drawn but with thinner gray bonds. Hydrogen bonds are shown as dotted white lines, and a tightly bound, conserved water molecule is represented by a pink sphere. The pattern of hydrogen bonding donors and acceptors on the protein would prevent cytosine from binding in the active site pocket. The conserved Tyr side chain sterically excludes ribose's 2′ OH group. [Courtesy of John Tainer, The Scripps Research Institute, La Jolla, California.]

Figure 28-19 Catalytic mechanism of thymidylate synthase. The methyl group is supplied by N^5,N^{10}-methylene-THF, which is concomitantly oxidized to dihydrofolate.

ternary covalent complex analogous to that after Step 2 in Fig. 28-19. Indeed, X-ray structural analysis by William Montfort revealed that crystals of *E. coli* TS that had been soaked in a solution containing FdUMP and N^5,N^{10}-methylene-THF contain precisely this complex (Fig. 28-20). Enzymatic inhibitors such as FdUMP, which inactivate an enzyme only after undergoing part or all of its normal catalytic reaction, are called **mechanism-based inhibitors** (alternatively, **suicide substrates** because they cause the enzyme to "commit suicide"). *Mechanism-based inhibitors, being targeted for particular enzymes, are among the most powerful, specific, and therefore useful enzyme inactivators.*

The strategic position of thymidylate synthase in DNA biosynthesis has led to the clinical use of FdUMP as an antitumor agent. Rapidly proliferating cells, such as cancer cells, require a steady supply of dTMP in order to survive and are therefore killed by treatment with FdUMP. In contrast, most normal mammalian cells, which grow slowly if at all, have a lesser requirement for dTMP, so that they are

Figure 28-20 The X-ray structure of the *E. coli* thymidylate synthase in covalent complex with FdUMP and THF. The active site region of one subunit of this dimeric enzyme is shown in ribbon form with its polypeptide chain colored in rainbow order from its N-terminus (*blue*) to its C-terminus (*red*). The FdUMP and N^5,N^{10}-methylene-THF are drawn in stick form with FdUMP C cyan, N^5,N^{10}-methylene-THF C green, N blue, O red, F purple, and P orange. The C5 and C6 atoms of FdUMP form covalent bonds with the CH_2 group substituent to N5 of THF and the S atom of Cys 146, whose side chain is drawn in stick form with C green and S yellow. [Based on an X-ray structure by William Montfort, University of Arizona, PDBid 1TSN.]

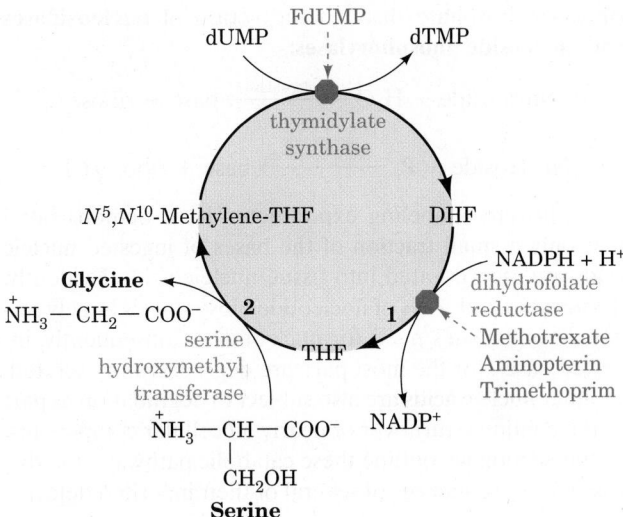

Figure 28-21 **Regeneration of N^5,N^{10}-methylenetetrahydrofolate.** The DHF product of the thymidylate synthase reaction is converted back to N^5,N^{10}-methylene-THF by the sequential actions of (1) dihydrofolate reductase and (2) serine hydroxymethyltransferase. Thymidylate synthase is inhibited by FdUMP, whereas dihydrofolate reductase is inhibited by the antifolates methotrexate, aminopterin, and trimethoprim.

relatively insensitive to FdUMP (some exceptions are the bone marrow cells that comprise the blood-forming tissues and much of the immune system, the intestinal mucosa, and hair follicles). **5-Fluorouracil** and **5-fluorodeoxyuridine** are also effective antitumor agents since they are converted to FdUMP through salvage reactions.

d. N^5,N^{10}-Methylene-THF Is Regenerated in Two Reactions

The thymidylate synthase reaction is biochemically unique in that it oxidizes THF to DHF; no other enzymatic reaction employing a THF cofactor alters this coenzyme's net oxidation state. The DHF product of the thymidylate synthase reaction is recycled to the enzyme's N^5,N^{10}-methylene-THF cofactor through two sequential reactions (Fig. 28-21):

1. DHF is reduced to THF by NADPH as catalyzed by **dihydrofolate reductase (DHFR;** Section 26-4D). Although, in most organisms, DHFR is a monomeric monofunctional enzyme, in protozoa and at least some plants, DHFR and TS occur on the same polypeptide chain to form a bifunctional enzyme that has been shown to channel DHF from its TS to its DHFR active sites.

2. Serine hydroxymethyltransferase (Section 26-3Bb) transfers the hydroxymethyl group of serine to THF, yielding N^5,N^{10}-methylene-THF and glycine.

e. Antifolates Are Anticancer Agents

Inhibition of DHFR quickly results in all of a cell's limited supply of THF being converted to DHF by the thymidylate synthase reaction. Inhibition of DHFR therefore not only prevents dTMP synthesis (Fig. 28-21), but also blocks all other THF-dependent biological reactions such as the

synthesis of purines (Section 28-1A), methionine (Section 26-5Ba), and, indirectly, histidine (Section 26-5Be). DHFR (Fig. 28-22) therefore offers an attractive target for chemotherapy.

Methotrexate (amethopterin), aminopterin, and **trimethoprim**

R = H   **Aminopterin**
R = CH$_3$ **Methotrexate (amethopterin)**

Trimethoprim

are DHF analogs that competitively although all but irreversibly bind to DHFR with an ~1000-fold greater affinity than does DHF. These **antifolates** (substances that interfere with the action of folate cofactors) are effective anticancer agents, particularly against childhood leukemias. In fact, a successful chemotherapeutic strategy is to treat a cancer

Figure 28-22 **X-ray structure of human dihydrofolate reductase in complex with folic acid.** The polypeptide is colored in rainbow order from its N-terminus (*blue*) to its C-terminus (*red*). The folic acid is drawn in space-filling form with C green, N blue, and O red. [Based on an X-ray structure by Joseph Kraut, University of California at San Diego. PDBid 1DHF.]

See Interactive Exercise 29.

victim with a lethal dose of methotrexate and some hours later "rescue" the patient (but hopefully not the cancer) by administering massive doses of 5-formyl-THF and/or thymidine. A low dose of methotrexate is also effective in the treatment of rheumatoid arthritis, inhibiting immune system activity and thus decreasing inflammation. Trimethoprim, which was discovered by George Hitchings and Gertrude Elion, binds much more tightly to bacterial DHFRs than to those of mammals and is therefore a clinically useful antibiotic.

4 NUCLEOTIDE DEGRADATION

Most foodstuffs, being of cellular origin, contain nucleic acids. Dietary nucleic acids survive the acid medium of the stomach; they are degraded to their component nucleotides, mainly in the duodenum, by pancreatic nucleases and intestinal phosphodiesterases. These ionic compounds, which cannot pass through cell membranes, are then hydrolyzed to nucleosides by a variety of group-specific nucleotidases and nonspecific phosphatases. Nucleosides may be directly absorbed by the intestinal mucosa or first undergo further degradation to free bases and ribose or

ribose-1-phosphate through the action of **nucleosidases** and **nucleoside phosphorylases:**

$$\text{Nucleoside} + \text{H}_2\text{O} \xrightarrow{\text{nucleosidase}} \text{base} + \text{ribose}$$

$$\text{Nucleoside} + \text{P}_i \xrightarrow[\text{phosphorylase}]{\text{nucleoside}} \text{base} + \text{ribose-1-P}$$

Radioactive labeling experiments have demonstrated that only a small fraction of the bases of ingested nucleic acids are incorporated into tissue nucleic acids. Evidently, the *de novo* pathways of nucleotide biosynthesis largely satisfy an organism's need for nucleotides. Consequently, ingested bases, for the most part, are degraded and excreted. Cellular nucleic acids are also subject to degradation as part of the continual turnover of nearly all cellular components. In this section we outline these catabolic pathways and discuss the consequences of several of their inherited defects.

A. Catabolism of Purines

The major pathways of purine nucleotide and deoxynucleotide catabolism in animals are diagrammed in Fig. 28-23. Other organisms may have somewhat different

Figure 28-23 **Major pathways of purine catabolism in animals.** The various purine nucleotides and deoxynucleotides are all degraded to uric acid.

(a)

(b)

Adenosine

Tetrahedral intermediate

NH₃

Inosine (enol tautomer)

Figure 28-24 **X-ray structure and mechanism of adenosine deaminase.** (*a*) A ribbon diagram of murine adenosine deaminase in complex with its transition state analog 6-hydroxy-1,6-dihydropurine ribonucleoside (HDPR). The polypeptide is drawn in ribbon form colored according to its secondary structure (helices cyan, β strands magenta, and loops salmon) and viewed approximately down the axis of the enzyme's α/β barrel from the N-terminal ends of its β strands. The HDPR is shown in stick form with its C, N, and O atoms green, blue, and red. The enzyme-bound Zn²⁺ ion, which is coordinated by HDPR's 6-hydroxyl group, is represented by a silver sphere. [Based on an X-ray structure by Florante Quiocho, Baylor College of Medicine. PDBid 1ADA.] (*b*) The proposed catalytic mechanism of adenosine deaminase. A Zn²⁺-polarized H₂O molecule (Section 15-1Cb) nucleophilically attacks C6 of the enzyme-bound adenosine molecule in a process that is facilitated by His 238 acting as a general base, Glu 217 acting as a general acid, and Asp 295 acting to orient the water molecule via hydrogen bonding. The resulting tetrahedral intermediate decomposes by the elimination of ammonia in a reaction that is aided by the now imidazolium and carboxyl side chains of His 238 and Glu 217 acting as a general acid and a general base, respectively. This yields inosine in its enol tautomeric form, which, on its release from the enzyme, largely assumes its dominant keto form. The Zn²⁺ is coordinated by three His side chains that are not shown. [After Wilson, D.K. and Quiocho, F.A., *Biochemistry* **32**, 1692 (1993).] 🖱️ **See Interactive Exercise 30**

pathways among these various intermediates (including adenine), but all of these pathways lead to uric acid. Of course, the intermediates in these processes may instead be reused to form nucleotides via salvage reactions. In addition, ribose-1-phosphate, a product of the reaction catalyzed by **purine nucleoside phosphorylase (PNP),** is isomerized by **phosphoribomutase** to the PRPP precursor ribose-5-phosphate.

Adenosine and deoxyadenosine are not degraded by mammalian PNP. Rather, adenine nucleosides and nucleotides are deaminated by **adenosine deaminase (ADA)** and **AMP deaminase** to their corresponding inosine derivatives, which, in turn, may be further degraded. The X-ray structure of murine ADA that was crystallized in the presence of its inhibitor **purine ribonucleoside** was determined by Florante Quiocho (Fig. 28-24*a*). The enzyme forms an

eight-stranded α/β barrel with its active site in a pocket at the C-terminal end of the β barrel, as occurs in nearly all known α/β barrel enzymes (Section 8-3Bh). Purine ribonucleoside binds to ADA in a normally rare hydrated form, **6-hydroxy-1, 6-dihydropurine ribonucleoside (HDPR)**,

Purine ribonucleoside   **6-Hydroxy-1,6-dihydropurine ribonucleoside (HDPR)**

a nearly ideal transition state analog of the ADA reaction. Although it had been previously reported that ADA does not require a cofactor, its X-ray structure clearly reveals that a zinc ion is bound in the deepest part of the active site pocket, where it is pentacoordinated by three His side chains, a carboxyl oxygen of Asp 295, and the O6 atom of HDPR. ADA's active site complex suggests a catalytic mechanism (Fig. 28-24b) reminiscent of that of carbonic anhydrase (Section 15-1Cb): His 238, which is properly positioned to act as a general base, abstracts a proton from a bound Zn^{2+}-activated water molecule, which nucleophilically attacks the adenine C6 atom to form a tetrahedral intermediate. Products are then formed by the elimination of ammonia.

a. Genetic Defects in ADA Result in Severe Combined Immunodeficiency Disease

Abnormalities in purine nucleoside metabolism arising from rare genetic defects in ADA selectively kill **lymphocytes** (a type of white blood cell). Since lymphocytes mediate much of the immune response (Section 35-2A), ADA deficiency results in **severe combined immunodeficiency disease (SCID)** that, without special protective measures, is invariably fatal in infancy due to overwhelming infection. The mutations in all eight known ADA variants obtained from SCID patients appear to structurally perturb the active site of ADA.

Biochemical considerations provide a plausible explanation of SCID's etiology (causes). In the absence of active ADA, deoxyadenosine is phosphorylated to yield levels of dATP that are 50-fold greater than normal. This high concentration of dATP inhibits ribonucleotide reductase (Section 28-3Ad), thereby preventing the synthesis of the other dNTPs, choking off DNA synthesis and thus cell proliferation. The tissue-specific effect of ADA deficiency on the immune system may be explained by the observation that lymphoid tissue is particularly active in deoxyadenosine phosphorylation.

SCID caused by ADA defects does not respond to treatment by the intravenous injection of ADA because the liver clears this enzyme from the bloodstream within minutes. If, however, several molecules of the biologically inert polymer **polyethylene glycol (PEG)**

$$HO \text{---} [CH_2 \text{---} CH_2 \text{---} O]_n H$$
Polyethylene glycol

are covalently linked to surface groups on ADA, the resulting **PEG–ADA** remains in the blood for 1 to 2 weeks, thereby largely resuscitating the SCID victim's immune system. The protein-linked PEG only reduces the catalytic activity of ADA by ~40% but, evidently, masks it from the receptors that filter it out of the blood. SCID can therefore be treated effectively by PEG–ADA. This treatment, however, is expensive and not entirely satisfactory. Consequently, ADA deficiency was selected as one of the first genetic diseases to be treated by gene therapy (Section 5-5Hb): Lymphocytes were extracted from the blood of an ADA-deficient child and grown in the laboratory, had a normal ADA gene inserted into them via genetic engineering techniques (Section 5-5), and were then returned to the child. After 12 years, 20 to 25% of the patient's lymphocytes contained the introduced ADA gene. However, ethical considerations have mandated that the patient continue receiving injections of PEG–ADA so that the efficacy of this gene therapy protocol is unclear.

b. The Purine Nucleotide Cycle

The deamination of AMP to IMP, when combined with the synthesis of AMP from IMP (Fig. 28-4, *left*), has the effect of deaminating aspartate to yield fumarate (Fig. 28-25). John Lowenstein demonstrated that this **purine nucleotide cycle** has an important metabolic role in skeletal muscle. An increase in muscle activity requires an increase in the activity of the citric acid cycle. This process usually occurs through the generation of additional citric acid cycle intermediates (Section 21-4). Muscles, however, lack most of the enzymes that catalyze these anaplerotic (filling up) reactions in other tissues. Rather, muscle replenishes its citric acid cycle intermediates as fumarate generated in the purine nucleotide cycle. The importance of the purine nucleotide cycle in muscle metabolism is indicated by the observation that the activities of the three enzymes involved are all severalfold higher in muscle than in other tissues. In fact, individuals with an inherited deficiency in muscle AMP deaminase (**myoadenylate deaminase deficiency**) are easily fatigued and usually suffer from cramps after exercise.

c. Xanthine Oxidase Is a Mini-Electron-Transport Protein

Xanthine oxidase (XO) converts hypoxanthine to xanthine, and xanthine to uric acid (Fig. 28-23, *bottom*). In mammals, this enzyme occurs mainly in the liver and the small intestinal mucosa. XO is a homodimer of ~1330-residue subunits, each of which binds a variety of electron-transfer agents: an FAD, two spectroscopically distinct [2Fe–2S] clusters, and a **molybdopterin complex (Mo-pt)**

Molybdopterin complex (Mo-pt)

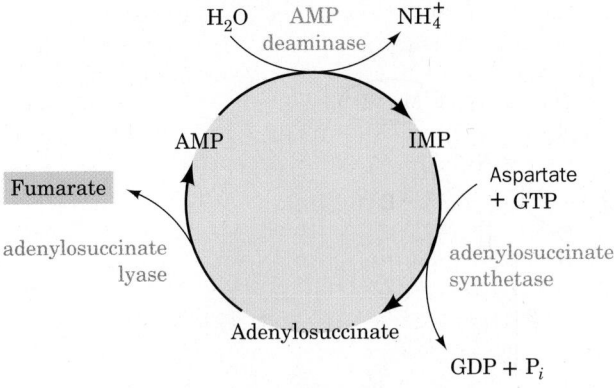

$$\text{Net: } H_2O + \text{Aspartate} + GTP \longrightarrow NH_4^+ + GDP + P_i + \text{fumarate}$$

Figure 28-25 **The purine nucleotide cycle.** This pathway functions, in muscle, to prime the citric acid cycle by generating fumarate.

in which the Mo atom cycles between its Mo(VI) and Mo(IV) oxidation states. The final electron acceptor is O_2, which is converted to H_2O_2, a potentially harmful oxidizing agent that is subsequently disproportionated to H_2O and O_2 by catalase (Section 1-2Ad). In XO, the polypeptide has been proteolytically cleaved into three segments (the uncleaved enzyme, which is known as **xanthine dehydrogenase,** preferably uses NAD^+ as its electron acceptor, whereas XO does not react with NAD^+).

The X-ray structure of XO from cow's milk in complex with the competitive inhibitor salicylic acid (Fig. 25-74), determined by Emil Pai, reveals that the FAD and the molybdopterin complex are interposed by the two [2Fe–2S] clusters to form a mini-electron-transport chain (Fig. 28-26). Each of its three peptide segments forms a separate domain with the N-terminal domain binding the two [2Fe–2S] clusters, the central domain binding the FAD, and the C-terminal domain binding the Mo-pt complex. Although the salicylic acid does not contact the Mo-pt complex, it binds to XO in a way that blocks the approach of substrates to the metal center.

XO hydroxylates xanthine at its C8 position (and hypoxanthine at its C2 position), yielding uric acid in its enol form that tautomerizes to the more stable keto form:

Uric acid (enol tautomer)   **Uric acid (keto tautomer)**

$pK = 5.4$

Urate

Figure 28-26 **X-ray structure of xanthine oxidase from cow's milk in complex with salicylic acid.**
(*a*) Ribbon diagram of its 1332-residue subunit in which the N-terminal domain (residues 2–165) is cyan, the central domain (residues 224–528) is gold, and the C-terminal domain (residues 571–1315) is lavender. The enzyme's redox cofactors and bound salicylic acid are shown in space-filling form with C green, N blue, O red, S yellow, P magenta, Fe orange, and Mo light blue. The ~50-residue peptide segments spanning domains are disordered and are apparently highly flexible. (*b*) The enzyme's redox cofactors and salicylic acid (Sal) drawn in stick form with their S, Fe, and Mo atoms represented by spheres. The atoms are colored as in Part *a* and viewed from the same direction but with greater magnification. [Based on an X-ray structure by Emil Pai, University of Toronto, Toronto, Ontario, Canada. PDBid 1FIQ.]

Figure 28-27   Mechanism of xanthine oxidase. The reduced enzyme is subsequently reoxidized by O_2, yielding H_2O_2.

(its enol form ionizes with a pK of 5.4; hence, the name *uric acid*). ^{18}O-labeling experiments have demonstrated that the C8 keto oxygen of uric acid is derived from H_2O, whereas the oxygen atoms of H_2O_2 come from O_2. Chemical and spectroscopic studies suggest that the enzyme has the following mechanism (Fig. 28-27):

1. The reaction is initiated by the attack of an enzyme nucleophile, X, on the C8 position of xanthine.

2. The C8—H atom is eliminated as a hydride ion that combines with the Mo(VI) complex, thereby reducing it to the Mo(IV) state.

3. Water displaces the enzyme nucleophile producing uric acid.

In the second stage of the reaction, the now reduced enzyme is reoxidized to its original Mo(VI) state by reaction with O_2. This complex process, not surprisingly, is but poorly understood. EPR measurements indicate that electrons are funneled from the Mo(IV) through the two [2Fe–2S] clusters to the flavin and ultimately to O_2, yielding H_2O_2 and regenerated enzyme.

B. *Fate of Uric Acid*

In humans and other primates, the final product of purine degradation is uric acid, which is excreted in the urine. The same is true of birds, terrestrial reptiles, and many insects, but these organisms, which do not excrete urea, also catabolize their excess amino acid nitrogen to uric acid via purine biosynthesis. This complicated system of nitrogen excretion has a straightforward function: *It conserves water.*

Uric acid is only sparingly soluble in water, so that its excretion as a paste of uric acid crystals is accompanied by very little water. In contrast, the excretion of an equivalent amount of the much more water-soluble urea osmotically sequesters a significant amount of water.

In all other organisms, uric acid is further processed before excretion (Fig. 28-28). Mammals other than primates oxidize it to their excretory product, **allantoin,** in a reaction catalyzed by the Cu-containing enzyme **urate oxidase.** A further degradation product, **allantoic acid,** is excreted by teleost (bony) fish. Cartilaginous fish and amphibia further degrade allantoic acid to urea prior to excretion. Finally, marine invertebrates decompose urea to their nitrogen excretory product, NH_4^+.

a. Gout Is Caused by an Excess of Uric Acid

Gout is a disease characterized by elevated levels of uric acid in body fluids. Its most common manifestation is excrutiatingly painful arthritic joint inflammation of sudden onset, most often in the big toe (Fig. 28-29), caused by deposition of nearly insoluble crystals of sodium urate. Sodium urate and/or uric acid may also precipitate in the kidneys and ureters as stones, resulting in renal damage and urinary tract obstruction. Gout, which affects ~3 per 1000 persons, predominantly males, has been traditionally, although inaccurately, associated with overindulgent eating and drinking. The probable origin of this association is that in previous centuries, when wine was often contaminated with lead during its manufacture and storage, heavy drinking resulted in chronic lead poisoning, which, among other things, decreases the kidney's ability to excrete uric acid.

Uric acid

Excreted by
{ Primates
Birds
Reptiles
Insects }

$2H_2O + O_2$ ⟶ urate oxidase ⟶ $CO_2 + H_2O_2$

Allantoin

{ Other mammals }

H_2O ⟶ allantoinase

Allantoic acid

{ Teleost fish }

H_2O ⟶ allantoicase ⟶ COOH / CHO

Glyoxylic acid

$2H_2N—\overset{O}{\underset{}{C}}—NH_2$

Urea

{ Cartilaginous fish
Amphibia }

$2H_2O$ ⟶ urease ⟶ $2CO_2$

$4NH_4^+$

{ Marine
invertebrates }

Figure 28-28 **Degradation of uric acid to ammonia.** The process is arrested at different stages in the indicated species and the resulting nitrogen-containing product is excreted.

Figure 28-29 ***The Gout***, **a cartoon by James Gillray (1799).** [Yale University Medical Historical Library.]

caused by glucose-6-phosphatase deficiency (von Gierke's glycogen storage disease; Section 18-4): The increased availability of glucose-6-phosphate stimulates the pentose phosphate pathway (Section 23-4), increasing the rate of ribose-5-phosphate production and consequently that of PRPP, which in turn stimulates purine biosynthesis.

Gout may be treated by administration of the xanthine oxidase inhibitor **allopurinol,** a hypoxanthine analog with interchanged N7 and C8 positions.

Allopurinol   **Hypoxanthine**

Xanthine oxidase hydroxylates allopurinol, as it does hypoxanthine, yielding **alloxanthine,**

Alloxanthine

which remains tightly bound to the reduced form of the enzyme, thereby inactivating it. Allopurinol consequently alleviates the symptoms of gout by decreasing the rate of uric acid production while increasing the levels of the more soluble hypoxanthine and xanthine. Although allopurinol controls the gouty symptoms of Lesch–Nyhan syndrome, it has no effect on its neurological symptoms.

The most prevalent cause of gout is impaired uric acid excretion (although usually for other reasons than lead poisoning). Gout may also result from a number of metabolic insufficiencies, most of which are not well characterized. One well-understood cause is HGPRT deficiency (Lesch–Nyhan syndrome in severe cases), which leads to excessive uric acid production through PRPP accumulation (Section 28-1D). Uric acid overproduction is also

Figure 28-30 **Major pathways of pyrimidine catabolism in animals.** The amino acid products of these reactions are taken up in other metabolic processes. UMP and dTMP are degraded by the same enzymes; the pathway for dTMP degradation is given in parentheses.

C. Catabolism of Pyrimidines

Animal cells degrade pyrimidine nucleotides to their component bases (Fig. 28-30, *top*). These reactions, like those of purine nucleotides, occur through dephosphorylation, deamination, and glycosidic bond cleavages. The resulting uracil and thymine are then broken down in the liver through reduction (Fig. 28-30, *middle*) rather than by oxidation, as

occurs in purine catabolism. The end products of pyrimidine catabolism, **β-alanine** and **β-aminoisobutyrate,** are amino acids and are metabolized as such. They are converted, through transamination and activation reactions, to malonyl-CoA and methylmalonyl-CoA (Fig. 28-30, *bottom left*) for further utilization (Sections 25-4A and 25-2Ea).

5 BIOSYNTHESIS OF NUCLEOTIDE COENZYMES

In this section we outline the assembly, in animals, of the nucleotide coenzymes NAD^+ and $NADP^+$, FMN and FAD, and coenzyme A, from their vitamin precursors. These vitamins are synthesized *de novo* only by plants and microorganisms.

A. Nicotinamide Coenzymes

The nicotinamide moiety of the nicotinamide coenzymes (NAD^+ and $NADP^+$) is derived, in humans, from dietary nicotinamide, nicotinic acid, or the essential amino acid tryptophan (Fig. 28-31). **Nicotinate phosphoribosyltransferase,**

Figure 28-31 **Pathways for the biosynthesis of NAD$^+$ and NADP$^+$.** These nicotinamide coenzymes are synthesized from their vitamin precursors, nicotinate and nicotinamide, and from the tryptophan degradation product, quinolinate.

Figure 28-32 Biosynthesis of FMN and FAD from the vitamin precursor riboflavin.

which occurs in most mammalian tissues, catalyzes the formation of **nicotinate mononucleotide** from nicotinate and PRPP. This intermediate may also be synthesized from **quinolinate,** a degradation product of tryptophan (Section 26-3G), in a reaction mediated by **quinolinate phosphoribosyltransferase,** which occurs mainly in liver and kidney. A poor diet, nevertheless, may result in pellagra (nicotinic acid deficiency; Section 13-3), since, under such conditions, tryptophan will be almost entirely utilized in protein biosynthesis. Nicotinate mononucleotide is joined via a pyrophosphate linkage to an ATP-derived AMP residue by **NAD⁺ pyrophosphorylase** to yield **nicotinate adenine dinucleotide (desamido NAD⁺).** Finally, **NAD⁺ synthetase** converts this intermediate to NAD⁺ by a transamidation reaction in which glutamine is the NH₂ donor.

NAD⁺ may also be synthesized from nicotinamide. This vitamin is converted to **nicotinamide mononucleotide (NMN)** by **nicotinamide phosphoribosyltransferase,** a widely occurring enzyme distinct from nicotinate phosphoribosyltransferase. However, NAD⁺ is synthesized from NMN and ATP by NAD pyrophosphorylase, the same enzyme that synthesizes nicotinate adenine dinucleotide.

NADP⁺ is formed via the ATP-dependent phosphorylation of the NAD⁺ adenosine residue's C2′ OH group by **NAD⁺ kinase.**

B. Flavin Coenzymes

FAD is synthesized from riboflavin in a two-reaction pathway (Fig. 28-32). First, the 5′-OH group of riboflavin's ribityl side chain is phosphorylated by **flavokinase,** yielding flavin mononucleotide (FMN; not a true nucleotide since its ribityl residue is not a true sugar). FAD may then be formed by the coupling of FMN and ATP-derived AMP in a pyrophosphate linkage in a reaction catalyzed by **FAD pyrophosphorylase.** Both of these enzymes are widely distributed in nature.

C. Coenzyme A

Coenzyme A is synthesized in mammalian cells according to the pathway diagrammed in Fig. 28-33. Pantothenate, an essential vitamin, is phosphorylated by **pantothenate kinase** and then coupled to cysteine, the future business end of CoA, by **phosphopantothenoylcysteine synthetase.** After decarboxylation by **phosphopantothenoylcysteine decarboxylase,** the resulting **4′-phosphopantethiene** is coupled to AMP in a pyrophosphate linkage by **dephospho-CoA pyrophosphorylase** and then phosphorylated at its adenosine 3′ OH group by **dephospho-CoA kinase** to form CoA. The latter two enzymatic activities occur on a single protein.

$$\text{HO-CH}_2-\underset{\underset{\text{CH}_3}{|}}{\overset{\overset{\text{H}_3\text{C}}{|}}{\text{C}}}-\underset{\underset{}{|}}{\overset{\overset{\text{OH}}{|}}{\text{CH}}}-\overset{\overset{\text{O}}{||}}{\text{C}}-\text{NH-CH}_2-\text{CH}_2-\text{COO}^-$$

Pantothenate

ATP ⟶ pantothenate kinase

ADP ⟵

$$^{-2}\text{O}_3\text{P-O-CH}_2-\underset{\underset{\text{CH}_3}{|}}{\overset{\overset{\text{H}_3\text{C}}{|}}{\text{C}}}-\underset{}{\overset{\overset{\text{OH}}{|}}{\text{CH}}}-\overset{\overset{\text{O}}{||}}{\text{C}}-\text{NH-CH}_2-\text{CH}_2-\text{COO}^-$$

4′-Phosphopantothenate

ATP + Cysteine ⟶ phosphopantothenoylcysteine synthetase

P_i + ADP ⟵

$$^{-2}\text{O}_3\text{P-O-CH}_2-\underset{\underset{\text{CH}_3}{|}}{\overset{\overset{\text{H}_3\text{C}}{|}}{\text{C}}}-\underset{}{\overset{\overset{\text{OH}}{|}}{\text{CH}}}-\overset{\overset{\text{O}}{||}}{\text{C}}-\text{NH-CH}_2-\text{CH}_2-\overset{\overset{\text{O}}{||}}{\text{C}}-\text{NH}-\underset{}{\overset{\overset{\text{COO}^-}{|}}{\text{CH}}}-\text{CH}_2-\text{SH}$$

4′-Phosphopantothenoylcysteine

phosphopantothenoylcysteine decarboxylase

CO_2 ⟵

$$^{-2}\text{O}_3\text{P-O-CH}_2-\underset{\underset{\text{CH}_3}{|}}{\overset{\overset{\text{H}_3\text{C}}{|}}{\text{C}}}-\underset{}{\overset{\overset{\text{OH}}{|}}{\text{CH}}}-\overset{\overset{\text{O}}{||}}{\text{C}}-\text{NH-CH}_2-\text{CH}_2-\overset{\overset{\text{O}}{||}}{\text{C}}-\text{NH}-\underset{}{\overset{\overset{\text{H}}{|}}{\text{CH}}}-\text{CH}_2-\text{SH}$$

4′-Phosphopantetheine

ATP ⟶ dephospho-CoA pyrophosphorylase

PP_i ⟵

$$\text{Adenosine-O}-\overset{\overset{\text{O}}{||}}{\underset{\underset{\text{O}^-}{|}}{\text{P}}}-\text{O}-\overset{\overset{\text{O}}{||}}{\underset{\underset{\text{O}^-}{|}}{\text{P}}}-\text{O-CH}_2-\underset{\underset{\text{CH}_3}{|}}{\overset{\overset{\text{H}_3\text{C}}{|}}{\text{C}}}-\underset{}{\overset{\overset{\text{OH}}{|}}{\text{CH}}}-\overset{\overset{\text{O}}{||}}{\text{C}}-\text{NH-CH}_2-\text{CH}_2-\overset{\overset{\text{O}}{||}}{\text{C}}-\text{NH-CH}_2-\text{CH}_2-\text{SH}$$

Dephosphocoenzyme A

ATP ⟶ dephospho-CoA kinase

ADP ⟵

$$\text{CH}_2-\text{O}-\overset{\overset{\text{O}}{||}}{\underset{\underset{\text{O}^-}{|}}{\text{P}}}-\text{O}-\overset{\overset{\text{O}}{||}}{\underset{\underset{\text{O}^-}{|}}{\text{P}}}-\text{O-CH}_2-\underset{\underset{\text{CH}_3}{|}}{\overset{\overset{\text{H}_3\text{C}}{|}}{\text{C}}}-\underset{}{\overset{\overset{\text{OH}}{|}}{\text{CH}}}-\overset{\overset{\text{O}}{||}}{\text{C}}-\text{NH-CH}_2-\text{CH}_2-\overset{\overset{\text{O}}{||}}{\text{C}}-\text{NH-CH}_2-\text{CH}_2-\text{SH}$$

Adenine

O

H H

H   H

O   OH

PO_3^{2-}

Coenzyme A (CoA)

Figure 28-33   Biosynthesis of coenzyme A from pantothenate, its vitamin precursor.

CHAPTER SUMMARY

1 Synthesis of Purine Ribonucleotides Almost all cells synthesize purine nucleotides *de novo* via similar metabolic pathways. The purine ring is constructed in an 11-step reaction sequence that yields IMP. AMP and GMP are then synthesized from IMP in separate pathways. Nucleoside diphosphates and triphosphates are sequentially formed from these products via phosphorylation reactions. The rates of synthesis of these various nucleotides are interrelated through feedback inhibition mechanisms that monitor their concentrations. Purine nucleotides may also be synthesized from free purines salvaged from nucleic acid degradation processes. The importance of these salvage reactions is demonstrated, for example, by the devastating and bizarre consequences of Lesch–Nyhan syndrome.

2 Synthesis of Pyrimidine Ribonucleotides Cells also synthesize pyrimidines *de novo* but, in this six-step process, a free base is formed before it is converted to a nucleotide, UMP. UTP is then formed by phosphorylation of UMP, and CTP is synthesized by the amination of UTP. Pyrimidine biosynthesis is regulated by feedback inhibition as well as by the concentrations of purine nucleotides.

3 Formation of Deoxyribonucleotides Deoxyribonucleotides are formed by reduction of the corresponding ribonucleotides. Three classes of ribonucleotide reductase (RNR) have been characterized: Class I RNR, which occurs in nearly all eukaryotes and many prokaryotes, contains an $Fe(III)$—O^{2-}—$Fe(III)$ group and a tyrosyl free radical; Class II and III RNRs, which occur only in prokaryotes, contain, respectively, a coenzyme B_{12} cofactor, and a [4Fe–4S] cluster together with a glycyl radical. All of them catalyze free radical–based reductions. The substrates for Class I and II RNRs are NDPs, whereas those for Class III RNRs are NTPs. Class I RNR has three independent regulatory sites that control its substrate specificity and its catalytic activity in part via its oligomerization state, thereby generating deoxynucleotides in the amounts required for DNA synthesis. The *E. coli* Class I RNR is reduced to its original state by electron-transport chains involving either thioredoxin, thioredoxin reductase, and NADPH; or glutaredoxin, glutathione, glutathione reductase, and NADPH. Thymine is synthesized by the methylation of dUMP by thymidylate synthase to form dTMP. The reaction's methyl source, N^5,N^{10}-methylene-THF, is oxidized in the reaction to yield dihydrofolate. N^5,N^{10}-Methylene-THF is subsequently regenerated through the sequential actions of dihydrofolate reductase and serine hydroxymethyltransferase. Since this sequence of reactions is required for DNA biosynthesis, it presents an excellent target for chemotherapy. FdUMP, a mechanism-based inhibitor of thymidylate synthase, and methotrexate, an antifolate that essentially irreversibly inhibits dihydrofolate reductase, are both highly effective anticancer agents.

4 Nucleotide Degradation Purine nucleotides are catabolized to yield uric acid. Depending on the species, the uric acid is either directly excreted or first degraded to simpler nitrogen-containing substances. Overproduction or underexcretion of uric acid in humans causes gout. Pyrimidines are catabolized in animal cells to amino acids.

5 Biosynthesis of Nucleotide Coenzymes The nucleotide coenzymes NAD^+ and $NADP^+$, FMN and FAD, and coenzyme A are synthesized in animals from vitamin precursors.

REFERENCES

General

Nyhan, W.L., Disorders of purine and pyrimidine metabolism, *Mol. Genet. Metab.* **86,** 25–33 (2005).

Valle, D. (Ed.), *The Online Metabolic & Molecular Bases of Inherited Disease,* http://www.ommbid.com/. [Part 11 contains chapters on defects in purine and pyrimidine metabolism.]

Purine Nucleotide Biosynthesis

Almassey, R.J., Janson, C.A., Kan, C.-C., and Hostomska, Z., Structures of the apo and complexed *Escherichia coli* glycinamide ribonucleotide transformylase, *Proc. Natl. Acad. Sci.* **89,** 6114–6118 (1992).

Eriksen, T.A., Kadziola, A., Bentsen, A.-K., Harlow, K.W., and Larsen, S., Structural basis for the function of *Bacillus subtilis* phosphoribosylpyrophosphate synthetase, *Nature Struct. Biol.* **7,** 303–308 (2000).

Greasley, S.E., Horton, P., Ramcharan, J., Beardsley, G.P., Benkovic, S.J., and Wilson, I.A., Crystal structure of a bifunctional transformylase and cyclohydrolase enzyme in purine biosynthesis, *Nature Struct. Biol.* **8,** 402–406 (2001).

Kappock, T.J., Ealick, S.E., and Stubbe, J., Modular evolution of the purine biosynthetic pathway, *Curr. Opin. Chem. Biol.* **4,** 567–572 (2000).

Levdikov, V.M., Barynin, V.V., Grebenko, A.I., Melik-Adamyan, W.R., Lamzin, V.S., and Wilson, K.S., The structure of SAICAR synthase: An enzyme in the *de novo* pathway of purine biosynthesis, *Structure* **6,** 363–376 (1998).

Li, C., Kappock, T.J., Stubbe, J., Weaver, T.M., and Ealick, S.E., X-ray crystal structure of aminoimidazole ribonucleotide synthetase (PurM) from the *Escherichia coli* purine biosynthetic pathway at 2.5 Å resolution, *Structure* **7,** 1155–1166 (1999).

Löffler, M., Fairbanks, L.D., Zameitat, E., Marinaki, A.M., and Simmonds, H.A., Pyrimidine pathways in health and disease, *Trends Mol. Med.* **11,** 430–437 (2005).

Mathews, I.I., Kappock, T.J., Stubbe, J., and Ealick, S.E., Crystal structure of *Escherichia coli* PurE, an unusual mutase in the purine biosynthetic pathway, *Structure* **7,** 1395–1406 (1999).

Poland, B.W., Fromm, H.J., and Honzatko, R.B., Crystal structures of adenylosuccinate synthetase from *Escherichia coli* complexed with GDP, IMP, hadacidin, NO_3^-, and Mg^{2+}, *J. Mol. Biol.* **264,** 1013–1027 (1996).

Sintchak, M.D., Fleming, M.A., Futer, O., Raybuck, S.A., Chambers, S.P., Caron, P.R., Murcko, M.A., and Wilson, K.P., Structure and mechanism of inosine monophosphate dehydrogenase in complex with the immunosuppressant mycophenolic acid, *Cell* **85,** 921–930 (1996).

Smith, J.L., Glutamine PRPP amidotransferase: Snapshots of an enzyme in action, *Curr. Opin. Struct. Biol.* **8,** 686–694 (1998).

Tesmer, J.J., Klem, T.J., Deras, M.L., Davisson, V.J., and Smith, J.L., The crystal structure of GMP synthetase reveals a novel catalytic triad and is a structural paradigm for two enzyme families, *Nature Struct. Biol.* **3,** 74–86 (1996).

Thoden, J.B., Kappock, T.J., Stubbe, J.A., and Holden, H.M., Three-dimensional structure of N^5-carboxyaminoimidazole ribonucleotide synthetase: A member of the ATP grasp protein superfamily, *Biochemistry* **38,** 15480–15492 (1999). [X-ray structure of PurK.]

Toth, E.A. and Yeates, T.O., The structure of adenylosuccinate lyase, an enzyme with dual activity in the *de novo* purine biosynthetic pathway, *Structure* **8,** 163–174 (2000).

Wang, W., Kappock, T.J., Stubbe, J.A., and Ealick, S.E., X-ray structure of glycinamide ribonucleotide synthetase from *Escherichia coli*, *Biochemistry* **37,** 15647–15662 (1998).

Zalkin, H. and Dixon, J.E., *De novo* purine nucleotide biosynthesis, *Prog. Nucleic Acid Res. Mol. Biol.* **42,** 259–285 (1992).

Pyrimidine Nucleotide Biosynthesis

Begley, T.P., Appleby, T.C., and Ealick, S.E., The structural basis for the remarkable catalytic proficiency of orotidine 5′-monophosphate decarboxylase, *Curr. Opin. Struct. Biol.* **10,** 711–718 (2000).

Evans, D.R. and Guy, H.I., Mammalian pyrimidine biosynthesis: Fresh insights into an ancient pathway, *J. Biol. Chem.* **279,** 33035–33038 (2004).

Jones, M.E., Orotidylate decarboxylase of yeast and man, *Curr. Top. Cell Regul.* **33,** 331–342 (1992).

Liu, S., Neidhardt, E.A., Grossman, T.H., Ocain, T., and Clardy, J., Structures of human dihydroorotate dehydrogenase in complex with antiproliferative agents, *Structure* **8,** 25–33 (1999).

Miller, B.G. and Wolfenden, R., Catalytic proficiency: The unusual case of OMP decarboxylase, *Annu. Rev. Biochem.* **71,** 847–885 (2002).

Scapin, G., Ozturk, D.H., Grubmeyer, C., and Sacchettini, J.C., The crystal structure of the orotate phosphoribosyltransferase complexed with orotate and α-D-5-phosphoribosyl-1-pyrophosphate, *Biochemistry* **34,** 10744–10754 (1995).

Thoden, J.B., Phillips, G.N., Jr., Neal, T.M., Raushel, F.M., and Holden, H.M., Molecular structure of dihydroorotase: A paradigm for catalysis through the use of a binuclear center, *Biochemistry* **40,** 6989–6997 (2001).

Traut, T.W. and Jones, M.E., Uracil metabolism—UMP synthesis from orotic acid or uridine and conversion of uracil to β-alanine: Enzymes and cDNAs, *Prog. Nucleic Acid Res. Mol. Biol.* **53,** 1–78 (1996).

Synthesis of Deoxynucleotides

Carreras, C.W. and Santi, D.V., The catalytic mechanism and structure of thymidylate synthase, *Annu. Rev. Biochem.* **64,** 721–762 (1995).

Eriksson, M., Uhlin, U., Ramaswamy, S., Ekberg, M., Regnström, K., Sjöberg, B.-M., and Eklund, H., Binding of allosteric effectors to ribonucleotide reductase protein R1: Reduction of active site cysteines promotes substrate binding, *Structure* **5,** 1077–1092 (1997).

Finer-Moore, J.S., Santi, D.V., and Stroud, R.M., Lessons and conclusions from dissecting the mechanism of a bisubstrate enzyme: thymidylate synthase mutagenesis, function, and structure, *Biochemistry* **42,** 248–256 (2003).

Kashlan, O.B., Scott, C.P., Lear, J.D., and Cooperman, B.S., A comprehensive model for the allosteric regulation of mammalian ribonuclease reductase. Functional consequences of ATP- and dATP-induced oligomerization of the large subunit, *Biochemistry* **41,** 462–474 (2002).

Knighton, D.R., Kan, C.-C., Howland, E., Janson, C.A., Hostomska, Z., Welsh, K.M., and Matthews, D.A., Structure of and kinetic channeling in bifunctional dihydrofolate reductase-thymidylate synthase, *Nature Struct. Biol.* **1,** 186–194 (1994).

Lennon, B.W., Williams, J.R., Jr., and Ludwig, M.L., Twists in catalysis: Alternating conformations in *Escherichia coli* thioredoxin reductase, *Science* **289,** 1190–1194 (2000).

Logan, D.T., Andersson, J., Sjöberg, B.-M., and Nordlund, P., A glycyl radical site in the crystal structure of a Class III ribonucleotide reductase, *Science* **283,** 1499–1504 (1999).

Matthews, D.A., Villafranca, J.E., Janson, C.A., Smith, W.W., Welsh, K., and Freer, S., Stereochemical mechanisms of action for thymidylate synthase based on the X-ray structure of the covalent inhibitory ternary complex with 5-fluoro-2′-deoxyuridylate and 5,10-methylenetetrahydrofolate, *J. Mol. Biol.* **214,** 937–948 (1990); *and* Hyatt, D.C., Maley, F., and Montfort, W.R., Use of strain in a stereospecific catalytic mechanism: Crystal structure of *Escherichia coli* thymidylate synthase bound to FdUMP and methylenetetrahydrofolate, *Biochemistry* **36,** 4585–4594 (1997).

Mol, C.D., Harris, J.M., McIntosh, E.M., and Tainer, J.A., Human dUTP pyrophosphatase: Uracil recognition by a β hairpin and active sites formed by three separate subunits, *Structure* **4,** 1077–1092 (1996).

Nordlund, P. and Eklund, H., Structure and function of the *Escherichia coli* ribonucleotide reductase protein R2, *J. Mol. Biol.* **232,** 123–164 (1993).

Nordlund, P. and Reichard, P., Ribonucleotide reductases, *Annu. Rev. Biochem.* **75,** 681–706 (2006).

Powis, G. and Montfort, W.R., Properties and biological activities of thioredoxins, *Annu. Rev. Biophys. Biomol. Struct.* **30,** 421–455 (2001).

Sintchak, M.D., Arjara, G., Kellog, B.A., Stubbe, J., and Drennan, C.L., The crystal structure of class II ribonucleotide reductase reveals how an allosterically regulated monomer mimics a dimer, *Nature Struct. Biol.* **9,** 293–300 (2002).

Stubbe, J. and Riggs-Gelasco, P., Harnessing free radicals: Formation and function of the tyrosyl radical in ribonucleotide reductase, *Trends Biochem. Sci.* **23,** 438–443 (1998).

Stubbe, J., Ge, J., and Yee, C.S., The evolution of ribonucleotide reduction revisited, *Trends Biochem. Sci.* **26,** 93–99 (2001); *and* Stubbe, J., Ribonucleotide reductases: The link between an RNA and a DNA world, *Curr. Opin. Struct. Biol.* **10,** 731–736 (2000).

Uhlin, U. and Eklund, H., Structure of ribonucleotide reductase protein R1, *Nature* **370,** 533–539 (1994).

Nucleotide Degradation

Enroth, C., Eger, B.T., Okamoto, K., Nishino, T., Nishino, T., and Pai, E., Crystal structure of bovine milk xanthine dehydrogenase and xanthine oxidase: Structure based mechanism of conversion, *Proc. Natl. Acad. Sci.* **97,** 10723–10728 (2000).

Parkman, R., Weinberg, K., Crooks, G., Nolta, I., Kapoor, N., and Kohn, D., Gene therapy for adenosine deaminase deficiency, *Annu. Rev. Med.* **51,** 33–47 (2000).

Wilson, D.K., Rudolph, F.B., and Quiocho, F.A., Atomic structure of adenosine deaminase complexed with a transition-state analog: Understanding catalysis and immunodeficiency mutations, *Science* **252,** 1278–1284 (1991); Wilson, D.K. and Quiocho, F.A., A pre-transition-state mimic of an enzyme: X-ray structure of adenosine deaminase with bound 1-deazaadenosine and zinc-activated water, *Biochemistry* **32,** 1689–1694 (1993); *and* Crystallographic observation of a trapped tetrahedral intermediate in a metalloenzyme, *Nature Struct. Biol.* **1,** 691–694 (1994).

Biosynthesis of Nucleotide Coenzymes

Belenky, P., Bogan, K.L., and Brenner, C., NAD^+ metabolism in health and disease, *Trends Biochem. Sci.* **32,** 12–19 (2007).

PROBLEMS

1. Azaserine (*O*-diazoacetyl-L-serine) and **6-diazo-5-oxo-L-norleucine (DON)**

$$\bar{N}=\overset{+}{N}=CH-\overset{O}{\overset{\|}{C}}-O-H_2C-\overset{NH_3^+}{\underset{COO^-}{\overset{|}{C}H}}$$

Azaserine

$$\bar{N}=\overset{+}{N}=CH-\overset{O}{\overset{\|}{C}}-CH_2-CH_2-\overset{NH_3^+}{\underset{COO^-}{\overset{|}{C}H}}$$

6-Diazo-5-oxo-L-norleucine (DON)

are glutamine analogs. They form covalent bonds to nucleophiles at the active sites of enzymes that bind glutamine, thereby irreversibly inactivating these enzymes. Identify the nucleotide biosynthesis intermediates that accumulate in the presence of either of these glutamine antagonists.

2. Suggest a mechanism for the AIR synthetase reaction (Fig. 28-2, Reaction 6).

*3. What is the energetic price, in ATPs, of synthesizing the hypoxanthine residue of IMP from CO_2 and NH_4^+?

4. Why is deoxyadenosine toxic to mammalian cells?

5. Indicate which of the following substances are mechanism-based inhibitors and explain your reasoning. (a) Tosyl-L-phenylalanine chloromethylketone with chymotrypsin (Section 15-3Ab). (b) Trimethoprim with bacterial dihydrofolate reductase. (c) The δ-lactone analog of $(NAG)_4$ with lysozyme (Section 15-2Cb). (d) Allopurinol with xanthine oxidase.

6. Why do individuals who are undergoing chemotherapy with cytotoxic (cell killing) agents such as FdUMP or methotrexate temporarily go bald?

7. Normal cells die in a nutrient medium containing thymidine and methotrexate that supports the growth of mutant cells defective in thymidylate synthase. Explain.

8. FdUMP and methotrexate, when taken together, are less effective chemotherapeutic agents than when either drug is taken alone. Explain.

9. Some microorganisms lack DHFR activity, but their thymidylate synthase has an FAD cofactor. What is the function of the FAD?

10. Why is gout more prevalent in populations that eat relatively large amounts of meat?

11. Gout resulting from the *de novo* overproduction of purines can be distinguished from gout caused by impaired excretion of uric acid by feeding a patient ^{15}N-labeled glycine and determining the distribution of ^{15}N in his or her excreted uric acid. What isotopic distributions are expected for each type of defect?

12. 6-Mercaptopurine,

6-Mercaptopurine

after its conversion to the corresponding nucleotide through salvage reactions, is a potent competitive inhibitor of IMP in the pathways for AMP and GMP biosynthesis. It is therefore a clinically useful anticancer agent. The chemotherapeutic effectiveness of 6-mercaptopurine is enhanced when it is administered with allopurinol. Explain the mechanism of this enhancement.

Schematic diagram of the eukaryotic preinitiation complex that is required for the transcription of DNA to messenger RNA. The TATA-box binding protein is shown in orange.

PART **V**

EXPRESSION AND TRANSMISSION OF GENETIC INFORMATION

PART V

Nucleic Acid Structures

Core subdomain II
Core subdomain I
C-terminal domain

CHAPTER 29

*There are two classes of nucleic acids, **deoxyribonucleic acid (DNA)** and **ribonucleic acid (RNA)**. DNA is the hereditary molecule in all cellular life-forms, as well as in many viruses.* It has but two functions:

1. To direct its own **replication** during cell division.

2. To direct the **transcription** of complementary molecules of RNA.

RNA, in contrast, has more varied biological functions:

1. The RNA transcripts of DNA sequences that specify polypeptides, **messenger RNAs (mRNAs),** direct the ribosomal synthesis of these polypeptides in a process known as **translation.**

2. The RNAs of ribosomes, which are about two-thirds RNA and one-third protein, have functional as well as structural roles.

3. During protein synthesis, amino acids are delivered to the ribosome by molecules of **transfer RNA (tRNA).**

4. Certain RNAs are associated with specific proteins to form **ribonucleoproteins** that participate in the post-transcriptional processing of other RNAs.

5. A variety of short RNAs participate in the control of eukaryotic gene expression and in protection against viruses, a phenomenon known as **RNA interference (RNAi).**

6. In many viruses, RNA, not DNA, is the carrier of hereditary information.

The structure and properties of DNA are introduced in Section 5-3. In this chapter we extend this discussion with emphasis on DNA; the structures of RNAs are detailed in Sections 31-4A and 32-2B. Methods of purifying, sequencing, and chemically synthesizing nucleic acids are discussed in Sections 6-6, 7-2, and 7-6, and recombinant DNA techniques are discussed in Section 5-5. Bioinformatics, as it concerns nucleic acids, is outlined in Section 7-4, and the Nucleic Acid Database is described in Section 8-3Cb.

1 DOUBLE HELICAL STRUCTURES

See Guided Exploration 23: DNA structures Double helical DNA has three major helical forms, B-DNA, A-DNA, and Z-DNA, whose structures are depicted in Fig. 29-1. In this section we discuss the major characteristics of each of these helical forms as well as those of double helical RNA and DNA–RNA hybrid helices.

A. B-DNA

The structure of **B-DNA** (Fig. 29-1, middle panels), the biologically predominant form of DNA, is described in Section 5-3A. To recapitulate (Table 29-1), B-DNA consists of a right-handed double helix whose two antiparallel sugar–phosphate chains wrap around the periphery of the helix. Its aromatic bases (A, T, G, and C), which occupy the core of the helix, form complementary A · T and G · C Watson–Crick base pairs (Fig. 5-12), whose planes are nearly perpendicular to the axis of the double helix. Neighboring base pairs, whose aromatic rings are 3.4 Å thick, are stacked in van der Waals contact, with the helix axis passing through the middle of each base pair. B-DNA is ~20 Å in diameter and has two deep grooves between its sugar–phosphate chains: the relatively narrow **minor groove,** which exposes that edge of the base pairs from which the glycosidic bonds (the bonds from the base N to the ribose C1′) extend (toward the bottom of Fig. 5-12), and the relatively wide **major groove,** which exposes the opposite edge of each base pair (toward the top of Fig. 5-12). Canonical (ideal) DNA has a helical twist of 10 base pairs (bp) per turn and hence a pitch (rise per turn) of 34 Å.

The Watson–Crick base pairs in either orientation are structurally interchangeable, that is, A · T, T · A, G · C, and

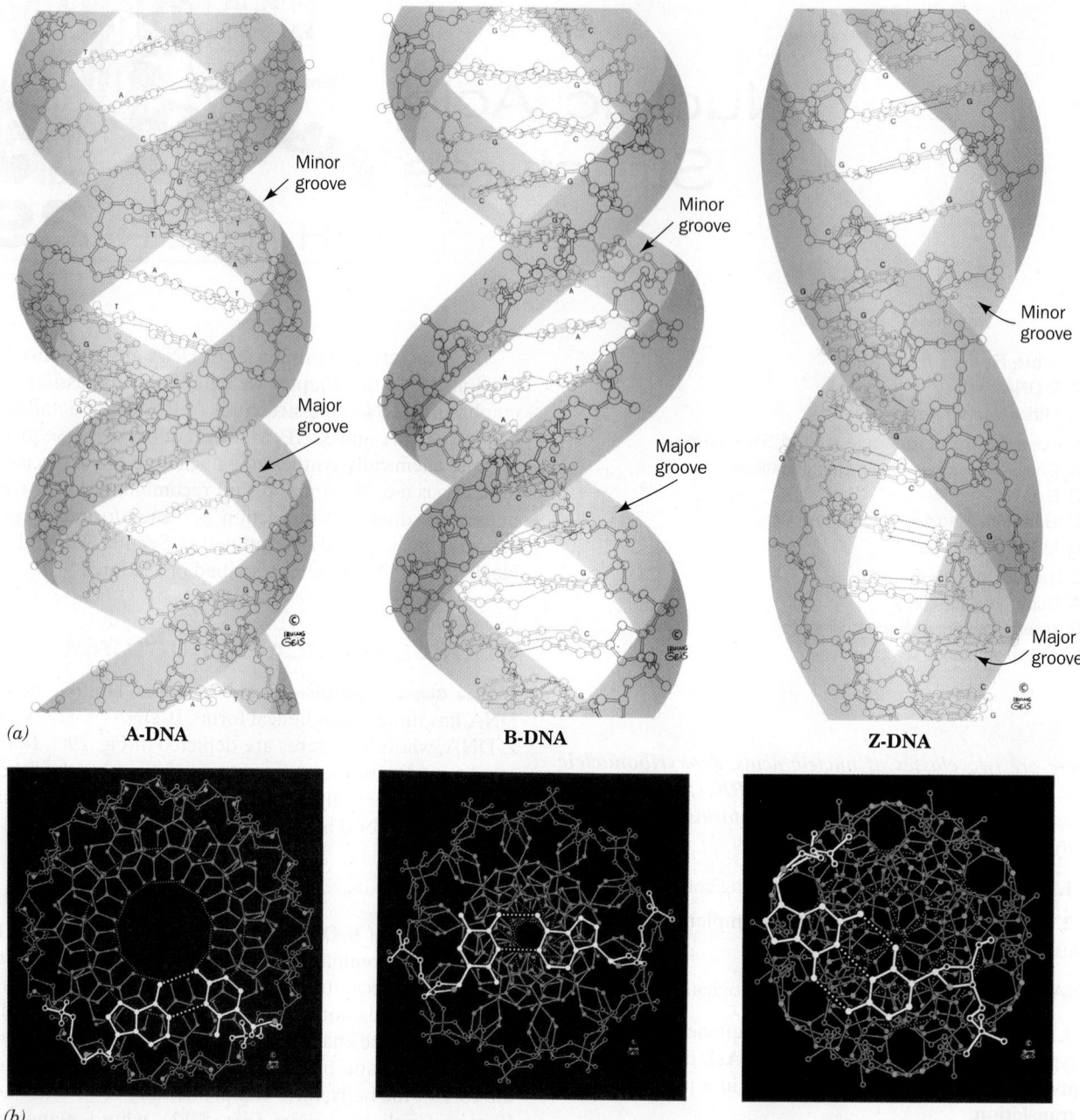

Figure 29-1   Structures of A-, B-, and Z-DNAs. (*a*) Ball-and-stick drawings viewed perpendicular to the helix axis. The sugar–phosphate backbones, which wind about the periphery of each molecule, are outlined by green ribbons, and the bases, which occupy its core, are red. Note that the two sugar–phosphate chains in each helix run in opposite directions so as to form right-handed double helices in A- and B-DNAs and a left-handed double helix in Z-DNA. (*b*) Views along the helix axis. The ribose ring O atoms are red and the nucleotide pair closest to the viewer is white. Note that the helix axis passes far "above" the major groove of A-DNA, through the base pairs of B-DNA, and through the edge of the minor groove of Z-DNA. Consequently, A-DNA has a hollow core whereas B- and Z-DNAs have solid cores. Also note that the deoxyribose residues in A- and B-DNAs have the same conformation in each helix, but those in Z-DNA have two different conformations so that alternate ribose residues

C · G can replace each other in the double helix without altering the positions of the sugar–phosphate backbones' C1' atoms. In contrast, any other combination of bases would significantly distort the double helix since the formation of a non-Watson–Crick base pair would require considerable reorientation of the DNA's sugar–phosphate backbones.

a. Real DNA Deviates from the Ideal Watson–Crick Structure

The DNA samples that were available when James Watson and Francis Crick formulated the Watson–Crick structure in 1953 were extracted from cells and hence consisted of molecules of heterogeneous lengths and base sequences.

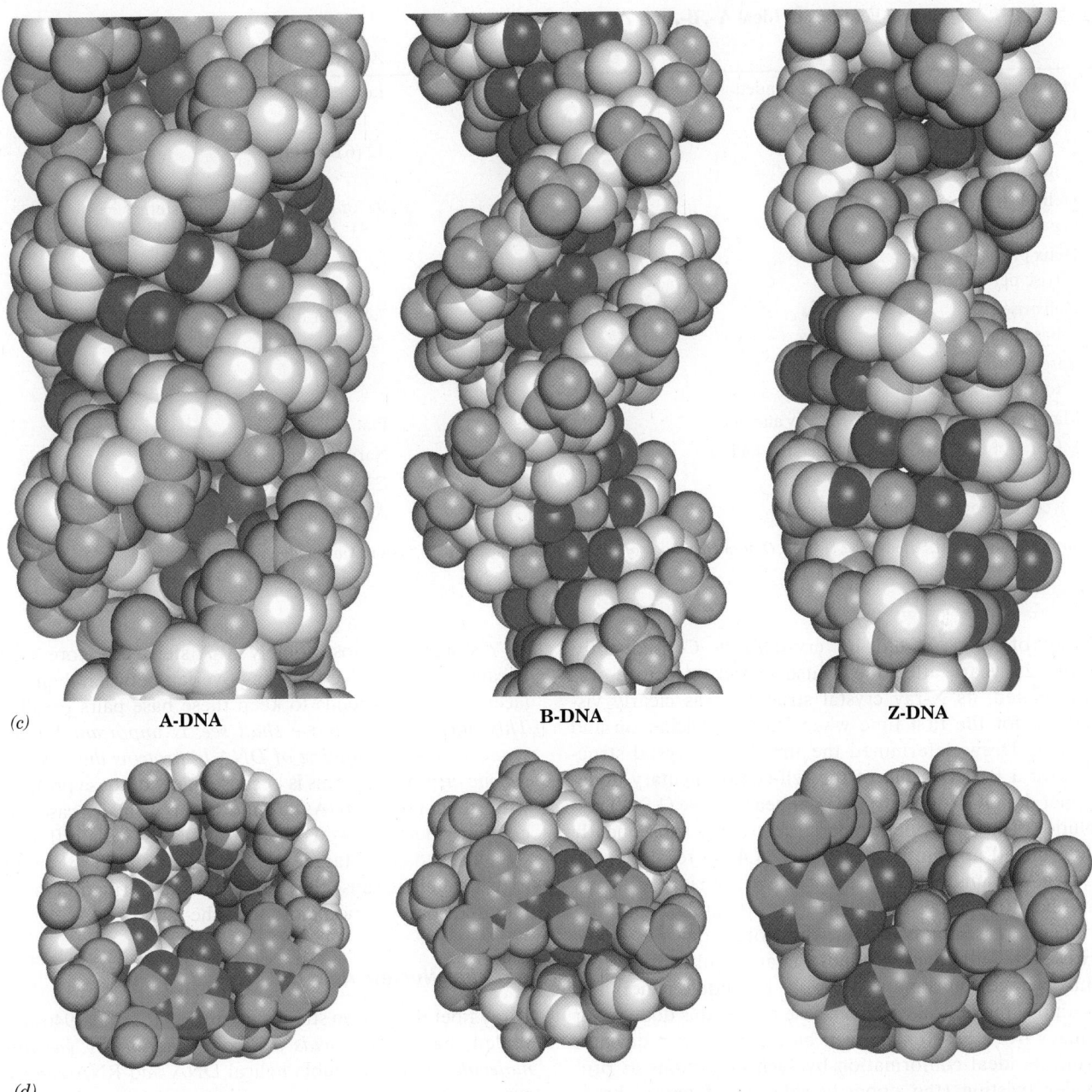

(c)   **A-DNA**   **B-DNA**   **Z-DNA**

(d)

lie at different radii. (*c*) Space-filling models viewed perpendicular to the helix axis and colored according to atom type (C white, N blue, O red, and P orange). (*d*) Space-filling models viewed along the helix axis and colored as in Part *c* but with the C atoms of the nucleotide pair closest to the viewer green. H atoms in Parts *b, c,* and *d* have been omitted for clarity. [Based on X-ray structures by the following: A-DNA, Olga Kennard, Dov Rabinovitch, Zippora Shakked, and Mysore Viswamitra,

Cambridge University, U.K. Nucleic Acid Database ID ADH010; B-DNA, Richard Dickerson and Horace Drew, Caltech. PDBid 1BNA; and Z-DNA, Andrew Wang and Alexander Rich, MIT. PDBid 2DCG. Illustration, Irving Geis. Image from the Irving Geis Collection, Howard Hughes Medical Institute. Reprinted with permission. Model coordinates for Parts *c* and *d* generated by Helen Berman, Rutgers University.] 🖊 **See Kinemage Exercises 17-1, 17-4, 17-5, and 17-6**

Such elongated molecules do not crystallize, but can be drawn into threadlike fibers in which the helix axes of the DNA molecules are all approximately parallel to the fiber axis but are poorly aligned, if at all, in any other way. The X-ray diffraction patterns of such fibers provide only crude, low-resolution images in which the base pair electron den-

sity is the average electron density of all the base pairs in the fiber. The Watson–Crick structure was based, in part, on the X-ray fiber diffraction pattern of B-DNA (Fig. 5-10).

By the late 1970s, advances in nucleic acid chemistry permitted the synthesis and crystallization of ever longer oligonucleotides of defined sequences (Section 7-6A),

Table 29-1 **Structural Features of Ideal A-, B-, and Z-DNA**

	A-DNA	B-DNA	Z-DNA
Helical sense	Right-handed	Right-handed	Left-handed
Diameter	~26 Å	~20 Å	~18 Å
Base pairs per helical turn	11.6	10	12 (6 dimers)
Helical twist per base pair	31°	36°	9° for pyrimidine–purine steps; 51° for purine–pyrimidine steps
Helix pitch (rise per turn)	34 Å	34 Å	44 Å
Helix rise per base pair	2.9 Å	3.4 Å	7.4 Å per dimer
Base tilt normal to the helix axis	20°	6°	7°
Major groove	Narrow and deep	Wide and deep	Flat
Minor groove	Wide and shallow	Narrow and deep	Narrow and deep
Sugar pucker	C3′-endo	C2′-endo	C2′-endo for pyrimidines; C3′-endo for purines
Glycosidic bond	Anti	Anti	Anti for pyrimidines; syn for purines

Source: Mainly Arnott, S., *in* Neidle, S. (Ed.), *Oxford Handbook of Nucleic Acid Structure*, p. 35, Oxford University Press (1999).

many of which could be crystallized. Consequently, some 25 years after the Watson–Crick structure was formulated, its X-ray crystal structure was clearly visualized for the first time when Richard Dickerson and Horace Drew determined the first X-ray crystal structure of a B-DNA, that of the self-complementary dodecamer d(CGCGAATTCGCG), at near-atomic (1.9 Å) resolution. This molecule, whose structure was subsequently determined at significantly higher (1.4 Å) resolution by Loren Williams, has an average rise per residue of 3.3 Å and has 10.1 bp per turn (a helical twist of 35.5° per bp), values that are nearly equal to those of canonical B-DNA. However, individual residues depart significantly from this average conformation (Fig. 29-1*a*, middle panel). For example, the helical twist per base pair in this dodecamer ranges from 26° to 43°. Each base pair further deviates from its ideal conformation by such distortions as propeller twisting (the opposite rotation of paired bases about the base pair's long axis; in the 1.4-Å resolution structure, this quantity ranges from −23° to −7°) and base pair roll (the tilting of a base pair as a whole about its long axis; this quantity ranges from −14° to 17°).

X-ray and NMR studies of numerous other double helical DNA oligomers have amply demonstrated that *the conformation of DNA, particularly B-DNA, is irregular in a sequence-specific manner,* although the rules specifying how sequence governs conformation have proved to be surprisingly elusive. This is because *base sequence does not so much confer a fixed conformation on a double helix as it establishes the deformability of the helix.* Thus, 5′-R–Y-3′ steps (where R and Y are the abbreviations for purines and pyrimidines, respectively) in B-DNA are easily bent because they exhibit relatively little ring–ring overlap between adjacent base pairs. In contrast, both Y–R steps and R–R steps (the latter, due to base pairing, are equivalent to

Y–Y steps), and most notably A–A steps, are more rigid because the extensive ring–ring overlap between their adjacent base pairs tends to keep these base pairs parallel. *This phenomenon, as we shall see, is important for the sequence-specific binding of DNA to proteins that process genetic information.* This is because many of these proteins wrap their target DNAs around them, in many cases by bending them by well over 90°. DNAs with different sequences than the target DNA would not bind so readily to the protein because they would resist deformation to the required conformation more than the target DNA.

B. *Other Nucleic Acid Helices*

X-ray fiber diffraction studies, starting in the mid-1940s, revealed that *nucleic acids are conformationally variable molecules.* Indeed, double helical DNA and RNA can assume several distinct structures that vary with such factors as the humidity and the identities of the cations present, as well as with base sequence. For example, fibers of B-DNA form in the presence of alkali metal ions such as Na⁺ when the relative humidity is 92%. In this subsection, we describe the other major conformational states of double-stranded DNA as well as those of double-stranded RNA and RNA–DNA hybrid helices.

a. **A-DNA's Base Pairs Are Inclined to the Helix Axis**

When the relative humidity is reduced to 75%, B-DNA undergoes a reversible conformational change to the so-called A form. Fiber X-ray studies indicate that *A-DNA forms a wider and flatter right-handed helix than does B-DNA* (Fig. 29-1, left panels; Table 29-1). A-DNA has 11.6 bp per turn and a pitch of 34 Å, which gives A-DNA an axial hole (Fig. 29-1*b, d*, left panels). A-DNA's most striking feature, however, is that the planes of its base pairs are tilted 20° with respect to the helix axis. Since its helix axis passes "above" the major groove side of the base pairs (Fig. 29-1*b*, *d*, left panels) rather than through them as in B-DNA, A-DNA has a deep major groove and a very shallow minor groove; it can be described as a flat ribbon wound around a 6-Å-diameter cylindrical hole. Most self-complementary oligonucleotides of <10 base pairs, for example, d(GGCCGGCC) and d(GGTATACC), crystallize in the A-DNA conformation. Like B-DNA, these molecules exhibit considerable sequence-specific conformational variation although the degree of variation is less than that in B-DNA.

A-DNA has, so far, been observed in only three biological contexts: at the cleavage center of topoisomerase II (Section 29-3Cd), at the active site of DNA polymerase (Section 30-2Ae), and in certain Gram-positive bacteria that have undergone **sporulation** (the formation, under environmental stress, of resistant although dormant cell types known as **spores;** a sort of biological lifeboat). Such spores contain a high proportion (20%) of **small acid-soluble spore proteins (SASPs).** Some of these SASPs induce B-DNA to assume the A form, at least *in vitro.* The DNA in bacterial spores exhibits a resistance to UV-induced damage that is abolished in mutants that lack these SASPs. This occurs because the B → A conformation change inhibits the UV-induced covalent cross-linking of pyrimidine bases (Section 30-5Aa), in part by increasing the distance between successive pyrimidines.

b. Z-DNA Forms a Left-Handed Helix

Occasionally, a seemingly well-understood or at least familiar system exhibits quite unexpected properties. Over 25 years after the discovery of the Watson–Crick structure, the crystal structure determination of the self-complementary hexanucleotide d(CGCGCG) by Andrew Wang and Alexander Rich revealed, quite surprisingly, *a left-handed double helix (Fig. 29-1, right panels; Table 29-1).* A similar helix is formed by d(CGCATGCG). *This helix, which has been dubbed* **Z-DNA,** *has 12 Watson–Crick base pairs per turn, a pitch of 44 Å, and, in contrast to A-DNA, a deep minor groove and no discernible major groove* (its helix axis passes "below" the minor groove side of its base pairs; Fig. 29-1*b,d,* right panels). Z-DNA therefore resembles a left-handed drill bit in appearance. The base pairs in Z-DNA are flipped 180° relative to those in B-DNA (Fig. 29-2) through conformational changes discussed in Section 29-2A. As a consequence, the repeating unit of Z-DNA is a dinucleotide, d(XpYp), rather than a single nucleotide as it is in the other DNA helices. The line joining successive phosphorus atoms on a polynucleotide strand of Z-DNA therefore follows a zigzag path around the helix (Fig. 29-1*a,c,* right panels; hence the name Z-DNA) rather than a smooth curve as it does in A- and B-DNAs (Fig. 29-1*a,c,* left and middle panels).

Fiber diffraction and NMR studies have shown that complementary polynucleotides with alternating purines and pyrimidines, such as poly d(GC) · poly d(GC) and poly

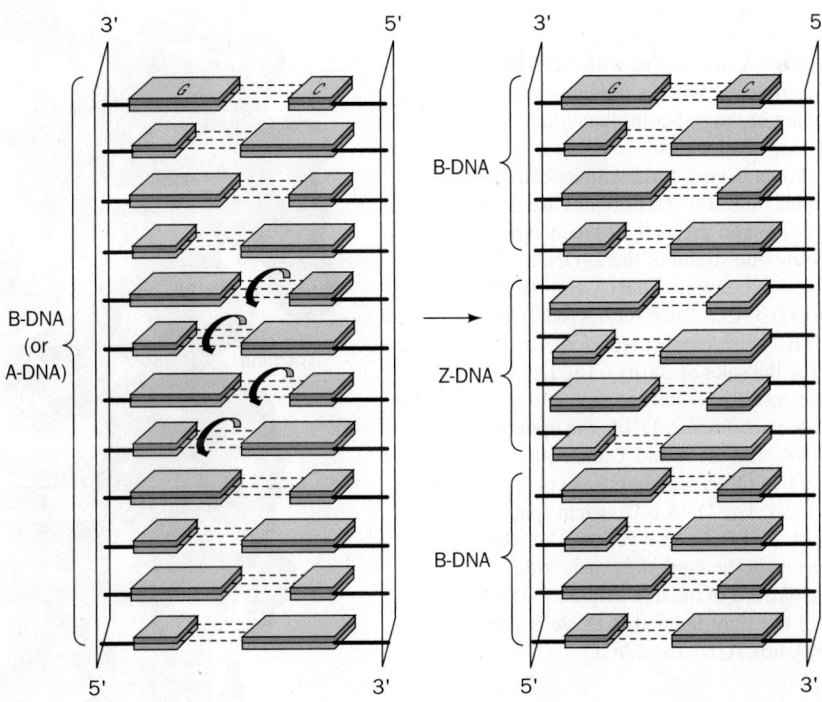

Figure 29-2 **Conversion of B-DNA to Z-DNA.** The conversion, here represented by a 4-bp DNA segment, involves a 180° flip of each base pair (*curved arrows*) relative to the sugar–phosphate chains (compare the base pair orientations of B- and Z-DNAs in Fig. 29-1*b, d*). Here, the different faces of the base pairs are colored red and green. Note that if the drawing on the left is taken as looking into the minor groove of unwound A- or B-DNA, then in the drawing on the right, we are looking into the major groove of the unwound Z-DNA segment. [After Rich, A., Nordheim, A., and Wang, A.H.-J., *Annu. Rev. Biochem.* **53,** 799 (1984).]

d(AC) · poly d(GT), take up the Z-DNA conformation at high salt concentrations. Evidently, *the Z-DNA conformation is most readily assumed by DNA segments with alternating purine–pyrimidine base sequences (for structural reasons explained in Section 29-2A).* A high salt concentration stabilizes Z-DNA relative to B-DNA by reducing the otherwise increased electrostatic repulsions between closest approaching phosphate groups on opposite strands (8 Å in Z-DNA vs 12 Å in B-DNA). The methylation of cytosine residues at C5, a common biological modification (Section 30-7), also promotes Z-DNA formation since a hydrophobic methyl group in this position is less exposed to solvent in Z-DNA than it is in B-DNA.

Does Z-DNA have any biological function? Rich has proposed that the reversible conversion of specific segments of B-DNA to Z-DNA under appropriate circumstances acts as a kind of switch in regulating genetic expression, and there are indications that it transiently forms behind actively transcribing RNA polymerase (Section 31-

4As). It was nevertheless surprisingly difficult to prove the *in vivo* existence of Z-DNA. A major difficulty was demonstrating that a particular probe for detecting Z-DNA, for example, a Z-DNA-specific antibody, does not in itself cause what would otherwise be B-DNA to assume the Z conformation—a kind of biological uncertainty principle (the act of measurement inevitably disturbs the system being measured). However, Rich has discovered several proteins that specifically bind Z-DNA, including a family of Z-DNA-binding protein domains named **Zα.** The existence of these proteins strongly suggests that Z-DNA does, in fact, exist *in vivo*.

The X-ray structure of the 81-residue Zα domain from the RNA editing enzyme **ADAR1** (Section 31-4As) in complex with d(TCGCGCG) has been determined (Fig. 29-3a). The CGCGCG segment of this heptanucleotide is self-complementary, and therefore forms a 2-fold symmetric, 6-bp segment of Z-DNA with an overhanging dT at the 5′ end of each strand (although these dT's are disordered in the

(a)

(b)

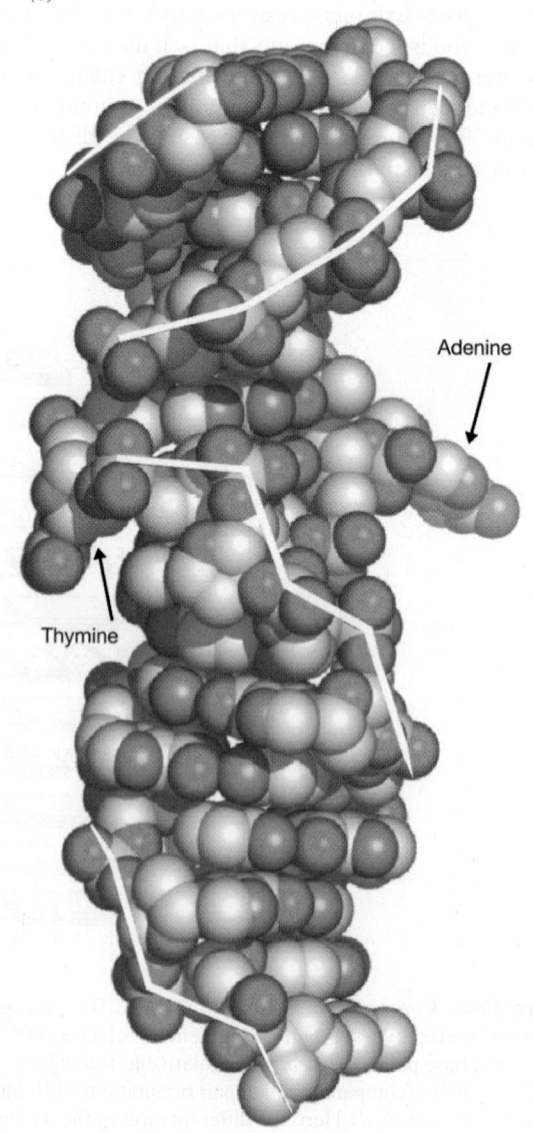

Figure 29-3   Structures of DNA in complex with ADAR1 Zα. (*a*) X-ray structure of the complex of two Zα domains with Z-DNA consisting of a duplex of the self-complementary hexamer d(CGCGCG) as viewed along its 2-fold axis of symmetry. The Z-DNA is shown in stick form with its backbones red and its remaining portions pink. The Zα domains are drawn in ribbon form with helices blue and sheets light green. Note that each Zα domain contacts only one strand of the Z-DNA. (*b*) X-ray structure of a junction between Z-DNA and B-DNA. The complex consists of d(GTCGCGCGCCATAAACC) and d(ACGGTTTATGGCGCGCG) in complex with four Zα domains (not shown here for the sake of clarity). The DNA forms a duplex with a 2-nucleotide overhang at each end in which the lower 8 nucleotide pairs form Z-DNA, the upper 6 nucleotide pairs form B-DNA, and the A and T nucleotides at the interface between these two DNA segments have been extruded from the double helix. The DNA is drawn in space-filling form with C gray, N blue, O red, and P yellow. The white lines connect adjacent P atoms in the same polynucleotide, thereby showing the zigzag pattern of the left-handed Z-DNA and the smoother pattern of the right-handed B-DNA. [Courtesy of Alexander Rich, MIT. PDBids 1QBJ and 2ACJ.]

X-ray structure). A monomeric unit of Zα binds to each strand of the Z-DNA, out of contact with the Zα that binds to the opposite strand. The protein primarily interacts with Z-DNA via hydrogen bonds and salt bridges between polar and basic protein side chains and the Z-DNA's sugar–phosphate backbone. Note that none of the DNA's bases participate in these associations. The protein's DNA-binding surface, which is complementary in shape to the Z-DNA, is positively charged, as is expected for a protein that interacts with several closely spaced, anionic phosphate groups. It is postulated that ADAR1's Zα domain targets it to the Z-DNA upstream of actively transcribing genes (for reasons discussed in Section 31-4As).

In an effort to visualize the structure of the junction between B-DNA and Z-DNA, Rich and Kyeong Kyu Kim cocrystallized Zα with a duplex DNA that has two overhanging nucleotides at each end and whose two 17-nucleotide strands have the sequences d(GTCGCGCGC-CATAAACC) and d(ACGGTTTATGGCGCGCG). The X-ray structure of this complex reveals that 8 nucleotide pairs at one end of the double helix are Z-DNA, 6 nucleotide pairs at the other end of the helix are B-DNA, and the nucleotides that would otherwise form an A · T base pair at the junction between these segments have been expelled from the double helix (Fig. 29-3b). The base pairs of the B- and Z-DNA segments form a continuous stack, which stabilizes the structure (Section 29-2C). Four Zα's are bound to the Z-DNA segment, two per polynucleotide strand, in a manner closely similar to that in the foregoing Zα–d(TCGCGCG) complex (for clarity, the Zα's are not shown in Fig. 29-3b) and the B-and Z-DNA segments adopt their standard conformations. Evidently, under the proper conditions, the handedness of duplex DNA can be reversed by breaking one base pair and ejecting its nucleotides from the duplex.

c. RNA-11 and RNA–DNA Hybrids Have an A-DNA-Like Conformation

Double helical RNA is unable to assume a B-DNA-like conformation because of steric clashes involving its 2′-OH groups. Rather, it usually assumes a conformation resembling A-DNA (Fig. 29-1, left panels) known as **A-RNA** or **RNA-11,** which ideally has 11.0 bp per helical turn, a pitch of 30.9 Å, and its base pairs inclined to the helix axis by 16.7°. Many RNAs, for example, transfer and ribosomal RNAs (whose structures are detailed in Sections 32-2B and 32-3A), contain complementary sequences that form double helical stems.

Hybrid double helices, which consist of one strand each of DNA and RNA, are also predicted to have A-RNA-like conformations. In fact, the X-ray structure, by Nancy Horton and Barry Finzel, of a 10-bp complex of the DNA oligonucleotide d(GGCGCCCGAA) with the complementary RNA oligonucleotide r(UUCGGGCGCC) reveals (Fig. 29-4) that it forms a double helix with A-RNA-like character (Table 29-1) in that it has 10.9 bp per turn, a pitch of 31.3 Å, and its base pairs are, on average, inclined to the helix axis by 13.9°. Nevertheless, this hybrid helix also has B-DNA-like qualities in that the width of its minor groove

Figure 29-4 X-ray structure of a 10-bp RNA–DNA hybrid helix consisting of d(GGCGCCCGAA) in complex with r(UUCGGGCGCC). The structure is shown in stick form with RNA C atoms cyan, DNA C atoms green, N blue, O red except for RNA O2′ atoms, which are magenta, and P yellow. [Based on an X-ray structure by Nancy Horton and Barry Finzel, Pharmacia & Upjohn, Inc., Kalamazoo, Michigan. PDBid 1FIX.]
🔗 See Interactive Exercise 31

(9.5 Å) is intermediate between those for canonical B-DNA (7.4 Å) and A-DNA (11 Å) and in that some of the ribose rings of its DNA strand have conformations characteristic of B-DNA (Section 29-2A), whereas others have conformations characteristic of A-RNA. Note that this structure is of biological significance because short segments of RNA · DNA hybrid helices occur in both the transcription of RNA on DNA templates (Section 31-2Ba) and in the initiation of DNA replication by short lengths of RNA (Section 30-1D). The RNA component of this helix is a substrate for **RNase H,** which specifically hydrolyzes the RNA strands of RNA · DNA hybrid helices *in vivo* (Section 30-4C).

2 FORCES STABILIZING NUCLEIC ACID STRUCTURES

Double-stranded DNA does not exhibit the structural complexity of proteins because it has only a limited repertoire of secondary structures and no comparable tertiary or quaternary structures (although see Section 29-3). This is perhaps to be expected since there is a far greater range of chemical and physical properties among the 20 amino acid residues of proteins than there is among the four DNA bases. However, many RNAs have well-defined tertiary structures (Sections 31-4A, 32-4Ca, 32-2B, and 32-3A).

In this section we examine the forces that give rise to the structures of nucleic acids. These forces are, of course, much the same as those that are responsible for the structures of proteins (Section 8-4) but, as we shall see, the way they combine gives nucleic acids properties that are quite different from those of proteins.

A. Sugar–Phosphate Chain Conformations

The conformation of a nucleotide unit, as Fig. 29-5 indicates, is specified by the six torsion angles of the sugar–phosphate backbone and the torsion angle describing the orientation of the base about the glycosidic bond. It would seem that these seven degrees of freedom per nucleotide would render polynucleotides highly flexible. Yet, as we shall see, these torsion angles are subject to a variety of internal constraints that greatly restrict their conformational freedom.

a. Torsion Angles About Glycosidic Bonds Have Only One or Two Stable Positions

The rotation of a base about its glycosidic bond is greatly hindered, as is best seen by the manipulation of a space-filling molecular model. Purine residues have two sterically permissible orientations relative to the sugar known as the **syn** (Greek: with) and **anti** (Greek: against) conformations (Fig. 29-6). For pyrimidines, only the anti conformation is easily formed because, in the syn conformation, the sugar residue sterically interferes with the pyrimidine's C2 substituent. In most double helical nucleic acids, all bases are in the anti conformation (Fig. 29-1a,c, left and middle panels). The exception is Z-DNA (Section 29-1Bb), in which the alternating pyrimidine and purine residues are anti and syn (Fig. 29-1a,c, right panels). *This explains Z-DNA's pyrimidine–purine alternation.* Indeed, the base pair flips that convert B-DNA to Z-DNA (Fig. 29-2) are brought about by rotating each purine base about its gly-

Figure 29-5   **The conformation of a nucleotide unit is determined by the seven indicated torsion angles.**

cosidic bond from the anti to syn conformation, whereas it is the sugars that rotate in the pyrimidine nucleotides, thereby maintaining them in their anti conformations.

b. Sugar Ring Pucker Is Largely Limited to Only a Few of Its Possible Arrangements

The ribose ring has a certain amount of flexibility that significantly affects the conformation of the sugar–phosphate backbone. The vertex angles of a regular pentagon are 108°, a value quite close to the tetrahedral angle (109.5°), so that one might expect the ribofuranose ring to be nearly flat. However, the ring substituents are eclipsed when the ring is planar. To relieve the resultant crowding, which even occurs between hydrogen atoms, the ring **puckers;** that is, it becomes slightly nonplanar, so as to reorient the ring substituents (Fig. 29-7; this is readily observed by the manipulation of a skeletal molecular model).

syn-Adenosine   anti-Adenosine   anti-Cytidine

Figure 29-6   **The sterically allowed orientations of purine and pyrimidine bases with respect to their attached ribose units.**

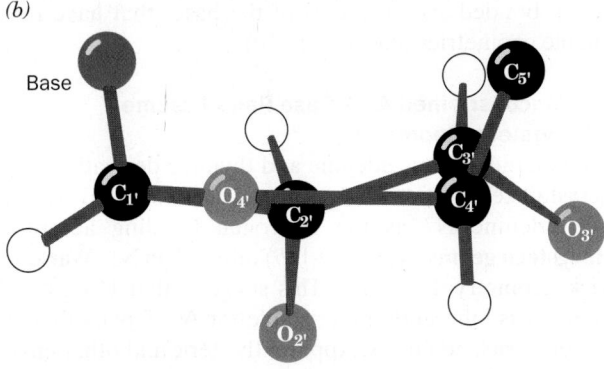

Figure 29-7   Ribose ring pucker. The substituents to (a) a planar ribose ring (here viewed down the C3′—C4′ bond) are all eclipsed. The resulting steric strain is partially relieved by ring puckering such as in (b), a half-chair conformation in which C3′ is the out-of-plane atom.

One would, in general, expect only three of a ribose ring's five atoms to be coplanar since three points define a plane. Nevertheless, in the great majority of the >50 nucleoside and nucleotide crystal structures that have been reported, four of the ring atoms are coplanar to within a few hundredths of an angstrom and the remaining atom is out of this plane by several tenths of an angstrom (the **half-**

chair conformation). If the out-of-plane atom is displaced to the same side of the ring as atom C5′, it is said to have the **endo** conformation (Greek: *endon,* within), whereas displacement to the opposite side of the ring from C5′ is known as the **exo** conformation (Greek: *exo,* out of). In the great majority of known nucleoside and nucleotide structures (molecules that are subject to few of the conformational constraints of double helices), the out-of-plane atom is either C2′ or C3′ (Fig. 29-8). C2′-*endo* is the most frequently occurring ribose pucker with C3′-*endo* and C3′-*exo* also being common. Other ribose conformations are rare.

The ribose pucker is conformationally important in nucleic acids because it governs the relative orientations of the phosphate substituents to each ribose residue (Fig. 29-8). For instance, it is difficult to build a regularly repeating model of a double helical nucleic acid unless the sugars are either C2′-*endo* or C3′-*endo*. In fact, canonical B-DNA has the C2′-*endo* conformation, whereas canonical A-DNA and RNA-11 are C3′-*endo*. In canonical Z-DNA, the purine nucleotides are all C3′-*endo* and the pyrimidine nucleotides are C2′-*endo*, which is another reason that the repeating unit of Z-DNA is a dinucleotide. The sugar puckers observed in the X-ray structures of A-DNA are, in fact, almost entirely C3′-*endo*. However, those of B-DNAs, although predominantly C2′-*endo*, exhibit significant variation including C4′-*exo*, O4′-*endo*, C1′-*exo*, and C3′-*exo*. This variation in B-DNA's sugar pucker is probably indicative of its greater flexibility relative to other types of DNA helices.

c. The Sugar–Phosphate Backbone Is Conformationally Constrained

If the torsion angles of the sugar–phosphate chain (Fig. 29-5) were completely free to rotate, there could probably be no stable nucleic acid structure. However, the comparison, by Muttaiya Sundaralingam, of some 40 nucleoside and nucleotide crystal structures has revealed that these angles are really quite restricted. For example, the torsion angle about the C4′—C5′ bond (γ in Fig. 29-5) is rather narrowly distributed such that O4′ usually has a gauche

Figure 29-8   Nucleotide sugar conformations. (a) The C3′-*endo* conformation (on the same side of the sugar ring as C5′), which occurs in A-DNA and RNA-11. (b) The C2′-*endo* conformation, which occurs in B-DNA. The distances between adjacent P atoms

in the sugar–phosphate backbone are indicated. [After Saenger, W., *Principles of Nucleic Acid Structure, p.* 237, Springer-Verlag (1983).] **See Kinemage Exercise 17-3**

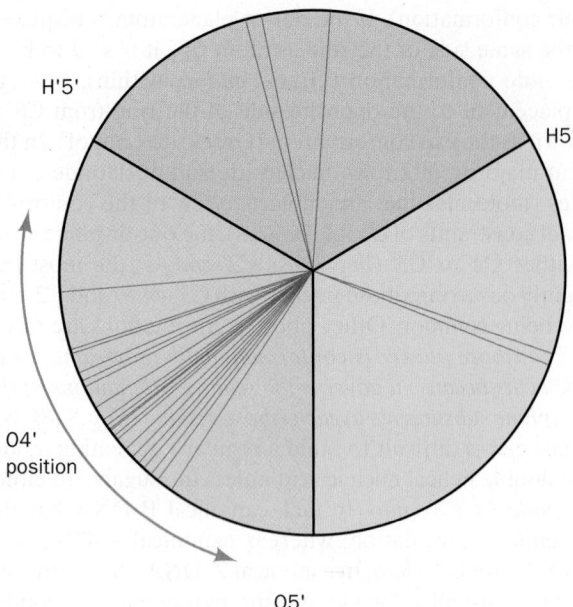

Figure 29-9 **Conformational wheel showing the distribution of the torsion angle about the C4′—C5′ bond.** The torsion angle (γ in Fig. 29-5) was measured in 33 X-ray structures of nucleosides, nucleotides, and polynucleotides. Each radial line represents the position of the C4′—O4′ bond in a single structure relative to the substituents of C5′ as viewed from C5′ to C4′. Note that most of the observed torsion angles fall within a relatively narrow range. [After Sundaralingam, M., *Biopolymers* **7**, 838 (1969).]

conformation (having a torsion angle of ~60°) with respect to O5′ (Fig. 29-9). This is because the presence of the ribose ring together with certain noncovalent interactions of the phosphate group stiffens the sugar–phosphate chain by restricting its range of torsion angles. These restrictions are even greater in polynucleotides because of steric interference between residues.

The sugar–phosphate conformational angles of the various double helices are all reasonably strain free. *Double helices are therefore conformationally relaxed arrangements of the sugar–phosphate backbone.* Nevertheless, the sugar–phosphate backbone is by no means a rigid structure, so, on strand separation, it assumes a random coil conformation.

B. Base Pairing

Base pairing is apparently a "glue" that holds together double-stranded nucleic acids. Only Watson–Crick pairs occur in the crystal structures of self-complementary oligonucleotides. It is therefore important to understand how Watson–Crick base pairs differ from other doubly hydrogen bonded arrangements of the bases that have reasonable geometries (e.g., Fig. 29-10).

a. Unconstrained A · T Base Pairs Assume Hoogsteen Geometry

When monomeric adenine and thymine derivatives are cocrystallized, the A · T base pairs that form invariably have adenine N7 as the hydrogen bonding acceptor **(Hoogsteen geometry;** Fig. 29-10*b*) rather than N1 (Watson–Crick geometry; Fig. 5-12). This suggests that Hoogsteen geometry is inherently more stable for A · T pairs than is Watson–Crick geometry. Apparently steric and other environmental influences make Watson–Crick geometry the preferred mode of base pairing in double helices. A · T pairs with Hoogsteen geometry are nevertheless of biological importance; for example, they help stabilize the tertiary structures of tRNAs (Section 32-2Ba). In contrast, monomeric G · C pairs always cocrystallize with Watson–Crick geometry as a consequence of their triply hydrogen bonded structures.

b. Non-Watson–Crick Base Pairs Are of Low Stability

The bases of a double helix, as we have seen (Section 5-3A), associate such that any base pair position may interchangeably be A · T, T · A, G · C, or C · G without affecting the conformations of the sugar–phosphate chains. One might reasonably suppose that this requirement of **geometric complementarity** of the Watson–Crick base pairs, A with T and G with C, is the only reason that other

Figure 29-10 **Some non-Watson–Crick base pairs.** (*a*) The pairing of adenine residues in the crystal structure of 9-methyladenine. (*b*) Hoogsteen pairing between adenine and thymine residues in the crystal structure of 9-methyladenine · 1-methylthymine. (*c*) A hypothetical pairing between cytosine and thymine residues. Compare these base pairs with the Watson–Crick base pairs in Fig. 5-12.

Figure 29-11 **The infrared spectra, in the N—H stretch region, of guanine, cytosine, and adenine derivatives.** The derivatives were analyzed both separately and in the indicated mixtures. The solvent, CDCl₃, does not hydrogen bond with the bases and is relatively transparent in the frequency range of interest. (*a*) G + C. The brown curve in the lower panel, which is the sum of the spectra in the two upper panels, is the calculated spectrum of G + C for noninteracting molecules. The band near 3500 cm⁻¹ in the observed G + C spectrum (*purple*) is indicative of a specific hydrogen bonding association between G and C. (*b*) G + A. The close match between the calculated and observed spectra of the G + A mixture indicates that G and A do not significantly interact. [After Kyogoku, Y., Lord, R.C., and Rich, A., *Science* **154,** 5109 (1966).]

base pairs do not occur in a double helical environment. In fact, this was precisely what was believed for many years after the DNA double helix was discovered.

Eventually, the failure to detect pairs of different bases in nonhelical environments other than A with T (or U) and G with C led Richard Lord and Rich to demonstrate, through spectroscopic studies, that *only the bases of Watson–Crick pairs have a high mutual affinity.* Figure 29-11*a* shows the infrared (IR) spectrum in the N—H stretch region of guanine and cytosine derivatives, both separately and in a mixture. The band in the spectrum of the G + C mixture that is not present in the spectra of either of its components is indicative of a specific hydrogen bonding interaction between G and C. Such an association, which can occur between like as well as unlike molecules, may be described by ordinary mass action equations.

$$B_1 + B_2 \rightleftharpoons B_1 \cdot B_2 \qquad K = \frac{[B_1 \cdot B_2]}{[B_1][B_2]} \qquad [29.1]$$

From analyses of IR spectra such as Fig. 29-11, the values of K for the various base pairs have been determined. The self-association constants of the Watson–Crick bases are given in the top of Table 29-2 (the hydrogen bonded association of like molecules is indicated by the appearance of new IR bands as the concentration of the molecule is increased). The bottom of Table 29-2 lists the association constants of the Watson–Crick pairs. Note that each of these latter quantities is larger than the self-association constants of either of their component bases, so that Watson–Crick

base pairs preferentially form from their constituents. In contrast, the non-Watson–Crick base pairs, A · C, A · G, C · U, and G · U, whatever their geometries, have association constants that are negligible compared with the self-pairing association constants of their constituents (e.g., Fig. 29-11*b*). *Evidently, a second reason that non-Watson–Crick base pairs do not occur in DNA double helices is that they have relatively little stability.* Conversely, the exclusive presence of Watson–Crick base pairs in DNA results, in part, from an **electronic complementarity** matching A to T and G to C. The theoretical basis of this electronic complementarity, which is an experimental observation, is obscure.

Table 29-2 **Association Constants for Base Pair Formation**

Base Pair	$K\ (M^{-1})^a$
Self-Association	
A · A	3.1
U · U	6.1
C · C	28
G · G	10^3–10^4
Watson–Crick Base Pairs	
A · U	100
G · C	10^4–10^5

[a]Data measured in deuterochloroform at 25°C.

Source: Kyogoku, Y., Lord, R.C., and Rich, A., *Biochim. Biophys. Acta* **179,** 10 (1969).

This is because the approximations inherent in theoretical treatments make them unable to accurately account for the minor (few $kJ \cdot mol^{-1}$) energy differences between specific and nonspecific hydrogen bonding associations. The double helical segments of many RNAs, however, contain occasional non-Watson–Crick base pairs, most often G · U, which have functional as well as structural significance (e.g., Sections 32-2Ba and 32-2Db).

c. Hydrogen Bonds Only Weakly Stabilize DNA

It is clear that hydrogen bonding is required for the specificity of base pairing in DNA that is ultimately responsible for the enormous fidelity required to replicate DNA with almost no error (Section 30-3D). Yet, as is also true for proteins (Section 8-4Ba), *hydrogen bonding contributes little to the stability of the double helix.* For instance, adding the relatively nonpolar ethanol to an aqueous DNA solution, which strengthens hydrogen bonds, destabilizes the double helix, as is indicated by its decreased melting temperature (T_m; Section 5-3Ca). This is because hydrophobic forces, which are largely responsible for DNA's stability (Section 29-2C), are disrupted by nonpolar solvents. In contrast, *the hydrogen bonds between the base pairs of native DNA are replaced in denatured DNA by energetically nearly equivalent hydrogen bonds between the bases and water.* This accounts for the thermodynamic observation that hydrogen bonding contributes only 2 to 8 kJ/mol to base pairing stability.

C. Base Stacking and Hydrophobic Interactions

Purines and pyrimidines tend to form extended stacks of planar parallel molecules. This has been observed in the structures of nucleic acids (e.g., Fig. 29-1) and in the several hundred reported X-ray crystal structures that contain nucleic acid bases. The bases in these structures are usually partially overlapped (e.g., Fig. 29-12). In fact, crystal structures of chemically related bases often exhibit similar stacking patterns. Apparently stacking interactions, which in the solid state are a form of van der Waals interaction (Section 8-4Ab), have some specificity, although certainly not as much as base pairing.

a. Nucleic Acid Bases Stack in Aqueous Solution

Bases aggregate in aqueous solution, as has been demonstrated by the variation of osmotic pressure with concentration. The van't Hoff law of osmotic pressure is

$$\pi = RTm \qquad [29.2]$$

where π is the osmotic pressure, m is the molality of the solute (mol solute/kg solvent), R is the gas constant, and T is the temperature. The molecular mass, M, of an ideal solute can be determined from its osmotic pressure since $M = c/m$, where $c = $ g solute/kg solvent.

If the species under investigation is of known molecular mass but aggregates in solution, Eq. [29.2] must be rewritten:

$$\pi = \phi RTm \qquad [29.3]$$

Figure 29-12   Stacking of adenine rings in the crystal structure of 9-methyladenine. The partial overlap of the rings is typical of the association between bases in crystal structures and in double helical nucleic acids. [After Stewart, R.F. and Jensen, L.H., *J. Chem. Phys.* **40,** 2071 (1964).]

where ϕ, the **osmotic coefficient,** indicates the solute's degree of association. ϕ varies from 1 (no association) to 0 (infinite association). The variation of ϕ with m for nucleic acid bases in aqueous solution (e.g., Fig. 29-13) is consistent with a model in which the bases aggregate in successive steps:

$$A + A \rightleftharpoons A_2 + A \rightleftharpoons A_3 + A \rightleftharpoons \cdots \rightleftharpoons A_n$$

where n is at least 5 (if the reaction goes to completion, $\phi = 1/n$). This association cannot be a result of hydrogen bonding since N^6, N^6-**dimethyladenosine,**

$$N^6, N^6\text{-Dimethyladenosine}$$

which cannot form interbase hydrogen bonds, has a greater degree of association than does adenosine (Fig. 29-13). Apparently *the aggregation arises from the formation of stacks of planar molecules.* This model is corroborated by proton NMR studies: The directions of the aggregates' chemical shifts are compatible with a stacked but not a hydrogen bonded model. The stacking associations of monomeric bases are not observed in nonaqueous solutions.

Single-stranded polynucleotides also exhibit stacking interactions. For example, poly(A) shows a broad increase of

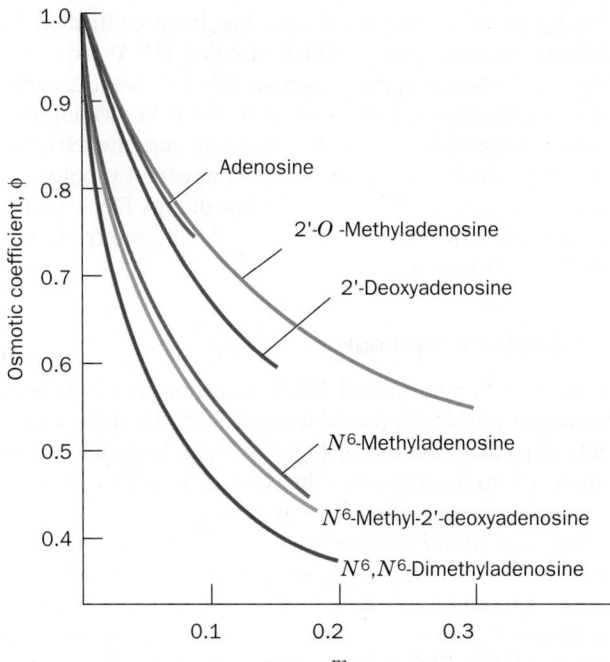

Figure 29-13 **Variation of the osmotic coefficient φ with the molal concentrations *m* of adenosine derivatives in H₂O.** The decrease of φ with increasing *m* indicates that these derivatives aggregate in solution. [After Broom, A.D., Schweizer, M.P., and Ts'o, P.O.P., *J. Am. Chem. Soc.* **89**, 3613 (1967).]

Figure 29-14 **Melting curves for poly(A) and ApA.** The broad temperature range of hyperchromic shifts at 258 nm of (*a*) poly(A) and (*b*) ApA is indicative of noncooperative conformational changes in these substances. Compare this figure with Fig. 5-16. [After Leng, M. and Felsenfeld, G., *J. Mol. Biol.* **15**, 457 (1966).]

UV absorbance with temperature (Fig. 29-14*a*). This hyperchromism (which is indicative of nucleic acid denaturation; Section 5-3Ca) is independent of poly(A) concentration, so that it cannot be a consequence of intermolecular disaggregation. Likewise, it is not due to a reduction in intramolecular hydrogen bonding because poly(N^6,N^6-dimethyladenosine) exhibits a greater degree of hyperchromism than does poly(A). The hyperchromism must therefore arise from some sort of stacking associations within a single strand that melt out with increasing temperature. This is not a very cooperative process, as is indicated by the broadness of the melting curve and the observation that short polynucleotides, including dinucleoside phosphates such as ApA, exhibit similar melting curves (Fig. 29-14*b*).

b. Nucleic Acid Structures Are Stabilized by Hydrophobic Forces

Stacking associations in aqueous solutions are largely stabilized by hydrophobic forces. One might reasonably suppose that hydrophobic interactions in nucleic acids are similar in character to those that stabilize protein structures. However, closer examination reveals that these two types of interactions are qualitatively different in character. Thermodynamic analysis of dinucleoside phosphate melting curves in terms of the reaction

Dinucleoside phosphate (*unstacked*)
$$\rightleftharpoons \text{dinucleoside phosphate}(stacked)$$

(Table 29-3) indicates that *base stacking is enthalpically*

driven and entropically opposed. Thus the hydrophobic interactions responsible for the stability of base stacking associations in nucleic acids are diametrically opposite in character to those that stabilize protein structures (which are enthalpically opposed and entropically driven; Section 8-4C). This is reflected in the differing structural properties of these interactions. For example, the aromatic side chains of proteins are almost never stacked and the crystal structures of aromatic hydrocarbons such as benzene, which resemble these side chains, are characteristically devoid of stacking interactions.

Hydrophobic forces in nucleic acids are poorly understood. The observation that they are different in character from the hydrophobic forces that stabilize proteins is nevertheless not surprising because the nitrogenous bases are considerably more polar than the hydrocarbon residues of proteins that participate in hydrophobic interactions. There is, however, no theory available that adequately explains the nature of hydrophobic forces in nucleic acids (our understanding of hydrophobic forces in proteins, it will be recalled, is similarly incomplete). They are complex interactions of which base stacking is probably a significant component. Whatever their origins, hydrophobic forces are of central importance in determining nucleic acid structures.

Table 29-3   **Thermodynamic Parameters for the Reaction**

Dinucleoside phosphate ⇌ dinucleoside phosphate
(*unstacked*)   (*stacked*)

Dinucleoside Phosphate	$\Delta H_{stacking}$ (kJ · mol^{-1})	$-T\,\Delta S_{stacking}$ (kJ · mol^{-1} at 25°C)
ApA	−22.2	24.9
ApU	−35.1	39.9
GpC	−32.6	34.9
CpG	−20.1	21.2
UpU	−32.6	36.2

Source: Davis, R.C. and Tinoco, I., Jr., *Biopolymers* **6**, 230 (1968).

D. *Ionic Interactions*

Any theory of the stability of nucleic acid structures must take into account the electrostatic interactions of their charged phosphate groups. Polyelectrolyte theory approximates the electrostatic interactions of DNA by considering the anionic double helix to be a homogeneously charged line or cylinder. We shall not discuss the details of this theory here, but note that it is often in reasonable agreement with experimental observations.

The melting temperature of duplex DNA increases with the cation concentration because these ions bind more tightly to duplex DNA than to single-stranded DNA due to the duplex DNA's higher anionic charge density. An increased salt concentration therefore shifts the equilibrium toward the duplex form, thus increasing the DNA's T_m. The observed relationship for Na^+ is

$$T_m = 41.1 X_{G+C} + 16.6 \log[Na^+] + 81.5 \quad [29.4]$$

where X_{G+C} is the mole fraction of $G \cdot C$ base pairs (recall that T_m increases with the $G + C$ content; Fig. 5-17); the equation is valid in the ranges $0.3 < X_{G+C} < 0.7$ and $10^{-3} M < [Na^+] < 1.0M$. Other monovalent cations such as Li^+ and K^+ have similar nonspecific interactions with phosphate groups. Divalent cations, such as Mg^{2+}, Mn^{2+}, and Co^{2+}, in contrast, specifically bind to phosphate groups, so that *divalent cations are far more effective shielding agents for nucleic acids than are monovalent cations.* For example, an Mg^{2+} ion has an influence on the DNA double helix comparable to that of 100 to 1000 Na^+ ions. Indeed, enzymes that mediate reactions with nucleic acids or just nucleotides (e.g., ATP) usually require Mg^{2+} for activity. Moreover, Mg^{2+} ions play an essential role in stabilizing the complex structures assumed by many RNAs such as transfer RNAs (tRNAs; Section 32-2B) and ribosomal RNAs (rRNAs; Section 32-3A).

3 SUPERCOILED DNA

♏ **See Guided Exploration 24: DNA supercoiling** Genetic analyses indicate that numerous viruses and bacteria have circular genetic maps, which implies that their chromosomes are

likewise circular. This conclusion has been confirmed by electron micrographs in which circular DNAs are seen (Fig. 29-15). Some of these circular DNAs have a peculiar twisted appearance, a phenomenon that is known equivalently as **supercoiling, supertwisting,** and **superhelicity.** Supercoiling arises from a biologically important topological property of covalently closed circular duplex DNA that is the subject of this section. It is occasionally referred to as DNA's tertiary structure.

A. *Superhelix Topology*

Consider a double helical DNA molecule in which both strands are covalently joined to form a circular duplex molecule as is diagrammed in Fig. 29-16 (each strand can be joined only to itself because the strands are antiparallel). *A geometric property of such an assembly is that the number of times one strand wraps about the other cannot be altered without first cleaving at least one of its polynucleotide strands.* You can easily demonstrate this to yourself with a buckled belt in which each edge of the belt represents a strand of DNA. The number of times the belt is twisted before it is buckled cannot be changed without unbuckling or cutting the belt (cutting a polynucleotide strand).

This phenomenon, as James White proved mathematically in 1969, is expressed

$$L = T + W \quad [29.5]$$

in which:

1. L, the **linking number** (also symbolized Lk), is the number of times that one DNA strand winds about the other. This integer quantity is most easily counted when the molecule's duplex axis is constrained to lie in a plane (see below). However, *the linking number is invariant no matter how the circular molecule is twisted or distorted so long as both its polynucleotide strands remain covalently intact; the linking number is therefore a topological property of the molecule.*

2. T, the **twist** (also symbolized Tw), is the number of complete revolutions that one polynucleotide strand makes about the duplex axis in the particular conformation

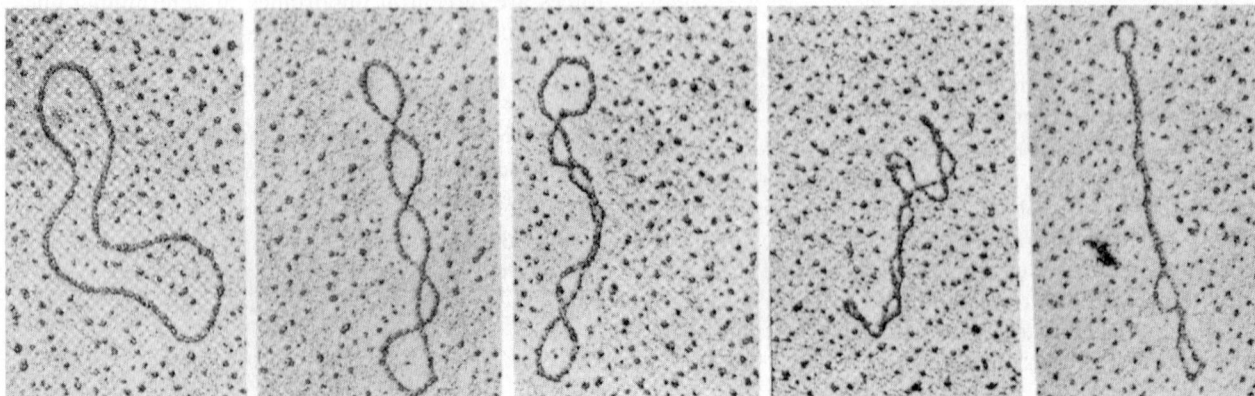

Figure 29-15 **Electron micrographs of circular duplex DNAs.** Their conformations vary from no supercoiling (*left*) to tightly supercoiled (*right*). [Electron micrographs by Laurien Polder.

From Kornberg, A. and Baker, T.A., *DNA Replication* (2nd ed.), p. 36, W.H. Freeman (1992). Used with permission.]

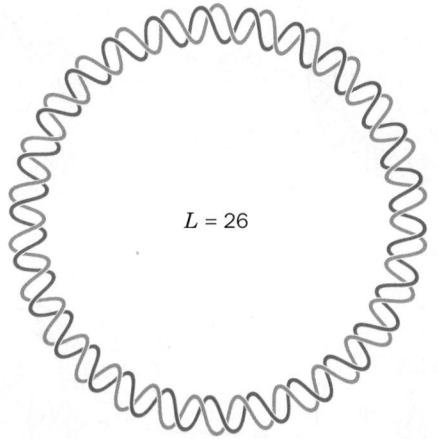

Figure 29-16 **Schematic diagram of covalently closed circular duplex DNA that has 26 double helical turns.** Its two polynucleotide strands are said to be **topologically bonded** to each other because, although they are not covalently linked, they cannot be separated without breaking covalent bonds.

perhelix axis in the conformation of interest. Its value is readily determined by projecting the DNA onto a plane and counting the number of times the duplex axis crosses itself (Fig. 29-17). *The writhing number is a measure of the DNA's superhelicity.* The difference between writhing and twisting is illustrated by the familiar example in Fig. 29-18. $W = 0$ when the DNA's duplex axis is constrained to lie in a plane (e.g., Fig. 29-16); then $L = T$, so L may be evaluated by counting the DNA's duplex turns.

The two DNA conformations diagrammed on the right of Fig. 29-19 are topologically equivalent; that is, they have the same linking number, L, but differ in their twists and writhing numbers. Note that T and W need not be integers (at least mathematically), only L. Although, strictly speaking, superhelicity is only defined for covalently closed circular duplex DNA, a linear segment of duplex DNA that is mechanically constrained from rotating at both ends (e.g., by protein anchors) has identical topological properties.

under consideration. By convention, T is positive for right-handed duplex turns, so that, for B-DNA in solution, the twist is normally the number of base pairs divided by 10.5 (the number of base pairs per turn of the B-DNA double helix under physiological conditions; see Section 29-3Bc).

3. W, the **writhing number** (also symbolized Wr), is the number of turns that the duplex axis makes about the su-

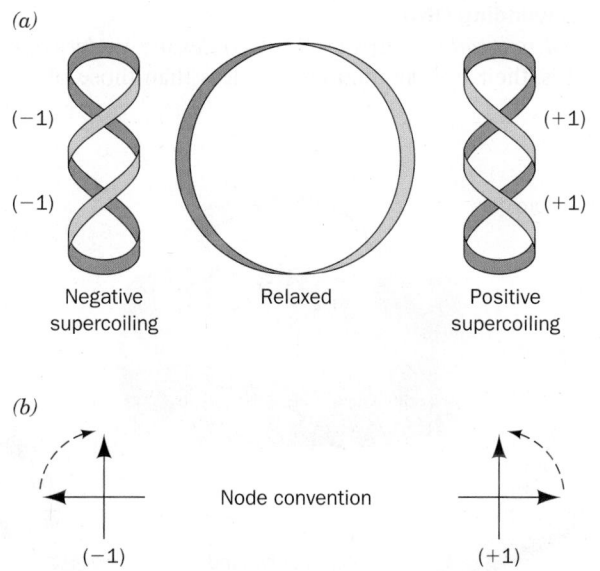

Figure 29-17 **Topological relationships in covalently closed duplex DNA.** (*a*) DNA molecules are represented by circular ribbons. DNA with no torsional strain is said to be relaxed. Underwinding or overwinding produces negative (-1) or positive ($+1$) supercoils. (*b*) A negative writhe has a crossover in which a clockwise rotation of the front segment of <180° aligns it over the back segment, whereas a positive writhe has the corresponding counterclockwise rotation. [After Deweese, J.E., Osheroff, M.A., and Osheroff, N., *Biochem. Mol. Biol. Educ.* **37,** 2 (2009).]

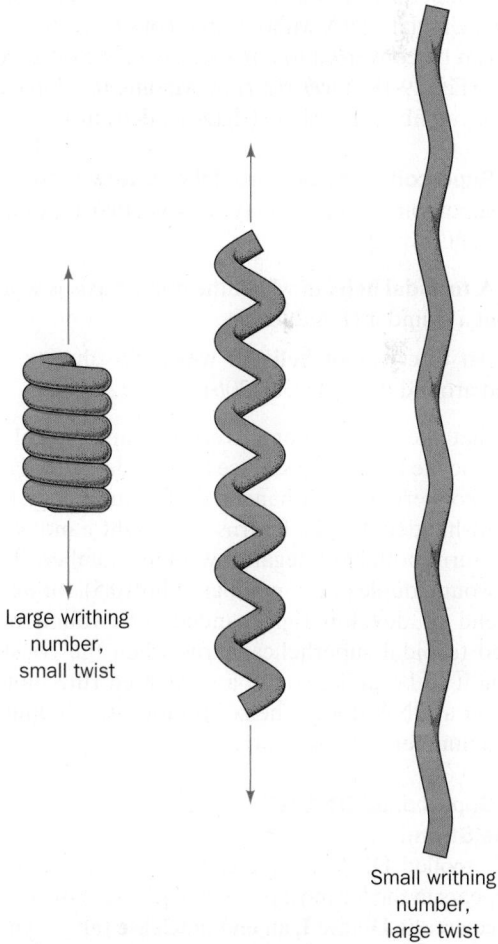

Figure 29-18 **The difference between writhing and twist as demonstrated by a coiled telephone cord.** Here the cord represents a double helical DNA molecule. In its relaxed state (*left*), the cord is in a helical form that has a large writhing number and a small twist. As the coil is pulled out (*middle*) until it is nearly straight (*right*), its writhing number becomes small as its twist becomes large.

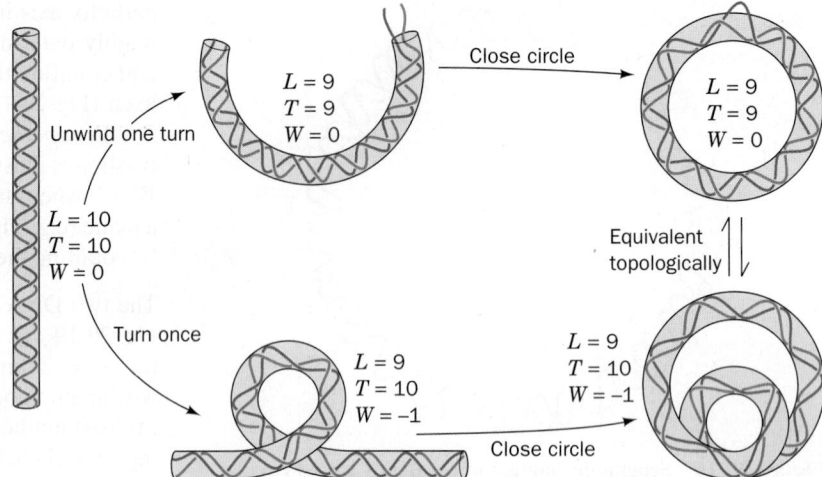

Figure 29-19   Two ways of introducing one supercoil into a DNA with 10 duplex turns. The two closed circular forms shown (*right*) are topologically equivalent; that is, they are interconvertible without breaking any covalent bonds. The linking number *L*, twist *T*, and writhing number *W* are indicated for each form. Strictly speaking, the linking number is only defined for a covalently closed circle.

Since L is constant in an intact duplex DNA circle, for every new double helical twist, ΔT, *there must be an equal and opposite superhelical twist, that is,* $\Delta W = -\Delta T$. For example, a closed circular DNA without supercoils (Fig. 29-19, *upper right*) can be converted to a negatively supercoiled conformation (Fig. 29-18, *lower right*) by winding the duplex helix the same number of positive (right-handed) turns.

a. Supercoils May Be Toroidal or Interwound

A supercoiled duplex may assume two topologically equivalent forms:

1. A toroidal helix in which the duplex axis is wound as if about a cylinder (Fig. 29-20*a*).

2. An interwound helix in which the duplex axis is twisted around itself (Fig. 29-20*b*).

Note that these two interconvertible superhelical forms have opposite handedness. Since left-handed toroidal turns may be converted to left-handed duplex turns (e.g., Fig. 29-19), left-handed toroidal turns and right-handed interwound turns both have negative writhing numbers. Thus an underwound duplex (*T* < number of bp/10.5), for example, will tend to develop right-handed interwound or left-handed toroidal superhelical turns when the constraints causing it to be underwound are released (the molecular forces in a DNA double helix promote its winding to its normal number of helical turns).

b. Supercoiled DNA Is Relaxed by Nicking One Strand

Supercoiled DNA may be converted to **relaxed circles** (as appears in the leftmost panel of Fig. 29-15) by treatment with **pancreatic DNase I**, an **endonuclease** (an enzyme that cleaves phosphodiester bonds within a polynucleotide strand) that cleaves only one strand of a duplex DNA. *One single-strand nick is sufficient to relax a supercoiled DNA.* This is because the sugar–phosphate chain opposite the nick is free to swivel about its backbone bonds (Fig. 29-5) so as to change the molecule's linking number and thereby alter its superhelicity. Supercoiling builds up elastic strain in a

DNA circle, much as it does in a rubber band. This is why the relaxed state of a DNA circle is not supercoiled.

B. *Measurements of Supercoiling*

Supercoiled DNA, far from being just a mathematical curiosity, has been widely observed in nature. In fact, its discovery in polyomavirus DNA by Jerome Vinograd stimulated the elucidation of the topological properties of superhelices rather than *vice versa*.

a. Intercalating Agents Control Supercoiling by Unwinding DNA

All naturally occurring DNA circles are underwound; that is, their linking numbers are less than those of their

Figure 29-20   Toroidal and interwound supercoils. A rubber tube that has been (*a*) toroidally coiled in a left-handed helix around a cylinder with its ends joined such that it has no twist jumps to (*b*) an interwound helix with the opposite handedness when the cylinder is removed. Neither the linking number, the twist, nor the writhing number are changed in this transformation.

Figure 29-21 **Sedimentation rate of underwound closed circular duplex DNA as a function of ethidium bromide concentration.** The intercalation of ethidium between the base pairs locally untwists the double helix (Fig. 29-22), which, since the linking number of the circle is constant, is accompanied by an equivalent increase in the writhing number. As the negatively coiled superhelix untwists, it becomes less compact and hence sediments more slowly. At the low point on the curve, the DNA circles have bound sufficient ethidium to become fully relaxed. As the ethidium concentration is further increased, the DNA supercoils in the opposite direction, yielding a positively coiled superhelix. The supertwisted appearances of the depicted DNAs have been verified by electron microscopy. [After Bauer, W.R., Crick, F.H.C., and White, J.H., *Sci. Am.* **243**(1), 129 (1980). Copyright © 1981 by Scientific American, Inc.]

corresponding relaxed circles. This phenomenon has been established by observing the effect of ethidium ion binding on the sedimentation rate of circular DNA (Fig. 29-21). Intercalating agents such as ethidium (a planar aromatic cation; Section 6-6Ca) alter a circular DNA's degree of superhelicity because they cause the DNA double helix to unwind (untwist) by ~26° at the site of the intercalated molecule (Fig. 29-22). $W < 0$ in an unconstrained underwound circle because of the tendency of a duplex DNA to maintain its normal twist of 1 turn per 10.5 bp. The titra-

Figure 29-22 **X-ray structure of a complex of ethidium with 5-iodo-UpA.** Ethidium (*red*) intercalates between the base pairs (*green with I purple*) of the double helically paired dinucleoside phosphate and thereby provides a model for the binding of ethidium to duplex DNA. [After Tsai, C.-C., Jain, S.C., and Sobell, H.M., *Proc. Natl. Acad. Sci.* **72,** 629 (1975).]

tion of a DNA circle by ethidium unwinds the duplex (decreases T), which must be accompanied by a compensating increase in W. This, at first, lessens the superhelicity of an underwound circle. However, as the circle binds more and more ethidium, its value of W passes through zero (relaxed circles) and then becomes positive, so that the circle again becomes superhelical. Thus the sedimentation rate of underwound DNAs, which is a measure of their compactness and therefore their superhelicity, passes through a minimum as the ethidium concentration increases. This is what is observed with native DNAs (Fig. 29-21). In contrast, the sedimentation rate of an overwound circle would only increase with increasing ethidium concentration.

b. DNAs Are Separated According to Their Linking Number by Gel Electrophoresis

Gel electrophoresis (Sections 6-4 and 6-6C) also separates similar molecules on the basis of their compactness, so that the rate of migration of a circular duplex DNA increases with its degree of superhelicity. The agarose gel electrophoresis pattern of a population of chemically identical DNA molecules with different linking numbers therefore consists of a series of discrete bands (Fig. 29-23). The molecules in a given band all have the same linking number and differ from those in adjacent bands by $\Delta L \pm 1$.

Comparison of the electrophoretic band patterns of **simian virus 40 (SV40)** DNA that had been enzymatically relaxed to varying degrees and then resealed (Fig. 29-23) reveals that 26 bands separate native from fully relaxed SV40 DNAs. Native SV40 DNA therefore has $W = -26$ (although it is somewhat heterogeneous in this quantity). Since SV40 DNA consists of 5243 bp, it has 1 negative superhelical turn per ~19 duplex turns. Such a **superhelix density** (W/T) is typical of circular DNAs from various biological sources.

c. DNA in Physiological Solution Has 10.5 Base Pairs per Turn

The insertion, using genetic engineering techniques (Section 5-5C), of an additional x base pairs into a superhelical DNA with a given linking number will increase the DNA's twist and hence decrease its writhing number by $x/h°$, where $h°$ is the number of base pairs per duplex turn. Such an insertion shifts the position of each band in the DNA's gel electrophoretic pattern by $x/h°$ of the spacing between bands. By measuring the effects of several such insertions, James Wang established that $h° = 10.5 \pm 0.1$ bp for B-DNA in solution under physiological conditions.

C. Topoisomerases

The normal biological functioning of DNA occurs only if it is in the proper topological state. In such basic biological processes as RNA transcription and DNA replication, the recognition of a base sequence requires the local separation of complementary polynucleotide strands. The negative supercoiling of naturally occurring DNAs results in a

Figure 29-23   Agarose gel electrophoresis pattern of SV40 DNA. Lane 1 contains the negatively supercoiled native DNA (*lower band;* the DNA was applied to the top of the gel). In lanes 2 and 3, the DNA was exposed for 5 and 30 min, respectively, to an enzyme, known as a type IA topoisomerase (Section 29-3Cb), that relaxes negative supercoils one at a time by increasing the DNA's linking number (L). The DNAs in consecutively higher bands of a given gel have successively increasing linking numbers ($\Delta L = +1$). [From Keller, W., *Proc. Natl. Acad. Sci.* **72,** 2553 (1975).]

torsional strain that promotes such separations since it tends to unwind the duplex helix (an increase in T must be accompanied by a decrease in W). *If DNA lacks the proper superhelical tension, the above vital processes (which themselves supercoil DNA; Sections 30-2C and 31-2Ca) occur quite slowly, if at all.*

*The supercoiling of DNA is controlled by a remarkable group of enzymes known as **DNA topoisomerases** (or simply **topoisomerases**).* They are so named because they alter the topological state (linking number) of circular DNA but not its covalent structure. There are two classes of topoisomerases:

1. Type I topoisomerases, the first of which was discovered by James Wang in 1971, act by creating transient single-strand breaks in DNA. Type I enzymes are subclassified into **types IA, IB,** and **IC topoisomerases** on the basis of their amino acid sequences and reaction mechanisms (see below). Type I topoisomerases are denoted by odd Roman numerals (e.g., topoisomerase I, III, etc.).

2. Type II topoisomerases, the first of which was discovered by Martin Gellert in 1976, act by making transient double-strand breaks in DNA with the accompanying hydrolysis of ATP to ADP + P_i. Type II enzymes are subclassified into **types IIA** and **IIB topoisomerases** on the basis of their amino acid sequences. Type II topoisomerases are denoted by even Roman numerals (e.g., topoisomerase II, IV, etc.).

a. Type I Topoisomerases Incrementally Relax Supercoiled DNA

Type I topoisomerases catalyze the relaxation of supercoils in DNA by changing their linking number in increments of one turn until the supercoil is entirely relaxed. In most organisms, type IA enzymes, which are present in all cells, relax only negatively supercoiled DNA, whereas type IB enzymes, which are present in all eukaryotes and many prokaryotes (but not *E. coli*), relax both negatively and positively coiled DNA. However, many hyperthermophiles, both eubacteria and archaea, have a type IA

topoisomerase known as **reverse gyrase** that induces positive supercoiling in DNA through the ATP-driven action of a **helicase** domain that is fused to the N-terminus of the topoisomerase domain (helicases are discussed in Section 30-2C). This suggests that positive supercoiling, which tightens the DNA double helix, protects DNA from thermal denaturation. Although types IA and IB topoisomerases are both monomeric, ~100-kD enzymes, they share no apparent sequence or structural similarities and function, as we shall see, via different enzymatic mechanisms.

A clue to the mechanism of type IA topoisomerase was provided by the observation that it reversibly **catenates** (interlinks) single-stranded circles (Fig. 29-24*a*). Apparently the enzyme operates by cutting a single strand, passing a single-strand loop through the resulting gap, and then resealing the break (Fig. 29-24*b*), thereby twisting double helical DNA by one turn. In support of this **strand passage** mechanism, the denaturation of type IA enzyme that has been incubated with single-stranded circular DNA yields a linear DNA that has its 5′-terminal phosphoryl group linked to the enzyme via a phosphoTyr diester linkage.

(a)

(b)

1   2   3

Duplex DNA (*n* turns)   Duplex DNA (*n* − 1 turns)

Figure 29-24 **Type IA topoisomerase action.** By cutting a single-stranded DNA, passing a loop of a second strand through the break, and then resealing the break, a type IA topoisomerase can (*a*) catenate two single-stranded circles or (*b*) unwind duplex DNA by one turn.

Type IA topoisomerase

In contrast, denatured type IB enzyme is linked to the 3′ end of DNA via a phosphoTyr linkage. *By forming such covalent enzyme–DNA intermediates, the free energy of the cleaved phosphodiester bond is preserved, so that no energy input is required to reseal the nick.*

b. Type IA Topoisomerase Functions via a Strand Passage Mechanism

Cells of *E. coli* contain two type IA topoisomerases named **topoisomerase I** (also called ω **protein**) and **topoisomerase III.** Topoisomerase III's Tyr 328 is the active site residue that forms a 5′-phosphoTyr linkage with the cleaved DNA. The X-ray structure of the inactive Y328F mutant of topoisomerase III in complex with the single-stranded octanucleotide d(CGCAACTT), determined by Alfonso

Figure 29-25 **X-ray structure of the Y328F mutant of *E. coli* topoisomerase III, a type IA topoisomerase, in complex with the single-stranded octanucleotide d(CGCAACTT).** The two views shown are related by a 90° rotation about a vertical axis. The protein's four domains are drawn in different colors. The DNA is drawn in space-filling form with C white, N blue, O red, and P yellow. The enzyme's active site is marked by the side chain of Phe 328, which is shown in space-filling form in yellow-green. [Based on an X-ray structure by Alfonso Mondragón, Northwestern University. PDBid 1I7D.]

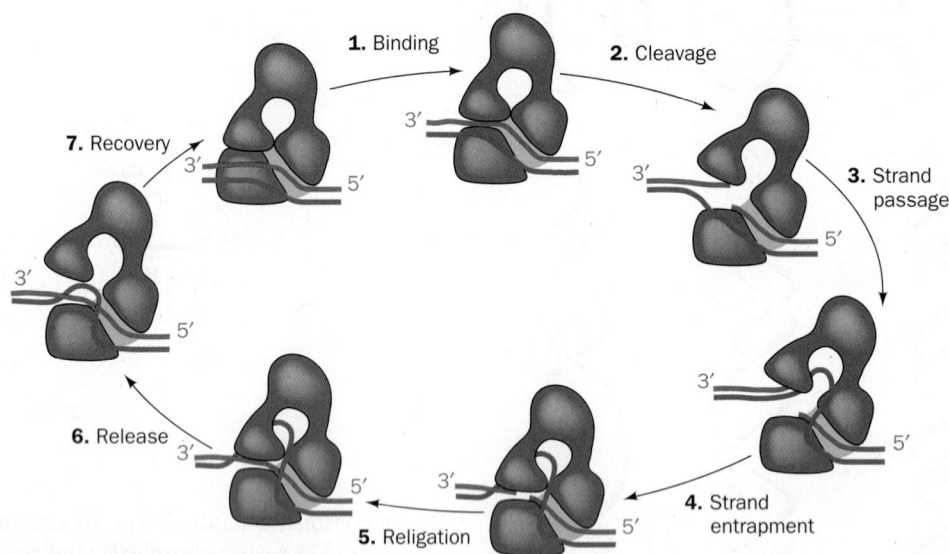

Figure 29-26 **Proposed mechanism for the strand passage reaction catalyzed by type IA topoisomerases.** The enzyme is shown in blue with the yellow patch representing the binding groove for single-stranded (ss) DNA. The two DNA strands, which are drawn in red and green, could represent the two strands of a covalently closed circular duplex or two ss circles. **(1)** The protein recognizes a ss region of the DNA, here the red strand, and binds it in its binding groove. This is followed by or occurs simultaneously with the opening of a gap between domains I and III. **(2)** The DNA is cleaved with the newly formed 5′ end becoming covalently linked to the active site Tyr and the segment with the newly formed 3′ end remaining tightly but noncovalently bound in the binding groove. **(3)** The unbroken (*green*) strand is passed through the opening or gate formed by the cleaved (*red*) strand to enter the protein's central hole. **(4)** The unbroken strand is trapped by the partial closing of the gap. **(5)** The two cleaved ends of the red strand are rejoined in what is probably a reversal of the cleavage reaction. **(6)** The gap between domains I and III reopens to permit the escape of the green strand, yielding the reaction product in which the green strand has been passed through a transient break in the red strand. **(7)** The enzyme returns to its initial state. If the two strands form a negatively supercoiled duplex DNA, its linking number, *L,* has increased by 1; if they are separate ss circles, they have been catenated or decatenated. For duplex DNA, this process can be repeated until all of its supercoils have been removed ($W = 0$). [After a drawing by Alfonso Mondragón, Northwestern University.]

Mondragón (Fig. 29-25), reveals that this 659-residue monomer folds into four domains which enclose an ~20 by 28 Å hole that is large enough to contain a duplex DNA and which is lined with numerous Arg and Lys side chains. The octanucleotide binds in a groove that is also lined with Arg and Lys side chains with its sugar–phosphate backbone in contact with the protein and with most of its bases exposed for possible base pairing. Curiously, this single-stranded DNA assumes a B-DNA-like conformation even though its complementary strand would be sterically excluded from the groove. The DNA strand is oriented with its 3' end near the active site, where, if the mutant Phe 328 were the wild-type Tyr, its side chain would be properly positioned to nucleophilically attack the phosphate group bridging the DNA's C6 and T7 to form a 5'-phosphoTyr linkage with T7 and release C6 with a free 3'-OH. This structure and that of the homologous and structurally similar *E. coli* topoisomerase I suggest the mechanism for the type IA topoisomerase-catalyzed strand passage reaction that is diagrammed in Fig. 29-26.

c. Type IB Topoisomerase Functions via a Controlled Rotation Mechanism

Human **topoisomerase I** is a 765-residue type IB topoisomerase (and hence is unrelated to *E. coli* topoisomerase I). It mediates the transient cleavage of one strand of a duplex DNA through the nucleophilic attack of Tyr 723 on a DNA P atom to yield a 3'-linked phosphoTyr diester bond and a free 5'-OH group on the succeeding nucleotide. Limited proteolysis studies revealed that topoisomerase I consists of four major regions: its N-terminal, core, linker, and C-terminal domains. The ~210-residue, highly polar, N-terminal domain, which is poorly conserved, contains several nuclear targeting signals and is dispensable for enzymatic activity.

The X-ray structure of the catalytically inactive Y723F mutant of topoisomerase I lacking its N-terminal 214

residues and in complex with a 22-bp palindromic duplex DNA was determined by Wim Hol (Fig. 29-27). The core domain of this bilobal protein is wrapped around the DNA in a tight embrace. If the mutant Phe 723 were the wild-type Tyr, its OH group would be colinear with the scissile P—O5' bond and hence ideally positioned to nucleophilically attack this P atom so as to form a covalent linkage with the 3' end of the cleaved strand. As expected, the protein interacts with the DNA in a largely sequence independent manner: Of the 41 direct contacts that the protein makes to the DNA, 37 are protein–phosphate interactions and only one is base-specific. The protein interacts to a much greater extent with the five base pairs of the DNA's upstream segment (which would contain the cleaved strand's newly formed 5' end; 29 of the 41 contacts) than it does with the base pairs of the DNA's downstream segment (to which Tyr 723 would be covalently linked; 12 of the 41 contacts).

Topoisomerase I does not seem sterically capable of unwinding supercoiled DNA via the strand passage mechanism that type IA topoisomerases appear to follow (Fig. 29-26). Rather, as is diagrammed in Fig. 29-28, it is likely that topoisomerase I relaxes DNA supercoils by permitting the cleaved duplex DNA's loosely held downstream segment to rotate relative to the tightly held upstream segment. This rotation can only occur about the sugar–phosphate bonds in the uncleaved strand (α, β, γ, ε, and ζ in Fig. 29-5) that are opposite the cleavage site because the cleavage frees these bonds to rotate. In support of this mechanism, the protein region surrounding the downstream segment contains 16 conserved, positively charged residues that form a ring about this duplex DNA, which would presumably hold the DNA in the ring but not in any specific orientation. Nevertheless, the downstream segment is unlikely to rotate freely because the cavity containing it is shaped so as to interact with the downstream segment during some portions of its rotation. Hence,

Figure 29-27   **X-ray structure of the N-terminally truncated, Y723F mutant of human topoisomerase I in complex with a 22-bp duplex DNA.** The protein's various domains and subdomains are drawn in different colors. The DNA's uncleaved strand is cyan, and the upstream and downstream portions of the scissile strand are purple and pink, respectively. [Courtesy of Wim Hol, University of Washington. PDBid 1A36.]

Figure 29-28   Controlled rotation mechanism for type IB topoisomerases. A highly negatively supercoiled DNA (*red, with a right-handed writhe*) is converted, via stages (*a*) through (*g*), to a less supercoiled form (*green*). Topoisomerase I is drawn as a bilobal space-filling structure, in which the cyan lobe is formed by core subdomains I and II (Fig. 29-27) and the purple lobe is formed by core subdomain III, the linker domain, and the C-terminal domain. The structure shown in (*d*), which is expanded by a factor of 2, shows the downstream portion of the rotating DNA (that containing the cleaved strand's new 5′ end) at 30° intervals, all differently colored. Since the enzyme is not always in direct contact with the rotating DNA, small rocking motions of the protein (*small curved arrows*) may accompany the controlled rotation. [Courtesy of Wim Hol, University of Washington.]

Controlled rotation

type IB topoisomerases are said to mediate a **controlled rotation** mechanism in relaxing supercoiled DNA. This unwinding is driven by the superhelical tension in the DNA and hence requires no other energy input. Eventually, the DNA is religated by a reversal of the cleavage reaction and the now less supercoiled DNA is released.

Type IC topoisomerase, whose only known family member, **topoisomerase V,** occurs exclusively in archaea, resembles type IB topoisomerases (which do not occur in archaea) in that it forms 3′-phosphoTyr intermediates and appears to function via a controlled rotation mechanism. However, it has no sequence or structural resemblance to type IB topoisomerases.

d. Type II Topoisomerases Function via a Strand Passage Mechanism

Bacteria have two types of type IIA topoisomerases: **DNA gyrase** (or just **gyrase**) and **topoisomerase IV,** both of which are A_2B_2 heterotetramers. Eukaryotic type IIA topoisomerases, which are named **topoisomerase II,** are homologous to bacterial type IIA topoisomerases but with

their A and B subunits fused so that they are homodimers. The type IIB topoisomerase family has only one member, **topoisomerase VI,** an A_2B_2 heterotetramer that occurs mainly in archaea (although some archaea express both type IIA and IIB topoisomerases). The A subunits of types IIA and IIB topoisomerases share a common ATPase module, but their B subunits are unrelated.

Gyrase is unique among topoisomerases in that it generates negative supercoils in DNA. All other topoisomerases but reverse gyrase only relax supercoils (DNA supercoiling in eukaryotes is generated differently from that in prokaryotes; Section 34-1Ba). It's A and B subunits are named **GyrA** and **GyrB.**

Type II topoisomerases can also catenate and decatenate double-stranded circles as well as tie and untie knots in them. The importance of this function can be seen as follows. The ~6.1 billion base pairs of DNA in a diploid human cell have an aggregate length of ~ 2 m and are confined to a nucleus that is 5 to 10 μm in diameter. Imagine that the ~20-Å-wide DNA was expanded by a factor of 5 million to the width of a 1-cm-diameter rope. It would then be ~10,000 km long (the distance from San Francisco to Rome) and confined to an expanded nucleus that is 25 to 50 m in diameter. Then imagine trying to manipulate the ropes in this system without generating tangles or knots. Yet such difficulties normally occur in cells during DNA replication, repair, and recombination (Chapter 30), as well as the segregation of the daughter chromosomes in dividing cells (Section 1-4B). It is the function of type II topoisomerases to untangle the DNA during these processes.

Bacterial but not eukaryotic type IIA topoisomerases are inhibited by a variety of substances including **novobiocin,** a member of the *Streptomyces*-derived **coumarin** family of antibiotics, and **ciprofloxacin** (trade name **Cipro**), a member of the synthetically generated **quinolone** family of antibiotics (their coumarin and quinolone groups are drawn in red):

Novobiocin

Ciprofloxacin

These agents profoundly inhibit bacterial DNA replication and RNA transcription, thereby demonstrating the importance of properly supercoiled DNA in these processes. Studies using *E. coli* gyrase mutants resistant to these substances have demonstrated that ciprofloxacin associates with GyrA and novobiocin binds to GyrB.

The gel electrophoretic pattern of duplex circles that have been exposed to gyrase shows a band pattern in which the linking numbers differ by increments of 2 rather than 1, as occurs with type I topoisomerases. *Evidently, gyrase acts by cutting both strands of a duplex, passing the duplex through the break, and resealing it* (Fig. 29-29). This hypothesis is corroborated by the observation that when gyrase is incubated with DNA and ciprofloxacin, and subsequently denatured with guanidinium chloride, a GyrA subunit remains covalently linked to the 5′ end of each of the two cut strands through a phosphoTyr linkage. These cleavage sites are staggered by 4 bp, thereby yielding sticky ends.

Saccharomyces cerevisiae (baker's yeast) topoisomerase II is a homodimer of 1428-residue subunits whose N- and C-terminal segments are homologous to *E. coli*'s GyrB (804 residues) and GyrA (878 residues) subunits, respectively. The breakage/reunion domain, which encompasses residues 410 to 1202, can, by itself, cleave duplex DNA but cannot transport it through the break without the action of the enzyme's ATPase domain (residues 1–409). However, the C-terminal segment (residues 1203–1428), which is poorly conserved, appears to be dispensable.

Although the structure of a full length type IIA topoisomerase is as yet unknown, those of its ATPase and breakage/reunion domains from both *E. coli* gyrase and yeast topoisomerase II have been determined. The X-ray structure of the homodimeric topoisomerase II ATPase in complex with the nonhydrolyzable ATP analog AMPPNP, determined by James Berger, consists of two domains (Fig.

Figure 29-29 A demonstration, in which DNA is represented by a ribbon, that cutting a duplex circle, passing the double helix through the resulting gap, and then resealing the break changes the linking number by 2. Separating the resulting single strands (slitting the ribbon along its length; *right*) indicates that one single strand makes two complete revolutions about the other.

(a)

(b)

(c)

Figure 29-30 **X-ray structures of *S. cerevisiae* topoisomerase II.** (*a*) Structure of a homodimer of the N-terminal ATPase domain (residues 7–406) in complex with AMPPNP as viewed with its 2-fold axis vertical. The protein is displayed in ribbon form with one subunit gray and the other colored in rainbow order from its N-terminus (*blue*) to its C-terminus (*red*). The bound AMPPNP molecules are shown in space-filling form with C green, N blue, O red, and P orange. (*b*) Structure of a homodimer of the DNA breakage/reunion domain (residues 419–1177) in complex with a doubly nicked 34-bp DNA. The structure is viewed with its 2-fold axis vertical. The DNA, which describes a 150° arc, is drawn in stick form with C green, N blue, O red, and P orange embedded in its semitransparent molecular surface (*orange*). The protein is displayed in ribbon form with one subunit gray and the other colored in rainbow order from its N-terminus (*blue*) to its C-terminus (*red*). The side chains of the two active site Tyr residues (Y782) are shown in space-filling form with C magenta and O red. The polypeptides in Parts *a* and *b*, which in the intact protein are joined by an 11-residue linker (represented by the dotted lines), presumably share the same 2-fold axis as drawn, although their relative orientation about this axis is unknown. (*c*) Same as Part *b* but viewed from below. Note that the Y782 side chain O atoms are positioned to link to opposite DNA strands at sites separated by 4 base pairs. [Based on X-ray structures by James Berger, University of California at Berkeley. PDBids 1PVG and 2RGR.] **See Interactive Exercise 32**

29-30*a*). The N-terminal domain binds AMPPNP and the C-terminal domains form the walls of a large hole through the dimer, which in the X-ray structure of the structurally similar *E. coli* GyrB is 20 Å across, the same width as the B-DNA double helix.

Berger also determined the X-ray structure of the yeast topoisomerase breakage/reunion domain in complex a 15-bp DNA that has a self-complementary 4-nucleotide overhang on the 5′ end of one its strands. The DNA thereby forms a 2-fold symmetric 34-bp duplex with nicks on opposite strands separated by 4 bp (Fig. 29-30*b,c*). These are precisely the sites at which the enzyme would cleave an intact DNA by linking its newly formed 5′-ending strands to the active site Tyr 782 residues. The protein binds the DNA in a positively charged groove that spans the width of the dimer and, in doing so, bends it through an arc of 150° (we shall see in later chapters that DNA-binding proteins often deform their bound DNA, although such an extreme deformation is unusual). Interestingly, the DNA between the two cleavage sites is essentially in the A form. There are almost no direct contacts between the protein and the DNA bases, as is expected for a protein with little sequence specificity. Note also that the C-terminal portions of the protein come together to enclose a large centrally located empty space.

Consideration of the foregoing two structures and those of the corresponding portions of *E. coli* gyrase suggests a type of strand passage model for the mechanism of type IIA topoisomerases (Fig. 29-31) in which the DNA duplex to be cleaved, the so-called G-segment (G for gate). binds in the above-described groove across the top of the cleavage/re-union domain. ATP binding to the ATP-binding domain then induces a sequence of conformational changes in which the G-segment is cleaved and the resulting two fragments are spread apart by at least 20 Å through the action of the protein. This permits the passage of the DNA's so-called T-segment (T for transported) through the break in the DNA and breakage/reunion domain's upper gate (which may also contain portions of the ATPase domain) into its central hole, thereby incrementing the DNA's linking number by 2. Then, in a process that is accompanied by ATP hydrolysis, the upper gate closes to reseal the cleaved DNA, and the T-segment passes through the breakage/reunion domain's bottom gate. Finally, the resulting ADP and P_i are released and the bottom gate closes to yield recycled enzyme. Many of these enzymatic states have been observed in the several known X-ray structures of type IIA topoisomerase components.

e. Topoisomerase Inhibitors Are Effective Antibiotics and Cancer Chemotherapy Agents

Coumarin derivatives such as novobiocin, and quinolone derivatives such as ciprofloxacin, specifically inhibit gyrase and are therefore antibiotics. In fact, ciprofloxacin is the most efficacious oral antibiotic against gram-negative bacteria presently in clinical use (novo-biocin's adverse side effects and the rapid generation of bacterial resistance to it have resulted in the discontinuation of its use in the treatment of human infections). A number of substances, including **doxorubicin** (also called **adriamycin;** a product of *Streptomyces peucetius*) and **etoposide** (a synthetic derivative),

Doxorubicin (Adriamycin)

Etoposide

Figure 29-31 **Model for the enzymatic mechanism of type II topoisomerases.** The protein's ATPase domain and the upper and lower portions of the breakage/reunion domain are colored yellow, red, and purple, respectively, and the DNA's G- and T-segments are colored gray and green, respectively. In **1,** the G-segment binds to the enzyme, thereby inducing the conformational change drawn in **2.** The binding of ATP (represented by asterisks) and a T-segment (**3**) induces a series of conformational changes in which the G-segment is cleaved as the upper gate opens. The ATPase domains dimerize, and the T-segment is transported through the break into the central hole (**4**). The DNA transport step is shown as proceeding through the hypothetical intermediate in square brackets. The G-segments are then resealed and the T-segment is released through the lower gate (**5**). This gate then closes as the ATP is hydrolyzed and the resulting ADP and P_i released to yield the enzyme in its starting state (**2**). [Courtesy of James Wang, Harvard University.]

inhibit eukaryotic type IIA topoisomerases and are therefore widely used in cancer chemotherapy. Indeed, ~50% of cancer chemotherapy regimens contain at least one drug targeted to type IIA topoisomerases.

Type IIA topoisomerase inhibitors act in either of two ways. Many of them, including novobiocin, inhibit their target enzyme's ATPase activity (novobiocin is a competitive inhibitor of ATP because it tightly binds to GyrB in a way that prevents the binding of ATP's adenine ring). They therefore kill cells by blocking topoisomerase activity, which results in the arrest of DNA replication and RNA transcription. However, other substances, including ciprofloxacin, doxorubicin, and etoposide, enhance the rate at which their target type IIA topoisomerases cleave double-stranded DNA and/or reduce the rate at which these enzymes reseal these breaks. Consequently, these agents induce higher than normal levels of transient protein-bridged breaks in the DNA of treated cells. These protein bridges are easily ruptured by the passage of the replication and transcription machinery, thereby rendering the breaks permanent. Although all cells have elaborate enzymatic systems to repair damaged DNA (Section 30-5), a sufficiently high level of DNA damage overwhelms these systems and hence results in cell death. Consequently, since rapidly replicating cells such as cancer cells have elevated levels of type IIA topoisomerases, they are far more likely to incur lethal DNA damage through the poisoning of their type IIA topoisomerases than are slow-growing or quiescent cells.

Type IB topoisomerases are specifically inhibited by the quinoline-based alkaloid **camptothecin**

Camptothecin

(a product of the Chinese yew tree, *Camptotheca acuminata*) and its derivatives, which act by stabilizing the covalent topoisomerase I–DNA complex. These compounds, the only known naturally occurring topoisomerase IB inhibitors, are potent anticancer agents.

CHAPTER SUMMARY

1 Double Helical Structures B-DNA consists of a right-handed double helix of antiparallel sugar–phosphate chains with ~10 bp per turn of 34 Å and with its bases nearly perpendicular to the helix axis. Bases on opposite strands hydrogen-bond in a geometrically complementary manner to form A · T and G · C Watson–Crick base pairs. At low humidity, B-DNA undergoes a reversible transformation to a wider, flatter right-handed double helix known as A-DNA. Z-DNA, which is formed at high salt concentrations by polynucleotides of alternating purine and pyrimidine base sequences, is a left-handed double helix. Double helical RNA and RNA · DNA hybrids have A-DNA-like structures. The conformation of DNA, particularly that of B-DNA, varies with its base sequence largely because DNA's flexibility varies with its base sequence.

2 Forces Stabilizing Nucleic Acid Structures The orientations about the glycosidic bond and the various torsion angles in the sugar–phosphate chain are sterically constrained in nucleic acids. Likewise, only a few of the possible sugar pucker conformations are commonly observed. Watson–Crick base pairing is both geometrically and electronically complementary. Yet hydrogen bonding interactions do not greatly stabilize nucleic acid structures. Rather, the structures are largely stabilized by hydrophobic interactions. Nevertheless, the hydrophobic forces in nucleic acids are qualitatively different in character from those that stabilize proteins. Electrostatic interactions between charged phosphate groups are also important structural determinants of nucleic acids.

3 Supercoiled DNA The linking number (L) of a covalently closed circular DNA is topologically invariant. Consequently, any change in the twist (T) of a circular duplex must be balanced by an equal and opposite change in its writhing number (W), which indicates its degree of supercoiling. Supercoiling can be induced by intercalation agents. The gel electrophoretic mobility of DNA increases with its degree of superhelicity. Naturally occurring DNAs are all negatively supercoiled and must be so in order to participate in DNA replication and RNA transcription.

Type IA topoisomerases relax negatively supercoiled DNAs via a strand passage mechanism in which they cleave a single strand of DNA to form a 5'-phosphoTyr bond, pass a single-strand DNA segment through the gap, and then reseal the gap. Type IB topoisomerases relax both negatively and positively supercoiled DNAs via a controlled rotation mechanism involving a single-strand cleavage in which a transient phosphoTyr bond is formed with the newly generated 3' end. Type II topoisomerases relax duplex DNA in increments of two supertwists at the expense of ATP hydrolysis by making a double-strand scission in the DNA so as to form two transient 5'-phosphoTyr linkages, passing the duplex through the break, and resealing it. DNA gyrase also generates negative supertwists in an ATP-dependent manner. Topoisomerases are the targets of several antibiotics and chemotherapeutic agents.

REFERENCES

General

Bloomfield, V.A., Crothers, D.M., and Tinoco, I., Jr., *Nucleic Acids: Structures, Properties, and Functions,* University Science Books (2000).

Creighton, T.E., *The Biophyical Chemistry of Nucleic Acids & Proteins,* Chapters 2–6, Helvetian Press (2010).

Neidle, S. (Ed.), *Oxford Handbook of Nucleic Acid Structure,* Oxford University Press (1999).

Saenger, W., *Principles of Nucleic Acid Structure,* Springer-Verlag (1984).

The double helix—50 years, *Nature* **421,** 395–453 (2003). [A supplement containing a series of articles on the historical, cultural, and scientific influences of the DNA double helix celebrating the fiftieth anniversary of its discovery.]

Structures and Stabilities of Nucleic Acids

Arnott, S., Historical article: DNA polymorphism and the early history of the double helix, *Trends Biochem. Sci.* **31,** 349–364 (2006).

Bacolla, A. and Wells, R.D., Non-B DNA conformations, genomic rearrangements, and human disease, *J. Biol. Chem.* **46,** 47411–47414 (2004).

Dickerson, R.E., Sequence-dependent B-DNA conformation in crystals and in protein complexes, *in* Sarma, R.H. and Sarma, M.H. (Eds.), *Structure, Motion, Interaction and Expression in Biological Molecules,* pp. 17–35, Adenine Press (1998); *and* DNA bending: the prevalence of kinkiness and the virtues of normality, *Nucleic Acids Res.* **26,** 1906–1926 (1998).

Fairhead, H., Setlow, B., and Setlow, P., Prevention of DNA damage in spores and *in vitro* by small, acid-soluble proteins from *Bacillus* species, *J. Bacteriol.* **175,** 1367–1374 (1993).

Ha, C.H., Lowenhaupt, K., Rich, A., Kim, Y.-G., and Kim, K.K., Crystal structure of a junction between B-DNA and Z-DNA reveals two extruded bases, *Nature* **437,** 1183–1186 (2005).

Horton, N.C. and Finzel, B.C., The structure of an RNA/DNA hybrid: A substrate of the ribonuclease activity of HIV-1 reverse transcriptase, *J. Mol. Biol.* **264,** 521–533 (1996).

Rich, A., The double helix: tale of two puckers, *Nature Struct. Biol.* **10,** 247–249 (2003). [A historical review.]

Schwartz, T., Rould, M.A., Lowenhaupt, K., Herbert, A., and Rich, A., Crystal structure of the Zα domain of the human editing enzyme ADAR1 bound to left-handed Z-DNA, *Science* **284,** 1841–1845 (1999).

Wing, R., Drew, H., Takano, T., Broka, C., Tanaka, S., Itakura, K., and Dickerson, R.E., Crystal structure analysis of a complete turn of B-DNA, *Nature* **287,** 755–758 (1980); and Shui, X., McFail-Isom, L., Hu, G.G., and Williams, L.D., The B-DNA decamer at high resolution reveals a spine of sodium, *Biochemistry* **37,** 8341–8355 (1998). [The Dickerson dodecamer at its original 2.5 Å resolution and at its later-determined 1.4 Å resolution.]

Supercoiled DNA

Bates, A.D. and Maxwell, A., *DNA Topology* (2nd ed.) Oxford University Press (2005).

Champoux, J.J., DNA topoisomerases: Structure, function, and mechanism, *Annu. Rev. Biochem.* **70,** 369–413 (2001).

Changela, A., DiGate, R., and Mondragón, A., Crystal structure of a complex of a type IA DNA topoisomerase with a single-stranded DNA, *Nature* **411,** 1077–1081 (2001); *and* Mondragón, A. and DiGate, R., The structure of *Escherichia coli* DNA topoisomerase III, *Structure* **7,** 1373–1383 (1999).

Classen, S., Olland, S., and Berger, J., Structure of the topoisomerase II ATPase region and its mechanism of inhibition by the chemotherapeutic agent ICRF-187, *Proc. Natl. Acad. Sci.* **100,** 10629–10634 (2003).

Deweese, J.E., Osheroff, M.A., and Osheroff, N., DNA topology and topoisomerases, *Biochem. Mol. Biol. Educ.* **37,** 2–10 (2009).

Dong, K.C. and Berger, J.M., Structural basis for the gate-DNA recognition and bending by type IIA topoisomerases. *Nature* **450,** 1201–1205 (2007). [X-ray structure of the yeast topoisomerase II breakage/reunion domain in complex with DNA.]

Dong, K.C. and Berger, J.M., Structure and function of DNA topoisomerases, *in* Rice, P.A. and Correll, C.C. (Eds.), *Protein–Nucleic Acid Interactions,* RSC Publishing (2008); *and* Corbett, K.D. and Berger, J.M., Structure, molecular mechanisms, and evolutionary relationships in DNA topoisomerases, *Annu. Rev. Biophys. Biomol. Struct.* **33,** 95–118 (2004).

Gadelle, D., Filée, J., Buhler, C., and Forterre, P., Phylogenomics of type II DNA topoisomerases, *Bioessays* **25,** 232–242 (2003).

Lebowitz, J., Through the looking glass: The discovery of supercoiled DNA, *Trends Biochem. Sci.* **15,** 202–207 (1990). [An informative eyewitness account of how DNA supercoiling was discovered.]

Li, T.-K. and Liu, L.F., Tumor cell death induced by topoisomerase-targeting drugs, *Annu. Rev. Pharmacol. Toxicol.* **41,** 53–77 (2001).

Maxwell, A., DNA gyrase as a drug target, *Biochem. Soc. Trans.* **27,** 48–53 (1999).

Redinbo, M.R., Stewart, L., Kuhn, P., Champoux, J.J., and Hol, W.G.J., Crystal structures of human topoisomerase I in covalent and noncovalent complexes with DNA, *Science* **279,** 1504–1513 (1998); Stewart, L., Redinbo, M.R., Qiu, X., Hol, W.G.J., and Champoux, J.J., A model for the mechanism of human topoisomerase I, *Science* **279,** 1534–1541 (1998); *and* Redinbo, M.R., Champoux, J.J., and Hol, W.G.J., Structural insights into the function of type IB topoisomerases, *Curr. Opin. Struct. Biol.* **9,** 29–36 (1999).

Taneja, B., Patel, A., Slesarev, A., and Mondraón, A., Structure of the N-terminal fragment of topoisomerase V reveals a new family of topoisomerases, *EMBO J.* **25,** 398–408 (2006).

Wang, J.C., Moving one DNA double helix through another by a type II DNA topoisomerase: The story of a simple molecular machine, *Q. Rev. Biophys.* **31,** 107–144 (1998).

Wang, J.C., Cellular roles of DNA topoisomerases: a molecular perspective, *Nature Rev. Mol. Cell Biol.* **3,** 430–440 (2002).

PROBLEMS

1. A · T base pairs in DNA exhibit greater variability in their propeller twisting than do G · C base pairs. Suggest the structural basis of this phenomenon.

*2. At Na^+ concentrations $>5M$, the T_m of DNA decreases with increasing $[Na^+]$. Explain this behavior. (*Hint:* Consider the solvation requirements of Na^+.)

*3. Why are the most commonly observed conformations of the ribose ring those in which either atom C2′ or atom C3′ is out of the plane of the other four ring atoms? (*Hint:* In puckering a planar ring such that one atom is out of the plane of the other four, the substituents about the bond opposite the out-of-plane atom remain eclipsed. This is best observed with a ball-and-stick model.)

4. Polyomavirus DNA can be separated by sedimentation at neutral pH into three components that have sedimentation coefficients of 20, 16, and 14.5S and that are known as Types I, II, and III DNAs, respectively. These DNAs all have identical base sequences and molecular masses. In 0.15M NaCl, both Types II and III DNA have melting curves of normal cooperativity and a T_m of 88°C. Type I DNA, however, exhibits a very broad melting curve and a T_m of 107°C. At pH 13, Types I and III DNAs have sedimentation coefficients of 53 and 16S, respectively, and Type II separates into two components with sedimentation coefficients of 16 and 18S. How do Types I, II, and III DNAs differ from one another? Explain their different physical properties.

5. When the helix axis of a closed circular duplex DNA of 2310 bp is constrained to lie in a plane, the DNA has a twist (T) of 207. When released, the DNA takes up its normal twist of 10.5 bp per turn. Indicate the values of the linking number (L), writhing number (W), and twist for both the constrained and unconstrained conformational states of this DNA circle. What is the superhelix density, σ, of both the constrained and unconstrained DNA circles?

6. A covalently closed circular duplex DNA has a 100-bp segment of alternating C and G residues. On transfer to a solution containing a high salt concentration, this segment undergoes a transition from the B conformation to the Z conformation. What is the accompanying change in its linking number, writhing number, and twist?

7. You have discovered an enzyme secreted by a particularly virulent bacterium that cleaves the C2′—C3′ bond in the deoxyribose residues of duplex DNA. What is the effect of this enzyme on supercoiled DNA?

8. A bacterial chromosome consists of a protein–DNA complex in which its single DNA molecule appears to be supercoiled, as demonstrated by ethidium bromide titration. However, in contrast to the case with naked circular duplex DNA, the light single-strand nicking of chromosomal DNA does not abolish this supercoiling. What does this indicate about the structure of the bacterial chromosome, that is, how do its proteins constrain its DNA?

9. Although types IA and IIA topoisomerases exhibit no significant sequence similarity, it has been suggested that they are distantly related based on the similarities of certain aspects of their enzymatic mechanisms. What are these similarities?

10. Draw the mechanism of DNA strand cleavage and rejoining mediated by topoisomerase IA.

Fingers   Thumb

3'

Palm   5'

DNA Replication, Repair, and Recombination

CHAPTER **30**

Here we begin a three-chapter series on the basic processes of gene expression: DNA replication (this chapter), transcription (Chapter 31), and translation (Chapter 32). These processes have been outlined in Section 5-4. We shall now discuss them in greater depth with an emphasis on how we have come to know what we know.

1 DNA REPLICATION: AN OVERVIEW

Watson and Crick's seminal paper describing the DNA double helix ended with the statement: "It has not escaped our notice that the specific pairing we have postulated immediately suggests a possible copying mechanism for the genetic material." In a succeeding paper they expanded on this rather cryptic remark by pointing out that a DNA strand could act as a template to direct the synthesis of its complementary strand. Although Meselson and Stahl demonstrated, in 1958, that DNA is, in fact, semiconservatively replicated (Section 5-3B), it was not until some 20 years later that the mechanism of DNA replication in prokaryotes was understood in reasonable detail. This is because, as we shall see in this chapter, the DNA replication process rivals translation in its complexity but is mediated by often loosely associated protein assemblies that are present in only a few copies per cell. *The surprising intricacy of DNA replication compared to the chemically similar transcription process (Section 31-2) arises from the need for extreme accuracy in DNA replication so as to preserve the integrity of the genome from generation to generation.*

A. Replication Forks

*DNA is replicated by enzymes known as **DNA-directed DNA polymerases** or simply **DNA polymerases**.* These enzymes utilize single-stranded DNA as templates on which to catalyze the synthesis of the complementary strand from the appropriate deoxynucleoside triphosphates (Fig. 30-1). The incoming nucleotides are selected by their ability to form Watson–Crick base pairs with the template DNA so that the newly synthesized DNA strand forms a double helix with the template strand. *Nearly all known DNA polymerases can only add a nucleotide donated by a nucleoside triphosphate to the free 3'-OH group of a base paired polynucleotide so that DNA chains are extended only in the 5' → 3' direction.* DNA polymerases are discussed further in Sections 30-2A, 30-2B, and 30-4B.

a. Duplex DNA Replicates Semiconservatively at Replication Forks

John Cairns obtained the earliest indications of how chromosomes replicate through the autoradiography of replicating DNA. Autoradiograms of circular chromosomes grown in a medium containing [³H]thymidine show

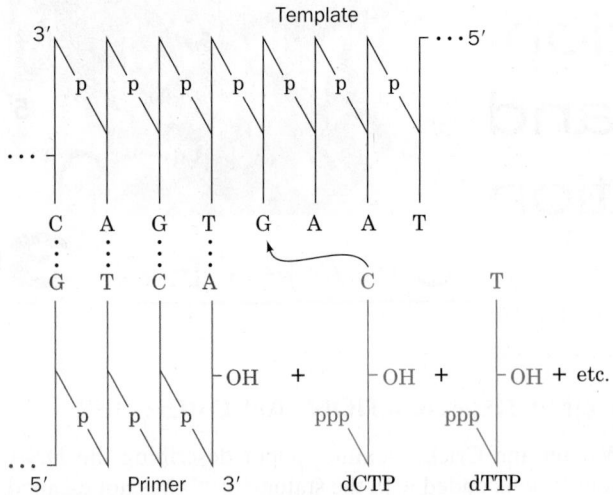

Figure 30-1 **Action of DNA polymerase.** DNA polymerases assemble incoming deoxynucleoside triphosphates on single-stranded DNA templates such that the growing strand is elongated in its $5' \rightarrow 3'$ direction.

the presence of replication "eyes" or "bubbles" (Fig. 30-2). These so-called **θ structures** (after their resemblance to the Greek letter theta) indicate that *double-stranded DNA (dsDNA) replicates by the progressive separation of its two parental strands accompanied by the synthesis of their complementary strands to yield two semiconservatively replicated duplex daughter strands* (Fig. 30-3). DNA replication involving θ structures is known as **θ replication.**

A branch point in a replication eye at which DNA synthesis occurs is called a **replication fork.** A replication bubble may contain one or two replication forks **(unidirectional** or **bidirectional replication).** Autoradiographic studies have demonstrated that θ replication is almost always bidirectional (Fig. 30-4). Moreover, such experiments, together with genetic evidence, have established that prokaryotic and bacteriophage DNAs have but one **replication origin** (point where DNA synthesis is initiated).

Figure 30-2 **Autoradiogram and its interpretive drawing of a replicating *E. coli* chromosome.** The bacterium had been grown for somewhat more than one generation in a medium containing [³H]thymidine, thereby labeling the subsequently synthesized DNA so that it appears as a line of dark grains in the photographic emulsion (*red lines in the interpretive drawing*). The size of the replication eye indicates that the circular chromosome is about one-sixth duplicated in the present round of replication. [Courtesy of John Cairns, Cold Spring Harbor Laboratory, New York.]

B. *Role of DNA Gyrase*

The requirement that the parent DNA unwind at the replication fork (Fig. 30-3) presents a formidable topological obstacle. For instance, *E. coli* DNA is replicated at a rate of ~1000 nucleotides/s. If its 1300-μm-long chromosome were linear, it would have to flail around within the confines of a 3-μm-long *E. coli* cell at ~100 revolutions/s

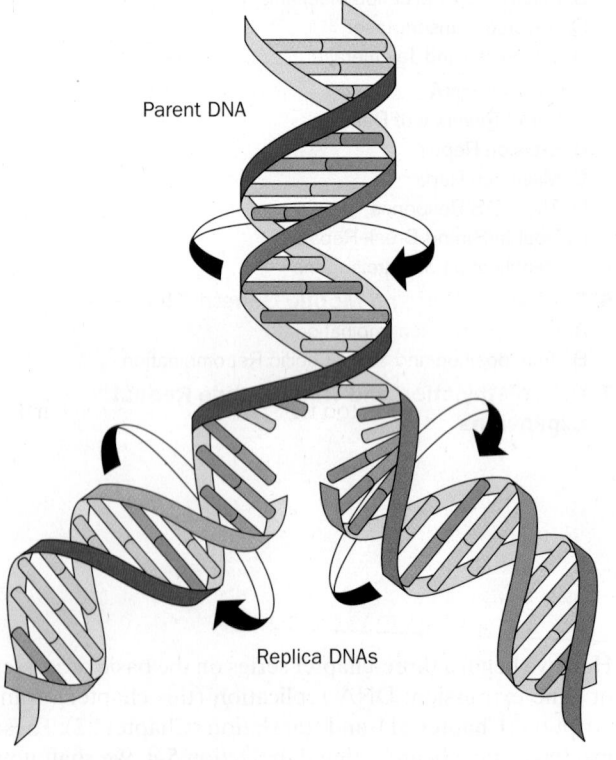

Figure 30-3 **Replication of DNA.**

(a)

Heavily labeled DNA — Unidirectional replication — Lightly labeled DNA

Bidirectional replication

(b)

Figure 30-4 Autoradiographic differentiation of unidirectional and bidirectional θ replication of DNA. (*a*) An organism is grown for several generations in a medium that is lightly labeled with [³H]thymidine so that all of its DNA will be visible in an autoradiogram. A large amount of [³H]thymidine is then added to the medium for a few seconds before the DNA is isolated **(pulse labeling)** in order to label only those bases near the replication fork(s). Unidirectional DNA replication will exhibit only one heavily labeled branch point (*above*), whereas bidirectional DNA replication will exhibit two such branch points (*below*). (*b*) An autoradiogram of *E. coli* DNA so treated, demonstrating that it is bidirectionally replicated. [Courtesy of David M. Prescott, University of Colorado.]

C. *Semidiscontinuous Replication*

The low-resolution images provided by autoradiograms such as Figs. 30-2 and 30-4*b* suggest that dsDNA's two antiparallel strands are simultaneously replicated at an advancing replication fork. Yet, all known DNA polymerases can only extend DNA strands in the $5' \rightarrow 3'$ direction. How, then, does DNA polymerase copy the parent strand that extends in the $5' \rightarrow 3'$ direction past the replication fork? This question was answered in 1968 by Reiji Okazaki through the following experiments. If a growing *E. coli* culture is pulse-labeled for 30 s with [³H]thymidine, much of the radioactive and hence newly synthesized DNA has a sedimentation coefficient in alkali of 7S to 11S. These so-called **Okazaki fragments** evidently consist of only 1000 to 2000 nucleotides (**nt;** 100–200 nt in eukaryotes). If, however, following the 30 s [³H]thymidine pulse, the *E. coli* are transferred to an unlabeled medium (a **pulse–chase** experiment), the resulting radioactively labeled DNA sediments at a rate that increases with the time that the cells had grown in the unlabeled medium. The Okazaki fragments must therefore become covalently incorporated into larger DNA molecules.

Okazaki interpreted his experimental results in terms of the **semidiscontinuous replication** model (Fig. 30-5). The two parent strands are replicated in different ways. *The newly synthesized DNA strand that extends $5' \rightarrow 3'$ in the direction of replication fork movement, the so-called **leading strand,** is essentially continuously synthesized in its $5' \rightarrow 3'$ direction as the replication fork advances. The other newly synthesized strand, the **lagging strand,** is also synthesized in its $5' \rightarrow 3'$ direction but discontinuously as Okazaki fragments. The Okazaki fragments are only covalently joined together sometime after their synthesis in a reaction catalyzed by the enzyme **DNA ligase** (Section 30-2D).*

The semidiscontinuous model of DNA replication is corroborated by electron micrographs of replicating DNA showing single-stranded regions on one side of the

(recall that B-DNA has ~10 bp per turn). But since the *E. coli* chromosome is, in fact, circular, even this could not occur. Rather, the DNA molecule would accumulate +100 supercoils/s (see Section 29-3A for a discussion of supercoiling) until it became too tightly coiled to permit further unwinding. Naturally occurring DNA's negative supercoiling promotes DNA unwinding but only to the extent of ~5% of its duplex turns (recall that naturally occurring DNAs are typically underwound by one supercoil per ~19 duplex turns; Section 29-3Bb). In prokaryotes, however, negative supercoils may be introduced into DNA through the action of a type IIA topoisomerase (DNA gyrase; Section 29-3Cd) at the expense of ATP hydrolysis. This process is essential for prokaryotic DNA replication as is demonstrated by the observation that DNA gyrase inhibitors, such as novobiocin, arrest DNA replication except in mutants whose DNA gyrase does not bind these antibiotics.

Figure 30-5 Semidiscontinuous DNA replication. In DNA replication, both daughter strands (*leading strand red, lagging strand blue*) are synthesized in their $5' \rightarrow 3'$ directions. The leading strand is synthesized continuously, whereas the lagging strand is synthesized discontinuously.

Figure 30-6 **Electron micrograph of a replication eye in** ***Drosophila melanogaster* DNA.** Note that the single-stranded regions (*arrows*) near the replication forks have the trans configuration consistent with the semidiscontinuous model of DNA replication. [From Kreigstein, H.J. and Hogness, D.S., *Proc. Natl. Acad. Sci.* **71**, 173 (1974).]

replication fork (Fig. 30-6). In bidirectionally replicating DNA, moreover, the two single-stranded regions occur, as expected, on diagonally opposite sides of the replication bubble.

D. *RNA Primers*

DNA polymerases' all but universal requirement for a free 3'-OH group to extend a DNA chain poses a question that was emphasized by the establishment of the semidiscontinuous model of DNA replication: How is DNA synthesis initiated? Careful analysis of Okazaki fragments revealed

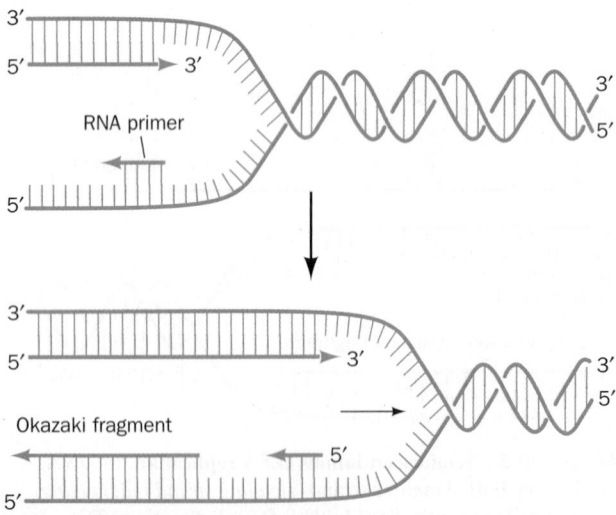

Figure 30-7 **Priming of DNA synthesis by short RNA segments.**

that *their 5' ends consist of RNA segments of 1 to 60 nt (a length that is species dependent) that are complementary to the template DNA chain* (Fig. 30-7). *E. coli* has two enzymes that can catalyze the formation of these **RNA primers: RNA polymerase,** the ~459-kD multisubunit enzyme that mediates transcription (Section 31-2), and the much smaller primase (60 kD), the monomeric product of the *dnaG* gene.

Primase is insensitive to the RNA polymerase inhibitor **rifampicin** (Section 31-2Bb). The observation that rifampicin inhibits only leading strand synthesis therefore indicates that *primase initiates the Okazaki fragment primers.* The initiation of leading strand synthesis in *E. coli,* a much rarer event than that of Okazaki fragments, can be mediated *in vitro* by either RNA polymerase or primase alone but is greatly stimulated when both enzymes are present. It is therefore thought that these enzymes act synergistically *in vivo* to prime leading strand synthesis.

Mature DNA does not contain RNA. The RNA primers are eventually removed and the resulting single-strand gaps are filled in with DNA by a mechanism described in Section 30-2Aj.

2 ENZYMES OF REPLICATION

DNA replication is a complex process involving a great variety of enzymes. It requires, to list only its major actors in their order of appearance: (1) DNA topoisomerases, (2) enzymes known as helicases that separate the DNA strands at the replication fork, (3) proteins that prevent them from reannealing before they are replicated, (4) enzymes that synthesize RNA primers, (5) a DNA polymerase, (6) an enzyme to remove the RNA primers, and (7) an enzyme to covalently link successive Okazaki fragments. In this section, we describe the properties and functions of many of these proteins.

A. *DNA Polymerase I*

In 1957, Arthur Kornberg reported that he had discovered an enzyme that catalyzes the synthesis of DNA in extracts of *E. coli* through its ability to incorporate the radioactive label from [^{14}C]thymidine triphosphate into DNA. This enzyme, which has since become known as **DNA polymerase I** or **Pol I,** consists of a monomeric 928-residue polypeptide.

Pol I couples deoxynucleoside triphosphates on DNA templates (Fig. 30-1) *in a reaction that occurs through the nucleophilic attack of the growing DNA chain's 3'-OH group on the α-phosphoryl of an incoming nucleoside triphosphate. The reaction is driven by the resulting elimination of PP$_i$ and its subsequent hydrolysis by inorganic pyrophosphatase.* The overall reaction resembles that catalyzed by RNA polymerase (Fig. 5-23) but differs from it by the strict requirement that the incoming nucleoside be linked to a free 3'-OH group of a polynucleoside that is base paired to the template (RNA polymerase initiates transcription by linking together two ribonucleoside triphosphates on a DNA template; Section 31-2C). The complementarity between the product DNA and the template was at first inferred through base composition and hybridization studies but was eventually directly established by base sequence determinations. The error rate of Pol I in copying the template is extremely low, as was first demonstrated by its *in vitro* replication of the 5386-nt DNA from bacteriophage **φX174** to yield fully infective phage DNA. In fact, its measured error rate is around one wrong base per 10 million.

Pol I is said to be **processive** in that it catalyzes a series of successive polymerization steps, typically 20 or more, without releasing the template (the opposite of "processive" is **"distributive"**). Pol I can, of course, work in reverse by degrading DNA through pyrophosphorolysis. This reverse reaction, however, probably has no physiological significance because of the low *in vivo* concentration of PP_i resulting from the action of inorganic pyrophosphatase.

a. Pol I Recognizes the Incoming dNTP According to the Shape of the Base Pair It Forms with the Template DNA

The specificity of Pol I for an incoming base arises from the requirement that it form a Watson–Crick base pair with the template rather than direct recognition of the incoming base (recall that the four base pairs, A · T, T · A, G · C, and C · G, have nearly identical shapes; Fig. 5-12). Thus, as Eric Kool demonstrated, when the "base" **2,4-difluorotoluene (F),**

2,4-Difluorotoluene base (F)   Thymine (T)

which is isosteric with (has the same shape as) thymine but does not accept hydrogen bonds, is synthetically inserted into a template DNA, Pol I incorporates A opposite the F with a similar rate of mismatches as it incorporates A opposite T. Likewise, dFTP is incorporated opposite template A with a similar fidelity as is dTTP. Yet the incorporation of F opposite an A in DNA destabilizes the double helix by 15 kJ/mol relative to T opposite this A. Evidently, *Pol I selects an incoming dNTP largely according to its ability to form a Watson–Crick-shaped pair with the template base but with little regard for its hydrogen bonding properties.* Indeed, the NMR structure of a 12-bp DNA containing a centrally located F opposite an A reveals that it assumes a B-DNA conformation in which the F–A pair closely resembles a T · A base pair in the same position of an otherwise identical DNA.

b. Pol I Can Edit Its Mistakes

In addition to its polymerase activity, Pol I has two independent hydrolytic activities:

1. It can act as a $3' \to 5'$ exonuclease.

2. It can act as a $5' \to 3'$ exonuclease.

The $3' \to 5'$ exonuclease reaction differs chemically from the pyrophosphorolysis reaction (the reverse of the polymerase reaction) only in that H_2O rather than PP_i is the nucleotide acceptor. Kinetic and crystallographic studies, however, indicate that these two catalytic activities occupy separate active sites (see below). The $3' \to 5'$ exonuclease function is activated by an unpaired 3'-terminal nucleotide with a free OH group. If Pol I erroneously incorporates a wrong (unpaired) nucleotide at the end of a growing DNA chain, the polymerase activity is inhibited and the $3' \to 5'$ exonuclease excises the offending nucleotide (Fig. 5-36). The polymerase activity then resumes DNA replication. *Pol I therefore has the ability to proofread or edit a DNA chain as it is synthesized so as to correct its mistakes.* This explains the great fidelity of DNA replication by Pol I: The overall fraction of bases that the enzyme misincorporates, $\sim 10^{-7}$, is the product of the fraction of bases that its polymerase activity misincorporates and the fraction of misincorporated bases that its $3' \to 5'$ exonuclease activity fails to excise. The price of this high fidelity is that ~3% of correctly incorporated nucleotides are also excised.

The Pol I $5' \to 3'$ exonuclease binds to dsDNA at single-strand nicks with little regard to the character of the 5' nucleotide (5'-OH or phosphate group; base paired or not). It cleaves the DNA in a base paired region beyond the nick such that the DNA is excised as either mononucleotides or oligonucleotides of up to 10 residues (Fig. 5-33). In contrast, the $3' \to 5'$ exonuclease removes only unpaired mononucleotides with 3'-OH groups.

c. Pol I's Polymerase and Two Exonuclease Functions Each Occupy Separate Active Sites

The $5' \to 3'$ exonuclease activity of Pol I is independent of both its $3' \to 5'$ exonuclease and its polymerase activities. In fact, as we saw in Section 7-2A, proteases such as subtilisin or trypsin cleave Pol I into two fragments: a larger C-terminal or **Klenow fragment** (residues 324–928), which contains both the polymerase and the $3' \to 5'$ exonuclease activities; and a smaller, N-terminal fragment (residues 1–323), which contains the $5' \to 3'$ exonuclease activity. Thus Pol I contains three active sites on a single polypeptide chain.

d. The X-Ray Structure of Klenow Fragment Indicates How It Binds DNA

The X-ray structure of Klenow fragment in complex with dsDNA, determined by Thomas Steitz, reveals that

Fingers

Thumb

3'

5'

Palm

3',5'-Exonuclease

Figure 30-8 **X-ray structure of *E. coli* DNA polymerase I Klenow fragment in complex with a dsDNA.** The protein is drawn in ribbon form colored in rainbow order from its N-terminus (*blue*) to its C-terminus (*red*) and embedded in its semitransparent molecular surface. The DNA is shown in stick form with the C atoms of its 10-nt template strand cyan, the C atoms of its 13-nt primer strand magenta, N blue, O red, and P orange, and with successive P atoms in each polynucleotide strand connected by orange rods. A Zn^{2+} ion (*purple sphere*) marks the $3' \rightarrow 5'$-exonuclease active site. [Based on an X-ray structure by Thomas Steitz, Yale University. PDBid 1KLN.]

See Interactive Exercise 33

the enzyme consists of two domains (Fig. 30-8). The smaller domain (residues 324–517) contains the $3' \rightarrow 5'$ exonuclease site, as was demonstrated by the absence of this function but not polymerase activity in a genetically engineered Klenow fragment mutant that lacks the divalent metal ion–binding sites known to be essential for $3' \rightarrow 5'$ exonuclease activity but which otherwise has a normal structure. The larger domain (residues 521–928) contains the polymerase active site at the bottom of a prominent cleft, a surprisingly large distance ($\sim$25 Å) from the $3' \rightarrow 5'$ exonuclease site. The cleft, which is lined with positively charged residues, has the appropriate size ($\sim$22 Å wide by $\sim$30 Å deep) and shape to bind a B-DNA molecule in a manner resembling a right hand grasping a rod. Consequently, Klenow fragment's polymerase domains are known as its "fingers," "thumb," and "palm" (Fig. 30-8). Indeed, all DNA polymerases of known structure have similar shapes with a fingers domain that binds the incoming nucleotide and the

templating base, a thumb domain that guides the newly formed dsDNA as it leaves the active site, and a palm domain that contains the active site at the bottom of the cleft between the fingers and thumb domains (Sections 30-2Ba, 30-4Ba, and 30-4Ca).

e. DNA Polymerase Distinguishes Watson–Crick Base Pairs via Sequence-Independent Interactions That Induce Domain Movements

DNA polymerase I from the thermophile *Thermus aquaticus* (Taq), which is 51% identical to its *E. coli* homolog, lacks a functional $3' \rightarrow 5'$ exonuclease. Its X-ray structure (Fig. 30-9a), also determined by Steitz, reveals that its C-terminal domain closely resembles Klenow fragment (Fig. 30-8), although the metal ion–binding carboxylate residues that are essential for the $3' \rightarrow 5'$ exonuclease function of Klenow fragment are absent in Taq polymerase. Its N-terminal $5' \rightarrow 3'$ exonuclease domain, which appears to be only loosely tethered to the C-terminal polymerase domain, contains a conserved cluster of metal ions at the bottom of a cleft that is $\sim$70 Å from the polymerase active site. It is therefore is unclear how the polymerase and $5' \rightarrow 3'$ exonuclease functions work in concert, as we shall see (Section 30-2Ai), to produce dsDNA with a single nick.

Gabriel Waksman crystallized the C-terminal domain of Taq DNA polymerase I [**Klentaq1**; which is often used in PCR experiments (Section 5-5F)] in complex with an 11-bp dsDNA that has a GGAAA-5' overhang at the 5' end of its template strand, and the crystals were incubated with 2′,3′-dideoxy-CTP (ddCTP; which lacks a 3′-OH group). The X-ray structure of these crystals (Fig. 30-9b) reveals that a ddC residue had been covalently linked to the 3' end of the primer and formed a Watson–Crick pair with the template overhang's 3' G. Moreover, a second ddCTP molecule (to which the primer's new 3'-terminal ddC residue is incapable of forming a covalent bond) occupies the enzyme's active site where it forms a Watson–Crick pair with the template's next G that is largely out of contact with the surrounding aqueous solution. Clearly, Klentaq1 retains its catalytic activity in this crystal.

A DNA polymerase must distinguish correctly paired bases from mismatches and yet do so via sequence-independent interactions with the incoming dNTP. The foregoing X-ray structure reveals that this occurs through an active site pocket that is complementary in shape to Watson–Crick base pairs. The pocket is formed by the stacking of a conserved Tyr side chain on the template base, as well as by van der Waals interactions with the protein and with the preceding base pair. In addition, although the dsDNA is mainly in the B conformation, the 3 base pairs nearest the active site assume the A conformation, as has also been observed in the X-ray structures of several other DNA polymerases in their complexes with DNA. The resulting wider and shallower minor groove (Section 29-1Ba) permits protein side chains to form hydrogen bonds with the otherwise inaccessible N3 atoms of the purine bases and O2 atoms of the pyrimidine bases. The positions of these hydrogen bond acceptors are sequence-

(a)

(b)

(c)

Figure 30-9 **X-ray structures of *T. thermophilus* DNA polymerase I.** (*a*) X-ray structure of the entire protein drawn in ribbon form. The N-terminal 5' → 3' exonuclease domain is colored in rainbow order from its N-terminus (*blue*) to its C-terminus (*red*). The Zn^{2+} ion at its active site is represented by a purple sphere. The C-terminal Klentaq1 domain, which is oriented similarly to the Klenow fragment in Fig. 30-8, is light green. (*b*) The X-ray structures of Klentaq1 in its closed conformation (with bound ddCTP) and (*c*) in its open conformation (ddCTP depleted). The protein, which is viewed similarly to Fig. 30-8, is drawn in worm form with its N-terminal, palm, fingers, and thumb domains colored yellow, magenta, green, and blue, respectively. The DNA is shown in stick form, whereas the primer strand's 3'-terminal ddC residue is drawn in space-filling form as is the ddCTP in Part *b* that is base paired with a template G residue in the enzyme's active site pocket. The atoms are colored according to type with template strand C cyan, primer strand C green, ddCTP C yellow, N blue, O red, and P orange. [Part *a* based on an X-ray structure by Thomas Steitz, Yale University. PDBid 1TAQ. Parts *b* and *c* based on X-ray structures by Gabriel Waksman, Washington University School of Medicine. PDBids 3KTQ and 2KTQ.]

independent as can be seen from an inspection of Fig. 5-12 [in contrast, the positions of the hydrogen bonding acceptors in the major groove vary with both the identity (A · T vs G · C) and the orientation (e.g., A · T vs T · A) of the base pair]. However, with a non-Watson–Crick pairing, these hydrogen bonds would be greatly distorted if not completely disrupted. The protein also makes extensive sequence-independent hydrogen bonding and van der Waals interactions with the DNA's sugar–phosphate backbone.

The above Klentaq1 · DNA · ddCTP crystals were partially depleted of ddCTP by soaking them in a stabilizing solution that lacks ddCTP. The X-ray structure of the ddCTP-depleted crystals (Fig. 30-9*c*) reveals that Klentaq1 assumes a so-called open conformation, which differs significantly from that in the so-called closed conformation described above by a 46° hingelike motion of the fingers domain away from the polymerase active site (Fig. 30-9*c*). Evidently, the formation of a Watson–Crick base pair at the polymerase active site triggers the formation of a productive ternary complex that buries the incoming nucleotide. In particular, the foregoing Tyr side chain, which extends from the rightmost helix of the fingers domain (Fig. 30-9*b*), stacks on the incoming base, an interaction that is absent in the open conformation. Moreover, a Lys and an Arg side chain, that also extend from the rightmost helix of the fingers domain, form salt bridges with the α- and β-phosphate groups of the incoming dNTP. These observations are consistent with kinetic measurements on Pol I indicating that the binding of the correct dNTP to the enzyme induces a rate-limiting conformational change that yields a tight ternary complex. It therefore appears that the enzyme rapidly samples the available dNTPs in its open conformation but only when it binds the correct dNTP in a Watson–Crick pairing with the template base does it form the catalytically competent closed conformation. The subsequent reaction steps then rapidly yield the product complex which, following a second conformational change, releases the product PP_i. Finally, the DNA is translocated in the active site, probably via a linear diffusion mechanism, so as to position it for the next reaction cycle.

The comparison of the above X-ray structures with that of Klentaq1 alone indicates that on binding DNA, the thumb domain moves to wrap around the DNA. It is likely that this conformational change is largely responsible for Pol I's processivity. In both Klentaq1 · DNA structures, neither the dsDNA nor the single-stranded DNA **(ssDNA)** passes through the cleft between the thumb and fingers domain as the shape and position of the cleft suggest. Rather, the template strand makes a sharp bend at the first unpaired base, thereby unstacking this base and positioning this ssDNA on the same side of the cleft as the dsDNA. Similar arrangements have been observed in X-ray structures of other DNA polymerases in their complexes with DNA.

f. The DNA Polymerase Catalytic Mechanism Involves Two Metal Ions

The X-ray structures of a variety of DNA polymerases suggest that they share a common catalytic mechanism for nucleotidyl transfer (Fig. 30-10). Their active sites all contain two metal ions, usually Mg^{2+}, that are liganded by two invariant Asp side chains in the palm domain. Metal ion B in Fig. 30-10 is liganded by all three phosphate groups of the bound dNTP, whereas metal ion A bridges the α-phosphate group of this dNTP and the primer's 3′-OH group. Metal ion A presumably activates the primer's 3′-OH group for an in-line nucleophilic attack on the α-phosphate group (Fig. 16-6*b*), whereas metal ion B functions to orient its bound triphosphate group and to electrostatically shield their negative charges as well as the additional negative charge on the transition state leading to the release of the PP_i ion (Section 16-2B).

Figure 30-10 **Schematic diagram for the nucleotidyl transferase mechanism of DNA polymerases.** A and B represent enzyme-bound metal ions that usually are Mg^{2+}. Atoms are colored according to type (C gray, N blue, O red, and P yellow) and metal ion coordination is represented by dotted lines. Metal ion A activates the primer's 3′-OH group for in-line nucleophilic attack on the incoming dNTP's α-phosphate group (*arrow*), whereas metal ion B acts to orient and electrostatically stabilize the negatively charged triphosphate group. [Courtesy of Tom Ellenberger, Harvard Medical School.]

g. Editing Complexes Contain the Primer Strand in the 3′ → 5′ Exonuclease Site

The complex of Klenow fragment with DNA shown in Fig. 30-8 contains a 13-nt primer strand [d(GCCTCGCG-GCGGC)] and a 10-nt template strand [d(GCCGC-GAGGC)] that is complementary to the 10-nt segment at the 5′ end of the primer strand. The 3′-terminal nucleotide of the primer strand (the last one that an active polymerase would have added) is bound at the 3′ → 5′ exonuclease active site. This arrangement is made possible by the opening up of the G · C base pair that would otherwise be formed by the 5′ nucleotide of the template strand, which remains bound near the entrance of the polymerase active site. Evidently, the Klenow fragment has bound the primer strand in an "editing" complex rather than in the polymerase cleft.

In *E. coli* Pol I, how does the 3′ end of the primer strand transfer between the polymerase active site and the 3′ → 5′ exonuclease active site? This appears to occur through the competition of these sites for the 3′ end of the primer strand, which base pairs to form dsDNA in the polymerase site and binds as a single strand to the exonuclease site. Thus, the formation of a Watson–Crick base pair facilitates the binding of the primer strand to the polymerase site preparatory for the next round of chain extension, whereas a mismatched base pair greatly slows the polymerase reaction while promoting the binding of the primer strand to the exonuclease site. Comparison of the editing complex with those of the Klentaq1 · DNA complexes suggests that the transfer of the primer strand from the polymerase to the editing sites of Klenow fragment requires that the dsDNA translocate backward (toward the 3′ end of the template strand) by several angstroms along the helix axis.

h. Pol I Functions Physiologically to Repair DNA

For some 13 years after Pol I's discovery, it was generally assumed that this enzyme was *E. coli*'s DNA replicase because no other DNA polymerase activity had been detected in *E. coli*. This assumption was made untenable by Cairns and Paula De Lucia's isolation, in 1969, of a mutant *E. coli* whose extracts exhibit <1% of the normal Pol I activity (although it has nearly normal levels of the 5′ → 3′ exonuclease activity) but which nevertheless reproduce at the normal rate. This mutant strain, however, is highly susceptible to the damaging effects of UV radiation and **chemical mutagens** (substances that chemically induce mutations; Section 32-1A). *Pol I evidently plays a central role in the repair of damaged (chemically altered) DNA.*

Damaged DNA, as we discuss in Section 30-5, is detected by a variety of DNA repair systems. Many of them endonucleolytically cleave the damaged DNA on the 5′ side of the lesion, thereby activating Pol I's 5′ → 3′ exonuclease. While excising this damaged DNA, Pol I simultaneously fills in the resulting single-strand gap through its polymerase activity. In fact, its 5′ → 3′ exonuclease activity increases 10-fold when the polymerase function is active. Perhaps the simultaneous excision and polymerization activities of Pol I protect DNA from the action of cellular nucleases that would further damage the otherwise gapped DNA.

i. Pol I Catalyzes Nick Translation

Pol I's combined 5′ → 3′ exonuclease and polymerase activities can replace the nucleotides on the 5′ side of a single-strand nick on otherwise undamaged DNA. These reactions, in effect, translate (move) the nick toward the DNA strand's 3′ end without otherwise changing the molecule (Fig. 30-11). This **nick translation** process, in the presence of labeled deoxynucleoside triphosphates, is synthetically employed to prepare highly radioactive DNA (the required nicks may be generated by treating the DNA with a small amount of pancreatic **Dnase I**).

j. Pol I's 5′ → 3′ Exonuclease Functions Physiologically to Excise RNA Primers

Pol I's 5′ → 3′ exonuclease also removes the RNA primers at the 5′ ends of newly synthesized DNA while its DNA polymerase activity fills in the resulting gaps (Fig. 5-34). The importance of this function was demonstrated by the isolation

Figure 30-11   **Nick translation as catalyzed by Pol I.**

of temperature-sensitive *E. coli* mutants that are neither viable nor exhibit any 5′ → 3′ exonuclease activity at the restrictive temperature of ~43°C (the low level of polymerase activity in the Pol I mutant isolated by Cairns and De Lucia is apparently sufficient to carry out this essential gap-filling process during chromosome replication). Thus Pol I has an indispensable role in *E. coli* DNA replication although a different one than was first supposed.

B. *DNA Polymerase III*

The discovery of normally growing *E. coli* mutants that have very little Pol I activity stimulated the search for an additional DNA polymerizing activity. This effort was rewarded by the discovery of two more enzymes, designated, in the order they were discovered, **DNA polymerase II (Pol II)** and **DNA polymerase III (Pol III)**. The properties of these enzymes are compared with that of Pol I in Table 30-1. Pol II and Pol III had not previously been detected because their combined activities in the assays used are normally <5% that of Pol I.

A mutant *E. coli* lacking measurable Pol II activity grows normally. However, Pol II has been implicated as a participant in repairing DNA damage via the **SOS response** (Section 30-5D), as have two additional *E. coli* enzymes that

Table 30-1   **Properties of *E. coli* DNA Polymerases**

	Pol I	Pol II	Pol III
Mass (kD)	103	90	130
Molecules/cell	400	?	10–20
Turnover number[a]	600	30	9000
Structural gene	*polA*	*polB*	*polC*
Conditionally lethal mutant	+	−	+
Polymerization: 5′ → 3′	+	+	+
Exonuclease: 3′ → 5′	+	+	+
Exonuclease: 5′ → 3′	+	−	−

[a]Nucleotides polymerized min^{-1} · molecule^{-1} at 37°C.

Source: Kornberg, A. and Baker, T.A., *DNA Replication* (2nd ed.), p. 167, Freeman (1992).

were more recently discovered: **DNA polymerase IV (Pol IV)** and **DNA polymerase V (Pol V)** (Section 30-5Db).

a. Pol III Is *E. coli's* DNA Replicase

The cessation of DNA replication in temperature-sensitive *polC* mutants above the restrictive (high) temperature demonstrates that *Pol III is E. coli's DNA replicase.* Its **Pol III core** has the subunit composition αεθ where α, the *polC* gene product (Table 30-2), contains the polymerase function. The catalytic properties of Pol III core resemble those of Pol I (Table 30-1) except for Pol III core's inability to replicate primed ssDNA or nicked dsDNA. Rather, Pol III core acts *in vitro* at single-strand gaps of <100 nucleotides, a situation that probably resembles the state of DNA at the replication fork. The Pol III $3' \rightarrow 5'$ exonuclease function, which resides on the enzyme's ε subunit, is DNA's primary editor during replication; it enhances the enzyme's replication fidelity by up to 200-fold. However, the Pol III $5' \rightarrow 3'$ exonuclease acts only on single-stranded DNA, so it cannot catalyze nick translation. θ is an accessory protein that stimulates the editing function of ε.

The X-ray structure of residues 1 to 917 of the 1160-residue *E. coli* Pol III α subunit, determined Mike O'Donnell and John Kuriyan, reveals that this protein has the expected thumb, fingers, and palm domains (Fig. 30-12). Nevertheless, it exhibits no significant sequence similarity and a different fold from all but two other DNA polymerases of known structure (both from gram-positive bacteria). In addition, Pol III α has an N-terminal PHP (for *p*olymerases and *h*istidinol *p*hosphatase) domain that buttresses both the palm and thumb domains.

*Pol III core (αεθ) functions in vivo as part of a complicated and labile multisubunit enzyme, the **Pol III holoenzyme**, which consists of at least 10 types of subunits (Table*

Figure 30-12 **X-ray structure of the *E. coli* Pol III α subunit.** The protein is drawn in ribbon form with its thumb, PHP, palm, and finger domains green, blue, magenta, and orange, respectively. Note the handlike shape of the protein but its entirely different fold from that of Klentaq1 (Fig. 30-9). [Based on an X-ray structure by Mike O'Donnell, The Rockefeller University, and John Kuriyan, University of California at Berkeley. PDBid 2HQA.]

30-2). The latter 7 subunits in Table 30-2 act to modulate Pol III core's activity. For example, Pol III core has a processivity of 10 to 15 residues; it can only fill in short single-stranded regions of DNA. However, Pol III core is rendered processive by association with the **β subunit** in the presence of the 7-subunit **γ complex** ($\gamma\tau_2\delta\delta'\chi\Psi$). Assembly of the processive enzyme is a two-stage process in which the γ complex transfers the β subunit to the primed template in an ATP-dependent reaction followed by the assembly of Pol III core with the β subunit on the DNA (Section 30-3Cc). The β subunit confers essentially unlimited processivity (>5000 residues) on the core enzyme even if the γ complex is subsequently removed. In fact, the β subunit is very strongly bound to the DNA, although it can freely slide along it.

b. The β Subunit Forms a Ringlike Sliding Clamp

The observation that a β subunit clamped to a cut circular DNA slides to the break and falls off suggests that the β subunit forms a closed ring around the DNA, thereby preventing its escape. Kuriyan and O'Donnell determined the X-ray structure of the β subunit in complex with a primer–template DNA (dsDNA with a single-stranded extension on the 5' end of one of its strands, the template strand). The protein forms a homodimer of C-shaped, 366-residue monomer units that associate to form an ~80-Å-diameter doughnut-shaped structure (Fig. 30-13*a*) that is equivalently known as the **sliding clamp** and the **β clamp.** The sliding clamp's central hole is ~35 Å in diameter, which is larger than the 20- and 26-Å diameters of B- and A-DNAs (recall that the hybrid helices which RNA

Table 30-2 **Components of *E. coli* DNA Polymerase III Holoenzyme**

Subunit	Mass (kD)	Structural Gene
α[a]	130	*polC (dnaE)*
ε[a]	27.5	*dnaQ*
θ[a]	10	*holE*
τ[b]	71	*dnaX*[c]
γ[b]	45.5	*dnaX*[c]
δ[b]	35	*holA*
δ'[b]	33	*holB*
χ[b]	15	*holC*
ψ[b]	12	*holD*
β	40.6	*dnaN*

[a]Components of the Pol III core.
[b]Components of the γ complex.
[c]The γ and τ subunits are encoded by the same gene sequence; the γ subunit comprises the N-terminal end of the τ subunit.

Sources: Kornberg, A. and Baker, T.A., *DNA Replication* (2nd ed.), p. 169, Freeman (1992); *and* Baker, T.A. and Wickner, S.H., *Annu. Rev. Genet.* **26,** 450 (1992).

(a)

(b)

Figure 30-13   X-ray structure of the β subunit of *E. coli* Pol III holoenzyme in complex with DNA. (*a*) The homodimeric sliding clamp is drawn in ribbon form embedded in its semitransparent surface diagram and viewed along its 2-fold axis with one subunit magenta and the other colored in rainbow order from its N-terminus (*blue*) to its C-terminus (*red*). The DNA, which consists of a 10-bp double-stranded segment with a 4-nt single-stranded extension at the 5′ end of one of the strands, is drawn in stick form with the C atoms of the template strand cyan, those of the primer strand green, N blue, O red, and P orange, and with an orange rod connecting successive P atoms in each strand. (*b*) Cutaway diagram of the structure in Part *a* rotated 90° about the horizontal axis. The DNA, which is shown in space-filling form, is inclined by ~22° to the protein's 2-fold axis, which is vertical in this diagram. Note that in a replicating system, the primer strand would be extended toward the top of the diagram. [Based on an X-ray structure by John Kuriyan, University of California at Berkeley, and Mike O'Donnell, The Rockefeller University. PDBid 3BEP.]

primers make with DNA have A-DNA-like conformations; Section 29-1Bc). Each β subunit consists of six tandem βαββ motifs of identical topology, which associate in pairs to form three pseudo-2-fold symmetric domains of very similar structures (although with <20% sequence identity). The dimeric ring therefore has the shape of a 6-pointed star in which the 12 helices line the central hole and the β strands associate in six β sheets that form the protein's outer surface. Electrostatic calculations indicate that the interior surface of the ring is positively charged, whereas its outer surface is negatively charged.

The α helices lining the protein's central hole are all approximately perpendicular to their radially adjacent segments of the sugar–phosphate backbone. These helices therefore span the major and minor grooves of the DNA rather than entering into them as do many helices that make sequence-specific interactions with dsDNA (e.g., Section 31-3Da). Since A- and B-DNAs have 11 and 10.5 bp per turn, whereas the sliding clamp has a pseudo-12-fold symmetry, it appears that the sliding clamp largely minimizes its associations with its threaded DNA, which facilitates the unencumbered passage of the DNA through the sliding clamp. Nevertheless, the primer-template DNA's helix axis is inclined to the homodimeric protein's 2-fold axis by ~22° such that its ssDNA segment is in van der Waals contact with a specific portion of the β clamp's inner wall (Fig. 30-13*b*; in previous model building studies based on the structure of the sliding clamp alone, these axes were assumed to be coincident). In fact, primer–template DNA binds to the sliding clamp ~4-fold more tightly than does dsDNA and the mutation of the residues that interact with the ssDNA segment significantly reduces the efficiency of DNA replication. Indeed, as is explained in Section 30-3Cc, the interaction between the ssDNA and the sliding clamp is physiologically significant.

C. Unwinding DNA: Helicases and Single-Strand Binding Protein

Pol III holoenzyme, unlike Pol I, cannot unwind dsDNA. Rather, *three proteins,* DnaB *protein (the product of the dnaB gene; proteins may be assigned the name of the gene specifying them but in roman letters with the first letter capitalized), Rep helicase, and single-strand binding protein*

Table 30-3 **Unwinding and Binding Proteins of *E. coli* DNA Replication**

Protein	Subunit Structure	Subunit Mass (kD)
DnaB protein	Hexamer	50
SSB	Tetramer	19
Rep protein	Monomer	68
PriA protein	Monomer	76

Source: Kornberg, A. and Baker, T.A., *DNA Replication* (2nd ed.), p. 366, Freeman (1992).

Figure 30-14   Unwinding of DNA by the combined action of DnaB and SSB proteins. The hexameric DnaB protein moves along the lagging strand template in the 5′ → 3′ direction. The resulting separated DNA strands are prevented from reannealing by SSB binding.

(SSB) (Table 30-3), work in concert to unwind the DNA before an advancing replication fork (Fig. 30-14) in a process that is driven by ATP hydrolysis.

a. Hexameric Helicases Mechanically Separate the Strands of dsDNA by Climbing Up One Strand

Access to the genetic information encoded in a double helical nucleic acid requires that its base-paired strands be separated. The proteins that do so, which are known as **helicases,** form a diverse group of enzymes that facilitate a variety of functions including DNA replication, recombination, and repair, as well as transcription termination (Section 31-2Da), RNA splicing, and RNA editing (Section 31-4A). Indeed, all forms of life contain helicases, 12 varieties of which occur in *E. coli*. A helicase functions by translocating along one strand of a double helical nucleic acid so as to separate the strands in its path. This, of course, requires free energy, and hence helicases are driven by NTP hydrolysis. Helicases have been classified into six superfamilies that vary in their characteristics, including their direction of translocation along their bound single strand (5′ → 3′ or 3′ → 5′) and whether they function as hexameric rings or dimers.

E. coli **DnaB** protein, a **hexameric helicase** of identical 471-residue subunits, separates the strands of dsDNA by translocating along the lagging strand template in the 5′ → 3′ direction while hydrolyzing ATP (it can also use GTP and CTP but not UTP). Electron microscopy and X-ray studies reveal that DnaB forms a hexameric ring that, depending on conditions, exhibits C_3 or C_6 symmetry and which encloses an ~30-Å-diameter central channel. Hexameric DnaB binds three primase molecules via the latter's helicase binding domain (Section 30-2E). Similarly, the bacteriophage **T7 gene 4 helicase/primase** (bacteriophage T7 infects *E. coli*) forms a two-tiered hexagonal ring (Fig. 30-15) whose N-terminal domains (residues 1–271) contain its primase activity and whose C-terminal domains (residues 272–566) carry out its helicase function. T7 gene 4 helicase/primase (also called **T7 gp4;** gp for *gene product*) translocates along ssDNA in the 5′ → 3′ direction while preferentially hydrolyzing dTTP (but also hydrolyzes dATP and ATP).

Leemor Joshua-Tor determined the only known X-ray structure of a hexameric helicase in complex with DNA, that of the **E1 protein** of bovine papillomavirus, which translocates along ssDNA in the 3′ → 5′ direction (the opposite direction of DnaB and T7 gp4). The protein in the structure, which contains the C-terminal 274 residues of the 605-residue E1 protein, consists of two domains: a 74-residue N-terminal oligomerization domain and a 200-residue C-terminal **AAA+ domain** (AAA+ for *A*TPases *a*ssociated with cellular *a*ctivities; a functionally diverse protein family). There are two families of hexameric helicases, the RecA family and the AAA+ family. RecA-family hexameric helicases (RecA catalyzes homologous recombination; Section 30-6Ab), such as DnaB and T7 gp4,

Figure 30-15   Electron microscopy–based image reconstruction of T7 gene 4 helicase/primase. In this two-tiered hexameric ring (*yellow*), the smaller lobe of each subunit forms the N-terminal primase domain and the larger lobe forms the C-terminal helicase domain. The protein is postulated to interact with DNA as is depicted by this model of a DNA fork consisting of a 30-bp duplex segment and two 25-nt single-stranded segments with the 5′ tail threaded through the hexameric ring. The way in which the 3′ tail interacts with the protein, if at all, is unknown. [Courtesy of S.S. Patel and K.M. Picha, University of Medicine and Dentistry of New Jersey.]

Figure 30-16   X-ray structure of bovine papillomavirus E1 helicase in complex with poly(dT) and ADP. (*a*) The protein is drawn in ribbon form viewed along the homohexamer's pseudo-6-fold axis with each protein subunit differently colored and with the oligomerization domain of each subunit more lightly shaded than its AAA+ domain. The protein loops extending radially inward from each subunit to interact with the DNA's phosphate groups are purple. The poly(dT), 6 nt of which are visible, is drawn in stick form with C gray, N blue, O red, and P orange and with successive P atoms joined by orange rods. Its 5′ end is closest to the viewer. The ADP is shown in space-filling form with C green, N blue, O red, and P orange. (*b*) Side view of the protein related to that in Part *a* by a 90° rotation about the horizontal axis. The blue and green AAA+ domains in Part *a*, except for their DNA-interacting loops, have been deleted to expose the DNA, which is drawn in space-filling form. [Based on an X-ray structure by Leemor Joshua-Tor, Cold Spring Harbor Laboratory, New York. PDBid 2GXA.]

translocate in the 5′ → 3′ direction and occur mainly in eubacteria and their phages, whereas AAA+ hexameric helicases, such as the E1 protein, translocate in the 3′ → 5′ direction and occur mainly in archaea, eukaryotes, and their viruses.

The E1 structure reveals that this helicase, which was crystallized with ADP and a 13-nt poly(dT) (although only 6 nt are visible in the X-ray structure), forms a two-layered hexagonal ring in which the oligomerization domains form a rigid collar with nearly perfect 6-fold symmetry. In contrast, the AAA+ domains deviate significantly from this symmetry (Fig. 30-16a). An ADP is bound at a radially peripheral site between each neighboring pair of AAA+ domains. The poly(dT) forms a right-handed helix that binds in the minimally ~13-Å-diameter central channel of the AAA+ domain hexamer (which is too narrow to admit dsDNA) with its 5′ end toward the top of the hexamer in Fig. 30-16. The DNA's phosphate groups each interact with a positively charged loop (residues 505–508) that extends radially inward from each AAA+ domain and hence these loops form an arrangement that resembles a right-handed spiral staircase that tracks the ssDNA's sugar–phosphate backbone. Apparently, the protein steps through a series of ATP-driven conformational changes that, via interactions with the loops, pushes the ssDNA through the channel from bottom to top in Fig. 30-16b. During this process, each loop maintains its grip on the same phosphate group. ATP

hydrolysis occurs toward the bottom of the spiral staircase and ADP release occurs between subunits located toward its top. A new ATP then binds to this site, which causes the topmost loop to drop to the bottom of the staircase, where it binds the next available phosphate group and repeats the catalytic cycle. Thus the E1 helicase mechanically separates the strands of dsDNA by pulling itself along the groove of one strand in its 3′ → 5′ direction but without turning relative to the DNA.

b. Rep Helicase Dimers Separate the Strands of dsDNA via an "Active, Rolling" Mechanism

Two other helicases, **Rep helicase** and **PriA protein,** have been implicated in the replication of various *E. coli* phage DNAs (Section 30-3B) and also participate in certain aspects of *E. coli* DNA replication (Section 30-3C). Both proteins translocate along DNA in the 3′ → 5′ direction (and hence along the opposite strand from DnaB) while hydrolyzing ATP. Rep helicase is not essential for *E. coli* DNA replication but the rate at which *E. coli* replication forks propagate is reduced ~2-fold in *rep⁻* mutants.

Rep helicase is a 673-residue monomer in solution but dimerizes on binding to DNA. Both subunits of the Rep dimer bind to ssDNA or dsDNA such that DNA binding to one subunit strongly inhibits DNA binding to the other (negative cooperativity). This observation led Timothy Lohman to propose the "active, rolling" mechanism for

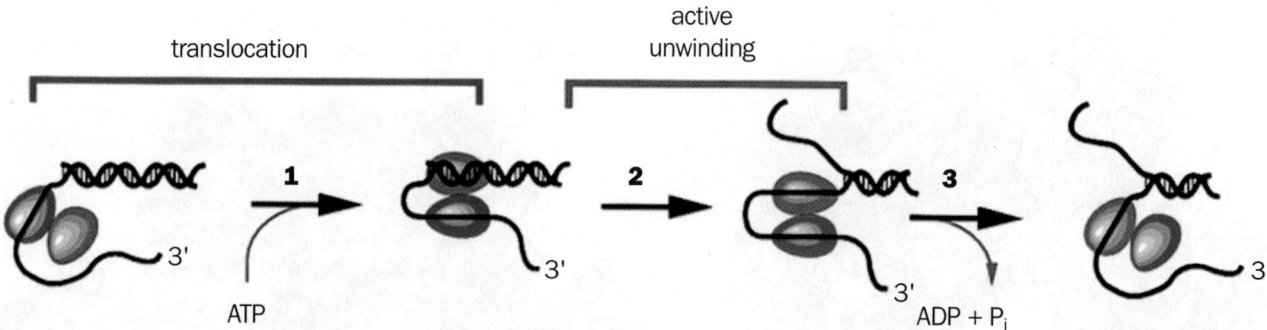

translocation

active
unwinding

ATP

ADP + P_i

3'

3'

3'

3'

Figure 30-17   Active, rolling mechanism for DNA unwinding by Rep helicase. (1) The subunit of dimeric Rep helicase that is not bound to ssDNA binds to dsDNA accompanied by ATP binding. **(2)** The subunit bound to dsDNA unwinds the double strand and remains bound to the 3'-ending strand. **(3)** In a process that is accompanied by the release of the ATP hydrolysis products, the subunit closer to the 3' end of the bound ssDNA releases it preparatory for a new cycle of dsDNA unwinding. [Courtesy of Gabriel Waksman, Washington University School of Medicine.]

Rep-mediated DNA unwinding in which the two subunits of the dimer alternate in binding dsDNA and the 3' end of the ssDNA at the ssDNA/dsDNA junction (Fig. 30-17). The two subunits then "walk" up the DNA while unwinding it in an ATP-dependent manner via a subunit switching mechanism in which the helicase subunit that is bound to the dsDNA displaces its 5'-starting strand while remaining bound to its 3'-starting strand. Release of the other subunit from the 3'-starting ssDNA then permits this subunit to bind to and unwind the new end of the dsDNA, thereby continuing the cycle.

The X-ray structure of *E. coli* Rep helicase in complex with the short ssDNA dT(pT)$_{15}$ and ADP (Fig. 30-18), determined by Lohman and Waksman, reveals that the relatively straight ssDNA molecule binds two contacting Rep monomers. A Rep monomer consists of two domains, 1 and 2, each of which is formed by two subdomains, A and B, with the two N-terminal subdomains (1A and 2A) homologous to each other. In the two Rep monomers that are bound to the same ssDNA, subdomain 2B exhibits strikingly different orientations with respect to the other three subdomains (Fig. 30-18). The Rep monomer that is bound to the 5' end of the ssDNA (which it contacts between bases 1 and 8) assumes the "open" conformation in which the four subdomains form an assembly that is reminiscent of a crab claw with one pincer (subdomain 2B) larger than the other (subdomain 1B). The DNA is bound at the bottom of the resulting cleft, whose floor is formed by subdomains 1A and 2A. In the Rep monomer that binds to the 3' end of the ssDNA (which it contacts between bases 9 and 16), subdomain 2B has reoriented relative to the other subdomains via a 130° rotation about a hinge region between subdomains 2A and 2B, thereby closing the cleft about the DNA to form the "closed" conformation. This conformation change is consistent with the active, rolling mechanism even though the way in which two Rep monomers form the dimer observed in solution remains unknown. The ADP binds to Rep between its subdomains 1A and 2A in close proximity to the DNA, suggesting that conformation changes at the ATP-binding site arising from ATP hydroly-

sis are transmitted to the DNA-binding site via the secondary structural elements that contact both sites. The way in which Rep separates the two strands of dsDNA is, as yet, unknown.

Figure 30-18   Superposition of the X-ray structures of Rep helicase in its open and closed forms in complex with dT(pT)$_{15}$ and ADP. The monomer in the open conformation is drawn in ribbon form colored according to secondary structure (helices magenta, β sheets yellow, and coil cyan) with its bound ssDNA segment and ADP drawn in stick form in blue and in red. In the closed conformation, subdomain 2B (transparent green ribbon) has rotated via a 130° hinge motion relative to subdomains 1A, 1B, and 2A so as to close over the ssDNA. [Courtesy of Gabriel Waksman, Washington University School of Medicine. PDBid 1UAA.]

c. Single-Strand Binding Protein Prevents ssDNA from Reannealing

If left to their own devices, the separated DNA strands behind an advancing helicase would rapidly reanneal to re-form dsDNA. What prevents them from doing so is the binding of **single-strand binding protein (SSB).** It also prevents ssDNA from forming fortuitous intramolecular secondary structures (helical stems) and protects it from nucleases. Numerous copies of SSB cooperatively coat ssDNA, thereby maintaining it in an unpaired state. Note, however, that ssDNA must be stripped of SSB before it can be replicated by Pol III holoenzyme.

E. coli SSB is a homotetramer of 177-residue subunits. SSB binds ssDNA in several distinct modes referred to as $(SSB)_n$, which differ by the number of nucleotides (n)

Figure 30-19 **X-ray structure of *E. coli* SSB in complex with dC(pC)$_{34}$.** The homotetramer, which has D_2 symmetry, is viewed along one of its 2-fold axes with its other 2-fold axes horizontal and vertical. Each of its subunits (which include the N-terminal 134 residues of the 177-residue polypeptide) are differently colored. Its two bound ssDNA molecules are drawn in space-filling form colored according to atom type with the upper strand C cyan, lower strand C green, N blue, O red, and P orange. (The lower strand is partially disordered and hence appears to consist of two fragments.) [Based on an X-ray structure by Timothy Lohman and Gabriel Waksman, Washington University School of Medicine. PDBid 1EYG.]

bound to each tetramer. The two major modes are $(SSB)_{35}$, in which only two of the tetramer's subunits strongly interact with the ssDNA, and $(SSB)_{65}$, in which all four subunits interact with the ssDNA. The $(SSB)_{35}$ mode displays unlimited cooperativity in that it forms extended strings of contacting tetramers along the length of a bound ssDNA, whereas the $(SSB)_{65}$ mode has limited cooperativity in that it forms beaded clusters on ssDNA that consist of only a few contacting tetramers.

Proteolysis studies have shown that SSB's ssDNA-binding site is contained within its 115 N-terminal residues. The X-ray structure of *E. coli* SSB's chymotryptic fragment (residues 1–135) in complex with dC(pC)$_{34}$, determined by Lohman and Waksman, reveals that the tetrameric protein has D_2 symmetry and binds two molecules of dC(pC)$_{34}$ (Fig. 30-19). For one of these 35-mers, 28 nucleotides (residues 3–30) were visible and these assumed the shape of an elongated horseshoe that wrapped around two SSB subunits with approximate 2-fold symmetry and with its apex contacting a third subunit. The other bound ssDNA was partially disordered such that only two segments were visible, one with 14 nt (residues 3–16) and the other with 9 nt (residues 19–27). The paths of the ss-DNA segments along the surface of the SSB suggested models that rationalize the different properties of $(SSB)_{35}$ and $(SSB)_{65}$. In the $(SSB)_{65}$ model, the two ends of a 65-nt segment emerge from the same side of the tetramer, which would limit the number of SSB tetramers that can bind to contiguous 65-nt segments of ssDNA. However, in the $(SSB)_{35}$ model, the two ends of a 35-nt segment emerge from opposite ends of the tetramer, thereby permitting an unlimited series of SSB tetramers to interact end-to-end along the length of a ssDNA.

D. DNA Ligase

Pol I, as we saw in Section 30-2A, replaces the Okazaki fragments' RNA primers with DNA through nick translation. *The resulting single-strand nicks between adjacent Okazaki fragments, as well as the nick on circular DNA after leading strand synthesis, are sealed in a reaction catalyzed by* **DNA ligase**. The free energy required by this reaction is obtained, in a species-dependent manner, through the coupled hydrolysis of either NAD^+ to NMN^+ + AMP or ATP to PP_i + AMP. The *E. coli* enzyme, which is also known as **LigA,** is a 671-residue monomer that utilizes NAD^+ and catalyzes a three-step reaction (Fig. 30-20):

1. The adenylyl group of NAD^+ is transferred to the ε-amino group of an enzyme Lys residue to form an unusual phosphoamide adduct that is, nevertheless, readily isolated.

2. The adenylyl group of this activated enzyme is transferred to the 5′-phosphoryl terminus of the nick to form an adenylylated DNA. Here, AMP is linked to the 5′-nucleotide via a pyrophosphate rather than the usual phosphodiester bond.

Figure 30-20 **The reactions catalyzed by *E. coli* DNA ligase.** In eukaryotic and T4 DNA ligases, NAD$^+$ is replaced by ATP so that PP$_i$ rather than NMN$^+$ is eliminated in the first reaction step. The numbered steps are described in the text.

3. DNA ligase catalyzes the formation of a phosphodiester bond by attack of the 3'-OH on the 5'-phosphoryl group, thereby sealing the nick and releasing AMP.

ATP-requiring DNA ligases, such as those of all eukaryotes and bacteriophage T4, release PP$_i$ in the first step of the reaction rather than NMN$^+$. T4 ligase is also noteworthy in that, at high DNA concentrations, it can link together two duplex DNAs **(blunt end ligation)** in a reaction that is a boon to genetic engineering (Section 5-5C).

The X-ray structure of *E. coli* DNA ligase in complex with a singly nicked 26-bp dsDNA and AMP was determined by Stewart Shuman. The complex was formed by reacting the protein with NAD$^+$ in the presence of Mg^{2+} (thus forming the phosphoamide product of step 1 of the DNA ligase reaction; Fig. 30-20), removing the Mg^{2+}, and then adding the nicked dsDNA. The X-ray structure of

crystals of this complex revealed that the protein forms a C-shaped clamp that encircles a 19-bp segment of the DNA centered on the nick (Fig. 30-21). Moreover, the complex had progressed through step 2 of the reaction, that is, the adenylyl group had formed a pyrophosphate linkage with the 5'-phosphate group at the nick. The reason that the enzyme did not complete its catalytic cycle by sealing the nick is presumably due to the absence of Mg^{2+}.

Residues 587 to 671 form a domain that is not visible in this X-ray structure although it is poorly resolved in the X-ray structure of DNA ligase from *Thermus filiformis*. Apparently, this domain has high mobility, which suggests that it folds out to allow the enzyme's nicked dsDNA substrate to bind to the active site and then folds back to help immobilize the DNA.

E. *Primase*

The primases from bacteria and several bacteriophages track the moving replication fork in close association with its DNA helicase. Thus, the N-terminal domain of T7 gene 4 helicase/primase forms its primase function (Fig. 30-15), whereas *E. coli* primase **(DnaG)** forms a noncovalent complex with DnaB. Since these DNA helicases translocate along the lagging strand template DNA in its 5' → 3' direction (Fig. 30-14), the primase must reverse its direction of travel in order to synthesize an RNA primer in its 5' → 3' direction. DnaG, which is held to the RNA primed site by its association with SSB, can synthesize up to 60-nt primers *in vitro*, although *in vivo*, primers have the length of 11 ± 1 nt. Since a replication fork in *E. coli* moves at ~1000 nt per second and Okazaki fragments are ~1000 nt in length, about one RNA primer must be synthesized per second at each replication fork. Primases tend to initiate synthesis at specific 3-nt sequences on the template. In *E. coli* this sequence is GTA, which is overrepresented in templates for lagging strand synthesis.

DnaG is a 581-residue monomeric protein. Proteolysis studies have shown that it consists of three domains: an N-terminal Zn^{2+}-binding domain (residues 1–110), which tetrahedrally ligands a Zn^{2+} ion via three Cys residues and a His residue and is implicated in recognizing ssDNA; a central RNA polymerase domain (residues 111–433) that carries out primer synthesis; and a C-terminal helicase binding domain (residues 434–581) that interacts with DnaB. Isolated DnaG is only weakly active *in vitro*; it synthesizes primers at a maximum rate of three per hour. However, in the presence of DnaB, it synthesizes primers at the rate observed *in vivo*. Since *E. coli* have 50 to 100 DnaG molecules per cell, this presumably limits primer synthesis to the replication fork. The importance of this function is underscored by T7 gp4, whose helicase and primase functions reside on the same polypeptide (Fig. 30-15).

The X-ray structure of the DnaG catalytic domain in complex with a 15-nt ssDNA (Fig. 30-22), determined by James Berger, reveals a cashew-shaped protein whose fold is unrelated to those of any other DNA or RNA polymerases. It contains an ~100-residue segment that is similar in both sequence and structure to segments in types IA

(a)   (b)

Figure 30-21   X-ray structure of *E. coli* DNA ligase in complex with a singly nicked 26-bp dsDNA and AMP. (*a*) The enzyme is drawn in ribbon form with its four domains colored, from N- to C-terminus, blue, aquamarine, yellow, and red. The dsDNA is shown in ladder form with the sugar–phosphate backbone of its 26-nt strand orange and those of its complementary two 13-nt strands magenta and purple. The AMP, which is covalently bound in pyrophosphate linkage to the phosphate group at the 5′-end of the purple strand, is drawn in space-filling form with C green, N blue, O red, and P orange. The side chain of Lys 115, which forms a phosphamide adduct in step 1

of the DNA ligase reaction (Fig. 30-20) is shown in stick form with C magenta and N blue. A Zn^{2+} ion, represented by a cyan sphere, is tetrahedrally liganded by four Cys residues. It is distant from the active site and therefore appears to have structural rather than catalytic function. (*b*) The 4 bp of nicked DNA in the boxed area of Part *a* in pyrophosphate linkage with the AMP together with the side chain of Lys 115 are all shown in stick form. The structure is viewed and colored as in Part *a* but with DNA C gray. The arrow points to the DNA's single-strand nick. [Based on an X-ray structure by Stewart Shuman, Sloan-Kettering Institute, New York, New York. PDBid 2OWO.]

Figure 30-22   X-ray structure of *E. coli* primase in complex with ssDNA. The protein is represented by its molecular surface colored according to its electrostatic potential with red negative, white nearly neutral, and blue positive. The DNA is drawn in stick form with C and P yellow, N blue, and O red. Note the strongly basic character of the DNA binding groove and the highly acidic nature of the active site region. [Courtesy of James Berger, University of California at Berkeley. PDBid 3B39.]

and IIA topoisomerases (Section 29-3C) and has therefore been named the **Toprim fold** (for *to*poisomerase and *pri-m*ase). The Toprim fold consists of a 4-stranded parallel β sheet flanked by three helices that resembles the nucleotide-binding (Rossmann) fold (Section 8-3Bi). The active site is marked by several residues that are highly conserved in DnaG-type primases, and in particular, a Glu and two Asp residues, which are invariant in all known Toprim folds and which, in the X-ray structure of a type IIA topoisomerase, coordinate an Mg^{2+} ion.

The ssDNA in the structure, only 5 nt of which are visible, occupies a positively charged groove on the surface of the catalytic subunit that feeds into its active site. The protein makes only a few hydrogen bonding and van der Waals interactions with the DNA's sugar–phosphate backbone and no specific interactions with its bases. Apparently, this DNA binding groove functions to nonspecifically capture a DNA template strand and direct it to the enzyme's active site.

3 PROKARYOTIC REPLICATION

Bacteriophages are among the simplest biological entities and their DNA replication mechanisms reflect this fact. Much of what we know about how DNA is replicated therefore stems from the study of this process in various phages. In this section we examine DNA replication in the **coliphages** (bacteriophages that infect *E. coli*) **M13** and φX174 and then consider DNA replication in *E. coli* itself. Eukaryotic DNA replication is discussed in Section 30-4.

A. *Bacteriophage M13*

Bacteriophage M13 carries a 6408-nt single-stranded circular DNA known as its **viral** or (+) **strand**. On infecting an *E. coli* cell, this strand directs the synthesis of its complementary or (−) strand to form the circular duplex **replicative form (RF),** which may be either nicked **(RF II)** or supercoiled **(RF I).** This replication process (Fig. 30-23)

1. DNA replication by Pol III
2. RNA excision and gap filling by Pol I
3. Gap sealing by DNA ligase
4. Supercoiling by DNA gyrase

Figure 30-23 **The synthesis of the M13 (−) strand DNA on a (+) strand template to form M13 RF I DNA.**

may be taken as a paradigm for leading strand synthesis in duplex DNA.

As the M13 (+) strand enters the *E. coli* cell, it becomes coated with SSB except at a palindromic 57-nt segment that forms a hairpin. RNA polymerase commences primer synthesis 6 nt before the start of the hairpin and extends the RNA 20 to 30 residues to form a segment of RNA–DNA hybrid duplex. The DNA that is displaced from the hairpin becomes coated with SSB so that when RNA polymerase reaches it, primer synthesis stops. Pol III holoenzyme then extends the RNA primer around the circle to form the (−) strand. The primer is removed by Pol I-catalyzed nick translation, thereby forming RF II, which is converted to RF I by the sequential actions of DNA ligase and DNA gyrase.

B. *Bacteriophage φX174*

Bacteriophage φX174, as does M13, carries a small (5386 nt) single-stranded circular DNA. Curiously, the *in vivo* conversion of the φX174 viral DNA to its replicative form is a much more complex process than that for M13 DNA in that φX174 replication requires the participation of a nearly 600-kD protein assembly known as a **primosome** (Table 30-4).

a. φX174 (−) Strand Replication Is a Paradigm for Lagging Strand Synthesis

φX174 (−) strand synthesis occurs in a six-step process (Fig. 30-24):

1. The reaction sequence begins in the same way as that for M13: The (+) strand is coated with SSB except for a 44-nt hairpin. A 70-nt sequence containing this hairpin, known as *pas* (for *primosome assembly site*), is then recognized and bound by the PriA, **PriB,** and **PriC** proteins.

2. DnaB and **DnaC** proteins in the form of a DnaB$_6$ · DnaC$_6$ complex add to the DNA with the help of **DnaT protein** in an ATP-requiring process. DnaC protein is then released yielding the **preprimosome.** The preprimosome, in turn, binds primase yielding the primosome.

Table 30-4 **Proteins of the Primosome**[a]

Protein	Subunit Structure	Subunit Mass (kD)
PriA	Monomer	76
PriB	Dimer	11.5
PriC	Monomer	23
DnaT	Trimer	22
DnaB	Hexamer	50
DnaC[b]	Monomer	29
Primase (DnaG)	Monomer	60

[a]The complex of all primosome proteins but primase is known as the preprimosome.
[b]Not part of the preprimosome or the primosome.

Source: Kornberg, A. and Baker, T.A., *DNA Replication* (2nd ed.), pp. 286–288, Freeman (1992).

1. Recognition

PriA, PriB, PriC

PriA

Single-strand
binding protein
(SSB)

$(DnaB)_6 \cdot (DnaC)_6$,
DnaT + ATP

Primase

2. Assembly

+ DnaC

Primasome

3. Migration

ATP

ADP + P_i

4. Priming

ATP
+ 4rNTPs

ADP

DNA polymerase III
holoenzyme, 4dNTPs

5'

3'

5. Elongation

RFII

Pol I, ligase,
gyrase, 4dNTPs,
NAD$^+$, ATP

**6. Excision,
Gap filling,
ligation,
supercoiling**

RFI

Figure 30-24 **The synthesis of the φX174 (−) strand on a (+)
strand template to form φX174 RF I DNA.** [After Arai, K., Low,
R., Kobori, J., Schlomai, J., and Kornberg, A., *J. Biol. Chem.* **256,**
5280 (1981).]

Figure 30-25 **Electron micrograph of a primosome bound to a
φX174 RF I DNA.** Such complexes always contain a single
primosome with one or two associated small DNA loops.
[Courtesy of Jack Griffith, Lineberger Cancer Research Center,
University of North Carolina.]

expense of ATP hydrolysis. This motion, which displaces
the SSB in its path, is opposite in direction to that of template reading during DNA chain propagation.

4. At randomly selected sites, the primosome reverses
its migration while primase synthesizes an RNA primer.
The initiation of primer synthesis requires the participation
of DnaB protein which, through concomitant ATP hydrolysis, is thought to alter template DNA conformation in a
manner required by primase.

5. Pol III holoenzyme extends the primers to form
Okazaki fragments.

6. Pol I excises the primers and replaces them by DNA.
The fragments are then joined by DNA ligase and supercoiled by DNA gyrase to form the φX174 RF I.

The primosome remains complexed with the DNA (Fig. 30-25)
where it participates in (+) strand synthesis (see below).

b. φX174 (+) Strand Replication Serves as a Model
for Leading Strand Synthesis

One strand of a circular duplex DNA may be synthe-
sized via the **rolling circle** or **σ-replication** mode (so called
because of the resemblance of the replicating structure to
the Greek letter sigma; Fig. 30-26). *The φX174 (+) strand is
synthesized on an RF I template by a variation on this
process, the **looped rolling circle** mode* (Fig. 30-27):

1. (+) strand synthesis begins with the primosome-
aided binding of the phage-encoded 513-residue enzyme
gene A protein to its ~30-bp recognition site. There, gene
A protein cleaves a specific phosphodiester bond on the
(+) strand nucleotide (near the beginning of gene A) by
forming a covalent bond between a Tyr residue and the
DNA's 5'-phosphoryl group, thereby conserving the
cleaved bond's energy.

2. Rep helicase (Section 30-2Cb) subsequently attaches
to the (−) strand at the gene A protein and, with the aid of

3. The primosome is propelled in the 5' → 3' direction
along the (+) strand by the PriA and DnaB helicases at the

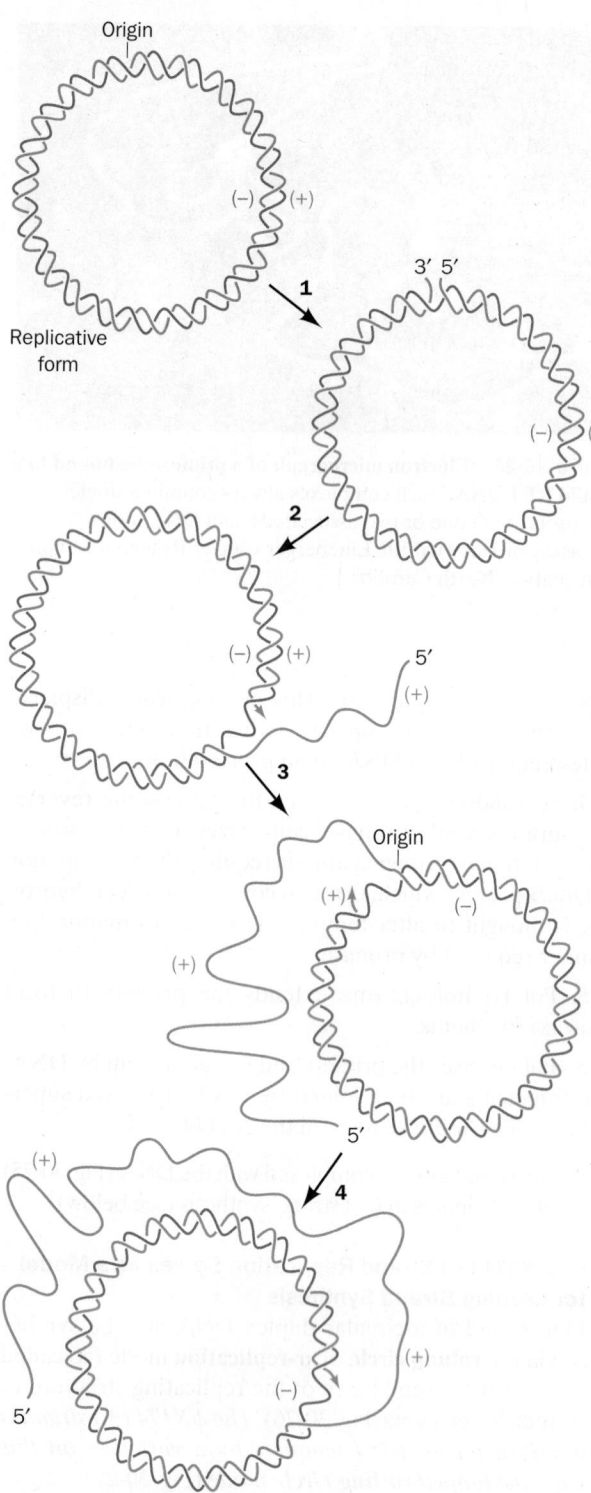

Figure 30-26 **The rolling circle mode of DNA replication.** The (+) strand being synthesized is extended from a specific cut made at the replication origin **(1)** so as to strip away the old (+) strand **(2 and 3).** The continuous synthesis of the (+) strand on a circular (−) strand template produces a series of tandemly linked (+) strands **(4),** which may later be separated by a specific endonuclease.

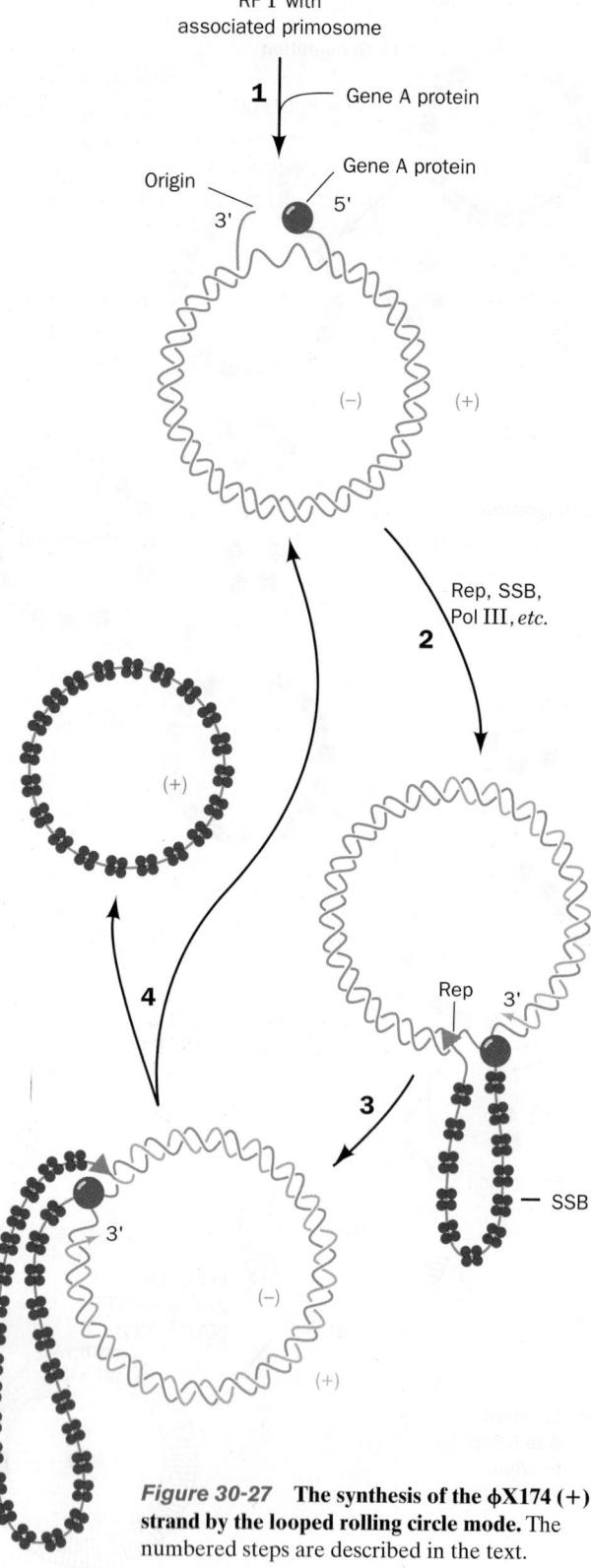

Figure 30-27 **The synthesis of the φX174 (+) strand by the looped rolling circle mode.** The numbered steps are described in the text.

the primosome still associated with the (+) strand, commences unwinding the duplex DNA from the (+) strand's

5′ end. The displaced (+) strand is coated with SSB, which prevents it from reannealing to the (−) strand. Rep helicase is essential for the replication of φX174 DNA, but not for the *E. coli* chromosome, as is demonstrated by the inability of φX174 to multiply in *rep⁻ E. coli.* Pol III holoenzyme extends the (+) strand from its free 3′-OH group.

3. The extension process generates a **looped rolling circle** structure in which the 5′ end of the old (+) strand remains linked to the gene A protein at the replication fork. It is thought that as the old (+) strand is peeled off the RF, the primosome synthesizes the primers required for the later generation of a new (−) strand.

4. When it has come full circle around the (−) strand, the gene A protein again makes a specific cut at the replication origin so as to form a covalent linkage with the new (+) strand's 5′ end. Simultaneously, the newly formed 3′-terminal OH group of the old, looped-out (+) strand nucleophilically attacks its 5′-phosphoryl attachment to the gene A protein, thereby liberating a covalently closed (+) strand. This is possible because the gene A protein has two closely spaced Tyr residues that alternate in their attachment to the 5′ ends of successively synthesized (+) strands.

The replication fork continues its progress about the duplex circle, producing new (+) strands in a manner reminiscent of linked sausages being pulled off a reel.

In the intermediate stages of a ϕX174 infection, each newly synthesized (+) strand directs the synthesis of the (−) strand to form RF I as described above. In the later stages of infection, however, the newly formed (+) strands are packaged into phage particles.

C. *Escherichia coli*

See Guided Exploration 25. The replication of DNA in E. coli *The E. coli chromosome replicates by the bidirectional θ mode from a single replication origin (Section 30-1Aa).* The most plausible model for events at the *E. coli* replication fork (Fig. 30-28) is largely derived from studies on the simpler and more experimentally accessible DNA replication mechanisms of coliphages such as M13 and ϕX174. Duplex DNA is unwound by DnaB helicase on the lagging strand template,

Figure 30-28 The replication of *E. coli* DNA. (*a*) The *E. coli* DNA replisome, which contains two DNA polymerase III holoenzyme complexes, synthesizes both the leading and the lagging strands. The lagging strand template must loop around to permit the holoenzyme to extend the primosome-primed lagging strand. Although not shown here, the DnaB helicase binds to τ₂ and hence moves with the replisome. (*b*) The holoenzyme releases the lagging strand template when it encounters the previously synthesized Okazaki fragment. This possibly signals the primosome to initiate the

synthesis of lagging strand RNA primer. (*c*) The holoenzyme rebinds the lagging strand template and extends the RNA primer to form a new Okazaki fragment. Note that in this model, leading strand synthesis is always ahead of lagging strand synthesis.

where it is joined by the primosome. The separated single strands are immediately coated by SSB. Leading strand synthesis is catalyzed by Pol III holoenzyme, as is that of the lagging strand after priming by primosome-associated primase. Both leading and lagging strand syntheses occur on a single ~900-kD multisubunit particle, the **replisome,** which contains two Pol III cores ($\alpha\epsilon\theta$) that are joined together by a dimer of τ subunits that bridges the α subunits. Hence, the lagging strand template must be looped around (Fig. 30-28). The τ_2 dimer also binds the DnaB helicase (an interaction that is not indicated in Fig. 30-28), thereby stimulating its helicase action while holding it to the replication fork. After completing the synthesis of an Okazaki fragment, the lagging strand holoenzyme relocates to a new primer near the replication fork, the primer heading the previously synthesized Okazaki fragment is excised by Pol I-catalyzed nick translation, and the nick is sealed by DNA ligase. Since lagging strand synthesis is more complex and hence more time-consuming than leading strand synthesis, the replisome functions to coordinate these two processes.

a. *E. coli* DNA Replication Is Initiated at *oriC* in a Process Mediated by DnaA Protein

The replication origin of the *E. coli* chromosome consists of a unique 250-bp segment known as the *oriC* locus. This sequence, segments of which are highly conserved among gram-negative bacteria, supports the bidirectional replication of the various plasmids into which it has been inserted. The *oriC* locus contains five highly conserved 9-bp segments with consensus sequence 5′-TTATCCACA-3′ known as **DnaA boxes** because they are specifically bound by **DnaA protein** (Fig. 30-29a). These are interspersed with several so-called **I-sites** that deviate from this consensus sequence and are bound by DnaA with lesser affinity. In addition, the "left" boundary region of *oriC* contains three tandemly repeated, 13-bp, AT-rich segments (consensus sequence 5′-GATCTNTTNTTTT-3′ where N marks nonspecific positions) that are known as **DNA unwinding elements (DUEs).**

DnaA (467 residues in *E. coli*) consists of four domains that are, from N- to C-terminus, a helicase interaction domain that mediates interactions with DnaB helicase (see below), a

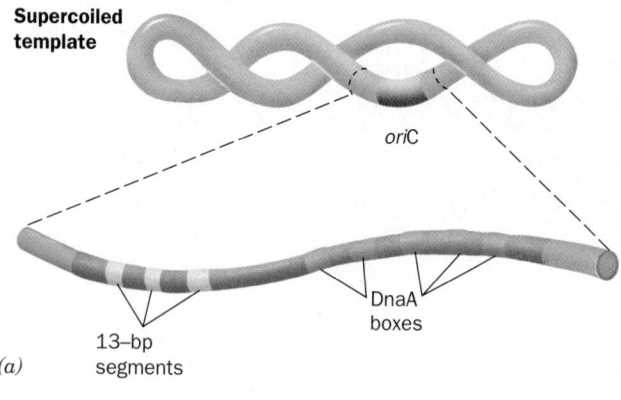

Supercoiled template

oriC

13–bp segments

DnaA boxes

(a)

(b)

Figure 30-29 **DNA replication initiation at** *oriC.* (*a*) Diagram of oriC showing the relative positions of its DnaA boxes (*green*) and its DNA unwinding elements (DUEs; *yellow*). (*b*) X-ray structure of the right-handed helical filament formed by the two C-terminal domains of *A. aeolicus* DnaA. It has eight subunits per turn and a pitch of 178 Å. Twelve subunits are shown, from right to left, in the alternating colors red, green, yellow, and blue. The right subunits are drawn as surface diagrams and the left subunits are represented by their polypeptide back bones with helices in tube form. (*c*) Model for initiation at *oriC.* The green ovals represent the N-terminal three domains of DnaA and the yellow ovals represent the associated C-terminal DNA-binding domains. See the text for an explanation. [Parts *b* and *c* courtesy of James Berger, University of California at Berkeley. PDBid 2HCB.]

ori binding

ATP-DnaA

ori melting

(−)

(+)

(superhelical strain?)

(AAA+ interactions?)

helicase loading

DnaB   ATP-DnaC

ADP-DnaC

(c)

flexible and poorly conserved linker, an ATPase domain that is a member of the AAA+ family (Section 30-2Ca), and a DNA binding domain. The X-ray structure of the C-terminal two domains of DnaA from *Aquifex aeolicus* (a thermophilic bacterium), determined by Berger, unexpectedly revealed that it forms a multisubunit right-handed helix (Fig. 30-29*b*).

Experiments with *oriC*-containing plasmids, pioneered by Kornberg, together with the X-ray structure of DnaA, indicate that replication initiation in *E. coli* occurs via the following process (Fig. 30-29*c*):

1. In the presence of ATP, DnaA, which is normally bound to three of *oriC*'s five DnaA boxes throughout *E. coli*'s lifetime, recruits additional DnaA subunits to the remaining DnaA boxes and to the I-sites so as to form a right-handed helix of DnaA subunits that is bound to the DNA. This generates local positive supercoils in the DNA. The superhelical strain resulting from the compensating negative supercoils (recall that the linking number of a covalently closed circular DNA such as an *E. coli* chromosome is invariant; Section 29-3A) melts the DUE-containing segment [Fig. 30-29*c*, *middle left*; recall that bacterial chromosomes are normally already negatively supercoiled (Section 29-2Bb)]. Alternatively, or in addition, the DnaA's ATPase domains may actively unwind the DNA (Fig. 30-29*c*, *middle right*). This process is facilitated by two homologous DNA-binding proteins, **HU** and **integration host factor (IHF),** that induce DNA bending (IHF is discussed in Section 33-3Ca).

2. The *oriC*–DnaA complex recruits two DnaB$_6$ · DnaC$_6$ complexes to opposite ends of the melted region to form the **prepriming complex.** DnaC, an ATPase that is a homolog of DnaA, functions to facilitate the loading of the DnaB hexamers onto the DNA. Its, X-ray structure, also determined by Berger, shows that it forms a helical assembly similar to that of DnaA. The AAA+ domains of DnaA and DnaC interact in an ATP-dependent manner to recruit and properly position the DnaB helicases, following which the DnaC is released.

In the presence of SSB and gyrase, DnaB helicase further unwinds the DNA in the prepriming complex in both directions so as to permit the entry of primase and RNA polymerase. The participation of both these enzymes in leading strand primer synthesis (Section 30-1D), together with the limitation of this process to the *oriC* site, suggests that the RNA polymerase activates primase to synthesize the primer. This perhaps explains the similarity of *oriC*'s DUEs to RNA polymerase's transcriptional promoters (Section 31-2Ba). The stage is thereby set for bidirectional DNA replication by Pol III holoenzyme as described above.

b. The Initiation of *E. coli* DNA Replication Is Strictly Regulated

Chromosome replication in E. coli occurs only once per cell division, so this process must be tightly controlled. The doubling (cell generation) time of *E. coli* at 37°C varies with growth conditions from <20 min to ~10 h. Yet the constant ~1000 nt/s rate of movement of each replication fork fixes the 4.6×10^6-bp *E. coli* chromosome's replication time, *C*, at ~40 min. Moreover, the segregation of cellular components and the formation of a septum between them, which must precede cell division, requires a constant time, *D* = 20 min, after the completion of the corresponding round of chromosome replication. *Cells with doubling times less than C + D = 60 min must consequently initiate chromosome replication before the end of the preceding cell division cycle.* This results in the formation of **multiforked chromosomes** as is diagrammed in Fig. 30-30 for a cell division time of 35 min.

Even in cells that contain multiple *oriC* sites, DNA replication is initiated at each such site once and only once per cell generation. However, after initiation has occurred, chain elongation proceeds at a uniform, largely uncontrolled rate. This suggests that a post-initiation *oriC* site is somehow sequestered from (prevented from interacting with) the replication initiation machinery, a phenomenon called **sequestration.** There is extensive morphological evidence, such as shown in Fig. 30-31, that the *E. coli* chromosome is associated with the cell membrane. This attachment

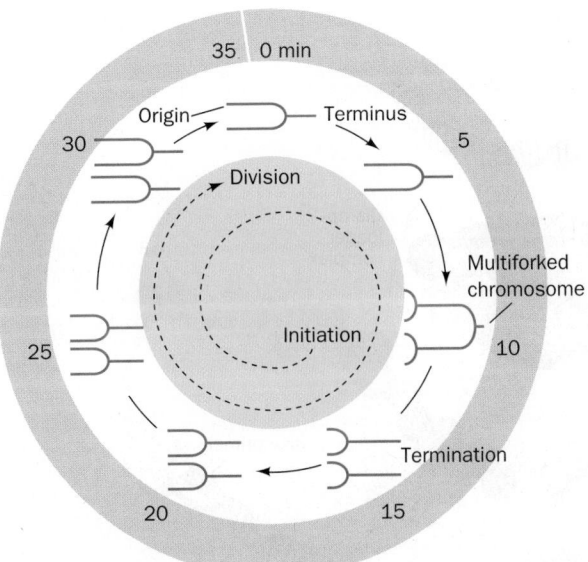

Figure 30-30   Multiforked chromosomes in *E. coli*. In cells that are dividing every 35 min, the fixed 60-min interval between the initiation of replication and cell division results in the production of multiforked chromosomes. [After Lewin, B., *Genes VII*, p. 370, Oxford University Press (2000).]

Figure 30-31   Electron micrograph of an intact and supercoiled *E. coli* chromosome attached to two fragments of the cell membrane. [From Delius, H. and Worcel, A., *J. Mol. Biol.* **82,** 108 (1974).]

would help explain how replicated chromosomes are segregated into different cells during cell division. But what is the mechanism of sequestration?

The sequence most commonly methylated in *E. coli* is the palindrome GATC, which is methylated at N6 of both its A bases by **Dam methyltransferase** (Section 30-7). GATC occurs 11 times in *oriC,* including at the beginning of all four of its 13-bp DUEs (see above). Newly replicated GATC segments are hemimethylated, that is, the GATC sequences on the newly synthesized strand are unmethylated. Although Dam methyltransferase begins methylating most hemimethylated GATC segments immediately after their synthesis (within ~1.5 min), those on *oriC* remain hemimethylated for around one-third of a cell generation. Consequently, the observation that membranes bind hemimethylated *oriC,* but not unmethylated or fully methylated *oriC,* suggests that hemimethylated *oriC* is bound to the membrane in a way that makes it inaccessible to both the initiation machinery and Dam methyltransferase.

The association of hemimethylated *oriC* with membrane requires the presence of the 181-residue **SeqA** protein, the product of *seqA* gene. Thus in *seqA⁻* cells: (1) the time to fully methylate hemimethylated GATC sites in *oriC* is reduced to 5 min, whereas the time to do so for other GATC sites is unaffected; (2) the synchrony of initiation of multiple *oriC* sites is lost; and (3) in the absence of functional Dam methyltransferase, fully methylated *oriC*-containing plasmids are replicated numerous times per cell generation, whereas in the presence of SeqA they are replicated only once. Evidently, sequestration occurs via the SeqA-mediated binding of hemimethylated *oriC* to the membrane. The hemimethylated promoter of the *dnaA* gene is similarly sequestered so as to repress its transcription, thereby providing an additional mechanism for preventing promiscuous initiation of DNA replication.

c. The Clamp Loader Loads the Sliding Clamp onto the DNA

Extensive investigations in many laboratories have led to the model of the *E. coli* replisome drawn in Fig. 30-32. The sliding clamp, which is responsible for Pol III's high processivity, is a ring-shaped dimer of β subunits through which the DNA strand being replicated is threaded (Section 30-2Bb). The two tightly associated β subunits ($K_D <$ 50 nM) that form the sliding clamp dissociate with a half-life of ~100 min at 37°C. Yet, since each replisome synthesizes around one Okazaki fragment per second, a sliding clamp must be loaded onto the lagging strand template at this frequency. This loading function is carried out in an ATP-dependent process by the γ complex (γτ₂δδ′χΨ). The τ and γ subunits are both encoded by the *dnaX* gene with τ (643 residues) the full-length product and γ (431 residues) its C-terminally truncated form; the C-terminal 122 residues of τ are known as τc. The γ complex, of which only τ₂ is diagrammed in Fig. 30-28, bridges the replisome's two Pol III cores via its τc segments, which also bind the DnaB helicase (Fig. 30-32). The χ and Ψ subunits form a heterodimer in which χ competes with primase for its binding site on SSB and hence functions to accelerate the dissociation of primase from the RNA primer it synthesized as well as link the γ complex to SSB. However, χ and Ψ are not essential participants in the clamp loading process and, therefore, we shall refer to the γτ₂δδ′ complex as the **clamp loader.** How does the clamp loader do its job?

Of the clamp loader's five subunits, only δ is capable of binding to and opening up the sliding clamp on its own. Kuriyan and O'Donnell determined the X-ray structure of the δ subunit in 1:1 complex with a β subunit that had two residues in its dimerization interface mutated so as to prevent its dimerization. The structure reveals (Fig. 30-33) that δ, which consists of three domains, inserts its β interaction element, a hydropho-

Figure 30-32 **Architecture of the *E. coli* replisome.** See the text for details. Compare this to Fig. 30-28. [Courtesy of Charles Richardson, Harvard Medical School.]

Figure 30-33   X-ray structure of the β–δ complex. A second β subunit taken from the X-ray structure of the sliding clamp (Fig. 30-13), the "Reference β monomer," is drawn in gray. The view is along the edge of the β ring. The δ subunit's β interaction element (*yellow*) consists largely of the α4 helix and two hydrophobic residues, Leu 73 and Phe 74, whose side chains are drawn in stick form. [Courtesy of John Kuriyan, University of California at Berkeley. PDB 1JQJ.]

bic plug that forms the tip of its N-terminal domain, into a hydrophobic pocket on one face of β. Comparison of δ in this structure with that in the γ₃δδ′ complex (see below) reveals that the β interaction element undergoes a dramatic conformational change on binding to β in which its α4 helix rotates by 45° and translates by 5.5 Å. Moreover, in forming the β–δ

complex, the β subunit increases its radius of curvature relative to that in the β dimer (Fig. 30-13) such that the β–δ interaction would induce the opening of one of the sliding clamp's β–β interfaces by ~15 Å. Such a gap is large enough to permit the passage of ssDNA but not dsDNA. Apparently, the clamp loader functions by trapping one β subunit of the sliding clamp in a conformation that prevents ring closure rather than actively pulling apart the two halves of the ring. This is corroborated by molecular dynamics simulations (Section 9-4a) suggesting that a β₂ dimer has a stable conformation but that an isolated β subunit with the conformation it has in the β₂ dimer rapidly (in ~1.5 ns) converts to a conformation resembling that in the β–δ complex. Thus, the conformational change of the δ subunit's β interaction element on binding to a β subunit is reminiscent of the action of a plumber's wrench in unlatching the nearby β–β interface so as to allow the sliding clamp to spring open.

The X-ray structure of the γ₃δδ′ complex (the clamp loader with both its τ subunits lacking τc; γ and τ are interchangeable in terms of their clamp loading functions) in complex with a primer–template DNA and ADP · BeF₃ (an ATP analog), also determined by Kuriyan and O'Donnell, suggests how the clamp loader functions. The γ, δ, and δ′ subunits all have similar folds; their N-terminal domains are all members of the widely distributed AAA+ family (DnaA and DnaC proteins are also members of this family) even though only the γ (and τ) subunits bind and hydrolyze ATP. The conserved regions of AAA+ proteins consist of two domains, an N-terminal ATP-binding domain and a smaller domain composed of a 3-helix bundle, whose relative orientations vary with ATP binding. The γ₃δδ′ pentamer's C-terminal domains form a ring-shaped collar (Fig. 30-34a) in which the subunits are arranged in clockwise order δ′–γ1–γ2–γ3–δ (Fig. 30-34b). The AAA+ domains are arranged in a right-handed spiral that tracks the minor groove of the dsDNA. Nevertheless, the clamp

Figure 30-34   X-ray structure of the γ₃δδ′ clamp loader in complex with a primer–template DNA and ADP · BeF₃. (*a*) View between the δ′ and δ subunits approximately perpendicular to the dsDNA's helical axis. The protein is drawn in tube-and-arrow form with its subunits colored as indicated and with the C-terminal domain of each subunit a lighter shade. The DNA consists of 10 bp with a 5′ overhang of 5 nt and, together with the ADP · BeF₃, is drawn in space-filling form with primer C cyan, template C green, ADP C magenta, N blue, O red, and P orange,

Be light green, and F light blue. (*b*) View rotated 90° about the horizontal axis relative to Part *a*. The C-terminal domain of each subunit has been deleted for clarity. Note how the single-stranded portion of the template strand turns by ~90° to avoid colliding with the collar formed by the clamp loader's C-terminal domains. The stiffness of dsDNA makes it unlikely that it could make such a turn. [Based on an X-ray structure by Mike O'Donnell, The Rockefeller University, and John Kuriyan, University of California at Berkeley. PDBid 3GLF.]

loader associates with the DNA almost entirely through contacts with the phosphate groups of the template strand alone. Thus, this structure is reminiscent of that of the E1 helicase in complex with ssDNA (Section 30-2Ca) with one of its six subunits missing.

The clamp loader must tightly bind the sliding clamp prior to its loading on the template DNA but must subsequently release the clamp to avoid interfering with its binding to the Pol III core ($\alpha\varepsilon\theta$). The structures of the clamp loader and the β–δ complex, together with a variety of biochemical evidence, suggest a model of how this might occur (Fig. 30-35): The binding of ATP to $\gamma1$ (the γ subunit that contacts δ') results in a conformational change that exposes the otherwise occluded ATP-binding site of $\gamma2$; ATP binding to $\gamma2$ likewise exposes $\gamma3$; and ATP binding to $\gamma3$ exposes the δ subunit's β interaction element, thereby permitting it to bind to a β subunit so as to spring open the sliding clamp. Primer–template DNA then inserts itself through the resulting gap in the sliding clamp. This process is facilitated by the gap between the AAA+ domains of δ and δ' subunits, which permits the clamp loader to track the template strand while avoiding contact with the primer strand. Eventually, β- and DNA-stimulated hydrolysis of the bound ATPs releases the β subunit from the clamp loader, whereon the sliding clamp closes around the DNA.

The departure of the clamp loader permits the Pol III core to bind to the sliding clamp. However, when the synthesis of an Okazaki fragment has been completed, the Pol III core must dissociate from the sliding clamp so that it can initiate the synthesis of the next Okazaki fragment. How does this occur?

Pol III's α subunit binds to the same hydrophobic pocket on the β_2 sliding clamp as does the δ subunit. This was shown by the observations that the phosphorylation of a kinase recognition sequence that had been engineered into the C-terminal segment of β is inhibited by both α and δ. The β subunit has an ~30-fold greater affinity for the γ complex in the presence of ATP than it has for the Pol III core. However, when primer–template DNA is also present, this order of affinity is reversed with β preferring to bind to the Pol III core (possibly due to the additional con-

tacts between the core and the DNA). Thus once the sliding clamp has been loaded onto the primer–template, the clamp loader is replaced by the Pol III core, which thereby blocks the clamp loader from unloading the clamp. Instead, the clamp loader loads a new clamp onto the lagging strand template in association with the primer that the primosome had synthesized in preparation for the next round of Okazaki fragment synthesis (Fig. 30-28b).

The X-ray structure of the sliding clamp in complex with primer–template DNA (Fig. 30-13) indicates that its ssDNA segment binds to the same site as do the α and δ subunits. This may serve to attract the primer–template DNA to the inside of the open clamp, which in turn may facilitate the release of the clamp loader and hence the closure of the sliding clamp. The binding of ssDNA to the sliding clamp may also prevent it from sliding away before it can be bound by the α subunit.

When the Pol III core has completed its synthesis of the Okazaki fragment, that is, when the gap between the two successively synthesized Okazaki fragments has been reduced to a nick, it releases the DNA and the sliding clamp. The Pol III core then binds to the newly primed template and its associated clamp (displacing the clamp loader), where it commences the synthesis of the next Okazaki fragment. Thus, a series of switches that are activated by ATP and DNA structure ensure the vectorial progression of lagging strand replication. Throughout this process, the Pol III holoenzyme is held at the replication fork by the leading strand Pol III core, which remains tethered to the DNA by its associated sliding clamp.

The sliding clamp that remains around the completed Okazaki fragment probably functions to recruit Pol I and DNA ligase so as to replace the RNA primer on the previously synthesized Okazaki fragment with DNA and seal the remaining nick. However, the sliding clamp must eventually be recycled. It was initially assumed that this was the job of the clamp loader. However, it is now clear that the release of the sliding clamp from its associated DNA is largely carried out by free δ subunit (the "wrench" in the clamp loader that cracks apart the β subunits forming the sliding clamp), which is synthesized in 5-fold excess over that required to populate the cell's few clamp loaders.

Figure 30-35 Schematic diagram of the clamp loader cycle. This speculative model is based on a combination of structural and biochemical information. The "arginine finger," an Arg side chain that interacts with the γ-phosphate group of an ATP bound to a neighboring subunit, is a common feature of AAA+ ATPases that form ringlike structures. [Modified from a drawing by Mike O'Donnell, The Rockefeller University, and John Kuriyan, University of California at Berkeley.]

d. Replication Termination Is Facilitated by Tus Protein

The *E. coli* replication terminus is a large (350 kb) region flanked by ten nearly identical nonpalindromic ~23-bp terminator sites, **TerH, TerI, TerE, TerD,** and **TerA** on one side and **TerJ, TerG, TerF, TerB,** and **TerC** on the other (Fig. 30-36; note that *oriC* is directly opposite the terminus region on the *E. coli* chromosome). A replication fork traveling counterclockwise as drawn in Fig. 30-36 passes through *TerJ, TerG, TerF, TerB,* and *TerC* but stops on encountering either *TerA, TerD, TerE, TerI,* or *TerH* (*TerD, TerE, TerI,* and *TerH* are presumably backup sites for *TerA*). Similarly, a clockwise-traveling replication fork transits *TerH, TerI, TerE, TerD,* and *TerA* but halts at *TerC* or, failing that, *TerB* or *TerF* or *TerG* or *TerI*. Thus, these termination sites act as one-way valves that allow replication forks to enter the termination region but not to leave it. This arrangement guarantees that the two replication forks generated by bidirectional initiation at *oriC* will meet in the replication terminus even if one of them arrives there well ahead of its counterpart.

The arrest of replication fork motion at *Ter* sites requires the action of **Tus** protein, a 309-residue monomer that is the product of the *tus* gene (for *t*erminator *u*tilization *s*ubstance). Tus specifically binds to a *Ter* site, where it prevents strand displacement by DnaB helicase, thereby arresting replication fork motion. The X-ray structure of Tus in complex with a 15-bp *Ter* sequence-containing DNA with a single T overhang at each 5′ end, determined by Kosuke Morikawa, reveals that Tus consists of two domains that form a deep positively charged cleft that largely envelops the bound DNA (Fig. 30-37). A 5-bp segment of the DNA near the side of Tus that permits the passage of the replication fork (the lower side of Fig. 30-37) is deformed and underwound relative to canonical (ideal) B-DNA such that its major groove becomes deeper and its minor groove is significantly expanded. The protein makes polar contacts with more than two-thirds of the phosphate groups in a 13-bp region and its interdomain β sheet penetrates the deepened major groove to make sequence-specific contacts with the exposed bases. The importance of this interdomain region for Tus function is demonstrated by the observation that most single residue mutations that reduce the ability of Tus to arrest replication occur in this interdomain region.

When Tus is fused to another DNA-binding protein, replication is inhibited at the other protein's binding site. This suggests that Tus does not act as a simple DNA-binding clamp, but interacts with DnaB helicase, the leading component of a replication fork (Fig. 30-32), to inhibit its helicase action. Apparently, Tus prevents the progress of DnaB in unwinding DNA from one side of Tus but not the other. Indeed, the encounter of DnaB with a Tus–*Ter* complex in the permissive direction causes Tus to rapidly dissociate from the DNA, whereas such an encounter from the nonpermissive direction generates a so-called locked Tus–*Ter* complex. Nevertheless, the way Tus and DnaB interact is unknown. Curiously, however, this termination system is not essential for termination. When the replication

Figure 30-36 **Map of the *E. coli* chromosome showing the positions of the *Ter* sites and the *oriC* site.** The *TerJ, TerG, TerF, TerB,* and *TerC* sites, in combination with Tus protein, allow a counterclockwise-moving replisome to pass but not a clockwise-moving replisome. The opposite is true of the *TerH, TerI, TerE, TerD,* and *TerA* sites. Consequently, two replication forks that initiate bidirectional DNA replication at *oriC* will meet between the oppositely facing *Ter* sites.

Figure 30-37 **X-ray structure of *E. coli* Tus protein in complex with a 15-bp *Ter*-containing DNA.** The protein is drawn in ribbon form colored in rainbow order from its N-terminus (*blue*) to its C-terminus (*red*). The DNA is shown in stick form with C gray, N blue, O red, and P orange and with successive P atoms in the same strand joined by orange rods. [Based on an X-ray structure by Kosuke Morikawa, Protein Engineering Research Institute, Osaka, Japan. PDBid 1ECR.] **See Interactive Exercise 34**

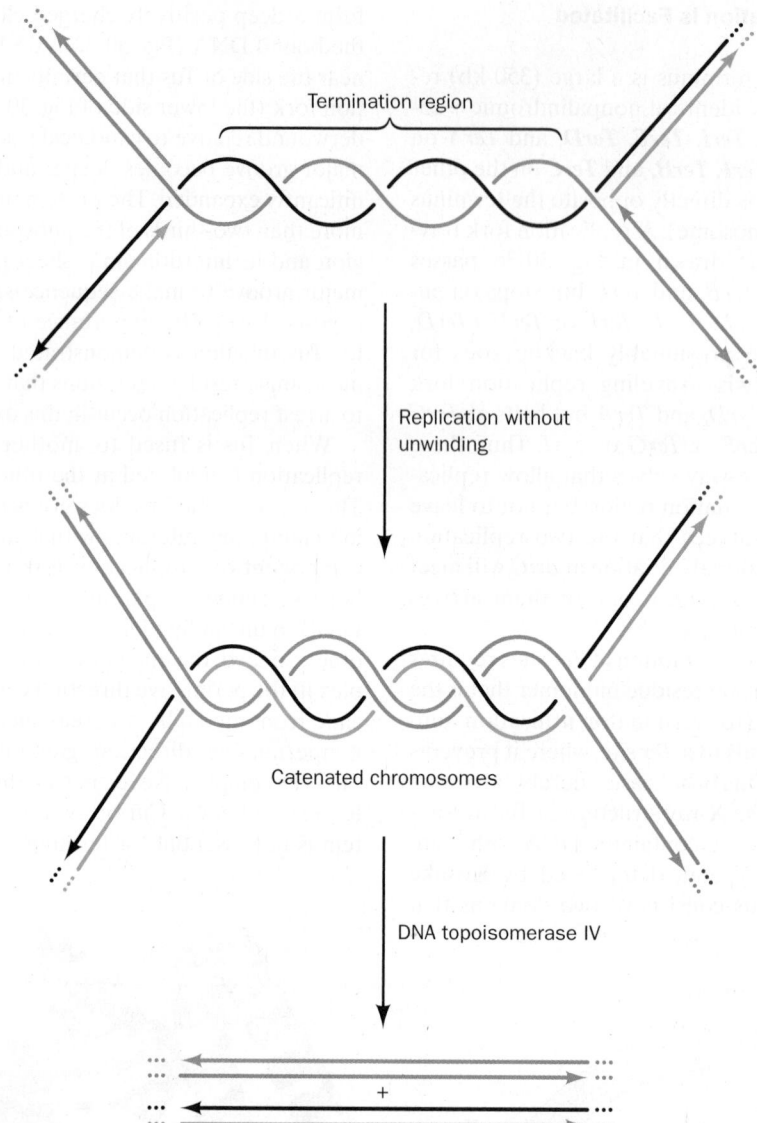

Figure 30-38   The formation and separation of catenated dsDNAs at the replication termination site. The parental strands are red and black and the daughter strands are green and blue. For clarity, the double helical character of the newly formed dsDNA molecules is not shown.

terminus is deleted, replication simply stops, apparently through the collision of opposing replication forks. Nevertheless, this termination system is highly conserved in gram-negative bacteria.

As two oppositely moving replication forks collide at the termination site, the newly synthesized strands become covalently linked to yield two covalently closed double-stranded chromosomes. However, since the parental DNA strands remain wound about each other by several turns (presumably, DNA gyrase cannot gain access to the DNA when the colliding replication forks closely approach each other), the product dsDNA strands must be wound about each other by the same number of turns (Fig. 30-38). The resulting catenated circular dsDNAs must be separated so that each can be passed to a different daughter cell. This is the job of the type II topoisomerase named topoisomerase IV (Section 29-3Cd).

D. *Fidelity of Replication*

Since a single polypeptide as small as the Pol I Klenow fragment can replicate DNA by itself, why does *E. coli* maintain a battery of >20 intricately coordinated proteins to replicate its chromosome? The answer apparently *is to ensure the nearly perfect fidelity of DNA replication required to preserve the genetic message's integrity from generation to generation.*

The rates of reversion of mutant *E. coli* or T4 phage to the wild type indicates that only one mispairing occurs per 10^8 to 10^{10} base pairs replicated. This corresponds to ~1 error per 1000 bacteria per generation. Such high replication accuracy arises from four sources:

1. Cells maintain balanced levels of dNTPs through the mechanism discussed in Section 28-3Ad. This is an important aspect of replication fidelity because a dNTP present been significant progress in our understanding of this essential process. In this section, we outline what is known about DNA replication in eukaryotes. We also discuss two DNA polymerases that are peculiar to eukaryotic systems: reverse transcriptase and telomerase.

A. *The Cell Cycle*

The **cell cycle,** the general sequence of events that occur during the lifetime of a eukaryotic cell, is divided into four distinct phases (Fig. 30-40):

1. Mitosis and cell division occur during the relatively brief **M phase** (for *m*itosis).

2. This is followed by the G_1 **phase** (for *g*ap), which covers the longest part of the cell cycle. This is the main period of cell growth.

3. G_1 gives way to the **S phase** (for *s*ynthesis), which in contrast to events in prokaryotes, *is the only period in the cell cycle when DNA is synthesized.*

4. During the relatively short G_2 **phase,** the now tetraploid cell prepares for mitosis. It then enters M phase once again and thereby commences a new round of the cell cycle.

The cell cycle for cells in culture typically occupies a 16- to 24-h period. In contrast, cell cycle times for the different types of cells of a multicellular organism may vary from as little as 8 h to >100 days. Most of this variation occurs in the G_1 phase. Moreover, many terminally differentiated cells, such as neurons or muscle cells, never divide; they assume a quiescent state known as the G_0 **phase.**

A cell's irreversible "decision" to proliferate is made during G_1. Quiescence is maintained if, for example, nutrients are in short supply or the cell is in contact with other cells **(contact inhibition).** Conversely, DNA synthesis may be induced by various agents such as carcinogens or tumor viruses, which trigger uncontrolled cell proliferation (can-

cer; Sections 19-3B and 34-4C); by the surgical removal of a tissue, which results in its rapid regeneration; or by proteins known as **mitogens,** which bind to cell-surface receptors and induce cell division (Section 34-4D).

a. The Cell Cycle Is Controlled by Cyclins and Cyclin-Dependent Protein Kinases

The progression of a cell through the cell cycle is regulated by proteins known as **cyclins** and **cyclin-dependent protein kinases (Cdks).** Cyclins are so named because they are synthesized during one phase of the cell cycle and completely degraded during a succeeding phase (protein degradation is discussed in Section 32-6). A particular cyclin specifically binds to and thereby activates its corresponding Cdk(s) to phosphorylate its target proteins, thus activating these proteins to carry out the processes comprising that phase of the cell cycle. In order to enter a new phase in the cell cycle, a cell must satisfy a corresponding **checkpoint,** which monitors whether the cell has satisfactorily completed the preceding phase [e.g., the attachment of all chromosomes to the mitotic spindle must precede mitosis (Section 1-4Aa); if this were not the case for even one chromosome, one daughter cell would lack this chromosome and the other would have two, both deleterious if not lethal conditions]. If the cell has not met the criteria of the checkpoint, the cell cycle is slowed or even arrested until it does so. We further discuss cell cycle control in Section 34-4C.

B. *Eukaryotic Replication Mechanisms*

Much of what we know about eukaryotic DNA replication has been learned from studies on budding yeast (*Saccharomyces cerevisiae*) and fission yeast (*Schizosaccharomyces pombe*), the simplest eukaryotes, and on simian virus 40 (SV40), which has a 5243-bp circular DNA chromosome that has only one replication origin. However, studies of DNA replication in the cells of **metazoa** (multicellular animals), particularly *Drosophila, Xenopus laevis* (an African clawed toad, whose eggs are easily studied), and humans, have also led to important advances in our knowledge.

a. Eukaryotic Cells Contain Numerous DNA Polymerases

The many known DNA polymerases can be classified into six families based on phylogenetic relationships. Members of families A (e.g., *E. coli* Pol I), B (e.g., *E. coli* Pol II), and C (e.g., *E. coli* Pol III) encompass all replicative polymerases as well as some repair polymerases, family D occurs only in archaea where its functions are poorly understood, and families X and Y participate in DNA repair. The fingers and thumb domains have structures that are unique to each family, whereas the catalytic residue–containing palm domains are similar in families A, B, and Y. Animal cells express at least four distinct types of DNA polymerases that are implicated in DNA replication (Table 30-5). They are designated, in the order of their discovery, DNA polymerases (pols) α, γ, δ, and ε (alternatively, POLA, POLG, POLD1, and POLE), of which pol γ is a member of family A and the others are members of family B.

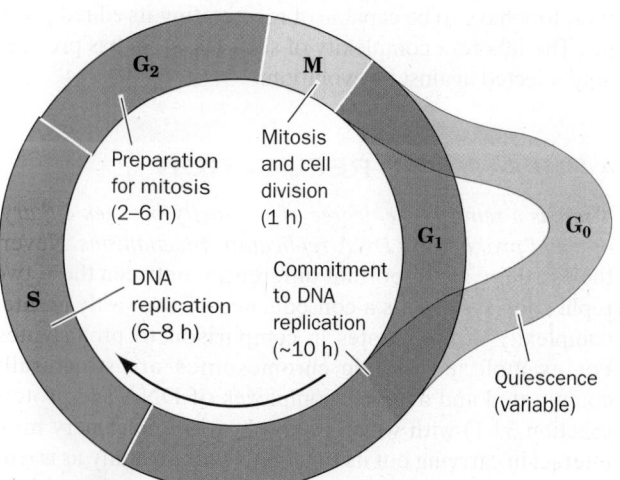

Figure 30-40 **The eukaryotic cell cycle.** Cells in G_1 may enter a quiescent phase (G_0) rather than continuing about the cycle.

Figure 30-39 **Chemical consequences if a DNA polymerase could synthesize DNA in its 3′ → 5′ direction.** (*a*) The coupling of each nucleoside triphosphate to the growing chain would be driven by the hydrolysis of the previously appended nucleoside triphosphate. (*b*) The editorial removal of an incorrect 5′-terminal nucleoside triphosphate would render the DNA incapable of further extension.

at aberrantly high levels is more likely to be misincorporated and, conversely, one present at low levels is more likely to be replaced by the dNTPs present at higher levels.

2. The polymerase reaction itself has extraordinary fidelity. This is because, as we have seen (Section 30-2Ae), the polymerase reaction occurs in two stages: (1) a binding step in which the incoming dNTP base-pairs with the template while the enzyme is in an open conformation that cannot catalyze the polymerase reaction; and (2) a catalysis step in which the polymerase forms a closed conformation about the newly formed base pair, which properly positions its catalytic residues (induced fit). Since the formation of the closed conformation requires that the incoming dNTP form a Watson–Crick-shaped base pair with the template, the conformation change constitutes a double check for correct base pairing.

3. The 3′ → 5′ exonuclease functions of Pol I and Pol III detect and eliminate the occasional errors made by their polymerase functions. In fact, mutations that increase a DNA polymerase's proofreading exonuclease activity decrease the rates of mutation of other genes.

4. A remarkable battery of enzyme systems, contained in all cells, function to repair residual errors in the newly synthesized DNA as well as any damage that it may incur after its synthesis through chemical and/or physical insults. We discuss these DNA repair systems in Section 30-5.

In addition, *the inability of a DNA polymerase to initiate chain elongation without a primer is a feature that increases DNA replication fidelity.* The first few nucleotides of a chain to be coupled together are those most likely to be mispaired because of the cooperative nature of base pairing interactions (Section 29-2). The editing of a short duplex oligonucleotide is similarly an error-prone process. The use of RNA primers eliminates this source of error since the

RNA is eventually replaced by DNA under conditic permit accurate base pairing to be achieved.

One might wonder why cells have evolved the cc system of discontinuous lagging strand synthesis than a DNA polymerase that could simply extend chains in their 3′ → 5′ direction. Consideration of the istry of DNA chain extension also leads to the con that this system promotes high-fidelity replication. Th ing of 5′-deoxynucleotide triphosphates in the 3′ → 5′ tion would require the retention of the growing cha terminal triphosphate group to drive the next couplir (Fig. 30-39*a*). On editing a mispaired 5′-terminal nucl (Fig. 30-39*b*), this putative polymerase would—in ar with Pol I, for example—excise the offending nucle leaving either a 5′-OH or a 5′-phosphate group. Neith these terminal groups is capable of energizing further extension. A proofreading 3′ → 5′ DNA polymerase v therefore have to be capable of reactivating its edited uct. The inherent complexity of such a system has pre ably selected against its evolution.

4 EUKARYOTIC REPLICATION

There is a remarkable degree of similarity between euk otic and prokaryotic DNA replication mechanisms. Ne theless, there are important differences between these replication systems as a consequence of the vastly gre complexity of eukaryotes in comparison to prokaryc For example, eukaryotic chromosomes are structur complicated and dynamic complexes of DNA and prot (Section 34-1) with which the replication machinery m interact in carrying out its function. Consequently, as is t of most aspects of biochemistry, our knowledge of h DNA is replicated in eukaryotes has lagged well behi that for prokaryotes, although in recent years there h

Table 30-5   Properties of Eukaryotic DNA Polymerases That Participate in DNA Replication

	α	γ	δ	ε
Location	Nucleus	Mitochondrion	Nucleus	Nucleus
Subunit masses (kD)[a]	167, 79, 62, 48	144	125, 55, 40	256, 78, 23, 22
	(166, 66, 59, 50)	(140, 55)	(124, 51, 51)	(262, 60, 17, 12)
Family	B	A	B	B

[a]Yeast *S. cerevisiae* (human cells).

Source: Mainly Johnson, A. and O'Donnell, M., *Annu. Rev. Biochem.* **74**, 283 (2005).

Pol α occurs only in the cell nucleus where it participates in the replication of chromosomal DNA. This function was largely established through the use of its specific inhibitor **aphidicolin**

Aphidicolin

and by the observation that pol α activity varies with the rate of cellular proliferation. Pol α, as do all DNA polymerases, replicates DNA by extending a primer $5' \rightarrow 3'$ under the direction of a single-stranded DNA template. This heterotetramer, which lacks exonuclease activity, consists of a 167-kD polymerase subunit, a 48-kD primase subunit, a 62-kD subunit that is required for full primase activity, and a 79-kD subunit that is implicated in the regulation of initiation, all of which are collectively known as **pol α/primase.**

Pol δ is a heterotrimer whose 125-kD catalytic subunit lacks an associated primase but contains a proofreading $3' \rightarrow 5'$ exonuclease domain. The X-ray structure of the yeast pol δ catalytic subunit (also called pol δ), determined by Aneel Aggarwal, reveals that this enzyme consists of five domains arranged around a central hole that is near its polymerase active site (Fig. 30-41). It has the right-hand-like architecture first seen in A-family DNA polymerases (Figs. 30-8 and 30-9), and its palm domain has a structurally similar core that contains the two invariant Asp residues implicated in the nucleotidyl transfer mechanism (Fig. 30-10). However, there are major differences between A-family polymerases (e.g., Fig. 30-8) and pol δ, which is representative of B-family polymerases. Most notably, in pol δ:

1. The fingers domain, which consists of only a pair of antiparallel helices, is rotated by ~60° relative to that in A-family polymerases.

2. The exonuclease domain projects from the top of the fingers domain rather than from the bottom of the palm domain, as it does in A-family polymerases.

3. There is a large N-terminal domain that A-family polymerases lack.

4. The template strand enters the active site from a cleft between the N-terminal and exonuclease domains, whereas in A-family polymerases, it does so from the fingers domain.

5. The newly formed dsDNA nearest the active site has a B-DNA-like conformation instead of the A-DNA-like conformation observed in A-family polymerases (Section 30-2Ae).

Figure 30-41   X-ray structure of yeast DNA polymerase δ (pol δ) in complex with primer–template DNA and dCTP. The protein is drawn in ribbon form with its five domains differently colored as indicated. The DNA, whose primer and template strands consist of 12 and 16 nt, together with the incoming dCTP, is drawn in stick form with template C green, primer C cyan, dCTP C magenta, N blue, O red, and P orange and with successive P atoms in each DNA strand connected by orange rods. The Ca^{2+} ions at the polymerase (Pol) active site are represented by dark green spheres as is the Ca^{2+} ion at the exonuclease (Exo) active site. [Based on an X-ray structure by Aneel Aggarwal, Mount Sinai School of Medicine, New York, New York. PDBid 3IAY.]

In contrast to pol α, which exhibits only moderate processivity (~100 nucleotides), that of pol δ is essentially unlimited (replicates the entire length of a template), but only when it is in complex with a protein named **proliferating cell nuclear antigen** (**PCNA;** so named because it occurs only in the nuclei of proliferating cells and reacts with antibodies produced by a subset of patients with the autoimmune disease systemic lupus erythematosus). The X-ray structure of PCNA (Fig. 30-42), determined by Kuriyan, reveals that it forms a trimeric ring with almost identical structure (and presumably function) as the *E. coli* β$_2$ sliding clamp (Fig. 30-13). Thus, each PCNA subunit consists of four rather than six of the structurally similar βαβββ motifs from which the *E. coli* β subunit is constructed. Intriguingly, PCNA and the β subunit exhibit no significant sequence identity, even when their structurally similar portions are aligned. Archaea also have sliding clamps with pseudohexagonal symmetry.

Pol δ in complex with PCNA is required for lagging strand synthesis. In contrast, pol α/primase functions to synthesize ~12-nt RNA primers, which it extends by an additional ~20 nt of DNA. Then, in a process called **polymerase switching,** the eukaryotic counterpart of the *E. coli* clamp loader (Section 30-3Cc), **replication factor C (RFC),** displaces the pol α and loads PCNA on the template DNA near the primer strand, following which pol δ binds to the PCNA and processively extends the DNA strand.

RFC, like the *E. coli* clamp loader, is a heteropentamer of AAA+ family subunits, but whereas the clamp loader contains three identical subunits (γ$_3$δδ′), RFC consists of five different subunits, RFC-A through RFC-E (alternatively, RFC1–RFC5). The X-ray structure of yeast RFC in complex with PCNA (Fig. 30-43), determined by O'Donnell and Kuriyan, reveals that RFC's A, B, and C subunits bind in conserved hydrophobic grooves on the face of PCNA on which its C-terminal residues are located (since PCNA's three identical subunits are linked head-to-tail, its two faces are different). In fact, this so-called C side of PCNA, which faces away from the direction of polymerase motion, binds many of the proteins that participate in replicative processes, including most DNA polymerases, and hence PCNA plays a major role in

Figure 30-43   **X-ray structure of yeast RFC in complex with PCNA, ADP, and ATPγS.** Both proteins are drawn in tube-and-arrow form with the RFC subunits colored as their homologs in Fig. 30-34a (with the RFC-A, B, C, D, and E subunits corresponding to the *E. coli* clamp loader's δ, γ3, γ2, γ1, and δ′ subunits, respectively) and the PCNA colored as in Fig. 30-42. The view is similar to that of the *E. coli* clamp loader in Fig. 30-34a. ADP, which binds to only RFC-E, and ATPγS, which binds to the other four subunits, are shown in space-filling form with ATPγS C green, ADP C magenta, N blue, O red, P orange, and S yellow. Note that all five RFC subunits have a nucleotide binding site, whereas the *E. coli* δ and δ′ subunits do not. [Based on an X-ray structure by Mike O'Donnell, The Rockefeller University, and John Kuriyan, University of California at Berkeley. PDBid 1SXJ.]

Figure 30-42   **X-ray structure of human PCNA.** Its three subunits, which form a 3-fold symmetric ring, are drawn in ribbon form embedded in their semitransparent surface diagram. One of these subunits is colored in rainbow order from its N-terminus (*blue*) to its C-terminus (*red*), another is pink, and the third is light green. A space-filling model of B-DNA viewed along its helix axis has been drawn in the center of the PCNA ring. Compare this structure with that of the β$_2$ sliding clamp of the *E. coli* Pol III holoenzyme (Fig. 30-13). [Based on an X-structure by John Kuriyan, University of California at Berkeley. PDBid 1AXC.] **See Interactive Exercise 35** recruiting the components of the various types of replication forks. The C-terminal domains of each RFC subunit associate to form a ring-shaped collar, as do the C-terminal domains of the *E. coli* clamp loader (Fig. 30-34). Likewise, RFC's AAA+ domains are arranged in a right-handed spiral that matches the helical path of the sugar–phosphate backbone of B-DNA. Presumably, prokaryotic and eukaryotic clamp loaders interact with primer–template DNA and their corresponding sliding clamps in similar ways.

Pol ε, a heterotetrameric nuclear enzyme, is the most enigmatic participant in DNA replication. Pol ε is highly processive in the absence of PCNA and has a $3' \rightarrow 5'$ exonuclease activity that degrades single-stranded DNA to 6- or 7-residue oligonucleotides rather than to mononucleotides, as does that of pol δ. Although pol ε is necessary for the viability of yeast, its essential function can be carried out by only the noncatalytic C-terminal half of its 256-kD catalytic subunit, which is unique among B-family DNA polymerases. This suggests that the C-terminal half of the pol ε catalytic subunit is required for the assembly of the replication complex. Nevertheless, Thomas Kunkel has shown that pol ε is probably the leading strand replicase, although it may also contribute to lagging strand synthesis. Moreover, pol δ may also participate in leading strand synthesis.

Pol γ, a monomer, occurs exclusively in the mitochondrion, where it presumably replicates the mitochondrial DNA. Chloroplasts contain a similar enzyme.

Eukaryotic cells contain batteries of DNA polymerases. These include the DNA polymerases that participate in chromosomal DNA replication (pols α, δ, and ε) and several that take part in DNA repair processes (Section 30-5) including **pols β, η, ι, κ,** and ζ (alternatively, POLB, POLH, POLI, POLK, and POLZ). Pol β, an X-family enzyme, is remarkable for its small size (a 335-residue monomer in humans).

b. Eukaryotic Chromosomes Consist of Numerous Replicons

Eukaryotic and prokaryotic DNA replication systems differ most obviously in that eukaryotic chromosomes have multiple replication origins in contrast to the single replication origin of prokaryotic chromosomes. Eukaryotic cells replicate DNA at the rate of ~50 nt/s (~20 times slower than does *E. coli*) as was determined by autoradiographically measuring the lengths of pulse-labeled sections of eukaryotic chromosomes. Since a eukaryotic chromosome typically contains 60 times more DNA than those of prokaryotes, its bidirectional replication from a single origin would require ~1 month to complete. Electron micrographs such as Fig. 30-44, however, reveal that eukaryotic chromosomes contain multiple origins, one every 3 to 300 kb depending on both the species and the tissue, so that S phase usually occupies only a few hours.

Cytological observations indicate that the various chromosomal regions are not all replicated simultaneously; rather, clusters of 20 to 80 adjacent **replicons**

Figure 30-44 **Electron micrograph of a fragment of replicating *Drosophila* DNA.** The arrows indicate its multiple replication eyes. [From Kreigstein, H.J. and Hogness, D.S., *Proc. Natl. Acad. Sci.* **71,** 136 (1974).]

(replication units; DNA segments that are each served by a replication origin) are activated simultaneously. New replicons are activated throughout S phase until the entire chromosome has been replicated. During this process, replicons that have already been replicated are distinguished from those that have not; that is, *a cell's chromosomal DNA is replicated once and only once per cell cycle.*

c. The Assembly of the Eukaryotic Initiation Complex Occurs in Two Stages

The once-and-only-once replication of eukaryotic DNA per cell cycle is conferred by a type of binary switch. A **pre-replicative complex (pre-RC)** is assembled at each replication origin during the G₁ phase of the cell cycle. This is the only period of the cell cycle during which the pre-RC can form and hence this process is known as **licensing.** However, a licensed pre-RC cannot initiate DNA replication. Rather, it must be activated to do so, a process that occurs only during S phase. *This temporal separation of pre-RC assembly and origin activation ensures that a new pre-RC cannot assemble on an origin that has already "fired" (commenced replication) so that an origin can only fire once per cell cycle.* How does this occur?

The licensing process and how the pre-RC is activated to form an initiation complex are still incompletely understood. Thus, although it appears that most of the proteins forming these complexes have been identified, their structures, interactions, and, in many cases, their functions are largely unknown. Keeping this in mind, let us consider what is known about these processes.

Replication origins are surprisingly variable among species, often within the same organism, and even vary with a given organism's developmental stage. Thus, whereas *S. cerevisiae* origins, which are known as **autonomously replicating sequences (ARS),** contain a highly conserved 11-bp AT-rich sequence within a less well defined ~125-bp region, some metazoan origins are dispersed over 10 to 50 kb "initiation zones" that contain multiple origins and, in some cases, require no specific DNA sequence at all. Despite this disparity, the proteins that participate in eukaryotic DNA replication are highly conserved from yeast to humans.

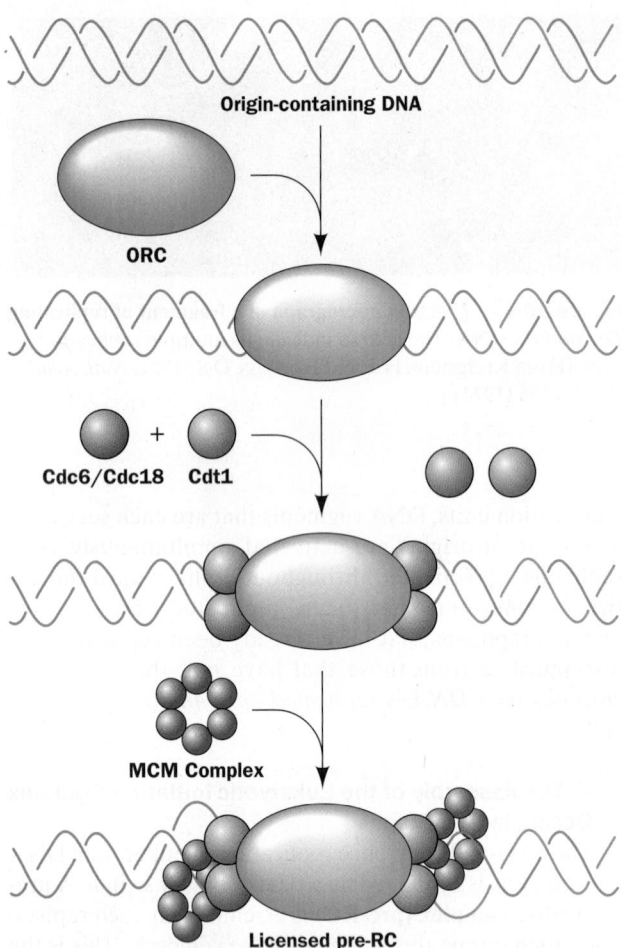

Figure 30-45   Schematic diagram for the assembly of the eukaryotic pre-replicative complex (pre-RC). The actual stoichiometries, positions, and interactions of its various components are largely unknown. The pre-RC only forms during the G_1 phase of the cell cycle.

The assembly of the pre-RC (Fig. 30-45) begins late in M phase or early in G_1 phase with the binding of the **origin recognition complex (ORC),** a hexamer of related proteins **(Orc1** through **Orc6),** to the origin, where it remains bound during most or all of the cell cycle. ORC, the functional analog of DnaA protein in *E. coli* replication initiation (Section 30-3Ca), then recruits two proteins, **Cdc6** in *S. cerevisiae* (**Cdc18** in *S. pombe*; Cdc for *cell division cycle*) and **Cdt1.** These proteins then cooperate with the ORC to load the **MCM complex** [named for its *m*inichromosome (plasmid) *m*aintenance functions], a hexamer of related subunits (**Mcm2** through **Mcm7**), onto the DNA to yield the licensed pre-RC. The MCM complex, a ring-shaped ATP-driven helicase, is the analog of *E. coli* DnaB helicase, whereas Cdc6/Cdc18 together with Cdt1 appears to be an analog of *E. coli* DnaC (which facilitates DnaB loading). With the exception of Cdt1, all of these proteins, Orc1 through Orc6, Cdc6/Cdc18, Mcm2 through Mcm7, as well as *E. coli* DnaA, DnaB, and DnaC, are AAA+ ATPases.

The conversion of a licensed pre-RC to an active initiation complex requires the addition of pol α/primase, pol ε, and several accessory proteins, which only occurs at the onset of S phase. This process begins with addition of **Mcm10** protein (which shares no sequence similarity with any of the subunits of the MCM complex) to the pre-RC, which probably displaces Cdt1. This is followed by the addition of at least two protein kinases, a Cdk and **Ddk,** the latter being a heterodimer of the protein kinase **Cdc7** with its activating subunit **Dbf4** (Ddk stands for *D*bf4-*d*ependent *k*inase). Ddk acts to phosphorylate five of the six MCM subunits (all but Mcm2) so as to activate the MCM complex as a helicase. In contrast, the way in which Cdks activate the pre-RC is poorly understood, although several ORC and MCM proteins as well as Cdc6/Cdc18 are phosphorylated by Cdks. Ddk together with a Cdk also recruits **Cdc45** to the growing initiation complex. Cdc45, in turn, is required for the assembly of the initiating synthetic machinery at the replication fork, including pol α/primase, pol ε, PCNA, and **replication protein A (RPA),** the heterotrimeric eukaryotic counterpart of SSB, thereby forming an active initiation complex.

d. Re-Replication Is Prevented through the Actions of Cdks and Geminin

Once initiation (priming) has occurred, the initiation complex is joined by RFC and pol δ and, as is described above, is converted to an active replicative complex by polymerase switching. DNA replication then proceeds bidirectionally until each replication fork has collided with an oppositely traveling replication fork, thereby completing the replication of the replicon. An active replication fork will destroy any licensed pre-RCs and unfired initiation complexes in its path, thereby preventing the DNA at such sites from being replicated twice. Eukaryotes appear to lack termination sequences and proteins analogous to the *Ter* sites and Tus protein in *E. coli*.

Several redundant mechanisms ensure that a pre-RC can initiate DNA synthesis only once. Cdks are active from late G_1 phase through late M phase. These elevated Cdk levels, which are required to activate initiation, also prevent reinitiation. The Cdk-mediated phosphorylation of Cdc6/Cdc18, which occurs late in G_1 after the pre-RCs have formed, causes Cdc6/Cdc18 to be proteolytically degraded in yeast and exported from the nucleus in mammalian cells. Evidently, Cdc6/Cdc18 is only required for the assembly of the pre-RC, not its activation. The helicase activity of the MCM complex is inhibited by phosphorylation, at least *in vitro*. Moreover, MCM proteins are exported from the nucleus in G_2 and M phases, a process that is interrupted by Cdk inactivation. However, the function of Cdk-mediated phosphorylation of ORC proteins is unclear.

Metazoan cells have yet another mechanism to prevent the assembly of a licensed pre-RC on already replicated DNA. High levels of a protein named **geminin** appear in S phase and continue to accumulate until late M phase, when geminin is degraded. Geminin associates with Cdt1 (which together with Cdc6/Cdc18 loads the MCM complex onto the ORC) so as to inhibit the assembly of the pre-RC. This inhibition can be reversed by the addition of excess Cdt1. It therefore seems likely that the presence of geminin provides protection against DNA re-replication under conditions when Cdks are inhibited by checkpoint activation. In addition, re-replication is also prevented by the degradation of Cdt1 after the origin with which it is associated has fired. The requirement for DNA replication in this process is indicated by its blockage by the DNA polymerase inhibitor aphidicolin.

Finally, cells that have shifted to the G_0 (quiescent) phase of the cell cycle (Fig. 30-40)—the majority of cells in the human body—cease making DNA. Such cells are characterized by the absence of Cdk activity. In proliferating cells, this would permit the re-replication of DNA. However, cells in G_0 also lack the proteins of the MCM complex and are therefore incapable of assembling licensed pre-RCs. Since cancerous cells are characterized by being in a state of rapid proliferation (Section 19-3B), the presence of MCM complex proteins in what should be quiescent cells is a promising diagnostic marker for cancer.

e. Primers Are Removed by RNase H1 and Flap Endonuclease-1

In lagging strand synthesis, when pol δ reaches the previously synthesized Okazaki fragment, it partially displaces its RNA primer through DNA synthesis, thereby generating an RNA flap. The primer is then removed through the actions of two enzymes: **RNase H1** removes most of the RNA leaving only a 5′ ribonucleotide adjacent to the DNA, which is then removed by **flap endonuclease-1 (FEN1).** However, as we have seen, pol α/primase extends the RNA primers it has made by ~20 nt of DNA before it is displaced by pol δ. Since pol α lacks proofreading ability, this primer extension is more likely to contain errors than the DNA synthesized by pol δ. However, FEN1, which is recruited to the replication fork by its binding to the C side of PCNA, provides what is, in effect, pol α's proofreading function: It is also an endonuclease that excises mismatch-containing oligonucleotides up to 15 nt long from the 5′ end of an annealed DNA strand. Moreover, FEN1 can make several such excisions in succession to remove more distant mismatches. The excised segment is later replaced by pol δ as it synthesizes the succeeding Okazaki fragment.

f. Mitochondrial DNA Is Replicated via RNA Okazaki Fragments

Mammalian mitochondria contain two to ten copies of their ~16-kb circular chromosome. The pol γ–mediated replication of this chromosome occurs unidirectionally from a single origin. In this process, as Ian Holt showed, the lagging strand Okazaki fragments are entirely synthesized as RNA (Fig. 30-46). The RNA is then replaced by DNA, although the way this occurs is poorly understood. One possibility is that this RNA is excised in much the same way as the primers in the nucleus, that is, through the actions of RNase H1 and FEN1, followed by the synthesis of lagging strand DNA by pol γ.

Figure 30-46 Replication of the mammalian mitochondrial chromosome. This circular chromosome is unidirectionally replicated from a single origin by a process in which the Okazaki fragments are entirely RNA.

C. Reverse Transcriptase

The **retroviruses,** which are RNA-containing eukaryotic viruses such as certain tumor viruses and human immunodeficiency virus (HIV), contain an **RNA-directed DNA polymerase (reverse transcriptase).** This enzyme, which was independently discovered in 1970 by Howard Temin and David Baltimore, acts much like Pol I in that it synthesizes DNA in the 5′ → 3′ direction from primed templates. In the case of reverse transcriptase, however, RNA is the template. The discovery of reverse transcriptase caused a mild sensation in the biochemical community because it was perceived by some as being heretical to the central dogma of molecular biology (Section 5-4). There is, however, no thermodynamic prohibition to the reverse transcriptase reaction; in fact, under certain conditions, Pol I can likewise copy RNA templates.

Reverse transcriptase transcribes the retrovirus's single-stranded RNA genome to a double-stranded DNA as follows (Fig. 30-47):

1. The retroviral RNA acts as a template for the synthesis of its complementary DNA (RNA-directed DNA polymerase activity), yielding an RNA–DNA hybrid helix. The DNA synthesis is primed by a host cell tRNA whose 3′ end partially unfolds to base pair with a complementary segment of the viral RNA.

2. The RNA strand is then nucleolytically degraded (**RNase H** activity; H for *hybrid*).

3. The DNA strand acts as a template for the synthesis of its complementary DNA (DNA-directed DNA polymerase activity), yielding double-stranded DNA.

The DNA is then integrated into a host cell chromosome.

Reverse transcriptase has been a particularly useful tool in genetic engineering because of its ability to transcribe mRNAs to complementary strands of DNA (cDNA). In transcribing eukaryotic mRNAs, which have poly(A) tails (Section 31-4Ab), the primer can be oligo(dT). cDNAs have been used, for example, as probes in Southern blotting (Section 5-5D) to identify the genes coding for their

Figure 30-47 **The reactions catalyzed by reverse transcriptase.**

corresponding mRNAs. An RNA's base sequence can be easily determined by sequencing its cDNA (Section 7-2A).

a. X-Ray Structure of HIV-1 Reverse Transcriptase

HIV-1 reverse transcriptase (RT) is a dimeric protein whose subunits are synthesized as identical 66-kD polypeptides, known as **p66** (p for *protein*), that each contain a polymerase domain and an RNase H domain. However, the RNase H domain of one of the two subunits is proteolytically excised, thereby yielding a 51-kD polypeptide named **p51.** Thus, RT is dimer of p66 and p51.

The first drugs to be clinically approved to treat AIDS, **3'-azido-3'-deoxythymidine (AZT; zidovudine), 2',3'-dideoxyinosine (ddI; didanosine), 2',3'-dideoxycytidine (ddC; zalcitabine), and 2',3'-didehydro-3'-deoxythymidine (stavudine),**

3'-Azido-3'-deoxythymidine (AZT; zidovudine)

2',3'-Dideoxyinosine, (ddI; didanosine)

2',3'-Dideoxycytidine (ddC; zalcitabine)

2',3'-Didehydro-3'-deoxythymidine (stavudine)

are RT inhibitors. Unfortunately, resistant strains of HIV-1 arise quite rapidly because RT lacks a proofreading exonuclease function and hence is highly error prone. Thus, as we have seen (Section 15-4Cd), effective long-term anti-HIV therapy requires the concurrent administration of at least one RT inhibitor and an HIV protease inhibitor.

Edward Arnold determined the X-ray structure of RT complexed to a 21-bp primer–template DNA with a 5-nt overhang at the 5' end of its template strand, a dideoxy-dG (ddG) residue at the 3' end of its primer strand (which prevents its further extension), and in complex with dATP (Fig. 30-48). The polymerase domains of p66 and p51 each contain four subdomains, which, because of their collective resemblance in p66 to DNA polymerases, are named, from N- to C-terminus, fingers, palm, thumb, and connection. Indeed, reverse transcriptases form a separate family of DNA polymerases, family RT. In p66, the RNase H domain follows the connection.

p51 has undergone a remarkable conformational change relative to p66: The connection has rotated by 155° and translated by 17 Å to bring it from a position in p66 in which it contacts only the RNase H domain (Fig. 30-48*a*) to one in p51 in which it contacts all three other polymerase subdomains (Fig. 30-48*b*). This permits p66 and p51 to bring different surfaces of their connections into juxtaposition to form, in part, RT's DNA-binding groove. Thus, the chemically identical polymerase domains of p66 and p51 are not related by 2-fold molecular symmetry (a rare but not unprecedented phenomenon), but instead, associate in a sort of head-to-head and tail-to-tail arrangement (Fig. 30-48*c*). Consequently, RT has only one polymerase active site and one RNase H active site. This is an example of viral genetic economy: HIV-1, with its limited genome size, has

(a)

(b)

Figure 30-48 **X-ray structure of HIV-1 reverse transcriptase in complex with primer–template DNA and dATP.** (*a*) A ribbon diagram of the p66 subunit in which the N-terminal fingers domain is blue, the palm is red, the thumb is green, the connection is yellow, and the RNase H domain is magenta. (*b*) The p51 subunit with its palm domain oriented similarly to that of p66. Note the different orientations of its four domains relative to p66. (*c*) A ribbon diagram of the HIV-1 RT p66/p51 heterodimer in complex with DNA and dATP. The domains of p66 and p51 are colored as in Parts *a* and *b* (the labels indicate subunit and domain; e.g., 51F and 66R denote the p51 finger domain and the p66 RNase H domain). The DNA is drawn in ladder form. The complex is oriented with its p66 polymerase subunit toward the top of the figure and viewed toward the protein's template–primer binding cleft (whose floor is largely composed of the connection domains of p66 and p51). [Based on an X-ray structure by Edward Arnold, Rutgers University. PDBid 3JYT.] See Interactive Exercise 36

(c)

succeeded in using a single polypeptide for what are essentially two different functions.

The dATP binds at the 3′ end of the primer strand, near p66's three catalytically essential Asp side chains, where it pairs with a template dT base. The DNA assumes a conformation that, near the polymerase active site, resembles A-DNA (note the A-DNA-like tilt of the bases with respect to the helix axis below the dATP in Fig. 30-48*c*), but elsewhere more closely resembles B-DNA (in which the bases are nearly perpendicular to the helix axis), a phenomenon that also has been observed in several structures of A-family DNA polymerases in their complexes with DNA (Section 30-2Ae). Most of the protein–DNA interactions involve the DNA's sugar–phosphate backbone.

The RT active site region contains the few sequence motifs that are conserved among the various polymerases. Indeed, this region of p66 has a striking structural resemblance to DNA polymerases of known structure (Sections 30-2A and 30-4Ba). This suggests that other DNA polymerases are likely to bind DNA in a similar manner.

D. *Telomeres and Telomerase*

The ends of linear chromosomes cannot be replicated by any of the mechanisms we have yet considered. This is because the RNA primer at the 5′ end of a completed lagging strand cannot be replaced with DNA; the primer required to do this would have no place to bind. How, then, are the DNA sequences at the ends of eukaryotic chromosomes, the **telomeres** (Greek: *telos,* end), replicated?

Telomeric DNA has an unusual sequence: It consists of up to several thousand tandem repeats of a simple, species-dependent, G-rich sequence concluding the 3′-ending strand of each chromosomal terminus. For example, the ciliated protozoan *Tetrahymena* has the repeating telomeric sequence TTGGGG, whereas in all vertebrates it is

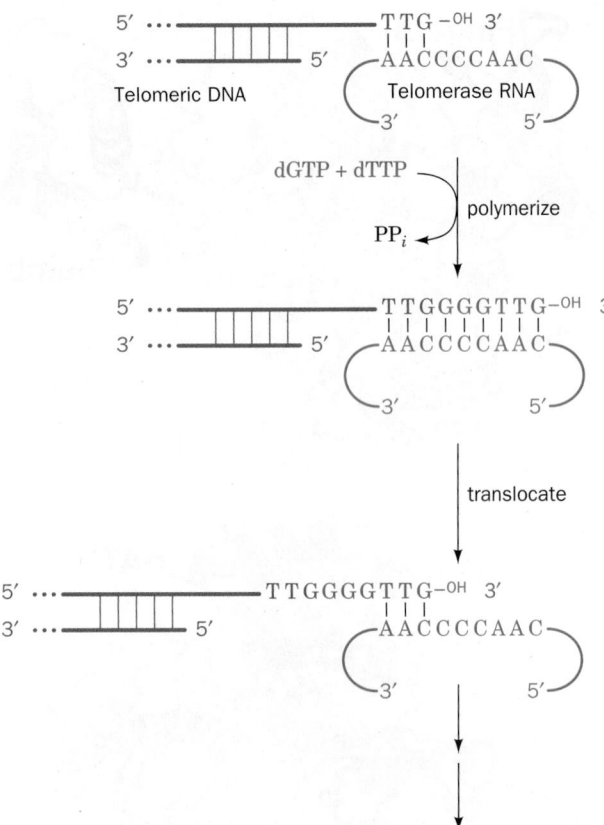

Figure 30-49   Proposed mechanism for the synthesis of telomeric DNA by *Tetrahymena* telomerase. The telomere's 5'-ending strand is later extended by normal lagging strand synthesis. [After Greider, C.W. and Blackburn, E.H., *Nature* **337**, 336 (1989).]

TTAGGG. Moreover, this strand ends with an overhang that varies from ~20 nt in yeast to ~200 bp in humans.

Elizabeth Blackburn, Carol Greider, and Jack Szostak have shown that telomeric DNA is synthesized by a novel mechanism. The enzyme that synthesizes the G-rich strand of telomeric DNA is named **telomerase.** *Tetrahymena* telomerase, for example, adds tandem repeats of the telomeric sequence TTGGGG to the 3' end of any G-rich telomeric oligonucleotide independently of any exogenously added template. A clue as to how this occurs came from the discovery that telomerases are ribonucleoproteins whose RNA components contain a segment that is complementary to the repeating telomeric sequence. This sequence apparently acts as a template in a kind of reverse transcriptase reaction that synthesizes the telomeric sequence, translocates to the DNA's new 3' end, and repeats the process (Fig. 30-49). This hypothesis is confirmed by the observation that mutationally altering the telomerase RNA gene segment complementary to telomere DNA results in telomere DNA with the corresponding altered sequence. In fact, telomerase's highly conserved protein component, which is named **TERT,** is homologous to known reverse transcriptases (its RNA component is called **TER).** The DNA strand complementary to the telomere's G-rich

strand is apparently synthesized by the normal cellular machinery for lagging strand synthesis, thereby accounting for the 3' overhang of the G-rich strand.

a. TERT Resembles Other DNA Polymerases

The X-ray structure of the 596-residue TERT from the red flour beetle *Tribolium castaneum,* determined by Emmanuel Skordalakes, reveals that this subunit contains the familiar fingers–palm–thumb domain organization of other DNA polymerases (Fig. 30-50) and, in particular, resembles the corresponding domains of the HIV-1 reverse transcriptase p66 subunit (Fig. 30-48). In addition, TERT has an N-terminal RNA-binding domain named TRBD (for *telomere repeat binding domain*). TRBD closes the gap between the thumb and fingers, thereby yielding a ringlike protein with a hole that is ~26 Å wide and ~21 Å deep. This is sufficient to accommodate an ~8-bp segment of double-stranded nucleic acid.

b. Telomeres Must Be Capped

Without the action of telomerase, a chromosome would be shortened at both ends by 50 to 100 nt with every cycle of DNA replication and cell division. It was therefore initially assumed that, in the absence of active telomerase, essential genes located near the ends of chromosomes would eventually be lost, thereby killing the descendents of the originally affected cells. However, it is now evident that telomeres serve a vital chromosomal function that is compromised before this can happen. Free DNA ends, which are subject to nuclease degradation, trigger DNA damage repair systems that normally function to rejoin the ends of broken chromosomes (as well as cell cycle arrest until this has happened). Thus exposed telomeric DNA would result in the end-to-end fusion of chromosomes, a process that leads to chromosomal instability and eventual cell death [fused chromosomes often break in mitosis (their two centromeres may cause them to be pulled in opposite directions), activating DNA damage checkpoints]. However, in a process known as **capping,** telomeric DNA is specifically bound by proteins that sequester the DNA ends. There is mounting evidence that capping is a dynamic process in which the probability of a telomere spontaneously upcapping increases as telomere length decreases. Since most somatic cells in multicellular organisms have very low levels of telomerase activity, this explains why such cells in culture can only undergo a limited number of doublings (20–60) before they reach senescence (a stage in which they cease dividing) and eventually die (Section 19-3B). Indeed, otherwise immortal *Tetrahymena* cultures with mutationally impaired telomerases exhibit characteristics reminiscent of senescent mammalian cells before dying off. Apparently, *the loss of telomerase function in somatic cells is a basis for aging in multicellular organisms.*

c. Telomere Length Correlates with Aging

There is strong experimental evidence in support of this theory of aging. The analysis of cultured human fibroblast

TRBD

Fingers

Palm

Thumb

Figure 30-50 **X-ray structure of TERT from *Tribolium castaneum.*** The protein is shown ribbon form with its TRBD, fingers, palm, and thumb colored orange, blue, red, and green, respectively, and embedded in its like-colored semitransparent surface diagram. Compare this structure with that of p66 subunit of HIV-1 reverse transcriptase (Fig. 30-48*a*). [Based on an X-ray structure by Emmanuel Skordalakes, The Wistar Institute, Philadelphia, Pennsylvania. PDBid 3DU5.]

from a number of donors between 0 and 93 years old indicates that there is only a weak correlation between the proliferative capacity of a cell culture and the age of its donor. There is, however, a strong correlation, valid over the entire donor age range, between the initial telomere length in a cell culture and its proliferative capacity. Thus, cells that initially have relatively short telomeres undergo significantly fewer doublings than cells with longer telomeres. Moreover, fibroblasts from individuals with **progeria** (a rare disease characterized by rapid and premature aging resulting in childhood death) have short telomeres, an observation that is consistent with their known reduced proliferative capacity in culture. In contrast, sperm (which, being germ cells, are in effect immortal) from donors ranging in age from 19 to 68 years had telomeres that did not vary in length with donor age, which indicates that telomerase is active at some stage of germ cell growth. Likewise, those few cells in a culture that become immortal (capable of unlimited proliferation) exhibit an active telomerase and a telomere of stable length, as do the cells of unicellular eukaryotes (which are also immortal). It therefore appears that telomere erosion is a significant cause of cellular senescence and hence aging. Indeed transgenic mice that constitutively (at a constant rate) express telomerase have increased lifespans (although, in contrast to humans, mice in which telomerase has been knocked out survive without significant problems for several generations before they become infertile).

d. Cancer Cells Have Active Telomerases

What advantage might multicellular organisms gain by eliminating telomerase activity in their somatic cells? An intriguing possibility is that cellular senescence is a mechanism that protects multicellular organisms from cancer. The two defining characteristics of cancer cells are that they are immortal and grow uncontrollably (Sections 19-3B and 34-4C). If mammalian cells were normally immortal, the incidence of cancer would probably be far greater than it is since immortalization, which requires an active telomerase, is a major step toward **malignant transformation** (cancer formation), which requires several independent genetic changes (Section 19-3B). Indeed, nearly all human cancers exhibit high telomerase activity. Moreover, as Robert Weinberg demonstrated, human fibroblasts in culture can be malignantly transformed by the acquisition of only three genes, those encoding: (1) TERT (its TER subunit is normally expressed in somatic cells), (2) an oncogenic variant of H-Ras (an essential participant in intracellular signal transduction pathways; Section 19-3C), and (3) the SV40 **large-T antigen** [SV40 is a tumor virus whose large-T antigen binds and functionally inactivates the tumor suppressor proteins known as **Rb** and **p53** (Section 34-4C; it also functions as a helicase in viral DNA replication)]. The age-related decline in telomere length in humans does not occur in mice, which suggests that telomere loss evolved to suppress tumor formation in long-lived animals, such as humans, but not in short-lived animals, such as mice. Thus, telomerase inhibitors may be effective antitumor agents.

e. Telomeric DNA Can Dimerize via G-Quartets

It has long been known that guanine forms strong Hoogsteen-type base pairs (Table 29-2) that can further associate to form cyclic tetramers known as **G-quartets** (Fig. 30-51a). Indeed, G-rich polynucleotides are notoriously difficult to work with because of their propensity to aggregate. The G-rich overhanging strands of telomeres dimerize to form stable complexes in solution, presumably via the formation of G-quartet–containing structures.

The 3′-terminal telomeric overhang of the ciliated protozoan *Oxytricha nova* has the sequence $d(T_4G_4)_2$, which resembles the repeating telomeric sequences of other organisms. The NMR structure of the dodecamer $d(G_4T_4G_4)$, determined by Juli Feigon (Fig. 30-51b), reveals that each oligonucleotide folds back on itself to form a hairpin, two of which associate in an antiparallel fashion to form a structure that contains four stacked G-quartets, with the T_4 sequences forming the loops at the ends of each stack.

The **telomere end binding protein (TEBP)** of *O. nova* is a heterodimeric capping protein that binds to and protects the foregoing 3′ overhang. The X-ray structure of TEBP in complex with $d(G_4T_4G_4)$, determined by Steve Schultz, reveals that the DNA binds in a deep cleft between the protein's α and β subunits, where it adopts an irregular nonhelical conformation (Fig. 30-52). In addition, two other $d(G_4T_4G_4)$ molecules form a G-quartet–linked dimer with the same conformation they adopt in solution (Fig. 30-51). The G-quartet assembly fits snugly into a small positively

Figure 30-51   NMR structure of the telomeric oligonucleotide d(GGGGTTTTGGGG). (*a*) The base-pairing interactions in the G-quartet at the end of the quadruplex in solution. (*b*) Schematic diagram of the NMR solution structure, in which the strand directions are indicated by arrows. The nucleotides are numbered 1 to 12 in one strand and 1* to 12* in the symmetry-related strand. Guanine residues G1 to G4 are represented by dark blue rectangles, G8 to G12 are light blue, G1* to G4* are red, and G9* to G12* are pink. [After Schultze, P., Smith, F.W., and Feigon, J., *Structure* **2**, 227 (1994). PDBid 156D.]

***Figure 30-52*   X-ray structure of *Oxytricha nova* telomere end binding protein (TEBP) in complex with d(G₄T₄G₄).** The TEBP is drawn in ribbon form with its α and β subunits magenta and cyan. The DNA is drawn in stick form with its bases yellow, the sugar–phosphate backbone of the single strand that binds in a cleft between the protein's α and β subunits blue, and the backbones of two strands that form a G-quartet–linked dimer red and green. The G-quartet–linked dimer binds in a cavity formed by the N-terminal domains of three symmetry-related α chains, although only one of them is shown here. [Based on an X-ray structure by Steve Schultz, University of Colorado. PDBid 1JB7.]

charged cavity formed by the N-terminal domains of three symmetry-related (in the crystal) α subunits at sites distinct from their ssDNA binding sites. The α subunit (TEBPα) contains three so-called **OB folds** (OB for *o*ligomer *b*inding), common oligonucleotide/oligosaccharide binding motifs that each contain a characteristic 5-stranded β barrel. Two of these OB folds participate in DNA binding and a third interacts with TEBPβ, which also contains an OB fold. The presence of both the ssDNA and the G-quartet assembly in the X-ray structure supports the hypothesis that multiple DNA structures and, in particular, G-quartets, play a role in telomere biology.

Although TEBP is not present in yeast or vertebrates, both humans and fission yeast express a telomere end-binding protein named **POT1** (for *p*rotection *o*f *t*elomeres-*1*) and its binding partner **TPP1** (so named because it had been previously called *T*INT1, *P*TOP, and *P*IP*1*), which bind to the single-stranded overhang at the ends of telomeres. POT1 consists mainly of two OB folds in which the N-terminal OB fold is homologous to that of TEBPα. The deletion of POT1 causes rapid loss of telomeric DNA and chromosomal end-joining. TPP1 is structurally similar to TEBPβ, despite their only 11% sequence identity, and hence the POT1–TPP1 complex appears to be a homolog of TEBP.

f. Telomeres Form T-Loops

Mammalian telomeric DNA is also capped by two related proteins, **TRF1** and **TRF2** (TRF for *t*elomere *r*epeat-binding *f*actor). Jack Griffith and Titia de Lange have shown through electron microscopy (EM) studies that, in the presence of TRF2, otherwise linear telomeric DNA forms large duplex end loops named **T-loops** (T for *t*elom-

ere; Fig. 30-53*a*). Moreover, the EM of DNA from mammalian telomeres, whose strands had been chemically cross-linked to preserve their structural relationships on deproteination, likewise revealed the presence of abundant T-loops of varying sizes. These observations suggest that T-loops are formed by the TRF2-induced invasion of the 3′ telomeric overhang into the repeating telomeric dsDNA (Fig. 30-53*b*) to form a **D-loop** (D for *d*isplacement; a segment of dsDNA whose two strands are separated). T-loops have also been observed in protozoa, suggesting that T-loops are a conserved feature of eukaryotic telomeres. TRF1 is implicated in controlling telomeric length, presumably by somehow limiting the number of TRF1 molecules that can bind to a telomere.

POT1, which binds to the single-stranded 3′ extension at chromosome ends, and TRF1 and TRF2, which bind double-stranded telomere DNA, are bridged by TPP1 and **TIN2** (for *T*RF1 *in*teracting protein *2*). A sixth protein, **RAP1** (for *r*epressor *a*ctivator protein *1*), binds mainly to TRF2. The complex formed by these six proteins, which is known as **shelterin,** apparently functions both to protect the mammalian telomere from being mistaken for a broken chromosome and hence being subject to DNA repair, and to limit telomere length by preventing the extension of its bound telomere by telomerase.

5 REPAIR OF DNA

DNA is by no means the inert substance that might be supposed from naive consideration of genome stability. Rather, the reactive environment of the cell, the presence

(a)

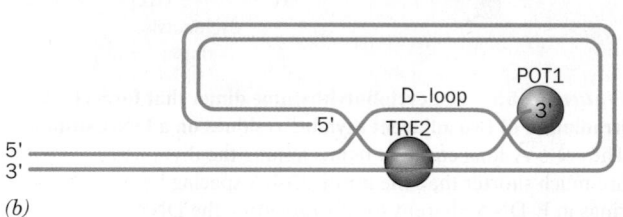

(b)

Figure 30-53   The telomeric T-loop. (*a*) An electron micrograph of a dsDNA consisting of a 3-kb unique sequence followed by ~2 kb of the repeating sequence TTAGGG on the strand that ends with a 150- to 200-nt 3′ overhang. This model telomeric DNA was then incubated with human TRF2. [Courtesy of Jack Griffith, University of North Carolina at Chapel Hill.] (*b*) The proposed structure of a T-loop. In a process that is mediated by TRF2, the repeating TTAGGG sequence in the DNA's 3′ overhang displaces a portion of the same strand (*blue*) in the double-stranded region of the DNA to form a duplex segment with the complementary strand (*red*), thereby generating a D-loop. The telomere end-binding protein POT1 specifically binds to the end of the 3′ overhang.

Figure 30-54 The types and sites of chemical damage to which DNA is normally susceptible *in vivo*. Red arrows indicate sites subject to oxidative attack, blue arrows indicate sites subject to spontaneous hydrolysis, and green arrows indicate sites subject to nonenzymatic methylation by *S*-adenosylmethionine. The width of an arrow is indicative of the relative frequency of the reaction. [After Lindahl, T., *Nature* **362,** 709 (1993).]

of a variety of toxic substances, and exposure to UV or ionizing radiation subjects it to numerous chemical insults that excise or modify bases and alter sugar–phosphate groups (Fig. 30-54). Indeed, some of these reactions occur at surprisingly high rates. For example, under normal physiological conditions, the glycosidic bonds of ~20,000 of the 6 billion purine nucleotides in each diploid human cell hydrolyze spontaneously each day.

Any DNA damage must be repaired if the genetic message is to maintain its integrity. Such repair is possible because of duplex DNA's inherent information redundancy. The biological importance of DNA repair is indicated by the identification of at least 130 genes in the human genome that participate in DNA repair and by the great variety of DNA repair pathways possessed by even relatively simple organisms such as *E. coli.* In fact, *the major DNA repair processes in eukaryotic cells and E. coli are chemically quite similar.* These processes are outlined in this section.

The two redundant copies of genetic information carried by dsDNA ideally suit it for the repair of damage to one of its strands: The damaged strand can be repaired under the direction of the undamaged strand. The importance of dsDNA to the storage of genetic information is indicated by the fact that only a few small viruses carry single-stranded DNA or RNA as their genetic material (e.g., φX174 and HIV). The DNA repair systems discussed below do not operate on single-stranded nucleic acids and hence these viruses have very high rates of mutation. Thus it appears that only organisms with very small genomes can afford the economy of not encoding their genomes on dsDNA.

A. Direct Reversal of Damage

a. Pyrimidine Dimers Are Split by DNA Photolyase

UV radiation of 200 to 300 nm promotes the formation of a cyclobutyl ring between adjacent thymine residues on the same DNA strand to form an intrastrand **thymine dimer** (Fig. 30-55). Similar cytosine and thymine–cytosine dimers are likewise formed but at lesser rates. Such **cyclobutane pyrimidine dimers (CPDs)** locally distort DNA's base-paired structure such that it can be neither transcribed nor replicated. Indeed, a single thymine dimer, if unrepaired, is sufficient to kill an *E. coli.*

Pyrimidine dimers may be restored to their monomeric forms through the action of light-absorbing enzymes named **photoreactivating enzymes** or **DNA photolyases** that are present in many prokaryotes and eukaryotes (including goldfish, rattlesnakes, and marsupials, but not placental mammals). These enzymes are 55- to 65-kD monomers that bind to a pyrimidine dimer in DNA, a process that can occur in the dark. A noncovalently bound chromophore, in

Figure 30-55 The cyclobutylthymine dimer that forms on UV irradiation of two adjacent thymine residues on a DNA strand. The ~1.6-Å-long covalent bonds joining the thymine rings (*red*) are much shorter than the normal 3.4-Å spacing between stacked rings in B-DNA, thereby locally distorting the DNA.

some species an N^5,N^{10}-methenyltetrahydrofolate (**MTHF;** Fig. 26-49) and in others a **5-deazaflavin,**

8-Hydroxy-7,8-didemethyl-5-deazariboflavin

then absorbs 300- to 500-nm light and transfers the excitation energy to a noncovalently bound FADH⁻, which in turn transfers an electron to the pyrimidine dimer, thereby splitting it. Finally, the resulting pyrimidine anion re-reduces the FADH· and the now unblemished DNA is released, thereby completing the catalytic cycle. DNA photolyases bind either dsDNA or ssDNA with high affinity but without regard to base sequence.

Thomas Carell and Lars-Oliver Essen determined the X-ray structure of the 474-residue DNA photolyase from the cyanobacterium *Anacystis nidulans* in complex with a 9-bp dsDNA containing a synthetic thymine dimer whose bridging phosphate group was replaced by a —O—CH₂—O— group (which does not affect the enzyme's ability to split the dimer). The DNA binds to a highly positively charged surface on the protein with its thymine dimer flipped out of the double helix and bound in a deep cavity (Fig. 30-56). This flip-out is probably facilitated by the relatively weak base pairing interactions of the thymine dimer and the distortions it imposes on the double helix. In the following discussions we shall see that this so-called **base flipping** (really nucleotide flipping since entire nucleotides flip out of the double helix) is by no means an unusual process for enzymes that perform chemistry on the bases of dsDNA. The DNA outside of the thymine dimer assumes the B conformation but at the thymine dimer is bent by 50° away from the protein, thereby unstacking the adenine bases complementary to the dimerized thymine bases. The "hole" in the DNA helix left by the flipped-out thymine dimer is partially occupied by an irregular protein ridge.

In the X-ray structure, the thymine dimer's C5—C5 and C6—C6 bonds are broken. Yet, in the crystal, the enzyme-bound thymine dimer is stable in the dark for at least a year. Apparently, the X-rays used to generate the diffraction data mimic the effects of the light that normally ruptures these bonds. Moreover, the FAD's isoalloxazine ring exhibits a 9° "butterfly" bend about its N5–N10 axis (the isoalloxazine ring's atomic numbering scheme is givien in Fig. 16-8), which is indicative that it is in the fully reduced FADH⁻ form. The isoalloxazine ring and adenine ring of the FAD⁻, which has a folded conformation, are in van der Waals contact with one or the other bases of the thymine dimer and the isoalloxazine ring is ~10 Å distant from the

Figure 30-56 **X-ray structure of *A. nidulans* DNA photolyase in complex with dsDNA containing a synthetic thymine dimer.** The protein is drawn in worm form embedded in its semitransparent surface diagram. The DNA, in which the phosphate group bridging the nucleotides of the thymine dimer is replaced by a —O—CH₂—O— group, is drawn mainly in stick form but with the bases of the thymine dimer in space-filling form, all colored according atom type (C green, N blue, O red, and P orange) and with successive P atoms in each polynucleotide chain connected by orange rods. The FAD and MTHF are drawn in stick form with their flavin and flavinlike rings in space-filling form and with FAD C magenta and MTHF C yellow. [Based on an X-ray structure by Thomas Carell, Ludwig Maximilians University, Munich, Germany, and Lars-Oliver Essen, Philipps University, Marburg, Germany. PDBid 1TEZ.]

flavin-like ring of the MTHF. This permits the observed efficient energy transfer in the photolyase reaction (which has a quantum yield of ~0.9).

b. Alkyltransferases Dealkylate Alkylated Nucleotides

The exposure of DNA to alkylating agents such as ***N*-methyl-*N'*-nitro-*N*-nitrosoguanidine (MNNG)**

***N*-Methyl-*N'*-nitro-*N*-nitrosoguanidine (MNNG)**

O^6-Methylguanine residue

yields, among other products, **O^6-alkylguanine** residues. The formation of these derivatives is highly mutagenic because on replication, they frequently cause the incorporation of thymine instead of cytosine.

O^6-**Methylguanine** and O^6-**ethylguanine** lesions of DNA in all species tested are repaired by O^6-**alkylguanine–DNA alkyltransferase,** which directly transfers the offending alkyl group to one of its own Cys residues. The reaction inactivates this protein, which therefore cannot be strictly classified as an enzyme. The alkyltransferase reaction has elicited considerable attention because carcinogenesis induced by methylating and ethylating agents is correlated with deficient repair of O^6-alkylguanine lesions.

The *E. coli* O^6-alkylguanine–DNA alkyltransferase activity occurs on the 178-residue C-terminal segment of the 354-residue **Ada protein** (the product of the ***ada*** gene). Its X-ray structure (Fig. 30-57*a*), determined by Eleanor Dodson and Peter Moody, reveals, unexpectedly, that its active site Cys residue, Cys 321, is buried inside the protein. Apparently, the protein must undergo a significant conformational change on DNA binding in order to effect the methyl transfer reaction.

Ada protein's 92-residue N-terminal segment has an independent function: It repairs methyl phosphotriesters in DNA (methylated phosphate groups) by irreversibly transferring the offending methyl group to its Cys 69. The NMR structure of Ada's N-terminal domain (Fig. 30-57*b*), determined by Gregory Verdine and Gerhard Wagner, reveals

that Cys 69, together with three other Cys residues, tetrahedrally coordinates a Zn^{2+} ion. This presumably stabilizes the thiolate form of Cys 69 over its thiol form, thereby facilitating its nucleophilic attack on the methyl group.

Intact Ada protein that is methylated at its Cys 69 binds to a specific DNA sequence, which is located upstream of the *ada* gene and several other genes encoding DNA repair proteins, thereby inducing their transcription. Evidently, Ada also functions as a chemosensor of methylation damage.

B. *Excision Repair*

Cells employ two types of excision repair mechanisms: (1) **nucleotide excision repair (NER),** which functions to repair relatively bulky DNA lesions; and (2) **base excision repair (BER),** which repairs nonbulky lesions involving a single base.

a. Nucleotide Excision Repair

NER is a DNA repair mechanism found in all cells that eliminates damage to dsDNA by excising an oligonucleotide containing the lesion and filling in the resulting single-strand

(b)

(a)

Figure 30-57 **Structure of the *E. coli* Ada protein.** (*a*) The X-ray structure of Ada's 178-residue C-terminal segment, which contains its O^6-alkylguanine–DNA alkyltransferase function. The protein is drawn in ribbon form colored in rainbow order from its N-terminus (*blue*) to its C-terminus (*red*). The side chain of Cys 146 (Cys 321 in the intact protein), to which the methyl group is irreversibly transferred, is shown in space-filling form with C green and S yellow. Note that this residue is almost entirely buried within the protein. [Based on an X-ray structure by Eleanor Dodson and Peter Moody, University of York, U.K. PDBid 1SFE.] (*b*) The NMR structure of Ada's 92-residue,

N-terminal segment, which mediates its methyl phosphotriester repair function. The protein is drawn in ribbon form colored in rainbow order from its N-terminus (*blue*) to its C-terminus (*red*). Its bound Zn^{2+} ion is represented by a purple sphere and its four tetrahedrally coordinating Cys side chains are shown in ball-and-stick form, with C green and S yellow except for the orange S atom of Cys 69, which becomes irreversibly methylated when the protein encounters a methylated phosphate group on DNA. The coordinate bonds to the Zn^{2+} ion are represented by gray sticks. [Based on an NMR structure by Gregory Verdine and Gerhard Wagner, Harvard University. PDBid 1ADN.]

gap. NER repairs lesions that are characterized by the displacement of bases from their normal positions, such as pyrimidine dimers, or by the addition of a bulky substituent to a base. This system appears to be activated by a helix distortion rather than by the recognition of any particular group. In humans, NER is the major defense against two important carcinogens, sunlight and cigarette smoke. The mechanism of NER in prokaryotes is similar to that in eukaryotes. However, prokaryotic NER employs 3 subunits, whereas eukaryotic NER involves the actions of 16 subunits. The eukaryotic proteins are conserved from yeast to humans but none of them exhibit any sequence similarity to the prokaryotic proteins, suggesting that the two NER systems arose by convergent evolution.

In *E. coli,* NER is carried out in an ATP-dependent process through the actions of the **UvrA, UvrB,** and **UvrC** proteins (the products of the *uvrA, uvrB,* and *uvrC* genes). This system, which is often referred to as the **UvrABC endonuclease** (although, as we shall see, there is no complex that contains all three subunits), cleaves the damaged DNA strand at the seventh and at the third or fourth phosphodiester bonds from the lesion's 5′ and 3′ sides, respectively (Fig. 30-58; this system is also known as an **excinuclease** to indicate that it *exci*ses a DNA segment rather than cleaving it in only one place as do most endonucleases). The excised 11- or 12-nt oligonucleotide is displaced by the binding of **UvrD** (also called **helicase II**) and replaced through the actions of Pol I and DNA ligase.

The mechanism of prokaryotic NER was elucidated mainly by Aziz Sancar. It begins with the damage recognition step in which a (UvrA)₂UvrB heterotrimer binds

tightly although nonspecifically to dsDNA, which it probes for damage according to its local propensity for bending and unwinding. The presence of a lesion activates the helicase function of UvrB to unwind 5 bp around the lesion in an ATP-driven process. This conformation change induces the dissociation of the UvrA from the complex, which allows the binding of UvrC. UvrB then makes the incision on the 3′ side of the lesion following which UvrC makes the incision on its 5′ side. UvrD binds to the resulting nicks in the DNA, which displaces UvrC and the lesion-containing oligomer. This makes the 5′ incision site accessible to Pol I, which fills in the gap and displaces UvrB. Finally, DNA ligase seals the remaining nick yielding refurbished DNA.

b. Xeroderma Pigmentosum and Cockayne Syndrome Are Caused by Genetically Defective NER

In humans, the rare inherited disease **xeroderma pigmentosum (XP;** Greek: *xeros,* dry + *derma,* skin) is mainly characterized by the inability of skin cells to repair UV-induced DNA lesions. Individuals suffering from this autosomal recessive condition are extremely sensitive to sunlight. During infancy they develop marked skin changes such as dryness, excessive freckling, and keratoses (a type of skin tumor; the skin of these children is described as resembling that of farmers with many years of sun exposure), together with eye damage, such as opacification and ulceration of the cornea. Moreover, they develop often fatal skin cancers at a 2000-fold greater rate than normal and internal cancers at a 10 to 20-fold increased rate. Curiously, many individuals with XP also have a bewildering variety of seemingly unrelated symptoms including progressive neurological degeneration and developmental deficits.

Cultured skin fibroblasts from individuals with xeroderma pigmentosum are defective in the NER of pyrimidine dimers. Cell-fusion experiments with cultured cells taken from various patients have demonstrated that this disease results from defects in any of 8 complementation groups (Section 1-4Cc), indicating that there must be at least 8 gene products, **XPA** through **XPG** and **XPV,** involved in this clearly important UV damage repair pathway.

What is the biochemical basis for the diverse group of symptoms associated with impaired NER? The reactive oxygen species (ROS; Section 22-4Cg) produced by oxidative metabolism readily damages DNA. Some of these oxidative lesions are repaired via NER. Since neurons have high rates of respiration and are long-lived nondividing cells, it seems likely that they would be particularly susceptible to oxidative damage in the absence of NER. This explains the progressive neurological deterioration in XP.

The need to repair a DNA lesion is particularly urgent if the damaged gene is being transcribed because RNA polymerase cannot transcribe through damaged DNA. Cells therefore recruit their DNA repair machinery to such genes in a process known as **transcription-coupled repair (TCR),** which operates only on the DNA strand that is being transcribed (lesions on the complementary strand are repaired at normal rates). For example, pyrimidine dimers are more rapidly removed from transcribed portions of DNA than from unexpressed sequences.

Figure 30-58   The mechanism of nucleotide excision repair (NER) of pyrimidine photodimers.

Cockayne syndrome (CS) is an inherited disease caused by defective TCR. Individuals with CS are hypersensitive to UV radiation (although they have a normal incidence of skin cancer) and exhibit stunted growth as well as neurological dysfunction due to neuron demyelination leading to death in childhood. CS is most often caused by mutations in two complementation groups, **CSA** and **CSB,** although certain defects in **XPB, XPD**, and XPG can also cause CS in addition to XP.

The retarded development typical of XPB and XPD defects and perhaps the demyelination that occurs in CS appear to be due more to impaired transcription than to defective NER. This is explained by the fact that the DNA helicases XPB and XPD are subunits of the ten-subunit eukaryotic transcription factor **TFIIH,** whose proper functioning is required for transcription initiation by RNA polymerase II (Section 34-3Bb) as well as NER and TCR. A eukaryotic RNA polymerase that is stalled at a DNA lesion is recognized by CSB, which then recruits the other TCR components. CSA interacts with both CSB and the **p44** subunit of TFIIH. Once the damage has been repaired, the RNA polymerase resumes transcription. Thus, in CS, RNA polymerase molecules become permanently stalled on DNA lesions.

TCR also occurs in bacteria. For example, in *E. coli,* the protein **TCRF** (for *t*ranscription *r*epair *c*ouping *f*actor; also known as **Mfd** protein, for *m*utation *f*requency *d*ecline) is an ATP-powered DNA translocase that displaces stalled RNA polymerase from the damaged template strand, following which TCRF recruits the proteins of the UvrABC system to the damage site. The repaired gene is then transcribed from its beginning. Since prokaryotic genes are much shorter than eukaryotic genes (most of which contain several long noncoding segments known as introns; Section 31-4Ac), this is a more efficient use of resources than generating the complex machinery necessary to restart transcription as occurs in eukaryotes.

c. Base Excision Repair

DNA bases are modified by reactions that occur under normal physiological conditions as well as through the action of environmental agents. For example, adenine and cytosine residues spontaneously deaminate at finite rates to yield hypoxanthine and uracil residues, respectively. *S*-Adenosylmethionine (SAM), a common metabolic methylating agent (Section 26-3Ea), occasionally nonenzymatically methylates a base to form derivatives such as 3-methyladenine and 7-methylguanine residues (Fig. 30-54). Ionizing radiation can promote ring opening reactions in bases. Such changes modify or eliminate base pairing properties.

DNA containing a damaged base may be restored to its native state through base excision repair (BER). Cells contain a variety of **DNA glycosylases** that each cleave the glycosidic bond of a corresponding specific type of altered nucleotide (Fig. 30-59), thereby leaving a deoxyribose residue in the backbone. Such **apurinic** or **apyrimidinic (AP) sites** (also called **abasic sites**) are also generated under normal physiological conditions by the spontaneous hydrolysis of a glycosidic bond. The deoxyribose residue is then cleaved on one side by an **AP endonuclease,** the deoxyribose and several adjacent residues are removed by the action of a

Figure 30-59   Action of DNA glycosylases. These enzymes hydrolyze the glycosidic bond of their corresponding altered base (*red*) to yield an AP site.

cellular exonuclease (possibly associated with a DNA polymerase), and the gap is filled in and sealed by a DNA polymerase and DNA ligase.

d. Uracil in DNA Would Be Highly Mutagenic

For some time after the essential functions of nucleic acids had been elucidated, there seemed no apparent reason for nature to go to the considerable metabolic effort of using thymine in DNA and uracil in RNA when these substances have virtually identical base pairing properties. This enigma was solved by the discovery of cytosine's penchant for conversion to uracil by deamination, either via spontaneous hydrolysis (Fig. 30-54), which is estimated to occur ~120 times per day in each human cell, or by reaction with nitrites (Section 32-1Aa). If U were the normal DNA base, the deamination of C would be highly mutagenic because there would be no indication of whether the resulting mismatched G · U base pair had originally been G · C or A · U. *Since T is DNA's normal base, however, any U in DNA is almost certainly a deaminated C.* U's that occur in DNA are efficiently excised by **uracil–DNA glycosylase [UDG; also called uracil *N*-glycosylase (UNG)]** and then replaced by C through BER.

UDG also has an important function in DNA replication. dUTP, an intermediate in dTTP synthesis, is present in all cells in small amounts (Section 28-3Ba). DNA polymerases do not discriminate well between dUTP and dTTP (recall that DNA polymerases select a base for incorporation into DNA according to its ability to form a Watson–Crick-shaped base pair with the template; Section 30-2Aa) so that, despite the low dUTP level that cells maintain, newly synthesized DNA contains an occasional U. These U's are rapidly replaced by T through BER. However, since excision occurs more rapidly than repair, all newly synthesized DNA is fragmented. When Okazaki fragments were first discovered (Section 30-1C), it therefore seemed that all DNA was synthesized discontinuously. This ambiguity was resolved with the discovery of *E. coli* defective in UDG. In these *ung*⁻ mutants, only about half of the newly synthesized DNA is fragmented, strongly suggesting that DNA's leading strand is synthesized continuously.

e. Uracil–DNA Glycosylase Induces Uridine Nucleotides to Flip Out

The X-ray structure of human UDG in complex with a 10-bp DNA containing a U · G mismatch (which can form a doubly hydrogen-bonded base pair whose shape differs from that of Watson–Crick base pairs; Section 32-2Db), determined by John Tainer, reveals that the UDG has bound the DNA with its uridine nucleotide flipped out of the ds-DNA (Fig. 30-60). Moreover, the enzyme has hydrolyzed uridine's glycosidic bond yielding the free uracil base and an AP site on the DNA, although both remain bound to the enzyme. The cavity in the DNA's base stack that would otherwise be occupied by the flipped-out uracil is filled by the side chain of Arg 272, which intercalates into the DNA from its minor groove side. The X-ray structure of a similar complex in which the U · G mismatch was replaced by a U · A base pair contained essentially identical features. However, when the U in the U · A-containing complex was replaced by **pseudouridine** (in which the "glycosidic" bond is made to uracil's C5 atom rather than to N1),

Pseudouridine

the uracil remained covalently linked to the DNA because the UDG could not hydrolyze its now C—C "glycosidic" bond.

How does UDG detect a base-paired uracil in the center of DNA and how does it discriminate so acutely between uracil and other bases, particularly the closely similar thymine? The above X-ray structures indicate that the phosphate groups flanking the flipped-out nucleotide are 4 Å closer together than they are in B-DNA (8 Å vs 12 Å), which causes the DNA to kink by ~45° in the direction parallel to the view in Fig. 30-60. These distortions arise from the binding of three rigid protein loops to the DNA, which would be unable to simultaneously bind to undistorted B-DNA. This led Tainer to formulate the "pinch–push–pull" mechanism for uracil detection in which he postulated that UDG rapidly scans a DNA for uracil by periodically binding to it so as to compress and thereby slightly bend the DNA's backbone (pinch). The DNA's presumed low resistance to bending at a uracil-containing site (a U · G base pair is smaller than C · G and hence leaves a space in the base stack, whereas a U · A base pair is even weaker than T · A) permits the enzyme to flip out the uracil by intercalating Arg 272 into the minor groove (push), thereby fully bending and kinking the DNA.

Figure 30-60 **X-ray structure of human uracil–DNA glycosylase (UDG) in complex with a 10-bp DNA containing a U · G base pair.** The protein (the C-terminal 223 residues of the 304-residue monomer) is represented by its ribbon diagram embedded in its transparent molecular surface. The DNA, viewed looking into its major groove, is drawn in stick form colored according to atom type (C green, N blue, O red, P orange). The **U · G base pair**'s uridine nucleotide has flipped out of the double helix (to the right of the DNA) and has been hydrolyzed to yield an AP nucleotide (stick form with C cyan) and uracil (space-filling form with C cyan), which remains bound in the enzyme's binding pocket. The side chain of Arg 272 (space-filling form with C yellow) has intercalated into the DNA base stack to fill the space vacated by the flipped-out uracil base. [Based on an X-ray structure by John Tainer, The Scripps Research Institute, La Jolla, California. PDBid 4SKN.]

This process is aided by the tight binding of the flipped-out uracil to the enzyme (pull). The exquisite specificity of this binding pocket for uracil prevents the binding and hence hydrolysis of any other base that the enzyme may have induced to flip out. Thus the overall shapes of adenine and guanine exclude them from this pocket, whereas thymine's 5-methyl group is sterically blocked by the rigidly held side chain of Tyr 147. Cytosine, which has approximately the same shape as uracil, is excluded through a set of hydrogen bonds emanating from the protein that mimic those made by adenine in a Watson–Crick A · U base pair.

AP sites in DNA are highly cytotoxic because they irreversibly trap mammalian topoisomerase I in its covalent complex with DNA (Section 29-3Ca). Moreover, since the ribose at the AP site lacks a glycosidic bond, it can readily convert to its linear form (Section 11-1B), whose reactive aldehyde group can cross-link to other cell components. This rationalizes why AP sites remain tightly bound to UDG in solution as well as in crystals. UDG activity is enhanced by AP endonuclease, the next enzyme in the BER pathway, but the two enzymes do not interact in the absence of DNA. This suggests that UDG remains bound to an AP site it generated until it is displaced by the more tightly binding AP endonuclease, thereby protecting the cell from the AP site's cytotoxic effects. It seems likely that other damage-specific DNA glycosylases function similarly.

C. *Mismatch Repair*

Any replicational mispairing that has eluded the editing functions of the various participating DNA polymerases may still be corrected by a process known as **mismatch repair (MMR).** For example, *E. coli* Pol I and Pol III have error rates of 10^{-6} to 10^{-7} per base pair replicated but the observed mutational rates in *E. coli* are 10^{-9} to 10^{-10} per base pair replicated. In addition, the MMR system can correct insertions or deletions of up to 4 nt (which arise from the slippage of one strand relative to the other in the active site of DNA polymerase). The importance of MMR is indicated by the fact that defects in the human MMR system result in a high incidence of cancer, most notably **hereditary nonpolyposis colorectal cancer (HNPCC;** which affects several organs and may be the most common inherited predisposition to cancer).

If an MMR system is to correct errors in replication rather than perpetuate them, it must distinguish the parental DNA, which has the correct base, from the daughter strand, which has an incorrect although normal base. In *E. coli,* as we have seen (Section 30-3C), this is possible because newly replicated GATC palindromes remain hemimethylated until the Dam methyltransferase has had sufficient time to methylate the daughter strand.

E. coli mismatch repair, which was elucidated in large part by Paul Modrich, requires the participation of three proteins and occurs as follows (Fig. 30-61):

1. MutS (853 residues) binds to a mismatched base pair or unpaired bases as a homodimer.

2. The MutS–DNA complex binds **MutL** (615 residues), also as a homodimer.

3. The MutS–MutL complex translocates along the DNA in both directions, thereby forming a loop in the DNA. The translocation appears to be driven by the ATPase function of MutS.

4. On encountering a hemimethylated GATC palindrome, the MutS–MutL complex recruits **MutH** (228 residues) and activates this single-strand endonuclease to make a nick on the 5′ side of the unmethylated GATC. This GATC may be located on either side of the mismatch and over 1000 bp distant from it, although repair efficiency decreases with the distance between the nick and the mismatch.

5. MutS–MutL recruits UvrD helicase, which in concert with an exonuclease separates the strands and degrades the nicked strand from the nick to beyond the mismatch. If the nick is on the 3′ side of the mismatch as shown, the exonuclease is **exonuclease I** (a 3′ → 5′ exonuclease), whereas if the nick is on the 5′ side of the mismatch, the exonuclease can be either **RecJ** or **exonuclease VII** (both 5′ → 3′ exonucleases).

The resulting gap is filled in by Pol III and sealed by DNA ligase, thereby correcting the mismatch. MutL is also an ATPase, which, it is postulated, functions to coordinate the various steps of mismatch repair.

Eukaryotic MMR systems are, not surprisingly, more complicated than those of *E. coli.* Eukaryotes express six homologs of MutS and five homologs of MutL that form

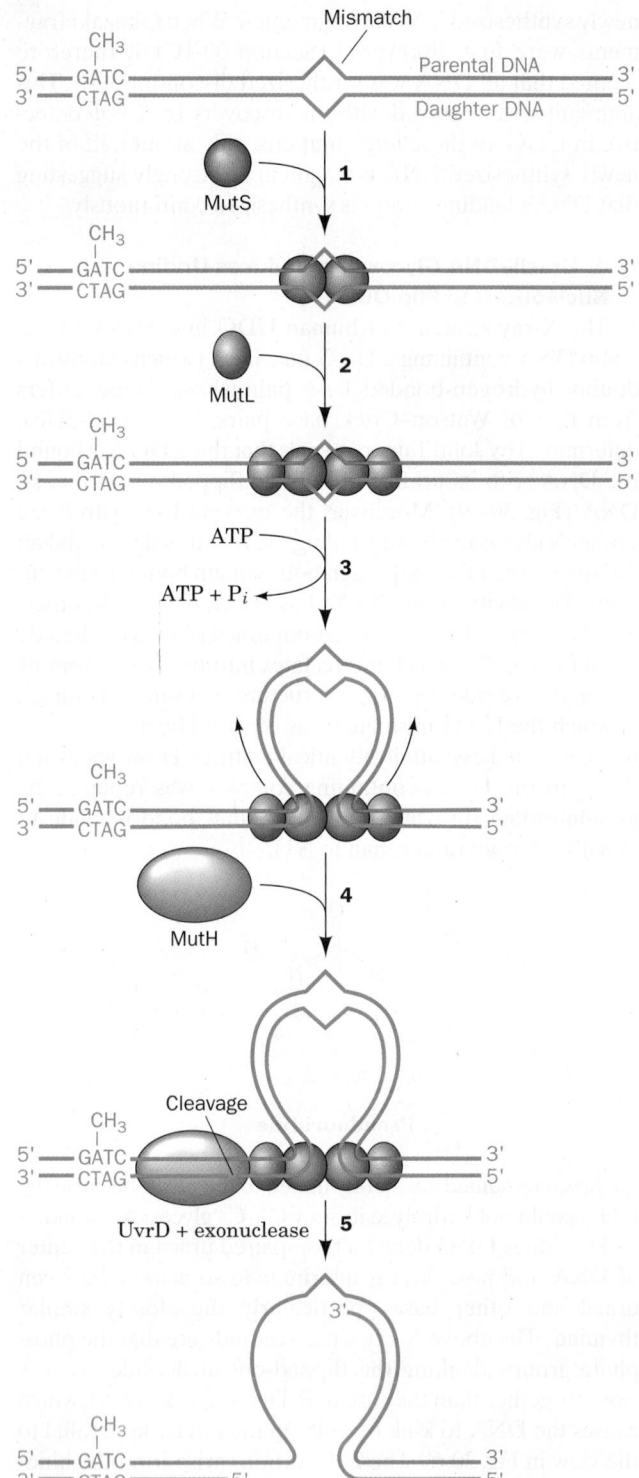

Figure 30-61 **The mechanism of mismatch repair in *E. coli.***

heterodimers on mismatched DNA. However, homologs of MutH only occur in gram-negative bacteria. Eukaryotes must have some other way of differentiating the parental and daughter DNA strands. Perhaps a newly synthesized daughter strand is identified by its as-yet unsealed nicks. DNA resynthesis is probably mediated by pol δ.

D. *The SOS Response*

Agents that damage DNA, such as UV radiation, alkylating agents, and cross-linking agents, induce a complex system of cellular changes in *E. coli* known as the **SOS response.** *E. coli so treated cease dividing and increase their capacity to repair damaged DNA.*

a. LexA Protein Represses the SOS Response

Clues as to the nature of the SOS response were provided by the observations that *E. coli* with mutant *recA* or *lexA* genes have their SOS response permanently switched on. **RecA,** a 353-residue protein that coats DNA as a multimeric helical filament, plays a central role, as we shall see, in homologous recombination (Section 30-6Ab). When *E. coli* are exposed to agents that damage DNA or inhibit DNA replication, their RecA specifically mediates the proteolytic cleavage

of **LexA** (202 residues) between its Asp 84 and Gly 85. RecA is activated to do so on binding to ssDNA (it was initially assumed that RecA catalyzes the proteolysis of LexA but subsequent experiments by John Little indicate that activated RecA stimulates LexA to cleave itself). Further investigations indicated that LexA functions as a repressor of 43 genes that participate in DNA repair and the control of cell division, including *recA, lexA, uvrA,* and *uvrB.* DNA sequence analyses of the LexA-repressible genes revealed that they are all preceded by a homologous 20-nt sequence, the so-called **SOS box,** that has the palindromic symmetry characteristic of operators (control sites to which repressors bind so as to interfere with transcriptional initiation by RNA polymerase; Section 5-4A). Indeed, LexA has been shown to specifically bind the SOS boxes of *recA* and *lexA.*

The preceding observations suggest a model for the regulation of the SOS response (Fig. 30-62). During normal

Figure 30-62 **Regulation of the SOS response in *E. coli.*** In a cell with undamaged DNA (*above*), LexA largely represses the synthesis of LexA, RecA, UvrA, UvrB, and other proteins involved in the SOS response. When there has been extensive DNA damage (*below*), RecA is activated by binding to the resulting single-stranded DNA to stimulate LexA self-cleavage. The consequent synthesis of the SOS proteins results in the repair of the DNA damage.

growth, LexA largely represses the expression of the SOS genes, including the *lexA* gene, by binding to their SOS boxes so as to inhibit RNA polymerase from initiating the transcription of these genes. When DNA damage has been sufficient to produce postreplication gaps, however, this ss-DNA binds to RecA so as to stimulate LexA cleavage. The LexA-repressible genes are consequently released from repression and direct the synthesis of SOS proteins including that of LexA (although this repressor continues to be cleaved through the influence of RecA). When the DNA lesions have been eliminated, RecA ceases stimulating LexA's autoproteolysis. The newly synthesized LexA can then function as a repressor, which permits the cell to return to normality.

b. SOS Repair Is Error Prone

The *E. coli* Pol III holoenzyme is unable to replicate through a variety of lesions such as AP sites and thymine dimers. On encountering such lesions, the replisome stalls and disassembles by releasing its Pol III cores, a process that is called replication fork "collapse." Cells have two general modes for restoring collapsed replication forks, **recombination repair** and **SOS repair.** Recombination repair circumvents the damaged template by using a homologous chromosome as its template DNA in a process known as **homologous recombination,** which also functions to generate genetic diversity. Hence we shall postpone our discussion of recombination repair until after our consideration of homologous recombination in Section 30-6A. In the following paragraphs we discuss SOS repair.

In SOS repair, the Pol III core lost from the collapsed replication fork is replaced by one of two so-called **bypass DNA polymerases,** whose synthesis is induced by the SOS response: **DNA polymerase IV (Pol IV,** the 336-residue product of the *dinB* gene) or **DNA polymerase V [Pol V;** the heterotrimeric product of the *umuD* and *umuC* genes, **UmuD′$_2$C** (umu for *UV mu*tagenesis), where UmuD′ is produced by the RecA-assisted self-cleavage of the 139-residue **UmuD** to remove its N-terminal 24 residues, and UmuC consists of 422 residues]. Both of these enzymes are Y-family DNA polymerases, all of whose members lack $3' \rightarrow 5'$ proofreading exonuclease activity and replicate undamaged DNA with poor fidelity and low processivity and hence are also known as **error-prone DNA polymerases.**

Translesion synthesis (TLS) by Pol V, which was characterized in large part by O'Donnell and Myron Goodman, requires the simultaneous presence of the β$_2$ sliding clamp, the γ complex (clamp loader), and SSB, together with a RecA filament in complex with the ssDNA arising from the action of helicase on the dsDNA ahead of the stalled replication fork. This so-called **Pol V mutasome** tends to incorporate G about half as often as A opposite thymine dimers and AP sites, with pyrimidines being installed infrequently. This process is, of course, highly mutagenic. But even in replicating undamaged DNA, Pol V is at least 1000-fold more error prone than is Pol I or Pol III holoenzyme. However, after synthesizing ~7 nt, the Pol V mutasome is replaced by Pol III holoenzyme, which commences normal

DNA replication after the now bypassed lesion. Pol II, a TLS participant that accurately replicates DNA, is also induced by the SOS response but it is synthesized well before Pol V appears (see below). The role of Pol II appears to be the mediation of error-free TLS, and only if this process fails is it replaced by Pol V to carry out error-prone TLS.

There are numerous types of DNA lesions besides AP sites and thymine dimers that interfere with normal DNA replication. Depending on the type of lesion, Pol IV, which is also error prone, may instead be recruited to carry out TLS. With many lesions, TLS may skip over the altered nucleotide, resulting in deletion of one or two bases in the daughter strand opposite the lesion (yielding a **frameshift mutation,** so called because it would change a structural gene's reading frame from that point onward; Section 5-4Bd). Moreover, Pol IV is prone to generating frameshift mutations even when replicating undamaged DNA.

The Y-family DNA polymerase **Dpo4** from the archaebacterium *Sulfolobus solfataricus* P2, a homolog of *E. coli* Pol IV and Pol V, misincorporates ~1 base per 500 replicated nucleotides. The X-ray structure of a complex of Dpo4 with a primer–template DNA that had been incubated with ddATP (which is complementary to the template base), determined by Wei Yang, reveals the structural basis for this low fidelity (Fig. 30-63). The 352-residue protein contains the fingers, palm, and thumb domains

Figure 30-63 X-ray structure of the bypass DNA polymerase Dpo4 from *Sulfolobus solfataricus* P2 in complex with a primer–template DNA and ddADP. The protein is drawn in ribbon form with its fingers, palm, thumb, and little finger domains blue, red, green, and purple, respectively. The DNA is gold with its backbones drawn as ribbons and its bases represented by rods. The ddADP, which is base-paired to a template T in the enzyme's active site, is shown in ball-and-stick form colored according to atom type (C pink, N blue, O red, and P magenta). [Courtesy of Wei Yang, NIH, Bethesda, Maryland. PDBid 1JX4.]

common to all known DNA polymerases (although their orders differ in the sequences of the different families of DNA polymerases) and, in addition, has a C-terminal domain unique to Y-family DNA polymerases that has been dubbed the "little finger" domain. The enzyme, as expected, has incorporated a ddA residue at the 3' end of the primer and, in addition, binds a ddADP in base-paired complex to the new template T. The little finger domain binds in the major groove of the DNA. However, the fingers and thumb domains are small and stubby compared to those of replicative DNA polymerases such as Klentaq1 (Fig. 30-9) and pol δ (Fig. 30-41), and the residues that contact the base pair in the active site are all Gly and Ala rather than the Phe, Tyr, and Arg that mainly do so in the replicative DNA polymerases. Moreover, the bound DNA is entirely in the B form rather than in the A form at the active site as occurs in many replicative DNA polymerases. Since the minor groove is more accessible in A-DNA than in B-DNA (Section 29-1B), this suggests that error-prone DNA polymerases have relatively little facility to monitor the base-pairing fidelity of the incoming nucleotide. This accounts for the ability of error-prone DNA polymerases to accommodate distorted template DNA as well as non-Watson–Crick base pairs at their active sites.

SOS repair is an error-prone and hence mutagenic process. It is therefore a process of last resort that is only initiated ~50 min after SOS induction if the DNA has not already been repaired by other means. Yet, DNA damage that normally activates the SOS response is nonmutagenic in the *recA⁻ E. coli* that survive. This is, as we saw, because bypass DNA polymerases will replicate over a DNA lesion even when there is no information as to which bases were originally present. Indeed, *most mutations in E. coli arise from the actions of the SOS repair system,* which is therefore a testimonial to the proposition that survival with a chance of loss of function (and the possible gain of new ones) is advantageous, in the Darwinian sense, over death, although only a small fraction of cells actually survive this process. It has therefore been suggested that, under conditions of environmental stress, the SOS system functions to increase the rate of mutation so as to increase the rate at which the *E. coli* adapt to the new conditions. Finally, it should be noted that the eukaryotic pols η, ι, and κ, all Y-family members, and pol ζ, an X-family member, are implicated in TLS and that pol η, the product of the *XPV* gene, is defective in the XPV form of xeroderma pigmentosum (Section 30-5Bb).

E. Double-Strand Break Repair

Double-strand breaks **(DSBs)** in DNA are produced when a replication fork encounters a nick and by the reactive oxygen species (ROS) by-products of oxidative metabolism and ionizing radiation (which also produces ROS). In fact, around 5 to 10% of dividing cells in culture exhibit at least one chromosome break at any given time. Moreover, DSBs are normal intermediates in certain specialized cellular processes such as recombination during meiosis (Section 1-4Ab) and **V(D)J recombination** in lymphoid cells, which helps generate the vast diversity of antigen-binding sites in antibodies and T-cell receptors (Section 35-2C). Unrepaired or misrepaired DSBs can be lethal to cells or cause chromosomal aberrations that may lead to cancer. Hence, the efficient repair of DSBs is essential for cell viability and genomic integrity.

Cells have two general modes to repair DSBs: recombination repair, which only occurs during the late S and G₂ phases of the cell cycle (when sister chromatids are present to serve as templates), and **nonhomologous end-joining (NHEJ)**, which functions throughout the cell cycle. Here we discuss NHEJ, a process which, as its name implies, directly rejoins DSBs. The recombination repair of DSBs is discussed in Section 30-6Ag.

In NHEJ, the broken ends of the DSB must be aligned, its frayed ends trimmed and/or filled in, and their strands ligated. The core NHEJ machinery in eukaryotes includes the DNA end-binding protein **Ku** (a heterodimer of homologous 70- and 83-kD subunits, **Ku70** and **Ku80**), **DNA ligase IV,** and the accessory protein **Xrcc4.** Ku, an abundant nuclear protein, binds to a DSB, whether blunt or with an overhang, and hence appears to be the cell's primary DSB sensor. The X-ray structure of Ku in complex with a 14-bp DNA, determined by Jonathan Goldberg, reveals that the protein cradles the dsDNA segment along its entire length and encircles its central ~3 bp segment (Fig. 30-64). The protein ring is also present in the closely

Figure 30-64   X-ray structure of human Ku protein in complex with DNA containing 14 bp. The subunits of Ku70 (*red helices and yellow strands*) and Ku80 (*blue helices and green strands*) are viewed along the pseudo-2-fold axis relating them. The DNA, viewed with its DSB pointing upward, is drawn in space-filling form with its sugar–phosphate backbone dark gray and its base pairs light gray. Note that the DNA is surrounded by a ring of protein. [Courtesy of John Tainer, The Scripps Research Institute, La Jolla, California. Based on an X-ray structure by Jonathan Goldberg, Memorial Sloan-Kettering Cancer Center, New York, New York. PDBid 1JEY.]

similar X-ray structure of Ku alone, thereby explaining why Ku that is bound to a dsDNA, which is then circularized, becomes permanently associated with it. Ku makes no specific contacts with the DNA's bases and few with its sugar–phosphate backbone, but instead fits snugly into the DNA's major and minor grooves so as to precisely orient it.

Ku–DNA complexes have been shown to dimerize so as to align the members of a DSB, both blunt ended and with short (1–4 bp) complementary single strands, for ligation as is diagrammed in Fig. 30-65. The DNA ends are exposed along one face of each Ku–DNA complex, presumably making them accessible to polymerases that fill in gaps and to nucleases that trim excess and inappropriate ends preparatory for ligation by DNA ligase IV in complex with Xrcc4. Nucleotide trimming, which of course generates mutations, appears to be carried out in

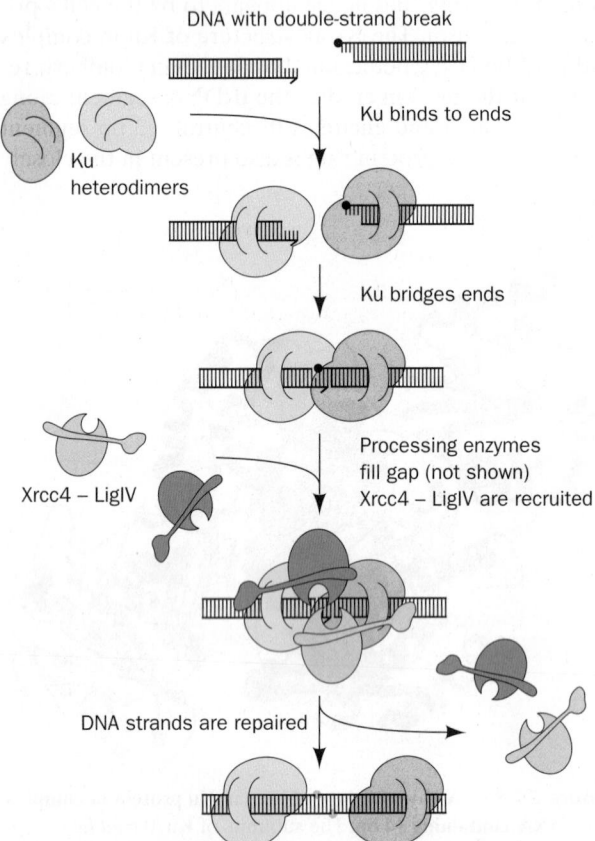

Figure 30-65 Schematic diagram of nonhomologous end-joining (NHEJ). The left dsDNA fragment is missing a base and the right fragment is blocked by a nonligatable group (*filled black circle*). The two Ku heterodimers are drawn in two shades of yellow and the Xrcc4–DNA ligase IV complexes are drawn in two shades of blue. The newly repaired links in the DNA are represented by pink circles. [After Jones, J.M., Gellert, M., and Yang, W., *Structure* **9**, 881 (2001).]

an ATP-dependent manner by the evolutionarily conserved **Mre11 complex,** which consists of two **Mre11** nuclease subunits and two **Rad50** ATPase subunits. Ku is eventually released from the rejoined DNA, perhaps by proteolytic cleavage.

The reason that the mutations generated by NHEJ are usually not unacceptably deleterious is that only a small fraction of the mammalian genome is expressed (Section 34-2A). In fact, the genome in a somatic cell of a 70-year-old human typically contains ~2000 "scars" caused by NHEJ.

F. *Identification of Carcinogens*

*Many forms of cancer are known to be caused by exposure to certain chemical agents that are therefore known as **carcinogens.*** It has been estimated that as much as 80% of human cancer arises in this fashion. There is considerable evidence that the primary event in carcinogenesis is often damage to DNA (carcinogenesis is discussed in Section 34-4C). Carcinogens are consequently also likely to induce the SOS response in bacteria and thus act as indirect mutagenic agents. In fact, there is a high correlation between carcinogenesis and mutagenesis (recall, e.g., the progress of xeroderma pigmentosum; Section 30-5Bb).

There are presently over 80,000 man-made chemicals of commercial importance and ~1000 new ones are introduced each year. The standard animal tests for carcinogenesis, exposing rats or mice to high levels of the suspected carcinogen and checking for cancer, are expensive and require ~3 years to complete. Thus, relatively few substances have been tested in this manner.

a. The Ames Test Assays for Probable Carcinogenicity

Bruce Ames devised a rapid and effective bacterial assay for carcinogenicity that is based on the high correlation between carcinogenesis and mutagenesis. He constructed special tester strains of *Salmonella typhimurium* that are *his⁻* (cannot synthesize histidine so that they are unable to grow in its absence), have cell envelopes that lack the lipopolysaccharide coating that renders normal *Salmonella* impermeable to many substances (Section 11-3Bc), and have inactivated excision repair systems. Mutagenesis in these tester strains is indicated by their reversion to the *his⁺* phenotype.

In the **Ames test,** ~10⁹ tester strain bacteria are spread on a culture plate that contains only a small amount of histidine to permit the bacteria to initially grow and mutate. Usually a mixture of several *his⁻* strains is used so that mutations due to both base changes and nucleotide insertions or deletions can be detected. A mutagen placed in the culture medium causes some of these *his⁻* bacteria to revert to the *his⁺* phenotype, which is detected by their growth into visible colonies after 2 days at 37°C (Fig. 30-66). The mutagenicity of a substance is scored as the number of such colonies less the few spontaneous revertant colonies that occur in the absence of the mutagen.

Figure 30-66   The Ames test for mutagenesis. A filter paper disk containing a mutagen, in this case the alkylating agent ethyl methanesulfonate, is centered on a culture plate containing *his⁻* tester strains of *Salmonella typhimurium* in a medium that initially contained only a small amount histidine. A dense halo of revertant bacterial colonies appears around the disk from which the mutagen diffused. The larger colonies distributed about the culture plate are spontaneous revertants. The bacteria near the disk have been killed by the toxic mutagen's high concentration. [Courtesy of Raymond Devoret, Institut Curie, Orsay, France.]

Many noncarcinogens are converted to carcinogens in the liver or in other tissues via a variety of detoxification reactions (e.g., those catalyzed by the cytochromes P450; Section 15-4Bc). A small amount of rat liver homogenate is therefore included in the Ames test medium in an effort to approximate the effects of mammalian metabolism.

b. Both Man-Made and Naturally Occurring Substances Can Be Carcinogenic

There is an ~80% correspondence between the compounds determined to be carcinogenic by animal tests and those found to be mutagenic by the Ames test. Dose–response curves, which are generated by testing a given compound at a number of concentrations, are almost always linear and extrapolate back to zero, indicating that *there is no threshold concentration for mutagenesis.* Several compounds to which humans have been extensively exposed that were found to be mutagenic by the Ames test were later found to be carcinogenic in animal tests. These include tris(2,3-dibromopropyl)phosphate, which was used as a flame retardant on children's sleepwear in the mid-1970s and can be absorbed through the skin; and furylfuramide, which was used in Japan in the 1960s and 1970s as an antibacterial additive in many prepared foods (and had passed two animal tests before it was found to be mutagenic). Carcinogens are not confined to man-made compounds but also occur in nature. For example, carcinogens are contained in many plants that are

common in the human diet, including alfalfa sprouts. **Aflatoxin B₁,**

Aflatoxin B₁

one of the most potent carcinogens known, is produced by fungi that grow on peanuts and corn. Charred or browned food, such as occurs on broiled meats and toasted bread, contains a variety of DNA-damaging agents. Thus, with respect to carcinogenesis, as Ames has written, "Nature is not benign."

6 RECOMBINATION AND MOBILE GENETIC ELEMENTS

The chromosome is not just a simple repository of genetic information. If this were so, the unit of mutation would have to be an entire chromosome rather than a gene because there would be no means of separating a mutated gene from the other genes of the same chromosome. Chromosomes would therefore accumulate deleterious mutations until they became nonviable.

It has been known from some of the earliest genetic studies that pairs of allelic genes may exchange chromosomal locations by a process known as **genetic recombination** (Section 1-4Cb). Mutated genes can thereby be individually tested, since their propagation is then not absolutely dependent on the propagation of the genes with which they had been previously associated. In this section, we consider the mechanisms by which genetic elements can move, both between chromosomes and within them.

A. Homologous Recombination

*Homologous recombination (also called **general recombination**) is defined as the exchange of homologous segments between two DNA molecules.* Both genetic and cytological studies have long indicated that such a crossing-over process occurs in higher organisms during meiosis (Fig. 1-27). Bacteria, which are normally haploid, likewise have elaborate mechanisms for the interchange of genetic information. They can acquire foreign DNA through transformation (Section 5-2A), through a process called **conjugation** (mating) in which DNA is directly transferred from one cell to another via a cytoplasmic bridge (Section 31-1Ac), and via **transduction** in which a defective bacteriophage that has erroneously acquired a segment of bacterial DNA rather than the viral chromosome transfers this DNA to another bacterial cell. In all of these processes, the foreign DNA is installed in the recipient's chromosome or plasmid through homologous recombination (to be propagated, a DNA segment must be part of a replicon; that is, be associated with a replication origin such as occurs in a chromosome, a plasmid, or a virus).

a. Recombination Occurs via a Crossed-Over Intermediate

The prototypical model for homologous recombination (Fig. 30-67) was proposed by Robin Holliday in 1964 on the basis of genetic studies on fungi. The corresponding strands of two aligned homologous DNA duplexes are nicked, and the nicked strands cross over to pair with the nearly complementary strands of the homologous duplex after which the nicks are sealed (Fig. 30-67*a–e*), thereby yielding a four-way junction known as a **Holliday junction** (Fig. 30-67*e*). A Holliday junction has, in fact, been observed in the X-ray structure of d(CCGGTACCGG), determined Shing Ho (Fig. 30-68), in which, perhaps unexpectedly, all the bases

form normal Watson–Crick base pairs without any apparent strain. The crossover point can move in either direction, often thousands of nucleotides, in a process known as **branch migration** (Fig. 30-67*e, f*) in which the four strands exchange base-pairing partners.

A Holliday junction can be resolved into two duplex DNAs in two equally probable ways (Fig. 30-67*g–l*):

1. The cleavage of the strands that did not cross over (right branch of Fig. 30-67*j–l*) exchanges the ends of the original duplexes to form, after nick sealing, the traditional recombinant DNA (Fig. 1-27*b*).

2. The cleavage of the strands that crossed over (left branch of Fig. 30-67*j–l*) exchanges a pair of homologous single-stranded segments.

The recombination of circular duplex DNAs results in the types of structures diagrammed in Fig. 30-69. Electron microscopic evidence for the existence of the postulated

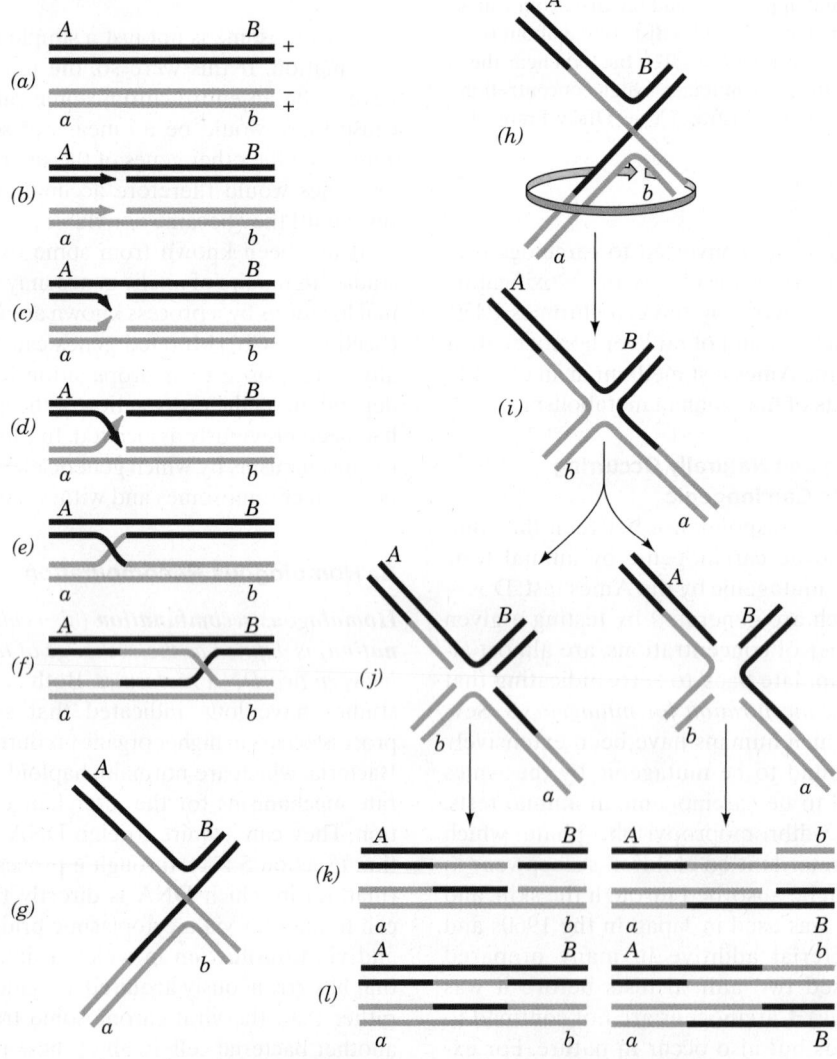

Figure 30-67 **The Holliday model of homologous recombination between homologous DNA duplexes.**
See the Animated Figures

(a)

(b)

Figure 30-68   X-ray structure of the self-complementary decameric DNA d(CCGGTACCGG).
(a) The secondary structure of the four-stranded Holliday junction formed by this sequence in which
the four strands, A, B, C, and D, are differently colored, their nucleotides are numbered 1 to 10 from
their 5' to 3' termini, and Watson–Crick base-pairing interactions are represented by black dashes.
The 2-fold axis relating the two helices of this so-called **stacked-X conformation** is represented by the
black lenticular symbol. (b) The observed three-dimensional structure of the Holliday junction, as
viewed along its 2-fold axis, in which the oligonucleotides are represented in stick form with their
backbones traced by ribbons, all colored as in Part a. With the exception of the backbones of strands
B and D at the crossovers, the two arms of this structure each form an undistorted B-DNA helix,
including the stacking of the base pairs flanking the crossovers. The two helices are inclined to each
other by 41°. Note that Fig. 30-67g is a schematic representation of the stacked-X conformation as
viewed perpendicular to both helices (from the side in this drawing and hence having the projected
appearance of the letter X). A Holliday junction can also assume a so-called **open-X conformation,**
which is represented by Fig. 30-67i. [Courtesy of Shing Ho, Oregon State University. PDBid 1DCW.]

**Figure 30-69   Homologous recombination between two
circular DNA duplexes.** This process can result either in two
circles of the original sizes or in a single composite circle.

(a)

(b)

Figure 30-70 **Electron micrographs of intermediates in the homologous recombination of two plasmids.** (*a*) A figure-8 structure. This corresponds to Fig. 30-69*d*. (*b*) A chi structure that results from the treatment of a figure-8 structure with a restriction endonuclease. Note the thinner single-stranded connections in the crossover region. [Courtesy of Huntington Potter, University of South Florida, and David Dressler, Oxford University, U.K.]

"figure-8" structures is shown in Fig. 30-70*a*. These figure-8 structures were shown not to be just twisted circles by cutting them with a restriction endonuclease to yield **chi structures** (after their resemblance to the Greek letter χ) such as that pictured in Fig. 30-70*b*.

b. Homologous Recombination in *E. coli* Is Catalyzed by RecA

The observation that *recA⁻ E. coli* have a 10^4-fold lower recombination rate than the wild-type indicates that *RecA protein has an important function in recombination.* Indeed, RecA greatly increases the rate at which complementary strands renature *in vitro.* This versatile protein (recall it also stimulates the autoproteolysis of LexA to trigger the SOS response and is an essential participant in the translesion synthesis of DNA; Section 30-5D) polymerizes cooperatively without regard to base sequence on ssDNA or on dsDNA that has a single-stranded gap. The resulting filaments, which may contain up to several thousand RecA monomers, specifically bind the homologous dsDNA, and, in an ATP-dependent reaction, catalyze strand exchange.

EM studies by Edward Egelman revealed that RecA filaments bound to ssDNA or dsDNA form a right-handed helix with ~6.2 RecA monomers per turn and a pitch (rise per turn) of 95 Å. The DNA in these filaments binds to the protein with 3 nt (or bp) per RecA monomer and hence is underwound with ~18.5 nt (or bp) per turn (vs 10 bp per turn for cannonical B-DNA).

The formation of RecA–DNA filaments is highly cooperative; it requires five or six RecA protomers to form a stable assembly. Consequently, attempts to crystallize RecA–DNA filaments over many years were unsuccessful. Nikola Pavletich ingeniously solved this conundrum by linking five or six *E. coli* RecA genes (each corresponding to residues 1–335 of this 353-residue protein) in tandem via 14-residue linkers and mutating the first and last RecA so as to prevent them from forming longer filaments. These fusion proteins, which had DNA-dependent ATPase and strand-exchange activities comparable to that of monomeric RecA, formed crystals containing both ssDNA and dsDNA.

The X-ray structure of the $RecA_5–(ADP–AlF_4^-)_5–$ $(dT)_{15}–(dA)_{12}$ complex (Fig. 30-71; $ADP–AlF_4^-$ is a nonhy-

Figure 30-71 **X-ray structure of the $RecA_5–(ADP–AlF_4^-)_5–$ $(dT)_{15}–(dA)_{12}$ complex viewed with its filament axis vertical.** The RecA units RecA¹ (the N-terminal unit) through RecA⁴ are colored green, cyan, magenta, and gray, respectively, with the C-terminal unit, RecA⁵, colored in rainbow order from its N-terminus (*blue*) to its C-terminus (*red*). The DNA and ADP–AlF₄⁻ are drawn in space-filling form with DNA C gray, ADP C green, N blue, O red, P orange, F light blue, and Al purple. [Based on an X-ray structure by Nikola Pavletich, Memorial Sloan-Kettering Cancer Center, New York, New York. PDBid 3CMX.]

drolyzable ATP analog) exhibits a straight filament axis with overall helical parameters that are closely similar to those derived from EM studies. Each RecA unit consists of a largely helical 30-residue N-terminal segment, a 240-residue α/β ATPase core, and a 64-residue C-terminal globular domain. The linkers connecting adjacent RecA units are disordered. Each RecA unit makes extensive contacts with its nearest neighbors so as to form a filament with a deep helical groove that exposes the DNA bound in its interior (Fig. 30-71 is viewed looking into this groove).

The two DNA strands, which lie close to the filament axis, form a complete set of Watson–Crick base pairs. However, rather than being smoothly stretched out, as had been expected, the dsDNA assumes an irregular conformation in which each 3-bp segment that is bound to a RecA unit closely resembles B-DNA with the steps between successive base pairs in this triplet having an axial rise of ~3.4 Å and a helical twist of ~30° (vs 3.4 Å and 36° for canonical B-DNA; Table 29-1). In contrast, the step between successive base pair triplets has an axial rise of 8.4 Å and a helical twist of −4°, thereby forming a 5-Å-high gap between successive triplets that is partially filled by the side chain of the conserved Ile 199. The sugar–phosphate backbone of the DNA strand furthest from the viewer in Fig. 30-71 [the $(dT)_{15}$]

makes extensive contacts with RecA. In contrast, the other strand [the $(dA)_{12}$] makes few contacts with the protein; it is held in place almost entirely by base pairing with the first strand. The $ADP–AlF_4^-$ is sandwiched between adjacent α/β ATPase cores, where it is completely buried.

The X-ray structure of the ssDNA-containing $RecA_6–(ADP–AlF_4^-)_6–(dT)_{18}$ complex closely resembles that of the foregoing dsDNA-containing complex but with the absence of the DNA strand closest to the viewer in Fig. 30-71. Thus, each RecA unit binds a $(dT)_3$ segment that is held in a B-DNA-like conformation with successive $(dT)_3$ segments separated by a 7.8-Å axial rise.

How does RecA mediate DNA strand exchange between single-stranded and duplex DNAs? On encountering a dsDNA with a strand that is complementary to its bound ssDNA, RecA partially unwinds the duplex and, in a reaction driven by RecA-catalyzed ATP hydrolysis, exchanges the ssDNA with the corresponding strand on the duplex. *This process tolerates only a limited degree of mispairing and requires that one of the participating DNA strands have a free end.* The assimilation (exchange) of a single-stranded circle with a strand on a linear duplex (Fig. 30-72) cannot proceed past the 3' end of a highly mismatched segment in the complementary strand. *The invasion of the single strand must*

Assimilation of 3' end of homologous DNA

Homologous DNA

Heterologous DNA

5'

3'

No assimilation of noncomplementary DNA

No assimilation

3'

5'

Assimilation stops at heterologous segment

Figure 30-72 **The RecA-catalyzed assimilation of a single-stranded circle by a dsDNA can occur only if the dsDNA has a 3' end that can base-pair with the circle (*red strand*).** Strand assimilation cannot proceed through a noncomplementary segment (*purple and orange strands*).

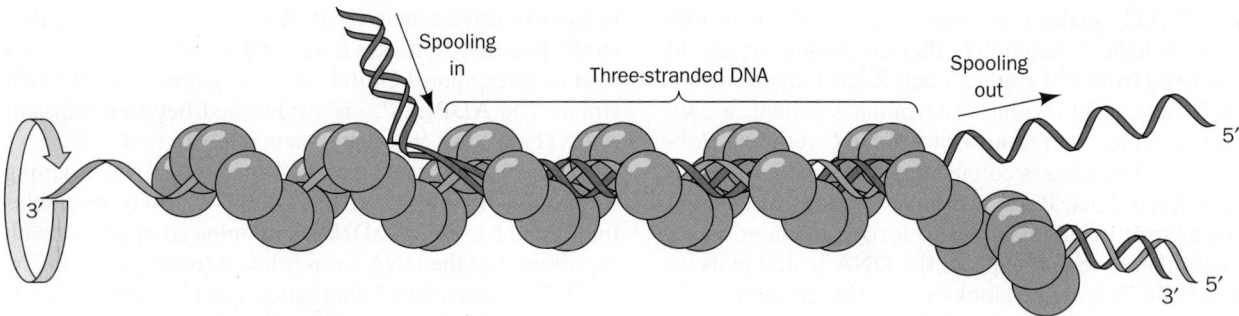

Figure 30-73 Hypothetical model for the RecA-mediated strand exchange reaction. Homologous DNA molecules are paired in advance of strand exchange in a three-stranded helix. The ATP-driven rotation of the RecA filament about its helix axis would cause duplex DNA to be "spooled in" to the filament, right to left as drawn. [After West, S.C., *Annu. Rev. Biochem.* **61,** 617 (1992).]

therefore begin at its 5′ end. A model for the consequent branch migration process is diagrammed in Fig. 30-73. Of course, two such strand exchange processes must occur simultaneously in a Holliday junction (Figs. 30-67 and 30-69).

The above structures suggest that the fidelity of homologous recombination arises from the B-DNA-like conformation that RecA imposes on the otherwise flexible bound ssDNA strand, which would exclude non-Watson–Crick base pairs. Strand exchange, of course, requires the separation of the two strands of the incoming dsDNA to permit one of its strands to sample base pairing with the ssDNA substrate. The above structures suggest that this is facilitated by the disruption of base stacking between base pair triplets in the RecA–DNA complex. However, the structure of the triple helical DNA intermediate in the strand exchange reaction is, as yet, unknown.

c. Eukaryotes Have RecA-Like Proteins

Yeast **RAD51** (339 residues) functions in the ATP-dependent repair and recombination of DNA in much the same way as does the 30% homologous *E. coli* RecA protein. The electron micrograph–based image reconstruction of RAD51 in complex with double-stranded DNA is nearly identical to that of RecA at low resolution: Both complexes form helical filaments in which the DNA has an ~5.1-Å rise per base pair and 18.6 bp per turn. Since RAD51 homologs occur in chickens, mice, and humans, it is very likely that such filaments universally mediate DNA repair and recombination.

d. RecBCD Initiates Recombination by Making Single-Strand Nicks

The single-strand nicks to which RecA binds are made by the **RecBCD** protein, the 330-kD heterotrimeric product of the SOS genes *recB, recC,* and *recD*. RecB is both a 3′ → 5′ helicase and a nuclease, whereas RecD is a 5′ → 3′ helicase. The formation of a RecA binding site begins with RecBCD binding to the end of a dsDNA and then unwinding it via its two ATP-driven helicase functions (Fig. 30-74). As it does so, RecB nucleolytically degrades the unwound single strands behind it, with the 3′-ending strand being

cleaved more often and hence broken down to smaller fragments than the 5′-ending strand. However, on RecC encountering the sequence GCTGGTGG from its 3′ end (the so-called **Chi sequence,** which occurs about every ~5 kb in the *E. coli* genome), the enzyme pauses and ceases its cleavage of the 3′-ending strand but increases the rate at which it cleaves the 5′-ending strand. RecBCD then helps load RecA onto the 3′-ending strand before dissociating from the DNA.

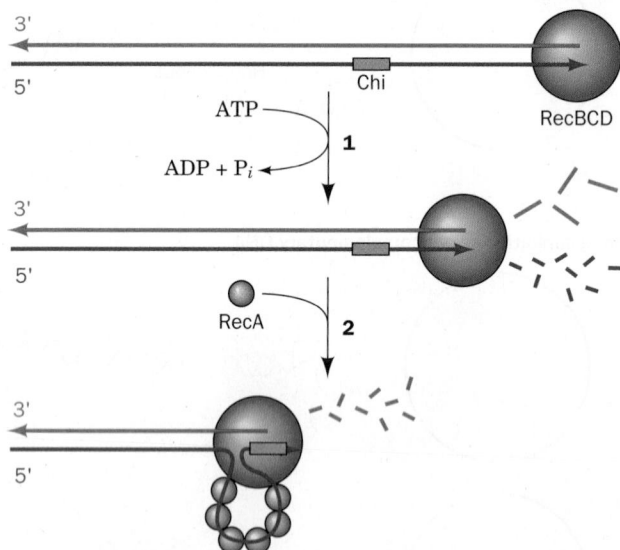

Figure 30-74 The generation of a 3′-ending single-strand DNA segment by RecBCD to initiate recombination. (1) RecBCD binds to a free end of a dsDNA and, in an ATP-driven process, advances along the helix, unwinding the DNA and degrading the resulting single strands behind it, with the 3′-ending strand cleaved more often than the 5′-ending strand. **(2)** When RecBCD encounters a properly oriented Chi sequence, it binds it and thus stops cleaving the 3′-ending strand but increases the frequency at which it cleaves the 5′-ending strand. This generates the potentially invasive 3′-ending strand segment to which RecA binds.

Figure 30-75 **X-ray structure of *E. coli* RecBCD in complex with a 51-nt DNA capable of forming a 21-bp hairpin loop.** The protein is drawn in semitransparent ribbon form with RecB yellow, RecC cyan, and RecD magenta. Note how the RecB nuclease domain is linked to the rest of the subunit by an extended polypeptide tether. The DNA is shown in space-filling form with C gray, N blue, O red, and P orange. A loop from RecC, which is drawn in space-filling form in green, is situated so as to wedge apart the two strands of the incoming dsDNA with the 3′-ending strand binding to the 3′ → 5′ helicase of RecB and the 5′-ending strand passing through RecC to bind to the 5′ → 3′ helicase of RecD. [Based on an X-ray structure by Dale Wigley, The London Research Institute, Herts, U.K. PDBid 3K70.]

Dale Wigley determined the X-ray structure of *E. coli* RecBCD in complex with a 51-nt DNA that could form a hairpin loop containing an up to 21-bp dsDNA stem (Fig. 30-75). The structure shows that RecB (1180 residues) and RecC (1122 residues) are intimately intertwined with RecB's C-terminal nuclease domain connected to the rest of the subunit by an extended 21-residue polypeptide tether. A 15-bp segment of dsDNA enters the protein through a tunnel between RecB and RecC. There it encounters a loop from RecC that appears to wedge the two strands apart, with the 6-nt 3′-ending single strand of the DNA binding to RecB and the 10-nt 5′-ending single strand binding to RecD (608 residues; the 5-nt loop connecting the two strands of the dsDNA at the top of Fig. 30-75 is disordered). The structure explains the different rates of cleavage of the two DNA strands. The 3′-ending strand emerges from a tunnel through RecC in the vicinity of the RecB nuclease domain, which is positioned to processively cleave it. The 5′-ending strand competes with the 3′-ending strand for the nuclease site, but since the 5′-ending strand is less favorably located, it is cleaved less frequently. However, after RecC has bound a Chi sequence,

the 3′-ending strand is no longer available for cleavage, which permits the nuclease to cleave the 5′-ending strand more frequently.

RecBCD can only commence unwinding DNA at a free duplex end. Such ends are not normally present in *E. coli,* which has a circular genome, but become available during such recombinational processes as bacterial transformation, conjugation, and viral transduction, as well as at collapsed replication forks.

e. RuvABC Mediates the Branch Migration and the Resolution of the Holliday Junction

The branch migration of the RecA-generated Holliday junction (Fig. 30-67*e,f*) requires the breaking and reforming of base pairs as the bases exchange partners in passing from one double helical stem to the other. Since $\Delta G = 0$ for this process, it was initially assumed that it occurs spontaneously. However, such a process would move forward and backward at random and, moreover, would be blocked by as little as a single mismatched base pair. In *E. coli,* and most other bacteria, branch migration is an ATP-driven unidirectional process that is mediated by two proteins whose synthesis is induced by the SOS response (Section 30-5D): **RuvB** (336 residues; Ruv for *r*epair of *UV* damage), an ATP-powered pump that drives branch migration but binds only weakly to DNA; and **RuvA** (203 residues), which binds to both a Holliday junction and to RuvB, thereby targeting **RuvB** to the DNA.

The X-ray structure of *Mycobacterium leprae* (the cause of leprosy) RuvA in complex with a synthetic and immobile Holliday junction (Fig. 30-76*a*), determined by Morikawa, reveals that RuvA forms a homotetramer to which the Holliday junction binds in its open-X conformation (Fig. 30-76*b*). The RuvA tetramer, which has the appearance of a four-petaled flower (it has C_4 symmetry rather than the D_2 symmetry of the vast majority of homotetramers), is relatively flat (80 × 80 × 45 Å) with one square face concave and the other convex. The concave face (that facing the viewer in Fig. 30-76*b*), which is highly positively charged and is studded with numerous conserved residues, has four symmetry-related grooves that bind the Holliday junction's four arms. This face's centrally located projection or "pin" is formed by the side chains of Glu 55 and Asp 56 from each subunit, and hence the repulsive forces between them and the Holliday junction's anionic phosphate groups probably facilitate the separation of the single-stranded DNA segments and guide them from one double helix to another.

RuvB is a member of the AAA+ family of ATPases (Section 30-2Ca). The X-ray structure of *Thermus thermophilus* RuvB crystallized in the presence of both ADP and AMPPNP, determined by Morikawa, reveals two molecules of RuvB with somewhat different conformations: one binding ADP and the other binding AMPPNP. Each RuvB molecule consists of three consecutive domains arranged in a crescentlike configuration with the adenine nucleotides binding at the interface between its N-terminal and middle domains. EM studies indicate that, in the presence of dsDNA, RuvB oligomerizes to form a hexamer (Fig. 30-77*a*),

(a)

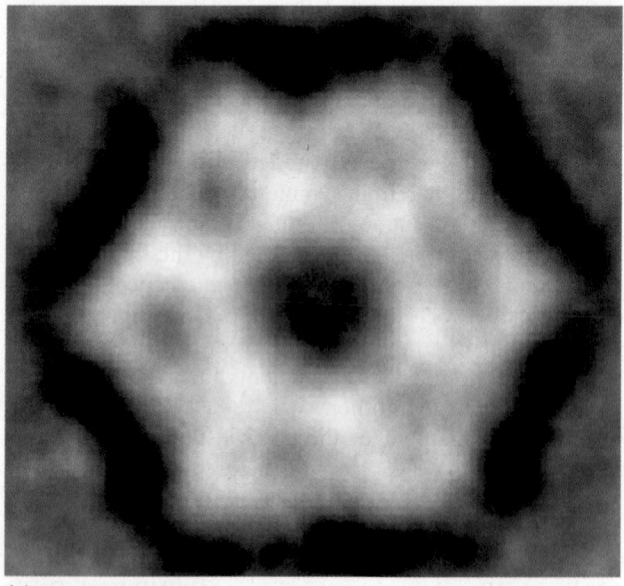

(a)

Figure 30-76 **X-ray structure of a RuvA tetramer in complex with a Holliday junction.** (*a*) A schematic drawing of the synthetic and immobile Holliday junction in this structure showing its base sequence. The two A · T base pairs that are disrupted at the crossover (and which, if the Holliday junction consisted of two homologous dsDNAs, as it normally does, would exchange base pairing partners) are magenta. (*b*) The RuvA–Holliday junction complex as viewed along the protein

tetramer's 4-fold axis. The protein is represented as its molecular surface with its subunits alternately colored pink and light blue. The DNA is drawn in space-filling form colored according to atom type with C atoms in different chains in different colors, N blue, O red, and P orange. [Part *a* courtesy of and Part *b* based on an X-ray structure by Kosuke Morikawa, Biomolecular Engineering Research Institute, Osaka, Japan. PDBid 1C7Y.]

as do most other AAA+ family members, including the D2 domain of NSF (Fig. 12-78). A hexameric model of RuvB (Fig. 30-77*b*), constructed by superimposing the N-terminal

domain of the RuvB monomer on the ATPase domains of the NSF D2 hexamer, agrees well with the EM-based image and contains no serious steric clashes. This 130-Å-diameter

(b)

Figure 30-77 **Proposed structure of the *T. thermophilus* RuvB hexamer.** (*a*) An EM-based image reconstruction of RuvB complexed with a 30-bp DNA (not visible) as viewed along its 6-fold axis. The image resolution is 30 Å. (*b*) A model of the RuvB hexamer that was constructed from the X-ray structure of RuvB monomers by superimposing their N-terminal domains on

the homologous ATPase domains of the NSF D2 homohexamer (Fig. 12-78). The N-terminal, middle, and C-terminal domains are blue, yellow, and green, respectively, and its bound AMPPNP is drawn in stick form in red. [Courtesy of Kosuke Morikawa, Biomolecular Engineering Research Institute, Osaka, Japan. PDBid 1HQC.]

hexameric model contains a 30-Å-diameter hole through which a single dsDNA can readily be threaded (see below). Moreover the six β hairpins, one per monomer, that have been implicated in binding to RuvA are located on the top face of the hexamer (as pictured in Fig. 30-77*b*).

The EM images of the RuvAB–Holliday junction complex indicate that RuvA binds two oppositely located RuvB hexamers. This has led to the model of their interaction depicted in Fig. 30-78 in which RuvA binds the Holliday junction and helps load the RuvB hexameric rings onto two opposing arms of the Holliday junction. The two hexameric rings are postulated to counter-rotate, each in the anticlockwise direction looking toward the center of the junction, so as to screw the horizontal DNA strands through the center of the junction and into the top and bottom double helices, thereby effecting branch migration (although rather than actually rotating relative to RuvA, a RuvB hexamer might pull the dsDNA through its central hole by "walking" up its grooves in a manner resembling that postulated for hexagonal helicases; Section 30-2Ca). The direction of branch migration depends on which pair of arms the RuvB hexamers are loaded.

The final stage of homologous recombination is the resolution of the Holliday junction into its two homologous dsDNAs. This process is carried out by **RuvC,** a homodimeric endonuclease of 173-residue subunits whose X-ray structure indicates that its active sites are located ∼30 Å apart on the same face of the protein. This suggests that RuvC sits down on the open face of the RuvAB–Holliday junction complex, that facing the viewer in Fig. 30-78, to cleave oppositely located strands at the Holliday junction. The resulting single-strand nicks in the now resolved dsDNAs are sealed by DNA ligase.

The X-ray structure of RuvC in complex with DNA has not been determined, although model building studies suggest that it binds Holliday junction DNA in its stacked-X conformation. However, Dietrich Suck determined the X-ray structure of bacteriophage **T4 endonuclease VII** in complex with a Holliday junction in the stacked-X conformation (Fig. 30-79). RuvC and the 157-residue T4 endonuclease VII exhibit no structural similarity but both are homodimers of relatively small subunits that have similar

Figure 30-78 Model of the RuvAB–Holliday junction complex. The model is based on electron micrographs such as that in the inset. The proteins are represented by their surface diagrams with the RuvA tetramer, as seen in its X-ray structure, green and the two oppositely oriented RuvB hexamers white. The DNA of the Holliday junction is drawn in space-filling form with its homologous blue and pink strands complementary to its light blue and red strands. The complex is postulated to drive branch migration via the ATP-driven counter-rotation of the RuvB hexamers relative to the RuvA tetramer. This pumps (screws) the horizontal dsDNAs through the RuvB hexamers to the center of the Holliday junction, where their strands separate and then base-pair with their homologs to form new dsDNAs, which are pumped out vertically. [Courtesy of Peter Artymiuk, University of Sheffield, U.K.]

Figure 30-79 X-ray structure of bacteriophage T4 endonuclease VII resolving a Holliday junction as viewed along its pseudo-2-fold axis. The Holliday junction DNA is drawn in ladder form with each of its four different 24-nt strands differently colored. The protein, a homodimer of 157-residue subunits, is shown in ribbon form embedded in its semitransparent molecular surface with one subunit red and the other blue. The arrows indicate the symmetrically located DNA cleavage sites. Compare the DNA in this structure to that in Fig. 30-68*b*. [Based on an X-ray structure by Dietrich Suck, European Molecular Biology Laboratory, Heidelberg, Germany. PDBid 2QNC.]

functions: the resolution of Holliday junctions into two duplex DNAs by introducing symmetrically placed nicks in equivalent strands (Fig. 30-67*j*).

The forgoing model of the **RuvABC resolvosome** provides a satisfying mechanism for branch migration and Holliday junction resolution. However, there is a fly in this particular ointment. The X-ray structure of an *M. leprae* RuvA–Holliday junction complex crystallized under conditions different from that in Fig. 30-76, determined by Laurence Pearl, resembles the complex in Fig. 30-76*b* but with a second RuvA tetramer in face-to-face contact with the concave (DNA-binding) side of the first. Hence, the Holliday junction is contained in two intersecting tunnels running through the resulting RuvA octamer. Are both RuvA–Holliday junction structures biologically relevant, or is one an artifact of crystallization? Pearl argues that the extensive complementary contacts between the two RuvA tetramers, which are strongly conserved, are unlikely to be artifactual and that a single RuvA tetramer is unlikely to withstand the torque exerted by the two (in effect) counter-rotating RuvB hexamers. However, if the RuvA octamer is biologically relevant, one of its tetramers would at some point have to dissociate in order to allow RuvC access to the Holliday junction. Yet, modeling studies indicate that the RuvC dimer cannot properly contact the RuvB tetramer-bound Holliday junction without it changing from its open-X to its stacked-X conformation. Further investigations are necessary to resolve these inconsistencies.

f. Recombination Repair Reconstitutes Damaged Replication Forks

Transformation, transduction, and conjugation are such rare events that the vast majority of bacterial cells never participate in these processes. Similarly, the only place in the metazoan life cycle at which gene shuffling through homologous recombination occurs is in meiosis (Section 1-4A). Why then do nearly all cells have elaborate systems for mediating homologous recombination? It is because damaged replication forks occur at a frequency of at least once per bacterial cell generation and perhaps 10 times per eukaryotic cell cycle. The DNA lesions that damage the replication forks can be circumvented via homologous recombination in a process named **recombination repair** [translesion synthesis, which is highly mutagenic, is a process of last resort (Section 30-5Db)]. Indeed, the rates of synthesis of RuvA and RuvB are greatly enhanced by the SOS response. Thus, as Michael Cox pointed out, *the primary function of homologous recombination is to repair damaged replication forks.* In what follows, we describe recombination repair as it occurs in *E. coli*.

Recombination repair is called into play when a replication fork encounters an unrepaired single-strand lesion (Fig. 30-80):

1. DNA replication is arrested at the lesion but continues on the opposing undamaged strand for some distance before the replisome fully collapses (Section 30-5Db).

2. The replication fork regresses to form a type of Holliday junction dubbed a "chicken foot." This process may occur spontaneously as driven by the positive supercoiling that has built up ahead of the replication fork, it may be mediated by RecA, or it may be promoted by **RecG,** an

ATP-driven helicase that catalyzes branch migration at DNA junctions with three or four branches.

3. The single-strand gap at the collapsed replication fork, now an overhang, is filled in by Pol I.

4. Reverse branch migration mediated by RuvAB or RecG yields a reconstituted replication fork, which supports replication restart (see below).

Note that this process does not actually repair the single-strand lesion that has caused the problem but instead reconstructs the replication fork in a way that permits the previously discussed DNA repair systems (Section 30-5) to eventually eliminate the lesion.

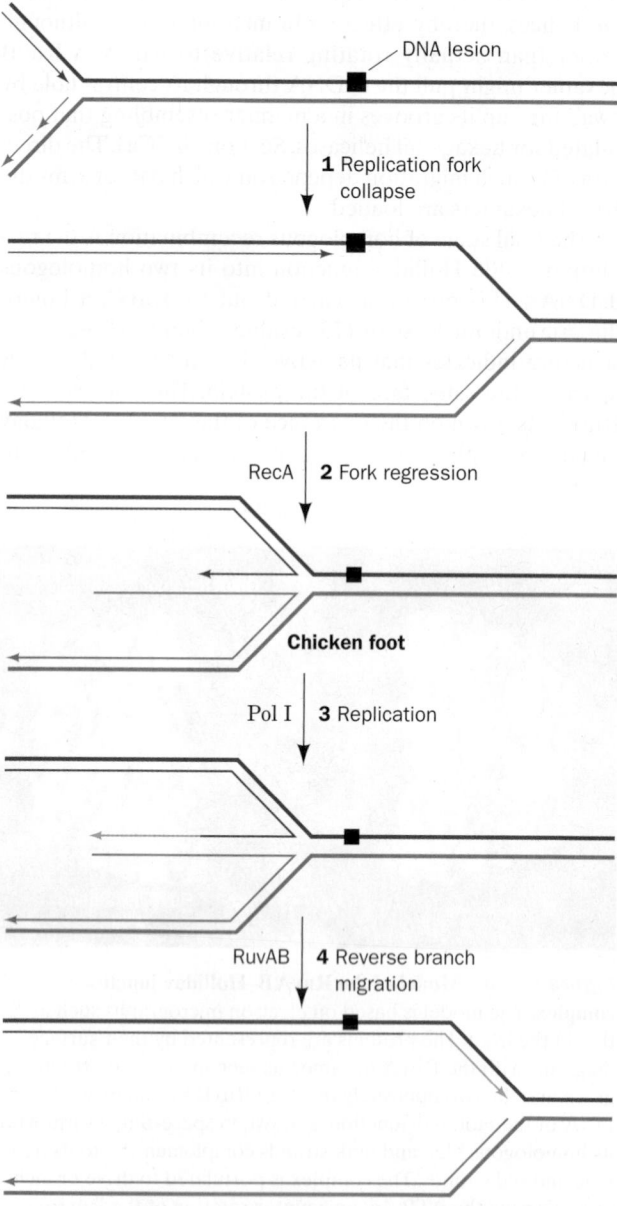

Figure 30-80   The recombination repair of a replication fork that has encountered a single-strand lesion. Thick lines indicate parental DNA, thin lines indicate newly synthesized DNA, the cyan lines indicate DNA that was synthesized by Pol I, and the arrows point in the 5′ → 3′ direction. [After Cox, M.M., *Annu. Rev. Genet.* **35,** 53 (2001).]

A second situation that requires recombination repair is the encounter of a replication fork with an unrepaired single-strand nick (Fig. 30-81):

1. When a single-strand nick is encountered, the replication fork collapses.

2. The repair process begins via the RecBCD plus RecA–mediated invasion of the newly synthesized and undamaged 3'-ending strand into the homologous dsDNA starting at its broken end.

3. Branch migration, as mediated by RuvAB, then yields a Holliday junction, which exchanges the replication fork's 3'-ending strands.

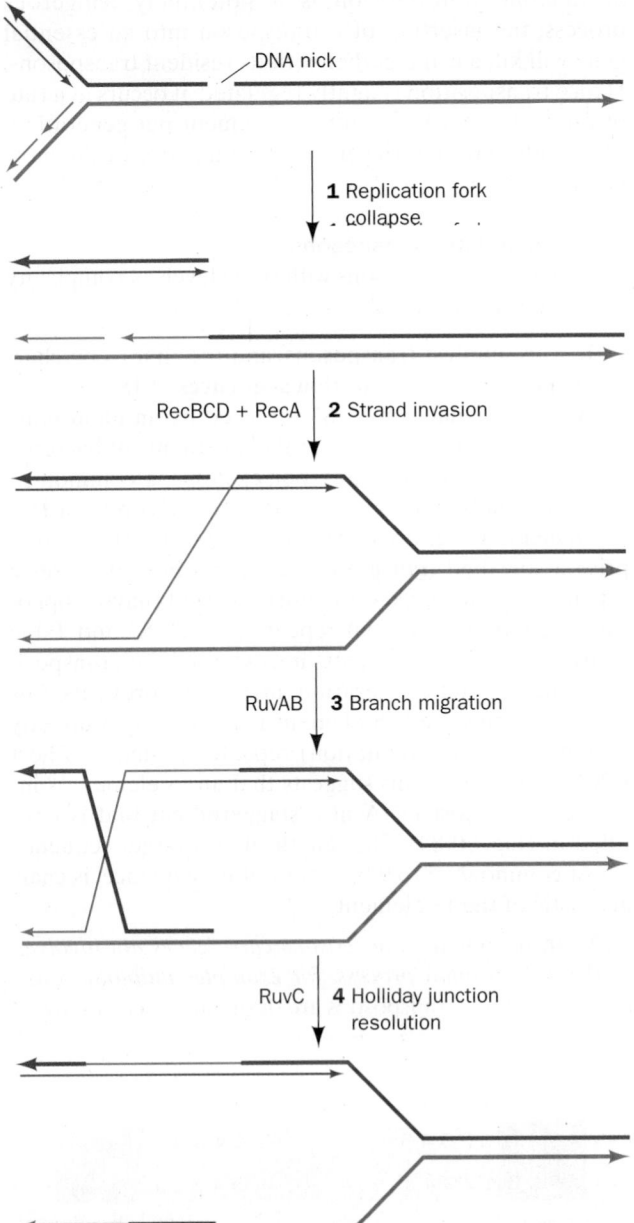

Figure 30-81 **The recombination repair of a replication fork that has encountered a single-strand nick.** Thick lines indicate parental DNA, thin lines indicate newly synthesized DNA, and the arrows point in the 5' → 3' direction. [After Cox, M.M., *Annu. Rev. Genet.* **35,** 53 (2001).]

4. RuvC then resolves the Holliday junction yielding a reconstituted replication fork ready for replication restart.

Thus, the 5'-ending strand of the nick has, in effect, become the 5' end of an Okazaki fragment.

The final step in the recombination repair process is the restart of DNA replication. This process is, of necessity, distinct from the replication initiation that occurs at *oriC* (Section 30-3Ca). **Origin-independent replication restart** is mediated by the same seven-protein primosome that initiates the minus strand replication of bacteriophage ϕX174 (Table 30-4), which has therefore been named the **restart primosome.**

g. Recombination Repair Reconstitutes Double-Strand Breaks

We have seen that double-strand breaks (DSBs) in DNA can be rejoined, often mutagenically, by nonhomologous end-joining (NHEJ; Section 30-5E). DSBs may also be nonmutagenically repaired through a recombination repair process known as homologous end-joining, which occurs via two Holliday junctions (Fig. 30-82):

1. The DSB's double-stranded ends are resected to produce single-stranded ends. One of the 3'-ending strands invades the corresponding sequence of a homologous

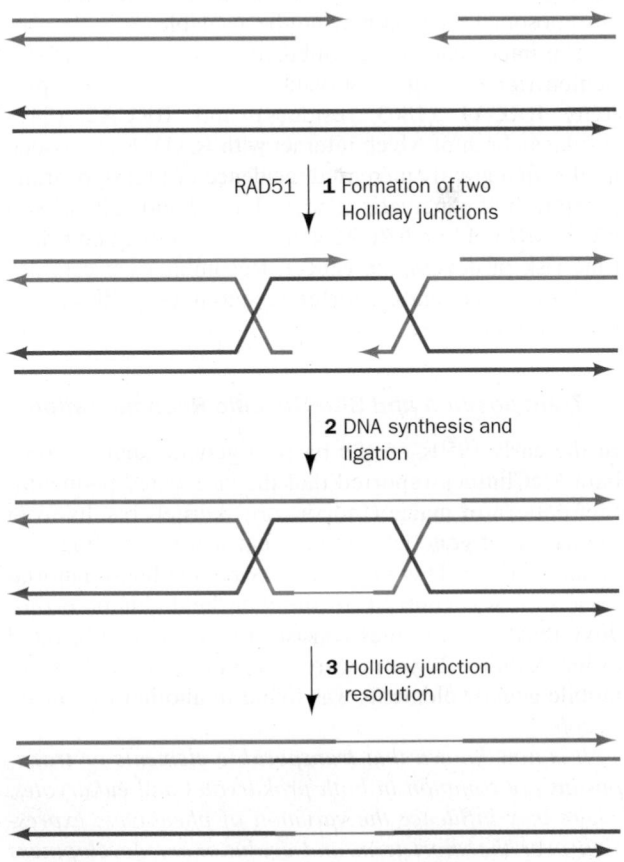

Figure 30-82 **The repair of a double-strand break in DNA by homologous end-joining.** Thick lines indicate parental DNA, thin lines indicate newly synthesized DNA, and the arrows point in the 5' → 3' direction. [After Haber, J.E., *Trends Genet.* **16,** 259 (2000).]

Table 30-6 **Properties of Some Insertion Elements**

Insertion Element	Length (bp)	Inverted Terminal Repeat (bp)	Direct Repeat at Target (bp)	Number of Copies in *E. coli* Chromosome
IS1	768	23	9	5–8
IS2	1327	41	5	5
IS4	1428	18	11–13	5
IS5	1195	16	4	1–2

Source: Mainly Lewin, B., *Genes IX, p.* 524, Oxford University Press (2008).

chromosome to form a Holliday junction, a process that, in eukaryotes, is mediated by the RecA homolog RAD51. The other 3′-ending strand pairs with the displaced strand segment on the homologous chromosome to form a second Holliday junction.

2. DNA synthesis and ligation fills in the gaps and seals the joints.

3. Both Holliday junctions are resolved to yield two intact double strands.

Thus, the sequences that may have been expunged in the formation of the DSB are copied from the homologous chromosome. Of course, a limitation of homologous end-joining, particularly in haploid cells, is that a homologous chromosomal segment may not be available.

The importance of recombination repair in humans is demonstrated by the observation that defects in the proteins **BRCA1** (1863 residues) and **BRCA2** (3418 residues), both of which interact with RAD51, are associated with a greatly increased incidence of breast, ovarian, prostate, and pancreatic cancers. Indeed, individuals with mutant *BRCA1* or *BRCA2* genes have up to an 80% lifetime risk of developing cancer. Recombination can also function to elongate shortened telomeres without the need for telomerase.

B. *Transposition and Site-Specific Recombination*

In the early 1950s, on the basis of genetic analysis, Barbara McClintock reported that the variegated pigmentation pattern of maize (Indian corn) kernels results from the action of genetic elements that can move about the maize genome. This proposal was resoundingly ignored because it was contrary to the then held genetic orthodoxy that chromosomes consist of genes linked in fixed order. Another 20 years were to pass before evidence of mobile genetic elements was found in another organism, *E. coli.*

It is now known that ***transposable elements*** *or* ***transposons*** *are common in both prokaryotes and eukaryotes, where they influence the variation of phenotypic expression over the short term and evolutionary development over the long term. Each transposon codes for the enzymes that specifically insert it into the recipient DNA.* This process has been described as **illegitimate recombination** because it requires no homology between donor and

recipient DNAs. Since the insertion site is chosen largely at random, transposition is a potentially dangerous process; the insertion of a transposon into an essential gene will kill a cell together with its resident transposons. Hence transposition is tightly regulated; it occurs at a rate of only 10^{-5} to 10^{-7} events per element per generation. The conditions that trigger transposition are, for the most part, unknown.

a. Prokaryotic Transposons

Prokaryotic transposons with three levels of complexity have been characterized:

1. The simplest transposons, and the first to be characterized, are named **insertion sequences** or **IS elements.** They are designated by "IS" followed by an identifying number. IS elements are normal constituents of bacterial chromosomes and plasmids. For example, a common *E. coli* strain has eight copies of **IS1** and five copies of **IS2.** IS elements generally consist of <2000 bp. These comprise a so-called **transposase** gene, and in some cases a regulatory gene, flanked by short inverted (having opposite orientation) terminal repeats (Fig. 30-83 and Table 30-6). The inverted repeats are essential for transposition; their genetic alteration invariably prevents this process. An inserted IS element is flanked by a directly (having the same orientation) repeated segment of host DNA (Fig. 30-83). This suggests that an IS element is inserted in the host DNA at a staggered cut that is later filled in (Fig. 30-84). The length of this target sequence (most commonly 5 to 9 bp), but not its sequence, is characteristic of the IS element.

2. *More complex transposons carry genes not involved in the transposition process, for example, antibiotic resistance genes.* Such transposons are designated "Tn" followed

Figure 30-83 **Structure of IS elements.** These and other transposons have inverted terminal repeats (*numerals*) and are flanked by direct repeats of host DNA target sequences (*letters*).

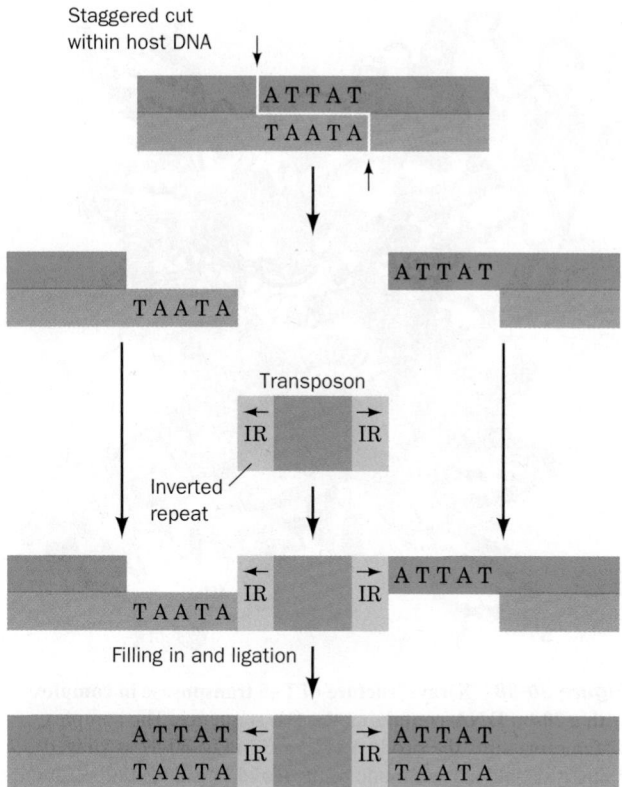

Staggered cut
within host DNA

Transposon

Inverted
repeat

IR   IR

Filling in and ligation

Figure 30-84 **A model for the generation of direct repeats of the target sequence by transposon insertion.**

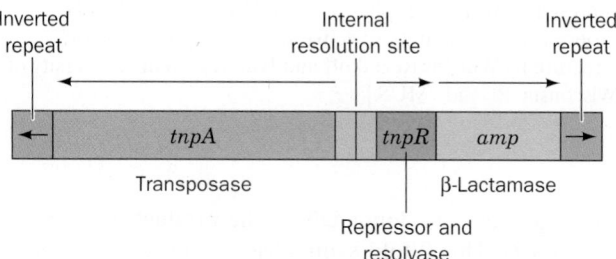

Inverted
repeat

Internal
resolution site

Inverted
repeat

tnpA

tnpR

amp

Transposase

β-Lactamase

Repressor and
resolvase

Figure 30-85 **A map of transposon Tn3.**

(a)

IS-like
module

IS-like
module

Central region

(b)

Central region

Inverted
repeats

Inverted
repeats

Figure 30-86 **A composite transposon.** This element consists of two identical or nearly identical IS-like modules (*green*) flanking a central region carrying various genes. The IS-like modules may have either (*a*) direct or (*b*) inverted relative orientations.

by an identifying number. For example, **Tn3** (Fig. 30-85) consists of 4957 bp and has inverted terminal repeats of 38 bp each. The central region of Tn3 codes for three proteins: (1) a 1015-residue transposase named **TnpA;** (2) a 185-residue protein known as **TnpR,** which mediates the **site-specific recombination** reaction necessary to complete the transposition process (see below) and also functions as a repressor for the expression of both *tnpA* and *tnpR;* and (3) a **β-lactamase** that inactivates ampicillin (Section 11-3Bb). The site-specific recombination occurs in an AT-rich region known as the **internal resolution site** that is located between *tnpA* and *tnpR.*

3. The so-called **composite transposons** (Fig. 30-86) consist of a gene-containing central region flanked by two identical or nearly identical IS-like modules that have either the same or an inverted relative orientation. It therefore seems that composite transposons arose by the association of two originally independent IS elements. Since the IS-like modules are themselves flanked by inverted repeats, the ends of either type of composite transposon must also be inverted repeats. Experiments demonstrate that composite transposons can transpose any sequence of DNA in their central region.

There are two modes of transposition: (1) **direct** or **simple transposition,** in which the transposon, as the name implies, physically moves from one DNA site to another; and (2) **replicative transposition,** in which the transposon remains at its original site and a copy of it is inserted at a target site. The two modes, as we shall see, have similar mechanistic features and, indeed, some transposons can move by either mode.

b. Direct Transposition of Tn5 Occurs by a Cut-and-Paste Mechanism

Tn5 is a 5.8-kb composite transposon that contains the gene encoding the 476-residue **Tn5 transposase** together with three antibiotic resistance genes. It is flanked by inverted IS-like modules ending in 19-bp sequences called **outside end (OE) sequences.** Tn5 undergoes direct transposition via a "cut-and-paste" mechanism that was elucidated in large part by William Reznikoff (Fig. 30-87):

1. Each of Tn5's two OE sequences on the donor DNA is bound by a monomer of Tn5 transposase.

2. The transposase dimerizes to form a catalytically active **synaptic complex** in which the transposon is held between the two transposase subunits.

3. Each transposase subunit activates a water molecule to nucleophilically attack the outermost nucleotide of its bound OE sequence, yielding a free 3'-OH group. This 3'-OH group is then activated to attack the opposite strand on the DNA to form a hairpin structure, thereby excising the transposon from the DNA. The hairpin is then hydrolyzed to yield a blunt-ended dsDNA at each end of the transposon, thus completing the "cut" portion of the transposition mechanism.

4. The synaptic complex binds to the target DNA.

Figure 30-87   The cut-and-paste transposition mechanism catalyzed by Tn5 transposase. The reactions comprising Steps 3 and 5 are indicated beside the braces to the right of these steps. [After Davies, D.R., Goryshin, I.Y., Reznikoff, W.S., and Rayment, I., *Science* **289**, 77 (2000).]

Figure 30-88   X-ray structure of Tn5 transposase in complex with a 20-bp DNA containing the OE sequence. The complex, which represents the product of Step 3 in Fig. 30-87, is viewed along its 2-fold axis with one of its two identical subunits colored in rainbow order from N-terminus (*blue*) to C-terminus (*red*) and the other subunit pink. The DNA is drawn in stick form with C green, N blue, O red, and P orange and with successive P atoms on the same polynucleotide connected by orange rods. The bound Mn^{2+} ions, which mark the enzyme's active site, are represented by purple spheres. The DNAs' reactive 3'-OH groups are located at these active sites. [Based on an X-ray structure by William Reznikoff and Ivan Rayment, University of Wisconsin. PDBid 1MUS.]

5. The transposon's 3'-OH groups nucleophilically attack the target DNA on opposite strands spaced 9 bp apart, thereby installing the transposon at the target site. Remarkably, this reaction and the three preceding lytic reactions are all mediated by the same catalytic site. The repair of the oppositely located single-strand gaps (Fig. 30-84) completes the "paste" portion of the mechanism.

Although, strictly speaking, not part of the transposition process, the double-strand break in the donor DNA left by the excision of the transposon must be repaired if the donor DNA is to be propagated (in bacteria, the donor DNA is often a plasmid so that its loss has little effect on the cell since plasmids are generally present in multiple copies).

The X-ray structure of a Tn5 synaptic complex (Fig. 30-88), determined by Reznikoff and Ivan Rayment, provides a model of the synaptic complex at the stage following its

cleavage from the donor DNA (the product of Step 3 in Fig. 30-87). This 2-fold symmetric complex consists of a dimer of Tn5 transposase subunits binding two 20-bp DNA segments containing the Tn5 transposon's 19-bp OE sequence with the outer end of each OE sequence bound to the protein (and whose opposite ends would, *in vivo*, be connected by the looped around transposon; Fig. 30-87). Both transposase subunits extensively participate in binding each DNA segment, thereby explaining why the individual subunits cannot cleave their bound DNA segments before forming the synaptic complex. The protein holds the DNA in a distorted B-DNA conformation with its two end pairs of nucleotides no longer base paired. Indeed, the penultimate base on the nontransferred strand is flipped out of the double helix and binds in a hydrophobic pocket. The transferred strand's free 3'-OH group, which occupies the active site, is bound in the vicinity of a cluster of three catalytically essential acidic residues, the so-called **DDE motif,** which is shared with other transposases. In the X-ray structure the DDE motif binds two Mn^{2+} ions, although physiologically it probably binds two Mg^{2+} ions. This suggests that transposases employ a metal-activated catalytic mechanism similar to that of the DNA polymerases (Section 30-2Af). The facing surface of the protein in Fig. 30-88 is positively charged with a prominent groove running from upper left to lower right that forms the apparent binding site for the target DNA.

Wild-type Tn5 transposase has such low catalytic activity that it is undetectable *in vitro*. However, that in the X-ray structure is a hyperactive mutant form that contains the mutations E54K and L372P (an unusual circumstance in that it is far more common to mutationally inhibit an enzyme under crystallographic study so as to trap it at some specific stage along its reaction pathway). Lys 54 is hydrogen bonded to O4 of a thymine base on the transferred strand. In the wild-type transposase, Glu 54 would probably have an unfavorable charge–charge repulsion with a nearby phosphate group, thus providing a structural basis for the increased activity of the E54K mutant. The L372P mutation disorders the peptide segment between residues 373 and 391 (it is ordered in the X-ray structure of wild-type Tn5 transposase lacking its N-terminal 55 residues), thereby suggesting that this mutation facilitates a conformational change required for substrate binding.

c. Replicative Transposition Occurs via Cointegrates

If a plasmid carrying a transposon resembling Tn3 is introduced into a bacterial cell carrying a plasmid that lacks the transposon, in some of the progeny cells both types of plasmid will contain the transposon (Fig. 30-89). Evidently, *such transposition involves the replication of the transposon into the recipient plasmid rather than its transfer from donor to recipient.*

Two plasmids, one containing a replicative transposon, will occasionally fuse to form a so-called **cointegrate** containing like-oriented copies of the transposon at both junctions of the original plasmids (Fig. 30-90). Yet, some of the progeny of a cointegrate-containing cell lack the cointegrate and instead contain both original plasmids, each with one copy of the transposon (Fig. 30-89). The cointegrate must therefore be an intermediate in the transposition process.

Although the mechanism of replicative transposition has not been fully elucidated, a plausible model for this process (and there are several) that accounts for the foregoing observations consists of the following steps (Fig. 30-91):

1. A pair of staggered single-strand cuts, such as is diagrammed in Fig. 30-84, is made by the transposon-encoded transposase at the target sequence of the recipient plasmid so as to liberate 3′-OH ends. Similarly, single-strand cuts are made on opposite strands to either side of the transposon. Note that these reactions resemble those catalyzed by Tn5 transposase (Fig. 30-87).

2. Each of the transposon's free ends is ligated to a protruding single strand at the insertion site. This forms a replication fork at each end of the transposon.

3. The transposon is replicated, thereby yielding a cointegrate.

4. Through a site-specific recombination between the internal resolution sites of the two transposons, the cointe-

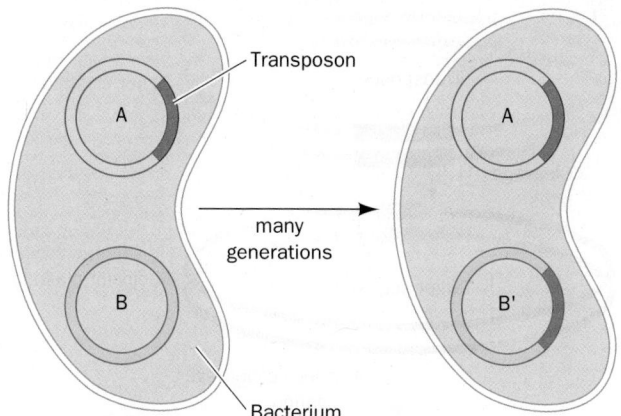

Figure 30-89 **Replicative transposition.** This type of transposition inserts a copy of the transposon at the target site while another copy remains at the donor site.

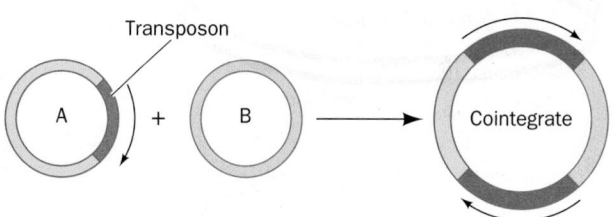

Figure 30-90 **A cointegrate.** This structure forms by the fusion of two plasmids, one carrying a transposon, such that both junctions of the original plasmid are spanned by transposons with the same orientation (*arrows*).

grate is resolved into the two original plasmids, each of which contains a transposon. This crossover process is catalyzed by transposon-encoded **recombinases** (TnpR in Tn3; also known as **resolvases**) rather than RecA; transposition proceeds normally in *recA⁻* cells (although RecA will resolve a cointegrate containing a transposon with a mutant resolvase and/or an altered internal resolution site, albeit at a much reduced rate).

Site-specific recombinases fall into only two protein families, **serine recombinases** and **tyrosine recombinases,** which are named after the amino acid residue that forms a transient covalent linkage to the DNA during the recombinase reaction. As we shall, these two types of recombinases function via different mechanisms.

d. γδ Resolvase Catalyzes Site-Specific Recombination

The γδ **resolvase,** a serine recombinase that forms a homodimer in solution, is a TnpR homolog that is encoded by the γδ **transposon** (a member of the Tn3 family of replicative transposons; Fig. 30-85). It catalyzes a site-specific recombination event in which a cointegrate containing two copies of the γδ transposon is resolved, via double-strand DNA cleavage, strand exchange, and

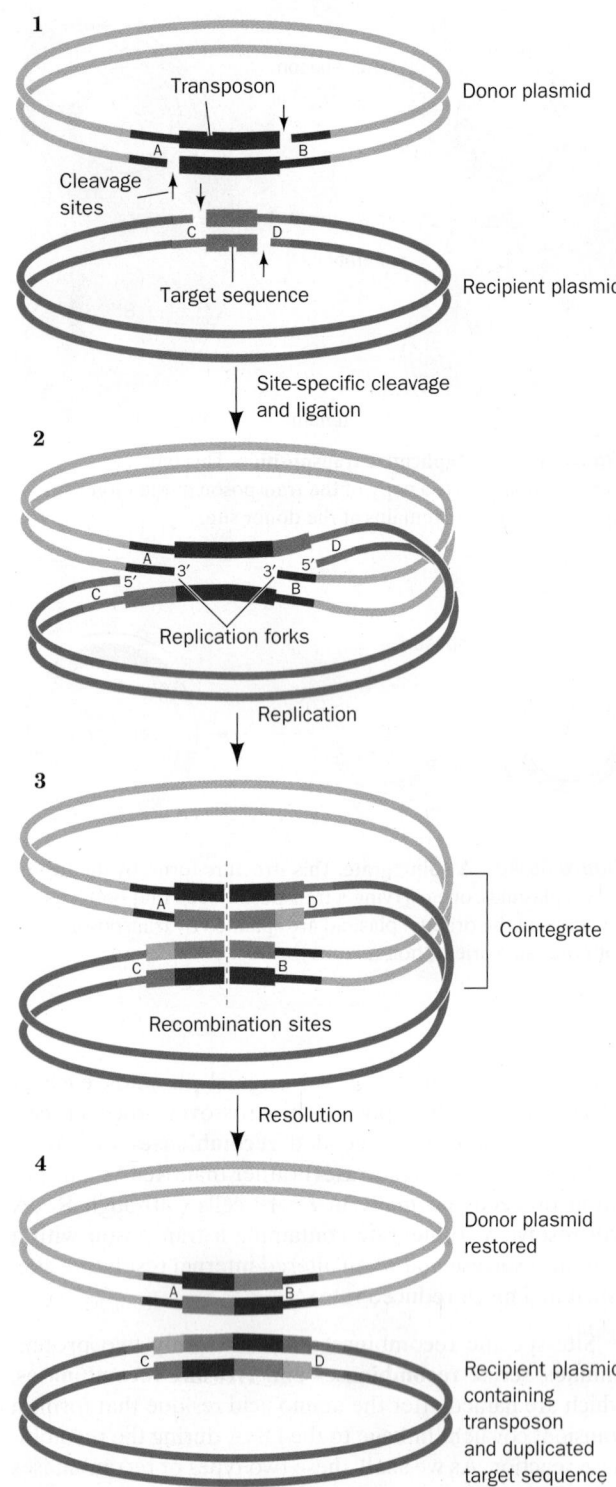

1
Transposon
Donor plasmid
A   B
Cleavage
sites
C   D
Target sequence
Recipient plasmid

Site-specific cleavage
and ligation

2
A   3'   3'   D
5'   5'
C   B
Replication forks

Replication

3
A   D
Cointegrate
C   B
Recombination sites

Resolution

4
Donor plasmid
restored
A   B

C   D
Recipient plasmid
containing
transposon
and duplicated
target sequence

**Figure 30-91   A model for transposition involving the
intermediacy of a cointegrate.** Here more lightly shaded bars
represent newly synthesized DNA. [After Shapiro, J.A., *Proc.
Natl. Acad. Sci.* **76,** 1934 (1979).]

res site that includes three binding sites for γδ resolvase
dimers, each of which contains an inverted repeat of the γδ
resolvase's 12-bp recognition sequence. The resolution of
the cointegrate involves the binding of a γδ resolvase homodimer to all six of these binding sites in the cointegrate
(three from each of its two transposons) as is diagrammed
in Fig. 30-92. The reaction proceeds via the formation of a
transient phosphoSer bond between Ser 10 and the 5'-
phosphate at each cleavage site.

The X-ray structure of the γδ resolvase homotetramer
in complex with two 34-bp palindromic dsDNA segments
containing an inverted repeat of the 12-bp recognition sequence separated by an 4-bp spacer (Fig. 30-93), determined by Nigel Grindley and Steitz, reveals that this synaptic tetramer has D_2 symmetry. Each 183-residue resolvase
monomer consists of an N-terminal catalytic domain
(residues 1–120) and a C-terminal DNA-binding domain
(residues 148–183) connected by an extended arm
(residues 121–147). Both dsDNAs, which are located at the
periphery of the protein core, have been cleaved with each
of the resulting four 5' ends in phosphoSer linkage to the
resolvase. This preserves the free energy of the cleaved
phosphodiester bond so that it can later be reformed with
a different partner, much as occurs with topoisomerases
(Section 29-3C).

Each centrally located catalytic domain approaches its
bound DNA from its minor groove side with its C-terminal
helix (helix E) binding over the minor groove (the segment
of the E helix that contacts the DNA is disordered in the
absence of the DNA). Each C-terminal domain binds over
the major groove of its recognition sequence on the opposite side of the DNA from its attached catalytic domain
with the extended arm that connects them running more or
less along the DNA's minor groove. The two C-terminal domains of the resolvase subunits labeled L and R (and the
symmetry-related L' and R' subunits) in Fig. 30-93 are
thereby separated by two helical turns along the cleaved
DNA, the segments of which closely assume the B-DNA
conformation. Each C-terminal helix binds in the DNA's
major groove and, together with its preceding helix, forms
a **helix–turn–helix (HTH) motif,** a common sequence-specific DNA-binding motif that occurs mainly in prokaryotic
transcriptional repressors and activators (Section 31-3Da).

The structure of the L–R dimer closely resembles that in
the X-ray structure of the dimer bound to uncleaved site I
DNA (the presynaptic dimer). This, and the short (17 Å)
distance between the free 3'-OH group in the L subunit
and the phosphoSer bond in the R subunit compared to
other such distances in the complex (L–L' and L–R'), indicates that the L–R and L'–R' dimers correspond to the initial site I–bound dimers soon after cleavage or just before
religation (Fig. 30-92). Consequently, the interface between
the L–R and L'–R' dimers must be the newly formed
synaptic interface.

How are the DNA strands in the synaptic complex
exchanged, that is, how is the DNA bound to the L subunit ligated to the DNA bound to either the L' subunit or
the R' subunit (and R to either R' or L')? In either case,
the free 3'-OH group on each subunit is ~50 Å from the

religation (the last step in Fig. 30-91), into two catenated
(linked) dsDNA circles that each contain one copy of the
γδ transposon (it also serves as its own transcriptional repressor as does TnpR). The γδ transposon contains a 114-bp

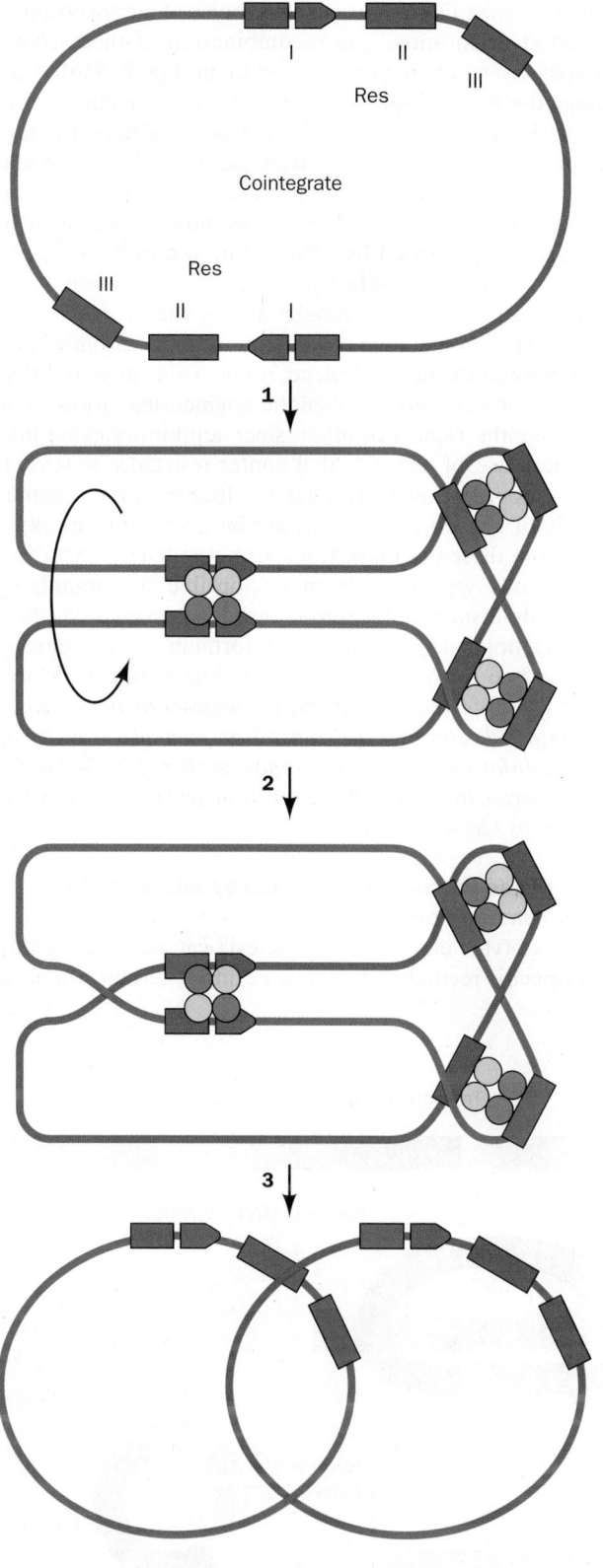

phosphoSer group that it would have to nucleophilically attack in the religation step and the space between them is filled with protein. Clearly, the synaptic complex must undergo a dramatic structural change to accomplish the religation step.

Figure 30-92 **A model for the resolution of a cointegrate containing two γδ transposons to form two catenated dsDNA circles. (1)** The γδ resolvase binds as six homodimers to its binding sites, I, II, and III, in each of the cointegrate's two *res* sites (yellow and green circles represent the γδ resolvase monomers initially bound to the red and blue *res* sites, respectively), which then associate to form synaptic tetramers. Although not shown as such, the synaptic tetramers bound to sites I associate with the synaptic tetramers bound to sites II and III to form, as seen in the electron microscope, a compact globule of unknown structure known as a **synaptosome. (2)** The dsDNA at sites I both undergo staggered (by 2 bp) double-strand scissions via the transient formation of phosphoSer bonds between Ser 10 and the 5′-phosphates at the cleavage sites. The cleaved strands then exchange places (cross over) in a process that apparently requires the rotation of one of the pairs of resolvase monomers with respect to the other and are then ligated. **(3)** The dissociation of the synaptosome yields the catenated dsDNA circles. [Courtesy of Gregory Mullen, University of Connecticut Health Center.]

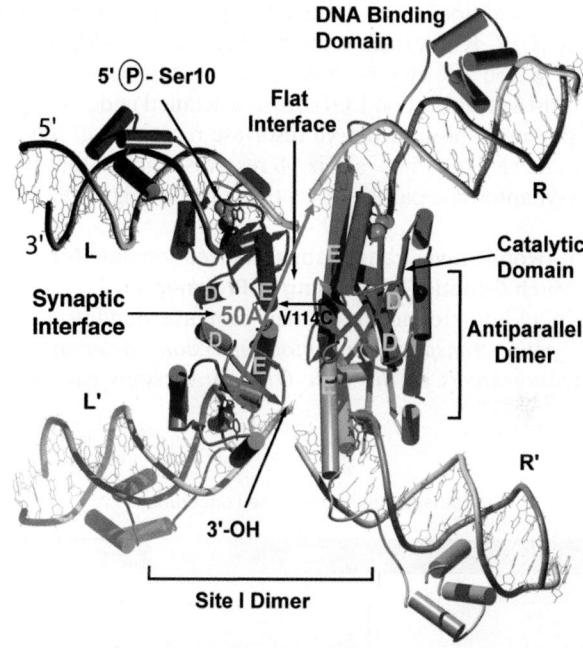

Figure 30-93 **X-ray structure of a γδ resolvase synaptic tetramer in complex with two 34-bp palindromic, site I–containing dsDNAs.** The D_2-symmetric complex is viewed with one of its 2-fold axes horizontal (the other 2-fold axes lie in the vertical plane labeled "Flat interface"). The DNAs (dark and light green and yellow and orange) have been cleaved into half-sites labeled L, R, L′, and R′ through the nucleophilic attack of the Ser 10 side chain (drawn in space-filling form with C blue and O red) on the 5′-phosphate group of A20 (drawn in stick form in green). The subunits of the resolvase tetramer, whose helices are drawn in tube form, are colored blue, green, red, and pink. The intact DNAs initially bind to the L–R and L′–R′ dimers, the so-called site I dimers. The L–L′ dimer and the symmetrically equivalent R–R′ dimer form so-called antiparallel dimers whose D and E helices associate as four-helix bundles. In addition, a Val 114 → Cys mutation (V114C) disulfide-cross-links the E helices of the L–R and L′–R′ dimers across the so-called flat interface. [Courtesy of Thomas Steitz, Yale University. PDBid 1ZR4.]

The observation that the interface between the L–L′ and R–R′ dimers is largely hydrophobic and unusually flat (Fig. 30-93) strongly suggests that the religation occurs after a 180° rotation of these dimers with respect to one another (about the horizontal 2-fold axis in Fig. 30-93), thus exchanging the positions of the R and R′ subunits with respect to the L and L′ subunits. The rotation is presumably driven by the superhelical tension in the naturally negatively supercoiled cointegrate. This model is supported by energy calculations and the observation that mutating Val 114 of the E helix to Cys, which, under oxidizing conditions, disulfide-links the E helices on the L and R (and L′ and R′) subunits (a mutation that was present in the foregoing structure), yields a complex that can form the covalent intermediate in Fig. 30-93 but cannot carry out the religation step—presumably because the disulfide bonds prevent the above rotation. However, reducing these disulfide bonds restores the complex's recombinational activity. Furthermore, mutation of Lys 136, an E helix residue located at the flat interface, to Cys prevents religation of the cleaved DNA when subjected to oxidizing conditions, even though it requires a 75° rotation of the L–L′/R–R′ interface in the structure in Fig. 30-93 to bring the Cys 136 side chains on opposing subunits close enough to form a disulfide bond. Of course, a detailed understanding of the mechanism of the γδ resolvase reaction will require the knowledge of how all six γδ resolvase dimers that form the synaptosome participate in the reaction (Fig. 30-92).

e. Replicative Transposons Are Responsible for Much Genetic Remodeling in Prokaryotes

In addition to mediating their own insertion into DNA, *replicative transposons promote inversions, deletions, and rearrangements of the host DNA.* Inversions can occur when the host DNA contains two copies of a transposon in inverted orientation. The recombination of these transposons inverts the region between them (Fig. 30-94a). If, instead, the two transposons have the same orientation, the resolution of this cointegrate-like structure deletes the segment between the two transposons (Fig. 30-94b; if the deleted segment lacks a replication origin, it will not be propagated). The deletion of a chromosomal segment in this manner, followed by its integration into the chromosome at a different site by a separate recombinational event, results in chromosomal rearrangement.

Transposition appears to be important in chromosomal and plasmid evolution. Indeed, it has been suggested that transposons are nature's genetic engineering "tools." For example, the rapid evolution, since antibiotics came into common use, of plasmids that confer resistance to several antibiotics (Section 5-5Ba) has resulted from the accumulation of the corresponding antibiotic-resistance transposons in these plasmids. Transposon-mediated rearrangements may well have been responsible for organizing originally distant genes into coordinately regulated operons (Section 5-4Aa) as well as for forming new proteins by linking two formerly independent gene segments. Moreover, *the occurrence of identical transposons in unrelated bacteria indicates that the transposon-mediated transfer of genetic information between organisms is not limited to related species, in contrast to genetic transfers mediated by homologous recombination.*

f. Phase Variation Is Mediated by Site-Specific Recombination

Phenotypic expression in bacteria can be regulated by site-specific recombination. For example, certain strains of

(a)

(b)

Figure 30-94 **Chromosomal rearrangement via recombination.** (*a*) The inversion of a DNA segment between two identical transposons with inverted orientations. (*b*) The deletion of a DNA segment between two identical transposons with the same orientation. This process parcels one transposon each to the resulting two DNA segments.

(a) **Phase 2**

H2–rh1
promoter

(b) **Phase 1**

**Figure 30-95   The mechanism of phase variation in
Salmonella.** (*a*) In Phase 2 bacteria, the *H2–rh1* promoter is
oriented so that H2 flagellin and repressor are synthesized.
Repressor binds to the *H1* gene, thereby preventing its expression.
(*b*) In Phase 1 bacteria, the segment preceding the *H2–rh1*
transcription unit has been inverted relative to its orientation in
Phase 2 bacteria. Hence this transcription unit cannot be
expressed because it lacks a promoter. This releases *H1* from
repression and results in the synthesis of H1 flagellin. The
inversion of the segment preceding the *H2–rh1* transcription unit
is mediated by the Hin protein, which is expressed in either
orientation by the *hin* gene.

Salmonella typhimurium make two antigenically distinct
versions of the protein **flagellin** (the major component of
the whiplike flagella with which bacteria propel themselves; Section 35-3I) that are designated **H1** and **H2**. Only
one of these proteins is expressed by any particular cell but
about once every 1000 cell divisions, in a process known as
phase variation, a cell switches the type of flagellin it synthesizes. It is thought that phase variation helps *Salmonella*
evade its host's immunological defenses.

What is the mechanism of phase variation? The two flagellin genes reside on different parts of the bacterial chromosome. *H2* is linked to the *rh1* gene that encodes a repressor of H1 expression (Fig. 30-95; *rh1*, *H2*, and *H1* are also
known as *fljA*, *fljB*, and *fljC*, respectively). Hence, when the
H2–rh1 transcription unit is expressed, H1 synthesis is

repressed; otherwise H1 is synthesized. Melvin Simon has
shown that the expression of the *H2–rh1* unit is controlled
by the orientation of a 995-bp segment that lies upstream of
H2 (Fig. 30-95) and that contains the following elements:

1. A promoter for *H2–rh1* expression.

2. The *hin* gene, which encodes the 190-residue **Hin
DNA invertase.** Hin, a serine recombinase, mediates the inversion of the DNA segment in a manner similar to that diagrammed in Fig. 30-94*a*. In fact, Hin is ~40% identical in
sequence with the γδ resolvase, which strongly suggests
that these proteins have similar structures.

3. Two closely related 26-bp sites, *hixL* and *hixR*, that
form the boundaries of the segment and hence contain its
cleavage sites. They each consist of two imperfect 12-bp inverted repeats separated by 2 nt.

In the Phase 2 orientation (Fig. 30-95*a*), the properly oriented promoter is just upstream of *H2*, so this gene and *rh1*
are coordinately expressed, thereby repressing H1 synthesis. In Phase 1 bacteria (Fig. 30-95*b*), however, this segment
has the opposite orientation. Consequently, neither *H2* nor
rh1, which then lacks a promoter, is expressed so that H1 is
synthesized.

g. Cre-Mediated Site-Specific Recombination Occurs via 3′-PhosphoTyr Intermediates

Bacteriophages, as we have seen (Fig. 1-31), replicate
themselves within their host bacterial cells which, in most
cases, they then lyse to release the progeny phage, a
lifestyle that is therefore known as the **lytic** mode. However, certain bacteriophages can assume an alternative,
nondestructive lifestyle, the **lysogenic** mode, in which they
install their DNA, usually in the host chromosome via site-
specific recombination, so that the phage DNA is passively
replicated with the host DNA. However, if the bacterial
host encounters conditions in which it is unlikely to survive, the phage DNA is excised from the bacterial chromosome via a reversal of the site-specific recombination reaction and it reenters the lytic mode so as to escape the
doomed host. We discuss the genetic factors that maintain
the balance between the lytic and lysogenic lifestyles in
bacteriophage λ in Section 33-3.

The enzymes that mediate the foregoing site-specific
recombination reactions are members of the **λ integrase
(λ Int;** alternatively, tyrosine recombinase) family, whose
~1000 known members also occur in prokaryotes and eukaryotes. These include the **XerC** and **XerD** proteins of *E.
coli* which, operating in concert, function to decatenate the
two linked circular dsDNA products of homologous recombination (Fig. 30-69*g, left*), as well as type IB topoisomerases (Section 29-3Cc).

The structurally best characterized member of the λ integrase family is the **Cre recombinase** of *E. coli* **bacteriophage P1.** In its lysogenic state, bacteriophage P1 is a sin-
gle-copy circular plasmid (rather than being inserted in the
host chromosome as is bacteriophage λ), but in the phage
head (the lytic mode), P1 DNA is a linear dsDNA that has
a 34-bp *loxP* site at each end. The main function of Cre,

Figure 30-96   The circularization of linear bacteriophage P1 DNA. This occurs through the Cre-mediated site-specific recombination between its two terminally located *loxP* sites (*red and green*) to yield its lysogenic plasmid.

which is encoded by bacteriophage P1, is to mediate the site-specific recombination between these two *loxP* sites so as to circularize the linear DNA (Fig. 30-96).

The *loxP* site is palindromic except for its central 8-bp crossover region, which confers directionality on the site. In carrying out the recombination reaction, the 343-residue Cre subunits form a homotetramer that binds two *loxP* sites in an antiparallel orientation, with each Cre subunit binding half of a *loxP* site. Then, as is diagrammed in Fig. 30-97, oppositely located Cre subunits catalyze single-strand scissions on the 5′ side of the crossover region on one strand of each of the two dsDNAs. This occurs through the nucleophilic attack of each of these active Cre subunit's conserved Tyr 324 residues on the DNA's scissile phosphoester bond to yield a 3′-phosphoTyr intermediate on one side of the cleaved bond and a free 5′-OH group on the other side (as similarly occurs in the reactions catalyzed by type IB topoisomerases; Section 29-3Cc). Each of the liberated 5′-OH groups then nucleophilically attacks the 3′-phospho-Tyr group on the opposite duplex to form a Holliday junction, thereby releasing the Tyr residues. The Holliday

Figure 30-97   The mechanism of Cre–*loxP* site-specific recombination. The dashed lines represent the nonpalindromic crossover regions of the *loxP* sites. The green and magenta Cre subunits are active for cleavage in the top and bottom parts of the diagram, respectively, with their roles being switched by the isomerization step. Note that the mechanism does not require branch migration of the Holliday junction intermediate. [Courtesy of Gregory Van Duyne, University of Pennsylvania School of Medicine.]

junction is resolved into two recombined dsDNAs when the two Cre subunits that had not yet participated in the reaction mediate the same cleavage and strand exchange reactions on the two heretofore unreacted single strands. This latter process must be preceded by a structural rearrangement (isomerization) of the Cre tetramer that positions the catalytic Tyr residues in the latter pair of subunits to participate in the reaction while those in the former pair of subunits are similarly removed from the scene of the action. Note that this mechanism differs from that mediated by serine recombinases in that the latter cleave all four DNA strands prior to initiating their exchange and hence do not have a Holliday junction intermediate (Section 30-6Bd).

The X-ray structures of Cre tetramers in their complexes with several *loxP* model DNAs, determined by Gregory Van Duyne, have helped elucidate its mechanism. When the DNA had a single-strand nick past the second nucleotide from the 5′ end of the crossover region, Cre-catalyzed strand scission yielded a free nucleotide (a CMP) that diffused away. Since this nucleotide contained the otherwise reactive 5′-OH group, the 3′-phosphoTyr intermediate was irreversibly trapped, that is, Cre could not carry out the strand exchange reaction in Fig. 30-97 (this nicked DNA is a suicide substrate for Cre; Section 28-3Bc). The X-ray structure of the Cre complex of this nicked DNA confirmed the presence of the 3′-phosphoTyr intermediate and indicated, through model building, that the 5′-OH group on the missing CMP residue would be well positioned to nucleophilically attack the 3′-phosphoTyr bond on the opposite strand (Fig. 30-98*a*). Note that this complex is only 2-fold symmetric although its four Cre subunits and much of the DNA are related by pseudo-4-fold symmetry. When the DNA was, instead, an immobile Holliday junction (Fig. 30-98*b*), the complex was also pseudo-4-fold symmetric with the single strands that had crossed over noticeably kinked at their centers. These structures revealed that the conformational changes necessary to carry out the strand exchange and isomerization reactions (Fig. 30-97) required surprisingly small movements on the part of the Cre subunits and that only the sugar–phosphate backbones of the strand-exchanged nucleotides needed to move in order to form the Holliday junction.

h. Most Transpositions in Eukaryotes Involve RNA Intermediates

Transposons similar to those in prokaryotes also occur in eukaryotes, including yeast, maize, *Drosophila,* and humans. In fact, ~3% of the human genome consists of DNA-based transposons although, in most cases, their sequences have mutated so as to render them inactive, that is, these transposons are evolutionary fossils. However, many eukaryotic transposons exhibit little similarity to those of prokaryotes. Rather, their base sequences resemble those of retroviruses (see below), which suggests that these transposons are degenerate retroviruses. The transposition of these so-called **retrotransposons** occurs via a pathway that resembles the replication of retroviral DNA (Section 15-4C): (1) their transcription to RNA, (2) the reverse transcriptase–mediated copying of this RNA to cDNA (Section 30-4C), and (3) the largely random insertion of this DNA into the host

Figure 30-98   X-ray structures of the Cre homotetramer in its complexes with model *loxP* DNAs. (*a*) Two identical dsDNAs that were nicked past the second nucleotide from the 5′ end of their crossover regions; and (*b*) an immobile Holliday junction. The left panels show the Cre–DNA complexes as viewed along their exact 2-fold and pseudo-4-fold axes, with the active and inactive subunits green and magenta, respectively (as in Fig. 30-97), and with the DNA gold. The right panels show only the DNAs in the X-ray structures as viewed from below the left panels. In the right panel of Part *a*, the active site Tyr that is covalently linked

to the 3′-OH group of the cleaved DNA strand is shown in stick form (*red*) and the modeled-in position of the cleaved CMP's 5′-OH group is shown positioned to nucleophilically attack the 3′-phosphoTyr group on the opposite dsDNA (*curved arrows*). In the right panel of Part *b*, the three base pairs that form as a consequence of strand exchange are indicated. Note that the vertical strands in the crossovers but not the horizontal strands are distinctly kinked at their centers. [Courtesy of Gregory Van Duyne, University of Pennsylvania School of Medicine. PDBids 2CRX, 3CRX, 4CRX, and 5CRX.]

organism's genome as mediated by enzymes known as **integrases** (which catalyze reactions similar to and structurally resemble cut-and-paste DNA transposases).

The involvement of RNA in retrotransposon-mediated transposition was ingeniously shown by Gerald Fink through his remodeling of **Ty1,** the most common transposable element in budding yeast (which has ~35 copies of this 6.3-kb element comprising ~13% of its 1700 kb genome; Ty stands for *Transposon yeast*), so that it contained a yeast intron (a sequence that is excised from an RNA transcript and hence is absent in the mature RNA; Section 5-4Ac) and was preceded by a galactose-sensitive yeast promoter. The transposition rate of this remodeled Ty1 element varied with the galactose concentration in the medium and the transposed elements all lacked the intron, thereby demonstrating the participation of an RNA intermediate.

A retroviral genome (Fig. 30-99*a*) is flanked by direct long terminal repeats (**LTRs**) of 250 to 600 bp and typically contains the genes encoding three polyproteins: **gag,** which is

Figure 30-99   Gene sequences of (*a*) retroviruses and (*b*) the Ty1 retrotransposon.

cleaved to the proteins comprising the viral core (Fig. 15-34); **pol,** which is cleaved to the above-mentioned reverse transcriptase and integrase, as well as the protease that catalyzes these cleavages; and **env,** which is cleaved to viral outer envelope proteins. Ty1 (Fig. 30-99*b*) is likewise flanked by LTRs (of 330 bp) but expresses only two polyproteins: **TYA** and **TYB,** the counterparts of gag and pol. Moreover, TYA and TYB, together with Ty1 RNA, form viruslike particles in the yeast cytoplasm. However, Ty1 lacks a counterpart of the retroviral *env* gene. Hence Ty1 is an "internal virus" that can only replicate within a genome, albeit at an extremely low rate compared to that of real retroviral infections.

Figure 30-100   Proposed mechanism for the transposition of nonviral retrotransposons. (1) The retrotransposon-encoded reverse transcriptase/endonuclease nicks one strand of the target DNA and then recruits the RNA transcript of the retrotransposon to this site. **(2)** The DNA-primed reverse transcription of the retrotransposon RNA. **(3)** The RNA is degraded, and the second DNA strand is synthesized using the first strand as its template (normal reverse transcriptase reactions; Section 30-4C), followed by the insertion of the resulting nonviral retrotransposon into the target DNA via a poorly understood process.

Copia (Latin for abundance), the most abundant retrotransposon in the *Drosophila* genome (which contains 20–60 copies of copia), resembles Ty1.

The LTRs in retroviruses and retrotransposons such as Ty1 and *copia* are essential elements for their transcription and hence for their transposition. Yet, vertebrate genomes also contain retrotransposons that lack LTRs and hence cannot be transcribed analogously to retroviruses. A common family of these **nonviral retrotransposons,** the 1- to 7-kb **long interspersed nuclear elements (LINEs),** each contain two open reading frames: *ORF1,* which contains sequences similar to those in *gag;* and *ORF2,* which contains sequences similar to those encoding reverse transcriptase. A proposed mechanism for the transposition of LINEs is diagrammed in Fig. 30-100.

Different types of transposons, DNA-only, retroviral, and nonviral, predominate in different organisms. Thus bacteria, as we have seen, contain nearly exclusively DNA-only transposons, yeast have mainly retroviral retrotransposons, *Drosophila* have all three types, and in humans LINEs predominate. In fact, the human genome contains an estimated 1.4 million LINEs or LINE fragments that comprise ~20% of the 3.0-billion-bp human genome (genomic organization is discussed in Section 34-2). The great majority of these molecular parasites have mutated to the point of inactivity but a few still appear capable of further transposition. Indeed, several hereditary diseases are caused by the insertion of a LINE into a gene. Several other types of retrotransposons also comprise significant fractions of the human genome as we shall see in Section 34-2.

7 DNA METHYLATION AND TRINUCLEOTIDE REPEAT EXPANSIONS

The A and C residues of DNA may be methylated, in a species-specific pattern, to form N^6-**methyladenine (m^{6}A),** N^4-**methylcytosine (m^{4}C),** and **5-methylcytosine (m^{5}C)** residues, respectively.

N^6-**Methyladenine (m^{6}A)**
residue

N^4-**Methylcytosine (m^{4}C)**
residue

5-Methylcytosine (m^{5}C)
residue

These are the only types of modifications to which DNA is subjected in cellular organisms (although all the C residues of T-even phage DNAs are converted to **5-hydroxymethyl-cytosine** residues,

5-Hydroxymethylcytosine residue

which may, in turn, be glycosylated). These methyl groups project into B-DNA's major groove, where they can interact with DNA-binding proteins. In most cells, only a few percent of the susceptible bases are methylated, although this figure rises to >30% of the C residues in some plants.

Bacterial DNAs are methylated at their own particular restriction sites, thereby preventing the corresponding restriction endonucleases from degrading the DNA (Section 5-5A). These restriction–modification systems, however, account for only part of the methylation of bacterial DNAs. In *E. coli*, most DNA methylation is catalyzed by the products of the *dam* and *dcm* genes. The **Dam methyltransferase (Dam MTase)** methylates the A residue in all GATC sequences, whereas the **Dcm MTase** methylates both C residues in CC^A_TGG at their C5 positions. Note

that both of these sequences are palindromic. We have seen that *E. coli* uses Dam Mtase–mediated methylation to differentiate parental from newly synthesized DNA in mismatch repair (Section 30-5C) and in limiting *oriC*-based DNA replication initiation to once per cell generation via sequestration (Section 30-3Cb).

a. The MTase Reaction Occurs via a Covalent Intermediate in Which the Target Base Is Flipped Out

The Dam and Dcm MTases, as do all known DNA MTases, use *S*-adenosylmethionine (SAM) as their methyl donor. Indeed, all m^5C-MTases share a set of conserved sequence motifs. Daniel Santi has proposed that the catalytic mechanism of these m^5C-MTases (Fig. 30-101) is similar to that of thymidylate synthase (Fig. 28-19) in that both types of enzymes transfer methyl groups to pyrimidine C5 atoms via a reaction that is initiated by the nucleophilic attack of a Cys thiolate group on the pyrimidine's C6 position. The pyrimidine's C5 atom is thereby activated as a resonance-stabilized carbanion that nucleophilically attacks the methyl donor's methyl group (which in thymidylate synthase is donated by N^5,N^{10}-methylene-THF rather than SAM) to yield a covalent intermediate. This intermediate subsequently decomposes to products through the enzymatic abstraction of the proton substituent to C5 and elimination of the enzyme. The Cys thiolate nucleophile is a component of a Pro-Cys dipeptide that is invariant in all known m^5C-MTases and thymidylate synthases.

Figure 30-101 The catalytic mechanism of 5-methylcytosine methyltransferases (m^5C-MTases). The methyl group is supplied by SAM, which thereby becomes *S*-adenosylhomocysteine. In M.HhaI, the DNA MTase from *Haemophilus haemolyticus,* the active site thiolate group, $^-$S—E, is on Cys 81, the enzyme general acid, E—A, is Glu 119, and the enzyme general base, E—B, has not been identified. [After Verdine, G.L., *Cell* **76,** 198 (1994).]

This mechanism is supported by the observation that the action of m⁵C-MTases on a **5-fluorocytosine (f⁵C)** residue

5-Fluorocytosine (f⁵C) residue

irreversibly traps the covalent intermediate (and hence inactivates the enzyme) because the enzyme cannot abstract fluorine, the most electronegative element, as an F^+ ion (5-fluorodeoxyuridylate is likewise a suicide substrate for thymidylate synthase; Section 28-3Bc). Stereochemical principles dictate that the enzyme's Cys thiolate group can nucleophilically attack cytosine's C5 position only from above or below the ring. This is possible because, as we shall see below, the enzyme induces its cytosine target to flip out of the DNA double helix.

The DNA MTase from *Haemophilus haemolyticus* **(M.HhaI),** a 327-residue monomer, is a component of this bacterium's restriction–modification system. M.HhaI methylates its recognition sequence, 5'-GCGC-3' in double-stranded DNA, to yield 5'-G-m⁵C-GC-3'. Richard Roberts and Xiaodong Cheng determined the X-ray structure of the inactivated M.HhaI–DNA complex formed by incubating the enzyme with the self-complementary sequence d(TGATA**G-f⁵C-GC**TATC) (in which the enzyme's recognition sequence is in bold) in the presence of SAM. The DNA binds to the enzyme in a large cleft between its two unequally sized domains (Fig. 30-102). The structure's most striking feature is that the f⁵C nucleotide has flipped out of the minor groove in the otherwise largely undistorted B-DNA helix and has inserted into the enzyme's active site.

There, the f⁵C has reacted with SAM so as to yield adenosylhomocysteine (SAM without its methyl group) and the methylated intermediate covalently linked to Cys 81. The side chain of Gln 237 fills the cavity in the DNA double helix left by the departure of the f⁵C by hydrogen-bonding to the opposing G base. Comparison of this structure with that of M.HhaI in complex only with SAM indicates that on binding the DNA the protein's so-called active site loop (residues 80–99) swings around to contact the DNA, a movement of up to 25 Å. Nearly all base-specific interactions are made in the major groove by two Gly-rich loops (residues 233–240 and 250–257), the so-called recognition loops. The protein also makes extensive sequence-nonspecific contacts with DNA phosphate groups.

Base flipping was first observed in the above X-ray structure. However, as is now clear from the structures of other MTases as well as those of variety of DNA repair enzymes (e.g., Sections 30-5Aa and 30-5Be), *base flipping is a common mechanism through which enzymes gain access to the bases in dsDNA on which they perform chemistry.*

b. DNA Methylation in Eukaryotes Functions in Gene Regulation

5-Methylcytosine is the only methylated base in most eukaryotic DNAs, including those of vertebrates. This modification occurs largely in the CG dinucleotide of various palindromic sequences. CG is present in the vertebrate genome at only about one-fifth its randomly expected

Figure 30-102   X-ray structure of the M.HhaI DNA methyltransferase in complex with *S*-adenosylhomocysteine and a dsDNA containing a methylated 5-fluorocytosine base at the enzyme's target site. The protein is represented by a semitransparent ribbon. The DNA is drawn in stick form colored according to atom type (C green, N blue, O red, and P orange). Its methylated 5-fluorocytosine residue (C atoms cyan) has swung out of the DNA helix into the enzyme's active site pocket, where its C6 forms a covalent bond with the S atom of an enzyme Cys residue (C atoms magenta and S yellow). The methyl group and a fluorine atom at C5 (which prevents the methylation reaction from going to completion) are represented by magenta and yellow-green spheres, respectively. The position of the flipped-out cytosine base in the DNA double helix is occupied by the side chain of a Gln residue (shown in space-filling form with C yellow), which hydrogen bonds to the "orphaned" guanine base. The *S*-adenosylmethionine, which has given up its methyl group, is drawn in stick form with its C atoms pink. [Based on an X-ray structure by Richard Roberts, New England Biolabs, Beverly, Massachusetts, and Xiaodong Cheng, Cold Spring Harbor Laboratory, New York. PDBid 1MHT.]

frequency. The upstream regions of many genes, however, have normal CG frequencies and are therefore known as **CpG islands.**

The degree of eukaryotic DNA methylation and its pattern are conveniently assessed by comparing the Southern blots (Section 5-5D) of DNA cleaved by the restriction endonucleases HpaII (which cleaves CCGG but not C-m⁵C-GG) and MspI (which cleaves both). Such studies indicate that eukaryotic DNA methylation varies with the species, the tissue, and the position along a chromosome.

The m^5C residues in a given DNA segment can be identified through **bisulfite sequencing,** in which the DNA is reacted with **bisulfite ion** (HSO_3^-), which selectively deaminates C (but not m^5C) residues to U, followed by PCR amplification (Section 5-5F), which copies these U's to T's and the m^5C's to C's. Comparison of the sequences of the amplified DNA with that of untreated DNA reveals which C's in the untreated DNA are methylated.

There is clear evidence that *DNA methylation switches off eukaryotic gene expression, particularly when it occurs in the promoter regions upstream of a gene's transcribed sequences.* For example, globin genes are less methylated in erythroid cells than they are in nonerythroid cells and, in fact, the specific methylation of the control region in a recombinant globin gene inhibits its transcription in transfected cells. In further support of the inhibitory effect of DNA methylation is the observation that **5-azacytosine (5-azaC),**

NH₂ structure

5-Azacytosine
(5-azaC)

a base analog that cannot be methylated at its N5 position and that inhibits DNA MTases, stimulates the synthesis of several proteins and changes the cellular differentiation patterns of cultured eukaryotic cells. The observation that repetitive intragenic parasites such as LINEs are highly methylated in somatic tissues has led to the hypothesis that CpG methylation in mammals arose to prevent the spurious transcriptional initiation of these retrotransposons.

The way in which DNA methylation prevents gene expression is poorly understood. One possibility is that DNA methylation can directly block the binding of transcriptional activators to their target sequences. However, in many cases, DNA methylation is recognized by a family of proteins that contain a conserved **methyl-CpG binding domain (MBD).** Since the methyl groups of m^5C residues extend into dsDNA's major groove, MBDs can bind to them without perturbing DNA's double helical structure. MBD-containing proteins inhibit the transcription of their bound promoter-methylated genes by recruiting protein complexes that induce the alteration of the local chromo-

some structure in a way that prevents the transcription of the associated gene (eukaryotic chromosome structure is discussed in Section 34-1). Another possibility has been raised by the observation that the methylation of synthetic poly(GC) stabilizes its Z-DNA conformation. Quite possibly, the formation of Z-DNA, which has been detected *in vivo* (Section 29-1Bb), acts as a conformational switch to turn off local gene expression.

c. DNA Methylation in Eukaryotes Is Self-Perpetuating

The palindromic nature of DNA methylation sites in eukaryotes permits the methylation pattern on a parental DNA strand to direct the generation of the same pattern in its daughter strand (Fig. 30-103). This **maintenance methylation** would result in the stable "inheritance" of a methylation pattern in a cell line and hence cause these cells to all have the same differentiated phenotype. Such changes to the genome are described as being **epigenetic** (Greek: *epi,*

Figure 30-103 **Maintenance methylation.** The pattern of methylation on a parental DNA strand induces the corresponding methylation pattern in the complementary strand. In this way, a stable methylation pattern may be maintained in a cell line.

upon or beside) because they provide an additional layer of information that specifies when and where specific portions of the otherwise fixed genome are expressed (an epigenetic change that we have already encountered is the lengthening of telomeres in germ cells; Section 30-4D). Epigenetic characteristics, as we shall see, are not bound by the laws of Mendelian inheritance.

There is considerable experimental evidence favoring the existence of maintenance methylation, including the observation that artificially methylated viral DNA, on transfection into eukaryotic cells, maintains its methylation pattern for at least 30 cell generations. Maintenance methylation in mammals appears to be mediated mainly by the protein **DNMT1** (for *DNA* methyl*transferase 1*), which has a strong preference for methylating hemimethylated substrate DNAs. In contrast, prokaryotic DNA MTases such as M.HhaI do not differentiate between hemimethylated and unmethylated substrate DNAs. The importance of maintenance methylation is demonstrated by the observation that mice homozygous for deletion of the *DNMT1* gene die early in embryonic development.

The interaction energy provided a single methyl group in the major groove of DNA seems insufficient to permit a protein to reliably differentiate between an m^5C and a C residue. Nevertheless, the **SRA** (for *SET* and *RING-associated*) domain of the protein **UHRF1** (for *ubiquitin*-like, containing *PHD* and *RING* finger domains *1*) does so and thereon recruits DNMT1 the site. How does the SRA domain distinguish a single m^5C residue from unmethylated C?

The X-ray structure of the 210-residue SRA domain in complex with a 12-bp DNA containing a centrally located hemimethylated CpG was independently determined by Masahiro Shirakawa and Sirano Dhe-Paganon. It reveals that the DNA's m^5C residue has flipped out from the minor groove side of the largely straight B-form DNA to tightly bind deep within in a protein pocket (Fig. 30-104). Two loops reach around the resulting gap in the DNA helix from the major and minor grooves to interact with the other three bases of the hemimethylated CpG site. The m^5C base is replaced in the double helix by the side chain of an Arg residue extending from the minor groove loop, which hydrogen-bonds to the "orphaned" G residue without greatly disturbing its conformation.

DNMT1 has both a catalytic domain and a UHRF1-binding domain. The catalytic domain is thought to have a similar structure and mode of DNA binding as bacterial MTases such as M.HhaI (Fig. 30-102). However, a model of the DNMT1 catalytic domain–DNA–SRA domain complex based on the structures of the M.HhaI–DNA and SRA–DNA complexes indicates that it is unlikely that both the m^5C and C residues of a double helical CpG island could be simultaneously flipped out of the DNA. This suggests that the binding of the UHRF1-binding domain of DNMT1 to the SRA–DNA complex causes the m^5C to flip back into the DNA double helix while the C to be methylated flips out to bind to the catalytic domain of DNMT1.

Figure 30-104   **X-ray structure of the SRA domain of mouse UHRF1 in complex with a hemimethylated 12-bp DNA.** The protein is drawn in ribbon form (*green*) embedded in its semitransparent molecular surface. The DNA is shown in stick form colored according to atom type (DNA C gray except for the m^5C residue C which is magenta, N blue, O red, and P orange) with successive P atoms in the same strand connected by orange rods. The methyl group of the m^5C residue is represented by a magenta sphere. An Arg side chain (C cyan) fills the space in the DNA double helix vacated by the flipped-out m^5C. [Based on an X-ray structure by Masahiro Shirakawa, Kyoto University, Japan. PDBid 2ZDK.]

The pattern of DNA methylation in mammals varies in early embryological development. DNA methylation levels are high in mature gametes (sperm and ova) but are nearly eliminated by the time a fertilized ovum has become a **blastocyst** (a hollow ball of cells, the stage at which the embryo implants into the uterine wall; embryonic development is discussed in Section 34-4A). After this stage, however, the embryo's DNA methylation levels globally rise until, by the time the embryo has reached the developmental stage known as a **gastrula,** its DNA methylation levels have risen to adult levels, where they remain for the lifetime of the animal. This *de novo* (new) methylation appears to be mediated by two DNA MTases distinct from DNMT1 named **DNMT3a** and **DNMT3b.** An important exception to this remethylation process is that the CpG islands of germline cells (cells that give rise to sperm or ova) remain unmethylated. This ensures the faithful transmission of the CpG islands to the succeeding generation in the face of the strong mutagenic pressure of m^5C deamination (which yields T, a mutation that mismatch repair occasionally fails to correct).

The change in DNA methylation levels (epigenetic reprogramming) during embryonic development suggests that the pattern of genetic expression differs in embryonic and somatic cells. This explains the observed high failure rate in cloning mammals (sheep, mice, cattle, etc.) by transferring the nucleus of an adult cell into an enucleated oocyte (immature ovum). Few of these animals survive to birth, many of those that do so die shortly thereafter, and most of the ~1% that do survive have a variety of abnormalities, most prominently an unusually large size. However, the survival of any embryos at all is indicative that the oocyte has the remarkable capacity to epigenetically reprogram somatic chromosomes (although it is rarely entirely successful in doing so) and that mammalian embryos are relatively tolerant of epigenetic abnormalities. Presumably, the reproductive cloning of humans from adult nuclei would result in similar abnormalities and for this reason (in addition to social and ethical prohibitions) should not be attempted.

d. Genomic Imprinting Results from Differential DNA Methylation

It has been known for thousands of years that maternal and paternal inheritance can differ. For example, a mule (the offspring of a mare and a male donkey) and a hinny (the offspring of a stallion and a female donkey) have obviously different physical characteristics, a hinny having shorter ears, a thicker mane and tail, and stronger legs than a mule. This is because, in mammals only, certain maternally and paternally supplied genes are differentially expressed, a phenomenon termed **genomic imprinting.** The genes that are subject to genomic imprinting are, as Rudolph Jaenisch has shown, differentially methylated in the two parents during gametogenesis and the resulting different methylation patterns are resistant to the wave of demethylation that occurs during the formation of the blastocyst and to the wave of *de novo* methylation that occurs thereafter.

The importance of genomic imprinting is demonstrated by the observation that an embryo derived from the transplantation of two male or two female pronuclei into an ovum fails to develop (pronuclei are the nuclei of mature sperm and ova before they fuse during fertilization). Inappropriate imprinting is also associated with certain diseases. For example, **Prader-Willi syndrome (PWS),** which is characterized by the failure to thrive in infancy, small hands and feet, marked obesity, and variable mental retardation, is caused by a >5000-kb deletion in a specific region of the paternally inherited chromosome 15. In contrast, **Angelman syndrome (AS),** which is manifested by severe mental retardation, a puppetlike ataxic (uncoordinated) gait, and bouts of inappropriate laughter, is caused by a deletion of the same region from the maternally inherited chromosome 15. These syndromes are also exhibited by those rare individuals who inherit both their chromosomes 15 from their mothers for PWS and from their fathers for AS. Evidently, certain genes on the deleted chromosomal region must be paternally inherited to avoid PWS and others must be maternally inherited to avoid AS. Several other human diseases are also associated with either maternal or paternal inheritance or lack thereof.

e. DNA Methylation Is Associated with Cancer

The mutation of an m^5C residue to T (with its associated G to A mutation on the complementary strand) is, by far, the most prevalent mutational change in human cancers. Such mutations usually convert proto-oncogenes to oncogenes (Section 19-3B) or inactivate tumor suppressors (Section 34-4Ca). In addition, the hypomethylation of proto-oncogenes and the hypermethylation of genes encoding tumor suppressors are associated with cancers, although it is unclear whether these are initiating or consolidating events for malignancies.

f. Several Neurological Diseases Are Associated with Trinucleotide Repeat Expansions

Fragile X syndrome, whose major symptoms include mental retardation and a characteristic long, narrow face with large ears, afflicts 1 in 4500 males and 1 in 9000 females. Fragile X syndrome is so named because, in affected individuals, the tip of the X chromosome's long arm is connected to the rest of the chromosome by a slender thread that is easily broken. The genetics of this condition are bizarre. The maternal grandfathers of individuals having fragile X syndrome may be asymptomatic, both clinically and cytogenetically. Their daughters are likewise asymptomatic, but these daughters' children of either sex may have the syndrome. Evidently, the fragile X defect is activated by passage through a female. Moreover, the **penetrance** of the disease, that is, the probability of a child having fragile X syndrome, and its severity, increase with each succeeding generation, a phenomenon termed **genetic anticipation.**

The affected gene in fragile X syndrome, *FMR1* (for *f*ragile X *m*ental *r*etardation *1*), encodes a 632-residue RNA-binding protein named **FMRP** (for *FMR* protein), which apparently functions in the transport of certain mRNAs from the nucleus to the cytoplasm (Section 34-3C), where it probably regulates their translation. FMRP, which is highly conserved in vertebrates, is expressed in most tissues but most heavily in brain neurons, where a variety of evidence indicates that its participation is required for the proper formation and/or function of synapses.

In the general population, the 5′ untranslated region of *FMR1* contains a polymorphic $(CGG)_n$ sequence with n ranging from 6 to 60 and often punctuated by one or two AGG interruptions. However, in certain asymptomatic individuals, n has increased from 60 to 200, a so-called premutation that males transmit in unchanged form to their daughters (they transmit a Y rather than an X chromosome to their sons). In the daughters' children, however, ~80% of the individuals inheriting a premutant *FMR1* gene exhibit an astonishing expansion (amplification) of the triplet repeat with n ranging from >200 to several thousand, as well as the symptoms of the disease, a so-called full mutation. These triplet repeats differ in size among siblings and often exhibit heterogeneity within an individual, suggesting that they are somatically generated.

These **dynamic mutations,** which expand more often than they contract, perhaps arise through slippage of the DNA during replication. One way that slippage might occur

(a)

(b)

Figure 30-105   The loop-out mechanism for the alteration of the number of consecutive triplet repeats in DNA through its replication. Here the template strand is red and its nascent (newly synthesized) daughter strand is blue. With long tracts of repeating sequences, the probability of a loop-out occurring increases because its neighboring sequences will remain base-paired. The loop-out of a $(CNG)_n$ repeat is stabilized by its ability to form a partially based-paired structure. (*a*) If the daughter strand loops out, the number of repeats increases. (*b*) If the template strand loops out, the number of repeats decreases.

is through the formation of loop-outs (Fig. 30-105) on either the newly synthesized strand (causing expansions) or on the template strand (causing contractions). Such $(CNG)_n$ loops are likely stabilized by their ability to form imperfectly base-paired stems. As expected, the frequency of slippage increases with the number of repeats.

The peculiar genetics of fragile X syndrome is a consequence of genomic imprinting through methylation. The *FMR1* gene is unmethylated in normal individuals. However, it is hypermethylated in individuals with a maternally transmitted full mutation. This maintains the *FMR1* gene in a transcriptionally silent (inactive) state, thereby accounting for the symptoms of the disease. The lesser frequency and severity of fragile X syndrome in females are accounted for by the fact that females have two X chromosomes, one of which is unlikely to be mutated.

Over 30 other pathological instances of the expansion of DNA repeats, most of which are also neurological diseases, are known including the following:

1. Myotonic dystrophy (DM), the most common adult form of **muscular dystrophy** (its estimated incidence is 1 in 8000). It is a multisystem autosomal dominant disorder that is mainly characterized by progressive muscle weakness and wasting, the severity of which increases with successive generations while the age of onset decreases (genetic anticipation). Its most severe form, congenital DM, is

exclusively maternally transmitted. DM arises from a trinucleotide expansion in the 3' untranslated region of the gene encoding **myotonic dystrophy protein kinase (MDPK),** which is expressed in the neurons affected by DM. The repeating triplet, $(CAG)_n$, is present in between 5 and 30 copies in the MDPK gene of normal individuals but expands from at least 50 repeats in minimally affected individuals to ~2000 repeats in severely affected individuals.

2. Huntington's disease (HD; previously called **Huntington's chorea),** a devastating neurodegenerative disorder characterized by progressively choreic (disordered) movements, cognitive decline, and emotional disturbances over an average 18-year course that is inevitably fatal. This dominant autosomal disorder, which affects ~1 in 10,000 individuals and has an average age of onset of ~40 years, is a consequence of the selective loss of certain groups of neurons in the brain. The *HD* gene, which encodes a widely expressed 3145-residue protein of unknown function named **huntingtin,** contains a polymorphic trinucleotide repeat, $(CAG)_n$, within its polypeptide coding sequence. The *HD* genes from 150 independent families with HD all contained between 37 and 86 repeat units, whereas those from normal individuals had 11 to 34 repeats. Moreover, the *HD* repeat length is unstable: >80% of meiotic transmissions show either increases or decreases, with the largest increases occurring in paternal transmissions (genomic imprinting). The number of repeats in afflicted individuals is inversely correlated with the age of onset of HD.

CAG is the codon for Gln (Table 5-3) and hence mutant huntingtin contains a long poly(Gln) tract. Synthetic poly(Gln) aggregates as β sheets that are linked by hydrogen bonds involving both their main chain and side chain amide groups. Indeed, the nuclei of HD-affected neurons contain inclusions that presumably consist of aggregates of huntingtin or its proteolytic products. It is these inclusions, as Max Perutz pointed out, that apparently kill the neurons in which they are contained, although the mechanism of how they do so is unknown. The long incubation period before the symptoms of HD become evident is attributed to the lengthy nucleation time for aggregate formation, much as occurs in the formation of amyloid fibrils (Section 9-5A).

3. Spinocerebellar ataxia (SCA) type 1, a progressive neurodegenerative disease whose age of onset is typically in the third or fourth decade, although it exhibits genetic anticipation. Like HD, it is caused by selective neuronal loss and is associated with an expansion of a CAG repeat in a coding region, in this case of a neuronal protein named **ataxin-1.** There it expands from ~28 to between 43 and 81 copies, thereby yielding a poly(Gln) tract of increased length (and tendency to aggregate). Four similar diseases, **SCA types 2, 3, 6,** and **7,** are caused by $(CAG)_n$ expansions in different neuronal proteins.

CHAPTER SUMMARY

1 and 2   DNA Replication   DNA is replicated in the 5′ → 3′ direction by the assembly of deoxynucleoside triphosphates on complementary DNA templates. Replication is initiated by the generation of short RNA primers, as mediated in *E. coli* by primase and RNA polymerase. The DNA is then extended from the 3′ ends of the primers through the action of a DNA polymerase (Pol III in *E. coli*). The leading strand at a replication fork is synthesized continuously, whereas the lagging strand is synthesized discontinuously by the formation of Okazaki fragments. RNA primers on newly synthesized DNA are excised and replaced by DNA through Pol I-catalyzed (in *E. coli*) nick translation. The single-strand nicks are then sealed by DNA ligase. Mispairing errors during DNA synthesis are corrected by the 3′ → 5′ exonuclease functions of both Pol I and Pol III. The Klenow fragment of Pol I and other DNA polymerases of known structure have a right-hand-like structure with the active site located in the palm domain. Pol I recognizes the incoming nucleotide according to the shape of the base pair it forms with the template base and catalyzes the formation of a phosphodiester bond via a mechanism involving two metal ions. DNA synthesis by *E. coli* Pol III requires the participation of many auxiliary proteins including DNA gyrase, DnaB helicase, single-strand binding protein (SSB), primase, the β_2 sliding clamp, and DNA ligase.

3   Prokaryotic Replication   DNA synthesis commences from specific sites known as replication origins. In the synthesis of the bacteriophage M13 (−) strand on the (+) strand template, the origin is recognized and primer synthesis is initiated by RNA polymerase. The analogous process in bacteriophage φX174, as well as in *E. coli*, is mediated by a complex primase-containing particle known as the primosome. φX174 (+) strands are synthesized according to the looped rolling circle mode of DNA replication on (−) strand templates of the replicative form in a process that is directed by the virus-specific gene *A* protein.

The *E. coli* chromosome is bidirectionally replicated in the θ mode from a single origin, *oriC*. It is recognized by DnaA protein, which together with DnaC protein, loads the DnaB helicase onto the developing replication forks. Leading strand synthesis is probably primed by RNA polymerase and primase working together, whereas Okazaki fragments are primed by primase in the primosome. The uncontrolled initiation of DNA replication is prevented by the sequestration of newly synthesized and hence hemimethylated *oriC* by membrane-associated SeqA protein, which prevents the *oriC* from becoming fully methylated at its multiple GATC sites. The β_2 sliding clamp, which is responsible for Pol III's processivity, is loaded onto the DNA by the $\gamma_3\delta\delta'$ clamp loader in an ATP-driven process. The δ subunit, when unmasked by ATP binding to the γ subunits, acts as a molecular "wrench" to spring open the sliding clamp, thereby permitting the entry of a single-stranded template DNA. Replication termination is facilitated by Tus protein which, on binding to an appropriately oriented *Ter* site, arrests the motion of a replication fork by binding to DnaB helicase. The great complexity of the DNA replication process functions to ensure the enormous fidelity necessary to maintain genome integrity.

4   Eukaryotic Replication   Progression through the eukaryotic cell cycle is mediated by cyclins complexed to their cognate cyclin-dependent protein kinases (Cdks). Chromosomal DNA replication is initiated by pol α/primase, which synthesizes a primer followed by a short length of DNA. Then, via polymerase switching mediated by replication factor C (RFC), the eukaryotic clamp loader, pol δ processively synthesizes both the lagging and leading strands in complex with PCNA, the eukaryotic sliding clamp.

Eukaryotic chromosomal DNA is synthesized in multiple origin-containing segments known as replicons. Nevertheless, chromosomal DNA is synthesized once and only once per cell cycle. The re-replication of DNA is prevented because replication initiation is licensed only in the G_1 phase of the cell cycle by the formation of the pre-replicative complex (pre-RC) but DNA is synthesized only in S phase by the activation of the pre-RC. The pre-RC is assembled in G_1 phase by the binding of the origin recognition complex (ORC) to an origin, which recruits Cdc6/Cdc18 and Cdt1 followed by the MCM complex, the replicative helicase. The activation of the pre-RC begins in S phase with the addition of Mcm10 followed by the phosphorylation of many of the pre-RC's subunits by Cdks and Ddk. Cdc45 then binds followed by pol α/primase, pol ε, PCNA, and the replication protein A (RPA), the SSB counterpart, to yield the active initiation complex. Re-replication is prevented through the actions of Cdks, which cause the elimination of Cdc6/Cdc18 and inhibit the helicase activity of the MCM complex. In metazoan cells, re-replication is also prevented by the binding of geminin to Cdt1. Mitochondrial DNA is replicated by polymerase γ via a process in which the Okazaki fragments are entirely RNA.

Retroviruses produce DNA on RNA templates in a reaction sequence catalyzed by reverse transcriptase. Telomeric DNA, a G-rich repeating octamer on the 3′-ending strand, is synthesized by the RNA-containing enzyme telomerase. Telomerase is active in germ cells but not somatic cells, a phenomenon that is at least in part responsible for cellular senescence and aging. The observation that telomerase is active in nearly all cancer cells suggests that telomerase inactivation is a defense against the development of cancer. The free DNA ends of telomeres are capped to prevent them from triggering DNA damage checkpoints. The *O. nova* telomere heterodimer TEBP binds both single strands of telomere DNA and a G–quartet-containing dimer. Its counterparts in humans and yeast are POT1 and TPP1. Telomeric DNA forms T-loops which are formed by the TRF2-mediated invasion of the 3′ telomeric overhang into repeating telomeric dsDNA to form a D-loop. POT1, TPP1, TRF1, together with TRF2, TIN2, and RAP1, form a complex on the telomere called shelterin that functions to protect telomeres from DNA repair processes and limit the size of the telomere.

5   Repair of DNA   Cells have a great variety of DNA repair mechanisms. DNA damage may be directly reversed such as in the photoreactivation of UV-induced pyrimidine dimers or in the repair of O^6-alkylguanine lesions by the transfer of the offending alkyl group to a repair protein. Pyrimidine dimers, as well as many other types of DNA lesions, may also be removed by nucleotide excision repair (NER), which in *E. coli* involves the UvrABC system. Xeroderma pigmentosum, an inherited human disease characterized by marked UV-induced skin changes and a greatly increased incidence of cancer, is caused by defects in any of seven complementation groups that participate in NER. Cockayne syndrome is associated with defects in the proteins responsible for transcription-coupled repair (TCR), which repairs lesions in actively transcribing genes. In base excision repair (BER), DNA glycosylases specifically remove the corresponding chemically altered bases, including uracil, through mechanisms that involve base flip-outs to form AP sites. The AP sites are cleaved on one side by an AP endonuclease, removed together with adjacent residues by an exonuclease, and replaced through the actions of a DNA polymerase and a DNA ligase. In mismatch repair (MMR), base-pairing mismatches arising from replication errors are corrected. In *E. coli* MMR, MutS and MutL bind to the mismatch and then identify the daughter strand, which contains the error, according to which strand of the nearest hemimethylated GATC palindrome is unmethylated. MutH then cleaves this strand, which is excised past the mismatch and replaced.

DNA damage in *E. coli* induces the SOS response, a LexA- and RecA-mediated process in which the error-prone bypass DNA polymerases Pol IV and Pol V replicate a damaged template DNA even if it provides no information as to which base to incorporate. Double-strand break (DSB) repair by nonhomologous end-joining (NHEJ) is facilitated by Ku protein, which holds two dsDNA ends together for ligation by DNA ligase IV in complex with Xrcc4. The high correlation between mutagenesis and carcinogenesis permits the detection of carcinogens by the Ames test.

6 Recombination and Mobile Genetic Elements Genetic information may be exchanged between homologous DNA sequences through homologous recombination, a process that occurs according to the Holliday model. In *E. coli*, strand invasion to form Holliday junctions is mediated by RecA after the RecBCD-mediated generation of the single-strand nicks to which RecA binds. Branch migration is mediated by RuvAB, which consists of a homotetramer (or a homooctamer) of RuvA, which binds both a Holliday junction and two oppositely located RuvB hexamers. In an ATP-driven process, the RuvB hexamers (in effect) counter-rotate to pump the dsDNA stems into the center of the RuvA-bound Holliday junction, where each of its single strands exchange base-pairing partners to form new dsDNA stems, which are translocated toward the periphery of the complex. The Holliday junction is eventually resolved to its component dsDNAs by RuvC and sealed by DNA ligase.

The primary function of homologous recombination is to repair damaged replication forks resulting from the encounters of replisomes with unrepaired single-strand lesions or breaks. DSBs may be rejoined via a recombination repair process called homologous end-joining.

Chromosomes and plasmids may be rearranged through the action of transposons. These DNA segments carry the genes that encode the proteins that mediate the transposition process as well as other genes. Tn5 transposase catalyzes the cut-and-paste transposition of the Tn5 transposon. Replicative transposition proceeds via the intermediacy of cointegrates, which are resolved through the action of enzymes such as the γδ resolvase. Transposition may be important in chromosomal and plasmid evolution and has been implicated in the control of phenotypic expression such as phase alternation in *Salmonella*, a process that is catalyzed by the Hin DNA invertase, a homolog of the γδ resolvase. Members of the λ integrase family of proteins, such as Cre recombinase, insert dsDNA segments into their target sites via a Holliday junction intermediate in which transient covalent bonds are formed between active site Tyr side chains and the 3′-OH groups at the cleavage sites. Retrotransposons undergo transposition through an RNA intermediate. Many retrotransposons, such as yeast Ty1, are "internal" retroviruses that can only replicate within a genome. Nonviral retrotransposons, such as LINEs, the dominant transposons in the human genome, have a different transpositional mechanism.

7 DNA Methylation and Trinucleotide Repeat Expansions Prokaryotic DNA may be methylated at its A or C bases. This prevents the action of restriction endonucleases and permits the correct mismatch repair of newly replicated DNA. In most eukaryotes, DNA methylation, which occurs, mainly at CpG islands, through the formation of m^5C, has been implicated in the control of gene expression and, via maintenance methylation, in genomic imprinting.

Over 30 inherited, mostly neurological diseases, including fragile X syndrome, myotonic dystrophy, and Huntington's disease, are characterized by the genetically bizarre expansion of segments of $(CNG)_n$ triplets. If an expanded triplet repeat occurs in an upstream noncoding region of a gene, its aberrant methylation, perhaps through genomic imprinting, may lead to the gene's transcriptional silencing; if the expanded repeat is instead manifested as a poly(Gln) tract in a protein, the resulting protein aggregates may kill the neurons in which it occurs.

REFERENCES

General

Alberts, B., Johnson, A., Lewis, J., Raff, M., Roberts, K., and Walter, P., *The Molecular Biology of the Cell* (5th ed.), Chap. 5, Garland Science (2008).

Kornberg, A., *For Love of Enzymes: The Odyssey of a Biochemist,* Harvard University Press (1989). [A scientific autobiography.]

Kornberg, A. and Baker, T.A., *DNA Replication* (2nd ed.), Freeman (1992). [A compendium of information about DNA replication whose first author is the founder of the field.]

Watson, J.D., Baker, T.A., Bell, S.P., Gann, A., Levine, M., and Losick, R., *Molecular Biology of the Gene* (6th ed.), Chapters 8–11, Cold Spring Harbor Laboratory Press (2008).

Prokaryotic DNA Replication

Bailey, S., Eliason, W.K., and Steitz, T.A., Structure of hexameric DnaB helicase and its complex with a domain of DnaG protein, *Science* **318,** 459–463 (2007).

Beese, L.S., Derbyshire, V., and Steitz, T.A., Structure of DNA polymerase I Klenow fragment bound to duplex DNA, *Science* **260,** 352–355 (1993).

Benkovic, J.J., Valentine, A.M., and Salinas, F., Replisome-mediated DNA replication, *Annu. Rev. Biochem.* **70,** 181–208 (2001).

Carr, K.M. and Kaguni, J.M., Stoichiometry of DnaA and DnaB protein in initiation at the *Escherichia coli* chromosomal origin, *J. Biol. Chem.* **276,** 44919–44925 (2001).

Caruthers, J.M. and McKay, D.B., Helicase structure and mechanism, *Curr. Opin. Struct. Biol.* **12,** 123–133 (2002).

Corn, J.E. and Berger, J.M., Regulation of bacterial priming and daughter strand synthesis through helicase–primase interactions, *Nucleic Acids Res.* **34,** 4082–4087 (2006).

Corn, J.E., Pelton, J.G., and Berger, J.M., Identitication of a DNA primase template tracking site redefines the geometry of primer synthesis, *Nature Struct. Mol. Biol.* **15,** 16–169 (2008).

Davey, M.J., Jeruzalmi, D., Kuriyan, J., and O'Donnell, M., Motors and switches: AAA$^+$ machines within the replisome, *Nature Rev. Mol. Cell Biol.* **3,** 1–10 (2002).

Doublié, S., Sawaya, M.R., and Ellenberger, T., An open and closed case for all polymerases, *Structure* **7,** R31–R35 (1999). [Reviews the mechanisms of DNA polymerases.]

Duerstadt, K.E. and Berger, J.M., AAA + ATPases in the initiation of DNA replication, *Crit. Rev. Biochem. Mol. Biol.* **43,** 163–187 (2008).

Enemark, E.J. and Joshua-Tor, L., Mechanism of DNA translocation in a replicative hexameric helicase, *Nature* **442,** 270–275 (2006).

Frick, D.N. and Richardson, C.C., DNA primases, *Annu. Rev. Biochem.* **70,** 39–80 (2001).

Georgescu, R.E., Kim, S.-S., Yurieva, O., Kuriyan, J., Kong, X.-P., and O'Donnell, M., Structure of a sliding clamp on DNA, *Cell* **132,** 43–54 (2008).

Hamdan, S.M. and Richardson, C.C., Motors, switches, and contacts in the replisone, *Annu. Rev. Biochem.* **78,** 205–243 (2009).

Jeruzalmi, D., O'Donnell, M., and Kuriyan, J., Clamp loaders and sliding clamps, *Curr. Opin. Struct. Biol.* **12,** 217–224 (2002).

Kaguni, J.M., DnaA: Controlling the initiation of bacterial DNA replication and more, *Annu. Rev. Microbiol.* **60,** 351–371 (2006).

Kamada, K., Horiuchi, T., Ohsumi, K., Shimamoto, N., and Morikawa, K., Structure of a replication–terminator protein complexed with DNA, *Nature* **383,** 598–603 (1996).

Keck, J.L., Roche, D.D., Lynch, A.S., and Berger, J.M., Structure of the RNA polymerase domain of *E. coli* primase, *Science* **287,** 2482–2486 (2000); *and* Podobnik, M., McInerney, P., O'Donnell, M., and Kuriyan, J., A TOPRIM domain in the crystal structure of the catalytic core of *Escherichia coli* primase confirms a structural link to DNA topoisomerases, *J. Mol. Biol.* **300,** 353–362 (2000).

Kim, Y., Eom, S.H., Wang, J., Lee, D.-S., Suh, S.W., and Steitz, T.A., Crystal structure of *Thermus aquaticus* DNA polymerase, *Nature* **376,** 612–616 (1995).

Kong, X.-P., Onrust, R., O'Donnell, M., and Kuriyan, J., Three-dimensional structure of the β subunit of *E. coli* DNA polymerase III holoenzyme: A sliding DNA clamp, *Cell* **69,** 425–437 (1992).

Kool, E.T., Active site tightness and substrate fit in DNA replication, *Annu. Rev. Biochem.* **71,** 191–219 (2002); *and* Hydrogen-bonding, base stacking, and steric effects in DNA replication, *Annu. Rev. Biophys. Biomol. Struct.* **30,** 1–2 (2001).

Korolev, S., Hsieh, J., Gauss, G.H., Lohman, T.M., and Waksman, G., Major domain swiveling revealed by the crystal structures of complexes of *E. coli* Rep helices bound to single-stranded DNA and ADP, *Cell* **90,** 635–647 (1997).

Kunkel, T.A. and Bebenek, K., DNA replication fidelity, *Annu. Rev. Biochem.* **69,** 497–529 (2000).

Lamers, M.H., Georgescu, R.E., Lee, S.-G., O'Donnell, M., and Kuriyan, J., Crystal structure of the catalytic α subunit of *E. coli* replicative DNA polymerase III, *Cell* **126,** 881–892 (2006); *and* Bailey, S., Wing, R.A., and Steitz, T.A., The structure of *T. aquaticus* polymerase III is distinct from eukaryotic replicative DNA polymerases, *Cell* **126,** 893–904 (2006).

Lee, J.Y., Chang, C., Song, H.K., Moon, J., Yang, J.K., Kim, H.-K., Kwon, S.-T., and Suh, S.W., Crystal structure of NAD$^+$-dependent DNA ligase: modular architecture and functional implications, *EMBO J.* **19,** 1119–1129 (2000).

Li, Y., Korolev, S., and Waksman, G., Crystal structures of open and closed forms of binary and ternary complexes of the large fragment of *Thermus aquaticus* DNA polymerase I: structural basis for nucleotide incorporation, *EMBO J.* **17,** 7514–7525 (1998).

Mott, M.L. and Berger, J.M., DNA replication initiation: mechanisms and regulation in bacteria, *Nature Rev. Microbiol.* **5,** 343-354 (2007).

Mott, M.L., Erzberger, J.P., Coons, M.M., and Berger, J.M., Structural synergy and molecular crosstalk between bacterial helicase loaders and replication initiators, *Cell* **134,** 623–634 (2008). [The X-ray structure of DnaC protein.]

Mulcair, M.D., Schaeffer, P.M., Oakley, A.J., Cross, H.F., Neylon, C., Hill, T.M., and Dixon, N.E., A molecular mousetrap determines polarity of termination of DNA replication in *E. coli,* *Cell* **125,** 1309–1319 (2006).

Nandakumar, J., Nair, P.A., and Shuman, S., Last stop on the road to repair: Structure of *E. coli* DNA ligase bound to nicked DNA-adenylate, *Mol. Cell* **26,** 257–271 (2007); *and* Shuman, S., DNA ligase: progress and prospects, *J. Biol. Chem.* **284,** 17365–17369 (2009).

O'Donnell, M., Replisome architecture and dynamics in *E. coli, J Biol. Chem.* **281,** 10653–10656 (2006); *and* Johnson, A. and O'Donnell, M., Cellular DNA replicases: Components and dynamics at the replication fork, *Annu. Rev. Biochem.* **74,** 283–315 (2005).

Patel, S.S. and Donmez, I., Mechanisms of helicases, *J. Biol. Chem.* **281,** 18265–18268 (2006); *and* Patel, S.S. and Picha, K.M., Structure and function of hexameric helicases, *Annu. Rev. Biochem.* **69,** 651–697 (2000).

Raghunathan, S., Kozlov, A.G., Lohman, T.M., and Waksman, G., Structure of the DNA binding domain of *E. coli* SSB bound to ssDNA, *Nature Struct. Biol.* **7,** 648–652 (2000).

Rothwell, P.J. and Waksman, G., Structure and mechanism of DNA polymerases, *Adv. Prot. Chem.* **71,** 401–440 (2005).

Simonetta, K.R., Kazmirski, S.L., Goedken, E.R., Cantor, A.J., Kelch, B.A., McNally, R., Seyedin, S.N., Makino, D.L., O'Donnell, M., and Kuriyan, J., The mechanism of ATP-dependent primer-template recognition by a clamp loader complex, *Cell* **137,** 659–671 (2009).

Singleton, M.R., Dillingham, M.S., and Wigley, D.B., Structure and mechanism of helicases and nucleic acid translocases, *Annu. Rev. Biochem.* **76,** 23–50 (2007).

Singleton, M.R., Sawaya, M.R., Ellenberger, T., and Wigley, D.B., Crystal structure of T7 gene 4 ring helicase indicates a mechanism for sequential hydrolysis of nucleotides, *Cell* **101,** 589–600 (2000).

Soultanas, P. and Wigley, D.B., Unwinding the 'Gordian knot' of helicase action, *Trends Biochem. Sci.* **26,** 47–54 (2001).

Watson, J.D. and Crick, F.H.C., Genetical implications of the structure of deoxyribonucleic acid, *Nature* **171,** 964–967 (1953). [The paper in which semiconservative DNA replication was first postulated.]

Eukaryotic DNA Replication

Allsopp, R.C., Vaziri, H., Patterson, C., Goldstein, S., Younglai, E.V., Futcher, A.B., Greider, C.W., and Harley, C.B., Telomere length predicts replicative capacity of human fibroblasts, *Proc. Natl. Acad. Sci.* **89,** 10114–10118 (1992).

Arezi, B. and Kuchta, R.D., Eukaryotic DNA primase, *Trends Biochem. Sci.* **25,** 572–576 (2000).

Armanios, M., Syndromes of telomere shortening, *Annu. Rev. Genomics Hum. Genet.* **10,** 45–61 (2009).

Autexier, C. and Lue, N.F., The structure and function of telomerase reverse transcriptase, *Annu. Rev. Biochem.* **75,** 493–517 (2006).

Bell, S.P. and Dutta, A., DNA replication in eukaryotic cells, *Annu. Rev. Biochem.* **71,** 333–374 (2002).

Blackburn E.H., Switching and signaling at the telomere, *Cell* **106,** 661–673 (2001); *and* Telomere states and cell fates, *Nature* **408,** 53–56 (2000).

Blackburn E.H., Greider, C., and Szostak, J.W., Telomeres and telomerase: the path from maize, *Tetrahymena* and yeast to human cancer and aging, *Nature Med.* **12,** 1133–1138 (2006). [A scientific memoir.]

Blow, J.J. and Dutta, A., Preventing re-replication of chromosomal DNA, *Nature Rev. Mol. Cell Biol.* **6,** 476–486 (2005).

Bowman, G.D., O'Donnell, M., and Kuriyan, J., Structural analysis of a eukaryotic sliding clamp–clamp loader complex, *Nature* **429,** 724–730 (2004).

Burger, P.M.J., Polymerase dynamics at the eukaryotic replication fork, *J. Biol. Chem.* **284,** 4041–4045 (2009). [Discusses the roles of pol δ, pol ε and FEN1.]

Cech, T.R., Life at the end of the chromosome: Telomeres and telomerase, *Angew. Chem.* **39,** 34–43 (2000).

Croy, J.E. and Wuttke, D.S., Themes in ssDNA recognition by telomere-end protection proteins, *Trends Biochem. Sci.* **31,** 516–525 (2006).

de Lange, T., Lundblad, V., and Blackburn, E. (Eds.), *Telomeres* (2nd ed.), Cold Spring Harbor Laboratory Press (2006).

DePamphilis, M.L., Replication origins in metazoan chromosomes: fact or fiction, *BioEssays* **21,** 5–16 (1999).

Diffley, J.F.X., DNA replication: Building the perfect switch, *Curr. Biol.* **11,** R367–R370 (2001).

Ding, J., Das, K., Hsiou, Y., Sarafianos, S.G., Clark, A.D., Jr., Jacobo-Molina, A., Tantillo, C., Hughes, S.H., and Arnold, E., Structure and functional implications of the polymerase active site region in a complex of HIV-1 RT with a double-stranded DNA template-primer and an antibody Fab fragment at 2.8 Å resolution, *J. Mol. Biol.* **284,** 1095–1111 (1998).

Franklin, M.C., Wang, J., and Steitz, T.A., Structure of the replicating complex of a pol α family DNA polymerase, *Cell* **105,** 657–667 (2001).

Ellenberger, T. and Tomkinson, A.E., Eukaryotic DNA ligases: structural and functional insights, *Annu. Rev. Biochem.* **77,** 313–338 (2008).

Garg, P. and Burgers, P.M.J., DNA polymerases that propagate the eukaryotic replication fork, *Crit. Rev. Biochem. Mol. Biol.* **40,** 115–128 (2005).

Gilbert, D.M., Making sense out of eukaryotic DNA replication origins, *Science* **294,** 96–100 (2001).

Gillis, A.J., Schuller, A.P., and Skordalakes, E., Structure of the *Tribolium castaneum* telomerase catalytic subunit TERT, *Nature* **455,** 633–637 (2008).

Gilson, E. and Géli, V., How telomeres are replicated, *Nature Rev. Mol. Cell Biol.* **8,** 825–838 (2007).

Griffith, J.D., Comeau, L., Rosenfield, S., Stansel, R.M., Bianchi, A., Moss, H., and de Lange, T., Mammalian telomeres end in a large duplex loop, *Cell* **97,** 503–514 (1999).

Hahn, W.C., Counter, C.M., Lundberg, A.S., Beijersbergen, R.L., Brooks, M.W., and Weinberg, R.A., Creation of human tumour cells with defined genetic elements, *Nature* **400,** 464–468 (1999).

Harrison, J.C. and Haber, J.E., Surviving the breakup: the DNA damage checkpoint, *Annu. Rev. Genet.* **40,** 209–235 (2006).

Holt, I.J., Mitochondrial DNA replication and repair: all a flap, *Trends Biochem. Sci.* **34,** 358–365 (2009).

Horvath, M.P. and Schultz, S.C., DNA G-quartets in a 1.86 Å resolution structure of an *Oxytricha nova* telomeric protein–DNA complex, *J. Mol. Biol.* **310,** 367–377 (2001).

Hübscher, U., Maga, G., and Spadari, S., Eukaryotic DNA polymerases, *Annu. Rev. Biochem.* **71,** 133–163 (2002).

Johansson, E. and MacNeil, S.A., The eukaryotic replicative DNA polymerase takes shape, *Trends Biochem. Sci.* **35,** 339–347 (2010).

Kelleher, C., Teixeira, M.T., Förstemann, K., and Lingner, J., Telomerase: Biochemical considerations for enzyme and substrate, *Trends Biochem. Sci.* **27,** 572–579 (2002).

Kelly, T.J. and Brown, G.W., Regulation of chromosome replication, *Annu. Rev. Biochem.* **69,** 829–880 (2000).

Lansdorp, P.M., Major cutbacks at chromosome ends, *Trends Biochem. Sci.* **30,** 388–395 (2005) [Reviews mechanisms of telomere erosion.]

Liu, Y., Kao, H.-I., and Bambara, R.A., Flap endonuclease 1: A central component of DNA metabolism, *Annu. Rev. Biochem.* **73,** 589–615 (2004).

Machida, Y.J., Hamlin, J.L., and Dutta, A., Right place, right time, and only once: Replication initiation in metazoans, *Cell* **123,** 13–24 (2005); *and* Machida, Y.J. and Dutta, A., Cellular checkpoint mechanisms monitoring proper initiation of DNA replication, *J. Biol. Chem.* **280,** 6253–6256 (2005).

Masai, H., Matsumoto, S., You, Z., Yoshizawa-Sugata, N., and Oda, M., Eukaryotic chromosome DNA replication: Where, when, and how? *Annu. Rev. Biochem.* **79,** 89–130 (2010).

McCulloch, S.D. and Kunkel, T.A., The fidelity of DNA synthesis by eukaryotic replicative and translesion synthesis polymerases, *Cell Res.* **18,** 148–161 (2008).

McEachern, M.J., Krauskopf, A., and Blackburn, E.H., Telomeres and their control, *Annu. Rev. Genet.* **34,** 331–358 (2000).

Moldovan, G.-L., Pfander, B., and Jentsch, S., PCNA, the maestro of the replication fork, *Cell* **129,** 665–679 (2007).

Neidle, S. and Parkinson, G., Telomere maintenance as a target for anticancer drug discovery, *Nature Rev. Drug Discov.* **1,** 383–393 (2002); *and* The structure of telomeric DNA, *Curr. Opin. Struct. Biol.* **13,** 275 (2003).

Osterhage, J.L. and Friedman, K.L., Chromosome end maintenance by telomerase, *J. Biol. Chem.* **284,** 16061–16065 (2009).

O'Sullivan, R.J. and Karlseder, J., Telomeres: protecting chromosomes against genome instability, *Nature Rev. Mol. Cell Biol.* **11,** 171–181 (2010).

Pursell, Z.F., Isoz, I., Lundström, E.-B., Johansson, E., and Kunkel, T.A., Yeast DNA polymerase ε participates in leading-strand replication, *Science* **317,** 127–130 (2007).

Riethman, H., Human telomere structure and biology, *Annu. Rev. Genomics Hum. Genet.,* **9,** 1–19 (2008).

Schultze, P., Smith, F.W., and Feigon, J., Refined solution structure of the dimeric quadruplex formed from the *Oxytricha* telomeric oligonucleotide d(GGGGTTTTGGGG), *Structure* **2,** 221–233 (1994).

Smogorzewska, A., and de Lange, T., Regulation of telomerase by telomeric proteins, *Annu. Rev. Biochem.* **73,** 177–208 (2004).

Swan, M.K., Johnson, R.E., Prakash, L., Prakash, S., and Aggarwal, A.K., Structural basis of high-fidelity DNA synthesis by yeast DNA polymerase δ, *Nature Struct. Mol. Biol.* **16,** 979–986 (2009).

Tye, B.K. and Sawyer, S., The hexameric eukaryotic MCM helicase: building symmetry from nonidentical parts, *J. Biol. Chem.* **275,** 34833–34836 (2000); *and* Tye, B.K., MCM proteins in DNA replication, *Annu. Rev. Biochem.* **68,** 649–686 (1999).

Urquidi, V., Tarin, D., and Goddison, S., Role of telomerase in cell senescence and oncogenesis, *Annu. Rev. Med.* **51,** 65–79 (2000).

Verdun, R.E. and Karlseder, J., Replication and protection of telomeres, *Nature* **447,** 924–931 (2007).

Repair of DNA

Ames, B.N., Identifying environmental chemicals causing mutations and cancer, *Science* **204,** 587–593 (1979).

Beckman, K.B. and Ames, B.N., Oxidative decay of DNA, *J. Biol. Chem.* **272,** 19633–19636 (1997).

Broyde, S., Wang, L., Rechkoblit, O., Geacintov, N.E., and Patel, D.J., Lesion processing: high-fidelity versus lesion-bypass DNA polymerases, *Trends Biochem. Sci.* **33,** 209–219 (2008).

David, S.S., O'Shea, V.L., and Kundu, S., Base excision repair of oxidative DNA damage, *Nature* **447,** 941–950 (2007).

Deaconescu, A.M., Savery, N., and Darst, S.A., The bacterial transcription repair coupling factor, *Curr. Opin. Struct. Biol.* **17,** 96–102 (2007).

Friedberg, E.C., Wagner, R., and Radman, M., Specialized DNA polymerases, cellular survival, and the genesis of mutations, *Science* **296,** 1627–1630 (2002).

Friedberg, E.C., Walker, G.C., Siede, W., Wood, R.D., Schultz, R.A., and Ellenberger, T., *DNA Repair and Mutagenesis* (2nd ed.), ASM Press (2006).

Garber, P.M., Vidanes, G., and Toczyski, D.P. Damage in transition, *Trends Biochem. Sci.* **30,** 63–66 (2005). [Describes how broken DNA can be repaired by NHEJ or recombination.]

Goodman, M.F., Error-prone repair DNA polymerases in prokaryotes and eukaryotes, *Annu. Rev. Biochem.* **71,** 17–50 (2002).

Harfe, B.D. and Jinks-Robertson, S., DNA mismatch repair and genetic instability, *Annu. Rev. Genet.* **34,** 359–399 (2000).

Heller, R.C. and Marians, K.J., Replisome assembly and the direct restart of replication forks, *Nature Rev. Mol. Cell Biol.* **7,** 932–943 (2006).

Hopfner, K.-P., Putnam, C.D., and Tainer, J.A., DNA double-strand break repair from head to tail, *Curr. Opin. Struct. Biol.* **12,** 115–122 (2002).

Kunkel, T.A. and Erie, D.A., DNA mismatch repair, *Annu. Rev. Biochem.* **74,** 681–710 (2005).

Lieber, M.R., The mechanism of double-strand DNA break repair by the nonhomologous DNA end-joining pathway, *Annu. Rev. Biochem.* **79,** 181–211 (2010).

Ling, H., Boudsocq, F., Woogate, R., and Yang, W., Crystal structure of a Y-family DNA polymerase in action: A mechanism for error-prone and lesion-bypass replication, *Cell* **107,** 91–102 (2001).

Mees, A., Klar, T., Gnau, P., Hennecke, U., Eker, A.P.M., Carell, T., and Essen, L.-O., Crystal structure of a photolyase bound to a CPD-like DNA lesion after in situ repair, *Science* **306,** 1789–1793 (2004).

McCullough, A.K., Dodson, M.L., and Lloyd, R.S., Initiation of base excision repair: glycosylase mechanism and structures, *Annu. Rev. Biochem.* **68,** 255–285 (1999).

Modrich, P., Mechanisms of eukaryotic mismatch repair, *J. Biol. Chem.* **281,** 30305–30309 (2006).

Mol, C.D., Parikh, S.S., Putnam, C.D., Lo, T.P., and Tainer, J.A., DNA repair mechanism for the recognition and removal of damaged DNA bases, *Annu. Rev. Biophys. Biomol. Struct.* **28,** 101–128 (1999).

Moore, M.H., Gulbis, J.M., Dodson, E.J., Demple, B., and Moody, P.C.E., Crystal structure of a suicidal DNA repair protein: Ada O^6-methylguanine-DNA methyltransferase from *E. coli,* *EMBO J.* **13,** 1495–1501 (1994).

Myers, L.C., Verdine, G.L., and Wagner, G., Solution structure of the DNA methyl triester repair domain of *Escherichia coli* Ada, *Biochemistry* **32,** 14089–14094 (1993).

Parikh, S.S., Mol, C.D., Slupphaug, G., Bharati, S., Krokan, H.E., and Tainer, J.A., Base excision repair initiation revealed by crystal structures and binding kinetics of human uracil–DNA glycosylase with DNA, *EMBO J.* **17,** 5214–5226 (1998).

Pham, P., Rangarajan, S., Woodgate, R., and Goodman, M.F., Roles of DNA polymerases V and II in SOS-induced error-prone and error-free repair in *Escherichia coli, Proc. Natl. Acad. Sci.* **98,** 8350–8354 (2001); *and* Goodman, M.F., Coping with replication 'train wrecks' in *Escherichia coli* using Pol V, Pol II, and RecA proteins, *Trends Biochem. Sci.* **25,** 189–195 (2000).

Prakash, S., Johnson, R.E., and Prakash, L., Eukaryotic translesion synthesis DNA polymerases: Specificity of structure and function, *Annu. Rev. Biochem.* **74,** 317–353 (2005).

Sancar, A., Lindsey-Bolz, L.A., Ünsal-Kaçmaz, K., and Linn, S., Molecular mechanisms of mammalian DNA repair and the DNA damage checkpoints, *Annu. Rev. Biochem.* **73,** 39–85 (2005).

Sarker, A.H., et al., Recognition of RNA polymerse II and transcription bubbles by XPG, CSB, and TFIIH: Insights for transcription-coupled repair and Cockayne syndrome, *Mol. Cell* **20,** 187–198 (2005).

Sutton, M.D., Smith, B.T., Godoy, V.G., and Walker, G.C., The SOS response: recent insights into *umuDC*-dependent mutagenesis and DNA damage tolerance, *Annu. Rev. Genet.* **34,** 479–497 (2000).

Tainer, J.A. and Friedberg, E.C. (Eds.), Biological Implications from Structures of DNA Repair Proteins, *Mutat. Res.* **460,** 139–335 (2000). [A series of authoritative reviews.]

Valle, D. (Ed.), *The Online Metabolic & Molecular Bases of Inherited Disease,* http://www.ommbid.com/. [Chapters 28 and 32 contain discussions of xeroderma pigmentosum, Cockayne syndrome, and hereditary nonpolyposis colorectal cancer.]

Walker, J.R., Corpina, R.A., and Goldberg, J., Structure of the Ku heterodimer bound to DNA and its implications for double-strand break repair, *Nature* **412,** 607–614 (2001).

Wyman, C. and Kanaar, R., DNA double-strand break repair: All's well that ends well, *Annu. Rev. Genet.* **40,** 363–383 (2006).

Yang, W. (Ed.), DNA Replication and Repair, *Adv. Prot. Chem.* **69** (2004).

Recombination and Mobile Genetic Elements

Ariyoshi, M., Nishino, T., Iwasaki, H., Shinagawa, H., and Morikawa, K., Crystal structure of the Holliday junction DNA in complex with a single RuvA tetramer, *Proc. Natl. Acad. Sci.* **97,** 8257–8262 (2000).

Biertümpfel, C., Yang, W., and Suck, D., Crystal structure of T4 endonuclease VII resolving a Holliday junction; *and* Hadden, J.M., Déclais, A.-C., Carrr, S.B., Lilley, D.M., and Phillips, S.E.V., The structural basis of Holliday junction resolution by T7 endonuclease I, *Nature* **449,** 616–620 *and* 621–625 (2007).

Changela, A., Perry, K., Taneja, B., and Mondragón, A., DNA manipulators: caught in the act, *Curr. Opin. Struct. Biol.* **13,** 15–22 (2003).

Chen, Z., Yang, H., and Pavletich, N.P., Mechanism of homologous recombination from the RecA–ssDNA/dsDNA structures, *Nature* **453,** 489–494 (2008).

Cox, M.M., Motoring along with the bacterial RecA protein, *Nature Rev. Mol. Cell Biol.* **8,** 127–138 (2007); *and* Regulation of bacterial RecA protein function, *Crit. Rev. Biochem. Mol. Biol.* **42,** 41–63 (2007).

Craig, N.L., Craigie, R., Gellert, M., and Lambowitz, A.M. (Eds.), *Mobile DNA II,* ASM Press (2002). [A compendium of authoritative articles.]

Grindley, N.G.F., Whiteson, K.L., and Rice, P.A., Mechanisms of site-specific recombination, *Annu. Rev. Biochem.* **75,** 567–605 (2006).

Haber, J.E., Partners and pathways. Repairing a double-strand break, *Trends Genet.* **16,** 259–264 (2000).

Ho, P.S. and Eichman, B.F., The crystal structures of Holliday junctions, *Curr. Opin. Struct. Biol.* **11**, 302–308 (2001).

Kuzminov, A., Recombinational repair of DNA damage in *Escherichia coli* and bacteriophage λ, *Microbiol. Mol. Biol. Rev.* **63**, 751–813 (1999).

Li, W., Kamtekar, S., Xiong, Y., Sarkis, G.J., Grindley, N.D.F., and Steitz, T.A., Structure of a synaptic γδ resolvase tetramer covalently linked to two cleaved DNAs, *Science* **309**, 1210–1215 (2005); *and* Yang, W. and Steitz, T.A., Crystal structure of the site-specific recombinase γδ resolvase complexed with a 34 bp cleavage site, *Cell* **82**, 193–207 (1995).

Lusetti, S.L. and Cox, M.M., The bacterial RecA protein and the recombinational DNA repair of stalled replication forks, *Annu. Rev. Biochem.* **71**, 71–100 (2002).

Marians, K.J., PriA-directed replication fork restart in *Escherichia coli, Trends Biochem. Sci.* **25**, 185–189 (2000).

Reznikoff, W. S., Transposon Tn5, *Annu. Rev. Genet.* **42**, 269–286 (2008).

Rice, P.A. and Baker, T.A., Comparative architecture of transposase and integrase complexes, *Nature Struct. Biol.* **8**, 302–307 (2001).

Roe, S.M., Barlow, T., Brown, T., Oram, M., Keeley, A., Tsaneva, I.R., and Pearl, L.H., Crystal structure of an octameric RuvA–Holliday junction complex, *Mol. Cell* **2**, 361–372 (1998).

Saikrishnan, K., Griffiths, S.P., Cook, N., Court, R., and Wigley, D.B., DNA binding to RecD: role of the 1B domain in SF1B helicase activity, *EMBO J.* **27**, 2222–2229 (2008).

Simon, M., Zieg, J., Silverman, M., Mandel, G., and Doolittle, R., Phase variation: evolution of a controlling element, *Science* **209**, 1370–1374 (1980).

Singleton, M.R., Dillingham, M.S., Gaudier, M., Kowalczykowski, S.C., and Wigley, D.B., Crystal structure of RecBCD enzyme reveals a machine for processing DNA breaks, *Nature* **432**, 187–193 (2004).

Sung, P. and Klein, H., Mechanism of homologous recombination: mediators and helicases take on regulatory functions, *Nature Rev. Mol. Cell Biol.* **7**, 739–750 (2006).

Van Duyne, G.D., A structural view of Cre–*loxP* site-specific recombination, *Annu. Rev. Biophys. Biomol. Struct.* **30**, 87–104 (2001).

Yamada, K., Ariyoshi, M., and Morikawa, K., Three-dimensional structural views of branch migration and resolution in DNA homologous recombination, *Curr. Opin. Struct. Biol.* **14**, 130–137 (2004).

Yamada, K., Kunishima, N., Mayanagi, K., Ohnishi, T., Nishino, T., Iwasaki, H., Shinagawa, H., and Morikawa, K., Crystal structure of the Holliday junction migration motor protein RuvB from *Thermus thermophilus* HB8, *Proc. Natl. Acad. Sci.* **98**, 1442–1447 (2001).

DNA Methylation and Trinucleotide Repeat Expansions

Arita, K., Ariyoshi, M., Tochio, H., Nakamura, Y., and Shirakawa, M., Recognition of hemi-methylated DNA by the SRA protein UHRF1 by a base-flipping mechanism; *and* Avvakumov, G.V., Walker, J.R., Xue, S., Li, Y., Duan, S., Bronner, C., Arrowsmith, C.H., and Dhe-Paganon, S., Structural basis for recognition of hemi-methylated DNA by the SRA domain of human UHRF1, Nature **455**, 818–821 *and* 822–825 (2008).

Bowater, R.P. and Wells, R.D., The intrinsically unstable life of DNA repeats associated with human hereditary disorders, *Prog. Nucleic Acid Res. Mol. Biol.* **66**, 159–202 (2001).

Castel, A.L., Cleary, J.D., and Pearson, C.E., Repeat instability as the basis for human diseases and as a potential target for therapy, *Nature Rev. Mol. Cell Biol.* **11**, 165–170 (2010).

Cummings, C.J. and Zoghbi, H.Y., Trinucleotide repeats: mechanisms and pathophysiology, *Annu. Rev. Genomics Hum. Genet.* **1**, 281–328 (2002).

Goodman, J. and Watson, R.E., Altered DNA methylation: a secondary mechanism involved in carcinogenesis, *Annu. Rev. Pharmacol. Toxicol.* **42**, 501–525 (2002).

Jones, P.A. and Baylin, S.B., The fundamental role of epigenetic events in cancer, *Nature Rev. Genet.* **3**, 415–428 (2002); *and* Jones, P.A. and Takai, D., The role of DNA methylation in mammalian epigenetics, *Science* **293**, 1068–1070 (2001).

Klimasauskas, S., Kumar, S., Roberts, R.J., and Cheng, X., HhaI methyltransferase flips its target base out of the DNA helix, *Cell* **76**, 357–369 (1994).

Klose, R.J. and Bird, A.P., Genomic DNA methylation: the mark and its mediators, *Trends Biochem. Sci.* **31**, 89–96 (2006).

Mirkin, S.M., Expandable DNA repeats and human disease, *Nature* **447**, 932–940 (2007).

O'Donnell, W.T. and Warren, S.T., A decade of molecular studies of fragile X syndrome, *Annu. Rev. Neurosci.* **25**, 315–338 (2002).

Perutz, M.F. and Windle, A.H., Causes of neural death in neurodegenerative diseases attributable to expansion of glutamine repeats, *Nature* **12**, 143–144 (2001); *and* Perutz, M.F., Glutamine repeats and neurodegenerative diseases: molecular aspects, *Trends Biochem. Sci.* **24**, 58–63 (1999).

Reik, W., Dean, W., and Walter, J., Epigenetic reprogramming in mammalian development, *Science* **293**, 1089–1093 (2001).

Rideout, W.M., III, Eggan, K., and Jaenisch, R., Nuclear cloning and epigenetic reprogramming of the genome, *Science* **293**, 1093–1098 (2001).

Shubert, H.L., Blumenthal, R.M., and Cheng, X., Many paths to methyltransfer: a chronicle of convergence, *Trends Biochem. Sci.* **28**, 329–335 (2003).

Szyf, M. and Detich, N., Regulation of the DNA methylation machinery and its role in cellular transformation, *Prog. Nucleic Acid Res. Mol. Biol.* **69**, 47–79 (2001).

Valle, D. (Ed.), *The Online Metabolic & Molecular Bases of Inherited Disease*, http://www.ommbid.com/. [Chaps. 64, 223, and 226 discuss fragile X syndrome, Huntington disease, and the spinocerebellar ataxias.]

PROBLEMS

1. Explain how certain mutant varieties of Pol I can be nearly devoid of DNA polymerase activity but retain almost normal levels of 5′ → 3′ exonuclease activity.

2. Why haven't Pol I mutants been found that completely lack 5′ → 3′ activity at all temperatures?

3. Why aren't type I topoisomerases necessary in *E. coli* DNA replication?

*4. The 3′ → 5′ exonuclease activity of Pol I excises only unpaired 3′-terminal nucleotides from DNA, whereas this enzyme's pyrophosphorolysis activity removes only properly paired 3′-terminal nucleotides. Discuss the mechanistic significance of this phenomenon in terms of the polymerase reaction.

5. You have isolated *E. coli* with temperature-sensitive mutations in the following genes. What are their phenotypes above their restrictive temperatures? Be specific. (a) *dnaB*, (b) *dnaE*, (c) *dnaG*, (d) *lig*, (e) *polA*, (f) *rep*, (g) *ssb*, and (h) *recA*.

6. About how many Okazaki fragments are synthesized in the replication of an *E. coli* chromosome?

***7.** What are the minimum and maximum number of replication forks that occur in a contiguous chromosome of an *E. coli* that is dividing every 25 min; every 80 min?

8. To put the *E. coli* replication system on a human scale, let us imagine that the 20-Å-diameter B-DNA was expanded to 1 m in diameter. If everything were proportionally expanded, then each DNA polymerase III holoenzyme would be about the size of a medium-sized truck. In such an expanded system: (a) How fast would each replisome be moving? (b) How far would each replisome travel during a complete replication cycle? (c) What would be the length of an Okazaki fragment? (d) What would be the average distance a replisome would travel between each error it made? Provide your answers in km/hr and km.

9. Why can't linear duplex DNAs, such as occur in bacteriophage T7, be fully replicated by only *E. coli*-encoded proteins?

***10.** What is the half-life of a particular purine base in the human genome assuming that it is subject only to spontaneous depurination? What fraction of the purine bases in a human genome will have depurinated in the course of a single generation (assume 25 years)? The DNAs of ~4000-year-old Egyptian mummies have been sequenced. Assuming that mummification did not slow the rate of DNA depurination, what fraction of the purine bases originally present in the mummy would still be intact today.

11. Why is the methylation of DNA to form O^6-methylguanine mutagenic?

12. A replication fork encountering a single-strand lesion may either dissociate or leave a single-strand gap. The latter process is more likely to occur during lagging strand synthesis than during leading strand synthesis. Explain.

13. The *E. coli* genome contains 1009 Chi sequences. Do these sequences occur at random, and, if not, how much more or less frequently than random do they occur?

14. *Deinococcus radiodurans*, which the *Guinness Book of World Records* has dubbed the world's toughest bacterium, can tolerate doses of ionizing radiation ~3000-fold greater than those that are lethal to humans (it was first discovered growing in a can of ground meat that had been "sterilized" by radiation). It appears to have several strategies to repair radiation damage to its DNA (which large doses of ionizing radiation fragment to many pieces) including a particularly large number of genes encoding proteins involved in DNA repair and 4 to 10 copies per cell of its genome, which consists of two circular chromosomes and two circular plasmids. Yet, these strategies, alone, do not account for *D. radiodurans'* enormously high radiation resistance. However, in an addi-

tional strategy, it organizes its multiple identical dsDNA circles into stacks in which, it is thought, the identical genes in the neighboring circles are aligned side by side. How would this latter strategy help *D. radiodurans* efficiently repair its fragmented DNA?

15. CpG islands occur in eukaryotic genomes at about one-fifth their expected random frequency. Suggest an evolutionary (mutational) process that eliminates CpG islands.

16. Explain why the brief exposure of a cultured eukaryotic cell line to 5-azacytosine results in permanent phenotypic changes to these cells.

17. Explain why chi structures, such as that shown in Fig. 30-70*b*, have two pairs of equal length arms.

***18.** Single-stranded circular DNAs containing a transposon have a characteristic stem-and-double-loop structure such as that shown in Fig. 30-106. What is the physical basis of this structure?

Figure 30-106 **Electron micrograph of a single-stranded circular DNA containing a transposon.** [Courtesy of Stanley N. Cohen, Stanford University School of Medicine.]

19. A composite transposon integrated in a circular plasmid occasionally transposes the DNA comprising the original plasmid rather than the transposon's central region. Explain how this is possible.

***20.** Cre recombinase has an additional function to that of circularizing the linear P1 dsDNA (Fig. 30-96). It is also required to resolve the circular dimers of P1 plasmids that result from their recombinational repair during replication, thereby permitting both daughter cells to receive a copy of the P1 plasmid. Using simple line diagrams, outline how these plasmids become dimerized and how Cre resolves them to circular monomers.

−35 element

σ₄

σ₂

σ₃

5′

−10 element

Transcription

CHAPTER **31**

There are three major classes of RNA, all of which participate in protein synthesis: **ribosomal RNA (rRNA), transfer RNA (tRNA),** and **messenger RNA (mRNA).** All of these RNAs are synthesized under the direction of DNA templates, a process known as **transcription.**

RNA's involvement in protein synthesis became evident in the late 1930s through investigations by Torbjörn Caspersson and Jean Brachet. Caspersson, using microscopic techniques, found that DNA is confined almost exclusively to the eukaryotic cell nucleus, whereas RNA occurs largely in the cytosol. Brachet, who had devised methods for fractionating cellular organelles, came to similar conclusions based on direct chemical analyses. He found, in addition, that the cytosolic RNA-containing particles are also protein rich. Both investigators noted that the concentration of these RNA–protein particles (which were later named ribosomes) is correlated with the rate at which a cell synthesizes protein, implying a relationship be-

tween RNA and protein synthesis. Indeed, Brachet even suggested that *the RNA–protein particles are the site of protein synthesis.*

Brachet's suggestion was shown to be valid when radioactively labeled amino acids became available in the 1950s. A short time after injection of a rat with a labeled amino acid, most of the label that had been incorporated in proteins was associated with ribosomes. This experiment also established that *protein synthesis is not immediately directed by DNA because, at least in eukaryotes, DNA and ribosomes are never in contact.*

In 1958, Francis Crick summarized the then dimly perceived relationships among DNA, RNA, and protein by what he called the **central dogma** of molecular biology: *DNA directs its own replication and its transcription to RNA which, in turn, directs its translation to proteins (Fig. 5-21).* The peculiar use of the word "dogma," one definition of which is a religious doctrine that the true believer cannot doubt, stemmed from a misunderstanding. When Crick formulated the central dogma, he was under the impression that dogma meant "an idea for which there was no reasonable evidence."

We begin this chapter by discussing experiments that led to the elucidation of mRNA's central role in protein synthesis. We then study the mechanism of transcription and its control in prokaryotes. Finally, in the last section, we consider post-transcriptional processing of RNA in both prokaryotes and eukaryotes. Translation is the subject of Chapter 32. Note that these subjects were outlined in Section 5-4. Here we shall delve into much greater detail.

1 THE ROLE OF RNA IN PROTEIN SYNTHESIS

The idea that proteins are specified by mRNA and synthesized on ribosomes arose from the study of **enzyme induction,** a phenomenon in which bacteria vary the synthesis rates of specific enzymes in response to environmental changes. In this section, we discuss the classic experiments that explained the basis of enzyme induction and revealed the existence of mRNA. We shall see that *enzyme induction occurs as a consequence of the regulation of mRNA synthesis by proteins that specifically bind to the mRNA's DNA templates.*

A. *Enzyme Induction*

E. coli can synthesize an estimated ~4300 different polypeptides. There is, however, enormous variation in the amounts of these different polypeptides that are produced. For instance, the various ribosomal proteins may each be present in over 10,000 copies per cell, whereas certain regulatory proteins (see below) normally occur in <10 copies per cell. Many enzymes, particularly those involved in basic cellular "housekeeping" functions, are synthesized at a more or less constant rate; they are called **constitutive enzymes.** Other enzymes, termed **adaptive** or **inducible enzymes,** are synthesized at rates that vary with the cell's circumstances.

a. Lactose-Metabolizing Enzymes Are Inducible

Bacteria, as has been recognized since 1900, adapt to their environments by producing enzymes that metabolize certain nutrients, for example, lactose, only when those substances are available. *E. coli* grown in the absence of lactose are initially unable to metabolize this disaccharide. To do so they require the presence of two proteins: **β-galactosidase,** which catalyzes the hydrolysis of lactose to its component monosaccharides,

and **galactoside permease** (also known as **lactose permease;** Section 20-4B), which transports lactose into the cell. *E. coli* grown in the absence of lactose contain only a few (<5) molecules of these proteins. Yet, a few minutes after lactose is introduced into their medium, *E. coli* increase the rate at which they synthesize these proteins by ~1000-fold (such that β-galactosidase can account for up to 10% of their soluble protein) and maintain this pace until lactose is no longer available. The synthesis rate then returns to its miniscule **basal level** (Fig. 31-1). *This ability to produce a series of proteins only when the substances they metabolize are present permits bacteria to adapt to their environment without the debilitating need to continuously synthesize large quantities of otherwise unnecessary substances.*

Lactose or one of its metabolic products must somehow trigger the synthesis of the above proteins. Such a substance is known as an **inducer.** The physiological inducer of the lactose system, the lactose isomer **1,6-allolactose,**

1,6-Allolactose

arises from lactose's occasional transglycosylation by β-galactosidase. Most experimental studies of the lactose system use **isopropylthiogalactoside (IPTG),**

Isopropylthiogalactoside (IPTG)

a potent inducer that structurally resembles allolactose but that is not degraded by β-galactosidase.

Lactose system inducers also stimulate the synthesis of **thiogalactoside transacetylase,** an enzyme that, *in vitro,* transfers an acetyl group from acetyl-CoA to the C6-OH group of a β-thiogalactoside such as IPTG. Since lactose fermentation proceeds normally in the absence of thiogalactoside transacetylase, however, this enzyme's physiological role is unknown.

b. *lac* System Genes Form an Operon

The genes specifying wild-type β-galactosidase, galactoside permease, and thiogalactoside transacetylase are designated Z^+, Y^+, and A^+, respectively. Genetic mapping of the defective mutants Z^-, Y^-, and A^- indicated that these

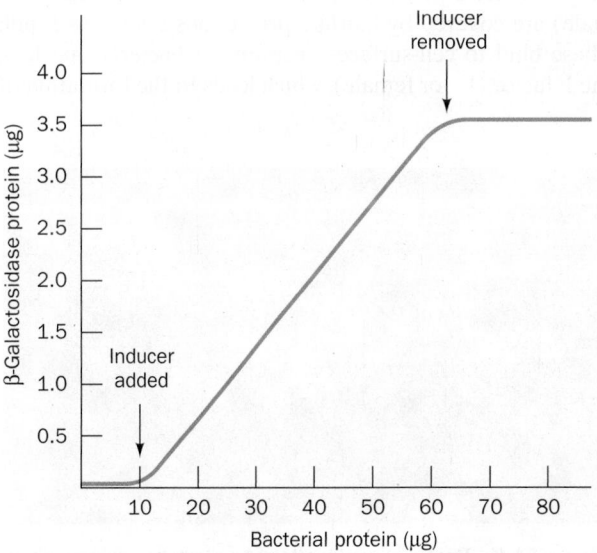

Figure 31-1 The induction kinetics of β-galactosidase in *E. coli.* [After Cohn, M., *Bacteriol. Rev.* **21,** 156 (1957).]

Figure 31-2 **Genetic map of the *E. coli lac* operon.** The map shows the genes encoding the proteins mediating lactose metabolism and the genetic sites that control their expression. The *Z*, *Y*, and *A* genes, respectively, encode β-galactosidase, galactoside permease, and thiogalactoside transacetylase.

lac **structural genes** (genes that specify polypeptides) are contiguously arranged on the *E. coli* chromosome (Fig. 31-2). *These genes, together with the control elements P and O, form a genetic unit called an **operon**, specifically the **lac operon**.* The nature of the control elements is discussed below. The role of operons in prokaryotic gene expression is examined in Section 31-3.

c. Bacteria Can Transmit Genes via Conjugation

An important clue as to how *E. coli* synthesizes protein was provided by a mutation that causes the proteins of the *lac* operon to be synthesized in large amounts in the absence of inducer. This so-called **constitutive mutation** occurs in a gene, designated *I*, that is distinct from although closely linked to the genes specifying the *lac* enzymes (Fig. 31-2). What is the nature of the *I* gene product? This riddle was solved in 1959 by Arthur Pardee, François Jacob, and Jacques Monod through an ingenious experiment that is known as the **PaJaMo experiment.** To understand this experiment, however, we must first consider **bacterial conjugation.**

Bacterial conjugation is a process, discovered in 1946 by Joshua Lederberg and Edward Tatum, through which some bacteria can transfer genetic information to others. The ability to conjugate ("mate") is conferred on an otherwise indifferent bacterium by a plasmid named **F factor** (for *fertility*). Bacteria that possess an F factor (designated F^+ or male) are covered by hairlike projections known as **F pili.** These bind to cell-surface receptors on bacteria that lack the F factor (F^- or female), which leads to the formation of

a cytoplasmic bridge between these cells (Fig. 31-3). The F factor then replicates and, as the newly replicated single strand is formed, it passes through the cytoplasmic bridge to the F^- cell where the complementary strand is synthesized (Fig. 31-4). This converts the F^- cell to F^+ so that the F factor is an infectious agent (a bacterial venereal disease?).

On very rare occasions, the F factor spontaneously integrates into the chromosome of the F^+ cell. In the resulting

Figure 31-4 **Diagram showing how an F^- cell acquires an F factor from an F^+ cell.** A single strand of the F factor is replicated, via the rolling circle mode (Section 30-3Bb), and is transferred to the F^- cell where its complementary strand is synthesized to form a new F factor.

Figure 31-3 **Bacterial conjugation.** An electron micrograph in false color showing an F^+ (*left*) and an F^- (*right*) *E. coli* engaged in sexual conjugation. [Dennis Kunkel/Phototake.]

Hfr (for *h*igh *f*requency of *r*ecombination) cells, the F factor behaves much as it does in the autonomous state. Its replication commences at a specific internal point in the F factor, and the replicated section passes through a cytoplasmic bridge to the F⁻ cell, where its complementary strand is synthesized. In this case, however, the replicated chromosome of the Hfr cell is also transmitted to the F⁻ cell (Fig. 31-5). *Bacterial genes are transferred from the Hfr cell to the F⁻ cell in fixed order.* This is because the F factor in a given Hfr strain is integrated into the bacterial chromosome at a specific site and because only a particular strand of the Hfr chromosomal DNA is replicated and transferred

to the F⁻ cell. Usually, only part of the Hfr bacterial chromosome is transferred during sexual conjugation because the cytoplasmic bridge almost always breaks off sometime during the ~90 min required to complete the transfer process. In the resulting **merozygote** (a partially diploid bacterium), the chromosomal fragment, which lacks a complete F factor, neither transforms the F⁻ cell to Hfr nor is subsequently replicated. However, the transferred chromosomal fragment recombines with the chromosome of the F⁻ cell (Section 30-6A), thereby permanently endowing the F⁻ cell with some of the traits of the Hfr strain.

The integrated F factor in an Hfr cell occasionally undergoes spontaneous excision to yield an F⁺ cell. In rare instances, the F factor is aberrantly excised such that a portion of the adjacent bacterial chromosome is incorporated in the subsequently autonomously replicating F factor. Bacteria carrying such a so-called **F′ factor** are permanently diploid for its bacterial genes.

d. *lac* Repressor Inhibits the Synthesis of *lac* Operon Proteins

In the PaJaMo experiment, Hfr bacteria of genotype I^+Z^+ were mated to an F⁻ strain of genotype I^-Z^- in the absence of inducer while the β-galactosidase activity of the culture was monitored (Fig. 31-6). At first, as expected,

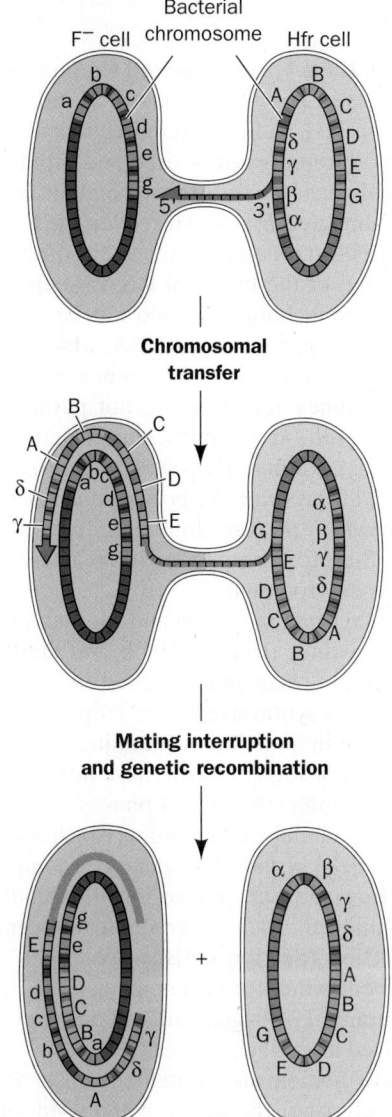

Chromosomal transfer

Mating interruption and genetic recombination

Figure 31-5 **Transfer of the bacterial chromosome from an Hfr cell to an F⁻ cell and its subsequent recombination with the F⁻ chromosome.** Here, Greek letters represent F factor genes, uppercase Roman letters represent bacterial genes from the Hfr cell, and lowercase Roman letters represent the corresponding alleles in the F⁻ cell. Since chromosomal transfer, which begins within the F factor, is rarely complete, the entire F factor is seldom transferred. Hence the recipient cell usually remains F⁻.

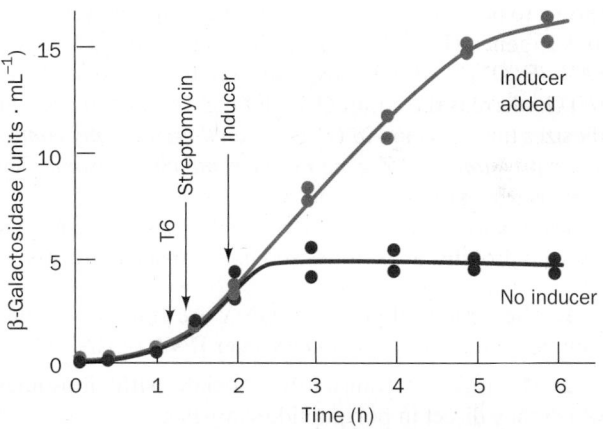

Figure 31-6 **The PaJaMo experiment.** This experiment demonstrated the existence of the *lac* repressor through the appearance of β-galactosidase in the transient merozygotes (partial diploids) formed by mating I^+Z^+ Hfr donors with I^-Z^- F⁻ recipients. The F⁻ strain was also resistant to both **bacteriophage T6** and **streptomycin,** whereas the Hfr strain was sensitive to these agents. Both types of cells were grown and mated in the absence of inducer. After sufficient time had passed for the transfer of the *lac* genes, the Hfr cells were selectively killed by the addition of T6 phage and streptomycin. In the absence of inducer (*lower curve*), β-galactosidase synthesis commenced at around the time at which the *lac* genes had entered the F⁻ cells but stopped after ~1 h. If inducer was added shortly after the Hfr donors had been killed (*upper curve*), enzyme synthesis continued unabated. This demonstrates that the cessation of β-galactosidase synthesis in uninduced cells is not due to the intrinsic loss of the ability to synthesize this enzyme but to the production of a repressor specified by the I^+ gene. [After Pardee, A.B., Jacob, F., and Monod, J., *J. Mol. Biol.* **1**, 173 (1959).]

there was no β-galactosidase activity because the Hfr donors lacked inducer and the F⁻ recipients were unable to produce active enzyme (only DNA passes through the cytoplasmic bridge connecting mating bacteria). About 1 h after conjugation began, however, when the I^+Z^+ genes had just entered the F⁻ cells, β-galactosidase synthesis began and only ceased after about another hour. The explanation for these observations is that the donated Z^+ gene, on entering the cytoplasm of the I^- cell, directs the synthesis of β-galactosidase in a constitutive manner. Only after the donated I^+ gene has had sufficient time to be expressed is it able to repress β-galactosidase synthesis. *The I^+ gene must therefore give rise to a diffusible product, the **lac repressor**, which inhibits the synthesis of β-galactosidase (and the other lac proteins).* Inducers such as IPTG temporarily inactivate *lac* repressor, whereas I^- cells constitutively synthesize *lac* enzymes because they lack a functional repressor. *Lac* repressor, as we shall see in Section 31-3B, is a protein.

B. *Messenger RNA*

The nature of the *lac* repressor's target molecule was deduced in 1961 through a penetrating genetic analysis by Jacob and Monod. A second type of constitutive mutation in the lactose system, designated O^c (for **operator constitutive**), which complementation analysis (Section 1-4Cc) has shown to be independent of the I gene, maps between the I and Z genes (Fig. 31-2). In the partially diploid F′ strain $O^cZ^-/F\ O^+Z^+$, β-galactosidase activity is inducible by IPTG, whereas the strain $O^cZ^+/F\ O^+Z^-$ constitutively synthesizes this enzyme. *An O^+ gene can therefore only control the expression of a Z gene on the same chromosome.* The same is true with the Y^+ and A^+ genes.

Jacob and Monod's observations led them to conclude that the proteins are synthesized in a two-stage process:

1. The structural genes on DNA are transcribed onto complementary strands of **messenger RNA (mRNA).**

2. The mRNAs transiently associate with ribosomes, which they direct in polypeptide synthesis.

This hypothesis explains the behavior of the *lac* system that we previously outlined in Section 5-4Ab (Fig. 5-25; See **Guided Exploration 2: Regulation of gene expression by the lac repressor system**). *In the absence of inducer, the lac repressor specifically binds to the O gene (the **operator**) so as to prevent the enzymatic transcription of mRNA. On binding inducer, the repressor dissociates from the operator, thereby permitting the transcription and subsequent translation of the lac enzymes.* The operator–repressor–inducer system thereby acts as a molecular switch so that the *lac* operator can only control the expression of *lac* enzymes on the same chromosome. The O^c mutants constitutively synthesize *lac* enzymes because they are unable to bind repressor. The **coordinate** (simultaneous) expression of all three *lac* enzymes under the control of a single operator site arises, as Jacob and Monod theorized, from the

transcription of the *lac* operon as a single **polycistronic mRNA** which directs the ribosomal synthesis of each of these proteins (the term **cistron** is a somewhat archaic synonym for gene). This transcriptional control mechanism is further discussed in Section 31-3. [DNA sequences that are on the same DNA molecule are said to be "in cis" (Latin: on this side of), whereas those on different DNA molecules are said to be "in trans" (Latin: across). Control sequences such as the O gene, which are only active on the same DNA molecule as the genes they control, are called **cis-acting elements.** Genes such as *lacI*, which specify the synthesis of diffusible products and can therefore be located on a different DNA molecule from the genes they control, are said to direct the synthesis of **trans-acting factors.**]

a. mRNAs Have Their Predicted Properties

The kinetics of enzyme induction, as indicated, for example, in Figs. 31-1 and 31-6, requires that the postulated mRNA be both rapidly synthesized and rapidly degraded. An RNA with such quick turnover had, in fact, been observed in T2-infected *E. coli*. Moreover, the base composition of this RNA fraction resembles that of the viral DNA rather than that of the bacterial RNA (keep in mind that base sequencing techniques would not be formulated for another ~15 years). Ribosomal RNA, which comprises up to 90% of a cell's RNA, turns over much more slowly than mRNA. Ribosomes are therefore not permanently committed to the synthesis of a particular protein (a once popular hypothesis). Rather, *ribosomes are nonspecific protein synthesizers that produce the polypeptide specified by the mRNA with which they are transiently associated.* A bacterium can therefore respond within a few minutes to changes in its environment.

Evidence favoring the Jacob and Monod model rapidly accumulated. Sydney Brenner, Jacob, and Matthew Meselson carried out experiments designed to characterize the RNA that *E. coli* synthesized after T4 phage infection. *E. coli* were grown in a medium containing ^{15}N and ^{13}C so as to label all cell constituents with these heavy isotopes. The cells were then infected with T4 phages and immediately transferred to an unlabeled medium (which contained only the light isotopes ^{14}N and ^{12}C) so that cell components synthesized before and after phage infection could be separated by equilibrium density gradient ultracentrifugation in CsCl solution (Section 6-5Bb). No "light" ribosomes were observed, which indicates, in agreement with the above-mentioned T2 phage results, that no new ribosomes are synthesized after phage infection.

The growth medium also contained either ^{32}P or ^{35}S so as to radioactively label the newly synthesized and presumably phage-specific RNA and protein, respectively. Much of the ^{32}P-labeled RNA was associated, as was postulated for mRNA, with the preexisting "heavy" ribosomes (Fig. 31-7). Likewise, the ^{35}S-labeled proteins were transiently associated with, and therefore synthesized by, these ribosomes.

Sol Spiegelman developed the RNA–DNA hybridization technique (Section 5-3Cb) in 1961 to characterize the

Figure 31-7 **The distribution, in a CsCl density gradient, of** **^{32}P-labeled RNA that had been synthesized by** *E. coli* **after T4** **phage infection.** Free RNA, being relatively dense, bands at the bottom of the centrifugation cell (*left*). Much of the RNA, however, is associated with the ^{15}N- and ^{13}C-labeled "heavy" ribosomes that had been synthesized before the phage infection. The predicted position of unlabeled "light" ribosomes, which are not synthesized by phage-infected cells, is also indicated. [After Brenner, S., Jacob, F., and Meselson, M., *Nature* **190,** 579 (1961).]

Figure 31-8 **The hybridization of ^{32}P-labeled RNA produced** **by T2-infected** *E. coli* **with ^{3}H-labeled T2 DNA.** On radioactive decay, ^{32}P and ^{3}H emit β particles (electrons) with characteristically different energies so that these isotopes can be independently detected. Although free RNA (*left*) in a CsCl density gradient is denser than DNA, much of the RNA bands with the DNA (*right*). This indicates that the two polynucleotides have hybridized and are therefore complementary in sequence. [After Hall, B.D. and Spiegelman, S., *Proc. Natl. Acad. Sci.* **47,** 141 (1961).]

RNA synthesized by T2-infected *E. coli.* He found that this phage-derived RNA hybridizes with T2 DNA (Fig. 31-8) but does not hybridize with DNAs from unrelated phage nor with the DNA from uninfected *E. coli.* This RNA must therefore be complementary to T2 DNA in agreement with Jacob and Monod's prediction; that is, the phage-specific RNA is a messenger RNA. Hybridization studies have likewise shown that mRNAs from uninfected *E. coli* are complementary to portions of *E. coli* DNA. In fact, other RNAs, such as transfer RNA and ribosomal RNA, have corresponding complementary sequences on DNA from the same organism. Thus, *all cellular RNAs are transcribed from DNA templates.*

2 RNA POLYMERASE

RNA polymerase (RNAP), *the enzyme responsible for the DNA-directed synthesis of RNA, was discovered independently in 1960 by Samuel Weiss and Jerard Hurwitz. The enzyme couples together the ribonucleoside triphosphates ATP, CTP, GTP, and UTP on DNA templates in a reaction that is driven by the release and subsequent hydrolysis of PP_i:*

$$(\text{RNA})_{n \text{ residues}} + \text{NTP} \rightleftharpoons (\text{RNA})_{n+1 \text{ residues}} + \text{PP}_i$$

All cells contain RNAP. In bacteria, one species of this enzyme synthesizes all of the cell's RNA except the RNA primers employed in DNA replication (Section 30-1D). Various bacteriophages encode RNAPs that synthesize only phage-specific RNAs. Eukaryotic cells contain four or five RNAPs that each synthesize a different class of RNA. In this section we first consider the properties of the bacterial RNAPs and then consider the eukaryotic enzymes.

E. coli RNAP's so-called **holoenzyme** is an ~459-kD protein with subunit composition $\alpha_2\beta\beta'\omega\sigma$ (Table 31-1) in which the β and β′ subunits contain several colinearly arranged homologous segments. Once RNA synthesis has been initiated, however, the σ subunit (also called **σ factor** or **σ^{70}** since its molecular mass is 70 kD) dissociates from the **core enzyme,** $\alpha_2\beta\beta'\omega$, which carries out the actual polymerization process (see below).

Table 31-1 **Components of *E. coli* RNA Polymerase** **Holoenzyme**

Subunit	Number of Residues	Structural Gene
α	329	*rpoA*
β	1342	*rpoB*
β′	1407	*rpoC*
ω	91	*rpoZ*
σ^{70}	613	*rpsD*

Figure 31-9 **An electron micrograph of *E. coli* RNA polymerase (RNAP) holoenzyme attached to various promoter sites on bacteriophage T7 DNA.** RNAP is one of the largest known soluble enzymes. [From Williams, R.C., *Proc. Natl. Acad. Sci.* **74,** 2313 (1977).]

Electron micrographs (Fig. 31-9) clearly indicate that RNAP, which has a characteristic large size, binds to DNA as a protomer. This large size is presumably a consequence of the holoenzyme's several complex functions including (1) template binding, (2) RNA chain initiation, (3) chain elongation, and (4) chain termination. We discuss these various functions below.

A. Template Binding

RNA synthesis is normally initiated only at specific sites on the DNA template. This was first demonstrated through hybridization studies of bacteriophage φX174 DNA with the RNA produced by φX174-infected *E. coli.* Bacteriophage φX174 carries a single strand of DNA known as the (+) strand. On its injection into *E. coli,* the (+) strand directs the synthesis of the complementary (−) strand with which it combines to form a circular duplex DNA known as the replicative form (Section 30-3Ba). The RNA produced by φX174-infected *E. coli* does not hybridize with DNA from intact phages but does so with the replicative form. Thus only the (−) strand of φX174 DNA, the so-called **antisense strand,** is transcribed, that is, acts as a template; the (+) strand, the **sense strand** (or **coding strand;** so called because it has the same sequence as the transcribed RNA), does not do so. Similar studies indicate that in larger phages, such as T4 and λ, the two viral DNA strands are the antisense (template) strands for different sets of genes. The same is true of cellular organisms.

a. Holoenzyme Specifically Binds to Promoters

RNA polymerase binds to its initiation sites through base sequences known as **promoters** *that are recognized by the corresponding σ factor.* The existence of promoters was originally suggested by mutations that enhance or diminish the transcription rates of certain genes, including those of the *lac* operon. *Genetic mapping of such mutations indicated that the promoter consists of an ~40-bp sequence that is located on the 5′ side of the transcription start site.* [By convention, the sequence of template DNA is represented by its sense (nontemplate) strand so that it will have the same directionality as the transcribed RNA. A base pair in a promoter region is assigned a negative or positive number that indicates its position, upstream or downstream in the direction of RNAP travel, from the first nucleotide that is transcribed to RNA; this start site is +1 and there is no 0.] RNA, as we shall see, is synthesized in the 5′ → 3′ direction (Section 31-2C). Consequently, the promoter lies on the "upstream" side of the RNA's starting nucleotide. Sequencing studies indicate that the *lac* promoter (*lacP*) overlaps the *lac* operator (Fig. 31-2).

The holoenzyme forms tight complexes with promoters (dissociation constant $K \approx 10^{-14}\,M$) and thereby protects the bound DNA segments from digestion by DNase I. The region from about −20 to +20 is protected against exhaustive DNase I degradation. The region extending upstream to about −60 is also protected but to a lesser extent, presumably because it binds holoenzyme less tightly.

Sequence determinations of the protected regions from numerous *E. coli* and phage genes have revealed the consensus sequence of *E. coli* promoters (Fig. 31-10). *Their most conserved sequence is a hexamer centered at about the −10 position, the so-called* **Pribnow box** (named after David Pribnow, who pointed out its existence in 1975). It has a consensus sequence of TATAAT in which the leading TA and final T are highly conserved. *Upstream sequences around position −35 also have a region of sequence similarity,* TTGACA, which is most evident in efficient promoters. The sequence of the segment between the −10 and the −35 sites is unimportant but its length is critical; it ranges from 16 to 19 bp in the great majority of promoters. The initiating (+1) nucleotide, which is nearly always A or G, is centered in a poorly conserved CAT or CGT sequence. Most promoter sequences vary considerably from the consensus sequence (Fig. 31-10). Nevertheless, a mutation in one of the partially conserved regions can greatly increase or decrease a promoter's initiation efficiency. In addition, Richard Gourse discovered that certain highly expressed genes contain an A + T–rich segment between positions −40 and −60, the **upstream promoter (UP) element,** which binds to the C-terminal domain of RNAP's α subunits. The UP element-containing genes include those encoding the ribosomal RNAs, the *rrn* genes (e.g., Fig. 31-10), which collectively account for 60% of the RNA synthesized by *E. coli. The rates at which genes are transcribed, which span a range of at least 1000, vary directly with the rate at which their promoters form stable initiation complexes with the holoenzyme.* Promoter mutations that increase or decrease the rate at which the associated gene is transcribed are known as **up mutations** and **down mutations.**

Operon	−35 region	−10 region (Pribnow box)	Initiation site (+1)
lac	ACCCCAGGCTTTACACTTTATGCTTCCGGCTCGTATGTTGTGTGGAATTGTGAGCGG
lacI	CCATCGAATGGCGCAAAACCTTTCGCGGTATGGCATGATAGCGCCCGGAAGAGAGTC
galP2	ATTTATTCCATGTCACACTTTTCGCATCTTTGTTATGCTATGGTTATTTCATACCAT
araBAD	GGATCCTACCTGACGCTTTTTATCGCAACTCTCTACTGTTTCTCCATACCCGTTTTT
araC	GCCGTGATTATAGACACTTTTGTTACGCGTTTTTGTCATGGCTTTGGTCCCGCTTTG
trp	AAATGAGCTGTTGACAATTAATCATCGAACTAGTTAACTAGTACGCAAGTTCACGTA
bioA	TTCCAAAACGTGTTTTTTGTTGTTAATTCGGTGTAGACTTGTAAACCTAAATCTTTT
bioB	CATAATCGACTTGTAAACCAAATTGAAAAGATTTAGGTTTACAAGTCTACACCGAAT
tRNA^{Tyr}	CAACGTAACACTTTACAGCGGCGCGTCATTTGATATGATGCGCCCCGCTTCCCGATA
rrnD1	CAAAAAAATACTTGTGCAAAAAATTGGGATCCCTATAATGCGCCTCCGTTGAGACGA
rrnE1	CAATTTTTCTATTGCGGCCTGCGGAGAACTCCCTATAATGCGCCTCCATCGACACGG
rrnA1	AAAATAAATGCTTGACTCTGTAGCGGGAAGGCGTATTATGCACACCCCGCGCCGCTG

Consensus sequence:

−35 region: **T T G A C A** (69 79 61 56 54 54) … 16–19 bp …
−10 region: **T A T A A T** (77 76 60 61 56 82) … 5–8 bp …
Initiation site: A 51, C 55, T 48, G 42

Figure 31-10 **The sense (nontemplate) strand sequences of selected *E. coli* promoters.** A 6-bp region centered around the −10 position (*red shading*) and a 6-bp sequence around the −35 region (*blue shading*) are both conserved. The transcription initiation sites (+1), which in most promoters occur at a single purine nucleotide, are shaded in green. The bottom row shows the consensus sequence of 298 *E. coli* promoters with the number below each base indicating its percentage occurrence. The downstream portions of the *rrn* genes' UP elements can be seen. [After Rosenberg, M. and Court, D., *Annu. Rev. Genet.* **13**, 321–323 (1979). Consensus sequence from Lisser, S. and Margalit, H., *Nucleic Acids Res.* **21**, 1512 (1993).]

b. Initiation Requires the Formation of an Open Complex

The promoter regions in contact with the holoenzyme were identified by determining where the enzyme alters the susceptibility of the DNA to alkylation by agents such as dimethyl sulfate (DMS), a procedure named **DMS footprinting** (Section 34-3Bh). These experiments demonstrated that the holoenzyme contacts the promoter mainly around its −10 and −35 regions. These protected sites are both on the same side of the B-DNA double helix as the initiation site, which suggests that holoenzyme binds to only one face of the promoter.

DMS methylates G residues at N7, A residues at N1 and N3, and C residues at N3. Since N1 on A and N3 on C participate in base pairing interactions, however, they can only react with DMS in single-stranded DNA. This differential methylation of single- and double-stranded DNAs provides a sensitive test for DNA strand separation or "melting." Such chemical footprinting studies indicate that the binding of holoenzyme "melts out" the promoter in a region of ~14 bp extending from the middle of the −10 region to just past the initiation site, thereby forming a so-called **transcription bubble.** The need to form this **open complex** explains why promoter efficiency tends to decrease with the number of G · C base pairs in the −10 region; this presumably increases the difficulty in opening the double helix as is required for chain initiation (recall that G · C pairs are more stable than A · T pairs).

Core enzyme, which does not specifically bind promoter (except when it has an UP element), tightly binds duplex DNA (the complex's dissociation constant is $K \approx 5 \times 10^{-12}$ M and its half-life is ~60 min). Holoenzyme, in contrast, binds to nonpromoter DNA comparatively loosely ($K \approx 10^{-7}$ M and a half-life >1 s). Evidently, the σ subunit allows holoenzyme to move rapidly along a DNA strand in search of the σ subunit's corresponding promoter. Once transcription has been initiated and the σ subunit jettisoned, the tight binding of core enzyme to DNA apparently stabilizes the ternary enzyme–DNA–RNA complex.

B. Chain Initiation

The 5′-terminal base of prokaryotic RNAs is almost always a purine with A occurring more often than G. The initiating reaction of transcription is simply the coupling of two nucleoside triphosphates in the reaction

$$\text{pppA} + \text{pppN} \rightleftharpoons \text{pppApN} + \text{PP}_i$$

and hence, unlike DNA replication, does not require a primer. Bacterial RNAs therefore have 5′-triphosphate groups as was demonstrated by the incorporation of radioactive label into RNA when it was synthesized with [γ-^{32}P]ATP. Only the 5′ terminus of the RNA can retain the label because the internal phosphodiester groups of RNA are derived from the α-phosphate groups of nucleoside triphosphates.

RNAP has a curious behavior: It frequently releases its newly synthesized RNA after only ~10 nt have been polymerized, a process known as **abortive initiation.** When RNAP initiates transcription, it keeps its grip on the promoter (which is on the DNA's nontemplate/sense strand). Consequently, conformational tension builds up as the template/antisense strand is pulled through the RNAP's active site, a process called **scrunching** because the resulting increased size of the transcription bubble in the downstream direction must somehow be accommodated within the RNAP. In abortive initiation, the RNAP fails to escape the promoter and instead relieves the conformational tension by releasing the newly synthesized RNA fragment, thereby letting the transcription bubble relax to its normal size. The RNAP then reinitiates transcription from the +1 position. In successful initiation, the strain eventually provides sufficient energy to strip the promoter from the RNAP, which then commences the processive (continuous) transcription of the template. This process requires the dissociation of the σ factor from the core–DNA–RNA complex to form the elongation complex, although recent experiments indicate that this process often occurs stochastically (randomly) over several nucleotide additions. The σ factor can then join with another core to form a new initiation complex as was demonstrated by a burst of RNA synthesis on addition of core enzyme to a transcribing reaction mixture that initially contained only holoenzyme.

a. Bacterial RNAP Has a Highly Complex Structure

The X-ray structure of *E. coli* RNAP has not been determined. However, Seth Darst and Dmitry Vassylyev independently determined the X-ray structures of the closely similar *Thermus aquaticus* (Taq) and *Thermus thermophilus* (Tth) RNAP core enzymes and holoenzymes. The structure of the Tth core enzyme in complex with DNA and RNA, in agreement with EM studies of *E. coli* RNAP, has the overall shape of a crab claw whose two "pincers" are formed by the β and β′ subunits (Fig. 31-11). The protein is ~150 Å long (parallel to the pincers), ~115 Å high, and ~110 Å deep, with the tunnel between the two pincers ~27 Å wide. The β and β′ subunits extensively interact with one another, particularly at the base of the tunnel (also called the main channel) where an active site Mg^{2+} ion is located, which is also where their homologous segments converge. The β′ subunit binds two Zn^{2+} ions, each via four Cys residues that are invariant in prokaryotes but not in eukaryotes. The outer surface of the RNAP is almost uniformly negatively charged, whereas those surfaces that interact with nucleic acids are positively charged.

The downstream dsDNA occupies the main channel, which directs the template strand to the active site. There it base-pairs with the incoming NTP (not present in this structure) at the so-called $i + 1$ site near the Mg^{2+} ion. The 3′ end of the RNA forms a 9-bp hybrid helix with the 5′ end of the template DNA strand and then exits the protein through a channel between the β and β′ subunits (the RNA exit channel) in which it adopts a conformation similar to that of a single strand within an RNA double helix. Thus the structure resembles that of a post-translocated elongation complex, although the paths taken by the template and nontemplate DNA strands to rejoin at the end of the transcription bubble are unclear.

The X-ray structure of the Tth holoenzyme indicates that its σ subunit (σ70) has three flexibly linked, largely α

Figure 31-11   **X-ray structure of Tth core RNAP in complex with a 23-nt template DNA, a 14-nt nontemplate DNA, and a 16-nt RNA.** The protein is drawn in ribbon form with its two α subunits yellow and green, its β subunit cyan, its β′ subunit pink, and its ω subunit gray. The bound Mg^{2+} and Zn^{2+} ions are represented by red and orange spheres, respectively. The DNA and RNA are shown in ladder form with template DNA green, nontemplate DNA blue, and RNA red. Note that residues 208 to 390 of the β′ subunit, which extend from the tip of its pincer, are disordered and hence not visible, as are the 86-residue C-terminal domains of both α subunits. [Based on an X-ray structure by Dmitry Vassylyev, University of Alabama at Birmingham. PDBid 5O5I.]

helical domains, σ2, σ3, and σ4, that extend across the top of the holoenzyme (Fig. 31-12; σ1 is not visible). The holoenzyme's pincers are ~10 Å farther apart than in the elongation complex. The binding cavities for the downstream dsDNA and the transcription bubble are partially occupied by the σ subunit's σ4 domain and its σ3–4 linker. This partially accounts for the above-described nucleic acid scrunching that precedes the transition from the initiation complex to the elongation complex. Thus, the release of only these segments of the σ subunit from RNAP is compatible with the formation of an elongation complex, thereby accounting for the above-mentioned stochastic release of the σ subunit from a successfully initiated RNAP complex.

A low (6.5 Å) resolution X-ray structure of Taq holoenzyme in complex with a dsDNA segment containing the promoter's −10 and −35 elements reveals that the DNA lies across one face of the holoenzyme, completely outside of the main channel (Fig. 31-13). All sequence-specific contacts that the holoenzyme makes with the −10 and −35 elements as well as with the so-called extended −10 region just upstream of the −10 element are mediated by the σ subunit via conserved residues. This structure presumably resembles the so-called **closed complex** in which the DNA has not yet entered the main channel to form a transcription bubble. The mechanism through which this occurs is largely unknown.

Figure 31-12 **X-ray structure of Tth RNAP holoenzyme viewed similarly to Fig. 31-11.** The subunits of the core enzyme are represented by their partially transparent molecular surface colored as in Fig. 31-11. The σ subunit is drawn with its α helices as cylinders and colored in rainbow order from its N-terminus (*blue*) to its C-terminus (*red*). [Based on an X-ray structure by Dmitry Vassylyev, University of Alabama at Birmingham. PDBid 1IW7.]

Figure 31-13 **Low (6.5 Å) resolution X-ray structure of Taq RNAP holoenzyme in complex with a promoter-containing dsDNA viewed as in Fig. 31-12.** The subunits of the core enzyme are represented by their partially transparent molecular surface colored as in Fig. 31-11 (which appears striated due to the structure's low resolution, which permits only the polypeptide backbones to be visualized). The σ subunit is drawn in worm form colored in rainbow order from its N-terminus (*blue*) to its C-terminus (*red*). The DNA's sugar–phosphate backbone is drawn in cartoon form with the template DNA strand blue and the nontemplate DNA's −10 element red, its −35 element purple, and its remaining portions orange. [Based on an X-ray structure by Seth Darst, The Rockefeller University. PDBid 1L9Z.]

b. Rifamycins Inhibit Prokaryotic Transcription Initiation

Two related antibiotics, **rifamycin B,** which is produced by *Streptomyces mediterranei,* and its semisynthetic derivative **rifampicin,**

Rifamycin B $R_1 = CH_2COO^-$; $R_2 = H$

Rifampicin $R_1 = H$; $R_2 = CH=N-N$⟨   ⟩$N-CH_3$

specifically inhibit transcription by prokaryotic, but not eukaryotic, RNAPs. This selectivity and their high potency (bacterial RNAP is 50% inhibited by 2×10^{-8} *M* rifampicin) made them medically useful bacteriocidal agents against gram-positive bacteria and tuberculosis. Indeed, few other antibiotics are effective against tuberculosis, which has reached epidemic levels in some parts of the world.

The finding that the β subunits of rifamycin-resistant mutants have altered electrophoretic mobilities first demonstrated that this subunit contains the rifamycin-binding site. Rifamycins inhibit neither the binding of RNAP to the promoter nor the formation of the first phosphodiester bond, but they prevent further chain elongation. The inactivated RNAP remains bound to the promoter, thereby blocking its initiation by uninhibited enzymes. Once RNA chain initiation has occurred, however, rifamycins have no effect on the subsequent elongation process. The rifamycins are therefore useful research tools because they permit the transcription process to be dissected into its initiation and its elongation phases.

The X-ray structure of Taq core enzyme in complex with rifampicin reveals how this antibiotic inhibits RNAP. Rifampicin binds with close complementary fit but little conformational change in a pocket in the β subunit that is located within the main channel, ~12 Å distant from the active site Mg^{2+} ion. Model building indicates that the bound rifampicin would sterically interfere with the RNA transcript at positions −2 to −5 in the transcription bubble. Thus, as is observed, rifampicin would not interfere with the initiation of transcription but would mechanically block the extension of the RNA transcript. The residues lining the pocket in which rifampicin binds are highly conserved among prokaryotes but not in eukaryotes, thereby explaining why rifamycins inhibit only bacterial RNAPs.

C. *Chain Elongation*

The direction of RNA chain elongation; that is, whether it occurs by the addition of incoming nucleotides to the 3' end of the nascent (growing) RNA chain ($5' \rightarrow 3'$ growth; Fig. 31-14a) or by their addition to its 5' terminus ($3' \rightarrow 5'$ growth; Fig. 31-14b), was established by determining the rate at which the radioactive label from [γ-^{32}P]GTP is incorporated into RNA. For $5' \rightarrow 3'$ elongation, the 5' γ-P is permanently labeled and, hence, the chain's level of radioactivity would not change on replacement of the labeled GTP with unlabeled GTP. However, for $3' \rightarrow 5'$ elongation, the 5' γ-P is replaced with the addition of every new nucleotide so that, on replacement of labeled with unlabeled GTP, the nascent RNA chains would lose their radioactivity. The former was observed. *Chain growth must therefore occur in the $5' \rightarrow 3'$ direction (Fig. 31-14a)*, the same direction as DNA is synthesized. This conclusion is corroborated by the observation that the antibiotic **cordycepin,**

Cordycepin
(3'-deoxyadenosine)

an adenosine analog that lacks a 3'-OH group, inhibits bacterial RNA synthesis. Its addition to the 3' end of RNA, as is expected for $5' \rightarrow 3'$ growth, prevents the RNA chain's

further elongation. Cordycepin would not have this effect if chain growth occurred in the opposite direction because it could not be appended to an RNA's 5' end.

a. Transcription Supercoils DNA

RNA chain elongation requires that the double-stranded DNA template be opened up at the point of RNA synthesis so that the template strand can be transcribed to its complementary RNA strand. In doing so, the RNA chain only transiently forms a short length of RNA–DNA hybrid duplex, as is indicated by the observation that transcription leaves the template duplex intact and yields single-stranded RNA. The unpaired transcription bubble of the DNA in the open initiation complex apparently travels along the DNA with the RNAP. There are two ways this might occur (Fig. 31-15):

1. If the RNAP followed the template strand in its helical path around the DNA, the DNA would build up little supercoiling because the DNA duplex would never be unwound by more than about a turn. However, the RNA transcript would wrap around the DNA, once per duplex turn. This model is implausible since it is unlikely that its DNA and RNA could be readily untangled: The RNA would not spontaneously unwind from the long and often circular DNA in any reasonable time, and no known topoisomerase can accelerate this process.

2. If the RNAP moves in a straight line while the DNA rotates, the RNA and DNA will not become entangled. Rather, the DNA's helical turns are pushed ahead of the advancing transcription bubble so as to more tightly wind the DNA ahead of the bubble (which promotes positive supercoiling), and the DNA behind the bubble becomes equivalently unwound (which promotes negative supercoiling, although note that the linking number of the entire DNA remains unchanged). This model is supported by the observa-

Figure 31-14 The two possible modes of RNA chain growth. Growth may occur (*a*) by the addition of nucleotides to the 3' end and (*b*) by the addition of nucleotides to the 5' end. RNA polymerase catalyzes the former reaction.

(a)

Figure 31-15 **RNA chain elongation by RNA polymerase.** In the region being transcribed, the DNA double helix is unwound by about a turn to permit the DNA's sense strand to form a short segment of DNA–RNA hybrid double helix with the RNA's 3′ end. As the RNAP advances along the DNA template (here to the right), the DNA unwinds ahead of the RNA's growing 3′ end and rewinds behind it, thereby stripping the newly synthesized RNA from the template (antisense) strand. *(a)* One way this might occur is by the RNAP following the path of the template strand about the DNA double helix, in which case the transcript would become wrapped about the DNA once per duplex turn.

(b) A second and more plausible possibility is that the RNA moves in a straight line while the DNA rotates beneath it. In this case the RNA would not wrap around the DNA but the DNA would become overwound ahead of the advancing transcription bubble and unwound behind it (consider the consequences of placing your finger between the twisted DNA strands in this model and pushing toward the right). The model presumes that the ends of the DNA, as well as the RNAP, are prevented from rotating by attachments within the cell (*black bars*). [After Futcher, B., *Trends Genet.* **4**, 271, 272 (1988).]

tions that the transcription of plasmids in *E. coli* causes their positive supercoiling in gyrase mutants (which cannot relax positive supercoils; Section 29-3Cd) and their negative supercoiling in topoisomerase I mutants (which cannot relax negative supercoils; Section 29-3Ca). In fact, by tethering RNAP to a glass surface and allowing it to transcribe DNA that had been fluorescently labeled at one end, Kazuhiko Kinosita demonstrated, through fluorescence microscopy (using techniques similar to those showing that the F_1F_0–ATPase is a rotary engine; Section 22-3Ce), that single DNA molecules rotated in the expected direction during transcription.

Inappropriate superhelicity in the DNA being transcribed halts transcription (Section 29-3C). Quite possibly the torsional tension in the DNA generated by negative superhe-

licity behind the transcription bubble is required to help drive the transcriptional process, whereas too much such tension prevents the opening and maintenance of the transcription bubble.

b. Transcription Occurs Processively and Rapidly

The *in vivo* rate of transcription is 20 to 70 nucleotides per second. Once an RNAP molecule has initiated transcription and moved away from the promoter, another RNAP can follow suit. The synthesis of RNAs that are needed in large quantities, ribosomal RNAs, for example, is initiated as often as is sterically possible, about once per second (Fig. 31-16). Processivity is accomplished without an obvious clamplike structure such as the sliding clamp of

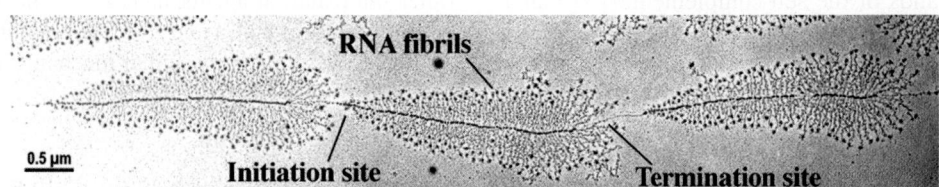

Figure 31-16 **An electron micrograph of three contiguous ribosomal genes from oocytes of the salamander *Pleurodeles waltl* undergoing transcription.** The "arrowhead" structures result from the increasing lengths of the nascent RNA chains as

the RNAP molecules synthesizing them move from the initiation site on the DNA to the termination site. [Courtesy of Ulrich Scheer, University of Würzburg, Germany.]

E. coli DNA polymerase III (Fig. 30-14). However, the RNAP itself apparently functions as a sliding clamp by binding tightly but flexibly to the DNA–RNA complex. In experiments in which the RNAP was immobilized and a magnetic bead was attached to the DNA, the bead was observed to undergo up to 180 rotations (representing nearly 2000 base pairs at 10.4 bp per turn) before the polymerase slipped.

c. Intercalating Agents Inhibit Both RNA and DNA Polymerases
Actinomycin D,

Actinomycin D

Figure 31-17 **NMR structure of actinomycin D in complex with a dsDNA of self-complementary sequence d(GAAGCTTC).** The actinomycin D is drawn in space-filling form and the DNA is drawn in stick form with successive P atoms on the same strand connected by orange rods, all colored according to atom type with actinomycin C green, DNA C cyan, H white, N blue, O red, and P orange. The complex is viewed toward the DNA's major groove. The actinomycin D's two cyclic depsipeptides are tightly wedged into the DNA's minor groove and the actinomycin D's phenoxazone ring system is intercalated between the DNA's central G · C base pairs. [Based on an NMR structure by Andrew Wang, University of Illinois. PDBid 1DSC.]

a useful antineoplastic (anticancer) agent produced by *Streptomyces antibioticus,* tightly binds to duplex DNA and, in doing so, strongly inhibits both transcription and DNA replication, presumably by interfering with the passage of RNA and DNA polymerases. The NMR structure of actinomycin D in complex with a duplex DNA composed of two strands of the self-complementary octamer d(GAAGCTTC) reveals that the DNA assumes a B-like conformation in which the actinomycin's **phenoxazone** ring system, as had previously been shown, is intercalated between the DNA's central G · C base pairs (Fig. 31-17). Consequently, the DNA helix is unwound by ~30° at the intercalation site and the central G · C base pairs are separated by ~7 Å. The DNA helix is severely distorted from the normal B-DNA conformation such that it is bent toward its major groove by ~30° and its minor groove is wide and shallow in a manner resembling that of A-DNA. Actinomycin D's two chemically identical cyclic **depsipeptides** (having both peptide bonds and ester linkages) extend in opposite directions from the intercalation site along the minor groove of the DNA. The complex is stabilized through the formation of base–peptide and phenoxazone–sugar–phosphate backbone hydrogen bonds, as well as by hydrophobic interactions, in a way that explains the preference of actinomycin D to bind to DNA with its phenoxazone ring intercalated between the base pairs of a 5′-GC-3′ sequence. Several other intercalation agents, including ethidium and acridine orange (Sections 6-6Ca and 29-3Ba), also inhibit nucleic acid synthesis, presumably by similar mechanisms.

D. *Chain Termination*

Electron micrographs such as Fig. 31-16 suggest that DNA contains specific sites at which transcription is terminated. In this section we discuss how transcription is terminated in bacteria. The eukaryotic process is discussed in Section 31-4Ab.

a. The RNA at Intrinsic Terminators Has an Oligo(U) Tract Preceded by a G + C–Rich Stem

Around half the transcriptional termination sites in *E. coli* are **intrinsic** or **spontaneous terminators,** that is, they induce termination without assistance. The sequences of these terminators share two common features (Fig. 31-18):

1. A tract of 7 to 10 consecutive A · T's with the A's on the template strand, sometimes interrupted by one or more different base pairs. The transcribed RNA is terminated in or just past this sequence.

2. A G + C–rich segment with a palindromic (2-fold symmetric) sequence that is immediately upstream of the series of A · T's.

The RNA transcript of this region can therefore form a self-complementary "hairpin" structure that is terminated by several U residues (Fig. 31-18).

The stability of a terminator's G + C–rich hairpin and the weak base pairing of its oligo(U) tail to template DNA are important factors in ensuring proper chain termination. In fact, model studies have shown that oligo(dA · rU) forms a particularly unstable hybrid helix although oligo(dA · dT) forms a helix of normal stability. In fact, oligo(dA · rU) tracts as long as 8 or 9 bp are unstable at room temperature when not bound to RNAP. The formation of the G + C–rich hairpin causes RNAP to pause for several seconds at the termination site. Mutations in the termination site that decrease the strengths of these associations reduce the efficiency of chain termination (the fraction of transcripts that are terminated at that site) and often eliminate it. Termination efficiency is similarly diminished when *in vitro* transcription is carried out with GTP replaced by **inosine triphosphate (ITP):**

I · C pairs are weaker than G · C pairs because the hypoxanthine base of I, which lacks the 2-amino group of G, can only make two hydrogen bonds to C, thereby decreasing the hairpin's stability.

Despite the foregoing, experiments by Michael Chamberlin in which segments of highly efficient terminators were swapped via recombinant DNA techniques indicate that the RNA terminator hairpin and U-rich 3′ tail do not function independently of their corresponding DNA's upstream and downstream flanking regions. Indeed, terminators that lack a U-rich segment can be highly efficient when joined to the appropriate sequence immediately downstream from the termination site.

These and other observations have led to three not necessarily mutually exclusive models to explain how intrinsic terminators work:

1. The forward translocation model, in which hairpin formation pushes the RNAP forward without the concomitant elongation of the RNA transcript. This would shorten the RNA–DNA hybrid by as much as several base pairs, thereby destabilizing it.

2. The RNA pullout model, in which hairpin formation mechanically pulls the RNA out of the RNA–DNA hybrid.

3. The allosteric model, in which hairpin formation induces a conformational change in the RNAP that permits the upstream nontemplate DNA strand to displace the weakly bound oligo(U) tail from the template DNA strand.

In an effort to differentiate these models, Robert Landick and Steven Block used **optical traps** to exert a pulling force on one or the other end of the DNA that a single RNAP molecule was transcribing or on its RNA transcript. An optical trap consists of a highly focused laser beam that is typically generated by sending it through a microscope objective lens. The resulting strong electric field gradient across the constricted region of the beam attracts dielectric (insulating) particles such as submicrometer sized polystyrene beads to the center of the beam where the electric field is strongest. The force on the particle varies directly with its displacement from the center of the beam. By attaching a single macromolecule of interest to such a bead in an optical trap and fixing the macromolecule's other end or attaching it to a bead in a second optical trap,

Inosine triphosphate (ITP)

5'-CAAAGCCCGCCGAAAGGCGGGCTTTTCT
3'-GTTTCGGGCGGCTTTCCG

5'-GACGCAGGCCAAA

Figure 31-18 **An *E. coli* intrinsic terminator.** Its transcription yields an RNA (*red*) with a self-complementary G + C–rich segment that forms a base-paired hairpin immediately followed by a sequence of 4 to 10 consecutive U's that base-pair with the template A's in the transcription bubble. The oval symbol represents the binding site for an incoming NTP. [After a drawing by Park, J.-S. and Roberts, J.W., Cornell University.]

a force can be exerted on the molecule by laterally displacing the beam as little as subnanometer distances. Such a device is known as an **optical tweezers.**

In the optical tweezers diagrammed in Fig. 31-19*a*, pulling apart the two optical traps would assist the RNAP in translocating along the DNA, whereas attaching the other end of the DNA to a bead would hinder this process. The application of either an assisting or hindering force to the DNA does not significantly affect the termination efficiencies of any of the three terminators shown in Fig. 31-19*c*. Evidently, forward translocation is not a general feature of intrinsic termination. However, the efficiency of the t500 terminator with a mutation in its hairpin varies with the force on the DNA, which indicates that forward translocation occurs with some terminators.

In the optical tweezers diagrammed in Fig. 31-19*b*, pulling on the RNA with sufficient force to disrupt the first 2 or 3 base pairs of the terminator hairpin reduces the ter-

mination efficiencies of all three terminators. If the force is greater than that required to fully unfold the hairpin, transcription efficiency is indistinguishable from that of the corresponding terminator containing only its oligo(U) tract. This suggests that the formation of the hairpin base pairs disrupts the adjacent RNA–DNA hybrid as predicted by the RNA pullout model.

Curiously, a force weaker than that required to disrupt hairpin base pairs increases termination efficiency. However, the presence of ssDNA complementary to the transcript eliminates this latter effect. Evidently, the RNA upstream of the terminator forms weakly base-paired secondary structures that compete with the formation of the terminator hairpin. In this way, the RNA sequence upstream of an intrinsic terminator modulates the efficiency of termination.

Since RNAP stabilizes the RNA–DNA hybrid, allosteric changes to RNAP by hairpin formation may also

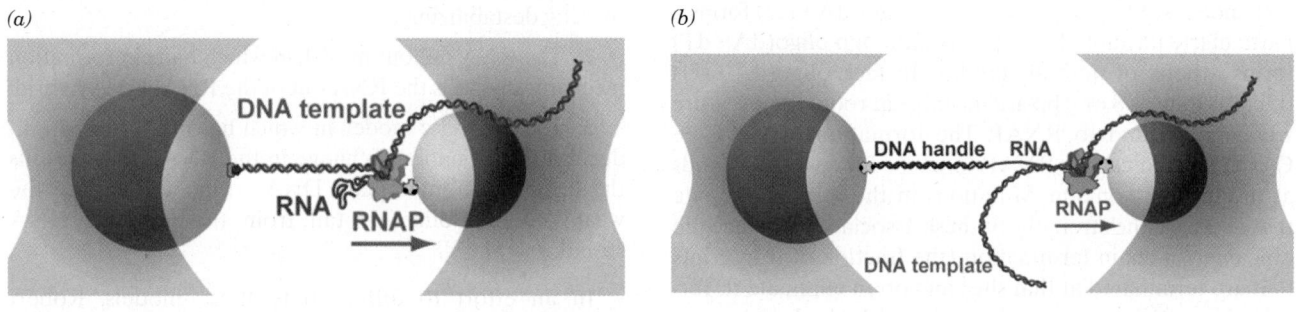

(a)

(b)

(c)

his terminator	t500 terminator	tR2 terminator

Figure 31-19 Apparatus for single molecule pulling assays on RNAP elongation complexes (not to scale). (*a*) A DNA-pulling assay. The RNAP (*green;* its direction of translocation is indicated by the green arrow) in an elongation complex was attached to an avidin-coated polystyrene bead (*light blue*) via a biotin–avidin linkage (*yellow and black;* Section 22-3Ce), and either the upstream end of the template DNA (*dark blue*) as shown or the downstream end was attached to a somewhat larger polystyrene bead via a digoxigenin–antidigoxigenin linkage [*purple and orange;* digoxigenin is a steroid related to digitalin (Fig. 20-21*b*) that has high antigenicity and antidigoxigenin is an antibody to which it specifically binds]. The RNA product of the complex (*red*) was untethered. The two beads were then placed in separate optical traps (*pink*) and the beads were pulled apart

while the elongation complex synthesized RNA. The horizontal displacement of a bead from the center of its optical trap is indicative of the pulling force on the bead. (*b*) An RNA-pulling assay. As in Part *a* but with the template DNA untethered and the RNA emerging from the RNAP attached to the second bead via a DNA handle (which had a 25-nt 3′ overhang complementary to the 5′ end of the RNA) tethered to the larger bead via a biotin–avidin linkage. (*c*) Structures of the three intrinsic terminators investigated in this study showing their hairpins and poly(U) tracts. The underlined bases are the transcript termination sites. The termination efficiencies of the *his,* t500, and tR2 terminators are normally 77%, 98%, and 46%, respectively. [Courtesy of Steven Block, Stanford University.]

influence termination efficiency. Indeed, mutations in the β subunit of RNAP can both increase and decrease termination efficiency. However, the way in which hairpin formation induces allosteric changes to RNAP is as yet unknown.

b. Many Bacterial Terminators Require the Assistance of Rho Factor

Around half the termination sites in *E. coli* lack any obvious similarities and are unable to form strong hairpins; *they require the participation of a protein known as **Rho factor** to terminate transcription.* Rho factor was discovered through the observation that *in vivo* transcripts are often shorter than the corresponding *in vitro* transcripts. Rho factor, a RecA family hexameric helicase (Section 30-2Ca) of identical 419-residue subunits, enhances the termination efficiency of spontaneously terminating transcripts as well as inducing the termination of nonspontaneously terminating transcripts.

Several key observations have led to a model of Rho-dependent termination:

1. Rho unwinds RNA–DNA and RNA–RNA double helices by translocating along a single strand of RNA in its $5' \rightarrow 3'$ direction. This process is powered by the hydrolysis of NTPs to NDPs + P_i with little preference for the identity of the base. NTPase activity is required for Rho-dependent termination as is demonstrated by its *in vitro* inhibition when the NTPs are replaced by their β,γ-imido analogs,

β,γ-Imido nucleoside triphosphate

substances that are RNAP substrates but cannot be hydrolyzed by Rho.

2. Genetic manipulations indicate that Rho-dependent termination requires the presence of a specific recognition sequence on the newly transcribed RNA upstream of the termination site. The recognition sequence must be on the nascent RNA rather than the DNA as is demonstrated by Rho's inability to terminate transcription in the presence of pancreatic RNase A. The essential features of this termination site have not been fully elucidated; the construction of synthetic termination sites indicates that it consists of 80 to 100 nucleotides that lack a stable secondary structure and contain multiple regions that are rich in C and poor in G.

These observations suggest that Rho attaches to nascent RNA at its recognition sequence [named ***rut*** (for *Rho ut*ilization), a C-rich segment of at least 40 nt] and then translocates along the RNA in the $5' \rightarrow 3'$ direction until it encounters an RNAP paused at the termination site (without the pause, Rho might not be able to overtake the RNA

polymerase). There, as Jeffrey Roberts has shown, Rho pushes the RNAP forward in a way that partially rewinds its dsDNA helix at the transcription bubble while unwinding the RNA–DNA hybrid helix (forward translocation), thus releasing the RNA. Rho-terminated transcripts have $3'$ ends that typically vary over a range of $\sim$50 nucleotides. This suggests that Rho pries the RNA away from the template DNA rather than "pushing" an RNA release "button." TCRF (alternatively, Mfd), which functions during transcription-coupled repair in *E. coli* to release a stalled RNAP from a damaged template by stripping away its bound RNA (Section 30-5Bb), is an ATP-powered DNA translocase that is thought to mechanically act on RNAP in much the same way as Rho.

Each Rho subunit consists of two domains that can be separated by proteolysis: Its N-terminal domain binds single-stranded polynucleotides and its C-terminal domain, which is homologous to the α and β subunits of the F_1–ATPase (Section 22-3Cb), binds an NTP. The X-ray structure of Rho in complex with AMPPNP and an 8-nt RNA, r(UC)$_4$ (Fig. 31-20*a*), determined by James Berger, reveals that Rho forms a hexameric lock washer–shaped helix that is 120 Å in diameter with an $\sim$30-Å-diameter central hole and whose first and sixth subunits are separated by a 12-Å gap and a rise of 45 Å along the helix axis. The RNAs, only a single UC unit of which is visible in each chain, bind along the top of the helix to the so-called primary RNA binding sites on the N-terminal domains, whereas AMPPNP binds to the C-terminal domains (which are further from the viewer in Fig. 31-20*a* than the N-terminal domains) at the interface between subunits. This X-ray structure represents an open state that has bound to *rut* site mRNA and is poised to bind additional mRNA upon its entry into the central cavity through the gap.

In the X-ray structure of Rho in complex with rU$_{12}$ and the ATP mimic ADP · BeF$_3$ (Fig. 31-20*b*), also determined by Berger, the helicase's six subunits have formed a closed ring in which each subunit has a different conformation. The RNA, only 6 nt of which are visible, assumes a right-handed helical conformation with its $5'$ end closest to the viewer in Fig. 31-20*b* and binds to Rho's N-terminal domains in the helicase's central channel, the so-called secondary RNA binding site. Protein loops that extend from the walls of the central channel to interact with the RNA are helically arranged like the steps of a right-handed spiral staircase such that they track the RNA's sugar–phosphate backbone, much like the central loops of E1 protein (an AAA+ family hexagonal helicase) tracks its centrally bound ssDNA (Section 30-2Ca). The different conformations of Rho's six subunits indicate that they sequentially undergo a series of six NTP-driven conformational changes and that these changes are allosterically coupled so that they progress around the hexamer in a wavelike manner. Since each of the foregoing loops maintains its grip on the same nucleotide during this process, the helicase translocates along its bound RNA, in much the same way that E1 protein translocates along its bound ssDNA. Why, then, do Rho and E1 protein move in opposite directions? Comparison of the structures of Rho and E1 protein

(a)

(b)

Figure 31-20 **X-ray structures of Rho factor.** (*a*) Rho in complex with r(UC)₄ (only one UC unit of which is visible) and AMPPNP. Each of the protein's six subunits are drawn in tube-and-arrow form in different colors with the upper right subunit colored in rainbow order from its N-terminus (*blue*) to its C-terminus (*red*). The UC units and the AMPPNP are shown in space-filling form with UC C green, AMPPNP C gray, N blue, O red, and P orange. The hexamer has a lock washer–like shape with the blue subunit ~45 Å closer to the viewer than the pink subunit. (*b*) Rho in complex with rU₁₂ (only 6 nt of which are visible) and ADP · BeF₃. The protein is drawn and colored as in Part *a*. The RNA and ADP · BeF₃ are shown in stick and space-filling form, respectively, with RNA C green, ADP C gray, N blue, O red, P orange, Be light green, and F light blue. Note that each of the Rho subunits has a different conformation. [Based on X-ray structures by James Berger, University of California at Berkeley. PDBid 1PVO and 3ICE.]

indicates that the relative order of the conformational states around the hexamer for Rho is opposite that for E1 protein. Evidently, the "firing order" of Rho's NTPase sites around the hexamer is the reverse of that of E1 protein, thus accounting for their differing directions of translocation. Presumably, other RecA and AAA+ family hexagonal helicases otherwise have similar mechanisms.

E. *Eukaryotic RNA Polymerases*

Eukaryotic nuclei, as Robert Roeder and William Rutter discovered, contain three distinct types of RNAPs that differ in the RNAs they synthesize:

1. RNA polymerase I (**RNAP I;** also called **Pol I** and **RNAP A**), which is located in the nucleoli (dense granular bodies in the nuclei that contain the ribosomal genes; Section 31-4Bb), synthesizes precursors of most ribosomal RNAs (rRNAs).

2. RNA polymerase II (**RNAP II;** also called **Pol II** and **RNAP B**), which occurs in the nucleoplasm, synthesizes mRNA precursors.

3. RNA polymerase III (**RNAP III;** also called **Pol III** and **RNAP C**), which also occurs in the nucleoplasm, syn-

thesizes the precursors of 5S ribosomal RNA, the tRNAs, and a variety of other small nuclear and cytosolic RNAs.

Eukaryotic nuclear RNAPs have considerably greater subunit complexity than those of prokaryotes. These enzymes have molecular masses of up to 600 kD and, as is indicated in Table 31-2, each contains two nonidentical "large" (>120 kD) subunits comprising ~65% of its mass that are homologs of the prokaryotic RNAP β′ and β subunits and up to 12 additional "small" (<50 kD) subunits, two of which are homologs of prokaryotic RNAP α, and one of which is a homolog of prokaryotic RNAP ω. Of these small subunits, five are identical in all three eukaryotic RNAPs and two others (the RNAP α homologs) are identical in RNAPs I and III. Two of the RNAP II subunits, Rbp4 and Rbp7, are not essential for activity and, in fact, are present in RNAP II in less than stoichiometric amounts. (Curiously, Rbp7 has a 102-residue segment that is 30% identical to a portion of σ⁷⁰, the predominant *E. coli* σ factor.) Thus 10 of the 12 RNAP II subunits are either identical or closely similar to subunits of RNAPs I and III (Table 31-2). Moreover, the sequences of these subunits are highly conserved (~50% identical) across species from yeast to humans (and to a lesser extent between eukaryotes and bac-

Table 31-2   **RNA Polymerase Subunits**[a]

S. cerevisiae RNAP I (14 subunits)	*S. cerevisiae* RNAP II (12 subunits)	*S. cerevisiae* RNAP III (15 subunits)	*E. coli* RNAP Core (5 subunits)	Class[b]
Rpa1 (A190)	Rbp1 (B220)	Rpc1 (C160)	β′	Core
Rpa2 (A135)	Rbp2 (B150)	Rpc2 (C128)	β	Core
Rpc5 (AC40)	Rpb3 (B44.5)	Rpc5 (AC40)	α	Core
Rpc9 (AC19)	Rpb11 (B13.6)	Rpc9 (AC19)	α	Core
Rbp6 (ABC23)	Rbp6 (ABC23)	Rpb6 (ABC23)	ω	Core/common
Rpb5 (ABC27)	Rpb5 (ABC27)	Rpb5 (ABC27)		Common
Rpb8 (ABC14.4)	Rpb8 (ABC14.4)	Rpb8 (ABC14.4)		Common
Rbp10 (ABC10β)	Rpb10 (ABC10β)	Rpb10 (ABC10β)		Common
Rbp12 (ABC10α)	Rpb12 (ABC10α)	Rpb12 (ABC10α)		Common
Rpa9 (A12.2)	Rpb9 (B12.6)	Rpc12 (C11)		
Rpa8 (A14)[c]	Rpb4 (B32)	—		
Rpa4 (A43)[c]	Rpb7 (B16)	Rpc11 (C25)		
+2 others[d]		+4 others[d]		

[a]Homologous subunits occupy the same row. In the alternative subunit names in parentheses, the letter(s) indicates the RNAPs in which the subunit is a component (A, B, and C for RNAPs I, II, and III) and the numbers indicate its approximate molecular mass in kilodaltons.
[b]Core: sequence partially homologous in all RNAPs; common: shared by all eukaryotic RNAPs.
[c]Potential homologs of Rbp4 and Rbp7.
[d]Rpa3 (A49) and Rpa5 (A34.5) in RNAP I and Rpc3 (C74), Rpc4 (C53), Rpc6 (C34), and Rpc8 (C31) in RNAP III.

Source: Mainly Cramer, P., *Curr. Opin. Struct. Biol.* **12,** 89 (2002).

teria). In fact, in all ten cases tested, a human RNAP II subunit could replace its counterpart in yeast without loss of cell viability.

Rpb1, the β′ homolog in RNAP II, has an extraordinary C-terminal domain (**CTD**). In mammals, it contains 52 highly conserved repeats of the heptad PTSPSYS (26 repeats in yeast with other eukaryotes having intermediate values). Five of the seven residues in these particularly hydrophilic repeats bear hydroxyl groups and at least 50 of them, predominantly those on the second Ser residue in each heptad, are subject to reversible phosphorylation by **CTD kinases** and **CTD phosphatases.** RNAP II initiates transcription only when the CTD is unphosphorylated but commences elongation only after the CTD has been phosphorylated, which suggests that this process triggers the conversion of RNAP II's initiation complex to its elongation complex. Charge–charge repulsions between nearby phosphate groups probably cause a highly phosphorylated CTD to project as far as 500 Å from the globular portion of RNAP II. Indeed, as we shall see, the phosphorylated CTD provides the binding sites for numerous auxiliary factors that have essential roles in the transcription process.

In contrast to the somewhat smaller prokaryotic RNAP holoenzymes, eukaryotic RNAPs do not independently bind their target DNAs. Rather, as we shall see in Section 34-3B, they are recruited to their target promoters through the mediation of complexes of transcription factors and their ancillary proteins that, in the case of RNAP II–transcribed genes, are so large and complicated that they collectively dwarf RNAP II.

In addition to the foregoing nuclear enzymes, eukaryotic cells contain separate mitochondrial and (in plants) chloroplast RNAPs. These small (~100 kD) single-subunit RNAPs, which resemble those encoded by certain bacteriophages, are much simpler than the nuclear RNAPs although they catalyze the same reaction.

a. X-Ray Structures of Yeast RNAP II Reveal a Transcribing Complex

In a crystallographic tour de force, Roger Kornberg determined the X-ray structure of yeast (*S. cerevisiae*) RNAP II that lacks its nonessential Rpb4 and Rpb7 subunits (Fig. 31-21). This enzyme, as expected, resembles Tth RNAP (Fig. 31-11) in its overall crab claw–like shape and in the positions and core folds of their homologous subunits although, of course, RNAP II is somewhat larger than and has several subunits that have no counterpart in bacterial RNAPs. RNAP II binds two Mg^{2+} ions at its active site in the vicinity of five conserved acidic residues (although one of these Mg^{2+} ions appears to be weakly bound and hence is only faintly visible in the X-ray structure; it apparently accompanies the incoming NTP). This suggests that RNAPs catalyze RNA elongation via a two-metal ion mechanism similar to that employed by DNA polymerases (Section 30-2Af). As is the case with bacterial RNAPs, the surface of RNAP II is almost entirely negatively charged except for its main channel and the region about the active site, which are positively charged.

Although, as mentioned above, RNAP II does not normally initiate transcription by itself, Kornberg found that it

(a)   *(b)*

Figure 31-21   **The X-ray structure of yeast RNAP II that lacks its Rpb4 and Rpb7 subunits.** (*a*) The enzyme is oriented similarly to Tth RNAP in Fig. 31-11 and its subunits are colored as is indicated in the accompanying diagram, with the subunits homologous to those of Tth RNAP given the same colors. The strongly bound Mn^{2+} ion (physiologically Mg^{2+}) that marks the active site is shown as a red sphere and the enzyme's 8 bound Zn^{2+} ions are shown as orange spheres. The Rpb1 C-terminal domain (CTD) is not visible due to disorder. In the accompanying diagram, the area of each numbered ellipsoid is proportional to the corresponding subunit's size and the width of each gray line connecting a pair of subunits is proportional to the surface area of their interface. (*b*) View of the enzyme from the right in Part *a* looking into its DNA-binding main channel. The black circle has the approximate diameter of B-DNA. [Based on an X-ray structure by Roger Kornberg, Stanford University. PDBid 1I50.]

will do so on a dsDNA bearing a 3′ single-stranded tail at one end. Consequently, incubating yeast RNAP II with the DNA shown in Fig. 31-22*a* and all NTPs but UTP yielded the DNA–RNA hybrid helix diagrammed in Fig. 31-22*a* bound to RNAP II. The X-ray structure of this paused transcribing complex revealed, as expected, that the dsDNA had bound in the enzyme's main channel (Fig. 31-22*b,c;* transcription resumed on soaking the crystals in UTP, thereby demonstrating that the crystalline complex was active). In comparison with the X-ray structure of RNAP II alone, a massive (~50 kD) portion of Rpb1 and Rpb2 named the "clamp" has swung down over the DNA to trap it in the main channel, in large part accounting for the enzyme's essentially infinite processivity. The mainly rigid motion of the clamp is mediated by conformational changes at five so-called switch regions at the base of the clamp in which three of these switches, which are disordered in the structure of RNAP II alone, become ordered in the transcribing complex.

The DNA unwinds by three bases before entering the active site (which is contained on Rpb1). Past this point, however, a portion of Rpb2 dubbed the "wall" directs the template strand out of the cleft in an ~90° turn. As a consequence, the template base at the active (*i* + 1) site points toward the floor of the cleft where it can be read out by the active site. This base is paired with the ribonucleotide at the 3′ end of the RNA, which is positioned above a 12-Å-diameter pore at the end of a funnel to the protein exterior (also called the secondary channel) through which NTPs presumably gain access to the otherwise sealed off active site. The RNA–DNA hybrid helix adopts a nonstandard conformation intermediate between those of A- and B-DNAs, which is underwound relative to that in the X-ray structure of an RNA–DNA hybrid helix alone (Fig. 29-4). Nearly all contacts that the RNAP makes with the RNA and DNA are with their sugar–phosphate backbones; none are with the edges of their bases. The specificity of the enzyme for a ribonucleotide rather than a deoxyribonucleotide is attributed to the enzyme's recognition of both the incoming ribose sugar and the RNA–DNA hybrid helix. After about one turn of hybrid helix, a loop extending from the clamp called the "rudder" separates the RNA and template DNA strands, thereby permitting the DNA double helix to reform as it exits the enzyme (although the unpaired 5′ tail of

```
                    G  3'
                    G
              5'    G
        +20   C-G
              T-A
              T-A      Downstream
              C-G      DNA duplex
              C-G
        +15   G-C
              C-G
              A-T      Non-template
              A-T      strand
              C-G
        +10   A-T
              A-T
              G-C
              A-T
              A-T
         +5   A-TTTTACGGACCAGAA  5'
              A
              A
Pause         A         RNA 3' end
site    +1    T-A
        -1    G-C • Mg²⁺
              C-G
              C-G
              T-A
DNA/RNA  -5   G-C     RNA
hybrid        G-C
              T-A       ↑
              C-G     RNAP II
              T  A
        -10   T  A
              C  G
              C  G     Transcription
              C (G) 5' start site
              C
(a)           CCCCCCCC  3'  Tail
```

Figure 31-22 X-ray structure of an RNAP II elongation complex. (*a*) The RNA · DNA complex in the structure with the template DNA cyan, the nontemplate DNA green, and the newly synthesized RNA red. The magenta dot marked Mg²⁺ represents the strongly bound active site metal ion. The black box encloses those portions of the complex that are clearly visible in the structure; the double-stranded portion of the DNA marked "Downstream DNA duplex" is poorly ordered, and the remaining portions of the complex are disordered. (*b*) View of the transcribing complex from the bottom of Fig. 31-21*a* in which portions of Rpb2 that form the near side of the cleft have been removed to expose the bound RNA · DNA complex. The protein is represented by its backbone in which the clamp, which is closed over the downstream DNA duplex, is yellow, the bridge helix is green, and the remaining portions of the protein are gray. The DNA and RNA are colored as in Part *a* with their well-ordered portions drawn in ladder form and their less-ordered portions drawn in backbone form. The active site Mg²⁺ ion is represented by a magenta sphere. (*c*) Cutaway schematic diagram of the transcribing complex in Part *b* in which the cut surfaces of the protein are light gray, its remaining surfaces are darker gray, and several of its functionally important structural features are labeled. The DNA, RNA, and active site Mg²⁺ ion

are colored as in Part *a* with portions of the DNA and RNA that are not visible in the X-ray structure represented by dashed lines. The α-amanitin binding site is marked by an orange circle. [Modified from diagrams by Roger Kornberg, Stanford University. PDBid 1I6H.] 🔗 **See Interactive Exercise 37**

the nontemplate strand and the 3′ tail of the template strand are disordered in the X-ray structure).

How does RNAP translocate its bound RNA–DNA assembly in preparation for a new round of synthesis? The highly conserved helical segment of Rpb1, dubbed the "bridge" because it bridges the two pincers forming the enzyme's cleft (Figs. 31-21 and 31-22), nonspecifically contacts

the template DNA base at the *i* + 1 position. Although this helix is straight in all X-ray structures of RNAP II yet determined, it is bent in that of Taq core RNAP. If the bridge helix, in fact, alternates between its straight and bent conformations, it would move by 3 to 4 Å. Kornberg has therefore speculated that translocation occurs through the bending of the bridge helix so as to push the paired nucleotides at posi-

(a)

(b)

Figure 31-23 **Proposed transcription cycle and translocation mechanism of RNAP.** (*a*) The nucleotide addition cycle in which the enzyme active site is marked by its strongly bound Mg^{2+} ion (*magenta*). The translocation of the transcribing RNA · DNA complex is proposed to be motivated by a conformational change of the bridge helix from straight (*gray circle*) to bent (*violet circle*). The relaxation of the bridge helix back to its straight form would complete the cycle by yielding an empty NTP binding site at the active ($i + 1$) site. (*b*) The RNA · DNA complex in RNAP II viewed and colored as in Fig. 31-22*b*. The RNAP II bridge helix is gray and the superimposed (and bent) Taq polymerase bridge helix is violet. The side chains extending from the bent helix would sterically clash with the hybrid base pair at position $i + 1$. [Courtesy of Roger Kornberg, Stanford University.]

tion $i + 1$ to position $i - 1$ (Fig. 31-23). The recovery of the bridge helix to its straight conformation would then yield an empty site at position $i + 1$ for entry of the next NTP, thereby preparing the enzyme for a new round of nucleotide addition. The reversal of this process is presumably prevented by the binding of the next substrate NTP and hence this mechanism is that of a **Brownian ratchet** [in which otherwise random thermal (Brownian) back-and-forth fluctuations are converted to coherent forward motion by inhibiting the backward motion; Section 12-4Bg].

RNAP II selects its substrate ribonucleotide through a two-stage process. The incoming NTP gains access to the active site through the funnel and pore (secondary channel) diagrammed in Fig. 31-22c. There it first binds to the so-called E (for *entry*) site (Fig. 31-24), which exhibits no selectivity for the identity of its base. The NTP then pivots to enter the A (for *addition*) site, which only accepts an NTP that forms a Watson–Crick base pair with the template base in the $i + 1$ position. This process is mediated by the Rbp1 subunit's so-called trigger loop, which swings in beneath the correctly base-paired NTP in the A site to form an extensive hydrogen-bonded network involving both the NTP and other portions of the RNAP, interactions that acutely discriminate against dNTPs.

b. Amatoxins Specifically Inhibit RNA Polymerases II and III

The poisonous mushroom *Amanita phalloides* (**death cap**), which is responsible for the majority of fatal mushroom poisonings, contains several types of toxic substances,

including a series of unusual bicyclic octapeptides known as **amatoxins**. α-Amanitin,

α-Amanitin

which is representative of the amatoxins, forms a tight 1:1 complex with RNAP II ($K = 10^{-8} M$) and a looser one with RNAP III ($K = 10^{-6} M$). Its binding slows an RNAP's rate of RNA synthesis from several thousand to only a few nucleotides per minute. α-Amanitin is therefore a useful tool for mechanistic studies of these enzymes. RNAP I as well as mitochondrial, chloroplast, and bacterial RNAPs are insensitive to α-amanitin.

The X-ray structure of RNAP II in complex with α-amanitin, also determined by Kornberg, reveals that α-amanitin binds in the funnel beneath the protein's bridge helix (Fig. 31-22c), where it interacts with residues of the bridge helix and the trigger loop. The α-amanitin binding site is too far away from the enzyme active site to directly interfere with NTP entry or RNA synthesis, consistent with the observation that α-amanitin does not influence the affinity of RNAP II for NTPs, although it reduces its selectivity. Mutation of Rbp1 His 1085, an invariant member of the trigger loop, to Tyr mimics the effects of α-amanitin. Moreover, this mutation renders RNAP II highly resistant to α-amanitin, in agreement with an X-ray structure indicating that α-amanitin interacts with the side chain of His 1085 so as to lock the trigger loop in a previously unobserved conformation. Evidently, α-amanitin interferes with the conformational change of the trigger loop postulated to promote catalysis (Fig. 31-24), which further supports this mechanism.

Despite the amatoxins' high toxicity (5–6 mg, which occurs in ~40 g of fresh mushrooms, is sufficient to kill a human adult), they act slowly. Death, usually from liver dysfunction, occurs no earlier than several days after mushroom ingestion (and after recovery from the effects of other mushroom toxins). This, in part, reflects the slow turnover of eukaryotic mRNAs and proteins.

c. RNAPs Can Correct Their Mistakes

RNAPs cannot read through a damaged template strand and consequently stall at the damage site. Moreover, if a deoxynucleotide or a mispaired ribonucleotide is mistakenly incorporated into RNA, the DNA–RNA hybrid helix becomes distorted, which also causes the RNAP

Figure 31-24 The A and E sites and the trigger loop in RNA polymerase II. A cutaway view of the transcribing complex viewed as in Fig. 31-22*b*. Its bound nucleic acids and nucleotides are colored differently with template DNA cyan, nontemplate DNA green, newly transcribed RNA red, GTP in the A site orange, and ATP in the E site blue. The trigger loop is magenta, the bridge helix is green, and the two Mg²⁺ ions at the active site are represented by magenta spheres. The RNAP II surface is gray. [Courtesy of Dong Wang and Roger Kornberg, Stanford University. PDBid 2E2H.]

to stall. How, then, do RNAPs avoid accumulating at damaged or mispaired sites, which, if it occurred on an essential gene, would be lethal?

RNAPs do not monotonically move forward along the template DNA. Instead, they frequently backtrack such that the RNA's penultimate nucleotide, which was in the $i - 1$ position, has re-entered the $i + 1$ position and the 3′-nucleotide, now in the $i + 2$ position, enters the secondary channel where it binds in the so-called P (for *p*roofreading) site. If the forward movement of the RNA is impeded by damage to the template or by mispairing, further backtracking becomes favored so that several more ribonucleotides enter the secondary channel. The backtracking of only one or a few nucleotides is reversible. Otherwise, transcription is arrested until the RNA is hydrolytically cleaved at the active site. In *E. coli,* this requires the assistance of the homologous proteins **GreA** and/or **GreB,** whereas with RNAP II, this function is carried out by the unrelated protein **TFIIS.** These proteins induce the RNAP active site to hydrolyze the phosphodiester bond between the ribonucleotides in the $i + 1$ and $i - 1$ positions (a reaction that is not the reverse of the polymerase reaction since this would be pyrophosphorolysis). In this way, RNAP can correct its mistakes and resume RNA synthesis. RNAP I and RNAP III also efficiently correct their mistakes.

Kornberg determined the X-ray structure of RNAP II in complex with TFIIS, a 28-nt template DNA, a 14-nt nontemplate DNA, and a 13-nt RNA that is complementary to 5′ end of the template DNA except for the last two residues at the RNA's 3′ end, which are mismatched (Fig. 31-25). The RNA–DNA hybrid has backtracked such that these latter residues occupy the $i + 1$ and $i + 2$ positions. The C-terminal domain of TFIIS is bound in the RNAP's funnel with one of its loops insinuated through the pore to interact with the RNAP's active site residues. There it presum-

Figure 31-25 X-ray structure of backtracked RNA polymerase II in complex with DNA, RNA, and TFIIS. The RNAP, which is represented by its semitransparent surface diagram, is viewed as in Figs. 31-22*b* and 31-24 with its subunits colored as in Fig. 31-21*a*. The DNA and RNA are shown in ladder form with the RNA (13 nt) red, the template DNA (28 nt) green, and the nontemplate DNA (14 nt) blue. The base pairs at the $i + 1$ and $i + 2$ positions of the RNA · DNA hybrid helix are mismatched. An Mg²⁺ ion, which is represented by a red sphere near the $i + 1$ position, marks the RNAP's active site. TFIIS is drawn in cartoon form in magenta. Its C-terminal domain is inserted into the RNAP's funnel (Fig. 31-22*c*) with a loop occupying the pore, where it is in proximity to the RNAP's active site. [Based on an X-ray structure by Roger Kornberg, Stanford University. PDBid 3GTM.]

ably facilitates the hydrolytic reaction, perhaps by liganding the active site Mg^{2+} ion that normally accompanies the incoming NTP (Section 31-2Ea) and/or positioning a hydrolytic water molecule. The trigger loop in this structure is in the "open" conformation. Interestingly, the cryo-EM–based structure of *E. coli* RNAP in complex with GreB indicates that GreB likewise inserts an extended protein finger in the RNAPs active site via its secondary channel, even though the structures of GreB and TFIIS are unrelated.

Despite the foregoing, transcription is less accurate than DNA replication: RNAPs incorporate one incorrect base for every $\sim 10^4$ transcribed, whereas, for example, *E. coli* Pol I incorporates one incorrect base in $\sim 10^7$ (Section 30-2Ab). Cells can tolerate the former rate because most genes are repeatedly transcribed. In contrast, errors in DNA synthesis alter all the affected gene's transcripts in the cell in which the error occurred and all of its progeny.

d. Mammalian RNA Polymerase I Has a Bipartite Promoter

Since, as we shall see in Section 31-4B, the numerous rRNA genes in a given eukaryotic cell have essentially identical sequences, its RNAP I only recognizes one promoter. Yet, in contrast to the case for RNAPs II and III, RNAP I promoters are species specific, that is, an RNAP I only recognizes its own promoter and those of closely related species. This is because only closely related species exhibit recognizable sequence identities near the transcriptional start sites of their rRNA genes. RNAP I promoters were therefore identified by determining how the transcription rate of an rRNA gene is affected by a series of increasingly longer deletions approaching its start site from either its upstream or its downstream sides. Such studies have indicated, for example, that mammalian RNAPs I require the presence of a so-called **core promoter element,** which spans positions -31 to $+6$ and hence overlaps the transcribed region. However, efficient transcription additionally requires an **upstream promoter element,** which is located between residues -187 and -107. These elements, which are G + C-rich and $\sim 85\%$ identical, are bound by specific transcription factors which then recruit RNAP I to the transcription start site.

e. RNA Polymerase II Promoters Are Complex and Diverse

The promoters recognized by RNAP II are considerably longer and more diverse than those of prokaryotic genes but have not yet been fully described. The structural genes expressed in all tissues, the so-called **housekeeping genes,** which are thought to be constitutively transcribed, have one or more copies of the sequence GGGCGG or its complement (the **GC box**) located upstream from their transcription start sites. The analysis of deletion and point mutations in eukaryotic viruses such as SV40 indicates that GC boxes function analogously to prokaryotic promoters. On the other hand, structural genes that are selectively expressed in one or a few types of cells often lack these GC-rich sequences. Rather, *many contain a conserved AT-rich sequence located 25 to 30 bp upstream from their transcription start sites* (Fig. 31-26). Note that this so-called **TATA box** resembles the -10 region of prokaryotic promoters (TATAAT), although they differ in their locations relative to the transcription start site (-27 vs -10). The functions of these two promoter elements are not strictly analogous, however, since the deletion of the TATA box does not necessarily eliminate transcription. Rather, TATA box deletion or mutation generates heterogeneities in the transcriptional start site, thereby indicating that the TATA box participates in selecting this site.

The gene region extending between about -50 and -110 also contains promoter elements. For instance, many eukaryotic structural genes, including those encoding the various globins, have a conserved sequence of consensus CCAAT (the **CCAAT box**) located between about -70 and -90 whose alteration greatly reduces the gene's transcription rate. Globin genes have, in addition, a conserved **CACCC box** upstream from the CCAAT box that has also been implicated in transcriptional initiation. Evidently, the promoter sequences upstream of the TATA box form the initial DNA-binding sites for RNA polymerase II and the other proteins involved in transcriptional initiation (see below).

f. Enhancers Are Transcriptional Activators That Can Have Variable Positions and Orientations

Perhaps the most surprising aspect of eukaryotic transcriptional control elements is that some of them need not

---

Figure 31-26   The promoter sequences of selected eukaryotic structural genes. The homologous segment, the TATA box, is shaded in red with the base at position -27 underlined and the initial nucleotide to be transcribed ($+1$) shaded in green. The bottom row indicates the consensus sequence of several such promoters with the subscripts indicating the percent occurrence of the corresponding base. [After Gannon, F., et al., *Nature* **278,** 433 (1978).]

Chicken ovalbumin

GAGGCTATATA**T**TCCCCAGGGCTCAGCCAGTGTCTGTA**C**A

Adenovirus late

GGGGCTATA**A**AAGGGGGTGGGGGCGCGTTCGTCCTC**A**CTC

Rabbit β globin

TTGGGCATAA**A**AGGCAGAGCAGGGCAGCTGCTGCTA**A**CACT

Mouse β globin major

GAGCATATA**A**GGTGAGGTAGGATCAGTTGCTCCTC**A**CATTT

$T_{82}A_{97}T_{93}A_{85}\,{}_{T_{37}}^{A_{63}}\,A_{83}\,{}_{T_{37}}^{A_{50}}$

have fixed positions and orientations relative to their corresponding transcribed sequences. For example, the SV40 genome, in which such elements were first discovered, contains two repeated sequences of 72 bp each that are located upstream from the promoter for early gene expression. Transcription is unaffected if one of these repeats is deleted but is nearly eliminated when both are absent. The analysis of a series of SV40 mutants containing only one of these repeats demonstrated that its ability to stimulate transcription from its corresponding promoter is all but independent of its position and orientation. Indeed, transcription is unimpaired when this segment is several thousand base pairs upstream or downstream from the transcription start site. Gene segments with such properties are named **enhancers** to indicate that they differ from promoters, with which they must be associated in order to trigger site-specific and strand-specific transcription initiation (although the characterization of numerous promoters and enhancers indicates that their functional properties are similar). Enhancers occur in both eukaryotic viruses and cellular genes.

Enhancers are required for the full activities of their cognate promoters. It was originally thought that enhancers somehow acted as entry points on DNA for RNAP II (perhaps by altering DNA's local conformation or through a lack of binding affinity for the histones that normally coat eukaryotic DNA; Section 34-1B). However, it is now clear that *enhancers are recognized by specific transcription factors that stimulate RNA polymerase II to bind to the corresponding but distant promoter.* This requires that the DNA between the enhancer and promoter loop around so that the transcription factor can simultaneously contact the enhancer and the RNAP II and/or its associated proteins at the promoter. Most cellular enhancers are associated with genes that are selectively expressed in specific tissues. It therefore seems, as we discuss in Section 34-3B, that *enhancers mediate much of the selective gene expression in eukaryotes.*

g. RNA Polymerase III Promoters Can Be Located Downstream from Their Transcription Start Sites

The promoters of genes transcribed by RNAP III can be located entirely within the genes' transcribed regions. Donald Brown established this through the construction of a series of deletion mutants of a *Xenopus borealis* 5S RNA gene. Deletions of base sequences that start from outside one or the other end of the transcribed portion of the 5S gene only prevent transcription if they extend into the segment between nucleotides +40 and +80. Indeed, a fragment of the 5S RNA gene consisting of only nucleotides 41 to 87, when cloned in a bacterial plasmid, is sufficient to direct specific initiation by RNAP III at an upstream site. This is because, as was subsequently demonstrated, the sequence contains the binding site for transcription factors that stimulate the upstream binding of RNAP III. Further studies have shown, however, that the promoters of other RNAP III–transcribed genes lie entirely upstream of their start sites. These upstream sites also bind transcription factors that recruit RNAP III.

3 CONTROL OF TRANSCRIPTION IN PROKARYOTES

Prokaryotes respond to sudden environmental changes, such as the influx of nutrients, by inducing the synthesis of the appropriate proteins. This process takes only minutes because transcription and translation in prokaryotes are closely coupled: *Ribosomes commence translation near the 5′ end of a nascent mRNA soon after it is extruded from RNA polymerase* (Fig. 31-27). Moreover, *most prokaryotic mRNAs are enzymatically degraded within 1 to 3 min of their synthesis,* thereby eliminating the wasteful synthesis of unneeded proteins after a change in conditions (protein degradation is discussed in Section 32-6). In fact, the 5′ ends of some mRNAs are degraded before their 3′ ends have been synthesized.

In contrast, the induction of new proteins in eukaryotic cells frequently takes hours or days, in part because transcription takes place in the nucleus and the resulting mRNAs must be transported to the cytoplasm, where translation occurs. However, eukaryotic cells, particularly those of multicellular organisms, have relatively stable environments; major changes in their transcription patterns usually occur only during cell differentiation.

Figure 31-27 **An electron micrograph and its interpretive drawing showing the simultaneous transcription and translation of an *E. coli* gene.** RNA polymerase molecules are transcribing the DNA from right to left while ribosomes are translating the nascent RNAs (mostly from bottom to top). [Courtesy of Oscar L. Miller, Jr. and Barbara Hamkalo, University of Virginia.]

In this section we examine some of the ways in which prokaryotic gene expression is regulated through transcriptional control. Eukaryotes, being vastly more complex creatures than are prokaryotes, have a correspondingly more complicated transcriptional control system whose general outlines are beginning to come into focus. We therefore defer discussion of eukaryotic transcriptional control until Section 34-3B, where it can be considered in light of what we know about the structure and organization of the eukaryotic chromosome.

A. *Promoters*

In the presence of high concentrations of inducer, the *lac* operon (Section 31-1Ab) is rapidly transcribed. In contrast, the *lacI* gene is transcribed at such a low rate that a typical *E. coli* cell contains <10 molecules of the *lac* repressor. Yet, the *I* gene has no repressor. Rather, it has such an inefficient promoter (Fig. 31-10) that it is transcribed an average of about once per bacterial generation. *Genes that are transcribed at high rates have efficient promoters.* In general, the more efficient a promoter, the more closely its sequence resembles that of the corresponding consensus sequence.

a. Gene Expression Can Be Controlled by a Succession of σ Factors

The processes of development and differentiation involve the temporally ordered expression of sets of genes according to genetically specified programs. Phage infections are among the simplest examples of developmental processes. Typically, only a subset of the phage genome, often referred to as *early* genes, are expressed in the host immediately after phage infection. As time passes, *middle* genes start to be expressed, and the *early* genes as well as the bacterial genes are turned off. In the final stages of phage infection, the *middle* genes give way to the *late* genes. Of course some phage types express more than three sets of genes and some genes may be expressed in more than one stage of an infection.

One way in which families of genes are sequentially expressed is through "cascades" of σ factors. In the infection of *Bacillus subtilis* by bacteriophage SP01, for example, the *early* gene promoters are recognized by the bacterial RNAP holoenzyme. Among these *early* genes is gene 28, whose gene product is a new σ subunit, designated σ^{gp28}, that displaces the bacterial σ subunit from the core enzyme. The reconstituted holoenzyme recognizes only the phage *middle* gene promoters, which all have similar −35 and −10 regions but bear little resemblance to the corresponding regions of bacterial and phage *early* genes. The *early* genes therefore become inactive once their corresponding mRNAs have been degraded. The phage *middle* genes include genes 33 and 34, which together specify yet another σ factor, $\sigma^{gp33/34}$, which, in turn, permits the transcription of only *late* phage genes.

Most bacteria, including *E. coli* and *B. subtilis,* likewise have several different σ factors (*E. coli* has seven). These are not necessarily utilized in a sequential manner. Rather,

those that differ from the predominant or primary σ factor (σ^{70} in *E. coli*) control the transcription of coordinately expressed groups of special purpose genes, whose promoters are quite different from those recognized by the primary σ factor. For example, in *E. coli,* the alternative σ factor σ^{32} is the master regulator of the heat shock response (Section 9–2C), whereas σ^{54} directs the expression of proteins involved in nitrogen assimilation. Likewise, sporulation in *B. subtilis,* a process in which the bacterial cell is asymmetrically partitioned into two compartments, the **forespore** (which becomes the **spore,** a germline cell from which subsequent progeny arise) and the **mother cell** (which synthesizes the spore's protective cell wall and is eventually discarded), is governed by five σ factors in addition to that of the **vegetative** (nonsporulating) cell: one that is active before cell partition occurs, two that are sequentially active in the forespore, and two that are sequentially active in the mother cell. Cross-regulation of the compartmentalized σ factors permits the forespore and mother cell to tightly coordinate this differentiation process.

B. *lac Repressor I: Binding*

In 1966, Benno Müller-Hill and Walter Gilbert isolated *lac* repressor on the basis of its ability to bind ^{14}C-labeled IPTG (Section 31-1Aa) and demonstrated that it is a protein. This was an exceedingly difficult task because *lac* repressor comprises only ~0.002% of the protein in wild-type *E. coli.* Now, however, *lac* repressor is available in quantity via molecular cloning techniques (Section 5-5G).

a. *lac* Repressor Finds Its Operator by Sliding Along DNA

The *lac* repressor is a tetramer of identical 360-residue subunits, each of which binds one IPTG molecule with a dissociation constant of $K = 10^{-6}\,M$. In the absence of inducer, the repressor tetramer nonspecifically binds duplex DNA with a dissociation constant of $K \approx 10^{-4}\,M$. However, it specifically binds to the *lac* operator with far greater affinity: $K \approx 10^{-13}\,M$. Limited proteolysis of *lac* repressor with trypsin reveals that each subunit consists of two functional domains: Its 58-residue N-terminal peptide binds DNA but not IPTG, whereas the remaining "core tetramer" binds only IPTG.

The observed rate constant for the binding of *lac* repressor to *lac* operator is $k_f \approx 10^{10}\,M^{-1}\,s^{-1}$. This "on" rate is much greater than that calculated for the diffusion-controlled process in solution: $k_f \approx 10^7\,M^{-1}\,s^{-1}$ for molecules the size of *lac* repressor. Since it is impossible for a reaction to proceed faster than its diffusion-controlled rate, the *lac* repressor must not encounter operator from solution in a random three-dimensional search. Rather, *it appears that lac repressor finds operator by nonspecifically binding to DNA and diffusing along it in a far more efficient one-dimensional search.*

b. *lac* Operator Has a Nearly Palindromic Sequence

The availability of large quantities of *lac* repressor made it possible to characterize the *lac* operator. *E. coli* DNA

Figure 31-28 The base sequence of the *lac* operator. The symmetry related regions (*red*) comprise 28 of its 35 bp. A "+" denotes positions at which repressor binding enhances methylation by dimethyl sulfate (which methylates G at N7 and A at N3) and a "−" indicates where this footprinting reaction is inhibited. The bottom row indicates the positions and identities of different point mutations that prevent *lac* repressor binding (*O^c* mutants). Those in red increase the operator's symmetry. [After Sobell, H.M., *in* Goldberger, R.F. (Ed.), *Biological Regulation and Development*, Vol. 1, *p.* 193, Plenum Press (1979).]

that had been sonicated to small fragments was mixed with *lac* repressor and passed through a nitrocellulose filter. Protein, with or without bound DNA, sticks to nitrocellulose, whereas duplex DNA, by itself, does not. The DNA was released from the filter-bound protein by washing it with IPTG solution, recombined with *lac* repressor, and the resulting complex treated with DNase I. The DNA fragment that *lac* repressor protects from nuclease degradation consists of a run of 26 bp that is embedded in a nearly 2-fold symmetric sequence of 35 bp (Fig. 31-28, *top*). *Such palindromic symmetry is a common feature of DNA segments that are specifically bound by proteins* (recall, for example, that restriction endonuclease recognition sites are also palindromic; Section 5-5Aa).

Palindromic DNA sequences, as we have seen, bind to proteins that have matching 2-fold symmetry. However, methylation protection experiments on the *lac* repressor–

operator system do not fully support this model: There is an asymmetric pattern of differences between free and repressor-bound operator in the susceptibility of its bases to reaction with DMS (Fig. 31-28). Furthermore, point mutations that render it operator constitutive (*O^c*), and that invariably weaken the binding of repressor to operator, may increase as well as decrease the operator's 2-fold symmetry (Fig. 31-28).

c. *lac* Repressor Prevents RNA Polymerase from Forming a Productive Initiation Complex

Operator occupies positions −7 through +28 of the *lac* operon relative to the transcription start site (Fig. 31-29). Nuclease protection studies, it will be recalled, indicate that, in the initiation complex, RNA polymerase tightly binds to the DNA between positions −20 and +20 (Section 31-2Aa). Thus, *the lac operator and promoter sites overlap.* It was therefore widely assumed for many years that *lac* repressor simply physically obstructs the binding of RNA polymerase to the *lac* promoter. However, the observation that *lac* repressor and RNA polymerase can simultaneously bind to the *lac* operon indicates that *lac* repressor must act by somehow interfering with the initiation process. Closer investigation of this phenomenon revealed that, in the presence of bound *lac* repressor, RNA polymerase holoenzyme still abortively synthesizes oligonucleotides, although they tend to be shorter than those made in the absence of repressor. Evidently, *lac repressor acts by somehow increasing the already high kinetic barrier for RNA polymerase to generate the open complex and commence processive elongation.*

We discuss the *lac* repressor structure and further aspects of *lac* operator organization in Section 31-3F.

C. *Catabolite Repression: An Example of Gene Activation*

Glucose is E. coli's metabolite of choice; the availability of adequate amounts of glucose prevents the full expression of >100 genes that encode proteins involved in the fermentation

Figure 31-29 The nucleotide sequence of the *E. coli lac* promoter–operator region. The region extends from the C-terminal portion of *lacI* (*left*) to the N-terminal portion of *lacZ* (*right*). The palindromic sequences of the operator and the CAP-binding site (Section 31-3C) are overscored or underscored. [After Dickson, R.C., Abelson, J., Barnes, W.M., and Reznikoff, W.A., *Science* **187**, 32 (1975).]

Figure 31-30 **The kinetics of *lac* operon mRNA synthesis following its induction with IPTG, and of its degradation after glucose addition.** *E. coli* were grown on a medium containing glycerol as their only carbon-energy source and ³H-labeled uridine. IPTG was added to the medium at the beginning of the experiment to induce the synthesis of the *lac* enzymes. After 3 min, glucose was added to stop the synthesis. The amount of ³H-labeled *lac* RNA was determined by hybridization with DNA containing the *lacZ* and *lacY* genes. [After Adesnik, M. and Levinthal, C., *Cold Spring Harbor Symp. Quant. Biol.* **35,** 457 (1970).]

of numerous other catabolites, including lactose (Fig. 31-30), *arabinose, and galactose, even when these metabolites are present in high concentrations.* This phenomenon, which is known as **catabolite repression,** prevents the wasteful duplication of energy-producing enzyme systems.

a. cAMP Signals the Lack of Glucose

The first indication of the mechanism of catabolite repression was the observation that, in *E. coli,* the level of cAMP, which was known to be a second messenger in animal cells (Section 18-3Cb), is greatly diminished in the presence of glucose. This observation led to the finding that the addition of cAMP to *E. coli* cultures overcomes catabolite repression by glucose. Recall that, in *E. coli,* adenylate cyclase is activated by the phosphorylated enzyme EIIA^{glc} (or possibly inactivated by dephospho-EIIA^{glc}), which is dephosphorylated on the transport of glucose across the cell membrane (Section 20-3D). *The presence of glucose, therefore, normally lowers the cAMP level in E. coli.*

b. CAP–cAMP Complex Stimulates the Transcription of Catabolite Repressed Operons

Certain *E. coli* mutants, in which the absence of glucose does not relieve catabolite repression, are missing a cAMP-

binding protein that is synonymously named **catabolite gene activator protein (CAP)** and **cAMP receptor protein (CRP).** CAP is a homodimer of 209-residue subunits that undergoes a large conformational change on binding cAMP. Its function was elucidated by Ira Pastan, who showed that *CAP–cAMP complex, but not CAP itself, binds to the lac operon (among others) and stimulates transcription from its otherwise low-efficiency promoter in the absence of lac repressor.* CAP is therefore a **positive regulator** (turns on transcription), in contrast to *lac* repressor, which is a **negative regulator** (turns off transcription).

The X-ray structure, by Thomas Steitz, of CAP–cAMP in complex with a palindromic 30-bp segment of duplex DNA whose sequence resembles that of the CAP binding sequence (Fig. 31-29) reveals that the DNA is bent by ~90° around the protein (Fig. 31-31*a*). The bend arises from two ~45° kinks in the DNA between the fifth and sixth bases out from the complex's 2-fold axis in both directions. This distortion results in the closing of the major groove and an enormous widening of the minor groove at each kink.

Why is the CAP–cAMP complex necessary to stimulate the transcription of its target operons? And how does it do so? The *lac* operon has a weak (low efficiency) promoter; its −10 and −35 sequences (TATGTT and TTTACA; Fig. 31-10) differ significantly from the corresponding consensus sequences of strong (high-efficiency) promoters (TATAAT and TTGACA; Fig. 31-10). Such weak promoters evidently require some sort of help for efficient transcriptional initiation.

Richard Ebright has shown that CAP interacts directly with RNAP via the C-terminal domain of its 85-residue α subunit (αCTD) in a way that stimulates RNAP to initiate transcription from a nearby promoter. The αCTD also binds dsDNA nonspecifically but does so with higher affinity at A + T–rich sites such as those of UP elements (Section 31-2Aa). It is flexibly linked to the rest of the α subunit and hence is not seen in the X-ray structure of Tth RNAP (Fig. 31-11) due to disorder.

Three classes of the over one hundred CAP-dependent promoters have been characterized:

1. Class I promoters, such as that of the *lac* operon, require only CAP–cAMP for transcriptional activation. The CAP binding site on the DNA can be located at various distances from the promoter provided that CAP and RNAP bind to the same face of the DNA helix. Thus, CAP–cAMP activates the transcription of the *lac* operon if its DNA binding site is centered near positions −62 (its wild-type position; Fig. 31-29), −72, −83, −93, or −103, all of which are one helical turn apart. For the latter sites, this requires that the DNA loop around to permit CAP–cAMP to contact the αCTD. Such looping is likely to be facilitated by the bending of the DNA around CAP–cAMP.

2. Class II promoters also require only CAP–cAMP for transcriptional activation. However, in class II promoters, the CAP binding site only occupies a fixed position that overlaps the RNAP binding site, apparently by replacing the promoter's −35 promoter region. CAP then interacts

Figure 31-31 X-ray structures of CAP–cAMP–dsDNA complexes. The dsDNA and cAMP in these 2-fold symmetric complexes are colored according to atom type with DNA C white, cAMP C green, N blue, O red, and P orange. (*a*) CAP–cAMP in complex with a palindromic 30-bp self-complementary DNA viewed with its 2-fold axis horizontal. The protein is drawn in ribbon form with its identical subunits pink and blue and with their C-terminal domains in darker shades. The DNA is shown in stick form with successive P atoms in the same strand connected by orange rods and with the cAMP drawn in space-filling form. (*b*) CAP–cAMP in complex with a 44-bp palindromic DNA and four αCTD subunits. The DNA, CAP, and cAMP are viewed as in

Part *a* with the DNA drawn in space-filling form. The αCTD subunits are drawn in ribbon form with the αCTD^{CAP,DNA} green and the αCTD^{DNA} dark green. (*c*) The same structure as in Part *a* showing the binding of the CAP dimer's two helix–turn–helix (HTH) motifs in successive major grooves of the DNA. The view is rotated 45° about the vertical axis relative to Part *a*. Note how CAP's F (recognition) helix is inserted into the DNA's major groove, as can also be seen in Parts *a* and *b*. [Parts *a* and *c* based on an X-ray structure by Thomas Steitz, Yale University. PDBid 1CGP. Part *b* based on an X-ray structure by Helen Berman and Richard Ebright, Rutgers University. PDBid 1LB2.]
🔊 **See Interactive Exercise 38**

with RNAP via interactions with both the αCTD and the α subunit's N-terminal domain.

3. Class III promoters require multiple activators to maximally stimulate transcription. These may be two or more CAP–cAMP complexes or a CAP–cAMP complex acting in concert with promoter-specific activators as occurs in the *araBAD* operon (Section 31-3E).

The X-ray structure of CAP–cAMP in complex with the *E. coli* αCTD and a 44-bp palindromic DNA containing the 22-bp CAP–cAMP binding site and 5′-AAAAAA-3′ at each end, determined by Helen Berman and Ebright, reveals how these components interact (Fig. 31-31*b*). The 2-fold symmetric CAP–cAMP–αCTD complex contains two differently located pairs of αCTDs. Each member of the pair designated αCTD^{CAP,DNA} binds to both CAP and to the DNA. CAP and

αCTD^{CAP,DNA} interact over a surprisingly small surface area involving only six residues on each protein that mutagenesis experiments had previously implicated. αCTD^{CAP,DNA} also interacts with the minor groove of a 6-bp segment of the DNA (5′-AAAAAG-3′) centered 19 bp from the center of the DNA. Each member of the other pair of αCTDs, designated αCTD^{DNA}, interacts with the minor groove of an UP element-like sequence (5′-GAAAAA-3′) that is fortuitously present in the DNA but it makes no contacts with other protein molecules. The common portions of the two CAP complexes pictured in Fig. 31-31*a,b* are closely superimposable, thereby indicating that the conformation of CAP and its interaction with DNA are not significantly altered by its association with the αCTD. Evidently, CAP–cAMP transcriptionally activates RNAP via a simple "adhesive" mechanism that facilitates and/or stabilizes its interaction with the promoter DNA. The structures of $\alpha CTD^{CAP \cdot DNA}$ and αCTD^{CAP} and their interactions with DNA are nearly identical, thereby suggesting that they are representative of the interaction of an αCTD with an UP element.

D. Sequence-Specific Protein–DNA Interactions

Since genetic expression is controlled by proteins such as CAP and *lac* repressor, an important issue in the study of gene regulation is how these proteins recognize their target base sequences on DNA. Sequence-specific DNA-binding proteins generally do not disrupt the base pairs of the duplex DNA to which they bind. Consequently, these proteins can only discriminate among the four base pairs $(A \cdot T, T \cdot A, G \cdot C, \text{ and } C \cdot G)$ according to the functional groups of these base pairs that project into DNA's major and minor grooves. An inspection of Fig. 5-12 reveals that the groups exposed in the major groove have a greater variation in their types and arrangements than do those that are exposed in the minor groove. Indeed, the positions of the hydrogen bonding acceptors in the major groove vary with both the identity and orientation of the base pair, whereas in the minor groove they are largely sequence independent. Moreover, the ~5-Å-wide and ~8-Å-deep minor groove of canonical (ideal) B-DNA is too narrow to admit protein structural elements such as an α helix, whereas its ~12-Å-wide and ~8-Å-deep major groove can do so. Thus, in the absence of major conformational changes to B-DNA, it would be expected that proteins could more readily differentiate base sequences from its major groove than from its minor groove. We shall see below that this is, in fact, the case.

a. The Helix–Turn–Helix Motif Is a Common DNA Recognition Element in Prokaryotes

See Guided Exploration 30: Transcription factor–DNA interactions
The CAP dimer's two symmetrically disposed F helices protrude from the protein surface in such a way that they fit into successive major grooves of B-DNA (Fig. 31-31). *CAP's E and F helices form a **helix–turn–helix (HTH) motif** (supersecondary structure) that conformationally resembles analogous HTH motifs in numerous other prokaryotic repressors of known X-ray and NMR structure,* including the *lac* repressor, the *E. coli* **trp repressor** (Section 31-3G), and the **cI repressors** and **Cro proteins** from **bacteriophages** λ and **434** (Section 33-3D). HTH motifs are ~20-residue polypeptide segments that form two α helices which cross at ~120° (Fig. 31-31*c*). They occur as components of domains that otherwise have widely varying structures, although all of them bind DNA. Note that HTH motifs are structurally stable only when they are components of larger proteins.

The X-ray and NMR structures of a number of protein–DNA complexes (see below) indicate that *DNA-binding proteins containing an HTH motif associate with their target base pairs mainly via the side chains extending from the second helix of the HTH motif, the so-called **recognition helix** (helix F in CAP, E in *trp* repressor, and α3 in the phage proteins). Indeed, replacing the outward-facing

residues of the 434 repressor's recognition helix with the corresponding residues of the related **bacteriophage P22** yields a hybrid repressor that binds to P22 operators but not to those of 434. Moreover, the HTH motifs in all these proteins have amino acid sequences that are similar to each other and to polypeptide segments in numerous other prokaryotic DNA-binding proteins, including *lac* repressor. Evidently, *these proteins are evolutionarily related and bind their target DNAs in a similar manner.*

How does the recognition helix recognize its target sequence? Since each base pair presents a different and presumably readily differentiated constellation of hydrogen bonding groups in DNA's major groove, it seemed likely that there would be a simple correspondence, analogous to Watson–Crick base pairing, between the amino acid residues of the recognition helix and the bases they contact in forming sequence-specific associations. The above X-ray structures, however, indicate this idea to be incorrect. Rather, base sequence recognition arises from complex structural interactions. For instance:

1. The X-ray structures of the 48% identical N-terminal domain of 434 repressor (residues 1–69) and the entire 71-residue 434 Cro protein in their complexes with the identical 20-bp target DNA (the expression of phage 434 is regulated through the differential binding of these proteins to the same DNA segments; Section 33-3Db) were both determined by Stephen Harrison. Both homodimeric proteins, as seen for CAP (Fig. 31-31), associate with the DNA in a 2-fold symmetric manner with their recognition helices bound in successive turns of the DNA's major groove (Figs. 31-32 and 31-33). In both complexes, the protein closely conforms to the DNA surface and interacts with its paired bases and sugar–phosphate chains through elaborate networks of hydrogen bonds, salt bridges, and van der Waals contacts. Nevertheless, the detailed geometries of these associations are significantly different. In the repressor–DNA complex (Fig. 31-32), the DNA bends around the protein in an arc of radius ~65 Å which compresses the minor groove by ~2.5 Å near its center (between the two protein monomers) and widens it by ~2.5 Å toward its ends. In contrast, the DNA in complex with Cro (Fig. 31-33), although also bent, is nearly straight at its center and has a less compressed minor groove (compare Figs. 31-32*a* and 31-33*a*). This explains why the simultaneous replacement of three residues in the repressor's recognition helix with those occurring in Cro does not cause the resulting hybrid protein to bind DNA with Cro-like affinity: *The different conformations of the DNA in the repressor and Cro complexes prevents any particular side chain from interacting identically with the DNA in the two complexes.*

2. Paul Sigler determined the X-ray structure of *E. coli* *trp* repressor in complex with a DNA containing an 18-bp palindrome (TGT<u>ACTAGTTA</u><u>ACTAGT</u>AC, where the *trp* repressor's target sequence is underlined) that closely resembles the *trp* operator (Section 31-3G). The dimeric protein's recognition helices bind, as expected, in successive major grooves of the DNA, each in contact with an

(a)

(b)

Figure 31-32   X-ray structure of the 69-residue N-terminal domain of 434 phage repressor in complex with a 20-bp dsDNA containing its target sequence. One strand of the DNA has the sequence d(TATACAAGAAAGTTTGTACT). (*a*) The complex viewed with the homodimeric protein's 2-fold axis horizontal. The protein is drawn in ribbon form with one of its two identical subunits blue and the other red and with their helix–turn–helix (HTH) motifs in darker shades. The DNA is drawn in stick form with C white, N blue, O red, and P orange, and with successive P atoms in the same chain connected by orange rods. (*b*) A

schematic drawing indicating how the HTH motif, which encompasses helices α2 and α3, interacts with its target DNA. Short bars emanating from the polypeptide chain represent peptide NH groups, hydrogen bonds are represented by dashed lines, and DNA phosphates are represented by numbered circles. The small circle is a water molecule. [Part *a* based on an X-ray structure by and Part *b* courtesy of Aneel Aggarwal, John Anderson, and Stephen Harrison, Harvard University. PDBid 2OR1.] **See Interactive Exercise 39 and Kinemage Exercise 18-1**

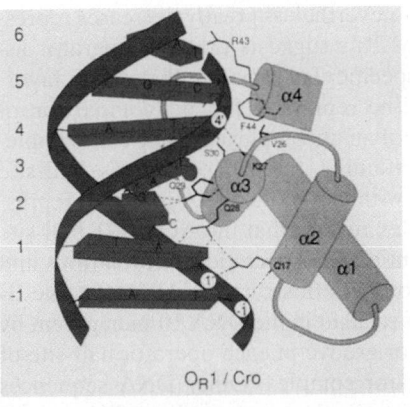

(a)

(b)

Figure 31-33   X-ray structure of the 71-residue 434 Cro protein in complex with the same 20-bp DNA shown in Fig. 31-32. Parts *a* and *b* correspond to those in Fig. 31-32. Note the close but not identical correspondence between the two structures and, in particular, the difference in the widths of the

minor groove between the two subunits in each structure. [Part *a* based on an X-ray structure by and Part *b* courtesy of Alfonso Mondragón, Cynthia Wolberger, and Stephen Harrison, Harvard University. PDBid 3CRO.]

Figure 31-34 **X-ray structure of an** *E. coli trp* **repressor–operator–tryptophan complex.** The complex is viewed with its molecular 2-fold axis horizontal. The protein's two identical subunits are drawn in ribbon form colored pink and blue with their HTH motifs (helices D and E) more deeply colored. The 18-bp-containing self-complementary dsDNA is shown in stick form with C white, N blue, O red, P orange, and with successive P atoms in the same chain connected by orange rods. The *trp* repressor binds its operator only when L-tryptophan, drawn in space-filling form with C green, is simultaneously bound. Note that the protein's recognition helices (E) bind, as expected, in successive major grooves of the DNA but extend approximately perpendicular to the DNA duplex axis. In contrast, the recognition helices of 434 repressor and Cro proteins are nearly parallel to the major grooves of their bound DNAs (Figs. 31-32 and 31-33), whereas those of CAP assume an intermediate orientation (Fig. 31-31). [Based on an X-ray structure by Paul Sigler, Yale University. PDBid 1TRO.]
🐭 **See Interactive Exercise 40**

operator half-site (<u>ACTAGT</u>; Fig. 31-34). There are numerous hydrogen bonding contacts between the *trp* repressor and its bound DNA's nonesterified phosphate oxygens. Astoundingly, however, *there are no direct hydrogen bonds or nonpolar contacts that can explain the repressor's specificity for its operator. Rather, all but one of the side chain–base hydrogen bonding interactions are mediated by bridging water molecules* (the one direct interaction involves a base that can be mutated without greatly affecting repressor binding affinity). Such buried water molecules have therefore been described as "honorary" protein side chains. In addition, the operator contains several base pairs that are not in contact with the repressor but whose mutation nevertheless greatly decreases repressor binding affinity. This suggests that the operator assumes a sequence-specific conformation that makes favorable contacts with the repressor. Indeed, comparison of the X-ray structure of an uncomplexed 10-bp self-complementary DNA containing the *trp* operator's half-site (CC<u>ACTAGT</u>GG) with that of the DNA in the *trp* repressor–operator complex reveals that the ACTAGT half-site assumes nearly identical idiosyncratic conformations and patterns of hydration in both structures. However, the B-DNA helix, which is straight in the DNA 10-mer, is bent by 15° toward the major groove in each operator half-site of the repressor–operator complex. Other DNA sequences could conceivably assume the repressor-bound operator's conformation but at too high an energy cost to form a stable complex with repressor (*trp* repressor's measured 10^4-fold preference for its operator over other DNAs implies an ~23 kJ · mol^{-1} difference in their binding free energies). This phenomenon, in which a protein senses the base sequence of DNA through the DNA's backbone conformation and/or flexibility, is referred to as **indirect readout.** 434 repressor apparently also employs indirect readout: Replacing the central A · T base pair of the operator shown in Fig. 31-32 with G · C reduces repressor binding affinity by 50-fold even though 434 repressor does not contact this region of the DNA.

It therefore appears that *there are no simple rules governing how particular amino acid residues interact with bases. Rather, sequence specificity results from an ensemble of mutually favorable interactions between a protein and its target DNA.*

b. The *met* Repressor Contains a Two-Stranded Antiparallel β Sheet That Binds in Its Target DNA's Major Groove

The *E. coli* **met repressor (MetJ),** when complexed with *S*-adenosylmethionine (SAM; Fig. 26-18), represses the transcription of its own gene and those encoding enzymes involved in the synthesis of methionine (Fig. 26-60) and SAM. The X-ray structure of the *met* repressor–SAM–operator complex (Fig. 31-35), determined by Simon Phillips, reveals a symmetric dimer of intertwined homodimers that lacks an HTH motif. Rather, *met* repressor binds to its palindromic target DNA sequence through two symmetry-related pairs of symmetrical two-stranded antiparallel β sheets (called **β ribbons)** that are inserted in successive major grooves of the DNA. Each β ribbon makes sequence-specific contacts with its target DNA sequence via hydrogen bonding and, probably, indirect readout.

(a)

(b)

Figure 31-35 **X-ray structure of the *E. coli met* repressor–SAM–operator complex.** (*a*) The overall structure of the complex as viewed along its 2-fold axis of symmetry. The self-complementary 18-bp DNA is drawn in stick form, and SAM, which must be bound to the repressor for it to also bind DNA, is shown in space-filling form with the DNA C white, SAM C green, N blue, O red, P orange, and S yellow. The DNA binds four identical 104-residue repressor subunits. Pairs of subunits (*light cyan and lavender*) form symmetric dimers in which each subunit donates one strand of the 2-stranded antiparallel β ribbon that is inserted in the DNA's major groove (*upper left and lower right*). Two such dimers pair across the complex's 2-fold axis via their antiparallel N-terminal helices, which contact one another over the DNA's minor groove. (*b*) Detailed view of the lower half of Part *a* showing the 2-stranded antiparallel β ribbon (residues 21–29) inserted into the DNA's major groove, as viewed along its local 2-fold axis (rotated relative to Part *a* by 50° about the vertical axis). The DNA is shown in space-filling form and the polypeptide chains are drawn in stick form with C green. [Based on an X-ray structure by Simon Phillips, University of Leeds, U.K. PDBid 1CMA.]
See Interactive Exercise 41

Phillips first determined the X-ray structure of *met* repressor in the absence of DNA. Model building studies aimed at elucidating how *met* repressor binds to its palindromic target DNA assumed that the 2-fold rotation axes of both molecules would be coincident, as they are in all prokaryotic protein–DNA complexes of known structure. There were, consequently, two reasonable choices: (1) The protein could dock to the DNA with the above pairs of β ribbons entering successive major grooves; or (2) a symmetry-related pair of protruding α helices on the opposite face of the protein could do so in a manner resembling the way in which the recognition helices of HTH motifs interact with DNA. A variety of structural criteria suggested that the α helices make significantly better contacts with the DNA than do the β ribbons. Thus, the observation that it is, in fact, the β ribbons that bind to the DNA provides an important lesson: *The results of model building studies must be treated with utmost caution.* This is because our imprecise understanding of the energetics of intermolecular interactions (Sections 8-4 and 29-2) prevents us from reliably predicting how associating macromolecules conform to one another. In the case of the *met* repressor, unpredicted mutual structural accommodations of the protein and DNA yielded a significantly more extensive interface than

had been predicted by simply docking the uncomplexed Met repressor to canonical B-DNA.

The numerous prokaryotic transcriptional regulators of known structure either contain an HTH motif or pairs of β ribbons like the *met* repressor (although numerous prokaryotic DNA-binding proteins, including CAP, contain an elaboration of the HTH motif known as the **winged helix** motif in which two protein loops, one of which contacts the DNA's minor groove, flank the HTH recognition helix like the wings of a butterfly). Moreover, most of these proteins are homodimers that bind to palindromic or pseudopalindromic DNA target sequences. However, eukaryotic transcription factors, as we shall see in Section 34-3B, employ a much wider variety of structural motifs to bind their target DNAs, many of which lack symmetry.

E. *araBAD Operon: Positive and Negative Control by the Same Protein*

Humans neither metabolize nor intestinally absorb the plant sugar L-arabinose. Hence, the *E. coli* that normally inhabit the human gut are periodically presented with a banquet of this pentose. Three of the five *E. coli* enzymes that

Figure 31-36 **A genetic map of the *E. coli araC* and *araBAD* operons.** The map indicates the proteins these operons encode and the reactions in which these proteins participate. The permease system, which transports arabinose into the cell, is the product of the *araE* and *araF* genes, which occur in two independent operons. The pathway product, xylulose-5-phosphate, is converted, via the pentose phosphate pathway, to the glycolytic intermediates fructose-6-phosphate and glyceraldehyde-3-phosphate (Section 23-4). [After Lee, N., *in* Miller, J.H. and Rezinkoff, W.S. (Eds.), *The Operon,* p. 390, Cold Spring Harbor Laboratory Press (1979).]

metabolize arabinose are products of the catabolite repressible **araBAD operon** (Fig. 31-36).

The *araBAD* operon, as Robert Schleif has shown, contains, moving upstream from its transcriptional start site, the *araI*, *araO*$_1$, and *araO*$_2$ control sites (Fig. 31-37*a*). The *araI* site (*I* for inducer) consists of two closely similar 17-bp half-sites, *araI*$_1$ and *araI*$_2$, that are direct repeats separated by 4 bp and are oriented such that *araI*$_2$, which overlaps the -35 region of the *araBAD* promoter, is downstream of *araI*$_1$. Likewise, *araO*$_1$ consists of two directly repeating half-sites, O_{1L} and O_{1R}. Intriguingly, however, *araO*$_2$ consists of a single half-site that is located in a noncoding upstream region of the *araC* gene (see below), at position -270 relative to the *araBAD* start site.

The transcription of the *araBAD* operon is regulated by both CAP–cAMP and the arabinose-binding protein **AraC.** Each 292-subunit of the homodimeric AraC consists of an N-terminal, arabinose-binding, dimerization domain (residues 1–170) connected via a flexible linker to a C-terminal DNA-binding domain (residues 178–292). Regulation of the *araBAD* operon occurs as follows (Fig. 31-37):

1. In the absence of AraC, RNA polymerase initiates transcription of the *araC* gene in the direction away from its upstream neighbor, *araBAD*. The *araBAD* operon is expressed at a low basal level.

2. When AraC is present, but neither arabinose nor CAP–cAMP (high glucose), AraC binds to *araO*$_2$ and *araI*$_1$ via two HTH motifs in each of its subunits. The binding of AraC to *araI*$_1$ prevents RNAP from initiating transcription of the *araBAD* operon (negative control). A series of deletion mutations indicate that the presence of *araO*$_2$ is also required for the repression of *araBAD*. The remarkably large 210-bp separation between *araO*$_2$ and *araI*$_1$ therefore strongly suggests that the DNA between them is looped such that a dimeric molecule of AraC protein simultaneously binds to both *araO*$_2$ and *araI*$_1$. This is corroborated by the observation that the level of repression is greatly diminished by the insertion of 5 bp (half a turn) of DNA between these two sites, thereby transferring *araO*$_2$ to the opposite face of the DNA helix relative to *araI*$_1$ in the putative loop. Yet, the insertion of 11 bp (one turn) of DNA has no such effect. Moreover, looping does not readily occur unless the DNA is supercoiled, which presumably drives the looping process. The AraC dimer also binds to *araO*$_1$, the operator of the *araC* gene, so as to block the transcription of *araC* but only at high concentrations. Thus, it is likely that DNA looping itself represses the transcription of *araC*. In either case, the expression of *araC* is autoregulatory.

3. When arabinose is present, it allosterically induces the AraC subunit bound to *araO*$_2$ to instead bind to *araI*$_2$. This activates RNAP to transcribe the *araBAD* genes (positive control). When the cAMP level is high (low glucose), CAP–cAMP, whose presence is required to achieve the maximum level of transcriptional activation, binds to a site between *araO*$_1$ and *araI*$_1$, where it functions to help break the loop between *araO*$_2$ and *araI*$_1$ and hence to increase the affinity of AraC for *araI*$_2$. The orientation of *araO*$_1$ with respect to *araC* is opposite to that of *araI* with respect to *araBAD*, and hence the binding of AraC–arabinose at *araO*$_1$ blocks RNAP binding at the *araC* promoter, that is, it represses the expression of AraC.

If the *araI*$_2$ subsite is mutated so as to increase AraC's affinity for it, arabinose is no longer required for transcriptional activation. This suggests that arabinose does not conformationally transform AraC to an activator but, rather, weakens its binding affinity for *araO*$_2$. If the *araI* site is *(a)* **When AraC is absent,** *araC* **is transcribed and** *araBAD* **is transcribed at a basal level**

araO₂ | araC | araO$_{1L}$ | araO$_{1R}$ | CAP | araI$_1$ | araI$_2$

RNA polymerase

araC mRNA

araBAD

araBAD mRNA (basal level)

(b) **When cAMP and L-arabinose are low, AraC represses** *araBAD* **transcription**

C-terminal DNA-binding domain

N-terminal arm

Linker

Arabinose-binding pocket

N-terminal dimerization domain

araC

araO₂

AraC

araO$_{1L}$ | araO$_{1R}$ | CAP | araI$_1$ | araI$_2$

araBAD

(c) **When cAMP and L-arabinose are abundant,** *araBAD* **transcription is activated**

CAP–cAMP   AraC–arabinose

araO₂ | araC | araO$_{1L}$ | araO$_{1R}$ | CAP | araI$_1$ | araI$_2$

RNA polymerase

araBAD

araBAD mRNA

Figure 31-37   The mechanism of *araBAD* **regulation.** (*a*) In the absence of AraC, RNAP initiates the transcription of *araC*. *araBAD* is also expressed but at a low basal level. (*b*) When AraC is present, but not L-arabinose or cAMP, AraC links together *araO₂* and *araI₁* to form a DNA loop, thereby repressing both *araC* and *araBAD*. (*c*) When AraC and L-arabinose are turned around or if it is moved upstream so that *araI₂* does not overlap the *araBAD* promoter, AraC cannot stimulate transcription. Evidently, *AraC activates RNAP through specific and relatively inflexible protein–protein interactions.*

both present and cAMP is abundant, the resulting AraC–arabinose complex releases *araO₂* and instead binds *araI₂*, thereby activating *araBAD* transcription. This process is facilitated by the binding of CAP–cAMP. *araC* is repressed by the binding of AraC–arabinose to *araO₁*.

The X-ray structures of the N-terminal domain of AraC (residues 2–178), in both the presence and the absence of arabinose, were determined by Schleif and Cynthia Wolberger. In the presence of arabinose, this domain consists of an 8-stranded β barrel followed by two antiparallel α helices (Fig. 31-38). Two such domains associate via an antiparallel coiled coil between each of their C-terminal helices to form the protein's dimerization interface. An arabinose molecule binds in a pocket of each β barrel via a network of direct and water-mediated hydrogen bonds with side chains that line the pocket. Residues 7 to 18 of the N-terminal arm lie across the mouth of the sugar-binding pocket (residues 2–6 are disordered), thereby fully enclosing the arabinose. The structure of the N-terminal domain in the absence of arabinose is largely superimposable on that in the complex with arabinose, with the exception that

Figure 31-38   X-ray structure of *E. coli* **AraC in complex with** L-arabinose. The homodimeric protein is viewed along its 2-fold axis with each of its subunits colored in rainbow order from N-terminus (*blue*) to C-terminus (*red*). The arabinose is drawn in space-filling form with C green and O red. [Based on an X-ray structure by Robert Schleif and Cynthia Wolberger, Johns Hopkins University. PDBid 2ARC.]

the N-terminal arm is disordered, a not unexpected observation considering that it interacts with bound arabinose via a series of hydrogen bonds.

How does arabinose binding induce the AraC subunit bound at *araO₂* to instead bind to *araI₂*? Several lines of evidence indicate that, in the absence of arabinose, AraC's N-terminal arm binds to its DNA-binding domain in a way that favors loop formation: (1) the deletion of the N-terminal arm beyond its sixth residue makes AraC act as if arabinose is present; (2) mutations to surface residues on the DNA-binding domain that presumably eliminate its binding of the N-terminal arm also constitutively activate AraC; and (3) mutations in the DNA-binding domain that weaken the binding of arabinose to the protein, presumably by strengthening the binding of the N-terminal arm, can be suppressed by a second mutation in the N-terminal arm or by the deletion of its five N-terminal residues. Evidently, *the binding of the N-terminal arms to the DNA-binding domains in the absence of arabinose rigidifies the AraC dimer such that it cannot simultaneously bind to the directly repeated araI₁ and araI₂ and hence induce the transcription of araBAD.* This is corroborated by the observations that (1) joining two AraC DNA-binding domains by flexible polypeptide linkers yields proteins that behave like AraC in the presence of arabinose, and (2) a construct consisting of two double-stranded *araI₁* half-sites flexibly connected by a 24-nt segment of ssDNA binds wild-type AraC with an affinity that is unaffected by arabinose.

F. *lac* Repressor II: Structure

Here we continue our discussions of the *lac* repressor, but now in terms of the concepts learned in Sections 31-3C–E.

a. Loop Formation Is Important in the Expression of the *lac* Operon

DNA loop formation, which is now known to occur in numerous bacterial and eukaryotic systems, apparently permits several regulatory proteins and/or regulatory sites on one protein to simultaneously influence transcription initiation by RNAP. In fact, *the lac repressor has three binding sites on the lac operon:* the primary operator (Fig. 31-28), now known as O_1, and two so-called pseudo-operators (previously thought to be nonfunctional evolutionary fossils), O_2 and O_3, which are located 401 bp downstream and 92 bp upstream of O_1 (within the *lacZ* gene and overlapping the CAP binding site, respectively). Müller-Hill determined the relative contributions of these various operators to the repression of the *lac* operon through the construction of a set of eight plasmids: Each contained the *lacZ* gene under the control of the natural *lac* promoter as well as the three *lac* operators (O_1, O_2, and O_3), which were either active or mutagenically inactive in all possible combinations. When all three operators are active, *lacZ* expression is repressed 1300-fold relative to when all three operators are inactive. The inactivation of only O_1 results in almost complete loss of repression whereas the inactivation of only O_2 or O_3 causes only a ~2-fold loss in repression. However, when O_2 and O_3 are both inactive, repres-

sion is decreased ~70-fold. These results suggest that efficient repression requires the formation of a DNA loop between O_1 and either O_2 or O_3. Indeed, such loop formation, and/or the cooperativity of repressor binding arising from it, appears to be a greater contributor to repression than repressor binding to O_1 alone, which provides only 19-fold repression.

b. The *lac* Repressor Is a Dimer of Dimers

Ponzy Lu and Mitchell Lewis determined the X-ray structures of the *lac* repressor alone, in its complex with IPTG, and in its complex with a 21-bp duplex DNA segment whose sequence is a palindrome of the left half of O_1 (Fig. 31-28). Each repressor subunit consists of five functional units (Fig. 31-39): (1) an N-terminal DNA-binding domain (residues 1–49) which is known as the "headpiece" because it is readily proteolytically cleaved away from the remaining still tetrameric "core" protein; (2) a hinge helix (residues 50–58) that also binds to the DNA; (3 and 4) a sugar-binding domain (residues 62–333) that is divided into an N-subdomain and a C-subdomain; and (5) a C-terminal tetramerization helix (residues 340–360).

Figure 31-39 **X-ray structure of the *lac* repressor subunit.** The DNA-binding domain (the headpiece), which contains an HTH motif, is red, the DNA-binding hinge helix is yellow, the N-subdomain of the sugar-binding domain is light blue, its C-subdomain is dark blue, and the tetramerization helix is purple. [Courtesy of Ponzy Lu and Mitchell Lewis, University of Pennsylvania. PDBid 1LBI.]

The *lac* repressor has an unusual quaternary structure (Fig. 31-40*a*). Whereas nearly all homotetrameric nonmembrane proteins of known structure have D_2 symmetry (three mutually perpendicular 2-fold axes; Fig. 8-65*b*), *lac* repressor is a V-shaped protein that has only 2-fold symmetry. Each leg of the V consists of a locally symmetric dimer of closely associated repressor subunits. Two such dimers

(a)

(b)

Figure 31-40 **The structure of the *lac* repressor in complex with DNA.** (*a*) The X-ray structure of the *lac* repressor tetramer bound to two 21-bp segments of symmetric *lac* operator DNA. The protein subunits are shown in ribbon form in yellow, cyan, green, and orange and the dsDNA segments are drawn in space-filling form with C white, N blue, O red, and P orange. [Courtesy of Ponzy Lu and Mitchell Lewis with coordinates generated by Benjamin Weider, University of Pennsylvania. PDBid 1LBG.] (*b*) The NMR structure of the 23-bp O_1 *lac* operator DNA in complex with two identical segments of the *lac* repressor consisting of its DNA-binding domain and its hinge helix. Each of the protein subunits is drawn in ribbon form colored in rainbow order from its N-terminus (*blue*) to its C-terminus (*red*). The DNA is represented in stick form with C white, N blue, O red, and P orange and with successive P atoms in the same chain connected by orange rods. The complex is viewed with its 2-fold axis vertical. Note that the protein dimer's two HTH motifs are inserted in successive major grooves at the periphery of the complex and that the insertion of the two centrally located hinge helices into the DNA's minor groove greatly widens and flattens the minor groove at this point and kinks the DNA in an upward bend. [Based on an NMR structure by Robert Kaptein and Rolf Boelens, Utrecht University, The Netherlands. PDBid 2KEI.]

associate rather tenuously, but with 2-fold symmetry, at the base (point) of the V to form a dimer of dimers.

In the structures of *lac* repressor alone and that of its IPTG complex, the DNA-binding domain is not visible, apparently because the hinge region that loosely tethers it to the rest of the protein is disordered. However, in the DNA complex, in which one DNA duplex binds to each of the two dimers forming the repressor tetramer, the DNA domain forms a compact globule containing three helices, the first two of which form a helix–turn–helix (HTH) motif. The two DNA-binding domains extending from each repressor dimer (at the top of each leg of the V) bind in successive major grooves of a DNA molecule via their HTH motifs, much as is seen, for example, in the complexes of 434 phage repressor and *trp* repressor with their target DNAs (Figs. 31-32 and 31-34). The binding of the *lac* repressor distorts the operator DNA such that it bends away from the DNA-binding domain with an ~60 Å radius of curvature due to an ~45° kink at the center of the operator that widens the DNA's minor groove to over 11 Å and reduces its depth to less than 1 Å. These distortions permit the now ordered hinge helix to bind in the minor groove so as to contact the identically bound hinge helix from the other subunit of the same dimer. NMR structures by Robert Kaptein and Rolf Boelens reveal that the DNA-binding domain, when cleaved from the repressor, binds to the *lac* operator without distorting the DNA, but that the DNA-binding domain together with the hinge helix forms a complex with the *lac* operator in which the hinge helix binds in the DNA's distorted minor groove (Fig. 31-40*b*) as in the X-ray structure. Thus, the binding of the two hinge helices to the *lac* operator appears necessary for DNA distortion. The two DNA duplexes that are bound to each repressor tetramer are ~25 Å apart and do not interact.

The sugar-binding domain consists of two topologically similar subdomains that are bridged by three polypeptide segments (Fig. 31-39). The two sugar-binding domains of a dimer make extensive contacts (Fig. 31-40*a*). IPTG binds to each sugar-binding domain between its subdomains. This does not significantly change the conformations of these subdomains, but it changes the angle between them. Although the hinge helix is not visible in the IPTG complex, model building indicates that, since the dimer's two hinge helices extend from its sugar-binding domains, this conformation change levers apart these hinge helices by 3.5 Å such that they and their attached HTH motifs can no longer simultaneously bind to their operator half-sites. Thus, inducer binding, which is allosteric within the dimer (has a positive homotropic effect; Section 10-4), greatly loosens the repressor's grip on the operator.

The C-terminal helices from each subunit, which are located on the opposite end of each subunit from the DNA-binding portion (at the point of the V), associate to form a bundle of four parallel helices that holds together the two repressor dimers, thereby forming the tetramer (Fig. 31-40*a*). The allosteric effects of inducer binding within each dimer are apparently not transmitted between dimers.

Moreover, the *E. coli* **purine repressor (PurR),** which is homologous to the *lac* repressor but lacks its C-terminal helix, crystallizes as a dimer whose X-ray structure closely resembles that of the *lac* repressor dimer. What then is the function of *lac* repressor tetramerization?

Model building suggests that when the *lac* repressor tetramer simultaneously binds to both the O_1 and O_3 operators, the 93-bp DNA segment containing them forms a loop ~80 Å in diameter (Fig. 31-41). Furthermore, the CAP–cAMP binding site is exposed on the inner surface of the loop. Adding the CAP–cAMP at its proper position to this model reveals that the ~90° curvature which CAP–cAMP binding imposes on DNA (Fig. 31-31) has the correct direction and magnitude to stabilize the DNA loop, thereby stabilizing this putative CAP–cAMP–*lac* repressor–DNA complex. It may seem paradoxical that the binding of CAP–cAMP, a transcriptional activator, stabilizes the repressor–DNA complex. However, when both glucose and lactose are in short supply, it is important that the bacterium lower its basal rate of *lac* operon expression in order to conserve energy. The binding site (promoter) for RNAP is also located on the inner surface of the loop. Thus, the large size of the RNAP molecule would prevent

Figure 31-41   Model of the 93-bp DNA loop formed when *lac* repressor binds to O_1 and O_3. The proteins are represented by their C_α backbones and the DNA is drawn in stick form with its sugar–phosphate backbones traced by helical ribbons. The model was constructed from the X-ray structure of the *lac* repressor (*magenta*) in complex with two 21-bp operator DNA segments (*red*) and the X-ray structure of CAP–cAMP (*blue*) in complex with its 30-bp target DNA (*cyan;* Fig. 31-28). The remainder of the DNA loop was generated by applying a smooth curvature to canonical B-DNA (*white*) with the −10 and −35 regions of the *lac* promoter highlighted in green. [Courtesy of Ponzy Lu and Mitchell Lewis, University of Pennsylvania.]

it from fully engaging the promoter in this looped complex, thereby maximizing repression.

c. Combining Genetic and Structural Studies of the *lac* Repressor Reveals Its Allosterically Important Residues

The phenotypes of 4042 point mutations of the *lac* repressor, which encompass nearly all of its 360 residues (making the *lac* repressor the most exhaustively mutationally characterized protein known) have been mapped onto its X-ray structure. Mutations with an I^- phenotype (*lac* repressors that fail to bind to the *lac* operator, so that β-galactosidase is constitutively synthesized) are located at the *lac* repressor's DNA-binding interface, at its dimer interface, or at internal residues of its inducer-binding core domain. Residues whose mutations result in the I^S phenotype (S for super-repressed; *lac* repressors that, in the presence of inducer, continue to repress the synthesis of β-galactosidase) appear to be of two types: (1) residues that are in direct contact with the inducer, whose alteration therefore interferes with inducer binding; and (2) residues at the dimer interface that are >8 Å from (not in direct contact with) the inducer-binding site. These latter mutations reveal which residues mediate the *lac* repressor's allosteric mechanism rather than directly binding the inducer or the DNA. Most of the allosterically important residues are located at the dimer interface and are members of the N-subdomain of the core domain, which links the inducer-binding sites to the operator DNA-binding sites. This is consistent with the observation that inducer binding causes a relative twist and translation of the N-subdomain, a movement which is propagated to the hinge helix and DNA-binding domain. This study demonstrates the power of combining genetic analysis with structural studies to elucidate structure–function relationships.

G. *trp* Operon: Attenuation

We now discuss a sophisticated transcriptional control mechanism named **attenuation** through which bacteria regulate the expression of certain operons involved in amino acid biosynthesis. This mechanism was discovered through the study of the *E. coli* **trp operon** (Fig. 31-42), which encodes five polypeptides comprising three enzymes that mediate the synthesis of tryptophan from chorismate (Section 26-5Bc). Charles Yanofsky established that the *trp* operon genes are coordinately expressed under the control of the *trp* repressor, a dimeric protein of identical 107-residue subunits that is the product of the *trpR* gene (which forms an independent operon). *The trp repressor binds* L-*tryptophan, the pathway's end product, to form a complex that specifically binds to trp operator (trpO,* Fig. 31-43) *so as to reduce the rate of trp operon transcription 70-fold.* The X-ray structure of the *trp* repressor–operator complex (Section 31-3Da) indicates that tryptophan binding allosterically orients *trp* repressor's two symmetry related helix–turn– helix "DNA reading heads" so that they can simultaneously bind to *trpO* (Fig. 31-34). Moreover, the bound tryptophan forms a hydrogen bond to a DNA phosphate group, thereby

Figure 31-42   A genetic map of the *E. coli trp* operon indicating the enzymes it specifies and the reactions they catalyze. The gene product of *trpC* catalyzes two sequential reactions in the synthesis of tryptophan. [After Yanofsky, C., *J. Am. Med. Assoc.* **218**, 1027 (1971).]

strengthening the repressor–operator association. Tryptophan therefore acts as a **corepressor;** its presence prevents what is then superfluous tryptophan biosynthesis (SAM similarly functions as a corepressor with the *met* repressor; Fig. 31-35*a*). The *trp* repressor also controls the synthesis of at least two other operons: the **trpR operon** and the **aroH operon** (which encodes one of three isozymes that catalyze the initial reaction of chorismate biosynthesis; Section 26-5Bc).

a. Tryptophan Biosynthesis Is Also Regulated by Attenuation

The *trp* repressor–operator system was at first thought to fully account for the regulation of tryptophan biosynthesis in *E. coli*. However, the discovery of *trp* deletion mutants located downstream from *trpO* that increase *trp* operon expression 6-fold indicated the existence of an additional transcriptional control element. Sequence analysis established that *trpE,* the *trp* operon's leading structural gene, is preceded by a 162-nucleotide **leader sequence** (*trpL*). Genetic analysis indicated that the new control element is located in *trpL,* ~30 to 60 nucleotides upstream of *trpE* (Fig. 31-42).

When tryptophan is scarce, the entire 6720-nucleotide polycistronic *trp* mRNA, including the *trpL* sequence, is synthesized. As the tryptophan concentration increases, the rate of *trp* transcription decreases as a result of the *trp* repressor–corepressor complex's consequent greater abundance. Of the *trp* mRNA that is transcribed, however, an increasing proportion consists of only a 140-nucleotide segment corresponding to the 5′ end of *trpL*. *The availability of tryptophan therefore results in the premature termination of trp operon transcription.* The control element responsible for this effect is consequently termed an **attenuator.**

b. The *trp* Attenuator's Transcription Terminator Is Masked when Tryptophan Is Scarce

What is the mechanism of attenuation? The attenuator transcript contains four complementary segments that can form one of two sets of mutually exclusive base-paired

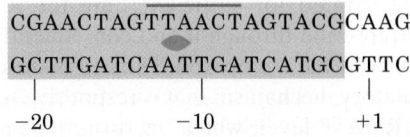

Figure 31-43   The base sequence of the *trp* operator. The nearly palindromic sequence is boxed and its −10 region is overscored.

Figure 31-44 **The alternative secondary structures of** *trpL* **mRNA.** The formation of the base paired 2 · 3 (antiterminator) hairpin (*right*) precludes the formation of the 1 · 2 and 3 · 4 (terminator) hairpins (*left*) and vice versa. Attenuation results in the premature termination of transcription immediately after hairpins (Fig. 31-44). *Segments 3 and 4 together with the succeeding residues comprise a normal intrinsic transcription terminator (Section 31-2Da):* a G + C–rich sequence that can form a self-complementary hairpin structure followed by several sequential U's (compare with Fig. 31-18). *Transcription rarely proceeds beyond this termination site unless tryptophan is in short supply.*

nucleotide 140 when the 3 · 4 hairpin is present. The arrow indicates the mRNA site past which RNA polymerase pauses until approached by an active ribosome. [After Fisher, R.F. and Yanofsky, C., *J. Biol. Chem.* **258,** 8147 (1983).]

A section of the leader sequence, which includes segment 1 of the attenuator, is translated to form a 14-residue polypeptide that contains two consecutive Trp residues (Fig. 31-44, *left*). The position of this particularly rare dipeptide segment (1.1% of the residues in *E. coli* proteins are Trp; Table 4-1) provided an important clue to the mechanism of attenuation. An additional essential aspect of this mechanism is that ribosomes commence the translation of a prokaryotic mRNA shortly after its 5′ end has been synthesized.

The above considerations led Yanofsky to propose the following model of attenuation (Fig. 31-45). An RNA polymerase that has escaped repression initiates *trp* operon transcription. Soon after the ribosomal initiation site of the *trpL* gene has been transcribed, a ribosome attaches to it and begins translation of the leader peptide. When tryptophan is abundant, so that there is a plentiful supply of **tryptophanyl–tRNA^{Trp}** (the transfer RNA specific for Trp with

an attached Trp residue; Section 32-2C), the ribosome follows closely behind the transcribing RNA polymerase so as to sterically block the formation of the 2 · 3 hairpin. Indeed, RNA polymerase pauses past position 92 of the transcript and only continues transcription on the approach of a ribosome, thereby ensuring the proximity of these two entities at this critical position. The prevention of 2 · 3 hairpin formation permits the formation of the 3 · 4 hairpin, the transcription terminator pause site, which results in the termination of transcription (Fig. 31-45*a*). When tryptophan is scarce, however, the ribosome stalls at the tandem UGG codons (which specify Trp; Table 5-3) because of the lack of tryptophanyl–tRNA^{Trp}. As transcription continues, the newly synthesized segments 2 and 3 form a hairpin because the stalled ribosome prevents the otherwise competitive formation of the 1 · 2 hairpin (Fig. 31-45*b*). The formation of the transcriptional terminator's 3 · 4 hairpin is thereby pre-empted for sufficient time for RNA polymerase to transcribe through it and consequently through the remainder of the *trp* operon. The cell is thus provided with a regulatory mechanism that is responsive to the tryptophanyl–tRNA^{Trp} level, which, in turn, depends on the protein synthesis rate as well as on the tryptophan supply.

There is considerable evidence supporting this model of attenuation. The *trpL* transcript is resistant to limited

(a) **High tryptophan**

Leader peptide

Transcription terminator

3 4

+

"Terminated" RNA polymerase

1 2

trpL mRNA

Ribosome transcribing the leader peptide mRNA

(b) **Low tryptophan**

Antiterminator

2 3

1

4

Transcribing RNA polymerase

trp operon mRNA

Ribosome stalled at tandem Trp codons

DNA encoding *trp* operon

Figure 31-45   Attenuation in the *trp* operon. (*a*) When tryptophanyl–tRNA^{Trp} is abundant, the ribosome translates *trpL* mRNA. The presence of the ribosome on segment 2 prevents the formation of the base-paired 2 · 3 hairpin. The 3 · 4 hairpin, an essential component of the transcriptional terminator, can thereby form, thus aborting transcription. (*b*) When tryptophanyl– tRNA^{Trp} is scarce, the ribosome stalls on the tandem Trp codons of segment 1. This situation permits the formation of the 2 · 3 hairpin which, in turn, precludes the formation of the 3 · 4 hairpin. RNA polymerase therefore transcribes through this unformed terminator and continues *trp* operon transcription.

RNase T1 digestion, indicating that it has extensive secondary structure. The significance of the tandem Trp codons in the *trpL* transcript is corroborated by their presence in *trp* leader regions of several other bacterial species. Moreover, the leader peptides of the five other amino acid–biosynthesizing operons known to be regulated by attenuation (most exclusively so) are all rich in their corresponding amino acid residues (Table 31-3). For example, the *E. coli* **his operon,** which specifies enzymes synthesizing histidine (Fig. 26-65), has seven tandem His residues in its leader peptide whereas the **ilv operon,** which specifies enzymes participating in isoleucine, leucine, and valine biosynthesis

(Fig. 26-61), has five Ile's, three Leu's, and six Val's in its leader peptide. Finally, the leader transcripts of these operons resemble that of the *trp* operon in their capacity to form two alternative secondary structures, one of which contains a trailing termination structure.

H. *Riboswitches Are Metabolite-Sensing RNAs*

We have just seen how the formation of secondary structure in a growing RNA transcript can regulate gene expression though attenuation. The conformational flexibility of mRNA also allows it to regulate genes by directly interacting

Table 31-3   Amino Acid Sequences of Some Leader Peptides in Operons Subject to Attenuation

Operon	Amino Acid Sequence[a]
trp	Met-Lys-Ala-Ile-Phe-Val-Leu-Lys-Gly-TRP-TRP-Arg-Thr-Ser
pheA	Met-Lys-His-Ile-Pro-PHE-PHE-PHE-Ala-PHE-PHE-PHE-Thr-PHE-Pro
his	Met-Thr-Arg-Val-Gln-Phe-Lys-HIS-HIS-HIS-HIS-HIS-HIS-HIS-Pro-Asp
leu	Met-Ser-His-Ile-Val-Arg-Phe-Thr-Gly-LEU-LEU-LEU-LEU-Asn-Ala- Phe-Ile-Val-Arg-Gly-Arg-Pro-Val-Gly-Gly-Ile-Gln-His
thr	Met-Lys-Arg-ILE-Ser-THR-THR-ILE-THR-THR-THR-ILE-THR-ILE- THR-THR-Gln-Asn-Gly-Ala-Gly
ilv	Met-Thr-Ala-LEU-LEU-Arg-VAL-ILE-Ser-LEU-VAL-VAL-ILE-Ser- VAL-VAL-VAL-ILE-ILE-ILE-Pro-Pro-Cys-Gly-Ala-Ala-Leu-Gly-Arg- Gly-Lys-Ala

[a]Residues in uppercase are synthesized in the pathway catalyzed by the operon's gene products.

Source: Yanofsky, C., *Nature* **289,** 753 (1981).

with certain cellular metabolites, thereby eliminating the need for sensor proteins such as the *lac* repressor, CAP, and the *trp* repressor.

In *E. coli*, the biosynthesis of thiamine pyrophosphate (TPP; Section 17-3Ba) requires the action of several proteins whose levels vary according to the cell's need for TPP. In at least two of the relevant genes the untranslated regions at the 5′ end of the mRNA include a highly conserved sequence called the *thi* **box.** The susceptibility of the *thi* box to chemical or enzymatic cleavage, as Ronald Breaker showed, differs in the presence and absence of TPP, suggesting that the RNA changes its secondary structure when TPP binds to it (the binding of a metabolite by RNA is not unprecedented; synthetic oligonucleotides known as **aptamers** bind specific molecules with high specificity and affinity; Section 7-6C). The TPP-sensing mRNA element has been dubbed a **riboswitch.**

The predicted secondary structure of the TPP-sensing riboswitch and its proposed mechanism are shown in Fig. 31-46*a*. In the absence of TPP, the mRNA assumes a con-

formation that allows a ribosome to begin translation. In the presence of TPP, an alternative secondary structure masks the sequence that identifies its translational initiation site to the ribosome (its so-called **Shine–Dalgarno sequence;** Section 32-3Cb) so that the ribosome cannot initiate the mRNA's translation. Thus, *the concentration of a metabolite can regulate the expression of genes required for its synthesis.* The X-ray structure of the 80-nt TPP-binding domain from the *E. coli* TPP-sensing riboswitch, determined by Breaker and Dinshaw Patel, reveals an intricately folded RNA that binds TPP in an extended conformation (Fig. 31-46*b*).

Over 20 classes of riboswitches have as yet been identified, including those that regulate the expression of enzymes involved in the metabolism of coenzyme B_{12} (Fig. 25-21), riboflavin (Fig. 16-8), *S*-adenosylmethionine (SAM; Fig. 26-18), lysine, and adenine. In general, they consist of two components, an aptamer that binds an effector and a so-called **expression platform** that transduces effector binding to a change in gene expression. In some cases, the

(a)   (b)

Figure 31-46 Structure of the TPP-sensing riboswitch from *E. coli.* (*a*) The predicted secondary structure of a 165-residue segment at the 5′ end of the *thiM* gene is shown in the absence (*left*) and presence (*right*) of TPP. The TPP-binding conformation masks the Shine–Dalgarno sequence (*orange*) required by the ribosome to initiate translation at the AUG start codon (*red*) just downstream. [After Winkler, W., Nahvi, A., and Breaker, R.R., *Nature* **419,** 952 (2002).] (*b*) The X-ray structure of the riboswitch's 90-nt TPP-sensing domain. The RNA is drawn in

cartoon form with its sugar–phosphate backbone represented by an orange rod and its bases represented by paddles with C green, N blue, and O red. The TPP is drawn in space-filling form with C cyan, N blue, O red, and S yellow. Mg^{2+} ions are represented by lavender spheres. [Based on an X-ray structure by Ronald Breaker, Yale University, and Dinshaw Patel, Memorial Sloan-Kettering Cancer Center, New York, New York. PDBid 2GDI.]

activated expression platform forms an intrinsic transcription termination site (Section 31-2Da) so that transcription beyond this site proceeds only when the effector is absent. In others, the activated expression platform cleaves itself, thereby inactivating the mRNA (the ability of RNA to act as an enzyme is discussed in Section 31-4Ae).

Riboswitches collectively regulate >2% of the genes in certain bacteria. Plants and fungi also contain riboswitches (although the above TTP-sensing riboswitch could not function in eukaryotes because eukaryotic ribosomes do not bind Shine–Dalgarno sequences; Section 32-3Cd). The fact that the interaction of riboswitches with their effectors does not require the participation of proteins suggests that they are relics of the RNA world (Section 1-5Ca) and hence among the oldest regulatory systems.

I. Regulation of Ribosomal RNA Synthesis: The Stringent Response

E. coli cells growing under optimal conditions divide every 20 min. Such cells contain up to 20,000 ribosomes and hence must synthesize ~10,000 ribosomes per cell division cycle. Yet RNAP can initiate the transcription of an rRNA gene no faster than about once per second. If *E. coli* contained only one copy of each of the three types of rRNA genes (those specifying the so-called 23S, 16S, and 5S rRNAs; Section 32-3A), fast-growing cells could synthesize no more than ~1200 ribosomes during their cell division cycle. However, *the E. coli genome contains seven separately located rRNA operons, all of which contain one nearly identical copy of each type of rRNA gene.* Moreover, rapidly growing cells contain multiple copies of their replicating chromosomes (Section 30-3Cb), thereby accounting for the observed rRNA synthesis rate.

Cells have the remarkable ability to coordinate the rates at which their thousands of components are synthesized. For example, *E. coli* adjust their ribosome content to match the rate at which they can synthesize proteins under the prevailing growth conditions. The rate of rRNA synthesis is therefore proportional to the rate of protein synthesis. One mechanism by which this occurs is known as the **stringent response:** *A shortage of any species of amino acid–charged tRNA (usually a result of "stringent" or poor growth conditions) that limits the rate of protein synthesis triggers a sweeping metabolic readjustment.* A major facet of this change is an abrupt 10- to 20-fold reduction in the rate of rRNA and tRNA synthesis. This **stringent control,** moreover, depresses numerous metabolic processes (including DNA replication and the biosynthesis of carbohydrates, lipids, nucleotides, proteoglycans, and glycolytic intermediates) while stimulating others (such as amino acid biosynthesis). The cell is thereby prepared to withstand nutritional deprivation.

a. (p)ppGpp Mediates the Stringent Response

*The stringent response is correlated with a rapid intracellular accumulation of two unusual nucleotides, **ppGpp** and **pppGpp** [known collectively as **(p)ppGpp**], and their prompt decay when amino acids become available.* The ob-

servation that mutants, designated *relA⁻*, which do not exhibit the stringent response (they are said to have **relaxed control**) lack (p)ppGpp suggests that these substances mediate the stringent response. This idea was corroborated by *in vitro* studies demonstrating, for example, that (p)ppGpp inhibits the transcription of rRNA genes but stimulates the transcription of the *trp* and *lac* operons as does the stringent response *in vivo*. Apparently, (p)ppGpp acts by somehow altering RNAP's promoter specificity at stringently controlled operons, a hypothesis that is supported by the isolation of RNAP mutants that exhibit reduced responses to (p)ppGpp. In addition, (p)ppGpp causes an increased frequency of pausing in RNAPs engaged in elongation, thereby reducing the rate of transcription.

The protein encoded by the wild-type *relA* gene, named **stringent factor (RelA),** catalyzes the reaction

$$ATP + GTP \rightleftharpoons AMP + pppGpp$$

and, to a lesser extent,

$$ATP + GDP \rightleftharpoons AMP + ppGpp$$

However, several ribosomal proteins convert pppGpp to ppGpp so that ppGpp is the stringent response's usual effector. Stringent factor is only active in association with a ribosome that is actively engaged in translation. (p)ppGpp synthesis occurs when a ribosome binds its mRNA-specified but uncharged tRNA (lacking an appended amino acid residue). The binding of a specified and charged tRNA greatly reduces the rate of (p)ppGpp synthesis. *The ribosome apparently signals the shortage of an amino acid by stimulating the synthesis of (p)ppGpp which, acting as an intracellular messenger, influences the rates at which a great variety of operons are transcribed.*

(p)ppGpp degradation is catalyzed by the *spoT* gene product. The *spoT⁻* mutants show a normal increase in (p)ppGpp level on amino acid starvation but an abnormally slow decay of (p)ppGpp to basal levels when amino acids again become available. The *spoT⁻* mutants therefore exhibit a sluggish recovery from the stringent response. *The (p)ppGpp level is apparently regulated by the countervailing activities of stringent factor and the spoT gene product.*

4 POST-TRANSCRIPTIONAL PROCESSING

The immediate products of transcription, the **primary transcripts,** are not necessarily functional entities. In order to acquire biological activity, many of them must be specifically altered in several ways: (1) by the exo- and endonucleolytic removal of polynucleotide segments; (2) by appending nucleotide sequences to their 3′ and 5′ ends; and (3) by the modification of specific nucleosides. The three major classes of RNAs, mRNA, rRNA, and tRNA, are altered in different ways in prokaryotes and in eukaryotes. In this section we shall outline these **post-transcriptional modification** processes.

A. *Messenger RNA Processing*

In prokaryotes, most primary mRNA transcripts function in translation without further modification. Indeed, as we have seen, ribosomes in prokaryotes usually commence translation on nascent mRNAs. In eukaryotes, however, mRNAs are synthesized in the cell nucleus, whereas translation occurs in the cytosol. Eukaryotic mRNA transcripts can therefore undergo extensive post-transcriptional processing while still in the nucleus.

a. Eukaryotic mRNAs Are Capped

*Eukaryotic mRNAs have a peculiar enzymatically appended **cap structure** consisting of a **7-methylguanosine** (m^7G) residue joined to the transcript's initial (5') nucleoside via a 5'–5' triphosphate bridge* (Fig. 31-47). This m^7 cap, which is added to the growing transcript before it is ~30 nucleotides long, defines the eukaryotic translational start site (Section 32-3Cd). A cap may be $O^{2'}$-methylated at the transcript's leading nucleoside (**cap-1,** the predominant cap in multicellular organisms), at its first two nucleosides (**cap-2**), or at neither of these positions (**cap-0,** the predominant cap in unicellular eukaryotes). If the leading nucleoside is adenosine (it is usually a purine), it may also be N^6-methylated.

Capping involves several enzymatic reactions: (1) the removal of the leading phosphate group from the mRNA's 5' terminal triphosphate group by an **RNA triphosphatase;** (2) the guanylation of the mRNA by **capping enzyme,** which requires GTP and yields the 5'–5' triphosphate bridge and PP_i; (3) the methylation of guanine by **guanine-7-methyltransferase** in which the methyl group is supplied by *S*-adenosylmethionine (SAM); and possibly (4) the O2' methylation of the mRNA's first and perhaps its second nucleotide by a SAM-requiring **2'-O-methyltransferase.** Both the capping enzyme and the guanine-7-methyltransferase bind to RNAP II's phosphorylated CTD (Section 31-2E). Hence it is likely that capping marks the completion of RNAP II's switch from transcription initiation to elongation.

b. Eukaryotic mRNAs Have Poly(A) Tails

Eukaryotic mRNAs, unlike those of prokaryotes, are invariably monocistronic. Moreover, in contrast to the case in bacteria (Section 31-2D), no eukaryotic transcriptional termination sequence has been identified (but see below). In fact, the eukaryotic termination process is imprecise; that is, the primary transcripts of a given structural gene have heterogeneous 3' sequences. Nevertheless, mature eukaryotic mRNAs have well-defined 3' ends; *almost all of them in mammals have 3'-poly(A) tails of ~250 nucleotides (~80 in yeast).* The poly(A) tails are enzymatically appended to the primary transcripts in two reactions that are mediated by a 500- to 1000-kD complex that consists of at least six proteins:

1. A transcript is cleaved to yield a free 3'-OH group at a specific site that is 15 to 25 nucleotides past an AAUAAA sequence and within 50 nucleotides before a

Figure 31-47   The structure of the 5' cap of eukaryotic mRNAs. It is known as cap-0, cap-1, or cap-2, respectively, if it has no further modifications, if the leading nucleoside of the transcript is $O^{2'}$-methylated, or if its first two nucleosides are $O^{2'}$-methylated.

U-rich or G + U–rich sequence. The AAUAAA sequence is highly conserved in higher eukaryotes (but not yeast) in which its mutation abolishes cleavage and polyadenylation. The precision of the cleavage reaction has apparently eliminated the need for accurate transcriptional termination. Nevertheless, the identity of the endonuclease that cleaves the RNA is uncertain although **cleavage factors I and II (CFI and CFII)** are required for this process.

2. The poly(A) tail is subsequently generated from ATP through the stepwise action of **poly(A) polymerase (PAP).** This enzyme, which by itself only weakly binds RNA, is recruited by **cleavage and polyadenylation specificity factor (CPSF)** on this heterotetramer's recognition of the AAUAAA sequence, which it does with almost no tolerance for sequence variation. The downstream G + U–rich element is recognized by the heterotrimeric **cleavage stimulation factor (CstF),** which increases the affinity with which CPSF binds the AAUAAA sequence. However, once the poly(A) tail has grown to ~10 residues, the AAUAAA sequence is no longer required for further chain elongation. This suggests that CPSF becomes disengaged from its recognition site in a manner reminiscent of the way σ factor is released from the transcriptional initiation site once the elongation of prokaryotic mRNA is under way (Section 31-2B). The final length of the poly(A) tail is controlled by **poly(A)-binding protein II (PAB II),** multiple copies of which bind to successive segments of poly(A). PAB II also increases the processivity of PAP.

Both CPSF and CstF bind to the phosphorylated RNAP II CTD (Section 31-2E); deleting the CTD inhibits polyadenylation. Evidently, the CTD couples polyadenylation to transcription.

PAP is a template-independent RNA polymerase that elongates an mRNA primer with a free 3'-OH group. The X-ray structure of the 530-residue D154A mutant form of yeast PAP (D154 is a catalytically essential active site residue) in complex with A_5 and ATP, determined by Andrew Bohm, reveals that this monomeric protein consists of three domains that form a prominent U-shaped cleft (Fig. 31-48). Hence it has the handlike domain arrangement of template-directed polymerases (Section 30-2Ad). Indeed, PAP's N-terminal domain, which contains the enzyme's active site, is homologous to the palm domain of DNA polymerase β, although it forms the thumb side of the cleft rather than its base. PAP's central domain, which forms the base of the cleft, is functionally but not structurally analogous to the fingers domain of template-directed polymerases in that it interacts with the β and γ phosphates of the incoming ATP. However, the C-terminal domain shows no resemblance to a fingers domain. Rather, it is topologically similar to the **RNA-recognition motif [RRM;** also known as the **RNA-binding domain (RBD)]** that occurs in >200 different RNA-binding proteins (see below). The A_5 binds in the cleft in an extended conformation such that, in contrast to the nucleic acids bound to template-dependent polymerases, its bases are not in contact. However, the 3'-terminal base of the A_5 stacks on the base of the ATP.

In comparison to the X-ray structure of yeast PAP in complex with 3'-dATP, also determined by Bohm, the N-terminal domain in the PAP–A_5–ATP structure has undergone an ~23° hingelike motion toward the C-terminal domain to form a closed conformation resembling that observed in the structure of Taq DNA polymerase I in complex with its substrates (Section 30-2Ae). This motion forms several new interactions, both direct and water-mediated, between PAP and its ATP substrate that helps differentiate adenine from other bases. In contrast, in template-dependent polymerases, the incoming base only makes sequence-specific contacts with the template base (Section 30-2Ae).

In vitro studies indicate that a poly(A) tail is not required for mRNA translation. Rather, the observations that an mRNA's poly(A) tail shortens as it ages in the cytosol and that unadenylated mRNAs have abbreviated cytosolic lifetimes suggest that poly(A) tails have a protective role. In fact, the only mature mRNAs that lack poly(A) tails, those of histones (which, with few exceptions, lack the AAUAAA cleavage–polyadenylation signal), have lifetimes of <30 min in the cytosol, whereas most other mRNAs last hours or days. The poly(A) tails are specifically complexed in the cytosol by **poly(A) binding protein (PABP;** not related to PAB II), which organizes poly(A)-bearing mRNAs into ribonucleoprotein particles. PABP is thought to protect mRNA from degradation as is suggested, for example, by the observation that the addition of PABP to a cell-free system containing mRNA and mRNA-degrading nucleases greatly reduces the rate at which the mRNAs are degraded and the rate at which their poly(A) tails are shortened.

All known PABPs contain four tandem and highly conserved RNA-recognition motifs (RRMs) followed by a less conserved Pro-rich C-terminal segment of variable length. A variety of evidence suggests that PABP's first two RRMs support most of the biochemical functions of full-length PABP. The X-ray structure of the first two RRMs of human PABP (RRM1/2; the N-terminal 190 residues of this 636-residue protein) in complex with A_{11}, determined by Stephen Burley, reveals that RRM1/2 forms a continuous trough-shaped surface in which the poly(A) binds in an extended conformation via interactions with conserved

Figure 31-48   X-ray structure of the D154A mutant of yeast poly(A) polymerase (PAP) in complex with A_5 and ATP. PAP is drawn in ribbon form with its N-terminal domain lavender, its central domain yellow-green, and its C-terminal domain light blue. The A_5 and ATP are drawn in stick form with A_5 C white, ATP C cyan, N blue, O red, and P orange. [Based on an X-ray structure by Andrew Bohm, Tufts University School of Medicine. PDBid 2Q66.]

Figure 31-49 **X-ray structure of the N-terminal two RNA-recognition motifs (RRMs) of human PABP in complex with A$_{11}$.** RRM1 is cyan, RRM2 is gold, and their linking segment is lavender. The poly(A), only nine of whose nucleotides are observed, is drawn in stick form with C green, N blue, O red, and P magenta. [Based on an X-ray structure by Stephen Burley, The Rockefeller University. PDBid 1CVJ.]

residues (Fig. 31-49). Each RRM, as also seen in the structures of a variety of other RNA-binding proteins, consists of a compact globule made of a 4-stranded antiparallel sheet that forms the RNA-binding surface backed by two helices.

The cleavage of a transcript past its AAUAAA sequence does not, in itself, terminate transcription. However, in yeast, the protein **Rtt103,** which binds to the phosphorylated CTD of RNAP II, recognizes the AAUAAA sequence and recruits the $5' \rightarrow 3'$ exonuclease known as **Rat1** (**Xrn2** in humans). Then, in what is termed the **torpedo model,** the highly processive Rat1/Xrn2 loads onto the newly liberated 5' end of the still nascent RNA and rapidly degrades it until it intercepts the RNAP and induces it to terminate RNA synthesis. It has been hypothesized that this occurs in much the same way as Rho factor terminates bacterial transcription (Section 31-2Db). This frees the RNAP to initiate a new round of transcription.

c. Eukaryotic Genes Consist of Alternating Expressed and Unexpressed Sequences

The most striking difference between eukaryotic and prokaryotic structural genes is that the coding sequences of

most eukaryotic genes are interspersed with unexpressed regions. Early investigations of eukaryotic structural gene transcription found, quite surprisingly, that primary transcripts are highly heterogeneous in length (from ~2000 to well over 20,000 nucleotides) and are much larger than was expected from the known sizes of eukaryotic proteins. Rapid labeling experiments demonstrated that little of this so-called **heterogeneous nuclear RNA (hnRNA)** is ever transported to the cytosol; most of it is quickly turned over (degraded) in the nucleus. Yet, the hnRNA's 5' caps and 3' tails eventually appear in cytosolic mRNAs. *The straightforward explanation of these observations, that **pre-mRNAs** are processed by the excision of internal sequences, seemed so bizarre that it came as a great surprise in 1977 when Phillip Sharp and Richard Roberts independently demonstrated that this is actually the case.* In fact, mammalian premRNAs typically contain eight noncoding **intervening sequences (introns)** whose aggregate length averages 4 to 10 times that of their flanking **expressed sequences (exons).** This situation is graphically illustrated in Fig. 31-50, which is an electron micrograph of chicken **ovalbumin** mRNA hybridized to the antisense strand of the ovalbumin gene (ovalbumin is the major protein component of egg white).

Figure 31-50 **An electron micrograph and its interpretive drawing of a hybrid between the antisense strand of the chicken ovalbumin gene and its corresponding mRNA.** The complementary segments of the DNA (*purple line in the drawing*) and mRNA (*red dashed line*) have annealed to reveal the exon positions (*L,* 1–7). The looped-out segments (I–VII), which have no complementary sequences in the mRNA, are the introns. [From Chambon, P., *Sci. Am.* **244**(5), 61 (1981).]

Figure 31-51 **The sequence of steps in the production of mature eukaryotic mRNA as shown for the chicken ovalbumin gene.** Following transcription, the primary transcript is capped and polyadenylated. The introns are then excised and the exons spliced together to form the mature mRNA. However, splicing may also occur cotranscriptionally.

Exons have lengths that range up to 17,106 nt (in the gene encoding the 34,350-residue muscle protein **titin,** the largest known single-chain protein; Section 35-3Ae) but with most <300 nt (and averaging 150 nt in humans). Introns, in contrast, are usually much longer, with lengths averaging ~3500 nt and as high as ~800,000 nt [in the gene encoding the muscle protein **dystrophin** (Section 35-3Ae), whose length is ~2400 kb and hence is the largest human gene] with no obvious periodicity. Moreover, the corresponding introns from genes in two vertebrate species can vary extensively in both length and sequence so as to bear little resemblance to one another. The number of introns in a gene averages 7.8 in the human genome and varies from none to 364 (with the latter number occurring in the gene encoding titin).

The formation of eukaryotic mRNA begins with the transcription of an entire structural gene, including its introns, to form pre-mRNA (Fig. 31-51). Then, following capping, the introns are excised and their flanking exons are connected, a process called **gene splicing** or just **splicing,** that often occurs cotranscriptionally. *The most striking aspect of gene splicing is its precision; if one nucleotide too few or too many were excised, the resulting mRNA could not be translated properly (Section 32-1B). Moreover, exons are never shuffled; their order in the mature mRNA is exactly the same as that in the gene from which it is transcribed.*

d. Exons Are Spliced in a Two-Stage Reaction

Sequence comparisons of exon–intron junctions from a diverse group of eukaryotes indicate that they have a high degree of homology (Fig. 31-52), including, as Richard Breathnach and Pierre Chambon first pointed out, *an invariant GU at the intron's 5' boundary and an invariant AG at its 3' boundary. These sequences are necessary and sufficient to define a splice junction:* Mutations that alter the sequences interfere with splicing, whereas mutations that change a nonjunction to a consensus-like sequence can generate a new splice junction.

Figure 31-52 **The consensus sequence at the exon–intron junctions of vertebrate pre-mRNAs.** The subscripts indicate the percentage of pre-mRNAs in which the specified base(s) occurs. Note that the 3' splice site is preceded by a tract of 11 predominantly pyrimidine nucleotides. [Based on data from Padgett, R.A., Grabowski, P.J., Konarska, M.M., Seiler, S.S., and Sharp, P.A., *Annu. Rev. Biochem.* **55,** 1123 (1986).]

Investigations of both cell-free and *in vivo* splicing systems by Argiris Efstratiadis, Tom Maniatis, Michael Rosbash, and Sharp established that intron excision occurs via two transesterification reactions that are remarkably similar from yeast to humans (Fig. 31-53):

1. The formation of a 2′,5′-phosphodiester bond between an intron adenosine residue and its 5′-terminal phosphate group with the concomitant liberation of the 5′ exon's 3′-OH group. *The intron thereby assumes a novel **lariat structure.*** The adenosine residue at the lariat branch has been identified in yeast as the last A in the highly conserved sequence UACUAAC and in vertebrates as the A in the equivalent but more permissive sequence YNCURAY [where R represents purines (A or G), Y represents pyrimidines (C or U), and N represents any nucleotide]. In yeast and vertebrates, the branch point A occurs ~50 and 18 to 40 residues upstream of the associated 3′ splice site, respectively. In yeast, which have relatively few introns, mutations that change this branch point A residue abolish splicing at that site. However, in higher eukaryotes, the mutation or deletion of a branch site often activates a so-called **cryptic branch site** that is also near the 3′ splice site. Evidently, the branch site functions to identify the nearest 3′ splice site as a target for linkage to the 5′ splice site.

2. The now free 3′-OH group of the 5′ exon forms a phosphodiester bond with the 5′-terminal phosphate of the 3′ exon yielding the spliced product and releasing the intron lariat with a free 3′-OH group. The intron lariat is then debranched (linearized) and, *in vivo,* is rapidly degraded. Mutations that alter the conserved AG at the 3′ splice site block this second step, although they do not interfere with lariat formation.

Note that the splicing process proceeds without free energy input; its transesterification reactions preserve the free energy of each cleaved phosphodiester bond through the concomitant formation of a new one.

The sequences required for splicing are the short consensus sequences at the 3′ and 5′ splice sites and at the branch site. Nevertheless, these sequences are poorly conserved. However, other short sequence elements within exons and introns that are known as **exonic** and **intronic splicing enhancers (ESEs and ISEs)** and **silencers (ESSs and ISSs)** also play important roles in splice site selection although their characteristics are poorly understood (even highly sophisticated computer programs are only ~50% successful in predicting actual splice sites over apparently equally good candidates that are not). In contrast, large portions of most introns can be deleted without impeding splicing.

e. Some Eukaryotic Genes Are Self-Splicing

It is now recognized that there are eight distinct types of introns, seven of which occur in eukaryotes (Table 31-4). **Group I introns** occur in the nuclei, mitochondria, and chloroplasts of diverse eukaryotes (but not vertebrates), and even in some bacteria. Thomas Cech's study of how

Figure 31-53 **The sequence of transesterification reactions that splice together the exons of eukaryotic pre-mRNAs.** The exons and introns are drawn in blue and orange, and R and Y represent purine and pyrimidine residues. **(1)** The 2′-OH group of a specific intron A residue nucleophilically attacks the 5′-phosphate at the 5′ intron boundary to yield an unusual 2′,5′-phosphodiester bond and thus form a lariat structure. **(2)** The liberated 3′-OH group forms a 3′,5′-phosphodiester bond with the 5′ terminal residue of the 3′ exon, thereby splicing the two exons together and releasing the intron in lariat form with a free 3′-OH.

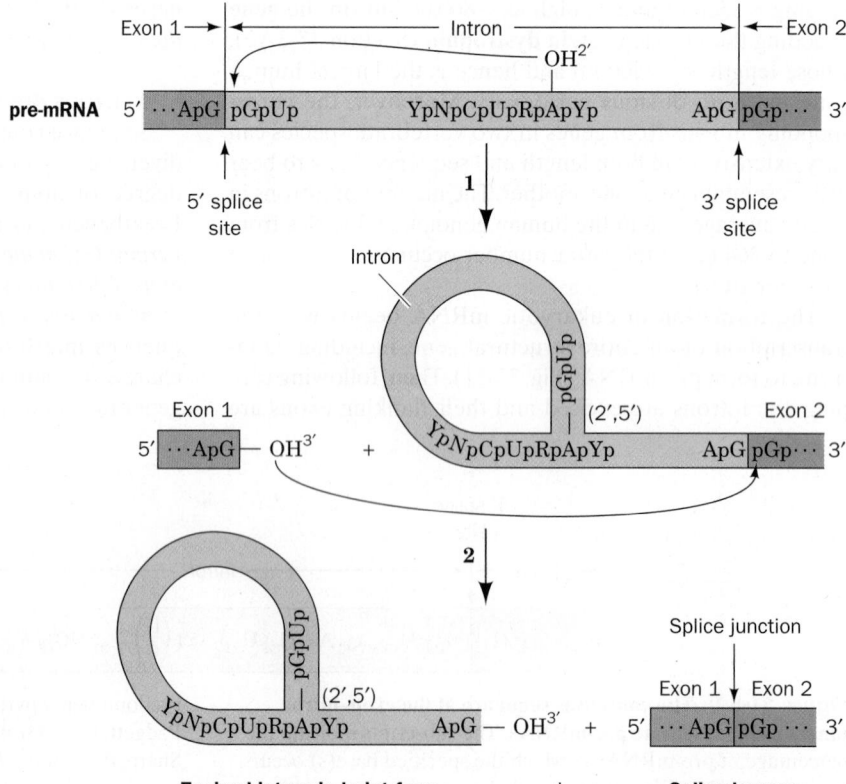

Table 31-4   Types of Introns

Intron Type	Where Found
GU–AG introns	Eukaryotic nuclear pre-mRNA
AU–AC introns	Eukaryotic nuclear pre-mRNA
Group I	Eukaryotic nuclear pre-mRNA, organelle RNAs, a few bacterial RNAs
Group II	Organelle RNAs, a few prokaryotic RNAs
Group III	Organelle RNAs
Twintrons (composites of two and/or more group II or III introns)	Organelle RNAs
Pre-tRNA introns	Eukaryotic nuclear pre-tRNAs
Archaeal introns	Various RNAs

Source: Brown, T.A., *Genomes* (3rd ed.), Garland Science, p. 355 (2007).

group I introns are spliced in the ciliated protozoan *Tetrahymena thermophila* led to an astonishing discovery: *RNA can act as an enzyme. When the isolated pre-rRNA of this organism is incubated with guanosine or a free guanine nucleotide (GMP, GDP, or GTP), but in the absence of protein, its single 421-nucleotide intron excises itself and splices together its flanking exons; that is, this pre-rRNA is self-splicing.* The three-step reaction sequence of this process (Fig. 31-54) resembles that of mRNA splicing:

1. The 3′-OH group of the guanosine forms a phosphodiester bond with the intron's 5′ end, liberating the 5′ exon.

2. The 3′-terminal OH group of the newly liberated 5′ exon forms a phosphodiester bond with the 5′-terminal phosphate of the 3′ exon, thereby splicing together the two exons and releasing the intron.

3. The 3′-terminal OH group of the intron forms a phosphodiester bond with the phosphate of the nucleotide 15 residues from the intron's 5′ end, yielding the 5′-terminal fragment with the remainder of the intron in cyclic form.

This self-splicing process consists of a series of transesterifications and therefore does not require free energy input. Cech further established the enzymatic properties of the

Figure 31-54   The sequence of reactions in the self-splicing of *Tetrahymena* **group I intron. (1)** The 3′-OH group of a guanine nucleotide attacks the intron's 5′-terminal phosphate so as to form a phosphodiester bond and release the 5′ exon. **(2)** The newly generated 3′-OH group of the 5′ exon attacks the 5′-terminal phosphate of the 3′ exon, thereby splicing the two exons and releasing the intron. **(3)** The 3′-OH group of the intron attacks the phosphate of the nucleotide that is 15 residues from the 5′ end so as to cyclize the intron and release its 5′-terminal fragment. Throughout this process, the RNA maintains a folded, internally hydrogen bonded conformation that permits the precise excision of the intron.

Tetrahymena intron, which stem from its three-dimensional structure, by demonstrating that it catalyzes the *in vitro* cleavage of poly(C) with an enhancement factor of 10^{10} over the rate of spontaneous hydrolysis. Indeed, this RNA catalyst even exhibits Michaelis–Menten kinetics (K_M = 42 μM and k_{cat} = 0.033 s^{-1} for C$_5$). Such RNA enzymes have been named **ribozymes.**

Although the idea that an RNA can have enzymatic properties may seem unorthodox, *there is no fundamental reason why an RNA, or any other macromolecule, cannot have catalytic activity* (recall that it was likewise once generally accepted that nucleic acids lack the complexity to carry hereditary information; Section 5-2). Of course, in order to be an efficient catalyst, a macromolecule must be able to assume a stable structure but, as we shall see below and in Sections 32-2B and 32-3Ae, RNAs, including tRNAs and rRNAs, can do so. In fact, the *Tetrahymena* intron undergoes a series of well-defined conformational changes during its reaction sequence. [Synthetic ssDNAs are also known to have catalytic properties although such **deoxyribozymes** are unknown in biology.]

The **group II introns,** which occur in the mitochondria of fungi and plants and comprise the majority of the introns in chloroplasts, are also self-splicing. They generally employ an internal A residue as their initial attacking nucleophile (instead of an external G) to form a lariat intermediate, a process that resembles the splicing of nuclear pre-mRNAs (Fig. 31-53). We shall see below that nuclear pre-mRNA splicing is mediated by complex ribonucleoprotein particles known as **spliceosomes.** The chemical similarities of the pre-mRNA and group II intron splicing reactions therefore suggest that *spliceosomes are ribozymal systems whose RNA components have evolved from primordial self-splicing RNAs and that their protein components serve mainly to fine-tune ribozymal structure and function.* Similarly, the RNA components of ribosomes, which are two-thirds RNA and one-third protein, clearly have a catalytic function in addition to the structural and recognition roles traditionally attributed to them (Section 32-3). Thus, the observations that nucleic acids but not proteins can direct their own synthesis, that cells contain batteries of protein-based enzymes for manipulating DNA but relatively few for processing RNA, and that many coenzymes are ribonucleotides (e.g., ATP, NAD$^+$, and CoA), led to the hypothesis that *RNAs were the original biological catalysts in precellular times (the RNA world) and that the chemically more versatile proteins were relative latecomers in macromolecular evolution (Section 1-5Ca).*

f. The X-Ray Structures of a Group I Ribozyme

Group I introns are the most abundant self-splicing introns, with >2000 such sequences known. The sequence of the 413-nt *Tetrahymena* group I intron, together with phylogenetic comparisons, indicates that it contains nine double helical segments that are designated P1 through P9 (Fig. 31-55a; P for base-*p*aired segment). Such analysis further indicates that the conserved catalytic core of group I introns consists of sets of coaxially stacked helices inter-spersed with internal loops that are organized into two domains, the P4-P5-P6 domain (also called P4-P6) and the P3-P7-P8-P9 domain (also called P3-P9).

Cech designed a 247-nt RNA (Fig. 31-55a) that encompasses both the P4-P6 and P3-P9 domains of the *Tetrahymena* group I intron (it lacks the P1-P2 domain and the attached exons), with the addition of a 3′ G (ωG), which functions as an internal guanosine nucleophile. This RNA is catalytically active; it binds the P1-P2 domain via tertiary interactions and, with the assistance of ωG, cleaves P1 in a manner similar to the intact intron.

The X-ray structure of this RNA (Fig. 31-55b,c) reveals that it is largely composed of three coaxially stacked sets of A-RNA-like helices with P4-P6 consisting of two pseudocontinous and straight parallel helices connected by a sharp bend and P3-P9 consisting of a curved helix that wraps around one side of P4-P6 through extensive interdomain interactions that form the ribozyme's active site. Of particular note are its so-called A-rich bulge, a 7-nt sequence about halfway along the short arm of the U-shaped P4-P6, and the 6-nt sequence at the tip of the short arm of the U, whose central GAAA assumes a characteristic conformation known as a **tetraloop.** In both of these substructures, the bases are splayed outward so as to stack on each other and to associate in the minor groove of specific segments of the long arm of the U via hydrogen bonding interactions involving ribose residues as well as bases. In many such interactions, the close packing of phosphate groups is mediated by hydrated Mg^{2+} ions. Throughout this structure, the defining characteristic of RNA, its 2′-OH group, is both a donor and an acceptor of hydrogen bonds to phosphates, bases, and other 2′-OH groups. Interestingly, although this overall fold is highly conserved among group I introns, their sequences are poorly conserved with the exception of a few crucial active site residues.

In the initial reaction catalyzed by group I ribozymes, the 3′-OH group of ωG nucleophilically attacks the phosphate group linking the 5′ exon to the ribozyme (Fig. 31-54). But how is only this 3′-OH group activated as a nucleophile? The binding site for the ωG substrate is composed of four coplanar base triples with the ωG–G264–C311 triple sandwiched by three other base triples (Fig. 31-56). Consequently, the base of ωG is stacked between those of A261 and C262 (Fig. 31-56a), which stabilizes the binding of ωG to this site through base stacking.

Divalent metal ions, usually Mg^{2+}, are often required for both the structural stability and the catalytic activity of ribozymes. Unfortunately, the relatively low (3.8 Å) resolution of the ribozyme X-ray structure precluded the direct observation of Mg^{2+} ions (which have the same number of electrons as water molecules). However, there is good evidence that the heavy metal ions (Eu and Ir) used to solve the X-ray structure occupied many of the same positions in the ribozyme as do Mg^{2+} ions and moreover, several of these sites were observed to contain Mg^{2+} ions in other group I introns whose X-ray structures are known. In particular, an Mg^{2+} ion is in contact with the 2′-OH group of ωG as well as being liganded by the phosphate groups of

(a)

Figure 31-55 **The group I intron from *Tetrahymena thermophila*.** (*a*) The secondary structure of the 414-nt ribozyme. Its P4-P6 and P3-P9 domains are shaded in blue and green, respectively, with the catalytically active 3′ ωG residue shaded in yellow, the base triples of the P7 domain shaded in red, and the A-rich bulge, the tetraloop, and the tetraloop receptor on the P4-P6 domain shaded in brown, pink, and cyan, respectively. Watson–Crick and non-Watson–Crick base pairing interactions are represented by short horizontal lines and small filled circles, whereas interdomain interactions are indicated by magenta lines. Every tenth residue is marked by an outwardly pointing dash. The positions of five residues that have been mutated to stabilize the ribozyme structure are circled and those of the seven mutations that facilitated crystallization are bracketed (this mutant form retains its catalytic activity). (*b*) The X-ray structure of the ribozyme, drawn in cartoon form with its bases shown as paddles, and colored as in Part *a*. The inferred positions of Mg^{2+} ions are represented by orange spheres. (*c*) As in Part *b* but rotated 140° about the vertical axis to better show the A-rich bulge and the interaction between the tetraloop and the tetraloop receptor. [Part *a* modified from a drawing by and Parts *b* and *c* based on an X-ray structure by Thomas Cech, University of Colorado. PDBid 1X8W.] 🔗 **See Interactive Exercise 42**

(b)

(c)

three surrounding nucleotides. This both orients the ribose group of ωG and nucleophilically activates its 3′-OH group. Biochemical studies indicate that a second Mg^{2+} ion, which accompanies the phosphate group of the RNA

substrate, also participates in the catalytic reaction. Note that two Mg^{2+} ions similarly participate in the phosphoryl-transfer reactions catalyzed by protein enzymes such as DNA polymerase (Section 30-2Af).

Figure 31-56 Structure of the guanosine binding site in the *Tetrahymena* group I intron. (*a*) Side view of the four coplanar base triples at the ribozymal active site in P7 (colored in red in Fig. 31-55). The sugar–phosphate backbone is drawn in stick form and the bases are drawn in paddle form. The nucleotides at each level have the same color with the exception of the C413–ωG dinucleotide, which is yellow. (*b*) Top view of the base triple interactions between ωG and the G264 · C311 base pair. (*c*) Top view of the base triple of C262, A263, and G312, which is located above that containing ωG. (*d*) Top view of the base triple of A261, A265, and U310, which is located below that containing ωG. [Courtesy of Thomas Cech, University of Colorado. PDBid 1X8W.]

g. Hammerhead Ribozymes Catalyze an In-Line Nucleophilic Attack

One of the best-characterized types of ribozymes is the **hammerhead ribozyme,** so called because of the superficial resemblance of its secondary structure, as it was originally laid out, to a hammer. This minimally ~40-nt RNA participates in the replication of certain viruslike RNAs that infect plants and also occurs in schistosomes (species of parasitic flatworms). The hammerhead ribozyme catalyzes the site-specific cleavage of one of its own phosphodiester bonds with an ~10^7-fold rate enhancement. However, it is not a true catalyst because it cannot return to its original state.

The secondary structure of the 63-nt hammerhead ribozyme from *Schistosoma mansoni* has three duplex stems and an active site core of two nonhelical segments (Fig. 31-57*a*). This ribozyme cleaves itself between its C-17 and C-1.1 nucleotides to yield a cyclic 2′,3′-phosphodiester on C-17 with inversion of configuration about the P atom, together with a free 5′-OH on C-1.1, much like the intermediate product in the RNA hydrolysis reaction catalyzed by RNase A (Section 15-1Ab). This suggests that the reaction proceeds via an "in-line" mechanism such as that diagrammed in Fig. 16-6*b* with the transition state forming a trigonal bipyramidal intermediate in which the attacking

nucleophile, the 2′-OH group (Y in Fig. 16-6*b*), and the leaving group, which forms the free 5′-OH group (X in Fig. 16-6*b*), occupy the axial positions.

The X-ray structure of the *S. mansoni* hammerhead ribozyme, determined by William Scott, reveals that its stem II, stem III and the terminal end of stem I are coaxially stacked. The remainder of stem I curves around in a manner that more closely resembles the handle of a suitcase than that of a hammer to form a junction with stems II and III that contains the ribozyme's active site core (Fig. 31-57*b*). The nucleotides in the helical stems mainly form normal Watson–Crick base pairs, whereas the nucleotides of the active site core participate in non-Watson–Crick base pairs. This explains the observations that most helical positions can be occupied by any Watson–Crick base pair but that few core bases can be changed without significantly reducing ribozymal activity.

The bases of the active site core participate in a hydrogen bonded network (Fig. 31-57*c*). This helps position C-17 such that its O2′ atom is properly oriented for an in-line

Figure 31-57 **Structure of the *Schistosoma mansoni* hammerhead ribozyme.** (*a*) The sequence and schematic structural organization of the ribozyme colored to match the X-ray structure drawn in Part *b*. Base pairs and tertiary interactions are represented by dashes and dotted lines, respectively. Nucleotides are labeled according to the universal numbering system for hammerhead ribozymes. (*b*) X-ray structure of the ribozyme drawn in paddle form with C atoms the same color as in Part *a* except that those of G-12 are light green and those of C-1.1 are light blue, N blue, O red, and P orange. Adjacent P atoms in the same strand are connected by thin orange rods. The P atom of the scissile phosphate group (bridging C-17 and C-1.1) and the nucleophile (O2′ of C-17) are represented by small spheres. (*c*) The ribozyme's active site residues, which are drawn in stick form with C green, O red, N blue, and P orange. Hydrogen bonds are represented by dashed black lines. The proposed reaction mechanism is indicated by the curved arrows with the dotted black line marking the in-line trajectory taken by atom O2′ of C-17 in nucleophilically attacking the P atom of the scissile phosphate group. [Part *a* courtesy of and Parts *b* and *c* based on an X-ray structure by William Scott, University of California at Santa Cruz. PDBid 2GOZ.] 🐌 **See Interactive Exercise 43**

nucleophilic attack on the P atom that links atom O3′ of C-17 to atom O5′ of C-1.1. The N1 of the invariant G-12 when deprotonated and the 2′-OH of the invariant G-8 appear to be properly positioned to respectively act as base and acid catalysts in this reaction (Fig. 31-57*c*), which strongly suggests that the reaction occurs via a concerted acid–base catalyzed mechanism. This reaction mechanism does not involve the participation of metal ions and none are observed in the ribozyme's catalytic core. However, in solution, the presence of divalent metal ions provides an ~50-fold enhancement rate over the presence of only monovalent metal ions. Perhaps divalent metal ions stabilize the negative charge on the trigonal bipyramidal intermediate (Fig. 16-6*b*) and/or they may help position and orient reactive groups.

h. Splicing of Pre-mRNAs Is Mediated by snRNPs in the Spliceosome

How are the splice junctions of pre-mRNAs recognized and how are the two exons to be joined brought together in the splicing process? Part of the answer to this question was established by Joan Steitz going on the assumption that one nucleic acid is best recognized by another. The eukaryotic nucleus, as has been known since the 1960s, contains numerous copies of several highly conserved 60- to 300-nucleotide RNAs called **small nuclear RNAs**

(snRNAs), which form protein complexes termed **small nuclear ribonucleoproteins (snRNPs;** pronounced "snurps"). Steitz recognized that the 5′ end of one of these snRNAs, **U1-snRNA** (so called because it is a member of a U-rich subfamily of snRNAs), is partially complementary to the consensus sequence of the 5′ splice site. The consequent hypothesis, that *U1-snRNA recognizes the 5′ splice site,* was

Figure 31-58 **An electron micrograph of spliceosomes in action.** A *Drosophila* gene that is ~6 kb long enters from the upper left of the micrograph and exits at the lower left. Transcription initiates near the point marked by an asterisk. The growing RNA chains appear as fibrils of increasing lengths that emanate from the DNA. The transcripts are undergoing cotranslational splicing as revealed by the progressive formation and loss of intron loops near the 5′ ends of the RNA transcripts (*arrows*). The beads at the base of each intron loop as well as elsewhere on the transcripts are the spliceosomes. The large arrow points to a transcript near the 3′ end of the gene that is no longer attached to the DNA template and hence appears to have recently been terminated and released. The bar is 200 nm long. [Courtesy of Ann Beyer and Yvonne Osheim, University of Virginia.]

corroborated by the observations that splicing is inhibited by the selective destruction of the U1-snRNA sequences that are complementary to the 5′ splice site or by the presence of anti-U1-snRNP antibodies (produced by patients suffering from **systemic lupus erythematosus,** an often fatal autoimmune disease). Three other snRNPs are also implicated in splicing: **U2-snRNP, U4–U6-snRNP** (in which the **U4-** and **U6-snRNAs** associate via base pairing), and **U5-snRNP.**

*Splicing takes place in an as yet poorly characterized ~2700 kD particle dubbed the **spliceosome*** (Fig. 31-58). The spliceosome brings together a pre-mRNA, the foregoing four snRNPs, and a variety of pre-mRNA binding proteins. Note that the spliceosome, which consists of 5 RNAs and ~150 polypeptides, is comparable in size and complexity to the ribosome (which in *E. coli* consists of 3 RNAs and 52 polypeptides with an aggregate mass of ~2500 kD; Section 32-3A).

In addition to its size and complexity, the spliceosome is a highly dynamic entity, with its various components associating and dissociating during specific stages of the splicing reaction (Fig. 31-59), while undergoing a variety of ATP-driven conformational changes. For example, to carry out the first transesterification reaction yielding the lariat structure (Fig. 31-53), the spliceosome undergoes a complex series of rearrangements that are schematically diagrammed in Fig. 31-60. Similarly extensive rearrangements are required to carry out the second transesterification reaction and to recycle the spliceosome for subsequent splicing reactions.

Although spliceosomal transesterification reactions were initially assumed to be mediated by protein catalysts, their chemical resemblance to the reactions carried out by the self-splicing group II introns suggests, as is noted above, that it is really the snRNAs that catalyze the splicing of pre-mRNAs (pre-mRNA introns have such varied sequences outside of their splice and branch sites that they are unlikely to play an active role in splicing). In fact, James Manley has shown that, in the absence of protein, segments of human U2- and U6-snRNAs catalyze an Mg^{2+}-dependent reaction in an intron branch site sequence–containing RNA that resembles splicing's first transesterification reaction.

i. Splicing Also Requires the Participation of Splicing Factors

Around 170 different proteins known as **splicing-associated factors** that are extrinsic to spliceosomes also participate in splicing, with individual assembly intermediates (e.g., complexes A, B, and C in Fig. 31-59) each associated with ~125 such proteins. Among them are **branch point–binding protein [BBP;** also known as **splicing factor 1 (SF1)]** and **U2-snRNP auxiliary factor (U2AF),** which cooperate to select the intron's branch point. U2AF binds to the polypyrimidine tract upstream of the 3′ splice site (Fig. 31-52), whereas BBP recognizes the nearby branch point sequence (Figs. 31-53 and 31-60). The NMR structure of the 131-residue RNA-binding segment of the 638-residue BBP in complex with an 11-nt RNA containing a branch

Figure 31-59 **The spliceosomal assembly/disassembly cycle.** The sequential actions of the spliceosomal snRNPs (*colored circles*), but not the non-snRNP proteins, are diagrammed in the process of excising an intron from a pre-RNA containing two exons (*blue*). Here 5′SS, BP, and 3′SS stand for the pre-mRNA's 5′ splice site, its branch point, and its 3′ splice site, respectively.

Eight conserved DExD/H box–containing **RNA-dependent RNA ATPases/helicases** as well as the GTPase **Snu114** act in specific steps of the splicing cycle to motivate RNA–RNA rearrangements and RNP remodeling reactions. [Courtesy of Reinhard Lührmann, Max-Planck-Institut für biophysikalische Chemie, Göttingen, Germany.]

Figure 31-60 **Schematic diagram of six rearrangements that the spliceosome undergoes in mediating the first transesterification reaction in pre-mRNA splicing.** The RNA is color coded to indicate segments that become base-paired. The black and green lines represent snRNA and pre-mRNA, and BBP stands for branch point–binding protein. U5, which participates in the second transesterification reaction, has been omitted for clarity. **(1)** Exchange of U1 for U6 in base pairing to the intron's 5′ splice site. **(2)** Exchange of BBP for U2 in binding to the intron's branch site. **(3)** Intramolecular rearrangement in

U2. **(4)** Disruption of a base-paired stem between U4 and U6 to form a stem–loop in U6. **(5)** Disruption of a second stem between U4 and U6 to form a stem between U2 and U6. **(6)** Disruption of a stem–loop in U2 to form a second stem between U2 and U6. The order of these rearrangements is unclear. The transesterification reaction is represented by the arrow from the A in the yellow segment of the pre-mRNA (*right panel*) to the 3′ end of the 5′ exon. [Adapted from Staley, J.P. and Guthrie, C., *Cell* **92**, 315 (1998).]

point sequence, determined by Michael Sattler, reveals that the RNA assumes an extended conformation and is largely buried in a groove that is lined with both aliphatic and basic residues (Fig. 31-61). The branch point adenosine, whose mutation abolishes BBP binding, is deeply buried and binds to BBP via hydrogen bonds that mimic Watson–Crick base pairing with uracil.

Other splicing factors include **SR proteins** and several members of the **heterogeneous nuclear ribonucleoprotein (hnRNP)** family. SR proteins each have one or more RRMs (RNA recognition motifs) near their N-terminus and a distinctive C-terminal **RS domain** that contains numerous Ser-Arg (SR) repeats and which participates in protein–protein interactions. SR proteins, when appropriately phosphorylated on their RS domains, specifically bind to their corresponding exonic splicing enhancers (ESEs) via their RRMs and thereby recruit the splicing machinery to the flanking 5' and 3' splice sites. The hnRNP proteins, which are highly abundant RNA-binding proteins, lack RS domains and hence cannot recruit the splicing machinery. Instead, they bind to their corresponding ESSs and ISSs (exonic and intronic splicing silencers) so as to block the binding of the splicing machinery at the flanking splice sites.

A simplistic interpretation of Fig. 31-53 suggests that any 5' splice site could be joined with any following 3' splice site, thereby eliminating all the intervening exons together with the introns joining them. However, such **exon skipping** does not normally take place (but see below). Rather, all of a pre-mRNA's introns are individually excised in what appears to be a largely fixed order that more or less proceeds in the 5' → 3' direction. This occurs, at least in part, because splicing takes place cotranscriptionally (Fig. 31-58). Thus, as a newly synthesized exon emerges

from an RNAP II, it is bound by splicing factors that are also bound to the RNAP II's highly phosphorylated C-terminal domain (CTD; Section 31-2E). This tethers the exon and its associated spliceosome to the CTD so as to ensure that splicing occurs when the next exon emerges from the RNAP II.

j. Spliceosomal Structures

All four snRNPs involved in pre-mRNA splicing contain the same so-called **snRNP core protein,** which consists of seven **Sm proteins** (so called because they react with autoantibodies of the Sm serotype from patients with systemic lupus erythematosus), which are named **B, D1, D2, D3, E, F,** and **G proteins.** Each of these Sm proteins contains two conserved segments, Sm1 and Sm2, that are separated by a linker of variable length. The seven Sm proteins collectively bind to a conserved RNA sequence, the **Sm RNA motif,** which occurs in U1-, U2-, U4-, and U5-snRNAs and which has the single-stranded sequence AAUU-UGUG. However, in the absence of a U-snRNA, the Sm proteins form three stable complexes D1–D2, D3–B, and E–F–G. None of these complexes alone bind U-snRNA. However, the D1–D2 and E–F–G complexes form a stable subcore snRNP with U-snRNA, to which D3–B binds to form the complete **Sm core domain.**

The X-ray structures of the D3–B and D1–D2 heterodimers, determined by Reinhard Lührmann and Kiyoshi Nagai, reveals that these four proteins share a common fold which consists of an N-terminal helix followed by a 5-stranded antiparallel β sheet that is strongly bent so as to form a hydrophobic core (Fig. 31-62a). The subunits of both dimers associate in a similar manner with the β5 strands of D3 and D1 binding to the β4 strands of B and D2, respectively, so as to join their β

(b)

Figure 31-62 **X-ray structures of Sm proteins.** (*a*) The structure of D3 protein. The N-terminal helix and the β strands of its Sm1 domain are red and blue and the β strands of its Sm2 domain are yellow. The B, D1, and D2 Sm proteins have similar structures with their L4 loops and N-terminal segments, including helix A, comprising their most variable portions. Several highly conserved residues are shown in stick form (with C gray, N blue, and O red), and a conserved hydrogen bonding network is represented by green dotted lines. (*b*) The D3–B dimer with D3 gold and B blue. The β5 strand of D3 associates with the β4 strand of B to form a continuous antiparallel β sheet. Note that their corresponding loops extend in similar directions. [Courtesy of Kiyoshi Nagai, MRC Laboratory of Molecular Biology, Cambridge, U.K. PDBid 1D3B.]

sheets (Fig. 31-62*b*). This, together with biochemical and mutagenic experiments, indicates that the Sm proteins form a closed heteroheptameric ring whose subunits are arranged in the order –B–D3–G–E–F–D2–D1–. This model is corroborated by the X-ray structure of an Sm-like protein from the hyperthermophilic archeon *Pyrobaculum aerophilum,* determined by David Eisenberg, that forms a homoheptameric ring that is structurally similar to the heteroheptameric model. This structure also supports the hypothesis that the seven eukaryotic Sm proteins arose through a series of duplications of an archaeal Sm-like protein gene.

Mammalian U1-snRNP consists of U1-snRNA and ten proteins, the seven Sm proteins that are common to all U-snRNPs as well as three that are specific to U1-snRNP: **U1-70K, U1-A,** and **U1-C** (437, 282, and 159 residues, respectively, in humans). The predicted secondary structure of the 165-nt U1-snRNA contains five double helical stems, four of which come together at a 4-way junction (Fig. 31-63*a*). U1-70K and U1-A bind directly to RNA stem–loops 1 and 2 (SL1 and SL2), respectively, whereas U1-C is bound by other proteins.

Nagai determined the X-ray structure of human U1 snRNP at 5.5 Å resolution. At this low resolution the major and minor grooves of the double-stranded RNA stems are visible. Moreover, the helices and sheets of the proteins are apparent, which allows the placement of protein folds of known structure. SL2 of the U1-snRNA had been altered and shortened to promote crystallization. This eliminated the binding site for U1-A. However, U1 snRNP in which U1-A is depleted is active in a splicing assay.

SL1 and SL2 are stacked coaxially as are SL3 and helix H, and these two stacked helices cross at an angle of ~90°. The Sm proteins form the predicted heteroheptameric ring, which is ~70 Å in diameter (Fig. 31-63*b*). The Sm RNA motif together with SL4 are threaded through the ring's funnel-shaped central hole such the Sm RNA motif interacts with the Sm proteins (Fig. 31-63*c*). U1-C associates with the Sm ring via an interaction between its zinc finger domain and the D3 subunit (Fig. 31-63*c*). An RRM in the C-terminal segment of U1-70K interacts with the loop of SL1 and this protein's N-terminal segment is draped across the external face of the Sm ring.

Fortuitously, the 5′ ends of two neighboring U1-snRNAs in the crystal form a double helical segment. Since, in the spliceosome, the 5′ end of U1-snRNA base-pairs with the 5′ splice site of the pre-RNA (Fig. 31-60), this interaction provides a model of how the spliceosome recognizes the 5′ splice site. The zinc finger domain of U1-C interacts with this double helix (left side of Fig. 31-63*c*) and presumably stabilizes it. This is consistent with the observation that mutants of U1-C in this region cannot initiate spliceosome formation (the first reaction in Fig. 31-59).

Difficulties in obtaining homogeneous preparations of spliceosomal subassemblies and their poor stabilities have hindered their crystallization and limited the resolution of their cryo-EM–based images. Nevertheless, the cryo-EM–based structures of a variety of spliceosomal components have been reported. For example, the ~40-Å resolution structure of human complex B after it has released U1-snRNP but before it has released U4-snRNP (Fig. 31-59) was determined by Lührmann and Holger

(a)

(b)

(c)

Figure 31-63 **X-ray structure of human U1-snRNP.** (*a*) The predicted secondary structure of U1-snRNA with the RNA segments at which the proteins U1-70K and U1-A bind indicated. Ψ is the symbol for pseudouridine residues (Section 30-5Be). (*b*) View of the X-ray structure along the axis of the Sm ring. The proteins are drawn in ribbon form in different colors and the RNA is shown in paddle form in gray. SL4 has been deleted for clarity. The yellow spheres represent the selenium atoms in selenoMet residues (Met with its S atom replaced by Se), which were mutagenically inserted into the proteins to aid in solving the structure. Their known positions in the polypeptide chains helped trace the polypeptides' paths. (*c*) The structure as viewed from above in Part *a*. The Zn^{2+} ion bound to the zinc finger motif of U1-C is represented by a green sphere. The 5′ terminal portion of the U1-snRNA forms a double helical segment with the 5′ end of a neighboring U1-snRNA, which is drawn in paddle form in purple. [Part *a* modified from a drawing by, and Parts *b* and *c* courtesy of, Kiyoshi Nagai, MRC Laboratory of Molecular Biology, Cambridge, U.K. PDBid 3CW1.]

Stark (Fig. 31-64). This $370 \times 270 \times 170$ Å particle, which is named BΔU1, has a roughly triangular body connected to a head domain. Other splicesome components are equally irregular.

k. The Significance of Gene Splicing

The analysis of the large body of known DNA sequences reveals that introns are rare in prokaryotic structural genes, uncommon in lower eukaryotes such as yeast (which has a total of 239 introns in its ~6600 genes and, with two exceptions, only one intron per polypeptide), and abundant in higher eukaryotes (the only known vertebrate structural genes lacking introns are those encoding histones and the antiviral proteins known as interferons). PremRNA introns, as we have seen, can be quite long and many genes contain large numbers of them. Consequently, unexpressed sequences constitute ~80% of a typical vertebrate structural gene and ~99% of a few of them.

Figure 31-64 **Cryo-EM–based structure of the human spliceosome at its BΔU1 stage of assembly/disassembly at 40 Å resolution.** The bar represents 100 Å. [Courtesy of Holger Stark, Max-Planck Institute for Biophysical Chemistry, Göttingen, Germany.]

The argument that introns are only molecular parasites **(junk DNA)** seems untenable since it would then be difficult to rationalize why the evolution of complex splicing machinery offered any selective advantage over the elimination of the split genes. What then is the function of gene splicing? Although, since its discovery, the significance of gene splicing has been often vehemently debated, two important roles for it have emerged: (1) It is an agent for rapid protein evolution; and (2) through **alternative splicing,** it permits a single gene to encode several (sometimes many) proteins that may have significantly different functions. In the following paragraphs, we discuss these aspects of gene splicing.

l. Many Eukaryotic Proteins Consist of Modules That Also Occur in Other Proteins

The 839-residue LDL receptor is a plasma membrane protein that functions to bind low-density lipoprotein (LDL) to coated pits for transport into the cell via endocytosis (Section 12-5Bc). LDL receptor's 45-kb gene contains 18 exons, most of which encode specific functional domains of the protein. *Moreover, 13 of these exons specify polypeptide segments that are homologous to segments in other proteins:*

1. Five exons encode a 7-fold repeat of a 40-residue sequence that occurs once in **complement C9** (an immune system protein; Section 35-2F).

2. Three exons each encode a 40-residue repeat similar to that occurring four times in **epidermal growth factor**

(EGF; Section 19-3A) and once each in three blood clotting system proteins: **factor IX, factor X,** and **protein C** (Section 35-1).

3. Five exons encode a 400-residue sequence that is 33% identical with a polypeptide segment that is shared only with EGF.

Evidently, the LDL receptor gene is modularly constructed from exons that also encode portions of other proteins. Numerous other eukaryotic proteins are similarly constituted including, as we have seen, many of the proteins involved in signal transduction (e.g., those containing SH2 and SH3 domains; Section 19-3C). Moreover, many exons encode complete domains that frequently have independent functions. *It therefore appears that the genes encoding these modular proteins arose by the stepwise collection of exons that were assembled by (aberrant) recombination between their neighboring introns.*

m. Alternative Splicing Greatly Increases the Number of Proteins Encoded by Eukaryotic Genomes

The expression of numerous cellular genes is modulated by the selection of alternative splice sites. Thus, certain exons in one type of cell may be introns in another. For example, a single rat gene encodes seven tissue-specific isoforms (splice variants) of the muscle protein **α-tropomyosin** (Section 35-3Ca) through the selection of alternative splice sites (Fig. 31-65).

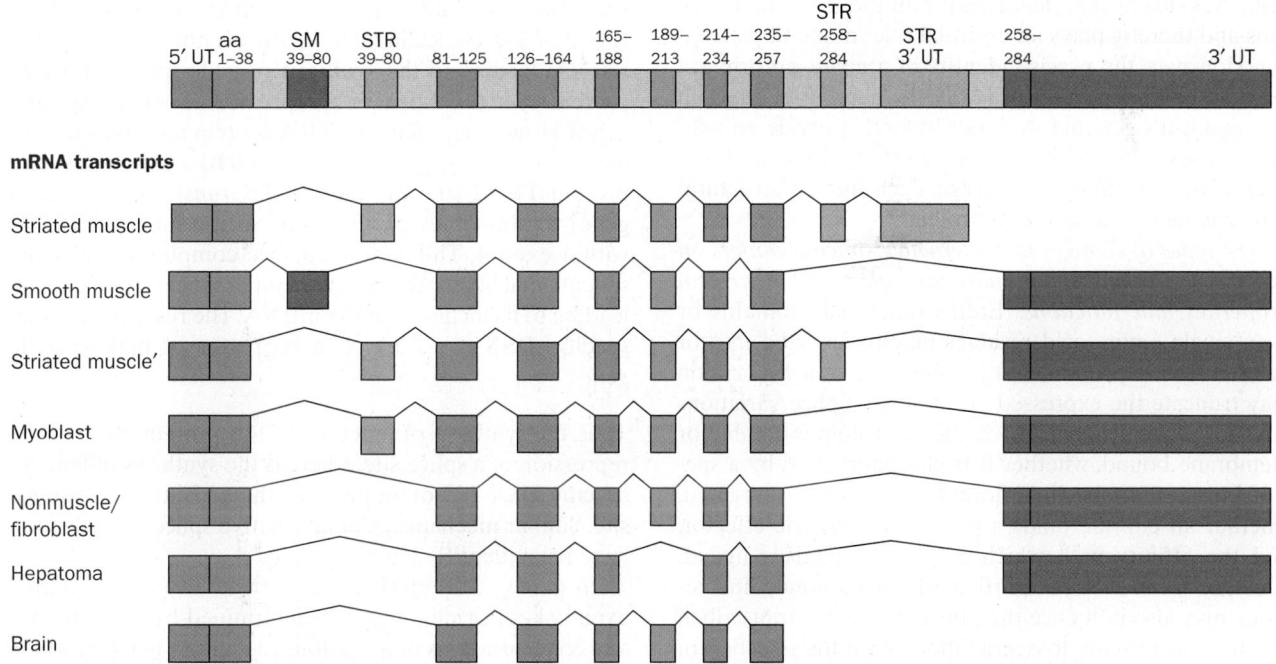

Figure 31-65   The organization of the rat α-tropomyosin gene and the seven alternative splicing pathways that give rise to cell-specific α-tropomyosin isoforms. The thin kinked lines indicate the positions occupied by the introns before they are spliced out to form the mature mRNAs. Tissue-specific exons are indicated together with the amino acid (aa) residues they encode: "constitutive" exons (those expressed in all tissues) are green, those expressed only in smooth muscle (SM) are brown, those expressed only in striated muscle (STR) are purple, and those variably expressed are yellow. Note that the smooth and striated muscle exons encoding amino acid residues 39 to 80 are mutually exclusive; likewise, there are alternative 3'-untranslated (UT) exons. [After Breitbart, R.E., Andreadis, A., and Nadal-Ginard, B., *Annu. Rev. Biochem.* **56,** 481 (1987).]

Alternative splicing occurs in all metazoa and is especially prevalent in vertebrates. In fact, microarray-based comparisons of the cDNAs obtained from various tissues indicate that ~95% of human structural genes are subject to at least one alternative splicing event. This rationalizes the discrepancy between the ~23,000 genes identified in the human genome (Section 7-2Bc) and earlier estimates that it contains 50,000 to 140,000 structural genes.

The variation in mRNA sequence can take several different forms: Exons can be retained in an mRNA or they can be skipped; introns may be excised or retained; and the positions of 5′ and 3′ splice sites can be shifted to make exons shorter or longer. Alterations in the transcriptional start site and/or the polyadenylation site can further contribute to the diversity of the mRNAs that are transcribed from a single gene. In a particularly striking example, the *Drosophila* protein **Dscam** (for *D*own *s*yndrome *c*ell-*a*dhesion *m*olecule), which functions in neuronal development, is encoded by 24 exons of which there are 12 mutually exclusive variants of exon 4, 48 of exon 6, 33 of exon 9, and 2 of exon 17 (which are therefore known as **cassette exons**) for total of 38,016 possible isoforms of this protein (compared to ~14,000 identified genes in the *Drosophila* genome). Although it is unknown if all possible Dscam isoforms are produced, experimental evidence suggests that the *Dscam* gene expresses many thousands of them. [Dscam is a membrane-anchored cell-surface protein of the immunoglobulin superfamily. The specific isoform expressed in a given neuron binds to itself but rarely to other isoforms. This permits the neuron to distinguish its own processes (axons and dendrites) from those of other neurons and thereby plays an essential role in neural patterning. However, the precise identity of a given isoform appears to be unimportant.] Clearly, the number of genes in an organism's genome does not by itself provide an adequate assessment of its protein diversity. Indeed, it has been estimated that, on average, each human structural gene encodes three different proteins.

The types of changes that alternative splicing confers on expressed proteins spans the entire spectrum of protein properties and functions. Entire functional domains or even single amino acid residues may be inserted into or deleted from a protein, and the insertion of a stop codon may truncate the expressed polypeptide. Splice variations may, for example, control whether a protein is soluble or membrane bound, whether it is phosphorylated by a specific kinase, the subcellular location to which it is targeted, whether an enzyme binds a particular allosteric effector, and the affinity with which a receptor binds a ligand. Changes in an mRNA, particularly in its noncoding regions, may also influence the rate at which it is transcribed and its susceptibility to degradation. Since the selection of alternative splice sites is both tissue- and developmental stage-specific, splice site choice must be tightly regulated in both space and time. In fact, it is estimated that from ~15% to 50% of human genetic diseases are caused by point mutations that result in pre-mRNA splicing defects. Some of these mutations delete functional splice sites, thereby activating nearby pre-existing **cryptic splice sites.** Others generate new splice sites that are used instead of the normal ones and yet others are in the genes encoding components of the splicing machinery. In addition, tumor progression is correlated with changes in levels of proteins implicated in alternative splice site selection.

How are alternative splice sites selected? Well-understood examples of such processes occur in the pathway responsible for sex determination in *Drosophila,* two of which we discuss here:

1. Exon 2 of *transformer (tra)* pre-mRNA contains two alternative 3′ splice sites (which succeed the excised intron), with the proximal (close; to exon 1) site used in males and the distal (far) site used in females (Fig. 31-66*a*). The region between these two sites contains a Stop codon (UAG). In males, the splicing factor U2AF binds to the proximal 3′ splice site to yield an mRNA containing this premature stop codon, which thereby directs the synthesis of truncated and hence nonfunctional **TRA** protein. In females, however, the proximal 3′ splice site is bound by the female-specific **SXL** protein, the product of the *sex-lethal (sxl)* gene (which is only expressed in females), so as to block the binding of U2AF, which then binds to the distal 3′ splice site, thereby excising the UAG and inducing the expression of functional TRA protein (U2AF and TRA both contain RS domains but not RRMs so that neither is an SR protein).

2. In *doublesex (dsx)* pre-mRNA, the first three exons are constitutively spliced in both males and females. However, the branch site immediately upstream of exon 4 has a suboptimal pyrimidine tract to which U2AF does not bind (Fig. 31-66*b*). Hence in males, exon 4 is not included in *dsx* mRNA, leading to the synthesis of male-specific **DSX-M** protein that functions as a repressor of female-specific genes. However, in females, TRA protein promotes the cooperative binding of the SR protein **RBP1** and the SR-like protein **TRA2** [the product of the *transformer 2 (tra-2)* gene] to six copies of an exonic splice enhancer (ESE) within exon 4. This heterotrimeric complex recruits the splicing machinery to the upstream 3′ splice site of exon 4, leading to its inclusion in *dsx* mRNA. The resulting female-specific **DSX-F** protein is a repressor of male-specific genes.

Thus, the synthesis of functional TRA protein involves the repression of a splice site, whereas the synthesis of female-specific DSX-F protein involves the activation of a splice site. Similar mechanisms of alternative splice site selection have been identified in vertebrates.

In general, the decision as to whether an alternative exon is kept or eliminated is determined by the activities and concentrations of its various regulators, many of which are SR proteins and hnRNPs. Hence the tissue-specific expression of these regulators and the phosphorylation state of the SR proteins are important contributors to the complex regulation of mRNA splicing. Moreover, extensive analysis of the sequences of numerous alternative splice sites has revealed the existence of a "splicing code" that uses combinations of over 200 RNA features that are

Figure 31-66 Mechanisms of alternative splice site selection in the *Drosophila* sex-determination pathway as described in the text. In all panels, exons are represented by colored rectangles and introns are shown as pale gray lines. (*a*) Alternative splicing in *tra* pre-mRNA. UAG is a Stop codon. (*b*) Alternative splicing in *dsx* pre-mRNA. The six ESEs (exonic splice enhancers) in exon 4 are indicated by green rectangles and S represents the splicing machinery. In females, polyadenylation (pA) of *dsx* mRNA occurs downstream of exon 4, whereas in males, it occurs downstream of exon 6. [After a drawing by Maniatis, T. and Tasic, B., *Nature* **418,** 236 (2002).]

present in both introns and exons and which are recognized by the foregoing regulators.

Riboswitches (Section 31-3H) have been implicated in controlling alternative splicing in eukaryotes. For example, as Breaker has shown, in the *NMT1* gene of the bread mold *Neurospora crassa* [which expresses an enzyme that participates in the metabolism of TPP (thiamine pyrophosphate)], a TPP-sensing riboswitch is contained in an intron that is located upstream of the mRNA's normal AUG translational start codon. This intron contains one 3' splice site and two 5' splice sites with two AUG start codons between them. When the concentration of TPP is low so that it does not bind to the riboswitch, one strand of the riboswitch's P4–P5 segment base-pairs with the second (downstream) splice site so as to inactivate it (Fig. 31-67*a*). The spliceosome then efficiently excises the entire intron yielding an mRNA that is readily translated (I-3; Fig. 31-67*b*). However, when the concentration of TPP is high so that it binds to the riboswitch, the riboswitch assumes a conformation that activates the second splice site but occludes the branch point A (Fig. 31-67*c*). Consequently, the spliceosome inefficiently excises only the downstream portion of the intron. The two upstream AUG codons, which are present in both the unspliced mRNA (I-1) and in the spliced mRNA containing only the upstream portion of the intron (I-2), compete for ribosomes with mRNA's normal AUG start codon and thereby repress the mRNA's translation (Fig. 31-67*c*).

n. AU–AC Introns Are Excised by a Novel Spliceosome

A small fraction of introns (~0.3%) have AU rather than GU at their 5' ends and AC rather than AG at their 3' ends, but are nevertheless excised via a lariat structure to an internal intron A. These so-called **AU–AC introns** (alternatively, **AT–AC introns** after their DNA sequences), which occur in organisms as diverse as *Drosophila*, plants, and humans, are excised by a novel so-called **AU–AC spliceosome** (alternatively, an **AT–AC spliceosome**) that has one snRNP, U5, in common with the major (GU–AG) spliceosome, and three others, **U11, U12,** and **U4atac–U6atac,** which are distinct from but structurally and functionally analogous to U1, U2, and U4–U6. Curiously, all genes known to contain AU–AC introns also contain multiple major class introns. Moreover, AU–AC introns are not conserved in either length or position in their host genes. Thus, the functional and evolutionary significance of the AU–AC spliceosome and introns is obscure.

o. Trans-Splicing

The types of splicing we have so far considered occur within single RNA molecules and hence are known as

← Activating   ⊢ Inhibiting

Figure 31-67   **Control of translation by the *N. crassa* TPP-sensing riboswitch through alternative splicing.** (*a*) The predicted secondary structure of the TPP aptamer, which resides in the 5′ untranslated region of the *NMT1* mRNA. The sequence of one strand of the P4-P5 stem is complementary (*orange shading*) to a segment that overlaps the second 5′ splice site. (*b*) At low TPP concentrations, the aptamer inhibits (*red tee*) splicing from the second 5′ splice site while activating (*green arrow*) the branch point A so that the spliceosome excises the RNA between the first 5′ splice site and the 3′ splice site, yielding an ORF (open reading frame; I-3) that is normally translated from its AUG start codon. (*c*) At high TPP concentrations, the binding of TPP to the aptamer activates the second 5′ splice site but occludes the branch point A. The spliceosome therefore inefficiently excises the intron's downstream portion yielding an mRNA (I-2) that contains two uORFs (*u*pstream ORFs) that compete with the translation of the primary ORF. The unspliced mRNA (I-1) also contains the two uORFs and hence is likewise inefficiently translated. [Courtesy of Ronald Breaker, Yale University.]

cis-splicing. The chemistry of the spliceosomal cis-splicing reaction, however, is the same as would occur if the two exons to be joined initially resided on two different RNA molecules, a process called **trans-splicing.** This, in fact, occurs in trypanosomes (kinetoplastid protozoa; the cause of African sleeping sickness). Trypanosomal mRNAs all have the same 35-nt noncoding leader sequence, although this leader sequence is not present in the corresponding genes. Rather, this sequence is part of a so-called **spliced leader (SL) RNA** that is transcribed from an independent gene. The 5′ splice site that succeeds the SL RNA leader sequence, and the branch site and 3′ splice site that precede the exon sequence have the same consensus sequences as occur in the RNAs spliced by the major spliceosome. Consequently, the SL RNA leader and the pre-mRNA are joined in a trans-splicing reaction that resembles the spliceosomal cis-splicing reaction (Fig. 31-53) with the exception that the product of the first transesterification reaction is necessarily Y-shaped rather than lariat-shaped (Fig. 31-68). Trypanosomes, whose pre-mRNAs lack introns, nevertheless have U2- and U4–U6-snRNPs but lack U1- and U5-snRNPs. However, the SL RNA, which is predicted to fold into three stem–loops and a single-stranded Sm RNA–like motif as does U1-snRNA (Fig. 31-63*a*), apparently carries out the functions of U1-snRNA in the trans-splicing reaction.

Trans-splicing has been shown to occur in nematodes (roundworms; e.g., *C. elegans*) and flatworms. These organisms also carry out cis-splicing and, indeed, perform both types of splicing on the same pre-mRNA. There are also several reports that trans-splicing occurs in higher eukaryotes such as *Drosophila* and vertebrates, but if it does occur, it does so in only a few pre-mRNAs and at a very low level.

p. mRNA Is Methylated at Certain Adenylate Residues

During or shortly after the synthesis of vertebrate pre-mRNAs, ~0.1% of their A residues are methylated at their N6 atoms. These m⁶A's tend to occur in the sequence RRm⁶ACX, where X is rarely G. Although the functional significance of these methylated A's is unknown, it should be noted that a large fraction of them are components of the corresponding mature mRNAs.

q. RNA Can Be Edited by the Insertion or Deletion of Specific Nucleotides

Certain mRNAs from a variety of eukaryotic organisms have been found to differ from their corresponding genes in several unexpected ways, including C → U and U → C changes, the insertion or deletion of U residues, and the insertion of multiple G or C residues. The most extreme examples of this phenomenon, which occur in the mitochon-

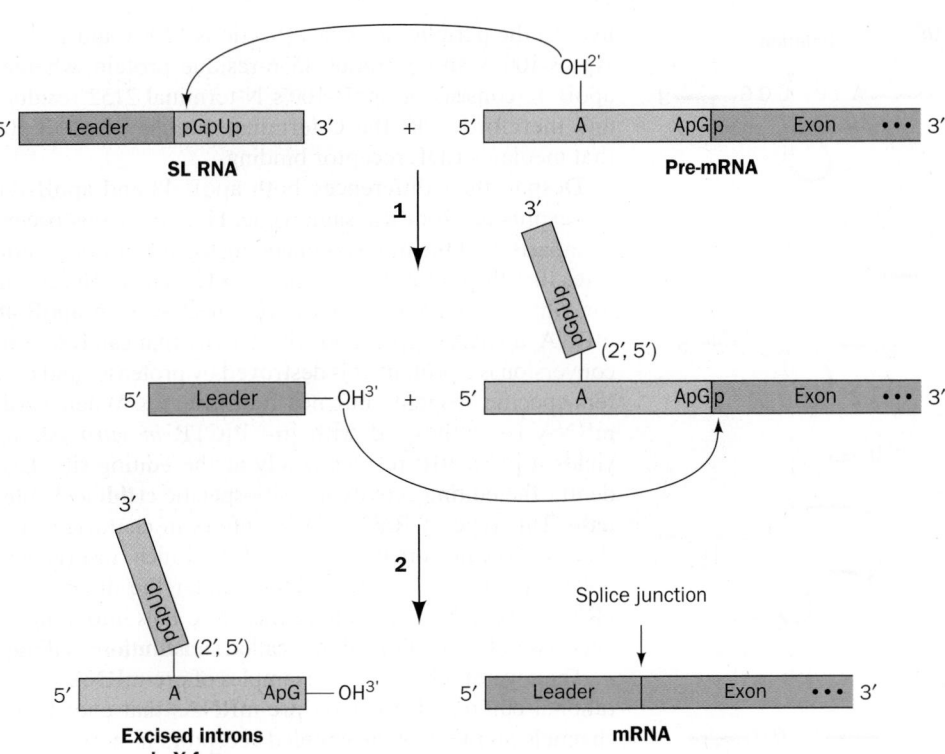

Figure 31-68 **The sequence of transesterification reaction that occurs in trans-splicing.** The chemistry is closely similar to that of pre-mRNA cis-splicing (Fig. 31-53).

dria of trypanosomes (whose DNA encodes only 20 genes), involve the addition and removal of up to hundreds of U's to and from 12 otherwise untranslatable mRNAs. The process whereby a transcript is altered in this manner is called **RNA editing** because it originally seemed that the required enzymatic reactions occurred without the direction of a nucleic acid template and hence violated the central dogma of molecular biology (Fig. 5-21). Eventually, however, a new class of trypanosomal mitochondrial transcripts called **guide RNAs (gRNAs)** was identified. gRNAs, which consist of 40 to 80 nucleotides, have 3' oligo(U) tails, an internal segment that is precisely complementary

to the edited portion of the pre-edited mRNA (if G · U pairs, which are common in RNAs, are taken to be complementary), and a 10- to 15-nt so-called anchor sequence near the 5' end that is largely complementary in the Watson–Crick sense to a segment of the mRNA that is not edited.

An unedited transcript presumably associates with the corresponding gRNA via its anchor sequence (Fig. 31-69). Then, in a process mediated by the appropriate enzymatic machinery in an ~20S RNP named the **editosome,** the gRNA's internal segment is used as a template to "correct" the transcript, thereby yielding the edited mRNA. Inser-

```
        U
5' −G−C−A  A−G−G−U−C−A−G−C−U−A−U−C−A−  3'   Pre-edited mRNA
3' −C  G−U−U  C−C  A−G  U−C−G−A−U−A−G−U−  5'   gRNA
    G  G      A    G G  A−A
    A  G           G−A
     ˋGˊ
```

```
5'  G−U−U−U−U−U−C−A ^ A−U−G−G−U−U−U−U−U−C−U−U−A−G−C−U−A−U−C−A   3'   Edited mRNA
3'  C−G−A−G−G−G−G−U−U−A−C−C−G−G−A−G−A−G−A−A−U−C−G−A−A−A−G−U   5'   gRNA
```

Figure 31-69 **A schematic diagram indicating how gRNAs direct the editing of trypanosomal pre-edited mRNAs.** The red U's in the edited mRNA are insertions and the triangle (Δ) marks a deletion. Several gRNAs may be necessary to direct the editing of consecutive segments of a pre-edited mRNA. [After Bass, B.L., *in* Gesteland, R.F. and Atkins, J.F. (Eds.), *The RNA World, p.* 387, Cold Spring Harbor Laboratory Press (1993).]

Figure 31-70 **Trypanosomal RNA editing pathways.** The RNAs being edited (*black*) are shown base-paired to the gRNAs (*blue*) with the U's that are (*a*) inserted by TUTase or (*b*) deleted by 3′-U-exo drawn in red. The arrowheads indicate the positions that are cleaved by the endonuclease. [After Madison-Antenucci, S., Grams, J., and Hajduk, S.L., *Cell* **108**, 435 (2002).]

tion editing requires at least three enzymatic activities that, somewhat surprisingly, are encoded by nuclear genes (Fig. 31-70*a*): (1) an endonuclease at a mismatch between the gRNA and the pre-edited mRNA to cleave the pre-edited mRNA on the 5′ side of the insertion point; (2) **terminal uridylyltransferase (TUTase)** to insert the new U(s); and (3) an **RNA ligase** to reseal the RNA. Deletion requires similar enzymatic apparatus with the exceptions that the endonuclease cleaves the RNA being edited on the 3′ side of the U(s) to be deleted and TUTase is replaced by **3′-U-exonuclease (3′-U-exo),** which excises the U(s) at the deletion site (Fig. 31-70*b*). A single gRNA mediates the editing of a block of 1 to 10 sites. Thus, the genetic information specifying an edited mRNA is derived from two or more genes. The functional advantage of this complicated process, either presently or more likely in some ancestral organism, is obscure.

r. RNA Can Be Edited by Base Deamination

Humans express two forms of **apolipoprotein B (apoB): apoB-48,** which is made only in the small intestine and functions in chylomicrons to transport triacylglycerols from the intestine to the liver and peripheral tissues; and **apoB-100,** which is made only in the liver and functions in VLDL, IDL, and LDL to transport cholesterol from the

liver to the peripheral tissues (Sections 12-5A and 12-5B). ApoB-100 is an enormous 4536-residue protein, whereas apoB-48 consists of apoB-100's N-terminal 2152 residues and therefore lacks the C-terminal domain of apoB-100 that mediates LDL receptor binding.

Despite their differences, both apoB-48 and apoB-100 are expressed from the same gene. How does this occur? Comparison of the mRNAs encoding the two proteins indicates that they differ by a single $C \rightarrow U$ change: The codon for Gln 2153 (CAA) in apoB-100 mRNA is, in apoB-48 mRNA, a UAA Stop codon. The activity that catalyzes this conversion is a protein: It is destroyed by proteases and protein-specific reagents but not by nucleases. When apoB mRNA is synthesized with $[\alpha\text{-}^{32}P]CTP$, *in vitro* editing yields a $[^{32}P]UMP$ residue solely at the editing site. Evidently, the editing activity is a site-specific **cytidine deaminase.** This type of RNA editing differs in character from that in trypanosomal mitochondria, which inserts and deletes multiple U's into mRNAs under the direction of gRNAs. ApoB mRNA editing therefore falls into a different class of RNA editing that is called **substitutional editing.**

The several other known examples of pre-mRNA substitutional editing all occur on pre-mRNAs that encode ion channels and G protein–coupled receptors in nerve tissue. Among them is vertebrate brain **glutamate receptor** pre-mRNA, which undergoes an $A \rightarrow I$ deamination [where I is inosine (guanosine lacking its 2-amino group), which the translational apparatus reads as G] that transforms a Gln codon (CAG) to that of a functionally important Arg (CIG; normally CGG). The vertebrate enzymes that catalyze such $A \rightarrow I$ RNA editing of pre-mRNAs, **ADAR1** (1200 residues), **ADAR2** (729 residues), and **ADAR3** (739 residues; ADAR for *adenosine deaminases acting on RNA*), have the curious requirement that their target A residues must be members of RNA double helices that are formed between the editing site and a complementary sequence that is usually located in a downstream intron (Fig. 31-71). Hence, ADAR-mediated editing must precede splicing.

Substitutional editing may contribute to protein diversity. For example, *Drosophila cacophony* pre-mRNA that encodes a voltage-gated Ca^{2+} channel subunit contains 10 different substitutional editing sites and hence has the potential of generating 1000 different isoforms in the absence of alternative splicing.

Substitutional editing can also generate alternative splice sites. For example, rat ADAR2 edits its own pre-mRNA by converting an intronic AA dinucleotide to AI, which mimics the AG normally found at 3′ splice sites (Fig. 31-53). The consequent new splice site adds 47 nucleotides near the 5′ end of the *ADAR2* mRNA so as to generate a new translational initiation site. The resulting ADAR2 isozyme is catalytically active but is produced in smaller amounts than that from unedited transcripts, perhaps due to a less efficient translational initiation site. Thus, rat ADAR2 appears to regulate its own rate of expression.

ADAR1 contains an N-terminal Z-DNA–binding domain, Zab, that is composed of two subdomains, Zα and Zβ. We have seen that in the X-ray structure of Zα in complex with Z-DNA (Fig. 29-3), Zα binds Z-DNA via

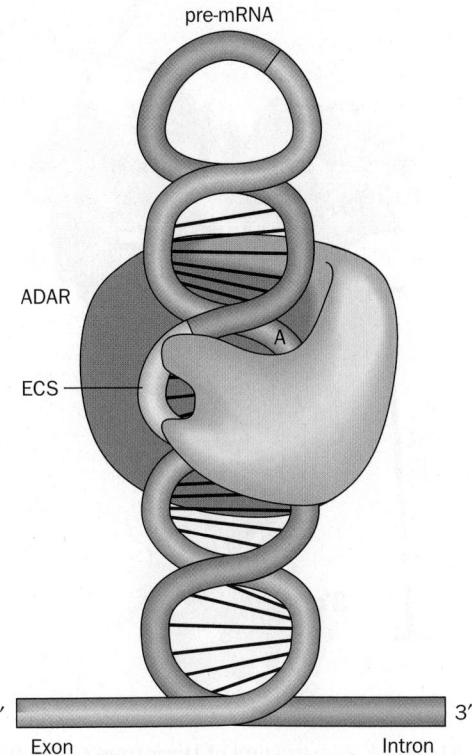

Figure 31-71 The recognition of ADAR editing sites. Both ADAR1 and ADAR2 bind to a 9- to 15-bp double-stranded RNA that is formed between the editing site (*orange*) on a pre-mRNA exon and a so-called **editing site complementary sequence (ECS;** *pink)* that is often located in a downstream intron (*brown*). A represents the adenosine that the ADAR (*green*) converts to inosine. [After Keegan, L.P., Gallo, A., and O'Connell, M.A., *Nature Rev. Genet.* **2,** 869 (2001).]

sequence-independent complementary surfaces (Section 29-1Bb). What is the function of Zab? Alexander Rich has proposed that since the negative supercoiling of the DNA immediately behind actively transcribing RNAP (Section 31-2Ca) stimulates the transient formation of Z-DNA (recall that Z-DNA has a left-handed helix), Zab targets ADAR1 to genes that are undergoing transcription. This would facilitate rapid A → I editing, which must take place before the next splicing reaction occurs.

s. RNA Interference Degrades mRNAs

Since the 1990s it has become increasingly clear that noncoding RNAs have important roles in controlling gene expression. One of the first indications of this phenomenon occurred in Richard Jorgensen's attempt to genetically engineer more vividly purple petunias by introducing extra copies of the gene that directs the synthesis of the purple pigment. Surprisingly, the resulting transgenic plants had variegated and often entirely white flowers. Apparently, the purple-making genes somehow switched each other off. Similarly, it is well known that **antisense RNA** (RNA that is complementary to at least a portion of an mRNA) prevents the translation of the corresponding mRNA because ribosomes cannot translate double-stranded RNA.

Yet, injecting **sense RNA** (RNA with the same sequence as an mRNA) into the nematode *Caenorhabditis elegans* also blocks protein production. Since the added RNA somehow interferes with gene expression, this phenomenon is known as **RNA interference (RNAi).** RNAi is now known to occur in all eukaryotes investigated except baker's yeast.

The mechanism of RNAi began to come to light in 1998 when Andrew Fire and Craig Mello showed that double-stranded RNA **(dsRNA)** was substantially more effective in causing RNAi in *C. elegans* than were either of its component strands alone. RNAi is induced by only a few molecules of dsRNA per affected cell, suggesting that RNAi is a catalytic rather than a stoichiometric effect. Further investigations, in large part in *Drosophila*, have led to the elucidation of the following pathway mediating RNAi (Fig. 31-72):

1. The trigger dsRNA, as Phillip Zamore discovered, is chopped up into ~21- to 25-nt-long double-stranded fragments known as **small interfering RNAs (siRNAs),** each of

Figure 31-72 A mechanism for RNA interference (RNAi). See the text for details. ATP is required for Dicer-catalyzed cleavage of RNA and for RISC-associated helicase unwinding of double-stranded RNA. Depending on the species, the mRNA may not be completely degraded.

whose strands has a 2-nt overhang at its 3′ end and a 5′ phosphate. This reaction is mediated by an ATP-dependent RNase named **Dicer,** a homodimer of ~1900-residue subunits in animals that is a member of the **RNase III** family of double-strand–specific RNA endonucleases.

2. An siRNA is transferred to a 250- to 500-kD multisubunit complex known as **RNA-induced silencing complex (RISC).** RISC has at least four protein components, one of which is an ATP-dependent RNA helicase that separates the two strands of the siRNA. The strand whose 5′ end has the lower free energy of binding, the **guide RNA,** is bound by the RISC whereas its complementary strand, the **passenger RNA,** is cleaved and discarded. In some species, but apparently not in humans, the original siRNA signal is amplified by the action of an **RNA-dependent RNA polymerase (RdRP).**

3. The guide RNA recruits the RISC complex to an mRNA with the complementary sequence.

4. An RNase III component of RISC known as **Argonaute (AGO;** also called **Slicer)** cleaves the mRNA opposite the bound guide RNA. The cleaved mRNA is then further degraded by cellular nucleases, thereby preventing its translation.

The X-ray structure of Dicer from the parasitic protozoan *Giardia intestinalis,* determined by Jennifer Doudna, reveals that its shape resembles that of a hatchet, with its two RNase III domains forming the blade and its PAZ domain (named for three proteins in which it is contained, *P*IWI, *A*rgonaute, and *Z*wille) forming the base of its handle (Fig. 31-73; Dicers from higher eukaryotes additionally contain an N-terminal DExD/H box helicase domain and a C-terminal dsRNA-binding domain). The two RNase III domains form an internal heterodimer that resembles the homodimeric structure of bacterial RNase III. Four conserved acidic residues in each RNase III domain bind two Mg^{2+} ions and hence are postulated to cleave an RNA strand via a two-metal-ion mechanism (Section 30-2Af). The two RNase III active sites are 17.5 Å apart, the width of dsRNA's major grove, and thus appear positioned to cleave the two strands of a bound dsRNA. The PAZ domain specifically binds dsRNA ends that have a 3′ two-nt overhang. The distance between this binding site and its closest RNase III domain active site is 65 Å, the length of a 25-bp dsRNA. This explains how Dicer cleaves an ~25-bp segment from the end of dsRNA.

Argonaute proteins consist of four domains: an N-terminal (N), a PAZ, a middle (Mid), and a PIWI (for *P*-element *i*nduced *w*impy testis) domain. The X-ray structures of several bacterial Argonaute proteins reveal that they have a bilobal architecture with the N and PAZ domains forming one lobe and the Mid and PIWI domains forming the other. The PIWI domain has an RNase H fold (RNase H cleaves the RNA strand of an RNA · DNA hybrid helix), which strongly suggests that it mediates Argonaute's "slicer" activity (bacterial Argonautes preferentially bind **guide DNA** over RNA). The X-ray structure of *T. thermophilus* Argonaute in ternary complex with a 21-nt

Figure 31-73 **X-ray structure of Dicer from *G. intestinalis.*** The protein is represented by its molecular surface colored according to its surface charge with red negative, blue positive, and white neutral. Bound Mg^{2+} ions, which are represented by green spheres, mark the active site of each of the protein's two RNase III domains. A dsRNA has been modeled into the structure with its 3′ overhang entering the PAZ domain's binding pocket (*asterisk*). The white arrows point to the dsRNA's scissile phosphate groups. Note that much of the surface to which the anionic dsRNA is presumably bound is positively charged (the calculated surface charge does not take into account the bound Mg^{2+} ions). [Courtesy of Jennifer Doudna, University of California at Berkeley. PDBid 2FFL.]

guide DNA and a 19-nt target RNA (Fig. 31-74), determined by Patel, reveals that the DNA · RNA hybrid helix binds in the cleft between Argonaute's two lobes with the phosphate group bridging RNA nucleotides 10 and 11 positioned for cleavage at the PIWI active site. The comparison of this structure with those of similar complexes that lack the target RNA or in which the target RNA has 12 or 15 nucleotides indicates that the guide DNA (and presumably the guide RNA in eukaryotes) initially binds to Argonaute with its 3′ end in the PAZ binding pocket, but as the hybrid helix lengthens beyond one turn, this 3′ end is released, which facilitates further winding of the hybrid helix.

t. RNAi Defends against Viral Infection and Regulates Gene Expression

What is the physiological function of RNAi? Since many eukaryotic viruses store and replicate their genomes as RNA (Chapter 33), it seems likely that RNAi arose as a defense against viral infections. Indeed, many plant viruses contain genes that suppress various steps of RNAi and which are essential for pathogenesis. RNAi has also been

Figure 31-74 **X-ray structure of *T. thermophilus* Argonaute in ternary complex with a 21-nt guide DNA and a 19-nt target RNA.** The protein is shown in ribbon form with its N, PAZ, Mid, and PIWI domains cyan, magenta, orange, and green, respectively, and with the linkers connecting these domains gray. The guide DNA and target RNA are drawn in stick form in red and blue with their P atoms yellow. Only DNA nucleotides 1 to 16 and RNA nucleotides 2 to 16 are visible. [Courtesy of Dinshaw Patel, Memorial-Sloan Kettering Cancer Center, New York, New York. PDBid 3HK2.]

shown to inhibit the intragenomic spread of retrotransposons (Section 30-6Bh).

A wide variety of eukaryotes, including plants, nematodes, flies, fish, and mammals, use RNAi to control gene expression. Certain mRNAs expressed by these organisms contain ~70-nt, imperfectly base-paired, stem–loop structures that are excised by a 1374-residue RNase III named **Drosha** (Fig. 31-75). The stem–loops are exported from the nucleus to the cytosol where they are cleaved by Dicer to liberate ~22-bp dsRNAs known as **microRNAs (miRNAs;** so called to differentiate these endogenous RNAs from exogenous siRNAs). The transcripts from which miRNAs are derived are known as **pri-miRNAs** (pri for *pri*mary), whereas the stem–loops are called a **pre-miRNAs** (pre for *pre*cursor). Pre-miRNAs can be located within both the introns and, less commonly, the exons of a pri-miRNA. *The miRNAs bind to RISC in which they function to identify the tens to hundreds of mRNAs containing segments that are partially complementary to the miRNA.*

The RISC-bound miRNA binds to its targets site, which is usually in the 3′ untranslated region (3′UTR) of an

Figure 31-75 **The generation of miRNAs from pri-miRNAs and their RISC-mediated binding to target mRNAs.** See the text for details.

mRNA. A lack of perfect complementarity prevents Argonaute from cleaving the mRNA (Argonaute's PIWI domain catalyzes slicing only if there is perfect complementarity to the miRNA's so-called **seed sequence,** which consists of nucleotides 2–8 from its 5′ end, and in fact, some species of Argonaute lack the catalytic residues to do so. Instead, miRNA-mediated silencing is thought to occur through the removal of its target mRNA's poly(A) tail or its m⁷G cap, which leads to the mRNA's degradation (Section 31-4Av), and/or the RISC-mediated repression of the target mRNA's translation by interfering with ribosomal initiation (Section 32-3Cd) and sequestering or degrading the mRNA in cytoplasmic granules known as **P bodies** (P for *p*rocessing).

In 1993, Victor Ambros discovered the first known miRNA, which is encoded by the *lin-4* gene of *C. elegans* (Fig. 31-76*a*). The *lin-4* gene was known to control the timing of larval development, although it was thought that it encoded a protein that repressed the expression of the *lin-14* gene. In fact, the *lin-4* miRNA is complementary to seven sites on the 3′UTR of the *lin-14* gene, which had previously been shown to mediate the repression of *lin-14* by the *lin-4* gene product. A puzzling observation at the time was that this regulation greatly reduces the amount of LIN-14 protein produced without altering the level of *lin-14* mRNA. These findings were eventually followed by the discovery that the *C. elegans let-7* gene encodes what is now known to be an miRNA (Fig. 31-76*b*) that controls the transition from larval to adult stages of development. Subsequently, *let-7* homologs were identified in the *Drosophila* and human genomes and *let-7* RNA was detected in these organisms as well as in numerous other animals.

Both the *lin-4* and *let-7* miRNAs were discovered by genetic analyses. However most of the nearly 10,000 miRNAs in plants and animals that are now known, including those in Fig. 31-76*c*, were identified through bioinformatic approaches (Section 7-4). The known miRNAs are catalogued in the miRBase database (http://www.mirbase.org/). Nearly all miRNAs are conserved among closely related animals (e.g., mice and humans) and many are more broadly conserved throughout animal lineages (e.g., over one-third of the 174 *C. elegans* miRNAs have homologs in humans). *The significance of miRNAs is indicated by the fact that humans express over 720 miRNAs that participate in regulating ~30% of their protein-coding genes.*

RNAi has become the method of choice for "knocking out" specific genes in plants and invertebrates. For example, in *C. elegans*, RNAi has been used to systematically inactivate over 16,000 of its ~19,000 protein-coding genes in an attempt to assign a function to each gene. *C. elegans* is particularly amenable to the RNAi approach, since these worms eat *E. coli* cells, and it is relatively easy to genetically engineer the bacterial cells to express double-stranded RNA that becomes part of the worms' diet. One limitation of the RNAi method is that it permits only the effects of gene inactivation—not gene activation—to be examined.

RNAi is of similar use in mammalian systems, even though mammals lack the mechanisms that amplify silencing in plants and nonvertebrates so that the effects of RNAi in mammals are transient. The exquisite specificity of RNAi may make it possible to prevent viral infections and to silence disease-causing mutant genes such as oncogenes. In fact, experiments have demonstrated that it is possible to use RNAi to block the liver's inflammatory response to a hepatitis virus, at least in mice, and to prevent HIV replication in cultured human cells. One challenge for the future is to devise protocols for more specific and longer-lasting gene silencing that would make it possible to prevent viral infections or to block the effects of disease-causing mutant genes.

u. Mature Eukaryotic mRNAs Are Actively Transported from the Nucleus to the Cytoplasm

The translation of prokaryotic mRNAs is often initiated, as we have seen in Fig. 31-27, before their synthesis is

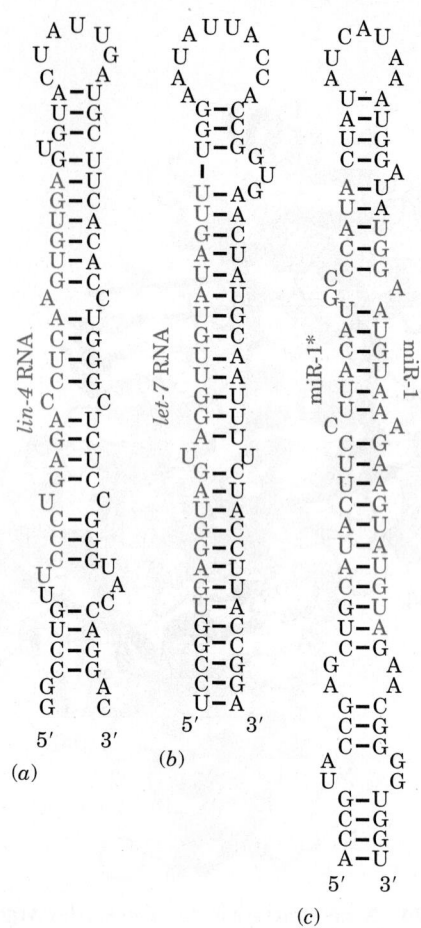

Figure 31-76 The predicted stem–loops of some pre-miRNAs. The miRNAs contained in these pre-mRNAs, all of which are from *C. elegans*, are red. (*a*) lin-4, (*b*) let-7, and (*c*) miR-1 and miR-1* (*in blue*), which are largely complementary to each other.

complete. This cannot occur in eukaryotes because the transcription and post-transcriptional processing of eukaryotic mRNAs occurs in the nucleus but their translation takes place in the cytosol. Consequently, mature mRNAs must be transported from the nucleus to the cytoplasm. This is a highly selective process because mature mRNAs comprise only a small fraction of the RNAs present in the nucleus, the remainder being pre-mRNAs, excised introns (which are usually much larger than the exons from which they were liberated), rRNAs, tRNAs, snRNAs, and a variety of RNAs that participate in the processing of rRNAs and tRNAs (Sections 31-4B and 31-4C). Indeed, only ~5% of the RNA that is synthesized ever leaves the nucleus.

How are mature mRNAs recognized and transported? As we have seen, throughout their residency in the nucleus, pre-mRNAs are continually associated with numerous proteins, including those that participate in synthesizing their m7G caps and poly(A) tails, and in splicing out their introns. In addition, the **exon junction complex (EJC),** which consists of four core proteins and several transiently associating proteins, is deposited onto mRNA during the splicing process at a site that is 20 to 24 nt upstream of the splice junction without regard to its sequence. The population of proteins bound to an mRNA changes as the mRNA is processed but some of the proteins, including SR proteins, hnRNPs (Section 31-4Ai), and EJCs remain associated with mature mRNAs in the nucleus. However, it appears that it is its entire collection of bound proteins rather than any individual protein that serves to identify an mRNA to the nuclear export machinery.

The eukaryotic nucleus (Fig. 1-5) is a double membrane-enveloped organelle that in animals is penetrated by an average of ~3000 pores. These are formed by **nuclear pore complexes (NPCs),** which are massive (~120,000 kD), 8-fold symmetric assemblies of ~30 different proteins known as **nucleoporins.** NPCs, which have inner diameters of ~90 Å (although this may be expandable to as much as 260 Å), allow the free diffusion of molecules of up to 50 kD, but most macromolecules, including mRNAs in their complexes with proteins, require an active transport process to pass through an NPC. Some of the proteins associated with mature mRNAs bear nuclear export signals that are recognized by a protein receptor that in yeast is named **Dbp5.** This 482-residue DExD/H box protein (Dbp5 stands for *DExD/H box protein 5*) is an ATP-driven RNA helicase that also binds to the NPC. This permits Dbp5 to pull the mRNA out into the cytosol while simultaneously stripping away many of its bound proteins. These proteins are later recycled by returning them to the nucleus through the NPCs.

v. mRNA Degradation Is Elaborately Controlled

The synthesis and maturation of mRNAs, as we have seen, are subject to multiple controls. The same is true of their degradation. Indeed, the range of mRNA stability in eukaryotic cells, measured in half-lives, varies from a few minutes to many hours or days. The mRNA molecules themselves contain elements that dictate their decay rates. These elements include the 3' poly(A) tail and the 5' m⁷G cap, which protect against exonucleases, as well as sequences that are located within the coding region.

A major route for mRNA degradation begins with the progressive removal of its poly(A) tail, a process catalyzed by **deadenylases** that are located throughout the cytosol. When the residual poly(A) tail is less than ~10 nt long and hence no longer capable of binding poly(A) binding protein (Section 31-4Ab), the mRNA becomes a substrate for a **decapping enzyme,** which hydrolytically excises the mRNA's m⁷G cap. This is possible because the translational initiation factor **eIF4G** interacts with both poly(A) binding protein and cap binding protein (Section 32-3Cd), thereby circularizing the mRNA so that events at its 3' end can be coupled to events at its 5' end. The decapped and deadenylated mRNA is then degraded by exonucleases, mainly the 1706-residue 5' → 3' exonuclease **Xrn1** and the 3' → 5' exonuclease complex named the **exosome.** A decapping enzyme, 5' → 3' exonucleases, and accessory proteins form P bodies (Section 31-4At) that function to either degrade mRNA or store it in an inactive form.

Proteins that bind to **AU-rich elements (AREs)** in the 3' untranslated region of mRNAs also appear to increase or decrease the rate of mRNA degradation, although their exact action is poorly understood. RNA secondary structure and RNA-binding proteins, which may be susceptible to modification by cellular signaling pathways, are thought to play a role in regulating mRNA stability.

The eukaryotic core exosome consists of single copies of nine different subunits. Its X-ray structure (Fig. 31-77), de-

termined by Christopher Lima, reveals that six of these subunits, **Rpr41** (Rpr for *rRNA processing;* the exosome was discovered as an activity that processed the 3' ends of rRNAs), **Rrp42, Mtr3, Rrp43, Rrp46,** and **Rrp45,** form a six-membered ring with the remaining three subunits, **Rrp4, Csl4,** and **Rrp40,** bound to the same face of this ring. These subunits are arranged such that the core exosome contains an ~9-Å-wide central channel that allows the entrance of only single-stranded RNAs.

The archeal exosome appears to be a simpler version of the eukaryotic core exosome. Its six-membered ring consists of only two types of subunits, Rrp41 and Rrp42, that alternate around the ring, with three copies of Rrp4 bound to the same face of the ring. Only Rrp41 contains an active site although Rrp42 is required for activity. Not surprisingly, eukaryotic Rrp4, Mtr3, and Rrp46 are homologs of archeal Rrp41, eukaryotic Rrp42, Rrp43, and Rrp45 are homologs of archeal Rrp42, and eukaryotic Rrp4, Csl4, and Rrp40 are homologs of archeal Rrp4. Nevertheless, despite the fact that each of its core subunits are essential for viability, eukaryotic core exosomes, from yeast to humans, are catalytically inactive due to changes in active site residues relative to their archeal homologs. However, the core exosome associates with two 3'-exonucleases, **Rrp6** and **Rrp44,** whose catalytically inactive mutants are individually viable in yeast but lethal in combination. Moreover, the core exosome interacts with numerous mostly multisubunit cofactors that carry out a variety of RNA processing activities in both the nucleus and the cytosol. Thus, the eukaryotic core exosome appears to be a structural platform upon which many RNA processing enzymes can be mounted.

Figure 31-77 X-ray structure of the human core exosome. The protein complex is drawn in ribbon form embedded in its semitransparent molecular surface, with each of its nine different subunits separately colored. The view is toward the face of the six-membered ring of subunits opposite that to which the three other subunits bind. [Based on an X-ray structure by Christopher Lima, Sloan-Kettering Institute, New York, New York. PDBid 2NN6.]

B. *Ribosomal RNA Processing*

The seven *E. coli* rRNA operons all contain one (nearly identical) copy of each of the three types of rRNA genes (Section 32-3A). Their polycistronic primary transcripts, which are ~5500 nucleotides in length, contain 16S rRNA at their 5' ends followed by the transcripts for 1 or 2 tRNAs, 23S rRNA, 5S rRNA, and, in some rRNA operons, 1 or 2 more tRNAs at the 3' end (Fig. 31-78). The steps in processing these primary transcripts to mature rRNAs were elucidated with the aid of mutants defective in one or more of the processing enzymes.

The initial processing, which yields products known as **pre-rRNAs,** commences while the primary transcript is still being synthesized. It consists of specific endonucleolytic cleavages by **RNase III, RNase P, RNase E,** and **RNase F** at the sites indicated in Fig. 31-78. The base sequence of the primary transcript suggests the existence of several base-paired stems. The RNase III cleavages occur in a stem consisting of complementary sequences flanking the 5' and 3' ends of the 23S segment (Fig. 31-79) as well as that of the 16S segment. Presumably, certain features of these stems constitute the RNase III recognition site.

The 5' and 3' ends of the pre-rRNAs are trimmed away in secondary processing steps (Fig. 31-78) through the action of **RNases D, M16, M23,** and **M5** to produce the mature rRNAs. These final cleavages only occur after the pre-rRNAs become associated with ribosomal proteins.

a. Ribosomal RNAs Are Methylated

During ribosomal assembly, the 16S and 23S rRNAs are methylated at a total of 24 specific nucleosides. The methylation reactions, which employ *S*-adenosylmethionine (Section 26-3Ea) as a methyl donor, yield N^6,N^6-dimethyladenine and $O^{2'}$-methylribose residues. $O^{2'}$-methyl groups may protect adjacent phosphodiester bonds from degradation by intracellular RNases (the mechanism of RNase hydrolysis involves utilization of the free 2'-OH group of ribose to eliminate the

substituent on the 3'-phosphoryl group via the formation of a 2',3'-cyclic phosphate intermediate; Figs. 5-3 and 15-3). However, the function of base methylation is unknown.

b. Eukaryotic rRNA Processing Is Guided by snoRNAs

The eukaryotic genome typically has several hundred tandemly repeated copies of rRNA genes that are contained in small, dark-staining nuclear bodies known as **nucleoli** (the site of rRNA transcription and processing and ribosomal subunit assembly; Fig. 1-5; note that nucleoli are not membrane enveloped). The primary rRNA transcript is an ~7500-nucleotide 45S RNA that contains, starting from its 5' end, the 18S, 5.8S, and 28S rRNAs separated by spacer sequences (Fig. 31-80). In the first stage of its processing, 45S RNA is specifically methylated at numerous sites (106 in humans) that occur mostly in its rRNA sequences. About 80% of these modifications yield $O^{2'}$-methylribose residues and the remainder form methylated bases such as N^6,N^6-dimethyladenine and 2-methylguanine. In addition, many pre-rRNA U's (95 in humans) are converted to pseudouridines (Ψ's) (Section 30-5Be), which may contribute to the rRNA's tertiary stability through hydrogen bonding involving its newly acquired ring NH group. The subsequent cleavage and trimming of the 45S RNA superficially resembles that of prokaryotic rRNAs. In fact, enzymes exhibiting RNase III–and RNase P–like activities occur in eukaryotes. The 5S eukaryotic rRNA is separately processed in a manner resembling that of tRNA (Section 31-4C).

The methylation sites in eukaryotic rRNAs occur exclusively within conserved domains that are therefore likely to participate in fundamental ribosomal processes. Indeed, the methylation sites generally occur in invariant sequences among yeast and vertebrates although the methylations themselves are not always conserved. These methylation sites do not appear to have a consensus structure

Figure 31-78 The post-transcriptional processing of *E. coli* rRNA. The transcriptional map is shown approximately to scale. The labeled arrows indicate the positions of the various nucleolytic cuts and the nucleases that generate them. [After Apiron, D., Ghora, B.K., Plantz, G., Misra, T.K., and Gegenheimer, P., *in* Söll, D., Abelson, J.N., and Schimmel P.R. (Eds.), *Transfer RNA: Biological Aspects, p.* 148, Cold Spring Harbor Laboratory Press (1980).]

Figure 31-79   **The stem-and-giant-loop secondary structure in the 23S region of the** ***E. coli*** **primary rRNA transcript.** The RNase III cleavage sites are indicated. [After Young, R.R., Bram, R.J., and Steitz, J.A., *in* Söll, D., Abelson, J.N., and Schimmel, P.R. (Eds.), *Transfer RNA: Biological Aspects*, p. 102, Cold Spring Harbor Laboratory Press (1980).]

The snoRNA nucleotide that pairs with the nucleotide to be O2′-methylated always precedes box D by exactly 5 nt. Evidently, each of these so-called **box C/D snoRNAs** act to guide the methylation of a single site. In fact, in those cases in which two adjacent ribose residues are methylated, two box C/D snoRNAs with overlapping sequences occur. The methylation is mediated by a complex of at least four nucleolar proteins, including **fibrillarin** (~325 residues; so called because it is located in the dense fibrillar region of the nucleolus), the likely methyltransferase, which together with a box C/D snoRNA form **snoRNPs.** The conversion of specific rRNA U's to Ψ's is similarly mediated by a different subgroup of snoRNAs, the **box H/ACA snoRNAs,** so called because they contain the sequence motifs ACANNN at the snoRNA's 3′ end and box H (ANANNA) at its 5′ end, so as to flank a sequence that partially base-pairs to the pre-rRNA segment containing the U to be converted to Ψ. Archaea also modify their rRNAs via RNA-guided methylations and U to Ψ conversions but, interestingly, the analogous reactions in eubacteria are mediated by protein enzymes that lack RNA.

C. *Transfer RNA Processing*

tRNAs, as we discuss in Section 32-2A, consist of ~80 nucleotides that assume a secondary structure with four base-paired stems known as the **cloverleaf structure** (Fig. 31-81). All tRNAs have a large fraction of modified bases (whose structures and functions are discussed in Section 32-2Aa) and each has the 3′-terminal sequence —CCA to which the

that might be recognized by a single methyltransferase. How, then, are these methylation sites targeted?

An important clue as to how the methylation sites on rRNA are selected came from the observation that pre-rRNA interacts with the members of a large family of **small nucleolar RNAs (snoRNAs;** ~100 in yeast and ~200 in mammals). The snoRNAs, whose lengths vary from 70 to 100 nt, contain segments of 10 to 21 nt that are precisely complementary to segments of the mature rRNAs that contain the O2′-methylation sites. These snoRNA sequences are located between the conserved sequence motifs known as box C (RUGAUGA) and box D (CUGA), which are respectively located on the 5′ and 3′ sides of the complementary segments. In intron-rich organisms such as vertebrates, most snoRNAs are encoded by the introns of structural genes so that not all excised introns are discarded.

Figure 31-80   **The organization of the 45S primary transcript of eukaryotic rRNA.**

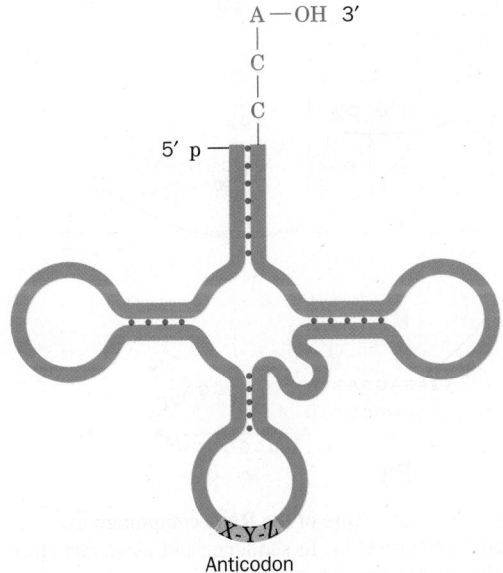

Figure 31-81   **A schematic diagram of the tRNA cloverleaf secondary structure.** Each dot indicates a base pair in the hydrogen bonded stems. The position of the anticodon triplet and the 3′-terminal —CCA are indicated.

corresponding amino acid is appended in the amino acid–charged tRNA. The **anticodon** (which is complementary to the codon specifying the tRNA's corresponding amino acid) occurs in the loop of the cloverleaf structure opposite the stem containing the terminal nucleotides.

The *E. coli* chromosome contains ~60 tRNA genes. Some of them are components of rRNA operons (Section 31-4B); the others are distributed, often in clusters, throughout the chromosome. The primary tRNA transcripts, which contain from one to as many as four or five identical tRNA copies, have extra nucleotides at the 3' and 5' ends of each tRNA sequence. The excision and trimming of these tRNA sequences resemble those for *E. coli* rRNAs (Section 31-4B) in that both processes employ some of the same nucleases.

a. RNase P Is a Ribozyme

RNase P, which generates the 5' ends of tRNAs (Fig. 31-78), is a particularly interesting enzyme because it has, in *E. coli,* a 377-nucleotide RNA component (~125 kD vs 14 kD for its 119-residue protein subunit) that is essential for its enzymatic activity. The enzyme's RNA was, quite understandably, first proposed to function in recognizing the sub-

strate RNA through base pairing and to thereby guide the protein subunit, which was presumed to be the actual nuclease, to the cleavage site. However, Sidney Altman demonstrated that *the RNA component of RNase P is, in fact, the enzyme's catalytic subunit* by showing that protein-free RNase P RNA catalyzes the cleavage of substrate RNA at high salt concentrations. RNase P protein, which is basic, evidently functions at physiological salt concentrations to electrostatically reduce the repulsions between the polyanionic ribozyme and substrate RNAs. The argument that trace quantities of RNase P protein are really responsible for the RNase P reaction was disposed of by showing that catalytic activity is exhibited by RNase P RNA that has been transcribed in a cell-free system. RNase P activity occurs in eukaryotes (nuclei, mitochondria, and chloroplasts) as well as in prokaryotes although eukaryotic nuclear RNase P's have 9 or 10 protein subunits, none of which are related to the bacterial protein. Indeed, RNase P

Figure 31-82 **Structure of the RNA component of** *T. maritima* **RNase P.** (*a*) Its sequence and secondary structure. The various segments (P for paired region, J for joining region, and L for loop) are shown in different colors. The black lines indicate major interactions that are observed in the X-ray structure, dashes indicate Watson–Crick base pairs, and small filled circles represent non-Watson–Crick base pairs. (*b*) Its X-ray structure, which is colored as in Part *a*. Of its 338 nucleotides, 309 are visible. The lower view is related to the upper view by a 90° rotation about the horizontal axis. [Courtesy of Alfonso Mondragón, Northwestern University. PDBid 2A2E.]

mediates one of the two ribozymal activities that occur in all cellular life, the other being associated with ribosomes (Section 32-3Dg).

The X-ray structures of the RNA components of RNase P from *Thermotoga maritima* (338 nt) and *Bacillus stearothermophilus* (417 nt), which were independently determined by Alfonso Mondragón and Norman Pace, reveal that these ribozymes consist mainly of stacked helical stems with overall compact structures typical of protein enzymes (Fig. 31-82). Biochemical studies and modeling indicate that the RNase active site lies in a cleft between the P2/P3 region (cyan in Fig. 31-82) and the P1/P4/P5 region (dark blue in Fig. 31-82). That portion of the structure encompassing P1 through P4, P9 through P11, J11-12 and J12-11 is present in all known RNase P's and hence is known as the universal minimum consensus structure. This structure was presumably present in the primordial RNase P.

b. Many Eukaryotic Pre-tRNAs Contain Introns

Eukaryotic genomes contain from several hundred to several thousand tRNA genes. Many eukaryotic primary tRNA transcripts, for example, that of yeast tRNA^{Tyr} (Fig. 31-83), contain a small intron adjacent to their anticodons as well as extra nucleotides at their 5′ and 3′ ends. Note that this intron is unlikely to disrupt the tRNA's cloverleaf structure.

c. The –CCA Ends of Eukaryotic tRNAs Are Post-Transcriptionally Appended

Eukaryotic tRNA transcripts lack the obligatory —CCA sequence at their 3′ ends. This is appended to the immature tRNAs by the enzyme **CCA-adding polymerase,** which sequentially adds two C's and an A to tRNA using CTP and ATP as substrates. This enzyme also occurs in prokaryotes, although, at least in *E. coli,* the tRNA genes all encode a —CCA terminus. The *E. coli* CCA-adding polymerase is therefore likely to function in the repair of degraded tRNAs.

Figure 31-83 The post-transcriptional processing of yeast tRNA^{Tyr}. A 14-nucleotide intervening sequence (*red*) and a 19-nucleotide 5′-terminal sequence (*green*) are excised from the primary transcript, a —CCA (*blue*) is appended to the 3′ end, and several of the bases are modified (their symbols are defined in Fig. 32-13) to form the mature tRNA. The anticodon is shaded. [After DeRobertis, E.M. and Olsen, M.V., *Nature* **278,** 142 (1989).]

CHAPTER SUMMARY

1 The Role of RNA in Protein Synthesis The central dogma of molecular biology states that "DNA makes RNA makes protein" (although RNA can also "make" DNA). There is, however, enormous variation among the rates at which the various proteins are made. Certain enzymes, such as those of the *lac* operon, are synthesized only when the substances whose metabolism they catalyze are present. The *lac* operon consists of the control sequences *lacP* and *lacO* followed by the tandemly arranged genes for β-galactosidase *(lacZ)*, galactoside permease *(lacY)*, and thiogalactoside transacetylase *(lacA)*. In the absence of inducer, physiologically allolactose, the *lac* repressor, the product of the *lacI* gene, binds to operator *(lacO)* so as to prevent the transcription of the *lac* operon by RNA polymerase. The binding of inducer causes the repressor to release the operator that allows the *lac* structural genes to be transcribed onto a single polycistronic mRNA. The mRNAs transiently associate with ribosomes so as to direct them to synthesize their encoded polypeptides.

2 RNA Polymerase The holoenzymes of bacterial RNA polymerases (RNAPs) have the subunit structure $\alpha_2\beta\beta'\omega\sigma$. They initiate transcription on the antisense (template) strand of a gene at a position designated by its promoter. In *E. coli* the most conserved region of the promoter is centered at about the -10 position and has the consensus sequence TATAAT. The -35 region is also conserved in efficient promoters. DMS footprinting studies indicate that the RNAP holoenzyme forms an open initiation complex with the promoter in which the template and nontemplate DNA strands are separated to form an ~14-nt transcription bubble. RNAPs have the shape of a crab claw. In the elongation complex, the template strand in the transcription bubble passes through a tunnel in the core enzyme to the active site, where it pairs with incoming ribonucleotides. The RNA product exits the enzyme through a channel between its β and β' subunits. In the closed complex of the bacterial RNAP holoenzyme, the σ subunit, which extends along the "top" of the holoenzyme, makes all the sequence-specific contacts with the promoter. After the initiation of RNA synthesis, the σ subunit dissociates from the core enzyme, which then autonomously catalyzes chain elongation in the $5' \rightarrow 3'$ direction. RNA synthesis is terminated by a segment of the transcript that forms a G + C–rich hairpin with an oligo(U) tail that spontaneously dissociates from the DNA. Termination sites that lack these sequences require the assistance of Rho factor for proper chain termination.

In the nuclei of eukaryotic cells, RNAPs I, II, and III, respectively, synthesize rRNA precursors, mRNA precursors, and tRNAs + 5S RNA. The structure of yeast RNAP II resembles that of bacterial RNAPs but is somewhat larger and has more subunits. The structure of its transcribing complex reveals a one-turn segment of RNA · DNA hybrid helix at the active site, which is in contact with the solvent via a pore leading into a funnel through which NTPs presumably pass. RNAPs can hydrolytically correct their mistakes with the aid of TFIIS in eukaryotes and GreA and GreB in bacteria. The minimal RNA polymerase I promoter extends between nucleotides -31 and $+6$. Many RNA polymerase II promoters contain a conserved TATAAAA sequence, the TATA box, located around position -27. Enhancers are transcriptional activators that can have variable positions and orientations relative to the transcription start site. RNA polymerase III promoters are located within the transcribed regions of their gene between positions $+40$ and $+80$.

3 Control of Transcription in Prokaryotes Prokaryotes can respond rapidly to environmental changes, in part because the translation of mRNAs commences during their transcription and because most mRNAs are degraded within 1 to 3 min of their synthesis. The expression of specific sets of genes is controlled, in most bacteria and some bacteriophages, by σ factors. The *lac* repressor is a tetrameric protein of identical subunits that, in the absence of inducer, nonspecifically binds to duplex DNA but binds much more tightly to *lac* promoter. The promoter sequence that *lac* repressor protects from nuclease digestion has nearly palindromic symmetry. Yet, methylation protection and mutational studies indicate that repressor is not symmetrically bound to promoter. *lac* repressor prevents RNA polymerase from properly initiating transcription at the *lac* promoter.

The presence of glucose represses the transcription of operons specifying certain catabolic enzymes through the mediation of cAMP. On binding cAMP, which accumulates only in the absence of glucose, catabolite gene activator protein (CAP) binds at or immediately upstream of the promoters of these operons, including the *lac* operon, thereby activating their transcription through the binding to the C-terminal domain of the associated RNAP's α subunit (αCTD). CAP's two symmetry equivalent DNA-binding domains each bind in the major groove of their target DNA via a helix–turn–helix (HTH) motif that also occurs in numerous prokaryotic repressors. The binding between these repressors and their target DNAs is mediated by mutually favorable associations between these macromolecules rather than any specific interactions between particular base pairs and amino acid side chains analogous to Watson–Crick base pairing. Sequence-specific interactions between the *met* repressor and its target DNA occur through a 2-fold symmetric antiparallel β ribbon that this protein inserts into the DNA's major groove. *araBAD* transcription is controlled by the levels of L-arabinose and CAP–cAMP through a remarkable complex of the control protein AraC to two binding sites, $araO_2$ and $araI_1$, that forms an inhibitory DNA loop. On binding L-arabinose and when CAP–cAMP is adjacently bound, AraC releases $araO_2$ and instead binds $araI_2$, thereby releasing the loop and activating RNA polymerase to transcribe the *araBAD* operon. The expression of the *lac* operon is also in part controlled by DNA loop formation. The *lac* repressor is a dimer of homodimers, one of which binds to the operator $lacO_1$ and the other to $lacO_2$ or $lacO_3$ to form a DNA loop that may interfere with RNAP binding to the *lac* promoter. The binding of an inducer such as IPTG to a *lac* repressor dimer core domain alters the angle between its two attached DNA-binding domains such that they cannot simultaneously bind to the *lac* operator, thereby weakening the repressor's grip on the DNA.

The expression of the *E. coli trp* operon is regulated by both attenuation and repression. On binding tryptophan, its corepressor, *trp* repressor binds to the *trp* operator, thereby blocking *trp* operon transcription. When tryptophan is available, much of the *trp* transcript that has escaped repression is prematurely terminated in the *trpL* sequence because its transcript contains a segment that forms a normal intrinsic terminator. When tryptophanyl–tRNA^{Trp} is scarce, ribosomes stall at the transcript's two tandem Trp codons. This permits the newly synthesized RNA to form a base-paired stem and loop that prevents the formation of the terminator structure. Several other operons are similarly regulated by attenuation. Riboswitches are mRNA components that regulate gene expression by specifically binding metabolites. The stringent response is another mechanism by which *E. coli* match the rate of transcription to charged tRNA availability. When a specified charged tRNA is scarce, stringent factor on active ribosomes synthesizes ppGpp, which inhibits the transcription of rRNA and some mRNAs while stimulating the transcription of other mRNAs.

4 Post-Transcriptional Processing Most prokaryotic mRNA transcripts require no additional processing. However, eukaryotic mRNAs have an enzymatically appended 5′ cap and, in most cases, an enzymatically generated poly(A) tail. Moreover, the introns of eukaryotic mRNA primary transcripts (hnRNAs) are precisely excised via lariat intermediates and their flanking exons are spliced together. Group I and group II introns are self-splicing, that is, their RNAs function as ribozymes (RNA enzymes). Ribozymes, such as the *Tetrahymena* pre-rRNA and hammerhead ribozymes, have complex structures containing several base-paired stems. Pre-mRNAs are spliced by large and complex particles named spliceosomes that consist of four different small nuclear ribonucleoproteins (snRNPs) and which are assisted by the participation of a variety of protein splicing factors. Many eukaryotic proteins consist of modules that also occur in other proteins and hence appear to have evolved via the stepwise collection of exons through recombination events. The alternative splicing of pre-mRNAs greatly increases the variety of proteins expressed by eukaryotic genomes. Certain mRNAs are subject to RNA editing, either by the replacement, insertion, or deletion of specific bases in a process that is directed by guide RNAs (gRNAs), or by substitutional editing mediated by cytidine deaminases or adenosine deaminases.

In RNA interference (RNAi), dsRNA is cleaved by the endoribonuclease Dicer to small interfering RNAs (siRNAs) that guide the hydrolytic cleavage of the complementary mRNAs by the Argonaute component of the RNA-induced silencing complex (RISC), thereby preventing the mRNAs' transcription. MicroRNAs (miRNAs) are generated through the excision of imperfectly base paired stem–loop structures from pri-miRNAs through the actions of Drosha and Dicer. The guide RNA strand of miRNAs binds to RISC and directs it to partially complementary sequences on 3′ untranslated regions (3′UTRs) of its target mRNAs, thus inhibiting the expression of the mRNAs and providing a major although only recently recognized mechanism for the regulation of gene expression. Mature eukaryotic mRNAs are selectively and actively transported from the nucleus to the cytosol via the nuclear pore complex. The degradation of mRNAs, which is elaborately controlled, is mediated in part by exosomes.

The primary transcript of *E. coli* rRNAs contains all three rRNAs together with some tRNAs. These are excised and trimmed by specific endonucleases and exonucleases. The eukaryotic 18S, 5.8S, and 28S rRNAs are similarly transcribed as a 45S precursor, which is processed in a manner resembling that of *E. coli* rRNAs. Eukaryotic rRNAs are modified by the methylation of specific nucleosides, as are prokaryotic rRNA, and by the conversion of certain U's to pseudouridines (Ψ's). These processes are guided by small nucleolar RNAs (snoRNAs). Prokaryotic tRNAs are excised from their primary transcripts and trimmed in much the same way as are rRNAs. In RNase P, one of the enzymes mediating this process, the catalytic subunit is an RNA. Eukaryotic tRNA transcripts also require the excision of a short intron and the enzymatic addition of a 3′-terminal —CCA to form the mature tRNA.

REFERENCES

General

Brown, T.A., *Genomes* (3rd ed.), Chapter 12, Wiley-Liss (2007).

Gesteland, R.F., Cech, T.R., and Atkins, J.F. (Eds.), *The RNA World* (3rd ed.), Cold Spring Harbor Laboratory (2006). [A series of authoritative articles on the nature of the prebiotic "RNA world" as revealed by the RNA "relics" in modern organisms.]

Hodgson, D.A. and Thomas, C.M. (Eds.), *Signals, Switches, Regulons and Cascades: Control of Bacterial Gene Expression*, Cambridge University Press (2002).

Watson, J.D., Baker, T.A., Bell, S.P., Gann, A., Levine, M., and Losick, R., *Molecular Biology of the Gene* (6th ed.), Chapters 12, 13, 16, and 18, Cold Spring Harbor Laboratory Press (2008).

The Genetic Role of RNA

Brachet, J., Reminiscences about nucleic acid cytochemistry and biochemistry, *Trends Biochem. Sci.* **12**, 244–246 (1987).

Brenner, S., Jacob, F., and Meselson, M., An unstable intermediate carrying information from genes to ribosomes for protein synthesis, *Nature* **190**, 576–581 (1960). [The experimental verification of mRNA's existence.]

Crick, F., Central dogma of molecular biology, *Nature* **227**, 561–563 (1970).

Hall, B.D. and Spiegelman, S., Sequence complementarity of T2-DNA and T2-specific RNA, *Proc. Natl. Acad. Sci.* **47**, 137–146 (1964). [The first use of RNA–DNA hybridization.]

Jacob, F. and Monod, J., Genetic regulatory mechanisms in the synthesis of proteins, *J. Mol. Biol.* **3**, 318–356 (1961). [The classic paper postulating the existence of mRNA and operons and explaining how the transcription of operons is regulated.]

Pardee, A.B., Jacob, F., and Monod, J., The genetic control and cytoplasmic expression of "inducibility" in the synthesis of β-galactosidase by *E. coli*, *J. Mol. Biol.* **1**, 165–178 (1959). [The PaJaMo experiment.]

Thieffry, D., Forty years under the central dogma, *Trends Biochem. Sci.* **23**, 312–316 (1998).

RNA Polymerase and mRNA

Campbell, E.A., Korzheva, N., Mustaev, A., Murakami, K., Nair, S., Goldfarb, A., and Darst, S.A., Structural mechanism for rifampicin inhibition of bacterial RNA polymerase, *Cell* **104**, 901–912 (2001).

Cramer, P., et al., Structure of eukaryotic RNA polymerases, *Annu. Rev. Biophys.* **37**, 337–352 (2008).

Cramer, P. and Arnold, A., Proteins: how RNA polymerases work, *Curr. Opin. Struct. Biol.* **19**, 680–682 (2009). [The editorial overview of a series of authoritative reviews on numerous aspects of RNA polymerases. See, in particular, the reviews beginning on pages 691, 701, 708, and 732.]

Cramer, P., Bushnell, D.A., and Kornberg, R.D., Structural basis of transcription: RNA polymerase at 2.8 Å resolution, *Science* **292,** 1863–1876 (2001); *and* Gnatt, A.L., Cramer, P., Fu, J., Bushnell, D.A., and Kornberg, R.D., Structural basis of transcription: An RNA polymerase II elongation complex at 3.3 Å resolution, *Science* **292,** 1876–1882 (2001).

Dahmus, M.E., Reversible phosphorylation of the C-terminal domain of RNA polymerase II, *J. Biol. Chem.* **271,** 19009–19012 (1996).

Darst, S.A., Bacterial RNA polymerase, *Curr. Opin. Struct. Biol.* **11,** 155–162 (2001).

Estrem, S.T., Gaal, T., Ross, W., and Gourse, R.L., Identification of an UP element consensus sequence for bacterial promoters, *Proc. Natl. Acad. Sci.* **95,** 9761–9766 (1998).

Gilmour, D.S. and Fan, R., Derailing the locomotive: transcription termination, *J. Biol. Chem.* **283,** 661–664 (2008).

Harada, Y., Ohara, O., Takatsuki, A., Itoh, H., Shimamoto, N., and Kinosita, K., Jr., Direct observation of DNA rotation during transcription by *Escherichia coli* RNA polymerase, *Nature* **409,** 113–115 (2001).

Kadonaga, K.T., The RNA polymerase II core promoter, *Annu. Rev. Biochem.* **72,** 449–479 (2003).

Kaplan, C.D., Larsson, K.-M., and Kornberg, R.D., The RNA polymerase II trigger loop functions in substrate selection and is directly targeted by α-amanitin, *Mol. Cell* **30,** 547–556 (2008).

Larson, M.H., Greenleaf, W.J., Landick, R., and Block, S.M., Applied force reveals mechanistic and energetic details of transcription termination, *Cell* **132,** 971–982 (2008); *and* Brueckner, F. and Cramer, P., Structural basis for transcription inhibition by α-amanitin and implications for RNA polymerase II translocation, *Nature Struct. Mol. Biol.* **15,** 811–818 (2008).

Lian, C., Robinson, H., and Wang, A.H.-J., Structure of actinomycin D bound with (GAAGCTTC)$_2$ and (GATGCTTC)$_2$ and its binding to the (CAG)$_n$:(CTG)$_n$ triplet sequence as determined by NMR analysis, *J. Am. Chem. Soc.* **118,** 8791–8801 (1996).

Mooney, R.A., Darst, S.A., and Landick, R., Sigma and RNA polymerase: An on-again, off-again relationship? *Mol. Cell.* **20,** 335–345 (2006).

Murikami, K.S., Masuda, S., and Darst, S., Structural basis of transcription initiation: RNA polymerase at 4 Å resolution, *Science* **296,** 1280–1284 (2002); Murikami, K.S., Masuda, S., Campbell, E.A., Muzzin, O., and Darst, S., Structural basis of transcription initiation: An RNA polymerase holoenzyme-DNA complex, *Science* **296,** 1285–1290 (2002); *and* Murakami, K.S. and Darst, S.A., Bacterial RNA polymerases: the whole story, *Curr. Opin. Struct. Biol.* **13,** 31–39 (2003).

Nudler, E., RNA polymerase active center: the molecular engine of transcription, *Annu. Rev. Biochem.* **78,** 335–361 (2009).

Revyakin, A., Liu, C., Ebright, R.H., and Strick, T.R., Abortive initiation and productive initiation by RNA polymerase involve DNA scrunching; *and* Kapanidis, A.N., Margeat, E., Ho, S.O., Kortkhonjia, E., Weiss, S, and Ebright, R.H., Initial transcription by RNA polymerase proceeds through DNA-scrunching mechanism, *Science* **314,** 1139–1143 *and* 1144–1150 (2006).

Reynolds, R. and Chamberlin, M.J., Parameters affecting transcription termination by *Escherichia coli* RNA. II. Construction and analysis of hybrid terminators, *J. Mol. Biol.* **224,** 53–63 (1992).

Roberts, J.W., Shankar, S., and Filter, J.J., RNA polymerase elongation factors, *Annu. Rev. Microbiol.* **62,** 211–233 (2008); *and*

Park, J.-S. and Roberts, J.W., Role of DNA bubble rewinding in enzymatic transcription termination, *Proc. Natl. Acad. Sci.* **103,** 4870–4875 (2006).

Shilatifard, A., Conway, R.C., and Conway, J.W., The RNA polymerase II elongation complex, *Annu. Rev. Biochem.* **72,** 693–715 (2003).

Skordalakes, E. and Berger, J.M., The structure of Rho transcription terminator: Mechanism of mRNA recognition and helicase loading, *Cell* **114,** 135–146 (2003).

Svejstrup, J.Q., Contending with transcriptional arrest during RNAPII transcript elongation, *Trends Biochem. Sci.* **32,** 165–171 (2007).

Thompson, N.D. and Berger, J.M., Running in reverse: The structural basis for translocation polarity in hexameric helicases, *Cell* **139,** 523–534 (2009).

Vassylyev, D.G., Sekine, S.-I., Laptenko, O., Lee, J., Vassylyeva, M.N., Borukhov, S., and Yokoyama, S., Crystal structure of a bacterial RNA polymerase holoenzyme at 2.6 Å resolution, *Nature* **417,** 712–718 (2002).

Vassylyev, D.G., Vassylyeva, M.N., Perederina, A., Tahirov, T.H., and Artsimovitch, I., Structural basis for transcription elongation by bacterial RNA polymerase; *and* Vassylyev, D.G., Vassylyeva, M.N., Zhang, J., Palangat, M., Artsimovitch, I., and Landick, R., Structural basis for substrate loading in bacterial RNA polymerase, *Nature* **448,** 157–162 *and* 163-168 (2007).

Wang, D., Bushnell, D.A., Huang, X., Westover, K.D., Levitt, M., and Kornberg, R.D., Structural basis of transcription: Backtracked RNA polymerase II at 3.4 Å resolution, *Science* **324,** 1203–1206 (2009); *and* Sydow, J.F., Bruekner, F., Cheung, A.C.M., Damsma, G.E., Dengl, S., Lehmann, E., Vassylyev, D., and Cramer, P., Structural basis of transcription: Mismatch-specific fidelity mechanisms and paused RNA polymerase II with frayed RNA, *Mol. Cell* **34,** 710–721 (2009).

Wang, D., Bushnell, D.A., Westover, K.D., Kaplan, C.D., and Kornberg, R.D., Structural basis of transcription: Role of the trigger loop in substrate specificity and catalysis, *Cell* **127,** 941–954 (2006).

Zhang, G., Campbell, E.A., Minakhin, L., Richter, C., Severinov, K., and Darst, S.A., Crystal structure of *Thermus aquaticus* core RNA polymerase at 3.3 Å resolution, *Cell* **98,** 811–824 (1999).

Control of Transcription in Prokaryotes

Anderson, J.E., Ptashne, M., and Harrison, S.C., The structure of the repressor–operator complex of bacteriophage 434, *Nature* **326,** 846–852 (1987).

Gilbert, W. and Müller-Hill, B., Isolation of the lac repressor, *Proc. Natl. Acad. Sci.* **56,** 1891–1898 (1966).

Harmor, T., Wu, M., and Schleif, R., The role of rigidity in DNA looping–unlooping by AraC, *Proc. Natl. Acad. Sci.* **98,** 427–431 (2001).

Huffman, J.L. and Brennan, R.G., Prokaryotic transcriptional regulators: more than just the helix-turn-helix motif, *Curr. Opin. Struct. Biol.* **12,** 98–106 (2002).

Kolter, R. and Yanofsky, C., Attenuation in amino acid biosynthetic operons, *Annu. Rev. Genet.* **16,** 113–134 (1982).

Lamond, A.I. and Travers, A.A., Stringent control of bacterial transcription, *Cell* **41,** 6–8 (1985).

Lawson, C.L., Swigon, D., Murakami, K.S., Darst, S.A., Berman, H.M., and Ebright, R.H., Catabolite activator protein: DNA binding and transcription activation, *Curr. Opin. Struct. Biol.* **14,** 10–20 (2004); *and* Bennoff, B., Yang, H., Lawson, C.L., Parkinson, G., Liu, J., Blatter, E., Ebright, Y.W., Berman, H.M., and Ebright, R.H., Structural basis of transcription activation:

The CAP-αCTD-DNA complex, *Science* **297**, 1562–1566 (2002).

Lewis, M., The *lac* repressor, *C. R. Biologies* **328**, 521–548 (2005).

Lewis, M., Chang, G., Horton, N.C., Kercher, M.A., Pace, H.C., Schumacher, M.A., Brennan, R.G., and Lu, P., Crystal structure of the lactose operon repressor and its complexes with DNA and inducer, *Science* **271**, 1247–1254 (1996).

Mondragón, A. and Harrison, S.C., The phage 434 Cro/O$_R$1 complex at 2.5 Å resolution, *J. Mol. Biol.* **219**, 321–334 (1991); *and* Wolberger, C., Dong, Y., Ptashne, M., and Harrison, S.C., Structure of phage 434 Cro/DNA complex, *Nature* **335**, 789–795 (1988).

Montange, R.K. and Batey, R.T., Riboswitches: emerging themes in RNA structure and function, *Annu. Rev. Biophys.* **387**, 117–133 (2007).

Mooney, R.A., Darst, S.A., and Landick, R., Sigma and RNA polymerase: an on-again, off-again relationship, *Mol. Cell* **20**, 335–345 (2005).

Oehler, S., Eismann, E.R., Krämer, H., and Müller-Hill, B., The three operators of the *lac* operon cooperate in repression, *EMBO J.* **9**, 973–979 (1990).

Pace, H.C., Kercher, M.A., Lu, P., Markiewicz, P., Miller, J.H., Chang, G., and Lewis, M., *Lac* repressor genetic map in real space, *Trends Biochem. Sci.* **22**, 334–339 (1997).

Reeder, T. and Schleif, R., AraC protein can activate transcription from only one position and when pointed in only one direction, *J. Mol. Biol.* **231**, 205–218 (1993).

Romanuka, J., Folkers, G.E., Biris, N., Tishchenko, E., Wienk, H., Bonvin, A.M.J.J., Kaptein, R., and Boelens, R., Specificity and affinity of lac repressor for the auxiliary operators *O2* and *O3* are explained by the structures of their protein–DNA complexes, *J. Mol. Biol.* **390**, 478–489 (2009).

Roth, A. and Breaker, R.R., The structural and functional diversity of metabolite-binding riboswitches, *Annu. Rev. Biochem.* **78**, 305–334 (2009).

Schleif, R., AraC protein: a love–hate relationship, *BioEssays* **25**, 274–282 (2003); *and* Regulation of the L-arabinose operon of *Escherichia coli*, *Trends Genet.* **16**, 559–565 (2000).

Schultz, S.C., Shields, G.C., and Steitz, T.A., Crystal structure of a CAP-DNA complex: The DNA is bent by 90°, *Science* **253**, 1001–1007 (1991).

Serganov, A., Polonskaia, A., Phan, A.T., Breaker, R.R., and Patel, D.J., Structural basis for gene regulation by a thiamine pyrophosphate-sensing riboswitch, *Nature* **441**, 1167–1171 (2006).

Shakked, Z., Guzikevich-Guerstein, G., Frolow, F., Rabinovich, D., Joachimiak, A., and Sigler, P.B., Determinants of repressor/operator recognition from the structure of the *trp* operator binding site, *Nature* **368**, 469–473 (1994).

Soisson, S.M., MacDougall-Shackleton, B., Schleif, R., and Wolberger, C., Structural basis for ligand-regulated oligomerization of AraC, *Science* **276**, 421–425 (1997). [The X-ray structure of AraC alone and in complex with arabinose.]

Somers, W.S. and Phillips, S.E.V., Crystal structure of the *met* repressor-operator complex at 2.8 Å resolution reveals DNA recognition by β-strands, *Nature* **359**, 387–393 (1992).

Yanofsky, C., Transcription attenuation, *J. Biol. Chem.* **263**, 609–612 (1988); *and* Attenuation in the control of expression of bacterial operons, *Nature* **289**, 751–758 (1981).

Post-transcriptional Processing

Bachellerie, J.-P. and Cavaillé, J., Guiding ribose methylation of rRNA, *Trends Biochem. Sci.* **22**, 257–261 (1997).

Balbo, P.B. and Bohm, A., Mechanism of poly(A) polymerase: structure of the enzyme-MgATP-RNA ternary complex and kinetic analysis, *Structure* **15**, 1117–2232 (2007).

Barash, Y., Calarco, J.A., Gao, W., Pan, Q., Wang, X., Shai, O., Blencoe, B.J., and Frey, B.J., Deciphering the splicing code, *Nature* **465**, 53–59 (2010).

Bartel, D.P., MicroRNAs: target recognition and regulatory function, *Cell* **136**, 215–233 (2009); *and* MicroRNAs: genomics, biogenesis, mechanism, and function, *Cell* **116**, 281–297 (2004).

Bass, B.L., RNA editing by adenosine deaminases that act on RNA, *Annu. Rev. Biochem.* **71**, 817–846 (2002).

Blencowe, B.J., Alternative splicing: new insights from global analyses, *Cell* **126**, 37–47 (2006).

Boehringer, D., Makarov, E.M., Sander, B., Makarova, O.V., Kastner, B., Lührmann, R., and Stark H., Three-dimensional structure of pre-catalytic human splicesomal complex B, *Nature Struct. Biol.* **11**, 463–468 (2004).

Brantl, S., Antisense regulation and RNA interference, *Biochim. Biophys. Acta* **1575**, 15–25 (2002).

Cheah, M.T., Wachter, A., Sudarsan, N., and Breaker, R.R., Control of alternative RNA splicing and gene expression by eukaryotic riboswitches, *Nature* **447**, 497–500 (2007).

Chen, M. and Manley, J.L., Mechanisms of alternative splicing regulation: insights from molecular and genomics approaches, *Nature Rev. Mol. Cell Biol.* **10**, 741–754 (2009).

Cole, C.N. and Scarcelli, J.J., Transport of messenger RNA from the nucleus to the cytoplasm, *Curr. Opin. Cell. Biol.* **18**, 299–306 (2006).

Davis, R.E., Spliced leader RNA *trans*-splicing in metazoa, *Parasitology Today* **12**, 33–40 (1996).

Decatur, W.A. and Fournier, M.J., RNA-guided nucleotide modification of ribosomal and other RNAs, *J. Biol. Chem.* **278**, 695–698 (2003).

Evans, D., Marquez, S.M., and Pace, N.R., RNase P: interface of the RNA and protein worlds, *Trends Biochem. Sci.* **31**, 333–341 (2006).

Fedor, M.J., Comparative enzymology and structural biology of RNA self-cleavage, *Annu. Rev. Biophys.* **38**, 271–299 (2009).

Filipowicz, W., Bhattacharyya, S.N., and Sonenberg, N., Mechanisms of post-transcriptional regulation by microRNAs: are the answers in sight? *Nature Rev. Genet.* **9**, 102–114 (2008).

Gannan, F., O'Hare, K., Perrin, F., LePennec, J.P., Benoist, C., Cochet, M., Breathnach, R., Royal, A., Garapin, A., Cami, B., and Chambon, P., Organization and sequences of the 5′ end of a cloned complete ovalbumin gene, *Nature* **278**, 428–434 (1979).

Garneau, N.L., Wilusz, J., and Wilusz, C.J., The highways and byways of mRNA decay, *Nature Rev. Mol. Cell Biol.* **8**, 113–126 (2007).

Gerber, A.P. and Keller, W., RNA editing by base deamination: more enzymes, more targets, new mysteries, *Trends Biochem. Sci.* **26**, 376–384 (2001).

Glisovic, T., Bachorik, J.L., Yong, J., and Dreyfuss, G., RNA-binding proteins and post-transcriptional gene regulation, *FEBS Lett.* **582**, 1977–1986 (2008).

Gott, J.M. and Emeson, R.B., Functions and mechanisms of RNA editing, *Annu. Rev. Genet.* **34**, 499–531 (2000).

Gu, M. and Lima, C.D., Processing the message: structural insights into capping and decapping mRNA, **15**, 99–106 (2005).

Guo, F., Gooding, A.R., and Cech, T.R., Structure of the *Tetrahymena* ribozyme: base triple sandwich and metal ion at the active site, *Mol. Cell* **16**, 351–362 (2004).

Jinek, M. and Doudna, J.A., A three-dimensional view of the molecular machinery of RNA interference, *Nature* **457**, 405–412 (2009).

Kambach, C., Walke, S., Young, R., Avis, J.M., de la Fortelle, E., Raker, V.A., Lührmann, R., and Nagai, K., Crystal structures of two Sm protein complexes and their implications for the assembly of the spliceosomal snRNPs, *Cell* **96**, 375–387 (1999).

Kawamata, T. and Tomari, Y., Making RISC, *Trends Biochem. Sci.* **35**, 368–376 (2010).

Keegan, L.P., Gallo, A., and O'Connell, M.A., The many roles of an RNA editor, *Nature Rev. Genet.* **2**, 869–878 (2001).

Krummel, D.A.P., Oubridge, C., Leung, A.K.W., Li, J., and Nagai, K., Crystal structure of human spliceosomal U1 snRNP at 5.5 Å resolution, *Nature* **458**, 475–480 (2009).

Li, Y. and Breaker, R.R., Deoxyribozymes: new players in an ancient game of biocatalysis, *Curr. Opin. Struct. Biol.* **9**, 315–323 (1999).

Liu, Q., Greimann, J.C. and Lima, C.D., Reconstitution, activities, and structure of the eukaryotic RNA exosome, *Cell* **127**, 1223–1237 (2006); *and* Errata, *Cell* **131**, 188-190 (2007).

Maas, S., Rich, A., and Nishikura, K., A-to-I RNA editing: Recent news and residual mysteries, *J. Biol. Chem.* **278**, 1391–1394 (2003); *and* Blanc, V. and Davidson, N.O., C-to-U RNA editing: Mechanisms leading to genetic diversity, *J. Biol. Chem.* **278**, 1395–1398 (2003).

MacRae, I.J., Zhou, K., Li, F., Repic, A., Brooks, A.N., Cande, W.Z., Adams, P.D., and Doudna, J.A., Structral basis for double-stranded RNA processing by Dicer, *Science* **311**, 195–198 (2006).

Madison-Antenucci, S., Grams, J., and Hajduk, S.L., Editing machines: the complexities of trypanosome editing, *Cell* **108**, 435–438 (2002).

Maniatis, T. and Tasic, B., Alternative pre-mRNA splicing and proteome expansion in metazoans, *Nature* **418**, 236–243 (2002).

Martick, M. and Scott, W.G., Tertiary contacts distant from the active site prime a ribozyme for catalysis, *Cell* **126**, 309–320 (2006). [The X-ray structure of *S. mansoni* hammerhead ribozyme.]

Mura, C., Cascio, D., Sawaya, M.R., and Eisenberg, D.S., The crystal structure of a heptameric archaeal Sm protein: Implication for the eukaryotic snRNP core, *Proc. Natl. Acad. Sci.* **98**, 5532–5537 (2001).

Nishikura, K., Functions and regulation of RNA editing by ADAR deaminases, *Annu. Rev. Biochem.* **79**, 321–349 (2010); *and* Hundley, H.A. and Bass, B.L., ADAR editing in double-stranded UTRs and other noncoding sequences, *Trends Biochem. Sci.* **35**, 377–383 (2010).

Pratt, A.J. and MacRae, I.J., The RNA-induced silencing complex: a versatile gene-silencing machine, *J. Biol. Chem.* **284**, 17897–17901 (2009).

Proudfoot, N.J., Furger, A., and Dye, M.J., Integrating mRNA processing with transcription, *Cell* **108**, 501–512 (2002).

Schmid, M. and Jensen, T.H., The exosome: a multipurpose RNA-decay machine, *Trends Biochem. Sci.* **33**, 501–510 (2008).

Serganov, A. and Patel, D.J., Ribozymes, riboswitches and beyond: regulation of gene expression without proteins, *Nature Rev. Genet.* **8**, 776–790 (2007).

Sharp, P.A., Split genes and RNA splicing, *Cell* **77**, 805–815 (1994).

Smith, H.C. (Ed.), *RNA and DNA Editing*, Wiley (2008).

Stahley, M.R. and Strobel, S.A., RNA splicing: group I crystal structures reveal the basis of splice site selection and metal ion catalysis, *Curr. Opin. Struct. Biol.* **16**, 319–326 (2006).

Stark, H. and Lührmann, R., Cryo-electron microscopy of spliceosomal components, *Annu. Rev. Biophys. Biomol. Struct.* **35**, 435–457 (2006).

Stuart, K.D., Schnaufer, A., Ernst, N.E., and Panigrahi, A.K., Complex management: RNA editing in trypanosomes, *Trends Biochem. Sci.* **30**, 97–105 (2005).

Tanaka Hall, T.M., Poly(A) tail synthesis and regulation: recent structural insights, *Curr. Opin. Struct. Biol.* **12**, 82–88 (2002).

Tarn, W.-Y. and Steitz, J.A., Pre-mRNA splicing: the discovery of a new spliceosome doubles the challenge, *Trends Biochem. Sci.* **22**, 132–137 (1997).

Toor, N., Keating, K.S., and Pyle, A.M., Structural insights into RNA splicing, *Curr. Opin. Struct. Biol.* **19**, 260–266 (2009).

Torres-Larios, A., Swinger, K.K., Krasilnikov, A.S., Pan, T., and Mondragón, A., Crystal structure of the RNA component of bacterial ribonuclease P, *Nature* **437**, 584–587 (2005); *and* Kazantsev, A.V., Krivenko, A.A., Harrington, D.J., Holbrook, S.R., Adams, P.D., and Pace, N.R., Crystal structure of a bacterial ribonuclease P RNA, *Proc. Natl. Acad. Sci.* **102**, 13392–13397 (2005).

Vicens, Q. and Cech, T.A., Atomic level architecture of group I introns revealed, *Trends Biochem. Sci.* **31**, 41–51 (2006).

Wahl, M.C., Will, C.L., and Lührmann, R., The spliceosome: design principles in a dynamic RNP machine, *Cell* **136**, 701–718 (2009). [A detailed review.]

Wang, Y., Juranek, S., Li, H., Sheng, G., Wardle, G.S., Tuschl, T., and Patel, D.J., Nucleation, propagation and cleavage of target RNAs in Ago silencing complexes, *Nature* **461**, 754–762 (2009). [X-ray structures of Argonaute–DNA · RNA complexes.]

Weinstein, L.B. and Steitz, J.A., Guided tours: from precursor to snoRNA to functional snoRNP, *Curr. Opin. Cell Biol.* **11**, 378–384 (1999).

Xiao, S., Scott, F., Fierke, C.A., and Enelke, D.R., Eukaryotic ribonuclease P: A plurality of ribonucleoprotein enzymes, *Annu. Rev. Biochem.* **71**, 165–189 (2002).

BIOINFORMATICS EXERCISES

Bioinformatics Exercises are available at www.wiley.com/college/voet.

Chapter 31

New Developments in RNA. Explore RNA web sites.

PROBLEMS

1. Indicate the phenotypes of the following *E. coli lac* partial diploids in terms of inducibility and active enzymes synthesized.

   a. $I^-P^+O^+Z^+Y^-/I^+P^-O^-Z^+Y^+$

   b. $I^-P^+O^cZ^+Y^-/I^+P^+O^+Z^-Y^+$

   c. $I^-P^+O^cZ^+Y^+/I^-P^+O^+Z^+Y^+$

   d. $I^+P^-O^cZ^+Y^+/I^-P^+O^cZ^-Y^-$

2. Superrepressed mutants, I^S, encode *lac* repressors that bind operator but do not respond to the presence of inducer. Indicate the phenotypes of the following genotypes in terms of inducibility and enzyme production.

   a. $I^SO^+Z^+$ b. $I^SO^cZ^+$ c. $I^+O^+Z^+/I^SO^+Z^+$

3. Why do *lacZ⁻ E. coli* fail to show galactoside permease activity after the addition of lactose in the absence of glucose? Why do *lac Y⁻* mutants lack β-galactosidase activity under the same conditions?

4. What is the experimental advantage of using IPTG instead of 1,6-allolactose as an inducer of the *lac* operon?

5. Describe the probable genetic defect that abolishes the sensitivity of the *lac* operon to the absence of glucose when other metabolic operons continue to be sensitive to the absence of glucose.

6. Indicate the −10 region, the −35 region, and the initiating nucleotide on the sense strand of the *E. coli* tRNA^Tyr promoter shown below.

5′ CAACGTAACACTTTACAGCGGCGCGTCATTTGATATGATGCGCCCCGCTTCCCGATA 3′

3′ GTTGCATTGTGAAATGTCGCCGCGCAGTAAACTATACTACGCGGGGCGAAGGGCTAT 5′

7. Why are *E. coli* that are diploid for rifamycin resistance and rifamycin sensitivity (*rif*^R/*rif*^S) sensitive to rifamycin?

8. Why does promoter efficiency tend to decrease with the number of G · C base pairs in the −10 region of a prokaryotic gene?

9. A eukaryotic ribosome contains 4 different rRNA molecules and ~82 different proteins. Why does a cell contain many more copies of the rRNA genes than the ribosomal protein genes?

10. What is the probability that the 4026-nucleotide DNA sequence coding for the β subunit of *E. coli* RNA polymerase will be transcribed with the correct base sequence? Perform the calculations for the probabilities of 0.0001, 0.001, and 0.01 that each base is incorrectly transcribed.

11. If an enhancer is placed on one plasmid and its corresponding promoter is placed on a second plasmid that is catenated (linked) with the first, initiation is almost as efficient as when the enhancer and promoter are on the same plasmid. However, initiation does not occur when the two plasmids are unlinked. Explain.

12. What is the probability that the symmetry of the *lac* operator is merely accidental?

13. Why does the inhibition of DNA gyrase in *E. coli* inhibit the expression of catabolite-sensitive operons?

14. Describe the transcription of the *trp* operon in the absence of active ribosomes and tryptophan.

15. Why can't eukaryotic transcription be regulated by attenuation?

16. Predict the effect of deleting the leader peptide sequence on regulation of the *trp* operon.

17. Charles Yanofsky and his associates have synthesized a 15-nucleotide RNA that is complementary to segment 1 of *trpL* mRNA (but only partially complementary to segment 3). What is its effect on the *in vitro* transcription of the *trp* operon? What is its effect if the *trpL* gene contains a mutation in segment 2 that destabilizes the 2 · 3 stem and loop?

18. Why are *relA*^− mutants defective in the *in vivo* transcription of *his* and *trp* operons?

19. Why aren't primary rRNA transcripts present in wild-type *E. coli*?

20. Why can't hammerhead ribozymes catalyze the cleavage of ssDNA?

21. Explain why the active site of poly(A) polymerase is much narrower than that of DNA and RNA polymerases.

22. Would you expect spliceosome-catalyzed intron removal to be reversible in a highly purified *in vitro* system and *in vivo*? Explain.

23. Introns in eukaryotic protein-coding genes may be quite large, but almost none are smaller than about 65 bp. What is the reason for this minimum intron size?

24. Infection with certain viruses inhibits snRNA processing in eukaryotic cells. Explain why this favors the expression of viral genes in the host cell.

25. Explain why RNAi would be a less efficient mechanism for regulating the expression of specific genes if Dicer hydrolyzed double-stranded RNA every 11 bp rather than every 22 bp.

P-site tRNA
50S
mRNA
30S
E-site tRNA
Polypeptide tunnel

Translation

CHAPTER 32

In this chapter we consider **translation,** the mRNA-directed biosynthesis of polypeptides. Although peptide bond formation is a relatively simple chemical reaction, the complexity of the translational process, which involves the coordinated participation of over 100 macromolecules, is mandated by the need to link 20 different amino acid residues accurately in the order specified by a particular mRNA. Note that we previewed this process in Section 5-4B.

We begin by discussing the **genetic code,** the correspondence between nucleic acid sequences and polypeptide sequences. Next, we examine the structures and properties of **tRNAs,** the amino acid–bearing entities that mediate the translation process. Following this, we consider the structure and functions of **ribosomes,** the complex molecular machines that catalyze peptide bond formation between the mRNA-specified amino acids. Peptide bond formation, however, does not necessarily yield a functional protein; many polypeptides must first be post-translationally modified as we discuss in the subsequent section. Finally, we study how cells degrade proteins, a process that must balance protein synthesis.

1 THE GENETIC CODE

How does DNA encode genetic information? According to the one gene–one polypeptide hypothesis, the genetic message dictates the amino acid sequences of proteins. Since the base sequence of DNA is the only variable element in this otherwise monotonously repeating polymer, the amino acid sequence of a protein must somehow be specified by the base sequence of the corresponding segment of DNA.

A DNA base sequence might specify an amino acid sequence in many conceivable ways. With only 4 bases to code for 20 amino acids, a group of several bases, termed a **codon,** is necessary to specify a single amino acid. A triplet code, that is, one with 3 bases per codon, is minimally required since there are $4^3 = 64$ different triplets of bases, whereas there can be only $4^2 = 16$ different doublets, which is insufficient to specify all the amino acids. In a triplet code, as many as 44 codons might not code for amino acids. On the other hand, many amino acids could be specified by more than one codon. Such a code, in a term borrowed from mathematics, is said to be **degenerate.**

Another mystery was, how does the polypeptide synthesizing apparatus group DNA's continuous sequence of bases into codons? For example, the code might be overlapping; that is, in the sequence

$$ABCDEFGHIJ\cdots$$

ABC might code for one amino acid, BCD for a second, CDE for a third, and so on. Alternatively, the code might be nonoverlapping, so that ABC specifies one amino acid, DEF a second, GHI a third, and so on. The code might also contain internal "punctuation" such as in the nonoverlapping triplet code

$$ABC,DEF,GHI,\cdots$$

5-Bromouracil (5BU)   **5BU**
  (keto tautomer)   **(enol tautomer)**   **Guanine**

Figure 32-1   5-Bromouracil. Its keto form (*left*) is its most common tautomer. However, it frequently assumes the enol form (*right*), which base pairs with guanine.

in which the commas represent particular bases or base sequences. A related question is, how does the genetic code specify the beginning and the end of a polypeptide chain?

The genetic code is, in fact, a nonoverlapping, comma-free, degenerate, triplet code. How this was determined and how the genetic code dictionary was elucidated are the subjects of this section.

A. *Chemical Mutagenesis*

The triplet character of the genetic code, as we shall see below, was established through the use of **chemical mutagens,** substances that chemically induce mutations. We therefore precede our study of the genetic code with a discussion of these substances. There are two major classes of mutations:

1. Point mutations, in which one base pair replaces another. These are subclassified as

(a) Transitions, in which one purine (or pyrimidine) is replaced by another.
(b) Transversions, in which a purine is replaced by a pyrimidine or vice versa.

2. Insertion/deletion mutations, in which one or more nucleotide pairs are inserted in or deleted from DNA.

A mutation in any of these three categories may be reversed by a subsequent mutation of the same but not another category.

a. Point Mutations Are Generated by Altered Bases

Point mutations can result from the treatment of an organism with base analogs or with substances that chemically alter bases. For example, the base analog **5-bromouracil (5BU)** sterically resembles thymine (5-methyluracil) but, through the influence of its electronegative Br atom, frequently assumes a tautomeric form that base pairs with guanine instead of adenine (Fig. 32-1). Consequently, when 5BU is incorporated into DNA in place of thymine, as it usually is, it occasionally induces an $A \cdot T \rightarrow G \cdot C$ transition in subsequent rounds of DNA replication. Occasionally, 5BU is also incorporated into DNA in place of cytosine, which instead generates a $G \cdot C \rightarrow A \cdot T$ transition.

The adenine analog **2-aminopurine (2AP),** normally base pairs with thymine (Fig. 32-2*a*) but occasionally forms an undistorted but singly hydrogen bonded base pair with cytosine (Fig. 32-2*b*). Thus 2AP generates $A \cdot T \rightarrow G \cdot C$ transitions.

In aqueous solutions, **nitrous acid** (HNO_2) oxidatively deaminates aromatic primary amines so that it converts cytosine to uracil (Fig. 32-3*a*) and adenine to the guaninelike **hypoxanthine** (which forms two of guanine's three hydrogen

2-Aminopurine (2AP)   **Thymine**

2AP   **Cytosine**

Figure 32-2   Base pairing by the adenine analog 2-aminopurine. It normally base pairs with thymine (*a*) but occasionally also does so with cytosine (*b*).

Cytosine   **Uracil**   **Adenine**

Adenine   **Hypoxanthine**   **Cytosine**

Figure 32-3   Oxidative deamination by nitrous acid. (*a*) Cytosine is converted to uracil, which base pairs with adenine. (*b*) Adenine is converted to hypoxanthine, a guanine derivative (it lacks guanine's 2-amino group) that base pairs with cytosine.

Figure 32-4   **Reaction with hydroxylamine converts cytosine to a derivative that base pairs with adenine.**

bonds with cytosine; Fig. 32-3*b*). Hence, treatment of DNA with nitrous acid, or compounds such as **nitrosamines**

Nitrosamines

that react to form nitrous acid, results in both A · T → G · C and G · C → A · T transitions. **Hydroxylamine** (NH_2OH) also induces G · C → A · T transitions by specifically reacting with cytosine to convert it to a compound that base pairs with adenine (Fig. 32-4).

   Nitrite, the conjugate base of nitrous acid, has long been used as a preservative of prepared meats such as frankfurters. However, the observation that many mutagens are also carcinogens (Section 30-5Fa) suggests that the consumption of nitrite-containing meat is harmful to humans. Proponents of nitrite preservation nevertheless argue that to stop it would result in far more fatalities. This is because lack of such treatment would greatly increase the incidence of **botulism,** an often fatal form of food poisoning caused by the ingestion of protein neurotoxins secreted by the anaerobic bacterium *Clostridium botulinum* (Section 12-4Dd).

   The use of alkylating agents such as dimethyl sulfate, **nitrogen mustard,** and **ethylnitrosourea**

Nitrogen mustard   **Ethylnitrosourea**

often generates transversions. The alkylation of the N7 position of a purine nucleotide causes its subsequent depurination. The resulting gap in the sequence is filled in by an error-prone repair system (Section 30-5D). Transversions arise when the missing purine is replaced by a pyrimidine. The repair of DNA that has been damaged by UV radiation may also generate transversions.

b. Insertion/Deletion Mutations Are Generated by Intercalating Agents

*Insertion/deletion mutations (also called **indels**), may arise from the treatment of DNA with intercalating agents such as acridine orange (Section 6-6Ca) or **proflavin.***

Proflavin

The distance between two consecutive base pairs is doubled by the intercalation of such a molecule between them. The replication of such a distorted DNA occasionally results in the insertion or deletion of one or more nucleotides in the newly synthesized polynucleotide. (Insertions and deletions of large DNA segments generally arise from aberrant crossover events; Section 34-2De.)

B. Codons Are Triplets

In 1961, Francis Crick and Sydney Brenner, through genetic investigations into the previously unknown character of proflavin-induced mutations, determined the triplet character of the genetic code. In bacteriophage T4, a particular proflavin-induced mutation, designated *FC*0, maps in the *rIIB* cistron (Section 1-4Eb). The growth of this mutant phage on a permissive host (*E. coli* B) resulted in the occasional spontaneous appearance of phenotypically wild-type phages as was demonstrated by their ability to grow on a restrictive host [*E. coli* K12(λ); recall that *rIIB* mutants form characteristically large plaques on *E. coli* B but cannot lyse *E. coli* K12(λ); Section 1-4Eb]. Yet, these doubly mutated phages are not genotypically wild type; the simultaneous infection of a permissive host by one of them and true wild-type phage yielded recombinant progeny that have either the *FC*0 mutation or a new mutation designated *FC*1. Thus the phenotypically wild-type phage is a double mutant that actually contains both *FC*0 and *FC*1. *These two genes are therefore **suppressors** of one another; that is, they cancel each other's mutant properties.* Furthermore, since they map together in the *rIIB* cistron, they are mutual **intragenic suppressors** (suppressors in the same gene).

   The treatment of *FC*1 in a manner identical to that described for *FC*0 provided similar results: the appearance of a new mutant, *FC*2, that is an intragenic suppressor of *FC*1. By proceeding in this iterative manner, Crick and Brenner collected a series of different *rIIB* mutants, *FC*3, *FC*4, *FC*5, etc., in which each mutant *FC*(*n*) is an intragenic suppressor of its predecessor, *FC*(*n* − 1). Recombination studies showed, moreover, that odd-numbered mutations are intragenic suppressors of even-numbered mutations, but neither pairs of different odd-numbered mutations nor pairs of different even-numbered mutations suppress each other. However, recombinants containing three odd-numbered mutations or three even-numbered mutations all are phenotypically wild type.

   Crick and Brenner accounted for these observations by the following set of hypotheses:

   1. The proflavin-induced mutation *FC*0 is either an insertion or a deletion of one nucleotide pair from the *rIIB* cistron. If it is a deletion then *FC*1 is an insertion, *FC*2 is a deletion, and so on, and vice versa.

2. *The code is read in a sequential manner starting from a fixed point in the gene.* The insertion or deletion of a nucleotide shifts the **frame** (grouping) in which succeeding nucleotides are read as codons (insertions or deletions of nucleotides are therefore also known as **frameshift mutations**). Thus the code has no internal punctuation that indicates the reading frame; that is, *the code is comma free.*

3. *The code is a triplet code.*

4. All or nearly all of the 64 triplet codons code for an amino acid; that is, *the code is degenerate.*

These principles are illustrated by the following analogy. Consider a sentence (gene) in which the words (codons) each consist of three letters (bases).

THE BIG RED FOX ATE THE EGG

(Here the spaces separating the words have no physical significance; they are only present to indicate the reading frame.) The deletion of the fourth letter, which shifts the reading frame, changes the sentence to

THE IGR EDF OXA TET HEE GG

so that all words past the point of deletion are unintelligible (specify the wrong amino acids). An insertion of any letter, however, say an X in the ninth position,

THE IGR EDX FOX ATE THE EGG

restores the original reading frame. Consequently, only the words between the two changes (mutations) are altered. As in this example, such a sentence might still be intelligible (the gene could still specify a functional protein), particularly if the changes are close together. Two deletions or two insertions, no matter how close together, would not suppress each other but just shift the reading frame. However, three insertions, say X, Y, and Z in the fifth, eighth, and twelfth positions, respectively, would change the sentence to

THE BXI GYR EDZ FOX ATE THE EGG

which, after the third insertion, restores the original reading frame. The same would be true of three deletions. As before, if all three changes were close together, the sentence might still retain much of its meaning.

Crick and Brenner did not unambiguously demonstrate that the genetic code is a triplet code because they had no proof that their insertions and deletions involved only single nucleotides. Strictly speaking, they showed that a codon consists of $3r$ nucleotides where r is the number of nucleotides in an insertion or deletion. Although it was generally assumed at the time that $r = 1$, proof of this assertion had to await the elucidation of the genetic code (Section 32-1C).

C. Deciphering the Genetic Code

The genetic code could, in principle, be determined by simply comparing the base sequence of an mRNA with the amino acid sequence of the polypeptide it specifies. In the 1960s, however, techniques for isolating and sequencing mRNAs had not yet been developed. The elucidation of the genetic code dictionary therefore proved to be a difficult task.

a. UUU Specifies Phe

The major breakthrough in deciphering the genetic code came in 1961 when Marshall Nirenberg and Heinrich Matthaei established that UUU is the codon specifying Phe. They did so by demonstrating that the addition of poly(U) to a cell-free protein synthesizing system stimulates only the synthesis of poly(Phe). The cell-free protein synthesizing system was prepared by gently breaking open *E. coli* cells by grinding them with powdered alumina and centrifuging the resulting cell sap to remove the cell walls and membranes. This extract contained DNA, mRNA, ribosomes, enzymes, and other cell constituents necessary for protein synthesis. When fortified with ATP, GTP, and amino acids, the system synthesized small amounts of proteins. This was demonstrated by the incubation of the system with ^{14}C-labeled amino acids followed by the precipitation of its proteins by the addition of trichloroacetic acid. The precipitate proved to be radioactive.

A cell-free protein synthesizing system, of course, produces proteins specified by the cell's DNA. On addition of DNase, however, protein synthesis stops within a few minutes because the system can no longer synthesize mRNA, whereas the mRNA originally present is rapidly degraded. Nirenberg found that crude mRNA-containing fractions from other organisms were highly active in stimulating protein synthesis in a DNase-treated protein synthesizing system. This system is likewise responsive to synthetic mRNAs.

The synthetic mRNAs that Nirenberg used in subsequent experiments were synthesized by the *Azotobacter vinelandii* enzyme **polynucleotide phosphorylase.** This enzyme, which was discovered by Severo Ochoa and Marianne Grunberg-Manago, links together nucleotides in the reaction

$$(RNA)_n + NDP \rightleftharpoons (RNA)_{n+1} + P_i.$$

In contrast to RNA polymerase, however, polynucleotide phosphorylase does not utilize a template. Rather, it randomly links together the available NDPs so that the base composition of the product RNA reflects that of the reactant NDP mixture.

Nirenberg and Matthaei demonstrated that poly(U) stimulates the synthesis of poly(Phe) by incubating poly(U) and a mixture of 1 radioactive and 19 unlabeled amino acids in a DNase-treated protein synthesizing system. Significant radioactivity appeared in the protein precipitate only when phenylalanine was labeled. *UUU must therefore be the codon specifying Phe.* In similar experiments using poly(A) and poly(C), it was found that poly(Lys) and poly(Pro), respectively, were synthesized. Thus *AAA specifies Lys and CCC specifies Pro.* [Poly(G) cannot function as a synthetic mRNA because, even under denaturing conditions, it aggregates to form a four-stranded helix (Section 30-4De). An mRNA must be single stranded to direct its translation; Section 32-2D.]

Table 32-1 **Amino Acid Incorporation Stimulated by a Random Copolymer of U and G in Mole Ratio 0.76 : 0.24**

Codon	Probability of Occurrence	Relative Incidence[a]	Amino Acid	Relative Amount of Amino Acid Incorporated
UUU	0.44	100	Phe	100
UUG	0.14	32	Leu	36
UGU	0.14	32	Cys	35
GUU	0.14	32	Val	37
UGG	0.04	9	Trp	14
GUG	0.04	9		
GGU	0.04	9	Gly	12
GGG	0.01	2		

[a]Relative incidence is defined here as 100 × probability of occurrence/0.44.

Source: Matthaei, J.H., Jones, O.W., Martin, R.G., and Nirenberg, M., *Proc. Natl. Acad. Sci.* **48,** 666 (1962).

Nirenberg and Ochoa independently employed ribonucleotide copolymers to further elucidate the genetic code. For example, in a poly(UG) composed of 76% U and 24% G, the probability of a given triplet being UUU is 0.76 × 0.76 × 0.76 = 0.44. Likewise, the probability of a particular triplet consisting of 2U's and 1G, that is, UUG, UGU, or GUU, is 0.76 × 0.76 × 0.24 = 0.14. The use of this poly(UG) as an mRNA therefore indicated the base compositions, but not the sequences, of the codons specifying several amino acids (Table 32-1). Through the use of copolymers containing two, three, and four bases, the base compositions of codons specifying each of the 20 amino acids were inferred. Moreover, *these experiments demonstrated that the genetic code is degenerate since, for example, poly(UA), poly(UC), and poly(UG) all direct the incorporation of Leu into a polypeptide.*

b. The Genetic Code Was Elucidated through Triplet Binding Assays and the Use of Polyribonucleotides with Known Sequences

In the absence of GTP, which is necessary for protein synthesis, trinucleotides but not dinucleotides are almost as effective as mRNAs in promoting the ribosomal binding of specific tRNAs. This phenomenon, which Nirenberg and Philip Leder discovered in 1964, permitted the various codons to be identified by a simple binding assay. Ribosomes, together with their bound tRNAs, are retained by a nitrocellulose filter but free tRNA is not. The bound tRNA was identified by using charged tRNA mixtures in which only one of the pendent amino acid residues was radioactively labeled. For instance, it was found, as expected, that UUU stimulates the ribosomal binding of only Phe tRNA. Likewise, UUG, UGU, and GUU stimulate the binding of Leu, Cys, and Val tRNAs, respectively. Hence

UUG, UGU, and GUU must be codons that specify Leu, Cys, and Val, respectively. In this way, the amino acids specified by some 50 codons were identified. For the remaining codons, the binding assay was either negative (no tRNA bound) or ambiguous.

The genetic code dictionary was completed and previous results confirmed through H. Gobind Khorana's synthesis of polyribonucleotides with specified repeating sequences (Section 7-6A). In a cell-free protein synthesizing system, UCUCUCUC···, for example, is read

UCU CUC UCU CUC UCU C···

so that it specifies a polypeptide chain of two alternating amino acid residues. In fact, it was observed that this mRNA stimulated the production of

Ser — Leu — Ser — Leu — Ser — Leu — ···

which indicates that either UCU or CUC specifies Ser and the other specifies Leu. This information, together with the tRNA-binding data, permitted the conclusion that UCU codes for Ser and CUC codes for Leu. These data also prove that codons consist of an odd number of nucleotides, thereby relieving any residual suspicions that codons consist of six rather than three nucleotides.

Alternating sequences of three nucleotides, such as poly(UAC), specify three different homopolypeptides because ribosomes may initiate polypeptide synthesis on these synthetic mRNAs in any of the three possible reading frames (Fig. 32-5). Analyses of the polypeptides specified by various alternating sequences of two and three nucleotides confirmed the identity of many codons and filled out missing portions of the genetic code.

c. mRNAs Are Read in the 5′ → 3′ Direction

The use of repeating tetranucleotides indicated the reading direction of the code and identified the chain

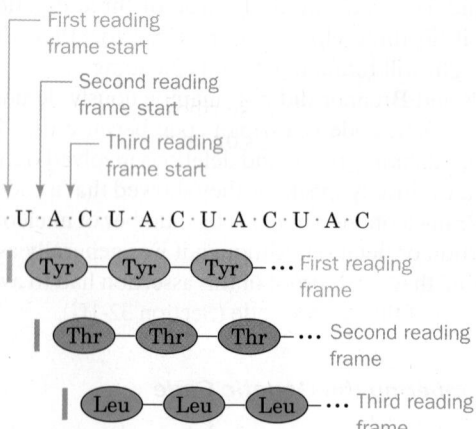

Figure 32-5 **The three potential reading frames of an mRNA.** Each reading frame would yield a different polypeptide.

termination codons. Poly(UAUC) specifies, as expected, a polypeptide with a tetrapeptide repeat:

$$5'\ \text{UAU CUA UCU AUC UAU CUA} \cdots 3'$$

$$\text{Tyr — Leu — Ser — Ile — Tyr — Leu — } \cdots$$

The amino acid sequence of this polypeptide indicates that the mRNA's 5′ end corresponds to the polypeptide's N-terminus; that is, *mRNA is read in the 5′ → 3′ direction.*

d. UAG, UAA, and UGA Are Stop Codons

In contrast to the above results, poly(AUAG) yields only dipeptides and tripeptides. This is because *UAG is a signal to the ribosome to terminate protein synthesis:*

$$\text{AUA GAU AGA UAG AUA GAU} \cdots$$

$$\text{Ile — Asp — Arg\quad Stop\quad Ile — Asp — } \cdots$$

Likewise, poly(GUAA) yields dipeptides and tripeptides because UAA is also a chain termination signal:

$$\text{GUA AGU AAG UAA GUA AGU} \cdots$$

$$\text{Val — Ser — Lys\quad Stop\quad Val — Ser — } \cdots$$

UGA is a third stop signal. These **Stop codons,** whose existence was first inferred from genetic experiments, are known, somewhat inappropriately, as **nonsense codons** because they are the only codons that do not specify amino acids. UAG, UAA, and UGA are sometimes referred to as *amber, ochre,* and *opal* codons. [They were so named as the result of a laboratory joke: The German word for amber is Bernstein, the name of an individual who helped discover *amber* mutations (mutations that change some other codon to UAG); *ochre* and *opal* are puns on *amber.*]

e. AUG and GUG Are Chain Initiation Codons

The codons AUG, and less frequently GUG, form part of the chain initiation sequence (Section 32-3Ca). However, they also specify the amino acid residues Met and Val, respectively, at internal positions of polypeptide chains. (Nirenberg and Matthaei's discovery that UUU specifies Phe was only possible because ribosomes indiscriminately initiate polypeptide synthesis on an mRNA when the Mg^{2+} concentration is unphysiologically high as it was, serendipitously, in their experiments.)

D. *The Nature of the Code*

The genetic code dictionary, as elucidated by the above methods, is presented in Table 32-2 as well as in Table 5-3. Examination of the table indicates that the genetic code has several remarkable features:

1. *The code is highly degenerate.* Three amino acids, Arg, Leu, and Ser, are each specified by six codons, and

Table 32-2 **The "Standard" Genetic Code**[a]

First position (5′ end)	Second position				Third position (3′ end)
	U	**C**	**A**	**G**	
U	UUU Phe / UUC	UCU / UCC Ser	UAU Tyr / UAC	UGU Cys / UGC	U / C
	UUA Leu / UUG	UCA / UCG	UAA Stop / UAG Stop	UGA Stop / UGG Trp	A / G
C	CUU / CUC Leu	CCU / CCC Pro	CAU His / CAC	CGU / CGC Arg	U / C
	CUA / CUG	CCA / CCG	CAA Gln / CAG	CGA / CGG	A / G
A	AUU / AUC Ile	ACU / ACC Thr	AAU Asn / AAC	AGU Ser / AGC	U / C
	AUA / AUG Met[b]	ACA / ACG	AAA Lys / AAG	AGA Arg / AGG	A / G
G	GUU / GUC Val	GCU / GCC Ala	GAU Asp / GAC	GGU / GGC Gly	U / C
	GUA / GUG	GCA / GCG	GAA Glu / GAG	GGA / GGG	A / G

[a]Nonpolar amino acid residues are tan, basic residues are blue, acidic residues are red, and nonpolar uucharged residues are purple.
[b]AUG forms part of the initiation signal as well as coding for internal Met residues.

most of the rest are specified by either four, three, or two codons. Only Met and Trp, two of the least common amino acids in proteins (Table 4-1), are represented by a single codon. Codons that specify the same amino acid are termed **synonyms.**

2. *The arrangement of the code table is nonrandom.* Most synonyms occupy the same box in Table 32-2; that is, they differ only in their third nucleotide. The only exceptions are Arg, Leu, and Ser, which, having six codons each, must occupy more than one box. XYU and XYC always specify the same amino acid; XYA and XYG do so in all but two cases. Moreover, changes in the first codon position tend to specify similar (if not the same) amino acids, whereas codons with second position pyrimidines encode mostly hydrophobic amino acids (tan in Table 32-2), and those with second position purines encode mostly polar amino acids (blue, red, and purple in Table 32-2). Apparently *the code evolved so as to minimize the deleterious effects of mutations.*

Many of the mutations causing amino acid substitutions in a protein can be rationalized, according to the genetic code, as single point mutations. *As a consequence of the genetic code's degeneracy, however, many point mutations at a third codon position are phenotypically silent; that is, the mutated codon specifies the same amino acid as the wild type.* Degeneracy may account for as much as 33% of the 25 to 75% range in the G + C content among the DNAs of different organisms (Section 5-1Ba). The frequent occurrence of Arg, Ala, Gly, and Pro also tends to give a high G + C content, whereas Asn, Ile, Lys, Met, Phe, and Tyr contribute to a low G + C content.

a. Some Phage DNA Segments Contain Overlapping Genes in Different Reading Frames

Since any nucleotide sequence may have three reading frames, it is possible, at least in principle, for a polynucleotide to encode two or even three different polypeptides. This idea was never seriously entertained, however, because it seemed that the constraints on even two overlapping genes in different reading frames would be too great for them to evolve so that both could specify sensible proteins. It therefore came as a great surprise, in 1976, when Frederick Sanger reported that the 5386-nucleotide DNA of bacteriophage φX174 (which, at the time, was the largest DNA to have been sequenced) contains two genes that are completely contained within larger genes of different reading frames (Fig. 32-6). Moreover, the end of the overlapping D and E genes contains the control sequence for the ribosomal initiation of the J gene so that this short DNA segment performs triple duty. Bacteria

also exhibit such coding economy; the ribosomal initiation sequence of one gene in a polycistronic mRNA often overlaps the end of the preceding gene. Nevertheless, completely overlapping genes have only been found in small single-stranded DNA phages, which presumably must make maximal use of the little DNA that they can pack inside their capsids.

b. The "Standard" Genetic Code Is Widespread but Not Universal

For many years it was thought that the "standard" genetic code (that given in Table 32-2) is universal. This assumption was, in part, based on the observations that one kind of organism (e.g., *E. coli*) can accurately translate the genes from quite different organisms (e.g., humans). This phenomenon is, in fact, the basis of genetic engineering. Once the "standard" genetic code had been established, presumably during the time of prebiotic evolution (Section 1-5B), any mutation that would alter the way the code is translated would result in numerous, often deleterious, protein sequence changes. Undoubtedly there is strong selection against such mutations.

Despite the foregoing, DNA sequencing studies in 1981 revealed that *the genetic codes of certain mitochondria (mitochondria contain their own genes and protein synthesizing systems but produce only a few mitochondrial proteins; Section 12-4E) are variants of the "standard" genetic code* (Table 32-3). For example, in mammalian mitochondria, AUA, as well as the standard AUG, is a Met/initiation codon; UGA specifies Trp rather than "Stop"; and AGA and AGG are "Stop" rather than Arg. Note that all mitochondrial genetic codes except those of plants simplify the "standard" code by increasing its degeneracy. For example, in the mammalian mitochondrial code, each amino acid is specified by at least two codons that differ only in their third nucleotide. Apparently the constraints preventing alterations of the genetic code are eased by the small sizes of mitochondrial genomes. More recent

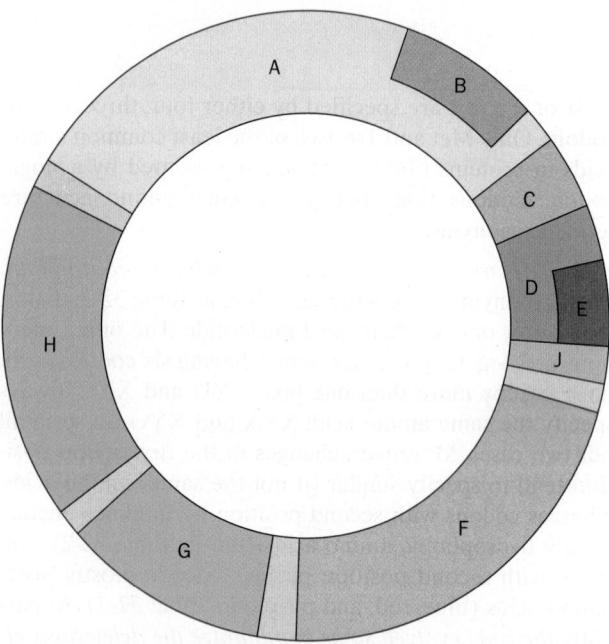

Figure 32-6 **Genetic map of bacteriophage φX174 as determined by DNA sequence analysis.** Genes are labeled A, B, C, etc. Note that gene B is wholly contained within gene A and gene E is wholly contained within gene D. These pairs of genes are read in different reading frames and therefore specify unrelated proteins. The unlabeled regions correspond to untranslated control sequences.

Table 32-3 **Mitochondrial Deviations from the "Standard" Genetic Code**

Mitochondrion	UGA	AUA	CUN[a]	AG$_G^A$	CGG
Mammalian	Trp	Met[b]		Stop	
Baker's yeast	Trp	Met[b]	Thr		
Neurospora crassa	Trp				
Drosophila	Trp	Met[b]		Ser[c]	
Protozoan	Trp				
Plant					Trp
"Standard" code	Stop	Ile	Leu	Arg	Arg

[a]N represents any of the four nucleotides.
[b]Also acts as part of an initiation signal.
[c]AGA only; no AGG codons occur in *Drosophila* mitochondrial DNA.

Source: Mainly Breitenberger, C.A. and RajBhandary, U.L., *Trends Biochem. Sci.* **10**, 481 (1985).

studies, however, have revealed that in ciliated protozoa, the codons UAA and UAG specify Gln rather than "Stop." Perhaps UAA and UAG were sufficiently rare codons in a primordial ciliate (which molecular phylogenetic studies indicate branched off very early in eukaryotic evolution) to permit the code change without unacceptable deleterious effects. At any rate, *the "standard" genetic code, although very widely utilized, is not universal.* Indeed, as we shall see in Section 32-2De, under the proper context in mRNA, certain codons can specify "nonstandard" amino acids.

Figure 32-7   The adaptor hypothesis. It postulates that the genetic code is read by molecules that recognize a particular codon and carry the corresponding amino acid.

2 TRANSFER RNA AND ITS AMINOACYLATION

The establishment of the genetic function of DNA led to the realization that cells somehow "translate" the language of base sequences into the language of polypeptides. Yet, nucleic acids originally appeared unable to bind specific amino acids [more recently RNA aptamers for specific amino acids have been generated; aptamers are nucleic acids that have been selected for their ability to bind specific ligands (Section 7-6C)]. In 1955, Crick, in what became known as the **adaptor hypothesis,** postulated that translation occurs through the mediation of "adaptor" molecules. Each adaptor was postulated to carry a specific enzymatically appended amino acid and to recognize the corresponding codon (Fig. 32-7). Crick suggested that these adaptors contain RNA because codon recognition could then occur by complementary base pairing. At about this time, Paul Zamecnik and Mahlon Hoagland discovered that in the course of protein synthesis, ^{14}C-labeled amino acids became transiently bound to a low molecular mass fraction of RNA. Further investigations indicated that these RNAs, which at first were called "soluble RNA" or "sRNA" but are now known as **transfer RNA (tRNA),** are, in fact, Crick's putative adaptor molecules.

A. *Primary and Secondary Structures of tRNA*

See Guided Exploration 26: The structure of tRNA In 1965, after a 7-year effort, Robert Holley reported the first known base sequence of a biologically significant nucleic acid, that of yeast **alanine tRNA (tRNA^{Ala};** Fig. 32-8). To do so Holley had to overcome several major obstacles:

1. All organisms contain many species of tRNAs (usually at least one for each of the 20 amino acids) which, because of their nearly identical properties (see below), are not easily separated. Preparative techniques had to be developed to provide the gram or so of pure yeast tRNA^{Ala} Holley required for its sequence determination.

2. Holley had to invent the methods that were initially used to sequence RNA (Section 7-2).

3. Ten of the 76 bases of yeast tRNA^{Ala} are modified (see below). Their structural formulas had to be elucidated

although they were never available in more than milligram quantities.

Since 1965, the techniques for tRNA purification and sequencing have vastly improved. A tRNA may now be sequenced in a few hours' time with only ~1 μg of material. Presently, the base sequences of many thousands of tRNAs from nearly 800 organisms are known, most from genomic sequences (they are compiled at the Genomic

Figure 32-8   Base sequence of yeast tRNA^{Ala} drawn in the cloverleaf form. The symbols for the modified nucleosides (*color*) are explained in Fig. 32-10.

tRNA Database, http://gtrnadb.ucsc.edu/). They vary in length from 54 to 100 nucleotides (18–28 kD) although most have ~76 nucleotides.

Almost all known tRNAs, as Holley first recognized, may be schematically arranged in the so-called cloverleaf secondary structure (Fig. 32-9). Starting from the 5′ end, they have the following common features:

1. A 5′-terminal phosphate group.

2. A 7-bp stem that includes the 5′-terminal nucleotide and that may contain non-Watson–Crick base pairs such as G · U. This assembly is known as the **acceptor** or **amino acid stem** because the amino acid residue carried by the tRNA is appended to its 3′-terminal OH group (Section 32-2C).

3. A 3- or 4-bp stem ending in a loop that frequently contains the modified base **dihydrouridine** (**D**; see below). This stem and loop are therefore collectively termed the **D arm.**

4. A 5-bp stem ending in a loop that contains the **anticodon,** the triplet of bases that is complementary to the codon specifying the tRNA. These features are known as the **anticodon arm.**

5. A 5-bp stem ending in a loop that usually contains the sequence TψC (where ψ is the symbol for **pseudouridine;** see below). This assembly is called the **TψC** or **T arm.**

6. All tRNAs terminate in the sequence CCA with a free 3′-OH group. The CCA may be genetically specified or enzymatically appended to immature tRNA (Section 31-4Cc).

7. There are 15 invariant positions (always have the same base) and 8 **semi-invariant** positions (only a purine or only a pyrimidine) that occur mostly in the loop regions. These regions also contain **correlated invariants,** that is, pairs of nonstem nucleotides that are base paired in all tRNAs. The purine on the 3′ side of the anticodon is invariably modified. The structural significance of these features is examined below.

The site of greatest variability among the known tRNAs occurs in the so-called **variable arm.** It has from 3 to 21 nucleotides and may have a stem consisting of up to 7 bp. The D loop also varies in length from 5 to 7 nucleotides.

a. Transfer RNAs Have Numerous Modified Bases

One of the most striking characteristics of tRNAs is their large proportion, up to 25%, of post-translationally modified or hypermodified bases. Nearly 80 such bases, found at >60 different tRNA positions, have been characterized. A few of them, together with their standard abbreviations, are indicated in Fig. 32-10. Hypermodified nucleosides, such as i[6]A, are usually adjacent to the anticodon's 3′ nucleotide when it is A or U. Their low polarities probably strengthen the otherwise relatively weak pairing associations of these bases with the codon, thereby increasing translational fidelity. Conversely, certain methylations block base pairing and hence prevent inappropriate structures from forming. Some of these modifications form important recognition elements for the enzyme that attaches the correct amino acid to a tRNA (Section 32-2Cb). However, none of them are essential for maintaining a tRNA's structural integrity (see below) or for its proper binding to the ribosome. Nevertheless, mutant bacteria unable to form certain modified bases compete poorly against the corresponding normal bacteria.

B. *Tertiary Structure of tRNA*

🔎 **See Guided Exploration 26: The structure of tRNA** The earliest physicochemical investigations of tRNA indicated that it has a well-defined conformation. Yet, despite numerous hydrodynamic, spectroscopic, and chemical cross-linking studies, its three-dimensional structure remained an enigma until 1974. In that year, the 2.5-Å resolution X-ray crystal structure of yeast **tRNA[Phe]** was separately elucidated by Alexander Rich in collaboration with Sung-Hou Kim and, in a different crystal form, by Aaron Klug. *The molecule assumes an L-shaped conformation in which one leg of the L is formed by the acceptor and T stems folded into a continuous A-RNA-like double helix (Section 29-1Ba) and the other leg is similarly composed of the D and*

Figure 32-9 Cloverleaf secondary structure of tRNA. Filled circles connected by dots represent Watson–Crick base pairs, and open circles in the double-helical regions indicate bases involved in non-Watson–Crick base pairing. Invariant positions are indicated: R and Y represent invariant purines and pyrimidines and ψ signifies pseudouridine. The starred nucleosides are often modified. The dashed regions in the D and variable arms contain different numbers of nucleotides in the various tRNAs.

Uracil derivatives

Pseudouridine (ψ)   Dihydrouridine (D)   Ribothymidine (T)   4-Thiouridine (s⁴U)

Cytosine derivatives

3-Methylcytidine (m³C)   N⁴-Acetylcytidine (ac⁴C)   Lysidine (L)

Adenine derivatives

1-Methyladenosine (m¹A)   N⁶-Isopentenyladenosine (i⁶A)   Inosine (I)

Guanine derivatives

N⁷-Methylguanosine (m⁷G)   N²,N²-Dimethylguanosine (m₂²G)   R = H   Wyosine (Wyo)

R = CH₂CH₂CHCOCH₃ Y

Figure 32-10   A selection of the modified nucleosides that occur in tRNAs together with their standard abbreviations. Note that although inosine chemically resembles guanosine, it is biochemically derived from adenosine. Nucleosides may also be methylated at their ribose 2′ positions to form residues symbolized, for instance, by Cm, Gm, and Um.

(a)

○ Constant nucleotide

◌ Constant purine
   or pyrimidine

Figure 32-11 Structure of yeast tRNA^Phe. (*a*) The base sequence drawn in cloverleaf form. Tertiary base pairing interactions are represented by thin red lines connecting the participating bases. Bases that are invariant or semi-invariant in all tRNAs are circled by solid and dashed lines, respectively. The 5' terminus is colored bright green, the acceptor stem is yellow, the D arm is white, the anticodon arm is light green, the variable arm is orange, the TψC arm is cyan, and the 3' terminus is red. (*b*) The X-ray structure drawn to show how its base paired stems form an L-shaped molecule. The tRNA is drawn in stick form with C atoms colored as in Part *a*, N blue, and O red. Adjacent P atoms are connected by rods colored as in Part *a*. [Based on an X-ray structure by Sung-Hou Kim, PDBid 6TRNA.] 🕮 **See Kinemage Exercises 19-1 and 19-2**

anticodon stems (Fig. 32-11). Each leg of the L is ~60 Å long and the anticodon and amino acid acceptor sites are at opposite ends of the molecule, some 76 Å apart. The narrow 20- to 25-Å width of native tRNA is essential to its biological function: During protein synthesis, three RNA molecules must simultaneously bind in close proximity at adjacent codons on mRNA (Section 32-3Ae).

a. tRNA's Complex Tertiary Structure Is Maintained by Hydrogen Bonding and Stacking Interactions

The structural complexity of yeast tRNA^Phe is reminiscent of that of a protein. Although only 42 of its 76 bases occur in double helical stems, *71 of them participate in stacking associations* (Fig. 32-12). The structure also contains 9 base pairing interactions that cross-link its tertiary structure (Figs. 32-11*a* and 32-12). Remarkably, all but one of these tertiary interactions, which appear to be the mainstays of the molecular structure, are non-Watson–Crick associations. Moreover, most of the bases involved in these interactions are either invariant or semi-invariant, which strongly suggests that all tRNAs have similar conformations (see below). The structure is also stabilized by several unusual hydrogen bonds between bases and either phosphate groups or the 2'-OH groups of ribose residues.

The compact structure of yeast tRNA^Phe results from its large number of intramolecular associations, which renders most of its bases inaccessible to solvent. The most notable exceptions to this are the anticodon bases and those of the amino acid–bearing—CCA terminus. Both of these groupings must be accessible in order to carry out their biological functions.

The observation that the molecular structures of yeast tRNA^Phe in two different crystal forms are essentially identical lends much credence to the supposition that its crystal structure closely resembles its solution structure. Transfer RNAs other than yeast tRNA^Phe have, unfortunately, been notoriously difficult to crystallize. As yet, the X-ray structures of only three other uncomplexed tRNAs have been reported (although the X-ray structures of numerous tRNAs in complex with the enzymes that append their corresponding amino acids and with ribosomes have been elucidated; Sections 32-2C and 32-3D). The major structural differences among them result from an apparent flexibility in the anticodon loop and the—CCA terminus as well as from a

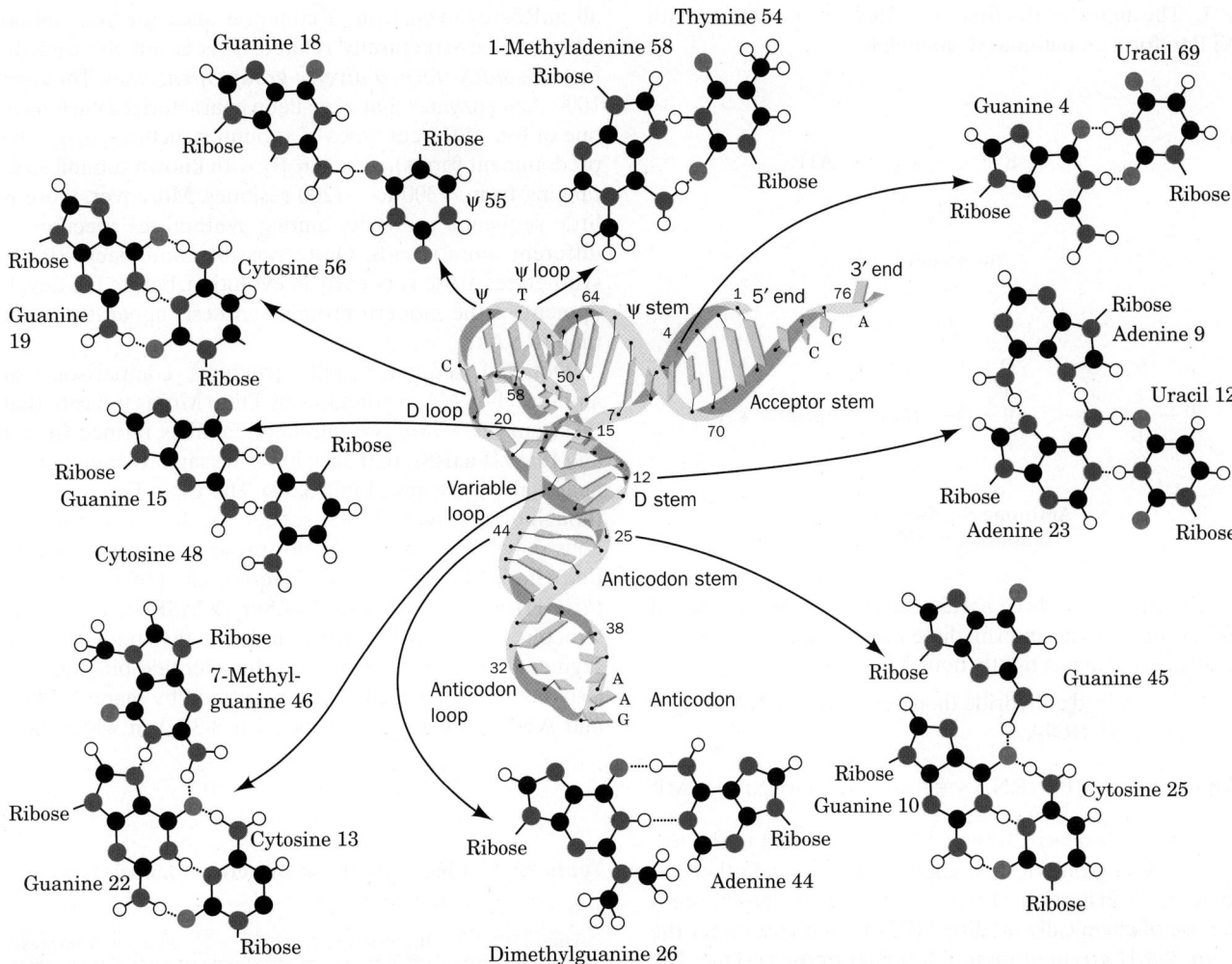

Figure 32-12   Tertiary base pairing interactions in yeast tRNA^Phe. Note that all but one of these nine interactions involve non-Watson–Crick pairs and that they are all located near the corner of the L. [After Kim, S.H., *in* Schimmel, P.R., Söll, D., and Abelson, J.N. (Eds.), *Transfer RNA: Structure, Properties and Recognition*, p. 87, Cold Spring Harbor Laboratory Press (1979). Illustration, Irving Geis. Image from the Irving Geis Collection, Howard Hughes Medical Institute. Reprinted with permission.]

See Kinemage Exercise 19-3

hingelike mobility between the two legs of the L that gives, for instance, yeast **tRNA^Asp** a boomerang-like shape. Such observations are in accord with the expectation that all tRNAs fit into the same ribosomal cavities.

C. Aminoacyl–tRNA Synthetases

See Guided Exploration 27: The structures of aminoacyl–tRNA synthetases and their interactions with tRNAs *Accurate translation requires two equally important recognition steps: (1) the choice of the correct amino acid for covalent attachment to a tRNA; and (2) the selection of the amino acid–charged tRNA specified by mRNA.* The first of these steps, which is catalyzed by amino acid–specific enzymes known as **aminoacyl–tRNA synthetases (aaRSs)**, appends an amino acid to the 3′-terminal ribose residue of its cognate tRNA to form an aminoacyl–tRNA (Fig. 32-13). This otherwise unfavorable process is driven by the hydrolysis of ATP in two sequential reactions that are catalyzed by a single enzyme.

Aminoacyl–tRNA

Figure 32-13   An aminoacyl–tRNA. The amino acid residue is esterified to the tRNA's 3′-terminal nucleoside at either its 3′-OH group, as shown here, or its 2′-OH group.

1. The amino acid is first "activated" by its reaction with ATP to form an **aminoacyl–adenylate**

$$
\begin{array}{c}
\text{H}\quad\;\;\text{O} \\
\;\;\;\;|\quad\;\;\;\;\| \\
\text{R}-\text{C}-\text{C} \\
\;\;\;\;|\qquad\quad\backslash \\
\;\;\;\text{NH}_3^+\qquad\text{O}^-
\end{array}
\quad +\;\text{ATP}
$$

Amino acid

$$
\begin{array}{c}
\text{H}\quad\;\;\text{O}\qquad\quad\;\;\text{O} \\
\;\;\;|\quad\;\;\;\|\qquad\quad\;\; \| \\
\text{R}-\text{C}-\text{C}-\text{O}-\text{P}-\text{O}-\text{Ribose}-\text{Adenine}\; +\;\text{PP}_i \\
\;\;\;\;|\qquad\qquad\quad\;| \\
\;\;\text{NH}_3^+\qquad\qquad\text{O}^-
\end{array}
$$

Aminoacyl–adenylate
(aminoacyl–AMP)

which, with all but three aaRSs, can occur in the absence of tRNA. Indeed, this intermediate may be isolated although it normally remains tightly bound to the enzyme.

2. This mixed anhydride then reacts with tRNA to form the aminoacyl–tRNA

$$\text{Aminoacyl–AMP} + \text{tRNA} \rightleftharpoons \text{aminoacyl–tRNA} + \text{AMP}$$

Some aaRSs exclusively append an amino acid to the terminal 2'-OH group of their cognate tRNAs, and others do so at the 3'-OH group. This selectivity was established with the use of chemically modified tRNAs that lack either the 2'- or 3'-OH group of their 3'-terminal ribose residue. The use of these derivatives was necessary because, in solution, the aminoacyl group rapidly equilibrates between the 2' and 3' positions.

The overall aminoacylation reaction is

$$
\begin{aligned}
\text{Amino acid} + \text{tRNA} + \text{ATP} &\rightleftharpoons \\
&\text{aminoacyl–tRNA} + \text{AMP} + \text{PP}_i
\end{aligned}
$$

These reaction steps are readily reversible because the free energies of hydrolysis of the bonds formed in both the aminoacyl–adenylate and the aminoacyl–tRNA are comparable to that of ATP hydrolysis. The overall reaction is driven to completion by the inorganic pyrophosphatase-catalyzed hydrolysis of the PP$_i$ generated in the first reaction step. Amino acid activation therefore chemically resembles fatty acid activation (Section 25-2A); the major difference between these two processes, which were both elucidated by Paul Berg, is that tRNA is the acyl acceptor in amino acid activation, whereas CoA performs this function in fatty acid activation.

a. There Are Two Classes of Aminoacyl–tRNA Synthetases

Most cells have one aaRS for each of the 20 amino acids. The similarity of the reactions catalyzed by these enzymes and the structural resemblance of all tRNAs suggests that all aaRSs evolved from a common ancestor and should therefore be structurally related. This is not the case. In fact, *the aaRSs form a diverse group of enzymes*. The over 1000 such enzymes that have been characterized each have one of four different types of subunit structures, α, α_2 (the predominant forms), α_4, and $\alpha_2\beta_2$, with known subunit sizes ranging from ~300 to ~1200 residues. Moreover, there is little sequence similarity among synthetases specific for different amino acids. Quite possibly, aminoacyl–tRNA synthetases arose very early in evolution, before the development of the modern protein synthesis apparatus other than tRNAs.

Detailed sequence and structural comparisons of aminoacyl–tRNA synthetases by Dino Moras indicate that these enzymes form two unrelated families, termed **Class I** and **Class II aaRSs,** that each have the same 10 members in nearly all organisms (Table 32-4). The Class I enzymes, although of largely dissimilar sequences, share two homologous polypeptide segments, not present in other proteins, that have the consensus sequences His-Ile-Gly-His (HIGH) and Lys-Met-Ser-Lys-Ser (KMSKS). The X-ray structures of Class I enzymes indicate that both of these segments are components of a dinucleotide-binding fold (Rossmann fold, which is also possessed by many NAD$^+$- and ATP-binding proteins; Section 8-3Bi) in which they

Table 32-4 **Characteristics of Bacterial Aminoacyl–tRNA Synthetases**

Amino Acid	Quaternary Structure	Number of Residues
Class I		
Arg	α	577
Cys	α	461
Gln	α	553
Glu	α	471
Ile	α	939
Leu	α	860
Met	α, α_2	676
Trp	α_2	325
Tyr	α_2	424
Val	α	951
Class II		
Ala	α, α_4	875
Asn	α_2	467
Asp	α_2	590
Gly	$\alpha_2\beta_2$	303/689
His	α_2	424
Lys	α_2	505
Pro	α_2	572
Phe	$\alpha_2\beta_2$, α	327/795
Ser	α_2	430
Thr	α_2	642

Source: Mainly Carter, C.W., Jr., *Annu. Rev. Biochem.* **62,** 715 (1993).

participate in ATP binding and are implicated in catalysis. The Class II synthetases lack the foregoing sequences but have three other sequences in common. Their X-ray structures reveal that these sequences occur in a so-called signature motif, a fold found only in Class II enzymes that consists of a 7-stranded antiparallel β sheet with three flanking helices, which forms the core of their catalytic domains.

Many Class I aaRSs require anticodon recognition to aminoacylate their cognate tRNAs. In contrast, several Class II enzymes, including **AlaRS** and **SerRS,** do not interact with their bound tRNA's anticodon. Indeed, several class II aaRSs accurately aminoacylate "microhelices" derived from only the acceptor stems of their cognate tRNAs. Another difference between Class I and Class II synthetases is that all Class I enzymes aminoacylate their bound tRNA's 3′-terminal 2′-OH group, whereas Class II enzymes, with the exception of **PheRS,** all charge the 3′-OH group. The amino acids for which the Class I synthetases are specific tend to be larger and more hydrophobic than those used by Class II synthetases. Finally, as Table 32-4 indicates, Class I aaRSs are mainly monomers, whereas most Class II aaRSs are homodimers.

LysRS has been classified as a Class II aaRS. However, a search of the genome sequences of *Methanococcus jannaschii* and *Methanobacterium thermoautotrophicum* failed to reveal the presence of such a LysRS. This led to the discovery that the LysRSs expressed by these archaebacteria are Class I rather than Class II enzymes. This raises the interesting question of how Class I LysRS evolved.

Prokaryotic aaRSs occur as individual protein molecules. However, in many higher eukaryotes (e.g., *Drosophila* and mammals), 9 aaRSs, some of each class, associate to form a multienzyme particle in which the glutamyl and prolyl synthetase functions are fused into a single polypeptide named **GluProRS.** The advantages of these systems are unknown.

b. The Structural Features Recognized by Aminoacyl-tRNA Synthetases May Be Quite Simple

As we shall see in Section 32-2D, ribosomes select aminoacyl–tRNAs only via codon–anticodon interactions, not according to the identities of their aminoacyl groups. *Accurate translation therefore requires not only that each tRNA be aminoacylated by its cognate aaRS but that it not be aminoacylated by any of its 19 noncognate aaRSs.* Moreover, since most cells express only one aaRS for each amino acid, each aaRS must aminoacylate all of the several, if not many, **isoaccepting tRNAs** (different tRNAs that are specific for the same amino acid) in each cell.

Considerable effort has therefore been expended, notably by LaDonne Schulman, Paul Schimmel, Olke Uhlenbeck, and John Abelson, in elucidating how aaRSs manage this feat, despite the close structural similarities of nearly all tRNAs. The experimental methods employed involved the use of specific tRNA fragments, mutationally altered tRNAs, chemical cross-linking agents, computerized sequence comparisons, and X-ray crystallography. The most common synthetase contact sites on tRNA occur on the inner (concave) face of the L. Other than that, there appears to be little regularity in how the various tRNAs are recognized by their cognate synthetases. Indeed, as we shall see, some aaRSs recognize only their cognate tRNA's acceptor stem, whereas others also interact with its anticodon region. Additional tRNA regions may also be recognized.

Genetic manipulations by Schimmel revealed that the tRNA features recognized by at least one type of aaRS are surprisingly simple. Numerous sequence alterations of *E. coli* tRNA^{Ala} do not appreciably affect its capacity to be aminoacylated with alanine. Yet, most base substitutions in the G3 · U70 base pair located in the tRNA's acceptor stem (Fig. 32-14*a*) greatly diminish this reaction. Moreover, the introduction of a G · U base pair into the analogous

Figure 32-14 Major identity elements in four tRNAs. Each base in the tRNA is represented by a filled circle. Red circles indicate positions that have been shown to be identity elements for the recognition of the tRNA by its cognate aminoacyl–tRNA synthetase. The anticodon bases that are identity elements are underlined. In each case, additional identity elements may yet be discovered. The base at position 73, which is an identity element in all four tRNAs shown here, is known as the **discriminator base.**

Figure 32-15   Three-dimensional model of *E. coli* tRNA^Ala^. This model is based on the X-ray structure of yeast tRNA^Phe^ (Fig. 32-11*b*) in which the nucleotides that are different in *E. coli* tRNA^Cys^ are highlighted in cyan and the G3 · U70 base pair is highlighted in ivory. [Courtesy of Ya-Ming Hou, MIT.]

position of **tRNA^Cys^** and tRNA^Phe^ causes them to be aminoacylated with alanine even though there are few other sequence identities between these mutant tRNAs and tRNA^Ala^ (e.g., Fig. 32-15). In fact, *E. coli* AlaRS even efficiently aminoacylates a 24-nt "microhelix" derived from only the G3 · U70–containing acceptor stem of *E. coli* tRNA^Ala^. Since the only *E. coli* tRNAs that normally have a G3 · U70 base pair are the tRNA^Ala^, and this base pair is also present in the tRNA^Ala^ from many organisms including yeast (Fig. 32-8), the foregoing observations strongly suggest that *the G3 · U70 base pair is a major feature recognized by AlaRSs.* These enzymes presumably recognize the distorted shape of the G · U base pair (Fig. 32-12), a hypothesis corroborated by the observation that base changes at G3 · U70 which least affect the acceptor identity of tRNA^Ala^ yield base pairs that structurally resemble G · U.

The elements of three other tRNAs, which are recognized by their cognate tRNA synthetases, are indicated in Fig. 32-14. As with tRNA^Ala^, these identity elements appear to comprise only a few bases. Note that the anticodon forms an identity element in two of these tRNAs. In another example of an anticodon identifier, the *E. coli* **tRNA^Ile^** specific for the codon AUA has the anticodon LAU, where L is **lysidine,** a modified cytosine whose 2-keto group is replaced by the amino acid lysine (Fig. 32-10). The L in this context pairs with A rather than G, a rare instance of base modification altering base pairing specificity. The replacement of this L with unmodified C, as expected, yields a tRNA that recognizes the Met codon AUG (codons bind anticodons in an antiparallel fashion). Surprisingly, how-

ever, this altered tRNA^Ile^ is also a much better substrate for **MetRS** than it is for **IleRS.** Thus, both the codon and the amino acid specificity of this tRNA are changed by a single post-transcriptional modification. The N^1-methylation of G37 in yeast tRNA^Asp^ (Fig. 32-14*b*) provides another example of a base modification forming an identity element. In the absence of this N^1-methyl group, tRNA^Asp^ is recognized by **ArgRS,** largely via its C36 and G37, whereas ArgRS normally recognizes only **tRNA^Arg^,** mainly via its C35 and U36.

The available experimental evidence has largely located the various tRNA identifiers in the acceptor stem and the anticodon loop (Fig. 32-16). The X-ray structures of several aaRS · tRNA complexes, which we consider next, have structurally rationalized some of these observations.

c. The X-Ray Structure of GlnRS · tRNA^Gln^, a Class I Complex

The X-ray structures of all 20 different amino acid–specific aaRSs from numerous organisms have been determined, many of which are in complex with ATP, their cognate amino acids, or their analogs. These structures reveal that the active sites of these enzymes bind the ATP and target amino acid in optimal positions for in-line nucleophilic displacement (Section 16-2B) during amino acid

Figure 32-16   Experimentally observed identity elements of tRNAs. The tRNA backbone is cyan and each of its nucleotides is represented by a yellow circle whose diameter is proportional to the fraction of the 20 tRNA acceptor types for which the nucleotide is an observed determinant. [Courtesy of William McClain, University of Wisconsin.]

activation and that the specificity of an aaRS for its target amino acid is determined by idiosyncratic contacts with the side chain of the amino acid.

The X-ray structures of 16 different aaRSs in their complexes with their cognate tRNAs (all but those of Gly, Ala, Lys, and His) have so far been reported. The first of them to be elucidated, that of *E. coli* **GlnRS**, a Class I synthetase, in its complex with **tRNA^{Gln}** and ATP (Fig. 32-17), was determined by Thomas Steitz. The tRNA^{Gln} assumes an L-shaped conformation that resembles those of tRNAs of known structures (e.g., Fig. 32-11*b*). GlnRS, a 553-residue monomeric protein that consists of four domains arranged to form an elongated molecule, interacts with the tRNA along the entire inside face of the L such that the anticodon is bound near one end of the protein and the acceptor stem is bound near its other end.

Genetic and biochemical data indicate that the identity elements of tRNA^{Gln} are largely clustered in its anticodon loop and acceptor stem (Fig. 32-14*c*). The anticodon loop of tRNA^{Gln} is extended by two novel non-Watson–Crick base pairs (2′-*O*-methyl-U32 · ψ38 and U33 · m^{2}A37), thereby causing the bases of the anticodon to unstack and splay outward in different directions so as to bind in separate recognition pockets of GlnRS. These structural features suggest that GlnRS uses all seven bases of the anticodon loop to discriminate among tRNAs. Indeed, changes to any one of the bases of residues C34 through ψ38 yield tRNAs with decreases in k_{cat}/K_M for aminoacylation by GlnRS by factors ranging from 70 to 28,000.

The GCCA at the 3′ end of the tRNA^{Gln} makes a hairpin turn toward the inside of the L rather than continuing helically onward (as does the ACCA at the 3′ end in the X-ray structure of tRNA^{Phe}; Fig. 32-11*b*). This conformation change is facilitated by the insinuation of a Leu side chain between the 5′ and 3′ ends of the tRNA so as to disrupt the first base pair of the acceptor stem (U1 · A72). The GlnRS reaction is therefore relatively insensitive to base changes in these latter two positions except when base pairing is strengthened by their conversion to G1 · C72. The GCCA end of the tRNA^{Gln} plunges deeply into a protein pocket

(*a*)

(*b*)

Figure 32-17   X-ray structure of *E. coli* GlnRS · tRNA^{Gln} · ATP. (*a*) The tRNA is drawn in stick form colored as the ATP but with the C atoms of the anticodon (UCG) and the 3′-CCA end yellow. An orange rod links its successive P atoms. The ATP bound in the protein's active site is drawn in space-filling form with C green, N blue, O red, and P orange. The protein is represented by a semitransparent light blue surface diagram that reveals the buried portions of the tRNA and ATP. Note that both the 3′ end of the tRNA (*top right*) and its anticodon bases (*bottom*) are inserted into deep pockets in the protein. (*b*) The complex viewed as in Part *a*. The tRNA's sugar–phosphate backbone is represented by an orange worm and the bases forming its identity elements (Fig. 32-14*c*) are drawn in stick form colored according to atom type with C magenta, N blue, and O red. The ATP is drawn as in Part *a*. The protein is shown in ribbon form colored in rainbow order from its N-terminus (*blue*) to its C-terminus (*red*). [Based on an X-ray structure by Thomas Steitz, Yale University. PDBid 1GTR.] **See Kinemage Exercise 20**

that also binds the enzyme's ATP and glutamine substrates. Three protein "fingers" are inserted into the minor groove of the acceptor stem to make sequence-specific interactions with base pairs G2 · C71 and G3 · C70 [recall that double helical RNA has an A-DNA-like structure (Section 29-1Bc) whose wide minor groove readily admits protein but whose major groove is normally too narrow to do so].

The GlnRS domain that binds glutamine, ATP, and the GCCA end of tRNA^{Gln}, the so-called catalytic domain, contains, as we previously discussed, a dinucleotide-binding fold. Much of this domain is nearly superimposable with and thus evolutionarily related to the corresponding domains of other Class I aaRSs.

d. The X-Ray Structure of AspRS · tRNA^{Asp}, a Class II Complex

Yeast **AspRS,** a Class II synthetase, is an α_2 dimer of 557-residue subunits. Its X-ray structure in complex with tRNA^{Asp}, determined by Moras, reveals that the protein symmetrically binds two tRNA molecules (Fig. 32-18). Like GlnRS, AspRS principally contacts its bound tRNA both at the end of its acceptor stem and in its anticodon region. The contacts in these two enzymes are, nevertheless, quite different in character (Fig. 32-19): Although both tRNAs approach their cognate synthetases along the inside of their L shapes, tRNA^{Gln} does so toward the direction of the minor groove of its acceptor stem, whereas tRNA^{Asp} does so toward the direction of its major groove. The GCCA at the 3' end of tRNA^{Asp} thereby continues its helical track as it plunges into AspRS's catalytic site, whereas, as we saw, the GCCA end of tRNA^{Gln} bends backward into a hairpin

turn that opens up the first base pair (U1 · A72) of its acceptor stem. Although the deep major groove of an A-RNA helix is normally too narrow to admit groups larger than water molecules (Section 29-1Bc), the major groove at the end of the acceptor stem in AspRS · tRNA^{Asp} is sufficiently widened for its base pairs to interact with a protein loop.

The anticodon arm of tRNA^{Asp} is bent by as much as 20 Å toward the inside of the L relative to that in the X-ray structure of uncomplexed tRNA^{Asp} and its anticodon bases are unstacked. The hinge point for this bend is a G30 · U40 base pair in the anticodon stem which, in nearly all other species of tRNA, is a Watson–Crick base pair. The anticodon bases of tRNA^{Gln} are also unstacked in contacting GlnRS but with a backbone conformation that differs from that in tRNA^{Asp}. Evidently, the conformation of a tRNA in complex with its cognate synthetase appears to be dictated more by its interactions with the protein (induced fit) than by its sequence.

Structural analyses of complexes of AspRS · tRNA^{Asp} with ATP and aspartic acid, and of GlnRS · tRNA^{Gln} with ATP, have permitted models of the aminoacyl–AMP complexes of these enzymes to be independently formulated. Comparison of these models reveals that the 3'-terminal A residues of tRNA^{Gln} and tRNA^{Asp} (to which the aminoacyl groups are appended; Fig. 32-13) are positioned on opposite sides of the enzyme-bound aminoacyl–AMP intermediate (Fig. 32-20). The 3'-terminal ribose residues are puckered C2'-*endo* for tRNA^{Asp} and C3'-*endo* for tRNA^{Gln} (see Fig. 29-8) such that the 2'-hydroxyl group of tRNA^{Gln} (Class I) is stereochemically positioned to attack the

Figure 32-18 X-ray structure of yeast AspRS · tRNA^{Asp} · ATP. (*a*) The homodimeric enzyme with its two symmetrically bound tRNAs is viewed with its 2-fold axis approximately vertical. The tRNAs are drawn in skeletal form colored according to atom type with the C atoms of the anticodon (GUC) and the 3'-CCA end yellow, the remaining C atoms green, N blue, O red, and P orange. An orange rod connects its successive P atoms. The two protein subunits are represented by semitransparent pink and light blue surface diagrams that reveal the buried portions of

the tRNAs. (*b*) A ribbon diagram of the AspRS · tRNA^{Asp} protomer. The tRNA's sugar–phosphate backbone is represented by an orange worm and the bases forming its identity elements (Fig. 32-14*b*) are drawn in stick form colored according to atom type with C magenta, N blue, and O red. The protein is shown in ribbon form colored in rainbow order from its N-terminus (*blue*) to its C-terminus (*red*). [Based on an X-ray structure by Dino Moras, CNRS/INSERM/ULP, Illkirch Cédex, France. PDBid 1ASY.]

(a)

Figure 32-19 **Comparison of the modes by which GlnRS and AspRS bind their cognate tRNAs.** The proteins and tRNAs are represented by blue and red spheres centered on their C_α and P atom positions. Note how GlnRS (*a*), a Class I synthetase, binds tRNA^{Gln} from the minor groove side of its acceptor stem so as to bend its 3' end into a hairpin conformation. In contrast, AspRS

(b)

(b), a Class II synthetase, binds tRNA^{Asp} from the major groove side of its acceptor stem so that its 3' end continues its helical path on entering the active site. [After drawings by Dino Moras, CNRS/INSERM/ULP, Illkirch Cédex, France. PDBids 1GTR abd 1ASY.]

aminoacyl–AMP's carboxyl group, whereas for tRNA^{Asp} (Class II), only the 3'-hydroxyl group is situated to do so. This clearly explains the different aminoacylation specificities of the Class I and Class II aaRSs.

A76 tRNA^{Asp} (Class II)   O3'   O2'   A76 tRNA^{Gln} (Class I)

Aminoacyl–AMP

e. Proofreading Enhances the Fidelity of Amino Acid Attachment to tRNA

The charging of a tRNA with its cognate amino acid is a remarkably accurate process: aaRSs display an overall error rate of about 1 in 10,000. We have seen that aaRSs bind only their cognate tRNAs through an intricate series of specific contacts. But how do they discriminate among the various amino acids, some of which are quite similar?

Experimental measurements indicate, for example, that IleRS transfers as many as 40,000 isoleucines to **tRNA^{Ile}** for every valine it so transfers. Yet, as Linus Pauling first pointed out, *there are insufficient structural differences between Val and Ile to permit such a high degree of discrimination in the direct generation of aminoacyl–tRNAs.* The X-ray structure of *Thermus thermophilus* IleRS, a monomeric

Figure 32-20 **Comparison of the stereochemistries of aminoacylation by Class I and Class II aaRSs.** The positions of the 3' terminal adenosine residues (A76) of AspRS (Class II, *left*) and GlnRS (Class I, *right*) are drawn relative to that of the enzyme-bound aminoacyl–AMP (*below*; only the carbonyl group of its aminoacyl residue is shown). Note how only O3' of tRNA^{Asp} and O2' of tRNA^{Gln} are suitably positioned to attack the aminoacyl residue's carbonyl group and thereby transfer the aminoacyl residue to the tRNA. [After Cavarelli, J., Eriani, G., Rees, B., Ruff, M., Boeglin, M., Mitschler, A., Martin, F., Gangloff, J., Thierry, J.-C., and Moras, D., *EMBO J.* **13,** 335 (1994).]

Class I aaRS, in complex with isoleucine, determined by Shigeyuki Yokoyama and Schimmel, indicates that isoleucine fits snugly into its binding site in the enzyme's Rossmann fold domain and hence that this binding site would sterically exclude leucine as well as larger amino acids. However, valine, which differs from isoleucine by only the lack of a single methylene group, fits into this isoleucine-binding site. The binding free energy of a methylene group is estimated to be ~ 12 kJ $\cdot$ mol^{-1}. Equation [3.17] indicates that the ratio f of the equilibrium constants, K_1 and K_2, with which two substances bind to a given binding site is given by

$$f = \frac{K_1}{K_2} = \frac{e^{-\Delta G_1^{\circ\prime}/RT}}{e^{-\Delta G_2^{\circ\prime}/RT}} = e^{-\Delta\Delta G^{\circ\prime}/RT} \qquad [32.1]$$

where $\Delta\Delta G^{\circ\prime} = \Delta G_1^{\circ\prime} - \Delta G_2^{\circ\prime}$ is the difference between the free energies of binding of the two substances. It is therefore estimated that isoleucyl–tRNA synthetase could discriminate between isoleucine and valine by no more than a factor of ~ 100.

Berg resolved this apparent paradox by demonstrating that, in the presence of tRNA^{Ile}, IleRS catalyzes the nearly quantitative hydrolysis of valyl–adenylate to valine + AMP rather than forming Val–tRNA^{Ile}. Moreover, the few Val–tRNA^{Ile} molecules that do form are hydrolyzed to valine + tRNA^{Ile}. Thus, *IleRS subjects both aminoacyl–adenylate and aminoacyl–tRNA^{Ile} to a* **proofreading** *or* **editing** *step that occurs at a separate catalytic site.* This site binds Val residues but excludes the larger Ile residues. *The enzyme's overall selectivity is therefore the product of the selectivities of its synthesis and proofreading steps, thereby accounting for the high fidelity of aminoacylation. Note that in this so-called* **double-sieve** *mechanism, editing occurs at the expense of ATP hydrolysis, the thermodynamic price of high fidelity (increased order).*

The X-ray structure of *Staphylococcus aureus* IleRS in complex with tRNA^{Ile} and the clinically useful antibiotic **mupirocin**

Figure 32-21   X-ray structure of *S. Aureus* isoleucyl–tRNA synthetase in complex with tRNA^{Ile} and mupirocin. The tRNA is white, the protein is colored by domain, and the mupirocin is shown in stick form in pink. [Courtesy of Thomas Steitz, Yale University. PDBid 1QU2.]

Mupirocin

(a product of *Pseudomonas fluorescens* that acts by specifically binding to bacterial IleRS so as to inhibit bacterial protein synthesis), determined by Steitz, suggests how IleRS carries out its editing process. The X-ray structure (Fig. 32-21) reveals that this complex resembles the GlnRS $\cdot$ tRNA^{Gln} $\cdot$ ATP complex (Fig. 32-17) but with IleRS having an additional editing domain (also called CP1 for connective peptide 1) inserted in its Rossmann fold domain. The two 3′ terminal residues of the tRNA^{Ile}, C75 and A76, are disordered but, when modeled so as to continue the acceptor stem's stacked A-form helix, extend into a cleft in the editing domain that has been implicated as its hydrolytic site (Fig. 32-22a, *left*). Thus, this IleRS complex ap-

pears to resemble an "editing complex" instead of a "transfer complex" as seen in the GlnRS structure. However, a transfer complex would form if the 3′ ending segment of the tRNA^{Ile} assumes a hairpin conformation (Fig. 32-22a, *right*) similar to that in the GlnRS structure (Figs. 32-17b and 32-19a; recall that IleRS and GlnRS are both Class I aaRSs). Steitz has therefore postulated that the aminoacyl group is shuttled between the IleRS's aminoacylation site and its editing site by such a conformational change (Fig. 32-22b). This process functionally resembles the way in which DNA polymerase I edits its newly synthesized strand (Section 30-2Ag), which Steitz also elucidated.

ValRS is a monomeric Class I aaRS that resembles IleRS. The X-ray structure of the complex of *T. thermophilus* ValRS, **tRNA^{Val}**, and the nonhydrolyzable **valyl–adenylate** analog **5′-*O*-[*N*-(L-valyl)sulfamoyl]adenosine (Val-AMS)**,

5′-*O*-[*N*-(L-Valyl)sulfamoyl]adenosine (Val-AMS)

(a)

Figure 32-22 Comparison of the putative aminoacylation and editing modes of IleRS · tRNA^{Ile}. (*a*) The superposition of tRNA^{Ile} in these two binding modes on the solvent-accessible surface of IleRS (*green*). The acceptor strand of tRNA^{Ile} in the editing mode observed in the X-ray structure of IleRS · tRNA^{Ile} · mupirocin (Fig. 32–21) is drawn in ribbon form in white with the modeled positions of C75 and A76 in red. This places the tRNA's 3′ end in the editing site. In contrast, the three 3′ terminal residues of tRNA^{Ile}, as positioned through homology modeling based on the X-ray structure of GlnRS · tRNA^{Gln} · ATP (Fig. 32-17) and drawn in ball-and-stick form with C yellow, N blue, O red, and P magenta, places the tRNA's 3′ end in the synthetic (aminoacylation) site, 34 Å distant from its position in the editing site. Note that there is a cleft running between the editing and synthetic sites and that the 3′ end of the tRNA continues its A-form helical path in the editing mode but assumes a hairpin conformation in the synthetic mode. (*b*) A cartoon comparing the positions of the 3′ end of tRNA^{Ile} in its complex with IleRS in its synthetic mode (*left*) and in its editing mode (*right*). [Part *a* courtesy of and Part *b* based on a drawing by Thomas Steitz, Yale University.]

Synthetic mode   **Editing mode**

(b)

Valyl-adenylate

determined by Yokoyama, reveals that the Val-AMS is bound in the aminoacylation pocket in the Rossmann fold domain, which accommodates the isosteric Val and Thr moieties but sterically excludes Ile. Modeling studies based on the IleRS · tRNA^{Ile} · mupirocin structure indicate that

the Thr side chain would fit into the ValRS editing pocket with its side chain hydroxyl group hydrogen bonded to the side chain of Asp 279 of ValRS, which protrudes into the pocket in contrast to the corresponding Asp 328 of IleRS, which does not. Consequently, a Val side chain would be excluded from the ValRS editing pocket because it cannot form such a hydrogen bond, thereby explaining why this editing pocket hydrolyzes **threonyl–adenylate** and Thr–tRNA^{Val} but not the corresponding Val derivatives. The ValRS · tRNA^{Val} structure also indicates that ValRS and tRNA^{Val} together form a tunnel connecting the ValRS's aminoacylation pocket with its editing pocket. Improperly formed threonyl–adenylate is proposed to be channeled through this tunnel for hydrolysis in the editing pocket, thereby explaining why tRNA^{Val} must be bound to ValRS for this pretransfer editing reaction to occur. Valyl–adenylate is presumably channeled through the similar IleRS · tRNA^{Ile} complex for its hydrolysis.

ThrRS, a Class II homodimer, has the opposite problem of ValRS: It must synthesize **Thr–tRNA^{Thr}** but not Val–tRNA^{Thr}. The X-ray structure of *E. coli* ThrRS that lacks its N-terminal domain but remains catalytically active in a complex with either threonine or the threonyl–adenylate analog **Thr-AMS,** determined by Moras, reveals that ThrRS's aminoacylation pocket contains a Zn^{2+} ion that is coordinated by the side chain hydroxyl and amino groups of the threonyl group as well as by three protein side chains. The isosteric valine could not coordinate the Zn^{2+} ion in this way and hence does not undergo adenylylation by ThrRS. However, what prevents ThrRS from synthesizing Ser–tRNA^{Thr}? In fact, the truncated ThrRS synthesizes Ser–tRNA^{Thr} at more than half the rate it synthesizes Thr–tRNA^{Thr}, thereby indicating that the N-terminal domain of wild-type ThrRS contains the enzyme's editing site. Mutational analysis of ThrRS has localized this editing site to a cleft in the N-terminal domain of wild-type ThrRS, whose X-ray structure in complex with tRNA^{Thr} was also determined by Moras. In this latter structure, the tRNA's 3′ end follows a regular helical path similar to that seen in the X-ray structure of AspRS · tRNA^{Asp} · ATP (Fig. 32-18) so as to enter the aminoacylation site. However, if the 3′ end of the bound tRNA^{Thr} assumed a hairpin conformation similar to that seen in X-ray structure of tRNA^{Gln} in complex with the Class I enzyme GlnRS and ATP (Fig. 32-17), its covalently linked aminoacyl group would enter the editing site. This indicates an intriguing "mirror symmetry" (Fig. 32-23): In Class I aaRSs that mediate a double-sieve editing mechanism, the 3′ end of the bound cognate tRNA assumes a hairpin conformation when it enters the aminoacylation site and a helical conformation when it enters the editing site, whereas the converse holds for Class II aaRSs. Finally, ThrRS does not appear to mediate pretransfer editing (does not hydrolyze **seryl–adenylate**), and, in fact, the ThrRS · tRNA^{Thr} complex lacks a channel connecting its aminoacylation and editing sites such as is seen in the ValRS · tRNA^{Val} complex.

Synthetases that have adequate selectivity for their corresponding amino acid lack editing functions. Thus, for example, the TyrRS aminoadenylylation site discriminates between tyrosine and phenylalanine through hydrogen bonding with the tyrosine —OH group. The cell's other amino acids, standard as well as nonstandard, have even less resemblance to tyrosine, which rationalizes why TyrRS lacks an editing site.

f. Gln–tRNA^{Gln} May Be Formed via an Alternative Pathway

Although it was long believed that each of the 20 standard amino acids is covalently linked to a tRNA by its corresponding aaRS, it is now clear that gram-positive bacteria, archaebacteria, cyanobacteria, mitochondria, and chloroplasts all lack GlnRS. Rather glutamate is linked to tRNA^{Gln} by the same GluRS that synthesizes **Glu–tRNA^{Glu}.** The resulting **Glu–tRNA^{Gln}** is then transamidated to Gln–tRNA^{Gln} by the enzyme **Glu–tRNA^{Gln} amidotransferase (Glu-AdT)** in an ATP-requiring reaction in which glutamine is the

Figure 32-23 Schematic diagram of the aminoacylation and editing mechanisms of Class I and Class II aaRSs emphasizing the "mirror symmetry" of their overall mechanisms. With Class I aaRSs (*left;* e.g., IleRS), the 3′ end of the bound tRNA's acceptor stem assumes a hairpin conformation in the synthetic mode and a helical conformation in the editing mode, whereas the converse occurs with Class II aaRSs (*right;* e.g., ThrRS). [Courtesy of Dino Moras, CNRS/INSERM/ULP, Illkirch Cedex, France.]

amide donor. Some microorganisms use a similar transamidation pathway for the synthesis of Asn–tRNA^{Asn} from **Asp–tRNA^{Asn}.**

The overall reaction catalyzed by Glu-AdT occurs in three stages (Fig. 32-24): (1) Glutamine is hydrolyzed to glutamate and the resulting NH_3 sequestered; (2) ATP reacts with the Glu side chain of Glu–tRNA^{Gln} to yield an activated acylphosphate intermediate and ADP; and (3) the acylphosphate intermediate reacts with the NH_3 to yield Gln–tRNA^{Gln} + P_i. Glu-AdT from *Bacillus subtilis,* which was characterized by Dieter Söll, is a heterotrimeric protein, none of whose subunits exhibit significant sequence similarity to GlnRS. The genes encoding these subunits, *gatA, gatB,* and *gatC,* form a single operon whose disruption is lethal, thereby demonstrating that *B. subtilis* has no alternative pathway for Gln–tRNA^{Gln} production. The **GatA** subunit of Glu-AdT appears to catalyze the activation of the side chain carboxyl of glutamic acid via a reaction resembling that catalyzed by carbamoyl phosphate synthetase (Section 26-2A). Nevertheless, GatA exhibits no sequence similarity with other known glutamine amidotransferases (members of the triad or Ntn families; Section 26-5Aa). The **GatB** subunit may be used to select the correct tRNA substrate. The role of the **GatC** subunit is unclear, although the observation that its presence is necessary for the expression of GatA in *E. coli* suggests that it participates in the modification, folding, and/or stabilization of GatA.

Since Glu is not misincorporated into *B. subtilis* proteins in place of Gln, the Glu–tRNA^{Gln} product of the above aminoacylation reaction must not be transported to the ribosome in the same way as Gln–tRNA^{Gln}. It is likely that this occurs because, as has been shown in chloroplasts, **EF-Tu,** the elongation factor that binds and transports most bacterial aminoacyl–tRNAs to the ribosome in a GTP-dependent process (Section 32-3D), does not bind Glu–tRNA^{Gln}. It is unclear why two independent routes have evolved for the synthesis of Gln–tRNA^{Gln}.

Figure 32-24 **The Glu-AdT–mediated synthesis of Gln–tRNA^{Gln} from Glu–tRNA^{Gln}.** The reaction involves the ATP-activated transfer of a glutamine-derived NH$_3$ to the glutamate moiety of Glu–tRNA^{Gln}.

g. Some Archaebacteria Lack a Separate CysRS

The genomes of certain archaebacteria such as *M. jannaschii* lack an identifiable gene for CysRS. This is because the enzyme responsible for synthesizing **Pro–tRNA^{Pro}** in these organisms also synthesizes **Cys–tRNA^{Cys}**. Interestingly, this enzyme, which is named **ProCysRS,** does not synthesize Pro–tRNA^{Cys} or Cys–tRNA^{Pro}. Although ProCysRS synthesizes **cysteinyl–adenylate** only in the presence of tRNA^{Cys}, it synthesizes **prolyl–adenylate** in the absence of tRNA^{Pro}. The binding of tRNA^{Cys} to ProCysRS blocks the activation of proline so that only cysteine can be activated. Conversely, the activation of proline facilitates the binding of tRNA^{Pro} while preventing the binding of tRNA^{Cys}. However, the mechanism through which ProCysRS carries out these mutually exclusive syntheses is unknown. In any case, it appears that some organisms can get by with as few as 17 different aaRSs; they may lack GlnRS, AspRS, and a separate CysRS.

D. Codon–Anticodon Interactions

In protein synthesis, the proper tRNA is selected only through codon–anticodon interactions; the aminoacyl group does not participate in this process. This phenomenon was demonstrated as follows. Cys–tRNA^{Cys}, in which the Cys residue was ^{14}C labeled, was reductively desulfurized with Raney nickel so as to convert the Cys residue to Ala:

The resulting ^{14}C-labeled hybrid, Ala–tRNA^{Cys}, was added to a cell-free protein synthesizing system extracted from rabbit reticulocytes. The product hemoglobin α chain's only radioactive tryptic peptide was the one that normally contains the subunit's only Cys. No radioactivity was found in the peptides that normally contain Ala but no Cys.

Evidently, *only the anticodons of aminoacyl–tRNAs partic-ipate in codon recognition.*

a. Genetic Code Degeneracy Is Largely Due to Variable Third Position Codon–Anticodon Interactions

One might naively guess that each of the 61 codons specifying an amino acid would be read by a different tRNA. Yet, even though most cells contain several groups of isoaccepting tRNAs, *many tRNAs bind to two or three of the codons specifying their cognate amino acids.* For example, yeast tRNA[Phe], which has the anticodon GmAA, recognizes the codons UUC and UUU (remember that the anticodon pairs with the codon in an antiparallel fashion),

$$
\begin{array}{cccc}
 & 3' & & 5' \\
\text{Anticodon:} & -\text{A}-\text{A}-\text{Gm}- \\
 & \vdots & \vdots & \vdots \\
 & 5' & & 3' \\
\text{Codon:} & -\text{U}-\text{U}-\text{C}-
\end{array}
\qquad
\begin{array}{cccc}
 & 3' & & 5' \\
 & -\text{A}-\text{A}-\text{Gm}- \\
 & \vdots & \vdots & \vdots \\
 & 5' & & 3' \\
 & -\text{U}-\text{U}-\text{U}-
\end{array}
$$

and yeast tRNA[Ala], which has the anticodon IGC, recognizes the codons GCU, GCC, and GCA.

$$
\begin{array}{cccc}
 & 3' & & 5' \\
\text{Anticodon:} & -\text{C}-\text{G}-\text{I}- \\
 & \vdots & \vdots & \vdots \\
 & 5' & & 3' \\
\text{Codon:} & -\text{G}-\text{C}-\text{U}-
\end{array}
\qquad
\begin{array}{cccc}
 & 3' & & 5' \\
 & -\text{C}-\text{G}-\text{I}- \\
 & \vdots & \vdots & \vdots \\
 & 5' & & 3' \\
 & -\text{G}-\text{C}-\text{C}-
\end{array}
$$

$$
\begin{array}{cccc}
 & 3' & & 5' \\
\text{Anticodon:} & -\text{C}-\text{G}-\text{I}- \\
 & \vdots & \vdots & \vdots \\
 & 5' & & 3' \\
\text{Codon:} & -\text{G}-\text{C}-\text{A}-
\end{array}
$$

It therefore seems that non-Watson–Crick base pairing can occur at the third codon–anticodon position (the anti-codon's first position is defined as its 3' nucleotide), the site of most codon degeneracy (Table 32-2). Note also that the third (5') anticodon position commonly contains a modified base such as Gm or I.

b. The Wobble Hypothesis Structurally Accounts for Codon Degeneracy

By combining structural insight with logical deduction, Crick proposed, in what he named the **wobble hypothesis,** how a tRNA can recognize several degenerate codons. He assumed that the first two codon–anticodon pairings have normal Watson–Crick geometry. The structural constraints that this places on the third codon–anticodon pairing ensure that its conformation does not drastically differ from that of a Watson–Crick pair. Crick then proposed that there could be a small amount of play or "wobble" in the third codon position which allows limited conformational adjustments in its pairing geometry. This permits the formation of several non-Watson–Crick pairs such as U · G and I · A (Fig. 32-25a). The allowed "wobble" pairings are indicated in Fig. 32-25b. Then, by analyzing the known pattern of codon–anticodon pairing, Crick deduced the most plausible sets of pairing combinations in the third codon–anticodon position (Table 32-5). Thus, an anticodon

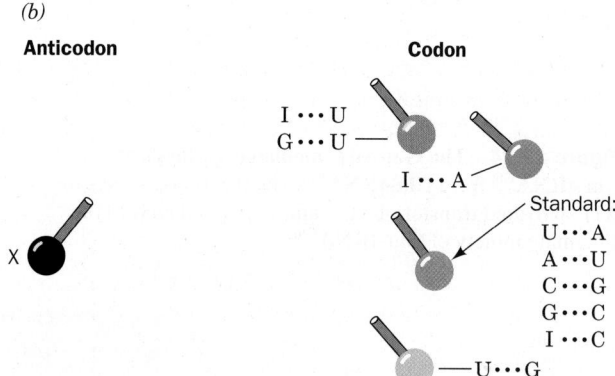

Figure 32-25 **Wobble pairing.** (*a*) U · G and I · A wobble pairs. Both have been observed in X-ray structures. (*b*) The geometry of wobble pairing. The spheres and their attached bonds represent the positions of ribose C1' atoms with their accompanying glycosidic bonds. X (*left*) designates the nucleoside at the 5' end of the anticodon (tRNA). The positions on the right are those of the 3' nucleoside of the codon (mRNA) in the indicated wobble pairings. [After Crick, F.H.C., *J. Mol. Biol.* **19**, 552 (1966).]

with C or A in its third position can only pair with its Watson–Crick complementary codon. If U, G, or I occupies the third anticodon position, two, two, or three codons are recognized, respectively.

Table 32-5 **Allowed Wobble Pairing Combinations in the Third Codon–Anticodon Position**

5'-Anticodon Base	3'-Codon Base
C	G
A	U
U	A or G
G	U or C
I	U, C, or A

No prokaryotic or eukaryotic cytoplasmic tRNA is known to participate in a nonwobble pairing combination. There is, however, no known instance of such a tRNA with an A in its third anticodon position, which suggests that the consequent A · U pair is not permitted. The structural basis of wobble pairing is poorly understood, although it is clear that it is influenced by base modifications.

A consideration of the various wobble pairings indicates that at least 31 tRNAs are required to translate all 61 coding triplets of the genetic code (there are 32 tRNAs in the minimal set because translational initiation requires a separate tRNA; Section 32-3Ca). Most cells have >32 tRNAs, some of which have identical anticodons. In fact, mammalian cells have >150 tRNAs. Nevertheless, *all isoaccepting tRNAs in a cell are recognized by a single aminoacyl–tRNA synthetase.*

c. Some Mitochondrial tRNAs Have More Permissive Wobble Pairings than Other tRNAs

The codon recognition properties of mitochondrial tRNAs must reflect the fact that mitochondrial genetic codes are variants of the "standard" genetic code (Table 32-3). For instance, the human mitochondrial genome, which consists of only 16,569 bp, encodes 22 tRNAs (together with 2 ribosomal RNAs and 13 proteins). Fourteen of these tRNAs each read one of the synonymous pairs of codons indicated in Tables 32-2 and 32-3 (MNX, where X is either C or U or else A or G) according to normal G · U wobble rules: The tRNAs have either a G or a modified U in their third anticodon position that, respectively, permits them to pair with codons having X = C or U or else X = A or G. The remaining 8 tRNAs, which, contrary to wobble rules, each recognize one of the groups of four synonymous codons (MNY, where Y = A, C, G, or U), all have anticodons with a U in their third position. Either this U can somehow pair with any of the four bases or these tRNAs read only the first two codon positions and ignore the third. Thus, not surprisingly, many mitochondrial tRNAs have unusual structures in which, for example, the GTψCRA sequence (Fig. 32-9) is missing, or, in the most bizarre case, a tRNA^Ser lacks the entire D arm.

d. Frequently Used Codons Are Complementary to the Most Abundant tRNA Species

The analysis of the base sequences of several highly expressed structural genes of *S. cerevisiae* has revealed a remarkable bias in their codon usage. Only 25 of the 61 coding triplets are commonly used. *The preferred codons are those that are most nearly complementary, in the Watson–Crick sense, to the anticodons in the most abundant species in each set of isoaccepting tRNAs.* Furthermore, codons that bind anticodons with two consecutive G · C pairs or three A · U pairs are avoided so that the preferred codon–anticodon complexes all have approximately the same binding free energies. A similar phenomenon occurs in *E. coli*, although several of its 22 preferred codons differ from those in yeast. The degree with which the preferred codons occur in a given gene is strongly correlated, in both organisms, with the gene's level of expression (the measured rates of aminoacyl–tRNA selection in *E. coli* span a

25-fold range). This, it has been proposed, permits the mRNAs of proteins that are required in high abundance to be rapidly and smoothly translated.

e. Selenocysteine and Pyrrolysine Are Carried by Specific tRNAs

Although it is widely stated, even in this text, that proteins are synthesized from the 20 "standard" amino acids, that is, those specified by the "standard" genetic code, some organisms, as Theresa Stadtman discovered, use a twenty-first amino acid, **selenocysteine (Sec; alternatively SeCys)**, in synthesizing a few of their proteins:

$$
\begin{array}{l}
| \\
NH \\
| \\
CH\!-\!CH_2\!-\!Se\!-\!H \\
| \\
C\!=\!O \\
|
\end{array}
$$

The selenocysteine (Sec) residue

Selenium, a biologically essential trace element, is a component of several enzymes in both prokaryotes and eukaryotes. These include thioredoxin reductase (Section 28-3Ae) and the **thyroid hormone deiodinases** (which participate in thyroid hormone synthesis; Section 19-1D) in mammals and three forms of **formate dehydrogenases** in *E. coli*, all of which contain Sec residues. The Sec residues are ribosomally incorporated into these proteins by a unique tRNA, **tRNA^Sec,** bearing a UCA anticodon that is specified by a particular (in the mRNA) UGA codon (normally the *opal* Stop codon). The Sec–tRNA^Sec is synthesized by the aminoacylation of tRNA^Sec with L-serine by the same SerRS that charges tRNA^Ser, followed by the enzymatic selenylation of the resulting Ser residue.

How does the ribosomal system differentiate a Sec-specifying UGA codon from a normal opal Stop codon? As we saw to be the case with Glu–tRNA^Gln (Section 32-2Cf), EF-Tu, the elongation factor that conducts most aminoacyl–tRNAs to the ribosome in a GTP-dependent process, does not bind Sec–tRNA^Sec. Instead it is bound by a specific elongation factor named **SELB**, a homolog of EF-Tu, that in its complex with GTP is recruited to a ribosomally bound mRNA stem–loop structure in the selenoprotein coding region on the 3′ side of the UGA codon specifying Sec.

Certain methanogenic (methane-producing) archaea express the enzyme **methylamine methyltransferase,** which contains the amino acid residue **pyrrolysine (Pyl),** a Lys with its ε-nitrogen in amide linkage to a pyrroline group:

$$
\begin{array}{l}
| \\
NH \\
| \\
CH\!-\!CH_2\!-\!CH_2\!-\!CH_2\!-\!CH_2\!-\!NH\!-\!C\!\!=\!\!O \cdots CH_3 \\
| \\
C\!=\!O \\
|
\end{array}
$$

The pyrrolysine (Pyl) residue

Unlike post-translationally modified Lys residues, such as 5-hydroxylysine (Hyl; Section 8-2B) and ε-*N*-acetyllysine

(Section 4-3B), Pyl is directly incorporated into proteins during translation. Pyl is specified by the codon UAG (normally the *amber* Stop codon). Pyl is carried to the ribosome by **tRNA^Pyl**, which contains a CUA anticodon and differs from typical tRNAs in having a D loop with five rather than eight residues, an anticodon stem with six rather than five base pairs, and a TψC loop that lacks the sequence TψC. A specific aminoacyl–tRNA synthetase, **PylRS**, that differs from known LysRSs, charges tRNA^Pyl with pyrrolysine in an ATP-dependent reaction, the first known example in nature of the direct aminoacylation of a tRNA with a "nonstandard" amino acid. Unlike the case for Ser–tRNA^Sec, Pyl–tRNA^Pyl is delivered to the ribosome by EF-Tu. This suggests that the mRNA contains a signal that causes UAG to be read as a Pyl codon rather than as a Stop codon. A conserved stem–loop structure located on the 3′ side of UAG codons specifying Pyl may comprise this signal. Alternatively, a Pyl–tRNA^Pyl may occasionally read a UAG codon and is therefore a type of nonsense suppressor (see below).

E. Nonsense Suppression

Nonsense mutations are usually lethal when they prematurely terminate the synthesis of an essential protein. An organism with such a mutation may nevertheless be "rescued" by a second mutation on another part of the genome. For many years after their discovery, the existence of such **intergenic suppressors** was quite puzzling. It is now known, however, that they usually arise from mutations in a tRNA gene that causes the tRNA to recognize a nonsense codon. Such a **nonsense suppressor** tRNA appends its amino acid (which is the same as that carried by the corresponding wild-type tRNA) to a growing polypeptide in response to the recognized Stop codon, thereby preventing chain termination. For example, the *E. coli* amber suppressor known as **su3** is a tRNA^Tyr whose anticodon has mutated from the wild-type GUA (which reads the Tyr codons UAU and UAC) to CUA (which recognizes the amber Stop codon UAG). An *su3*^+ *E. coli* with an otherwise lethal *amber* mutation in a gene coding for an essential protein would be viable if the replacement of the wild-type amino acid residue by Tyr does not inactivate the protein.

There are several well-characterized examples of *amber* (UAG), *ochre* (UAA), and *opal* (UGA) suppressors in *E. coli* (Table 32-6). Most of them, as expected, have mutated anticodons. **UGA-1** tRNA, however, differs from the wild type only by a G → A mutation in its D stem, which changes a G · U pair to a stronger A · U pair. This mutation apparently alters the conformation of the tRNA's CCA anticodon so that it can form an unusual wobble pairing with UGA as well as with its normal codon, UGG. Nonsense suppressors also occur in yeast.

a. Suppressor tRNAs Are Mutants of Minor tRNAs

How do cells tolerate a mutation that both eliminates a normal tRNA and prevents the termination of polypeptide synthesis? They survive because the mutated tRNA is usually a minor member of a set of isoaccepting tRNAs and be-

Table 32-6 **Some *E. coli* Nonsense Suppressors**

Name	Codon Suppressed	Amino Acid Inserted
*su*1	UAG	Ser
*su*2	UAG	Gln
*su*3	UAG	Tyr
*su*4	UAA, UAG	Tyr
*su*5	UAA, UAG	Lys
*su*6	UAA	Leu
*su*7	UAA	Gln
UGA-1	UGA	Trp
UGA-2	UGA	Trp

Source: Körner, A.M., Feinstein, S.I., and Altman, S., *in* Altman, S. (Ed.), *Transfer RNA,* p. 109, MIT Press (1978).

cause nonsense suppressor tRNAs must compete for Stop codons with the protein factors that mediate the termination of polypeptide synthesis (Section 32-3F). Consequently, the rate of suppressor-mediated synthesis of active proteins with either UAG or UGA nonsense mutations rarely exceeds 50% of the wild-type rate, whereas mutants with UAA, the most common termination codon, have suppression efficiencies of <5%. Many mRNAs, moreover, have two tandem Stop codons so that even if their first Stop codon were suppressed, termination could occur at the second. Nevertheless, many suppressor-rescued mutants grow relatively slowly because they cannot make an otherwise prematurely terminated protein as efficiently as do wild-type cells.

Other types of suppressor tRNAs are also known. **Missense suppressors** act similarly to nonsense suppressors but substitute one amino acid in place of another. **Frameshift suppressors** have eight nucleotides in their anticodon loops rather than the normal seven. They read a four-base codon beyond a base insertion thereby restoring the wild-type reading frame.

3 RIBOSOMES AND POLYPEPTIDE SYNTHESIS

Ribosomes were first seen in cellular homogenates by dark-field microscopy in the late 1930s by Albert Claude who referred to them as "microsomes." It was not until the mid-1950s, however, that George Palade observed them in cells by electron microscopy, thereby disposing of the contention that they were merely artifacts of cell disruption. The name ribosome derives from the fact that these particles in *E. coli* consist of approximately two-thirds RNA and one-third protein. (**Microsomes** are now defined as the artifactual vesicles formed by the endoplasmic reticulum on cell disruption. They are easily isolated by differential centrifugation and are rich in ribosomes.) The correlation between the amount of RNA in a cell and the rate at which it synthesizes protein led to the suspicion that ribosomes are the site of protein synthesis. This hypothesis was confirmed in 1955 by Paul Zamecnik, who demonstrated that ^14C-labeled amino acids are transiently associated with

Table 32-7 **Components of *E. coli* Ribosomes**

	Ribosome	Small Subunit	Large Subunit
Sedimentation coefficient	70S	30S	50S
Mass (kD)	2520	930	1590
RNA			
Major		16S, 1542 nucleotides	23S, 2904 nucleotides
Minor			5S, 120 nucleotides
RNA mass (kD)	1664	560	1104
Proportion of mass	66%	60%	70%
Proteins		21 polypeptides	31 polypeptides
Protein mass (kD)	857	370	487
Proportion of mass	34%	40%	30%

ribosomes before they appear in free proteins. Further research showed that ribosomal polypeptide synthesis has three distinct phases: (1) chain initiation, (2) chain elongation, and (3) chain termination.

In this section we examine the structure of the ribosome and then outline the ribosomal mechanism of polypeptide synthesis. In doing so we shall compare the properties of ribosomes from prokaryotes with those of eukaryotes.

A. *Ribosome Structure*

The *E. coli* ribosome, which has a particle mass of ~2.5 × 10^6 D and a sedimentation coefficient of 70S, is a spheroidal particle that is ~250 Å across in its largest dimension. It may be dissociated, as James Watson discovered, into two unequal subunits (Table 32-7). The small (30S) subunit consists of a 16S rRNA molecule and 21 different polypeptides, whereas the large (50S) subunit contains a 5S and a 23S rRNA together with 31 different polypeptides. The up to 20,000 ribosomes in an *E. coli* cell account for ~80% of its RNA content and ~10% of its protein.

Structural studies of the ribosome through electron microscopy began soon after its discovery. Three-dimensional (3D) structures of the ribosome and its subunits at low (~50 Å) resolution first became available in the 1970s through image reconstruction techniques, pioneered by Klug, in which electron micrographs of a single particle or ordered sheets of particles taken from several directions are combined to yield its 3D image. The small subunit is a roughly mitten-shaped particle, whereas the large subunit is spheroidal with three protuberances on one side (Fig. 32-26).

Figure 32-26 A low resolution model of the *E. coli* ribosome. The small subunit (*top; red*) associates with the large subunit (*middle; green*) to form the intact ribosome (*bottom*). Two perpendicular views of each particle are provided. These models are based on transmission electron micrographs of negatively stained particles (in which the particle being imaged is embedded in electron-absorbing heavy metal salts, thereby providing contrast between the relatively electron-transparent particle and the background).

(a) 16S RNA   (b)   **23S RNA**

Figure 32-27 **Secondary structures of the *E. coli* ribosomal RNAs.** (*a*) 16S RNA and (*b*) 23S and 5S RNAs. The rRNAs are colored by domain with short lines spanning a stem representing Watson–Crick base pairs, small dots representing G · U base pairs, and large dots representing other non-Watson–Crick base pairs. Note the flowerlike series of stems and loops forming each domain. [Courtesy of Venki Ramakrishnan, MRC Laboratory of Molecular Biology, Cambridge, U.K., and Peter Moore, Yale University. Adapted from diagrams in http://www.rna.ccbb.utexas.edu/.]

a. Ribosomal RNAs Have Complicated Secondary Structures

The *E. coli* 16S rRNA, which was sequenced by Harry Noller, consists of 1542 nucleotides. A computerized search of this sequence for stable double helical segments yielded many plausible but often mutually exclusive secondary structures. However, the comparison of the sequences of 16S rRNAs from several prokaryotes, under the assumption that their structures have been evolutionarily conserved, led to the flowerlike secondary structure for 16S rRNA seen in Fig. 32-27*a*. This four-domain structure is 54% base paired. Its double helical stems tend to be short (<8 bp) and many of them are imperfect. Intriguingly, electron micrographs of the 16S rRNA resemble those of the complete 30S subunit, thereby suggesting that the 30S subunit's overall shape is largely determined by the 16S rRNA. The large ribosomal subunit's 5S and 23S rRNAs, which consist of 120 and 2904 nucleotides, respectively, have also been sequenced. As with the 16S rRNA, they have extensive secondary structures (Fig. 32-27*b*).

b. Ribosomal Proteins Have Been Partially Characterized

Ribosomal proteins are difficult to separate because most of them are insoluble in ordinary buffers. By convention, ribosomal proteins from the small and large subunits are designated with the prefixes S and L, respectively, followed by a number indicating their position, from upper left to lower right, on a two-dimensional gel electrophoretogram (roughly in order of decreasing molecular mass; Fig. 32-28). Only protein S20/L26 appears to be common to both subunits. One of the large subunit proteins is partially acetylated at its N-terminus so that it gives rise to two elec-

trophoretic spots (L7/L12). Four copies of this protein, a dimer of dimers, are present in the large subunit. Moreover, these four copies of L7/L12 aggregate with L10 to form a stable complex that was initially thought to be a unique protein, "L8." All the other ribosomal proteins occur in only one copy per subunit.

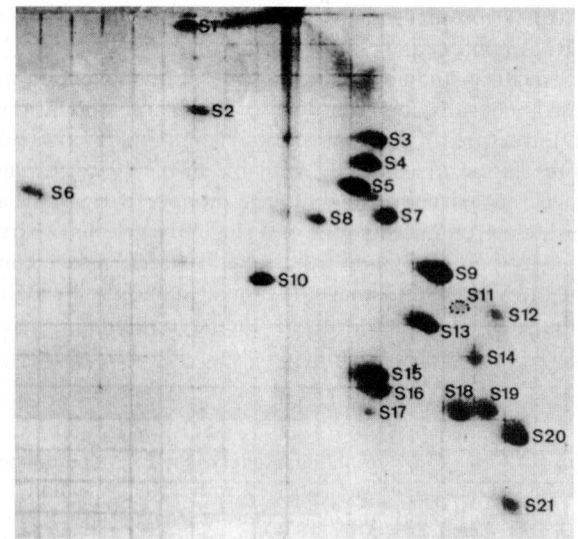

Figure 32-28 **Two-dimensional gel electrophoretogram of *E. coli* small ribosomal subunit proteins.** First dimension (*vertical*): 8% acrylamide, pH 8.6; second dimension (*horizontal*): 18% acrylamide, pH 4.6. [From Kaltschmidt, E. and Wittmann, H.G., *Proc. Natl. Acad. Sci.* **67,** 1277 (1970).]

The amino acid sequences of all 52 *E. coli* ribosomal proteins were elucidated, mainly by Heinz-Günter Wittmann and Brigitte Wittmann-Liebold. They range in size from 46 residues for L34 to 557 residues for S1. Most of these proteins, which exhibit little sequence similarity with one another, are rich in the basic amino acids Lys and Arg and contain few aromatic residues as is expected for proteins that are closely associated with polyanionic RNA molecules.

The X-ray and NMR structures of around half of the ribosomal proteins or their fragments have been independently determined. These proteins have a wide variety of structural motifs although most of their folds occur in other proteins of known structure. Around one-third of these ribosomal proteins contain the **RNA-recognition motif (RRM;** Section 31-4Ab), which occurs in >200 RNA-binding proteins including rho factor (the transcriptional termination protein, which contains four such motifs; Section 31-2Db), poly(A) polymerase, poly(A)-binding protein (PABP), several proteins involved in gene splicing (Section 31-4A), and the translational initiation factor **eIF4B** (Section 32-3Cd). All of these proteins presumably evolved from an ancient RNA-binding protein.

c. Ribosomal Subunits Are Self-Assembling

Ribosomal subunits form, under proper conditions, from mixtures of their numerous macromolecular components. *Ribosomal subunits are therefore self-assembling entities.* Masayasu Nomura determined the order in which this occurs through partial reconstitution experiments. If one macromolecular component is left out of an otherwise self-assembling mixture of proteins and RNA, the other components that fail to bind to the resulting partially assembled subunit must somehow interact with the omitted component. Through the analysis of a series of such partial reconstitution experiments, Nomura constructed an assembly map of the small (30S) subunit (Fig. 32-29). This map indicates that the initial steps in small subunit assembly are the independent binding to naked 16S rRNA of six so-called primary (1°) binding proteins (S4, S7, S8, S15, S17, and S20). The resulting assembly intermediates provide the molecular scaffolding for binding secondary (2°) binding proteins, which after a significant conformational change, form the attachment sites for tertiary (3°) binding proteins. An analogous assembly map for the large subunit was elucidated by Knud Nierhaus. The observation that similar assembly intermediates occur *in vivo* and *in vitro* suggests that *in vivo* and *in vitro* assembly processes are much alike.

In the cell, the 16S RNA folds in an ordered manner such that each domain is folded before the next domain is transcribed. The assembly of the small ribosomal subunit is then facilitated by a variety of **assembly factors,** proteins that bind to immature complexes but not to mature subunits. Many assembly factors associate with segments of the 16S RNA that change conformation during the latter stages of assembly. Presumably, the assembly of the large ribosomal subunit follows a similar course.

d. The Atomic Structure of the Prokaryotic Ribosome Has Been Long in Coming

The elucidation of the ribosome's atomic structure was a tortuous affair extending over four decades in which slow incremental improvements were occasionally punctuated by significant technical gains. The process began in the 1960s with shadowy transmission electron micrographs that provided only rough 2D shapes. This was followed in the 1970s by image reconstruction techniques that generated 3D models although still at low resolution (Fig. 32-26). Later in the 1970s, the sites of many of the ribosome's proteins were determined by James Lake and Georg Stöffler through **immune electron microscopy,** a technique in which antibodies raised against a particular ribosomal protein are used to mark its position in electron micrographs of the antibody complexed to a ribosomal subunit. These results were improved and extended in the 1980s by neutron scattering experiments conducted by Donald Engelman and Peter Moore on the 30S subunit, which indicated the distances between the centers of mass of its component proteins and hence their three-dimensional distribution. These structural studies were supplemented by a variety of chemical cross-linking and fluorescence transfer studies that demonstrated the proximity of various ribosomal components.

The molecular structure of the prokaryotic ribosome began to come into focus in the mid-1990s through the development of **cryoelectron microscopy (cryo-EM).** In this technique, the sample is cooled to near liquid N_2 temperatures ($-196°C$) so rapidly (in a few milliseconds) that the water in the sample does not have time to crystallize but,

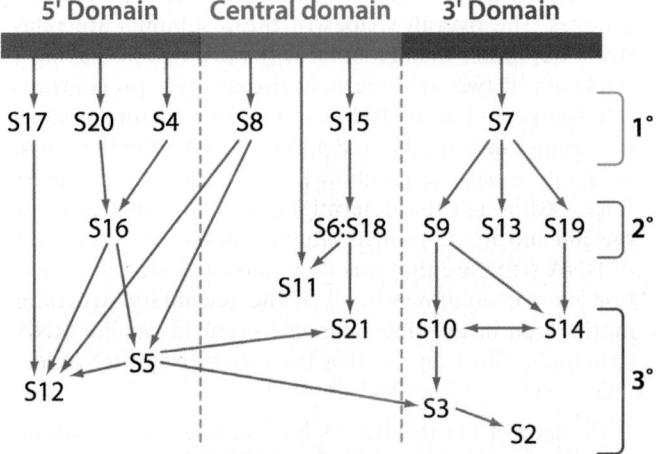

Figure 32-29 **Assembly map of the *E. coli* small subunit.** The map is organized according to the domains of the 16S RNA (Fig. 32-27) with arrows indicating the facilitation of binding. For example, the arrow from the 5′ domain of the 16S rRNA to S20 indicates that S20 binds directly to the 16S rRNA in the absence of other proteins and is therefore a primary (1°) binding protein; the arrow from S20 to S16 indicates that S20 facilitates the binding of S16, which is therefore a secondary (2°) binding protein; and the arrows from S16 to S5 and S12 indicate that S5 and S12 are tertiary (3°) binding proteins. [Courtesy of James Williamson, The Scripps Research Institute, La Jolla, California.]

Figure 32-30 **Cryoelectron microscopy–based image of the *E. coli* ribosome.** The 30S subunit (*yellow*) is on the left and the 50S subunit (*blue*) is on the right. The tRNAs that occupy the A, P, and E sites (Section 32-3B) are colored magenta, green, and brown. A portion of the 50S subunit has been cut away to reveal the polypeptide exit tunnel. A segment of mRNA (*5′ end brown and 3′ end lavender*) and the nascent polypeptide chain (*yellow*) have been modeled into the structure. [Courtesy of Joachim Frank, State University of New York at Albany.]

rather, assumes a vitreous (glasslike) state. Consequently, the sample remains hydrated and hence retains its native shape to a greater extent than in conventional electron microscopy (in which the sample is vacuum dried). Studies, carried out in large part by Joachim Frank, revealed the positions where tRNAs and mRNA as well as various soluble protein factors bind to the ribosome (Fig. 32-30). The highest resolution achieved by cryo-EM of ribosomes has gradually improved over the years to ~8 Å.

Ribosomal subunits were first crystallized by Ada Yonath in 1980 although they diffracted X-rays poorly. Over the course of several years, however, the quality of these crystals were incrementally improved until, in 1991, Yonath reported crystals of the 50S subunit that diffracted X-rays to 3-Å resolution. It was not until later in the 1990s, however, that technology was up to the task of determining the X-ray structures of these gargantuan molecular complexes. In 2000, the *annus mirabilis* (miracle year) of ribosomology, Moore and Steitz reported the X-ray structure of the 50S ribosomal subunit of the halophilic (salt-loving) bacterium *Haloarcula marismortui* at atomic (2.4-Å) resolution and Venki Ramakrishnan and Yonath independently reported the X-ray structure of the 30S subunit of *T. thermophilus* at ~3-Å resolution. In 2001, Noller reported the 5.5-Å resolution structure of the entire *T. thermophilus* ribosome, which was gradually improved to 2.8 Å. In addition, the structures of the *E. coli* ribosome, by Jamie Cate, and the large subunit from *Deinococcus radiodurans*, by Yonath, have been determined. In the following paragraphs we discuss the properties of these groundbreaking structures. We consider their functional implications starting in Section 32-3C.

e. Ribosomal Architecture

Several generalizations can be made about ribosomal architecture based on the structures of the 30S and 50S subunits:

1. Both the 16S and 23S rRNAs are assemblies of coaxially stacked helical elements connected by loops, most of which are irregular extensions of helices (Fig. 32-31). These structures, which are in close accord with previous secondary structure predictions (Fig. 32-27), are stabilized by interactions between helices such as minor groove to minor groove packing, which also occur in the structures of the group I intron and RNase P (Sections 31-4Af and 31-4Ca; recall that A-form RNA has a very shallow minor groove); the insertion of a phosphate ridge into a minor groove; and adenines that are distant in sequence but often highly conserved that are inserted into minor grooves. The overall shapes of these subunits are relatively flat rather than globular with most regions having a thickness of two or three helical diameters [in contrast, other large (>100 nt) RNAs of known structure, such as the group I intron, RNase P, and the RNA from the signal recognition particle (Section 12-4Bb), are only one helix thick]. Although the determination of the structures of the 30S and 50S ribosomal subunits increased the amount of RNA structure that was then known at atomic resolution by ~10-fold, nearly all of the secondary structural motifs seen in the ribosome also occur in smaller RNA structures. This suggests that the repertoire of RNA secondary structural motifs is limited.

2. Each of the 16S RNA's four domains, which extend out from a central junction (Fig. 32-27a), forms a morphologically distinct portion of the 30S subunit (Fig. 32-31a): The 5′ domain forms most of the body (Fig. 32-26), the central domain forms the platform, the 3′ major domain forms the entire head, and the 3′ minor domain, which consists of just two helices, is located at the interface between the 30S and 50S subunits. In contrast, the 23S RNA's six domains (Fig. 32-27b) are intricately intertwined in the 50S subunit (Fig. 32-31b). Since the ribosomal proteins are embedded in the RNA (see below), this suggests that the domains of the 30S subunit can move relative to one

Interface View   **Back View**

Figure 32-31 Tertiary structures of the ribosomal RNAs. (*a*) The 16S rRNA of *T. thermophilus.* (*b*) The 23S rRNA of *H. marismortui.* The rRNAs are colored according to domain as in Fig. 32-27. The interface view of a ribosomal subunit (*left*) is toward its surface that associates with the other subunit in the whole ribosome and the back view (*right*) is from the opposite (solvent-exposed) side. Note that the secondary structure domains of the 16S rRNA fold as separate tertiary structure domains, whereas in the 23S rRNA the secondary structure domains are convoluted together. [Courtesy of Venki Ramakrishnan, MRC Laboratory of Molecular Biology, Cambridge, U.K., and Peter Moore, Yale University. PDBids 1J5E and 1JJ2.]

another during protein synthesis, whereas the 50S subunit appears to be rigid.

3. The distribution of the proteins in the two ribosomal subunits is not uniform (Fig. 32-32). The vast majority of the ribosomal proteins are located on the back and sides of their subunits. In contrast, the face of each subunit that forms the interface between the two subunits, particularly those regions that bind the tRNAs and mRNA (see below), is largely devoid of proteins.

4. Most ribosomal proteins consist of a globular domain, which is, for the most part, located on a subunit surface (Fig. 32-32), and a long segment that is largely devoid of secondary structure and unusually rich in basic residues that infiltrates between the RNA helices into the subunit interior (Fig. 32-33). Indeed, a few ribosomal proteins lack a globular domain altogether (e.g., L39e in Fig. 32-33*b*). Ribosomal proteins make far fewer base-specific interactions than do other known RNA-binding proteins. They tend to interact with the RNA through salt bridges between their positively charged side chains and the RNAs' negatively charged phosphate oxygen atoms, thereby neutralizing the repulsive charge–charge interactions between nearby RNA segments. This is consistent with the hypothesis that

Interface view   **Back view**

Figure 32-32 Distribution of protein and RNA in the ribosomal subunits. (*a*) The 30S subunit of *T. thermophilus.* (*b*) The 50S subunit of *H. marismortui.* The subunits are drawn in space-filling form with their RNAs gray and their proteins in various colors. Note that the interface side of each subunit is largely free of protein, particularly in its regions that interact with mRNA and tRNAs. [Part *a* based on an X-ray structure by Venki Ramakrishnan, MRC Laboratory of Molecular Biology, Cambridge, U.K. Part *b* based on an X-ray structure by Peter Moore and Thomas Steitz, Yale University. PDBids 1J5E and 1JJ2.]

(a)

(b)

Figure 32-33 Gallery of ribosomal protein structures. Proteins from (*a*) the 30S subunit and (*b*) the 50S subunit. The proteins are represented by their backbones with their globular portions green and their highly extended segments red. The globular portions are exposed on the surface of their associated subunit (Fig. 32-32), whereas the extended segments are largely buried in the RNA. The Zn^{2+} ions bound by L37e and L44e are represented by magenta spheres. [Courtesy of Venki Ramakrishnan, MRC Laboratory of Molecular Biology, Cambridge, U.K., and Peter Moore, Yale University. PDBid 1J5E.]

the primordial ribosome consisted entirely of RNA (the RNA world) and that the proteins that were eventually acquired stabilized its structure and fine-tuned its function.

The X-ray structure of the entire *T. thermophilus* ribosome in complex with three tRNAs and a 11-nt mRNA segment was determined by Ramakrishnan (Fig. 32-34; ribosomes, as we shall see in Section 32-3B, have three functionally distinct tRNA-binding sites known as the A, P, and E sites). The structures of the 30S and 50S subunits in this enormous molecular machine closely resemble those of the isolated subunits although there are several regions at the subunit interface that exhibit significant conformational shifts, which suggests that these changes occur as a consequence of subunit association. In addition, several disordered portions of the isolated *H. marismortui* 50S subunit are ordered in the intact *T. thermophilus* ribosome, although this may be a consequence of the latter's greater thermal stability.

The ribosome binds all three tRNAs in a similar manner with their anticodon stem–loops bound to the 30S subunit and their remaining portions, the D stem, elbow, and acceptor stem, bound to the 50S subunit. These interactions, which mainly consist of RNA–RNA contacts, are made to the tRNAs' universally conserved segments, thereby permitting the ribosome to bind different species of tRNAs in the similar ways.

The small and large ribosomal subunits contact each other at 12 positions via RNA–RNA, protein–protein, and

RNA–protein bridges. These intersubunit bridges have a distinct distribution: The RNA–RNA bridges are centrally located adjacent to the three bound tRNAs, whereas the protein–protein and RNA–protein bridges are peripherally located away from the ribosome's functional sites. The RNA–RNA contacts consist mainly of minor groove–minor groove interactions although major groove, loop, and backbone contacts also occur. In the RNA–protein bridges, the proteins contact nearly all types of RNA features including major groove, minor groove, backbone, and loop elements.

We discuss the path of the mRNA and how it interacts with the tRNAs in Section 32-3D. There we shall see that *the large subunit is mainly involved in mediating biochemical tasks such as catalyzing the reactions of polypeptide elongation, whereas the small subunit is the major actor in ribosomal recognition processes such as mRNA and tRNA binding* (although, as we have seen, the large subunit also participates in tRNA binding). We shall also see that *rRNA has the major functional role in ribosomal processes* (recall that RNA has demonstrated catalytic properties; Sections 31-4Ae and 31-4Ca).

f. Eukaryotic Ribosomes Are Larger and More Complex than Prokaryotic Ribosomes

Although eukaryotic and prokaryotic ribosomes resemble one another in both structure and function, they differ in nearly all details. Eukaryotic ribosomes have particle masses in the range 3.9 to 4.5×10^6 D and have a nominal

(a)

16S RNA · 5S RNA · 23S RNA

(b)

P · E · A · mRNA · 16S RNA

(c)

5S RNA · E P · A · 23S RNA · **Peptidyl transferase center**

(d)

A · P · E · mRNA

Figure 32-34   **X-ray structure of the *T. thermophilus* ribosome in complex with tRNA and mRNA at 3.5 Å resolution.** The E site binds tRNA^Phe and the A and P sites bind Phe–tRNA^Phe (in which the O atom forming an ester linkage from Phe to O3′ of the 3′ terminal A has been replaced by an NH group to prevent the hydrolysis of the Phe—tRNA linkage). (*a*) The RNA components of the ribosomal complex (its proteins are omitted for clarity) drawn in cartoon form except for the 11-residue mRNA, which is shown in space-filling form. The 16S RNA is cyan, the 23S RNA is yellow, the 5S RNA is blue, the tRNAs in the A, P, and E sites are green, magenta, and red, and the mRNA, which is largely occluded by the 16S RNA, is colored according to atom type with C pink, N blue, O red, and P orange. (*b*) The 16S RNA in interface view with its bound tRNAs and mRNA, all

represented as in Part *a*. (*c*) The 23S RNA in interface view (rotated 180° about the vertical direction relative to Part *b*) with its bound tRNAs all represented as in Part *a*. (*d*) The interactions of the tRNAs with the mRNA. This assembly is drawn in cartoon form with A-site C green, P-site C magenta, E-site C red, mRNA C pink, N blue, and O red and with successive P atoms connected by orange rods. The Phe residues appended to the A- and P-site tRNAs are represented in space-filling form. Note the close approach of these Phe residues and that the tRNAs in the A and P sites, but not that in the E site, form based paired codon–anticodon interactions with the mRNA. [Based on an X-ray structure by Venki Ramakrishnan, MRC Laboratory of Molecular Biology, Cambridge, U.K. PDBids 2WDK and 2WDL.] **See Interactive Exercise 44.**

Table 32-8 **Components of Rat Liver Cytoplasmic Ribosomes**

	Ribosome	Small Subunit	Large Subunit
Sedimentation coefficient	80S	40S	60S
Mass (kD)	4220	1400	2820
RNA			
Major		18S, 1874 nucleotides	28S, 4718 nucleotides
Minor			5.8S, 160 nucleotides
			5S, 120 nucleotides
RNA mass (kD)	2520	700	1820
Proportion of mass	60%	50%	65%
Proteins		33 polypeptides	49 polypeptides
Protein mass (kD)	1700	700	1000
Proportion of mass	40%	50%	35%

sedimentation coefficient of 80S. They dissociate into two unequal subunits that have compositions that are distinctly different from those of prokaryotes (Table 32-8; compare with Table 32-7). The small **(40S)** subunit of the rat liver cytoplasmic ribosome, which together with the yeast ribosome is the most well-characterized eukaryotic ribosome, consists of 33 unique polypeptides and an **18S rRNA.** Its large **(60S)** subunit contains 49 different polypeptides and three rRNAs of **28S, 5.8S,** and **5S.** The additional complexity of the eukaryotic ribosome relative to its prokaryotic counterpart is presumably due to the eukaryotic ribosome's additional functions: Its mechanism of translational initiation is more complex (Section 32-3Cd); it must be transported from the nucleus, where it is formed, to the cy-toplasm, where translation occurs; and the machinery with which it participates in the secretory pathway is more complicated (Section 12-4B).

Sequence comparisons of the corresponding rRNAs from various species indicates that evolution has conserved their secondary structures rather than their base sequences (Figs. 32-27a and 32-35). For example, a G · C in a base paired stem of E. coli 16S rRNA has been replaced by an A · U in the analogous stem of yeast 18S rRNA. The **5.8S rRNA,** which occurs in the large eukaryotic subunit in base paired complex with the **28S rRNA,** is homologous in sequence to the 5′ end of prokaryotic 23S rRNA. Apparently 5.8S RNA arose through mutations that altered rRNA's post-transcriptional processing, producing a fourth rRNA.

Figure 32-35 **Predicted secondary structures of evolutionarily distant 16S-like rRNAs.** (*a*) Archaebacteria (*Halobacterium volcanii*), (*b*) eukaryotes (*S. cerevisiae*), and (*c*) mammalian mitochondria (bovine). Compare them with Fig. 32-27a, the secondary structure of 16S RNA from eubacteria (*E. coli*). Note the close similarities of these assemblies; they differ mostly by insertions and deletions of stem-and-loop structures. The 23S-like rRNAs from a variety of species likewise have similar secondary structures. [After Gutell, R.R., Weiser, B., Woese, C.R., and Noller, H.F., *Prog. Nucleic Acid Res. Mol. Biol.* **32,** 183 (1985).]

(a)

(b)

(c)

Figure 32-36 Cryo-EM–based image of the yeast 80S ribosome at 15 Å resolution. (*a*) The ribosome shown in side view analogous to Fig. 32-30 of the *E. coli* ribosome. The small (40S) subunit is yellow, the large (60S) subunit is cyan, and the tRNA that is bound in the ribosomal P site is green. Portions of this ribosome that are not homologous to the RNA or proteins of the *E. coli* ribosome are shown in gold for the small subunit and magenta for the large subunit. (*b*) The computationally isolated small subunit shown in interface view analogous to the left panel of Fig. 32-32*a*. (*c*) The computationally isolated large subunit shown in interface view analogous to the left panel of Fig. 32-32*b*. [Courtesy of Joachim Frank, State University of New York at Albany.] 🖱 **See the Interactive Exercise.**

The cryo-EM–based image of the yeast 80S ribosome (Fig. 32-36), determined at 15 Å resolution by Andrej Sali, Günter Blobel, and Frank, reveals that there is a high degree of structural conservation between eukaryotic and prokaryotic ribosomes. Although the yeast 40S subunit (which consists of a 1798-nt 18S rRNA and 32 proteins) contains an additional 256 nt of RNA and 11 proteins relative to the *E. coli* 30S subunit (Table 32-8; 15 of the *E. coli* proteins are homologous to those of yeast), both exhibit a similar division into head, neck, body, and platform (Fig. 32-36*b* vs Figs. 32-32*a* and 32-34*b*). Many of the differences between these two small ribosomal subunits are accounted for by the 40S subunit's additional RNA and proteins, although their homologous portions exhibit several distinct conformational differences. Similarly, the yeast 60S subunit (Fig. 32-35*c*; which consists of an aggregate of 3671 nt and 45 proteins) structurally resembles the considerably smaller (Table 32-7) prokaryotic 50S subunit (Fig. 32-32*b*). The yeast ribosome exhibits 16 intersubunit bridges, 12 of which match the 12 that were observed in the X-ray structure of the *T. thermophilus* ribosome, a remarkable evolu-

tionary conservation that indicates the importance of these bridges. Moreover, the tRNA that occupies the P site of the yeast ribosome has a conformation that more closely resembles that of the P-site tRNA in the *T. thermophilus* ribosome than that of free tRNA[Phe].

B. *Polypeptide Synthesis: An Overview*

Before we commence our detailed discussion of polypeptide synthesis, it will be helpful to outline some of its major features.

a. Polypeptide Synthesis Proceeds from N-Terminus to C-Terminus

The direction of ribosomal polypeptide synthesis was established, in 1961 by Howard Dintzis, through radioactive labeling experiments. He exposed reticulocytes that were actively synthesizing hemoglobin to ^{3}H-labeled leucine for times less than that required to make an entire polypeptide. The extent to which the tryptic peptides from the soluble (completed) hemoglobin molecules were labeled increased

Figure 32-37 **Demonstration that polypeptide synthesis proceeds from the N-terminus to the C-terminus.** Rabbit reticulocytes were incubated with [^{3}H]leucine. The curves show the distribution of [^{3}H]Leu among the tryptic peptides from the β subunit of soluble rabbit hemoglobin after the indicated incubation times. The numbers on the horizontal axis are peptide identifiers arranged from N-terminus to C-terminus. [After Dintzis, H.M., *Proc. Natl. Acad. Sci.* **47**, 255 (1961).]

with their proximity to the C-terminus (Fig. 32-37). Incoming amino acids must therefore be appended to a growing polypeptide's C-terminus; that is, *polypeptide synthesis proceeds from N-terminus to C-terminus.*

b. Ribosomes Read mRNA in the 5′ → 3′ Direction

The direction in which the ribosome reads mRNAs was determined through the use of a cell-free protein synthesizing system in which the mRNA was poly(A) with a 3′-terminal C.

$$5'\ A-A-A-\cdots-A-A-A-C\ 3'$$

Such a system synthesizes a poly(Lys) that has a C-terminal Asn.

$$\overset{+}{H_3N}-Lys-Lys-Lys-\cdots-Lys-Lys-Asn-COO^-$$

This, together with the knowledge that AAA and AAC code for Lys and Asn and the polarity of polypeptide synthesis, indicates that *the ribosome reads mRNA in the 5′ → 3′ direction.* Since mRNA is synthesized in the 5′ → 3′ direction, this accounts for the observation that, in prokaryotes, ribosomes initiate translation on nascent mRNAs (Section 31-3).

c. Active Translation Occurs on Polyribosomes

Electron micrographs, as Rich discovered, reveal that ribosomes engaged in protein synthesis are tandemly arranged on mRNAs like beads on a string (Fig. 31-27) to

form assemblies known as **polyribosomes (polysomes).** Polysomes arise because once an active ribosome has cleared its initiation site on an mRNA, a second ribosome can initiate at that site.

Cryo-EM–based image reconstruction of *E. coli* polysomes, by Ulrich Hartl and Wolfgang Baumeister, has revealed that adjacent ribosomes are densely packed along the mRNA and have staggered or pseudohelical arrangements (Fig. 32-38). The mRNA is located along the inside of these assemblies, the tRNA entrance sites are accessible, and the polypeptide exit tunnel on each ribosome faces the cytosol. Modeling suggests that such an arrangement maximizes the distance between the nascent polypeptide chains exiting neighboring ribosomes. This reduces the ability of these incompletely folded polypeptides to nonspecifically aggregate and hence increases the yield of natively folded protein.

d. Chain Elongation Occurs by the Linkage of the Growing Polypeptide to the Incoming tRNA's Amino Acid Residue

During polypeptide synthesis, amino acid residues are sequentially added to the C-terminus of the nascent, ribosomally bound polypeptide chain. If the growing polypeptide is released from the ribosome by treatment with high salt concentrations, its C-terminal residue is invariably esterified to a tRNA molecule as a **peptidyl–tRNA:**

Peptidyl–tRNA

The nascent polypeptide must therefore grow by being transferred from the peptidyl–tRNA to the incoming aminoacyl–tRNA to form a peptidyl–tRNA with one more

Figure 32-38 **Cryo–EM–based image of an *E. coli* polysome.**
The mRNA (which is mostly occluded) is represented by a red
line, the small ribosomal subunits are yellow, the large subunits
are blue-gray, and the red cones point to the polypeptide exit
tunnel on each large subunit. The assembly shown has a pseudohelical arrangement of ribosomes in which the center-to-center
distance between adjacent ribosomes averages ~230 Å.
Polysomes with somewhat different although equally densely
packed arrangements of ribosomes have also been observed.
[Courtesy of Ulrich Hartl and Wolfgang Baumeister, Max Planck
Institute of Biochemistry, Martinsreid, Germany.]

residue (Fig. 32-39). Apparently, the ribosome has at least
two tRNA-binding sites: the so-called **peptidyl** or **P site,**
which binds the peptidyl–tRNA, and the **aminoacyl** or **A
site,** which binds the incoming aminoacyl–tRNA (Fig. 32-
39). Consequently, after the formation of a peptide bond,
the newly deacylated P-site tRNA must be released and replaced by the newly formed peptidyl–tRNA from the A
site, thereby permitting a new round of peptide bond for-

mation. The finding by Knud Nierhaus that each ribosome
can bind up to three deacylated tRNAs but only two
aminoacyl–tRNAs indicates, however, that the ribosome
has a third tRNA-binding site, the **exit** or **E site,** which
transiently binds the outgoing deacylated tRNA. All three
sites, as we have seen (Fig. 32-34), extend over both ribosomal subunits.

The details of the chain elongation process are discussed in Section 32-3D. Chain initiation and chain termination, which are special processes, are examined in Sections 32-3C and 32-3F, respectively. In all of these sections
we shall first consider the process of interest in *E. coli* and
then compare it with the analogous eukaryotic activity.

C. *Chain Initiation*

a. fMet Is the N-Terminal Residue of Prokaryotic Polypeptides

The first indication that the initiation of translation requires a special codon, since identified as AUG (and, in
prokaryotes, occasionally GUG), was the observation that
almost half of the *E. coli* proteins begin with the otherwise
uncommon amino acid Met. This was followed by the discovery of a peculiar form of Met–tRNA^{Met} in which the
Met residue is *N*-formylated:

$$
\begin{array}{c}
\text{S}-\text{CH}_3 \\
| \\
\text{CH}_2 \\
| \\
\text{O} \quad\quad \text{CH}_2 \quad \text{O} \\
\| \quad\quad\quad | \quad\quad \| \\
\text{HC}-\text{NH}-\text{CH}-\text{C}-\text{O}-\text{tRNA}_f^{\text{Met}}
\end{array}
$$

***N*-Formylmethionine–tRNA$_f^{\text{Met}}$
(fMet–tRNA$_f^{\text{Met}}$)**

Peptidyl–tRNA   Aminoacyl–tRNA   Uncharged tRNA   Peptidyl–tRNA

Figure 32-39 **Ribosomal peptidyl transferase
reaction forming a peptide bond.** The ribosome
catalyzes the nucleophilic attack of the amino
group of the aminoacyl–tRNA in the A site on
the peptidyl–tRNA ester in the P site, thereby
forming a new peptide bond and transferring
the nascent polypeptide to the A-site tRNA,
while displacing the P-site tRNA.

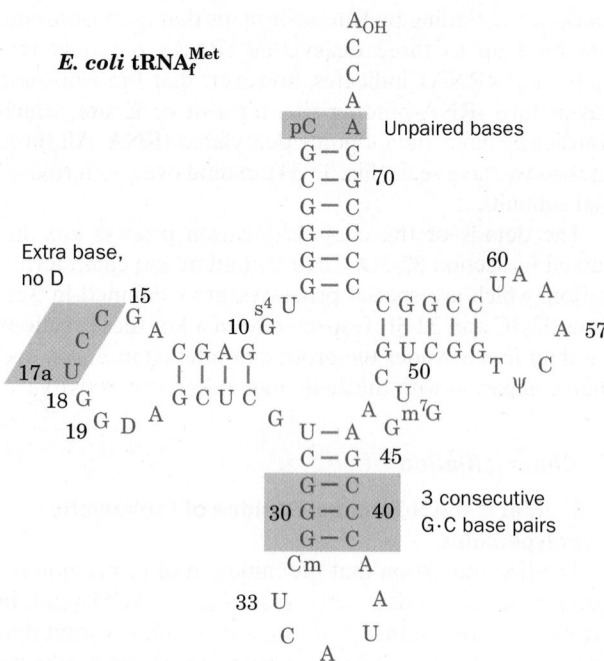

E. coli tRNA$_f^{Met}$

Unpaired bases

Extra base, no D

3 consecutive G·C base pairs

Figure 32-40 **Nucleotide sequence of *E. coli* tRNA$_f^{Met}$ shown in cloverleaf form.** The shaded boxes indicate the significant differences between this initiator tRNA and noninitiator tRNAs such as yeast tRNA^{Ala} (Fig. 32-8). [After Woo, N.M., Roe, B.A., and Rich, A., *Nature* **286**, 346 (1980).]

The ***N*-formylmethionine** residue **(fMet)** already has an amide bond and can therefore only be the N-terminal residue of a polypeptide. In fact, polypeptides synthesized in an *E. coli*–derived cell-free protein synthesizing system always have a leading fMet residue. *fMet must therefore be E. coli's initiating residue.*

The tRNA that recognizes the initiation codon, **tRNA$_f^{Met}$** (Fig. 32-40), differs from the tRNA that carries internal Met residues, **tRNA$_m^{Met}$**, although they both recog-

nize the same codon. In *E. coli,* uncharged (deacylated) tRNA$_f^{Met}$ is first aminoacylated with methionine by the same MetRS that charges tRNA$_m^{Met}$. The resulting **Met–tRNA$_f^{Met}$** is specifically *N*-formylated to yield **fMet–tRNA$_f^{Met}$** in an enzymatic reaction that employs N^{10}-formyltetrahydrofolate (Section 26-4D) as its formyl donor. The formylation enzyme does not recognize Met–tRNA$_m^{Met}$. The X-ray structures of *E. coli* tRNA$_f^{Met}$ and yeast tRNA^{Phe} (Fig. 32-11*b*) are largely similar but differ conformationally in their acceptor stems and anticodon loops. Perhaps these structural differences permit tRNA$_f^{Met}$ to be distinguished from tRNA$_m^{Met}$ in the reactions of chain initiation and elongation (see Section 32-3D).

*E. coli proteins are post-translationally modified by a **deformylase**, which hydrolytically deformylates their fMet residue, and, in many proteins, by the subsequent removal of the resulting N-terminal Met.* This processing usually occurs on the nascent polypeptide, which accounts for the observation that mature *E. coli* proteins all lack fMet.

b. Base Pairing between mRNA and the 16S rRNA Helps Select the Translational Initiation Site

AUG codes for internal Met residues as well as the initiating Met residue of a polypeptide. Moreover, mRNAs usually contain many AUGs (and GUGs) in different reading frames. Clearly, *a translational initiation site must be specified by more than just an initiation codon.* This occurs in two ways: (1) the masking of AUGs that are not initiation codons by mRNA secondary structure; and (2) interactions between the mRNA and the 16S rRNA that select the initiating AUG as we now discuss.

The 16S rRNA contains a pyrimidine-rich sequence at its 3′ end. This sequence, as John Shine and Lynn Dalgarno pointed out in 1974, is partially complementary to a purine-rich tract of 3 to 10 nucleotides, the **Shine–Dalgarno sequence,** that is centered ~10 nucleotides upstream from the start codon of nearly all known prokaryotic mRNAs (Fig. 32-41). *Base pairing interactions between an mRNA's*

Figure 32-41 **Some translational initiation sequences recognized by *E. coli* ribosomes.** The mRNAs are aligned according to their initiation codons (*blue shading*). Their Shine–Dalgarno sequences (*red shading*) are complementary, counting G · U pairs, to a portion of the 16S rRNA's 3′ end

(*green shading; below*). [After Steitz, J.A., *in* Chambliss, G., Craven, G.R., Davies, J., Davis, K., Kahan, L., and Nomura, M. (Eds.), *Ribosomes. Structure, Function and Genetics,* pp. 481–482, University Park Press (1979).]

Shine–Dalgarno sequence and the 16S rRNA apparently permit the ribosome to select the proper initiation codon. Thus ribosomes with mutationally altered anti-Shine–Dalgarno sequences often have greatly reduced ability to recognize natural mRNAs, although they efficiently translate mRNAs whose Shine–Dalgarno sequences have been made complementary to the altered anti-Shine–Dalgarno sequences. Moreover, treatment of ribosomes with the bactericidal protein **colicin E3** (produced by *E. coli* strains carrying the E3 plasmid), which specifically cleaves a 49-nucleotide fragment from the 3′ terminus of 16S rRNA, yields ribosomes that cannot initiate new polypeptide synthesis but can complete the synthesis of a previously initiated chain.

The X-ray structure of the 70S ribosome reveals, in agreement with Fig. 32-30, that a 27-nt segment of the mRNA is wrapped in a groove that encircles the neck of the 30S subunit (Fig. 32-42). The mRNA codons in the A and P sites are exposed on the interface side of the 30S subunit (as in Fig. 32-34*b*), whereas its 5′ and 3′ ends are bound in tunnels composed of RNA and protein. The mRNA's

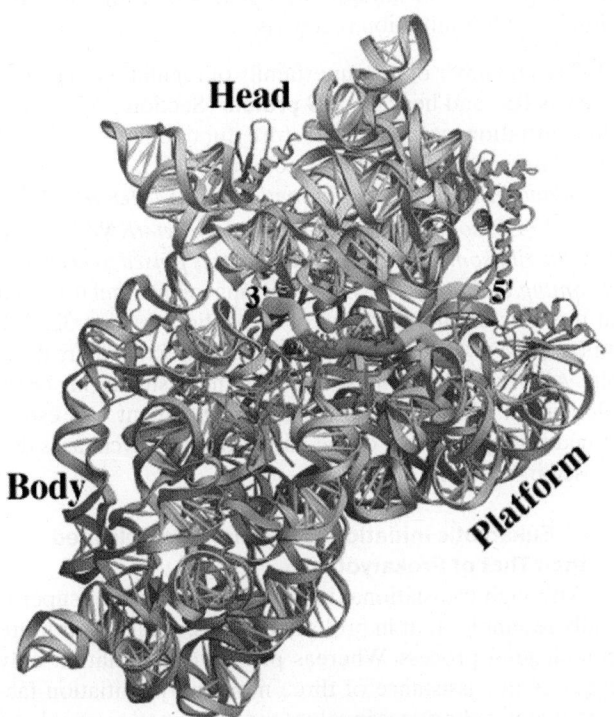

Figure 32-42 Path of mRNA through the ribosomal 30S subunit as viewed from its interface side. The 16S RNA is cyan, and the 27-nt mRNA is represented in worm form with its A- and P-site codons orange and red, the Shine–Dalgarno helix (which includes a segment of 16S RNA) magenta, and its remaining segments yellow. The S3, S4, and S5 proteins are green, the S7, S11, and S12 proteins are purple, and the remaining ribosomal proteins have been omitted for clarity. The S3, S4, and S5 proteins, which in part form the tunnel through which the mRNA enters the ribosome, may function as a helicase to remove secondary structure from the mRNA that would otherwise interfere with tRNA binding. [Courtesy of Gloria Culver, Iowa State University. Based on an X-ray structure by Harry Noller, University of California at Santa Cruz. PDBid 1JGO.]

Shine–Dalgarno sequence, which is located near its 5′ end, is base paired, as expected, with the 16S RNA's anti-Shine–Dalgarno sequence, which is situated close to the E site. The resulting double helical segment is accommodated in a cleft formed by both RNA and protein elements of the 16S subunit's head, neck, and platform (Fig. 32-26).

c. Prokaryotic Initiation Is a Three-Stage Process That Requires the Participation of Soluble Protein Initiation Factors

See Guided Exploration 28: Translational initiation Intact ribosomes do not directly bind mRNA so as to initiate polypeptide synthesis. Rather, *initiation is a complex process in which the two ribosomal subunits and fMet–tRNA$_f^{Met}$ assemble on a properly aligned mRNA to form a complex that is competent to commence chain elongation. This assembly process also requires the participation of protein **initiation factors** that are not permanently associated with the ribosome.* Initiation in *E. coli* involves three initiation factors designated **IF-1, IF-2,** and **IF-3** (Table 32-9). Their existence was discovered when it was found that washing small ribosomal subunits with 1*M* ammonium chloride solution,

Table 32-9 **The Soluble Protein Factors of *E. coli* Protein Synthesis**

Factor	Number of Residues[a]	Function
Initiation Factors		
IF-1	71	Assists IF-3 binding
IF-2	890	Binds initiator tRNA and GTP
IF-3	180	Releases mRNA and tRNA from recycled 30S subunit and aids new mRNA binding
Elongation Factors		
EF-Tu	393	Binds aminoacyl–tRNA and GTP
EF-Ts	282	Displaces GDP from EF-Tu
EF-G	703	Promotes translocation through GTP binding and hydrolysis
Release Factors		
RF-1	360	Recognizes UAA and UAG Stop codons
RF-2	365	Recognizes UAA and UGA Stop codons
RF-3	528	Stimulates RF-1/RF-2 release via GTP hydrolysis
RRF	185	Together with EF-G, induces ribosomal dissociation to small and large subunits

[a]All *E. coli* translational factors are monomeric proteins.

which removes the initiation factors but not the "permanent" ribosomal proteins, prevents initiation.

The initiation sequence in *E. coli* ribosomes has three stages (Fig. 32-43):

1. On completing a cycle of polypeptide synthesis, the 30S and 50S subunits are separated (Section 32-3Fa). IF-3 then binds to the 30S subunit so as to prevent the reassociation of the 50S subunit. The X-ray structure of the 30S subunit in complex with the C-terminal domain of IF-3 (which by itself prevents the association of the 30S and 50S subunits), determined by Yonath and François Franceschi, indi-

cates that IF-3 binds to the upper end of the platform (Fig. 32-26) on its solvent (back) side. Hence IF-3 does not function by physically blocking the binding of the 50S subunit.

2. mRNA and IF-2 in a ternary complex with GTP and fMet–tRNA$_f^{Met}$ that is accompanied by IF-1 subsequently bind to the 30S subunit in either order. Hence, fMet–tRNA$_f^{Met}$ recognition must not be mediated by a codon–anticodon interaction. This interaction, nevertheless, helps bind fMet–tRNA$_f^{Met}$ to the ribosome. IF-1 binds in the A site where it may function to prevent the inappropriate or premature binding of a tRNA. IF-3 also functions in this stage of the initiation process: it destabilizes the binding of tRNAs that lack the three G · C pairs in the anticodon stem of RNA$_f^{Met}$ (Fig. 32-40) and helps discriminate between matched and mismatched codon–anticodon interactions.

3. Last, in a process that is preceded by IF-1 and IF-3 release, the 50S subunit joins the 30S initiation complex in a manner that stimulates IF-2 to hydrolyze its bound GTP to GDP + P$_i$. This irreversible reaction conformationally rearranges the 30S subunit and releases IF-2 for participation in further initiation reactions.

IF-2 is a member of the superfamily of regulatory GTPases such as Ras and hence is a **G protein** (Section 19-2A). The 30S initiation complex therefore functions as its **GAP** (GTPase-activating protein; Section 19-2Ca).

Initiation results in the formation of an fMet–tRNA$_f^{Met}$ · mRNA · ribosome complex in which the fMet–tRNA$_f^{Met}$ occupies the ribosome's P site while its A site is poised to accept an incoming aminoacyl–tRNA (an arrangement similar to that at the conclusion of a round of elongation; Section 32-3D). In fact, tRNA$_f^{Met}$ is the only tRNA that directly enters the P site. All other tRNAs must do so via the A site during chain elongation (Section 32-3D). This arrangement was established through the use of the antibiotic puromycin as is discussed in Section 32-3Df.

d. Eukaryotic Initiation Is Far More Complicated than That of Prokaryotes

Although translational initiation in eukaryotes superficially resembles that in prokaryotes, it is, in fact, a far more complicated process. Whereas prokaryotic initiation only requires the assistance of three monomeric initiation factors, that in eukaryotes involves the participation of at least 12 initiation factors (designated eIFn; "e" for eukaryotic) that consist of at least 26 polypeptide chains. Eukaryotic initiation occurs as follows (Fig. 32-44):

1. The process begins with the binding of **eIF3** (which in mammals consists of 13 different subunits) and **eIF1A** (a monomer and homolog of bacterial IF-1) to the 40S subunit in the inactive 80S ribosome (which had terminated elongation in its previous elongation cycle) so that it releases the 60S subunit.

2. The ternary complex of **eIF2** (a heterotrimer), GTP, and **Met–tRNA$_i^{Met}$** binds to the 40S ribosomal subunit accompanied by **eIF1** (a monomer) to form the so-called **43S**

Figure 32-43 **Translational initiation pathway in *E. coli*.** The E site, which is unoccupied during this process, has been omitted for clarity.

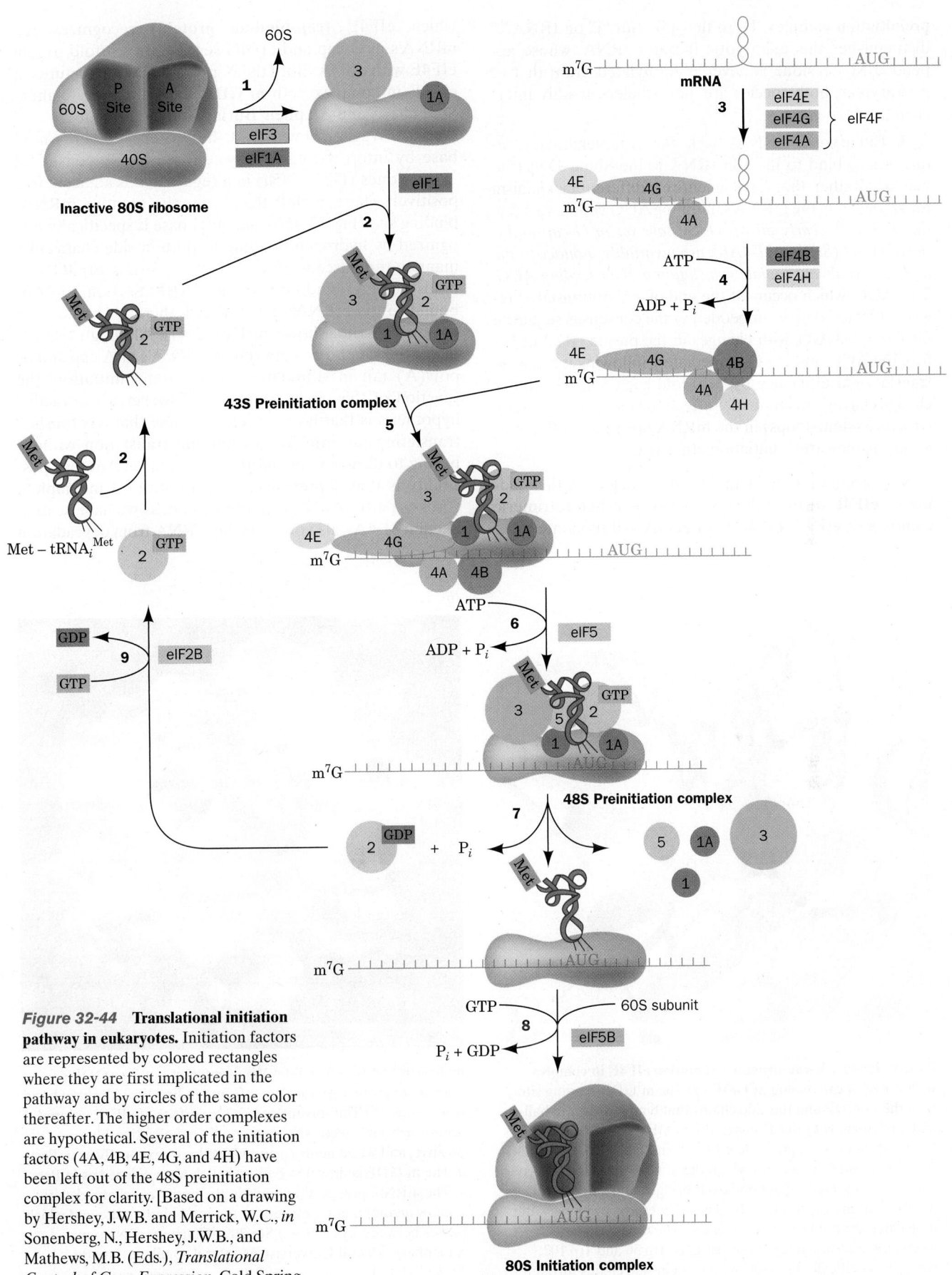

Figure 32-44 Translational initiation pathway in eukaryotes. Initiation factors are represented by colored rectangles where they are first implicated in the pathway and by circles of the same color thereafter. The higher order complexes are hypothetical. Several of the initiation factors (4A, 4B, 4E, 4G, and 4H) have been left out of the 48S preinitiation complex for clarity. [Based on a drawing by Hershey, J.W.B. and Merrick, W.C., *in* Sonenberg, N., Hershey, J.W.B., and Mathews, M.B. (Eds.), *Translational Control of Gene Expression*, Cold Spring Harbor Laboratory Press (2000).]

preinitiation complex. Here the subscript "i" on tRNA$_i^{Met}$ distinguishes this eukaryotic initiator tRNA, whose appended Met residue is never *N*-formylated, from that of prokaryotes; both species are, nevertheless, readily interchangeable *in vitro.*

3. Eukaryotic mRNAs lack the complementary sequences to bind to the 18S rRNA in the Shine–Dalgarno manner. Rather, they have an entirely different mechanism for recognizing the mRNA's initiating AUG codon. *Eukaryotic mRNAs, nearly all of which have an m^{7}G cap and a poly(A) tail (Section 31-4Ab), are invariably monocistronic and almost always initiate translation at their leading AUG.* This AUG, which occurs at the end of a 5'-untranslated region of 50 to 70 nt, is embedded in the consensus sequence GCCRCC**AUG**G, with changes in the purine (R) 3 nt before the AUG and the G immediately following it reducing translational efficiency by ~10-fold each and with other changes having much smaller effects. In addition, secondary structure (stem–loops) in the mRNA upstream of the initiation site may affect initiation efficiency.

The recognition of the initiation site begins by the binding of **eIF4F** to the m^{7}G cap. eIF4F is a heterotrimeric complex of **eIF4E, eIF4G,** and **eIF4A** (all monomers), in which eIF4E **(cap-binding protein)** recognizes the mRNA's m^{7}G cap and eIF4G serves as a scaffold to join eIF4E with eIF4A. Both the X-ray and NMR structures of eIF4E in complex with **m^{7}GDP,** determined by Nahum Sonenberg and Stephen Burley and by Sonenberg and Gerhard Wagner, reveal that the protein binds the m^{7}G base by intercalating it between two highly conserved Trp residues (Fig. 32-45*a*) in a region that is adjacent to a positively charged cleft that forms the putative mRNA-binding site (Fig. 32-45*b*). The m^{7}G base is specifically recognized by hydrogen bonding to protein side chains in a manner reminiscent of G · C base pairing. eIF4G also binds poly(A)-binding protein (PABP; Section 31-4Ab) bound to the mRNA's poly(A) tail, thereby circularizing the mRNA (not shown in Fig. 32-44). Although this explains the synergism between an mRNA's m^{7}A cap and its poly(A) tail in stimulating translational initiation, the function of this circle is unclear. However, an attractive hypothesis is that it enables a ribosome that has finished translating the mRNA to reinitiate translation without having to disassemble and then reassemble. Another possibility is that it prevents the translation of incomplete (broken) mRNAs. This circularization, as we have seen in Section 31-4Av, also protects the mRNA from degradation

(a)

(b)

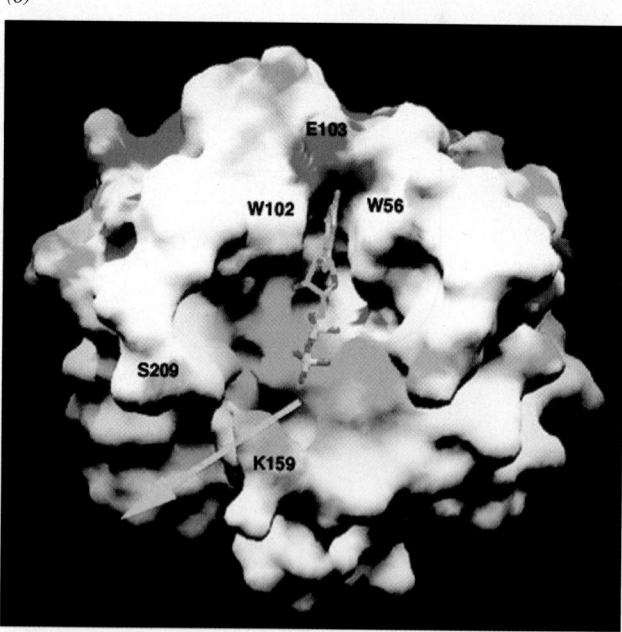

Figure 32-45 X-ray structure of murine eIF4E in complex with the m^{7}G cap analog m^{7}GDP. (*a*) The m^{7}GDP-binding site with the m^{7}GDP and the side chains that bind it drawn in ball-and-stick form with the atoms of the m^{7}GDP colored according to type (C green, N dark yellow, O red, and P bright yellow) and the protein side chains with which the m^{7}GDP interacts drawn in various colors. Hydrogen bonds, salt bridges, and van der Waals interactions are represented by dashed lines and bridging water molecules are drawn as black spheres. The m^{7}G base is intercalated between the indole rings of Trp 56 and Trp 102, where it specifically interacts with protein side chains through hydrogen bonds and van der Waals interactions. The GDP's phosphate groups interact directly and indirectly with three basic side chains. (*b*) The solvent-accessible surface of eIF4E colored according to its electrostatic potential (red negative, blue positive, and white neutral) and viewed approximately as in Part *a*. The m^{7}GDP is drawn in ball-and-stick form colored as in Part *a*. The mRNA presumably binds in the positively charged cleft (*yellow arrow*) that is adjacent to the m^{7}G binding site and which passes between Lys 159 and Ser 209. [Courtesy of Nahum Sonenberg, McGill University, Montréal, Québec, Canada. PDBid 1EJ1.]

by preventing the action of decapping enzyme until the mRNA's poly(A) tail has been shortened to the point that it can no longer bind PABP.

4. eIF4B (an RRM-containing homodimer) and **eIF4H** (a monomer) join the eIF4F–mRNA complex where they stimulate the RNA helicase activity of eIF4A to unwind the mRNA's helical segments in an ATP-dependent process. This presumably also strips away the proteins that are bound to the mRNA (Section 31-4Au). eIF4A is the prototype of the so-called **DEAD-box family** of proteins (also known as DExD/H family proteins; Section 31-4Au), which is named after one of the sequence motifs shared by the diverse members of this family, all of which have NTPase activity.

5. The eIF4F–mRNA–eIF4B–eIF4H complex joins the 43S preinitiation complex through a protein–protein interaction between eIF4G and the 40S subunit-bound eIF3. This differs substantially from the corresponding prokaryotic process (Fig. 32-43) in which the mRNA is bound to the 30S ribosomal subunit via associations between RNA molecules (involving the Shine–Dalgarno sequence and the codon–anticodon interaction).

6. eIF5 (a monomer) joins the growing assembly. The 43S preinitiation complex then translocates along the mRNA, an ATP-dependent process called **scanning,** until it encounters the mRNA's AUG initiation codon, which is optimally in the sequence GCC(A/G)CC**AUG**G. This yields the **48S preinitiation complex.** The recognition of the AUG occurs mainly through base pairing with the CUA anticodon on the bound Met–tRNA$_i^{Met}$, as was demonstrated by the observation that mutating this anticodon results in the recognition of the new cognate codon instead of AUG. This explains why the initiator tRNA must bind to the small subunit before the mRNA.

7. The formation of the 48S preinitiation complex induces eIF2 to hydrolyze its bound GTP to GDP + P$_i$, which results in the release of all the initiation factors, thereby leaving the Met–tRNA$_i^{Met}$ in the small subunit's P site. The hydrolysis reaction is stimulated by eIF5, acting as a GAP (Section 19-2Ca).

8. The 60S subunit then joins the mRNA-bound Met–tRNA$_i^{Met}$–40S subunit complex in a GTPase reaction mediated by **eIF5B** (a monomer and homolog of bacterial IF-2), thereby yielding the 80S ribosomal initiation complex. Thus eukaryotic translation initiation consumes two GTPs versus one for prokaryotic initiation (Fig. 32-43).

9. What remains is to recycle the eIF2 · GDP complex by exchanging its GDP for GTP. This reaction is mediated by **eIF2B** (a heteropentamer), which therefore functions as eIF2's **GEF** (guanine nucleotide exchange factor; Section 19-2Ca).

Many eukaryotic initiation factors are subject to phosphorylation/dephosphorylation and are therefore likely to participate in the control of eukaryotic translation, a subject we discuss in Section 32-4.

Although the initiation sites on most eukaryotic mRNAs are identified by the above-described scanning mechanism, a few mRNAs have an **internal ribosome entry site (IRES)** to which the 40S subunit can directly bind in a process reminiscent of prokaryotic initiation. However, little is yet known about the mechanism of IRES-based initiation. Indeed, IRESs lack clearly identifiable consensus sequences.

D. Chain Elongation

See Guided Exploration 29: Translational elongation *Ribosomes elongate polypeptide chains in a three-stage reaction cycle that adds amino acid residues to a growing polypeptide's C-terminus* (Fig. 32-46):

1. Decoding, in which the ribosome selects and binds an aminoacyl–tRNA, whose anticodon is complementary to the mRNA codon in the A site.

2. Transpeptidation, in which the peptidyl group on the P-site tRNA is transferred to the aminoacyl group in the A site through the formation of a peptide bond (Fig. 32-39).

3. Translocation, in which A-site and P-site tRNAs are respectively transferred to the P site and E site accompanied by their bound mRNA; that is, the mRNA, together with its base paired tRNAs, is ratcheted through the ribosome by one codon.

Translational elongation, which occurs at a rate of 10 to 20 residues/s, involves the participation of several nonribosomal proteins known as **elongation factors** (Table 32-9). We describe these processes in the following paragraphs.

a. Decoding
In the decoding stage of the *E. coli* elongation cycle, a binary complex of GTP with the elongation factor **EF-Tu** (also called **EF1A**) combines with an aminoacyl–tRNA. The resulting ternary complex binds to the ribosome, and, in a reaction that hydrolyzes the GTP to GDP + P$_i$, the aminoacyl–tRNA is bound in a codon–anticodon complex to the ribosomal A site and EF-Tu · GDP + P$_i$ is released. In the remainder of this stage, the bound GDP is replaced by GTP in a reaction mediated the elongation factor **EF-Ts** (also called **EF1B**). EF-Tu, as are several other GTP-binding ribosomal factors, is a G-protein, and hence the ribosome functions as its GAP and EF-Ts is its GEF.

Aminoacyl–tRNAs can bind to the ribosomal A site without the mediation of EF-Tu but at a rate too slow to support cell growth. The importance of EF-Tu is indicated by the fact that it is the most abundant *E. coli* protein; it is present in ~100,000 copies per cell (>5% of the cell's protein), which is approximately the number of tRNA molecules in the cell. Consequently, *the cell's entire complement of aminoacyl–tRNAs is essentially sequestered by EF-Tu.*

Figure 32-46   Elongation cycle in *E. coli* ribosomes. The E site, to which discharged tRNAs are transferred before being released to solution, is not shown. Eukaryotic elongation follows a similar cycle but EF-Tu and EF-Ts are replaced by a single multisubunit protein, eEF1, and EF-G is replaced by eEF2.

b. EF-Tu Is Sterically Prevented from Binding Initiator tRNA

The X-ray structure of the Phe–tRNA^{Phe} · EF-Tu · GMPPNP ternary complex (GMPPNP is a nonhydrolyzable GTP analog; Section 19-3Ch), determined by Brian Clark and Jens Nyborg, reveals that these two macromolecules associate to form a corkscrew-shaped complex in which the EF-Tu and the tRNA's acceptor stem form a knoblike handle and the tRNA's anticodon helix forms the screw (Fig. 32-47). The conformation of the tRNA^{Phe} closely resembles that of the uncomplexed molecule (Fig. 32-11*b*). The EF-Tu folds into three distinct domains that are connected by flexible peptides, rather like beads on a string. The N-terminal domain 1, which binds guanine nucleotides and catalyzes GTP hydrolysis, structurally resembles other known G-proteins.

The two macromolecules associate rather tenuously via three major regions: (1) the CCA—Phe segment at the 3′ end of the Phe–tRNA^{Phe} binds in a cleft between EF-Tu's domains 1 and 2 (the blue and green mainly helical domain and the yellow β sheet domain in Fig. 32-47) that ends in a pocket large enough to accommodate all amino acid residues; (2) the 5′-phosphate of the tRNA binds in a depression at the junction of EF-Tu's three domains; and (3) one side of the tRNA's TψC stem contacts the exposed main chain and side chains of EF-Tu's C-terminal domain 3 (the orange and red β barrel–containing domain in Fig. 32-47). The tight association of the aminoacyl group with EF-Tu appears to greatly increase the affinity of EF-Tu for the otherwise loosely bound tRNA, which explains why EF-Tu does not bind uncharged elongator tRNAs.

EF-Tu binds neither formylated aminoacyl–tRNAs nor unformylated Met–tRNA$_f^{Met}$, which is why the initiator tRNA never reads internal AUG or GUG codons. The first base pair of tRNA$_f^{Met}$ is mismatched (C · A; Fig. 32-40) and hence this initiator tRNA has a 3′ overhang of 5 nt vs 4 nt for an elongator tRNA. It seems likely that this mismatch, together with the formyl group, prevents fMet–tRNA$_f^{Met}$ from binding to EF-Tu. Indeed, EF-Tu binds to *E. coli* tRNA$_f^{Met}$ whose 5′-terminal C residue has been deaminated

Figure 32-48   Comparison of the X-ray structures of EF-Tu in its complexes with GDP and GMPPNP. The protein is represented by its C_α backbone with domain 1, its GTP-binding domain, purple in the GDP complex and red in the GMPPNP complex. Domain 2 and domain 3, which have the same orientation in both complexes, are green and cyan. The bound GDP and GMPPNP are shown in stick form with C yellow, N blue, O red, and P green. [Courtesy of Morten Kjeldgaard and Jens Nyborg, University of Aarhus, Århus, Denmark. PDBid 1EFT.] **See Interactive Exercise 45**

Figure 32-47   X-ray structure of the ternary complex of yeast Phe–tRNA^{Phe}, *Thermus aquaticus* EF-Tu, and GMPPNP. The EF-Tu is drawn in ribbon form colored in rainbow order from its N-terminus (*blue*) to its C-terminus (*red*). The tRNA is shown in stick form colored according to atom type with C green, N blue, O red, and P orange and with orange rods linking successive P atoms. The tRNA's appended aminoacyl–Phe residue and the GMPPNP that is bound to the EF-Tu are drawn in space-filling form with C atoms cyan and yellow, respectively. Two bound Mg^{2+} ions are represented by magenta spheres. [Based on an X-ray structure by Jens Nyborg, University of Aarhus, Århus, Denmark. PDBid 1TTT.]

by bisulfite treatment (Section 30-7b), which reestablishes the "missing" base pair as U · A. Similarly, Sec–tRNA^{Sec}, which is also not bound by EF-Tu (but rather by SELB; Section 32-2De), has 8 bp in its acceptor stem vs 7 bp in those of other elongator tRNAs. However, initiator tRNAs from several sources have fully base paired acceptor stems, and the U1 · A72 base pair of tRNA^{Gln} is opened up on binding to GlnRS (Section 32-2Cc).

c. EF-Tu Undergoes a Major Conformational Change on Hydrolyzing GTP

Morten Kjeldgaard and Nyborg determined the X-ray structures of *T. aquaticus* EF-Tu (405 residues) in com-

plex with GMPPNP and the 70% identical *E. coli* EF-Tu (393 residues) in complex with GDP (Fig. 32-48). The conformation of EF-Tu in its complex with only GMPPNP closely resembles that in its ternary complex with Phe–tRNA^{Phe} and GMPPNP (Fig. 32-47). However, comparison of the GMPPNP and GDP complexes indicates that, on hydrolyzing its bound GTP, EF-Tu undergoes a major structural reorganization. Its greatest local conformational changes occur in the Switch I and Switch II regions of domain 1, which in all G-proteins signal the state of the bound nucleotide to interacting partners (Section 19-2Cb; here domains 2 and 3): Switch I converts from a β hairpin to a short α helix and the α helix of Switch II shifts toward the C-terminus by 4 residues. As a consequence, this latter helix reorients by 42°, which results in domain 1 rigidly changing its orientation with respect to domains 2 and 3 by a dramatic 91° rotation. The tRNA binding site is thereby eliminated.

d. EF-Ts Disrupts the Binding of GDP to EF-Tu

EF-Tu has a 100-fold higher affinity for GDP than GTP. Hence, replacement of the EF-Tu–bound GDP by GTP must be facilitated by the interaction of EF-Tu with EF-Ts (Fig. 32-46, *top*). The X-ray structure of the EF-Tu · EF-Ts complex, determined by Stephen Cusack and Reuben

Figure 32-49   X-ray structure of the *E. coli* EF-Tu · EF-Ts complex. Domains 1, 2, and 3 of EF-Tu are magenta, green, and cyan, respectively, and EF-Ts is orange. [Based on an X-ray structure by Stephen Cusack and Reuben Leberman, EMBL, Grenoble Cedex, France. PDBid 1EFU.]

Leberman, reveals that the EF-Tu has a conformation resembling that of its GDP complex (Fig. 32-49) but with its domains 2 and 3 swung away from domain 1 by ~18°. EF-Ts is an elongated molecule that binds along the right side of EF-Tu as shown in Fig. 32-49, where it contacts EF-Tu's domains 1 and 3. The intrusive interactions of EF-Ts side chains with the GDP binding pocket on EF-Tu domain 1 disrupts the Mg^{2+} ion binding site. This reduces the affinity of EF-Tu for GDP, thereby facilitating its exchange for GTP (after EF-Ts has dissociated), which has a 10-fold higher concentration in the cell than does GDP (the GEF-containing segment of Sos similarly interferes with Mg^{2+} binding and hence guanine nucleotide binding by Ras; Sec-

tion 19-3Cf). EF-Tu's subsequent binding of a charged elongator tRNA increases its affinity for GTP.

e. Transpeptidation

In the transpeptidation stage of the elongation cycle (Fig. 32-46), the peptide bond is formed through the nucleophilic displacement of the P-site tRNA by the amino group of the 3'-linked aminoacyl–tRNA in the A site (Fig 32-39). The nascent polypeptide chain is thereby lengthened at its C-terminus by one residue and transferred to the A-site tRNA. The reaction occurs without the need of activating cofactors such as ATP because the ester linkage between the nascent polypeptide and the P-site tRNA is a "high-energy" bond. The **peptidyl transferase** center that catalyzes peptide bond formation is located entirely on the large subunit as is demonstrated by the observation that in high concentrations of organic solvents such as ethanol, the large subunit alone catalyzes peptide bond formation. The organic solvent apparently distorts the large subunit in a way that mimics the effect of small subunit binding.

f. Puromycin Is an Aminoacyl–tRNA Analog

The ribosomal elongation cycle was originally characterized through the use of the antibiotic **puromycin** (Fig. 32-50). This product of *Streptomyces alboniger*, which resembles the 3' end of Tyr–tRNA, causes the premature termination of polypeptide chain synthesis. Puromycin, in competition with the mRNA-specified aminoacyl–tRNA but without the need of elongation factors, binds to the ribosomal A site which, in turn, catalyzes a normal transpeptidation reaction to form peptidyl–puromycin. Yet, the ribosome cannot catalyze the transpeptidation reaction in the next elongation cycle because puromycin's "amino acid residue" is linked to its "tRNA" via an amide rather than an ester bond. Polypeptide synthesis is therefore aborted and the peptidyl–puromycin is released.

In the absence of the elongation factor EF-G (see below), an active ribosome cannot bind puromycin because its A site is at least partially occupied by a peptidyl–tRNA. A newly initiated ribosome, however, violates this rule; it

Figure 32-50   Puromycin. This antibiotic (*left*) resembles the 3'-terminus of tyrosyl–tRNA (*right*).

Puromycin   **Tyrosyl–tRNA**

catalyzes fMet–puromycin formation. *These observations demonstrated the functional existence of the ribosomal P and A sites and established that fMet–tRNA$_f^{Met}$ binds directly to the P site, whereas other aminoacyl–tRNAs must first enter the A site.*

g. Transpeptidation: The Ribosome Is a Ribozyme

What is the nature of the peptidyl transferase center, that is, does it consist of RNA, protein, or both? Since all proteins, including those associated with ribosomes, are ribosomally synthesized, the primordial ribosome must have preceded the primordial proteins and hence consisted entirely of RNA. Despite this (in hindsight) obvious evolutionary argument, the idea that rRNA functions catalytically was not seriously entertained until after it had been discovered that RNA can, in fact, act as a catalyst (Section 31-4Ae). Several other observations further indicate that the ribosome is a ribozyme:

1. The absence from the 50S subunit of any one of its proteins but L2, L3, and L4 does not abolish its peptidyl transferase function.

2. rRNAs are more highly conserved throughout evolution than are ribosomal proteins.

3. Most mutations that confer resistance to antibiotics that inhibit protein synthesis occur in genes encoding rRNAs rather than ribosomal proteins.

Nevertheless, the unambiguous demonstration that rRNA functions catalytically in polypeptide synthesis proved to be surprisingly elusive. Noller succeeded in showing that the *T. thermophilus* large ribosomal subunit from which ~95% of the protein had been removed by treatment with SDS and **proteinase K** followed by phenol extraction (which denatures proteins; Section 6-6A) maintained >80% of its peptidyl transferase activity in a model reaction. Moreover, this activity was abolished by RNase treatment. However, since the remaining protein was due to several intact ribosomal proteins (which are presumably sequestered within the 23S RNA), it could be argued that these proteins are essential for ribosomal catalytic function, a reasonable expectation in light of the >3.5 billion years over which ribosomal proteins and RNAs have coevolved.

Steitz and Moore unequivocally determined the nature of the peptidyl transferase center through its identification in the X-ray structure of the 50S subunit. Peptide bond formation presumably resembles the reverse of peptide bond hydrolysis such as that catalyzed by serine proteases (Section 15-3C). The ribosomal reaction's tetrahedral intermediate (Fig. 32-51a) is mimicked by a compound synthesized by Michael Yarus that consists of the trinucleotide CCdA linked to puromycin via a phosphoramidite group (Fig. 32-51b). This compound, which is named **CCdA-p-Puro,** binds tightly to the ribosome so as to inhibit its peptidyl transferase activity. The X-ray structure of the 50S subunit in complex with CCdA-p-Puro reveals that the inhibitor binds to domain V of the 23S RNA (Fig. 32-27b) at the entrance to the ~100-Å-long polypeptide exit tunnel that runs through to the back of the subunit (Figs. 32-30 and 32-34c). There, *the inhibitor is completely enveloped in RNA*

Figure 32-51   Ribosomal tetrahedral intermediate and its analog. (*a*) The chemical structure of the tetrahedral intermediate (*red C*) in ribosomally mediated peptide bond formation in which the A-site aminoacyl residue is Tyr. (*b*) CCdA-p-Puro, the transition state analog of the tetrahedral intermediate in Part *a* produced by linking the 3′-OH group of CCdA to the amino group of puromycin's *O*-methyltyrosine residue via a phosphoryl group.

with no protein side chain approaching closer than ~18 Å to the inhibitor's phosphoramidite group and with the nearest Mg^{2+} ion 8.5 Å away—both too far away to participate in catalysis. Moreover, all the nucleotides that contact the CCdA-p-Puro are >95% conserved among all three kingdoms of life. Clearly, *the ribosomal transpeptidase reaction is catalyzed by RNA.*

Despite the foregoing, the X-ray structure of the *T. thermophilus* ribosome in complex with tRNA and mRNA (Fig. 32-34) reveals that the N-terminal tail of L27 interacts with the phosphate group of residue C75 of the A-site tRNA. Moreover, L16 interacts with the elbow region of the A-site tRNA via interactions between conserved Arg residues and phosphate and 2'-OH groups in the elbow region of the tRNA. These observations confirm previous mutational studies indicating that L16 and L27 help stabilize the binding of the A-site tRNA. However, some organisms lack L27, which indicates that its binding to A-site tRNA is not part of an evolutionarily conserved mechanism.

h. Peptide Bond Formation Does Not Occur via Acid–Base Catalysis

The ribosomal peptidyl transferase reaction occurs ~10^7-fold faster than the uncatalyzed reaction. How does the ribosome catalyze this reaction? Peptide bond formation is naively expected to proceed via the nucleophilic attack of the amino group on the carbonyl group of an ester to form a tetrahedral intermediate that collapses to an amide and an alcohol (Fig. 32-39). However, in the physiological pH range, the attacking amino group is predominantly in its ammonium form (RNH_3^+), and hence lacks the lone pair necessary to undertake a nucleophilic attack. This suggests that the peptidyl transferase reaction is catalyzed in part by a general base that abstracts a proton from the ammonium group to generate the required free amino group (RNH_2).

Inspection of the peptidyl transferase center in *H. marismortui* reveals that the only basic group within 5 Å of the inferred position of the attacking amino group is atom N3 of the invariant rRNA base A2486 (A2451 in *E. coli*). It is ~3 Å from and hence hydrogen bonded to the attacking amino group (Fig. 32-52). This further suggests that the protonated A2486-N3 electrostatically stabilizes the oxyanion of the tetrahedral reaction intermediate and then donates the proton to the leaving group of the P-site tRNA to yield a 3'-OH group (general acid catalysis). However, in order for A2486-N3 to act as a general base in abstracting the proton from an ammonium group (whose pK is ~10), it must have a pK of at least 7 (recall that proton transfers between hydrogen-bonded groups occur at physiologically significant rates only when the pK of the proton donor is no more than 2 or 3 pH units greater than that of the proton acceptor; Section 15-3Dd). Yet, the pK of N3 in AMP is <3.5. Moreover, several lines of evidence indicate that A2486 does not function as an acid–base catalyst including (1) the model displayed in Fig. 32-52 indicates that the tetrahedral intermediate's oxyanion would point away from and hence could not be stabilized by protonated

Figure 32-52 Model of the substrate complex of the 50S ribosomal subunit. Atoms are colored according to type with the A-site substrate C and P purple, P-site substrate C and P green, 23S rRNA C and P orange, N blue, and O red. The attacking amino group of the A-site aminoacyl residue is held in position for nucleophilic attack (*cyan arrow*) on the carbonyl C of the P-site aminoacyl ester through hydrogen bonds (*dashed black lines*) to A2486-N3 and the 2'-O of the P-site A76. [Courtesy of Peter Moore and Thomas Steitz, Yale University.]

A2486-N3; and (2) mutagenic replacement of A2486 by any other base does not greatly alter the ribosome's catalytic rate.

i. The Ribosome Is an Entropy Trap

If the peptidyl transfer reaction does not involve acid–base catalysis, what is the origin of the ribosome's catalytic power? Marina Rodnina and Richard Wolfenden noted that the uncatalyzed reaction of esters with amines to form amides occurs quite facilely in aqueous solution. They therefore measured the rates of both uncatalyzed peptide bond formation by model compounds and peptidyl transfer by the ribosome at several different temperatures. This provided values of $\Delta\Delta H_{cat}^{\ddagger}$ and $\Delta\Delta S_{cat}^{\ddagger}$, the reaction's change in the enthalpy and entropy of activation by the ribosome relative to the uncatalyzed reaction. Here $\Delta\Delta H_{cat}^{\ddagger} - T\Delta\Delta S_{cat}^{\ddagger} = \Delta\Delta G_{cat}^{\ddagger} = \Delta G_{cat}^{\ddagger}(\text{uncat}) - \Delta G^{\ddagger}(\text{cat})$, where $\Delta\Delta G_{cat}^{\ddagger}$ is the change in the reaction's free energy of activation by the ribosome, $\Delta G^{\ddagger}(\text{cat})$, relative to that of the uncatalyzed reaction, $\Delta G^{\ddagger}(\text{uncat})$ (Section 14-1C). The measured value of $\Delta\Delta H_{cat}^{\ddagger}$ is -19 kJ · mol^{-1}, a quantity that would be positive, not negative, if the ribosomal reaction had a significant component of chemical catalysis such as acid–base catalysis and/or the formation of new hydrogen bonds. In contrast, the value of $T\Delta\Delta S_{cat}^{\ddagger}$ is 52 kJ · mol^{-1}, which indicates that the Michaelis complex in the ribosomal reaction is significantly more ordered relative to the transition state than is the uncatalyzed reaction. This value of $T\Delta\Delta S_{cat}^{\ddagger}$ largely accounts for the observed ~10^7-fold rate enhancement of the ribosomal reaction relative to

Figure 32-53 **The mechanism of ribosome-catalyzed peptidyl transfer.** The nucleophilic attack of the α-amino group of the aminoacyl–tRNA (*red*) on the carbonyl C of the peptidyl–tRNA (*blue*) occurs in concert with a proton shuttle that involves the O3′ and 2′-OH of the P-site A76 together with the α-amino group of the aminoacyl–tRNA. The reaction proceeds through a transition state (*center; enclosed by square brackets*) that contains a six-membered ring of partially bonded atoms and which collapses to the reaction products drawn on the right.

the uncatalyzed reaction (the rate enhancement by the ribosome is given by $e^{\Delta\Delta G^{\ddagger}_{cat}/RT}$; Section 14-1Cd). Evidently, *the ribosome enhances the rate of peptide bond formation by properly positioning and orienting its substrates and/or excluding water from the preorganized electrostatic environment of the active site (a form of reactant ordering) rather than by chemical catalysis.*

The X-ray structures of the large ribosomal subunit in complex with aminoacyl–tRNA and peptidyl–tRNA indicate that the ribosome uses an induced fit mechanism, as occurs in enzymes such as hexokinase (Section 17-2Aa). Conformational changes in the 23S rRNA, presumably triggered by proper binding of the aminoacyl–tRNA in the A site, orient the ester group of the peptidyl–tRNA for nucleophilic attack. The hydrogen bond between the 2′-OH of the P-site A76 and the attacking amino group (Fig. 32-52) is crucial in doing so. In fact, replacing this 2′-OH group with H or F reduces the reaction rate by at least 10^6. This suggests that the peptidyl transferase reaction occurs via the substrate-assisted proton shuttle mechanism diagrammed in Fig. 32-53. Although the peptidyl transferase reaction is relatively sluggish compared to the reactions catalyzed by many protein enzymes, it is sufficiently fast to keep up with the other ribosomal processes (which collectively link together 10–20 residues/s). Apparently, the ribosome's peptidyl transferase function is a molecular fossil from the RNA world.

In the absence of a tRNA in the A site, the ester bond linking the peptidyl group to the P-site tRNA is shielded by U2585 of the 23S RNA from nucleophilic attack by water, which would otherwise release the peptidyl group from the ribosome.

j. Translocation: The Ribosome Moves to the Next Codon

In the translocation stage of the elongation cycle, the now uncharged P-site tRNA (at first tRNA$_f^{Met}$ but subsequently an elongator tRNA) is transferred to the E site (not shown in Fig. 32-46), its former occupant having been previously expelled (see below). Simultaneously, the peptidyl–tRNA in the A site, together with its bound mRNA, is moved to the P site. This prepares the ribosome for the next elongation cycle. The maintenance of the peptidyl–tRNA's codon–anticodon association is no longer necessary for amino acid specification. Rather, it acts as a place-keeper that permits the ribosome to precisely step off the three nucleotides along the mRNA required to preserve the reading frame. Indeed, the observation that frameshift suppressor tRNAs induce a four-nucleotide translocation (Section 32-2Ea) indicates that mRNA movement is directly coupled to tRNA movement. An Mg^{2+}-stabilized kink in the mRNA between the A and P codons apparently helps prevent slippage.

k. EF-G Structurally Mimics the EF-Ts · tRNA Complex

The translocation process requires the participation of elongation factor **EF-G** (also called **EF2**), which binds to the ribosome together with GTP and is only released on hydrolysis of the GTP to GDP + P_i (Fig. 32-46). EF-G release is a prerequisite for beginning the next elongation cycle because EF-G and EF-Tu bind to the same site of the ribosome and hence their binding is mutually exclusive.

The X-ray structure of *T. thermophilus* EF-G · GMPPNP, determined by Anders Liljas and Derek Logan, reveals a tadpole-shaped monomeric protein that consists of five

Figure 32-54 **X-ray structure of EF-G from *T. thermophilus* in complex with GMPPNP.** The protein is drawn in ribbon form colored in rainbow order from its N-terminus (*blue*) to its C-terminus (*red*). The GMPPNP is drawn in space-filling form colored according to atom type with C yellow, N blue, O red, and P orange. An Mg^{2+} ion that is bound to the GMPPNP is represented by a magenta sphere. Portions of the structure are not visible. Note the remarkable resemblance in shape between this structure and that of Phe–tRNA^Phe · EF-Tu · GMPPNP (Fig. 32-47). [Based on an X-ray structure by Anders Liljas and Derek Logan, Lund University, Lund, Sweden. PDBid 2BV3.]

domains (Fig. 32-54). The first two domains closely resemble those in EF-Tu · GMPPNP rather than those in EF-Tu · GDP (Fig. 32-48). This, it is argued, is because the two elongation factors have reciprocal functions with EF-Tu · GTP facilitating the conversion of the ribosome from its post- to its pre-translocational state and EF-G · GTP promoting the reverse transition. This idea is supported by the intriguing observation that the Phe–tRNA^Phe · EF-Tu · GMPPNP and EF-G · GMPPNP complexes are almost identical in appearance: EF-G's three C-terminal domains (yellow through red in Fig. 32-54), which have no counterparts in EF-Tu, closely resemble the EF-Tu–bound tRNA in shape, a remarkable case of **macromolecular mimicry.** Indeed, EF-G occupies the same ribosomal site as does EF-Tu (Fig. 32-55).

EF-G is unusual among G-proteins in that it has no corresponding GEF. However, its N-terminal guanine nu-

(a)

(b)

Figure 32-55 **X-ray structure of the *T. thermophilus* ribosome in its complexes with (*a*) EF-G and (*b*) EF-Tu.** The ribosome is drawn in ribbon form with its 23S RNA orange, its 50S subunit proteins brown, its 16S RNA cyan, and its 30S proteins blue. Its bound tRNAs, mRNA, EF-G, and EF-Tu are represented by their surface diagrams with E-site tRNA yellow, P-site tRNA green, A-site tRNA magenta, mRNA black, and both EF-G and EF-Tu red. In Part *b*, the A-site tRNA bound to EF-Tu is in the A/T conformation (see below). The positions of the decoding center (**DC**) and the peptidyl transferase center (**PTC**) are indicated. [Courtesy of Martin Schmeing and Venki Ramakrishnan, MRC Laboratory of Molecular Biology, Cambridge, U.K. PDBids 2WRI, 2WRJ, 2WRN, and 2WRO.]

cleotide–binding domain contains a unique α helical insert (green in Fig. 32-54) that contacts the domain's conserved core at sites analogous to those in EF-Tu that interact with EF-Ts. This suggests that this subdomain acts as an internal GEF.

I. Translocation Occurs via Intermediate States

Chemical footprinting studies (Section 31-2Ab) by Noller revealed that certain bases in the 16S rRNA are protected by tRNAs bound in the ribosomal A and P sites and that certain bases in the 23S rRNA are protected by tRNAs in the A, P, and E sites. Almost all of these protected bases are absolutely conserved in evolution and many of them have been implicated in ribosomal function through biochemical or genetic studies.

Variations in chemical footprinting patterns during the elongation cycle together with the more recently determined X-ray and cryo-EM structures indicate that the translocation of tRNA occurs in several discrete steps (Fig. 32-56):

1. Let us start with the ribosome in its **post-translocational state:** a deacylated tRNA bound to the E subsites of both the 30S and 50S subunits (the E/E binding state), a peptidyl–tRNA bound in the P subsites of both subunits

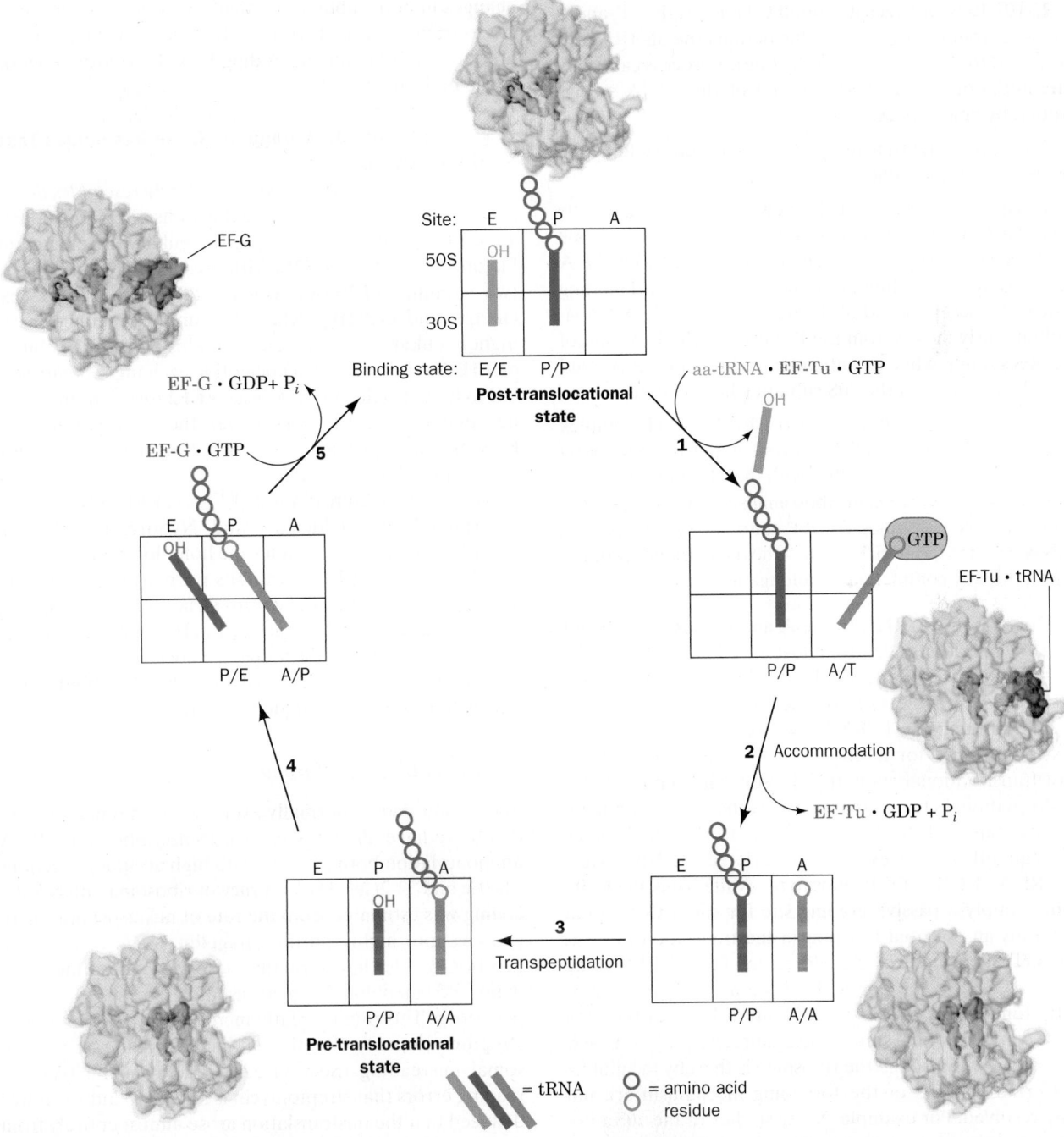

Figure 32-56   Ribosomal binding states in the elongation cycle. Note how this scheme elaborates the classic elongation cycle diagrammed in Fig. 32-46. The drawings are accompanied by 17-Å-resolution cryo-EM–based images of the *E. coli* 70S ribosome in the corresponding binding states in which the 30S subunit is transparent yellow, the 50S subunit is transparent blue, and the tRNAs and elongation factors are colored as in the drawing they accompany. [Cryo-EM images courtesy of Knud Nierhaus, Max-Planck-Institut für Molekulare Genetik, Berlin, Germany, and Joachim Frank, Wadsworth Center, State University of New York at Albany.]

(the P/P state), and the A site empty. An aminoacyl–tRNA (aa–tRNA) in ternary complex with EF-Tu and GTP binds to the A site accompanied by the release of the E-site tRNA (but see below). This yields a complex in which the incoming aa–tRNA is bound in the 30S subunit's A subsite via a codon–anticodon interaction (recall that the mRNA is bound to the 30S subunit) but with the EF-Tu preventing the entry of the tRNA's aminoacyl end into the 50S subunit's A subsite, an arrangement termed the A/T state (T for EF-*T*u).

2. EF-Tu hydrolyzes its bound GTP to GDP + P$_i$ and is released from the ribosome. This permits the aa–tRNA to fully bind to the A site (the A/A state), a process called **accommodation** in which the 3' end of the tRNA swings around by nearly 70 Å.

3. The peptidyl transferase reaction occurs, yielding the **pre-translocational state.**

4. The acceptor end of the new peptidyl–tRNA shifts from the A subsite of the 50S subunit to its P-subsite, while the tRNA's anticodon end remains associated with the A subsite of the 30S subunit (yielding the A/P hybrid binding state). The acceptor end of the newly deacylated tRNA simultaneously moves from the P subsite to the E subsite of the 50S subunit while its anticodon end remains associated with the P subsite of the 30S subunit (the P/E state).

5. The ribosomal binding of the EF-G · GTP complex and the subsequent GTP hydrolysis impel the anticodon ends of these tRNAs, together with their bound mRNA, to move relative to the small ribosomal subunit such that the peptidyl–tRNA assumes the P/P state and the deacylated tRNA assumes the E/E state (the post-translocational state), thereby completing the elongation cycle.

The binding of tRNA to the A and E sites, as Nierhaus has shown, exhibits negative allosteric cooperativity. In the pre-translocational state, the E site binds the newly deacylated tRNA with high affinity (the E site is sterically unable to bind an aminoacyl–tRNA), whereas the empty A site has low affinity for aminoacyl–tRNA. However, in the post-translocational state, the ribosome has undergone a conformational change that converts the A site to a high-affinity state and the E site to a low-affinity state, which consequently releases the deacylated tRNA when aa–tRNA · EF-Tu · GTP binds to the A site. Thus, the E site is not simply a passive holding site for spent tRNAs but performs an essential function in the translation process. The GTP hydrolysis by the elongation factors EF-Tu and EF-G as well as the peptidyl transferase reaction apparently function to reduce the activation barriers between these conformational states. The unidirectional A → P → E flow of tRNAs through the ribosome is thereby facilitated.

Certain aspects of the foregoing mechanism are not fully resolved. For example, X-ray studies of the 70S ribosome in complex with three tRNAs (e.g., Fig. 32-34) suggest that the E-site tRNA is not released from the ribosome until Step 2 of Fig. 32-56. However, Nierhaus and Frank argue that these complexes were crystallized in presence of unphysiologically high tRNA concentrations.

Whatever the case, it is clear that the changes in binding states result in large-scale tRNA movements, in some instances >50 Å. Moreover, cryo-EM studies indicate that on binding EF-G · GDP(CH$_2$)P (like GMPPNP but with a CH$_2$ group rather than an NH group bridging its β and γ phosphates), the 30S subunit rotates with respect to the 50S subunit by 6° clockwise when viewed from the 30S subunit's solvent side, which results in a maximum displacement of ~19 Å at the periphery of the ribosome. This rotation is accompanied by many smaller conformational changes in both subunits, particularly in the regions about the entrance and exit to the mRNA channel. Clearly, we are far from fully understanding how the ribosome works at the molecular level.

m. The Eukaryotic Elongation Cycle Resembles That of Prokaryotes

The eukaryotic elongation cycle closely resembles that of prokaryotes. In eukaryotes, the functions of EF-Tu and EF-Ts are respectively assumed by the eukaryotic elongation factors **eEF1A** and **eEF1B,** with yeast eEF1B consisting of two subunits: **eEF1Bα,** which catalyzes nucleotide exchange, and **eEF1Bγ,** which has unknown function (in higher eukaryotes, eEF1B contains a third subunit, **eEF1Bβ,** that possesses a nucleotide exchange activity similar to that of eEF1Bα). Likewise, **eEF2** functions in a manner analogous to EF-G. However, the corresponding eukaryotic and prokaryotic elongation factors are not interchangeable.

The X-ray structure of yeast eEF1A · eEF1Bα (Fig. 32-57), determined by Kjeldgaard and Nyborg, reveals that eEF1A structurally resembles the homologous EF-Tu (Fig. 32-49), whereas eEF1Bα exhibits no resemblance to EF-Ts, either in sequence or in structure. Nevertheless, eEF1Bα functionally interacts with eEF1A much as EF-Ts interacts with EF-Tu: Both GEFs associate with their corresponding G-protein so as to disrupt the Mg^{2+} binding site associated with its bound guanine nucleotide.

E. Translational Accuracy

The genetic code is normally expressed with remarkable fidelity. We have already seen that transcription and tRNA aminoacylation both proceed with high accuracy (Sections 31-2Ec and 32-2Ce). The accuracy of ribosomal mRNA decoding was estimated from the rate of misincorporation of [^{35}S]Cys into highly purified **flagellin,** an *E. coli* protein (Section 35-3I) that normally lacks Cys. These measurements indicated that the mistranslation rate is ~10^{-4} errors per codon. This rate is greatly increased in the presence of **streptomycin,** an antibiotic that increases the rate of ribosomal misreading (Section 32-3Ga). From the types of reading errors that streptomycin is known to induce, it was deduced that the mistranslation arose almost entirely from the confusion of the Arg codons CGU and CGC for the Cys codons UGU and UGC. The above error rate is therefore largely caused by mistakes in ribosomal decoding.

An aminoacyl–tRNA is selected by the ribosome only according to its anticodon. Yet the binding energy loss

Figure 32-57 X-ray structure of yeast eEF1A · eEF1Bα.
Domains 1, 2, and 3 of eEF1A are magenta, green, and cyan,
respectively, and eEF1Bα is orange. The complex is oriented so
as to emphasize the structural resemblance between eEF1A and
the similarly colored EF-Tu in its complex with EF-Ts (Fig.
32-49). Note the lack of resemblance between eEF1Bα and
EF-Ts. [Based on an X-ray structure by Morten Kjeldgaard and
Jens Nyborg, University of Aarhus, Århus, Denmark. PDBid 1F60.]

arising from a single base mismatch in a codon–anticodon interaction is estimated to be ~12 kJ · mol^{-1}, which, according to Eq. [32.1], cannot account for a ribosomal decoding
accuracy of less than ~10^{-2} errors per codon. Moreover,
the base pairing interaction between the UUU codon for
Phe and the GAA anticodon of tRNA^{Phe} would be naively
expected to be less stable than the incorrect pairing be-

tween the UGC codon for Ser and the GCG anticodon of
tRNA^{Arg}. This is because both interactions have one G · U
base pair and the former correct interaction's remaining
two A · U base pairs are weaker than the latter incorrect
interaction's remaining two G · C base pairs. Evidently, the
ribosome has some sort of proofreading mechanism that
increases its overall decoding accuracy.

a. The Ribosome Monitors the Formation of a Correct Codon–Anticodon Complex

As we have seen (Figs. 32-55*b* and 32-56), the aminoacyl–tRNA · EF-Tu · GTP ternary complex initially binds to
the ribosome with the tRNA in the A/T binding state. The
tRNA only assumes the fully bound A/A state (accommodation) after the GTP has been hydrolyzed and the EF-Tu ·
GDP complex has been released from the ribosome. These
two states presumably permit the ribosome to double-
check (proofread) the codon–anticodon complex that the
mRNA makes with the incoming tRNA.

The X-ray structure of the *T. thermophilus* 30S subunit
in complex with a U$_6$ hexanucleotide mRNA and a 17-nt
RNA consisting of the tRNA^{Phe} anticodon stem–loop
(Fig. 32-11, although its nucleotides are unmodified), determined by Ramakrishnan, revealed how an mRNA-
specified tRNA initially binds to the ribosome. The
codon–anticodon association is stabilized by its interactions with three universally conserved ribosomal bases,
A1492, A1493, and G530 (Fig. 32-58):

1. The first codon–anticodon base pair, that between
mRNA U1 and tRNA A36, is stabilized by the binding of
the rRNA A1493 base in the base pair's minor groove (Fig.
32-58*a*).

2. The second codon–anticodon base pair, that between
U2 and A35, is bolstered by A1492 and G530, which both
bind in this base pair's minor groove (Fig. 32-58*b*).

3. The third codon–anticodon base pair (the wobble
pair; Section 32-2Db), that between U3 and G34, is reinforced through minor groove binding by G530 (Fig. 32-58*c*).

Figure 32-58 Codon–anticodon interactions in the ribosome.
The (*a*) first, (*b*) second, and (*c*) third codon–anticodon base
pairs as seen in the X-ray structure of the *T. thermophilus* 30S
subunit in complex with U$_6$ (a model mRNA) and the 17-nt
anticodon stem–loop of tRNA^{Phe} (whose anticodon is GAA).
The structures are drawn in ball-and-stick form embedded in

their semitransparent van der Waals surfaces. Codons are purple,
anticodons are yellow, and rRNA is brown or gray with non-C
atoms colored according to type (N blue, O red, and P green).
Protein C atoms are gray and Mg^{2+} ions are represented by
magenta spheres. [Courtesy of Venki Ramakrishnan, MRC
Laboratory of Molecular Biology, Cambridge, U.K. PDBid 1IBM.]

(a)

(b)

Figure 32-59   **Ribosomal decoding site.** The X-ray structures of *T. thermophilus* 30S subunit (*a*) alone and (*b*) in its complex with U$_6$ and the 17-nt anticodon stem–loop of tRNA^{Phe}. The RNAs are drawn as ribbons with their nucleotides in paddle form with tRNA gold, A-site mRNA purple, rRNA gray, and nucleotides that undergo conformational changes red. Protein S12 is tan and Mg^{2+} ions are represented by red spheres. Compare Part *b* with Fig. 32-58. [Courtesy of Venki Ramakrishnan, MRC Laboratory of Molecular Biology, Cambridge, U.K. PDBids (*a*) 1FJF and (*b*) 1IBM.]

This latter interaction appears to be less stringent than those in the first and second codon–anticodon positions, which is consistent with the need for the third codon–anticodon pairing to tolerate non-Watson–Crick base pairs (Section 32-2D).

Comparison of this structure with that of the 30S subunit alone reveals that the foregoing rRNA nucleotides undergo conformational changes on the formation of a codon–anticodon complex (Fig. 32-59). In the absence of tRNA, the bases of A1492 and A1493 stack in the interior of an RNA loop but flip out of this loop to form the codon–anticodon complex, whereas the G530 base switches from the syn to the anti conformation (Section 29-2Aa). These interactions enable the ribosome to monitor whether an incoming tRNA is cognate to the codon in the A site; a non-Watson–Crick base pair could not bind these ribosomal bases in the same way. Indeed, any mutation of A1492 or A1493 is lethal because pyrimidines in these positions could not reach far enough to interact with the codon–anticodon complex or G530 and because a G in either position would be unable to form the required hydrogen bonds and its N2 would be subjected to steric collisions. An incorrect codon–anticodon provides insufficient free energy to bind the tRNA to the ribosome and it therefore dissociates from it, still in its ternary complex with EF-Tu and GTP.

b. GTP Hydrolysis by EF-Tu Is a Thermodynamic Prerequisite to Ribosomal Proofreading

A proofreading step must be entirely independent of the initial selection step. Only then can the overall probability of error be equal to the product of the probabilities of error of the individual selection steps. We have seen that DNA polymerases and aminoacyl–tRNA synthetases maintain the independence of their two selection steps by carrying them out at separate active sites (Sections 30-2Ac and 32-2Ce). Yet the ribosome only recognizes the incoming aminoacyl–tRNA according to its anticodon's complementarity to the codon in the A site. Consequently, the ribosome must somehow examine this codon–anticodon interaction in two separate ways.

The formation of a correct codon–anticodon complex induces EF-Tu to hydrolyze its bound GTP, although how this occurs is unclear (note that EF-Tu's GTPase domain is bound in the 50S subunit which, together with the observation that GTP hydrolysis requires an intact tRNA, suggests that the hydrolysis signal is at least in part transmitted through the tRNA). The resulting conformational change in EF-Tu (Fig. 32-48) swings its bound tRNA into the A/A state (accommodation), a process that moves the 3' end of the tRNA by nearly 70 Å. This, it is hypothesized, subjects the codon–anticodon interaction to a strain that only a correct pairing can withstand. The codon–anticodon interaction is thereby subjected to a second screening that only permits a cognate aminoacyl–tRNA to enter the peptidyl transferase center. The irreversible GTPase reaction must precede this proofreading step because otherwise the dissociation of a noncognate tRNA (the release of its anticodon from the codon) would simply be the reverse of the initial binding step, that is, it would be part of the initial selection step rather than proofreading. *GTP hydrolysis therefore provides the second context necessary for proofreading; it is the entropic price the system must pay for accurate tRNA selection.*

F. Chain Termination

Polypeptide synthesis under the direction of synthetic mRNAs such as poly(U) terminates with a peptidyl–tRNA in association with the ribosome. However, *the translation of natural mRNAs, which contain the Stop codons UAA, UGA, or UAG, results in the release of free polypeptides.*

Accurate termination is essential, not only because it prevents the wasteful synthesis of nonfunctional polypeptides, but also because prematurely terminated polypeptides may be toxic.

a. Prokaryotic Termination

In *E. coli,* chain termination has several stages (Fig. 32-60):

1. The termination codons, the only codons that normally have no corresponding tRNAs, are recognized by class I **release factors** (Table 32-9): **RF-1** recognizes UAA and UAG, whereas the 39% identical **RF-2** recognizes UAA and UGA. Swapping a conserved PXT tripeptide in RF-1 with a conserved SPF tripeptide in RF-2 interchanges their Stop codon specificities, which suggests that these tripeptides mimic anticodons.

2. *On binding to their corresponding Stop codon, RF-1 and RF-2 induce the transfer of the peptidyl group from*

Figure 32-60 **Termination pathway in *E. coli* ribosomes.** RF-1 recognizes the Stop codons UAA and UAG, whereas RF-2 (not shown) recognizes UAA and UGA. Eukaryotic termination follows an analogous pathway but requires only a single class I release factor, eRF1, that recognizes all three Stop codons.

Figure 32-61   Ribosome-catalyzed hydrolysis of peptidyl–tRNA to form a polypeptide and free tRNA.

tRNA to water rather than to an aminoacyl–tRNA, thereby releasing the completed polypeptide (Fig 32-61). This occurs with an error rate of 10^{-5} without proofreading. The class I release factors act at the ribosomal A site as is indicated by the observations that they compete with suppressor tRNAs for termination codons and that they cannot bind to the ribosome simultaneously with EF-G. A GGQ tripeptide that is universally conserved in all class I release factors is implicated in catalyzing the hydrolysis of the peptidyl–tRNA ester linkage (see below).

3. Once the newly synthesized polypeptide has been released from the ribosome, the class II release factor **RF-3,** in its complex with GDP, binds to the ribosome at the same site as do EF-Tu and EF-G. In fact, the X-ray structure of RF-3 · GDP resembles that of EF-Tu · GMPPNP (Fig. 32-47). Free RF-3 has a greater affinity for GDP than GTP but on binding to the ribosome–RF-1/2 complex, it exchanges its bound GDP for GTP. The resulting change in the conformation of RF-3, as seen in cryo-EM studies, causes it to bind more tightly to the ribosome and expel the RF-1/2. RF-3 is not required for cell viability although it is necessary for maximum growth rate; RF-3 only accelerates the dissociation of RF-1/2 from the ribosome by ~5-fold.

4. The interaction of RF-3 · GTP with the ribosome stimulates it to hydrolyze its bound GTP, much as occurs with EF-Tu · GTP and EF-G · GTP. The resulting RF-3 · GDP then dissociates from the ribosome. Subsequently, **ribosomal recycling factor (RRF)** binds in the ribosomal A site followed by EF-G · GTP. RRF, which was discovered by Akira Kaji, is essential for cell viability.

5. EF-G hydrolyzes its bound GTP, which causes RRF to be translocated to the P site and the tRNAs previously in the P and E sites (the latter not shown in Fig. 32-60) to be released. Finally, the small and large ribosomal subunits separate, a process that is facilitated by the binding of IF-3 (Section 32-3Cc), and RRF, EF-G · GDP, and mRNA are released. The ribosomal subunits can then participate in a new round of initiation (Fig. 32-43).

b. Eukaryotic Termination

Chain termination in eukaryotes resembles that in prokaryotes, but it has only one class I release factor, **eRF1,** that recognizes all three Stop codons. It is unrelated in sequence to RF-1 and RF-2. However, the eukaryotic class II release factor, **eRF3,** resembles RF-3 in both sequence and function. Nevertheless, eRF3 is essential for eukaryotic cell viability.

c. The Ribosome Binds RF-1 and RF-2 in a Conformation That Catalyzes Peptide Release

The X-ray structures of the *T. thermophilus* ribosome with RF-1, an mRNA containing a UAA Stop codon, and deacylated tRNAs in its P and E sites was determined by Noller (Fig. 32-62*a*), and the closely similar structures containing the tRNAs, RF-2, and an mRNA with a UAA or a UGA Stop codon, were respectively determined by Noller and Ramakrishnan. These are all product complexes since they lack peptidyl groups on their P-site tRNAs. The structurally similar RF-1 (Fig. 32-62*b*) and RF-2 each consist of four domains with domains 2 and 4 occupying the ribosome's decoding center (DC), where they contact the mRNA's Stop codon, and with domain 3 occupying the peptidyl transferase center (PTC), where it interacts with the ribose residue of the P-site tRNA's A76 to which a peptidyl group would be linked in a substrate complex. The deletion of domain 1 does not affect peptide release activity but is required for the RF-3–facilitated dissociation of RF-1/2 from the ribosome (see below).

(a)

(b)

***Figure 32-62*   X-ray structure of the *T. thermophilus* ribosome in complex with RF-1, a UAA Stop codon–containing mRNA, and deacylated tRNAs in its P and E sites.** (*a*) The overall structure with proteins shown in ribbon form and all RNAs shown in ladder form but the mRNA, which is drawn in worm form in green. The ribosome is semitransparent with its 23S RNA gray, its 5S RNA light blue, its 16S RNA cyan, the 50S subunit's proteins magenta, and the 30S subunit's proteins purple. The tRNAs occupying the P and E sites are orange and red, and the RF-1, which in part occupies the ribosomal A site, is yellow. (*b*) Close-up of the interactions between the P-site tRNA, mRNA, and RF-1. The ribosome, tRNA, and mRNA are drawn as in Part *a* and the RF-1 is colored with its domains 1, 2, 3, and 4 green, yellow, blue, and magenta, respectively. The so-called switch–loop, which connects domains 3 and 4 and undergoes a major conformational rearrangement between the free and ribosome-bound RF-1, is orange. The PVT tripeptide implicated in Stop codon recognition and the GGQ tripeptide implicated in catalyzing the ester hydrolysis reaction are drawn in stick form in red. [Courtesy of Harry Noller, University of California at Santa Cruz. PDBids 3D5A and 3D5B.]

The binding of RF-1 or RF-2 in the DC causes A530 and A1492 of the 16S RNA to flip out from their resting state (Fig. 32-59*a*) as occurs with the binding of a tRNA to its cognate codon (Fig. 32-59*b*). However, A1493 does not flip out because in doing so it would clash with domain 2 of either release factor. Instead, it stacks on A1913 of the 23S RNA. The Stop codons are recognized by hydrogen bonding and van der Waals interactions with the similarly located PXT and SPF tripeptides on domain 2 of RF-1 and RF-2. However, the observation that mutations in RF-2 distant from its SQF motif result in altered specificity suggests that Stop codon recognition arises from a subtle balance of binding energy and conformational changes as we have seen to be the case for codon recognition by tRNA (Section 32-3Ea).

In the PTC, the GGQ tripeptide on domain 3 of both RF-1 and RF-2 contacts the 3′-terminal ribose residue (A76) of the deacylated P-site tRNA. Both Gly residues adopt backbone conformations that are forbidden for other amino acid residues, which accounts for the observations that the mutation of either residue results in an up to 10^4-fold reduction in the rate of peptide release. The main chain NH group of the Gln residue is hydrogen bonded to the 3′-OH group of A76, which, it is hypothesized, positions it to also hydrogen bond to and thereby stabilize the tran-

siton state oxyanion in the hydrolysis reaction. In agreement with this hypothesis, the mutation of the Gln residue to Pro, which lacks a main chain NH group, abolishes the hydrolysis reaction. The side chain of the Gln residue is pointed away from the ribose (top of Fig. 32-62*b*) where, it is proposed, it helps position a water molecule for an in-line nucleophilic attack on the scissile ester bond. In addition, the observation that peptide release is nearly abolished by the removal of the 2′-OH from the P-site tRNA's 3′ terminal residue suggests that peptide release utilizes a substrate-assisted proton shuttle mechanism similar to that of peptide bond formation (Section 32-3Di). Finally, the binding of RF-1/2 shifts U2585 so as to expose the otherwise protected scissile ester bond to nucleophilic attack (Section 32-3Di). Nevertheless, the formulation of a definitive mechanism for peptide release must await the X-ray structure of a ribosome in complex with both a release factor and a peptidyl–tRNA, that is, a substrate complex.

How does the binding of a release factor to a Stop codon in the DC induce the ~75-Å distant PTC to hydrolyze the scissile ester bond? In the X-ray structures of RF-1 or RF-2 alone, their PXT/SPF and GGQ motifs are only ~23 Å apart due to a change in conformation of their switch–loop segments (Fig. 32-62*b*) relative to that in the ribosomal complexes. The conformation of the switch–loop observed in the ribosomal complexes is only possible when a Stop codon is recognized. This is because the flipping out of A1493, which only occurs when a tRNA binds its cognate codon in the DC (Fig. 32-59), or the failure of A1913 to stack on A1493, alters the binding pocket for the ribosomally bound switch–loop. Noller has therefore proposed that the binding of a Stop codon by its corresponding release factor and the consequent rearrangement of both its switch–loop and the DC cooperatively permit the GGQ motif to bind to the PTC in a way that catalyzes peptide release.

d. RRF Binds in the Ribosomal A Site

The X-ray structure of *T. thermophilus* ribosomal recycling factor (RRF), determined by Yoshikazu Nakamura, reveals it to be a two-domain structure that resembles tRNA in its overall shape (Fig. 32-63). The comparison of this structure with those of several other bacterial RRFs indicates that the two linkers connecting the RRF domains are flexible such that domain II can rotate about the axis of the three-helix bundle forming domain I.

The X-ray structure of the *T. thermophilus* ribosome in complex with RRF in its A site, the anticodon stem–loop (ASL) of tRNA^{Phe} in its P site, tRNA$_f^{Met}$ in its E site, and an mRNA with a UAG Stop codon in the A site was determined by Ramakrishnan (Fig. 32-64). Domain I of the RRF spans the A and P sites of the 50S ribosome, a position in which the tip of its domain I would clash with a tRNA in the P site (and which rationalizes why a ribosomal complex of RRF and a tRNA in the P site has not been crystallized). This suggests that RRF binding forces a tRNA bound in the P site into the P/E hybrid binding state (Section 32-3D1). Previous structural studies suggested that RRF binding induced changes in the bridges connecting the small and large subunits (Section 32-3Ae). However, no such changes are observed in the above structure. Perhaps they occur in the P/E hybrid state.

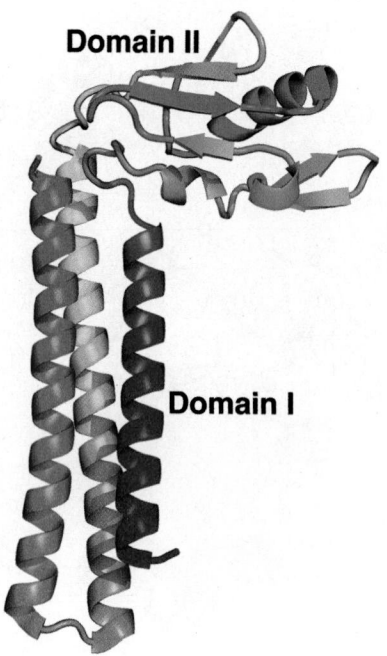

Figure 32-63   **X-ray structure of *T. thermophilus* RRF.** This monomeric protein is drawn in ribbon form colored in rainbow order from its N-terminus (*blue*) to its C-terminus (*red*). [Based on an X-ray structure by Yoshikazu Nakamura, The University of Tokyo, Japan. PDBid 1EH1.]

e. GTP Hydrolysis Speeds Up Ribosomal Processes

What is the role of the GTP hydrolysis reactions mediated by the various ribosomally associated G proteins (IF-2, EF-Tu, EF-G, and RF-3 in bacteria)? Translation occurs in the absence of GTP, albeit slowly, so that the free energy of the peptidyl transferase reaction is sufficient to drive the entire translational process. Moreover, none of the GTP

Figure 32-64   **X-ray structure of the *T. thermophilus* ribosome in complex with RRF in its A site, the anticodon stem–loop (ASL) of tRNA^{Phe} in its P site, tRNA$_f^{Met}$ in its E site, and an mRNA with a UAG Stop codon in the A site.** The ribosomal RNAs are shown as semitransparent surface diagrams with 23S RNA light blue, 5S RNA blue-green, and 16S RNA yellow. The ribosomal proteins are drawn in ribbon form with 50S subunit proteins blue and 30S subunit proteins tan. The RRF, tRNAs, and mRNA are represented by their surface diagrams with domains I and II of RRF red and blue, the mRNA magenta, the P-site ASL purple, and the E-site tRNA$_f^{Met}$ gray. [Courtesy of Venki Ramakrishnan, MRC Laboratory of Molecular Biology, Cambridge, U.K. PDBids 2V46 and 2V47.]

hydrolysis reactions yields a "high-energy" covalent intermediate as does, say, ATP hydrolysis in numerous biosynthetic reactions. Instead, the ribosomal binding of a G-protein induces it to hydrolyze its bound GTP to GDP resulting a conformational change that causes the ribosome to carry out a particular process (fMet–tRNA$_f^{Met}$ binding for IF-2, accommodation for EF-Tu, translocation for EF-G, and RF-1/2 release for RF-3) and release the resulting G-protein · GDP complex. *The high rate and irreversibility of the GTP hydrolysis reaction ensures that the various complex ribosomal processes to which it is coupled, initiation, elongation, and termination, will themselves be fast and irreversible.* In essence, G-protein · GTP complexes act as Maxwell's demons to trap the ribosome in functionally productive conformations. Hence, as we saw to be the case for ribosomal proofreading (Section 32-3Eb), the ribosome utilizes the free energy of GTP hydrolysis to gain a more ordered (lower entropy) state rather than a higher energy state as often occurs in ATP-dependent processes.

G. *Protein Synthesis Inhibitors: Antibiotics*

Antibiotics are bacterially, fungally, or synthetically produced substances that inhibit the growth of microorganisms. Antibiotics are known to inhibit a variety of essential biological processes, including DNA replication (e.g., ciprofloxacin; Section 29-3Cd), transcription (e.g., rifamycin B; Section 31-2Bb), and bacterial cell wall synthesis (e.g., penicillin; Section 11-3Bb). However, *the majority of known antibiotics, including a great variety of medically useful substances, block translation.* This situation is presumably a consequence of the translational machinery's enormous complexity, which makes it vulnerable to disruption in many ways. Antibiotics have also been useful in an-alyzing ribosomal mechanisms because, as we have seen for puromycin (Section 32-3Df), the blockade of a specific function often permits its biochemical dissection into its component steps. Table 32-10 and Fig. 32-65 present several medically significant and/or biochemically useful translational inhibitors. We study the mechanisms of a few of the best characterized of them below.

a. Streptomycin

Streptomycin, which was discovered in 1944 by Selman Waksman, is a medically important member of a family of antibiotics known as **aminoglycosides** that inhibit prokaryotic ribosomes in a variety of ways. At low concentrations, streptomycin induces the ribosome to characteristically misread mRNA: One pyrimidine may be mistaken for the other in the first and second codon positions and either pyrimidine may be mistaken for adenine in the first position. This inhibits the growth of susceptible cells but does not kill them. At higher concentrations, however, streptomycin prevents proper chain initiation and thereby causes cell death.

Certain streptomycin-resistant mutants (str^R) have ribosomes with an altered protein S12 compared with streptomycin-sensitive bacteria (str^S). Intriguingly, a change in base C912 of 16S rRNA (which lies in its central domain; Fig. 32-27*a*) also confers streptomycin resistance. (Some mutant bacteria are not only resistant to streptomycin but dependent on it; they require it for growth.) In partial diploid bacteria that are heterozygous for streptomycin resistance (str^R/str^S), streptomycin sensitivity is dominant. This puzzling observation is explained by the finding that, in the presence of streptomycin, str^S ribosomes remain bound to initiation sites, thereby excluding str^R ribosomes from these sites. Moreover, the mRNAs in these blocked complexes are degraded after a few

Table 32-10 **Some Ribosomal Inhibitors**

Inhibitor	Action
Chloramphenicol	Inhibits peptidyl transferase on the prokaryotic large subunit
Cycloheximide	Inhibits peptidyl transferase on the eukaryotic large subunit
Erythromycin	Inhibits translocation by the prokaryotic large subunit
Fusidic acid	Inhibits elongation in prokaryotes by binding to EF-G · GDP in a way that prevents its dissociation from the large subunit
Paromomycin	Increases the ribosomal error rate
Puromycin	An aminoacyl–tRNA analog that causes premature chain termination in prokaryotes and eukaryotes
Streptomycin	Causes mRNA misreading and inhibits chain initiation in prokaryotes
Tetracycline	Inhibits the binding of aminoacyl–tRNAs to the prokaryotic small subunit
Diphtheria toxin	Catalytically inactivates eEF2 by ADP-ribosylation
Ricin/abrin/α-sarcin	**Ricin** and **abrin** are poisonous plant glycosidases that catalytically inactivate the eukaryotic large subunit by hydrolytically depurinating a specific highly conserved A residue of the 28S RNA, which is located on the so-called **sarcin–ricin loop** that forms a critical part of the ribosomal factor–binding center; α-**sarcin** is a fungal protein that cleaves a specific phosphodiester bond in the sarcin–ricin loop

Chloramphenicol

Erythromycin

Streptomycin

Cycloheximide

Fusidic acid

Tetracycline

Paromomycin

Figure 32-65 **Selection of antibiotics that act as translational inhibitors.**

minutes, which allows the *str*^S ribosomes to bind to newly synthesized mRNAs as well.

b. Chloramphenicol

Chloramphenicol, the first of the "broad-spectrum" antibiotics, inhibits the peptidyl transferase activity on the large subunit of prokaryotic ribosomes. However, its clinical uses are limited to only severe infections because of its toxic side effects, which are caused, at least in part, by the chloramphenicol sensitivity of mitochondrial ribosomes. The 23S RNA is implicated in chloramphenicol binding by the observation that some of its mutants are chloramphenicol resistant. Indeed, X-ray studies indicate that chloramphenicol binds in the large subunit's polypeptide exit tunnel in the vicinity of the A site. This explains why chloramphenicol competes for binding with the 3′ end of aminoacyl–tRNAs as well as with puromycin (whose ribosomal binding site overlaps that of chloramphenicol) but not with peptidyl–tRNAs. These observations suggest that chloramphenicol inhibits peptidyl transfer by interfering with the interactions of ribosomes with A site–bound aminoacyl–tRNAs.

c. Paromomycin

Paromomycin, a clinically useful aminoglycoside antibiotic, increases the ribosomal error rate. The X-ray structure of the 30S subunit in complex with paromomycin (Fig. 32-66) reveals that it binds to the interior of the RNA loop in which the bases of A1492 and A1493 are normally stacked (Fig. 32-59a). This causes these bases to flip out of the loop and assume a conformation resembling that in the codon–anticodon–30S subunit complex (Fig. 32-59b). Indeed, this codon–anticodon–30S subunit complex is not significantly disturbed by the binding of paromomycin. As we have seen in Section 32-3Ea, the 30S subunit employs A1492 and A1493 to ascertain whether the first two codon–anticodon base pairs are Watson–Crick base pairs, that is, whether the incoming tRNA is cognate to the codon in the A site. Noncognate tRNAs normally have insufficient codon–anticodon binding energy to flip A1492 and A1493 out of the loop and consequently are rejected by the ribosome. However, the binding of paromomycin to the 30S subunit pays the energetic price of these base flips. This facilitates the ribosomal acceptance (stabilizes the binding) of near-cognate aminoacyl–tRNAs and hence the erroneous incorporation of their amino acid residues into the polypeptide being synthesized.

d. Tetracycline

Tetracycline and its derivatives are broad-spectrum antibiotics that bind to the small subunit of prokaryotic ribosomes, where they inhibit aminoacyl–tRNA binding. An X-ray structure of tetracycline in complex with the 30S subunit reveals that tetracycline mainly binds in a crevice comprised of only the 3′ major domain of 16S RNA (Fig. 32-27a) and which is located in the neck of the 30S subunit just above its A site. This permits the initial screening of the aminoacyl–tRNA to proceed but physically blocks its accommodation into the peptidyl transferase (A/A) site after

Figure 32-66 **X-ray structure of the 30S ribosome in complex with the antibiotic paromomycin.** The view and coloring are the same as those in Fig. 32-59 with the paromomycin (PAR) drawn in stick form in yellow-green. [Courtesy of Venki Ramakrishnan, MRC Laboratory of Molecular Biology, Cambridge, U.K. PDBid 1IBK.]

EF-Tu–catalyzed GTP hydrolysis has occurred, resulting in the release of the tRNA. Hence, in addition to preventing protein synthesis, tetracycline binding causes the unproductive hydrolysis of GTP which, since this occurs every time a cognate aminoacyl–tRNA binds to the ribosome, poses an enormous energetic drain on the cell. The nucleotides forming the tetracycline binding site are poorly conserved in eukaryotic ribosomes, thereby accounting for tetracycline's bacterial specificity.

Tetracycline also blocks the stringent response (Section 31-3I) by inhibiting (p)ppGpp synthesis. This indicates that deacylated tRNA must bind to the A site in order to activate stringent factor.

Tetracycline-resistant bacterial strains have become quite common, thereby precipitating a serious clinical problem. Resistance is often conferred by a decrease in bacterial cell membrane permeability to tetracycline rather than any alteration of ribosomal components.

e. Diphtheria Toxin

Diphtheria is a disease resulting from bacterial infection by *Corynebacterium diphtheriae* that harbor the bacteriophage **corynephage β**. Diphtheria was a leading cause of childhood death until the late 1920s when immunization became prevalent. Although the bacterial infection is usually confined to the upper respiratory tract, the bacteria secrete a phage-encoded protein, known as **diphtheria toxin (DT),** which is responsible for the disease's lethal effects. *Diphtheria toxin specifically inactivates the eukaryotic elongation factor eEF2, thereby inhibiting eukaryotic protein synthesis.*

The pathogenic effects of diphtheria are prevented, as was discovered in the 1880s, by immunization with **toxoid** (formaldehyde-inactivated toxin). Individuals who have contracted diphtheria are treated with antitoxin from horse serum, which binds to and thereby inactivates DT, as well as with antibiotics to combat the bacterial infection.

DT is a member of the family of bacterial toxins that includes cholera toxin (CT) and pertussis toxin (PT; Section 19-2Ce). It is a monomeric 535-residue protein that is readily cleaved past its Arg residues 190, 192, and 193 by trypsin and trypsinlike enzymes. This hydrolysis occurs around the time diphtheria toxin encounters its target cell, yielding two fragments, A and B, which, nevertheless, remain linked by a disulfide bond. The B fragment's C-terminal domain binds to a specific receptor on the plasma membrane of susceptible cells, thereby inducing DT's uptake into the endosome (Fig. 12-91) via receptor-mediated endocytosis (Section 12-5Bc; free fragment A is devoid of toxic activity). The endosome's low pH of 5 triggers a conformational change in the B fragment's N-terminal domain, which then inserts into the endosomal membrane so as to facilitate the entry of the A fragment into the cytoplasm. The disulfide bond linking the A and B subunits is then cleaved by the cytoplasm's reducing environment.

Within the cytosol, the A fragment catalyzes the **ADP-ribosylation** of eEF2 by NAD^+,

$$eEF2 + NAD^+$$
$$(active)$$

$$\downarrow diphtheria\ toxin$$

$$ADP\text{-ribosyl-eEF2} + Nicotinamide + H^+$$
$$(inactive)$$

thereby inactivating this elongation factor. Since the A fragment acts catalytically, *one molecule is sufficient to ADP-ribosylate all of a cell's eEF2s, which halts protein synthesis and kills the cell.* Only a few micrograms of diphtheria toxin are therefore sufficient to kill an unimmunized individual.

Diphtheria toxin specifically ADP-ribosylates a modified His residue on eEF2 known as **diphthamide:**

ADP-Ribosylated diphthamide

Diphthamide occurs only in eEF2 (not even in its bacterial counterpart, EF-G), which accounts for the specificity of diphtheria toxin in exclusively modifying eEF2 (recall that CT ADP-ribosylates a specific Arg residue on $G_{s\alpha}$ and PT ADP-ribosylates a specific Cys residue on G_{ia}; Section 19-2C). Since diphthamide occurs in all eukaryotic eEF2s, it probably is essential to eEF2 activity. Yet, certain mutant cultured animal cells, which have unimpaired capacity to synthesize proteins, lack the enzymes that post-translationally modify His to diphthamide (although mutating the diphthamide His to Asp, Lys, or Arg inactivates translation). Perhaps the diphthamide residue has a control function.

4 CONTROL OF EUKARYOTIC TRANSLATION

The rates of ribosomal initiation on prokaryotic mRNAs differ by factors of up to 100, a variation that is largely a consequence of their different Shine–Dalgarno sequences. Moreover, the genes forming an operon are often expressed in decreasing molar amounts from the operon's 5′ end to its 3′ end. For example, the proteins specified by the *E. coli lac* operon (Section 31-1Ab), β-galactosidase, galactose permease, and thiogalactoside transacetylase, are produced in molar ratios of 10:5:2. Such **polarity** may arise when the initiation codon of a gene that lacks a Shine–Dalgarno sequence is very near the Stop codon of its upstream gene, a situation that occurs most often when the Stop codon overlaps the initiation codon as in the sequence AUGA. The translation of the upstream gene will then be required for the translation of the downstream gene, a phenomenon termed **translational coupling.** The polarity arises because a ribosome often dissociates from the mRNA on encountering the upstream gene's Stop codon. Alternatively, an mRNA may fold in a way that masks an internal Shine–Dalgarno sequence, for example, by the base pairing of a segment adjacent to the Shine–Dalgarno sequence to a downstream element of the preceding gene. Such Shine–Dalgarno sequences only become available when a ribosome that is translating the preceding gene disrupts the folded structure.

Genetic expression in prokaryotes is largely transcriptionally controlled (Section 31-3). This is apparently because prokaryotic mRNAs have lifetimes of only a few minutes and, hence, it is a more efficient use of resources to control their transcription. Nevertheless, the expression of certain prokaryotic genes is translationally controlled, most notably those encoding the ribosomal proteins (which comprise 10% of cellular proteins), which must be produced in equimolar amounts. The production of ribosomal proteins is controlled, in part, through a process in which a ribosomal protein binds to the mRNA of the operon encoding it in the vicinity of a translational start site located near the mRNA's 5′ end so as to inhibit its translational initiation. However, each such protein binds more tightly to an rRNA in forming the ribosome. Consequently, only when there is an excess of that protein will it inhibit its own translation as well as those of other proteins encoded by its operon.

Eukaryotic cells, whose mRNAs have lifetimes of hours or days, respond to many of their needs through translational control. In this section, we examine how eukaryotic translation is regulated via the phosphorylation/dephosphorylation of eIF2 and eIF4E. We then consider translational control by mRNA masking and cytoplasmic polyadenylation and end by discussing the uses of antisense oligonucleotides.

A. *Regulation of eIF2*

Four important pathways for the regulation of translation in eukaryotes involve the phosphorylation of the conserved Ser 51 on the α subunit of eIF2 (**eIF2α;** recall that eIF2 is an $\alpha\beta\gamma$ trimer that conducts Met–tRNA$_i^{Met}$ to the 40S ribosomal subunit, and the resulting complex scans the bound mRNA for the initiating AUG codon to form the 48S preinitiation complex; Section 32-3Cd). The so-called **eIF2α kinases** that do so share a conserved kinase domain but have unique regulatory domains.

a. Heme Availability Controls Globin Translation

Reticulocytes synthesize protein, almost exclusively hemoglobin, at an exceedingly high rate and are therefore a favorite subject for the study of eukaryotic translation. Hemoglobin synthesis in fresh reticulocyte lysates proceeds normally for several minutes but then abruptly stops because of the inhibition of translational initiation and the consequent polysome disaggregation. This process is prevented by the addition of heme [a mitochondrial product (Section 26-4A) that this *in vitro* system cannot synthesize], thereby indicating that *globin synthesis is regulated by heme availability*. The inhibition of globin translational initiation is also reversed by the addition of the eukaryotic initiation factor eIF2 and by high levels of GTP.

In the absence of heme, reticulocyte lysates accumulate an eIF2α kinase named **heme-regulated inhibitor [HRI;** also called heme-controlled repressor **(HCR)]**. HRI is a homodimer whose 629-residue subunits each contain two heme-binding sites. When heme is plentiful, both of these sites are occupied and the protein, which is autophosphorylated at several Ser and Thr residues, is inactive. However, when heme is scarce, one of these sites loses its bound heme, thereby activating HRI to autophosphorylate itself at several additional sites and to phosphorylate Ser 51 of eIF2α.

Phosphorylated eIF2 can participate in the ribosomal initiation process in much the same way as unphosphorylated eIF2. This puzzling observation was clarified by the discovery that GDP does not dissociate from phosphorylated eIF2 at the completion of the initiation process as it normally does through a process facilitated by eIF2B acting as a GEF (Fig. 32-44). This is because phosphorylated eIF2 forms a much tighter complex with eIF2B than does unphosphorylated eIF2. This sequesters eIF2B (Fig. 32-

Figure 32-67 **Model for heme-controlled protein synthesis in reticulocytes.**

67), which is present in lesser amounts than eIF2, thereby preventing the regeneration of the eIF2 · GTP required for translational initiation. The presence of heme reverses this process by inhibiting HRI, whereon the phosphorylated eIF2 molecules are reactivated through the action of **eIF2 phosphatase,** which is unaffected by heme. The reticulocyte thereby coordinates its synthesis of globin and heme.

b. Interferons Protect against Viral Infection

Interferons are cytokines that are secreted by virus-infected vertebrate cells. On binding to surface receptors of other cells, interferons convert them to an antiviral state, which inhibits the replication of a wide variety of RNA and DNA viruses. Indeed, the discovery of interferons in the 1950s arose from the observation that virus-infected individuals are resistant to infection by a second type of virus.

There are three families of interferons: **type α** or **leukocyte interferon** (165 residues; leukocytes are white blood cells), the related **type β** or **fibroblast interferon** (166 residues; fibroblasts are connective tissue cells), and **type γ** or **lymphocyte interferon** (146 residues; lymphocytes are immune system cells). *Interferon synthesis is induced by the double-stranded RNA (dsRNA) that is generated during infection by both DNA and RNA viruses, as well as by the synthetic dsRNA poly(I) · poly(C).* Interferons are effective antiviral agents in concentrations as low as $3 \times 10^{-14} M$, which makes them among the most potent biological substances known. Moreover, they have far wider specificities than antibodies raised against a particular virus. They have therefore elicited great medical interest, particularly since some cancers are virally induced (Section 19-3B). Indeed, they are in clinical use against certain tumors and viral infections. These treatments are made possible by the production of large quantities of these otherwise quite scarce proteins through recombinant DNA techniques (Section 5-5G).

Interferons prevent viral proliferation largely by inhibiting protein synthesis in infected cells (lymphocyte interferon also modulates the immune response). They do so in two independent ways (Fig. 32-68):

1. Interferons induce the production of an eIF2α kinase, **double-stranded RNA–activated protein kinase [PKR;** also known as **double-stranded RNA–activated inhibitor (DAI);** 551 residues], which on binding dsRNA, dimerizes and autophosphorylates itself. This activates PKR to phosphorylate eIF2α at its Ser 51, thereby inhibiting ribosomal initiation and hence the proliferation of viruses in virus-infected cells. The importance of PKR to cellular antiviral defense is indicated by the observation that many viruses express inhibitors of PKR.

(a)

Inhibition of Translation

(b)

mRNA Degradation

Figure 32-68 **The action of interferon.** In interferon-treated cells, the presence of dsRNA, which normally results from a viral infection, causes (*a*) the inhibition of translational initiation and (*b*) the degradation of mRNA, thereby blocking translation and preventing virus replication.

2. Interferons also induce the synthesis of **(2′,5′)-oligoadenylate synthetase (2,5A synthetase).** In the presence of dsRNA, this enzyme catalyzes the synthesis from ATP of the unusual oligonucleotide **pppA(2′p5′A)$_n$** where $n = 1$ to 10. *This compound,* **2,5-A,** *activates a preexisting endonuclease,* ***RNase L,*** *to degrade mRNA, thereby inhibiting protein synthesis.* 2,5-A is itself rapidly degraded by an enzyme named **(2′,5′)-phosphodiesterase** so that it must be continually synthesized to maintain its effect.

The independence of the 2,5-A and PKR systems is demonstrated by the observation that the effect of 2,5-A on protein synthesis is reversed by added mRNA but not by added eIF2. [Recall that RNA interference (RNAi; Section 31-4At) constitutes an alternative dsRNA-based antiviral defense.]

c. PERK Prevents the Buildup of Unfolded Proteins in the ER

PKR-like endoplasmic reticulum kinase (PERK), a 1087-residue transmembrane protein, resides in the endoplasmic reticulum (ER) membrane of all multicellular eukaryotes. It is repressed by its binding to the ER-resident chaperone BiP (Section 12-4Bf). When the ER contains an excessive amount of unfolded proteins (caused by various forms of stress such as high temperatures), BiP dissociates from PERK, thereby activating PERK to phosphorylate eIF2α at its Ser 51 and hence inhibit translation. Thus PERK functions to protect the cell from the irreversible damage caused by the accumulation of unfolded proteins in the ER.

Wolcott–Rallison syndrome is a genetic disease characterized mainly by insulin-dependent (type I) diabetes that develops in early infancy (type I diabetes usually first appears in childhood; Section 27-4B). It is caused by mutations in the catalytic domain of PERK. This results in the death of pancreatic β cells, in which PERK is particularly abundant. Multiple systemic disorders subsequently occur including **osteoporosis** (reduction in the quantity of bone) and growth retardation.

d. GCN2 Regulates Amino Acid Biosynthesis

GCN2 (1590 residues), the sole eIF2α kinase in yeast, is a transcriptional activator of the gene encoding **GCN4,** a transcriptional activator of numerous yeast genes, many of which encode enzymes that participate in amino acid biosynthetic pathways. The C-terminal domain of GCN2, which resembles histidyl–tRNA synthetase (HisRS), preferentially binds uncharged tRNAs (whose presence is indicative of an insufficient supply of amino acids). The binding of an uncharged tRNA to this HisRS-like domain activates the adjacent eIF2α kinase domain and thereby inhibits translational initiation, although at only a modest level.

Despite this inhibition of yeast protein synthesis, activated GCN2 induces the expression of GCN4. This seemingly paradoxical property of GCN2, as Alan Hinnebusch explained, arises from the fact that GCN4 mRNA contains four short so-called **upstream open reading frames**

(uORFs), uORF1 to uORF4, in its 5′ leader that precedes the sequence encoding GCN4. Under the normal nutrient conditions in which GCN2 is inactive, the ribosome binds to the mRNA near its 5′ cap and scans for the nearest AUG initiation codon (which is in uORF1), where it forms the 48S preinitiation complex (Fig. 32-44) and commences the translation of uORF1 (Section 32-3Cd). On terminating translation at uORF1's Stop codon, the presence of the surrounding A + U–rich sequences causes the ribosome to resume scanning for the next AUG codon, where it initiates the translation of uORF2. This process repeats until the ribosome terminates at the end of uORF4, where its Stop codon's surrounding G + C–rich sequences induce the ribosome to disengage from the mRNA. Hence GCN4 is only expressed at a low basal level. However, under the low nutrient conditions in which GCN2 phosphorylates eIF2α at its Ser 51, the resulting reduced level of the eIF2 · Met–tRNA$_i^{Met}$· GTP ternary complex causes the 40S subunit to scan longer distances before it can form the 48S preinitiation complex. Consequently, ~50% of the ribosomes scan past uORF2, uORF3, and uORF4 and only initiate translation at the *GCN4* AUG codon, which is therefore translated at a high level (uORF2 and uORF3 can be mutationally eliminated without significantly affecting translational control).

Mammalian homologs of GCN2 are activated under conditions of amino acid starvation. This suggests that the foregoing process has been conserved throughout eukaryotic evolution.

B. *Regulation of eIF4E*

eIF4E (cap-binding protein) binds to the m⁷G cap of eukaryotic mRNAs and thereby participates in translational initiation by helping to identify the initiating AUG codon (Section 32-3Cd). When mammalian cells are treated with hormones, cytokines, **mitogens** (substances that induce mitosis), and/or growth factors, Ser 209 of human eIF4E is phosphorylated via a Ras-activated MAP kinase cascade (Sections 19-3C and 19-3D), thereby increasing eIF4E's affinity for capped mRNA and hence stimulating translational initiation. Ser 209 occupies a surface position on eIF4E adjacent to the binding site for the β phosphate group of the m⁷GDP and flanking the putative binding cleft for mRNA (Fig. 32-45*b*). The structure of eIF4E suggests that the phosphoryl group of phosphorylated Ser 209 forms a salt bridge with Lys 159, which occupies the other side of the putative mRNA-binding cleft, so as to form a clamp that would help stabilize the bound mRNA. The importance of regulating eIF4E activity is indicated by the observations that the overexpression of eIF4E causes the malignant transformation of rodent cell lines and that eIF4E expression is elevated in several human cancers.

The homologous ~120-residue proteins known as **4E-BP1, 4E-BP2,** and **4E-BP3** (BP for *b*inding *p*rotein; the first two are also known as **PHAS-I** and **PHAS-II**) inhibit cap-dependent translation. They do so by binding on the opposite side of eIF4E from its mRNA-binding site, presumably to a patch of seven highly conserved surface residues, and hence do not prevent eIF4E from binding the m⁷G cap. Rather, they block eIF4E from binding to eIF4G and thereby interfere with the formation of the eIF4F complex that positions the 40S ribosomal subunit-bound Met–tRNA$_i^{Met}$ on the mRNA's initiating AUG codon (Section 32-3Cd). In fact, the 4E-BPs and eIF4G all possess the sequence motif YXXXXLϕ (where ϕ is an aliphatic residue, most often L but also M or F) through which they bind to eIF4E.

The treatment of responsive cells with insulin or any of several protein growth factors causes the 4E-BPs to dissociate from eIF4E. This is because the presence of these hormones induces the phosphorylation of the 4E-BPs at six Ser/Thr residues via the signal transduction pathway involving PI3K, PKB, and mTOR (Fig. 19-67). Evidently, the phosphorylation of eIF4E and the 4E-BPs have similar if not synergistic effects in the hormonal regulation of translation in eukaryotes.

C. *mRNA Masking and Cytoplasmic Polyadenylation*

It has been known since the nineteenth century that early embryonic development in animals such as sea urchins, insects, and frogs is governed almost entirely by information present in the oocyte (egg) before fertilization. Indeed, sea urchin embryos exposed to sufficient actinomycin D (Section 31-2Cc) to inhibit RNA synthesis without blocking DNA replication develop normally through their early stages without a change in their protein synthesis program. This is in part because an unfertilized egg contains large quantities of mRNA that is "masked" by associated proteins to form ribonucleoprotein particles, thereby preventing the mRNAs' association with the ribosomes that are also present. On fertilization, this mRNA is "unmasked" in a controlled fashion, quite possibly by the dephosphorylation of the associated proteins, and commences directing protein synthesis. Development of the embryo can therefore start immediately on fertilization rather than wait for the synthesis of paternally specified mRNAs. Thus, gene expression in the early stages of development is entirely translationally controlled; transcriptional control only becomes important when transcription is initiated.

a. Cytoplasmic Polyadenylation

Another mechanism of translational control in oocytes and early embryos involves the polyadenylation of mRNAs in the cytoplasm (polyadenylation usually occurs in the nucleus, following which the mRNA is exported to the cytoplasm; Section 31-4Ab). A substantial number of maternally supplied mRNAs in oocytes have relatively short poly(A) tails (20–40 nt versus a usual length of ~250 nt). The 3′ untranslated region of these mRNAs contains both the AAUAAA polyadenylation signal (which is required for polyadenylation in the nucleus; Section 31-4Ab) together with a so-called **cytoplasmic polyadenylation element (CPE),** which has the consensus sequence UUUUUAU. The CPE is recognized by **CPE-binding protein (CPEB),** which contains two RNA recognition motifs (RRMs) as well as a **zinc finger** motif (Section 34-3Bl) that contribute to its binding to the mRNA. Joel Richter discovered that CPEB recruits a 931-residue protein named **maskin** which, in turn, binds the eIF4E (cap-binding protein) that is bound to the mRNA's 5′ cap (Fig. 32-69a). Maskin contains the same YXXXXLϕ motif through which the 4E-BPs and eIF4G bind to eIF4E (Section 32-4B), thereby blocking the binding of eIF4G to eIF4E and hence preventing the formation of the 48S preinitiation complex (Fig. 32-44).

In the maturation of *Xenopus laevis* oocytes, a process that precedes fertilization and is stimulated by the steroid hormone progesterone (Section 19-1Gb), a variety of mRNAs, including those encoding several cyclins (which participate in cell cycle control; Section 34-4Da) are translationally activated. Soon after exposure to progesterone, a protein kinase named **aurora** phosphorylates the mRNA-bound CPEB at its Ser 174. This increases CPEB's affinity for cleavage and polyadenylation specificity factor (CPSF; Section 31-4Ab), which then binds to the mRNA's AAUAAA sequence, where it recruits poly(A) polymerase (PAP) to lengthen the mRNA's poly(A) tail (Fig. 32-69b).

Translational initiation and cytoplasmic polyadenylation occur simultaneously, which suggests that these processes are linked. Indeed, Richter has shown that this occurs through the binding to poly(A) of poly(A)-binding protein (PABP; Section 31-4Ab), which as we saw (Section 32-3Cd), also binds to eIF4G to circularize the mRNA. The eIF4G in this complex displaces maskin from eIF4E, thereby permitting the formation of the 48S preinitiation complex and hence the mRNA's translation (Fig. 32-69b).

Mammalian cells also exhibit cell cycle–dependent cytoplasmic polyadenylation of mRNAs. This suggests that translational control by polyadenylation is a general feature in animal cells.

D. *Antisense Oligonucleotides*

Since ribosomes cannot translate double-stranded RNA or DNA–RNA hybrid helices, the translation of a given mRNA can be inhibited by a segment of its complementary RNA or DNA, that is, an **antisense RNA** or an **antisense oligodeoxynucleotide,** which are collectively known as **antisense oligonucleotides.** Moreover, endogenous RNase H's (enzymes that cleave the RNA strand of an RNA–DNA duplex; Section 31-4C) cleave an mRNA–oligodeoxynucleotide duplex on its mRNA strand, leaving the antisense oligodeoxynucleotide intact for binding to another mRNA.

Since the human genome consists of ~3 billion bp, an ~15-nt oligonucleotide (which is easily synthesized; Section 7-6Aa) should ideally be able to target any segment of the human genome. This exquisite specificity provides the delivery of an antisense oligonucleotide to, or its expression in, a selected tissue or organism with enormous biomedical and biotechnological potential. However, care must be taken that an antisense oligonucleotide does not also eliminate nontarget mRNAs.

Methods for the delivery of a therapeutically useful antisense oligonucleotide to a target tissue are as yet in their infancy. This is in large part because oligonucleotides are readily degraded by the many nucleases present in an organism and because they do not readily pass through cell membranes. Moreover, a target mRNA is likely to be associated with cellular proteins and hence not available for binding to other molecules. The nuclease resistance of oligonucleotides can be increased by derivatizing them, for example, by replacing a nonbridging oxygen at each phosphate group with a methyl group or an S atom so as to yield **methylphosphonate** or **phosphorothioate oligonucleotides,** although this reduces their antisense activity. The expression of antisense oligonucleotides in the specified tissues would, of course, circumvent the delivery problem but has all the difficulties associated with gene therapy (Section 5-5H).

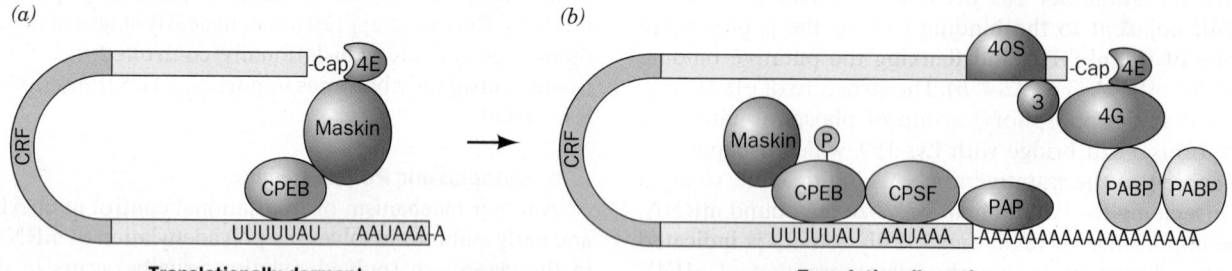

(a) Translationally dormant

(b) Translationally active

Figure 32-69 **CPEB-mediated translational control.** (*a*) In immature *Xenopus* oocytes, an mRNA containing the CPE (UUUUUAU) is bound by CPEB, which binds maskin, which then binds eIF4E so as to prevent it from binding eIF4G, thereby maintaining the mRNA in a translationally dormant (masked) state. (*b*) In the maturation process, CPEB is phosphorylated by an aurora protein kinase. The phosphorylated CPEB binds CPSF, which recruits PAP to extend the mRNA's heretofore short poly(A) tail. PABP binds to the newly lengthened poly(A) tail and simultaneously binds to eIF4G so as to displace maskin. This permits the 48S preinitiation complex to assemble and hence the translation of the mRNA to proceed. [Based on a drawing by Mendez, R. and Richer, J.D., *Nature Rev. Mol. Cell Biol.* **2,** 521 (2001).]

Despite the foregoing, antisense technology is beginning to show success. **Fomivirsen** (trade name **Vitravene**), a 21-nt phosphorothioate oligonucleotide that is complementary to an mRNA expressed by **cytomegalovirus (CMV),** is effective in the treatment of retinitis (inflammation of the retina) caused by CMV infection in individuals with AIDS (CMV is an opportunistic pathogen that rarely infects individuals with normally functioning immune systems). It was approved for human use in 1998 by the FDA, the first antisense drug so approved. A number of antisense oligonucleotides that are mainly targeted against genes that are overexpressed in specific cancers and autoimmune diseases as well as other viral infections are in clinical trials (Section 15-4Bb), although additional antisense drugs have not yet been approved for human use.

Antisense technology has also had some success in the arena of biotechnology. For example, in tomatoes and other fruits, the enzyme **polygalacturonase (PG),** which is expressed during ripening, depolymerizes the pectin (mainly polygalacturonic acid) in the cell wall. This results in a softening of tomatoes to the point that vine-ripened (and hence better tasting) tomatoes are unable to withstand the rigors of shipping and hence must be picked before they are ripe. The introduction into a tomato, via genetic engineering techniques, of a gene expressing antisense PG RNA yielded the so-called Flavr Savr tomato that had substantially reduced PG expression and hence remained firm after vine ripening.

5 POST-TRANSLATIONAL MODIFICATION

To become mature proteins, polypeptides must fold to their native conformations, their disulfide bonds, if any, must form, and, in the case of multisubunit proteins, the subunits must properly combine. Moreover, as we have seen throughout this text, many proteins are modified in enzymatic reactions that proteolytically cleave certain peptide bonds and/or derivatize specific residues. In this section we shall review some of these **post-translational modifications.**

A. *Proteolytic Cleavage*

Proteolytic cleavage is the most common type of post-translational modification. Probably all mature proteins have been so modified, if by nothing else than the proteolytic removal of their leading Met (or fMet) residue shortly after it emerges from the ribosome. Many proteins, which are involved in a wide variety of biological processes, are synthesized as inactive precursors that are activated under proper conditions by limited proteolysis. Some examples of this phenomenon that we have encountered are the conversion of trypsinogen and chymotrypsinogen to their active forms by tryptic cleavages

of specific peptide bonds (Section 15-3E), and the formation of active insulin from the 84-residue proinsulin by the excision of its internal 33-residue C chain (Section 9-1Aa). Inactive proteins that are activated by removal of polypeptides are called **proproteins,** whereas the excised polypeptides are termed **propeptides.**

a. Propeptides Direct Collagen Assembly

Collagen biosynthesis is illustrative of many facets of post-translational modification. Recall that collagen, a major extracellular component of connective tissue, is a fibrous triple-helical protein whose polypeptides each contain the amino acid sequence $(Gly-X-Y)_n$ where X is often Pro, Y is often 4-hydroxyproline (Hyp), and $n \approx 340$ (Section 8-2B). The polypeptides of **procollagen** (Fig. 32-70) differ from those of the mature protein by the presence of both N-terminal and C-terminal propeptides of ~100 residues whose sequences, for the most part, are unlike those of mature collagen. The procollagen polypeptides rapidly assemble, *in vitro* as well as *in vivo,* to form a collagen triple helix. In contrast, polypeptides extracted from mature collagen will reassemble only over a period of days, if at all. *The collagen propeptides are apparently necessary for proper procollagen folding.*

The N- and C-terminal propeptides of procollagen are respectively removed by **amino-** and **carboxyprocollagen**

Figure 32-70 **Electron micrograph of procollagen aggregates that have been secreted by fibroblasts into the extracellular medium.** [Courtesy of Jerome Gross, Massachusetts General Hospital, Harvard Medical School.]

Figure 32-71 **Schematic representation of the procollagen molecule.** Gal, Glc, GlcNAc, and Man, respectively, denote galactose, glucose, *N*-acetylglucosamine, and mannose residues. Note that the N-terminal propeptide has intrachain disulfide bonds while the C-terminal propeptide has both intrachain and interchain disulfide bonds. [After Prockop, D.J., Kivirikko, K.I., Tuderman, L., and Guzman, N.A., *New Engl. J. Med.* **301,** 16 (1979).]

peptidases (Fig. 32-71), which may also be specific for the different collagen types. An inherited defect of aminoprocollagen peptidase in cattle and sheep results in a bizarre condition, **dermatosparaxis,** that is characterized by extremely fragile skin. An analogous disease in humans, **Ehlers–Danlos syndrome VII,** is caused by a mutation in one of the procollagen polypeptides that inhibits the enzymatic removal of its aminopropeptide. Collagen molecules normally spontaneously aggregate to form collagen fibrils (Figs. 8-31 and 8-32). However, electron micrographs of dermatosparaxic skin show sparse and disorganized collagen fibrils. *The retention of collagen's aminopropeptides apparently interferes with proper fibril formation.* (The dermatosparaxis gene was bred into some cattle herds because heterozygotes produce tender meat.)

b. Signal Peptides Are Removed from Nascent Proteins by a Signal Peptidase

Many transmembrane proteins or proteins that are destined to be secreted are synthesized with an N-terminal **signal peptide** of 13 to 36 predominantly hydrophobic residues. As we saw in our discussion of the **secretory pathway** (Section 12-4B), a signal peptide is recognized by a **signal recognition particle (SRP).** The SRP binds a ribosome synthesizing a signal peptide to a protein pore known as the **translocon** that is embedded in the membrane [the rough endoplasmic reticulum (RER) in eukaryotes and the plasma membrane in bacteria] and conducts the signal peptide and its following nascent polypeptide through the translocon.

Proteins bearing a signal peptide are known as **preproteins** or, if they also contain propeptides, as **preproproteins.**

Once the signal peptide has passed through the membrane, it is specifically cleaved from the nascent polypeptide by a membrane-bound **signal peptidase.** Both insulin and collagen are secreted proteins and are therefore synthesized with leading signal peptides in the form of **preproinsulin** and **preprocollagen.** These and many other proteins are therefore subject to three sets of sequential proteolytic cleavages: (1) the deletion of their initiating Met residue, (2) the removal of their signal peptides, and (3) the excision of their propeptides.

c. Polyproteins

Some proteins are synthesized as segments of **polyproteins,** polypeptides that contain the sequences of two or more proteins. Examples include many polypeptide hormones (Section 34-3C); the proteins synthesized by many viruses, including those causing polio (Section 33-2C) and AIDS (Section 15-4Cb); and **ubiquitin,** a highly conserved eukaryotic protein involved in protein degradation (Section 32-6Bb). Specific proteases post-translationally cleave polyproteins to their component proteins, presumably through the recognition of the cleavage site sequences. Some of these proteases are conserved over remarkable evolutionary distances. For instance, ubiquitin is synthesized as several tandem repeats (**polyubiquitin**) that *E. coli* properly cleave even though prokaryotes lack ubiquitin. Other proteases have more idiosyncratic cleavage sequences. This has allowed medicinal chemists to design inhibitors of **HIV protease** (which catalyzes an essential step in the viral life cycle) that have been highly effective in attenuating if not preventing the progression of AIDS (Section 15-4Cd).

B. *Covalent Modification*

Proteins are subject to specific chemical derivatizations, both at the functional groups of their side chains and at their terminal amino and carboxyl groups. Over 150 different types of side chain modifications, involving all side chains but those of Ala, Gly, Ile, Leu, Met, and Val, are known (Section 4-3A). These include acetylations, glycosylations, hydroxylations, methylations, nucleotidylations, phosphorylations, and ADP-ribosylations as well as numerous "miscellaneous" modifications.

Some protein modifications, such as the phosphorylation of glycogen phosphorylase (Section 18-1A) and the ADP-ribosylation of eEF2 (Section 32-3Ge), modulate protein activity. Several side chain modifications covalently bond cofactors to enzymes, presumably to increase their catalytic efficiency. Examples of linked cofactors that we have encountered are N^ε-lipoyllysine in dihydrolipoyl transacetylase (Section 21-2Ac) and 8α-histidylflavin in succinate dehydrogenase (Section 21-3F). The attachment of complex carbohydrates, which occur in almost infinite variety, alter the structural properties of proteins and form recognition markers in various types of targeting and cell–cell interactions (Sections 11-3C, 12-3E, and 23-3B). Modifications that cross-link proteins, such as occur in collagen (Section 8-2Bc), stabilize supramolecular aggregates. The functions of most side chain modifications, however, remain enigmatic.

a. Collagen Assembly Requires Chemical Modification

Collagen biosynthesis (Fig. 32-72) is illustrative of protein maturation through chemical modification. As the nascent procollagen polypeptides pass into the RER of the fibroblasts that synthesized them, the Pro and Lys residues are hydroxylated to Hyp, 3-hydroxy-Pro, and 5-hydroxy-Lys (Hyl). The enzymes that do so are sequence specific: **Prolyl 4-hydroxylase** and **lysyl hydroxylase** act only on the Y residues of the Gly-X-Y sequences, whereas **prolyl 3-hydroxylase** acts on the X residues but only if Y is Hyp. Glycosylation, which also occurs in the RER, subsequently attaches sugar residues to Hyl residues (Section 8-2Bb). The folding of three polypeptides into the collagen triple helix must follow hydroxylation and glycosylation because the hydroxylases and glycosyl transferases do not act on helical substrates. Moreover, the collagen triple helix denatures below physiological temperatures unless stabilized by hydrogen bonding interactions involving Hyp residues (Section 8-2B). Folding is also preceded by the formation of specific interchain disulfide bonds between the carboxylpropeptides. This observation bolsters the previously discussed conclusion that collagen propeptides help select and align the three collagen polypeptides for proper folding.

The procollagen molecules pass into the Golgi apparatus where they are packaged into **secretory vesicles** (Sections 12-4C and 12-4D) and secreted into the extracellular spaces of connective tissue. The aminopropeptides are excised just after procollagen leaves the cell and the carboxypropeptides are removed sometime later. The collagen molecules then spontaneously assemble into fibrils, which suggests that an important propeptide function is to prevent intracellular fibril formation. Finally, after the action of the extracellular enzyme lysyl oxidase, the collagen molecules in the fibrils spontaneously cross-link (Fig. 8-33).

C. *Protein Splicing: Inteins and Exteins*

Protein splicing is a post-translational modification process in which an *in*ternal pro*tein* segment (an **intein**) excises itself from a surrounding *ex*ternal pro*tein*, which it ligates to form the mature **extein.** The portions of the unspliced

Figure 32-72   **Schematic representation of procollagen biosynthesis.** Saccharides are represented as in Fig. 32-71. The diagram does not indicate the removal of signal peptides. [After Prockop, D.J., Kivirikko, K.I., Tuderman, L., and Guzman, N.A., *New Engl. J. Med.* **301,** 18 (1979).]

Figure 32-73 **Series of reactions catalyzed by inteins to splice themselves out of a polypeptide chain.**
See the text for details.

extein on the N- and C-terminal sides of the intein are called the **N-extein** and the **C-extein.** Over 500 putative inteins, ranging in length from 100 to 1650 residues, have so far been identified in archaebacteria, eubacteria, single-celled eukaryotes, and viruses (and are registered in the Intein Database at http://www.neb.com/neb/inteins.html). The various exteins in which these inteins are embedded exhibit no significant sequence similarity and, in fact, can be replaced by other polypeptides, thereby indicating that exteins do not contain the catalytic elements that mediate protein splicing. In contrast, the ~130-residue splicing elements of inteins exhibit significant sequence similarity. All of them have four conserved splice-junction residues: (1) a Ser/Thr/Cys at the intein's N-terminus; and (2 and 3) a His–Asn/Gln dipeptide at the intein's C-terminus; which is immediately followed by (4) a Ser/Thr/Cys at the N-terminus of the C-extein.

Protein splicing occurs via a reaction sequence that involves four successive nucleophilic displacements, the first three of which are mediated by the intein (Fig. 32-73):

1. Attack by the N-terminal intein residue (Ser, Thr, or Cys; shown in Fig. 32-73 as Ser) on its preceding carbonyl group, yielding a linear (thio)ester intermediate.

2. A transesterification reaction in which the —OH or —SH group on the C-extein's N-terminal residue (shown in Fig. 32-73 as Ser) attacks the above (thio)ester linkage, thereby yielding a branched intermediate in which the N-extein has been transferred to the C-extein.

3. Cleavage of the amide linkage connecting the intein to the C-extein by cyclization of the intein's C-terminal Asn or Gln (shown in Fig. 32-73 as Asn). The succinimide ring of the excised intein then spontaneously hydrolyzes to regenerate Asn (or iso-Asn).

4. Spontaneous rearrangement of the (thio)ester linkage between the ligated exteins to yield the more stable peptide bond.

The X-ray structure of the 198-residue **GyrA intein** from *Mycobacterium xenopi,* determined by James Sacchetini, indicates how this intein catalyzes the foregoing splicing reactions. This intein's N-terminal residue, Cys 1, was replaced by an Ala–Ser dipeptide with the expectation that the mutant protein would resemble the intein's presplicing state (the new N-terminal residue, Ala 0, presumably represents the C-terminal residue of the N-extein). The X-ray structure reveals that this monomeric protein consists primarily of β strands, two of which curve about the periphery of the entire protein to give it the shape of a flattened horseshoe (Fig. 32-74). The intein's catalytic site is located at the bottom of a broad and shallow cleft near the center of this so-called **β-horseshoe,** where the intein's N-terminal and C-terminal residues are in close proximity. The Ala 0—Ser 1 peptide bond, the bond cleaved in Reaction 1 of the protein splicing process (Fig. 32-73) assumes the cis conformation (Fig. 8-2), a rare high-energy conformation (except when the peptide bond is followed by Pro) that destabilizes this bond. Its amide nitrogen atom is hydrogen bonded to the side chain of the highly conserved

Figure 32-74   X-ray structure of the *M. xenopi* Gyr A intein in which Cys 1 was replaced by an Ala 0–Ser 1 dipeptide. The protein is drawn in ribbon form with its N-terminal Ala 0–Ser 1 dipeptide and its C-terminal His 197–Asn 198 dipeptide as well as the side chains of residues 72 through 75 drawn in stick form colored according to atom type (C of residues 0–1 magenta, C of residues 72–75 green, C of residues 197–198 cyan, N blue, and O red). Hydrogen bonds are represented by thin gray bonds. [Based on an X-ray structure by James Sacchetini, Texas A&M University. PDBid 1AM2.]

His 75. Hence His 75 is well positioned to donate a proton that would promote the breakdown of the tetrahedral intermediate in Reaction 1. The side chains of Thr 72 and Asn 74 appear well positioned to stabilize this tetrahedral intermediate in a manner resembling that of the oxyanion hole in serine proteases (Section 15-3Db). The position of Ser 1 and a modeled Thr at the intein's C-terminus is consistent with Reaction 2 of the splicing process. The side chain of the invariant His 197 is hydrogen bonded to the carboxylate of the C-terminal Asn 198 and hence is positioned to protonate the peptide bond cleaved in Reaction 3.

a. Most Inteins Encode a Homing Endonuclease

What is the biological function of inteins? Nearly all inteins contain polypeptide inserts forming so-called **homing endonucleases.** These are site-specific endonucleases that make a double-strand break in genes that are homologous to their corresponding extein but which lack inteins. The break initiates the double-strand break repair of the DNA via recombination (Section 30-6Ag). Since the intein-containing gene is likely to be the only other gene in the cell containing extein-like sequences, the intein gene is copied into the break. Thus, most inteins mediate a highly specific transposition or "homing" of the genes that insert them in similar sites. The intein's protease and endonuclease activities appear to have a symbiotic relationship: The protease activity excises the intein from the host protein, thereby preventing deleterious effects on the host, whereas the endonuclease activity assures the mobility of the intein gene. Thus intein genes appear to be molecular parasites (junk DNA) that only function to propagate themselves. Indeed, homing endonucleases are also encoded by certain types of introns.

6 PROTEIN DEGRADATION

The pioneering work of Henry Borsook and Rudolf Schoenheimer around 1940 demonstrated that the components of living cells are constantly turning over. For example, adult humans normally turn over ~2% of their proteins per day. Proteins have lifetimes that range from as short as a few minutes to weeks or more. In any case, *cells continuously synthesize proteins from and degrade them to their component amino acids.* The function of this seemingly wasteful process is 2-fold: (1) to eliminate abnormal proteins whose accumulation would be harmful to the cell, and (2) to permit the regulation of cellular metabolism by eliminating superfluous enzymes and regulatory proteins. Indeed, since the level of an enzyme depends on its rate of degradation as well as its rate of synthesis, *controlling a protein's rate of degradation is as important to the cellular economy as is controlling its rate of synthesis.* In this section we consider the processes of intracellular protein degradation and their consequences.

A. Degradation Specificity

Cells selectively degrade abnormal proteins. For example, hemoglobin that has been synthesized with the valine analog **α-amino-β-chlorobutyrate**

α-Amino-β-chlorobutyrate   **Valine**

has a half-life in reticulocytes of ~10 min, whereas normal hemoglobin lasts the 120-day lifetime of the red cell (which makes it perhaps the longest lived cytoplasmic protein). Likewise, unstable mutant hemoglobins are degraded soon after their synthesis, which, for reasons explained in Section 10-3A, results in the hemolytic anemia characteristic of these molecular disease agents. Bacteria also selectively degrade abnormal proteins. For instance, *amber* and *ochre* mutants of β-galactosidase have half-lives in *E. coli* of only a few minutes, whereas the wild-type enzyme is almost indefinitely stable. Most abnormal proteins, however, probably arise from the chemical modification and/or spontaneous denaturation of these fragile molecules in the cell's reactive environment rather than by mutations or the rare errors in transcription or translation. *The ability to eliminate damaged proteins selectively is therefore an essential recycling mechanism that prevents the buildup of substances that would otherwise interfere with cellular processes.*

Normal intracellular proteins are eliminated at rates that depend on their identities. A given protein is elimi-

Table 32-11   **Half-Lives of Some Rat Liver Enzymes**

Enzyme	Half-Life (h)
Short-Lived Enzymes	
Ornithine decarboxylase	0.2
RNA polymerase I	1.3
Tyrosine aminotransferase	2.0
Serine dehydratase	4.0
PEP carboxylase	5.0
Long-Lived Enzymes	
Aldolase	118
GAPDH	130
Cytochrome *b*	130
LDH	130
Cytochrome *c*	150

Source: Dice, J.F. and Goldberg, A.L., *Arch. Biochem. Biophys.* **170**, 214 (1975).

nated with first-order kinetics, indicating that the molecules being degraded are chosen at random rather than according to their age. The half-lives of different enzymes in a given tissue vary substantially as is indicated for rat liver in Table 32-11. Remarkably, *the most rapidly degraded enzymes all occupy important metabolic control points, whereas the relatively stable enzymes have nearly constant catalytic activities under all physiological conditions. The susceptibilities of enzymes to degradation have evidently evolved together with their catalytic and allosteric properties so that cells can efficiently respond to environmental changes and metabolic requirements.* The criteria through which native proteins are selected for degradation are considered in Section 32-6B.

The rate of protein degradation in a cell also varies with its nutritional and hormonal state. Under conditions of nutritional deprivation, cells increase their rate of protein degradation so as to provide the necessary nutrients for indispensable metabolic processes. The mechanism that increases degradative rates in *E. coli* is the stringent response (Section 31-3I). A similar mechanism may be operative in eukaryotes since, as happens in *E. coli,* increased rates of degradation are prevented by antibiotics that block protein synthesis.

B. Degradation Mechanisms

Eukaryotic cells have dual systems for protein degradation: lysosomal mechanisms and ATP-dependent cytosolically based mechanisms. We consider both mechanisms below.

a. Lysosomes Mostly Degrade Proteins Nonselectively

Lysosomes are membrane-encapsulated organelles (Section 1-2Ad) that contain ~50 hydrolytic enzymes, including a variety of proteases known as **cathepsins.** The lysosome maintains an internal pH of ~5 and its enzymes have acidic pH optima. This situation presumably protects the cell against accidental lysosomal leakage since lysosomal enzymes are largely inactive at cytosolic pH's.

Lysosomes recycle intracellular constituents by fusing with membrane-enclosed bits of cytoplasm known as **autophagic vacuoles** and subsequently breaking down their contents. They similarly degrade extracellular substances that the cell takes up via endocytosis (Section 12-5Bc). The existence of these processes has been demonstrated through the use of lysosomal inhibitors. For example, the antimalarial drug **chloroquine**

Chloroquine

is a weak base that, in uncharged form, freely penetrates the lysosome where it accumulates in charged form, thereby increasing the intralysosomal pH and inhibiting lysosomal function. The treatment of cells with chloroquine reduces their rate of protein degradation. Similar effects arise from treatment of cells with cathepsin inhibitors such as the polypeptide antibiotic **antipain.**

Antipain

Lysosomal protein degradation in well-nourished cells appears to be nonselective. Lysosomal inhibitors do not affect the rapid degradation of abnormal proteins or short-lived enzymes. Rather, they prevent the acceleration of nonselective protein breakdown on starvation. However, the continued nonselective degradation of proteins in starving cells would rapidly lead to an intolerable depletion of essential enzymes and regulatory proteins. Lysosomes therefore also have a selective pathway, which is activated only after a prolonged fast, that takes up and degrades proteins containing the pentapeptide Lys-Phe-Glu-Arg-Gln (KFERQ) or a closely related sequence. Such KFERQ proteins are selectively lost in fasting animals from tissues that atrophy in response to fasting (e.g., liver and kidney) but not from tissues that do not do so (e.g., brain and testes). KFERQ proteins are specifically bound in the cytosol and delivered to the lysosome by a 73-kD **peptide recognition protein (prp73),** a member of the 70-kD heat shock protein (Hsp70) family (Section 9-2C).

Both normal and pathological processes are associated with increased lysosomal activity. **Diabetes mellitus** (Section 27-4B) stimulates the lysosomal breakdown of proteins. Similarly, muscle wastage caused by disuse, denervation, or traumatic injury arises from increased lysosomal activity. The regression of the uterus after childbirth, in which this muscular organ reduces its mass from 2 kg to 50 g in 9 days, is a striking example of this process. Many chronic inflammatory diseases, such as **rheumatoid arthritis,** involve the extracellular release of lysosomal enzymes that break down the surrounding tissues.

b. Ubiquitin Marks Proteins Selected for Degradation

It was initially assumed that protein degradation in eukaryotic cells is primarily a lysosomal process. Yet, reticulocytes, which lack lysosomes, selectively degrade abnormal proteins. The observation that protein breakdown is inhibited under anaerobic conditions led to the discovery of a cytosolically based ATP-dependent proteolytic system that is independent of the lysosomal system. This phenomenon was thermodynamically unexpected since peptide hydrolysis is an exergonic process.

The analysis of a cell-free rabbit reticulocyte system demonstrated that **ubiquitin** (Fig. 32-75) is required for ATP-dependent protein degradation. *This 76-residue monomeric protein, so named because it is ubiquitous as well as abundant in eukaryotes, is the most highly conserved protein known:* It is identical in such diverse organisms as humans, toad, trout, and *Drosophila* and differs in only three residues between humans and yeast. Evidently, ubiquitin is all but uniquely suited to an essential cellular function.

Proteins that are selected for degradation are so marked by covalently linking them to ubiquitin. This process, which is reminiscent of amino acid activation (Section 32-2C),

Figure 32-75 X-ray structure of human ubiquitin. The polypeptide is drawn in ribbon form colored in rainbow order from its N-terminus (*blue*) to its C-terminus (*red*) with the side chain of Lys 48 shown in stick form with C green and N blue. This α + β architecture is known as a **β-Grasp** fold because its β sheets appear to grasp its α helix. [Based on an X-ray structure by Charles Bugg, University of Alabama at Birmingham. PDBid 1UBQ.] **See Interactive Exercise 46.**

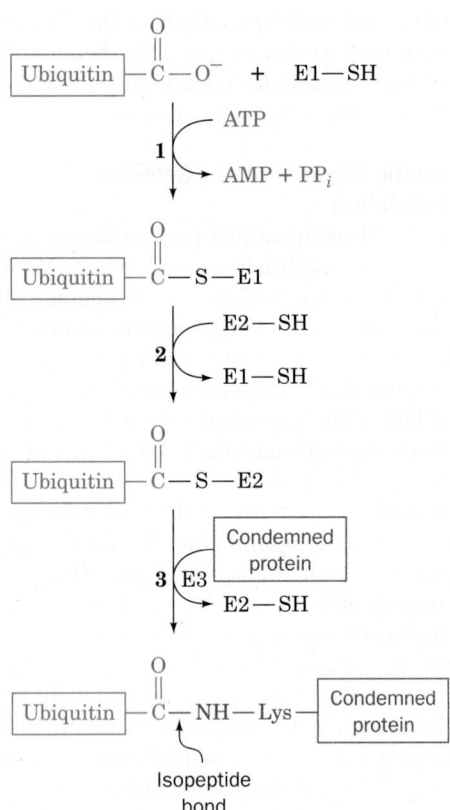

Figure 32-76 **Reactions involved in the attachment of ubiquitin to a protein.** In the first part of the process, ubiquitin's terminal carboxyl group is joined, via a thioester linkage, to E1 in a reaction driven by ATP hydrolysis. The activated ubiquitin is subsequently transferred to a sulfhydryl group of an E2 and then, in a reaction catalyzed by an E3, to a Lys ε-amino group on a condemned protein, thereby flagging the protein for proteolytic degradation by the 26S proteasome.

occurs in a three-step pathway that was elucidated notably by Avram Hershko, Aaron Ciechanover, and Irwin Rose (Fig. 32-76):

1. In an ATP-requiring reaction, ubiquitin's terminal carboxyl group is conjugated, via a thioester bond, to **ubiquitin-activating enzyme (E1),** a homodimer of ~1050-residue subunits. In this process, the substrate protein's terminal carboxyl group is initially adenylylated and then transferred to the E1 Cys—SH group with the elimination of AMP. Most organisms, including yeast and humans, have only one type of E1.

2. The ubiquitin is then transferred to a specific Cys sulfhydryl group on one of numerous proteins named **ubiquitin-conjugating enzymes (E2s;** 11 in yeast and over 20 in mammals). The various E2's are characterized by ~150-residue catalytic cores containing the active site Cys that exhibit at least 25% sequence identities and which mainly vary by the presence or absence of N- and/or C-terminal extensions that exhibit little sequence identity to each other. The X-ray and NMR structures of several species of E2 reveal that their catalytic cores all assume closely similar α/β structures (e.g., Fig. 32-77) in which

Figure 32-77 **X-ray structure of an E2 protein from** *Arabidopsis thaliana.* The protein is drawn in ribbon form colored in rainbow order from its N-terminus (*blue*) to its C-terminus (*red*). The side chain of Cys 88, to which ubiquitin is covalently linked, is shown in ball-and-stick form with C green and S yellow. [Based on an X-ray structure by William Cook, University of Alabama at Birmingham. PDBid 2AAK.]

most of the identical residues are clustered on one surface near the ubiquitin-accepting Cys residue, where they presumably interact with ubiquitin and E1.

3. **Ubiquitin–protein ligase (E3)** *transfers the activated ubiquitin from E2 to a Lys ε-amino group of its target protein, thereby forming an* **isopeptide bond.** Each of the many E3s present in eukaryotic cells mediates the ubiquitination (alternatively, ubiquitylation) of a specific set of proteins and thereby marks them for degradation. Each E3 is served by one or a few specific E2s. Most E3s are members of two unrelated families, those containing a **HECT** domain (HECT for *h*omologous to *E*6AP *C-t*erminus) and those containing a so-called **RING** domain (also called a **RING-finger** domain; RING for *r*eally *i*nteresting *n*ew *g*ene), although some E2s react well with members of both families. The human genome contains 28 HECT genes and 616 RING genes, more than its number of protein kinase genes (518), which is indicative of the specialized and varied functions of E3s (although not all RING domain–containing proteins are E3s). HECT domain E3s are modularly constructed with a unique N-terminal domain that interacts with its target proteins via their so-called **ubiquitination signals** (usually short polypeptide segments; see below) and an ~350-residue HECT domain that mediates E2 binding and catalyzes the ubiquitination reaction. RING domains, which are implicated in recognizing a substrate protein's ubiquitination signal, are 40- to 60-residue motifs that each bind two structurally but not catalytically implicated Zn^{2+} ions via a total of 8 Cys and His residues in a characteristic consensus sequence (much like the zinc finger motifs in certain DNA-binding proteins, Section 34-3B1). RING domain–containing E3s may consist of a single subunit or may be multisubunit proteins in which the RING domain is contained in one subunit. HECT E3–mediated ubiquitination occurs via the transfer of ubiquitin from E2 to a conserved Cys residue on the HECT domain followed by its transfer to the substrate protein Lys side chain. In contrast, RING domain E3s act

(a)

(b)

Figure 32-78 **X-ray structure of human tetraubiquitin.** (*a*) A ribbon drawing in which the isopeptide bonds connecting successive ubiquitin molecules, together with the Lys side chains making them, are drawn in stick form with C green, N blue, and O red. However, since the C-terminal three residues of ubiquitin 2 are disordered, the isopeptide bond connecting it to ubiquitin 3 is represented by a thin gray bond (this isopeptide bond nevertheless exists, as was demonstrated by SDS–PAGE of dissolved crystals). The monomer units in a multiubiquitin chain of any length are likely to be arranged with the repeating symmetry of the tetraubiquitin structure, although the weak interactions between adjacent ubiquitin units suggests that this chain is conformationally flexible. (*b*) A surface diagram, viewed as in Part *a*, in which basic residues (Arg, Lys, His) are blue, acidic residues (Asp, Glu) are red, uncharged polar residues (Gly, Ser, Thr, Asn, Gln) are purple, and hydrophobic residues (Ile, Leu, Val, Ala, Met, Phe, Tyr, Pro) are tan (ubiquitin lacks Cys and Trp residues). Note the unusually large solvent-exposed surface occupied by the hydrophobic residues. [Based on an X-ray structure by William Cook, University of Alabama at Birmingham, and Cecile Pickart, Johns Hopkins University. PDBid 1TBE.]

as adapters that position the reactive E2–ubiquitin thioester bond for the direct transfer of the ubiquitin to the substrate protein Lys side chain.

In order for a target protein to be efficiently degraded, it must be linked to a chain of at least four tandemly linked ubiquitin molecules in which Lys 48 of each ubiquitin forms an isopeptide bond with the C-terminal carboxyl group of the succeeding ubiquitin (Fig. 32-78). These **polyubiquitin (polyUb)** chains, which can reach lengths of 50 or more ubiquitin molecules, are generated by the E3s, although how they switch from transferring a ubiquitin to the target protein to processively synthesizing a polyubiquitin chain is unknown.

c. Ubiquitinated Proteins Are Hydrolyzed in the Proteasome

*A ubiquitinated protein is proteolytically degraded to short peptides in an ATP-dependent process mediated by a large (2000 kD, 26S) multisubunit protein complex named the **26S proteasome** (sometimes spelled "proteosome") that electron micrographic studies reveal has the shape of a bi-capped hollow barrel (Fig. 32-79).* Proteolysis occurs inside the barrel, which permits this process to be extensive and processive, while preventing nonspecific proteolytic damage to other cellular components. PolyUb chains are the signals that target a protein to the proteasome; the identity of the target protein has little effect on the efficiency with which it is degraded by the proteasome. Nevertheless, the proteasome does not degrade ubiquitin molecules; they are returned to the cell. The size and functional complexity of this entire proteolytic system, which occurs in the nucleus as well as the cytosol, rivals that of the ribosome (Section 32-3) and the spliceosome (Section 31-4A) and hence is

Figure 32-79 **Electron microscopy–based image of the** ***Drosophila melanogaster*** **26S proteasome.** The complex is around 450 × 190 Å. The central portion of this 2-fold symmetric multiprotein complex (*yellow*), the **20S proteasome,** consists of four stacked 7-membered rings of subunits that form a hollow barrel in which the proteolysis of ubiquitin-linked proteins occurs. The **19S caps** (*cyan*), which may attach to one or both ends of the 20S proteasome, control the access of condemned proteins to the 20S proteasome (see below). [Courtesy of Wolfgang Baumeister, Max Planck Institute of Biochemistry, Martinsreid, Germany.]

indicative of the importance of properly managing protein degradation. Indeed, ~5% of the proteins expressed by yeast participate in protein degradation. We discuss the structure and function of the 26S proteasome below.

d. Many E3s Have Elaborate Modular Structures

The proto-oncogene product **c-Cbl** (906 residues) is a single-subunit, RING domain–containing E3 that functions to ubiquitinate certain activated receptor tyrosine kinases (RTKs; Section 19-3A), thereby terminating their signaling. Nikola Pavletich determined the X-ray structure of the N-terminal half of c-Cbl (residues 47–447) in its ternary complex with the E2 protein **UbcH7** (which consists of little more than the ~150-residue E2 catalytic core) and a 9-residue peptide containing the ubiquitination signal from a nonreceptor tyrosine kinase (NRTK) named **ZAP-70** (Fig. 19-44). The structure (Fig. 32-80) reveals that UbcH7 and c-Cbl's RING domain and SH2-containing tyrosine kinase–binding (TKB) domains interact with one another across multiple interfaces to form a compact and apparently rigid structure. The RING domain consists of a 3-stranded β sheet, an α helix, and two large loops that are held together by two tetrahedrally coordinated Zn^{2+} ions. UbcH7 adopts the characteristic α/β fold of other E2s of known structure (e.g., Fig. 32-77). The ZAP-70 peptide is bound on the opposite side of the TKB domain from the UbcH7 active site Cys residue (Cys 86) and is ~60 Å distant from it.

SCF complexes are multisubunit RING E3s that consist of **Cul1** (a member of the **cullin** family; 776 residues), **Rbx1** (which contains the complex's RING domain; 108 residues), **Skp1** (163 residues), and a member of the **F-box protein** family (~430 to >1000 residues; SCF for *S*kp1–*c*ullin–*F*-box protein). Rbx1 and Cul1 form the complex's catalytic core that binds E2; F-box proteins consist of an ~40-residue **F-box** that binds Skp1 followed by protein–protein interaction modules such as **leucine-rich repeats (LRRs)** or WD40 repeats (Section 19-2C) that bind substrate protein; and Skp1 functions as an adapter that links the F-box to Cul1. Cells contain numerous different F-box proteins (at least 38 in humans) that presumably permit the specific ubiquitination of a diverse variety of protein substrates (see below).

Pavletich has also determined the X-ray structures of two segments of the **SCF^Skp2** complex (where the superscript identifies the complex's F-box protein, here **Skp2**, 436 residues). The structure of the Skp1–Skp2 complex (Fig. 32-81) reveals that it has the shape of a sickle with the Skp1 and the 3-helix F-box of Skp2 forming the handle and its 10 LRRs (~26 residues each) forming the curved blade. The structure of the Cul1–Rbx1–Skp1–F-box^Skp2 quaternary complex (Fig. 32-82) shows that Cul1 is an elongated protein that consists of a long stalk formed by three repeats of a novel five-helix motif known as a cullin repeat followed by a globular domain that binds Rbx1. Apparently Cul1 acts like a rigid scaffold that organizes the Skp1–F-box^Skp2 complex and Rbx1 so as to hold them over 100 Å

Figure 32-80 **X-ray structure of the human c-Cbl–UbcH7–ZAP-70 peptide ternary complex drawn in tube-and-arrow form.** UbcH7, an E2 that consists almost entirely of the E2 catalytic core, is colored orange with the side chain of its active site Cys 86 shown in stick form with C green and S yellow. c-Cbl (residues 47–447 of the 903-residue protein), a monomeric RING E3, is colored according to domain with its TKB domain green, its RING domain red, and the linker joining them yellow. The RING domain's two bound Zn^{2+} ions are represented by cyan spheres. The 9-residue ubiquitination site of the RTK ZAP-70, whose fourth residue is phospho-Tyr, is drawn in stick form with C magenta, N blue, and O red, and P orange. [Based on an X-ray structure by Nikola Pavletich, Memorial Sloan-Kettering Cancer Center, New York, New York. PDBid 1FBV.]

Figure 32-81 **X-ray structure of the human Skp1–Skp2 complex.** Skp1 and Skp2 are drawn in tube-and-arrow form in blue and magenta. Skp2 consists of an N-terminal F-box that forms three helices, followed by 3 noncanonical so-called linker leucine-rich repeats (LRRs) that are contiguous with 7 LRRs that were predicted from their amino acid sequences for a total of 10 LRRs. After the tenth LRR, Skp2's ~30-residue C-terminal tail extends back past the first LRR by packing under the concave surface of the LRR domain. [Based on an X-ray structure by Nikola Pavletich, Memorial Sloan-Kettering Cancer Center, New York, New York. PDBid 1FQV.]

Figure 32-82   X-ray structure of the human Cul1–Rbx1–Skp1–F-box^{Skp2} quaternary complex. Cul1, Rbx1, Skp1, and the Skp2 F-box are drawn in tube-and-arrow form and respectively colored green, red, blue, and magenta. The three cullin repeats of Cul1 are indicated. The three Zn^{2+} ions bound to Rbx1 are represented by cyan spheres. [Based on an X-ray structure by Nikola Pavletich, Memorial Sloan-Kettering Cancer Center, New York, New York. PDBid 1LDK.]

apart. The Rbx1 RING domain contains a 20-residue insert that forms the binding site for a third tetrahedrally liganded Zn^{2+} ion.

The apparent rigidity of the foregoing three structures has enabled Pavletich to construct a model of the intact SCF^{Skp2}–E2 complex by superimposing Skp1–Skp2 on Cul1–Rbx1–Skp1–F-box^{Skp2} and docking the E2 UbcH7 onto the Rbx1 RING domain based on the c-Cbl–UbcH7 structure (Fig. 32-83). The model indicates that E2 and the LRR-containing domain of Skp2 are on the same side of the SCF complex but separated by a distance of ~50 Å.

Figure 32-83   Model of the SCF^{Skp2}–E2 complex. This model, which is based on the X-ray structures in Figs. 32-80, 32-81, and 32-82, is colored and viewed as in Fig. 32-82. E2 is yellow with its active site Cys residue, to which ubiquitin would be covalently linked, drawn in space-filling form in cyan. The Zn^{2+} ions associated with the Rbx1 RING domain are represented by yellow spheres. The gray arrow indicates the 50-Å gap between the tip of the Skp2 LRR domain and the E2 active site. [Courtesy of Nikola Pavletich, Memorial Sloan-Kettering Cancer Center, New York, New York.]

This suggests that Cul1's long stalk functions to separate the complex's substrate-binding and catalytic sites so that substrates with different sizes and various distances between their ubiquitinated Lys residues and their ubiquitination signals can be accommodated.

e. The Ubiquitin System Has Both Housekeeping and Regulatory Functions

Until the mid-1990s, it appeared that the ubiquitin system functioned mainly in a "housekeeping" capacity to maintain the proper balance among metabolic proteins and to eliminate damaged proteins. Indeed, as Alexander Varshavsky discovered, *the half-lives of many cytoplasmic proteins vary with the identities of their N-terminal residues (Table 32-12).* Thus, in a selection of 208 cytoplasmic proteins known to be long lived, all have a "stabilizing" residue, Met, Ser, Ala, Thr, Val, or Gly, at their N-termini. This so-called **N-end rule** is true for both eukaryotes and prokaryotes, which suggests the system that selects proteins for degradation is conserved in eukaryotes and prokaryotes, even though prokaryotes lack ubiquitin. The N-end rule results from the actions of the single-subunit, RING E3 named **E3α** (~1950 residues; also known as **Ubr1**) whose ubiquitination signals are the destabilizing N-terminal residues in Table 32-12.

Similarly, it has long been known that proteins with segments rich in Pro (P), Glu (E), Ser (S), and Thr (T), the so-called **PEST proteins,** are rapidly degraded. This is because these PEST elements often contain phosphorylation sites that target their proteins for ubiquitination.

Table 32-12   Half-Lives of Cytoplasmic Enzymes as a Function of Their N-Terminal Residues

N-Terminal Residue	Half-Life
Stabilizing	
Met	>20 h
Ser	
Ala	
Thr	
Val	
Gly	
Destabilizing	
Ile	~30 min
Glu	
Tyr	~10 min
Gln	
Highly Destabilizing	
Phe	~3 min
Leu	
Asp	
Lys	
Arg	~2 min

Source: Bachmair, A., Finley, D., and Varshavsky, A., *Science* **234,** 180 (1986).

It is now clear, however, that the ubiquitin system is far more sophisticated than a simple garbage disposal system. Thus, the known E3s each respond to certain ubiquitination signals that often occur on a quite limited range of target proteins, many of which have regulatory functions. For example, *the ubiquitination system has an essential function in cell cycle progression.* The cell cycle, as we have seen in Section 30-4Aa and will further discuss in Section 34-4D, is regulated by a series of proteins known as cyclins. A given cyclin, which is expressed immediately preceding and/or during a specific phase of the cell cycle, binds to a corresponding **cyclin-dependent protein kinase (Cdk),** which then phosphorylates its target proteins so as to activate them to carry out the processes of that phase of the cell cycle. Moreover, many cyclins also inhibit the transition to the subsequent phase of the cell cycle (e.g., DNA replication or mitosis). Consequently, for a cell to progress from one phase of the cell cycle to the next, the cyclin(s) governing that phase must be eliminated. This occurs via the specific ubiquitination of the cyclin, thereby condemning it to be destroyed by the proteasome. The E3s responsible for this process are the SCF complexes containing F-box proteins targeted to a corresponding cyclin and a multisubunit complex known as the **anaphase-promoting complex (APC;** alternatively the **cyclosome;** Section 34-4Da). APC, an ~1500-kD RING domain–containing particle that in yeast consists of 11 subunits, specifically ubiquitinates proteins that contain the 9-residue consensus sequence RTALGDIGN, the so-called **destruction box,** near their N-termini.

The transcription factor **NF-κB,** which plays a central role in immune and inflammatory responses (Section 34-3Bs), is maintained in an inactive state in the cytosol through its binding to the inhibitor **IκBα** in a way that occludes the short internal basic sequence that directs NF-κB's import into the nucleus (its **nuclear localization signal; NLS**). However, the stimulation of cell-surface receptors by proinflammatory cytokines such as **tumor necrosis factor-α (TNFα;** Section 19-3Db) and **interleukin-1 (IL-1;** Section 19-3Eb) initiate a signal transduction pathway (Section 19-3D) that phosphorylates IκBα bound to NF-κB at both Ser residues in the sequence DSGLDS. This phosphorylated sequence is the ubiquitination signal for the SCF complex containing the F-box protein **β-TrCP** (605 residues), which mediates the ubiquitination of the phosphorylated IκBα. The consequent destruction of IκBα exposes the NLS of NF-κB, which is then translocated to the nucleus where it activates the transcription of its target genes (Section 34-3Bs).

Some viruses usurp the ubiquitin system. Oncogenic forms of **human papillomavirus (HPV),** the cause of nearly all cervical cancers (a leading cause of death of women in developing countries), encode the ~150-residue **E6 protein,** which combines with the 875-residue cellular protein named **E6-associated protein** (**E6AP;** the first E3 known to contain a HECT domain) to ubiquitinate **p53,** thereby marking it for destruction. This latter protein is a transcription factor that monitors genome integrity and hence is important in preventing malignant transformation and the proliferation of cancer cells (Section 34-4Ca), that is, it is a **tumor suppressor** (a protein whose loss of function is a cause of cancer). Consequently, HPV provokes the uncontrolled growth of the cells it infects and hence its own proliferation. E6AP normally functions to ubiquitinate certain members of the Src family of protein tyrosine kinases (Section 19-3Ba), including Src itself. The deletion of the segment of chromosome 15 that contains the E6AP gene causes Angelman syndrome, which as we have seen (Section 30-7d) is characterized by severe mental retardation and is exclusively maternally inherited due to genomic imprinting.

The foregoing are only a few examples of the numerous cellular processes that are regulated by the ubiquitin-mediated proteolysis system. Not surprisingly, therefore, many pathological conditions in humans, including inflammatory, neurodegenerative, and muscle-wasting diseases, are attributable to malfunctioning ubiquitination systems.

f. The 20S Proteasome Catalyzes Proteolysis Inside a Hollow Barrel

The 26S proteasome (Fig. 32-79) is an ~2100-kD multisubunit protein that catalyzes the ATP-dependent hydrolysis of ubiquitin-linked proteins. This yields oligopeptides with lengths of 4 to 25 residues and averaging 7 to 9 residues that are subsequently degraded to their component amino acids by cytosolic exopeptidases. The 26S proteasome consists of a **20S proteasome** (~670 kD), the barrel-shaped catalytic core of the 26S proteasome, and its **19S caps** (~700 kD; also known as **PA700** and the **19S regulator**), which associate with the ends of the 20S proteasome and stimulate its activity (PA for *proteasome activator*). The 20S proteasome only hydrolyzes unfolded proteins in an ATP-independent manner; the 19S caps function to identify and unfold the ubiquitinated protein substrates.

The 20S proteasome occurs in the nuclei and cytosol of all eukaryotic cells and in all archaebacteria yet examined. However, the only eubacteria in which it occurs are those of the class Actinobacteria, which suggests that they obtained it via horizontal gene transfer from some other organism.

The 20S proteasome of *Thermoplasma acidophilum* (an archaebacterium) consists of 14 copies each of α and β subunits (233 and 203 residues) that electron microscopy studies revealed form a 150-Å long and 110-Å-diameter barrel in which the subunits are arranged in four stacked rings (as is evident in the central portion of the 26S proteasome seen in Fig. 32-79). The α and β subunits are 26% identical in sequence except for an ~35-residue N-terminal tail of the α subunit, which the β subunit lacks. Eukaryotic 20S proteasomes are more complex in that they consist of 7 different α-like and 7 different β-like subunits versus only one of each type for the *T. acidophilum* 20S proteasome.

The X-ray structure of the *T. acidophilum* 20S proteasome, determined by Baumeister and Robert Huber, reveals that its two inner rings each consist of 7 β subunits and its two outer rings each consist of 7 α subunits, all arranged with D_7 symmetry (Fig. 32-84). Thus the overall

(a)

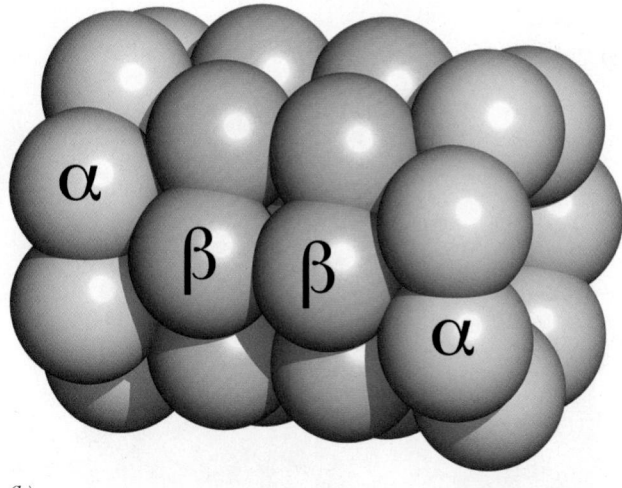

(b)

Figure 32-84 **X-ray structure of the *T. acidophilum* 20S proteasome.** (*a*) Ribbon diagram, viewed with its 7-fold axis tipped toward the viewer by 20°, in which the α subunits are pink and the β subunits are light blue except for one α subunit (*left*) and a similarly oriented β subunit (*middle*), which are both colored in rainbow order from their N-termini (*blue*) to their C-termini (*red*). Note their close structural resemblance. (*b*) Diagram, viewed as in Part *a*, in which the subunits are represented by equal-sized spheres with α subunits pink and β subunits light blue. (*c*) Surface diagram, viewed along a 2-fold axis, with the subunits nearest the viewer removed to expose the proteasome's triple-chambered internal cavity, which is maximally ~100 Å long and ~60 Å in diameter. The active sites on the β subunits are marked by the bound inhibitor, LLnL, which is drawn in stick form in yellow. [Part *a* based on an X-ray structure by and Part *c* courtesy of Robert Huber, Max-Planck-Institut für Biochemie, Martinsried, Germany. PDBid 1PMA.]

(c)

structure of the 20S proteasome superficially resembles that of the unrelated molecular chaperone GroEL (Section 9-2Ca). The structures of the α and β subunits are remarkably similar (Fig. 32-84*a*) except, of course, for the α subunit's N-terminal tail (blue in Fig. 32-84*a*), which extends radially inward to contact the N-terminal tail of a neighboring α subunit. This accounts for the observation that α subunits alone spontaneously assemble into 7-membered rings (a capacity that is abolished by the deletion of their N-terminal 35 residues), whereas β subunits alone remain monomeric.

The central cavity of the *T. acidophilum* 20S proteasome consists of three large chambers (Fig. 32-84*c*): Two are located at the interfaces between adjoining rings of α and β subunits, with the third, larger chamber centrally located between the two rings of β subunits. Unfolded polypeptide substrates enter the central chamber of the barrel (where the proteasome's active sites are located; see below) through ~13-Å-diameter axially located apertures in the α rings that are lined with hydrophobic residues. This allows only unfolded

proteins to enter the central chamber, thereby protecting properly folded proteins from indiscriminant degradation by this omnivorous protein-dismantling machine.

The X-ray structure of the yeast 20S proteasome, determined by Huber, demonstrates that its outer and inner rings respectively consist of seven different α-type subunits and seven different β-type subunits, all of which are uniquely arranged (Fig. 32-85). The α-like subunits have folds that are similar to one another as well as to that of the *T. acidophilum* 20S proteasome and likewise for the β-like subunits. Consequently, this 28-subunit, 6182-residue protein complex has exact 2-fold rotational symmetry relating its two pairs of rings but only pseudo-7-fold rotational symmetry relating the subunits within each ring. The narrow axial apertures in the α rings through which unfolded polypeptides enter the hydrolytic chamber (Fig. 32-85*d*) are occluded in the closed state (Fig. 32-85*c*) by a plug formed by the interdigitation of its α subunits' N-terminal tails. This indicates that the 19S caps of the 26S proteasome, which have been shown to activate the 20S proteasome,

(a)

α6   α7   α1   α2

β7   β1   β2

β2'   β1'   β7'

α2'   α1'   α7'   α6'

(b)

Channel
entry port

α

β

β5'   β2   β1

Proteolytic
sites

β'

α'

(c)

α4

α5   α3

α6

α7   α1   α2

(d)

α4

α5   α3

α6

α7   α1   α2

■ α-pockets of
   the core particle

□ Bound PA26 C termini

Figure 32-85 (*Opposite*) **X-ray structure of the yeast 20S proteasome.** (*a*) Surface diagram viewed along the 28-mer's 2-fold axis. Each pair of identical subunits has the same color except for β1 and the symmetry related β1', which are colored differently. (*b*) Cutaway surface diagram sliced along the complex's pseudo-7-fold axis. The slice surface is green and the active sites of the β1, β2, and β5' subunits are marked in red. The brackets indicate the approximate positions of the channel entry ports as seen in the open state. (*c*) End view of the 20S proteasome showing its α-ring in its closed state represented as in Part *a*. The pockets where PA26 binds (see below) are highlighted in blue. (*d*) The open state represented as in Part *c* and with the C-termini of its bound PA26 subunits shown in worm form in yellow. [Courtesy of Daniel Finley, Harvard Medical School. Based on X-ray structures by Robert Huber, Max-Planck-Institut für Biochemie, Martinsried, Germany (*closed state*), and Christopher Hill, University of Utah (*open state*). PDBids 1RYP and 1FNT.]

control the access to it by inducing conformational changes in its α rings (see below). The X-ray structure of the bovine 20S proteasome, determined by Tomitake Tsukihara, reveals that its arrangement of seven α-type and seven β-type subunits is similar to that in yeast.

g. The Proteasome Catalyzes Peptide Hydrolysis via a Novel Mechanism

The X-ray structure of the *T. acidophilum* 20S proteasome in complex with the aldehyde inhibitor **acetyl-Leu-Leu-norleucinal (LLnL)**

$$CH_3-\overset{O}{\overset{\|}{C}}-Leu-Leu-NH-\overset{(CH_2)_3}{\overset{|}{CH}}-\overset{O}{\overset{\|}{CH}}$$ with CH_3 atop

Acetyl-Leu-Leu-norleucinal (LLnL)

reveals that its active sites are on the inner surfaces of its rings of β subunits, with the aldehyde function of the LLnL close to the side chain of the highly conserved Thr 1β. Deletion of this Thr or its mutation to Ala yields properly assembled 20S proteasomes that are completely inactive. Evidently, 20S proteasomes catalyze peptide hydrolysis by a novel mechanism in which the hydroxyl group of its Thr 1β is the attacking nucleophile. This, as yet, poorly understood mechanism, in which the amino group at the N-terminus and possibly a bound water molecule act to nucleophilically activate the hydroxyl side chain, is now known to be employed by other hydrolases (e.g., glutamate synthase; Section 26-5Aa), which are collectively known as the **N-terminal nucleophile (Ntn)** family of hydrolases. The *T. acidophilum* β subunits preferably cleave polypeptides after hydrophobic residues. However, in the yeast and the bovine 20S proteasomes, only subunits β1, β2, and β5 are catalytically active. Their respective preferences for cleavage after acidic (caspase-like), basic (trypsin-like), and hydrophobic (chymotrypsin-like) residues are explained by

the respective basic, acidic, and nonpolar characters of their pockets that bind the side chain of the residue preceding the scissile peptide bond, although this specificity is relatively low. The functions of the four different catalytically inactive β subunits are unknown, although mutagenically modifying an inactive β subunit can abolish the catalytic activity of an active β subunit.

h. The 19S Caps Control the Access of Ubiquitinated Proteins to the 20S Proteasome

The 20S proteasome probably does not exist alone *in vivo;* it is most often in complex with two 19S caps that function to recognize ubiquitinated proteins, unfold them, and feed them into the 20S proteasome in an ATP-dependent manner (it may also associate with other regulatory complexes; see below). The 19S cap, which consists of ~18 different subunits, is poorly characterized due in large part to its low intrinsic stability. Its so-called base complex consists of 9 different subunits, 6 of which are ATPases that form a ring that abuts the α ring of the 20S proteasome (Fig. 32-79). Each of these ATPases contains an ~230-residue ATPase module that is a member of the AAA+ family (Section 30-2Ca). Cecile Pickart demonstrated via cross-linking experiments that one of these ATPases, named **S6'** (alternatively **Rpt5**), contacts the polyUb signal that targets a condemned protein to the 26S proteasome. This suggests that the recognition of the polyUb chain as well as substrate protein unfolding are ATP-driven processes. Moreover, the ring of ATPases must function to open (gate) the otherwise closed axial aperture of the 20S proteasome so as to permit the entry of the unfolded substrate protein.

Eight additional subunits form the so-called lid complex, the portion of the 19S cap that is more distal to (distant from) the 20S proteasome. The functions of the lid subunits are largely unknown, although a truncated 26S proteasome that lacks the lid subunits is unable to degrade polyubiquitinated substrates. Several other subunits may be transiently associated with the 19S cap and/or with the 20S proteasome.

i. Deubiquitinating Enzymes Have Several Functions

The enzymes that hydrolytically cleave the isopeptide bonds linking successive ubiquitin units in polyUb are known as **deubiquitinating enzymes (DUBs).** Cells contain a surprisingly large number of DUBs (at least 17 in yeast and ~100 in humans). Nearly all known DUBs are **cysteine proteases,** enzymes whose catalytic mechanism resembles that of serine proteases (Section 15-3C) but whose attacking nucleophile is Cys—S⁻ rather than Ser—OH.

DUBs may release entire polyUb chains from a condemned protein or sequentially release ubiquitin units from the chain terminus. It has been proposed that this latter process functions as a clock to time the protein degradation process. If a polyUb chain is trimmed to less than four ubiquitin units before degradation begins, then its attached protein is likely to escape destruction. This would spare proteins that had been inappropriately tagged with only short polyUb chains.

The mammalian 19S lid subunit known as **POH1** (**Rpn11** for the 65% identical yeast subunit) appears to be responsible for the deubiquitination of target proteins prior to their degradation; its inactivation prevents target protein degradation. Curiously, this DUB is a Zn^{2+}-dependent protease (as is carboxypeptidase A; Fig. 15-42) rather than a cysteine protease.

Certain DUBs function to dismember polyUb chains that have been released from substrate proteins by sequentially removing ubiquitin units from the end of the chain that is nearest to the substrate protein (that with a free C-terminus). Consequently, these DUBs cannot remove ubiquitin units from polyUb chains that are still attached to substrate proteins, thereby preventing their premature removal.

Cells express ubiquitin as polyproteins containing several ubiquitin units (Section 32-5Ac) or with ubiquitin fused to certain ribosomal subunits (there is no gene that encodes a single ubiquitin unit). These polyproteins are rapidly processed by certain DUBs to yield free ubiquitin.

j. The 11S Activator Forms a Heptameric Barrel That Opens the 20S Proteasome

Higher eukaryotes contain an **11S activator** (alternatively, **11S regulator**) that functions to open the channel into the 20S proteasome in an ATP-independent manner so as to permit the entrance of polypeptides (but not folded proteins). The mammalian 11S activator, which functions in the generation of peptides for presentation to the immune system (Section 35-2E), is named **REG** (alternatively **PA28**). It is a heteroheptameric complex of two ~245-residue subunits, **REGα** and **REGβ,** that exhibit ~50% sequence identity except for a highly variable internal 18-residue segment that is thought to confer subunit-specific properties. Indeed REGα alone forms a heptamer whose biochemical properties are similar to that of REG (although both subunits must be present *in vivo*).

The trypanosome *Trypanosoma brucei*, which lacks 19S caps, expresses a homoheptameric 11S activator named **PA26** that is only 14% identical to human REGα. Nevertheless, the various 11S activators activate 20S proteasomes from widely divergent species. Thus, rat 20S proteasome is activated by PA26 and the yeast 20S proteasome is activated by human REGα despite the fact that yeast lacks 11S activators.

(a)

(b)

Figure 32-86 **X-ray structure of *T. brucei* PA26 in complex with the yeast 20S proteasome.** (*a*) The PA26 heptamer in ribbon form viewed with its 7-fold axis vertical. Each of its subunits are differently colored with that closest to the viewer colored in rainbow order from its N-terminus (*blue*) to its C-terminus (*red*). (*b*) Cutaway diagram of the entire complex drawn in worm form and viewed with its 7-fold axis vertical. The PA26 is yellow, the α and β subunits of the 20S proteasome are magenta and blue, its α-annulus is green, and its N-terminal segments that are ordered and partially disordered are red and pink. [Part *a* based on an X-ray structure by and Part *b* courtesy of Christopher Hill, University of Utah. PDBid 1FNT.]

The X-ray structure of PA26 in complex with the yeast 20S proteasome, determined by Christopher Hill, reveals that each PA26 monomer consists of an up–down–up–down 4-helix bundle. These monomers form a 7-fold symmetric heptameric barrel that is 90 Å in diameter, 70 Å long, and has a 33-Å-diameter central pore (Fig. 32-86a) and which closely resembles the previously determined X-ray structure of human REGα. Two PA26 barrels associate coaxially with the 20S proteasome, one at each end (Fig. 32-86b). The conformation of the 20S proteasome in this complex, for the most part, is closely similar to that of the 20S proteasome alone (Fig. 32-85). However, the C-terminal tails of the PA26 subunits insert into pockets on the 20S proteasome's α subunits in a way that induces conformational changes in its N-terminal tails that clear the 20S proteasome's otherwise blocked central aperture (Fig. 32-85c,d), thus permitting unfolded polypeptides to enter the proteasome's central chamber.

k. Bacteria Contain a Variety of Self-Compartmentalized Proteases

Nearly all eubacteria lack 20S proteasomes. Nevertheless they have ATP-dependent proteolytic assemblies that share the same barrel-shaped architecture and carry out similar functions. For example, in *E. coli,* two proteins known as **Lon** and **Clp** mediate up to 80% of the bacterium's protein degradation, with additional contributions from at least three other proteins including **heat shock locus UV (HslUV).** Thus, *all cells appear to contain proteases whose active sites are only available from the inner cavity of a hollow particle to which access is controlled.* These so-called **self-compartmentalized proteases** appear to have arisen early in the history of cellular life, before the advent of eukaryotic membrane-bound organelles such as the lysosome, which similarly carry out degradative processes in a way that protects the cell contents from indiscriminant destruction.

Clp protease consists of two components, the proteolytically active **ClpP** and one of several ATPases, which in *E. coli* are **ClpA** and **ClpX.** The X-ray structure of ClpP, determined by John Flanagan, reveals that it oligomerizes to form an ~90-Å-long and -wide hollow barrel that consists of two back-to-back 7-fold symmetric rings of 193-residue subunits (Fig. 32-87) and thereby has the same D_7 symmetry as does the 20S proteasome. Nevertheless, the ClpP subunit has a novel fold that is different from that of the 20S proteasome's homologous α and β subunits. The ClpP active site, which is only exposed on the inside of the barrel, contains a catalytic triad composed of Ser 97, His 122, and Asp 171, and hence is a serine protease (Section 15-3Ab).

HslUV protease appears to be a hybrid of Clp and the 26S proteasome. Its **HslV** subunits in *Haemophilus influenzae* (174 residues) are 18% identical to the β subunits of the *T. acidophilum* 20S proteasome, whereas its regulatory **HslU** caps (444 residues) have ATPase activity and are homologous to *E. coli* ClpX. The X-ray structure of *H. influenzae* HslUV, determined by David McKay, indicates that HslV forms a dimer of hexameric rather than

(a)

(b)

Figure 32-87   X-ray structure of *E. coli* ClpP. (*a*) View of the heptameric complex along its 7-fold axis, drawn in ribbon form in which the lower ring is pale cyan and the upper ring is pink with one subunit colored in rainbow order from its N-terminus (*blue*) to its C-terminus (*red*). (*b*) View along the complex's 2-fold axis (rotated 90° about a horizontal axis with respect to Part *a*). [Based on an X-ray structure by John Flanagan, Brookhaven National Laboratory, Upton, New York. PDBid 1TYF.]

HslU

HslV

HslV

HslU

Figure 32-88 **X-ray structure of *H. influenzae* HslVU in complex with ATP.** The 821-kD complex is drawn in ribbon form viewed along a 2-fold axis with its 6-fold axis vertical. The D_6 symmetric dodecamer of HslV subunits is coaxially bound at both ends by C_6 symmetric HslU hexamers to yield a complex with overall D_6 symmetry. The subunits of the lower HslU hexamer are alternately orange and cyan, whereas those of the upper HslU hexamer are pale orange except for one subunit, which is colored in rainbow order from its N-terminus (*blue*) to its C-terminus (*red*). The subunits of the lower HslV hexamer are alternately green and magenta, whereas those of the upper hexamer are pale blue except for one subunit, which is colored in rainbow order. The ATPs, which are bound at the interfaces between neighboring HslU subunits, are drawn in space-filling form with C green, N blue, O red, and P orange. [Based on an X-ray structure by David McKay, Stanford University School of Medicine. PDBid 1G3I.]

heptameric rings (Fig. 32-88). A hexameric ring of HslU subunits binds to both ends of the HslV dodecamer to form a 24-subunit assembly with D_6 symmetry, rather than the D_7 symmetry of the 26S proteasome. Nevertheless, both the fold and the intersubunit contacts of the HslV subunits are closely similar to those of the 20S proteasome β subunits. In addition, both have N-terminal Thr residues. Thus, HslV can be regarded as the eubacterial homolog of archaebacterial and eukaryotic 20S proteasomes.

Thermoplasma acidophilum contains another large proteolytic complex, which is unrelated to the proteasome. The X-ray structure of this protease (Fig. 32-89), determined by Huber, indicates that it forms a 730-kD toroidal hexameric ring with D_3 symmetry that has a peculiar triangular shape reminiscent of a tricorn (a hat whose brim is turned up on three sides) and hence was named **tricorn protease.** Cryo-EM studies indicate that 20 of these tricorn hexamers associate to form a 14,600-kD hollow icosahedron (Fig. 32-89*c*; an icosahedron is shown in Fig. 8-65*c*), making it by far the largest homooligomeric enzyme complex known (it is even larger than some virus particles, many of which also have icosahedral symmetry; Section 33-2Aa).

I. Ubiquitination Has Multiple Proteasome-Independent Functions

Proteins may be monoubiquitinated or polyubiquitinated or even monoubiquitinated on more than one Lys residue. Moreover, ubiquitin has seven Lys residues so that seven types of polyubiquitin chains are possible.

Many types of ubiquitination mediate processes other than directing their associated proteins to the proteasome. This occurs through mechanisms reminiscent of protein phosphorylation, but instead of being recognized by specialized phosphoprotein-binding motifs such as SH2 (Section 19-3Cb), ubiquitinated proteins are recognized by conserved **ubiquitin-binding domains (UBDs).** Moreover, whereas protein phosphorylation is reversible through the action of protein phosphatases (Section 19-3F), ubiquitination is reversible through the agency of DUBs.

Ubiquitination participates in regulating such diverse cellular processes as endocytosis, protein trafficking, DNA repair, intracellular signaling, and transcription. For example, during S phase of the cell cycle, the monoubiquitination of PCNA (the sliding clamp associated with the eukaryotic DNA replication fork; Section 30-4Ba) at its Lys 164 recruits an error-prone translesion DNA polymerase to the replication fork at a DNA damage site, whereas the polyubiquitination of the same site with a Lys 63–linked chain recruits DNA polymerases that mediate error-free lesion repair. In another example, in the signal transduction pathway that activates NF-κB (Section 32-6Be), IκBα is phosphorylated by **IκB kinase (IKK),** which has a regulatory subunit named **NEMO** (for *NF*-κB *e*ssential *mo*dulator). The binding of cytokines such as TNFα and IL-1 to their transmembrane receptors in this pathway activates the receptor to Lys 63–linked polyubiquitinate NEMO, which in turn activates IKK to phosphorylate IκBα. This, as we saw, induces the formation of the Lys 48–linked polyubiquitination signal that leads to the destruction of IκBα and hence the translocation of NF-κB to the nucleus. *Yersinia pestis,* the bacterium that causes bubonic plague (Section 19-3Fc), produces a virulence factor named **YopJ,** which functions as a DUB that prevents the activation of NF-κB (which is an

(a)

(b)

(c)

Figure 32-89 **Structure of the tricorn protease.** (*a*) The X-ray structure of its D_3 symmetric hexamer drawn in ribbon form and viewed along its 3-fold axis. Its subunits are differently colored with that on the lower right colored in rainbow order from its N-terminus (*blue*) to its C-terminus (*red*). (*b*) View along the hexameric complex's 2-fold axis (rotated 90° about a horizontal axis with respect to Part *a*). (*c*) A cryo-EM–based image of the icosahedral complex. Each of its component "plates" represents a hexameric complex such as those drawn in Parts *a* and *b*. Note how neighboring plates are related by both 3-fold axes (which are coincident with those of the hexameric complexes) and 5-fold axes. The white scale bar is 50 nm long. [Parts *a* and *b* are based on an X-ray structure by Robert Huber, Max-Planck-Institut für Biochemie, Martinsried, Germany. PDBid 1K32. Part *c* is courtesy of Wolfgang Baumeister, Max-Planck-Institut für Biochemie, Martinsried, Germany.]

important immune system activator). We shall see in Section 34-3Bff that the monoubiquitination of certain histones functions to regulate transcription.

m. Ubiquitinlike Modifiers Participate in a Variety of Regulatory Processes

Eukaryotic cells express several proteins that have the same β-Grasp fold as ubiquitin and are similarly conjugated to other proteins, although their surface residues and charge distributions differ significantly. These **ubiquitinlike modifiers (Ubls),** which participate in a variety of fundamental cellular processes, each have a corresponding activating enzyme (E1), at least one conjugating enzyme (E2), and one or more ligases (E3s), that function to link the Ubl to its target protein(s) in a manner closely resembling that of ubiquitin.

Two of the most extensively studied Ubls are **SUMO** (*s*mall *u*biquitin-related *mo*difier; 18% identical to ubiquitin) and **RUB1** (*r*elated-to-*ub*iquitin *1*; called **NEDD8** in

vertebrates; 50% identical to ubiquitin), proteins that are highly conserved from yeast to humans. The sumoylation of PCNA at the same Lys residue at which it is ubiquitinated promotes normal DNA replication during S phase. The sumoylation of IκBα at the same residue (Lys 21) at which it is ubiquitinated blocks its ubiquitination and subsequent degradation and thereby prevents the translocation of NF-κB to the nucleus. Evidently, there is a complex regulatory interplay between the ubiquitination and sumoylation of both PCNA and IκBα. SUMO also modifies two mammalian glucose transporters, GLUT1 and GLUT4 (Section 20-2Eb), and in doing so, increases the availability of GLUT4 but decreases that of GLUT1.

All known RUB1 targets are cullins, all of which are subunits of SCF complexes, the multisubunit RING E3s (Section 32-6Bd). In fact, β-TrCP, the E3 that directs the ubiquitination of IκBα, must be conjugated to RUB1 before it can do so, thereby adding further complexity to the control of NF-κB.

CHAPTER SUMMARY

1 The Genetic Code   Point mutations are caused by either base analogs that mispair during DNA replication or by substances that react with bases to form products that mispair. Insertion/deletion (frameshift) mutations arise from the association of DNA with intercalating agents that distort the DNA structure. The analysis of a series of frameshift mutations that suppress one another established that the genetic code is an unpunctuated triplet code. In a cell-free protein synthesizing system, poly(U) directs the synthesis of poly(Phe), thereby demonstrating that UUU is the codon specifying Phe. The genetic code was elucidated through the use of polynucleotides of known composition but random sequence, by the ability of defined triplets to promote the ribosomal binding of tRNAs bearing specific amino acids, and through the use of synthetic mRNAs of known alternating sequences. The latter investigations have also demonstrated that the 5' end of mRNA corresponds to the N-terminus of the polypeptide it specifies and have established the sequences of the Stop codons. Degenerate codons differ mostly in the identities of their third base. Small single-stranded DNA phages such as φX174 contain overlapping genes in different reading frames. The genetic code used by mitochondria differs in several codons from the "standard" genetic code.

2 Transfer RNA and Its Aminoacylation   Transfer RNAs consist of 54 to 100 nucleotides that can be arranged in the cloverleaf secondary structure. As many as 10% of a tRNA's bases may be modified. Yeast tRNA^{Phe} forms a narrow, L-shaped, three-dimensional structure that resembles that of other tRNAs. Most of the bases are involved in stacking and base pairing associations including nine tertiary interactions that appear to be essential for maintaining the molecule's native conformation. Amino acids are appended to their cognate tRNAs in a two-stage reaction catalyzed by the corresponding aminoacyl–tRNA synthetase (aaRS). There are two classes of aaRSs, each containing 10 members. Class I aaRSs have two conserved sequence motifs that occur in the Rossmann fold common to the catalytic domain of these enzymes. Class II aaRSs have three conserved sequence motifs that occur in the 7-stranded antiparallel β sheet–containing fold that forms the core of their catalytic domains. In binding only their cognate tRNAs, aaRSs recognize only an idiosyncratic but limited number of bases (identity elements) that are, most often, located at the anticodon and in the acceptor stem. The great accuracy of tRNA charging arises from the proofreading of the bound amino acid by certain aminoacyl–tRNA synthetases via a double-sieve mechanism and at the expense of ATP hydrolysis.

Many organisms and organelles lack a GlnRS and instead synthesize Gln–tRNA^{Gln} by the GluRS-catalyzed charging of tRNA^{Gln} with glutamate followed by its transamidation using glutamine as the amido group source in a reaction mediated by Glu–tRNA^{Gln} amidotransferase (Glu-AdT). Ribosomes select tRNAs solely on the basis of their anticodons. Sets of degenerate codons are read by a single tRNA through wobble pairing. The UGA codon, which is normally the *opal* Stop codon may, depending on its context in mRNA, specify a selenoCys (Sec) residue, which is carried by a specific tRNA (tRNA^{Sec}), thereby forming a selenoprotein. Nonsense mutations may be suppressed by tRNAs whose anticodons have mutated to recognize a Stop codon.

3 Ribosomes and Polypeptide Synthesis   The ribosome consists of a small and a large subunit whose complex shapes have been revealed by cryoelectron microscopy and X-ray crystallography. The three RNAs and 52 proteins comprising the *E. coli* ribosome self-assemble under proper conditions. Both ribosomal subunits consist of an RNA core in which the proteins are embedded, mainly as globular domains on the back and sides of the particle, with long basic polypeptide segments that infiltrate between the RNA helices so as to neutralize their anionic charges. Eukaryotic ribosomes are larger and more complex than those of prokaryotes.

Ribosomal polypeptide synthesis proceeds by the addition of amino acid residues to the C-terminal end of the nascent polypeptide. The mRNAs are read in the $5' \rightarrow 3'$ direction. mRNAs are usually simultaneously translated by several ribosomes in the form of polysomes. The ribosome has three tRNA-binding sites: the A site, which binds the incoming aminoacyl–tRNA; the P site, which binds the peptidyl–tRNA; and the E site, which transiently binds the outgoing deacylated tRNA. During polypeptide synthesis, the nascent polypeptide is transferred to the aminoacyl–tRNA, thereby lengthening the nascent polypeptide by one residue. The newly deacylated tRNA is translocated to the E site and the new peptidyl–tRNA, with its associated codon, is translocated to the P site. In prokaryotes, the initiation sites on mRNA are recognized through their Shine–Dalgarno sequences and by their initiating codon. Prokaryotic initiating codons specify fMet–tRNA$_f^{Met}$. Initiation involves the participation of three initiation factors that induce the assembly of the ribosomal subunits with fMet–tRNA$_f^{Met}$ in the P site and mRNA. Eukaryotic initiation is a far more complicated process that requires the participation of at least 11 initiation factors. The system binds the mRNA's 5' cap and scans along the mRNA until it finds its AUG initiation codon, usually the mRNA's first AUG, through codon–anticodon interactions with the initiating tRNA, Met–tRNA$_i^{Met}$.

Polypeptides are elongated in a three-part cycle, consisting of aminoacyl–tRNA decoding, transpeptidation, and translocation, that requires the participation of elongation factors and is vectorially driven by GTP hydrolysis. EF-Tu, which functions to escort aminoacyl–tRNAs into the ribosomal A site, undergoes a major conformational change on hydrolyzing its bound GTP. The X-ray structure of the 50S subunit clearly shows that the ribosomal peptidyl transferase center is distant from any protein and hence that the ribosome is a ribozyme. Peptide bond formation is catalyzed by a substrate-assisted mechanism in which the ribosome functions as an entropy trap. Translocation is motivated through the EF-G–catalyzed hydrolysis of GTP. EF-G · GDP, which binds to the same ribosomal site as aminoacyl–tRNA · EF-Tu · GTP, is a macromolecular mimic of this complex. Translocation occurs via intermediate states, the A/P and P/E states, in which the newly formed peptidyl–tRNA and the newly deacylated tRNA are respectively bound to the A and P subsites of the 30S subunit and to the P and E subsites of the 50S subunit, following which EF-G hydrolyzes its bound GTP and shifts these tRNAs to the P/P and E/E states. The ribosome initially selects an aminoacyl–tRNA whose anticodon is cognate to its A-site–bound codon through interactions involving three universally conserved 30S subunit bases while the tRNA is in the A/T binding state. The codon–anticodon interaction is then proofread in an independent process that follows the hydrolysis of the EF-Tu–bound GTP and which occurs when the tRNA has shifted to the A/A binding state, a process called accommodation.

Termination codons bind release factors that induce the hydrolysis of the peptidyl–tRNA bond. Eukaryotic elongation and termination resemble those of prokaryotes. Ribosomal inhibitors, many of which are antibiotics, are medically important and biochemically useful in elucidating ribosomal function. Streptomycin causes mRNA misreading and inhibits prokaryotic chain initiation, chloramphenicol inhibits prokaryotic peptidyl transferase, paromomycin causes codon misreading, tetracycline inhibits aminoacyl–tRNA binding to the prokaryotic 30S subunit, and diphtheria toxin ADP-ribosylates eEF2.

4 Control of Eukaryotic Translation Several mechanisms of translational control have been elucidated in eukaryotes. eIF2α kinases catalyze the phosphorylation of eIF2α, which then tightly binds eIF2B so as to prevent it from exchanging eIF2-bound GDP for GTP and hence inhibits translational initiation. These eIF2α kinases include heme-regulated inhibitor (HRI), which functions to coordinate globin synthesis with heme availability; double-stranded RNA-activated protein kinase (PKR), an interferon-induced protein that functions to inhibit viral proliferation; and PKR-like endoplasmic reticulum kinase (PERK), which functions to protect the cell from the irreversible damage caused by the accumulation of unfolded proteins in the ER. GCN2, in contrast, is an eIF2α kinase that, when amino acids are scarce, stimulates the translation of the transcriptional activator GCN4 by causing the 40S ribosomal subunit to scan across four upstream open reading frames (uORFs) in the *GCN4* mRNA, thereby permitting the ribosome to initiate translation at the GCN4 coding sequence. The phosphorylation of eIF4E (cap-binding protein) by a MAP kinase cascade increases eIF4E's affinity for capped mRNA and thereby stimulates translational initiation. The binding of 4E-BPs to eIF4E blocks its binding of eIF4G and hence prevents initiation. However, the insulin-induced phosphorylation of the 4E-BPs causes them to dissociate from eIF4E. The mRNAs in certain animal oocytes are masked by the binding of proteins, which prevents their translation.

Many oocyte mRNAs have short poly(A) tails that are preceded by a cytoplasmic polyadenylation element (CPE) that is bound by CPE-binding protein (CPEB). CPEB binds maskin which binds eIF4E, thereby inhibiting translational initiation. However, when CPEB is phosphorylated, it recruits poly(A) polymerase (PAP), which extends the mRNA's poly(A) tail such that it is bound by poly(A)-binding protein (PABP). PABP then binds to eIF4G, which in turn displaces maskin from eIF4E, thereby permitting the translation of the mRNA. Antisense oligonucleotides can be used to inhibit the translation of their complementary mRNAs. Although the delivery of antisense oligonucleotides to their sites of action has proved to be a difficult problem, their use is starting to show some medical and biotechnological successes.

5 Post-Translational Modification Proteins may be post-translationally modified in a variety of ways. Proteolytic cleavages, usually by specific peptidases, activate proproteins. The signal peptides of preproteins are removed by signal peptidases. Covalent modifications alter many types of side chains in a variety of ways that modulate the catalytic activities of enzymes, provide recognition markers, and stabilize protein structures.

Protein splicing occurs via the intein-catalyzed self-excision between an N-extein and a C-extein accompanied by the ligation of the N- and C-exteins via a peptide bond. Most inteins contain a homing endonuclease that makes a double-strand nick in a gene similar to that encoding the corresponding extein, thereby triggering a recombinational double-strand DNA repair process that copies the gene encoding the intein into the break. Inteins therefore appear to be molecular parasites.

6 Protein Degradation Proteins in living cells are continually turning over. This controls the level of regulatory enzymes and disposes of abnormal proteins that would otherwise interfere with cellular processes. Proteins are degraded by lysosomes via a nonspecific process as well via a process specific for KFERQ proteins that is stimulated during starvation. A cytosolically based ATP-dependent system degrades normal as well as abnormal proteins in a process that flags these proteins by the covalent attachment of Lys 48–linked polyubiquitin chains to their Lys residues. This process is mediated by three consecutively acting enzymes: ubiquitin-activating enzyme (E1), ubiquitin-conjugating enzyme (E2), and ubiquitin–protein ligase (E3). Most cells have one species of E1, several species of E2, and numerous species of E3, each of which is served by one or a few E2s. The polyubiquitinated protein is proteolytically degraded in the 26S proteasome.

E3s can have complicated modular structures, each having different specificities for target proteins. SCF complexes, one of whose several subunits contains a RING domain, are particularly elaborate. The RING E3 known as E3α functions to ubiquitinate proteins that satisfy the N-end rule. The transcription factor NF-κB is activated through the ubiquitination and subsequent destruction of its otherwise bound inhibitor IκBα by the SCF β-TrCP, which is activated through phosphorylation via a signal transduction cascade. Cyclins, which mediate the cell cycle, are destroyed in a programmed manner through ubiquitination by their cognate E3s, one of which is anaphase-promoting complex (APC).

The 26S proteasome consists of a hollow protein barrel formed by two rings of seven α subunits flanking two rings of seven β subunits known as the 20S proteasome, which is bound at each end by 19S caps that each consist of ~18 subunits. The active sites of the β subunits, which are members of the N-terminal nucleophile (Ntn) family of hydrolases, are inside the barrel. Ubiquitinated proteins are selected by the 19S caps, which unfold them in an ATP-dependent matter and then feed them into the 20S proteasome via its axial channel.

The polyubiquitin (polyUb) chains are excised from the condemned protein by proteasome-associated deubiquitinating enzymes (DUBs), while other DUBs dismember the polyUb chains to their component ubiquitin units, thereby recycling them. The 11S activator is a heptameric complex that, on binding to one end of a 20S proteasome, opens its axial channel in an ATP-independent manner, thereby permitting polypeptides, but not folded proteins, to enter the 20S proteasome. Eubacteria, nearly all of which lack proteasomes, nevertheless express a variety of self-compartmentalized proteases, including ClpP, heat shock locus UV (HslUV), and tricorn protease, that function to proteolytically dispose of their cellular proteins. Monoubiquitination and polyubiquitination with chains linked through ubiquitin residues other than Lys 48 regulate a variety of cellular processes. Ubiquitinlike modifiers (Ubls), such as SUMO and RUB1, also participate in numerous regulatory processes.

REFERENCES

General

Watson, J.D., Baker, T.A., Bell, S.P., Gann, A., Levine, M., and Losick, R., *Molecular Biology of the Gene* (6th ed.), Chap. 14 and 15, Cold Spring Harbor Laboratory Press (2008).

The Genetic Code

Attardi, G., Animal mitochondrial DNA: an extreme example of genetic economy, *Int. Rev. Cytol.* **93,** 93–145 (1985).

Benzer, S., The fine structure of the gene, *Sci. Am.* **206**(1), 70–84 (1962).

Crick, F.H.C., The genetic code, *Sci. Am.* **207**(4), 66–74 (1962) [The structure of the code as determined by phage genetics]; *and* The genetic code: III, *Sci. Am.* **215**(4), 55–62 (1966). [A description of the nature of the code after its elucidation was almost complete.]

Crick, F.H.C., Burnett, L., Brenner, S., and Watts-Tobin, R.J., General nature of the genetic code for proteins, *Nature* **192,** 1227–1232 (1961).

Fox, T.D., Natural variation in the genetic code, *Annu. Rev. Genet.* **21,** 67–91 (1987).

Judson, H.F., *The Eighth Day of Creation,* Expanded Edition, Part II, Cold Spring Harbor Laboratory Press (1996). [A fascinating historical narrative on the elucidation of the genetic code.]

Khorana, H.G., Nucleic acid synthesis in the study of the genetic code, *Nobel Lectures in Molecular Biology, 1933–1975,* pp. 303–331, Elsevier (1977).

Knight, R.D., Freeland, S.J., and Landweber, L.F., Selection, history and chemistry: the three faces of the genetic code, *Trends Biochem. Sci.* **24,** 241–247 (1999).

Nirenberg, M., The genetic code, *Nobel Lectures in Molecular Biology, 1933–1975,* pp. 335–360, Elsevier (1977).

Nirenberg, M., Historical review: deciphering the genetic code—a personal account, *Trends Biochem. Sci.* **29,** 46–54 (2004).

Nirenberg, M. and Leder, P., RNA code words and protein synthesis, *Science* **145,** 1399–1407 (1964). [The determination of the genetic code by the ribosomal binding of tRNAs using specific trinucleotides.]

Nirenberg, M.W. and Matthaei, J.H., The dependence of cell-free protein synthesis in *E. coli* upon naturally occurring or synthetic polyribonucleotides, *Proc. Natl. Acad. Sci.* **47,** 1588–1602 (1961). [The landmark paper reporting the finding that poly(U) stimulates the synthesis of poly(Phe).]

Singer, B. and Kusmierek, J.T., Chemical mutagenesis, *Annu. Rev. Biochem.* **51,** 655–693 (1982).

The Genetic Code, Cold Spring Harbor Symp. Quant. Biol. **31** (1966). [A collection of papers describing the establishment of the genetic code. See especially the articles by Crick, Nirenberg, and Khorana.]

Yanofsky, C., Establishing the triplet nature of the genetic code, *Cell* **128,** 815–818 (2007).

Yarus, M., Caporaso, J.G., and Knight, R., Origins of the genetic code: the escaped triplet theory, *Annu. Rev. Biochem.* **74,** 179–198 (2005).

Transfer RNA and Its Aminoacylation

Alexander, R.W. and Schimmel, P., Domain–domain communication in aminoacyl–tRNA synthetases, *Prog. Nucleic Acid Res. Mol. Biol.* **69,** 317–349 (2001).

Ambrogelly, A., Palioura, S., and Söll, D., Natural expansion of the genetic code, *Nature Chem. Biol.* **3,** 29–35 (2007).

Björk, G.R., Ericson, J.U., Gustafsson, C.E.D., Hagervall, T.G.,

Jösson, Y.H., and Wikström, P.M., Transfer RNA modification, *Annu. Rev. Biochem.* **56,** 263–287 (1987).

Böck, A., Forschhammer, K., Heider, J., and Baron, C., Selenoprotein synthesis: an expansion of the genetic code, *Trends Biochem. Sci.* **16,** 463–467 (1991).

Crick, F.H.C., Codon–anticodon pairing: the wobble hypothesis, *J. Mol. Biol.* **19,** 548–555 (1966).

Fukai, S., Nureki, O., Sekine, S., Shimada, A., Tao, J., Vassylyev, D.G., and Yokoyama, S., Structural basis for double-sieve discrimination of L-valine from L-isoleucine and L-threonine by the complex of tRNA^{Val} and valyl–tRNA synthetase, *Cell* **103,** 793–803 (2000).

Ibba, M. and Söll, D., Aminoacyl–tRNA synthesis, *Annu. Rev. Biochem.* **69,** 617–650 (2000).

Ibba, M.A., Francklyn, C., and Cusack, S. (Eds.), *The Aminoacyl-tRNA Synthetases,* Landes Bioscience/Eurekah.com (2005). [Contains reviews on each of the twenty aminoacyl–tRNA synthetases.]

Jacquin-Becker, C., Ahel, I., Ambrogelly, A., Ruan, B., Söll, D., and Stathopoulos, C., Cysteinyl–tRNA formation and prolyl–tRNA synthetase, *FEBS Lett.* **514,** 34–36 (2002).

Kim, S.H., Suddath, F.L., Quigley, G.J., McPherson, A., Sussman, J.L., Wang, A.M.J., Seeman, N.C., and Rich, A., Three-dimensional tertiary structure of yeast phenylalanine transfer RNA, *Science* **185,** 435–440 (1974); *and* Robertus, J.D., Ladner, J.E., Finch, J.T., Rhodes, D., Brown, R.S., Clark, B.F.C., and Klug, A., Structure of yeast phenylalanine tRNA at 3 Å resolution, *Nature* **250,** 546–551 (1974). [The landmark papers describing the high-resolution structure of a tRNA.]

Krzycki, J.A., The direct genetic encoding of pyrrolysine, *Curr. Opin. Microbiol.* **8,** 706–712 (2005); *and* Nozawa, K., O'Donoghue, P., Gundllapalli, S., Araiso, Y., Ishitani, R., Umehara, T., Söll, D., and Nureki, O., Pyrolysyl–tRNA synthetase–tRNA^{Pyl} structure reveals the molecular basis of orthogonality, *Nature* **457,** 1163–1167 (2009).

Ling, J., Reynolds, N., and Ibba, M., Aminoacyl–tRNA synthesis and translational control, *Annu. Rev. Microbiol.* **63,** 61–78 (2009).

Nureki, O., Vassylyev, D.G., Tateno, M., Shimada, A., Nakama, T., Fukai, S., Konno, M., Hendrickson, T.L., Schimmel, P., and Yokoyama, S., Enzyme structure with two catalytic sites for double-sieve selection of substrate, *Science* **280,** 578–582 (1998). [The X-ray structures of IleRS in complexes with isoleucine and valine.]

Rould, M.A., Perona, J.J., and Steitz, T.A., Structural basis of anti-codon loop recognition by glutaminyl–tRNA synthetase, *Nature* **352,** 213–218 (1991).

Ruff, M., Krishnaswamy, S., Boeglin, M., Poterszman, A., Mitschler, A., Podjarny, A., Rees, B., Thierry, J.C., and Moras, D., Class II aminoacyl transfer RNA synthetases: Crystal structure of yeast aspartyl–tRNA synthetase complexed with tRNA^{Asp}, *Science* **252,** 1682–1689 (1991).

Schimmel, P. and Beebe, K., Aminoacyl tRNA synthetases: from the RNA to the theater of proteins. *In* Gesteland, R.F., Cech, T.R., and Atkins, J.F. (Eds.), *The RNA World* (3rd ed.), pp. 227–255, Cold Spring Harbor Laboratory Press (2006).

Silvian, L.F., Wang, J., and Steitz, T.A., Insights into editing from an Ile-tRNA synthetase structure with tRNA^{Ile} and mupirocin, *Science* **285,** 1074–1077 (1999).

Söll, D. and RajBhandary, U.L. (Eds.), *tRNA: Structure, Biosynthesis, and Function,* ASM Press (1995).

Stadtman, T.C., Selenocysteine, *Annu. Rev. Biochem.* **65,** 83–100 (1996).

Ribosomes and Polypeptide Synthesis

Aitken, C.E., Petrov, A., and Puglisi, J.D., Single ribosome dynamics and the mechanism of translation, *Annu. Rev. Biophys.* **39**, 491–513 (2010).

Ban, N., Nissen, P., Hansen, J., Moore, P.B., and Steitz, T., The complete atomic structure of the large ribosomal subunit at 2.4 Å resolution, *Science* **289**, 905–920 (2000).

Bashan, A. and Yonath, A., Correlating ribosome function with high-resolution structures, *Trends Microbiol.* **16**, 326–335 (2008).

Bell, C.E. and Eisenberg, D.E., Crystal structure of diphtheria toxin bound to nicotinamide adenine dinucleotide, *Biochemistry* **35**, 1137–1149 (1996).

Brandt, F., Etchells, S.A., Ortiz, J.O., Elcock, A.H., Hartl, F.U., and Baumeister, W., The native 3D organization of bacterial polysomes, *Cell* **136**, 261–271 (2009).

Brodersen, D.E., Clemons, W.M., Jr., Carter, A.P., Morgan-Warren, R.J., Wimberly, B.T., and Ramakrishnan, V., The structural basis for the action of the antibiotics tetracycline, pactamycin, and hygromycin B on the 30S ribosomal subunit, *Cell* **103**, 1143–1154 (2000).

Czworkowski, J., Wang, J., Steitz, J.A., and Moore, P.B., The crystal structures of elongation factor G complexed with GDP, at 2.7 Å resolution; *and* Ævarsson, A., Brazhnikov, E., Garber, M., Zheltonosova, J., Chirgadze, Yu., Al-Karadaghi, S., Svensson, L.A., and Liljas, A., Three-dimensional structure of the ribosomal translocase: elongation factor G from *Thermus thermophilus*, *EMBO J.* **13**, 3661–3668 *and* 3669–3677 (1994).

Dintzis, H.M., Assembly of the peptide chains of hemoglobin, *Proc. Natl. Acad Sci.* **47**, 247–261 (1961); *and* The wandering pathway to determining N to C synthesis of proteins: Some recollections concerning protein structure and biosynthesis, *Biochem. Mol. Biol. Educ.* **34**, 241–246 (2006). [The determination of the direction of polypeptide biosynthesis.]

Dunkle, J.A. and Cate, J.H.D., Ribosome structure and dynamics during translocation and termination, *Annu. Rev. Biophys.* **39**, 227–244 (2010).

Frank, J., Single-particle imaging of macromolecules by cryo-electron microscopy, *Annu. Rev. Biophys. Biomol. Struct.* **31**, 303–319 (2002).

Frank, J. and Agrawal, R.K., A ratchet-like inter-subunit reorganization of the ribosome during translocation, *Nature* **406**, 318–322 (2000).

Frank, J. and Gonzalez, R.L., Jr., Structure and dynamics of a processive Brownian motor: the translating ribosome, *Annu. Rev. Biochem.* **79**, 381–412 (2010).

Gao, H., et al., RF3 induces ribosomal conformational changes responsible for dissociation of class I release factors, *Cell* **129**, 929–941 (2007).

Gingras, A.-C., Raught, B., and Sonnberg, N., eIF4 initiation factors: effectors of mRNA recruitment to ribosomes and regulators of translation, *Annu. Rev. Biochem.* **68**, 913–963 (1999).

Held, W.A., Ballou, B., Mizushima, S., and Nomura, M., Assembly mapping of 30S ribosomal proteins from *Escherichia coli, J. Biol. Chem.* **249**, 3103–3111 (1974).

Holbrook, S.R., Structural principles from large RNAs, *Annu. Rev. Biophys.* **37**, 445–464 (2008).

Jackson, R.J., Hellen, C.U.T., and Pestova, T.V., The mechanism of eukaryotic translation and principles of its regulation, *Nature Rev. Mol. Cell Biol.* **10**, 113–127 (2010).

Kawashima, T., Berthet-Colominas, C., Wulff, M., Cusack, S., and Leberman, R., The structure of the *Escherichia coli* EF-Tu · EF-Ts complex at 2.5 Å resolution, *Nature* **379**, 511–518 (1996).

Kiel, M.C., Kaji, H., and Kaji, A., Ribosome recycling, *Biochem. Mol. Biol. Educ.* **35**, 40-44 (2007); *and* Hirokawa, G., Demesh-kina, N., Iwakura, N., Kaji, H., and Kaji, A., The ribosome-recycling step: consensus or controversy? *Trends Biochem. Sci.* **31**, 143–149 (2006).

Kjeldgaard, M. and Nyborg, J., Refined structure of elongation factor EF-Tu from *Escherichia coli, J. Mol. Biol.* **223**, 721–742 (1992); *and* Kjeldgaard, M., Nissen, P., Thirup, S., and Nyborg, J., The crystal structure of elongation factor EF-Tu from *Thermus aquaticus* in the GTP conformation, *Structure* **1**, 35–50 (1993).

Korostelev, A. and Noller, H.F., The ribosome in focus: new structures bring new insights, *Trends Biochem. Sci.* **32**, 434–441 (2007); *and* Korostelev, A., Trakhanov, S., Laurberg, M., and Noller, H.F., Crystal structure of the 70S ribosome–tRNA complex reveals functional interactions and rearrangements, *Cell* **126**, 1065–1077 (2006).

Laurberg, M., Asahara, H., Korostelev, A., Zhu, J., Trakhanov, S., and Noller, H.F., Structural basis for translation termination on the 70S ribosome, *Nature* **454**, 852–857 (2008); Korostelev, A., Asahara, H., Lancaster, L., Laurberg, M., Hirschi, A., Zhu, J., Trakhanov, S., Scott, W.G., and Noller, H.F., Crystal structure of a translation termination complex formed with release factor RF2, *Proc. Natl. Acad. Sci.* **105**, 19684–19689 (2008); *and* Weixl-baumer, A., Jin, H., Neubauer, C., Voorhees, R.M., Petry, S., Kelley, A.C., and Ramakrishnan, V., Insights into translational termination from the structure of RF2 bound to the ribosome, *Science* **322**, 953–956 (2008).

Marcotrigiano, J., Gingras, A.-C., Sonenberg, N., and Burley, S.K., Cocrystal structure of the messenger RNA 5' cap-binding protein (eIF4E) bound to 7-methyl-GDP, *Cell* **89**, 951–961 (1997).

Moazed, D. and Noller, H.F., Intermediate states in the movement of transfer RNA in the ribosome, *Nature* **342**, 142–148 (1989).

Moore, P.B. and Steitz, T.A., The involvement of RNA in ribosome function, *Nature* **418**, 229–235 (2002); *and* The structural basis of large ribosomal subunit function, *Annu. Rev. Biochem.* **72**, 813–850 (2003).

Munro, J.B., Sanbonmatsu, K.Y., Spahn, C.M.T., and Blanchard, S.C., Navigating the ribosome's metastable energy landscape, *Trends Biochem. Sci.* **31**, 390–399 (2009).

Nissen, P., Kjeldgaard, M., Thirup, S., Polekhina, G., Reshetnikova, L., Clark, B.F.C., and Nyborg, J., Crystal structure of the ternary complex of Phe–tRNA^Phe, EF-Tu, and a GTP analog, *Science* **270**, 1464–1472 (1995).

Noller, H.F., Hoffarth, V., and Zimniak, L., Unusual resistance of peptidyl transferase to protein extraction procedures, *Science* **256**, 1416–1419 (1992); *and* Noller, H.F., Peptidyl transferase: protein, ribonucleoprotein, or RNA? *J. Bacteriol.* **175**, 5297–5300 (1993).

Ogle, J.M., Brodersen, D.E., Clemons, W.M., Jr., Tarry, M.J., Carter, A.P., and Ramakrishnan, V., Recognition of cognate transfer RNA by the 30S ribosomal subunit, *Science* **292**, 897–902 (2001); *and* Ogle, J.M., Carter, A.P., and Ramakrishnan, V., Insights into the decoding mechanism from recent ribosome structures, *Trends Biochem. Sci.* **28**, 259–266 (2003).

Pioletti, M., et al., Crystal structure of complexes of the small ribosomal subunit with tetracycline, edeine and IF3, *EMBO J.* **20**, 1829–1839 (2001).

Rodnina, M.V., Beringer, M., and Wintermeyer, W., How the ribosome makes peptide bonds, *Trends Biochem. Sci.* **32**, 20–26 (2007).

Schluenzen, F., Tocilj, A., Zarivach, R., Harms, J., Gluehmann, M., Janell, D., Bashan, A., Bartels, H., Agmon, I., Franceschi, F., and Yonath, A., Structure of functionally activated small ribosomal subunit at 3.3 Å resolution, *Cell* **102**, 615–623 (2000).

Schmeing, T.M., and Ramakrishnan, V., What recent ribosome structures have revealed about the mechanism of translation,

Nature **461,** 1234–1242 (2009); Simonovic, M. and Steitz, T.A., A structural view of the mechanism of the ribosome-catalyzed peptide bond mechanism, *Biochim. Biophys. Acta* **1789,** 612–623 (2009); *and* Zimmerman, E. and Yonath, A., Biological implications of the ribosome's stunning stereochemistry, *ChemBioChem* **10,** 63–72 (2009). [Authoritative reviews.]

Schmeing, T.M., Voorhees, R.M., Kelley, A.C., Gao, Y.-G., Murphy, F.V., IV, Weir, J.R., and Ramakrishnan, V., The crystal structure of the ribosome bound to EF-Tu and aminoacyl–tRNA; *and* Gao, Y.-G., Selmer, M., Dunham, M., Weixlbaumer, A., Kelley, A.C., and Ramakrishnan, V., The structure of the ribosome with elongation factor G trapped in the post-translocational state, *Science* **326,** 688–693 *and* 694–699 (2009).

Selmer, M., Dunham, C.M., Murphy, F.V., IV, Weixlbaumer, A., Petry, S., Kelley, A.C., Weir, J.R., and Ramakrishnan, V., Structure of the 70S ribosome complexed with mRNA and tRNA, *Science* **313,** 1935–1942 (2006); *and* Voorhees, R.M., Weixlbaumer, A., Loakes, D., Kelley, A.C., and Ramakrishnan, V., Insights into substrate stabilization from snapshots of the peptidyl transferase center of the intact 70S ribosome, *Nature Struct. Mol. Biol.* **16,** 528–533 (2009).

Sievers, A., Beringer, M., Rodnina, M.V., and Wolfenden, R., The ribosome as an entropy trap. *Proc. Natl. Acad. Sci.* **101,** 7897–7901 (2004).

Sonenberg, N., eIF4E, the mRNA cap-binding protein: from basic discovery to translational research, *Biochem. Cell Biol.* **86,** 178–183 (2008).

Sonenberg, N. and Dever, T.E., Eukaryotic translation initiation factors and regulators, *Curr. Opin. Struct. Biol.* **13,** 56–63 (2003).

Spahn, C.M.T., Beckmann, R., Eswar, N., Penczek, P.A., Sali, A., Blobel, G., and Frank, J., Structure of the 80S ribosome from *Saccharomyces cerevisiae*—tRNA-ribosome and subunit-subunit interactions, *Cell* **107,** 373–386 (2001).

Steitz, J.A. and Jakes, K., How ribosomes select initiator regions in mRNA: base pair formation between the 3′ terminus of 16S RNA and the mRNA during initiation of protein synthesis in *Escherichia coli, Proc. Natl. Acad. Sci.* **72,** 4734–4738 (1975).

Sykes, M.T. and Williamson, J.R., A complex assembly landscape for the 30S ribosomal subunit, *Annu. Rev. Biophys.* **38,** 197–215 (2009); *and* Woodson, S.A., RNA folding and ribosome assembly, *Curr. Opin. Struct. Biol.* **12,** 667–673 (2008).

Vestergaard, B., Van, L.B., Andersen, G.R., Nyborg, J., Buckingham, R.H., and Kjeldgaard, M., Bacterial polypeptide release factor RF2 is structurally distinct from eukaryotic eRF1, *Mol. Cell* **8,** 1375–1382 (2001).

Weixlbaumer, A., Petry, S., Dunham, C.M., Selmer, M., Kelley, A.C., and Ramakrishnan, V., Crystal structure of the ribosome recycling factor bound to the ribosome, *Nature Struct. Mol. Biol.* **14,** 733–737 (2007).

Wimberly, B.T., Broderson, D.E., Clemons, W.M., Jr., Morgan-Warren, R., von Rhein, C., Hartsch, T., and Ramakrishnan, V., Structure of the 30S ribosomal subunit, *Nature* **407,** 327–339 (2000); *and* Broderson, D.E., Clemons, W.M., Jr., Carter, A.P., Wimberly, B.T., and Ramakrishnan, V., Crystal structure of the 30S ribosomal subunit from *Thermus thermophilus:* Structure of the proteins and their interactions with 16S RNA, *J. Mol. Biol.* **316,** 725–768 (2002).

Yonath, A., The search and its outcome: High resolution structures of ribosomal particles from mesophilic, thermophilic, and halophilic bacteria at various functional states, *Annu. Rev. Biophys. Biomol. Struct.* **31,** 257–273 (2002).

Yonath, A., Antibiotics targeting ribosomes: resistance, selectivity, synergism, and cellular regulation, *Annu. Rev. Biochem.* **74,** 649–679 (2005).

Youngman, E.M., McDonald, M.E., and Green, R., Peptide release on the ribosome: mechanism and implications for translational control, *Annu. Rev. Microbiol.* **62,** 33–373 (2008).

Yusupova, G.Z., Yusupov, M.M., Cate, J.D.H., and Noller, H.F., The path of messenger RNA through the ribosome, *Cell* **106,** 233–241 (2001).

Control of Eukaryotic Translation

Branch, A.D., A good antisense molecule is hard to find, *Trends Biochem. Sci.* **23,** 45–50 (1998).

Calkhoven, C.F., Müller, C., and Leutz, A., Translational control of gene expression and disease, *Trends Mol. Med.* **8,** 577–583 (2002).

Chen, J.-J., and London, I.M., Regulation of protein synthesis by heme-regulated eIF-2α kinase, *Trends Biochem. Sci.* **20,** 105–108 (1995).

Clemens, M.J., PKR—A protein kinase regulated by double-stranded RNA, *Int. J. Biochem. Cell Biol.* **29,** 945–949 (1997).

Dever, T.E., Gene-specific regulation by general translation factors, *Cell* **108,** 545–556 (2002).

Gray, N.K. and Wickens, M., Control of translation initiation in animals, *Annu. Rev. Cell Dev. Biol.* **14,** 399–458 (1998).

Lawrence, J.C., Jr. and Abraham, R.T., PHAS/4E-BPs as regulators of mRNA translation and cell proliferation, *Trends Biochem. Sci.* **22,** 345–349 (1997).

Lebedeva, I. and Stein, C.A., Antisense oligonucleotides: promise and reality, *Annu. Rev. Pharmacol. Toxicol.* **41,** 403–419 (2001).

Matthews, M.B., Sonnenberg, N., and Hershey, J.W.B. (Eds.), *Translational Control in Biology and Medicine,* Cold Spring Harbor Laboratory Press (2007).

Mendez, R. and Richter, J.D., Translational control by CPEB: a means to the end, *Nature Rev. Mol. Cell Biol.* **2,** 521–529 (2001).

Phillips, M.I. (Ed.), *Antisense Technology: Part A. General Methods, Methods of Delivery, and RNA Studies;* and *Part B. Applications, Methods Enzymol.* **313** *and* **314** (2000).

Sen, G.C. and Lengyel, P., The interferon system, *J. Biol. Chem.* **267,** 5017–5020 (1992).

Sheehy, R.E., Kramer, M., and Hiatt, W.R., Reduction of polygalacturonase activity in tomato fruit by antisense RNA, *Proc. Natl. Acad. Sci.* **85,** 8805–8809 (1988).

Sonenberg, N. and Hinnebusch, A.G., Regulation of translation initiation in eukaryotes: mechanisms and biological targets, *Cell* **136,** 731–745 (2009).

Tafuri, S.R. and Wolffe, A.P., Dual roles for transcription and translation factors in the RNA storage particles of *Xenopus* oocytes, *Trends Cell. Biol.* **3,** 94–98 (1993).

Tamm, I., Dörken, B., and Hartmann, G., Antisense therapy in oncology: new hope for an old idea? *Lancet* **358,** 489–497 (2001).

Post-Translational Modification

Gogarten, J.P., Senejani, A.G., Zhaxybayeva, O., Olendzenski, L., and Hilario, E., Inteins: structure, function, and evolution, *Annu. Rev. Microbiol.* **56,** 263–287 (2002).

Harding, J.J., and Crabbe, M.J.C. (Eds.), *Post-Translational Modifications of Proteins,* CRC Press (1992).

Klabunde, T., Sharma, S., Telenti, A., Jacobs, W.R., Jr., and Sacchetini, J.C., Crystal structure of Gyr A protein from *Mycobacterium xenopi* reveals structural basis of splicing, *Nature Struct. Biol.* **5,** 31–36 (1998).

Liu, X.-Q., Protein-splicing intein: genetic mobility, origin, and evolution, *Annu. Rev. Genet.* **34,** 61–76 (2000).

Saleh, L. and Perler, F.B., Protein splicing *in cis* and *in trans, Chem. Rec.* **6,** 183–193 (2006).

Wold, F., *In vivo* chemical modification of proteins, *Annu. Rev. Biochem.* **50,** 783–814 (1981).

Wold, F. and Moldave, K. (Eds.), *Posttranslational Modifications, Parts A and B, Methods Enzymol.* **106** and **107** (1984). [Contains extensive descriptions of the amino acid "zoo."]

Protein Degradation

Bochtler, M., Ditzel, L., Groll, M., Hartmann, C., and Huber, R., The proteasome, *Annu. Rev. Biophys. Biomol. Struct.* **28**, 295–317 (1999).

Brandstetter, H., Kim, J.-S., Groll, M., and Huber, R., Crystal structure of the tricorn protease reveals a protein disassembly line, *Nature* **414**, 466–470 (2001); Walz, J., Tamura, T., Tamura, N., Grimm, R., Baumeister, W., and Koster, A.J., Tricorn protease exists as an icosahedral supermolecule *in vivo, Mol. Cell* **1**, 59–65 (1997); *and* Walz, J., Koster, A.J., Tamura, T., and Baumeister, W., Capsids of tricorn protease studied by cryomicroscopy, *J. Struct. Biol.* **128**, 65–68 (1999).

Cook, W.J., Jeffrey, L.C., Kasperek, E., and Pickart, C.M., Structure of tetraubiquitin shows how multiubiquitin chains can be formed, *J. Mol. Biol.* **236**, 601–609 (1994).

Cook, W.J., Jeffrey, L.C., Sullivan, M.L., and Vierstra, R.D., Three-dimensional structure of a ubiquitin-conjugating enzyme (E2), *J. Biol. Chem.* **267**, 15116–15121 (1992).

DeMartino, G.N. and Gillette, T.G., Proteasomes: machines for all reasons, *Cell* **129**, 659–662 (2007).

Deshaies, R.J. and Joazeiro, C.A.P., RING domain E3 ubiquitin ligases, *Annu. Rev. Biochem.* **78**, 399–434 (2009).

Dye, B.T. and Schulman, B.A., Structural mechanisms underlying posttranslational modification by ubiquitin-like proteins, *Annu. Rev. Biophys. Biomol. Struct.* **36**, 131–150 (2007).

Finley, D., Recognition and processing of ubiquitin–protein conjugates by the proteasome, *Annu. Rev. Biochem.* **78**, 477–513 (2009).

Glickman, M.H. and Ciechanover, A., The ubiquitin-proteasome proteolytic pathway: destruction for the sake of construction, *Physiol. Rev.* **82**, 373–428 (2002).

Hershko, A. and Ciechanover, A., The ubiquitin system, *Annu. Rev. Biochem.* **67**, 425–479 (1998).

Jentsch, S. and Pyrowalakis, G. Ubiquitin and its kin: how close are the family ties, *Trends Cell Biol.* **10**, 335–342 (2003). [Discusses Ubls.]

Liu, F. and Walters, K.J., Multitasking with ubiquitin through multivalent interactions, *Trends Biochem. Sci.* **35**, 352–360 (2010).

Löwe, J., Stock, D., Jap, B., Zwicki, P., Baumeister, W., and Huber, R., Crystal structure of the 20S proteasome from the archeon *T. acidophilum* at 3.4 Å resolution, *Science* **268**, 533–539 (1995); *and* Groll, M., Ditzel, L., Löwe, J., Stock, D., Bochtler, M., Bartunik, H.D., and Huber, R., Structure of 20S proteasome from yeast at 2.4 Å resolution, *Nature* **386**, 463–471 (1997).

Manchado, E., Eguren, M., and Malumbres, M., The anaphase-promoting complex/cyclosome (APC/C): cell-cycle-dependent and -independent functions, *Biochem. Soc. Trans.* **38**, 65–71 (2010).

Mukhopadhyay, D. and Riezman, H., Proteosome-independent functions of ubiquitin in endocytosis and signaling, *Science* **315**, 201–205 (2007).

Navon, A. and Ciechanover, A., The 26S proteosome: from basic mechanism to drug targeting, *J. Biol. Chem.* **284**, 33713–33718 (2009).

Page, A.M. and Hieter, P., The anaphase-promoting complex: new subunits and regulators, *Annu. Rev. Biochem.* **68**, 583–609 (1999).

Pickart, C.M., Back to the future with ubiquitin, *Cell* **116**, 181–190 (2004); *and* Mechanisms underlying ubiquitination, *Annu. Rev. Biochem.* **70**, 503–533 (2001).

Schwartz, A.L. and Ciechanover, A., Targeting proteins for destruction by the ubiquitin system: implication for human pathobiology, *Annu. Rev. Pharmacol. Toxicol.* **49**, 73–96 (2009).

Senahdi, V.-J., Bugg, C.E., Wilkinson, K.D., and Cook, W.J., Three-dimensional structure of ubiquitin at 2.8 Å resolution, *Proc. Natl. Acad. Sci.* **82**, 3582–3585 (1985).

Skaug, B., Jiang, X., and Chen, Z.J., The role of ubiquitin in NF-κB regulatory pathways, *Annu. Rev. Biochem.* **78**, 769–796 (2009).

Sousa, M.C., Trame, C.B., Tsuruta, H., Wilbanks, S.M., Reddy, V.J., and McKay, D.B., Crystal and solution structures of an HslUV protease–chaperone complex, *Cell* **103**, 633–643 (2000).

Unno, M., Mizushima, T., Morimoto, Y., Tomisugi, Y., Tanaka, K., Yasuoka, N., and Tsukihara, T., The structure of the mammalian proteasome at 2.75 Å resolution, *Structure* **10**, 609–618 (2002).

VanDemark, A.P. and Hill, C.P., Structural basis of ubiquitylation, *Curr. Opin. Struct. Biol.* **12**, 822–830 (2002).

Varshavsky, A., Regulated protein degradation, *Trends Biochem. Sci.* **30**, 283–286 (2005).

Varshavsky, A., Turner, G., Du, F., and Xie, Y., The ubiquitin system and the N-end rule, *Biol. Chem.* **381**, 779–789 (2000).

Voges, D., Zwickl, P., and Baumeister, W., The 26S proteasome: a molecular machine designed for controlled proteolysis, *Annu. Rev. Biochem.* **68**, 1015–1068 (1999).

Wang, J., Hartling, J.A., and Flanagan, J.M., The structure of ClpP at 2.3 Å resolution suggests a model for ATP-dependent proteolysis, *Cell* **91**, 447–456 (1997).

Whitby, F.G., Masters, E.I. Kramer, L., Knowlton, J.R., Yao, Y., Wang, C.C., and Hill, C.P., Structural basis for the activation of 20S proteasomes by 11S regulators, *Nature* **408**, 115–120 (2000); *and* Förster, A., Whitby, F.G., and Hill, C.P., The pore of activated 20S proteasomes has an ordered 7-fold symmetric conformation, *EMBO J.* **22**, 4356–4354 (2003).

Zheng, N., et al., Structure of the Cul1–Rbx1–Skp1–F-box^{Skp2} SCF ubiquitin ligase complex, *Nature* **41**, 703–709 (2002); Schulman, B.A., et al., Insights into SCF ubiquitin ligases from the structure of the Skp1–Skp2 complex, *Nature* **408**, 381–386 (2000); *and* Zheng, N., Wang, P., Jeffrey, P.D., and Pavletich, N.P., Structure of a c-Cbl–UbcH7 complex: RING domain function in ubiquitin-protein ligases, *Cell* **102**, 533–539 (2000).

Zwickl, P., Seemüller, E., Kapelari, B., and Baumeister, W., The proteasome: A supramolecular assembly designed for controlled proteolysis, *Adv. Prot. Chem.* **59**, 187–222 (2002).

PROBLEMS

1. What is the product of the reaction of guanine with nitrous acid? Is the reaction mutagenic? Explain.

2. What is the polypeptide specified by the following DNA antisense strand? Assume translation starts at the first initiation codon.

5′-TCTGACTATTGAGCTCTCTGGCACATAGCA-3′

*3. The fingerprint of a protein from a phenotypically revertant mutant of bacteriophage T4 indicates the presence of an altered tryptic peptide with respect to the wild type. The wild-type and mutant peptides have the following sequences:

Wild type   Cys-Glu-Asp-His- Val-Pro-Gln-Tyr-Arg
Mutant   Cys-Glu-Thr- Met-Ser- His-Ser- Tyr-Arg

Indicate how the mutant could have arisen and give the base sequences, as far as possible, of the mRNAs specifying the two peptides. Comment on the function of the peptide in the protein.

4. Explain why the various classes of mutations can reverse a mutation of the same class but not a different class.

5. Which amino acids are specified by codons that can be changed to an *amber* codon by a single point mutation?

6. The mRNA specifying the α chain of human hemoglobin contains the base sequence

···UCCAAUACCGUUAAGCUGGA···

The C-terminal tetrapeptide of the normal α chain, which is specified by part of this sequence, is

-Ser-Lys-Tyr-Arg

In hemoglobin Constant Spring, the corresponding region of the α chain has the sequence

-Ser-Lys-Tyr-Arg-Gln-Ala-Gly-···

Specify the mutation that causes hemoglobin Constant Spring.

7. Explain why a minimum of 32 tRNAs are required to translate the "standard" genetic code.

8. Draw the wobble pairings not in Fig. 32-25*a*.

9. A colleague of yours claims that by exposing *E. coli* to HNO_2 she has mutated a $tRNA^{Gly}$ to an *amber* suppressor. Do you believe this claim? Explain.

***10.** Deduce the anticodon sequences of all suppressors listed in Table 32-6 except UGA-1 and indicate the mutations that caused them.

11. How many different types of macromolecules must be minimally contained in a cell-free protein synthesizing system from *E. coli*? Count each type of ribosomal component as a different macromolecule.

12. Why do oligonucleotides containing Shine–Dalgarno sequences inhibit translation in prokaryotes? Why don't they do so in eukaryotes?

13. Why does m^7GTP inhibit translation in eukaryotes? Why doesn't it do so in prokaryotes?

14. What would be the distribution of radioactivity in the completed hemoglobin chains on exposing reticulocytes to ^3H-labeled leucine for a short time followed by a chase with unlabeled leucine?

15. Design an mRNA with the necessary prokaryotic control sites that codes for the octapeptide Lys-Pro-Ala-Gly-Thr-Glu-Asn-Ser.

***16.** Indicate the translational control sites in and the amino acid sequence specified by the following prokaryotic mRNA.

5′-CUGAUAAGGAUUUAAAUUAUGUGUCAAUCACGA-

AUGCUAAUCGAGGCUCCAUAAUAACACUU CGAC-3′

17. What is the energetic cost, in ATP equivalents, for the *E. coli* synthesis of a polypeptide chain of 100 residues starting from amino acids and mRNA? Assume that no losses are incurred as a result of proofreading.

18. It has been suggested that Gly–tRNA synthetase does not require an editing mechanism. Why?

19. Explain why prokaryotic ribosomes can translate a circular mRNA molecule, whereas eukaryotic ribosomes normally cannot, even in the presence of the required cofactors.

20. EF-Tu binds all aminoacyl–tRNAs with approximately equal affinity so that it can deliver them to the ribosome with the same efficiency. Based on the experimentally determined binding constants for EF-Tu and correctly charged and mischarged aminoacyl–tRNAs (see table), explain how the tRNA–EF-Tu recognition system could prevent the incorporation of the wrong amino acid during translation.

Aminoacyl–tRNA	Dissociation Constant (n*M*)
Ala–tRNA^{Ala}	6.2
Gln–tRNA^{Ala}	0.05
Gln–tRNA^{Gln}	4.4
Ala–tRNA^{Gln}	260

Source: LaRiviere, F.J., Wolfson, A.D., and Uhlenbeck, O.C., *Science* **294**, 167 (2001).

21. All cells contain an enzyme called **peptidyl–tRNA hydrolase,** and cells that are deficient in this enzyme grow very slowly. What is the probable function of this enzyme and why is it necessary?

22. An antibiotic named fixmycin, which you have isolated from a fungus growing on ripe passion fruit, is effective in curing several types of sexually transmitted diseases. In characterizing fixmycin's mode of action, you have found that it is a bacterial translational inhibitor that binds exclusively to the large subunit of *E. coli* ribosomes. The initiation of protein synthesis in the presence of fixmycin results in the generation of dipeptides that remain associated with the ribosome. Suggest a mechanism of fixmycin action.

23. Heme inhibits protein degradation in reticulocytes by allosterically regulating ubiquitin-activating enzyme (E1). What physiological function might this serve?

24. Genbux Inc., a biotechnology firm, has cloned the gene encoding an industrially valuable enzyme into *E. coli* such that the enzyme is produced in large quantities. However, since the firm wishes to produce the enzyme in ton quantities, the expense of isolating it would be greatly reduced if the bacterium could be made to secrete it. As a high-priced consultant, what general advice would you offer to solve this problem?

INDEX

Page references in **bold face** refer to a major discussion of the entry. Positional and configurational designations in chemical names (e.g., 3-, α, *N*-, *p*-, *trans*, D-, *sn*-) are ignored in alphabetizing. Numbers and Greek letters are otherwise alphabetical as if they were spelled out.

Note: Chapters 33–35, which are only available on our book companion website, www.wiley.com/college/voet, are separately indexed at this website.

Dipole moment, 39, 43T
Dipyrromethane, 1053
Directional cloning, 118, 118F
Direct transposition, 1237–1239
Disaccharides, **367**
Disc electrophoresis, 148, 148F, 149F
Discontinuous pH electrophoresis, 148, 148F
Discriminator base, 1351F
Disorder, entropy and, 54–57
Dissociation constant (K_a):
    and acid strength, 45–47, 46T
    and pH, 47
    of polyprotic acids, 49–50
Distal histidine, in hemoglobin, 332, 340
Distance-based tree-building, 205
Distance matrix, 204–205
Distributive enzymes, 1177
Disulfide bonds:
    cysteine, 71, 71F
    insulin, 165F
    interchange reaction, 280F
    keratin, 235
    in native proteins, 292
    in protein sequencing, 168–169, 172, 184
    and protein stability, 264–265
Divergent evolution, 254, **317**
Dixon, J., 736
DL3, 714
D-loops, 1213
DM (myotonic dystrophy), 1252
DMB (5,6-dimethylbenzimidazole), 954
DMF (*N,N*-dimethylformamide), 134, 262
DMS (dimethyl sulfate), 1267
DMSO (dimethyl sulfoxide), 43T, 134, 262
DNA (deoxyribonucleic acid), 3, 18, 1145.
    See also related topics, e.g.: DNA repair
    A-DNA, 1146F, 1147F, **1148–1149,**
        1148T, 1151
    B-DNA, 1145–1148, *see* B-DNA
    as carrier of genetic information, **85–88**
    and central dogma of molecular biology,
        95, 1260
    chemical structure, **84–85**
    cloning, *see* Molecular cloning
    complementary base pairing, 18
    decay of, 116–117
    denaturation and renaturation, 90–93,
        92F, 93F
    double helical structure, **1145–1148**
    fractionation, 156–159
    frictional ratio, 155
    junk, 1317
    mutationos of, *see* Mutations
    in nucleus, 8–9
    replication, 18F, 90, 101–104, 102F
    sedimentation coefficients, 154F
    semiconservative replication, 1173–1174
    semidiscontinuous replication,
        1175–1176, 1175F
    size, 94–95
    structure, 18F
    supercoiled, **1158–1170**
    transcription, 95–98
    X-ray diffraction, 88F
    Z-DNA, 1146F, 1147F, 1148T, **1149–1151**
DnaA boxes, 1194, 1195
dnaA gene, 1196
DnaA protein, 1194–1195, 1206
dnaB gene, 1182T
DnaB helicase, 1191, 1193–1194
DNA-binding domain, 705
DnaB protein, 1206
    in prokaryotic replication, 1190, 1190T,
        1191, 1195, 1199
    in unwinding of DNA, 1183, **1184,** 1184T

DNA chips, *see* DNA microarrays
DnaC protein, 1190, 1190T, 1195, 1206
DNA Data Bank of Japan (DDBJ), 195T
DNA-directed DNA polymerases, *see* DNA
    polymerases
DNA-directed polymerases, 1173. *See also*
    DNA polymerases
dnaG gene, 1176
DNA glycosylases, 1218
DnaG protein, 1190T
DNA gyrase, 1166–1169, **1174–1175,**
    1190, 1195
DnaJ, 293
DnaK, 293
DNA ligase, 103, 103F, 1176
    in DNA replication, **1187–1188,**
        1188F, 1189F
    in mismatch repair, 1220
DNA ligase IV, 1224
DNA methylation, **1246–1252**
DNA methyltransferase, 104, 105
DNA microarrays, **211–213,** 211F, 212F,
    213F
dnaN gene, 1182T
DNA photolyases, 1214–1215, 1215F
DNA polymerase, 101–103, 102F,
    1173–1174, 1174F
    bypass, 1222
    error-prone, 1222
    eukaryotic, 1202–1205, 1203T
    intercalating agents inhibit, 1272
    nucleotidyl transferase mechanism, 1180F
    in PCR, 114–115
    properties of selected animal, 1203T
    replication fidelity, 1200–1201, 1201F
    RNA-directed, 95F, 116
    and semidiscontinuous replication,
        1175–1176
    and TERT, 1210
T7 DNA polymerase, 179
DNA polymerase α (pol α), 1202–1203
DNA polymerase β (pol β), 1205
DNA polymerase γ, *see* Pol γ
DNA polymerase δ, *see* Pol δ
DNA polymerase ε, *see* Pol ε
DNA polymerase ζ, *see* Pol ζ
DNA polymerase η, *see* Pol η
DNA polymerase ι, *see* Pol ι
DNA polymerase κ, *see* Pol κ
DNA polymerase I (Pol I), 103, 103F,
    1176–1181
    in bacteriophage φX174 replication, 1191
    in bacteriophage M13 replication, 1190F
    error correction, 103–104
    error rate, 1220
    as A family polymerase, 1202
    in gene manipulation, 109
    role in recombination repair, 1234
    in site-directed mutation, 119
    T. thermophilus, 1179F
    use in nucleic acid sequencing, 177–179
DNA polymerase II (Pol II), **1181,** 1181T
    as B family polymerase, 1202
    role in SOS repair, 1222
DNA polymerase III (Pol III), 103, **1182–1183**
    as C family polymerase, 1202
    error correction, 103–104
    role in mismatch repair, 1220
DNA polymerase IV (Pol IV), 1182, 1222
DNA polymerase V (Pol V), 1182, 1222
DNA polymerase A family, 1202, 1205
DNA polymerase B family, 1202
DNA polymerase C family, 1202
DNA polymerase D family, 1202
DNA polymerase X family, 1202

DNA polymerase Y family, 1202
dnaQ gene, 1182T
DNA recombination, *see* Recombination
DNA repair, **1213–1225**
    and carcinogen identification, 1224–1225
    direct damage reversal, 1214–1216
    double-strand break repair, **1223–1224**
    excision repair, **1216–1219**
    mismatch repair, **1220**
    and Pol I, 1181
    and replication fidelity, 1200–1201
    and SOS response, 1221–1223
DNA replicase, 1182
DNA replication, 18–19, 1145, **1173–1213.**
    See also related topics, e.g.: DNA ligase
    enzymes for, 1176–1189
    in eukaryotes, **1201–1213**
    in meiosis, 21F
    in mitosis, 20F
    overview, 1173–1176
    in prokaryotes, 1173, **1190–1201,** 1200F
    semiconservative nature of, 90, 91F,
        1173–1174
    semidiscontinuous, 1175–1176, 1175F
    telomeres, 1209–1213
    unwinding, 1183–1187
    and viruses, 27
DNA–RNA hybrids, 96, 1151, 1151F
DNase I, 1181
DNA sequencing:
    automated, 179–180
    genome sequencing, 180–182
    Sanger method, 176–180
    sequence databases, 194–195
DNA topoisomerases, 1176
DnaT protein, 1190, 1190T
DNA unwinding elements (DUEs), 1194
dnaX gene, 1182T
DNMT1, 1250
DNMT3a, 1250
DNMT3b, 1250
DNP (2,4-dinitrophenol), 860
dNTPs (deoxynucleoside triphosphates):
    in chain-terminator method, 177–179
    in DNA replication, 101, 1170, 1174F
    and replication fidelity, 1200–1201
Docking proteins, 420, 710
4,7,10,13,16,19-Docosahexenoic acid,
    see DHA
Dodecyltriethylammonium bromide
    (DTAB), 399F
Dodson, E., 1216
Dogs:
    fatty acid experiments with, 945
    number of chromosomes, 19T
Dolichol, 883, 883F
Dolichol-P-mannose, 884
Dolichol-PP-oligosaccharide synthesis,
    884F, 885
Domains:
    of life, 6
    of proteins, **247,** 281
Domain swapping, 718
Dominant traits, 20–22
DON (6-Diazo-5-oxo-L-norleucine), 1142
Donnan equilibrium, 787
Donohue, J., 88
Dopamine, 80, 80F, 783, 1058–1060
Dopamine β-hydroxylase, 1060F
Dopamine receptors, 689
Dot matrices, sequence alignment using,
    196–197, 197F
Dot plot, 196
Double blind tests, 542–543
Double-displacement reactions, 499

Double helical structure, of DNA, **88–90**
Double-reciprocal plot, 490, 490F, 494F–496F, 499F, 500F
doublesex (*dsx*) pre-mRNA (*Drosophila melanogaster*), 1318
Double-sieve mechanism, 1356
Double-strand breaks, *see* DSBs
Double-stranded DNA, *see* dsDNA
Double-stranded RNA, *see* dsRNA
Double-stranded RNA-activated inhibitor (DAI), 1400
Double-stranded RNA-activated protein kinase (PKR), 1400
Doublié, S., 1303
Doubly wound sheets, 254, 255F
Douce, R., 1032
Doudna, J., 1324
Down mutations, 1266
Down's syndrome, 312
Doxorubicin, 1169, 1170
DPG (2,3-diphosphoglycerate), **329**
dp gene (fruit fly), 25F
Dpo4, 1222F, 1222F
Drew, H., 89, 1148
Driving force, 587
drk protein, 709
DRMs (detergent-resistant membranes), 411
Drosha, 1325
Drosophila melanogaster, 23F
  alternative splicing, 1318, 1319F
  cacophony pre-mRNA, 1322
  chromosome number, 19T
  complementation test, 25F
  DNA, 94, 94T, 95F
  Dscam protein, 1318
  genetic experiments, 23–26
  genetic map, 25F
  genome sequencing, 113, 177T, 182
  mitochondria, 1344T
  per gene, 120F
  26S proteasome, 1411F
  replication, 1176F, 1202, 1205F
  RNA interference, 1324
  spliceosomes, 1312F
  transposons, 1244, 1246
  ubiquitin, 1409
Drugs, 78, 396, 968
Drug design, **539**
  cytochromes P450, 543–545
  discovery techniques, **539–541**
  HIV protease and its inhibitors, **545–551**
  pharmacology, **542–545**
  structure-based, 541
Drug discovery, **539–541**
Drug–drug interactions, 544
Druker, B., 719
DSBs (double-strand breaks), 1223–1224, 1235F
Dscam protein, 1318
dsDNA (double-stranded DNA), 18F
  detection by electrophoresis, 158, 158F
  Klentaq1, 1180
  replication, 102–103, 1174
  from semiconservative replication, 90
dsRNA (double-stranded RNA), 85, 1323
DSX-F protein, 1318
DSX-M protein, 1318
dsx (*doublesex*) pre-mRNA (*Drosophila melanogaster*), 1318
DT, *see* Diphtheria toxin
DTAB (dodecyltriethylammonium bromide), 399F
dTMP, *see* Deoxythymidine monophosphate
DUBs (deubiquitinating enzymes), 1417–1418, 1420–1421

Ducruix, A., 709
DUEs (DNA unwinding elements), 1194
Dunathan, H., 1033
Dunn, M., 1078
Duplex DNA, 90, 1159F. *See also* dsDNA (double-stranded DNA)
dUTP, 1218
dUTP diphosphohydrolase (dUTPase; dUTP pyrophosphate), 1126
Dutton, L., 841
Dutzler, R., 755
du Vigneaud, V., 205
Duysens, L., 909
Dwarfism, 685
Dyad axes, 89F
Dynamic mutations, 1251–1252
Dynamin, 436
Dynamite, 743
Dynein, 440
Dystrophin, 1305

E
E1, *see* Pyruvate dehydrogenase; Ubiquitin-activating enzyme
E1b, 799
EI (Enzyme I), of PTS system, 765
E1o, *see* α-Ketoglutarate dehydrogenase
E1p, 799
E1 protein:
  of bovine papillomavirus, 1184–1185, 1185F
  and Rho factor, 1275–1276
E2, *see* Dihydrolipoyl transacetylase; Ubiquitin-conjugating enzyme
EIIA, of PTS system, 765, 767F
EIIB, of PTS system, 765
EIIC, of PTS system, 765
EII, of PTS system, 765
E2o (dihydrolipoyl transsuccinylase), 810
E3, *see* Ubiquitin-protein ligase
E3α, 1413, 1414
E3b, 799
E3 binding protein (E3BP), 798
E3o, 799
E3p, 799
E4P, *see* Erythrose-4-phosphate
E6-associated protein (E6AP), 1414
E6 protein, 1414
E. coli, 4–5, 4F. *See also* Bacteriophages; *specific genes*
  χ[1776] strain, 123
  Ada protein, 1216
  ampicillin resistant, 26
  aspartokinase isoenzymes, 1075T
  BAC replication in, 109
  and bacteriophages, 27F, 28, 86F, 87, 87F, 1340
  biological structures, 14, 15F
  carbamoyl phosphate synthetase, 1027–1028
  catabolite repression, 1285–1288
  CCA-adding polymerase, 1331
  chromosome replication, 1195
  cloning of, 104
  DNA conservation experiments, 90, 91F
  DNA ligase, 1188, 1188F, 1189F
  DNA methylation, 1247
  DNA repair, 1214
  DNA replication, 102F–104F, 103, 104
  DNA size, 94T
  elongation cycle, 1380F
  enzyme induction, 1260–1264
  fMet, 1374
  genome sequencing, 113, 177T
  GroEL/ES system, 294–302

heat shock proteins, 293, 294
homologous recombination, 1228–1230
inclusion bodies, 117, 117F
internal structure, 5F
isocitrate dehydrogenase, 817
lac operon, 97, 97F
membrane of, 398T
mismatch repair, 1220, 1220F
molecular composition, 5T
multiforked chromosomes, 1195, 1195F
nonsense suppressors, 1362, 1362T
nucleotide excision repair, 1217
Pol I, 1176–1181, 1181T
Pol I Klenow fragment, 1178F
Pol II, 1181, 1181T
Pol III, 1181T, 1182–1184, 1182T, 1183F
Pol IV, 1182
Pol V, 1182
pre-mRNA splicing, 1312
primase, 1188–1189, 1189F
promoters, 1266, 1267F
pUC18 cloning vector, 107, 107F, 111
pyrimidine synthesis regulation, 1118F
pyruvate dehydrogenase complex, 792, 793F
RecBCD, 1231, 1231F
replication, 1174F, **1192–1200**, 1193F
replication in, 1183–1187, 1184T
replication rate, 1174–1175
replisome, 1196F
Rho factor, 1276F
ribosome complements, 1363T
ribosomes, 367F, 1363–1365, 1363F–1366F, 1370–1371
RNAP, 1265–1266
RNA primers, 1176
RNAP subunits, 1277T
RuvABC in, 1231
semidiscontinuous replication, 1175
shuttle vectors, 118
σ factors, 1284
simultaneous transcription and translation, 1283F
as source of proteins for purification, 130
termination pathway in ribosomes, 1391F
Ter sites and *oriC* locus, 1199F
thymine dimer effect, 1214–1215
topoisomerase III Y328F mutant, 1163–1165, 1164F
TPP-sensing riboswitch, 1300, 1300F
transcription in, 1271–1273, 1273F, 1284–1301
translation in, 1375F, 1376F
tRNA genes, 1330
two-dimensional gel electrophoresis of, 152F
type II restriction enzymes, 105T
UV absorbance spectra of nucleic acids and DNA, 92F
E54K mutation, 1239
Ealick, S., 1116
Early endosome antigen 1 (EEA 1), 734, 735F
early genes, 1284
Eaton, W., 346, 353
4E-BP1, 1401
4E-BP2, 1401
4E-BP3, 1401
Ebright, R., 1286, 1287
EBS (epidermolysis bullosa simplex), 235
Echinoderms, 13F
Eck, M., 716, 721
Eclipsed conformation, 222, 223F
EcoRI, 105, 105F, 105T, 106
EcoRII, 105T

EcoRV, 105–106, 105F, 105T, 109
ECS (editing site complementary sequence), 1323F
Ectodomains, 453, 684
ED$_{50}$, 539
Edema factor (EF), anthrax, 715
Edidin, M., 408
Editing:
of DNA chain, 1170, 1201
of tRNA, 1355–1358
Editing site complementary sequence (ECS), 1323F
Editosome, 1321–1322
Edman, P., 165, 171
Edman degradation, 165, 167F, 171, 172, 174
Edman's reagent, 165, 167F
EDTA (ethylenediaminetetraacetic acid), 157
EEA 1, *see* Early endosome antigen 1
eEF1A, 1388, 1389F
eEF1B, 1388
eEF1B α, 1388, 1389F
eEF1B β, 1388
eEF2, 1388, 1398, 1405
EF, *see* Edema factor; Elongation factor
EF1A, 1379
EF1B, 1379
EF2, 1385–1386
Effectors, 348, 442, 625T, 688
EF-G, 1375T, **1385–1388,** 1386F
EF hands, 656
Efstradiadis, A., 1306
EF-Ts, 1375T, 1379, 1381–1382, 1382F
EF-Tu, 1375T
in chain initiation, 1382F, 1386, 1386F, 1388
and Gln-tRNA^{Gln}, 1358
and initiator tRNA, 1380–1381, 1381F
and translational accuracy, 1390–1391
Egelman, E., 1228
EGF, *see* Epidermal growth factor
EGF receptor, 699
EH, *see* Enoyl-CoA hydratase
E Helix, 242F
EHK (epidermolytic hyperkeratosis), 235
Ehlers-Danlos syndromes, 240, 1404
Eicosanoid metabolism, 993–997
for leukotrienes and lipoxins, 996F, **1000–1004**
for prostaglandins, prostacyclins, and thromboxanes, 996F, **997–1000**
Eicosanoid receptors, 689
5,8,11,14,17-Eicosapentaenoic acid, *see* EPA
8,11,14-Eicosatrienoic acid, 994
eIF1A, 1376
eIF2, 1376, 1399–1401
eIF2α, 1399–1401
eIF2B, 1379, 1399
eIF2α kinases, 1399
eIF2 phosphatase, 1399
eIF3, 1376
eIF4A, 1378
eIF4B, 1365, 1379
eIF4E, 1378, 1399, 1401
eIF4F, 1378
eIF4G, 1327, 1378
eIF4H, 1379
eIF5, 1379
eIF5B, 1379
eIF*n*, 1376, 1378
80S subunit (ribosome), 1370T, 1371, 1371F
Einstein (unit), 903
Eisenberg, D., 310, 929, 1068, 1315
EJC (exon junction complex), 1326–1327
Eklund, H., 934, 1119, 1122
Elastase, 525, 525T, 529T

gene duplication, 194
proenzymes, 538
specificity, 169T
X-ray structure, 528F, 530, 534–535, 535F
Elasticity coefficient, 622
Elastin, 240, 529
Electrical synapses, 778
Electric fish, 779–780
Electroblotting, 111
Electrochemical cells, 584, 584F
Electrochemical potential, 745
Electromotive force (emf), 584
Electron acceptor, 584
Electron density, 241
Electron density maps, 241–242, 242F
Electron donor, 584
Electronic complementarity, Watson–Crick base pairs, 1155–1156
Electron pairs, acids/bases and, 45
Electron paramagnetic resonance (EPR), 911
Electron paramagnetic spectroscopy, 911
Electron spin resonance (ESR) spectroscopy, 911
Electron-transfer flavoprotein (ETF), 948
Electron-transfer reactions, 586
Electron transport, 188, 386, 823, **828–845.** *See also* Oxidative phosphorylation
and acyl-CoA reoxidation, 948
Coenzyme Q:Cytochrome *c* oxidoreductase, 832T, 840–841, 841F
components of, **833–845**
cytochrome *c* oxidase, 832T, 841–845, 842F
inhibitor studies, 831, 831F
in mitochondria, **823–828**
NADH:Coenzyme Q reductase, 832T, 834–835
net reaction, 824F, 829–830, 829F
P/O ratio, 831–833, 833F
in purple photosynthetic bacteria, 909–913, 913F
reduction potentials of components in testing mitochondria, 832T
rigid coupling of phosphorylation and oxidation, 831–833
sequence of, **829–833**
thermodynamics, **828–829**
two-center, in photosynthesis, 913–925
Electron tunneling, 841
Electrophiles, 563–564, 563F
Electrophilic catalysis, 510
Electrophoresis, 132T
capillary, 152
gel, 147–150
isolectric focusing, 150–152
nucleic acids, **158–159**
paper, 146–147
proteins, **146–152**
SDS–PAGE, 150
Electrophoretic mobility, 146
Electroplaques, 779, 779F
Electrospray ionization (ESI), 172
Electrostatic forces, protein stability and, 259–261
Elements, periodic table, 31F
Elemental composition, humans, 29T
Elementary reactions, **483**
11S regulator, 1418–1419
Elimination reactions, **566–567,** 567F
Elion, G., 1130
ELISA, *see* Enzyme-linked immunosorbent assay
Ellis, J., 293
Elongases, **969–971**

Elongation factor (EF), 422–423, 1375T, **1379**
Elution, in ion exchange chromatography, 136
Embden, G., 595
Embden–Meyerhoff–Parnas pathway, 595. *See also* Glycolysis
EMBL Nucleotide Sequence Database, 195T
Embryogenesis, 1250–1251, 1401
Embryological development, 14F
Embryonic stem cells, 121
Emergent properties, 738
Emf (electromotive force), 584
Enantiomers, 74–76, 74F
Endergonic processes, 57
End groups, amino acid, 165–168
Endo, T., 446
Endo conformation, 1153, 1153F
Endocrine glands, 671, 672F
Endocrine hormones, 671
Endocrine system, 672F
Endocytosis, 8, 433, 454F, 986, 986F
Endoglycosidases, 366
Endonucleases, 1160, 1407–1408. *See also* Restriction endonucleases
Endopeptidases, 169–170, 169T, 194
Endopeptidase Arg-C, 169T
Endopeptidase Asp-N, 169T
Endopeptidase Glu-C, 169T
Endopeptidase Lys-C, 169T
Endoplasmic reticulum (ER), 7F, 9, 11F
metabolic functions, 562T
rough, *see* Rough endoplasmic reticulum
secretory pathway in, 420–424
smooth, *see* Smooth endoplasmic reticulum
β-Endorphin, 673T, 685
Endosomes, 455F
Endothelial NOS (eNOS), 686–688
Endothelium, 686
Endotoxins, 688
δ-Endotoxins, 122
End point, of titration, 51
Enediolate anion, 567
Energy, **53**
conservation of, 52–54
and evolution, 34
metabolic processes, 17
for prokaryotes, 5–6
Energy barrier, to free rotation of polypeptide backbone, 222–223
Energy coupling, 845
Energy metabolism strategies, 1088–1090
Energy-rich bonds, 580
Energy-rich compounds, *see* High-energy compounds
Energy surface, in landscape theory, 287
Energy transduction, 587, 845
Engelman, D., 1365
Englander, W., 285
Englemann, T., 901
Enhancers, 1282–1283, 1337
Enolase, 567, 596F, **612–613,** 613F
Enolate, 568, 569F
Enol forms, of nucleotide bases, 88
Enol phosphates, 580
5-Enolpyruvylshikimate-3-phosphate synthase (ESPS-synthase), 1075
eNOS (endothelial NOS), 686–688
Enoyl-ACP reductase (ER), 965
Enoyl-CoA hydratase (EH), 947, 947F, 948, 948F
Enoyl-CoA isomerase, 950
3,2-Enoyl-CoA isomerase, 952
Enteropeptidase, 537

Enthalpy, **53–54**
Entner–Doudoroff pathway, 636–637
Entropy, **55–57**
of peptidyl transferase reaction, 1384–1385
and protein folding, 284, 287
and temperature, 56F
unitary entropy change, 263T
env, 1246
Enzymatic interconversion, *see* Covalent modification
Enzymes, 4, 163, **469–479**. *See also specific enzymes*
activity control, 474–479
allosteric control, 476–479
in assays, 131
catalytic activity of, 243
chemical catalysts vs., 469
coenzymes, **473–474**
for DNA replication, 1176–1189
DNA replication error repairing, 103–104
feedback inhibition control for, 475
geometric specificity, **472–473**
as globular proteins, 241
historical perspective, 469–470
for liposome delivery, 396
nomenclature, **479**
one gene-one enzyme maxim, 26
processive and distributive, 1177
selective catalysis by, 588–589
stereospecificity, **470–472**
substrate specificity, **470–473**, 470F
Enzyme I (EI), of PTS system, 765
Enzyme catalysis. *See also* Drug design
acid–base catalysis, **506–510**
covalent catalysis, **510–511**
electrostatic catalysis, **512**
lysozyme, **517–525**
metal ion catalysis, **511–512**
and preferential transition state binding, **515–516,** 516F
and proximity/orientation effects, **512–515**
serine proteases, **525–537**
zymogens, **537–538**
Enzyme Commission, 470
Enzyme induction, 1260–1264
Enzyme inhibition, **493–496**
Enzyme kinetics, 482, 487–492
bisubstrate reactions, **497–501**
competitive inhibition, **493–494**
data analysis, **490–491**
and flux control coefficient, 621–622
Haldane relationship, **491–492**
Michaelis–Menten kinetics, **488–491,** 496–497, 501–503
mixed inhibition, **495–496**
pH effects, **496–497**
reversible reaction, **491–492**
uncompetitive inhibition, **495**
Enzyme-linked immunosorbent assay (ELISA), **131–132,** 132F
Enzyme-substrate complex, 470F, **488**
Enzymology, 469
EPA (5,8,11,14,17-eicosapentaenoic acid), 387T, 971, 994, 1002–1003
Epidermal growth factor (EGF), 453–454, 997, 1317
Epidermal growth factor (EGF) receptor, 699
Epidermis, 16F
Epidermolysis bullosa simplex (EBS), 235
Epidermolytic hyperkeratosis (EHK), 235
Epigenetic genome changes, 1249–1250
Epigenetic reprogramming, 1250–1251
Epimers, 361
Epimerization, 567
Epinephrine, 673T, **679–680**

in fatty acid metabolism regulation, 973
in glycogen metabolism regulation, 660
synthesis, 1058–1060
Epithelial cells, 373, 410
Epitopes, 205
EPO, *see* Erythropoietin
Epoxy-activated agarose, 142, 143F
2′,3′-Epoxy-ATP, 1180
EPR (electron paramagnetic resonance), 911
E protein, 1314
Equatorial groups, in sugars, 363
Equilibrium, 55, **58–61,** 587
Equilibrium constants, 45–47, 58–59, 58T
Equilibrium density gradient ultracentrifugation, 90, 91F, 155, 156, 159
Equilibrium thermodynamics, 587–588
Equivalence point, 48
ER, *see* Endoplasmic reticulum; Enoyl-ACP reductase
Erabutoxin, 781
ERAD process, 427
Erbitux, 720
eRF1, 1392
eRF3, 1392
Ergocalciferol, 678
Ergosterol, 393, 678
ERKs (extracellular-signal-regulated kinases), 713
Ernst, O., 694
ERp57, 886
Error catastrophe, 587
Error correction:
in DNA replication, 103–104
in RNA replication, 1281, 1281F
Error-prone DNA polymerases, 1222
Erythomycin, 1395T, 1396F
Erythrocruorins, 325
Erythrocytes, 324–325
heme synthesis regulation, 1055–1056
and sickle-cell anemia, 185–188, 185F, **343–347**
Erythrocyte ghosts, 411
Erythrocyte glucose transporter, 746–748, 746F, 747F
Erythrocyte membranes, **411–414,** 411F
and blood groups, **414–415**
composition, 398T
freeze-etch micrograph, 410F
glycophorin A, 401F
Erythromycin A, 968, 969F
Erythropoietic protoporphyria, 1055
Erythropoietin (EPO), 119, 123, 717
D-Erythrose, 360F
Erythrose-4-phosphate (E4P), 636, 896, 929
D-Erythrulose, 361F
Escherich, T., 4
Escherichia coli, see E. coli
ESEs, *see* Exonic splicing enhancers
ESI (electrospray ionization), 172
ESI-MS, 173
E (exit) site, 1373, 1385, 1387
ESPS-synthase (5-enolpyruvylshikimate-3-phosphate synthase), 1075
ESR (electron spin resonance) spectroscopy, 911
Essen, L.-O., 1215
Essential amino acids, 1019, 1065T, **1072–1078**
Essential fatty acids, 971
ESSs (exonic splicing silencers), 1306
Esters, rates of reaction and motional freedom, 515T
β-Estradiol, 681
Estrogens, 458, 673T, 680, 991

ESTs (expressed sequence tags), 181
ETF (electron-transfer flavoprotein), 948
ETF:ubiquinone oxidoreductase, 837, 948
Ethanol, 783. *See also* Yeast alcohol dehydrogenase
dielectric constant and dipole moment, 43T
from Entner–Doudoroff pathway, 636, 637F
from fermentation, 616
prochiral atoms of, 77F
protein solubility in, 134
Ethanolamine, 389T, 390, 1004
Ethics, of gene therapy, 123–124
Ethidium ion, 157, 1161–1162, 1162F
Ethyl-CoA, 874
Ethylenediaminetetraacetic acid (EDTA), 157
Ethylene glycol, 265
O^6-Ethylguanine residues, 1216
N-Ethylmaleimide (NEM), 440–441
Ethylnitrosurea, 1340
Etoposide, 1169, 1170
Eubacteria, 6, 192
Euchromatin, 182
Eukarya, 6, 7F
Eukaryotes, 3, **6–14**
cellular architecture, 8–12
chromosome number, 19T
citric acid cycle in, 789
DNA, 86, 88, 94T
DNA methylation, 1248–1251
DNA replication, **1201–1213**
evolution of, 13F, 34
genome sequencing, 176
glucose transporters, 751
introns of, 1316
metabolic pathway location, 562
mismatch repair, 1220
mRNA posttranscriptional modification, 1302–1327
nuclear pores of, 1327
nucleotide excision repair, 1217
phylogeny and differentiation, 12–14
polypeptide synthesis, 1376–1379, 1388, 1392
RecA-like proteins, 1230
ribosomes, 1369–1371
rRNA processing, 1328–1329
secretary proteins, 421F
translational in, 1376–1379, 1377F
translation in, 1398–1402
transposons, 1244–1246
Eukaryotic gene expression:
in bacteria vs. eukaryotic cells, 118–119
and colony hybridization, 114
and mRNA lifetime, 101
posttranscriptional modification, 97–98
RNAP, 96, 1276–1283
transcriptional initiation, 96
European corn borer, 122
Evans, M., 262
Evans, R., 1104
Even-chain fatty acids, 945F
Evolution, **29,** 30F. *See also* Chemical evolution; Mutations
of citric acid cycle, 814–815
convergent and divergent, 252, 254, **317,** 531
and divergence of kingdoms, 192
of protein structures, 316–318
and transposition, 1242
Evolutionary distance, 189
Evolutionary tree, 13F
Excinuclease, 1217
Excision repair, **1216–1219**
Excitatory synapses, 778
Exciton transfer, 905

Schizosaccharomyces pombe, 1202, 1206–1207
Schleiden, M., 3
Schleif, R., 1292, 1293
Schlessinger, J., 729
Schoenheimer, R., 1408
Schofield, C., 534
Schulman, L., 1351
Schultz, S., 1212
Schulz, G., 403, 628, 803, 983
Schwann, T., 3
Schwann cell, 778F
SCID, *see* Severe combined immunodeficiency disease
SCID-X1, 123
SciFinder Scholar, 35
Scintillation counter, 149
SCIPs, 735
Scissile peptide bond, 169
SCOP, *see* Structural Classification of Proteins
Scorpion venom, 777
Scott, W., 1310
Scrapie, 312–315, 313F
Screens, for drugs, 539
Scrunching, 1268
Scurvy, 236, 364
SDS (sodium dodecyl sulfate), 150, 169, 171, 399F
SDS-PAGE (polyacrylamide gel electrophoresis), **150,** 151F, 169, 411F. *See also* PAGE (polyacrylamide gel electrophoresis)
Sec (selenocysteine), 1361
Sec1, 441
Sec7, 435
Sec12p, 437
Sec13/31, 430, 438–439, 438F
Sec23/24, 430
Sec61, 424
Sec62/Sec63 complex, 428
SecA, 428
SecB, 428
Secondary lysosome, 455F
Secondary structure, 163, **221–233**
    helical, 225–229, 225F
    nonrepetitive, 230, 232
    peptide group in, 221–225
    predicting, for proteins, **302–304**
    proteins, **221–233**
    tRNA, 1345–1346, 1346F
Second law of thermodynamics, **54–57**
Second messengers, 661, **725–727**
Second-order rate equation, 484
Second-order reactions, 483, 484F
Secretary proteins, 421F
β-Secretase, 312, 547
γ-Secretase, 312
Secretin, 673T, 675
Secretory pathway, **420–428**
Secretory vesicles, 429F, 1405
SecY, 424–425, 425F
SeCys (selenocysteine), 1361
Sedimentation, **152–153**
Sedimentation coefficient, 153, 154F
Sedoheptulose-7-phosphate (S7P), 636, 894
Sedoheptulose bisphosphatase (SBPase), 929, 933F
Seeberger, P., 383
Seed sequence, 1325
Séguin, A., 823
SELB, 1361
Selectable markers, 111
Selenocysteine (Sec) (SeCys), 1361

SELEX (Systematic Evolution of Ligands by EXponential enrichment), **213–214**
Self-assembly, 17, 278, 302
Self-compartmentalized proteases, 1419–1420
Self organization, 29, 30F
Self-splicing genes, 1306–1308
Sem-5, 709
Semiconservative replication, 90
Semiconservative replication, of DNA, 90, **1173–1174**
Semidiscontinuous replication, of DNA, **1175–1176,** 1175F
Semi-invariant position, 1346
Sendai virus, 408, 409F
Senebier, J., 901
Senescence, 703, 1210
Sense RNA, 1323
Sense strand, 203, **1266,** 1267F
Sephadex gels, 139
Sepharose gels, 139
Septic shock, 688
seqA gene, 1196
SeqA protein, 1196
Sequence databases, 194–195
Sequence-tagged connectors (STCs), 181
Sequence-tagged sites (STSs), 181
Sequencing, *see* Nucleic acid sequencing; Protein sequencing
Sequencing gel, 177–178, 178F, 179F
Sequential model, of allosteric interactions, 351–352
Sequential reactions, 498–501, 500F
Sequestration, 1195
SER, *see* Smooth endoplasmic reticulum
Ser, *see* Serine
SERCA complex, 762–764, 762F, 763F
Series-5 leukotrienes, 1003
Serine:
    amino group, 208F
    biosynthesis, 1071–1072
    degradation, 1030–1034
    genetic codes for, 100T, 1343T
    in globular proteins, 246–247
    in glycerophospholipids, 389T
    half-life, 1413T
    α helix/β sheet propensity, 302T
    inserted by *E. coli* nonsense suppressor, 1362T
    in native unfolded proteins, 283
    as nonessential amino acid, 1065T
    in PEST proteins, 1413
    side chains, 71, 208F, 264T
    structure and general properties, 69T
Serine carboxypeptidase II, 531, 531F
Serine dehydratase, 1031, 1408T
Serine hydroxymethyltransferase, 1031, 1033–1034, 1129
Serine palmitoyltransferase, 1008
Serine proteases, 525T
    catalytic mechanism, **531–537,** 532F
    evolutionary relationships among, 530–531
    kinetics and catalytic groups, **525–527**
    specificity pockets, 529F
    transition state stabilization, 534F
    X-ray structures, **527–531**
    zymogens, **537–538**
Serine recombinase, 1239
Serotonin, **783,** 1046, 1059
Serotonin receptors, 689
Serpentine receptors, 689
SerRS, 1351
Serum, disc electrophoresis of, 149F
Serum albumin, 134T, 140F, 944–945, 945F

Serum albumin C, 133F
Serum albumin dimer, 140F
Seryl-adenylate, 1358
Severe combined immunodeficiency disease (SCID), 122, 123, 1132
Sex chromosomes, 22–23
sex-lethal (sxl) gene *(Drosophila melanogaster),* 1318
Sex linked genes, 24
SF1 (splicing factor 1), 1312
S form (*RS* system), 76–77
SH1 domains, 706
SH2-containing inositol- 5-phosphatases (SHIPs), 735
SH2 domains, 706–707, 1317
SH3 domains, 708, 1317
Sharon, N., 523
Sharp, K., 259
Sharp, P., 1304, 1306
Shc, 707, 709–710
Shear degradation, DNA, 95
Sheep, scrapie in, 312–315
Shelterin, 1213
Shemin, D., 574, 1048
Shi, Y., 722
Shimomura, O., 120
Shine, J., 1374
Shine–Dalgarno sequences, 1300, 1301, 1374–1375, 1398
SHIP2, 735
Shipley, G., 696
SHIPs (SH2-containing inositol-5-phosphatases), 735
Shirakawa, M., 1250
Shoelson, S., 721
Shoolingin-Jordan, P., 1053
Short, strong hydrogen bonds, 536
Short tandem repeats (STRs), 115–116
Short-term regulation, of fatty acid metabolism, 973
Shotgun cloning, 113
Shotgun sequencing, 181
Shotton, D., 529
SHP-2, 721–722
Shulman, G., 1103
Shuman, S., 1188
Shuttle vectors, 118
Sialic acid, 1010
Sialic acids, 365, 392F. *See also* NANA
Sialidase, 381
Sickle-cell anemia, 185–188, **343–347**
    gene therapy, 123
    Southern blot for diagnosis or prenatal detection, 112–113
Sickle-cell hemoglobin, 81. *See also* Hemoglobin S
Sickle-cell trait, 187–188
Side chains, of amino acids, 70–71, 208F, 264T, 511
SIDS (sudden infant death syndrome), 948
si face, 77, 77F
Sigler, P., 294, 692, 729, 942, 1288
σ^{32}, 1284
σ^{54}, 1284
σ^{gp28}, 1284
$\sigma^{gp33/34}$, 1284
σ^{70}, 1265, 1284
σ factors, 1265, 1266, 1284
σ-Replication, 1191–1193, 1192F
Sigmoidal shape, oxygen-binding curve, 326, 326F, 339
Signal-anchor sequence, 426
Signal-gated channels, 772
Signal peptidase, 422, 1404
Signal peptides, 420, 1404

Synchrotrons, 241
Syn conformation, 1152F
Synechocystis sp., 177T
Synge, R., 144
Synonyms, in genetic code, 99, 1343
Syntaxin, 433F, 441, 441F, 443, 444
2,5A synthetase ([2′,5′]-oligoadenylate synthetase), 1400
Syphilis, arsenicals for, 799
SYPRO dyes, 149
Systematic Evolution of Ligands by EXponential enrichment (SELEX), **213–214**
Systematic names, enzymes, 479
Systemic lupus erythematosus, 1204, 1312
Systems, 52–53
Systems biology, **576–578,** 576F
Szent-Györgyi, A., 791
Szostak, J., 1210

T

T (ribothymidine), 1347F
T1 G6P translocase, 665F
T2 transporter, 665F
T3 transporter, 665F
T4 Endonuclease VII, 1233–1234, 1233F
T4 Lysozyme, 282, 304
T7 gene 4 helicase/primase, 1184, 1184F, 1188
T7 gp4, 1184, 1184F
Tabun, 782
Tacrolimus, 724
Tagamet, 764
D-Tagatose, 361F
Tail spike protein, 289–290
Tainer, J., 686, 1126, 1219
D-Talose, 360F
Tandem mass spectrometry (MS/MS), 173–175, 174F, 175F
Tangier disease, 458
Taq DNA polymerase, 114, 179
TaqI, 105T
Taq RNAP, 1268, 1269, 1269F
T arm, 1346
Tarui's disease, 667
TATA box, 1282F
Tatum, E., 26, 570, 1262
Tau protein, 312
Taurine conjugates, 993, 993F
Taxonomy, 6, 12, 189
Taylor, S., 653, 654
Tay–Sachs disease, 123, 392, **1011, 1013,** 1013F
TBSV (tomato bushy stunt virus), 277
TψC arm, 1346
TC10, 738
TCA (tricarboxylic acid) cycle, 789. *See also* Citric acid cycle
T cell receptor, 724
T cells, 123, 292–293, 671
TCR (transcription-coupled repair), 1217
TD₅₀, 539
TDF (testes-determining factor), 682
TE (palmitoyl thioesterase), 965
TEBP (telomere end-binding protein), 1212–1213, 1212F
Teeth, 235
Teichoic acids, 378, 378F
Telomerase, **1209–1213**
Telomere end-binding protein, *see* TEBP
Telomere repeat-binding factor 1 (TRF1), 1213
Telomere repeat-binding factor 2 (TRF2), 1213
Telomeres, 109, **1209–1213**
Telomeric T-loops, 1213, 1213F
Telophase, 20F, 21F
Temin, H., 545, 1207

Temperature:
 denaturation, 57
 and entropy, 56F
 and lipid bilayer fluidity, **397–399**
Temperature-jump, 284
Template binding, RNAP, 1266–1267, 1266F
Template strand, 96, 102F
Tendons, 232, 233, 235, 240T
Tensin, 736
TER, 1210
TerA, 1199
TerB, 1199
TerC, 1199
TerD, 1199
TerE, 1199
TerF, 1199
TerG, 1199
TerH, 1199
TerI, 1199
TerJ, 1199
Terminal deoxynucleotidyl transferase, 109, 109F
Terminal desaturases, 970–971
Terminal transferase, 109, 109F
Terminal uridylyltransferase (TUTase), 1322
Termites, 369
Termolecular reactions, 483
Terpenoids, 976–977
Ter sites, 1199, 1206
TERT (telomerase subunit), 1210, 1211, 1211F
Tertiary structure, 163
 α/β barrels, 252, 254
 domains, 247–249, 251–253, 253F
 globular proteins, 245–256
 α helices and β sheets, 246
 open β sheets, 254–256
 predicting, for proteins, 304–305
 and protein folding, 287
 side chain polarity, 246–247
 supersecondary structures, 249–251
 tRNA, 1348–1349
Testes, 672F
Testes-determining factor (TDF), 682
Testicular feminization, 681
Testosterone, 681
Tetanus, 442
Tethering factors, 441
Tetracycline, 128, 1395T, 1396F, **1397**
Tetracycline-resistant bacterial strains, 1397
Tetrahedral intermediate, 531–533
Tetrahedral symmetry, proteins, 268, 268F
5,6,7,8-Tetrahydrobiopterin, *see* BH₄
Tetrahydrofolate, 473T, 492
Tetrahydrofolate (THF), **1062–1064**
Tetrahydrofuran-2-carboxylate, 516
Tetrahymena thermophila:
 Group I introns, 1307–1310, 1307F, 1309F, 1310F
 telomerase, 1209–1210
Tetraloop, 1308
2,3,4,6-*O*-Tetramethyl-α-D-glucose, 508
Tetramethyl-*p*-phenylenediamine (TMPD), 831
Tetranucleotide hypothesis, 85
Tetrapeptides, 73F
Tetraubiquitin, 1411F
Tetrazole, 211
Tetrodotoxin, 777
TeTx, 442
TFIIH, 1218
TFIIS protein, 1281–1282
TG, *see* Thapsigargin
TGN, *see* Trans Golgi network

Thalassemias, 219, 342
Thalidomide, 78, 78F
Thapsigargin (TG), 763, 763F
ThDP (thiamin diphosphate), 616
Theoretical plate, in chromatography, 135
Therapeutic index, 539
Thermal stability, proteins, 266
Thermatoga maritima, 1394F
Thermoacidophiles, 6
Thermocycler, 114
Thermodynamics, **52–61.** *See also* Equilibrium
 chemical equilibria, 58–61
 defined, 52
 electron-transport chain, **828–829**
 first law, **52–54**
 free energy, **57–58,** 61
 glycogen metabolism, 643–644
 of life, **586–589**
 of membrane transport, **744–745**
 of phosphate compounds, **578–583**
 second law, **54–57**
 transition state, 486
 units and constants, 53T
Thermogenesis, 629, 860–861, **1099–1001**
Thermogenin, 861, 1101
Thermolysin, 169T
Thermoplasma acidophilum, 1415, 1415F
Thermoproteus, 7F
Thermosomes, 301
Thermostable proteins, 266
Thermus aquaticus, 179, 466
 DNA polymerase I (Klentaq1), 1178–1180
 EF-Tu, 1381F
 in PCR, 114
 RNAP, 1268, 1269, 1269F
Thermus thermophilus:
 30S subunit, 1389–1390
 Argonaute, 1324, 1325F
 EF-G complex with GDP, 1385–1386, 1386F
 proteinase K, 1383
 ribosomal complex, 1384, 1392–1394, 1393F, 1394F
 ribosomal subunits, 1366–1368, 1371, 1386F
 RNAP, 1268, 1268F–1269F
 RuvB, 1231–1233, 1232F
θ replication, 1174, 1175F
θ structures, 1174
THF, *see* Tetrahydrofolate
Thiamin diphosphate (ThDP), 616
Thiamine, 474T, 618
Thiamine pyrophosphate, *see* TPP
Thiazolidinediones (TZDs), 1103–1104
Thiazolinone, 165
Thiazolium ring, 616
thi box, 1300
Thin layer chromatography (TLC), 145
Thiogalactoside transacetylase, 1261
Thiohemiacetal, 608
Thiokinases, 945
Thiolase, 947, 958, 976
Thiolase reaction, 949–950
Thioredoxin (Trx), 291, 933–934, 934F, 1124
Thioredoxin reductase (TrxR), 1125
4-Thiouridine (s⁴U), 1347F
30S subunit (ribosome), 363, 1363T, 1368, 1368F, 1369F, **1370**
 path of mRNa through, 1375, 1376F
Thirumalai, D., 300
Thr, *see* Threonine
Thr-AMS, 1358
Threading (fold recognition), 305

Wlodawer, A., 549
Wobble pairing, 1360–1361, 1360F, 1360T
WOC (water-oxidizing center), 916
Woese, C., 6, 8
Wolberger, C., 1293
Wolcott–Rallison syndrome, 1400
Wolfenden, R., 515, 1384
Wolman's disease, 466
Wolynes, P., 287
Woodward, R. B., 975
Wool, 233, 235
Work, 53–54, 57–58
World Wide Web, 195T, 256T
Writhing number, of superhelix, 1159–1160, 1159F
Wunsch, C., 199
Wüthrich, K., 243, 314
WW domain, 708
Wyman, J., 349

X
Xanthine, 1112, 1130F
Xanthine dehydrogenase, 1133
Xanthine oxidase (XO), 1130F, 1132–1133
Xanthomas, 457
Xanthosine, 1130F
Xanthosine monophosphate, *see* XMP
X chromosome, 22–23
Xenobiotics, 542
Xenopus borealis, 1283
Xenopus laevis, 1163
Xenotransplantation, 121–122
XerC, 1243
XerD, 1243
Xeroderma pigmentosum (XP), 1217–1218, 1223, 1224
X-gal (5-bromo-4-chloro-3-indolyl-β-D-galactoside), 110
XhoI, 105T
X-linked phosphorylase kinase deficiency, 667
XMP (xanthosine monophosphate), 1112, 1130F
XO, *see* Xanthine oxidase
XP, *see* Xeroderma pigmentosum
XPA, 1217
XPB, 1217, 1218
XPC, 1217
XPD, 1217, 1218
XPE, 1217

XPF, 1217
XPG, 1217, 1218
XPV, 1217
X-ray crystallography:
  and double helix of DNA, 88, 88F
  of enzyme–substrate complexes, 470
  interpretation of, 241–245
  and NMR, 244
  protein dynamics studies, 306
Xrcc4, 1223–1224
Xrn1 exonuclease, 1327
Xrn2 exonuclease, 1304
Xu, W., 722
Xu5P, *see* Xylulose-5-phosphate
Xylitol, 365
XylNAc residue (*N*-acetylxylosamine), 523
Xylose, 597–598
D-Xylose, 360F, 361
D-Xylulose, 361
Xylulose-5-phosphate (Xu5P), 892–894, 929

Y
YACs (yeast artificial chromosomes), 108–109, 113, 114, 180–181
YADH, *see* Yeast alcohol dehydrogenase
Yalow, R., 674
Yang, W., 1222
Yanoffsky, C., 1296, 1298, 1337
Yarus, M., 1383
Y chromosome, 23
Yeast, 13F. *See also Saccharomyces cerevisiae* (Baker's yeast)
  20S proteasome, 1416F–1417F
  chromosome number, 19T
  cloning of, in host organism, 104. *See also* YACs [yeast artificial chromosomes]
  DNA size, 94T
  eEF1A-eEF1Bb, 1389F
  exon splicing, 1305–1306
  fermentation, 593–594, 616–619
  GCN2, 1400–1401
  genome sequencing, 176, 177T, 182
  introns in, 1316
  prions in, 315–316
  protein disulfide isomerase, 291–292, 291F
  RAD51, 1230
  Rpn11, 1418
  80S rRNA, 1371F
  rRNA methylation sites, 1328–1329

shuttle vectors, 118
snoRNAs, 1329
transposons, 1244
tRNA^Ala, 1345F, 1346
tRNA^Asp, 1349
tRNA^Phe, 1346, 1346F, 1348
vesicle fusion, 440
Yeast alanine tRNA, 176
Yeast alcohol dehydrogenase (YADH), 618, 619
  geometric specificity, 472–473
  molecular mass, 140F
  stereospecificity, 470–472
Yeast artificial chromosomes, *see* YACs
Yellow mosaic virus protein, 153T
Yersinia pestis, 1420–1421
Y^+ gene, 1261–1262, 1264
Ylid, 616
-*yl* suffix, 73
Yokoyama, S., 1355–1357
Yonath, A., 1365, 1366, 1376
Yonetani, T., 354
YopJ, 1420
Yoshida, M., 857
Yoshikawa, S., 841
Young, W., 594

Z
Zα, 1150–1151, 1150F
Zalcitabine, 1208
Zamecnik, P., 1345, 1362
Zamore, Phillip, 1323
ZAP-70, 1412, 1412F
Z-DNA, 1146F, 1147F, 1148T, **1149–1151**
Zero order reactions, 488
Z^+ gene, 1261–1262, 1264
Zidovudine, *see* 3'-Azido-3'-deoxythymidine
α Zigzag, 433
Zimm, B., 94
Zinc, 473, 511–513
Zinc fingers, 1402
Zipping, 443F
Zocor, 990, 991F
Zonal ultracentrifugation, 155, 156F, 159
Zone electrophoresis, 146
Z-scheme, 914, 915F, 917F
Zwitterions, 67F, 70
Zygote, 20
Zymase, 594
Zymogens, **537–538**

One- and Three-Letter Symbols for the Amino Acids[a]

A	Ala	Alanine
B	Asx	Asparagine or aspartic acid
C	Cys	Cysteine
D	Asp	Aspartic acid
E	Glu	Glutamic acid
F	Phe	Phenylalanine
G	Gly	Glycine
H	His	Histidine
I	Ile	Isoleucine
K	Lys	Lysine
L	Leu	Leucine
M	Met	Methionine
N	Asn	Asparagine
P	Pro	Proline
Q	Gln	Glutamine
R	Arg	Arginine
S	Ser	Serine
T	Thr	Threonine
V	Val	Valine
W	Trp	Tryptophan
Y	Tyr	Tyrosine
Z	Glx	Glutamine or glutamic acid

[a]The one-letter symbol for an undetermined or nonstandard amino acid is X.

Thermodynamic Constants and Conversion Factors

Joule (J)
$1 \text{ J} = 1 \text{ kg} \cdot \text{m}^2 \cdot \text{s}^{-2}$ $1 \text{ J} = 1 \text{ C} \cdot \text{V}$ (coulomb volt)
$1 \text{ J} = 1 \text{ N} \cdot \text{m}$ (newton · meter)

Calorie (cal)
1 cal heats 1 g of H_2O from 14.5 to 15.5°C
$1 \text{ cal} = 4.184 \text{ J}$

Large calorie (Cal)
$1 \text{ Cal} = 1 \text{ kcal}$ $1 \text{ Cal} = 4184 \text{ J}$

Avogadro's number (N)
$N = 6.0221 \times 10^{23}$ molecules · mol^{-1}

Coulomb (C)
$1 \text{ C} = 6.241 \times 10^{18}$ electron charges

Faraday ($\mathscr{F}$)
$1 \mathscr{F} = N$ electron charges
$1 \mathscr{F} = 96,485 \text{ C} \cdot \text{mol}^{-1} = 96,485 \text{ J} \cdot \text{V}^{-1} \cdot \text{mol}^{-1}$

Kelvin temperature scale (K)
0 K = absolute zero   273.15 K = 0°C

Boltzmann constant (k_B)
$k_B = 1.3807 \times 10^{-23} \text{ J} \cdot \text{K}^{-1}$

Gas constant (R)
$R = Nk_B$ $R = 1.9872 \text{ cal} \cdot \text{K}^{-1} \cdot \text{mol}^{-1}$
$R = 8.3145 \text{ J} \cdot \text{K}^{-1} \cdot \text{mol}^{-1}$ $R = 0.08206 \text{ L} \cdot \text{atm} \cdot \text{K}^{-1} \cdot \text{mol}^{-1}$

The Standard Genetic Code

First Position (5′ end)	Second Position				Third Position (3′ end)
	U	C	A	G	
U	UUU Phe	UCU Ser	UAU Tyr	UGU Cys	U
	UUC Phe	UCC Ser	UAC Tyr	UGC Cys	C
	UUA Leu	UCA Ser	UAA Stop	UGA Stop	A
	UUG Leu	UCG Ser	UAG Stop	UGG Trp	G
C	CUU Leu	CCU Pro	CAU His	CGU Arg	U
	CUC Leu	CCC Pro	CAC His	CGC Arg	C
	CUA Leu	CCA Pro	CAA Gln	CGA Arg	A
	CUG Leu	CCG Pro	CAG Gln	CGG Arg	G
A	AUU Ile	ACU Thr	AAU Asn	AGU Ser	U
	AUC Ile	ACC Thr	AAC Asn	AGC Ser	C
	AUA Ile	ACA Thr	AAA Lys	AGA Arg	A
	AUG Met[a]	ACG Thr	AAG Lys	AGG Arg	G
G	GUU Val	GCU Ala	GAU Asp	GGU Gly	U
	GUC Val	GCC Ala	GAC Asp	GGC Gly	C
	GUA Val	GCA Ala	GAA Glu	GGA Gly	A
	GUG Val	GCG Ala	GAG Glu	GGG Gly	G

[a]AUG forms part of the initiation signal as well as coding for internal Met residues.

Some Common Biochemical Abbreviations[a]

A	adenine	E4P	erythrose-4-phosphate
aa	amino acid	EPR	electron paramagnetic resonance
aaRS	aminoacyl–tRNA synthetase	ER	endoplasmic reticulum
ACAT	acyl-CoA:cholesterol acyltransferase	ESI	electrospray ionization
ACh	acetylcholine	EST	expressed sequence tag
AChE	acetylcholinesterase	ETF	electron-transfer flavoprotein
ACP	acyl-carrier protein	FAD	flavin adenine dinucleotide, oxidized form
ADA	adenosine deaminase	FADH·	flavin adenine dinucleotide, radical form
ADH	alcohol dehydrogenase	FADH$_2$	flavin adenine dinucleotide, reduced form
AdoCbl	5′-deoxyadenosylcobalamin	FAS	fatty acid synthase
AdoMet	adenosylmethionine	FBP	fructose-1,6-bisphosphate
ADP	adenosine diphosphate	FBPase	fructose-1,6-bisphosphatase
ADPNP	adenosine-5′-(β,γ-imido)triphosphate	Fd	ferredoxin
AIDS	acquired immunodeficiency syndrome	FGF	fibroblast growth factor
AKAP	A-kinase anchoring protein	FH	familial hypercholesterolemia
ALA	δ-aminolevulinic acid	fMet	N-formylmethionine
AMP	adenosine monophosphate	FMN	flavin mononucleotide
AMPK	AMP-dependent protein kinase	FNR	ferredoxin-NADP$^+$ reductase
AMPPNP	adenosine-5′-(β,γ-imido)triphosphate	F1P	fructose-1-phosphate
ARS	autonomously replicating sequence	F2,6P	fructose-2,6-bisphosphate
ATCase	aspartate transcarbamoylase	F6P	fructose-6-phosphate
ATP	adenosine triphosphate	G	guanine
BAC	bacterial artificial chromosome	GABA	γ-aminobutyric acid
BChl	bacteriochlorophyll	Gal	galactose
BCKDH	branched chain α-keto acid dehydrogenase	GalNAc	N-acetylgalactosamine
BH$_4$	5,6,7,8-tetrahydrobiopterin	GAP	glyceraldehyde-3-phosphate
bHLH	basic helix–loop–helix	GAP	GTPase activating protein
bp	base pair	GAPDH	glyceraldehyde-3-phosphate dehydrogenase
BPG	D-2,3-bisphosphoglycerate	GDH	glutamate dehydrogenase
BPheo	bacteriopheophytin	GDP	guanosine diphosphate
BPTI	bovine pancreatic trypsin inhibitor	GEF	guanine nucleotide exchange factor
C	cytosine	GK	glucokinase
CaM	calmodulin	Glc	glucose
CAM	crassulacean acid metabolism	GlcNac	N-acetylglucosamine
cAMP	3′,5′-cyclic AMP	GLUT	glucose transporter
CAP	catabolite gene activator protein	GMP	guanosine monophosphate
CCV	clathrin-coated vesicle	gp	gene product
CD	circular dichroism	G1P	glucose-1-phosphate
CDK	cyclin-dependent protein kinase	G6P	glucose-6-phosphate
cDNA	complementary DNA	G6PD	glucose-6-phosphate dehydrogenase
CDP	cytidine diphosphate	GPI	glycosylphosphatidylinositol
CDR	complementarity-determining region	gRNA	guide RNA
CE	capillary electrophoresis	GSH	glutathione
cGMP	3′5′-cyclic GMP	GSSG	glutathione disulfide
CGN	cis Golgi network	GTF	general transcription factor
Chl	chlorophyll	GTP	guanosine triphosphate
CM	carboxymethyl	HA	hemagglutinin
CMP	cytidine monophosphate	Hb	hemoglobin
CoA or CoASH	coenzyme A	HDL	high density lipoprotein
COP	coat protein	HGPRT	hypoxanthine-guanine phosphoribosyl transferase
CoQ	coenzyme Q (ubiquinone)	HIV	human immunodeficiency virus
COX	cyclooxygenase or cytochrome c oxidase	HMG-CoA	β-hydroxy-β-methylglutaryl-CoA
CPS	carbamoyl phosphate synthetase	hnRNA	heterogeneous nuclear RNA
CTD	C-terminal domain	HPETE	hydroperoxyeicosatetraenoic acid
CTP	cytidine triphosphate	HPLC	high-performance liquid chromatography
D	dalton	Hsp	heat shock protein
d	deoxy	HTH	helix–turn–helix
DAG	1,2-diacylglycerol	Hyl	5-hydroxylysine
dd	dideoxy	Hyp	4-hydroxyproline
DEAE	diethylaminoethyl	IDL	intermediate density lipoprotein
DG	sn-1,2-diacylglycerol	IF	initiation factor
DHAP	dihydroxyacetone phosphate	IgG	immunoglobulin G
DHF	dihydrofolate	IHP	inositol hexaphosphate
DHFR	dihydrofolate reductase	IMP	inosine monophosphate
DMF	N,N-dimethylformamide	IP$_3$	inositol-1,4,5-trisphosphate
DMS	dimethyl sulfate	IPTG	isopropylthiogalactoside
DNA	deoxyribonucleic acid	IR	infrared
DNP	2,4-dinitrophenol	IS	insertion sequence
dNTP	2′-deoxynucleotide triphosphate	ISP	iron–sulfur protein
Dol	dolichol	ITP	inosine triphosphate
L-DOPA	L-3,4-dihydroxyphenylalanine	JAK	Janus kinase
dsDNA	double-stranded DNA	K_M	Michaelis constant
DUB	deubiquitinating enzyme	kb	kilobase pair
EF	elongation factor	kD	kilodalton
EGF	epidermal growth factor	KF	Klenow fragment
ELISA	enzyme-linked immunosorbent assay	LCAT	lecithin:cholesterol acyl transferase
EM	electron microscopy	LDH	lactate dehydrogenase
emf	electromotive force	LDL	low density lipoprotein

[a]The three-letter and one-letter abbreviations for the "standard" amino acid residues are given in Table 4-1.

(table continued on following page)